Standard Atomic Masses of the Elements 2013, IUPAC

Based on Relative Atomic Mass of ^{12}C = 12, where ^{12}C is a neutral atom in its nuclear and electronic ground state.[1]

Name	Symbol	Atomic Number	Atomic Weight
Actinium[2]	Ac	89	(227)
Aluminum	Al	13	26.981 5386(8)
Americium[2]	Am	95	(243)
Antimony	Sb	51	121.760(1)
Argon	Ar	18	39.948(1)
Arsenic	As	33	74.921 60(2)
Astatine[2]	At	85	(210)
Barium	Ba	56	137.327(7)
Berkelium[2]	Bk	97	(247)
Beryllium	Be	4	9.012 182(3)
Bismuth	Bi	83	208.980 40(1)
Bohrium[2]	Bh	107	(270)
Boron	B	5	10.811(7)
Bromine	Br	35	79.904(1)
Cadmium	Cd	48	112.411(8)
Calcium	Ca	20	40.078(4)
Californium[2]	Cf	98	(251)
Carbon	C	6	12.0107(8)
Cerium	Ce	58	140.116(1)
Cesium	Cs	55	132.905 4519(2)
Chlorine	Cl	17	35.453(2)
Chromium	Cr	24	51.9961(6)
Cobalt	Co	27	58.933 195(5)
Copper	Cu	29	63.546(3)
Curium[2]	Cm	96	(247)
Darmstadtium[2]	Ds	110	(281)
Dubnium[2]	Db	105	(268)
Dysprosium	Dy	66	162.500(1)
Einsteinium[2]	Es	99	(252)
Erbium	Er	68	167.259(3)
Europium	Eu	63	151.964(1)
Fermium[2]	Fm	100	(257)
Fluorine	F	9	18.998 4032(5)
Francium[2]	Fr	87	(223)
Gadolinium	Gd	64	157.25(3)
Gallium	Ga	31	69.723(1)
Germanium	Ge	32	72.64(1)
Gold	Au	79	196.966 569(4)
Hafnium	Hf	72	178.49(2)
Hassium[2]	Hs	108	(277)
Helium	He	2	4.002 602(2)
Holmium	Ho	67	164.930 32(2)
Hydrogen	H	1	1.00794(7)
Indium	In	49	114.818(3)
Iodine	I	53	126.904 47(3)
Iridium	Ir	77	192.217(3)
Iron	Fe	26	55.845(2)
Krypton	Kr	36	83.798(2)
Lanthanum	La	57	138.905 47(7)
Lawrencium[2]	Lr	103	(262)
Lead	Pb	82	207.2(1)
Lithium	Li	3	6.941
Lutetium	Lu	71	174.9668(1)
Magnesium	Mg	12	24.3050(6)
Manganese	Mn	25	54.938 045(5)
Meitnerium[2]	Mt	109	(276)
Mendelevium[2]	Md	101	(258)
Mercury	Hg	80	200.59(2)
Molybdenum	Mo	42	95.96(2)
Neodymium	Nd	60	144.242(3)
Neon	Ne	10	20.1797(6)
Neptunium[2]	Np	93	(237)
Nickel	Ni	28	58.6934(4)
Niobium	Nb	41	92.906 38(2)
Nitrogen	N	7	14.0067(2)
Nobelium[2]	No	102	(259)
Osmium	Os	76	190.23(3)
Oxygen	O	8	15.9994(3)
Palladium	Pd	46	106.42(1)
Phosphorus	P	15	30.973 762(2)
Platinum	Pt	78	195.084(9)
Plutonium[2]	Pu	94	(244)
Polonium[2]	Po	84	(209)
Potassium	K	19	39.0983(1)
Praseodymium	Pr	59	140.907 65(2)
Promethium[2]	Pm	61	(145)
Protactinium[2]	Pa	91	231.035 88(2)
Radium[2]	Ra	88	(226)
Radon[2]	Rn	86	(222)
Rhenium	Re	75	186.207(1)
Rhodium	Rh	45	102.905 50(2)
Roentgenium[2]	Rg	111	(280)
Rubidium	Rb	37	85.4678(3)
Ruthenium	Ru	44	101.07(2)
Rutherfordium[2]	Rf	104	(265)
Samarium	Sm	62	150.36(2)
Scandium	Sc	21	44.955 912(6)
Seaborgium[2]	Sg	106	(271)
Selenium	Se	34	78.96(3)
Silicon	Si	14	28.0855(3)
Silver	Ag	47	107.8682(2)
Sodium	Na	11	22.989 769 28(2)
Strontium	Sr	38	87.62(1)
Sulfur	S	16	32.065(5)
Tantalum	Ta	73	180.947 88(2)
Technetium[2]	Tc	43	(98)
Tellurium	Te	52	127.60(3)
Terbium	Tb	65	158.925 35(2)
Thallium	Tl	81	204.3833(2)
Thorium[2]	Th	90	232.038 06(2)
Thulium	Tm	69	168.934 21(2)
Tin	Sn	50	118.710(7)
Titanium	Ti	22	47.867(1)
Tungsten	W	74	183.84(1)
Uranium[2]	U	92	238.028 91(3)
Vanadium	V	23	50.9415(1)
Xenon	Xe	54	131.293(6)
Ytterbium	Yb	70	173.054(5)
Yttrium	Y	39	88.905 85(2)
Zinc	Zn	30	65.38(2)
Zirconium	Zr	40	91.224(2)
Copernicium[2]	Cn	112	(285)
—[2,3]		113	(284)
Flerovium[2]	Fl	114	(289)
—[2,3]		115	(288)
Livermorium[2]	Lv	116	(293)
—[2,3]		117	(294)
—[2,3]		118	(294)

1. The atomic masses of many elements vary depending on the origin and treatment of the sample. This is particularly true for Li; commercially available lithium-containing materials have Li atomic masses in the range of 6.939 and 6.996. Uncertainties are given in parentheses following the last significant figure to which they are attributed.
2. Elements with no stable nuclide; the value given in parentheses is the atomic mass number of the isotope of longest known half-life. However, three such elements (Th, Pa, and U) have a characteristic terrestrial isotopic composition, and the atomic weight is tabulated for these.
3. Not yet named.

PRINCIPLES OF MODERN CHEMISTRY

EIGHTH EDITION

DAVID W. OXTOBY
Pomona College

H.P. GILLIS
University of California—Los Angeles

LAURIE J. BUTLER
The University of Chicago

CENGAGE Learning

Australia • Brazil • Mexico • Singapore • United Kingdom • United States

CENGAGE Learning

Principles of Modern Chemistry,
Eighth Edition
David W. Oxtoby, H.P. Gillis, Laurie J. Butler

Product Director: Mary Finch
Product Manager: Krista Mastroianni
Managing Developer: Peter McGahey
Content Developer: Ed Dodd
Content Developer: Elizabeth Woods
Product Assistant: Morgan Carney
Senior Content Developer: Lisa Weber
Marketing Manager: Julie Schuster
Content Project Manager: Teresa L. Trego
Art Director: Andrei Pasternak
Manufacturing Planner: Judy Inouye
Production Service: Graphic World Inc.
Photo Researcher: Lumina Datamatics
Text Researcher: Lumina Datamatics
Copy Editor: Graphic World Inc.
Illustrator: Graphic World Inc.
Text Designer: Ellen Pettengell Design
Cover Designer: Delgado and Company
Cover Image: Dr. Eric Heller
Compositor: Graphic World Inc.

Library of Congress Control Number: 2014953673

ISBN: 978-1-305-07911-3

Cengage Learning
20 Channel Center Street
Boston, MA 02210
USA

Cengage Learning is a leading provider of customized learning solutions with office locations around the globe, including Singapore, the United Kingdom, Australia, Mexico, Brazil, and Japan. Locate your local office at **www.cengage.com/global**.

Cengage Learning products are represented in Canada by Nelson Education, Ltd.

To learn more about Cengage Learning Solutions, visit **www.cengage.com**.

Purchase any of our products at your local college store or at our preferred online store **www.cengagebrain.com**.

Printed in the United States of America
Print Number: 01 Print Year: 2015

IN APPRECIATION OF

Karl F. Freed

William M. Gelbart

Yuan T. Lee

our Ph.D. advisers

for their distinguished careers in scientific research and education

The search for truth is in one way hard and in another easy, for it is evident that no one can master it fully or miss it completely.
But each adds a little to our knowledge of nature, and from all the facts assembled there arises a certain grandeur.

(Greek inscription, taken from Aristotle, on the facade of the National Academy of Sciences building in Washington, D.C.)

BRIEF CONTENTS

CONTENTS

UNIT III Kinetic Molecular Description of the States of Matter 368

UNIT IV Equilibrium in Chemical Reactions 478

APPLICATIONS

PREFACE

The eighth edition of *Principles of Modern Chemistry* is written for students in honors and upper-mainstream general chemistry courses who seek to understand and interpret chemical events at the molecular level. The relation of molecular structure to function and properties requires the introduction of molecular structure early in the course and the use of structural arguments in presenting the remaining topics. Moreover, students will soon be introduced to the predictive power of chemical computations and simulations, for which a solid background in the description of molecular structure is essential.

The organization of the eighth edition is fundamentally the same as that of the seventh edition, and presents the material from a unified, molecular point of view that continues to emphasize the central role of structure, but with greater focus on the electronic structure of molecules as a unifying theme. This edition also deepens and broadens the introduction of chemical principles in several chapters. Recognizing that in the top colleges and universities that use this text, over 90% of students taking introductory chemistry have taken a single-variable calculus course in high school, or are taking calculus as a co-requisite for the course, we have improved several portions of the text. These revisions include a more quantitative treatment of the forces and potential energy in molecules (Chapter 3), dipole moments (Chapter 3), molecular collisions (Chapters 9 and 18), and the Maxwell-Boltzmann distribution of molecular speeds (Chapter 9). The eighth edition also includes major revisions in the chapters on quantum mechanics and molecular structure (Chapter 6), electrochemistry (Chapter 17), nuclear chemistry (Chapter 19), and molecular spectroscopy and photochemistry (Chapter 20).

Significant Changes in This Edition

- ***OWLv2 for General Chemistry***—Our online learning solution OWLv2 has been revised to offer more flexible student and instructor functionality, new personalized study tools, enhanced integration with learning management systems, and improved analytics. OWLv2 includes Quick Prep, new iPad-compatible visualizations, tutorials, and simulations, and new Adaptive Study Plans. Tutors, Visualizations, and Simulations have been converted from Flash to HTML, allowing for easier navigation and streamlined grading.
- ***Revised Writing Style without Loss of Rigor***—The language is more modern and less formal. We have introduced a more conversational writing style, designed to engage our students as active participants in developing the presentation. We have examined every sentence in the book to simplify and lighten the language without compromising intellectual integrity.

- *Greater Flexibility in Topic Coverage*—In response to comments by students, faculty, and reviewers, greater modularity and flexibility have been built into the text to make it compatible with alternative sequences of topics. While keeping the discussion of bonding and structure at the beginning of the book, we have been careful to maintain the option to follow the "macro-to-micro" approach used in previous editions. Selecting alternative approaches is facilitated by the unit structure of the book; we offer several suggestions in the **Teaching Options** section.
- *Streamlined Topic Coverage*—In response to suggestions by both users and reviewers of the text, we have taken care in this revision to retain only those topics that were felt to be critical for students. Chapters 19 and 20, especially, were shortened after extensive surveying of professors who teach the course; what remains (we think) are chapters that are more readily covered within the framework of a conventional two-semester course.
- *New Problems*—We've added approximately 30 new and revised problems throughout the book. These follow the unique tradition established in previous editions that all problems are based on actual experimental data measured on real chemical systems. We intend the problems to guide our students in developing intuition for chemical results and the magnitudes of chemical quantities, as well as facility in numerical calculations.

Organizational Approach for Selected Chapters and Major Changes to Content for Eighth Edition

Chapter 1: The Atom in Modern Chemistry

This chapter describes the physical structure of the atom, as determined from the classic experiments of Thomson, Millikan, and Rutherford. For this edition, we added an introduction of the electrostatic force on negatively and positively charged particles (see Section 1.4.3), improved the coverage of the J.J. Thompson experiment measuring the charge-to-mass ratio of the electron (see Section 1.4.3), and incorporated coverage of the mole, counting molecules by weighing (which was Section 2.1 in the seventh edition; it is Section 1.6 here in the eighth edition).

Chapter 3: Atomic Shells and Classical Models of Chemical Bonding

This chapter provides a substantial introduction to molecular structure by coupling experimental observation with interpretation through simple classical models. Today, the tools of classical bonding theory—covalent bonds, ionic bonds, polar covalent bonds, electronegativity, Lewis electron dot diagrams, and VSEPR theory—have all been explained by quantum mechanics. It is a matter of preference whether to present the classical theory first and then gain deeper insight from the quantum explanations, or to cover the quantum theory first and then see the classical theory as a limiting case. Our experience has been that presenting the classical description first enables our students to bring considerably greater sophistication to their first encounter with quantum mechanics and therefore to develop a deeper appreciation for that subject. We have seen that this approach offers definitive pedagogical advantages by enabling students to

- learn the language and vocabulary of the chemical bond starting from familiar physical concepts.

- become familiar with the properties of a broad array of real molecules *before* attempting to explain these results using quantum mechanics.
- develop experience in using physical concepts and equations to describe the behavior of atoms and molecules.

We have organized this chapter to meet these goals. Changes for the eighth edition include the following:

- The chapter now presents attractive and repulsive forces between nuclei and electrons, and between atoms in a diatomic molecule, using simple vectors in two dimensions, so that students may correctly calculate the direction of attractive and repulsive forces between particles (see Section 3.3). We emphasize the physical insight gained from the new quantitative treatment.
- We have streamlined and improved the discussion of polar covalent bonding, removing the application of the virial theorem but retaining the basic description of Coulomb forces between electrons and nuclei (see Sections 3.7 and 3.9).
- We introduce the direction of a molecular dipole moment, in addition to its magnitude (see Section 3.9.4).

Chapter 4: Introduction to Quantum Mechanics

This chapter presents a significant introduction to the concepts and vocabulary of quantum mechanics through very careful choice of language, illustrations with experimental data, interpretation with aid of simple models, and extensive use of graphical presentations.

For the eighth edition we improved the treatment of the particle-in-a-box with examples (Example 4.7 and the solution to Problem 37) that show students how to use the mathematical solutions to answer questions about the probability of the electron being found in different regions of space. We chose the chemical system of 1,3-butadiene for the example. The added material (mainly in Section 4.6) is appropriate both for students who have had integration in their single-variable calculus course and for students who would simply estimate the area under a curve graphically.

Chapter 5: Quantum Mechanics and Atomic Structure

This chapter provides a comprehensive introduction to the hydrogen atomic orbitals, the Hartree orbitals, the shell model of the atom as explained by the Hartree orbitals, and the relation of the shell model to experimental measurements such as photoelectron spectroscopy and the periodic properties of atoms. In this edition, recognizing that students' quantitative understanding of electron density in molecular systems begins with their introduction to atomic wave functions, we added graphical and analytical examples (Examples 5.2 and 5.3) to help students navigate the three-dimensional coordinate system in atoms (see Sections 5.1.2 and 5.1.3). Equation 5.7 reduces the integration over three dimensions to an integral that students may solve using single-variable calculus (or estimate graphically) to aid their understanding of the graphs of $r^2\psi^2$ and r^2R^2 presented in the chapter.

Chapter 6: Quantum Mechanics and Molecular Structure

This chapter provides a gentle ramp-up, starting from a qualitative overview of the quantum picture of the chemical bond and its relation to the potential energy curve for a molecule. The discussion proceeds through molecular orbital theory (MO), then valence bond theory (VB), then the combined use of MO and VB, and ends with a comparison of MO with VB. Changes for this edition include:

- The pedagogy in the introductory section titled Quantum Picture of the Chemical Bond (Section 6.1) has been shortened and improved.
- The graphics and equations presented in the treatment of LCAO-MO theory (see Sections 6.5 and 6.6) have been revised to represent constructive and destructive interference using the same wave functions that are presented in Chapter 5. The linear combinations of atomic orbitals used to construct hybrid atomic orbitals (Section 6.9) have been revised and the pedagogy is shortened and improved.
- The treatment of polyatomic molecules has been improved, including delocalized π bonding in NO_2, 1,3-butadiene, and benzene.
- The summary and comparison of LCAO-MO theory and the valence bond model was shortened and improved (Section 6.12).

Chapter 7: Bonding in Organic Molecules

The purpose of this chapter is to describe the bonding and nomenclature in alkanes, alkenes, alkynes, aromatics, and conjugated hydrocarbons and in the major functional groups. Our main goal is to illustrate the bonding theories from Chapter 6 with examples from organic chemistry that can be used in conjunction with Chapter 6. For the new edition we updated the nomenclature to give IUPAC names (retaining the common names where advisable) and moved the graphics related to delocalized π molecular orbitals in 1,3-butadiene and benzene to Chapter 6.

Chapter 8: Bonding in Transition Metal Compounds and Coordination Complexes

We present a comprehensive introduction to bonding in transition metal compounds and coordination complexes using MO and VB theory as developed in Chapter 6. Our goal is to demonstrate that MO theory is not limited to the first- and second-period diatomic molecules and that it provides the most satisfactory method for describing bonding in coordination complexes. The material covered in this chapter now provides a self-contained introduction to structure and bonding in inorganic chemistry that should provide sound preparation for an advanced inorganic chemistry course. For the new edition, we have revised and improved the figures related to chiral molecules (see Section 8.3.3) and crystal field theory (see Section 8.4.2).

Chapter 9: The Gaseous State

In this chapter we improved the treatment of the Maxwell-Boltzmann distribution of molecular speeds (see Section 9.5.2) to help students estimate the fraction of molecules with speed in a given range and to understand why the rate of a chemical reaction might increase with temperature. We also introduced the concept of relative velocity in the treatment of molecular collisions (see Section 9.7.2) so that students have a basis for deriving (in Chapter 18) the expression for the rate of an elementary bimolecular reaction in terms of the product of the concentrations of the two reacting species.

Chapter 10: Solids, Liquids, and Phase Transitions

For the eighth edition, we have introduced calculus to the treatment of compressibility and thermal expansion (see Sections 10.1.2 and 10.1.3).

Chapter 12: Thermodynamic Processes and Thermochemistry

The treatment in this and prior editions gives the instructor flexibility on how to cover traditional thermochemistry in conjunction with direct applications of the laws of thermodynamics. Stronger students may be encouraged to solve problems starting from the axioms and deriving the needed formulas from first principles.

Chapter 13: Spontaneous Processes and Thermodynamic Equilibria

The chapter presents a molecular statistical interpretation of entropy alongside the traditional macroscopic description involving the reversible transfer of heat. It carefully develops the conditions under which one may use the change in Gibbs free energy to assess the spontaneity of a reaction.

Chapter 14: Chemical Equilibrium

To provide flexibility for instructors, this chapter is written to allow thermodynamics to be taught either before or after equilibrium. Each topic is introduced first from the empirical point of view, and then followed immediately with the thermodynamic treatment of the same topic. Instructors who prefer to treat thermodynamics first can use the chapter as written, whereas those who prefer the empirical approach can skip appropriate sections, and then come back and pick up the thermo-based equilibrium sections after they cover basic thermodynamics. "Signposts" are provided in each section to guide these two groups of readers; the options are clearly marked. Specific examples of this flexible approach are:

- Section 14.2 provides a thorough discussion of procedures for writing the empirical law of mass action for gas-phase, solution, and heterogeneous reactions, with specific examples for each.
- Section 14.3 follows with the thermodynamic prescription for calculating equilibrium constants from tabulated Gibbs free energy values for gas-phase, solution, and heterogeneous reactions, with specific examples for each. For students who have covered Chapters 12 and 13, it gives a succinct derivation of $\Delta G = \Delta G° + RT \ln Q$ and the relationship between Q and K.
- Sections 14.4 and 14.5 present a variety of equilibrium calculations based on the empirical law of mass action.
- Section 14.6 discusses direction of change in terms of the empirical reaction quotient Q, with illustrations in gas-phase, solution, and heterogeneous reactions.
- Section 14.7 discusses direction of change from the point of view of thermodynamics, relating Q to the Gibbs free energy change and the equilibrium constant.

Chapter 15: Acid–Base Equilibria

The examples establishing the molecular-level interpretation of acid-base equilibria (see Section 15.8) have been carefully vetted and improved.

Chapter 17: Electrochemistry

This chapter provides a molecular level interpretation of electrochemical processes unique in an undergraduate textbook, as far as we are aware, to complement the standard thermodynamic treatment of the subject. We introduce the idea that a redox potential (a free energy) can be associated with an orbital energy level, which allows us to use energy-level diagrams to help students visualize electron transfer processes pictorially. This approach also allows us to introduce an electrostatic driving force for electrochemical processes and connect it to the thermodynamic driving force. We explicitly identify the conditions under which this approximation is valid (outer sphere electron transfer processes, negligible entropic contribution to the Gibbs free energy) so that our students can use this description with confidence.

For the eighth edition we streamlined and improved the introductory material on electric potential versus potential energy (see Section 17.1), moved the essential material on the molecular interpretation of electrochemical processes to later sections in the chapter to better integrate the pedagogy and applications, and deleted the long subsection on alternative reference electrodes in Section 17.2, moving the essential material to earlier in Section 17.2. We revised the applications in the molecular electrochemistry section (which was Section 17.5 in the seventh edition and is now Section 17.4 in the eighth edition) to shorten the presentation and improve clarity, including the discussion of using semiconductors in solar cells for generating H_2 as an alternative fuel.

Chapter 18: Chemical Kinetics

For the eighth edition we revised and improved the treatment of collision theory (see Section 18.6.1) to allow students to solve problems of importance in atmospheric chemistry, such as the reaction of Cl with methane. The concepts of relative velocity and impact parameter are developed with new graphics and simple examples (Examples 18.9, 18.10, and 18.11) and the connection with molecular collisions (as introduced in Chapter 9) are firmly established. We also expanded the material on molecular beam experiments (see Section 18.6.4). The seventh edition subsection on mechanisms of enzyme-catalyzed reactions is now presented as a *Connection to Biology.*

Chapter 19: Nuclear Chemistry

Chapter 19 has been extensively reorganized and streamlined, with the focus now on balancing nuclear reactions, mass–energy relationships, and predicting the spontaneity of nuclear reactions (with improved notation in the eighth edition), and the kinetics of radioactive decay. This material is followed by a detailed description of selected key applications. The first set of applications, covered under the heading, Radiation in Biology and Medicine (see Section 19.4), retains the *Connection to Medicine* presented in the seventh edition. The chapter ends with applications in nuclear fission, including nuclear power reactors (see Section 19.5.1), and fusion (in the sun and other stars and in proposed energy applications).

Chapter 20: Molecular Spectroscopy and Photochemistry

Chapter 20 has been reorganized in the eighth edition to provide an introduction to molecular energy levels first, including a discussion of how one might probe them using traditional absorption spectroscopy or more modern laser spectroscopy (specifically laser-induced fluorescence). Other changes include:

- A streamlined discussion of direct absorption methods and Raman spectroscopy (see Section 20.2), which included removing the detailed description of Fourier transform methods that was in the middle of the vibrational spectroscopy section in the last edition.
- A reorganized order of topics, so that rotational spectroscopy in diatomic molecules is now treated first, followed by vibrations in diatomics and then infrared spectroscopy of diatomic molecules involving a change in both the vibrational and the rotational quantum number (see consecutive subsections in Section 20.3.1). Rotations and vibrations in polyatomic molecules follow (see Section 20.3.2), ending with material from the seventh edition on characteristic vibrational frequencies of organic functional groups.
- A simplified treatment of nuclear magnetic resonance (NMR) spectroscopy, making it more physically intuitive (see Section 20.4), while retaining material on Fourier-transform NMR from the seventh edition.
- The revision of several figures in the chapter with a focus on scientific accuracy.
- A shortened, streamlined section titled Introduction to Atmospheric Chemistry (see Section 20.6) that retains the discussion of the chemistry of ozone but removes, in response to reviewer feedback, much of the last edition's coverage of climate change. (The full treatment that was found in the seventh edition will be made available on the instructors' website.)
- A shortened, streamlined *A Deeper Look* section (Section 20.8), that retains the coverage on lasers but omits the coverage of the Einstein radiation relations.

Chapter 21: Structure and Bonding in Solids

For the eighth edition we improved the discussion regarding symmetries of the cube primitive (see Section 21.1) and improved the accuracy of several figures.

Chapter 22: Inorganic Materials

For the eighth edition we improved graphics in the section on silicates (see Section 22.1.1).

Appendices

For the eighth edition we added to the appendix on mathematical background for general chemistry a subsection that introduces vectors (see Section C.7). We also improved the notation and pedagogy in portions of the appendix on physics background for general chemistry (see Section B.2).

Teaching Options

The text is structured and written to give instructors significant flexibility in choosing the order in which topics are presented. We suggest several such possibilities here. In all cases we recommend starting with Chapter 1 to provide a contemporary introduction to the structure and properties of the atom, as well as to help our students understand how we came to acquire this understanding. Our own students report that this early introduction to the scientific method, following these historical examples, has been helpful to them in subsequent courses. We then recommend working through the material in Chapter 2 to establish a secure foundation in "chemical accounting methods" that is necessary for studying all the remaining chapters. Particularly well-prepared students can skip Chapter 2, especially if diagnostics are available to ascertain satisfactory background.

Classical Bonding before Introduction to Quantum Theory

Chapters 1–6; selections from Chapters 7 and 8; Chapters 9–23

This is the sequence we have found most effective overall all for two reasons: (1) introducing the classical description before tackling quantum mechanics helps our students see the need to understand the latter approach, and (2) it enables our students to bring substantially greater maturity to their first exposure to quantum theory. This leads to deeper and quicker mastery of quantum theory and its applications to atomic and molecular structure. Instructors who wish to introduce molecular spectroscopy earlier can easily cover Sections 20.1 through 20.4 immediately after Chapter 6.

Introduction to Quantum Theory before Bonding

Chapters 1, 2, 4, 5, 3, 6; selections from Chapters 7 and 8; Chapters 9–23

These sequences are appropriate for instructors who prefer to establish a background in quantum theory before discussing ionic and covalent bonding, Lewis diagrams, and VSEPR theory. Instructors who prefer to cover these classical bonding topics after quantum mechanics but before MO and VB theory would cover Chapter 3 before Chapter 6. Those who want to present the full quantum story first and then present the classical description as the limiting case would cover Chapter 3 after Chapter 6. We recommend that both of these sequences cover Section 3.3 (force and potential energy in atoms) before Chapter 4 to give students a good physical feeling for Rutherford's planetary model of the atom in preparation for the quantum theory. Instructors who wish to introduce molecular spectroscopy earlier can easily cover Sections 20.1 through 20.4 immediately after Chapter 6.

Traditional "Macro-to-Micro" Approach

Chapters 1, 2, 9–19, 3–8, 20–23

This sequence covers fully the macroscopic descriptions of chemical phenomena and then begins to interpret them in terms of molecular structure. Instructors could choose either of the two bonding approaches suggested earlier for the specific order of Chapters 3 through 6 late in this course. This sequence represents a rather pure form of the "macro-to-micro" approach that was followed in the first three editions. Alternatively, they could cover Chapter 3 between Chapter 2 and Chapter 9, as was done in the fourth and fifth editions. This approach has the advantage of building a substantial foundation in structure—and a complete discussion of chemical nomenclature—as the basis for the macroscopic descriptions, while leaving the quantum theory of bonding to come later in the course.

Thermodynamics before Chemical Equilibrium

Chapters 12–17

This is the sequence we have found to be the most effective. If students first have a good understanding for the physical basis of equilibrium, then the facts and trends of chemical equilibrium quickly begin to form patterns around molecular structure. The equilibrium state is determined by the changes in entropy and bond energies associated with each chemical reaction.

Empirical Chemical Equilibrium before Thermodynamics

Chapter 14 (omit Sections 14.3, 14.7); Chapters 15, 16, 12, 13; Sections 14.3, 14.7; Chapter 17

Perhaps to provide background for quantitative laboratory work, others may wish to present chemical equilibrium earlier in the course in a more empirical fashion, before the presentation of thermodynamics. Chapter 14 is clearly marked with "signposts" to facilitate this sequence.

Modified "Macro-to-Micro" Approach

Chapters 1–3, 9, 10, 12–14, 11, 15, 4–7, 20, 16–19, 8, and 21; selections from Chapters 22 and 23

The syllabus currently in use at The University of Chicago follows the approach of thermodynamics before chemical equilibrium, and covers most of quantum mechanics in the winter months (so students co-enrolled in calculus are up to speed by then). Molecular spectroscopy (Chapter 20) and applications to organic chemistry (selected from Chapter 7) immediately follow, or are interleaved with, the coverage of quantum mechanics and molecular structure during the winter. Chapters 16 to 19 and Chapter 21 follow in the spring with Chapter 8 immediately following Chapter 19 in our regular course and Chapter 7 in our honors course. This allows students to switch between our honors and regular course in the winter if they desire. For an honors course populated primarily by students who have scored a 4 or 5 in AP chemistry, we retain the student's interest by covering only selected sections of Chapters 1, 3, 9 (focusing on the material in Section 9.5 and later), 10, and 11, and we skip Chapter 2. We then focus on a first-principles development of thermodynamics and chemical equilibrium in Chapters 12 to 14, immediately followed by chemical applications in Chapter 15 and selected portions of Chapter 16. This is followed by the comprehensive treatment of quantum mechanics, Chapters 4 to 6, the last with added material on molecular orbitals for a variety of polyatomics covered in Chapter 7 along with some radical species such as the allyl radical. Chapters 8 and 20 are typically covered then, followed by Chapters 16 through 21 in the spring and special topics selected by the instructor.

General Aspects of Flexibility

Certain topics may be omitted without loss of continuity. For example, a principles-oriented course might cover the first 20 chapters thoroughly and then select one or two specific topics in the last chapters for close attention. A course with a more descriptive orientation might omit the sections entitled *A Deeper Look . . .*, which are more advanced conceptually and mathematically than the sections in the main part of the book, and cover the last three chapters more systematically. Additional suggestions are given in the *Instructor's Manual* that accompanies the book.

Features

Mathematical Level

This book presupposes a solid high school background in algebra and coordinate geometry. The concepts of slope and area are introduced in the physical and chemical contexts in which they arise, and differential and integral notation is used in parallel with these concepts. Thus, students with a single-variable calculus background are engaged fully, while students without that background will not be at a disadvantage. Simple vectors are introduced to aid in discussing force and velocity, with an elementary section in Appendix C covering vectors in Cartesian coordinates. The book is fully self-contained in its use of mathematical methods. Methods are introduced at "point of use," and Appendix C provides a more comprehensive introduction (or review) of the material as needed.

Key equations in the text are highlighted in color and numbered on the right side of the text column. Students should practice using them for chemical calculations. The steps to derive each of these equations are presented in the text, so students do not conclude they must simply memorize these equations. Rather they are taught how to develop these relationships from first principles.

Updated Design and New Illustrations and Photographs

This eighth edition features a modern design, whose elements have been carefully arranged for maximum clarity and whose aesthetics should engage today's visually oriented students. We have selected photographs and illustrations to amplify and illuminate concepts in the narrative text. All illustrations of atomic and molecular orbitals, charge density, and electrostatic potential energy maps were generated expressly for this textbook. Most orbitals and charge densities were calculated by Mr. Hatem Helal (Cambridge University, UK) in the Materials Simulation Center at the California Institute of Technology, directed by Professor William A. Goddard III. Dr. Kelly Gaither (Director, Visualization and Data Analysis group, Texas Advanced Computing Center) plotted the images using state-of-the-art software at the Scientific Visualization Laboratory at The University of Texas at Austin. The colors, lighting effects, and viewing angles were chosen to display three-dimensional objects with maximum clarity and to provide chemical insight. In many cases quantitative contour plots accompany the three-dimensional isosurfaces representing orbitals to help our students understand how the appearances of isosurfaces depend on choices made by scientists and that these isosurfaces are not unique.

Molecular Art

The sixth edition introduced an art program in which molecular shapes are rendered with quantitative accuracy and in modern graphical style. Most of the illustrations of atomic and molecular orbitals, charge density, and electrostatic potential energy maps were generated from accurate quantum chemistry calculations carried out at the California Institute of Technology. Orbitals were plotted using state-of-the-art visualization software at the Texas Advanced Computing Center, University of Texas at Austin. The colors, lighting effects, and viewing angles were chosen to display three-dimensional objects with maximum clarity and to provide chemical insight.

Worked Examples

This textbook includes worked examples that demonstrate the methods of reasoning applied in solving chemical problems. The examples are inserted immediately after the presentation of the corresponding principles, and cross-references are made to related problems appearing at the end of the chapter.

A Deeper Look

Sections entitled *A Deeper Look . . .* provide students with a discussion of the physical origins of chemical behavior. The material that they present is sometimes more advanced mathematically than that in the main parts of the book. The material provided in these sections allows instructors to more easily tailor the breadth and depth of their courses to meet their specific objectives.

Key Terms

Key terms appear in boldface where they are first introduced. Definitions for all key terms are also included in the Index/Glossary for ready reference.

"Connections to..."

A number of essays provide "Connections to..." other branches of science, engineering, and medicine. Coupled with the interdisciplinary Cumulative Exercises that have long been a hallmark of *Principles of Modern Chemistry,* these "Connections" give a substantial sampling of applications of the principles of chemistry.

Chapter Summary

Immediately at the end of each chapter is a summary that ties together the main themes of the chapter in a retrospective narrative. This complements the introductory passage at the beginning of the chapter in a manner that conveys the importance of the chapter. The summary is the first in a set of four end-of-chapter features that constitute a comprehensive set of tools for organizing, studying, and evaluating mastery of the chapter.

Cumulative Exercise

At the end of each of Chapters 2 through 21 is a cumulative exercise, a unique feature of *Principles* since its inception that focuses on a problem of chemical interest and draws on material from the entire chapter for its solution. Working through a chapter's cumulative exercise provides a useful review of material in the chapter, helps our students put principles into practice, and prepares them to solve the problems that follow.

Concepts and Skills

Each chapter concludes with a list of concepts and skills (task-oriented) for each section in the chapter for review by our students. Included in this list are cross-references to the section in which the topic was covered, a concise review of material essential to that topic, the key equations for each topic, and cross-references to end-of-chapter problems that help test mastery of the particular skill involved. Our own students report that this feature has been very helpful to them for self-testing and review of material.

Problems

Problems are grouped into three categories. Answers to odd-numbered "paired problems" (which have red problem numbers) are provided in Appendix G; they enable students to check the answer to the first problem in a pair before tackling the second problem. *Additional Problems,* which are unpaired, illustrate further applications of the principles developed in the chapter. *Cumulative Problems* integrate material from the chapter with topics presented earlier in the book. We integrate more challenging problems throughout the problems sets and identify them with asterisks.

Appendices

Appendices A, B, and C are important pedagogically. Appendix A discusses experimental error and scientific notation. Appendix B introduces the SI system of units used throughout the book and describes the methods used for converting units. Appendix B also provides a brief review of some fundamental principles in physics, which may be particularly helpful to students in understanding topics covered in Chapters 3, 4, 5, 6, 9, 10, 12, 13, 17, 18, 19, and 20. Appendix C provides a review of mathematics for general chemistry. Appendices D, E, and F are compilations of thermodynamic, electrochemical, and physical data, respectively.

Index/Glossary

The Index/Glossary at the back of the book provides brief definitions of all boldfaced key terms, as well as cross-references to the pages on which the terms appear.

Supporting Materials

Student Resources

Please visit http://www.cengage.com/chemistry/oxtoby/POMC8e for information about student resources for this text.

Instructor Resources

Please visit http://www.cengage.com/chemistry/oxtoby/POMC8e for information about instructor resources for this text.

Acknowledgments

In preparing the eighth edition, we have benefited greatly from the comments of students who used the first seven editions over the years. We would also like to acknowledge the many helpful suggestions of colleagues at Pomona College, The University of Chicago, the University of California–Los Angeles, and other colleges and universities who have taught from this book. We are particularly grateful to Professors Robin Garrell, Ken Houk, Herb Kaesz, and Thomas Mason of UCLA, Professor Greg Engel of The University of Chicago, Professor Michael Topp of the University of Pennsylvania, Professor Gina Frey of Washington University, and Professors Eric Anslyn, Al Bard, Ray Davis, Brad Holliday, Simon Humphrey, Brent Iverson, Richard Jones, Peter Rossky, Jason Shear, John Stanton, Keith Stevenson, David Vanden Bout, Grant Willson, and Robert Wyatt of The University of Texas–Austin, Professor Samir Anz of California Polytechnic State University–Pomona, and Professor Andrew J. Pounds of Mercer University for their helpful comments and advice.

We extend special thanks to the following professors who offered comments on the seventh edition or reviewed manuscript for the eighth edition:

Joseph BelBruno, Dartmouth College
Kenneth Brown, Georgia Institute of Technology
Lynne M. Butler, Haverford College
Justin Ferman, University of Massachusetts–Amherst
Regina F. Frey, Washington University of St. Louis
Henry C. Griffin, University of Michigan
Philippe Guyot-Sionnest, The University of Chicago
Paul J. Karol, Carnegie Mellon University
Walter Loveland, Oregon State University
Doug Loy, University of Arizona
Tom Mason, University of California–Los Angeles
Donald R. Neu, St. Cloud State University
Andrew J. Pounds, Mercer University
Robert C. Scarrow, Haverford College
George Schatz, Northwestern University
Michael L. Stein, The University of Chicago
Bozhi Tian, The University of Chicago
Michael R. Topp, University of Pennsylvania
Lauren Webb, University of Texas at Austin
Steven W. Yates, University of Kentucky

We are grateful to Dr. Melita Morton Balch for her critical corrections to the manuscript, and to Dr. Justin Fermann for his very careful attention to detail as accuracy reviewer of the eighth edition.

We are much indebted to our longtime friend Professor Eric J. Heller of Harvard University for the beautiful and striking images that grace the covers of the sixth, seventh, and eighth editions of our book. Professor Heller's work demonstrates that images of great beauty can arise from scientific research and that artistic renderings effectively convey the meaning of scientific results. We are certain this image will entice readers to peek between the covers of our book, and we hope they find scientific beauty on the inside as well as on the cover!

We are particularly grateful to friends and colleagues who provided original scientific illustrations for the book. They are Professor Wilson Ho (University of California–Irvine), Dr. Gilberto Medeiros-Ribeiro and Dr. R. Stanley Williams (Hewlett-Packard Research Laboratories), Professor Leonard Fine (Columbia University), Professor Andrew J. Pounds (Mercer University) and Dr. Mark Iken (Scientific Visualization Laboratory, Georgia Institute of Technology), Dr. Stuart Watson and Professor Emily Carter (Princeton University), Professor Nathan Lewis (California Institute of Technology), Dr. Don Eigler (IBM Almaden Research Center), Dr. Gerard Parkinsen and Mr. William Gerace (OMICRON Vakuumphysik), Dr. Richard P. Muller and Professor W.A. Goddard III (California Institute of Technology), Professor Moungi Bawendi and Ms. Felice Frankel (Massachusetts Institute of Technology), Professor Graham Fleming (University of California–Berkeley), Professors Donald Levy, Philippe Guyot-Sionnest, and David Mazziotti (The University of Chicago), Professor W.E. Moerner (Stanford University), Dr. Jane Strouse (University of California–Los Angeles), Professor James Speck and Professor Stephen Den Baars (University of California–Santa Barbara), Professor John Baldeschwieler (California Institute of Technology), and Dr. Thomas Koop (Bielefeld University).

We are especially grateful to Mr. Hatem H. Helal (California Institute of Technology and Cambridge University, UK), who carried out the quantum chemistry calculations for most of the orbital illustrations in Chapters 4, 5, 6, and 8, and to Dr. Kelly P. Gaither (Texas Advanced Computing Center, The University of Texas at Austin), who generated these illustrations from the results of the calculations. Our longtime friend and colleague Professor William A. Goddard III (California Institute of Technology) very generously made his computational facilities available for these calculations and provided much good advice as we selected and prepared these illustrations. Sarah Chandler (The University of Texas at Austin) was very helpful in generating a number of graphs and two-dimensional surfaces.

We are also indebted to Professor Charles M. Knobler of the University of California–Los Angeles; Professor Jurg Waser, formerly of the California Institute of Technology; and Mrs. Jean T. Trueblood (widow of the late Professor Kenneth N. Trueblood of the University of California–Los Angeles) for permission to incorporate selected problems from their distinguished textbook *ChemOne,* Second Edition, McGraw-Hill, New York (1980).

On a personal note, it gives us genuine pleasure to dedicate this eighth edition of our textbook to our own Ph.D. research advisers Professors Bill Gelbart (Oxtoby), Karl Freed (Gillis), and Yuan T. Lee (Butler). They showed us the excitement of doing scientific research and the joy of transmitting scientific knowledge to the next generation. Their legacy inspires our work with our own students in the laboratory, in the classroom, and in the pages of this textbook.

The staff members at Cengage Learning have been most helpful in preparing this eighth edition. In particular, we gratefully acknowledge the superb work of content developer Ed Dodd, who made sure all the changes to the text and graphics for the eighth edition were done perfectly. We also acknowledge the role of our associate product manager, chemistry, Krista M. Mastroianni. Content developer Elizabeth Woods coordinated production of the ancillary materials. Senior content developer Lisa Weber handled the media products. Senior content project manager, Teresa L. Trego, at Cengage, and Cassie Carey, senior production editor at Graphic World Publishing Services, kept the schedule moving smoothly. We acknowledge the contributions of art director Maria Epes, and John Sarantakis, project manager, intellectual property acquisition. We are grateful to marketing director Nicole Hamm for helping us obtain valuable comments from users and reviewers. We gratefully acknowledge the continuing support of product director Mary Finch.

David W. Oxtoby
Pomona College

H.P. Gillis
University of California–Los Angeles

Laurie J. Butler
The University of Chicago

November 2014

ABOUT THE AUTHORS

David W. Oxtoby

David W. Oxtoby became the ninth president of Pomona College on July 1, 2003. An internationally noted chemist, he previously served as dean of physical sciences at The University of Chicago. At Pomona, he holds a coterminous appointment as president and professor of chemistry. Before coming to Pomona, he was associated with the University of Chicago for nearly three decades, with brief interludes to serve as a visiting professor at such places as the University of Paris; the University of Bristol in Great Britain; and the University of Sydney in Australia. Oxtoby is a fellow of the American Physical Society and a member of the American Chemical Society and the American Association for the Advancement of Science. After earning his bachelor's degree, summa cum laude, from Harvard University, he went on to earn his Ph.D. at the University of California, Berkeley. As a research chemist, he is author or co-author of more than 165 scientific articles on such subjects as light scattering, chemical reaction dynamics, and phase transitions. In addition to co-authoring *Principles of Modern Chemistry* and *Chemistry: Science of Change,* he has received fellowships from the Guggenheim, von Humboldt, Dreyfus, Sloan, Danforth, and National Science foundations.

H.P. Gillis

H.P. Gillis conducts experimental research in the physical chemistry of electronic materials, emphasizing phenomena at solid surfaces and interfaces. Dr. Gillis received his B.S. (Chemistry and Physics) at Louisiana State University and his Ph.D. (Chemical Physics) at The University of Chicago. After postdoctoral research at the University of California–Los Angeles and 10 years with the technical staff at Hughes Research Laboratories in Malibu, California, Dr. Gillis joined the faculty of Georgia Institute of Technology. Dr. Gillis moved to University of California–Los Angeles, where he currently serves as adjunct professor of materials science and engineering. He has taught courses in general chemistry, physical chemistry, quantum mechanics, surface science, and materials science at UCLA and at Georgia Institute of Technology.

Laurie J. Butler

Laurie J. Butler uses molecular beam and laser spectroscopic techniques, along with associated computational modeling, to study the chemical reaction dynamics of radical intermediates and of electronically excited molecules. She received her B.S. (Chemistry) at the Massachusetts Institute of Technology and her Ph.D. (Physical Chemistry) at the University of California, Berkeley (the latter under the direction of Nobel Laureate Y.T. Lee). After a short postdoctoral position at University of Wisconsin–Madison, she accepted a faculty position at The University of Chicago. Fellow of the AAAS, the American Chemical Society, and the American Physical Society, she has co-authored over a hundred scientific papers. She teaches introductory and physical chemistry to the brilliant undergraduates at Chicago, for which she received the University's Quantrell teaching award, as well as graduate-level courses in quantum mechanics and in chemical reaction dynamics.

Introduction to the Study of Modern Chemistry

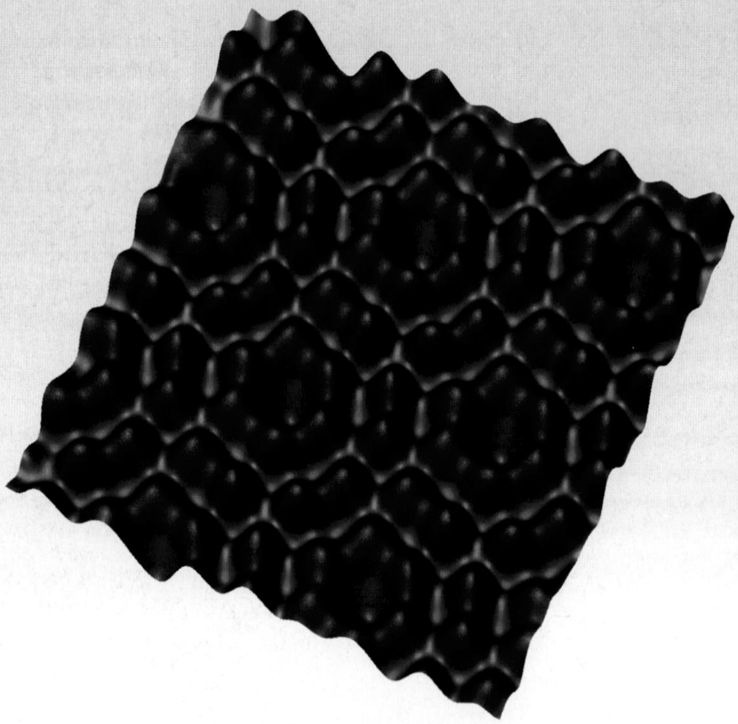

The surface of a silicon crystal imaged using a scanning tunneling microscope. Individual silicon atoms appear as purple protrusions above the background. The surface was cleaned in an ultrahigh vacuum to remove all impurity atoms and the image was taken at very low temperatures (−220°C) to obtain the high resolution shown here. There are two kinds of surface silicon atoms shown in this image: "corner" silicon atoms that form hexagonal rings around a hole in the surface layer and "center" silicon atoms that appear as pairs arranged around the hexagonal rings.

UNIT I

Modern chemistry explores the world of atoms and molecules, seeking to explain not only their bonding, structures, and properties, but also how these very structures are transformed in chemical reactions. The search for atoms and molecules began with the speculations of ancient philosophers and—stimulated by the classic experiments of the 18th and 19th centuries—led to John Dalton's famous atomic hypothesis in 1808. The quest continues. The invention of the scanning tunneling microscope (STM) in the 1980s, along with other microscopic and spectroscopic techniques, has enabled contemporary scientists to detect and manipulate individual atoms and molecules.

Unit Chapters

CHAPTER 1

The Atom in Modern Chemistry

CHAPTER 2

Chemical Formulas, Equations, and Reaction Yields

Unit Goals

- To describe the key experiments and the underlying physical models that justify the central role of the atom in modern chemistry
  - Chemical evidence for the existence and properties of atoms and molecules
  - Physical evidence for the existence and properties of atoms and molecules
  - The modern, planetary model of the atom
- To introduce the established quantitative procedures that describe chemical reactions as rearrangements of atoms, forming products from reactants
  - The mole concept that relates weighing and counting of molecules and atoms
  - Balanced chemical equations that relate moles of reactants to moles of products

CHAPTER 1

The Atom in Modern Chemistry

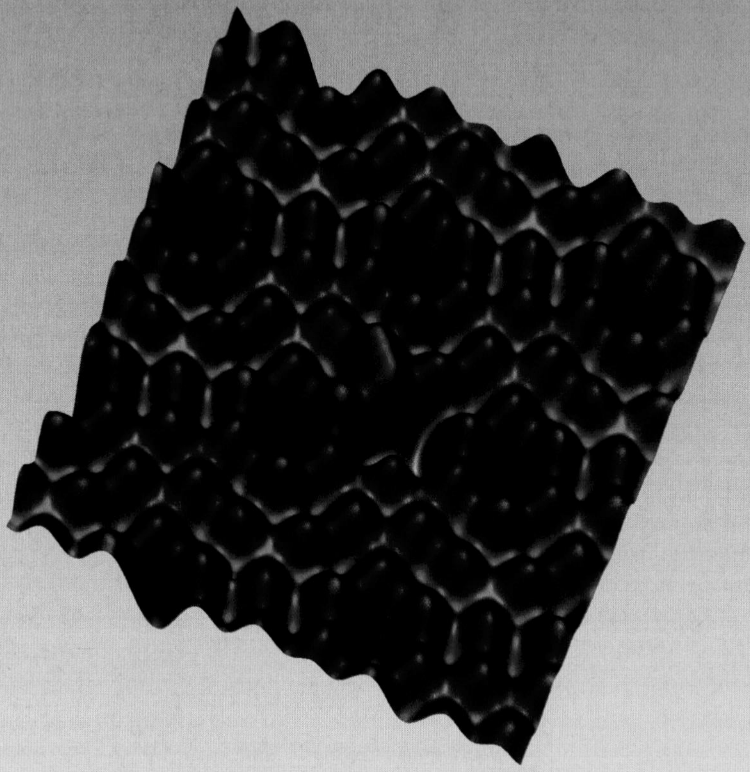

Photo courtesy of Wilson Ho, University of California, Irvine. Reprinted by permission of Physical Review Letters. 79, 4397–4400. (1997).

Reversible single atom transfer using the scanning tunneling microscope. This image was taken under the same conditions as the one shown opposite page 1. One of the "center" silicon atoms (imaged in red) has been transferred halfway to another center atom site by the scanning tunneling microscope tip. The atom is stable in this position at low temperatures but returns to its home site as the temperature is raised above −100°C.

1.1 The Nature of Modern Chemistry

Chemists study the properties of substances and the reactions that transform substances into other substances. Chemists are particularly interested in understanding how and why specific chemical reactions occur, in order to tailor the properties of existing substances to meet particular needs—and to create entirely new substances designed to have specific properties. Chemistry has improved agricultural production, helped prevent and cure many diseases, increased the efficiency of energy production, and reduced environmental pollution, to cite just a few advances. A particularly exciting challenge for modern chemical research is to understand the dynamics of these chemical transformations, because they govern phenomena as diverse as the evolution of small carbon-containing molecules in interstellar space, changes in terrestrial atmospheric and climatic patterns caused by pollutants, and the unfolding of life processes in living organisms.

Chemistry influences almost every area of science and technology; advances in chemistry inform disciplines as different as solid-state physics and molecular biology, and the synthetic methods and analytical techniques developed by chemists support research and manufacturing in important areas like medicine and microelectronics. Perhaps no other science covers as broad a range of topics as does chemistry. Within a single modern chemistry department, you're likely to find chemists creating new materials and developing strategies for sustainable energy, devising synthetic routes that conserve all of the atoms in starting materials, detecting and identifying single molecules, designing new molecules for therapeutic purposes and translating those developments into clinical trials, and developing highly selective integrated sensors for a variety of applications in science and technology. Despite the diversity of these areas of scientific inquiry, they are all unified by a single set of fundamental scientific principles, which we will introduce in this textbook.

Painting The Alchemist by Hendrick Heerschop, 1671. Courtesy of Dr. Alfred Bader

FIGURE 1.1 Alchemists searched in vain for procedures that would turn base metals into gold. Their apparatus foreshadowed equipment in modern chemical laboratories.

Chemistry is a relatively young science, and its foundations weren't established until the last quarter of the 18th century. Before that, most chemists were known as *alchemists*—early entrepreneurs who sought to transform the properties of materials for economic gain (Fig. 1.1). For centuries their obsession was to transform "base" metals, such as lead, into gold. They boldly assumed that the properties of one material could somehow be extracted and transferred to another. If the essential properties—such as yellow color, softness, and ductility—could be assembled from various inexpensive sources, then gold could be created at great profit.

The alchemists persisted in their efforts for more than a thousand years. Although they collected many useful, empirical results that have since been incorporated into modern chemistry, they never transformed base metals into gold. Many scientists had begun to challenge the basic assumptions of the alchemists by the middle of the 17th century. These doubts culminated with the publication of *The Sceptical Chymist* by Robert Boyle in England in the 1660s, one of the pivotal events that marked the beginning of modern chemistry. Another century was required to establish the conceptual foundations of modern chemistry, a field that flourished throughout the 19th and 20th centuries and remains vibrant today.

The error of the alchemists is obvious to modern observers: they did not follow the scientific method. A new idea is accepted only temporarily in the scientific method, in the form of a **hypothesis**. It is then subjected to rigorous testing, in carefully controlled experiments. A hypothesis is elevated to a **scientific law** only after it has survived many such tests. Concepts or ideas that have earned the status of scientific laws by direct and repeated testing then can be applied with confidence in new situations. A scientific law must be predictive, in addition to being explanatory; failure to accurately predict the results of a new experiment is sufficient to invalidate a scientific law. Had a proper set of tests been made in separate, independent experiments, the alchemists would have recognized that the properties of a material are, in fact, intrinsic, inherent characteristics of that material and cannot be extracted from it.

The history of the alchemists shows the origin of a certain duality in the nature of modern chemistry, which persists to the present. Because chemistry contributes to the foundations of numerous professions and industries, we see the urge to apply established chemical knowledge for profit. But we also see the urge to create new chemical knowledge, driven by both intellectual curiosity and by the desire to have reliable information for applications. Many scientists and engineers from different disciplines contribute to both basic and applied chemical research and development. Irrespective of the specific context, conducting chemical research requires scrupulous adherence to the scientific method, in which the new knowledge generated is subjected to rigorous scrutiny before it earns the confidence of the scientific community.

Most students who study chemistry will apply what they have learned during the course of their professional careers rather than conduct research in chemistry. Still, a useful strategy for learning to think like an experienced chemist is to assume that you are personally responsible for establishing the scientific foundations of chemistry for the very first time. Upon encountering a new topic, try this: imagine that you are the first person ever to see the laboratory results on which it is based. Imagine that you must construct the new concepts and explanations to interpret these results, and that you will present and defend your conclusions before the scientific community. Be suspicious. Cross-check everything. Demand independent confirmations. Always remain, with Boyle, the "skeptical chemist." Follow the scientific method in your acquisition of knowledge, even from textbooks. In this way, you will make the science of chemistry your own, and you will experience the intellectual joys of discovery and interpretation. Most important, you will recognize that chemistry is hardly a closed set of facts and formulas. Quite the contrary, it is a living, growing method for investigating all aspects of human experience that depend on changes in the composition of substances. Finally, learning to examine critically what we call "the nature of the evidence" will help you make better decisions as consumers and citizens in a world where science and technology continue to become increasingly important parts of modern life.

1.1.1 Conservation of Matter and Energy

The science of chemistry rests on two well-established principles: the conservation of matter and the conservation of energy. The total amount of matter involved in chemical reactions is conserved—that is, it remains constant during the course of every chemical reaction. Matter is neither created nor destroyed in chemical reactions; the components of the reactants are simply rearranged to form products.

These rearrangements are inevitably accompanied by changes in energy, which brings us to the second principle. The total amount of chemical energy stored in the reactants is almost always different than that stored in the products; this difference manifests itself in the form of thermal, electrical, or mechanical energy required for, or produced by, chemical reactions. But energy is neither created nor destroyed during chemical reactions; it has always been found to be conserved.

These two core principles must be modified slightly for nuclear reactions, which occur at energies so high that matter and energy can be converted into one another through Einstein's relation, $E = mc^2$. The *sum* of mass and energy is conserved in nuclear reactions.

1.1.2 Macroscopic Methods and Nanoscopic Models

Chemical reasoning, as used both in applications and in basic research, resembles a detective story in which tangible clues lead to a mental picture of events never directly witnessed by the detective. Chemical experiments are conducted in laboratories equipped with beakers, flasks, analytical balances, pipettes, optical and infrared spectrometers, lasers, vacuum pumps, pressure gauges, mass spectrometers, centrifuges, and other apparatus. Each of these devices exists on the *macroscopic* scale—that is, it is perceptible by ordinary human senses. Macroscopic objects range in size from about 1 meter (m) down to about 1 millimeter (mm), which is 1×10^{-3} m. But the actual chemical transformation events occur in the *nanoscopic* world of atoms and molecules—objects far too small to be detected by the naked eye, even with the aid of a first-class microscope. One nanometer (nm) is 1×10^{-9} m. So our modern laboratory instruments are the bridge between these worlds, giving us the means not only to influence the actions of

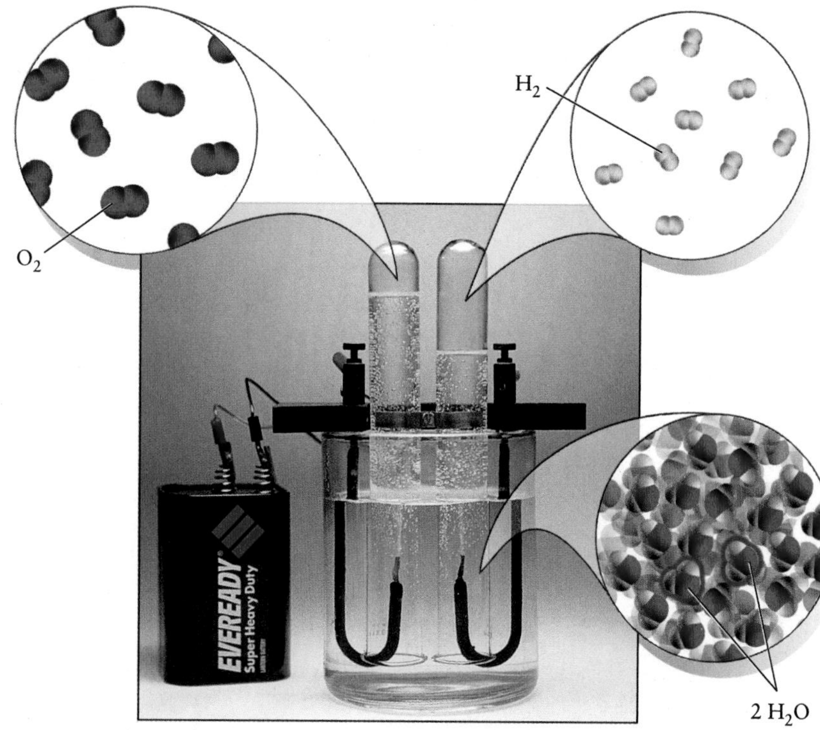

FIGURE 1.2 Hydrogen and oxygen gas are produced in the ratio 2:1 when an electrical current is passed through water that contains dissolved sulfuric acid. The insert illustrates how chemists view this macroscopic chemical reaction as arising from the rearrangement of atoms on the nanoscale.

the atoms and molecules but also to measure their response. Figure 1.2 shows views of both worlds simultaneously. In illustrating the chemical decomposition of water into gaseous hydrogen and oxygen by electrolysis, the figure shows the relationship between events on the macroscale and on the nanoscale. Chemists *think* in the highly visual nanoscopic world of atoms and molecules, but they *work* in the tangible world of macroscopic laboratory apparatus. These two approaches to the chemical sciences cannot be divorced, and we emphasize their interplay throughout this textbook. Students of chemistry must master not only the fascinating concepts of chemistry, which describe the nanoscopic world of atoms and molecules, but also the macroscopic procedures of chemistry on which those concepts are founded.

1.2 Elements: The Building Blocks of Matter

Chemists study how sets of pure substances transform into other sets of pure substances in chemical reactions. These investigations apply two modes of reasoning and physical procedures—analysis (taking things apart) and synthesis (putting things together)—that go back to early Greek philosophers, who sought to identify the constituents of all matter as four elements: air, earth, fire, and water. Contemporary chemists classify matter using a very different set of fundamental building blocks, but the analysis and synthesis steps are basically unchanged.

1.2.1 Substances and Mixtures

Investigating chemical reactions can be greatly complicated and often obscured by the presence of extraneous materials. The first step is to learn how to analyze and classify materials to ensure that you are working with *pure* substances before initiating any reactions (Fig. 1.3). Suppose you take a sample of a

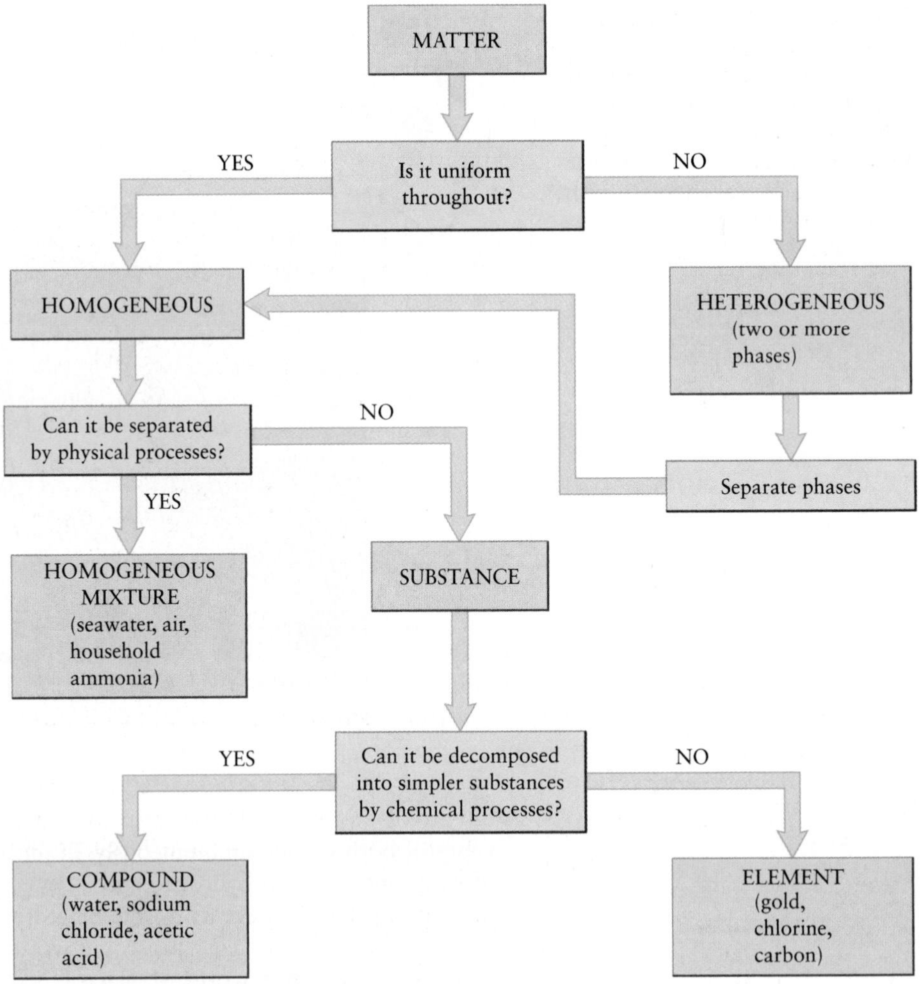

FIGURE 1.3 Process flowchart for analyzing matter.

material—a gas, liquid, or solid—and examine its various properties or distinguishing characteristics, such as its color, odor, or density. How uniform are those properties throughout the sample? Different regions of a piece of wood, for example, have different properties, such as variations in color and texture. Wood, then, is said to be **heterogeneous.** Other materials, such as air or mixtures of salt and water, are classified as **homogeneous** materials because their properties are the same throughout a given sample. But these mixtures, although homogeneous, cannot be called pure substances. Thus, call a substance a **mixture** when it is possible to separate it into components by ordinary physical means such as melting, freezing, boiling, or dissolving in solvents (Fig. 1.4). These operations provide

(a) (b) (c)

FIGURE 1.4 (a) A solid mixture of blue $Cu(NO_3)_2 \cdot 6H_2O$ and yellow CdS is added to water. (b) Although the $Cu(NO_3)_2 \cdot 6H_2O$ dissolves readily and passes through the filter, the CdS remains largely undissolved and is retained by the filter. (c) Evaporation of the solution leaves nearly pure crystals of $Cu(NO_3)_2 \cdot 6H_2O$.

ways of separating materials from one another by their properties: freezing points, boiling points, and solubilities. Air, for example, is a mixture of several components—oxygen, nitrogen, argon, and various other gases. If air is liquefied and then warmed slowly, the gases with the lowest boiling points evaporate first, leaving behind in the liquid those with higher boiling points. Such a separation would not be perfect, but the processes of liquefaction and evaporation could be repeated to improve the purity of the component gases to any required degree.

If all these physical procedures (and many others) fail to separate matter into portions that have different properties, the material is said to be a **pure substance.** What about the common material sodium chloride, which we call table salt? Is it a substance? The answer is yes if we use the term *sodium chloride* (NaCl), but no if we use the term *table salt*. Table salt is a mixture of mostly sodium chloride but also small amounts of sodium iodide (needed by the thyroid gland) and magnesium carbonate (needed to prevent the salt from caking). Even if these two components were not added, table salt contains other impurities that had not been removed in its preparation, so to that extent, table salt is a mixture. In contrast, when we refer to sodium chloride, we imply that it is a pure substance that contains only NaCl. The pure substance sodium chloride is a compound, not an element.

Nothing is absolutely pure in practice, so the word **substance** is an idealization. Among the purest materials ever prepared are silicon (Fig. 1.5) and germanium. These elements are used in electronic devices and solar cells, and their electronic properties require either high purity or else precisely controlled concentrations of deliberately added impurities. Meticulous chemical and physical methods have enabled scientists to prepare germanium and silicon with concentrations of impurities that are less than one part per billion. Any higher concentrations of certain impurities would alter the electrical properties of these materials.

1.2.2 Elements

Literally millions of substances have so far been either discovered or synthesized and formally identified. Are these the fundamental building blocks of matter? Happily not, for their classification alone would pose an insurmountable task. In fact, all of these substances are merely combinations of much smaller numbers of building blocks called **elements.** Elements are substances that cannot be decomposed into two or more simpler substances by ordinary physical or chemical means. The word *ordinary* excludes the processes of radioactive decay, whether natural or artificial, and high-energy nuclear reactions that *do* transform elements into one another. When a substance contains two or more chemical elements, we call it a **compound.** For example, hydrogen and oxygen are elements because no further chemical separation is possible, whereas water is a

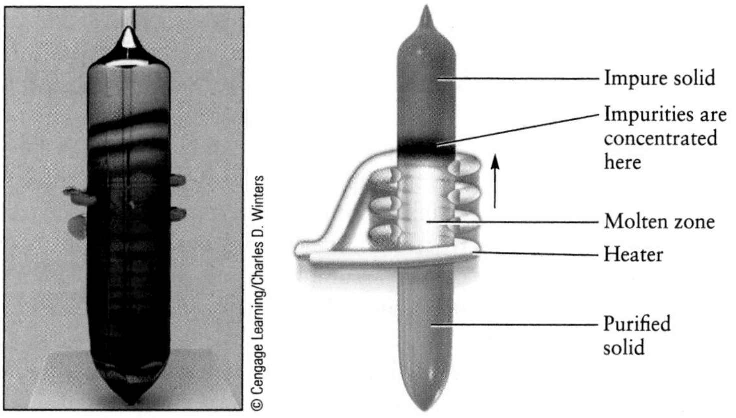

FIGURE 1.5 Nearly pure elemental silicon is produced by pulling a 10-inch-long solid cylinder (called a boule) out of the melt, leaving most of the impurities behind.

compound because it can be separated into hydrogen and oxygen by passing an electric current through it (see Fig. 1.2). *Binary* compounds are substances, such as water, that contain two elements, *ternary* compounds contain three elements, *quaternary* compounds contain four elements, and so on.

At present, scientists have identified some 118 chemical elements. A few have been known since before recorded history, principally because they occur in nature as elements rather than in combination with one another in compounds. Gold, silver, lead, copper, and sulfur are the most common of these. Gold is found in streams in the form of little granules (placer gold) or nuggets in loosely consolidated rock. Sulfur is associated with volcanoes, and copper often can be found in its native state in shallow mines. Iron occurs (only rarely) in its elemental state (in meteorites); it usually is combined with oxygen or other elements. Ancient metallurgists, in the second millennium B.C., somehow learned to reduce iron oxide to metallic iron using charcoal in forced-draft fires, and the Iron Age was born.

The names of the chemical elements and the symbols that designate them have a fascinating history. The symbols for many elements come from their Latin names, such as gold (aurum, symbol Au), copper (cuprum, Cu), and iron (ferrum, Fe). Some elements have names that describe their characteristic reactions or source; hydrogen (H), for example, means "water former" and potassium (kalium, K) takes its common name from potash (potassium carbonate), a useful chemical obtained in early times by leaching the ashes of wood fires with water. Many elements take their names from Greek and Roman mythology: cerium (Ce) from Ceres, goddess of plenty; tantalum (Ta) from Tantalus, who was condemned in the afterlife to an eternity of hunger and thirst while close to water and fruit that were always tantalizingly just out of reach; niobium (Nb) from Niobe, daughter of Tantalus; and mercury (hydrargyrum, Hg), which means silver water, named after the quickly moving god. Some elements are named for continents: europium (Eu) and americium (Am). Other elements are named after countries: germanium (Ge), francium (Fr), and polonium (Po). Cities provided the names of other elements: holmium (Stockholm, Ho), ytterbium (Ytterby, Yb), and berkelium (Berkeley, Bk). Some are named for the planets: uranium (U), plutonium (Pu), and neptunium (Np). Other elements take their names from colors: praseodymium (green, Pr), rubidium (red, Rb), and cesium (sky blue, Cs). Still others honor great scientists: curium (Marie Curie, Cm), mendelevium (Dmitri Mendeleev, Md), fermium (Enrico Fermi, Fm), einsteinium (Albert Einstein, Es), and seaborgium (Glenn Seaborg, Sg).

1.3 Indirect Evidence for the Existence of Atoms: Laws of Chemical Combination

How did we acquire the chemical evidence for the existence of atoms and the scale of relative atomic masses? It is an instructive story, both in its own right and as an illustration of how science progresses.

We may know the elements to be the most fundamental substances, and we may know that they can be combined chemically to form compound substances, but that knowledge provides us no information about the nanoscopic structure of matter or how that nanoscopic structure controls and is revealed by chemical reactions. Ancient philosophers dealt with these fascinating questions by proposing assumptions, or *postulates,* about the structure of matter. The Greek philosopher Democritus (c. 460–370 B.C.) postulated the existence of unchangeable *atoms* of the elements, which he imagined to undergo continuous random motion in the vacuum, a remarkably modern point of view. It follows from this postulate that matter is not divisible without limit; there is a lower limit to which a compound can be divided before it becomes separated into atoms of the elements from which it is made. Lacking both experimental capabilities and the essentially modern scientific view that theories must be tested and refined by experiment, the Greek philosophers were content to leave their views in the form of assertions.

Experimental measurements showed that the ratios of the masses of compounds that react to form other compounds is also fixed. These results could be interpreted only by inferring that the smallest indivisible units of the elements (atoms) combined to form the smallest indivisible units of the compounds (molecules). The definite mass ratios involved in reactions suggested a convenient method for counting the number of atoms of each element participating in the reaction. These results, summarized as the **laws of chemical combination,** provided overwhelming, if indirect, evidence for the existence of atoms and molecules.

For more than a century, we have become so accustomed to speaking of atoms that we rarely stop to consider the experimental evidence for their existence that was collected in the 18th and 19th centuries. Twentieth-century science developed a number of sophisticated techniques to measure the properties of single atoms and powerful microscopes to observe them (see *Connection to Nanotechnology*). But long before single atoms were detected, chemists could speak with confidence about their existence and the ways in which they combine to form molecules. Moreover, although the absolute masses of single atoms of oxygen and hydrogen were not measured until the early 20th century, chemists could assert (correctly) some 50 years earlier that the *ratio* of their masses was close to 16 : 1.

1.3.1 Law of Conservation of Mass

The first steps toward formulating the laws of chemical composition were taken during the 18th century in the course of studies of heat and combustion. It had been observed that an organic material, such as wood, left a solid residue of ash when burned; similarly, a metal heated in air was transformed into a "calx," which we now call an oxide. The popular explanation for these phenomena in the early 18th century was that a property called *phlogiston* was driven out of wood or metal by the heat of a fire. From the modern perspective, this seems absurd, because the ash weighed less than the original wood, whereas the calx weighed more than the metal. The principle of conservation of mass had not yet been established at that time, and people saw no reason why the mass of a material should not change when heated.

Further progress could be made only by carefully measuring the changes in mass[1] that occur in chemical reactions. The balance had been known since antiquity, but it had been used principally as an assayer's tool and for verifying the masses of coins or commodities in commerce. The analytical balance developed in the 18th century was accurate to perhaps 1 part in 10,000, enabling much more accurate measurements of mass changes accompanying chemical reactions than had been possible previously. French chemist Antoine Lavoisier used the analytical balance (see the photo on page 35) to demonstrate that the sum of the masses of the products of a chemical reaction equals the sum of the masses of the reactants to the high degree of accuracy provided by the instrument. Lavoisier heated mercury in a sealed flask that contained air. After several

[1]Chemists sometimes use the term *weight* in place of *mass*. Strictly speaking, weight and mass are not the same. The mass of a body is an invariant quantity, but its weight is the force exerted on it by gravitational attraction (usually by the Earth). Newton's second law relates the two ($weight = m \times g$, where g is the acceleration due to gravity). As g varies from place to place on the Earth's surface, so does the weight of a body. In chemistry, we deal mostly with ratios, which are the same for masses and weights. In this textbook we use the term *mass* exclusively, but *weight* is still in colloquial chemical use.

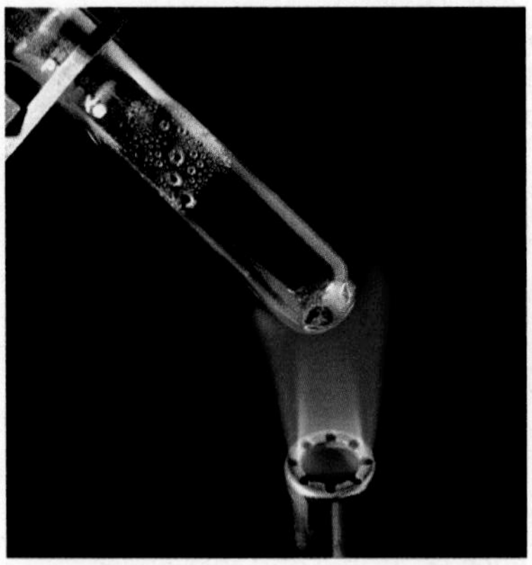

FIGURE 1.6 The red solid, mercury oxide, decomposes into liquid mercury and oxygen gas, when heated. Note the drops of liquid mercury condensing on the side of the test tube.

days, a red substance, an oxide of mercury, was produced. The gas remaining in the flask was reduced in mass and could no longer support life or combustion; a candle was extinguished by it, and animals suffocated when forced to breathe it. We now know that this residual gas was nitrogen, and that the oxygen in the air had reacted with the mercury. Lavoisier then took a carefully weighed amount of the red oxide of mercury and heated it to a very high temperature (Fig. 1.6). He weighed both the mercury and the gas that were produced and showed that their combined mass was the same as that of the mercury oxide with which he had started. After further experiments, Lavoisier was able to state the **law of conservation of mass:**

Mass is neither created nor destroyed in chemical reactions; it is conserved.

Lavoisier was the first to observe that a chemical reaction is analogous to an algebraic equation. We would write his second reaction as

$$2\ \mathrm{HgO} \longrightarrow 2\ \mathrm{Hg} + \mathrm{O_2}$$

although during Lavoisier's lifetime, the identity of the gas (oxygen) was only just being discovered by Joseph Priestley.

1.3.2 Law of Definite Proportions

Rapid progress ensued as chemists began to make accurate determinations of the masses of reactants and products. A controversy arose between two schools of thought, led by a pair of French chemists, Claude Berthollet and Joseph Proust. Berthollet believed that the proportions (by mass) of the elements in a particular compound were not fixed, but could actually vary over a certain range. Water, for example, might contain more or less than 11.1% hydrogen by mass, the generally accepted value. Proust disagreed, arguing that any apparent variation was due to impurities and experimental errors. He also stressed the difference between homogeneous mixtures and chemical compounds. In 1794, Proust published the fundamental **law of definite proportions:**

In a given chemical compound, the proportions by mass of the elements that compose it are fixed, independent of the origin of the compound or its mode of preparation.

Pure sodium chloride contains 60.66% chlorine by mass, whether we obtain it from salt mines, crystallize it from waters of the oceans or inland salt seas, or synthesize it from its elements, sodium and chlorine.[2]

The law of definite proportions was a crucial step in the development of modern chemistry, and Proust's conclusions had become widely accepted by the time Dalton published his atomic theory. We now recognize that this law is not strictly true in all cases. Although all gaseous compounds obey Proust's law, certain solids have compositions that vary over small ranges; such a solid is called a **nonstoichiometric compound.** An example is wüstite, which has the nominal chemical formula FeO (with 77.73% iron by mass), but the composition of which, in fact, ranges continuously from $\mathrm{Fe_{0.95}O}$ (with 76.8% iron) down to $\mathrm{Fe_{0.85}O}$ (74.8% iron), depending on the method of preparation. Such compounds are called **berthollides,** in honor of Berthollet. We now know, on the atomic level, why they are nonstoichiometric (see discussion in Section 21.4).

The development of the law of definite proportions provides an excellent example of how science progresses. Measurements of the compositions of a large number of compounds supported the law of definite proportions, but later,

[2]This statement needs some qualification. As explained in the next section, many elements have several *isotopes,* which are species whose atoms have almost identical chemical properties but different masses. Natural variation in isotope abundance leads to small variations in the mass proportions of elements in a compound, and larger variations can be induced by artificial isotopic enrichment.

more precise measurements uncovered exceptions to the general principle that had been established. The following explanation of the exceptions leads to a deeper understanding.

1.3.3 Dalton's Atomic Theory

English scientist John Dalton was by no means the first person to propose the existence of atoms; as we have seen, speculations about them date back to ancient Greek times (the word *atom* is derived from Greek *a-* ["not"] plus *tomos* ["cut"], meaning "not divisible"). Dalton's major contribution to chemistry was to marshal the evidence for the existence of atoms. He showed that the mass relationships found by Lavoisier and Proust could be interpreted most simply by postulating the existence of atoms of the various elements.

In 1808, Dalton published *A New System of Chemical Philosophy,* in which the following five postulates comprise the **atomic theory of matter**:

1. Matter consists of indivisible atoms.
2. All the atoms of a given chemical element are identical in mass and in all other properties.
3. Different chemical elements have different kinds of atoms; in particular, their atoms have different masses.
4. Atoms are indestructible and retain their identities in chemical reactions.
5. Atoms of the elements combine with each other in small integer ratios to form compounds.

Dalton's fourth postulate clearly is related to the law of conservation of mass. The fifth aims to explain the law of definite proportions. Perhaps Dalton's reasoning went something like this: Suppose you reject the atomic theory and believe instead that compounds are subdivisible without limit. What, then, ensures the constancy of composition of a substance such as sodium chloride? Nothing! But if each sodium atom in sodium chloride is matched by one chlorine atom, then the constancy of composition can be understood. So in this argument for the law of definite proportions, it does not matter how small the atoms of sodium and chlorine are. It is important merely that there be some lower bound to the subdivisibility of matter, because the moment we put in such a lower bound, arithmetic steps in. Matter becomes countable, and the units of counting are simply atoms. Believing in the law of definite proportions as an established experimental fact, Dalton *postulated* the existence of the atom.

1.3.4 Law of Multiple Proportions

The composition of a compound is shown by its **chemical formula.** The symbol H_2O for water specifies that there are two atoms of hydrogen for each atom of oxygen in one unit of water. We now know that the two hydrogen atoms are strongly bound to the oxygen atom in a discrete unit called a **molecule** and that the atoms in one water molecule are not strongly bound to the atoms of any other water molecule. Many solid substances do not exist as molecules but as infinite arrays of ions or atoms bonded to each other in ways that do not allow us to identify a particular molecule uniquely; they are called ionic and covalent solids. The chemical formula for these pure substances gives the relative ratios of the numbers of their constituent atoms. How do we know that these are the true proportions? The determination of chemical formulas (and the accompanying determination of relative atomic masses), building on the atomic hypothesis of Dalton, was a major accomplishment of 19th-century chemistry.

The simplest compounds made from two different elements is a **diatomic molecule** that contains one atom of each element. Eighteenth- and nineteenth-century chemists knew, however, that two elements often combine in different proportions, suggesting the existence of multiple compounds that are more complicated than diatomic molecules.

Carbon (C) and oxygen (O), for example, combine under different conditions to form two different compounds, which we will call A and B. Analysis shows that A contains 1.333 grams (g) of oxygen per 1.000 g of carbon, and B contains 2.667 g of oxygen per 1.000 g of carbon. Although at this point we know nothing about the chemical formulas of the two oxides of carbon, we can say immediately that molecules of compound A contain half as many oxygen atoms per carbon atom as do molecules of compound B. The evidence for this is that the ratio of the masses of oxygen in A and B, for a fixed mass of carbon in each compound, is 1.333:2.667, or 1:2. If the formula of compound A were CO, then the formula of compound B would have to be CO_2, C_2O_4, C_3O_6, or some other multiple of CO_2. We cannot say from these data which of these (or an infinite number of other possibilities) are the true formulas of the molecules of compounds A and B, but we do know this: The number of oxygen atoms per carbon atom in the two compounds is the *quotient of integers.*

Let's consider another example. Arsenic (As) and sulfur (S) combine to form two sulfides, A and B, in which the masses of sulfur per 1.000 g of arsenic are 0.428 and 0.642 g, respectively. The ratio of these masses is 0.428:0.642 = 2:3. We conclude that *if* the formula of compound A is a multiple of AsS, then the formula of compound B must be a multiple of As_2S_3.

These two examples illustrate the **law of multiple proportions:**

When two elements form a series of compounds, the masses of one element that combine with a fixed mass of the other element are in the ratio of small integers to each other.

In the first example, the ratio of the masses of oxygen in the two compounds, for a given mass of carbon, was 1:2. In the second example, the ratio of the masses of sulfur in the two compounds, for a given mass of arsenic, was 2:3. Today, we know that the carbon oxides are CO (carbon monoxide) and CO_2 (carbon dioxide), and the arsenic sulfides are As_4S_4 and As_2S_3. Dalton could not have known this, however, because he had no information from which to decide how many atoms of carbon and oxygen are in one molecule of the carbon–oxygen compounds or how many atoms of arsenic and sulfur are in the arsenic–sulfur compounds.

EXAMPLE 1.1

Chlorine (Cl) and oxygen form four different binary compounds. Analysis gives the following results:

Compound	Mass of O Combined with 1.0000 g Cl
A	0.22564 g
B	0.90255 g
C	1.3539 g
D	1.5795 g

(a) Show that the law of multiple proportions holds for these compounds.

(b) If the formula of compound A is a multiple of Cl_2O, then determine the formulas of compounds B, C, and D.

Solution

(a) Form ratios by dividing each mass of oxygen by the smallest, which is 0.22564 g:

0.22564 g:0.22564 g = 1.0000 for compound A

0.90255 g:0.22564 g = 4.0000 for compound B

1.3539 g:0.22564 g = 6.0003 for compound C

1.5795 g:0.22564 g = 7.0001 for compound D

The ratios are whole numbers to a high degree of precision, and the law of multiple proportions is satisfied.

(Continued)

(b) If compound A has a formula that is some multiple of Cl_2O, then compound B is Cl_2O_4 (or ClO_2, or Cl_3O_6, and so forth) because it is four times richer in oxygen than is compound A. Similarly, compound C, which is six times richer in oxygen than compound A, is Cl_2O_6 (or ClO_3, or Cl_3O_9, and so forth), and compound D, which is seven times richer in oxygen than compound A, is Cl_2O_7 (or a multiple thereof).

Related Problems: 7, 8, 9, 10

Dalton made a sixth assumption, which was incorrect, to resolve the dilemma of the absolute number of atoms present in a molecule; he called it the "rule of greatest simplicity." If two elements A and B form only one compound then the chemical formula for that compound will be the simplest possible: AB. Thus, he assumed that when hydrogen and oxygen combine to form water, the reaction is

$$H + O \longrightarrow HO$$

Dalton was wrong, however, as we now know, and the correct reaction is

$$2\ H_2 + O_2 \longrightarrow 2\ H_2O$$

1.3.5 Law of Combining Volumes

French chemist Joseph Gay-Lussac conducted some important experiments, at about the same time as Dalton, on the relative volumes of gases that react completely with one another to form new gases. He discovered the **law of combining volumes:**

The ratio of the volumes of any pair of gases in a gas phase chemical reaction (at the same temperature and pressure) is the ratio of simple integers.

Here are three examples:

2 volumes of hydrogen + 1 volume of oxygen ⟶ 2 volumes of water vapor

1 volume of nitrogen + 1 volume of oxygen ⟶ 2 volumes of nitrogen oxide

3 volumes of hydrogen + 1 volume of nitrogen ⟶ 2 volumes of ammonia

1.3.6 Avogadro's Hypothesis

Gay-Lussac did not interpret his experimental findings theoretically but, shortly after their publication in 1811, the Italian chemist Amedeo Avogadro used them to formulate an important postulate that became known as Avogadro's hypothesis:

Equal volumes of different gases at the same temperature and pressure contain equal numbers of particles.

The question immediately arose: Are "particles" of the elements the same as Dalton's atoms? Avogadro believed that they were not; rather, he proposed that elements could exist as diatomic molecules. Avogadro's hypothesis could explain Gay-Lussac's law of combining volumes (Fig. 1.7). Thus, the reactions we wrote out in words become

$$2\ H_2 + O_2 \longrightarrow 2\ H_2O$$

$$N_2 + O_2 \longrightarrow 2\ NO$$

$$3\ H_2 + N_2 \longrightarrow 2\ NH_3$$

The coefficients in the above reactions are proportional to the volumes of the reactant and product gases in Gay-Lussac's experiments, and the chemical

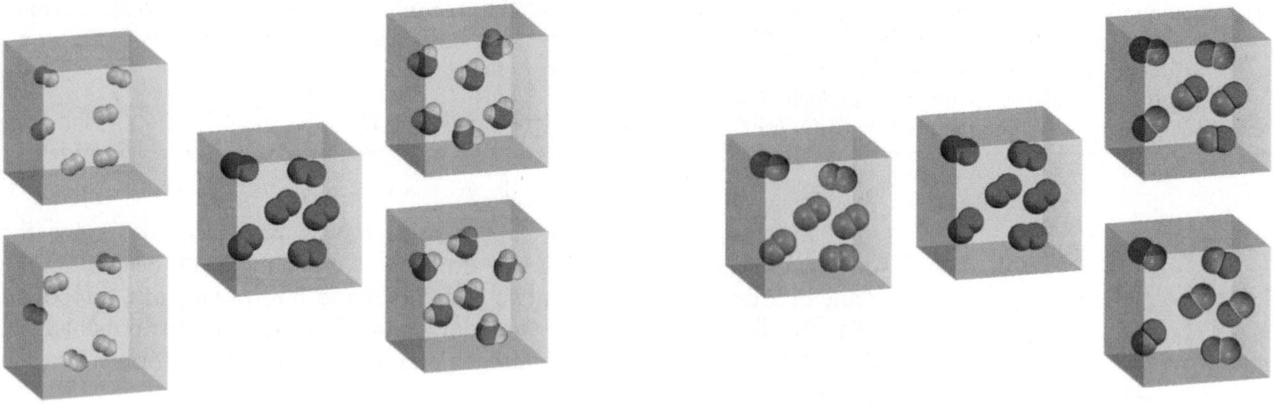

$$2\,H_2 + O_2 \longrightarrow 2\,H_2O \qquad\qquad N_2 + O_2 \longrightarrow 2\,NO$$

FIGURE 1.7 The cubes shown contain equal volumes of different gases at the same temperature and pressure. The combining volumes that Gay-Lussac observed for the two reactions can be understood if each cube contains the same number of molecules (Avogadro's hypothesis), and if hydrogen, oxygen, and nitrogen exist as diatomic molecules, as shown.

formulas of the reactions agree with modern results. Dalton, on the other hand, would have written those reactions as

$$H + O \longrightarrow OH$$

$$N + O \longrightarrow NO$$

$$H + N \longrightarrow NH$$

The combining volumes predicted by Avogadro's hypothesis for all three of these reactions as written would be

$$1 \text{ volume} + 1 \text{ volume} \longrightarrow 1 \text{ volume}$$

which did not agree with the results of Gay-Lussac's experiments.

Avogadro's hypothesis not only predicted the correct molecular formulas but also the correct relative atomic masses of the elements. Chemical analysis during the 18th century had demonstrated that 1 g of hydrogen gas reacts completely with 8 g of oxygen gas to produce 9 g of water. If Dalton's formula for water, HO, were correct, then an atom of oxygen would weigh 8 times as much as an atom of hydrogen; that is, Dalton's assumption requires the **relative atomic mass** of oxygen to be 8 on a scale where the relative atomic mass of hydrogen had been chosen to be 1. Avogadro's hypothesis predicted, however, that each water molecule has twice as many atoms of hydrogen as oxygen; therefore, to explain the observed experimental mass relation, the relative mass for oxygen must be 16, a result consistent with modern measurements.

We might expect that Dalton would have welcomed Avogadro's brilliant hypothesis, but he did not. Dalton and others insisted that elements could not exist as diatomic molecules. One reason for their belief was the then-popular idea that a force called *affinity* held molecules together. Affinity expressed the attraction of opposites, just as we think of the attraction between positive and negative electric charges. Why should two atoms of the same type be held together in a molecule if the affinity theory were true? Moreover, if atoms of the same element did bond together somehow in pairs, why shouldn't they aggregate further to form molecules with three, four, or six atoms, and so forth? Avogadro's reasoning did not attract the attention it deserved because so many chemists continued to believe in the affinity theory, resulting in great confusion due to the adoption of different chemical formulas for the same molecule. A textbook published in 1861 by the German chemist August Kekulé titled *Lehrbuch der Organischen Chemie* gave 19 different chemical formulas for acetic acid!

In 1860, 50 years after Avogadro's work, Italian chemist Stanislao Cannizzaro presented a paper at the First International Chemical Congress in Karlsruhe, Germany, that convinced others to accept Avogadro's approach. Cannizzaro had analyzed many gaseous compounds and was able to show that their chemical formulas could be established with a consistent scheme that used Avogadro's hypothesis and avoided any extra assumptions about molecular formulas. Gaseous hydrogen, oxygen, and nitrogen (as well as fluorine, chlorine, bromine, and iodine), did indeed turn out to be diatomic molecules under ordinary conditions, thus vindicating Avogadro and his hypothesis.

1.4 The Physical Structure of Atoms

The laws of chemical combination that culminated in Dalton's atomic theory and Avogadro's hypothesis reinforced the original Greek concept that the atom was the ultimate and indivisible building block of matter. By the end of the 19th century, new experimental results forced scientists to abandon this view and to conclude that atoms themselves were composed of a number of smaller, *elementary* particles. It took about 40 years to identify and measure the properties of these sub-atomic particles. This was a fascinating period in the development of modern science. Physicists built upon the advances of chemists to develop a deeper understanding of the fundamental structure of matter. Knowledge of the components of the atom and of the forces that hold them together stimulated entirely new fields of basic science and technology that continue to the present. Three themes characterize the research in this period: (1) how electric and magnetic fields (at that time only recently understood and described by the empirical laws of electricity and magnetism) became the dominant probes of the structure of matter, (2) the importance of clearly formulated hypotheses that could be tested experimentally, and (3) how the results we describe for several key experiments led to the general conclusions and insights that were drawn.

We suggest that you review the relevant sections of Appendix B before continuing because interpreting the results of experiments described in the following paragraphs requires an elementary understanding of electricity and magnetism.

1.4.1 Electrolysis and the Existence of Ions

The story begins around the last quarter of the 19th century, by which time Faraday's laws of electrolysis had been generally accepted. Faraday had studied electrochemical reactions extensively; reactions in which the passage of an electric charge produces chemical changes (see Section 17.1). The electrolysis of water, shown in Figure 1.2, is an example of an electrochemical reaction—the electric current passing through the water decomposes it into the elements hydrogen and oxygen. Faraday discovered that the amount of water decomposed, and the amount of hydrogen and oxygen produced at the electrodes, were directly proportional to the quantity of charge passed. He observed the same result for many different electrochemical reactions, which led him to suggest that there must exist a fixed unit of electricity that is transferred in all these reactions. This conclusion was perhaps the first suggestion that all matter is composed of some electrical component that can be stripped away and transferred from reactants to products. Faraday named the electrically charged entities **ions**, the Greek word for "wanderer." He measured the amount of charge necessary to produce 1 g of hydrogen from the electrolysis of water and thus calculated the **charge-to-mass ratio** for the hydrogen ion, which was about 10^8 C kg^{-1} in modern units.

Faraday's measurement was important, for nothing was known about the detailed structure or even the mass of the hydrogen ion at the time. Knowing the charge-to-mass ratio made it possible to investigate the properties of ions through their response to electric and magnetic fields. These studies would have to be carried out in gases at reduced pressure, where the charged particles could travel long distances without collisions, and their deflections by electric and magnetic fields could be measured. Also, Faraday's measurement established the order of magnitude for this quantity against which the charge-to-mass ratios of other particles would later be compared.

1.4.2 Glow Discharges and Cathode Rays

The electrical components of gaseous substances were discovered in studies of the *glow discharge,* a process in which soft and beautiful light is emitted when electricity is forced to pass through a gas at low pressure. This phenomenon created great excitement among physicists and stimulated intense research projects to explain the nature of the discharge and the origin of the light. An early example of a glow discharge tube is the *Crookes tube,* an apparatus (Figure 1.8) invented by William Crookes to study the principles of electric discharges. Modern neon signs are examples of glow discharge tubes. The color of the glow depends upon the gas; neon produces a red emission, argon is purple, and mercury is blue, as shown in Figure 1.8. The glass tube has inside two conducting plates separated from each other, and when a large electrical potential difference is applied (imagine hooking up the negative terminal of a battery to the plate called the "cathode" and the positive terminal to the plate called the "anode"), current passes through the gas and the gas glows. The glass on the right end of the tube is coated with a phosphor that glows when illuminated by radiation. Figure 1.8b shows a faint blue glow discharge that becomes weak as the gas pressure is reduced and eventually disappears at very low pressures. The phosphor at the end of the tube continues to glow, however, even when all of the gas has been evacuated. It appeared to be illuminated by some mysterious rays being emitted from the cathode and these rays were observed to travel in straight lines. Because these mysterious rays persist even when there is no gas in the tube, they must be emitted from the cathode and so were named **cathode rays**. Several decades of research revealed that these cathode rays could be deflected by both magnetic and electric fields and could heat a piece of metal foil in the tube until it glowed. Influenced by Maxwell's recent electromagnetic theory of light, one school of physicists believed the cathode rays to be a strange form of invisible light, whereas another group considered them a stream of negatively charged particles. The latter group was right, as demonstrated by the definitive experiments of J. J. Thompson described in the next section.

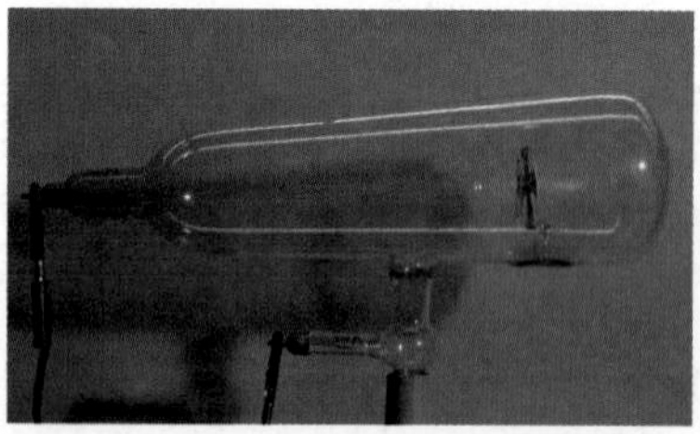

D-Kuru/Wikimedia Commons

(a)

D-Kuru/Wikimedia Commons

(b)

FIGURE 1.8 A Crookes tube. (a) The Crookes tube is a glass cylinder that is evacuated and then filled with different gases to study the properties of glow discharges. A discharge is established when voltages are applied between the cathode (the electrode on the left) and the anode (the electrode in the arm below the main tube). The glass face on the right side of the tube is coated with a phosphor that glows when irradiated. (b) A weak blue glow discharge that disappears at low pressure is shown. The phosphorescence from the end of the tube persists even when no gas is present, providing evidence for the existence of cathode rays.

1.4.3 Negative Charge in the Atom: Electrons

In 1897, British physicist J. J. Thomson conducted a set of experiments that established that the cathode rays are negatively charged particles, now called **electrons**. His experiments showed that these negatively charged particles were deflected by an electric field and that the direction of the displacement was consistent with their negative charge.

Figure 1.9 shows the force exerted on a positively charged particle (+) and on a negatively charged particle (−), when the particles are in a uniform electric field $\vec{E}$. The upper metal plate has an excess negative charge on it, while the lower plate has an excess positive charge on it; this establishes an electric field $\vec{E}$ between the plates that points upward, in the $+y$ direction, as shown by the blue **vector** (arrow) in the figure. This electric field exerts a force $\vec{F}$ on a charged

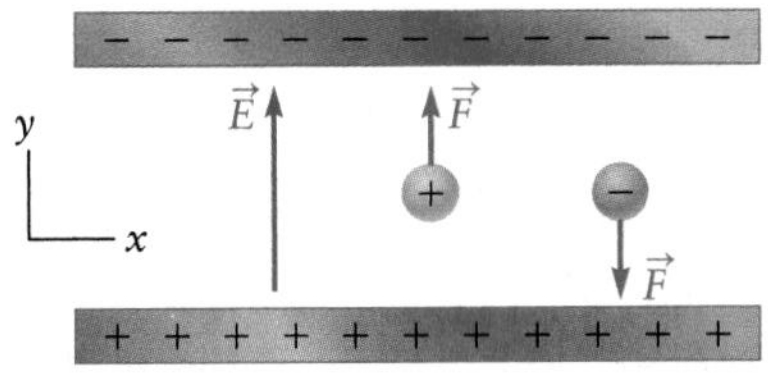

FIGURE 1.9 The direction of the electric field between two charged metal plates, and the resulting direction of the electrostatic force exerted on a negatively charged particle and on a positively charged particle.

particle given by $\vec{F} = q\vec{E}$, where q is the charge of the particle. Thus, when q is positive the force $\vec{F}$ points in the same direction as $\vec{E}$ (upward, in the $+y$ direction), as shown by the red vector $\vec{F}$ in the figure, but when q is negative the force $\vec{F}$ points downward (in the $-y$ direction), as shown by the green arrow in Figure 1.9. In the key experiment by J. J. Thompson described in the next paragraph, the negatively charged particles are moving in the $+x$ direction when they experience this downward force, in the $-y$ direction, so the component of the force in the y direction, F_y, is a negative number (downward), while the component of the force in the x direction, F_x, is zero.

Charge-to-Mass Ratio of the Electron

The cathode ray tube used to measure the charge-to-mass ratio of the electron is shown schematically in Figure 1.10. The glass tube is evacuated to very low pressures so that electrons can travel the length of the tube without being scattered by residual gases. The glass face on the right side of the tube is coated with a phosphor that emits light when irradiated. Electrons emitted from the cathode are accelerated toward the anode, and those that pass through the hole in the anode form a collimated beam that lights up the phosphor where it hits. The electrical deflection plates bend the beam down, and the displacement is measured directly from the positions of the phosphorescent spots on the screen. The magnetic field is then turned on, and the field strength required to restore the beam to its original position is recorded. The charge-to-mass ratio of the electron can be calculated by measuring the displacement and knowing the strengths of the electric and magnetic fields applied.

The voltage applied to the deflection plates establishes a uniform electric field between them that exerts a constant downward force on electrons in the region that is given by

$$F_y = -eE = m_e a_y \qquad [1.1]$$

where $-e$ is the charge on the electron and E is the strength of the electric field. The constant acceleration that results from this force is calculated using Newton's second law,

$$a_y = \frac{-eE}{m_e} \qquad [1.2]$$

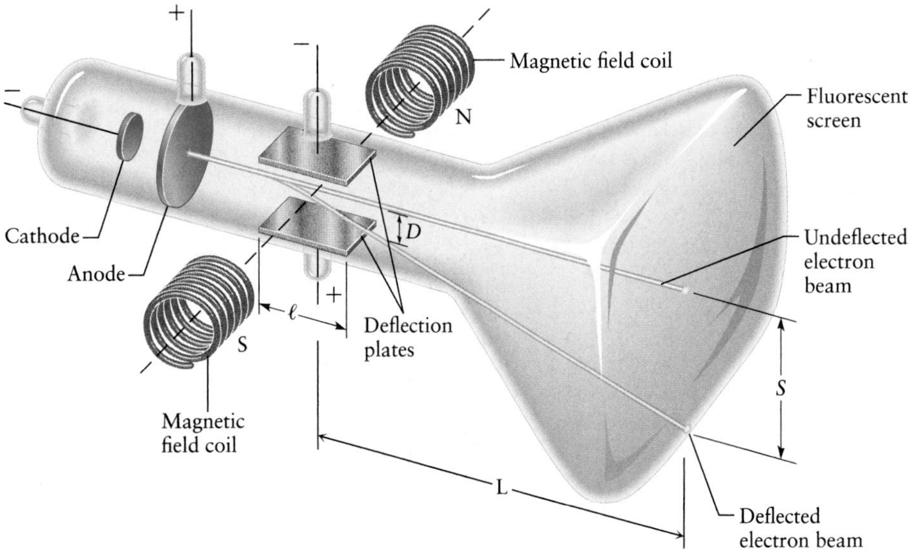

FIGURE 1.10 Thomson's apparatus, used to measure the charge-to-mass ratio, e/m_e, of the electron. Electrons emitted from the cathode (cathode rays) travel across the tube from left to right. An electric field deflects the beam down and a magnetic field deflects the beam up. The deflection S, due to the electric field alone, was measured. The beam was then restored to its original position by applying the magnetic field. The charge-to-mass ratio for the electron was determined from these two measurements, as discussed in the text. (ℓ is the length of the electric deflection plates.)

which results in a displacement in the downward direction with magnitude given by

$$D = \frac{1}{2}|a_y|\, t^2 \qquad [1.3]$$

where t is the time required to travel the distance ℓ, the length of the plates. Substituting $t = \ell/|v_x|$, for electrons traveling through the plates at constant speed $v_x = dx/dt$, and $|a_y|$ (the magnitude of the downward acceleration due to the electric field) from Equation 1.2 into Equation 1.3 gives:

$$D = \frac{1}{2}|a_y|\, t^2 = \frac{1}{2}\left(\frac{e}{m_e}\right)\left(\frac{\ell}{|v_x|}\right)^2 E \qquad [1.4]$$

the magnitude of the downward displacement. The deflected electrons travel in straight lines after they leave the region between the plates, because there are no additional forces acting on them, and they strike the fluorescent screen at distance S below the position of the undeflected beam. This extra path length "magnifies" the displacement by the factor $2L/\ell$, where L is the distance from the center of the plates (a distance of $\ell/2$ from the left edge) to the screen, as shown in Figure 1.11.

Electrons arriving at the screen have been displaced by

$$S = \left(\frac{2L}{\ell}\right) D = \left(\frac{e}{m_e}\right)\left(\frac{\ell}{|v_x|}\right)^2 \left(\frac{L}{\ell}\right) E \qquad [1.5]$$

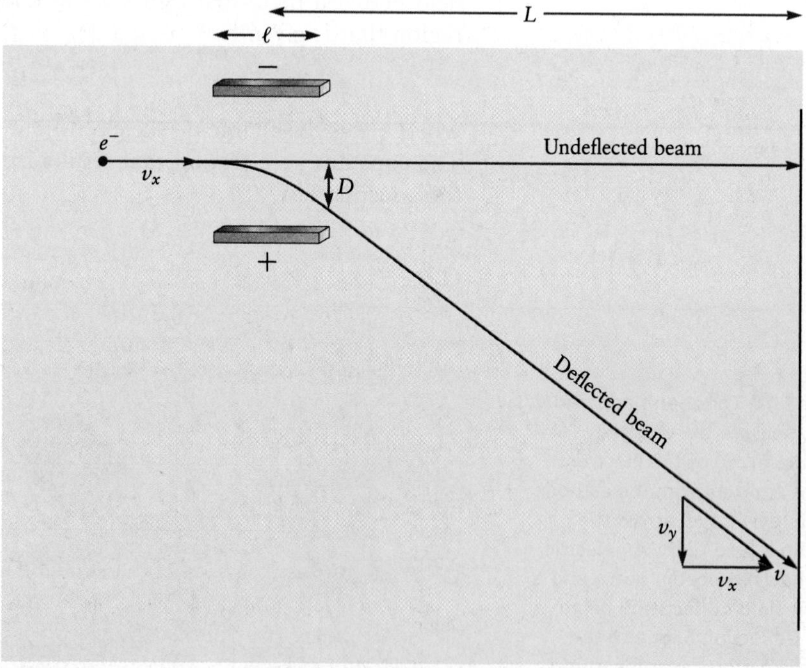

FIGURE 1.11 Path followed by an electron beam subject to a force in the downward direction between two electrostatic plates. No force is exerted in the horizontal direction, so the component of velocity in the $+x$ direction, v_x, is constant, but between the plates of length ℓ the force is in the downward direction, so v_y, the y component of the velocity, is negative (downward) and increases in magnitude until the electron reaches the right edge of the plates. After that the electrons travel with constant v_x and v_y, so the path is straight. The average v_y between the plates is half of the final value of v_y at the right edge of the plates, so if one extrapolates the trajectory backwards from the straight portion of the deflected beam, it intersects the undeflected path at $x = \ell/2$. This gives the magnification ratio $S/D = L/(\ell/2) = 2L/\ell$.

Equation 1.5 could be solved for the charge-to-mass ratio, if the velocity of the electrons could be determined, because all of the other quantities could be measured directly in this experiment.

Thomson made an ingenious modification to his experiment to measure the velocity of the electrons directly and thereby determine e/m_e. He established a magnetic field in the same region as the electric field by passing an electric current through a pair of coils, as shown in Figure 1.10. The magnetic field was oriented perpendicular to both the electric field and to the flight path of the electrons, and it exerted an upward force on them. Thomson could return the deflected beam to its original position by varying the strengths of the two fields so that the net force on the electrons was zero. The force due to the electric field E was

$$F_{y\ (due\ to\ electric\ field)} = -eE \qquad [1.6]$$

and the force due to the magnetic field B was

$$F_{y\ (due\ to\ magnetic\ field)} = e|v_x|B \qquad [1.7]$$

so the velocity can be calculated by noting that when the two fields are adjusted so the beam is not deflected, the sum of the two forces must be zero. Thus, $-eE + e|v_x|B = 0$ gives:

$$|v_x| = \frac{E}{B} \qquad [1.8]$$

Substituting this result for the absolute value of v_x into Equation 1.5 gives the deflection observed under the influence of the electric field alone as

$$S = \left(\frac{e}{m_e}\right)\left(\frac{\ell B}{E}\right)^2\left(\frac{L}{\ell}\right)E \qquad [1.9]$$

which can be solved to find the magnitude of the charge-to-mass ratio of the electron.

$$\frac{e}{m_e} = \frac{SE}{\ell L B^2} \qquad [1.10]$$

Each of the quantities on the right-hand side of Equation 1.10 was measured directly in Thomson's experiment, allowing him to determine the magnitude of the charge-to-mass ratio for the electron for the first time. The currently accepted value is $e/m_e = 1.758820 \times 10^{11}\ \text{C kg}^{-1}$, with the charge and mass measured in the SI units coulombs and kilograms, respectively. (See Appendix B for a full discussion of units of measure.)

Charge of the Electron

Thomson was able to measure only the charge-to-mass *ratio* of the electron in his experiment, and so an independent measurement of either the charge or the mass was required to determine the values of both of these fundamental physical quantities. The American physicist Robert Millikan and his student H. A. Fletcher measured the charge of the electron in 1906, in an elegant experiment that is illustrated schematically in Figure 1.12. An atomizer, like one used to spray perfume, injects very small (about 1 μm) droplets of oil into a chamber above a pair of electrically charged plates that are separated by an insulator. From time to time a drop falls through a hole in the upper plate and enters the region between the plates; the droplets pick up a charge during the spraying process or by collisions with ions in the chamber air. The voltage applied to the plates establishes a constant electric field in this region that pushes negatively charged particles up, countering the downward pull of gravity. Millikan and his student observed hundreds to thousands of single droplets for extended periods

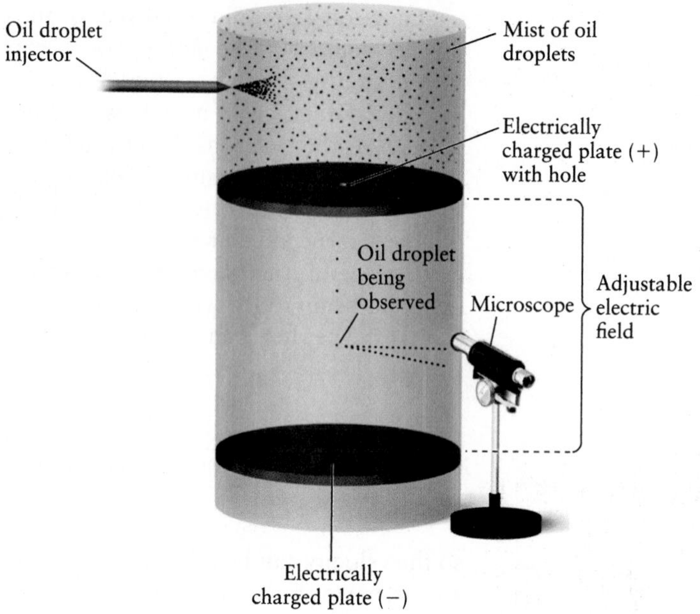

FIGURE 1.12 Millikan's apparatus, used to measure the charge of the electron, −e. Individual drops are suspended by adjusting the electric field to provide an electrostatic force that opposes the gravitational force; the charge on the electron was determined by equating these two forces as discussed in the text.

of time (as long as many hours), and they were able to control the motions of individual droplets by controlling the applied voltage. Individual droplets could be suspended for long periods of time by balancing the gravitational force by an opposing electrical force. Once suspended, individual droplets were sometimes observed to suddenly jump either upward or downward as they acquired negative or positive charges from ions in the air; they could be returned to their stable positions by adjusting the voltages on the plates.

The net force on stationary droplets is zero; balancing the electrical and gravitational forces gives $Mg = QE$, where M is the mass of the droplet, g is the acceleration due to gravity, Q is the magnitude of the charge on the droplet, and E is the electric field strength. The masses of individual droplets were calculated by multiplying their densities by their volumes (determined by measuring their diameters using the microscope). The gravitational acceleration g is a known constant, and the electric field strength was controlled and measured in each experiment, allowing Millikan to calculate the Q for each individual drop. The charges observed varied significantly from droplet to droplet but they were always an integral multiple of the one smallest unit of charge. Millikan pointed out for the first time that the magnitude of the fundamental unit of positive charge was exactly the same as the magnitude of the fundamental unit of negative charge. The value that Millikan reported for the magnitude of the charge of the electron was 1.59×10^{-19} C. The currently accepted value is $e = 1.602176565 \times 10^{-19}$ C which, along with the currently accepted value for the e/m_e ratio gives $m_e = 9.10938291 \times 10^{-31}$ kg for the electron mass.

1.4.4 Positive Charge in the Atom: The Nucleus

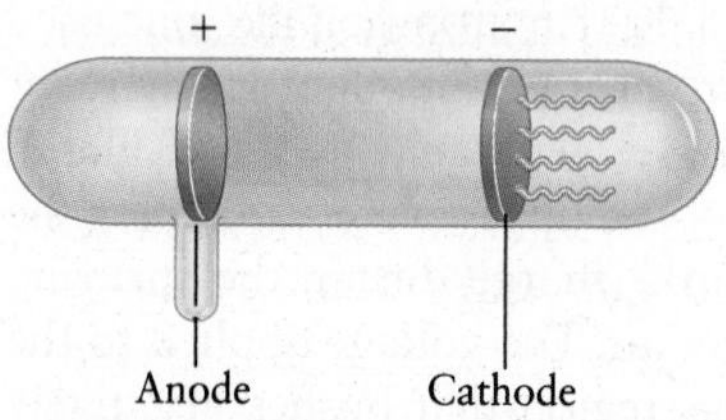

FIGURE 1.13 Modified Crookes tube.

Careful observation of the discharges in Crookes tubes showed that there was a glow that extended behind the cathode; a special version of the Crookes tube, shown in Figure 1.13 (based upon an original sketch by Thomson), was used to study the origin of this glow. The holes in the cathode (called canals) allowed part of the discharge to pass through and create pencils of light that were called **canal rays.** The German physicist Wilhelm Wien proceeded to study their properties by measuring their deflections in both electric and magnetic fields. He drew the following conclusions from these measurements:

1. The canal rays passing through the cathode were deflected by magnetic fields in a direction that established them as positively charged particles.
2. The electric and magnetic fields required to deflect these particles were much larger than those used in Thomson's experiments, proving that the particles were much more massive than the electron, at least as massive as the hydrogen atom.
3. The electric and magnetic fields required to deflect particles from different gases by the same amount were different, proving that the positively charged particles associated with different gases had different masses.

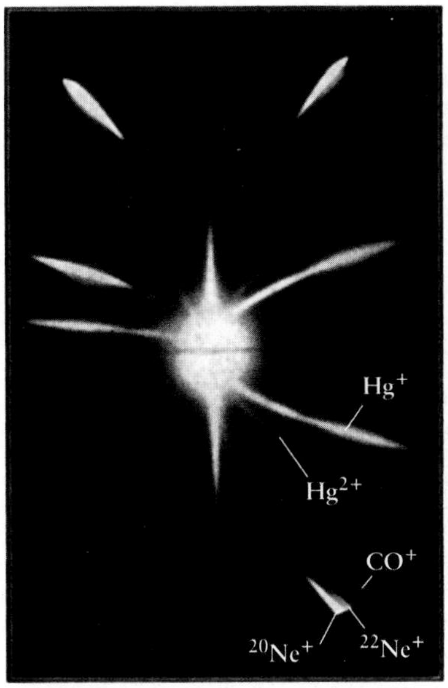

FIGURE 1.14 Traces of parabolic paths produced by positive ions in Thomson's experiments on the "positive rays." The ions giving rise to particular trajectories in the lower right quadrant of the figure are identified.

Thomson took up the further study of these "rays of positive electricity," as he called them, using a variation of his cathode ray tube apparatus. Imagine a version of the apparatus shown in Figure 1.10 with three modifications: (1) the voltages on the anode (the plate with the hole) and cathode are reversed so that positively charged particles pass through the hole and strike the front of the tube; (2) the magnetic field coils are rotated by 90° so that the electric and magnetic fields are now parallel to one another; and (3) the front of the tube is fitted with a photographic plate. The magnetic field deflects the particles horizontally while the electric field deflects them vertically, as before, and the film shows parabolic traces that depend upon the charge-to-mass ratio of the ions and on their velocities. Figure 1.14 shows one such set of traces, the identity of the particles being established by calculating the trajectories expected from their charge-to-mass ratios and the known field strengths. The particles are recognized as positively charged *ions* formed by removal of one or more electrons from an atom or molecule, usually by collisions with high-energy electrons. Mercury ions (from the vacuum pump) are clearly identified, a set with one unit of positive charge and a second set with two units of positive charge. CO^+ and neon ions with two different masses were identified (see later). Other experiments (not shown) established the utility of the method for identifying the individual components of complex mixtures of gases. Thomson did not fail to point out the importance of this new method for chemical analysis, which forms the basis for mass spectrometry, one of our most important methods for determining molecular weights and structures.

Thomson and Wien had discovered two new, and quite different, types of electrically charged particles that comprise matter: a light particle with a negative charge that appeared to be a common constituent of all atoms, and a number of much heavier, positively charged particles whose relative masses depended on the elements from which they were produced. Although it was generally then agreed that these particles were the building blocks of atoms, it was not at all clear how they were assembled. That piece of the puzzle remained unsolved until a stunning discovery was made in the laboratory of Ernest Rutherford in 1911, which relied upon results in earlier studies of radioactivity.

Discovery of the Atomic Nucleus

Radioactivity from natural sources was discovered in 1889 by the French physicists Henri Becquerel and Marie and Pierre Curie. Becquerel showed that rays emitted from uranium salts and uranium metal darkened nearby photographic plates that had been shielded from light by black paper. The Curies discovered the radioactive elements thorium, polonium, and radium. They demonstrated that while emitting radiation these elements were transformed into other elements by *radioactive decay*. This apparent violation of one of the key postulates of Dalton's atomic theory stimulated intense interest in discovering

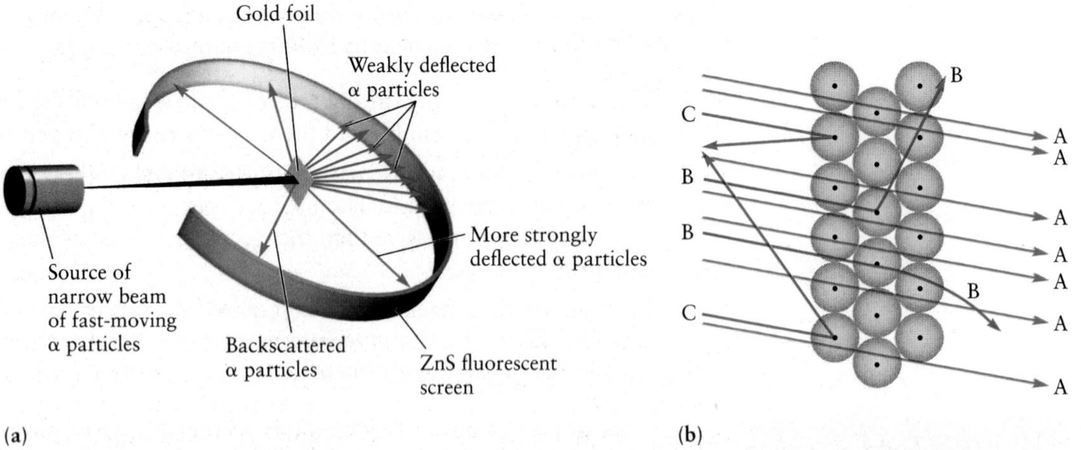

FIGURE 1.15 Schematic of Rutherford's experiment on the scattering of particles by thin (600 nm) gold foils. (a) Flashes of light mark the arrival of alpha particles at the fluorescent screen. More than 100,000 alpha particles per minute were weakly deflected, and only about 20 alpha particles per minute were deflected backwards, under typical conditions. (b) Interpretation of the Rutherford experiment: Most of the alpha particles pass through the space between nuclei and are deflected only slightly (A). A few pass close to a nucleus and are more strongly deflected (B). Some are even scattered backward (C). The size of the nucleus, relative to the size of a gold atom, is much smaller than the diagram suggests.

the mechanism of radioactive decay. Chapter 19 describes this very important process from a modern point of view, and reassures us that elements are not transmuted one into another by ordinary chemical reactions. Radioactive decay is not a revival of medieval alchemy! Our interest here is to use the emitted radiation as another probe of atomic structure.

By 1911, New Zealander Ernest Rutherford and his students at the University of Manchester had been investigating radioactive decay for a number of years. They had determined that the emitted radiation has at least two components, which they labeled α and β on the basis of their relative ability to penetrate solid materials. They had shown that α particles are doubly charged He atoms, by measuring their charge-to-mass ratios and by physically trapping and identifying the He gas produced. Now they wanted to study their interactions with matter to see if they could serve as probes of atomic structure. Rutherford chose to work with very thin gold foils (600 nm or about 2000 gold atoms thick) so the α particles could pass through the sample and their properties be measured after they exited the foil. Deflections of a collimated beam of alpha particles by a gold foil were measured by observing the scintillations they produced on a fluorescent ZnS screen (Fig. 1.15). Almost all of the alpha particles passed straight through the foil, but a few were deflected through large angles. Some particles even scattered backwards, in a few rare events! Rutherford was astounded, because the alpha particles were relatively massive and fast moving. In his words, "It was almost as incredible as if you fired a 15-inch shell at a piece of tissue paper and it came back and hit you." He and his students studied the frequency with which such large deflections occurred. They concluded that most of the mass in the gold foil was concentrated in dense, extremely small, positively charged particles that they called nuclei. By analyzing the trajectories of the particles scattered by the foil, they estimated the radius of the gold nucleus to be less than 10^{-14} m and the positive charge on each nucleus to be approximately $+100e$ (the actual value is $+79e$).

1.4.5 Rutherford's Planetary Model of the Atom

Based on these experimental results, Rutherford proposed a model of the atom in which the charge on the nucleus is $+Ze$, with Z electrons surrounding the nucleus out to a distance of about 10^{-10} m (0.1 nm). This integer Z is called the atomic number of the element. Atomic numbers are given on the inside front cover of this book. The Rutherford model for a gold atom has 79 electrons (each with charge $-1e$) arranged about a nucleus of charge $+79e$. The electrons occupy nearly the entire volume of the atom, whereas nearly all its mass is concentrated in the nucleus.

The model of the atom Rutherford proposed is often called the "planetary model" of the atom because he envisioned the electrons occupying most of the atomic volume (like the planets in a solar system), centered on the small, dense nucleus (like the sun). The Rutherford model has become the universally accepted picture of the structure of the atom. The properties of a given chemical element arise from the charge $+Ze$ on its nucleus and the presence of Z electrons around the nucleus. Although the planetary model has been extremely successful in explaining many of the properties of atoms, it has an inherent flaw that was recognized later: it couldn't possibly exist according to the laws of classical physics! We discuss this and other failures of classical physics to explain the properties of atoms in Chapter 4.

1.5 Mass Spectrometry, Isotopes, and the Measurement of Relative Mass

Mass spectrometry developed rather quickly after Thomson had shown how relative atomic and molecular masses could be measured directly by observing the deflections of ions in electric and magnetic fields. The most startling new result of his studies was the appearance of a new ion with a relative mass of 22 (see Fig. 1.14) that did not correspond to any known element; the doubly charged ion appeared at mass 11, leaving little doubt that this was a new species. This result was obtained with every sample of neon gas studied, including those of exceptional purity, and led Thomson to propose that neon was in fact not a simple gas but a mixture of two gases, the more abundant component having a relative mass of 20 and the less abundant component a relative mass of 22. It appeared, therefore, that the atomic mass of neon, determined by chemical means to be 20.2, was somehow an average of the masses of the two kinds of neon atoms. This was an astonishing result that seemed inconsistent with what was known about the elements at the time. The sample was chemically pure, and there was no room in the periodic table for a new element of mass 22 whose chemical properties were identical to those of neon. This interpretation was later confirmed using more advanced mass spectrometric methods, and this experiment provided one of the earliest clues for the existence of what became known as an **isotope** from the Greek *isos-* ["equal"] plus *topos* ["place"].

Refinements of Thomson's technique made it possible to measure relative masses extremely accurately and led to the discovery of the isotopes of many elements. A simple mass spectrometer, like the one shown schematically in Figure 1.16, consists of three elements: a region in which atoms or molecules are ionized and accelerated, one in which their flight paths are bent by magnetic forces, and a detector that measures the ion current. Positive ions are created by electron impact, the electrons being emitted from a heated filament and accelerated by an applied voltage as shown. The ions are then accelerated by an electric field created by a voltage applied to the plates shown, and enter the magnetic field region where they experience a magnetic force with magnitude given by

$$F_B = qvB$$

where q is the positive charge of the ion and v is its speed.

The deflection due to this force can be calculated using the same approach we described for Thomson's experiments, but for the present purpose it is

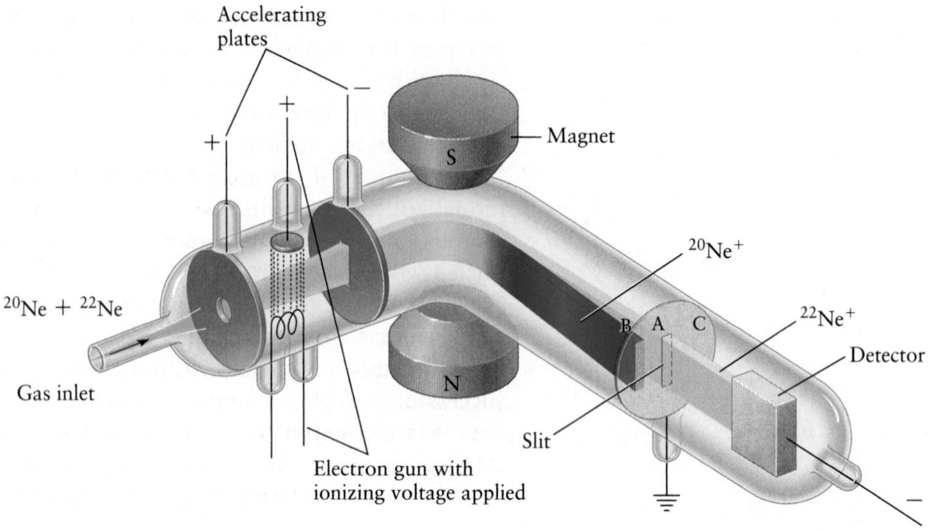

FIGURE 1.16 A simplified representation of a modern mass spectrometer. A gas mixture containing the isotopes ^{20}Ne and ^{22}Ne is introduced through the gas inlet. Atoms are ionized by electron impact and then accelerated by the electric field established between the accelerating plates. The ion beam passes into a magnetic field, where it is separated into components on the basis of the ions' charge-to-mass ratios. The conditions chosen for this illustration allow the heavier $^{22}Ne^+$ ions to pass through the slit in the plate (A) and reach the detector while the lighter $^{20}Ne^+$ ions strike the plate at position B and are not detected. Different conditions would move the $^{22}Ne^+$ beam to position C and allow the $^{20}Ne^+$ ions to pass through the slit and reach the detector.

sufficient to note that the lighter ions are deflected more than the heavier ions for a given magnetic field strength. The figure shows, for example, that ^{20}Ne has been deflected more than ^{22}Ne, which is passing through the slit and being detected under this set of conditions. A mass spectrum is acquired by scanning the magnetic field and measuring the ion current as a function of the magnetic field strength. The mass spectrum of neon, for example, consists of two peaks, as shown schematically in Figure 1.17; the more intense peak is ^{20}Ne and the less intense peak is ^{22}Ne, the relative intensities reflecting their relative natural abundances.

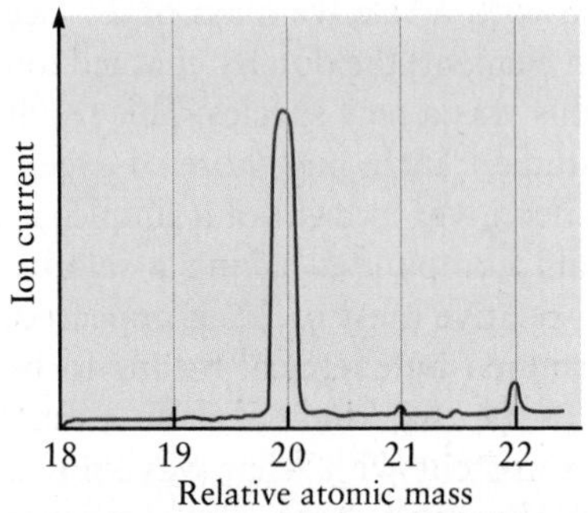

FIGURE 1.17 A sketch of the mass spectrum of Ne, showing only the two isotopes ^{20}Ne and ^{22}Ne.

Mass spectrometry confirmed the existence of isotopes for many elements and established a physical method for determining relative atomic masses. One consequence of this development was the realization that the relative atomic masses of the elements determined from the laws of chemical combination are actually averages of the relative atomic masses of its isotopes weighted by their relative natural abundances, averages that could now be calculated from its mass spectrum as follows. Suppose, for example, that there existed only the two neon isotopes identified so far and that 91% of the neon atoms were ^{20}Ne and 9% were ^{22}Ne. The average relative atomic mass would then be $(0.91)(20) + (0.09)(22) = 20.18$, which is close to the value reported in the table of relative atomic masses of the elements on the inside front cover of the book. Similarly, chlorine has two isotopes, about 75% being ^{35}Cl and 25% being ^{37}Cl, from which we calculate an average mass of 35.5, which is also close to the value reported in the table just cited. The average relative atomic mass of an element comprised of n isotopes with relative atomic masses A_i and relative fractional abundances p_i is given by

$$A = p_1A_1 + p_2A_2 + \cdots + p_nA_n = \sum_{1}^{n} p_iA_i$$

The discovery of isotopes created a major problem for chemists and physicists trying to establish a relative mass scale that was acceptable to both communities. Prior to the discovery of isotopes, and especially the isotopes of oxygen, the masses of the elements were determined from the laws of chemical composition using oxygen as the reference, whose relative atomic mass was defined to be exactly 16. Oxygen was a logical choice for the reference mass because it forms many compounds with the elements through combustion reactions. Having discovered a way to measure the relative atomic masses of the isotopes of all of the elements, the physicists argued that it made more sense, on physical grounds, to set the relative atomic mass of the ^{16}O isotope to 16, exactly. The problem with this choice, from the point of view of the chemists, was that reference data that had been generated for many decades would have to be revised because the error introduced by making this choice would be unacceptably high. A compromise was reached in 1961 in which the mass of the ^{12}C isotope was defined to have a relative atomic mass of 12, exactly. This choice was appealing to the physicists because it was based upon a physical measurement of a particular isotope and it was acceptable to the chemists because a fortuitous distribution of the isotopes of both carbon and oxygen in natural abundance averaged out the errors and reduced them to an acceptable level.

Relative atomic masses have no units because they are ratios of two masses measured in whatever units we choose (grams, kilograms, pounds, and so forth). The **relative molecular mass** of a compound is the sum of the relative atomic masses of the elements that constitute it, each one multiplied by the number of atoms of that element in a molecule. For example, the formula of water is H_2O, so its relative molecular mass is

$$2 \text{ (relative atomic mass of H)} + 1 \text{ (relative atomic mass of O)} = 2(1.0079) + 1(15.9994) = 18.0152$$

EXAMPLE 1.2

Calculate the relative atomic mass of carbon, taking the relative atomic mass of ^{13}C to be 13.003354 on the ^{12}C scale.

Solution

Set up the following table:

Isotope	Isotopic Mass × Abundance
^{12}C	12.000000 × 0.98892 = 11.867
^{13}C	13.003354 × 0.01108 = 00.144

Chemical relative atomic mass = 12.011

Related Problems: 15, 16, 17, 18

The number of significant figures in a table of chemical or natural relative atomic masses (see the inside front cover of this book) is limited not only by the accuracy of the mass spectrometric data but also by any variability in the natural abundances of the isotopes. If lead from one mine has a relative atomic mass of 207.18 and lead from another has a mass of 207.23, there is no way a result more precise than 207.2 can be obtained. In fact, geochemists now use small variations in the $^{16}O:^{18}O$ ratio as a "thermometer" to deduce the temperatures at which different oxygen-containing rocks were formed in the Earth's crust over geological time scales. They also find anomalies in the oxygen isotopic compositions of certain meteorites, implying that their origins may lie outside our solar system. Temperature variations over the past million years of the earth's history have been established by measuring the $^{16}O:^{18}O$ and $^{2}H:^{1}H$ ratios as a function of depth in the Greenland and Antarctic ice cores and correlated with atmospheric CO_2 concentration, providing important clues as to the origins of climate change (see Section 20.6).

1.5.1 Structure of the Nucleus: Protons, Neutrons, and Isotopes

The experiments described earlier led to the discovery of electrons and nuclei and to the planetary model of the atom, but they did not provide sufficient evidence to establish a model for the structure of the nucleus. Subsequent experiments, of the same general type we have discussed, identified the two other elementary particles, the proton and the neutron, whose properties we describe here. A full discussion of the nature of those experiments is beyond the scope of this textbook. The smallest and simplest nucleus is that of the hydrogen atom—the proton (from the Greek *protos* ["first"])—so named because it is a fundamental component of all nuclei. The proton has a positive charge of $+e$, where e is the elementary charge, but its mass is 1.67262×10^{-27} kg, which is 1836 times greater than the electron mass. Nuclei of the other elements contain an integral number of protons, which is given by their **atomic number** Z; thus, the charge on the nucleus is $+Ze$. The existence of isotopes, elements with the same atomic number but different masses, implied that there was another nuclear particle whose mass was the same as the proton mass but with no charge—that particle is the neutron. The **mass number** A of a particular isotope is the sum of the numbers of protons and neutrons for the isotope. $A = Z + N$. We write chemical symbols for the isotopes (nuclides) as follows: ${}^{A}_{Z}X$, where X is the chemical symbol for the element. The isotopes of hydrogen, for example, are written as: ${}^{1}_{1}H$, ${}^{2}_{1}H$, and ${}^{3}_{1}H$ for hydrogen, deuterium, and tritium, respectively. It is customary, though redundant, to include the atomic number in addition to the symbol for the element, primarily because it is helpful when balancing nuclear chemical reactions (see Chapter 19).

EXAMPLE 1.3

Radon-222 (${}^{222}Rn$) has recently received publicity because its presence in basements may increase the number of cancer cases in the general population, especially among smokers. State the number of electrons, protons, and neutrons that make up an atom of ${}^{222}Rn$.

Solution

From the table on the inside front cover of the book, the atomic number of radon is 86; thus, the nucleus contains 86 protons and $222 - 86 = 136$ neutrons. The atom has 86 electrons to balance the positive charge on the nucleus.

Related Problems: 19, 20, 21, 22

1.6 The Mole: Counting Molecules by Weighing

The laws of chemical combination assert that chemical reactions occur in such a way that the number of atoms of a given type is conserved in every chemical reaction, except nuclear reactions. It is impractical to count the numbers of atoms or molecules in laboratory or industrial scale reactions, however, so we must find a way to relate the masses of the reactants and products in those reactions to the numbers of atoms or molecules involved. Chemists established a scale of relative atomic masses in the 19th century, while developing the laws of chemical combination; the accuracy of that scale was grealy improved upon in the 20th century using mass spectrometry. That relative atomic mass scale must be converted to a macroscopic scale that allows us to count atoms and molecules by weighing. The concept and methods that allow us to do this are developed in this section.

1.6.1 Relation between Atomic and Macroscopic Masses: Avogadro's Number

Laboratory or industrial chemical reactions are carried out with quantities that range from milligrams to tons, so we must be able to relate the relative atomic mass scale to the macroscopic scales used in practice. The link between the two scales is provided by **Avogadro's number** (N_A), defined as the number of atoms in exactly 12 g of ^{12}C, the currently accepted value of which is

$$N_A = 6.02214129 \times 10^{23}$$

The mass of a single ^{12}C atom is then found by dividing exactly 12 g carbon (C) by N_A:

$$\text{Mass of a } ^{12}\text{C atom} = \frac{12.00000 \text{ g}}{6.02214129 \times 10^{23}} = 1.9926467 \times 10^{-23} \text{ g}$$

The masses of individual atoms are truly small and any macroscopic quantity of mass contains an amazingly large number of atoms.

Avogadro's number is defined relative to the ^{12}C atom because that isotope has been chosen by international agreement to form the basis for the modern scale of relative atomic masses. We can find the masses of Avogadro's number of the other elements by simply taking ratios, as follows. Consider sodium, which has a relative atomic mass of 22.98977. A sodium atom is 22.98977/12 times as heavy as a ^{12}C atom. If the mass of N_A atoms of ^{12}C is 12 g, then the mass of N_A atoms of sodium must be

$$\frac{22.98977}{12}\left(12 \text{ g}\right) = 22.98977 \text{ g}$$

The mass (in grams) of N_A atoms of *any* element is numerically equal to the relative atomic mass of that element. The same conclusion applies to molecules. From the relative molecular mass of water determined earlier, the mass of N_A molecules of water is 18.0152 g.

EXAMPLE 1.4

One of the heaviest atoms found in nature is ^{238}U. Its relative atomic mass is 238.0508 on a scale in which 12 is the atomic mass of ^{12}C. Calculate the mass (in grams) of one ^{238}U atom.

Solution

Because the mass of N_A atoms of ^{238}U is 238.0508 g and N_A is $6.02214129 \times 10^{23}$, the mass of one ^{238}U atom must be

$$\frac{238.0508 \text{ g}}{6.02214129 \times 10^{23}} = 3.952926 \times 10^{-22} \text{ g}$$

Related Problems: 23, 24

1.6.2 The Mole

Because the masses of atoms and molecules are so small, laboratory scale chemical reactions must involve large numbers of atoms and molecules. It is convenient to group atoms or molecules in counting units of $N_A = 6.02214129 \times 10^{23}$ to measure the **number of moles** of a substance. One of these counting units is called a **mole** (abbreviated mol, whether singular or plural, derived from Latin *moles,* meaning "heap" or "pile"). One mole of a substance is the amount that contains Avogadro's number of atoms, molecules, or other entities. That is, 1 mol of ^{12}C contains N_A ^{12}C atoms, 1 mol of water contains N_A water molecules, and so forth. We must be careful in some cases, because a phrase such as "1 mol of oxygen" is ambiguous. We should refer instead to "1 mol of O_2" if there are N_A oxygen *molecules,* and "1 mol of O" if there are N_A oxygen *atoms.* Henceforth, for *any* species we use "number of moles of a particular species" to describe the number of moles in a sample of that species.

We define the **molar mass** of an element (often called the atomic mass or the atomic weight) as the mass of one mole of that element in grams; it is determined by taking the ratio of the relative atomic mass of the element to that of ^{12}C and multiplying the result by 12 g. The same procedure is used to calculate the molar masses (often called the molecular weights) for molecules. Thus, the relative molecular mass of water is 18.0152, and its molar mass is 18.0152 g mol^{-1}.

To determine the number of moles of a given substance, we use the chemist's most powerful tool, the laboratory balance. If a sample of iron weighs 8.232 g, then

$$\begin{aligned}\text{moles of iron} &= \frac{\text{number of grams of iron}}{\text{molar mass of iron}}\\ &= \frac{8.232 \text{ g Fe}}{55.847 \text{ g mol}^{-1}}\\ &= 0.1474 \text{ mol Fe}\end{aligned}$$

where the molar mass of iron was obtained from the periodic table of the elements or a table of relative atomic masses (see the inside front and back covers of this book). The calculation can be turned around as well. Suppose a certain amount, for example, 0.2000 mol, of water is needed in a chemical reaction. We have

$$\left(\text{moles of water}\right) \times \left(\text{molar mass of water}\right) = \text{mass of water}$$

$$\left(0.2000 \text{ mol H}_2\text{O}\right) \times \left(18.015 \text{ g mol}^{-1}\right) = 3.603 \text{ g H}_2\text{O}$$

We simply weigh 3.603 g water to get the 0.2000 mol needed for the reaction. In both cases, the molar mass is the conversion factor between the mass of the substance and the number of moles of the substance.

Although the number of moles in a sample is generally determined by weighing, it is still preferable to think of a mole as a fixed number of atoms or molecules (Avogadro's number) rather than as a fixed mass. The term *mole* is thus analogous to a term such as *dozen*: the mass of one dozen pennies is 26 g, which is substantially less than the mass of one dozen nickels, 60 g; each group contains 12 coins. Figure 1.18 shows mole quantities of several

FIGURE 1.18 One mole of a number of different substances. Clockwise from top: graphite (C), potassium permanganate ($KMnO_4$), copper sulfate pentahydrate ($CuSO_4 \cdot 5\,H_2O$), copper (Cu), sodium chloride (NaCl), and potassium dichromate ($K_2Cr_2O_7$). Antimony (Sb) is at the center.

substances. A mole of most common household substances (water, sugar, salt) is about a tablespoon.

EXAMPLE 1.5

Nitrogen dioxide (NO_2) is a major component of urban air pollution. For a sample containing 4.000 g NO_2, calculate (a) the number of moles of NO_2 and (b) the number of molecules of NO_2.

Solution

(a) From the tabulated molar masses of nitrogen (14.007 g mol^{-1}) and oxygen (15.999 g mol^{-1}), the molar mass of NO_2 is

$$14.007 \text{ g mol}^{-1} + (2 \times 15.999 \text{ g mol}^{-1}) = 46.005 \text{ g mol}^{-1}$$

The number of moles of NO_2 is then

$$\text{mol of NO}_2 = \frac{4.000 \text{ g NO}_2}{46.005 \text{ g mol}^{-1}} = 0.08695 \text{ mol NO}_2$$

(b) To convert from moles to number of molecules, multiply by Avogadro's number:

$$\text{molecules of NO}_2 = \left(0.08695 \text{ mol NO}_2\right) \times 6.0221 \times 10^{23} \text{ mol}^{-1}$$
$$= 5.236 \times 10^{22} \text{ molecules NO}_2$$

Related Problems: 29, 30

1.6.3 Density and Molecular Size

The **mass density** of a substance, hereafter density, is the ratio of its mass to its volume:

$$\text{density} = \frac{\text{mass}}{\text{volume}} \qquad [1.11]$$

The base unit of mass in the International System of Units (SI; see discussion of base units in Appendix B) is the kilogram (kg), but when discussing solubility or mass density the gram often is used instead; moreover, the gram is used in tables of molar masses. Several units for volume are in frequent use. The base SI unit of the cubic meter (m^3) is unwieldy for laboratory purposes (1 m^3 water weighs 1000 kg, or 1 metric ton). We use the gram (g) for mass and the liter (1 L = 10^{-3} m^3) or the cubic centimeter (cm^3 or milliliter, mL) for volume (1 cm^3 = 1 mL = 10^{-3} L = 10^{-6} m^3). Table 1.1 lists the densities of some substances in units of grams per cubic centimeter.

The density of a substance is not a fixed, invariant property of the substance; its value depends on the pressure and temperature at the time of measurement. For some substances (especially gases and liquids), the volume may be more convenient to measure than the mass, and the density provides the conversion factor between volume and mass. For example, the density of liquid benzene (C_6H_6) is 0.8765 g cm^{-3} near room temperature. Suppose that we wanted to find the mass of benzene contained in a volume that measured 0.2124 L. We simply multiply the volume by the density as follows:

$$m = \rho V$$

TABLE 1.1

Densities of Selected Substances

Substance	Density (g cm^{-3})
Hydrogen	0.000082
Oxygen	0.00130
Water	1.00
Magnesium	1.74
Sodium chloride	2.16
Quartz	2.65
Aluminum	2.70
Iron	7.86
Copper	8.96
Silver	10.5
Lead	11.4
Mercury	13.5
Gold	19.3
Platinum	21.4

These densities were measured at room temperature and at average atmospheric pressure near sea level.

where m is the mass, ρ is the density, and V is the volume. Therefore, the value of the mass of benzene is

$$m = 0.2124 \text{ L} \times (1 \times 10^3 \text{ cm}^3 \text{ L}^{-1}) \times (0.8765 \text{ g cm}^{-3}) = 186.2 \text{ g}$$

If we wanted to know the number of moles of benzene in that sample we would simply divide the mass by the molar mass of benzene (78.114 g mol^{-1}) to get 2.384 mol.

Knowing the density and molar mass of a substance, we can readily compute its **molar volume, *V*/*n*,** that is, the volume occupied by one mole of a substance:

$$\frac{V}{n} = \frac{\text{molar mass } \left(\text{g mol}^{-1}\right)}{\text{density } \left(\text{g cm}^{-3}\right)} = \text{molar volume } \left(\text{cm}^3 \text{ mol}^{-1}\right)$$

For example, near 0°C, ice has a density of 0.92 g cm^{-3}; thus, the molar volume of solid water under these conditions is

$$\frac{V}{n} = \frac{18.0 \text{ g mol}^{-1}}{0.92 \text{ g cm}^{-3}} = 20 \text{ cm}^3 \text{ mol}^{-1}$$

The molar volume of a gas is much larger than that of either a liquid or a solid. For O_2 under room conditions, the data in Table 1.1 give a molar volume of 24,600 $\text{cm}^3 \text{ mol}^{-1}$ = 24.6 L mol^{-1}, which is more than 1000 times larger than the molar volume just computed for ice under the same conditions of temperature and pressure. (You should remember this ratio because you will use it often later on, when making estimates of the properties of gases, liquids, and solids.) How can we interpret this fact on a microscopic level? We also note that the volumes of liquids and solids do not shift much with changes in temperature or pressure, but that the volumes of gases are quite sensitive to these changes. One hypothesis that would explain these observations is that the molecules in liquids and solids are close enough to touch one another, but that they are separated by large distances in gases. If this hypothesis is correct (as has been well established by further study), then the sizes of the molecules themselves can be estimated from the volume occupied per molecule in the liquid or solid state. The volume per molecule is the molar volume divided by Avogadro's number; for ice, this gives

$$\text{Volume per H}_2\text{O molecule} = \frac{20 \text{ cm}^3 \text{ mol}^{-1}}{6.02 \times 10^{23} \text{ mol}^{-1}} = 3.3 \times 10^{-23} \text{ cm}^3$$

This volume corresponds to that of a cube with edges about 3.2×10^{-8} cm (0.32 nm) on a side. We conclude from this and other density measurements that the characteristic size of atoms and small molecules is about 10^{-8} cm, or about 0.1 nm. This length, 0.1 nm or 1×10^{-10} m, occurs so frequently in chemistry that it has been given a special name, the ångström (Å), in honor of the Swedish physicist Anders Ångström. Avogadro's number provides the link between the length and mass scales of laboratory measurements and the masses and volumes of single atoms and molecules.

CONNECTION TO NANOTECHNOLOGY

Imaging Atoms, Molecules, and Chemical Reactions by Scanning Tunnelling Microscopy

The laws of chemical combination provided indirect evidence for the existence of atoms. The experiments of Thomson, Wien, and Rutherford provided direct physical evidence for the existence of the elementary particles that make up the atom. We conclude this chapter by describing an experimental method that allows us not only to image individual atoms and molecules but also to observe and control a chemical reaction at the single molecule level—a feat only dreamed of as recently as the mid-1980s. The ability to manipulate single atoms, molecules, and nanoparticles has developed into a set of technologies that allows scientists and engineers to create two-dimensional arrays of these entities with applications in materials science, biology, and medicine.

Microscopy began with the fabrication of simple magnifying glasses; the development of the compound microscope in the late 17th century made it possible to observe single biological cells, enabling the study of biology on the cellular level for the first time. The invention of the electron microscope in the 1930s opened the way to observe objects with dimensions much smaller than the wavelength of light, the resolution limit of optical microscopes. Unfortunately, however, sample damage caused by the high energy of the electron beam required is still a limitation of the technique for certain applications. Gerd Binnig and Heinrich Rohrer developed the scanning tunneling microscope (STM), which images atoms using low-energy electrons, for which they received the 1986 Nobel Prize in Physics. The STM measures small currents produced by electrons that tunnel between a very sharp conducting tip and the sample at very small tip-surface separations, as shown in Figure C1.1. The tunneling current is an exponential function of the tip-surface separation. A feedback loop monitors the tunneling current, which is fed to the piezoelectric element to keep the tip-surface separation constant as the tip is scanned across the sample. The three-dimensional STM images shown are plots of the tip height as a function of the tip position in the plane of the sample. Individual atoms of the single crystal metal surface are clearly resolved, as are some missing atoms (defects) shown in the inset.

Scanning tunneling microscope images have visually confirmed many properties of solids, surfaces, and adsorbed species, such as the sizes of atoms and the distances between them, which are already known from other techniques. But, much new information has been obtained as well. The STM images have shown the positions and shapes of molecules undergoing chemical reactions on surfaces, which helps guide the search for new ways of carrying out such reactions. They have also revealed the shape of the surface of the molecules of the nucleic acid DNA, which plays a central role in genetics.

Images of Single Molecules in Chemical Reactions

The STM has been used to image the surfaces of materials since the mid-1980s, but only recently has it been used to image single molecules and to initiate chemical reactions at the single molecule level, as we illustrate with the

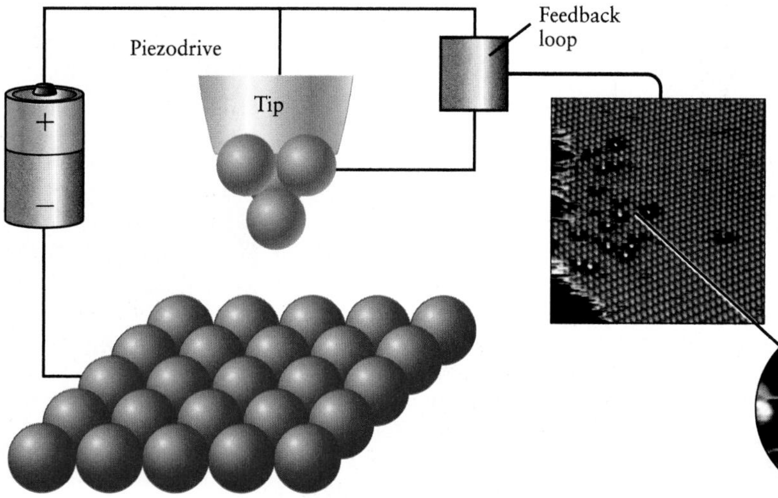

FIGURE C1.1 Schematic of STM experiment. The distance from the atoms on the STM tip (blue) and the atoms on the surface being studied (green) is typically a few angstroms. The electrostatic potential difference between the tip and the surface is controlled, in this figure, by a battery, but normally one would use a tunable voltage supply.

following example. Although the STM can be used to image objects in air, the experiments described here were conducted in ultrahigh vacuum (extremely low pressure) to ensure that only the reactants of interest were present on the surface. Figure C1.2 shows an STM tip hovering over a silver (Ag) surface on which an iron (Fe) atom and a carbon monoxide (CO) molecule have been chemically bonded (adsorbed), and a series of steps that leads to the formation of a product molecule Fe(CO).

Figure C1.2(a) shows the tip positioned over a single carbon monoxide (CO) molecule, ready to pluck it from the silver (Ag) surface; in part (b) CO is adsorbed onto the tip, bonded via the carbon (C) atom, and the tip is then translated across the surface to a region near an iron (Fe) atom; part (c) shows CO being transferred to the Fe atom and forming an Fe-CO bond; and finally, in part (d) the tip is withdrawn with the product molecule, FeCO, remaining bound to the Ag surface. Bonding between the carbon monoxide and the tip is controlled by the tip voltage and current. CO is plucked from the surface when electrons flow from the tip to the surface and it is released to the surface when electrons flow in the opposite direction.

The schematic serves as a guide to the eye for interpreting the real STM images shown in Figure C1.3. Each image represents an area of the surface that measures 6.3 × 6.3 nm. The false color scale reflects the height of the objects above the plane of the silver surface atoms; the red end of the scale represents protrusions, whereas the purple end represents depressions. The identity of each chemical species was established by the nature of the image and also by the way in which the current varied with the applied voltage. That variation provides a chemical signature.

In part A of Figure C1.3, five Fe atoms and five CO molecules are clearly seen; the red arrow identifies one Fe atom that is a bit difficult to see otherwise. The curved white arrow shows a CO molecule in close proximity to an Fe atom. Part B shows the FeCO molecule formed as a result of the transfer of that CO molecule to the Fe atom by the tip, as well as another potentially reactive pair identified by the white curved arrow. From the shape of the resulting image in part C, we can see that another FeCO molecule has been formed. The white curved

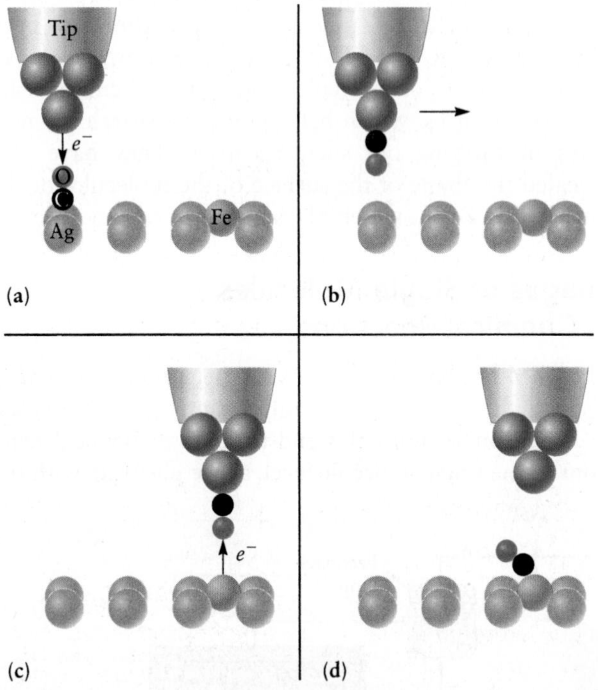

FIGURE C1.2 Moving a CO molecule adsorbed to a surface with an STM tip. In (a) the electrostatic potential difference (provided by a "battery" in the schematic in Fig C1.1) is biased so that an electron tunnels from the tip to the adsorbed CO molecule. In (b) this results in the CO molecule being picked up by the STM tip (note that the CO molecule flips direction), and the experimenter then moves the STM tip toward another part of the surface where an iron atom resides. In (c), changing the bias of the electrostatic potential (by reversing the direction of the "battery") allows an electron to tunnel from the iron atom adsorbed on the surface to the CO molecule on the STM tip; finally in (d) CO transfers from the STM tip back to the iron atom on the surface and they react with each other to make FeCO.

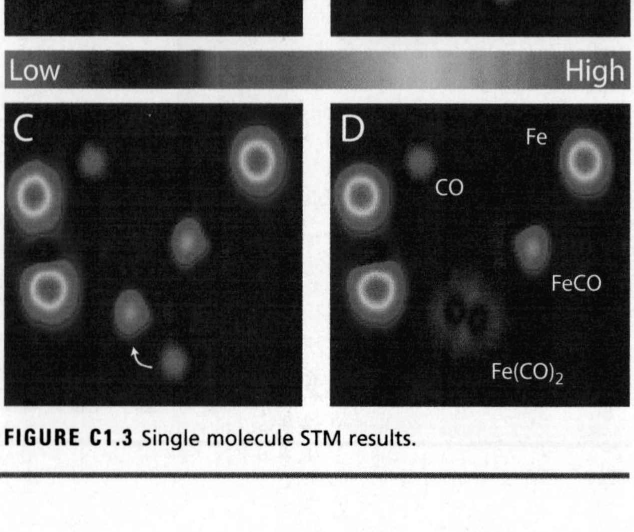

FIGURE C1.3 Single molecule STM results.

arrow suggests the possibility of adding an additional CO molecule to the first FeCO synthesized to form $Fe(CO)_2$, which, indeed, occurs as shown in part D. This remarkable sequence of images shows clearly the synthesis of a pair of distinct Fe(CO) molecules, as well as an $Fe(CO)_2$ molecule, from the reactants Fe and CO adsorbed onto a silver surface. These syntheses were accomplished by manipulating single CO molecules to place them sufficiently close to Fe atoms to initiate a chemical reaction, demonstrating our ability to observe and control chemical reactions at the single molecule level.

The ability to manipulate single molecules and nanoparticles has led to the recent development of **dip-pen lithography**, a promising approach for synthesizing two-dimensional arrays of a wide variety of materials of interest. The method relies upon the forces between molecules and surfaces (the basis of **atomic force microscopy**) instead of tunneling to assemble arrays of inorganic materials for catalysis, sensing or microelectronics applications as well as arrays of biomolecules for rapid throughput screening.

CHAPTER SUMMARY

We have come a long way since the attempts of the alchemists to turn base metals into gold, to transmute one element into another. Through the early chemical experiments of Dalton, Gay-Lussac, and Avogadro, we have learned that matter is ultimately indivisible, at least as far as its physical and chemical properties are concerned. The experiments of Thomson, Wien, and Rutherford confirmed, from the results of physical measurements, the existence of the atom. These experiments also identified and characterized the elementary particles from which the atom is made, and this led to the modern model of the atom as an object with a small, dense nucleus surrounded by a much larger volume occupied by electrons. Physicists in the 21st century have developed tools of unprecedented power with which to analyze and synthesize single molecules, an achievement that has already led to exciting new applications in almost every area of modern science and engineering. We have also shown you how chemists count molecules by weighing macroscopic quantities of substances. Avogadro's number connects the nanoscopic world of atoms and molecules to the macroscopic world of the laboratory: 1 mol = 6.022×10^{23} atoms or molecules.

CONCEPTS AND SKILLS

Section 1.1 – The Nature of Modern Chemistry

State and interpret the laws of conservation of energy and conservation of mass and distinguish between macroscopic and nanoscopic length scales.

- Energy and mass are separately conserved in ordinary (not nuclear) chemical reactions. The total mass of the products equals the total mass of the reactants. The total amount of energy contained in the products is equal to the total amount of energy contained in the reactants. Chemical reactions occur on the scale of nanometers with masses of the order 10^{-26} kg, but we observe them in the laboratory on a scale of grams and centimeters.

Section 1.2 – Elements: The Building Blocks of Matter

Describe in operational terms how to distinguish among mixtures, compounds, and elements (Problems 1–4).

- Mixtures may be separated into simpler substances by *physical* processes like filtration or distillation. Substances that can be separated into simpler substances by chemical reactions are called compounds; those that cannot are called elements.

Section 1.3 – Indirect Evidence for the Existence of Atoms: Laws of Chemical Combination

Outline Dalton's atomic theory of matter and describe its experimental foundation.

- Dalton asserted that matter consisted of indivisible atoms and that all atoms of a given element were identical in mass and all other physical and chemical properties. Atoms of different elements have different characteristic physical and chemical properties, such as mass, that distinguish them from each other. Atoms are indestructible and retain their identities in chemical reactions. Compounds are formed from atoms in the elements combined in small whole-number ratios. Dalton's atomic theory was developed to explain the law of conservation of mass and the law of definite proportions.

Describe how chemical methods lead to the determination of chemical formulas (Problems 7–10).

- The laws of chemical combination, which include the law of multiple proportions and Avogadro's hypothesis, provided the basis for developing a procedure with which to determine chemical formulas. The ratios of the masses of elements in particular compounds were always fixed, and the ratios of the masses of the elements in compounds that contained the same elements were themselves also a ratio of small integers. Empirical (simplest) chemical formulas were established on the basis of these ratios.

Section 1.4 – The Physical Structure of Atoms

Describe the experiments that led to the discovery and characterization of electrons, nuclei, and isotopes (Problems 15–16).

- Cathode rays, emitted from the cathodes in glow discharge tubes, were shown to be negatively charged particles whose charge-to-mass ratio was first measured by Thomson. Millikan measured the charge on these particles, thus determining both the charge and the mass of the electron. Canal rays were shown to be massive, positively charged particles whose charge-to-mass ratio depended on the gas used in the discharge. Radioactive decay suggested, and mass spectroscopy confirmed, the existence of isotopes of the elements, with similar physical and chemical properties but different masses.

Describe the experiments that led to the planetary model of the atom and discuss its features.

- The scattering of α particles by thin gold films established the planetary model of the atom, with a small, dense nucleus surrounded by electrons that occupied most of the volume of the atom.

Section 1.5 – Mass Spectrometry, Isotopes, and Relative Masses

State the numbers of protons, neutrons, and electrons for any atom in the periodic table (Problems 19–22).

Section 1.6 – The Mole: Counting Molecules by Weighing

Interconvert mass, number of moles, number of molecules, and (using density) the molar volume of a substance (Problems 23–34).

- Avogadro's number (6.02×10^{23}) is the conversion factor between moles and numbers of molecules. The molar mass is the conversion factor between moles and mass. Setting up calculations to convert any of these quantities can be done by inspection, using dimensional analysis to guide you. Molar volumes are calculated by dividing the molar mass by the density.

Describe in a general way the operation of the scanning tunneling microscope (STM) and the kinds of information it provides.

PROBLEMS

Answers to problems whose numbers are red appear in Appendix G. Problems that are more challenging are indicated with asterisks.

1.2 Elements: The Building Blocks of Matter

1. Classify the following materials as substances or mixtures: table salt, wood, mercury, air, water, seawater, sodium chloride, and mayonnaise. If they are mixtures, subclassify them as homogeneous or heterogeneous; if they are substances, subclassify them as compounds or elements.

2. Classify the following materials as substances or mixtures: absolute (pure) alcohol, milk (as purchased in a store), copper wire, rust, barium bromide, concrete, baking soda, and baking powder. If they are mixtures, subclassify them as homogeneous or heterogeneous; if they are substances, subclassify them as compounds or elements.

3. A 17th-century chemist wrote of the "simple bodies which enter originally into the composition of mixtures and into which these mixtures resolve themselves or may be finally resolved." What is being discussed?

4. Since 1800, almost 200 sincere but erroneous reports of the discovery of new chemical elements have been made. Why have mistaken reports of new elements been so numerous? Why is it relatively easy to prove that a material is not a chemical element, but difficult to prove absolutely that a material is an element?

1.3 Indirect Evidence for the Existence of Atoms: Laws of Chemical Combination

5. A sample of ascorbic acid (vitamin C) is synthesized in the laboratory. It contains 30.0 g carbon and 40.0 g oxygen. Another sample of ascorbic acid, isolated from lemons (an excellent source of the vitamin), contains 12.7 g carbon. Compute the mass of oxygen (in grams) in the second sample.

6. A sample of a compound synthesized and purified in the laboratory contains 25.0 g hafnium and 31.5 g tellurium. The identical compound is discovered in a rock formation. A sample from the rock formation contains 0.125 g hafnium. Determine how much tellurium is in the sample from the rock formation.

7. Nitrogen (N) and silicon (Si) form two binary compounds with the following compositions:

Compound	Mass % N	Mass % Si
1	33.28	66.72
2	39.94	60.06

   (a) Compute the mass of silicon that combines with 1.0000 g of nitrogen in each case.

   (b) Show that these compounds satisfy the law of multiple proportions. If the second compound has the formula Si_3N_4, what is the formula of the first compound?

8. Iodine (I) and fluorine (F) form a series of binary compounds with the following compositions:

Compound	Mass % I	Mass % F
1	86.979	13.021
2	69.007	30.993
3	57.191	42.809
4	48.829	51.171

   (a) Compute in each case the mass of fluorine that combines with 1.0000 g iodine.

   (b) By figuring out small whole-number ratios among the four answers in part (a), show that these compounds satisfy the law of multiple proportions.

9. Vanadium (V) and oxygen (O) form a series of compounds with the following compositions:

Mass % V	Mass % O
76.10	23.90
67.98	32.02
61.42	38.58
56.02	43.98

   What are the relative numbers of atoms of oxygen in the compounds for a given mass of vanadium?

10. Tungsten (W) and chlorine (Cl) form a series of compounds with the following compositions:

Mass % W	Mass % Cl
72.17	27.83
56.45	43.55
50.91	49.09
46.36	53.64

   If a molecule of each compound contains only one tungsten atom, what are the formulas for the four compounds?

11. A liquid compound containing only hydrogen and oxygen is placed in a flask. Two electrodes are dipped into the liquid and an electric current is passed between them. Gaseous hydrogen forms at one electrode and gaseous oxygen at the other. After a time, 14.4 mL hydrogen has evolved at the negative terminal, and 14.4 mL oxygen has evolved at the positive terminal.

   (a) Assign a chemical formula to the compound in the cell.

   (b) Explain why more than one formula is possible as the answer to part (a).

12. A sample of liquid N_2H_4 is decomposed to give gaseous N_2 and gaseous H_2. The two gases are separated, and the nitrogen occupies 13.7 mL at room conditions of pressure and temperature. Determine the volume of the hydrogen under the same conditions.

13. Pure nitrogen dioxide (NO_2) forms when dinitrogen oxide (N_2O) and oxygen (O_2) are mixed in the presence of a certain catalyst. What volumes of N_2O and oxygen are needed to produce 4.0 L NO_2 if all gases are held at the same conditions of temperature and pressure?

14. Gaseous methanol (CH_3OH) reacts with oxygen (O_2) to produce water vapor and carbon dioxide. What volumes of water vapor and carbon dioxide will be produced from 2.0 L methanol if all gases are held at the same temperature and pressure conditions?

1.4 The Physical Structure of Atoms

15. In J. J. Thompson's experiment depicted in Figures 1.10 and 1.11, assume that the electrons are initially traveling at a speed of 8.00×10^6 m s^{-1} in the $+x$ direction and the electrostatic plates have a length of $\ell = 0.10$ m, between which the electrons are deflected toward the $-y$ direction (downward) by an electric field of magnitude 200 V m^{-1}. (Note that 1 volt = 1 V = 1 J C^{-1}.) Calculate S, the magnitude of the downward displacement, if the fluorescent screen is located 0.90 m beyond the exit of the region between the two electrostatic plates. No magnetic field is applied to counteract the electrostatic force. (Caution: L is not 1.0 m nor 0.9 m; see Figure 1.11.)

16. In the problem 15 above, what is v_y, the y-component of the electron's velocity, when it has reached the right edge of the plates? Note that $a_y = \dfrac{\Delta v_y}{\Delta t}$ (be careful with the sign). How is the ratio S/L related to v_x and v_y, if v_y is calculated at the right edge of the plates? The distances S and L are shown in Figure 1.11.

1.5 Mass Spectrometry, Isotopes, and the Measurement of Relative Mass

17. The natural abundances and isotopic masses of the element silicon (Si) relative to $^{12}C = 12.00000$ are

Isotope	% Abundance	Isotopic Mass
^{28}Si	92.21	27.97693
^{29}Si	4.70	28.97649
^{30}Si	3.09	29.97376

Calculate the atomic mass of naturally occurring silicon.

18. The natural abundances and isotopic masses of the element neon (Ne) are

Isotope	% Abundance	Isotopic Mass
^{20}Ne	90.00	19.99212
^{21}Ne	0.27	20.99316
^{22}Ne	9.73	21.99132

Calculate the atomic mass of naturally occurring neon.

19. Only two isotopes of boron (B) occur in nature; their atomic masses and abundances are given in the following table. Complete the table by computing the relative atomic mass of ^{11}B to four significant figures, taking the tabulated relative atomic mass of natural boron as 10.811.

Isotope	% Abundance	Atomic Mass
^{10}B	19.61	10.013
^{11}B	80.39	?

20. More than half of all the atoms in naturally occurring zirconium are ^{90}Zr. The other four stable isotopes of zirconium have the following relative atomic masses and abundances:

Isotope	% Abundance	Atomic Mass
^{91}Zr	11.27	90.9056
^{92}Zr	17.17	91.9050
^{94}Zr	17.33	93.9063
^{96}Zr	2.78	95.9083

Compute the relative atomic mass of ^{90}Zr to four significant digits, using the tabulated relative atomic mass 91.224 for natural zirconium.

21. The isotope of plutonium used for nuclear fission is ^{239}Pu. Determine (a) the ratio of the number of neutrons in a ^{239}Pu nucleus to the number of protons, and (b) the number of electrons in a single plutonium atom.

22. The last "missing" element from the first six periods was promethium, which was finally discovered in 1947 among the fission products of uranium. Determine (a) the ratio of the number of neutrons in a ^{145}Pm nucleus to the number of protons, and (b) the number of electrons in a single promethium atom.

23. The americium isotope ^{241}Am is used in smoke detectors. Describe the composition of a neutral atom of this isotope for protons, neutrons, and electrons.

24. In 1982, the production of a single atom of $^{266}_{109}Mt$ (meitnerium-266) was reported. Describe the composition of a neutral atom of this isotope for protons, neutrons, and electrons.

1.6 The Mole: Counting Molecules by Weighing

25. Compute the mass (in grams) of a single iodine atom if the relative atomic mass of iodine is 126.90447 on the accepted scale of atomic masses (based on 12 as the relative atomic mass of ^{12}C).

26. Determine the mass (in grams) of exactly 100 million atoms of fluorine if the relative atomic mass of fluorine is 18.998403 on a scale on which exactly 12 is the relative atomic mass of ^{12}C.

27. Compute the relative molecular masses of the following compounds on the ^{12}C scale:
(a) P_4O_{10} **(b)** BrCl
(c) $Ca(NO_3)_2$ **(d)** $KMnO_4$
(e) $(NH_4)_2SO_4$

28. Compute the relative molecular masses of the following compounds on the ^{12}C scale:
(a) $[Ag(NH_3)_2]Cl$ **(b)** $Ca_3[Co(CO_3)_3]_2$
(c) OsO_4 **(d)** H_2SO_4
(e) $Ca_3Al_2(SiO_4)_3$

29. Suppose that a person counts out gold atoms at the rate of one each second for the entire span of an 80-year life. Has the person counted enough atoms to be detected with an ordinary balance? Explain.

30. A gold atom has a diameter of 2.88×10^{-10} m. Suppose the atoms in 1.00 mol of gold atoms are arranged just touching their neighbors in a single straight line. Determine the length of the line.

31. The vitamin A molecule has the formula $C_{20}H_{30}O$, and a molecule of vitamin A_2 has the formula $C_{20}H_{28}O$. Determine how many moles of vitamin A_2 contain the same number of atoms as 1.000 mol vitamin A.

32. Arrange the following in order of increasing mass: 1.06 mol SF_4; 117 g CH_4; 8.7×10^{23} molecules of Cl_2O_7; and 417×10^{23} atoms of argon (Ar).

33. Mercury is traded by the "flask," a unit that has a mass of 34.5 kg. Determine the volume of a flask of mercury if the density of mercury is 13.6 g cm^{-3}.

34. Gold costs \$400 per troy ounce, and 1 troy ounce = 31.1035 g. Determine the cost of 10.0 cm^3 gold if the density of gold is 19.32 g cm^{-3} at room conditions.

35. Aluminum oxide (Al_2O_3) occurs in nature as a mineral called corundum, which is noted for its hardness and resistance to attack by acids. Its density is 3.97 g cm^{-3}. Calculate the number of atoms of aluminum in 15.0 cm^3 corundum.

36. Calculate the number of atoms of silicon (Si) in 415 cm^3 of the colorless gas disilane at 0°C and atmospheric pressure, where its density is 0.00278 g cm^{-3}. The molecular formula of disilane is Si_2H_6.

ADDITIONAL PROBLEMS

37. Soft wood chips weighing 17.2 kg are placed in an iron vessel and mixed with 150.1 kg water and 22.43 kg sodium hydroxide. A steel lid seals the vessel, which is then placed in an oven at 250°C for 6 hours. Much of the wood fiber decomposes under these conditions; the vessel and lid do not react.
(a) Classify each of the materials mentioned as a substance or mixture. Subclassify the substances as elements or compounds.
(b) Determine the mass of the contents of the iron vessel after the reaction.

38. In a reproduction of the Millikan oil-drop experiment, a student obtains the following values for the charges on nine different oil droplets.

6.563×10^{-19} C	13.13×10^{-19} C	19.71×10^{-19} C
8.204×10^{-19} C	16.48×10^{-19} C	22.89×10^{-19} C
11.50×10^{-19} C	18.08×10^{-19} C	26.18×10^{-19} C

(a) Based on these data alone, what is your best estimate of the number of electrons on each of the above droplets? (*Hint:* Begin by considering differences in charges between adjacent data points, and see into what groups these are categorized.)
(b) Based on these data alone, what is your best estimate of the charge on the electron?
(c) Is it conceivable that the actual charge is half the charge you calculated in (b)? What evidence would help you decide one way or the other?

39. A rough estimate of the radius of a nucleus is provided by the formula $r = kA^{1/3}$, where k is approximately 1.3×10^{-13} cm and A is the mass number of the nucleus. Estimate the density of the nucleus of ^{127}I (which has a nuclear mass of 2.1×10^{-22} g) in grams per cubic centimeter. Compare with the density of solid iodine, 4.93 g cm^{-3}.

40. In a neutron star, gravity causes the electrons to combine with protons to form neutrons. A typical neutron star has a mass half that of the sun, compressed into a sphere of radius 20 km. If such a neutron star contains 6.0×10^{56} neutrons, calculate its density in grams per cubic centimeter. Compare this with the density inside a ^{232}Th nucleus, in which 142 neutrons and 90 protons occupy a sphere of radius 9.1×10^{-13} cm. Take the mass of a neutron to be 1.675×10^{-24} g and that of a proton to be 1.673×10^{-24} g.

41. Dalton's 1808 version of the atomic theory of matter included five general statements (see Section 1.3). According to modern understanding, four of those statements require amendment or extension. List the modifications that have been made to four of the five original postulates.

42. Naturally occurring rubidium (Rb) consists of two isotopes: ^{85}Rb (atomic mass 84.9117) and ^{87}Rb (atomic mass 86.9092). The atomic mass of the isotope mixture found in nature is 85.4678. Calculate the percentage abundances of the two isotopes in rubidium.

43. A sample of a gaseous binary compound of boron and chlorine weighing 2.842 g occupies 0.153 L. This sample is decomposed to give 0.664 g solid boron and enough gaseous chlorine (Cl_2) to occupy 0.688 L at the same temperature and pressure. Determine the number of atoms of boron and the number of atoms of chlorine in a molecule of this compound.

CHAPTER 2

Chemical Formulas, Equations, and Reaction Yields

An "assay balance of careful construction" of the type used by Lavoisier before 1788. This balance became the production model that served as a general, all-purpose balance for approximately 40 years. Users of this type of balance included Sir Humphrey Davy and his young assistant Michael Faraday.

Chapter 1 explained how chemical and physical methods are used to establish chemical formulas and relative atomic and molecular masses. We also developed the concept of the mole, which allows us to count molecules by weighing macroscopic quantities of matter. This chapter begins our study of chemical reactions. We examine the balanced chemical equations that summarize these reactions and show how to relate the masses of substances consumed to the masses of substances produced. This is an immensely practical and important subject. The questions that ask how much of a substance will react with a given amount of another substance and then how much product will be generated are central to all chemical processes, whether industrial, geological, or biological.

2.1 Empirical and Molecular Formulas

According to the laws of chemical combination, each substance may be described by a chemical formula that specifies the relative numbers of atoms of the elements in that substance. We now distinguish between two types of formulas: the molecular formula and the empirical formula. The **molecular formula** of a substance specifies the number of atoms of each element in one molecule of that substance. Thus, the molecular formula of carbon dioxide is CO_2; each molecule of carbon dioxide contains 1 atom of carbon and 2 atoms of oxygen. The molecular formula of glucose is $C_6H_{12}O_6$; each glucose molecule contains 6 atoms of carbon, 6 atoms of oxygen, and 12 atoms of hydrogen. Molecular formulas can be defined for all gaseous substances and for those liquids or solids that, like glucose, possess well-defined molecular structures.

In contrast, the **empirical formula** of a compound is the simplest formula that gives the correct relative numbers of atoms of each kind in a compound. For example, the empirical formula for glucose is CH_2O, indicating that the numbers of atoms of carbon, hydrogen, and oxygen are in a ratio of 1:2:1. Molecular formulas, when known, are clearly preferable to empirical formulas because they provide more detailed information. In some solids and liquids, however, distinct small molecules do not exist, and the only meaningful chemical formula is an empirical one. Solid cobalt(II) chloride, which has the empirical formula $CoCl_2$, is an example. There are strong attractive forces between a cobalt atom and two adjoining chlorine (Cl) atoms in solid cobalt(II) chloride, but it is impossible to distinguish the forces *within* such a "molecule" of $CoCl_2$ from those operating *between* it and a neighbor; the latter are equally strong. Consequently, cobalt(II) chloride is represented with an empirical formula and referred to by a **formula unit** of $CoCl_2$, rather than by "a molecule of $CoCl_2$." Many solids can be represented only by their formula units because it is not possible to identify a molecular unit in a unique way; other examples include sodium chloride (NaCl), the major component in table salt, and silicon dioxide (SiO_2), the major component of sand (see Figure 3.3). In some cases, small molecules are incorporated into a solid structure, and the chemical formula is written to show this fact explicitly. Thus, cobalt and chlorine form not only the anhydrous salt $CoCl_2$ mentioned earlier but also the hexahydrate $CoCl_2 \cdot 6\ H_2O$, in which six water molecules are incorporated per $CoCl_2$ formula unit (Fig. 2.1). The dot in this formula is used to set off a well-defined molecular component of the solid, such as water.

FIGURE 2.1 When cobalt(II) chloride crystallizes from solution, it brings with it six water molecules per formula unit, producing a red solid with the empirical formula $CoCl_2 \cdot 6\ H_2O$. This solid melts at 86°C; it loses some of the water when heated above 110°C to form a lavender solid with the empirical formula $CoCl_2 \cdot 2\ H_2O$.

2.2 Chemical Formula and Percentage Composition

The empirical formula H_2O specifies that for every atom of oxygen in water, there are two atoms of hydrogen. Equivalently, one mole of H_2O contains two moles of hydrogen atoms and one mole of oxygen atoms. The number of atoms and the number of moles of each element are present in the same ratio, namely, 2:1. The empirical formula for a substance is clearly related to the percentage composition by mass of that substance. This connection can be used in various ways.

2.2.1 Empirical Formula and Percentage Composition

The empirical formula of a compound can be simply related to the mass percentage of its constituent elements using the mole concept. For example, the empirical formula for ethylene (molecular formula C_2H_4) is CH_2. Its composition by mass is calculated from the masses of carbon and hydrogen in 1 mol of CH_2 formula units:

$$\text{mass of C} = 1 \text{ mol C} \times \left(12.011 \text{ g mol}^{-1}\right) = 12.011 \text{ g}$$

$$\text{mass of H} = 2 \text{ mol H} \times \left(1.00794 \text{ g mol}^{-1}\right) = 2.0159 \text{ g}$$

Adding these masses together gives a total mass of 14.027 g. The mass percentages of carbon and hydrogen in the compound are then found by dividing each of their masses by this total mass and multiplying by 100%, giving 85.628% C and 14.372% H by mass, respectively.

2.2.2 Determination of Empirical Formula from Mass Composition

We can reverse the procedure just described and determine the empirical formula from the elemental analysis of a compound, as illustrated by Example 2.1.

EXAMPLE 2.1

A 60.00-g sample of a dry-cleaning fluid was analyzed and found to contain 10.80 g carbon, 1.36 g hydrogen, and 47.84 g chlorine. Determine the empirical formula of the compound using a table of atomic masses.

Solution

The amounts of each element in the sample are

$$\text{carbon: } \frac{10.80 \text{ g C}}{12.011 \text{ g mol}^{-1}} = 0.8992 \text{ mol C}$$

$$\text{hydrogen: } \frac{1.36 \text{ g H}}{1.008 \text{ g mol}^{-1}} = 1.35 \text{ mol H}$$

$$\text{chlorine: } \frac{47.84 \text{ g Cl}}{35.453 \text{ g mol}^{-1}} = 1.349 \text{ mol Cl}$$

The ratio of the amount of carbon to that of chlorine (or hydrogen) is 0.8992:1.349 = 0.6666, which is close to 2:3. The numbers of moles form the ratio 2:3:3; therefore, the empirical formula is $C_2H_3Cl_3$. Additional measurements would be necessary to find the actual molecular mass and the correct *molecular* formula from among $C_2H_3Cl_3$, $C_4H_6Cl_6$, or any higher multiples $(C_2H_3Cl_3)_n$.

Related Problems: 7, 8, 9, 10, 11, 12

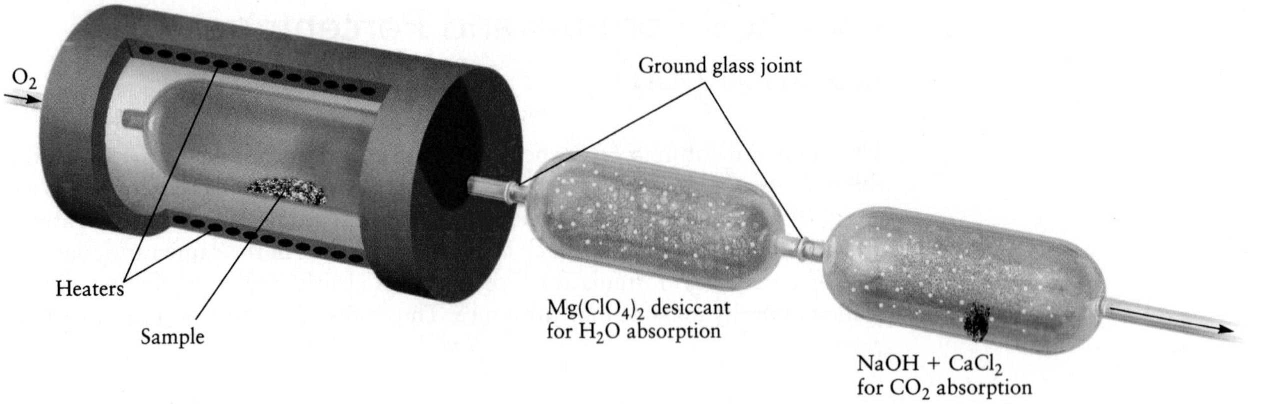

FIGURE 2.2 A combustion train used to determine the relative amounts of carbon and hydrogen in hydrocarbons. A sample of known weight is burned in a flow of oxygen to produce water and carbon dioxide. These combustion products pass over a drying agent, such as magnesium perchlorate, $Mg(ClO_4)_2$, which absorbs the water. The carbon dioxide then passes through to the second stage where it is absorbed on finely divided particles of sodium hydroxide, NaOH, mixed with calcium chloride, $CaCl_2$. The masses of the water and carbon dioxide produced in the reaction are determined by weighing the absorption tubes before and after the reaction.

2.2.3 Empirical Formula Determined from Elemental Analysis by Combustion

A **hydrocarbon** is a compound that contains only carbon and hydrogen. Its empirical formula can be determined by using the combustion train shown in Figure 2.2. In this device, a known mass of the hydrocarbon is burned completely in oxygen, producing carbon dioxide and water whose masses are measured. The empirical formula for the compound is then calculated from these data by the procedure illustrated in Example 2.2.

EXAMPLE 2.2

A certain compound, used as a welding fuel, contains only carbon and hydrogen. Burning a small sample of this fuel completely in oxygen produces 3.38 g CO_2, 0.692 g water, and no other products. What is the empirical formula of the compound?

Solution

We first compute the amounts of CO_2 and H_2O. Because all the carbon has been converted to CO_2 and all the hydrogen to water, the amounts of C and H in the unburned gas can be determined:

$$\text{mol of C} = \text{mol of CO}_2 = \frac{3.38 \text{ g}}{44.01 \text{ g mol}^{-1}} = 0.0768 \text{ mol}$$

$$\text{mol of H} = 2(\text{mol of H}_2\text{O}) = 2\left(\frac{0.692 \text{ g}}{18.02 \text{ g mol}^{-1}}\right) = 0.0768 \text{ mol}$$

Because each water molecule contains two hydrogen atoms, it is necessary to multiply the number of moles of water by 2 to find the number of moles of hydrogen atoms. Having found that the compound contains equal numbers of moles of carbon and hydrogen, we have determined that its empirical formula is CH. Its molecular formula may be CH, C_2H_2, C_3H_3, and so on.

Related Problems: 13, 14

2.2.4 Connection between Empirical and Molecular Formulas

The molecular formula is some whole-number multiple of the empirical formula. To determine the molecular formula, you must know the approximate molar mass of the compound under study. From Avogadro's hypothesis, the ratio of molar masses of two gaseous compounds is the same as the ratio of their densities, provided that those densities are measured at the same temperature and pressure. (This is true because a given volume contains the same number of molecules of the two gases.) The density of the welding gas from Example 2.2 is 1.06 g L^{-1} at 25°C and atmospheric pressure. Under the same conditions, the density of gaseous oxygen (which exists as diatomic O_2 molecules with molar mass of 32.0 g mol^{-1}) is 1.31 g L^{-1}. The approximate molar mass of the welding gas is, therefore,

$$\text{molar mass of welding gas} = \frac{1.06 \text{ g L}^{-1}}{1.31 \text{ g L}^{-1}}\left(32.0 \text{ g mol}^{-1}\right) = 25.9 \text{ g mol}^{-1}$$

The molar mass corresponding to the *empirical* formula CH is 13.0 g mol^{-1}. Because 25.9 g mol^{-1} is approximately twice this value, there must be two CH units per molecule; therefore, the molecular formula is C_2H_2. The gas is acetylene.

2.3 Writing Balanced Chemical Equations

Chemical reactions combine elements into compounds, decompose compounds back into elements, and transform existing compounds into new compounds. Because atoms are indestructible in chemical reactions, the same number of atoms (or moles of atoms) of each element must be present before and after any ordinary (as opposed to nuclear) chemical reaction. The conservation of matter in a chemical change is represented in a balanced chemical equation for that process. The study of the relationships between the numbers of reactant and product molecules is called **stoichiometry** (derived from the Greek *stoicheion,* meaning "element," and *metron,* meaning "measure"). Stoichiometry is fundamental to all aspects of chemistry.

An equation can be balanced using stepwise reasoning. Consider the decomposition of ammonium nitrate (NH_4NO_3) when heated gently to produce dinitrogen oxide (N_2O) and water. An *unbalanced* equation for this reaction is

$$NH_4NO_3 \longrightarrow N_2O + H_2O$$

Substances on the left side of the arrow are called reactants, and those on the right side are called products. We read chemical equations just like we read sentences; reactions progress from left to right. This equation is unbalanced because there are 3 mol of oxygen atoms on the left side of the equation (and 4 mol of hydrogen atoms), but only 2 mol of oxygen atoms and 2 mol of hydrogen atoms on the right side. To balance the equation, begin by assigning 1 as the coefficient of one species, usually the species that contains the most elements—in this case, NH_4NO_3. Next, look for elements in that substance that appear only once elsewhere in the equation and assign coefficients to balance the number of moles of that element on both sides. The only other species in this reaction that contains nitrogen is N_2O, and assigning a coefficient of 1 for the N_2O ensures that there are 2 mol of nitrogen atoms on each side of the equation.

Hydrogen appears in H_2O; thus, its coefficient is 2 to balance the 4 mol of hydrogen atoms on the left side. This gives

$$NH_4NO_3 \longrightarrow N_2O + H_2O$$

Finally, verify that the last element, oxygen, is also balanced by noting that there are 3 mol of oxygen atoms on each side. The coefficients of 1 in front of the NH_4NO_3 and N_2O are omitted by convention.

As a second example, consider the reaction in which butane (C_4H_{10}) is burned in oxygen to form carbon dioxide and water:

$$\_\, C_4H_{10} + \_\, O_2 \longrightarrow \_\, CO_2 + \_\, H_2O$$

Spaces have been left for the coefficients that specify the number of moles of each reactant and product. Begin with 1 mol of butane, C_4H_{10}. It contains 4 mol of carbon atoms and must produce 4 mol of carbon dioxide molecules to conserve the number of carbon atoms in the reaction. Therefore, the coefficient for CO_2 is 4. In the same way, the 10 mol of hydrogen *atoms* must form 5 mol of water *molecules,* because each water molecule contains 2 hydrogen atoms; thus, the coefficient for the H_2O is 5:

$$C_4H_{10} + \_\, O_2 \longrightarrow 4\, CO_2 + 5\, H_2O$$

Four moles of CO_2 contain 8 mol of oxygen atoms, and 5 mol of H_2O contain 5 mol of oxygen atoms, resulting in a total of 13 mol of oxygen atoms. Thirteen moles of oxygen atoms are equivalent to $\frac{13}{2}$ moles of oxygen molecules; therefore, the coefficient for O_2 is $\frac{13}{2}$. The balanced equation is

$$C_4H_{10} + \tfrac{13}{2}\, O_2 \longrightarrow 4\, CO_2 + 5\, H_2O$$

There is nothing wrong with fractions such as $\frac{13}{2}$ in a balanced equation, because fractions of moles are perfectly meaningful. It is often customary, however, to eliminate such fractions because the equation can be interpreted on the molecular level as well as on the macroscopic level. Your instructor will very likely have a preference on this matter, so make sure that you know what it is and balance equations accordingly. In this case, multiplying all coefficients in the equation by 2 gives

$$2\, C_4H_{10} + 13\, O_2 \longrightarrow 8\, CO_2 + 10\, H_2O$$

The procedure may be summarized as follows:

1. Assign 1 as the coefficient of one species. The best choice is the most complicated species; that is, the species with the largest number of elements.
2. Identify, in sequence, elements that appear in only one chemical species, the coefficient of which has not yet been determined. Choose that coefficient to balance the number of moles of atoms of that element. Continue until all coefficients have been identified.
3. If desired, multiply the whole equation by the smallest integer that will eliminate any fractions.

This method of balancing equations "by inspection" works in many, but not all, cases. Section 11.4 (the number before the period refers to the chapter number) presents techniques for balancing certain more complex chemical equations.

Once the reactants and products are known, balancing chemical equations is a routine, mechanical process of accounting. We often tell students something like "if you have three dimes, two nickels, and a penny before the reaction, make sure that you have three dimes, two nickels, and a penny after the reaction" to give them a visual image. The difficult part (and the part where chemistry comes in) is to know which substances will react with each other

CONNECTION TO CHEMICAL ENGINEERING

Sulfuric Acid Manufacturing

Chemistry is big business. The chemical industry in the United States alone generates nearly a trillion dollars per year in annual sales—a significant component of the U.S. gross domestic product. It is impossible to overstate the importance of the chemical industry to our modern way of life; we depend on chemicals for energy, materials, agriculture, and health. The chemical and petrochemical industries transform raw ingredients, such as minerals and petroleum, into a variety of products on truly large scales. The process by which sulfuric acid is manufactured illustrates the practical importance of the topics developed in this chapter.

Sulfuric acid is produced in greater quantity than any other chemical; about 200 million metric tonnes (10^3 kg) were manufactured worldwide in 2010. Most sulfuric acid plants are located near their product's point of use—phosphate fertilizer manufacturing plants, nickel ore leaching plants, and petroleum refineries—because it is less expensive to transport elemental sulfur (the starting material) than it is to transport sulfuric acid (the product). Sulfuric acid is also used in a variety of industrial chemical processes that include the manufacture of commodity and specialty chemicals, polymers, pharmaceuticals, soaps, and detergents and in the pulp and paper industry. The lead acid storage battery in your car is one of the few consumer products that actually contains sulfuric acid as such.

Elemental sulfur mined from historically important locations like Sicily, and later Indonesia, Chile, and Japan, used to be the dominant source of the raw material. Most sulfur occurs as fine powders or in polycrystalline form, but brightly colored yellow single crystals of sulfur are occasionally found. Virtually all of the sulfur used to manufacture sulfuric acid today was removed from petroleum fuels or feedstocks by a process called hydrosulfurization, because sulfur poisons the catalysts used in petroleum refining and other processes. Canada has become the world's largest exporter of elemental sulfur. Figure C2.1 shows a stockpile of bright yellow sulfur awaiting shipment.

© Gunter Marx Photography/CORBIS

FIGURE C2.1 A stockpile of yellow sulfur.

Most sulfuric acid produced today is made by the contact process, which got its name because the second step in the reaction is carried out in contact with a catalyst, typically vanadium pentoxide (V_2O_5) mixed with proprietary (secret) additives. The main steps of the overall reaction sequence are:

$$S(s) + O_2(g) \rightleftharpoons SO_2(g)$$

$$SO_2(g) + \tfrac{1}{2}O_2(g) \longrightarrow SO_3(g)$$

$$SO_3(g) + H_2O(\ell) \longrightarrow H_2SO_4(\ell)$$

The first step (the double arrows indicating an equilibrium) is the combustion of elemental sulfur in air to produce sulfur dioxide, which is further oxidized catalytically in the second reaction to produce sulfur trioxide. The second step is the key reaction of the process, and considerable effort has been devoted to optimizing the conditions under which this reaction takes place. Sulfuric acid is produced by the reaction between sulfur trioxide and water in the final step of the sequence. As you will see later in this textbook, optimizing the yields and rates of industrial chemical processes requires careful consideration of both the thermodynamics (Chapter 13) and the kinetics (Chapter 18) of the reactions. As is often the case in industrial chemical processes, thermodynamic considerations favor low temperatures, whereas kinetic considerations favor high temperatures, so the operating temperatures of sulfuric acid plants represent a compromise between those competing factors. The catalyst (something that speeds up a chemical reaction) is required for the conversion of SO_2 to SO_3, because the reaction rate would be unacceptably slow without it.

The schematic in Figure C2.2 shows two stages of a typical modern multi-stage sulfuric acid plant that is designed to achieve maximum conversion yields at acceptable rates; these plants typically convert 99.7% of the elemental sulfur into sulfuric acid. The high yields are due to several important design features. First, and most important, the temperature of the SO_2 produced by the initial combustion reaction is lowered prior to subsequent reaction with molecular oxygen. This step is taken to enhance the conversion of SO_2 to SO_3, and it would be repeated several times in a multistage plant. Second, unreacted SO_3 is absorbed by sulfuric acid in one or more absorbing towers, which allows any unreacted SO_2 to make one or more passes over the catalytic reactor to increase the amount converted. Finally, most modern plants have scrubbers that remove trace amounts of SO_2 from the exhaust gases, primarily for environmental reasons (see Section 20.6), but also to recover a marketable product.

Understanding reaction stoichiometries is clearly essential in order to design and operate chemical plants efficiently. The masses of raw materials necessary to produce the desired quantity of products is calculated using basic stoichiometry relationships, taking into account the limiting reactant(s) involved. Many practical reactions are run with an excess of one or more of the reactants to ensure complete conversion. Designing the air handling system to provide sufficient oxygen for the combustion and subsequent oxidation of sulfur requires quantitative understanding of the reaction stoichiometry.

To give you a feel for the amount of material converted in a modern sulfuric acid plant, let's calculate the amount of sulfur needed as raw material for a new plant under construction in Saudi Arabia in 2010. This plant, when complete, will produce 13,500 tons of sulfuric acid *per day* with the entire output devoted to phosphate fertilizer manufacture. H_2SO_4 is 32.7% S by weight, so we need (.327) × (13,500 tons) = 4400 *tons* of sulfur *per day.* A typical rail car holds 100 tons, so the owners of this plant need to provide 44 rail cars of sulfur per day, or a full trainload of sulfur every three days—truly a staggering amount!

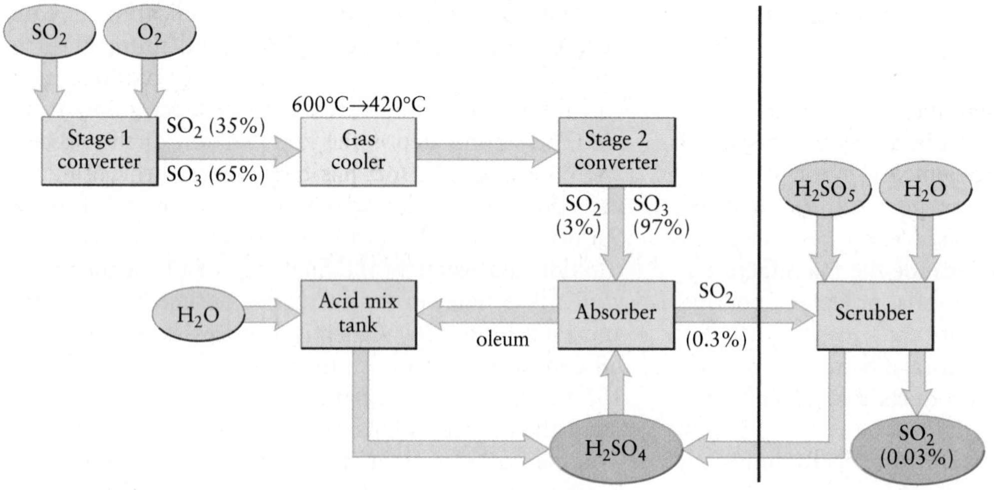

FIGURE C2.2 Schematic showing the contact process for producing sulfuric acid.

and to determine which products are formed. We return to this question many times throughout this book.

EXAMPLE 2.3

Hargreaves process is an industrial procedure for making sodium sulfate (Na_2SO_4) for use in papermaking. The starting materials are sodium chloride (NaCl), sulfur dioxide (SO_2), water, and oxygen. Hydrogen chloride (HCl) is generated as a by-product. Write a balanced chemical equation for this process.

Solution

The unbalanced equation is

$$\_NaCl + \_SO_2 + \_H_2O + \_O_2 \longrightarrow \_Na_2SO_4 + \_HCl$$

Begin by assigning a coefficient of 1 to Na_2SO_4 because it is the most complex species, composed of 3 different elements. There are 2 mol of sodium atoms on the right;

(Continued)

therefore, the coefficient for NaCl must be 2. Following the same argument, the coefficient for SO_2 must be 1 to balance the 1 mol of sulfur on the right. This gives

$$2\ NaCl + SO_2 + \_H_2O + \_O_2 \longrightarrow Na_2SO_4 + \_HCl$$

Next, we note that there are 2 mol of Cl atoms on the left (reactant) side; therefore, the coefficient for HCl must be 2. Hydrogen is the next element to balance, with 2 mol on the right side, and therefore a coefficient of 1 for the H_2O:

$$2\ NaCl + SO_2 + H_2O + \_O_2 \longrightarrow Na_2SO_4 + 2\ HCl$$

Finally, the oxygen atoms must be balanced. There are 4 mol of oxygen atoms on the right side, but there are 2 mol from SO_2 and 1 mol from H_2O on the left side; therefore, 1 mol of oxygen *atoms* must come from O_2. Therefore, the coefficient for O_2 is $\frac{1}{2}$:

$$2\ NaCl + SO_2 + H_2O + \tfrac{1}{2}O_2 \longrightarrow Na_2SO_4 + 2\ HCl$$

Multiplying all coefficients in the equation by 2 gives

$$4\ NaCl + 2\ SO_2 + 2\ H_2O + O_2 \longrightarrow 2\ Na_2SO_4 + 4\ HCl$$

In balancing this equation, oxygen was considered last because it appears in several places on the left side of the equation.

Related Problems: 19, 20

2.4 Mass Relationships in Chemical Reactions

A balanced chemical equation makes a quantitative statement about the relative masses of the reacting substances. The chemical equation for the combustion of butane,

$$2\ C_4H_{10} + 13\ O_2 \longrightarrow 8\ CO_2 + 10\ H_2O$$

can be interpreted as either

$$2 \text{ molecules of } C_4H_{10} + 13 \text{ molecules of } O_2 \longrightarrow 8 \text{ molecules of } CO_2 + 10 \text{ molecules of } H_2O$$

or

$$2 \text{ mol of } C_4H_{10} + 13 \text{ mol of } O_2 \longrightarrow 8 \text{ mol of } CO_2 + 10 \text{ mol of } H_2O$$

Multiplying the molar mass of each substance in the reaction by the number of moles represented in the balanced equation gives

$$116.3 \text{ g } C_4H_{10} + 416.0 \text{ g } O_2 \longrightarrow 352.1 \text{ g } CO_2 + 180.2 \text{ g } H_2O$$

The coefficients in a balanced chemical equation relate the amounts of substances consumed in or produced by a chemical reaction. If 6.16 mol butane react according to the preceding equation, the amounts of O_2 consumed and CO_2 generated are

$$\text{mol } O_2 = 6.16 \text{ mol } C_4H_{10} \times \left(\frac{13 \text{ mol } O_2}{2 \text{ mol } C_4H_{10}}\right) = 40.0 \text{ mol } O_2$$

$$\text{mol } CO_2 = 6.16 \text{ mol } C_4H_{10} \times \left(\frac{8 \text{ mol } CO_2}{2 \text{ mol } C_4H_{10}}\right) = 24.6 \text{ mol } CO_2$$

For most practical purposes we are interested in the *masses* of reactants and products, because those are the quantities that are directly measured. In this case, the molar masses (calculated from a table of atomic masses) are used to convert the number of moles of a substance (in moles) to its mass (in grams), as illustrated by Example 2.4. Sometimes, however, we are also interested in knowing the number of molecules in a sample. The mole concept allows us to convert easily from mass to numbers of molecules as follows:

$$\textit{mass} \overset{\textit{molar mass}}{\longleftrightarrow} \textit{moles} \overset{N_A}{\longleftrightarrow} \textit{number of molecules}$$

Mass and moles are related by the molar mass; Avogadro's number N_A relates number and moles. You should practice using these relationships to calculate any desired quantity from any given quantity. You can use dimensional analysis to help you figure out whether to divide or multiply in any given problem.

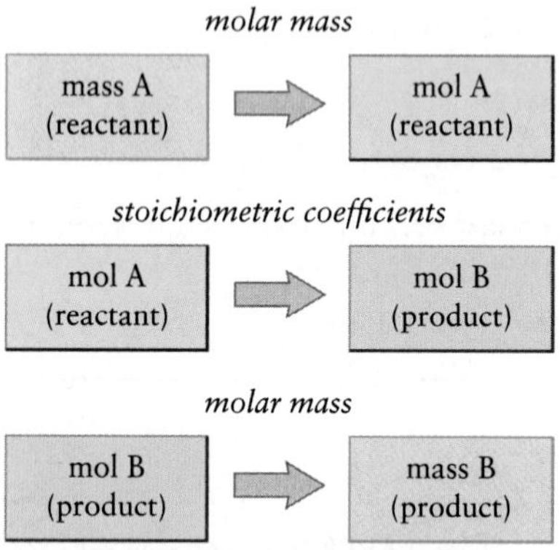

FIGURE 2.3 The steps in a stoichiometric calculation. In a typical calculation, the mass of one reactant or product is known and the masses of one or more other reactants or products are to be calculated using the balanced chemical equation and a table of relative atomic masses.

EXAMPLE 2.4

Calcium hypochlorite, $Ca(OCl)_2$, is used as a bleaching agent. It is produced from sodium hydroxide, calcium hydroxide, and chlorine according to the following overall equation:

$$2\ NaOH + Ca(OH)_2 + 2\ Cl_2 \longrightarrow Ca(OCl)_2 + 2\ NaCl + 2\ H_2O$$

How many grams of chlorine and sodium hydroxide react with 1067 g $Ca(OH)_2$, and how many grams of calcium hypochlorite are produced?

Solution

The amount of $Ca(OH)_2$ consumed is

$$\frac{1067\ \text{g Ca(OH)}_2}{74.09\ \text{g mol}^{-1}} = 14.40\ \text{mol Ca(OH)}_2$$

where the molar mass of $Ca(OH)_2$ has been obtained from the molar masses of calcium, oxygen, and hydrogen as

$$40.08\ \text{g mol}^{-1} + 2(15.999\ \text{g mol}^{-1}) + 2(1.0079\ \text{g mol}^{-1}) = 74.09\ \text{g mol}^{-1}$$

According to the balanced equation, 1 mol $Ca(OH)_2$ reacts with 2 mol NaOH and 2 mol Cl_2 to produce 1 mol $Ca(OCl)_2$. If 14.40 mol of $Ca(OH)_2$ reacts completely, then

$$\text{mol NaOH} = 14.40\ \text{mol Ca(OH)}_2 \left(\frac{2\ \text{mol NaOH}}{1\ \text{mol Ca(OH)}_2}\right)$$

$$= 28.80\ \text{mol NaOH}$$

$$\text{mol Cl}_2 = 14.40\ \text{mol Ca(OH)}_2 \left(\frac{2\ \text{mol Cl}_2}{1\ \text{mol Ca(OH)}_2}\right)$$

$$= 28.80\ \text{mol Cl}_2$$

$$\text{mol Ca(OCl)}_2 = 14.40\ \text{mol Ca(OH)}_2 \left(\frac{1\ \text{mol Ca(OCl)}_2}{1\ \text{mol Ca(OH)}_2}\right)$$

$$= 14.40\ \text{mol Ca(OCl)}_2$$

From the number of moles and molar masses of reactants and products, the following desired masses are found:

$$\text{Mass NaOH reacting} = (28.80\ \text{mol})(40.00\ \text{g mol}^{-1}) = 1152\ \text{g}$$

$$\text{Mass Cl}_2\ \text{reacting} = (28.80\ \text{mol})(70.91\ \text{g mol}^{-1}) = 2042\ \text{g}$$

$$\text{Mass Ca(OCl)}_2\ \text{produced} = (14.40\ \text{mol})(142.98\ \text{g mol}^{-1}) = 2059\ \text{g}$$

Related Problems: 21, 22, 23, 24

In calculations such as the one illustrated in Example 2.4, we are given a known mass of one substance and are asked to calculate the masses of one or more of the other reactants or products. Figure 2.3 summarizes the three-step process used. With experience, it is possible to write down the answers in a shorthand form so that all three conversions are conducted at the same time. The amount of NaOH reacting in the preceding example can be written as

$$\left(\frac{1067 \text{ g Ca(OH)}_2}{74.10 \text{ g mol}^{-1}}\right) \times \left(\frac{2 \text{ mol NaOH}}{1 \text{ mol Ca(OH)}_2}\right) \times 40.00 \text{ g mol}^{-1} = 1152 \text{ g NaOH}$$

At first, however, it is better to follow a stepwise procedure for such calculations.

2.5 Limiting Reactant and Percentage Yield

In the cases we have considered so far, the reactants were present in the exact ratios necessary for them all to be completely consumed in forming products. This is not the usual case, however. It is necessary to have methods for describing cases in which one of the reactants may not be present in sufficient amount and in which conversion to products is less than complete.

2.5.1 Limiting Reactant

Suppose arbitrary amounts of reactants are mixed and allowed to react. The one that is used up first is called the **limiting reactant** (limiting reagent in some texts); some quantity of the other reactants remains after the reaction has gone to completion. These other reactants are present **in excess.** An increase in the amount of the limiting reactant leads to an increase in the amount of product formed. This is not true of the other reactants. In an industrial process, the limiting reactant is often the most expensive one, to ensure that none of it is wasted. For instance, the silver nitrate used in preparing silver chloride for photographic film by the reaction

$$AgNO_3 + NaCl \longrightarrow AgCl + NaNO_3$$

is far more expensive than the sodium chloride (ordinary salt). Thus, it makes sense to perform the reaction with an excess of sodium chloride to ensure that as much of the silver nitrate as possible reacts to form products.

There is a systematic method to find the limiting reactant and determine the maximum possible amounts of products. Take each reactant in turn, assume that it is used up completely in the reaction, and calculate the mass of one of the products that will be formed. Whichever reactant gives the *smallest* mass of this product is the limiting reactant. Once it has reacted fully, no further product can be formed.

EXAMPLE 2.5

Sulfuric acid (H_2SO_4) forms in the chemical reaction

$$2\ SO_2 + O_2 + 2\ H_2O \longrightarrow 2\ H_2SO_4$$

Suppose 400 g SO_2, 175 g O_2, and 125 g H_2O are mixed and the reaction proceeds until one of the reactants is used up. Which is the limiting reactant? What mass of H_2SO_4 is produced, and what masses of the other reactants remain?

(Continued)

Solution

The number of moles of each reactant originally present is calculated by dividing each mass by the corresponding molar mass:

$$\frac{400 \text{ g } SO_2}{64.06 \text{ g mol}^{-1}} = 6.24 \text{ mol } SO_2$$

$$\frac{175 \text{ g } O_2}{32.00 \text{ g mol}^{-1}} = 5.47 \text{ mol } O_2$$

$$\frac{125 \text{ g } H_2O}{18.02 \text{ g mol}^{-1}} = 6.94 \text{ mol } H_2O$$

If all the SO_2 reacted, it would give

$$6.24 \text{ mol } SO_2 \times \left(\frac{2 \text{ mol } H_3SO_4}{2 \text{ mol } SO_2}\right) = 6.24 \text{ mol } H_2SO_4$$

If all the O_2 reacted, it would give

$$5.47 \text{ mol } O_2 \times \left(\frac{2 \text{ mol } H_2SO_4}{1 \text{ mol } O_2}\right) = 10.94 \text{ mol } H_2SO_4$$

Finally, if all the water reacted, it would give

$$6.94 \text{ mol } H_2O \times \left(\frac{2 \text{ mol } H_2SO_4}{2 \text{ mol } H_2O}\right) = 6.94 \text{ mol } H_2SO_4$$

In this case, SO_2 is the limiting reactant because the computation based on its amount produces the smallest amount of product (6.24 mol H_2SO_4). Oxygen and water are present in excess. After reaction, the amount of each reactant that remains is the original amount minus the amount reacted:

$$\text{mol } O_2 = 5.47 \text{ mol } O_2 - \left(6.24 \text{ mol } SO_2 \times \frac{1 \text{ mol } O_2}{2 \text{ mol } SO_2}\right)$$

$$= 5.47 - 3.12 \text{ mol } O_2 = 2.35 \text{ mol } O_2$$

$$\text{mol } H_2O = 6.94 \text{ mol } H_2O - \left(6.24 \text{ mol } SO_2 \times \frac{2 \text{ mol } H_2O}{2 \text{ mol } SO_2}\right)$$

$$= 6.94 - 6.24 \text{ mol } H_2O = 0.70 \text{ mol } H_2O$$

The masses of reactants and products after the reaction are

$$\text{mass } H_2SO_4 \text{ produced} = (6.24 \text{ mol})(98.07 \text{ g mol}^{-1}) = 612 \text{ g}$$

$$\text{mass } O_2 \text{ remaining} = (2.35 \text{ mol})(32.00 \text{ g mol}^{-6}) = 75 \text{ g}$$

$$\text{mass } H_2O \text{ remaining} = (0.70 \text{ mol})(18.02 \text{ g mol}^{-1}) = 13 \text{ g}$$

The total mass at the end is 612 g + 13 g + 75 g = 700 g, which is, of course, equal to the total mass originally present, 400 g + 175 g + 125 g = 700 g, as required by the law of conservation of mass.

Related Problems: 35, 36

2.5.2 Percentage Yield

The amounts of products calculated so far have been **theoretical yields**, determined by assuming that the reaction goes cleanly and completely. The **actual yield** of a product (that is, the amount present after separating it from other products and reactants and purifying it) is less than the theoretical yield. There are several possible reasons for this. The reaction may stop short of completion, so reactants remain unreacted. There may be competing reactions that give other products, and therefore reduce the yield of the desired one. Finally, in the process of separation and purification, some of the product is invariably lost, although that amount can be reduced by careful experimental techniques. The ratio of the actual yield to the theoretical yield (multiplied by 100%) gives the **percentage yield** for that product in the reaction.

EXAMPLE 2.6

The sulfide ore of zinc (ZnS) is reduced to elemental zinc by "roasting" it (heating it in air) to give ZnO, and then heating the ZnO with carbon monoxide. The two reactions can be written as

$$ZnS + \tfrac{3}{2}O_2 \longrightarrow ZnO + SO_2$$

$$ZnO + CO \longrightarrow Zn + CO_2$$

Suppose 5.32 kg ZnS is treated in this way and 3.30 kg pure Zn is obtained. Calculate the theoretical yield of zinc and its actual percentage yield.

Solution

From the molar mass of ZnS (97.46 g mol^{-1}), the number of moles of ZnS initially present is

$$\frac{5320 \text{ g ZnS}}{97.46 \text{ g mol}^{-1}} = 54.6 \text{ mol ZnS}$$

Because each mole of ZnS gives 1 mol of ZnO in the first chemical equation, and each mole of ZnO then gives 1 mol of Zn, the theoretical yield of zinc is 54.6 mol. In grams, this is

$$54.6 \text{ mol Zn} \times 65.39 \text{ g mol}^{-1} = 3570 \text{ g Zn}$$

The ratio of actual yield to theoretical yield, multiplied by 100%, gives the percentage yield of zinc:

$$\% \text{ yield} = \left(\frac{3.30 \text{ kg}}{3.57 \text{ kg}}\right) \times 100\% = 92.4\%$$

Related Problems: 37, 38

It is clearly desirable to achieve the highest percentage yield of product possible to reduce the consumption of raw materials. In some synthetic reactions (especially in organic chemistry), the final product is the result of many successive reactions. In such processes, the yields in the individual steps must be quite high if the synthetic method is to be a practical success. Suppose, for example, that ten consecutive reactions must be performed to reach the product, and that each has a percentage yield of only 50% (a fractional yield of 0.5). The overall yield is the product of the fractional yields of the steps:

$$\underbrace{(0.5) \times (0.5) \times \ldots \times (0.5)}_{10 \text{ terms}} = (0.5)^{10} = 0.001$$

This overall percentage yield of 0.1% makes the process useless for synthetic purposes. If all the individual percentage yields could be increased to 90%, however, the overall yield would then be $(0.9)^{10} = 0.35$, or 35%. This is a much more reasonable result, and one that might make the process worth considering.

CHAPTER SUMMARY

We have shown you how the relative number of atoms in a molecule or solid is given by its empirical or its molecular formula, and how these formulas are determined experimentally. The principle of conservation of mass has been sharpened a bit in our discussion of balancing chemical reactions. Not only is the total mass conserved in ordinary (as opposed to nuclear) chemical reactions, but the total number of atoms of every element is also conserved. Balancing a chemical reaction requires nothing more than assuring that the same numbers of atoms (or moles of atoms) of each element appear on each side of the balanced equation. Because chemists weigh macroscopic quantities of reactants and products, it is important to understand how mass ratios relate to mole ratios in chemical reactions. Finally, we point out that not every reactant is completely consumed in a chemical reaction, and that the limiting reactant determines the maximum theoretical yield; the percentage yield may be somewhat less.

CONCEPTS AND SKILLS

Section 2.1 – Empirical and Molecular Formulas

Distinguish between empirical and molecular formulas.

Section 2.2 – Chemical Formula and Percentage Composition

Given the percentages by mass of the elements in a compound, determine its empirical formula and vice versa (Problems 1–12).

- Determine the number of moles of each element in a compound by dividing the mass of each element (from mass percentages) by the molar mass of the element and then take ratios, reducing them to the smallest ratio of integers.

Use the masses of products obtained in combustion train measurements to determine empirical formulas (Problems 12 and 13).

Use ratios of gas densities to estimate molar mass and determine molecular formulas (Problems 15–18).

- The ratio of the density of an unknown gas to that of a known gas (using Avogadro's hypothesis) gives an approximate molar mass.

Section 2.3 – Writing Balanced Chemical Equations

Balance simple chemical equations (Problems 19 and 20).

- Choose the substance with the most elements and assign it a stoichiometric coefficient of 1.
- Choose an element that appears in only one other substance in the reaction and balance it with respect to the first substance. Continue in this way until all of the elements have been balanced.

Section 2.4 – Mass Relationships in Chemical Reactions

Given the mass of a reactant or product in a chemical reaction, use a balanced chemical equation to calculate the masses of other reactants consumed and other products formed (Problems 21–34).

- Use the flowchart (Figure 2.3).
- Convert masses of reactants to moles of reactants using their molar masses.
- Convert moles of reactants to moles of products using stoichiometry.
- Convert moles of products to masses of products using their molar masses.

Section 2.5 – Limiting Reactant and Percentage Yield

Given a set of initial masses of reactants and a balanced chemical equation, determine the limiting reactant and calculate the masses of reactants and products after the reaction has gone to completion (Problems 35 and 36).

- Calculate the number of moles of each reactant and determine which one would produce the smallest amount of product if the reaction went to completion.
- Calculate the number of moles of reactants consumed and products produced using the number of moles of this limiting reactant and the reaction stoichiometry.

CUMULATIVE EXERCISE

Wolfgang Kumm/Corbis

FIGURE 2.4 A jet engine fan blade made of a single-crystal titanium alloy.

Titanium in Industry

Metallic titanium and its alloys (especially those with aluminum and vanadium) combine the advantages of high strength and lightweight and are therefore used widely in the aerospace industry for the bodies and engines of airplanes (Figure 2.4). The major natural source for titanium is the ore rutile, which contains titanium dioxide (TiO_2).

(a) An intermediate in the preparation of elemental titanium from TiO_2 is a volatile chloride of titanium (boiling point 136°C) that contains 25.24% titanium by mass. Determine the empirical formula of this compound.

(b) At 136°C and atmospheric pressure, the density of this gaseous chloride is 5.6 g L^{-1}. Under the same conditions, the density of gaseous nitrogen (N_2, molar mass 28.0 g mol^{-1}) is 0.83 g L^{-1}. Determine the molecular formula of this compound.

(c) The titanium chloride dealt with in parts (a) and (b) is produced by the reaction of chlorine with a hot mixture of titanium dioxide and coke (carbon), with carbon dioxide generated as a by-product. Write a balanced chemical equation for this reaction.

(d) What mass of chlorine is needed to produce 79.2 g of the titanium chloride?

(e) The titanium chloride then reacts with liquid magnesium at 900°C to give titanium and magnesium chloride ($MgCl_2$). Write a balanced chemical equation for this step in the refining of titanium.

(f) Suppose the reaction chamber for part (e) contains 351 g of the titanium chloride and 63.2 g liquid magnesium. Which is the limiting reactant? What maximum mass of titanium could result?

(g) Isotopic analysis of the titanium from a particular ore gave the following results:

Isotope	Relative Mass	Abundance (%)
^{46}Ti	45.952633	7.93
^{47}Ti	46.95176	7.28
^{48}Ti	47.947948	73.94
^{49}Ti	48.947867	5.51
^{50}Ti	49.944789	5.34

Calculate the mass of a single ^{48}Ti atom and the *average* mass of the titanium atoms in this ore sample.

Answers

(a) $TiCl_4$

(b) $TiCl_4$

(c) $TiO_2 + C + 2\ Cl_2 \longrightarrow TiCl_4 + CO_2$

(d) 59.2 g

(e) $TiCl_4 + 2\ Mg \longrightarrow Ti + 2\ MgCl_2$

(f) Mg; 62.3 g

(g) 7.961949×10^{-23} g; 7.950×10^{-23} g

PROBLEMS

Answers to problems whose numbers are red appear in Appendix G. Problems that are more challenging are indicated with asterisks.

2.2 Chemical Formula and Percentage Composition

1. A newly synthesized compound has the molecular formula $ClF_2O_2PtF_6$. Compute, to four significant figures, the mass percentage of each of the four elements in this compound.

2. Acetaminophen is the generic name of the pain reliever in Tylenol and some other headache remedies. The compound has the molecular formula $C_8H_9NO_2$. Compute, to four significant figures, the mass percentage of each of the four elements in acetaminophen.

3. Arrange the following compounds from left to right in order of increasing percentage by mass of hydrogen: H_2O, $C_{12}H_{26}$, N_4H_6, LiH.

4. Arrange the following compounds from left to right in order of increasing percentage by mass of fluorine: HF, C_6HF_5, BrF, UF_6.

5. "Q-gas" is a mixture of 98.70% helium and 1.30% butane (C_4H_{10}) by mass. It is used as a filling for gas-flow Geiger counters. Compute the mass percentage of hydrogen in Q-gas.

6. A pharmacist prepares an antiulcer medicine by mixing 286 g Na_2CO_3 with water, adding 150 g glycine ($C_2H_5NO_2$), and stirring continuously at 40°C until a firm mass results. The pharmacist heats the mass gently until all the water has been driven away. No other chemical changes occur in this step. Compute the mass percentage of carbon in the resulting white crystalline medicine.

7. Zinc phosphate is used as a dental cement. A 50.00-mg sample is broken down into its constituent elements and gives 16.58 mg oxygen, 8.02 mg phosphorus, and 25.40 mg zinc. Determine the empirical formula of zinc phosphate.

8. Bromoform is 94.85% bromine, 0.40% hydrogen, and 4.75% carbon by mass. Determine its empirical formula.

9. Fulgurites are the products of the melting that occurs when lightning strikes the earth. Microscopic examination of a sand fulgurite shows that it is a globule with variable composition that contains some grains of the definite chemical composition Fe 46.01%, Si 53.99%. Determine the empirical formula of these grains.

10. A sample of a "suboxide" of cesium gives up 1.6907% of its mass as gaseous oxygen when gently heated, leaving pure cesium behind. Determine the empirical formula of this binary compound.

11. Barium and nitrogen form two binary compounds containing 90.745% and 93.634% barium, respectively. Determine the empirical formulas of these two compounds.

12. Carbon and oxygen form no fewer than five different binary compounds. The mass percentages of carbon in the five compounds are as follows: A, 27.29; B, 42.88; C, 50.02; D, 52.97; and E, 65.24. Determine the empirical formulas of the five compounds.

13. A sample of 1.000 g of a compound containing carbon and hydrogen reacts with oxygen at elevated temperature to yield 0.692 g H_2O and 3.381 g CO_2.
    (a) Calculate the masses of C and H in the sample.
    (b) Does the compound contain any other elements?
    (c) What are the mass percentages of C and H in the compound?
    (d) What is the empirical formula of the compound?

14. Burning a compound of calcium, carbon, and nitrogen in oxygen in a combustion train generates calcium oxide (CaO), carbon dioxide (CO_2), nitrogen dioxide (NO_2), and no other substances. A small sample gives 2.389 g CaO, 1.876 g CO_2, and 3.921 g NO_2. Determine the empirical formula of the compound.

15. The empirical formula of a gaseous fluorocarbon is CF_2. At a certain temperature and pressure, a 1-L volume holds 8.93 g of this fluorocarbon, whereas under the same conditions, the 1-L volume holds only 1.70 g gaseous fluorine (F_2). Determine the molecular formula of this compound.

16. At its boiling point (280°C) and at atmospheric pressure, phosphorus has a gas density of 2.7 g L^{-1}. Under the same conditions, nitrogen has a gas density of 0.62 g L^{-1}. How many atoms of phosphorus are there in one phosphorus molecule under these conditions?

17. A gaseous binary compound has a vapor density that is 1.94 times that of oxygen at the same temperature and pressure. When 1.39 g of the gas is burned in an excess of oxygen, 1.21 g water is formed, removing all the hydrogen originally present.
    (a) Estimate the molecular mass of the gaseous compound.
    (b) How many hydrogen atoms are there in a molecule of the compound?
    (c) What is the maximum possible value of the atomic mass of the second element in the compound?
    (d) Are other values possible for the atomic mass of the second element? Use a table of atomic masses to identify the element that best fits the data.
    (e) What is the molecular formula of the compound?

18. A gaseous binary compound has a vapor density that is 2.53 times that of nitrogen at 100°C and atmospheric pressure. When 8.21 g of the gas reacts with $AlCl_3$ at 100°C, 1.62 g gaseous nitrogen is produced, removing all of the nitrogen originally present.
    (a) Estimate the molecular mass of the gaseous compound.
    (b) How many nitrogen atoms are there in a molecule of the compound?
    (c) What is the maximum possible value of the atomic mass of the second element?
    (d) Are other values possible for the atomic mass of the second element? Use a table of atomic masses to identify the element that best fits the data.
    (e) What is the molecular formula of the compound?

2.3 Writing Balanced Chemical Equations

19. Balance the following chemical equations:

(a) $H_2 + N_2 \longrightarrow NH_3$
(b) $K + O_2 \longrightarrow K_2O_2$
(c) $PbO_2 + Pb + H_2SO_4 \longrightarrow PbSO_4 + H_2O$
(d) $BF_3 + H_2O \longrightarrow B_2O_3 + HF$
(e) $KClO_3 \longrightarrow KCl + O_2$
(f) $CH_3COOH + O_2 \longrightarrow CO_2 + H_2O$
(g) $K_2O_2 + H_2O \longrightarrow KOH + O_2$
(h) $PCl_5 + AsF_3 \longrightarrow PF_5 + AsCl_3$

20. Balance the following chemical equations:

(a) $Al + HCl \longrightarrow AlCl_3 + H_2$
(b) $NH_3 + O_2 \longrightarrow NO + H_2O$
(c) $Fe + O_2 + H_2O \longrightarrow Fe(OH)_2$
(d) $HSbCl_4 + H_2S \longrightarrow Sb_2S_3 + HCl$
(e) $Al + Cr_2O_3 \longrightarrow Al_2O_3 + Cr$
(f) $XeF_4 + H_2O \longrightarrow Xe + O_2 + HF$
(g) $(NH_4)_2Cr_2O_7 \longrightarrow N_2 + Cr_2O_3 + H_2O$
(h) $NaBH_4 + H_2O \longrightarrow NaBO_2 + H_2$

2.4 Mass Relationships in Chemical Reactions

21. For each of the following chemical reactions, calculate the mass of the underlined reactant that is required to produce 1.000 g of the underlined product.

(a) $\underline{Mg} + 2\ HCl \longrightarrow \underline{H_2} + MgCl_2$
(b) $\underline{2\ CuSO_4} + 4\ KI \longrightarrow 2\ CuI + \underline{I_2} + 2\ K_2SO_4$
(c) $\underline{NaBH_4} + 2\ H_2O \longrightarrow NaBO_2 + \underline{4\ H_2}$

22. For each of the following chemical reactions, calculate the maximum mass of the underlined product that could be produced from 1.000 g of the underlined reactant.

(a) $\underline{CaCO_3} + H_2O \longrightarrow \underline{Ca(OH)_2} + CO_2$
(b) $\underline{C_3H_8} + 5\ O_2 \longrightarrow \underline{3\ CO_2} + 4\ H_2O$
(c) $\underline{2\ MgNH_4PO_4} \longrightarrow \underline{Mg_2P_2O_7} + 2\ NH_3 + H_2O$

23. An 18.6-g sample of K_2CO_3 was treated in such a way that all of its carbon was captured in the compound $K_2Zn_3[Fe(CN)_6]_2$. Compute the mass (in grams) of this product.

24. A chemist dissolves 1.406 g pure platinum (Pt) in an excess of a mixture of hydrochloric and nitric acids and then, after a series of subsequent steps involving several other chemicals, isolates a compound of molecular formula $Pt_2C_{10}H_{18}N_2S_2O_6$. Determine the maximum possible yield of this compound.

25. Disilane (Si_2H_6) is a gas that reacts with oxygen to give silica (SiO_2) and water. Calculate the mass of silica that would form if 25.0 cm^3 disilane (with a density of 2.78×10^{-3} g cm^{-3}) reacted with excess oxygen.

26. Tetrasilane (Si_4H_{10}) is a liquid with a density of 0.825 g cm^{-3}. It reacts with oxygen to give silica (SiO_2) and water. Calculate the mass of silica that would form if 25.0 cm^3 tetrasilane reacted completely with excess oxygen.

27. Cryolite (Na_3AlF_6) is used in the production of aluminum from its ores. It is made by the reaction

$$6\ NaOH + Al_2O_3 + 12\ HF \longrightarrow 2\ Na_3AlF_6 + 9\ H_2O$$

Calculate the mass of cryolite that can be prepared by the complete reaction of 287 g Al_2O_3.

28. Carbon disulfide (CS_2) is a liquid that is used in the production of rayon and cellophane. It is manufactured from methane and elemental sulfur via the reaction

$$CH_4 + 4\ S \longrightarrow CS_2 + 2\ H_2S$$

Calculate the mass of CS_2 that can be prepared by the complete reaction of 67.2 g sulfur.

29. Potassium nitrate (KNO_3) is used as a fertilizer for certain crops. It is produced through the reaction

$$4\ KCl + 4\ HNO_3 + O_2 \longrightarrow 4\ KNO_3 + 2\ Cl_2 + 2\ H_2O$$

Calculate the minimum mass of KCl required to produce 567 g KNO_3. What mass of Cl_2 will be generated as well?

30. Elemental phosphorus can be prepared from calcium phosphate via the overall reaction

$$2\ Ca_3(PO_4)_2 + 6\ SiO_2 + 10\ C \longrightarrow 6\ CaSiO_3 + P_4 + 10\ CO$$

Calculate the minimum mass of $Ca_3(PO_4)_2$ required to produce 69.8 g P_4. What mass of $CaSiO_3$ is generated as a by-product?

31. An element X has a dibromide with the empirical formula XBr_2 and a dichloride with the empirical formula XCl_2. The dibromide is completely converted to the dichloride when it is heated in a stream of chlorine according to the reaction

$$XBr_2 + Cl_2 \longrightarrow XCl_2 + Br_2$$

When 1.500 g XBr_2 is treated, 0.890 g XCl_2 results.

(a) Calculate the atomic mass of the element X.
(b) By reference to a list of the atomic masses of the elements, identify the element X.

***32.** An element A has a triiodide with the formula AI_3 and a trichloride with the formula ACl_3. The triiodide is quantitatively converted to the trichloride when it is heated in a stream of chlorine, according to the reaction

$$AI_3 + \tfrac{3}{2}\ Cl_2 \longrightarrow ACl_3 + \tfrac{3}{2}\ I_2$$

If 0.8000 g AI_3 is treated, 0.3776 g ACl_3 is obtained.

(a) Calculate the atomic mass of the element A.
(b) Identify the element A.

***33.** A mixture consisting of only sodium chloride (NaCl) and potassium chloride (KCl) weighs 1.0000 g. When the mixture is dissolved in water and an excess of silver nitrate is added, all the chloride ions associated with the original mixture are precipitated as insoluble silver chloride (AgCl). The mass of the silver chloride is found to be 2.1476 g. Calculate the mass percentages of sodium chloride and potassium chloride in the original mixture.

***34.** A mixture of aluminum and iron weighing 9.62 g reacts with hydrogen chloride in aqueous solution according to the parallel reactions

$$2\ Al + 6\ HCl \longrightarrow 2\ AlCl_3 + 3\ H_2$$

$$Fe + 2\ HCl \longrightarrow FeCl_2 + H_2$$

A 0.738-g quantity of hydrogen is evolved when the metals react completely. Calculate the mass of iron in the original mixture.

2.5 Limiting Reactant and Percentage Yield

35. When ammonia is mixed with hydrogen chloride (HCl), the white solid ammonium chloride (NH_4Cl) is produced. Suppose 10.0 g ammonia is mixed with the same mass of hydrogen chloride. What substances will be present after the reaction has gone to completion, and what will their masses be?

36. The poisonous gas hydrogen cyanide (HCN) is produced by the high-temperature reaction of ammonia with methane (CH_4). Hydrogen is also produced in this reaction.

(a) Write a balanced chemical equation for the reaction that occurs.

(b) Suppose 500.0 g methane is mixed with 200.0 g ammonia. Calculate the masses of the substances present after the reaction is allowed to proceed to completion.

37. The iron oxide Fe_2O_3 reacts with carbon monoxide (CO) to give iron and carbon dioxide:

$$Fe_2O_3 + 3\ CO \longrightarrow 2\ Fe + 3\ CO_2$$

The reaction of 433.2 g Fe_2O_3 with excess CO yields 254.3 g iron. Calculate the theoretical yield of iron (assuming complete reaction) and its percentage yield.

38. Titanium dioxide, TiO_2, reacts with carbon and chlorine to give gaseous $TiCl_4$:

$$TiO_2 + 2\ C + 2\ Cl_2 \longrightarrow TiCl_4 + 2\ CO$$

The reaction of 7.39 kg titanium dioxide with excess C and Cl_2 gives 14.24 kg titanium tetrachloride. Calculate the theoretical yield of $TiCl_4$ (assuming complete reaction) and its percentage yield.

ADDITIONAL PROBLEMS

39. Human parathormone has the impressive molecular formula $C_{691}H_{898}N_{125}O_{164}S_{11}$. Compute the mass percentages of all the elements in this compound.

40. A white oxide of tungsten is 79.2976% tungsten by mass. A blue tungsten oxide also contains exclusively tungsten and oxygen, but it is 80.8473% tungsten by mass. Determine the empirical formulas of white tungsten oxide and blue tungsten oxide.

41. A dark brown binary compound contains oxygen and a metal. It is 13.38% oxygen by mass. Heating it moderately drives off some of the oxygen and gives a red binary compound that is 9.334% oxygen by mass. Strong heating drives off more oxygen and gives still another binary compound, which is only 7.168% oxygen by mass.

(a) Compute the mass of oxygen that is combined with 1.000 g of the metal in each of these three oxides.

(b) Assume that the empirical formula of the first compound is MO_2 (where M represents the metal). Give the empirical formulas of the second and third compounds.

(c) Name the metal.

42. A binary compound of nickel and oxygen contains 78.06% nickel by mass. Is this a stoichiometric or a nonstoichiometric compound? Explain.

43. Two binary oxides of the element manganese contain, respectively, 30.40% and 36.81% oxygen by mass. Calculate the empirical formulas of the two oxides.

44. A possible practical way to eliminate oxides of nitrogen (such as NO_2) from automobile exhaust gases uses cyanuric acid, $C_3N_3(OH)_3$. When heated to the relatively low temperature of 625°F, cyanuric acid converts to gaseous isocyanic acid (HNCO). Isocyanic acid reacts with NO_2 in the exhaust to form nitrogen, carbon dioxide, and water, all of which are normal constituents of the air.

(a) Write balanced equations for these two reactions.

(b) If the process described earlier became practical, how much cyanuric acid (in kilograms) would be required to absorb the 1.7×10^{10} kg NO_2 generated annually in auto exhaust in the United States?

45. Aspartame (molecular formula $C_{14}H_{18}N_2O_5$) is a sugar substitute in soft drinks. Under certain conditions, 1 mol of aspartame reacts with 2 mol of water to give 1 mol of aspartic acid (molecular formula $C_4H_7NO_4$), 1 mol of methanol (molecular formula CH_3OH), and 1 mol of phenylalanine. Determine the molecular formula of phenylalanine.

46. 3′-Methylphthalanilic acid is used commercially as a "fruit set" to prevent premature drop of apples, pears, cherries, and peaches from the tree. It is 70.58% carbon, 5.13% hydrogen, 5.49% nitrogen, and 18.80% oxygen. If eaten, the fruit set reacts with water in the body to produce an innocuous product, which contains carbon, hydrogen, and oxygen only, and *m*-toluidine ($NH_2C_6H_4CH_3$), which causes anemia and kidney damage. Compute the mass of the fruit set that would produce 5.23 g *m*-toluidine.

47. Aluminum carbide (Al_4C_3) reacts with water to produce gaseous methane (CH_4). Calculate the mass of methane formed from 63.2 g Al_4C_3.

48. Citric acid ($C_6H_8O_7$) is made by fermentation of sugars such as sucrose ($C_{12}H_{22}O_{11}$) in air. Oxygen is consumed and water generated as a by-product.

(a) Write a balanced equation for the overall reaction that occurs in the manufacture of citric acid from sucrose.

(b) What mass of citric acid is made from 15.0 kg sucrose?

49. A sample that contains only $SrCO_3$ and $BaCO_3$ weighs 0.800 g. When it is dissolved in excess acid, 0.211 g carbon dioxide is liberated. What percentage of $SrCO_3$ did the sample contain? Assume all the carbon originally present is converted to carbon dioxide.

50. A sample of a substance with the empirical formula XBr_2 weighs 0.5000 g. When it is dissolved in water and all its bromine is converted to insoluble AgBr by addition of an excess of silver nitrate, the mass of the resulting AgBr is found to be 1.0198 g. The chemical reaction is

$$XBr_2 + 2\ AgNO_3 \longrightarrow 2\ AgBr + X(NO_3)_2$$

(a) Calculate the molecular mass (that is, formula mass) of XBr_2.

(b) Calculate the atomic mass of X and give its name and symbol.

51. A newspaper article about the danger of global warming from the accumulation of greenhouse gases such as carbon dioxide states that "reducing driving your car by 20 miles a week would prevent release of over 1000 pounds of CO_2 per year into the atmosphere." Is this a reasonable statement? Assume that gasoline is octane (molecular formula C_8H_{18}) and that it is burned completely to CO_2 and H_2O in the engine of your car. Facts (or reasonable guesses) about your car's gas mileage, the density of octane, and other factors will also be needed.

52. In the Solvay process for producing sodium carbonate (Na_2CO_3), the following reactions occur in sequence:

$$NH_3 + CO_2 + H_2O \longrightarrow NH_4HCO_3$$

$$NH_4HCO_3 + NaCl \longrightarrow NaHCO_3 + NH_4Cl$$

$$2\ NaHCO_3 \xrightarrow{\text{heat}} Na_2CO_3 + H_2O + CO_2$$

How many metric tons of sodium carbonate would be produced per metric ton of NH_3 if the process were 100% efficient (1 metric ton = 1000 kg)?

53. A yield of 3.00 g $KClO_4$ is obtained from the (unbalanced) reaction

$$KClO_3 \longrightarrow KClO_4 + KCl$$

when 4.00 g of the reactant is used. What is the percentage yield of the reaction?

54. An industrial-scale process for making acetylene consists of the following sequence of operations:

$$\underset{\text{limestone}}{CaCO_3} \longrightarrow \underset{\text{lime}}{CaO} + \underset{\text{carbon dioxide}}{CO_2}$$

$$CaO + 3\ C \longrightarrow \underset{\text{calcium carbide}}{CaC_2} + \underset{\text{carbon monoxide}}{CO}$$

$$CaC_2 + 2\ H_2O \longrightarrow \underset{\text{calcium hydroxide}}{Ca(OH)_2} + \underset{\text{acetylene}}{C_2H_2}$$

What is the percentage yield of the overall process if 2.32 metric tons C_2H_2 is produced from 10.0 metric tons limestone (1 metric ton = 1000 kg)?

55. Silicon nitride (Si_3N_4), a valuable ceramic, is made by the direct combination of silicon and nitrogen at high temperature. How much silicon must react with excess nitrogen to prepare 125 g silicon nitride if the yield of the reaction is 95.0%?

Chemical Bonding and Molecular Structure

Courtesy of Dr. Richard P. Muller and Professor William A. Goddard III, California Institute of Technology.

The electron density in a delocalized three-center bond for H_3^+ calculated by quantum mechanics.

UNIT II

Unit II is a journey into the nanoscopic realm of atoms and molecules where the fundamental events of chemistry take place. A molecule is a tiny physical object that has a specific structure, with dimensions in the range 0.1 nm to about 2 nm, held together by a delicate balance of electrical forces. In polymeric and biological materials, molecular dimensions may be as large as micrometers or even millimeters. The structure takes a characteristic three-dimensional shape, usually of a type already familiar in geometry, and frequently with a high degree of symmetry. Molecular attributes—structure, electron density, shape—can be predicted by theory and observed in experiments. They determine the physical properties and chemical reactivity of molecules in situations ranging from pure chemistry to materials science to biology.

Unit Chapters

Unit Goals

- To introduce the classical theory of chemical bonding as a tool for describing the structures and shapes of molecules
- To convey the basic concepts and methods of quantum mechanics that describe the restrictions on the energy and the physical motions of microscopic systems
- To develop an intuition for the behavior of quantum systems and an appreciation for the magnitudes of physical properties in the quantum regime
- To use quantum mechanics to:
  - Describe the allowed energies and electron densities in atoms
  - Explain the structure of the periodic table and periodic trends in the properties of atoms
  - Describe covalent bond formation and the structures of diatomic and small polyatomic molecules
  - Describe covalent bond formation and the structures of organic molecules
  - Describe bonding in more complex structures that include transition metal complexes

3 CHAPTER

Atomic Shells and Classical Models of Chemical Bonding

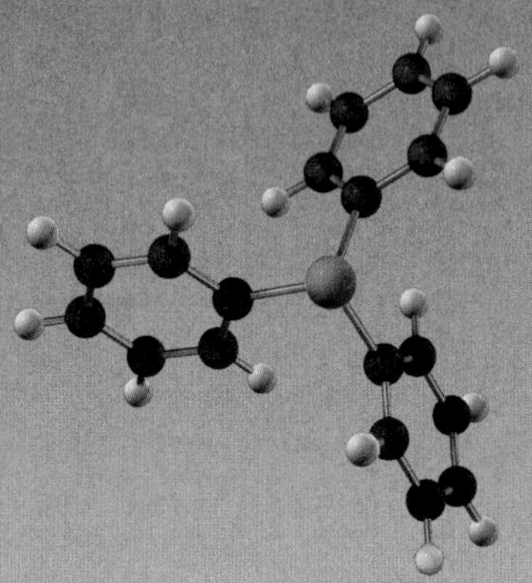

(a)

(b)

(a) The shape of the molecule triphenyl-phosphine, $(C_6H_5)_3P$, is determined by locating the valence shell electron pairs in those positions that minimize the overall energy of the molecule. (b) The space-filling representation aids the analysis and understanding of the steric environment responsible for the specific molecular geometry observed.

The previous chapters showed how the laws of conservation of mass and conservation of atomic identity, together with the concept of the mole, determine quantitative mass relationships in chemical reactions. That discussion assumed prior knowledge of the chemical formulas of the reactants and products in each equation. The far more open-ended questions of which compounds are found in nature (or which can be made in the laboratory) and what types of reactions they undergo now arise. Why are some elements and compounds violently reactive and others inert? Why are there compounds with chemical formulas H_2O and NaCl, but never H_3O or $NaCl_2$? Why are helium and the other noble gases monatomic, but molecules of hydrogen and chlorine diatomic? All of these questions can be answered by examining the formation of chemical bonds between atoms.

When two atoms come sufficiently close together, the electrons of each atom experience the additional attractive force of the nucleus in the other atom, the electrons repel each other, and the positively charged nuclei repel each other. A stable chemical bond between two atoms in the gas phase is formed only when the total energy of the resulting molecule is lower than that of the two isolated atoms. **Quantum mechanics**—the fundamental branch of physics that describes the properties, interactions, and motions of atomic and subatomic particles—is necessary to explain the re-distribution of electrons that leads to formation of chemical bonds between atoms.

Before the quantum explanation of bonding was available, chemists had developed a powerful set of concepts and tools based on classical mechanics—covalent bonds, ionic bonds, polar covalent bonds, electronegativity, Lewis electron dot diagrams, and valence shell electron-pair repulsion (VSEPR) theory—that rationalized a great deal of information about the structure of molecules and patterns of chemical reactivity. This set of concepts and tools constitutes the *classical description of the chemical bond,* and it is part of the daily vocabulary of every working chemist, especially in organic and biological chemistry. These tools are the foundation of chemical intuition, by which we mean the ability to explain and even predict chemical phenomena. Intuition is judgment informed by experience. Extensive practice in applying these tools to interpret factual information is essential to develop your own chemical intuition.

The classical theory of chemical bonding and molecular shapes starts with conceptual models of the chemical bond which can be understood on the basis of simple electrostatics—the forces between and energies of systems of stationary charged particles. Chemical bonds form by sharing or transferring electrons between atoms. Chemists generally identify two extreme cases. In a **covalent bond**, the electrons are shared more or less equally between the two atoms comprising the bond. In an **ionic bond**, one or more electrons is completely transferred from one atom to the other, and the dominant contribution to the strength of the bond is the electrostatic attraction between the resulting positive and negative ions. Although many real chemical bonds are well described by these idealized models, most bonds are neither completely ionic nor completely covalent and are best described as having a mixture of ionic and covalent character. In a **polar covalent bond**, a partial transfer of charge from one atom to the other occurs. **Electronegativity,** the tendency of an atom in a molecule to attract electrons from other atoms, explains whether a given pair of atoms forms an ionic, covalent, or polar covalent bond.

Two simple tools are used to implement the classical theory of bonding and structure. First, the **Lewis electron dot diagram** shows the number of **valence** (outermost) electrons associated with each atom in a molecule and indicates whether they are bonding (shared) or nonbonding. These diagrams are useful in predicting the structural formula—that is, which atoms are bonded to each other in polyatomic molecules—but they do not describe the three-dimensional shapes of molecules. Second, **VSEPR (valence shell electron pair repulsion) theory** predicts molecular shapes, based on the electrostatic argument that electron pairs in a molecule will arrange themselves to be as far apart as possible.

We start with a brief survey of the images used to represent molecular structure, then we describe the **periodic table,** a list of the elements arranged to display at a glance patterns of their physical properties and chemical reactivity. Relating bond formation between a pair of atoms to their positions in the periodic table reveals trends that build up chemical intuition. Next, we invoke Rutherford's planetary model of the atom and show how electrical forces control the gain or loss of electrons by the atom. We then examine the electrical forces within molecules and show how they lead to the ionic and covalent models of the chemical bond. The use of Lewis diagrams to describe bond formation and the VSEPR theory to describe molecular shapes complete the classical theory of bonding. We conclude with a brief survey of the procedures for assigning proper names to chemical compounds.

The foundations of classical bonding theory have been successfully explained by quantum mechanics. Today it is largely a matter of preference whether you first learn the classical theory and then gain deeper insight from the quantum explanations, or you first learn the quantum theory and then see the classical theory as a limiting case. We prefer to present the classical description first, and this chapter is devoted to that subject. That way, we establish the language and vocabulary of the chemical bond and allow you to become familiar with the properties of a broad array of real molecules *before* attempting to explain these results using quantum mechanics in Chapters 4, 5, and 6. Your instructor may prefer the opposite sequence, in which case you will read Chapters 4 and 5 before this chapter. We wrote this textbook to accommodate either approach.

3.1 Representations of Molecules

We discussed the distinction between empirical and molecular formulas for compounds in Chapter 2 and introduced you to combustion analysis, a classical method for the determination of empirical formulas. Compounds may either be discrete molecules or extended solids. A **molecule** is a collection of atoms bonded together, with the elements in fixed proportions and with a (typically) well-defined three-dimensional structure. Molecules are stable in the gas phase and can be condensed to form liquids and solids while preserving their identities. Individual molecules of molecular liquids or solids can be uniquely identified; Figure 1.2 shows two water molecules outlined by closed lines that separate them from other water molecules. Molecules are characterized by their molecular formulas, which we discuss here. It is not possible, however, to identify a "molecule" in extended solids (see later in this section) and we can only characterize them by their empirical formulas.

The molecular formula specifies the number of atoms of each element present in one molecule of a compound. The molecular formula for glucose is $C_6H_{12}O_6$; each molecule of glucose contains 6 atoms of carbon, 12 atoms of hydrogen, and 6 atoms of oxygen. The molecular formula is the starting point for predicting the existence of particular molecules, explaining their structures and shapes, and describing their reactions.

How do we determine a molecular formula? In order to prove that a substance is a compound, we must identify the elements present and confirm that the relative composition satisfies the law of definite proportions. Historically this proof was accomplished by combustion analysis but today is done primarily by mass spectrometry (see *Connection to Instrumental Analysis*). From this information we deduce the empirical formula, which lists the atoms present and gives their relative numbers. But the empirical formula alone does not uniquely describe a molecule. For example, the empirical formula CH_2O characterizes not only glucose $C_6H_{12}O_6$ but also acetic acid $C_2H_4O_2$ and formaldehyde CH_2O. To determine the molecular formula, we measure the molar mass of the compound under study from its gas-law behavior or by mass spectrometry. We take the ratio of that molar mass to the molar mass of the empirical formula, and obtain the molecular formula as a simple integral multiple of the empirical formula. (See Sections 2.1 and 2.2 for a more detailed discussion.)

Once we have determined that a molecule has a specific and well-defined composition, with a known mass, we can begin to ask interesting and important questions about its behavior. For example, why does it contain only the particular numbers of atoms stated in its formula? How are these atoms held together in the molecule? What is the three-dimensional structure and the shape of the molecule? We study each of these questions in the chapters in Unit II.

The first step in understanding the properties of a molecule is to find its **condensed structural formula**, which specifies which atoms are bonded to each other and by what types of bonds. For example, the molecular formula for ethanol is C_2H_6O and its condensed structural formula is CH_3CH_2OH. The condensed structural formula tells us (by convention) that the two carbon atoms are bonded to one another, and that the second carbon atom is also bonded to the oxygen atom. Three hydrogen atoms are bonded to the first carbon atom, two hydrogen atoms are bonded to the second carbon atom, and one hydrogen atom is bonded to the oxygen atom. We also represent structural formulas as simple two-dimensional line drawings, as shown in Figure 3.1a. We use the Lewis model to find structural formulas; this model defines each bond as a pair of electrons localized between two particular atoms and represents structural formulas using Lewis dot diagrams. Figure 3.1a shows examples of Lewis dot diagrams and their elaboration as two-dimensional simple line drawings for molecules with the condensed structural formulas CH_4 (methane), NH_3 (ammonia), and H_2O (water).

Despite its name, the structural formula fails to convey the detailed structure of a molecule as a three-dimensional object in space. For this purpose we use ball-and-stick models in which the balls represent the atoms and the sticks represent the bonds between them. The lengths of the bonds are represented by the lengths of the sticks and the bond angles are the angles between the sticks attached to particular spheres (see Figure 3.1b). Ball-and-stick figures can be drawn on paper or they can be constructed as three-dimensional structural models. In either case they provide a clear picture of the geometrical relationship between bonds at each atom, but they do not provide any detail about the sizes of the atoms. The bond lengths and bond angles represented in these structures can be measured experimentally, and they can be predicted at various levels of detail by the different theories of chemical bonding.

The shapes of molecules influence their behavior and function, especially the ease with which they can fit into various guest-host configurations important in biology and biochemistry. We describe the shapes of molecules by recognizing

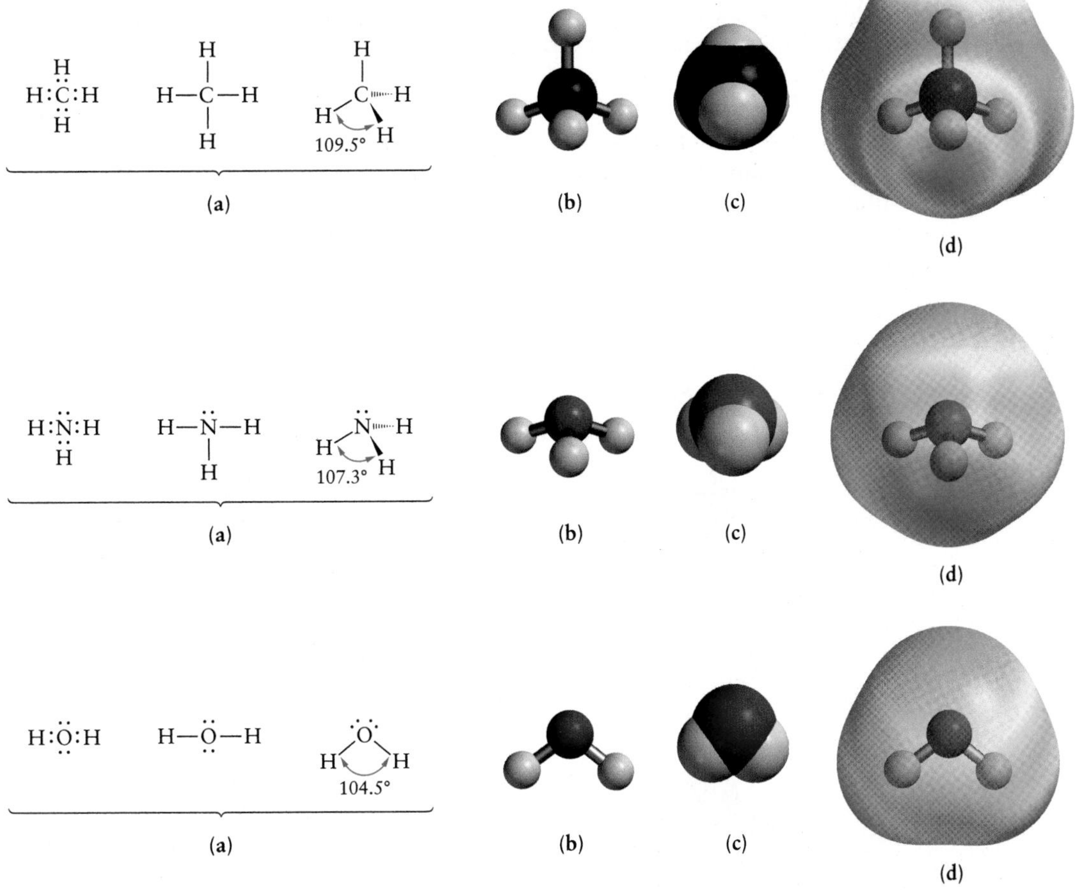

FIGURE 3.1 Different molecular representations of methane, ammonia, and water. (a) Lewis dot diagrams, line structures, and line angle representations. (b) Ball-and-stick models. (c) Space-filling models. (d) Electrostatic potential energy diagrams (elpots), shown, for clarity, over a larger volume than the space-filling models in (c).

that the positions of the atoms and the bonds define polyhedra or other shapes familiar to us from geometry. Frequently these shapes show a high degree of symmetry. Molecular shapes can be determined from experiment, and predicted by bonding theories.

Molecules have size, or volume, and occupy space due to the volume occupied by the electrons in the atoms of the molecule. This important property is represented by space-filling models, which show that the atoms have specific sizes that physically contact each other in molecules (see Fig. 3.1c). The images in that figure are constructed from computer calculations of the three-dimensional distribution of electron density around the molecule, made possible by quantum mechanical theories of chemical bonding.

The final representation we introduce to display molecular properties is the **electrostatic potential energy diagram** (sometimes called "elpot" diagram). This diagram displays the electrostatic potential energy that a small positive "test charge" would experience at every position on the electron density surface that defines the space-filling model. The results are displayed by color-coding the space-filling model to represent the local variations in electrostatic potential measured by the test charge (see Fig. 3.1d). Elpot diagrams give us a feel for the spatial distribution of electrons in molecules; these images are generated by computer calculations based on quantum mechanical theories of chemical bonding. The elpot diagrams are extremely helpful for evaluating the chemical reactivity of different sites on molecules.

We use all these representations throughout the chapters in Unit II as we develop the theoretical and experimental methods to explain the sizes, structures, and shapes of molecules. We encourage you to use them to develop the skill to visualize molecules as three-dimensional objects.

It frequently occurs that different compounds found in nature or in the laboratory have the same molecular formula but different molecular structures and therefore different properties. Such compounds, called **isomers**, are named and characterized independently. The structural formula, detailed molecular structure, and shape must be worked out for each isomer. The existence of isomers illustrates dramatically that the properties of compounds are determined by the structures of their molecules (see Fig. 3.2). Several different classes of isomeric

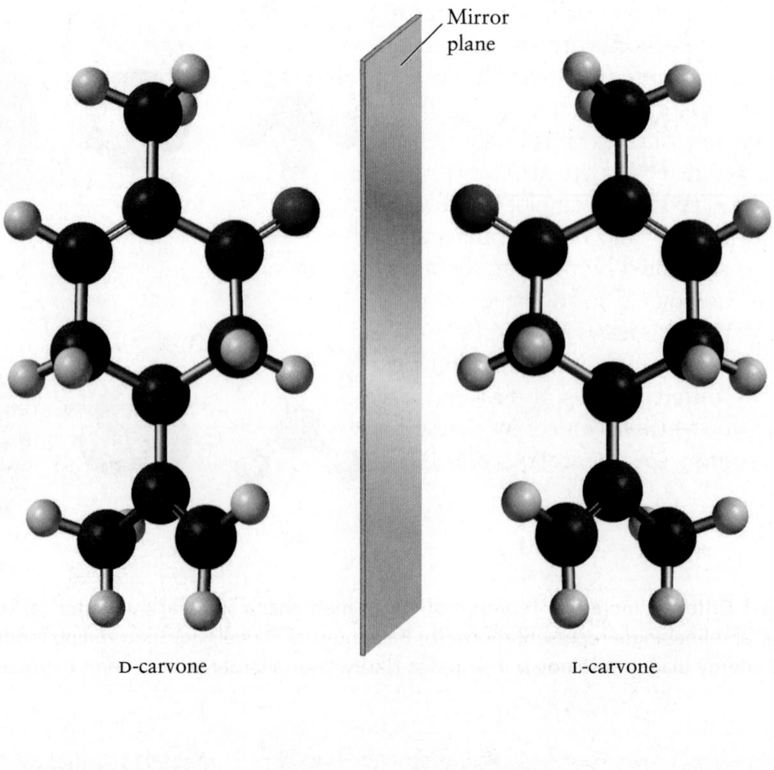

FIGURE 3.2 Structures of the optical isomers of carvone. D-carvone is on the left and L-carvone is on the right. Their mirror images are not superimposable. D-carvone has the odor of spearmint oil and L-carvone has the odor of caraway and dill seed oil.

CONNECTION TO INSTRUMENTAL ANALYSIS

Mass Spectrometry

We introduced you to mass spectrometry in Section 1.4, in which we showed how atoms of different masses could be separated from each another on the basis of their mass-to-charge ratios. J. J. Thomson immediately recognized the potential of mass spectrometry for chemical analysis (J.J. Thomson, *Rays of Positive Electricity and their Application to Chemical Analyses* [London, New York: Longmans, Green and Co., 1913]) but it was not until commercial instruments became available in the 1950s that mass spectrometry began to be used routinely for that purpose. Modern mass spectrometry is undoubtedly our most important technique for the accurate determination of molecular masses but it also provides us with valuable information about chemical formulas and molecular structure, including, for example, the determination of the sequences of amino acids in proteins.

The schematic of a mass spectrometer shown in Figure 1.16 shows the three elements common to all mass spectrometers: a source region, a mass analyzer, and a detector. Modern instruments vary considerably in the means by which ions are produced and masses are analyzed, with specialized instruments having been designed for specific purposes. It is beyond the scope of this book to discuss these instruments in any detail so we introduce you to the basic elements, point out the special features of a couple of the most widely used specialized spectrometers, and introduce you to the interpretation of mass spectra. Mass spectrometers are operated under high to ultrahigh vacuum conditions to minimize contamination and to ensure that ions do not suffer collisions in the mass analyzer. Samples are introduced into the source by a variety of methods that depend upon their physical state and vapor pressure; the gases are ionized and enter the mass analyzer, which separates them spatially. The ion current is measured by a detector that can count single ions and the mass spectrum produced is a plot of the ion current versus the mass-to-charge ratio m/z. (In mass spectrometry, the ratio m/z is calculated by dividing the mass m of the ion in atomic mass units by z, an integer corresponding to the total charge on the species divided by $e = 1.602 \times 10^{-19}$ C.) We introduce you to three different ionization methods and two different levels of mass resolution, the kinds of analyses for which each of these methods are used, and how the resulting spectra are interpreted.

Ionization Methods

Electron impact (EI) ionization has historically been the most important ionization method in which electrons emitted from a hot filament are accelerated to give them sufficient kinetic energy to ionize molecules in the source. 70 eV has been chosen as the standard kinetic energy because it is sufficiently high to ionize any molecule and also because it provides enough excess kinetic energy to cause fragmentation, which provides important information about molecular structure. Fragmentation patterns at specified ionization energies are characteristic of a particular molecule, enabling the identification of unknowns in a mixture (after separation by gas chromatography, see Figure 14.17) by comparing their spectra with those in mass spectral databases. It is often desirable, on the other hand, to use soft ionization techniques to prevent fragmentation and to ensure that the parent molecular ion dominates the mass spectrum. Matrix-assisted laser desorption ionization (MALDI) is among the most popular and powerful of the soft ionization techniques because of its ability to handle molecules with large molar masses and provide high mass resolution. Compounds of interest (the analyte) are dissolved in a solution that also contains molecules that absorb laser radiation of a particular wavelength (the matrix). Evaporating the solvent leaves the analyte trapped in the solid matrix, which is then introduced into the mass spectrometer. A pulsed laser deposits energy into the matrix, ionizing and ejecting the analyte into the gas phase, with the singly charged parent molecular ion dominating. Electrospray (ES) ionization is particularly useful for analyzing high molecular weight polymers, including biological macromolecules such as proteins and nucleic acids. A solution containing the analyte is sprayed through a pair of coaxial needles with a large voltage difference between them. The solvent evaporates, producing charged analytes that enter the mass analyzer.

Mass Resolution

Mass analyzers are generally classified into two broad categories based upon their mass resolution: low-resolution instruments generally require that two masses differ by more than 0.1% to be separated whereas high-resolution instruments are capable of mass resolution of one part in 10^4 or better. Low-resolution instruments are generally used for the identification of compounds in mixtures by comparison to standard reference spectra, or to determine the chemical formulas and molecular structures of newly synthesized compounds. The low-resolution EI mass spectrum of the neurotransmitter dopamine is shown in Figure C3.1.

The most intense (base) peak, from a fragment of the molecular ion, appears at $m/z = 124$, with a relative abundance of 100. The fragmentation mechanism that

[redacted]

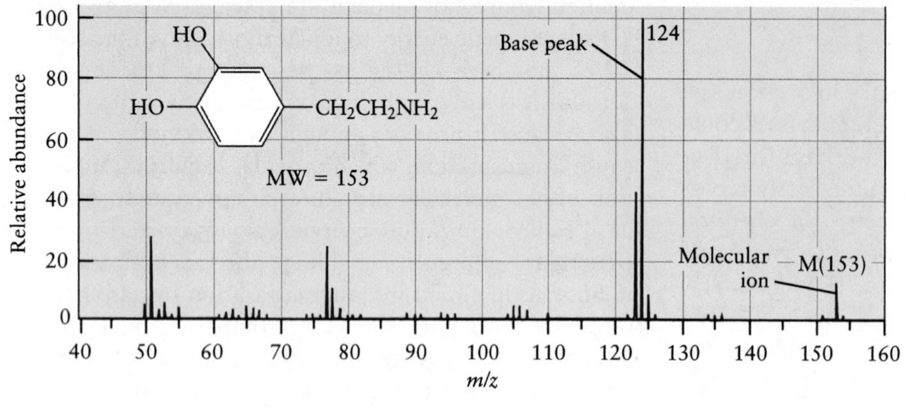

FIGURE C3.1 Low-resolution EI mass spectrum of dopamine. (Adapted from *Organic Chemistry* 5th ed. W. H. Brown, C. S. Foote, B. L. Iverson, and E. Anslyn. Brooks Cole/Cengage, 2009, Chapter 14, Figure 14.2, p. 523.)

produces this ion is hard to explain but the peak at 123 amu is easy to explain; it is due to the loss of the CH_2NH_2 group. Characteristic losses like these are used to help identify molecular structures. The spectra of linear alkanes (see Section 7.2), for example, have a series of peaks separated by 14 amu due to the loss of CH_2 groups. Common fragmentation patterns for cyclic alkanes include the loss of side groups, for example methyl at 15 amu, and the elimination of CH_2CH_2 groups as ethylene, producing a peak at M − 28, where M is the mass of the molecular ion. A common fragmentation mechanism for alcohols (see Section 7.6) is the elimination of water to produce an M − 18 peak. Finally, the mass spectra of molecules with π bonds often show prominent molecular ion peaks because the bond in the resulting ion is still quite strong and the molecular ion quite stable.

The electrospray mass spectrum of a small polypeptide (see Section 7.6) is shown in Figure C3.2. Note the simplicity of this spectrum compared with that of dopamine; much less fragmentation is observed. Polypeptides typically fragment by cleaving peptide bonds as shown by the dashed lines in the figure. The peak at M − 17 is due to the loss of either —OH or —NH_3 and the peaks at lower masses arise from cleavage of one or more peptide bonds. The M + 1 and M + 23 peaks arise from proton or sodium ion attachment.

High-resolution mass spectrometry provides not only accurate molecular weights but, perhaps more importantly, molecular formulas. The molecules CO, N_2, CH_2N and C_2H_4 all have nominal masses of 28 amu and are indistinguishable at low resolution. The precise masses of $^{12}C^{16}O$, $^{14}N_2$, $^{12}C^1H_2{}^{14}N$ and $^{12}C_2{}^1H_4$ (see Table 19.1) are actually 27.9949, 28.0062, 28.0187 and 28.0312 amu, respectively, are readily separated at high resolution (1 part per 10^4 − 10^6 is common). More complicated molecules have significantly more possibilities than this but measuring exact masses alone at high resolution is sufficient to determine molecular formulas uniquely, supplanting chemical methods such as combustion analysis (see Figure 2.2) which is now used primarily to establish the purity of newly synthesized compounds.

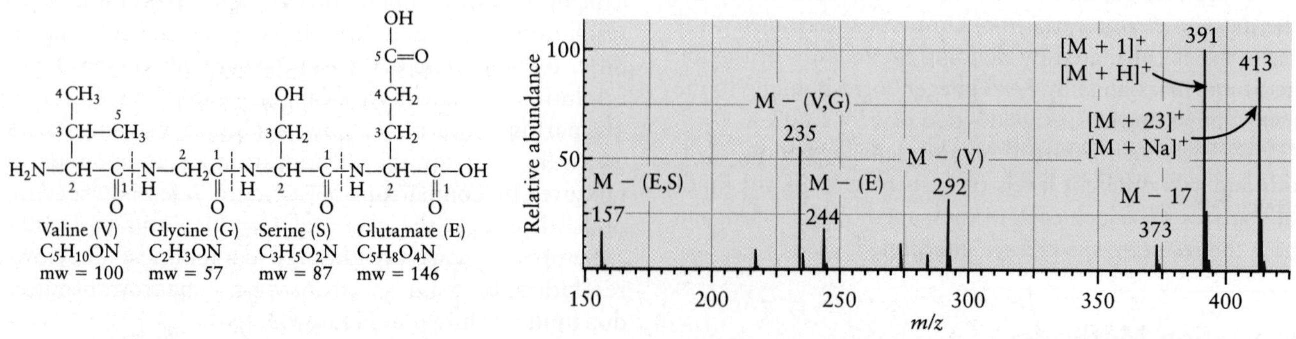

FIGURE C3.2 Electrospray mass spectrum of a small polypeptide. (Adapted from *Spectrometric Identification of Organic Compounds* 7th ed. Robert M. Silverstein, Francis X. Webster, and David Kiemle. John Wiley & Sons, New York, 2005, p. 8.)

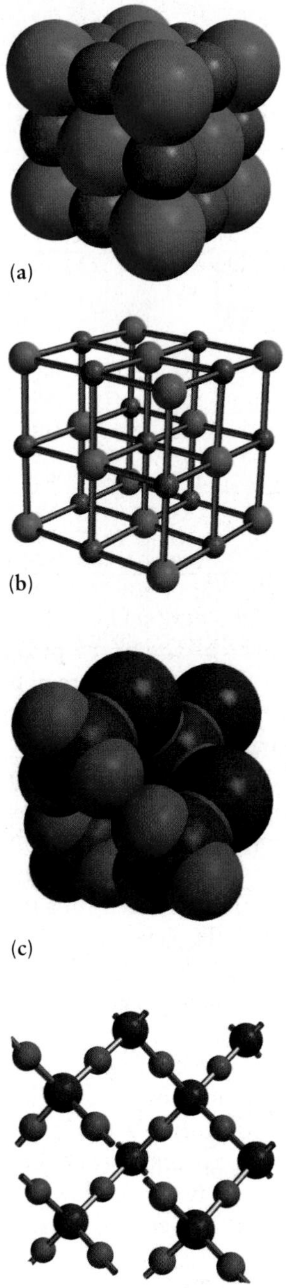

FIGURE 3.3 (a) Space-filling and (b) ball-and-stick models of NaCl. (c) Space-filling and (d) ball-and-stick models of the mineral cristobalite, one of many crystalline forms of SiO_2. The NaCl and SiO_2 models are not scaled.

compounds appear in organic, inorganic, and biological chemistry where an appreciation of the three-dimensional structures of molecules is essential for explaining the behavior of compounds.

The representations just described are appropriate for compounds formed by covalent bonding primarily between non-metallic elements. These compounds have discrete molecules which are stable in the gas phase. These gaseous molecules can be condensed to liquids and sometimes also into solids while maintaining their identities.

By contrast, ionic bonding between metallic and non-metallic atoms produces extended solid-state ionic compounds, which exist at room temperature, in which each ion is surrounded by a group of ions of opposite charge. The formula unit (see Section 2.1) of such a solid reflects the composition given by the empirical formula for the solid. The formula unit is also the smallest collection of ions that is electrically neutral. Thus solid sodium chloride contains not "a molecule of NaCl" but rather a formula unit of NaCl. Similarly, solid silicon dioxide contains a formula unit of SiO_2 (see Fig. 3.3).

3.2 The Periodic Table

The number of known chemical compounds is already huge, and it continues to increase rapidly as the result of significant investments in chemical research. An unlimited number of chemical reactions are available among these compounds. The resultant body of chemical knowledge, viewed as a collection of facts, is overwhelming in its size, range, and complexity. It has been made manageable by the observation that the properties of the elements naturally display certain regularities. These regularities enable the classification of the elements into families whose members have similar chemical and physical properties. When the elements are arranged in order of increasing atomic number, Z, remarkable patterns emerge. Families of elements with similar chemical properties are easily identified by their locations in this arrangement. This discovery is summarized concisely by the **periodic law:**

The chemical properties of the elements are periodic functions of the atomic number Z.

Consequently, the elements listed in order of increasing Z can be arranged in a chart called the periodic table, which displays, at a glance, the patterns of chemical similarity. The periodic table then permits systematic classification, interpretation, and prediction of all chemical information.

The modern periodic table (Fig. 3.4 and the inside front cover of this book) places elements in both a **group** (arranged vertically) and a **period** (arranged horizontally).

There are eight groups of **representative elements,** or "main-group" elements. In addition to the representative elements, there are ten groups (and three periods) of **transition-metal elements,** a period of elements with atomic numbers 57 through 71 called the rare-earth or **lanthanide elements,** and a period of elements from atomic numbers 89 through 103 called the **actinide elements,** all of which are unstable and most of which must be produced artificially. The lanthanide and actinide elements are usually placed below the rest of the table to conserve space. The groups of representative elements are numbered (using Roman numerals) from I to VIII, with the letter A sometimes added to differentiate them from the transition-metal groups, which are labeled from IB to VIIIB. This book uses group numbers exclusively for the representative elements (dropping the A) and refers to the transition-metal elements by the first element in the corresponding group. For example, the

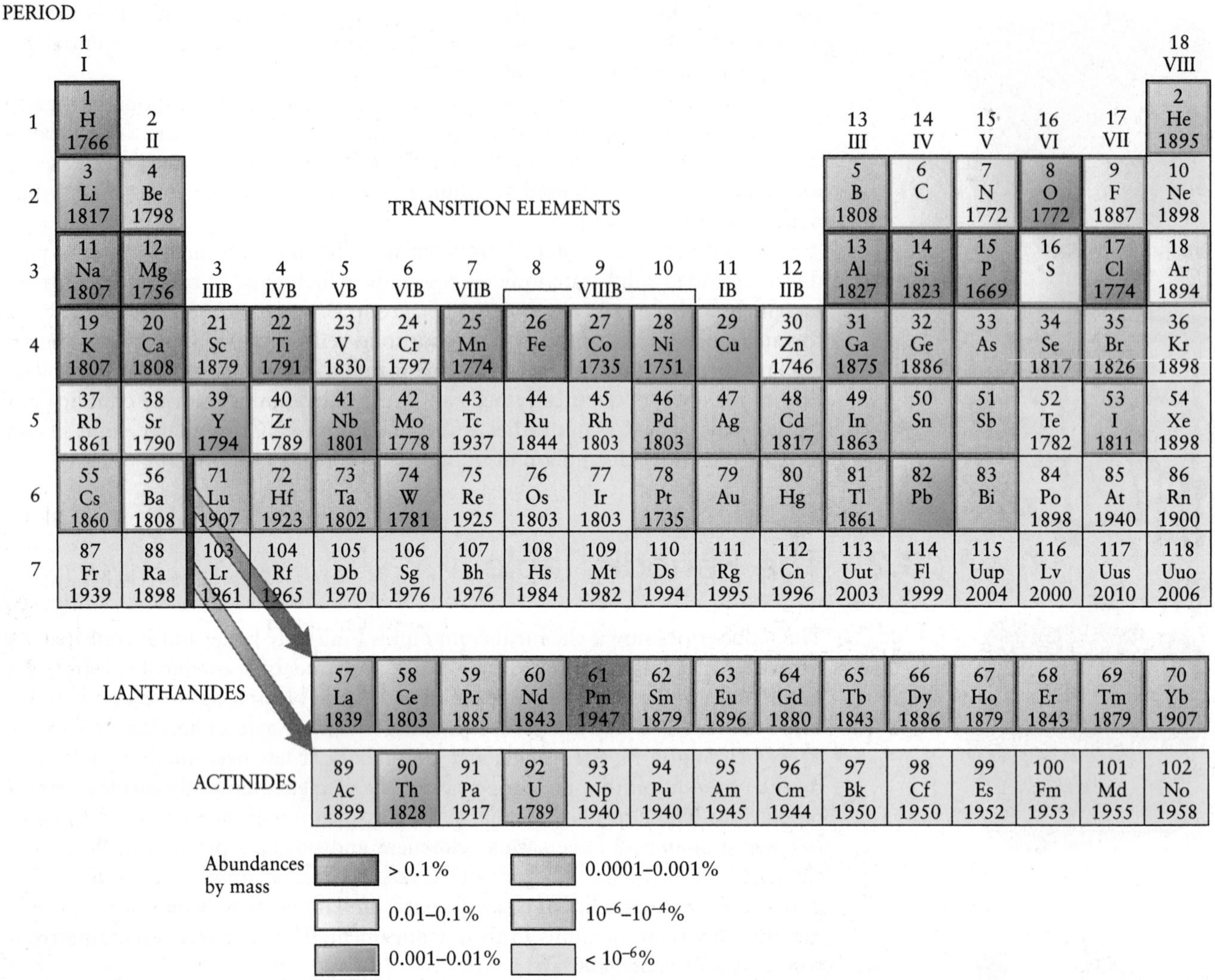

FIGURE 3.4 The modern periodic table of the elements. Below each symbol is the year in which that element was discovered; elements with no dates have been known since ancient times. Above each symbol is the atomic number. The color coding indicates the relative abundance by mass of the elements in the world (the atmosphere, oceans and fresh water bodies, and the Earth's crust to a depth of 40 km). Oxygen alone comprises almost 50% of the mass, and silicon comprises more than 25%.

elements in the carbon group (C, Si, Ge, Sn, Pb) are designated as Group IV, and the elements chromium (Cr), molybdenum (Mo), and tungsten (W) as the chromium group.[1]

3.2.1 Survey of Physical and Chemical Properties: The Representative Elements

Lothar Meyer and Dmitri Mendeleev independently and simultaneously organized the elements into tables based on their atomic weights. Meyer was more interested in the periodic variation in the physical properties of the elements, whereas Mendeleev was more interested in patterns of chemical reactivity. Therefore, most chemists consider Mendeleev to be responsible for the creation of the modern periodic table.

[1]We do not use the group numbers 1-18 recommended by the International Union of Pure and Applied Chemistry in their 1990 publication *Nomenclature of Inorganic Chemistry IUPAC Recommendations 1990.* An update of this reference (also called the Red Book) was published in 2005.

The elements have been classified empirically based on similarities in their physical or chemical properties. **Metals** and **nonmetals** are distinguished by the presence (or absence) of a characteristic metallic luster, good (or poor) ability to conduct electricity and heat, and malleability (or brittleness). Certain elements (boron, silicon, germanium, arsenic, antimony, and tellurium) resemble metals in some respects and nonmetals in others, and are therefore called **metalloids** (or sometimes **semimetals**). Their ability to conduct electricity, for example, is much worse than metals, but is not essentially zero like the nonmetals.

The empirical formulas of the binary compounds formed by the elements with chlorine (their *chlorides*), with oxygen (their *oxides*), and with hydrogen (their *hydrides*) show distinct periodic trends.

Group I, the **alkali metals** (lithium, sodium, potassium, rubidium, and cesium), are all relatively soft metals with low melting points that form 1:1 compounds with chlorine, with chemical formulas such as NaCl and RbCl. The alkali metals react with water to liberate hydrogen; potassium, rubidium, and cesium liberate enough heat upon reaction to ignite the hydrogen. Group II, the **alkaline-earth metals** (beryllium, magnesium, calcium, strontium, barium, and radium), react in a 1:2 atomic ratio with chlorine, producing compounds such as $MgCl_2$ and $CaCl_2$.

Of the nonmetallic elements, Group VI, the **chalcogen elements** (oxygen, sulfur, selenium, and tellurium), forms 1:1 compounds with the alkaline-earth metals (such as CaO and BaS) but 2:1 compounds with the alkali metals (such as Li_2O and Na_2S). Members of Group VII, the **halogen elements** (fluorine, chlorine, bromine, and iodine), differ significantly in their physical properties (fluorine and chlorine are gases at room temperature, bromine is a liquid, and iodine a solid), but their chemical properties are similar. Any alkali metal will combine with any halogen in 1:1 proportion to form a compound such as LiF or RbI, which is called an **alkali halide**.

The remaining elements fall into three additional groups whose chemical and physical properties are somewhat less clearly delineated than those already discussed. Group III includes a metalloid (boron) and four metals (aluminum, gallium, indium, and thallium). The metals form 1:3 chlorides (such as $GaCl_3$) and 2:3 oxides (such as Al_2O_3). Group IV comprises the elements carbon, silicon, germanium, tin, and lead. All of these elements form 1:4 chlorides (such as $SiCl_4$), 1:4 hydrides (such as GeH_4), and 1:2 oxides (such as SnO_2). Tin and lead are metals with low melting points, and silicon and germanium are **semiconductors**. Although we classified silicon and germanium as metalloids earlier, their electrical properties can be finely tuned by incorporating small amounts of impurities. These two elements form the basis for the modern semiconductor industry, which manufactures computer chips and other solid-state devices. Several different allotropes of elemental carbon exist (for example, graphite, diamond, and the recently discovered fullerenes). *Allotropes* are modifications of an element with differing atomic arrangements that lead to different physical and chemical properties. For example, ozone (O_3) and ordinary diatomic oxygen (O_2) are also allotropes. Group V includes nitrogen, phosphorus, arsenic, antimony, and bismuth. These elements form binary compounds with hydrogen and oxygen that have empirical formulas such as PH_3 and N_2O_5. The hydrides become increasingly unstable as their molar masses increase, and BiH_3 is stable only below 45°C. A similar trend exists for the oxides, and Bi_2O_5 has never been obtained in pure form. The lighter members of this group are clearly nonmetals (nitrogen and phosphorus), bismuth is clearly a metal, and arsenic and antimony are classified as semimetals.

Group VIII, the **noble gases** (helium, neon, argon, krypton, xenon, and radon), are sometimes called the **inert gases** because of their relative inertness toward chemical combination. They are all monatomic, in contrast with the other elements that exist as gases at room temperature and atmospheric pressure (hydrogen, oxygen, nitrogen, fluorine, chlorine), which are diatomic molecules.

Systematic trends in both the physical and chemical properties of the elements give important clues as to the structure of the atom. In addition to the properties that distinguish metals from nonmetals (electrical and thermal conductivity, malleability, luster, and ductility), there are a number of other physical properties that show clear periodic trends; these properties include melting and boiling points, densities, atomic sizes, and the energy changes that occur when an electron is added to or removed from a neutral atom. Numerical values for most of these properties are tabulated in Appendix F. In general, the elements on the left side of the table (especially in the later periods) are metallic solids and good conductors of electricity. On the right side (especially in earlier periods), they are poor conductors of electricity and are generally gases at room temperature and atmospheric pressure. The semimetals occupy a region of the periodic table located between the metals and the nonmetals that is represented by a diagonal band (see the inside front cover of this textbook).

Patterns in chemical reactivity of the elements correlate with patterns in the physical structure of the atom; they are both periodic functions of Z. Reading across the periodic table (horizontally) shows that each main-group element (Groups I-VIII) in Period 3 has exactly 8 more electrons than the element immediately above it in Period 2. Similarly, each main-group element in Periods 4 and 5 has exactly 18 more electrons than the corresponding element in the period above. The sequence of numbers, 8, 8, 18, 18, and so forth, that organize the periodic table into groups (columns), whose elements have similar physical and chemical properties, arises from the quantum theory of atomic structure (see discussion in Chapter 5).

3.3 Forces and Potential Energy in Atoms

The atom arose in the domain of chemistry, its existence inferred indirectly from the laws of chemical combination. Beginning with the work of Thomson and Rutherford, understanding the atom also fell within the province of physics, which sought to explain its structure and behavior as consequences of the electrical forces between the electrons and the nucleus. Modern chemistry combines these themes to explain chemical behavior in terms of the electrical forces within the atom. The purpose of this section is to give you an appreciation for the nature and magnitudes of these forces, and the associated potential energy, in preparation for your studies of chemical bond formation. It is essential that you understand and learn to use potential energy diagrams for atoms. We suggest that you review the background material on force, work, potential energy, potential energy diagrams, and electricity and magnetism in Appendix B.2 before continuing to study this section. Appendix C.7 gives an introduction to vectors.

Rutherford's planetary model describes the atom as a dense, central nucleus of positive charge $+Ze$ surrounded by a total of Z electrons, each in motion like the planets around the sun. The attractive force between each electron and the nucleus, and the repulsive force between each pair of the electrons, are all determined by Coulomb's law. Figure 3.5 shows an example of an atom with just one electron; the vector $\vec{r}$ points from the nucleus to the instantaneous position of the electron. The distance from the nucleus to the electron is the length of the vector $\vec{r}$, which we will denote as r without the over-arrow. (Mathematics texts would give the symbol $\|r\|$ for the length, or norm, and denote the vector by bolding it.)

Looking at Figure 3.5, when the electron is at the top of the circular orbit the velocity $\vec{v}$ is in the $+x$ direction ($\vec{v} = (v_x, v_y) = (v_x, 0)$, where $v_x = dx/dt$ is positive at this point in the orbit) and the force attracting the electron is

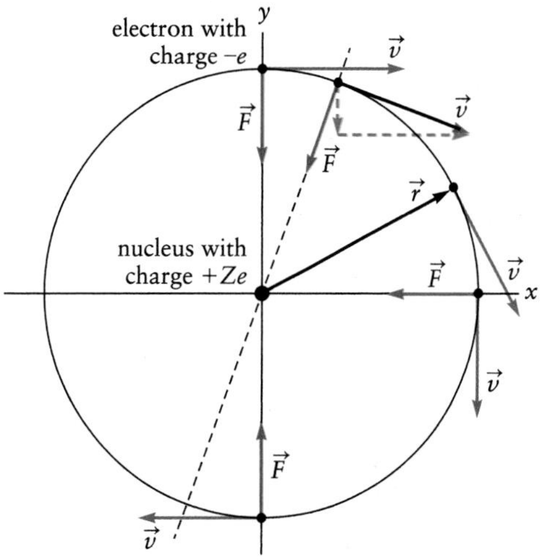

FIGURE 3.5 An electron, with charge $-e$, undergoing a clockwise circular orbit around a nucleus with charge $+Ze$. The red arrows show the force that the electron feels at each point in its trajectory; the force felt by the electron changes direction as the electron moves from one position to the next (four positions on the circular orbit are shown), but the force felt by the electron always points toward the nucleus. The velocity vector (blue) is always tangent to the circular orbit.

downward, in the $-y$ direction. Thus the force $\vec{F}$ does not slow the electron down, it changes the *direction* of the velocity. When the electron reaches the point in the circular orbit at about 20°, you can see that the attractive force has tilted the velocity vector downward, so it now has a velocity vector with a negative component v_y, but the speed of the electron has not changed. The kinetic energy of the electron, $\frac{1}{2}mv^2$, is thus constant as the electron orbits around the nucleus. The force $\vec{F}$ (shown by the red vectors in Figure 3.5) always points from the electron to the nucleus because the electrostatic potential energy is lower when two particles with opposite charges (one positive and one negative) get closer to each other. Thus the Coulomb force $\vec{F}$ on the electron, always attracted to the nucleus, points in the opposite direction of $\vec{r}$, the vector from the nucleus to the electron.

In general, for two particles, one with charge q_1 and the other with charge q_2, separated by distance r, the electrostatic potential energy given by Coulomb's law is

$$V(r) = \frac{q_1 q_2}{4\pi\epsilon_0 r} \qquad \textbf{[3.1]}$$

where ϵ_0, called the *permittivity of vacuum*, is a proportionality constant with a numerical value of $8.854 \times 10^{-12}\ C^2\ J^{-1}\ m^{-1}$ in the International System of Units (SI) described in Appendix B.1. Charge is measured in coulombs (C), distance in meters (m), and force in newtons (N). By convention, $V(r) \to 0$ as $r \to \infty$. This is a logical choice for the zero of potential energy because there is no interaction between the particles at such large distances. If the charges have the same sign, the potential energy, as expressed by Equation 3.1, is positive and it decreases as r increases. Thus dV/dr is negative if the two charges have the same sign, and the potential energy decreases if the charged particles move farther away from each other. (Like a bicycle that starts at the top of a hill [high potential energy] with zero kinetic energy and then rolls downhill, if you hold two protons close to each other and then let them loose, they will repel each other and start moving away from each other, gaining kinetic energy as the distance between them increases and the potential energy decreases.) The upper curve in Figure 3.6, a calculation using Equation 3.1 of the potential energy between two protons as a function of the distance r between them, shows that the potential energy $V(r)$ decreases when the distance between the protons increases. But what about a proton, with charge $+e$ and an electron with charge $-e$? These two charges have opposite signs, so using Equation 3.1, the potential energy is negative when the electron is near the proton (your bicycle is in a deep

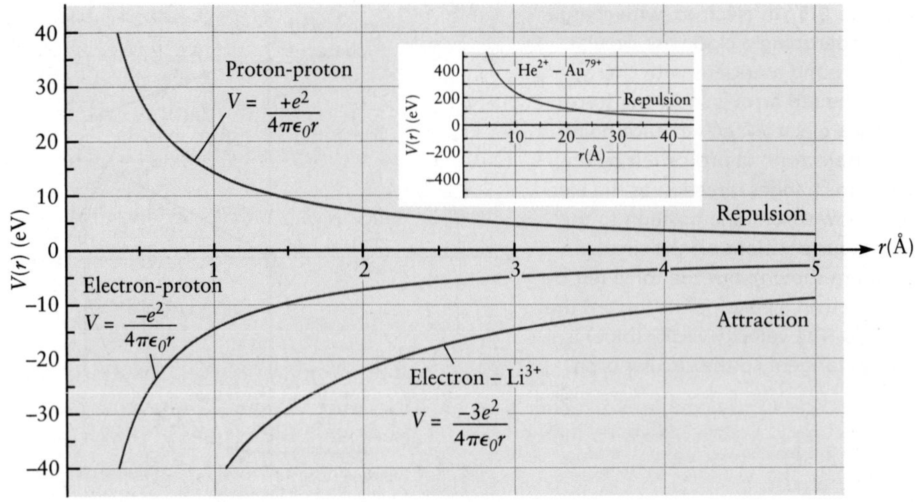

FIGURE 3.6 Potential energy curves for pairs of charged particles interacting according to Coulomb's law.

hole or "potential energy well"!) and it increases (becomes less negative) as r increases. Thus for an electron interacting with a proton, dV/dr is positive; the potential energy gets higher (less negative) as you try to move an electron farther from the proton. This is a reflection of the fact that the electron feels an attractive force from the proton.

Typical distances between nuclei and electrons in atoms are about 10^{-10} m. Figure 3.6 thus gives the distance between the charged particles in a distance unit called an angstrom (Å), where 1 Å = 10^{-10} m. The potential energy of the hydrogen atom ($Z = 1$) when the proton and electron are separated by 1 Å is

$$V(1\text{Å}) = -\frac{(1.602 \times 10^{-19}\ \text{C})^2}{4\pi(8.854 \times 10^{-12}\ \text{C}^2\ \text{J}^{-1}\ \text{m}^{-1})(1 \times 10^{-10}\ \text{m})} \quad [3.2]$$

$$= -\frac{(8.988 \times 10^{9})(1.602 \times 10^{-19})^2}{(1 \times 10^{-10})}\ \text{J}$$

$$V(1\text{Å}) = -2.307 \times 10^{-18}\ \text{J}$$

Energy values in this range appear in many branches of science because of the small quantities of electrical charge found in individual atoms and molecules, so it is convenient to define a new energy unit for these applications. An **electron volt** (eV) is the kinetic energy gained by an electron when it is accelerated through a potential difference of magnitude 1 volt (V), where 1 V = 1 J/C. Thus 1 eV = $(1.602177 \times 10^{-19}$ C) (1 V) = 1.602177×10^{-19} J. Using this conversion factor between joules and electron volts, the potential energy of a proton and electron separated by 1 Å in a hydrogen atom is -2.307×10^{-18} J = -14.40 eV. This is the energy unit used on the vertical axis in Figure 3.6. The potential energy scale is defined so that $V(r) \to 0$ as $r \to \infty$, where the particles do not interact with one another. (If you need to define the potential energy at large r as other than zero, as we do later in this chapter when considering the Coulomb interaction between K^+ and F^-, you need only add the appropriate energy offset to the expression in Equation 3.1.) Figure 3.6 also shows the potential energy function for an electron interacting with a lithium nucleus ($Z = 3$); at the same distance r the potential energy is lower because the attractive forces between the electron and the Li nucleus as they were approaching each other were stronger. Using Equation 3.1, the electrostatic potential energy between a nucleus of charge $+Ze$ and an electron (charge $-e$) is

$$V(r) = -\frac{Ze^2}{4\pi\epsilon_0 r} \quad [3.3]$$

Once we know the potential energy function $V(r)$, we can use it to predict the motions of the particles. The direction of the force exerted by one particle on another, either attractive or repulsive, is determined by the slope (the derivative) of the potential energy function with respect to the distance between the two particles (see Appendix B.2). If $\frac{dV}{dr}$ is positive at a particular distance r between the two particles, the two particles have an attractive force between them, and if $\frac{dV}{dr}$ is negative, the two particles have a repulsive force between them. The resulting equations are particularly simple if the particle exerting the force is at the origin of your coordinate system, as in Figure 3.5, and the vector $\vec{r}$ points from the origin to the particle experiencing the force. However this will not generally be the case—even diatomic molecules have two nuclei and many electrons, all of which are experiencing Coulomb forces. So let's consider how to calculate the force on both particles involved, where particle number 1 is at the position given by $\vec{r}_1 = (x_1, y_1, z_1)$ and particle number 2 is at the position given by $\vec{r}_2 = (x_2, y_2, z_2)$. The relationship between the force $\vec{F}_2$ on the electron (calling the electron particle number 2 and calling the nucleus with charge $+Ze$ particle number 1) is given by

$$\vec{F}_2(r) = -\frac{dV}{dr}\hat{r}_{12} = -\frac{d\left(\frac{q_1 q_2}{4\pi\epsilon_0 r}\right)}{dr}\hat{r}_{12} = -\frac{d\left(\frac{(+Ze)(-e)}{4\pi\epsilon_0 r}\right)}{dr}\hat{r}_{12} = -\frac{Ze^2}{4\pi\epsilon_0 r^2}\hat{r}_{12} \quad [3.4]$$

where $\vec{r}_{12} = \vec{r}_2 - \vec{r}_1$ is the vector pointing from particle 1 to particle 2 and $\hat{r}_{12}$ is the unit vector pointing the same direction ($\hat{r}_{12} = \vec{r}_{12}/r_{12}$). Thus when dV/dr is positive (the potential energy increases if the two particles move farther from each other), as it is for an electron and a proton, the force on the electron is in the opposite direction as the vector pointing from the proton to the electron. (You can see this clearly in Figure 3.5; the red force vector $\vec{F}$ on the electron is opposite in direction to the vector pointing from the positively charged nucleus to the electron.) Note that if you wish to calculate the force on the proton, particle 1, it is just $\vec{F}_1(r) = -\frac{dV}{dr}\hat{r}_{21}$, but now the unit vector $\hat{r}_{21}$ $(= -\hat{r}_{12})$ is pointing from the electron to the proton, so $\vec{F}_2(r) = -\vec{F}_1(r)$ and you recover Newton's second law. Here the magnitude of both $\vec{r}_{12} = \vec{r}_2 - \vec{r}_1$ and $\vec{r}_{21} = \vec{r}_1 - \vec{r}_2$ is given simply as the distance r, but in equations later in this chapter, when we have several particles, we'll be careful to put subscripts on the distance r to indicate the two particles it is a distance between.

EXAMPLE 3.1

Assume in Figure 3.6 that some sort of clamp is applied to hold an electron fixed at the distance of 2.5 Å from a lithium nucleus Li^{3+}, which is also clamped in its position.

(a) Calculate the potential energy of the system of these two fixed particles.

(b) If the lithium atom is at the coordinates $(x_1, y_1) = (1.5\text{ Å}, 0\text{ Å})$ and the electron is located at $(x_2, y_2) = (-1.0\text{ Å}, 0)$, calculate the direction of the force $\vec{F}_2$ on the electron by giving its components F_x and F_y. Does the electron move toward or away from the nucleus? Is the force attractive or repulsive? Then calculate the force $\vec{F}_1$ on the nucleus and give its magnitude.

(c) Make a plot like Figure 3.5 to show the position of the electron and the position of the nucleus, and superimpose on your plot the vector force $\vec{F}_2$ on the electron as we did for the electron in Figure 3.5. Also show the vector force $\vec{F}_1$ on the nucleus.

(Continued)

Solution

(a) We use Equation 3.1 or Equation 3.3 to calculate the potential energy as follows:

$$V(2.5\ \text{Å}) = \frac{(+3e)(-e)}{4\pi(8.854 \times 10^{-12}\ \text{C}^2\ \text{J}^{-1}\ \text{m}^{-1})(2.5 \times 10^{-10}\ \text{m})}$$

$$= -\frac{3(1.602 \times 10^{-19}\ \text{C})^2}{(2.782 \times 10^{-20}\ \text{C}^2\ \text{J}^{-1})}$$

$$V(2.5\ \text{Å}) = -2.768 \times 10^{-18}\ \text{J} = -17.28\ \text{eV}$$

The potential energy at 2.5 Å is negative (i.e., lower than the value at infinite separation) because these oppositely charged particles attract one another.

(b) First calculate $\vec{r}_{12}$, the vector pointing from particle number 1, the Li^{3+} nucleus, to particle number 2, the electron. $\vec{r}_{12} = \vec{r}_2 - \vec{r}_1 = (-1\ \text{Å}, 0) - (1.5\ \text{Å}, 0\ \text{Å}) = (-2.5\ \text{Å}, 0\ \text{Å})$. The magnitude of $\vec{r}_{12} = \sqrt{(-2.5)^2 + (0)^2}\ \text{Å} = 2.5\ \text{Å}$, so the distance between the two particles, r, is 2.5 Å. The unit vector $\hat{r}_{12} = (-2.5\ \text{Å}, 0\ \text{Å})/2.5\ \text{Å} = (-1, 0)$, so this vector is indeed a unit vector pointing from the Li^{3+} nucleus toward the electron. Using the expression for Coulomb's law in Equation 3.4 to calculate the force on particle number 2, the electron (exerted by particle number 1, the Li^{3+} nucleus).

$$\vec{F}_2 = \frac{(+3e)(-e)}{4\pi\left(8.854 \times 10^{-12}\ \text{C}^2\ \text{J}^{-1}\ \text{m}^{-1}\right)\left(2.5 \times 10^{-10}\ \text{m}\right)^2}(-1, 0)$$

$$= \left(-\frac{3\left(1.602 \times 10^{-19}\right)^2}{\left(1.113 \times 10^{-10}\right)\left(2.5 \times 10^{-10}\right)^2}\ \text{J m}^{-1}\right)(-1, 0)$$

$$\vec{F}_2(2.5\ \text{Å}) = (-1.1 \times 10^{-8}\ \text{N})(-1, 0) = (1.1 \times 10^{-8}\ \text{N}, 0\ \text{N})$$

The SI unit of force is the **newton** (N); $1\ \text{N} = 1\ \text{kg m s}^{-2} = 1\ \text{J m}^{-1}$. The x-component of the force, $F_x = 1.1 \times 10^{-8}$ N, so the force on the electron is in the $+x$ direction, toward the nucleus. The electron is attracted toward the nucleus. This force accelerates the electron in the $+x$ direction, as $F_x = m_e a_x = m_e dv_x/dt$. The y-component of the force, F_y, is 0 N. The force on the nucleus is $\vec{F}_1 = -\vec{F}_2$, so $\vec{F}_1 = (-1.1 \times 10^{-8}\ \text{N}, 0\ \text{N})$. The nucleus thus feels a force in the $-x$ direction, and so would be accelerated toward the electron, but its mass is more than 10,000 times the mass of an electron, so it barely moves. For both particles the force is attractive (see Figure 3.7), because the slope of the potential energy curve in Figure 3.6, dV/dr, is positive at all values of r for oppositely charged particles.

(c) See Figure 3.7. The positions of the electron and the Li^{3+} nucleus given in the problem are plotted on the x-axis, and the force on the electron $\vec{F}_2$ is shown in red while the force on the nucleus, $\vec{F}_1$, is shown in green. Because we are superimposing force in newtons on a coordinate system that shows positions of the particles in distance units (Å), the lengths of the force vectors shown are arbitrary. The directions are important, however, as is the fact that the magnitude of $\vec{F}_1$ is equal to the magnitude of $\vec{F}_2$, 1.1×10^{-8} N.

Related Problems: 5, 6, 7, 8

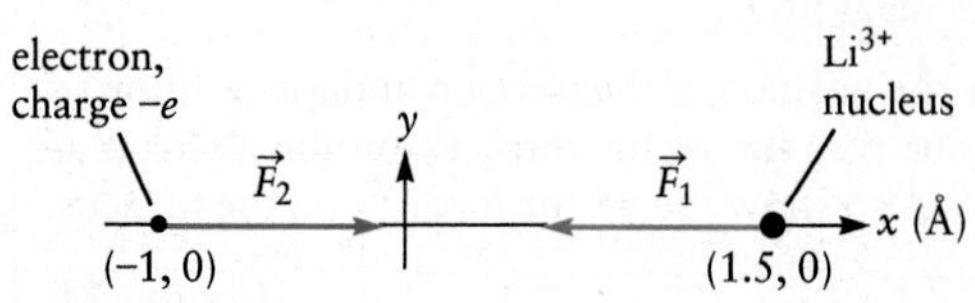

FIGURE 3.7 An electron at the location $(x_2, y_2) = (-1.0\ \text{Å}, 0\ \text{Å})$ experiencing a force $\vec{F}_2$ (red vector) due to the electrostatic attraction to the Li^{3+} nucleus, with charge +3e, located at the point $(x_1, y_1) = (1.5\ \text{Å}, 0\ \text{Å})$. The force the electron feels (red vector) is equal in magnitude to, but opposite in direction to, the force $\vec{F}_1$ that the nucleus feels (green vector).

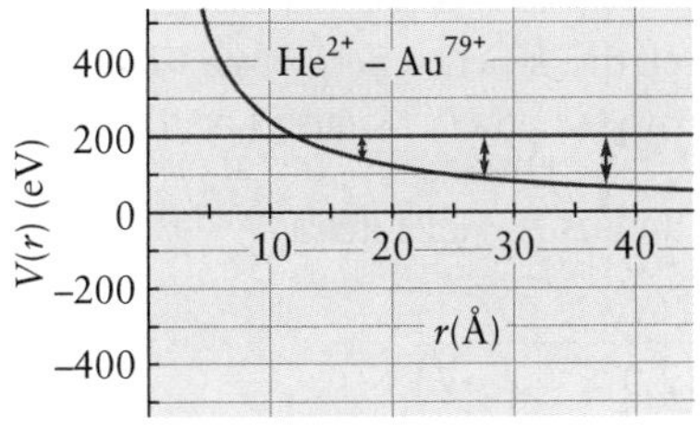

FIGURE 3.8 Potential energy curve for He^{2+} ions interacting with Au^{79+} ions. The initial kinetic energy of the alpha particles is 200 eV. The vertical arrows represent their decreasing kinetic energy as the alpha particles are slowed by repulsive force.

We can use the inset in Figure 3.6 to describe Rutherford's experiment (see Section 1.4) if we make the approximation that the alpha particles, He^{2+} collide with bare Au^{79+} nuclei in the gold foil, thus neglecting the electrons in the gold atoms. Imagine that we aim a beam of alpha particles at the foil, each particle moving with kinetic energy of 200 eV. With the stated approximation, we can represent one of these incoming alpha particles by drawing a solid line on the inset across the energy graph at the value +200 eV (see Figure 3.8).

In Figure 3.8, the red line represents the total energy of the alpha particle, because at infinite distance from the foil all of the energy of the particle is kinetic energy. As the alpha particle comes closer to the foil, it will lose some of its kinetic energy because the repulsive Coulomb force exerted by a gold nucleus slows down the particle. The particle has to give up some of its energy to overcome the repulsive force from the gold nuclei and keep moving toward the foil. If you think of an alpha particle moving directly toward a single gold nucleus, at any position along the r-axis the kinetic energy of the alpha particle is the difference between 200 eV and the value of the potential energy at that point: kinetic energy $\mathcal{T}$ at distance $r = 200\text{ eV} - V(r)$. We can represent the changes in the kinetic energy of the alpha particle by drawing a series of vertical arrows from the potential energy curve to the horizontal red line at total $E = 200$ eV at several different positions for the alpha particle. These arrows become progressively shorter as the alpha particle comes closer to the gold nucleus, because the particle is slowed more and more. Somewhere near 12 Å, the potential energy curve intersects the line for 200 eV. At this point the kinetic energy goes to zero, the alpha particle stops, turns around, and heads back away from the foil because of the repulsive force exerted by the gold nucleus. The position where the particle stops is called the *turning point,* and its value can be calculated by setting the potential energy equal to 200 eV. If we do this experiment again with higher kinetic-energy alpha particles, the turning point gets closer to the gold nucleus. In Rutherford's experiment, the kinetic energy of the alpha particles was 2.0×10^6 eV, abbreviated as 2.0 MeV. Example 3.2 calculates the turning point in Rutherford's experiment to get some sense of how close his alpha particle could get to a gold nucleus if it was initially headed straight toward that nucleus. (If the alpha particle were not headed straight toward the nucleus, but instead underwent a glancing collision, its trajectory would be a curved one; the speed would never become zero, but it would be deflected.)

EXAMPLE 3.2

Calculate the turning point of an alpha particle ($Z = 2$) directed with initial kinetic energy of 2.0 MeV, if the trajectory of the alpha particle was headed directly toward a single gold nucleus ($Z = 79$) (the nuclei in the gold foil in Rutherford's experiment). How close could the alpha particle in Rutherford's experiment get to the center of a gold nucleus?

Solution

At the turning point r_{tp} the alpha particle stops for an instant before it turns back. Total energy is conserved, so if the total energy of the two particles when they were far apart was Kinetic Energy + Potential Energy = 20 MeV + 0 MeV (since $V(r)$ is zero for large r), then at the separation r_{tp}, the kinetic energy of the alpha particle is equal to zero, so the total energy, 20 MeV = 0 + $V(r_{tp})$. Thus $V(r_{tp})$ = 20 MeV; all of the initial alpha particle's kinetic energy has been converted to potential energy at $r = r_{tp}$. (Here we assume that the gold nucleus is so much more massive than the alpha particle that we can assume it does not move; in reality the equation below is

(Continued)

only correct if the initial kinetic energy were the relative kinetic energy (see Chapter 18).) From Equation 3.1 we obtain the relation $V(r_{tp}) = \frac{q_1 q_2}{4\pi\epsilon_0 r_{tp}} = 20$ MeV.

Solving this for r_{tp} and substituting $q_1 = +2e$ and $q_2 = +79e$ gives

$$r_{tp} = \frac{(+2e)(+79e)}{4\pi\left(8.854 \times 10^{-12}\ \text{C}^2\ \text{J}^{-1}\ \text{m}^{-1}\right)\left(2.0 \times 10^6\ \text{eV}\right)}$$

$$r_{tp} = \frac{(2)(79)\left(1.602 \times 10^{-19}\right)^2\ \text{J m}}{\left(1.114 \times 10^{-10}\right)\left(2.0 \times 10^6\ \text{eV}\right)\left(1.602 \times 10^{-19}\ \text{J eV}^{-1}\right)}$$

$$r_{tp} = 11.4 \times 10^{-14}\ \text{m} = 11.4 \times 10^{-12}\ \text{cm} = 11.4 \times 10^{-4}\ \text{Å}$$

The initial kinetic energy was large enough that at the turning point the alpha particle came close to, but did not penetrate, the gold nucleus. (Experimental and theoretical studies suggest that the radius of a nucleus is on the order of 10^{-14} m.)

Related Problems: 5, 6, 7, 8

Now let's examine the motion of the electron in the hydrogen atom, which has only one proton and one electron. The total energy (kinetic and potential) of the electron in the atom is

$$E = \frac{1}{2}m_e v^2 - \frac{Ze^2}{4\pi\epsilon_0 r} \qquad [3.7]$$

The total energy E of the atom is fixed (assuming no external forces or radiative processes). It is informative to represent this condition by a horizontal line on the potential energy curve and show the kinetic energy $\mathcal{T}$ at each point as a vertical arrow connecting V to E. (See Appendix B.2 for background.)

Figure 3.9a shows the total energy set to +10.0 eV. The graph shows that the electron has significant kinetic energy everywhere, so the case of positive total E corresponds to *unbound motion* in which the electron approaches the proton but passes by without becoming trapped or attached.

Figure 3.9b shows the total energy set to −13.6 eV. The kinetic energy is large at small values of r and decreases to zero at the point where this line intersects V. For values of r larger than this, the kinetic energy is negative, which is not allowed in Newtonian mechanics. Thus, at this total energy, the motion of the electron is *bound*; the electron is said to be "trapped within a potential well centered on the proton," and its motions are limited to the range between $r = 0$ and the distance at which the total energy $E = V(r)$. We describe the hydrogen atom using quantum mechanics in Chapter 5, where we see that only certain specific values of the bound state energy are allowed, one of which is −13.6 eV. (Note that in this simple example we do not have the electron circling the nucleus, as it would in Rutherford's planetary model of the atom. Then conservation of angular momentum keeps the orbit at even smaller distances than the distance at which the total energy $E = V(r)$, because the kinetic energy is never zero during the bound motion.)

EXAMPLE 3.3

Assume that a hydrogen atom has bound state energy $E = -13.6$ eV. Calculate the maximum distance r_{max} of the electron from the nucleus allowed by Newtonian mechanics.

(Continued)

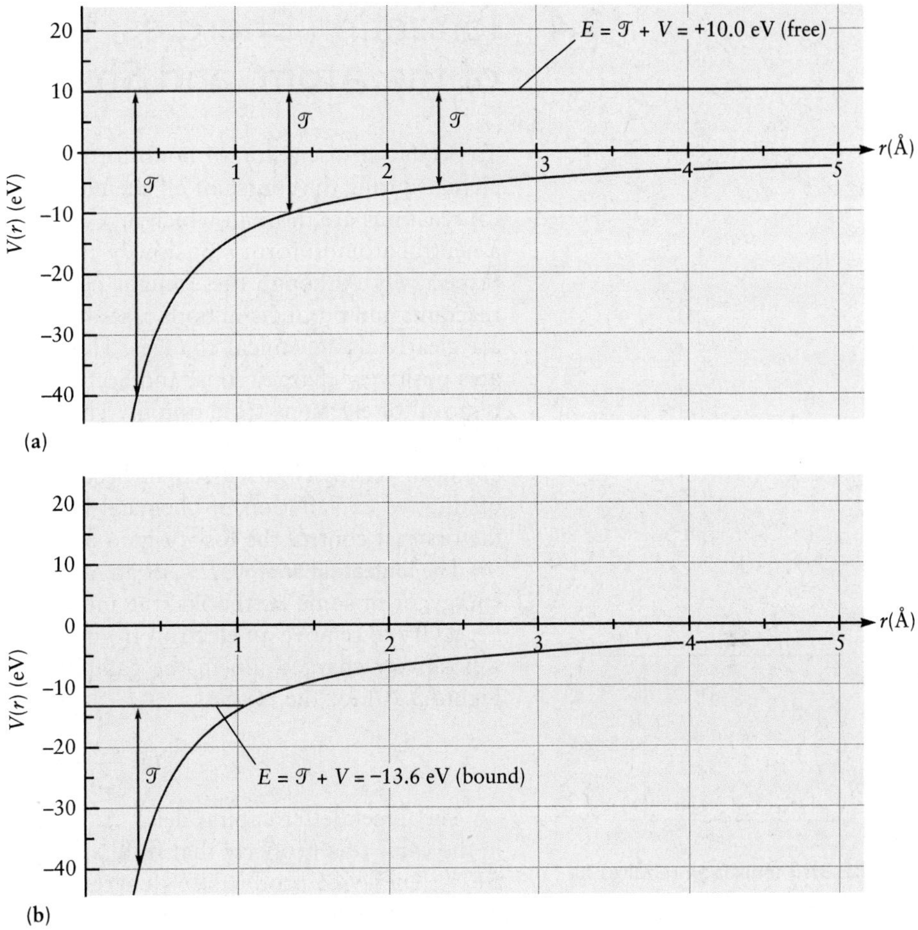

FIGURE 3.9 Potential energy, total energy, and kinetic energy for interaction of an electron with a proton. When the total energy is fixed, the kinetic energy at each point is represented by a vertical arrow from the potential energy curve to the value of the total energy shown by the red horizontal line. (a) If the potential energy is 0 at large *r*, a total energy $E > 0$ corresponds to unbound motion, characterized by significant kinetic energy at all positions. (b) Total energy $E < 0$ corresponds to bound motion where the electron is confined to distances smaller than the point at which the potential and total energy are equal and the kinetic energy is 0.

Solution

When the electron is located at r_{max} the potential energy is equal to the bound state energy of the atom. From Equation 3.3 we obtain the following relations:

$$V(r_{max}) = -\frac{e^2}{4\pi\epsilon_0 r_{max}} = -13.6 \text{ eV}$$

$$r_{max} = \frac{e^2}{4\pi\epsilon_0(13.6 \text{ eV})}$$

$$r_{max} = \frac{(1.602 \times 10^{-19}\text{C})^2}{4\pi(8.854 \times 10^{-12}\text{C}^2\text{ J}^{-1}\text{ m}^{-1})(13.6 \text{ eV})(1.602 \times 10^{-19}\text{ J eV}^{-1})}$$

$$r_{max} = 1.06 \times 10^{-10}\text{m} = 1.06 \text{ Å}$$

It is tempting to interpret r_{max} as a measure of the "size" of the hydrogen atom, but that idea has to be refined when we describe the hydrogen atom by quantum mechanics. In Chapter 5 we see that quantum mechanics describes the position of the electron around the nucleus by a probability function and that 99% of the probability lies within a sphere of radius 2.2Å centered on the nucleus when the bound state energy is −13.6 eV. Quantum mechanics allows small particles such as electrons to penetrate into regions where classical mechanics forbids them to go.

Related Problems: 5, 6, 7, 8

3.4 Ionization Energies, the Shell Model of the Atom, and Shielding

To begin our discussion of bond formation, we note that electron distributions change during the course of all chemical reactions. The simplest possible chemical reactions are those in which an electron is either removed from or added to a neutral atom to form a positively charged **cation** or negatively charged **anion,** respectively. Although these might be considered to be physical processes, the reactants and products in both cases have different chemical properties, so these are clearly also chemical changes. This section focuses on the process that creates positively charged ions, and Section 3.5 discusses formation of anions and removal of electrons from anions. The energy changes associated with each of these processes show clear periodic trends that correlate with the trends in chemical reactivity discussed in Section 3.2. This correlation suggests that a qualitative explanation of chemical bonding may begin by understanding the factors that control the loss or gain of electrons by atoms.

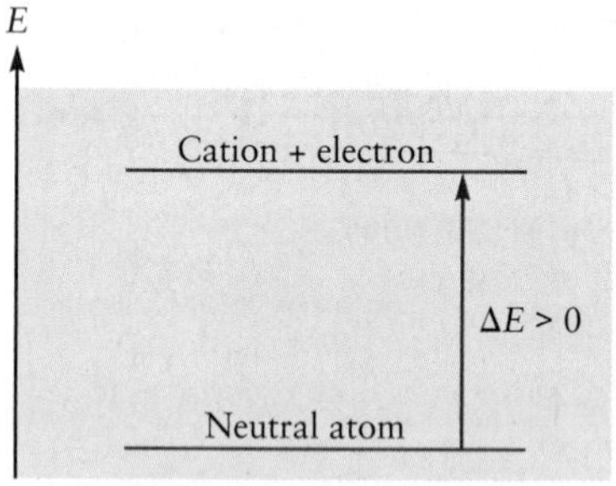

FIGURE 3.10 Ionization requires sufficient energy to enable the electron to escape from the potential energy well of the atom.

The **ionization energy,** IE_1, of an atom (also referred to as the first ionization energy, or in some textbooks, the ionization potential) is the minimum energy necessary to remove an electron from a neutral atom in the gas phase and form a positively charged ion in the gas phase. It is the change in energy, ΔE (see Figure 3.10) for the process

$$X(g) \longrightarrow X^+(g) + e^- \qquad \Delta E = IE_1$$

The Greek letter capital delta, Δ, is widely used to symbolize the difference in the value of a property that results from a physical or chemical process. Here ΔE = [energy of products] − [energy of reactants]; it is positive when energy must be provided for the process to occur, and it is negative if the process liberates energy. To achieve the ionization of X(g) to form the products $X^+(g) + e^-$, it is necessary to supply energy to the neutral atom X(g). The energy added enables the electron to escape from the potential energy well that holds it in the atom. Therefore, the energy of the final state [the electron e^- and the ion $X^+(g)$] is greater than that of the initial state (the neutral atom).

ΔE for ionization reactions is always positive. The ionization energy is a measure of the stability of the free atom. Those atoms with larger ionization energies are more stable than those atoms with smaller ionization energies because their electrons must be removed from deeper potential energy wells.

Figure 3.11 shows the measured ionization energies of the elements plotted as a function of their atomic numbers. The left vertical scale is the energy in eV per atom and the right vertical scale is the energy in kJ per *mole* of atoms. Recall from the discussion in Section 3.2 that 1 eV per *atom* equals 96.48 kJ per mole of atoms.

The values generally increase moving across a period (from left to right), becoming large for each noble gas atom, and then fall abruptly for the alkali atom at the beginning of the next period. The large values for the noble gas atoms demonstrate that their electron configurations are extremely stable, and that considerable energy is required to liberate their electrons. Moreover, the electron configurations of the noble gas atoms are more stable than those of the atoms immediately before and after them in the periodic table.

Ionization energy is thus a periodic property of the elements (see Section 3.2). The small local increases and decreases superimposed on the generally increasing trend across a period are explained in detail by the quantum description of atomic structure in Chapter 5. Our primary objective here is to demonstrate the experimental result that the energy required to remove the first

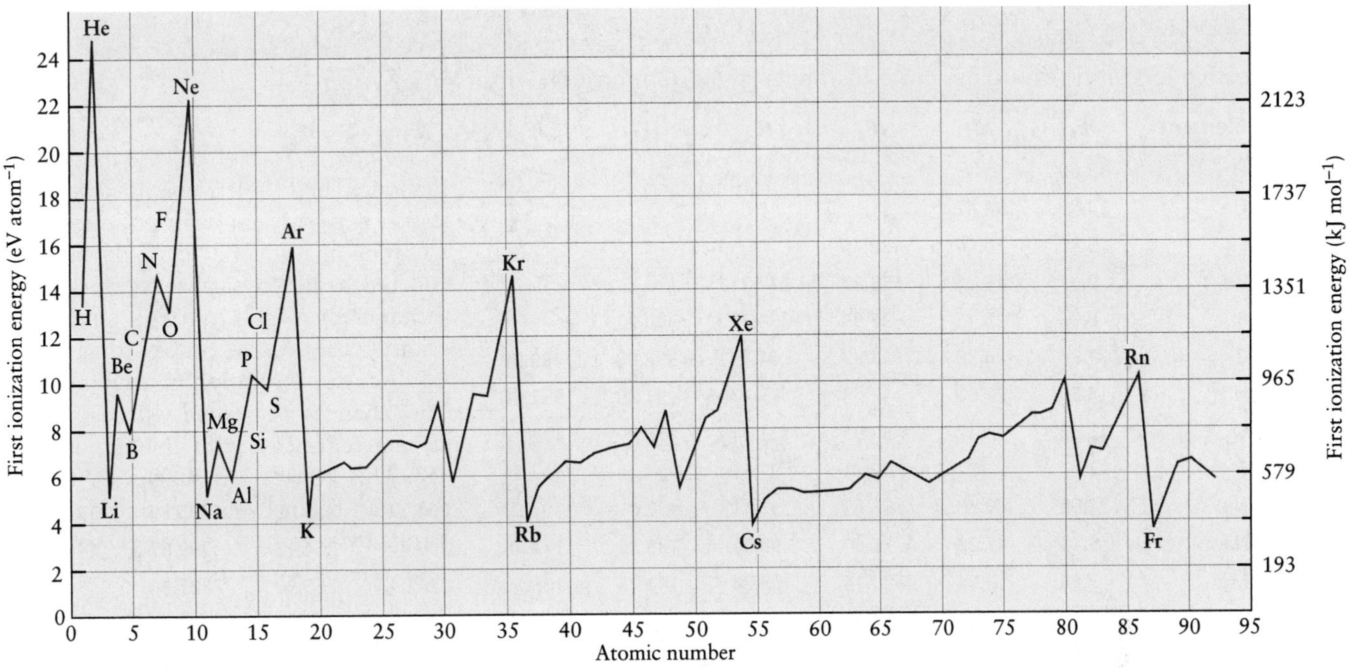

FIGURE 3.11 First ionization energy plotted versus atomic number shows periodic behavior. Symbols for the noble gases are shown in red; those for alkali metals are shown in blue.

electron is periodic in the atomic number Z. It is not necessary to consider the small local variations observed in the data for this purpose.

The *second* ionization energy, IE_2, is the minimum energy required to remove a second electron, as represented by the process

$$X^+(g) \longrightarrow X^{2+}(g) + e^- \qquad \Delta E = IE_2$$

(Thus the reactant in this chemical equation is the cation formed from removing the first electron from the neutral atom.)

The third, fourth, and higher ionization energies are defined in an analogous fashion. Successive ionization energies always increase due to the greater electrostatic attraction of the electron to the product ions, which have increasingly greater positive charges.

Examination of successive ionization energies suggests the idea that the electrons in an atom are organized in a very interesting structure. This hypothesis arises from patterns revealed in Table 3.1, which shows the first ten ionization energies for the elements H through Ar. Ionization energies are only expressed in eV (per atom) in this table, rather than in kJ mol^{-1} (see Figure 3.11) to make it easier to display them in tabular form.

Let's first consider He. IE_1 for He is 24.59 eV, which is much greater than that of H (13.60 eV) or Li (5.39 eV). The electronic structure of He is thus much more stable than that of either H or Li. Further disruption of the stable He structure by removing a second electron requires $IE_2 = 54.42$ eV.

Next, let's consider Li, for which IE_1 is 5.39 eV and IE_2 is 75.64 eV. This is a far greater difference between IE_1 and IE_2 than for He. One electron is removed easily from Li to form Li^+, which has two electrons that are more tightly held than those in neutral He. Note that the difference $IE_3 - IE_2$ for Li is about 1.5 times the difference $IE_2 - IE_1$ for He. The last electron to be removed from Li is more strongly bound than the last electron removed from He.

As we proceed across the 2nd period, an interesting pattern develops. The ionization energies for Be show a large jump between IE_2 and IE_3, demonstrating that Be easily loses two electrons to form Be^{2+}, which has two electrons and

TABLE 3.1

Successive Ionization Energies of the Elements Hydrogen through Argon (in eV Atom^{-1})

Z	Element	IE_1	IE_2	IE_3	IE_4	IE_5	IE_6	IE_7	IE_8	IE_9	IE_{10}
1	H	13.60									
2	He	24.59	54.42								
3	Li	5.39	75.64	122.45							
4	Be	9.32	18.21	153.89	217.71						
5	B	8.30	25.15	37.93	259.37	340.22					
6	C	11.26	24.38	47.89	64.49	392.08	489.99				
7	N	14.53	29.60	47.45	77.47	97.89	552.06	667.03			
8	O	13.62	35.12	54.93	77.41	113.90	138.12	739.32	871.39		
9	F	17.42	34.97	62.71	87.14	114.24	157.16	185.18	953.89	1103.08	
10	Ne	21.56	40.96	63.45	97.11	126.21	157.93	207.27	239.09	1195.79	1362.16
11	Na	5.14	47.29	71.64	98.91	138.39	172.15	208.47	264.18	299.87	1465.10
12	Mg	7.65	15.04	80.14	109.24	141.26	186.50	224.94	265.90	327.94	367.53
13	Al	5.99	18.83	28.45	119.99	153.71	190.47	241.43	284.59	330.21	398.57
14	Si	8.15	16.35	33.49	45.14	166.77	205.05	246.52	303.17	351.10	401.43
15	P	10.49	19.73	30.18	51.37	65.02	220.43	263.22	309.41	371.73	424.50
16	S	10.36	23.33	34.83	47.30	72.68	88.05	280.93	328.23	379.10	447.10
17	Cl	12.97	23.81	39.61	53.46	67.8	97.03	114.19	348.28	400.03	455.62
18	Ar	15.76	27.63	40.74	59.81	75.02	91.01	124.32	143.46	422.43	478.68
19	K	4.34	31.63	45.72	60.91	82.66	100.0	117.56	154.86	175.82	503.44
20	Ca	6.11	11.87	50.91	67.10	84.41	108.78	127.7	147.24	188.54	211.27
21	Sc	6.54	12.80	24.76	73.47	91.66	111.1	138.0	158.7	180.02	225.32

displays helium-like stability. Boron ionization energies have a large jump between IE_3 and IE_4, showing that three electrons are easily removed, and carbon has a large jump between IE_4 and IE_5, showing that four electrons are easily removed. The pattern continues, showing that fluorine has seven electrons that can be removed more easily than the last two electrons, and neon has eight. This pattern is shown in Table 3.1 by highlighting the ionization energies for the more easily removed electrons in each atom. These results suggest that the electrons in each atom of the 2nd period are arranged in two groups: two electrons exist in a stable, helium-like configuration, while the others are less tightly bound. It is reasonable to assume that the two most tightly bound electrons are located close to the nucleus, where the attractive Coulomb force is very strong, and the others are much farther away from the nucleus. We hypothesize from these experimental results that electrons in the atoms of 2nd period are organized as a stable, helium-like *inner core,* surrounded by less tightly bound electrons whose number increases from one to eight as the atomic number increases from three to ten.

Examination of the atoms in the 3rd period, Na through Ar, reveals a similar pattern of relatively more easily removed electrons outside a stable core, which resembles the Ne atom. This pattern is shown in Table 3.1 by highlighting the ionization energy values for the more easily removed electrons in each atom in the 3rd period. (The beginning of this pattern in the 4th period is shown by highlighting the ionization energy values for the easily removed electrons in K, Ca, and Sc.) Na appears to have a single weakly bound electron outside

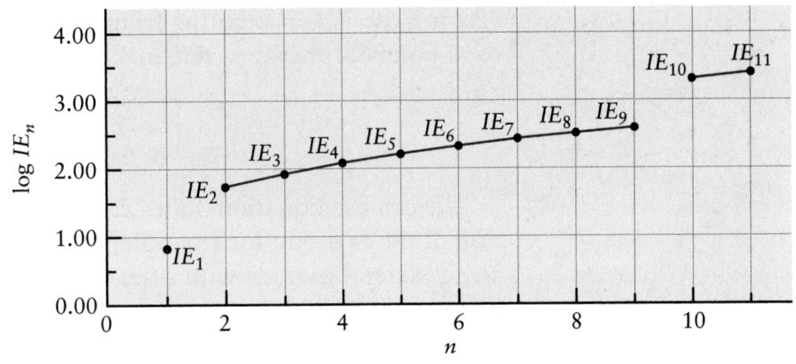

FIGURE 3.12 Logarithm of successive ionization energies for Na versus number of electrons removed suggests a three-shell electronic structure.

a neon-like core. Also note the large difference between IE_9 and IE_{10} for Na. Further insight into the physical origin of this arrangement is obtained by plotting the successive ionization energies of Na versus n, the total number of electrons that have been removed at each step. It is convenient to plot the logarithm of ionization energy versus n to compress the vertical scale. The result (Fig. 3.12) suggests that the electrons in the Na atom are arranged in three groups. The first electron is easily removed to produce Na^+, which has neon-like stability as indicated by the large jump between IE_1 and IE_2. Electrons 2 through 9 are in the second group, and all of them are more tightly bound than the first electron. A big jump between IE_9 and IE_{10} suggests that the last two electrons occupy a third group, the electrons of which are the most tightly bound of all. This is the helium-like stable core identified in our analysis of the 2nd period.

We summarize these results by proposing the **shell model for atomic structure.** The electrons are grouped into shells based upon the energy required to remove them from the atom. The shells correspond to the periods of the periodic table: the first shell contains at most 2 electrons, the second shell contains at most 8 electrons, and the third shell contains at most 8 electrons. We assume that these shells are spherical and concentric, centered at the nucleus. We also assume that the ionization energies of the electrons in each shell decrease as we move outward from the nucleus. The shell model predicts that the sulfur atom (S, $Z = 16$) would have 6 electrons in its third (outer) shell, 8 electrons in its second (middle) shell, and 2 electrons in its first (innermost) shell.

3.4.1 The Shell Model of the Atom

Examining the ionization process as a simple chemical reaction leads us to conclude that electrons occupy a set of shells that surround the nucleus. This is a remarkable experimental result. All of the electrons in an atom are identical, and they all interact with the same nucleus. Why should they be arranged in shells, and what determines the number of electrons that can occupy a given shell?

We begin to seek an explanation by considering the forces that act within many-electron atoms, and the potential energy functions associated with these forces. Consider Li, for which $Z = 3$.

Each of the electrons is attracted to the nucleus and repelled by the other electrons via the Coulomb interaction. If we put the Li nucleus at the origin of our coordinate system, the vector $\vec{r}_1$ shown in Figure 3.13 points from the nucleus to electron number 1, labeled e_1 in the figure. (The vector $\vec{r}_1$ gives the position of that electron by its Cartesian coordinates $(x_1, y_1, z_1) = \vec{r}_1$.) The distance from the nucleus to that electron is then the length r_1 of the vector $\vec{r}_1$. (Thus $r_1 = +\sqrt{x_1^2 + y_1^2 + z_1^2}$ according to the Pythagorean theorem.) The distance from electron number 1 to electron number 2 is r_{12}, the length of the

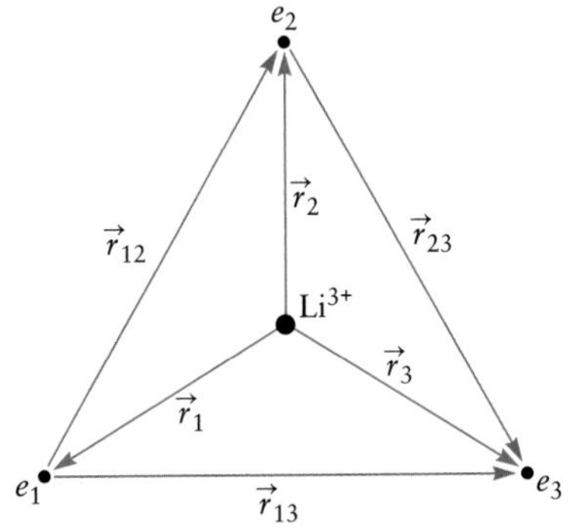

FIGURE 3.13 Li atom with three electrons.

vector $\vec{r}_{12} = \vec{r}_2 - \vec{r}_1$ pointing from electron number 1 to electron number 2. The total potential energy V due to Coulomb interactions is then given by

$$V = \frac{e^2}{4\pi\epsilon_0}\left(-\frac{Z}{r_1} - \frac{Z}{r_2} - \frac{Z}{r_3} + \frac{1}{r_{12}} + \frac{1}{r_{13}} + \frac{1}{r_{23}}\right) \qquad [3.8]$$

We can use Equation 3.8 to calculate the potential energy for any configuration of the Li atom; for example, we could place the electrons at the vertices of an equilateral triangle with sides of a given length, centered on the nucleus (see Figure 3.13). Unlike the case of the hydrogen atom, there is no simple way to relate the potential energy in this example to the bound motion of all of the electrons in the atom and to use it to calculate the energy required to remove one of them from the atom.

To make progress, let's invent a simple one-electron model by assuming that each electron moves under the influence of forces that can be derived from an **effective potential energy** function that takes into account both the attractive electron-nuclear forces and the average of the repulsive force among the electrons. To see how this idea works out, let's assume that electron 2 is located between electron 1 and the nucleus at some particular time (see Figure 3.14). The real effect of electron 2 is to *screen* or *shield* electron 1 from the full strength of the Coulomb force exerted by the nucleus with $Z = 3$. It is useful to think of the Coulomb interaction as strictly "line of sight," so intrusion by another electron will reduce its strength. In effect, the charge of the nucleus, as seen by electron 1, has been reduced by the presence of the other electrons. Suppose electrons 2 and 3 get between electron 1 and the nucleus. The extent of screening is even greater, and the magnitude of Z_{eff} is smaller. We can set $Z_{eff} = Z - S$, where the screening constant S measures the extent to which the other electrons screen out the nuclear charge felt by electron 1. We interpret S physically as the average number of electrons spending time between the nucleus and the electron we have labeled as 1. By making estimates for S, we can then find the effective potential energy using the formula for the Coulomb potential energy but with an *effective charge* Z_{eff} on the nucleus:

$$V_{eff}(r) = -\frac{Z_{eff}e^2}{4\pi\epsilon_0 r} \qquad [3.9]$$

Let's make some simple estimates of the effective potential energy to illustrate the effects of screening, beginning with the lithium atom. The simplest model predicts $Z_{eff} = 3 - 2 = 1$; this is very close to $Z_{eff} = 1.3$, which can be extracted from Equation 3.9 and the experimental ionization energy for Li. Electrons in the innermost shell each contribute about $-0.85\ e$ to screen the outer electrons from the full nuclear charge. The amount of shielding experienced by an electron in more complicated atoms depends on its position in the atom and the arrangement of the other electrons with respect to that electron

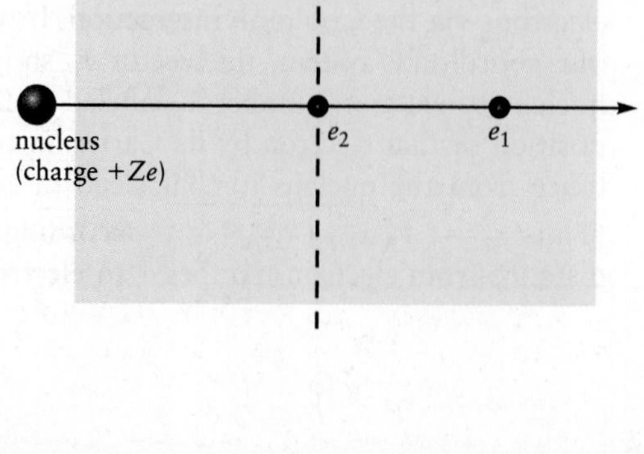

FIGURE 3.14 Electron 1 is partially shielded from the full force of the nuclear attraction when electron 2 comes between electron 1 and the nucleus.

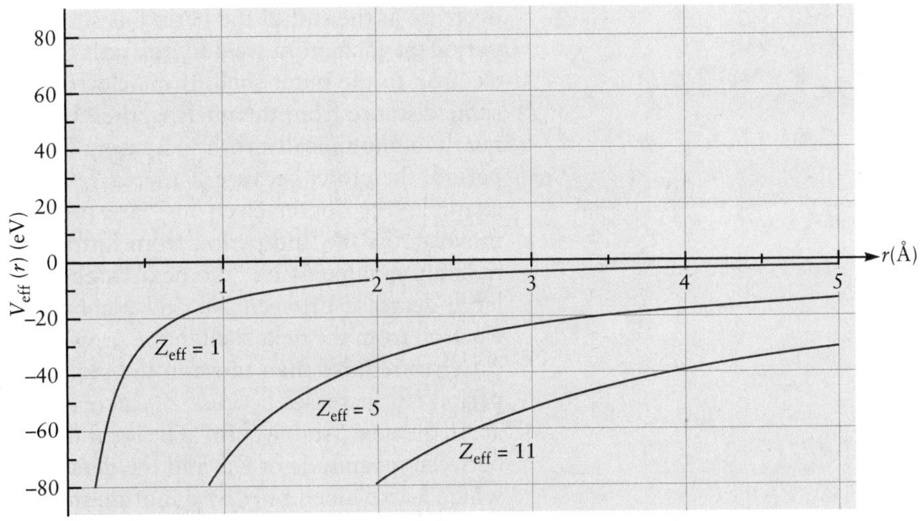

FIGURE 3.15 Curves for the effective potential energy $V_{eff}(r)$ for electrons in Na ($Z = 11$) when $Z_{eff} = 1, 5, 11$. An electron at any location is more strongly bound in the atom as the value of Z_{eff} increases.

and the nucleus. An electron that spends most of its time very near the nucleus will be screened only very slightly, so its screening constant will be small and it will experience an effective nuclear charge Z_{eff} that is very nearly equal to the actual nuclear charge Z of the atom. An electron that spends most of its time in the outermost regions of the atom experiences Z_{eff} in the range 1–2, because most of the other $(Z - 1)$ electrons provide significant amounts of screening. An electron that spends most of its time in the intermediate region of the atom is partly screened and experiences an intermediate value of Z_{eff}. Consider Na, for which $Z = 11$. Let's examine cases of significant screening ($Z_{eff} = 1$), intermediate screening ($Z_{eff} = 5$), and no screening ($Z = 11$). Figure 3.15 shows plots of V_{eff} curves for $Z_{eff} = 1, 5, 11$. Clearly, those electrons that experience the lower values of Z_{eff} are more weakly bound than those with higher values of Z_{eff}, as can be seen by comparing the values of the potential energy for the three curves at particular values of *r*. Each shell in an atom is characterized by a value of Z_{eff} that varies only slightly for the electrons in that shell; the value of Z_{eff} is dramatically different for electrons in different shells.

The shell model (see Figure 3.16) is a refinement of Rutherford's planetary model because it describes the Z electrons as being organized into a series of concentric shells based on their ionization energies. The shell model is inspired by patterns in experimental measurements of successive ionization energies. Our one-electron approximation based on Z_{eff} and simple classical electrostatics provides a qualitative physical interpretation of the shell model. Altogether, this is a very physical and intuitively appealing picture of the electronic structure within the atom. We have great confidence in this picture and use it to organize and interpret vast amounts of chemical information.

In Chapter 5 we treat Rutherford's model by quantum mechanics and see that the shell structure is completely justified by quantum mechanics. We see why each specific shell contains 2, 8, or 18 electrons. We generate values for the screening constant S and Z_{eff} systematically, and see that they range from 1 to the full, unscreened nuclear charge Z for an atom.

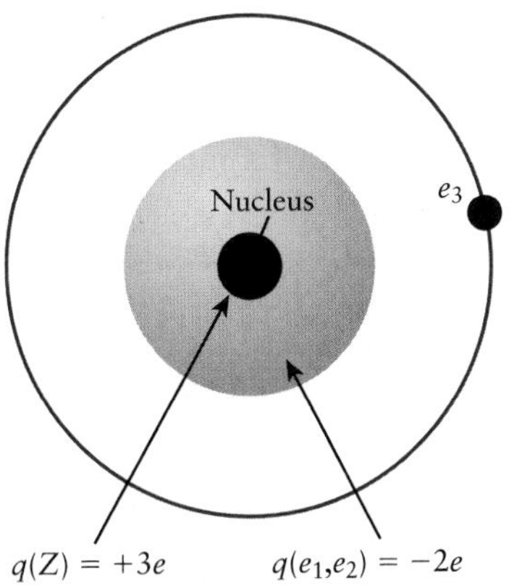

FIGURE 3.16 Shell model of the atom. The outer electron is partially shielded from the nuclear attraction by the electrons in the inner shell.

3.4.2 The Shell Model of the Atom and Periodic Behavior in Chemical Bonding

Examining the ionization process as the prototype of a simple chemical reaction led us to the shell model. Now we use the shell model and the concept of Z_{eff} to explain the general increase of IE_1 across each period, followed by an abrupt decrease at the end of the period as shown in Figure 3.11. As we move across a period, at each atom we add one unit of positive charge to the nucleus and one electron to the outer shell. If the electrons in a shell are located at roughly the same distance from the nucleus, then the attractive electrostatic forces increase nearly monotonically with increasing Z. Z_{eff} always increases moving across a period, therefore, because Z increases and there is never perfect shielding. For example, the "outer electrons" are progressively more strongly bound as we move across the 2nd period from Li to F, and IE_1 increases until it achieves its maximum value at Ne, the next "stable core." The ionization energy shows a large decrease between Ne and Na because at Na we are now removing an electron from the next shell in the progression. We know that these electrons are easier to remove than those in the second shell. As we continue across the 3rd period IE_1 increases because Z_{eff} increases until we achieve the maximum value at Ar, the next "stable core." The local fluctuations result from a delicate balance between magnitude of Z_{eff} and the details of the electron distributions in atoms, which is explained only by quantum mechanics.

There is a strong correlation between the ease of removing electrons from an atom and the chemistry of that atom. The electrons that are easiest to remove are most likely to participate in chemical bond formation and in chemical reactions. We now know, for every atom, that there is a very large energy difference in the energies of the electrons in the outermost shell and those in the inner shells. This difference is displayed dramatically in Table 3.1 as the right-hand end of the highlighted IE values for each atom. Electrons in the inner shells (called **core electrons**) do not participate significantly in chemical reactions because they are quite difficult to remove from the atom. The outermost, partially filled shell (called the **valence shell**) contains the electrons involved in chemical bonding, the **valence electrons**. The number of valence electrons in a neutral atom of a main-group element (those in Groups I–VIII) of the second and third periods is equal to the group number of the element in the periodic table. However, the main-group elements that follow a series of transition-metal elements require some special attention. Atoms of bromine, for example, have 17 more electrons than atoms of argon, the preceding noble gas, but only 7 are considered to be valence electrons. This is true for two reasons. First, in the fourth, fifth, and sixth rows, the 10 electrons added to complete the transition metal series (although they are important for the bonding of those elements) have become *core* electrons by the time the end of the transition-metal series is reached. They are closer to the nucleus, on average, than the electrons that fill the rest of the shells of those periods, and it might be useful to visualize them as occupying a subshell. Second, and more importantly, the chemical properties of the main group elements in this part of the periodic table are characteristic of the group to which they belong. The bonding properties of an element such as bromine, for example, resemble those of the lighter elements in its group.

Progressing through the elements in order of increasing atomic number along a period, we see that stability increases from left to right, as indicated by increasing values of the ionization energy. As we progress across a period, the valence shell of that period becomes a filled shell, which is the stable configuration of the noble gases helium, neon, argon, and so on. Atoms with filled shells are extremely stable chemically, as shown by the large values of their ionization energies. The 2nd period ends with a filled valence shell of 8 electrons in Ne, and the 3rd period ends with a filled valence shell of 8 electrons in Ar. These two configurations, called the *stable octets,* play a key role in theories of chemical bond formation. To a large extent the bond forming possibilities for atoms in the 2nd and 3rd periods can be categorized by their ability to lose, gain or share electrons in order to achieve stable octets in their valence shells, and therefore the great stability of a noble gas configuration.

The American chemist G. N. Lewis introduced a useful notation that describes the electronic structure of atoms. The Lewis model represents the valence electrons as dots arranged around the chemical symbol for an atom; the core electrons are not shown. The first four dots are arranged individually around the four sides of the symbol for each element. If an atom has more than four valence electrons, dots representing them are then paired with those already present. The result is a **Lewis dot symbol** for that atom. The Lewis notation for the elements of the first two periods is

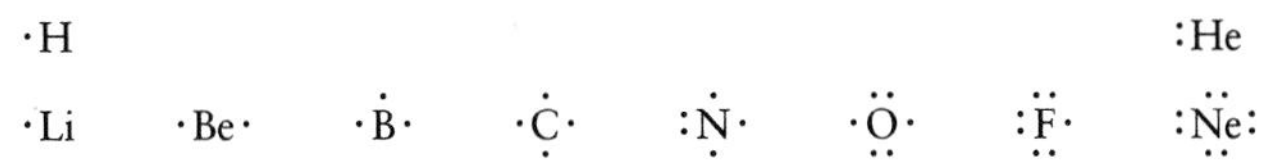

The Lewis symbols vividly display how the valence shell structure changes across a period and suggest at a glance the bonding possibilities that would establish stable octets about each atom.

3.5 Electron Affinity

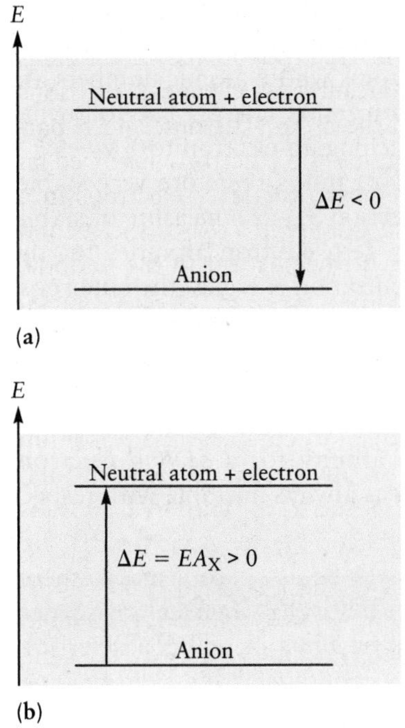

FIGURE 3.17 (a) When an electron attaches to a neutral atom to form a stable anion (as shown), the electron gets trapped in the potential well of the atom, and the energy of the products is lower than the energy of the reactants. Thus the ΔE for the reaction $X(g) + e^- \rightarrow X^-(g)$ is negative; when you add an electron to a neutral atom such as Cl that has a positive electron affinity, energy is released. This $\Delta E = -EA_X$ (b) ΔE for the reaction $X^-(g) \rightarrow X(g) + e^-$; this $\Delta E = EA_X$.

Section 3.4 describes the prototypical chemical reaction in which an electron is removed from a neutral atom in the gas phase to form a cation. Ionization energy (IE) measures the difficulty with which an atom gives up an electron to form a cation. The present section describes a different prototypical reaction, in which an electron is attached to a neutral gas phase atom to form an anion. The ease with which an atom accepts an extra electron to form an anion is measured by the electron affinity (*EA*) of the atom, to be defined shortly.

An anion is formed by the electron attachment reaction,

$$X(g) + e^- \longrightarrow X^-(g),$$

for which the energy change ΔE (see definition in Section 3.4) is called the **electron attachment energy.** This reaction is represented on a plot that shows the potential energy of the atom $X(g)$ and the electron versus the distance between them, similar to those shown in Figure 3.15. Initially, the atom and the electron are very far apart and do not interact, at which point the potential energy is defined to be zero. They approach one another and form the anion $X^-(g)$, whose energy is usually lower than that of the separated atom and electron. When X is, for example, a halogen, the energy of the products is lower than the energy of the reactants, ΔE is negative, energy is usually released in the reaction, and the anion is stable (see Figure 3.17a). Electron attachment is an example of a reaction type called an **exothermic reaction,** which means energy is released. The anion is stable because the neutral atom can accommodate an extra electron, which is strongly bound by the effective potential V_{eff}. The energy change, ΔE, for the reverse reaction, in which the electron is removed from $X^-(g)$ to give the neutral atom $X(g)$, is positive because energy must be supplied to overcome V_{eff} that holds the electron in the anion (see Figure 3.17b).

It is difficult to measure the electron attachment energy directly. It is easier to investigate the reverse reaction: start with the gaseous anion and measure the energy required to remove the electron in the same way that ionization energies are measured in Section 3.3. For example, the reaction

$$Cl^-(g) \longrightarrow Cl(g) + e^-$$

requires $\Delta E = +349 \text{ kJ mol}^{-1}$ to remove the electron from Cl^-.

The **electron affinity** EA_X of atom X is defined as the energy required to detach the electron from the anion X^- and give the neutral atom:

$$X^-(g) \longrightarrow X(g) + e^- \qquad \Delta E = EA_X$$

Thus, EA for Cl is 349 kJ mol^{-1}. The electron affinity is a property of the neutral atom, but it is measured directly by removing an electron from the anion. The EA value defined and measured this way is positive because it is the energy that must be invested to remove an electron from a stable species, the anion. Removing an electron from a stable species is always an endothermic reaction, as seen in Section 3.4.

Values of electron affinity for selected elements are shown in Table 3.2. We choose not to include "negative" electron affinities, which you might see tabulated in other textbooks or reference data collections. An atom with a negative electron affinity would require some external force to hold the electron on the atom. While this situation could provide interesting opportunities to study the forces between electrons and atoms, such an "anion" would be unstable and therefore not particularly useful as a prototypical chemical reaction product in our view.

The periodic trends in electron affinity largely parallel those in ionization energy, increasing across a period to become large for the halogens, then decreasing abruptly to essentially zero for the noble gases. A notable difference between the trends in ionization energy and electron affinity is that the dramatic decreases in electron affinities occur between atoms whose atomic numbers are one lower than the corresponding breaks in ionization energy. The following examples illustrate and explain this point. Attaching an electron to F gives F^-, which has the same electron arrangement as Ne, and is therefore very stable. (Recall the discussion of ionization energy in Section 3.3 as a measure of stability.) Similarly, the noble gases have essentially zero electron affinities for the same reason that the alkali metals have small ionization energies; the outermost electron in Ne^- or Xe^- would reside in a new shell, and the resulting ion would be less stable than the neutral parent atom. Indeed, Ne^- has a nuclear charge of $+10e$ while Na has a nuclear charge of $+11e$, so Ne^- is not a stable anion at all. Chlorine has the highest electron affinity, and that of Ne is zero.

No gaseous atom has a positive electron affinity for a *second* electron, because a gaseous ion with a net charge of $-2e$ is always unstable with respect

TABLE 3.2

Electron Affinity of Selected Atoms (in kJ mol^{-1})

H							**He**
73							*
Li	**Be**	**B**	**C**	**N**	**O**	**F**	**Ne**
60	*	27	122	*	141	328	*
Na	**Mg**	**Al**	**Si**	**P**	**S**	**Cl**	**Ar**
53	*	42	134	72	200	349	*
K	**Ca**	**Ga**	**Ge**	**As**	**Se**	**Br**	**Kr**
48	2	41	119	79	195	325	*
Rb	**Sr**	**In**	**Sn**	**Sb**	**Te**	**I**	**Xe**
47	5	29	107	101	190	295	*
Cs	**Ba**	**Tl**	**Pb**	**Bi**	**Po**	**At**	**Rn**
46	14	19	35	91	183	270	*

*No stable anion A^- exists for this element in the gas phase.

to ionization. Attaching a second electron means bringing it close to a species that is already negatively charged. The two repel each other, and the potential energy of the system increases. In crystalline environments, however, doubly negative ions such as O^{2-} can be stabilized by electrostatic interactions with neighboring positive ions.

EXAMPLE 3.4

Consider the elements selenium (Se) and bromine (Br). Without consulting tables of data, predict which one should have the higher value of ionization energy IE_1. Predict which should have the higher value of electron affinity EA.

Solution

These elements are in adjacent groups in the 4th period of the periodic table. Progressing from selenium to bromine adds one unit of positive charge, which increases Z_{eff}. At the same time we add one more electron into the 4th shell like the valence electrons in Se. Since the nuclear attraction is greater for the added electron in bromine and its distance from the nucleus is about the same, it will be more tightly bound than the valence electrons in selenium. Therefore bromine will have the larger value of IE_1.

If we place an additional electron in the 4th shell of each atom to form an anion, the additional electron on bromine will experience a larger value of Z_{eff}, and the Br^- anion will have greater stability than the Se^- anion. Moreover, Br^- has a stable octet configuration whereas Se^- does not. So Br has the larger value of EA.

Related Problems: 9, 10, 11, 12, 13, 14

In other textbooks you may see electron affinity defined as the energy change that occurs when an electron is added to a gaseous atom, which is the same as the electron attachment energy defined above. That definition gives EA values with signs opposite from the definition we have adopted. To avoid possible confusion, be sure to check whether the data source defines electron affinity as the electron attachment energy for formation of an anion or as the electron detachment energy for removing the electron from an anion. We prefer the definition adopted here because it is the original definition of electron affinity in the scientific literature, and it is still used in most data tabulations and current research papers. Moreover, because the word "affinity" in common language implies an attractive or favorable relationship between two entities, understanding technical usage is easier if "affinity" is a positive number whose magnitude increases as the interaction becomes more favorable.

3.6 Electronegativity: The Tendency of Atoms to Attract Electrons in Molecules

Classical models of the chemical bond generally identify three categories of bonds formed between a pair of atoms, based upon the degree of charge separation in the bond. Ionic bonds are those that result from the complete transfer of one or more valence electrons from one atom to the other atom, creating a positive ion and a negative ion that are bound to one another by electrostatic forces. Covalent bonds are those in which the electrons are shared evenly between the atoms with no (or very little) charge separation. Polar covalent bonds represent the intermediate situation, in which the degree of charge separation is expressed by imagining that a fraction of an elementary charge is transferred between the atoms to establish a partial charge separation or polarization along the bond axis, but not to produce ions. Each of these models is defined in more detail and illustrated later in this chapter.

Let's figure out how to qualitatively predict whether a pair of atoms might form (1) a covalent bond, with electrons equally shared between the two atoms, or (2) a polar covalent bond with unequally shared electrons, or (3) an ionic bond. In homonuclear diatomics like H_2 or Cl_2, the shared electrons are equally shared, so the bond is a covalent one and the resulting molecule is non-polar (the electrons in the molecule are distributed equally between the two atoms). At the other extreme are bonds between an atom with very high electron affinity, such as a Cl atom, and an atom with a low ionization energy, such as a K atom. Then an electron may be easily transferred from the K atom to the Cl atom, so an ionic bond results; it almost looks like an electron has been fully transferred from the K atom to the Cl atom, then the K^+ and the Cl^- are drawn together by electrostatic attraction. This is most likely to be the case when a Group I atom (with low ionization energy) is bonded to a Group VII atom, with high electron affinity. Intermediate between these two cases are atoms that form a polar covalent bond in which the electron affinity of one atom is modest and the ionization energy of the other atom is modest, so one imagines that an electron is not fully transferred from one to the other, but rather, on average, there is a partial negative charge on one atom and a partial positive charge on the other atom. A bond between N and O is expected to be a polar covalent bond. These ideas lead to the concept of **electronegativity** developed below.

3.6.1 Mulliken's Electronegativity Scale

The American physicist Robert Mulliken proposed, in 1934, that the ionization energy and the electron affinity of atoms could be used to define a new quantity called the **electronegativity** that would measure their tendency to attract electrons. He observed that elements located in the lower left corner of the periodic table have both small ionization energies and small electron affinities. This means that they give up electrons readily (to form positive ions) but do not readily accept electrons (to form negative ions). They tend to act as electron *donors* when forming bonds with other elements. In contrast, elements in the upper right corner of the periodic table have large ionization energies and also (except for the noble gases) large electron affinities. As a result, these elements accept electrons easily but give them up only reluctantly; they act as electron *acceptors* when forming bonds with other elements. Mulliken simply *defined* electronegativity, based on these observations, as a quantity that is proportional to the average of the ionization energy and the electron affinity:

$$\text{EN (Mulliken)} = \frac{1}{2}C(IE_1 + EA) \qquad [3.10]$$

Electronegativity is defined to be a pure number, so the proportionality constant C must have dimensions of $(\text{energy})^{-1}$. Once C has been evaluated, we can calculate the EN for any atom using the values for *IE* and *EA* in Tables 3.1 and 3.2. Electron acceptors (such as the halogens) have both large ionization energies and large electron affinities; they are highly **electronegative.** Electron donors (such as the alkali metals) have small ionization energies and small electron affinities, and therefore low electronegativities; they are **electropositive.** The noble gases rarely participate in chemical bonding. Their large ionization energies and small (essentially zero) electron affinities mean that they are neither electron donors nor acceptors and are not predicted to bond to other elements. Electronegativities therefore are not generally assigned to the noble gases. Mulliken's procedure is intuitively very appealing because it is based upon the simple shell model of the atom. Unfortunately, reliable measurements of the electron affinities of the elements were very difficult to obtain at the time, with data being available for only about 20 elements. This situation greatly limited use of Mulliken's definition in chemistry. Chemists rely on Mulliken's method for fundamental understanding of the EN concept and use other measures of EN in practical applications.

3.6.2 Pauling's Electronegativity Scale

The American chemist Linus Pauling had proposed a different electronegativity scale two years before Mulliken's work was first published. Pauling's method was based on a comparison of the bond dissociation energies of a large number of homonuclear and heteronuclear bonds in which one atom of the bond was the same (HF, HCl, and HBr compared with H_2, F_2, Cl_2 and Br_2 for example). Pauling observed that bonds formed between elements from opposite sides of the periodic table were stronger than those between identical elements or even those located close to one another. He suggested that this extra stability was provided by partial charge separation in the bond, called **ionic character.** This separation enabled the bonding electrons to be pulled closer toward one nucleus, experience its effective nuclear charge more strongly, and make an *ionic* contribution to the bond strength. Pauling constructed an empirical formula to calculate electronegativties of the elements that was based on the additional stabilization energy provided by the ionic character of the bond, which he calculated by the following procedure.

Let the bond dissociation energy of an A—A bond be symbolized by ΔE_{AA}, and that of a B—B bond by ΔE_{BB}; both bonds are covalent because the two atoms are identical. Pauling proposed that the *covalent* contribution to the bond dissociation energy of an A—B bond is the (geometric) mean of the two bond dissociation energies, $\sqrt{\Delta E_{AA}\Delta E_{BB}}$. He reasoned that the A—B bond must also include some ionic character, due to partial charge transfer between the atoms, that contributes to the strength of the bond and that this **excess bond energy Δ** could be calculated as follows

$$\Delta = \Delta E_{AB} - \sqrt{\Delta E_{AA}\Delta E_{BB}} \qquad [3.11]$$

Finally, Pauling defined the electronegativity difference as

$$\chi_A - \chi_B = 0.102\Delta^{1/2} \qquad [3.12]$$

where χ_A and χ_B (Greek letter small chi) are the EN values of A and B, respectively, to account for the ionic contribution to the bond strength. The coefficient 0.102 is appropriate when Δ is measured in kilojoules per mole (kJ mol^{-1}). The choice of the geometric mean to estimate the covalent contribution was originally inspired by elementary valence bond theory (see Section 6.8) arguments, but Pauling states explicitly in his published work that he retained it simply because it gave better agreement with experiment than the arithmetic mean. Note that the formula, as defined, only allows the calculation of electronegativity differences but identifies the more electronegative element as the one with the larger value of χ. Modern electronegativity (now abbreviated EN) scales, like the one shown in Figure 3.18, were developed by assigning an arbitrary value to an element (fluorine is the current choice) and stating electronegativities of the other elements relative to that choice.

Eighty years later, Pauling electronegativities are still among the chemist's most powerful guides for predicting the charge distributions and the nature of the bond between two atoms. Bonds between elements with EN differences of greater than 2 are generally considered to be largely ionic whereas bonds between elements with EN differences close to zero are considered to be covalent. Those with EN differences in the range between 0.2 and 2 are considered to be polar covalent bonds. We present a more quantitative discussion in Section 3.9 in which we connect the predictions based on EN values to experimental measurements of the degree of charge separation, the dipole moment of the molecule.

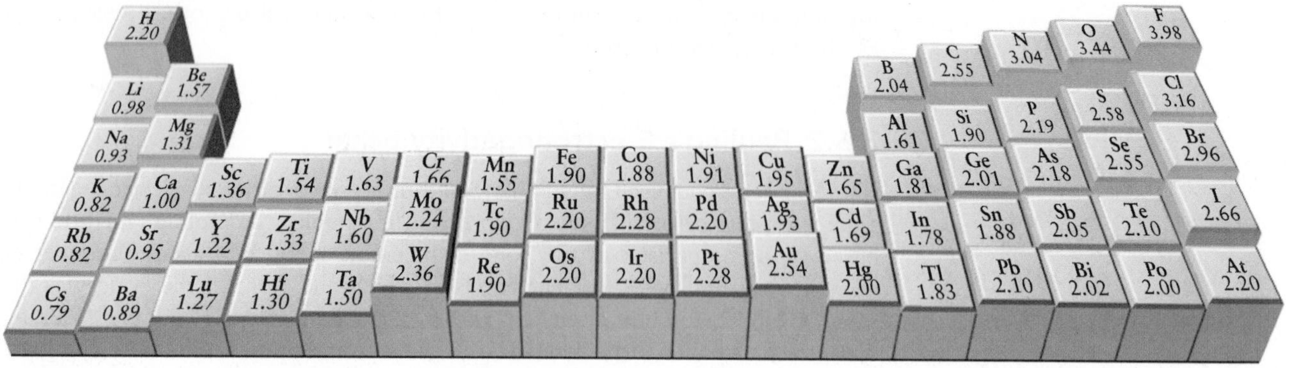

FIGURE 3.18 Average electronegativity of atoms, computed with the method that Linus Pauling developed. Electronegativity values have no units.

EXAMPLE 3.5

Without consulting tabulated values of electronegativity and guided only by the periodic table, arrange these atoms in order of increasing electronegativity: H, F, Al, O. Give a brief explanation of your answer.

Solution

The sequence of increasing electronegativity values is Al < H < O < F. Al is a metal and therefore electropositive, with a low value of electronegativity. In general, electronegativity increases from left to right in the periodic table. As the effective nuclear charge increases, the ability to accommodate an extra electron also increases.

Related Problems: 15, 16

It is informative to compare Mulliken's and Pauling's approaches to the development of their electronegativity scales. Mulliken's approach was based on the physical properties of isolated, individual atoms, and could not take into account any effects attributable to the properties of the other atom of the bond. In contrast, Pauling's method took into account both the covalent and ionic contributions to the character of the bond to arrive at a scale that, in some sense, represented the average electronegativity of an element, when bonded to a wide range of other elements. By explicitly separating the ionic and covalent contributions that determine the character of the chemical bond, Pauling provided the framework for our contemporary understanding of the polar covalent bond, in particular, and its effect on the measurable properties of molecules, including their dipole moments. These considerations, which also determine the nature of the forces between molecules, are discussed in Section 3.9.

Despite the differences between their two approaches, Mulliken's and Pauling's scales produce EN values that are very nearly proportional to one another. Pauling's scale has been more widely adopted simply because the data on which it is based are more complete and considered to be more reliable. Electronegativity as a concept has been criticized over the years because it is not measurable and must be constructed indirectly. Nonetheless, it is extremely useful because it succinctly summarizes the most characteristic chemical properties of an atom in a single parameter. Research to refine the concept continues to this day, relying on advanced methods for calculating molecular electronic structure and advanced laser spectroscopy for high-precision measurements of *EA* values through electron detachment from anions.

3.7 Forces and Potential Energy in Molecules: Formation of Chemical Bonds

In the previous sections we investigated the energetics of the loss and gain of electrons as prototype chemical reactions. Here, we want to define the energetics of more general types of bond formation. Then we will have all the necessary background to describe in detail the formation of ionic bonds, covalent bonds, and polar covalent bonds in the following sections.

What determines the stability of a chemical bond? Why is the H_2 molecule more stable than a pair of separated hydrogen atoms in the gas phase at normal temperatures and pressures?

Experience shows that physical objects move spontaneously toward configurations that reduce their potential energy. A car rolls downhill, converting its gravitational potential energy into kinetic energy. We have already learned in Section 3.2 that microscopic charged particles move in directions that will reduce their electrostatic (Coulomb) potential energy. But formation of a chemical bond is more subtle than ordinary motion of charged particles under Coulomb's force law, because it involves a special event. Two atoms flying toward each other have a certain total energy that includes contributions from their internal structure, their potential energy relative to each other, and their kinetic energy. To enter into what the distinguished chemist and author George C. Pimentel has called "the blissful state of bondedness"—in which the atoms fly together as a bonded pair forever after—they must give up some of their total energy. A diatomic molecule is more stable than the separated atoms from which it was formed because its *total energy* is less than that of the two separated atoms. You can reach the same conclusion by examining the reverse process; dissociation of a diatomic molecule requires input of energy.

Let's interpret this fact in terms of the potential energy changes in formation of the molecule, illustrated with the specific example of H_2 shown in Figure 3.19. The nuclei are labeled A and B, and the electrons are labeled 1 and 2. The distance between each electron and each proton (r_{1A}, r_{1B}, r_{2A}, r_{2B}) is shown in blue in Figure 3.19, while the distance between the two electrons and the distance between the two protons (R_{AB}) are shown in red. The potential energy of the molecule is most conveniently expressed in terms of these distances.

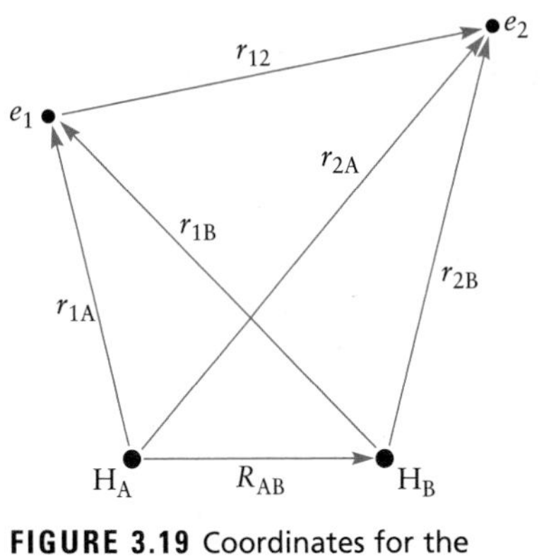

FIGURE 3.19 Coordinates for the hydrogen molecule. The nuclei are assumed stationary at fixed positions separated by the distance R_{AB}. The distance of electron 1 from nuclei A and B is given by r_{1A}, r_{1B}; the distance of electron 2 from nuclei A and B is given by r_{2A}, r_{2B}; the distance between the electrons is given by r_{12}. Note that all the blue vectors pointing from the nuclei to the electrons are of the form $\vec{r}_{A1} = \vec{r}_1 - \vec{r}_A = -\vec{r}_{1A}$, but because the length of $\vec{r}_{A1}$ equals the length of $\vec{r}_{1A}$ ($r_{A1} = r_{1A}$), the labels on the picture showing the distances are correct.

$$V = -\frac{e^2}{4\pi\epsilon_0}\left(\frac{1}{r_{1A}} + \frac{1}{r_{2A}} + \frac{1}{r_{1B}} + \frac{1}{r_{2B}}\right) + \frac{e^2}{4\pi\epsilon_0}\left(\frac{1}{r_{12}}\right) + \frac{e^2}{4\pi\epsilon_0}\left(\frac{1}{R_{AB}}\right) \qquad [3.13]$$

$$V = V_{en} + V_{ee} + V_{nn}$$

The first four terms in Equation 3.13 represent the attractions between the electrons and the nuclei, V_{en}, and all are negative. The last two terms represent the repulsions between the pair of electrons, V_{ee}, and the pair of protons, V_{nn}, and both are positive. The value of V can be calculated for any internuclear distance R_{AB}, if the positions of the electrons were known. Then the total energy would be the sum of the kinetic energies of all the electrons, the relative kinetic energy of the nuclei, and the potential energies in Equation 3.13. However, when a diatomic molecule such as H_2 dissociates to H + H, the electrons change their kinetic energies and positions, one electron going with each H atom, as R_{AB} increases! Chemists are typically interested in the motion of the A and B nuclei relative to each other (for example, could the A-B bond stretch and break?), so we define a new effective potential energy function, V_{eff}, that includes the change in electron kinetic energies as the bond stretches. We define this V_{eff} so that the total energy

E of the non-rotating H-H molecule is the sum of only two terms, the relative kinetic energy of the two nuclei and V_{eff}. Thus V_{eff} includes all the terms in Equation 3.13 *and* the sum the kinetic energies of all the electrons, arranged in orbits around the separating nuclei, that readjust as the nuclei move apart or together. Thus V_{eff} must depend on R_{AB}, which tracks the transition from a diatomic molecule to two separated atoms. Since V_{eff} includes kinetic energies of the electrons and, for H_2, one electron is close to each H atom at large R_{AB}, the zero of V_{eff} is not naturally 0 at large R_{AB} (it is -27.2 eV for large R_{AB} for H + H). Two conventions are common, adding an offset to the potential energy at all R_{AB} so that $V_{eff} = 0$ at large R_{AB} (see Figure 3.20) or adding a larger offset so that $V_{eff} = 0$ at $R_{AB} = R_e$, the equilibrium internuclear distance for the molecule (see Figure 20.15 and Equation 20.12). At large distances the slope $dV/dR_{AB} = 0$ because the isolated atoms do not interact. Assuming you set $V_{eff} = 0$ at large R_{AB}, as R_{AB} decreases, V_{eff} must become negative because the atoms begin to attract each other, thanks to the two electrons interacting with both nuclei. At very small distances, V_{eff} must become positive and large as $V_{eff} \rightarrow \infty$ due to the repulsion between the protons. Therefore, at some intermediate internuclear separation that we label as R_e the decreasing potential function must reach a minimum and change its slope as it heads upward toward positive values. Figure 3.20 is a sketch of a generic V_{eff} versus R_{AB} that shows all these features.

The slope dV/dR_{AB} of the effective potential energy curve in Figure 3.20 is negative at small R_{AB}, so the two atoms in the diatomic molecule feel a repulsive force between them. The positively charged protons are too close together for the surrounding electrons to keep them together. At the minimum of $V(R_{AB})$, at $R_{AB} = R_e$, the atoms in the diatomic molecule neither attract or repel each other, the force is 0, so the molecule is most stable at that internuclear distance. For internuclear distances $R_{AB} > R_e$, the slope $\Delta V/\Delta R_{AB}$ is positive, so the atoms feel an attractive force between them. Therefore, we identify R_e as the *equilibrium bond length* of the molecule. If you start with the atoms in the diatomic molecule at the separation R_e, where V_{eff} is a minimum, and if the relative kinetic energy at that point is small, you must add energy to separate the atoms (pulling against this attractive force). The energy you must add to fully separate the atoms in the molecule, or break the bond, is called the **bond dissociation energy,** ΔE_d. Figure 3.20 shows that the deeper the well in V_{eff}, the more energy you must add to break the bond, and the larger the bond dissociation energy. Thus very stable molecules have deep potential energy wells, effectively holding the atoms together.

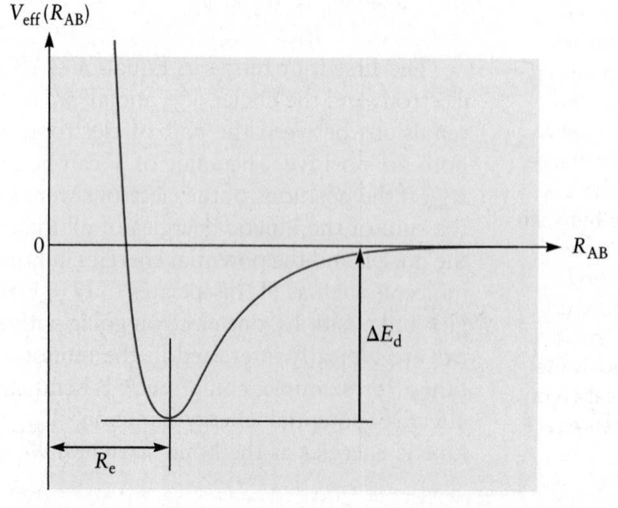

FIGURE 3.20 Dependence of the effective potential energy curve V_{eff} for a diatomic molecule on the internuclear separation R_{AB}. The location of the minimum corresponds to the equilibrium bond length. If the well is low, or deep, relative to the energy of the separated atoms, then the AB molecule is particularly stable, as it would require more energy to break the A-B bond.

At this point, you might wonder how a chemical bond is ever formed. since if two atoms collide with each other, by virtue of conservation of energy they have enough total energy to separate again, even if they would have the lowest potential energy at $R_{AB} = R_e$. When the two atoms are close together during the collision, they must transfer energy to another molecule or atom in the surroundings. In a gas phase system, such as a collision between a Cl atom in the atmosphere and an NO_2 radical, another molecule, such as N_2, must collide with the Cl-NO_2 complex when the internuclear distance is small enough that the potential energy between the Cl and NO_2 is low. Then some of the energy in the complex can be transferred to the N_2 molecule, and the $ClNO_2$ molecule is then confined in the potential energy well. The energy lost by the Cl-NO_2 complex is transferred to the random thermal motion of the surrounding molecules. This process is very efficient if the reaction is happening in solution; the solvent molecules take away energy from the A-B system as the A-B bond is formed. Energy is thus released to the surrounding molecules as the bond is formed between atom A and atom B.

3.8 Ionic Bonding

An **ionic bond** forms between atoms with large differences in electronegativity, such as sodium and fluorine. A practical definition of an ionic bond is one in which the dominant contribution to the strength of the bond is the electrostatic attraction between the ions. Conceptually, the formation of an ionic bond from neutral gas-phase atoms can be thought of as the result of two sequential processes. The more electropositive ion transfers an electron to the more electronegative atom, forming an ion pair that is then drawn together by the attractive electrostatic force. Although we focus our discussion on ionic bonding in a gaseous diatomic molecule where we can clearly identify the forces responsible, most ionic compounds are solids under normal conditions. In an ionic solid, ions of one charge are surrounded by an orderly array of ions of the opposite charge, resulting in extremely large Coulomb stabilization energies. They generally have high melting and boiling points (for example, NaCl melts at 801°C and boils at 1413°C) and can form large crystals. Solid ionic compounds usually conduct electricity poorly, but their melts (molten salts) conduct well.

The creation of positive and negative ions is represented using the Lewis dot symbols for the valence shell of atoms introduced at the end of Section 3.4, by removing or adding dots and also by writing the net electric charge of the ion as a right superscript. For example:

Na·	Na^+	·Ca·	Ca^{2+}
Sodium atom	Sodium ion	Calcium atom	Calcium ion
:F̤̈·	:F̤̈:$^-$	·S̤̈·	:S̤̈:$^{2-}$
Fluorine atom	Fluoride ion	Sulfur atom	Sulfide ion

As demonstrated in Section 3.4, special stability results when an atom, by either losing or gaining electrons, forms an ion whose outermost shell has the same number of electrons as the outermost shell of a noble-gas atom. Except for hydrogen and helium, whose valence shells are completed with two electrons, atoms of the first few periods of the periodic table have a maximum of eight electrons in their valence shells. We say that a chlorine ion ($:\ddot{\underset{..}{\mathrm{Cl}}}:^-$) or an argon atom ($:\ddot{\underset{..}{\mathrm{Ar}}}:$) has a completed octet in its valence shell.

The tendency of atoms to achieve valence octets describes and organizes a large amount of chemical reaction data. Atoms of elements in Groups I and II achieve an octet by losing electrons to form cations; atoms of elements in Groups VI and VII do so by gaining electrons to form anions. Reactions of the metallic elements on the left side of the periodic table with the nonmetallic elements on the right side always transfer just enough electrons to form ions with completed octets. The following equations use Lewis symbols to show the formation first of a cation and an anion and then of an **ionic compound**.

$$\mathrm{Na}\cdot \longrightarrow \mathrm{Na}^+ + e^- \qquad \text{Loss of a valence electron}$$

$$e^- + :\ddot{\underset{..}{\mathrm{Cl}}}\cdot \longrightarrow :\ddot{\underset{..}{\mathrm{Cl}}}:^- \qquad \text{Gain of a valence electron}$$

$$\mathrm{Na}^+ + :\ddot{\underset{..}{\mathrm{Cl}}}:^- \longrightarrow \mathrm{Na}^+\ :\ddot{\underset{..}{\mathrm{Cl}}}:^- \qquad \text{Combination to form the compound NaCl}$$

Another example is the formation of $CaBr_2$:

$$\cdot\mathrm{Ca}\cdot + 2\ :\ddot{\underset{..}{\mathrm{Br}}}\cdot \longrightarrow \mathrm{Ca}^{2+} + 2\ :\ddot{\underset{..}{\mathrm{Br}}}:^- \longrightarrow \mathrm{Ca}^{2+}(\mathrm{Br}^-)_2$$

Neutral atoms not having octets	Positive ion with octet	Negative ions with octets	Ionic compound

The total charge on the compound must be zero, thus the stoichiometry is determined by charge balance. The model predicts a 1:1 compound between Na and Cl and a 1:2 compound between Ca and Br, in agreement with experiment. By invoking the octet rule, the Lewis electron dot model predicts the formula for ionic compounds formed between atoms from Groups I and II and atoms from Groups VI and VII.

Ionic compounds are named by a systematic procedure. Elemental cations retain the name of the parent element, whereas the suffix *-ide* is added to the root name of the element that forms the anion. For example, Cl^- is chloride, and the compound it forms with Na^+ is sodium chloride, the major ingredient in table salt. For this reason, ionic solids are often called salts. Simple, binary ionic compounds are easily named by inspection. Calcium bromide is a binary ionic compound with two Br^- anions per Ca^{2+} cation, to maintain electrical neutrality. Thus the formula unit of this ionic solid is $CaBr_2$. A more comprehensive discussion of inorganic nomenclature is presented in Section 3.13.

Let's consider the energetics of formation of an ionic bond from two neutral gas-phase atoms, potassium and fluorine to be specific. When the atoms are infinitely far apart, their interactions are negligible and we assign their potential energy of interaction as zero (see discussion in Section 3.5). Ionizing potassium requires energy, whereas attaching an electron to fluorine releases energy. The relevant reactions and their energy changes are

$$\mathrm{K} \longrightarrow \mathrm{K}^+ + e^- \qquad \Delta E = IE_1 = +419 \text{ kJ mol}^{-1}$$

and

$$\mathrm{F} + e^- \longrightarrow \mathrm{F}^- \qquad \Delta E = -EA = -328 \text{ kJ mol}^{-1}$$

The total energy cost for the creation of this ion pair, when the parent atoms are infinitely far apart, is

$$\Delta E_\infty = IE_1(\mathrm{K}) - EA(\mathrm{F}) = +91 \text{ kJ mol}^{-1}$$

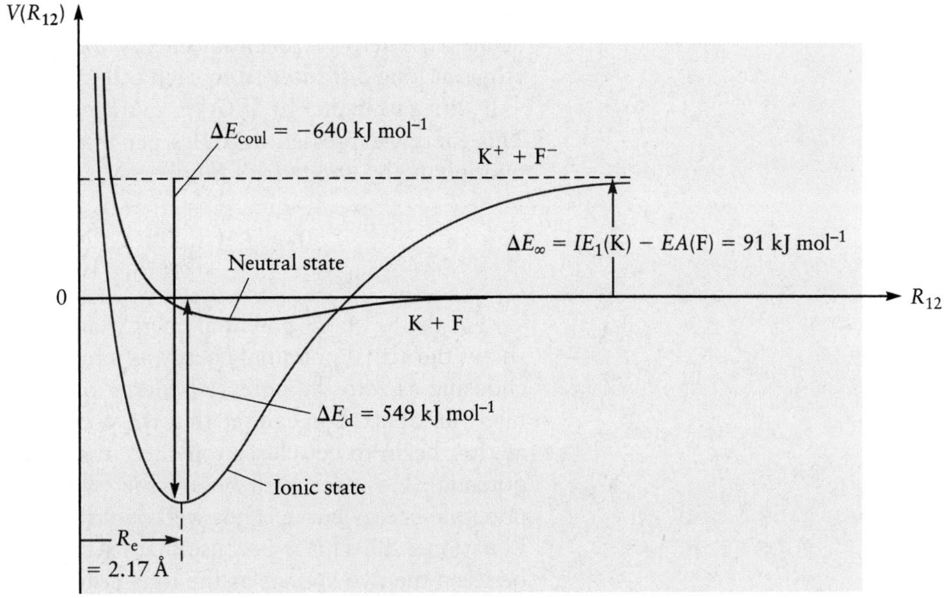

FIGURE 3.21 The potential energy of the ions K^+ and F^-, and the atoms K and F, as a function of their internuclear separation R_{12}.

Note that even for this case, in which one element is highly electronegative and the other is highly electropositive, it still *costs* energy to transfer an electron from a potassium atom to a fluorine atom. This is always true. Because the smallest ionization energy of any element (Cs, 376 kJ mol^{-1}) is larger than the largest electron affinity of any element (Cl, 349 kJ mol^{-1}), creating an ion pair from neutral atoms always requires energy. This fact is illustrated by the higher energy of the red curve as compared to the blue curve at large internuclear distances in Figure 3.21. The blue curve in Figure 3.21 shows how the potential energy of interaction between neutral F and K atoms depends on their separation. When the atoms are very far apart (right side of the figure), they have essentially no interaction, and their potential energy is by convention set to zero. As the atoms approach one another more closely, their potential energy becomes negative due to attractive forces. When the atoms are very close, their potential energy becomes positive due to repulsive forces. The minimum value in the curve occurs at the separation where attractive and repulsive forces exactly balance. The nature of the short-range attractive and repulsive forces between neutral atoms that do not share electrons to form a covalent bond is explained in Section 10.2. The minimum value of the blue curve in Figure 3.21 represents a very modest reduction in potential energy compared with the noninteracting free atoms.

Now let's start from an ion pair separated by a large distance and see what interaction and mechanism will reduce the potential energy of the system to form an ionic bond. The ions are attracted to one another (because they have opposite charges) by the electrostatic force, and the potential energy of the system is described by Coulomb's law:

$$V(R_{12}) = \frac{q_1 q_2}{4\pi\epsilon_0 R_{12}} \text{(J per ion pair)} + V(\infty) \qquad [3.19]$$

where q_1 and q_2 are the charges on the ions, R_{12} is the separation between the ions, and ϵ_0 is defined in Equation 3.1. We have added the potential energy offset to the Coulomb potential of Equation 3.1 because we have already defined the zero of potential energy to be the potential energy of the neutral atoms at long distances from each other. The potential energy of the ion pair $A^+ + B^-$ must be higher by $IE(A) - EA(B)$ when A^+ and B^- are far from each other. This energy, expressed in joules per ion pair, can be converted to kJ mol^{-1} by multiplying by Avogadro's number, N_A, and dividing by 10^3 to get

$$V(R_{12}) = \frac{q_1 q_2}{4\pi\epsilon_0 R_{12}} \frac{N_A}{10^3} (\text{kJ mol}^{-1}) + V(\infty) \qquad [3.20]$$

For the $K^+ + F^-$ potential energy curve, $V(\infty) = +91$ kJ mol^{-1}. Figure 3.21 shows the actual potential energy as a function of the distance between the ions, choosing as zero the potential energy of the neutral atoms when they are infinitely far apart. You can see that the potential energy between K^+ and F^- drops as they begin to get closer together, in agreement with the attractive Coulomb potential. However, well before the two ions reach $R_{12} = R_e$ the ionic (red) potential energy curve is not well described by the simple Coulomb potential in Equation 3.20. This is because there is some sharing of the transferred electron between the two species as the ions near each other.

Both the neutral potential energy curve shown in blue and the ionic potential energy curve shown in red have a repulsive region at small R_{12}; at these short distances the two positively charged nuclei (and the inner shells of electrons) are too close to each other for the shared electrons in the covalent bond or the Coulomb potential, respectively, to counterbalance this repulsion, so the potential energy goes up at small internuclear distances. The equilibrium bond distance R_e on the ionic (red) potential energy curve is shorter than that on the covalent (blue) potential energy curve because en route from large internuclear distances to $R_{12} = R_e$ there has been a strong electrostatic attraction between the two ions, lowering the potential energy as R decreases from large to near R_e.

Starting at the right side of the curve in Figure 3.21, the potential energy of the pair of ions is greater than that of the neutral atoms by

$$\Delta E_\infty = IE_1(\text{K}) - EA(\text{F}) = +91 \text{ kJ mol}^{-1}$$

Moving toward the left side of Figure 3.21, the potential energy decreases rapidly as one follows the red curve to smaller R_{12}, due largely to the attractive *Coulomb stabilization energy* and reaches a minimum value at the equilibrium bond length, $R_e = 2.17$ Å. If we were to try to force the ions to move closer together, we would encounter the resistance depicted by the steep repulsive wall on the left side of the curve, which arises from the repulsive interactions between the core electrons of each of the ions. The equilibrium bond length, R_e, is determined by the balance between the attractive and repulsive forces.

At last, we can very roughly estimate the stabilization of an ionic bond such as KF relative to the neutral atoms, which is the key to formation of an ionic bond from two neutral atoms. From Figure 3.21, if the red ionic curve followed a Coulomb potential with $q_1 = +e$ and $q_2 = -e$ all the way from large R_{12} to $R_{12} = R_e$,

$$\Delta E_d \approx -\frac{q_1 q_2}{4\pi\epsilon_0 R_e} \frac{N_A}{10^3} - \Delta E_\infty \qquad [3.22]$$

where $\Delta E_\infty = IE_1(\text{K}) - EA(\text{F})$. This energy measures the strength of the ionic bond and is approximately equal to the bond dissociation energy, which is the energy required to break the ionic bond and liberate neutral atoms.

EXAMPLE 3.6

Estimate the energy of dissociation to neutral atoms for KF, which has a bond length of 2.17×10^{-10} m. For KF, $\Delta E_\infty = IE_1(\text{K}) - EA(\text{F}) = 91\ \text{kJ mol}^{-1}$.

Solution

$$\Delta E_d \approx -\frac{q_1 q_2}{4\pi\epsilon_0 R_e}\frac{N_A}{10^3} - \Delta E_\infty$$

$$= -\frac{-(1.602 \times 10^{-19}\,\text{C})^2(6.022 \times 10^{23}\,\text{mol}^{-1})}{(4)(3.1416)(8.854 \times 10^{-12}\,\text{C}^2\,\text{J}^{-1}\,\text{m}^{-1})(2.17 \times 10^{-10}\,\text{m})(10^3\,\text{J kJ}^{-1})}$$

$$-91\ \text{kJ mol}^{-1}$$

$$= 640\ \text{kJ mol}^{-1} - 91\ \text{kJ mol}^{-1}$$

$$= 549\ \text{kJ mol}^{-1}$$

This estimate compares fairly well with the experimentally measured dissociation energy of 498 kJ mol^{-1}.

Related Problems: 23, 24, 25, 26

As shown in Example 3.6, our simple model for ionic bonding in KF predicts a bond dissociation energy ΔE_d (the energy required to dissociate the molecule into neutral atoms, starting from the equilibrium bond length R_e) of 549 kJ mol^{-1}, which agrees reasonably well with the experimental value of 498 kJ mol^{-1}. We can conclude that the bonding is predominantly ionic, and that the driving force for the formation of the bond is indeed the reduction of the potential energy of the system, relative to that of the separated atoms. The formation of the ions is a key intermediate step between the separated atoms and the stable ionic bond.

There are several reasons why this simple model does not do a better job in calculating the bond energy. First, all bonds have some degree of covalent character, which involves electron sharing between the atoms. Second, we have assumed that each ion is a point charge. In reality, the distribution of electrons around the fluoride ion is distorted by presence of the potassium ion; this distortion is called **polarization**. The effect of the nonsymmetric shape of the charge distribution on the bond energy is accounted for in more detailed calculations.

The mechanism by which an ionic bond forms from gas-phase atoms is interesting. The explanation is provided by comparing the potential energy curves for the ionic and neutral states in Figure 3.21 to determine which has the lower value at each separation R. For large separations the neutral atoms are always more stable than the ions. As the atoms draw closer together, their potential energy decreases slowly following the blue curve, until it approaches the red curve for ionic interactions at a particular distance. At distances shorter than this intersection point, the ionic species are favored because the Coulomb attraction between the ions more than compensates for the energy required to transfer the electron between the neutral atoms, and an ionic bond forms. At these shorter distances, the atom pair become ions and follow the red curve because it provides much lower potential energy. At very short distances the electrons of the two ions begin to repel each other and their potential energy rises steeply. The equilibrium bond length R_e of an ionic bond is determined by a balance of attractive and repulsive forces at the minimum of the red curve.

The Canadian chemist and Nobel Laureate John Polanyi proposed the following intriguing possibility. Electron transfer takes place when the neutral K and F atoms get close enough to each other, reaching the point at which the red and blue curves in Figure 3.21 intersect. The strong Coulombic attraction of the newly created ion pair then rapidly pulls the reactants together, where they form an ionic bond. The metal has sent its electron to "harpoon" the halogen, pulling it in with the "rope" of the Coulomb interaction. This **harpoon mechanism** has been studied extensively for a variety of systems and is generally agreed to provide a satisfactory semiquantitative description of the formation of gas-phase alkali halide molecules.

You should keep in mind that although gas-phase molecules with predominantly ionic bonding can be prepared and are stable at high temperatures and low pressures, most ionic bonds occur in ionic solids. In an ionic solid, ions of one charge are surrounded by an orderly array of ions of the opposite charge, resulting in extremely large Coulomb stabilization energies. See Figure 3.3 in Section 3.1.

3.9 Covalent and Polar Covalent Bonding

We have discussed how ionic bonding results from electron transfer and Coulomb stabilization of the resulting ions and that the propensity of a pair of atoms to form an ionic bond is determined by the difference in their electronegativities. What kinds of bonds are formed between elements of identical or comparable electronegativities to give compounds such as H_2 or CO?

For example, carbon (electronegativity 2.55) and hydrogen (electronegativity 2.20) react to form methane (CH_4). Unlike ionic compounds, methane is a gas at room temperature, not a solid. Therefore, it is possible to prepare isolated methane molecules and determine their structure by experiment. Cooling methane to low temperatures condenses it to a solid in which the CH_4 molecules do not strongly interact, so retain their identities. Methane dissolves in water to a slight extent, but it does not ionize. Thus, it is not useful to think of methane as an ionic substance made up of C^{4-} and H^+ ions (or C^{4+} and H^- ions). Methane molecules are formed by covalent, not ionic, bonding.

What is the driving force for covalent bond formation from separated atoms in the gas phase? Section 3.7 provides a general argument for how a chemical bond forms from a pair of isolated atoms by arriving at an arrangement of electrons around the nuclei, which reduces the energy of the molecule below that of the isolated atoms. We have used that argument to show that for ionic compounds the driving force is the Coulomb stabilization of the ion pair. We present a plausibility argument to suggest why a covalent bond might form, by focusing on the *forces* acting on the nuclei due to the electrons.

Let's consider the simplest possible molecule, H_2^+ (the hydrogen molecule ion), which has only one electron. In Figure 3.22, H_A identifies the position of nucleus A, and H_B that of nucleus B. The distance between the two nuclei is R_{AB}, represented by the dashed black line. The distances between the electron and each nucleus are r_{Ae} and r_{Be}, respectively, and are represented by the dashed red lines.

Let's identify all the forces between the three particles. There is the internuclear repulsive force, $F_{AB} \propto (+Z_A e)(+Z_B e)/R_{AB}^2$, and two electron-nuclear attractive forces, $F_{Ae} \propto (-e)(+Z_A e)/r_{Ae}^2$ and $F_{Be} \propto (-e)(+Z_B e)/r_{Be}^2$. The internuclear repulsive force always opposes formation of a chemical bond (Fig. 3.22a), so we must identify some force that overcomes this repulsion. We need only consider the attractive forces and ask, "Over what region in space does the electron exert

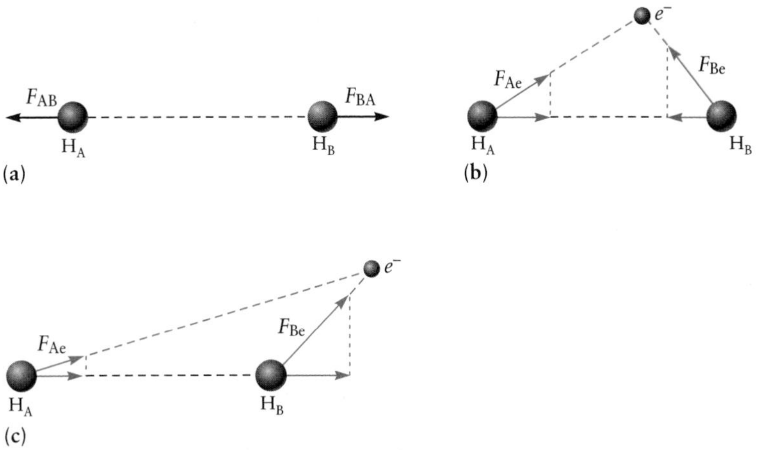

FIGURE 3.22 The forces between the particles in H_2^+. (a) The internuclear repulsion always opposes bonding the nuclei together. (b) An electron positioned in a region that will tend to bond the nuclei together, as shown by the force (red) that the electron exerts on each nucleus, and the component (blue) of each force along the internuclear axis. (c) An electron positioned in a region that will tend to pull the nuclei apart. (Adapted from G.C. Pimentel and R.D. Spratley, *Chemical Bonding Clarified through Quantum Mechanics,* Holden-Day Inc., San Francisco. 1969. Page 74.)

forces on the nuclei that will tend to pull them together?" Only the component of the attractive force directed along the internuclear axis is effective in drawing them together. In Figure 3.22b, the component of the attractive force that the electron exerts on each nucleus is shown in blue. When an electron is located in a region "between" the nuclei, as in Figure 3.22b, these blue components of the forces act to pull the nuclei together. In contrast, when the electron is "outside" the internuclear region (for example, see Fig. 3.22c), it exerts a greater force on the nearer nucleus than the farther, pulling the nuclei apart. The shared electrons in a covalent chemical bond might be postulated to be found at positions like that depicted in Figure 3.22b, not Figure 3.22c. Then the component of the force along the internuclear axis exerted on each nucleus by the electron is attractive, encouraging bonding. This simple model is supported by experimental data for H_2^+. Its equilibrium bond length R_e is 1.06 Å, and its bond dissociation energy ΔE_d is 255.5 kJ mol^{-1}, which is characteristic of a stable covalent bond.

This picture of covalent bonding in the H_2^+ molecular ion can be applied to other molecules. For example, the H_2 molecule is quite stable (its bond dissociation energy is 432 kJ mol^{-1}), yet it consists of two identical atoms. There is no possibility of a net charge transfer from one to the other to form an ionic bond. The stability of H_2 arises from the sharing of electrons between atoms in a covalent bond.

Any classical theory of chemical bond formation must explain certain properties of the chemical bond, explain trends observed in bonding, and most importantly, predict likely bonding properties of molecules not yet made. The most important classical descriptors of the chemical bond are the bond length, energy, order, and polarity.

3.9.1 Bond Lengths

For a diatomic molecule, the only relevant structural parameter is the bond length, that is, the distance between the nuclei of the two atoms. Table 3.3 lists the equilibrium bond lengths R_e of a number of diatomic molecules, expressed in units of angstroms (1 Å = 10^{-10} m). Certain systematic trends are immediately obvious. Among the members of a group in the periodic table, bond lengths usually increase with increasing atomic number Z. The I_2 bond is longer than the F_2 bond, for example, and those of Cl_2 and Br_2 fall in line, as they should. A particularly significant result from experiment is that the length of a bond of the same type (see later) between a given pair of atoms changes little from one molecule to another. For example, C—H bond lengths in acetylsalicylic acid (aspirin, $C_9H_8O_4$) are about the same as they are in methane (CH_4), although

TABLE 3.3

Properties of Diatomic Molecules

Molecule	Bond Length (Å)	Bond Energy (kJ mol^{-1})
H_2	0.751	432
N_2	1.100	942
O_2	1.211	495
F_2	1.417	155
Cl_2	1.991	240
Br_2	2.286	190
I_2	2.669	148
HF	0.926	565
HCl	1.284	429
HBr	1.424	363
HI	1.620	295
ClF	1.632	252
BrF	1.759	282
BrCl	2.139	216
ICl	2.324	208
NO	1.154	629
CO	1.131	1073

the molecules have different structures and physical and chemical properties. Table 3.4 shows that the lengths of O—H, C—C, and C—H bonds in a number of molecules are constant to within a few percent.

3.9.2 Bond Energies

The stability of a molecule is determined by the energy required to dissociate the molecule into its constituent atoms. The greater the energy required, the more stable the molecule. The **bond energy**, also called the bond dissociation energy, is the energy required to break one mole of the particular bond under

TABLE 3.4

Reproducibility of Bond Lengths

Bond	Molecule	Bond Length (Å)
O—H	H_2O	0.958
	H_2O_2	0.960
	HCOOH	0.95
	CH_3OH	0.956
C—C	Diamond	1.5445
	C_2H_6	1.536
	CH_3CHF_2	1.540
	CH_3CHO	1.50
C—H	CH_4	1.091
	C_2H_6	1.107
	C_2H_4	1.087
	C_6H_6	1.084
	CH_3Cl	1.11
	CH_2O	1.06

Bond lengths measured experimentally by rotational spectroscopy.

CONNECTION TO INSTRUMENTAL ANALYSIS

Molecular Spectroscopy

Much, if not most, of what we know about the structures and properties of molecules has been provided by molecular spectroscopy (see Chapter 20). We introduce you here to the interactions between electromagnetic radiation and matter and show how rotational and vibrational spectroscopy provide the experimental values of bond lengths (see Tables 3.3–3.6) as well as bond force constants (bond stiffness). Electromagnetic radiation may be absorbed, emitted, or scattered by molecules at frequencies that are characteristic of particular molecular motions. We ask you to simply accept, for now, three assertions about quantum mechanics: 1) that only certain values of the energy associated with these motions are allowed; 2) that electromagnetic radiation can be thought of as a beam of particles called photons, each of which has energy $E = h\nu$, where h is Planck's constant and the frequency ν is the speed of light c divided by its wavelength (see Section 4.1), and 3) that absorbing, emitting, or scattering photons causes transitions between these "energy levels." We can then introduce simple classical models that allow us to extract molecular properties from spectra.

Molecules in the gas phase rotate about their centers of mass at frequencies of 10^{10}–10^{12} Hz (s^{-1}); absorbing microwave radiation at these frequencies excites them to higher energy rotational levels. Chemical bonds can be thought of as springs connecting atoms that are oscillating back and forth. Absorbing infrared radiation (10^{12}–10^{13} Hz) forces these oscillators to vibrate with more energy. Let's consider the mechanisms by which absorption and scattering induce molecular motion. Electromagnetic radiation (see Section 4.1) consists of oscillating electric and magnetic fields oriented perpendicular to one another and to the direction in which they are travelling; the interactions between the electric fields and molecules are the most important. We showed earlier that charges are unevenly distributed in chemical bonds if the electronegativities of the atoms are different. Electric fields exert forces on charged particles, moving the positive charges in the direction of the field and the negative charges in the opposite direction. Figure C3.3 shows the forces exerted by an electric field on a heteronuclear diatomic molecule, with partial charges δ^+ and δ^- on the two atoms. Figure C3.3a shows how an oscillating electric field induces rotation and Figure C3.3b shows how it induces vibration. Atoms of opposite charge move in opposite directions under the influence of the field; these directions reverse when the field direction reverses as it oscillates. Inelastic light scattering by molecules (Raman scattering) also induces molecular

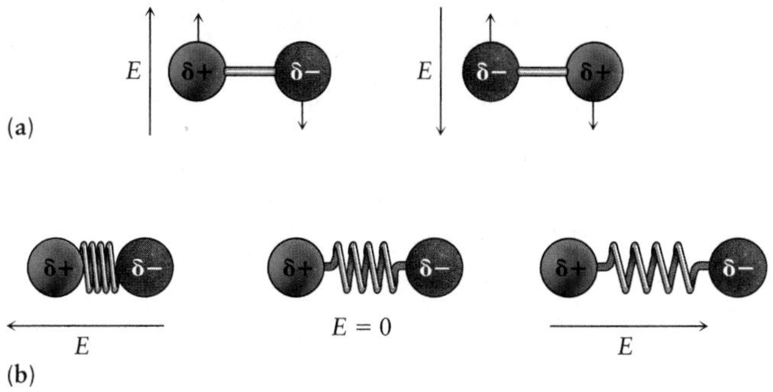

FIGURE C3.3 In (a), the applied electric would cause a molecule nonzero dipole moment to rotate, if you change the direction of the field in sync with the rotational period. In (b) changing the direction of the applied electric field during the vibrational motion of the polar diatomic molecule can induce it to vibrate with more energy. Thus the frequency ν at which the applied electric field oscillates must be equal to the natural vibrational frequency of the diatomic molecule.

motion, even in nonpolar, homonuclear diatomic molecules. The energy difference between the incident and scattered photons is equal to the difference between the molecular energy levels.

Rotational Spectroscopy: Bond Lengths and Bond Angles

Rotations of diatomic molecules are analyzed using the rigid rotor model (see Section 20.3) in which the atoms are represented as point masses attached to one another by a massless rigid rod with moment of inertia $I = \mu R_e^2$, where $\mu = m_1 m_2/(m_1 + m_2)$ is the reduced mass and R_e is the equilibrium bond length. Rotational spectra, obtained by microwave absorption or Raman spectroscopy, comprise a series of lines separated by $2\tilde{B} = h/8\pi^2 cI$ where $\tilde{B}$ (cm^{-1}) is the rotational constant, which is proportional to energy. Each line in the microwave absorption spectrum of carbon monoxide shown in Figure C3.4 corresponds to the absorption of microwave radiation that excites molecules from one level to the next higher energy level. Many such transitions are allowed, as shown in Figure C3.4, but we are only interested in the value of $\tilde{B}$ for now.

Rotational constants for H_2, N_2, O_2, and CO are 59.2, 1.99, 1.44, and 1.9 cm^{-1}, respectively, from which you can calculate the bond lengths shown in Table 3.3 using the formulas introduced above. Rotational spectra of polyatomic molecules are much more complicated than those of diatomic molecules but they provide not only bond lengths but also bond angles and thus complete three-dimensional structures of gas-phase polyatomic molecules.

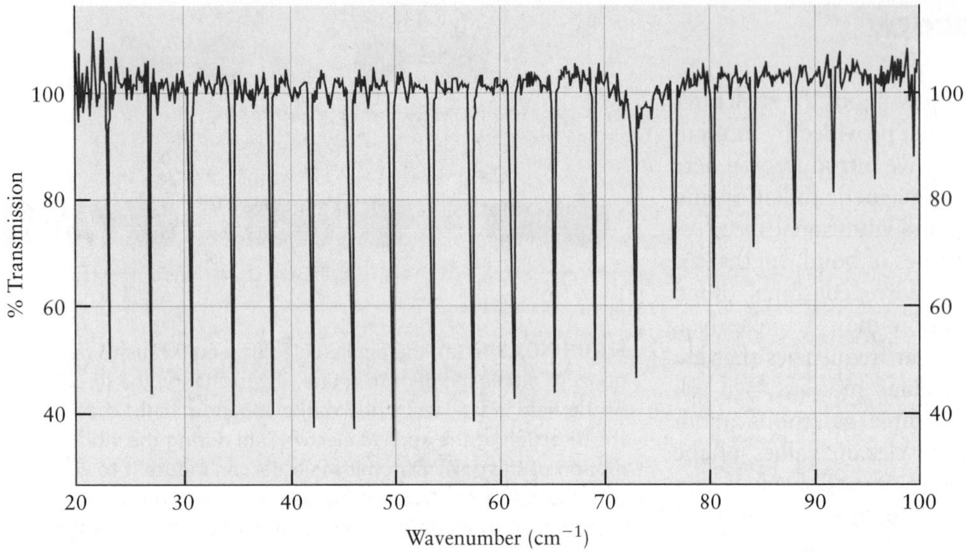

FIGURE C3.4 Microwave absorption spectrum of carbon monoxide (CO). The horizontal axis gives the inverse of the wavelength of the light that is absorbed (multiply that by $hc = 2 \times 10^{-23}$ J cm to obtain the energy of the photon absorbed).

Vibrational Spectroscopy: Force Constants and Molecular Structure

The harmonic oscillator model (see Section 20.3) is used to interpret vibrational spectra. Diatomic molecules are represented by two atomic masses, of reduced mass μ, connected by a spring with stiffness (force constant) k. The fundamental frequency of a harmonic oscillator is given by $\nu = \frac{1}{2\pi}\sqrt{\frac{k}{\mu}}$. The energy levels of a harmonic oscillator are equally spaced with the separation between adjacent levels being given by $\Delta E = h\nu$, where $h = 6.626 \times 10^{-34}$ J s. Only transitions between adjacent levels are allowed, as for rotational spectroscopy, but they all occur at the same energy, resulting in spectra that are much simpler to interpret than rotational spectra. Figure C3.5 shows the Raman spectra of H_2, N_2, O_2, and CO to illustrate the ability of the technique to measure vibrational spectra of both homonuclear and heteronuclear diatomic molecules. Of these molecules, infrared absorption spectroscopy could measure only CO, because the mechanism discussed earlier requires that there be charge separation in the molecule to interact with the electric field of the radiation.

The spectra clearly reveal the dependence of the vibrational frequencies observed on both the reduced mass and the force constant. The vibrational frequency of H_2 (4160 cm^{-1}) is the highest of the group because it is the lightest molecule and has only a single bond. The correlation between bond order and vibrational frequencies among a series of molecules with comparable reduced masses is clearly shown by the spectra of N_2, CO, and O_2, with vibrational frequencies of 2331, 2143, and 1550 cm^{-1}, respectively. The vibrational frequencies of N_2 and CO are comparable and much higher than that of O_2, providing solid evidence for the greater strength of the triple bond in those molecules compared with the double bond in O_2. The ratio of the frequencies is very nearly 3/2, the ratio of the bond orders.

Infrared spectroscopy is also widely used in chemical analysis to identify and characterize polyatomic molecules on the basis of distinctive functional group (see Section 7.4) frequencies. For example, the infrared spectrum of an amide shows the characteristic N—H and C=O stretching vibrations of the amide group as well as the C—H stretching vibrations of the methyl (CH_3) and methylene (CH_2) groups.

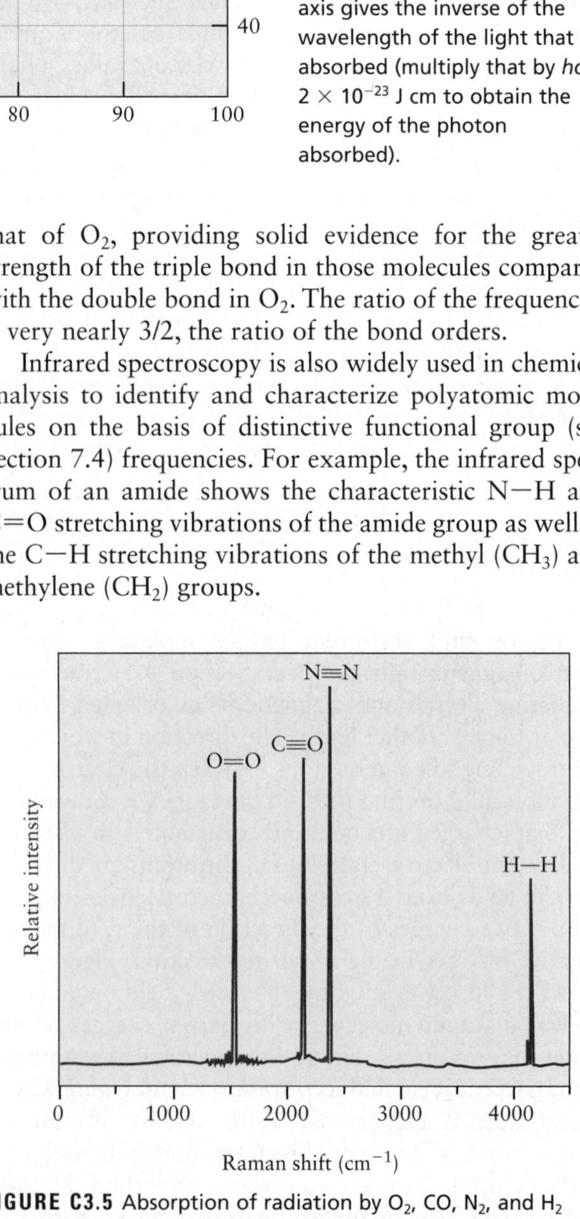

FIGURE C3.5 Absorption of radiation by O_2, CO, N_2, and H_2 molecules as a function of the frequency of the radiation. The frequency shown on the horizontal axis is ν/c.

discussion (see Section 3.7). The bond energy is denoted by ΔE_d ("d" stands for *dissociation* here) and is measured directly as the energy change for the dissociation reaction, for example

$$H_2(g) \longrightarrow 2H(g) \qquad \Delta E = \Delta E_d$$

Bond energy is measured as part of the science of thermochemistry, described in Section 12.6. In essence, one mole of the compound is placed in a closed container and heated sufficiently to dissociate it completely into atoms. The amount of energy required is measured in units of kJ mol^{-1}. Table 3.3 lists bond energies for selected diatomic molecules. Again, certain systematic trends with changes in atomic number are evident. Bonds generally grow weaker with increasing atomic number, as shown by the decrease in the bond energies of the hydrogen halides in the order HF > HCl > HBr > HI. Note, however, the unusual weakness of the bond in the fluorine molecule, F_2. Its bond energy is significantly *smaller* than that of Cl_2 and comparable with that of I_2. Bond strength decreases dramatically in the diatomic molecules from N_2 (942 kJ mol^{-1}) to O_2 (495 kJ mol^{-1}) to F_2 (155 kJ mol^{-1}). What accounts for this behavior? A successful theory of bonding must explain both the general trends and the reasons for particular exceptions.

Bond energies, like bond lengths, are fairly reproducible (within about 10%) from one compound to another. It is therefore possible to tabulate *average* bond energies from measurements on a series of compounds. The energy required to break the O—H bond, for example, in each of the compounds shown in Table 3.4, is roughly 465 kJ mol^{-1}.

3.9.3 Bond Order

Sometimes, the length and energy of the bond between two specific kinds of atoms are *not* comparable among different compounds, but rather are sharply different. Table 3.5 shows the great differences in bond lengths and bond energies of carbon–carbon bonds in ethane (H_3CCH_3), ethene (H_2CCH_2), and ethyne (HCCH, commonly called acetylene). Carbon–carbon bonds from many other molecules all fit into one of the three classes given in the table (that is, some carbon-carbon bond lengths are close to 1.54 Å, others are close to 1.34 Å, and still others are close to 1.20 Å). This observation confirms the existence of not one, but three types of carbon–carbon bonds. We classify these as single, double, and triple bonds, respectively, based on their bond lengths and bond dissociation energies. The longest and weakest (as in ethane) is a single bond represented by C—C; that of intermediate strength (as in ethene) is a double bond, C=C; and the shortest and strongest (as in acetylene) is a triple bond, C≡C. We define the **bond order** as the number of shared electron pairs between the two atoms. Bond order is predicted by models of covalent bond formation.

Even these three types do not cover all the carbon–carbon bonds found in nature, however. In benzene (C_6H_6), the experimental carbon–carbon bond length is 1.397 Å, and its bond dissociation energy is 505 kJ mol^{-1}. This bond is intermediate between a single bond and a double bond (its bond order is $1\frac{1}{2}$). In fact, the bonding in compounds such as benzene differs from that in many other

TABLE 3.5

Three Types of Carbon–Carbon Bonds

Bond	Molecule	Bond Length (Å)	Bond Energy (kJ mol^{-1})
C—C	C_2H_6 (or H_3CCH_3)	1.536	345
C=C	C_2H_4 (or H_2CCH_2)	1.337	612
C≡C	C_2H_2 (or HCCH)	1.204	809

TABLE 3.6

Average Bond Lengths (in Å)

C—C	1.54	N—N	1.45	C—H	1.10
C=C	1.34	N=N	1.25	N—H	1.01
C≡C	1.20	N≡N	1.10	O—H	0.96
C—O	1.43	N—O	1.43	C—N	1.47
C=O	1.20	N=O	1.18	C≡N	1.16

compounds (see Chapter 7). Although many bonds have properties that depend primarily on the two atoms that form the bond (and thus are similar from one compound to another), bonding in benzene and related molecules, and a few other classes of compounds, depends on the nature of the whole molecule.

Multiple bonds occur between atoms other than carbon and even between unlike atoms. Some representative bond lengths are listed in Table 3.6.

The bond order is not measured experimentally. To test various theoretical predictions of bond order, we compare with a quantity called the **bond force constant**, k, which is measured experimentally by the technique of molecular vibration spectroscopy (see *Connection to Instrumental Analysis*). Experimental results show that the force constant increases with bond order. Comparing this result with Tables 3.5 and 3.6 shows that increased force constant correlates with higher bond energy and smaller bond length: stiffer bonds are stronger and shorter if the identity of the two atoms bonded together is unchanged. Comparing the theoretical quantity, bond order, with the experimental quantity, force constant, is one of the main ways to evaluate the accuracy of theoretical models of covalent bond formation.

3.9.4 Polar Covalent Bonding: Electronegativity and Dipole Moments

Laboratory measurements show that most real bonds are neither fully ionic nor fully covalent, but instead possess a mixture of ionic and covalent character. A bond in which there is a partial transfer of charge is called a **polar covalent bond.** This section provides an approximate description of the polar covalent bond based on the relative ability of each atom to attract the electron pair toward its nucleus. This ability is estimated by comparing the electronegativity values for the two atoms.

On the Pauling scale (see Fig. 3.18 and Appendix F), electronegativities range from 3.98 (for fluorine) to 0.79 (for cesium). These numerical values are useful for exploring periodic trends and for making semiquantitative comparisons. They represent the average tendency of an atom to attract electrons within a molecule, based on the properties of the bond it makes in a large range of compounds.

The absolute value of the difference in electronegativity of two bonded atoms tells the degree of *polarity* in their bond. A large difference (greater than about 2.0) means that the bond is ionic and that an electron has been transferred completely or nearly completely to the more electronegative atom. A small difference (less than about 0.4) means that the bond is largely covalent, with electrons in the bond shared fairly evenly. Intermediate values of the difference signify a polar covalent bond with intermediate character. These suggested dividing points between bond types are not hard and fast rules, and your instructor may suggest alternative values.

EXAMPLE 3.7

Using Figure 3.10, arrange the following bonds in order of decreasing polarity: H—C, O—O, H—F, I—Cl, Cs—Au.

Solution

The differences in electronegativity among the five pairs of atoms (without regard to sign) are 0.35, 0.00, 1.78, 0.50, and 1.75, respectively. The order of decreasing polarity is the same as the order of decrease in this difference: H—F, Cs—Au, I—Cl, H—C, and O—O. The last bond in this listing is nonpolar.

Related Problems: 31, 32, 33, 34

3.9.5 Dipole Moments and Percent Ionic Character

A bond that is almost purely ionic, such as that of KF, can be thought of as arising from the nearly complete transfer of one electron from the electropositive to the electronegative species. KF can be described fairly accurately as K^+F^-, with charges $+e$ and $-e$ on the two ions. However, characterizing the charge distribution for a molecule such as HF, which has significant covalent character, is more complex. If we wish to approximate the bond by its ionic character, it is best described as $H^{\delta+}F^{\delta-}$, where some fraction, δ of the full charge, e, is on each nucleus. A useful measure of the ionic character of a bond, arising from electronegativity differences, especially for diatomic molecules, is the dipole moment of the molecule. If two charges of equal magnitude and opposite sign, $+q$ and $-q$, are separated by a distance R, the **dipole moment** $\vec{\mu}$ (Greek letter lowercase mu) of that charge distribution is

$$\vec{\mu} = q\vec{R} \qquad [3.23]$$

$\vec{R}$ is the vector from the atom with partial negative charge to the atom with partial positive charge. In SI units the magnitude μ of the dipole moment is measured in Coulomb meters, an inconveniently large unit for discussing molecules. The unit most often used is the debye (D), which is related to SI units by

$$1 \text{ D} = 3.336 \times 10^{-30} \text{ C m}$$

(This apparently peculiar definition arises from the transition from electrostatic units to SI units.) The debye can also be defined as the dipole moment of two charges $\pm e$ separated by 0.2082 Å. If δ is the fraction of a unit charge on each atom in a diatomic molecule ($q = \delta e$) and R is the equilibrium bond length, then

$$\mu(\text{D}) = [R(\text{Å})/0.2082 \text{ Å}]\ \delta \qquad [3.24]$$

This equation can, of course, be inverted to determine the fraction ionic character from the experimental value of the dipole moment. Dipole moments are measured experimentally by electrical and spectroscopic methods and provide useful information about the nature of bonding. In HF, for example, the value of δ calculated from the dipole moment ($\mu = 1.82$ D) and bond length ($R = 0.917$ Å) is 0.41, substantially less than the value of 1 for a purely ionic bond. We convert δ to a "percent ionic character" by multiplying by 100% and say that the bond in HF is 41% ionic. Deviations from 100% ionic bonding occur for two reasons: (1) covalent contributions lead to electron sharing between atoms, and (2) the electronic charge distribution around one ion may be distorted by the electric field of the other ion (polarization). When polarization is extreme, regarding the ions as point charges is no longer a good approximation, and a more accurate description of the distribution of electric charge is necessary.

Table 3.7 provides a scale of ionic character for diatomic molecules, based on the definition of δ. The degree of ionic character inferred from the dipole moment is reasonably well correlated with the Pauling electronegativity

TABLE 3.7

Dipole Moments of Diatomic Molecules

Molecule	Bond Length (Å)	Dipole Moment (D)	% Ionic Character (100 δ)
H_2	0.751	0	0
CO	1.131	0.112	2
NO	1.154	0.159	3
HI	1.620	0.448	6
ClF	1.632	0.888	11
HBr	1.424	0.828	12
HCl	1.284	1.109	18
HF	0.926	1.827	41
CsF	2.347	7.884	70
LiCl	2.027	7.129	73
LiH	1.604	5.882	76
KBr	2.824	10.628	78
NaCl	2.365	9.001	79
KCl	2.671	10.269	82
KF	2.176	8.593	82
LiF	1.570	6.327	84
NaF	1.931	8.156	88

differences (Fig. 3.23). A great deal of ionic character usually corresponds to a large electronegativity difference, with the more electropositive atom carrying the charge $+\delta e$. There are exceptions to this general trend, however. Carbon is less electronegative than oxygen, so one would predict a charge distribution of $C^{\delta+}O^{\delta-}$ in the CO molecule. In fact, the measured dipole moment is quite small in magnitude and is oriented in the opposite direction: $C^{\delta-}O^{\delta+}$ with $\delta = 0.02$. The discrepancy arises because of the lone-pair electron density on the carbon atom (which is reflected in the formal charge of -1 carried by that atom, as discussed in Section 3.10).

In summary, the properties of a covalent chemical bond are often quite similar in a variety of compounds, but we must be alert for exceptions that may signal new types of chemical bonding.

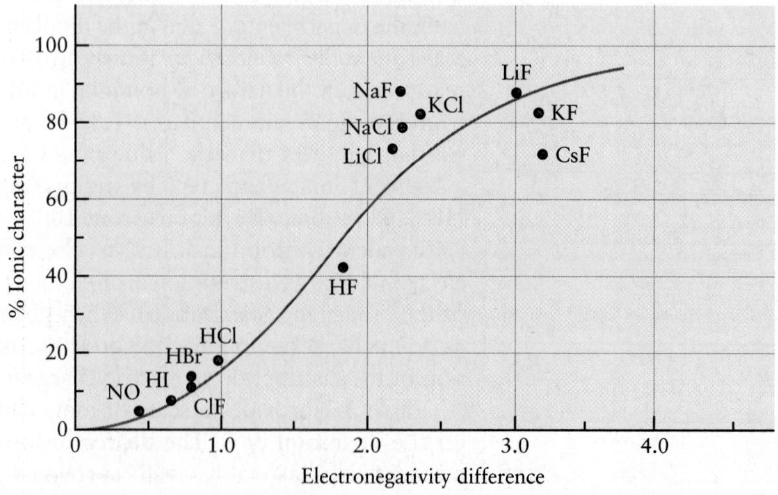

FIGURE 3.23 Two measures of ionic character for diatomic molecules are the electronegativity difference (from Fig. 3.18) and the percent ionic character 100 δ, calculated from the observed dipole moment and bond length. The curve shows that the two correlate approximately but that there are many exceptions.

3.10 Electron Pair Bonds and Lewis Diagrams for Molecules

The goal of the Lewis model is to predict the *structural formula* for a molecule whose molecular formula is already known. In what sequence are the atoms bonded to each other, and by what kinds of bonds? The Lewis model approaches this goal by considering covalent bonds to be shared pairs of valence electrons positioned between two nuclei, where they can help attract the nuclei together and to contribute to the bond strength through the mechanism described in Fig. 3.14b. This bonding model is visually described with the Lewis electron dot symbols for the valence electrons in atoms introduced at the end of Section 3.4.

The Lewis model for covalent bonding starts with the recognition that electrons are not transferred from one atom to another in a nonionic compound, but rather are *shared* between atoms to form covalent bonds. Hydrogen and chlorine combine, for example, to form the **covalent compound** hydrogen chloride. This result can be indicated with a *Lewis diagram* for the molecule of the product, in which the valence electrons from each atom are redistributed so that one electron from the hydrogen atom and one from the chlorine atom are now shared by the two atoms. The two dots that represent this electron pair are placed between the symbols for the two elements:

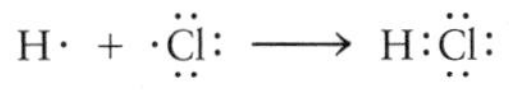

The basic rule that governs Lewis diagrams is the **octet rule**: Whenever possible, the electrons in a covalent compound are distributed in such a way that each main-group element (except hydrogen) is surrounded by eight electrons (an *octet* of electrons). Hydrogen has two electrons in such a structure. When the octet rule is satisfied, the atom attains the special stability of a noble-gas shell. As a reminder, the special stability of the noble-gas configuration arises from the fact that electrons in a filled shell experience the maximum electron–nuclear attraction possible. In the structure for HCl shown earlier, the H nucleus has two valence electrons in its shell (like the noble gas, He), and Cl has eight (like Ar). Electrons that are shared between two atoms are counted as contributing to the filling of the valence shells of both atoms.

A shared pair of electrons is more typically represented by a short line (–)

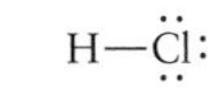

The unshared electron pairs around the chlorine atom in the Lewis diagram are called **lone-pair electrons**, and they make no contribution to the bond between the atoms. Lewis diagrams of some simple covalent compounds are

NH_3 H_2O CH_4

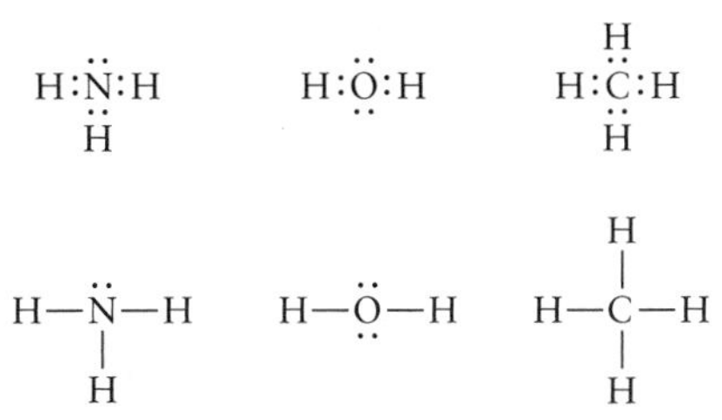

Lewis diagrams show how bonds connect the atoms in a molecule, but they do not show the spatial geometry of the molecule. The ammonia molecule is not planar, but pyramidal, with the nitrogen atom at the apex. The water molecule

(a)

(b)

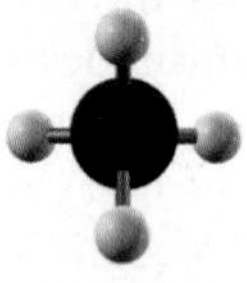

(c)

FIGURE 3.24 Molecules of three familiar substances, drawn in ball-and-stick fashion. The sizes of the balls have been reduced somewhat to show the bonds more clearly, but the relative sizes of the balls are correct. (a) Ammonia, NH_3. (b) Water, H_2O. (c) Methane, CH_4.

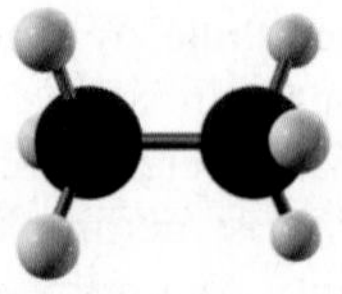

FIGURE 3.25 Ethane, C_2H_6, can be burned in oxygen as a fuel, and if strongly heated, it reacts to form hydrogen and ethylene.

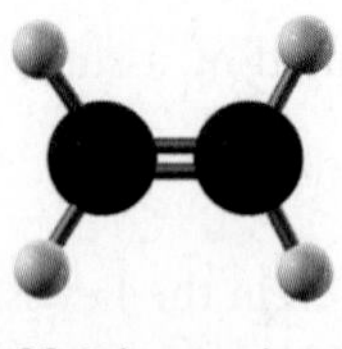

FIGURE 3.26 Ethene, also called ethylene, C_2H_4, is the largest volume organic (carbon-containing) chemical produced.

FIGURE 3.27 Ethyne, or acetylene, C_2H_2, has a triple bond that makes it highly reactive.

is bent rather than straight. Three-dimensional geometries can be represented by ball-and-stick models (such as those shown in Fig. 3.24).

More than one pair of electrons may be shared by two atoms in a bond. For example, in the oxygen molecule, each atom has six valence electrons. Thus, for each to achieve an octet configuration, *two* pairs of electrons must be shared, making a **double bond** between the atoms:

```
 ..  ..          ..  ..
 O: :O     or    O==O
 ..  ..          ..  ..
```

Similarly, the N_2 molecule has a **triple bond,** involving three shared electron pairs:

```
:N:::N:     or     :N≡N:
```

In contrast, the F_2 molecule has only a single bond. The number of shared electron pairs in a bond determines the order of the bond, which has already been connected with bond energy and bond length in Section 3.9. The decrease in bond order from 3 to 2 to 1 explains the dramatic reduction in the bond energies of the sequence of diatomic molecules N_2, O_2, and F_2. A carbon–carbon bond can involve the sharing of one, two, or three electron pairs. A progression from single to triple bonding is found in the hydrocarbons ethane (C_2H_6, Fig. 3.25), ethene (C_2H_4, Fig. 3.26), and ethyne (C_2H_2, Fig. 3.27):

```
  H H        H  H
  .. ..      ..  ..
H:C:C:H      C::C       H:C:::C:H
  .. ..      ..  ..
  H H        H  H
```

This progression corresponds to the three types of carbon–carbon bonds with properties that are related to bond order in Section 3.9 and are summarized in Tables 3.4 and 3.5.

Multiple bonding to attain an octet most frequently involves the elements carbon, nitrogen, oxygen, and, to a lesser degree, sulfur. Double and triple bonds are shorter than a single bond between the same pair of atoms (see illustrative examples in Table 3.5).

3.10.1 Formal Charges

In a **homonuclear diatomic molecule** (in which both atoms are the same, as in H_2 and Cl_2), the electrons are shared equally between the two atoms, and the covalency is nearly ideal for such molecules.

Consider, however, a molecule of carbon monoxide (CO). Its Lewis diagram has a triple bond

```
:C:::O:
```

that uses the ten valence electrons (four from the C and six from the O) and gives each atom an octet. If the six bonding electrons were shared equally, the carbon atom would own five valence electrons (one *more* than its group number) and the oxygen atom would own five electrons (one *less* than its group number). Equal sharing implies that, formally, the carbon atom must gain an electron and the oxygen atom must lose an electron. This situation is described by assigning a **formal charge** to each atom, defined as the charge an atom in a molecule would have if the electrons in its Lewis diagram were divided equally among the atoms that share them. Thus, in CO, C has a formal charge of −1 and O has a formal charge of +1:

```
 −1   +1
:C:::O:
```

FIGURE 3.28 Carbon monoxide, CO, is a colorless, odorless, and toxic gas produced by the incomplete burning of hydrocarbons in air. It is used in the production of elemental metals from their oxide ores.

Carbon monoxide (Fig. 3.28) is a covalent compound, and the assignment of formal charges does not make it ionic. Formal charges are not real charges, and they are not measured. They appear only in the context of drawing Lewis diagrams, where they guide us in deciding whether a proposed diagram represents a reasonable structural formula for a real compound.

We emphasize that the idea of equal sharing of electrons in the bond of a **heteronuclear diatomic molecule**, such as CO, has no experimental basis. It is merely assumed as a starting point for estimating the possibility for separation of positive and negative charge that would arise in a proposed molecular structure solely from the difference in valence (group number) of the participating atoms. If the formal charges in a proposed Lewis diagram differ significantly from what we see in ordinary chemical experience, that diagram should be considered a poor candidate for describing a real molecule, and other possible Lewis diagrams must be examined. There are no hard and fast rules, but here are some general guidelines. Diagrams with like formal charges on adjacent atoms would correspond to unstable molecules and should be discarded. Diagrams with opposite formal charges on adjacent atoms are good so long as the negative charge is on the more electronegative atom and the positive charge is on the more electropositive atom. (The example of CO above shows that this guideline cannot always be achieved.) Diagrams with positive formal charges on the more electronegative atom and negative charges on the more electropositive atom are less favorable. Opposite formal charges are more favorable if they are located on atoms close together than if they are located on atoms farther apart. Diagrams with low formal charges (0, +1, −1) on atoms are favorable.

The formal charge on an atom in a Lewis diagram is simple to calculate. If the valence electrons were removed from an atom, it would have a positive charge equal to its group number in the periodic table (elements in Group VI, the chalcogens, have six valence electrons, and therefore a charge of +6 when those electrons are removed). From this positive charge, subtract the number of lone-pair valence electrons possessed by the atom in the Lewis diagram, and then subtract half of the number of bonding electrons it shares with other atoms:

formal charge = number of valence electrons − number of electrons in lone pairs $-\frac{1}{2}$ (number of electrons in bonding pairs)

EXAMPLE 3.8

Compute the formal charges on the atoms in the following Lewis diagram, which represents the azide ion (N_3^-).

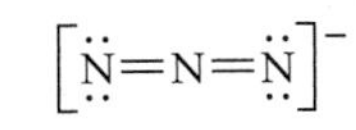

Solution

Nitrogen is in Group V. Hence, each N atom contributes 5 valence electrons to the bonding, and the negative charge on the ion contributes 1 more electron. The Lewis diagram correctly represents 16 electrons.

Each of the terminal nitrogen atoms has four electrons in lone pairs and four bonding electrons (which comprise a double bond) associated with it. Therefore,

$$\text{formal charge}_{(\text{terminal N})} = 5 - 4 - \tfrac{1}{2}(4) = -1$$

The nitrogen atom in the center of the structure has no electrons in lone pairs. Its entire octet comprises the eight bonding electrons:

$$\text{formal charge}_{(\text{central N})} = 5 - 0 - \tfrac{1}{2}(8) = +1$$

The sum of the three formal charges is −1, which is the true overall charge on this polyatomic ion. Failure of this check indicates an error either in the Lewis diagram or in the arithmetic.

Related Problems: 39, 40, 41, 42

3.10.2 Drawing Lewis Diagrams

When drawing Lewis diagrams, we shall assume that the molecular "skeleton" (that is, a plan of the bonding of specific atoms to other atoms) is known. In this respect, it helps to know that hydrogen and fluorine bond to only one other atom and are always terminal atoms in Lewis diagrams. A systematic procedure for drawing Lewis diagrams can then be followed, as expressed by the following guidelines:

1. Count the total number of valence electrons available by first using the group numbers to add the valence electrons from all the atoms present. If the species is a negative ion, *add* additional electrons to achieve the total charge. If it is a positive ion, *subtract* enough electrons to result in the total charge.
2. Calculate the total number of electrons that would be needed if each atom had its *own* noble-gas shell of electrons around it (two for hydrogen, eight for carbon and heavier elements).
3. Subtract the number in step 1 from the number in step 2. This is the number of shared (or bonding) electrons present.
4. Assign two bonding electrons (one pair) to each bond in the molecule or ion.
5. If bonding electrons remain, assign them in pairs by making some double or triple bonds. In some cases, there may be more than one way to do this. In general, double bonds form only between atoms of the elements C, N, O, and S. Triple bonds are usually restricted to C, N, or O.
6. Assign the remaining electrons as lone pairs to the atoms, giving octets to all atoms except hydrogen.
7. Determine the formal charge on each atom and write it next to that atom. Check that the formal charges add up to the correct total charge on the molecule or polyatomic ion. (This step not only guides you to the better diagrams, it also provides a check for inadvertent errors such as the wrong number of dots).
8. If more than one diagram is possible, choose the one with the smallest magnitudes of formal charges (0, +1, −1) and with any negative formal charges placed on the most electronegative atoms.

The use of these guidelines is illustrated by Example 3.9.

EXAMPLE 3.9

Write a Lewis electron dot diagram for phosphoryl chloride, $POCl_3$ (Fig. 3.29). Assign formal charges to all the atoms.

Solution

The first step is to calculate the total number of valence electrons available in the molecule. For $POCl_3$, it is

$$5 \text{ (from P)} + 6 \text{ (from O)} + [3 \times 7 \text{ (from Cl)}] = 32$$

Next, calculate how many electrons would be necessary if each atom were to have its own noble-gas shell of electrons around it. Because there are 5 atoms in the present case (none of them hydrogen), 40 electrons would be required. From the difference of these numbers (40 − 32 = 8), each atom can achieve an octet only if 8 electrons are shared between pairs of atoms. The 8 electrons correspond to 4 electron pairs, so each of the four linkages in $POCl_3$ must be a single bond. (If the number of shared electron pairs were *larger* than the number of bonds, double or triple bonds would be present.)

FIGURE 3.29 Phosphoryl chloride, $POCl_3$, is a reactive compound used to introduce phosphorus into organic molecules in synthesis reactions. Experimental studies show that the P—O bond is more like a double bond than a single bond. Lewis diagrams rationalizing the existence of the P=O double bond can be constructed as an example of valence shell expansion. See the discussion in Section 3.10.4.

(Continued)

The other 24 valence electrons are assigned as lone pairs to the atoms in such a way that each achieves an octet configuration. The resulting Lewis diagram is

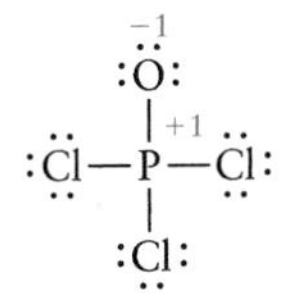

Formal charges are already indicated in blue in this diagram. Phosphorus has the group number V, and it shares eight electrons with no lone-pair electrons, so

$$\text{formal charge on P} = 5 - 4 = +1$$

Oxygen has the group number VI with six lone-pair electrons and two shared electrons, so

$$\text{formal charge on O} = 6 - 6 - \tfrac{1}{2}(2) = -1$$

All three chlorine atoms have zero formal charge, computed by

$$\text{formal charge on Cl} = 7 - 6 - \tfrac{1}{2}(2) = 0$$

Related Problems: 43, 44, 45, 46, 47, 48, 49, 50

FIGURE 3.30 Ozone, O_3, is a pale blue gas with a pungent odor. It condenses to a dark blue liquid below −112°C.

3.10.3 Resonance Forms

For certain molecules or molecular ions, two or more equivalent Lewis diagrams can be written. An example is ozone (O_3, Fig. 3.30), for which there are two possible Lewis diagrams:

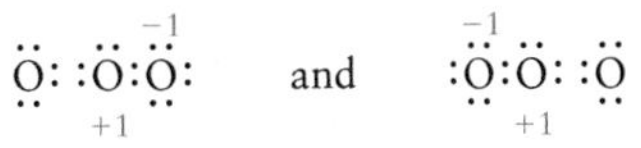

These diagrams suggest that one O—O bond is a single bond and the other one is a double bond, so the molecule would be asymmetric. In fact, the two O—O bond lengths are found experimentally to be identical, with a bond length of 1.28 Å, which is intermediate between the O—O single bond length in H_2O_2 (1.49 Å) and the O=O double bond length in O_2 (1.21 Å).

The Lewis model fails to predict the correct structural formula for O_3. The Lewis model handles such cases by saying that the actual bonding in O_3 is represented as a **resonance hybrid** of the two Lewis diagrams in each of which one of the bonds is intermediate between a single bond and a double bond. This hybrid is represented by connecting the diagrams with a double-headed arrow:

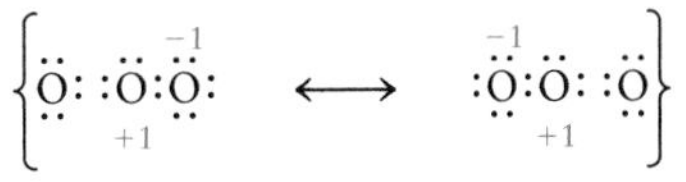

The term *resonance* does not mean that the molecule physically oscillates back and forth from one of these bonding structures to the other. Rather, within the limitations of the Lewis dot model of bonding, the best representation of the actual bonding is a hybrid diagram that includes features of each of the acceptable individual diagrams. This concept of resonance arises because the Lewis model very strictly defines a chemical bond as a pair of electrons localized between two nuclei. In cases like O_3 where the electrons are delocalized over the entire molecule, we seek another bonding model to overcome this limitation in the Lewis model. The need for resonance is avoided by using molecular orbital theory to describe chemical bonding (see Chapter 6).

EXAMPLE 3.10

Draw three resonance forms for the nitrate ion NO_3^-, (Fig. 3.31), and estimate the bond lengths.

Solution

NO_3^- contains 24 valence electrons. For each atom to have its own octet, $4 \times 8 = 32$ electrons would be required. Therefore, $32 - 24 = 8$ electrons must be shared between atoms, implying a total of four bonding pairs. These can be distributed in one double and two single bonds, leading to the equivalent resonance diagrams:

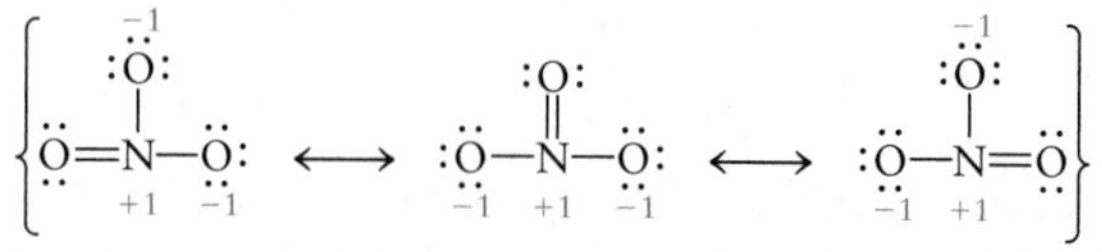

The two singly bonded oxygen atoms carry formal charges of −1, and the charge of the nitrogen atom is +1. The bond lengths should all be equal and should lie between the values given in Table 3.6 for N—O (1.43 Å) and N=O (1.18 Å). The experimentally measured value is 1.24 Å.

Related Problems: 51, 52, 53, 54, 55, 56

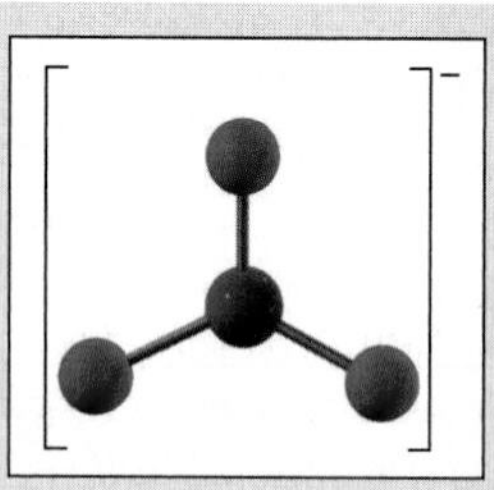

FIGURE 3.31 The nitrate ion, NO_3^- has a symmetric planar structure.

3.10.4 Breakdown of the Octet Rule

Lewis diagrams and the octet rule are useful tools for predicting the types of molecules that will be stable under ordinary conditions of temperature and pressure. For example, we can write a simple Lewis diagram for water (H_2O),

H:Ö:H

in which each atom has a noble-gas configuration. It is impossible to do this for OH or for H_3O, which suggests that these species are either unstable or highly reactive.

There are several situations in which the octet rule is *not* satisfied.

Case 1: Odd-Electron Molecules

The electrons in a Lewis diagram that satisfies the octet rule must occur in pairs—bonding pairs or lone pairs. Any molecule that has an odd number of electrons cannot satisfy the octet rule. Most stable molecules have even numbers of electrons, but a few have odd numbers. An example is nitrogen oxide (NO), a stable (although reactive) molecule that is an important factor in air pollution. Nitrogen oxide has 11 electrons, and the best electron dot diagram for it is

:Ṅ: :Ö

in which only the oxygen atom has a noble-gas configuration. The stability of NO contradicts the octet rule.

FIGURE 3.32 Boron trifluoride, BF_3, is a highly reactive gas that condenses to a liquid at −100°C. Its major use is in speeding up a large class of reactions that involve carbon compounds.

Case 2: Octet-Deficient Molecules

Some molecules are stable even though they have too few electrons to achieve an octet. For example, the standard rules for BF_3 (Figure 3.32) would lead to the Lewis diagram

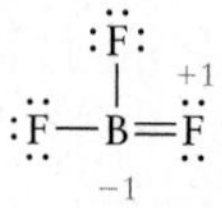

but experimental evidence strongly suggests there are no double bonds in BF_3 (fluorine never forms double bonds).

Moreover, the placement of a positive formal charge on fluorine is highly non-favored. The following diagram avoids both problems:

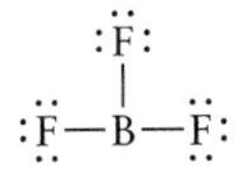

Although this Lewis diagram denies an octet to the boron atom, it does at least assign zero formal charges to all atoms.

Case 3: Valence Shell Expansion

FIGURE 3.33 Sulfur hexafluoride, SF_6, is an extremely stable, dense, and nonreactive gas. It is used as an insulator in high-voltage generators and switches.

Lewis diagrams become more complex in compounds formed from elements in the third and subsequent periods. For example, sulfur forms some compounds that are readily described by Lewis diagrams that give closed shells to all atoms. An example is hydrogen sulfide (H_2S), which is analogous to water in its Lewis diagram. In sulfur hexafluoride (SF_6) (Figure 3.33), however, the central sulfur is bonded to six fluorine atoms.

This molecule cannot be described by a Lewis diagram unless more than eight electrons are allowed around the sulfur atom, a process called **valence shell expansion**. The resulting Lewis diagram is

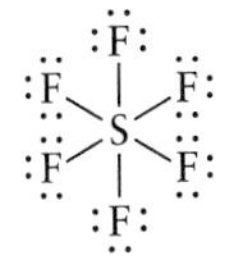

The fluorine atoms have octets, but the central sulfur atom shares a total of 12 electrons.

In the standard procedure for writing Lewis diagrams, the need for valence shell expansion is signaled when the number of shared electrons is not sufficient to place a bonding pair between each pair of atoms that are supposed to be bonded. In SF_6, for example, 48 electrons are available, but 56 are needed to form separate octets on 7 atoms. This means that $56 - 48 = 8$ electrons would be shared. Four electron pairs are not sufficient to make even single bonds between the central S atom and the 6 terminal F atoms. In this case, we still follow rule 4 (assign one bonding pair to each bond in the molecule or ion; see earlier), even though it requires that we use more than 8 electrons. Rule 5 becomes irrelevant because there are no extra shared electrons. Rule 6 is now replaced with a new rule:

Rule 6′: Assign lone pairs to the terminal atoms to give them octets. If any electrons still remain, assign them to the central atoms as lone pairs if the central atom is not one from the 2nd period.

The effect of rule 6′ is to abandon the octet rule for the central atom but preserve it for the terminal atoms.

EXAMPLE 3.11

Write a Lewis diagram for I_3^-, the linear (tri-iodide) ion.

Solution

There are 7 valence electrons from each iodine atom plus 1 from the overall ion charge, giving a total of 22. Because $3 \times 8 = 24$ electrons would be needed in separate octets, $24 - 22 = 2$ are shared according to the original rules. Two electrons are not sufficient to make two different bonds, however, so valence expansion is necessary.

(Continued)

A pair of electrons is placed in each of the two bonds, and rule 6′ is used to complete the octets of the two terminal I atoms. This leaves

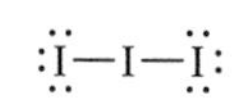

At this stage, 16 valence electrons have been used. The remaining 6 are placed as lone pairs on the central I atom. A formal charge of −1 then resides on this atom:

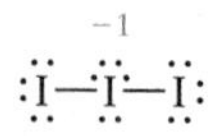

Note the valence expansion on the central atom: It shares or owns a total of 10 electrons, rather than the 8 required by adherence to the octet rule.

Related Problems: 57, 58

The Lewis model represents covalent bonds as shared valence-electron pairs positioned between two nuclei, where they presumably are involved in net attractive interactions that pull the nuclei together and contribute to the strengthening of the bond. The bonding mechanism cannot be explained by classical physics, and is examined through quantum mechanics in Chapter 6.

Despite its uncertainties and limitations, the Lewis model is still the best starting point for a systematic exploration of the bonding and structure in a molecule. Most working chemists draw a Lewis diagram to propose a structural formula as the first step of this exploration described in Section 3.1.

3.11 The Shapes of Molecules: Valence Shell Electron-Pair Repulsion Theory

When two molecules approach one another to begin a chemical reaction, the probability of a successful encounter can depend critically on the three-dimensional shapes and the relative orientation of the molecules, as well as on their chemical identities. Shape is especially important in biological and biochemical reactions, in which molecules must fit precisely onto specific sites on membranes and templates; drug and enzyme activity are important examples. Characterization of molecular shape is therefore an essential part of the study of molecular structure.

The structure of a stable molecule is defined by the three-dimensional arrangement of its constituent atoms. Pictorial representations of molecular structure—for example, the familiar "ball-and-stick" models and sketches—show the positions of the nuclei of the constituent atoms, but not the positions of their electrons. The electrons are responsible for the chemical bonds that hold the atomic nuclei together as the molecule. Several properties characterize the three-dimensional structure of molecules (Fig. 3.34). The bond length measures the distance between the atomic nuclei in a particular bond; summing bond lengths projected along the three Cartesian axes provides a measure of the size and shape of the molecule. Bond angles, defined as the angle between the axes of adjacent bonds, provide a more detailed view of the three-dimensional structures of molecules. Finally, the relationships between two planes, each defined by three atoms with the planes having two atoms in common (the angle between these planes is the *dihedral angle*) provide additional insights into the topology of simple molecules. However, molecules are not rigid entities, with structures that are precisely defined by the coordinates of their nuclei. Their atoms vibrate about their equilibrium positions, albeit with relatively small displacements. Average bond lengths and angles are measured by spectroscopic techniques (see Chapter 20) and X-ray diffraction (see Section 21.1).

Molecular shape or geometry is governed by energetics; a molecule assumes the geometry that gives it the lowest potential energy. Sophisticated quantum

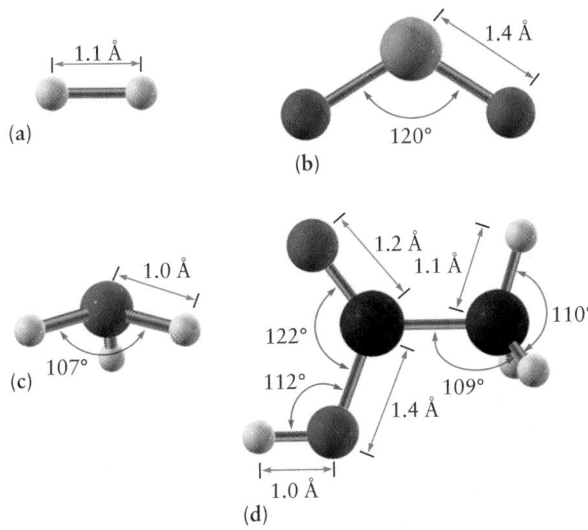

FIGURE 3.34 Three-dimensional molecular structures of (a) H_2, (b) SO_2, (c) NH_3, and (d) $C_2H_4O_2$, showing bond lengths and angles. (Courtesy of Prof. Andrew J. Pounds, Mercer University, Macon, GA, and Dr. Mark A. Iken, Scientific Visualization Laboratory, Georgia Institute of Technology, Atlanta, GA.)

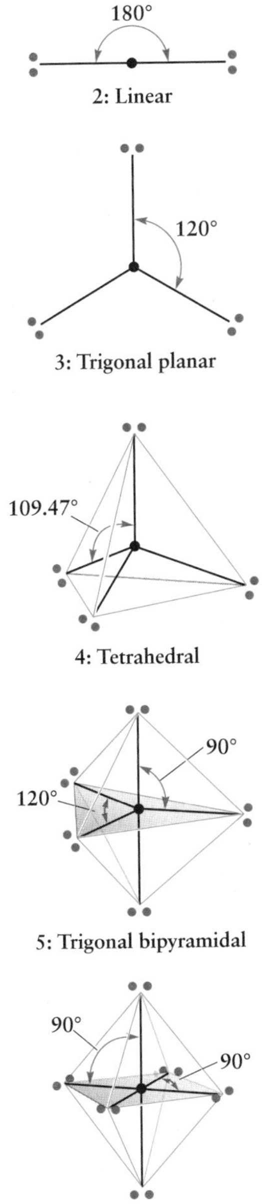

FIGURE 3.35 The positions of minimum energy for electron pairs on a sphere centered on the nucleus of an atom. The angles between the electron pairs are indicated. For two, three, and four electron pairs they are 180°, 120°, and 109.47°, respectively.

mechanical calculations consider numerous possible geometrical arrangements for a molecule, calculate the total potential energy of the molecule for each arrangement, and identify the arrangement that gives the lowest potential energy. This procedure can be mimicked within the approximate classical model described in this chapter by considering numerous possible arrangements of bond angles and then identifying the one that corresponds to the lowest potential energy of the molecule. Because a covalent bond is formed by the sharing of a pair of electrons between two atoms (as described by the Lewis model in Section 3.10), changes in bond angles change the relative positions of the electron pairs around a given central atom. Electron pairs tend to repel each other through the electrostatic (Coulomb) repulsion between like charges and through quantum mechanical effects. Consequently, it is desirable in terms of energy for electrons to avoid each other. The VSEPR theory provides procedures for predicting molecular geometry by minimizing the potential energy due to electron-pair repulsions.

3.11.1 The Valence Shell Electron-Pair Repulsion Theory

The VSEPR theory starts with the fundamental idea that electron pairs in the valence shell of an atom repel each other. These include both lone pairs, which are localized on the atom and are not involved in bonding, and bonding pairs, which are shared covalently with other atoms. The electron pairs position themselves as far apart as possible to minimize their repulsions. The molecular geometry, which is defined by the positions of the *nuclei,* is then traced from the relative locations of the electron pairs.

The arrangement that minimizes repulsions naturally depends on the number of electron pairs. Figure 3.35 shows the minimum energy configuration for two to six electron pairs located around a central atom. Two electron pairs place themselves on opposite sides of the atom in a linear arrangement, three pairs form a trigonal planar structure, four arrange themselves at the corners of a tetrahedron, five define a trigonal bipyramid, and six define an octahedron. To find which geometry applies, we determine the **steric number,** SN, of the central atom, which is defined as

$$\text{SN} = \begin{pmatrix}\text{number of atoms}\\ \text{bonded to central atom}\end{pmatrix} + \begin{pmatrix}\text{number of lone pairs}\\ \text{on central atom}\end{pmatrix}$$

The steric number of an atom in a molecule can be determined by inspection from the Lewis diagram of the molecule.

EXAMPLE 3.12

Calculate steric numbers for iodine in IF_4^- and for bromine in BrO_4^-. These molecular ions have central I or Br surrounded by the other four atoms.

Solution

The central I^- atom has eight valence electrons.

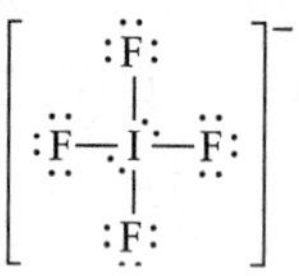

Each F atom has seven valence electrons of its own and needs to share one of the electrons from I^- to achieve a noble-gas configuration. Thus, four of the I^- valence electrons take part in covalent bonds, leaving the remaining four to form two lone pairs. The steric number is given by

$$SN = 4 \text{ (bonded atoms)} + 2 \text{ (lone pairs)} = 6$$

In BrO_4^-, each oxygen atom needs to share two of the electrons from Br^- to achieve a noble-gas configuration.

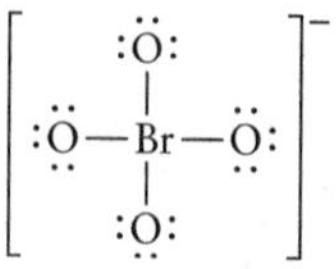

Because this assignment accounts for all eight of the Br^- valence electrons, there are no lone pairs on the central atom and

$$SN = 4 \text{ (bonded atoms)} + 0 \text{ (lone pairs)} = 4$$

Double-bonded or triple-bonded atoms count the same as single-bonded atoms in determining the steric number. In CO_2, for example, two double-bonded oxygen atoms are attached to the central carbon and there are no lone pairs on that atom, so SN = 2.

The steric number is used to predict molecular geometries. In molecules XY_n, in which there are no lone pairs on the central atom X (the simplest case),

$$SN = \text{number of bonded atoms} = n$$

The n bonding electron pairs (and therefore the outer atoms) position themselves (see Fig. 3.35) to minimize electron-pair repulsion. Thus, CO_2 is predicted (and found experimentally) to be linear, BF_3 is trigonal planar, CH_4 is tetrahedral, PF_5 is trigonal bipyramidal, and SF_6 is octahedral.

When lone pairs are present, the situation changes slightly. There can now be three different types of repulsions: (1) bonding pair against bonding pair, (2) bonding pair against lone pair, and (3) lone pair against lone pair. Consider the ammonia molecule (NH_3), which has three bonding electron pairs and one lone pair (Fig. 3.36a). The steric number is 4, and the electron pairs arrange themselves into an approximately tetrahedral structure. The lone pair is not identical to the three bonding pairs, however, so there is no reason for the electron-pair structure to be *exactly* tetrahedral. It is found that lone pairs tend to occupy more space than bonding pairs (because the bonding pairs are held closer to the central atom), so the angles of bonds opposite to them are reduced. The geometry of the *molecule*, as distinct from that of the electron pairs, is named for the sites occupied by actual atoms. The description of the molecular geometry makes no reference to lone pairs that may be present on the central atom, even

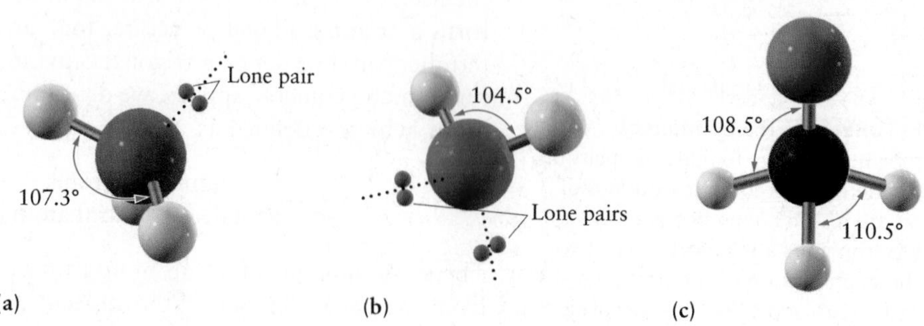

FIGURE 3.36 (a) Ammonia (NH_3) has a pyramidal structure in which the bond angles are less than 109.5°. (b) Water (H_2O) has a bent structure with a bond angle less than 109.5° and smaller than that of NH_3. (c) CH_3Cl has a distorted tetrahedral structure.

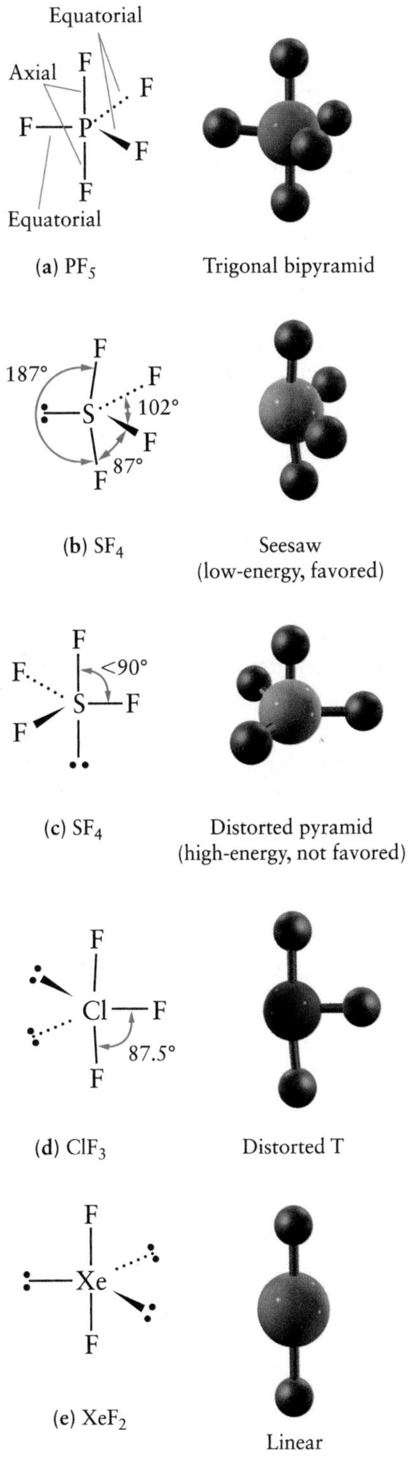

FIGURE 3.37 Molecules with steric number 5. The molecular geometry is named for the sites occupied by the *atoms,* not the underlying trigonal-bipyramidal structure of electron pairs.

though their presence affects that geometry. The structure of the ammonia molecule is thus predicted to be a trigonal pyramid in which the H—N—H bond angle is smaller than the tetrahedral angle of 109.5°. The observed structure has an H—N—H bond angle of 107.3°. The H—O—H bond angle in water, which has two lone pairs and two bonding pairs, is still smaller at 104.5° (see Fig. 3.36b).

A similar distortion takes place when two types of outer atoms are present. In CH_3Cl, the bonding electron pair in the C—Cl bond is not the same as those in the C—H bonds, so the structure is a distorted tetrahedron (see Fig. 3.36c). Because Cl is more electronegative than H, it tends to attract electrons away from the central atom, reducing the electron-pair repulsion. This allows the Cl—C—H bond angles to become 108.5°, which is smaller than tetrahedral, whereas the H—C—H angles become 110.5°, which is larger than tetrahedral. In effect, electropositive substituents repel other substituents more strongly than do electronegative substituents.

The fluorides PF_5, SF_4, ClF_3, and XeF_2 all have steric number 5 but have different numbers of lone pairs (0, 1, 2, and 3, respectively). What shapes do their molecules have? We have already mentioned that PF_5 is trigonal bipyramidal. Two of the fluorine atoms occupy **axial sites** (see Fig. 3.37a), and the other three occupy **equatorial sites.** Because the two kinds of sites are not equivalent, there is no reason for all of the P—F bond lengths to be equal. Experiment shows that the equatorial P—F bond length is 1.534 Å, which is shorter than the 1.577 Å axial P—F lengths.

SF_4 has four bonded atoms and one lone pair. Does the lone pair occupy an axial or an equatorial site? In the VSEPR theory, when electron pairs form a 90° angle to the central atom, they repel each other much more strongly than when the angle is larger. A single lone pair therefore finds a position that minimizes the number of 90° repulsions it has with bonding electron pairs. It occupies an equatorial position with two 90° repulsions (see Fig. 3.37b), rather than an axial position with three 90° repulsions (see Fig. 3.37c). The axial S—F bonds are bent slightly away from the lone pair; consequently, the molecular structure of SF_4 is a distorted seesaw. A second lone pair (in ClF_3, for example) also takes an equatorial position, leading to a distorted T-shaped molecular structure (see Fig. 3.37d). A third lone pair (in XeF_2 or I_3^-, for example) occupies the third equatorial position, and the molecular geometry is linear (see Fig. 3.37e). Table 3.8 summarizes the molecular shapes predicted by VSEPR and provides examples of each.

EXAMPLE 3.13

Predict the geometry of the following molecules and ions: (a) ClO_3^+, (b) ClO_2^+, (c) SiH_4, (d) IF_5.

Solution

(a) The central Cl atom has all of its valence electrons (six, because the ion has a net positive charge) involved in bonds to the surrounding three oxygen atoms and has no lone pairs. Its steric number is 3. In the molecular ion, the central Cl should be surrounded by the three O atoms arranged in a trigonal planar structure.

(b) The central Cl atom in this ion also has a steric number of 3, comprising two bonded atoms and a single lone pair. The predicted molecular geometry is a bent molecule with an angle somewhat less than 120°.

(c) The central Si atom has a steric number of 4 and no lone pairs. The molecular geometry should consist of the Si atom surrounded by a regular tetrahedron of H atoms.

(Continued)

FIGURE 3.38 The structure of IF_5. Note the distortions of F—I—F bond angles from 90° because of the lone pair at the bottom (not shown).

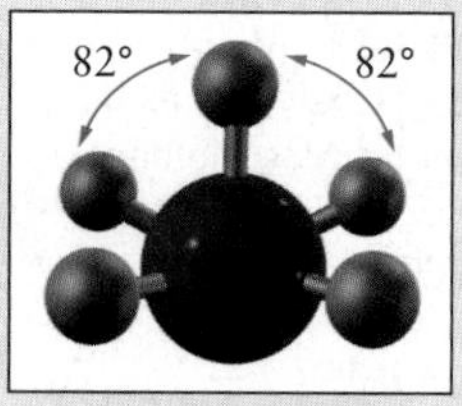

(d) Iodine has seven valence electrons, of which five are shared in bonding pairs with F atoms. This leaves two electrons to form a lone pair, so the steric number is 5 (bonded atoms) + 1 (lone pair) = 6. The structure will be based on the octahedron of electron pairs from Figure 3.35, with five F atoms and one lone pair. The lone pair can be placed on any one of the six equivalent sites and will cause the four F atoms to bend away from it toward the fifth F atom, giving the distorted structure shown in Figure 3.38.

Related Problems: 59, 60, 61, 62, 63, 64

The VSEPR theory is a simple but remarkably powerful model for predicting the geometries and approximate bond angles of molecules that have a central atom. Its success is even more remarkable when we realize that the VSEPR theory is based purely on empirical arguments, not theoretical calculations.

TABLE 3.8

Molecular Shapes Predicted by the Valence Shell Electron-Pair Repulsion Theory

Molecule	Steric Number	Predicted Geometry		Example
AX_2	2	Linear		CO_2
AX_3	3	Trigonal planar	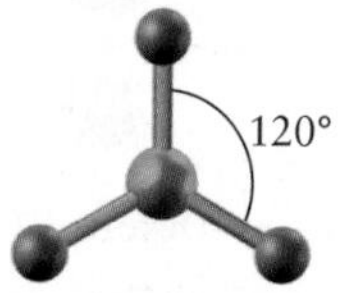	BF_3
AX_4	4	Tetrahedral	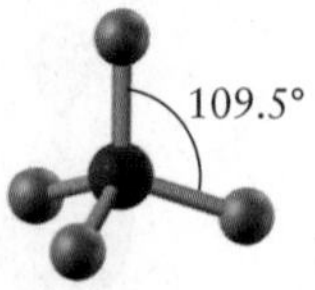	CF_4
AX_5	5	Trigonal bipyramidal	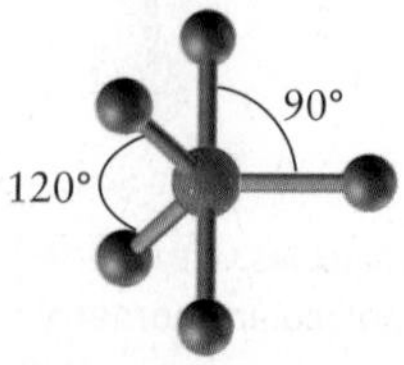	PF_5
AX_6	6	Octahedral	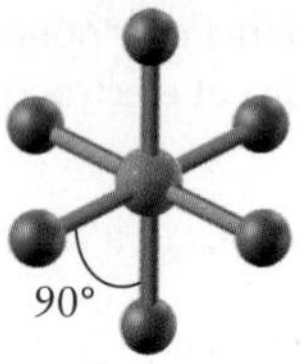	SF_6

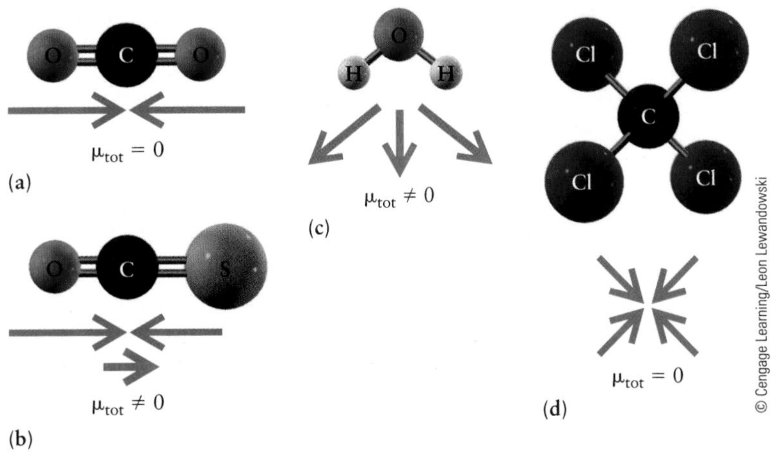

FIGURE 3.39 The total dipole moment of a molecule is obtained by vector addition of its bond dipoles. The net dipole moment vector is shown in red. (a) CO_2. (b) OCS. (c) H_2O. (d) CCl_4.

However, it has its limitations. The VSEPR theory does not account for the fact that observed bond angles in the Group V and VI hydrides H_2S (92°), H_2Se (91°), PH_3 (93°), and AsH_3 (92°) are so far from tetrahedral (109.5°) and so close to right angles (90°).

3.11.2 Dipole Moments of Polyatomic Molecules

Polyatomic molecules, like the diatomic molecules considered in the previous section, may have dipole moments. Often, a good approximation is to assign a dipole moment (a vector, shown as an arrow pointing from the negative charge to the positive charge) to each bond, and then obtain the total dipole moment by carrying out a vector sum of the bond dipoles. In CO_2, for example, each C—O has a bond dipole (Fig. 3.39a). However, because the dipoles are equal in magnitude and point in opposite directions, the total molecular dipole moment vanishes. In OCS, which is also linear with a central carbon atom, the C—O and C—S bond dipole moments have different magnitudes, leaving a net non-zero dipole moment (see Fig. 3.39b). In the water molecule (Fig. 3.39c), the two bond dipoles add vectorially to give a net molecular dipole moment. The more symmetric molecule CCl_4, on the other hand, has no net dipole moment (see Fig. 3.39d). Even though each of the four C—Cl bonds (pointing from the four corners of a tetrahedron) has a dipole moment, their vector sum is zero. Molecules such as H_2O and OCS, with nonzero dipole moments, are **polar**; those such as CO_2 and CCl_4, with no net dipole moment, are **nonpolar**, even though they contain polar bonds. Intermolecular forces differ between polar and nonpolar molecules (see discussion in Chapter 10), a fact that significantly affects their physical and chemical properties.

EXAMPLE 3.14

Predict whether the molecules NH_3 and SF_6 will have dipole moments.

Solution

The NH_3 molecule has a dipole moment because the three N—H bond dipoles add to give a net dipole pointing downward from the N atom to the base of the pyramid that defines the NH_3 structure. The SF_6 molecule has no dipole moment because each S—F bond dipole is balanced by one of equal magnitude pointing in the opposite direction on the other side of the molecule.

Related Problems: 65, 66

3.12 Oxidation Numbers

The terms *ionic* and *covalent* have been used in this chapter as though they represent the complete transfer of one or more electrons from one atom to another or the equal sharing of pairs of electrons between atoms in a molecule. In reality, both types of bonding are idealizations that rarely apply exactly. To account for the transfer of electrons from one molecule or ion to another in oxidation–reduction reactions and to name different binary compounds of the same elements, it is not necessary to have detailed knowledge of the exact electron distributions in molecules, whether they are chiefly ionic or covalent. Instead, we can assign a convenient fictitious charge to each atom in a molecule and call each the **oxidation number** of the corresponding atom, making certain that the law of charge conservation is strictly observed. Oxidation numbers are chosen so that in ionic compounds the sum of the oxidation numbers on all the atoms equals the charge on the ion. The following simple conventions are useful:

1. The oxidation numbers of the atoms in a neutral molecule must add up to zero, and those in an ion must add up to the charge on the ion.
2. Alkali-metal atoms are assigned the oxidation number $+1$, and the alkaline-earth atoms $+2$, in their compounds.
3. Fluorine is always assigned oxidation number -1 in its compounds. The other halogens are generally assigned oxidation number -1 in their compounds, except those containing oxygen and other halogens in which the halogen can have a positive oxidation number.
4. Hydrogen is assigned oxidation number $+1$ in its compounds, except in metal hydrides such as LiH, where convention 2 takes precedence and the oxidation number of hydrogen is -1.
5. Oxygen is assigned oxidation number -2 in nearly all compounds. However, there are two exceptions: In compounds with fluorine, convention 3 takes precedence, and in compounds that contain O—O bonds, conventions 2 and 4 take precedence. Thus, the oxidation number for oxygen in OF_2 is $+2$; in peroxides (such as H_2O_2 and Na_2O_2), it is -1. In superoxides (such as KO_2), the oxidation number of oxygen is $-\frac{1}{2}$.

Convention 1 is fundamental because it guarantees charge conservation: The total number of electrons must remain constant in chemical reactions. This rule also makes the oxidation numbers of the neutral atoms of all elements zero. Conventions 2 to 5 are based on the principle that in ionic compounds the oxidation number should equal the charge on the ion. Note that fractional oxidation numbers, although uncommon, are allowed and, in fact, are necessary to be consistent with this set of conventions.

With the preceding conventions in hand, chemists can assign oxidation numbers to the atoms in most compounds. Apply conventions 2 through 5 as listed previously, noting the exceptions given, and then assign oxidation numbers to the other elements in such a way that convention 1 is always obeyed. Note that convention 1 applies not only to free ions, but also to the components that make up ionic solids. Chlorine has oxidation number -1 not only as a free Cl^- ion, but in the ionic solid AgCl and in covalent CH_3Cl. It is important to recognize common ionic species (especially molecular ions) and to know the total charges they carry. Table 3.9 lists the names and formulas of many common anions. Inspection of the table reveals that several elements exhibit different oxidation numbers in different compounds. In Ag_2S, sulfur appears as the sulfide ion and

TABLE 3.9

Formulas and Names of Some Common Anions

F^-	Fluoride	CO_3^{2-}	Carbonate
Cl^-	Chloride	HCO_3^-	Hydrogen carbonate
Br^-	Bromide	NO_2^-	Nitrite
I^-	Iodide	NO_3^-	Nitrate
H^-	Hydride	SiO_4^{4-}	Silicate
O^{2-}	Oxide	PO_4^{3-}	Phosphate
S^{2-}	Sulfide	HPO_4^{2-}	Hydrogen phosphate
O_2^{2-}	Peroxide	$H_2PO_4^-$	Dihydrogen phosphate
O_2^-	Superoxide	SO_3^{2-}	Sulfite
OH^-	Hydroxide	SO_4^{2-}	Sulfate
CN^-	Cyanide	HSO_4^-	Hydrogen sulfate
CNO^-	Cyanate	ClO^-	Hypochlorite
SCN^-	Thiocyanate	ClO_2^-	Chlorite
MnO_4^-	Permanganate	ClO_3^-	Chlorate
CrO_4^{2-}	Chromate	ClO_4^-	Perchlorate
$Cr_2O_7^{2-}$	Dichromate		

has oxidation number −2, but in Ag_2SO_4, it appears as part of a sulfate (SO_4^{2-}) ion. In this case,

$$(\text{oxidation number of S}) + [4 \times (\text{oxidation number of O})] = \text{total charge on ion}$$

$$x + [4(-2)] = -2$$

$$\text{oxidation number of S} = x = +6$$

A convenient way to indicate the oxidation number of an atom is to write it directly above the corresponding symbol in the formula of the compound:

$$\overset{+1\,-2}{N_2O} \qquad \overset{+1\,-1}{LiH} \qquad \overset{0}{O_2} \qquad \overset{+6\,-2}{SO_4^{2-}}$$

EXAMPLE 3.15

Assign oxidation numbers to the atoms in the following chemical compounds and ions: NaCl, ClO^-, $Fe_2(SO_4)_3$, SO_2, I_2, $KMnO_4$, CaH_2.

Solution

$\overset{+1\,-1}{NaCl}$	From conventions 2 and 3.
$\overset{+1\,-2}{ClO^-}$	From conventions 1 and 5.
$\overset{+3\,\,+6\,-2}{Fe_2(SO_4)_3}$	From conventions 1 and 5. This is solved by recognizing the presence of sulfate SO_4^{2-} groups.
$\overset{+4\,-2}{SO_2}$	From conventions 1 and 5.
$\overset{0}{I_2}$	From convention 1 (I_2 is an element).
$\overset{+1\,+7\,-2}{KMnO_4}$	From conventions 1, 2, and 5.
$\overset{+2\,-1}{CaH_2}$	From conventions 1 and 2 (metal hydride case).

Related Problems: 71, 72

Oxidation numbers must not be confused with the formal charges on Lewis dot diagrams (see Section 3.10). They resemble formal charges to the extent that both are assigned, by arbitrary conventions, to symbols in formulas for specific purposes. The purposes differ, however. Formal charges are used solely to identify preferred Lewis diagrams. Oxidation numbers are used in nomenclature, in identifying oxidation–reduction reactions, and in exploring trends in chemical reactivity across the periodic table. Oxidation numbers are often not the same

FIGURE 3.40 Several oxides of manganese. They are arranged in order of increasing oxidation number of the Mn, counterclockwise from the bottom left: MnO, Mn_3O_4, Mn_2O_3, and MnO_2. A compound of still higher oxidation state, Mn_2O_7, is a dark red liquid that explodes easily.

as the true charges on atoms. In $KMnO_4$ or Mn_2O_7, for example, Mn has the oxidation number +7, but the actual net charge on the atom is much less than +7. A high oxidation state usually indicates significant covalent character in the bonding of that compound. The oxides of manganese with lower oxidation states (Fig. 3.40) have more ionic character.

Figure 3.41 shows the most common oxidation states of the elements of the main groups. The strong diagonal lines for these elements reflect the stability of closed electron octets. For example, the oxidation states +6 and −2 for sulfur would correspond to losing or gaining enough electrons to attain a noble gas configuration, in this case that of Ar. The octet-based Lewis model is not useful for compounds of the transition elements (see Chapter 8 for further discussion).

3.13 Inorganic Nomenclature

Sections 3.8 and 3.9 used the periodic table and the Lewis model to describe the transfer and sharing of electrons in chemical compounds. We conclude this chapter by discussing the systematic nomenclature of inorganic compounds. The definitive source for the naming of inorganic compounds is *Nomenclature of Inorganic Chemistry—IUPAC Recommendations 2005* (N. G. Connelly and T. Damhus, Sr., eds. Royal Society of Chemistry, 2005). Of the wide variety of inorganic compounds, we discuss only simple covalent and ionic compounds and emphasize current usage.

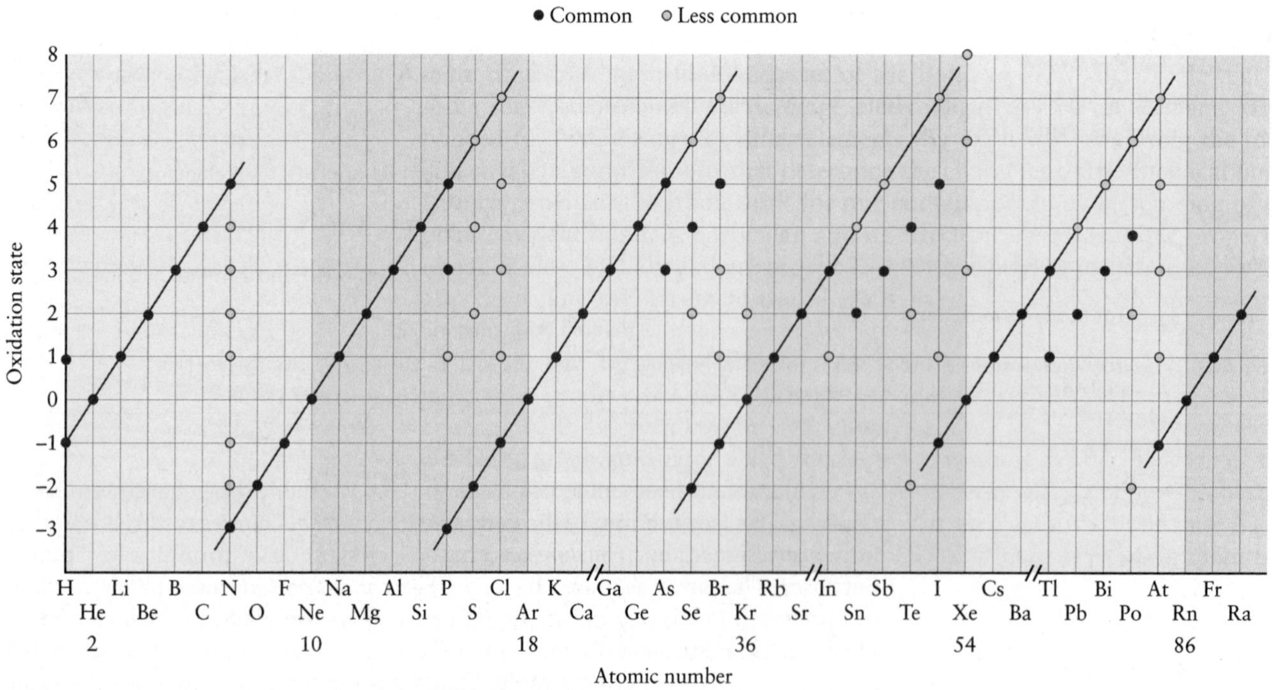

FIGURE 3.41 Important oxidation states of the main-group elements.

3.13.1 Names and Formulas of Ionic Compounds

Ionic compounds are made from ions, bound by electrostatic forces such that the total charge of the compound is zero. The stoichiometry of an ionic compound is determined by inspection, to satisfy charge neutrality. As mentioned earlier, gas-phase ionic compounds exist, but most ionic compounds are solids in which each ion is strongly bound to a number of the nearest neighbor ions of the opposite sign. An ionic compound is named by listing the name of the cation followed by that of the anion. Ions can either be monatomic or polyatomic; the latter are also referred to as molecular ions.

A monatomic cation bears the name of the parent element. We have already encountered the sodium ion (Na^+) and the calcium ion (Ca^{2+}); ions of the other elements in Groups I and II are named in the same way. The transition metals and the metallic elements of Groups III, IV, and V differ from the Group I and II metals in that they often form several stable ions in compounds and in solution. Although calcium compounds never contain Ca^{3+} (always Ca^{2+}), the element iron forms both Fe^{2+} and Fe^{3+}, and thallium forms both Tl^+ and Tl^{3+}. When a metal forms ions of more than one charge, we distinguish them by placing a Roman numeral in parentheses after the name of the metal, indicating the charge on the ion:

Cu^+	copper(I) ion	Fe^{2+}	iron(II) ion	Sn^{2+}	tin(II) ion
Cu^{2+}	copper(II) ion	Fe^{3+}	iron(III) ion	Sn^{4+}	tin(IV) ion

An earlier method for distinguishing between such pairs of ions used the suffixes *-ous* and *-ic* added to the root of the (usually Latin) name of the metal to indicate the ions of lower and higher charge, respectively. Thus, Fe^{2+} was called the ferrous ion and Fe^{3+} the ferric ion. This method, although still used occasionally, is not recommended for systematic nomenclature and is not used in this book.

A few polyatomic cations are important in inorganic chemistry. These include the ammonium ion, NH_4^+ (obtained by adding H^+ to ammonia); the hydronium ion, H_3O^+ (obtained by adding H^+ to water); and the particularly interesting molecular ion formed by mercury, Hg_2^{2+} the mercury(I) ion. This species must be carefully distinguished from Hg^{2+}, the mercury(II) ion. The Roman numeral I in parentheses means, in this case, that the average charge on each of the two mercury atoms is +1. Compounds with the empirical formulas HgCl and HgBr have molecular formulas Hg_2Cl_2 and Hg_2Br_2, respectively.

A monatomic anion is named by adding the suffix *-ide* to the first portion of the name of the element. Thus, chlor*ine* becomes the chlor*ide* ion, and ox*ygen* becomes the ox*ide* ion. The other monatomic anions of Groups V, VI, and VII are named similarly. Many polyatomic anions exist, and the naming of these species is more complex. The names of the oxoanions (each contains oxygen in combination with a second element) are derived by adding the suffix *-ate* to the stem of the name of the second element. Some elements form two oxoanions. The *-ate* ending is then used for the oxoanion with the larger number of oxygen atoms (such as NO_3^-, nitr*ate*), and the ending *-ite* is added for the name of the anion with the smaller number (such as NO_2^- nitr*ite*). For elements such as chlorine, which form more than two oxoanions, we use the additional prefixes *per-* (largest number of oxygen atoms) and *hypo-* (smallest number of oxygen atoms). An oxoanion that contains hydrogen as a third element includes that word in its name. For example, the HCO_3^- oxoanion is called the hydrogen carbonate ion in preference to its common (nonsystematic) name, "bicarbonate ion," and HSO_4^-, often called "bisulfate ion," is better designated as the hydrogen sulfate ion. Table 3.9 lists some of the most important anions. It is important to be able to recognize and name the ions from that table, bearing in mind that the electric charge is an essential part of the formula.

It is customary (and recommended) to name ionic compounds (salts) using the Roman numeral notation, based on group numbers or oxidation states. $CuSO_4$ is copper(I) sulfate, $Fe(NO_3)_2$ is iron(II) nitrate and $KMnO_4$ is potassium permanganate, the Roman numeral being omitted in the last example because potassium cations always carry only +1 charge.

The composition of an ionic compound is determined by overall charge neutrality. The total positive charge on the cations must exactly balance the total negative charge on the anions. The following names and formulas of ionic compounds illustrate this point:

Tin(II) bromide	One 2+ cation, two 1− anions	$SnBr_2$
Potassium permanganate	One 1+ cation, one 1− anion	$KMnO_4$
Ammonium sulfate	Two 1+ cations, one 2− anion	$(NH_4)_2SO_4$
Iron(II) dihydrogen phosphate	One 2+ cation, two 1− anions	$Fe(H_2PO_4)_2$

EXAMPLE 3.16

Give the chemical formulas of (a) calcium cyanide and (b) copper(II) phosphate.

Solution

(a) Calcium cyanide is composed of Ca^{2+} and CN^-. For the overall charge to be 0, there must be two CN^- for each Ca^{2+}. Thus, the chemical formula of calcium cyanide is $Ca(CN)_2$.

(b) The ions present in copper(II) phosphate are Cu^{2+} and PO_4^{3-}. To ensure charge neutrality, there must be three Cu^{2+} (total charge +6) and two PO_4^{3-} (total charge −6) per formula unit. Thus, the chemical formula of copper(II) phosphate is $Cu_3(PO_4)_2$.

Related Problems: 73, 74, 75, 76, 77, 78

3.13.2 Naming Binary Covalent Compounds

How do we name covalent (nonionic) compounds? If a pair of elements forms only one compound, begin with the name of the element that appears first in the chemical formula, followed by the second element, with the suffix *-ide* added to its root. This is analogous to the naming of ionic compounds. Just as NaBr is sodium bromide, so the following names designate typical covalent compounds:

HBr	hydrogen bromide
$BeCl_2$	beryllium chloride
H_2S	hydrogen sulfide
BN	boron nitride

Many well-established nonsystematic names continue to be used (even the strictest chemist does not call water "hydrogen oxide"!) They include:

H_2O	water
NH_3	ammonia
N_2H_4	hydrazine
PH_3	phosphine
AsH_3	arsine
$COCl_2$	phosgene

TABLE 3.10

Prefixes Used for Naming Binary Covalent Compounds

Number	Prefix
1	*mono-*
2	*di-*
3	*tri-*
4	*tetra-*
5	*penta-*
6	*hexa-*
7	*hepta-*
8	*octa-*
9	*nona-*
10	*deca-*

If a pair of elements forms more than one compound, two methods can be used to distinguish between them:

1. Use Greek prefixes (Table 3.10) to specify the number of atoms of each element in the molecular formula of the compound (*di-* for two, *tri-* for three, and so forth). If the compound is a solid without well-defined molecules, name the empirical formula in this way. The prefix for one (*mono-*) is omitted, except in the case of carbon monoxide.
2. Write the oxidation number of the first-named element in Roman numerals and place it in parentheses after the name of that element.

Applying the two methods to the oxides of nitrogen gives

N_2O	dinitrogen oxide	nitrogen(I) oxide
NO	nitrogen oxide	nitrogen(II) oxide
N_2O_3	dinitrogen trioxide	nitrogen(III) oxide
NO_2	nitrogen dioxide	nitrogen(IV) oxide
N_2O_4	dinitrogen tetraoxide	nitrogen(IV) oxide
N_2O_5	dinitrogen pentaoxide	nitrogen(V) oxide

The first method has some advantages over the second and is recommended. It distinguishes between NO_2 (nitrogen dioxide) and N_2O_4 (dinitrogen tetraoxide), two distinct compounds that would both be called nitrogen(IV) oxide under the second system of nomenclature. Two of these oxides have common (nonsystematic) names that may be encountered elsewhere: N_2O is often called nitrous oxide, and NO is called nitric oxide.

CHAPTER SUMMARY

The classical description of chemical bonding provides the conceptual framework and language used by all chemists in their daily work. The classical description is based largely on simple electrostatics. We can understand a great deal about the nature of the chemical bond by examining the forces and potential energy of interaction between the electrons and the nuclei, which are governed by Coulomb's law. Potential energy diagrams will help you understand the interactions between and among electrons and nuclei that determine a wide variety of properties of atoms and molecules. We encourage you to learn to interpret and use these diagrams, because they will appear over and over again in this text and in subsequent chemistry courses.

The periodic table organizes the elements in a way that shows similar physical and chemical properties of groups and the variations of these properties across periods. Examination of periodic trends in ionization energies suggests that the electrons in an atom are organized in a series of concentric shells around the nucleus. These trends can be explained by the decreased electron–nuclear attraction in successive shells, as well as the screening or shielding of the outer electrons from the full nuclear charge by the inner electrons. Electron affinity is a measure of the tendency of an atom to form a stable anion. Periodic trends in electron affinities are explained in the same way as those observed for ionization energies.

Chemical bonds are generally classified according to the amount of charge separation in the bond. The character of a particular chemical bond—ionic, covalent, or polar covalent—can be predicted by comparing the electronegativities of the atoms involved. The electronegativity of an element is a measure of its relative propensity to attract electrons in a chemical bond. Elements with large ionization energies and large electron affinities tend to attract electrons, whereas those with small ionization energies and small electron affinities tend to donate electrons. Bonds formed between two atoms with large electronegativity differences tend to be ionic, whereas those formed between those atoms with nearly the same electronegativities tend to be covalent. Most bonds are somewhere in between—that is, they are polar covalent.

The arrangements of atoms in a molecule and the three-dimensional shape of a molecule can be rationalized or predicted using the Lewis dot model and the VSEPR theory, respectively. The Lewis model predicts the most likely arrangement of atoms in a molecule, the existence of multiple bonds, and charge distributions using the idea of formal charges. It is based on the idea that the representative elements (except hydrogen) are most stable with a filled octet of valence electrons (hydrogen needs only two). This stability can be understood from the empirical fact that eight electrons is the maximum each valence shell can hold, and that each electron added to a shell decreases the energy of the atom via the Coulomb interaction with the nucleus. Each element is represented by its chemical symbol, with its valence electrons arranged as dots. The atoms are combined in a way that creates the maximum number of filled octets around each representative element and a pair of electrons around each hydrogen atom. The VSEPR theory predicts the three-dimensional structures of molecules by arranging electron pairs to minimize their mutual repulsion.

Oxidation numbers are used to track the gain and loss of electrons in chemical reactions and are used in the systematic naming of inorganic compounds.

The concepts introduced in this chapter provide a sound basis for understanding a great deal of chemistry. A firm understanding of these concepts is essential for you to conduct and interpret experiments, read the chemical literature, and develop a deeper understanding based on quantum mechanics, which we develop in Chapters 4–6.

CONCEPTS AND SKILLS

Section 3.2 – The Periodic Table

Describe the structure of the periodic table, and predict chemical and physical properties of an element based on those of others in its group and period (Problems 1–4).

- The periodic table organizes the elements into eighteen groups, with the main-group elements labeled I to VIII in this text, with similar physical and chemical properties based on the experimental observations that these properties are periodic functions of the atomic number Z of the elements.
- The dependence of properties on Z suggests that periodic behavior originates in the electronic structure of the atom, since Z is the number of electrons in the atom and also the magnitude of the positive charge on the nucleus.
- Experiments (summarized below) that measure gain or loss of electrons by atoms exhibit periodic behavior and thereby confirm that periodicity originates in the electronic structure of the atom.
- Quantum mechanics (in Chapter 5) connects periodic behavior of physical and chemical properties directly to electronic structure and provides a complete theoretical explanation for the organization of the periodic table.

Section 3.3 – Forces and Potential Energy in Atoms

Calculate the force and potential energy between a pair of charged particles, and predict the direction of relative motion of the particles (Problems 5–8).

- The force on a charged particle, number 2, with charge q_2, exerted by particle number 1, with charge q_1, separated by distance r is calculated from Coulomb's law as $\vec{F}_2(r) = -\frac{dV}{dr}\hat{r}_{12} = \frac{q_1 q_2}{4\pi\epsilon_0 r^2}\hat{r}_{12}$
- The potential energy of the system of two charged particles is calculated from Coulomb's law as $V(r) = \frac{q_1 q_2}{4\pi\epsilon_0 r}$ [3.2]
- The force between the particles is obtained by calculating the slope of the potential energy function $\frac{dV}{dr}$. When the slope of the potential energy is negative the force is repulsive and pushes the particles apart. When the slope of the potential energy is positive, the force is attractive and pulls the particles together.
- The physical structure of the atom, as determined by experiments, is summarized in the planetary model. In a neutral atom with atomic number Z, there are Z electrons moving around a dense nucleus, which has positive charge $+Ze$.
- The physical interactions inside the atom can be understood using Coulomb's law, which describes the forces between each electron and the nucleus and the forces between the electrons in many-electron atoms.
- The stable physical structure of the atom can be explained only by quantum mechanics (Chapter 5). Classical physics predicts that such a structure would not be stable.

Section 3.4 – Ionization Energies, the Shell Model of the Atom, and Shielding

Describe the trends in ionization energy across the periodic table (Problems 9–10).

- The ionization energy IE_1 is the minimum energy required to remove the first electron from an atom and place it infinitely far away with zero kinetic energy.
- The value of IE_1 increases across each period and reaches a maximum value at the noble gas element. It falls back to a minimum for the alkali element at the beginning of the next period, and increases across this period to reach a maximum at the noble gas element. Small variations appear within this general increasing trend across each period.

Describe the shell structure of the atom, and represent valence shell electrons of an atom by its Lewis electron dot symbol (Problems 11–12).

- Periodic trends in successive ionization energies suggest the shell model of the atom in which electrons occupy concentric shells located at increasing distances from the nuclei.
- Valence shell electrons are represented by dots (from 1 to 8 in number) arranged around the chemical symbol for the element.
- Each electron is shielded from the full positive charge of the nucleus as other electrons come between it and the nucleus. Consequently each electron experiences an effective nuclear charge $Z_{eff} = Z - S$, where the shielding constant S is essentially the average number of electrons that intervene. The amount of shielding is larger, and the magnitude of Z_{eff} is smaller, for electrons in outer shells than for those in inner shells.

- Periodic trends in Z_{eff} explain the periodic trends in the ionization energy across the periodic table. Passing across a period, at each element the nuclear charge Z increases by one unit and one more electron is added to the outer shell. The added electron therefore feels a stronger Z. Thus at each element in a given period the value of Z_{eff} increases, the added electron is more tightly bound, and the value of IE_1 increases to reach a maximum at the noble gas element. The large value of IE_1 at the noble gas element reflects the special stability of the filled octet electronic structure in the valence shell of that atom.

Section 3.5 to 3.6 – Electron Affinity and Electronegativity: The Tendency of Atoms to Attract Electrons in Molecules

Describe the trends in electron affinity and electronegativity across the periodic table (Problems 13–16).

- The electron affinity of an atom is the energy required to detach an electron from the stable anion of that atom.
- The electron affinity increases across a period for the same reason IE_1 increases; at each element the increase in Z_{eff} binds the electron added to form the anion more strongly. The increasing pattern lags behind that for IE_1 because adding one more electron to form the anion fills the shell one step sooner. Electron affinity reaches its maximum at the halogen element because the anion has the special stability of the filled octet electronic structure.
- Electronegativity is the relative tendency of atoms in a chemical bond to attract electrons. Mulliken's scale is based on a simple physical model for a gas-phase atom that averages the ionization energy and the electron affinity for a single element. Pauling's scale averages this tendency over a variety of bonding partners.
- Electronegativity increases across a period because the increase in Z_{eff} makes it energetically more difficult to form a positive ion and energetically more favorable to form an anion.

Section 3.7 – Forces and Potential Energy in Molecules: Formation of Chemical Bonds

Correlate the shape of the effective potential energy function between the nuclei with trends in the bond length and bond energies of diatomic molecules (Problems 17–20).

- The bond energy of a diatomic molecule is the amount of energy you must add to a molecule, initially at the minimum of the effective potential energy function between the nuclei, to bring it up to the effective potential energy of the isolated atoms (which is often defined to be zero).
- For ionic bond formation, the reduction in potential energy relative to the free atoms is calculated easily using Coulomb's law.
- The formation of a covalent chemical bond can be rationalized in terms of classical physics by examining the forces between the electrons and the nuclei to identify bonding and antibonding regions in the locations of the electrons.
- The formation of a covalent bond is fully explained only by quantum mechanics, including the effects of the kinetic energy of electrons in the bond (Chapter 6).

Section 3.8 – Ionic Bonding

Use the principle of charge balance and Lewis diagrams to write chemical formulas for ionic compounds (Problems 21–22).

- The composition of an ionic compound is determined by overall charge neutrality. The total positive charge on the cations must exactly balance the total negative charge on the anions.

Calculate the energy of dissociation of gaseous diatomic ionic compounds into neutral atoms and ions (Problems 23–26).

- Calculate the Coulomb stabilization energy of the ions as the difference between the minimum in their potential energy curve at R_e and the potential energy of the ions at infinity. Subtract the energy investment ΔE_∞ required to create the two ions at infinite separation. The difference is the bond dissociation energy ΔE_d. The method is illustrated in Example 3.6.

Section 3.9 – Covalent and Polar Covalent Bonding

Describe the relations among bond order, length, and energy (Problems 27–32).

- Chemists generally distinguish among three types of chemical bonds based on the electronegativity differences between the atoms: covalent (nearly identical electronegativities), ionic (greatly different electronegativities), and polar covalent (anything in between these extremes). The percent ionic character can be determined by measuring experimentally the dipole moments of molecules.
- Important measurable properties of the chemical bond include its length, dissociation energy, stiffness (vibrational force constant), and dipole moment.
- Increased force constant correlates with higher bond energy and smaller bond length: stiffer bonds are stronger and shorter.
- Theoretical models of bond formation predict the bond order, the number of electron pairs participating in the bond. Comparing the theoretical quantity bond order with measured values of force constant is one of the best ways to evaluate the accuracy of theoretical models.
- Bond lengths and bond energies of the same type (for example, CH single bonds) are remarkably similar in all compounds in which they appear. Bond lengths decrease and bond dissociation energies increase for double and triple bonds formed between the same pairs of atoms.

Estimate the percent ionic character of a bond from its dipole moment (Problems 33–38).

- The percent ionic character can be determined by measuring experimentally the dipole moments of molecules in units of Debye and the bond length in Å. The calculated result is % Ionic Character $= 100\ \delta = 0.2082\ \mu/R$.

Section 3.10 – Electron Pair Bonds and Lewis Diagrams for Molecules

Given a molecular formula, draw a Lewis diagram for the molecule and assign a formal charge (Problems 39–50).

- Lewis dot diagrams are a tool for predicting the most likely connectivity, or arrangement of bonds between atoms, in a molecule. They are also useful for predicting the existence of multiple bonds and for determining qualitatively the distribution of charges in a molecule.
- An atom in a molecule is assigned a formal charge by assuming that all the electrons in the Lewis diagram are divided equally among the atoms that share them.
- Formal charges have no reality. They are simply calculational tools to assist in evaluating the best Lewis diagrams for a specific molecule. The procedures and useful guidelines are illustrated in Examples 3.8 and 3.9.

Assign formal charges and identify resonance diagrams for a given Lewis diagram (Problems 51–58).

- Two or more equivalent Lewis diagrams for a molecule are called resonance diagrams. Taken together as a "resonance hybrid" they suggest that the actual bond properties lie somewhere between the equivalent diagrams.

Section 3.11 – The Shapes of Molecules: Valence Shell Electron-Pair Repulsion Theory

Predict the geometries of molecules by use of the VSEPR model (Problems 59–64).

- The VSEPR theory predicts the three-dimensional shapes of molecules. It is based on simple electrostatics: Electron pairs in a molecule will arrange themselves in such a way as to minimize their mutual repulsion. The steric number determines the geometry of the electron pairs (linear, trigonal pyramidal, tetrahedral, and so forth), whereas the molecular geometry is determined by the arrangement of the nuclei and may be less symmetric than the geometry of the electron pairs.

Determine whether a polyatomic molecule is polar or nonpolar (Problems 65–70).

- Identify the dipole moment for each polar bond in the molecule, and add up all these dipoles as a vector sum.
- If the sum is zero, the molecule is nonpolar. If the sum is not equal to zero, the molecule is polar.

Section 3.12 – Oxidation Numbers

Assign oxidation numbers to atoms in compounds (Problems 71–72).

- Oxidation numbers are assigned to elements to name inorganic compounds, to keep track of electrons in electron transfer (oxidation–reduction) reactions, and to explore trends in chemical reactivity across the periodic table.

Section 3.13 – Inorganic Nomenclature

Name inorganic compounds, given their chemical formulas, and write chemical formulas for inorganic compounds (Problems 73–84).

- Systematic procedures have been agreed upon for naming inorganic compounds based on their molecular and structural formulas. It is essential to translate smoothly and quickly between the name and the structural formula, and vice versa. The guidelines are summarized in Section 13.13 and the procedures are illustrated in Example 3.16.

CUMULATIVE EXERCISE

Structure and Bonding in Metal Oxides and Peroxides

Consider the three compounds KO_2, BaO_2, and TiO_2. Each contains two oxygen atoms per metal atom, but the oxygen occurs in different forms in the three compounds.

(a) The oxygen in TiO_2 occurs as O^{2-} ions. What is the Lewis dot symbol for this ion? How many valence electrons does this ion have? What is the chemical name for TiO_2?

(b) Recall that Group II elements form stable 2+ ions. Using Table 3.9, identify the oxygen-containing ion in BaO_2 and give the name of the compound. Draw a Lewis diagram for the oxygen-containing ion, showing formal charge. Is the bond in this ion a single or a double bond?

(c) Recall that Group I elements form stable 1+ ions. Using Table 3.9, identify the oxygen-containing ion in KO_2 and give the name of the compound. Show that the oxygen-containing ion is an odd-electron species. Draw the best Lewis diagram you can for it.

Answers

(a) The ion $:\underset{\cdot\cdot}{\ddot{O}}:^{2-}$ has eight valence electrons (an octet). TiO_2 is titanium(IV) oxide.

(b) The ion in BaO_2 must be the peroxide ion (O_2^{2-}), and the compound is barium peroxide. The Lewis diagram for the peroxide ion is

and the O—O bond is a single bond.

(c) The ion in KO_2 must be the superoxide ion (O_2^{-}), and the compound is potassium superoxide. The superoxide ion has 13 valence electrons. The best Lewis diagram is a pair of resonance diagrams

 and 

in which only one of the oxygen atoms attains an octet electron configuration.

PROBLEMS

Answers to problems whose numbers are red appear in Appendix G. Problems that are more challenging are indicated with asterisks.

3.2 The Periodic Table

1. Before the element scandium was discovered in 1879, it was known as "eka-boron." Predict the properties of scandium from averages of the corresponding properties of its neighboring elements in the periodic table. Compare your predictions with the observed values in Appendix F.

Element	Symbol	Melting Point (°C)	Boiling Point (°C)	Density (g cm^{-3})
Calcium	Ca	839	1484	1.55
Titanium	Ti	1660	3287	4.50
Scandium	Sc	?	?	?

2. The element technetium (Tc) is not found in nature but has been produced artificially through nuclear reactions. Use the data for several neighboring elements in the table below to estimate the melting point, boiling point, and density of technetium. Compare your predictions with the observed values in Appendix F.

Element	Symbol	Melting Point (°C)	Boiling Point (°C)	Density (g cm^{-3})
Manganese	Mn	1244	1962	7.2
Molybdenum	Mo	2610	5560	10.2
Rhenium	Re	3180	5627	20.5
Ruthenium	Ru	2310	3900	12.3

3. Use the group structure of the periodic table to predict the empirical formulas for the binary compounds that hydrogen forms with the elements antimony, bromine, tin, and selenium.

4. Use the group structure of the periodic table to predict the empirical formulas for the binary compounds that hydrogen forms with the elements germanium, fluorine, tellurium, and bismuth.

3.3 Forces and Potential Energy in Atoms

5. An electron is located at the origin of the coordinates, and a second electron is brought to a position 2 Å from the origin in the $+y$ direction.

(a) Calculate the force on the electron at the origin, giving the components F_x and F_y.

(b) Calculate the potential energy of the two electrons.

6. A gold nucleus is located at the origin of coordinates, and an electron is brought to a position 2 Å from the origin in the $+y$ direction.
(a) Calculate the force on the gold nucleus exerted by the electron giving its components F_x and F_y.
(b) Calculate the potential energy of the gold nucleus and the electron.

7. The electron in a hydrogen atom is initially at a distance 2.12 Å from the proton, and then moves in the $+x$ direction to a distance 0.529 Å from the proton.
(a) Calculate the magnitude of the force on the electron at each separation.
(b) Calculate the change in the potential energy between the proton and the electron.
(c) Calculate the change in the speed of the electron if the electron was initially stationary.

8. A gold nucleus is located at the origin of coordinates, and a helium nucleus initially at the coordinates (0, 2 Å) moves to the position (0, 1 Å).
(a) Calculate the magnitude of the force on the electron at each separation.
(b) Calculate the change in the potential energy of the two nuclei.
(c) Calculate the change in the kinetic energy of the helium nucleus assuming its value at 2 Å is 2.0 MeV if you assume the gold nucleus is so massive that it stays at the origin.

3.4 Ionization Energies, the Shell Model of the Atom, and Shielding

9. For each of the following pairs of atoms, state which you expect to have the higher first ionization energy: **(a)** Rb or Sr; **(b)** Po or Rn; **(c)** Xe or Cs; **(d)** Ba or Sr.

10. For each of the following pairs of atoms, state which you expect to have the higher first ionization energy: **(a)** Bi or Xe; **(b)** Se or Te; **(c)** Rb or Y; **(d)** K or Ne.

11. Use the data in Table 3.1 to plot the logarithm of ionization energy versus the number of electrons removed for Be. Describe the electronic structure of the Be atom.

12. Use the data in Table 3.1 to plot the logarithm of ionization energy versus the number of electrons removed for Ne. Describe the electronic structure of the Ne atom.

3.5 Electron Affinity
3.6 Electronegativity: The Tendency of Atoms to Attract Electrons in Molecules

13. For each of the following pairs of atoms, state which you expect to have the greater electron affinity: **(a)** Xe or Cs; **(b)** Pm or F; **(c)** Ca or K; **(d)** Po or At.

14. For each of the following pairs of atoms, state which you expect to have the higher electron affinity: **(a)** Rb or Sr; **(b)** I or Rn; **(c)** Ba or Te; **(d)** Bi or Cl.

15. Ignoring tables of electronegativity values and guided only by the periodic table, arrange these atoms in order of increasing electronegativity: O, F, S, Si, K. Briefly explain your reasoning.

16. Ignoring tables of electronegativity values and guided only by the periodic table, arrange these atoms in order of increasing electronegativity: S, Cl, Sb, Se, In. Briefly explain your reasoning.

3.7 Forces and Potential Energy in Molecules: Formation of Chemical Bonds

17. H_2 has equilibrium bond length of 0.751 Å and bond dissociation energy of 432 kJ mol^{-1}, whereas F_2 has equilibrium bond length of 1.417 Å and bond dissociation energy of 155 kJ mol^{-1}. On the same graph show qualitative sketches of the effective potential energy curve V_{eff} for an H_2 and an F_2 molecule. In your solution, show the conversion from kJ mol^{-1} to the energy of a single molecule.

18. N_2 has equilibrium bond length of 1.100 Å and bond dissociation energy of 942 kJ mol^{-1}, whereas O_2 has equilibrium bond length of 1.211 Å and bond dissociation energy of 495 kJ mol^{-1}. On the same graph show qualitative sketches of the effective potential energy curve V_{eff} for an N_2 and an O_2 molecule. In your solution, show the conversion from kJ mol^{-1} to the energy of a single molecule.

19. HF has equilibrium bond length of 0.926 Å and bond dissociation energy of 565 kJ mol^{-1}. Compare the effective potential curve for HF with those for H_2 and F_2 in Problem 17.

20. NO has equilibrium bond length of 1.154 Å and bond dissociation energy of 629 kJ mol^{-1}. Compare the effective potential curve for NO with those for N_2 and O_2 in Problem 18.

3.8 Ionic Bonding

21. For each of the following atoms or ions, state the total number of electrons, the number of valence electrons, and the number of core electrons.
(a) Rn **(b)** Sr^+ **(c)** Se^{2-} **(d)** Sb^-

22. For each of the following atoms or ions, state the total number of electrons, the number of valence electrons, and the number of core electrons.
(a) Ra^{2+} **(b)** Br **(c)** Bi^{2-} **(d)** Ga^+

23. Use the data in Figure 3.11 and Table 3.2 to calculate the energy changes (ΔE) for the following pairs of reactions:
(a) $K(g) + Cl(g) \longrightarrow K^+(g) + Cl^-(g)$
$K(g) + Cl(g) \longrightarrow K^-(g) + Cl^+(g)$
(b) $Na(g) + Cl(g) \longrightarrow Na^+(g) + Cl^-(g)$
$Na(g) + Cl(g) \longrightarrow Na^-(g) + Cl^+(g)$

Explain why K^+Cl^- and Na^+Cl^- form in preference to K^-Cl^+ and Na^-Cl^+.

24. Use the data in Figure 3.11 and Table 3.2 to calculate the energy changes (ΔE) for the following pairs of reactions:
(a) $Na(g) + I(g) \longrightarrow Na^+(g) + I^-(g)$
$Na(g) + I(g) \longrightarrow Na^-(g) + I^+(g)$
(b) $Rb(g) + Br(g) \longrightarrow Rb^+(g) + Br^-(g)$
$Rb(g) + Br(g) \longrightarrow Rb^-(g) + Br^+(g)$

Explain why Na^+I^- and Rb^+Br^- form in preference to Na^-I^+ and Rb^-Br^+.

25. In a gaseous KCl molecule, the equilibrium internuclear distance is 2.67×10^{-10} m. Using data from Appendix F and assuming that a Coulomb potential is a good approximation of the potential from $K^+ + Cl^-$ to 2.67×10^{-10} m, estimate the dissociation energy of gaseous KCl into K and Cl atoms (in kJ mol^{-1}). (Note: Table 3.7 shows that a Coulomb potential is only a good approximation to the potential between two charged atoms at long inter-atomic distances; the percent ionic character of KCl at equilibrium is only 82%.)

26. In a gaseous RbF molecule, the bond length is 2.274×10^{-10} m. Using data from Appendix F and making the same oversimplified assumption as in the prior problem on the shape of the potential curve from $Rb^+ + F^-$ to an internuclear separation of 2.274×10^{-10} m, calculate the energy in kJ mol^{-1} required to dissociate RbF to neutral atoms.

3.9 Covalent and Polar Covalent Bonding

27. The bond lengths of the X—H bonds in NH_3, PH_3, and SbH_3 are 1.02, 1.42, and 1.71 Å, respectively. Estimate the length of the As—H bond in AsH_3, the gaseous compound that decomposes on a heated glass surface in Marsh's test for arsenic. Which of these four hydrides has the weakest X—H bond?

28. Arrange the following covalent diatomic molecules in order of the lengths of the bonds: BrCl, ClF, IBr. Which of the three has the weakest bond (the smallest bond energy)?

29. The bond length in H—I (1.62 Å) is close to the sum of the atomic radii of H (0.37 Å) and I (1.33 Å). What does this fact indicate about the polarity of the bond?

30. The bond length in F_2 is 1.417 Å, instead of twice the atomic radius of F, which is 1.28 Å. What can account for the unexpected length of the F—F bond?

31. Use electronegativity values to arrange the following bonds in order of decreasing polarity: N—O, N—N, N—P, and C—N.

32. Use electronegativity values to rank the bonds in the following compounds from least ionic to most ionic in character: IF, ICl, ClF, BrCl, and Cl_2.

33. Ionic compounds tend to have higher melting and boiling points and to be less volatile (that is, have lower vapor pressures) than covalent compounds. For each of the following pairs, use electronegativity differences to predict which compound has the higher vapor pressure at room temperature.
(a) CI_4 or KI
(b) BaF_2 or OF_2
(c) SiH_4 or NaH

34. For each of the following pairs, use electronegativity differences to predict which compound has the higher boiling point.
(a) $MgBr_2$ or PBr_3
(b) OsO_4 or SrO
(c) Cl_2O or Al_2O_3

35. Estimate the percent ionic character of the bond in each of the following diatomic molecules, based on the dipole moment.

	Bond Length (Å)	Dipole Moment (D)
ClO	1.573	1.239
KI	3.051	10.82
TlCl	2.488	4.543
InCl	2.404	3.79

36. Estimate the percent ionic character of the bond in each of the following species. All the species are unstable or reactive under ordinary laboratory conditions, but they can be observed in interstellar space.

	Bond Length (Å)	Dipole Moment (D)
OH	0.980	1.66
CH	1.131	1.46
CN	1.175	1.45
C_2	1.246	0

37. The percent ionic character of a bond can be approximated by the formula $16\Delta + 3.5\Delta^2$, where Δ is the magnitude of the difference in the electronegativities of the atoms (see Fig. 3.18). Calculate the percent ionic character of HF, HCl, HBr, HI, and CsF, and compare the results with those in Table 3.7.

38. The percent ionic character of the bonds in several interhalogen molecules (as estimated from their measured dipole moments and bond lengths) are ClF (11%), BrF (15%), BrCl (5.6%), ICl (5.8%), and IBr (10%). Estimate the percent ionic characters for each of these molecules, using the equation in Problem 37, and compare them with the given values.

3.10 Electron Pair Bonds and Lewis Diagrams for Molecules

39. Assign formal charges to all atoms in the following Lewis diagrams.

(a) SO_4^{2-}

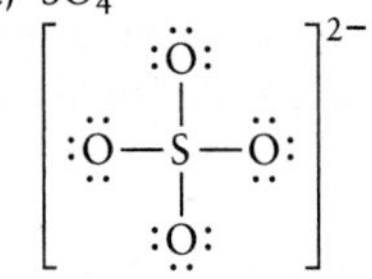

(b) $S_2O_3^{2-}$

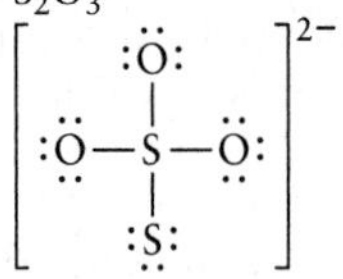

(c) SbF_3

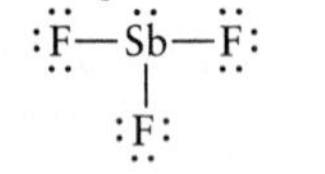

(d) SCN^-

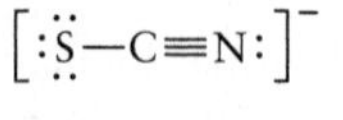

40. Assign formal charges to all atoms in the following Lewis diagrams.

(a) ClO_4^-

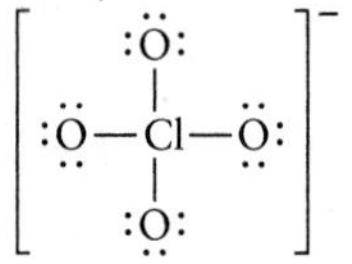

(b) SO_2

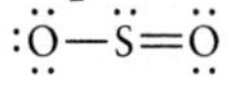

(c) BrO_2^-

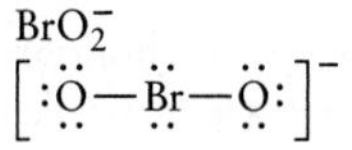

(d) NO_3^-

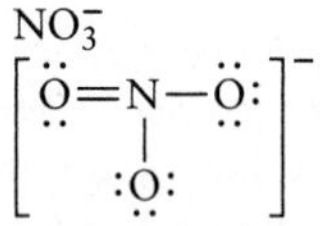

41. Determine the formal charges on all the atoms in the following Lewis diagrams.

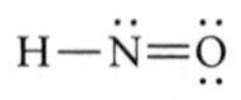 and 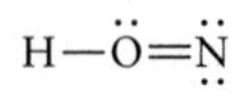

Which one would best represent bonding in the molecule HNO?

42. Determine the formal charges on all the atoms in the following Lewis diagrams.

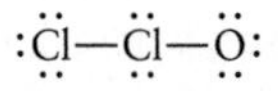 and 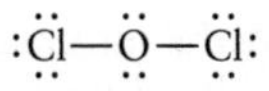

Which one would best represent bonding in the molecule Cl_2O?

43. In each of the following Lewis diagrams, Z represents a main-group element. Name the group to which Z belongs in each case and give an example of such a compound or ion that actually exists.

(a)

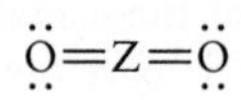

(b)

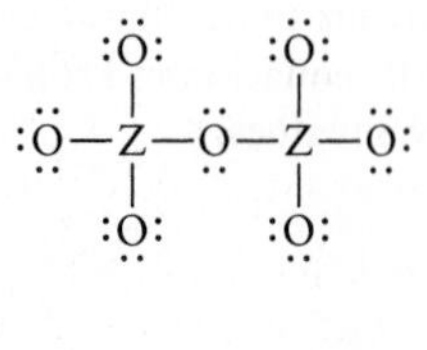

(c)

(d)

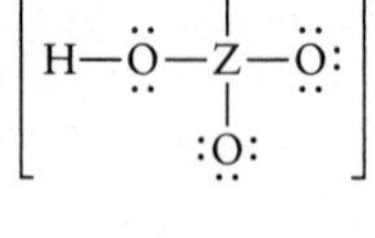

44. In each of the following Lewis diagrams, Z represents a main-group element. Name the group to which Z belongs in each case and give an example of such a compound or ion that actually exists.

(a)

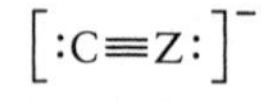

(b)

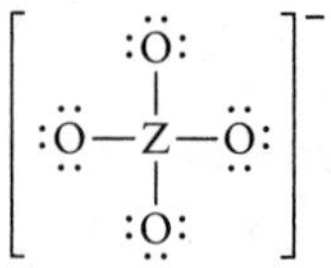

(c)

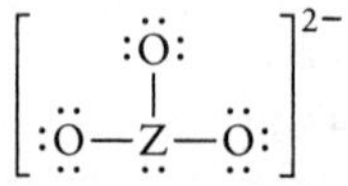

(d)

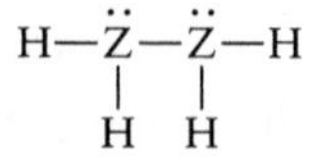

45. Draw Lewis electron dot diagrams for the following species: **(a)** AsH_3; **(b)** HOCl; **(c)** KrF^+; **(d)** $PO_2Cl_2^-$ (central P atom).

46. Draw Lewis electron dot diagrams for the following species: **(a)** methane; **(b)** carbon dioxide; **(c)** phosphorus trichloride; **(d)** perchlorate ion.

47. Urea is an important chemical fertilizer with the chemical formula $(H_2N)CO(NH_2)$. The carbon atom is bonded to both nitrogen atoms and the oxygen atom. Draw a Lewis diagram for urea and use Table 3.6 to estimate its bond lengths.

48. Acetic acid is the active ingredient of vinegar. Its chemical formula is CH_3COOH, and the second carbon atom is bonded to the first carbon atom and to both oxygen atoms. Draw a Lewis diagram for acetic acid and use Table 3.6 to estimate its bond lengths.

49. Under certain conditions, the stable form of sulfur consists of rings of eight sulfur atoms. Draw the Lewis diagram for such a ring.

50. White phosphorus (P_4) consists of four phosphorus atoms arranged at the corners of a tetrahedron. Draw the valence electrons on this structure to give a Lewis diagram that satisfies the octet rule.

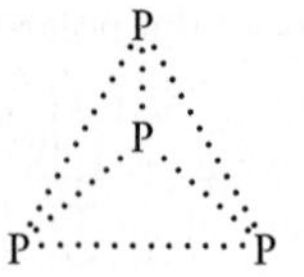

51. Draw Lewis electron dot diagrams for the following species, indicating formal charges and resonance diagrams where applicable.

(a) H_3NBF_3

(b) CH_3COO^- (acetate ion)

(c) HCO_3^- (hydrogen carbonate ion)

52. Draw Lewis electron dot diagrams for the following species, indicating formal charges and resonance diagrams where applicable.
(a) HNC (central N atom)
(b) SCN^- (thiocyanate ion)
(c) H_2CNN (the first N atom is bonded to the carbon and the second N)

53. Draw Lewis diagrams for the two resonance forms of the nitrite ion, NO_2^-. In what range do you expect the nitrogen–oxygen bond length to fall? (*Hint:* Use Table 3.6.)

54. Draw Lewis diagrams for the three resonance forms of the carbonate ion, CO_3^{2-}. In what range do you expect the carbon–oxygen bond length to fall? (*Hint:* Use Table 3.6.)

55. Methyl isocyanate, which was involved in the disaster in Bhopal, India, in 1984, has the chemical formula CH_3NCO. Draw its Lewis diagram, including resonance forms. (**Note:** The N atom is bonded to the two C atoms.)

56. Peroxyacetyl nitrate (PAN) is one of the prime irritants in photochemical smog. It has the formula $CH_3COOONO_2$, with the following structure:

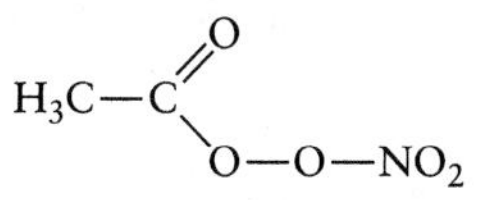

Draw its Lewis diagram, including resonance forms.

57. Draw Lewis diagrams for the following compounds. In the formula the symbol of the central atom is given first. (*Hint:* The valence octet may be expanded for the central atom.)
(a) PF_5 (b) SF_4 (c) XeO_2F_2

58. Draw Lewis diagrams for the following ions. In the formula the symbol of the central atom is given first. (*Hint:* The valence octet may be expanded for the central atom.)
(a) BrO_4^- (b) PCl_6^- (c) XeF_6^+

3.11 The Shapes of Molecules: Valence Shell Electron-Pair Repulsion Theory

59. For each of the following molecules, give the steric number and sketch and name the approximate molecular geometry. In each case, the central atom is listed first and the other atoms are all bonded directly to it.
(a) CBr_4 (b) SO_3
(c) SeF_6 (d) $SOCl_2$
(e) ICl_3

60. For each of the following molecules or molecular ions, give the steric number and sketch and name the approximate molecular geometry. In each case, the central atom is listed first and the other atoms are all bonded directly to it.
(a) PF_3 (b) SO_2Cl_2
(c) PF_6^- (d) ClO_2^-
(e) GeH_4

61. For each of the following molecules or molecular ions, give the steric number, sketch and name the approximate molecular geometry, and describe the directions of any *distortions* from the approximate geometry due to lone pairs. In each case, the central atom is listed first and the other atoms are all bonded directly to it.
(a) ICl_4^- (b) OF_2
(c) BrO_3^- (d) CS_2

62. For each of the following molecules or molecular ions, give the steric number, sketch and name the approximate molecular geometry, and describe the direction of any *distortions* from the approximate geometry due to lone pairs. In each case, the central atom is listed first and the other atoms are all bonded directly to it.
(a) TeH_2 (b) AsF_3
(c) PCl_4^+ (d) XeF_5^+

63. Give an example of a molecule or ion having a formula of each of the following types and structures.
(a) AB_3 (planar) (b) AB_3 (pyramidal)
(c) AB_2^- (bent) (d) AB_3^{2-} (planar)

64. Give an example of a molecule or ion having a formula of each of the following types and structures.
(a) AB_4^- (tetrahedral) (b) AB_2 (linear)
(c) AB_6^- (octahedral) (d) AB_3^- (pyramidal)

65. For each of the answers in Problem 59, state whether the species is polar or nonpolar.

66. For each of the answers in Problem 60, state whether the species is polar or nonpolar.

67. The molecules of a certain compound contain one atom each of nitrogen, fluorine, and oxygen. Two possible structures are NOF (O as central atom) and ONF (N as central atom). Does the information that the molecule is bent limit the choice to one of these two possibilities? Explain.

68. Mixing $SbCl_3$ and $GaCl_3$ in a 1:1 molar ratio (using liquid sulfur dioxide as a solvent) gives a solid ionic compound of empirical formula $GaSbCl_6$. A controversy arises over whether this compound is $(SbCl_2^+)(GaCl_4^-)$ or $(GaCl_2^+)(SbCl_4^-)$.
(a) Predict the molecular structures of the two anions.
(b) It is learned that the cation in the compound has a bent structure. Based on this fact, which formulation is more likely to be correct?

69. (a) Use the VSEPR theory to predict the structure of the NNO molecule.
(b) The substance NNO has a small dipole moment. Which end of the molecule is more likely to be the positive end, based only on electronegativity?

70. Ozone (O_3) has a nonzero dipole moment. In the molecule of O_3, one of the oxygen atoms is directly bonded to the other two, which are not bonded to each other.
(a) Based on this information, state which of the following structures are possible for the ozone molecule: symmetric linear, nonsymmetric linear (for example,

different O—O bond lengths), and bent. (**Note:** Even an O—O bond can have a bond dipole if the two oxygen atoms are bonded to different atoms or if only one of the oxygen atoms is bonded to a third atom.)

(b) Use the VSEPR theory to predict which of the structures of part (a) is observed.

3.12 Oxidation Numbers

71. Assign oxidation numbers to the atoms in each of the following species: $SrBr_2$, $Zn(OH)_4^{2-}$, SiH_4, $CaSiO_3$, $Cr_2O_7^{2-}$, $Ca_5(PO_4)_3F$, KO_2, CsH.

72. Assign oxidation numbers to the atoms in each of the following species: NH_4NO_3, $CaMgSiO_4$, $Fe(CN)_6^{4-}$, B_2H_6, BaH_2, $PbCl_2$, $Cu_2O(SO_4)$, $S_4O_6^{2-}$.

3.13 Inorganic Nomenclature

73. Give the name and formula of an ionic compound involving only the elements in each pair that follows. Write Lewis symbols for the elements both before and after chemical combination.
(a) Chlorine and cesium **(b)** Calcium and astatine
(c) Aluminum and sulfur **(d)** Potassium and tellurium

74. Give the name and formula of an ionic compound involving only the elements in each pair that follows. Write Lewis symbols for the elements both before and after chemical combination.
(a) Gallium and bromine **(b)** Strontium and polonium
(c) Magnesium and iodine **(d)** Lithium and selenium

75. Give systematic names to the following compounds:
(a) Al_2O_3 **(b)** Rb_2Se
(c) $(NH_4)_2S$ **(d)** $Ca(NO_3)_2$
(e) Cs_2SO_4 **(f)** $KHCO_3$

76. Give systematic names to the following compounds:
(a) KNO_2 **(b)** $Sr(MnO_4)_2$
(c) $MgCr_2O_7$ **(d)** NaH_2PO_4
(e) $BaCl_2$ **(f)** $NaClO_3$

77. Write the chemical formulas for the following compounds:
(a) Silver cyanide
(b) Calcium hypochlorite
(c) Potassium chromate
(d) Gallium oxide
(e) Potassium superoxide
(f) Barium hydrogen carbonate

78. Write the chemical formulas for the following compounds:
(a) Cesium sulfite
(b) Strontium thiocyanate
(c) Lithium hydride
(d) Sodium peroxide
(e) Ammonium dichromate
(f) Rubidium hydrogen sulfate

79. Trisodium phosphate (TSP) is a heavy-duty cleaning agent. Write its chemical formula. What would be the systematic name for this ionic compound?

80. Monoammonium phosphate is the common name for a compound made up of NH_4^+ and $H_2PO_4^-$; it is used as a flame retardant. (Its use for this purpose was first suggested by Gay-Lussac in 1821.) Write its chemical formula. What is the systematic chemical name of this compound?

81. Write the chemical formula for each of the following compounds:
(a) Silicon dioxide
(b) Ammonium carbonate
(c) Lead(IV) oxide
(d) Diphosphorus pentaoxide
(e) Calcium iodide
(f) Iron(III) nitrate

82. Write the chemical formula for each of the following compounds:
(a) Lanthanum(III) sulfide
(b) Cesium sulfate
(c) Dinitrogen trioxide
(d) Iodine pentafluoride
(e) Chromium(III) sulfate
(f) Potassium permanganate

83. Give the systematic name for each of the following compounds:
(a) Cu_2S and CuS **(b)** Na_2SO_4
(c) As_4O_6 **(d)** $ZrCl_4$
(e) Cl_2O_7 **(f)** Ga_2O

84. Give the systematic name for each of the following compounds:
(a) Mg_2SiO_4 **(b)** $Fe(OH)_2$ and $Fe(OH)_3$
(c) As_2O_5 **(d)** $(NH_4)_2HPO_4$
(e) SeF_6 **(f)** Hg_2SO_4

ADDITIONAL PROBLEMS

85. Refer to Figure 3.18 and compute the difference in electronegativity between the atoms in LiCl and those in HF. Based on their physical properties (see below), are the two similar or different in terms of bonding?

	LiCl	HF
Melting point	605°C	83.1°C
Boiling point	1350°C	19.5°C

86. Ordinarily, two metals, when mixed, form alloys that show metallic character. If the two metals differ sufficiently in electronegativity, they can form compounds with significant ionic character. Consider the solid produced by mixing equal chemical amounts of Cs and Rb, compared with that produced by mixing Cs and Au. Compute the electronegativity difference in each case, and determine whether either mixture has significant ionic character. If either compound is ionic or partially ionic, which atom carries the net negative charge? Are there alkali halides with similar or smaller electronegativity differences?

87. At large interatomic separations, an alkali halide molecule MX has a lower energy as two neutral atoms, M + X; at short separations, the ionic form (M+)(X−) has a lower energy. At a certain distance, R_c, the energies of the two forms become equal, and it is near this distance that the electron will jump from the metal to the halogen atom during a collision. Because the forces between neutral atoms are weak at large distances, a reasonably good approximation can be made by ignoring any variation in potential $V(R)$ for the neutral atoms between R_c and $R = \infty$. For the ions in this distance range, $V(R)$ is dominated by their Coulomb attraction.

(a) Express R_c for the first ionization energy of the metal M and the electron affinity of the halogen X.

(b) Calculate R_c for LiF, KBr, and NaCl using data from Appendix F.

88. Use the data in Appendix F to compute the energy changes (ΔE) of the following pairs of reactions:

(a) $Na(g) + I(g) \longrightarrow Na^+(g) + I^-(g)$ and
$Na(g) + I(g) \longrightarrow Na^-(g) + I^+(g)$

(b) $K(g) + Cl(g) \longrightarrow K^+(g) + Cl^-(g)$ and
$K(g) + Cl(g) \longrightarrow K^-(g) + Cl^+(g)$

Explain why Na^+I^- and K^+Cl^- form in preference to Na^-I^+ and K^-Cl^+.

89. The carbon–carbon bond length in C_2H_2 is 1.20 Å, that in C_2H_4 is 1.34 Å, and that in C_2H_6 is 1.53 Å. Near which of these values would you predict the bond length of C_2 to lie? Is the experimentally observed value, 1.31 Å, consistent with your prediction?

90. Two possible Lewis diagrams for sulfine (H_2CSO) are

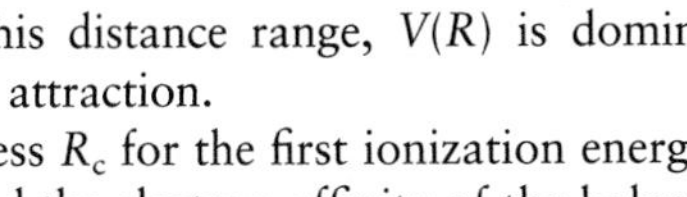

(a) Compute the formal charges on all atoms.

(b) Draw a Lewis diagram for which all the atoms in sulfine have formal charges of zero.

91. There is persuasive evidence for the brief existence of the unstable molecule OPCl.

(a) Draw a Lewis diagram for this molecule in which the octet rule is satisfied on all atoms and the formal charges on all atoms are zero.

(b) The compound OPCl reacts with oxygen to give O_2PCl. Draw a Lewis diagram of O_2PCl for which all formal charges are equal to zero. Draw a Lewis diagram in which the octet rule is satisfied on all atoms.

92. The compound SF_3N has been synthesized.

(a) Draw the Lewis diagram of this molecule, supposing that the three fluoride atoms and the nitrogen atom surround the sulfur atom. Indicate the formal charges. Repeat, but assume that the three fluorine atoms and the sulfur atom surround the nitrogen atom.

(b) From the results in part (a), speculate about which arrangement is more likely to correspond to the actual molecular structure.

93. In nitryl chloride (NO_2Cl), the chlorine atom and the two oxygen atoms are bonded to a central nitrogen atom, and all the atoms lie in a plane. Draw the two electron dot resonance forms that satisfy the octet rule and that together are consistent with the fact that the two nitrogen–oxygen bonds are equivalent.

94. The molecular ion $S_3N_3^-$ has the cyclic structure

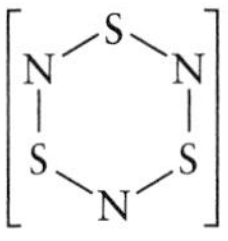

All S—N bonds are equivalent.

(a) Give six equivalent resonance hybrid Lewis diagrams for this molecular ion.

(b) Compute the formal charges on all atoms in the molecular ion in each of the six Lewis diagrams.

(c) Determine the charge on each atom in the polyatomic ion, assuming that the true distribution of electrons is the *average* of the six Lewis diagrams arrived at in parts (a) and (b).

(d) An advanced calculation suggests that the actual charge resident on each N atom is −0.375 and on each S atom is +0.041. Show that this result is consistent with the overall +1 charge on the molecular ion.

***95.** The two compounds nitrogen dioxide and dinitrogen tetraoxide are introduced in Section 3.13.

(a) NO_2 is an odd-electron compound. Draw the best Lewis diagrams possible for it, recognizing that one atom cannot achieve an octet configuration. Use formal charges to decide whether that should be the (central) nitrogen atom or one of the oxygen atoms.

(b) Draw resonance forms for N_2O_4 that obey the octet rule. The two N atoms are bonded in this molecule.

96. Although magnesium and the alkaline-earth metals situated below it in the periodic table form ionic chlorides, beryllium chloride ($BeCl_2$) is a covalent compound.

(a) Follow the usual rules to write a Lewis diagram for $BeCl_2$ in which each atom attains an octet configuration. Indicate formal charges.

(b) The Lewis diagram that results from part (a) is an extremely unlikely one because of the double bonds and formal charges it shows. By relaxing the requirement of placing an octet on the beryllium atom, show how a Lewis diagram without formal charges can be written.

97. (a) The first noble-gas compound, prepared by Neil Bartlett in 1962, was an orange-yellow ionic solid that consisted of XeF^+ and PtF_{11}^-. Draw a Lewis diagram for XeF^+.

(b) Shortly after the preparation of the ionic compound discussed in part (a), it was found that the irradiation of mixtures of xenon and fluorine with sunlight produced white crystalline XeF_2. Draw a Lewis diagram for this molecule, allowing valence expansion on the central xenon atom.

***98.** Represent the bonding in SF_2 (F—S—F) with Lewis diagrams. Include the formal charges on all atoms. The dimer of this compound has the formula S_2F_4. It was isolated in 1980 and shown to have the structure F_3S—SF. Draw a possible Lewis diagram to represent the bonding in the dimer, indicating the formal charges on all atoms. Is it possible to draw a Lewis diagram for S_2F_4 in which all atoms have valence octets? Explain why or why not.

***99.** A stable triatomic molecule can be formed that contains one atom each of nitrogen, sulfur, and fluorine. Three bonding structures are possible, depending on which is the central atom: NSF, SNF, and SFN.

(a) Write a Lewis diagram for each of these molecules, indicating the formal charge on each atom.

(b) Often, the structure with the least separation of formal charge is the most stable. Is this statement consistent with the observed structure for this molecule—namely, NSF, which has a central sulfur atom?

(c) Does consideration of the electronegativities of N, S, and F from Figure 3.18 help rationalize this observed structure? Explain.

100. The gaseous potassium chloride molecule has a measured dipole moment of 10.3 D, which indicates that it is a very polar molecule. The separation between the nuclei in this molecule is 2.67 Å. What would the dipole moment of a KCl molecule be if there were opposite charges of one fundamental unit (1.60×10^{-19} C) at the nuclei?

101. **(a)** Predict the geometry of the $SbCl_5^{2-}$ ion, using the VSEPR method.

(b) The ion $SbCl_6^{3-}$ is prepared from $SbCl_5^{2-}$ by treatment with Cl^-. Determine the steric number of the central antimony atom in this ion, and discuss the extension of the VSEPR theory that would be needed for the prediction of its molecular geometry.

102. The element xenon (Xe) is by no means chemically inert; it forms a number of chemical compounds with electronegative elements such as fluorine and oxygen. The reaction of xenon with varying amounts of fluorine produces XeF_2 and XeF_4. Subsequent reaction of one or the other of these compounds with water produces (depending on conditions) XeO_3, XeO_4, and H_4XeO_6, as well as mixed compounds such as $XeOF_4$. Predict the structures of these six xenon compounds, using the VSEPR theory.

103. Predict the arrangement of the atoms about the sulfur atom in F_4SPO, assuming that double-bonded atoms require more space than single-bonded atoms.

104. Draw Lewis diagrams and predict the geometries of the following molecules. State which are polar and which are nonpolar.

(a) ONCl **(b)** O_2NCl

(c) XeF_2 **(d)** SCl_4

(e) CHF_3

***105.** Suppose that any given kind of bond, such as O—H, has a characteristic electric dipole. That is, suppose that electric dipole moments can be assigned to bonds just as bond energies can be. Both are usefully accurate approximations. Consider the water molecule

Show that if μ_{OH} is the dipole moment of the OH bond, then the dipole moment of water is $\mu(H_2O) = 2\mu_{OH} \cos(\theta/2)$. What is the dipole moment μ_{OH} if $\mu(H_2O)$ is 1.86 D?

106. A good method of preparing pure oxygen on a small scale is the decomposition of $KMnO_4$ in a vacuum above 215°C:

$$2\ KMnO_4(s) \longrightarrow K_2MnO_4(s) + MnO_2(s) + O_2(g)$$

Assign an oxidation number to each atom and verify that the total number of electrons lost is equal to the total number gained.

107. Bismuth forms an ion with the formula Bi_5^{3+}. Arsenic and fluorine form a complex ion $[AsF_6]^-$, with fluorine atoms arranged around a central arsenic atom. Assign oxidation numbers to each of the atoms in the bright yellow crystalline solid with the formula $Bi_5(AsF_6)_3 \cdot 2SO_2$.

108. In some forms of the periodic table, hydrogen is placed in Group I; in others, it is placed in Group VII. Give arguments in favor of each location.

***109.** **(a)** Determine the oxidation number of lead in each of the following oxides: PbO, PbO_2, Pb_2O_3, Pb_3O_4.

(b) The only known lead ions are Pb^{2+} and Pb^{4+}. How can you reconcile this statement with your answer to part (A)?

110. There have been some predictions that element 114 will be relatively stable in comparison with many other elements beyond uranium in the periodic table. Predict the maximum oxidation state of this element. Based on the trends in the oxidation states of other members of its group, is it likely that this oxidation state will be the dominant one?

CUMULATIVE PROBLEMS

111. A certain element, M, is a main-group metal that reacts with chlorine to give a compound with the chemical formula MCl_2 and with oxygen to give the compound MO.

(a) To which group in the periodic table does element M belong?

(b) The chloride contains 44.7% chlorine by mass. Name the element M.

***112.** An ionic compound used as a chemical fertilizer has the composition (by mass) 48.46% O, 23.45% P, 21.21% N, 6.87% H. Give the name and chemical formula of the compound and draw Lewis diagrams for the two types of ions that make it up.

113. A compound is being tested for use as a rocket propellant. Analysis shows that it contains 18.54% F, 34.61% Cl, and 46.85% O.

(a) Determine the empirical formula for this compound.

(b) Assuming that the molecular formula is the same as the empirical formula, draw a Lewis diagram for this molecule. Review examples elsewhere in this chapter to decide which atom is most likely to lie at the center.

(c) Use the VSEPR theory to predict the structure of the molecule from part (b).

114. Many important fertilizers are ionic compounds that contain the elements nitrogen, phosphorus, and potassium because these are frequently the limiting plant-growth nutrients in soil.

(a) Write the chemical formulas for the following chemical fertilizers: ammonium phosphate, potassium nitrate, ammonium sulfate.

(b) Calculate the mass percentage of nitrogen, phosphorus, and potassium for each of the compounds in part (a).

4 CHAPTER

Introduction to Quantum Mechanics

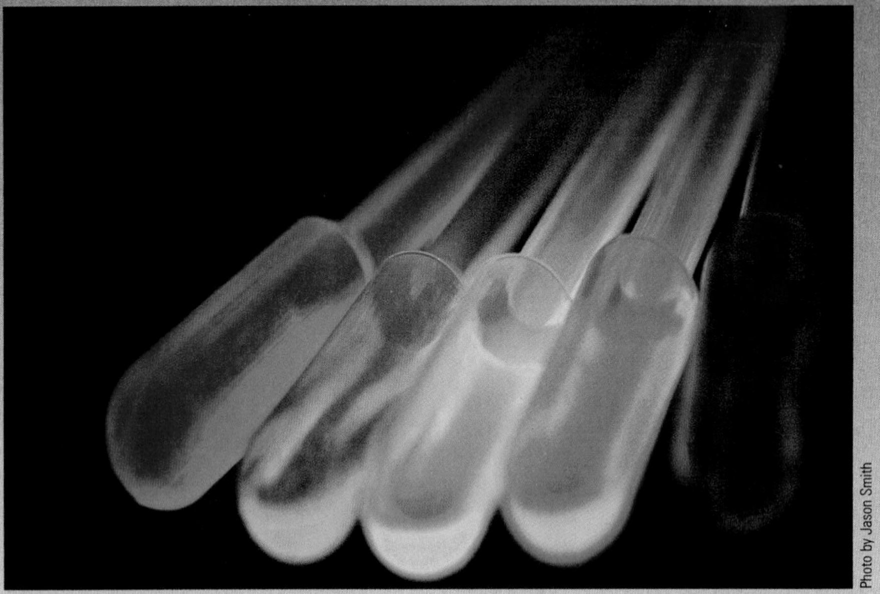

Colloidal solutions of CdSe/ZnSe core/shell nanoparticles are synthesized with different size distributions then excited with UV light. The nanocrystals in the different vials show different colors in visible light because the wavelength at which they emit is determined by their size. This size-dependent behavior in the nanometer regime can be explained only by quantum mechanics. Such nanocrystals, also called "quantum dots," are incorporated into new optical and electronic device designs.

Science can advance in different ways. Usually, the slow and steady accumulation of experimental results supports and refines existing models, which leads to a more satisfactory description of natural phenomena. Occasionally, however, the results of new experiments directly contradict previously accepted theories. In this case, a period of uncertainty ensues; it is resolved only through the eventual emergence of a new and more complete theory that explains both the previously understood results and the new experiments. This process is called a *scientific revolution.* In the first 25 years of the 20th century, a revolution in physics led to the development of the quantum theory, which also profoundly affected the science of chemistry.

One of the fundamental assumptions of early science was that nature is continuous; that is, nature does not make "jumps." On the macroscopic scale, this appears to be true enough. We can measure out an amount of graphite (carbon) of mass 9, or 8.23, or 6.4257 kg, and it appears that the mass can have any value provided that our balance is sufficiently accurate. On the atomic scale, however, this apparently continuous behavior breaks down. An analogy may be useful here. A sand beach from a distance appears smooth and continuous, but a close look reveals that it is made up of individual grains of sand. This same "graininess" is found in matter observed on the atomic scale. The mass of carbon (^{12}C) comes in "packets," each of which weighs 1.99265×10^{-26} kg. In principle, two, three, or any integral number of such packets can be "weighed out," but we cannot obtain $1\frac{1}{2}$ packets. Carbon is not a continuous material. It comes in chunks, each containing the minimum measurable mass of carbon—that of an atom. Similarly, electric charge comes in packets of size *e*, as shown in Section 1.4, and fractional charges are never observed in chemical reactions.

The central idea of quantum theory is that energy, like matter, is not continuous but it exists only in discrete packets. Discreteness of matter and charge on the microscopic scale seems entirely reasonable and familiar to us, based on the modern picture of atomic structure. But, the idea that energy also exists only in discrete chunks is contrary to our experience of the macroscopic world. The motions of a soccer ball rolling up and down the sides of a gully involve arbitrary amounts of kinetic and potential energy; nothing in ordinary human experience suggests that the energy of a system should change abruptly by "jumps." Understanding quantum mechanics requires that we develop a new kind of physical intuition, based on the results of experiments that are impossible to understand using classical mechanics. These results are completely divorced from ordinary human experience in the macroscopic world around us, and our physical intuition from the macroscopic world cannot be transferred to the quantum domain. We must resist the urge to interpret these quantum results in terms of ordinary experience.

To understand the far-reaching nature of the quantum revolution, you should consider the state of physics at the end of the 19th century. The 200 years that followed the seminal work (published in 1687) of Isaac Newton were the classical period in the study of mechanics, the branch of physics that predicts the motions of particles and the collections of particles that make up working mechanisms. By the end of that period, about 1900, physicists had achieved a deep understanding that successfully dealt with problems ranging from the motions of the planets in their orbits to the design of a bicycle. These achievements make up the field now called **classical mechanics**.

Classical mechanics can predict the future positions of a group of particles from their present positions and velocities if the forces among them are known. At the end of the 19th century, it was naturally thought that the motion of elementary particles—such as the recently discovered electron—could be described by classical mechanics. Once the correct force laws operating between the elementary particles were discovered, the properties of atoms and molecules could be predicted to any desired accuracy by solving Newton's equations of motion. It was believed that all the fundamental laws of physics had been discovered. At the dedication of the Ryerson Physics Laboratory at the University of Chicago in 1894, the American physicist A. A. Michelson said, "Our future discoveries must be looked for in the sixth decimal place." Little did he imagine the revolutionary changes that would shake physics and chemistry during the following 30 years.

Central to those changes was not only the recognition that energy is quantized but also the discovery that all particles display wavelike properties in their motions. The effects of wavelike properties are most pronounced for small, low-mass particles such as electrons in atoms. **Quantum mechanics** incorporated both the ideas of "wave–particle duality" and energy quantization into a single comprehensive theory that superseded classical mechanics to describe the properties of matter on the nanometer length scale. Quantum mechanics is one of the greatest intellectual achievements of the 20th century.

This chapter describes the origins of the quantum theory, summarizes its techniques, and demonstrates their application to simple model systems. Our goals are to help you become skilled and confident in using the language, concepts, and tools of quantum theory. With these skills, we will guide you to develop an intuitive understanding of the behavior of quantum systems—so foreign to our ordinary human experience—and the magnitudes of the observable quantities (energy, momentum, length) in the quantum domain. Chapter 5 shows how quantum mechanics explains the structure of atoms and the periodic table, and Chapter 6 shows how the quantum theory explains the formation of chemical bonds.

4.1 Preliminaries: Wave Motion and Light

Many kinds of waves are studied in physics and chemistry. Familiar examples include water waves stirred up by the winds over the oceans, set off by a stone dropped into a quiet pool, or created for teaching demonstrations by a laboratory water-wave machine. Sound waves are periodic compressions of the air that move from a source to a detector such as the human ear. Light waves, as discussed later in this chapter, consist of oscillating electric and magnetic fields moving through space. Even some chemical reactions occur in such a way that waves of color pass through the sample as the reaction proceeds. Common to all these wave phenomena is the oscillatory variation of some property with time at a given fixed location in space (Table 4.1). All of these waves are described by the same equations.

TABLE 4.1

Kinds of Waves

Wave	Oscillating Quantity
Water	Height of water surface
Sound	Density of air
Light	Electric and magnetic fields
Chemical	Concentrations of chemical species

A snapshot of a water wave (Fig. 4.1) records the crests and troughs present at some instant in time. The **amplitude,** A, of the wave is the maximum height or the maximum displacement of the water surface compared with the undisturbed level of the water; this undisturbed height is customarily chosen as the reference height and assigned the value zero. Positive vertical displacements increase the level of the water, whereas negative vertical displacements decrease the level of the water. We define the amplitude as either the height of a crest or the depth of a trough, and it is always given as an absolute value. The distance between two successive crests (or troughs) is called the **wavelength,** λ (Greek lambda), of the wave, provided that this distance is reproducible from peak to peak. The **frequency** of a water wave can be measured by counting the number of peaks or troughs observed moving past a fixed point in space per second. The frequency, ν (Greek symbol, pronounced nu), is measured in units of waves (or cycles) per second, or simply s^{-1}. The fundamental frequency unit one cycle per second has been named the **hertz (Hz)** in honor of the German physicist Heinrich Hertz. For example, if 12 water-wave peaks are observed to pass a certain point in 30 seconds, the frequency is

$$\text{frequency} = \nu = \frac{12}{30 \text{ s}} = 0.40 \text{ s}^{-1} = 0.40 \text{ Hz}$$

The wavelength and frequency of a wave are related through its speed—the rate at which a particular wave crest moves through the medium. In Figure 4.1, the crest at the left end of the horizontal black arrow will move forward exactly one wavelength in one cycle of the wave. By definition, the time required for the crest to travel this distance is the reciprocal of the frequency, $\tau = \nu^{-1}$, so the speed (the distance traveled divided by the time elapsed) is given by

$$\text{speed} = \frac{\text{distance traveled}}{\text{time elapsed}} = \frac{\lambda}{\nu^{-1}} = \lambda\nu$$

The speed of a wave is the product of its wavelength and its frequency.

The dependence of the vertical displacement, $h(x,t) - h_0$, of the water as a function of both position x and time t in Figure 4.1 for a wave traveling in the $+x$ direction may be summarized by the expression $h(x,t) - h_0 = A(\cos 2\pi x/\lambda - 2\pi t/\tau)$ where A is the amplitude of the wave.

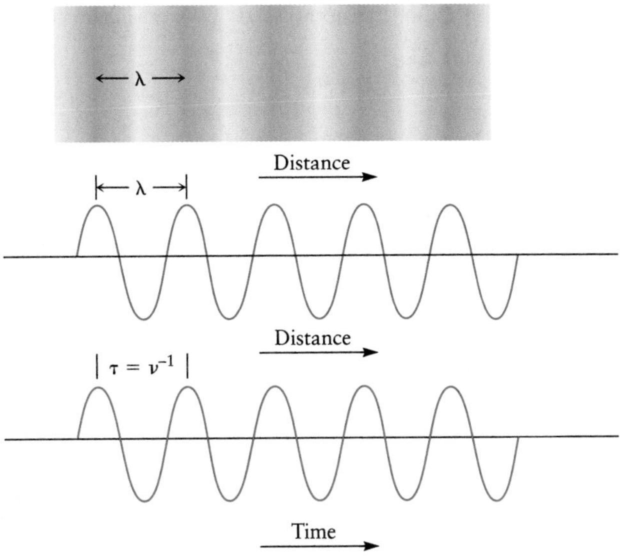

FIGURE 4.1 As a water wave moves across an otherwise calm tank, its amplitude and wavelength can be determined. Its speed is found as the ratio of the distance traveled by a particular wave crest to the time elapsed.

4.1.1 Electromagnetic Radiation

By the end of the 18th century, the behavior of light was well described by a wave model. The signature properties of light—diffraction, interference, and polarization—were understood as consequences of wave propagation. In 1865, the Scottish physicist James Clerk Maxwell proposed a theory that described visible light as a propagating wave of **electromagnetic radiation** that carries both energy and momentum. Unlike water and sound waves, electromagnetic waves are not sustained by some "propagating medium" such as water or air. Rather, a beam of light consists of oscillating *electric* and *magnetic fields* oriented perpendicular to one another and to the direction in which the light is propagating (Fig. 4.2). These fields are produced by the motion of charged particles in the source of the light. These oscillating fields can transfer energy and momentum to other charged particles that intercept the beam in some location that is remote from the source. Electromagnetic waves carry information from a broadcast source to a remote receiver in wireless communication. Indeed, one

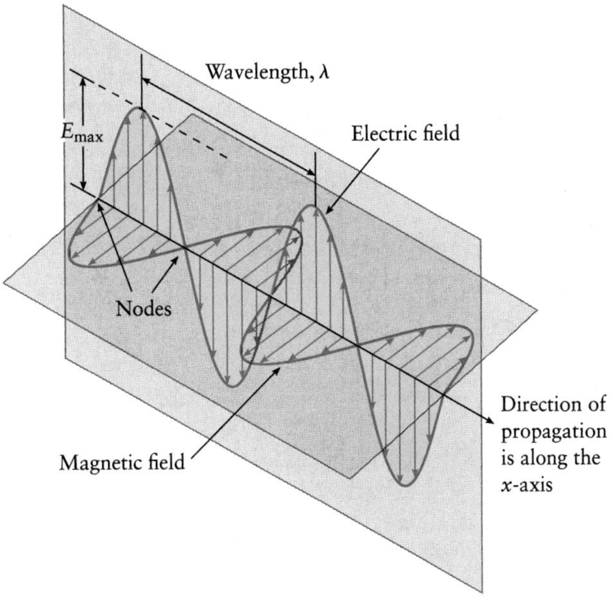

FIGURE 4.2 Light consists of waves of oscillating electric and magnetic fields that are perpendicular to each other and to the direction of propagation of the light.

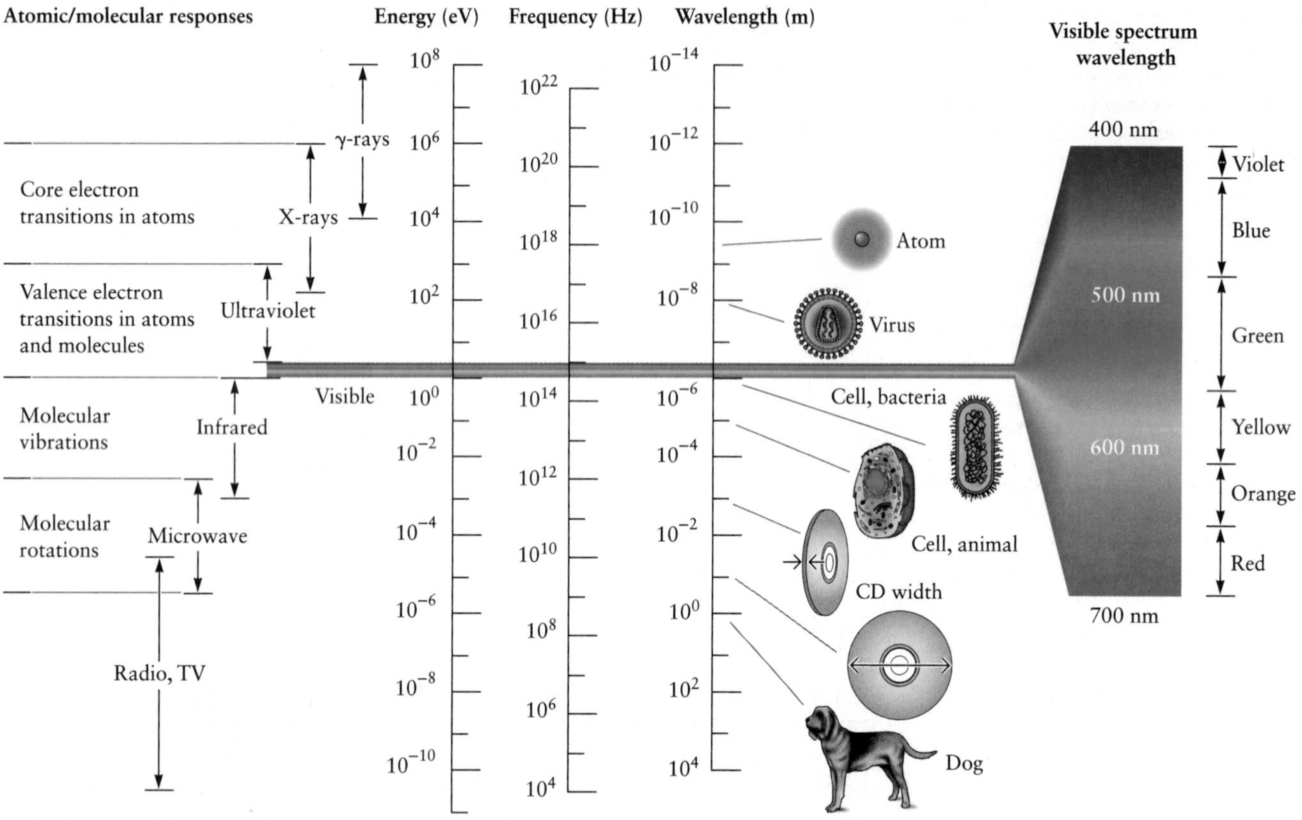

FIGURE 4.3 The electromagnetic spectrum. Note the small fraction that is visible to the human eye.

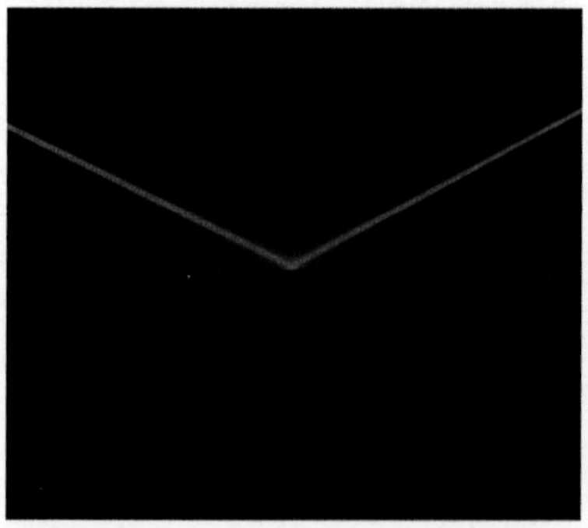

© Cengage Learning/Henry Leap and Jim Lehman

FIGURE 4.4 A laser emits a well-collimated beam of light with a narrow range of wavelengths. The direction of motion of a laser beam can be manipulated by inserting mirrors in the path of the beam.

of the early triumphs of Maxwell's theory was the development of the radio, based largely on the experimental work of Heinrich Hertz. We will see that electromagnetic radiation is both emitted and absorbed by atoms and molecules. It is, therefore, one of our most effective tools for probing the nature of atoms and molecules.

Electromagnetic waves are described by the equations introduced earlier. A detector located at any point x along the x-axis in Figure 4.2 will see the electric field $E(x,t)$ oscillate with time as the wave propagates along the x-axis according to the equation

$$E(x, t) = E_{\max} \cos[2\pi(x/\lambda - \nu t)]$$

where $E_{\max}$ is the amplitude of the wave. The speed, c, of light passing through a vacuum is equal to the product $\lambda\nu$, and its value is

$$c = \lambda\nu = 2.99792458 \times 10^8 \text{ m s}^{-1} \qquad [4.1]$$

The speed, c, is a universal constant; it is the same for all types of radiation in the electromagnetic spectrum (Fig. 4.3). Regions of the electromagnetic spectrum are characterized by different values of wavelength and frequency. The region visible to the eye, which is a small fraction of the entire spectrum, comprises bands of colored light that cover particular ranges of wavelength and frequency. The band of light we perceive as green is centered about 5.7×10^{14} Hz with wavelengths near 5.3×10^{-7} m (530 nm). Red light is characterized by a lower frequency and a longer wavelength than green light, and violet light is characterized by a higher frequency and shorter wavelength than green light. A laser, such as the one shown in Figure 4.4, emits nearly monochromatic light (light of a

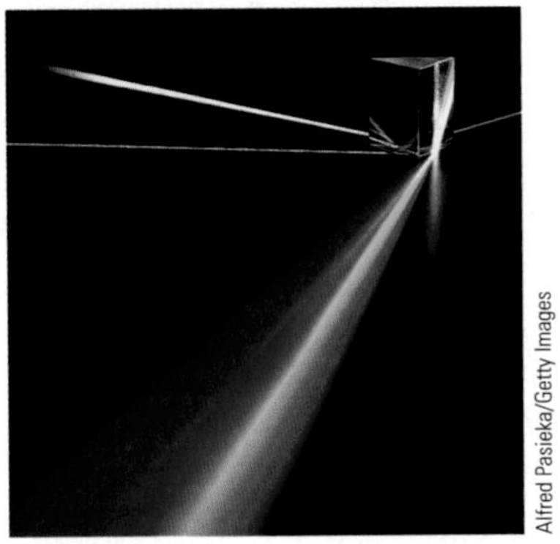

FIGURE 4.5 When white light is passed through slits to produce a narrow beam and then refracted in a glass prism, the various colors are dispersed, or separated from one another, because each wavelength is refracted through a different angle.

single frequency and wavelength). White light contains the full range of visible wavelengths; it can be resolved into its component wavelengths, or refracted, by passing it through a prism (Fig. 4.5).

Electromagnetic radiation that lies outside the visible region is also familiar to us (see Fig. 4.3). The warmth radiated from a stone pulled from a fire is largely due to infrared radiation, whose wavelength is longer than that of visible light. Microwave ovens use radiation whose wavelength is longer than infrared wavelengths, and radio communication uses still longer wavelengths. Radio stations are identified by their broadcast frequencies. FM stations typically broadcast at frequencies of tens to hundreds of megahertz (1 MHz $= 10^6$ s^{-1}), whereas AM stations broadcast at lower frequencies, from hundreds to thousands of kilohertz (1 kHz $= 10^3$ s^{-1}). You might check the frequencies of some of your favorite radio stations; ours include a classical music station broadcasting at 90.5 MHz (FM) and a sports station broadcasting at 1300 kHz (AM). Radiation with wavelengths shorter than that of visible light includes ultraviolet light, X-rays, and gamma rays; radiation in these regions of the electromagnetic spectrum (with wavelengths shorter than about 340 nm) can cause ionization and damage in biological tissue and are often collectively called **ionizing radiation.**

EXAMPLE 4.1

Almost all commercially available microwave ovens use radiation with a frequency of 2.45×10^9 Hz. Calculate the wavelength of this radiation.

Solution

The wavelength is related to the frequency as follows:

$$\lambda = \frac{c}{\nu} = \frac{3.00 \times 10^8 \text{ m s}^{-1}}{2.45 \times 10^9 \text{ s}^{-1}} = 0.122 \text{ m}$$

Thus, the wavelength is 12.2 cm.

Related Problems: 3, 4

The wave theory of light is based on experimental demonstrations of interference and diffraction, results that can only be explained as consequences of wave motion. When two light waves pass through the same region of space, they interfere to create a new wave called the *superposition* of the two. If the crests of the waves are aligned, the superposition wave is their sum and has greater amplitude than the original waves. This process is called **constructive interference.** If the crest of one wave is aligned with the trough of the other, the superposition wave is their difference, and the waves cancel through **destructive interference** (Fig. 4.6a, b). If the waves are not perfectly aligned for either constructive or destructive interference, the superposition wave has amplitude

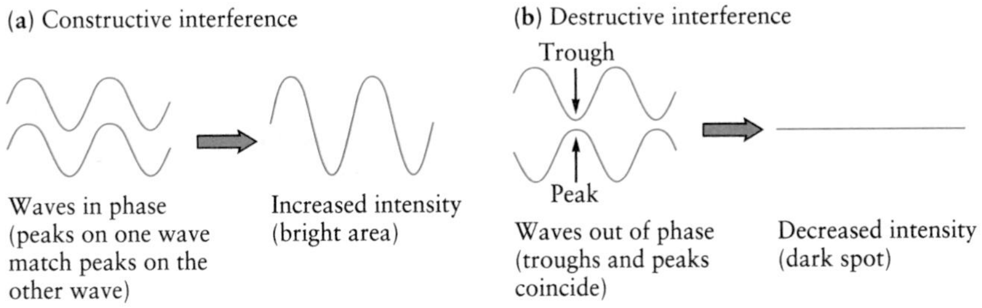

FIGURE 4.6 Interference of waves (a) Constructive interference. When two waves are in phase their crests and troughs add to produce a new wave of greater amplitude. (b) Destructive interference. When two waves are out of phase, their crests and troughs cancel each other and the net amplitude is 0. (Adapted from Zumdahl/DeCoste, *Chemical Principles, 6e*, Fig. 12.7; Cengage Learning.)

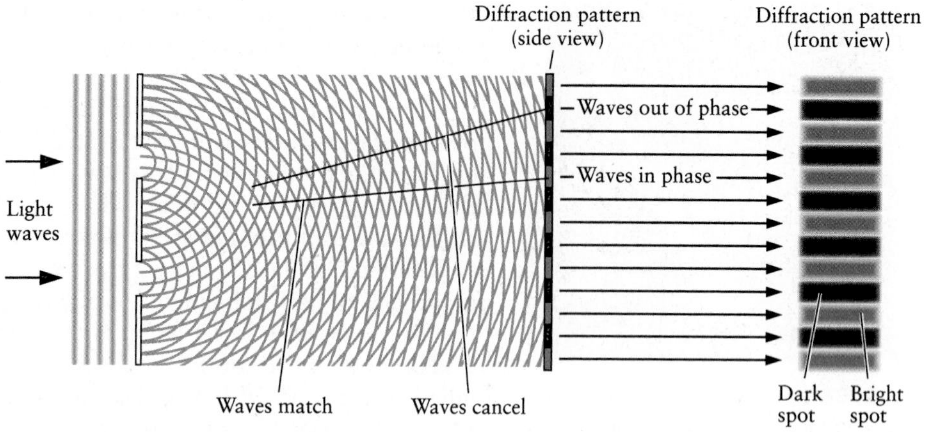

FIGURE 4.7 Diffraction of waves. The double slit experiment demonstrates diffraction as a consequence of constructive and destructive interference of waves arriving from the slits. The diffraction pattern on the screen behind the slits shows alternating bright regions (constructive intereference) and dark regions (destructive interference). (Adapted from Moore/Stanitski/Jurs, *Chemistry: The Molecular Science 4e*, Fig. 7.5, p. 278; Cengage Learning.)

intermediate between the two extreme results. When a monochromatic beam of light, such as that from a laser (see Fig. 4.4), passes through two open slits in a solid wall, each slit becomes the source of new light waves. These outgoing new waves interfere to produce a **diffraction pattern**, which has alternating bright and dark regions. The dark regions are the result of destructive interference, and the bright regions arise from constructive interference. The solid lines in Figure 4.7 trace one set of points in space where two outgoing waves undergo constructive interference and another set where they show destructive interference. We see later (Section 21.1) that electromagnetic waves interact with the atoms in a crystalline solid to generate outgoing waves that produce diffraction patterns. From the dimensions of these patterns we can determine the distances between atoms in the solid.

4.2 Evidence for Energy Quantization in Atoms

Rutherford's planetary model of the atom was inconsistent with the laws of classical physics (see discussion in Section 3.3). According to Maxwell's electromagnetic theory, accelerated charges must emit electromagnetic radiation. An electron in orbit around the nucleus is accelerating because its direction is constantly changing. It must, therefore, emit electromagnetic radiation, lose energy, and eventually spiral into the nucleus. The very existence of stable atoms was perhaps the most fundamental of all the conceptual challenges facing physicists in the early 1900s. The recognition that energy is quantized in atoms was the first step toward resolving the conceptual conflicts.

This section begins with a discussion of blackbody radiation, the experiment that introduced energy quantization into science. Next, two sets of experiments that demonstrated quantization of energy in free atoms in the gas phase are described. We describe these experiments using energy level diagrams, which represent the discrete energy states of the atom. Our goal here is to introduce you to the relationship between the experimental evidence for energy quantization and the energy-level diagrams used to describe these experiments. Later in the chapter, we use the quantum theory to explain how energy is quantized and to predict the allowed values of the energy for several model problems. We have organized our discussion to group key concepts together, for better coherence and to provide physical insight; it does not strictly follow the historical development of the field.

4.2.1 Blackbody Radiation and Planck's Quantum Hypothesis

We are about to discuss a monumental achievement in the development of modern science, which changed forever the way we look at the world. This was the recognition that objects cannot gain or lose energy in arbitrary or continuous amounts, but instead transfer energy only in discrete, discontinuous amounts that are multiples of some fundamental quantity of energy. The German physicist Max Planck achieved this insight in 1901 while trying to explain some puzzling new experimental measurements on the interaction of solid objects with radiant energy, which was known as **blackbody radiation**. You will shortly see that you are already familiar with blackbody radiation in various guises, and we relate the discussion closely to the experimental results so that you can always see the problem exactly as Planck saw it. We invite you to read and think along with Planck and to witness an important demonstration of how science advances. When experimental results do not agree with established scientific theories, the theories must be either modified or discarded and replaced with new ones, to account for both the new and the old experimental results. This process leads to the development of theories that provide a more fundamental understanding of a wider range of phenomena than their predecessors.

Every object emits energy from its surface in the form of thermal radiation. This energy is carried by electromagnetic waves; the distribution of the wavelengths of electromagnetic waves depends on the temperature of the object. At ordinary temperatures, thermal radiation falls within the infrared portion of the electromagnetic spectrum; images formed by this radiation are used to map the surface of Earth from satellites in space and for tracking the movement of people in darkness using "night vision" detectors. As objects are heated to higher temperatures, the total intensity of radiation emitted over all frequencies increases, and the frequency distribution of the intensity also changes. The solid curves in Figure 4.8 show how the measured radiation intensity depends on frequency and temperature. There are two important features of these curves. First, the maximum in the radiation intensity distribution moves to higher frequency (shorter wavelength) as the temperature increases. This phenomenon is observed in familiar objects such as the heating element on an electric kitchen

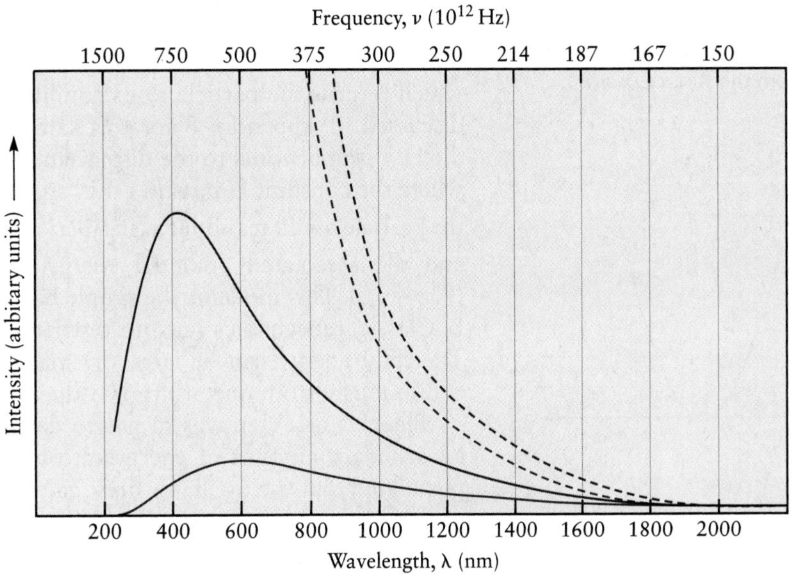

FIGURE 4.8 The dependence of the intensity of blackbody radiation on wavelength for two temperatures: 5000 K (red curve) and 7000 K (blue curve). The sun has a blackbody temperature near 5780 K, and its light-intensity curve lies between the two shown. The classical theory (dashed curves) disagrees with observation at shorter wavelengths.

range or the filament in an incandescent lightbulb. As these objects are heated, they first glow red, then orange, then yellow, and finally, white. It also explains the differences in color among stars; the hottest stars appear to be nearly white, whereas the colors of cooler stars can range from red to yellow. Second—and this is a key result—the radiation intensity falls to zero at extremely high frequencies for objects heated to any temperature.

The sources of blackbody radiation, according to classical physics, are oscillating electrical charges in the surfaces of these objects that have been accelerated by ordinary thermal motion. Each motion persists for a certain period, producing radiation whose frequency is inversely related to that period. A number of scientists used different methods to calculate the radiation intensity curves using this simplified model and arrived at the following result:

$$\rho_T(\nu) = \frac{8\pi k_B T \nu^2}{c^3} \qquad [4.2]$$

where $\rho_T(\nu)$ is the intensity of the radiation at the frequency ν; k_B is a fundamental constant called the Boltzmann constant, which is discussed in Sections 9.5 and 9.6; T is the temperature in degrees Kelvin (K) and c is the speed of light.

These calculated results, shown for 5000 and 7000 K by the dashed curves in Figure 4.8, agree well with experiment at lower frequency. But the theory does not predict a maximum in the intensity distribution, and even worse, it disagrees badly with the experimental results at high frequencies. This feature of the result was called the "ultraviolet catastrophe" because it predicts an infinite intensity at very short wavelengths, whereas the experimental intensities always remain finite and actually fall to zero at very short wavelengths (very high frequencies). The calculated result failed completely to explain the frequency distribution in blackbody radiation; yet, it is a direct consequence of the laws of classical physics. How could this conflict be resolved?

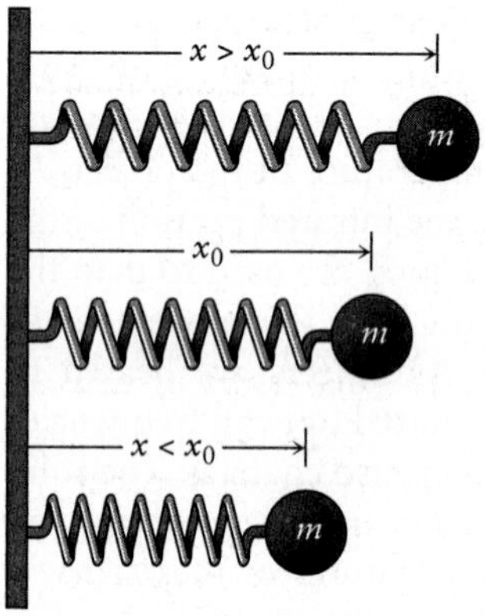

FIGURE 4.9 A charged particle of mass m bound to a solid surface by a spring is a model for the oscillatory motions of the surface atoms of a black body. The particle is shown at its rest position x_0, at a position closer to the surface ($x < x_0$), and at one farther from the surface ($x > x_0$).

Blackbody radiation was explained by Max Planck in 1901, but only by overthrowing the very foundations of classical mechanics. Planck reasoned that the very high-frequency oscillators must not be excited by the thermal energy of the hot body to the same degree as the lower frequency oscillators. This was a challenge to explain because classical mechanics allows an oscillator to have any energy. Planck's argument involved two steps, which are explained as follows.

For simplicity in following Planck's hypothesis, let us focus on just one of the oscillating charged particles and visualize it as a ball of mass, m, held in place by a spring (Fig. 4.9). As the particle moves in response to the thermal motion of the atoms in the hot body, the spring exerts a "restoring force," F, which returns the particle to its equilibrium position, which we will call x_0. As discussed in Appendix B for this same model problem, the restoring force is directly proportional to the displacement, and the force law is $F = -k(x - x_0)$, where the constant k measures the "stiffness" of the spring. The displacement of the particle oscillates about x_0 in a periodic motion of frequency $\nu = (1/2\pi)\sqrt{k/m}$, and the associated potential energy of the particle is $V(x) = \frac{1}{2}k(x - x_0)^2$. (Fig. 4.10). This model is the simple harmonic oscillator described in Appendix B. Classical mechanics puts no restrictions on the value of the total energy, E. The total energy can be large or small, and it can be changed smoothly and continuously from one arbitrary value to another.

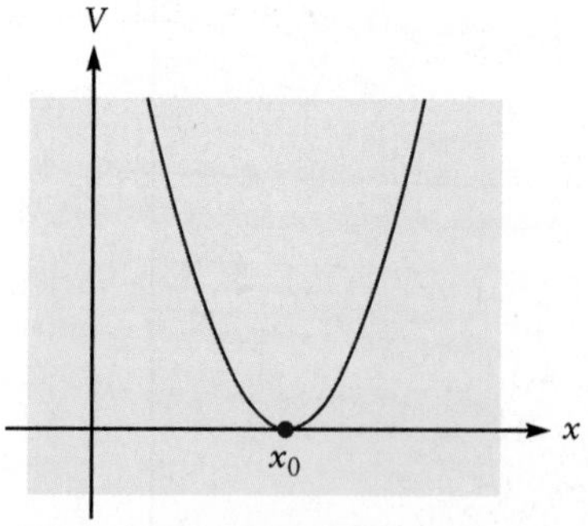

FIGURE 4.10 The potential energy curve for an oscillator is a consequence of the "restoring force" that always drives the oscillator toward its equilibrium position.

Planck's first step was to pose a daring hypothesis: It is *not* possible to put an arbitrary amount of energy into an oscillator of frequency ν. Instead, he postulated that the oscillator must gain and lose energy in "packets," or *quanta*, of magnitude $h\nu$, and that the total energy of an oscillator, ε_{osc}, can take only discrete values that are integral multiples of $h\nu$:

$$\varepsilon_{osc} = nh\nu \qquad n = 1, 2, 3, 4, \ldots \qquad [4.3]$$

In Planck's hypothesis, h was a constant with physical units *energy* × *frequency*$^{-1}$ = *energy* × *time*, but the value of which was yet to be determined.

You can easily visualize the consequences of Planck's hypothesis using the simple harmonic oscillator model. Replace the spring and ball with a rubber band stretched between your fingers. Experience shows that you can stretch the band to any arbitrary length by applying the right amount of energy (so long as you do not rupture the band). But under Planck's hypothesis, the band would accept only certain specific values of energy. The rubber band would behave as if it could be stretched only to certain specific positions. It simply would not respond to attempts to give it energy between these specific values. This fact is contrary to all ordinary human experience with tangible, macroscopic objects. And yet, this is how energy transfer operates in the microscopic world of atoms, electrons, and molecules.[1]

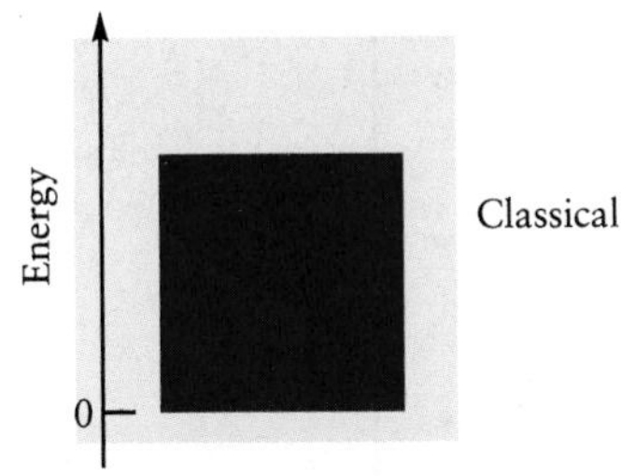

FIGURE 4.11 An oscillator obeying classical mechanics has continuous values of energy and can gain or lose energy in arbitrary amounts.

The dramatic contrast between the energy values allowed by classical mechanics and those that arise from Planck's postulate is illustrated using **energy-level diagrams**, in which a horizontal line represents an allowed energy value for a system. The height of each line above the zero of energy represents the total energy in that level. In macroscopic systems that are well described by classical mechanics, all energies are allowed; the upper energy level diagram in Figure 4.11 represents the continuum of energies that the rubber band can accept up to the point where it breaks. For the quantum oscillators that Planck proposed, only those levels shown on the lower energy level diagram are allowed (Fig. 4.12).

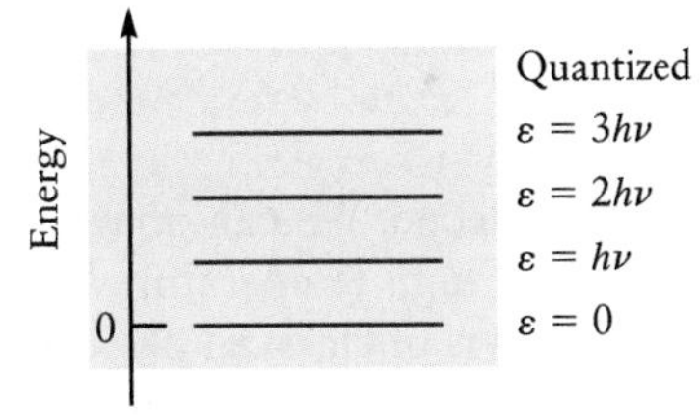

FIGURE 4.12 An oscillator described by Planck's postulate has discrete energy levels. It can gain or lose energy only in amounts that correspond to the difference between two energy levels.

Planck's second step was to predict the radiation intensity curves by calculating the average energy in these quantized oscillators at each frequency as a function of temperature. The key idea is that the excitation of a particular oscillator is an all-or-nothing event; there is either enough thermal energy to cause it to oscillate or there is not. According to Planck, the falloff in the intensity with frequency at a given temperature of the blackbody radiation is due to a diminishing probability of exciting the high-frequency oscillators. Planck's distribution for the intensity of radiation connected with the surface oscillators is

$$\rho_T(\nu) = \frac{8\pi h\nu^3}{c^3}\frac{1}{e^{h\nu/k_B T} - 1} \qquad [4.4]$$

All of the symbols in Equation 4.4 have been identified earlier in this chapter. The value of h was determined by finding the best fit between this theoretical expression and the experimental results. Figure 4.13 shows the fit for $T = 1646$ K, resulting in the value $h = 6.63 \times 10^{-34}$ J s. The value of h, a fundamental constant of nature, has been measured to very high precision over the years by a number of other techniques. It is referred to as **Planck's constant**, and the currently accepted value is

$$h = 6.62606957 \times 10^{-34} \text{ J s}$$

We ask you to accept that the second fraction on the right-hand side of Equation 4.4 is the probability that an oscillator of frequency ν is activated at a given temperature T. Chapter 12 presents the origin of this probability in the famous Boltzmann distribution, but in this chapter we want to use the result to demonstrate some additional consequences of Planck's hypothesis.

Before proceeding to explore the implications of the Planck distribution, we need to check whether it reduces to the classical expression under the appropriate conditions. It is always important to check whether new concepts can be matched with old concepts under appropriate conditions; this demonstrates that

[1]We will see later that energy transfer into macroscopic objects is also quantized. However, the discrete values are so closely spaced that they appear continuous, and the system behaves as if it obeys classical mechanics. This demonstrates the classical limit of quantum mechanics.

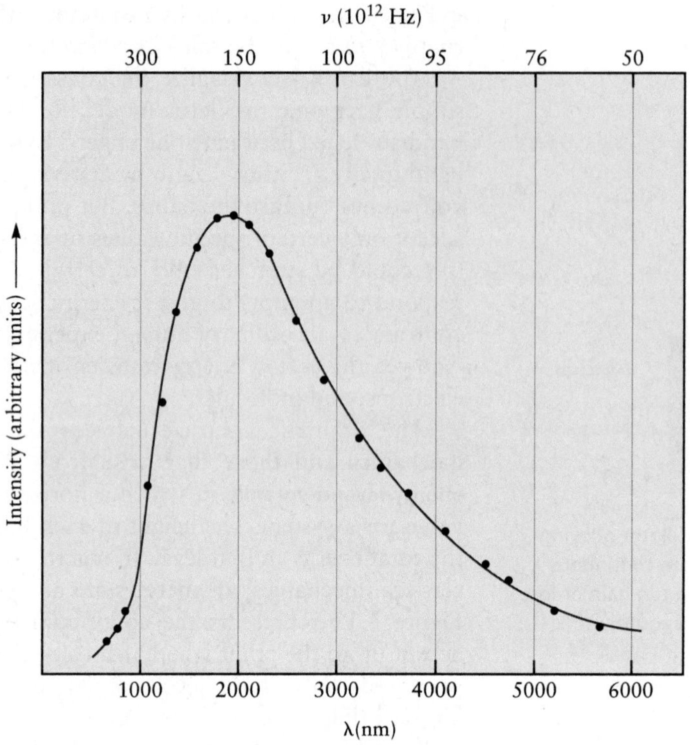

FIGURE 4.13 Experimental test of Planck's distribution for blackbody radiation. The dots represent experimental data acquired at $T = 1646$ K. The continuous curve represents Planck's predicted distribution, with the parameter $h = 6.63 \times 10^{-34}$ J s. Agreement between experiment and theory is spectacular, demonstrating the validity of Planck's theory and also determining the value of the previously unknown parameter h.

the new concepts represent an orderly advance in knowledge. We can imagine that all of the oscillators would be excited at sufficiently high temperatures, in which case, the system should behave according to the laws of classical physics. We express this condition mathematically by setting $h\nu/k_BT \ll 1$, a ratio that is nearly zero. You will soon learn in calculus that most functions can be represented by simpler forms as the argument of the function approaches zero. For the exponential function, $\exp(x) \approx 1 + x$ when x is nearly zero. Using this approximation, we obtain the high temperature limit of Planck's distribution

$$\rho_T(\nu) = \frac{8\pi h\nu^3}{c^3}\frac{1}{e^{h\nu/k_BT} - 1} \approx \frac{8\pi h\nu^3}{c^3}\frac{1}{([1 + h\nu/k_BT] - 1)} = \frac{8\pi k_BT\nu^2}{c^3} \quad [4.5]$$

which is valid as $T \longrightarrow \infty$. This is indeed the classical result quoted in Equation 4.2.

Physically, Planck's dramatic explanation of blackbody radiation includes three fundamentally new ideas:

1. The energy of a system can take only discrete values, which are represented on its energy-level diagram.
2. A quantized oscillator can gain or lose energy only in discrete amounts $\Delta\epsilon$, which are related to its frequency by $\Delta\epsilon = h\nu$.
3. To emit energy from higher energy states, the temperature of a quantized system must be sufficiently high to excite those states.

These three ideas have permeated all areas of science and technology. They are the basis for our understanding that energy (like matter) is discrete, not continuous, and that it can be transferred only in discrete chunks and not by arbitrary amounts. Every system has its own energy-level diagram that describes the allowed energy values and the possible values of energy transfers.

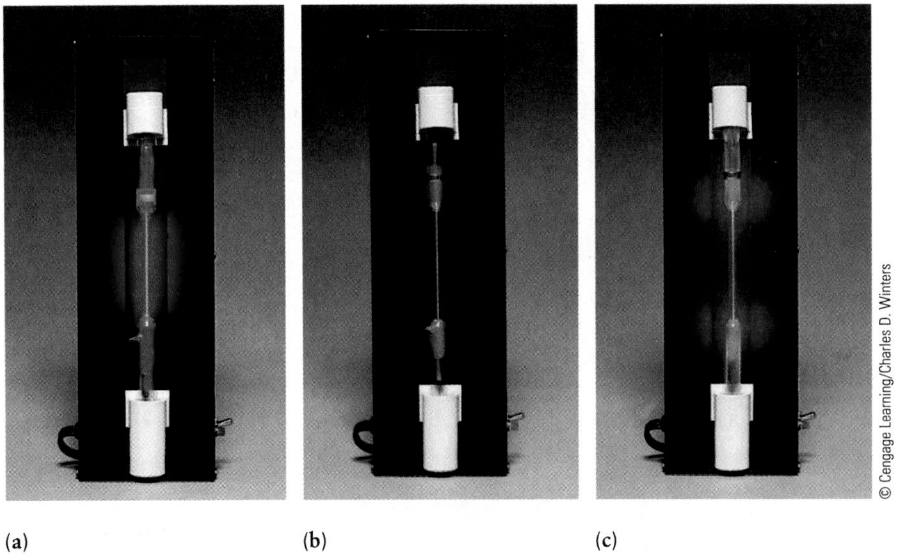

FIGURE 4.14 When a gas is excited in an electrical discharge, it glows as it emits light. The colors of the light emitted by three gases are shown: (a) neon, (b) argon, and (c) mercury. Each emission consists of several wavelengths of light, and the perceived color depends on which wavelength predominates.

4.2.2 Atomic Spectra and Transitions between Discrete Energy States

Light that contains a number of different wavelengths (see Fig. 4.5) can be resolved into its components by passing it through a prism, because each wavelength is refracted through a different angle. In the last quarter of the 19th century, experiments showed that elemental gases excited in electrical discharges emitted light in characteristic colors (Fig. 4.14). To understand this process, physicists sought to resolve this light into its component wavelengths. One instrument used for this purpose is called a **spectrograph** (Fig. 4.15a). The spectrograph is enclosed in a boxlike container to exclude stray light. The light to be analyzed enters through a narrow slit in the walls. The light passes to the prism, where it is dispersed into its components, and then falls on a photographic plate or other detector. The detector records the position and intensity of an image of the slit formed by each component wavelength. The recorded array of images is called the *spectrum* of the incoming light. If the incoming light contained all wavelengths, the spectrum would be continuous bands of dispersed colors as in Figure 4.5. Instead, the light emitted from gaseous atoms excited in flames or in electrical discharges gives discrete emission spectra, that is, a series of parallel lines. Each line is an image of the slit at a specific wavelength (see Fig. 4.15a), and the spectrum is unique for each element (Fig. 4.16). If white light is passed through a sample of gaseous atoms and the transmitted light is then sent into the spectrograph, the resulting absorption spectrum consists of dark slit images superimposed on the continuous spectrum of white light (see Fig. 4.15b). These experiments show that atoms emit and absorb light at a discrete set of frequencies characteristic of a particular element. For example, in 1885, J. J. Balmer discovered that hydrogen atoms emit a series of lines in the visible region, with frequencies that fit the following simple formula:

$$\nu = \left[\frac{1}{4} - \frac{1}{n^2}\right] \times 3.29 \times 10^{15}\ \text{s}^{-1} \qquad n = 3, 4, 5, \ldots \qquad [4.6]$$

The hydrogen atom lines shown in Figure 4.16 fit this equation with $n = 3$, 4, 5, and 6 corresponding to the red, green, blue, and violet lines in the spectrum, respectively. Trying to understand the existence of discrete line spectra

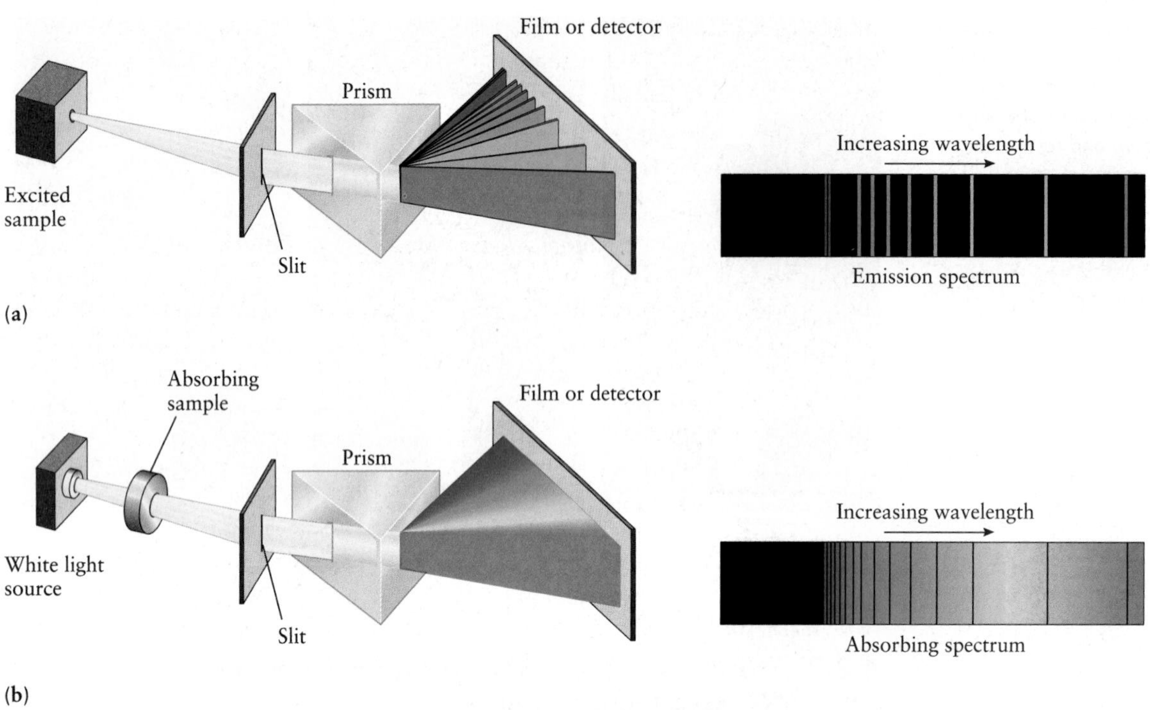

FIGURE 4.15 (a) The emission spectrum of atoms is measured by passing the light emitted from an excited sample through a prism to separate it according to wavelength, then recording the image on photographic film or an electronic detector. The spectrum consists of discrete bright lines against a dark background. (b) In absorption spectroscopy, white light from a source passes through the unexcited sample, which absorbs certain discrete wavelengths of light. Dark lines appear on a bright background.

and the various empirical equations that relate the frequencies of the lines challenged physicists for more than three decades.

The first explanation for these surprising experimental results was provided in 1913 by the Danish physicist Niels Bohr. He proposed a model of the hydrogen atom that allowed only discrete energy states to exist. He also proposed that light absorption resulted from a *transition* of the atoms between two of these states. The frequency of the light absorbed is connected to the energy of the

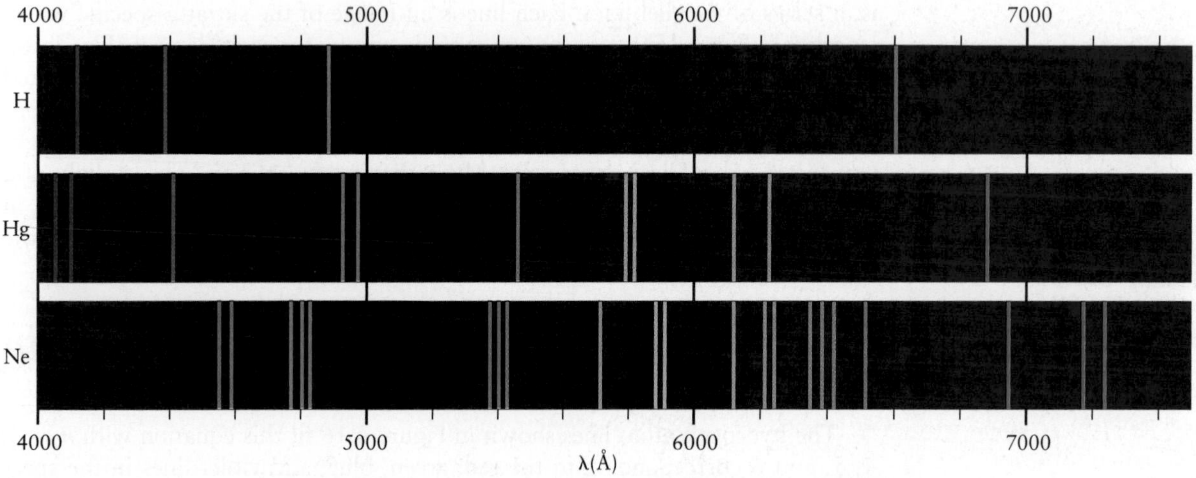

FIGURE 4.16 Atoms of hydrogen, mercury, and neon emit light at discrete wavelengths. The pattern seen is characteristic of the element under study. 1 Å = 10^{-10} m.

initial and final states by the expression $E_i + h\nu = E_f$, where $h\nu$ is the energy of the absorbed light (see Section 4.4.1). This gives

$$\nu = \frac{E_f - E_i}{h} \quad \text{or} \quad \Delta E = h\nu \text{ (absorption)} \qquad [4.7]$$

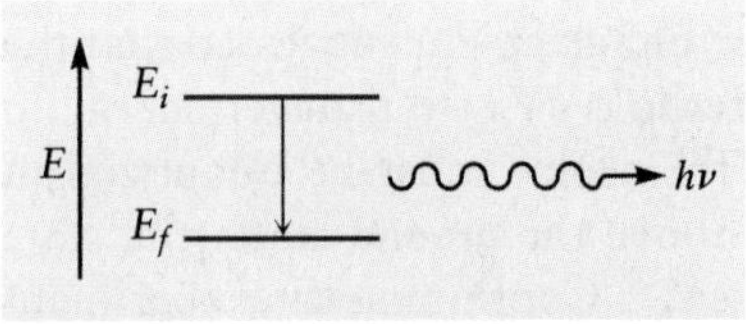

FIGURE 4.17 An atom makes a transition from state E_i to E_f and emits a photon of frequency $\nu = [E_i - E_f]/h$.

where h is Planck's constant. In absorption, the energy of the final state, E_f, is greater than that of the initial state so the signs work out correctly; ν is a positive number as it must be. For emission, however, $E_f < E_i$, and energy conservation gives $E_i = E_f + h\nu$. The energy of the light is now on the final (product) side of the equation, as the light is produced or emitted (not absorbed). This gives $\Delta E = E_f - E_i = -h\nu$ for emission (Fig. 4.17). Thus $\Delta E > 0$ for absorption, whereas $\Delta E < 0$ for emission. The Bohr model also accounts for the values of the discrete energy levels in the hydrogen atom (see Section 4.3).

The atoms of every element can be represented by an energy-level diagram in which the energy difference between two levels is related by Equation 4.7 to the frequency of a specific line in the experimental spectrum of the atom. Except for the simplest case of hydrogen, however, constructing the energy-level diagram from the experimental spectrum is difficult because numerous transitions are involved. Nonetheless, spectroscopists have assigned the atomic spectra of most of the elements in the periodic table, and extensive tabulations of the results are readily available.

4.2.3 The Franck–Hertz Experiment and the Energy Levels of Atoms

In 1914, the German physicists James Franck and Gustav Hertz (nephew of Heinrich Hertz) conducted an experiment to test Bohr's hypothesis that the energy of atoms is quantized by measuring the energy transferred to an atom in collisions with electrons. In their apparatus (Fig. 4.18), electrons of known energy collided with gaseous atoms, and the energy lost from the electrons was measured. Electrons were emitted from the heated cathode C and accelerated toward the anode A. Holes in the anode allowed electrons to pass toward the collector plate P with known kinetic energy controlled by the accelerating voltage between C and A. The apparatus was filled to a low pressure with the gas to be studied. The current arriving at P was studied as a function of the kinetic energy of the electrons by varying the accelerating voltage.

The experiment was started using a very low accelerating voltage, and the current was found to increase steadily as the accelerating voltage was increased.

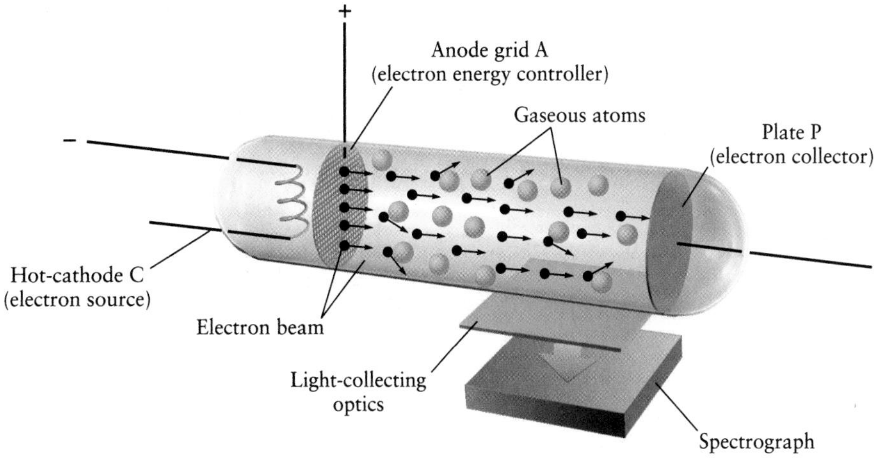

FIGURE 4.18 Apparatus of Franck and Hertz that demonstrates the quantization of energy in atoms. Gaseous atoms collide with electrons and gain energy by collisions only when the energy of the electron exceeds a certain threshold. The excited atom then emits a photon whose frequency is determined by the energy transferred to the atom during the collision.

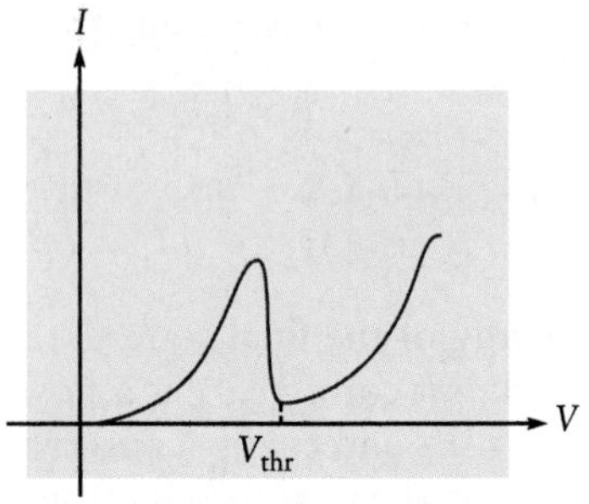

FIGURE 4.19 The current in the Franck–Hertz experiment shows a sharp change at a particular value of the accelerating voltage, corresponding to the threshold for energy transfer from the electron to a gaseous atom.

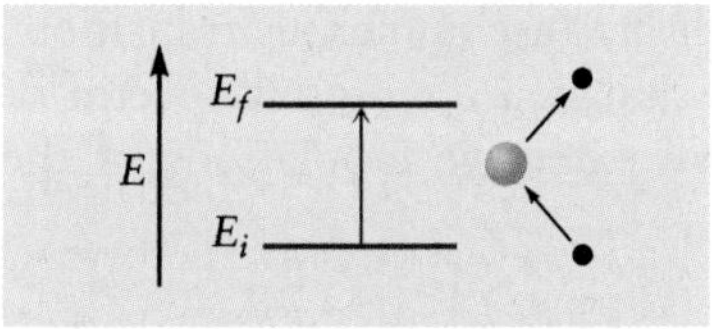

FIGURE 4.20 The Franck–Hertz experiment measures directly the separation between energy levels of the atom by measuring the energy lost by an electron colliding with the atom.

At a certain voltage, V_{thr}, the current dropped sharply, going nearly to zero (Fig. 4.19). This observation implied that most of the electrons had lost their kinetic energy in collisions with the gas atoms and were unable to reach the collector. As the voltage was increased above V_{thr}, the current rose again. This result indicated that electrons were reaccelerated after collisions and gained sufficient energy to reach the collector.

The abrupt fall in the plot of current versus voltage at V_{thr} suggested that the kinetic energy of the electrons must reach a threshold eV_{thr} to transfer energy to the gas atoms, suggesting that the energy of the atoms must be quantized in discrete states. The **first excited state** must lie above the **ground state** (the state with lowest energy) (Fig. 4.20) by the amount eV_{thr}. Continuing the experiment with higher values of accelerating voltage revealed additional energy thresholds corresponding to excited states with higher energies.

To confirm their interpretation, Franck and Hertz used a spectrograph to analyze light that was emitted by the excited atoms. When the accelerating voltage was below V_{thr}, no light was observed. When the accelerating voltage was slightly above V_{thr}, a single emission line was observed whose frequency was very nearly equal to

$$\nu = \frac{E_{\text{upper}} - E_{\text{lower}}}{h} = \frac{eV_{\text{thr}}}{h} \qquad [4.8]$$

At higher accelerating voltages, additional spectral emission lines appeared as each additional excitation energy threshold was reached.

EXAMPLE 4.2

The first two excitation voltage thresholds in the Franck–Hertz study of mercury vapor were found at 4.9 and 6.7 V. Calculate the wavelength of light emitted by mercury atoms after excitation past each of these thresholds.

Solution

The emitted wavelength at a particular value of V_{thr} is given by the following equation:

$$\lambda = \frac{hc}{E_{\text{upper}} - E_{\text{lower}}} = \frac{hc}{eV_{\text{thr}}}$$

$$= \frac{(6.6261 \times 10^{-34}\text{ J s})(2.9979 \times 10^{8}\text{ m s}^{-1})(1 \times 10^{9}\text{ nm m}^{-1})}{(1.6022 \times 10^{-19}\text{ C})(V_{\text{thr}})}$$

$$= \frac{1239.8\text{ nm J C}^{-1}}{V_{\text{thr}}}$$

The value of each emission wavelength is calculated by substituting in a particular value of V_{thr} expressed in units of volts (V). The units in the numerator simplify to nm J C^{-1} = nm V.

At V_{thr} = 4.9 V, the calculated wavelength is λ = 250 nm. The wavelength actually observed above this threshold was 253.7 nm.

At V_{thr} = 6.7 V, the calculated wavelength is λ = 180 nm. The wavelength actually observed above this threshold was 184.9 nm.

Energy differences measured by the Franck–Hertz method and by optical emission spectroscopy agree quite closely. The optical measurements are more precise. These results enable us to begin to construct the energy diagram for mercury, showing the location of the first two excited states relative to the ground state (Fig. 4.21).

Related Problems: 17, 18

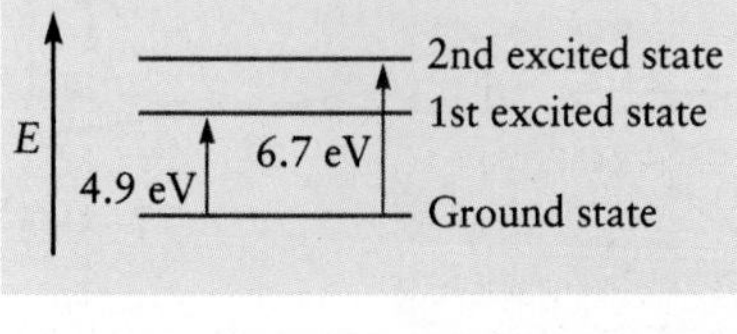

FIGURE 4.21 Simplified energy-level diagram for mercury.

The significance of the Franck–Hertz experiment in the development of modern science cannot be exaggerated. It demonstrated that atoms absorb energy in collisions with electrons only in discrete, quantized amounts. The energy is then released only in discrete, quantized amounts by light emission. The Franck–Hertz experiment provided dramatic confirmation of Bohr's hypothesis that the energy of atoms is quantized in discrete states. It also provided a direct electrical method for measuring the energy differences between these states and for constructing the energy-level diagram starting with the ground state. The technique continues to be used today to construct energy-level diagrams for molecules in the methods called "electron impact spectroscopy" or "electron energy loss spectroscopy."

4.3 The Bohr Model: Predicting Discrete Energy Levels in Atoms

Atomic spectra and Franck–Hertz experiments measure the differences between energy levels, and they enable the energy level diagrams to be constructed from the experimental data using Equations 4.7 and 4.8. In 1913, Niels Bohr developed the first theoretical model to predict the energy levels of the hydrogen atom and one-electron ions such as He^+, Li^{2+}, and Be^{3+}. You should review Section 3.3 before continuing further in this chapter. Pay careful attention to the definition of an absolute energy scale for atoms by choosing a reference state whose energy we set as zero. The logical choice, as discussed in Section 3.3, is the electron at rest located infinitely far from the nucleus.

The Bohr theory started from Rutherford's planetary model of the atom. Bohr supplemented Rutherford's planetary model with the assumption that an electron of mass m_e moves in a circular orbit of radius r about a fixed nucleus. The total energy of the hydrogen atom, kinetic plus potential, is given by Equation 3.7, which we reproduce and renumber here as Equation 4.9 for convenience:

$$E = \frac{1}{2}m_e v^2 - \frac{Ze^2}{4\pi\epsilon_0 r} \qquad [4.9]$$

The expression for the Coulomb force that attracts the electron to the nucleus is $\vec{F} = F_{\text{coul}}\hat{r}$, where $\hat{r}$ is a unit vector pointing from the nucleus to the electron and F_{coul} is the negative derivative of the potential energy with respect to the separation r: $F_{\text{coul}} = -Ze^2/4\pi\epsilon_0 r^2$. Newton's second law relating force and acceleration is $\vec{F} = m_e\vec{a}$, and for uniform circular motion, the acceleration, $\vec{a}$, of the electron has magnitude v^2/r. Combining these results gives the following relation for the *magnitude* of the force:

$$|F_{\text{Coulomb}}| = |m_e\vec{a}| \qquad [4.10a]$$

$$\frac{Ze^2}{4\pi\epsilon_0 r^2} = m_e\frac{v^2}{r} \qquad [4.10b]$$

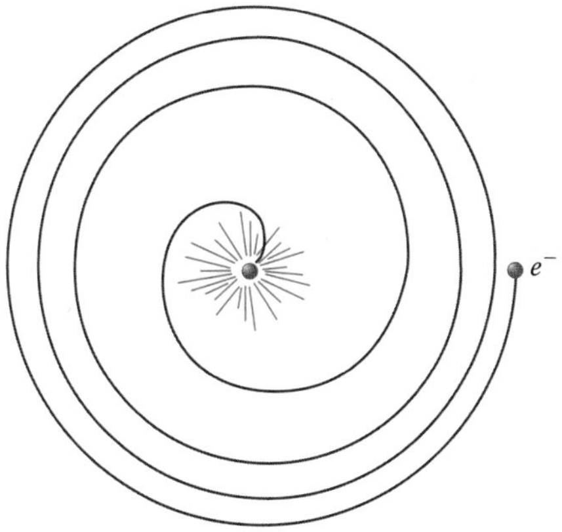

FIGURE 4.22 Classical theory states that atoms constructed according to Rutherford's nuclear model are not stable. The motion of electrons around the nucleus would cause them to radiate energy and quickly spiral into the nucleus.

As mentioned earlier, classical physics requires that an accelerated electron emit electromagnetic radiation, thereby losing energy and eventually spiraling into the nucleus (Fig. 4.22). Bohr avoided this conflict by simply *postulating* that only certain discrete orbits (characterized by radius r_n and energy E_n) are allowed, and that light is emitted or absorbed only when the electron "jumps" from one stable orbit to another. This bold assertion was Bohr's attempt to explain the existence of stable atoms, a well-established experimental fact. Faced with the contradiction between the experimental results and the requirements of classical electrodynamics, he simply discarded the latter in the formulation of his model.

The next step in the development of the Bohr model was his assertion that the angular momentum of the electron is quantized. This was an *ad hoc* assumption designed to produce stable orbits for the electron; it had no basis in either classical theory or experimental evidence. The magnitude of the **linear momentum** of an electron is the product of its mass and its speed, $m_e v$. The **angular momentum**, $\vec{L}$, is a different quantity that describes rotational motion about an axis. An introduction to angular momentum is provided in Appendix B. For the circular paths (see Fig. 3.5) of the Bohr model, the magnitude of the angular momentum of the electron is the product of its mass, its speed, and the radius of the orbit ($L = m_e vr$). Bohr postulated that the angular momentum is quantized in integral multiples of $h/2\pi$, where h is Planck's constant:

$$L = m_e vr = n\frac{h}{2\pi} \qquad n = 1, 2, 3, \ldots \qquad [4.11]$$

The existence of discrete orbits and quantized energies follows directly as a consequence of the quantization of angular momentum.

We can determine the properties of these discrete orbits as follows. Equations 4.9 and 4.10 contain two unknowns, v and r. Solving Equation 4.11 for v ($= nh/2\pi m_e r$), inserting it into Equation 4.10, and solving for r gives the allowed values for radius of the orbits:

$$r_n = \frac{\epsilon_0 n^2 h^2}{\pi Z e^2 m_e} = \frac{n^2}{Z} a_0 \qquad [4.12]$$

where a_0, the **Bohr radius**, has the numerical value 5.29×10^{-11} m = 0.529 Å. (The Bohr radius ($a_0 = \epsilon_0 h^2/\pi e^2 m_e$) is a convenient, fundamental unit of length in atomic physics that relieves us from the burden of carrying along all of the constants in Eq. 4.12.) This first prediction of the Bohr model is the existence of a series of orbits whose distances from the nucleus increase dramatically with increasing n. Substituting r_n from Equation 4.12 into Equation 4.11 allows us to calculate the speed v_n corresponding to the orbit with radius r_n.

$$v_n = \frac{nh}{2\pi m_e r_n} = \frac{Ze^2}{2\epsilon_0 nh} \qquad [4.13]$$

The results obtained for r_n and v_n can now be substituted into Equation 4.9 to give us the allowed values of the energy:

$$E_n = \frac{-Z^2 e^4 m_e}{8\epsilon_0^2 n^2 h^2} = -(2.18 \times 10^{-18}\ \text{J})\frac{Z^2}{n^2} \qquad n = 1, 2, 3, \ldots \qquad [4.14a]$$

That these energies are negative is a consequence of our choice for the zero of energy, as discussed in Section 3.3 and shown in Figure 3.6. For the same reason that we introduced the Bohr radius, it is convenient to express atomic energy levels using a unit known as a **rydberg,** where 1 rydberg = 2.18×10^{-18} J. The energy level expression then becomes

$$E_n = -\frac{Z^2}{n^2} \quad (\text{rydberg}) \qquad n = 1, 2, 3, \ldots \qquad [4.14b]$$

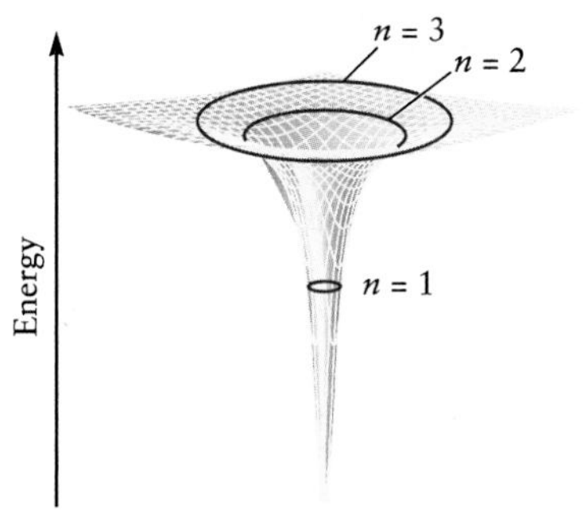

FIGURE 4.23 The potential energy of the electron and nucleus in the hydrogen atom has its lowest (most negative) value when the electron is closest to the nucleus. (Compare with the one-dimensional plot in Fig. 3.6.) The electron moving away from the nucleus can be seen as moving up the sides of a steep potential energy well. In the Bohr theory, the electron moves in a circular orbit which can have only specific allowed values of *r*, the radius, and *E*, the energy. The first three of these are shown qualitatively by rings.

The Bohr model thus predicts a discrete energy-level diagram for the one-electron atom (Figs. 4.23 and 4.24). The ground state is identified by $n = 1$, and the excited states have higher values of n (see Fig. 4.23).

The **ionization energy** is the minimum energy required to remove an electron from an atom (see Section 3.4). In the Bohr model, ionization involves a transition from the $n = 1$ state to the $n = \infty$ state, in which $E_n = 0$. The associated energy change is

$$\Delta E = E_{\text{final}} - E_{\text{initial}} = 0 - (-2.18 \times 10^{-18}\ \text{J}) = 2.18 \times 10^{-18}\ \text{J}$$

Multiplying this result by Avogadro's number gives the ionization energy, *IE*, per mole of atoms:

$$IE = (6.022 \times 10^{23}\ \text{atoms mol}^{-1})(2.18 \times 10^{-18}\ \text{J atom}^{-1}) = 1.31 \times 10^{6}\ \text{J mol}^{-1} = 1310\ \text{kJ mol}^{-1}$$

This prediction agrees with the experimentally observed ionization energy of hydrogen atoms and provides confidence in the validity of the Bohr model. Note that the allowed values of the energy E_n given in Equation 4.14 are the sum of the kinetic and potential energy, with the speed of the electron decreasing as n increases (see Eqn. 4.13). The Bohr model was the first physical theory that could predict ionization energies with remarkable accuracy.

EXAMPLE 4.3

Consider the $n = 2$ state of Li^{2+}. Using the Bohr model, calculate the radius of the electron orbit, the electron speed, and the energy of the ion relative to that of the nucleus and electron separated by an infinite distance.

Solution

Because $Z = 3$ for Li^{2+} (the nuclear charge is $+3e$) and $n = 2$, the radius is

$$r_2 = \frac{n^2}{Z} a_0$$

$$= \frac{4}{3} a_0 = \frac{4}{3}(0.529\ \text{Å}) = 0.705\ \text{Å}$$

The speed is

$$v_2 = \frac{nh}{2\pi m_e r_2}$$

$$= \frac{2(6.626 \times 10^{-34}\ \text{J s})}{2\pi(9.11 \times 10^{-31}\ \text{kg})(0.705 \times 10^{-10}\ \text{m})}$$

$$= 3.28 \times 10^{6}\ \text{m s}^{-1}$$

The energy is

$$E_2 = -\frac{(3)^2}{(2)^2}(2.18 \times 10^{-18}\ \text{J}) = -4.90 \times 10^{-18}\ \text{J}$$

Typically, atomic sizes fall in the range of angstroms, and atomic excitation energies in the range of 10^{-18} J. This is consistent with the calculations of Coulomb potential energies in electron volts and dimensions in angstroms in Section 3.3.

Related Problems: 19, 20

4.3.1 Atomic Spectra: Interpretation by the Bohr Model

When a one-electron atom or ion undergoes a transition from a state characterized by quantum number n_i to a state lower in energy with quantum number n_f ($n_i > n_f$), light of frequency ν is emitted to carry off the energy $h\nu$ lost by the atom. By conservation of energy, $E_i = E_f + h\nu$; thus,

$$h\nu = \frac{Z^2e^4m_e}{8\epsilon_0^2h^2}\left[\frac{1}{n_f^2} - \frac{1}{n_i^2}\right] \quad \text{(emission)} \qquad [4.15]$$

As n_i and n_f take on a succession of integral values, lines are seen in the emission spectrum (see Fig. 4.24) with frequencies

$$\nu = \frac{Z^2e^4m_e}{8\epsilon_0^2h^3}\left(\frac{1}{n_f^2} - \frac{1}{n_i^2}\right) = (3.29 \times 10^{15}\ \text{s}^{-1})Z^2\left(\frac{1}{n_f^2} - \frac{1}{n_i^2}\right) \qquad [4.16]$$

$$n_i > n_f = 1, 2, 3, \ldots \text{(emission)}$$

Conversely, an atom can *absorb* energy $h\nu$ from a photon as it undergoes a transition to a *higher* energy state ($n_f > n_i$). In this case, conservation of energy requires $E_i + h\nu = E_f$; thus, the absorption spectrum shows a series of lines at frequencies

$$\nu = (3.29 \times 10^{15}\ \text{s}^{-1})\ Z^2\left(\frac{1}{n_i^2} - \frac{1}{n_f^2}\right) \qquad [4.17]$$

$$n_f > n_i = 1, 2, 3, \ldots \text{(absorption)}$$

For hydrogen, which has an atomic number of $Z = 1$, the predicted emission spectrum with $n_f = 2$ corresponds to the series of lines in the visible region measured by Balmer and shown in Figure 4.24. A series of lines at higher frequencies (in the ultraviolet region) is predicted for $n_f = 1$ (the Lyman series), and other series are predicted at lower frequencies (in the infrared region) for $n_f = 3, 4, \ldots$ In fact, the predicted and observed spectra of hydrogen and one-electron ions are in excellent agreement—a major triumph of the Bohr theory.

Despite these successes, the Bohr theory has a number of shortcomings. Most important, it cannot predict the energy levels and spectra of atoms and ions with more than one electron. Also, more fundamentally, it was an uncomfortable hybrid of classical and nonclassical concepts. The postulate of quantized angular momentum—which led to only certain circular orbits being allowed—had no fundamental basis and was simply grafted onto classical physics to force the predictions of classical physics to agree with the experimental results. In 1926, the Bohr theory was replaced by modern quantum mechanics in which the quantization of energy and angular momentum arise as natural consequences of the basic postulates and require no additional assumptions. The circular orbits of the Bohr theory do not appear in quantum mechanics. The Bohr theory provided the conceptual bridge from classical theoretical physics to the new quantum mechanics. Its historical and intellectual importance cannot be exaggerated.

4.4 Evidence for Wave–Particle Duality

The Bohr theory provided a prescription for calculating the discrete energy levels of a one-electron atom or ion, but it did not explain the origin of energy quantization. A key step toward the development of modern quantum mechanics was

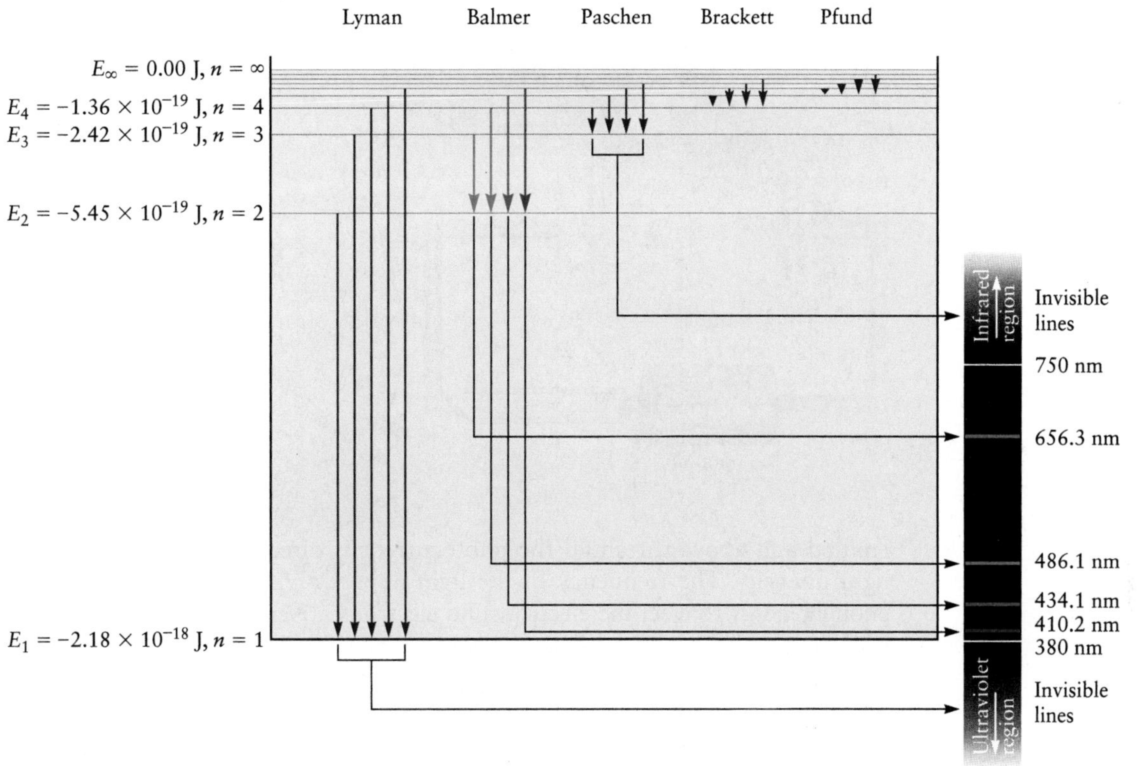

FIGURE 4.24 In the energy levels of the hydrogen atom, the separated electron and proton are arbitrarily assigned zero energy, and all other energies are more negative than that. Atoms emit light as they fall from higher to lower energy levels as indicated by the arrows. Each series of related transitions is named after the person who discovered it. The arrows identifying transitions in the Balmer series are color-coded to match the emission lines to which they correspond. These lines in the Balmer series are shown in the hydrogen emission spectrum in Figure 4.16.

the concept of **wave–particle duality**—the idea that particles sometimes behave as waves, and vice versa. Experiments were forcing physicists to recognize that physical systems could display either particle or wave characteristics, depending on the experimental conditions to which they were subjected. The physicist Albert Einstein introduced wave–particle duality to explain the photoelectric effect, in which light acted as a particle. French physicist Louis de Broglie suggested that particles could exhibit wavelike properties, and the stage was set for the new quantum mechanics to synthesize wave–particle duality and energy quantization into a comprehensive new theory.

4.4.1 The Photoelectric Effect

In addition to the conceptual problems with the planetary model of the atom and the difficulties with blackbody radiation, another conflict between experiment and classical theory arose in the early 20th century from the observation of the **photoelectric effect**. A beam of light shining onto a metal surface (called the *photocathode*) can eject electrons (called *photoelectrons*) and cause an electric current (called a *photocurrent*) to flow (Fig. 4.25). The photocurrent shows an extremely interesting dependence on the frequency and intensity of the incident light (Fig. 4.26). Regardless of the light intensity, no photocurrent flows until the frequency exceeds a particular threshold value ν_0, which is unique for each metal. Low-frequency (long wavelength; for example, red) light apparently cannot provide enough energy to eject the electrons, no matter how intense it is. When the frequency of the light is increased through the threshold value (corresponding, perhaps, to green or blue light), electrons are

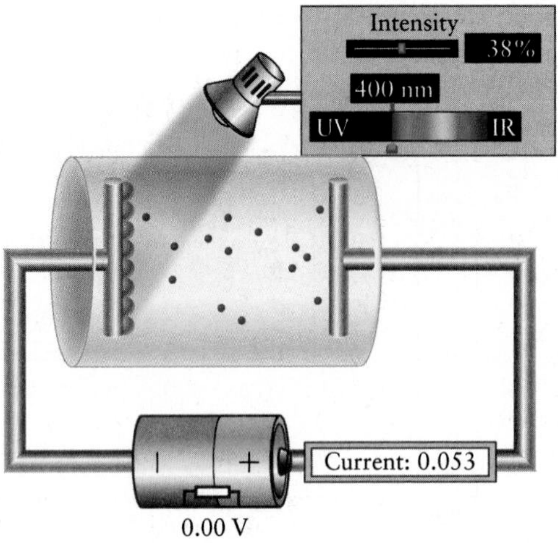

FIGURE 4.25 In a photoelectric cell (photocell), light strikes a metal surface in an evacuated space and ejects electrons. The electrons are attracted to a positively charged collector, and a current flows through the cell.

emitted and above threshold the photocurrent is directly proportional to the light intensity. The frequency of the light apparently is the key to delivering enough energy to eject the electrons; no electrons are emitted when the surface is excited by light whose frequency is below the threshold frequency, but electrons are readily emitted for all frequencies above the threshold frequency. These results could not be explained by classical physics. According to classical electromagnetic theory, the energy associated with electromagnetic radiation depends on only the intensity of the radiation, not on its frequency. Why, then, could a low-intensity (dim) beam of blue light (high frequency) eject electrons from sodium when a high-intensity (bright) beam of red light (low frequency) had no effect at all (Fig. 4.27a)? The key to the explanation is to relate the energy imparted by the light at the threshold frequency, ν_0, to the energy with which the photoelectrons are emitted from atoms in the metal.

The photoelectrons leaving the metal surface and traveling toward the detector have a range of kinetic energies. Let's assume that those photoelectrons arriving at the collector with E_{max} were emitted from atoms at the surface of the

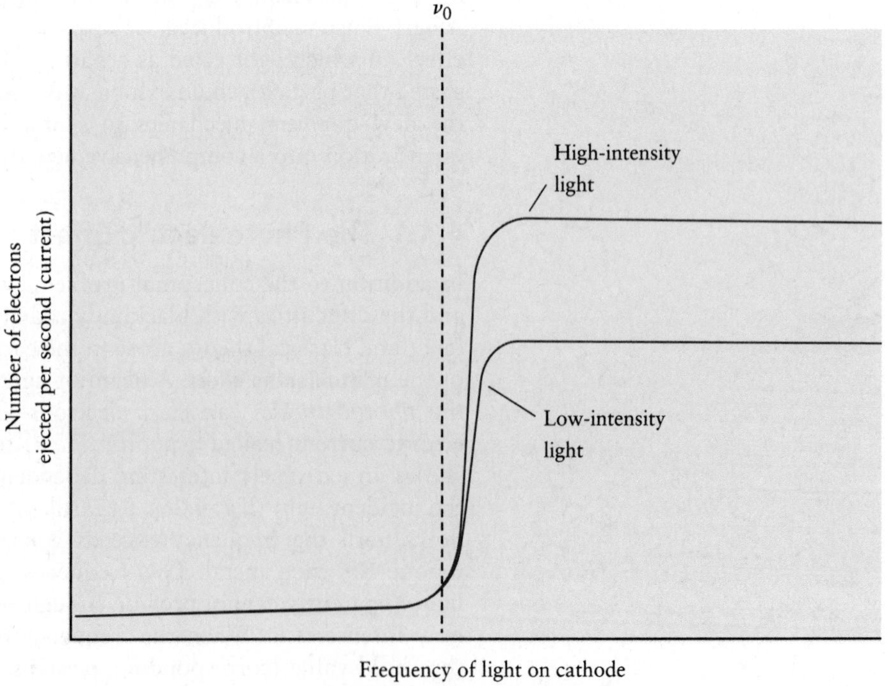

FIGURE 4.26 Frequency and intensity dependence of the photoelectric effect. Only light above the threshold frequency can eject photoelectrons from the surface. Once the frequency threshold has been passed, the total current of photoelectrons emitted depends on the intensity of the light, not on its frequency.

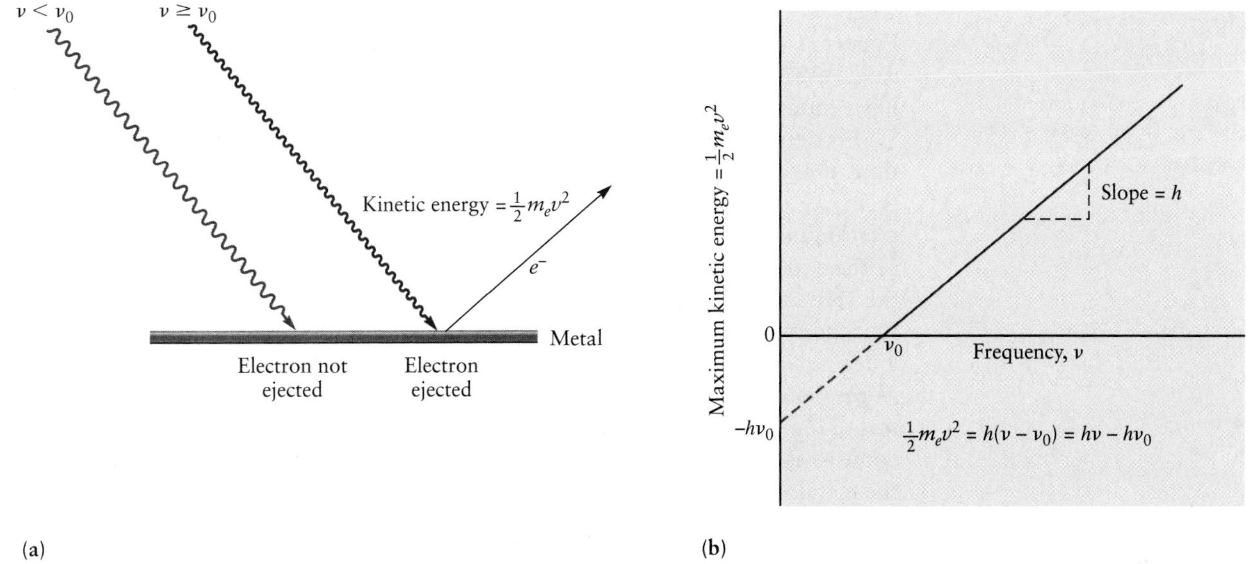

FIGURE 4.27 (a) Two key aspects of the photoelectric effect. Blue light is effective in ejecting electrons from the surface of this metal, but red light is not. (b) The maximum kinetic energy of the ejected electrons varies linearly with the frequency of light used.

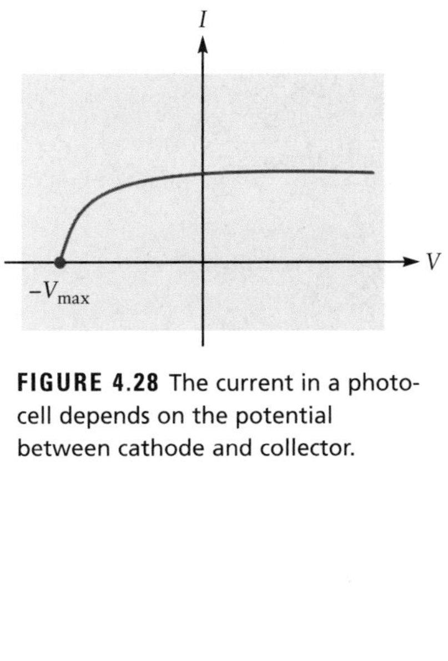

FIGURE 4.28 The current in a photocell depends on the potential between cathode and collector.

metal. Assume those arriving with lower kinetic energy were emitted deeper in the metal but lost some kinetic energy through collisions with other metal atoms before escaping from the surface. Then, the value of E_{max} should be directly related to the energy acquired by the photoelectron during the ejection process. We determine this maximum kinetic energy as follows. When the frequency and intensity of the beam are held constant, the magnitude of the photocurrent depends on the electrical potential (voltage) of the collector relative to the photocathode (Fig. 4.28). At sufficiently positive potentials, all of the photoelectrons are attracted to the collector and the current–voltage curve becomes flat, or saturated. As the potential of the collector is made more negative, photoelectrons arriving with kinetic energies less than the maximum are repelled and the photocurrent decreases. Only those photoelectrons with sufficient kinetic energy to overcome this repulsion reach the collector. As the collector is made still more negative, the photocurrent drops sharply to zero at $-V_{max}$, identifying the maximum in the kinetic energy of the photoelectrons: $E_{max} = eV_{max}$. The potential required to stop all of the electrons from arriving at the collector is thus a direct measure of their maximum kinetic energy, expressed in units of electron volts (eV) (see Section 3.3). But what was the connection between ν_0 and E_{max}?

In 1905, Einstein used Planck's quantum hypothesis to explain the photoelectric effect. First, he suggested that a light wave of frequency ν consists of quanta of energy (a quantum of light was later called a **photon** by G. N. Lewis), each of which carries energy, $E_{photon} = h\nu$. Second, Einstein assumed that, in the photoelectric effect, an electron in the metal absorbs a photon of light and thereby gains the energy required to escape from the metal. A photoelectron emitted from beneath the surface will lose energy E' in collisions with other atoms and Φ in escaping from the surface, after which it travels through the vacuum to the detector with kinetic energy E_{kin}. Conservation of energy leads to the relation $h\nu = E' + \Phi + E_{kin}$ for the process. Electrons with the maximum kinetic energy are emitted at the surface, so for them $E' = 0$. Therefore, Einstein's theory predicts that the maximum kinetic energy of photoelectrons emitted by light of frequency ν is given by

$$E_{max} = \frac{1}{2}mv_e^2 = h\nu - \Phi \qquad [4.18]$$

where $\Phi = h\nu_0$ is a constant characteristic of the metal. The key idea of Einstein's explanation is that the interaction of a photon with an electron is a single event and the result is all or nothing; either the photon does or does not have enough energy to overcome the forces that bind the electron to the solid.

Einstein's theory predicts that the maximum kinetic energy is a linear function of the frequency, which provides a means for testing the validity of the theory. Experiments conducted at several frequencies demonstrated that the relation between E_{max} and frequency is indeed linear (see Fig. 4.27b). The slope of the experimental data determined the numerical value of h to be identical to the value that Planck found by fitting the experimental data to his theoretical blackbody radiation intensity distribution. Einstein's interpretation also provided a means to obtain the value of the quantity Φ from the experimental data as the "energy intercept" of the linear graph. Φ, called the **work function** of the metal, represents the binding energy, or energy barrier, that electrons must overcome to escape from the metal surface after they have absorbed a photon inside the metal. Φ governs the extraction of electrons from metal surfaces by heat and by electric fields, as well as by the photoelectric effect, and it is an essential parameter in the design of numerous electronic devices.

EXAMPLE 4.4

Light with a wavelength of 400 nm strikes the surface of cesium in a photocell, and the maximum kinetic energy of the electrons ejected is 1.54×10^{-19} J. Calculate the work function of cesium and the longest wavelength of light that is capable of ejecting electrons from that metal.

Solution

The frequency of the light is

$$\nu = \frac{c}{\lambda} = \frac{3.00 \times 10^8 \text{ m s}^{-1}}{4.00 \times 10^{-7} \text{ m}} = 7.50 \times 10^{14} \text{ s}^{-1}$$

The binding energy $h\nu_0$ can be calculated from Einstein's formula:

$$E_{max} = h\nu - h\nu_0$$

$$1.54 \times 10^{-19} \text{ J} = (6.626 \times 10^{-34} \text{ J s})(7.50 \times 10^{14} \text{ s}^{-1}) - h\nu_0$$

$$= 4.97 \times 10^{-19} \text{ J} - h\nu_0$$

$$\Phi = h\nu_0 = (4.97 - 1.54) \times 10^{-19} \text{ J} = 3.43 \times 10^{-19} \text{ J}$$

The minimum frequency ν_0 for the light to eject electrons is then

$$\nu_0 = \frac{3.43 \times 10^{-19} \text{ J}}{6.626 \times 10^{-34}} = 5.18 \times 10^{14} \text{ s}^{-1}$$

From this, the maximum wavelength λ_0 is

$$\lambda_0 = \frac{c}{\nu_0} = \frac{3.00 \times 10^8 \text{ m s}^{-1}}{5.18 \times 10^{14} \text{ s}^{-1}} = 5.79 \times 10^{-7} \text{ m} = 579 \text{ nm}$$

Related Problems: 27, 28

Planck and Einstein proposed quantum hypotheses to explain experimental data in two very different areas of physics. The fact that data from both of these experiments led to the same value of h inspired great confidence in the validity of these hypotheses, despite their conflicts with established scientific principles. Einstein's bold assertion that light consisted of a stream of bundles of energy that

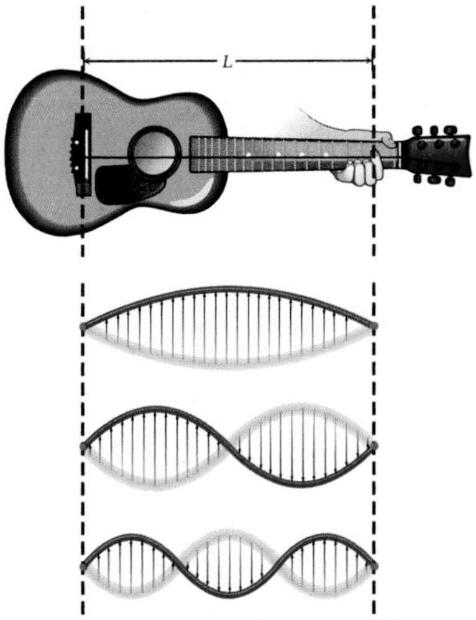

FIGURE 4.29 A guitar string of length L with fixed ends can vibrate in only a restricted set of ways. The position of the spring at the turning points of the motion are shown here. In standing waves such as these, the whole string is in motion except at its ends and at the nodes.

appeared to transfer their energy through collisions like those of material particles was completely at odds with the classical wave representation of light, which had already been amply confirmed by experimental studies. How could light be both a wave and a particle?

By 1930, these paradoxes had been resolved by quantum mechanics, which superseded Newtonian mechanics. The classical wave description of light is adequate to explain phenomena such as interference and diffraction, but the emission of light from matter and the absorption of light by matter are described by the particlelike photon picture. A hallmark of quantum, as opposed to classical, thinking is not to ask "What *is* light?" but instead "How does light behave under particular experimental conditions?" Thus, wave–particle duality is not a contradiction, but rather part of the fundamental nature of light and also of matter.

4.4.2 De Broglie Waves

Thus far, this chapter has considered only one type of wave, a **traveling wave.** Electromagnetic radiation (light, X-rays, and gamma rays) is described by such a traveling wave moving through space at speed c. Another type of wave is a **standing wave,** of which a simple example is a guitar string with fixed ends (an example of a physical *boundary condition*). A plucked string vibrates, but only certain oscillations of the string are possible. Because the ends are fixed, the only oscillations that can persist are those in which an integral number of half-wavelengths fits into the length of string, L (Fig. 4.29). The condition on the allowed wavelengths is

$$n\frac{\lambda}{2} = L \qquad n = 1, 2, 3, \ldots \qquad [4.19]$$

It is impossible to create a wave with any other value of λ if the ends of the string are fixed. The oscillation with $n = 1$ is called the **fundamental** or first harmonic, and higher values of n correspond to higher harmonics. At certain points on the standing wave, the amplitude of oscillation is zero; any of these points is called a **node.** (The fixed ends are not counted as nodes.) The higher the number of the harmonic n, the more numerous the nodes, the shorter the wavelength, the higher the frequency, and the higher the energy of the standing wave.

De Broglie realized that such standing waves are examples of quantization: Only certain discrete vibrational modes, characterized by the "quantum number" n, are allowed for the vibrating string. He suggested that the quantization of energy in a one-electron atom might have the same origin, and that the electron might be associated with a standing wave, in this case, a *circular* standing wave oscillating about the nucleus of the atom (Fig. 4.30). For the amplitude of the wave to be well defined (single valued and smooth), an integral number of wavelengths must fit into the circumference of the circle ($2\pi r$). The condition on the allowed wavelengths for standing circular waves is

$$n\lambda = 2\pi r \qquad n = 1, 2, 3, \ldots \qquad [4.20]$$

FIGURE 4.30 A circular standing wave on a closed loop. The state shown has $n = 7$, with seven full wavelengths around the circle.

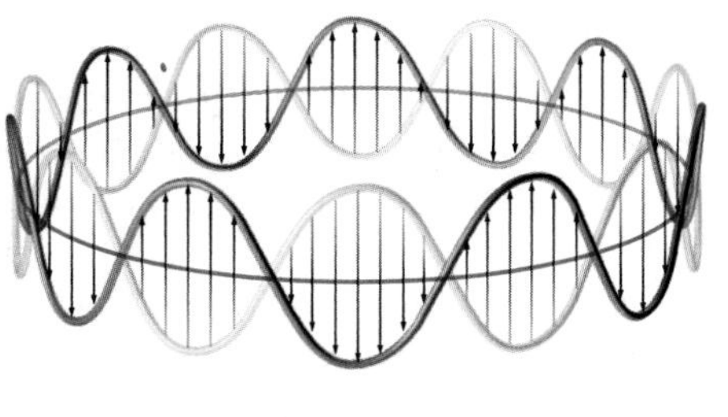

Bohr's assumption about quantization of the angular momentum of the electron was

$$m_e vr = n\frac{h}{2\pi} \qquad [4.21]$$

which can be rewritten as

$$2\pi r = n\left[\frac{h}{m_e v}\right] \qquad [4.22]$$

Comparison of de Broglie's equation (see Eq. 4.20) with Bohr's equation (see Eq. 4.22) shows that the wavelength of the standing wave is related to the magnitude of the linear momentum, p, of the electron by the following simple formula:

$$\lambda = \frac{h}{m_e v} = \frac{h}{p} \qquad [4.23]$$

De Broglie used the theory of relativity to show that exactly the same relationship holds between the wavelength and momentum of a *photon*. De Broglie therefore proposed as a generalization that any particle—no matter how large or small—moving with magnitude of linear momentum p has wavelike properties and a wavelength of $\lambda = h/p$ associated with its motion.

EXAMPLE 4.5

Calculate the de Broglie wavelengths of (a) an electron moving with speed 1.0×10^6 m s^{-1} and (b) a baseball of mass 0.145 kg, thrown with a speed of 30 m s^{-1}.

Solution

(a) $\lambda = \frac{h}{p} = \frac{h}{m_e v} = \frac{6.626 \times 10^{-34}\text{ J s}}{(9.11 \times 10^{-31}\text{ kg})(1.0 \times 10^6\text{ m s}^{-1})}$

$= 7.3 \times 10^{-10}\text{ m} = 7.3\text{ Å}$

(b) $\lambda = \frac{h}{mv} = \frac{6.626 \times 10^{-34}\text{ J s}}{(0.145\text{ kg})(30\text{ m s}^{-1})}$

$= 1.5 \times 10^{-34}\text{ m} = 1.5 \times 10^{-24}\text{ Å}$

The latter wavelength is far too small to be observed. For this reason, we do not recognize the wavelike properties of baseballs or other macroscopic objects, even though they are always present. However, on a microscopic level, electrons moving in atoms show wavelike properties that are essential for explaining atomic structure.

Related Problems: 31, 32

4.4.3 Electron Diffraction

Under what circumstances does the wavelike nature of particles become apparent? When two waves pass through the same region of space, they interfere with each other. Consider water waves as an example. When two crests meet, constructive interference occurs and a higher crest (greater amplitude) appears; where a crest of one wave meets a trough of the other, destructive interference (smaller or even zero amplitude) occurs (Fig. 4.6). Light waves passing through two slits in a solid barrier produce a diffraction pattern of alternating bright and

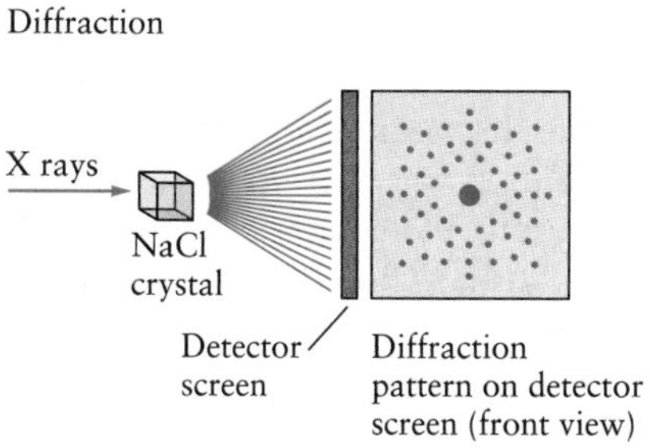

FIGURE 4.31 X-ray diffraction by a crystal produces a Laue diffraction pattern. (Adapted from Zumdahl/DeCoste, *Chemical Principles, 6e,* Fig. 12.7; Cengage Learning.)

dark spots as the result of constructive and destructive interference (Fig. 4.7). A way to test de Broglie's hypothesis is to see whether a beam of electrons will generate a diffraction pattern similar to a beam of light.

As discussed in more detail in Chapter 21, it was known by 1914 that X-rays diffract from the lattice planes of a single crystal solid because the spacing between successive planes of atoms is comparable to the wavelength of the X-rays used in the experiment (Fig. 4.31). As the high energy X-rays penetrate deeply into the solid, a new "wavelet" is generated at each atom, much as water waves colliding with the posts supporting a dock in a lake generate new waves originating at the posts. Because the spacing between the successive atomic scattering centers is comparable to the wavelength of the X-rays, these "wavelets" interfere constructively and destructively to generate a diffraction pattern in the outgoing wave.

If the de Broglie hypothesis is correct, particles whose de Broglie wavelengths are comparable with lattice spacings should also diffract from crystals. This result was demonstrated in 1927 by the American physicists C. Davisson and L. H. Germer. In their experiment, a beam of low-energy electrons was directed toward a single crystal nickel sample in vacuum. The kinetic energy of the electrons could be varied continuously over the range 20 eV to 200 eV to change their wavelengths. Let's confirm that the de Broglie wavelength of the electrons used in this experiment is comparable with atomic lattice spacings, as required for the electrons to diffract from the planes of atoms in the solid. The kinetic energy of an electron accelerated from rest to a final voltage V is $\mathcal{T} = eV$, where e is the magnitude of the charge on the electron. Recalling that $p = mv$ and $\mathcal{T} = \frac{1}{2}mv^2$, we can set $\mathcal{T} = p^2/2m_e$ and solve for the magnitude of the momentum of the electron to get $p = \sqrt{2m_e eV}$. The de Broglie wavelength of the electron is therefore $\lambda = h/\sqrt{2m_e eV}$. Calculations using these formulas must express the kinetic energy of the electron in joules ($1 \text{ eV} = 1.6 \times 10^{-19}$ J) in order to obtain the proper units for momentum and wavelength. An electron with kinetic energy of 50 eV (typical of the Davisson–Germer experiment) has a de Broglie wavelength of 1.73 Å, which is comparable to the spacing between atomic planes in metal crystals.

The geometry for electron diffraction is sketched in Figure 4.32. Because these low-energy electrons do not penetrate deeply into the solid, the diffraction pattern originates primarily at the surface atoms due to interference between "wavelets" originating from *adjacent* atoms. Figure 4.32a shows a side view of

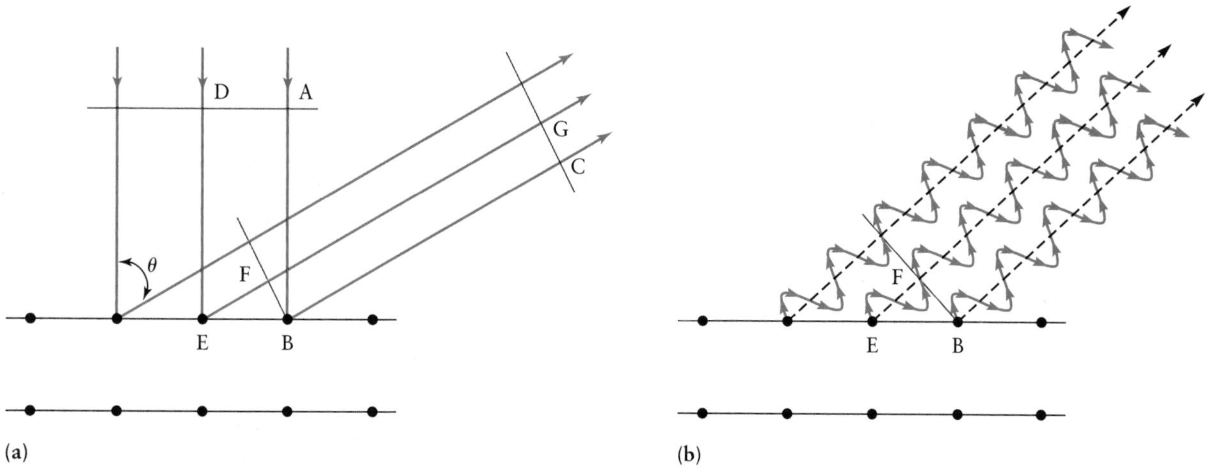

FIGURE 4.32 Geometry for diffraction of electrons at a solid surface. (a) Incident waves representing the incoming electron beam are shown as lines for simplicity. (b) Scattered waves after electrons collide with the surface illustrate constructive interference by which the diffracted beams are generated.

the crystal in which atoms in the top two planes are represented as dots. The distance between atoms in each plane is a. The electron beam comes in normal (perpendicular) to the surface, and the outgoing waves are at angle θ with respect to the surface normal. Line AD represents a wave front of electron waves that are in phase as they approach the crystal. To prevent clutter in the diagram we represent each wave as a straight line that shows its direction of motion. The wave that is scattered at atom B follows the path ABC, while the wave scattered at E follows the path DEFG. This second wave travels a greater distance than the first, and the difference in path length is EF. The construction in the figure shows that $EF = a \sin \theta$. In order for the outgoing waves to be in phase at the wave front CG, this extra distance must correspond to an integral multiple of the wavelength of the electron. This relationship between the two outgoing waves is shown in Figure 4.32b. The diffraction condition is thus

$$n\lambda = a \sin \theta \qquad [4.24]$$

For a particular choice of crystal sample (the value a) and X-ray wavelength λ, the diffracted intensity appears in the direction defined by θ. The two-dimensional surface may have a periodic spacing a between atoms along the x-axis and b between atoms along the y-axis. This situation gives two diffraction conditions that must be satisfied at the same time:

$$n_a\lambda_a = a \sin \theta_a \quad \text{and} \quad n_b\lambda_b = b \sin \theta_b$$

The interference of waves in two dimensions to produce diffracted beams departing the surfaces is suggested by the sketch in Figure 4.33.

A modern apparatus for demonstrating electron diffraction is sketched in Figure 4.34. The diffraction pattern on the phosphor screen is observed visually though the viewport. Some portion of the pattern may be blocked from view by

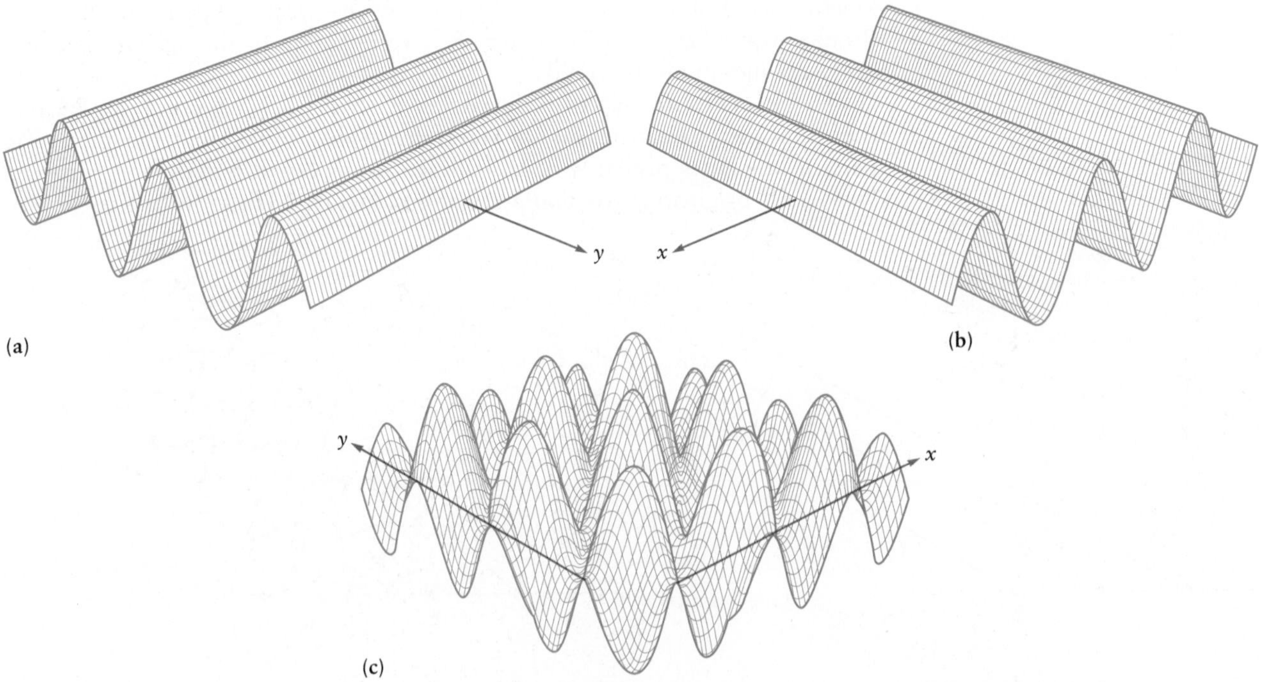

FIGURE 4.33 When two waves (a) and (b) moving at right angles to one another meet and overlap, their amplitudes add. Where two crests meet, there is constructive interference and the resultant amplitude is greater than that of either wave alone (shown by the peaks in part (c)). Where a crest meets a trough, destructive interference gives a value of zero for the sum of the two waves. At positions where the resulting wave oscillates in time with a large amplitude, you are likely to find a scattered electron.

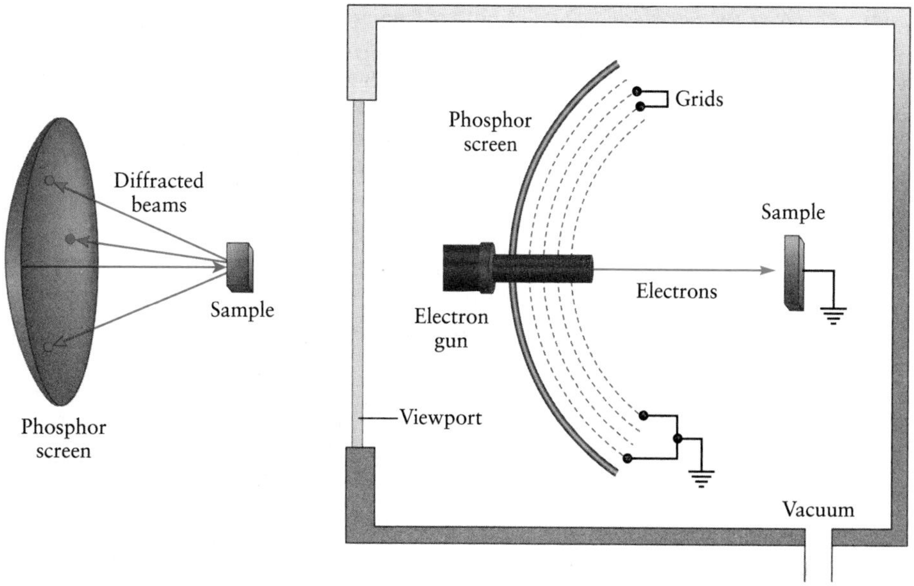

FIGURE 4.34 Apparatus for demonstrating diffraction of electrons by solid surfaces. The incoming electron beam is oriented perpendicular (normal) to the plane of the surface. Electrons are back-scattered after collisions with the surface. Voltages applied to the grids reject background electrons and allow only those electrons that have been diffracted by the surface to arrive at the phosphor screen. The diffracted beams stimulate emission from the screen and thereby display the diffraction pattern of the surface as an array of bright spots against a dark background. (Adapted from L. M. Feldman and J. M. Mayer, *Thin Film and Surface Analysis*, Fig. 7.6, p. 163, North Holland, 1984.)

the electron gun. Figure 4.35 shows the electron diffraction pattern for the surface of a crystal of silicon.

The original experiments of Davisson and Germer demonstrated electron diffraction, and thereby provided a striking confirmation of de Broglie's hypothesis about the wavelike nature of matter. The modern form of their experiments described here enables a technique called low-energy electron diffraction (LEED) that is now used widely to study the atomic structure of solid surfaces. Measurements of the diffraction angle for electrons with a specific value of kinetic energy gives the atomic spacing a by using Equation 4.24.

The Davisson-Germer experiments remind us that "waves" and "particles" are idealized models that describe objects found in nature, and that wave–particle duality is a fact of nature. Photons, electrons, and even helium atoms all have both wave and particle character; which aspect they display depends strongly on the conditions under which they are observed.

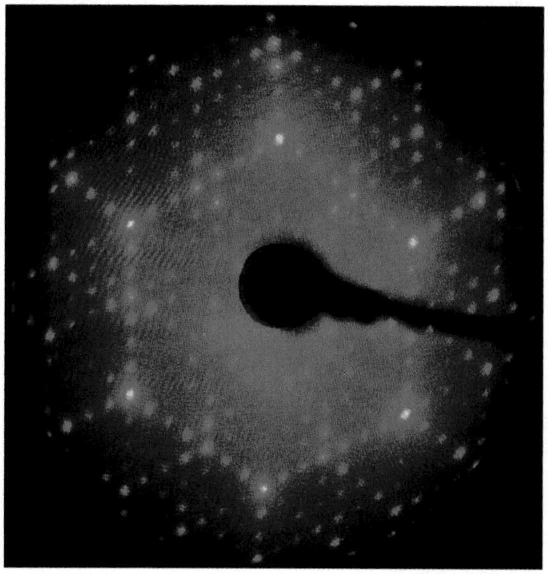

Courtesy of Dr. Gerard Parkinsen and Mr. William Gerace, OMICRON Vakuumphysik GMBH, Tanusstein, Germany

FIGURE 4.35 Low-energy electron diffraction pattern of the same silicon surface imaged by scanning tunneling microscopy in the figure opposite page 1 of this book.

4.4.4 Indeterminacy and Uncertainty: The Heisenberg Principle

An inevitable consequence of de Broglie's standing-wave description of an electron in an orbit around the nucleus is that the position and momentum of a particle cannot be known both precisely and simultaneously. The magnitude of the momentum of the circular standing wave shown in Figure 4.30 is given exactly by $p = h/\lambda$, but because the wave is spread uniformly around the circle, we cannot specify the angular position of the electron on the circle at all. We say the angular position is **indeterminate** because it has no definite value. This conclusion is in stark contrast with classical physics where the positions and momenta are all known precisely and the trajectories of particles are well defined. How was this paradox resolved?

In 1927, the German physicist Werner Heisenberg proposed that **indeterminacy** is a general feature of quantum systems. Indeterminacy presents a fundamental limit to the "knowability" of the properties of these systems that is intrinsic and not just a limitation of our ability to make more precise measurements. In particular, it influences which combinations of properties can be measured together. Heisenberg identified pairs of properties that *cannot* be measured together with complete precision, and estimated the best precision we can hope to obtain when we do measure them. For example, we cannot measure position and momentum simultaneously and obtain sharp, definite values for each. The same is true for energy and time. Notice that the combination of dimensions *length* × *momentum* is the same as the combination *energy* × *time*. (You should verify this by simple dimensional analysis.) Either combination is called *action*, and it has the same dimensions as Planck's constant. The **Heisenberg indeterminacy principle** states that when we measure two properties, A and B, the product of which has dimensions of action, we will obtain a spread in results for each identified by σ_A and σ_B that will satisfy the following condition:

$$(\sigma_A)(\sigma_B) \geq h/4\pi \qquad [4.25]$$

If we try to measure A precisely and make σ_A nearly zero, then the spread in σ_B will have to increase to satisfy this condition. Trying to determine the angular position of the orbiting electron described by the de Broglie wave discussed earlier is a perfect illustration. Indeterminacy is intrinsic to the quantum description of matter and applies to all particles no matter how large or small.

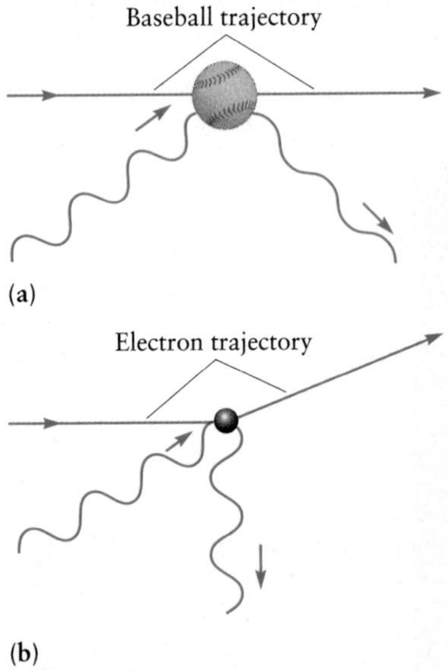

FIGURE 4.36 A photon, which has a negligible effect on the trajectory of a baseball (a), significantly perturbs the trajectory of the far less massive electron (b).

The practical consequence of indeterminacy for the outcome of measurements is best seen by applying the Heisenberg principle in specific cases. How are the position and momentum of a macroscopic object such as a baseball in motion determined? The simplest way is to take a series of snapshots at different times, with each picture recording the light (photons) scattered by the baseball. This is true for imaging any object; we must scatter something (like a photon) from the object and then record the positions of the scattered waves or particles (Fig. 4.36). Scattering a photon from a baseball does not change the trajectory of the baseball appreciably (see Fig. 4.36a) because the momentum of the photon is negligible compared with that of the baseball, as shown in Example 4.6. Scattering photons from an electron, however, is another thing altogether. To locate the *position* of any object, we must use light with a wavelength that is comparable with or shorter than the size of the object. Thus, to measure the position of an electron in an atom to a precision of, say, 1% of the size of the atom, we would need a probe with a wavelength of order 10^{-12} m. The magnitude of the momentum of such a wave, given by the de Broglie relation, is 6.625×10^{-22} kg m s^{-1}. The magnitude of the linear momentum of the electron in a ground state ($n = 1$) hydrogen atom using the Bohr model is 2.0×10^{-24} kg m s^{-1}. Trying to measure the position of an electron to a precision of 10^{-12} m with a photon of sufficiently short wavelength turns out to be roughly equivalent to trying to measure the position of a marble using a bowling ball as a probe! So on the length and mass scale of elementary particles, it is clear that we cannot measure simultaneously, to arbitrary precision, the values of the momentum and the position of a particle.

To make a rough estimate of the precision allowed by the indeterminacy principle, let us take as our spread of values in the position, σ_x, the wavelength of our probe λ. This choice means that we can locate the particle somewhere between two crests of the wave. Let us take as our estimate of the spread in the momentum, σ_p, the value of the momentum itself, p; that is, we know p to within $\pm p$. Their product is therefore $\sigma_x\sigma_p = h$, but because we have asserted that this is the *best* we can do, we write $\sigma_x\sigma_p \geq h$. If, instead of these estimates, you properly calculate the spread in both variables as one standard deviation or the root mean square deviation from a very large number of measurements, the result becomes

$$\sigma_x\sigma_p \geq h/4\pi \qquad [4.26]$$

which is in agreement with the Heisenberg principle.

At last, we can resolve the paradox between de Broglie waves and classical orbits, which started our discussion of indeterminacy. The indeterminacy principle places a *fundamental limit* on the precision with which the position and momentum of a particle can be known simultaneously. It has profound significance for how we think about the motion of particles. According to classical physics, the position and momentum are fully known simultaneously; indeed, we must know both to describe the classical trajectory of a particle. The indeterminacy principle forces us to abandon the classical concepts of trajectory and orbit. The most detailed information we can possibly know is the statistical spread in position and momentum allowed by the indeterminacy principle. In quantum mechanics, we think not about particle trajectories, but rather about the probability distribution for finding the particle at a specific location.

EXAMPLE 4.6

Suppose photons of green light (wavelength 5.3×10^{-7} m) are used to locate the position of the baseball from Example 4.5 with precision of one wavelength. Calculate the minimum spread in the *speed* of the baseball.

Solution

The Heisenberg relation,

$$\sigma_x\sigma_p \geq h/4\pi$$

with the very crude estimate of λ for σ_x gives

$$\sigma_p \geq \frac{h}{4\pi\sigma_x}$$

$$= \frac{6.626 \times 10^{-34}\ \text{J s}}{4\pi(5.3 \times 10^{-7}\ \text{m})} = 9.9 \times 10^{-29}\ \text{kg m s}^{-1}$$

Because the momentum is just the mass (a constant) times the speed, the spread in the speed is

$$\sigma_v = \frac{\sigma_p}{m} \geq \frac{9.9 \times 10^{-29}\ \text{kg m s}^{-1}}{(0.145\ \text{kg})}$$

$$= 6.8 \times 10^{-28}\ \text{m s}^{-1}$$

This is such a tiny fraction of the speed of the baseball ($30\ \text{m s}^{-1}$) that indeterminacy plays a negligible role in the measurement of the baseball's motion. Such is the case for all macroscopic objects.

Related Problems: 35, 36

Describing the indeterminacy principle presents certain challenges to language. Clearly, when a property of the system is indeterminate, measurements will produce a statistical spread of values for that property. In a colloquial sense, there will be *uncertainty* in the measurement, because its outcome is not precisely predictable, just as there is uncertainty in the outcome of playing a game of chance. In almost all English-language books on quantum mechanics, the spread in value of a property σ_A is called the *uncertainty in A*, and the relation in Equation 4.25 is called the *Heisenberg uncertainty principle*. A property is indeterminate if it has no definite value, whereas it is uncertain if it does have a definite value, but that value is not known to the experimenter. We prefer the phrase *indeterminacy principle* because it more accurately conveys that a fundamental limitation on measurements is being described, whereas *uncertainty principle* suggests that the uncertainty could be reduced by conducting more carefully designed experiments.

4.5 The Schrödinger Equation

De Broglie's work attributed wavelike properties to electrons in atoms, which inspired the Austrian physicist Erwin Schrödinger to think about how to describe electrons as waves. Schrödinger, a recognized authority on the theory of vibrations and the associated "quantization" of standing waves, reasoned that an electron (or any other particle with wavelike properties) might well be described by a wave function. A **wave function** maps out what the wave looks like in three dimensions; it may also be a function of time. Ocean waves have displacements from sea level that vary in both space and time. An electromagnetic wave has an electrical field whose component along its direction of polarization oscillates in both space and time. Schrödinger's wave function, symbolized by the Greek letter psi (ψ), characterizes the wave associated with a particle, at a position located by the coordinates x, y, z at time t. It is important to emphasize that a particle's wave function may oscillate in time, like an ordinary water wave or electromagnetic wave, and that at a particular time t the value of a particle's wave function may be positive, negative, or zero at different possible positions of the particle. We'll call the latter function of position $\psi(x)$ if the particle is restricted to lie along a line, or $\psi(x,y,z)$ if the particle is to be found in three dimensional coordinate space. The value of $\psi(x)$ may be positive or negative, and at coordinates where the value of $\psi(x)$ is zero (where it changes from positive to negative) one says there is a **node** in the wave function. The next section describes what the wave function for a particle $\psi(x)$ tells you physically about the particle's position and total energy. It also allows us to describe how two particles might interfere with each other as two waves might, constructively if the value of the wave function for both particles have the same sign at a particular position, and destructively if the wave function for both particles have opposite signs. This allows one to quantitatively describe diffraction experiments done with electrons, similar to diffraction of light in Figure 4.7. This constructive and destructive interference also plays an important role in how shared electrons form covalent bonds! After working a few simple examples to help you get used to wave functions for particles in this chapter, we'll examine wave functions for electrons in atoms (Chapter 5) and in covalent bonds (Chapter 6).

4.5.1 Origins of the Schrödinger Equation

Although it is beyond the scope of this text to explain the origins of the **Schrödinger equation**, it is nevertheless worthwhile to work through the logic that might have stimulated Schrödinger's thinking and, more importantly, to explore some of the properties of the mathematical form of the Schrödinger equation. Having been trained in the classical theory of waves and inspired by de Broglie's hypothesis, it was natural for Schrödinger to seek a wave equation that described the properties of matter on the atomic scale. Classical wave equations relate the second derivatives of the wave function with respect to position x to the second derivatives with respect to time (or in quantum mechanics, the first derivative with respect to time). For simplicity let's see if we can find a wave equation that relates the second derivative of a function with respect to position to the function itself, leaving the time dependence for more advanced work.

We begin by considering a particle moving freely in one dimension (along the x-axis) with classical momentum, p. Such a particle is associated with a wave of wavelength $\lambda = h/p$. Two "wave functions" that describe such a wave are

$$\psi(x) = A \sin \frac{2\pi x}{\lambda} \quad \text{and} \quad \psi(x) = B \cos \frac{2\pi x}{\lambda} \tag{4.27}$$

where A and B are constants. Choosing the sine function, for example, let's see what its second derivative with respect to x looks like. From differential calculus, the derivative (or slope) of the first $\psi(x)$ in Equation 4.27 is

$$\frac{d\psi(x)}{dx} = A \frac{2\pi}{\lambda} \cos \frac{2\pi x}{\lambda}$$

The slope of *this* function is given by the second derivative of ψ, written $\frac{d^2\psi(x)}{dx^2}$, which is equal to

$$\frac{d^2\psi(x)}{dx^2} = -A\left(\frac{2\pi}{\lambda}\right)^2 \sin \frac{2\pi x}{\lambda}$$

This is just a constant, $-(2\pi/\lambda)^2$, multiplied by the original wave function $\psi(x)$:

$$\frac{d^2\psi(x)}{dx^2} = -\left(\frac{2\pi}{\lambda}\right)^2 \psi(x)$$

This is an equation (called a *differential equation*) that is satisfied by the function $\psi(x) = A \sin (2\pi x/\lambda)$. It is easy to verify that this equation is also satisfied by the function $\psi(x) = B \cos (2\pi x/\lambda)$.

Let's now replace the wavelength λ with h divided by the momentum p from the de Broglie relation:

$$\frac{d^2\psi(x)}{dx^2} = -\left(\frac{2\pi}{h} p\right)^2 \psi(x) \tag{4.28}$$

We can rearrange this equation into a suggestive form by multiplying both sides by $-h^2/8\pi^2 m$, giving

$$-\frac{h^2}{8\pi^2 m} \frac{d^2\psi(x)}{dx^2} = \frac{p^2}{2m} \psi(x) = \mathcal{T}\psi(x)$$

where $\mathcal{T} = p^2/2m$ is the kinetic energy of the particle. This form of the equation suggests that there is a fundamental relationship between the second derivative of the wave function (also called its *curvature*) and the kinetic energy, $\mathcal{T}$.

If external forces are present, a potential energy term $V(x)$ (due to the presence of walls enclosing the particle or to the presence of fixed charges, for example) must be included. Writing the total (kinetic + potential) energy as E the analog to the previous equation is

$$-\frac{h^2}{8\pi^2 m} \frac{d^2\psi(x)}{dx^2} + V(x)\psi(x) = E\psi(x) \tag{4.29}$$

This is the Schrödinger equation for a particle moving in one dimension. The development provided here is not a derivation of this central equation of quantum mechanics; rather, it is a plausibility argument based on the idea that the motions of particles can be described by a wave function with the wavelength of the particle being given by the de Broglie relation.

4.5.2 The Validity of the Schrödinger Equation

The validity of any scientific theory must be tested by extensive comparisons of its predictions with a large body of experimental data. Although we have presented a plausibility argument that suggests how Schrödinger might have initially developed his equation, understanding the source of his inspiration is not nearly as important as evaluating the accuracy of the theory. It is the same for all great scientific discoveries; the story of Newton and the apple is not nearly as important as the fact that classical mechanics has been shown to describe the behavior of macroscopic systems to astonishingly high accuracy. Quantum mechanics superseded Newtonian mechanics because the latter failed to account for the properties of atoms and molecules. We believe that quantum mechanics is correct because its predictions agree with experiment, in many cases to better than $10^{-10}\%$. It is generally considered to be among the most accurate theories of nature because of this astonishingly good agreement. But even quantum mechanics began to fail as scientists were able to make more accurate measurements than those made in the early part of the 20th century. Relativistic corrections to Schrödinger's equations improved the situation dramatically, but only with the development of **quantum electrodynamics**—in which matter and radiation are treated completely equivalently—did complete agreement between theory and experiment occur. Quantum electrodynamics is an extremely active field of research today, and it continues to ask questions such as, "How does the system know that it is being measured?" and "How can we use quantum mechanics to make computers of unprecedented power?" The fundamental ideas of quantum mechanics—energy quantization and wave–particle duality—appear to be universally true in science. Although these properties of nature are less evident in the macroscopic world, the predictions of quantum mechanics agree well with those of classical mechanics on the relevant length and mass scales for macroscopic systems.

4.5.3 Interpretation of the Energy in the Schrödinger Equation

The Schrödinger equation can be solved exactly for any number of model problems and for a few real problems, notably the hydrogen atom. What do the solutions of this equation tell us about the energies and other properties of quantum systems? Or, to phrase the question slightly differently, how do we interpret ψ, and what information does it contain?

Let's focus initially on the energy. For all systems confined in space by a potential $V(x)$ (or, more generally $V(x, y, z)$) that does not vary in time, solutions of the Schrödinger equation given in Equation 4.29 can be found only for certain discrete values of the energy; energy quantization is a natural consequence of the Schrödinger equation. Any state described by these time-independent wave functions is called a **stationary state**. For a given system, there may be many states with different energies characterized by different wave functions. The solution that corresponds to the lowest energy is called the ground state (just as in the Bohr model), and higher energy solutions are called excited states.

4.5.4 Interpretation of the Wave Function in the Schrödinger Equation

What is the physical meaning of the wave function ψ? We have no way of measuring ψ directly, just as in classical wave optics we have no direct way of measuring the oscillating electric and magnetic fields that constitute the light wave (see Fig. 4.2). What *can* be measured in the latter case is the intensity of the light wave, which, according to the classical theory of electromagnetism, is proportional to the square of the amplitude of the electric field:

$$\text{intensity} \propto (E_{\text{max}})^2$$

However, if we view electromagnetic radiation as a collection of *particles* (photons), then the intensity is simply proportional to the density of photons in a region of space. Connecting the wave and particle views of the electromagnetic field suggests that the *probability* of finding a photon is proportional to the square of the amplitude of the electric field.

By analogy, we interpret the square of the wave function ψ^2 for a particle as a probability density for that particle. That is, $|\psi(x, y, z)|^2 \, dV$ is the probability that the particle will be found in a small volume $dV = dxdydz$ centered at the point (x, y, z). This probabilistic interpretation of the wave function, proposed by the German physicist Max Born, is now generally accepted because it provides a consistent picture of particle motion on a microscopic scale.

The probabilistic interpretation requires that any function must meet three mathematical conditions before it can be used as a wave function. The next section illustrates how these conditions are extremely helpful in solving the Schrödinger equation. To keep the equations simple, we will state these conditions for a single particle moving in only one dimension. All the conditions extend immediately to three dimensions when proper coordinates and notation are used. (You should read Appendix C6, which reviews probability concepts and language, before proceeding further with this chapter.)

When the possible outcomes of a probability experiment are continuous (for example, the position of a particle along the x-axis) as opposed to discrete (for example, flipping a coin), the distribution of results is given by the probability density function $P(x)$. The product $P(x)dx$ gives the probability that the result falls in the interval of width dx centered about the value x. The first condition, that the probability density must be **normalized,** ensures that probability density is properly defined (see Appendix C6), and that all possible outcomes are included. This condition is expressed mathematically as

$$\int_{-\infty}^{+\infty} P(x)\, dx = \int_{-\infty}^{+\infty} |\psi(x)|^2 \, dx = 1 \qquad [4.30]$$

The second and third conditions are subsidiary to the first, in that they must be satisfied to enable the first one to be satisfied. The second condition is that $P(x)$ must be continuous at each point x. At some specific point, call it x_a, the form of the probability density may change for physical reasons, but its value at x_a must be the same regardless of whether x_a is approached from the left or from the right. This translates into the condition that $\psi(x)$ and its first derivative $\psi'(x)$ are continuous at each point x (with the exception that $d\psi/dx$ is not required to be continuous where the slope of $V(x)$ approaches infinity). The third condition is that if the particle is in a bound state (due to a potential energy function $V(x)$), $\psi(x)$ must be bounded at large values of x. This is stated mathematically as

$$\psi \longrightarrow 0 \quad \text{as} \quad x \longrightarrow \pm\infty \qquad [4.31]$$

The requirement that $\psi(x)$ be continuous results in **boundary conditions**, which are restrictions that must be satisfied by the solutions to differential equations such as the Schrödinger equation. A differential equation does not completely define a physical problem until the equation is supplemented with boundary conditions. These conditions invariably arise from physical analysis, and they help to select from the long list of possible solutions to the differential equation those that apply specifically to the problem being studied.

We must acknowledge that our information about the location of a particle is limited, and that it is statistical in nature. So not only are we restricted by the indeterminacy principle as to what we can measure, but we must also come to grips with the fact that fundamental properties of quantum systems are unknowable, except in a statistical sense. If this notion troubles you, you are in good company. Many of the best minds of the 20th century, notably Einstein, never became comfortable with this central conclusion of the quantum theory.

4.5.5 Procedures for Solving the Schrödinger Equation

The application of quantum mechanics to solve for the properties of any particular system is straightforward in principle. You need only substitute the appropriate potential energy term for that system into the Schrödinger equation and solve the equation to obtain two principal results: the allowed energy values and the corresponding wave functions. You will find that solutions exist only for specific, discrete energy values. Energy quantization arises as a direct consequence of the boundary conditions imposed on the Schrödinger equation (see later discussion) with no need for extra assumptions to be grafted on. Each energy value corresponds to one or more wave functions; these wave functions describe the distribution of particles when the system has a specific energy value.

We illustrate this procedure in detail for a simplified model in the next section so that you will see how energy levels and wave functions are obtained. We also use the model problem to illustrate important general features of quantum mechanics including restrictions imposed on the form of the wave function by the Schrödinger equation and its physical interpretation.

4.6 Quantum Mechanics of Particle-in-a-Box Models

We are about to show you how to solve the Schrödinger equation for a simple but important model for which we can carry out every step of the complete solution using only simple mathematics. We will convert the equations into graphical form and use the graphs to provide physical interpretations of the solutions. It is important for you to develop a physical understanding from the graphical forms of the solution. Later in this textbook we present the solutions for more complex applications only in graphical form, and you will rely on the skills you develop here to see the physical interpretation for a host of important chemical applications of quantum mechanics. This section is, therefore, one of the most important sections in the entire textbook.

4.6.1 One-Dimensional Boxes

The simplest model problem for which the Schrödinger equation can be solved, and in which energy quantization appears, is the so-called particle in a box. It consists of a particle confined by potential energy barriers to a certain region of

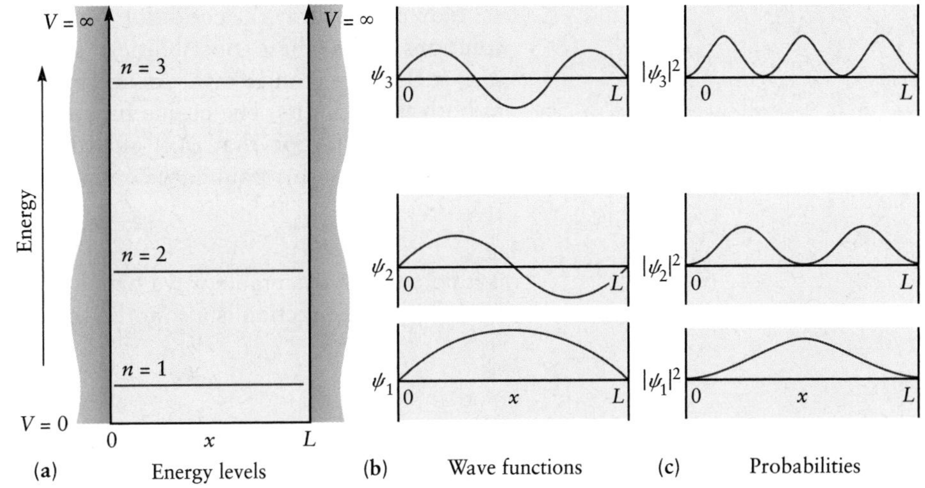

FIGURE 4.37 (a) The potential energy for a particle in a box of length L, with the first three energy levels marked. (b) Wave functions showing the ground state ψ_1 and the first two excited states. The more numerous the nodes, the higher the energy of the state. (c) The squares of the wave functions from (b), equal to the probability density for finding the particle at a particular point in the box.

space (the "box"). In one dimension, the model is visualized easily as a bead sliding along a wire between barriers at the ends of the wire. The particle is located on the x-axis in the interval between 0 and L, where L is the length of the box (Fig. 4.37a). If the particle is to be completely confined in the box, the potential energy $V(x)$ must rise abruptly to an infinite value at the two end walls to prevent even fast-moving particles from escaping. Conversely, inside the box, the motion of the particle is free, so $V(x) = 0$ everywhere inside the box. This means that the total energy, $E = \mathcal{T} + V$, must be positive at each point inside the box. We will determine the possible values of E by solving the Schrödinger equation. The solution for this model illustrates the general methods used for other more difficult potential energy functions.

A quick inspection of the potential energy function tells us the general nature of the solution. Wherever the potential energy V is infinite, the probability of finding the particle must be zero. Hence, $\psi(x)$ and $\psi^2(x)$ must be zero in these regions:

$$\psi(x) = 0 \text{ for } x \leq 0 \text{ and } x \geq L \qquad [4.32]$$

Inside the box, where $V = 0$, the Schrödinger equation has the following form:

$$-\frac{h^2}{8\pi^2 m}\frac{d^2\psi(x)}{dx^2} = E\psi(x)$$

(This is often written as $\mathcal{H}\psi = E\psi$ where $\mathcal{H}$ is called the **Hamiltonian operator.** $\mathcal{H}$ is neither a number or a function, it is an expression that tells you to take a second derivative of the function to the right of it, then multiply the result by a constant. Typically it would also have a $V(x)$ term added to it, but in this problem $V(x) = 0$ inside the box. The result of $\mathcal{H}\psi$ will be the total energy E multiplied by the same wave function ψ for every ψ that solves the Schrödinger equation.) Simplifying the equation above gives

$$\frac{d^2\psi(x)}{dx^2} = -\frac{8\pi^2 mE}{h^2}\psi(x)$$

As shown earlier, the sine and cosine functions are two possible solutions to this equation, because the second derivative of each function is the function itself multiplied by a (negative) constant.

Now let us apply the conditions defined in Section 4.5 to select the allowed solutions from these possibilities. The boundary conditions require that $\psi(x) = 0$ at $x = 0$ and $x = L$. An acceptable wave function must be continuous at both these points. The cosine function can be eliminated because it cannot satisfy the condition that $\psi(x)$ must be 0 at $x = 0$. The sine function does, however, satisfy this boundary condition since $\sin(0) = 0$ so

$$\psi(x) = A \sin kx \qquad [4.33]$$

is a potentially acceptable wave function.

If the wave function is also to be continuous at $x = L$, then we must have

$$\psi(L) = 0 \qquad [4.34]$$

or

$$\psi(L) = A \sin kL = 0$$

This can be true only if

$$kL = n\pi \qquad n = 1, 2, 3, \ldots$$

because $\sin(n\pi) = 0$. Thus, the combination of the boundary conditions and continuity requirement gives the allowed solutions as

$$\psi(x) = A \sin\left(\frac{n\pi x}{L}\right) \qquad n = 1, 2, 3, \ldots \qquad [4.35]$$

where the constant A is still to be determined. (Note that $n = 0$ is not a possibility as it would give $\psi = 0$ at all x axis values. Negative integers are not included as they give the same set of wave functions.) The restriction of the solutions to this form in which n is an integer quantizes the energy and the wave functions.

As explained in Section 4.5, the wave function must be normalized. This condition is not always satisfied by solutions to the Schrödinger equation; thus, we must see how we can enforce it. To normalize the wave function just obtained we set

$$A^2 \int_0^L \sin^2\left(\frac{n\pi x}{L}\right) dx = 1$$

and solve for A. Evaluating the definite integral gives $L/2$, so

$$A^2\left(\frac{L}{2}\right) = 1$$

and

$$A = \sqrt{\frac{2}{L}}$$

The normalized wave function for the one-dimensional particle in a box is

$$\psi_n(x) = \sqrt{\frac{2}{L}} \sin\left(\frac{n\pi x}{L}\right) \qquad n = 1, 2, 3, \ldots \qquad [4.36]$$

where n labels a particular allowed solution of the Schrödinger equation.

To find the energy E_n for a particle described by the wave function ψ_n, we calculate the second derivative:

$$\frac{d^2\psi_n(x)}{dx^2} = \frac{d^2}{dx^2}\left[\sqrt{\frac{2}{L}}\sin\left(\frac{n\pi x}{L}\right)\right]$$

$$= -\left(\frac{n\pi}{L}\right)^2\left[\sqrt{\frac{2}{L}}\sin\left(\frac{n\pi x}{L}\right)\right]$$

$$= -\left(\frac{n\pi}{L}\right)^2 \psi_n(x)$$

This must be equal to $-\dfrac{8\pi^2 mE_n}{h^2}\psi_n(x)$. Setting the coefficients equal to one another gives

$$\frac{8\pi^2 mE_n}{h^2} = \frac{n^2\pi^2}{L^2} \qquad \text{or}$$

$$E_n = \frac{n^2h^2}{8mL^2} \qquad n = 1, 2, 3, \ldots \qquad [4.37]$$

This solution of the Schrödinger equation demonstrates that the energy of a particle in the box is quantized. The energy, E_n, and wave function $\psi_n(x)$ are unique functions of the quantum number n, which must be a positive integer. These are the only allowed stationary states of the particle in a box. The allowed energy levels are plotted together with the potential energy function in Figure 4.37a. A system described by the particle-in-a-box model will have an emission or absorption spectrum that consists of a series of frequencies given by

$$h\nu = |E_n - E_{n'}| \qquad [4.38]$$

where n and n' are positive integers, and the E_n values are given by Equation 4.37.

The wave functions $\psi_n(x)$ plotted in Figure 4.37b for the quantum states n are the standing waves of Figure 4.29. The guitar string and the particle in the box are physically analogous. The boundary condition that the value of the wave function ψ must be zero at each end of the guitar string (where its amplitude of oscillation with time is zero) or at each wall of the box is responsible for the quantization of energy and the restriction on the motions allowed.

Following Born's interpretation that the probability of finding the particle near a particular position is proportional to the square of its wave function evaluated at that position, we can study the probability distributions for the particle in a box in various quantum states. Figure 4.37c shows the probability distributions for the first three states of the particle in a box. For the ground state ($n = 1$), we see that the most likely place to find the particle is in the middle of the box, with a small chance of finding it near either wall. In the first excited state ($n = 2$), the probability is a maximum when it is near $L/4$ and $3L/4$ and zero near 0, $L/2$, and L. And for the $n = 3$ state, the maxima are located at $L/6$, $3L/6$, and $5L/6$, with nodes located at $L/3$ and $2L/3$. Wherever there is a node in the wave function, the probability is zero that the particle will be found at that location. To be more precise, the probability of finding the particle between the position $x = x_1$ and $x = x_2$ is calculated with the following integral:

$$\text{Probability of finding particle between } x_1 \text{ and } x_2 = \int_{x_1}^{x_2} |\psi(x)|^2\, dx$$

where $\psi(x)$ is the wave function of the particle. If you are not yet comfortable with integrals, you can calculate the probability from the area under the curve in a plot of $|\psi(x)|^2$ versus x between x_1 and x_2. If the interval $\Delta x = x_2 - x_1$ is small, this integral has a value close to $|\psi(x_0)|^2\,\Delta x$, where x_0 is evaluated in the middle of the range from x_1 to x_2. Checking the units on this expression is helpful. The unit of $|\psi(x)|^2$ is m^{-1} (or $Å^{-1}$) and Δx has units of m (or Å), so the probability has no units; a probability of 0.1 means that there is a 10% chance of finding the particle in this range of positions.

EXAMPLE 4.7

The Lewis dot structure of 1,3-butadiene, $CH_2{=}CH{-}CH{=}CH_2$ has two double bonds. Four electrons make two of the single bonds into double bonds, the extra pair of bonding electrons on the first, and the extra pair on the third of the carbon–carbon bonds. In a simple particle-in-a-box model for these four electrons, they may move freely along all three carbon–carbon bonds in the molecule but not past the end C atoms. If you draw a line along the carbon–carbon–carbon backbone of this molecule and label the center of one end C atom as $x = 0$, this corresponds to a one-dimensional box of length equal to the sum of the carbon-carbon bond lengths: C=C bond + C—C bond + C=C bond = 1.34 Å + 1.54 Å + 1.34 Å = 4.22 Å. When two electrons are in the $n = 1$ state and two electrons are in the $n = 2$ state:

(a) Calculate the most likely and the least likely positions of one of the $n = 2$ electrons.

(b) Calculate the probability of finding this electron between the first two carbon atoms, between the second two carbon atoms, and between the third two carbon atoms.

Solution

(a) Using Equation 4.36 the wave function for an electron in this $n = 2$ state is

$$\psi_2 = \sqrt{\frac{2}{4.22\ Å}}\sin\left(\frac{2\pi x}{4.22\ Å}\right) \quad \text{between } x = 0\ Å \text{ and } x = 4.22\ Å$$

To find the probability density, you first square the wave function

$$|\psi_2|^2 = \frac{2}{4.22\ Å}\sin^2\left(\frac{2\pi x}{4.22\ Å}\right) \quad \text{between } x = 0\ Å \text{ and } x = 4.22\ Å$$

The maximum values of $|\psi_2|^2$ occur when the sin function is +1 and −1, so at $2\pi x/(4.22\ Å) = \pi/2$ and $3\pi/2$. Thus the most likely positions of this $n = 2$ electron are $x = 1.055$ Å and $x = 3.165$ Å. The minimum value of $|\psi_2|^2$ occurs at $x = 2.11$ Å, so this is the least likely position of the $n = 2$ electron.

(b) The probability of finding this electron between the first two carbon atoms equals the area under the curve of $|\psi(x)|^2$ vs. x from $x_1 = 0$ Å to $x_2 = 1.34$ Å; this is the shaded area in Figure 4.38.

$$\int_{0Å}^{1.34\ Å} \frac{2}{4.22\ Å}\sin^2\left(\frac{2\pi x}{4.22\ Å}\right)dx = 0.38 = 38\%$$

If you are not yet comfortable with integrals, you can estimate the shaded area curve as 190 rectangles multiplied by the area per rectangle, $(0.02\ Å^{-1})(0.1Å) = 0.002$. This gives a probability of 0.38 = 38%, the same result you get by integrating $|\psi(x)|^2$ from $x_1 = 0$ Å to $x_2 = 1.34$ Å.

The probability of finding the electron between the third and the fourth carbon is also 38%. The probability of finding the electron between the second and the third carbon should be much smaller, as you can see by estimating the area under the curve in Figure 4.38 from $x_1 = 1.34$ Å to $x_2 = 2.88$ Å. The integration

(Continued)

FIGURE 4.38 Integrating the probability density to calculate the probability of finding the particle between $x = 0$ and $x = 1.34$ Å.

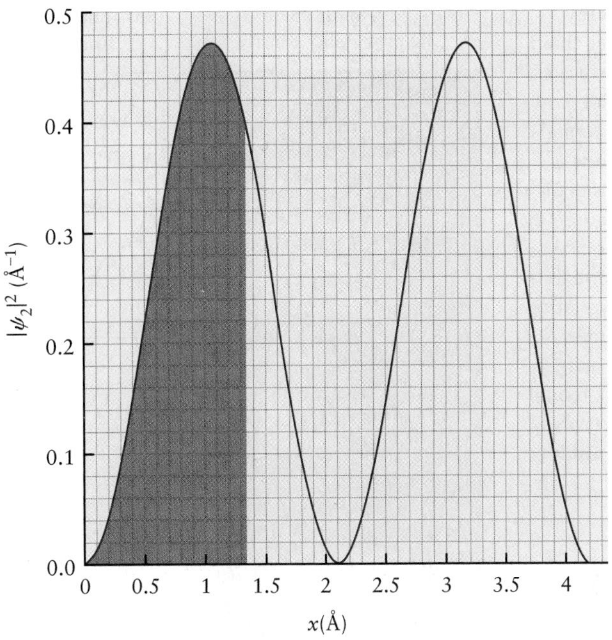

over this range gives a probability of $0.25 = 25\%$. If you sum all three probabilities they should sum to $1 = 100\%$ (We get 101% due to rounding errors.) This is because the wave function is normalized, meaning that if we look for the particle everywhere, there should be 100% chance of finding it.

Related Problems: 37, 38

The number of nodes in the wave function is important not only for helping us understand probability distributions but also because it provides an important, and perfectly general, criterion for ordering the energy levels in any quantum system. The wave function ψ_n has $n - 1$ nodes, and it is clear from Figure 4.37c that the number of nodes increases with the energy of each state. This is a general feature in quantum mechanics: For a given system, the relative ordering of the energy levels can be determined simply by counting the number of nodes. The particle-in-a-box model provides a useful interpretation for experimental systems in which electrons are free to move in one dimension over a relatively short distance between two barriers that constitute the walls of the "box." A very interesting example is shown in Figure 4.39. The scanning tunneling microscope (STM) was used to assemble a linear chain of 20 Pd atoms to form a "wire" of length 6.0 nm on the surface of a NiAl crystal as described in Chapter 1. The STM was then used to measure the probability density for locating electrons along the chain of atoms. Figure 4.39a shows two-dimensional images of this density for the first four quantum states. The number of nodes increases from 0 to 3. Figure 4.39b shows linear scans of the probability density along the chain for these same four quantum states. Figure 4.39c shows the probability density for the particle-in-a-box model from Figure 4.37c calculated for a box of length $L = 6.0$ nm.

Can a particle in a box have zero energy? Setting $n = 0$ and solving for E using the equation $E_n = n^2h^2/8mL^2$ gives $E_0 = 0$. But this is not possible. Setting n equal to zero in $\psi_n(x) = A \sin(n\pi x/L)$ makes $\psi_0(x)$ zero everywhere. In this case, $\psi_0^2(x)$ would be zero everywhere in the box, and thus there would be no particle in the box. The same conclusion comes from the indeterminacy principle. If the energy of the lowest state could be zero, the momentum of the particle would also be zero. Moreover, the uncertainty or spread in the particle

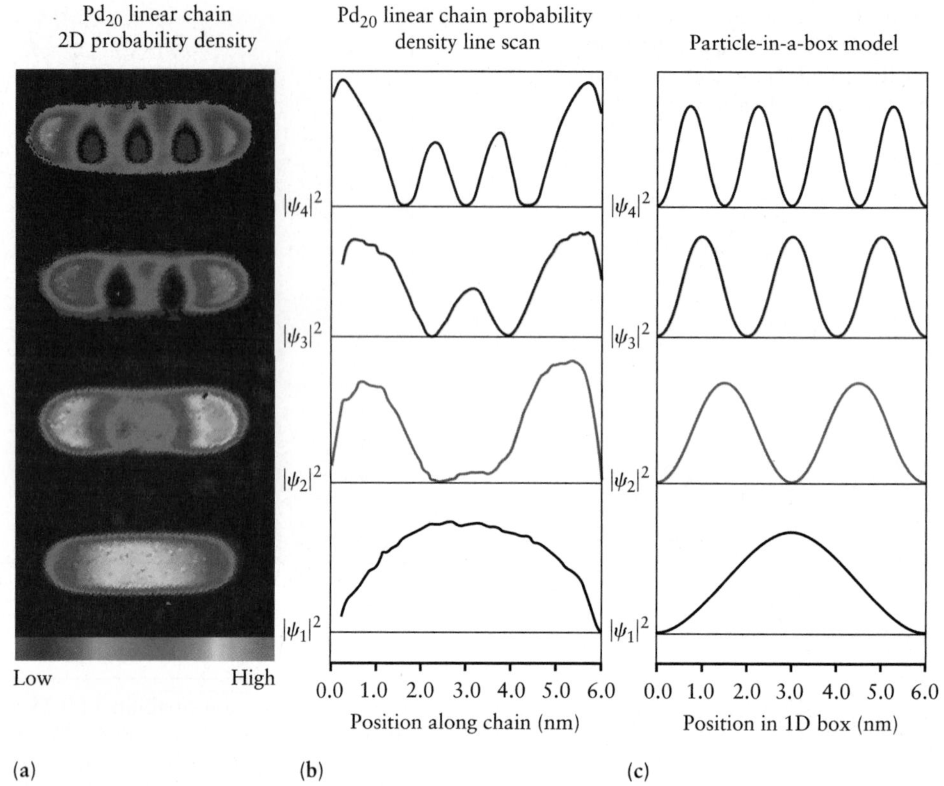

FIGURE 4.39 Quantized energy states for electrons in a linear chain of 20 atoms of Pd on the surface of a NiAl crystal. The atoms were assembled using a scanning tunneling microscope (STM). (a) These images show the two-dimensional probability distribution measured by the STM. (b) These curves are line scans across the measured probability densities, showing their variation along the chain. (c) These curves show the predictions of the one-dimensional particle-in-a-box model for this system. (Courtesy of Professor Wilson Ho, University of California, Irvine.)

momentum, σ_{p_x}, would also be zero, requiring that σ_x be infinite, which contradicts our assertion that the particle is confined to a box of length L. Even at the absolute zero of temperature, where classical kinetic theory would suggest that all motion ceases, a finite quantity of energy remains for a bound system. It is required by the indeterminacy principle and is called the **zero-point energy.**

The wave functions for a particle in a box illustrate another important principle of quantum mechanics: the correspondence principle. We have already stated earlier (and will often repeat) that all successful physical theories must reproduce the explanations and predictions of the theories that preceded them on the length and mass scales for which they were developed. Figure 4.40 shows the probability density for the $n = 5$, 10, and 20 states of the particle in a box. Notice how the probability becomes essentially uniform across the box, and at $n = 20$ there is little evidence of quantization. The **correspondence principle** requires that the results of quantum mechanics reduce to those of classical mechanics for large values of the quantum numbers, in this case, n.

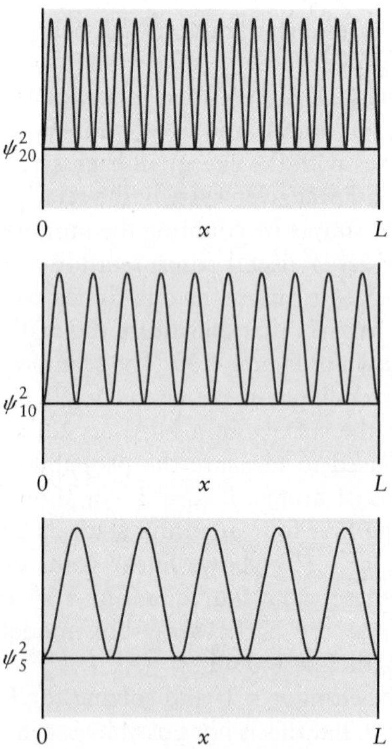

FIGURE 4.40 The probability distribution for a particle in a box of length L in the quantum states $n = 5$, 10, and 20. Compare these results with those shown in Figure 4.37c.

4.6.2 Energy Levels for Particles in Two- and Three-Dimensional Boxes

The Schrödinger equation is readily generalized to describe a particle in a box of two or three dimensions. A particle in a two-dimensional box can be visualized as a marble moving in the x-y plane at the bottom of a deep elevator shaft, with infinite potential walls confining its motion in the x and y directions. A particle in a three-dimensional rectangular box has infinite potential walls confining its motion in the x, y, and z directions. In both cases, the potential energy function is zero throughout the interior of the box. The wave function ψ and potential V now depend on as many as three coordinates (x, y, z), and derivatives with respect to each coordinate appear in the Schrödinger equation. Because the potential energy is constant in all directions inside the box, the motions in the x direction are independent of the motions in the y and z directions, and vice versa. For potential functions of this type, the Schrödinger equation can be solved by the method of *separation of variables,* and the results are quite interesting. The wave function is the product of the wave functions for independent motion in each direction, and the energy is the sum of the energies for independent motion in each direction. Therefore, we can immediately apply the results for the one-dimensional motions developed earlier to discuss, in turn, the energies and wave functions for multidimensional boxes.

The allowed energies for a particle in a three-dimensional rectangular box are

$$E_{n_x n_y n_z} = \frac{h^2}{8m}\left[\frac{n_x^2}{L_x^2} + \frac{n_y^2}{L_y^2} + \frac{n_z^2}{L_z^2}\right] \qquad \begin{cases} n_x = 1, 2, 3, \ldots \\ n_y = 1, 2, 3, \ldots \\ n_z = 1, 2, 3, \ldots \end{cases} \qquad [4.39]$$

where L_x, L_y, and L_z are the side lengths of the box. Here the state is designated by a set of three quantum numbers, (n_x, n_y, n_z). Each quantum number ranges independently over the positive integers. We can obtain the energy levels for a particle in a two-dimensional box in the x-y plane from Equation 4.39 by setting $n_z = 0$ and restrict the box to be a square by setting $L_x = L_y = L$. Similarly, we can specialize Equation 4.39 to a cubic box by setting $L_x = L_y = L_z = L$.

$$E_{n_x n_y n_z} = \frac{h^2}{8mL^2}\left[n_x^2 + n_y^2 + n_z^2\right] \qquad \begin{cases} n_x = 1, 2, 3, \ldots \\ n_y = 1, 2, 3, \ldots \\ n_z = 1, 2, 3, \ldots \end{cases} \qquad [4.40]$$

Figure 4.41 plots the first few energy levels from Equation 4.40. We see that certain energy values appear more than once because the squares of different sets of quantum numbers can add up to give the same total. Such energy levels, which correspond to more than one quantum state, are called **degenerate**. Degenerate energy levels appear in systems with potential energy functions that have symmetric features. In Chapters 5 and 6, we encounter many examples of degenerate energy levels in atomic and molecular systems, as consequences of symmetry. As an exercise, we suggest that you apply Equation 4.40 to determine the energy levels in a square box and examine their degeneracy.

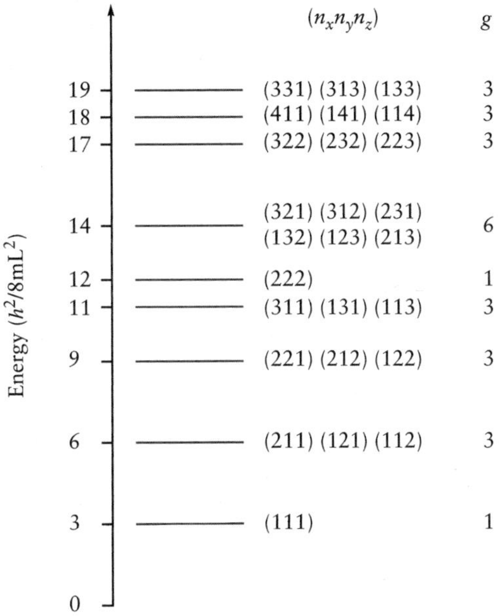

FIGURE 4.41 The energy levels for a particle in a cubic box. The quantum numbers identifying the quantum states and the degeneracy values are given for each energy level.

EXAMPLE 4.8

Consider the following two systems: (a) an electron in a one-dimensional box of length 1.0 Å and (b) a helium atom in a cube 30 cm on an edge. Calculate the energy difference between ground state and first excited state, expressing your answer in kJ mol^{-1}.

Solution

(a) For a one-dimensional box,

$$E_{\text{ground state}} = \frac{h^2}{8mL^2}(1^2)$$

$$E_{\text{first excited state}} = \frac{h^2}{8mL^2}(2^2)$$

Then, for one electron in the box,

$$\Delta E = \frac{3h^2}{8mL^2}$$

$$= \frac{3(6.626 \times 10^{-34}\text{ J s})^2}{8(9.11 \times 10^{-31}\text{ kg})(1.0 \times 10^{-10}\text{ m})^2}$$

$$= 1.8 \times 10^{-17}\text{ J}$$

Multiplying this result by 10^{-3} kJ J^{-1} and by $N_A = 6.022 \times 10^{23}$ mol^{-1} gives

$$\Delta E = 1.1 \times 10^4\text{ kJ mol}^{-1}$$

(b) For a three-dimensional cube, $L_x = L_y = L_z = L$, and

$$E_{\text{ground state}} = \frac{h^2}{8mL^2}(1^2 + 1^2 + 1^2)$$

$$E_{\text{first excited state}} = \frac{h^2}{8mL^2}(2^2 + 1^2 + 1^2)$$

In this case, the three states (2, 1, 1), (1, 2, 1), and (1, 1, 2) have the same energy.

$$\Delta E = \frac{3h^2}{8mL^2}$$

$$= \frac{3(6.626 \times 10^{-34}\text{ J s})^2}{8(6.64 \times 10^{-27}\text{ kg})(0.30\text{ m})^2}$$

$$= 2.8 \times 10^{-40}\text{ J}$$

$$= 1.7 \times 10^{-19}\text{ kJ mol}^{-1}$$

The energy levels are so close together in the latter case (due to the larger mass of the particle and the much larger dimensions of the box) that they appear continuous, and quantum effects play no role. The properties are almost those of a classical particle.

Related Problems: 39, 40

4.7 A DEEPER LOOK Wave Functions for Particles in Two- and Three-Dimensional Boxes

4.7.1 Wave Functions for Particles in Square Boxes

The wave function for a particle in a square box of length L on each side in the x-y plane, which we denote as Ψ, is given by

$$\Psi_{n_x n_y}(x, y) = \psi_{n_x}(x)\psi_{n_y}(y) = \frac{2}{L}\sin\left(\frac{n_x \pi x}{L}\right)\sin\left(\frac{n_y \pi y}{L}\right) \qquad [4.41]$$

as explained earlier. To generate a graphical representation, we calculate the value of Ψ at each point (x, y) in the plane and plot this value as a third dimension above the x-y plane. To make our graphs apply to square boxes of any size, we show them for dimensionless variables $\tilde{x} = x/L$ and $\tilde{y} = y/L$, which range from 0 to 1. We also plot the value of the wave function as a dimensionless function $\tilde{\Psi}$, defined as the ratio of the value of Ψ to its maximum value, $\tilde{\Psi}(\tilde{x}, \tilde{y}) = \Psi(\tilde{x}, \tilde{y})/\Psi_{\text{max}}$. The value of $\tilde{\Psi}$ ranges from -1 to 1. We show three examples in Figure 4.42.

The wave function for the ground state $\tilde{\Psi}_{11}(\tilde{x}, \tilde{y})$ (see Fig. 4.42a) has no nodes and has its maximum at the center of the box, as you would expect from the one-dimensional results in Figure 4.37 from which $\tilde{\Psi}_{11}(\tilde{x}, \tilde{y})$ is constructed.

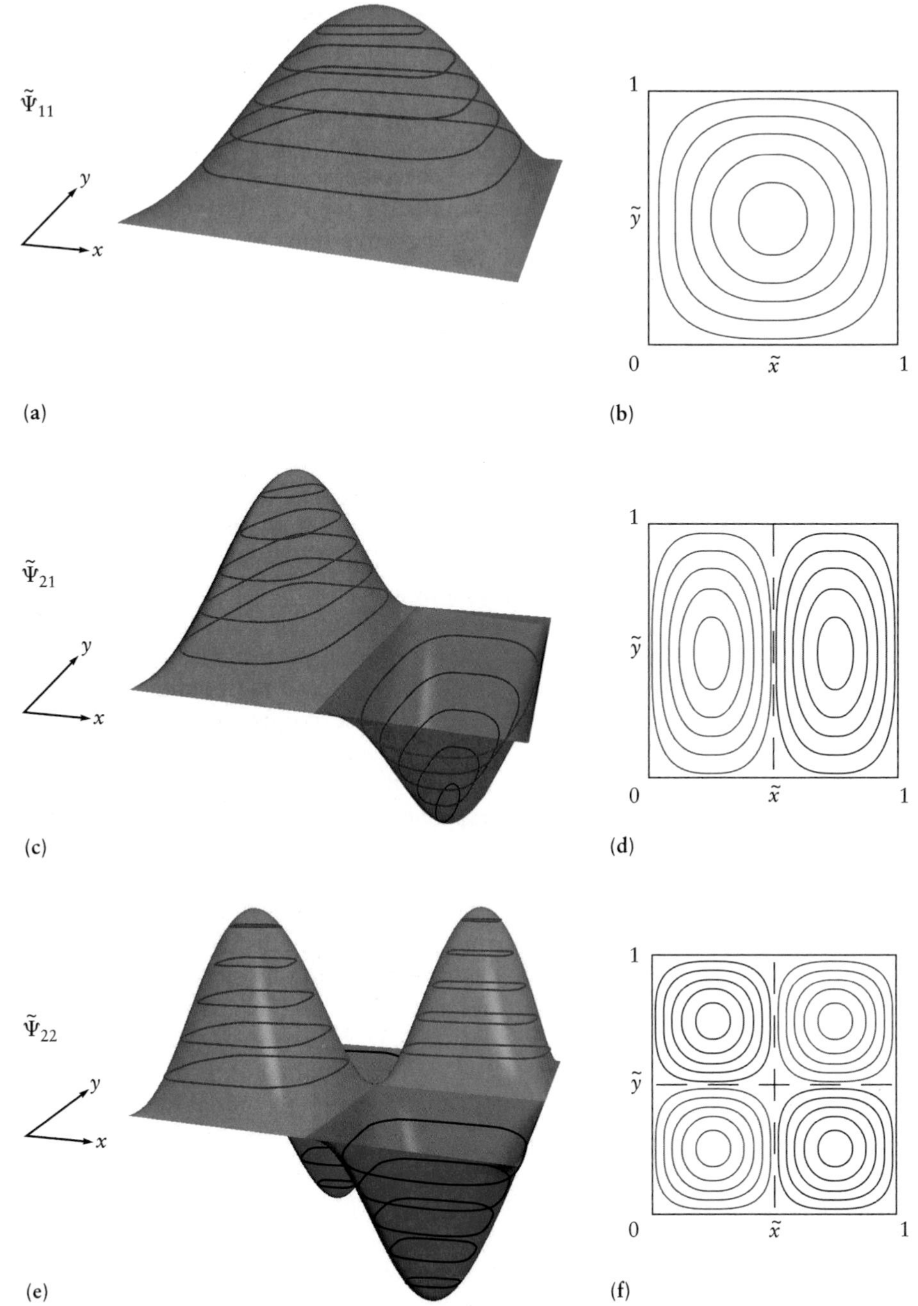

FIGURE 4.42 Wave function for a particle in a square box in selected quantum states. Dimensionless variables are used. (a) Three-dimensional plot for the ground state $\tilde{\Psi}_{11}(\tilde{x}, \tilde{y})$. (b) Contour plot for $\tilde{\Psi}_{11}(\tilde{x}, \tilde{y})$. (c) Three-dimensional plot for the first excited state $\tilde{\Psi}_{21}(\tilde{x}, \tilde{y})$. (d) Contour plot for $\tilde{\Psi}_{21}(\tilde{x}, \tilde{y})$. (e) Three-dimensional plot for the second excited state $\tilde{\Psi}_{22}(\tilde{x}, \tilde{y})$. (f) Contour plot for $\tilde{\Psi}_{22}(\tilde{x}, \tilde{y})$.

Figure 4.42b shows $\tilde{\Psi}_{11}(\tilde{x}, \tilde{y})$ as a contour plot in the x-y plane, generated by choosing a particular value of the wave function in Figure 4.42a, "slicing" its three-dimensional image at that value, and projecting each point on the edge of that slice down to the x-y plane to form a closed contour in that plane. The contour then defines all points in the x-y plane for which $\tilde{\Psi}_{11}(\tilde{x}, \tilde{y})$ has the particular value selected. The process is continued by selecting other values of $\tilde{\Psi}_{11}(\tilde{x}, \tilde{y})$ until the entire three-dimensional image has been collapsed into a set of concentric contours in the two-dimensional x-y plane. Mountain climbers throughout the world use this method to generate contour maps of mountain ranges. The outermost contour identifies points at which the wave function has 10% of its maximum value. The second contour identifies points with 30% of the maximum value, and so on to the innermost contour that identifies points with 90% of the maximum value. Note that the contours in Figure 4.42d, which correspond to uniform increases in amplitude, become much closer together as we approach the node. This indicates the value of the wave function is changing rapidly as it passes through zero. Notice in Figure 4.42b that the contours are circular at the large values of $\tilde{\Psi}_{11}(\tilde{x}, \tilde{y})$, but at lower values, they become squarish and approach perfect squares as $\tilde{\Psi}_{11}(\tilde{x}, \tilde{y}) \longrightarrow 0$, as required by the boundary conditions imposed by the square box.

The wave function for the first excited state $\tilde{\Psi}_{21}(\tilde{x}, \tilde{y})$ is shown in Figure 4.42c. It has a maximum (positive) at $\tilde{x} = 0.25, \tilde{y} = 0.50$ and a minimum (negative) at $\tilde{x} = 0.75, \tilde{y} = 0.50$. The wave function changes sign as it moves along $\tilde{x}$ for any value of $\tilde{y}$. There is a *nodal line* that lies along $\tilde{x} = 0.5$. The wave function does not change sign as it moves along $\tilde{y}$ for any value of $\tilde{x}$. Make sure you see how these characteristics trace back to the one-dimensional solutions in Figure 4.37. Be especially mindful that nodal points in one dimension have become nodal lines in two dimensions. Figure 4.42d shows the contour plot for $\tilde{\Psi}_{21}(\tilde{x}, \tilde{y})$. Note that the contours are nearly circular near the maximum and minimum values, and they become ellipsoidal at smaller values of the wave function. This asymmetry in shape occurs because the motion in the x dimension occurs at a higher level of excitation than that in the y direction. At still lower values of the wave function, the contours begin to resemble rectangles. They approach perfect rectangles as $\tilde{\Psi}_{21}(\tilde{x}, \tilde{y}) \longrightarrow 0$ to match the nodal line along $\tilde{x} = 0.5$ and the boundary conditions enforced by the box. To build up your expertise, construct and examine the wave function $\tilde{\Psi}_{12}(\tilde{x}, \tilde{y})$ in a square box. Convince yourself it is degenerate with $\tilde{\Psi}_{21}(\tilde{x}, \tilde{y})$, and that its plots are the same as those of $\tilde{\Psi}_{21}(\tilde{x}, \tilde{y})$ rotated by 90 degrees in the x-y plane. Give a physical explanation why the two sets of plots are related in this way.

Finally, we plot the wave function for the second excited state $\tilde{\Psi}_{22}(\tilde{x}, \tilde{y})$ (see Fig. 4.42e). It has two maxima (positive) and two minima (negative) located at the values 0.25 and 0.75 for $\tilde{x}$ and $\tilde{y}$. There are two nodal lines, along $\tilde{x} = 0.5$ and $\tilde{y} = 0.5$. They divide the x-y plane into quadrants, each of which contains a single maximum (positive) or minimum (negative) value. Make sure that you see how these characteristics trace back to the one-dimensional solutions in Figure 4.37. Figure 4.42f shows the contour plots for $\tilde{\Psi}_{22}(\tilde{x}, \tilde{y})$. As the magnitude (absolute value) of $\tilde{\Psi}_{22}(\tilde{x}, \tilde{y})$ decreases, the contours distort from circles to squares to match the nodal lines and boundary conditions of the box.

The pattern is now clearly apparent. You can easily produce hand sketches, in three dimensions and as contour plots, for any wave function for a particle in a square box. You need only pay attention to the magnitude of the quantum numbers n_x and n_y, track the number of nodes that must appear along the x and y axes, and convert these into nodal lines in the x-y plane.

The probability of locating the particle in a small element of area of size $d\tilde{x}d\tilde{y}$ centered on the point $(\tilde{x}, \tilde{y})$ is given by $\left|\tilde{\Psi}_{n_x n_y}(\tilde{x}, \tilde{y})\right|^2 d\tilde{x}d\tilde{y}$ when the system is in the quantum state described by n_x and n_y. For the wave functions shown in Figures 4.42a, c, and e, $\tilde{\Psi}^2$ will show 1, 2, and 4 peaks above the x-y plane, respectively. You should make hand sketches of these probability functions and also of their contour plots in the x-y plane. The physical interpretation is straightforward. In the ground state, the probability has a global maximum at the center of the box. In progressively higher excited states, the probability spreads out from the center into a series of local maxima, just as in the one-dimensional case in Figure 4.37. These local maxima are arranged in a pattern determined by the quantum numbers in the x and y directions. If $n_x = n_y$, the local maxima will form a highly symmetric arrangement with pairs separated by nodal lines. If $n_x \neq n_y$, the pattern will not be symmetric. As n_x and n_y take on larger values, the probability becomes more nearly uniform through the box, and the motion becomes more like that predicted by classical mechanics (see the one-dimensional case in

Fig. 4.41). As an exercise, we suggest that you determine the number of nodal lines and the number of local probability maxima for the highly excited state $\tilde{\Psi}_{20,20}(\tilde{x}, \tilde{y})$ and predict the nature of the motion of the particle.

4.7.2 Wave Functions for Particles in Cubic Boxes

The wave function for a particle in a cubic box of length L on each side, with one corner located at the origin of coordinates, is given by

$$\Psi_{n_x n_y n_z}(x, y, z) = \left(\frac{2}{L}\right)^{3/2} \sin\left(\frac{n_x \pi x}{L}\right) \sin\left(\frac{n_y \pi y}{L}\right) \sin\left(\frac{n_z \pi z}{L}\right) \qquad [4.42]$$

where each of the quantum numbers n_x, n_y, and n_z can be any of the positive integers. Graphical representation of these wave functions requires some care. Equation 4.42 tells us simply to go to the point (x, y, z), evaluate the wave function there, and draw a graph showing the results of visiting many such points. However, all three spatial dimensions have already been used up to define the location; thus, we would need a fourth dimension to display the value of the wave function. Alternatively, we could set up a table of numbers giving the value of the wave function at each point (x, y, z), but it would be difficult to develop any intuition about shapes and structures from this table. We will get around these problems by slicing up three-dimensional space into various two- and one-dimensional regions, evaluating the wave function from Equation 4.42 at each point in these restricted regions, and generating graphical representations over these restricted regions. From the behavior of the wave function in these regions, we draw inferences about its overall behavior, even though we cannot graphically display its overall behavior in complete detail. For example, we could evaluate Equation 4.42 only at points in the x-y plane, and thereby generate contour maps of these wave functions in the x-y plane similar to those for the two-dimensional case shown in Figures 4.42b, d, and f. We could repeat this operation at several other "cut planes" through the box, and the resulting series of contour plots would provide considerable insight into the characteristics of the wave function.

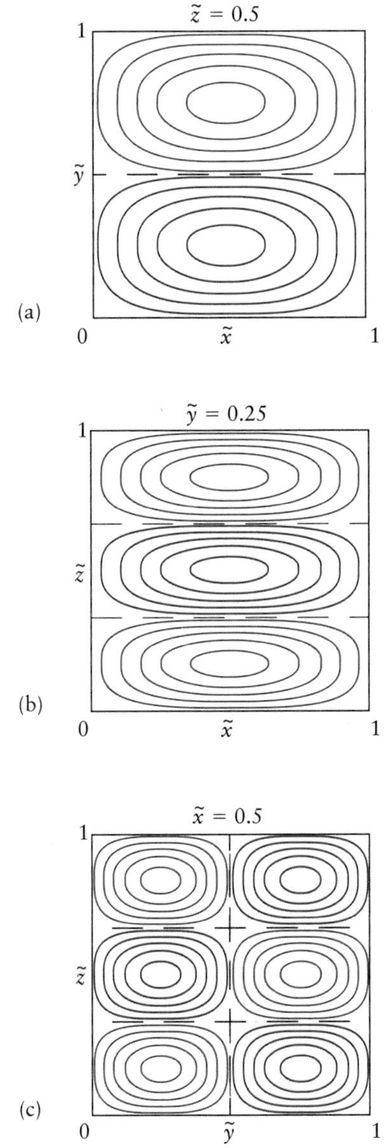

FIGURE 4.43 Contour plots for $\tilde{\Psi}_{123}(\tilde{x}, \tilde{y}, \tilde{z})$ for a particle in a cubic box. (a) Contours generated in a cut at $\tilde{z} = 0.5$. (b) Contours generated in a cut at $\tilde{y} = 0.25$. (c) Contours generated in a cut at $\tilde{x} = 0.5$. The location of nodal lines and shapes of the contours are explained in the text.

In Figure 4.43, we examine the behavior of $\tilde{\Psi}_{123}(\tilde{x}, \tilde{y}, \tilde{z})$ using dimensionless variables defined earlier. Figure 4.43a shows a contour plot generated in a cut plane parallel to the x-y axis at $\tilde{z} = 0.5$. It demonstrates two sets of ellipses separated by one nodal line, arising because $n_x = 1$, $n_y = 2$. Figure 4.43b shows a contour plot generated in a cut plane at $\tilde{y} = 0.25$ or 0.75 parallel to the x-z plane. It shows three sets of ellipses separated by two nodal lines, as a consequence of $n_x = 1$, $n_z = 3$. Figure 4.43c shows a contour plot in the cut plane at $\tilde{x} = 0.5$. It shows six circles and three nodal lines, due to $n_y = 2$, $n_z = 3$. All of this suggests that $\tilde{\Psi}_{123}(\tilde{x}, \tilde{y}, \tilde{z})$ is an interesting object indeed!

How can we get some sense of the three-dimensional shape of a wave function? We can do something similar to the level curves in the contour plots shown at the right of Figure 4.42 for the particle in a two-dimensional box. But of course for particle in a three-dimensional box, rather than level curves, these will be level surfaces, or surfaces plotted in three dimensions that show where the wave function has a certain value. The result is a set of concentric "blimp-shaped" surfaces in three dimensions. Each of them identifies a surface of points (x, y, z) at every one of which $\tilde{\Psi}_{123}(\tilde{x}, \tilde{y}, \tilde{z})$ has the same value. These surfaces are called *isosurfaces* because the wave function has constant value at each point on them. In practice, one generates these isosurfaces by evaluating $\tilde{\Psi}_{123}(\tilde{x}, \tilde{y}, \tilde{z})$ at many points in the cubic box and tracking in the computer all points that have, for example, the value $\tilde{\Psi} = \pm 0.9$. Then the computer plots the resulting

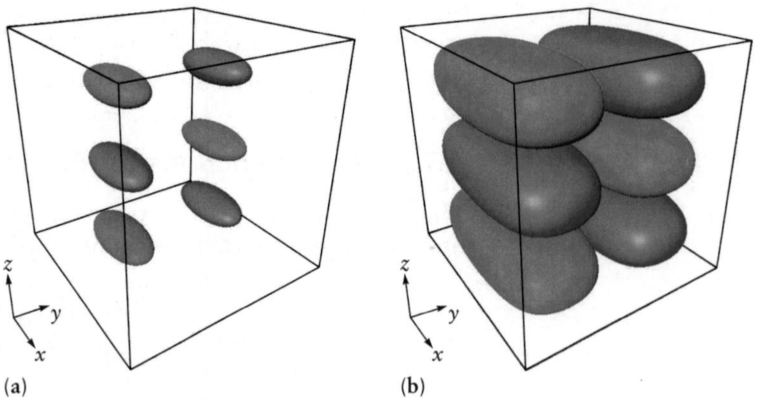

FIGURE 4.44 Isosurfaces for $\tilde{\Psi}_{123}(\tilde{x}, \tilde{y}, \tilde{z})$ for a particle in a cubic box. (a) Isosurfaces for wave function value $\tilde{\Psi}_{123} = \pm 0.8$. (b) Isosurfaces for wave function value $\tilde{\Psi}_{123} = \pm 0.2$. Each isosurface is shown in the same color as the corresponding contour in Figure 4.43.

isosurfaces in three dimensions. Figure 4.44 shows the isosurfaces for $\tilde{\Psi}_{123}(\tilde{x}, \tilde{y}, \tilde{z})$ at the values $\tilde{\Psi} = \pm 0.8, \pm 0.2$.

Figure 4.45 briefly summarizes key images for $\tilde{\Psi}_{222}(\tilde{x}, \tilde{y}, \tilde{z})$ for a particle in a cubic box. Figure 4.45a shows a contour plot in a cut plane at $\tilde{z} = 0.75$. Convince yourself that the contour plot in a cut at $\tilde{z} = 0.25$ would have the same pattern but each positive peak would become negative, and vice versa. Why should we not take a cut at $\tilde{z} = 0.5$? Be sure you understand the same concerns for cut planes perpendicular to the x-axis or the y-axis. Figures 4.45b and c shows the isosurfaces for the maxima and minima of $\tilde{\Psi}_{222}(\tilde{x}, \tilde{y}, \tilde{z})$ at the values $\tilde{\Psi} = \pm 0.9, \pm 0.3$.

Notice how in Figures 4.44 and 4.45 the shape depends on the value selected for the isosurface. This demonstrates an important point about plots of wave functions for a particle moving in three dimensions: It is not possible to show the shape of the wave function in three dimensions. You should be mindful that the appearance of the wave function in three-dimensional representations depends strongly on choices made by the illustrator. Be certain you understand these choices in each image you examine (or create!). These same issues appear in Chapter 5 when we discuss the wave functions for electrons in atoms, called *atomic orbitals.* Throughout this book, we have taken great care to generate accurate contour plots and isosurfaces for them from computer calculations to guide your thinking about the distribution of electrons in atoms and molecules.

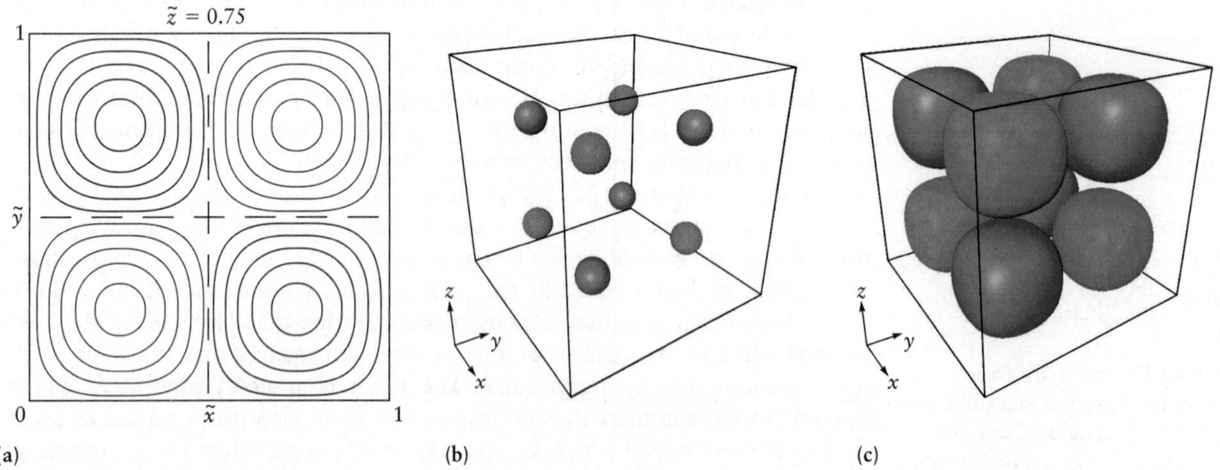

FIGURE 4.45 Representations of $\tilde{\Psi}_{222}(\tilde{x}, \tilde{y}, \tilde{z})$ for a particle in a cubic box. (a) Contour plots for a cut taken at $\tilde{z} = 0.75$. (b) Isosurfaces for wave function value $\tilde{\Psi}_{222} = \pm 0.9$. (c) Isosurfaces for wave function value $\tilde{\Psi}_{222} = \pm 0.3$. Each isosurface is shown in the same color as the corresponding contour in (a).

CHAPTER SUMMARY

The concepts and methods of quantum mechanics were developed to explain the behavior of matter on the nanometer-length scale. The results of a number of key experiments demanded the creation of a new physical theory in the first third of the 20th century because classical mechanics and electrodynamics failed completely to account for these new observations. These experimental facts included the wavelength and temperature dependence of blackbody radiation, the very existence of stable atoms and their discrete line spectra, the photoelectric effect, and electron diffraction. Taken together, these experiments demonstrated unequivocally the existence of energy quantization (blackbody radiation and atomic spectra) and wave–particle duality (photoelectric effect and electron diffraction), which would become the central concepts of the new theory quantum mechanics.

Explaining each of these experiments required scientists to abandon long-held concepts from classical physics. Particularly striking were Planck's explanation of blackbody radiation and Bohr's model of the hydrogen atom. Both scientists started afresh and made whatever assumptions were necessary to fit the experimental results, ignoring any conflicts with classical physics. Only after their models agreed so well with experiment did they and other scientists begin to consider the radical philosophical implications of quantum mechanics and develop a new way of thinking about nature on the nanometer-length scale. This was undoubtedly one of the most significant shifts in the history of science.

The key new concepts developed in quantum mechanics include the quantization of energy, a probabilistic description of particle motion, wave–particle duality, and indeterminacy. These ideas appear foreign to us because they are inconsistent with our experience of the macroscopic world. Nonetheless, we have accepted their validity because they provide the most comprehensive account of the behavior of matter and radiation and because the agreement between theory and the results of all experiments conducted to date has been impressively accurate.

Energy quantization arises for all systems whose motions are confined by a potential well. The one-dimensional particle-in-a-box model shows why quantization only becomes apparent on the atomic scale. Because the energy level spacing is inversely proportional to the mass and to the square of the length of the box, quantum effects become too small to be observed for macroscopic systems, like a ball in a squash court.

Wave–particle duality accounts for the probabilistic nature of quantum mechanics and for indeterminacy. Once we accept that particles can behave as waves, we can form analogies with classical electromagnetic wave theory to describe the motion of particles. For example, the probability of locating the particle at a particular location is the square the spatial part of its wavefunction (the square of the amplitude of the wave function plotted as a function of time, as it oscillates at a particular point in coordinate space). Zero-point energy is a consequence of the Heisenberg indeterminacy relation; all particles bound in potential wells have finite energy even at the absolute zero of temperature.

Particle-in-a-box models illustrate a number of important features of quantum mechanics. The energy-level structure depends on the nature of the potential, $E_n \propto n^2$, for the particle in a one-dimensional box, so the separation between energy levels increases as n increases. The probability density distribution is different from that for the analogous classical system. The most probable location for the particle-in-a-box model in its ground state is the center of the box, rather than uniformly over the box as predicted by classical mechanics. Normalization ensures that the probability of finding the particle at some position in the box, summed over all possible positions, adds up to 1. Finally, for large values of n, the probability distribution looks much more classical, in accordance with the correspondence principle.

Different kinds of energy level patterns arise from different potential energy functions, for example the hydrogen atom (see Section 5.1) and the harmonic oscillator (see Section 20.3).

These concepts and principles are completely general; they can be applied to explain the behavior of any system of interest. In the next two chapters, we use quantum mechanics to explain atomic and molecular structure, respectively. It is important to have a firm grasp of these principles because they are the basis for our comprehensive discussion of chemical bonding in Chapter 6.

CONCEPTS AND SKILLS

Section 4.1 – Preliminaries: Wave Motion and Light

Relate the frequency, wavelength, and speed of light waves. Do the same for other kinds of waves (Problems 1–8).

- Waves are oscillating disturbances in space and time which are characterized by their amplitude, wavelength, and frequency. The speed of a wave is given by the product $\lambda\nu$, where λ and ν are the wavelength and frequency, respectively. Light waves are propagating waves of electromagnetic radiation. For this case the speed is given by $c = \lambda\nu$, where c is the speed of light (3×10^8 m s^{-1}).

Section 4.2 – Evidence for Energy Quantization in Atoms

Describe blackbody radiation and discuss how related paradoxes of classical physics were resolved by quantum mechanics (Problems 9–10).

- Blackbody radiation emitted from hot sources has a characteristic frequency distribution that is temperature-dependent. The spectrum of cooler objects has a comparatively narrow band that peaks near the red end of the visible spectrum, whereas that of hotter objects has a much broader band that is shifted toward the blue.
- The peak observed in the frequency distribution of blackbody radiation is completely inconsistent with the predictions of classical electromagnetic theory. This failure of classical physics is called the ultraviolet catastrophe.
- The only way that Planck could fit the experimental spectrum was to postulate that:
    - the oscillating charges responsible for the radiation were restricted to discrete energies

$$\varepsilon_{\text{osc}} = nh\nu \; (n = 1, 2, 3, 4, \ldots), \qquad [4.3]$$

    - an oscillator was either excited or not
    - the probability of an oscillator being excited depends on the temperature.

Use experimental emission and absorption spectra to determine spacing between energy levels in atoms (Problems 11–16).

- Atoms emit and absorb energy in discrete amounts. The frequency ν of the light absorbed or emitted is related to the difference between two energy levels as

$$\nu = \frac{E_f - E_i}{h} \quad \text{or} \quad \Delta E = h\nu \text{ (for absorbed light)} \qquad [4.7]$$

Use the Franck–Hertz method to determine spacings between adjacent energy levels in atoms (Problems 17–18).

- Electrons can excite atoms from one quantum state to another by energy transferred during collisions. The threshold energy for excitation exactly matches the emission of light as the atom drops back down to the lower state, thus confirming the existence of quantized states and showing that they may be excited by either mechanical impact of electrons or absorption of photons.

$$\nu = \frac{E_{\text{upper}} - E_{\text{lower}}}{h} = \frac{eV_{\text{thr}}}{h} \qquad [4.8]$$

Section 4.3 – The Bohr Model: Predicting Discrete Energy Levels in Atoms

Use the Bohr model to calculate the energy levels of one-electron atoms and to find the frequencies and wavelengths of light emitted in transitions between energy levels (Problems 19–22).

- Bohr postulated quantization of the angular momentum:

$$L = m_e vr = n\frac{h}{2\pi} \qquad n = 1, 2, 3, \ldots \qquad [4.11]$$

- He used this relation with the classical equations of motion, and correctly predicted the energy levels of one-electron atoms:

$$E_n = \frac{-Z^2e^4m_e}{8\epsilon_0^2n^2h^2} = -(2.18 \times 10^{-18}\text{ J})\frac{Z^2}{n^2} \qquad n = 1, 2, 3, \ldots \qquad [4.14a]$$

- Bohr assumed that emission or absorption of light by atoms involved in transitions between energy levels of the atom and the frequency of the light was related to the energy difference between levels as $\Delta E = h\nu$. Combining this relation with his equation for the energy levels predicted the frequencies of light absorbed or emitted by the atom.

$$\nu = \frac{Z^2e^4m_e}{8\epsilon_0^2h^3}\left(\frac{1}{n_f^2} - \frac{1}{n_i^2}\right) = (3.29 \times 10^{15}\text{s}^{-1})Z^2\left(\frac{1}{n_f^2} - \frac{1}{n_i^2}\right) \qquad [4.16]$$

$n_i > n_f = 1, 2, 3, \ldots$ (emission)

$$\nu = (3.29 \times 10^{15}\text{ s}^{-1})Z^2\left(\frac{1}{n_i^2} - \frac{1}{n_f^2}\right) \qquad [4.17]$$

$n_f > n_i = 1, 2, 3, \ldots$ (absorption)

- Bohr's model could not account for the spectra of many-electron atoms.

Section 4.4 – Evidence for Wave–Particle Duality

Describe the photoelectric effect, and discuss how related paradoxes of classical physics were resolved by quantum mechanics (Problems 23–24).

- In the photoelectric effect, light shines on a metal surface in a vacuum. The kinetic energy and photocurrent (number of electrons per second) emitted are measured as a function of frequency and intensity of the light.
- Experimental results of the photoelectric effect:
    - No electrons are emitted below a threshold frequency ν_0 regardless of intensity.
    - Above the threshold, the photocurrent is proportional to the light intensity.

- Interpretation of the photoelectric effect: Light behaves like a stream of particles called photons, each with an energy $E = h\nu$. A photon of a given energy either does or does not provide enough energy to overcome the forces binding the electron to the metal.
- Above the threshold, the excess energy goes into the kinetic energy of the photoelectron $\frac{1}{2}mv^2 = h\nu - h\nu_0$, where h is Planck's constant 6.625×10^{-34} J sec^{-1}.
- The photoelectric effect demonstrated the particle nature of electromagnetic radiation (formerly described only as waves).

Using the law of conservation of energy, relate the work function of a metal to the wavelength of light used to eject electrons in the photoelectric effect and the kinetic energy of those electrons (Problems 25–28).

- Above the threshold, the maximum kinetic energy of the electron is $E_{\text{max}} = \frac{1}{2}mv_e^2 = h\nu - \Phi$, where h is Planck's constant 6.625×10^{-34} J sec^{-1}.
- $\Phi = h\nu_0$ is the work function of the metal, which is the energy barrier the electron must overcome to escape from the metal.

Discuss the de Broglie relation and use it to calculate the wavelengths associated with particles in motion (Problems 29–32).

- Wave–particle duality: De Broglie postulated that the motion of an electron in an atom could be described as a circular standing wave, which required an integral number of wavelengths to fit the circumference, $n\lambda = 2\pi r$. Combining this result with Bohr's quantization of angular momentum led to the de Broglie relation $\lambda = h/p$.
- The de Broglie relation asserts that a particle moving with magnitude of momentum p can behave as if it were a wave of wavelength $\lambda = h/p$. This relation is the origin of wave–particle duality, which means that matter can show either particlelike or wavelike behavior depending on the conditions of the observation.

Describe interference between wave functions for an electron. Explain how constructive and destructive interference influence the probability for finding the electron at a particular location (Problems 33–34).

- Diffraction of low energy electrons at solid surfaces confirms the wavelike nature of the electron. Electrons departing the surface after collisions with atoms in the surface are described by wave functions whose wavelengths depend on the energies of the electrons. The outgoing waves can interfere constructively to define the bright spots on the display, or destructively to produce no image on the display. The probability of finding an electron at a particular location is highest at locations where the wave functions interfere constructively.

State the Heisenberg indeterminacy principle and use it to establish bounds within which the position and momentum of a particle can be known (Problems 35–36).

- The Heisenberg indeterminacy relation $\sigma_x\sigma_p \geq h/4\pi$ is the quantitative statement of a fundamental limit on our ability to know simultaneously the values of two properties of a particle with an arbitrarily high precision.
- For example, if the position of a particle is known to fall within the range $(\Delta x)_0$, the uncertainty in the momentum cannot be smaller than $\sigma_p = h/[4\pi(\Delta x)_0]$.

Section 4.5 – The Schrödinger Equation

State the conditions a function must satisfy in order to be a solution of the Schrödinger equation. Explain how these conditions provide the probability interpretation of the wave function.

- The wave function and its first derivative must be continuous at each point in space.
- The wave function must be bounded at very large values of x (if the particle is not free). This means that

$$\psi \longrightarrow 0 \quad \text{as} \quad x \longrightarrow \pm\infty \qquad [4.31]$$

- The square of the wave function gives the probability of finding the particle at a particular position in space. Normalization of the wave function by requiring that $\int_{\text{all space}} \psi^2 dV = 1$ ensures that the particle will be found somewhere.

Section 4.6 – Quantum Mechanics of Particle-in-a-Box Models

Determine the energy levels for particles confined in boxes with rigid walls that are infinite potential barriers (Problems 37–40).

- In one-dimensional boxes the energy of the particle is characterized by a single quantum number, and the energy-level spacing is

$$E_n = \frac{h^2 n^2}{8mL^2} \qquad n = 1, 2, 3, \ldots \qquad [4.37]$$

- The wave functions for a particle in a one-dimensional box are given by

$$\psi_n(x) = \sqrt{\frac{2}{L}} \sin\left(\frac{n\pi x}{L}\right) \qquad n = 1, 2, 3, \ldots \qquad [4.36]$$

  In a one-dimensional problem, the probability of finding the particle between the position $x = x_1$ and $x = x_2$ is calculated with the following integral

$$\int_{x_1}^{x_2} |\psi(x)|^2 \, dx$$

  If the interval $\Delta x = x_2 - x_1$ is small, this integral has a value close to $|\psi(x_0)|^2 \Delta x$, where x_0 is evaluated in the middle of the range from x_1 to x_2.

- In three-dimensional boxes the energy of the particle is characterized by three quantum numbers, one each for motion in the x, y, and z directions. The energy levels in a cubic box are

$$E_{n_x n_y n_z} = \frac{h^2}{8mL^2}\left[n_x^2 + n_y^2 + n_z^2\right] \qquad \begin{cases} n_x = 1, 2, 3, \ldots \\ n_y = 1, 2, 3, \ldots \\ n_z = 1, 2, 3, \ldots \end{cases} \qquad [4.40]$$

Section 4.7 – A Deeper Look . . . Wave Functions for Particles in Two- and Three-Dimensional Boxes

Prepare hand-drawn sketches for contour diagrams and isosurfaces of the wave functions for particles in square and cubic boxes. Relate the number and locations of the nodes, maxima, and minima to the quantum numbers for the wave function (Problems 41–44).

- In square boxes, the quantum numbers n_x and n_y, give the number of nodes ($n_x - 1$ and $n_y - 1$ respectively) that appear along the x and y axes, and convert these into nodal lines in the x–y plane.
- In cubic boxes, show how the quantum numbers n_x, n_y, and n_z, are related to the number of nodes that must appear along the x, y, and z axes, and convert these into nodal planes in the x–y–z space.

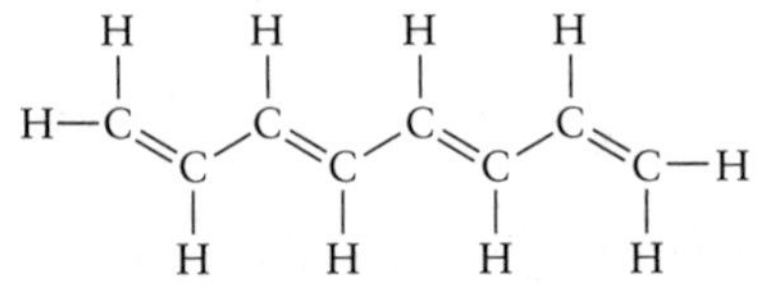

FIGURE 4.46 The conjugated molecule octatetraene.

CUMULATIVE EXERCISE

Conjugated Molecules in Dyestuffs and Biological Materials

An interesting class of carbon-containing molecules called *conjugated molecules* have structures that consist of a sequence of alternating single and double bonds (Fig. 4.46). These chainlike molecules are represented as zigzag structures in which the angle between adjacent segments is determined by the geometry of the C—C double bond. Various properties to be explored in later chapters indicate that the electrons forming the double bonds are "de-localized" over the entire chain. Such molecules absorb light in the visible and ultraviolet regions of the electromagnetic spectrum. Many dyestuffs and molecules with biological significance have these structures. The properties of these molecules can be described approximately by the particle-in-a-box model in which we assume there is no interaction between the electrons, the potential energy is constant along the chain, and the potential energy is infinite at the ends of the chain. Assume the length of the potential well is Nd, where N is the number of carbon atoms in the chain, and d is half the sum of the lengths of a C—C single bond and a C—C double bond. In a conjugated molecule of N carbon atoms in the chain, there will be N electrons involved in the double bonds.

(a) Write the equation for the energy levels of an electron in this potential well.

(b) Write the equation for the wave function of an electron in this potential well.

To describe the placement of N electrons in the energy levels, we anticipate a principle to be developed in Chapter 5 that requires that no more than two electrons can occupy a level. Therefore, levels will be occupied from the ground state up to level $n = N/2$. Absorption of light can cause one electron to move to the next level $n = (N/2) + 1$.

(c) Write the equation for the frequency of light that will cause this transition.

(d) The molecule butadiene has four carbon atoms with conjugated structure; thus, $N = 4$. Calculate the wavelength of light in the first transition of butadiene.

(e) The molecular structure of vitamin A is conjugated with $N = 10$. Calculate the wavelength of light in the first transition of vitamin A.

(f) The molecule β-carotene has $N = 22$. Calculate the wavelength of light in the first transition of β-carotene.

We see in part (c) that the frequency of light absorbed should be inversely proportional to the length of the chain. Short-chain conjugated molecules absorb in the ultraviolet, whereas longer chain molecules absorb in the visible. This qualitative trend is predicted by the simple particle-in-a-box model. Later chapters detail how these results can be improved.

Answers

(a) $E_n = \dfrac{n^2h^2}{8mN^2d^2}$

(b) $\psi_n(x) = A\sin\left(\dfrac{n\pi x}{Nd}\right)$

(c) $h\nu = E_{(N/2)+1} - E_{N/2} = \dfrac{h^2(N+1)}{8md^2N^2} \simeq \dfrac{h^2}{8md^2N}$

(d) $\lambda = 2050$ Å

(e) $\lambda = 3150$ Å

(f) $\lambda = 4410$ Å

PROBLEMS

Answers to problems whose numbers are red appear in Appendix G. Problems that are more challenging are indicated with asterisks.

4.1 Preliminaries: Wave Motion and Light

1. Some water waves reach the beach at a rate of one every 3.2 s, and the distance between their crests is 2.1 m. Calculate the speed of these waves.

2. The spacing between bands of color in a chemical wave from an oscillating reaction is measured to be 1.2 cm, and a new wave appears every 42 s. Calculate the speed of propagation of the chemical waves.

3. An FM radio station broadcasts at a frequency of $9.86 \times 10^7\ s^{-1}$ (98.6 MHz). Calculate the wavelength of the radio waves.

4. The gamma rays emitted by ^{60}Co are used in radiation treatment of cancer. They have a frequency of $2.83 \times 10^{20}\ s^{-1}$. Calculate their wavelength, expressing your answer in meters and in angstroms.

5. Radio waves of wavelength 6.00×10^2 m can be used to communicate with spacecraft over large distances.
   (a) Calculate the frequency of these radio waves.
   (b) Suppose a radio message is sent home by astronauts in a spaceship approaching Mars at a distance of 8.0×10^{10} m from Earth. How long (in minutes) will it take for the message to travel from the spaceship to Earth?

6. An argon ion laser emits light of wavelength of 488 nm.
   (a) Calculate the frequency of the light.
   (b) Suppose a pulse of light from this laser is sent from Earth, is reflected from a mirror on the moon, and returns to its starting point. Calculate the time elapsed for the round trip, taking the distance from Earth to the moon to be 3.8×10^5 km.

7. The speed of sound in dry air at 20°C is 343.5 m s^{-1}, and the frequency of the sound from the middle C note on a piano is 261.6 s^{-1} (according to the American standard pitch scale). Calculate the wavelength of this sound and the time it will take to travel 30.0 m across a concert hall.

8. Ultrasonic waves have frequencies too high to be detected by the human ear, but they can be produced and detected by vibrating crystals. Calculate the wavelength of an ultrasonic wave of frequency $5.0 \times 10^4\ s^{-1}$ that is propagating through a sample of water at a speed of 1.5×10^3 m s^{-1}. Explain why ultrasound can be used to probe the size and position of the fetus inside the mother's abdomen. Could audible sound with a frequency of 8000 s^{-1} be used for this purpose?

4.2 Evidence for Energy Quantization in Atoms

9. The maximum in the blackbody radiation intensity curve moves to shorter wavelength as temperature increases. The German physicist Wilhelm Wien demonstrated the relation to be $\lambda_{max} \propto 1/T$. Later, Planck's equation showed the maximum to be $\lambda_{max} = 0.20\ hc/kT$. In 1965, scientists researching problems in telecommunication discovered "background radiation" with maximum wavelength 1.05 mm (microwave region of the EM spectrum) throughout space. Estimate the temperature of space.

10. Use the data in Figure 4.8 to estimate the ratio of radiation intensity at 10,000 Å (infrared) to that at 5000 Å (visible) from a blackbody at 5000 K. How will this ratio change with increasing temperature? Explain how this change occurs.

11. Excited lithium atoms emit light strongly at a wavelength of 671 nm. This emission predominates when lithium atoms are excited in a flame. Predict the color of the flame.

12. Excited mercury atoms emit light strongly at a wavelength of 454 nm. This emission predominates when mercury atoms are excited in a flame. Predict the color of the flame.

13. Barium atoms in a flame emit light as they undergo transitions from one energy level to another that is 3.6×10^{-19} J lower in energy. Calculate the wavelength of light emitted and, by referring to Figure 4.3, predict the color visible in the flame.

14. Potassium atoms in a flame emit light as they undergo transitions from one energy level to another that is 4.9×10^{-19} J lower in energy. Calculate the wavelength of light emitted and, by referring to Figure 4.3, predict the color visible in the flame.

15. The sodium D-line is actually a pair of closely spaced spectroscopic lines seen in the emission spectrum of sodium atoms. The wavelengths are centered at 589.3 nm. The intensity of this emission makes it the major source of light (and causes the yellow color) in the sodium arc light.
    (a) Calculate the energy change per sodium atom emitting a photon at the D-line wavelength.
    (b) Calculate the energy change per mole of sodium atoms emitting photons at the D-line wavelength.
    (c) If a sodium arc light is to produce 1.000 kilowatt (1000 J s^{-1}) of radiant energy at this wavelength, how many moles of sodium atoms must emit photons per second?

16. The power output of a laser is measured by its wattage, that is, the number of joules of energy it radiates per second (1 W = 1 J s^{-1}). A 10-W laser produces a beam of green light with a wavelength of 520 nm (5.2×10^{-7} m).
    (a) Calculate the energy carried by each photon.
    (b) Calculate the number of photons emitted by the laser per second.

17. In a Franck–Hertz experiment on sodium atoms, the first excitation threshold occurs at 2.103 eV. Calculate the wavelength of emitted light expected just above this threshold. (**Note:** Sodium vapor lamps used in street lighting emit spectral lines with wavelengths 5891.8 and 5889.9 Å.)

18. In a Franck–Hertz experiment on hydrogen atoms, the first two excitation thresholds occur at 10.1 and 11.9 eV. Three optical emission lines are associated with these levels. Sketch an energy-level diagram for hydrogen atoms

based on this information. Identify the three transitions associated with these emission lines. Calculate the wavelength of each emitted line.

4.3 The Bohr Model: Predicting Discrete Energy Levels in Atoms

19. Use the Bohr model to calculate the radius and the energy of the B^{4+} ion in the $n = 3$ state. How much energy would be required to remove the electrons from 1 mol of B^{4+} in this state? What frequency and wavelength of light would be emitted in a transition from the $n = 3$ to the $n = 2$ state of this ion? Express all results in SI units.

20. He^+ ions are observed in stellar atmospheres. Use the Bohr model to calculate the radius and the energy of He^+ in the $n = 5$ state. How much energy would be required to remove the electrons from 1 mol of He^+ in this state? What frequency and wavelength of light would be emitted in a transition from the $n = 5$ to the $n = 3$ state of this ion? Express all results in SI units.

21. The radiation emitted in the transition from $n = 3$ to $n = 2$ in a neutral hydrogen atom has a wavelength of 656.1 nm. What would be the wavelength of radiation emitted from a doubly ionized lithium atom (Li^{2+}) if a transition occurred from $n = 3$ to $n = 2$? In what region of the spectrum does this radiation lie?

22. Be^{3+} has a single electron. Calculate the frequencies and wavelengths of light in the emission spectrum of the ion for the first three lines of each of the series that are analogous to the Lyman and the Balmer series of neutral hydrogen. In what region of the spectrum does this radiation lie?

4.4 Evidence for Wave–Particle Duality

23. Both blue and green light eject electrons from the surface of potassium. In which case do the ejected electrons have the higher average kinetic energy?

24. When an intense beam of green light is directed onto a copper surface, no electrons are ejected. What will happen if the green light is replaced with red light?

25. Cesium frequently is used in photocells because its work function (3.43×10^{-19} J) is the lowest of all the elements. Such photocells are efficient because the broadest range of wavelengths of light can eject electrons. What colors of light will eject electrons from cesium? What colors of light will eject electrons from selenium, which has a work function of 9.5×10^{-19} J?

26. Alarm systems use the photoelectric effect. A beam of light strikes a piece of metal in the photocell, ejecting electrons continuously and causing a small electric current to flow. When someone steps into the light beam, the current is interrupted and the alarm is triggered. What is the maximum wavelength of light that can be used in such an alarm system if the photocell metal is sodium, with a work function of 4.41×10^{-19} J?

27. Light with a wavelength of 2.50×10^{-7} m falls on the surface of a piece of chromium in an evacuated glass tube. If the work function of chromium is 7.21×10^{-19} J, determine (a) the maximum kinetic energy of the emitted photoelectrons and (b) the speed of photoelectrons that have this maximum kinetic energy.

28. Calculate the maximum wavelength of electromagnetic radiation if it is to cause detachment of electrons from the surface of metallic tungsten, which has a work function of 7.29×10^{-19} J. If the maximum speed of the emitted photoelectrons is to be 2.00×10^6 m s^{-1}, what should the wavelength of the radiation be?

29. A guitar string with fixed ends has a length of 50 cm.
    - **(a)** Calculate the wavelengths of its fundamental mode of vibration (that is, its first harmonic) and its third harmonic.
    - **(b)** How many nodes does the third harmonic have?

30. Suppose we picture an electron in a chemical bond as being a wave with fixed ends. Take the length of the bond to be 1.0 Å.
    - **(a)** Calculate the wavelength of the electron wave in its ground state and in its first excited state.
    - **(b)** How many nodes does the first excited state have?

31. Calculate the de Broglie wavelength of the following:
    - **(a)** an electron moving at a speed of 1.00×10^3 m s^{-1}
    - **(b)** a proton moving at a speed of 1.00×10^3 m s^{-1}
    - **(c)** a baseball with a mass of 145 g, moving at a speed of 75 km hr^{-1}

32. Calculate the de Broglie wavelength of the following:
    - **(a)** electrons that have been accelerated to a kinetic energy of 1.20×10^7 J mol^{-1}
    - **(b)** a helium atom moving at a speed of 353 m s^{-1} (the root-mean-square speed of helium atoms at 20 K)
    - **(c)** a krypton atom moving at a speed of 299 m s^{-1} (the root-mean-square speed of krypton atoms at 300 K)

33. In a particular Low Energy Electron Diffraction (LEED) study of a solid surface, electrons at 45 eV were diffracted at $\theta = 53°$.
    - **(a)** Calculate the crystal spacing a.
    - **(b)** Calculate the diffraction angle for 90 eV electrons on this same surface.

34. What electron energy is required to obtain the diffraction pattern for a surface with crystal spacing of 4.0 Å?

35. **(a)** The position of an electron is known to be within 10 Å (1.0×10^{-9} m). What is the minimum uncertainty in its velocity component in this direction?
    - **(b)** Repeat the calculation of part (a) for a helium atom.

36. No object can travel faster than the speed of light, so it would appear evident that the uncertainty in the speed of any object is at most 3×10^8 m s^{-1}.
    - **(a)** What is the minimum uncertainty in the position of an electron, given that we know nothing about its speed except that it is slower than the speed of light?
    - **(b)** Repeat the calculation of part (a) for the position of a helium atom.

4.6 Quantum Mechanics of Particle-in-a-Box Models

37. **(a)** Using Equation 4.36, make a graph of the $n = 3$ wave function and the square of this wave function for a particle in a box with the edges at $x = 0$ and $x = 7$ Å. Estimate the probability of finding the particle between

$x_1 = 3$ Å and $x_2 = 4$ Å. You may either do an integral or find the area under the curve by counting the number of rectangular boxes (roughly estimating the fractional boxes), and then multiply by the area of each box, watching your units. Show your calculation with units for credit.

(b) Verify that the $n = 3$ wave function in (a) is an allowed quantum state (sometimes called an *eigenfunction*) of $\mathcal{H}$ by substituting it for ψ in the left side of the equation for $\mathcal{H}\psi = E\psi$ immediately following Eq. 4.32. Calculate the derivatives to see if your result $E\psi$ is a constant multiplied by the original $n = 3$ wavefunction. What does the constant you get tell you about the energy of this eigenstate? Is it in agreement with the energy for the $n = 3$ level you would calculate from Equation 4.37?

38. Using a simple particle-in-a-box model for the multiple bonding in 1,2-butadiene (see Example 4.7) and the $n = 2$ wave function for the weakest bound electron, calculate the probability of finding the electron in an 0.1 interval centered midway between the two inner carbon atoms (that is, the center is at $x = 2.11$ Å, so this interval is from $x_1 = 2.06$ Å to $x_2 = 2.16$ Å). Then calculate the probability of finding the electron in an 0.1 interval centered midway between an end carbon atom and the carbon atom that is double-bonded to it in the Lewis dot structure. (You may calculate the appropriate integrals or estimate the relevant areas under the curve graphically.) Then recalculate the probabilities by approximating the integral as $|\psi(x_0)|^2\,\Delta x$, where x_0 is evaluated in the middle of the range from x_1 to x_2. Finally, explain why you can't approximate the integral needed for Problem 37(a) in this way.

39. Chapter 3 introduced the concept of a double bond between carbon atoms, represented by C=C, with a length near 1.34 Å. The motion of an electron in such a bond can be treated crudely as motion in a one-dimensional box. Calculate the energy of an electron in each of its three lowest allowed states if it is confined to move in a one-dimensional box of length 1.34 Å. Calculate the wavelength of light necessary to excite the electron from its ground state to the first excited state.

40. When metallic sodium is dissolved in liquid sodium chloride, electrons are released into the liquid. These dissolved electrons absorb light with a wavelength near 800 nm. Suppose we treat the positive ions surrounding an electron crudely as defining a three-dimensional cubic box of edge L, and we assume that the absorbed light excites the electron from its ground state to the first excited state. Calculate the edge length L in this simple model.

4.7 Wave Functions for Particles in Two- and Three-Dimensional Boxes

41. Write the wave function $\tilde{\Psi}_{12}(\tilde{x}, \tilde{y})$ for a particle in a square box.

(a) Convince yourself it is degenerate with $\tilde{\Psi}_{21}(\tilde{x}, \tilde{y})$.

(b) Convince yourself that its plots are the same as those of $\tilde{\Psi}_{21}(\tilde{x}, \tilde{y})$ rotated by 90 degrees in the x-y plane.

(c) Give a physical explanation why the two sets of plots are related in this way.

42. Write the wave function for the highly excited state $\tilde{\Psi}_{100,100}(\tilde{x}, \tilde{y})$ for a particle in a square box.

(a) Determine the number of nodal lines and the number of local probability maxima for this state.

(b) Describe the motion of the particle in this state.

43. Consider the wave function $\tilde{\Psi}_{222}(\tilde{x}, \tilde{y}, \tilde{z})$ for a particle in a cubic box. Figure 4.45a shows a contour plot in a cut plane at $\tilde{z} = 0.75$.

(a) Convince yourself that the contour plot in a cut at $\tilde{z} = 0.25$ would have the same pattern, but each positive peak would become negative, and vice versa.

(b) Describe the shape of this wave function in a plane cut at $\tilde{y} = 0.5$.

44. Consider the wave function $\tilde{\Psi}_{222}(\tilde{x}, \tilde{y}, \tilde{z})$ for a particle in a cubic box. Figure 4.45a shows a contour plot in a cut plane at $\tilde{z} = 0.75$.

(a) Describe the shape of this wave function in a cut plane at $\tilde{x} = 0.5$.

(b) Describe the shape of this wave function in a cut plane at $\tilde{y} = 0.5$.

ADDITIONAL PROBLEMS

45. A piano tuner uses a tuning fork that emits sound with a frequency of 440 s^{-1}. Calculate the wavelength of the sound from this tuning fork and the time the sound takes to travel 10.0 m across a large room. Take the speed of sound in air to be 343 m s^{-1}.

46. The distant galaxy called Cygnus A is one of the strongest sources of radio waves reaching Earth. The distance of this galaxy from Earth is 3×10^{24} m. How long (in years) does it take a radio wave of wavelength 10 m to reach Earth? What is the frequency of this radio wave?

47. Hot objects can emit blackbody radiation that appears red, orange, white, or bluish white, but never green. Explain.

48. Compare the energy (in joules) carried by an X-ray photon (wavelength $\lambda = 0.20$ nm) with that carried by an AM radio wave photon ($\lambda = 200$ m). Calculate the energy of 1.00 mol of each type of photon. What effect do you expect each type of radiation to have for inducing chemical reactions in the substances through which it passes?

49. The maximum in Planck's formula for the emission of blackbody radiation can be shown to occur at a wavelength $\lambda_{max} = 0.20\, hc/kT$. The radiation from the surface of the sun approximates that of a blackbody with $\lambda_{max} =$ 465 nm. What is the approximate surface temperature of the sun?

50. Photons of wavelength 315 nm or less are needed to eject electrons from a surface of electrically neutral cadmium.

(a) What is the energy barrier that electrons must overcome to leave an uncharged piece of cadmium?

(b) What is the maximum kinetic energy of electrons ejected from a piece of cadmium by photons of wavelength 200 nm?

(c) Suppose the electrons described in (b) were used in a diffraction experiment. What would be their wavelength?

51. When ultraviolet light of wavelength 131 nm strikes a polished nickel surface, the maximum kinetic energy of ejected electrons is measured to be 7.04×10^{-19} J. Calculate the work function of nickel.

52. Express the speed of the electron in the Bohr model in terms of the fundamental constants (m_e, e, h, ϵ_0), the nuclear charge Z, and the quantum number n. Evaluate the speed of an electron in the ground states of He^+ ion and U^{91+}. Compare these speeds with the speed of light c. As the speed of an object approaches the speed of light, relativistic effects become important. In which kinds of atoms do you expect relativistic effects to be greatest?

53. Photons are emitted in the Lyman series as hydrogen atoms undergo transitions from various excited states to the ground state. If ground-state He^+ are present in the same gas (near stars, for example), can they absorb these photons? Explain.

54. Name a transition in C^{5+} that will lead to the absorption of green light.

55. The energies of macroscopic objects, as well as those of microscopic objects, are quantized, but the effects of the quantization are not seen because the difference in energy between adjacent states is so small. Apply Bohr's quantization of angular momentum to the revolution of Earth (mass 6.0×10^{24} kg), which moves with a speed of 3.0×10^4 m s^{-1} in a circular orbit (radius 1.5×10^{11} m) about the sun. The sun can be treated as fixed. Calculate the value of the quantum number n for the present state of the Earth–sun system. What would be the effect of an increase in n by 1?

56. Sound waves, like light waves, can interfere with each other, giving maximum and minimum levels of sound. Suppose a listener standing directly between two loudspeakers hears the same tone being emitted from both. This listener observes that when one of the speakers is moved 0.16 m farther away, the perceived intensity of the tone decreases from a maximum to a minimum.

(a) Calculate the wavelength of the sound.

(b) Calculate its frequency, using 343 m s^{-1} as the speed of sound.

57. **(a)** If the kinetic energy of an electron is known to lie between 1.59×10^{-19} J and 1.61×10^{-19} J, what is the smallest distance within which it can be known to lie?

(b) Repeat the calculation of part (a) for a helium atom instead of an electron.

58. It has been suggested that spacecraft could be powered by the pressure exerted by sunlight striking a sail. The force exerted on a surface is the momentum p transferred to the surface per second. Assume that photons of 6000 Å light strike the sail perpendicularly. How many must be reflected per second by 1 cm^2 of surface to produce a pressure of 10^{-6} atm?

59. It is interesting to speculate on the properties of a universe with different values for the fundamental constants.

(a) In a universe in which Planck's constant had the value $h = 1$ J s, what would be the de Broglie wavelength of a 145-g baseball moving at a speed of 20 m s^{-1}?

(b) Suppose the speed of the ball from part (a) is known to lie between 19 and 21 m s^{-1}. What is the smallest distance within which it can be known to lie?

(c) Suppose that in this universe the mass of the electron is 1 g and the charge on the electron is 1 C. Calculate the Bohr radius of the hydrogen atom in this universe.

60. The normalized wave function for a particle in a one-dimensional box in which the potential energy is zero is $\psi(x) = \sqrt{2/L}\,\sin(n\pi x/L)$, where L is the length of the box (with the left wall at $x = 0$). What is the probability that the particle will lie between $x = 0$ and $x = L/4$ if the particle is in its $n = 2$ state?

61. A particle of mass m is placed in a three-dimensional rectangular box with edge lengths $2L$, L, and L. Inside the box the potential energy is zero, and outside it is infinite; therefore, the wave function goes smoothly to zero at the sides of the box. Calculate the energies and give the quantum numbers of the ground state and the first five excited states (or sets of states of equal energy) for the particle in the box.

5

CHAPTER

Quantum Mechanics and Atomic Structure

Fireworks above Paris; La Grande Arche is in the foreground. Many of the colors in fireworks are produced from atomic emission: red from strontium, orange from calcium, yellow from sodium, green from barium, and blue from copper. The sharp lines observed in the emission spectra of atoms can only be explained using the quantum theory of atomic structure.

The atom is the most fundamental concept in the science of chemistry. The prior chapters have given many examples of chemical reactions in which bonds between atoms in the reactants are broken, and new bonds are formed between atoms in the products. Although the Bohr model for Rutherford's planetary atom, rooted in the laws of classical physics, gives an excellent prediction for the emission and absorption spectra of one-electron atoms, it fails even for helium. Although simple electrostatics motivates why a positive Na^+ ion might be drawn to form a bond with a Cl^- ion, it cannot explain why the effective potential energy between two H atoms goes down as the two atoms near each other and the electrons rearrange themselves around the two nuclei. The stability of atoms and molecules could not be explained until quantum mechanics was developed.

The goal of this chapter is to describe the structure and properties of atoms using quantum mechanics. We couple the physical insight into the atom developed in Sections 3.3, 3.4, 3.5, and 3.6 with the quantum methods of Chapter 4 to develop a quantitative description of atomic structure.

We begin with the hydrogen atom, for which the Schrödinger equation can be solved exactly because it has only one electron. The method of solving the Schrödinger equation for the hydrogen atom is the same as that used for the particle-in-a-box models in Section 4.6. Because the mathematics is more complicated, we do not show the details here, but we present the solutions in both analytic and graphical form. We obtain exact expressions for the energy levels and the wave functions. Each wave function that solves the Schrödinger equation for the H atom is called a **hydrogen atomic orbital**. The square of a wave function gives the probability per unit volume of locating the electron in a small volume located at a particular distance from the nucleus and in a given direction. This reveals the sizes and shapes of hydrogen atomic orbitals; these orbitals are the starting points for understanding multielectron atoms and covalent bonding in polyatomic molecules.

There is no exact solution for any other atom, so we must develop approximate solutions. We treat each electron in a many-electron atom as if it were moving in an *effective force field* that results from averaging its interactions with all the other electrons and the nucleus. The effective field was introduced by purely physical arguments in Section 3.4. Here, we develop this concept systematically and from it obtain approximate one-electron wave functions called **Hartree atomic orbitals** (to honor the English physicist Douglas Hartree who pioneered the method), which include the effect of all the other electrons in the atom. The shapes of the Hartree orbitals are similar to those of the hydrogen atomic orbitals, but their sizes and their energy level patterns are quite different. We use the Hartree orbitals to explain the shell model of the atom, the structure of the periodic table, and the periodic behavior of atomic properties. The result is a comprehensive, approximate quantum description of atomic structure, which serves as the starting point for the quantum description of the chemical bond in Chapter 6.

Your primary objective in this chapter should be to understand the shapes and structures of the hydrogen atomic orbitals and the Hartree orbitals from these graphical representations. You should be able to predict how the probability distribution for the electrons depends on the properties of the orbitals, as specified by their quantum numbers. Always keep in mind the distinction between the hydrogen atomic orbitals and the Hartree orbitals. The former apply only to the hydrogen atom, and the latter only to many-electron atoms. Be aware of their differences, as well as their similarities.

5.1 The Hydrogen Atom

The hydrogen atom is the simplest example of a one-electron atom or ion; other examples are He^{+}, Li^{2+}, and other ions in which all but one electron have been stripped off. They differ only in the charge $+Ze$ on the nucleus, and therefore in the magnitude of the attractive force experienced by the electron.

The potential energy for the one-electron atom, discussed in the context of the planetary model in Section 3.3 and of the Bohr model in Section 4.3, depends only on the distance of the electron from the nucleus; it does not depend on angular orientation (see Fig. 4.23). Solution of the Schrödinger equation is most easily carried out in coordinates that reflect the natural symmetry of the potential energy function. For an isolated one-electron atom or ion,

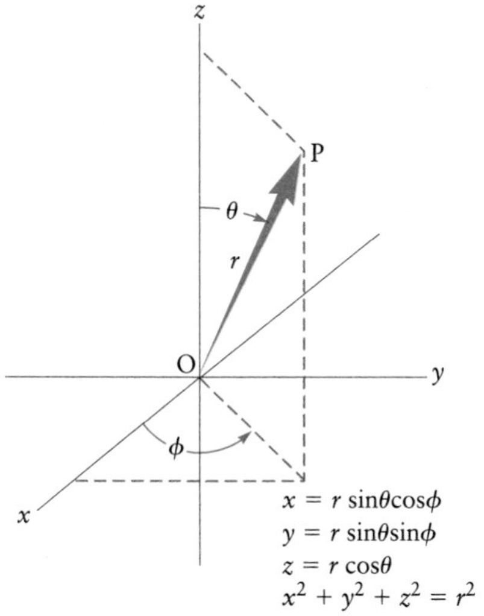

FIGURE 5.1 The relationship between spherical coordinates (r, θ, ϕ) and Cartesian coordinates (x, y, z). Here, θ is the angle with respect to the Cartesian z-axis, which ranges from 0 to π, and ϕ is the azimuthal angle (the angle between the x-axis and the projection onto the x-y plane of the arrow from the origin to P), which ranges from 0 to 2π. Here, r is the distance of the electron from the nucleus at the origin, and ranges from 0 to ∞.

spherical coordinates are more appropriate than the more familiar Cartesian coordinates. Spherical coordinates are defined in Figure 5.1: r is the distance of the electron at P from the nucleus at O, and the angles θ and ϕ are similar to those used to locate points on the surface of the globe; θ is related to the latitude, and ϕ is related to the longitude.

The Schrödinger equation for H is similar to that in Equation 4.29, except in three dimensions. The potential energy $V(x)$ in Equation 4.29 is replaced by the Coulomb potential $V(r)$ given in Equation 3.1. The variable r is the distance from the nucleus to the electron, or the magnitude of the red vector $\vec{r}$ shown in Figure 5.1. After rewriting the second derivatives in the Schrödinger equation in spherical coordinates (r, θ, ϕ), using the relationships in the caption of Figure 5.1, the equation looks impressively complicated, but it can be solved using the method outlined in Section 4.5. The solution must be continuous in all three coordinates, and the radial portion must satisfy the boundary condition: $\psi \longrightarrow 0$ as $r \longrightarrow \infty$. This procedure leads to quantization of the energy and to equations for the associated wave functions just as we saw with the particle-in-a-box models in Section 4.6. We describe these parts of the solution in turn in the remainder of this section.

5.1.1 Energy Levels

Solutions of the Schrödinger equation for the one-electron atom exist only for particular values of the energy[1]:

$$E = E_n = -\frac{Z^2 e^4 m_e}{8\epsilon_0^2 n^2 h^2} \qquad n = 1, 2, 3, \ldots \qquad [5.1a]$$

In energy units of rydbergs (1 rydberg $= 2.18 \times 10^{-18}$ J), this equation becomes (see Eqs. 4.14a and 4.14b):

$$E_n = -\frac{Z^2}{n^2} \quad \text{(rydberg)} \qquad n = 1, 2, 3, \ldots \qquad [5.1b]$$

The integer n, called the **principal quantum number**, indexes the individual energy levels. These are identical to the energy levels predicted by the Bohr theory. Quantization arises because of the physical boundary conditions imposed on the solutions to the Schrödinger equation rather than from making arbitrary assumptions about the angular momentum.

The energy of a one-electron atom depends only on the principal quantum number n, because the potential energy depends only on the radial distance. The Schrödinger equation also quantizes L^2, the square magnitude of the angular momentum, as well as L_z, the projection of the angular momentum along the z-axis. (A review of elementary aspects of angular momentum in Appendix B provides useful background for this discussion.) Quantization of the square of the angular momentum as well as its projection along the z-axis requires two additional quantum numbers. The **angular momentum quantum number** ℓ may take on any integral value from 0 to $n - 1$, and the angular momentum projection quantum number m may take on any integral value from $-\ell$ to ℓ. The quantum number m is referred to as the **magnetic quantum number** because its value helps determine how the energy of the atom would shift when placed in

[1]Strictly speaking, the electron mass m_e in the expressions for the energy levels of one-electron atoms should be replaced by the *reduced mass* μ, equal to $m_e m_N/(m_e + m_N)$, where m_N is the nuclear mass; μ differs from m_e by less than 0.1%.

an external magnetic field. The allowed values of the square of the angular momentum L and its z projection are given by

$$\text{Allowed values of } L^2 = \ell(\ell+1)\frac{h^2}{4\pi^2} \qquad \ell = 0, 1, \ldots, n-1 \qquad [5.2a]$$

$$\text{Allowed values of } L_z = m\frac{h}{2\pi} \qquad m = -\ell, -\ell+1, \ldots, 0, \ldots, \ell-1, \ell \qquad [5.2b]$$

For $n = 1$ (the ground state), the only allowed values for the angular momentum quantum numbers are ($\ell = 0, m = 0$). For $n = 2$, there are $n^2 = 4$ allowed sets of quantum numbers:

$$(\ell = 0, m = 0), (\ell = 1, m = 1), (\ell = 1, m = 0), (\ell = 1, m = -1)$$

The restrictions on ℓ and m give rise to n^2 sets of quantum numbers for every value of n. Each set (n, ℓ, m) identifies a specific **quantum state** of the atom in which the electron has energy equal to E_n, angular momentum equal to $\sqrt{\ell(\ell+1)}\, h/2\pi$, and z-projection of angular momentum equal to $mh/2\pi$. When $n > 1$, a total of n^2 specific quantum states correspond to the single energy level E_n; consequently, this set of states is said to be **degenerate**.

It is conventional to label specific states by replacing the angular momentum quantum number with a letter; we signify $\ell = 0$ with *s*, $\ell = 1$ with *p*, $\ell = 2$ with *d*, $\ell = 3$ with *f*, $\ell = 4$ with *g*, and on through the alphabet. Thus, a state with $n = 1$ and $\ell = 0$ is called a 1*s* state, one with $n = 3$ and $\ell = 1$ is a 3*p* state, one with $n = 4$ and $\ell = 3$ is a 4*f* state, and so forth. The letters *s*, *p*, *d*, and *f* derive from early (pre–quantum mechanics) spectroscopy, in which certain spectral lines were referred to as sharp, principal, diffuse, and fundamental. These terms are not used in modern spectroscopy, but the historical labels for the values of the quantum number ℓ are still followed. Table 5.1 summarizes the allowed combinations of quantum numbers.

TABLE 5.1

Allowed Values of Quantum Numbers for One-Electron Atoms

n	1	2		3		
ℓ	0	0	1	0	1	2
m	0	0	−1, 0, +1	0	−1, 0, +1	−2, −1, 0, +1, +2
Number of degenerate states for each ℓ	1	1	3	1	3	5
Number of degenerate states for each n	1	4		9		

EXAMPLE 5.1

Give the names of all the orbitals with $n = 4$, and state how many m values correspond to each type of orbital.

Solution

The quantum number ℓ may range from 0 to $n - 1$; thus, its allowed values in this case are 0, 1, 2, and 3. The labels for the groups of orbitals are then:

$\ell = 0$	4*s*	$\ell = 2$	4*d*
$\ell = 1$	4*p*	$\ell = 3$	4*f*

The quantum number m ranges from $-\ell$ to $+\ell$; thus, the number of m values is $2\ell + 1$. This gives one 4*s* orbital, three 4*p* orbitals, five 4*d* orbitals, and seven 4*f* orbitals for a total of $16 = 4^2 = n^2$ orbitals with $n = 4$. They all have the same energy, but they differ in shape.

Related Problems: 3, 4

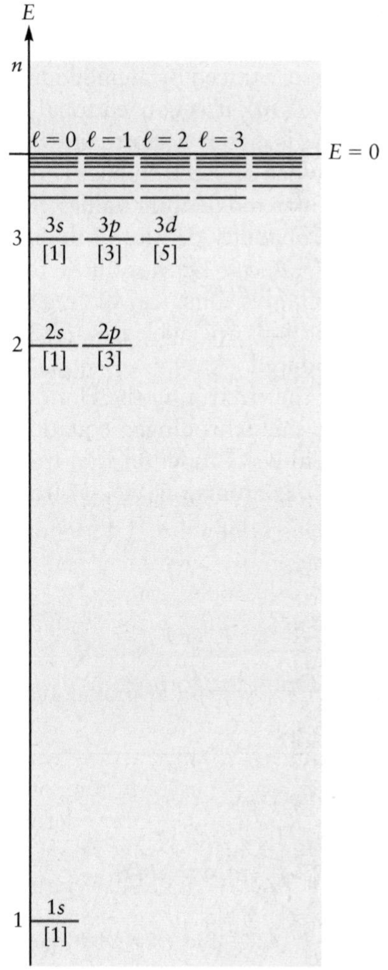

FIGURE 5.2 The energy-level diagram of the H atom predicted by quantum mechanics is arranged to show the degeneracy of the energy levels. At each value of n there are n states with different values of ℓ. At each value of ℓ there are $2\ell + 1$ states with different values of m. Altogether, each value of n corresponds to n^2 degenerate states.

The energy levels, including the degeneracy due to m and with states labeled by the spectroscopic notation, are conventionally displayed on a diagram as shown in Figure 5.2. Compare this diagram with the energy level diagram for the hydrogen atom obtained by the Bohr model in Figure 4.24 which shows only the dependence on the principal quantum number n.

5.1.2 Wave Functions

For each quantum state (n, ℓ, m), solution of the Schrödinger equation provides a wave function of the form

$$\psi_{n\ell m}(r, \theta, \phi) = R_{n\ell}(r)Y_{\ell m}(\theta, \phi) \quad [5.3]$$

in which the total wave function is the product of a radial part, $R_{n\ell}(r)$, and an angular part, $Y_{\ell m}(\theta, \phi)$. This product form is a consequence of the spherically symmetric potential energy function, and it enables separate examination of the angular and radial contributions to the wave function. The functions $Y_{\ell m}(\theta, \phi)$ are called **spherical harmonics**. They appear in many physical problems with spherical symmetry and were already well known before the advent of the Schrödinger equation. The angular motions of the electron described by θ and ϕ influence the *shape* of the wave function through the angular factor $Y_{\ell m}$, even though they do not influence the energy.

The wave function itself is not measured. It is to be viewed as an intermediate step toward calculating the physically significant quantity ψ^2, which is the probability density for locating the electron at a particular point in the atom. More precisely,

$$(\psi_{n\ell m})^2 dV = [R_{n\ell}(r)]^2[Y_{\ell m}(\theta, \phi)]^2 dV \quad [5.4]$$

gives the probability of locating the electron within a small three-dimensional volume, dV, located at the position (r, θ, ϕ) when it is known that the atom is in the state n, ℓ, m. Specific examples are presented in succeeding paragraphs. The volume element $dV = dxdydz$ (using volume = length × width × height) in Cartesian coordinates may be converted to spherical coordinates (see Figure 5.3) to give

$$dV = dxdydz = r^2 \sin\theta\, drd\theta d\phi \quad [5.5]$$

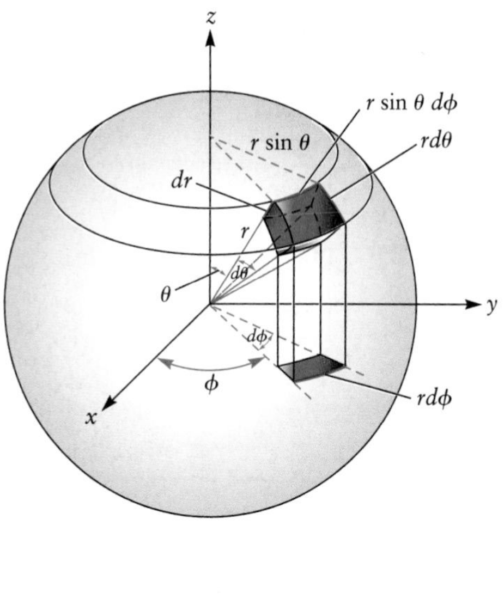

FIGURE 5.3 The differential volume element in spherical polar coordinates.

A wave function $\psi_{n\ell m}(r, \theta, \phi)$ for a one-electron atom in the state (n, ℓ, m) is called an **orbital**. This term recalls the circular orbits of the Bohr atom, but there is no real resemblance. An orbital is *not* a trajectory traced by an individual electron. When the one-electron atom is in state (n, ℓ, m), it is conventional to say the electron is "in an (n, ℓ, m) orbital." This phrase is merely a shorthand way of making the precise but cumbersome statement: "When an electron has energy, total angular momentum, and z component of angular momentum values corresponding to the quantum numbers n, ℓ, m, the probability density of finding the electron at the point (r, θ, ϕ) is given by $\psi^2_{n\ell m}(r, \theta, \phi)$." Do not allow this verbal shorthand to mislead you into thinking an orbital is some sort of "region in space" inside which the electron is confined. The orbitals are labeled 1*s*, 2*s*, 2*p*, 3*s*, ... by the spectroscopic notation previously introduced.

Table 5.2 gives the allowed stationary states for the electron in the H atom, or any one-electron atom, obtained from solving the Schrödinger equation, where each $\psi(r, \theta, \phi)$ is obtained by choosing a radial wave function $R(r)$ from the right column and multiplying it by an angular wave function $Y(\theta, \phi)$ from the left column. (But keep in mind the rules in Table 5.1; e.g., if $n = 1$ you can

TABLE 5.2

Angular and Radial Parts of Wave Functions for One-Electron Atoms

Angular Part $Y(\theta, \phi)$	Radial Part $R_{n\ell}(r)$
$\ell = 0$: $Y_s = \left(\frac{1}{4\pi}\right)^{1/2}$	$R_{1s} = 2\left(\frac{Z}{a_0}\right)^{3/2} \exp(-\sigma)$
	$R_{2s} = \frac{1}{2\sqrt{2}}\left(\frac{Z}{a_0}\right)^{3/2}(2-\sigma)\exp(-\sigma/2)$
	$R_{3s} = \frac{2}{81\sqrt{3}}\left(\frac{Z}{a_0}\right)^{3/2}(27-18\sigma+2\sigma^2)\exp(-\sigma/3)$
$\ell = 1$: $Y_{p_x} = \left(\frac{3}{4\pi}\right)^{1/2}\sin\theta\cos\phi$	$R_{2p} = \frac{1}{2\sqrt{6}}\left(\frac{Z}{a_0}\right)^{3/2}\sigma\exp(-\sigma/2)$
$Y_{p_y} = \left(\frac{3}{4\pi}\right)^{1/2}\sin\theta\sin\phi$	$R_{3p} = \frac{4}{81\sqrt{6}}\left(\frac{Z}{a_0}\right)^{3/2}(6\sigma-\sigma^2)\exp(-\sigma/3)$
$Y_{p_z} = \left(\frac{3}{4\pi}\right)^{1/2}\cos\theta$	
$\ell = 2$: $Y_{d_{z^2}} = \left(\frac{5}{16\pi}\right)^{1/2}(3\cos^2\theta-1)$	$R_{3d} = \frac{4}{81\sqrt{30}}\left(\frac{Z}{a_0}\right)^{3/2}\sigma^2\exp(-\sigma/3)$
$Y_{d_{xz}} = \left(\frac{15}{4\pi}\right)^{1/2}\sin\theta\cos\theta\cos\phi$	
$Y_{d_{yz}} = \left(\frac{15}{4\pi}\right)^{1/2}\sin\theta\cos\theta\sin\phi$	
$Y_{d_{xy}} = \left(\frac{15}{16\pi}\right)^{1/2}\sin^2\theta\sin 2\phi$	
$Y_{d_{x^2-y^2}} = \left(\frac{15}{16\pi}\right)^{1/2}\sin^2\theta\cos 2\phi$	

$$\sigma = \frac{Zr}{a_0} \qquad a_0 = \frac{\epsilon_0 h^2}{\pi e^2 m_e} = 0.529 \times 10^{-10}\ \text{m}$$

only use the $\ell = 0$ angular function, Y_s, in Table 5.2, making the 1*s* orbital. A 1*p* orbital ($n = 1, \ell = 1$) does not exist, since for a given n, ℓ is an integer ranging from 0 to $n - 1$ only.) Note that each of the angular functions $Y(\theta, \phi)$ in Table 5.2 has a specific ℓ quantum number, given by the s, p, or d subscript, but the functions are linear combinations (see Section 5.1.3) of two or more of the spherical harmonics $Y_{\ell m}$ in Equation 5.3. This results in those orbitals not having a well-defined quantum number m, but they are the familiar orbitals like $2p_x$, $2p_y$, and $2p_z$ commonly used in introductory texts.

The electronic wave function for an atom or molecule allows you to calculate where the negative charge density is largest; you simply multiply the charge of an electron, -1.6×10^{-19} C, by the electron probability density ψ^2, where ψ^2 is evaluated at a particular location in the atom or molecule. This allows you to predict the dipole moment of a molecule and the location of unpaired electron density in a molecule. For example, if you integrate the unpaired electron density of the allyl radical, CH_2CHCH_2, over a volume centered near the first carbon atom, or centered near the middle carbon atom, or centered near the third carbon atom, you find that there is a 50% probability of finding the unpaired electron near each end carbon, and 0% probability of finding unpaired electron on the center carbon atom. The example below shows you how to integrate the electron density over a small volume for an H atom in its ground electronic state, and for an H atom that has been electronically excited by absorbing a 121-nm photon polarized in the z direction. The method is similar to Example 4.7b, where we integrated ψ^2 between x_1 and x_2 to calculate the probability of finding the electron in that range of x values. However, in three dimensions we have to integrate ψ^2 (with units of m^{-3}) over a volume with units of m^3 to find the probability of finding the electron in that volume. Formally, this is the triple integral, but if you are searching for the electron in a small volume centered at a position $\vec{r}_0 = (x_0, y_0, z_0)$, it reduces to the following simple expression:

$$\int_V |\psi|^2 dV = \int_{z_1}^{z_2}\int_{y_1}^{y_2}\int_{x_1}^{x_2} |\psi(x, y, z)|^2 dxdydz$$

$$\approx |\psi(x_0, y_0, z_0)|^2 \Delta x \Delta y \Delta z = |\psi(x_0, y_0, z_0)|^2 \times \text{Volume searched} \quad [5.6]$$

Here we have used the fact that as long as the numerical value of the wave function does not change much over the volume searched you can approximate the integral by treating the wave function as a constant at the value $\psi(x_0, y_0, z_0)$. The product $\Delta x \Delta y \Delta z$ is simply the volume over which you are searching for the electron. Example 5.2 demonstrates how to use Equation 5.6, one that also shows how to go between Cartesian and spherical coordinates using the relations given in Figure 5.1.

EXAMPLE 5.2

(a) What is the probability of finding the electron in the ground state of the H atom in a small cubical volume = 0.5 pm^3 centered at a distance of 20 pm from the nucleus in the $-y$ direction?

(b) If you move the volume searched closer to the nucleus, to only 10 pm from the nucleus, does the probability of finding the electron increase?

(c) If the electron in your H atom is in the $2p_z$ orbital, an excited electronic state of the H atom, what is the probability now?

(Continued)

Solution

(a) In Cartesian coordinates (x, y, z), the volume is centered at $\vec{r}_0 = (0, -20 \text{ pm}, 0)$. The distance from the nucleus is $r = \sqrt{x^2 + y^2 + z^2}$, so you are searching for the electron in a volume centered at a distance of $\sqrt{0^2 + (-20 \text{ pm})^2 + 0^2} = 20$ pm from the nucleus. The volume of 0.5 pm^3 is very small (think of trying to trap the electron in a little cage with length = width = height = $\sqrt[3]{0.5 \text{ pm}^3} = 0.79$ pm, so the value of ψ is roughly constant in the entire volume searched. The approximation in Equation 5.6 is thus good, so the probability = $|\psi(x_0, y_0, z_0)|^2 \times$ Volume searched, = $|\psi(x_0, y_0, z_0)|^2 \times 0.5$ pm^3. To calculate ψ^2, note that ground state H atoms have the lowest energy, so $n = 1$. The electron is thus in the 1s orbital ($n = 1$, $\ell = 0$) with $\psi = R_{1s}(r)Y_s$. Using these functions from Table 5.2 with $Z = 1$ gives, at $r = 20$ pm,

$$\psi = 2\left(\frac{1}{52.9 \text{ pm}}\right)^{\frac{3}{2}} e^{-\frac{20 \text{ pm}}{52.9 \text{ pm}}} \sqrt{\frac{1}{4\pi}} = 0.0010 \text{ pm}^{-3/2}$$

Squaring this and multiplying by the volume searched gives the probability of finding the electron = $(0.0000010 \text{ pm}^{-3})(0.5 \text{ pm}^3) = 5 \times 10^{-7}$.

(b) There is a higher probability of finding the electron in the same size volume but closer to the nucleus, because ψ^2 is larger at $r = 10$ pm.

(c) If the electron is in the 2p_z orbital, the wave function $R(r)Y(\theta, \phi)$ has angular part $Y(\theta, \phi) = (3/4\pi)^{1/2}\cos\theta$. To evaluate this, note that the volume is in which you are searching for the electron is centered at $\vec{r}_0 = (0, -20 \text{ pm}, 0)$. So the corresponding red vector in Figure 5.1 would be pointing from the origin to this point, in the negative y direction. Any point in the x-y plane is at an angle of 90 degrees = $\pi/2$ from the z-axis, so $\theta = \pi/2$. To point over in the negative y direction, the ϕ angle must be $3\pi/2$. (In the spherical coordinates defined in Figure 5.1, the range of θ is 0 to π and the range of ϕ is 0 to 2π, so that you don't double count any points in space.) Thus the angular part of the wave function at $\theta = \pi/2$ is $(3/4\pi)^{1/2}\cos(\pi/2) = 0$. This is a node in the wave function (indeed $\psi = 0$ in the entire xy plane), so the probability of finding the electron near $\vec{r}_0 = (0, -20 \text{ pm}, 0)$ is approximately 0; the 2p_z wave function is 0 at the center of the volume searched and the volume searched is small.

Related Problems: 5, 6

5.1.3 Sizes and Shapes of Orbitals

The sizes and shapes of the hydrogen atom orbitals are important in chemistry because they provide the foundations for the quantum description of chemical bonding and the molecular shapes to which it leads. Sizes and shapes of the orbitals are revealed by graphical analysis of the wave functions, of which the first few are given in Table 5.2. Note that the radial functions are written in terms of the dimensionless variable σ, which is the ratio of Zr to a_0, the Bohr radius (Eq. 4.12). For $Z = 1$, $\sigma = 1$ at the radius of the first Bohr orbit of the hydrogen atom.

Graphical representation of the orbitals requires some care. Equation 5.3 tells us simply to go to the point (r, θ, ϕ), evaluate the wave function there, and draw a graph showing the results of visiting many such points. But all three spatial dimensions have already been used to define the location; thus we would need a fourth dimension to display the value of the wave function at that point. Alternatively, we could create a table of numbers giving the value of the wave function at each point (r, θ, ϕ), but it would be difficult to develop intuition about shapes and structures from this table.

One way we get around these problems is by slicing up three-dimensional space into various two- and one-dimensional regions and examining the value

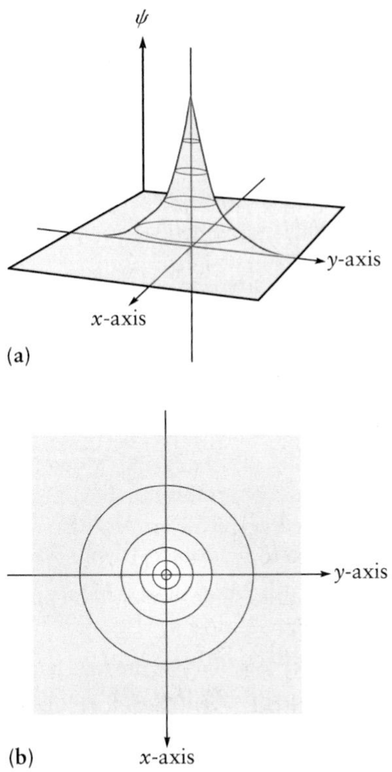

FIGURE 5.4 Representations of the hydrogen 1*s* orbital for points in the *x-y* plane. (a) The value of the orbital at each point in the *x-y* plane (where $\theta = 0$, see Fig. 5.1) is plotted above the point. (b) Slices through the image in (a) define contours of the image in the *x-y* plane.

of the wave function at each point in these regions. This means we look at the wave function when the motion of the electron is restricted to the specific regions we select. For example, suppose in Figure 5.1 we look only at points in the *x-y* plane and evaluate the wave function at each point and plot this value as a third dimension above the *x-y* plane. We obtain a mountain-like shape above the plane. Figure 5.4a shows this result for the 1*s* orbital, $\psi_{100}(r, \theta, \phi)$. Figure 5.4b shows this function as a contour plot in the *x-y* plane, generated by choosing a particular value of the wave function in Figure 5.4a, "slicing" its three-dimensional image perpendicular to the *z*-axis at that value, and projecting each point on the edge of that slice down to the *x-y* plane to form a closed contour in that plane. The contour then defines all points in the *x-y* plane for which $\psi_{100}(r, \theta, \phi)$ has the particular value selected. The process is continued by selecting other values of $\psi_{100}(r, \theta, \phi)$ until the entire three-dimensional image has been collapsed into a set of concentric contours in the two-dimensional *x-y* plane. Mountain climbers throughout the world use this method to generate contour maps of mountain ranges. The outermost contour in Fig. 5.4b identifies points at which the wave function has 10% of its maximum value. The second contour identifies points with 30% of the maximum value, and so on to the innermost contour that identifies points with 90% of the maximum value. Note that these contours, which correspond to uniform increases in the value of ψ, become much closer together as we approach the origin of the coordinates. This indicates that the value of the wave function is increasing rapidly as the electron gets closer to the proton. The contours are circles because $\psi_{100}(r, \theta, \phi)$ does not depend on the angles, as we see from its defining equation. We can generate contour maps in any other "cut planes" in the same way. We rely on contour plots to display the angular shapes of wave functions. As we examine other orbitals that do have angular dependence, the contours will no longer be circular. Displaying the contours in different "cut planes" gives great insight into the shapes of such orbitals, and we exploit that technique throughout the chapter.

A second approach is to look only at the radial behavior. We start at the origin in Figure 5.1 and move out along the specific direction (r, θ, ϕ), holding (θ, ϕ) constant, and plot the wave function calculated from Table 5.2 at each value of r. Figure 5.5 shows the result for the 1*s* orbital. This image is a "vertical slice" through the plot shown in Figure 5.4a. The radial behavior of $\psi_{100}(r, \theta, \phi)$ is an exponential decrease as r increases, as shown by its defining equation. We rely on such two-dimensional graphs of the wave function versus r to display radial behavior of each of the hydrogen orbitals.

We use the same techniques to generate graphical representations of $(\psi_{n\ell m})^2$, where, $|\psi_{n\ell m}|^2 dV$ is the probability of finding the electron in a small volume element dV (see Eq. 5.4). The electron density for the 1*s* orbital is spherical, because the wave function does not depend on the angles, and its contours in the *x-y* plane are circles. The radial dependence of $(\psi_{100})^2$ is shown in Figure 5.6.a. We can combine the radial dependence and the contours to generate a three-dimensional image of the electron density by heavily shading the regions where ψ^2 is largest and lightly shading the regions where ψ^2 is smallest. Figure 5.6b shows a cross-section of this three-dimensional image in the *x-y* plane.

In the remainder of this section we use the methods just described to examine systematically the shapes of the hydrogen orbitals and the probability densities they define. Images that combine these angular and radial effects display size and shape in three-dimensional space. We always state the conditions and limitations of such three-dimensional images. Because three dimensions have already been used up to specify the location of the electron, the appearance of the wave function in three-dimensional representations depends strongly on choices made by the illustrator. Be certain you understand these choices in each image you examine (or create!).

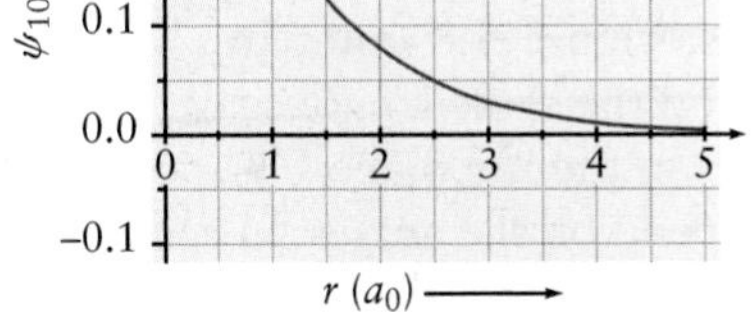

FIGURE 5.5 Plot of the hydrogen 1*s* orbital against distance from the nucleus. The distance is measured in units of the Bohr radius a_0.

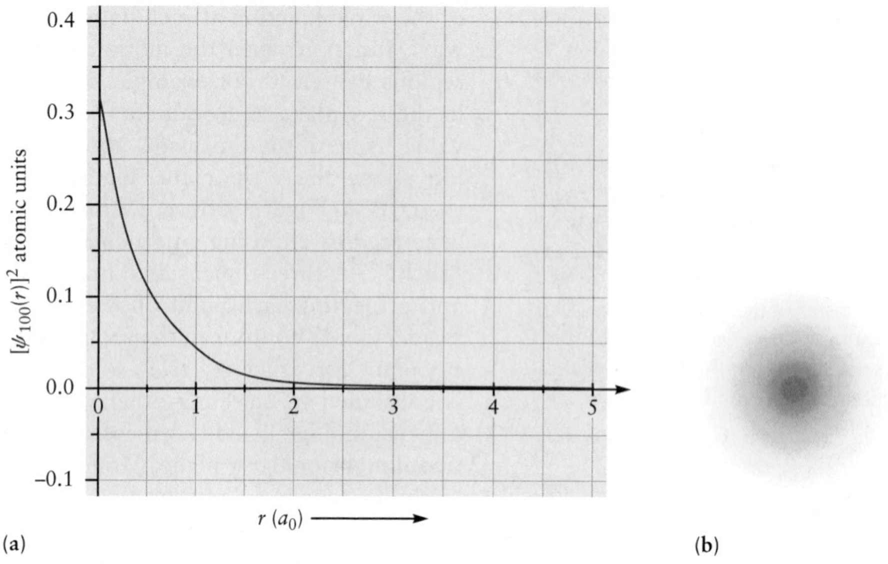

FIGURE 5.6 Representations of the probability density for the hydrogen 1s orbital. (a) Plot of $[\psi_{100}(r)]^2$ against distance from the nucleus. A contour map (not shown here) can be generated in the *x-y* plane just as in Figure 5.4. (b) In three dimensions the probability density can be represented by varying the intensity of a color. The color is most intense where the probability is highest. A sphere drawn around the nucleus with radius of 4.2 a_0 will enclose 99% of the probability density.

s Orbitals

Let's begin with the **s orbital**, corresponding to $\psi_{n\ell m}$ with $\ell = 0$ (therefore, $m = 0$ as well). For all *s* orbitals, the angular part *Y* is a constant (see Table 5.2). Because ψ does not depend on either θ or ϕ, all *s* orbitals are spherically symmetric about the nucleus. This means that the value of the wave function for an *s* orbital (and therefore also the probability of finding the electron near some point in space) depends only on its distance, *r*, from the nucleus and not on its direction in space. There are several ways to visualize the *ns* orbitals with $n = 1, 2, 3, \ldots$, and the probability density they describe.

Figure 5.7 displays the relationships between the radial dependence of ψ and ψ^2 and the electron density representation for the 1*s*, 2*s*, and 3*s* orbitals. These schematic images display key features of these functions, but the radial dependence shown is not quantitative. The 2*s* and 3*s* orbitals have nodes, values of *r* at which the value of the wave function is 0, with lobes of positive phase and negative phase separated by the nodes. See Figure 5.7b. This behavior is contained in their equations in Table 5.2. The radial function R_{2s} has a node where $\sigma = 2$ which corresponds to $r = 2a_0$. The function R_{3s} has a node where $27 - 18\sigma + 2\sigma^2 = 0$ which occurs at $r = 1.9a_0$ and $r = 7.1a_0$. Figure 5.7c shows that ψ^2 passes through a local maximum value after *r* crosses each node, and then continues to fall off after the final node as *r* approaches very large values. From this behavior we generate the electron density representations in Figure 5.7a, in which the intensity of shading indicates the magnitude of the electron density in each region. The shading will be greatest at the local maximum values in ψ^2 and must fade to white at the nodes.

Having established the key features of these orbitals and their probability functions, we examine them once more to demonstrate *quantitatively* their dependence on *r*. Quantitative information is essential for comparing behavior of these three orbitals in chemical applications. Figure 5.8 compares various representations of the 1*s*, 2*s*, and 3*s* orbitals all on the same length scale. Figure 5.8a shows contour plots for these three orbitals in the *x-y* plane. The contours are circles because the wave functions do not depend on the angles. In each plot, the outermost contour identifies points at which the amplitude of ψ is 5% of its maximum value. The second circle identifies points with ψ at 10% of the maximum, and so on in steps of 20% to the innermost contour, which identifies points with ψ at 90% of the maximum. Contours with positive phase (positive sign for the wave function) are shown in red, whereas negative phase (negative

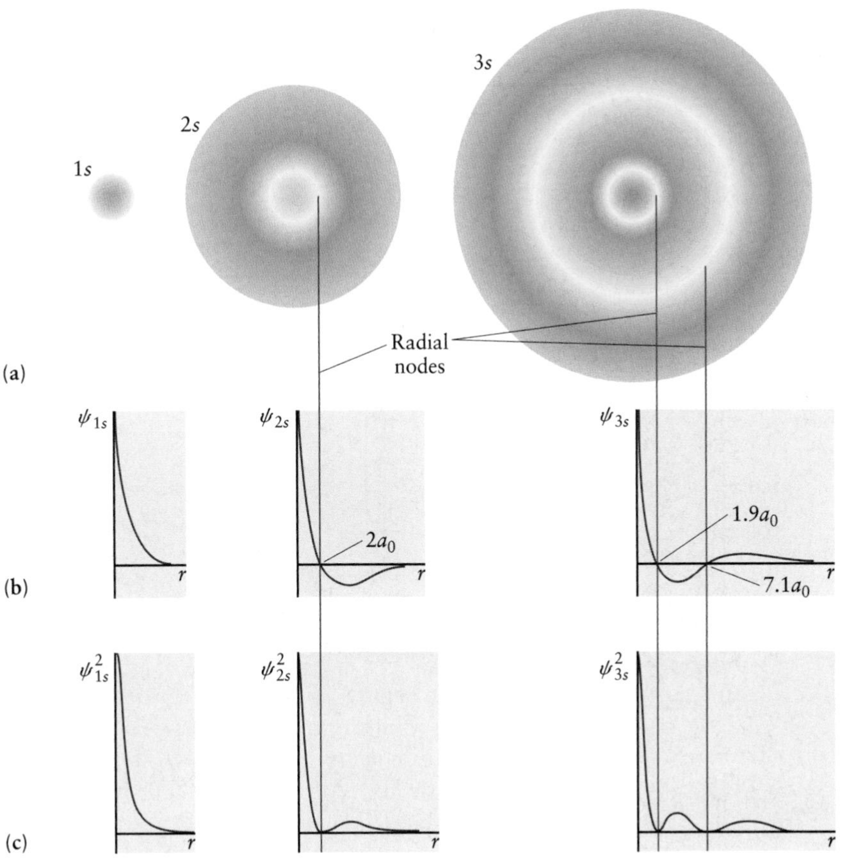

FIGURE 5.7 Qualitative representations of hydrogen s orbitals. (a) An electron probability density representation of a hydrogen atom in its 1s, 2s, and 3s states. The spheres are cut off at a radius that encloses 90% of the probability density. (b) The radial wave functions plotted against distance from the nucleus, r. (c) Dependence of the probability density on the distance from the nucleus, given by the square of the wave function (prior to integrating over a thin spherical shell a distance r from the nucleus).

sign for the wave function) is represented in blue. Note that these contours, which identify uniform increases in magnitude, become much closer together as we approach the origin. The numerical value of the wave function increases rapidly as we approach the nucleus. The x-y plane corresponds to $\theta = \pi/2$ in Figure 5.1. Because the wave function does not depend on angles, this same contour plot could be obtained by tilting the x-y plane to any value of θ. It could also be obtained by starting with the x-z plane and the y-z plane and tilting either of them to any value of θ. Therefore, we can rotate the contour plot in Figure 5.8a to generate a set of concentric spheres in three dimensions. Each sphere identifies a surface of points in (r, θ, ϕ) at each of which the wave function has the same value. These spheres (not shown here) are called *isosurfaces*.

Figure 5.8b shows the radial portions of the 1s, 2s, and 3s wave functions $\psi_{n00} \propto R_{n0}(r)$ directly. These curves quantitatively represent how ψ changes as a function of the distance from the nucleus and identify the locations of the nodes.

Figure 5.8c plots the **radial probability density** $r^2[R_{n0}(r)]^2$, which we introduce here for the first time. This is the probability density of finding the electron at any point in space at a distance r from the nucleus, after integrating over all angles θ and ϕ. More precisely, the product $r^2[R_{n0}(r)]^2\,dr$ gives the probability of finding the electron anywhere within a thin spherical shell of thickness dr, located at distance r from the nucleus. This spherical shell is easily visualized with the aid of Figure 5.3. As the angle ϕ runs through its entire range from 0 to 2π, a circular annulus of width dr located between r and $r + dr$ is traced out in the x-y plane. This annulus will become a spherical shell as the angle θ runs through its range from 0 to π. The factor r^2 in front accounts for the increasing volume of spherical shells at greater distances from the nucleus. The radial probability distribution is small near the nucleus, where the shell volume (proportional to r^2) is small, and reaches its maximum value at the distance where

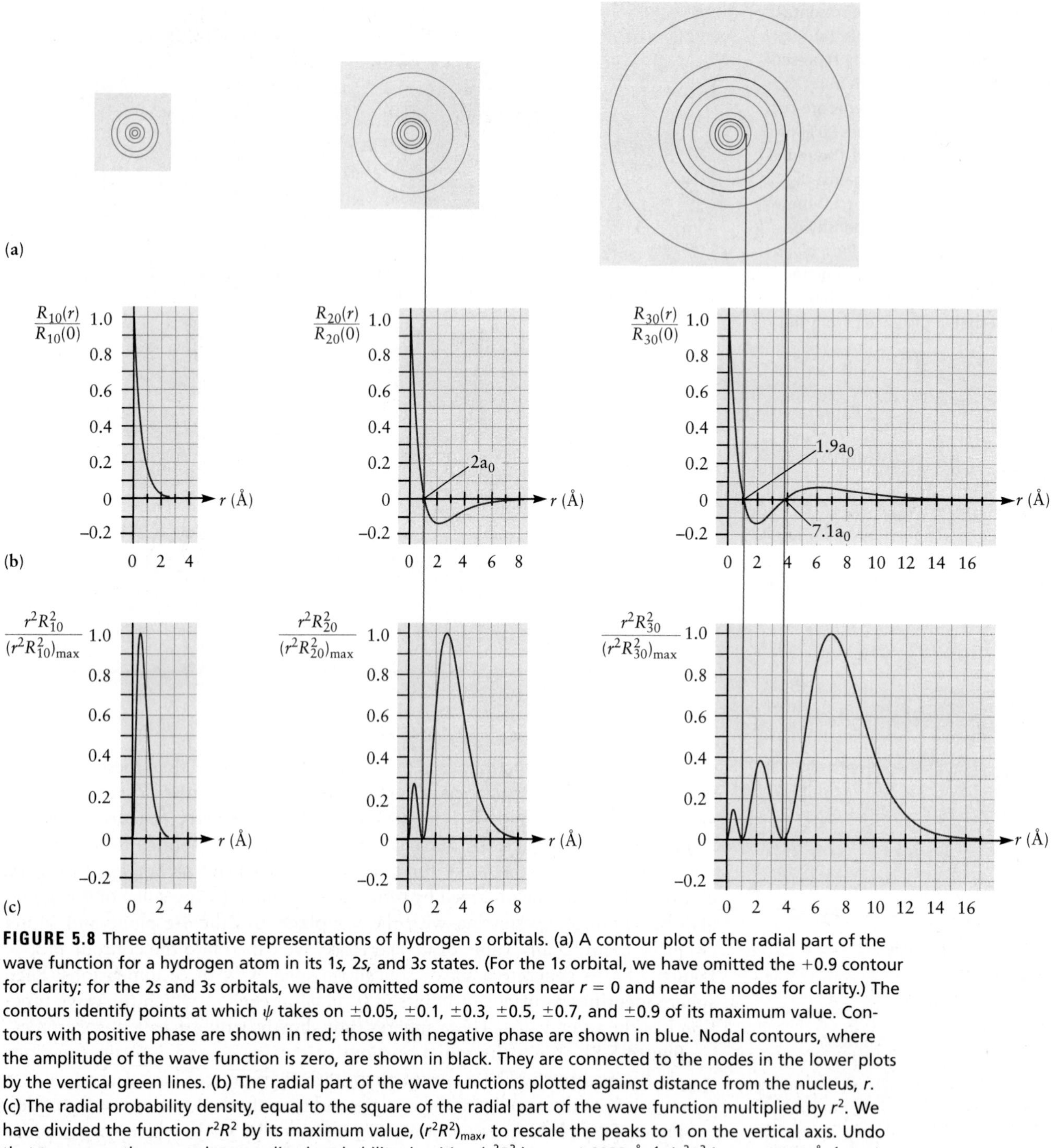

FIGURE 5.8 Three quantitative representations of hydrogen *s* orbitals. (a) A contour plot of the radial part of the wave function for a hydrogen atom in its 1*s*, 2*s*, and 3*s* states. (For the 1*s* orbital, we have omitted the +0.9 contour for clarity; for the 2*s* and 3*s* orbitals, we have omitted some contours near $r = 0$ and near the nodes for clarity.) The contours identify points at which ψ takes on ±0.05, ±0.1, ±0.3, ±0.5, ±0.7, and ±0.9 of its maximum value. Contours with positive phase are shown in red; those with negative phase are shown in blue. Nodal contours, where the amplitude of the wave function is zero, are shown in black. They are connected to the nodes in the lower plots by the vertical green lines. (b) The radial part of the wave functions plotted against distance from the nucleus, *r*. (c) The radial probability density, equal to the square of the radial part of the wave function multiplied by r^2. We have divided the function r^2R^2 by its maximum value, $(r^2R^2)_{max}$, to rescale the peaks to 1 on the vertical axis. Undo that to recover the properly normalized probability densities. $(r^2R_{10}^2)_{max} = 1.0233$ Å^{-1}, $(r^2R_{20}^2)_{max} = 0.361$ Å^{-1}, and $(r^2R_{30}^2)_{max} = 0.192$ Å^{-1}.

the electron is most likely to be found. This maximum arises from two contributions. The probability density for finding the electron at a specific location decreases as we move out from the nucleus (see Fig. 5.8b). But the volume of the spherical shell inside which we are adding up probability density increases as *r* increases. These two competing contributions determine the maximum in the radial probability function at some distance out from the nucleus.

Finally, we want to describe the size of the orbital. What is meant by the *size* of an orbital? Strictly speaking, the wave function of an electron in an atom stretches out to infinity, so an atom has no clear boundary. We define the size of an atom as the extent of a "balloon skin" inside which 90% of the probability density of the electron is contained. Calculations show that spheres containing 90% of the probability density for an electron in the 1*s*, 2*s*, and 3*s* orbitals of the hydrogen atom have radii equal to 1.41 Å, 4.83 Å, and 10.29 Å, respectively. These results show that the size of an orbital increases with increasing quantum number n. A 3*s* orbital is larger than a 2*s* orbital, which, in turn, is larger than a 1*s* orbital. This is the quantum analog of the increase in radius of the Bohr orbits with increasing n.

Let's see how to verify, using the functions plotted in Figure 5.8c, that when you integrate all the 2*s* orbital's electron density over a sphere of radius 4.83 Å, that it encloses 90% of the electron density. You might be wondering why the function plotted in Figure 5.8c does not have the angular part of the wave function $Y_s(\theta, \phi) = (1/(4\pi))^{1/2}$ in it. This is easy to see from a combination of Equations 5.5 and 5.6, replacing dV by $r^2\sin\theta dr d\theta d\phi$ instead of $dxdydz$. Because you are now looking for the electron at all values of θ and ϕ, you integrate from $\phi = 0$ to 2π and from $\theta = 0$ to π as r ranges from 0 to 4.83 Å. This gives

$$\int_V |\psi_{2s}|^2 dV = \int_{\phi=0}^{2\pi} \int_{\theta=0}^{\pi} \int_{r=0}^{r=4.83\text{Å}} |R_{2s}(r)Y_s(\theta,\phi)|^2 r^2 \sin\theta dr d\theta d\phi$$

$$= \int_{\phi=0}^{2\pi} \int_{\theta=0}^{\pi} \frac{1}{4\pi} \sin\theta \, d\theta \, d\phi \left[\int_{r=0}^{r=4.83\text{Å}} |R_{2s}(r)|^2 \, r^2 \, dr \right] \qquad [5.7]$$

$$= (2\pi)(2)\left(\frac{1}{4\pi}\right)\left[\int_{r=0}^{r=4.83\text{Å}} |R_{2s}(r)|^2 \, r^2 \, dr \right]$$

$$= \int_{r=0}^{r=4.83\text{Å}} |R_{2s}(r)|^2 \, r^2 \, dr$$

So all you need to do is to integrate the function $r^2|R_{2s}(r)|^2$, or estimate the area under the curve in Figure 5.8c from $r = 0$ to $r = 4.83$ Å. (You'll need to multiply the numbers on the vertical axis by 0.361 A^{-1} because the figure plots a rescaled function $(r^2R_{2s}{}^2)/(r^2R_{2s}{}^2)_{max}$.) This gives the integral as 0.90, or 90%. (Note that all of the $Y(\theta, \phi)$ functions in Table 5.2 have constants in them, which ensures that the integral of $|Y(\theta, \phi)|^2\sin\theta d\theta d\phi$, from $\phi = 0$ to 2π and $\theta = 0$ to π, equals 1.)

EXAMPLE 5.3

What is the probability of finding the electron in the ground state of the H atom closer to the nucleus than $r = a_0 = 0.529$ Å? Compare this with the probability of finding the electron in the ground state of the He^+ closer to the nucleus than $r = a_0 = 0.529$ Å.

Solution

Using the result of Equation 5.7, the probability is

$$\int_V |\psi_{1s}|^2 dV = \int_{r=0}^{r=a_0} |R_{1s}(r)|^2 r^2 \, dr$$

For the H atom,

$$R_{1s} = 2\left(\frac{1}{0.529\text{ Å}}\right)^{\frac{3}{2}} e^{-\frac{r}{0.529\text{ Å}}}$$

Substituting this into the integral gives

$$\int_{r=0}^{r=0.529\text{Å}} 4\left(\frac{1}{0.529\text{ Å}}\right)^3 e^{-\frac{2r}{0.529\text{ Å}}} r^2 \, dr = 0.32 = 32\%$$

(Use an integrator app or a table of integrals to evaluate this. Alternatively try integrating by parts, or estimating the area under the curve in Figure 5.8c, but remember to rescale the values on the vertical axis with the $(r^2R^2)_{max}$ given in the figure caption.)

For the He^+ ion,

$$R_{1s} = 2\left(\frac{2}{0.529\text{ Å}}\right)^{\frac{3}{2}} e^{-\frac{2r}{0.529\text{ Å}}}$$

(Continued)

Substituting this into the integral gives

$$\int_{r=0}^{r=0.529\,\text{Å}} 4\left(\frac{2}{0.529\,\text{Å}}\right)^3 e^{-\frac{4r}{0.529\,\text{Å}}} r^2\,dr = 0.76 = 76\%$$

The higher nuclear charge draws the 1*s* electron closer to the nucleus, so there is a higher probability of finding it within a distance of 0.529 Å from the nucleus.

Related Problem: 48

Another measure of the size of an orbital is the most probable distance of the electron from the nucleus in that orbital. Figure 5.8c shows that the most probable location of the electron is progressively farther from the nucleus in *ns* orbitals for larger *n*. Nonetheless, there is a finite probability density for finding the electron at the nucleus in both 2*s* and 3*s* orbitals. This happens because electrons in *s* orbitals have no angular momentum ($\ell = 0$), and thus can approach the nucleus along the radial direction. The ability of electrons in *s* orbitals to "penetrate" close to the nucleus has important consequences in the structure of many-electron atoms and molecules (see later).

Finally, note that an *ns* orbital has $n - 1$ radial nodes; a **radial node** is a spherical surface about the nucleus on which ψ and ψ^2 are 0. These spherical surfaces are the analogues of the nodal planes in the wave functions for a particle in a cubic box (Fig. 4.43). The more numerous the nodes in an orbital, the higher the energy of the corresponding quantum state of the atom. Just as for the particle in a box, the energies of orbitals increase as the number of nodes increases.

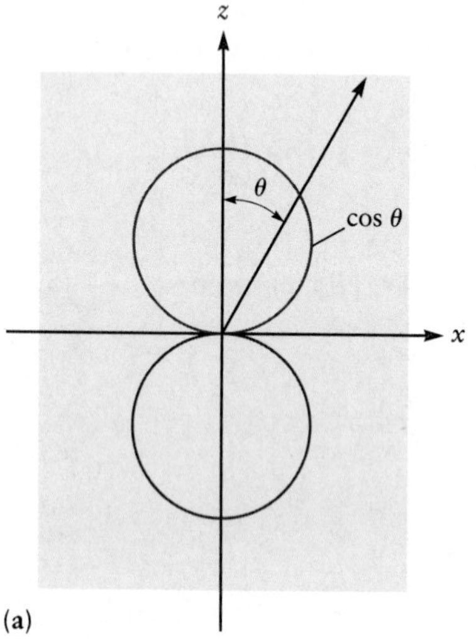

(a)

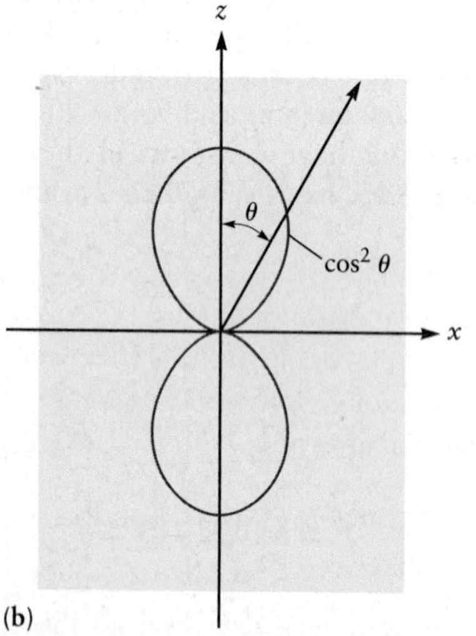

(b)

FIGURE 5.9 Two aspects of hydrogen *p* orbitals. (a) The angular wave function for the p_z orbital. The p_x and p_y orbitals are the same, but are oriented along the *x*- and *y*-axis, respectively. (b) The square of the angular wave function for the p_z orbital. Results for the p_x and p_y orbitals are the same, but are oriented along the *x*- and *y*-axis, respectively.

p Orbitals

Orbitals with angular quantum numbers different from 0 are not spherically symmetric. Interesting angular effects arise from the quantization of angular momentum. The angular wave function $Y_{\ell m}(\theta, \phi)$ has separate lobes with positive and negative phase, with a node between them. Equation 5.2b specifies $2\ell + 1$ projections of these angular momentum values along the *z*-axis. The three angular wave functions with $\ell = 1$, for which the allowed *m* values are -1, 0, and $+1$, lead to three orbitals (each called a ***p* orbital**) with the same shapes but different orientations in space.

The angular wave function $Y_{10}(\theta, \phi)$ with the combination $\ell = 1, m = 0$ is called the angular portion of the p_z orbital here because it is oriented along the *z*-axis. The wave function Y_{p_z} for the p_z orbital (see Table 5.2) is proportional to $\cos\theta$. From the relation between spherical and Cartesian coordinates illustrated in Figure 5.1, you can see that $\cos\theta \propto z$; thus, the maximum probability density for this orbital (Figure 5.9b) is along the *z*-axis (where $\theta = 0$ or π). There is a node in the *x*-*y* plane (where $\theta = \pi/2$, so $\cos\theta = 0$) (Fig. 5.9a). This nodal plane is called an **angular node**; the wave function changes sign across it. The p_z orbital therefore points along the *z*-axis, with its positive phase (red in Fig. 5.9a) on the side of the *x*-*y* plane where the *z*-axis is positive and negative phase (blue in Fig. 5.9a) on the side where the *z*-axis is negative.

The angular wave functions $Y_{11}(\theta, \phi)$ for $\ell = 1, m = 1$ and $Y_{1,-1}(\theta, \phi)$ for $\ell = 1, m = -1$ do not have a simple geometrical interpretation. However, their sum and their difference, which are also allowed solutions of the Schrödinger equation for the hydrogen atom, do have simple interpretations. Therefore, we form two new angular wave functions:

$$\begin{aligned} Y_{p_x} &= c_1(Y_{11} + Y_{1,-1}) \\ Y_{p_y} &= c_2(Y_{11} - Y_{1,-1}) \end{aligned} \qquad [5.8]$$

where c_1 and c_2 are appropriate constants. The resulting expressions for Y_{p_x} and Y_{p_y} are given in Table 5.2. A comparison of these expressions with Figure 5.1 shows that Y_{p_x} lies along the *x*-axis and Y_{p_y} lies along the *y*-axis, but otherwise

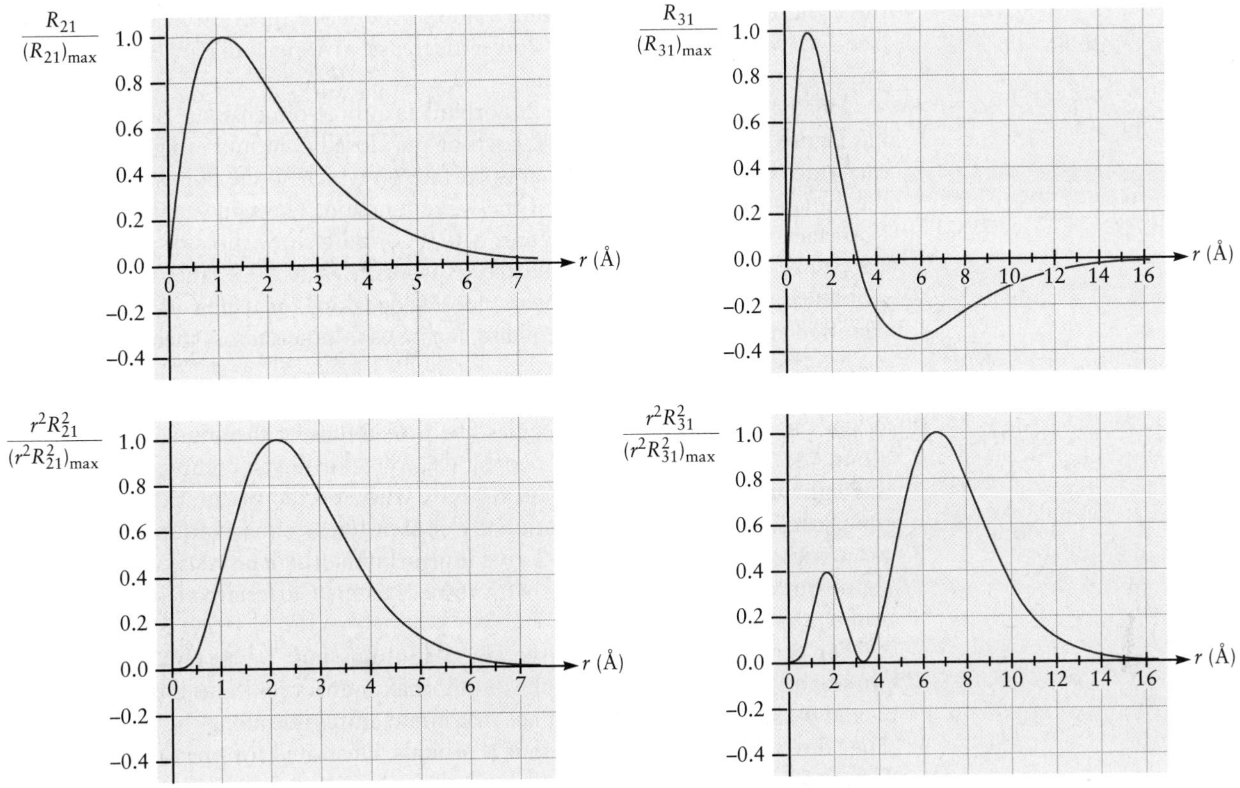

FIGURE 5.10 Radial wave functions $R_{n\ell}$ for *np* orbitals and the corresponding radial probability densities $r^2R^2_{n\ell}$.

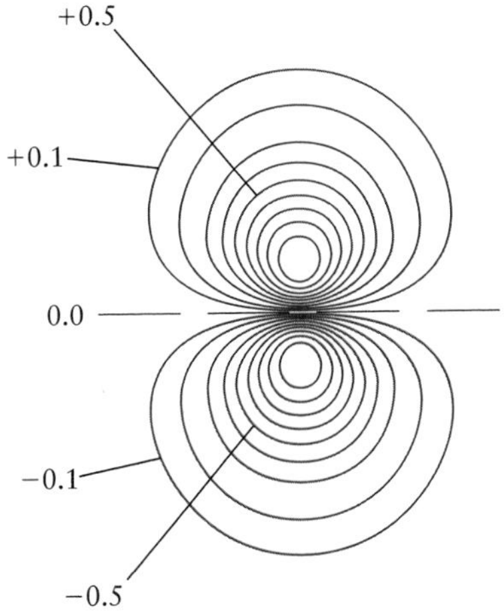

FIGURE 5.11 Contour plot of ψ for the $2p_z$ orbital for the hydrogen atom. This plot lies in the *x-z* plane. The *z*-axis (not shown) would be vertical in this figure, and the *x*-axis (not shown) would be horizontal. The lobe with positive phase is shown in red, and the lobe with negative phase in blue. The *x-y* nodal plane is shown as a dashed black line. Compare with Figure 5.9a.

have the same shape as Y_{p_z}. They have nodes at the *y-z* and *x-z* planes, respectively.

The radial parts of the *np* wave functions (represented by $R_{n\ell}$) are illustrated in Figure 5.10. The *p* orbitals, like the *s* orbitals, may have radial nodes, at which the probability density vanishes at certain distances from the nucleus regardless of direction. From Figure 5.10, the R_{21} wave function has no radial nodes, and the R_{31} function has one radial node; the R_{41} (not shown) function has two radial nodes. The $R_{n\ell}$ wave functions have n-ℓ-1 radial nodes. Because the angular part of the *np* wave function always has a nodal plane, the total wave function has $n - 1$ nodes ($n - 2$ radial and 1 angular), which is the same number as an *s* orbital with the same principal quantum number. The $R_{21}(r)$ function (that is, R_{2p}) in Table 5.2 contains the factor σ, which is proportional to r ($\sigma = Zr/a_0$) and causes it to vanish at the nucleus. This is true of all the radial wave functions except the *ns* functions, and it means that the probability density is zero for the electron to be at the nucleus for all wave functions with $\ell > 0$ (*p, d, f*, . . .). Physically, electrons with angular momentum are moving around the nucleus, not toward it, and cannot "penetrate" toward the nucleus.

Finally, we combine the angular and radial dependence to get a sense of the shape of the complete orbital, $\psi_{n\ell m} = R_{n\ell}Y_{\ell m}$. Let's examine the $2p_z$ orbital at points (r, θ, ϕ) confined to the *x-z* plane. At each point, we calculate the value of R_{21} (as in Fig. 5.10) and the value of Y_{2p_z} (as in Fig. 5.9a). Then we multiply these values together to obtain the value of ψ_{2p_z} at that point. We continue this process and generate a contour plot for ψ_{2p_z} in the *x-z* plane. The results are shown in Figure 5.11. Contours identify points at which ψ_{2p_z} takes on ±0.1, ±0.2, ±0.3, ±0.4, ±0.5, ±0.6, ±0.7, ±0.8, ±0.9 of its maximum value. Contours with positive phase are shown in red; blue contours represent negative phase. The radial wave function from Figure 5.10 has dramatically changed the circular angular wave function from Figure 5.9a. The contours are not concentric, but rather bunch together on the sides nearest the *x-y* plane. This reflects the rapid decrease in the absolute value of the wave function near the nucleus (see Fig. 5.10) and the much slower decrease at longer distances beyond the maximum in the radial function.

Finally, we can represent the $2p_z$ orbital as a three-dimensional object by rotating Figure 5.11 about the z-axis. Each of the closed contours in Figure 5.11 will then trace out a three-dimensional isosurface on which all the points (r, θ, ϕ) have the same amplitude and phase of the wave function. The same analysis generates isosurfaces for $2p_x$ and $2p_y$. Figures 5.12a, b, and c show plots of all three, with the phases and nodal planes indicated, as isosurfaces at ± 0.2 times the maximum amplitude. Each $2p$ orbital appears, loosely speaking, as a pair of flattened and distorted spheres, with opposite phase, facing each other across their nodal plane.

d Orbitals

When $\ell = 2$ Equation 5.2b specifies five projections of the angular momentum along the z-axis. As with the p orbitals, we take linear combinations of the angular wave functions to obtain orbitals with specific orientations relative to the Cartesian axes. The conventionally chosen linear combinations of the solutions with $m = -2, -1, +1, +2$ give four orbitals (the ***d* orbitals**) with the same shape but different orientations with respect to the Cartesian axes: d_{xy}, d_{yz}, d_{xz}, and $d_{x^2-y^2}$ (Fig. 5.13).

For example, a d_{xy} orbital (see Fig. 5.13a) has four lobes, two with positive phase and two with negative phase; the maximum amplitude is at 45° to the x- and y-axes. The $d_{x^2-y^2}$ orbital has maximum amplitude along the x- and y-axes. The "down-axis" view of these four d orbitals, illustrated for one of them (d_{xy}) in Figure 5.13f, shows that they all have the same shape when viewed down the appropriate axis. The fifth orbital, d_{z^2}, which corresponds to $m = 0$, has a different shape from the rest, with maximum amplitude along the z-axis and a little "doughnut" in the x-y plane. Each d orbital has two angular nodes (for example, the d_{xy} orbital has the x-z and y-z planes as its nodal surfaces). The radial functions, $R_{n2}(r)$, have $n - 3$ radial nodes, giving once again $n - 1$ total nodes.

The wave functions for f orbitals and orbitals of higher ℓ can be calculated, but they play a smaller role in chemistry than do the s, p, and d orbitals.

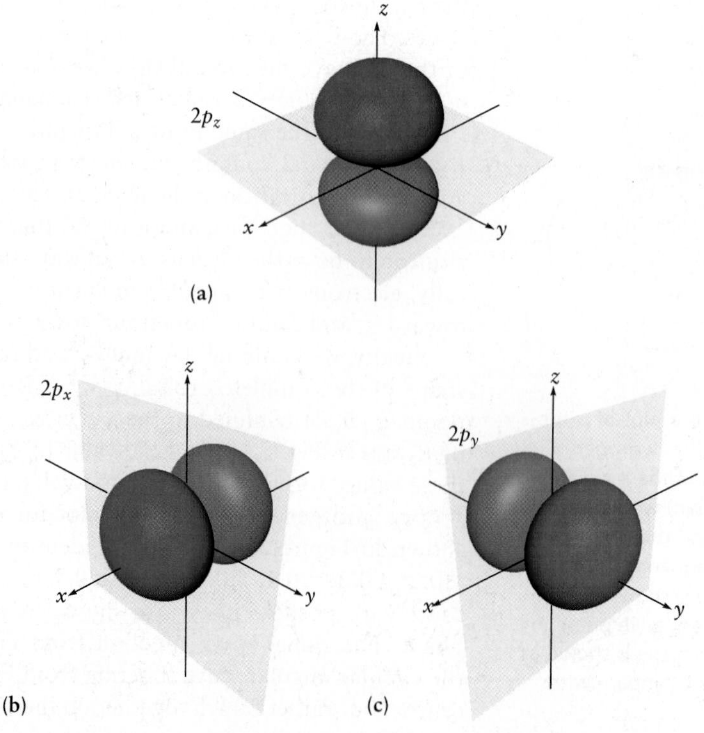

FIGURE 5.12 The shapes of the three $2p$ orbitals, with phases and nodal planes indicated. The isosurfaces in (a), (b), and (c) identify points where value of each wave function is ± 0.2 of its maximum value. (a) $2p_z$ orbital. (b) $2p_x$ orbital. (c) $2p_y$ orbital.

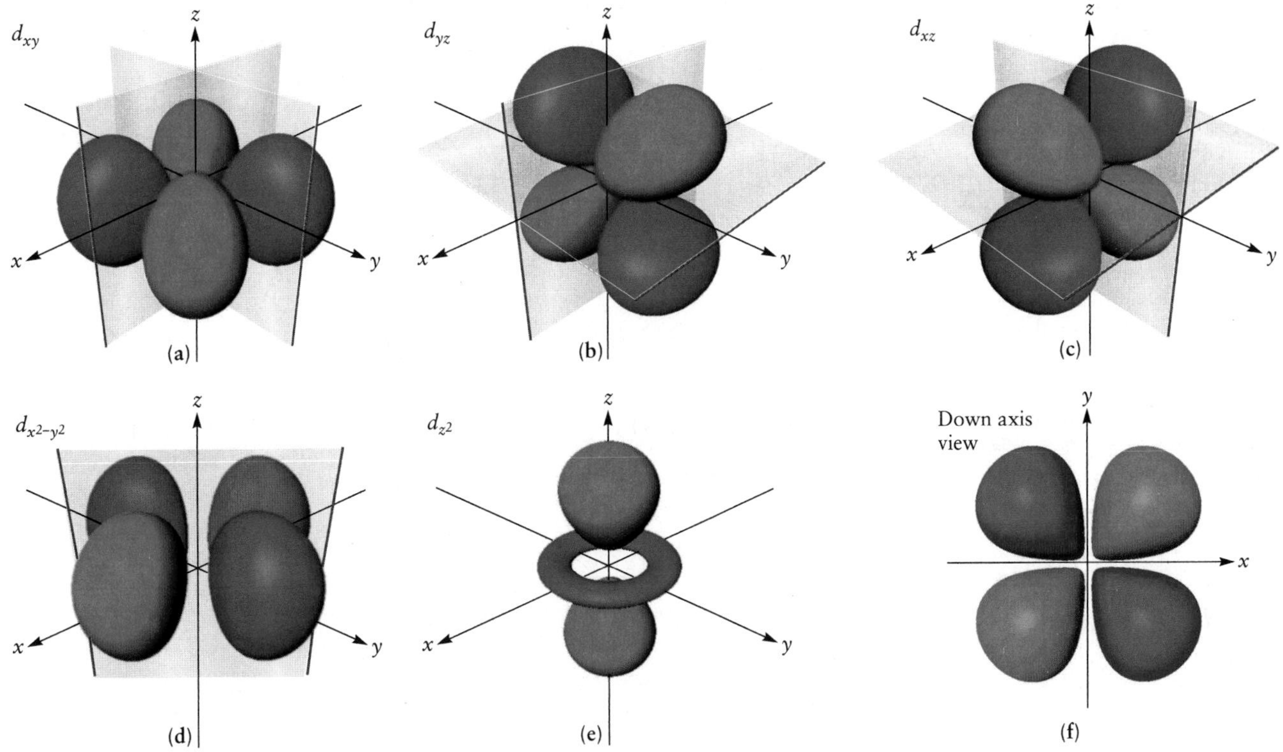

FIGURE 5.13 The shapes of the five 3*d* orbitals, with phases and nodal surfaces indicated. The "down-axis" view shows the shapes of the first four orbitals (a)–(d) when viewed down the appropriate axis; the specific example shown here is the d_{xy} orbital viewed down the *z*-axis.

We summarize the important features of orbital shapes and sizes as follows:

1. For a given value of ℓ, an increase in n leads to an increase in the average distance of the electron from the nucleus, and therefore in the size of the orbital (see Figs. 5.8 and 5.10).
2. An orbital with quantum numbers n and ℓ has ℓ angular nodes and $n - \ell - 1$ radial nodes, giving a total of $n - 1$ nodes. An angular node appears as a nodal plane in plots of p orbitals. The value of the radial or angular part of the wave function changes sign as you cross a radial or angular node respectively. For a one-electron atom or ion, the energy depends only on the number of nodes—that is, on n but not on ℓ or m. The energy increases as the number of nodes increases.
3. As r approaches 0, $\psi(r, \theta, \phi)$ vanishes for all orbitals except s orbitals; thus, only an electron in an s orbital can "penetrate to the nucleus," that is, have a finite probability density right at the nucleus.

The next section shows that these general statements are important for determining the electronic structure of many-electron atoms even though they are deduced from the one-electron case.

The characteristics of the orbitals of a one-electron atom (or ion) are especially well displayed by a quantitative plot showing s, p, and d orbitals all on the same scale (Fig. 5.14). The best quantitative measure of the size of an orbital is $\bar{r}_{n\ell}$, the average value of the distance of the electron from the nucleus in that orbital. Quantum mechanics calculates $\bar{r}_{n\ell}$ as

$$\bar{r}_{n\ell} = \frac{n^2 a_0}{Z}\left\{1 + \frac{1}{2}\left[1 - \frac{\ell(\ell+1)}{n^2}\right]\right\} \qquad [5.9]$$

FIGURE 5.14 Dependence of radial probability densities on distance from the nucleus for one-electron orbitals with n = 1, 2, 3. The small arrow above each curve locates the value of $\bar{r}_{n\ell}$ for that orbital. The distance axis is expressed in the same dimensionless variable introduced in Table 5.1. The value 1 on this axis is the first Bohr radius for the hydrogen atom. Because the radial probability density has dimensions (length)$^{-1}$, the calculated values of $r^2[R_{n\ell}(r)]^2$ are divided by $(a_0)^{-1}$ to give a dimensionless variable for the probability density axis. (Adapted from Eisberg, R. *Fundamentals of Modern Physics*, page 306, Figure 10-2, Wiley, 1961.)

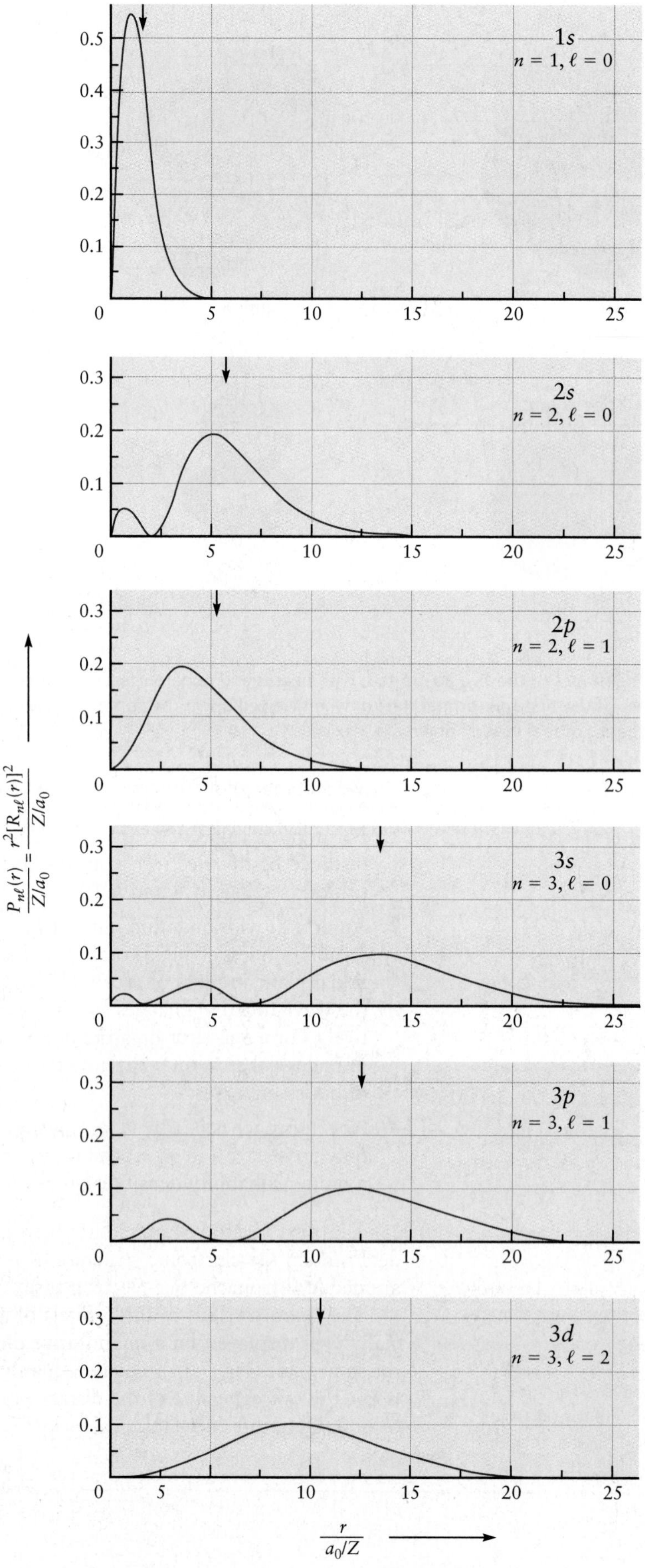

The leading term of this expression is the radius of the nth Bohr orbit (see Eq. 4.12). In Figure 5.14, the small arrow on each curve locates the value of $\bar{r}_{n\ell}$ for that orbital.

EXAMPLE 5.4

Compare the 3p and 4d orbitals of a hydrogen atom with respect to the (a) number of radial and angular nodes and (b) energy of the corresponding atom.

Solution

(a) The 3p orbital has a total of $n - 1 = 3 - 1 = 2$ nodes. Of these, one is angular ($\ell = 1$) and one is radial. The 4d orbital has $4 - 1 = 3$ nodes. Of these, two are angular ($\ell = 2$) and one is radial.

(b) The energy of a one-electron atom depends only on n. The energy of an atom with an electron in a 4d orbital is higher than that of an atom with an electron in a 3p orbital, because $4 > 3$.

Related Problems: 7, 8

5.1.4 Electron Spin

If a beam of hydrogen atoms in their ground state (with $n = 1, \ell = 0, m = 0$) is sent through a magnetic field whose intensity increases in the plane perpendicular to the flight of the beam, it splits into two beams, each containing half of the atoms (Fig. 5.15). The pioneering experiment of this type is called the Stern–Gerlach experiment after the German physicists who performed it, Otto Stern and Walther Gerlach.

Recall that a magnet, unlike a single electric charge, has two poles, and that a magnetic dipole moment can be used to describe the interactions of a magnet with a magnetic field in the same way that an electric dipole moment is used to describe the interaction of a pair of charges with an electric field. You may recall from classical physics (or your own experience) that a small bar magnet will rotate to orient itself in the presence of an external magnetic field. If the magnetic field changes strength along a particular direction, then a force will be exerted on the bar magnet that will cause it to *move* in the direction of the changing field, and not just rotate to a new direction. If the magnetic dipole moments of the hydrogen atoms were randomly oriented in space (as predicted by classical physics), then the beam would be smeared out at the detector to reflect all possible orientations of the magnetic moment. That the original beam is split into only two well-defined beams in this experiment demonstrates the unexpected fact that the *orientation* of the magnetic moment of the electron is quantized. The result of this experiment is explained by introducing a *fourth* quantum number, m_s, which can take on two values, conventionally chosen to

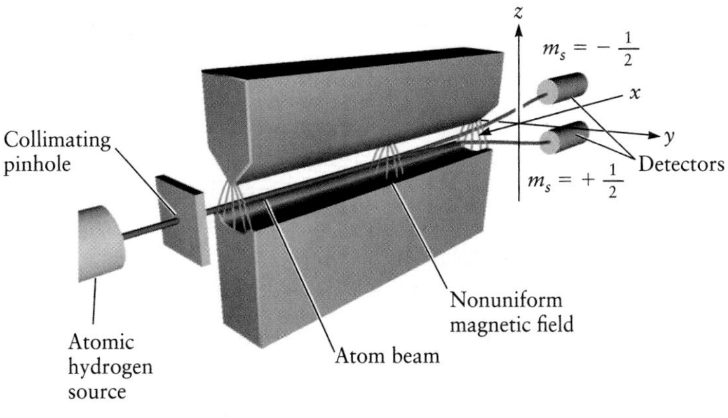

FIGURE 5.15 A beam of hydrogen atoms is split into two beams when it traverses a magnetic field, the value of which is not constant in the plane perpendicular to the path of the beam. The nonconstant field is created by the specially shaped cross section of the north and south poles of the magnet in the z-y plane. The green curved lines trace the pattern over which the field is varied. Regions where the green lines are closer together are regions of greater magnetic field. Atoms with spin quantum number $m_s = +\frac{1}{2}$ follow one trajectory, and those with $m_s = -\frac{1}{2}$ follow another.

be $+\frac{1}{2}$ and $-\frac{1}{2}$. For historical reasons, the fourth quantum number is referred to as the **spin quantum number**. When $m_s = +\frac{1}{2}$, the electron spin is said to be "up," and when $m_s = -\frac{1}{2}$, the spin is "down." The spin quantum number arises from relativistic effects that are not included in the Schrödinger equation. For most practical purposes in chemistry, it is sufficient simply to solve the ordinary Schrödinger equation, and then associate with each electron a spin quantum number $m_s = +\frac{1}{2}$ or $-\frac{1}{2}$ which does not affect the spatial probability distribution of the electron. Including the spin doubles the number of allowed quantum states with principal quantum number n, from n^2 to $2n^2$. This fact will assume considerable importance when considering the many-electron atoms in the next section.

5.2 Shell Model for Many-Electron Atoms

As we move from one-electron to many-electron atoms, both the Schrödinger equation and its solutions become increasingly complicated. The simplest many-electron atom, helium (He), has two electrons and a nuclear charge of $+2e$. The positions of the two electrons in a helium atom can be described using two sets of Cartesian coordinates, (x_1, y_1, z_1) and (x_2, y_2, z_2), relative to the same origin. The wave function ψ depends on all six of these variables: $\psi = \psi(x_1, y_1, z_1, x_2, y_2, z_2)$. Its square, $\psi^2(x_1, y_1, z_1, x_2, y_2, z_2)$, is the probability density of finding the first electron at point (x_1, y_1, z_1) and, simultaneously, the second electron at (x_2, y_2, z_2). The Schrödinger equation is now more complicated, and an explicit solution for helium is not possible. Nevertheless, modern computers have enabled us to solve this equation numerically with high accuracy, and the predicted properties of helium are in excellent agreement with experiment.

Although these numerical calculations demonstrate conclusively the usefulness of the Schrödinger equation for predicting atomic properties, they suffer from two defects. First, they are somewhat difficult to interpret physically, and second, they become increasingly difficult to solve, even numerically, as the number of electrons increases. As a result, approximate approaches to the many-electron Schrödinger equation have been developed.

5.2.1 Hartree Orbitals

The **self-consistent field (SCF) orbital approximation method** developed by Hartree is especially well suited for applications in chemistry. Hartree's method generates a set of approximate one-electron orbitals, φ_α, and associated energy levels, ε_α, reminiscent of those for the H atom. The subscript α represents the appropriate set of quantum numbers (see later in this chapter for a definition). The electronic structure of a neutral atom with atomic number Z is then "built up" by placing Z electrons into these orbitals in accordance with certain rules (see later in this chapter for descriptions of these rules).

In this section, we introduce Hartree's method and use it to describe the electron arrangements and energy levels in many-electron atoms. Later sections detail how this approximate description rationalizes periodic trends in atomic properties and serves as a starting point for descriptions of chemical bond formation.

For any atom, Hartree's method begins with the exact Schrödinger equation in which each electron is attracted to the nucleus and repelled by all the other electrons in accordance with the Coulomb potential. The following three simplifying assumptions are made immediately:

1. Each electron moves in an *effective field* created by the nucleus and all the other electrons, and the effective field for electron i depends only on its position r_i.
2. The effective field for electron i is obtained by averaging its Coulomb potential interactions with each of the other electrons over all the positions of the other electrons so that r_i is the only coordinate in the description.
3. The effective field is spherically symmetric; that is, it has no angular dependence.

Under the first assumption, each electron moves as an independent particle and is described by a one-electron orbital similar to those of the hydrogen atom. The wave function for the atom then becomes a product of these one-electron orbitals, which we denote $\varphi_\alpha(r_i)$. For example, the wave function for lithium (Li) has the form $\psi_{\text{atom}} = \varphi_\alpha(r_1)\varphi_\beta(r_2)\varphi_\gamma(r_3)$. This product form is called the **orbital approximation for atoms.** The second and third assumptions in effect convert the exact Schrödinger equation for the atom into a set of simultaneous equations for the unknown effective field and the unknown one-electron orbitals. These equations must be solved by iteration until a self-consistent solution is obtained. (In spirit, this approach is identical to the solution of complicated algebraic equations by the method of iteration described in Appendix C.) Like any other method for solving the Schrödinger equation, Hartree's method produces two principal results: energy levels and orbitals.

These Hartree orbitals resemble the atomic orbitals of hydrogen in many ways. Their angular dependence is identical to that of the hydrogen orbitals, so quantum numbers ℓ and m are associated with each atomic orbital. The radial dependence of the orbitals in many-electron atoms differs from that of one-electron orbitals because the effective field differs from the Coulomb potential, but a principal quantum number n can still be defined. The lowest energy orbital is a 1*s* orbital and has no radial nodes, the next lowest *s* orbital is a 2*s* orbital and has one radial node, and so forth. Each electron in an atom has associated with it a set of four quantum numbers (n, ℓ, m, m_s). The first three quantum numbers describe its spatial distribution and the fourth specifies its spin state. The allowed quantum numbers follow the same pattern as those for the hydrogen atom. However, the number of states associated with each combination of (n, ℓ, m) is twice as large because of the two values for m_s.

5.2.2 Sizes and Shapes of Hartree Orbitals

The spatial properties of Hartree orbitals are best conveyed through a specific example. We present the results for argon (Ar), taken from Hartree's original work. The ground state of the argon atom has 18 electrons in the 1*s*, 2*s*, 2*p*, 3*s*, and 3*p* Hartree orbitals (see later). Figure 5.16 shows the radial probability density distributions for these five occupied orbitals as calculated by Hartree's method. The probability density distribution shown for the 2*p* level is the sum of the distributions for the $2p_x$, $2p_y$, and $2p_z$ orbitals; similarly, the 3*p* probability density distribution includes the $3p_x$, $3p_y$, and $3p_z$ orbitals. Comparing Figure 5.16 with Figure 5.14 shows that each Hartree orbital for argon is "smaller" than the corresponding orbital for hydrogen in the sense that the region of maximum probability density is closer to the nucleus. This difference occurs because the argon nucleus ($Z = 18$) exerts a much stronger attractive force on

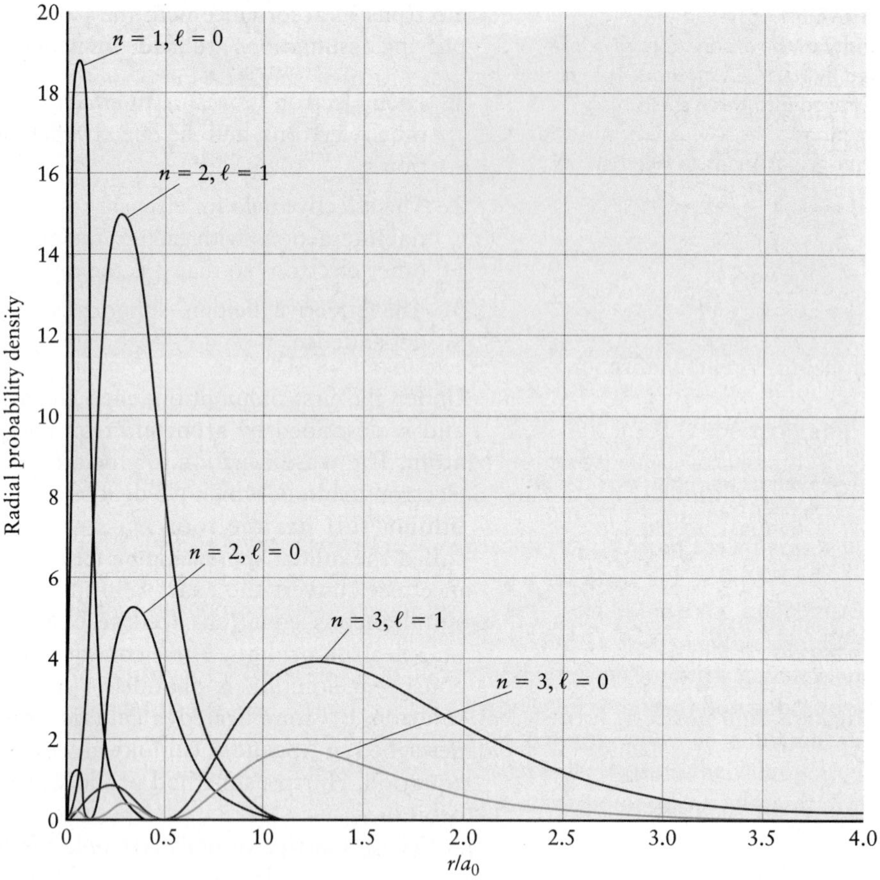

FIGURE 5.16 Dependence of radial probability densities on distance from the nucleus for Hartree orbitals in argon with $n = 1, 2, 3$. The results were obtained from self-consistent calculations using Hartree's method. Distance is plotted in the same dimensionless variable used in Figure 5.14 to facilitate comparison with the results for hydrogen. The fact that the radial probability density for all orbitals with the same value of n have maxima very near one another suggests that the electrons are arranged in "shells" described by these orbitals. (Adapted from Eisberg, R. *Fundamentals of Modern Physics*, page 399, Figure 13-2, Wiley, 1961.)

electrons than does the hydrogen nucleus ($Z = 1$). We develop a semiquantitative relation between orbital size and Z in the next subsection.

The fact that Hartree orbitals with the same value of n are large in the same narrow regions of space, despite their different values of ℓ, has interesting consequences. The total radial probability density function for a many-electron atom gives (upon integration) the probability of finding an electron near a distance r from the nucleus regardless of which orbital it occupies. We obtain this function by summing up the radial probability density functions of all the occupied orbitals. The resulting probability function is proportional to the *radial charge density distribution function* $\rho(r)$ for the atom. If the radial probability density functions in Figure 5.16 are all added together, the result reflects the contributions of electrons to the charge density $\rho(r)$ in a thin spherical shell of radius r, regardless of the orbital to which the electron belongs. A plot of $\rho(r)$ on the same scale as Figure 5.16 shows three peaks at r values of approximately 0.1, 0.3, and 1.2 in units of a_0 (Fig. 5.17) The total electron density of the argon atom thus is concentrated in three concentric shells, where a **shell** is defined as all electrons with the same value of n. Each shell has a radius determined by the principal quantum number n. The shell model summarizes the coarse features of the electron density of an atom by averaging over all those local details not described by the principal quantum number n. Within each shell, a more detailed picture is provided by the subshells; a **subshell** is a group of orbitals that have the same value of both n and ℓ.

The subshells (see Fig. 5.16) determine the structure of the periodic table and the formation of chemical bonds. In preparation for a discussion of these connections, it is necessary to describe the energy values for Hartree orbitals.

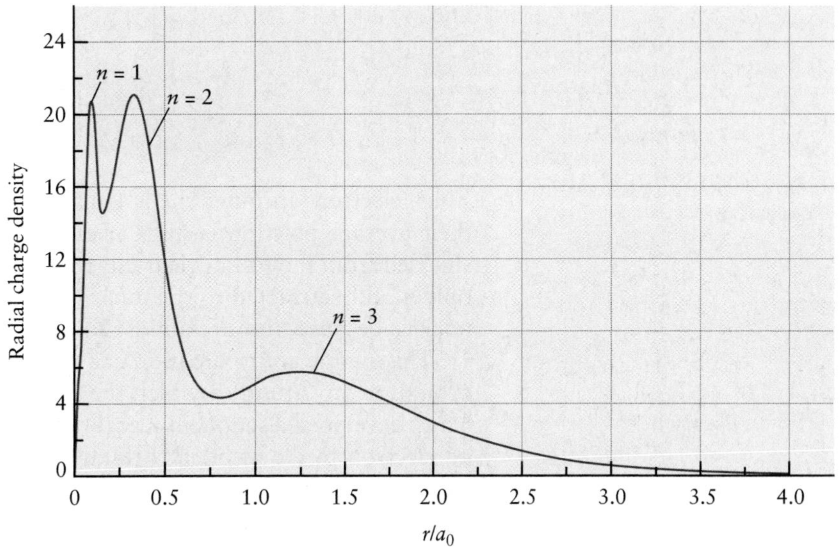

FIGURE 5.17 The radial charge density in the argon atom as calculated by Hartree's method. The charge is arrayed into three shells corresponding to the values 1, 2, and 3 for the principal quantum number *n*.

5.2.3 Shielding Effects: Energy Sequence of Hartree Orbitals

The energy-level diagrams calculated for many-electron atoms by Hartree's method resemble the diagram for the hydrogen atom (see Fig. 5.2), but differ in two important respects. First, orbitals with the same n but different ℓ, such as the 2s and the 2p orbital, no longer have the same energy. Because the effective field in Hartree's method is different from the Coulomb field in the hydrogen atom, the energy levels of Hartree orbitals depend on both n and ℓ. Second, the energy values are distinctly shifted from the values of corresponding hydrogen orbitals because of the stronger attractive force exerted by nuclei with $Z > 1$.

These two effects can be explained qualitatively by a highly simplified one-electron model. Assume each of the electrons in shell n is moving in a Coulomb potential given approximately by

$$V_n^{\text{eff}}(r) \approx -\frac{Z_{\text{eff}}(n)e^2}{r} \qquad [5.10]$$

where $Z_{\text{eff}}(n)$ is the **effective nuclear charge** in that shell. To understand the origin and magnitude of $Z_{\text{eff}}(n)$, consider a particular electron e_1 in an atom. Inner electrons near the nucleus *shield* e_1 from the full charge Z of the nucleus by effectively canceling some of the positive nuclear charge. $Z_{\text{eff}}(n)$ is thus the net reduced nuclear charge experienced by a particular electron, due to the presence of the other electrons. The effective nuclear charge $Z_{\text{eff}}(n)$ is given by $(Z-S)$ where Z is the nuclear charge and S is the shielding or screening constant for an orbital with principal quantum number n. (See Section 3.4.) For a neutral atom, $Z_{\text{eff}}(n)$ can range from a maximum value of Z near the nucleus (no screening) to a minimum value of 1 far from the nucleus (complete screening by the other $Z - 1$ electrons). The Hartree calculations for argon described above show that $Z_{\text{eff}}(1) \sim 16$, $Z_{\text{eff}}(2) \sim 8$, and $Z_{\text{eff}}(3) \sim 2.5$. The effect of shielding on the energy and radius of a Hartree orbital is easily estimated in this simplified picture by using the hydrogen atom equations with Z replaced by $Z_{\text{eff}}(n)$. We use ε_n to distinguish a Hartree orbital energy from the H atom orbital E_n. Thus,

$$\varepsilon_n \approx -\frac{[Z_{\text{eff}}(n)]^2}{n^2} \quad \text{(rydbergs)} \qquad [5.11]$$

and

$$\bar{r}_{n\ell} \approx \frac{n^2 a_0}{Z_{\text{eff}}(n)}\left\{1 + \frac{1}{2}\left[1 - \frac{\ell(\ell+1)}{n^2}\right]\right\} \quad [5.12]$$

Thus, electrons in inner shells (small n) are tightly bound to the nucleus, and their average position is quite near the nucleus because they are only slightly shielded from the full nuclear charge of Z protons. Electrons in outer shells are only weakly attracted to the nucleus, and their average position is far from the nucleus because they are almost fully shielded from the nuclear charge.

This simple approximation can be used to estimate the energies and radii of orbitals in any atom for which the effective nuclear charge values are available. Slater developed a set of empirical rules for estimating the shielding constant for an electron in an ns or np orbital in order to assign effective nuclear charge values:

1. Write the electronic configuration for the atom with terms in the order and grouped as follows: $(1s)$ $(2s, 2p)$ $(3s, 3p)$ $(3d)$ $(4s, 4p)$ $(4d)$ $(4f)$ $(5s, 5p)$ etc.
2. Electrons in any group to the right of a (ns, np) group do not shield the (ns, np) electrons and therefore make no contribution to the shielding constant.
3. Each of the other electrons in a (ns, np) group contributes 0.35 to the shielding constant. (This includes how valence electrons in main group elements shield each other.)
4. Each electron in the $(n - 1)$ shell contributes 0.85 to the shielding constant.
5. Each electron in shells $(n - 2)$ or lower fully shield the outer electrons, contributing 1.00 to the shielding constant.
6. For He $(1s)^2$, the shielding constant is 0.31.

Slater's rules in effect summarize and quantify the general properties of radial distributions functions presented above, and they provide great physical insight into the shielding process. Today, values of Z_{eff} for each element are obtained directly from advanced computer calculations that are more sophisticated than Hartree's method.

The results for the first two periods are shown in Table 5.3; the actual Z values are shown in parentheses after each atomic symbol.

EXAMPLE 5.5

Estimate the energy and the average value of r in the $1s$ orbital of argon. Compare the results with the corresponding values for hydrogen.

Solution

Using Equation 5.11 and the value $Z_{\text{eff}}(1) \sim 16$ leads to $\varepsilon_{1s} \sim -256$ rydbergs for argon. The Ar$(1s)$ electron is more strongly bound than the H$(1s)$ electron by a factor 256. (Compare Eq. 5.1b for the hydrogen atom.)

Using Equation 5.12 and the value $Z_{\text{eff}}(1) \sim 16$ leads to $\bar{r}_{1s} = \dfrac{3a_0}{2 \cdot 16}$ for argon. This is smaller by a factor of 16 than $\bar{r}_{1s}$ for hydrogen.

A comparison of Figure 5.16 with Figure 5.14 demonstrates that each Hartree orbital for argon is "smaller" than the corresponding orbital for hydrogen in the sense that the region of maximum probability is closer to the nucleus.

Related Problems: 11, 12, 13, 14, 15, 16

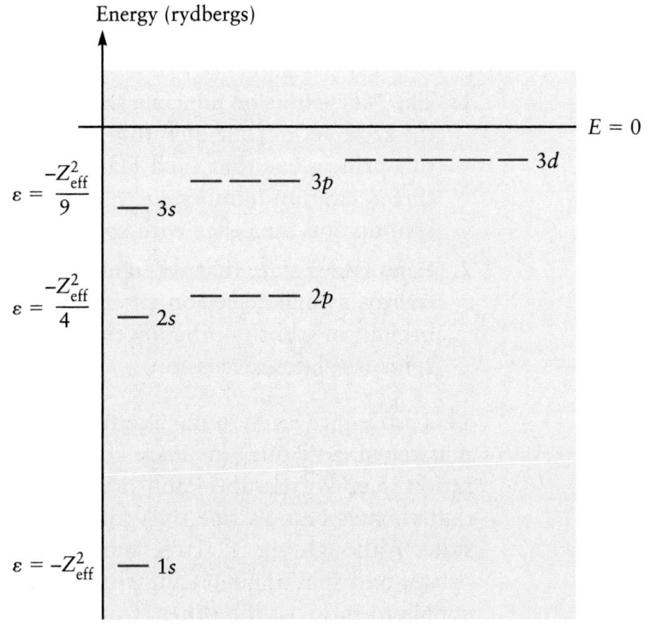

FIGURE 5.18 Approximate energy-level diagram for Hartree orbitals, estimated by incorporating values of Z_{eff}. Energy values are in units of rydbergs. The result of electron–electron repulsion is to remove the degeneracy of the hydrogen atom states with different ℓ values. Note that the value of Z_{eff} is different for each value of n (see Table 5.3).

TABLE 5.3

Z_{eff} for Selected Atoms

	H(1)							He(2)
1s	1.00							1.69
	Li(3)	**Be(4)**	**B(5)**	**C(6)**	**N(7)**	**O(8)**	**F(9)**	**Ne(10)**
1s	2.69	3.68	4.68	5.67	6.66	7.66	8.65	9.64
2s	1.28	1.91	2.58	3.22	3.85	4.49	5.13	5.76
2p			2.42	3.14	3.83	4.45	5.10	5.76

The dependence of the energy on ℓ in addition to n can be explained by comparing the extent of shielding in different subshells. Figures 5.8 through 5.14 show that only the *s* orbitals penetrate to the nucleus; both *p* and *d* orbitals have nodes at the nucleus. Consequently, the shielding will be smallest, and the electron most tightly bound, in *s* orbitals. Calculations show that

$$\varepsilon_{ns} < \varepsilon_{np} < \varepsilon_{nd}$$

The energy-level diagram for Hartree orbitals is shown qualitatively in Figure 5.18. Energy levels for *ns* and *np* orbitals, can be estimated by using values for Z_{eff} such as those in Table 5.3 in Equation 5.11.

How do we use these Hartree energy levels and orbitals to describe the electrons in an atom? The answer is the subject of the next section.

5.3 Aufbau Principle and Electron Configurations

The ground-state electronic configuration of an atom with atomic number Z is built up by arranging the Hartree atomic orbitals in order of increasing energy and adding one electron at a time, starting with the lowest energy orbital, until all Z electrons in the neutral atom are in place. The following additional restrictions are imposed at each step:

1. The **Pauli exclusion principle** states that no two electrons in an atom can have the same set of four quantum numbers (n, ℓ, m, m_s). Another way of stating this principle is that each Hartree atomic orbital (characterized by a set of three quantum numbers, n, ℓ, and m) holds at most two electrons, one with spin up and the other with spin down.
2. **Hund's rules** state that when electrons are added to Hartree orbitals of equal energy, a single electron enters each orbital before a second one enters any orbital. In addition, the lowest energy configuration is the one with parallel spins (see later discussion).

This procedure leads to the **electron configuration** for each atom, which gives the number of electrons "in" each Hartree orbital. Any proposed electron configuration that violates the Pauli principle is invalid. Any proposed configuration that violates Hund's rule will give an excited state of the atom, not its ground state. Although the Hartree model is commonly used and informative, the orbital and spin angular momenta of one electron in a multielectron atom can couple to those of the other electrons, resulting in quantum numbers different from the ℓ, m and m_s for each individual electron.

5.3.1 Building Up from Helium to Argon

Let's see how the aufbau principle works for the atoms from helium (He) through neon (Ne). The lowest energy orbital is always the 1*s* orbital; therefore, helium has two electrons (with opposite spins) in that orbital. The ground-state electron configuration of the helium atom is symbolized as $1s^2$ and is conveniently illustrated by a diagram (for example, Fig. 5.19). The 1*s* orbital in the helium atom is somewhat larger than the 1*s* orbital in the helium ion (He^+). In the ion, the electron experiences the full nuclear charge $+2e$, but in the atom, each electron partially screens or shields the other electron from the nuclear charge. The orbital in the helium atom can be described by the approximate equations given previously with an "effective" nuclear charge Z_{eff} of 1.69, which lies between +1 (the value for complete shielding by the other electron) and +2 (no shielding).

A lithium (Li) atom has three electrons. The third electron does not join the first two in the 1*s* orbital. It occupies a different orbital because, by the Pauli principle, at most two electrons may occupy any one orbital. The third electron goes into the 2*s* orbital, which is the next lowest in energy. The ground state of the lithium atom is therefore $1s^22s^1$, and $1s^22p^1$ is an excited state.

The next two elements present no difficulties. Beryllium (Be) has the ground-state configuration $1s^22s^2$ and the 2*s* orbital is now filled. Boron (B), with five electrons, has the ground-state configuration $1s^22s^22p^1$. Because the three 2*p* orbitals of boron have the same energy, there is an equal chance for the electron to be in each one.

With carbon (C), the sixth element, a new question arises. Will the sixth electron go into the same orbital as the fifth (for example, both into a $2p_x$ orbital with opposite spins), or will it go into the $2p_y$ orbital, which has equal energy? The answer is found in the observation that two electrons that occupy the same atomic orbital experience stronger electron–electron repulsion than they would if they occupied orbitals in different regions of space. Thus, putting the last two electrons of carbon into two different *p* orbitals, such as $2p_x$ and $2p_y$, in accordance with Hund's rule, leads to lower energy than putting them into the same *p* orbital. The electron configuration of carbon is then $1s^22s^22p_x^12p_y^1$, or more simply, $1s^22s^22p^2$. This configuration is shown in Figure 5.20.

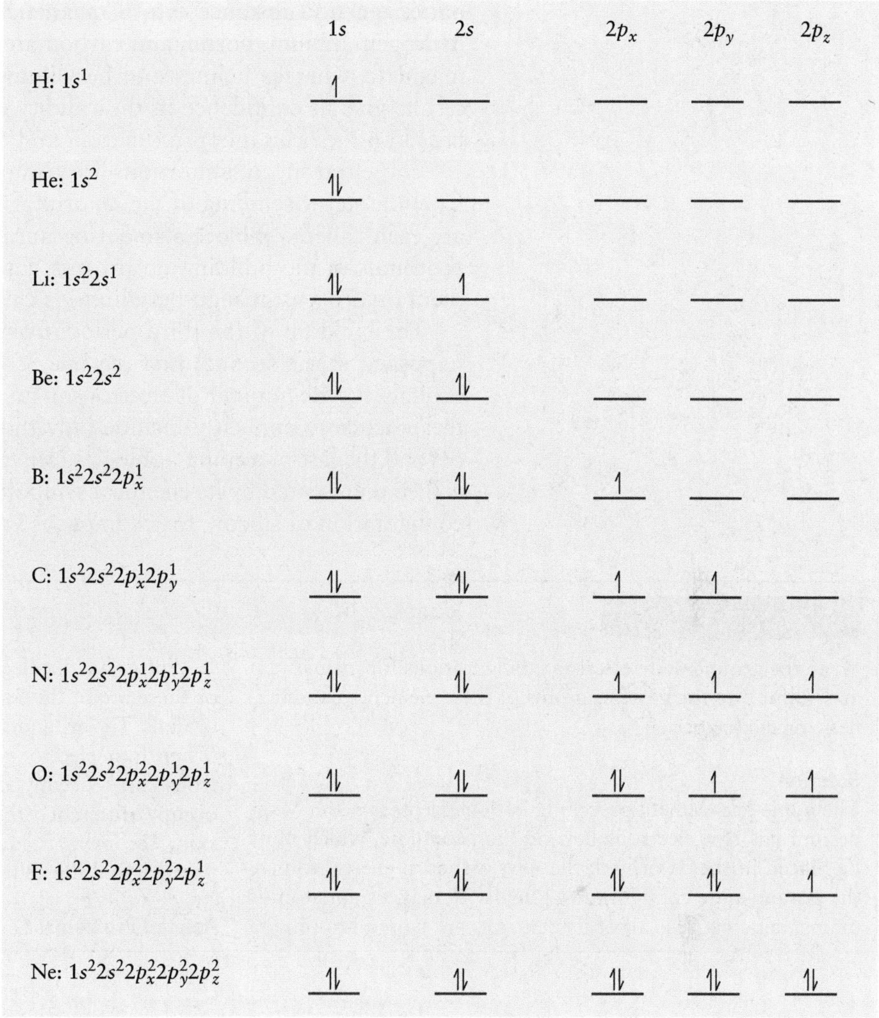

FIGURE 5.19 The ground-state electron configurations of first- and second-period atoms. Each horizontal line represents a specific atomic orbital. Arrows pointing up represent electrons with spin quantum number $m_s = +\frac{1}{2}$ and arrows pointing down represent electrons with spin quantum number $m_s = -\frac{1}{2}$.

The behavior of atoms in a magnetic field provides a test of their electron configuration. A substance is **paramagnetic** if it is attracted into a magnetic field. Any substance that has one or more unpaired electrons in the atoms, molecules, or ions that compose it is paramagnetic because a net magnetic moment arises from each of the unpaired electrons. (Recall the Stern–Gerlach experiment described in Section 5.1.) A substance in which all the electrons are paired is weakly **diamagnetic**: It is pushed *out* of a magnetic field, although the force it experiences is much smaller in magnitude than the force that pulls a typical

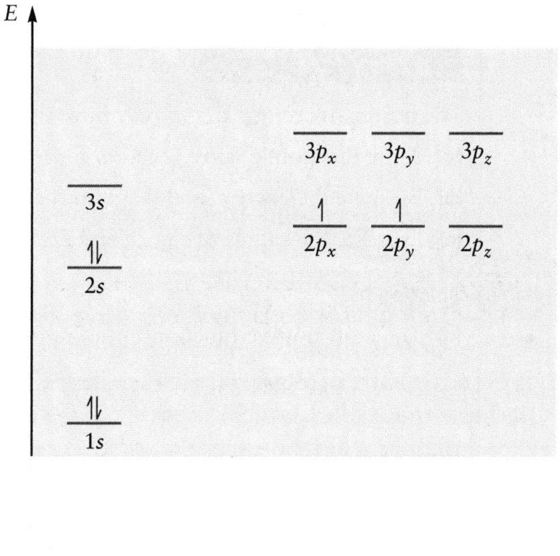

FIGURE 5.20 For a many-electron atom such as carbon, orbitals with different ℓ values (such as the 2*s* and 2*p* orbitals) have different energies (see Fig. 5.18). When two or more orbitals have the *same* energy (such as the three 2*p* orbitals here), electrons occupy different orbitals with parallel spins in the ground state.

paramagnetic substance into a magnetic field. Of the atoms discussed so far, hydrogen, lithium, boron, and carbon are known from experiments to be paramagnetic, whereas helium and beryllium are known to be diamagnetic. These results give us confidence in the validity of our description of atomic structure based on the orbital approximation and SCF calculations.

The electron configurations from nitrogen (N) through neon (Ne) follow from the stepwise filling of the $2p$ orbitals. The six elements from boron to neon are each called a ***p*-block element** because their configurations involve filling of p orbitals in the building-up process. Each of the four elements that precedes them (hydrogen through beryllium) is called an ***s*-block element**.

The build-up of the third period, from sodium to argon, is an echo of what happened in the second; first the one $3s$ orbital is filled, and then the three $3p$ orbitals. As the number of electrons in an atom reaches 15 or 20, it is frequently the practice to explicitly include only those electrons added in the building up beyond the last preceding noble-gas element. The configuration of that noble gas is then represented by its chemical symbol enclosed in brackets. The ground-state configuration of silicon, for example, is written $[\text{Ne}]3s^23p^2$ using this system.

EXAMPLE 5.6

Write the ground-state electron configurations for magnesium and sulfur. Are the gaseous atoms of these elements paramagnetic or diamagnetic?

Solution

The noble-gas element preceding both elements is neon. Magnesium has two electrons beyond the neon core, which must be placed in the $3s$ orbital, the next higher in energy, to give the ground-state configuration $[\text{Ne}]3s^2$. A magnesium atom is diamagnetic because all of its electrons are paired in orbitals.

Sulfur has six electrons beyond the neon core; the first two of these are in the $3s$ orbital, and the next four are in the $3p$ orbitals. The ground-state configuration of sulfur is $[\text{Ne}]3s^23p^4$. When four electrons are put into three p orbitals, two electrons must occupy one of the orbitals, and the other two occupy different orbitals to reduce electron–electron repulsion. According to Hund's rules, these two electrons' spins are parallel, and the sulfur atom is paramagnetic.

Related Problems: 17, 18, 19, 20

In summary, we remind you that the electron configuration for an atom is a concise, shorthand notation that represents a great deal of information about the structure and energy levels of the atom. Each configuration corresponds to an atomic wave function comprising a product of occupied Hartree orbitals. Each orbital has a well-defined energy (given by Eq. 5.11 and shown in Fig. 5.18) and average radius (given by Eq. 5.12). The orbitals are grouped into subshells that are characterized by radial distribution functions (see Fig. 5.16). Chapter 6 describes the formation of chemical bonds by starting with the electron configurations of the participating atoms. We encourage you to become expert with atomic electronic configurations and all the information that they summarize.

EXAMPLE 5.7

The boron atom with $Z = 5$ may have the electron configuration $\text{B}:(1s)^2(2s)^2(2p_x)^1$.

(a) Write the atomic wave function for a B atom.

(b) Estimate the energy level diagram for a B atom.

(c) Estimate the radius of the $2s$ and $2p_x$ orbitals.

Solution

(a) The atomic wave function for a B atom is

$$\psi_B(\vec{r}_1, \vec{r}_2, \vec{r}_3, \vec{r}_4, \vec{r}_5) = [\varphi_{1s}(\vec{r}_1)\varphi_{1s}(\vec{r}_2)][\varphi_{2s}(\vec{r}_3)\varphi_{2s}(\vec{r}_4)][\varphi_{2p_x}(\vec{r}_5)]$$

(Continued)

(b) Estimate the energy levels of a B atom.

$$\varepsilon_{1s} \approx -\frac{(4.68)^2}{1^2} = -21.90 \text{ Ry}$$

$$\varepsilon_{2s} \approx -\frac{(2.58)^2}{2^2} = -1.66 \text{ Ry}$$

$$\varepsilon_{2p} \approx -\frac{(2.42)^2}{2^2} = -1.46 \text{ Ry}$$

(c) Use Equation 5.12 and Z_{eff} values for boron from Table 5.3 to estimate orbital radii as follows:

$$\bar{r}_{2s} \approx \frac{2^2 a_0}{(2.58)}\left\{1+\frac{1}{2}\left[1-\frac{0(0+1)}{2^2}\right]\right\} = \frac{4a_0}{(2.58)}\left\{\frac{3}{2}\right\} = 2.33a_0$$

$$\bar{r}_{2p} \approx \frac{2^2 a_0}{(2.42)}\left\{1+\frac{1}{2}\left[1-\frac{1(1+1)}{2^2}\right]\right\} = \frac{4a_0}{(2.42)}\left\{\frac{5}{4}\right\} = 2.07a_0$$

Related Problems: 21, 22, 23, 24, 25, 26

5.3.2 Transition-Metal Elements and Beyond

After the $3p$ orbitals have been filled with six electrons, the natural next step is to continue the build-up process using the $3d$ subshell. Advanced calculations for elements 19 (K) through 30 (Zn) predict that ε_{3d} and ε_{4s} are very close, so the build-up process becomes rather subtle. For K and Ca, the calculations show that $\varepsilon_{4s} < \varepsilon_{3d}$. Optical spectroscopy confirms that the ground state of K is [Ar]$3d^04s^1$ and that of Ca is [Ar]$3d^04s^2$, as predicted by the sequence of calculated orbital energies. For Sc and the elements beyond, advanced calculations predict that $\varepsilon_{3d} < \varepsilon_{4s}$. Filling the $3d$ orbitals first would give the configurations [Ar]$3d^34s^0$ for Sc, [Ar]$3d^44s^0$ for Ti, and so on to [Ar]$3d^{10}4s^0$ for Ni. These configurations are inconsistent with numerous optical, magnetic, and chemical properties of these elements, so some consideration besides the energies of the individual orbitals must also influence the build-up process. Let's consider the alternative configuration [Ar]$3d^14s^2$ for Sc and compare its total energy with that of the [Ar]$3d^34s^0$ configuration. This comparison must add the electrostatic repulsion energy between the electrons to the sum of the energies of the occupied one-electron orbitals to find the total energy of the atom. Because the $3d$ orbital is much more localized than the $4s$ orbital, the much greater repulsion energy of the two electrons in the $3d$ orbital outweighs the fact that $\varepsilon_{3d} < \varepsilon_{4s}$ and the configuration with two d electrons has higher energy. Thus, the configuration [Ar]$3d^14s^2$ has the lower energy and is the ground state for Sc. The same reasoning—minimizing the energy of the atom as a whole—predicts ground state electron configurations from [Ar]$3d^14s^2$ for Sc to [Ar]$3d^{10}4s^2$ for Zn that agree with experimental results. Each of the 10 elements from scandium to zinc is called a ***d*-block element** because its configuration involves the filling of a d orbital in the building-up process.

Experimental evidence shows that chromium and copper do not fit this pattern. In its ground state, chromium has the configuration [Ar]$3d^54s^1$ rather than [Ar]$3d^44s^2$, and copper has the configuration [Ar]$3d^{10}4s^1$ rather than [Ar]$3d^94s^2$. Similar anomalies occur in the fifth period and others, such as the ground-state configuration [Kr]$4d^75s^1$ that is observed for ruthenium in place of the expected [Kr]$4d^65s^2$.

In the sixth period, the filling of the $4f$ orbitals (and the generation of the ***f*-block elements**) begins as the rare-earth (lanthanide) elements from lanthanum

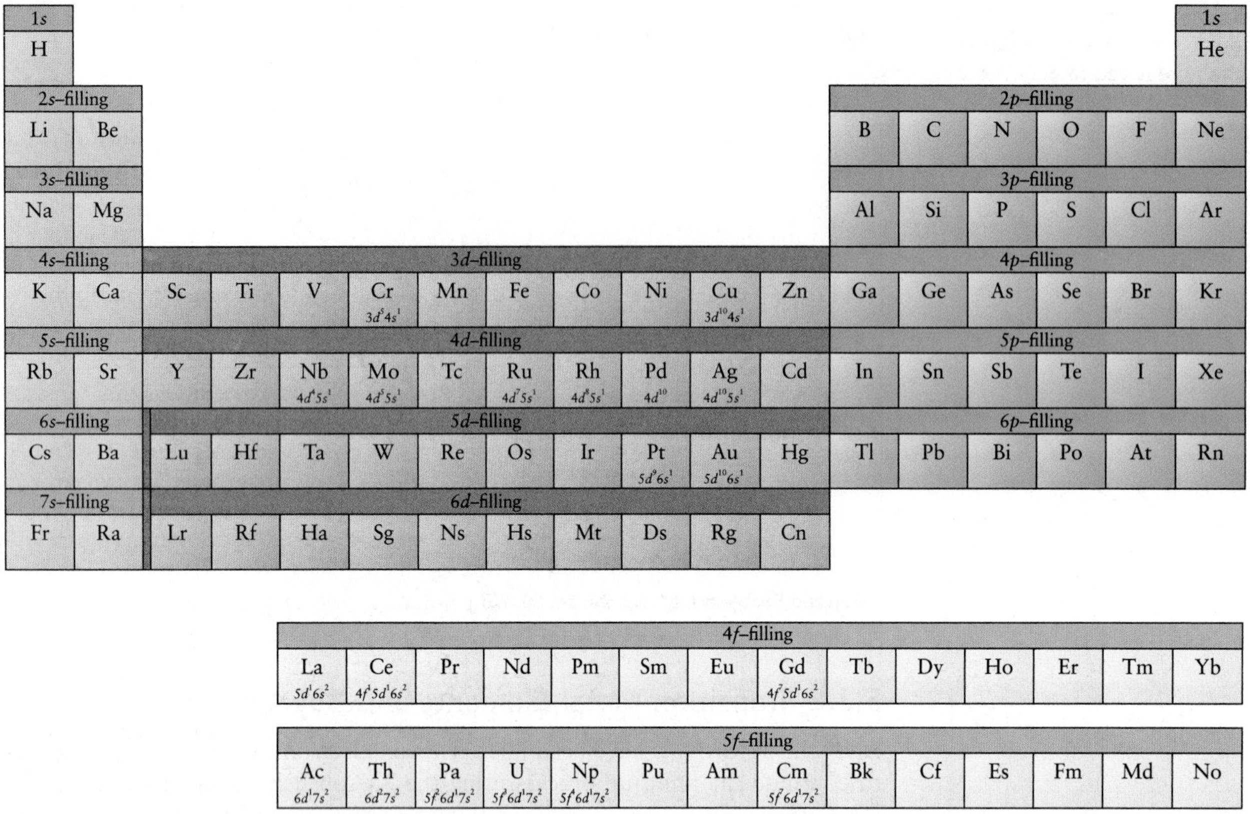

FIGURE 5.21 The filling of shells and the structure of the periodic table. Only the "anomalous" electron configurations are shown.

to ytterbium are reached. The configurations determined from calculations and experiment can be recalled as needed by *assuming* that the orbitals are filled in the sequence $1s \rightarrow 2s \rightarrow 2p \rightarrow 3s \rightarrow 3p \rightarrow 4s \rightarrow 3d \rightarrow 4p \rightarrow 5s \rightarrow 4d \rightarrow 5p \rightarrow 6s \rightarrow 4f \rightarrow 5d \rightarrow 6p \rightarrow 7s \rightarrow 5f \rightarrow 6d$. The energies of the 4*f*, 5*d*, and 6*s* orbitals are comparable over much of the sixth period, and thus their order of filling is erratic.

The periodic table shown in Figure 5.21 classifies elements within periods according to the subshell that is being filled as the atomic number increases. Configurations are given explicitly for exceptions to this "standard" order of filling.

5.4 Shells and the Periodic Table: Photoelectron Spectroscopy

Our discussion of electronic structure began in Section 3.4 by analyzing patterns in the successive ionization energies of the atoms; these patterns suggested that the electrons are arranged in shells within the atom. In Section 5.2, we demonstrated that quantum theory *predicts* the shell structure of the atom. A shell is defined precisely as a set of orbitals that have the same principal quantum number, reflecting the fact that the average positions of the electrons in each of these shells are close to each other, but far from those of orbitals with different n values (see Fig. 5.16). Now we can accurately interpret the results in

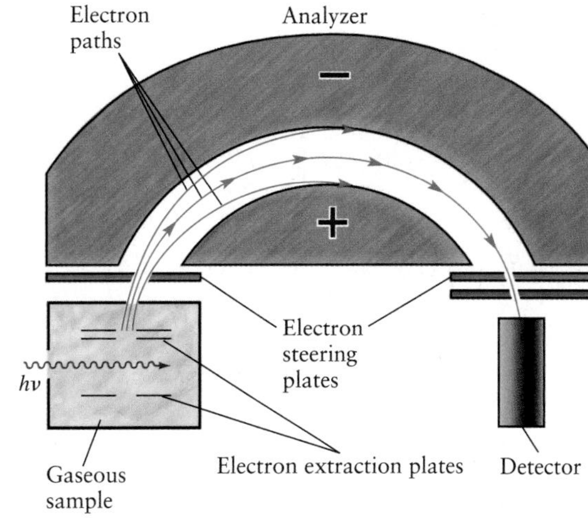

FIGURE 5.22 The energy of photoelectrons is determined by measuring the voltage required to deflect the electrons along a semicircular pathway between two charged metallic hemispherical plates in vacuum so that they arrive at the detector.

Figure 3.12 as showing that Na has two electrons in the $n = 1$ shell, eight in the $n = 2$ shell, and one in the $n = 3$ shell.

The shell structure shows that two elements in the same group (column) of the periodic table have related valence electron configurations. For example, sodium (configuration [Ne]$3s^1$) and potassium (configuration [Ar]$4s^1$) each have a single valence electron in an *s* orbital outside a closed shell; consequently, the two elements closely resemble each other in their chemical properties. A major triumph of quantum mechanics is its ability to account for the periodic trends discovered by chemists many years earlier and organized empirically in the periodic table by Mendeleev and others (see Section 3.2). The ubiquitous octets in the Lewis electron dot diagrams of second- and third-period atoms and ions in Chapter 3 arise from the eight available sites for electrons in the one *s* orbital and three *p* orbitals of the valence shell. The special properties of the transition-metal elements are ascribed to the partial filling of their *d* orbitals (see Chapter 8 for further discussion of this feature).

The shell structure is confirmed directly by an important experimental technique called **photoelectron spectroscopy,** or PES. Photoelectron spectroscopy determines the energy level of each orbital by measuring the ionization energy required to remove each electron from the atom. Photoelectron spectroscopy is simply the photoelectric effect of Section 4.4 applied not to metals, but instead to free atoms. If radiation of sufficiently high frequency ν (in the ultraviolet or X-ray region of the spectrum) strikes an atom, an electron will be ejected with kinetic energy $\frac{1}{2}m_e v^2$. The kinetic energy of the ejected electrons is measured by an *energy analyzer*, which records the voltage required to deflect the electrons around a semicircular pathway in vacuum to reach the detector (Fig. 5.22). As the voltage between the hemispherical plates is changed, electrons with different values of kinetic energy will be deflected to the detector, and the spectrum of kinetic energy values can be recorded. Measuring the kinetic energy by deflection is analogous to measuring energy in the photoelectric effect experiments, and the results are conveniently expressed in units of electron volts (eV). Then the ionization energy spectrum, *IE,* is calculated by the principle of conservation of energy (see Section 4.4),

$$IE = h\nu_{\text{photon}} - \frac{1}{2} m_e v^2_{\text{electron}} \quad [5.13]$$

Figure 5.23 shows the measured photoelectron spectrum for neon excited by X-rays with wavelength 9.890×10^{-10} m, and Example 5.8 shows how the

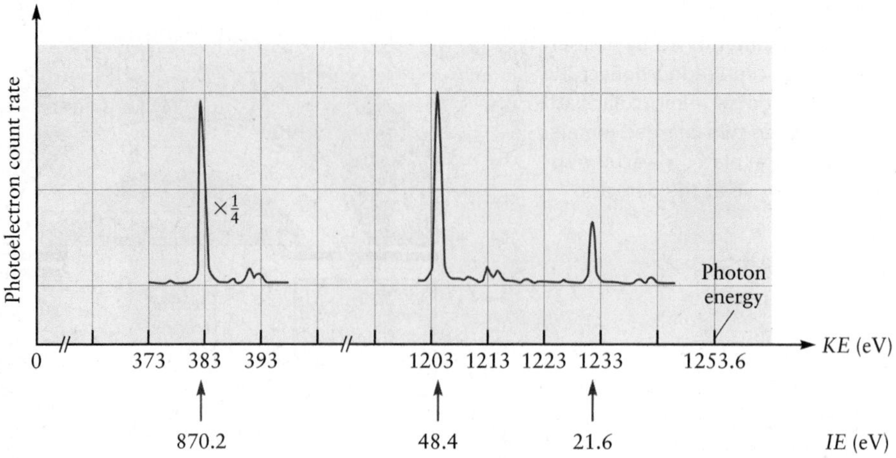

FIGURE 5.23 Photoelectron spectrum of neon. The spectrum shows three peaks, demonstrating that the electrons of neon are organized in three bonding states of distinct energy values. The peak at 383.4 eV has been reduced by a factor of 4 for display on the same scale as the other two.

spectrum is obtained and interpreted. Three peaks appear with kinetic energy values 383.4, 1205.2, and 1232.0 eV. The corresponding ionization energy is shown beneath each peak. (See Example 5.8 for details.) Note that ionization energy increases from right to left in Figure 5.23, opposite to kinetic energy. The peak at lowest ionization energy (highest kinetic energy) is produced by the most weakly bound electrons (see Eq. 5.13). This is the minimum amount of energy required to detach an electron from an atom and is the same as the ionization energy IE_1 introduced in Section 3.4. Clearly, there can be no signal in the photoelectron spectrum at ionization energies less than this value. Peaks at higher ionization energies correspond to electrons removed from more strongly bound states. This spectrum demonstrates that the ten electrons of neon are arranged in bonding states that produce three distinct, discrete energy levels.

These results are connected to the shell model by **Koopmans' approximation,** which asserts in a form suitable for our discussion that the ionization energy of an electron is the negative of the energy of the Hartree orbital of the electron:

$$IE_\alpha = -\varepsilon_\alpha \qquad [5.14]$$

FIGURE 5.24 The ionization energy $IE_{n,\ell}$ is the energy required to remove an electron from this atomic state and set it free from the atom.

The Hartree orbital energies are intrinsically negative because they represent the energy stabilization of an electron bound in an atom relative to the free electron and a positive ion. The ionization energy is positive because it must be supplied to liberate the electron from the atom (Fig. 5.24). Therefore, we should be able to read off the orbital energies directly from the measured spectrum of ionization energies. Koopmans' approximation is not strictly valid because it assumes the orbital energies are the same in the ion as in the parent atom, despite the loss of an electron. This is called the **frozen orbital approximation**. The theorem assumes no energy is lost to *relaxation* of the electronic structure during the ionization process. In fact, relaxation effects are usually no larger than 1 − 3 eV. (The ionization of He is an obvious exception.) They can be included with orbital energies calculated by the more advanced Hartree–Fock method. So Koopmans' approximation and PES provide a quantitative test for advanced theoretical models of electronic structure.

These experimental results for neon are consistent with the electron configuration Ne: $1s^22s^22p^6$ predicted by the aufbau principle. Ionization energies measured in this way are used to construct the energy-level diagram for atoms and to show explicitly the value of the ground-state energy.

EXAMPLE 5.8

Construct the energy-level diagram for neon from the data in Figure 5.23.

Solution

The ionization energy of each level is calculated as $IE = E_{photon} - \mathcal{T}_{electron}$. Because the measured kinetic energy values for the photoelectrons are reported in units of electron volts, the most convenient approach is to calculate the photon energy in electron volts and then subtract the kinetic energy values. The energy of the photon is given by

$$E_{photon} = \frac{hc}{\lambda} = \frac{(6.6261 \times 10^{-34}\ \mathrm{J\ s})(2.9979 \times 10^{8}\ \mathrm{m\ s^{-1}})}{(9.890 \times 10^{-10}\ \mathrm{m})(1.6022 \times 10^{-19}\ \mathrm{J\ eV^{-1}})} = 1253.6\ \mathrm{eV}$$

The calculated ionization energies for the peaks are summarized as follows:

Kinetic Energy (eV)	Ionization Energy (eV)
383.4	870.2
1205.2	48.4
1232.0	21.6

The energy-level diagram (Fig. 5.25) is drawn by showing the negative of each ionization energy value as the energy of an orbital, in accordance with Equation 5.14 and Koopmans' approximation.

Related Problems: 27, 28, 29, 30, 31, 32

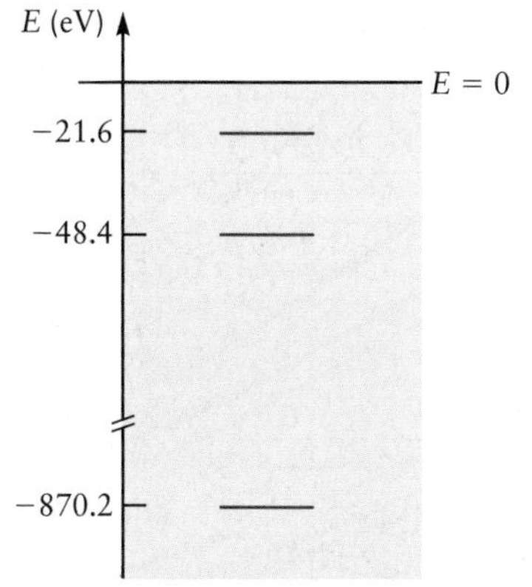

FIGURE 5.25 Energy-level diagram of neon as determined by photoelectron spectroscopy.

This method has been used to determine the energy levels for orbitals in most neutral atoms (Fig. 5.26). The energies are reported in units of rydbergs and plotted on a logarithmic scale. These data confirm the existence of subshells, which are grouped into shells having similar energies. However, there are significant exceptions. The $3d$ subshell for elements 21 through 29 (scandium through copper) lies substantially higher than $3s$ and $3p$ and only slightly lower than $4s$. This is consistent with the chemical observation that the $3d$ electrons are valence electrons in these transition metals. As Z goes above 30, the energy of the $3d$ subshell decreases rapidly, so the $3d$ electrons are not valence electrons for zinc and higher elements. The $4d$, $5d$, $4f$, and $5f$ subshells all behave similarly, so electrons in filled d and f subshells are not valence electrons. We can develop an approximate criterion for distinguishing valence and core electrons by examining the noble gases (elements 2, 10, 18, 36, 54, and 86), which participate poorly or not at all in chemical bonding. The highest-energy subshell for each of them lies below −1 rydberg. Therefore, −1 rydberg is a reasonable approximate boundary for the difference between valence and core electrons.

5.5 Periodic Properties and Electronic Structure

5.5.1 Sizes of Atoms and Ions

The sizes of atoms and ions influence how they interact in chemical compounds. Although atomic radius is not a precisely defined concept, these sizes can be estimated in several ways. If the electron density is known from theory or experiment, a contour surface of fixed electron density can be drawn, as demonstrated in Section 5.1 for one-electron atoms. Alternatively, if the atoms or

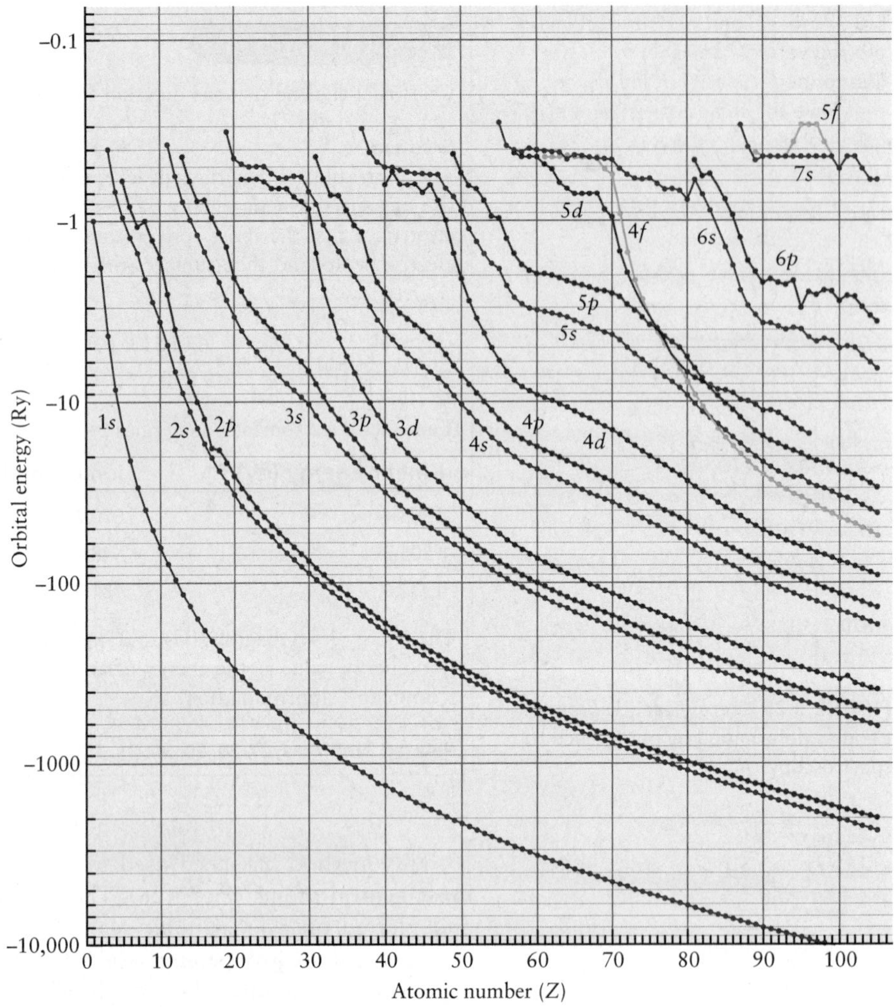

FIGURE 5.26 The energies of different subshells in the first 97 elements, as determined by photoelectron spectroscopy. Negative values on the vertical axis correspond to the bound state orbital energies. Subshells having the same principal quantum number *n*, such as 2*s* and 2*p*, have similar energies and are well separated from orbitals of different *n*. Significant exceptions do exist, as explained in the text. Note the logarithmic energy scale. One rydberg is 2.18×10^{-18} J.

ions in a crystal are assumed to be in contact with one another, a size can be defined from the measured distances between their centers (this approach is explored in greater detail in Chapter 21). These and other measures of size are reasonably consistent with each other and allow for the tabulation of sets of atomic and ionic radii, many of which are listed in Appendix F.

Certain systematic trends appear in these radii. For a series of elements or ions in the same group (column) of the periodic table, the radius usually increases with increasing atomic number. This occurs mainly because the Pauli exclusion principle effectively excludes added electrons from the region occupied by the core electrons, thus forcing an increase in size as more distant electron shells are occupied. By contrast, Coulomb (electrostatic) forces cause the radii of atoms to *decrease* with increasing atomic number across a period. As the nuclear charge increases steadily, electrons are added to the same valence shell and are ineffective in shielding each other from its attraction. This "incomplete shielding" of the added proton by the added electron as we go from atomic number Z to $Z + 1$ leads to an increase in Z_{eff} across a period.

Superimposed on these broad trends are some subtler effects that have significant consequences in chemistry. One dramatic example is shown in Figure 5.27. The radii of several sets of ions and atoms increase with atomic number in a given group (see earlier), but the *rate* of this increase changes considerably when the ions and atoms that contain the same number of electrons as argon are reached (S^{2-}, Cl^-, Ar, K^+, Ca^{2+}, Sc^{3+}, Ti^{4+}). For example, the change in size from Li^+ to Na^+ to K^+ is substantial, but the subsequent changes, to Rb^+ and Cs^+, are

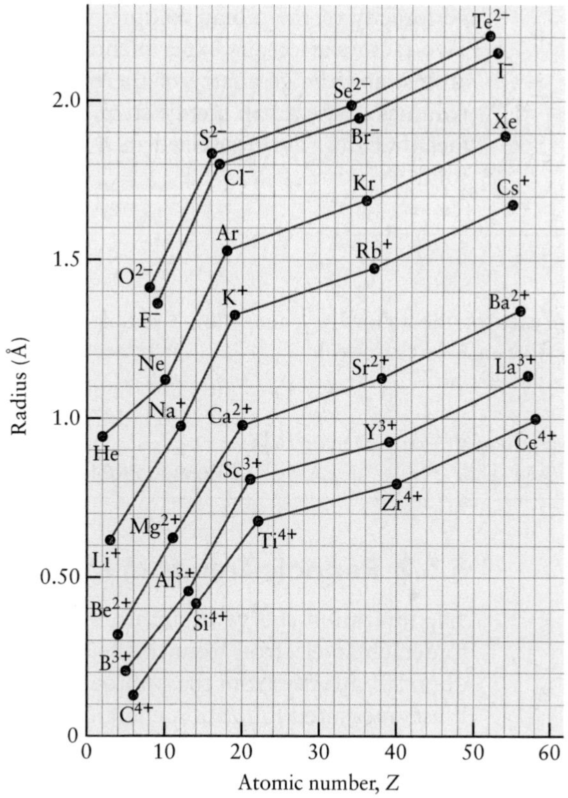

FIGURE 5.27 Ionic and atomic radii plotted versus atomic number. Each line connects a set of atoms or ions that have the same charge; all species have noble-gas configurations.

significantly smaller due to the filling of the *d* orbitals, which begins after K^+ is reached. Because atomic and ionic size decrease from left to right across a series of transition-metal elements (due to the increased effective nuclear charge), the radius of a main-group element is smaller than it would have been had the transition series not intervened. A similar phenomenon, called the **lanthanide contraction,** occurs during the filling of the 4*f* orbitals in the lanthanide series. Its effect on the sizes of transition-metal atoms is discussed in Section 8.1.

A different measure of atomic size is the volume occupied by a mole of atoms of the element in the solid phase. Figure 5.28 shows the pronounced

H 11.4																	He 21.0
Li 13.0	Be 4.85											B 4.39	C 3.42	N 13.5	O 17.4	F 11.2	Ne 13.2
Na 23.8	Mg 14.0											Al 10.0	Si 12.1	P 17.0	S 15.5	Cl 17.4	Ar 22.6
K 45.9	Ca 26.2	Sc 15.0	Ti 10.6	V 8.32	Cr 7.23	Mn 7.35	Fe 7.09	Co 6.67	Ni 6.59	Cu 7.11	Zn 9.16	Ga 11.8	Ge 13.6	As 13.0	Se 16.4	Br 19.8	Kr 28.0
Rb 55.8	Sr 33.9	Y 19.9	Zr 14.0	Nb 10.8	Mo 9.38	Tc 8.63	Ru 8.17	Rh 8.28	Pd 8.56	Ag 10.3	Cd 13.0	In 15.8	Sn 16.3	Sb 18.2	Te 20.5	I 25.7	Xe 35.9
Cs 70.9	Ba 38.2	Lu 17.8	Hf 13.4	Ta 10.9	W 9.47	Re 8.86	Os 8.42	Ir 8.52	Pt 9.09	Au 10.2	Hg 14.1	Tl 17.2	Pb 18.3	Bi 21.3	Po 23.0		Rn 50.5

FIGURE 5.28 The molar volumes (measured in $cm^3\ mol^{-1}$ of atoms) of some elements in their solid states. Note the large values for the alkali metals.

periodicity of the molar volume, with maxima occurring for the alkali metals. Two factors affect the experimentally measured molar volume: the "size" of the atoms, and the geometry of the bonding that connects them. The large molar volumes of the alkali metals stem both from the large size of the atoms and the fact that they are organized in a rather open, loosely packed structure in the solid.

EXAMPLE 5.9

Predict which atom or ion in each of the following pairs should be larger: (a) Kr or Rb, (b) Y or Cd, (c) F^- or Br^-.

Solution

(a) Rb should be larger because it has an extra electron in a 5*s* orbital beyond the Kr closed shell.

(b) Y should be larger because the effective nuclear charge increases through the transition series from Y to Cd.

(c) Br^- should be larger because the extra outer electrons are excluded from the core.

Related Problems: 33, 34, 35, 36

5.5.2 Periodic Trends in Ionization Energies

The ionization energy of an atom is defined as the minimum energy necessary to detach an electron from the neutral gaseous atom (see Section 3.4). It can be obtained directly from the photoelectron spectrum of an atomic gas. Appendix F lists measured ionization energies of the elements, and Figure 5.29 shows the periodic trends in first and second ionization energies with increasing atomic number.

Let's use insight from quantum mechanics to examine the periodic trends in the first ionization energy. We obtain deeper understanding of the stabilities of the various electron configurations using this approach than we did in our empirical discovery of shell structure in Section 3.4. There is a large reduction in IE_1 from helium to lithium for two reasons: (1) a 2*s* electron is much farther from the nucleus than a 1*s* electron, and (2) the 1*s* electrons screen the nucleus

FIGURE 5.29 First and second ionization energies of atoms of the first three periods.

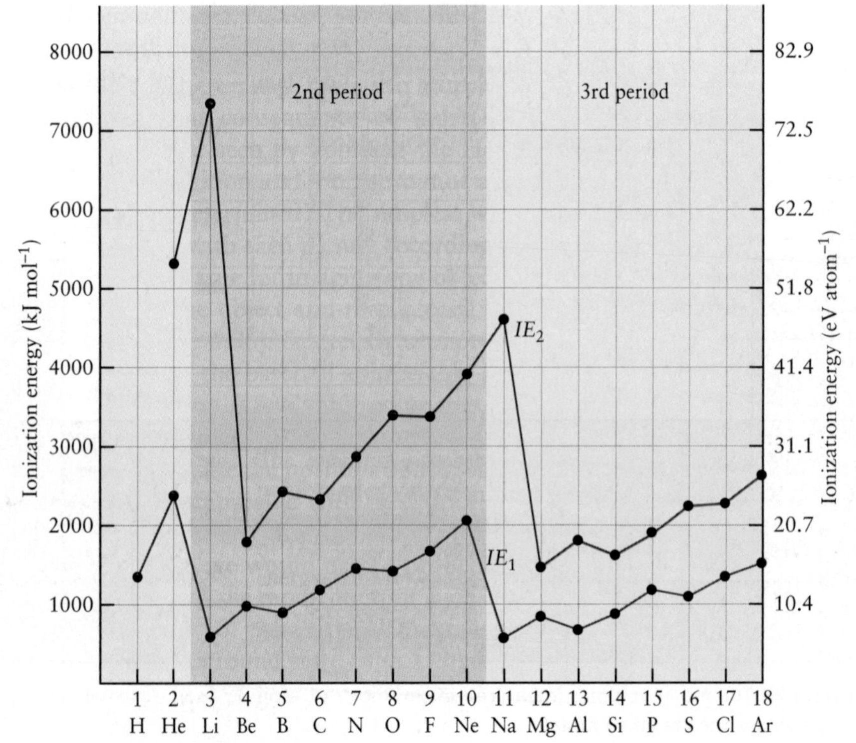

in lithium so effectively that the 2*s* electron "sees" a net positive charge close to +1, rather than the larger charge seen by the electrons in helium. Beryllium shows an increase in IE_1 compared with lithium because the effective nuclear charge has increased, but the electron being removed is still from a 2*s* orbital.

The IE_1 of boron is somewhat less than that of beryllium because the fifth electron is in a higher energy (and therefore less stable) 2*p* orbital. In carbon and nitrogen, the additional electrons go into 2*p* orbitals as the effective nuclear charge increases to hold the outer electrons more tightly; hence, IE_1 increases. The nuclear charge is higher in oxygen than in nitrogen, which would give it a higher ionization energy if this were the only consideration. However, oxygen must accommodate two electrons in the same 2*p* orbital, leading to greater electron–electron repulsion and diminished binding, thus more than compensating for the increased electron-nuclear interaction. Consequently, oxygen has a lower IE_1 than nitrogen. Fluorine and neon have successively higher first ionization energies because of increasing effective nuclear charge. The general trends of increasing ionization energy across a given period, as well as the dips that occur at certain points, can thus be rationalized through the orbital description of many-electron atoms.

The ionization energy tends to decrease down a group in the periodic table (for example, from lithium to sodium to potassium). As the principal quantum number increases, so does the distance of the outer electrons from the nucleus. There are some exceptions to this trend, however, especially for the heavier elements in the middle of the periodic table. For example, the first ionization energy of gold is higher than that of silver or copper. This fact is crucial in making gold a "noble metal"; that is, one that is resistant to attack by oxygen.

Similar trends are observed in the second ionization energies, but they are shifted higher in atomic number by one unit (see Fig. 5.29). Thus, IE_2 is large for lithium (because Li^+ has a filled $1s^2$ shell), but relatively small for beryllium (because Be^+ has a single electron in the outermost 2*s* orbital).

5.5.3 Electron Affinity

The **electron affinity,** *EA*, of an atom X is the energy required to detach an electron from the anion X^- and give the neutral atom X (see Section 3.5). Appendix F lists the electron affinities of the elements.

The periodic trends in electron affinity (Fig. 5.30) parallel those in ionization energy for the most part, except that they are shifted one unit *lower* in atomic

H 73																	He *
Li 60	Be *											B 27	C 122	N *	O 141	F 328	Ne *
Na 53	Mg *											Al 43	Si 134	P 72	S 200	Cl 349	Ar *
K 48	Ca 2	Sc 18	Ti 8	V 51	Cr 64	Mn *	Fe 16	Co 64	Ni 111	Cu 118	Zn *	Ga 29	Ge 116	As 78	Se 195	Br 325	Kr *
Rb 47	Sr 5	Y 30	Zr 41	Nb 86	Mo 72	Tc 53	Ru 99	Rh 110	Pd 52	Ag 126	Cd *	In 29	Sn 116	Sb 103	Te 190	I 295	Xe *
Cs 46	Ba 14	Lu 50	Hf *	Ta 31	W 79	Re 14	Os 106	Ir 151	Pt 214	Au 223	Hg *	Tl 19	Pb 35	Bi 91	Po 183	At 270	Rn *

FIGURE 5.30 Electron affinities (measured in kJ mol^{-1}) of gaseous atoms of the elements. An asterisk means that the element does not have a stable anion in the gas phase.

number. The reason is clear. Attaching an electron to F gives F^-, with the configuration $1s^22s^22p^6$, the same as that for neon. Fluorine has a large affinity for electrons because the added electron fills the $n = 2$ shell, rather than being forced into the higher energy 3*s* orbital. In contrast, the noble gases do not have well-defined electron affinities because the "extra" electron would reside in a new shell far from the nucleus and be almost totally screened from the nuclear charge.

EXAMPLE 5.10

Consider the elements selenium (Se) and bromine (Br). Which has the higher first ionization energy? Which has the higher electron affinity?

Solution

These two atoms are adjacent to each other in the periodic table. Bromine has one more electron in the 4*p* subshell, and this electron should be more tightly bound than the 4*p* electrons in selenium because of the incomplete shielding and the extra unit of positive charge on the nucleus. Thus, IE_1 should be greater for bromine.

Bromine has a greater electron affinity than selenium. Gaining the extra electron changes the Br atom into Br^-, which has a particularly stable closed-shell electron configuration (the same as that of the noble gas atom krypton [Kr]). No such configuration is created when a Se atom gains an additional electron.

Related Problems: 37, 38, 39, 40

CHAPTER SUMMARY

The planetary model of the atom provided the first physical picture of the structure of the atom. This model, however, is rooted in classical mechanics, a theory that failed to predict the results of experiments conducted in the period roughly 1890 to 1915, an era of great excitement in the development of modern science. For example, the Bohr model was only able to predict the hydrogen atom emission spectra discovered by Balmer in 1885, not the emission spectra of multi-electron atoms. The theoretical explanation of the structure, properties, and behavior of atoms requires quantum mechanics. Quantum mechanics tells us that an atom can have only specific, discrete amounts of energy. Indeed, the very existence of quantum states explains the stability of the atom, which was predicted to collapse according to classical physics. Quantum mechanics also demonstrates that the concept of planetary orbits is simply not applicable on the atomic scale. We cannot know the detailed trajectory of an electron in the intuitive, classical sense familiar to ordinary human perception. Instead, we describe the probability for finding the electron at a particular location in the atom, based on knowing the quantum state of the atom.

For the hydrogen atom, we can solve the Schrödinger equation exactly to obtain the allowed energy levels and the hydrogen atomic orbitals. The sizes and shapes of these orbitals tell us the probability distribution for the electron in each quantum state of the atom. We are led to picture this distribution as a smeared cloud of electron density (probability density) with a shape that is determined by the quantum state.

For all other atoms, we have to generate approximations to solve the Schrödinger equation. The Hartree orbitals describe approximately the spatial distribution for each electron in the atom, moving under an effective force obtained by averaging over the interactions with all the other electrons. The Hartree orbitals have shapes similar to those of the hydrogen atomic orbitals—but very different sizes and energy values—and thus guide us to view the probability distribution for each electron as a smeared cloud of electron density.

The Hartree orbitals are the foundation of the quantum explanation of atomic structure. They justify the shell model of the atom, they explain the structure of the periodic table, and they provide the starting point for the quantum explanation of chemical bond formation in the following chapter.

CONCEPTS AND SKILLS

Section 5.1 – The Hydrogen Atom

Define the quantum numbers that characterize one-electron atoms; discuss the shapes, sizes, and nodal properties of the corresponding orbitals; and calculate the probability of finding the electron in different regions of space around the nucleus. (Problems 1–8).

- Quantum mechanics explains the physical stability of the atom by predicting its allowed discrete energy levels and defining the wave functions (also called atomic orbitals) associated with each energy level. The orbitals determine the probability density for finding the electrons at particular locations in the atom when the electrons are in a specific quantum state.
- The allowed energy levels for a one-electron atom or ion with atomic number Z are given by the expression

$$E_n = -\frac{Z^2 e^4 m_e}{8\epsilon_0^2 n^2 h^2} \qquad n = 1, 2, 3, \ldots \qquad [5.1a]$$

  These values are negative numbers because they measure the energy of the bound states of the stable atom relative to a separated electron and cation, which is defined to be the zero of the energy scale. The energy of the ground state of the hydrogen atom is -13.6 eV, a number worth remembering.
- Because the potential energy in a one-electron atom depends only on r, the wave functions (atomic orbitals) have the product form $\psi_{n\ell m}(r, \theta, \phi) = R_{n\ell}(r)Y_{\ell m}(\theta, \phi)$. The quantum number ℓ describes quantization of the total angular momentum of the electron, and the quantum number m describes quantization of the component of angular momentum along the z-axis. Whereas the quantum number n determines the allowed energy levels, the quantum number ℓ determines the shapes of the orbitals, and m determines their orientation relative to the z-axis.
- The probability for finding an electron in the volume element $dV = r^2 \sin\theta \, dr d\theta d\phi$ when the electron is in the specific quantum state (n, ℓ, m) is given by

$$(\psi_{n\ell m})^2 dV = [R_{n\ell}(r)]^2 [Y_{\ell m}(\theta, \phi)]^2 dV \qquad [5.4]$$

  Examples 5.2 and 5.3 show how to use this equation for numerical calculations.
- Describe the shape, orientation, and size of the H atomic orbitals. Relate these properties to the following equation for the dependence of the size and shape on the nuclear charge:

$$\bar{r}_{n\ell} = \frac{n^2 a_0}{Z}\left\{1 + \frac{1}{2}\left[1 - \frac{\ell(\ell+1)}{n^2}\right]\right\} \qquad [5.9]$$

- The Stern–Gerlach experiment demonstrates that the electron has a property called *spin*, which leads to a magnetic dipole moment. Spin is quantized with only two allowed values described by the quantum number m_s. Complete determination of the quantum state of the electron requires values for all four quantum numbers (n, ℓ, m, m_s).

Section 5.2 – Shell Model for Many-Electron Atoms

Prepare an approximate energy-level diagram for an atom using values for Z_{eff} (Problems 11–16).

- Atoms with many electrons are described by Hartree's SCF method, in which each electron is assumed to move under the influence of an effective potential $V_{eff}(r)$ due to the spatial distribution of all the other electrons. This method generates a set of one-electron wave functions called the Hartree orbitals $\varphi_\alpha(r)$ with energy values ε_α, where α represents the assumed set of one-electron quantum numbers. Hartree orbitals bear close relation to the hydrogen atomic orbitals but are not the same functions.

$$V_n^{\text{eff}}(r) = -\frac{Z_{\text{eff}}(n)e^2}{r} \qquad [5.10]$$

- Energies of the Hartree orbitals are different from those of the corresponding hydrogen atomic orbitals. For an atom with atomic number Z they can be estimated as

$$\varepsilon_n \approx -\frac{[Z_{\text{eff}}(n)]^2}{n^2}\text{(rydbergs)} \qquad [5.11]$$

  where the effective nuclear charge experienced by each electron is determined by screening of that electron from the full nuclear charge by other electrons.

- The Hartree orbitals have the same shapes as the corresponding hydrogen atomic orbitals, but their sizes are quite different, as described by the expression

$$\bar{r}_{n\ell} \approx \frac{n^2 a_0}{Z_{\text{eff}}(n)}\left\{1+\frac{1}{2}\left[1-\frac{\ell(\ell+1)}{n^2}\right]\right\} \qquad [5.12]$$

Section 5.3 – Aufbau Principle and Electron Configurations

Use the aufbau principle to predict electron configurations of atoms and ions and to account for the structure of the periodic table (Problems 17–26).

- The electron configuration for an atom with atomic number Z is determined by arranging the Hartree orbitals in order of increasing energy, then placing at most two electrons in each orbital in accordance with the Pauli exclusion principle and Hund's rule until all Z electrons have been placed. The configuration consists of specifying the set of four quantum numbers (n, ℓ, m, m_s) for each electron in the atom.
- The Hartree orbitals and their electron configurations justify the shell model of the atom; that is, the electrons are grouped into shells of 2, 8, or 18 electrons arranged concentrically around the nucleus at increasing distances from the nucleus. As we move outward from the nucleus, the electrons in each successive shell are bound progressively less strongly to the nucleus. Electrons in the outermost shell, called the valence electrons, are the least strongly bound, and they participate in the formation of chemical bonds.
- The Hartree orbitals and their electron configurations explain the structure of the periodic table.

Section 5.4 – Shells and the Periodic Table: Photoelectron Spectroscopy

Construct the energy-level diagram for an atom using PES for orbital energies (Problems 27–32).

- Ionization energy is the amount of energy required to remove an electron from an atom and place it infinitely far away with zero kinetic energy. For any state n of a one-electron atom, the ionization energy IE_n is given by the relation $IE_n = E_\infty - E_n$. Ionization energy is intrinsically positive because the energy of the final state is higher than the energy of the initial state. The ionization energy of the ground state of the hydrogen atom is +13.6 eV.
- The shell model is verified experimentally by the technique of PES, in which ionization energy is measured for electrons in each shell:

$$IE = h\nu_{\text{photon}} - \frac{1}{2} m_e v^2_{\text{electron}} \qquad [5.13]$$

  The results are connected to the shell model by Koopmans' theorem, which asserts that the orbital energy is the negative of the ionization energy, $IE_\alpha = -\varepsilon_\alpha$.

Section 5.5 – Periodic Properties and Electronic Structure

Describe the trends in ionization energy and electron affinity across the periodic table and relate them to the electronic structure of atoms (Problems 35–40).

- The Hartree orbitals and the shell model explain periodic trends in ionization energy, electron affinity, and the radii of atoms and ions. Small changes in these properties within a period are further explained by detailed changes in Z_{eff} within that period.

CUMULATIVE EXERCISE

Atoms in Interstellar Space

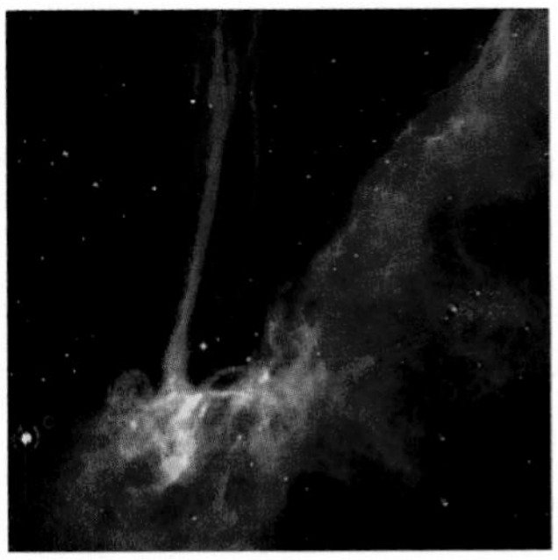
J. J. Hester (Arizona State University), and NASA

FIGURE 5.31 Clouds of gas surround hot stars in these galactic clusters. The red color arises from hydrogen radiation.

The vast stretches of space between the stars are by no means empty. They contain both gases and dust particles at very low concentrations (Fig. 5.31). Interstellar space extends so far that these low-density species significantly affect the electromagnetic radiation arriving from distant stars and other sources, which is detected by telescopes. The gas in interstellar space consists primarily of hydrogen atoms (either neutral or ionized) at a concentration of about one atom per cubic centimeter. The dust (thought to be mostly solid water, methane, or ammonia) is even less concentrated, with typically only a few dust particles (each 10^{-4} to 10^{-5} cm in radius) per cubic kilometer.

(a) The hydrogen in interstellar space near a star is largely ionized by the high-energy photons from the star. Such regions are called H II regions. Suppose a ground-state hydrogen atom absorbs a photon with a wavelength of 65 nm. Calculate the kinetic energy of the electron ejected. (**Note:** This is the gas-phase analog of the photoelectric effect for solids.)

(b) What is the de Broglie wavelength of the electron from part (a)?

(c) Free electrons in H II regions can be recaptured by hydrogen nuclei. In such an event, the atom emits a series of photons of increasing energy as the electrons cascade down through the quantum states of the hydrogen atom. The particle densities are so low that extremely high quantum states can be detected in interstellar space. In particular, the transition from the state

$n = 110$ to $n = 109$ for the hydrogen atom has been detected. What is the Bohr radius of an electron for hydrogen in the state $n = 110$?

(d) Calculate the wavelength of light emitted as an electron undergoes a transition from level $n = 110$ to $n = 109$. In what region of the electromagnetic spectrum does this lie?

(e) H II regions also contain ionized atoms that are heavier than hydrogen. Calculate the longest wavelength of light that will ionize a ground-state helium atom. Use data from Appendix F.

(f) The regions farther from stars are called H I regions. There, almost all of the hydrogen atoms are neutral rather than ionized and are in the ground state. Will such hydrogen atoms absorb light in the Balmer series emitted by atoms in H II regions?

(g) We stated in Section 5.1 that the energy of the hydrogen atom depends only on the quantum number n. In fact, this is not quite true. The electron spin (m_s quantum number) couples weakly with the spin of the nucleus, making the ground state split into two states of almost equal energy. The radiation emitted in a transition from the upper to the lower of these levels has a wavelength of 21.2 cm and is of great importance in astronomy because it allows the H I regions to be studied. What is the energy difference between these two levels, both for a single atom and for a mole of atoms?

(h) The gas and dust particles between a star and the earth scatter the star's light more strongly in the blue region of the spectrum than in the red. As a result, stars appear slightly redder than they actually are. Will an estimate of the temperature of a star based on its apparent color give too high or too low a number?

Answers

(a) 8.8×10^{-19} J

(b) 0.52 nm = 5.2 Å

(c) 6.40×10^{-7} m = 6400 Å

(d) 5.98×10^{-2} m = 5.98 cm, in the microwave region

(e) 50.4 nm

(f) No, because the lowest energy absorption for a ground-state hydrogen atom is in the ultraviolet region of the spectrum.

(g) 9.37×10^{-25} J; 0.564 J mol^{-1}

(h) Too low, because red corresponds to emitted light of lower energy. Experience with blackbody radiation curves would assign a lower temperature to a star emitting lower energy light.

PROBLEMS

Answers to problems whose numbers are red appear in Appendix G. Problems that are more challenging are indicated with asterisks.

5.1 The Hydrogen Atom

1. Which of the following combinations of quantum numbers are allowed for an electron in a one-electron atom? Which are not?
   (a) $n = 2, \ell = 2, m = 1, m_s = \frac{1}{2}$
   (b) $n = 3, \ell = 1, m = 0, m_s = -\frac{1}{2}$
   (c) $n = 5, \ell = 1, m = 2, m_s = \frac{1}{2}$
   (d) $n = 4, \ell = -1, m = 0, m_s = \frac{1}{2}$

2. Which of the following combinations of quantum numbers are allowed for an electron in a one-electron atom? Which are not?
   (a) $n = 3, \ell = 2, m = 1, m_s = 0$
   (b) $n = 2, \ell = 0, m = 0, m_s = -\frac{1}{2}$
   (c) $n = 7, \ell = 2, m = -2, m_s = \frac{1}{2}$
   (d) $n = 3, \ell = -3, m_s = -\frac{1}{2}$

3. Label the orbitals described by each of the following sets of quantum numbers:
   (a) $n = 4, \ell = 1$
   (b) $n = 2, \ell = 0$
   (c) $n = 6, \ell = 3$

4. Label the orbitals described by each of the following sets of quantum numbers:
   (a) $n = 3, \ell = 2$
   (b) $n = 7, \ell = 4$
   (c) $n = 5, \ell = 1$

5. Estimate the probability of finding an electron which is excited into the 2*s* orbital of the H atom, looking in a cubical box of volume 0.75×10^{-36} m^3 centered at the nucleus. Then estimate the probability of finding the electron if you move the volume searched to a distance of 105.8 pm from the nucleus in the positive *z* direction. (Note that since these volumes are small, it does not matter whether the volume searched is cubical or spherical.)

6. Using Table 5.2, write down the mathematical expression for the $2p_x$ wave function for an electronically excited H atom. Estimate the probability of finding the $2p_x$ electron if you look in a cubical box of volume of 0.8 (pm)^3 centered at a distance of 0.500×10^{-10} m in the $\theta = \pi/2$, $\phi = 0$ direction. Does this probability change as you change ϕ? At what ϕ angles is the probability of finding the electron smallest and at what ϕ angles is the probability the largest? (Note that $\phi = 2\pi$ is the same location as $\phi = 0$, so don't double count.)

7. How many radial nodes and how many angular nodes does each of the orbitals in Problem 3 have?

8. How many radial nodes and how many angular nodes does each of the orbitals in Problem 4 have?

9. Use the mathematical expression for the $2p_z$ wave function of a one-electron atom (see Table 5.2) to show that the probability of finding an electron in that orbital anywhere in the *x-y* plane is 0. What are the nodal planes for a d_{xz} orbital and for a $d_{x^2-y^2}$ orbital?

10. **(a)** Use the radial wave function for the 3*p* orbital of a hydrogen atom (see Table 5.2) to calculate the value of *r* for which a node exists.
    (b) Find the values of *r* for which nodes exist for the 3*s* wave function of the hydrogen atom.

5.2 Shell Model for Many-Electron Atoms

11. Calculate the average distance of the electron from the nucleus in a hydrogen atom when the electron is in the 2*s* orbital. Repeat the calculation for an electron in the 2*p* orbital.

12. The helium ion He^+ is a one-electron system whose wave functions and energy levels are obtained from those for H by changing the atomic number to $Z = 2$. Calculate the average distance of the electron from the nucleus in the 2*s* orbital and in the 2*p* orbital. Compare your results with those in Problem 11 and explain the difference.

13. Spectroscopic studies show that Li can have electrons in its 1*s*, 2*s*, and 2*p* Hartree orbitals, and that $Z_{eff}(2s) = 1.28$ for the $1s^2 2s^1$ configuration. Estimate the energy of the 2*s* orbital of Li. Calculate the average distance of the electron from the nucleus in the 2*s* orbital of Li.

14. Spectroscopic studies of Li also show that $Z_{eff}(2p) = 1.02$. Estimate the energy of the 2*p* orbital of Li. Calculate the average distance of the electron from the nucleus in the 2*p* orbital of Li. Comparing your results with those in Problem 13 shows that the energy values differ by about 50%, whereas the average distances are nearly equal. Explain this observation.

15. Spectroscopic studies show that Na can have electrons in its 1*s*, 2*s*, 2*p*, and 3*s* Hartree orbitals, and that $Z_{eff}(3s) = 1.84$. Using data from Problem 13, compare the energies of the Na 3*s* orbital, the Li 2*s* orbital, and the H 1*s* orbital.

16. Using data from Problems 13 and 15, calculate the average distance of the electron from the nucleus in the Na 3*s* orbital, the Li 2*s* orbital, and the H 1*s* orbital. Explain the trend in your results.

5.3 Aufbau Principle and Electron Configurations

17. Give the ground-state electron configurations of the following elements:
    (a) C **(b)** Se **(c)** Fe

18. Give the ground-state electron configurations of the following elements:
    (a) P **(b)** Tc **(c)** Ho

19. Write ground-state electron configurations for the ions Be^+, C^-, Ne^{2+}, Mg^+, P^{2+}, Cl^-, As^+, and I^-. Which do you expect will be paramagnetic due to the presence of unpaired electrons?

20. Write ground-state electron configurations for the ions Li^-, B^+, F^-, Al^{3+}, S^-, Ar^+, Br^+, and Te^-. Which do you expect to be paramagnetic due to the presence of unpaired electrons?

21. Identify the atom or ion corresponding to each of the following descriptions:
    (a) an atom with ground-state electron configuration $[Kr]4d^{10}5s^2 5p^1$
    (b) an ion with charge -2 and ground-state electron configuration $[Ne]3s^2 3p^6$
    (c) an ion with charge $+4$ and ground-state electron configuration $[Ar]3d^3$

22. Identify the atom or ion corresponding to each of the following descriptions:
    (a) an atom with ground-state electron configuration $[Xe]4f^{14}5d^6 6s^2$
    (b) an ion with charge -1 and ground-state electron configuration $[He]2s^2 2p^6$
    (c) an ion with charge $+5$ and ground-state electron configuration $[Kr]4d^6$

23. Predict the atomic number of the (as yet undiscovered) element in the seventh period that is a halogen.

24. **(a)** Predict the atomic number of the (as yet undiscovered) alkali-metal element in the eighth period.
(b) Suppose the eighth-period alkali-metal atom turned out to have atomic number 137. What explanation would you give for such a high atomic number (recall that the atomic number of francium is only 87)?

25. Suppose that the spin quantum number did not exist, and therefore only one electron could occupy each orbital of a many-electron atom. Give the atomic numbers of the first three noble-gas atoms in this case.

26. Suppose that the spin quantum number had three allowed values $\left(m_s = 0, +\frac{1}{2}, -\frac{1}{2}\right)$. Give the atomic numbers of the first three noble-gas atoms in this case.

5.4 Shells and the Periodic Table: Photoelectron Spectroscopy

27. Photoelectron spectra of mercury (Hg) atoms acquired with radiation from a helium lamp at 584.4 Å show a peak in which the photoelectrons have kinetic energy of 11.7 eV. Calculate the ionization energy of electrons in that level.

28. Quantum mechanics predicts that the energy of the ground state of the H atom is −13.6 eV. Insight into the magnitude of this quantity is gained by considering several methods by which it can be measured.
(a) Calculate the longest wavelength of light that will ionize H atoms in their ground state.
(b) Assume the atom is ionized by collision with an electron that transfers all its kinetic energy to the atom in the ionization process. Calculate the speed of the electron before the collision. Express your answer in meters per second (m s^{-1}) and miles per hour (miles h^{-1}).
(c) Calculate the temperature required to ionize a H atom in its ground state by thermal excitation. (*Hint:* Recall the criterion for thermal excitation of an oscillator in Planck's theory of blackbody radiation is that $h\nu \approx k_BT$.)

29. Photoelectron spectroscopy studies of sodium atoms excited by X-rays with wavelength 9.890×10^{-10} m show four peaks in which the electrons have speeds 7.992×10^6 m s^{-1}, 2.046×10^7 m s^{-1}, 2.074×10^7 m s^{-1}, and 2.009×10^7 m s^{-1}. (Recall that 1 J = 1 kg m^2 s^{-2}.)
(a) Calculate the ionization energy of the electrons in each peak.
(b) Assign each peak to an orbital of the sodium atom.

30. Photoelectron spectroscopy studies of silicon atoms excited by X-rays with wavelength 9.890×10^{-10} m show four peaks in which the electrons have speeds 2.097×10^7 m s^{-1}, 2.093×10^7 m s^{-1}, 2.014×10^7 m s^{-1}, and 1.971×10^7 m s^{-1}. (Recall that 1 J = 1 kg m^2 s^{-2}.)
(a) Calculate the ionization energy of the electrons in each peak.
(b) Assign each peak to an orbital of the silicon atom.

31. Photoelectron spectroscopy studies have determined the orbital energies for fluorine atoms to be

1*s*	−689 eV
2*s*	−34 eV
2*p*	−12 eV

Estimate the value of Z_{eff} for F in each of these orbitals.

32. Photoelectron spectroscopy studies have determined the orbital energies for chlorine atoms to be

1*s*	−2,835 eV
2*s*	−273 eV
2*p*	−205 eV
3*s*	−21 eV
3*p*	−10 eV

Estimate the value of Z_{eff} for Cl in each of these orbitals.

5.5 Periodic Properties and Electronic Structure

33. For each of the following pairs of atoms or ions, state which you expect to have the larger radius.
(a) Na or K **(b)** Cs or Cs^+
(c) Rb^+ or Kr **(d)** K or Ca
(e) Cl^- or Ar

34. For each of the following pairs of atoms or ions, state which you expect to have the larger radius.
(a) Sm or Sm^{3+} **(b)** Mg or Ca
(c) I^- or Xe **(d)** Ge or As
(e) Sr^+ or Rb

35. Predict the larger ion in each of the following pairs. Give reasons for your answers.
(a) O^-, S^{2-} **(b)** Co^{2+}, Ti^{2+}
(c) Mn^{2+}, Mn^{4+} **(d)** Ca^{2+}, Sr^{2+}

36. Predict the larger ion in each of the following pairs. Give reasons for your answers.
(a) S^{2-}, Cl^- **(b)** Tl^+, Tl^{3+}
(c) Ce^{3+}, Dy^{3+} **(d)** S^-, I^-

37. The first ionization energy of helium is 2370 kJ mol^{-1}, the highest for any element.
(a) Define *ionization energy* and discuss why for helium it should be so high.
(b) Which element would you expect to have the highest *second* ionization energy? Why?
(c) Suppose that you wished to ionize some helium by shining electromagnetic radiation on it. What is the maximum wavelength you could use?

38. The energy needed to remove one electron from a gaseous potassium atom is only about two-thirds as much as that needed to remove one electron from a gaseous calcium atom, yet nearly three times as much energy as that needed to remove one electron from K^+ as from Ca^+. What explanation can you give for this contrast? What do you expect to be the relation between the ionization energy of Ca^+ and that of neutral K?

39. Without consulting any tables, arrange the following substances in order and explain your choice of order:
(a) Mg^{2+}, Ar, Br^-, Ca^{2+} in order of increasing radius
(b) Na, Na^+, O, Ne in order of increasing ionization energy
(c) H, F, Al, O in order of increasing electronegativity

40. Both the electron affinity and the ionization energy of chlorine are higher than the corresponding quantities for sulfur. Explain why in terms of the electronic structure of the atoms.

41. The cesium atom has the lowest ionization energy, 375.7 kJ mol^{-1}, of all the neutral atoms in the periodic table. What is the longest wavelength of light that could ionize a cesium atom? In which region of the electromagnetic spectrum does this light fall?

42. Until recently, it was thought that Ca^- was unstable, and that the Ca atom therefore had a negative electron affinity. Some new experiments have now measured an electron affinity of +2.0 kJ mol^{-1} for calcium. What is the longest wavelength of light that could remove an electron from Ca^-? In which region of the electromagnetic spectrum does this light fall?

ADDITIONAL PROBLEMS

43. In the hydrogen atom, the transition from the $2p$ state to the $1s$ state emits a photon with energy 16.2×10^{-19} J. In an iron atom, the same transition emits X-rays with wavelength 0.193 nm. Calculate the energy difference between these two states in iron. Explain the difference in the $2p$-$1s$ energy level spacing in these two atoms.

44. The energy needed to ionize an atom of element X when it is in its most stable state is 500 kJ mol^{-1}. However, if an atom of X is in its lowest excited state, only 120 kJ mol^{-1} is needed to ionize it. What is the wavelength of the radiation emitted when an atom of X undergoes a transition from the lowest excited state to the ground state?

45. Suppose an atom in an excited state can return to the ground state in two steps. It first falls to an intermediate state, emitting radiation of wavelength λ_1, and then to the ground state, emitting radiation of wavelength λ_2. The same atom can also return to the ground state in one step, with the emission of radiation of wavelength λ. How are λ_1, λ_2, and λ related? How are the frequencies of the three radiations related?

46. For the Li atom, the energy difference between the ground state and the first excited state, in which the outermost electron is in a $2p$ orbital, is 2.96×10^{-19} J. In the Li^{2+} ion, the energy difference between the $2s$ and $2p$ levels is less than 0.00002 of this value. Explain this observation.

47. How does the $3d_{xy}$ orbital of an electron in O^{7+} resemble the $3d_{xy}$ orbital of an electron in a hydrogen atom? How does it differ?

48. The wave function of an electron in the lowest (that is, ground) state of the hydrogen atom is

$$\psi(r) = \left(\frac{1}{\pi a_0^3}\right)^{1/2} \exp\left(-\frac{r}{a_0}\right)$$

$$a_0 = 0.529 \times 10^{-10} \text{ m}$$

(a) What is the probability of finding the electron inside a sphere of volume 1.0 pm^3, centered at the nucleus (1 pm = 10^{-12} m)?
(b) What is the probability of finding the electron in a volume of 1.0 pm^3 at a distance of 52.9 pm from the nucleus, in a fixed but arbitrary direction?
(c) What is the probability of finding the electron in a spherical shell of 1.0 pm in thickness, at a distance of 52.9 pm from the nucleus?

49. An atom of sodium has the electron configuration $[Ne]6s^1$. Explain how this is possible.

50. (a) The nitrogen atom has one electron in each of the $2p_x$, $2p_y$, and $2p_z$ orbitals. By using the form of the angular wave functions, show that the total electron density, $\psi^2(2p_x) + \psi^2(2p_y) + \psi^2(2p_z)$, is spherically symmetric (that is, it is independent of the angles θ and ϕ). The neon atom, which has *two* electrons in each $2p$ orbital, is also spherically symmetric.
(b) The same result as in part (a) applies to d orbitals, thus a filled or half-filled subshell of d orbitals is spherically symmetric. Identify the spherically symmetric atoms or ions among the following: F^-, Na, Si, S^{2-}, Ar^+, Ni, Cu, Mo, Rh, Sb, W, Au.

51. Chromium(IV) oxide is used in making magnetic recording tapes because it is paramagnetic. It can be described as a solid made up of Cr^{4+} and O^{2-}. Give the electron configuration of Cr^{4+} in CrO_2, and determine the number of unpaired electrons on each chromium ion.

52. Use the data from Appendix F to graph the variation of atomic radius with atomic number for the rare-earth elements from lanthanum to lutetium.
(a) What is the general trend in these radii? How do you account for it?
(b) Which two elements in the series present exceptions to the trend?

53. Arrange the following seven atoms or ions in order of size, from smallest to largest: K, F^+, Rb, Co^{25+}, Br, F, Rb^-.

54. Which is higher, the third ionization energy of lithium or the energy required to eject a $1s$ electron from a Li atom in a PES experiment? Explain.

55. The outermost electron in an alkali-metal atom is sometimes described as resembling an electron in the corresponding state of a one-electron atom. Compare the first ionization energy of lithium with the binding energy of a 2*s* electron in a one-electron atom that has nuclear charge Z_{eff}, and determine the value of Z_{eff} that is necessary for the two energies to agree. Repeat the calculation for the 3*s* electron of sodium and the 4*s* electron of potassium.

56. In two-photon ionization spectroscopy, the combined energies carried by two different photons are used to remove an electron from an atom or molecule. In such an experiment, a K atom in the gas phase is to be ionized by two different light beams, one of which has a 650-nm wavelength. What is the maximum wavelength for the second beam that will cause two-photon ionization?

57. For the H atom, the transition from the 2*p* state to the 1*s* state is accompanied by the emission of a photon with an energy of 16.2×10^{-19} J. For an Fe atom, the same transition (2*p* to 1*s*) is accompanied by the emission of X-rays of 0.193-nm wavelengths. What is the energy difference between these states in iron? Comment on the reason for the variation (if any) in the 2*p*-1*s* energy-level spacing for these two atoms.

58. **(a)** Give the complete electron configuration ($1s^22s^22p\ldots$) of aluminum in the ground state.

(b) The wavelength of the radiation emitted when the outermost electron of aluminum falls from the 4*s* state to the ground state is about 395 nm. Calculate the energy separation (in joules) between these two states in the Al atom.

(c) When the outermost electron in aluminum falls from the 3*d* state to the ground state, the radiation emitted has a wavelength of about 310 nm. Draw an energy-level diagram of the states and transitions discussed here and in (b). Calculate the separation (in joules) between the 3*d* and 4*s* states in aluminum. Indicate clearly which has higher energy.

59. What experimental evidence does the periodic table provide that an electron in a 5*s* orbital is slightly more stable than an electron in a 4*d* orbital for the elements with 37 and 38 electrons?

Quantum Mechanics and Molecular Structure

6

CHAPTER

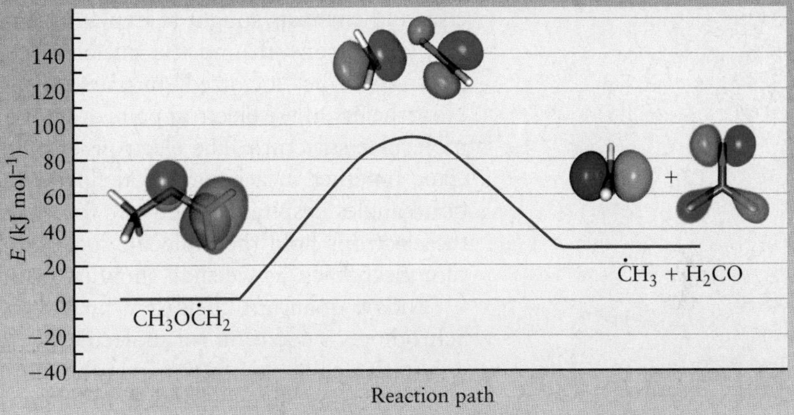

Potential energy diagram for the decomposition of the methyl methoxy radical, an important intermediate in the combustion of dimethyl ether. Highly accurate computational quantum chemistry methods were used to calculate the configurations and the relative energies of the species shown. (Courtesy of Prof. Emily Carter, Princeton)

Chemists seek to understand the atomic interactions that form molecules and extended solid structures, to understand intermolecular interactions, and, most important, to understand chemical reactivity. Quantum mechanics provides a firm conceptual foundation for understanding these phenomena, all of which involve the chemical bond.

The concept of the chemical bond as an agent for holding atoms together in molecules was formulated around the middle of the 19th century. In the 1860s, the chemists August Wilhelm Hofmann and Edward Frankland used three-dimensional arrays of colored wooden balls as models for molecular structure in their lectures in Berlin and London. In 1875 the Dutch chemist Jacobus van't Hoff popularized these developments in a book entitled *Chemistry in Space,* and by 1885 the three-dimensional representation of molecules was universally accepted.

Today, chemistry is a sophisticated science, fully integrated with modern computer technology, and chemists use a broad array of display tools aided by computer graphics to describe various aspects of molecular structure. Even so, chemists continue to use "ball-and-stick models," in which each ball represents an atom, centered on the position of the nucleus, and the sticks represent chemical bonds, to help them think about the geometry of molecular structure. Molecular geometry is defined by the lengths of the bonds between the nuclei and the angles between the bonds (see Figs. 3.1 and 3.24).

After J. J. Thomson discovered the electron in 1897, and especially after Ernest Rutherford formulated the planetary model of the atom in 1912, physicists and chemists sought to explain the chemical bond as special arrangements of the electrons around the nuclei. For example, G. N. Lewis considered the electrons to be "localized" in pairs to form covalent bonds between the nuclei. Nonetheless, these electron pairs are not explicitly included in the description of molecular structure. The electrons provide attractive forces that hold the molecule together in a geometry defined by a particular set of bond lengths and bond angles despite the repulsive forces that tend to push the nuclei apart. How the electrons hold the molecule together can only be explained by using quantum mechanics, as we shall show in Section 6.1 below.

Today, quantum chemistry provides very good approximate solutions to Schrödinger's equation for molecules. Perhaps more important, it provides new qualitative concepts for representing and describing chemical bonds, molecular structure, and chemical reactivity. The quantum description of the chemical bond justifies the electron pair model, shows the limits of its validity, and points the way to advances beyond the pair model. The quantum description forms the basis for all modern studies in structural chemistry.

This chapter begins with a description of the quantum picture of the chemical bond and then applies this picture to the simplest possible molecule, H_2^+, which contains only one electron. Schrödinger's equation for the motion of the electron in H_2^+ can be solved exactly, and we use its solutions to display the general features of each **molecular orbital (MO)**, the one-electron wave functions that describe the electronic structure of molecules. Just as we used the **atomic orbitals (AOs)** of the H atom to suggest approximate AOs for complex atoms, we let the MOs for H_2^+ guide us to develop approximations for the MOs of more complex molecules.

Guided by these insights, we then describe the two main ways to construct approximate molecular electronic wave functions from atomic orbitals, the **linear combination of atomic orbitals (LCAO)** method and the **valence bond (VB) method**. The LCAO method generates MOs that are *delocalized* over the entire molecule, and it builds up the electronic configurations of molecules using an aufbau principle just like the one for atoms. In contrast, the VB method describes electron pairs that are *localized* between a pair of atoms, and it provides a quantum mechanical foundation for the Lewis electron dot diagrams and for the valence shell electron pair repulsion (VSEPR) theory (see Ch. 3). We apply both of these methods to describe structure and bonding in a variety of molecules. We conclude the chapter by comparing the LCAO and VB methods and showing how each is the starting point for using modern methods for computational quantum chemistry. These are now sufficiently accurate and so easy to use that they are becoming part of every chemist's set of tools for both research and education.

The central conceptual goal of this chapter is for you to understand that as two atoms approach each other and form a bond, their electron densities begin to interpenetrate and to form a new electron density characteristic of the molecule. We want you to see how to use the wave functions for electrons initially located on different atoms to form new wave functions that represent chemical bonds in molecules.

6.1 Quantum Picture of the Chemical Bond

Visualize two H atoms (labeled A and B), initially quite far apart, as they approach one another with the possibility of forming a chemical bond. We know from Chapter 3 that a bond can form only if the resulting molecule has lower energy than the pair of isolated atoms. Let's track progress by plotting the potential energy of the atoms as the distance between them R_{AB} decreases, using methods developed in Sections 3.3 and 3.7, and let's display the results in Figure 6.1. (We have already introduced this curve in a general, intuitive way in Figure 3.20.) At very large separation we define the energy of interaction to be zero. As the atoms come closer together, the potential energy decreases (becomes negative) due to attractive forces between the atoms. At very close distances the potential energy becomes positive due to repulsive forces between the atoms. At some intermediate but small distance between the atoms the attractive interactions dominate the repulsion, and the potential energy is negative; it reaches a minimum value where the attractive and repulsive forces balance exactly. This minimum is the signature of chemical bond formation. The value of the potential energy at this minimum measures the extent to which the bond is more stable than the separated atoms. The difference between this minimum and the value for the separated atoms is the amount of energy that must be invested to break the bond and liberate the free atoms. This is a positive number called the bond dissociation energy and denoted either D_e or ΔE_d. The distance R_e at which the minimum occurs is the equilibrium length of the bond. For H_2, the bond dissociation energy D_e is 458 kJ mol^{-1}, or 4.75 eV molecule^{-1}, and experimental measurements show that the bond length is 0.74 Å.

The total energy of the molecule is quantized, so we can visualize a set of energy levels superposed on the potential energy curve. Just as shown for the particle-in-a-box model (see Section 4.6), the uncertainty principle requires that there is a **zero-point energy** for the molecule. This is the lowest allowed value of the total energy, and is represented by the line E_0. Because the zero-point energy can never be removed from the molecule, it provides a reference point for the amount of energy required to dissociate the molecule. Relative to the zero-point energy, the **bond dissociation energy** is defined as D_0. Although both D_0 and D_e are called dissociation energies, only the former is measurable experimentally as the energy needed to dissociate the molecule. D_e is useful as a parameter to construct model potentials and optimize geometry in calculations. The bond dissociation energy D_0 for H_2 is 432 kJ mol^{-1} (measured experimentally and calculated with quantum theory); it is smaller than D_e by the zero-point energy

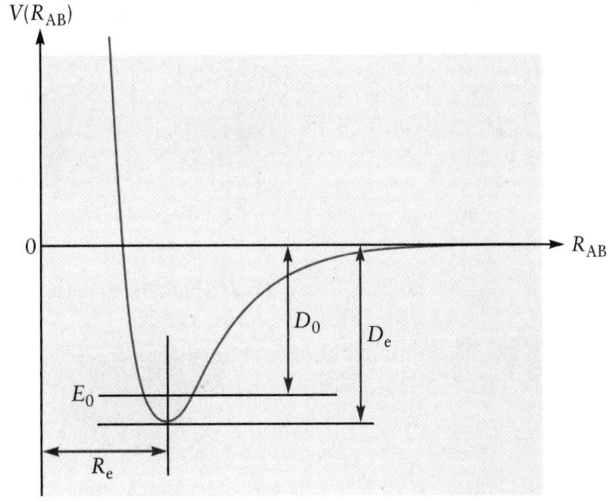

FIGURE 6.1 Schematic representation of the potential energy of two hydrogen atoms as a function of the distance between them. As distance decreases, the potential energy reaches a minimum value of −458 kJ mol^{-1} at a distance of 0.74 Å. The value of D_0 for H_2 is 432 kJ mol^{-1} or 4.48 eV molecule^{-1}.

of H_2 as shown in Figure 6.1. (If a polyatomic molecule dissociates, D_0 is measured from the lowest quantum state of the reactants, called the zero-point level, to the zero-point level of the products, so again equals the experimentally measured bond dissociation energy at 0 K.)

The meaning of this curve is considerably more subtle than the simple description we gave to introduce it. Each atom contains one proton and one electron. Each of these four particles interacts with each of the others by a force described by Coulomb's law (see Fig. 3.19), and all these contributions must be included in the potential energy of the two atoms. However, the coordinates in Figure 6.1 track only the distance between the two protons, and do not describe the electrons at all.

The sequence of events traced in Figure 6.2 shows us what has happened. As the two atoms draw near, their electron densities interpenetrate and form a new electron density characteristic of the molecule. Because electrons behave like waves, the two atomic electronic wave functions can interfere either constructively or destructively. If the interference is constructive, the electron density in the region between the nuclei increases. The net result of this increased negative charge density between the nuclei is to attract them toward each other and offset their mutual Coulomb repulsion so they arrive at a stable position.

In fact, the curve in Figure 6.1 plots the *effective potential energy function* for the two protons in the molecule, against the distance between the protons. This effective potential energy function includes several contributions: (a) Coulomb repulsion between the protons; (b) Coulomb repulsion between the electrons; (c) Coulomb attractions of both electrons to both protons; (d) kinetic energies of the electrons. The contributions from the electrons have been averaged out in some sense, and their coordinates do not appear in the effective potential energy curve. The effective potential energy function governs the formation of the molecule (it gives the depth of the potential well relative to the isolated atoms), the structure of the molecule (by defining the equilibrium bond length), and the vibrations of the molecule (by defining the force that determines the relative

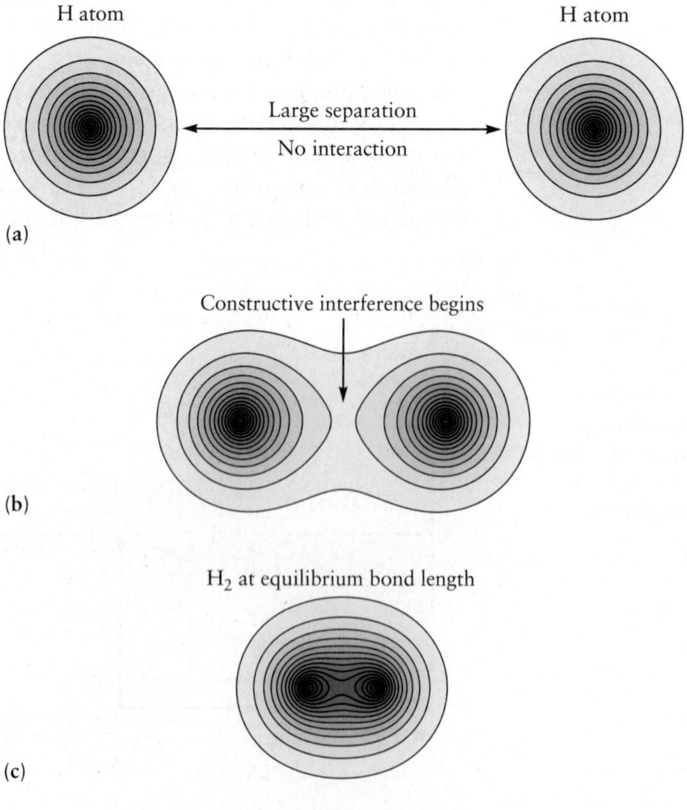

FIGURE 6.2 The probability density for finding an electron in the H_2 molecule, plotted at three internuclear distances: (a) $8a_0$, (b) $3.5a_0$, and (c) $1.4a_0$. At the equilibrium bond length (c) significant build up of electron density between the nuclei is visible. (*Courtesy of Chad Heaps, David A. Mazziotti, The University of Chicago*)

motions of the nuclei). If we can find a way to predict the effective potential energy curve as a function of the distance between the nuclei, we will know all the structural properties of the molecule.

The only justification for the effective potential energy curve for a molecule is provided by the quantum description of the chemical bond. Not only does this theory define the concept of the effective potential energy curve, it also provides an explicit procedure for calculating the curve. The quantum description of the chemical bond therefore has enormous predictive power. It can determine whether a particular molecule will exist, and if it does exist what will be its structure and properties.

The definition of the effective potential energy curve for a molecule, and the method for calculating it, arise naturally within the Born–Oppenheimer approximation, which is the foundation of molecular calculations in quantum mechanics. Now let's see how to obtain the effective potential energy curve and how to use it.

6.1.1 Born–Oppenheimer Approximation: Slow Nuclei, Fast Electrons

This approximation method was developed by the physicists Max Born (German) and J. Robert Oppenheimer (American) in 1927, just one year after Schrödinger's quantum solution for the H atom. Born and Oppenheimer recognized that because the nuclei are much more massive than the electrons (the mass of the proton is 1836 times the mass of the electron), the nuclei in molecules will move much more slowly than the electrons. This enables dividing the problem of describing the molecule in detail into two parts:

(A) Consider the nuclei to be fixed at a specific set of positions. (For the H_2 molecule in Figure 6.1 this means a specific fixed value of R_{AB}.) Then solve Schrödinger's equation for the electrons moving around the fixed nuclei and obtain the allowed quantized energy levels, E_{el}, and wave functions, ψ_{el}, for the electrons. (For H_2 this involves two electrons and six coordinates to track their positions in three-dimensional space.) Next we move the nuclei a bit, and repeat the calculation for the electrons at the new fixed location for the nuclei. The Schrödinger equation for this electronic problem includes the kinetic energies of the electrons, the Coulomb attraction of both electrons to both nuclei, the Coulomb repulsion between the electrons, and the Coulomb repulsion between the nuclei (at their fixed positions, which is the separation R, or R_{AB} in Fig. 6.1, for H_2), so the only term left out is the kinetic energy of the nuclei. Then (B) use the resulting energy, E_{el}, from solving $\mathcal{H}_{el}\psi_{el} = E_{el}\psi_{el}$, as the effective potential energy for the motion of the nuclei. The lowest energy solution for E_{el} at each fixed internuclear position gives the $V(R)$ shown in Figure 6.1, the effective potential energy as a function of internuclear distance for H_2. That gives us the bond dissociation energy and the equilibrium bond length. Using this potential energy function and the kinetic energy for the relative motion of the nuclei (that we left out previously), we construct and solve a Schrödinger equation for the nuclear motion. The resulting wave functions for the nuclei and their corresponding energies give the allowed energies for the vibrational motion (see Figure 12.23 and Section 20.3) and rotational motion (see Section 20.3) of the molecule.

The Born–Oppenheimer approximation invites a new interpretation of how electrons are distributed in molecules. Instead of the localized electron pairs visualized by Lewis, we visualize a group of electrons moving rapidly around the sluggish nuclei, to establish a distribution of electron density described by quantum mechanics. This distribution holds the molecule together with fixed bond lengths and bond angles.

The Born–Oppenheimer approximation is equally applicable to polyatomic molecules, but the details become complicated rapidly as the number of atoms

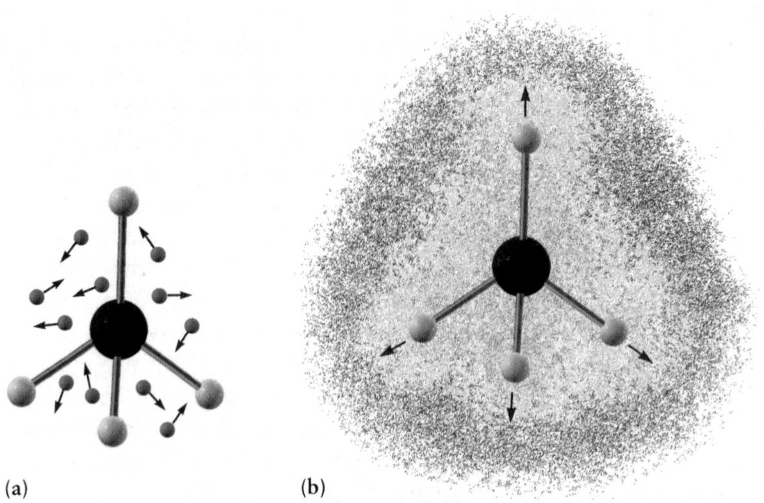

FIGURE 6.3 Schematic representation of the Born–Oppenheimer approximation for methane. (a) The nuclei are assumed to be fixed at positions corresponding to the equilibrium bond distances, while the electrons move rapidly around them. The wave functions and allowed energy values are calculated for the electrons at many different positions of the nuclei. (b) The allowed energy values calculated in (a), with the internuclear repulsions added in, are viewed as the effective potential energy functions that govern the relative motions of the nuclei in the molecule, as occurs in vibrations. The electron density calculated in (a) is shown, with shades of red indicating electron densities between 11×10^{-3} and 16.5×10^{-3} Å^{-3} turning to yellow for electron densities greater than 16.5×10^{-3} Å^{-3}.

increases and require multidimensional spaces for their description. We illustrate the concepts schematically for the methane molecule CH_4 (5 nuclei, 10 electrons) in Figure 6.3. During the calculations in Part A, 15 nuclear coordinates are held fixed at sequential positions, and 30 coordinates for the electrons are allowed to vary. All the electrons, both valence and core, in all the atoms are included in these calculations. That means the Schrödinger equation for Part A has 30 independent variables. The resulting ψ_{el}, when squared, gives the electron density in the methane molecule shown in Figure 6.3b. The corresponding energy E_{el} is the potential energy V, now a function of 3(5)-6 bond angles and internuclear distances in methane. This effective potential governs the nuclear motion (vibrations and bond-breaking).

Today chemists have very sophisticated computer programs to carry out calculations of this type, even for quite large molecules. The effective potential energy functions are represented as highly visual sections of multi-dimensional surfaces that, when used as the potential energy in the Schrödinger equation for the nuclear motion, allows you to predict not only vibrations of existing bonds but also the dynamics and kinetics of breaking old bonds and forming new bonds to describe chemical reactivity.

The Born–Oppenheimer approximation is a remarkable scientific achievement. It started out as a mathematical simplification for trying to solve the Schrödinger equation for a molecule, and turned out to provide the most important concept for describing a chemical bond. Even more remarkable, the approximation originates in the simple physical fact that protons are nearly 2000 times more massive than electrons, and therefore must move quite sluggishly while the electrons move very rapidly.

Most chemists need never work through the details of the Born–Oppenheimer approximation, but they must surely be aware of its scientific content and consequences: not only does it define the effective potential energy curve for a molecule, but it also defines the route for explicitly calculating that curve for any particular molecule.

In the remainder of Chapter 6 we introduce various methods to construct approximate wave functions that describe the distribution of electron density around the nuclei in a molecule. In each case we will obtain the effective potential energy curve and use it to interpret and predict molecular structure from the approximate wave function.

6.2 Exact Molecular Orbitals for the Simplest Molecule: H_2^+

Before we begin constructing approximate molecular orbitals for familiar molecules, it is useful to examine briefly the one molecule whose electronic wave function can be solved exactly by quantum mechanics.

The hydrogen molecular ion contains a single electron bound to two protons. It is a stable but highly reactive species produced by electrical discharge in H_2 gas. Its bond length is 1.06 Å, and its bond dissociation energy D_e to produce H and H^+ is 2.79 eV = 269 kJ mol^{-1}. The similarity of these values to those for more familiar molecules (see Chapter 3) suggests that the exact quantum solutions for H_2^+ will provide insights into chemical bonding that can be transferred to more complex molecules. The solutions for H_2^+ also introduce essential notation and terminology used in describing the covalent bonding in larger molecules.

The hydrogen molecular ion is sketched in Figure 6.4. The two nuclei, for convenience labeled A and B, are separated by the distance R_{AB} along the internuclear axis, chosen by convention to be the z-axis. The electron is located at distance r_A from nucleus A, and at distance r_B from nucleus B. The angle ϕ describes the electron's location about the internuclear axis. For a fixed value of R_{AB}, the position of the electron is more conveniently specified by the values of (r_A, r_B, ϕ) than by (x, y, z) since the former set reflects the natural symmetry of the system. The potential energy of the system is given by

$$V = -\frac{e^2}{4\pi\epsilon_0}\left(\frac{1}{r_A} + \frac{1}{r_B}\right) + \frac{e^2}{4\pi\epsilon_0}\left(\frac{1}{R_{AB}}\right) = V_{en} + V_{nn} \qquad [6.1]$$

The first two terms in Equation 6.1 represent the attractions between the electron and the two nuclei, V_{en}, and the last term represents the repulsion between the pair of protons, V_{nn}. For each combination of values for r_A and r_B

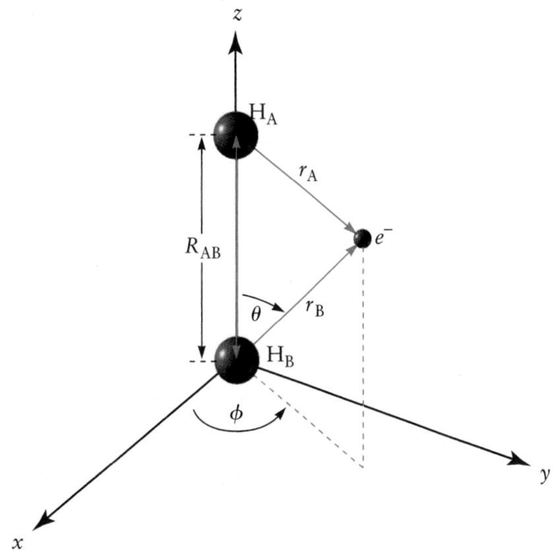

FIGURE 6.4 Coordinates for the H_2^+ molecular ion. The two nuclei are located along the z-axis, separated by the distance R_{AB}. The vectors $\vec{r}_A$ and $\vec{r}_B$ shown have lengths r_A and r_B, the distances of the electron from nuclei A and B, respectively; r_A and r_B range from 0 to ∞. The angle θ is determined from r_A, r_B, and R_{AB} by the law of cosines; it does not appear explicitly in the calculation. The angle ϕ varies from 0 to 2π.

the potential energy has the same value at all values of the angle ϕ. The potential energy has cylindrical symmetry around the R_{AB} axis, and the angle ϕ does not appear in the equation for potential energy.

The quantum treatment of this molecule appears to be a straightforward extension of that for the H atom (see Section 5.1 for the solutions). One more proton has been added, and the symmetry of the system has changed from spherical to ellipsoidal. The new feature is the internuclear distance R_{AB}, the bond length of the molecule. Since there is no exact solution that describes the motion of all three particles, let's use the Born–Oppenheimer approximation (described in Section 6.1). We assume that R_{AB} is held at the measured equilibrium bond length, and we find $\psi_{el}(r_A, r_B, \phi; R_{AB})$ the *electronic wave function* for the electron around the fixed nuclei. The semicolon inside the parentheses indicates that the nuclear coordinates are held fixed as a *parameter* (for now at the equilibrium bond length 1.06 Å) while the electronic coordinates range over all values as we seek the solution for ψ_{el}. Because the potential energy is the same at all values of the angle ϕ, the angular coordinate will influence the shape of the molecular orbital but not the energy levels of the electron, just as the angles (θ, ϕ) in Figure 5.1 contributed to the shape of the hydrogen orbitals but not the energy levels. Therefore we omit ϕ in subsequent equations for H_2^+.

The electronic wave function ψ_{el} lies at the heart of chemical bonding because its square gives the probability density for locating the electrons around the nuclei. To obtain ψ_{el} for H_2^+, we solve the Schrödinger equation for the electron by the same methods we applied to the H atom in Section 5.1. We require the solution to be smooth, single-valued, and finite in value in all regions of space so its square is a well-defined probability density function. We also enforce physical boundary conditions to guarantee that our solution describes a bound state of the electron (ψ_{el} must approach 0 as $r_A \rightarrow \infty$ and as $r_B \rightarrow \infty$). We find that solutions exist only when the total energy and the component of angular momentum directed along the internuclear axis are quantized. The complete set of quantum numbers is more extensive than that for the H atom, and will not be discussed here.

6.2.1 Electronic Wave Functions for H_2^+

We emphasize that the solutions for ψ_{el} are mathematically exact. Because of their ellipsoidal symmetry, they cannot be written as simple exponential and polynomial functions that are easily manipulated, as were the H atom solutions. Consequently we present and interpret these exact solutions in graphical form.

The first eight wave functions, starting with the ground state, are shown in Figure 6.5. These are plotted using the same coordinates as in Figure 6.4 (but with the axes reoriented so the z-axis is horizontal), where the two protons lie on the z-axis and the value of R_{AB} is the experimental bond length for H_2^+, which is 1.06 Å. Each wave function is shown as an isosurface comprising all those points in three-dimensional space where the wave function has the value equal to ± 0.1 of its maximum value. Regions where the wave function has a positive value are shown in red, and regions where the wave function has a negative value in blue, as in the display images of atomic orbitals in Chapter 5.

Each wave function is identified by four labels. By analogy with the quantum number subscripts on the hydrogen atomic orbitals, these labels summarize the energy and the shape of each wave function. Each provides insight into the shape and symmetry of the wave function and its corresponding probability density for locating the electron. They help us understand how the electron density has been shifted away from the spherical symmetry of the H atom to accommodate the second nucleus and the internuclear distance in the molecule.

First, the integer is an index that tracks the relative energies of the wave functions of each symmetry type. For example, $1\sigma_g$ is the first (lowest on the

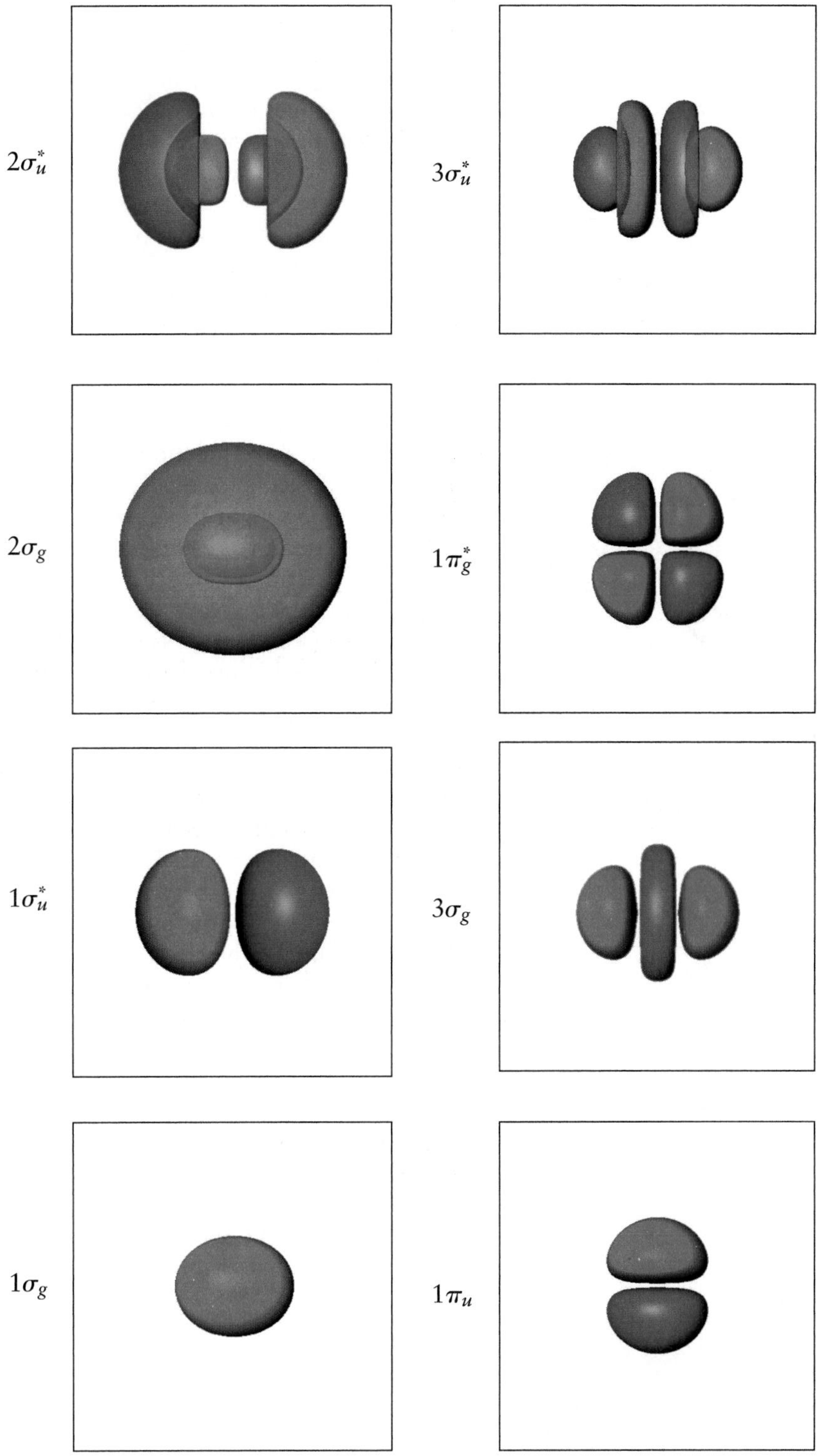

FIGURE 6.5 Wave functions for the first eight energy levels of the H_2^+ molecular ion, calculated exactly by quantum mechanics. The two nuclei lie along the *z*-axis, which is in the plane of the paper. Regions of positive and negative values of the wave function are shown in red and blue, respectively. The labels for each orbital are explained in the text. The images are isosurfaces corresponding to contours at $\pm$ 0.1 of the maximum value. The $1\sigma_g$ is the orbital the electron occupies when H_2^+ is in its ground electronic state, and the energies increase in the order $1\sigma_g < 1\sigma_u^* < 2\sigma_g < 2\sigma_u^* < 1\pi_u < 3\sigma_g < 1\pi_g^* < 3\sigma_u^*$.
(Courtesy of Mr. Hatem Helal and Professor William A. Goddard III, California Institute of Technology, and Dr. Kelly P. Gaither, University of Texas at Austin.)

energy scale) of the σ_g wave functions, while $2\sigma_u^*$ has the second lowest energy of the σ_u wave functions, and $1\pi_u$ has the lowest energy of the π_u wave functions. The energy index integer is analogous to the principal quantum number n for atomic orbitals. As the index increases within each symmetry type, the number of nodes along the internuclear axis increases.

Second, the Greek letter tells us how the amplitude of the wave function is distributed around the internuclear axis. The letter σ indicates the amplitude has cylindrical symmetry around the axis; a cross-section of the orbital in a plane perpendicular to the axis is a disc. The letter π signifies that the wave function has a nodal plane that contains the internuclear axis. A cross-section perpendicular to the axis shows a node at the axis with the wave function having positive values on one side of the axis and negative values on the other side. These properties originate in the angular momentum of the electron as it moves around the internuclear axis, and remind us of the *s* and *p* atomic orbitals. An electron in a σ wave function has no component of angular momentum along the axis and therefore can approach the axis. An electron described by a π wave function has a nonzero component of angular momentum along the axis and thus does not approach the axis.

Third, the subscript *g* or *u* describes how the wave function changes as we invert our point of observation through the center of the molecule. Imagine Cartesian coordinates with their origin at the center of the molecule, and compare the wave function at the point (x, y, z) and at the point $(-x, -y, -z)$. If the sign of the wave function is the same at these two points, it is called symmetric and labeled *g* for the German word *gerade* (even). If the sign of the wave function is opposite at these two points, it is called antisymmetric and labeled *u*, for the German word *ungerade* (odd). The sign of a wave function is critical because it determines whether, if there is an overlap in space with another atom's or molecule's wave function, the two waves interfere constructively or destructively (see Chapter 4) during the formation of a chemical bond.

Finally, we examine the behavior of the wave function when the point of observation is reflected through a plane perpendicular to the internuclear axis and located at the center of the molecule. If the wave function changes sign upon reflection, it receives the superscript *. If it does not change sign upon reflection, no symbol is added. Each wave function labeled * has a node on the internuclear axis at the center of the molecule, which means there is zero electron density at this point. We will see later that such nodes in the electronic wave function located between two atoms in a molecule have special significance in the description of chemical bonds. (Note that in a heteronuclear diatomic molecule, the electron density is not shared equally between the two atoms, so there is no plane of symmetry perpendicular to the internuclear axis between the two atoms; thus the nodal surface across which the wave function changes from positive to negative in value is closer to one atom, but a * is still used as a label.)

We call each of these exact one-electron wave functions a **molecular orbital,** just as we called the exact one-electron wave functions for the H atom atomic orbitals. These exact MOs play a fundamental role in the quantum description of chemical bonding. If you want to know more about these exact MOs, Figure 6.55 in Section 6.13, *A Deeper Look . . .*, provides several additional illustrations.

EXAMPLE 6.1

In Section 5.1 we classified the atomic orbitals of H according to their number of angular nodes (each defined by a plane) and their radial nodes (each defined by a spherical surface). How do we extend this analysis to the molecular orbitals of H_2^+?

Solution

As we move from atoms to diatomic molecules, the most important new feature is a unique choice for the *z*-axis of the coordinate system, defined along the internuclear axis of the molecule. We classify the molecular orbitals by their response to the symmetry operations of this cylindrical configuration. Rotating the molecule about this axis does not change the σ MOs, but does change the π MOs. Inverting the coordinates through the center of the molecule does not change the *g* MOs, but does change the *u* MOs. Reflecting the coordinates through a plane perpendicular to the axis at the center of the molecule changes the * MOs, but does not change the MOs not labeled *. All these changes are described in the preceding text and are revealed by examining cross sections perpendicular to the *z*-axis. The cited problems illustrate these effects for specific MOs.

Related Problems: 1, 2, 3, 4, 5, 6

6.2.2 Nature of the Chemical Bond in H_2^+

Our intuitive understanding of the chemical bond is that the nuclei and electrons arrange themselves in a manner that reduces their total energy to a value lower than the total energy of the isolated atoms. Achieving this arrangement requires that new attractive interactions come into play to overcome the internuclear repulsion as the bond is formed. Naively, we expect this to occur when the electrons are arranged so they spend most of their time "between" the nuclei where they would experience maximum attraction to all the nuclei, not just the nucleus of the parent atom. We saw in Section 3.7 that classical electrostatics could not explain this stabilization mechanism, so we relied on the qualitative Lewis model of electron-shared pair bonds. Now we want to see how these ideas are handled by quantum mechanics in the simplest case. We recommend that you review Section 3.7 at this point.

First, let's examine the electron density around the nuclei, assumed fixed at the equilibrium bond length, which is calculated by squaring each MO wave function in Figure 6.5. Detailed quantum calculations show that in the $1\sigma_g$ orbital the electron density between the two nuclei in the H_2^+ ion is greater than that of two noninteracting H atoms, and so is consistent with our expectations about chemical bonding. But in the $1\sigma_u^*$ orbital the electron density between the nuclei is less than that for two noninteracting H atoms. Indeed, the electron density in the $1\sigma_u^*$ orbital goes to zero halfway between the nuclei (clearly visible in the image of the orbital in Figure 6.5) and thus appears inconsistent with a chemical bond. If you want to know more about the electron density in these exact MOs, Figure 6.56 in Section 6.13, *A Deeper Look* . . . provides quantitative plots of electron density in each of the 8 MOs and some further discussion.

It appears that $1\sigma_g$ supports formation of the bond by increasing electron density between the nuclei whereas $1\sigma_u^*$ opposes bond formation by reducing electron density between the nuclei. Yet, both functions are part of the exact quantum solution for H_2^+. To understand the role each of them plays in forming the bond, we need to determine how the energy of these states compares to the energy of the isolated atom and proton. Reducing the energy relative to the separated particles is the key to bond formation.

To see how the energy depends on the internuclear separation, we simply fix the nuclei in space, solve Schrödinger's equation for the electron with the nuclei fixed at that position (with the Coulomb repulsion between the nuclei at fixed R_{AB} included in $\mathcal{H}_{el}$), then move the nuclei a bit, and repeat the calculation until we have covered all reasonable values of the nuclear positions. (This process is carried out systematically using the Born–Oppenheimer approximation which is described in Section 6.1.) This calculation gives the electronic energy $E_{el}(R_{AB})$ as a function of the positions of the nuclei so is the effective potential energy that governs the motion of the nuclei shown in Figure 6.6. When the electron is in the $1\sigma_u^*$ MO, the effective potential energy is greater than that of the separated atoms at all distances, and increases as the internuclear distance decreases. This energy represents a repulsive interaction, under which the nuclei would fly apart rather than remain close together.

In contrast, when the electron is in the $1\sigma_g$ MO the effective potential energy is lower than that for noninteracting H atoms, and indeed reaches a minimum between the nuclei. When the electron is in this orbital, the molecular ion is energetically stable with a bond length that corresponds to the minimum in the potential energy curve.

Considered together, the increased electron density between the nuclei when the electron is in the $1\sigma_g$ orbital and the resulting lowering of the potential energy at $R_{AB} = R_e$ (the equilibrium bond length) indicate formation of a chemical bond. Consequently, $1\sigma_g$ is called a **bonding molecular orbital**.

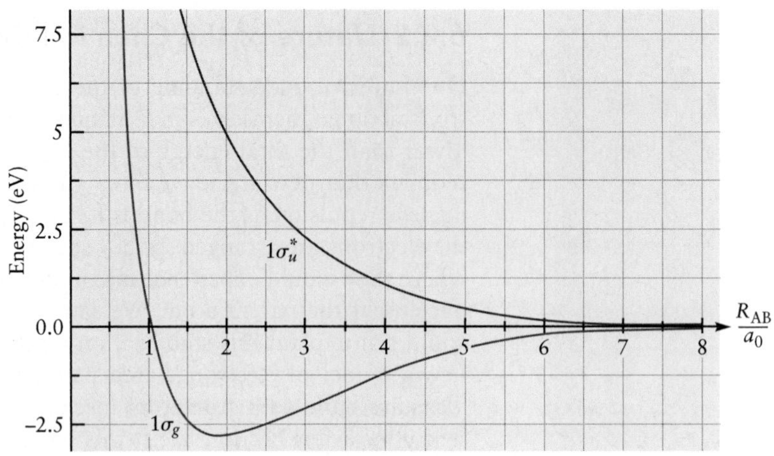

FIGURE 6.6 The total energy (excluding nuclear kinetic energy) of H_2^+ as a function of internuclear separation when the electron is in the $1\sigma_g$ orbital (red curve), and when the electron is in the $1\sigma_u^*$ orbital (blue curve), obtained in the exact solution of Schrödinger's equation for H_2^+. The internuclear separation is plotted in units of the Bohr radius. As explained in the text these curves are interpreted as the effective potential energy of the protons when the electron is in each of the respective orbitals. (Adapted from R. S. Berry, S. A. Rice, and J. Ross, *Physical Chemistry*, (2nd ed.), New York: Oxford, 2000, Figure 6.6.)

By contrast, when the electron is in the $1\sigma_u^*$ orbital the molecular ion shows zero electron density halfway between the nuclei and an increase in the effective potential energy (the total energy less the nuclear kinetic energy) as the atoms approach each other. Consequently this is called an **antibonding molecular orbital.** We can gain physical insight into the nature of the antibonding MO by realizing that an electron in this MO has significant probability density in regions that are not "between the protons." In the classical electrostatic picture depicted in Chapter 3, Figure 3.22, an electron in this orbital is still part of the molecule, but it spends most of its time tugging the protons apart rather than pulling them together. From this analysis we conclude that all the MOs labeled with * in Figure 6.5 are antibonding. Because each has a nodal plane between the two atoms perpendicular to the internuclear axis, the effective potential energy curve will be repulsive.

6.2.3 Summary: Key Features of the Quantum Picture of Chemical Bonding

Key features of the quantum picture of the chemical bond are revealed by our study of the exact molecular orbitals for the H_2^+ ion. In the remainder of this chapter, they guide the development of approximate molecular orbital methods for more complicated molecules. These features are

1. In representing molecular structures, we consider the nuclei to be fixed (more precisely, slowly vibrating around) at positions corresponding to the equilibrium geometry of the molecule while the electrons move rapidly around them (the Born–Oppenheimer approximation).
2. A molecular orbital is a one-electron wave function whose square describes the distribution of electron density around the nuclei in their equilibrium positions.
3. An electron in a bonding molecular orbital has increased electron density in the region between the nuclei when $R = R_e$, decreasing the effective potential energy relative to that of the separated atoms.
4. Each bonding molecular orbital is related to an antibonding molecular orbital which has zero amplitude on a surface between the nuclei and thus a node on the internuclear axis. An electron in an antibonding molecular orbital leads to effective potential energy higher than that of the separated atoms.

5. Bonding and anti-bonding molecular orbitals of type σ are those for which the electron density is distributed symmetrically about the internuclear axis. A σ orbital has cylindrical symmetry; cross-sections perpendicular to the internuclear axis are discs.
6. Bonding and anti-bonding molecular orbitals of type π are those for which the electron density has a nodal plane that contains the internuclear axis. Consequently, the amplitude is concentrated "off the axis" and the orbital does not have cylindrical symmetry about the internuclear axis.

6.3 Molecular Orbital Theory and the Linear Combination of Atomic Orbitals Approximation for H_2^+

The goal of this section is to introduce the first of the two main methods for generating approximate electronic wave functions for molecules. The LCAO method consists of selecting sums and differences (linear combinations) of atomic orbitals to generate the best approximation to each type of molecular orbital. The approximate MO constructed in this way is then used to calculate properties of the molecule using standard equations of quantum mechanics, by hand for simple cases and by computer for larger molecules.

The LCAO method is best explained with the help of a specific example, so we start with the hydrogen molecular ion H_2^+. We can evaluate the success of the method by comparing its results to the exact solution described in Section 6.2, and gain confidence in applying the method to more complex molecules.

The LCAO approximation is motivated by physical reasoning as follows. Consider the 1 σ_g MO shown in Figure 6.5. When the electron is close to nucleus A, it experiences a potential not very different from that in an isolated hydrogen atom. The ground-state MO in the region near A should therefore resemble a 1*s* atomic wave function φ_{1s}^{A}. Near B, the MO should resemble φ_{1s}^{B}. (These atomic orbitals were given the symbol ψ_{1s} in Chapter 5.) The 1*s* orbitals at A and B are identical (with the exception that the origin of each orbital's coordinate system is on a different atom); the labels are attached to emphasize the presence of two nuclei. Note that A and B are merely labels, not exponents. A simple way to construct a MO with these properties is to approximate the $1\sigma_g$ MO as a sum of the H 1*s* orbitals with adjustable coefficients. Similarly, we can approximate the other exact MOs in Figure 6.5 by forming linear combinations of the H atomic orbitals which they resemble when the electron is near A or B.

The general form for an approximate MO for H_2^+ is a linear combination of two atomic orbitals, obtained by adding or subtracting the two: with coefficients whose values depend on R_{AB}:

$$\psi_{MO} = C_A(R_{AB})\varphi_{1s}^{A} \pm C_B(R_{AB})\varphi_{1s}^{B}$$

For our purposes, it is adequate to ignore the dependence of the coefficients on R_{AB} and evaluate them by normalizing the wave function.

The coefficients C_A and C_B give the relative weights of the two atomic orbitals in determining the character of the MO. If C_A were greater in magnitude than C_B, the φ_{1s}^{A} orbital would be more heavily weighted and the electron would be more likely to be found near nucleus A and vice versa. But because the two nuclei in the H_2^+ ion are identical, the electron is just as likely to be found near one nucleus as the other. Therefore, the magnitudes of C_A and C_B must be equal, so either $C_A = C_B$ or $C_A = -C_B$. For both these choices $(\psi_{MO})^2$ does not change if the A and B nuclei are interchanged.

To maintain the distinctions among the various orbitals, we will use the following notation in describing the LCAO approximation. Atomic orbitals will be represented by φ and molecular orbitals by σ or π. Generic wave functions will be represented by ψ. Occasionally, ψ will represent some special wave function, in which case appropriate subscripts will be attached.

6.3.1 LCAO Molecular Orbitals for H_2^+

Proceeding as described above, we construct approximate molecular orbitals for the exact $1\sigma_g$ and $1\sigma_u^*$ MOs in Figure 6.5:

$$1\sigma_g \approx \sigma_{g1s} = C_g[\varphi_{1s}^{A} + \varphi_{1s}^{B}] \quad [6.2a]$$

$$1\sigma_u^* \approx \sigma_{u1s}^* = C_u[\varphi_{1s}^{A} - \varphi_{1s}^{B}] \quad [6.2b]$$

where C_g and C_u are chosen to ensure that the total probability of finding the electron *somewhere* is unity and we have ignored their dependence on the choice of R_{AB} at which we have fixed the nuclei. Notice that we have introduced new symbols σ_{g1s} and σ_{u1s}^* for the approximate MOs not only to distinguish them from the exact MOs but also to indicate explicitly the AOs from which they were constructed. The symbols for the exact and approximate MOs are summarized in Table 6.1. To simplify the notation, henceforth we omit the dependence of σ_{g1s} and σ_{u1s}^* on R_{AB}.

TABLE 6.1

Molecular Orbitals for Homonuclear Diatomic Molecules

Exact MO Notation	LCAO MO Notation
$1\sigma_g$	σ_{g1s}
$1\sigma_u^*$	σ_{u1s}^*
$2\sigma_g$	σ_{g2s}
$2\sigma_u^*$	σ_{u2s}^*
$1\pi_u$	π_{u2p_x}, π_{u2p_y}
$3\sigma_g$	σ_{g2p_z}
$1\pi_g^*$	$\pi_{g2p_x}^*, \pi_{g2p_y}^*$
$3\sigma_u^*$	$\sigma_{u2p_z}^*$

The distribution of electron probability density is obtained by squaring each of the approximate MOs:

$$[\sigma_{g1s}]^2 = C_g^2[(\varphi_{1s}^{A})^2 + (\varphi_{1s}^{B})^2 + 2\varphi_{1s}^{A}\varphi_{1s}^{B}] \quad [6.3a]$$

$$[\sigma_{u1s}^*]^2 = C_u^2[(\varphi_{1s}^{A})^2 + (\varphi_{1s}^{B})^2 - 2\varphi_{1s}^{A}\varphi_{1s}^{B}] \quad [6.3b]$$

These can be compared with the electron probability distribution for a noninteracting (n.i.) system (obtained by averaging the probabilities for $H_A + H_B^+$ and $H_A^+ + H_B$), which is

$$\psi_{n.i.}^2 = C_3^2[(\varphi_{1s}^{A})^2 + (\varphi_{1s}^{B})^2] \quad [6.4]$$

To describe the noninteracting system as one electron distributed over two possible sites, we set $C_3^2 = 0.5$. The interpretation of these approximate MOs in relation to the noninteracting system is best explained graphically. Figure 6.7 shows plots of these various wave functions (left side) and their squares (right side). Compared with the noninteracting pair of atoms, the system described by the approximate MO σ_{g1s} shows increased electron density between the nuclei and is therefore a bonding orbital as defined in Section 6.2. By contrast, the approximate MO σ_{u1s}^* shows reduced probability for finding an electron between the nuclei and so is an antibonding orbital. Note in Figure 6.7 that σ_{u1s}^* has a node between the nuclei and is antisymmetric with respect to inversion through the molecular center (see Section 6.2). Comparing Figure 6.7 with Figure 6.5 shows that the LCAO method has reproduced qualitatively the probability density in the first two exact wave functions for H_2^+.

A simple way to understand why a particular linear combination of atomic orbitals results in increasing or decreasing electron density between the nuclei is shown graphically in Figure 6.8. In the linear combination of atomic orbitals for the σ_{g1s} orbital in Figure 6.8a, the two atomic orbitals have the same sign (both positive) where they overlap between the nuclei. Thus, the electron waves

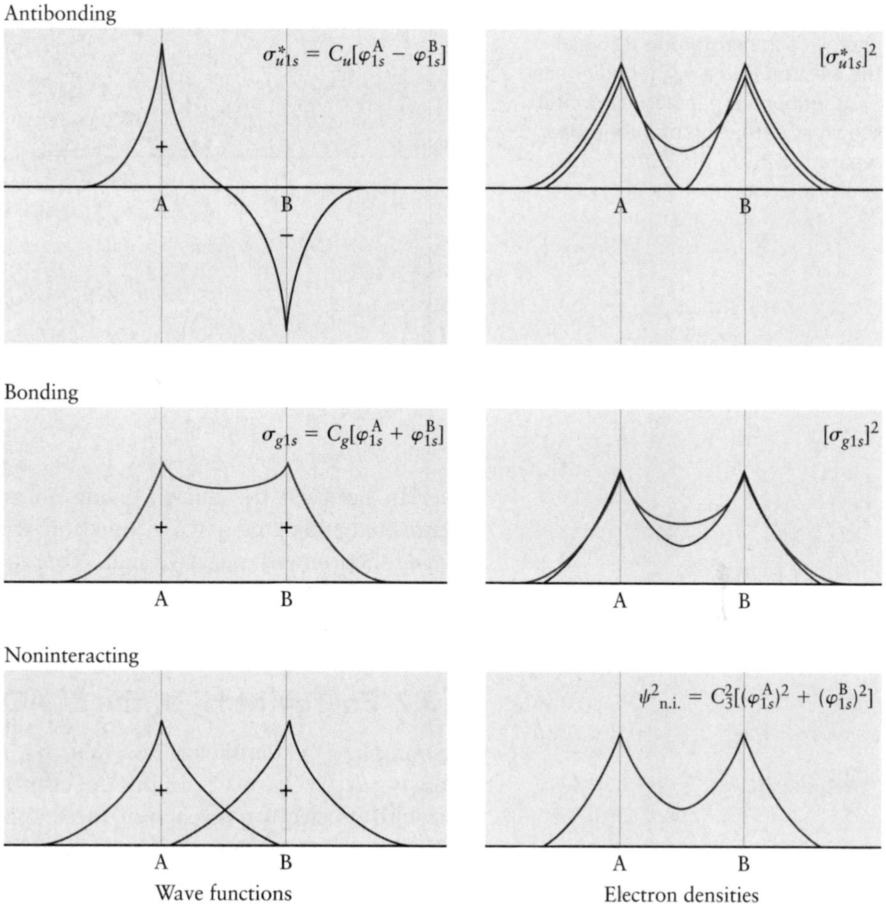

FIGURE 6.7 Antibonding and bonding molecular orbitals of H_2^+ along the internuclear axis in the linear combination of atomic orbitals (LCAO) approximation. For comparison, the green lines show the independent atomic orbitals and the electron probability distribution ψ^2(n.i.) for a noninteracting system. Compared with this reference system, the bonding orbital shows *increased* probability density between the nuclei, but the antibonding orbital shows *decreased* probability density in this region.

constructively interfere, giving more electron density in that region. The linear combination for the σ^*_{u1s} molecular orbital is $C_u[\varphi^A_{1s} + (-\varphi^B_{1s})]$, so a wave function with positive sign is added to a wave function of negative sign, $(-\varphi^B_{1s})$. Thus, as shown in Figure 6.8b, the waves destructively interfere where they

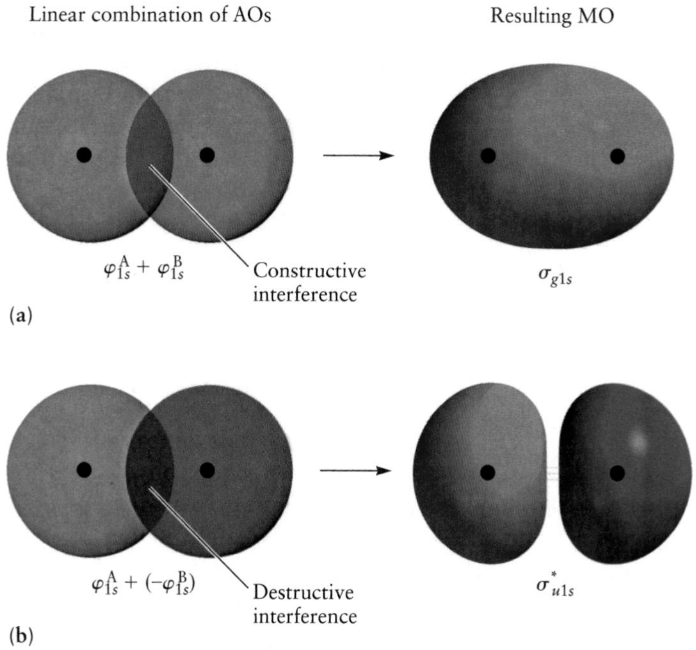

FIGURE 6.8 Graphical representation of the linear combination of atomic orbitals (AOs) resulting in a bonding molecular orbital, shown in (a), and an antibonding molecular orbital, shown in (b). The black dots show the approximate positions of the nuclei.

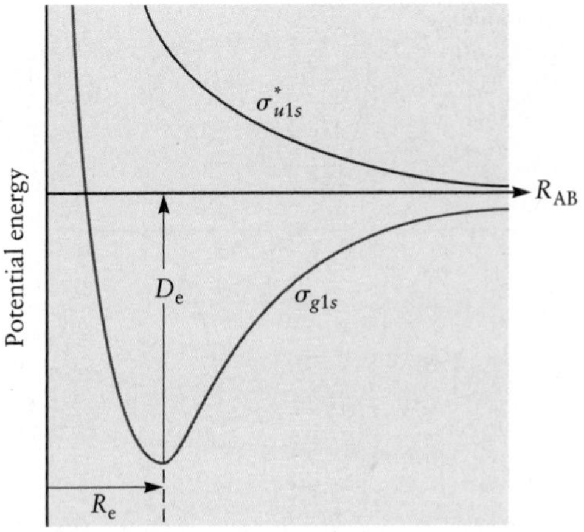

FIGURE 6.9 Effective potential energy of the protons in H_2^+ when the electron is in a σ_{g1s} (bonding) and σ^*_{u1s} (antibonding) molecular orbital, shown as a function of internuclear separation R_{AB} in the LCAO approximation.

overlap between the nuclei, giving an anti-bonding molecular orbital. In the figure, red indicates a wave function with positive sign and blue indicates a wave function with negative sign. Note that φ_{1s}^{B} is positive everywhere (see Table 5.2), so $(-\varphi_{1s}^{B})$ is negative everywhere; thus $(-\varphi_{1s}^{B})$ is shown in blue for the linear combination that gives the σ^*_{u1s} molecular orbital.

6.3.2 Energy of H_2^+ in the LCAO Approximation

To complete the demonstration that σ_{g1s} and σ^*_{u1s} are bonding and antibonding MOs respectively, we examine the effective potential energy curve for the H_2^+ ion with its electron in each of these approximate MOs. Figure 6.9 shows the effective potential energy of the H_2^+ ion in the LCAO approximation for the σ_{g1s} and σ^*_{u1s} MOs. In the ground electronic state shown in red, resulting from the σ_{g1s} electron probability density, the H_2^+ ion has a bound region of the potential, with the equilibrium internuclear distance R_e shown. In the excited electronic state shown in blue, there is no bound potential well, and a repulsive force acts to separate the transient molecular ion into H + H^+.

How well does the LCAO approximation describe the effective potential energy curve in H_2^+? We compare the exact and LCAO results in Figure 6.10, where the zero of energy at infinite separation is again taken to be the separated species H and H^+. The energy in σ_{g1s} has a minimum at R_{AB} = 1.32 Å, and the predicted energy to dissociate the ion to H and H^+ is D_e = 1.76 eV. These results compare reasonably well to the exact theoretical values R_{AB} = 1.06 Å and D_e = 2.79 eV, (see Section 6.2) and to the experimentally measured bond energy D_0 (after correcting for zero-point energy).

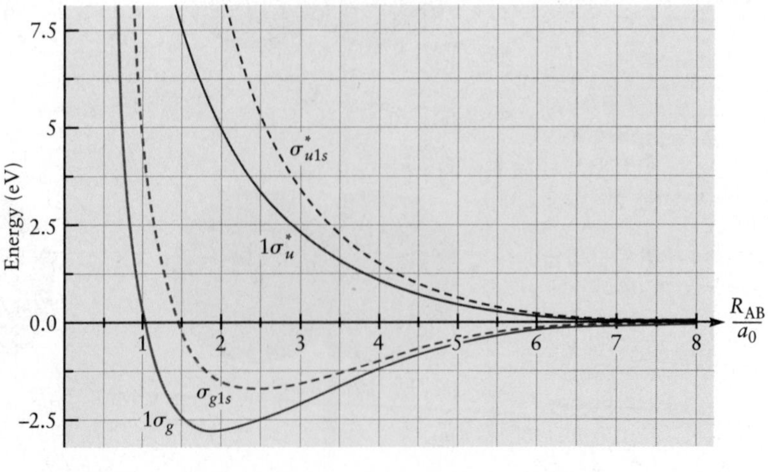

FIGURE 6.10 Comparison of the effective potential energy for the σ_{g1s} and σ^*_{u1s} orbitals of H_2^+ in the LCAO approximation (dashed lines) with the exact results (solid lines). The internuclear separation is plotted in units of the Bohr radius. (Adapted from R. S. Berry, S. A. Rice, and J. Ross, *Physical Chemistry*, 2nd ed., New York: Oxford, 2000, Figure 6.6.)

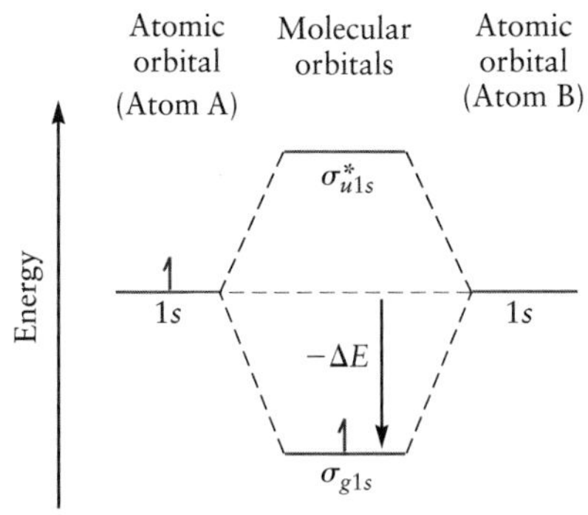

FIGURE 6.11 Correlation diagram for H_2^+ in the linear combination of atomic orbitals (LCAO) approximation. The bonding orbital is stabilized relative to the noninteracting system. $E_{1s} + (-\Delta E) = E_{\sigma_{g1s}}$ so ΔE is a positive number.

Let's put these results of the LCAO approximation in perspective. The results in Figure 6.7 and Figure 6.9 were obtained by working out the details of the approximation expressed in Equation 6.2ab. These LCAO results have captured qualitatively the results of the exact calculation. Therefore we can apply the LCAO method in other more complex cases and be confident we have included the qualitative essential features of bond formation. And, we can always improve the results by following up with a self-consistent computer calculation that produces optimized MOs. We will give some examples later in the chapter to illustrate how advanced calculations flow very easily from the simple LCAO theory introduced here.

The energy-level diagram within the LCAO approximation is given by a **correlation diagram** (Fig. 6.11), which shows that two 1*s* atomic orbitals have been combined to give a σ_{g1s} MO with energy lower than the atomic orbitals and a σ^*_{u1s} MO with higher energy than the atomic orbitals. This diagram is a purely qualitative representation of the same information contained in Figure 6.9. The actual energy level values will depend on the distance between the fixed nuclei (as shown in Figure 6.9) and must be determined from calculations. Even without the results shown in Figure 6.9, we would know that an electron in an antibonding orbital has higher energy than one in a bonding orbital because the antibonding orbital has a node. Consequently, in the ground state of H_2^+, the electron occupies the σ_{g1s} molecular orbital. By forming the bond in the molecular ion, the total system of two hydrogen nuclei and one electron becomes more stable than the separated atoms, where $E_{\sigma_{g1s}} = E_{1s} + (-\Delta E)$ and ΔE is positive (see Figure 6.11).

6.4 Homonuclear Diatomic Molecules: First-Period Atoms

We can combine the LCAO method for H_2^+ with an aufbau principle, analogous to that developed for atoms, to describe the electron configurations of more complex molecules. Electrons available from the two atoms are "fed" into the molecular orbitals, starting with the MO of lowest energy. At most, two electrons can occupy each molecular orbital. The ground-state H_2 molecule, therefore, accommodates two electrons with opposite spins in a σ_{g1s} bonding molecular orbital, as shown in Figure 6.12. The diatomic molecule is more stable than the isolated atoms by the energy difference $-2\Delta E$. The value of $-\Delta E$ here is different from that in Figure 6.11 because the present case involves the effects of electron–electron repulsion.

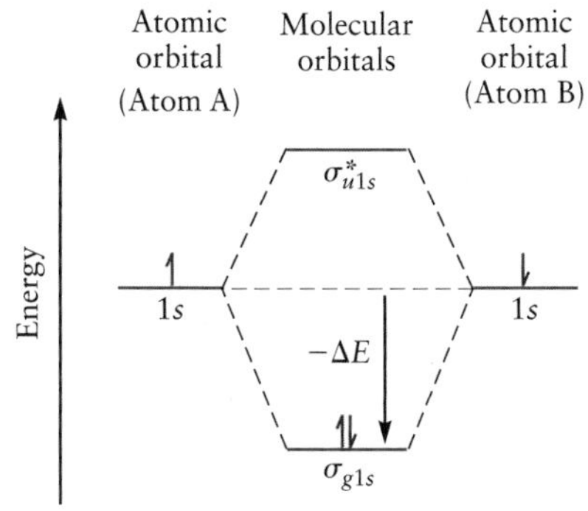

FIGURE 6.12 Correlation diagram for first-period diatomic molecules. Blue arrows indicate the electron filling for the H_2 molecule. All of the atomic electrons are pooled and used to fill the molecular orbitals following the aufbau principle. In the molecules, electrons are no longer connected to any particular atom.

The LCAO approximation can be applied in this same way to He_2^+ and He_2, with one change. The MOs must be generated as linear combinations of He 1*s* atomic orbitals, not H 1*s* orbitals. The reason is that when the electrons in He_2^+ and He_2 approach close to one of the nuclei, they experience a potential much closer to that in a He atom than in a H atom. Therefore, the equations for the MOs are

$$\sigma_{g1s} = C_g[\varphi^{A}_{He1s} + \varphi^{B}_{He1s}] \quad [6.5a]$$

$$\sigma^*_{u1s} = C_u[\varphi^{A}_{He1s} - \varphi^{B}_{He1s}] \quad [6.5b]$$

We rewrite Equations 6.5a and 6.5b using a simpler notation, which we adopt for the remainder of the text.

$$\sigma_{g1s} = C_g[1s^{A} + 1s^{B}] \quad [6.6a]$$

$$\sigma^*_{u1s} = C_u[1s^{A} - 1s^{B}] = C_u[1s^{A} + (-1s^{B})] \quad [6.6b]$$

The second equality shown in Equation 6.6b emphasizes that the minus sign changes the sign of the $1s$ wave function, from positive (red) to negative (blue), to help you understand the graphical representation of these linear combinations in Figure 6.8 and the other figures in this chapter.

In these equations the symbol for the atomic wave function φ has been dropped and the atomic orbitals are identified by their hydrogenic labels *1s, 2s, 2p,* etc. The superscripts A and B will be used to identify particular atoms of the same element in bonds formed from the same element in homonuclear diatomics. These helium MOs have the same general shapes as those shown in Figure 6.7 for the MOs constructed from H $1s$. They lead to potential energy curves as shown in Figure 6.9 and a correlation diagram similar to Figure 6.12. Quantitative calculations of electron density and energy (these calculations are not performed in this book) would produce different values for the two sets of MOs in Equations 6.2a and 6.2b and Equations 6.5a and 6.5b. Keep in mind that we construct the MOs as combinations of all the atomic orbitals required to accommodate the electrons in the ground states of the atoms forming the molecule. This set of atomic orbitals is called the *minimum basis set* for that specific molecule. Therefore, quantitative calculations for each molecule are influenced by the detailed properties of the atoms in the molecule.

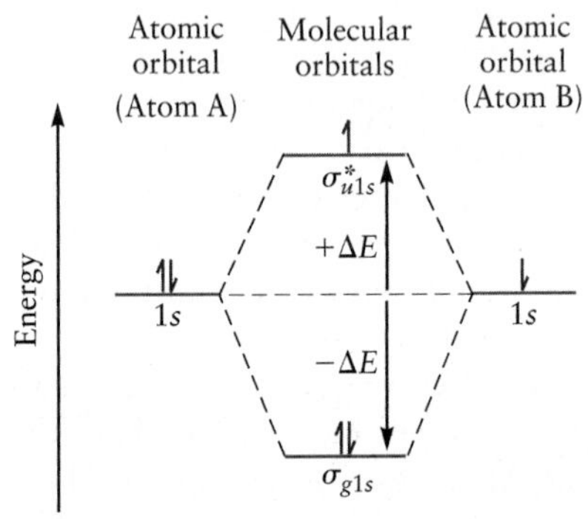

FIGURE 6.13 Correlation diagram for first-period diatomic molecules. Blue arrows indicate the electron filling for the He_2^+ molecule. The aufbau principle fills the bonding orbital with two electrons, so the third electron must go into the antibonding orbital, thus reducing the bond order compared with that in H_2.

Because He_2^+ and He_2 have more than two electrons, the aufbau principle requires them to have one or two electrons, respectively, in the σ^*_{u1s} antibonding orbital. The electrons in the antibonding orbital contribute a destabilization energy in the amount $+\Delta E$ relative to the separated He atoms, as shown in Figure 6.13. This effect competes with the stabilization energy of $-\Delta E$ that arises from the electrons in the σ_{g1s} bonding orbital, giving a weak bond in He_2^+ and no stable bond in He_2.

The general features of covalent bonding in the LCAO picture can be summarized as follows. Covalent bond formation arises from the presence of electrons (most often electron pairs with opposite spins) in *bonding* MOs. The average electron density between the nuclei is greater than between the noninteracting atoms, and tends to pull them together. Electrons in an *antibonding* MO tend to force the nuclei apart, reducing the bond strength. This competition is described by the **bond order,** defined as follows:

bond order = $\frac{1}{2}$ (number of electrons in bonding molecular orbitals − number of electrons in antibonding molecular orbitals)

In the LCAO molecular orbital description, the H_2 molecule in its ground state has a pair of electrons in a bonding molecular orbital and thus a single bond (that is, its bond order is one). Later on, as we describe more complex diatomic molecules in the LCAO approximation, we will see bond orders greater than one. This quantum mechanical definition of bond order generalizes the concept first developed in the Lewis theory of chemical bonding, that a shared pair of electrons corresponds to a single bond, two shared pairs to a double bond, and so forth.

EXAMPLE 6.2

Give the ground-state electron configuration and the bond order of the He_2^+ molecular ion.

Solution

The He_2^+ ion has three electrons, which are placed in molecular orbitals to give the ground-state configuration $(\sigma_{g1s})^2(\sigma^*_{u1s})^1$, indicating that the ion has a doubly

(Continued)

occupied σ_{g1s} orbital (bonding) and a singly occupied σ^*_{u1s} orbital (antibonding). The bond order is

$$\text{bond order} = \frac{1}{2}(2 \text{ electrons in } \sigma_{g1s} - 1 \text{ electron in } \sigma^*_{u1s}) = \frac{1}{2}$$

This should be a weaker bond than that in H_2.

Related Problems: 7, 8, 9, 10, 11, 12, 13, 14, 15, 16

TABLE 6.2

Electron Configurations and Bond Orders for First-Row Diatomic Molecules

Species	Electron Configuration	Bond Order	Bond Energy (kJ mol^{-1})	Bond Length (Å)
H_2^+	$(\sigma_{g1s})^1$	$\frac{1}{2}$	255	1.06
H_2	$(\sigma_{g1s})^2$	1	431	0.74
He_2^+	$(\sigma_{g1s})^2(\sigma^*_{u1s})^1$	$\frac{1}{2}$	251	1.08
He_2	$(\sigma_{g1s})^2(\sigma^*_{u1s})^2$	0	~0	Large

Table 6.2 lists the molecular orbital configurations of **homonuclear diatomic molecules** and molecular ions made from first-period elements. These configurations are simply the occupied molecular orbitals in order of increasing energy, together with the number of electrons in each orbital. Higher bond order corresponds to greater bond dissociation energies and shorter bond lengths. The species He_2 has bond order 0 and does not form a true chemical bond.

The preceding paragraphs have illustrated the LCAO approximation with specific examples and shown how the character of the chemical bond is determined by the difference in the number of electrons in bonding and antibonding MOs. The systematic procedure for applying the LCAO approximation to define the MOs for any diatomic molecule consists of three steps:

1. Form linear combinations of the minimum basis set of AOs (all of the AOs in the valence shell of each atom, whether or not these AOs are occupied in the ground state of each atom) to generate MOs. The total number of MOs formed in this way must equal the number of AOs used.
2. Arrange the MOs in order from lowest to highest energy.
3. Put in electrons (at most two electrons per MO), starting from the orbital of lowest energy. Apply Hund's rules when appropriate.

6.5 Homonuclear Diatomic Molecules: Second-Period Atoms

Now that we have some experience with the LCAO method, it is useful to collect several important insights before we proceed to more complicated molecular examples. The LCAO method extends to molecules the description developed for many-electron atoms in Section 5.2. We wrote the electron configuration for a many-electron atom by placing electrons in a set of single-particle AOs according to the Pauli exclusion principle. Here we will see how to write the electron configuration for a molecule by placing electrons in a set of single-particle MOs, guided by the Pauli exclusion principle.

How do we obtain the single-electron molecular orbitals? The LCAO method was invented to construct *approximate molecular orbitals* directly from the Hartree atomic orbitals for the specific atoms in the molecule, guided by molecular symmetry and chemical intuition. We start with the Hartree AOs (see Section 5.2) because they already include the effects of the atomic number Z and of shielding on the size and energy of the AOs in each atom. We thus have a quick check on, for example, the relative energies of the $2s$ and $2p$ AOs in a pair of different atoms, and from this we can quickly see the extent to which electrons in these AOs interact or compete to determine the size and energy of the resulting MO.

The essential new feature compared to the atomic case is that the (multi-center) approximate MOs are spread around all the nuclei in the molecule, so the electron density is *delocalized* over the entire molecule. The approximate MOs therefore differ considerably from the (single-center) atomic orbitals used in Section 5.2. Constructing the approximate MOs and recognizing how their electron density is distributed over the entire network of fixed nuclei are the key tools for describing molecular structure and bonding. Mastering these tools for progressively more complicated molecules is our objective in this Section and the next three after it.

In order to discuss bonding with atoms in the second period and beyond, we must generate approximate MOs to accommodate electrons from AOs higher than the $1s$ orbitals. Let's try to motivate these combinations by physical reasoning, as we did at the beginning of Section 6.3. Imagine that two Li atoms in their ground-state electronic configurations $(1s)^2(2s)^1$ approach each other. It is reasonable to expect that their $2s$ AOs will interact to produce MOs very similar in shape to the σ_{1s} bonding and anti-bonding MOs but larger in size, and that two electrons will go into this new bonding MO which we temporarily name σ_{2s}. What happens to the $1s$ electrons? We can use Figure 5.26 to estimate that the energy of the Li($1s$) AO is about 3.8 Ry (or 52 eV) below that of the Li($2s$) AO. Because this is a very significant difference, it is unlikely that the $1s$ and $2s$ electrons will interact, and it is reasonable to suppose the $1s$ orbitals will combine to form σ_{1s} bonding and anti-bonding MOs as described in the previous section, independent of the $2s$ electrons. If our thinking is correct, then the Li_2 molecule would have the electron configuration $(\sigma_{g1s})^2(\sigma^*_{u1s})^2(\sigma_{g2s})^2$. In a similar way we would expect Be_2 to have the configuration $(\sigma_{g1s})^2(\sigma^*_{u1s})^2(\sigma_{g2s})^2(\sigma^*_{u2s})^2$. Because of the very poor spatial overlap of the $1s$ AOs, however, it is typical to see those electrons listed as simply remaining in the two $1s$ AOs. They are often omitted from the LCAO-MO correlation diagram. In this case one should specify that you are constructing the MOs that result from the valence AOs only, and don't forget to subtract off the core electrons (two $1s$ electrons on each Li atom) from the total number of electrons to calculate the number of electrons to put in your LCAO-MO correlation diagram.

In the remainder of Period 2 from B to Ne we must consider the possibility that the $2p$ AOs will interact with each other and also with the $2s$ AOs. Because the $2p$ AOs are not spherical, we expect the relative orientation of the two orbitals will strongly influence the formation of MOs. Qualitatively, we expect that two dumbbell shapes interacting end-to-end will produce a different result than side-to-side or end-to-side. Moreover, the difference in phase (the sign of the wave function) of the two lobes will strongly influence the result. Careful geometrical analysis is required to identify the MOs formed from the $2p$ AOs.

Two conclusions from more advanced discussions of quantum theory justify the results of our physical reasoning and tell us how to construct the MOs for second-period atoms in a systematic way.

1. *Two atomic orbitals contribute significantly to bond formation only if their atomic energy levels are very close to one another.*

Consequently, we can ignore mixing between the core-shell 1*s* orbitals and the valence-shell 2*s* and 2*p* orbitals. Similarly, we can ignore mixing between the 2*s* and 2*p* orbitals except in special cases to be described below.

2. *Two atomic orbitals on different atoms contribute significantly to bond formation only if they overlap significantly.*

The term *overlap* is used in bonding theory to describe the extent to which orbitals on separate atoms "interact" or "inter-penetrate" as two atoms approach one another at close distance. Two orbitals overlap significantly if both have appreciable amplitudes over the same region of space. The net overlap may be positive, negative, or zero, depending on the relative amplitudes and phases of orbitals involved in the overlap region. Formally, one calculates an overlap integral, the integral of the product of φ^A with φ^B over all space. If this is zero, the two orbitals are said to have zero overlap. Thus, though they appear to have overlapping regions, the 2*s* orbital on atom A has zero overlap with the $2p_x$ or the $2p_y$ orbital on atom B (defining the *z* axis as the internuclear axis). Bonding molecular orbitals arise from positive net overlap (constructive interference between the wave functions for atomic orbitals), antibonding molecular orbitals result from negative net overlap (destructive interference between the wave functions for atomic orbitals). Nonbonding molecular orbitals can arise in two ways: a) there may be negligible overlap of the atomic orbitals because the separation between the nuclei is greater than the spatial extent of the orbitals or b) regions of positive overlap cancel regions of negative overlap to give zero net overlap. We give specific examples in the following paragraphs.

For *s* orbitals it is rather easy to guess the degree of overlap; the closer the nuclei, the greater the overlap. If the wave functions have the same phase, the overlap is positive; if they have opposite phases, the overlap is negative. For more complex cases, the overlap between participating atomic orbitals depends strongly on both the symmetry of the arrangement of the nuclei and on the phases of the orbitals. If the two orbitals are shaped so that neither has substantial amplitude in the region of interest, then their overlap is negligible. However, if they both have significant magnitude in the region of interest, it is important to know whether regions of positive overlap (where the two orbitals have the same phase) are canceled by regions of negative overlap (where the two orbitals have opposite phases). Such cancellation leads to negligible or zero overlap between the orbitals. Qualitative sketches that illustrate significant or negligible overlap in several common cases are shown in Figure 6.14. In particular, constructive interference and overlap between *s* and *p* orbitals is significant only in the case where an *s* orbital approaches a *p* orbital "end-on." The phase of the *p* orbital lobe pointing toward the *s* orbital must be the same as that of the *s* orbital. You should review the "sizes and shapes" of hydrogenic orbitals discussed in Section 5.1 and depicted in Figure 5.4. A great deal of qualitative insight into the construction of molecular orbitals can be gleaned from these considerations.

The two conclusions stated above justify the following approximate LCAO MOs for the second-period homonuclear diatomic molecules. As with the first-period atoms, we use the new labels in Table 6.1 to distinguish the approximate MOs from the exact H_2^+ MOs in Figure 6.5 and to indicate their atomic parentage.

We combine the 2*s* atomic orbitals of the two atoms in the same fashion as 1*s* orbitals, giving a σ_{g2s} bonding orbital and a σ_{u2s}^* antibonding orbital. The

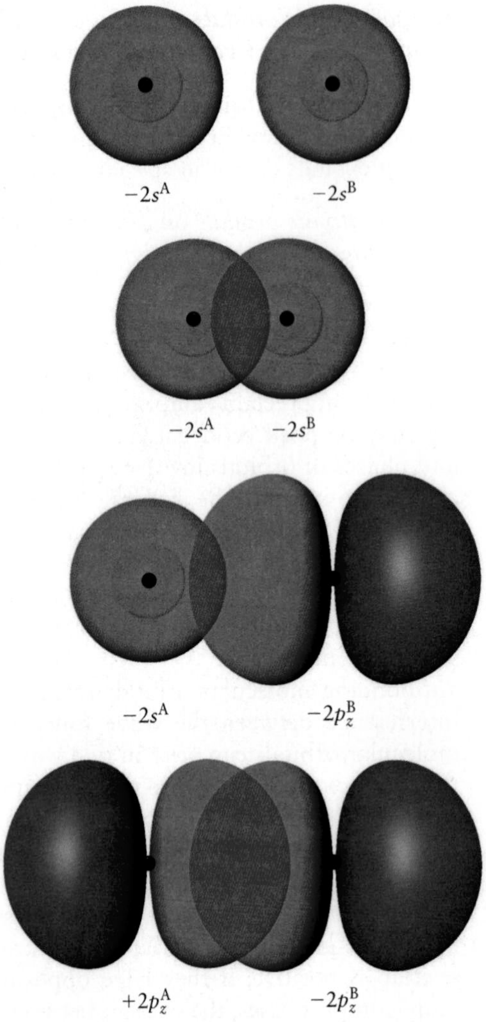

FIGURE 6.14 Overlap of orbitals in several common combinations. The magnitude of overlap can be estimated qualitatively from the relative size and symmetry of the two orbitals involved. (Note the radial nodes in the 2*s* orbitals, clearly visible in these images, and the minus sign in front of each. Table 5.2 and Figure 5.7 show the sign of the 2*s* wave function far from the nucleus is negative, but we are depicting the orbitals in these linear combinations as red. The red spherical orbital depicted is the 2*s* function in Table 5.2 multiplied by a −1, so the resulting −2*s* wave function is positive far from the nucleus.) For the $2p_z$ orbitals, the *z*-axis points from left to right along the internuclear axis. (Courtesy of Mr. Hatem Helal and Professor William A. Goddard III, California Institute of Technology, and Dr. Kelly P. Gaither, University of Texas at Austin.)

particular linear combinations that give the molecular orbitals depicted in Figure 6.18 are

$$\sigma_{g2s} = C_g[-2s^A - 2s^B] \quad [6.7a]$$

$$\sigma^*_{u2s} = C_u[-2s^A + 2s^B] \quad [6.7b]$$

(Note that after the wave function is squared to obtain the MO's electron density, the linear combination calculated by writing σ_{g2s} as either $2s^A + 2s^B$ or as $-2s^A - 2s^B$, give the same electron density, so correspond to the same molecular orbital. We are only being specific about the signs so that it is easy to match the linear combination with the graphical representations in Figures 6.14 and 6.18.)

The choice of appropriate combinations of the 2*p* orbitals is guided by the overlap arguments and by recalling that the bond axis is the *z*-axis. The 2*p* orbitals form different MOs depending on whether they are parallel or perpendicular to the internuclear (bond) axis. Consider first the $2p_z$ orbitals, which can be used to form two different kinds of σ orbitals. If the phase of the $2p_z$ orbital on atom B is reversed by multiplying it by a −1 (see Equation 6.8a), it then has the same phase in the internuclear region as the $2p_z$ orbital on atom A (see Figure 6.15a). They thus interfere constructively in the internuclear region, and a bonding σ_{g2p_z} orbital is formed. Conversely, if lobes of opposite phase overlap, they form an antibonding MO labeled $\sigma^*_{u2p_z}$ (see Figure 6.15b, corresponding to the linear combination shown in Equation 6.8b).

$$\sigma_{g2p_z} = C_g[2p_z^A - 2p_z^B] \qquad [6.8a]$$

$$\sigma^*_{u2p_z} = C_u[2p_z^A + 2p_z^B] \qquad [6.8b]$$

The bonding MO shows increased electron density between the nuclei, whereas the antibonding MO has a node.

The two $2p_x$ orbitals, whose lobes are perpendicular to the bond axis, can overlap "side by side" to form a bonding MO denoted π_{u2p_x} and an antibonding MO denoted $\pi^*_{g2p_x}$. The overlap leading to these MOs is shown in Fig. 6.16.

$$\pi_{u2p_x} = C_u[2p_x^A + 2p_x^B] \qquad [6.9a]$$

$$\pi^*_{g2p_x} = C_g[2p_x^A - 2p_x^B] \qquad [6.9b]$$

These orbitals have a nodal plane that contains the internuclear axis (in this case, the y-z plane perpendicular to the page). They are designated by π to signify their origin in AOs that have one unit of angular momentum along the internuclear axis. The π MOs do not have cylindrical symmetry about the internuclear axis. In the same way, π_{u2p_y} and $\pi^*_{g2p_y}$ orbitals can be formed from the $2p_y$ atomic orbitals. Their lobes project above and below the x-z nodal plane, which is the plane of the page, so they are shown in perspective in Figure 6.18.

$$\pi_{u2p_y} = C_u[2p_y^A + 2p_y^B] \qquad [6.10a]$$

$$\pi^*_{g2p_y} = C_g[2p_y^A - 2p_y^B] \qquad [6.10b]$$

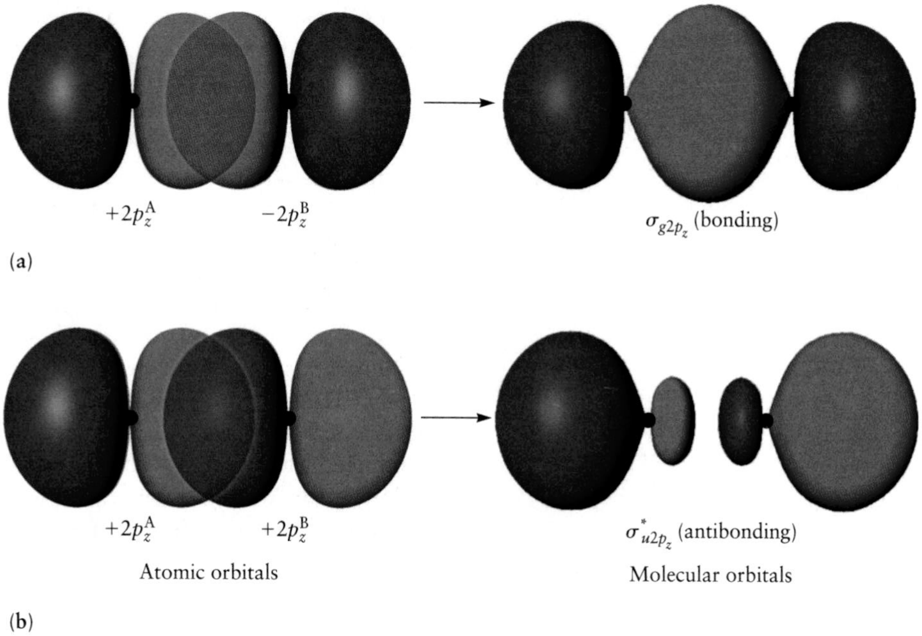

FIGURE 6.15 Formation of (a) σ_{g2p_z} bonding and (b) $\sigma^*_{u2p_z}$ antibonding molecular orbitals from $2p_z$ orbitals on atoms A and B. Regions where the wave function has positive sign are shown in red, and those with negative sign are shown in blue.

(Courtesy of Mr. Hatem Helal and Professor William A. Goddard III, California Institute of Technology, and Dr. Kelly P. Gaither, University of Texas at Austin.)

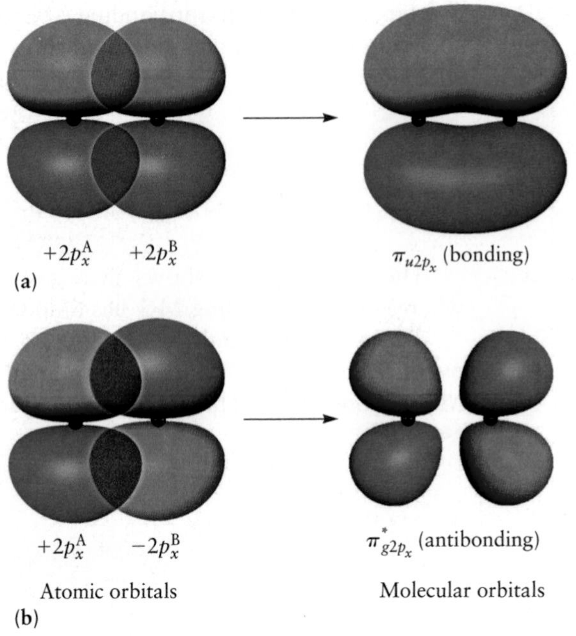

FIGURE 6.16 Formation of (a) π_{u2p_x} bonding and (b) $\pi^*_{g2p_x}$ antibonding molecular orbitals from $2p_x$ orbitals on atoms A and B. (Courtesy of Mr. Hatem Helal and Professor William A. Goddard III, California Institute of Technology, and Dr. Kelly P. Gaither, University of Texas at Austin.)

Like the $2p$ AOs from which they are constructed, the π_{2p} MOs are degenerate: π_{u2p_x} and π_{u2p_y} have the same energy, and $\pi^*_{g2p_x}$ and $\pi^*_{g2p_y}$ have the same energy. We expect π_{u2p_x} and π_{u2p_y} to be less effective than σ_{g2p_z} as bonding orbitals, because the overlap in the π orbitals is smaller and because it occurs off the internuclear axis and therefore has less tendency to increase the electron density between the nuclei and to pull them closer together.

The most important point to understand in constructing MOs and predicting their behavior by the overlap argument is that the relative phases of the two AOs determine whether the resulting MO is bonding or antibonding. Bonding orbitals form through the overlap of wave functions with the same phase, by constructive interference of "electron waves"; antibonding orbitals form through the overlap of wave functions with opposite phase, by destructive interference of "electron waves."

By considering the symmetry and relative energies of the participating AOs, we have generated a total of 8 approximate LCAO MOs to accommodate the electrons in the $2s$ and $2p$ AOs. Including the two MOs originating from the $1s$ AOs, we now have a total of 10 approximate MOs which can describe diatomic molecules containing up to 20 electrons, from H_2 to Ne_2. Each of these is a simple sum or difference of two AOs.

The next step is to determine the energy ordering of the MOs. In general, that step requires a calculation, as we did for the first-period diatomics in Figure 6.11 and Figure 6.12. The results for Li_2 through F_2 are shown in Figure 6.17. The electrons for each molecule have been placed in MOs according to the aufbau principle. We show only the MOs formed from the $2s$ and $2p$ orbitals. In second-period diatomic molecules, the $1s$ orbitals of the two atoms barely overlap. Because the σ_{g1s} bonding and σ^*_{u1s} antibonding orbitals are both doubly occupied, they have little net effect on bonding properties and need not be considered. Thus in Figure 6.18 we only show the MOs from the valence AOs, and we leave the electrons in the core $1s$ orbitals out of the diagram, filling the MOs generated from valence AOs with the valence electrons only.

There are two different energy orderings for diatomic molecules formed from second-period elements. The first (Fig. 6.18a) applies to the molecules with atoms Li through N (that is, the first part of the period) and their positive and negative ions. The second (Fig. 6.18b) applies to the later elements, O, F, and Ne, and their positive and negative ions. This difference is based on experimental measurements of molecular orbital energies by photoelectron spectroscopy (see Section 6.7) summarized in Figure 6.17. While the energy of the $\pi_{u2p_{x(y)}}$ molecular orbital remains nearly constant as we move across Period 2, the energy of the σ_{g2p_z} molecular orbital lies above $\pi_{u2p_{x(y)}}$ in the first part of the period and falls below it for the later elements O, F, and Ne. This behavior is borne out by advanced calculations of the energies for these two molecular orbitals. A simplified physical interpretation relates this result to the extent to which the 2*s* and $2p_z$ atomic orbitals can mix while contributing to the molecular orbital. The 2*s* AOs have the right symmetry to mix with $2p_z$ AOs to form a σ molecular orbital directed along the *z*-axis. In the first part of Period 2 the 2*s* and 2*p* are sufficiently close in energy that mixing occurs and increases the energy of the nominal σ_{g2p_z} orbital while decreasing the energy of the nominal σ_{g2s} orbital. In the latter part the energy separation of 2*p* and 2*s* AOs is too great for mixing to occur. Because the 2*s* and $2p_x$ and $2p_y$ AOs do not have the right symmetry to mix and form a π molecular orbital, the energy of the $\pi_{u2p_{x(y)}}$ MO remains nearly constant across the period even when the 2*s* and 2*p* AOs are close together in energy.

An important prediction comes from the molecular orbital energy levels in Figure 6.17. Hund's rules require that, in the ground state, the electrons occupy different orbitals and have parallel spins, so B_2 and O_2 are predicted to be paramagnetic. This paramagnetism is exactly what is found experimentally (Fig. 6.19). In contrast, in the Lewis electron dot diagram for O_2,

:Ö::Ọ:

all the electrons appear to be paired. The highly reactive nature of molecular oxygen as compared to N_2 can be rationalized as resulting from the readiness of the two π^* electrons, unpaired and in different regions of space, to find additional bonding partners in other molecules.

The electron configurations in Figure 6.17 allow us to calculate the bond order for each molecule and correlate it with other properties of the molecules.

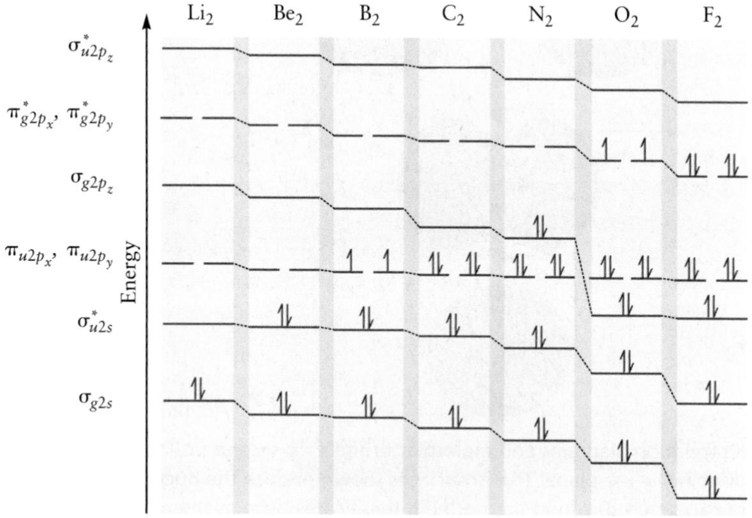

FIGURE 6.17 Energy levels for the homonuclear diatomics Li_2 through F_2. Notice how the highest occupied level changes with the number of valence electrons. Notice especially the change between N_2 and O_2.

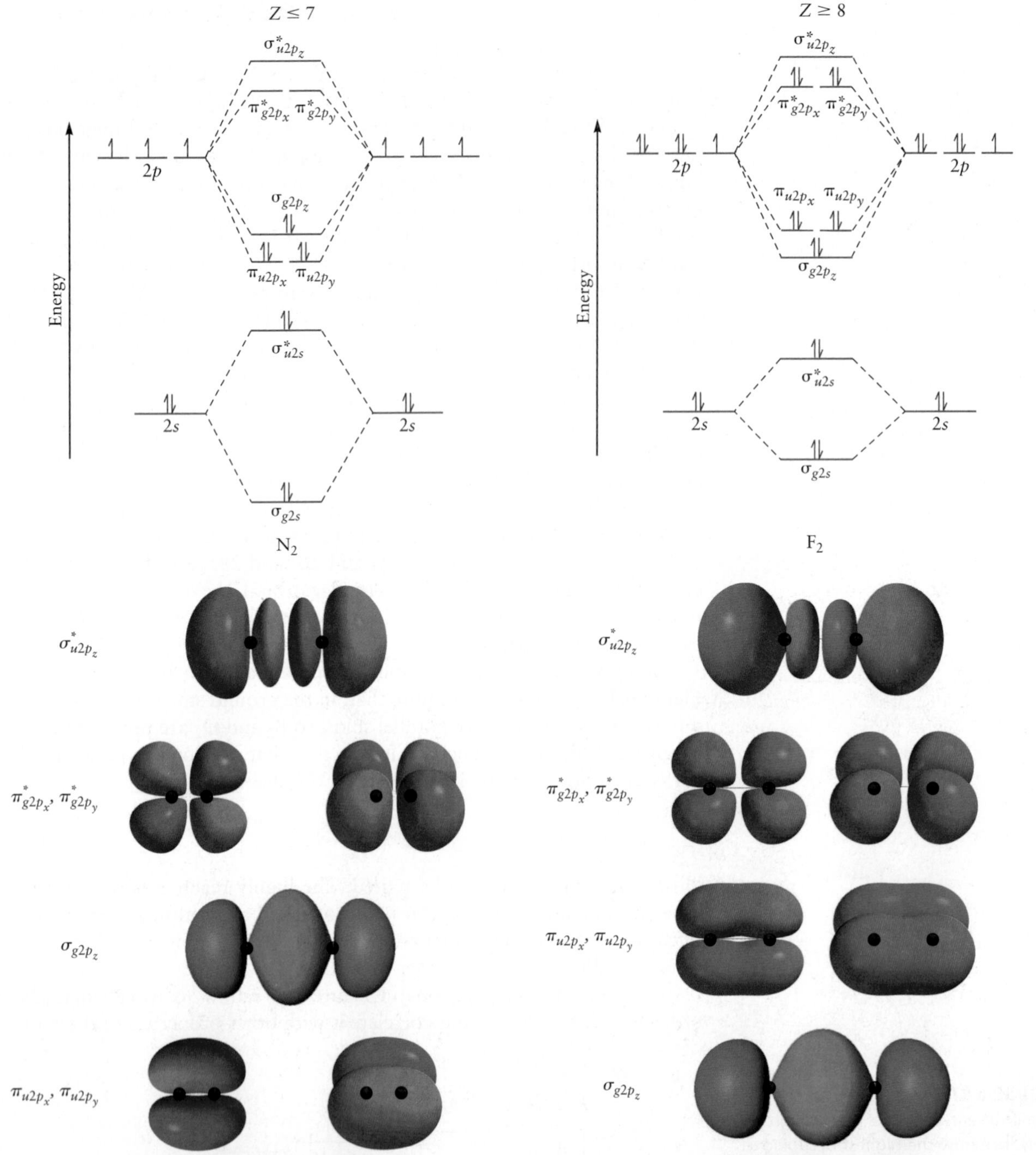

FIGURE 6.18 Correlation diagrams and molecular orbitals for second-period diatomic molecules. The black dots represent the positions of the nuclei in the *x–z* plane. The isosurfaces shown enclose the nuclei in some cases and it may be difficult to visualize the curved nodal surfaces that surround the nuclei in these cases, such as the $\sigma^*_{u2p_z}$ orbitals. (a) Correlation diagram and molecular orbitals calculated for N_2. (b) Correlation diagram and molecular orbitals calculated for F_2. (Courtesy of Mr. Hatem Helal and Professor William A. Goddard III, California Institute of Technology, and Dr. Kelly P. Gaither, University of Texas at Austin.)

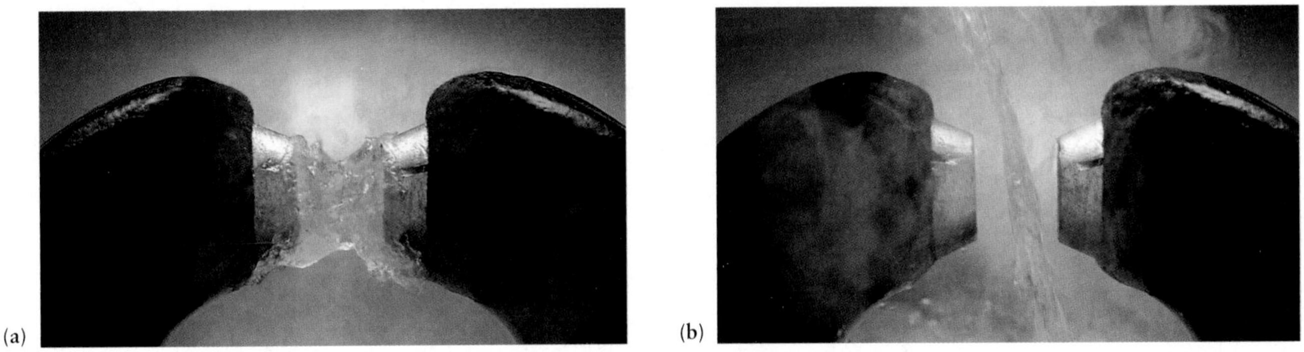

FIGURE 6.19 (a) Oxygen is paramagnetic; liquid oxygen (O_2) poured between the pole faces of a magnet is attracted and held there. (b) When the experiment is repeated with liquid nitrogen (N_2), which is diamagnetic, the liquid pours straight through.

EXAMPLE 6.3

Determine the ground-state electron configuration and bond order of the F_2 molecule.

Solution

Each atom of fluorine has 7 valence electrons, so 14 electrons are placed in the molecular orbitals constructed from the valence atomic orbitals (occupied and unoccupied) to represent bonding in the F_2 molecule. The correlation diagram of Figure 6.18 gives the electron configuration

$$(\sigma_{g2s})^2(\sigma_{u2s}^*)^2(\sigma_{g2p_z})^2(\pi_{u2p_x}, \pi_{u2p_y})^4(\pi_{g2p_x}^*, \pi_{g2p_y}^*)^4$$

Because there are 8 valence electrons in bonding orbitals and 6 in antibonding orbitals, the bond order is

$$\text{bond order} = \tfrac{1}{2}(8 - 6) = 1$$

and the F_2 molecule has a single bond.

Related Problems: 17, 18, 19, 20, 21, 22, 23, 24

Table 6.3 summarizes the properties of second-period homonuclear diatomic molecules. Note the close relationship among bond order, bond length, and bond energy and the fact that the bond orders calculated from the MOs agree completely with the results of the Lewis electron dot model. How these properties depend upon the number of electrons in the molecules is shown in Figure 6.20. The bond orders simply follow the filling of MOs in a given subshell, rising from

TABLE 6.3

Molecular Orbitals of Homonuclear Diatomic Molecules

Species	Number of Valence Electrons	Valence Electron Configuration	Bond Order	Bond Length (Å)	Bond Energy (kJ mol^{-1})
H_2	2	$(\sigma_{g1s})^2$	1	0.74	431
He_2	4	$(\sigma_{g1s})^2(\sigma_{u1s}^*)^2$	0		
Li_2	2	$(\sigma_{g2s})^2$	1	2.67	105
Be_2	4	$(\sigma_{g2s})^2(\sigma_{u2s}^*)^2$	0	2.45	9
B_2	6	$(\sigma_{g2s})^2(\sigma_{u2s}^*)^2(\pi_{u2p})^2$	1	1.59	289
C_2	8	$(\sigma_{g2s})^2(\sigma_{u2s}^*)^2(\pi_{u2p})^4$	2	1.24	599
N_2	10	$(\sigma_{g2s})^2(\sigma_{u2s}^*)^2(\pi_{u2p})^4(\sigma_{g2p_z})^2$	3	1.10	942
O_2	12	$(\sigma_{g2s})^2(\sigma_{u2s}^*)^2(\sigma_{g2p_z})^2(\pi_{u2p})^4(\pi_{g2p}^*)^2$	2	1.21	494
F_2	14	$(\sigma_{g2s})^2(\sigma_{u2s}^*)^2(\sigma_{g2p_z})^2(\pi_{u2p})^4(\pi_{g2p}^*)^4$	1	1.41	154
Ne_2	16	$(\sigma_{g2s})^2(\sigma_{u2s}^*)^2(\sigma_{g2p_z})^2(\pi_{u2p})^4(\pi_{g2p}^*)^4(\sigma_{u2p_z}^*)^2$	0		

π_{u2p} refers to the π orbitals constructed from the $2p_x$ and/or the $2p_y$ orbitals.

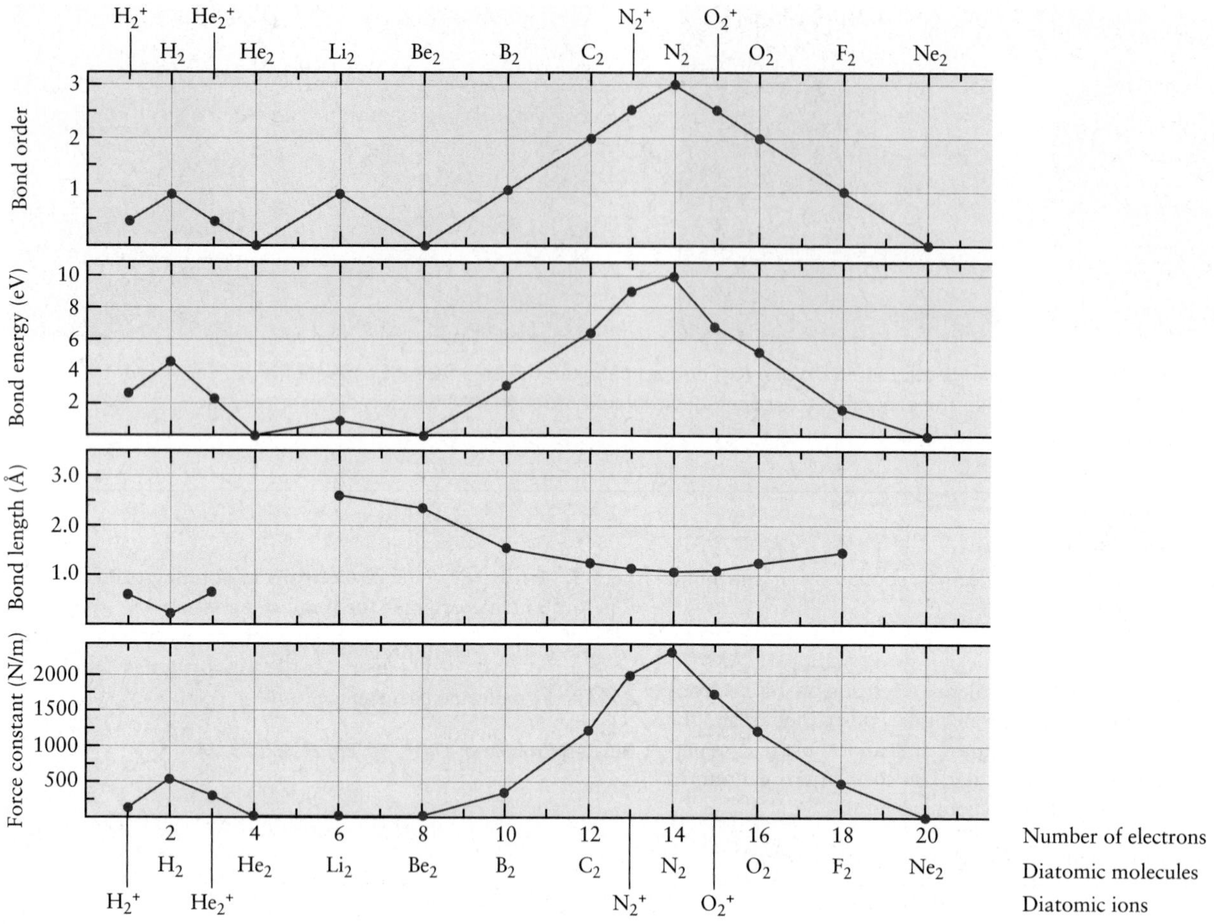

FIGURE 6.20 Trends in bond order, bond length, bond energy, and force constant with the number of valence electrons in the second-row diatomic molecules.

zero to one and then falling back to zero for the first-period diatomics and also for Li_2, Be_2, and their ions. The bond orders move in half-integral steps if we include the molecular ions; they increase as the σ orbitals are filled and decrease as the σ^* orbitals begin to fill. These MOs are all constructed from *s* orbitals and so there is no possibility for multiple bonds. Moving from Be_2 through Ne_2, the bond orders move again in half-integer increments from zero to three and back to zero as the π orbitals and then the π^* orbitals are filled. Both bond energies and force constants are directly correlated with the bond order, whereas the bond length varies in the opposite direction. This makes perfect sense; multiple bonds between atoms should be stronger and shorter than single bonds.

In summary, the simple LCAO method provides a great deal of insight into the nature of chemical bonding in homonuclear diatomic molecules and the trends in their properties that result. It is consistent with the predictions of simpler theories, like that of G. N. Lewis; but clearly more powerful and more easily generalized to problems of greater complexity.

6.6 Heteronuclear Diatomic Molecules

A diatomic molecule formed from atoms of two different elements (such as CO and NO) is called a **heteronuclear diatomic molecule.** We construct molecular orbitals for such molecules by following the procedure described earlier, with two changes. First, we use a different set of labels because heteronuclear diatomic molecules lack the inversion symmetry of homonuclear diatomic molecules. We therefore drop the *g* and *u* subscripts on the MO labels. Second, we recognize that the AOs on the participating atoms now correspond to different energies. For example, we combine the 2*s* atomic orbital of carbon and the 2*s* atomic orbital of oxygen to produce a bonding MO (without a node)

$$\sigma_{2s} = C_A(-2s^A) + C_B(-2s^B) \qquad [6.11a]$$

and an antibonding MO (with a node)

$$\sigma^*_{2s} = C'_A(-2s^A) + C_B(2s^B) \qquad [6.11b]$$

where A and B refer to the two different atoms in the molecule. In the homonuclear case we argued that $C_A = C_B$ and $C'_A = C'_B$ because the electron must have equal probability of being near each nucleus, as required by symmetry. When the two nuclei are different, this reasoning does not apply. If atom B is more electronegative than atom A, then $|C_B| > |C_A|$ for the bonding σ MO (and the electron spends more time on the electronegative atom); $|C'_A| > |C'_B|$ for the higher energy σ^* MO and it more closely resembles a $2s^A$ AO.

Molecular orbital correlation diagrams for heteronuclear diatomics start with the energy levels of the more electronegative atom displaced *downward* because that atom attracts valence electrons more strongly than does the less electronegative atom. Figure 6.21 shows the diagram appropriate for many heteronuclear diatomic molecules of second-period elements (where the electronegativity difference is not too great). This diagram has been filled with the valence electrons for the ground state of the molecule BO. Another example, NO, with 11 valence electrons (5 from N, 6 from O), has the ground-state configuration

$$(\sigma_{2s})^2(\sigma^*_{2s})^2(\pi_{2p_x}, \pi_{2p_y})^4(\sigma_{2p_z})^2(\pi^*_{2p_x}, \pi^*_{2p_y})^1$$

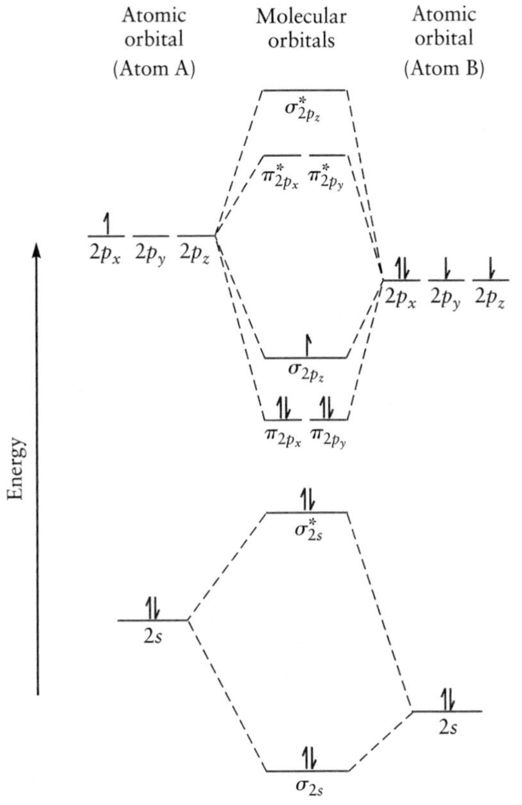

FIGURE 6.21 Correlation diagram for heteronuclear diatomic molecules, AB. The atomic orbitals for the more electronegative atom (B) are displaced downward because they have lower energies than those for A. The orbital filling shown is that for (boron monoxide) BO.

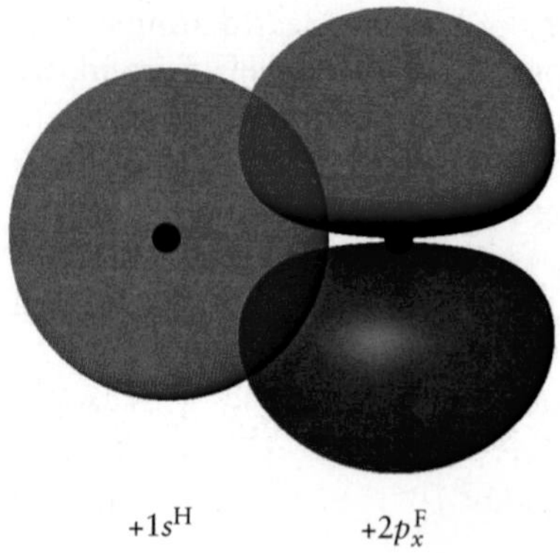

FIGURE 6.22 The H 1*s* and F $2p_x$ (or $2p_y$) orbitals have no net overlap in HF.

With eight electrons in bonding orbitals and three in antibonding orbitals, the bond order of NO is $\frac{1}{2}(8 - 3) = 2\frac{1}{2}$ and it is paramagnetic. The bond energy of NO should be smaller than that of CO, which has one fewer electron but a bond order of three; experiment agrees with this prediction.

We explained earlier that in homonuclear diatomics, atomic orbitals mix significantly to form molecular orbitals only if they are fairly close in energy (within 1 Ry or so) and have similar symmetries. The same reasoning is very helpful in constructing MOs for heteronuclear diatomics. For example, in the HF molecule, both the 1*s* and 2*s* orbitals of the F atom are far too low in energy to mix with the H 1*s* orbital. Moreover, the overlap between the H 1*s* and F 2*s* is negligible (Fig. 6.23a). The net overlap of the H 1*s* orbital with the $2p_x$ or $2p_y$ F orbital is zero (Fig. 6.22) because the regions of positive and

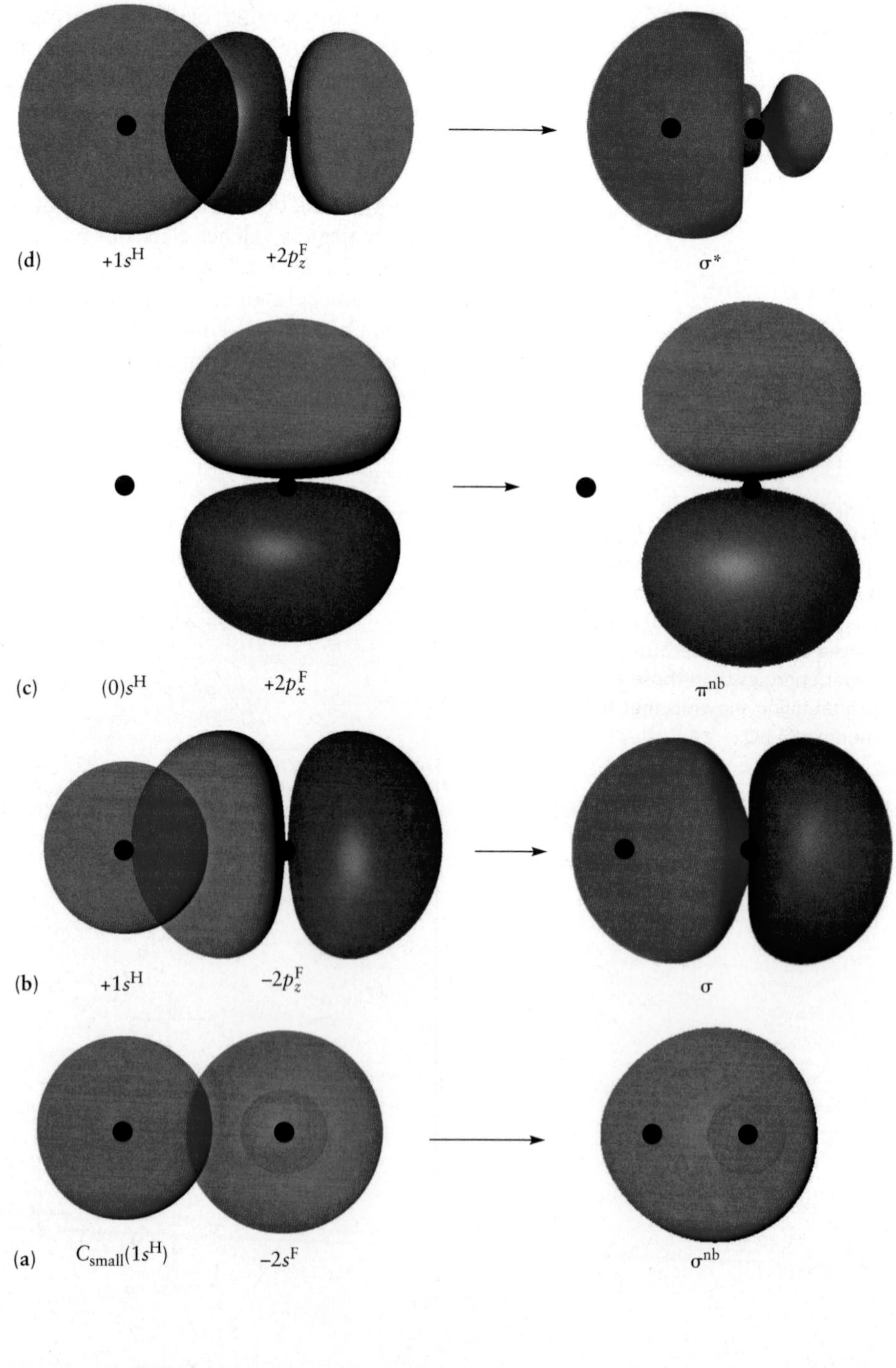

FIGURE 6.23 Overlap of atomic orbitals in HF. (**Note:** in the correlation diagram in Figure 6.24 it is assumed that lowest energy orbital is a nonbonding one that looks identical to a 2*s* orbital on the F atom, with no contribution from the H 1*s* orbital. In the graphical depiction of this orbital in (a) in this figure, a small coefficient multiplies the H 1*s* orbital in this linear combination; the overlap is negligible.) (Courtesy of Mr. Hatem Helal and Professor William A. Goddard III, California Institute of Technology, and Dr. Kelly P. Gaither, University of Texas at Austin.)

negative overlap sum to zero. This leaves only the $2p_z$ orbital of F to mix with the H 1*s* orbital to give both σ bonding and σ^* anti-bonding orbitals (see Figs. 6.23b and d). The correlation diagram for HF is shown in Figure 6.24. The 2*s*, $2p_x$, and $2p_y$ orbitals of fluorine do not mix with the 1*s* of hydrogen and therefore remain as atomic (nonbonding) orbitals denoted by σ^{nb} and π^{nb}. Electrons in these orbitals do not contribute significantly to chemical bonding. Because fluorine is more electronegative than hydrogen, its 2*p* orbitals lie below the 1*s* hydrogen orbital in energy. The σ orbital then contains more fluorine $2p_z$ character, and the σ^* orbital more closely resembles a hydrogen 1*s* AO. When the eight valence electrons are put in for HF, the result is the electron configuration:

$$(\sigma^{\text{nb}})^2(\sigma)^2(\pi_x^{\text{nb}}, \pi_y^{\text{nb}})^4$$

The net bond order is 1 because electrons in **nonbonding atomic orbitals** do not affect bond order. The electrons in the σ orbital are more likely to be found near the fluorine atom than near the hydrogen, so HF has the dipole moment $H^{\delta+}F^{\delta-}$.

If a more electropositive atom (such as Na or K) is substituted for H, the energy of its outermost *s* orbital will be higher than that of the H atom, because its ionization energy is lower. In this case, the σ orbital will resemble a fluorine $2p_z$ orbital even more (that is, the coefficient C_F of the fluorine wave function will be close to 1, and C_A for the alkali atom will be very small). In this limit the molecule can be described as having the valence electron configuration $(\sigma^{\text{nb}})^2(\sigma)^2(\pi_x^{\text{nb}}, \pi_y^{\text{nb}})^4$, which corresponds to the ionic species Na^+F^- or K^+F^-. The magnitudes of the coefficients in the molecular orbital wave function are thus closely related to the ionic–covalent character of the bonding and to the dipole moment.

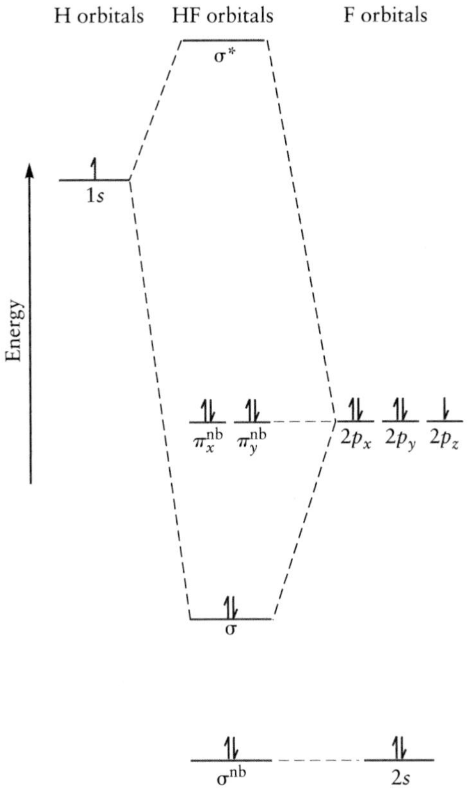

FIGURE 6.24 Correlation diagram for HF. The 2*s*, $2p_x$, and $2p_y$ atomic orbitals of fluorine do not mix with the 1*s* atomic orbital of hydrogen, and therefore remain nonbonding.

CONNECTION TO INSTRUMENTAL ANALYSIS

Photoelectron Spectroscopy

In photoelectron spectroscopy (PES), we illuminate a sample with high-frequency radiation (ultraviolet or X-ray) and measure the kinetic energy of the photoelectrons emitted from the sample (see Figs. 5.22 and 5.23). We used PES in Section 5.4 to confirm the shell structure of the atom predicted by quantum mechanics (see Fig. 5.26). For molecules, PES confirms the MO description of bonding and measures the energy, ε, for individual MOs. The bridge between PES results and MO theory is Koopmans's approximation (stated in Section 5.4). These three tools are used together to study the electronic structure of molecules in all branches of chemistry.

As a concrete example, suppose we illuminate a diatomic gaseous sample with He(I) radiation, which has energy of 21.22 eV and a wavelength of 58.43 nm, and we measure the kinetic energy of the emitted photoelectrons with an energy analyzer. The resulting PES spectrum shows a series of peaks, each of which we label with an index *i*. We subtract the measured kinetic energy from the photon energy, which is fixed in our experiment. Thus, by conservation of energy, we are measuring the *ionization energy*, IE_i, required to liberate those electrons that contribute to peak *i*. Koopmans's theorem states that the measured ionization energy is the negative of the energy of the orbital from which the photoelectrons were emitted: $IE_i = -\varepsilon_i$. (Recall that IE_i is positive because it must be provided to the system, and that ε_i is negative because it measures the amount by which the molecular orbital is stabilized relative to the cation plus a free electron.) Koopmans's theorem is only an approximation, because it assumes that the ion produced during photoemission has the same orbital energies as the parent neutral molecule. In addition to the relaxation of the resulting ion as observed for atoms (see discussion in Section 5.4), some of the energy provided by the photon can be used to excite vibrational states in the molecular ion, which requires an amount of energy, $E_i^{(\mathrm{vib})}$. Now the energy conservation equation is

$$h\nu_{\mathrm{photon}} - \frac{1}{2}m_e v^2 = -\varepsilon_i + E_i^{(\mathrm{vib})} = IE_i$$

As a result of the vibrational excitation, the peak *i* in the spectrum is actually a series of narrower peaks; the separation between adjacent peaks depends on the vibrational frequency of the diatomic ion. The result illustrated in Figure C6.1 shows the PES of hydrogen. The peak near 15.5 eV corresponds to the ionization energy for removing electrons with no vibrational excitation of the resulting molecular ion. As the energy increases along the axis toward 18 eV, the amount of vibrational excitation of the H_2^+ ion increases, and the spacing between vibrational levels becomes smaller. The H_2^+ ion is approaching its dissociation limit.

These vibrational "fine-structure" peaks on the PES data at first appear to be a nuisance, but in fact, they greatly aid in relating experimental data to particular MOs. The connection is made through the concept of bond order introduced in Section 6.4.

Case A: If the photoelectron is removed from a bonding MO, the bond order of the positive ion will be smaller than the bond order of the parent molecule. Consequently, the bond in the molecular ion will be less stiff, and its vibrational frequency (determined directly from the PES fine structure) will be lower than that of the parent molecule (determined by vibrational spectroscopy).

Case B: If the photoelectron is emitted from an antibonding orbital, the bond order of the positive ion will be larger than the bond order of the parent molecule. The bond in the diatomic molecular ion will be stiffer and will show a higher vibrational frequency.

Case C: If the photoelectron is emitted from a nonbonding orbital, there is no change in the bond order, and consequently little or no change in the vibrational frequency. The PES spectrum for the orbital will show few vibrational peaks, because the disturbance to the bond during photoemission is quite small. By contrast, the spectrum in Case A will show several vibrational fine structure peaks because removal of a bonding electron is a major disturbance that starts many vibrations of the bond. Case B is intermediate, with fewer vibrational

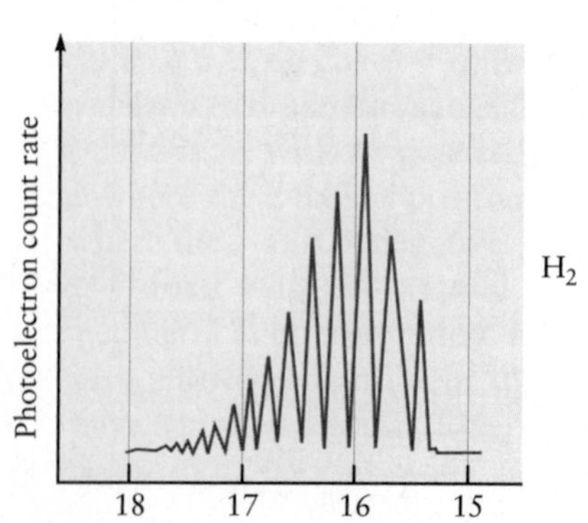

FIGURE C6.1 The photoelectron spectrum of H_2 shows a series of peaks corresponding to vibrational excitation of H_2^+.

subpeaks, because removing an antibonding electron disturbs the bond, but less so than in Case A.

The photoelectron spectra for N_2 and O_2 are shown in Figures C6.2 and C6.3. The experimental peaks have been assigned to orbitals by slightly more complex versions of the arguments used previously. Note that for N_2, the energy for the σ_{g2p_z} MO is lower than that for π_{u2p_x} and π_{u2p_y} whereas the order is switched for O_2, as indicated in Figure 6.17, Table 6.3, and the related text. This switch is due to interaction between the 2*s* and 2*p* AOs.

The photoelectron spectrum for NO is shown in Figure C6.4. The orbital assignments are based on the arguments summarized above. Note that the 1*s* core levels for both N and O appear at the same orbital energies as they do in their respective elemental gases, N_2 and O_2. This experimental result demonstrates clearly that the core levels do not participate in chemical bond formation and can be neglected in the MO analysis of bond formation.

These examples show that photoelectron spectroscopy is useful in testing theoretical models for bonding because it directly measures ionization energies that can be correlated with theoretical orbital energies through Koopmans's theorem. These methods are readily extended to polyatomic molecules (see Section 20.5).

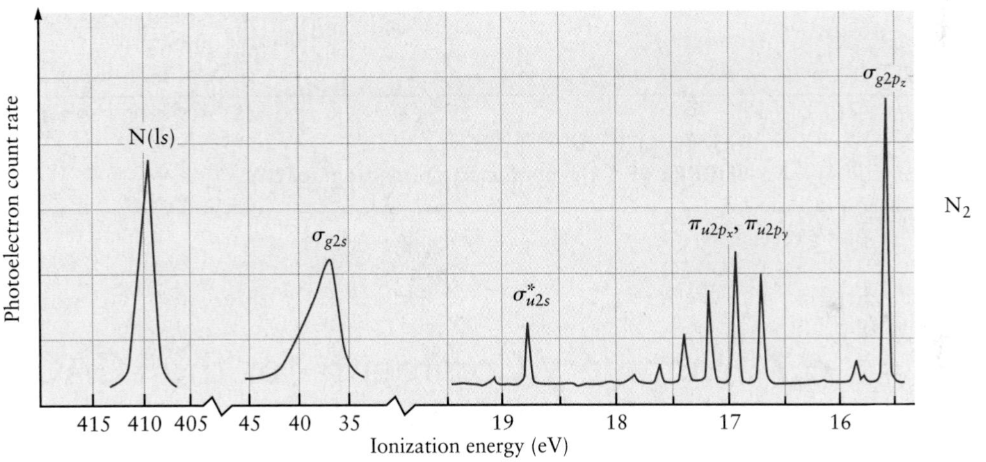

FIGURE C6.2 The photoelectron spectrum for N_2 shows the valence electrons in the occupied molecular orbitals and the N(1*s*) core electrons.

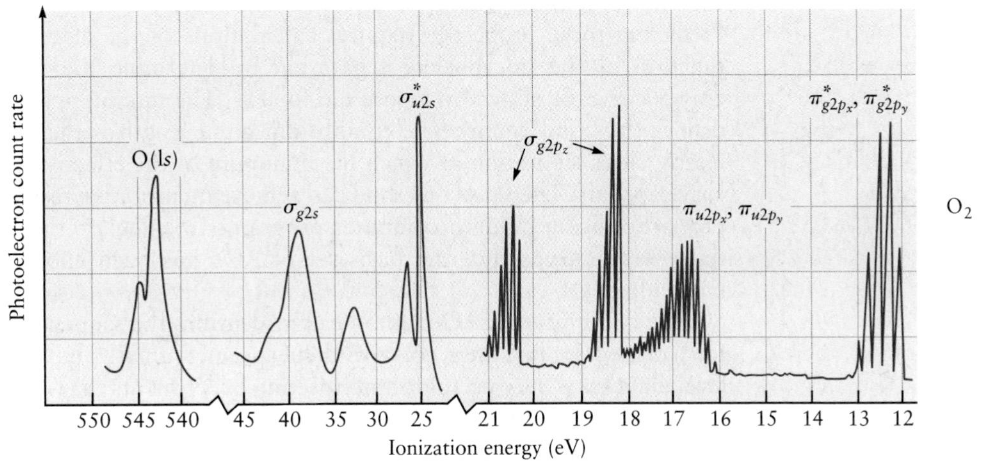

FIGURE C6.3 The photoelectron spectrum for O_2 shows valence electrons in the occupied molecular orbitals and the O(1*s*) core electrons.

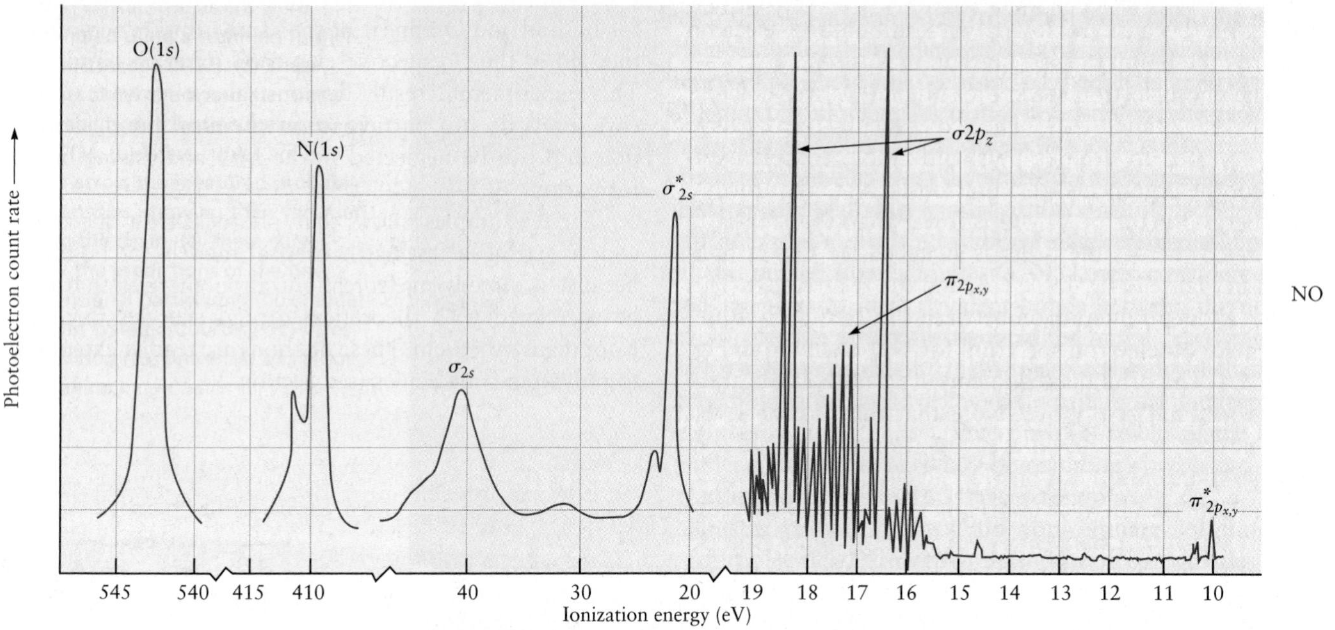

FIGURE C6.4 The photoelectron spectrum for NO.

6.7 Summary Comments for the LCAO Method and Diatomic Molecules

The qualitative LCAO method presented in Sections 6.4–6.6 can rationalize experimentally observed trends in bond length and bond energy in a group of molecules by relating both these properties to bond order, but it cannot predict the values of bond energy or molecular geometry for any specific molecule. Predicting these properties requires calculations of the electronic energy as a function of the internuclear separation to determine the effective potential energy curve, as shown in Figure 6.6 for H_2^+. The minimum value of this curve defines the bond energy. The equilibrium bond length is then identified as the internuclear separation at which the minimum of the effective potential energy curve appears. The LCAO method can achieve quantitative results for molecular properties using modern computer programs to calculate the effective potential energy curve, once the basis set of AOs has been chosen for a specific molecule.

The qualitative LCAO method easily identifies the sequence of energy levels for a molecule, but does not give their specific values. The qualitative energy level diagram is very useful for interpreting experiments that involve adsorption and emission of energy such as spectroscopy, ionization by electron removal, and electron attachment. Trends in such results for a group of molecules can be easily rationalized, but predicting specific results requires the quantitative calculations described in the preceding paragraph.

6.8 Valence Bond Theory and the Electron Pair Bond

The characteristics (bond length, bond energy, polarity, etc.) of specific chemical bonds such as O—H, C—C, and C—H do not differ much from molecule to molecule (see Section 3.9, especially Table 3.4). If the bonding electrons are spread out over the entire molecule, as described by the LCAO model, then why should the properties of a bond be nearly independent of the nature of the rest of the molecule? Would some other model that associates electrons directly with the bond provide a better description of chemical bonds?

The **valence bond (VB) theory** was developed to provide a quantum explanation and justification for the Lewis electron pair model in which the chemical bond is described as a pair of electrons localized between two atoms. The valence bond theory constructs a wave function for each individual chemical bond by assuming that each participating atom arrives with at least one unpaired electron in an AO. The VB wave function for the bond is a *product* of two one-electron AO wave functions, each describing an electron localized on one of the atoms. The spins of the electrons must be paired to satisfy the Pauli exclusion principle.

The VB description for H_2 was developed by the German physicists Walter Heitler and Fritz London in 1927, just one year after Schrödinger introduced wave mechanics to explain the structure of the H atom. The American physicist John C. Slater also made important contributions to developing the VB method. Establishing the VB method as one of the cornerstones of modern structural chemistry awaited the pioneering work of the American chemist Linus Pauling, who used it to describe structure and bonding in polyatomic molecules, starting in 1931. VB theory predates LCAO theory, and until modern computer calculations became available in the 1960s, VB theory was more widely used than LCAO because it could describe molecular structure without performing detailed calculations. VB theory is still very popular today for the same reason.

This section presents a qualitative version of VB theory, which easily describes the structure and geometry of bonds in polyatomic molecules by hand-drawn sketches. Because of this ease, qualitative VB theory is a very useful complement to qualitative LCAO theory. It is widely used in organic and inorganic chemistry and in biochemistry. Later in this chapter we compare VB and LCAO methods and see when one is preferable to the other. In some cases we see it is beneficial to use the two methods together.

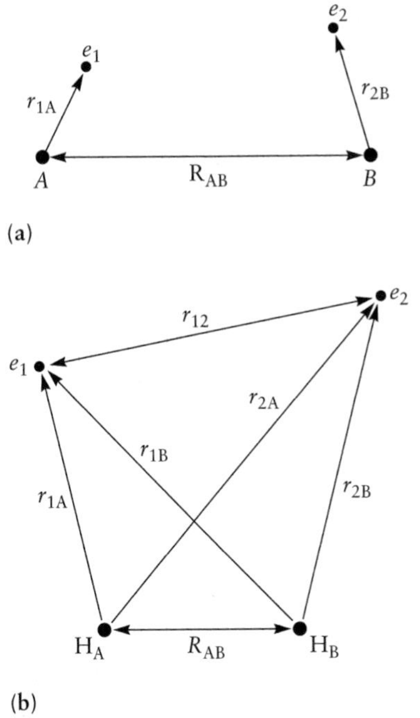

FIGURE 6.25 Two hydrogen atoms approach one another. The protons are separated by the distance R_{AB}. (a) At large values of R_{AB} each electron interacts only with the proton to which it is bound. (b) As the atoms approach closer, both electrons interact with both protons. The distance of electron 1 from nuclei A and B is given by r_{1A}, r_{1B}; the distance of electron 2 from nuclei A and B is given by r_{2A}, r_{2B}; the distance between the electrons is given by r_{12}.

6.8.1 Single Bonds

Consider that the hydrogen molecule, described by the Lewis structure H:H, is formed by combining two hydrogen atoms each with the electron configuration H: $(1s)^1$. The two atoms approach one another and begin to interact as shown in Figure 6.1, and in more detail in Figure 6.25. The protons are separated by the distance R_{AB}. At very large separations, each electron is bound to its own proton, and is located by coordinate r_{1A} or r_{2B}. At very large distances, the atoms are independent of each other, and the wave function describing the pair of them is $\varphi^A(r_{1A})\varphi^B(r_{2B})$. This result is a consequence of the form of the Schrödinger equation. If two objects are independent, their total energy is the sum of the individual energies, and the wave function for the two of them is the product of the individual wave functions. As the atoms approach closer together so that bond formation is a possibility, it is reasonable to propose an approximate wave function for the two of them would take the form

$$\psi^{el}(r_{1A}, r_{2B}; R_{AB}) = c(R_{AB})\varphi^{A}(r_{1A})\varphi^{B}(r_{2B}) \qquad [6.12]$$

in which the coefficient $c(R_{AB})$ compensates for the fact that the product form is strictly valid only at very large values of R_{AB}.

As the atoms begin to interact strongly, we cannot determine whether electron 1 arrived with proton A and electron 2 with proton B, or vice versa, because the electrons are indistinguishable. To allow for both possibilities we propose a wave function that is a sum of two of the product forms just discussed. We assign electron 1 to proton A and electron 2 to proton B in the first product form, and in the second product we switch each electron to the other proton. The result is

$$\psi^{el}(r_{1A}, r_{2B}; R_{AB}) = c_1(R_{AB})\varphi^{A}(r_{1A})\varphi^{B}(r_{2B}) + c_2(R_{AB})\varphi^{A}(r_{2A})\varphi^{B}(r_{1B}) \qquad [6.13]$$

Symmetry requires that $c_1 = c_2$ and $c_1 = -c_2$ be equally valid choices. We label these combinations *gerade (g)* and *ungerade (u)* respectively to show how each behaves under inversion symmetry, as in Sections 6.2 and 6.3. We must check both cases to determine whether they describe bond formation, using our familiar criteria of increased electron density between the nuclei and energy reduced below that of the separated atoms.

It requires some care to calculate the electron density for ψ_g^{el} and ψ_u^{el}. Unlike the one-electron wave functions we have seen earlier, these are examples of *two-electron functions,* which depend explicitly on the positions of both electrons. Their squares give the probability density for finding electron 1 at r_1 and electron 2 at r_2. To calculate the probability density for finding electron 1 at r_1, no matter where electron 2 is located, we must square the function and then average over all possible locations for electron 2. Similarly, we calculate the probability density for finding electron 2 at r_2 regardless of the location of electron 1. Adding these results together gives the total electron density at each point in space, as a function of the internuclear distance R_{AB}. The results for ψ_g^{el} and ψ_u^{el} are shown in Figure 6.26a. The wave function ψ_g^{el} shows increased electron density between the nuclei, whereas ψ_u^{el} shows reduced electron density between the nuclei. The shape of the electron density in three dimensions is more conveniently obtained from computer calculations using a quantitative version of the VB method which will be described later. Figure 6.26b shows a three-dimensional isosurface of the electron density for H_2 calculated from ψ_g^{el}.

It is a straightforward exercise in quantum mechanics—although beyond the scope of this book—to calculate the energy of the hydrogen molecule as a function of R_{AB} when the electrons are described by ψ_g^{el} and by ψ_u^{el}. The results give the effective potential energy curve for each value of R_{AB}. The two calculated effective potential energy curves (not shown here) are qualitatively similar to those in Figure 6.9. They show that ψ_g^{el} describes a state with lower energy than that of the separated atoms, whereas ψ_u^{el} describes a state whose energy is higher than that of the separated atoms for all values of R_{AB}.

Taken together, the reduced effective potential energy and increased electron density between the nuclei demonstrate that ψ_g^{el} describes a stable chemical bond, while ψ_u^{el} describes a state that is strictly repulsive everywhere and does not lead to bond formation. Therefore, we conclude that the correct wave function to describe an electron pair bond in the VB method is

$$\psi_g^{el} = C_1[1s^{A}(1)1s^{B}(2) + 1s^{A}(2)1s^{B}(1)] \qquad [6.14]$$

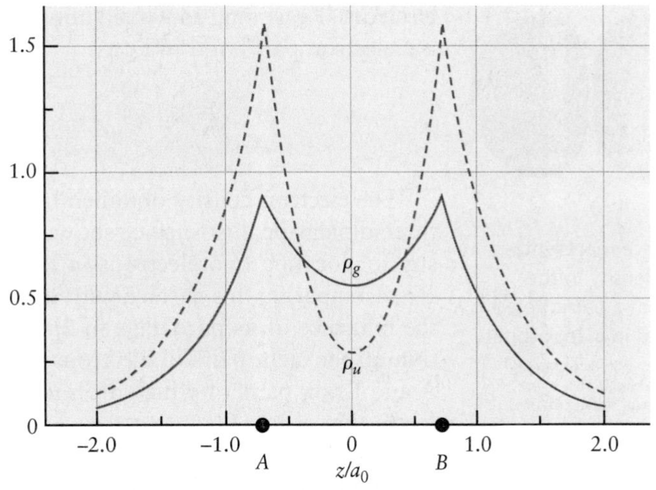

(a)

(b)

FIGURE 6.26 The electron density for the ψ_g^{el} and ψ_u^{el} wave functions in the simple valence bond model for H_2. (a) The electron density ρ_g for ψ_g^{el} and ρ_u for ψ_u^{el} calculated analytically as described in the text. (b) Three-dimensional isosurface of the electron density for the ψ_g^{el} wave function obtained by computer calculations. (Courtesy of Mr. Hatem Helal and Professor William A. Goddard III, California Institute of Technology, and Dr. Kelly P. Gaither, University of Texas at Austin.)

In this and the following equations we use the simplified notation for atomic orbitals introduced earlier in our discussion of the LCAO approximation in Section 6.3. In addition, we use "1" and "2" as shorthand notation for the coordinates locating electron 1 and electron 2, and C_1 is a normalization constant.

Equations 6.13 and 6.14 are the basic descriptions of the electron pair bond in VB theory. At first glance they may seem a bit daunting with the numerous subscripts and superscripts, but they provide a concise summary of a great deal of important information. They symbolically represent the elaborate verbal description of what happens when the electron densities of two H atoms begin to interpenetrate and form a new density characteristic of the H_2 molecule: "The wave function for the electron pair bond in a hydrogen molecule is constructed by multiplying a 1*s* orbital for electron 1 on atom A by a 1*s* orbital for electron 2 on atom B and adding the product to the result of multiplying a 1*s* orbital for electron 2 on atom A by a 1*s* orbital for electron 1 on atom B and finally normalizing the resulting sum." There is a beautiful and satisfying interplay between visualizing the physical events, describing them verbally, and describing them in simple equations. We use these equations repeatedly to describe several different types of electron pair bonds in VB theory. The atoms and the orbitals involved change from one case to another, but the verbal description of the process and the equations are always the same.

Now let's consider the formation of F_2, represented by its Lewis diagram, from two F atoms each with electron configuration F: $(1s)^2(2s)^2(2p_x)^2(2p_y)^2(2p_z)^1$. Suppose the two atoms labeled A and B approach each other along the *z*-axis so that their $2p_z$ orbitals with the same phase point toward each other. As the atoms draw close, these two orbitals can overlap to form a single bond with two electrons. Reasoning as we did above for H_2, we write the VB wave function for the electron pair bond in F_2 as

$$\psi_g^{\text{bond}} = C_1[2p_z^A(1)2p_z^B(2) + 2p_z^A(2)2p_z^B(1)] \qquad [6.15]$$

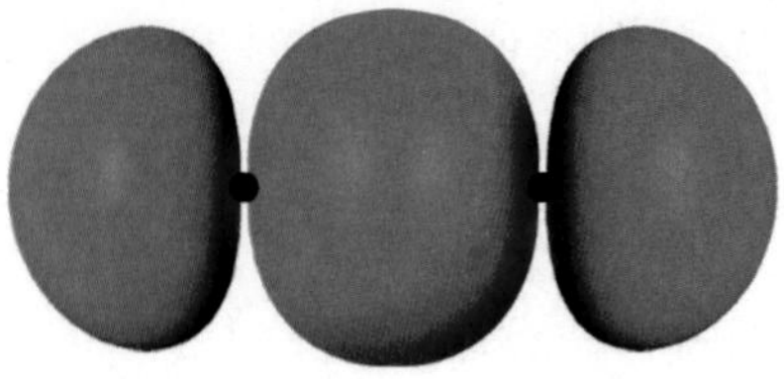

F F

FIGURE 6.27 Isosurface representation of the electron density in the F_2 σ bond formed from a pair of electrons initially localized in a $2p_z$ orbital on each F atom. (Courtesy of Mr. Hatem Helal and Professor William A. Goddard III, California Institute of Technology, and Dr. Kelly P. Gaither, University of Texas at Austin.)

The electron density obtained from this wave function is represented by the three-dimensional isosurface shown in Figure 6.27. It gives the probability density for locating two electrons in the F_2 molecule with internuclear separation R_{AB}. Comparing this surface with the one shown in Figure 6.26b clearly reveals the influence of its parentage in $2p$ AOs. This wave function gives no information on the eight pairs of electrons remaining in their atomic orbitals on atoms A and B, six pairs of which are shown as unshared pairs in the Lewis diagram for F_2 (Fig. 6.28).

:F̤̈:F̤̈:

FIGURE 6.28 Lewis diagram for F_2.

The VB model also describes bond formation in heteronuclear diatomics. We can combine the features of the two preceding examples to describe HF, which has one shared pair in a single bond produced by overlap of H $1s$ and F $2p_z$. To write a wave function for the electron pair bond, one might begin with

$$\psi^{\text{bond}} = C_1[2p_z^F(1)1s^H(2)] + C_2[2p_z^F(2)1s^H(1)] \qquad [6.16]$$

However, C_1 and C_2 are equal and should be omitted. The electron density isosurface is depicted in Figure 6.29. Because the wave function for the bond pair is symmetric (does not change sign) upon interchanging the spatial positions of the two electrons, the Pauli principle tells us that the spin part of the wave function should be antisymmetric (i.e. $\alpha(1)\beta(2) - \alpha(2)\beta(1)$, where α indicates $m_s = 1/2$ and β indicates $m_s = -1/2$).

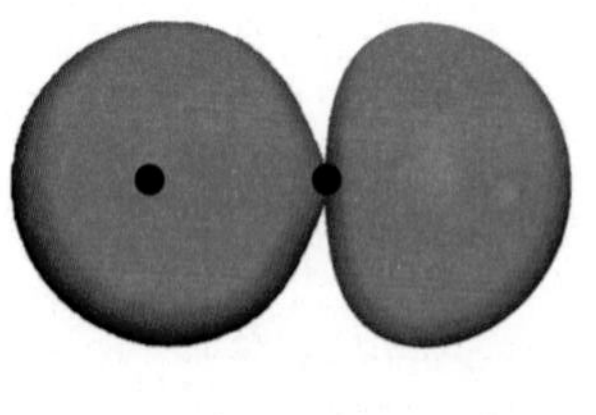

H F

FIGURE 6.29 Isosurface representation of the electron density in the HF σ bond formed from a pair of electrons initially localized in a $1s$ orbital on H and in a $2p_z$ orbital on F. (Courtesy of Mr. Hatem Helal and Professor William A. Goddard III, California Institute of Technology, and Dr. Kelly P. Gaither, University of Texas at Austin.)

The bond pair wave functions in Equations 6.14, 6.15, and 6.16 were specially constructed to describe two electrons localized between two atoms as a single chemical bond between the atoms. These wave functions should not be called MOs, because they are not single-electron functions and they are not delocalized over the entire molecule. The corresponding single bonds, shown in Figures 6.26, 6.27, and 6.29 are examples of a **sigma (σ) bond**, because their electron density is cylindrically symmetric about the bond axis. Because of this symmetry it is said to have zero angular momentum about the bond axis. Finally, electrons are not placed in these bonds by the aufbau principle. Rather, each bond is formed by overlap of two AOs, each of which is already half-filled with one electron. The electrons in the two participating AOs must have opposite spin, so the bond corresponds to an electron pair with opposite, or "paired" spins.

6.8.2 Multiple Bonds

:N:::N:

FIGURE 6.30 Lewis diagram for N_2.

To see how the VB method describes multiple bonds, let's examine N_2 (Fig. 6.30). Suppose two nitrogen atoms with electron configuration N: $(1s)^2(2s)^2(2p_x)^1(2p_y)^1(2p_z)^1$ approach one another along the z-axis. The two $2p_z$ orbitals can overlap and form a σ bond whose wave function is

$$\psi_\sigma^{\text{bond}} = C_1[2p_z^A(1)2p_z^B(2) + 2p_z^A(2)2p_z^B(1)] \qquad [6.17]$$

The $2p_x$ orbitals and the $2p_y$ orbitals on the two atoms do not approach head-on in this configuration, but rather side-to-side. Therefore, the positive lobes of the $2p_x$ orbitals can overlap laterally, as can the negative lobes. Together they form a **pi (π) bond**, which has a node through the plane containing the

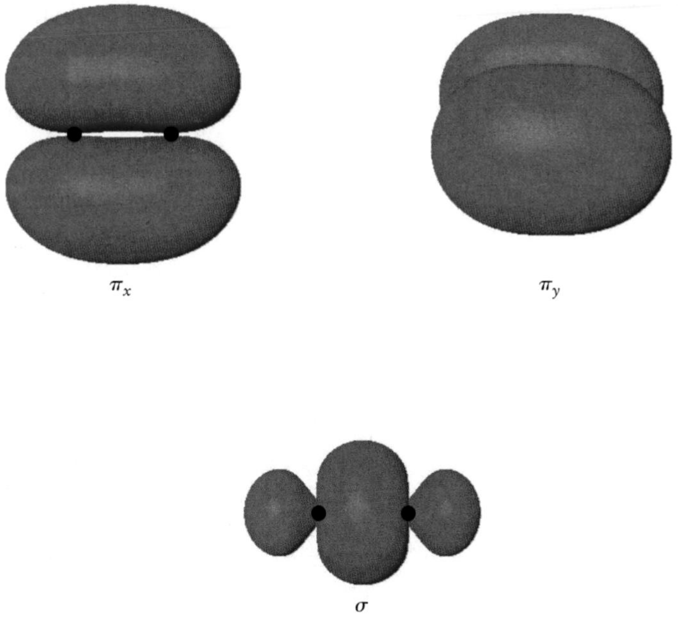

FIGURE 6.31 Isosurface representation of the electron densities in the σ bond and the two π bonds for nitrogen obtained by computer calculations. (Courtesy of Mr. Hatem Helal and Professor William A. Goddard III, California Institute of Technology, and Dr. Kelly P. Gaither, University of Texas at Austin.)

bond axis with amplitude above and below the plane. The wave function for the π_x bond is

$$\psi_{\pi_x}^{\text{bond}}(1, 2) = C_1[2p_x^A(1)2p_x^B(2)] + C_1[2p_x^A(2)2p_x^B(1)] \quad [6.18]$$

Similarly, the $2p_y$ orbitals on the two atoms can overlap to form a second π bond, the π_y bond in the y-z plane, for which the wave function is

$$\psi_{\pi_y}^{\text{bond}}(1, 2) = C_1[2p_y^A(1)2p_y^B(2) + 2p_y^A(2)2p_y^B(1)] \quad [6.19]$$

Again the two spins in each bond are opposite. The three bonds in N_2 are shown as isosurfaces of electron density in Figure 6.31.

The problems for Section 6.8 at the end of the chapter give practice in writing out the wave functions for electron pair bonds in several other diatomic molecules, and identifying the bond order in each molecule. You should prepare hand sketches of the electron density isosurfaces for each bond you describe.

6.8.3 Polyatomic Molecules

Describing the three-dimensional structure of polyatomic molecules requires that we include bond angles as well as bond lengths. Let's test the VB approximation on the second-period hydrides, whose structures we have already examined using VSEPR theory in Chapter 3, to see how well it describes bond angles and molecular shapes.

H:Be:H

FIGURE 6.32 Lewis diagram for BeH_2.

Beryllium hydride, BeH_2, has four valence electrons, two from Be and one each from the two H atoms, all of which appear in the Lewis diagram in Figure 6.32. In VSEPR theory, the steric number is 2, so the molecule is predicted to be linear, and this prediction is verified by experiment. The electron configuration of the central atom is Be: $(1s)^2(2s)^2$. There are no unpaired electrons to overlap with H(1s) orbitals, so the VB model fails to predict the formation of BeH_2.

H:B:H
H

FIGURE 6.33 Lewis diagram for BH_3.

Boron hydride, BH_3 (Fig. 6.33), has six valence electrons corresponding to steric number 3 and a trigonal planar structure. With the electron configuration

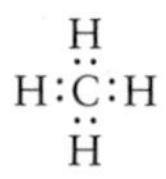

FIGURE 6.34 Lewis diagram for CH_4.

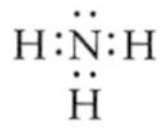

FIGURE 6.35 Lewis diagram for NH_3.

FIGURE 6.36 Lewis diagram for H_2O.

B: $(1s)^2(2s)^2(2p)^1$ on the central atom, the VB model cannot account for the formation of BH_3 and, in fact, predicts that BH is the stable molecule, which does not agree with experiment.

Methane, CH_4 (Fig. 6.34), has steric number 4, and VSEPR predicts a tetrahedral structure, which is confirmed by experiment. Starting with the electron configuration C: $(1s)^2(2s)^2(2p)^2$, the VB model cannot account for the formation of CH_4 and predicts that CH_2 would be the stable hydride, which is again contrary to the experimental results.

Ammonia, NH_3 (Fig. 6.35), has steric number 4 with three shared pairs and one unshared pair on the N atom. VSEPR predicts a trigonal pyramid structure, as a subcase of tetrahedral structure, with angles slightly less than 109.5° due to repulsion between the unshared pair and the three bonding pairs. Experiment verifies this structure with angles of 107°. The electron configuration N: $(1s)^2(2s)^2(2p_x)^1(2p_y)^1(2p_z)^1$ would permit the formation of three σ bonds by overlap of H(1*s*) orbitals with each of the 2*p* orbitals on N. Because these 2*p* orbitals are all mutually perpendicular, VB predicts a trigonal pyramid but one with bond angles of 90°.

Finally, water, H_2O (Fig. 6.36), has steric number 4 with two shared pairs and two unshared pairs on the O atom. VSEPR predicts a bent structure, as a subcase of tetrahedral structure, with angles significantly less than the tetrahedral value of 109° due to repulsion between the two unshared pairs and the bonding pairs. Experiment verifies this prediction with a bond angle of 104.5°. The electron configuration O: $(1s)^2(2s)^2(2p_x)^2(2p_y)^1(2p_z)^1$ would permit formation of two σ bonds by overlap of H(1*s*) orbitals with each of the 2*p* orbitals on O. VB predicts a bent structure for H_2O, but with bond angle of 90°.

These examples show that the VB model does not accurately describe bonding in the second-period hydrides. It predicts the wrong valence for atoms in Groups IIA through IVA, and the wrong structure for atoms in Groups VA and VIA. Clearly, the model had to be improved. Linus Pauling gave the answer in 1931 by introducing the concepts of *promotion* and *hybridization.*

Atoms such as Be, B, and C can have the correct valence for bonding by **promotion** of valence electrons from the ground state to excited states at higher energy. For example,

$$\text{Be: } (1s)^2(2s)^2 \rightarrow \text{Be: } (1s)^2(2s)^1(2p)^1$$

and

$$\text{C: } (1s)^2(2s)^2(2p)^2 \rightarrow \text{C: } (1s)^2(2s)^1(2p_x)^1(2p_y)^1(2p_z)^1$$

are ready to form BeH_2 and CH_4, respectively. These excited states are known from spectroscopy. The C excited state lies about 8.26 eV (190 kcal mol^{-1}) above the ground state; promoting an electron to create the excited state clearly requires energy. Pauling argued that this investment would be repaid by the energy released when the C—H bonds of methane form (about 100 kcal mol^{-1} for each bond).

Even though the valence is correct after promotion, the structure would still be wrong. Beryllium hydride would have two different kinds of bonds, and methane would have three identical bonds formed by overlap of H(1*s*) with the C(2*p*) orbitals and a different bond formed by H(1*s*) and C(2*s*). Pauling proposed that new atomic orbitals with the proper symmetry for bond formation could be formed by **hybridization** of 2*s* and 2*p* orbitals after promotion. The Be(2*s*) and Be($2p_z$) orbitals would combine to form two equivalent hybrid atomic orbitals oriented 180° apart. The C(2*s*) would hybridize with the three C(2*p*) orbitals to give four equivalent new atomic orbitals arranged as a tetrahedron around the C atom.

Pauling's achievements made it possible to describe polyatomic molecules by VB theory, and hybridization has provided the vocabulary and structural concepts for much of the fields of inorganic chemistry, organic chemistry, and biochemistry.

6.9 Orbital Hybridization for Polyatomic Molecules

Pauling developed the method of hybrid orbitals to describe the bonding in molecules containing second-period atoms with steric numbers 2, 3, and 4. Let's discuss these hybridization schemes in sequence, starting with BeH_2. In each case we will learn how to construct the hybrid orbitals and how to use them to describe the bonds and the three-dimensional structure of polyatomic molecules. We will use the lower case Greek letter "chi" χ to represent hybrid orbitals.

6.9.1 *sp* Hybridization

Because the BeH_2 molecule is known to be linear, we need two new atomic orbitals oriented 180° apart on the Be atom, each of which can overlap with one H atom. We can generate these new atomic orbitals in the following way. Let's define the z-axis of the coordinate system to lie along the H—Be—H bonds, and place the Be nucleus at the origin. We mix the $2s$ and $2p_z$ orbitals of Be to form two new orbitals on the Be atom. The linear combinations that correspond to Figure 6.37b (with the z-axis pointing to the right), are

$$\chi_1(r) = \frac{1}{\sqrt{2}}[-2s - 2p_z] \qquad [6.20a]$$

$$\chi_2(r) = \frac{1}{\sqrt{2}}[-2s + 2p_z] \qquad [6.20b]$$

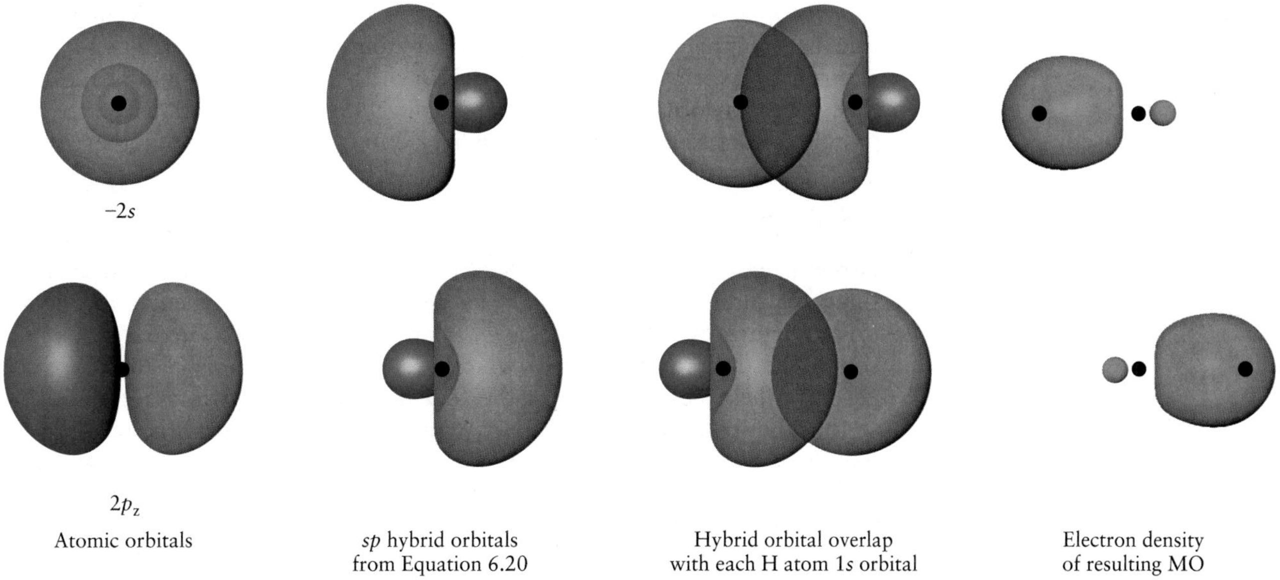

FIGURE 6.37 Formation, shapes, and bonding of the *sp* hybrid orbitals in the BeH_2 molecule. (a) The $-2s$ and $2p_z$ orbitals of the Be atom. (b) The two *sp* hybrid orbitals formed from the $2s$ and $2p_z$ orbitals on the beryllium atom. (c) The two σ bonds that form from the overlap of the *sp* hybrid orbitals with the hydrogen $1s$ orbitals, making two single bonds in the BeH_2 molecule. (d) Electron density in the two σ bonds obtained by computer calculations.
(Courtesy of Mr. Hatem Helal and Professor William A. Goddard III, California Institute of Technology, and Dr. Kelly P. Gaither, University of Texas at Austin.)

The coefficient $1/\sqrt{2}$ is a normalization constant. We call these **sp hybrid atomic orbitals** because they are formed as the sum or difference of one *s* orbital and one *p* orbital. Like the familiar *s* and *p* orbitals, a hybrid atomic orbital is a one-electron wave function that is defined at every point in space. The value of the wave function at each point is the sum or difference of the original AOs combined to form the hybrid. Its square at each point in three-dimensional space gives the probability density for finding the electron at that point, when the electron is in the hybrid orbital.

The formation, shapes, and orientation of the *sp* hybrid orbitals—and their participation in chemical bonds—are shown in Figure 6.37. The first column shows the standard orbitals on the Be atom before hybridization, and the second shows the hybrid atomic orbitals on Be resulting from the linear combination of the Be 2*s* and $2p_z$ atomic orbitals given in Equation 6.20. The hybrid atomic orbital shown on the top of the second column is the one with large electron density pointing in the negative *z* direction, from Equation 6.20a. It is the one that will have substantial overlap with the H atom coming in from the left as you form H-Be-H. As the two hydrogen atoms approach the Be atom from opposite directions along the *z*-axis, each shares its electron with the nearest hybrid orbital to form a localized σ bond (see Fig. 6.37c). The result is a pair of localized σ bonds, one on each side of the Be atom. The third column in Figure 6.37 illustrates these two σ bonds by locating the Be and H nuclei at distances apart equal to the experimental bond length of BeH_2, and placing an *sp* hybrid on the Be atom and a 1*s* AO on the H atom in order to show where these orbitals overlap. Since the advent of the VB method, chemists have used qualitative sketches similar to this third column to show the location, shape, and orientation of the σ bonds. With modern computational methods, detailed representations of the electron density can be obtained from the wave functions for the two bonding pairs. The fourth column in Figure 6.37 shows the electron density in the σ bonds of BeH_2 calculated in this way.

sp hybridization explains how an atom can have a pair of σ bonds at an angle of 180° apart and therefore be part of a linear molecule. This result is broadly applicable and can be combined with other aspects of molecular structure. Later in this section we shall see that atoms having more than two valence electrons can form linear molecules through *sp* hybridization and at the same time can also form multiple bonds through linear combinations of the two *p* orbitals not used in making the *sp* hybrid AOs with orbitals on other atoms.

6.9.2 sp^2 Hybridization

Experimental data suggest that the BH_3 molecule has a trigonal planar structure with three equivalent bonds, as predicted by VSEPR theory. To explain this structure, we need three new orbitals oriented 120° apart in a plane containing the B atom, each of which can overlap with one H atom. We can generate these new orbitals in the following way. Let's choose coordinates so that the structure lies in the *x*-*y* plane with the B atom at the origin. Promotion of one of the 2*s* electrons creates the excited-state configuration B: $2s^1 2p_x^1 2p_y^1$. The two occupied 2*p* AOs already lie in the desired plane. Geometry shows they can be mixed with the 2*s* AO to form three equivalent new orbitals separated by 120°. To calculate the sp^2 hybrid orbitals depicted in Figure 6.29b, if the *y*-axis is pointing vertically on the page while the *x*-axis points to the right, the necessary linear combinations of boron's AOs, for the top, middle and bottom sp^2 hybrid AOs, respectively, are

$$\chi_1(r) = (1/3)^{1/2}(-2s) - (1/6)^{1/2}(2p_x) + (1/2)^{1/2}(2p_y) \qquad [6.22a]$$

$$\chi_2(r) = (1/3)^{1/2}(-2s) + 2(1/6)^{1/2}(2p_x) \quad [6.22b]$$

$$\chi_3(r) = (1/3)^{1/2}(-2s) - (1/6)^{1/2}(2p_x) - (1/2)^{1/2}(2p_y) \quad [6.22c]$$

These are called ***sp*^{2} hybrid atomic orbitals** because they are generated from one *s* and two *p* orbitals. The formation, shape, and orientation of the sp^2 hybrids are shown in Figure 6.38. They lie in the *x*-*y* plane with an angle of 120° between them. After hybridization, the electron configuration of the atom is B: $(1s)^2(\chi_1)^1(\chi_2)^1(\chi_3)^1$. Each of the sp^2 hybrids can overlap with an H(1*s*) orbital to produce a σ bond. The third column shows the traditional sketches of the orbital overlap leading to each of these σ bonds in BH_3, and the fourth column shows the electron density in the resulting bonding MOs. Experimentally, BH_3 molecules turn out to be unstable and react rapidly to form B_2H_6 or other higher compounds called *boranes,* but BH_3 can be detected in a mass spectrometer as a fragment of the higher compounds. The closely related BF_3 molecule is stable, and experiments have shown its structure to be the trigonal planar geometry

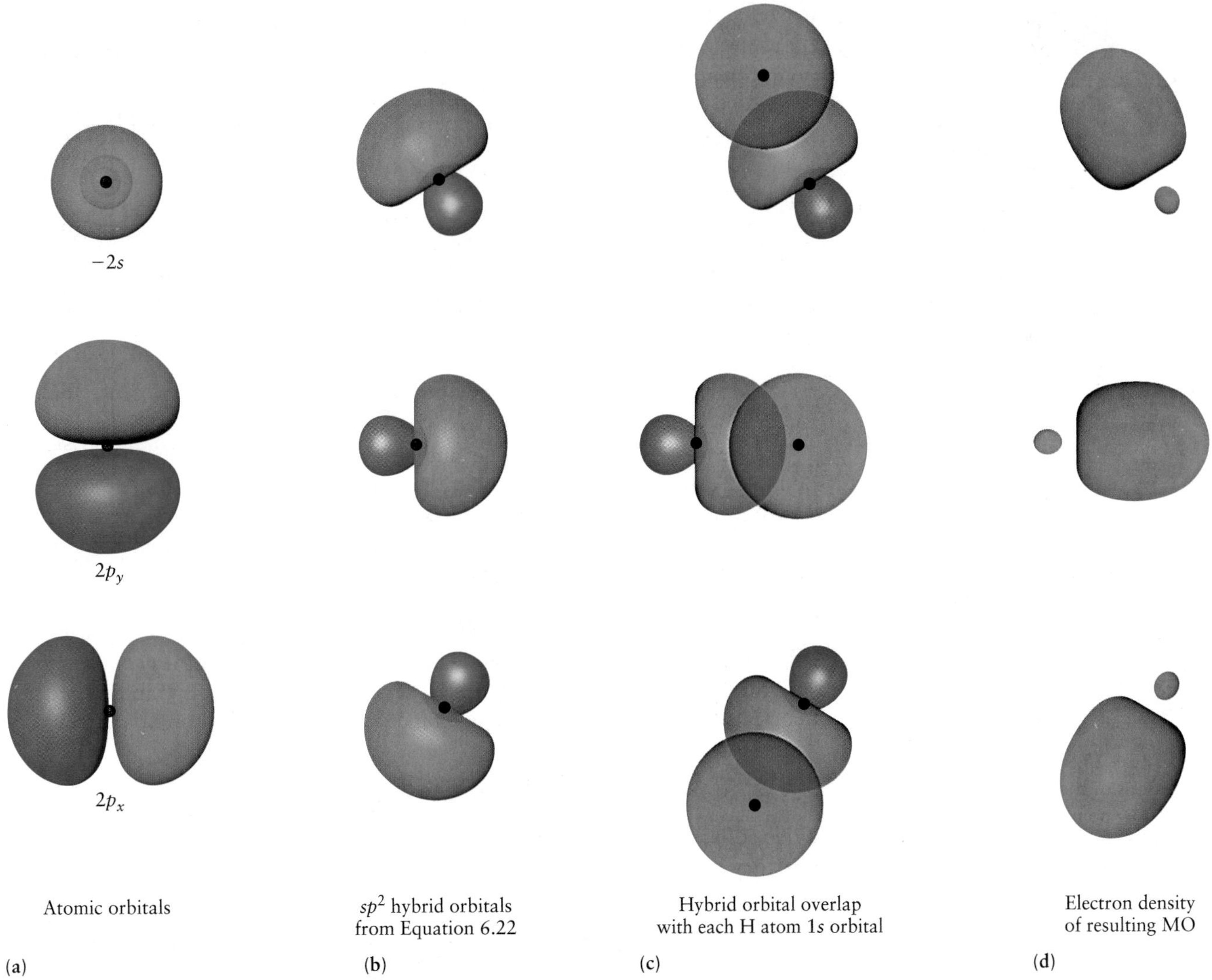

FIGURE 6.38 Formation, shapes, relative orientation, and bonding of the three sp^2 hybrid orbitals in the BH_3 molecule. (The *y*-axis is vertical here and the *x*-axis points to the right.) (a) The −2*s*, $2p_y$, and $2p_x$ atomic orbitals on a boron atom. (b) The three sp^2 hybrid orbitals on a boron atom. (c) Overlap of the sp^2 hybrid orbitals with hydrogen 1*s* orbitals to form three σ bonds in BH_3. (d) Electron density in the three σ bonds (nuclei not shown) obtained by computer calculations. (Courtesy of Mr. Hatem Helal and Professor William A. Goddard III, California Institute of Technology, and Dr. Kelly P. Gaither, University of Texas at Austin.)

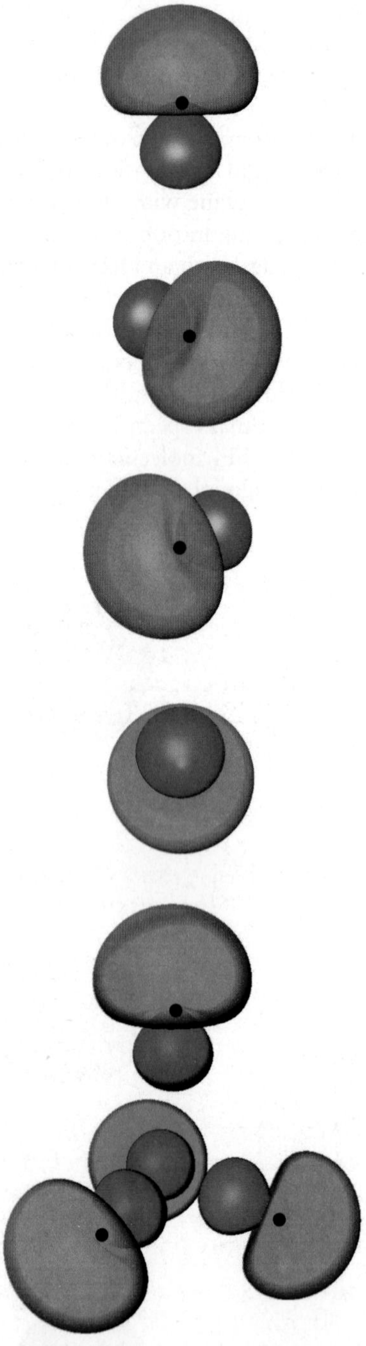

FIGURE 6.39 Shapes and relative orientations of the four sp^3 hybrid orbitals in CH_4 pointing at the corners of a tetrahedron with the carbon atom at its center. The "exploded view" at the bottom shows the tetrahedral geometry. (Courtesy of Mr. Hatem Helal and Professor William A. Goddard III, California Institute of Technology, and Dr. Kelly P. Gaither, University of Texas at Austin.)

characteristic of sp^2 hybridization. It forms three σ bonds by overlap of a $B(sp^2)$ hybrid with an $F(2p)$ orbital directed along each B-F bond in the x-y plane.

sp^2 hybridization explains how an atom can form three σ bonds at an angle of 120° in a plane and therefore be part of a trigonal planar molecule. This is a general result, which can be combined with other aspects of molecular structure. Later in this section we shall see that atoms having more than three valence electrons can form trigonal structures through sp^2 hybridization and at the same time can also form multiple bonds through linear combinations of each p orbital not used in making the three sp^2 hybrid AOs with orbitals on neighboring atoms.

6.9.3 sp^3 Hybridization

The structure of methane CH_4 has been shown by numerous experimental measurements to be tetrahedral with H—C—H bond angles of 109.5° and C—H bond length of 1.093 Å. To describe this structure for CH_4, we need four new equivalent orbitals on C oriented as a tetrahedron about the C atom. To construct these orbitals, we first promote C from C: $(1s)^2(2s)^2(2p_x)^1(2p_y)^1$ to C: $(1s)^2(2s)^1(2p_x)^1(2p_y)^1(2p_z)^1$. Then we combine the $2s$ and three $2p$ orbitals of the central carbon atom to form four equivalent **sp^3 hybrid atomic orbitals**, which point toward the vertices of a tetrahedron:

$$\chi_1(r) = -\frac{1}{2}\left[2s + 2p_x + 2p_y + 2p_z\right] \quad [6.23a]$$

$$\chi_2(r) = -\frac{1}{2}\left[2s - 2p_x - 2p_y + 2p_z\right] \quad [6.23b]$$

$$\chi_3(r) = -\frac{1}{2}\left[2s + 2p_x - 2p_y - 2p_z\right] \quad [6.23c]$$

$$\chi_4(r) = -\frac{1}{2}\left[2s - 2p_x + 2p_y - 2p_z\right] \quad [6.23d]$$

Figure 6.39 shows the shape and orientation of these four hybrid AOs, pointing toward the vertices of a tetrahedron, which has the C atom at its center. The bottom image in Figure 6.39 shows an "exploded view" in which the hybrid AOs have been displaced from one another to clearly reveal the tetrahedral geometry. Each hybrid AO can overlap a $1s$ orbital of one of the hydrogen atoms to give an overall tetrahedral structure for CH_4.

The key result of sp^3 hybridization is to explain how an atom can form four σ bonds at an angle of 109.5° and thereby generate tetrahedral geometry. This structural unit is a major building block in organic chemistry and polymeric materials based on C, and in semiconductor materials based on Si and Ge.

6.9.4 Summary of Hybridization Results

Table 6.4 summarizes the three types of hybrid orbitals and the molecular geometry they produce when there are no nonbonding electron pairs on atom A.

TABLE 6.4

Orbital Hybridization and Molecular Geometry

Molecule	Hybrid Orbitals on Central Atom	Molecular Geometry	Example
AX_2	sp	Linear	BeH_2
AX_3	sp^2	Trigonal planar	BH_3
AX_4	sp^3	Tetrahedral	CH_4

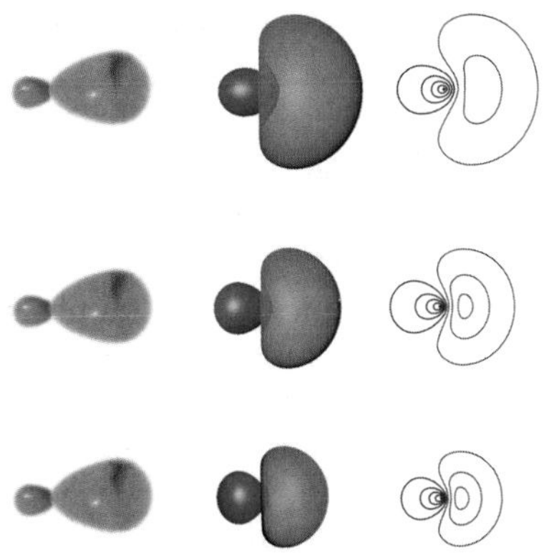

FIGURE 6.40 Exact and approximate representations of the hybrid orbital shapes. For each type of hybrid orbital shown, the left column shows typical chemists' sketches, the center column shows isosurfaces, and the right column shows contour plots. The top row is the *sp* hybrid orbitals, the middle row is the sp^2 hybrid orbitals, and the bottom row is the sp^3 hybrid orbitals. (Courtesy of Mr. Hatem Helal and Professor William A. Goddard III, California Institute of Technology, and Dr. Kelly P. Gaither, University of Texas at Austin.)

Because of their widespread use in chemistry, it is important to have a good sense of the sizes and shapes of hybrid orbitals. The shapes of the *sp* hybrid orbitals in Figure 6.37 are quantitatively correct and properly scaled in size relative to the other orbitals shown. Chemists tend to sketch these orbitals by hand like those shown in Figure 6.40, which gives the misleading impression that the *sp* hybrids are thin, cigar-like shapes with highly directional electron density concentrated right along the direction of the bonds. A contour map of χ_1 (from Equation 6.20b) shows that the orbital is rather diffuse and broadly spread out, despite its directional concentration (see Fig. 6.40). Because this plot is symmetrical about the z-axis, each of these contours can be rotated about the z-axis to produce a three-dimensional isosurface at a specified fraction of the total amplitude; the isosurfaces in Figure 6.37 were generated in just that way. These isosurfaces show the real effect of *sp* hybridization: The amplitudes of the Be AOs are now "pooched out" a bit in the $+z$ and $-z$ directions, but it has not been squeezed down into a thin tube. The $2p_x$ and $2p_y$ orbitals remain unchanged, oriented perpendicular to each other and to the *sp* hybrid orbitals.

The chemist's sketches typically drawn to emphasize directionality of the sp^2 orbitals and a contour plot of the actual shape are shown in the second row of Figure 6.40. Each of these contours can be rotated about the x-y plane to produce a three-dimensional isosurface at a value (+ or −) chosen to be a specific fraction of the maximum value of the wave function. These isosurfaces demonstrate that sp^2 hybridization give a hybrid AO wave function of the B atom that is "pooched out" at three equally spaced locations around the "equator" of the atom (see Fig. 6.38). The $2p_z$ orbital is not involved and remains perpendicular to the plane of the sp^2 hybrids. The electron density in each hybrid AO is more directional (as it is the square of the wave function).

The standard chemist's sketches of the sp^3 orbitals and a contour plot that shows the exact shape and directionality of each orbital are shown in the third row of Figure 6.40. The isosurfaces shown in Figure 6.39 were generated from these contour plots.

6.9.5 Hybridization and Lone Pairs

Let's return to our systematic study of bonding in hydrides of second-period atoms and see how hybridization can correctly explain the structure of the ammonia and water molecules. The Lewis diagram for ammonia shows the N atom has one unshared pair, giving steric number 4. Let's use sp^3 hybridization to describe the bonding of this atom. Of the eight valence electrons in NH_3, six are involved in σ bonds between nitrogen and hydrogen; the other two occupy the fourth sp^3 hybrid orbital as a lone pair (see Fig. 6.41a). This model predicts a trigonal pyramid with three equivalent bonds, in accordance with VSEPR theory. Placing the unshared pair in an sp^3 hybrid orbital predicts bond angles of 109.5°, reasonably close to the measured values of 107° for NH_3. The general conclusion is that sp^3 hybridization with a lone pair in one of the hybrid AOs explains the trigonal pyramid structure for AX_3 molecules, as a subcase of tetrahedral structure. Examples include the hydrides and halides of the Group V elements N, P, As, Sb, and Bi, all of which have the trigonal pyramid structure with bond angles ranging from 92° to 107°. Additional effects must be considered to explain the range of bond angle values in this group of molecules.

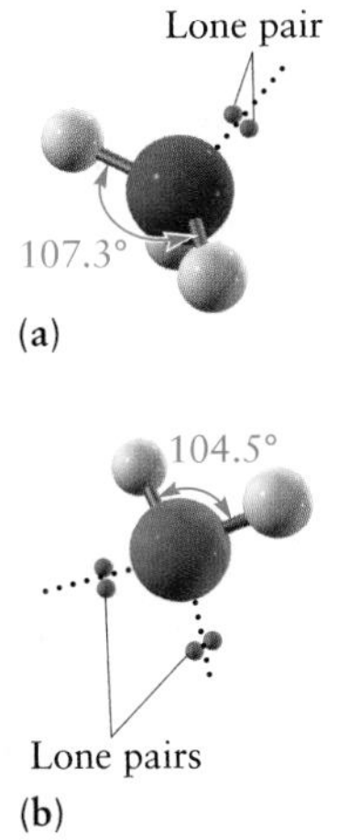

FIGURE 6.41 (a) Ammonia (NH_3) has a pyramidal structure in which the bond angles are less than 109.5°. (b) Water (H_2O) has a bent structure, with a bond angle less than 109.5° and smaller than that of NH_3.

Oxygen in H_2O likewise has steric number 4 and can be described with sp^3 hybridization, with two lone pairs in sp^3 orbitals (see Figure 6.41b). Placing the unshared pairs in an sp^3 hybrid orbital predicts bond angles of 109.5°, reasonably close to the measured values of 104.5° for H_2O. The general conclusion is that sp^3 hybridization with lone pairs in two of the hybrid AOs explains the bent or angular structure for AX_2 molecules, as a subcase of tetrahedral structure. Other examples besides water are H_2S, OF_2, and SF_2. Again, additional effects must be considered to explain the range of bond angle values in this group of molecules.

6.9.6 Hybridization and Multiple Bonds in Organic Carbon Compounds

Carbon (C) is unique among all the elements in its ability to form large numbers of molecules with great variety in their structures. Different hydrocarbons (molecules consisting of H and C atoms) have structures with single, double, and triple carbon–carbon bonds. This variety traces to the location of carbon in the periodic table. As a Group IV element, carbon can form four bonds, which we have already seen is more than the other elements in the second period. Because of its small size, carbon can form double and triple bonds, which are rare in the larger atoms (Ge, Si) in Group IV. With intermediate value of electronegativity, carbon forms covalent compounds with more electronegative elements such as nitrogen, oxygen, and the halogens and also with more electropositive elements such as hydrogen, mercury, and lead.

The chemistry of carbon constitutes the field of organic chemistry, to which Chapter 7 provides a brief systematic introduction organized around the bonding in organic molecules. Orbital hybridization plays a key role in explaining bonding and structure of organic molecules. The methane molecule CH_4 explained by sp^3 hybrid bonding is the simplest example of single bonds in organic chemistry. Here we give a brief introduction to double and triple bonds in organic chemistry based on sp^2 and sp hybridization, respectively. More extensive discussions of these multiple bonds and their implications for organic reactions are in Chapter 7.

We begin with the compound ethene C_2H_4 for which the Lewis diagram indicates a double bond between the C atoms and a single bond between each H and the adjacent C. Experiments show that ethene is a planar molecule with H—C—H bond angles of 117° and H—C—C bond angles of 120°. This structure can be explained with sp^2 hybrid orbitals on each carbon atom. To construct these orbitals, we first promote C from C: $(1s)^2(2s)^2(2p_x)^1(2p_y)^1$ to C: $(1s)^2(2s)^1(2p_x)^1(2p_y)^1(2p_z)^1$. Then we combine the $2s$ and two of the $2p$ orbitals to form three equivalent sp^2 hybrid atomic orbitals which lie 120° apart in a plane. After hybridization, each C atom has the electron configuration C: $(1s)^2(\chi_1)^1(\chi_2)^1(\chi_3)^1(2p_z)^1$. The new feature compared to hybridization in boron discussed earlier is that the nonhybridized $2p_z$ has one electron, and this orbital has its original orientation perpendicular to the plane containing the hybrids. Figure 6.42a shows the three sp^2 hybrid orbitals, and the nonhybridized

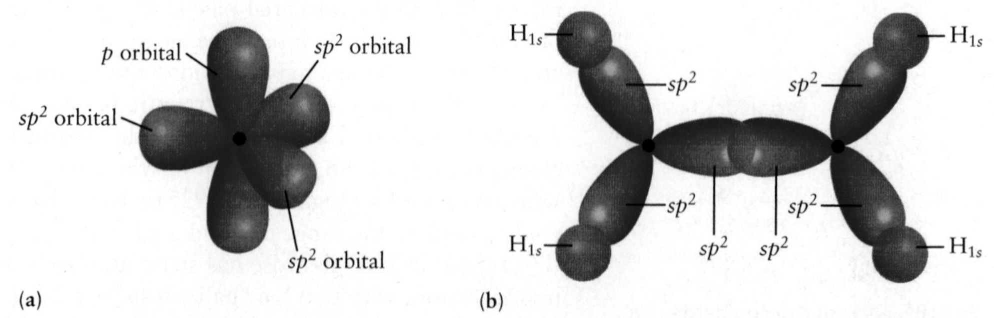

FIGURE 6.42 Formation of σ bonds in ethene. (a) The three sp^2 hybridized orbitals are oriented in a plane with their axes at angles of 120°. The non-hybridized $2p$ orbital is perpendicular to the plane containing the three sp^2 hybrid orbitals. (b) Top view showing formation of a C—C σ bond by overlap of two sp^2 hybrid orbitals and formation of four C—H σ bonds by overlap between four C sp^2 hybrid orbitals and four H $1s$ orbitals.

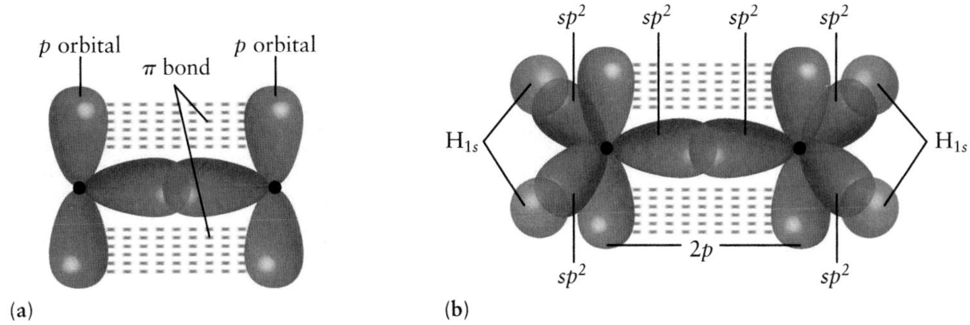

FIGURE 6.43 Formation of π bonds in ethene. (a) Overlap of parallel 2*p* orbitals forms a π bond. (b) The complete bonding scheme for ethene includes a C=C double bond and four C—H single bonds.

$2p_z$ orbital. A σ bond is formed by overlap of the sp^2 orbitals on each carbon atom, and the remaining four sp^2 orbitals form σ bonds with the four hydrogen atoms. Each carbon forms σ bonds with two hydrogen atoms (Fig. 6.42b). The nonhybridized $2p_z$ orbitals on the two carbon atoms are parallel to each other and overlap to form a π bond (see Fig. 6.43a), just as illustrated in Figure 6.31. The result is a double bond between the two carbon atoms (Fig. 6.43b). This description of double bonds is used for a large class of organic molecules called *alkenes.*

H:C:::C:H

FIGURE 6.44 Lewis diagram for C_2H_2.

Now we turn to acetylene (also called ethyne, the IUPAC name), C_2H_2, the Lewis diagram for which (Fig. 6.44) shows a triple bond between the carbon atoms. Acetylene is known to be a linear molecule. This structure can be explained by *sp* hybridization on each of the carbon atoms. To construct these orbitals, we define the *z*-axis to lie along the bond, and we promote each carbon atom from C: $(1s)^2(2s)^2(2p_x)^1(2p_y)^1$ to C: $(1s)^2(2s)^1(2p_x)^1(2p_y)^1(2p_z)^1$. On each carbon atom we form a pair of *sp* hybrid orbitals oriented 180° apart along the *z*-axis. After hybridization each carbon atom has the electron configuration C: $(1s)^2(\chi_1)^1(\chi_2)^1(2p_x)^1(2p_y)^1$. Figure 6.45a sketches the formation of the *sp* hybrid orbitals and shows these along with the $2p_x$ and $2p_y$ nonhybridized atomic orbitals. Sigma bond formation in acetylene is shown in Figure 6.45b. A σ bond between the two carbon atoms is formed by overlap of *sp* hybrids on each carbon, and each carbon forms a σ bond with one hydrogen atom using its other *sp* hybrid. The $2p_x$ and $2p_y$ non-hybridized atomic orbitals are parallel pairs on the two carbon atoms. Each pair overlaps to form a π bond, as shown in Figure 6.46a. The result is a triple bond in acetylene (see Fig. 6.46b), analogous to the triple bond in N_2 shown in Figure 6.31. This description of triple bonds is used for a large class of organic molecules called *alkynes.*

Finally, the concept of orbital hybridization has played a key role in leading us to the modern theory of the chemical bond, and the tools of hybridization are used to this day in every branch of chemistry. Hybridization has inspired a great deal of discussion over the years, some of it impassioned, on the meaning and significance of "promotion" and "return on the energy investment." Where

FIGURE 6.45 Formation of sigma bonds in acetylene. (a) The two *sp* hybridized orbitals are oriented in a plane with their axes at angles of 180°. Two nonhybridized 2*p* orbitals are oriented perpendicular to the plane of the two *sp* hybrid orbitals. (b) Top view showing formation of a C—C σ bond by overlap of two *sp* hybrid orbitals. The other two *sp* hybrid orbitals shown in (b) are available to form σ C—H bonds with each H atom to form HCCH.

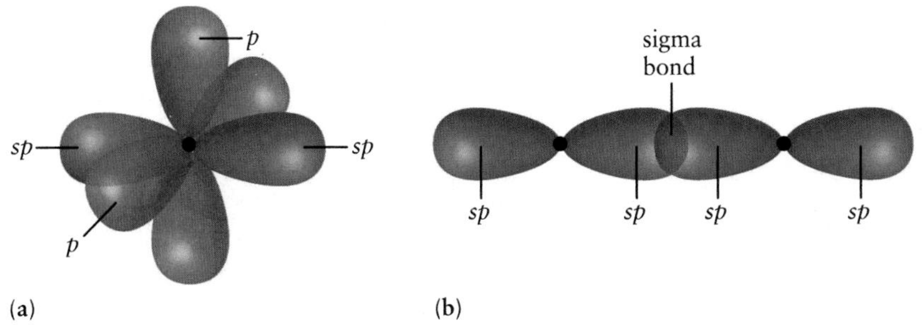

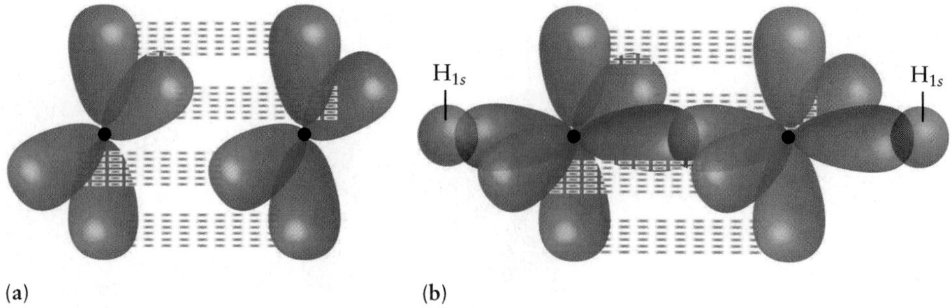

FIGURE 6.46 Formation of π bonds in acetylene. (a) Overlap of two sets of 2p orbitals forms two π bonds. (b) The complete bonding scheme for acetylene includes a C=C double bond and two C—H single bonds.

does the energy input for promotion come from? How is the bond formed? How does the energy released on bond formation compensate for promotion? Do these concerns cast doubt on the validity and usefulness of the hybrid orbital representations of the chemical bond? These may be legitimate concerns if one is trying to describe the dynamic events by which the bond is actually formed. However, these concerns are largely side issues for our main question: is hybridization a useful way to describe the *structure* of a chemical bond after it has been formed? Quantum mechanics with exact molecular orbital theory offers accurate predictions for molecular shapes without any assumptions inherent in VSEPR theory, but bonding by first constructing hybrid AOs continues to give a good qualitative picture for many organic molecules.

6.10 Predicting Molecular Structures and Shapes

The three-dimensional shape of a molecule strongly influences how a pair of molecules approach one another, and whether this pair will successfully encounter for a chemical reaction or join together in a "guest-host" arrangement. Molecular shape has enormous influence over the function of molecules to determine the properties of a material or to achieve specific biological activity. The "shape of a molecule" is a descriptive term such as "square planar" or "tetrahedral" that we apply once we know the detailed molecular structure, specified by the bond lengths and bond angles in the molecule. The term "shape" connotes a specific geometrical relationship between bonds, but does not imply specific quantitative dimensions. If we know the detailed molecular structure, we can determine the shape by inspection.

To what extent can we predict the shape of a particular molecule based on the information we have developed so far? The VSEPR theory in Section 3.11 can predict the idealized shape by counting up the total number of bonding and unshared electron pairs in the valence shell and finding the spatial arrangement of all the pairs that minimizes the potential energy. The location of the bonding pairs within this optimum arrangement determines the location of the bonds in the molecule. From this, the theory assigns the shape of the molecule to a particular geometrical label. The results are summarized in Table 3.8. Deviations from these perfect shapes are rationalized by ranking the importance of repulsions between bonding pairs and unshared pairs. Note, however, that orbital hybridization as described in this book has no predictive capabilities at all. Recall that we had to specify the bond angles that we wanted to achieve before starting each type of hybridization. Orbital hybridization provides a quantum justification or interpretation for a molecular shape that we already know or assume, but it does not predict the shape or the geometrical structure.

At this point, the best method we can provide to describe the structure and shape of a molecule is summarized by the following steps:

1. Determine the empirical formula for the substance from an elemental analysis based on combustion (see Section 2.2.3).
2. Determine the molecular formula for the substance by its behavior as a gas (see Section 2.2.4) or by mass spectrometry (see the *Connection* in Chapter 3).
3. Determine the structural formula for the molecule by writing a Lewis diagram based on the molecular formula (see Section 3.10).
4. Determine the molecular shape experimentally, e.g., by X-ray scattering, or predict the shape by using VSEPR theory (see Section 3.11).
5. Identify the hybridization scheme that best explains the shape predicted by VSEPR (see Section 6.9).

This method is illustrated in Example 6.4 for the compound hydrazine, which is used as a fuel in rocket propulsion.

EXAMPLE 6.4

Hydrazine is a potentially explosive gas used in rocket fuels. Elemental analysis shows its mass percent composition to be 87.419% nitrogen and 12.581% hydrogen. The density of hydrazine at 1 atm pressure and 25°C is 1.31 g L^{-1}. Determine the molecular formula for hydrazine. Predict the structure of hydrazine by writing down its Lewis diagram and using VSEPR theory. What is the hybridization of the nitrogen atoms?

Solution

The background for the first part of this example is described in Section 2.2 and Examples 2.1 and 2.2. The number of moles of nitrogen and hydrogen in a sample of 100.00 g of hydrazine are

$$\text{nitrogen}: \frac{87.42 \text{ g N}}{14.9967 \text{ g mol}^{-1}} = 6.2413 \text{ mol N}$$

$$\text{hydrogen}: \frac{12.58 \text{ g H}}{1.0079 \text{ g mol}^{-1}} = 12.4814 \text{ mol H}$$

$$\text{ratio}: \frac{\text{moles H}}{\text{moles N}} = \frac{12.4814}{6.2413} = 1.9998$$

The empirical formula is therefore close to NH_2. To determine the true molecular formula we estimate the molar mass of hydrazine using the ideal gas law (see Sec. 9.3). If m is the mass of the sample in grams and $\mathcal{M}$ is the molar mass in grams mol^{-1}, the molar mass $\mathcal{M}$ is determined from the density ρ and the gas properties as follows:

$$PV = nRT = \frac{m}{\mathcal{M}}RT$$

$$\mathcal{M} = \frac{mRT}{PV} = \left(\frac{m}{V}\right)\frac{RT}{P} = \rho\frac{RT}{P}$$

$$\mathcal{M} = \frac{(1.31 \text{ g L}^{-1})(0.0821 \text{ L atm mol}^{-1} \text{ K}^{-1})(298.15 \text{ K})}{1 \text{ atm}}$$

$$\mathcal{M} = 32.066 \text{ g mol}^{-1}$$

FIGURE 6.47 The structure of hydrazine, N_2H_4.

The molar mass corresponding to the empirical formula NH_2 is 16.0125 g mol^{-1}. Because the true molar mass is approximately twice this value the molecular formula for hydrazine is N_2H_4, which is frequently written as H_2NNH_2. The Lewis diagram is

```
    H
H:N:N:H
  H
```

Lewis diagram for N_2NNH_2.

Both nitrogen atoms have steric number 4 and are sp^3 hybridized, with H—N—H and H—N—N angles of approximately 109.5°. The extent of rotation about the N—N bond cannot be predicted from the VSEPR theory or the hybrid orbital model. Figure 6.47 shows the full three-dimensional structure of hydrazine.

Related Problems: 49, 50, 51, 52, 53, 54, 55, 56

TABLE 6.5

Molecular Shapes Predicted by the Valence Shell Electron-Pair Repulsion Theory and Rationalized by Orbital Hybridization

Molecule	Steric Number	Number of Lone Pairs	Orbital Hybridization	Predicted Geometry	Image	Example
AX_2	2	0	*sp*	Linear		BeH_2, CO_2
AX_3	3	0	sp^2	Trigonal planar	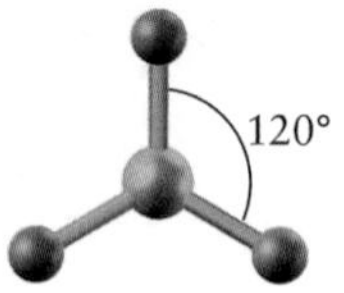	BF_3, SO_3
AX_2	3	1	sp^2	Bent		SO_2
AX_4	4	0	sp^3	Tetrahedral	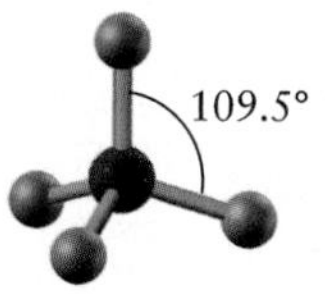	CF_4, SO_4^{2-}
AX_3	4	1	sp^3	Trigonal pyramidal		NH_3, PF_3, $AsCl_3$
AX_2	4	2	sp^3	Bent		H_2O, H_2S, SF_2

Table 6.5 summarizes the results of this procedure for molecules with steric numbers 2–4 and gives several examples of each type.

6.10.1 The Shapes of Molecules and Electrostatic Forces

The **electrostatic potential energy map** for a molecule combines information from two different sources into one representation. The size and shape of the molecule come from the spatial distribution of the electron density (from high-level electronic structure calculations) represented as an isosurface. The electrostatic potential energy that a positive test charge would experience is indicated at each point on that isosurface. Let's construct each of these pieces in turn.

Isosurfaces of electron density are obtained from the probability density isosurfaces for molecules described in Sections 6.4–6.9. These are surfaces in three-dimensional space that include all the points at which ψ^2 has a particular value. The value of electron density chosen to define the isosurface is selected by some definite, though arbitrary, criterion. There is broad acceptance of a standard density of $0.002\ e/(a_0)^3$, where a_0 is the Bohr radius. This value is thought to best represent the "sizes" and shapes of molecules because it corresponds to the van der Waals atomic radii discussed earlier in the context of atomic size. These are the same dimensions depicted in space-filling models of molecules. Figures 6.48a and 6.48b show, respectively, space-filling models and electron density isosurfaces plotted at $0.002\ e/(a_0)^3$ for water, ammonia, and methane. The electron densities plotted here include all of the electrons in the molecule. They are calculated using state-of-the-art quantum chemical methods.

To understand how a value of the electrostatic potential energy can be associated with each point on the electron density isosurface, we imagine a thought experiment in which a positive unit test charge moves over this isosurface, interacting with all of the electrons and the nuclei of the molecule as it visits every position on the isosurface. The magnitude of the test charge must be extremely small to avoid distorting and polarizing the electron density of the

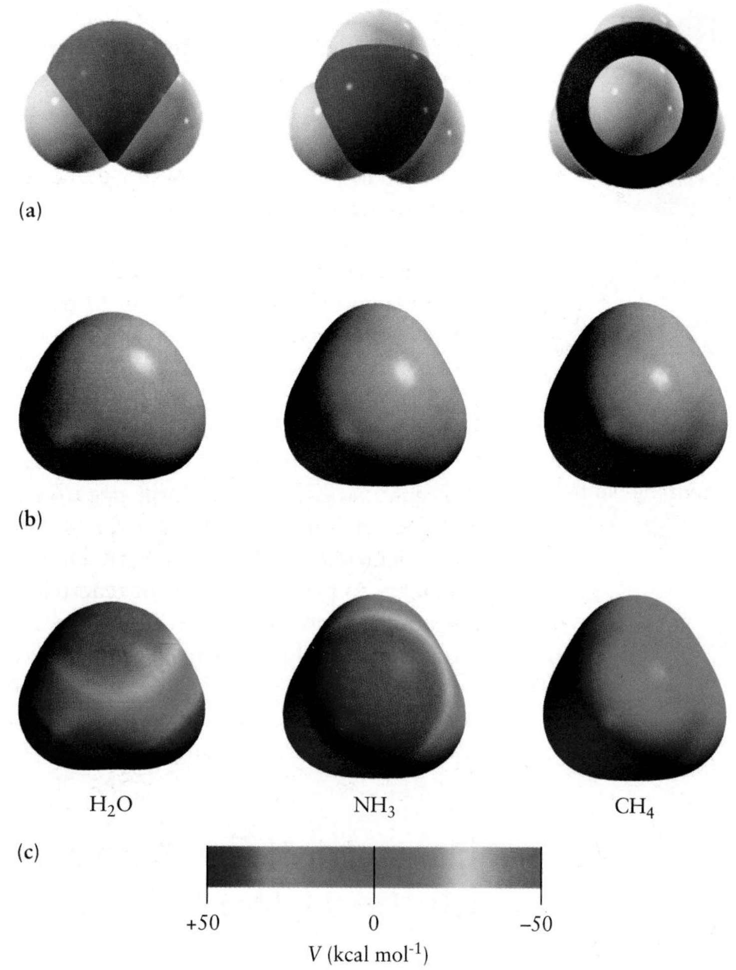

FIGURE 6.48 Representations of the shape, size, and electrostatic potential of molecules. (a) Space-filling models, (b) 0.002 $e/(a_0)^3$ electron density isosurfaces, and (c) electrostatic potential energy surfaces for water, ammonia, and methane. Note that the energy units are kcal mol^{-1}, which are still in common use in organic chemistry.

molecule. The test particle is attracted to the molecule at those points where it experiences negative electrostatic potential energy, and it is repelled away from the molecule at points where it experiences positive electrostatic potential energy. It is convenient to summarize these explorations by assigning color to each location on the isosurface on the basis of the sign and magnitude of the electrostatic potential energy experienced by the test charge. The conventional choice for the color map is based on the visible spectrum, in which the most negative potential is represented by red and the most positive potential is represented by blue. The colors assigned to each point on the electron density isosurface vary continuously between these extremes on the basis of the sign and magnitude of the electrostatic potential energy. The potential energy values near zero are represented by green.

This thought experiment is carried out as follows to obtain actual values of electrostatic potential energy through computer calculations. After the electron density isosurface has been calculated by quantum mechanics, a unit (elementary) positive test charge is moved around the molecule at locations corresponding to points on the isosurface. At each location the electrostatic potential energy is calculated using Coulomb's law to describe the interaction of the test charge with each nucleus and the distributed electron density in the molecule (with the total charge on the electrons equal to the number of electrons multiplied by $-e$). At each location, the value of the electrostatic potential energy represents the balance between attraction of the test charge by electrons and repulsion by the nuclei in the molecule. The value obtained at each location of the test particle is then mapped onto the corresponding point on the electron

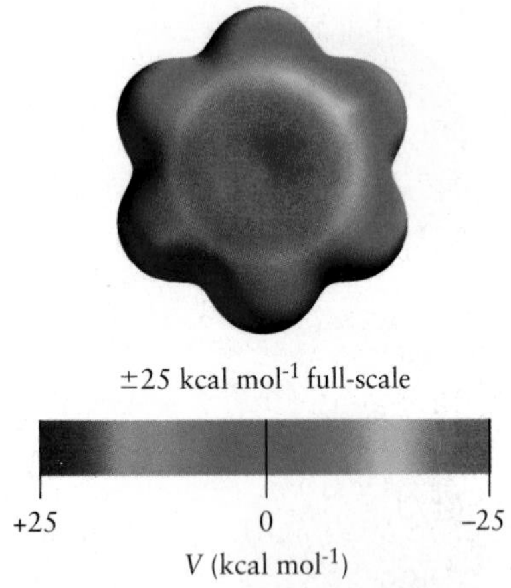

FIGURE 6.49 Electrostatic potential energy surface for benzene plotted at ± 25 kcal mol^{-1} full-scale.

density isosurface for the molecule by assigning color according to the color map. The electrostatic potential energy maps calculated for the molecules shown in Figures 6.48a and 6.48b are shown in Figure 6.48c.

The electrostatic potential energy map for a given molecule can be used to predict how the target molecule influences the motion of charged particles as they approach it. The images show at a glance which portions of the molecule are most likely to attract or repel a proton. For example, in Figure 6.48c the red region shows a region of low electrostatic potential energy, to which a proton will be attracted. Thus, positive ions will be attracted to the oxygen end of the water molecule and to the nitrogen end of the ammonia molecule; the opposite is true for negative ions. The interaction of a positive or a negative ion with methane is weak and shows no pronounced directionality. Comparing Figure 6.48c with Figure 6.49 shows that a positive ion experiences greater attraction to the oxygen end of a water molecule than to the center of a benzene molecule.

Electrostatic potential energy maps can be used to identify reactive sites on molecules. Locations with large negative values of the electrostatic potential are relatively rich in electron density, and those with large positive values are relatively depleted in electron density. These maps are now widely used in organic chemistry to predict patterns of reactivity for electrophilic (electron-loving) and nucleophilic (proton-loving) molecules and to explain how the presence of different functional groups in the molecule can affect these patterns. These methods are effective aids in identifying sites for chemical reactivity in more complicated molecules, including those of biological interest. They are widely used in molecular modeling simulations of drug design.

6.11 Using the LCAO and Valence Bond Methods Together

The LCAO and VB approaches are both good starting points for describing bonding and reactivity. You can apply either one to set up a purely qualitative description of the problem of interest, confident that you can move on to a high-level quantitative calculation as your needs demand. Which method you choose at the beginning depends primarily on the area of chemistry in which you are working and the broad class of problems you are investigating. LCAO theory is most often used to describe the electronic states of molecules in contexts that require knowledge of energy levels. Examples include molecular spectroscopy, photochemistry, and phenomena that involve ionization (such as electron-induced reactions and photoelectron spectroscopy). VB theory is more widely used to describe molecular structure, especially in pictorial ways.

Many chemists use a mixture of the two methods, where localized VB σ bonds describe the network holding the molecule together and de-localized LCAO π bonds describe the spread of electron density over the molecule. We will illustrate this combination for the case of linear and also bent triatomic molecules here, then discuss two cases that illustrate delocalized π bonding in polyatomic molecules, 1,3-butadiene and benzene.

6.11.1 Triatomic Nonhydrides

Many triatomic molecules and molecular ions can be formed from atoms of elements in the second and third periods. Some are linear (examples are CO_2, N_2O, OCS, and NO_2^+), and others are bent (examples include NO_2, O_3, NF_2, and NO_2^-).

Experiments show that triatomic nonhydrides with 16 or fewer valence electrons generally are linear, whereas those with 17 to 20 valence electrons generally are bent. This is consistent with the VSEPR theory, because the former have steric number 2 (giving linear molecules), whereas the latter have steric number 3 or 4 (giving bent molecules with one or two lone-pair orbitals on the central atom). These experimental results can be explained by the combined approach in which we use localized orbitals for the σ bonds and delocalized orbitals for the π bonds of these molecules.

6.11.2 Linear Triatomic Nonhydrides

For linear molecules (examples are CO_2, N_2O, OCS, and NO_2^+), it is logical to use *sp* hybridization to describe the central atom. This leaves two *p* orbitals (p_x and p_y) perpendicular to the bond axis. Hybridization of the outer atoms is not necessary to treat such molecules; the outer atoms are most simply described as having fully occupied *s* orbitals not involved in bonding, a p_z orbital that takes part in σ bonding, and p_x and p_y orbitals that form delocalized π bonds with the other atoms. The σ bonds (which extend along the bond axes) can be obtained by combining the p_z orbital of each of the outer atoms with the *sp* hybrid orbital of the central atom that points toward it to form a localized bonding orbital. These two localized σ bonds are filled by two pairs of electrons.

Linear combinations of the p_x and p_y orbitals of all three atoms make up the delocalized molecular orbitals involved in π bonding. How do three p_x orbitals combine to form three molecular orbitals? Two of the combinations are obvious. The lowest in energy of the three, the π_x molecular orbital (Fig. 6.50a) is fully bonding, with no nodes, and the highest in energy of the three, the π_x^* molecular orbital (Fig. 6.50c) is fully antibonding, with two nodes, one between each pair of atoms. The third linear combination is less obvious. Intermediate

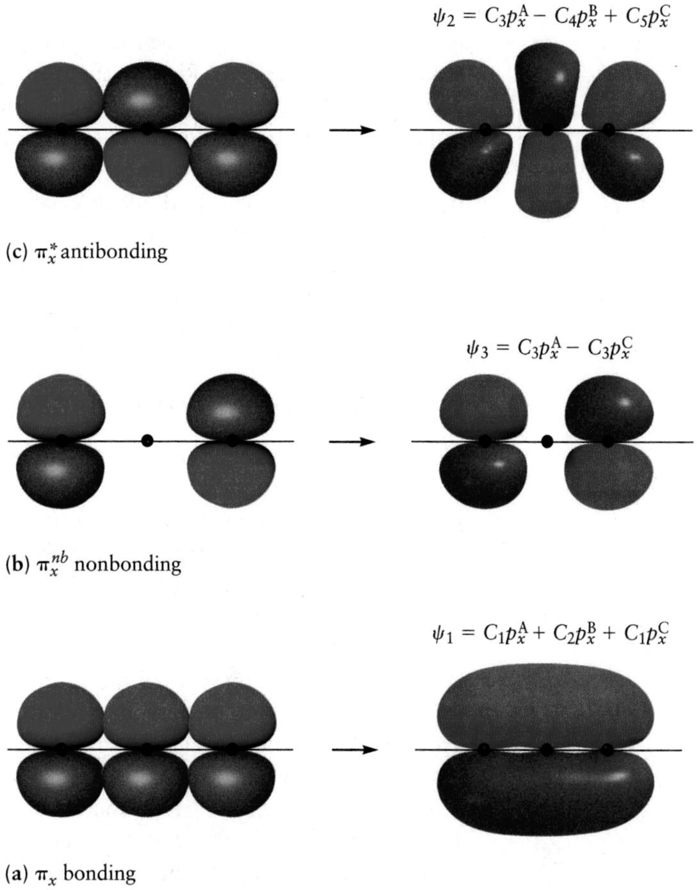

FIGURE 6.50 π bonding in linear triatomic molecules. From three *p* orbitals lying perpendicular to the bond axis, one bonding, one antibonding, and one nonbonding molecular orbital can be constructed. A second group of molecular orbitals can be constructed from the *p* orbitals that lie perpendicular to the plane of the page.

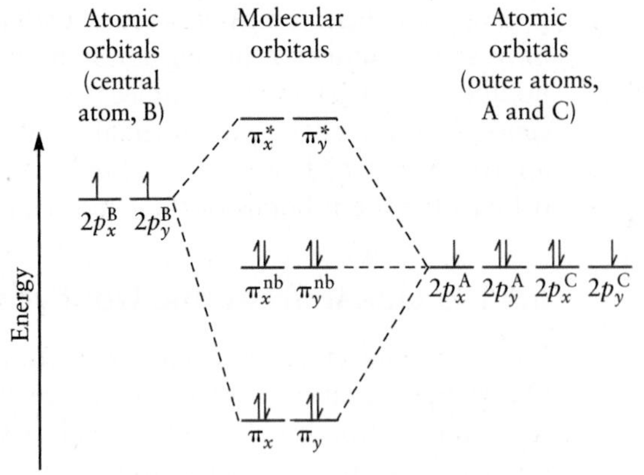

FIGURE 6.51 Correlation diagram for π electrons in linear triatomic molecules. The orbital filling shown is that for CO_2.

in energy, the π_x^{nb} orbital has the form of Figure 6.50b, with a single node and with the coefficient of the wave function on the central atom being zero. Because there is only one molecular orbital at this energy formed from $2p_x$ AOs, the electron density corresponding to this orbital must be evenly distributed, so an orbital with a node between the left and center atoms would not reflect the correct symmetry of the molecule. This orbital is nonbonding (π_x^{nb}) because there is neither a node nor a region of increased electron density between the central atom and its neighbors. The three p_y orbitals can be combined to give the same types of molecular orbitals as the p_x. Figure 6.51 is the resulting correlation diagram for the π levels only, with the nonbonding orbitals lying between the bonding and the antibonding orbitals in energy. As before, the energy increases with the number of nodes.

Consider the CO_2 molecule. It has 16 valence electrons (4 from the carbon and 6 from each oxygen). Two electrons on each oxygen atom are nonbonding electrons in 2*s* atomic orbitals. Of the remaining 12 electrons, 4 take part in localized σ bonds, 2 between each oxygen and the carbon. This leaves 8 electrons for the π system. Placing them in the correlation diagram of Figure 6.51 gives two electron pairs in π orbitals and two pairs in π^{nb} orbitals. The total bond order for the molecule is 4, because there are two pairs of electrons in σ bonding orbitals (from the *sp* hybrids) and two pairs in π bonding orbitals (the nonbonding π^{nb} and atomic oxygen 2*s* orbitals neither increase nor decrease the bond order). Each bond is then of order 2, in agreement with the Lewis diagram result for CO_2. The meaning is slightly different here, however, in that some of the bonding electrons (those in the π orbital system) are delocalized over the full molecule rather than shared by only two of the atoms.

6.11.3 Nonlinear Triatomic Molecules

Nonlinear triatomic molecules (examples include NO_2, O_3, NF_2, and NO_2^-) can be described through sp^2 hybridization of the central atom. If the molecule lies in the *x*-*y* plane, then the valence *s*, p_x, and p_y orbitals of the central atom can be combined to form three sp^2 hybrid orbitals with an angle close to 120° between each pair. One of these orbitals holds a lone pair of electrons, and the other two take part in σ bonds with the outer atoms. The fourth orbital, a p_z orbital, takes part in delocalized π bonding (Fig. 6.52a). On the outer atoms, the *p* orbital pointing toward the central atom takes part in a localized σ bond, and the p_z orbital takes part in π bonding; the third *p* orbital and the *s* orbital are AOs that do not participate in bonding; in the molecular orbital correlation diagram these would be represented by nonbonding molecular orbitals localized on each respective atom (with the same energy as they had in the atom). Finally, the three p_z AOs can be combined into bonding, nonbonding, and antibonding π orbitals (Fig. 6.52a) much as in the linear molecule case. Here

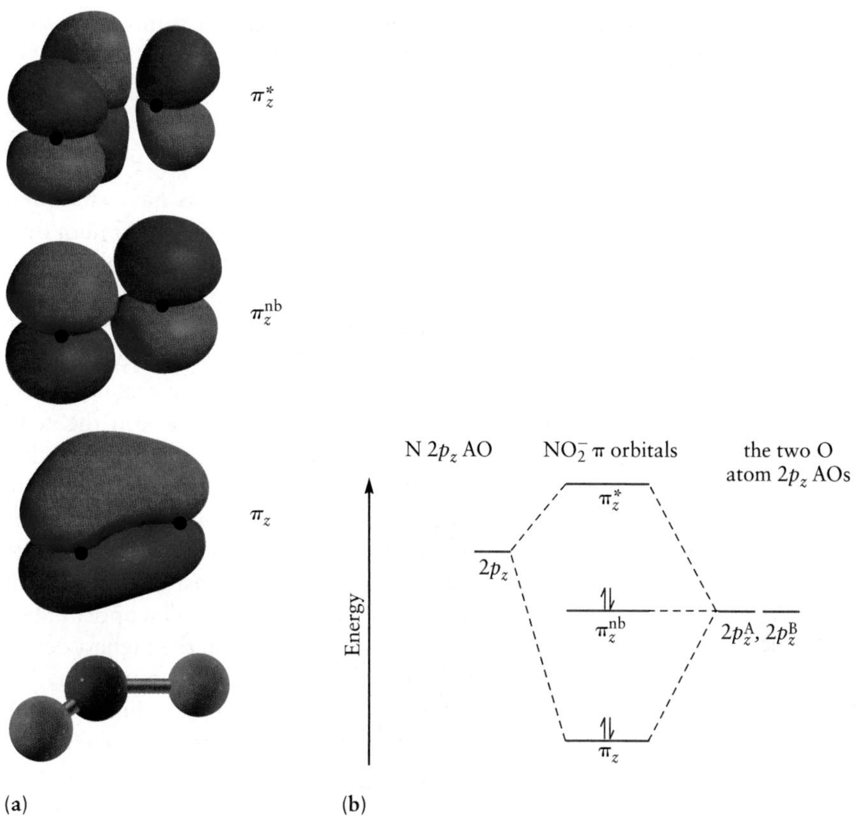

FIGURE 6.52 (a) Ball-and-stick model (bottom) and molecular orbitals for bent triatomic molecules. The central atom has three sp^2 hybrid orbitals (not shown) that would lie in the plane of the molecule. From the three p_z orbitals perpendicular to this plane, three π orbitals can be constructed. (b) Correlation diagram for the π orbitals.

there is only one of each type of orbital (π, π^{nb}, π^*) rather than two as for linear molecules.

Consider a specific example, NO_2^-, with 18 electrons. Two electrons are placed in each oxygen $2s$ orbital and 2 more in each nonbonding oxygen $2p$ orbital, so that a total of 8 electrons are localized on oxygen atoms. Two electrons also are placed as a lone pair in the third sp^2 orbital of the nitrogen atom. Of the remaining 8 electrons, 4 are involved in the two σ bonds between nitrogen and the oxygen atoms. The last 4 are placed into the π electron system: 2 into the bonding π orbital and 2 into the nonbonding π^{nb} orbital. Because a total of 6 electrons are in bonding orbitals and none are in antibonding, the net bond order for the molecule is 3, or $1\frac{1}{2}$ per bond. In the Lewis model, 2 resonance forms are needed to represent NO_2^-. The awkwardness of the multiple resonance forms required by the Lewis model is avoided by treating the electrons in the bonding π molecular orbital as delocalized over the three atoms in the molecule.

6.11.4 Organic Molecules with Delocalized Electrons

Organic chemists make extensive use of the combined approach, where localized VB σ bonds describe the network holding the molecule together and de-localized LCAO π bonds describe the spread of electron density over the molecule. In Chapter 7 we illustrate this approach in some detail for several molecules that belong to the classes substituted alkenes and alkynes (Section 7.3, Example 7.2). Here in this chapter, we now describe the delocalized π bonding in a polyene, 1,3-butadiene, and in benzene, a cyclic hydrocarbon with two resonance Lewis structures.

De-localized π bonding is important in many polyenes, the simplest of which is 1,3-butadiene ($CH_2CHCHCH_2$). Its Lewis dot structure, seen below, does not show resonance, yet the π molecular orbitals in 1,3-butadiene are de-localized over all four carbon atoms.

All four carbon atoms have steric number 3, so all are sp^2 hybridized. The remaining p_z orbitals have maximum overlap when the four carbon atoms lie in the same plane, so this molecule is predicted to be planar. From these four p_z atomic orbitals, four MOs can be constructed by combining their phases, as shown in Figure 6.53. The four electrons that remain after the σ orbitals are filled are placed in the two lowest π orbitals. The first of these is bonding among all four carbon atoms; the second is bonding between the outer carbon atom pairs, but antibonding between the central pair. Therefore, 1,3-butadiene has stronger and shorter bonds between the outer carbon pairs than between the two central carbon atoms. It is an example of a **conjugated π electron system**, in which two or more double or triple bonds alternate with single bonds. Such conjugated systems have lower energies than would be predicted from localized bond models and are best described with delocalized MOs extending over the entire π electron system. The optical absorption spectrum of 1,3-butadiene evidences a transition from the highest occupied π orbital to the lowest unoccupied π^* orbital. This requires a much smaller photon energy than in ethene, as the energies of the two middle orbitals in Figure 6.53 are much closer together than are the π and π^* orbitals of ethene.

Finally, we show the de-localized π orbitals of benzene in Figure 6.54. You can easily tell from the perspective view in Figure 6.54(b) what particular linear

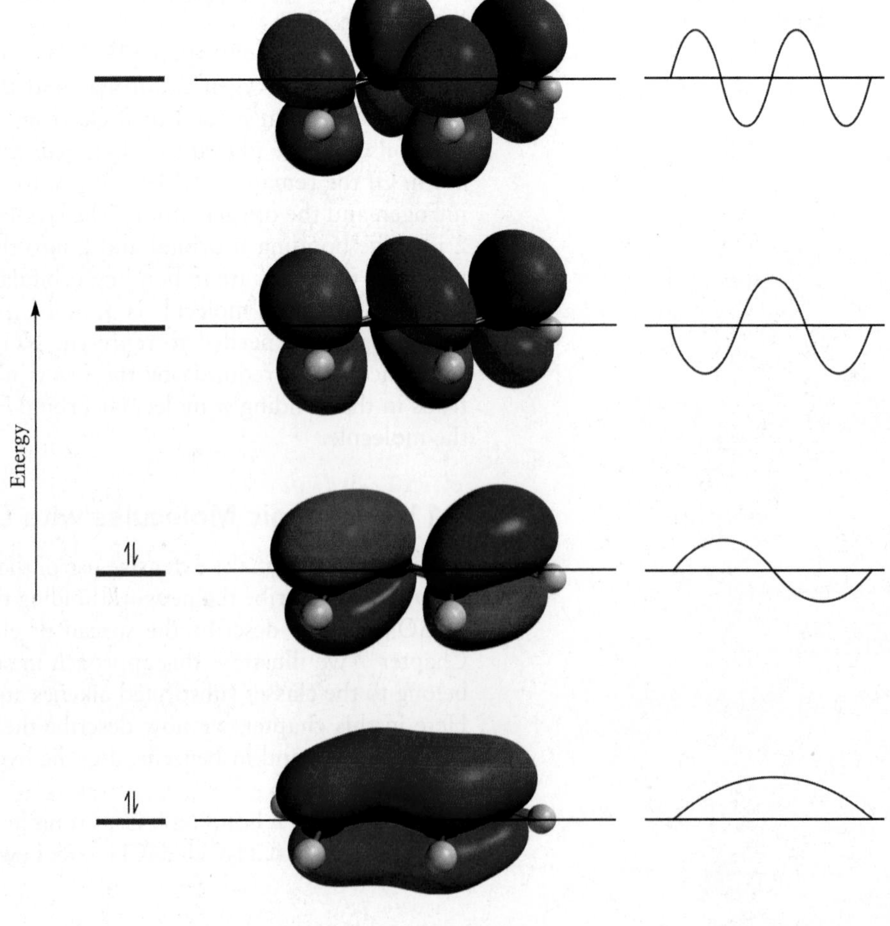

FIGURE 6.53 The four π molecular orbitals formed from four $2p_z$ atomic orbitals in 1,3-butadiene, viewed from the side. In all but the lowest energy orbital there are *y–z* nodal planes between some of the carbon atoms. The horizontal black line represents the *x–y* nodal plane of the π orbitals; it is also the molecular plane. Note the similarity in the *y–z* nodal patterns to those of the first four modes of a vibrating string or the first four wave functions of the one-dimensional particle in a box (right). Only the two lowest energy π orbitals are occupied in the ground state of 1,3-butadiene.

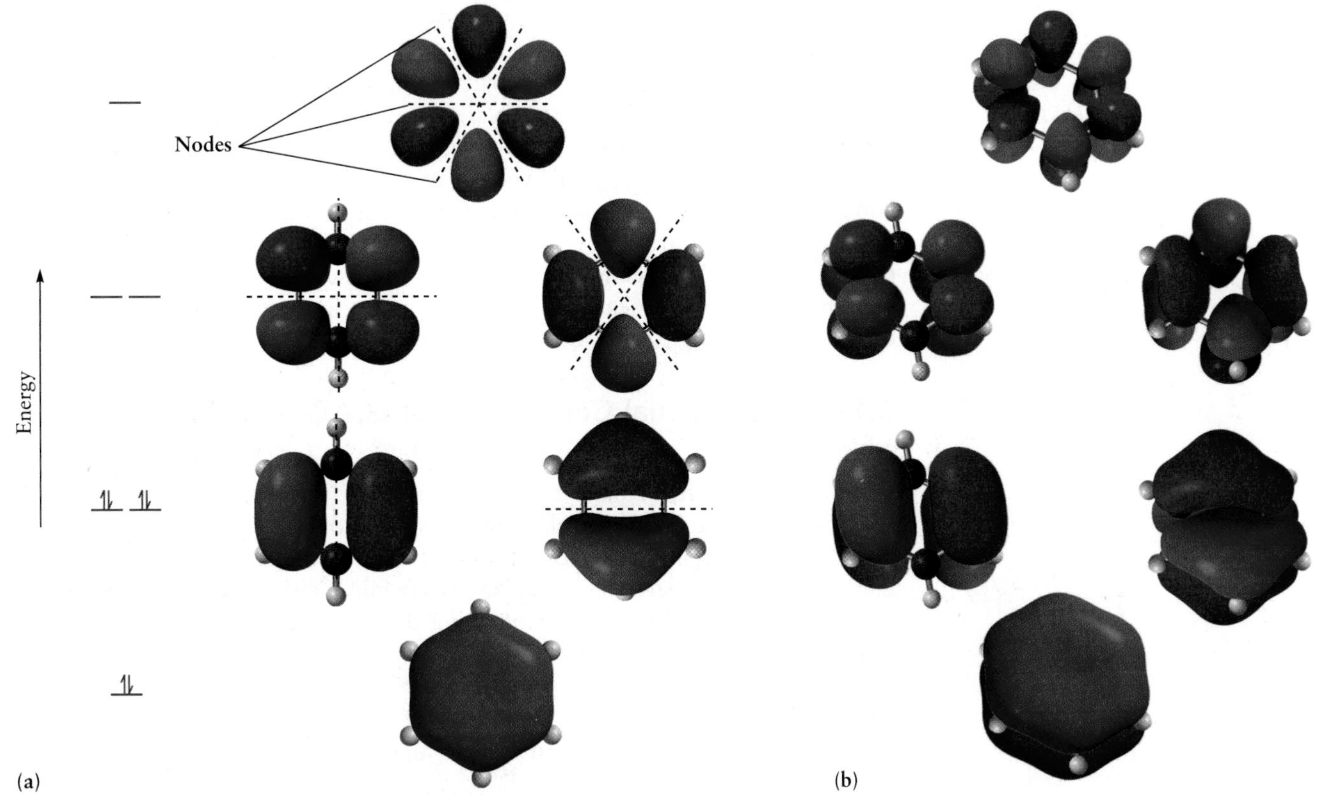

FIGURE 6.54 The six π molecular orbitals for benzene. In the view from the top in (a) you can more easily see the nodal planes indicated in dashed lines. In the view from the side in (b), you can see how each molecular orbital results from a linear combination of six $2p_z$ atomic orbitals oriented perpendicular to the plane of the molecule. Only the three lowest energy orbitals are occupied in molecules of benzene in the ground state.

combination of the six $2p$ orbitals perpendicular to the plane of the molecule were used to make each de-localized π orbital. The six π MOs are characterized by their energies and by the number of nodal planes perpendicular to the molecular plane. The lowest energy MO has no such nodal planes, the next highest energy orbitals are a degenerate pair with one nodal plane, the next highest in energy are also degenerate with two nodal planes, and the highest energy orbital has three nodal planes. The C_6H_6 molecule has 30 valence electrons, of which 24 occupy σ bonding orbitals formed from a linear combination of two sp^2 hybrids (for each C—C bond) or a linear combination of an sp^2 hybrid with an H 1s AO (for each C—H bond). When the six remaining valence electrons are placed in the three lowest energy π orbitals, the resulting electron distribution is the same in all six carbon–carbon bonds. As a result, benzene has six carbon–carbon bonds of equal length whose properties are intermediate between those of single and double bonds.

6.12 Summary and Comparison of the LCAO and Valence Bond Methods

Most of our understanding of chemical bonding, structure, and reactivity is based on the simple approaches to MO and VB theory presented in this chapter. They provide the foundation upon which our chemical intuition has been built. The concepts in these simple theories are central to all areas of modern chemistry. The LCAO and VB methods start with very different quantum mechanical approaches to describe chemical bonding. Each originated as an approximation method for constructing approximate wave functions for molecules, using very different initial models of the chemical bond. The two methods look very different at the beginning, and their results presented in this book look very different. Are the two methods equally valid? Are they equally applicable to a broad range of molecules? Can they both be used for quantitative calculations on large molecules? The answer is *yes* for each of these questions. This section compares the two methods to give deeper insight into their conceptual meanings and to provide a basis for choosing to use one over the other in a particular chemical investigation.

6.12.1 Conceptual Comparison of LCAO and VB

The accuracy of any quantum chemical method is judged by how well its predictions agree with the results of experiment. Predicting measurable values of molecular properties requires that we know the molecular electronic wave function for the molecule. Therefore, the best way to compare the LCAO and VB methods is to compare the molecular electronic wave functions generated by each. We illustrate this comparison using the molecular electronic wave functions for H_2 constructed using both methods.

The LCAO method for H_2 constructs a σ bonding MO as a linear combination of H 1*s* AOs centered on the two hydrogen atoms. This MO, delocalized over the entire molecule, is given by the equation

$$\sigma_{g1s} = C_g[\varphi_{1s}^{A} + \varphi_{1s}^{B}] \qquad [6.24]$$

Neglecting the normalization constant and using the simplified notation from Section 6.3 gives the form

$$\sigma_{g1s} = [1s^{A} + 1s^{B}] \qquad [6.25]$$

Both electrons occupy this bonding orbital, satisfying the condition of indistinguishability and the Pauli principle. Recall from Section 6.4 that the molecular electronic wave function in the LCAO approximation is the *product* of all of the occupied MOs, just as the wave function for an atom is the product of all the occupied Hartree orbitals. Thus we have the molecular wave function

$$\psi_{MO}^{el} = \sigma_{g1s}(1)\sigma_{g1s}(2) = [1s^{A}(1) + 1s^{B}(1)][1s^{A}(2) + 1s^{B}(2)] \qquad [6.26]$$

The VB model, in contrast, starts by assuming that a good approximation to the molecular electronic wave function for H_2 is the product of an H 1*s* orbital centered on atom A, occupied by electron 1, and another H 1*s* orbital centered on atom B, occupied by electron 2. As shown in Section 6.8, this molecular electronic wave function is given by

$$\psi_{VB}^{el}(r_{1A}, r_{2B}) = c_1\varphi_A(r_{1A})\varphi_B(r_{2B}) + c_2\varphi_A(r_{2A})\varphi_B(r_{1B}) \qquad [6.27]$$

which upon dropping the normalization factors and using the simplified notation introduced in Section 6.4 becomes

$$\psi_{VB}^{el} = 1s^{A}(1)1s^{B}(2) + 1s^{A}(2)1s^{B}(1) \qquad [6.28]$$

Now we can compare the LCAO and VB molecular electronic wave functions directly by multiplying out ψ_{MO}^{el} and rearranging terms to obtain

$$\psi_{MO}^{el} = [1s^{A}(1)1s^{B}(2) + 1s^{A}(2)1s^{B}(1)] + [1s^{A}(1)1s^{A}(2) + 1s^{B}(1)1s^{B}(2)] \qquad [6.29]$$

The first bracketed term in ψ_{MO}^{el} is identical to ψ_{VB}^{el}. It represents the purely covalent structure H–H because the two electrons are shared equally between atoms A and B. The second term may be labeled ψ_{ionic} because it is a mixture of the ionic states $H_A^- H_B^+$ and $H_A^+ H_B^-$, respectively. This can be seen by looking at the two terms in the second set of brackets; the first term puts both electrons on nucleus A (making it H^-) and the second term puts both electrons on nucleus B (making it H^-).

Thus, our comparison shows that the LCAO method includes an ionic contribution to the bond but the VB method does not. In fact, the simple MO approach suggests that the bond in H_2 is 50% covalent and 50% ionic, which is contrary both to experience and intuition. Because the electronegativities of the two atoms in a homonuclear diatomic molecule are the same, there is no reason to expect any ionic contribution to the bond, much less such a large one. The complete absence of ionic contributions in the VB molecular wave function suggests this method is not well suited for polar molecules like HF. Thus, the truth in describing the chemical bond and molecular structure appears to lie somewhere between the LCAO and VB methods.

It is also informative to compare how these methods describe chemical reactivity, which requires bonds to be broken. We already know that the VB wave function for H_2 correctly describes the long-distance limit as two separate H atoms. But, the LCAO wave function predicts that in the long-distance limit H_2 dissociates into ionic species as well as H atoms. Ionic products are not usually produced by dissociation under ordinary thermal conditions. Again, the best description must lie between the extremes defined by the simple LCAO and VB methods.

The simple form of LCAO and VB methods, as presented here, can be refined to provide more accurate wave functions for molecules and solids from which measurable properties can be calculated. Both methods have been improved significantly in a number of ways. We illustrate one approach for improving ψ_{VB}, not only because it is easy to understand but also because the method is generally applicable in many areas of quantum chemistry. The simple VB wave function can be improved by adding in (mixing) some ionic character. We write

$$\psi_{improved} = \psi_{VB} + \lambda\psi_{ionic} \qquad [6.30]$$

and then choose λ on the basis of some criterion (see discussion later). For example, the purely covalent description of the HF molecule in Equation 6.16 could be improved in this way to describe the bond polarity and the percent ionic character determined from measurements of the dipole moment (see Sec. 3.9 and Eq. 3.24). The most common way to do this is to adjust λ to minimize the energy. We calculate the energy, using the methods developed in Chapter 5, with λ as a parameter, and then differentiate the result with respect to λ to find the value of λ that minimizes the energy, just as in ordinary calculus. Using that special value of λ in Equation 6.30 gives a wave function that is a better approximation to the true wave function than the simple VB wave function. Moreover, the energy calculated for the ground state of the system using the "improved" wave function is guaranteed never to be lower than the true ground-state energy. These results are consequences of the **variational principle** in quantum mechanics, which gives a guideline for improving the accuracy of various approximations; lower energy is always better. The variational principle is also routinely used to improve predictions from modern versions of LCAO theory.

6.12.1 Elementary LCAO and VB Theories as Qualitative Predictive Tools

The introductory levels of the LCAO and VB methods presented in this chapter are appropriate for the hand-drawn sketches and "back of the envelope" calculations that occur in the early stages of any project in basic or applied chemistry. Depending on the molecules under investigation and the kinds of questions being asked, one method may be more useful than the other at this early stage. Either one can serve as the starting point for a later calculation.

Elementary LCAO theory offers qualitative predictions for processes in diatomic and polyatomic molecules that involve promoting a molecule to an excited electronic state. Using diagrams like that in Figure 6.18 and Figure 6.53, one can determine the relative energy ordering of occupied and unoccupied orbitals in a molecule. We can immediately use these diagrams to describe the absorption and emission of light, formation of cations by electron removal, formation of anions by electron attachment, and other processes that involve loss or gain of electronic energy.

These results are quite hard to achieve with elementary VB. Bond formation pairs up two electrons in half-filled AOs on two separate atoms, and ignores all the other electrons on both atoms. The model predicts the energy in the bonding and the antibonding states of this particular bond, and thereby guarantees that the bond has a stable ground state lower than the energy of the separated atoms. The model gives no additional information about energy levels.

Most of our understanding of chemical bonding, structure, and reactivity is based on the elementary approaches to MO and VB theory presented in this chapter. They provide the foundation upon which our chemical intuition has been built; the concepts introduced in this chapter are central to all areas of modern chemistry. Publicly available software packages enable computer-based calculations that remove the limitations on the elementary approaches. You will likely begin to use computational chemistry software in your second-year courses in college-level chemistry. Both the MO and VB elementary approaches serve as the starting point for such calculations, and the elementary concepts are essential for interpreting the results of the computational chemistry tools.

6.13 A DEEPER LOOK Properties of the Exact Molecular Orbitals for H_2^+

Here we give a more detailed analysis of the exact wave functions for H_2^+ presented in Section 6.2 and Figure 6.5, and for the associated electron density functions.

6.13.1 Wave Functions

The first eight wave functions, starting with the ground state, are shown in Figure 6.55 using representations that give additional insight into their structure beyond Figure 6.5. These are plotted using the same coordinates as in Figure 6.4, where the two protons lie on the z-axis and the value of R_{AB} is the experimental bond length for H_2^+, which is 1.06 Å. Each wave function is shown in three different representations: (a) an isosurface comprising all those points in three-dimensional space where the wave function has a value equal to 0.1 of its maximum value; (b) a contour plot in a plane containing the internuclear axis with contours shown for ± 0.1, ± 0.3, ± 0.5, ± 0.7, and ± 0.9 of the maximum value; (c) a plot of the amplitude along the internuclear axis, which amounts to a "line scan" across the contour plot. Part (a) is the same as Figure 6.5, with red isosurfaces representing positive values of the wave function representing negative values. In part (b) the contours have the same color coding, with nodal contours represented as dashed black lines. We encourage you to compare these representations to develop intuition for the shape and symmetry of each wave function and its corresponding probability density for locating the electron. It is important to develop a very quick feel for "How many nodes and where are they?" for each wave function. In developing these insights you will be guided by the four labels on each wave function; their meanings are explained in Section 6.2.

Here we give deeper insight into the meaning of the Greek letter in each label. The position of the electron in H_2^+ around the internuclear axis is given by the angle ϕ (see Fig. 6.4). Because the potential energy of H_2^+ is the same at all values of ϕ (see Eq. 6.1), the angular coordinate will influence the shape of the molecular orbital, but not the energy levels, just as the angles (θ, ϕ) in Figure 5.1 influence the shape of the atomic orbitals for hydrogen, but not the energy levels. The component of the electron's angular momentum along the internuclear axis in H_2^+ is quantized exactly the same as the component of angular momentum along the z-axis in the hydrogen atom (see Eq. 5.2b). This quantization condition is given by the expression,

$$\text{Allowed value of } L_z = m\frac{h}{2\pi} \qquad m = 0, \pm 1, \pm 2, \pm 3, \ldots$$

Because the total angular momentum of the electron in H_2^+ is not quantized, there is no analog of the atomic quantum number ℓ to define an upper limit to the allowed values of m. Greek letters are used to identify the component of angular momentum along the internuclear axis as follows:

$\sigma \longrightarrow$ angular momentum component $= 0$

$\pi \longrightarrow$ momentum component $= \pm h/2\pi$

$\delta \longrightarrow$ angular momentum component $= \pm 2h/2\pi$

$\varphi \longrightarrow$ angular momentum component $= \pm 3h/2\pi$

The Greek letter tells us how the electron probability density is distributed around the internuclear axis, as viewed in a plane perpendicular to the axis. The σ wave functions have finite amplitudes on the internuclear axis, and the probability of finding the electron on that axis is therefore finite. As you can deduce from Figure 6.55, in a cross-section perpendicular to the internuclear axis the σ wave functions are cylindrically symmetric. The π wave functions, in contrast, describe electron motion about the internuclear axis with angular momentum $+h/2\pi$ or $-h/2\pi$. Physically, these two cases correspond to clockwise and counter-clockwise motion of the electron around the internuclear axis. These two molecular orbitals do not have a simple geometrical representation, but we can form linear combinations of them just as we did to define the p_x and p_y atomic orbitals (see Eq. 5.6). This leads to two wave functions—π_{u2p_x} and π_{u2p_y}—with the same energy, one of which lies along the x-axis and the other along the y-axis (the internuclear axis is chosen to be the z-axis). Because the π_{u2p_x} and π_{u2p_y} wave functions have nodal planes that include the internuclear axis, there is zero probability of finding the electron anywhere along the z-axis just as there is zero probability of finding the electron at the nucleus in an atomic p orbital. Viewed perpendicular to the internuclear axis, the π_{u2p_x} and π_{u2p_y} wave functions do not have cylindrical symmetry about the axis. These results remind us of the s, p, d, f progression for angular momentum of the electron about the nucleus in the hydrogen atom and the fact that s orbitals have amplitude at the nucleus, whereas p orbitals have nodes at the nucleus.

6.13.2 Electron Probability Density Functions

To understand the nature of the chemical bond in H_2^+, we must calculate the electron density around the nuclei, which are assumed to be fixed at their equilibrium positions. In quantum mechanics, the electron probability density in a region of space is simply proportional to the probability density for an electron to be in that region. The probability density functions for locating the electron at each point in space are shown in Figure 6.56; they were calculated by

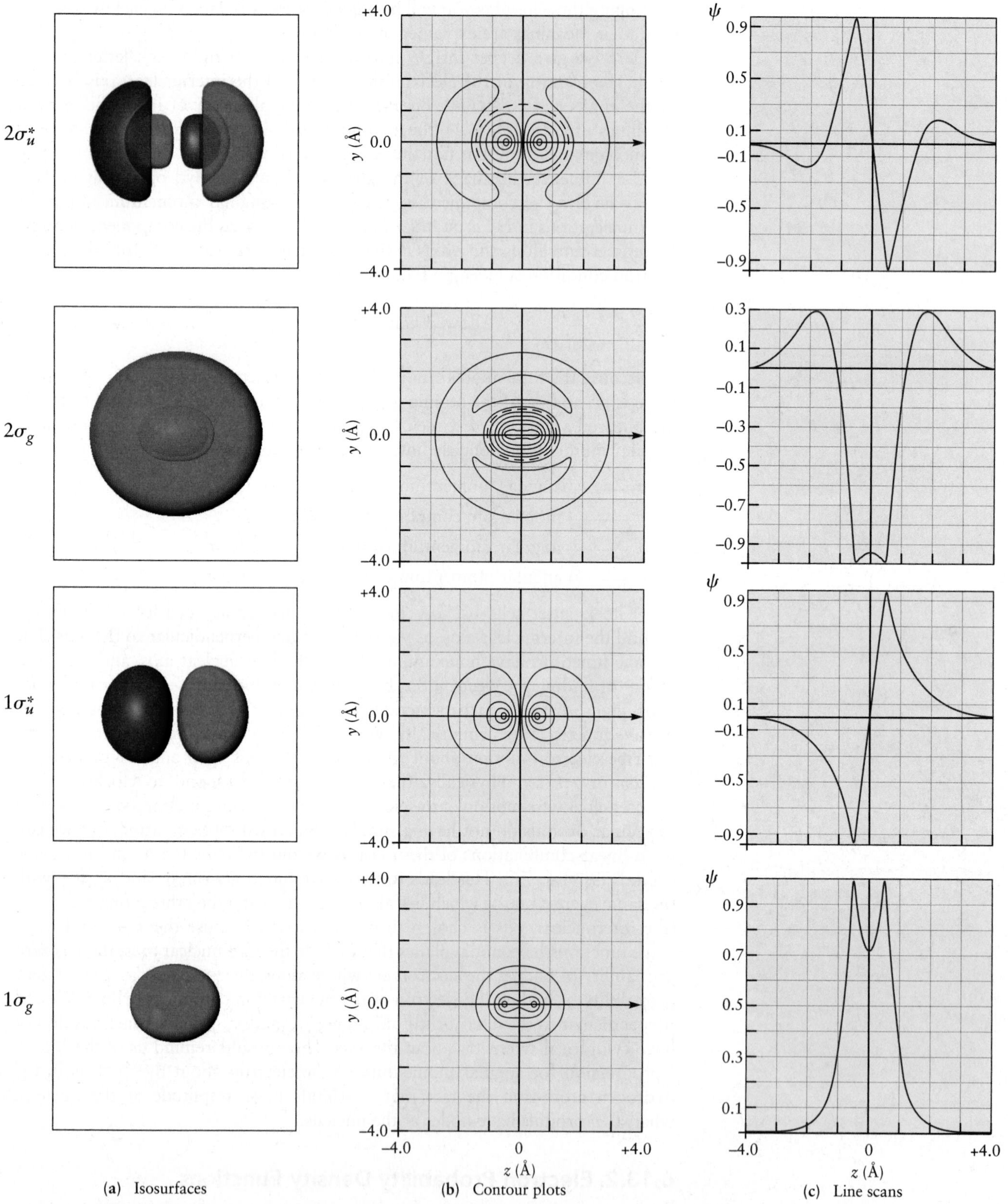

FIGURE 6.55 Wave functions for the first eight energy levels of the H_2^+ molecular ion, calculated exactly by quantum mechanics. The ground-state wave function is at the bottom of the figure; the others are arranged above it in order of increasing energy. The two nuclei lie along the *z*-axis, which is in the plane of the paper. Regions of positive and negative amplitude are shown in red and blue, respectively. The labels for each orbital are explained in the text. (a) Isosurfaces corresponding to contours at ± 0.1 of the maximum amplitude.
(Courtesy of Mr. Hatem Helal and Professor William A. Goddard III, California Institute of Technology, and Dr. Kelly P. Gaither, University of Texas at Austin.)

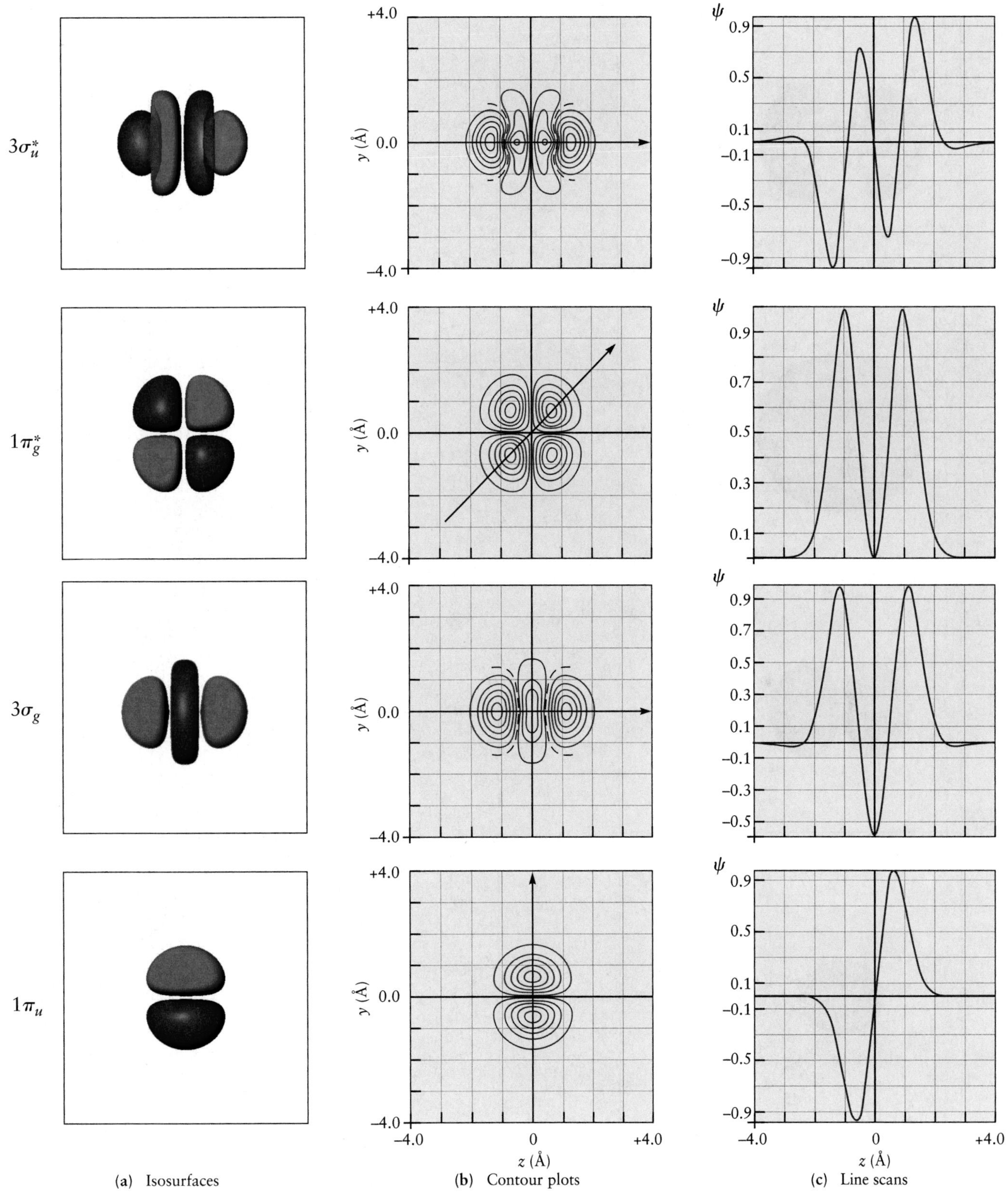

FIGURE 6.55 cont'd (b) Contours of constant amplitude in the *x-z* plane, at values ±0.1, ±0.3, ±0.5, ±0.7, and ±0.9 of the maximum amplitude. Nodes are represented by black dashed lines. (c) The amplitude along the *z*-axis, obtained as a "line scan" across the contour plot, along the direction indicated by the green arrow in (b).
(Courtesy of Mr. Hatem Helal and Professor William A. Goddard III, California Institute of Technology, and Dr. Kelly P. Gaither, University of Texas at Austin.)

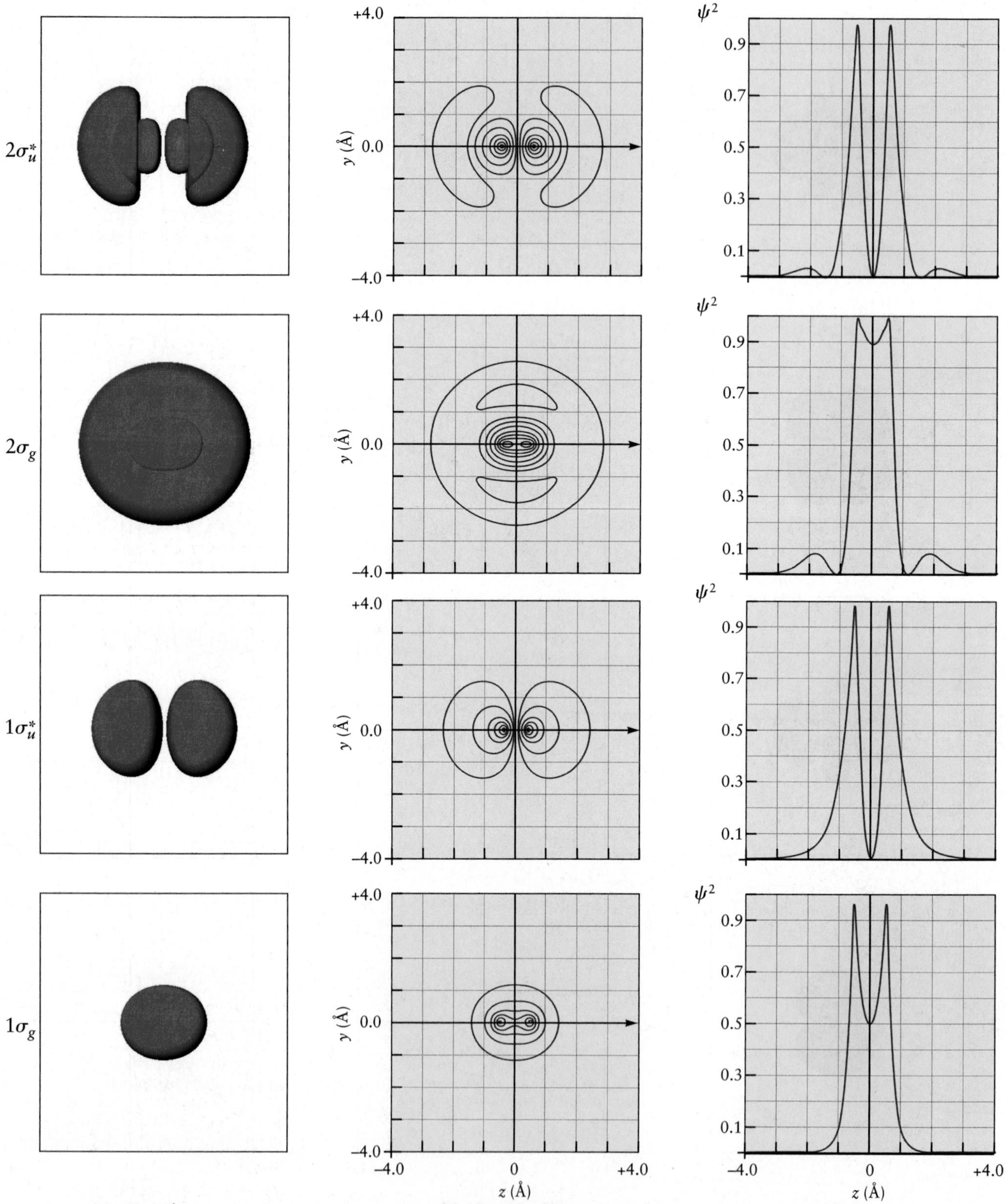

FIGURE 6.56 Probability density distributions for the first eight energy levels of the H_2^+ molecular ion, calculated exactly by quantum mechanics. (a) Isosurfaces comprising all points at which the probability density is 0.01 of its maximum value. (b) Contour plots in the *x–z* plane with contours at 0.01, 0.1, 0.3, 0.5, 0.7, and 0.9 of the maximum value. (Courtesy of Mr. Hatem Helal and Professor William A. Goddard III, California Institute of Technology, and Dr. Kelly P. Gaither, University of Texas at Austin.)

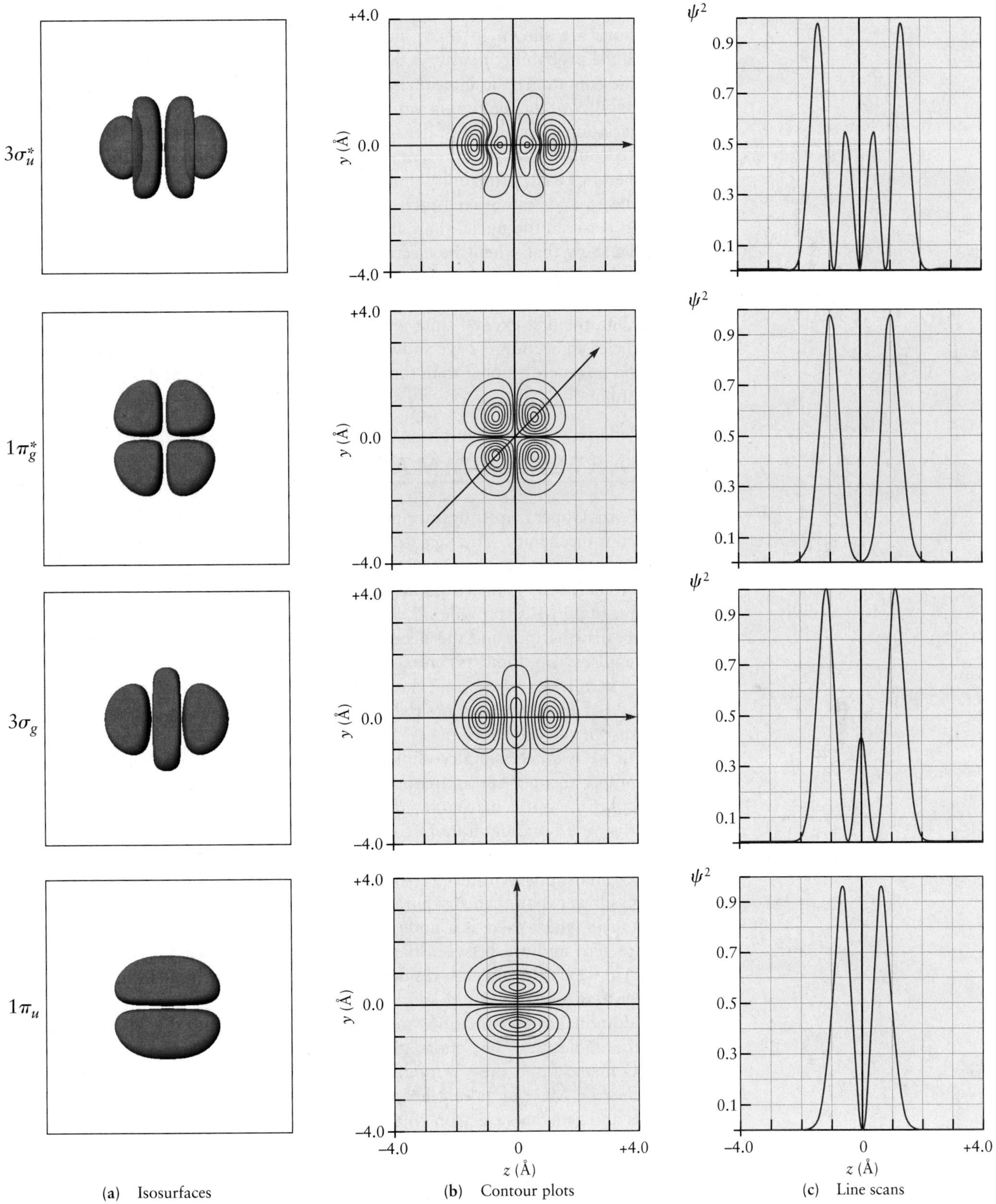

FIGURE 6.56 cont'd (c) Line scans across the contour plot, along the direction indicated by the green arrow in (b), showing the variation in probability density along the internuclear axis.
(Courtesy of Mr. Hatem Helal and Professor William A. Goddard III, California Institute of Technology, and Dr. Kelly P. Gaither, University of Texas at Austin.)

squaring each MO wave function shown in Figure 6.55. The probability density functions are shown in three views: (a) isosurfaces comprising all points at which the probability density is 0.01 of its maximum value; (b) contour plots in a plane containing the internuclear axis with contours at 0.05, 0.01, 0.3, 0.5, 0.7, and 0.9 of the maximum value; (c) line scans across the contour plot showing variation in probability density along the internuclear axis. We encourage you to compare these representations to develop intuition for the shape and symmetry of each probability density for locating the electron.

The ground-state wave function $1\sigma_g$ has much greater electron density in the region between the nuclei than at the extreme ends of the molecule. In Section 6.2 we show that when the electron is in the $1\sigma_g$ MO, the H_2^+ ion has energy lower than that of the separated atoms. These two effects together are consistent with our expectations about chemical bonding, and so $1\sigma_g$ is labeled a bonding MO. But, the first excited state wave function $1\sigma_u^*$ has a node halfway between the nuclei. In Section 6.2 we show that when the electron is in the $1\sigma_u^*$ MO, the H_2^+ ion has energy higher than that of the separated atoms, and so $1\sigma_u^*$ is labeled an antibonding MO.

CHAPTER SUMMARY

The Born–Oppenheimer approximation is the starting point for all of molecular quantum mechanics. The fact that electrons move so much faster than nuclei allows us to treat electronic motion independently of nuclear motion by solving an electronic Schrödinger equation for each value of the internuclear separation. The resulting MOs provide all of the information of interest, the probability density distributions and the electronic energies being the most important. With the repulsive Coulomb potential included in the electronic Hamiltonian at a fixed internuclear separation, the lowest energy solution to $\mathcal{H}_{el}\psi_{el} = E_{el}\psi_{el}$ as a function of geometry becomes the ground-state potential energy curve $V(R)$ that governs the nuclear motion.

The MOs for the simplest molecule, H_2^+, can be calculated exactly within the Born–Oppenheimer approximation. The results illustrate the general features of molecular quantum mechanics that form the basis for our understanding of structure and bonding in more complicated molecules. The orbitals are characterized by their symmetry, the number and nature of their nodes, and their energies. Orbitals in which the electron density increases between the nuclei lead to energies lower than that of separated H and H^+; these are called bonding MOs. Orbitals in which there is a node in the wave function and the electron density goes to zero midway between the nuclei are called antibonding orbitals. States of H_2^+ in which the electron resides in an antibonding orbital are unstable with respect to dissociation to H and H^+.

Additional approximations must be made to calculate the MOs for many-electron molecules. The most important approximation procedures are the LCAO method and the VB model. The LCAO method constructs de-localized one-electron MOs by taking linear combinations of AOs centered on different atoms and generates electron configurations by placing electrons in these MOs using an aufbau principle and invoking Hund's rules for the ground-state configuration. The VB model constructs a wave function for a localized pair bond starting with an "occupied" AO on each of the two atoms that form the bond. These two procedures provide the conceptual foundation and vocabulary for qualitative and even semiquantitative understanding of chemical bonding and molecular structure in contemporary chemistry. A variety of sophisticated computational methods have been developed using these procedures as starting points, and the results of these calculations are now sufficiently accurate to have both analytical and predictive value. Our goal has been to give you a comprehensive introduction to molecular quantum mechanics so that you can easily read more advanced treatments and begin to use commercially available software with intelligence and confidence.

CONCEPTS AND SKILLS

Section 6.1 – Quantum Picture of the Chemical Bond

Give a general description of the Born–Oppenheimer approximation and explain why it is the foundation for all molecular quantum mechanics. Describe the key features of the Born–Oppenheimer approximation (Section 6.1).

- Nuclei are so much heavier than electrons that they may be considered fixed in space while the electrons move rapidly around them.
- The Born–Oppenheimer approximation allows us to solve the electronic Schrödinger equation for H_2^+ for a fixed internuclear separation R_{AB}. The result is a one-electron MO, which is analogous to the one-electron hydrogen AO.
- With the repulsive Coulomb potential included in the electronic Hamiltonian at a fixed internuclear separation, the lowest energy solution to $\mathcal{H}_{el}\psi_{el} = E_{el}\psi_{el}$ as a function of the relative positions of the nuclei, R_{AB} for a diatomic molecule, is the ground-state potential energy curve $V(R_{AB})$ that governs the nuclear motion. Excited states are obtained from higher energy solutions.
- We write Schrödinger's equation with the kinetic energy of the nuclei and the potential energy described earlier to find the quantized ground-state energy of the molecule. The nuclei vibrate about the equilibrium bond length—this is the zero point motion required by the uncertainty principle. The energy required to dissociate the molecular ion from the ground state into a separated proton and hydrogen atom is the bond dissociation energy.

Section 6.2 – Exact Molecular Orbitals for the Simplest Molecule: H_2^+

Give a general description of the quantum picture of the chemical bond and how it differs from the classical picture. Describe the key features of the quantum picture, including the nature of bonding and antibonding MOs, symmetry of MOs, and the energy sequence of MOs (Problems 1–8).

- We illustrate graphically the first eight MOs for H_2^+ to show their shapes and to characterize them by their energies and symmetry, just as we characterized the atomic orbitals for the hydrogen atom. The MOs are characterized by the component of the angular momentum along the internuclear axis: by analogy to the hydrogen atom, these are called σ for orbitals with cylindrical symmetry and π for MOs with one nodal plane that contains the internuclear axis.

Sections 6.3–6.6 – Delocalized Bonds: Molecular Orbital Theory and the LCAO

Show how MOs can be constructed from the AOs of two atoms that form a chemical bond, and explain how the electron density between the atoms is related to the MO (Problems 9–16).

- A good approximation to the one-electron MOs for a diatomic molecule is the weighted sum or difference of AOs of the atoms of the molecule. The linear combination for which the contributing AOs overlap with the same phase (either both positive or both negative) leads to increased electron density between the nuclei and bonding; the linear combination for which the contributing AOs overlap with the opposite phase (one positive and the other negative) leads to a node between the nuclei and decreased electron density, and it is antibonding.

- Correlation diagrams show how pairs of AOs lead to bonding and antibonding pairs of MOs. An aufbau principle is used to build up electron configurations, just like for atoms. Hund's rules predict the lowest energy electron configurations and either paramagnetic or diamagnetic behavior.
- The bond order is found by counting the number of electrons in bonding orbitals, subtracting the number in antibonding orbitals, and dividing the result by 2. Electrons in antibonding orbitals effectively cancel the bonding capacity of those in bonding orbitals. This scheme explains the trends in bond length, bond stiffness, and bond dissociation energy of the first- and second-row diatomic molecules.
- The energy sequence of the MOs is slightly more complicated in second-row homonuclear diatomic molecules, because the *p* orbitals can overlap in two different ways. Moving from left to right across the row, the energy-level ordering changes at N_2 because the energy of the π orbital remains nearly constant, whereas that of the σ orbital drops rapidly (see Fig. 6.16). Therefore, two energy level diagrams are required to explain the bonding in the second-period diatomic molecules.
- The MOs for heteronuclear diatomic molecules are obtained by the same approach, with the AO energies of the more electronegative element placed lower (more stable) than those of the other element. If the difference in AO energies is small, the MO energy sequence is that given in Figure 6.21; for larger AO energy differences, the MOs are those described in Figure 6.24.

Connection to Instrumental Analysis: Photoelectron Spectroscopy Section 6.7 – Summary Comments for the LCAO Method and Diatomic Molecules

Relate photoelectron spectra to correlation diagrams for MOs (Problems 35–40).

- Photoelectron spectroscopy confirms the validity of the orbital approximation by measuring the ionization energies of the MOs directly. The ionization energy of the orbital is obtained as the difference in the energy of the photon used to ionize the molecule and the measured kinetic energy of the emitted electrons. Koopmans's approximation states that the orbital energy ε in the LCAO method is the negative of the ionization energy.
- In addition to the orbital energies, PES provides a great deal of information about the nature of the orbital (bonding, nonbonding, or antibonding) from the vibrational fine structure observed in the spectra.

Sections 6.8–6.10 – Localized Bonds: The Valence Bond Model

Use the VB method to construct wave functions for localized electron pair bonds, including multiple bonds, and predict the molecular geometry from these bonds (Problems 41–48).

- The VB model constructs wave functions to describe localized electron-pair bonds. The model describes bonding in diatomic molecules, including the formation of multiple (σ and π) bonds. It is most frequently applied to organize and correlate data on molecular structures, especially for molecules of the type AB_x, the geometries of which are described by VSEPR theory.
- The simple VB model is augmented with the concept of orbital hybridization to account for the valence of second-row atoms and the structures of their compounds. Hybrid orbitals are constructed by adding *s* and *p* orbitals with different coefficients (weights or percentage contributions) and phases. The number of hybrid orbitals produced equals the number of starting AOs; there are two *sp* hybrid orbitals, three sp^2 hybrid orbitals, and four sp^3 hybrid orbitals.

Sections 6.11–6.12 – Comparison of LCAO and Valence Bond Methods and Using Them Together

Compare the LCAO and VB approaches, and combine them to describe the molecular network and delocalized bonds in certain classes of molecules (Problems 57–62).

- Comparing the LCAO and VB treatments for the hydrogen molecule at the level of the *electronic wave function for the molecule* gives considerable insight into the differences between the methods and also suggests ways to improve each. The VB wave function predicts a purely covalent bond, whereas the LCAO wave function predicts a bond with an equal mixture of covalent and ionic character. Neither of these is the best representation of bonding in most diatomics so refinements of both approaches are necessary to produce results that are in better agreement with experiment.
- Many methods have been developed to improve both the simple LCAO and VB models; it is easiest to illustrate one approach for improving the VB model. Let $\psi_{\text{improved}} = \psi_{\text{VB}} + \lambda\psi_{\text{MO}}$, where λ is chosen so that the energy of the electronic state ψ for the energy of the ground electronic state is minimized. The variational principle ensures that the true energy is always lower than the energy calculated using an approximate wave function. This provides a well-defined criterion to judge improvement—lower energy is always better.
- Many chemists combine the LCAO and VB methods to describe bonding in polyatomic molecules. They use the VB model to describe the localized σ bonds that provide "connectivity" for the molecule structure and use the LCAO method to describe the de-localized π bonds that distribute electrons over the entire structure.
- In hydrocarbon molecules with alternating single and double bonds along a carbon backbone, such as 1,3-butadiene, the electron density distribution and energy levels of the delocalized π orbitals can be understood using particle-in-a-box models. The energies of a set of levels increases with the number of nodes in the wave function.

CUMULATIVE EXERCISE

Iodine in the Human Diet

FIGURE 6.57 Iodine sublimes from the bottom of the beaker and condenses on the bottom of the chilled round-bottom flask.

The shiny purple-black crystals of elemental iodine (Fig. 6.57) were first prepared in 1811 from the ashes of seaweed. Several species of seaweed concentrate the iodine that is present in low proportions in seawater, and for many years, seaweed was the major practical source of this element. Most iodine is now produced from natural brines via oxidation of iodide ion with chlorine.

(a) Iodine is an essential trace element in the human diet, and iodine deficiency causes goiter, the enlargement of the thyroid gland. Much of the salt intended for human consumption is "iodized" by the addition of small quantities of sodium iodide to prevent goiter. Calculate the electronegativity difference between sodium and iodine. Is sodium iodide an ionic or a covalent compound? What is its chemical formula?

(b) Iodine is an important reagent in synthetic organic chemistry because bonds between carbon and iodine form readily. Use electronegativities to determine whether the C—I bond is ionic, purely covalent, or polar covalent in character.

(c) Give the steric numbers for the iodine atom and identify the geometries of the following ions containing iodine and oxygen: IO_3^-, IO_6^{5-}, and IO_4^-.

(d) What is the ground-state electron configuration of the valence electrons of iodine molecules (I_2)? Is iodine paramagnetic or diamagnetic?

(e) What is the electron configuration of the I_2^+ molecular ion? Is its bond stronger or weaker than that in I_2? What is its bond order?

Answers

(a) The electronegativity difference is 1.73; thus, the compound is largely ionic, with formula NaI.

(b) The electronegativity difference is 0.11; thus, the C—I bond is largely covalent, with nearly equal sharing of electrons between the atoms.

(c) IO_3^-: $SN = 4$, structure is pyramidal; IO_6^{5-}: $SN = 6$, structure is octahedral; IO_4^-: $SN = 4$, structure is tetrahedral.

(d) $(\sigma_{g5s})^2(\sigma_{u5s}^*)^2(\sigma_{g5p_z})^2(\pi_{u5p_x}\pi_{u5p_y})^4(\pi_{g5p_x}^*\pi_{g5p_y}^*)^4$; iodine is diamagnetic.

(e) $(\sigma_{g5s})^2(\sigma_{u5s}^*)^2(\sigma_{g5p_z})^2(\pi_{u5p_x}\pi_{u5p_y})^4(\pi_{g5p_x}^*\pi_{g5p_y}^*)^3$; the bond is stronger; bond order is 3/2 versus 1.

PROBLEMS

Answers to problems whose numbers are red appear in Appendix G. Problems that are more challenging are indicated with asterisks.

6.1 Quantum Picture of the Chemical Bond

1. Determine the number of nodes along the internuclear axis for each of the σ molecular orbitals for H_2^+ shown in Figure 6.5.

2. Determine the number of nodes along the internuclear axis and the number of nodal planes for each of the π molecular orbitals for H_2^+ shown in Figure 6.5.

3. Sketch the shape of each of the σ molecular orbitals for H_2^+ shown in Figure 6.5 in a plane perpendicular to the internuclear axis located at the midpoint between the two nuclei. Repeat the sketches for a plane perpendicular to the internuclear axis located at a point one quarter of the distance between the two nuclei.

4. Sketch the shape of each of the π molecular orbitals for H_2^+ shown in Figure 6.5 in a plane perpendicular to the internuclear axis located at the midpoint between the two nuclei. Repeat the sketches for a plane perpendicular to the internuclear axis located at a point one quarter of the distance between the two nuclei.

5. Compare the electron density in the $1\sigma_g$ and $1\sigma_u^*$ molecular orbitals for H_2^+ shown in Figure 6.5 with the classical model for bonding for H_2^+ summarized in Figure 3.22b. Which of these molecular orbitals describes the bond in H_2^+?

6. Explain why $1\sigma_g$ is the ground state for H_2^+. By combining your answer with the answer to Problem 5, what conclusions can you draw about the molecular orbital description of the bond in H_2^+?

6.3 Molecular Orbital Theory and the Linear Combination of Atomic Orbitals Approximation for H_2^+

6.4 Homonuclear Diatomic Molecules: First-Period Atoms

7. The ground state of H_2 has the electron configuration $(\sigma_{g1s})^2$. There are excited states that have the following configurations
   (a) $(\sigma_{g1s})^1(\sigma_{u1s}^*)^1$
   (b) $(\sigma_{g1s})(\sigma_{u1s}^*)^2$

   Which state do you predict to have higher energy?

8. Predict the ground electronic state of the He_2^{2+} ion. What is the bond order? Will it be stable in the ground state?

9. Without consulting tables of data, predict which species has the larger bond energy, H_2 or He_2^+.

10. Without consulting tables of data, predict which species has the larger bond energy, H_2^+ or H_2.

11. Without consulting tables of data, predict which species has the greater bond length, H_2 or He_2^+.

12. Without consulting tables of data, predict which species has the greater bond length, H_2^+ or H_2.

13. Without consulting tables of data, on the same graph sketch the effective potential energy curves for H_2 and He_2^+.

14. Without consulting tables of data, on the same graph sketch the effective potential energy curves for H_2^+ and H_2.

15. Suppose we supply enough energy to H_2 to remove one of its electrons. Is the bond energy of the resulting ion larger or smaller than that of H_2? Is the bond length of the resulting ion larger or smaller than that of H_2?

16. Suppose we supply enough energy to He_2^+ to remove its most weakly bound electron. Is the bond energy of the resulting ion larger or smaller than that of He_2^+? Is the bond length of the resulting ion larger or smaller than that of He_2?

6.5 Homonuclear Diatomic Molecules: Second-Period Atoms

17. If an electron is removed from a fluorine molecule, an F_2^+ molecular ion forms.
   (a) Give the molecular electron configurations for F_2 and F_2^+ (for the MOs constructed from valence AOs).
   (b) Give the bond order of each species.
   (c) Predict which species should be paramagnetic.
   (d) Predict which species has the greater bond dissociation energy.

18. When one electron is added to an oxygen molecule, a superoxide ion (O_2^-) is formed. The addition of two electrons gives a peroxide ion (O_2^{2-}). Removal of an electron from O_2 leads to O_2^+.
   (a) Construct the correlation diagram for O_2^-.
   (b) Give the valence electron configuration for each of the following species: O_2^+, O_2, O_2^-, O_2^{2-}.
   (c) Give the bond order of each species.
   (d) Predict which species are paramagnetic.
   (e) Predict the order of increasing bond dissociation energy among the species.

19. Predict the valence electron configuration and the total bond order for the molecule S_2, which forms in the gas phase when sulfur is heated to a high temperature. Will S_2 be paramagnetic or diamagnetic?

20. Predict the valence electron configuration and the total bond order for the molecule I_2. Will I_2 be paramagnetic or diamagnetic?

21. For each of the following valence electron configurations of a homonuclear diatomic molecule or molecular ion, identify the element X, Q, or Z and determine the total bond order.
   (a) $X_2{:}(\sigma_{g2s})^2(\sigma_{u2s}^*)^2(\sigma_{g2p_z})^2(\pi_{u2p})^4(\pi_{g2p}^*)^4$
   (b) $Q_2^+{:}(\sigma_{g2s})^2(\sigma_{u2s}^*)^2(\pi_{u2p})^4(\sigma_{g2p_z})^1$
   (c) $Z_2^-{:}(\sigma_{g2s})^2(\sigma_{u2s}^*)^2(\sigma_{g2p_z})^2(\pi_{u2p})^4(\pi_{g2p}^*)^3$

22. For each of the following valence electron configurations of a homonuclear diatomic molecule or molecular ion, identify the element X, Q, or Z and determine the total bond order.
   (a) $X_2{:}(\sigma_{g2s})^2(\sigma_{u2s}^*)^2(\sigma_{g2p_z})^2(\pi_{u2p})^4(\pi_{g2p}^*)^2$
   (b) $Q_2^-{:}(\sigma_{g2s})^2(\sigma_{u2s}^*)^2(\pi_{u2p})^3$
   (c) $Z_2^{2+}{:}(\sigma_{g2s})^2(\sigma_{u2s}^*)^2(\sigma_{g2p_z})^2(\pi_{u2p})^4(\pi_{g2p}^*)^2$

23. For each of the electron configurations in Problem 21, determine whether the molecule or molecular ion is paramagnetic or diamagnetic.

24. For each of the electron configurations in Problem 22, determine whether the molecule or molecular ion is paramagnetic or diamagnetic.

6.6 Heteronuclear Diatomic Molecules

25. Following the pattern of Figure 6.21, work out the correlation diagram for the CN molecule, showing the relative energy levels of the atoms and the bonding and antibonding orbitals of the molecule. Indicate the occupation of the MOs with arrows. State the order of the bond and comment on the magnetic properties of CN.

26. Following the pattern of Figure 6.21, work out the correlation diagram for the BeN molecule, showing the relative energy levels of the atoms and the bonding and antibonding orbitals of the molecule. Indicate the occupation of the MOs with arrows. State the order of the bond and comment on the magnetic properties of BeN.

27. The bond length of the transient diatomic molecule CF is 1.291 Å; that of the molecular ion CF^+ is 1.173 Å. Explain why the CF bond shortens with the loss of an electron. Refer to the proper MO correlation diagram.

28. The compound nitrogen oxide (NO) forms when the nitrogen and oxygen in air are heated. Predict whether the nitrosyl ion (NO^+) will have a shorter or a longer bond than the NO molecule. Will NO^+ be paramagnetic like NO or diamagnetic?

29. What would be the electron configuration for a HeH^- molecular ion? What bond order would you predict? How stable should such a species be?

30. The molecular ion HeH^+ has an equilibrium bond length of 0.774 Å. Draw an electron correlation diagram for this ion, indicating the occupied MOs. Is HeH^+ paramagnetic? When HeH^+ dissociates, is a lower energy state reached by forming $He + H^+$ or $He^+ + H$?

31. Predict the ground state electronic configurations of CF, CH, CH^+, and CN^-. Do any of them have unpaired electrons?

32. Predict the ground state electronic configuration of HeBe. What is the bond order?

33. The bond dissociation energies for the species NO, CF^-, and CF^+ are ordered as $CF^+ > NO > CF^-$. Use MO theory to explain this ordering.

34. The ionization energy of CO is greater than that of NO. Explain this difference based on the electron configurations of these two molecules.

Connection to Instrumental Analysis: Photoelectron Spectroscopy

35. Photoelectron spectra were acquired from a sample of gaseous N_2 using He(I) light with energy 21.22 eV as the ionization source. Photoelectrons were detected with kinetic energy values 5.63 eV and also with 4.53 eV. Calculate the ionization energy for each group of electrons. Identify the MOs that were most likely the sources of these two groups of electrons.

36. Photoelectron spectra were acquired from a sample of gaseous O_2 using X-ray radiation with wavelength 0.99 nm and energy 1253.6 eV. The spectrum contained a large peak for photoelectrons with speed of 1.57×10^7 m s^{-1}. Calculate the ionization energy of these electrons. Identify the orbital from which they were most likely emitted.

37. From the $n = 0$ peaks in the photoelectron spectrum for N_2 shown in the figure in *Connections* box, prepare a quantitative energy level diagram for the molecular orbitals of N_2.

38. From the $n = 0$ peaks in the photoelectron spectrum for O_2 shown in the figure in *Connections* box, prepare a quantitative energy level diagram for the molecular orbitals of O_2.

39. The photoelectron spectrum of HBr has two main groups of peaks. The first has ionization energy 11.88 eV. The next peak has ionization energy 15.2 eV, and it is followed by a long progression of peaks with higher ionization energies. Identify the molecular orbitals corresponding to these two groups of peaks.

40. The photoelectron spectrum of CO has four major peaks with ionization energies of 14.5, 17.2, 20.1, and 38.3 eV. Assign these peaks of molecular orbitals of CO, and prepare a quantitative energy level correlation diagram for CO. The ionization energy of carbon atoms is 11.26 eV, and the ionization energy of oxygen atoms is 13.62 eV.

6.8 Valence Bond Theory and Electron Pair Bond

41. Write simple valence bond wave functions for the diatomic molecules Li_2 and C_2. State the bond order predicted by the simple VB model and compare with the LCAO predictions in Table 6.3.

42. Write simple valence bond wave functions for the diatomic molecules B_2 and O_2. State the bond order predicted by the simple VB model and compare with the LCAO predictions in Table 6.3.

43. Both the simple VB model and the LCAO method predict that the bond order of Be_2 is 0. Explain how each arrives at that conclusion.

44. Both the simple VB model and the LCAO method predict that the bond order of Ne_2 is 0. Explain how each arrives at that conclusion.

45. Write simple valence bond wave functions for formation of bonds between B atoms and H atoms. What B—H compound does the VB model predict? What geometry does it predict for the molecules?

46. Write simple valence bond wave functions for formation of bonds between C and H atoms. What C—H compound does the VB model predict? What geometry does it predict for the molecules?

47. Write simple valence bond wave functions for the bonds in NH_3. What geometry does the VB model predict for NH_3?

48. Write simple valence bond wave functions for the bonds in H_2O. What geometry does the VB model predict for H_2O?

6.9 Orbital Hybridization for Polyatomic Molecules
6.10 Predicting Molecular Structures and Shapes

49. Formulate a localized bond picture for the amide ion (NH_2^-). What hybridization do you expect the central nitrogen atom to have, and what geometry do you predict for the molecular ion?

50. Formulate a localized bond picture for the hydronium ion (H_3O^+). What hybridization do you expect the central oxygen atom to have, and what geometry do you predict for the molecular ion?

51. Draw a Lewis electron dot diagram for each of the following molecules and ions. Formulate the hybridization for the central atom in each case and give the molecular geometry.
(a) CCl_4
(b) CO_2
(c) OF_2
(d) CH_3^-
(e) BeH_2

52. Draw a Lewis electron dot diagram for each of the following molecules and ions. Formulate the hybridization for the central atom in each case and give the molecular geometry.
(a) BF_3
(b) BH_4^-
(c) PH_3
(d) CS_2
(e) CH_3^+

53. Describe the hybrid orbitals on the chlorine atom in the ClO_3^+ and ClO_2^+ molecular ions. Sketch the expected geometries of these ions.

54. Describe the hybrid orbitals on the chlorine atom in the ClO_4^- and ClO_3^- molecular ions. Sketch the expected geometries of these ions.

55. The sodium salt of the unfamiliar orthonitrate ion (NO_4^{3-}) has been prepared. What hybridization is expected on the nitrogen atom at the center of this ion? Predict the geometry of the NO_4^{3-} ion.

56. Describe the hybrid orbitals used by the carbon atom in N≡C—Cl. Predict the geometry of the molecule.

6.11 Using the LCAO and Valence Bond Methods Together

57. Describe the bonding in the bent molecule NF_2. Predict its energy level diagram and electron configuration.

58. Describe the bonding in the bent molecule OF_2. Predict its energy level diagram and electron configuration.

59. The azide ion (N_3^-) is a weakly bound molecular ion. Formulate its MO structure for localized σ bonds and de-localized π bonds. Do you expect N_3 and N_3^+ to be bound as well? Which of the three species do you expect to be paramagnetic?

60. Formulate the MO structure of (NO_2^+) for localized σ bonds and de-localized π bonds. Is it linear or nonlinear? Do you expect it to be paramagnetic? Repeat the analysis for NO_2 and for NO_2^-.

61. Discuss the nature of the bonding in the nitrite ion (NO_2^-). Draw the possible Lewis resonance diagrams for this ion. Use the VSEPR theory to determine the steric number, the hybridization of the central nitrogen atom, and the geometry of the ion. Show how the use of resonance structures can be avoided by introducing a de-localized π MO. What bond order does the MO model predict for the N—O bonds in the nitrite ion?

62. Discuss the nature of the bonding in the nitrate ion (NO_3^-). Draw the possible Lewis resonance diagrams for this ion. Use the VSEPR theory to determine the steric number, the hybridization of the central N atom, and the geometry of the ion. Show how the use of resonance structures can be avoided by introducing a de-localized π MO. What bond order is predicted by the MO model for the N—O bonds in the nitrate ion?

*__63.__ The pyridine molecule (C_5H_5N) is obtained by replacing one C—H group in benzene with a nitrogen atom. Because nitrogen is more electronegative than the C—H group, orbitals with electron density on nitrogen are lower in energy. How do you expect the π MOs and energy levels of pyridine to differ from those of benzene?

64. For each of the following molecules, construct the π MOs from the $2p_z$ atomic orbitals perpendicular to the plane of the carbon atoms.

(a) Cyclobutadiene

(b) Allyl radical

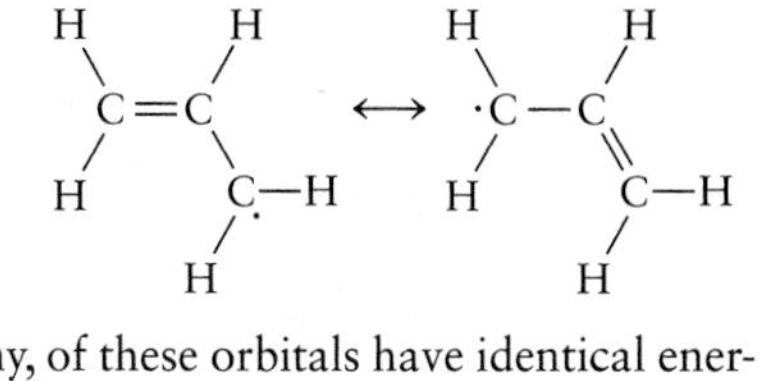

Indicate which, if any, of these orbitals have identical energies from symmetry considerations. Show the number of electrons occupying each π MO in the ground state, and indicate whether either or both of the molecules are paramagnetic. Assume that the C atoms in the allyl radical are all sp^2 hybridized.

ADDITIONAL PROBLEMS

65. **(a)** Sketch the occupied MOs of the valence shell for the N_2 molecule. Label the orbitals as σ or π orbitals, and specify which are bonding and which are antibonding.
(b) If one electron is removed from the highest occupied orbital of N_2, will the equilibrium N—N distance become longer or shorter? Explain briefly.

66. Calcium carbide (CaC_2) is an intermediate in the manufacturing of acetylene (C_2H_2). It is the calcium salt of the carbide (also called acetylide) ion (C_2^{2-}). What is the electron configuration of this molecular ion? What is its bond order?

67. The B_2 molecule is paramagnetic; show how this indicates that the energy ordering of the orbitals in this molecule is given by Figure 6.18a rather than 6.18b.

68. The Be_2 molecule has been detected experimentally. It has a bond length of 2.45 Å and a bond dissociation energy of 9.46 kJ mol^{-1}. Write the ground-state electron configuration of Be_2 and predict its bond order using the theory developed in the text. Compare the experimental bonding data on Be_2 with those recorded for B_2, C_2, N_2, and O_2 in Table 6.3. Is the prediction that stems from the simple theory significantly incorrect?

*__69.__ **(a)** The ionization energy of molecular hydrogen (H_2) is *greater* than that of atomic hydrogen (H), but that of molecular oxygen (O_2) is *lower* than that of atomic oxygen (O). Explain. (*Hint:* Think about the stability of the molecular ion that forms in relation to bonding and antibonding electrons.)
(b) What prediction would you make for the relative ionization energies of atomic and molecular fluorine (F and F_2)?

70. The molecular ion HeH^+ has an equilibrium bond length of 0.774 Å. Draw an electron correlation diagram for this molecule, indicating the occupied MOs. If the lowest energy MO has the form $C_1\psi_{1s}^{H} + C_2\psi_{1s}^{He}$, do you expect C_2 to be larger or smaller than C_1?

*__71.__ The MO of the ground state of a heteronuclear diatomic molecule AB is

$$\psi_{mol} = C_A\varphi^A + C_B\varphi^B$$

If a bonding electron spends 90% of its time in an orbital φ^A on atom A and 10% of its time in φ^B on atom B, what are the values of C_A and C_B? (Neglect the overlap of the two orbitals.)

72. The stable molecular ion H_3^+ is triangular, with H—H distances of 0.87 Å. Sketch the molecule and indicate the region of greatest electron density of the lowest energy MO.

***73.** According to recent spectroscopic results, nitramide

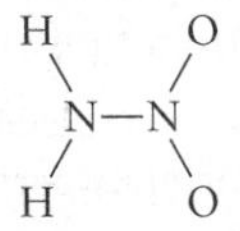

is a nonplanar molecule. It was previously thought to be planar.

(a) Predict the bond order of the N—N bond in the nonplanar structure.

(b) If the molecule really were planar after all, what would be the bond order of the N—N bond?

74. *trans*-tetrazene (N_4H_4) consists of a chain of four nitrogen atoms with each of the two end atoms bonded to two hydrogen atoms. Use the concepts of steric number and hybridization to predict the overall geometry of the molecule. Give the expected structure of *cis*-tetrazene.

7

CHAPTER

Bonding in Organic Molecules

Royalty-free/CORBIS

A petroleum refining tower.

Carbon (C) is unique among the elements in the large number of compounds it forms and in the variety of their structures. There are thousands of hydrocarbons (compounds that contain only hydrogen and carbon) that contain single, double, and triple bonds arranged as chains, rings, branched structures, and cages (Fig. 7.1). Hydrogen and oxygen, in contrast, form only two stable neutral compounds, water and hydrogen peroxide, and there are only six stable compounds that contain only nitrogen and oxygen.

The unique properties of carbon relate to its position in the periodic table. As a second-period element, carbon atoms are relatively small. Therefore, carbon can easily form the double and triple bonds that are rare in the compounds of related elements, such as silicon. Carbon can make four bonds, as a group IV element, which is more than the other second-period elements; the number of possible bonds plus the variety in their geometries (linear, trigonal planar, and

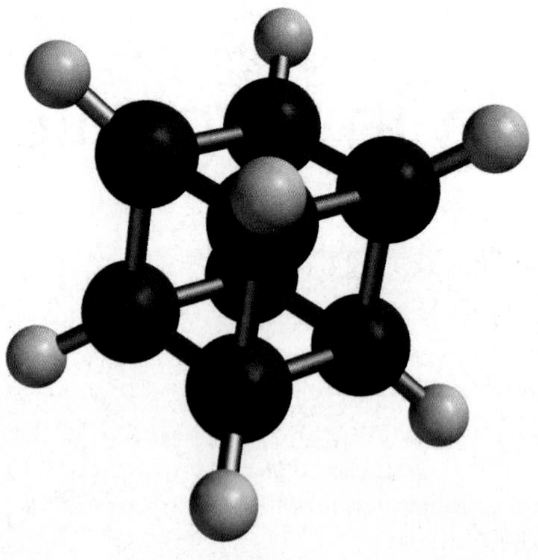

FIGURE 7.1 One simple and unusual hydrocarbon is cubane (C_8H_8), in which the eight carbon atoms are arranged at the corners of a cube. Recently, a derivative was made in which all eight hydrogen atoms were replaced by $-NO_2$ groups.

tetrahedral) means an enormous variety of possible structures. Finally, as an element of intermediate electronegativity, carbon forms covalent compounds both with relatively electronegative elements, such as oxygen, nitrogen, and the halogens, and with relatively electropositive elements, such as hydrogen and the heavy metals mercury and lead.

The study of the compounds of carbon is the discipline traditionally called **organic chemistry**, although the chemistry of carbon is intimately connected to that of the inorganic elements and to biochemistry. This chapter builds on the general principles of covalent bonding in carbon compounds presented in Chapter 6. The relationship between molecular structure and the properties of organic substances is illustrated by examining the composition, refining, and chemical processing of petroleum, the primary starting material for the production of hydrocarbons and their derivatives. The chapter continues with an introduction to the types of compounds produced when elements such as chlorine, oxygen, and nitrogen combine with carbon and hydrogen. It concludes with a brief introduction to some organic molecules important to agriculture and to medicine.

7.1 Petroleum Refining and the Hydrocarbons

The effects of petroleum on everyday life could not have been anticipated when the first oil well was drilled in 1859 near Titusville, Pennsylvania. Today, the petroleum and petrochemical industries span the world and influence nearly every aspect of our daily lives. In the early years of the 20th century, the development of the automobile, fueled by low-cost gasoline derived from petroleum, dramatically changed many people's lifestyles. The subsequent use of gasoline and oil to power trains and planes, tractors and harvesters, and pumps and coolers transformed travel, agriculture, and industry. Natural gas and heating oil warm most homes in the United States. Finally, the spectacular growth of the petrochemical industry since 1945 has led to the introduction of innumerable new products, ranging from pharmaceuticals to plastics and synthetic fibers. More than half of the chemical compounds produced in greatest volumes are synthesized from petroleum feedstocks.

Prospects for the continued availability of cheap petroleum and petrochemicals are clouded for the 21st century and beyond. Many wells have been drained, and the remaining petroleum is relatively difficult and costly to extract. Petroleum is not easy to make. It originated from the deposition and decay of organic matter (of animal or vegetable origin) in oxygen-poor marine sediments. Petroleum subsequently migrated to the porous sandstone rocks from which it is extracted today. Over the past 100 years, we have consumed a significant fraction of the petroleum accumulated in the earth over many millions of years. The imperative for the future is to save the remaining reserves for uses for which few substitutes are available (such as the manufacture of specialty chemicals) while finding other sources of heat and energy.

Although crude petroleum contains small amounts of oxygen, nitrogen, and sulfur, its major constituents are **hydrocarbons**—compounds of carbon and hydrogen. Isolating individual hydrocarbon substances from petroleum mixtures is an industrial process of central importance. Moreover, it provides a fascinating story that illustrates how the structures of molecules determine the properties of substances and the behavior of those substances in particular processes. The next three sections present a brief introduction to this story, emphasizing the structure–property correlations.

7.2 The Alkanes

TABLE 7.1

Straight-Chain Alkanes

Name	Formula
Methane	CH_4
Ethane	C_2H_6
Propane	C_3H_8
Butane	C_4H_{10}
Pentane	C_5H_{12}
Hexane	C_6H_{14}
Heptane	C_7H_{16}
Octane	C_8H_{18}
Nonane	C_9H_{20}
Decane	$C_{10}H_{22}$
Undecane	$C_{11}H_{24}$
Dodecane	$C_{12}H_{26}$
Tridecane	$C_{13}H_{28}$
Tetradecane	$C_{14}H_{30}$
Pentadecane	$C_{15}H_{32}$
.	.
.	.
.	.
Triacontane	$C_{30}H_{62}$

7.2.1 Normal Alkanes

The most prevalent hydrocarbons in petroleum are the **straight-chain alkanes** (also called normal alkanes, or *n*-alkanes), which consist of chains of carbon atoms bonded to one another by single bonds, with enough hydrogen atoms on each carbon atom to bring it to the maximum bonding capacity of four. These alkanes have the generic formula C_nH_{2n+2}; Table 7.1 lists the names and formulas of the first few alkanes. The ends of the molecules are methyl ($-CH_3$) groups, with methylene ($-CH_2-$) groups between them. We could write pentane (C_5H_{12}) as $CH_3CH_2CH_2CH_2CH_3$ to indicate the structure more explicitly or, in abbreviated fashion, as $CH_3(CH_2)_3CH_3$.

Bonding in the normal alkanes is explained by the valence bond (VB) model with orbital hybridization described in Section 6.8. The carbon atom in methane has four sp^3 hybridized orbitals, which overlap with hydrogen 1*s* orbitals to form four σ bonds pointing toward the vertices of a tetrahedron with the carbon atom at its center. These orbitals are represented in Figures 6.39 and 6.40; the methane molecule is shown in Figure 7.2a. The bonds in ethane are also described by sp^3 hybridization. One hybrid orbital on each carbon atom overlaps another hybrid orbital to form the C—C σ bond. The remaining three hybrids on each carbon overlap with hydrogen 1*s* orbitals to form σ bonds. The ethane molecule is shown in Figure 7.2b.

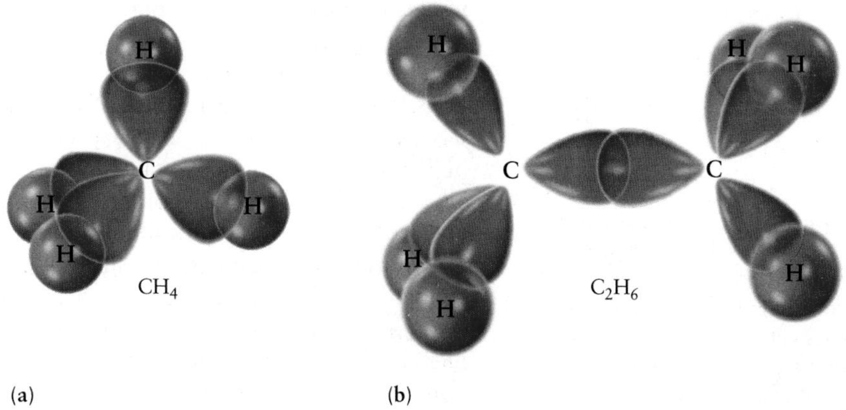

FIGURE 7.2 Bonding in the alkanes involves sp^3 hybridized orbitals on carbon. (a) Methane. (b) Ethane. The orbitals shown here are typical sketches used by organic chemists to describe bonding in organic molecules. Figure 6.40 compares these shapes to the actual shapes of the hybrid orbitals. (For clarity, the minor negative (blue) lobe of each hybrid orbital is not shown.)

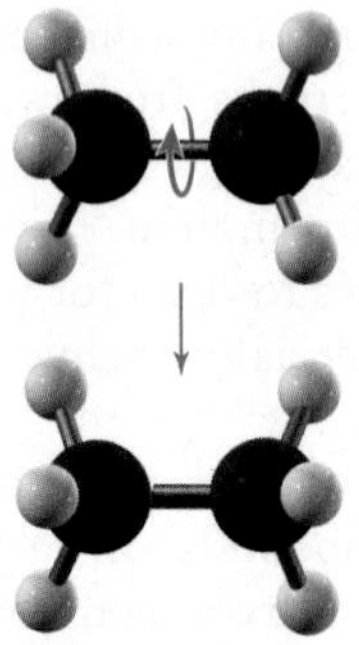

FIGURE 7.3 The two $-CH_3$ groups in ethane rotate easily about the bond that joins them.

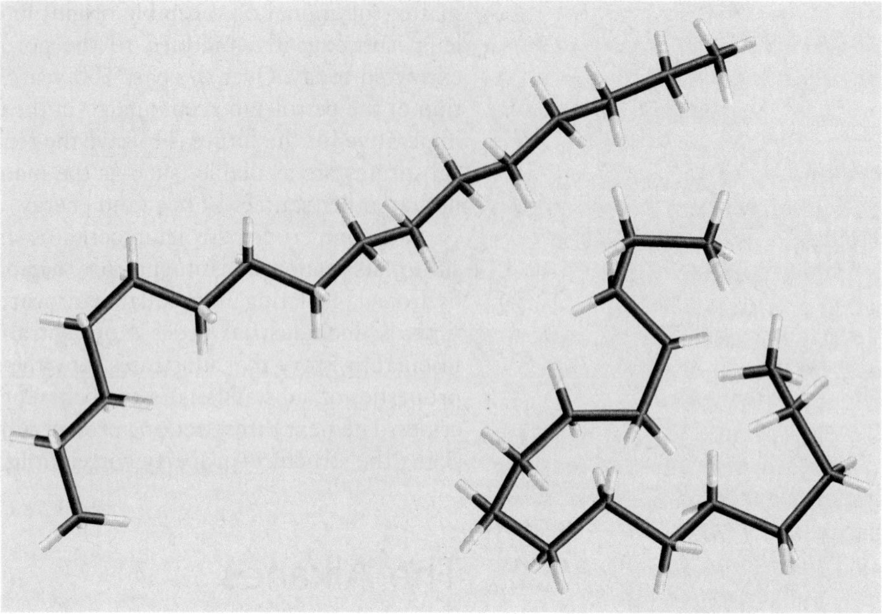

FIGURE 7.4 Two of the many possible conformations of the alkane $C_{16}H_{34}$. The carbon atoms are not shown explicitly, but they lie at the black intersections. Hydrogen atoms are at the white ends. Eliminating the spheres representing atoms in these tube (or Dreiding) models reveals the conformations more clearly.

FIGURE 7.5 The melting and boiling points of the straight-chain alkanes increase with chain length *n*. Note the alternation in the melting points: Alkanes with *n* odd tend to have lower melting points because they are more difficult to pack into a crystal lattice.

The same bonding scheme applies to the larger straight-chain alkanes. Alkanes with all their carbon atoms in a straight chain are called normal alkanes or *n*-alkanes. Two of the sp^3 hybrid orbitals on each carbon atom overlap those of adjacent atoms to form the backbone of the chain, and the remaining two bond to hydrogen atoms. Although molecular vibrations do not change bond lengths very much, internal rotation about C—C single bonds is quite easy (Fig. 7.3). Thus, a hydrocarbon molecule in a gas or liquid is constantly changing conformations which, for long chain hydrocarbons, can range from rather linear molecules like strands of spaghetti to more compact forms like balls of yarn. The term *straight chain* refers only to the bonding pattern in which each carbon atom is bonded to the next one in a sequence; it does not mean that the carbon atoms are positioned along a straight line. An alkane molecule with 10 to 20 carbon atoms looks quite different when "balled up" than when its bonds are extended to give a "stretched" molecule (Fig. 7.4). These two extreme conformations and many others interconvert rapidly at room temperature.

Figure 7.5 shows the melting and boiling points of the straight-chain alkanes, which both increase with the number of carbon atoms and thus with molecular mass. This is a consequence of the increasing strength of *dispersion forces* between these molecules (see discussion in Section 10.2). Methane, ethane, propane, and butane are all gases at room temperature, but the hydrocarbons that follow them in the alkane series are liquids. Alkanes beyond about $C_{17}H_{36}$ are waxy solids at 20°C, whose melting points increase with the number of carbon atoms present. Paraffin wax, a low-melting solid, is a mixture of alkanes with 20 to 30 carbon atoms per molecule. Petrolatum (petroleum jelly, or Vaseline) is a different mixture that is semisolid at room temperature.

Mixtures of hydrocarbons such as petroleum do not boil at single, sharply defined temperatures. Instead, as such a mixture is heated, the compounds with lower boiling points (the most volatile) boil off first, and as the temperature increases, more and more of the material vaporizes. The existence of a boiling-point range permits components of a mixture to be separated by distillation (see discussion in Section 11.6). The earliest petroleum distillation was a simple

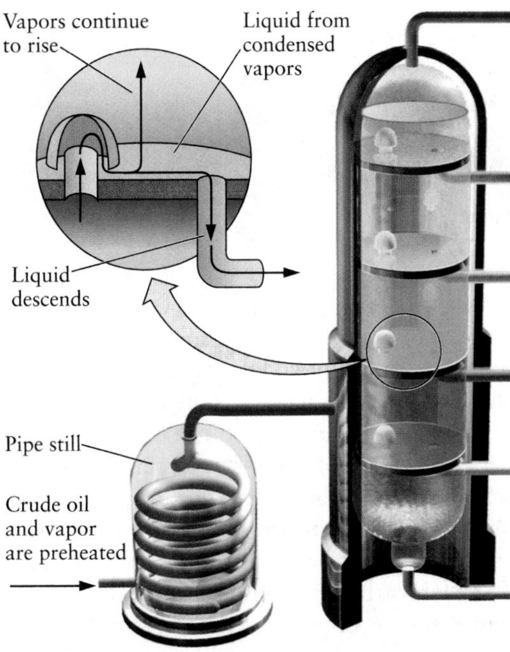

FIGURE 7.6 In the distillation of petroleum, the lighter, more volatile hydrocarbon fractions are removed from higher up the column and the heavier fractions from lower down.

batch process: The crude oil was heated in a still, the volatile fractions were removed at the top and condensed to gasoline, and the still was cleaned for another batch. Modern petroleum refineries use much more sophisticated and efficient distillation methods, in which crude oil is added continuously and fractions of different volatility (vapor pressure) are tapped off at various points up and down the distillation column (Fig. 7.6). Heat exchangers capture the heat liberated from condensation of the liquid products to save on energy costs.

Distillation separates hydrocarbons by their boiling points, and thus by molecular mass. A mixture of gases emerges from the top of the column, resembling the natural gas that collects in rock cavities above petroleum deposits. These gas mixtures contain ethane, propane, and butane, which can be separated further by redissolving them in a liquid solvent such as hexane. The methane-rich mixture of gases that remains is used for chemical synthesis or is shipped by pipeline to heat homes. The gases dissolved in hexane can be separated by redistilling, after which they can be used as starting materials in chemical processes. Propane and butane are also bottled under pressure and sold as liquefied petroleum gas, which is used for fuel in areas where natural gas is not available from pipelines. The next fraction to emerge from the petroleum distillation column after the gases is naphtha, which is used primarily to manufacture gasoline. Subsequent fractions of successively higher molecular mass are used for jet and diesel fuel, heating oil, and machine lubricating oil. The heavy, nonvolatile sludge that remains at the bottom of the distillation unit is pitch or asphalt, which is used for roofing and paving.

7.2.2 Cyclic Alkanes

In addition to the straight-chain alkanes, the cyclic alkanes also appear in petroleum. A **cycloalkane** consists of at least one chain of carbon atoms attached at the ends to form a closed loop. Two hydrogen atoms must be eliminated to form this additional C—C bond; thus, the general formula for cycloalkanes having one ring is C_nH_{2n} (Fig. 7.7). The cycloalkanes are named by adding the prefix *cyclo-* to the name of the straight-chain alkane that has the same number of

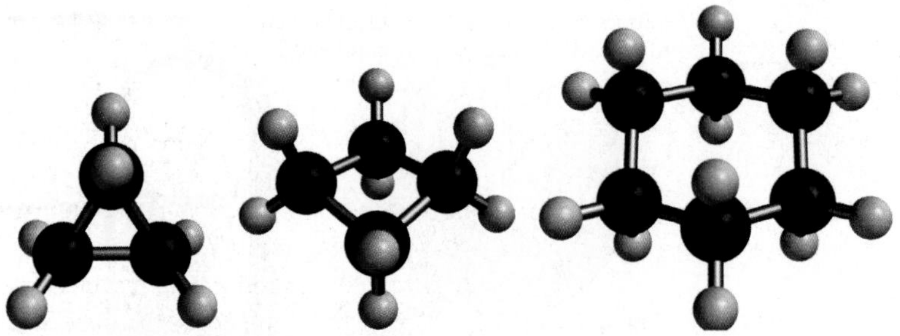

(a) (b) (c)

FIGURE 7.7 Three cyclic hydrocarbons. (a) Cyclopropane, C_3H_6. (b) Cyclobutane, C_4H_8. (c) Cyclohexane, C_6H_{12}.

carbon atoms as the ring compound. Bonding in the cycloalkanes involves sp^3 hybridization of the carbon atoms, just as in the straight-chain alkanes. But, coupling the tetrahedral angle of 109.5° with the restriction of a cyclic structure leads to two new interesting structural features that introduce *strain energy* in the cycloalkanes and influence the stability of their conformations.

It is easy to see from inspection of molecular models that two distinct conformations of cyclohexane can be formed when the tetrahedral angle is maintained at each carbon atom. These are called either the **boat conformation** or the **chair conformation** because of their resemblance to these objects (Fig. 7.8). The chair conformation has four carbon atoms in a plane with one carbon atom above and one carbon atom below that plane, located on opposite sides of the molecule. The boat conformation also has four carbon atoms in a plane, but both of the remaining atoms are located above this plane. Both conformations exist and appear to interconvert rapidly at room temperature through a sequence of rotations about single bonds (see Fig. 7.3). The chair conformation is significantly more stable than the boat configuration, because the hydrogen atoms can become quite close and interfere with one another in the boat conformation. When atoms that are not bonded to each other come sufficiently close in space to experience a repulsive interaction, this increase in potential energy reduces the stability of the molecule. Such interactions are called *steric strain,* and they play a significant role in determining the structure of polyatomic molecules. When the hydrogen atoms on cyclohexane are replaced with larger substituents, these effects can prevent interconversion between the boat and chair conformations. This effect is seen in many large molecules of biological significance, where the cyclohexane ring is an important structural unit, locked into one of its conformations.

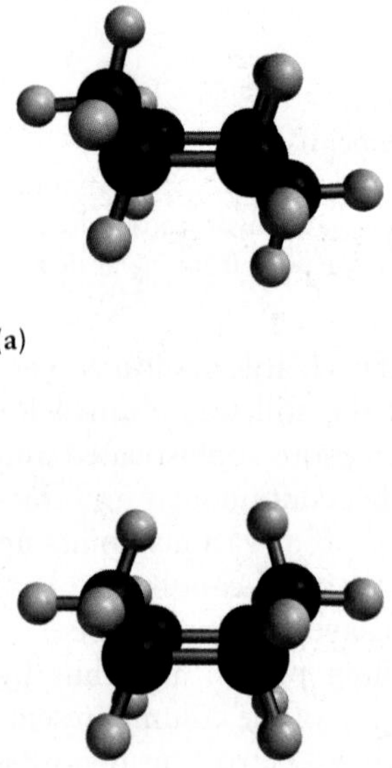

(a)

(b)

FIGURE 7.8 The conformations of cyclohexane. (a) Chair. (b) Boat.

Consider the possibility that cyclohexane could have a planar hexagonal structure. Then each C—C—C bond angle would be 120° resulting in angle strain of 10.5°. This distortion of the bond angle from the tetrahedral value increases the potential energy of the bond above its equilibrium value (which can be understood qualitatively using VSEPR; see Section 3.11), and the resulting **angle strain** energy reduces the stability of the molecule. Cyclohexane minimizes this effect through rotation about single bonds. The geometries of organic molecules are those that minimize the total strain, balancing the opposing contributions from angle and steric strain. The smallest cycloalkanes, namely, cyclopropane and cyclobutane, have much less freedom to rotate about single bonds. Consequently, they are strained compounds because the C—C—C bond angle is 60° (in C_3H_6) or 90° (in C_4H_8), which is far less than the normal tetrahedral angle of 109.5°. As a result, these compounds are more reactive than the heavier cycloalkanes or their straight-chain analogs, propane and butane.

7.2.3 Branched-Chain Alkanes and Isomerism

A **branched-chain alkane** is a hydrocarbon that contains only C—C and C—H single bonds, but in which the carbon atoms are no longer arranged in a

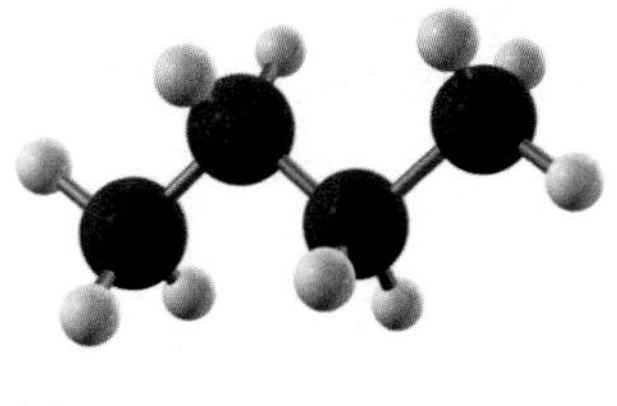

(a)

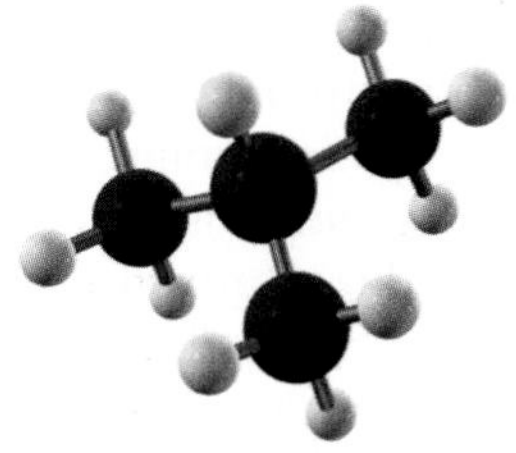

(b)

FIGURE 7.9 Two isomeric hydrocarbons with the molecular formula C_4H_{10}. (a) Butane. (b) 2-Methylpropane.

straight chain. One or more carbon atoms in each molecule is bonded to three or four other carbon atoms, rather than to only one or two as in the normal alkanes (*n*-alkanes) or cycloalkanes. The simplest branched-chain molecule (Fig. 7.9) is 2-methylpropane, sometimes referred to as isobutane. This molecule has the same molecular formula as butane (C_4H_{10}) but a different bonding structure in which the central carbon atom is bonded to three $-CH_3$ groups and only one hydrogen atom. The compounds butane and 2-methylpropane are each called a **geometrical isomer** of the other. The molecules have the same formula but different three-dimensional structures that can be interconverted only by breaking and re-forming chemical bonds.

The number of possible geometrical isomers increases rapidly with increasing numbers of carbon atoms. Butane and 2-methylpropane are the only two isomers with the chemical formula C_4H_{10}, but there are three isomers of C_5H_{12}, five isomers of C_6H_{14}, nine isomers of C_7H_{16}, and millions of isomers of $C_{30}H_{62}$. A systematic procedure for naming these isomers has been codified by the International Union of Pure and Applied Chemistry (IUPAC). The following set of rules is a part of that procedure:

1. Find the longest continuous chain of carbon atoms in the molecule. The molecule is named as a derivative of this alkane. In Figure 7.9b, a chain of three carbon atoms can be found; thus, the molecule is a derivative of propane.
2. The hydrocarbon groups attached to the chain are called alkyl groups. Their names are obtained by dropping the ending *-ane* from the corresponding alkane and replacing it with *-yl* (Table 7.2). The methyl group, CH_3, is derived from methane (CH_4), for example. Note that the isopropyl group attaches by its middle carbon atom. Alkyl side groups not bonded to a hydrocarbon chain, as represented in Table 7.2, are free radicals, neutral species with a single unpaired electron.
3. Number the carbon atoms along the chain identified in rule 1. Identify each alkyl group by the number of the carbon atom at which it is attached to the chain. The methyl group in the molecule in Figure 7.9b is attached to the second of the three carbon atoms in the propane chain; therefore, the molecule is called 2-methylpropane. The carbon chain is numbered from the end that gives the lowest number for the position of the first attached group.
4. If more than one alkyl group of the same type is attached to the chain, use the prefixes *di-* (two), *tri-* (three), *tetra-* (four), *penta-* (five), and so forth to specify the total number of such attached groups in the molecule. Thus, 2,2,3-trimethylbutane has two methyl groups attached to the second carbon atom and one to the third carbon atom of the four-atom butane chain. It is an isomer of heptane (C_7H_{16}).
5. If several types of alkyl groups appear, name them in alphabetical order. Ethyl is listed before methyl, which appears before propyl.

TABLE 7.2

Alkyl Side Groups

Name	Formula
Methyl	$-CH_3$
Ethyl	$-CH_2CH_3$
Propyl	$-CH_2CH_2CH_3$
Isopropyl	$-CH(CH_3)_2$
Butyl	$-CH_2CH_2CH_2CH_3$

EXAMPLE 7.1

Name the following branched-chain alkane:

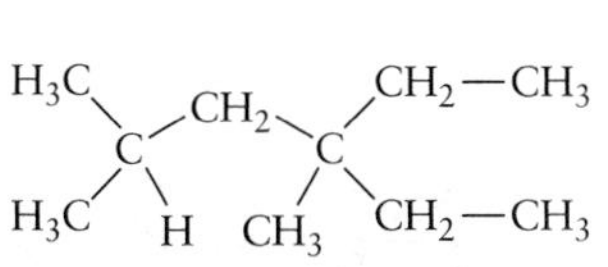

Solution

The longest continuous chain of carbon atoms is six, so this is a derivative of hexane. Number the carbon atoms starting from the left.

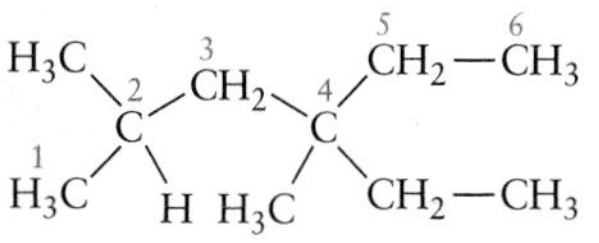

(Continued)

Methyl groups are attached to carbon atoms 2 and 4, and an ethyl group is attached to atom 4. The name is thus 4-ethyl-2,4-dimethylhexane. Note that if we had started numbering from the right, the higher number 3 would have appeared for the position of the first methyl group; therefore, the numbering from the left is preferred.

Related Problems: 7, 8, 9, 10, 11, 12

Optical isomerism, or **chirality**, is a second type of isomerism that is characteristic of organic molecules. A carbon atom that makes single bonds to four different atoms or groups of atoms can exist in two forms that are mirror images of each other but that cannot be interconverted without breaking and re-forming bonds (Fig. 7.10). If a mixture of the two forms is resolved into its optical isomers, the two forms rotate the plane of polarized light in different directions; therefore, such molecules are said to be "optically active." Although paired optical isomers have identical physical properties, their chemical properties can differ when they interact with other optically active molecules. As discussed in Section 23.4, proteins and other biomolecules are optically active. One goal of pharmaceutical research is to prepare particular optical isomers of carbon compounds for medicinal use. In many cases, one optical isomer is beneficial and the other is useless or even harmful.

The fraction of branched-chain alkanes in gasoline affects how it burns in an engine. Gasoline consisting entirely of straight-chain alkanes burns unevenly, causing "knocking" that can damage the engine. Blends that are richer in branched-chain and cyclic alkanes burn with much less knocking. Smoothness of combustion is rated quantitatively via the **octane number** of the gasoline, which was defined in 1927 by selecting as references one compound that causes large amounts of knocking and another that causes little to no knocking. Pure 2,2,4-trimethylpentane (commonly known as isooctane) burns smoothly and was assigned an octane number of 100. Of the compounds examined at the time, pure heptane caused the most knocking and was assigned octane number 0. Mixtures of heptane and isooctane cause intermediate amounts of knocking. Standard mixtures of these two compounds define a scale for evaluating the knocking caused by real gasolines, which are complex mixtures of branched- and straight-chain hydrocarbons. If a gasoline sample produces the same amount of knocking in a test engine as a mixture of 90% (by volume) 2,2,4-trimethylpentane and 10% heptane, it is assigned the octane number 90.

Certain additives increase the octane rating of gasoline. The least expensive of these is tetraethyllead, $Pb(C_2H_5)_4$, a compound that has weak bonds between the central lead atom and the ethyl carbon atoms. It readily releases ethyl radicals ($\cdot C_2H_5$) into the gasoline during combustion; these reactive species speed and smooth the combustion process, reducing knocking and giving better fuel performance. A **radical** is an atom or molecule that has one or more unpaired electrons; a radical that is electrically neutral is called a **free radical.** Chlorine atoms and oxygen molecules are both free radicals, for example, with one and

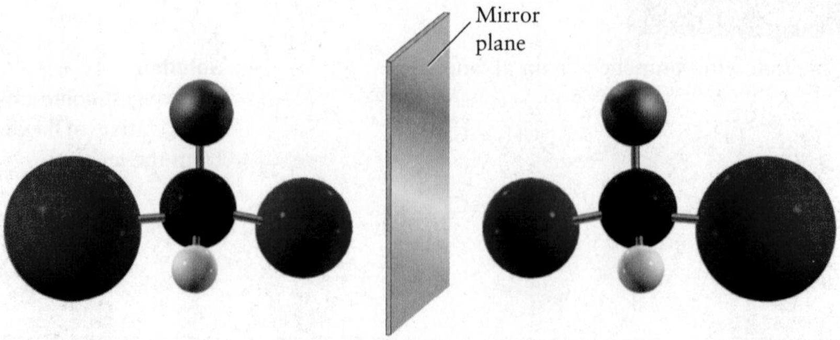

FIGURE 7.10 A molecule such as CHBrClF, which has four different atoms or groups of atoms bonded to a single carbon atom, exists in two mirror-image forms that cannot be superimposed by rotation of the molecule in space. Such pairs of molecules are optical isomers; the carbon atom is called a chiral center.

two unpaired electrons, respectively. The lead released into the atmosphere is a long-term health hazard and poisons catalytic converters (Section 18.8), rendering them ineffective. Leaded gasoline has been phased out and other low-cost additives have been developed to increase octane numbers. Chemical processing to make branched-chain compounds from straight-chain compounds is also used to control the octane number of gasoline.

7.3 The Alkenes and Alkynes

The hydrocarbons discussed so far in this chapter are referred to as **saturated,** because all the carbon–carbon bonds are single bonds. Any hydrocarbon that has double and triple carbon–carbon bonds is referred to as an **unsaturated hydrocarbon** (Fig. 7.11). Ethene (C_2H_4) has a double bond between its carbon atoms and is called an **alkene**. The simplest **alkyne** is acetylene (C_2H_2), which has a triple bond between its carbon atoms. In naming these compounds, the *-ane* ending of the corresponding alkane is replaced by *-ene* when a double bond is present and by *-yne* when a triple bond is present. Ethyne is thus the systematic name for acetylene, although we will continue to use its more common name. For any compound with a carbon backbone of four or more carbon atoms, it is necessary to specify the location of the double or triple bond. This is done by numbering the carbon–carbon bonds and putting the number of the lower numbered carbon involved in the multiple bond before the name of the alkene or alkyne. Thus, the two different isomeric alkynes with the formula C_4H_6 are

$$HC{\equiv}C{-}CH_2{-}CH_3 \qquad \textbf{1-butyne}$$

$$CH_3{-}C{\equiv}C{-}CH_3 \qquad \textbf{2-butyne}$$

Bonding in alkenes is described by the VB method with sp^2 hybrid orbitals on each carbon atom. (This method is described in Section 6.8 and shown in Figures 6.42 and 6.43. You should review that material before proceeding.) Figure 7.12a shows the three sp^2 hybrid orbitals and the nonhybridized $2p_z$ orbital. A σ bond is formed between the carbon atoms by overlap of one sp^2 orbital on each carbon atom, and the remaining four sp^2 orbitals form σ bonds with the four hydrogen atoms, as shown in Figure 7.12b. The nonhybridized $2p_z$ orbitals on the two carbon atoms are oriented parallel to one other and overlap to form a π bond (Fig. 7.12c). The overall result is the formation of a double bond between the two carbon atoms (Fig. 7.12d).

Bonding in alkynes is explained by sp hybridization (see description in Section 6.8 and illustrations in Figs. 6.44 and 6.45). Figure 7.13a shows two

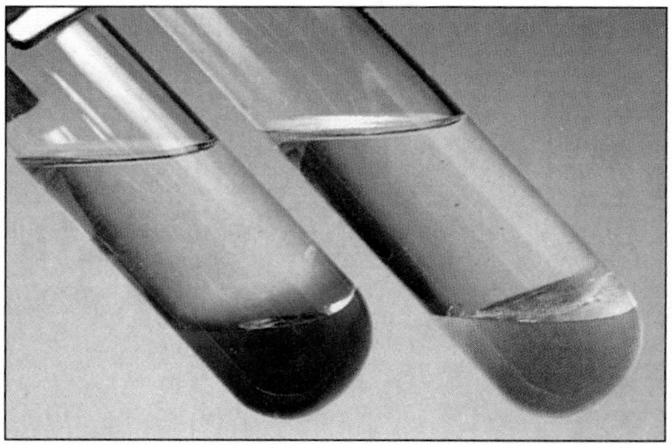

FIGURE 7.11 One way to distinguish alkanes from alkenes is by their reactions with aqueous $KMnO_4$. This strong oxidizing agent does not react with hexane and retains its purple color (left). But, when $KMnO_4$ is placed in contact with 1-hexene, a redox reaction occurs in which the brown solid MnO_2 forms (right) and $-OH$ groups are added to both sides of the double bond in the 1-hexene, giving a compound with the formula $CH_3(CH_2)_3CH(OH)CH_2OH$.

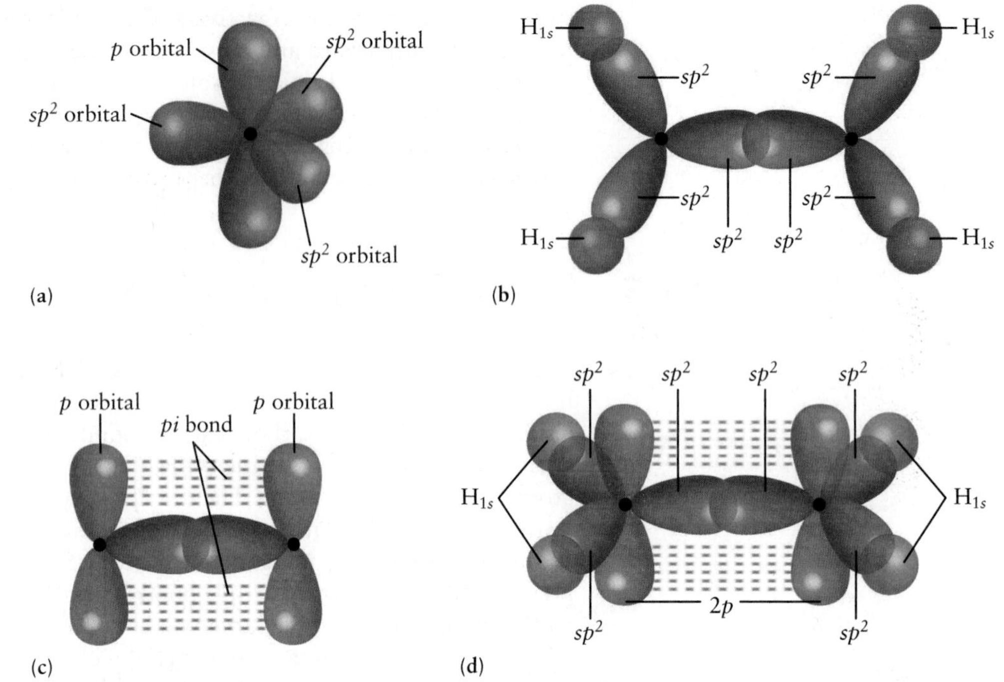

FIGURE 7.12 Bonding in ethene. (a) Three sp^2 hybrid orbitals and the nonhybridized 2*p* orbital on each carbon. (b) Top view showing formation of a C—C σ bond by overlap of two C sp^2 hybrid orbitals and four C—H σ bonds by overlap between four C sp^2 hybrid orbitals and four H 1*s* orbitals. (c) Overlap of parallel 2*p* orbitals to form a π bond. (d) The complete bonding scheme includes one C=C double bond and four C—H single bonds.

sp orbitals on each carbon atom as well as the nonhybridized $2p_x$ and $2p_y$ orbitals. A σ bond between the two carbon atoms is formed by overlap of *sp* hybrids on each carbon atom as shown in Figure 7.13b. The $2p_x$ and $2p_y$ nonhybridized atomic orbitals are parallel pairs on the two adjacent carbon atoms; each pair overlaps to form one π as shown in Figure 7.13c. The result is a triple bond in acetylene, analogous to the triple bond in N_2 shown in Figure 6.31.

FIGURE 7.13 Bonding in acetylene. (a) One *sp* orbital on each carbon atom with a nonhybridized $2p_x$ and $2p_y$ orbital. (b) Overlap of two sp^2 hybrid orbitals to form a C—C single bond. (c) Overlap of two mutually perpendicular pairs of nonhybridized *p* orbitals to form two perpendicular π bonds. (d) The complete bonding scheme includes a C≡C triple bond and two C—H single bonds.

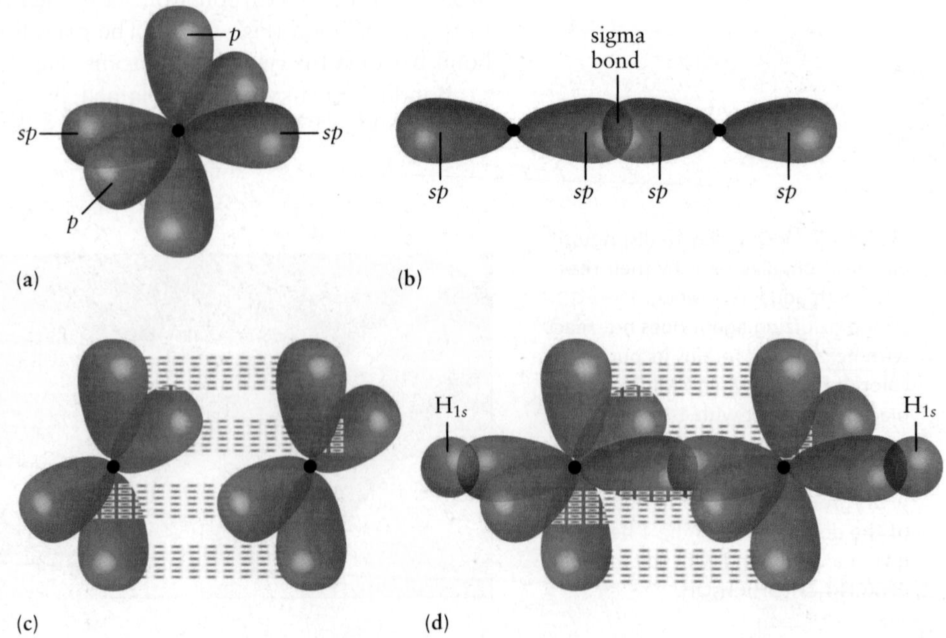

Figure 7.13d shows the complete bonding scheme with the formation of two C—H bonds formed by overlap of two carbon *sp* orbitals with two H 1*s* orbitals.

As explained later in this chapter, bond rotation does not occur readily about a carbon–carbon double bond. Many alkenes therefore exist in contrasting isomeric forms, depending on whether bonding groups are on the same (*cis*) or opposite (*trans*) sides of the double bond. There is only a single isomer of 1-butene but two of 2-butene, distinguished by the two possible placements for the outer CH_3 groups relative to the double bond:

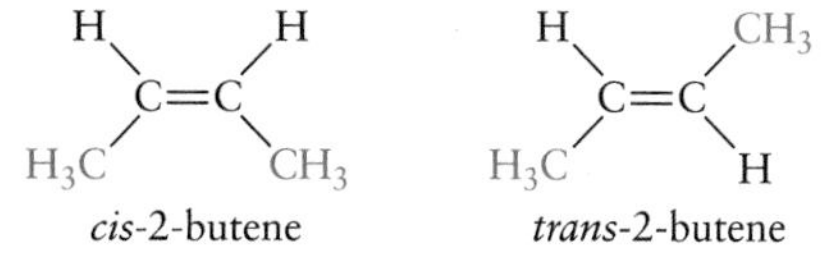

These compounds differ in melting and boiling points, density, and other physical and chemical properties.

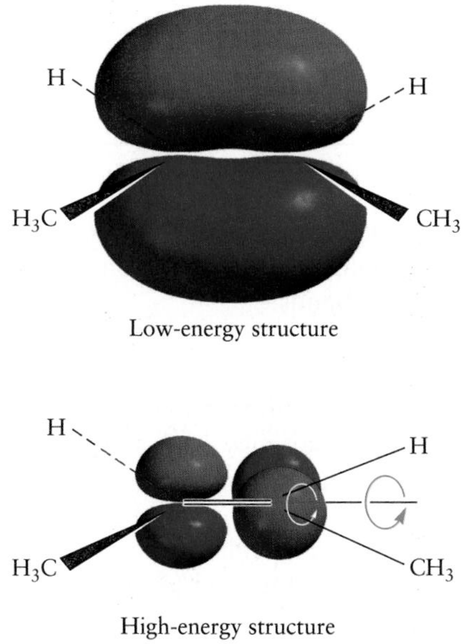

FIGURE 7.14 The overlap between the two *p* orbitals decreases when the 2-butene molecule is twisted about the C—C bond. The bonding MO becomes a nonbonding MO, and the energy of the molecule increases.

The structures and bonding in substituted alkenes and alkynes can be described through the combined molecular orbital (MO) and VB picture presented in Section 6.11, which uses localized VB bonds to describe the molecular framework and delocalized MOs to describe the π electrons. Let's apply this method to 2-butene ($CH_3CHCHCH_3$) and gain deeper understanding of the isomers discussed earlier. The Lewis diagram for 2-butene is

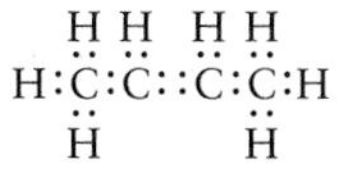

From the valence shell electron-pair repulsion (VSEPR) theory, the steric number of the two outer carbon atoms is 4 (so they are sp^3 hybridized), and that of the two central carbon atoms is 3 (sp^2 hybridized). The bonding around the outer carbon atoms is tetrahedral, and that about the central ones is trigonal planar. Each localized σ bond uses two electrons, resulting in a single bond between each pair of bonding atoms. In the case of 2-butene, these placements use 22 of the 24 available valence electrons, forming a total of 11 sigma bonds.

Next, the remaining *p* orbitals that were not involved in hybridization are combined to form π MOs. The p_z orbitals from the two central carbon atoms can be mixed to form a π (bonding) MO and a π^* (antibonding) MO. The remaining two valence electrons are placed into the π orbital, resulting in a double bond between the central carbon atoms. If the p_z orbital of one of these atoms is rotated about the central C—C axis, its overlap with the p_z orbital of the other carbon atom changes (Fig. 7.14). The overlap is greatest and the energy lowest when the two p_z orbitals are parallel to each other. In the most stable molecular geometry, the hydrogen atoms on the central carbon atoms lie in the same plane as the C—C—C—C carbon skeleton. This prediction is verified by experiment.

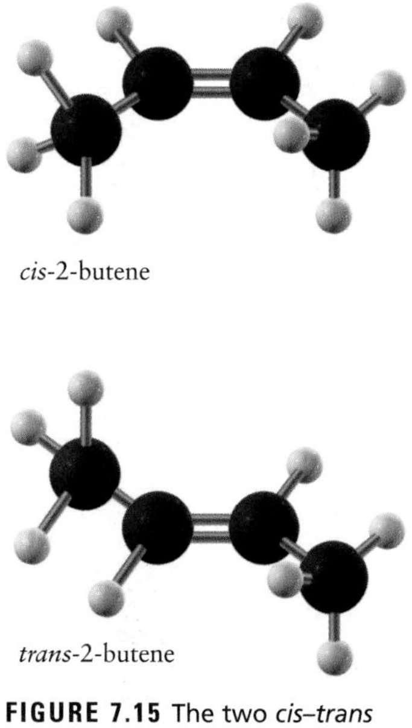

FIGURE 7.15 The two *cis–trans* isomers of 2-butene.

Figure 7.15 shows the structures of the isomers *cis*-2-butene and *trans*-2-butene. Converting one form to the other requires breaking the central π bond (by rotating the two p_z orbitals 180° with respect to each other as in Fig. 7.14), then re-forming it in the other configuration. Because breaking a π bond costs a significant amount of energy, both *cis* and *trans* forms are stable at room temperature, and interconversion between the two is slow. *Cis–trans* isomerization can be induced in photochemical reactions, however (see Section 20.6). Molecules such as *trans*-2-butene can absorb ultraviolet light, which excites an electron from a π to a π^* MO. In the excited electronic state of *trans*-2-butene, the carbon–carbon double bond is effectively reduced to a single bond, and one CH_3 group can rotate relative to the other to form *cis*-2-butene.

EXAMPLE 7.2

The Lewis diagram for propyne (CH_3CCH) is

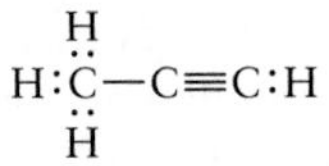

Discuss its bonding and predict its geometry.

Solution

The leftmost carbon atom in the structure is sp^3 hybridized, and the other two carbon atoms are sp hybridized. The atoms in the molecule are located on a single straight line, with the exception of the three hydrogen atoms on the leftmost carbon atom, which point outward toward three of the vertices of a tetrahedron. There is a σ bond between each pair of bonded atoms. The p_x and p_y orbitals on carbon atoms 2 and 3 combine to form two π orbitals and two π^* orbitals; only the former are occupied in the ground-state electron configuration.

Related Problems: 13, 14

Compounds with two double bonds are called *dienes,* those with three double bonds are called *trienes,* and so forth. The compound 1,3-pentadiene, for example, is a derivative of pentane with two double bonds:

$$CH_2{=}CH{-}CH{=}CH{-}CH_3$$

Each double bond in **polyenes** may lead to *cis* and *trans* conformations, depending on its neighboring groups; therefore, several isomers may have the same bonding patterns but different molecular geometries and physical properties.

When two or more double or triple bonds occur close to each other in a molecule, delocalized π bonding may best describe the bonding. See Section 6.11.4 for a detailed description of the delocalized π orbitals of 1,3-butadiene ($CH_2CHCHCH_2$).

Alkenes are not present to a significant extent in crude petroleum. They are essential starting compounds for the synthesis of organic chemicals and polymers, however, so their production from alkanes is of great importance. One way to produce alkenes is by **cracking** the petroleum using heat or catalysts. In **catalytic cracking,** the heavier fractions from the distillation column (compounds of C_{12} or higher) are passed over a silica–alumina catalyst at temperatures of 450°C to 550°C. Catalysts increase reaction rates without being consumed or produced by the reaction (see Section 18.8). Reactions such as

$$CH_3(CH_2)_{12}CH_3 \longrightarrow CH_3(CH_2)_4{-}CH{=}CH_2 + C_7H_{16}$$

occur to break the long chain into fragments. This type of reaction accomplishes two purposes. First, the shorter chain hydrocarbons have lower boiling points and can be added to gasoline. Second, the alkenes that result have higher octane numbers than the corresponding alkanes and perform better in the engine. Moreover, these alkenes can react with alkanes to give the more highly branched alkanes that are desirable in gasoline. **Thermal cracking** uses higher temperatures of 850°C to 900°C and no catalyst. It produces shorter chain alkenes, such as ethene and propene, through reactions such as

$$CH_3(CH_2)_{10}CH_3 \longrightarrow CH_3(CH_2)_8CH_3 + CH_2{=}CH_2$$

The short-chain alkenes are too volatile to be good components of gasoline, but they are among the most important starting materials for chemical synthesis.

7.4 Aromatic Hydrocarbons

Our final example of a group of hydrocarbons found in crude petroleum is the **aromatic hydrocarbon**, of which benzene is the simplest example. Benzene is a cyclic molecule with the formula C_6H_6. In the language of Chapter 3, benzene is represented as a resonance hybrid of two Lewis diagrams:

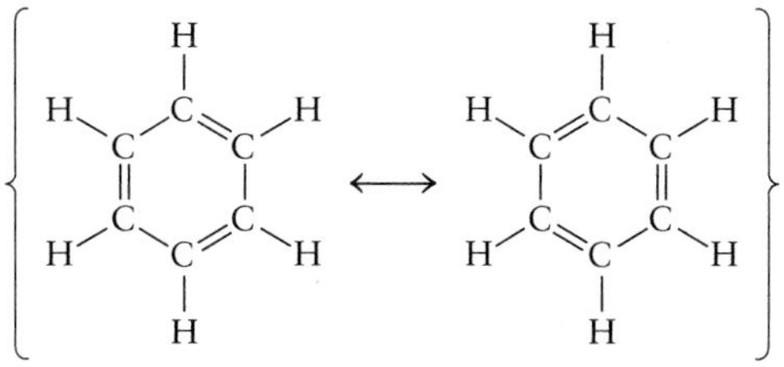

The modern view of resonant structures is that the molecule does not jump between two structures, but rather has a single, time-independent electron distribution in which the π bonding is described by delocalized MOs. Each carbon atom is sp^2 hybridized, and the remaining six p_z orbitals combine to give six MOs delocalized over the entire molecule. The π MOs are described in Section 6.11, and depicted in Figure 6.54 with their energy level diagram. In the ground electronic state of benzene, the three bonding π orbitals are occupied. As a result, benzene has six carbon–carbon bonds of equal length whose properties are intermediate between those of single and double bonds.

Benzene is sometimes represented by its chemical formula C_6H_6 and sometimes (to show structure) by a hexagon with a circle inside it:

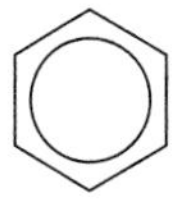

The six points of the hexagon represent the six carbon atoms, with the hydrogen atoms omitted for simplicity. The circle represents the de-localized π electrons, which are spread out evenly over the ring. The molecules of other aromatic compounds contain benzene rings with various side groups or two (or more) benzene rings linked by alkyl chains or fused side by side, as in naphthalene ($C_{10}H_8$):

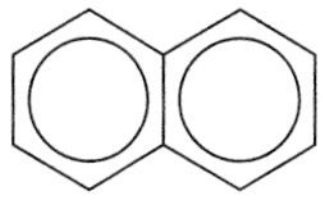

The most prevalent aromatic compounds in petroleum, other than benzene, are toluene, in which one hydrogen atom on the benzene ring is replaced by a methyl group, and the xylenes, in which two such replacements are made:

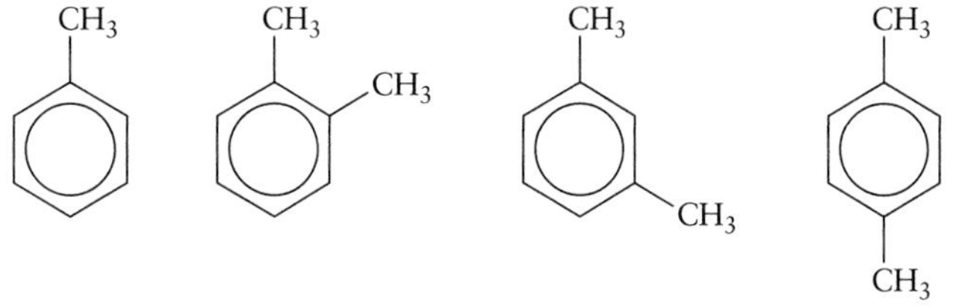

Toluene *o*-Xylene *m*-Xylene *p*-Xylene

This set of compounds is referred to as BTX (for *b*enzene-*t*oluene-*x*ylene). The BTX in petroleum is important to polymer synthesis (see Section 23.1). These components also significantly increase octane numbers and are used to make high-performance fuels with octane numbers above 100, as are required in modern aviation. There are ring compounds that contain alternating single and double bonds that are not aromatic. Only those ring compounds with $4n + 2$ π electrons

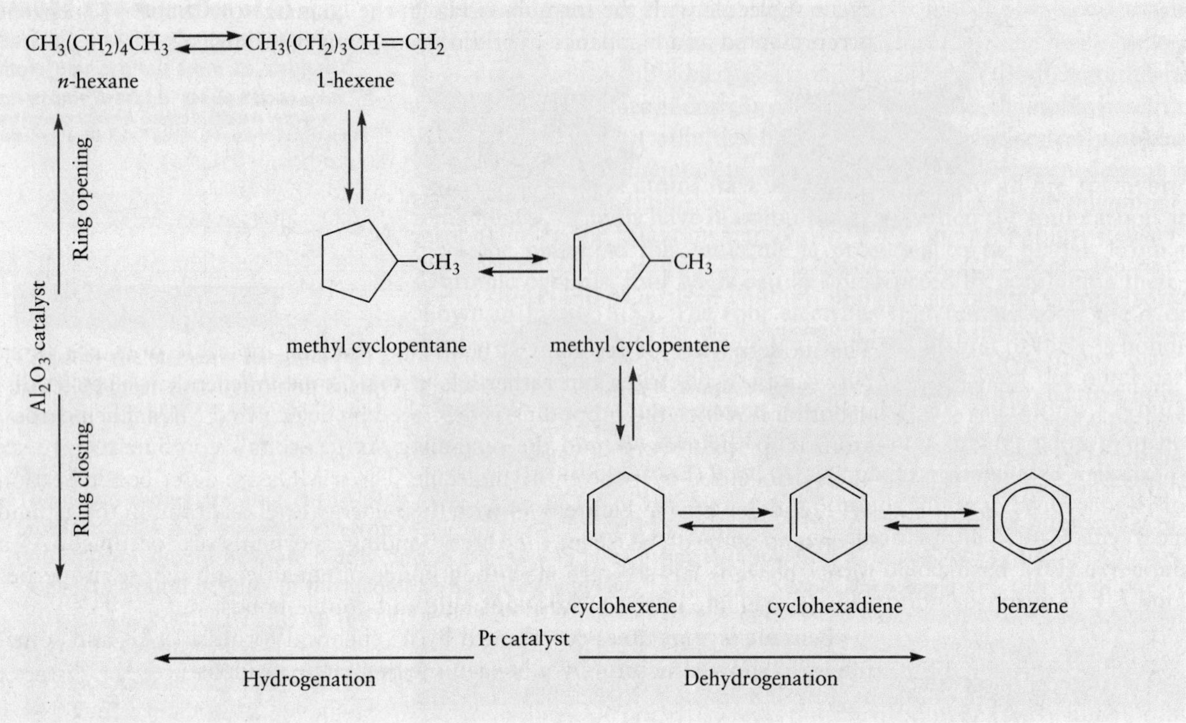

FIGURE 7.16 The reforming reaction that produces benzene from hexane uses a bifunctional catalyst of platinum metal supported on alumina. Platinum catalyzes hydrogenation/dehydrogenation reactions (blue arrows) while the alumina catalyzes ring opening and closing reactions (red arrows). It is common in reaction schemes of this type to omit most of the hydrogen atoms. The carbon atom at each vertex is assumed to have enough hydrogen atoms to satisfy the octet rule.

are considered to be aromatic, on the basis of an additional stability conferred by that electronic configuration.

A major advance in petroleum refining has been the development of **reforming reactions,** which produce BTX aromatics from straight-chain alkanes that contain the same numbers of carbon atoms. A fairly narrow distillation fraction that contains only C_6 to C_8 alkanes is taken as the starting material. The reactions use high temperatures and transition-metal catalysts such as platinum or rhenium on alumina supports, and their detailed mechanisms are not fully understood. Apparently, a normal alkane such as hexane is cyclized and dehydrogenated to give benzene as the primary product (Fig. 7.16). Heptane yields mostly toluene, and octane yields a mixture of xylenes. Toluene is replacing benzene as a solvent in industrial applications because tests on laboratory animals show it to be far less carcinogenic (cancer-causing) than benzene. Benzene is more important than toluene as a starting material for chemical synthesis; so a large fraction of the toluene produced is converted to benzene by **hydrodealkylation:**

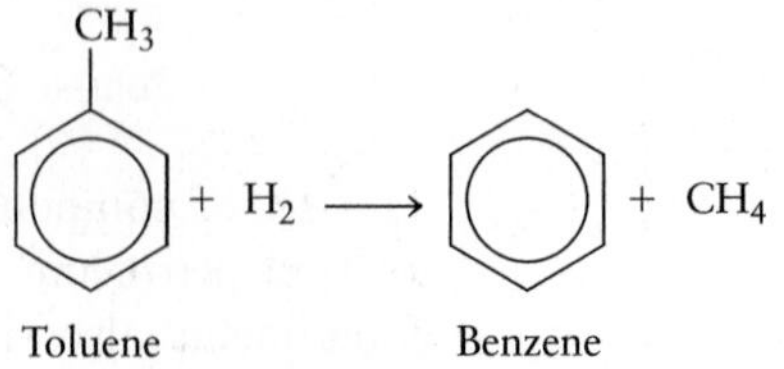

This reaction is conducted at high temperatures (550–650°C) and pressures of 40 to 80 atm.

7.5 Fullerenes

Among the most interesting conjugated π electron systems is a molecule discovered in 1985: buckminsterfullerene, C_{60}. Previously, only two forms of pure carbon (diamond and graphite) were known. In 1985, Harold Kroto, Robert Curl, and Richard Smalley were studying certain long-chain carbon molecules that had been discovered in the vicinity of red giant stars by radioastronomers using spectroscopy. They sought to duplicate the conditions near those stars by vaporizing a graphite target with a laser beam. Analysis of the products by mass spectrometry demonstrated not only the hoped-for molecules but a large proportion of molecules of molar mass 720 g mol^{-1}, which corresponds to the molecular formula C_{60}. Although the amounts of C_{60} present were far too small to isolate for direct determination of molecular structure, Kroto, Curl, and Smalley correctly suggested the cage structure shown in Figure 7.17a and named the molecule *buckminsterfullerene* after the architect Buckminster Fuller, the inventor of the geodesic dome, which the molecular structure of C_{60} resembles.

Molecules of C_{60} have a highly symmetric structure: 60 carbon atoms are arranged in a closed net with 20 hexagonal faces and 12 pentagonal faces. The pattern is exactly the design on the surface of a soccer ball (see Fig. 7.17b). Every carbon atom has a steric number of 3; all 60 atoms are sp^2 hybridized accordingly, although 1 of the 3 bond angles at each carbon atom must be distorted from the usual 120° sp^2 bond angle down to 108°. The π electrons of the double bonds are de-localized: The 60 p orbitals (1 from each carbon atom) mix to give 60 MOs with amplitude spread over both the inner and outer surfaces of the molecule. The lowest 30 of these MOs are occupied by the 60 π electrons.

Scientists succeeded in synthesizing C_{60} in gram quantities, in 1990, by striking an electric arc between two carbon rods held under an inert atmosphere. The carbon vapors condensed to form soot, which was extracted with an organic solvent. Using chromatography, the scientists could separate the C_{60}—first as a solution of a delicate magenta hue and, finally, as a crystalline solid—from the various impurities in the growth batch. C_{60} has subsequently been found in soot-forming flames when hydrocarbons are burned. Thus, the newest form of carbon has been (in the words of one of its discoverers) "under our noses since time immemorial." In 1994, the first buckminsterfullerene molecules were brought back from outer space in the form of the impact crater from a tiny meteorite colliding with an orbiting spacecraft.

Buckminsterfullerene is not the only new form of carbon to emerge from the chaos of carbon vapor condensing at high temperature. Synthesis of C_{60}

(a) (b)

FIGURE 7.17 (a) The structure of C_{60}, buckminsterfullerene. Note the pattern of hexagons and pentagons. (b) The design on the surface of a soccer ball has the same pattern as the structure of C_{60}.

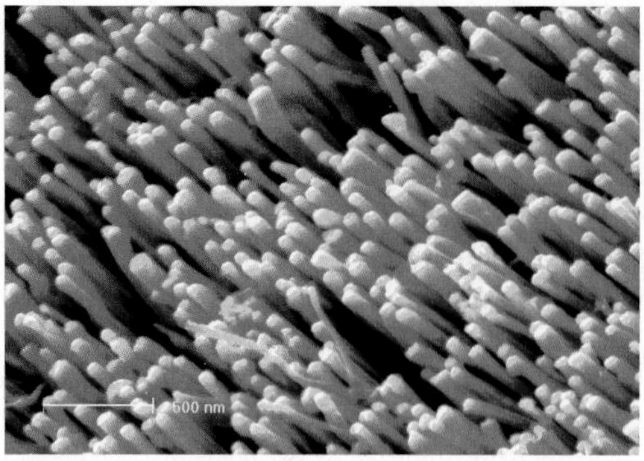

FIGURE 7.18 A bundle of carbon nanotubes, each about 1.4 nm in diameter. The bundle is 10 to 20 nm in thickness.

simultaneously produces a whole family of closed-cage carbon molecules called **fullerenes**. All the fullerenes have even numbers of atoms, with formulas ranging up to C_{400} and even higher. These materials offer exciting prospects for technical applications. For example, because C_{60} readily accepts and donates electrons to or from its π MOs, it has possible applications in batteries. It forms compounds (such as Rb_3C_{60}) that are superconducting (have zero resistance to the passage of an electric current) up to 30 K. Fullerenes also can encapsulate foreign atoms present during synthesis. If a graphite disk is soaked with a solution of $LaCl_3$, dried, and used as a laser target, the substance La@C_{60} forms, where the symbol @ means that the lanthanum atom is trapped within the 60-atom carbon cage.

Condensation of the carbon vapor under certain conditions favors the formation of *nanotubes*, which consist of seamless, cylindrical shells of thousands of sp^2 hybridized carbon atoms arranged in hexagons. The ends of the tubes are capped by pentagons inserted into the hexagonal network. These structures all have a de-localized π-electron system that covers the inner and outer surfaces of the cage or cylinder. Nanotubes also offer exciting prospects for material science and technological applications. For example, the mechanical properties of nanotubes suggest applications as high-strength fibers (Fig. 7.18).

7.6 Functional Groups and Organic Reactions

We discussed petroleum refining and the various classes of hydrocarbons produced in Sections 7.1–7.4. We now consider the structures, properties, and reactions of molecules formed by adding substituent atoms such as oxygen, nitrogen, and the halogens to the hydrocarbon backbone. In doing so, we shift our attention from the structures of entire molecules to the properties of **functional groups,** which are atoms or groups of atoms within molecules, with characteristic physical and chemical properties that are largely independent of the molecules of which they are a part. Organic chemists have identified functional groups on the basis of patterns of chemical reactivity. This fact permits us to regard an organic molecule as a hydrocarbon frame, which mainly governs size and shape, to which are attached functional groups that mainly determine the chemistry of the molecule. Table 7.3 shows some of the most important functional groups.

A few small molecules (methane, ethene, propene, benzene, and xylene) derived from petroleum or natural gas and their derivatives are the starting materials for chemicals produced in high-volume industrial processes. This section introduces the common functional groups, describes their bonding, and illustrates typical reactions in their synthesis or applications.

TABLE 7.3
Common Functional Groups

Functional Group†	Type of Compound	Examples
R—F, —Cl, —Br, —I	Alkyl or aryl halide	CH_3CH_2Br (bromoethane)
R—OH	Alcohol	CH_3CH_2OH (ethanol)
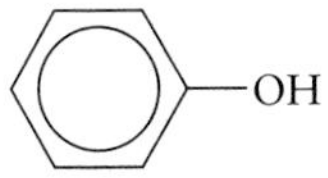	Phenol	C_6H_5—OH (phenol)
R—O—R′	Ether	H_3C—O—CH_3 (dimethyl ether)
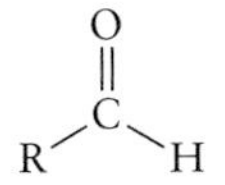	Aldehyde	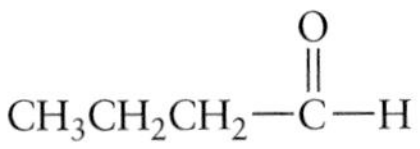 (butyraldehyde, or butanal)
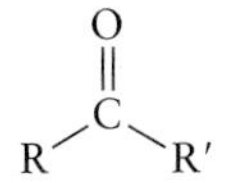	Ketone	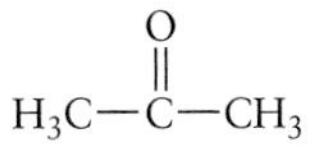 (propanone, or acetone)
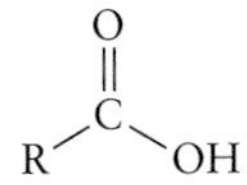	Carboxylic acid	CH_3COOH (acetic acid)
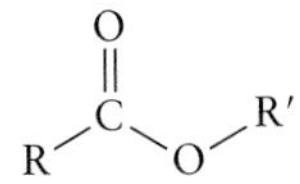	Ester	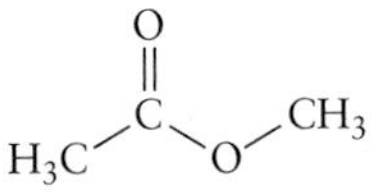 (methyl acetate)
R—NH_2	Amine	CH_3NH_2 (methylamine)
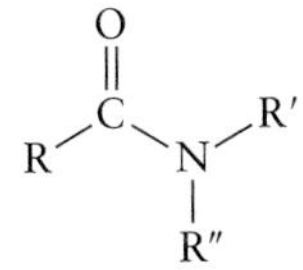	Amide	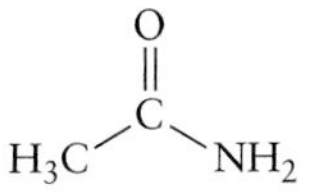 (acetamide)

†The symbols R, R′, and R″ stand for alkyl groups such as CH_3 or C_2H_5. In some cases, they may represent an H atom.

7.6.1 Halides

One of the simplest functional groups consists of a single halogen atom, which we take to be chlorine for illustrative purposes. The chlorine atom forms a σ bond to a carbon atom by overlap of its $3p_z$ orbital with a hybridized orbital on the carbon. The hybridized orbital may be sp^3, sp^2, or sp depending on the bonding in the hydrocarbon frame. An **alkyl halide** forms when mixtures of alkanes and halogens (except iodine) are heated or exposed to light.

$$\underset{\text{Methane}}{CH_4} + \underset{\text{Chlorine}}{Cl_2} \xrightarrow{250°C–400°C \text{ or light}} \underset{\text{Chloromethane}}{CH_3Cl} + \underset{\text{Hydrogen chloride}}{HCl}$$

The mechanism is a free radical chain reaction (see Section 18.4). Ultraviolet light initiates the reaction by dissociating a small number of chlorine molecules into highly reactive chlorine radicals. The chlorine radical abstracts a hydrogen atom from methane to form HCl and a methyl radical. Chlorine atoms and methyl radicals are both free radicals, which means that they are neutral species with one unpaired electron. Free radicals are often very reactive because they have incomplete octets and because the products formed by abstraction reactions are often very stable:

$$Cl\cdot + CH_4 \longrightarrow HCl + \cdot CH_3 \qquad \text{(propagation)}$$

$$\cdot CH_3 + Cl_2 \longrightarrow CH_3Cl + Cl\cdot \qquad \text{(propagation)}$$

The chlorine atoms and the methyl species are called free radicals, and are denoted by the dots next to the chemical symbols. A **free radical** is a chemical species that contains an odd (unpaired) electron; it is usually formed by breaking a covalent bond to form a pair of such species. They often appear as intermediates in reactions. Chloromethane (also called methyl chloride) is used in synthesis to add methyl groups to organic molecules. If sufficient chlorine is present, more highly chlorinated methanes form, providing a route for the industrial synthesis of dichloromethane (CH_2Cl_2, also called methylene chloride), trichloromethane ($CHCl_3$, chloroform), and tetrachloromethane (CCl_4, carbon tetrachloride). All three chloromethanes are used as solvents and their vapors have anesthetic or narcotic effects; environmental and health concerns have led to their production and use being highly regulated.

Adding chlorine to C=C bonds is a more important industrial route to alkyl halides than the free-radical reactions just described. Billions of kilograms of 1,2-dichloroethane (sometimes called by its old, now incorrect, name of ethylene dichloride) are manufactured each year, making this compound the largest organic chemical produced by volume. It is made by adding chlorine to ethene over an iron(III) oxide catalyst at moderate temperatures (40–50°C), either in the vapor phase or in a solution of 1,2-dibromoethane:

$$CH_2{=}CH_2 + Cl_2 \longrightarrow ClCH_2CH_2Cl$$

Almost all of the 1,2-dichloroethane produced is used to make chloroethene (vinyl chloride, $CH_2{=}CHCl$). This is accomplished by heating the 1,2-dichloroethane to 500°C over a charcoal catalyst to abstract HCl:

$$ClCH_2CH_2Cl \longrightarrow CH_2{=}CHCl + HCl$$

The HCl can be recovered and converted to Cl_2 for further production of 1,2-dichloroethane from ethene. Vinyl chloride has a much lower boiling point than 1,2-dichloroethane (−13°C compared with 84°C), so the two are easily separated by fractional distillation. Vinyl chloride is used in the production of polyvinyl chloride plastic (see Section 23.1).

7.6.2 Alcohols and Phenols

Alcohols have the —OH functional group attached to a tetrahedral carbon atom, that is, a carbon atom with single bonds to four other atoms. The carbon atom in the functional group is sp^3 hybridized. The oxygen atom is likewise sp^3 hybridized; two of the hybrid orbitals form σ bonds, whereas the other two hold lone pairs of electrons.

The simplest alcohol is methanol (CH_3OH), which is made from methane in a two-step process. The first is the **reforming reaction**

$$CH_4(g) + H_2O(g) \longrightarrow CO(g) + 3\,H_2(g)$$

conducted at high temperatures (750–1000°C) with a nickel catalyst. The gas mixture that results, called **synthesis gas**, reacts directly to form methanol at 300°C.

$$CO(g) + 2\,H_2(g) \longrightarrow CH_3OH(g)$$

The next higher alcohol, ethanol (CH_3CH_2OH) (Fig. 7.19), can be produced from the fermentation of sugars. Although fermentation is the major source of

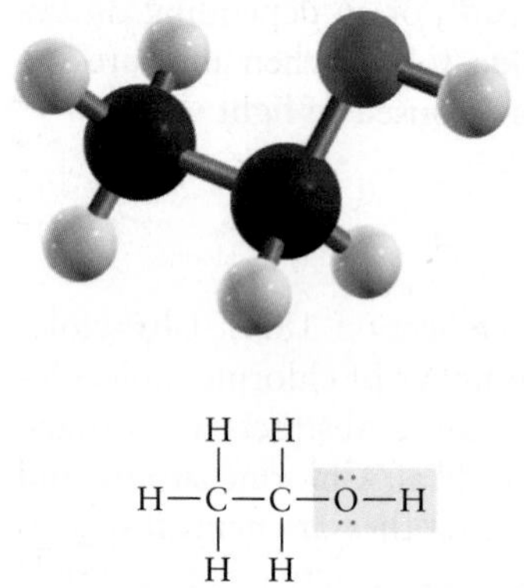

FIGURE 7.19 Ethanol (an alcohol).

ethanol for alcoholic beverages and for gasoline additives (like E10 or E85, where the number is the percent alcohol by volume), it is not significant for industrial production, which relies on the direct hydration of ethene:

$$CH_2{=}CH_2 + H_2O \longrightarrow CH_3CH_2OH$$

Temperatures of 300°C to 400°C and pressures of 60 to 70 atm are used with a phosphoric acid catalyst. Both methanol and ethanol are used widely as solvents and as intermediates for further chemical synthesis.

Two three-carbon alcohols exist, depending on whether the —OH group is attached to a terminal carbon atom or the central carbon atom. They are named 1-propanol and 2-propanol:

$$\underset{\text{1-Propanol}}{CH_3CH_2CH_2OH} \qquad \underset{\text{2-Propanol}}{\begin{array}{c} CH_3CHCH_3 \\ | \\ OH \end{array}}$$

1-Propanol and 2-propanol are commonly referred to as *n*-propyl alcohol and isopropyl alcohol, respectively. The systematic names of alcohols are obtained by replacing the *-ane* ending of the corresponding alkane with *-anol* and using a numeric prefix, when necessary, to identify the carbon atom to which the —OH group is attached.[1] Isopropanol is manufactured by the acid-catalyzed direct hydration of propene, the first step of which is

$$H_3C{-}CH{=}CH_2 + H^+ \longrightarrow H_3C{-}\overset{+}{C}H{-}CH_3$$

producing a transient charged species in which the positive charge is centered on the central carbon atom. Water reacts at this positive site to give the following intermediate:

$$\begin{array}{c} H_3C{-}CH{-}CH_3 \\ | \\ +OH_2 \end{array}$$

which loses a proton (the catalyst) and produces 2-propanol as follows:

$$H_2O + \begin{array}{c} H_3C{-}CH{-}CH_3 \\ | \\ +OH_2 \end{array} \longrightarrow \begin{array}{c} H_3C{-}CH{-}CH_3 \\ | \\ OH \end{array} + H_3O^+$$

2-propanol is produced preferentially because secondary carbon cations (with the positive charge localized on the carbon atom bonded to two carbon atoms) are more stable than primary carbon cations.

The compound 1-propanol is a *primary alcohol*: The carbon atom to which the —OH group is bonded has exactly one other carbon atom attached to it. The isomeric compound 2-propanol is a *secondary alcohol* because the carbon atom to which the —OH group is attached has two carbon atoms (in the two methyl groups) attached to it. The simplest *tertiary alcohol* (in which the carbon atom attached to the —OH group is also bonded to three other carbon atoms) is 2-methyl-2-propanol:

$$\begin{array}{c} OH \\ | \\ H_3C{-}CH{-}CH_3 \\ | \\ CH_3 \end{array}$$

Primary, secondary, and tertiary alcohols differ in chemical properties.

[1]Compare the names of alcohols with the corresponding names of alkyl halides. If the —OH group were replaced by a chlorine atom, the names of these compounds would be 1-chloropropane and 2-chloropropane.

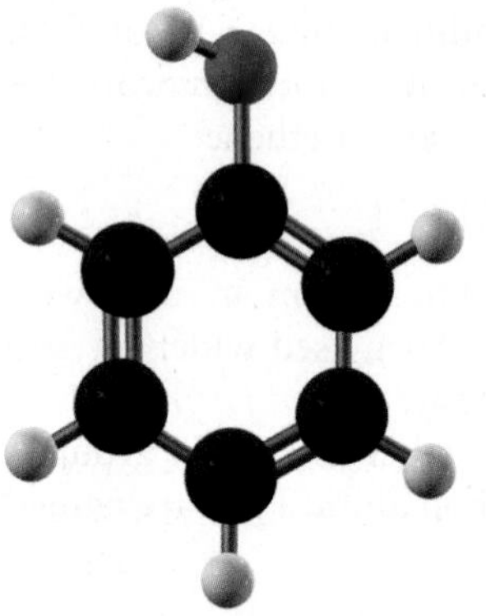

FIGURE 7.20 The structure of phenol, C_6H_5OH.

A **phenol** is a compound in which an —OH group is attached directly to an aromatic ring. The simplest example is phenol itself (C_6H_5OH, Fig. 7.20). As in the alcohols, the oxygen atom is sp^3 hybridized with two unshared pairs. The carbon atom, which is part of the aromatic ring, is sp^2 hybridized (see Section 7.4).

The manufacture of phenols uses quite different types of reactions from those used to make alcohols. One method, introduced in 1924 and still used to a small extent today, involves the chlorination of the benzene ring followed by reaction with sodium hydroxide:

$$C_6H_6 + Cl_2 \longrightarrow HCl + C_6H_5Cl$$

$$C_6H_5Cl + 2\ NaOH \longrightarrow C_6H_5O^-Na^+ + NaCl + H_2O$$

This approach illustrates a characteristic difference between the reactions of aromatics and alkenes. When chlorine reacts with an alkene, it *adds* across the double bond (as shown in the production of 1,2-dichloroethane). When an aromatic ring is involved, substitution of chlorine for hydrogen occurs instead and the aromatic π-bonding structure is preserved.

A different approach is used to make almost all phenol today. It involves, first, the acid-catalyzed reaction of benzene with propene to give cumene, or isopropyl benzene:

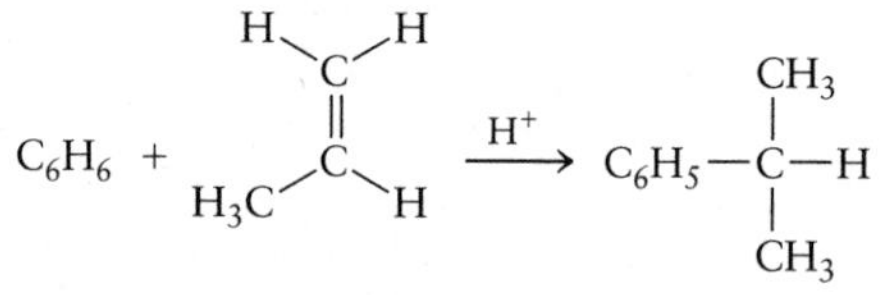

As in the production of 2-propanol, the first step is the addition of H^+ to propene to give $CH_3—CH^+—CH_3$ (see earlier in this section). This ion then attaches to the benzene ring through its central carbon atom to give the cumene and regenerate H^+. Subsequent reaction of cumene with oxygen (Fig. 7.21) gives phenol and acetone, an important compound that is discussed later in this section. The main use of phenol is in the manufacturing of polymers and aspirin.

7.6.3 Ethers

An **ether** is characterized by the —O— functional group, in which an oxygen atom provides a link between two separate alkyl or aromatic groups. Figure 7.22

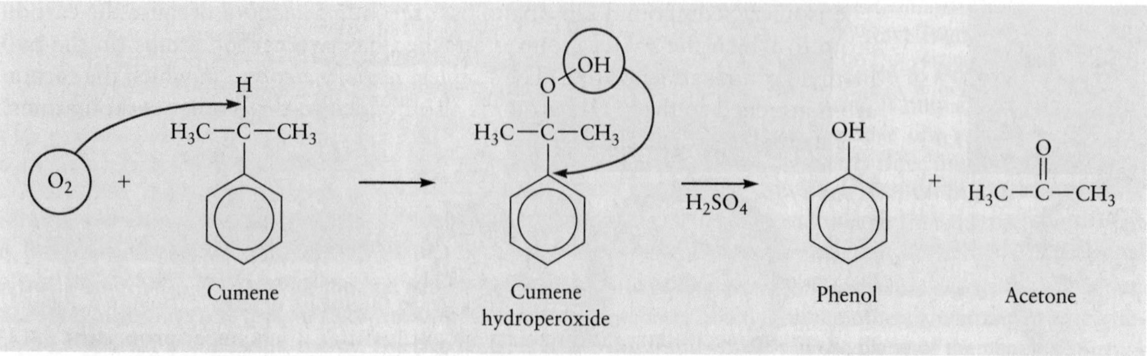

FIGURE 7.21 The synthesis of phenol and acetone from cumene is a two-step process that involves insertion of O_2 into a C—H bond to make a peroxide (R—O—O—R′), followed by acid-catalyzed migration of the —OH group to form the products.

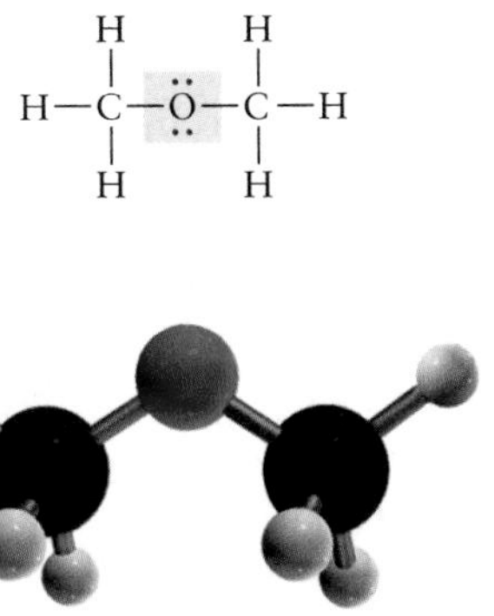

(a)

(b)

FIGURE 7.22 Structure of dimethyl ether. (a) Lewis diagram. (b) Ball-and-stick model.

shows the simplest ether, dimethyl ether. The oxygen atom is sp^3 hybridized. Two of these hybrid orbitals form σ bonds to the carbon atoms, whereas each of the other two holds an unshared pair. The C—O—C bond angle is 110.3°, which is close to the tetrahedral value 109.5° predicted by sp^3 hybridization.

One important ether is diethyl ether, often called simply ether, in which two ethyl groups are linked to the same oxygen atom:

$$C_2H_5-O-C_2H_5$$

Diethyl ether is a useful solvent for organic reactions, and was formerly used as an anesthetic. It can be produced by a **condensation reaction** (a reaction in which a small molecule such as water is eliminated to join [condense] two species) between two molecules of ethanol in the presence of concentrated sulfuric acid as a dehydrating agent:

$$CH_3CH_2OH + HOCH_2CH_3 \xrightarrow{H_2SO_4} CH_3CH_2-O-CH_2CH_3 + H_2O$$

Another ether of considerable importance is methyl *t*-butyl ether (MTBE):

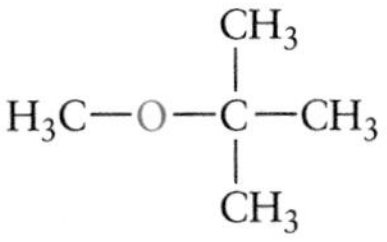

This compound appeared to be a successful replacement for tetraethyllead as an additive to increase the octane ratings of gasolines. The bond between the oxygen and the *t*-butyl group is weak, and it breaks to form radicals that assist the smooth combustion of gasoline. MTBE is readily soluble in water, however, and it has appeared in drinking water supplies through leaks from underground storage tanks for gasoline. Concern over possible health risks has caused MTBE to be phased out in various regions of the United States.

In a cyclic ether, oxygen forms part of a ring with carbon atoms, as in the common solvent tetrahydrofuran (Fig. 7.23). The smallest such ring has two carbon atoms bonded to each other and to the oxygen atom; it occurs in oxirane, commonly called ethylene oxide,

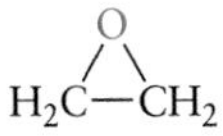

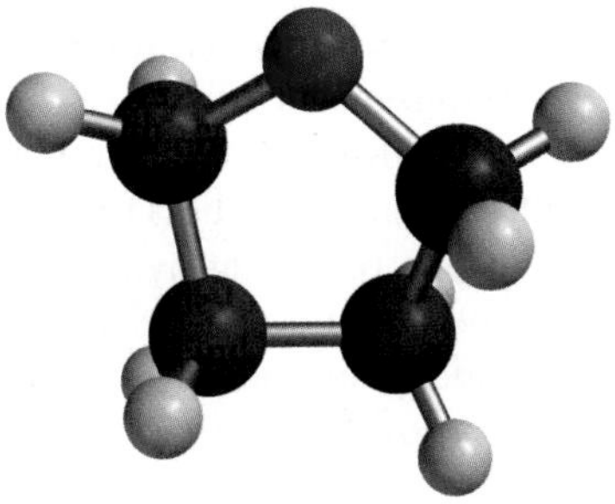

FIGURE 7.23 The structure of tetrahydrofuran, C_4H_8O.

which is made by direct oxidation of ethene over a silver catalyst:

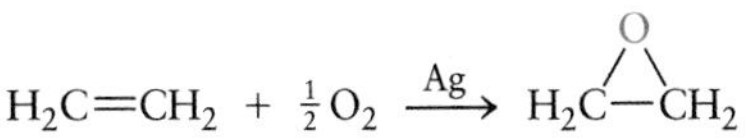

Any ether with three-membered rings is called an **epoxide**. The major use of oxirane is in the preparation of 1,2-dihydroxyethane (commonly called ethylene glycol):

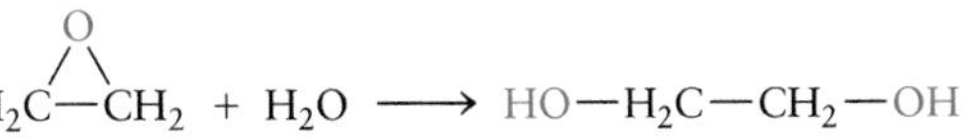

This reaction is conducted either at 195°C under pressure or at lower temperatures (50–70°C) with sulfuric acid as a catalyst. 1,2-dihydroxyethane is a dialcohol, or **diol**, in which two —OH groups are attached to adjacent carbon atoms. Its primary use is as a component in antifreeze to decrease the freezing point of water in automobile radiators (see the discussion of colligative properties in Section 11.5).

FIGURE 7.24 Acetaldehyde (an aldehyde).

7.6.4 Aldehydes and Ketones

An **aldehyde** contains the characteristic $-\mathrm{C(=O)-H}$ functional group (Fig. 7.24). Figure 7.26 shows the bonding in formaldehyde, which is the simplest organic molecule with a double bond between carbon and oxygen. The carbon is sp^2 hybridized and forms σ bonds to two hydrogen atoms. The oxygen is also sp^2 hybridized. Carbon and oxygen form one σ bond by overlap of sp^2 orbitals and one π bond by overlap of parallel nonhybridized $2p$ orbitals. This is the same bonding model used to describe C═C double bonds in Figure 7.12c.

Aldehydes can be prepared by the dehydrogenation of primary alcohols. Formaldehyde results from the dehydrogenation of methanol at high temperatures using an iron oxide–molybdenum oxide catalyst:

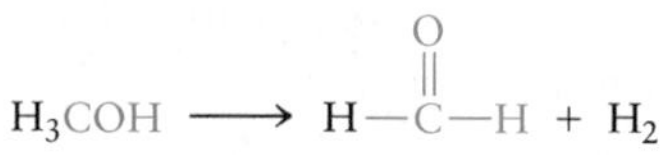

Another reaction that gives the same primary product is the oxidation reaction

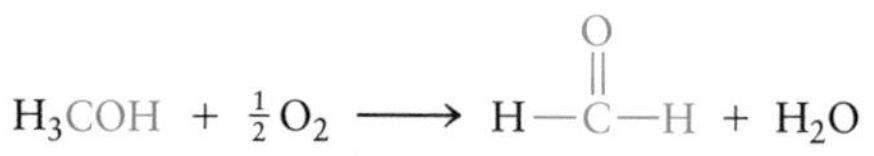

Formaldehyde is readily soluble in water, and a 40% aqueous solution of formaldehyde called *formalin* is used to preserve biological specimens. It is a component of wood smoke and helps to preserve smoked meat and fish, probably by reacting with nitrogen-containing groups in the proteins of attacking bacteria. Its major use is in making polymer adhesives and insulating foam.

The next aldehyde in the series is acetaldehyde, the structure of which is shown above. Industrially, acetaldehyde is produced not from ethanol but by the oxidation of ethene, using a $PdCl_2$ catalyst.

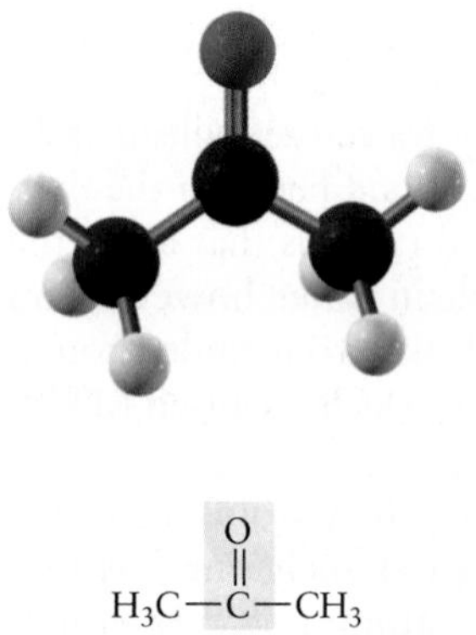

FIGURE 7.25 Acetone (a ketone).

A **ketone** has the $-\mathrm{C(=O)}-$ functional group in which a carbon atom forms a double bond to an oxygen atom and single bonds to two separate alkyl or aromatic groups. The simplest ketone is acetone (Fig. 7.25), in which two methyl groups are bonded to the central carbon. The bonding scheme is the same as that shown in Figure 7.26, with alkyl or aromatic groups replacing hydrogen atoms in the two single bonds. Such compounds can be prepared by dehydrogenation or oxidation of secondary alcohols, just as aldehydes come from primary alcohols. Acetone is made by the dehydrogenation of 2-propanol over a copper oxide or zinc oxide catalyst at 500°C:

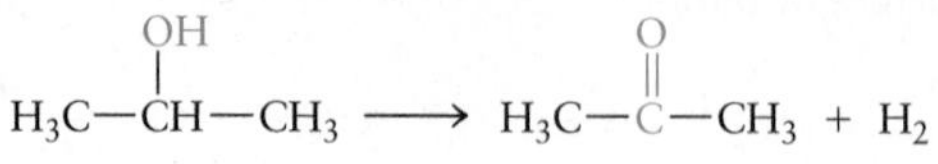

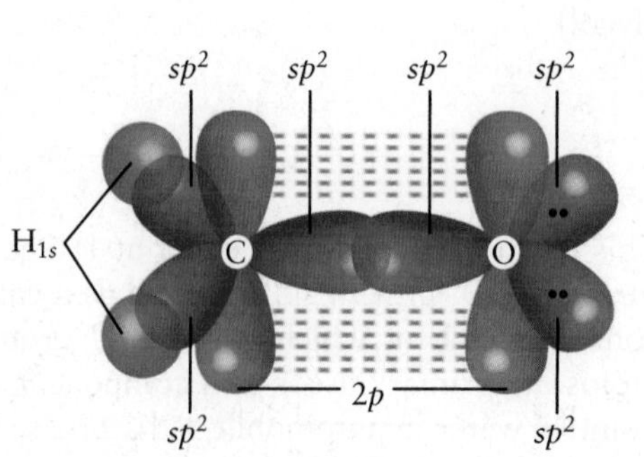

FIGURE 7.26 Bonding in formaldehyde involves sp^2 hybrid orbitals of both the carbon (C) and oxygen (O) atoms. The C—O σ bond is constructed by overlap of C and O sp^2 hybrid orbitals and the C—H bonds from overlap of C sp^2 orbitals with H $1s$ orbitals. Two lone pairs occupy the other O sp^2 hybrid orbitals. The π bond is formed by overlap of the parallel unhybridized $2p$ orbitals.

Acetone is also produced (in greater volume) as the coproduct with phenol of the oxidation of cumene (see earlier). It is a widely used solvent and is the starting material for the synthesis of a number of polymers.

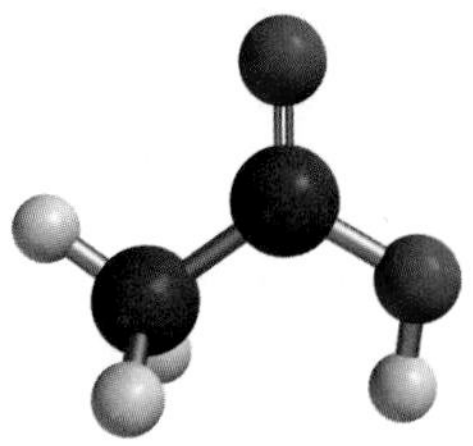

$$H_3C-C(=O)-O-H$$

FIGURE 7.27 Acetic acid (a carboxylic acid).

7.6.5 Carboxylic Acids and Esters

A **carboxylic acid** contains the $-\overset{\text{O}}{\overset{\|}{\text{C}}}-\text{OH}$ functional group (also written as $-COOH$). The bonding scheme is a variation of that shown in Figure 7.26, in which the doubly bonded oxygen atom is sp^2 hybridized and the singly bonded oxygen is sp^3 hybridized with unshared pairs in two of the hybrid orbitals. Carboxylic acids are the products of the oxidation of aldehydes, just as aldehydes are the products of the oxidation of primary alcohols. (The turning of wine to vinegar is a two-step oxidation leading from ethanol through acetaldehyde to acetic acid.) Acetic acid (Fig. 7.27) can be produced industrially by the air oxidation of acetaldehyde over a manganese acetate catalyst at 55°C to 80°C:

$$H_3C-C(=O)-H + \tfrac{1}{2}O_2 \xrightarrow{Mn(CH_3COO)_2} H_3C-C(=O)-OH$$

The reaction now preferred on economic grounds for acetic acid production is the combination of methanol with carbon monoxide (both derived from natural gas) over a catalyst that contains rhodium and iodine. The overall reaction is

$$CH_3OH(g) + CO(g) \xrightarrow{Rh,I_2} CH_3COOH(g)$$

and can be described as a **carbonylation**, or the insertion of CO into the methanol C—O bond.

Acetic acid is a member of a series of carboxylic acids with formulas $H-(CH_2)_n-COOH$. The simplest carboxylic acid is formic acid (HCOOH), with $n = 0$. This compound was first isolated from extracts of the crushed bodies of ants, and its name stems from the Latin word *formica,* meaning "ant." Formic acid is the strongest acid of the series, and acid strength decreases with increasing length of the hydrocarbon chain. (See Section 15.8.) Long chain carboxylic acids are called fatty acids. Sodium stearate, the sodium salt of stearic acid, $CH_3(CH_2)_{16}COOH$, is a typical component of soap. It cuts grime by simultaneously interacting with grease particles along its hydrocarbon tail and with water molecules at its carboxylate ion end group to make the grease soluble in water.

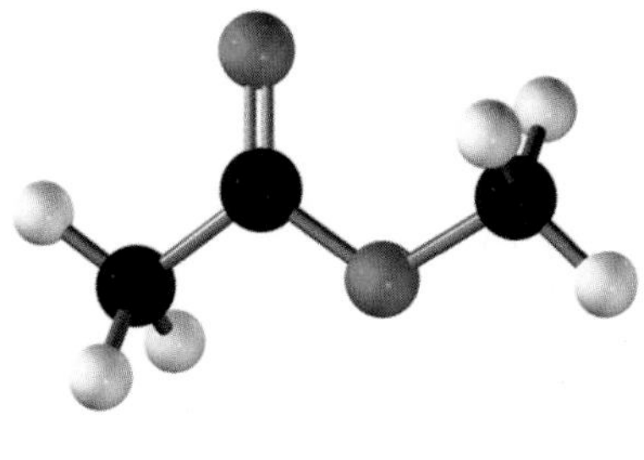

$$H_3C-C(=O)-O-CH_3$$

FIGURE 7.28 Methyl acetate (an ester).

An **ester** is created when a carboxylic acid reacts with an alcohol or a phenol, with water forming as the coproduct in a condensation reaction. The bonding scheme in esters is a variation of that shown in Figure 7.26, in which the singly bonded oxygen is sp^3 hybridized, with unshared pairs occupying two of the hybrid orbitals. An example is the condensation of acetic acid with methanol to give methyl acetate (Fig. 7.28):

$$H_3C-C(=O)-\boxed{OH + H}OCH_3 \longrightarrow H_3C-C(=O)-OCH_3 + H_2O$$

Esters are named by stating the name of the alkyl group of the alcohol (the methyl group in this case), followed by the name of the carboxylic acid with the ending *-ate* (acetate). One of the most important esters in commercial production is vinyl acetate, with the structure

$$H_3C-C(=O)-O-CH=CH_2$$

CONNECTION TO BIOLOGY

Functional Groups in Proteins

Proteins are natural polymers of amino acids linked by amide bonds to form long polyamide chains. Amides are functional groups produced by condensation reactions of amines with carboxylic acids, reactions that are analogous to those that form esters (see p. 315 for an introduction to amines and amides). The amide bond is generally referred to as a peptide bond in biochemistry and the polyamide chains are generally called polypeptides. Proteins serve a variety of biological functions: structural proteins are long fibers that make up the basic structures of hair, skin, and nails; enzymes are exquisitely selective catalysts for biochemical reactions, and hormones are chemical messengers that send signals to trigger specific biological responses. The structures and functions of proteins are, in large part, determined by the nature of the functional groups of the amino acids from which they are constructed and by the sequences in which they are assembled. An introduction to the structure and function of proteins provides an excellent opportunity to illustrate the utility of the functional group concept in chemistry and biology.

The general structures of the biologically important α-amino acids are represented by the reactants in the condensation reaction shown below.

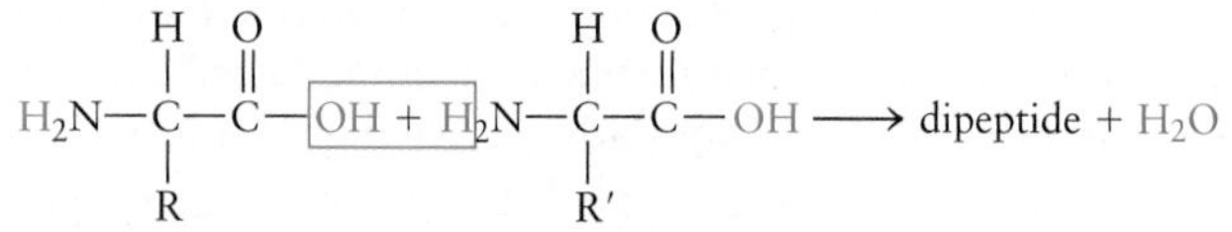

Condensation

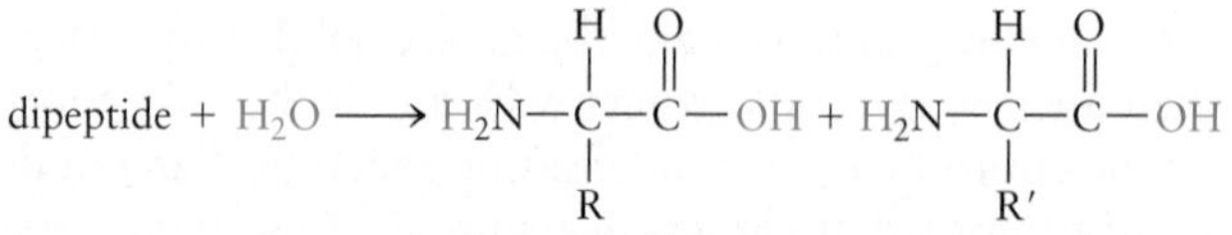

Hydrolysis

There are four different functional groups bonded to the central (α) carbon atom: a carboxylic acid group, an amine group, a hydrogen atom, and an R group, which is generally referred to as a side chain in biochemistry. Amino acids are bifunctional molecules, which allows them to form proteins via condensation reactions as shown. The dipeptide, whose structure is not shown, has a free amine group on the left side and a free carboxylic acid group on the right side, which are available to react with the complementary functional groups of other amino acids to form polypeptides of increasing size through condensation polymerization reactions of this type. The reverse hydrolysis reactions break up proteins into smaller polypeptides and even down to their constituent amino acids.

The 20 common amino acids differ only in the nature of the R groups attached to the α carbon atom, but those differences lead to a great diversity in the properties of proteins from which they are constructed. The side chains fall into several classes (see Table 23.2) that are defined primarily by their polarity but also by their acid/base character. The nonpolar alkane and aromatic hydrocarbon side chains are called hydrophobic (water avoiding) whereas the water-soluble side chains like alcohols, carboxylic acids, and amines are called hydrophilic (water seeking).

Proteins that consist of a single polypeptide chain are characterized by three levels of structure, whereas those that contain two or more polypeptide subunits, like hemoglobin (see the *Connection to Biology* in Chapter 14) have an additional level of organization. The primary structure is simply the protein's amino acid sequence. The three-dimensional structure of the enzyme chymotrypsin (Fig. C7.1) provides an excellent example of secondary and tertiary protein structures.

Secondary structures are determined primarily by the extent of inter-chain hydrogen bonding; the α-helix and β-pleated sheets (see Figs. 23.19 and 23.20) are examples of highly organized secondary structures that contrast with less highly organized structures called random coils. Chymotripsin's α-helical regions are represented by the coiled ribbons, β-pleated sheets by the broad

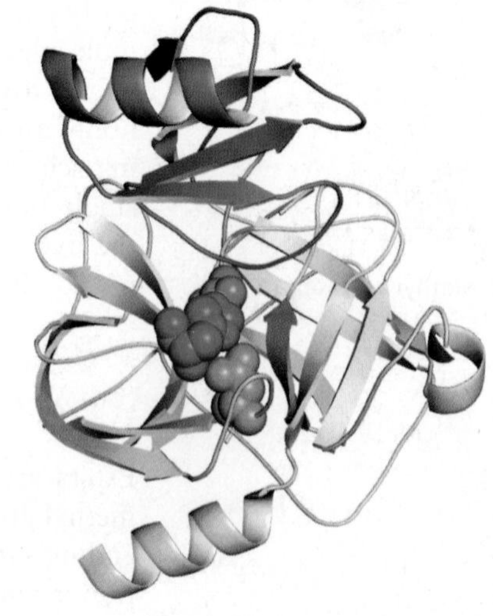

FIGURE C7.1 Three-dimensional structure of the enzyme chymotrypsin. (Adapted from Garrett/Grisham, *Biochemistry, 4e*, Figure 4.17, p. 438, Cengage Learning, 2009, © Cengage Learning)

arrows, and random coils by the thin tubes. Proteins fold to form a variety of global three-dimensional (tertiary) structures that are determined, to a large degree, by the locations of their hydrophobic and hydrophilic regions in the sequence. The hydrophobic regions of proteins tend to be located in the interior of folded proteins, whereas the hydrophilic regions tend to be found on the exterior, where they can interact favorably with water and solvated ions. The folded tertiary structure of chymotrypsin shown is the result of these interactions.

The active site of an enzyme is the site at which catalysis occurs. The active site in chymotrypsin comprises three polar amino acids (the catalytic triad) located in a hydrophobic pocket of the enzyme as shown by the colored space-filling models in the center of the enzyme structure in Fig. C7.1 and in the ball-and-stick models in the expanded view in Figure C7.2 below. Acidic and basic side chains, such as serine (Ser) and aspartic acid (Asp) help catalyze biochemical reactions by transferring protons, as exemplified by the hydrolysis reaction catalyzed by chymotrypsin.

Chymotrypsin is a digestive enzyme that catalyzes the hydrolysis of peptide bonds in proteins; it is a member of a class of proteins called serine proteases because the amino acid serine present in the active site plays a critical role in the reaction mechanism. The selectivity of different serine proteases for specific side chains is remarkable: the hydrophobic pocket in chymotrypsin can accommodate aromatic amino acids, as shown,

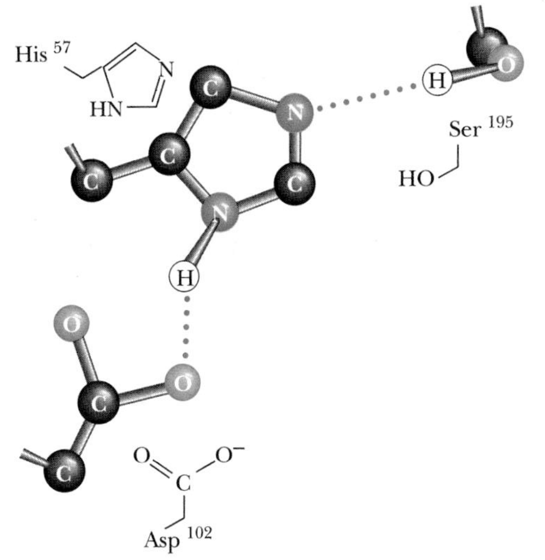

FIGURE C7.2 Expanded view of the acive site in chymotrypsin. (The ball-and-stick models omit H atoms except those in the hydrogen bonds between the three amino acids.)

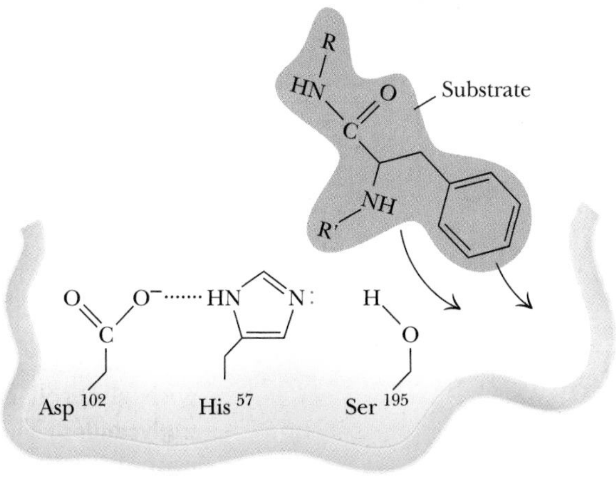

FIGURE C7.3 Binding of the amino acid phenylalanine in the active site of chymotrypsin. (Adapted from Garrett/Grisham, *Biochemistry, 4e*, Figure 4.18, p. 438, Cengage Learning, 2009, © Cengage Learning.)

whereas a negatively charged amino acid in the active site of the closely related enzyme trypsin binds positively charged amino acids such as arginine and histidine. Binding of the amino acid phenylalanine (the substrate) in the active site is illustrated in Figure C7.3.

The mechanism by which chymotrypsin hydrolyzes peptide bonds is described by the following abbreviated reaction sequence: 1) The aromatic amino acids phenylalanine (shown in the figure) or tyrosine bind in the active site with their aromatic side chains in the hydrophobic pocket. 2) A proton is transferred from the serine —OH group to a nitrogen atom of the nearby histidine residue, releasing a pair of electrons to form a covalent anionic intermediate. 3) Proton transfer from the histidine residue to the amino group of the substrate induces electron transfer from the oxygen lone pair on the anion to the amide C—N bond, cleaving that bond and releasing the amino-terminated fragment. 4) The carboxyl-terminated fragment covalently binds as an ester to serine. 5) Water finally hydrolyzes the ester bond, releasing the carboxylic acid product. This mechanism, one of the most well established in all of enzyme catalysis, shows the importance of: 1) hydrophobic interactions that allow the active site to selectively bind aromatic amino acids, and 2) spatial organization of the three amino acids in the active site that facilitates bond cleavage by proton transfer among both acidic and basic functional groups. Many enzyme-catalyzed reactions use acidic functional groups in this way to shuttle protons and catalyze the hydrolysis of amide and ester bonds in natural biopolymers.

Despite its name, vinyl acetate is not prepared by the reaction of acetic acid with an alcohol but rather with ethene and oxygen over a catalyst such as $CuCl_2$ and $PdCl_2$:

$$H_3C{-}\overset{\overset{\displaystyle O}{\|}}{C}{-}OH + CH_2{=}CH_2 + \tfrac{1}{2}O_2 \xrightarrow{CuCl_2} H_3C{-}\overset{\overset{\displaystyle O}{\|}}{C}{-}O{-}CH{=}CH_2 + H_2O$$

Esters are colorless, volatile liquids that often have pleasant odors. Many esters occur naturally in flowers and fruits. Isoamyl acetate (Fig. 7.29a) is generated in apples as they ripen and contributes to the flavor and odor of the fruit. Benzyl acetate, the ester formed from acetic acid and benzyl alcohol (see Fig. 7.29b), is a major component of oil of jasmine and is used in the preparation of perfumes.

Animal fats and vegetable oils are triesters of long-chain carboxylic acids with glycerol, $HOCH_2CH(OH)CH_2OH$, a trialcohol; they are referred to as **triglycerides.** These are energy-storage molecules of biological origin. A large proportion of sunflower oil is an oily liquid composed of molecules with the structural formula

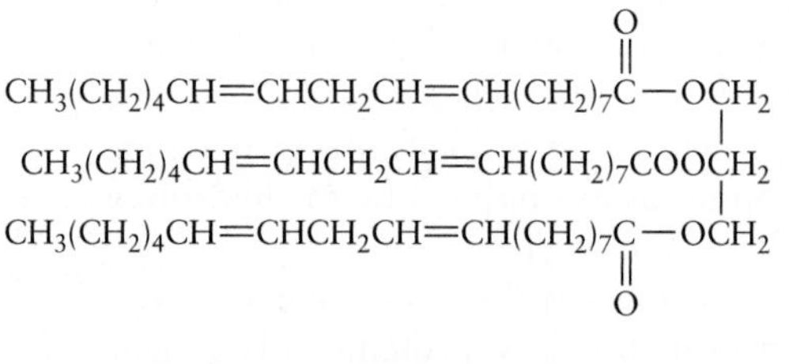

Molecules of this type are called polyunsaturated because they have more than one double bond. Butter is a mixture of triglycerides, many of which are **saturated** because their hydrocarbon chains contain no double bonds. Hydrogen is used in food processing to convert unsaturated liquid vegetable oils to saturated solids. **Hydrogenation** of sunflower oil with 6 mol H_2 in the presence of a catalyst saturates it, and the product has a high enough melting point to make it a solid at room conditions. The use of solid fats (or solidified oils) has advantages in food processing and preservation; therefore, "hydrogenated vegetable oil" is an ingredient in many foodstuffs. The distinction between fats and oils is a practical one: fats are solid at room temperature, and oils are liquids. Heavy consumption of saturated fats has been linked to diseases of the heart and circulatory system, and the presence of *un*hydrogenated (polyunsaturated) oils in foods is now extensively advertised. *Trans* fats, which have only one double

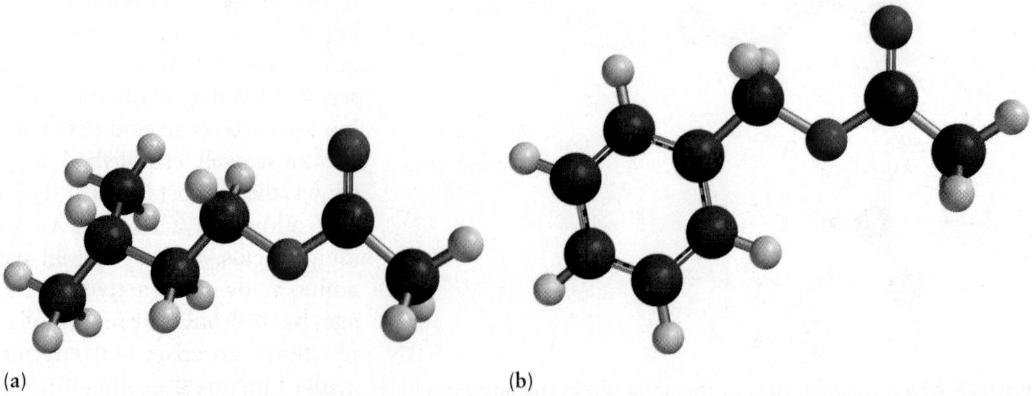

(a) (b)

FIGURE 7.29 The structures of (a) isoamyl acetate, $CH_3COO(CH_2)_2CH(CH_3)_2$, and (b) benzyl acetate, $CH_3COOCH_2C_6H_5$.

bond with hydrocarbon chains in the *trans* configuration, have recently been implicated in heart disease, and a ban on their use in food preparation is under consideration. When a triglyceride is hydrolyzed (by adding water to break bonds, the reverse of condensation) through addition of sodium hydroxide, the ester bonds are broken and glycerol and sodium salts of long-chain carboxylic acids are produced. This reaction is the basis for traditional soap making through the addition of lye (sodium hydroxide) to animal fats.

7.6.6 Amines and Amides

Amines are derivatives of ammonia with the general formula R_3N, where R can represent a hydrocarbon group or a hydrogen atom. If only one hydrogen atom of ammonia is replaced by a hydrocarbon group, the result is a *primary amine*. Examples include ethylamine and aniline:

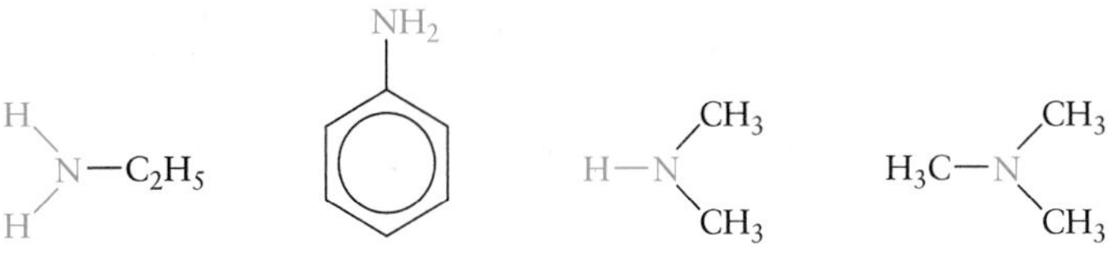

Ethylamine Aniline Dimethylamine Trimethylamine

If two hydrocarbon groups replace hydrogen atoms in the ammonia molecule, the compound is called a *secondary amine* (such as dimethylamine), and three replacements make a *tertiary amine* (trimethylamine). You should draw Lewis dot diagrams for several amines and recognize that the nitrogen atom is sp^3 hybridized and forms three σ bonds with one unshared pair occupying the fourth hybrid orbital. Amines are bases, like ammonia, because the lone pair on the nitrogen atom can act either as a hydrogen ion acceptor or as an electron pair donor (see Section 15.1).

Primary or secondary amines (or ammonia itself) can react with carboxylic acids to form amides, in condensation reactions that are analogous to the formation of esters from alcohols and carboxylic acids. An example of amide formation is

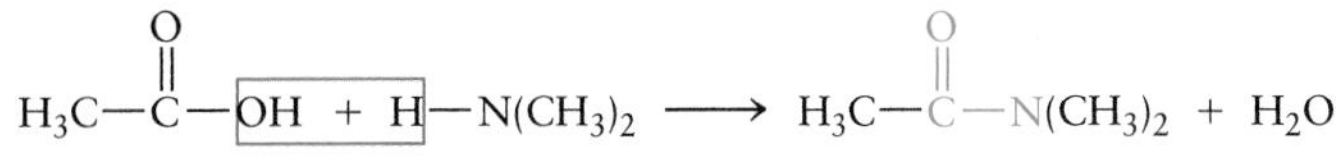

If ammonia is the reactant, an $-NH_2$ group replaces the $-OH$ group in the carboxylic acid as the amide is formed:

Amide linkages are present in the backbone of every protein molecule and are very important in biochemistry, where the structure of the molecule strongly influences its function (see Section 23.4). The bonding scheme in amides has two resonance structures; both are shown in Figure 7.30. In one resonance structure the doubly bonded carbon and oxygen atoms are sp^2 hybridized and the singly bonded nitrogen atom is sp^3 hybridized with an unshared pair occupying one of the hybrid orbitals. In the other, the N atom is double bonded to the C atom, and has no lone pair, so is sp^2 hybridized, while the CO bond is a single one. The resulting π bonding is delocalized, much like that in NO_2^- in Figure 6.52. The four bonds of the amide group lie in the same plane, as shown in Figure 7.30, a fact that has important consequences for the structures of proteins.

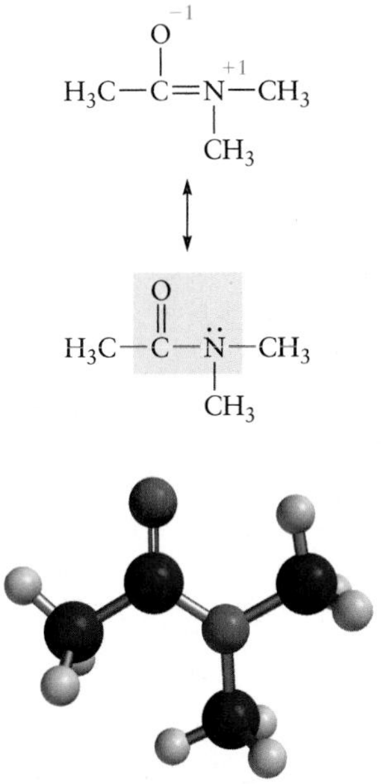

FIGURE 7.30 Bonding in dimethyl acetamide, an amide. The amide linkage is planar, in agreement with the expectation of delocalized π bonding.

7.7 Pesticides and Pharmaceuticals

Most of the organic compounds discussed so far in this chapter are relatively small molecules that are produced in large volume. Molecules like these are starting materials for the synthesis of numerous structurally more complex organic compounds with applications in agriculture, medicine, and consumer products. This section discusses a selection of these compounds, all of which are used in agriculture or as pharmaceuticals. Some of the structures and syntheses of these compounds are intricate; do not try to memorize them. Your goals, instead, should be to recognize the hydrocarbon frameworks of the molecules, to recognize functional groups, and to begin to appreciate the extremely diverse structures and properties of organic compounds.

Chemists have developed a shorthand notation to represent the structures of complex organic molecules. These line structures which focus attention to the most important structural elements, are illustrated in Figure 7.31. In this notation, the symbol "C" for a carbon atom is omitted, and only the C—C bonds are shown. A carbon atom is assumed to lie at each end of the line segments that represent bonds. In addition, symbols for the hydrogen atoms attached to carbon atoms are omitted. Terminal carbon atoms (those at the end of chains) and their associated hydrogen atoms are shown explicitly. To generate the full structure (and the molecular formula) from line structures, carbon atoms must be inserted at the end of each bond, and enough hydrogen atoms must be attached to each carbon atom to satisfy its valence of four.

7.7.1 Insecticides

The chemical control of insect pests dates back thousands of years. The earliest insecticides were inorganic compounds of copper, lead, and arsenic, as well as some naturally occurring organic compounds such as nicotine (Fig. 7.32a). Few of these "first-generation" insecticides are in use today because of their adverse side effects on plants, animals, and humans.

Controlled organic syntheses developed after World War II gave rise to a second generation of insecticides. The success of these agents led to rapid growth in the use of chemicals for insect control. The leading insecticide of the 1950s and 1960s was DDT (an abbreviation for dichlorodiphenyltrichloroethane; see Fig. 7.32b). DDT was extremely important worldwide in slowing the spread of typhus (transmitted by body lice) and malaria (transmitted by

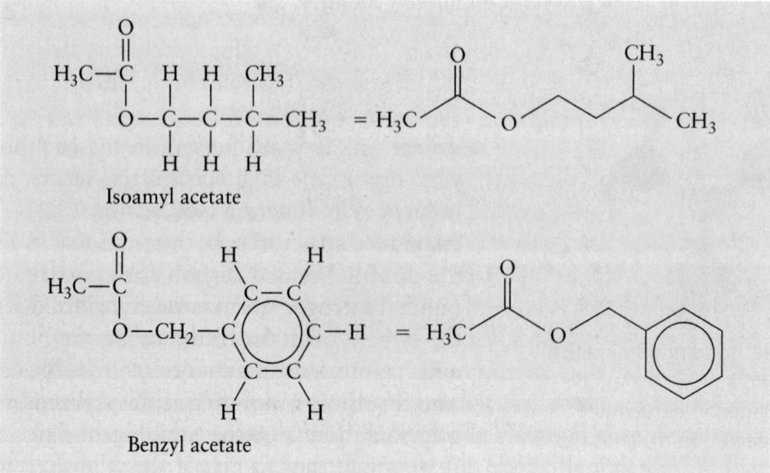

FIGURE 7.31 Extended molecular structures for two esters are shown on the left side of the figure with their associated line structures shown on the right. Carbon atoms are assumed to lie at the intersections of the lines representing bonds, and there are enough hydrogen atoms attached to each carbon atom to satisfy its valence. It is customary to show the carbon and hydrogen atoms of terminal methyl groups, however.

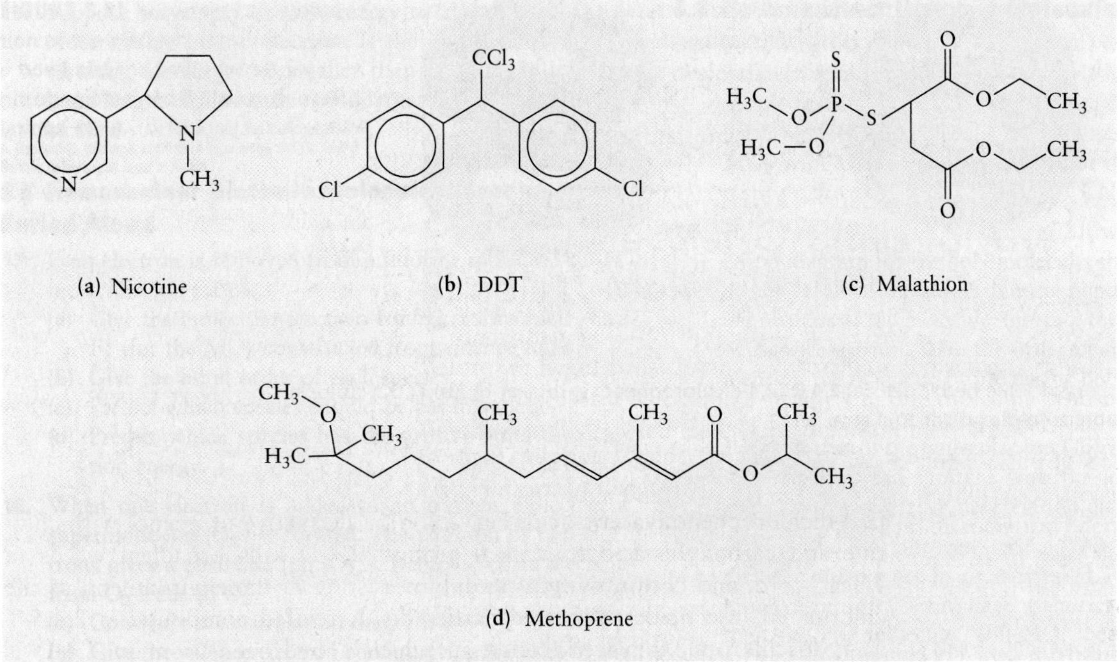

FIGURE 7.32 Structures of several insecticides: (a) nicotine; (b) dichlorodiphenyltrichloroethane (DDT); (c) malathion; and (d) methoprene.

mosquitoes). But mosquitoes developed strong resistance to DDT, and its use was banned in the United States in 1972 because of its adverse effects on birds, fish, and other life-forms that can accumulate DDT to high concentrations. Many other chlorine-substituted hydrocarbons are no longer used as insecticides for the same reason and organophosphorus compounds are used widely instead. The structure of the insecticide malathion, in which phosphorus appears with organic functional groups, is shown in Figure 7.32c. Note the two ester groups, the two kinds of sulfur atoms, and the "expanded octet" (see valence shell expansion in Section 3.10) on the central phosphorus atom that lets it form five bonds.

Second-generation insecticides frequently kill beneficial insects along with the pests unless they are applied at the right times and in properly controlled doses.Third-generation insecticides are more subtle. Many of these are based on sex attractants (to collect insects together in one place before exterminating them or to lead them to mate with sterile partners) or juvenile hormones (to prevent insects from maturing and reproducing). These compounds have the advantages of being specific against the pests and doing little or no harm to other organisms. Moreover, they can be used in small quantities, and they degrade rapidly in the environment. An example is the juvenile hormone methoprene (see Fig. 7.32d), which is used in controlling mosquitoes. It consists of a branched dialkene chain with a methyl ether (methoxy) and an isopropyl-ester functional group.

7.7.2 Herbicides

Chemical control of weeds, together with use of fertilizers, has contributed to the "green revolution" that began in the 1940s and during which agricultural productivity has increased dramatically throughout the world. The first herbicide of major importance, introduced in 1945 and still in use today, was 2,4-D

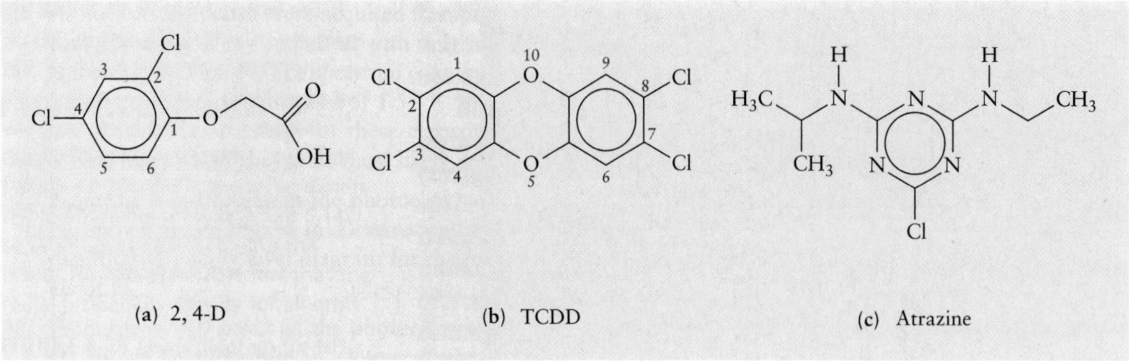

FIGURE 7.33 Structures of some herbicides. (a) 2,4-D (2,4-dichlorophenoxyacetic acid); (b) TCDD ("dioxin," or 2,3,7,8-tetrachlorodibenzo-*p*-dioxin); and (c) atrazine.

(2,4-dichlorophenoxyacetic acid; Fig. 7.33a), a derivative of phenol with chlorine and carboxylic acid functional groups. 2,4-D kills broadleaf weeds in wheat, corn, and cotton without unduly persisting in the environment, as the chlorinated insecticides discussed earlier do. A related compound is 2,4,5-T (2,4,5-trichlorophenoxyacetic acid), in which a hydrogen atom in 2,4-D is replaced by a chlorine atom. Much attention has been given recently to TCDD ("dioxin," or 2,3,7,8-tetrachlorodibenzo-*p*-dioxin; see Fig. 7.33b), which occurs as a trace impurity (10–20 ppb by mass) in 2,4,5-T and which, in animal tests, is the most toxic compound of low-to-moderate molar mass currently known. The use of 2,4,5-T as a defoliant (Agent Orange) during the Vietnam War led to a lawsuit by veterans who claimed that health problems arose from contact with the traces of TCDD present in the 2,4,5-T. Although such a direct connection has never been proved, the use of chlorinated phenoxy herbicides has decreased, and that of other herbicides, such as atrazine (see Fig. 7.33c), has grown.

7.7.3 Analgesics

Drugs that relieve pain are called analgesics, the oldest of which was probably a tea brewed from the bark of a willow tree and "prescribed" by the physician Hippocrates in the fourth century BC. The active ingredient in willow bark tea is salicylic acid, which is an effective analgesic but can cause severe intestinal bleeding, so chemists in the late 1800s sought to modify salicylic acid to preserve its desirable properties while minimizing its undesirable side effects. The German chemist, Fritz Hoffman, working for the Bayer chemical company, succeeded in making an ester of salicylic acid by reacting its alcohol functional group with acetic acid to make acetylsalicylic acid, the structure of which is shown in Figure 7.34a. You probably know this molecule by its generic name, aspirin, or by its trade name Bayer Aspirin.

Aspirin is not only an analgesic but also reduces fever (an antipyretic) and inflammation (an anti-inflammatory). Two related drugs, acetaminophen and ibuprofen, are shown in Figures 7.34b and 7.34c. This family of drugs provides an excellent example of structure-property relationships in organic chemistry and how subtle changes in the functional groups arranged around the hydrocarbon framework can dramatically alter biological function. All three of these molecules (called nonsteroidal anti-inflammatories or NSAIDs) work by binding to a site in a family of enzymes called cyclooxygenases (COX) that catalyze the synthesis of an important family of hormones called prostaglandins. Hormones

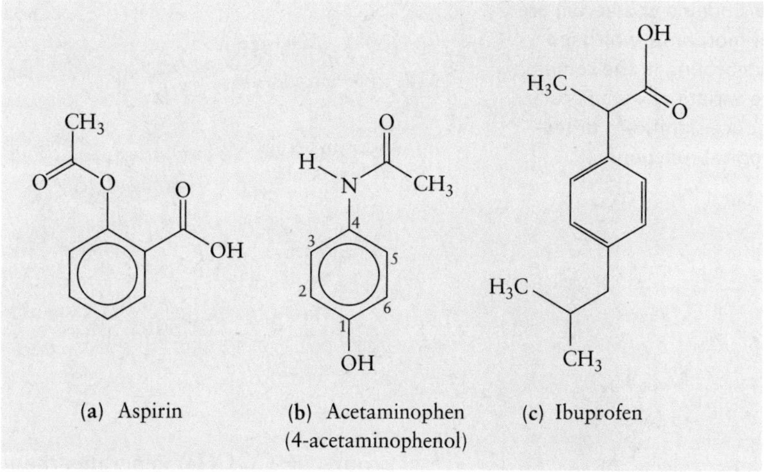

FIGURE 7.34 The molecular structures of some analgesics: (a) aspirin; (b) acetaminophen; and (c) ibuprofen.

are chemical messengers that instruct cells to carry out specific functions, for instance, to suppress inflammation or block the transmission of pain signals. These so-called COX inhibitors block prostaglandin synthesis by occupying the active site of these enzymes, inhibition being an important mode mechanism of drug action. Note the similarity of the hydrocarbon frameworks in these molecules; they contain the phenyl structural group, which is a hydrocarbon that is soluble in fatty cell membranes where the COX enzymes reside. The frameworks all have approximately the same three-dimensional geometries that allow them to fit in the binding site like a hand in a glove. As stated earlier, it is the framework that determines the three-dimensional structure of organic molecules but the functional groups that determine the chemistry. Ibuprofen is the best anti-inflammatory of the three drugs, with acetaminophen showing almost no anti-inflammatory activity at all. Acetaminophen (Fig. 7.35), however, does not inhibit clotting to the extent that aspirin does, which is why "more hospitals use Tylenol."

A much more powerful pain reliever, which is available only by prescription because of its addictive properties, is morphine, which acts on the central nervous system, apparently because its shape fits a receptor site on the nerve cell, and blocks the transmission of pain signals to the brain. Its structure contains five interconnected rings. A small change (the replacement of one —OH group by an $-OCH_3$ group, giving a methyl ether) converts morphine into codeine, a prescription drug used as a cough suppressant. Replacing *both* —OH groups by

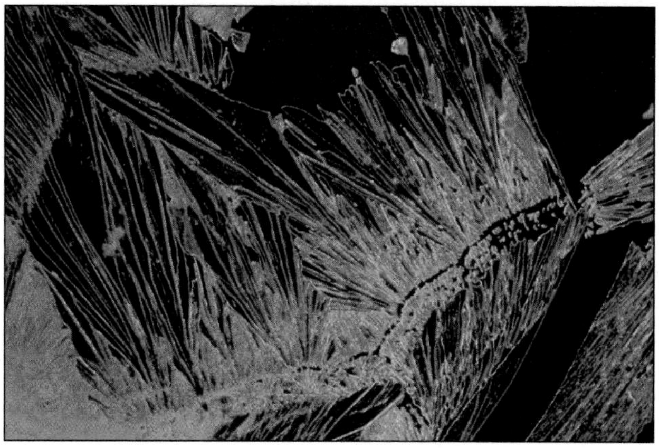

FIGURE 7.35 Crystals of 4-acetaminophenol (Tylenol) viewed under polarized light.

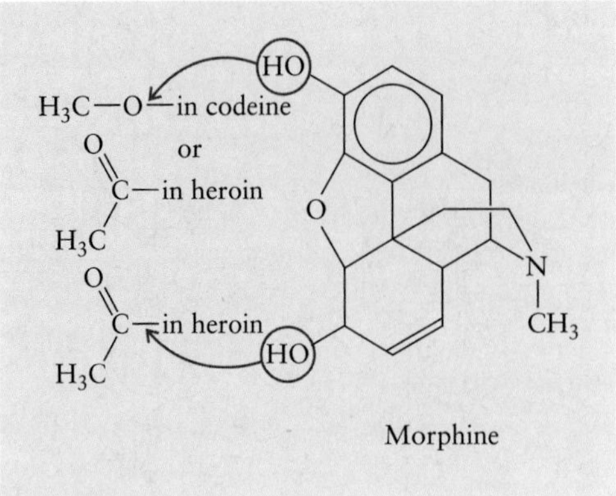

FIGURE 7.36 Codeine and heroin are derivatives of morphine, which is a natural product found in the common poppy. Subtle variations in molecular structure produce significant differences in biological function.

acetyl groups ($-COCH_3$) generates the notoriously addictive substance heroin (Fig. 7.36).

7.7.4 Antibacterial Agents

The advent of antibacterial agents changed the treatment of bacterial diseases such as tuberculosis and pneumonia dramatically beginning in the 1930s. The first "wonder drug" was sulfanilamide (Fig. 7.37a), a derivative of aniline. Other "sulfa drugs" are obtained by replacing one of the hydrogen atoms on the sulfonamide group by other functional groups. Bacteria mistake sulfanilamide for *p*-aminobenzoic acid, a molecule with a very similar shape but a carboxylic acid ($-COOH$) group in place of the $-SO_2NH_2$ group. The drug then interferes with the bacterium's synthesis of folic acid, an essential biochemical, so the organism dies. Mammals do not synthesize folic acid (they obtain it from their diet), so they are not affected by sulfanilamide.

The penicillin molecule (see Fig. 7.37b) contains an amide linkage that connects a substituted double ring (including sulfur and nitrogen atoms) to a benzyl (phenylmethyl) group. It is a natural product formed by certain molds. Although the total synthesis of penicillin was achieved in 1957, that chemical route is not competitive economically with biosynthesis via fermentation. The mold grows

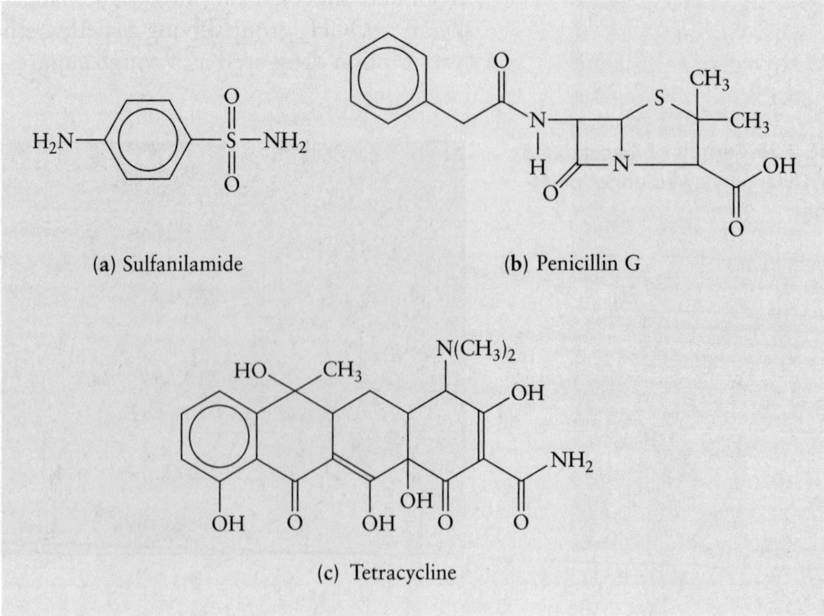

FIGURE 7.37 Molecular structures of some antibiotics: (a) sulfanilamide; (b) penicillin G; and (c) tetracycline.

FIGURE 7.38 Fermentation tanks used in modern penicillin production.

for several days in tanks that may hold up to 100,000 L of a fermentation broth (Fig. 7.38). The penicillin is later separated by solvent extraction. Penicillin functions by deactivating enzymes responsible for building cell walls in the bacteria. Derivatives of natural penicillin have been developed and are commercially available.

Finally we mention the tetracyclines, which are derivatives of the four-ring aromatic compound represented in Figure 7.37c. These drugs have the broadest spectrum of antibacterial activity found to date.

7.7.5 Steroids

The **steroids** are a family of naturally occurring compounds with a wide variety of functions. Most of them are synthesized from cholesterol. The structure of cholesterol (Fig. 7.39a) contains a group of four fused hydrocarbon rings (3 six-atom rings and 1 five-atom ring). All steroids possess this "steroid nucleus." Cholesterol itself is present in all tissues of the human body. When present in excess in the bloodstream, it can accumulate in the arteries, restricting the flow of blood and leading to heart attacks. Its derivatives have widely different functions. The hormone cortisone (see Fig. 7.39b), which is secreted by the adrenal glands, regulates the metabolism of sugars, fats, and proteins in all body cells. As a drug, cortisone reduces inflammation and moderates allergic responses. It often is prescribed to combat arthritic inflammation of the joints.

The human sex hormones are also derivatives of cholesterol. Here the resourcefulness of nature for building compounds with quite different functions from the same starting material is particularly evident. The female sex hormone progesterone (see Fig. 7.39c) differs from the male sex hormone testosterone only by replacing an acetyl ($-C(O)CH_3$) group by a hydroxyl ($-OH$) group. Oral contraceptives are synthetic compounds with structures that are closely related to progesterone.

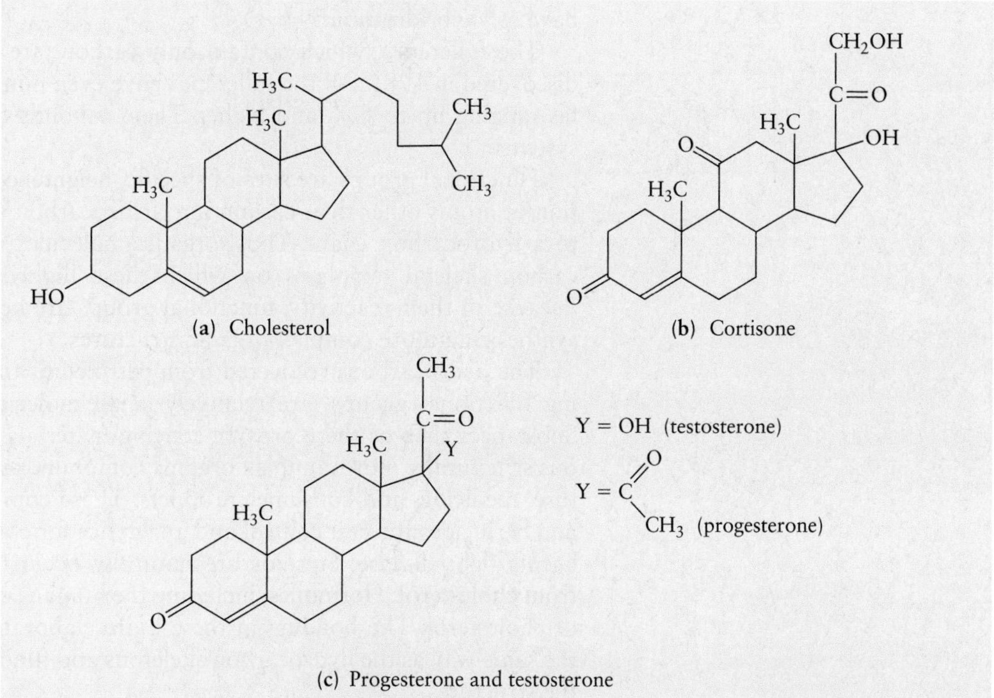

FIGURE 7.39 Molecular structures of some steroids: (a) cholesterol; (b) cortisone; and (c) progesterone and testosterone.

CHAPTER SUMMARY

The element carbon has a rich and varied chemistry because of its location in the periodic table. As a Group IV element, each carbon atom forms four covalent bonds, more than any other second-period element. In consequence of its intermediate value of electronegativity, carbon can bond with the more electronegative elements, such as oxygen, nitrogen, and the halogens, as well as the more electropositive elements such as hydrogen and some of the heavy metals, by virtue of its intermediate electronegativity. Carbon atoms also bond with other carbon atoms, forming single, double, and triple bonds.

The hydrocarbons—molecules that contain only carbon and hydrogen—are fundamental to organic chemistry because they provide the archetypes for models of bond formation and the starting point for synthesis of other organic molecules. The hydrocarbons fall naturally into three families, based on their chemical properties and types of bonds.

The alkanes, also called saturated hydrocarbons, have only single bonds. The bonds are described by sp^3 hybridization of the carbon atoms. When the carbon atoms are not bonded in a linear sequence, cases can occur in which two molecules with the same formula can have different structures, called geometrical isomers, with quite distinct properties.

Unsaturated hydrocarbons have double or triple bonds between carbon atoms. The alkenes have C=C double bonds, described by sp^2 hybridization of the carbon atoms. The alkynes have C≡C triple bonds, described by *sp* hybridization of the carbon atoms. Because bond rotation does not occur readily about a carbon–carbon double bond, many alkenes exist in contrasting isomeric forms, depending on whether bonding groups are on the same (*cis*) or opposite (*trans*) sides of the double bond. When two or more double or triple bonds are separated by one single bond, the *p* orbitals form a conjugated system, in which the de-localized π orbitals are best described by MO theory.

The aromatic hydrocarbons are conjugated cyclic structures in which the π bonding is described through de-localized MOs formed at carbon atoms that have sp^2 hybridization.

The fullerenes, which contain only carbon, are allotropic forms of carbon discovered in 1985. All the fullerenes have even numbers of atoms, with formulas ranging up to C_{400} and higher. Their π bonds form conjugated π electron systems.

Functional groups are sites of specific, heightened reactivity caused by insertion of atoms other than carbon into hydrocarbon structures or by attachment to a hydrocarbon chain. Thus, organic molecules are conveniently viewed as carbon skeletal templates on which these highly reactive sites are located. Because of their reactivity, functional groups are key elements in strategies for synthesizing more complex organic structures.

The hydrocarbons recovered from petroleum, and their derivatives containing functional groups, are relatively small molecules with simple structures. Substances such as these provide starting materials for the synthesis of numerous structurally more complex organic compounds with applications in agriculture, medicine, and consumer products. These compounds include insecticides and herbicides for pest control and analgesics for controlling pain. Antibacterial agents fight disease. Steroids are naturally occurring compounds that derive from cholesterol. Hormones, including the human sex hormones, are derivatives of cholesterol. The bonding in these more elaborate structures is explained in the same way as the hydrocarbon skeletons and functional groups that comprise the structures.

CONCEPTS AND SKILLS

Sections 7.1–7.4 – Petroleum Refining and the Hydrocarbons, Alkanes, Alkenes, Alkynes, and Aromatic Hydrocarbons

Identify important classes of hydrocarbons found in crude oil and natural gas, and describe their behavior in combustion reactions (Problems 1–4).

- Hydrocarbons are the principal components of crude oil and natural gas; they are used as fuels and as building blocks for the synthesis of organic molecules of industrial and commercial importance, such as polymers and pharmaceuticals.
- Alkanes are saturated hydrocarbons with the general formula C_nH_{2n+2}, in which every carbon atom makes four bonds either to other carbon atoms or to hydrogen.
- Alkenes have one or more C=C double bonds.
- Alkynes have one or more C≡C triple bonds.
- Aromatic hydrocarbons are planar ring structures with $4n + 2$ delocalized π electrons, with n being an integer.

Write names and structural formulas for hydrocarbons (Problems 5–12).

- Hydrocarbons are named using a combination of prefixes, which depend on the number of carbon atoms in a molecule, and suffixes that identify the hydrocarbon class to which it belongs. The molecule is named after its longest chain using the following rules:
- Hydrocarbons that contain 1–4 carbon atoms in a chain are named after the hydrocarbons from which they are derived: methane, ethane, propane, and butane, respectively.
- Hydrocarbons that contain five or more carbon atoms are named using the Greek prefixes *penta-*, *hexa-*, *hepta-*, etc., in which the Greek prefix identifies the number of carbon atoms in the molecule.
- The ending *-ane* is replaced by *-ene* for alkenes and *-yne* for alkynes.
- Branches are named by identifying the radical and the position at which it is located along the main chain: 2-methylpentane, for example, has a methyl group attached to the second carbon atom in the pentane chain.

Describe the bonding in hydrocarbons using the VB model with hybrid orbitals on the carbon atoms (Problems 13–14).

- Bonding in alkanes is described using sp^3 hybrid orbitals on each carbon atom to make four σ bonds either to other carbon atoms or to hydrogen.
- Bonding in alkenes is described using sp^2 hybrids orbitals on each carbon atom to make three σ bonds either to other carbon atoms or to hydrogen and one π bond to an adjacent carbon atom. There is no rotation about the double bonds in alkenes, which makes it possible to have geometric isomers.
- Bonding in alkynes is described using sp hybrid orbitals on each carbon atom to make two σ bonds either to other carbon atoms or to hydrogen and two π bonds to an adjacent carbon atom.

Sections 7.3–7.5 – Alkenes, Alkynes, Aromatic Hydrocarbons, and Fullerenes

Discuss the delocalization of the π electrons in organic molecules and fullerenes (Problems 15–16).

- Alternating single and double bonds in linear or cyclic alkenes can be described using resonance structures in which the electron density is delocalized over the entire molecule.

Section 7.6 – Functional Groups and Organic Reactions

Identify important functional groups and outline chemical processes by which important chemical compounds are synthesized (Problems 17–26).

- Functional groups are atoms or groups of atoms with similar physical and chemical properties that are largely independent of the molecules of which they are a part.
- Important organic functional groups include carboxylic acids (weak acids), amines (weak bases), alcohols, esters (produced by reactions between carboxylic acids and alcohols), amides (produced by reactions between carboxylic acids and amines) and ethers, which are not particularly reactive species.

Describe the bonding in important functional groups using the VB model with hybrid orbitals on carbon, oxygen, and nitrogen atoms (Problems 27–30).

- The bonding in alkyl functional groups is described using sp^3 hybridization on each carbon, as is the bonding for all functional groups that involve only single bonds, for example, alcohols, ethers, and amines. The bonding in all functional groups that contain double bonds is described using sp^2 hybridization on the two atoms connected by the double bond; examples include aldehydes, ketones, carboxylic acids, esters, and amides.

Section 7.7 – Pesticides and Pharmaceuticals

Recognize and describe the shapes and functional groups for molecules used as pesticides and pharmaceuticals (Problems 31–36).

- Pesticides and pharmaceuticals are molecules that include functional groups such as phenyl groups, alcohols, ethers, aldehydes, ketones, carboxylic acids, amines, and amides. Many of these structures have more than one ring, which gives them characteristic three-dimensional shapes that are important for their biological functions.

PROBLEMS

Answers to problems whose numbers are red appear in Appendix G. Problems that are more challenging are indicated with asterisks.

7.1 Petroleum Refining and the Hydrocarbons

1. Is it possible for a gasoline to have an octane number exceeding 100? Explain.

2. Is it possible for a motor fuel to have a negative octane rating? Explain.

3. A gaseous alkane is burned completely in oxygen. The volume of the carbon dioxide that forms equals twice the volume of the alkane burned (the volumes are measured at the same temperature and pressure). Name the alkane and write a balanced equation for its combustion.

4. A gaseous alkyne is burned completely in oxygen. The volume of the water vapor that forms equals the volume of the alkyne burned (the volumes are measured at the same temperature and pressure). Name the alkyne and write a balanced equation for its combustion.

5. **(a)** Write a chemical equation involving structural formulas for the catalytic cracking of decane into an alkane and an alkene that contain equal numbers of carbon atoms. Assume that both products have straight chains of carbon atoms.
   (b) Draw and name one other isomer of the alkene.

6. **(a)** Write an equation involving structural formulas for the catalytic cracking of 2,2,3,4,5,5-hexamethylhexane. Assume that the cracking occurs between carbon atoms 3 and 4.
   (b) Draw and name one other isomer of the alkene.

7. Write structural formulas for the following:
   (a) 2,3-Dimethylpentane
   (b) 3-Ethyl-2-pentene
   (c) Methylcyclopropane
   (d) 2,2-Dimethylbutane
   (e) 3-Propyl-2-hexene
   (f) 3-Methyl-1-hexene
   (g) 4-Ethyl-2-methylheptane
   (h) 4-Ethyl-2-heptyne

8. Write structural formulas for the following:
   (a) 2,3-Dimethyl-1-cyclobutene
   (b) 2-Methyl-2-butene
   (c) 2-Methyl-1,3-butadiene
   (d) 2,3-Dimethyl-3-ethylhexane
   (e) 4,5-Diethyloctane
   (f) Cyclooctene
   (g) Propadiene
   (h) 2-pentyne

9. Write structural formulas for *trans*-3-heptene and *cis*-3-heptene.

10. Write structural formulas for *cis*-4-octene and *trans*-4-octene.

11. Name the following hydrocarbons:
   (a) 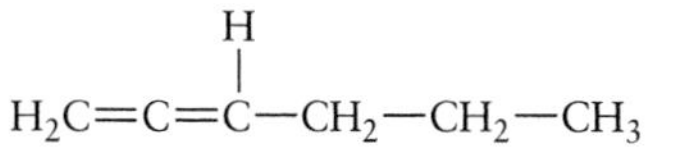
   (b) 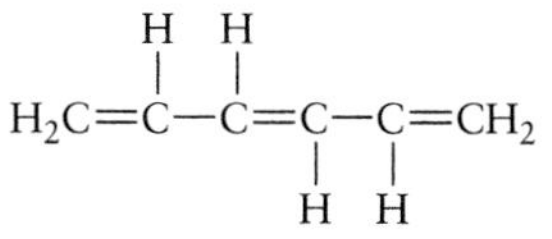
   (c) 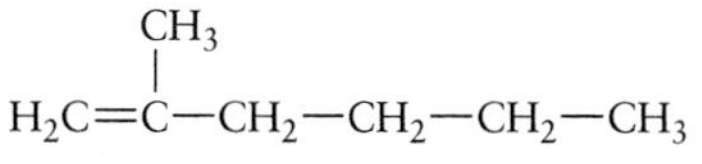
   (d) $CH_3-CH_2-C{\equiv}C-CH_2-CH_3$

12. Name the following hydrocarbons:
   (a) 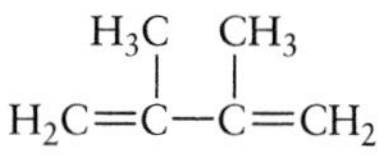
   (b) 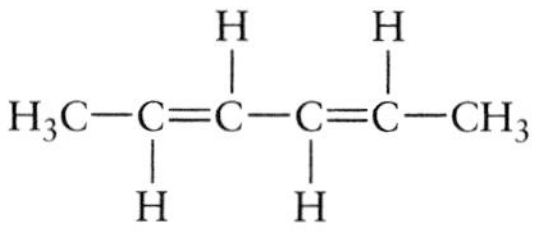
   (c) 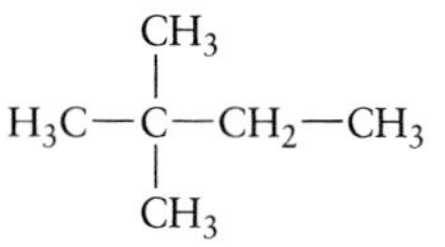
   (d) 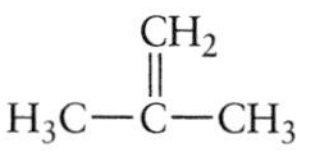

13. State the hybridization of each of the carbon atoms in the hydrocarbon structures in Problem 11.

14. State the hybridization of each of the carbon atoms in the hydrocarbon structures in Problem 12.

7.5 Fullerenes

15. To satisfy the octet rule, fullerenes must have double bonds. How many? Give a simple rule for one way of placing them in the structure shown in Figure 7.17a.

16. It has been suggested that a compound of formula $C_{12}B_{24}N_{24}$ might exist and have a structure similar to that of C_{60} (buckminsterfullerene).
   (a) Explain the logic of this suggestion by comparing the number of valence electrons in C_{60} and $C_{12}B_{24}N_{24}$.
   (b) Propose the most symmetric pattern of carbon, boron, and nitrogen atoms in $C_{12}B_{24}N_{24}$ to occupy the 60 atom sites in the buckminsterfullerene structure. Where could the double bonds be placed in such a structure?

7.6 Functional Groups and Organic Reactions

17. In a recent year, the United States produced 6.26×10^9 kg of 1,2-dichloroethane (sometimes still called by its old name of ethylene dichloride) and 15.87×10^9 kg ethene. Assuming that all significant quantities of 1,2-dichloroethane were produced from ethene, what fraction of the ethene production went into making 1,2-dichloroethane? What mass of chlorine was required for this conversion?

18. In a recent year, the United States produced 6.26×10^9 kg 1,2-dichloroethane and 3.73×10^9 kg vinyl chloride. Assuming that all significant quantities of vinyl chloride were produced from 1,2-dichloroethane, what fraction of the 1,2-dichloroethane production went into making vinyl chloride? What mass of hydrogen chloride was generated as a by-product?

19. Write balanced equations for the following reactions. Use structural formulas to represent the organic compounds.
    (a) The production of butyl acetate from butanol and acetic acid
    (b) The conversion of ammonium acetate to acetamide and water
    (c) The dehydrogenation of 1-propanol
    (d) The complete combustion (to CO_2 and H_2O) of heptane

20. Write balanced equations for the following reactions. Use structural formulas to represent the organic compounds.
    (a) The complete combustion (to CO_2 and H_2O) of cyclopropanol
    (b) The reaction of isopropyl acetate with water to give acetic acid and isopropanol
    (c) The dehydration of ethanol to give ethene
    (d) The reaction of 1-iodobutane with water to give 1-butanol

21. Outline, using chemical equations, the synthesis of the following from easily available petrochemicals and inorganic starting materials.
    (a) Vinyl bromide ($CH_2{=}CHBr$)
    (b) 2-Butanol
    (c) Acetone (CH_3COCH_3)

22. Outline, using chemical equations, the synthesis of the following from easily available petrochemicals and inorganic starting materials.
    (a) Vinyl acetate ($CH_3COOCH{=}CH_2$)
    (b) Formamide ($HCONH_2$)
    (c) 1,2-Difluoroethane

23. Write a general equation (using R to represent a general alkyl group) for the formation of an ester by the condensation of a tertiary alcohol with a carboxylic acid.

24. Explain why it is impossible to form an amide by the condensation of a tertiary amine with a carboxylic acid.

25. Calculate the volume of hydrogen at 0°C and 1.00 atm that is required to convert 500.0 g linoleic acid ($C_{18}H_{32}O_2$) to stearic acid ($C_{18}H_{36}O_2$). (See Sec. 9.1.)

26. A chemist determines that 4.20 L hydrogen at 298 K and a pressure of 1.00 atm is required to completely hydrogenate 48.5 g of the unsaturated compound oleic acid to stearic acid ($C_{18}H_{36}O_2$). How many units of unsaturation (where a unit of unsaturation is one double bond) are in a molecule of oleic acid? (See Sec. 9.1.)

27. Acetic acid can be made by the oxidation of acetaldehyde (CH_3CHO). Molecules of acetaldehyde have a $-CH_3$ group, an oxygen atom, and a hydrogen atom attached to a carbon atom. Draw the Lewis diagram for this molecule, give the hybridization of each carbon atom, and describe the π orbitals and the number of electrons that occupy each one. Draw the three-dimensional structure of the molecule, showing all angles.

28. Acrylic fibers are polymers made from a starting material called acrylonitrile, $H_2C(CH)CN$. In acrylonitrile, a $-C{\equiv}N$ group replaces a hydrogen atom on ethene. Draw the Lewis diagram for this molecule, give the hybridization of each carbon atom, and describe the π orbitals and the number of electrons that occupy each one. Draw the three-dimensional structure of the molecule, showing all angles.

29. Compare the bonding in formic acid (HCOOH) with that in its conjugate base formate ion ($HCOO^-$). Each molecule has a central carbon atom bonded to the two oxygen atoms and to a hydrogen atom. Draw Lewis diagrams, determine the steric numbers and hybridization of the central carbon atom, and give the molecular geometries. How do the π orbitals differ in formic acid and the formate molecular ion? The bond lengths of the C—O bonds in HCOOH are 1.23 (for the bond to the lone oxygen) and 1.36 Å (for the bond to the oxygen with a hydrogen atom attached). In what range of lengths do you predict the C—O bond length in the formate ion to lie?

30. Section 7.3 shows that the compound 2-butene exists in two isomeric forms, which can be interconverted only by breaking a bond (in that case, the central double bond). How many possible isomers correspond to each of the following chemical formulas? Remember that a simple rotation of an entire molecule does not give a different isomer. Each molecule contains a central C=C bond.
    (a) $C_2H_2Br_2$
    (b) C_2H_2BrCl
    (c) $C_2HBrClF$

7.7 Pesticides and Pharmaceuticals

31. **(a)** The insecticide methoprene (see Fig. 7.32d) is an ester. Write the structural formulas for the alcohol and the carboxylic acid that react to form it. Name the alcohol.
    (b) Suppose that the carboxylic acid from part (a) is changed chemically so that the OCH_3 group is replaced by a hydrogen atom and the COOH group is replaced by a CH_3 group. Name the hydrocarbon that would result.

32. **(a)** The herbicide 2,4-D (see Fig. 7.33a) is an ether. Write the structural formulas of the two alcohol or phenol compounds that, on condensation, would form this ether. (The usual method of synthesis does not follow this plan.)
(b) Suppose that hydrogen atoms replace the chlorine atoms and a $-CH_3$ group replaces the carboxylic acid group in the two compounds in part (a). Name the resulting compounds.

33. **(a)** Write the molecular formula of acetylsalicylic acid (see Fig. 7.34a).
(b) An aspirin tablet contains 325 mg acetylsalicylic acid. Calculate the number of moles of that compound in the tablet.

34. **(a)** Write the molecular formula of acetaminophen (see Fig. 7.34b).
(b) A tablet of Extra Strength Tylenol contains 500 mg acetaminophen. Calculate the chemical amount (in moles) of that compound in the tablet.

35. Describe the changes in hydrocarbon structure and functional groups that are needed to make cortisone from cholesterol (see Fig. 7.39).

36. Describe the changes in hydrocarbon structure and functional groups that are needed to make testosterone from cortisone (see Fig. 7.39).

ADDITIONAL PROBLEMS

37. *trans*-Cyclodecene boils at 193°C, but *cis*-cyclodecene boils at 195.6°C. Write structural formulas for these two compounds.

38. A compound $C_4H_{11}N$ is known from its reactivity and spectroscopic properties to have no hydrogen atoms attached directly to the nitrogen atom. Write all structural formulas consistent with this information.

39. A compound C_3H_6O has a hydroxyl group but no double bonds. Write a structural formula consistent with this information.

40. Consider the following proposed structures for benzene, each of which is consistent with the molecular formula C_6H_6.

(i)

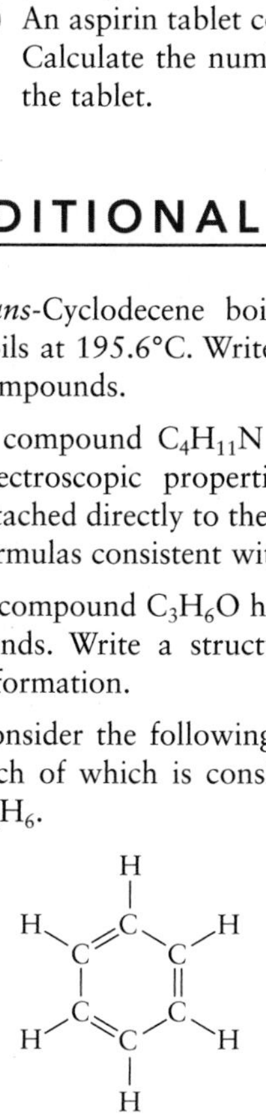

(ii)

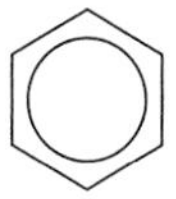

(iii)

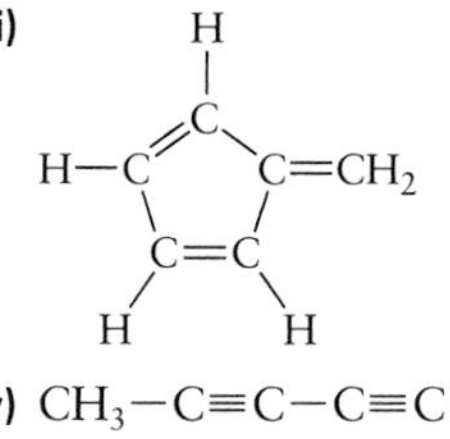

(iv) $CH_3-C\equiv C-C\equiv C-CH_3$
(v) $CH_2=CH-C\equiv C-CH=CH_2$

(a) When benzene reacts with chlorine to give C_6H_5Cl, only one isomer of that compound forms. Which of the five proposed structures for benzene are consistent with this observation?
(b) When C_6H_5Cl reacts further with chlorine to give $C_6H_4Cl_2$, exactly three isomers of the latter compound form. Which of the five proposed structures for benzene are consistent with this observation?

41. Acetyl chloride, CH_3COCl, reacts with the hydroxyl groups of alcohols to form ester groups with the elimination of HCl. When an unknown compound X with formula $C_4H_8O_3$ reacted with acetyl chloride, a new compound Y with formula $C_8H_{12}O_5$ was formed.
(a) How many hydroxyl groups were there in X?
(b) Assume that X is an aldehyde. Write a possible structure for X and a possible structure for Y consistent with your structure for X.

42. When an ester forms from an alcohol and a carboxylic acid, an oxygen atom links the two parts of each ester molecule. This atom could have come originally from the alcohol, from the carboxylic acid, or randomly from either. Propose an experiment using isotopes to determine which is the case.

43. Hydrogen can be added to a certain unsaturated hydrocarbon in the presence of a platinum catalyst to form hexane. When the same hydrocarbon is oxidized with $KMnO_4$, it yields acetic acid and butanoic acid. Identify the hydrocarbon and write balanced chemical equations for the reactions.

44. (a) It is reported that ethene is released when pure ethanol is passed over alumina (Al_2O_3) that is heated to 400°C, but diethyl ether is obtained at a temperature of 230°C. Write balanced equations for both of these dehydration reactions.
(b) If the temperature is increased well above 400°C, an aldehyde forms. Write a chemical equation for this reaction.

45. In what ways do the systematic developments of pesticides and of pharmaceuticals resemble each other, and in what ways do they differ? Consider such aspects as "deceptor" molecules, which are mistaken by living organisms for other molecules; side effects; and the relative advantages of a broad versus a narrow spectrum of activity.

46. The steroid stanolone is an androgenic steroid (a steroid that develops or maintains certain male sexual characteristics). It is derived from testosterone by adding a molecule of hydrogen across the C=C bond in testosterone.

(a) Using Figure 7.39c as a guide, draw the molecular structure of stanolone.

(b) What is the molecular formula of stanolone?

CUMULATIVE PROBLEMS

47. The structure of the molecule cyclohexene is

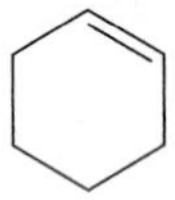

Does the absorption of ultraviolet light by cyclohexene occur at longer or at shorter wavelengths than the absorption by benzene? Explain.

48. The naphthalene molecule has a structure that corresponds to two benzene molecules fused together:

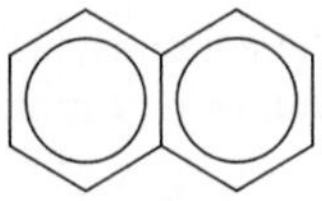

The π electrons in this molecule are delocalized over the entire molecule. The wavelength of maximum absorption in benzene is 255 nm. Will the corresponding wavelength in naphthalene be shorter or longer than 255 nm?

8 CHAPTER

Bonding in Transition Metal Compounds and Coordination Complexes

The colors of gemstones originate with transition-metal ions. Emerald is the mineral beryl (beryllium aluminum silicate $3BeO{\cdot}Al_2O_3{\cdot}6\ SiO_2$), in which some of the Al^{3+} ions have been replaced by Cr^{3+} ions. The spatial arrangement of the oxide ions around each Cr^{3+} ion breaks the degeneracy of the 3*d* orbitals into two sets of orbitals at different energies; the absorption of red and yellow light causes transitions between these levels. There is also another absorption band in the blue-violet region of the spectrum, and the color of emeralds is due to the transmission of the green light.

The partially filled *d*-electron shells of the transition-metal elements are responsible for a range of physical properties and chemical reactions quite different from those of the main-group elements. The presence of unpaired electrons in the transition-metal elements and their compounds, the availability of low-energy unoccupied orbitals, and the facility with which transition-metal oxidation states change are important factors that determine their rich and fascinating chemistry. Transition-metal complexes are characterized by a wide variety of geometric structures, variable and striking colors, and magnetic properties that depend on subtle details of their structure and bonding. This chapter begins with a descriptive overview of the systematic trends in the properties of these metals, and then presents a comprehensive introduction to the classical and quantum mechanical models that describe bond formation in their compounds.

8.1 Chemistry of the Transition Metals

Let's begin by surveying some of the most important physical and chemical properties of the transition-metal elements and interpreting the trends observed in those properties using the quantum theory of atomic structure developed in Chapter 5. We focus initially on the fourth-period elements, also called the first transition series (those from scandium through zinc in which the 3*d* shell is progressively filled). We then discuss the periodic trends in the melting points and atomic radii of the second and third transition series elements.

8.1.1 Physical Properties

Table 8.1 lists some of the most important physical properties of the elements of the first transition series, which were obtained primarily from Appendix F. The general trends in all of these properties can be understood by recalling that both the number of protons and the number of electrons increase as we move from left to right along a period. We showed in Chapter 5 that the trends in ionization energies for the main group elements could be explained by the shell model of the atom, in which electrons are added to the same shell, each being located at about the same distance from the nucleus. Increasing effective nuclear charge leads to rather smooth increases in the ionization energies as we move across each of the first three periods. First and second ionization energies generally increase as we move across the first transition series but the trend is not smooth. The energies of the 4*s* and 3*d* orbitals are so close to one another that the electron configurations of the neutral atoms and their ions are not easily predicted from the simplest model of atomic structure.

The increase in electron–nuclear attraction that results from increasing Z_{eff} causes atomic and ionic radii initially to decrease from left to right along each period, as shown in Table 8.1 and Figure 8.1, but then to increase toward the end of each series as electron–electron repulsion begins to dominate. Atoms of the second transition series (from yttrium to cadmium) are larger than those of the first transition series, as expected from the fact that the 4*d* orbitals are larger than the 3*d* orbitals. However, the atomic radii of the third transition series are not that much different from those of the second (see Fig. 8.1). This

TABLE 8.1

Properties of the Fourth-Period Transition Elements

Element	Sc	Ti	V	Cr	Mn	Fe	Co	Ni	Cu	Zn
IE_1 (kJ mol^{-1})	631	658	650	653	717	759	758	737	745	906
IE_2 (kJ mol^{-1})	1235	1310	1414	1592	1509	1562	1648	1753	1958	1733
Boiling point (°C)	2831	3287	3380	2672	1962	2750	2870	2732	2567	907
Melting point (°C)	1541	1660	1890	1857	1244	1535	1495	1453	1083	420
Atomic radius (Å)	1.61	1.45	1.31	1.25	1.37	1.24	1.25	1.25	1.28	1.34
M^{2+} Ionic radius (Å)	0.81	0.68	0.88	0.89	0.80	0.72	0.72	0.69	0.72	0.74
M^{2+} configuration	d^1	d^2	d^3	d^4	d^5	d^6	d^7	d^8	d^9	d^{10}
$\Delta H_{hyd}(M^{2+})^{\dagger}$ (kJ mol^{-1})				−2799	−2740	−2839	−2902	−2985	−2989	−2937

†Defined as $\Delta H_f^\circ(M^{2+}(aq)) - \Delta H_f^\circ(M^{2+}(g))$.

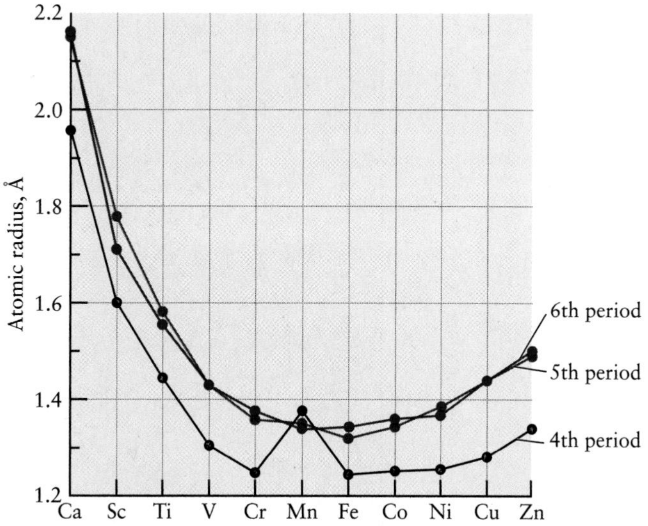

FIGURE 8.1 Variation of atomic radii through the fourth-, fifth-, and sixth-period transition-metal elements. The chemical symbols shown at the bottom of the graph refer to the fourth-period elements.

experimental fact is explained by the **lanthanide contraction.** The first and second transition series are separated by 18 elements, whereas the second and third are separated by 32 elements. The lanthanides have intervened, but their presence is not so obvious from the modern periodic table. Both the nuclear charge and the electron count have increased between the fifth and sixth period elements in a given group, but the *f* orbitals are much more diffuse than the *d* orbitals and much less effective in screening the nuclear charge. Elements in the sixth period experience much greater effective nuclear charges than expected and are therefore smaller. The atomic and ionic radii of hafnium, in the sixth period, are essentially the same as those of zirconium, in the fifth period. These two elements are difficult to separate from one another because they have the same valence, similar size, and, therefore, similar properties. (A similar effect, shown in Fig. 5.27 and discussed in Section 5.5 for main-group atoms and ions, arises from the filling of the 3*d* shell.)

The term *lanthanide contraction* is also used to identify another trend—the decrease in the atomic and ionic radii of the lanthanides from left to right along the sixth period. This trend has the same physical origin as that discussed previously in Chapter 5 for the second- and third-period elements and repeated earlier here: Nuclear charge is increasing while electrons are being added to the same subshell, in this case, the *f* shell. Using the same term to identify two different phenomena can cause confusion; therefore, we suggest you pay careful attention to the context when you see the term *lanthanide contraction.*

Metal–metal bond strengths first increase and then decrease going across each period, reaching a maximum in the middle. Evidence supporting this conclusion comes from the periodic variation in the melting and boiling points of the fourth-period elements shown in Table 8.1 and the corresponding trends in melting points for the three periods shown in Figure 8.2. The melting and boiling points are functions of the bond strengths between the atoms or ions in solids. Both of these properties correlate roughly with the number of unpaired electrons in the elements involved, which reaches a maximum in the middle of the corresponding series. We can rationalize this correlation by thinking about the number of covalent bonds a given metal atom can form with its neighbors; the larger the number of unpaired electrons available for bonding, the greater the number of potential bonds. Another way to understand this trend is to think about bonding in terms of the formation of molecular orbitals that are delocalized over the entire solid (see Section 21.3). These molecular orbitals are constructed from the outermost *d* orbitals on the metal atoms, and each of them can accommodate two electrons. Lower energy orbitals in the solid are

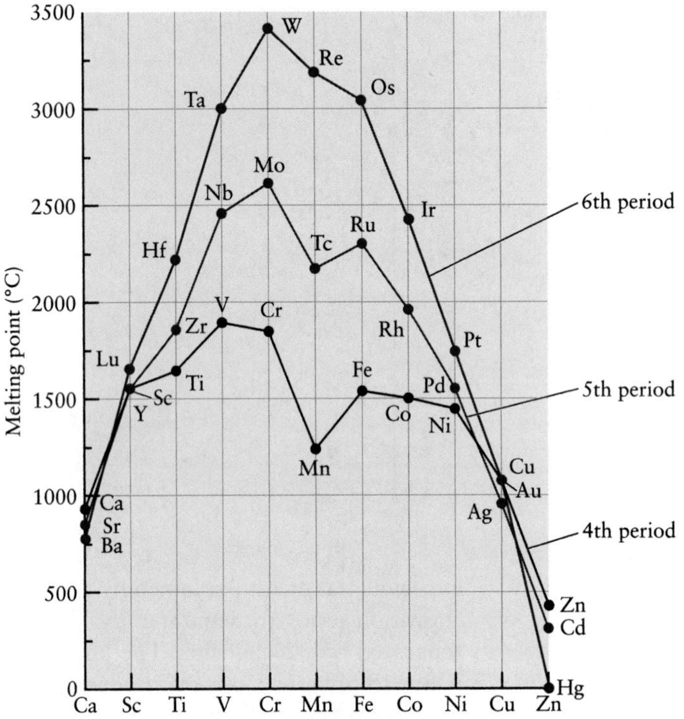

FIGURE 8.2 Variation in the melting points of the transition metals across three periods with the alkaline earth metals included for reference.

primarily bonding. As they are progressively filled through the first half of each transition series, the overall metal–metal bond strength increases. In the second half of each transition series, the higher energy antibonding orbitals become filled, and the bond strength decreases. Tungsten, near the middle of the sixth period, has a very high melting point (3410°C), which makes it useful in light-bulb filaments; mercury, at the end of the same period, has a melting point well below room temperature (−39°C).

Most reactions involving transition metal ions have historically been carried out in aqueous solution; it is important, therefore, to understand the enthalpy changes that occur when metal ions are hydrated. The enthalpy of reaction, ΔH_{rxn}, $= \Delta E + \Delta(PV)$, is close to the energy change for the reaction ΔE (for small $\Delta(PV)$). Thus the enthalpy of hydration is roughly equal to the energy change associated with the hydration of the gas phase ion

$$M^{2+}(g) \longrightarrow M^{2+}(aq)$$

Equivalently, if six water molecules comprise the solvation sphere around the aqueous metal ion, you might write

$$M^{2+}(g) + 6\,H_2O(\ell) \longrightarrow M(H_2O)_6{}^{2+}(aq)$$

Enthalpies of hydration for the M^{2+} ions of the first transition series show an interesting trend (see Table 8.1 and Fig. 8.3). Although we have not yet discussed how ions interact with water in aqueous solutions (see Section 10.2), we might expect the strength of the interactions to increase as the ionic radii decrease, allowing the water molecules to approach the ions more closely. A roughly linear trend that might be expected using this reasoning is shown as the red line in Figure 8.3. The experimental results shown as the black points connected by the blue line follow the same general trend, but clear deviations from linearity suggest that factors other than ionic radii are important. In particular, the experimental results show that ions with filled shells (Ca^{2+}[Ar] and Zn^{2+}[d^{10}]) or half-filled shells (Mn^{2+}[d^5]) follow the simple linear trend quite

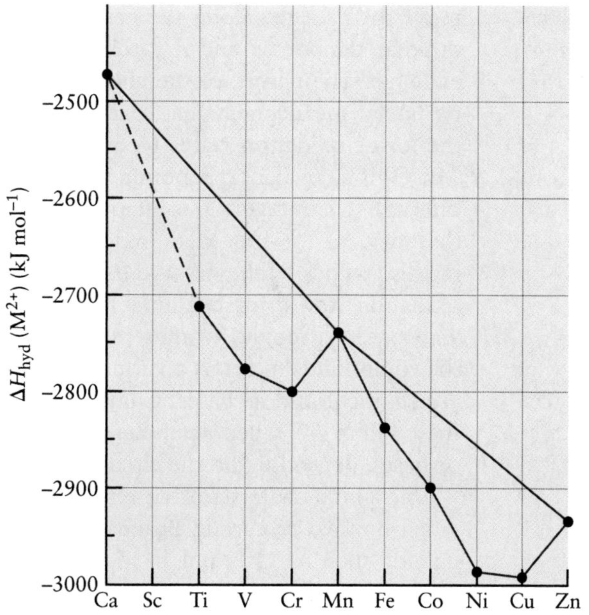

FIGURE 8.3 Enthalpies of hydration of M^{2+} ions, defined as $\Delta H_f^\circ(M^{2+}(aq)) - \Delta H_f^\circ(M^{2+}(g))$. The crystal field stabilization energy (discussed in Section 8.4) preferentially stabilizes certain ions, lowering ΔH_{hyd} from a line representing a roughly linear change with increasing atomic number (red line) to the experimental results (blue line).

closely. We connect this behavior to the electron configurations of the elements in Section 8.4. The anomalously low melting and boiling points and larger atomic radius of manganese, relative to its neighbors, as shown in Figures 8.1 and 8.2, also arises from its half-filled *d* shell.

8.1.2 Oxidation States of the Transition-Metal Elements

Figure 8.4 shows the characteristic oxidation states of the transition metals in common compounds, which you should compare with those of the main group elements shown in Figure 3.41. The maximum oxidation states of the early members of each period reflect, in a formal sense, the participation of all the outer *s* and *d* electrons in ionic or covalent bonding. The maximum oxidation state increases from +3 for Sc to +7 for Mn and then decreases as we continue to

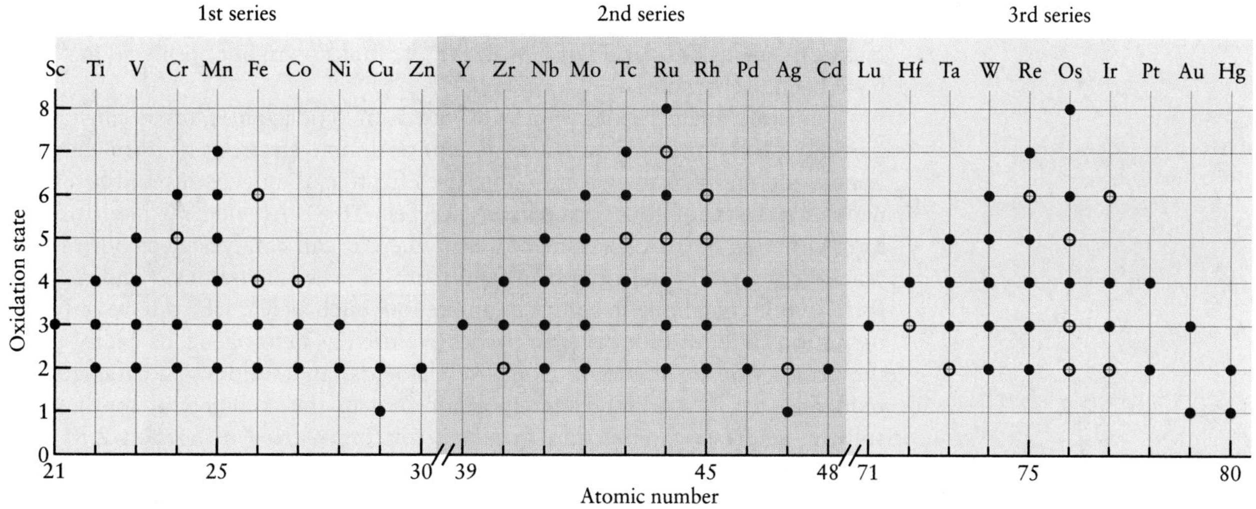

FIGURE 8.4 Some of the oxidation states found in compounds of the transition-metal elements. The more common oxidation states are represented by solid circles, and the less common ones are represented by open circles.

move to the right along the period. The electronic configuration of Sc^{3+} is the same as that of Ar and it forms ionic compounds with halogens like ScF_3, for example. The nature of the chemical bond varies with the oxidation state of a particular metal; the higher oxidation states form more covalent bonds, whereas the lower oxidation states form more ionic bonds. Oxides are good examples: Mn_2O_7 is a covalent compound that is a liquid at room temperature (crystallizing only at 6°C), but Mn_3O_4 is an ionic compound, containing both Mn^{2+} and Mn^{3+}, that melts at 1564°C. Metal oxides can be classified as acids or bases (see following discussion) by the size and oxidation state of the metal cation. Cations in high oxidation states, for example Ti^{4+}, attract electrons from the oxygen atom in water quite strongly, forming more covalent bonds, making it relatively easy for the compounds to act as proton donors. Compounds that readily accept protons are characterized as bases. Compounds whose behavior is intermediate between these limits are called **amphoteric,** which means that they can act either as acids or bases, depending on the circumstances.

Inorganic chemists often refer to higher oxidation states being "stabilized" when bonded to certain **ligands** (bonding partners). This generally means that ligands such as O^{2-} and F^- donate electrons to the metal cation in order to achieve a more neutral configuration overall. This tendency to transfer charge is an example of Pauling's principle of *electroneutrality*, which we used in Section 3.10 (without naming it as such) to help choose among different possible Lewis diagrams by minimizing the degree of charge separation. A covalent bond that results from electron pair donation is called a **dative bond**, and these are extremely important in the chemistry of the transition metals.

Higher oxidation states for metals in a given group are more commonly found for the heavier members of the group because the valence electrons are in the 5*s*4*d* and 6*s*5*d* shells, which are much further from the nucleus than those of the first transition series. The chemistry of iron is dominated by the +2 and +3 oxidation states, as found in the common oxides FeO and Fe_2O_3, but the +8 state, which is nonexistent for iron, has been observed for the heavier members of the iron group, ruthenium and osmium. The oxide OsO_4, for example, is a volatile yellow solid that melts at 41°C and boils at 131°C. It selectively oxidizes C=C double bonds to *cis* diols, which makes it useful in organic synthesis (a similar reaction is shown in Fig. 7.11) and as a biological stain, where it precipitates out as black osmium metal that is easily seen. The chemistry of nickel is almost entirely that of the +2 oxidation state, but the chemistry of the heavier elements in the nickel group, palladium and platinum, is dominated by the +4 state. For example, NiF_2 is the only stable fluoride of nickel, but both PdF_2 and PdF_4 exist. PtF_2 is not known, but both PtF_4 and PtF_6 have been prepared.

Transition metals exist in a wide range of oxidation states because their partially filled *d* orbitals can either accept or donate electrons to form chemical bonds, and many of their compounds are effective homogeneous and heterogeneous catalysts for this very reason (see Sec. 18.8). An element like iron can exist either as Fe^{2+} or Fe^{3+} in solution, where it can catalyze electron-transfer reactions by easily shuttling back and forth between its oxidized and reduced forms while remaining in solution, unlike ions such as K^+, which have only one oxidation state in solution. Iron oxides are used as heterogeneous catalysts for the production of ammonia from the elements, and solid V_2O_5 catalyzes the oxidation of SO_2 to SO_3, an important step in the commercial synthesis of sulfuric acid (see *Connection to Chemical Engineering* in Section 2.3). The active sites of many enzymes (biological catalysts) include transition-metal ions that catalyze a variety of important redox reactions such as the reduction of O_2 to H_2O during respiration. General mechanisms of enzyme catalysis are discussed in Section 18.8.

EXAMPLE 8.1

Assign oxidation states to the transition-metal ions in the following compounds: TiO_2, $ZrCl_4$, VCl_3, CrF_6, Mo_2S_3, FeO, $AuCl_3$. Explain, where appropriate, which ligands stabilize the higher oxidation states and why.

Solution

Recall from Section 3.12 that the most important oxidation states of the halogens and the chalcogenides (Group VI) are −1 and −2, respectively. In this chapter we assume that these are the only oxidation states of these elements, unless otherwise stated. The sum of the oxidation numbers for neutral species must equal zero so the oxidation states of the metal cations must be equal in magnitude but opposite in sign to the sum of the oxidation states of the halogens or the chalcogenides. These may be determined by inspection: Ti(IV) in TiO_2; Zr(IV) in $ZrCl_4$; V(III) in VCl_3; Cr(VI) in CrF_6; Mo(III) in Mo_2S_3; Fe(II) in FeO; Au(III) in $AuCl_3$. V(IV) and Cr(VI) are the highest oxidation states of those two elements; they are stabilized only by small ligands with high charge densities, O^{2-} and F^-.

Related Problems: 1, 2, 3, 4

8.1.3 Hard and Soft Acids and Bases

Inorganic compounds can be broadly characterized as acids or bases using the Lewis model (see Section 15.1), in which bases are defined as electron pair donors and acids are defined as electron pair acceptors. The American inorganic chemist Ralph Pearson extended this model by developing the additional categories "hard" and "soft," which has enabled chemists to organize the physical and chemical properties of these compounds in a simple way and to make predictions about their reactions. We introduce you to the Lewis model to provide the background necessary to develop the hard and soft acid–base concept.

You are probably familiar with the Brønsted-Lowry definition of acids and bases in which acids are defined as proton donors and bases are defined as proton acceptors, as illustrated by the following reaction:

$$HCl(g) + OH^-(aq) \longrightarrow H_2O(\ell) + Cl^-(aq)$$

A proton has been transferred from HCl(g) (the acid) to $OH^-(aq)$ (the base) to form the water $H_2O(\ell)$. This reaction can be interpreted equally well using the Lewis model by focusing on the sharing of one of the oxygen lone pairs with the proton to form a new OH bond. (In the Lewis jargon, one would say the lone pair on the O atom is *donated to* the proton, so the $OH^-(aq)$ is an electron pair donor, or a Lewis base.) The Lewis concept is more general than the Brønsted-Lowry concept because it allows us to treat systems that don't involve protons at all. For example, the reaction

$$Ag^+(aq) + Cl^-(aq) \longrightarrow AgCl(s)$$

can be considered to be an acid–base reaction in which a lone pair of electrons in $Cl^-(aq)$ (the Lewis base) is donated to $Ag^+(aq)$ (the Lewis acid) to form the solid precipitate AgCl. Most, if not all, of the compounds we discuss in this chapter can be identified either as Lewis acids or Lewis bases, and we can predict their most likely reactions using this concept. Bonding in inorganic compounds, and especially the coordination complexes discussed in Section 8.6, is also described largely in terms of the transfer of one or more pairs of electrons from electron donors (Lewis bases) to electron acceptors (Lewis acids).

Most of the elements in the periodic table are metals that tend to form positive ions because of their relatively low ionization energies and low electron affinities (or, equivalently, their low electronegativities). These metal cations are electron pair acceptors and are classified as Lewis acids. They may be classified as hard, soft, or borderline. Most metal ions are *hard acids* because of their low electronegativities and their propensity to accept electrons. Charge density is also a good measure of hardness; small, multiply charged ions like Be^{2+} and

Mg^{2+} are hard acids, for example, and the hardness of a given transition metal increases with the increasing oxidation number as the charge density of the cation increases. The *soft acids* are found in the lower-right region of the metallic elements in the periodic table; they have high electronegativities and relatively low charge densities because of their large size and generally low oxidation states. Finally, a few elements form borderline acids because their physical properties are intermediate between those of the hard and soft acids. We may discuss some examples in passing, but note that the classification of several of these ions changes with oxidation state. Cu^{+}, with its relatively low charge density, is a soft acid, whereas Cu^{2+}, with a somewhat higher charge density due to its 2+ charge, is a borderline acid. Representative examples are shown in Table 8.2.

There are significantly fewer bases than acids. The hard bases include F^{-}, polyatomic anions that bond to metals through oxygen, such as sulfate and nitrate, and the neutral Lewis bases, water and ammonia, which are lone pair donors. The soft bases include the large, less electronegative anions such as I^{-} and S^{2-}, as well as a few polyatomic species that may either be neutral (CO) or anionic (CN^{-}, SCN^{-}). Cl^{-} and Br^{-} are the most important of the very few borderline bases. These examples are also shown in Table 8.2.

Let's illustrate the utility of the hard/soft acid/base concept in understanding chemical reactivity before discussing how the model also provides insight into the nature of the bonding in these species. Predicting reactivity follows one simple rule: Hard acids tend to pair with hard bases, whereas soft acids tend to pair with soft bases. The outcomes of many possible reactions can be predicted by simply looking at the character of the species involved. For example, the outcome of the following gas phase reaction can be predicted using this concept.

$$HgF_2(g) + BeI_2(g) \longrightarrow BeF_2(g) + HgI_2(g)$$

The reactants include a soft acid paired with a hard base, and a hard acid paired with a soft base; they exchange partners to pair the hard acid with the hard base and vice versa, as predicted. The concept can also be used to predict the course of displacement reactions as illustrated by the reaction

$$AgBr(s) + I^{-}(aq) \longrightarrow AgI(s) + Br^{-}(aq)$$

because the soft acid Ag^{+} prefers to bond with the softer of the two bases, which is $I^{-}(aq)$. The concept can be used to quickly assess the outcomes of possible reactions as well as to predict trends in the solubility of inorganic compounds in aqueous solutions. For example, the solubilities of the sodium halides

TABLE 8.2

Classification of Lewis Acids and Bases†

	Hard	Borderline	Soft
Acids	H^{+}, Li^{+}, Na^{+}, K^{+}, Be^{2+}, Mg^{2+}, Ca^{2+}, Cr^{2+}, Cr^{3+}, Al^{3+}, SO_3, BF_3	Fe^{2+}, Co^{2+}, Ni^{2+}, Cu^{2+}, Zn^{2+}, Pb^{2+}, SO_2, BBr_3	Cu^{+}, Ag^{+}, Au^{+}, Tl^{+}, Hg^{+}, Pd^{2+}, Cd^{2+}, Pt^{2+}, Hg^{2+}, BH_3
Bases	F^{-}, OH^{-}, H_2O, NH_3, CO_3^{2-}, NO_3^{-}, O^{2-}, SO_4^{2-}, PO_4^{3-}, ClO_4^{-}	$\underline{N}O_2^{-}$, SO_3^{2-}, Br^{-}, N_3^{-}, N_2, C_6H_5N, $SC\underline{N}^{-}$	H^{-}, R^{-}, $\underline{C}N^{-}$, $\underline{C}O$, I^{-}, CN^{-}, R_3P, C_6H_6, R_2S

†The underlined element identifies the electron pair donor if there is more than one possible choice. (R indicates an alkyl group, such as CH_3.)

increase in the order $F^- < Cl^- < Br^- < I^-$, whereas the solubilities of the silver halides increase in the inverse order, reflecting the preference of the hard acid Na^+ to bond with the hard bases and the preference of the soft acid Ag^+ to bond with the soft bases. These preferences also tell us something about the nature of the bonding involved. Hard acid/hard base interactions are primarily ionic in nature, whereas soft acid/soft base interactions are primarily covalent.

EXAMPLE 8.2

Use the hard/soft acid/base concept to predict whether the following reactions will occur.

(a) $CaF_2(s) + CdI_2(s) \longrightarrow CaI_2(s) + CdF_2(s)$

(b) $Cr(CN)_2(s) + Cd(OH)_2(s) \longrightarrow Cd(CN)_2(s) + Cr(OH)_2(s)$

Solution

(a) No. Calcium is a hard acid that prefers to bond to the hard base F^-.

(b) Yes. Chromium is a hard acid that prefers to bond to the hard base OH^-.

Related Problems: 5, 6

8.2 Introduction to Coordination Chemistry

8.2.1 Formation of Coordination Complexes

The Alsatian-Swiss chemist Alfred Werner pioneered the field of coordination chemistry in the late 19th century. At that time, a number of compounds of cobalt(III) chloride with ammonia were known. They had the following chemical formulas and colors:

Compound 1: $CoCl_3 \cdot 6NH_3$	Orange-yellow
Compound 2: $CoCl_3 \cdot 5NH_3$	Purple
Compound 3: $CoCl_3 \cdot 4NH_3$	Green
Compound 4: $CoCl_3 \cdot 3NH_3$	Green

Treatment of these compounds with aqueous hydrochloric acid did not remove the ammonia, which suggested that it was somehow closely bound with the cobalt ions. Treatment with aqueous silver nitrate at 0°C, however, gave interesting results. With compound 1, all of the chloride present precipitated as solid AgCl. With compound 2, only two thirds of the chloride precipitated, and with compound 3, only one third precipitated. Compound 4 did not react at all with the silver nitrate. These results suggested that there were two different kinds of species associated with the cobalt ions, which Werner called *valences* (recall that the electron, the key player in the formation of the chemical bond, was just being characterized by J. J. Thomson). The primary, or ionizable, valences were anions like Cl^- in simple salts such as $CoCl_3$, whereas the secondary valences could be either simple anions or neutral molecules such as NH_3. Werner assumed that the primary valences were nondirectional, whereas the secondary valences were oriented along well-defined directions in space. The picture that emerged was that of a metal atom coordinated to the secondary valences (ligands) in the **inner coordination sphere** surrounded by the primary valences and solvent in the **outer coordination sphere**. The primary valences neutralize the positive charge on the complex ion. Werner accounted for experimental results described earlier by positing the existence of **coordination complexes** with six ligands (either chloride ions or ammonia molecules) attached to each Co^{3+} ion. Specifically, he wrote the formulas for compounds 1 through 4 as

Compound 1: $[Co(NH_3)_6]^{3+} (Cl^-)_3$

Compound 2: $[Co(NH_3)_5Cl]^{2+} (Cl^-)_2$

Compound 3: $[Co(NH_3)_4Cl_2]^+ (Cl^-)$

Compound 4: $[Co(NH_3)_3Cl_3]$

with the charge on each complex ion being balanced by an equal number of Cl^- ions in the outer coordination sphere. Only these chloride ions, which were *not* bonded directly to cobalt, could react with the silver ions in cold aqueous silver nitrate to form the AgCl precipitate.

Werner realized that he could test his hypothesis by measuring the electrical conductivity of aqueous solutions of the salts of these complex ions. Ions are the electrical conductors in aqueous solutions, and the conductivity is proportional to the ion concentration. If Werner's proposal was correct, then an aqueous solution of Compound 1, for example, should have a molar conductivity close to that of an aqueous solution of $Al(NO_3)_3$, which also forms four ions per formula unit when it dissociates completely in water (one 3+ ion and three 1− ions). His experiments confirmed that the conductivities of these two solutions were, indeed, similar. Furthermore the conductivity of aqueous solutions of compound 2 was close to those of $Mg(NO_3)_2$, and solutions of compound 3 conducted electricity about as well as those containing $NaNO_3$. Compound 4, in contrast, did not dissociate into ions when dissolved in water, producing a solution of very low electrical conductivity.

Werner and other chemists studied a variety of other coordination complexes, using both physical and chemical techniques. Their research showed that 6 is the most common coordination number by far, as in the cobalt complexes discussed earlier. Coordination numbers ranging from 1 to 16 have subsequently been discovered, however. The most common of these include coordination numbers 2 (as in $[Ag(NH_2)_2]^-$), 4 (as in $[PtCl_4]^{2-}$), and 5 (as in $[Ni(CN)_5]^{3-}$).

A second example illustrates the ability of transition metals to form complexes with small molecules and ions. Copper metal and hot concentrated sulfuric acid ("oil of vitriol") react to form solid copper(II) sulfate, commonly called "blue vitriol" by virtue of its deep blue color. There is more to this compound than copper and sulfate, however; it contains water as well. When the water is driven away by heating, the blue color vanishes, leaving greenish white anhydrous copper(II) sulfate (Fig. 8.5). The blue color of blue vitriol comes from a **coordination complex** in which H_2O molecules bond directly to Cu^{2+} ions to form coordination complexes with the formula $[Cu(OH_2)_4]^{2+}$. Bonding in this complex ion can be described qualitatively using the Lewis theory of acids and bases (discussed previously and in more detail in Section 15.1). The transfer of a pair of electrons from water to Cu^{2+} to form a dative bond can be thought of as an acid–base reaction. As a Lewis acid, the Cu^{2+} ion *coordinates* four water molecules into a group by accepting electron density from a lone pair on each. By acting as electron-pair donors and sharing electron density with the Cu^{2+} ion, the four water molecules are the ligands, occupying the inner coordination sphere of the ion. The chemical formula of blue vitriol is $[Cu(OH_2)_4]SO_4 \cdot H_2O$; the fifth water molecule is not coordinated directly to copper. Coordination complexes are those in which a central metal atom or ion is surrounded by a set of ligands. (Some textbooks use the term *coordination compound* to refer to a neutral coordination complex or to an ionic compound in which at least one of the ions is a coordination complex. The ion $[Cu(OH_2)_4]^{2+}$ would be referred to as a coordination complex, whereas blue vitriol would be referred to as a coordination compound. We often use the terms interchangeably.)

FIGURE 8.5 Hydrated copper(II) sulfate, $CuSO_4{\cdot}5H_2O$, is blue (left), but the anhydrous compound, $CuSO_4$, is greenish white (right). A structural study of the solid compound demonstrates that four of the water molecules are closely associated with the copper and the fifth is not. Thus, a better representation of the hydrated compound is $[Cu(OH_2)_4]SO_4{\cdot}H_2O$.

The positive ions of every metal in the periodic table accept electron density to some degree and can therefore coordinate surrounding electron donors, even if only weakly. The solvation of Na^+ by H_2O molecules in aqueous solution (see Fig. 10.6) is an example of weak coordination. The ability to make fairly strong, *directional* bonds by accepting electron pairs from neighboring molecules or ions is characteristic of the transition-metal elements. Coordination occupies a middle place energetically between the weak intermolecular attractions in molecular solids and liquids (see Chapter 10) and the stronger covalent and ionic bonding (see Chapters 3 and 6). Thus, heating blue vitriol disrupts the $Cu{-}OH_2$ bonds at temperatures well below those required to break the covalent bonds in the SO_4^{2-} group (here M represents a metallic element such as Cu). The energy (more precisely, the enthalpy) required to break a $M^{2+}{-}OH_2$ bond in a transition metal coordination complex falls in the range between 170 and 210 kJ mol^{-1}. This bond dissociation energy is far less than the bond energies of the strongest chemical bonds (e.g., 942 kJ mol^{-1} for N_2), but it is by no means small. The bond dissociation enthalpy increases as the charge density on the metal atom increases; Cu^{3+} binds water more strongly than Cu^{2+}, for example.

A wide variety of molecules and ions bond to metals as ligands; common ones include the halide ions (F^-, Cl^-, Br^-, I^-), ammonia (NH_3), carbon monoxide (CO), water, and a few other simple ligands listed in Table 8.3. Ligands that bond to a metal atom through a single point of attachment are called *monodentate* (derived from Latin *mono,* meaning "one," plus *dens,* meaning "tooth," indicating that they bind at only one point). More complex ligands can bond through two or more attachment points; they are referred to as *bidentate, tridentate,* and so forth. Ethylenediamine ($NH_2CH_2CH_2NH_2$) (abbreviated as "en" in chemical formulas), in which two NH_2 groups are connected by a hydrocarbon chain, is a particularly important bidentate ligand. Both nitrogen atoms in ethylenediamine have lone electron pairs to share. If all the nitrogen atoms in three ethylenediamine molecules bind to a single Co^{3+} ion, then that ion has a coordination number of 6 and the formula of the resulting complex is $[Co(en)_3]^{3+}$. Complexes in which a ligand coordinates via two or more donors to the same central atom are called **chelates** (derived from Greek *chele,* meaning "claw," because the ligand grabs onto the central atom like a pair of pincers). Figure 8.6 shows the structures of some important chelating ligands.

TABLE 8.3

Common Ligands and Their Names

Ligand†	Name
$:NO_2^-$	Nitro
$:OCO_2^{2-}$	Carbonato
$:ONO^-$	Nitrito
$:CN^-$	Cyano
$:SCN^-$	Thiocyanato
$:NCS^-$	Isothiocyanato
$:OH^-$	Hydroxo
$:OH_2$	Aqua
$:NH_3$	Ammine
$:CO$	Carbonyl
$:NO^+$	Nitrosyl

†The ligating atom is indicated by a pair of dots (:) to show a lone pair of electrons. In the CO_3^{2-} ligand, either one or both of the O atoms can donate a lone pair to a metal.

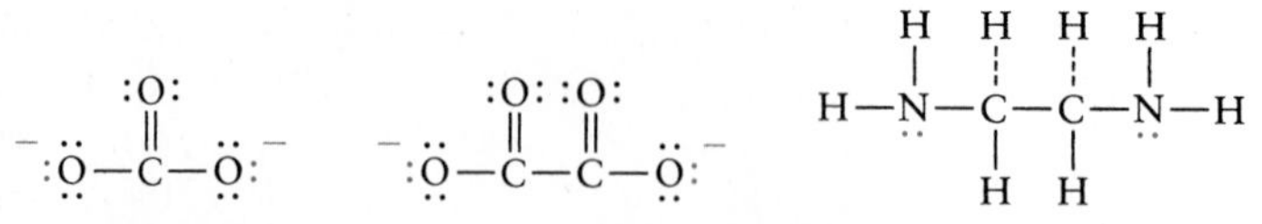

Carbonate ion, CO_3^{2-} — Oxalate ion, $C_2O_4^{2-}$ — Ethylenediamine, $NH_2CH_2CH_2NH_2$

FIGURE 8.6 Three bidentate ligands, each capable of donating two pairs of electrons.

Brackets are used to identify complex ions in modern notation. The charge of the ion in a coordination compound is not written explicitly but inferred from the charges of the counterions. In the formula $[Pt(NH_3)_6]Cl_4$, the portion in brackets represents a coordination complex in which Pt coordinates six NH_3 ligands. Four Cl^- ions surround it so the charge of the complex ion in brackets is +4. The brackets emphasize that a complex is a distinct chemical entity and within the brackets, the symbol of the central atom comes first. The electric charge on a coordination complex is the sum of the oxidation number of the metal ion and the charges of the ligands that surround it. Thus, the complex of copper(II) (Cu^{2+}) with four Br^- ions is an anion with a −2 charge, $[CuBr_4]^{2-}$ and the Pt in the above complex is in a +4 oxidation state.

EXAMPLE 8.3

Determine the oxidation state of the coordinated metal atom in each of the following compounds:

(a) $K[Co(NH_3)_2(CN)_4]$; **(b)** $Os(CO)_5$;
(c) $Na[Co(OH_2)_3(OH)_3]$.

Solution

(a) The oxidation state of K is known to be +1, so the complex in brackets is an anion with a −1 charge, $[Co(NH_3)_2(CN)_4]^-$. The charge on the two NH_3 ligands is 0, and the charge on each of the four CN^- ligands is −1. The oxidation state of the Co must then be +3, because 4×-1 (for the CN^-) + 2×0 (for the NH_3) + 3 (for Co) equals the required −1.

(b) The ligand CO has zero charge, and the complex has zero charge as well. Therefore, the oxidation state of the osmium is 0.

(c) There are three neutral ligands (the water molecules) and three ligands with −1 charges (the hydroxide ions). The Na^+ ion contributes only +1; thus, the oxidation state of the cobalt must be +2.

Related Problems: 9, 10

Coordination modifies the chemical and physical properties of both the central atom and the ligands. Consider the chemistry of aqueous cyanide (CN^-) and iron(II) (Fe^{2+}) ions. The CN^- ion reacts immediately with acid to generate gaseous HCN, a deadly poison, and Fe^{2+}, when mixed with aqueous base, instantly precipitates a gelatinous hydroxide. The reaction between Fe^{2+} and CN^- produces the complex ion $[Fe(CN)_6]^{4-}(aq)$, which undergoes neither of the two reactions just described nor any others considered characteristic of CN^- or Fe^{2+}. Ions or molecules may be present in multiple forms in the same compound. The two Cl^- ions in $[Pt(NH_3)_3Cl]Cl$ are chemically different, because one is coordinated and the other is not. Treatment of an aqueous solution of this substance with Ag^+ immediately precipitates the uncoordinated Cl^- as AgCl(*s*), but *not* the coordinated Cl^-, just as it did for Werner's complexes discussed earlier.

Ionic coordination complexes of opposite charges can combine with each other—just as any positive ion can combine with a negative ion—to form salts. For example, the cation $[Pt(NH_3)_4]^{2+}$ and the anion $[PtCl_4]^{2-}$ form an ionic compound with the formula $[Pt(NH_3)_4][PtCl_4]$. This compound and the following four compounds,

$$Pt(NH_3)_2Cl_2 \qquad [Pt(NH_3)_3Cl][Pt(NH_3)Cl_3]$$

$$[Pt(NH_3)_3Cl]_2[PtCl_4] \qquad [Pt(NH_3)_4][Pt(NH_3)Cl_3]_2$$

all contain Pt, NH_3, and Cl in the ratio of 1:2:2; that is, they have the same percentage composition. Two pairs even have the same molar mass. Yet, the five compounds differ in structure and in physical and chemical properties. The concept of coordination organizes an immense number of chemical compositions and patterns of reactivity by considering combinations of ligands linked in varied ratios with central metal atoms or ions.

8.2.2 Naming Coordination Compounds

So far, we have identified coordination compounds only by their chemical formulas, but names are also useful for many purposes. Some substances were named before their structures were known. Thus, $K_3[Fe(CN)_6]$ was called potassium ferricyanide, and $K_4[Fe(CN)_6]$ was called potassium ferrocyanide [these are complexes of Fe^{3+} (ferric) and Fe^{2+} (ferrous) ions, respectively]. These older names are still used conversationally, but systematic names are preferred to avoid ambiguity. The definitive source for the naming of inorganic compounds is *Nomenclature of Inorganic Chemistry: IUPAC Recommendations 2005* (N. G. Connelly and T. Damhus, Sr., Eds. Royal Society of Chemistry, 2005).

1. The name of a coordination complex is written as a single word built from the names of the ligands, a prefix before each ligand to indicate how many ligands of that kind are present in the complex, and the name of the central metal atom.
2. Compounds containing coordination complexes are named following the same rules as those for simple ionic compounds: The positive ion is named first, followed (after a space) by the name of the negative ion.
3. Anionic ligands are named by replacing the usual ending with the suffix *-o*. The names of neutral ligands are unchanged. Exceptions to the latter rule are aqua (for water), ammine (for NH_3), and carbonyl (for CO) (see Table 8.3).
4. Greek prefixes (*di-*, *tri-*, *tetra-*, *penta-*, *hexa-*) are used to specify the number of ligands of a given type attached to the central ion, if there is more than one. The prefix *mono-* (for one) is not used. If the name of the ligand itself contains a term such as *mono-* or *di-* (as in ethylenediamine), then the name of the ligand is placed in parentheses and the prefixes *bis-*, *tris-*, and *tetrakis-* are used instead of *di-*, *tri-*, and *tetra-*.
5. The ligands are listed in alphabetical order, without regard for the prefixes that tell how often each type of ligand occurs in the coordination sphere.
6. A Roman numeral, enclosed in parentheses placed immediately after the name of the metal, specifies the oxidation state of the central metal atom. If the complex ion has a net negative charge, the ending *-ate* is added to the stem of the name of the metal.

A few examples of complexes and their systematic names are shown in Table 8.4.

TABLE 8.4

Examples of Complexes and Their Systematic Names

Complex	Systematic Name
$K_3[Fe(CN)_6]$	Potassium hexacyanoferrate(III)
$K_4[Fe(CN)_6]$	Potassium hexacyanoferrate(II)
$Fe(CO)_5$	Pentacarbonyliron(0)
$[Co(NH_3)_5CO_3]Cl$	Penta-amminecarbonatocobalt(III) chloride
$K_3[Co(NO_2)_6]$	Potassium hexanitrocobaltate(III)
$[Cr(OH_2)_4Cl_2]Cl$	Tetra-aquadichlorochromium(III) chloride
$[Pt(NH_2CH_2CH_2NH_2)_3]Br_4$	Tris(ethylenediamine)platinum(IV) bromide
$K_2[CuCl_4]$	Potassium tetrachlorocuprate(II)

EXAMPLE 8.4

Interpret the names and write the formulas of these coordination compounds:

(a) sodium tricarbonatocobaltate(III)

(b) diamminediaquadichloroplatinum(IV) bromide

(c) sodium tetranitratoborate(III)

Solution

(a) In the anion, three carbonate ligands (with −2 charges) are coordinated to a cobalt atom in the +3 oxidation state. Because the complex ion thus has an overall charge of −3, three sodium cations are required, and the correct formula is $Na_3[Co(CO_3)_3]$.

(b) The ligands coordinated to a Pt(IV) include two ammonia molecules, two water molecules, and two chloride ions. Ammonia and water are electrically neutral, but the two chloride ions contribute a total charge of $2 \times (-1) = -2$ that sums with the +4 of the platinum and gives the complex ion a +2 charge. Two bromide anions are required to balance this; thus, the formula is $[Pt(NH_3)_2(OH_2)_2Cl_2]Br_2$.

(c) The complex anion has four −1 nitrate ligands coordinated to a central boron(III). This gives a net charge of −1 on the complex ion and requires one sodium ion in the formula $Na[B(NO_3)_4]$.

Related Problems: 11, 12

8.2.3 Ligand Substitution Reactions

We discuss a few ligand substitution reactions to give you a feel for the properties and reactions of coordination complexes. These simple reactions are aptly named; one or more ligands are simply substituted for one another. We have already discussed one example of a series of ligand substitution reactions: the exchange of NH_3 and Cl^- in Werner's cobalt complexes.

If the yellow crystalline solid nickel(II) sulfate is exposed to moist air at room temperature, it takes up six water molecules per formula unit. These water molecules coordinate the nickel ions to form a bright green complex:

$$\underset{\text{Yellow}}{NiSO_4(s)} + \underset{\text{Colorless}}{6\ H_2O(g)} \rightleftharpoons \underset{\text{Green}}{[Ni(OH_2)_6]SO_4(s)}$$

Heating the green hexaaquanickel(II) sulfate to temperatures well above the boiling point of water drives off the water and regenerates the yellow $NiSO_4$ in the reverse reaction. A different coordination complex forms when yellow $NiSO_4(s)$ is exposed to gaseous ammonia, $NH_3(g)$. This time, the product is a blue-violet complex:

$$\underset{\text{Yellow}}{NiSO_4(s)} + \underset{\text{Colorless}}{6\ NH_3(g)} \rightleftharpoons \underset{\text{Blue-violet}}{[Ni(NH_3)_6]SO_4(s)}$$

Heating the blue-violet product drives off ammonia, and the color of the solid returns to yellow. Given these facts, it is not difficult to explain the observation that a green $[Ni(OH_2)_6]^{2+}(aq)$ solution turns blue-violet when treated with $NH_3(aq)$ (Fig. 8.7). NH_3 must have displaced the H_2O ligands from the coordination sphere forming the blue-violet $[Ni(NH_3)_6]^{2+}$ complex.

$$\underset{\text{Green}}{[Ni(OH_2)_6]^{2+}(aq)} + \underset{\text{Colorless}}{6\ NH_3(aq)} \longrightarrow \underset{\text{Blue-violet}}{[Ni(NH_3)_6]^{2+}(aq)} + \underset{\text{Colorless}}{6\ H_2O}$$

A **labile complex** is one in which ligand substitution reactions proceed rapidly. One in which substitution proceeds slowly or not at all is called an **inert complex.** These terms are used to describe the different kinetics (rates) of the reactions (see Chapter 18) that are thermodynamically allowed (see Chapter 13). Large activation energy barriers for ligand substitution reactions of inert complexes make those reactions slow even though there may be a thermodynamic driving force for them to proceed. In the substitution reaction,

$$[Co(NH_3)_6]^{3+}(aq) + 6\ H_3O^+(aq) \rightleftharpoons [Co(OH_2)_6]^{3+}(aq) + 6\ NH_4^+(aq)$$

FIGURE 8.7 When ammonia is added to the green solution of nickel(II) sulfate on the left (which contains $[Ni(OH_2)_6]^{2+}$ ions), ligand substitution occurs to give the blue–violet solution on the right (which contains $[Ni(NH_3)_6]^{2+}$ ions).

the products are favored thermodynamically by an enormous amount, yet the inert $[Co(NH_3)_6]^{3+}$ complex ion lasts for weeks in acidic solution because there is no low-energy pathway for the reaction. We would say that the cobalt(III) ion, $[Co(NH_3)_6]^{3+}$, is thermodynamically unstable relative to $[Co(OH_2)_6]^{3+}$, but kinetically stable (inert). The closely related cobalt(II) complex, $[Co(NH_3)_6]^{2+}$, reacts with water in a matter of seconds:

$$[Co(NH_3)_6]^{2+}(aq) + 6\ H_3O^+(aq) \longrightarrow [Co(OH_2)_6]^{2+}(aq) + 6\ NH_4^+(aq)$$

The hexa-aminecobalt(II) complex is both thermodynamically unstable and kinetically labile.

Ligand substitution reactions proceed sequentially, and they can usually be stopped at intermediate stages by controlling the reaction conditions. The following series of stable complexes represents all possible four-coordinate complexes of Pt(II) with the two ligands NH_3 and Cl^-.

$$[Pt(NH_3)_4]^{2+},\ [PtNH_3Cl]^{+},\ [Pt(NH_3)_2Cl_2],\ [Pt(NH_3)Cl_3]^{-},\ \text{and}\ [PtCl_4]^{2-}$$

These mixed-ligand complexes provide wonderful examples of the variety and richness of coordination chemistry.

8.3 Structures of Coordination Complexes

8.3.1 Octahedral Geometries

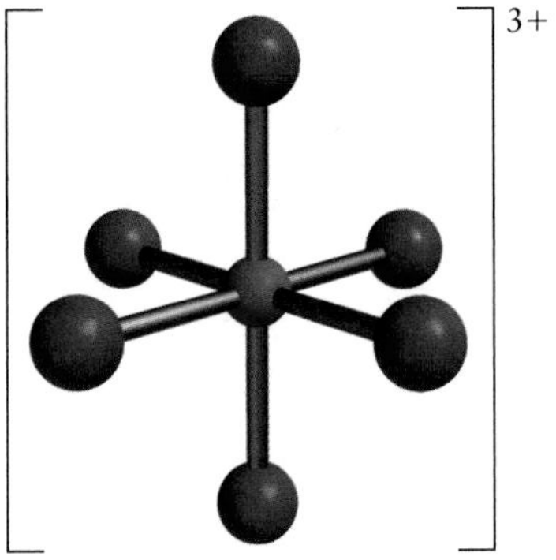

FIGURE 8.8 Ball-and-stick model of the octahedral complex $[Co(NH_3)_6]^{3+}$, with the hydrogen atoms not shown for clarity. All six ligands are equivalent.

What is the geometric structure of the complex $[Co(NH_3)_6]^{3+}$? This question naturally occurred to Werner, who suggested that the arrangement should be the simplest and most symmetric possible, with the ligands positioned at the six vertices of a regular octahedron (Fig. 8.8). Modern methods of X-ray diffraction (see Section 21.1) enable us to make precise determinations of atomic positions in crystals and have confirmed Werner's proposed octahedral structure for this complex. X-ray diffraction techniques were not available in the late 19th century, however, so Werner turned to a study of the properties of substituted complexes to test his hypothesis. Having a set of molecular ball-and-stick models at hand as you read this section will make it much easier for you to visualize the structures and transformations described. We also provide line structures for selected complexes to introduce you to the shorthand way inorganic chemists sketch these complexes. As defined in Section 3.8, equatorial ligands are those that lie in the horizontal plane and axial ligands are those that lie along the vertical line (see Fig. 8.8).

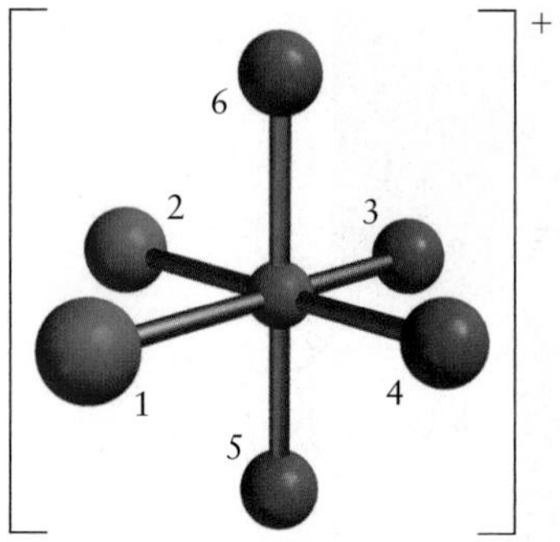

(a)

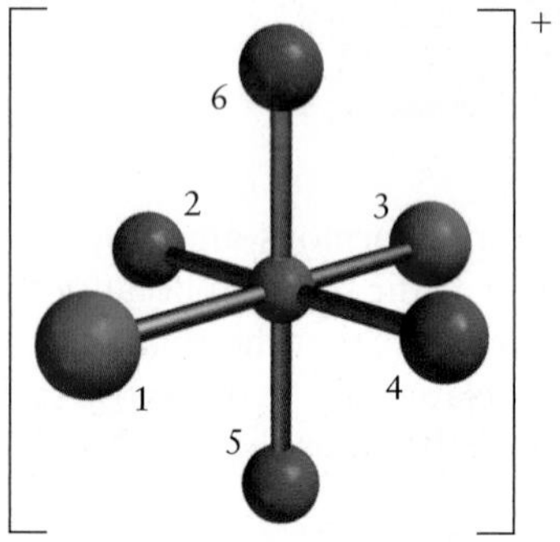

(b)

FIGURE 8.9 (a) The *cis*-$[Co(NH_3)_4Cl_2]^+$ and (b) *trans*-$[Co(NH_3)_4Cl_2]^+$ ions. The *cis* complex is purple in solution, but the *trans* complex is green.

Replacing one ammonia ligand by a chloride ion results in a complex with the formula $[Co(NH_3)_5Cl]^{2+}$, in which one vertex of the octahedron is occupied by Cl^- and the other five by NH_3. Only one structure of this type is possible, because all six vertices of a regular octahedron are equivalent, and the various singly substituted complexes $[MA_5B]$ (where $A = NH_3$, $B = Cl^-$, $M = Co^{3+}$, for example) can be superimposed on one another. Now, suppose a second NH_3 ligand is replaced by Cl^-. The second Cl^- can lie in one of the four equivalent positions closest to the first Cl^- (in the horizontal plane; see Fig. 8.9a) or in position labeled 3, on the opposite side of the central metal atom (see Fig. 8.9b). The first of these *geometric isomers*, in which the two Cl^- ligands are closer to each other, is called *cis*-$[Co(NH_3)_4Cl_2]^+$, and the second, with the two Cl^- ligands farther apart, is called *trans*-$[Co(NH_3)_4Cl_2]^+$. The octahedral structure model predicts that there can be only two different ions with the chemical formula $[Co(NH_3)_4Cl_2]^+$. You may have already encountered geometric isomers in Chapter 7. When Werner began his work, only the green *trans* (across) form was known, but by 1907, he had prepared the *cis* (near) isomer and shown that it differed from the *trans* isomer in color (it was violet rather than green) and other physical properties. The isolation of two, and only two, geometric isomers of this ion was good (although not conclusive) evidence that the octahedral structure was correct. Similar isomerism is displayed by the complex ion $[CoCl_2(en)_2]^+$, which also exists in a purple *cis* form and a green *trans* form (Fig. 8.10).

EXAMPLE 8.5

How many geometric isomers does the octahedral coordination compound $[Co(NH_3)_3Cl_3]$ have?

Solution

We begin with the two isomers of $[Co(NH_3)_4Cl_2]^+$ shown in Figure 8.9 and see how many different structures can be made by replacing one of the ammonia ligands with Cl^-.

Starting with the *trans* form (see Fig. 8.9b), it is clear that replacement of any of the four NH_3 ligands at site 2, 4, 5, or 6 gives a set of equivalent structures that can be superimposed on one another by rotation of the starting structure. Figure 8.11b shows the structure that results from substitution at the position 2. What isomers can be made from the *cis* form shown in Figure 8.9a? If either the ammonia ligand at site 3 or the one at site 4 (i.e., one of the two that are *trans* to existing Cl^- ligands) is replaced, the result is simply a rotated version of Figure 8.11b; therefore, these replacements do *not* give another isomer. Replacement of the ligand at site 5 or 6, however, gives a different structure (see Fig. 8.11a). (Note that Cl^- occupying sites 1, 2, 5 and 1, 2, 6 results in equivalent structures.)

We conclude that there are two, and only two, possible isomers of the octahedral complex structure MA_3B_3. In fact, only one form of $[Co(NH_3)_3Cl_3]$ has been prepared to date, presumably because the two isomers interconvert rapidly. However, two isomers are known for the closely related coordination complex $[Cr(NH_3)_3(NO_2)_3]$.

Related Problems: 17, 18

8.3.2 Square-Planar, Tetrahedral, and Linear Geometries

Complexes with coordination numbers of 4 are typically either tetrahedral or square planar. The tetrahedral geometry (Fig. 8.12a) predominates for four-coordinate complexes of the early transition metals (those toward the left side of the *d* block of elements in the periodic table). Geometric isomerism is not possible for tetrahedral complexes of the general form MA_2B_2, because all four tetrahedral sites are completely equivalent.

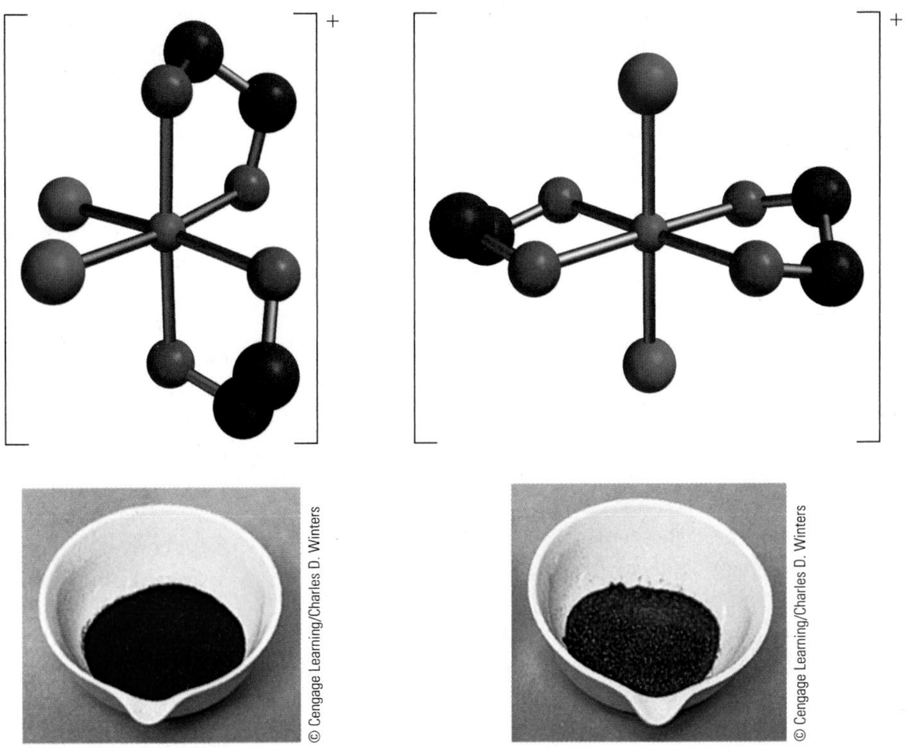

FIGURE 8.10 The complex ion $[CoCl_2(en)_2]^+$ is an octahedral complex that has *cis* and *trans* isomers, according to the relative positions of the two Cl^- ligands. Salts of the *cis* isomers are purple, and salts of the *trans* isomers are green. For clarity, hydrogen atoms are not shown.

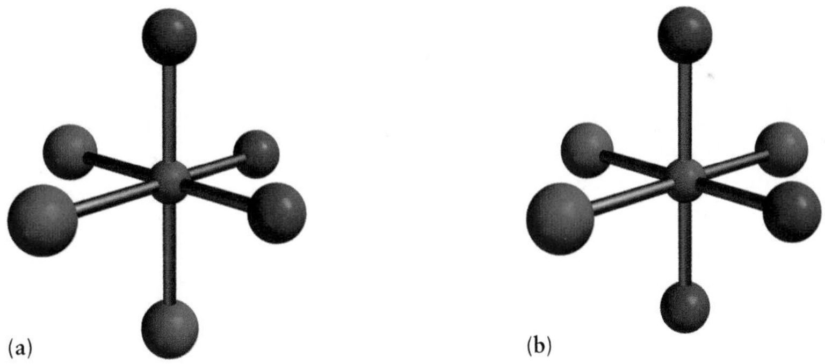

FIGURE 8.11 Structural isomers of $Co(NH_3)_3Cl_3$. (a) The *fac* isomer with identical ligands arranged cis to each other (b) the *mer* isomer with identical ligands lying in the same plane (with the central atom)

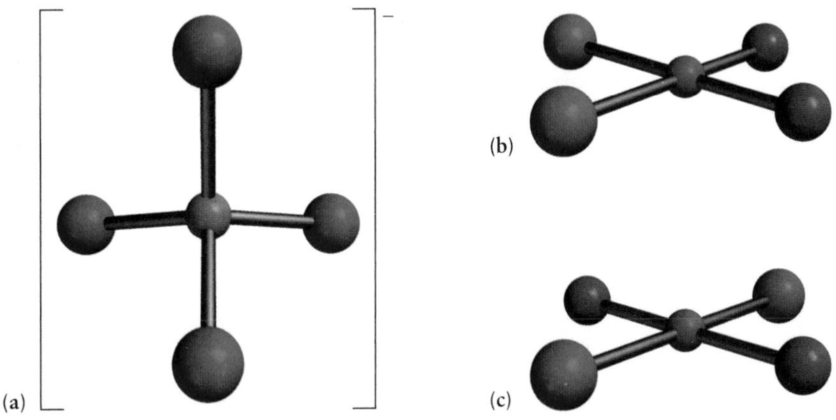

FIGURE 8.12 Four-coordinate complexes. (a) Tetrahedral, $[FeCl_4]^-$. (b, c) Square planar, illustrating (b) the *cis* and (c) the *trans* forms of $[Pt(NH_3)_2Cl_2]$.

CONNECTION TO BIOLOGY

Coordination Complexes in Heme Proteins

Coordination complexes of the transition metals play central roles in the chemistry of life processes; understanding the nature of these roles is an important goal of the field of bioinorganic chemistry. Nine transition-metal ions are classified as essential minerals that must be present in at least trace amounts to sustain life. These include the microminerals Fe, Zn, and Cu (daily requirement, 1–10 mg for adult humans), as well as the trace minerals V, Cr, Mn, Co, Ni, and Mo (daily requirement, few hundred μg for adult humans). Most of these metals are found in metalloenzymes, biological catalysts in which the metal ion plays a central role, but also in molecules like the chlorophylls (see Section 20.7) in which light absorption initiates a series of redox reactions that ultimately convert solar energy into chemical potential energy. Metalloenzymes are distinguished from the larger class of metalloproteins by their ability to bind the metal ion in the "resting state." Coordination complexes serve a variety of biological functions that include oxygen binding and transport (see Section 14.4), metal transport and storage, and electron transport and catalysis. Many of the most important complexes in nature are built from a class of chelating ligands called porphyrins. The figures show the parent porphyrin ligand, porphine, and two biologically important coordination complexes, heme and chlorophyll *a*. We briefly discuss the structures and functions of these two complexes and then provide an example of another complex of this type, cytochrome P-450, that illustrates how the variable oxidation states of the transition metals enable enzymes to efficiently catalyze redox reactions.

Porphine (Fig. C8.1) is a planar tetradentate ligand that coordinates to metal ions through the nitrogen lone pairs in the equatorial positions of an octahedral complex. This arrangement leaves one of the axial positions available to bind an amino acid residue in proteins and the other axial position available to bind substrates in enzyme-catalyzed reactions. Heme (Fig. C8.2) is an iron porphyrin complex that is part of *hemoglobin*, the protein responsible for oxygen transport in blood. The heme is bound to the globin protein through the fifth coordination site, leaving the sixth coordination site available to bind oxygen reversibly. Hemoglobin's cooperative binding equilibria are discussed in Section 14.4. Chlorophyll *a* (Fig. C8.3) is a magnesium porphyrin that absorbs light as the primary event in photosynthesis, creating excited states that initiate a series of redox reactions that ultimately convert CO_2 to carbohydrates as our primary energy source (see Section 20.7). There are a number of chlorophyll molecules found in photosynthetic systems that absorb light over different regions of the visible spectrum. The diversity of these molecules enables more efficient conversion of solar energy than could be achieved by a single kind of chlorophyll molecule.

Enzymes are proteins that catalyze biological reactions (see Section 18.8). Cytochromes are heme-based, membrane-bound proteins that catalyze a variety of oxidation-reduction reactions. The cytochrome P-450 family is an important set of enzymes that catalyze the addition of oxygen to hydrocarbon substrates as part of the body's defense against hydrophobic compounds such as pharmaceuticals, steroids, and pesticides. The hydroxylation of R—H to R—OH makes these compounds more soluble in aqueous solutions, facilitating their elimination from the body. Many, if not most, common pharmaceuticals are metabolized and eliminated by cytochrome P-450 in the liver. An iron porphyrin is bound to the cytochrome protein via a sulfur linkage to the amino acid cysteine on the fifth (axial) coordination site, with the sixth (axial) site available for coordination to the substrate. These cytochromes absorb visible light most strongly at 450 nm, when CO is the sixth ligand, thus giving them their name. Figure C8.4 shows the basic

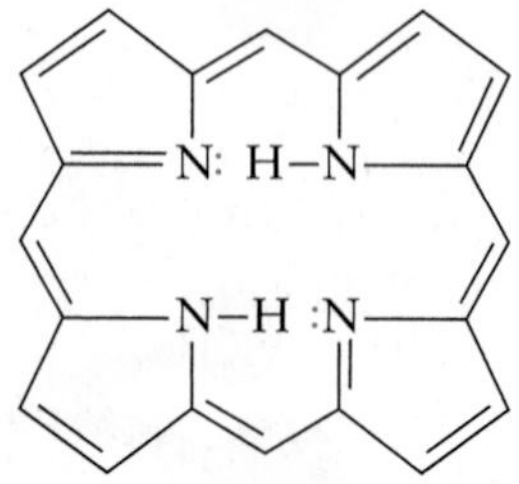

FIGURE C8.1 Porphine.

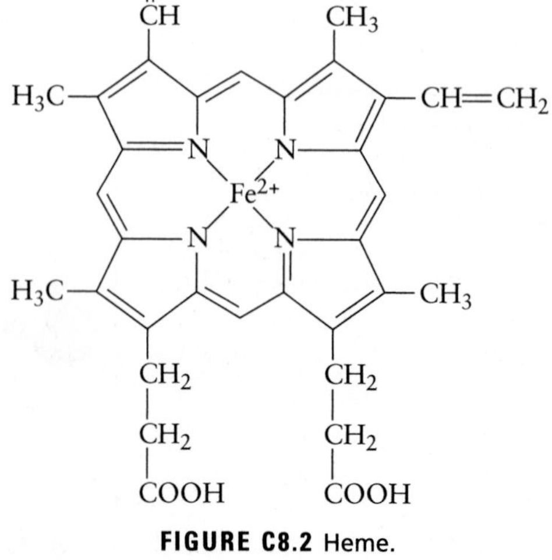

FIGURE C8.2 Heme.

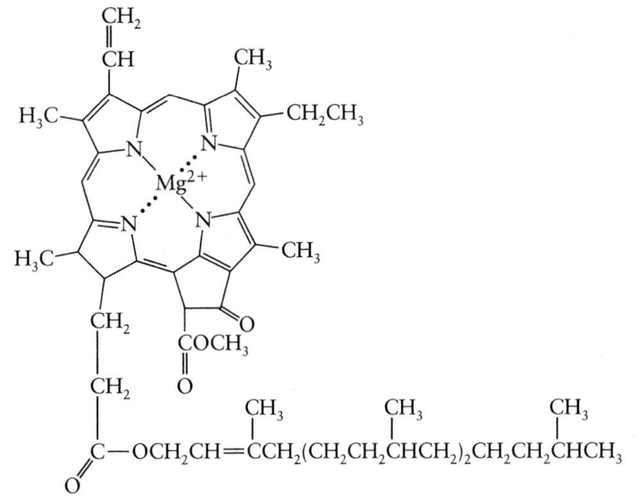

FIGURE C8.3 Chlorophyll *a*.

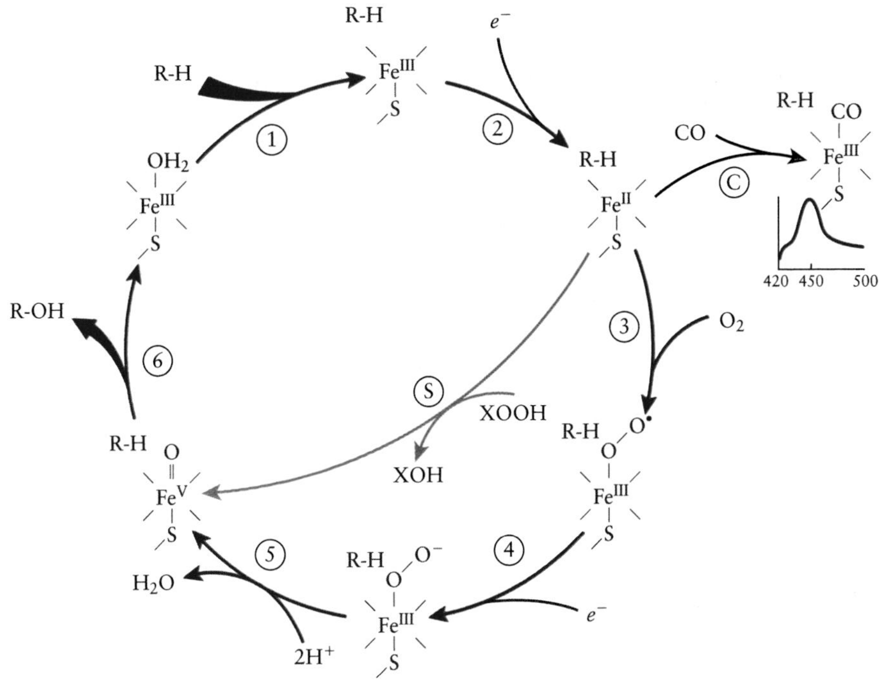

FIGURE C8.4 Cytochrome P-450 catalytic cycle.

catalytic cycle for cytochrome P-450, beginning with the enzyme in its resting state in which water is bound as the sixth ligand and iron is present in the Fe(III) oxidation state. Hydrocarbon (R—H) substrates bind to the active site, displacing water and changing the electronic configuration of iron from low-spin to high-spin (step 1). An electron transferred from an electron donor reduces Fe(III) to Fe(II), and O_2 binds to the sixth site as a free radical, oxidizing Fe(II) to Fe(III) in the process. The transfer of a second electron creates a short-lived peroxy (O_2^-) species that reacts with two protons to eliminate water and form a highly reactive Fe(V)-oxo species that is stabilized by the sulfur ligand. This highly reactive oxo ligand is then thought to insert into the R—H bond as a neutral atom, oxidizing the substrate to R—OH and reducing Fe(V) back to Fe(III) to begin the cycle all over again. The figure shows two alternate reactions that begin with the Fe(II) species. Exposure to CO at this point in the cycle produces the species that gives rise to the strong absorption at 450 nm, as shown in step C. The reaction labeled S shows an alternate route to the production of the Fe(V)-oxo species if stronger oxidizing agents like peroxide (XOOH) are available. This mechanism provides an excellent example of the way in which the variable oxidation states of transition metals in coordination complexes can effect catalytic reactions by electron transfer.

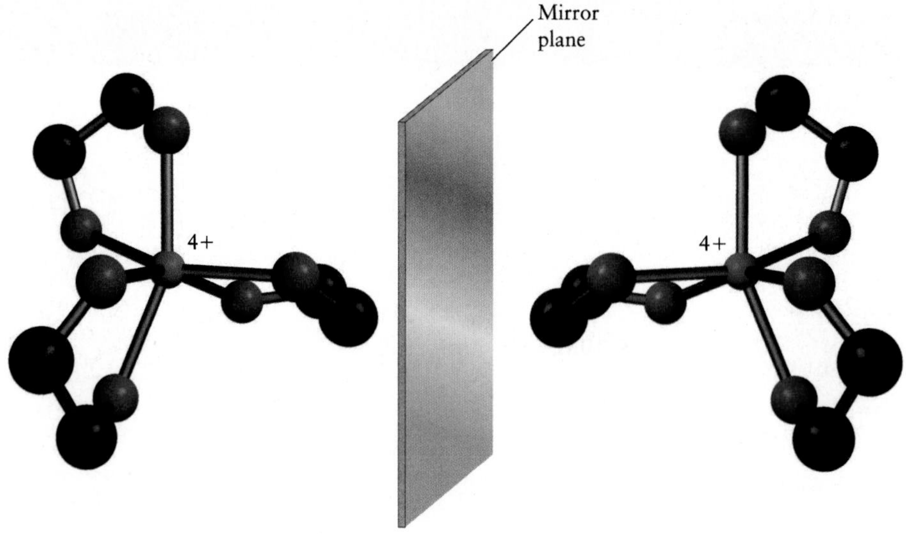

FIGURE 8.13 Enantiomers of the $[Pt(en)_3]^{4+}$ ion. Reflection through the mirror plane transforms one enantiomer into the other. The two cannot be superimposed by simple rotation.

The square-planar geometry (see Figs. 8.12b,c) is common for four-coordinate complexes of Au^{3+}, Ir^{+}, Rh^{+}, and especially common for ions with the d^8 valence electron configurations: Ni^{2+}, Pd^{2+}, and Pt^{2+}, for example. The Ni^{2+} ion forms a few tetrahedral complexes, but four-coordinate Pd^{2+} and Pt^{2+} are nearly always square planar. Square-planar complexes of the type MA_2B_2 can have isomers, as illustrated in Figures 8.12b and c for *cis*- and *trans*-$[Pt(NH_3)_2Cl_2]$. The *cis* form of this compound is a potent and widely used anticancer drug called cisplatin, but the *trans* form has no therapeutic properties.

Finally, linear complexes with coordination numbers of 2 are known, especially for ions with d^{10} configurations such as Cu^{+}, Ag^{+}, Au^{+}, and Hg^{2+}. The central Ag atom in a complex such as $[Ag(NH_3)_2]^{+}$ in aqueous solution strongly attracts several water molecules as well, however, so its actual coordination number under these circumstances may be greater than 2.

8.3.3 Chiral Structures

Molecules that rotate plane polarized light in opposite directions are called optical isomers (see Section 7.1, Fig. 7.10). They typically have chiral structures that cannot be superimposed on their mirror images by rotation. The two structures shown in Figure 8.13 for the complex ion $[Pt(en)_3]^{4+}$ are examples of such a mirror-image pair.

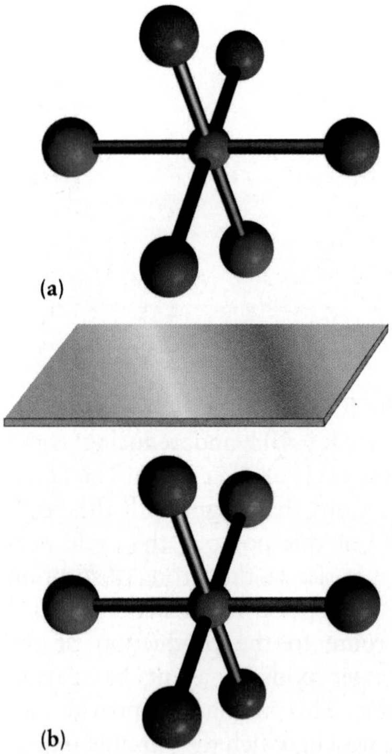

FIGURE 8.14 The structure of (a) the all-*cis*-$[Co(NH_3)_2(OH_2)_2Cl_2]^{+}$ complex ion, together with (b) its mirror image.

EXAMPLE 8.6

Suppose that the complex ion $[Co(NH_3)_2(OH_2)_2Cl_2]^{+}$ is synthesized with the two ammine ligands *cis* to each other, the two aqua ligands *cis* to each other, and the two chloro ligands *cis* to each other (Fig. 8.14a). Is this complex optically active?

Solution

We represent a mirror by a plane in Figure 8.14, shown as tilted for clarity, and create the mirror image by making each point in the image lie at the same distance from the plane as the corresponding point in the original structure. Comparing the original (see Fig 8.14a) with the mirror image (see Fig. 8.14b) shows that *cis*, *cis*-$[Co(NH_3)_2(OH_2)_2Cl_2]^{+}$ is chiral, because the two structures cannot be superimposed even after they are turned. Although many chiral complexes contain chelating ligands, this example proves that nonchelates can be chiral.

Related Problems: 19, 20

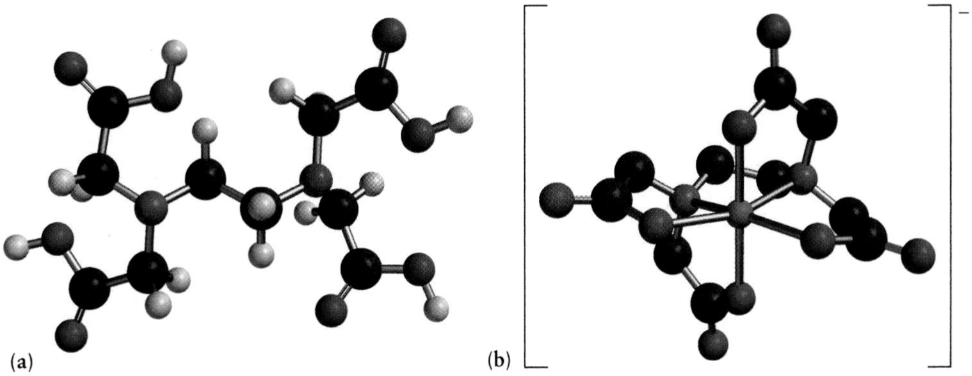

(a) (b)

FIGURE 8.15 (a) The neutral form of EDTA (showing four protons bonded to $EDTA^{4-}$ (the protons in each OH group are removed when EDTA binds to metal ions. (b) The chelation complex of EDTA with cobalt(III). Each EDTA ion has six donor sites at which it can bind (by donating lone-pair electrons) to the central metal ion (in (b), the H atoms on the CH_2 groups not shown for clarity).

The hexadentate ligand EDTA (ethylenediaminetetra-acetate ion) forms chiral complexes. Figure 8.15 shows the structure of this chelating ligand coordinated to a Co^{3+} ion. The central metal ion is literally "enveloped" by the ligand, forming six coordinate covalent bonds with two nitrogen atoms and four oxygen anions. A chelating agent like EDTA has a strong affinity for certain metal ions and can **sequester** them effectively in solution. EDTA solubilizes the scummy precipitates that Ca^{2+} ion forms with anionic constituents of soap by forming a stable complex with Ca^{2+}. In so doing it breaks up the main contributor to bathtub rings and it is a "miracle ingredient" in some bathtub cleaners. EDTA is also used to recover trace contaminants from water (some metal ions, especially heavy ones, are toxic). It has been used as an antidote for lead poisoning because of its great affinity for Pb^{2+} ions. Iron complexes of EDTA in plant foods permit a slow release of iron to the plant. EDTA also sequesters copper and nickel ions in edible fats and oils. Because these metal ions catalyze the oxidation reactions that turn oils rancid, EDTA preserves freshness.

8.4 Crystal Field Theory: Optical and Magnetic Properties

What is the nature of the bonding in coordination complexes of the transition metals that leads to their special properties? Why does Pt(IV) form only octahedral complexes, whereas Pt(II) forms square-planar ones, and under what circumstances does Ni(II) form octahedral, square-planar, and tetrahedral complexes? Can trends in the length and strength of metal–ligand bonds be understood? To answer these questions, we need a theoretical description of bonding in coordination complexes.

8.4.1 Crystal Field Theory

Crystal field theory, which is based on an ionic description of metal–ligand bonding, provides a simple and useful model for understanding the electronic structure, optical properties, and magnetic properties of coordination complexes. The theory was originally developed to explain these properties of ions in solids, for example, the red color of ruby, which arises from Cr^{3+} ions in an Al_2O_3 lattice. It was quickly applied to the related problem of understanding the bonding, structures, and other properties of coordination complexes. The theory treats the complex as a central metal ion perturbed by negatively charged ligands. In an octahedral complex, the six ligands are treated as negative point charges aligned along the $\pm x$, $\pm y$, and $\pm z$ coordinate axes with respect to a metal atom or ion at the origin. The energy of an electron is the same in any of the five degenerate *d* orbitals of a metal atom or ion in the gas phase. When external charges are present, however (Fig. 8.16), the energies of electrons in the various *d* orbitals change by different amounts because of the Coulomb repulsion between the external charges and the electrons in the *d* orbitals. The magnitude of the Coulomb repulsion depends inversely on separation between the charge densities of the ligands and of the electron densities in each of the orbitals, which is different for the various orbitals. An electron in a $d_{x^2-y^2}$ or d_{z^2} orbital on the central metal atom is most likely to be found along the coordinate axes, where it experiences a strong repulsive interaction with the electrons from the ligand, raising the energy of the orbital. In contrast, an electron in a d_{xy}, d_{yz}, or d_{xz} orbital of the metal is most likely to be found *between* the coordinate axes, and therefore experiences less repulsion from an octahedral array of charges; the energy of these orbitals is also raised but not by as much as that of the $d_{x^2-y^2}$ and d_{z^2} orbitals. The octahedral crystal field breaks the

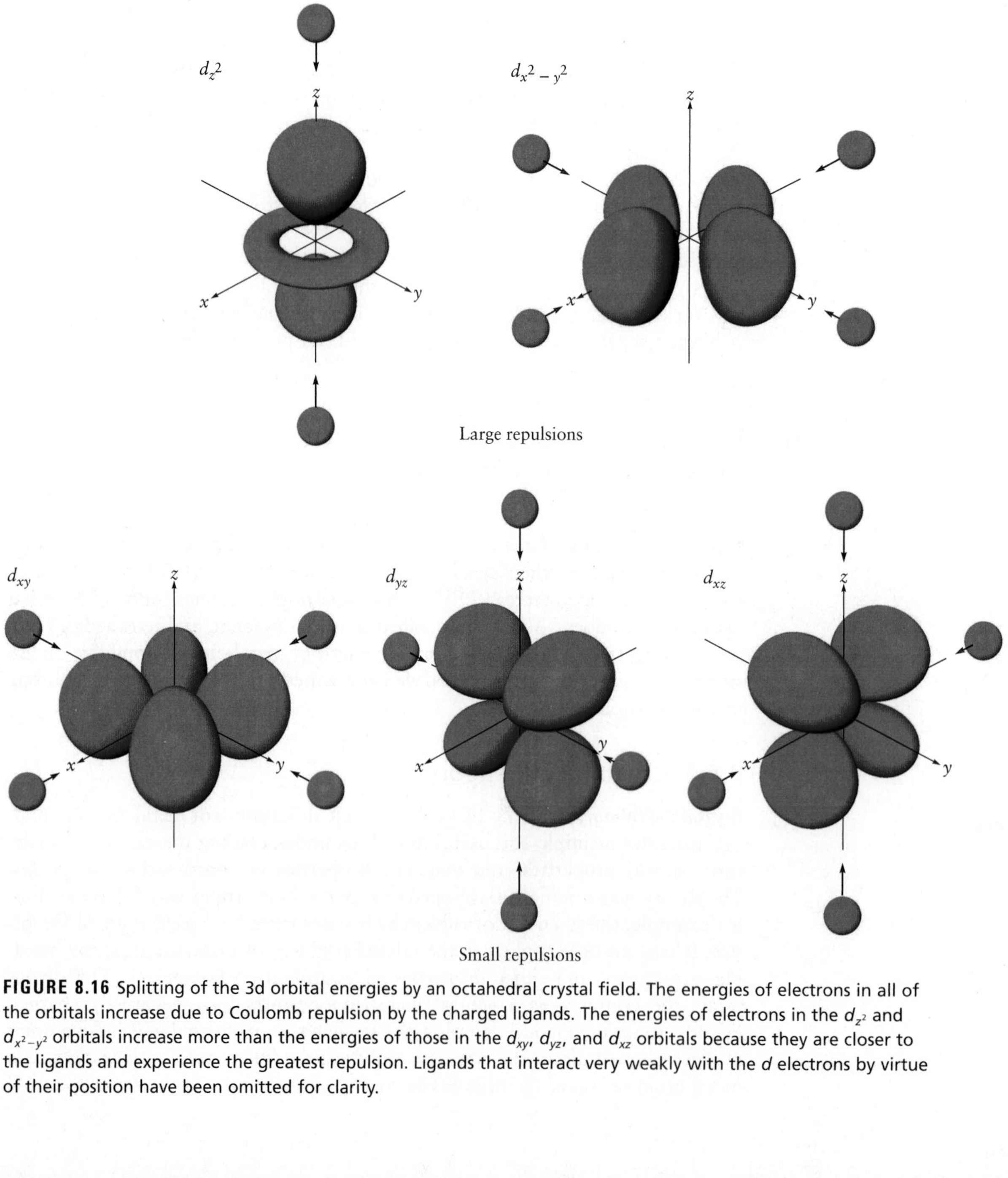

FIGURE 8.16 Splitting of the 3d orbital energies by an octahedral crystal field. The energies of electrons in all of the orbitals increase due to Coulomb repulsion by the charged ligands. The energies of electrons in the d_{z^2} and $d_{x^2-y^2}$ orbitals increase more than the energies of those in the d_{xy}, d_{yz}, and d_{xz} orbitals because they are closer to the ligands and experience the greatest repulsion. Ligands that interact very weakly with the *d* electrons by virtue of their position have been omitted for clarity.

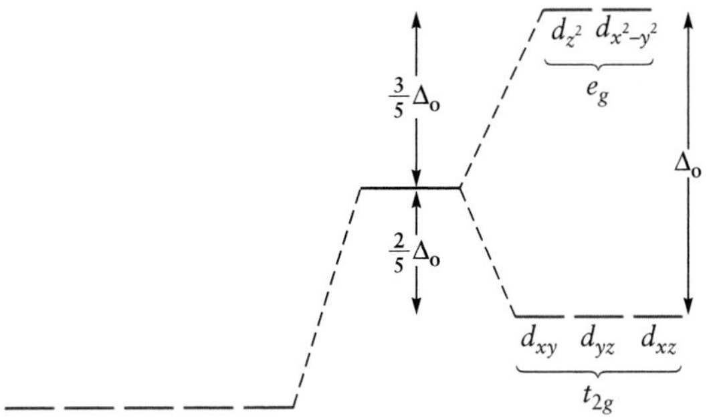

FIGURE 8.17 An octahedral crystal field increases the energies of all five *d* orbitals, but the increase is greater for the d_{z^2} and $d_{x^2-y^2}$ orbitals.

degeneracy of the *d* orbitals into two groups: a set of triply degenerate t_{2g} orbitals at lower energy and a set of doubly degenerate e_g orbitals at higher energy, as shown in Figure 8.17. The d_{xy}, d_{yz}, and d_{xz} orbitals of the isolated atom become the three t_{2g} orbitals in a complex, and the $d_{x^2-y^2}$ and d_{z^2} orbitals become the two e_g orbitals. The labels t_{2g} and e_g specify the symmetry and degeneracy (number of orbitals with the same energy) of each set of orbitals. *t* orbitals are threefold (triply) degenerate, whereas *e* orbitals are twofold (doubly) degenerate. The subscript *g* has its usual meaning; *g* orbitals are symmetric with respect to inversion of the coordinates (see Section 6.1). The difference in energy between the two levels is called the *crystal field splitting energy*, denoted by the symbol Δ_o for an octahedral crystal field. The crystal field splitting is typically measured using optical absorption spectroscopy because the energy level separation corresponds to the energy of photons in the visible region of the spectrum, which is why these complexes often show such vibrant colors.

Figure 8.17 shows the increase in the energies of all of the *d* orbitals expected for ions that are surrounded by spherical crystal fields, as well as the splitting caused by octahedral fields. The energy of the t_{2g} level is lower by $\frac{2}{5}\Delta_o$ than it would have been without splitting, and the energy of the e_g level is higher by $\frac{3}{5}\Delta_o$. Let's build up the *d*-electron configurations of the common ions in the first transition series using the Aufbau principle. Ti^{3+} is the first common ion we encounter that has partially filled *d* orbitals, with a single electron in one of the lower energy t_{2g} orbitals. Ti^{2+} and V^{3+} are isoelectronic with a d^2 configuration. Hund's rule tells us that the lower energy configuration is one in which there is a single electron in each of two t_{2g} orbitals with parallel spins. V^{2+} and Cr^{3+} form an isoelectronic pair with a d^3 configuration; the lowest energy state is the one in which a single electron occupies each of the t_{2g} orbitals with parallel spins. All of these ions have one or more unpaired electrons and are therefore paramagnetic. There are two possibilities for the d^4 configurations of Cr^{2+} and Mn^{3+}; the fourth electron can occupy one of the vacant e_g orbitals with parallel spin or it can be paired with one of the electrons in an occupied t_{2g} orbital. The lowest energy configuration is determined by comparing the energy cost of pairing an electron (the pairing energy) with the cost of promoting an electron to the higher energy e_g orbital. Both the crystal field splitting and the pairing energy depend on the metal and the ligands, and so it is not immediately obvious which configuration has the lowest energy in any particular case.

We have illustrated the two possibilities for the electronic configurations of Cr^{2+} and Mn^{3+} in Figure 8.18. The fourth electron will occupy an orbital in the lower level if the crystal field splitting is much larger than the pairing energy. Conversely, the fourth electron will occupy an orbital in the upper level if the crystal field splitting is much smaller than the pairing energy. The total spin (sum of the spins of the electrons) is lower in stronger crystal fields than in weaker crystal fields, and so the complexes are referred to as **low-spin complexes** with **low-spin configurations**, and **high-spin complexes** with **high-spin configurations**, respectively. Low spin configurations arise from strong crystal fields, so they are also called **strong-field configurations**, and the ligands that produce such fields are

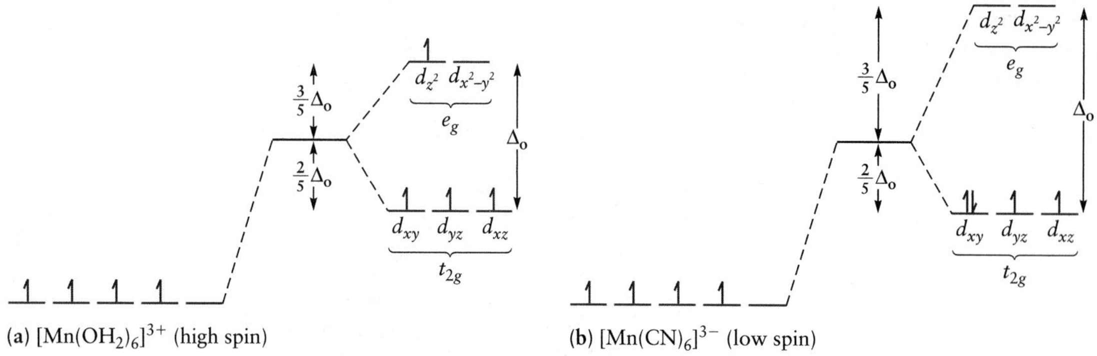

FIGURE 8.18 Electron configuration for (a) high-spin (small Δ_o) and (b) low-spin (large Δ_o) Mn(III) complexes.

called **strong-field ligands.** Conversely, high-spin configurations arise from weak crystal fields so they are also called **weak-field configurations** and the ligands that produce such fields are called **weak-field ligands.** These terms are used interchangeably (weak-field ⟷ high-spin; strong-field ⟷ low-spin); you should be alert, therefore, when reading inorganic texts or the scientific literature. These two configurations are only important for complexes with 4–7 *d* electrons. There is no ambiguity in assigning configurations for complexes having three or fewer *d* electrons or more than seven *d* electrons.

Table 8.5 summarizes the electron configurations possible for 10 electrons in an octahedral crystal field, provides specific examples of transition metal ions from the first transition series with these configurations, and tabulates the **crystal field stabilization energy (CFSE)** of each type of octahedral complex. The CFSE is the energy difference between electrons in an octahedral crystal field and those in the hypothetical spherical crystal field introduced in Figure 8.16. If these five orbitals are either fully occupied or half full (as in d^{10} or high-spin d^5 complexes), then the energy of the ion is predicted to be the same as in a spherical field: the CFSE is zero. If the lower energy orbitals are preferentially occupied, however, the configuration is stabilized. For example, in a low-spin d^4 complex, the energy of each of the four electrons in the t_{2g} orbitals is lowered by $\frac{2}{5}\Delta_o$, resulting in a total CFSE of $-\frac{8}{5}\Delta_o$.

The CFSE helps to explain the trends in the enthalpies of hydration of ions in the first transition series shown in Figure 8.3. If each measured value (blue curve) is adjusted by correcting for the CFSE of that complex ion, results quite close to the straight red lines are obtained. The relatively small magnitude of the enthalpy of hydration for Mn^{2+} arises from the high-spin d^5 configuration of Mn^{2+}, which results in a CFSE of zero. Negative CFSEs lower the enthalpies of hydration for ions on either side of Mn^{2+}.

8.4.2 Square-Planar and Tetrahedral Complexes

Crystal field theory applies to square-planar and tetrahedral complexes, as well as to octahedral complexes. Let's consider a square-planar complex in which four negative charges are arranged around a metal ion along the $\pm x$- and $\pm y$-axes, as shown in the upper right and lower left images in Figure 8.16. The relative *d*-orbital energy level ordering in a square-planar crystal field can be predicted using the same reasoning we applied to the octahedral case. The magnitude of the repulsive interaction between electrons in a given orbital and the electrons of the ligand depends on the degree to which the electron density is concentrated along the *x*- and *y*- axes. An electron in the $d_{x^2-y^2}$ orbital, which is oriented along these axes, experiences the greatest repulsion and lies at the highest energy. The energy of the d_{xy} orbital is lower than that of the $d_{x^2-y^2}$

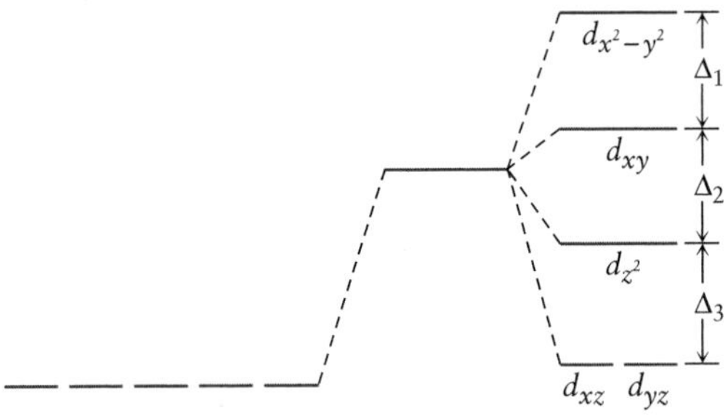

FIGURE 8.19 Energy-level structure of the 3d orbitals in a square planar crystal field, derived by withdrawing axial ligands from an octahedral field.

orbital because the lobes of this orbital are oriented at 45° to the axes. The d_{z^2} orbital energy level is lower still because its electron density is concentrated along the z-axis, with a small component in the x-y plane. Finally, the d_{xz} and d_{yz} orbitals experience the least repulsion. These orbitals are the most stable in a square-planar crystal field because they have nodes in the x-y plane. Figure 8.19 shows the resulting energy level diagram.

Tetrahedral complexes result from ligands located on four of the eight corners of an imaginary cube with the metal ion at its center. The $d_{x^2-y^2}$ and d_{z^2} orbitals point toward the centers of the cube faces, but the other three orbitals point toward the centers of the cube edges, which are closer to the corners occupied by the ligands. Electrons in the latter orbitals are therefore more strongly repelled than those in the former. Figure 8.20 shows the result; the energy level ordering is the reverse of that found for octahedral complexes. In addition, the magnitude of the splitting Δ_t is about half that of Δ_o.

Another way to think of the square-planar crystal field splittings is to consider what happens to the octahedral energy levels shown in Figure 8.17 as the two ligands on the $\pm z$ axis move away from the metal. Let's consider each of the two degenerate levels separately. As the ligands retreat, the energy of the d_{z^2} orbital falls because of the decreased repulsion; the energy of the $d_{x^2-y^2}$ orbital must increase to conserve the total energy of the e_g level. Similarly, the energies of the d_{xz} and d_{yz} orbitals of the t_{2g} level are stabilized as the ligands are pulled away and the energy of the d_{xy} orbital increases. So the octahedral crystal field energy levels distort smoothly into the square planar levels as the z axis ligands are pulled away. Some coordination complexes are described as having distorted octahedral structures, with the two z-axis ligands moved outward but not removed completely. The level splittings observed in these cases are intermediate between the octahedral and the square-planar patterns. Figure 8.21 tracks the energies of the orbitals for various crystal fields beginning with a spherical field in which the five d orbitals are degenerate. The tetrahedral crystal field and its associated energy level diagram is shown to the left, the octahedral crystal field is shown to the right, followed by the series of structures derived from the octahedral field by distorting and finally removing the axial ligands.

Octahedral complexes are the most common because the formation of six bonds to ligands, rather than four, confers greater stability. Square-planar arrangements are important primarily for complexes of d^8 ions with strong field ligands. The low-spin configuration that leaves the high-energy $d_{x^2-y^2}$ orbital vacant is the most stable. The low-spin configuration of a square planar d^8 complex is more stable than the corresponding configuration of an octahedral complex because the energy of the highest occupied orbitals in the square planar complex is lower than those in the octahedral complex. Finally, tetrahedral complexes are less stable than either octahedral or square planar complexes for two reasons. First, the crystal field splitting is smaller so the lower set of energy levels is stabilized less than in the other geometries. Second, because the lower-energy set of levels is only doubly degenerate, electrons must be placed into the upper level at an earlier stage in the filling process.

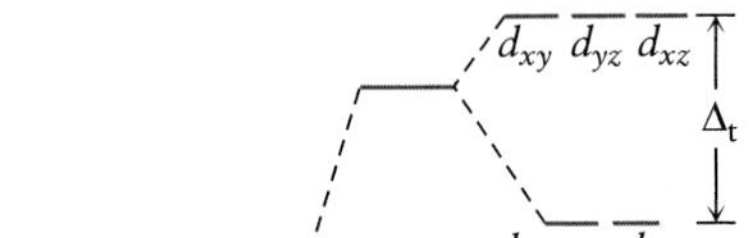

(a)

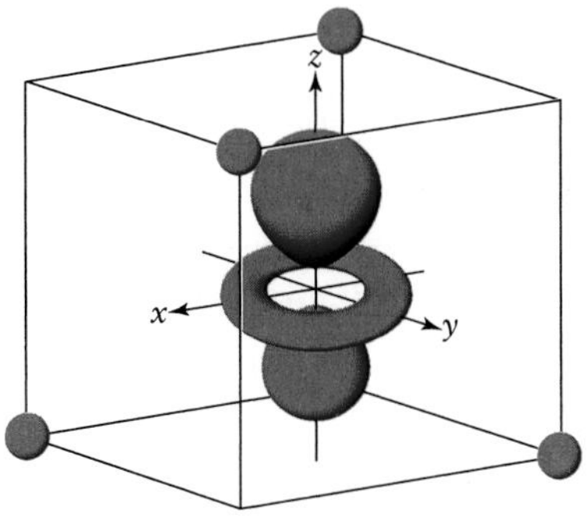

(b)

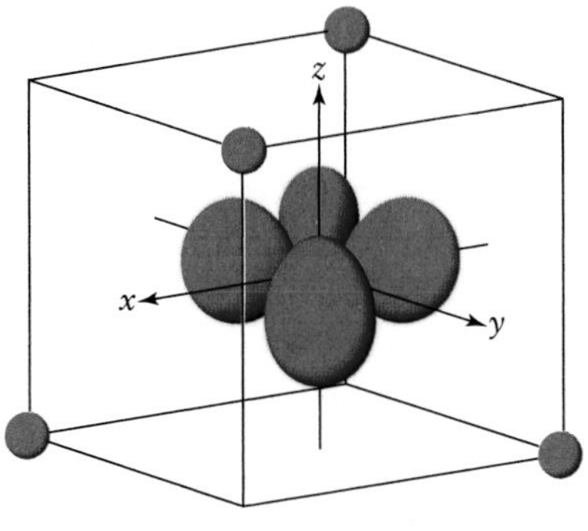

(c)

FIGURE 8.20 (a) Energy level diagram for a tetrahedral crystal field. (b) The energies of the d_{z^2} (shown) and $d_{x^2-y^2}$ orbitals lie lower in energy because they point to the cube faces, away from the ligands. (c) The energies of the d_{xy} (shown), d_{yz}, and d_{xz} orbitals lie higher in energy because they point to the cube edges, closer to the ligands.

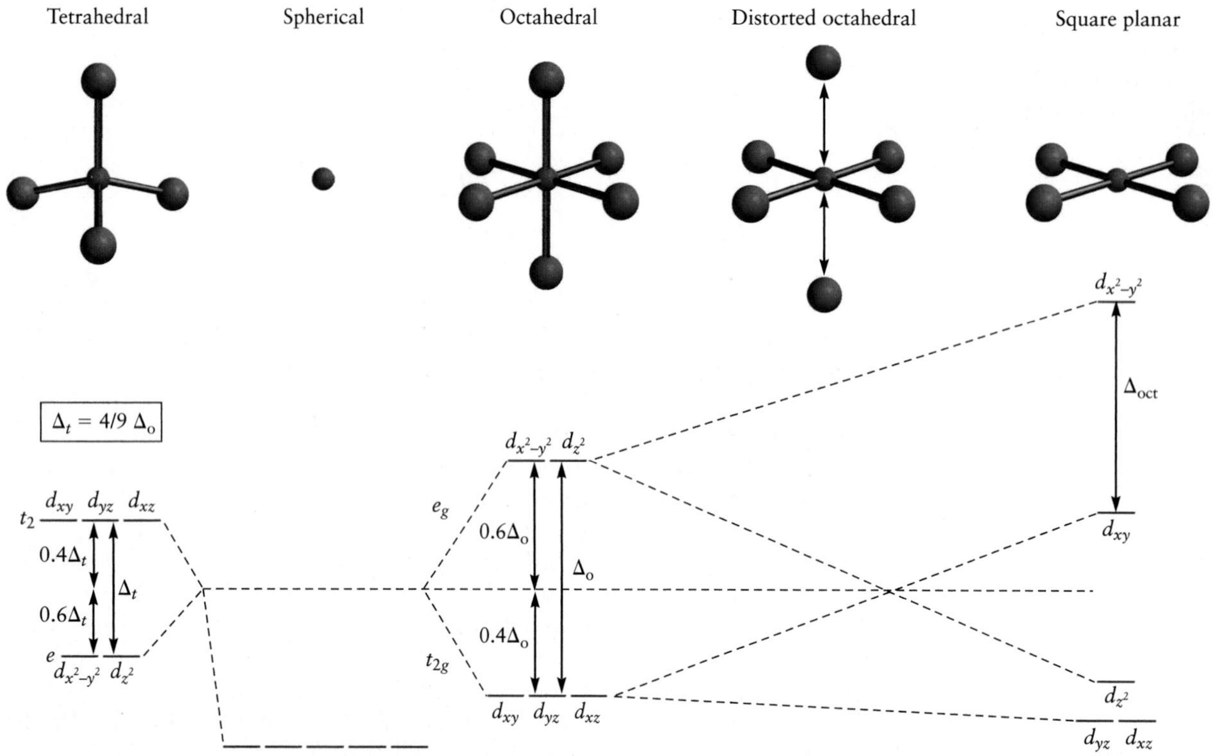

FIGURE 8.21 Correlation diagram showing the relationships among *d*-orbital energy levels in crystal fields of different symmetries. The five orbitals are degenerate in an isolated ion or atom and shift upwards in a spherical field. Splitting of the levels by an octahedral crystal field and those derived from it are shown to the right; the energy level diagram for a tetrahedral crystal field is shown on the left. Dashed lines track the energies of specific orbitals in the different crystal fields.

Crystal field theory allows us to predict, at least qualitatively, what geometries to expect for different coordination complexes, as determined by the identity and oxidation state of the metal and the characteristics of the ligands. Higher coordination numbers are more frequently observed for elements of the second and the third transition series than for those of the first transition series because the larger atomic and ionic radii of the former can accommodate more ligands. For a given metal in a given oxidation state, lower coordination numbers become favored as the ligands become more bulky. Steric repulsion between the ligands, a nonbonded interaction discussed in Section 7.2, overcomes the attractive interactions between the ligands and the central metal. Tetrahedral coordination is favored over octahedral coordination when the central atom is small and the ligands are bulky, for this reason. The oxoanions of metals in high oxidation states on the left side of the *d*-block, for example, generally have tetrahedral geometries, as do complexes of the heavier halides (Cl^-, Br^-, I^-) bound to the M^{2+} metals on the right side of the *d*-block. Common examples include $[VO_4]^{3-}$, $[CrO_4]^{2-}$, and $[MnO_4]^-$ for the oxoanions and $[FeCl_4]^{2-}$ and $[CuBr_4]^{2-}$ for the halides. Square planar complexes of metals in the first transition series are typically formed between metal ions with d^8 configurations and ligands that can accept electrons from the metal ion, such as CN^- (see Section 8.6), whereas d^8 metals in the second transition series are often square planar, irrespective of the nature of the ligand. Prominent examples include $[NiCN_4]^{2-}$ and $[Pt(NH_3)_4]^{2+}$.

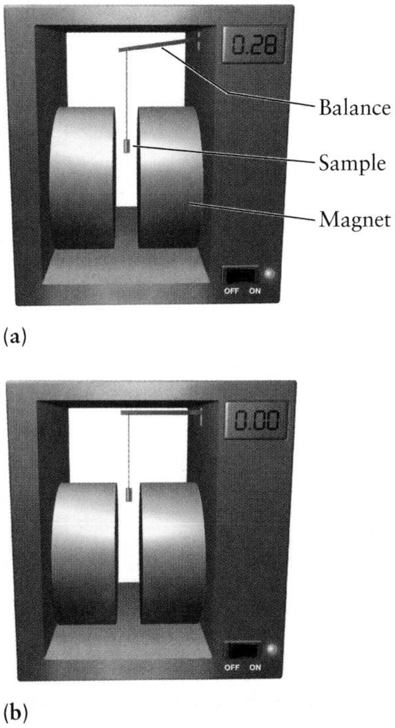

FIGURE 8.22 Samples are initially inserted about halfway into a magnetic field to create a field gradient that exerts a force on magnetic dipoles. (a) Paramagnetic compounds are attracted by the field and "weigh" more than they do with no field present. (b) Diamagnetic compounds are repelled slightly by the field and "weigh" less than they do with no field present.

8.4.3 Magnetic Properties

The existence of high- and low-spin configurations accounts for the magnetic properties of many different coordination compounds. As discussed in Section 6.5, substances are classified as paramagnetic or diamagnetic according to whether they are attracted into a magnetic field. Figure 8.22 shows a schematic of an experiment that demonstrates the universal *susceptibility* of substances to the influence of magnetic fields. A cylindrical sample is suspended between the poles of a powerful magnet whose field is not uniform throughout space. It is weighed accurately in the absence of a magnetic field and then again in the presence of the field. (There are several ways to do this; either the sample or the magnet may be moved, or the magnet may be switched off and then on.) The net force on the sample (apparent weight) is found to be different in the presence of the magnetic field. The measurements just described provide not only qualitative characterization of a sample's magnetic properties, but also a quantitative value for its **magnetic susceptibility,** that is, the strength of its interaction with a magnetic field. *Diamagnetic* compounds are slightly repelled by the magnetic field and weigh less, whereas *paramagnetic* compounds are attracted by the magnetic field and weigh more. Diamagnetic compounds have small, negative susceptibilities, whereas paramagnetic compounds have much larger, positive susceptibilities.

Paramagnetic compounds have one or more unpaired electrons, as discussed in Section 6.5, whereas all of the electrons in diamagnetic substances are paired. Magnetic susceptibility measurements not only identify substances as paramagnetic or diamagnetic, but also count the number of unpaired spins; these measurements were very important in validating crystal field theory and more refined theories that followed.

These facts emerge in connection with coordination complexes, because paramagnetism is prevalent among transition-metal complexes, whereas most other chemical substances are diamagnetic. Among complexes of a given metal ion, the number of unpaired electrons, as observed by magnetic susceptibility, varies with the identities of the ligands. Both $[Co(NH_3)_6]^{3+}$ and $[CoF_6]^{3-}$ have six ligands surrounding a central Co^{3+} ion; yet, the former is diamagnetic (because it is a strong-field, low-spin complex), and the latter is paramagnetic with four unpaired electrons (because it is a weak-field, high-spin complex). Similarly, $[Fe(CN)_6]^{4-}$ is diamagnetic, but $[Fe(OH_2)_6]^{2+}$ has four unpaired electrons; these complexes also correspond to the two d^6 configurations shown in Table 8.5.

TABLE 8.5

Electron Configurations and Crystal Field Stabilization Energies for High- and Low-Spin Octahedral Complexes

Configuration		d^1	d^2	d^3	d^4	d^5	d^6	d^7	d^8	d^9	d^{10}
Examples		Ti^{3+}	Ti^{2+}, V^{3+}	V^{2+}, Cr^{3+}	Cr^{2+}, Mn^{3+}	Mn^{2+}, Fe^{3+}	Fe^{2+}, Co^{3+}	Co^{2+}, Ni^{3+}	Ni^{2+}, Pt^{2+}	Cu^{2+}	Zn^{2+}
HIGH SPIN	e_g	— —	— —	— —	↑ —	↑ ↑	↑ ↑	↑ ↑	↑ ↑	↑↓ ↑	↑↓ ↑↓
	t_{2g}	↑ — —	↑ ↑ —	↑ ↑ ↑	↑ ↑ ↑	↑ ↑ ↑	↑↓ ↑ ↑	↑↓ ↑↓ ↑	↑↓ ↑↓ ↑↓	↑↓ ↑↓ ↑↓	↑↓ ↑↓ ↑↓
	CFSE	$-\frac{2}{5}\Delta_o$	$-\frac{4}{5}\Delta_o$	$-\frac{6}{5}\Delta_o$	$-\frac{3}{5}\Delta_o$	0	$-\frac{2}{5}\Delta_o$	$-\frac{4}{5}\Delta_o$	$-\frac{6}{5}\Delta_o$	$-\frac{3}{5}\Delta_o$	0
LOW SPIN	e_g				— —	— —	— —	↑ —			
	t_{2g}				↑↓ ↑ ↑	↑↓ ↑↓ ↑	↑↓ ↑↓ ↑↓	↑↓ ↑↓ ↑↓			
	CFSE	Same as high spin			$-\frac{8}{5}\Delta_o$	$-\frac{10}{5}\Delta_o$	$-\frac{12}{5}\Delta_o$	$-\frac{9}{5}\Delta_o$	Same as high spin		

CFSE, Crystal field stabilization energies.

EXAMPLE 8.7

The octahedral complex ions $[FeCl_6]^{3-}$ and $[Fe(CN)_6]^{3-}$ are both paramagnetic, but the former is high spin and the latter is low spin. Identify the *d*-electron configurations in these two octahedral complex ions. In which is the octahedral field splitting greater? How does the CFSE differ between the complexes?

Solution

The Fe^{3+} ion has five *d* electrons. A high-spin complex such as $[FeCl_6]^{3-}$ has five unpaired spins ($t_{2g}^3 e_g^2$); a low-spin complex such as $[Fe(CN)_6]^{3-}$ has one unpaired spin (t_{2g}^5). The splitting Δ_o must be greater for cyanide than for chloride ion ligands. The CFSE for the $[FeCl_6]^{3-}$ complex is zero, whereas that for the $[Fe(CN)_6]^{3-}$ complex is $-2\Delta_o$.

Related Problems: 27, 28

8.5 Optical Properties and the Spectrochemical Series

FIGURE 8.23 Several colored coordination compounds. (clockwise from top left) They are $Cr(CO)_6$ (white), $K_3[Fe(C_2O_4)_3]$ (green), $[Co(en)_3]I_3$ (orange), $[Co(NH_3)_5(OH_2)]Cl_3$ (red), and $K_3[Fe(CN)_6]$ (red-orange).

Transition-metal complexes are characterized by their rich colors, which are often deep, vibrant, and saturated (Fig. 8.23). The colors depend on the oxidation state of the metal ion, the number and nature of the ligands, and the geometry of the complex. Earlier figures have shown the color changes that accompany dehydration and ligand substitution reactions and also the different colors of a pair of geometric isomers. The following series of Co(III) complexes shows how the colors of coordination complexes of the same ion can vary with different ligands:

$[Co(NH_3)_6]^{3+}$	Orange
$[Co(NH_3)_4Cl_2]^{+}$	A green form and a violet form
$[Co(NH_3)_5(OH_2)]^{3+}$	Purple

Coordination complexes appear colored when they absorb visible light. Recall that atoms absorb light when the energy of an incident photon exactly matches the energy difference between two atomic energy levels (see Fig. 4.15). The missing wavelengths appear as dark lines against the spectral rainbow. The transmitted light still appears quite white to our eyes, however, because atomic absorption lines are so narrow. Only a small fraction of the visible light has been absorbed. Coordination complexes, on the other hand, absorb light over significant regions of the visible spectrum. What we see is the color that is *complementary* to the color that is most strongly absorbed (see Section 20.5). The $[Co(NH_3)_5Cl]^{2+}$ ion, for example, absorbs greenish yellow light, with the strongest absorption occurring near 530 nm. Only the red and blue components of white light are transmitted through an aqueous solution of this ion, which appears purple to us. Materials that absorb all visible wavelengths appear gray or black, and those that absorb visible light weakly or not at all appear colorless.

Crystal field theory was developed, in part, to explain the colors of transition-metal complexes. It was not completely successful, however. Its failure to predict trends in the optical absorption of a series of related compounds stimulated the development of ligand field and molecular orbital theories and their application in coordination chemistry. The colors of coordination complexes are due to the excitation of the *d* electrons from filled to empty *d* orbitals (*d-d* transitions). In octahedral complexes, the electrons are excited from occupied t_{2g} levels to empty e_g levels. The crystal field splitting Δ_o is measured directly from the optical absorption spectrum of the complexes. The wavelength of the strongest absorption is called λ_{max} and it is related to Δ_o as follows. $E = h\nu$, so

FIGURE 8.24 The colors of the hexa-aqua complexes of metal ions (from left) Mn^{2+}, Fe^{3+}, Co^{2+}, Ni^{2+}, Cu^{2+}, and Zn^{2+}, prepared from their nitrate salts. Note that the d^{10} Zn^{2+} complex is colorless. The green color of the Ni^{2+} is due to absorption of both red and blue light that passes through the solution. The yellow color of the solution containing $[Fe(OH_2)_6]^{3+}$ is caused by hydrolysis of that ion to form $[Fe(OH)(OH_2)_5]^{2+}$; if this reaction is suppressed, the solution is pale violet.

$\Delta_o = h\nu = hc/\lambda_{max}$. Because energy is inversely proportional to wavelength, compounds with small crystal field splittings absorb light with longer wavelengths, toward the red end of the visible spectrum, and those with large crystal field splittings absorb light with shorter wavelengths, toward the blue end of the spectrum.

In $[Co(NH_3)_6]^{3+}$, an orange compound that absorbs most strongly in the violet region of the spectrum, the crystal field splitting Δ_o is larger than in $[Co(NH_3)_5Cl]^{2+}$, a violet compound that absorbs most strongly at lower frequencies (longer wavelengths) in the yellow–green region of the spectrum. d^{10} complexes (like those of Zn^{2+} or Cd^{2+}) are colorless because all of the d levels (both t_{2g} and e_g) are filled. Because there are no empty orbitals available to accept an excited electron, the transition is not allowed, which means that the absorption is weak or nonexistent. High-spin d^5 complexes such as $[Mn(OH_2)_6]^{2+}$ and $[Fe(OH_2)_6]^{3+}$ also show only weak absorption bands because excitation of an electron from a filled t_{2g} level to an empty e_g level would require its spin to flip in order to satisfy the Pauli principle. (Recall that all unpaired spins are oriented parallel in high-spin complexes.) Light absorption rarely reverses the spin of an electron, so the optical absorption of these compounds is weak, as shown by the pale pink color of the hexa-aqua Mn^{2+} complex in Figure 8.24. Table 8.6 lists the crystal field splitting energies (CFSE) expressed in wave numbers $1/\lambda_{max}$ in units cm^{-1} and absorption wavelengths for a number of coordination complexes to give you a feel for their diversity.

Experimental measurements of the optical absorption spectra and magnetic properties of transition-metal complexes have provided a critical test of the validity of crystal field theory. The theory makes specific predictions about the strengths of

TABLE 8.6

Crystal Field Splitting Energies and Wavelengths of Maximum Absorption for Selected Octahedral Transition Metal Complexes

Complex	λ_{max} (nm)	CFSE (cm^{-1})	Complex	λ_{max} (nm)	CFSE (cm^{-1})
$[TiF_6]^{3-}$	588	17,006	$[Co(NH_3)_6]^{3+}$	437	22,883
$[Ti(OH_2)_6]^{3+}$	492	20,325	$[Co(CN)_6]^{3-}$	290	34,483
$[V(OH_2)_6]^{3+}$	560	17,857	$[Co(OH_2)_6]^{2+}$	1075	9,302
$[V(OH_2)_6]^{2+}$	806	12,407	$[Ni(OH_2)_6]^{2+}$	1176	8,503
$[Cr(OH_2)_6]^{3+}$	575	17,452	$[Ni(NH_3)_6]^{2+}$	926	10,799
$[Cr(NH_3)_6]^{3+}$	463	21,598	$[RhBr_6]^{3-}$	463	21,519
$[Cr(CN)_6]^{3-}$	376	26,596	$[RhCl_6]^{3-}$	439	22,780
$Cr(CO)_6$	311	32,154	$[Rh(NH_3)_6]^{3+}$	293	34,130
$[Fe(CN)_6]^{3-}$	310	32,258	$[Rh(CN)_6]^{3-}$	227	44,053
$[Fe(CN)_6]^{4-}$	296	33,784	$[IrCl_6]^{3-}$	362	26,724
$[Co(OH_2)_6]^{3+}$	549	18,215	$[Ir(NH_3)_6]^{3+}$	250	40,000

crystal fields produced by different ligands for a given metal cation in a specific oxidation state. It also makes specific predictions about trends in the crystal field splitting with the oxidation state of a particular metal for a series of complexes containing the same ligands. Simple ionic ligands such as the halides should produce much stronger crystal fields than neutral ligands due to the larger electrostatic repulsion between the negatively charged anion and the *d* electrons of the metal ion. Coulomb's law also predicts that these repulsive interactions should increase with decreasing ionic ligand radius, for a given metal cation, and with increasing oxidation state for a given metal, which allows closer approach of the ligands. Crystal field theory predicts splittings that increase in the order $I^- < Br^- < Cl^- < F^-$. A systematic ranking of the strength of various ligands was obtained by comparing the optical absorption spectra of a series of complexes with the general formula $[Co(III)(NH_3)_5X]^{n+}$. The strength of the interaction between a single ligand X and the Co^{3+} ion could be measured directly because all other interactions and the geometry of the complex remained constant. Ligands were ranked from weakest to strongest on the basis of their crystal field splittings in the **spectrochemical series** as follows (below, py = pyridine and Ph = phenyl):

$$I^- < Br^- < S_2^- < SCN^- < Cl^- < NO_3^- < N_3^- < F^- < OH^- <$$
$$OH_2 < NCS^- < py < NH_3 < en < NO_2^- < PPh_3 < CN^- \approx CO \approx NO^+$$

Weak-field ligands (high-spin) Intermediate-field ligands Strong-field ligands (low-spin)

Although this order is not followed for all metal ions, it is a useful generalization that helps us understand the properties of various coordination complexes. There is a similar spectrochemical series for metal ions that is largely independent of the ligands. It is $Mn^{2+} < Ni^{2+} < Co^{2+} < Fe^{2+} < V^{2+} < Fe^{3+} < Co^{3+} < Mn^{4+} < Mo^{3+}, Rh^{3+} < Ru^{3+} < Pd^{4+} < Ir^{3+} < Pt^{4+}$. These spectrochemical series illustrate the failure of crystal field theory to provide a satisfactory account of the factors that govern crystal field splitting. Neutral ligands with lone pairs, such as water and ammonia, produce larger splittings than any of the halides, and ligands with low-lying antibonding π molecular orbitals produce the largest splittings of all. Some of the trends in the series for the metal ions make sense—the increase in Δ_o with increasing oxidation number, for example—but others do not. The increase in Δ_o moving down a group is inconsistent with a simple electrostatic analysis because the larger ionic radii of the second and third transition series ions would be expected to reduce rather than increase the Coulomb repulsion between the metal *d* electrons and the ionic ligands. A more comprehensive theory is clearly required to explain the spectrochemical series. We apply the concepts and methods of molecular orbital theory developed in Chapter 6 to account for the trends observed in the spectrochemical series, as well as to provide a more satisfactory description of structure and bonding in coordination complexes and show how molecular orbital theory correctly accounts for the trend observed in the spectrochemical series.

EXAMPLE 8.8

Predict which of the following octahedral complexes has the shortest λ_{max}: $[FeF_6]^{3-}$, $[Fe(CN)_6]^{3-}$, $[Fe(OH_2)_6]^{3+}$.

Solution

$[Fe(CN)_6]^{3-}$ has the strongest field ligands of the three complexes; thus, its energy levels are split by the greatest amount. The frequency of the light absorbed should be greatest, and λ_{max} should be the shortest for this ion.

$[Fe(CN)_6]^{3-}$ solutions are red, which means that they absorb blue and violet light. Solutions of $[Fe(OH_2)_6]^{3+}$ are a pale violet due to the weak absorption of red light, and $[FeF_6]^{3-}$ solutions are colorless, indicating that the absorption lies beyond the long wavelength limit of the visible spectrum.

Related Problems: 37, 38, 39, 40

8.6 Bonding in Coordination Complexes

8.6.1 Valence Bond Theory

Valence bond (VB) theory is used widely in contemporary chemistry to describe structure and bonding in transition-metal compounds, especially coordination complexes. The VB model is intuitively appealing for this purpose for several reasons. Because transition-metal compounds, particularly coordination complexes, often comprise a central atom surrounded by ligands in a symmetric arrangement, forming hybrid orbitals on the central atom with the appropriate symmetry to bond to these ligands is a natural approach to the problem. Often, little interaction occurs among metal–ligand bonds, so the local description is reasonable. Participation of the *d* electrons enables a much more varied set of structures and hybrid orbitals than can be formed from only *s* and *p* orbitals. This section describes two sets of hybrid orbitals used to describe bonding in transition-metal compounds and complexes, and provides examples of each.

Hybridization is justified here for precisely the same reasons we laid out in Section 6.8. The lobes of the hybrid orbitals point toward the ligands and overlap the ligand orbitals more strongly than the standard atomic orbitals. The energy cost to promote electrons from lower energy atomic orbitals (e.g., 3*d* and 4*s*) to the highest energy orbital (4*p*) to form hybrid orbitals is more than offset by the energy gained by forming a stronger bond with the ligand.

We construct the first set of hybrid orbitals from one *s* atomic orbital, the three *p* atomic orbitals and the d_{z^2} atomic orbital; they are called sp^3d hybrid orbitals. The principal quantum numbers of the participating atomic orbitals depend on the particular metal atom under consideration; for Co, they would be the 3*d*, 4*s*, and 4*p* atomic orbitals. The sp^3d hybrid orbitals in the most general case are written out as

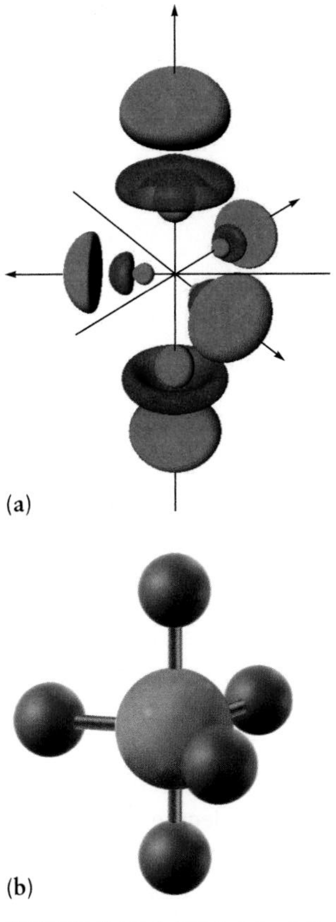

(a)

(b)

FIGURE 8.25 (a) Hybrid orbitals formed from linear combinations of d_{z^2}, s, p_x, p_y, and p_z orbitals. The pair of orbitals that point along the positive and negative *z*-axes are the same except for their orientation in space; they are called *axial* orbitals. The set of three orbitals in the *x-y* plane are equivalent to one another and are called *equatorial* orbitals. All five orbitals have been pulled apart from their proper positions at the origin for clarity. (b) This set of hybrid orbitals can be used to describe the bonding in PF_5, for example.

$$\chi_1 = \sqrt{\frac{1}{3}}\,[s + \sqrt{2}\,(p_x)]$$

$$\chi_2 = \sqrt{\frac{1}{3}}\left[s - \sqrt{\frac{1}{2}}\,(p_x) + \sqrt{\frac{3}{2}}\,(p_y)\right]$$

$$\chi_3 = \sqrt{\frac{1}{3}}\left[s - \sqrt{\frac{1}{2}}\,(p_x) - \sqrt{\frac{3}{2}}\,(p_y)\right]$$

$$\chi_4 = \sqrt{\frac{1}{2}}\,[p_z + d_{z^2}]$$

$$\chi_5 = \sqrt{\frac{1}{2}}\,[p_z - d_{z^2}]$$

As shown in Figure 8.25a, these orbitals point to the vertices of a trigonal bipyramid; there are three equivalent equatorial hybrids and two equivalent axial hybrids. Examples of molecules whose shapes are described by sp^3d hybridization include PF_5, which you have seen in Section 3.8, and $CuCl_5^{3-}$. PF_5 is shown in Figure 8.25b to show how VB theory using sp^3d hybrid orbitals can rationalize its structure.

The second set of hybrid orbitals we construct are the sp^3d^2 hybrids; these are six equivalent orbitals directed toward the vertices of an octahedron (Fig. 8.26a). They describe the structures and bonding in all of the octahedral

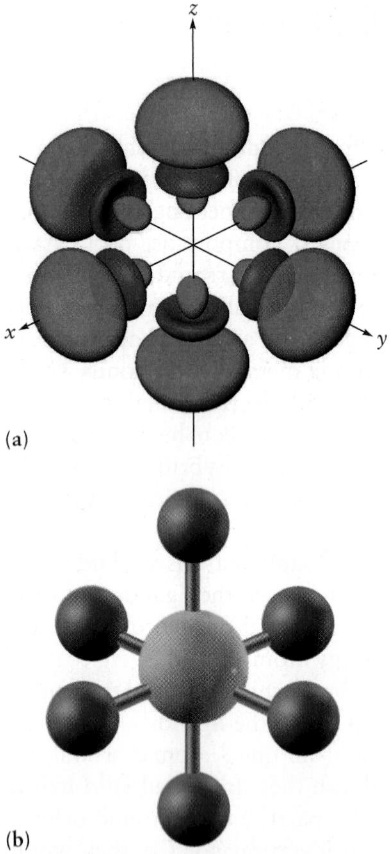

FIGURE 8.26 (a) Hybrid orbitals formed from linear combinations of d_{z^2}, $d_{x^2-y^2}$, p_x, p_y, and p_z orbitals. All six orbitals are equivalent except for their orientation in space. The six orbitals have been pulled apart from their proper positions at the origin for clarity. (b) This set of hybrid orbitals can be used to describe the bonding in SF_6, for example.

TABLE 8.7

Examples of Hybrid Orbitals and Bonding in Complexes

Coordination Number	Hybrid Orbital	Configuration	Examples
2	*sp*	Linear	$[Ag(NH_3)_2]^+$
3	sp^2	Trigonal planar	BF_3, NO_3^-, $[Ag(PR_3)_3]^+$
4	sp^3	Tetrahedral	$Ni(CO)_4$, $[MnO_4]^-$, $[Zn(NH_3)_4]^{2+}$
4	sp^2d	Square planar	$[Ni(CN)_4]^{2-}$, $[Pt(NH_3)_4]^{2+}$
5	sp^3d	Trigonal bipyramidal	TaF_5, $[CuCl_5]^{3-}$, $[Ni(PEt_3)_2Br_3]$
6	sp^3d^2	Octahedral	$[Co(NH_3)_6]^{3+}$, $[PtCl_6]^{2-}$

Adapted from G.E. Kimball. *J. Chem. Phys.* 1940, 8, 188.

coordination complexes discussed in Section 8.3, as well as that in SF_6, which we show in Figure 8.26b.

$$\chi_1 = \sqrt{\frac{1}{6}}\,[s + \sqrt{3}\,(p_z) + \sqrt{2}\,(d_{z^2})]$$

$$\chi_2 = \sqrt{\frac{1}{6}}\left[s + \sqrt{3}\,(p_z) - \sqrt{\frac{1}{2}}\,(d_{z^2}) + \sqrt{\frac{3}{2}}\,(d_{x^2-y^2})\right]$$

$$\chi_3 = \sqrt{\frac{1}{6}}\left[s + \sqrt{3}\,(p_z) - \sqrt{\frac{1}{2}}\,(d_{z^2}) - \sqrt{\frac{3}{2}}\,(d_{x^2-y^2})\right]$$

$$\chi_4 = \sqrt{\frac{1}{6}}\left[s - \sqrt{3}\,(p_z) - \sqrt{\frac{1}{2}}\,(d_{z^2}) + \sqrt{\frac{3}{2}}\,(d_{x^2-y^2})\right]$$

$$\chi_5 = \sqrt{\frac{1}{6}}\left[s - \sqrt{3}\,(p_z) - \sqrt{\frac{1}{2}}\,(d_{z^2}) - \sqrt{\frac{3}{2}}\,(d_{x^2-y^2})\right]$$

$$\chi_6 = \sqrt{\frac{1}{6}}\,[s - \sqrt{3}\,(p_z) + \sqrt{2}\,(d_{z^2})]$$

Table 8.7 shows the variety of hybrid orbitals that can be constructed from various combinations of *s*, *p*, and *d* orbitals, the shapes of the molecules that result, and selected examples.

VB theory with hybrid orbitals is widely used to rationalize the structures of coordination complexes. It complements classical valence shell electron-pair repulsion (VSEPR) theory by using methods of quantum mechanics to describe the geometry of coordination complexes. As with main-group elements, VB theory is better suited to rationalize structure and bonding after the fact than to predict structure. And by treating the bonds as local and equivalent, it fails completely to account for the colors and magnetic properties of coordination complexes. These shortcomings motivate the application of molecular orbital theory to describe structure and bonding in coordination chemistry.

8.6.2 Molecular Orbital Theory

The failure of crystal field theory and VB theory to explain the spectrochemical series stimulated the development of **ligand field theory**, which applies qualitative methods of molecular orbital theory to describe the bonding and structure of coordination complexes. The terms *ligand field theory* and *molecular orbital theory* are often used interchangeably when discussing bonding in modern inorganic chemistry; we prefer to use the term *molecular orbital (MO) theory* for consistency throughout this textbook.

Let's apply MO theory to describe bonding in octahedral coordination complexes using the same approach developed in Section 6.5. MOs are formed as linear combinations of atomic orbitals of the different atoms of the complex. We include the valence *d, s,* and *p* orbitals of the central metal, as well as the *s* and *p* orbitals of the ligands in our minimal basis set. We restrict our discussion to metals of the first transition series, so the valence orbitals of the metal are the 3*d*, 4*s*, and 4*p* orbitals. The relevant ligand orbitals are the 2*s*, 2*p*, 3*s*, and 3*p* orbitals for most of the ligands we discuss, for example, the 2*p* orbitals of F^- or OH^-, or the 3*p* orbitals of Cl^-. They may also be hybrid orbitals constructed from this set of atomic orbitals, for example, the sp^3 orbitals of the aqua (H_2O) or ammine (NH_3) ligands.

We begin by constructing a set of σ MOs using the metal orbitals and six of the ligand orbitals that point along the metal–ligand bonds, which are oriented along the Cartesian axes in octahedral complexes. Figure 8.27 illustrates the

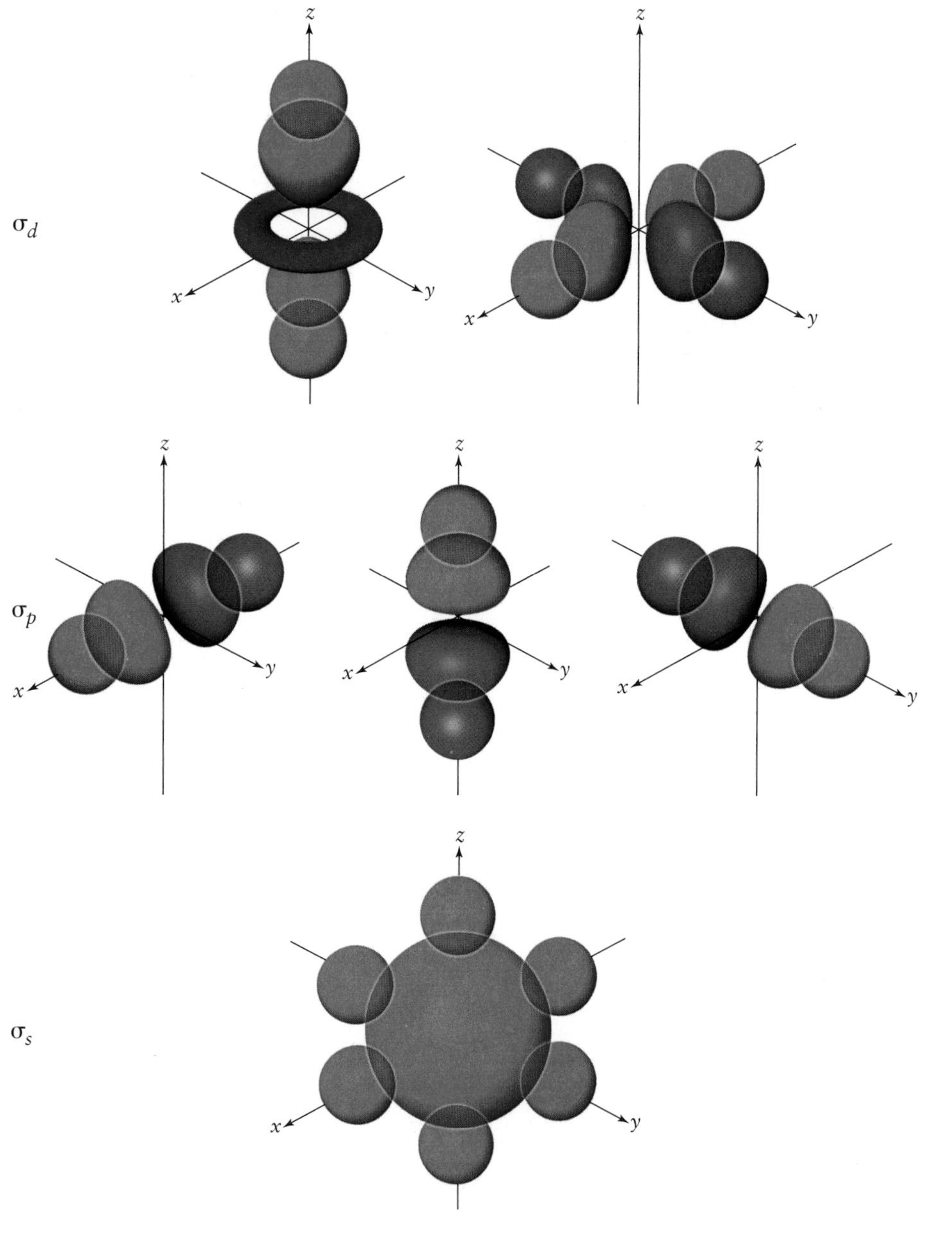

FIGURE 8.27 Overlap of metal orbitals with ligand orbitals to form σ bonds. The ligand orbitals can be either *p* orbitals or hybrid orbitals (for example, sp^3 for water), and thus they are represented only schematically.

bonding MOs; the corresponding antibonding MOs, as well as a set of nonbonding MOs, are not shown. A bonding σ orbital is formed by overlap of the metal 4*s* orbital with six ligand orbitals of the same phase. Overlap of the 4*s* metal orbital with six ligand orbitals of the opposite phase forms the corresponding antibonding σ^* MO. We label these orbitals σ_s and σ_s^*, respectively, to identify them by the number of radial nodes (angular momentum) and by their bonding and antibonding character, as well as to keep track of the atomic orbitals from which they were constructed, just as we did in Chapter 6. We then construct another set of bonding and antibonding σ orbitals from the three metal *p* orbitals and three ligand orbitals, which we denote σ_p and σ_p^*, respectively; each of these sets of orbitals is triply degenerate, differing only in their orientation in space. Only two of the five metal *d* orbitals are aligned along the bond axes in octahedral complexes. Overlap of the metal d_{z^2} and $d_{x^2-y^2}$ with ligand orbitals generates a pair of bonding and antibonding MOs that we label σ_d and σ_d^*, respectively. The d_{xy}, d_{yz}, and d_{xz} orbitals, whose lobes are oriented at 45° to the bond axes (see Fig. 5.13), have zero net overlap with the ligand orbitals, and are therefore nonbonding. We label these orbitals δ_{xy}^{nb}, δ_{yz}^{nb}, and δ_{zx}^{nb} by analogy to the way we labeled the nonbonding π orbital in HF shown in Figure 6.2; this scheme allows us to keep track of the nodal structure and angular momentum of the orbitals in a consistent way. We have constructed 15 MOs from our minimal basis set of 15 atomic orbitals; there are the six bonding σ orbitals shown in Figure 8.27, the six corresponding antibonding orbitals, and the three nonbonding metal δ^{nb} orbitals.

We have constructed the orbital correlation diagram shown in Figure 8.28 using the same procedure outlined in Section 6.6. The relative energies of the

FIGURE 8.28 Orbital correlation diagram for an octahedral ligand field, showing the energy-level filling for a $[CrCl_6]^{3-}$ ion. We have added an asterisk to the e_g label to explicitly identify this set of orbitals as antibonding.

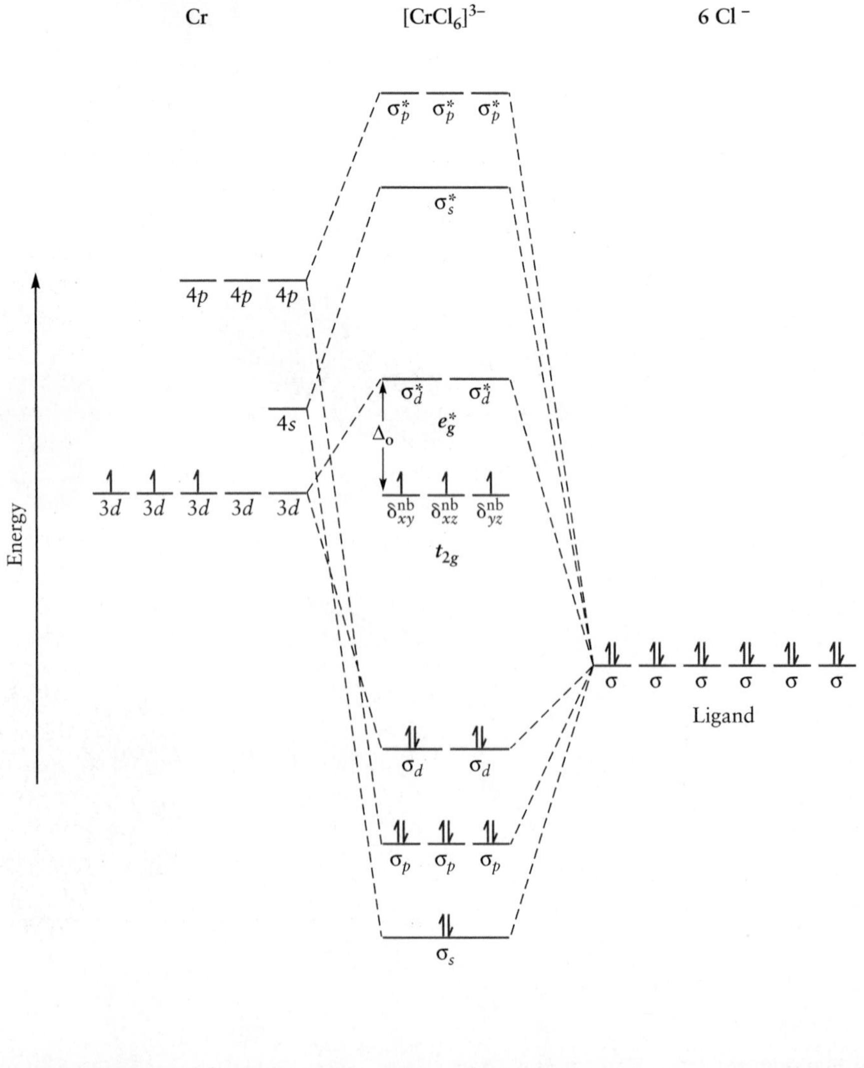

metal orbitals and the ligand orbitals were located in an approximate way using the ionization energies plotted in Figure 5.26. The metal orbitals lie at much higher energies than orbitals of the ligands used to construct σ bonds. The relative energies of the metal orbitals of the first transition series increase in the order $3d < 4s < 4p$, as shown. The molecular orbitals shown were constructed from the metal orbitals and the ligand orbitals by recalling that atomic orbitals interact strongly with one another to form MOs only when: (1) they are close to one another in energy and (2) they overlap significantly in space. The orbital correlation diagram shown is quite typical for transition metal complexes in which the bonds are primarily σ bonds, complexes that include ligands such as F^-, H_2O, and NH_3, in which the ligand orbitals lie much lower in energy than the metal orbitals.

Recall from our discussion of bonding in heteronuclear diatomic molecules (see Section 6.6) that the character of bonding MOs is predominately that of the more electronegative element. Metal fluoride σ orbitals, for example, are derived primarily from atomic fluorine orbitals, so we can think of the bonding in metal fluoride complexes as primarily ionic. The fluoride ion acts as a lone pair donor (Lewis base) that forms dative bonds with metal ions. The correlation diagram shows the six bonding MOs being populated by electron pair donation from the ligands, a bonding scheme called **ligand-to-metal** (L$\rightarrow$M) σ donation by inorganic chemists. MO theory predicts six bonds, just like VB theory, but each of these six σ bonds is delocalized over the entire complex, in contrast to the set of six localized M—L bonds predicted using VB theory.

Molecular orbital theory produces the same energy level structure for the center of the correlation diagram but the physical origin of the crystal field splitting is quite different, as discussed later. The same general conclusions and considerations apply, therefore, with respect to the optical and magnetic properties of coordination complexes. Orbitals are populated using an Aufbau principle with the metal electrons filling the t_{2g} and e_g^* levels as before. Low-spin configurations result when the crystal field splitting is larger than the pairing energy, and high-spin configurations result from the opposite situation. MO theory provides considerably more insight into the origin of the crystal field splitting, however, and a satisfactory explanation for the trends observed in the spectrochemical series that could not be provided by the earlier theories. MO theory identifies the t_{2g} orbitals as the δ_{xy}^{nb}, δ_{yz}^{nb}, and δ_{zx}^{nb} nonbonding MOs and the e_g^* orbitals as the antibonding σ_d^* orbitals; the crystal field splitting Δ_o is the energy difference between them. The energies of the δ^{nb} orbitals are unaffected by bonding, but the energies of the σ_d^* orbitals that lie above them are determined by the strength of the metal-ligand bonds. Recall from Section 6.5 that the energy of bonding MOs decreases with increasing bond strength while the energy of the corresponding antibonding MOs increases with increasing bond strength. The increase in crystal field splitting observed through the middle of the spectrochemical series correlates well with the strength of the metal-ligand σ bonds, but MO theory cannot account for the order observed throughout the rest of the series without considering π bonding.

Let's consider the effects of π bonding by first examining the nature of the bonds formed between a metal atom or ion and a single ligand. We then extend the treatment to examine the effects of π bonding on the energy level structure and properties of octahedral complexes. Figure 8.29a shows a bonding and an antibonding MO constructed from a metal d orbital and a ligand p orbital oriented "side-by-side." These MOs are typical for metals with few d electrons, such as Cr^{3+}, and for ligands with filled p orbitals that lie close to, but somewhat lower than, the energy of the metal d orbitals, for example, the $3p$ orbitals of Cl^-. The empty $4p$ ligand orbitals are not considered because they lie too high in energy. The bonding π orbital is primarily ligand in character, because the ligand orbitals are lower in energy than the metal orbitals; the antibonding

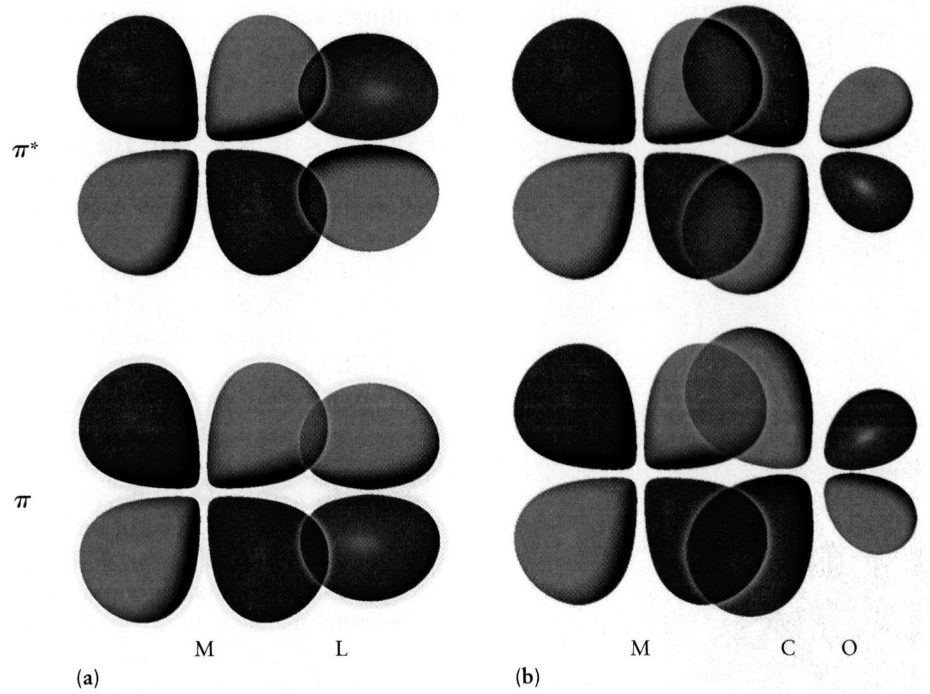

FIGURE 8.29 (a) Bonding and antibonding π orbitals formed by "side-by-side" overlap of a metal *d* orbital with an atomic ligand *p* orbital. (b) Bonding and antibonding π orbitals formed by "side-by-side" overlap of a metal *d* orbital with a molecular ligand π* orbital.

π* orbital is primarily metal in character for the same reason. The bonding orbital is populated by charge donation from the filled ligand orbitals; we call this kind of bonding (L→M) π donation, by analogy to that just described for (L→M) σ donation. Figure 8.29b illustrates the opposite situation in which the filled ligand orbitals lie at much lower energies than the metal orbitals and do not participate in bonding. There is an empty ligand π* orbital, however, that is sufficiently close in energy and of the correct symmetry that it can overlap with the metal *d* orbital to form a bonding and an antibonding MO by overlapping the metal orbital "side-by-side," as shown. The character of the bonding MO is largely determined by the character of the lower energy metal orbital, and it is filled by electrons transferred from the metal. Inorganic chemists call this bonding arrangement (M→L) π donation. It is important for metals with filled, or nearly filled, *d* orbitals and for ligands with π* orbitals that lie close in energy, CO and CN^- being among the most important examples. We see from the figure that the MO is bonding between the metal and one atom of the ligand, but antibonding between the ligand atoms. (M→L) π donation strengthens the M—C bond but weakens the C—O bond, restoring electron density on the oxygen atom; it is often called **backbonding** because it transfers charge back to the ligand, reducing excess negative charge on the metal due to (L→M) σ donation alone (in an attempt to achieve electroneutrality). (L→M) σ donation and (M→L) π donation are often called synergistic for these reasons.

We are now in a position to construct π MOs from the nonbonding metal orbitals oriented "side-by-side" with a set of four ligand *p* orbitals, an example of which is shown in Figure 8.30. Three pairs of bonding and antibonding MOs are generated in this way; they are labeled t_{2g} and t_{2g}^*, and they lie in the *x*–*y*, *y*–*z*, and *z*–*x* planes, respectively. These orbitals correspond to the π and π* orbitals just discussed for a single ligand. The character of these π orbitals is determined by the relative energies of the metal and ligand orbitals from which they are constructed, as well as the occupation of those levels before bonds are formed. Let's focus on the nature of the ligands by choosing a particular metal, say Cr^{3+}, and work out the effects of π bonding on the energy level structure for different classes of ligands. We first consider those ligands that have filled *p* orbitals with energies that lie slightly lower than the energies of the

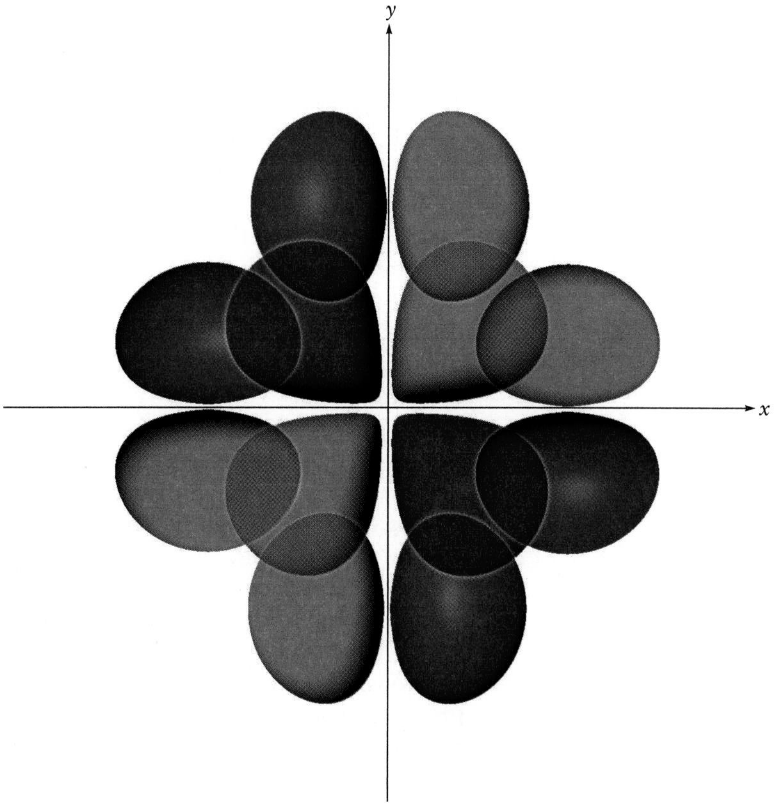

FIGURE 8.30 Bonding π MO formed by constructive overlap of a metal d_{xy} orbital with four ligand *p* orbitals. The corresponding π^* orbital is formed by overlap with *p* orbitals of the opposite phase. Equivalent sets of orbitals are formed using the metal d_{yz} and d_{zx} orbitals.

metal orbitals, as shown by the energy level diagram in Figure 8.31a. The overlap between the metal t_{2g} orbitals and the four ligand *p* orbitals creates a set of three new bonding t_{2g} MOs as well as a set of three new antibonding t_{2g}^* MOs that lie above and below the metal t_{2g} level. The bonding orbitals are mostly

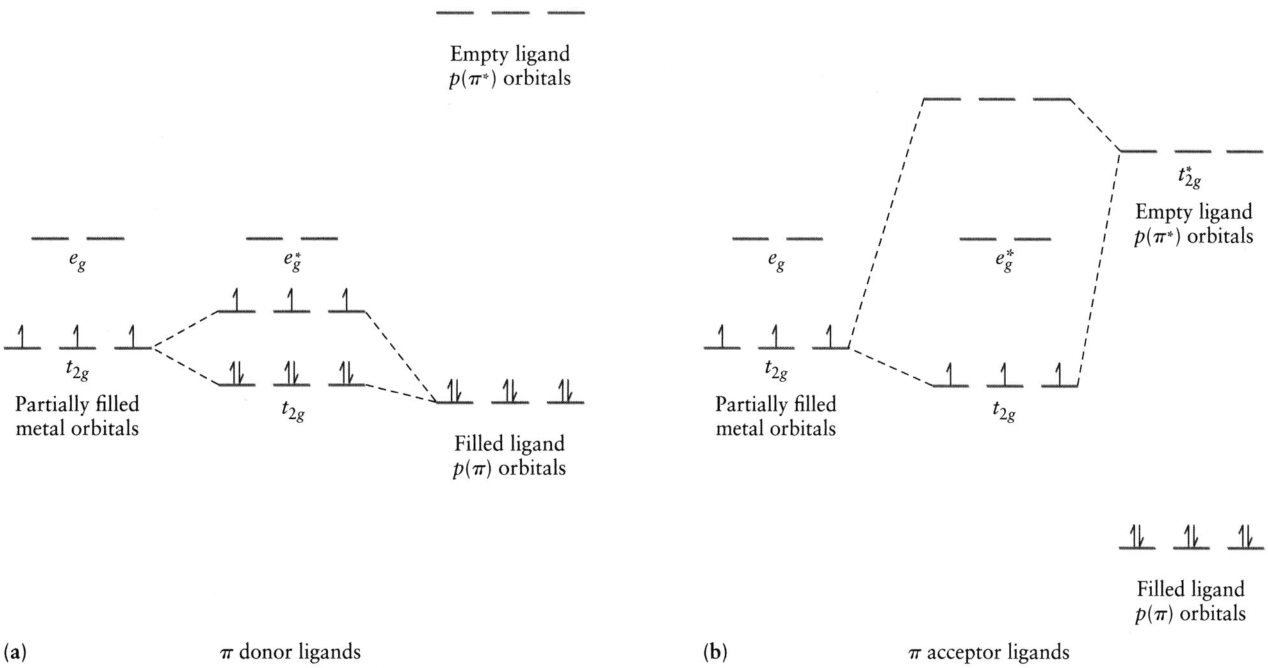

FIGURE 8.31 (a) Ligand-to-metal π donation showing the formation of bonding and antibonding MOs and a reduction in Δ_o compared with that from σ bonding alone. (b) Metal-to-ligand π donation showing the formation of bonding and antibonding MOs and an increase in Δ_o compared with that from σ bonding alone.

ligand in character, for the reasons discussed previously, and they have been completely filled by electrons transferred from the lone pairs of the six ligands. The *d* electrons occupy the new t_{2g}^* antibonding MO, which is located at higher energy than the metal t_{2g} level. The energy of the metal e_g^* level is unchanged, so the effect of this ligand-to-metal (L→M) π donation is to increase the energy of the highest occupied molecular orbital and decrease the crystal field splitting Δ_o. Strong π-donor ligands include I^- and Br^-, with occupied *p* orbitals whose energies lie close to the energies of the metal *d* orbitals (because they are 4*p* and 5*p* orbitals, respectively). The greater spatial extent of these more diffuse orbitals also overlaps strongly with the metal *d* orbitals resulting in the large splitting between the bonding and antibonding MOs observed. π donors account for the small crystal field splittings observed for ligands on the left side of the spectrochemical series.

Let's now consider the opposite situation in which the filled *p* orbitals of the ligands lie much lower in energy than the metal *d* orbitals, so much lower that they do not effectively contribute to the formation of MOs with the metal orbitals. Ligands on the right-hand side of the spectrochemical series have unoccupied π^* orbitals with energies sufficiently close to those of the metal *d* orbitals that they can effectively combine with the metal orbitals to form new MOs, as shown in Figure 8.31b. This interaction creates a set of bonding MOs with energy lower than the metal t_{2g} level that is populated by transfer of the metal electrons. This **metal-to-ligand (M→L) π donation** increases the crystal field splitting Δ_o and accounts for the splittings observed on the right side of the spectrochemical series. Ligands are classified according to their abilities to act as π donors, σ donors, or π acceptors, and this classification provides an explanation of the trends observed for the crystal field splitting through the spectrochemical series as follows.

$$\underbrace{I^- < Br^- < Cl^- < F^-}_{\pi \text{ donors}} < \underbrace{OH_2 < NH_3 < en}_{\sigma \text{ donors}} < \underbrace{CO, CN^-}_{\pi \text{ acceptors}}$$

EXAMPLE 8.9

Construct an MO correlation diagram for the complex ion $[Ni(NH_3)_6]^{2+}$ and populate the orbitals with electrons. Do you expect π bonding to be important for this complex? What features of this diagram correspond to those predicted using crystal field theory? What kind of bonding would VB theory predict for d^8 complexes, and how does MO theory account for the bonding more accurately, even qualitatively?

Solution

The correlation diagram looks like that shown in Figure 8.28, with the bonding sigma orbitals being populated through electron donation from the six ligand lone pairs. The eight metal *d* electrons fill the nonbonding t_{2g} orbitals, with one electron occupying each of the antibonding e_g^* orbitals. The electron spins are paired in the t_{2g} orbitals but are parallel to one another in the e_g^* orbitals, making the complex paramagnetic. There are no low energy π or π^* orbitals available on the ligands for π bonding. The center portion of the MO diagram is identical to that predicted by crystal field theory, as shown in Figure 8.17, and the magnetic properties are predicted to be the same by both methods. VB theory fails to describe bonding in this complex because it would need eight hybrid orbitals to accommodate the eight *d* electrons to form two-center two-electron bonds with the ligands. MO theory correctly accounts for the optical and magnetic properties of the complex.

Figure 8.32 summarizes the molecular orbital picture of bonding in octahedral coordination complexes, identifies the interactions that determine the splitting energy Δ_o, and provides guidance for predicting the bonding and energy-level structure for any coordination compound of interest, based on the oxidation state of the metal (number of *d* electrons) and the nature and occupancy of the ligand orbitals. The center portion of the figure shows the orbital splitting predicted using the simple molecular orbital picture that includes only σ bonding; the three *d* electrons shown could be found in complexes of V^{2+}, Cr^{3+}, or Mn^{4+}, for example. Crystal field theory would make the same prediction for the

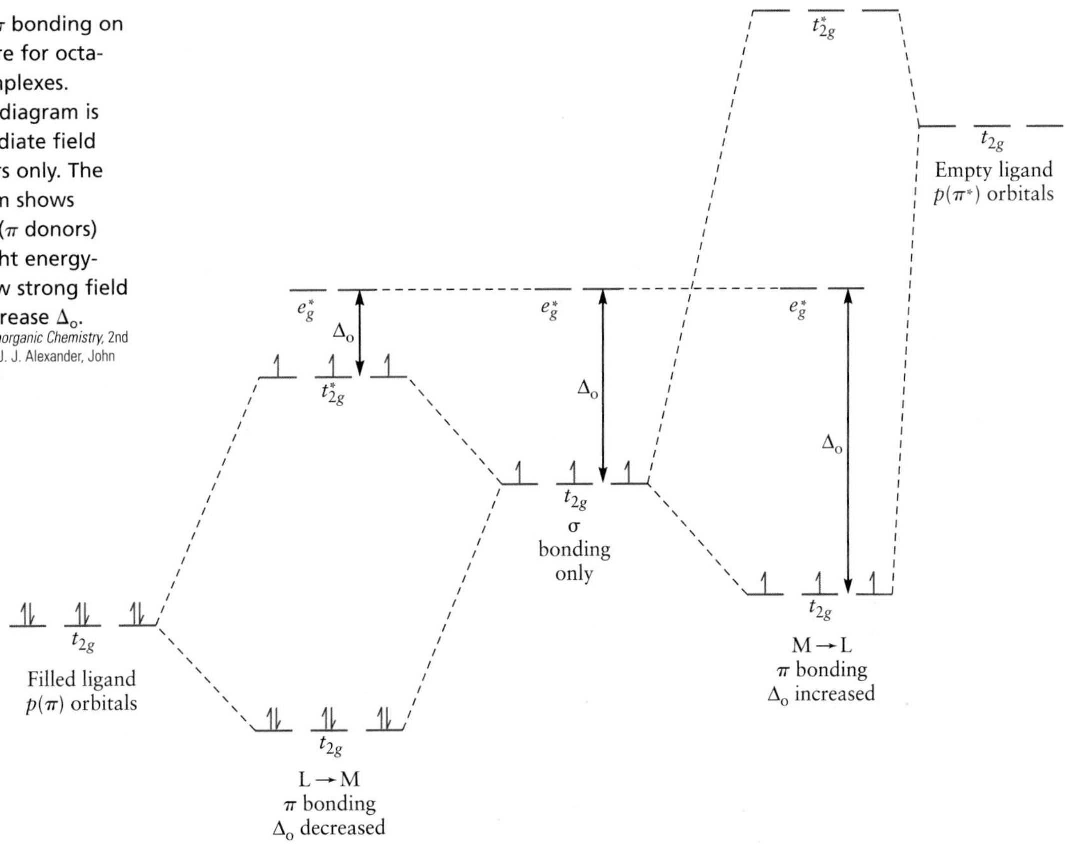

FIGURE 8.32 Effect of π bonding on the energy-level structure for octahedral coordination complexes. The center energy-level diagram is appropriate for intermediate field ligands that are σ donors only. The left energy-level diagram shows how weak field ligands (π donors) decrease Δ_o, and the right energy-level diagram shows how strong field ligands (π acceptors) increase Δ_o.
(Adapted from *Concepts and Models of Inorganic Chemistry*, 2nd edition, B. Douglas, D. H. McDaniel, and J. J. Alexander, John Wiley and Sons, New York, 1983, p. 293.)

energy-level structure of $[Cr(NH_3)_6]^{3+}$, and it can be considered a limiting case of molecular orbital theory.

The left side of Figure 8.32 could represent the bonding interactions and energy-level structure for $[CrCl_6]^{3-}$ in which a set of filled ligand orbitals (in this case, Cl 2*p* orbitals) donate electrons to a set of empty (or only partially filled) metal *d* orbitals (L→M π donation). The ligand orbitals are assumed to lie lower in energy than the metal orbitals, so the resulting t_{2g} molecular orbital is mostly ligand in character and lies at a relatively low energy. There is a corresponding t_{2g}^* orbital, which is mostly metal-like, at higher energy. This antibonding orbital is the highest occupied molecular orbital in this example; six of the nine available electrons go into the bonding t_{2g} orbital, leaving three to occupy the t_{2g}^* orbital. There are two important differences between this picture and the one in the center of Figure 8.32. First, the t_{2g} orbitals that were nonbonding metal *d* orbitals in the absence of π bonding have now become antibonding MOs by virtue of L→M π donation. Second, the energy-level splitting, Δ_o, has been reduced from its value without π bonding interactions. So, we now have an explanation for the mechanism by which weak field ligands lead to the smallest field splittings and a criterion with which to classify them. Weak field ligands have filled low-energy *p* orbitals that can donate electron density to empty metal *d* orbitals to produce an occupied antibonding molecular orbital at higher energy than the nonbonding metal *d* orbitals. All of the weak field ligands identified empirically in the spectrochemical series (the halides) fall into this class.

The right side of Figure 8.32 illustrates the case in which the metal *d* orbitals are nearly filled and the ligand has empty π* orbitals at higher energies. This situation leads to a bonding and antibonding pair of MOs as before, but in this case, the energy of the t_{2g}^* orbital is too high for it to be occupied. The highest occupied molecular orbital is the bonding t_{2g} orbital, and Δ_o is now the difference in energy between this orbital and the antibonding e_g^* orbital. In the language of inorganic chemistry, the t_{2g} orbital has been stabilized because of M→L π donation. These interactions are important for metals with filled or nearly filled t_{2g} orbitals (such as Fe, Co, and Ni) and ligands with empty, low-lying orbitals such as the *d* orbitals of P or S or the π^* orbitals of CO, CN^-, or NO^+.

Molecular orbital theory provides a simple yet comprehensive way to understand and predict bonding patterns and energy levels in coordination complexes. Although we have worked out only the octahedral geometry in detail, the same considerations apply for all of the other geometries. Ligands classified empirically by the strengths of their interactions as revealed by the spectrochemical series can now be classified by the bonding interactions responsible. Weak field ligands are **π donors**, intermediate field ligands are **sigma (σ) donors** with little to no π interactions at all, and strong field ligands are **π acceptors.** These classifications enable us to determine the nature of the ligand by simply looking for these characteristic features and to understand the bonding in coordination complexes by combining this knowledge with the oxidation state of the metal ion of interest.

CHAPTER SUMMARY

Transition-metal compounds and coordination complexes display a much wider variety of physical and chemical properties than the main-group elements in large part because of the participation of the *d* electrons in bonding and chemical reactions. Because the cost of transferring electrons to and from the *d* orbitals is low, these elements have several stable oxidation states and can make several kinds of chemical bonds. The ease with which electrons can be transferred to and from transition-metal compounds also makes them excellent catalysts by providing electron transfer reactions as alternate pathways for chemical reactions. Bonding in transition-metal compounds and coordination complexes is well described by molecular orbital theory. The overlap of the *d* orbitals of the metal with the σ and π orbitals of ligands forms bonding and antibonding molecular orbitals; these interactions are often described as ligand-to-metal (L→M) σ and π donation and metal-to-ligand (M→L) π donation, respectively.

Coordination complexes are molecules or ions in which a central metal atom is bound to one or more ligands in a symmetric arrangement. Linear, tetrahedral, square-planar, trigonal bipyramidal, and octahedral arrangements are all known; the octahedral geometry is by far the most common. Crystal field theory accounts for the colors and magnetic properties of coordination complexes by considering the strengths of the repulsive interactions between the electrons in the various *d* orbitals and the ligands, represented as point charges located along the Cartesian axes. For octahedral geometries, the degenerate set of *d* orbitals splits into two levels, one set of three at lower energy and a set of two at higher energy. The colors of coordination complexes, as well as their magnetic properties, are rationalized using this model. Large energy differences between the two sets of levels result in the absorption of light in the blue or ultraviolet portion of the spectrum, and the complexes appear red in color. These strong fields also favor electron configurations in which the electrons preferentially occupy the lower levels in pairs, the low-spin configuration. Weak fields lead to optical absorption in the red or yellow regions of the spectrum, and the compounds appear blue. The electrons occupy both sets of orbitals with their spins parallel, the high-spin configuration.

Crystal field theory is only partially successful in explaining the optical and magnetic properties of the coordination complexes; it cannot explain the relative strengths of the ligands in the spectrochemical series. Molecular orbital theory (an earlier version of which is called ligand field theory) provides a more complete and quantitative description of bonding, optical, and magnetic properties by allowing for the formation of both σ and π bonds between the central metal ion and the ligands. Molecular orbital theory provides qualitative insight by classifying ligands by their bonding types and by considering the oxidation states of the metal ion.

CONCEPTS AND SKILLS

Section 8.1 – Chemistry of the Transition Metals

Discuss the systematic variation of physical and chemical properties of the transition elements through the periodic table (Problems 1–6).

- The properties of transition metal compounds and coordination complexes are determined, in large part, by the participation of *d* electrons in bonding and chemical reactions. Most transition metals have a number of stable oxidation states that lead to different kinds of chemical bonds and facilitate electron transfer reactions.

Identify the common oxidation states of the transition metals, rationalize the periodic trends observed, and be able to determine oxidation states in inorganic compounds (Problems 7–10).

- The highest oxidation state observed for each element on the left side of each period corresponds to ionization to form a noble gas configuration. Higher oxidation states are more common for the heavier members of each group and all higher oxidation states are preferentially stabilized by small anionic ligands such as F^- and O^{2-}.

Use the concept of hard and soft acids and bases to predict the stability and reactivity of ionic compounds (Problems 11–14).

- Hard acids are small cations with large charge densities whereas soft acids are larger atoms or ions with small charge densities but large polarizabilities. Hard bases include F^- and neutral Lewis bases whereas soft bases include larger anions with smaller charge densities. Hard acids prefer to pair with hard bases and soft acids prefer to pair with soft bases.

Section 8.2 – Introduction to Coordination Chemistry

Define the terms coordination compound, coordination number, ligand, and chelation (Problems 15–18).

- Coordination complexes comprise a central metal atom or ion coordinated by dative bonds to a number of ligands in a symmetrical arrangement. Linear, tetrahedral, square planar, trigonal bipyramidal, and octahedral geometries are all known, with octahedral being by far the most common.

Name coordination compounds, given their molecular formulas (Problems 19–22).

Section 8.3 – Structures of Coordination Complexes

Draw geometric isomers of octahedral, tetrahedral, and square-planar complexes (Problems 25–28).

Describe several coordination complexes that have roles in biology.

- A number of coordination complexes are important in biology, many built around porphyrin ligands; these include the oxygen transport protein hemoglobin and the family of enzymes called cytochromes that catalyze many different biochemical reactions.

Section 8.4 – Crystal Field Theory: Optical and Magnetic Properties

Use crystal field theory to interpret the magnetic properties of coordination compounds in terms of the electron configurations of their central ions (Problems 29–33).

- Crystal field theory accounts for the optical and magnetic properties of coordination complexes by considering the electrostatic repulsion between the metal *d* electrons and the charges on ionic ligands. The set of five degenerate *d* orbitals split into several sets that depend upon the symmetry of each crystal field. Octahedral crystal fields, which are the most common, split the orbitals into a triply degenerate set at lower energy and a doubly degenerate set at higher energy.
- Magnetic properties are determined by the competition between the cost of promoting an electron to the higher level and the cost of pairing an electron with another electron in an occupied orbital.

Section 8.5 – Optical Properties and the Spectrochemical Spectrum

Relate the colors of coordination compounds to their crystal field splitting energies and CFSE (Problems 35–44).

- The crystal field splitting is measured by optical absorption spectroscopy, which results from *d–d* transitions between the two sets of levels.

Section 8.6 – Bonding in Coordination Complexes

Use molecular orbital theory to order the energy levels in coordination compounds and to account for the spectrochemical series.

- Molecular orbital theory that considers only σ bonding produces results similar to crystal field theory but with a different physical interpretation for the physical origin of the splitting.

CUMULATIVE EXERCISE

Platinum

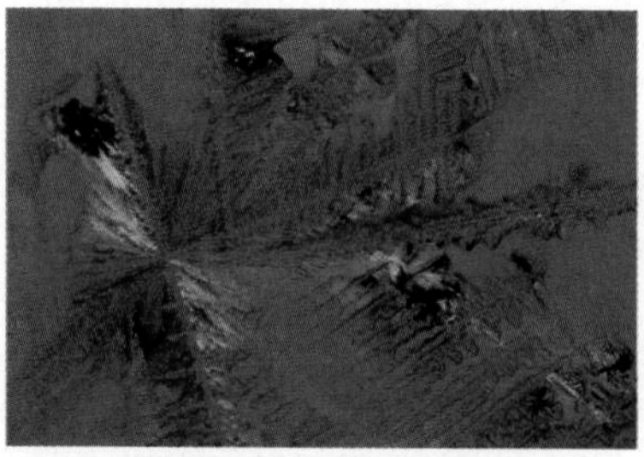

FIGURE 8.33 Cisplatin crystals.

The precious metal platinum was first used by South American Indians, who found impure, native samples of it in the gold mines of what is now Ecuador and used the samples to make small items of jewelry. Platinum's high melting point (1772°C) makes it harder to work than gold (1064°C) and silver (962°C), but this same property and a high resistance to chemical attack make platinum suitable as a material for high-temperature crucibles. Although platinum is a noble metal, in the +4 and +2 oxidation states it forms a variety of compounds, many of which are coordination complexes. Its coordinating abilities make it an important catalyst for organic and inorganic reactions.

(a) The anticancer drug cisplatin, *cis*-$[Pt(NH_3)_2Cl_2]$ (see Figs. 8.12b and 8.33), can be prepared from K_2PtCl_6 via reduction with N_2H_4 (hydrazine), giving K_2PtCl_4, followed by replacement of two chloride ion ligands with ammonia. Give systematic names to the three platinum complexes referred to in this statement.

(b) The coordination compound diamminetetracyanoplatinum(IV) has been prepared, but salts of the hexacyanoplatinate(IV) ion have not. Write the chemical formulas of these two species.

(c) Platinum forms organometallic complexes quite readily. In one of the simplest of these, Pt(II) is coordinated to two chloride ions and two molecules of ethylene (C_2H_4) to give an unstable yellow crystalline solid. Can this complex have more than one isomer? If so, describe the possible isomers.

(d) Platinum(IV) is readily complexed by ethylenediamine. Draw the structures of both enantiomers of the complex ion *cis*-$[Pt(Cl)_2(en)_2]^{2+}$. In this compound, the Cl^- ligands are *cis* to one another.

(e) In platinum(IV) complexes, the octahedral crystal field splitting Δ_o is relatively large. Is K_2PtCl_6 diamagnetic or paramagnetic? What is its *d* electron configuration?

(f) Is cisplatin diamagnetic or paramagnetic?

(g) The salt $K_2[PtCl_4]$ is red, but $[Pt(NH_3)_4]Cl_2 \cdot H_2O$ is colorless. In what regions of the spectrum do the dominant absorptions for these compounds lie?

(h) When the two salts from part (g) are dissolved in water and the solutions mixed, a green precipitate called Magnus's green salt forms. Propose a chemical formula for this salt and assign the corresponding systematic name.

Answers

(a) *cis*-Diamminedichloroplatinum(II), potassium hexachloroplatinate(IV), and potassium tetrachloroplatinate(II)

(b) $[Pt(NH_3)_2(CN)_4]$ and $[Pt(CN)_6]^{2-}$

(c) Pt(II) forms square-planar complexes. There are two possible forms, arising from *cis* and *trans* placement of the ethylene molecules. These are analogous to the two isomers shown in Figures 8.12b and c.

(d) The structures are the *cis* form shown in Figure 8.10 and its mirror image.

(e) The six *d* electrons in Pt(IV) are all in the lower t_{2g} level in a low-spin, large-Δ_o complex. All are paired; thus, the compound is diamagnetic.

(f) Diamagnetic, with the bottom four levels in the square-planar configuration (all but $d_{x^2-y^2}$) occupied

(g) Red transmission corresponds to absorption of green light by $K_2[PtCl_4]$. A colorless solution has absorptions at either higher or lower frequency than visible. Because Cl^- is a weaker field ligand than NH_3, the absorption frequency should be higher for the $Pt(NH_3)_4^{2+}$ complex, putting it in the ultraviolet region of the spectrum.

(h) $[Pt(NH_3)_4][PtCl_4]$, tetra-ammineplatinum(II) tetrachloroplatinate(II)

PROBLEMS

Answers to problems whose numbers are red appear in Appendix G. Problems that are more challenging are indicated with asterisks.

8.1 Chemistry of the Transition Metals

1. Of the compounds PtF_4 and PtF_6, predict (a) which is more soluble in water and (b) which is more volatile.

2. The melting point of $TiCl_4$ (−24°C) lies below those of TiF_4 (284°C) and $TiBr_4$ (38°C). Explain why by considering the covalent-ionic nature of these compounds and the intermolecular forces in each case.

3. The decavanadate ion is a complex species with chemical formula $V_{10}O_{28}^{6-}$. It reacts with an excess of acid to form the dioxovanadium ion, VO_2^+, and water. Write a balanced chemical equation for this reaction. In what oxidation state is vanadium before and after this reaction? What vanadium oxide has the same oxidation state?

4. What is the chemical formula of vanadium(III) oxide? Do you predict this compound to be more basic or more acidic than the vanadium oxide of Problem 3? Write a balanced chemical equation for the reaction of vanadium(III) oxide with a strong acid.

5. Titanium(III) oxide is prepared by reaction of titanium(IV) oxide with hydrogen at high temperature. Write a balanced chemical equation for this reaction. Which oxide do you expect to have stronger basic properties?

6. Treatment of cobalt(II) oxide with oxygen at high temperatures gives Co_3O_4. Write a balanced chemical equation for this reaction. What is the oxidation state of cobalt in Co_3O_4?

8.2 Introduction to Coordination Chemistry

7. What are the oxidation states of manganese in the following series of oxides? MnO, MnO_2, Mn_2O_3, Mn_2O_7, Mn_3O_4,

and $[MnO_4]^-$? Recall that you must include the total charge on the ion for ionic species.

8. What are the oxidation states of the transition metals in the following series of compounds? Nb_2O_5, MoS_2, $RuCl_3$, RhO_2, PdF_2, and Ag_2O.

9. Why is the highest oxidation state observed for $CrCl_4$ only the +4 oxidation state, in contrast to the +6 state observed in CrF_6?

10. OH^- tends to stabilize intermediate oxidation states. Why is this?

11. Nickel and copper ores are generally sulfides, whereas aluminum ore is an oxide. Explain this observation on the basis of the hard/soft acid/base concept.

12. Would you expect the following reaction to proceed in aqueous solution as written?

$$CuCl_2(aq) + S^{2-}(aq) \longrightarrow CuS(s) + 2\ Cl^-(aq)$$

13. What experimental evidence would allow you to determine whether $TiCl_4$ is an ionic compound or a covalent compound? What would you predict based upon oxidation state?

14. Which of the two common oxides of chromium, CrO_2 or Cr_2O_3, would you predict to have the higher melting point and why?

15. Will methylamine (CH_3NH_2) be a monodentate or a bidentate ligand? With which of its atoms will it bind to a metal ion?

16. Show how the glycinate ion ($H_2N{-}CH_2{-}COO^-$) can act as a bidentate ligand. (Draw a Lewis diagram if necessary.) Which atoms in the glycinate ion will bind to a metal ion?

17. Determine the oxidation state of the metal in each of the following coordination complexes: $[V(NH_3)_4Cl_2]$, $[Mo_2Cl_8]^{4-}$, $[Co(OH_2)_2(NH_3)Cl_3]^-$, $[Ni(CO)_4]$.

18. Determine the oxidation state of the metal in each of the following coordination complexes: $Mn_2(CO)_{10}$, $[Re_3Br_{12}]^{3-}$, $[Fe(OH_2)_4(OH)_2]^+$, $[Co(NH_3)_4Cl_2]^+$.

19. Give the chemical formula that corresponds to each of the following compounds:
- **(a)** Sodium tetrahydroxozincate(II)
- **(b)** Dichlorobis(ethylenediamine)cobalt(III) nitrate
- **(c)** Triaquabromoplatinum(II) chloride
- **(d)** Tetra-amminedinitroplatinum(IV) bromide

20. Give the chemical formula of each of the following compounds:
- **(a)** Silver hexacyanoferrate(II)
- **(b)** Potassium tetraisothiocyanatocobaltate(II)
- **(c)** Sodium hexafluorovanadate(III)
- **(d)** Potassium trioxalatochromate(III)

21. Assign a systematic name to each of the following chemical compounds:
- **(a)** $NH_4[Cr(NH_3)_2(NCS)_4]$
- **(b)** $[Tc(CO)_5]I$
- **(c)** $K[Mn(CN)_5]$
- **(d)** $[Co(NH_3)_4(OH_2)Cl]Br_2$

22. Give the systematic name for each of the following chemical compounds:
- **(a)** $[Ni(OH_2)_4(OH)_2]$
- **(b)** $[HgClI]$
- **(c)** $K_4[Os(CN)_6]$
- **(d)** $[FeBrCl(en)_2]Cl$

8.3 Structures of Coordination Complexes

23. Suppose 0.010 mol of each of the following compounds is dissolved (separately) in 1.0 L water: KNO_3, $[Co(NH_3)_6]Cl_3$, $Na_2[PtCl_6]$, $[Cu(NH_3)_2Cl_2]$. Rank the resulting four solutions in order of conductivity, from lowest to highest.

24. Suppose 0.010 mol of each of the following compounds is dissolved (separately) in 1.0 L water: $BaCl_2$, $K_4[Fe(CN)_6]$, $[Cr(NH_3)_4Cl_2]Cl$, $[Fe(NH_3)_3Cl_3]$. Rank the resulting four solutions in order of conductivity, from lowest to highest.

25. Draw the structures of all possible isomers for the following complexes. Indicate which isomers are enantiomer pairs.
- **(a)** Diamminebromochloroplatinum(II) (square-planar)
- **(b)** Diaquachlorotricyanocobaltate(III) ion (octahedral)
- **(c)** Trioxalatovanadate(III) ion (octahedral)

26. Draw the structures of all possible isomers for the following complexes. Indicate which isomers are enantiomer pairs.
- **(a)** Bromochloro(ethylenediamine)platinum(II) (square-planar)
- **(b)** Tetra-amminedichloroiron(III) ion (octahedral)
- **(c)** Amminechlorobis(ethylenediamine)iron(III) ion (octahedral)

27. Iron(III) forms octahedral complexes. Sketch the structures of all the distinct isomers of $[Fe(en)_2Cl_2]^+$, indicating which pairs of structures are mirror images of each other.

28. Platinum(IV) forms octahedral complexes. Sketch the structures of all the distinct isomers of $[Pt(NH_3)_2Cl_2F_2]$, indicating which pairs of structures are mirror images of each other.

8.4 Crystal Field Theory: Optical and Magnetic Properties

29. For each of the following ions, draw diagrams like those in Figure 8.18 to show orbital occupancies in both weak and strong octahedral fields. Indicate the total number of unpaired electrons in each case.
- **(a)** Mn^{2+}
- **(b)** Zn^{2+}
- **(c)** Cr^{3+}
- **(d)** Mn^{3+}
- **(e)** Fe^{2+}

30. Repeat the work of the preceding problem for the following ions:
- **(a)** Cr^{2+}
- **(b)** V^{3+}
- **(c)** Ni^{2+}
- **(d)** Pt^{4+}
- **(e)** Co^{2+}

31. Experiments can measure not only whether a compound is paramagnetic, but also the number of unpaired electrons. It is found that the octahedral complex ion $[Fe(CN)_6]^{3-}$ has fewer unpaired electrons than the octahedral complex ion $[Fe(OH_2)_6]^{3+}$. How many unpaired electrons are present in each species? Explain. In each case, express the CFSE in terms of Δ_o.

32. The octahedral complex ion $[MnCl_6]^{3-}$ has more unpaired spins than the octahedral complex ion $[Mn(CN)_6]^{3-}$. How many unpaired electrons are present in each species? Explain. In each case, express the CFSE in terms of Δ_o.

33. Explain why octahedral coordination complexes with three and eight *d* electrons on the central metal atom are particularly stable. Under what circumstances would you expect complexes with five or six *d* electrons on the central metal atom to be particularly stable?

34. Mn, Fe, and Co in the +2 and +3 oxidation states all form hexaaquacomplexes in acidic aqueous solution. The reduction reactions of the three species are represented schematically below, where the water ligands are not shown for simplicity. It is an experimental fact from electrochemistry that Mn^{2+} and Co^{2+} are more easily reduced than Fe^{3+}; that is, they will more readily accept an electron. Based on the electron configurations of the ions involved, explain why Fe^{3+} is harder to reduce than Mn^{2+} and Co^{2+}.

$$Mn^{3+} + e \longrightarrow Mn^{2+}$$

$$Fe^{3+} + e \longrightarrow Fe^{2+}$$

$$Co^{3+} + e \longrightarrow Co^{2+}$$

8.5 Optical Properties and the Spectrochemical Series

35. An aqueous solution of zinc nitrate contains the $[Zn(OH_2)_6]^{2+}$ ion and is colorless. What conclusions can be drawn about the absorption spectrum of the $[Zn(OH_2)_6]^{2+}$ complex ion?

36. An aqueous solution of sodium hexaiodoplatinate(IV) is black. What conclusions can be drawn about the absorption spectrum of the $[PtI_6]^{2-}$ complex ion?

37. Estimate the wavelength of maximum absorption for the octahedral ion hexacyanoferrate(III) from the fact that light transmitted by a solution of it is red. Estimate the crystal field splitting energy Δ_o (in kJ mol^{-1}).

38. Estimate the wavelength of maximum absorption for the octahedral ion hexa-aquanickel(II) from the fact that its solutions are colored green by transmitted light. Estimate the crystal field splitting energy Δ_o (in kJ mol^{-1}).

39. Estimate the CFSE for the complex in Problem 29. (**Note:** This is a high-field (low-spin) complex.)

40. Estimate the CFSE for the complex in Problem 30.

41. The chromium(III) ion in aqueous solution is blue–violet.
(a) What is the complementary color to blue–violet?
(b) Estimate the wavelength of maximum absorption for a $Cr(NO_3)_3$ solution.
(c) Will the wavelength of maximum absorption increase or decrease if cyano ligands are substituted for the coordinated water? Explain.

42. An aqueous solution containing the hexa-amminecobalt(III) ion is yellow.
(a) What is the complementary color to yellow?
(b) Estimate this solution's wavelength of maximum absorption in the visible spectrum.

43. **(a)** An aqueous solution of $Fe(NO_3)_3$ has only a pale color, but an aqueous solution of $K_3[Fe(CN)_6]$ is bright red. Do you expect a solution of $K_3[FeF_6]$ to be brightly colored or pale? Explain your reasoning.
(b) Would you predict a solution of $K_2[HgI_4]$ to be colored or colorless? Explain.

44. **(a)** An aqueous solution of $Mn(NO_3)_2$ is very pale pink, but an aqueous solution of $K_4[Mn(CN)_6]$ is deep blue. Explain why the two differ so much in the intensities of their colors.
(b) Predict which of the following compounds would be colorless in aqueous solution: $K_2[Co(NCS)_4]$, $Zn(NO_3)_2$, $[Cu(NH_3)_4]Cl_2$, $CdSO_4$, $AgClO_3$, $Cr(NO_3)_2$.

45. Predict the geometry and number of unpaired spins in the following four-coordinate complexes: $[AuBr_4]^-$, $[NiBr_4]^{2-}$. Explain your reasoning.

46. Predict the geometry and number of unpaired spins in the following four-coordinate complexes: $[Ni(CN)_4]^{2-}$, $[NiCl_4]^{2-}$. Explain your reasoning.

ADDITIONAL PROBLEMS

47. Of the ten fourth-period transition metal elements in Table 8.1, which one has particularly low melting and boiling points? How can you explain this in terms of the electronic configuration of this element?

***48.** Although copper lies between nickel and zinc in the periodic table, the reduction potential of Cu^{2+} is above that for both Ni^{2+} and Zn^{2+} (see Table 8.3). Use other data from Table 8.3 to account for this observation. (*Hint:* Think of a multistep process to convert a metal atom in the solid to a metal ion in solution. For each step, compare the relevant energy or enthalpy changes for Cu with those for Ni or Zn.)

49. A reference book lists five different values for the electronegativity of molybdenum, a different value for each oxidation state from +2 through +6. Predict which electronegativity is highest and which is lowest.

50. If *trans*-$[Cr(en)_2(NCS)_2]SCN$ is heated, it forms gaseous ethylenediamine and solid $[Cr(en)_2(NCS)_2][Cr(en)(NCS)_4]$. Write a balanced chemical equation for this reaction. What are the oxidation states of the Cr ions in the reactant and in the two complex ions in the product?

51. A coordination complex has the molecular formula $[Ru_2(NH_3)_6Br_3](ClO_4)_2$. Determine the oxidation state of ruthenium in this complex.

52. Heating 2.0 mol of a coordination compound gives 1.0 mol NH_3, 2.0 mol H_2O, 1.0 mol HCl, and 1.0 mol $(NH_4)_3[Ir_2Cl_9]$. Write the formula of the original (six-coordinate) coordination compound and name it.

53. Explain why ligands are usually negative or neutral in charge and only rarely positive.

54. Match each compound in the group on the left with the compound on the right that is most likely to have the same electrical conductivity per mole in aqueous solution.

(a) $[Fe(OH_2)_5Cl]CO_3$	HCN
(b) $[Mn(OH_2)_6]Cl_3$	$Fe_2(SO_4)_3$
(c) $[Zn(OH_2)_3(OH)]Cl]$	NaCl
(d) $[Fe(NH_3)_6]_2(CO_3)_3$	$MgSO_4$
(e) $[Cr(NH_3)_3Br_3]$	Na_3PO_4
(f) $K_3[Fe(CN)_6]$	$GaCl_3$

55. Three different compounds are known to have the empirical formula $CrCl_3 \cdot 6H_2O$. When exposed to a dehydrating agent, compound 1 (which is dark green) loses 2 mol water per mole of compound, compound 2 (light green) loses 1 mol water, and compound 3 (violet) loses no water. What are the probable structures of these compounds? If an excess of silver nitrate solution is added to 100.0 g of each of these compounds, what mass of silver chloride will precipitate in each case?

56. The octahedral structure is not the only possible six-coordinate structure. Other possibilities include a planar hexagonal structure and a triangular prism structure. In the latter, the ligands are arranged in two parallel triangles, one lying above the metal atom and the other below the metal atom with its corners directly in line with the corners of the first triangle. Show that the existence of two and only two isomers of $[Co(NH_3)_4Cl_2]^+$ is evidence against both of these possible structures.

57. Cobalt(II) forms more tetrahedral complexes than any other ion except zinc(II). Draw the structure(s) of the tetrahedral complex $[CoCl_2(en)]$. Could this complex exhibit geometric or optical isomerism? If one of the Cl^- ligands is replaced by Br^-, what kinds of isomerism, if any, are possible in the resulting compound?

58. Is the coordination compound $[Co(NH_3)_6]Cl_2$ diamagnetic or paramagnetic?

59. A coordination compound has the empirical formula $PtBr(en)(SCN)_2$ and is diamagnetic.

(a) Examine the *d*-electron configurations on the metal atoms, and explain why the formulation $[Pt(en)_2(SCN)_2]$ $[PtBr_2(SCN)_2]$ is preferred for this substance.

(b) Name this compound.

60. We used crystal field theory to order the energy-level splittings induced in the five *d* orbitals. The same procedure could be applied to *p* orbitals. Predict the level splittings (if any) induced in the three *p* orbitals by octahedral and square-planar crystal fields.

61. The three complex ions $[Mn(CN)_6]^{5-}$, $[Mn(CN)_6]^{4-}$, and $[Mn(CN)_6]^{3-}$ have all been synthesized and all are low-spin octahedral complexes. For each complex, determine the oxidation number of Mn, the configuration of the *d* electrons (how many t_{2g} and how many e_g), and the number of unpaired electrons present.

62. On the basis of the examples presented in Problem 51, can you tell whether $Mn^{2+}(aq)$ is more easily oxidized or more easily reduced? What can you conclude about the stability of $Mn^{2+}(aq)$?

63. The following ionic radii (in angstroms) are estimated for the +2 ions of selected elements of the first transition-metal series, based on the structures of their oxides: $Ca^{2+}(0.99)$, $Ti^{2+}(0.71)$, $V^{2+}(0.64)$, $Mn^{2+}(0.80)$, $Fe^{2+}(0.75)$, $Co^{2+}(0.72)$, $Ni^{2+}(0.69)$, $Cu^{2+}(0.71)$, $Zn^{2+}(0.74)$. Draw a graph of ionic radius versus atomic number in this series, and account for its shape. The oxides take the rock salt structure. Are these solids better described as high- or low-spin transition-metal complexes?

64. The coordination geometries of $[Mn(NCS)_4]^{2-}$ and $[Mn(NCS)_6]^{4-}$ are tetrahedral and octahedral, respectively. Explain why the two have the same room-temperature molar magnetic susceptibility.

65. The complex ion $CoCl_4^{2-}$ has a tetrahedral structure. How many *d* electrons are on the Co? What is its electronic configuration? Why is the tetrahedral structure stable in this case?

66. In the coordination compound $(NH_4)_2[Fe(OH_2)F_5]$, the Fe is octahedrally coordinated.

(a) Based on the fact that F^- is a weak-field ligand, predict whether this compound is diamagnetic or paramagnetic. If it is paramagnetic, tell how many unpaired electrons it has.

(b) By comparison with other complexes reviewed in this chapter, discuss the likely color of this compound.

(c) Determine the *d*-electron configuration of the iron in this compound.

(d) Name this compound.

67. The compound $Cs_2[CuF_6]$ is bright orange and paramagnetic. Determine the oxidation number of copper in this compound, the most likely geometry of the coordination around the copper, and the possible configurations of the *d* electrons of the copper.

68. In what directions do you expect the bond length and vibrational frequency of a free CO molecule to change when it becomes a CO ligand in a $Ni(CO)_4$ molecule? Explain your reasoning.

69. Give the number of valence electrons surrounding the central transition-metal ion in each of the following known organometallic compounds or complex ions: $[Co(C_5H_5)_2]^+$, $[Fe(C_5H_5)(CO)_2Cl]$, $[Mo(C_5H_5)_2Cl_2]$, $[Mn(C_5H_5)(C_6H_6)]$.

70. Molecular nitrogen (N_2) can act as a ligand in certain coordination complexes. Predict the structure of $[V(N_2)_6]$, which is isolated by condensing V with N_2 at 25 K. Is this compound diamagnetic or paramagnetic? What is the formula of the carbonyl compound of vanadium that has the same number of electrons?

71. What energy levels are occupied in a complex such as hexacarbonylchromium(0)? Are any electrons placed into antibonding orbitals that are derived from the chromium *d* orbitals?

*72. The compound $WH_2(C_5H_5)_2$ acts as a base, but $TaH_3(C_5H_5)_2$ does not. Explain.

73. Discuss the role of transition-metal complexes in biology. Consider such aspects as their absorption of light, the existence of many different structures, and the possibility of multiple oxidation states.

CUMULATIVE PROBLEMS

74. Predict the volume of hydrogen generated at 1.00 atm and 25°C by the reaction of 4.53 g scandium with excess aqueous hydrochloric acid.

*75. Mendeleev's early periodic table placed manganese and chlorine in the same group. Discuss the chemical evidence for these placements, focusing on the oxides of the two elements and their acid–base and redox properties. Is there a connection between the electronic structures of their atoms? In what ways are the elements different?

76. An orange–yellow osmium carbonyl compound is heated to release CO and leave elemental osmium behind. Treatment of 6.79 g of the compound releases 1.18 L CO(*g*) at 25°C and 2.00 atm pressure. What is the empirical formula of this compound?

Kinetic Molecular Description of the States of Matter

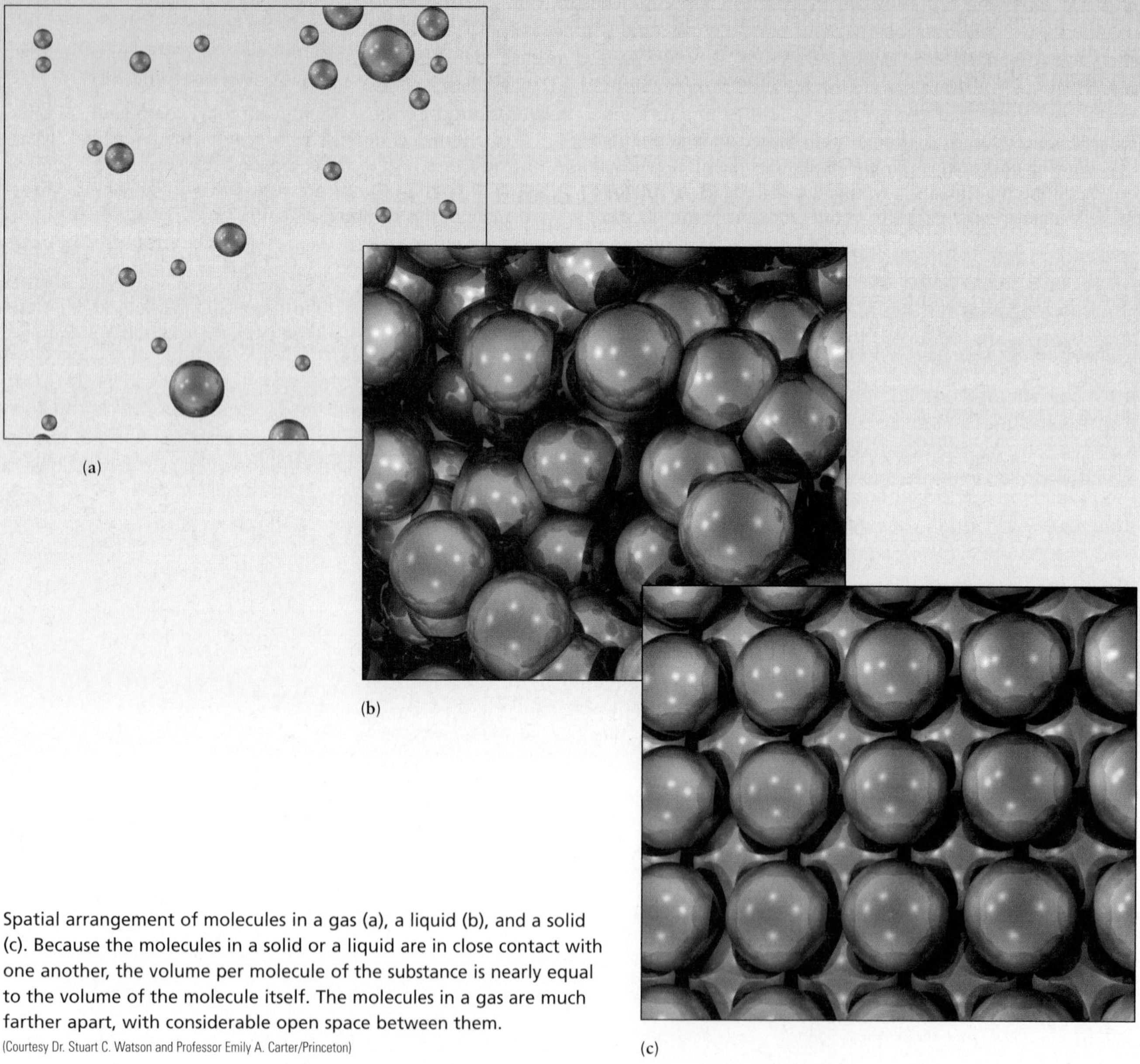

Spatial arrangement of molecules in a gas (a), a liquid (b), and a solid (c). Because the molecules in a solid or a liquid are in close contact with one another, the volume per molecule of the substance is nearly equal to the volume of the molecule itself. The molecules in a gas are much farther apart, with considerable open space between them.

(Courtesy Dr. Stuart C. Watson and Professor Emily A. Carter/Princeton)

UNIT III

Unit III relates the behavior of matter as perceived and measured on the scale of human experience to the structures and interactions of the molecules in the matter. We see that macroscopic properties are determined by nanoscopic structures and the forces between them. We recognize gases, liquids, and solids quite easily because we are surrounded by important examples of each: air, water, and earth. Gases and liquids are fluid, but solids are rigid. A gas expands to fill any container it occupies. A liquid has a fixed volume but flows to conform to the shape of its container. A solid has a fixed volume *and* a fixed shape, both of which resist deformation. These characteristics originate in the arrangements and motions of molecules, which are determined by the sizes and shapes of the molecules and the forces between them. In a solid or liquid, molecules (shown as spheres in the images on the previous page) are in very close contact. At these distances, molecules experience strong mutual forces of attraction. In a liquid, molecules can slide past one another, whereas in a solid, molecules oscillate around particular locations and occasionally hop to new locations. In a gas, the molecules are much farther apart, and the forces between them are strong only during collisions. The molecules in a gas can roam freely throughout the container, unlike those in solids and liquids.

Unit Chapters

Unit Goals

- To define the essential properties of solids, liquids, and gases
- To relate the magnitudes of these properties to the structures and motions of molecules and to the forces between molecules
- To define the properties of solutions and relate their magnitudes to composition of the solutions

9 CHAPTER

The Gaseous State

Lithium metal, in the spoon, reacts with drops of water (H_2O) to form LiOH and H_2, which is not visible. The steam observed is water having been evaporated by the heat produced from the reaction.

The *kinetic molecular theory of matter* asserts that the macroscopic properties of matter are determined by the structures of its constituent molecules and the interactions between them. Starting in the latter half of the 19th century, scientists have developed a magnificent theoretical structure called **statistical mechanics** to explain the connection between microscopic structure and macroscopic properties. Statistical mechanics provides this bridge for all types of matter, ranging from biological materials to solid-state integrated circuits. Today, every student of science must be familiar with the concepts that make this connection. We give a qualitative introduction to these concepts, and apply them to numerous cases in the next four chapters.

We start the discussion with gases, because gases provide the simplest opportunity for relating macroscopic properties to the structures and interactions of molecules. This is possible because gases are much less dense than solids or liquids. On the macroscopic level, gases are distinguished from liquids and solids by their much smaller values of *mass density* (conveniently measured in grams per cubic centimeter). On the microscopic level, the *number density* (number of molecules in 1 cm^3 of the sample) is smaller—and the distances between molecules are much greater—than in liquids and solids. Molecules with no net electrical charge exert significant forces on each other only when they are close. Consequently, in the study of gases it is a legitimate simplification to ignore interactions between molecules until they collide and then to consider collisions between only two molecules at a time.

In this chapter we develop two themes that are key to relating macroscopic behavior to molecular structure. First, we show how to define and measure the macroscopic properties of gases as temperature, pressure, and volume are changed. This discussion leads us to develop and apply the ideal gas law, which—as verified by experiment—accurately represents the bulk properties of gases at low density. Second, we interpret and explain the ideal gas law in terms of the structures and motions of individual molecules. We obtain refinements to the ideal gas law by analyzing the consequences of collisions between pairs of molecules. The results and insights we obtain are essential background for later study of the rate and extent of chemical reactions since chemical reactions occur through molecular collisions.

9.1 The Chemistry of Gases

The ancient Greeks considered air to be one of the four fundamental elements in nature. European scientists began to study the properties of air (such as its resistance to compression) as early as the 17th century. The chemical composition of air (Table 9.1) was unknown until late in the 18th century, when Lavoisier, Priestley, and others showed that it consists primarily of two substances: oxygen and nitrogen. Oxygen was characterized by its ability to

TABLE 9.1

Composition of Air

Constituent	Formula	Fraction by Volume
Nitrogen	N_2	0.78110
Oxygen	O_2	0.20953
Argon	Ar	0.00934
Carbon dioxide	CO_2	0.00038
Neon	Ne	1.82×10^{-5}
Helium	He	5.2×10^{-6}
Methane	CH_4	1.5×10^{-6}
Krypton	Kr	1.1×10^{-6}
Hydrogen	H_2	5×10^{-7}
Dinitrogen oxide	N_2O	3×10^{-7}
Xenon	Xe	8.7×10^{-8}

Air also contains other constituents, the abundances of which are quite variable in the atmosphere. Examples are water (H_2O), 0–0.07; ozone (O_3), 0–7×10^{-8}; carbon monoxide (CO), 0–2×10^{-8}; nitrogen dioxide (NO_2), 0–2×10^{-8}; and sulfur dioxide (SO_2), 0–1×10^{-6}.

support life. Once the oxygen in a volume of air had been used up (by burning a candle in a closed container, for example), the nitrogen that remained could no longer keep animals alive. More than 100 years elapsed before a careful reanalysis of the composition of air using multiple experimental techniques showed that oxygen and nitrogen account for only about 99% of the total volume, most of the remaining 1% being a new gas, given the name "argon." The other noble gases (helium, neon, krypton, and xenon) are present in air to lesser extents.

FIGURE 9.1 Carbon dioxide, CO_2, is dissolved in aqueous solution to form carbonated beverages. Carbon dioxide also reacts with water to produce $H_2CO_3(aq)$, which provides some of the acidity in soft drinks. Solid carbon dioxide (dry ice) is used for refrigeration.

Other gases are found on the surface of the Earth and in the atmosphere. Methane (CH_4), formerly known as "marsh gas," is produced by bacterial processes, especially in swampy areas. It is a major constituent of natural-gas deposits formed over many millennia by decay of plant matter beneath the surface of the Earth. Recovery of methane from municipal landfills for use as a fuel is now a commercially feasible process. Gases also form when liquids evaporate. The most familiar example is water vapor in the air from the evaporation of liquid water; it provides the humidity of air.

FIGURE 9.2 Sulfur dioxide, SO_2, is produced by burning sulfur in air as the first step in the production of sulfuric acid.

Gases are also formed by chemical reactions. Some solids decompose upon heating to give gaseous products. One famous example is the decomposition of mercury(II) oxide to mercury and oxygen (Fig. 1.6):

$$2\,HgO(s) \xrightarrow{\text{Heat}} 2\,Hg(\ell) + O_2(g)$$

Joseph Priestly discovered the element oxygen while investigating this reaction. Even earlier, in 1756, Joseph Black showed that marble, which consists primarily of calcium carbonate ($CaCO_3$), decomposes upon heating to give quicklime (CaO) and carbon dioxide:

$$CaCO_3(s) \xrightarrow{\text{Heat}} CaO(s) + CO_2(g)$$

FIGURE 9.3 Sulfur trioxide, SO_3, is a corrosive gas produced by the further oxidation of sulfur dioxide.

Ammonium chloride (NH_4Cl) decomposes under heat to produce two gases: ammonia and hydrogen chloride:

$$NH_4Cl(s) \xrightarrow{\text{Heat}} NH_3(g) + HCl(g)$$

Some gas-forming reactions proceed explosively. The decomposition of nitroglycerin is a detonation in which all the products are gases:

$$4\,C_3H_5(NO_3)_3(\ell) \longrightarrow 6\,N_2(g) + 12\,CO_2(g) + O_2(g) + 10\,H_2O(g)$$

Several elements react with oxygen to form gaseous oxides. Carbon dioxide (Fig. 9.1) forms during animal respiration and is also produced by burning coal, oil, and other materials that contain carbon compounds. The role of carbon dioxide in global warming is the subject of intense research and policy debates. Oxides of sulfur (see Figs. 9.2 and 9.3) are produced by burning elemental sulfur, and oxides of nitrogen arise from combustion of elemental nitrogen. Sulfur is a common impurity in fossil fuels, and nitrogen is burned in high-temperature environments like automobile engines:

$$S(s) + O_2(g) \longrightarrow SO_2(g)$$

$$2\,SO_2(g) + O_2(g) \longrightarrow 2\,SO_3(g)$$

$$N_2(g) + O_2(g) \longrightarrow 2\,NO(g)$$

$$2\,NO(g) + O_2(g) \longrightarrow 2\,NO_2(g)$$

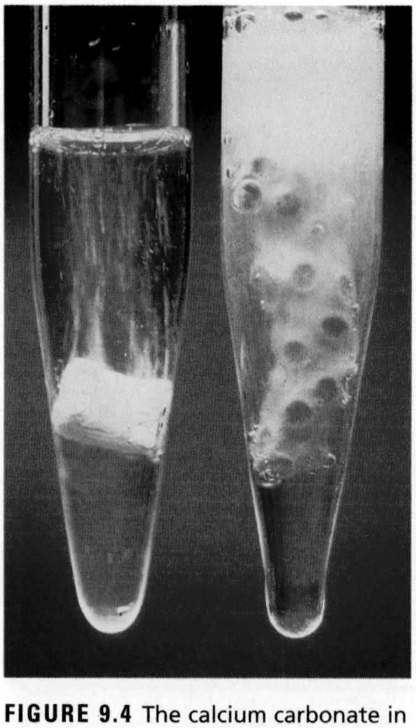

FIGURE 9.4 The calcium carbonate in a piece of chalk reacts with an aqueous solution of hydrochloric acid to produce bubbles of carbon dioxide.

The role of these compounds in air pollution is described in Section 20.6.

Gases can also be produced by the reactions of acids with ionic solids. Carbon dioxide is produced by the reaction of acids with carbonates (Fig. 9.4):

$$CaCO_3(s) + 2\,HCl(g) \longrightarrow CaCl_2(s) + CO_2(g) + H_2O(\ell)$$

Other examples of this type of reaction include:

$$Na_2S(s) + 2\ HCl(g) \longrightarrow 2\ NaCl(s) + H_2S(g)$$
$$K_2SO_3(s) + 2\ HCl(g) \longrightarrow 2\ KCl(s) + SO_2(g) + H_2O(\ell)$$
$$NaCl(s) + H_2SO_4(\ell) \longrightarrow NaHSO_4(s) + HCl(g)$$

In these reactions, the metal ions (sodium, calcium, and potassium) play no direct role; so, parallel reactions can be written for other metal ions.

The gases mentioned earlier vary greatly in their chemical properties. Some, such as HCl and SO_3, are reactive, acidic, and corrosive, whereas others, such as N_2O and N_2, are much less reactive. Whereas the chemical properties of gases vary significantly, their physical properties are quite similar and much simpler to understand. At sufficiently low densities, all gases behave physically in the same way. Their properties are summarized and interpreted by a model system called the "ideal" gas, which is the subject of the following sections.

9.2 Pressure and Temperature of Gases

The macroscopic behavior of a fixed mass of a gas is completely characterized by three properties: volume (V), pressure (P), and temperature (T). Volume is self-evident and needs no comment; as stated in Section 2.1, we use the liter as a unit of volume. The definition of pressure and temperature require a little more care.

9.2.1 Pressure and Boyle's Law

The force exerted by a gas on a unit area of the walls of its container is called the **pressure** of that gas. We do not often stop to think that the air around us exerts a pressure on us and on everything else at the surface of the Earth. Evangelista Torricelli (1608–1647), an Italian scientist who had been an assistant to Galileo, demonstrated this phenomenon in an ingenious experiment. He sealed a long glass tube at one end and filled it with mercury. He then covered the open end with his thumb, turned the tube upside down, and immersed the open end in a dish of liquid mercury (Fig. 9.5a), taking care that no air leaked in. The mercury in the tube fell, leaving a nearly perfect vacuum at the closed end, but it did not all run out of the tube. It stopped when its top was about 76 cm above

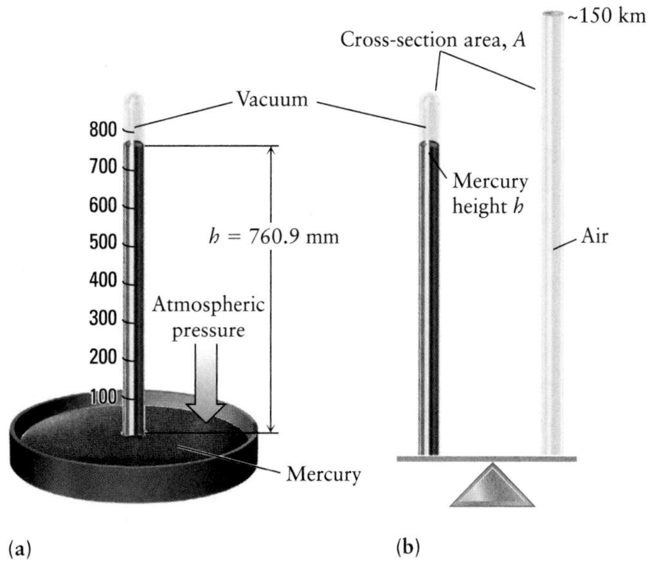

FIGURE 9.5 (a) In Torricelli's barometer, the top of the mercury in the tube is approximately 76 cm higher than that in the open beaker. (b) The mass of mercury in the column of height h exactly balances that of a column of air of the same diameter extending to the top of the atmosphere.

the level of the mercury in the dish. Torricelli showed that the exact height varied somewhat from day to day and from place to place.

This simple device, called a **barometer,** works like a balance, one arm of which is loaded with the mass of mercury in the tube and the other with a column of air of the same cross-sectional area that extends to the top of the Earth's atmosphere, approximately 150 km above the surface of the Earth (see Fig. 9.5b). The height of the mercury column adjusts itself so that its mass and that of the air column become equal. This means that the two forces on the surface of the mercury in the dish are balanced. Day-to-day changes in the height of the column occur as the force exerted by the atmosphere varies with the weather. Atmospheric pressure varies strongly with altitude; it is lower at higher altitudes because the column of air pressing down is shorter and, therefore, has less mass.

How Torricelli's invention measures the pressure of the atmosphere is explained by Newton's second law of motion, which states:

$$\text{force} = \text{mass} \times \text{acceleration}$$

$$F = ma$$

in which the acceleration of a body (a) is the rate at which its velocity changes. The gravitational field of Earth exerts an attractive force that accelerates all bodies toward Earth. The standard acceleration due to the Earth's gravitational field (usually denoted by g instead of a) is $g = 9.80665 \text{ m s}^{-2}$. Pressure is the force per unit area, or the total force, F, divided by the area, A:

$$P = \frac{F}{A} = \frac{mg}{A}$$

Because the volume of mercury in the barometer is $V = Ah$,

$$P = \frac{mg}{A} = \frac{mg}{V/h} = \frac{mgh}{V}$$

Writing the density as $\rho = m/V$ and substituting, we get

$$P = \rho g h \qquad [9.1]$$

We can use this equation to calculate the pressure exerted by the atmosphere. The density of mercury at 0°C, in SI units (see Appendix B for a description of SI units), is

$$\rho = 13.5951 \text{ g cm}^{-3} = 1.35951 \times 10^4 \text{ kg m}^{-3}$$

and the height of the mercury column under ordinary atmospheric conditions near sea level is close to 0.76 m (760 mm). Let's use exactly this value for the height in our computation:

$$P = \rho g h = (1.35951 \times 10^4 \text{ kg m}^{-3})(9.80665 \text{ m s}^{-2})(0.760000 \text{ m})$$
$$= 1.01325 \times 10^5 \text{ kg m}^{-1} \text{ s}^{-2}$$

Pressure is expressed in various units. The SI unit for pressure is the **pascal** (Pa), which is 1 kg m^{-1} s^{-2}. One **standard atmosphere** (1 atm) is defined as exactly 1.01325×10^5 Pa. The standard atmosphere is a useful unit because the pascal is inconveniently small and because "atmospheric pressure" is important as a standard of reference. We must express pressures in pascals when we perform calculations entirely in SI units.

For historical reasons, a number of different pressure units are commonly used in different fields of science and engineering. Although we will work primarily with the standard atmosphere, it is important that you recognize other units and be able to convert among them. For example, the atmospheric pressure (often called the barometric pressure) recorded in weather reports and forecasts is typically expressed as the height (in millimeters or inches) of the column of mercury it supports. One standard atmosphere supports a 760-mm

TABLE 9.2

Units of Pressure

Unit	Definition or Relationship
pascal (Pa)	1 kg m^{-1} s^{-2}
bar	1 × 10^5 Pa
atmosphere (atm)	101,325 Pa
torr	1/760 atm
760 mm Hg (at 0°C)	1 atm
14.6960 pounds per square inch (psi, lb in^{-2})	1 atm

column of mercury at 0°C; thus, we often speak of 1 atm pressure as 760 mm or 30 inches (of mercury [Hg]). Because the density of mercury depends slightly on temperature, for accurate work it is necessary to specify the temperature and make the proper corrections to the density. A more precise term is the "torr," defined as 1 torr = 1/760 atm (or 760 torr = 1 atm) at *any* temperature. Only at 0°C do the torr and the millimeters of mercury (mm Hg) coincide. These and other units of pressure are summarized in Table 9.2.

Robert Boyle, an English natural philosopher and theologian, studied the properties of confined gases in the 17th century. He noted that a gas tends to spring back to its original volume after being compressed or expanded. Such behavior resembles that of metal springs, which were being investigated by his collaborator Robert Hooke. Boyle published his experiments on the compression and expansion of air in the 1662 monograph titled "The Spring of the Air and Its Effects."

Boyle studied how the volume of a confined gas responded to changes in pressure while the temperature was held constant. Boyle worked with a simple piece of apparatus: a J-tube in which air was trapped at the closed end by a column of mercury (Fig. 9.6). If the difference in height, *h,* between the two

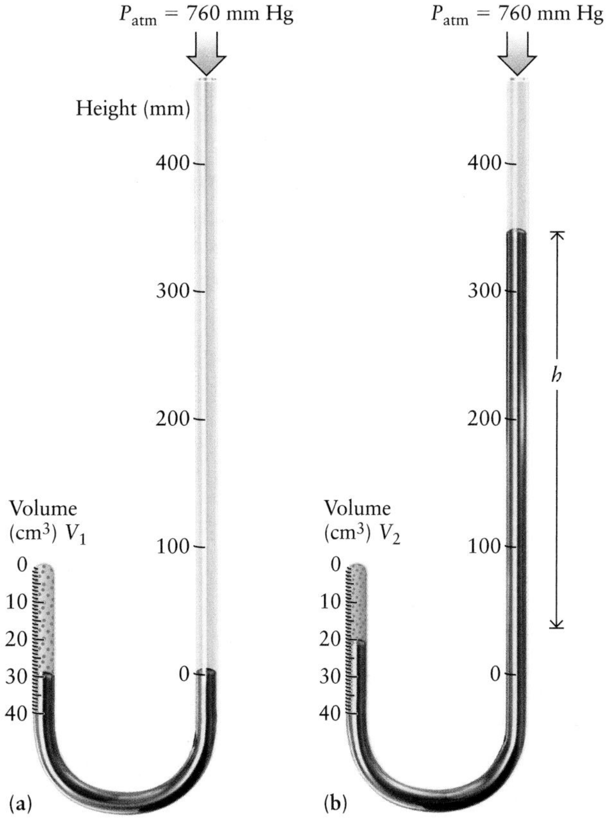

FIGURE 9.6 (a) Boyle's J-tube. When the heights of mercury on the two sides of the tube are the same, the pressure of the confined gas must equal that of the atmosphere, 1 atm or 760 mm Hg. (b) After mercury has been added, the pressure of the gas is increased by the number of millimeters of mercury in the height difference *h*. The compression of the gas causes it to occupy a smaller volume.

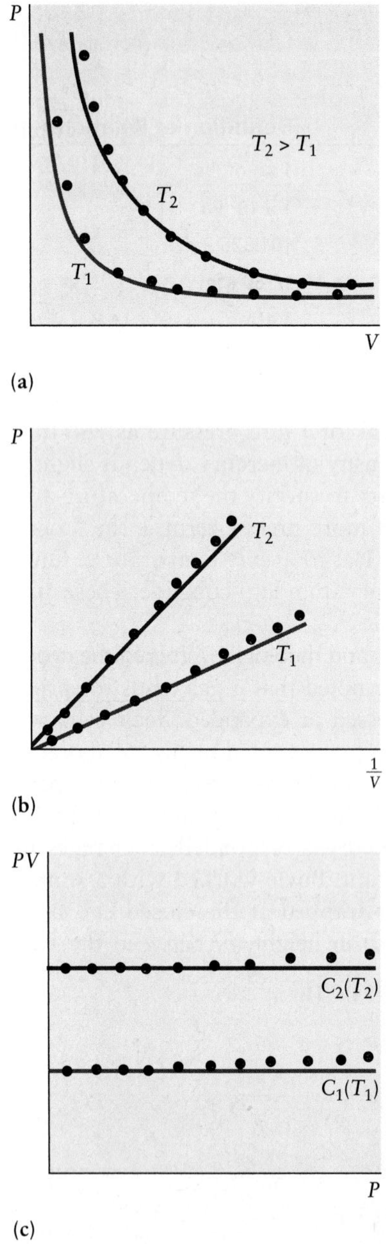

FIGURE 9.7 (a–c) Three ways of depicting Boyle's law. Small deviations from the law (especially apparent in b and c) arise for real gases at higher pressures or smaller volumes.

mercury levels in such a tube is 0, then the pressure of the air in the closed part exactly balances that of the atmosphere; its pressure, P, is 1 atm. Adding mercury to the open end of the tube increases the pressure applied to the confined air; its internal pressure increases by 1 mm Hg for every 1-mm difference between the levels of mercury on the open and closed sides of the tube. Expressed in atmospheres, the pressure in the closed end of the tube is

$$P = 1 \text{ atm} + \frac{h \text{ (mm)}}{760 \text{ mm atm}^{-1}}$$

The volume of the confined air can be read from the scale on the previously calibrated tube. The temperature is held constant by the ambient surroundings of the tube. Boyle discovered that the product of pressure and volume, PV, has the constant value C so long as temperature and the number of moles of confined gas remain fixed. The value of the constant C depends on the amount of gas and the temperature of the gas.

Boyle's data showed that P and V are inversely related, and *suggested* the relationship might be described by the equation $PV = C$, as shown in Figure 9.7a. Before the advent of graphing calculators and computers, it was not always easy to determine the mathematical function that best fit experimental data. With a limited amount of data, it is difficult to distinguish $P = C/V$ from $P = P^* \exp(-C/V)$, where P^* is some fixed reference value of pressure. How are we to know whether Boyle's experiments are best described by $PV = C$ or some more complicated equation? There are two ways of plotting the data to answer this question. The first is to rewrite Boyle's proposed equation in the form

$$P = \frac{C}{V} = C\left(\frac{1}{V}\right)$$

which makes P directly or linearly proportional to $1/V$. If Boyle's equation correctly fits his data, then replotting the data in the form P against $1/V$ (rather than V) should give a straight line passing through the origin with slope C (this is shown to be true in Fig. 9.7b). Alternatively, if $PV = C$ is correct, replotting the data in the form PV against P should give a straight line independent of P (the results are shown in Fig. 9.7c). Rewriting proposed equations to make them linear ($y = mx + b$) and plotting the data together with a proposed equation provides a good test of that equation's validity. The two plots studied here show convincingly that Boyle's equation accurately describes the relation between pressure and volume, at least over the range of temperatures and pressures measured.

$$PV = C \quad \text{(fixed temperature and fixed amount of gas)} \qquad [9.2]$$

This result is known as **Boyle's law.**

Keep in mind that the constant C depends on the temperature T and the amount (number of moles n) of gas in the closed container. For each combination of T and n, the limiting value of C at low P can be obtained by extrapolating plots (such as Fig. 9.7c) to zero pressure. For 1 mol gas (for example, 31.999 g O_2, 28.013 g N_2, or 2.0159 g H_2) at 0°C, extrapolation of PV to zero pressure (as shown by the red line in Fig. 9.7c) gives a limiting value of 22.414 L atm for C. Therefore, for these conditions, Boyle's law takes the special form:

$$PV = 22.414 \text{ L atm} \quad \text{(for 1 mol gas at 0°C)} \qquad [9.3]$$

If the pressure is 1.00 atm, the volume is 22.4 L; if P is 4.00 atm, V is $22.414/4.00 = 5.60$ L.

Boyle's law is an idealized expression that is satisfied exactly by all gases at *very* low pressures. For real gases near 1 atm pressure, small corrections may be necessary for highly accurate studies of P-V-T behavior. At pressures beyond 50–100 atm, substantial corrections are necessary.

EXAMPLE 9.1

The long cylinder of a bicycle pump has a volume of 1131 cm^3 and is filled with air at a pressure of 1.02 atm. The outlet valve is sealed shut and the pump handle is pushed down until the volume of the air is 517 cm^3. Compute the pressure inside the pump. Express its value in atmospheres and pounds per square inch.

Solution

Note that the temperature and amount of gas are not stated in this problem; thus, the value of 22.414 L atm cannot be used for the constant C. It is necessary only to assume that the temperature does not change as the pump handle is pushed down. If P_1 and P_2 are the initial and final pressures and V_1 and V_2 the initial and final volumes, then

$$P_1V_1 = P_2V_2$$

because the temperature and amount of air in the pump do not change. Substitution gives

$$(1.02 \text{ atm})(1131 \text{ cm}^3) = P_2(517 \text{ cm}^3)$$

which can be solved for P_2:

$$P_2 = 2.23 \text{ atm}$$

In pounds per square inch (see Table 9.2), this pressure is

$$P_2 = 2.23 \text{ atm} \times 14.696 \text{ psi atm}^{-1} = 32.8 \text{ psi}$$

Related Problems: 11, 12

9.2.2 Temperature and Charles's Law

Temperature is one of those elusive properties that we all think we understand but is, in fact, difficult to pin down in a quantitative fashion. We have an instinctive feeling (through the sense of touch) for *hot* and *cold*. Water at its freezing point is obviously colder than at its boiling point, so we assign it a lower temperature. Both the Celsius and Fahrenheit temperature scales were defined using the freezing and boiling points of water. Water freezes at 0°C (32°F) and boils at 100°C (212°F). Although both scales use the same reference points, it is interesting to see how the size of the degree was defined and the reference temperatures were determined. Fahrenheit initially chose as reference points the freezing point for a saturated saltwater solution, which he assigned as 0° (then thought to be as cold as possible), and normal body temperature (that of his wife), assigned to be 96°. His choice of 96° for body temperature is thought to have been stimulated by the earlier work of Newton, who had devised a similar scale by dividing the interval between the boiling and freezing points of water into 12 units. As Fahrenheit's thermometers got better, he increased their resolution by several factors of 2 until the number 96 was reached. Further calibration led to the modern definition based on the freezing and boiling points of pure water cited earlier. The Celsius scale (originally called the centigrade scale) is a bit more logical. Celsius chose the same end points but divided the range into 100 units for ease of calculation.

Assigning two fixed points in this way does not show how to define a temperature scale. Ether boils at atmospheric pressure somewhere between 0°C and 100°C, but what temperature should be assigned to its boiling point? Further arbitrary choices are certainly not the answer. The problem is that temperature is not a mechanical quantity like pressure; therefore, it is more difficult to define.

One way around this problem is to find some mechanical property that depends on temperature and use it to define a temperature scale. If we measure the value of this property of an object immersed in boiling water and again when it is immersed in boiling ether, we can quantitatively compare the boiling points of these two liquids. A number of mechanical properties depend on

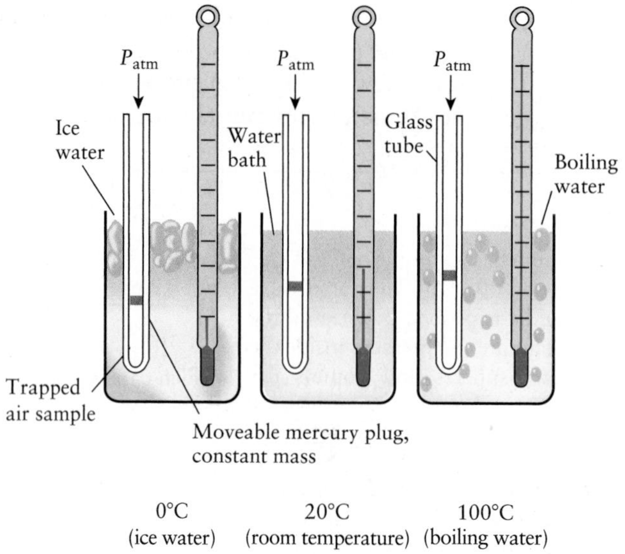

FIGURE 9.8 The volume of a gas confined at constant pressure increases as the temperature increases.

temperature. For example, as liquid mercury is heated from 0°C to 100°C, its volume increases by 1.82%. This change in volume could be used to define a temperature scale. If we *assume* that the volume of mercury is a linear function of temperature, then we can simply measure temperature by measuring the volume of mercury in a tube (a mercury thermometer). The problem is that this definition is tied to the properties of a single substance, mercury, and rests on the assumption that the volume of a sample of mercury is directly proportional to the temperature. Can we define a temperature scale that is more universal, one that does not depend on the properties of a specific material or the assumption of linearity? To answer this question, let's examine the behavior of gases upon heating or cooling.

Boyle observed that the product of the pressure and volume of a confined gas changes on heating, but the first quantitative experiments on the temperature dependence of the properties of gases were performed by the French scientist Jacques Charles more than a century later. Charles observed that *all* gases expand by the same relative amount between the same initial and final temperatures, when studied at sufficiently low pressures. For example, heating a sample of N_2 from the freezing point of water to the boiling point causes the gas to expand to 1.366 times its original volume (Fig. 9.8). The same 36.6% increase in volume occurs for O_2, CO_2, and other gases. (In contrast, liquids and solids vary widely in their thermal expansion.) This universal behavior suggests that temperature is a linear function of gas volume. We write this function as

$$t = c\left(\frac{V}{V_0} - 1\right)$$

where V is the volume of the gas at temperature t, V_0 is its volume at the freezing point of water, and c is a constant that is the same for all gases. We have written the linear equation in this form to ensure that the freezing point of water (when $V = V_0$) will be at $t = 0$, corresponding to the zero of the Celsius scale of temperature. From the measured fractional increase in V to the boiling point (taken to be 100°C), the value of c can be determined. In 1802, Gay-Lussac reported a value for c of 267°C. Subsequent experiments have refined this result to give $c = 273.15°\text{C}$. The definition of temperature (in degrees Celsius) is then

$$t = 273.15°\text{C}\left(\frac{V}{V_0} - 1\right)$$

The temperature of a gas sample at low pressure can be measured by comparing its volume with the volume it occupies at the freezing point of water. For many gases, atmospheric pressure is sufficiently low; but for highly accurate temperature determinations, it is necessary to use pressures below atmospheric or to apply small corrections.

With this definition of temperature in mind, let's return to mercury and measure its *actual* changes in volume with temperature. The result found is almost, but not quite, linear. If a mercury thermometer is calibrated to match the gas thermometer at 0°C and 100°C and if the scale in between is divided evenly into 100 parts to mark off degrees, a small error will result from using this thermometer. A temperature of 40.00°C on the gas thermometer will be read as 40.11°C on the mercury thermometer, because the volume of liquid mercury is not exactly a linear function of temperature.

We can rewrite the preceding equation to express the gas volume in terms of the temperature. The result is

$$V = V_0\left(1 + \frac{t}{273.15°\text{C}}\right) \qquad [9.4]$$

In words, the volume of a gas varies linearly with its temperature. This is the most common statement of **Charles's law**, but it is somewhat misleading because the linearity is built in through the definition of temperature. The key observation is the universal nature of the constant 273.15°C, which is the same for all gases at low pressures. Written in this form, Charles's law suggests that an interesting lower limit to the temperature exists. Negative temperatures on the Celsius scale correspond to temperatures below the freezing point of water and, of course, are meaningful. But what would happen as t approached −273.15°C? The volume would then approach zero (Fig. 9.9), and if t could go below this value, the volume would become negative, clearly an impossible result. We therefore surmise that t = −273.15°C is a fundamental limit below which the temperature cannot be lowered. All real gases condense to liquid or solid form before they reach this **absolute zero** of temperature, so we cannot check for the existence of this limit simply by measuring the volume of gases. More rigorous arguments show that *no* substance (gas, liquid, or solid) can be cooled below

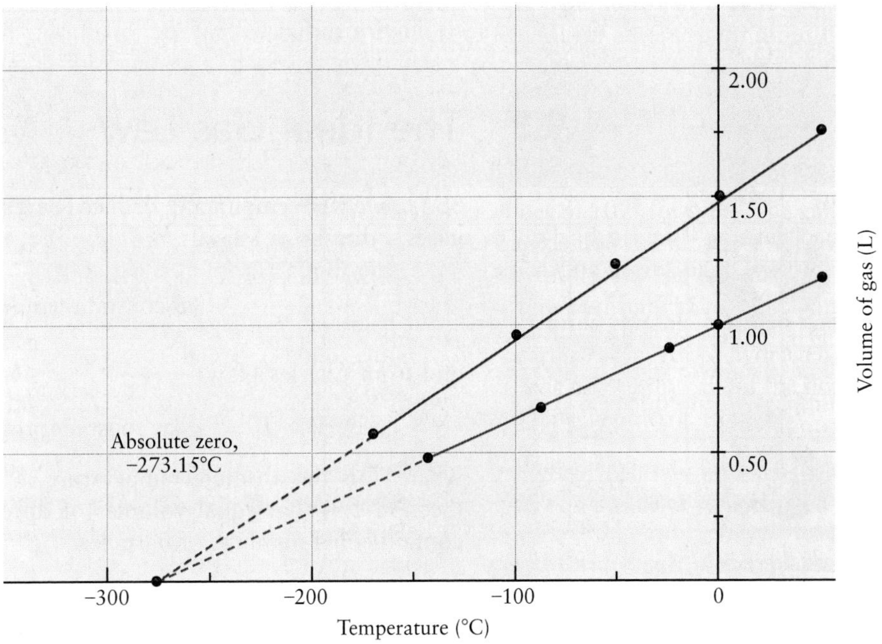

FIGURE 9.9 The volume of a sample of a gas is a function of temperature at constant pressure. The observed straight-line response of volume to temperature illustrates Charles's law. The volume of a particular sample of a gas (red line) is 1.0 L at a temperature of 0°C. Another sample of a gas (blue line) held at the same pressure takes up more volume at 0°C but shrinks faster as cooled. The *percentage* change in volume is the same as that of the first sample for every degree of temperature change. Extrapolation of the trends (dashed lines) predicts that the volumes of the samples go to zero at a temperature of −273.15°C. Similar observations are made regardless of the chemical identities of the gases.

−273.15°C. In fact, it becomes increasingly difficult to cool a substance as absolute zero is approached. The coldest temperatures reached to date are less than 1 nanodegree above absolute zero.

The absolute zero of temperature is a compellingly logical choice as the zero point of a temperature scale. The easiest way to create such a new scale is to add 273.15 to the Celsius temperature, which leads to the **Kelvin temperature scale:**

$$T \text{ (Kelvin)} = 273.15 + t \text{ (Celsius)} \qquad [9.5]$$

The capital T signifies that this is an absolute scale, the unit of which is *kelvin* (K). Thus, a temperature of 25.00°C corresponds to 273.15 + 25.00 = 298.15 K. (**Note:** The unit is the kelvin, not the °K.) On this scale, Charles's law takes the following form:

$$V \propto T \quad \text{(fixed pressure and fixed amount of gas)} \qquad [9.6]$$

where the proportionality constant is determined by the pressure and the amount (number of moles) of gas present. The ratios of volumes occupied at two different temperatures by a fixed amount of gas at fixed pressure are:

$$\frac{V_1}{V_2} = \frac{T_1}{T_2}$$

EXAMPLE 9.2

A scientist studying the properties of hydrogen at low temperature takes a volume of 2.50 L hydrogen at atmospheric pressure and a temperature of 25.00°C and cools the gas at constant pressure to −200.00°C. Predict the volume that the hydrogen occupies at the lower temperature.

Solution

The first step is always to convert temperatures to kelvins:

$$t_1 = 25.00\text{°C} \Rightarrow T_1 = 273.15 + 25.00 = 298.15 \text{ K}$$

$$t_2 = -200.00\text{°C} \Rightarrow T_2 = 273.15 - 200.00 = 73.15 \text{ K}$$

The ratio in Charles's law is

$$\frac{V_1}{T_1} = \frac{2.50 \text{ L}}{298.15 \text{ K}} = \frac{V_2}{T_2} = \frac{V_2}{73.15 \text{ K}}$$

$$V_2 = \frac{(73.15 \text{ K})(2.50 \text{ L})}{298.15 \text{ K}} = 0.613 \text{ L}$$

Related Problems: 13, 14, 15, 16, 17, 18

9.3 The Ideal Gas Law

So far, we have empirically deduced several relationships between properties of gases. From Boyle's law,

$$V \propto \frac{1}{P} \qquad \text{(at constant temperature, fixed amount of gas)}$$

and from Charles's law,

$$V \propto T \qquad \text{(at constant pressure, fixed amount of gas)}$$

where T is the absolute temperature in kelvins. From Avogadro's hypothesis (Section 1.3) that equal volumes of different gases held at the same T and P contain equal numbers of particles,

$$V \propto n \qquad \text{(at constant temperature and pressure)}$$

where n is the number of moles of substance. These three statements may be combined in the form

$$V \propto \frac{nT}{P}$$

A proportionality constant called R converts this proportionality to an equation:

$$V = R\frac{nT}{P} \quad \text{or} \quad PV = nRT \qquad [9.7]$$

Because Avogadro's hypothesis states that equal volumes of all gases contain the same number of molecules (or moles), R is a *universal* constant. Equation 9.7 states the **ideal gas law**, which holds approximately for all gases near atmospheric pressure and room temperature and becomes increasingly accurate at lower pressure and higher temperature. It is a limiting law that describes the behavior of gases at low *densities*. We obtained the ideal gas law empirically from experimental studies of the P-V-T behavior of gases, and we shall see in Section 9.5 that the same relation is predicted by a molecular model of gases. The ideal gas law is our first example of a direct connection between the experimentally observed macroscopic behavior of matter and the structure and interactions of its constituent molecules.

Situations frequently arise in which a gas undergoes a change that takes it from some initial condition (described by P_1, V_1, T_1, and n_1) to a final condition (described by P_2, V_2, T_2, and n_2). Because R is a constant,

$$\frac{P_1V_1}{n_1T_1} = \frac{P_2V_2}{n_2T_2} \qquad [9.8]$$

This is a useful alternative form of the ideal gas law.

EXAMPLE 9.3

A weather balloon filled with helium (He) has a volume of 1.0×10^4 L at 1.00 atm and 30°C. It rises to an altitude at which the pressure is 0.60 atm and the temperature is −20°C. What is the volume of the balloon then? Assume that the balloon stretches in such a way that the pressure inside stays close to the pressure outside.

Solution

Because the amount of helium does not change, we can set n_1 equal to n_2 and cancel it out of Equation 9.8, giving

$$\frac{P_1V_1}{T_1} = \frac{P_2V_2}{T_2}$$

Solving this for the only unknown quantity, V_2, gives

$$V_2 = V_1\left(\frac{P_1T_2}{P_2T_1}\right)$$

$$= 1.0 \times 10^4\ \text{L}\left(\frac{1.00\ \text{atm}}{0.60\ \text{atm}}\right)\left(\frac{253\ \text{K}}{303\ \text{K}}\right)$$

$$= 1.4 \times 10^4\ \text{L} = 14{,}000\ \text{L}$$

Remember that temperatures must always be converted to kelvins when using the ideal gas law.

Related Problems: 19, 20, 21, 22

The approach outlined in Example 9.3 can be applied when other combinations of the four variables (P, V, T, and n) remain constant (Fig. 9.10). When n and T remain constant, they can be canceled from the equation to give Boyle's law, which was used in solving Example 9.1. When n and P remain constant, this relation reduces to Charles's law, used in Example 9.2.

FIGURE 9.10 In a hot-air balloon, the volume remains nearly constant (the balloon is rigid) and the pressure is nearly constant as well (unless the balloon rises very high). Thus, *n* is inversely proportional to *T*; as the air inside the balloon is heated, its amount decreases and its density falls. This reduced density gives the balloon its lift.

The numerical value of *R* depends on the units chosen for *P* and *V*. At the freezing point of water ($T = 273.15$ K), the product *PV* for 1 mol of any gas approaches the value 22.414 L atm at low pressures; hence, *R* has the value

$$R = \frac{PV}{nT} = \frac{22.414 \text{ L atm}}{(1.000 \text{ mol})(273.15 \text{ K})} = 0.082057 \text{ L atm mol}^{-1} \text{ K}^{-1}$$

If *P* is measured in SI units of pascals (kg m^{-1} s^{-2}) and *V* in cubic meters, then *R* has the value

$$R = \frac{(1.01325 \times 10^5 \text{ kg m}^{-1} \text{ s}^{-2})(22.414 \times 10^{-3} \text{ m}^3)}{(1.0000 \text{ mol})(273.15 \text{ K})}$$

$$= 8.3145 \text{ kg m}^2 \text{ s}^{-2} \text{ mol}^{-1} \text{ K}^{-1}$$

Because 1 kg m^2 s^{-2} is defined to be 1 *joule* (the SI unit of energy, abbreviated J), the gas constant may also be expressed as

$$R = 8.3145 \text{ J mol}^{-1} \text{ K}^{-1}$$

EXAMPLE 9.4

What mass of helium is needed to fill the weather balloon from Example 9.3?

Solution

First, solve the ideal gas law for *n*:

$$n = \frac{PV}{RT}$$

If *P* is expressed in atmospheres and *V* is expressed in liters, then the value $R = 0.08206$ L atm mol^{-1} K^{-1} must be used.

$$n = \frac{(1.00 \text{ atm})(1.00 \times 10^4 \text{ L})}{(0.08206 \text{ L atm mol}^{-1} \text{ K}^{-1})(303.15 \text{ K})} = 402 \text{ mol}$$

Because the molar mass of helium is 4.00 g mol^{-1}, the mass of helium required is

$$(402 \text{ mol})(4.00 \text{ g mol}^{-1}) = 1610 \text{ g} = 1.61 \text{ kg}$$

Related Problems: 23, 24

9.3.1 Chemical Calculations for Gases

One of the most important applications of the gas laws in chemistry is to calculate the volumes of gases consumed or produced in chemical reactions. If the conditions of pressure and temperature are known, the ideal gas law can be used to convert between the number of moles and gas volume. Instead of working with the mass of each gas taking part in the reaction, we can then use its volume, which is easier to measure. This is illustrated by the following example.

FIGURE 9.11 When copper metal is immersed in concentrated nitric acid, the copper is oxidized and an aqueous solution of blue copper(II) nitrate forms. In addition, some of the nitrate ion is reduced to brown gaseous nitrogen dioxide, which bubbles off.

EXAMPLE 9.5

Concentrated nitric acid acts on copper to give nitrogen dioxide and dissolved copper ions (Fig. 9.11) according to the balanced chemical equation:

$$Cu(s) + 4\,H^+(aq) + 2\,NO_3^-(aq) \longrightarrow 2\,NO_2(g) + Cu^{2+}(aq) + 2\,H_2O(\ell)$$

(Continued)

Suppose that 6.80 g copper is consumed in this reaction, and that the NO_2 is collected at a pressure of 0.970 atm and a temperature of 45°C. Calculate the volume of NO_2 produced.

Solution

The first step (as in Fig. 2.3) is to convert from the mass of the known reactant or product (in this case, 6.80 g Cu) to the number of moles by using the molar mass of copper, 63.55 g mol^{-1}:

$$\frac{6.80 \text{ g Cu}}{63.55 \text{ g mol}^{-1}} = 0.107 \text{ mol Cu}$$

Next, the number of moles of NO_2 generated in the reaction is calculated using the stoichiometric coefficients in the balanced equation:

$$0.107 \text{ mol Cu} \times \left(\frac{2 \text{ mol NO}_2}{1 \text{ mol Cu}}\right) = 0.214 \text{ mol NO}_2$$

Finally, the ideal gas law is used to find the volume from the number of moles (remember that the temperature must first be expressed in kelvins by adding 273.15):

$$V = \frac{nRT}{P} = \frac{(0.214 \text{ mol})(0.08206 \text{ L atm mol}^{-1} \text{ K}^{-1})(273.15 + 45)\text{K}}{0.970 \text{ atm}} = 5.76 \text{ L}$$

Therefore, 5.76 L NO_2 is produced under these conditions. This calculation neglects the formation of $N_2O_4(g)$, dimers of $NO_2(g)$ (see Example 9.6).

Related Problems: 25, 26, 27, 28, 29, 30, 31, 32

9.4 Mixtures of Gases

Suppose a mixture of gases occupies a container at a certain temperature. How does each gas contribute to the total pressure of the mixture? We define the **partial pressure** of each gas as the pressure that gas would exert if it alone were present in the container. John Dalton concluded, from experiment, that the total pressure measured, P_{tot}, is the sum of the partial pressures of the individual gases (Fig. 9.12). This should come as no surprise, given the validity of Avogadro's hypothesis. Even so, it is an important result. **Dalton's law** holds under the same conditions as the ideal gas law itself: It is approximate at moderate pressures and becomes increasingly more accurate as the pressure is lowered.

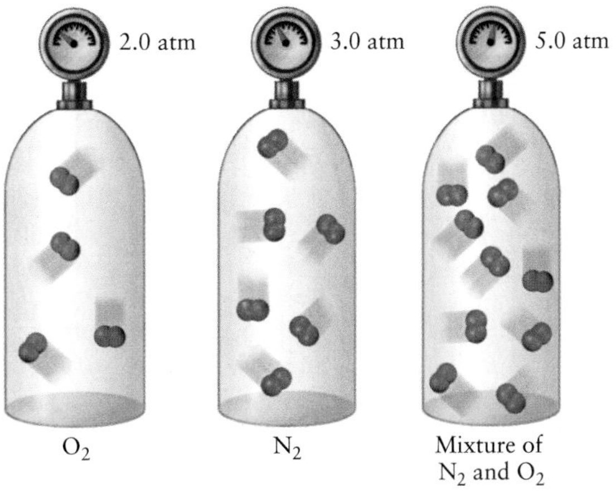

FIGURE 9.12 According to Dalton's law, the total pressure of a gas mixture is the sum of the pressures exerted by the individual gases. Note that the total volume is the same in all three containers.

For a gas mixture at low pressure, the partial pressure of one component, A, is

$$P_A = n_A \frac{RT}{V}$$

The total pressure is the sum of the partial pressures:

$$P_{tot} = P_A + P_B + P_C + \cdots = (n_A + n_B + n_C + \cdots)\frac{RT}{V} = n_{tot}\frac{RT}{V} \qquad [9.9]$$

where n_{tot} is the total number of moles in the gas mixture. Dividing the first equation by the second gives

$$\frac{P_A}{P_{tot}} = \frac{n_A}{n_{tot}} \quad \text{or} \quad P_A = \frac{n_A}{n_{tot}} P_{tot}$$

We define

$$X_A = \frac{n_A}{n_{tot}}$$

as the **mole fraction** of A in the mixture—that is, the number of moles of A divided by the total number of moles present. Then

$$P_A = X_A P_{tot} \qquad [9.10]$$

The partial pressure of any component in a mixture of ideal gases is the total pressure multiplied by the mole fraction of that component. Note that the volume fractions in Table 9.1 are the same as mole fractions.

EXAMPLE 9.6

When NO_2 is cooled to room temperature, some of it reacts to form a *dimer,* N_2O_4, through the reaction

$$2\ NO_2(g) \longrightarrow N_2O_4(g)$$

Suppose 15.2 g of NO_2 is placed in a 10.0-L flask at high temperature and the flask is cooled to 25°C. The total pressure is measured to be 0.500 atm. What partial pressures and mole fractions of NO_2 and N_2O_4 are present?

Solution

Initially, there is 15.2 g/46.01 g mol^{-1} = 0.330 mol NO_2, and therefore, the same number of moles of nitrogen atoms. If at 25°C there are n_{NO_2} moles NO_2 and $n_{N_2O_4}$ moles N_2O_4, then, because the total number of moles of nitrogen atoms is unchanged,

$$n_{NO_2} + 2n_{N_2O_4} = 0.330 \text{ mol} \qquad \text{(a)}$$

To find a second relation between n_{NO_2} and $n_{N_2O_4}$, use Dalton's law:

$$P_{NO_2} + P_{N_2O_4} = 0.500 \text{ atm}$$

$$\frac{RT}{V} n_{NO_2} + \frac{RT}{V} n_{N_2O_4} = 0.500 \text{ atm}$$

$$n_{NO_2} + n_{N_2O_4} = 0.500 \text{ atm} \frac{V}{RT}$$

$$= \frac{(0.500 \text{ atm})(10.0 \text{ L})}{(0.08206 \text{ L atm mol}^{-1}\text{ K}^{-1})(298 \text{ K})}$$

$$n_{NO_2} + n_{N_2O_4} = 0.204 \text{ mol} \qquad \text{(b)}$$

(Continued)

Subtracting Equation b from Equation a gives

$$n_{N_2O_4} = 0.126 \text{ mol}$$

and substituting that back into Equation a or Equation b gives

$$n_{NO_2} = 0.078 \text{ mol}$$

From these results, NO_2 has a mole fraction of 0.38 and a partial pressure of (0.38)(0.500 atm) = 0.19 atm, and N_2O_4 has a mole fraction of 0.62 and a partial pressure of (0.62)(0.500 atm) = 0.31 atm.

Related Problems: 33, 34, 35, 36, 37, 38

9.5 The Kinetic Theory of Gases

The ideal gas law summarizes certain physical properties of gases at low pressures. It is an empirical law, the consequence of experimental observations, but its simplicity and generality prompt us to ask whether it has some underlying microscopic explanation that involves the properties of atoms and molecules in a gas. Such an explanation would allow other properties of gases at low pressures to be predicted and would clarify why real gases deviate from the ideal gas law to small but measurable extents. Such a theory was developed in the 19th century, notably by the physicists Rudolf Clausius, James Clerk Maxwell, and Ludwig Boltzmann. The **kinetic theory of gases** is one of the great milestones of science, and its success provides strong evidence for the atomic theory of matter (see discussion in Chapter 1).

This section introduces a type of reasoning different from that used so far. Instead of proceeding from experimental observations to empirical laws, we begin with a model and use the basic laws of physics with mathematical reasoning to show how this model helps explain the measured properties of gases. In this way, the kinetic theory of gases provides a microscopic understanding of Boyle's law and also a microscopic mechanical definition of temperature as a measure of the average kinetic energy of the molecules in a gas.

The underlying assumptions of the kinetic theory of gases are simple:

1. A pure gas consists of a large number of identical molecules separated by distances that are great compared with their size.
2. The gas molecules are constantly moving in random directions with a distribution of speeds.
3. The molecules exert no forces on one another between collisions, so between collisions they move in straight lines with constant velocities.
4. The collisions of molecules with the walls of the container are *elastic;* no energy is lost during a collision.

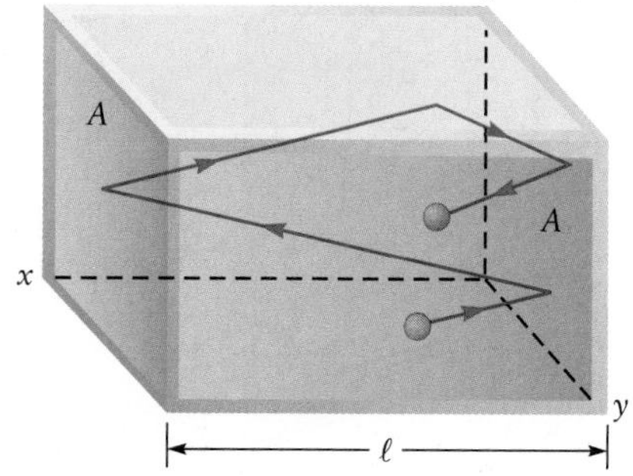

FIGURE 9.13 The path of a molecule in a box.

9.5.1 The Meaning of Temperature

We first use the kinetic theory of gases to find a relation among pressure, volume, and the motions of molecules in an ideal gas. Comparing the result obtained with the ideal gas law ($PV = nRT$) provides a deeper understanding of the meaning of temperature.

Suppose a container has the shape of a rectangular box of length ℓ, with end faces, each of which has area A (Fig. 9.13). A single molecule moving with speed v in some direction is placed in the box. It is important to distinguish between **speed** and **velocity.** The velocity of a molecule specifies both the rate at which it

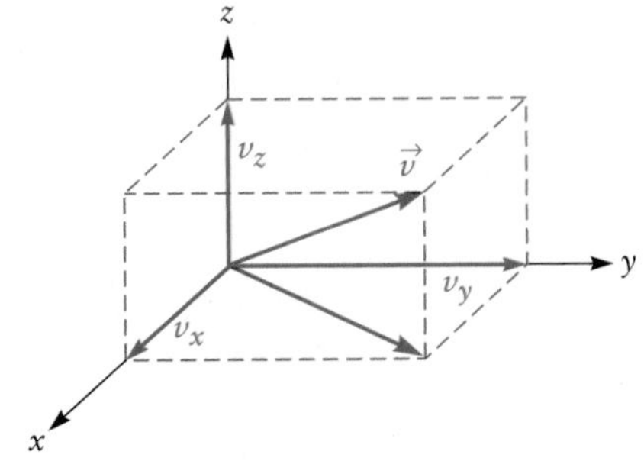

FIGURE 9.14 Velocity is shown by an arrow of length v. It can be separated into three components, v_x, v_y, and v_z, along the three Cartesian coordinate axes and projected into the x-y plane.

is moving (its speed, in meters per second) and the direction of motion. As shown in Figure 9.14, the velocity can be indicated by an arrow (vector) $\vec{v}$, which has a length equal to the speed v and which points in the direction of motion of the molecule. The velocity can also be represented by its components along three coordinate axes: v_x, v_y, and v_z. These are related to the speed, v, by the Pythagorean theorem:

$$v^2 = v_x^2 + v_y^2 + v_z^2$$

(If vectors and components are new to you, see Appendix C.7 for introductory examples.) The **momentum** of a molecule, $\vec{p}$, is its velocity multiplied by its mass. When the molecule collides elastically with a wall of the box, such as one of the end faces of area A, the y and z components of the velocity, v_y and v_z, are unchanged, but the x component (perpendicular to A) reverses sign (Fig. 9.15). Thus $v_{x,\text{f}} = -v_x$ where the subscript f indicates final (after the collision). The change in the x component of the momentum of the molecule, $\Delta p_{x,\text{mol}}$, is

$$\Delta p_{x,\text{mol}} = \text{final momentum} - \text{initial momentum} = m(-v_x) - mv_x = -2\,mv_x$$

The total momentum of the system (molecule plus box) must be conserved, so this momentum change of the molecule is balanced by an equal and opposite momentum change given to the wall:

$$\Delta p_{x,\text{wall}} = 2mv_x$$

After colliding with the wall, the molecule reverses direction, strikes the opposite face of the box, and then again approaches the original face. In between, it may strike the top, bottom, and sides. These collisions do not change the magnitude of v_x, so they do not affect the time between collisions with the original end face (see Fig. 9.13). The distance traveled in the x direction is 2ℓ, and the magnitude of the velocity component in this direction is the absolute value of v_x, so the time elapsed between collisions with this end face is

$$\Delta t = \frac{2\ell}{|v_x|}$$

The magnitude of the momentum transferred to the wall per second is the magnitude of the momentum change per collision divided by Δt:

$$\frac{|\Delta p_{x,\text{wall}}|}{\Delta t} = \frac{|2mv_x|}{2\ell/|v_x|} = \frac{mv_x^2}{\ell}$$

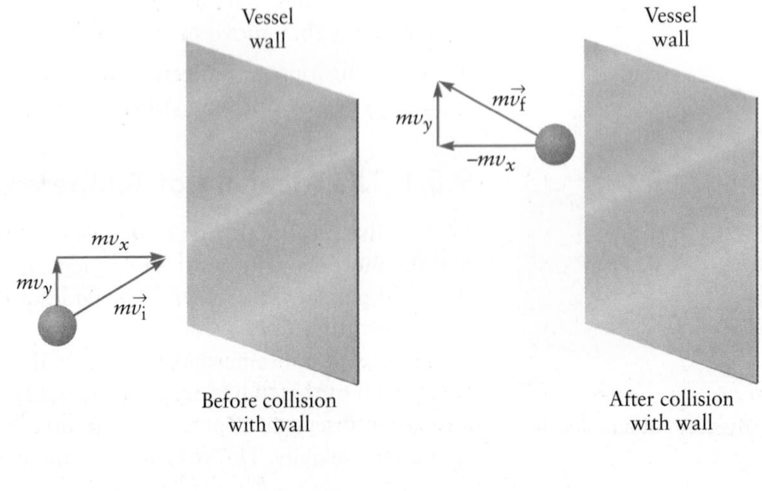

FIGURE 9.15 An elastic collision of a molecule with a wall. The component of the molecule's momentum perpendicular to the wall reverses sign, from mv_x to $-mv_x$. The component parallel to the wall, mv_y, is unchanged. The total momentum is shown by the red arrow. Although the direction of the red arrow is changed by the collision, its length, which represents the magnitude of the momentum, is not changed. Speeds of molecules are not affected by elastic collisions with the walls of the container.

From Newton's second law, the magnitude of the force exerted on the original face by repeated collisions of this molecule is:

$$f = ma = m\frac{|\Delta v|}{\Delta t} = \frac{|\Delta p|}{\Delta t} = \frac{mv_x^2}{\ell}$$

Suppose now that a large number, N, of molecules of mass m are moving independently in the box with x components of velocity, v_{x1}, v_{x2}, v_{x3}, and so forth. Then the total force exerted on the face by the N molecules is the sum of the forces exerted by the individual molecules:

$$F = \frac{mv_{x1}^2}{\ell} + \frac{mv_{x2}^2}{\ell} + \cdots + \frac{mv_{xN}^2}{\ell} = \frac{Nm}{\ell}\overline{v_x^2}$$

where

$$\overline{v_x^2} = \frac{1}{N}(v_{x1}^2 + v_{x2}^2 + \cdots + v_{xN}^2)$$

Here, $\overline{v_x^2}$ is the average of the square of the x component of the velocity of the N molecules, obtained by summing v_x^2 for the N molecules and dividing by N. The pressure is the total force on the wall divided by the area, A, so

$$P = \frac{F}{A} = \frac{Nm}{A\ell}\overline{v_x^2}$$

Because $A\ell$ is the volume, V, of the box, we conclude that

$$PV = Nm\overline{v_x^2}$$

There is no preferred direction of motion for the gas molecules; thus, $\overline{v_x^2}$, $\overline{v_y^2}$, and $\overline{v_z^2}$ should all be equal to one other. We therefore conclude that

$$\overline{v^2} = \overline{v_x^2} + \overline{v_y^2} + \overline{v_z^2} = 3\overline{v_x^2}$$

so

$$PV = \tfrac{1}{3}Nm\overline{v^2} \qquad [9.11]$$

where $\overline{v^2}$ is the **mean-square speed** of the gas molecules. From the ideal gas law,

$$PV = nRT$$

so, we conclude that

$$\tfrac{1}{3}Nm\overline{v^2} = nRT$$

We have achieved our major goal with the derivation of this equation: a relationship between the temperature and the speeds of molecules. It can be simplified to provide additional insights. The equation has the number of molecules, N, on the left and the number of moles, n, on the right. Because N is just n multiplied by Avogadro's number, N_A, we can divide both sides by n to find

$$\tfrac{1}{3}N_A m\overline{v^2} = RT \qquad [9.12]$$

Let's examine this equation in two ways. First, we note that the kinetic energy of a molecule of mass m moving at speed v is equal to $\frac{1}{2}mv^2$, so the *average* kinetic energy of N_A molecules (1 mol), which we denote by $\overline{E}$, is

$\frac{1}{2}N_A m\overline{v^2}$. This quantity is exactly the same as that in the left side of Equation 9.12, with the factor $\frac{1}{2}$ replacing $\frac{1}{3}$:

$$\bar{E} = \frac{1}{2}N_A m\overline{v^2} = \frac{3}{2}\times\left(\frac{1}{3}N_A m\overline{v^2}\right) = \frac{3}{2}RT \qquad [9.13]$$

We obtain the average kinetic energy per molecule, $\bar{\varepsilon}$, by dividing $\bar{E}$ by Avogadro's number:

$$\bar{\varepsilon} = \frac{3}{2}\,k_B T \qquad [9.14]$$

where k_B is **Boltzmann's constant** and is defined as R/N_A. The average kinetic energy of the molecules of a gas depends only on the temperature. It does not depend on the mass of the molecules or their number density in the gas. This relation is the most fundamental result of the kinetic theory of gases, and it is used in all branches of science.

A second way to look at the equation is to recall that if m is the mass of a single molecule, then $N_A m$ is the mass of 1 mol of molecules—the molar mass, abbreviated $\mathcal{M}$. Solving Equation 9.13 for the mean-square speed, we find that

$$\overline{v^2} = \frac{3RT}{\mathcal{M}} \qquad [9.15]$$

The mean-square speed of a gas molecule is proportional to temperature and inversely proportional to its mass. All molecules move faster at higher temperatures, and lighter molecules move faster than heavier ones at the same temperature.

9.5.2 Distribution of Molecular Speeds

One of the fundamental assumptions of the kinetic theory is that the molecules travel through the gas with a range of possible speeds. We would like to know how the molecules are distributed over the range of possible speeds.

As a first step, we can get some sense of the typical speeds in the gas by the following method. We define the **root-mean-square speed,** v_{rms}, as the square root of the mean-square speed $3RT/\mathcal{M}$:

$$v_{rms} = \sqrt{\overline{v^2}} = \sqrt{\frac{3RT}{\mathcal{M}}} \qquad [9.16]$$

To calculate v_{rms} in m s^{-1}, one uses T in kelvin units and the molar mass in kg mol^{-1} and the following value for the universal gas constant.

$$R = 8.3145 \text{ J mol}^{-1}\text{ K}^{-1} = 8.3145 \text{ kg m}^2\text{ s}^{-2}\text{ mol}^{-1}\text{ K}^{-1}$$

Note that molar masses $\mathcal{M}$ must be converted to *kilograms* per mole for use in the equation.

EXAMPLE 9.7

Calculate v_{rms} for (a) a helium atom, (b) an oxygen molecule, and (c) a xenon atom at 298 K.

(Continued)

Solution

Because the factor $3RT$ appears in all three expressions for v_{rms}, let's calculate it first:

$$3RT = (3)(8.3145 \text{ kg m}^2 \text{ s}^{-2} \text{ mol}^{-1} \text{ K}^{-1})(298 \text{ K}) = 7.43 \times 10^3 \text{ kg m}^2 \text{ s}^{-2} \text{ mol}^{-1}$$

The molar masses of He, O_2, and Xe are 4.00 g mol^{-1}, 32.00 g mol^{-1}, and 131.3 g mol^{-1}, respectively. Convert them to 4.00×10^{-3} kg mol^{-1}, 32.00×10^{-3} kg mol^{-1}, and 131.3×10^{-3} kg mol^{-1}, and insert them together with the value for $3RT$ into the equation for v_{rms}:

$$v_{rms}(\text{He}) = \sqrt{\frac{7.43 \times 10^3 \text{ kg m}^2 \text{ s}^{-2} \text{ mol}^{-1}}{4.00 \times 10^{-3} \text{ kg mol}^{-1}}} = 1360 \text{ m s}^{-1}$$

$$v_{rms}(\text{O}_2) = \sqrt{\frac{7.43 \times 10^3 \text{ kg m}^2 \text{ s}^{-2} \text{ mol}^{-1}}{32.00 \times 10^{-3} \text{ kg mol}^{-1}}} = 482 \text{ m s}^{-1}$$

$$v_{rms}(\text{Xe}) = \sqrt{\frac{7.43 \times 10^3 \text{ kg m}^2 \text{ s}^{-2} \text{ mol}^{-1}}{131.3 \times 10^{-3} \text{ kg mol}^{-1}}} = 238 \text{ m s}^{-1}$$

At the same temperature, the He, O_2, and Xe molecules all have the same average kinetic energy; lighter molecules move faster to compensate for their smaller masses. These rms speeds convert to 3050, 1080, and 532 mph, respectively. The average molecule moves along quite rapidly at room temperature!

Related Problems: 41, 42, 43, 44

It is useful to have a complete picture of the entire distribution of molecular speeds. This turns out to be important when we study chemical kinetics (see Chapter 18). There we'll develop an understanding of the rate of a chemical reaction in terms of collisions between reacting molecules. For example, the collision of a gaseous OH radical in the atmosphere with a methane molecule can produce CH_3 radicals. The reaction $OH + CH_4 \longrightarrow H_2O + CH_3$ will not occur unless the colliding molecules approach each other with very high relative speeds (like two fast cars on the highway undergoing a head-on collision) because the energetic barrier to this reaction is about 5 kcal/mol (= 21 kJ mol^{-1}). Thus, it is not the average relative speed of the colliding OH radicals and methane molecules that is important for this reaction, but rather it is the collisions with the fastest relative speeds that matter.

So how can one calculate the fraction of molecules in a gas with speed between v_1 and v_2, say the fraction of methane molecules with speeds between $v_1 = 1000$ and $v_2 = 1100$ m s^{-1}? You need a speed distribution $f(v)$ (see Eq. 9.19 below and Fig. 9.16) for the molecules to do this and you need the range $\Delta v = v_2 - v_1$. Then the fraction of molecules with speed between v_1 and v_2 is $f(v)\,\Delta v$ as long as the range Δv is small. The distribution of molecular speeds, $f(v)$ is shown in the upper frame of Figure 9.16 for methane molecules at 298 K. From this, you can calculate the fraction of methane molecules with speeds between 1000 and 1100 m s^{-1} to be (0.00041 s m^{-1}) (1100 m s^{-1} − 1000 m s^{-1}) = 0.041. Thus 4.1% of the methane molecules in a sample of gas at room temperature have speeds between 1000 and 1100 m s^{-1}. Note that we used a value for $f(v)$ in the middle of the speed range, $f(1050) = 0.00041$ s m^{-1}. If the range of speeds Δv is not small, then the correct formula for calculating the fraction of molecules with speed between v_1 and v_2 is:

$$\text{Fraction of molecules with speed between } v_1 \text{ and } v_2 = \int_{v_1}^{v_2} f(v)\,dv \qquad [9.17]$$

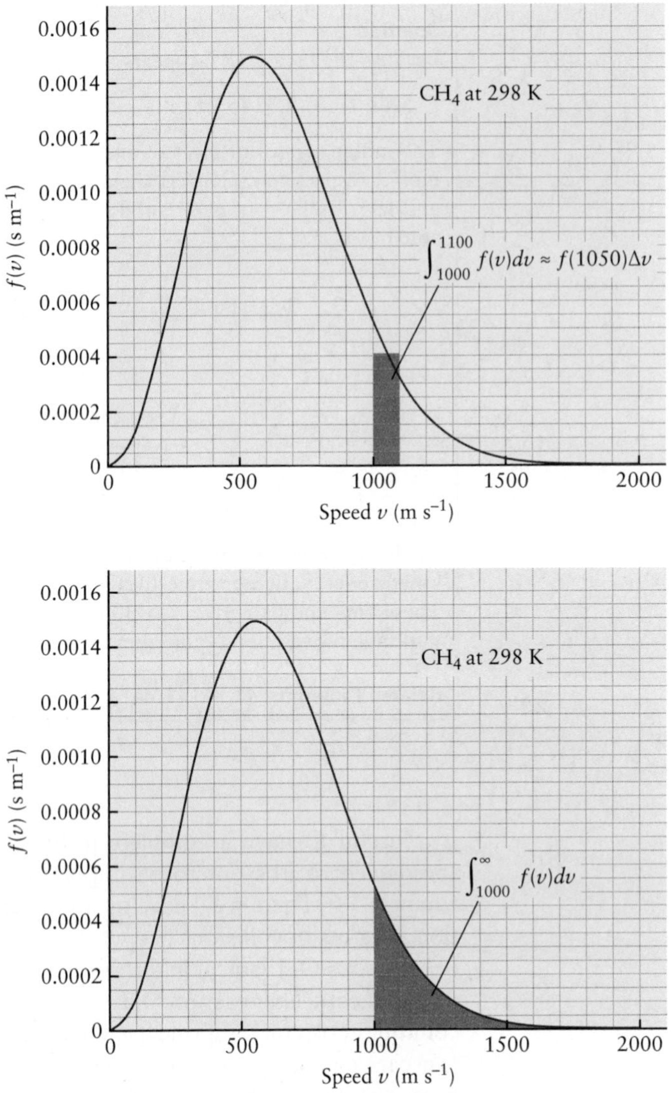

FIGURE 9.16 The distribution of molecular speeds of gaseous methane at room temperature.

This properly accounts for the fact that the function $f(v)$ changes with speed v. The lower frame of Figure 9.16 shows graphically how you would calculate the fraction of methane molecules with speed v greater than $v_1 = 1000 \text{ m s}^{-1}$

$$\text{Fraction of molecules with speed greater than } v_1 = \int_{v_1}^{\infty} f(v)\,dv \qquad [9.18]$$

The speed distribution $f(v)$ was predicted theoretically by Maxwell and Boltzmann about sixty years before it was first measured. It is called the **Maxwell–Boltzmann speed distribution** for a gas of molecules of mass m at temperature T:

$$f(v) = 4\pi\left(\frac{m}{2\pi k_B T}\right)^{\frac{3}{2}} v^2 e^{\frac{-mv^2}{2k_B T}} \qquad [9.19]$$

In this equation m is the mass of one molecule and k_B is Boltzmann's constant defined in Equation 9.14. (Alternatively, you can use the molar mass and the gas constant R.) You can measure this speed distribution experimentally with the apparatus sketched in Figure 9.17. The entire apparatus is enclosed in a large vacuum chamber. The molecules leak out of their container to form a *molecular beam* that passes into a speed analyzer. The analyzer consists of two rotating plates each with a notch in its edge, separated by a fixed distance L. Molecules

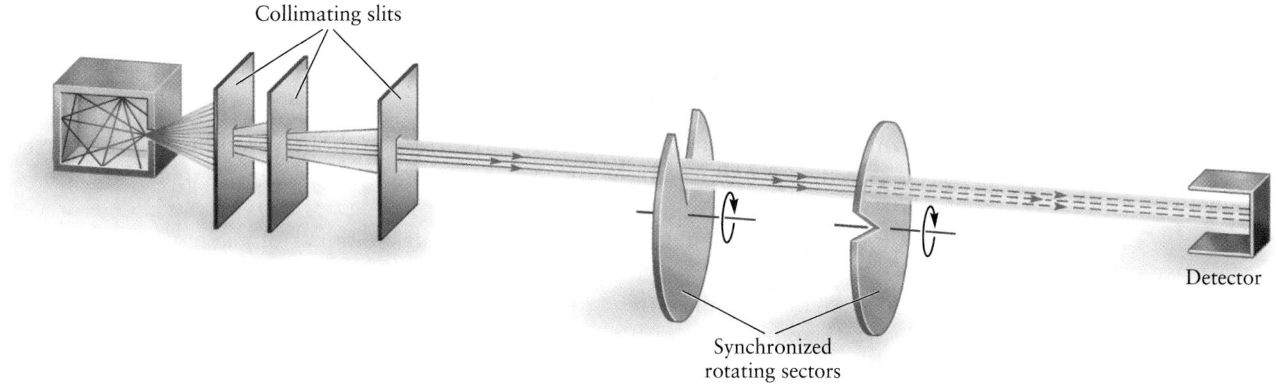

FIGURE 9.17 A device for measuring the distribution of molecular speeds. Only those molecules with the correct velocity to pass through both rotating sectors will reach the detector, where they will be counted. Changing the rate of rotation of the sectors allows the speed distribution to be determined.

with speed v that pass through the center of the notch in the first plate will get to the second plate in a time $t = \frac{L}{v}$. If the two notches are offset by 90 degrees, the molecules can only pass the second plate if it has rotated enough to bring the second notch up to the molecular beam in that same amount of time. If the plates are rotated at 400 Hz, the second notch moves up to the molecular beam in 6.25×10^{-4} s, so if the distance L between the plates is 0.2 m then the molecules with speeds near $v = \frac{L}{t} = 320 \text{ m s}^{-1}$ can pass through the notch to the detector and be counted. (If the notches are wider, molecules with a wider range of speeds Δv near $v = 320 \text{ m s}^{-1}$ can pass through the second slot.) To measure the number of molecules with speeds near 160 m s^{-1} one simply rotates the plates slower, at 200 Hz, so the flight time between the two plates is lengthened to 12.5×10^{-4} s. Then only the slower molecules get to the second plate when the notch in the second plate is crossing the molecular beam, the faster molecules hit the second plate before the notch moves into position. Thus, changing the rotation rate of the plates lets you count the number of molecules in each speed range.

Figure 9.18 shows how the speed distribution $f(v)$ changes as you change the temperature of the gas. The $f(v)$ for methane at 500 K is shown in red in Figure 9.18 with the prior 298 K distribution in blue for comparison. We use

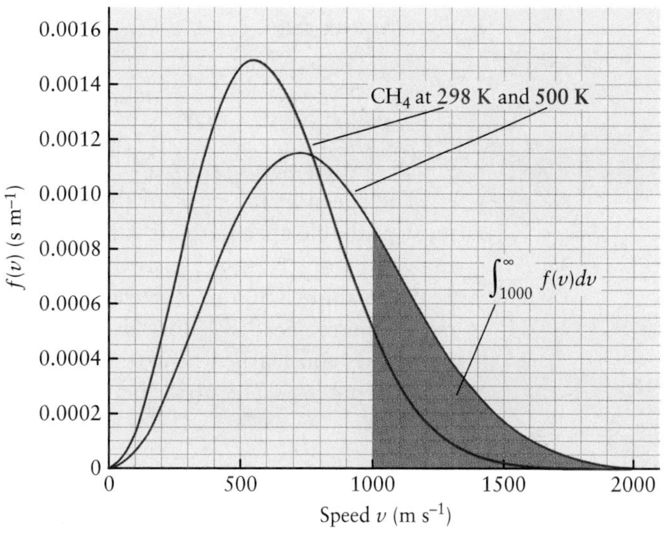

FIGURE 9.18 The distribution of molecular speeds of gaseous methane at room temperature and at 500 K.

Equation 9.18 to first calculate what fraction of methane molecules in a gas at 298 K have a speed greater than 1000 m s^{-1}. You can integrate this numerically with your calculator, or simply estimate the area under the curve shown in the lower frame of Figure 9.16. To estimate the area under the curve, count the number of rectangles in the shaded blue area (about 18 adding up the partial rectangles) and multiply by the area of one rectangle (height × width = 0.00005 s m^{-1} × 100 m s^{-1} = 0.005). This gives 0.09, so at 298 K about 9 percent of the methane molecules in your container of gas have speeds faster than 1000 m s^{-1}. If you pick a molecule out of the gas at random, there is a 9% chance, or probability, that it will have a speed greater than 1000 m s^{-1}. (Mathematically, the function $f(v)$ is one example of a "probability density," described further in Appendix C.6.) Now, if you raise the temperature to 500 K, the distribution of molecular speeds broadens and has a larger high-speed tail, as shown by the red distribution in Figure 9.18. The most probable speed v_{mp} (the speed at the peak of each distribution) has only gone up modestly, from 556 m s^{-1} at 298 K to 720 m s^{-1} at 500 K, but the fraction of molecules with speeds greater than 1000 m s^{-1} is now 27%, 3 times what it was at 298 K! These faster methane molecules are the ones that are more likely to react with OH in the atmosphere, so you can expect the rate of this reaction to increase as you increase the temperature.

From the Maxwell–Boltzmann speed distribution at a given temperature, you can calculate some other useful quantities, like the average speed of the gas molecules. The procedure is described in Appendix C.6 for probability densities in general. For the Maxwell–Boltzmann speed distribution this calculation gives the **average speed** $\bar{v}$, which is

$$\bar{v} = \sqrt{\frac{8k_BT}{\pi m}} = \sqrt{\frac{8RT}{\pi \mathcal{M}}} \qquad [9.20]$$

If a probability distribution is symmetrical about its maximum, like the familiar "bell curve," the most probable value and the average value are the same. The Maxwell–Boltzmann distribution is not symmetrical; the area under the curve to the right of the maximum is somewhat larger than the area under the curve to the left of the maximum. (The next paragraphs use the mathematical form of the distribution to explain this fact.) Consequently, $\bar{v}$ will be larger than the most probable value of v.

The root-mean-square value can be calculated from the probability distribution, as shown in Appendix C.6. For a symmetrical distribution, this would be equal to the average value. For the Maxwell–Boltzmann distribution, we have already seen that

$$v_{rms} = \sqrt{\frac{3k_BT}{m}} = \sqrt{\frac{3RT}{\mathcal{M}}}$$

which verifies that $\bar{v} < v_{rms}$.

The speed at the peak of the Maxwell–Boltzmann distribution is called the most probable speed, v_{mp}. It is the easiest to identify from a plot of the distribution. In Figure 9.16, the distribution of molecular speeds for methane at 298 K peaks at about 556 m s^{-1}. However it is not as physically important as the root mean square speed, v_{rms}, or the average speed $\bar{v}$. Earlier in this section, in Equation 9.13, you learned how to calculate the average kinetic energy of a sample of gas from v_{rms}. The ratios between them are

$$v_{mp} : \bar{v} : v_{rms} = 1.000 : 1.128 : 1.225$$

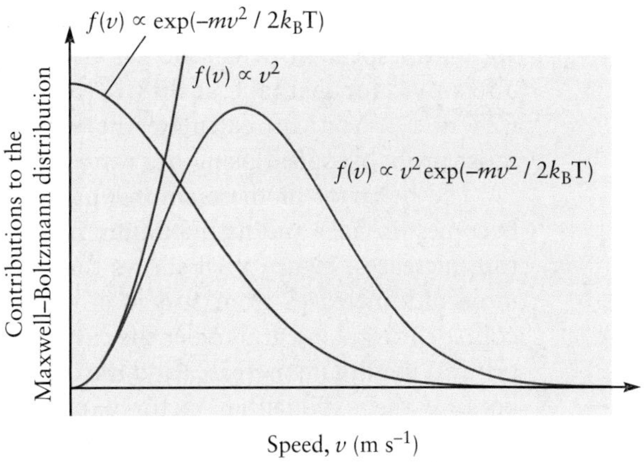

FIGURE 9.19 Mathematical form of the Maxwell–Boltzmann speed distribution. The factor v^2 cuts off the distribution at small values of v, whereas the exponential factor causes it to die off at large values of v. The competition between these effects causes the distribution to achieve its maximum value at intermediate values of v.

The Maxwell–Boltzmann distribution is not symmetrical because it has the mathematical form

$$f(v) \propto v^2 \exp(-mv^2/2k_BT)$$

which describes a competition between the two factors that depend on v^2. The competition arises because these factors behave oppositely, for physical reasons, as the value of v changes. We can get physical insight into the distribution by studying the behavior of these factors separately while T is held constant.

The exponential factor can be viewed graphically as the right half of a bell curve with its maximum at $v = 0$ (Fig. 9.19). At low values of v, this factor behaves as $\exp(-mv^2/2k_BT) \longrightarrow \exp(-0) = 1$. At very large values of v, this factor behaves as $\exp(-mv^2/k_BT) = 1/[\exp(mv^2/k_BT)] \longrightarrow 1/\infty = 0$. The role of this factor is to describe the statistical weight given to each value of v in relation to T. The limits we have just examined show this factor gives large statistical weight to small values of v, and increasingly small weight to large values of v, eventually forcing the distribution to fall off to zero at extremely high values of v. This is exactly what we expect on physical grounds.

The factor v^2 can be viewed as the right half of a parabola with its minimum at $v = 0$ (see Fig. 9.19). The value of this factor approaches zero as v decreases towards 0, and it grows without bound as v becomes extremely large. Although we do not provide all the details, the role of this factor is to count the number of different ways molecules in the gas can achieve a particular value of the speed, v. With Avogadro's number of molecules moving around the vessel, it is physically sensible that many different combinations of v_x, v_y, and v_z correspond to a given value of the speed. And, we expect the number of such combinations to increase as the value of the speed increases. The shape of this factor strongly favors molecules with large values of v and it rapidly cuts off the distribution for small values of v.

The net result of these two competing factors is to keep the probability small for both extremely large and extremely small values of v. The probability will have a maximum at some intermediate value of v where the increasing effect of v^2 is just balanced by the decreasing effect of the exponential factor (see Fig. 9.19). This is the *most probable* value of v, denoted by v_{mp}, and it can be identified by setting to zero the derivative of the curve with respect to v. Because v^2 approaches zero for small values of v more rapidly than the exponential factor approaches zero for large values of v, the probability is larger to the right side of the maximum. The area under the curve to the right of the maximum is somewhat larger than the area under the curve to the left of the maximum. Consequently, the average value of v denoted by $\bar{v}$ will be larger than the most probable value of v. Figure 9.18 shows that the most probable speed, the speed at which the Maxwell–Boltzmann distribution peaks, is about 556 m s^{-1} for methane at 298 K (black line), but the average speed is larger, 627 m s^{-1}. You can calculate the average speed with Equation 9.20 and the most probable speed using the ratio given earlier in this section.

The behavior of these competing factors also explains why the distribution becomes broader and its maximum moves to a higher value of v as the temperature increases. Figure 9.18 shows that the most probable speed of the methane molecules increases from 556 m s^{-1} at 298 K to 720 m s^{-1} at 500 K, (but of course the total integral under the curve is still 1). The speed at which the distribution is a maximum increases and the distribution broadens towards higher speeds, because the exponential factor $\exp(-mv^2/2k_BT)$ decreases more slowly with increasing v at high T than at low T. Figure 9.18 shows that the fraction of methane molecules with speed greater than 1000 m s^{-1} increases by a factor of 3 when the temperature is raised from 298 to 500 K.

The Maxwell–Boltzmann speed distribution defines temperature in the kinetic theory of gases as proportional to the average kinetic energy per molecule through Equation 9.14. Unless the molecular speed distribution for a given gas corresponds to the Maxwell–Boltzmann distribution, temperature has no meaning for the gas. Temperature describes a system of gaseous molecules only when their speed distribution is represented by the Maxwell–Boltzmann function. Consider a closed container filled with molecules whose speed distribution is not "Maxwellian." Such a situation is possible (for example, just after an explosion), but it cannot persist for long. Any distribution of molecular speeds other than a Maxwell–Boltzmann distribution quickly becomes Maxwellian through molecular collisions that exchange energy. Once attained, the Maxwell–Boltzmann distribution persists indefinitely (or at least until some new disturbance is applied). The gas molecules have come to **thermal equilibrium** with one another, and we can speak of a system as having a temperature only if the condition of thermal equilibrium exists.

9.6 Real Gases: Intermolecular Forces

The ideal gas law, $PV = nRT$, is a particularly simple example of an **equation of state**—an equation relating the pressure, temperature, number of moles, and volume to one another. Equations of state can be obtained from either theory or experiment. They are useful not only for ideal gases but also for real gases, liquids, and solids.

Real gases follow the ideal gas equation of state only at sufficiently low densities. Deviations appear in a variety of forms. Boyle's law, $PV = C$, is no longer satisfied at high pressures, and Charles's law, $V \propto T$, begins to break down at low temperatures. Deviations from the predictions of Avogadro's hypothesis appear for real gases at moderate pressures. At atmospheric pressure, the ideal gas law is quite well satisfied for most gases, but for some with polar molecules (like water vapor and ammonia), there are deviations of 1 to 2%. The easiest way to detect these deviations is to calculate the **compressibility factor** z from experimental P-V-T data:

$$z = \frac{PV}{nRT} \qquad [9.21]$$

When z differs from 1 (Fig. 9.20), the ideal gas law is inadequate, and a more accurate equation of state is necessary.

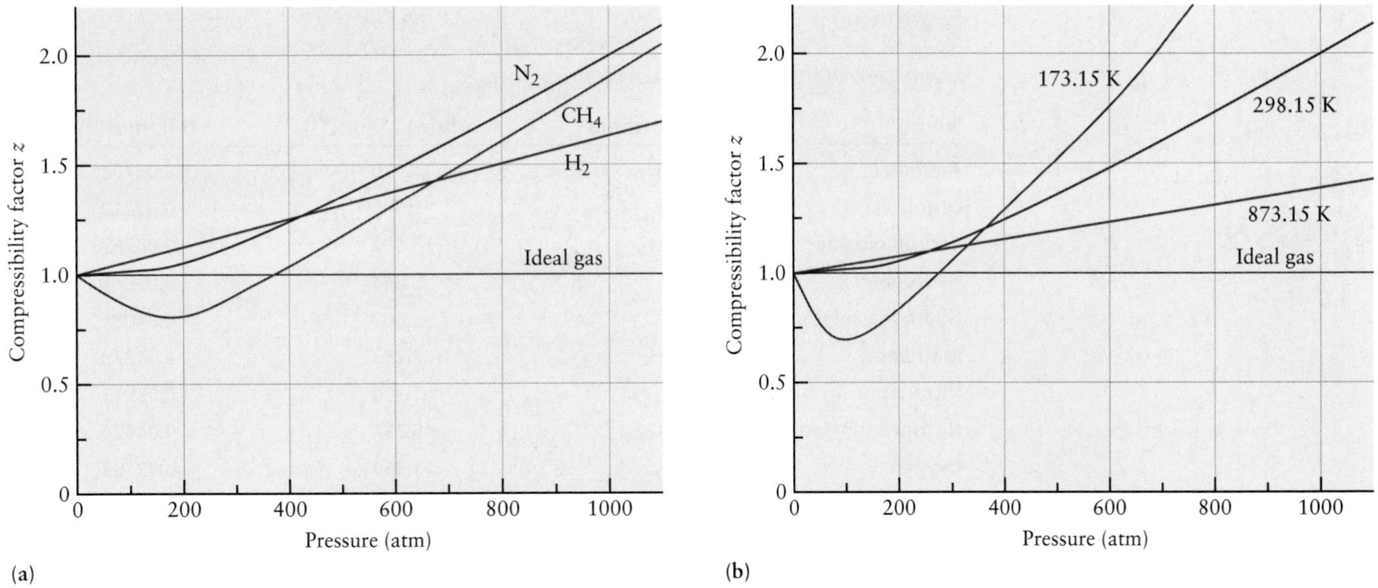

FIGURE 9.20 A plot of $z = PV/nRT$ against pressure shows deviations from the ideal gas law quite clearly. For an ideal gas, $z = 1$, is represented by a straight horizontal line. (a) Deviation of several real gases at 298.15 K. (b) Deviation of nitrogen at several temperatures.

9.6.1 The van der Waals Equation of State

One of the earliest and most important improvements on the ideal gas equation of state was proposed in 1873 by the Dutch physicist Johannes van der Waals. The **van der Waals equation of state** is:

$$\left(P + a\frac{n^2}{V^2}\right)(V - nb) = nRT \tag{9.22a}$$

$$P = \frac{nRT}{V - nb} - a\frac{n^2}{V^2} \tag{9.22b}$$

To obtain this equation, the ideal gas law—which ignores interactions between molecules—requires two modifications to describe the effects of the forces between molecules, which are repulsive at short distances and attractive at large distances. We know from Section 9.5 that pressure is determined by the product of the momentum transferred per collision with the walls of the container times the number of collisions per second. So, it is necessary to see how repulsive and attractive forces modify the collision rate away from the value it would have in the ideal gas. Because of repulsive forces, molecules cannot occupy the same space at the same time. They exclude other molecules from the volumes they occupy; in this way, the effective volume available to a given molecule is not V, but $V - nb$, where b is a constant describing the *excluded volume* per mole of molecules. This effect pushes the molecules away from each other and toward the walls, thereby increasing the rate of wall collisions. The result is a pressure higher than the ideal gas value, as shown in the first term of Equation 9.22b. Attractive forces hold pairs or groups of molecules together. Any tendency to cluster together reduces the effective number of independent molecules in the

TABLE 9.3

Van der Waals Constants of Several Gases

Name	Formula	a (atm L^2 mol^{-2})	b (L mol^{-1})
Ammonia	NH_3	4.170	0.03707
Argon	Ar	1.345	0.03219
Carbon dioxide	CO_2	3.592	0.04267
Hydrogen	H_2	0.2444	0.02661
Hydrogen chloride	HCl	3.667	0.04081
Methane	CH_4	2.253	0.04278
Nitrogen	N_2	1.390	0.03913
Nitrogen dioxide	NO_2	5.284	0.04424
Oxygen	O_2	1.360	0.03183
Sulfur dioxide	SO_2	6.714	0.05636
Water	H_2O	5.464	0.03049

gas and, therefore, reduces the rate of collisions with the walls of the container. Having fewer wall collisions reduces the pressure below the ideal gas law prediction. Because this reduction depends on attractions between *pairs* of molecules, van der Waals argued that it should be proportional to the *square* of the number of molecules per unit volume (N^2/V^2) or, equivalently, proportional to n^2/V^2. Compared with the ideal gas, this intermolecular attraction reduces the pressure by an amount $a(n/V)^2$, where a is a positive constant that depends on the strength of the attractive forces. This effect gives the second term in Equation 9.22b. Rearranging Equation 9.22b gives the standard form of the van der Waals equation shown in Equation 9.22a.

The constants a and b are obtained by fitting experimental P-V-T data for real gases to Equation 9.22a or 9.22b (Table 9.3). The units for these constants are

$$a: \quad \text{atm L}^2\text{ mol}^{-2}$$

$$b: \quad \text{L mol}^{-1}$$

when R has the units L atm mol^{-1} K^{-1}.

EXAMPLE 9.8

A sample of 8.00 kg gaseous nitrogen fills a 100-L flask at 300°C. What is the pressure of the gas, calculated from the van der Waals equation of state? What pressure would be predicted by the ideal gas equation?

Solution

The molar mass of N_2 is 28 g mol^{-1}, so

$$n = \frac{8.00 \times 10^3\text{ g}}{28.0\text{ g mol}^{-1}} = 286\text{ mol}$$

The temperature (in kelvins) is $T = 300 + 273 = 573$ K, and the volume, V, is 100 L. Using $R = 0.08206$ L atm mol^{-1} K^{-1} and the van der Waals constants for nitrogen given in Table 9.3, we calculate $P = 151 - 11 = 140$ atm. If the ideal gas law is used instead, a pressure of 134 atm is calculated. This illustrates the magnitude of deviations from the ideal gas law at higher pressures.

Related Problems: 47, 48, 49, 50

The effects of the two van der Waals parameters are clearly apparent in the compressibility factor for this equation of state:

$$z = \frac{PV}{nRT} = \frac{V}{V - nb} - \frac{a}{RT}\frac{n}{V} = \frac{1}{1 - bn/V} - \frac{a}{RT}\frac{n}{V} \qquad [9.23]$$

Repulsive forces (through b) increase z above 1, whereas attractive forces (through a) reduce z.

We illustrate the effects of a by comparing Equation 9.23 with the experimental data for the compressibility factor shown in Figure 9.20a. At lower pressures, for example 200 atm, the intermolecular forces reduce z for CH_4 to a value significantly below the ideal gas value. For N_2, the effect that decreases z is readily apparent, but it is smaller than the effect that increases z. For H_2, the effect that decreases z is completely dominated by the forces that increase z. These results are consistent with the a-parameter value for CH_4 being about twice that for N_2 and about 10 times that for H_2 (see Table 9.3). The values of a originate in the structure of the molecules and vary significantly between highly polar molecules such as H_2O and nonpolar molecules such as H_2.

The constant b is the volume excluded by 1 mol of molecules and should therefore be close to V_m, the volume per mole in the liquid state, where molecules are essentially in contact with each other. For example, the density of liquid nitrogen is 0.808 g cm^{-3}. One mole of N_2 weighs 28.0 g, so

$$V_m \text{ of } N_2(\ell) = \frac{28.0 \text{ g mol}^{-1}}{0.808 \text{ g cm}^{-3}} = 34.7 \text{ cm}^3 \text{ mol}^{-1}$$
$$= 0.0347 \text{ L mol}^{-1}$$

This is reasonably close to the van der Waals b parameter of 0.03913 L mol^{-1}, obtained by fitting the equation of state to P-V-T data for nitrogen. In Table 9.3, the values for b are all quite similar. All the molecules in Table 9.3 are about the same size and have similar values of molar volume in the liquid state.

9.6.2 Intermolecular Forces

Deeper understanding of the attractive force parameter, a, and the excluded volume per mole, b, comes from examination of the forces acting between the atoms or molecules in a gas. As a pair of molecules approaches one another, the forces between them generate potential energy, which competes with the kinetic energy associated with their speeds. This potential energy can increase the molar volume through intermolecular repulsions, or decrease the pressure by temporarily attracting molecules to form dimers and so reducing the rate of collisions with the walls. The potential energy is our means to describe systematically how intermolecular forces cause these two effects. The noble gases provide the simplest example. As two noble gas atoms approach one another, attractive forces dominate until the distance between their centers, R, becomes short enough for the repulsive forces to begin to become significant. If the atoms are forced still closer together, they repel each other with a strongly increasing force as the distance between them is reduced. These interactions can be described by a **potential energy curve** $V(R)$ (see Section 3.7, Section 6.1, and Appendix B) such as that shown for argon in Figure 9.21. If two molecules can lower their energy by moving closer together, then a net attractive force exists between them; if they can lower their energy only by moving apart, there is a net repulsive force. Graphically, this means that at a separation R where the slope dV/dR is positive, the force between the particles is attractive. At very small separations between the two particles, where dV/dR is negative, the force between the particles is repulsive. The force changes from attractive to repulsive at the minimum of $V(R)$, where the net force between atoms is zero. Potential energy curves for atoms are generated by fitting equations for $V(R)$ to measured properties of real gases. More accurate versions come from experiments in which beams of atoms collide with one another.

For many purposes, the detailed shape of the potential is less important than two characteristic parameters: the depth and location of the potential minimum.

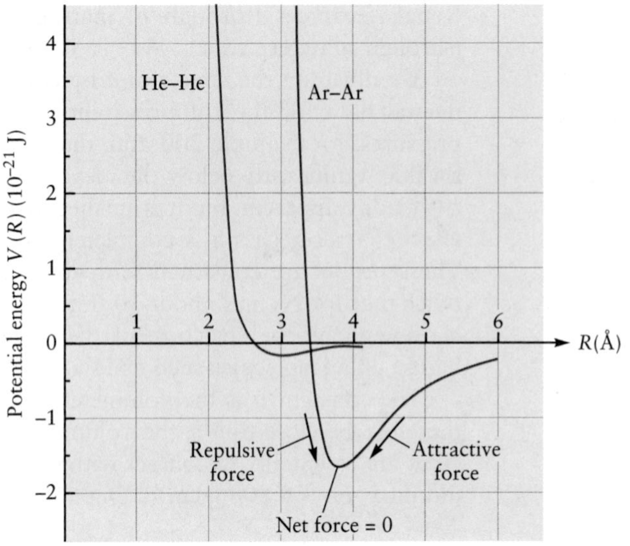

FIGURE 9.21 Potential energy curves $V(R)$ for pairs of helium atoms (red) and pairs of argon atoms (blue) obtained from atomic beam collision studies. At any point, the force between atoms is calculated from the negative of the slope of V (see Section 3.7, Section 6.1, and Appendix B). In regions where the slopes of the curves are negative, the atoms repel each other. In regions where the slopes are positive, the atoms attract one another. The greater well depth for Ar arises from stronger intermolecular attractions, and the location of the minimum for Ar at larger R correlates roughly with relative molecular size.

A simple expression frequently used to model these interactions between atoms is the **Lennard–Jones potential:**

$$V_{LJ}(R) = 4\varepsilon\left[\left(\frac{\sigma}{R}\right)^{12} - \left(\frac{\sigma}{R}\right)^{6}\right] \quad [9.24]$$

where ε is the depth and σ is the distance at which $V(R)$ passes through zero. This potential has an attractive part, proportional to R^{-6}, and a repulsive part, proportional to R^{-12}. The minimum is located at $2^{1/6}\sigma$, or 1.122σ. Table 9.4 lists Lennard–Jones parameters for a number of atoms. Note that the depth and range of the potential increase for the heavier noble-gas atoms, but the well depths are quite small compared to well depths for the potential between two atoms that may covalently bond. The interaction between two molecules such as N_2 that are nearly spherical, neutral, and non-polar can also be described approximately with Lennard–Jones potentials. The two parameters ε and σ in the Lennard–Jones potential, like the van der Waals parameters a and b, are simple ways of characterizing the interactions between molecules in real gases.

TABLE 9.4

Lennard–Jones Parameters for Atoms and Molecules

Substance	σ(m)	ε(J)
He	2.56×10^{-10}	1.41×10^{-22}
Ne	2.75×10^{-10}	4.92×10^{-22}
Ar	3.40×10^{-10}	1.654×10^{-21}
Kr	3.60×10^{-10}	2.36×10^{-21}
Xe	4.10×10^{-10}	3.06×10^{-21}
H_2	2.93×10^{-10}	5.11×10^{-22}
O_2	3.58×10^{-10}	1.622×10^{-21}
CO	3.76×10^{-10}	1.383×10^{-21}
N_2	3.70×10^{-10}	1.312×10^{-21}
CH_4	3.82×10^{-10}	2.045×10^{-21}

The Lennard–Jones potential and the Maxwell–Boltzmann distribution (see Section 9.5) together explain how deviations from ideal gas behavior depend on temperature. Qualitatively, the most important effect is the ratio of the well depth ε to k_BT. When T is low enough to make $k_BT \ll \varepsilon$, a pair of molecules remains close together sufficiently long to reduce the rate of wall collisions, and thereby reduce the pressure below its ideal gas value. When T is such that $k_BT \gg \varepsilon$, the molecules are primarily sensitive to the repulsive part of the Lennard–Jones potential, and the pressure is increased above its ideal gas value. These effects are illustrated in Figure 9.20, which shows that at low pressure the compressibility factor for N_2 is dominated by attractive forces at 173.15 K and by repulsive forces at 873.15 K.

At 300 K, the value of k_BT is 4.14×10^{-21} J, which is comparable to and slightly larger than the well depths for typical gases listed in Table 9.4.

9.7 A DEEPER LOOK Molecular Collisions and Rate Processes

The kinetic theory we have developed can be applied to several important properties of gases. A study of the rates at which atoms and molecules collide with a wall and with one another helps to explain phenomena ranging from isotope separation based on gaseous diffusion to gas-phase chemical kinetics.

9.7.1 Molecule–Wall Collisions

Let's call Z_w the rate of collisions of gas molecules with a section of wall of area A. A full mathematical calculation of Z_w requires integral calculus and solid geometry. We present instead some simple physical arguments to show how this rate depends on the properties of the gas.

First of all, Z_w should be proportional to the area A, because doubling the area will double the number of collisions with the wall. Second, Z_w should be proportional to the average molecular speed, $\bar{v}$, because molecules moving twice as fast will collide twice as often with a given wall area. Finally, the wall collision rate should be proportional to the number density, N/V, because twice as many molecules in a given volume will have twice as many collisions with the wall. All of these arguments are consistent with the kinetic theory of gases and are confirmed by the full mathematical analysis. We conclude that

$$Z_w \propto \frac{N}{V}\,\bar{v}A$$

Note that the units of both sides are s^{-1}, as required for these expressions to represent a rate. The proportionality constant can be calculated from a complete analysis of the directions from which molecules impinge on the wall; it turns out to have the value $\frac{1}{4}$. So the wall collision rate is

$$Z_w \propto \frac{1}{4}\frac{N}{V}\bar{v}A = \frac{1}{4}\frac{N}{V}\sqrt{\frac{8RT}{\pi\mathcal{M}}}A \qquad [9.25]$$

We have used the result for $\bar{v}$ given earlier in Equation 9.20. This simple equation has many applications. It sets an upper limit on the rate at which a gas may react with a solid. It is also the basis for calculating the rate at which molecules effuse through a small hole in the wall of a vessel.

EXAMPLE 9.9

Calculate the number of collisions that oxygen molecules make per second on 1.00 cm^2 of the surface of the vessel containing them if the pressure is 1.00×10^{-6} atm and the temperature is 25°C (298 K).

Solution

First compute the quantities that appear in the equation for Z_w:

$$\frac{N}{V} = \frac{N_A n}{V} = \frac{N_A P}{RT}$$

$$= \frac{(6.022 \times 10^{23} \text{ mol}^{-1})(1.00 \times 10^{-6} \text{ atm})}{(0.08206 \text{ L atm mol}^{-1} \text{ K}^{-1})(298 \text{ K})}$$

$$= 2.46 \times 10^{16} \text{ L}^{-1} = 2.46 \times 10^{19} \text{ m}^{-3}$$

$$A = 1.00 \text{ cm}^2 = 1.00 \times 10^{-4} \text{ m}^2$$

$$\bar{v} = \sqrt{\frac{8RT}{\pi \mathcal{M}}}$$

$$= \sqrt{\frac{8(8.3145 \text{ J mol}^{-1} \text{ K}^{-1})(298 \text{ K})}{\pi(32.00 \times 10^{-3} \text{ kg mol}^{-1})}}$$

$$= 444 \text{ m s}^{-1}$$

The collision rate is then

$$Z_w = \frac{1}{4}\frac{N}{V}\bar{v}A$$

$$= \tfrac{1}{4}(2.46 \times 10^{19} \text{ m}^{-3})(444 \text{ m s}^{-1})(1.00 \times 10^{-4} \text{ m}^2)$$

$$= 2.73 \times 10^{17} \text{ s}^{-1}$$

Related Problems: 51, 52

Equation 9.25 is the basis of explaining **Graham's law of effusion.** In 1846, Thomas Graham showed experimentally that the rate of effusion of a gas through a small hole into a vacuum (Fig. 9.22) is inversely proportional to the square root of its molar mass. Assuming that different gases are studied at the same temperature and pressure, their number density, N/V, is the same, and the rate of effusion of each gas depends only on the factor $1/\sqrt{\mathcal{M}}$ in Equation 9.25, exactly as observed by Graham. Explaining this experimental result is yet another success for the kinetic theory of gases.

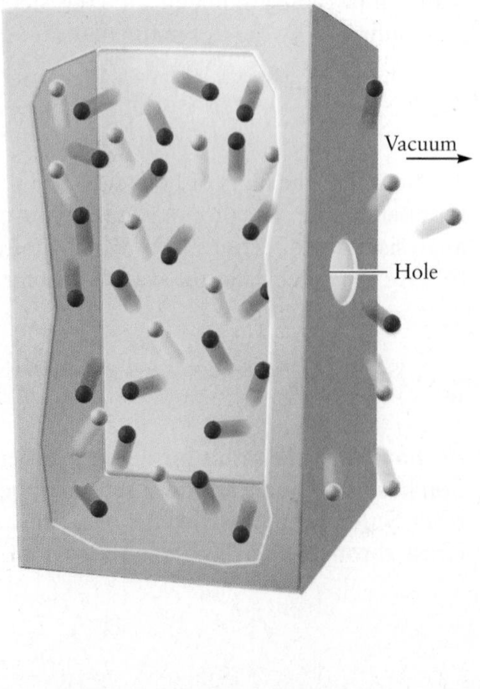

FIGURE 9.22 A small hole in the box permits molecules to effuse out into a vacuum. The less massive particles (here, helium atoms, red) effuse at greater rates than the more massive oxygen molecules (purple) because their speeds are greater on the average.

Graham's law also describes the effusion of a mixture of two gases through a small hole. The ratio of the rates of effusion of the two species, A and B, is

$$\frac{\text{rate of effusion of A}}{\text{rate of effusion of B}} = \frac{\frac{1}{4}\frac{N_A}{V}\sqrt{\frac{8RT}{\pi \mathcal{M}_A}}A}{\frac{1}{4}\frac{N_B}{V}\sqrt{\frac{8RT}{\pi \mathcal{M}_B}}A} = \frac{N_A}{N_B}\sqrt{\frac{\mathcal{M}_B}{\mathcal{M}_A}} \qquad [9.26]$$

This ratio is equal to the ratio of the numbers of molecules of the two species effusing through the hole in a short time interval. N_A and N_B are the number of molecules of species A and B, respectively. The emerging gas is enriched in the lighter component because lighter molecules effuse more rapidly than heavier ones. If B is heavier than A, the **enrichment factor** is $\sqrt{\mathcal{M}_B/\mathcal{M}_A}$ for the lighter species, A.

EXAMPLE 9.10

Calculate the enrichment factors from effusion for a mixture of $^{235}UF_6$ and $^{238}UF_6$, uranium hexafluoride with two different uranium isotopes. The atomic mass of ^{235}U is 235.04, and that of ^{238}U is 238.05. The atomic mass of fluorine is 19.00.

Solution

The two molar masses are

$$\mathcal{M}(^{238}UF_6) = 238.05 + 6(19.00) = 352.05 \text{ g mol}^{-1}$$

$$\mathcal{M}(^{235}UF_6) = 235.04 + 6(19.00) = 349.04 \text{ g mol}^{-1}$$

$$\text{enrichment factor} = \sqrt{\frac{\mathcal{M}(^{238}UF_6)}{\mathcal{M}(^{235}UF_6)}} = \sqrt{\frac{352.05 \text{ g mol}^{-1}}{349.04 \text{ g mol}^{-1}}} = 1.0043$$

Related Problems: 53, 54

Graham's law of effusion holds true only if the opening in the vessel is small enough and the pressure low enough that most molecules follow straight-line trajectories through the opening without colliding with one another. A related but more complex phenomenon is **gaseous diffusion through a porous barrier.** This differs from effusion in that molecules undergo many collisions with one another and with the barrier during their passage through it. Just as in effusion, the diffusion rate is inversely proportional to the square root of the molar mass of the gas. But the reasons for this dependence are not those outlined for the effusion process. If a mixture of gases is placed in contact with a porous barrier, the gas passing through is enriched in the lighter component, A, by a factor of $\sqrt{\mathcal{M}_B/\mathcal{M}_A}$, and the gas remaining behind is enriched in the heavier component.

In order to develop the atomic bomb, it was necessary to separate the more easily fissionable isotope ^{235}U from ^{238}U. Because the natural abundance of ^{235}U is only 0.7%, its isolation in nearly pure form was a daunting task. The procedure adopted was to react uranium with F_2 gas to form the relatively volatile compound UF_6 (boiling point, 56°C), which can be enriched in ^{235}U by passing it through a porous barrier. As shown in Example 9.11, each passage gives an enrichment factor of only 1.0043. So, successive passage through many such barriers is necessary to provide sufficient enrichment of the ^{235}U component. A multi-state gaseous diffusion chamber was constructed in a short time at the Oak Ridge National Laboratories, and the enriched ^{235}U was used in the first atomic bomb. Similar methods are still used today to enrich uranium for use as fuel in nuclear power plants (Fig. 9.23).

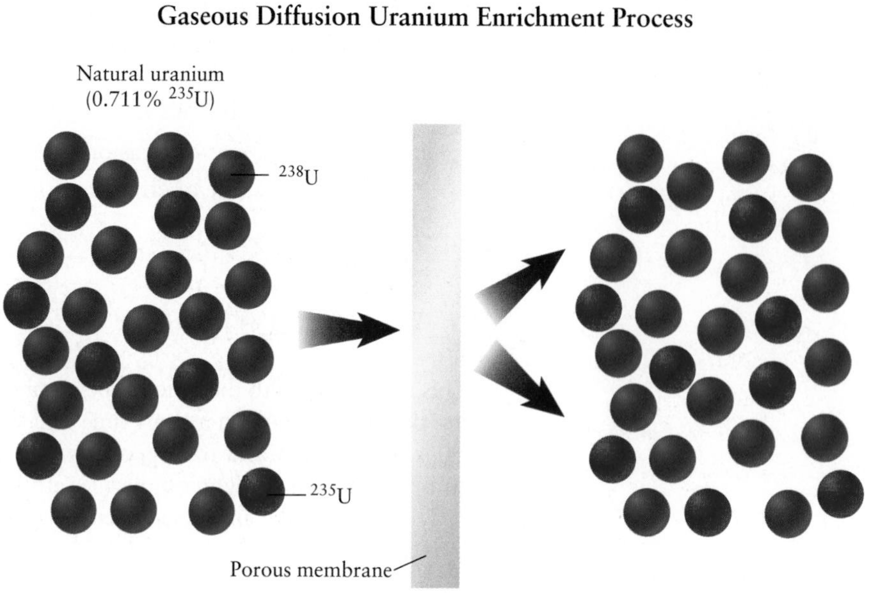

FIGURE 9.23 Schematic showing uranium isotope enrichment by diffusion through a porous membrane with pores a few nanometers in diameter. Isotope enrichment occurs because the lighter UF_6 gas molecules (with ^{235}U atoms) tend to diffuse through the barrier more rapidly than the heavier UF_6 gas molecules containing ^{238}U. The enrichment factor has been greatly exaggerated for purposes of illustration. The gas must pass through several hundred barriers in sequence to build up enough ^{235}U to be useful in reactor fuel, as illustrated in the following *Connection to Chemical Engineering*.

EXAMPLE 9.11

How many diffusion stages are required if ^{235}U is to be enriched from 0.70 to 7.0% by means of the gaseous UF_6 diffusion process?

Solution

From Example 9.10, the enrichment factor per stage is 1.0043; thus, the first stage is described by the relation

$$\left(\frac{N^{235}UF_6}{N^{238}UF_6}\right)_1 = \left(\frac{N^{235}UF_6}{N^{238}UF_6}\right)_0 (1.0043)$$

where the subscripts 0 and 1 denote the initial concentration and that after the first stage, respectively. The ratio of the numbers after n stages satisfies the equation

$$\left(\frac{N^{235}UF_6}{N^{238}UF_6}\right)_n = \left(\frac{N^{235}UF_6}{N^{238}UF_6}\right)_0 (1.0043)^n$$

$$\frac{7.0}{93.0} = \frac{0.70}{99.30}(1.0043)^n$$

$$10.677 = (1.0043)^n$$

This equation can be solved by taking logarithms of both sides (see Appendix C):

$$\log_{10}(10.677) = n \log_{10}(1.0043)$$

$$1.02845 = 0.0018635n$$

$$n = 552 \text{ stages}$$

Related Problems: 55, 56

9.7.2 Molecule–Molecule Collisions

Collisions lie at the very heart of chemistry, because chemical reactions can occur only when molecules collide with one another. The kinetic theory of gases provides methods for estimating the frequency of molecular collisions and the average distance traveled by a molecule between collisions, both important in understanding the rates of chemical reactions (See Chapter 18).

We assume molecules are approximately spherical, with diameters, d, on the order of 10^{-10} m (see Section 2.1). We initially suppose that a particular molecule moves through a gas of stationary molecules of the same diameter. Such a molecule "sweeps out" a cylinder with cross-sectional area, $A = \pi d^2$ (Fig. 9.24).

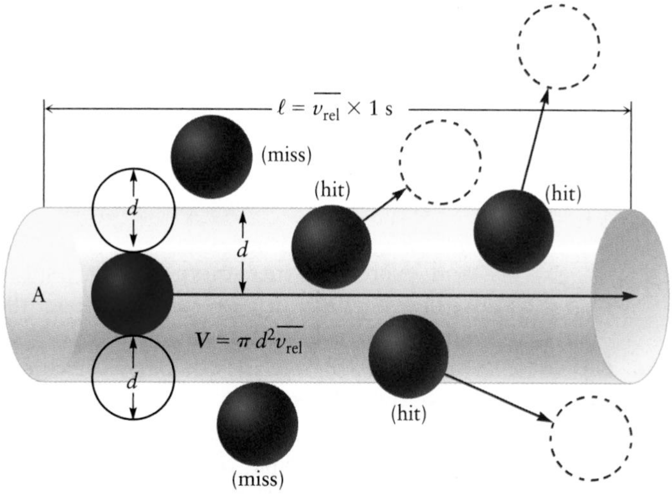

FIGURE 9.24 An average molecule (red) sweeps out a cylinder of volume $\pi d^2 \overline{v_{rel}}$ in 1 second. It will collide with any molecules whose centers lie within the cylinder. Using this construction, we can calculate the rate of collisions with other molecules.

This particular molecule collides with any molecule whose center lies inside the cylinder, within a distance, *d*, of the center of the moving molecule. In 1 second, the length of such a cylinder is $\bar{v}\Delta t = \bar{v} \times 1$ s, where $\bar{v}$ is the average molecular speed. It does not matter that the moving molecule is scattered in another direction when it collides with other molecules; these little cylinders are merely joined together, end to end, to define a cylinder whose length is $\bar{v} \times 1$ s.

In 1 second, the moving molecule therefore sweeps out the volume

$$V_{cyl} = \pi d^2 \bar{v} \qquad \text{(if blue molecules still)}$$

If N_2/V is the number of blue molecules per unit volume in the gas, then the number of collisions per second experienced by the moving red molecule is:

$$\text{collision rate} = Z_1 = \frac{N_2}{V} V_{cyl} = \frac{N_2}{V} \pi d^2 \overline{v_{rel}} \qquad [9.27]$$

Here we have specified the speed as the average *relative* speed between the red molecule and the blue molecules, rather than the average speed of the red molecule, as the blue molecules are also moving.

If the red and blue molecules do not have the same diameter, then *d* is simply the radius of the red molecule plus the radius of a blue molecule.

The average relative speed, $\overline{v_{rel}}$, between a red molecule and the blue molecules with which it is colliding (with both averaged over their respective Maxwell–Boltzmann distribution of molecular speeds) is very similar to the average speed in Equation 9.20, except one replaces m by $\mu = m_1 m_2/(m_1 + m_2)$. (See Chapter 18.) Using this in Equation 9.27 gives

$$Z_1 = \frac{N_2}{V} \pi d^2 \overline{v_{rel}} = \frac{N_2}{V} \pi d^2 \sqrt{\frac{8k_B T}{\pi \mu}} = \frac{N_2}{V} d^2 \sqrt{\frac{8\pi k_B T}{\mu}} \qquad [9.28]$$

You can use this to calculate the collision frequency of a single molecule in a pure gas (the number of collisions per second suffered by a single molecule in a pure gas). Using $\overline{v_{rel}} = \bar{v}\sqrt{2}$ (since $m_1 = m_2$), the expression becomes:

$$Z_1 \text{ in pure gas} = \frac{N}{V} \pi d^2 \overline{v_{rel}} = \frac{N}{V} \pi d^2 \bar{v}\sqrt{2} = 4\frac{N}{V} d^2 \sqrt{\frac{\pi k_B T}{m}} \qquad [9.29]$$

Now, if the number density of the red molecules is N_1/V, you can easily calculate the number of collisions between red molecules and blue molecules per second per unit volume of your sample by multiplying Equation 9.28 above by N_1/V:

$$\frac{\text{collisions of type 1 molecules with type 2}}{\text{m}^3\ \text{s}} = \frac{N_1}{V}\frac{N_2}{V}\pi d^2\overline{v_{rel}}$$
$$= \frac{N_1}{V}\frac{N_2}{V}\pi d^2\sqrt{\frac{8k_BT}{\pi\mu}} \qquad [9.30]$$

and expressing the volumes in m^3, speed in m s^{-1}, distances in m, and the reduced mass μ in kg (calculated from the mass of each of the colliding molecules in kg). Equation 9.30 is very important in expressing the reaction rate between two molecules, as two molecules can react when they collide with each other (see Chapter 18). Noting that in Equation 9.30 the term $N_1/V = P_1/k_BT$, one sees that the number of potentially reactive collisions between molecules of type 1 and molecules of type 2 is proportional to the product of the partial pressures, or the product of the concentrations, of the two molecules. We will use this result in Chapter 18.

Finally, if the gas has only one component, say type 1, the number of collisions between molecules in the gas per unit volume per unit time can be calculated from Equation 9.30, but you have to divide by a factor of 2 so that you don't double count one collision. Also, using $\overline{v_{rel}} = \overline{v}\sqrt{2}$, this gives the following:

$$\frac{\text{collisions between molecules in a pure gas}}{\text{m}^3\ \text{s}} = \frac{1}{2}\left(\frac{N}{V}\right)^2\pi d^2\overline{v_{rel}}$$
$$= \frac{\sqrt{2}}{2}\left(\frac{N}{V}\right)^2\pi d^2\overline{v}$$

EXAMPLE 9.12

Calculate the collision frequency for (**a**) a molecule in a sample of oxygen at 1.00 atm pressure and 25°C, and (**b**) a molecule of hydrogen in a region of interstellar space where the number density is 1.0×10^{10} molecules per cubic meter and the temperature is 30 K. Take the diameter of O_2 to be 2.92×10^{-10} m and that of H_2 to be 2.34×10^{-10} m.

Solution

(a) $$\frac{N}{V} = \frac{N_AP}{RT} = \frac{(6.022 \times 10^{23}\ \text{mol}^{-1})(1.00\ \text{atm})}{(0.08206\ \text{L atm mol}^{-1}\ \text{K}^{-1})(298\ \text{K})}$$
$$= 2.46 \times 10^{22}\ \text{L}^{-1} = 2.46 \times 10^{25}\ \text{m}^{-3}$$

In Example 9.9, we found $\overline{v}$ to be 444 m s^{-1}. Thus, using Equation 9.29, the collision frequency is

$$Z_1 = (2.46 \times 10^{25}\ \text{m}^{-3})(2.92 \times 10^{-10}\ \text{m})^2(444\ \text{m s}^{-1})\sqrt{2}\pi$$
$$= 4.14 \times 10^9\ \text{s}^{-1}$$

This is the average number of collisions experienced by each molecule per second.

(b) An analogous calculation gives

$$Z_1 = 1.4 \times 10^{-6}\ \text{s}^{-1}$$

In other words, the average molecule under these conditions waits 7.3×10^5 seconds, or 8.5 days, between collisions.

Related Problems: 88, 89

9.7.3 Mean Free Path and Diffusion

Z_1 is the rate at which a particular molecule collides with other molecules. Its inverse, Z_1^{-1}, therefore measures the average time between collisions. During this interval, a molecule travels an average distance $\overline{v}\,Z_1^{-1}$ which is called the **mean free path,** λ in a pure gas.

$$\lambda = \overline{v}Z_1^{-1} = \frac{\overline{v}}{\sqrt{2}\,(N/V)\,\pi d^2\overline{v}} = \frac{1}{\sqrt{2}\pi d^2 N/V} \qquad [9.31]$$

FIGURE 9.25 A gas molecule follows a straight-line path for only a short time before undergoing a collision, so its overall path is a zigzag one. The net displacement Δr of a particular molecule in time, Δt, is shown. The path taken by the molecules is traced out in blue, and the red arrow represents the net displacement during the period Δt. The mean-square net displacement $\overline{\Delta r^2}$ is equal to $6Dt$, where D is the diffusion constant of the molecules in the gas.

The mean free path, unlike the collision frequency, does not depend on the molar mass. The mean free path must be much larger than the molecular diameter for a gas to show ideal gas behavior. For a molecule of diameter 3×10^{-10} m, the mean free path at 25°C and atmospheric pressure is 1×10^{-7} m, which is 300 times larger.

Molecules in a gas move in straight lines only for rather short distances before they are deflected by collisions and change direction (Fig. 9.25). Because each molecule follows a zigzag course, they take more time to travel a particular net distance from their starting points than they would if there were no collisions. This fact helps explain why diffusion is slow in gases. Recall that at room temperature the speed of a molecule is on the order of 1×10^3 m s^{-1}. If molecules traveled in straight-line trajectories, a perfume released in one part of a room would be noticed across the room almost instantaneously. Instead, there is a time lapse because the molecules follow irregular paths that we call "random walks."

We can describe diffusion in a gas using averaged quantities such as the mean-square displacement, $\overline{\Delta r^2} = \overline{\Delta x^2} + \overline{\Delta y^2} + \overline{\Delta z^2}$, from the zig-zag motion. If there are no gas currents to perturb the motion of the gas molecules—a rather strenuous condition requiring strict isolation of the experiment from the surroundings—then $\overline{\Delta r^2}$ is found to be proportional to the time elapsed, t:

$$\overline{\Delta r^2} = 6Dt \qquad [9.32]$$

The proportionality constant is $6D$, where D is the **diffusion constant** of the molecules. So, the root-mean-square displacement $\sqrt{\overline{\Delta r^2}}$ is equal to $\sqrt{6Dt}$.

The diffusion constant has units of m^2 s^{-1}. It is proportional to the mean free path, λ, and to the mean molecular speed, $\bar{v}$, but the proportionality constant is difficult to calculate in general. For the simplest case, a single-component gas, the proportionality constant is $3\pi/16$, so

$$D = \frac{3\pi}{16}\lambda\bar{v} = \frac{3\pi}{16}\sqrt{\frac{8RT}{\pi\mathcal{M}}}\frac{1}{\sqrt{2}\pi d^2 N/V}$$
$$= \frac{3}{8}\sqrt{\frac{RT}{\pi\mathcal{M}}}\frac{1}{d^2 N/V} \qquad [9.33]$$

EXAMPLE 9.13

Calculate the mean free path and the diffusion constant for the molecules in Example 9.12.

Solution

(a) The mean free path is

$$\lambda = \frac{1}{\sqrt{2}\pi\left(2.92 \times 10^{-10}\ \text{m}\right)^2\left(2.46 \times 10^{25}\ \text{m}^{-1}\right)}$$
$$= 1.07 \times 10^{-7}\ \text{m}$$

The diffusion constant is

$$D = \frac{3\pi}{16}\lambda\bar{v} = \frac{3\pi}{16}\left(1.07 \times 10^{-7}\ \text{m}\right)\left(444\ \text{m s}^{-1}\right)$$
$$= 2.80 \times 10^{-5}\ \text{m}^2\ \text{s}^{-1}$$

(b) An analogous calculation for an average molecule of H_2 in interstellar space gives

$$\lambda = 4.1 \times 10^8\ \text{m}$$

For comparison, the distance from the Earth to the moon is 3.8×10^8 m. The diffusion constant is

$$D = 1.4 \times 10^{11}\ \text{m}^2\ \text{s}^{-1}$$

Related Problems: 57, 58

CONNECTION TO CHEMICAL ENGINEERING

Uranium Enrichment for Nuclear Reactor Fuel

Nuclear energy is an important source of power throughout the world today. In March 2010, the European Nuclear Society (ENS) reported that 437 nuclear power plants are in operation and 55 plants are under construction worldwide. Inside the nuclear reactor in a power plant, nuclei of the uranium isotope ^{235}U are bombarded with neutrons and break apart into lighter nuclides in a process called fission. (See Section 19.5.) Fission liberates enormous amounts of energy in the form of heat. The heat creates steam to drive turbines that generate electrical power. The isotopic abundance in naturally occurring uranium is 0.7% for ^{235}U and 99.3% for ^{238}U, which is not used in the fission process. In order to be used as fuel for nuclear reactors, uranium must have the weight percentage of ^{235}U lie between 3% and 5%. Increasing the ratio of ^{235}U to ^{238}U through an enrichment process is the essential step in preparing fuel for a nuclear reactor. The isotopes of uranium, ^{235}U and ^{238}U, are chemically identical, but differ in their physical properties, especially their masses. This small difference in mass allows limited separation of molecules containing these isotopes by the process of gaseous diffusion through a porous barrier. (See Section 9.7.) The uranium to be enriched is converted to uranium hexafluoride, UF_6, which is the only compound of uranium that becomes a vapor at a modest temperature (56° C). Because fluorine has only the single isotope ^{19}F, the mass difference between the molecules $^{235}UF_6$ and $^{238}UF_6$ is only three units. Consequently, very limited separation is achieved in a single pass through the porous barrier, and multiple stages are required to achieve the levels of enrichment necessary for use as nuclear fuel. (See Example 9.11 in Section 9.7.) This industrial process has been used for over 60 years to increase or enrich the percentage of ^{235}U.

Uranium Mining, Refining, and Conversion

Uranium is obtained by mining its ore to recover the stable oxide U_3O_8 or UO_3. To alleviate environmental concerns, open-pit mining has been replaced in the United States and Australia by *in situ* leaching of uranium ore from the earth. In this process, oxygenated acidic or basic solutions are circulated through the buried deposits of uranium ore via injection and recovery wells.

The uranium obtained in the solution is mostly uranyl sulfate, $UO_2(SO_4)_3^{4-}$, in acid leach conditions or uranyl carbonate, $UO_2(CO_3)_3^{4-}$ in a carbonate leach system. In the United States this solution is pumped to the treatment plant where the uranium is removed on an ion exchange resin (Fig. C9.1). In one such system, the resin is rinsed with sulfuric acid and hydrogen peroxide is added to precipitate uranyl peroxide.

After the precipitate dries at a low temperature, U_3O_8, called "yellowcake" because of its bright yellow color, is obtained. This stable oxide is the form in which uranium is marketed and exported.

A solvent extraction process is used to recover uranium from the acidic leaching solution in Australia. Organic amines dissolved in kerosene remove uranium oxide by forming molecular complexes that are soluble in kerosene. Cationic and ionic impurities are removed from this organic phase, after which the uranium complexes are stripped out in a solution of ammonium sulfate. Gaseous ammonia is added to neutralize this solution and to precipitate ammonium diuranate.

$$2NH_3 + 2UO_2(SO_4)_3^{4-} \longrightarrow (NH_4)_2U_2O_7 + 4SO_4^{2-}$$

The diuranate is treated by a process called dewatering and then roasted to yield U_3O_8 as the final product.

For uranium to be enriched via the gaseous diffusion process it must be in a gaseous form. This is achieved by converting uranium oxide to uranium hexafluoride (UF_6) (known as 'hex' in the industry), by the following processes.

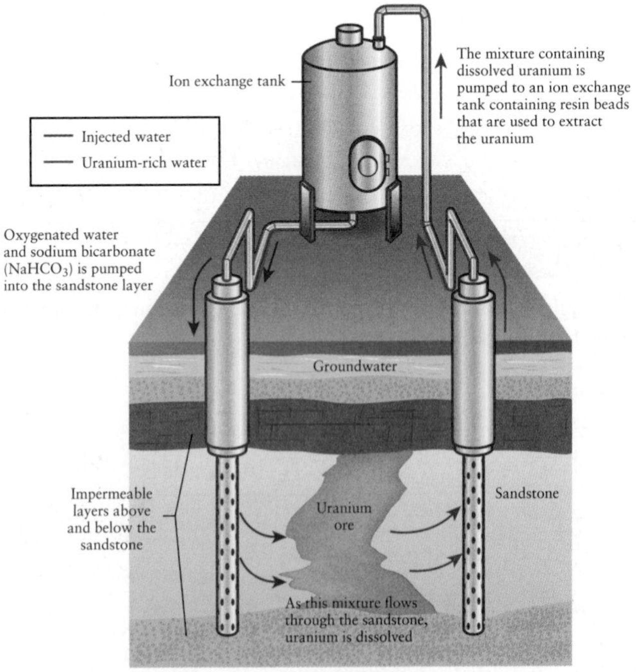

FIGURE C9.1 In-situ uranium mining process.

First, U_3O_8 is reduced to uranium dioxide (UO_2) by hydrogen in a kiln using the "dry process." Similarly UO_3 is reduced by the "wet process."

$$\text{Dry Process: } U_3O_8 + 2H_2 \longrightarrow 3UO_2 + 2H_2O$$

$$\text{Wet Process: } UO_3 + H_2 \longrightarrow UO_2 + H_2O$$

Next, uranium dioxide is reacted in another kiln with gaseous hydrogen fluoride HF (in the dry process) or aqueous HF (in the wet process) to form uranium tetrafluoride (UF_4).

$$UO_2 + 4HF \longrightarrow UF_4 + 2H_2O$$

Finally, uranium tetrafluoride (UF_4) is fed into a flame tower to react with gaseous fluorine (dry process), or into a fluidized bed reactor to react with aqueous fluorine (wet process), to produce UF_6.

$$UF_4 + F_2 \longrightarrow UF_6$$

The UF_6 product forms a highly corrosive liquid, which is stored in thick-walled steel cylinders. After cooling and pressurization, the UF_6 in the cylinder becomes a white crystalline solid.

Gaseous Diffusion

At the gaseous diffusion enrichment plant, solid UF_6 is heated in its container to a temperature above 56°C and becomes a vapor. The gas is fed into the plant's pipelines and arrives at special filters called diffusion barriers (semi-porous membranes). The barriers contain hundreds of millions of pores per square inch. The diameter of the pores is extremely small, with an average size on the order of 25 nm. The pores must be uniform in size. The barrier must be fabricated from a material that will not react with UF_6 gas (which is highly corrosive), and it must withstand high flow rates for successful operation. Typical barrier materials are aluminum oxide or nickel.

These pores are designed to be just large enough to allow UF_6 gas molecules to flow through. The kinetic energy of the uranium hexafluoride molecules depends on the temperature, which is held constant at each stage. Since the kinetic energy of a molecule is proportional to the product of its mass and the square of its speed, the lighter molecules ($^{235}UF_6$) have higher speeds than the heavy molecules ($^{238}UF_6$). Because they move at a faster speed, the lighter molecules hit the barrier more frequently and have a higher probability of passing through the pores. As the lighter $^{235}UF_6$ molecules diffuse through the barriers at a faster rate than the heavier $^{238}UF_6$ molecules, isotope enrichment occurs.

Each stage has an input where gas with a particular mole fraction of $^{235}UF_6$ is fed in. As the gas travels through each diffusion stage, the gas on the exit side of the barrier is enriched (contains more of the lighter gas) while the gas on the entrance side of the barrier is depleted (contains less of the lighter gas). As this process is repeated throughout multiple diffusion stages the concentration of $^{235}UF_6$ relative to $^{238}UF_6$ increases. The enrichment factor achieved in each diffusion stage is only 1.0043 due to the small mass difference between these two isotopes. (See Example 9.10.) Thus, many stages are required to achieve the desired level of isotope enrichment.

Figure C9.2 shows one of the 3122 diffusion stages in the gaseous diffusion plant at Oak Ridge National Laboratory. Gaseous UF_6 entering the left side of the stage is progressively enriched as it moves through the stage from left to right, as indicated by an increase in the number of red dots representing $^{235}UF_6$ molecules. There are three diffusion barriers in each stage and gases

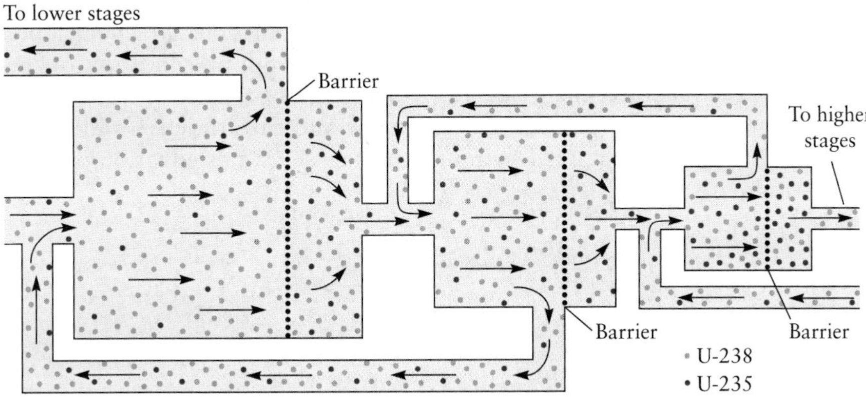

FIGURE C9.2 Schematic of a diffusion stage at the gaseous diffusion plant at Oak Ridge National Laboratory. The concentration of $^{235}UF_6$ (red dots) is increased at each stage by only a factor of 1.0043, so the enrichment per stage is much less than that shown.

are recycled through the barriers within a stage, as well as between stages, multiple times, as shown by the arrows flowing back to the left.

At the end of the gaseous diffusion process, the enriched UF_6 gas is condensed into a liquid that is cooled, pressurized, and solidified before being sent to a fuel fabrication facility for further processing. The enriched UF_6 is converted into uranium dioxide (UO_2) powder that is processed to form hard, ceramic pellets of enriched uranium with uniform size. These pellets are used to form fuel rods for the reactor.

After running in the reactor for some period of time, the rods are "spent" because the ^{235}U has been depleted by fission. Spent nuclear fuel is a highly radioactive, complex mixture of uranium oxide and the fission products of ^{235}U. Safe disposal and storage of spent nuclear material present considerable challenges due to its long-lived radioactivity. Exposure to radiation and to high temperature in the reactor will have altered the physical and chemical structure of the materials in ways that make subsequent processing quite difficult. Active research in this problem engages chemists, physicists, chemical engineers, and materials experts to devise proper methods. Each of these professions plays a key role at each step of the nuclear fuel cycle, which is the progression of nuclear material from the *front end* (preparation), through the *service period* (use during reactor operation), to the *back end* (safely manage, contain, and re-process or dispose of spent fuel).

Uranium enrichment for commercial applications originated in the United States with the gaseous diffusion process, which is now used worldwide and produces approximately 25% of the world's enriched uranium. Only one gaseous diffusion plant operates in the United States today, in Paducah, Kentucky. According to the United States Energy Commission (USEC), 70% of the production cost at the Paducah plant is due to electricity used for the uranium enrichment process. Due to energy requirements, gaseous diffusion is being phased out in favor of more efficient enrichment processes based on gas centrifuges and laser isotope separation technology.

CHAPTER SUMMARY

Gases provide the simplest opportunity for relating macroscopic properties of matter to the structure, motions, and interactions of molecules. Because molecules in gases are quite far apart most of the time, we can neglect intermolecular forces and represent the molecules as point masses that have only kinetic energy and collide with the walls of the container but not with each other. The simplest treatment of this physical model predicts the ideal gas law, which was discovered empirically. More elaborate mathematical treatments of the same model produce the full probability distribution for molecular speeds. From this distribution, various average quantities can be calculated and used to interpret numerous experimental phenomena in gases at low density. At higher density, intermolecular forces can no longer be neglected. Their effect is described systematically by the intermolecular potential energy function, which includes both the attractive and repulsive effects. The well depth and location of the minimum in the potential energy curve are very useful parameters for summarizing these effects. They provide deeper insight into the attractive and repulsive constants that are obtained by fitting the van der Waals equation of state to empirical data.

CONCEPTS AND SKILLS

Section 9.1 – The Chemistry of Gases

Write chemical equations for several reactions that lead to gas formation (Problems 1–4).

- In addition to the evaporation of liquids (such as water and gasoline), certain classes of chemical reactions lead to gaseous products that can escape into the atmosphere if not confined. Several examples are described in Section 9.1.

Section 9.2 – Pressure and Temperature of Gases

Describe how pressure and temperature are defined and measured (Problems 5–10).

- The pressure exerted on the walls of a container, either by gas enclosed within it or by the atmosphere outside it, is defined as the force exerted per unit area of wall. Therefore, in SI units pressure has dimensions of N m^{-2}, which has the special name pascal (Pa). Other units may be more convenient in particular applications; conversion factors are available in Table 9.2. Torricelli's barometer measures the pressure of the atmosphere quite directly as equivalent to the force per unit area exerted by a mercury column of height h: $P = \rho gh$. Boyle used the J-tube as a means to increase the pressure on a fixed amount of gas, and to measure the resulting decrease in volume. The results are summarized in Boyle's law, $PV = C$, where C is a constant for a fixed amount of gas and fixed temperature. For 1 mol of gas at 0°C, the value of the constant is 22.414 L atm.
- We cannot measure temperature directly because it is not a mechanical property. Instead, we measure changes in some mechanical property chosen because we believe its changes are proportional to changes in temperature, such as the length of a column of mercury inside a narrow tube, and calibrate these changes against selected reference points to define scales of temperature and practical thermometers. A better illustration is the "gas thermometer" which displays the change in volume of a gas as its temperature increases. (Here the mechanical property, the gas volume, changes more linearly with temperature than does the mercury thermometer volume.) The gas must be held at constant pressure inside a container whose walls can move as the gas expands. See Figure 9.8. These measurements suggest the existence of the absolute zero of temperature at $t = -273.15$°C and lead to the Kelvin temperature scale, T (Kelvin) = 273.15 + t(Celsius). The changes in volume of a gas as temperature is changed are summarized in Charles's law, $V \propto T$, but only when temperature is expressed in Kelvin. The proportionality constant depends on the amount of gas and on the fixed pressure at which the measurements are made.

Section 9.3 – The Ideal Gas Law

Use the ideal gas law to relate pressure, volume, temperature, and number of moles of an ideal gas and to do stoichiometric calculations involving gases (Problems 19–32).

- Boyle's law connects V and P at constant T and constant n (number of moles). Charles's law connects V and T at constant P and constant n. Avogadro's hypothesis connects V and n at constant T and P. Therefore, the experiments show that the four measurable properties V, T, P, and n are all interconnected. Furthermore, experiments show that for a pure gas, holding any three of them at a fixed value forces the remaining one to assume a specific fixed value. It is very convenient to summarize this vast range of experimental results in a single compact equation, the ideal gas law $PV = nRT$. Since any one of these can be considered the "dependent variable" and its value can be calculated as a result of changes in the remaining "independent variables," the equation can be rearranged into a variety of useful forms for specific applications. When solving any problem, it is a good idea to read the problem as a description of a laboratory measurement. That way you will know immediately which variables are independent (being manipulated directly by the experimenter) and dependent (responding to the changes induced by the experimenter) and you can immediately put the ideal gas law into the most useful form. Chemical reactions involving gaseous products or reactants are described by balanced equations that relate the number of moles of each reactant and product. When describing pure solid substances as in Chapter 2, we determine the number of moles of each participant as the ratio of its mass to its molecular weight. In gaseous reactions we relate the number of moles of each participant to the conditions of the gas through the equation $n = PV/RT$.

Section 9.4 – Mixtures of Gases

Use Dalton's law to calculate partial pressures in gas mixtures (Problems 33–38).

- When several gaseous species are present in a container with volume V and temperature T, each of the gases generates its own individual pressure, called the partial pressure, against the walls of the container. The sum of all the partial pressure contributions is the total pressure. Physically this arises because the frequency at which a gas molecule hits the walls of the container, and the velocity it has when it hits the wall, is largely unaffected by the other gaseous molecules present. The equations to describe these events are straightforward. For species A, we have two expressions for the partial pressure $P_A = n_A(RT/V)$ and $P_A = X_A P_{tot}$, both of which state that the partial pressure contribution of species A is proportional to the number of moles of species A that are present. Using either version, the sum of all the partial pressures adds up to the total pressure.

Section 9.5 – The Kinetic Theory of Gases

Use the Maxwell–Boltzmann distribution of molecular speeds to calculate root-mean-square, most probable, and average speeds of molecules in a gas. Integrate the distribution from a speed v_1 to a faster speed v_2 to calculate the fraction of molecules with speeds in that range. (Problems 41–46).

- The kinetic molecular theory views a vessel of gas as a collection of molecules and describes their behavior, in particular the range of speeds at which they move, using statistics and probability. Three standard statistical parameters indicate the typical behavior of a molecule: most probable speed v_{mp}, average speed $\bar{v}$, and root-mean-square speed v_{rms}. The magnitudes of these parameters are ranked in the order $v_{mp} < \bar{v} < v_{rms}$ because the Maxwell–Boltzmann distribution is not symmetrical; the area under the curve to the right of the maximum is slightly greater than that to the left of the maximum. Calculations of these quantities are straightforward; equations are provided using the mass of an individual molecule and also the molar mass of the substance. All these statistical indications of a typical molecular speed increase as T increases and decrease as the molecular mass increases. You can calculate the fraction of molecules with speeds between v_1 and v_2 by integrating the Maxwell–Boltzmann distribution numerically, or estimating the area graphically.

Describe the connection between temperature and the speeds or kinetic energies of the molecules in a gas (Problems 47–48).

- One of the great achievements of the kinetic theory of gases is to explain the meaning of temperature as proportional to the average kinetic energy of the molecules. This makes it quite clear why temperature cannot be represented as a macroscopic mechanical quantity. Two equations are very useful to describe this relation: $\bar{\varepsilon} = \frac{3}{2}k_BT$ for the average kinetic energy and $\overline{v^2} = 3RT/\mathcal{M}$ for the average of the square of the speed.

Section 9.6 – Real Gases: Intermolecular Forces

Use the van der Waals equation to relate the pressure, volume, temperature, and number of moles of a nonideal gas (Problems 49–52).

- The van der Waals equation can be viewed as a "correction" to the ideal gas law to account for the finite size of the molecules and attractive forces between them. In essence the attractions between molecules, represented by the a parameter, decrease the pressure the gas would exert on the walls from that expected for an ideal gas. The finite size of the molecules, represented by the b parameter, increases the volume the gas occupies (compared to the volume an ideal gas). The van der Waals equation describes all the same kinds of experiments as the ideal gas law, and is used in exactly the same way. Always try to interpret the results in terms of the effects of the a and b parameters described above and their magnitudes in Table 9.3.

Discuss how forces between atoms and molecules vary with distance.

- Examine the Lennard–Jones potential energy curve for a pair of atoms as presented in Figure 9.21, Equation 9.24, and Table 9.4. Visualize a sequence of curves plotted for the gases listed in Table 9.4 and identify from each curve the typical separations for which the force between the atoms is attractive or repulsive. Note that the minimum of the curves in Figure 9.21 occurs at the distance $R = 1.122\sigma$, where σ is the value of R at which $V(R) = 0$. Note the very small well depths for the He-He potential, and the Ar-Ar potential, as compared with well depths of covalently bound atoms.

Section 9.7 – A Deeper Look . . . Molecular Collisions and Rate Processes

Calculate the rate of collisions of molecules with a wall, and from that calculation determine the effusion rate of a gas through a small hole of known area (Problems 53–54).

- Three physical effects contribute to Z_w, the rate of collisions with the wall: the rate is proportional to the area A of the wall, proportional to the average speed of the molecules, and proportional to the number density of the molecules. This reasoning leads directly to Equation 9.25 for Z_w. If we take A to be the area of the small hole, all the molecules that collide with this area escape from the vessel, and the effusion rate is the same as Z_w.

Calculate the enrichment factor for lighter molecules when a gas consisting of a mixture of light and heavy molecules effuses through a small aperture in a vessel wall (Problems 55–58).

- If gas A is lighter than gas B, the escaping gas is richer in A because the A molecules effuse more rapidly. Taking the ratio Z_A/Z_B shows that the enrichment factor is $\sqrt{\mathcal{M}_B/\mathcal{M}_A}$. See Equation 9.26.

Calculate the collision frequency, mean free path, and diffusion constant for gases (Problems 57–58).

- The key physical idea behind all these applications is to calculate the rate (or frequency) with which one molecule collides with other molecules rather than with the wall. The reasoning is described in Figure 9.24, which shows the "collision cylinder" swept out by a molecule with diameter d during 1 sec as it moves through a gas with average relative speed $\overline{v_{rel}}$. This molecule collides with all other molecules that appear in its "collision cylinder." So Z_1, the rate (or frequency) of collisions with other molecules, is proportional to the volume of the collision cylinder and the number density of molecules in the gas. This reasoning leads directly to Equations 9.28 and 9.29. The mean free path λ is the distance traveled by a typical molecule between collisions as it moves with average speed $\overline{v}$. The time elapsed between collisions is the reciprocal of Z_1. Therefore, $\lambda = \overline{v}\, Z_1^{-1}$. Equation 9.31 gives the details for calculating λ. Over long periods of time a molecule experiences many collisions and therefore travels a zig-zag path. Its net displacement over a time interval is not given by the simple equation *distance* = *speed* × *time*, but must be described statistically by the root-mean-square displacement. The mean-square displacement is related to the elapsed time interval by the diffusion constant D, which is proportional to the mean free path and to the average speed. See Equation 9.33 for the method of calculating D.

CUMULATIVE EXERCISE

NASA

FIGURE 9.26 A space shuttle taking off.

Ammonium Perchlorate as a Rocket Fuel

Ammonium perchlorate (NH_4ClO_4) is a solid rocket fuel that was used in space shuttles (see Fig. 9.26). When heated above 200°C, it decomposes to a variety of gaseous products, of which the most important are N_2, Cl_2, O_2, and water vapor.

(a) Write a balanced chemical equation for the decomposition of NH_4ClO_4, assuming the products just listed are the only ones generated.

(b) The sudden appearance of hot gaseous products in a small initial volume leads to rapid increases in pressure and temperature, which give the rocket its thrust. What total pressure of gas would be produced at 800°C by igniting 7.00×10^5 kg NH_4ClO_4 (a typical charge of the booster rockets in the space shuttle) and allowing it to expand to fill a volume of 6400 m^3 (6.40×10^6 L)? Use the ideal gas law.

(c) Calculate the mole fraction of chlorine and its partial pressure in the mixture of gases produced.

(d) The van der Waals equation applies strictly to pure real gases, not to mixtures. For a mixture like the one resulting from the reaction of part (a), it may still be possible to define effective a and b parameters to relate total pressure, volume, temperature, and total number of moles. Suppose the gas mixture has $a = 4.00$ atm L^2 mol^{-2} and $b = 0.0330$ L mol^{-1}. Recalculate the pressure of the gas mixture in part (b) using the van der Waals equation. Why is the result smaller than that in part (b)?

(e) Calculate and compare the root-mean-square speeds of water and chlorine molecules under the conditions of part (b).

(f) The gas mixture from part (b) cools and expands until it reaches a temperature of 200°C and a pressure of 3.20 atm. Calculate the volume occupied by the gas mixture after this expansion has occurred. Assume ideal gas behavior.

Answers

(a) $2\ NH_4ClO_4(s) \longrightarrow N_2(g) + Cl_2(g) + 2\ O_2(g) + 4\ H_2O(g)$

(b) 328 atm

(c) $X_{Cl_2} = \frac{1}{8} = 0.125$ (exactly); $P_{Cl_2} = 41.0$ atm

(d) 318 atm. The real pressure is less than that calculated using the ideal gas law because half of the products are water molecules that have very strong intermolecular attractions, leading to the large value of the *a* parameters for the gas mixture.

(e) $v_{rms}(H_2O) = 1220 \text{ m s}^{-1}$; $v_{rms}(Cl_2) = 614 \text{ m s}^{-1}$

(f) $2.89 \times 10^5 \text{ m}^3$

PROBLEMS

Answers to problems whose numbers are red appear in Appendix G. Problems that are more challenging are indicated with asterisks.

9.1 The Chemistry of Gases

1. Solid ammonium hydrosulfide (NH_4HS) decomposes entirely to gases when it is heated. Write a chemical equation representing this change.

2. Solid ammonium carbamate ($NH_4CO_2NH_2$) decomposes entirely to gases when it is heated. Write a chemical equation representing this change.

3. Ammonia (NH_3) is an important and useful gas. Suggest a way to generate it from ammonium bromide (NH_4Br). Include a balanced chemical equation.

4. Hydrogen cyanide (HCN) is a poisonous gas. Explain why solutions of potassium cyanide (KCN) should never be acidified. Include a balanced chemical equation.

9.2 Pressure and Temperature of Gases

5. Suppose a barometer were designed using water (with a density of 1.00 g cm^{-3}) rather than mercury as its fluid. What would be the height of the column of water balancing 1.00 atm pressure?

6. A vessel that contains a gas has two pressure gauges attached to it. One contains liquid mercury, and the other an oil such as dibutyl phthalate. The difference in levels of mercury in the two arms of the mercury gauge is observed to be 9.50 cm. Given

$$\text{density of mercury} = 13.60 \text{ g cm}^{-3}$$
$$\text{density of oil} = 1.045 \text{ g cm}^{-3}$$
$$\text{acceleration due to gravity} = 9.806 \text{ m s}^{-2}$$

   (a) What is the pressure of the gas?
   (b) What is the difference in height of the oil in the two arms of the oil pressure gauge?

7. Calcium dissolved in the ocean is used by marine organisms to form $CaCO_3(s)$ in skeletons and shells. When the organisms die, their remains fall to the bottom. The amount of calcium carbonate that can be dissolved in seawater depends on the pressure. At great depths, where the pressure exceeds about 414 atm, the shells slowly redissolve. This reaction prevents all the Earth's calcium from being tied up as insoluble $CaCO_3(s)$ at the bottom of the sea. Estimate the depth (in feet) of water that exerts a pressure great enough to dissolve seashells.

8. Suppose that the atmosphere were perfectly uniform, with a density throughout equal to that of air at 0°C, 1.3 g L^{-1}. Calculate the thickness of such an atmosphere that would cause a pressure of exactly 1 standard atm at the Earth's surface.

9. The "critical pressure" of mercury is 172.00 MPa. Above this pressure mercury cannot be liquefied, no matter what the temperature. Express this pressure in atmospheres and in bars (1 bar $= 10^5$ Pa).

10. Experimental studies of solid surfaces and the chemical reactions that occur on them require very low gas pressures to avoid surface contamination. High-vacuum apparatus for such experiments can routinely reach pressures of 5×10^{-10} torr. Express this pressure in atmospheres and in pascals.

11. Some nitrogen is held in a 2.00-L tank at a pressure of 3.00 atm. The tank is connected to a 5.00-L tank that is completely empty (evacuated), and a valve is opened to connect the two tanks. No temperature change occurs in the process. Determine the total pressure in this two-tank system after the nitrogen stops flowing.

12. The Stirling engine, a heat engine invented by a Scottish minister, has been considered for use in automobile engines because of its efficiency. In such an engine, a gas goes through a four-step cycle: (1) expansion at constant *T*, (2) cooling at constant *V*, (3) compression at constant *T* to its original volume, and (4) heating at constant *V* to its original temperature. Suppose the gas starts at a pressure of 1.23 atm and the volume of the gas changes from 0.350 to 1.31 L during its expansion at constant *T*. Calculate the pressure of the gas at the end of this step in the cycle.

13. The absolute temperature of a 4.00-L sample of gas doubles at constant pressure. Determine the volume of the gas after this change.

14. The Celsius temperature of a 4.00-L sample of gas doubles from 20.0°C to 40.0°C at constant pressure. Determine the volume of the gas after this change.

15. The gill is an obscure unit of volume. If some $H_2(g)$ has a volume of 17.4 gills at 100°F, what volume would it have if the temperature were reduced to 0°F, assuming that its pressure stayed constant?

16. A gas originally at a temperature of 26.5°C is cooled at constant pressure. Its volume decreases from 5.40 L to 5.26 L. Determine its new temperature in degrees Celsius.

17. Calcium carbide (CaC_2) reacts with water to produce acetylene (C_2H_2), according to the following equation:

$$CaC_2(s) + 2\ H_2O(\ell) \longrightarrow Ca(OH)_2(s) + C_2H_2(g)$$

A certain mass of CaC_2 reacts completely with water to give 64.5 L C_2H_2 at 50°C and $P = 1.00$ atm. If the same mass of CaC_2 reacts completely at 400°C and $P = 1.00$ atm, what volume of C_2H_2 will be collected at the higher temperature?

18. A convenient laboratory source for high purity oxygen is the decomposition of potassium permanganate at 230°C:

$$2\ KMnO_4(s) \longrightarrow K_2MnO_4(s) + MnO_2(s) + O_2(g)$$

Suppose 3.41 L oxygen is needed at atmospheric pressure and a temperature of 20°C. What volume of oxygen should be collected at 230°C and the same pressure to give this volume when cooled?

9.3 The Ideal Gas Law

19. A bicycle tire is inflated to a gauge pressure of 30.0 psi at a temperature of $t = 0$°C. What will its gauge pressure be at 32°C if the tire is considered nonexpandable? (**Note:** The gauge pressure is the *difference* between the tire pressure and atmospheric pressure, 14.7 psi.)

20. The pressure of a poisonous gas inside a sealed container is 1.47 atm at 20°C. If the barometric pressure is 0.96 atm, to what temperature (in degrees Celsius) must the container and its contents be cooled so that the container can be opened with no risk for gas spurting out?

21. A 20.6-L sample of "pure" air is collected in Greenland at a temperature of −20.0°C and a pressure of 1.01 atm and is forced into a 1.05-L bottle for shipment to Europe for analysis.
   (a) Compute the pressure inside the bottle just after it is filled.
   (b) Compute the pressure inside the bottle as it is opened in the 21.0°C comfort of the European laboratory.

22. Iodine heptafluoride (IF_7) can be made at elevated temperatures by the following reaction:

$$I_2(g) + 7\ F_2(g) \longrightarrow 2\ IF_7(g)$$

Suppose 63.6 L gaseous IF_7 is made by this reaction at 300°C and a pressure of 0.459 atm. Calculate the volume this gas will occupy if heated to 400°C at a pressure of 0.980 atm.

23. According to a reference handbook, "The weight of one liter of $H_2Te(g)$ is 6.234 g." Why is this information nearly valueless? Assume that $H_2Te(g)$ is an ideal gas, and calculate the temperature (in degrees Celsius) at which this statement is true if the pressure is 1.00 atm.

24. A scuba diver's tank contains 0.30 kg oxygen (O_2) compressed into a volume of 2.32 L.
   (a) Use the ideal gas law to estimate the gas pressure inside the tank at 5°C, and express it in atmospheres and in pounds per square inch.
   (b) What volume would this oxygen occupy at 30°C and a pressure of 0.98 atm?

25. Hydrogen is produced by the complete reaction of 6.24 g sodium with an excess of gaseous hydrogen chloride.
   (a) Write a balanced chemical equation for the reaction that occurs.
   (b) How many liters of hydrogen will be produced at a temperature of 50.0°C and a pressure of 0.850 atm?

26. Aluminum reacts with excess aqueous hydrochloric acid to produce hydrogen.
   (a) Write a balanced chemical equation for the reaction. (*Hint:* Water-soluble $AlCl_3$ is the stable chloride of aluminum.)
   (b) Calculate the mass of pure aluminum that will furnish 10.0 L hydrogen at a pressure of 0.750 atm and a temperature of 30.0°C.

27. The classic method for manufacturing hydrogen chloride, which is still in use today to a small extent, is the reaction of sodium chloride with excess sulfuric acid at elevated temperatures. The overall equation for this process is

$$NaCl(s) + H_2SO_4(\ell) \longrightarrow NaHSO_4(s) + HCl(g)$$

What volume of hydrogen chloride is produced from 2500 kg sodium chloride at 550°C and a pressure of 0.97 atm?

28. In 1783, the French physicist Jacques Charles supervised and took part in the first human flight in a hydrogen balloon. Such balloons rely on the low density of hydrogen relative to air for their buoyancy. In Charles's balloon ascent, the hydrogen was produced (together with iron(II) sulfate) from the action of aqueous sulfuric acid on iron filings.
   (a) Write a balanced chemical equation for this reaction.
   (b) What volume of hydrogen is produced at 300 K and a pressure of 1.0 atm when 300 kg sulfuric acid is consumed in this reaction?
   (c) What would be the radius of a spherical balloon filled by the gas in part (b)?

29. Potassium chlorate decomposes when heated, giving oxygen and potassium chloride:

$$2\ KClO_3(s) \longrightarrow 2\ KCl(s) + 3\ O_2(g)$$

A test tube holding 87.6 g $KClO_3$ is heated, and the reaction goes to completion. What volume of O_2 will be evolved if it is collected at a pressure of 1.04 atm and a temperature of 13.2°C?

30. Elemental chlorine was first produced by Carl Wilhelm Scheele in 1774 using the reaction of pyrolusite (MnO_2) with sulfuric acid and sodium chloride:

$$4\ NaCl(s) + 2\ H_2SO_4(\ell) + MnO_2(s) \longrightarrow 2\ Na_2SO_4(s) + MnCl_2(s) + 2\ H_2O(\ell) + Cl_2(g)$$

Calculate the minimum mass of MnO_2 required to generate 5.32 L gaseous chlorine, measured at a pressure of 0.953 atm and a temperature of 33°C.

31. Elemental sulfur can be recovered from gaseous hydrogen sulfide (H_2S) through the following reaction:

$$2\ H_2S(g) + SO_2(g) \longrightarrow 3\ S(s) + 2\ H_2O(\ell)$$

(a) What volume of H_2S (in liters at 0°C and 1.00 atm) is required to produce 2.00 kg (2000 g) sulfur by this process?
(b) What minimum mass and volume (at 0°C and 1.00 atm) of SO_2 are required to produce 2.00 kg sulfur by this reaction?

32. When ozone (O_3) is placed in contact with dry, powdered KOH at −15°C, the red-brown solid potassium ozonide (KO_3) forms, according to the following balanced equation:

$$5\ O_3(g) + 2\ KOH(s) \longrightarrow 2\ KO_3(s) + 5\ O_2(g) + H_2O(s)$$

Calculate the volume of ozone needed (at a pressure of 0.134 atm and −15°C) to produce 4.69 g KO_3.

9.4 Mixtures of Gases

33. Sulfur dioxide reacts with oxygen in the presence of platinum to give sulfur trioxide:

$$2\ SO_2(g) + O_2(g) \longrightarrow 2\ SO_3(g)$$

Suppose that at one stage in the reaction, 26.0 mol SO_2, 83.0 mol O_2, and 17.0 mol SO_3 are present in the reaction vessel at a total pressure of 0.950 atm. Calculate the mole fraction of SO_3 and its partial pressure.

34. The synthesis of ammonia from the elements is conducted at high pressures and temperatures:

$$N_2(g) + 3\ H_2(g) \longrightarrow 2\ NH_3(g)$$

Suppose that at one stage in the reaction, 13 mol NH_3, 31 mol N_2, and 93 mol H_2 are present in the reaction vessel at a total pressure of 210 atm. Calculate the mole fraction of NH_3 and its partial pressure.

35. The atmospheric pressure at the surface of Mars is 5.92×10^{-3} atm. The Martian atmosphere is 95.3% CO_2 and 2.7% N_2 by volume, with small amounts of other gases also present. Compute the mole fraction and partial pressure of N_2 in the atmosphere of Mars.

36. The atmospheric pressure at the surface of Venus is 90.8 atm. The Venusian atmosphere is 96.5% CO_2 and 3.5% N_2 by volume, with small amounts of other gases also present. Compute the mole fraction and partial pressure of N_2 in the atmosphere of Venus.

37. A gas mixture at room temperature contains 10.0 mol CO and 12.5 mol O_2.
(a) Compute the mole fraction of CO in the mixture.
(b) The mixture is then heated, and the CO starts to react with the O_2 to give CO_2:

$$CO(g) + \tfrac{1}{2}O_2(g) \longrightarrow CO_2(g)$$

At a certain point in the heating, 3.0 mol CO_2 is present. Determine the mole fraction of CO in the new mixture.

38. A gas mixture contains 4.5 mol Br_2 and 33.1 mol F_2.
(a) Compute the mole fraction of Br_2 in the mixture.
(b) The mixture is heated above 150°C and starts to react to give BrF_5:

$$Br_2(g) + 5\ F_2(g) \longrightarrow 2\ BrF_5(g)$$

At a certain point in the reaction, 2.2 mol BrF_5 is present. Determine the mole fraction of Br_2 in the mixture at that point.

39. The partial pressure of water vapor in saturated air at 20°C is 0.0230 atm.
(a) How many molecules of water are in 1.00 cm^3 of saturated air at 20°C?
(b) What volume of saturated air at 20°C contains 0.500 mol water?

40. The partial pressure of oxygen in a mixture of oxygen and hydrogen is 0.200 atm, and that of hydrogen is 0.800 atm.
(a) How many molecules of oxygen are in a 1.500-L container of this mixture at 40°C?
(b) If a spark is introduced into the container, how many grams of water will be produced?

9.5 The Kinetic Theory of Gases

41. (a) Compute the root-mean-square speed of H_2 molecules in hydrogen at a temperature of 300 K.
(b) Repeat the calculation for SF_6 molecules in gaseous sulfur hexafluoride at 300 K.

42. Researchers recently reported the first optical atomic trap. In this device, beams of laser light replace the physical walls of conventional containers. The laser beams are tightly focused. They briefly (for 0.5 s) exert enough pressure to confine 500 sodium atoms in a volume of 1.0×10^{-15} m^3. The temperature of this gas is 0.00024 K, the lowest temperature ever reached for a gas. Compute the root-mean-square speed of the atoms in this confinement.

43. Compare the root-mean-square speed of helium atoms near the surface of the sun, where the temperature is approximately 6000 K, with that of helium atoms in an interstellar cloud, where the temperature is 100 K.

44. The "escape velocity" necessary for objects to leave the gravitational field of the Earth is 11.2 km s^{-1}. Calculate the ratio of the escape velocity to the root-mean-square speed of helium, argon, and xenon atoms at 2000 K. Does your result help explain the low abundance of the light gas helium in the atmosphere? Explain.

45. What is the probability that an O_2 molecule in a sample of oxygen at 300 K has a speed between 5.00×10^2 and 5.10×10^2 m s^{-1}? (*Hint:* Try approximating the area under the Maxwell–Boltzmann distribution by small rectangles.)

46. In a sample of argon gas at 298.15 K, what fraction of the Ar atoms have a speed between 600 and 1500 m/s? In a sample of neon gas at the same temperature, what fraction of the Ne atoms have a speed between 600 and 1500 m/s? Give your answers to two significant figures, so that you may estimate the area graphically or integrate numerically. Include a graph of the two Maxwell–Boltzmann distributions with your answer, both in the same plot, shading the relevant area under each curve.

47. Chlorine dioxide (ClO_2) is used for bleaching wood pulp. In a gaseous sample held at thermal equilibrium at a particular temperature, 35.0% of the molecules have speeds exceeding 400 m s^{-1}. If the sample is heated slightly, will the percentage of molecules with speeds in excess of 400 m s^{-1} then be greater than or less than 35%? Explain.

48. The ClO_2 described in Problem 47 is heated further until it explodes, yielding Cl_2, O_2, and other gaseous products. The mixture is then cooled until the original temperature is reached. Is the percentage of *chlorine* molecules with speeds in excess of 400 m s^{-1} greater than or less than 35%? Explain.

9.6 Real Gases: Intermolecular Forces

49. Oxygen is supplied to hospitals and chemical laboratories under pressure in large steel cylinders. Typically, such cylinders have an internal volume of 28.0 L and contain 6.80 kg oxygen. Use the van der Waals equation to estimate the pressure inside such cylinders at 20°C in atmospheres and in pounds per square inch.

50. Steam at high pressures and temperatures is used to generate electrical power in utility plants. A large utility boiler has a volume of 2500 m^3 and contains 140 metric tons (1 metric ton = 10^3 kg) of steam at a temperature of 540°C. Use the van der Waals equation to estimate the pressure of the steam under these conditions, in atmospheres and in pounds per square inch.

51. Using (a) the ideal gas law and (b) the van der Waals equation, calculate the pressure exerted by 50.0 g carbon dioxide in a 1.00-L vessel at 25°C. Do attractive or repulsive forces dominate?

52. When 60.0 g methane (CH_4) is placed in a 1.00-L vessel, the pressure is measured to be 130 atm. Calculate the temperature of the gas using (a) the ideal gas law and (b) the van der Waals equation. Do attractive or repulsive forces dominate?

9.7 A Deeper Look . . . Molecular Collisions and Rate Processes

53. A spherical bulb with a volume of 500 cm^3 is evacuated to a negligibly small residual gas pressure and then closed off. One hour later, the pressure in the vessel is found to be 1.00×10^{-7} atm because the bulb has a tiny hole in it. Assume that the surroundings are at atmospheric pressure, T = 300 K, and the average molar mass of molecules in the atmosphere is 28.8 g mol^{-1}. Calculate the radius of the hole in the vessel wall, assuming it to be circular.

54. A 200-cm^3 vessel contains hydrogen gas at a temperature of 25°C and a pressure of 0.990 atm. Unfortunately, the vessel has a tiny hole in its wall, and over a period of 1 hour, the pressure drops to 0.989 atm. What is the radius of the hole (assumed to be circular)?

55. Methane (CH_4) effuses through a small opening in the side of a container at the rate of 1.30×10^{-8} mol s^{-1}. An unknown gas effuses through the same opening at the rate of 5.42×10^{-9} mol s^{-1} when maintained at the same temperature and pressure as the methane. Determine the molar mass of the unknown gas.

56. Equal chemical amounts of two gases, fluorine and bromine pentafluoride, are mixed. Determine the ratio of the rates of effusion of the two gases through a small opening in their container.

57. Calculate the theoretical number of stages that would be needed to enrich ^{235}U to 95% purity by means of the barrier diffusion process, using $^{235}UF_6$ and $^{238}UF_6$ as the gaseous compounds. The natural abundance of ^{238}U is 99.27%, and that of ^{235}U is 0.72%. Take the relative atomic masses of ^{235}U and ^{238}U to be 235.04 and 238.05, respectively.

58. A mixture of H_2 and He at 300 K effuses from a tiny hole in the vessel that contains it. What is the mole fraction of H_2 in the original gas mixture if 3.00 times as many He atoms as H_2 molecules escape from the orifice in unit time? If the same mixture is to be separated by a barrier-diffusion process, how many stages are necessary to achieve H_2 of 99.9% purity?

59. At what pressure does the mean free path of krypton (Kr) atoms ($d = 3.16 \times 10^{-10}$ m) become comparable with the diameter of the 1-L spherical vessel that contains them at 300 K? Calculate the diffusion constant at this pressure.

60. At what pressure does the mean free path of Kr atoms ($d = 3.16 \times 10^{-10}$ m) become comparable with the diameter of a Kr atom if T = 300 K? Calculate the diffusion constant at this pressure. Assume that Kr obeys the ideal gas law even at these high pressures.

ADDITIONAL PROBLEMS

61. The Earth is approximately a sphere of radius 6370 km. Taking the average barometric pressure on the Earth's surface to be 730 mm Hg, estimate the total mass of the Earth's atmosphere.

***62.** After a flood fills a basement to a depth of 9.0 feet and completely saturates the surrounding earth, the owner buys an electric pump and quickly pumps the water out of the basement. Suddenly, a basement wall collapses, the structure is severely damaged, and mud oozes in. Explain this event by estimating the difference between the outside pressure at the base of the basement walls and the pressure inside the drained basement. Assume that the density of the mud is 4.9 g cm^{-3}. Report the answer both in atmospheres and in pounds per square inch.

63. The density of mercury is 13.5955 g cm^{-3} at 0.0°C, but only 13.5094 g cm^{-3} at 35°C. Suppose that a mercury barometer is read on a hot summer day when the temperature is 35°C. The column of mercury is 760.0 mm long. Correct for the expansion of the mercury and compute the true pressure in atmospheres.

64. When a gas is cooled at constant pressure, the volume decreases according to the following equation:

$$V = 209.4 \text{ L} + \left(0.456 \frac{\text{L}}{°\text{F}}\right) \times t_F$$

where t_F is the temperature in degrees Fahrenheit. From this relationship, estimate the absolute zero of temperature in degrees Fahrenheit.

65. Amonton's law relates pressure to absolute temperature. Consider the ideal gas law and then write a statement of Amonton's law in a form that is analogous to the statements of Charles's law and Boyle's law in the text.

66. The density of a certain gas is 2.94 g L^{-1} at 50°C and P = 1.00 atm. What is its density at 150°C? Calculate the molar mass of the gas, assuming it obeys the ideal gas law.

67. A lighter-than-air balloon contains 1005 mol helium at 1.00 atm and 25.0°C.

(a) Compute the difference between the mass of the helium it contains and the mass of the air it displaces, assuming the molar mass of air to be 29.0 g mol^{-1}.

(b) The balloon now ascends to an altitude of 10 miles, where the temperature is −80.0°C. The walls of the balloon are elastic enough that the pressure inside it equals the pressure outside. Repeat the calculation of part (a).

68. Baseball reporters say that long fly balls that would have carried for home runs in July "die" in the cool air of October and are caught. The idea behind this observation is that a baseball carries better when the air is less dense. Dry air is a mixture of gases with an effective molar mass of 29.0 g mol^{-1}.

(a) Compute the density of dry air on a July day when the temperature is 95.0°F and the pressure is 1.00 atm.

(b) Compute the density of dry air on an October evening when the temperature is 50.0°F and the pressure is 1.00 atm.

(c) Suppose that the humidity on the July day is 100%; thus, the air is saturated with water vapor. Is the density of this hot, moist air less than, equal to, or greater than the density of the hot, dry air computed in part (a)? In other terms, does high humidity favor the home run?

69. Sulfuric acid reacts with sodium chloride to produce gaseous hydrogen chloride according to the following reaction:

$$NaCl(s) + H_2SO_4(\ell) \longrightarrow NaHSO_4(s) + HCl(g)$$

A 10.0-kg mass of NaCl reacts completely with sulfuric acid to give a certain volume of HCl(g) at 50°C and P = 1.00 atm. If the same volume of hydrogen chloride is collected at 500°C and P = 1.00 atm, what mass of NaCl has reacted?

70. Exactly 1.0 lb Hydrone, an alloy of sodium with lead, yields (at 0.0°C and 1.00 atm) 2.6 ft^3 of hydrogen when it is treated with water. All the sodium reacts according to the following reaction:

$$2\ Na_{\text{in alloy}} + 2\ H_2O(\ell) \longrightarrow 2\ NaOH(aq) + H_2(g)$$

and the lead does not react with water. Compute the percentage by mass of sodium in the alloy.

71. A sample of limestone (calcium carbonate, $CaCO_3$) is heated at 950 K until it is completely converted to calcium oxide (CaO) and CO_2. The CaO is then all converted to calcium hydroxide by addition of water, yielding 8.47 kg of solid $Ca(OH)_2$. Calculate the volume of CO_2 produced in the first step, assuming it to be an ideal gas at 950 K and a pressure of 0.976 atm.

72. A gas exerts a pressure of 0.740 atm in a certain container. Suddenly, a chemical change occurs that consumes half of the molecules originally present and forms two new molecules for every three consumed. Determine the new pressure in the container if the volume of the container and the temperature are unchanged.

73. The following arrangement of flasks is set up. Assuming no temperature change, determine the final pressure inside the system after all stopcocks are opened. The connecting tube has zero volume.

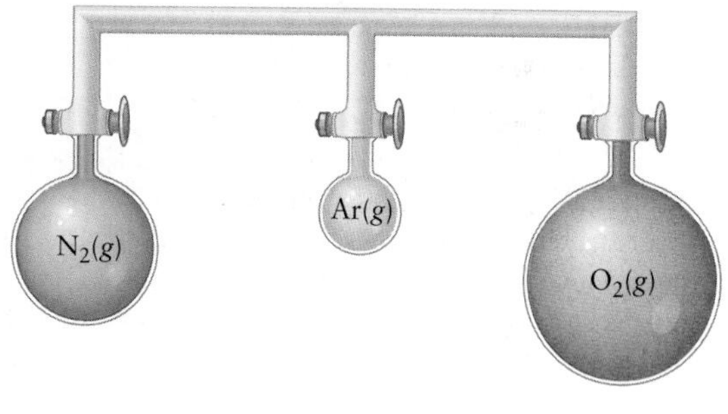

V = 4.00 L	V = 3.00 L	V = 5.00 L
P = 0.792 atm	P = 1.23 atm	P = 2.51 atm

***74.** A mixture of $CS_2(g)$ and excess $O_2(g)$ in a 10.0-L reaction vessel at 100.0°C is under a pressure of 3.00 atm. When the mixture is ignited by a spark, it explodes. The vessel successfully contains the explosion, in which all of the $CS_2(g)$ reacts to give $CO_2(g)$ and $SO_2(g)$. The vessel is cooled back to its original temperature of 100.0°C, and the pressure of the mixture of the two product gases and the unreacted $O_2(g)$ is found to be 2.40 atm. Calculate the mass (in grams) of $CS_2(g)$ originally present.

75. Acetylene reacts with hydrogen in the presence of a catalyst to form ethane according to the following reaction:

$$C_2H_2(g) + 2\ H_2(g) \longrightarrow C_2H_6(g)$$

The pressure of a mixture of acetylene and an excess of hydrogen decreases from 0.100 to 0.042 atm in a vessel of a given volume after the catalyst is introduced, and the temperature is restored to its initial value after the reaction reaches completion. What was the mole fraction of acetylene in the original mixture?

76. Refer to the atomic trap described in Problem 42.

(a) Assume ideal gas behavior to compute the pressure exerted on the "walls" of the optical bottle in this experiment.

(b) In this gas, the mean free path (the average distance traveled by the sodium atoms between collisions) is 3.9 m. Compare this with the mean free path of the atoms in gaseous sodium at room conditions.

77. Deuterium (^2H), when heated to sufficiently high temperature, undergoes a nuclear fusion reaction that results in the production of helium. The reaction proceeds rapidly at a temperature, T, at which the average kinetic energy of the deuterium atoms is 8×10^{-16} J. (At this temperature, deuterium molecules dissociate completely into deuterium atoms.)

(a) Calculate T in kelvins (atomic mass of $^2H = 2.015$).

(b) For the fusion reaction to occur with ordinary H atoms, the average energy of the atoms must be about 32×10^{-16} J. By what factor does the average speed of the ^1H atoms differ from that of the ^2H atoms of part (a)?

78. Molecules of oxygen of the following isotopic composition are separated in an oxygen enrichment plant: $^{16}O^{16}O$, $^{16}O^{17}O$, $^{16}O^{18}O$, $^{17}O^{17}O$, $^{17}O^{18}O$, $^{18}O^{18}O$.

(a) Compare the average translational kinetic energy of the lightest and heaviest molecular oxygen species at 200°C and at 400°C.

(b) Compare the average speeds of the lightest and heaviest molecular oxygen species at the same two temperatures.

***79.** Molecules in a spherical container make wall collisions in a great circle plane of the sphere. All paths traveled between collisions are equal in length.

(a) Relate the path length $\Delta\ell$ to the angle θ and the sphere radius r.

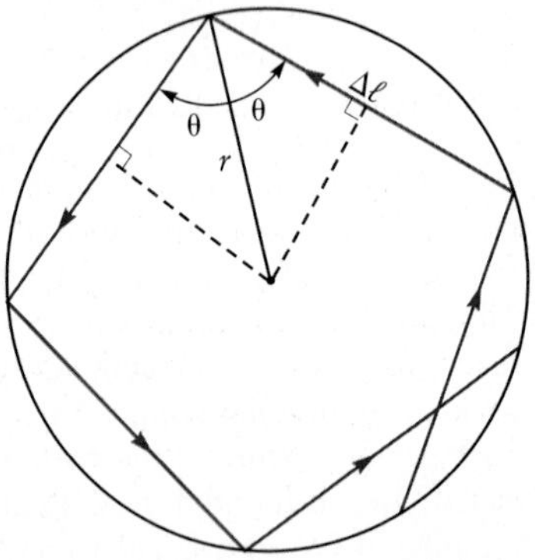

(b) Express the component of the molecular momentum transferred to the wall in a collision in terms of the molecular mass m, the speed v, and the angle θ.

(c) Using parts (a) and (b), calculate the average force exerted on the wall by one molecule.

(d) Derive the relation $PV = PV = \frac{1}{3}Nm\overline{v^2}$ for a spherical container. This is the same relation found for a rectangular box in Section 9.5.

***80.** The van der Waals constant b is related to the volume excluded per mole of molecules, so it should be proportional to $N_A\sigma^3$, where σ is the distance parameter in the Lennard–Jones potential.

(a) Make a plot of b against $N_A\sigma^3$ for Ar, H_2, CH_4, N_2, and O_2, using data from Tables 9.3 and 9.4. Do you see an overall correlation between the two?

(b) The van der Waals constant a has dimensions of pressure times the square of the molar volume. Rewrite the units of a in terms of energy, length, and number of moles, and suggest a relation between a and some combination of the constants ε, σ, and N_A. Make a plot of a against this combination of constants for the gases of part (a). Do you see an overall correlation in this case?

***81.** Take the derivative of the Lennard–Jones potential to express the force exerted on one atom by another as a function of the distance R between them. Calculate the forces (in joules per meter) on a pair of interacting argon atoms at distances of 3.0, 3.4, 3.8, and 4.2 $\times 10^{-10}$ m. To calculate the direction of the force on each atom, assume that one Ar atom is at the origin and the other argon atom is at the coordinates (−3.0 Å, 0), (−3.4 Å, 0), (−3.8 Å, 0), and (−4.2 Å, 0). Calculate the vector components F_x and F_y for the force $\vec{F} = (F_x, F_y)$ on the argon atom at the origin (particle 1 in the solutions) and the vector components for the force on the other atom (particle 2 in the solution), showing explicitly what you are using for the unit vector $\hat{r}_{21}$ or the unit vector $\hat{r}_{12}$. (*Hint:* Use Eq. 3.4, but with the correct $V(R)$.) Is the force attractive or repulsive at each of these distances?

82. A vessel with a small hole in its wall is filled with oxygen to a pressure of 1.00 atm at 25°C. In a 1.00-minute period, 3.25 g oxygen effuses out through the hole into a vacuum. The vessel is evacuated and filled with an unknown gas at the same pressure and volume. In this case, 5.39 g of the unknown gas effuses in 1.00 minute. Calculate the molar mass of the unknown gas.

83. A cylindrical storage tank for natural gas (mostly methane, CH_4) with a 20-ft radius and a 50-ft height is filled to a pressure of 2000 psi at 20°C. A small leak of 1.0-mm^2 area develops at one of the welds. Calculate the mass of CH_4 (in grams) that leaks out of the tank in one day. What fraction of the total gas escapes per day?

84. A thermos bottle (Dewar vessel) has an evacuated space between its inner and outer walls to diminish the rate of transfer of thermal energy to or from the bottle's contents. For good insulation, the mean free path of the residual gas (air; average molecular mass = 29) should be at least 10 times the distance between the inner and outer walls, which is about 1.0 cm. What should be the maximum residual gas pressure in the evacuated space if $T = 300$ K? Take an average diameter of $d = 3.1 \times 10^{-10}$ m for the molecules in the air.

85. A tanker truck carrying liquid ammonia overturns, releasing ammonia vapor into the air.

(a) Approximating ammonia, oxygen, and nitrogen as spheres of equal diameter (3×10^{-10} m), estimate the diffusion constant of ammonia in air at atmospheric pressure and 20°C.

(b) Calculate the time required for a 100-m root-mean-square displacement of ammonia from the truck, and express this time in everyday units (seconds, minutes, hours, days, or years). The actual time for the ammonia to travel this distance is far shorter because of the existence of air currents (even when there is no wind).

86. Molecules of UF_6 are approximately 175 times more massive than H_2 molecules; however, Avogadro's number of H_2 molecules confined at a set temperature exert the same pressure on the walls of the container as the same number of UF_6 molecules. Explain how this is possible.

87. The number density of atoms (chiefly hydrogen) in interstellar space is about 10 per cubic centimeter, and the temperature is about 100 K.

(a) Calculate the pressure of the gas in interstellar space, and express it in atmospheres.

(b) Under these conditions, an atom of hydrogen collides with another atom once every 1×10^9 seconds (that is, once every 30 years). By using the average speed, estimate the distance traveled by a H atom between collisions. Compare this distance with the distance from the Earth to the Sun (150 million km).

88. A sample of 2.00 mol argon is confined at low pressure in a volume at a temperature of 50°C. Describe quantitatively the effects of each of the following changes on the pressure, the average energy per atom in the gas, the root-mean-square speed, the rate of collisions with a given area of wall, the frequency of Ar–Ar collisions, and the mean free path:

(a) The temperature is decreased to −50°C.

(b) The volume is doubled.

(c) The amount of argon is increased to 3.00 mol.

89. By assuming that the collision diameter of a CH_4 molecule is given by its Lennard–Jones σ parameter (see Table 9.4), estimate the rate at which methane molecules collide with one another at 25°C and a pressure of (a) 1.00 atm and (b) 1.0×10^{-7} atm.

CUMULATIVE PROBLEMS

90. A gaseous hydrocarbon, in a volume of 25.4 L at 400 K and a pressure of 3.40 atm, reacts in an excess of oxygen to give 47.4 g H_2O and 231.6 g CO_2. Determine the molecular formula of the hydrocarbon.

91. A sample of a gaseous binary compound of boron and chlorine weighing 2.842 g occupies 0.153 L at 0°C and 1.00 atm pressure. This sample is decomposed to give solid boron and gaseous chlorine (Cl_2). The chlorine occupies 0.688 L at the same temperature and pressure. Determine the molecular formula of the compound.

92. A mixture of calcium carbonate, $CaCO_3$, and barium carbonate, $BaCO_3$, weighing 5.40 g reacts fully with hydrochloric acid, $HCl(aq)$, to generate 1.39 L $CO_2(g)$, measured at 50°C and 0.904 atm pressure. Calculate the percentages by mass of $CaCO_3$ and $BaCO_3$ in the original mixture.

93. A solid sample of Rb_2SO_3 weighing 6.24 g reacts with 1.38 L gaseous HBr, measured at 75°C and 0.953 atm pressure. The solid RbBr, extracted from the reaction mixture and purified, has a mass of 7.32 g.

(a) What is the limiting reactant?

(b) What is the theoretical yield of RbBr, assuming complete reaction?

(c) What is the actual percentage yield of product?

94. (a) A commercial 737 jet transporting 143 passengers and 5 crew members from Kansas City (MCI) to Baltimore (BWI) burned 11,800 lb (about 1700 gallons) of Jet A fuel en route. Jet A fuel is kerosine based, consisting primarily of C_nH_{2n+2} hydrocarbons, with $n = 6$ to 16, so the carbon:hydrogen ratio is close to 1:2. During this flight, how much CO_2 was released into the atmosphere? Assume the combustion of the fuel was complete, so all the fuel was burned to form CO_2 and H_2O. Give both the mass of CO_2 produced (in kg and in lbs) and the volume it would occupy at 298 K, 1 atm.

(b) How much CO_2 would be released into the atmosphere if those passengers and crew made the trip instead, in pairs, in hybrid cars at 40 miles per gallon. Assume the density of the gasoline is 0.75 kg L^{-1} and that carbon and hydrogen dominate the composition in a ratio of 1:2. The road trip is 1082 miles.

10

CHAPTER

Solids, Liquids, and Phase Transitions

Solid iodine is converted directly to a vapor (sublimes) when warmed. Here, purple iodine vapor is redeposited as a solid on the cooler upper surfaces of the vessel.

The bulk properties of gases, liquids, and solids—molar volume, density, compressibility, and thermal expansion, among others—differ widely, often by orders of magnitude. All of these properties depend on the temperature and pressure. The local structure—the arrangement of atoms or molecules on the nanometer length scale—is the key microscopic feature that distinguishes the three states of matter from one another and explains the differences in their bulk properties. The average separation between molecules, and the nature of the intermolecular forces, determine the local structure and therefore the properties of the bulk.

We begin with a brief survey of selected bulk properties and describe how the measured results depend on the number density of molecules and the strength of intermolecular forces. Then, we define the types of intermolecular forces and show how these forces originate in the structures of molecules. Finally, we relate the transitions between the states of matter to intermolecular forces. In the previous chapter we derived the ideal gas law using kinetic molecular arguments and neglecting all intermolecular forces in the gas. In this chapter we apply similar kinetic molecular arguments to solids and liquids. Because we cannot neglect the intermolecular forces, the discussion of solids and liquids is more qualitative than that for gases.

10.1 Bulk Properties of Gases, Liquids, and Solids: Molecular Interpretation

Each of the following macroscopic measurements clearly distinguishes gases, liquids, and solids from each other and also probes the strength of the intermolecular forces, albeit indirectly. Each shows that in gases at low densities, molecules are on average far apart and interact only weakly, whereas in condensed phases, molecules are closely packed together and interact quite strongly. These general conclusions apply almost universally to substances consisting of small, nearly rigid molecules such as N_2, CO_2, CH_4, and acetic acid $C_2H_4O_2$. In contrast, many biological materials and synthetic polymers contain complex chain-like molecules that can become strongly entangled. Clear-cut classification of these materials as solids or liquids becomes difficult.

Relating bulk properties qualitatively to microscopic properties requires only consideration of the long-range attractive forces and short-range repulsive forces between molecules; it is not necessary to take into account the details of molecular shapes. We have already shown one kind of potential energy function that describes these long-range and short-range forces, the Lennard–Jones 6–12 potential used in Section 9.6 to obtain corrections to the ideal gas law. In Section 10.2, we discuss more detailed aspects of intermolecular forces, most of which are derived from electrostatic (Coulomb) interactions that depend on molecular structure and shape.

10.1.1 Molar Volume

One mole of a typical solid or liquid occupies a volume of 10 to 100 cm^3 at room conditions, but the **molar volume** of a gas under the same conditions is about 24,000 cm^3 mol^{-1}. This large difference explains why solids and liquids are called the condensed states of matter. Because a mole of any substance contains Avogadro's number of molecules, the molar volume is inversely related to the **number density** (number of molecules per cubic centimeter) of the different phases. Liquids and solids have high number densities, and gases have very low number densities. On melting, most solids change volume by only 2% to 10%, showing that the solid and liquid states of a given substance are condensed, relative to the gaseous state, by roughly the same amount.

The similarity of the molar volumes of solid and liquid forms of the same substance suggests that the separation between neighboring molecules in the two states is approximately the same. Density measurements (see Section 1.6.3) show that the intermolecular contacts, the distances between the nuclei of atoms at the far edge of one molecule and the near edge of a neighbor, usually range from 3×10^{-10} m to 5×10^{-10} m in solids and liquids. At these distances, longer range attractive forces and shorter range repulsive forces just balance one another, resulting in a minimum in the potential energy (see Section 9.6). Although these intermolecular separations are significantly greater than chemical bond lengths (which range from 0.5–2.5 $\times 10^{-10}$ m), they are much shorter than the intermolecular separations in gases, which average about 30×10^{-10} m under room conditions. The distinction is shown schematically in Figure 10.1.

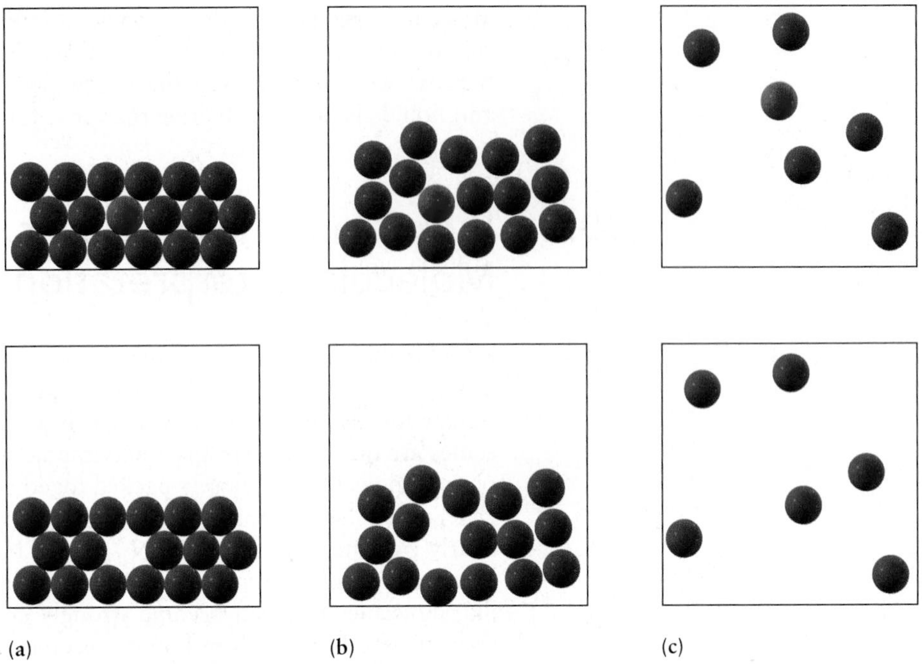

FIGURE 10.1 Intermolecular forces create structure in liquids and solids. If a single atom is removed from a snapshot of the atomic arrangement in a solid (a), it is easy to figure out exactly where to put it back in. For a liquid (b), the choices are limited. If a single atom is removed from a gas (c), no clue remains to tell where it came from.

10.1.2 Compressibility

The **compressibility** of a substance is defined as the fractional decrease in volume per unit increase in pressure. (This property is different from the compressibility factor Z [see Section 9.6], which measures non-ideality in gases.) The compressibility is usually denoted by the Greek letter kappa, κ, where $\kappa = -\frac{1}{V}\left(\frac{\partial V}{\partial P}\right)_T$.

The subscript T indicates that the temperature is held constant as you take the derivative of V with respect to P. For small changes in pressure and volume, this is approximately $\kappa = -(1/V)(\Delta V/\Delta P)$. A sample with volume V is subjected to a pressure increase, ΔP, and the resulting change in volume, ΔV, is measured. The ratio is divided by V; thus, the tabulated value depends only on the substance being measured and not on the geometry of the sample. Consequently, the unit of κ is inverse pressure, typically atm^{-1}. The minus sign is included in the definition to make κ positive because ΔV is negative. The measurements are always performed at fixed temperature to eliminate any thermal effects on the volume (see later). Consequently, κ is called the *isothermal compressibility,* and the temperature of measurement is quoted together with the tabulated values. Both solids and liquids are nearly incompressible, whereas gases are very compressible. According to Boyle's law, doubling the pressure exerted on an ideal gas from 1 to 2 atm reduces its volume to half its original volume (at constant temperature). The corresponding compressibility is pressure dependent, ranging from 1 atm^{-1} (or 10 MPa^{-1}) at the beginning of the compression to 0.5 atm^{-1} near the end of the compression. (Using the definition of the isothermal compressibility given, it is easy to show that the isothermal compressibility of an ideal gas is P^{-1}.) In contrast, doubling the pressure exerted on water or steel scarcely changes the volume at all; typical compressibilities for these materials are of the order 10^{-5} to 10^{-6} atm^{-1} (Table 10.1). Increasing the pressure on a liquid or solid by a factor of 2 changes its volume by 1% or less. The high compressibility of gases and the low compressibilities of solids and liquids suggest that in the gas phase there is substantial space between the molecules, but in the condensed states the molecules or atoms of a substance are in contact or nearly in contact.

TABLE 10.1

Isothermal Compressibility† and Thermal Expansion Coefficients

Compound	$\kappa/(10^{-6}\ \text{atm}^{-1})$	$\alpha/(10^{-4}\ K^{-1})$
Liquids		
Benzene	92.1	12.4
Ethanol	76.8	11.2
Mercury	38.7	1.82
Water	49.7	2.1
Solids		
Copper	0.735	0.501
Diamond	0.187	0.030
Iron	0.597	0.354
Lead	2.21	0.861

†Values at 20°C.

The much greater separation between molecules in the gas phase than in the condensed phases dramatically influences the effect of intermolecular forces. The repulsive force, although very strong, is very short-ranged and becomes significant only when molecules are quite close together. So, gases require only modest forces to compress them significantly, because the molecules can be pushed much closer together before experiencing repulsive forces. Liquids and solids require much greater forces to oppose the strongly repulsive forces already operating because the molecules are in contact (see Section 9.6).

10.1.3 Thermal Expansion

The **coefficient of thermal expansion** α is defined as the fractional increase in the volume of a substance per degree increase in temperature. It is defined by $\alpha = \frac{1}{V}\left(\frac{\partial V}{\partial T}\right)_P$, where the subscript P indicates that the pressure is held constant as you take the derivative of V with respect to T. For small changes in volume and temperature, this is approximately $\alpha = (1/V)(\Delta V/\Delta T)$. The measurements are performed at constant pressure; thus, α is called the *isobaric coefficient of thermal expansion,* and the pressure at which the measurements were taken is quoted together with tabulated values. Charles's law shows that this coefficient is the same for all gases and takes the value $1/273.15(°\text{C})^{-1}$ at 0°C. You can verify this by substituting $\Delta V/\Delta T = nR/P$ (from the ideal gas law) into the definition given and simplifying the expression. Increasing the temperature by 1°C thus causes a gas to expand by 1/273.15, or 0.366% of its original volume at 0°C, as long as the pressure is constant. The thermal expansion coefficients of liquids and solids are much smaller. Heating water from 20°C to 21°C increases its volume by only 0.0212%, and the volume of mercury goes up by only 0.0177% over the same temperature interval. The coefficients of thermal expansion of solids are mostly less than 0.02% per degree Celsius (see Table 10.1).

The difference in thermal expansion between condensed states and gases is explained by strong intermolecular forces (deep intermolecular potential wells) acting over short distances in the condensed states, but not in the gaseous state. Because of the much greater intermolecular separations, these forces are much weaker in gases. Increasing the volume in a solid or liquid requires that attractive forces between each molecule and its neighbors be partially overcome.

Because the intermolecular distances in a solid or liquid fall in the range where intermolecular attractive forces are strongest, relatively small expansion occurs when the temperature is increased. By contrast, molecules in a gas are so far apart that attractive forces are essentially negligible; the same temperature increase produces much greater expansion in a gas than in condensed phases.

10.1.4 Fluidity and Rigidity

The most characteristic property of gases and liquids is their **fluidity,** which contrasts with the **rigidity** of solids. Liquids possess definite volumes but keep no definite shapes of their own; they flow easily under stress (externally applied mechanical force). The resistance of a material to macroscopic flow is measured by its **shear viscosity.** On the microscopic level, shear viscosity arises from the resistance of one thin layer of molecules "dragged across" another thin layer. The shear viscosities of most liquids are about 16 orders of magnitude smaller than those of most solids, and those of gases are smaller yet. A *rigid* material retains its shape under stress; it manifests structural strength by resisting flow when stress is applied. The properties of **hardness** (resistance to indentation) and **elasticity** (capacity to recover shape when a deforming stress is removed) are closely related to rigidity, or high shear viscosity. Solids possess these properties in good measure; gases and liquids do not.

10.1.5 Diffusion

When two different substances are placed in contact—for example, a drop of red ink into a beaker of water—they start to mix. Molecules of one type migrate, or **diffuse,** into regions initially occupied only by the other type. Molecules dispersed in gases at room conditions diffuse at rates on the order of centimeters per second. If you have ever passed by a perfume counter, this rate should not surprise you. Molecules in liquids and solids diffuse far more slowly. The **diffusion constant** of a substance measures the rate of diffusive mixing. At room temperature and pressure, diffusion constants for the diffusion of liquids into liquids are about four orders of magnitude smaller than those for gases into gases; diffusion constants of solids into solids are many orders of magnitude smaller yet. Diffusion in solids is really quite slow. Values of the diffusion constant for selected materials are shown in Table 10.2.

Figure 10.1 shows a "snapshot" of a liquid and a solid, fixing the positions of the atoms at a particular instant in time. The paths followed by the molecules in these two states can also be examined over a short time interval (Fig. 10.2).

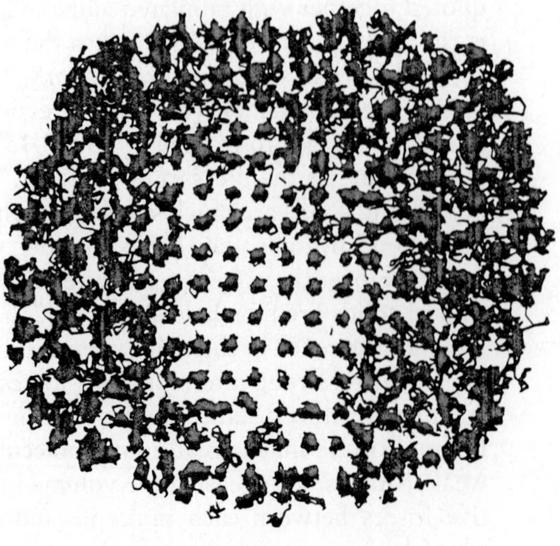

FIGURE 10.2 In this computer-simulated picture of the motions of atoms in a tiny melting crystal, the atoms at the center (in the solid) move erratically about particular sites. The atoms at the surface (in the liquid) move over much greater distances.

TABLE 10.2

Diffusion Constants

Diffusing Species	Host Material	Diffusion Constant ($m^2\ s^{-1}$)	Temperature (K)
Ar	Ar	2.3×10^{-6}	100
Ar	Ar	1.86×10^{-5}	300
N_2	N_2	2.05×10^{-5}	300
O_2	O_2	1.8×10^{-5}	273
CH_4	CH_4	2.06×10^{-5}	273
HCl	HCl	1.24×10^{-5}	295
Cu	Cu	4.2×10^{-19}	500
Al	Al	4.2×10^{-14}	500
Cu	Al	4.1×10^{-14}	500
Cu	Ni	1.3×10^{-22}	500
Fe	Fe	3.0×10^{-21}	500
Fe	Fe	1.8×10^{-15}	900
C	Fe	1.7×10^{-10}	900

In liquids, molecules are free to travel through the sample, changing neighbors constantly in the course of their diffusive motion. In a solid, the molecules constantly vibrate about their equilibrium positions, but remain quite close to those positions. The low shear viscosity of a liquid implies that its molecules can quickly change neighbors, finding new interactions as the liquid flows in response to an external stress. The rigidity of solids suggests, in contrast, a durable arrangement of neighbors about any given molecule. The durable arrangement of molecules in a solid, as opposed to the freedom of molecules to diffuse in a liquid at comparable packing density, is the critical difference between the solid and liquid states.

Short-range attractive intermolecular forces in liquids lead to well-defined local structures that persist for short periods. Individual molecules experience interactions with neighbors that lead, at any instant, to a local environment closely resembling that in a solid, but they quickly move on. Their trajectories consist of "rattling" motions in a temporary cage formed by neighbors and superimposed on erratic displacements over larger distances. In this respect, a liquid is intermediate between a gas and a solid. A gas (see Fig. 9.25) provides no temporary cages, so each molecule of a gas travels a longer distance before colliding with a second molecule. Consequently, the diffusion constant of a gas is larger than that of a liquid. In a solid, the cages are nearly permanent, thus diffusion is slow. Melting occurs as thermal energy increases the amplitude of vibration of the molecules around their equilibrium positions in a solid to such a degree that they are free to make major excursions. For these same reasons, liquids can dissolve substances much more rapidly than do solids. Individual molecules of a liquid quickly wander into contact with molecules of an added substance, and new attractions between the unlike molecules have an early chance to replace those existing originally in the pure liquid.

Liquids and gases may also mix through **convection,** as well as by diffusion. In convection, the net flow of a whole region of fluid with respect to another region leads to mixing at far greater rates than occurs through simple diffusion. Convection is the primary mechanism by which mixing occurs in the oceans and in the atmosphere. Convection is not observed in solids.

(a)

(b)

FIGURE 10.3 (a) Surface tension causes the spherical shape of the water droplet in this photograph, which was taken an instant after a drop of water hit the surface of a pool and bounced up, pulling with it a column of water. (b) Small droplets of mercury on a tabletop form nearly perfect spheres due to the high surface tension of mercury. Larger droplets are flattened by the force of gravity.

10.1.6 Surface Tension

Boundaries between phases have special importance in chemistry and biology. Each type of boundary has its own unique characteristics. The surface of water (or any liquid) in contact with air (or any gas) resists attempts to increase its area (Fig. 10.3a). This **surface tension** causes the surface to behave like a weak, elastic skin. Effects of surface tension are particularly apparent under zero gravity, where liquids float around as spherical drops because spheres contain the largest volume for the smallest surface area of any geometric shape. If two small drops encounter each other, they tend to coalesce into a larger drop because one large drop has a smaller surface area than two small drops. The surface tension of water is larger than that of most other liquids at room temperature, but it is about six times smaller than that for the liquid metal mercury, which has one of the highest values known for any liquid at room temperature and pressure (see Fig. 10.3b).

Surface tension results from the intermolecular attractions among the molecules in a liquid (Fig. 10.4). Increasing the surface area of a liquid requires redistributing some of the molecules that were originally buried in the interior to positions at the enlarged boundaries. Molecules at the edges have no neighbors on one side and experience attractions only from molecules in the bulk of the liquid. Their potential energy is greater than it would be if they were in the interior. Thus, energy is required to increase the surface area of a liquid. Liquids such as water and mercury with high values of surface tension have particularly strong intermolecular attractions, as confirmed by measurement of other properties.

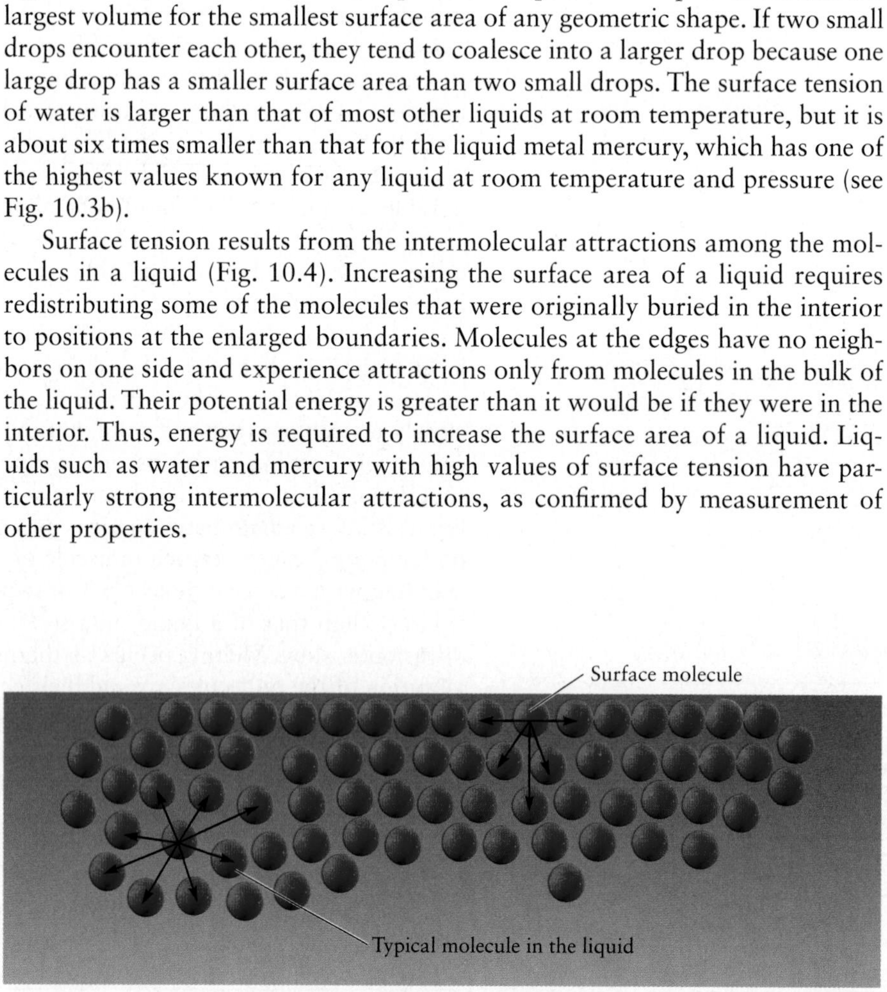

FIGURE 10.4 The intermolecular attractions acting on a molecule at the surface of a liquid pull it downward and to the sides but not upward. In the interior, a molecule is pulled more or less equally in all directions.

10.2 Intermolecular Forces: Origins in Molecular Structure

To provide a more quantitative explanation of the magnitudes of bulk properties, we must describe molecular interactions with more detail than the simple net repulsion and attraction summarized in the Lennard–Jones model potential. For this purpose we introduce other intermolecular forces in the following paragraphs and show how they arise from molecular structure. Intermolecular forces are distinct from intramolecular forces, which lead to the covalent chemical bonds that establish and maintain the structure of discrete molecules. (See Chapters 3 and 6.) Intermolecular forces differ from intramolecular forces in several important ways:

1. Intermolecular forces are generally weaker than covalent chemical bonds. For example, it takes 239 kJ to break 1 mol of Cl—Cl covalent bonds, but only 1.2 kJ to overcome 1 mol of Ar—Ar attractions.
2. Intermolecular forces are much less directional than covalent chemical bonds.
3. Intermolecular forces operate at longer range than covalent chemical bonds.

It is useful to distinguish different classes of intermolecular forces based on their strength, directionality, and range and to relate these aspects to the structure of the molecules.

10.2.1 Ion–Ion Forces

Ionic solids and liquids are made up of electrically charged entities, sometimes monatomic ions such as Na^+, Cl^-, and Ca^{2+}, and sometimes polyatomic ions such as NH_4^+ and SO_4^{2-}. The dominant interaction among these ions is the Coulomb force of electrostatic attraction or repulsion, which leads to the Coulomb potential described in Section 3.3. Ions of like charge repel one another, and ions of unlike charge attract one another. These **ion–ion forces** can be as strong as those in the covalent bond, and they are long ranged. The potential energy is proportional to R^{-1} and so approaches the value at infinite separation much less rapidly than other types of interactions. Ion–ion forces are not directional; each ion interacts equally strongly with neighboring ions on all sides. Ion–ion forces lead to the formation of ionic bonds through the Coulomb stabilization energy (see Section 3.8).

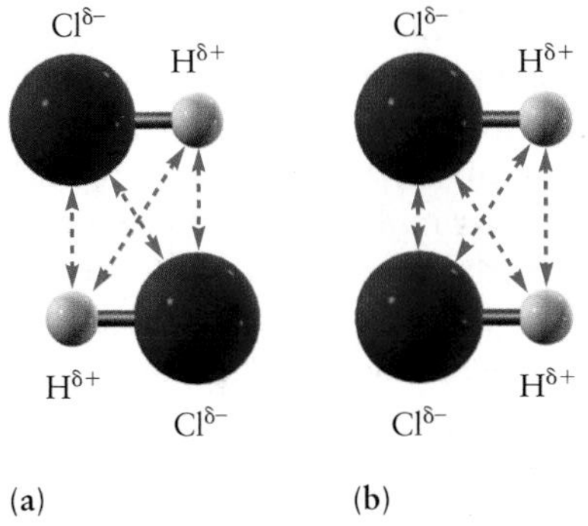

FIGURE 10.5 A molecule of HCl can be represented as having a small net negative charge on the Cl end, balanced by a small net positive charge on the H end. The forces between two HCl molecules depend on their orientations. (a) The oppositely charged ends (blue arrows) are closer than the ends with the same charge (red arrows). This gives a net attractive force. (b) Here, the opposite is true, and the net force is repulsive.

10.2.2 Dipole–Dipole Forces

The dominant force between polar molecules is the **dipole–dipole force.** This is a second example of electrostatic forces that arise from interactions between fixed charges, in this case the magnitudes of the permanent dipole moments of polar molecules. As shown in Figure 10.5, these forces also depend on the orientations of the two molecules and can be either attractive or repulsive or zero. Random motions of polar molecules in gases and liquids lead to a variety of energetically favorable temporary dipole–dipole orientations. The potential energy between dipoles separated by the distance R falls off as R^{-3}. Thus the attractive force gets weaker more rapidly as the separation is increased for a dipole-dipole interaction than for a Coulomb interaction. Increasing the distance between a pair of ions by a factor of 10 changes the Coulomb potential energy to be closer to 0 by only a factor of 10, whereas increasing the distance between a pair of dipoles

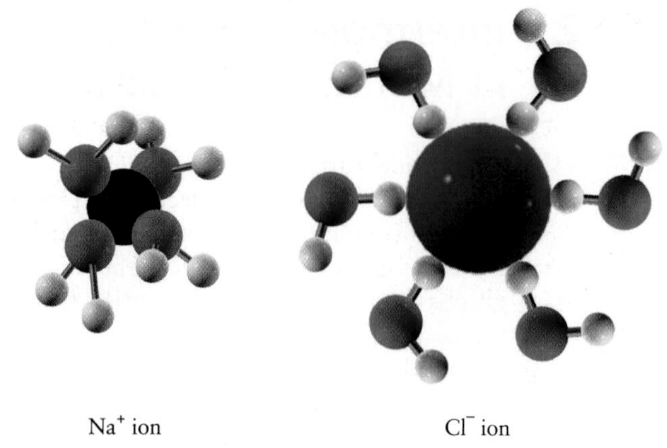

FIGURE 10.6 Solvation of ions in liquid water. The water molecules have dipole moments; thus, the oxygen (O) atoms bear small, negative charges, whereas the hydrogen (H) atoms bear small, positive charges. (a) Positive ions are attracted to neighboring water molecules in aqueous solution by ion–dipole forces. (b) Negative ions form hydrogen bonds with water, with a nearly linear bond from O to H to the anion.

by a factor of 10 changes the potential energy to be closer to 0 by a factor of 1000. In liquids, thermal energy can overcome dipole–dipole attractions and disrupt favorable orientations; dipole–dipole interactions are too weak to hold molecules in a liquid together in a nearly rigid arrangement. Nonetheless, they are sufficiently strong to influence many physical properties, including boiling points, melting points, and molecular orientations in solids.

10.2.3 Ion–Dipole Forces

A third example of electrostatic forces occurs when a polar molecule is near an ion. The interaction between a polar solvent molecule, such as water, and a dissolved ion is the most common case of ion–dipole interaction. Figure 10.6 shows dissolved Na^+ and Cl^- ions interacting with water dipoles. Positive ions are attracted by the negative end of the dipole and repelled by the positive end; thus, the ion is surrounded by a shell of water molecules whose oxygen (O) atoms are near the cation and whose hydrogen (H) atoms point outward into the solution. For many years it was believed that the opposite would be true for negative ions, that they would be surrounded by a shell of water dipoles whose H atoms were both near the anion. Since about 1980, neutron diffraction has been used to determine the distances between atoms in ionic aqueous solutions. A series of such studies has shown that the halide anion interacts with only one of the H atoms, and the atoms $O{-}H{-}Cl^-$ lie nearly in a straight line. The other H atom points in a direction determined by the geometry of the water molecule. The solvation of the anion is not governed by ion–dipole forces. Rather, the $O{-}H{-}Cl^-$ interaction is an example of the *hydrogen bond*, a special intermolecular force that occurs only in liquids. The hydrogen bond is discussed in the next section, and the solvation of ions is discussed more thoroughly in Chapter 11.

10.2.4 Charge-Induced Dipole Forces: Polarizability

The electrons in a nonpolar molecule or atom are distributed symmetrically, but the distribution can be distorted by an approaching electrical charge. An argon (Ar) atom has no dipole moment, but an approaching Na^+, with its positive charge, attracts the electrons on the side near it more strongly than those on the far side. By tugging on the nearby electrons harder, Na^+ induces a temporary dipole moment in the Ar atom (Fig. 10.7). The electron distribution of the nonpolar molecule is said to be *polarizable*, and the magnitude of the dipole moment induced measures the **polarizability** of the molecule. As long as the induced dipole is present, the interaction between molecules is similar to the ion–dipole case just described. **Induced dipole forces** also can be caused by a negative ion or by another dipole. These so-called induction forces differ from

FIGURE 10.7 As an ion approaches an atom or molecule, its electrostatic field distorts the distribution of the outer electrons. The effect of this distortion is to create a dipole moment that exerts an attractive force back on the ion.

the electrostatic forces between permanent fixed charges such as ions or dipoles. Rather, they arise from interactions between the permanent charges or moments on one molecule and the induced moments, or the polarizability, of another molecule. These interactions are weak and are effective only at short range. The induced dipole moment closely tracks the motion of the charge or dipole moment of the inducing molecule. The induced dipole is dynamically correlated with the motion of the inducing molecule. A good way to study induced dipole forces quantitatively is to collide a beam of Na^+ ions and a beam of Ar atoms in vacuum, measure the energy and direction of their deflections, and deduce the correct potential function to explain these deflections.

10.2.5 Induced Dipole–Induced Dipole Forces: London Dispersion Forces

Helium (He) atoms, like the atoms of the other noble gases, are electrically neutral and nonpolar, so none of the forces discussed so far explains the observed fact that there are attractions between He atoms. We know such attractions must exist, because helium becomes a liquid at 4.2 K and 1 atm. Attractions between neutral, nonpolar atoms or molecules arise from the **London dispersion forces** (often called van der Waals forces) that exist between all atoms and molecules. Dispersion forces are, in effect, a mutual interaction between the polarizable charge distributions on two separate molecules, and they are always attractive. Although the electron probability distribution around the molecule is described by the square of the wave function, dynamic motions of charge around the molecule can lead to an instantaneous, temporary dipole moment. Such a temporary dipole on one molecule will induce a temporary dipole in the other molecule. These transient, fluctuating dipoles attract one another in much the same way as do permanent dipoles. Figure 10.8 provides a simple view of the source of this interaction. The polarizability increases with the number of electrons in the atom or molecule. Heavier atoms or molecules interact more strongly by dispersion forces than do lighter ones because their outer electrons are located in shells farther from the nucleus. These electrons are less strongly bound than the outer electrons of the lighter elements, because they are shielded from the full attraction of the nucleus by intervening electrons (see Sections 3.4 and 5.2). Consequently, they are more easily distorted by external fields of neighboring dipoles. Dispersion forces are always attractive and fall off as R^{-6}. These interactions are short ranged, much more so than dipole–dipole forces. Dispersion forces provide the attractive term in the Lennard–Jones potential (see description in Section 9.6).

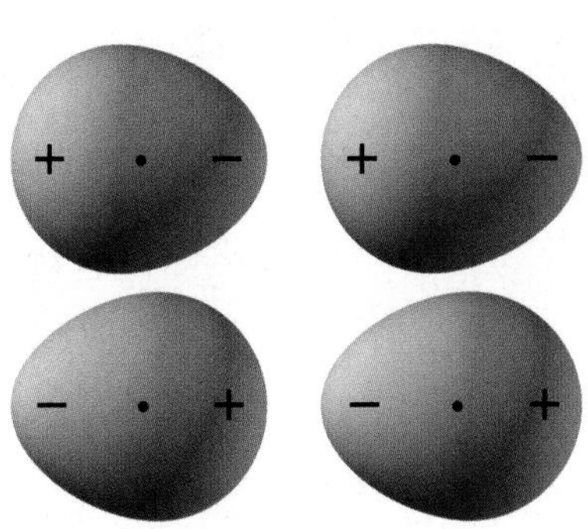

FIGURE 10.8 A fluctuation of the electron distribution on one atom induces a corresponding temporary dipole moment on a neighboring atom. The two dipole moments interact to give a net attractive force, called a "dispersion force."

10.2.6 Repulsive Forces

As atoms or molecules approach each other closely, a **repulsive force** comes into play and can overcome the attractive forces considered so far. The source of this force is the strong repulsion between the core (nonvalence) electrons when neighboring atoms are forced close to each other. This contribution is negligible until the distance between centers becomes small, at which point the potential energy increases rapidly as distance is reduced further. Two mathematical models are used to describe the repulsive region of the potential energy function, although neither has a simple physical foundation to guide the choice of the parameters involved. The exponential form $Ae^{-R/\rho}$ is successful when A and ρ are chosen to fit experimental data such as compressibility measurements. The inverse power form R^{-n}, where n is quite large, is also successful. The choice $n = 12$ is widely used because, when combined with the $n = 6$ term as shown in Equation 9.24, the resulting Lennard–Jones potential reproduces the trends in experimental data over significant ranges. Regardless of mathematical form, this steep, repulsive interaction at extremely small distances justifies modeling atoms as hard, nearly incompressible spheres with characteristic dimensions called the **van der Waals radii.** This label honors the early contributions of Johannes van der Waals to the study of non-bonded interactions between molecules and their influence on the properties of materials. The minimum distance between molecules in a condensed phase is determined by the sum of the van der Waals radii of their atoms. Space-filling models and drawings are usually designed to approximate the van der Waals surface of molecules, which represents the distance of closest approach by neighboring molecules. Van der Waals radii for atoms are typically a few angstroms (Å).

EXAMPLE 10.1

State which attractive intermolecular forces are likely dominant in the following substances:

(a) $F_2(s)$

(b) $HBr(\ell)$

(c) $NH_4Cl(s)$

Solution

(a) Molecules of F_2 are nonpolar, thus the predominant attractive forces between molecules in $F_2(s)$ come from dispersion.

(b) The HBr molecule has a permanent dipole moment. The predominant forces between molecules are dipole–dipole. Dispersion forces will also contribute to associations, especially because Br is a rather heavy atom.

(c) The ammonium ions are attracted to the chloride ions primarily by ion–ion forces.

Related Problems: 15, 16, 17, 18, 19, 20

10.2.7 Comparison of Potential Energy Curves

The relative strengths and effective ranges of several intermolecular forces are illustrated in Figure 10.9, which shows how the potential energy depends on the intermolecular separation (center-to-center distance) for several pairs of ions, atoms, or molecules. The potentials illustrated here include Coulomb (R^{-1}), dipole–dipole (R^{-3}), dispersion (R^{-6}), and repulsive (R^{-12}) potentials. The species shown in Figure 10.9 were chosen so that the interacting atoms, ions, or molecules have the same number of electrons (Ar, Cl^-, K^+, HCl). The ion–ion interaction of K^+ with Cl^- is the strongest, followed by the interaction between two HCl molecules (dipole–dipole and dispersion), and the Ar–Ar interaction (dispersion only). The key points illustrated here (and detailed in the caption for Fig. 10.9) are the dramatically different depths of the wells (several orders of magnitude), as well as the distinctly different distances at which the minima occur. For comparison, the potential energy curve for the covalent bond in Cl_2 is also shown. These graphs demonstrate that the intermolecular forces between atoms or molecules that are close together but not joined by chemical bonds are much weaker than those between two ions involved in an ionic bond, and between two atoms involved in a covalent bond. Because the kinetic energy of translational motions of molecules is on the order of 2–3 kJ mol^{-1}, molecules of KCl and Cl_2 will almost never be dissociated by collisions, whereas pairs of argon atoms or HCl molecules will break apart readily. The qualitative shape of the potential wells is the same for bonding and non-bonding interactions, but the depth of the well compared to thermal energy of molecules is the fundamental distinction between these two types.

10.2.8 The Shapes of Molecules and Electrostatic Forces

The potential energy diagrams in Figure 10.9 depend only on the distance between the two species. Interactions between complex molecules also depend strongly on their relative orientations, so we need a three-dimensional generalization of the

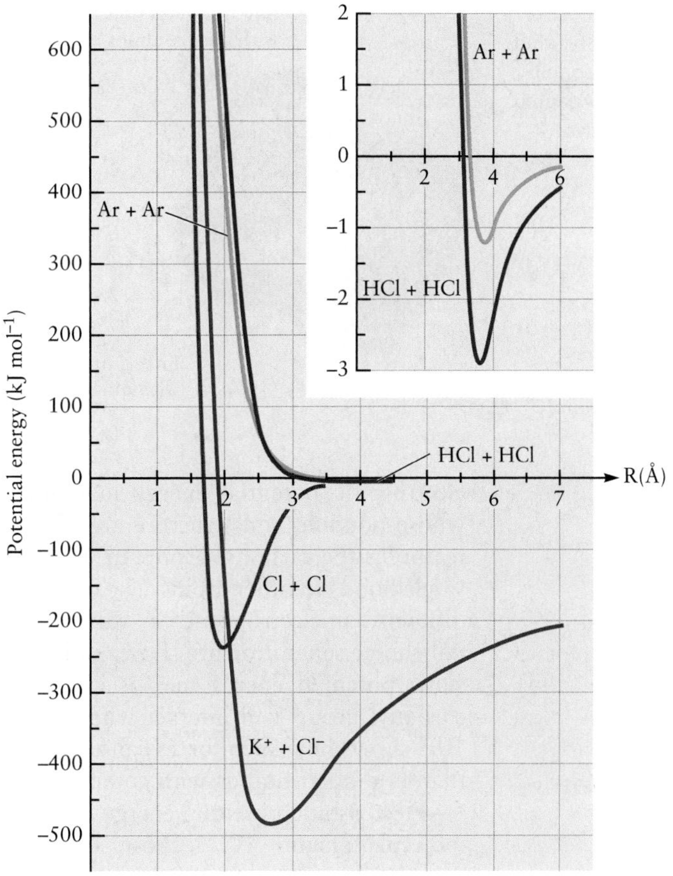

FIGURE 10.9 The potential energy of a pair of atoms, ions, or molecules depends on the distance between the members of the pair. Here, the potential energy at large separations (to the right side of the graph) is arbitrarily set to zero by convention (see Section 3.3). As pairs of particles approach each other, the potential energy *decreases* (becomes negative) because attractive forces come into effect. The lowest point in each curve occurs at the distance where attractive and repulsive forces exactly balance. The relative potential energy values at these minima measure the relative strength of the attractive forces in the various cases illustrated. Note the shallow potential energy minimum reached when two HCl molecules or two argon atoms near each other. The inset shows these same two curves with the vertical scale expanded by a factor of 100. (The HCl—HCl curve was computed for the relative orientations of Fig. 10.5a.)

potential energy diagram to describe these interactions more fully. This need could be met by constructing a potential energy *surface* where the interaction energy is plotted as a function of all three spatial coordinates that locate the center of one molecule relative to the center of the other. Instead of aiming for quantitative calculations of potential energy surfaces from quantum mechanics, we use approximate representations to describe the influence of shape and orientation as two molecules approach one another. One such approximate representation is the electrostatic potential energy map, which shows the shape and size of the molecule, as well as the sign and magnitude of the electrostatic potential at the "surface" of the molecule. These maps were introduced and the methods by which they are obtained were defined in Section 6.10.

The electrostatic potential energy map for a given molecule, called the "target" molecule, shows the spatial shape of the electrostatic field around the molecule, and so it can be used to predict how the target molecule influences the motion of charged particles as they approach it. Figure 10.10 shows the

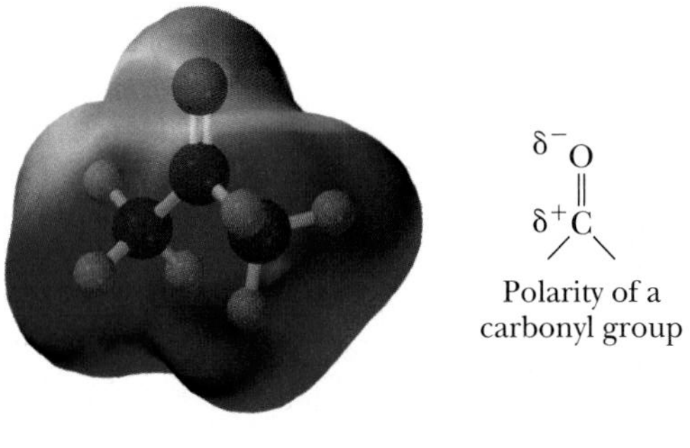

FIGURE 10.10 Electrostatic potential energy map for acetone. Note the charge separation indicated by the red (negative) region on the carbonyl oxygen and the blue (positive) region on the three carbon atoms. Because of this charge separation, acetone is a polar molecule. (© Cengage Learning; Adapted from Brown/Foote Iverson/Anslyn: *Organic Chemistry 5/e,* Ch. 16, p. 564)

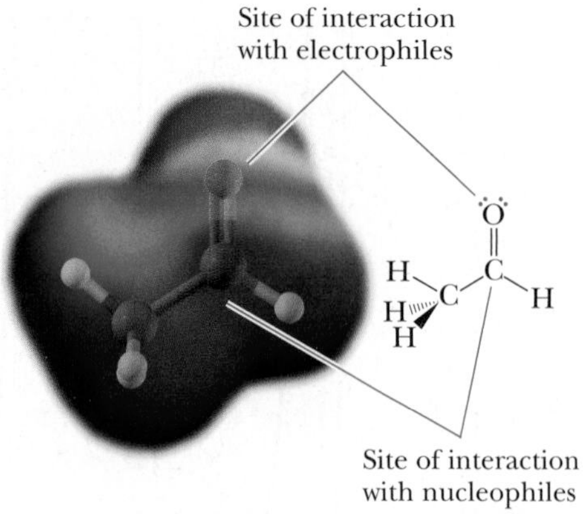

FIGURE 10.11 Electrostatic potential energy map for acetaldehyde showing sites for electrophilic and nucleophilic attack. (© Cengage Learning; Adapted from Brown/Foote Iverson/Anslyn: *Organic Chemistry 5/e,* Ch. 16, p. 566)

electrostatic potential energy map for the molecule acetone, $CH_3C(O)CH_3$, whose bonding and structure are discussed in Section 7.6. The key structural feature is the carbonyl group, in which the C has a double bond to O and two single bonds to other atoms. The carbonyl group is quite polar due to the strong difference in electronegativity values for C and O. The dipole moment and partial charge separation are shown in Figure 10.10, and their origin in the electrostatic potential energy map is clearly indicated. Molecules that contain the carbonyl group will interact with each other by dipole–dipole forces. Section 10.3 shows how such forces cause the pure liquids to have higher boiling points than non-polar liquids with comparable molecular weight.

Electrostatic potential energy maps can be used to identify reactive sites on molecules. Figure 10.11 shows the map for acetaldehyde (CH_3CHO) which contains the carbonyl group and is discussed in Section 7.6. Comparison with Figure 10.10 shows the consequences of replacing one of the terminal $-CH_3$ groups in acetone with a single $-H$. Locations with large negative values of the electrostatic potential, such as the red region around the O atom, are relatively rich in electron density and tend to attract reaction partners that lack electron density. Regions with large positive values, indicated in blue, are relatively depleted in electron density and tend to attract electron-rich reaction partners. These maps are now widely used in organic chemistry to predict patterns of reactivity for electrophilic (electron-loving) and nucleophilic (proton-loving) molecules and to explain how the presence of different functional groups in the molecule can affect these patterns. These methods are effective aids in identifying sites for chemical reactivity in more complicated molecules, including those of biological interest. They are widely used in molecular modeling simulations of drug design.

10.3 Intermolecular Forces in Liquids

The same intermolecular forces that make gases deviate from ideal behavior (see Sections 9.6 and 10.2) are responsible for the existence of solids and liquids. At very high temperatures, these forces are negligible because the high kinetic energy of the molecules disrupts all possible attractions; all materials are gaseous at sufficiently high temperatures. At lower temperatures, where materials are in the liquid state, molecules are close together and the details of the intermolecular potential energy determine their properties. Section 10.2 describes the influence of molecular structure on the intermolecular potential energy. This section surveys the correlation between the properties of liquids and the structure of their constituent molecules. Special attention is given to the unusual properties of water.

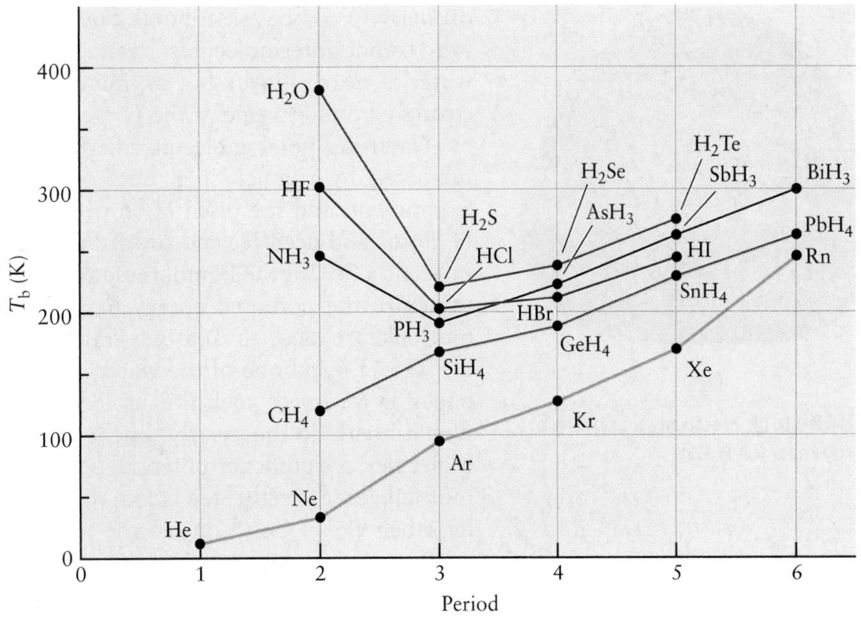

FIGURE 10.12 Trends in the boiling points of hydrides of some main-group elements and the noble gases.

Substances with strong attractive intermolecular forces tend to remain liquids at higher temperatures than those with weaker intermolecular forces; they have higher normal boiling points, T_b. Ionic liquids generally have the strongest attractions, because of the Coulomb interaction among charged ions, and thus have high boiling points. Molten NaCl, for example, boils at 1686 K under atmospheric pressure. At the opposite extreme, the boiling point of helium at 1 atm pressure is only 4.2 K. Within a series of related compounds, those of higher molar mass tend to have higher normal boiling points. This trend arises from the increased polarizability of the heavier compounds, not from the increased mass per se. Progressing from helium to xenon, normal boiling points increase (Fig. 10.12), as do the strengths of the attractive dispersion forces among the noble gases. These forces, represented by the well depth ε in Table 9.4, arise from the polarizability of the atoms.

Between the noble-gas and ionic liquids falls a class of liquids called **polar liquids**. In liquid HCl, the molecules arrange themselves to the greatest extent possible with neighboring dipoles oriented to minimize the dipole–dipole potential energy. As described in Section 10.2, the dipole–dipole intermolecular forces in such polar liquids are weaker than the ion–ion Coulombic forces in ionic liquids but stronger than the dispersion forces in **nonpolar liquids** such as N_2. These three forces operate respectively between molecules in which the bonding is polar covalent, fully ionic, and fully covalent. As shown in Figure 10.12, HCl has a higher boiling point than argon (a nonpolar fluid of atoms with nearly the same molar mass) because of its polar nature. The magnitudes of the intermolecular forces in HCl and argon are compared explicitly in Figure 10.9.

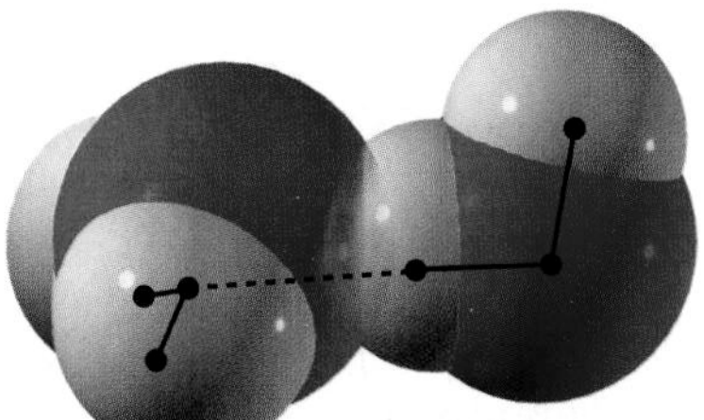

FIGURE 10.13 A single hydrogen bond between water molecules forms a dimer. This bond is far weaker than a covalent bond but still strong enough to resist dissociation at room temperature. The shared hydrogen (H) atom at the center approaches the neighboring oxygen (O) atom quite closely.

10.3.1 Hydrogen Bonds

Figure 10.12 shows the normal boiling points of several series of hydrides, in which the boiling point increases with increasing molar mass in a series of related compounds. The dramatic deviations from these systematic trends shown by HF, NH_3, and especially H_2O indicate the strength and importance of the special type of bond that is common to these cases, a **hydrogen bond**. Such a bond forms when an H atom bonded to an O, N, or F atom (highly electronegative atoms) also interacts with the lone electron pair of another such atom nearby. Figure 10.13 shows the interaction of a pair of water molecules to form a dimer in the gas phase. The hydrogen bond that forms is weaker than an

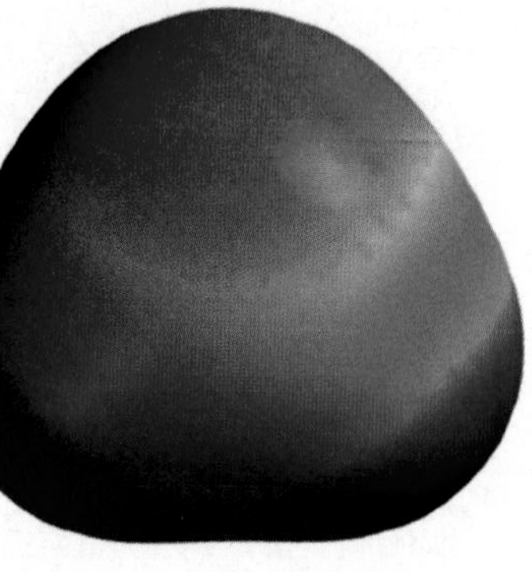

FIGURE 10.14 Electrostatic potential energy map for H_2O.

ordinary O—H covalent bond, but the interaction is significantly stronger than most other intermolecular interactions. Like most hydrogen bonds, that in water is nearly linear but asymmetric, with the H atom closer to and more strongly bound to one of the O atoms. It is indicated as O—H · · · O.

Water is a polar molecule, like HCl and H_2S. The water molecule is bent, as predicted by valence shell electron-pair repulsion theory and confirmed by experiment, and the orientation of its dipole moment (positive end toward the H atoms and negative end toward the O atom) has been related to its structure in Section 3.9. In the liquid, these molecules orient themselves in directions that minimize the potential energy between them; consequently, H atoms on one molecule are close to O atoms on neighboring molecules. In Figure 10.13, the H—O—H backbone of the water molecule on the right lies in the plane of the paper. If we try to rock this molecule in the plane of the paper in a way that would break up the (nearly) linear arrangement of O—H—O between the two molecules, a significant potential barrier opposes this movement. But the water molecule on the right side is free to rotate about the O—H—O bond and move its other O—H bond out of the plane of the paper to any other orientation consistent with the bent structure of the water molecule. The H atom in a bond such as O—H is surrounded by a relatively low density of negative charge (see Fig. 10.14) because, unlike all other elements, it has no electrons other than valence electrons. As a result, it can approach close to the lone-pair electrons on a neighboring O atom, causing a strong electrostatic (Coulomb) interaction between the two. In addition, a small amount of covalent bonding arises from the sharing of electrons between the two O atoms and the intervening H atom. These effects combine to make the interaction unusually strong. For the same reason, hydrogen bonds form with anions in aqueous solution (see Fig. 10.6).

10.3.2 Special Properties of Water

Water makes up about 0.023% of the total mass of the earth. About 1.4×10^{21} kg of it is distributed above, on, and below the earth's surface. The volume of this vast amount of water is about 1.4 billion km^3. Most of the earth's water (97.7%) is contained in the oceans, with about 1.9% in the form of ice or snow and most of the remainder (a small fraction of the total) available as freshwater in lakes, rivers, underground sources, and atmospheric water vapor. A small but important fraction is bound to cations in certain minerals, such as clays and hydrated crystalline salts. More than 80% of the surface of the earth is covered with water—as ice and snow near the poles, as relatively pure water in lakes and rivers, and as a salt solution in the oceans.

The unusual properties of water, which come from its network of hydrogen bonds, have profound effects on life on earth. Figure 10.12 compares the boiling points of water and hydrides that lack hydrogen bonds. An extrapolation of the trends from the latter compounds would give a boiling point for "water without hydrogen bonds" near 150 K (−123°C). Life as we know it would not be possible under these circumstances.

If all possible hydrogen bonds form in a mole (N_A molecules) of pure water, then every oxygen atom is surrounded by four H atoms in a tetrahedral arrangement: its own two and two from neighboring molecules. This tetrahedral arrangement forms a three-dimensional network with a structure similar to that of diamond or SiO_2. The result is an ordered array of six-membered rings of water molecules (Fig. 10.15) that manifests itself macroscopically in the characteristic sixfold symmetry of snowflakes.

The density of water reaches its maximum at 4°C (Fig. 10.16), and it expands on freezing. This unusual behavior, which is seen in few other liquids, also is caused by hydrogen bonds. When ice melts, some of the hydrogen bonds that maintain the open structure shown in Figure 10.15 break and the structure

FIGURE 10.15 The structure of ice is quite open. Each water molecule has only four nearest neighbors with which it interacts by means of hydrogen bonds (blue dashed lines).

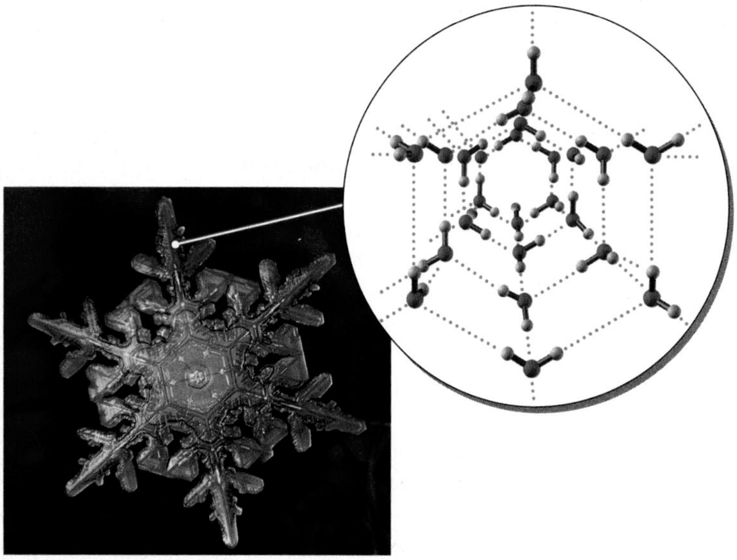

partially collapses, producing a liquid with a smaller volume (higher density). The reverse process, a sudden expansion of water on freezing, can cause bursting of water pipes and freeze/thaw cracking of concrete. Such expansion also has many beneficial effects. If ice were denser than water, the winter ice that forms at the surface of a lake would sink to the bottom and the lake would freeze from the bottom up. Instead, the ice remains at the surface, and the water near the bottom achieves a stable wintertime temperature near 4°C, which allows fish to survive.

EXAMPLE 10.2

Predict the order of increase in the normal boiling points of the following substances: F_2, HBr, NH_4Cl, and HF.

Solution

As an ionic substance, NH_4Cl should have the highest boiling point of the four (measured value: 520°C). HF should have a higher boiling point than HBr, because its molecules form hydrogen bonds (see Fig. 10.13) that are stronger than the dipolar interactions in HBr (measured values: 20°C for HF, −67°C for HBr). Fluorine, F_2, is nonpolar and contains light atoms, and thus should have the lowest boiling point of the four substances (measured value: −188°C).

Related Problems: 23, 24

FIGURE 10.16 The density of water rises to a maximum as it is cooled to 3.98°C, then starts to decrease slowly. Undercooled water (water chilled below its freezing point but not yet converted to ice) continues the smooth decrease in density. When liquid water freezes, the density drops abruptly.

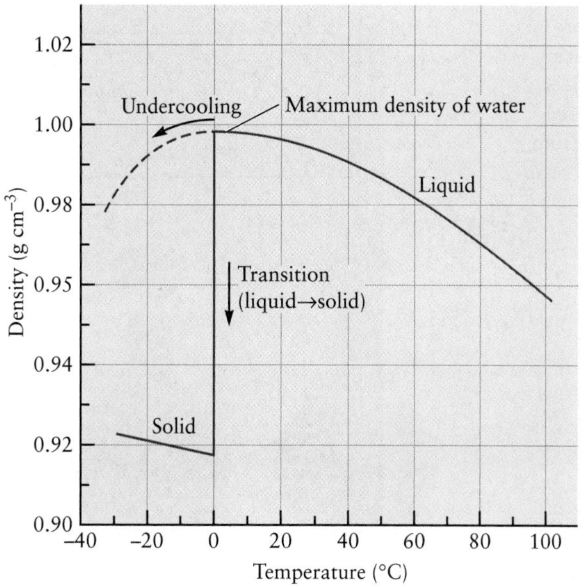

10.4 Phase Equilibrium

In the fall, if a warm day is followed by a chilly night, dew forms on grass as gaseous water molecules in the air condense to form droplets; H_2O has changed its **phase** from gaseous to liquid. Liquids and solids, like gases, are phases—samples of matter that are uniform throughout in both chemical constitution and physical state. Two or more phases can coexist. Suppose a small quantity of liquid water is put in an evacuated flask, with the temperature held at 25°C by placing the system in a constant-temperature bath. A pressure gauge is used to monitor changes in the pressure of water vapor inside the flask. Immediately after the water enters the flask, the pressure of water vapor begins to rise from zero. It increases with time and gradually levels off at a value of 0.03126 atm, which is the **vapor pressure** of water at 25°C. The contents of the flask have reached **equilibrium,** a condition in which no further changes in macroscopic properties occur as long as the system remains isolated. This passage toward equilibrium is a spontaneous process, occurring in a closed system without any external influence. If some of the water vapor that has formed is removed, additional water evaporates from the liquid to reestablish the same vapor pressure, $P_{vap}(H_2O) = 0.03126$ atm.

What is happening on a microscopic scale to cause this spontaneous movement of the system toward equilibrium? According to the kinetic theory, the molecules of water in the liquid are in a constant state of thermal motion. Some of those near the surface are moving fast enough to escape the attractive forces holding them in the liquid; this process of **evaporation** causes the pressure of the water vapor to increase. As the number of molecules in the vapor phase increases, the reverse process begins to occur: Molecules in the vapor strike the surface of the liquid, and some are captured, leading to **condensation.** As the pressure of the gas increases, the rate of condensation increases until it balances the rate of evaporation from the surface (Fig. 10.17). Once this occurs, there is no further net flow of matter from one phase to the other; the system has reached **phase equilibrium,** characterized by a particular value of the water vapor pressure. Water molecules continue to evaporate from the surface of the liquid,

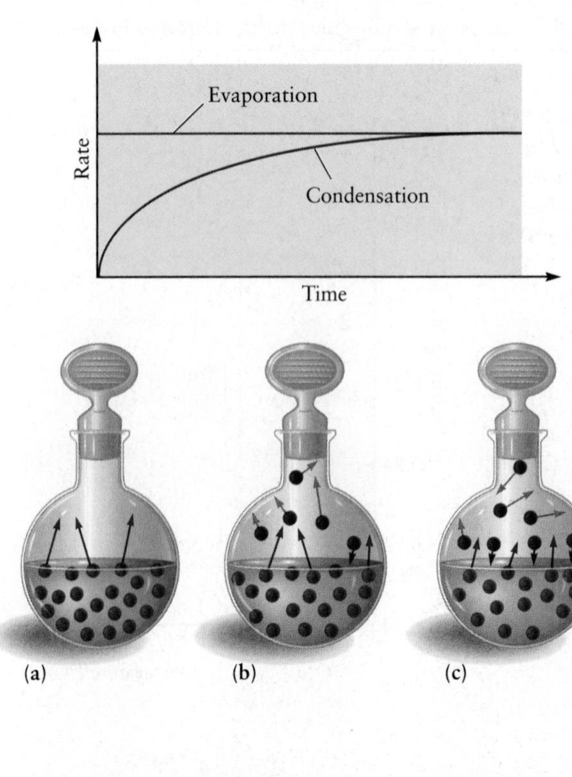

FIGURE 10.17 Approach to equilibrium in evaporation and condensation. Initially, the pressure above the liquid is very low, and many more molecules leave the liquid surface than return to it. As time passes, more molecules fill the gas phase until the equilibrium vapor pressure, P_{vap}, is approached; the rates of evaporation and condensation then become equal.

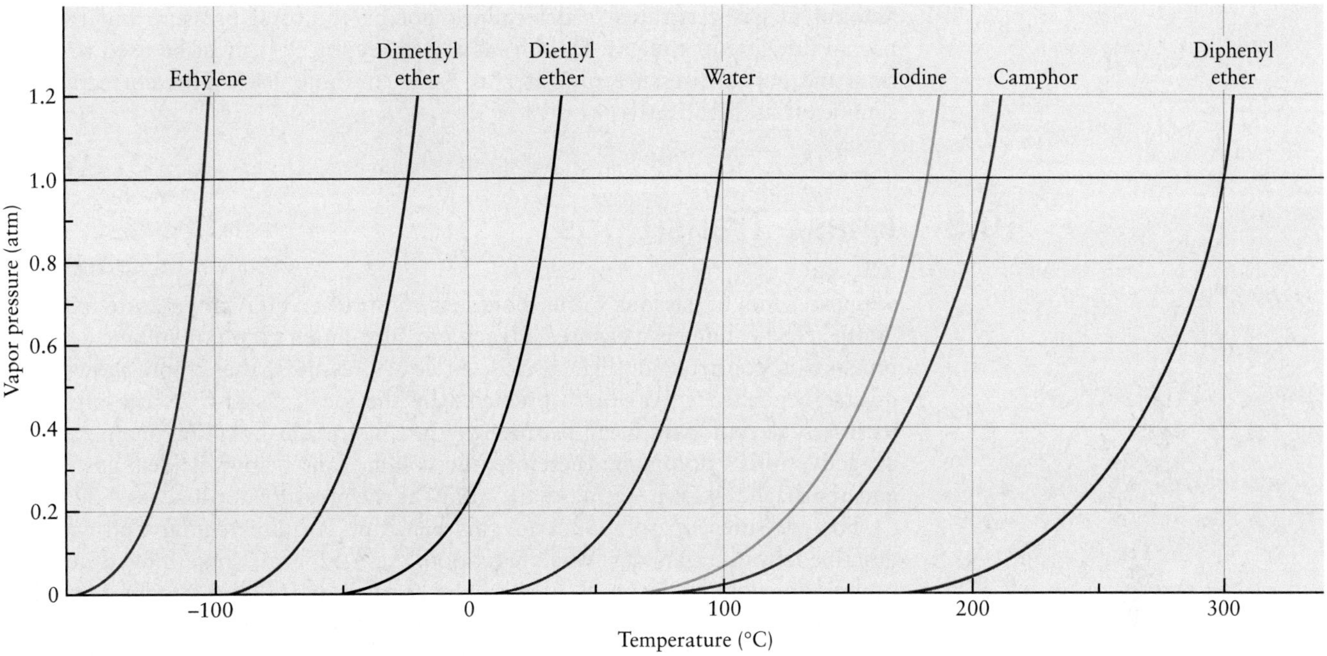

FIGURE 10.18 The vapor pressure of a solid or liquid depends strongly on temperature. The temperature at which the vapor pressure becomes 1 atm defines the normal boiling point of a liquid and the normal sublimation point of a solid.

TABLE 10.3

Vapor Pressure of Water at Various Temperatures

Temperature (°C)	Vapor Pressure (atm)
15.0	0.01683
17.0	0.01912
19.0	0.02168
21.0	0.02454
23.0	0.02772
25.0	0.03126
30.0	0.04187
50.0	0.12170

but other water molecules return to the liquid from the vapor at an equal rate. A similar phase equilibrium is established between an ice cube and liquid water at the freezing point.

The vapor pressure of the water is independent of the size and shape of the container. If the experiment is duplicated in a larger flask, then a greater *amount* of water evaporates on the way to equilibrium, but the final pressure in the flask at 25°C is still 0.03126 atm as long as some liquid water is present. If the experiment is repeated at a temperature of 30.0°C, everything happens as just described, except that the pressure in the space above the water reaches 0.04187 atm. A higher temperature corresponds to a larger average kinetic energy for the water molecules. A new balance between the rates of evaporation and condensation is struck, but at a higher vapor pressure. The vapor pressure of water, and of all other substances, increases with rising temperature (Fig. 10.18; Table 10.3).

Phase equilibrium is a *dynamic* process that is quite different from the static equilibrium achieved as a marble rolls to a stop after being spun into a bowl. In the equilibrium between liquid water and water vapor, the partial pressure levels off, not because evaporation and condensation stop, but because at equilibrium their rates become the same. The properties of a system at equilibrium are independent of the direction from which equilibrium is approached, a conclusion that can be drawn by observing the behavior of the liquid–vapor system. If we inject enough water vapor into the empty flask so that initially the pressure of the vapor is *above* the vapor pressure of liquid water, $P_{vap}(H_2O)$, then liquid water will condense until the same equilibrium vapor pressure is achieved (0.03126 atm at 25°C). Of course, if we do not use enough water vapor to exceed a pressure of 0.03126 atm, all the water will remain in the vapor phase, and two-phase equilibrium will not be reached.

The presence of water vapor above an aqueous solution has an important practical consequence. If a reaction in aqueous solution generates gases, these gases are "wet," containing water vapor at a partial pressure given by the equilibrium vapor pressure of water at the temperature of the experiment. The amount of gas generated is determined not by the total pressure but by the partial pressure of the gas. Dalton's law (see Section 9.4) must be used to subtract the partial pressure of water as listed in Table 10.3. This correction is significant in quantitative work.

10.5 Phase Transitions

Suppose 1 mol of gaseous sulfur dioxide is compressed at a temperature fixed at 30.0°C. The volume is measured at each pressure, and a graph of volume against pressure is constructed (Fig. 10.19). At low pressures, the graph shows the inverse dependence ($V \propto 1/P$) predicted by the ideal gas law. As the pressure increases, deviations appear because the gas is not ideal. At this temperature, attractive forces dominate; therefore, the volume falls below its ideal gas value and approaches 4.74 L (rather than 5.50 L) as the pressure approaches 4.52 atm.

For pressures up to 4.52 atm, this behavior is quite regular and can be described by the van der Waals equation. At 4.52 atm, something dramatic occurs: The volume decreases abruptly by a factor of 100 and remains small as the pressure is increased further. What has happened? The gas has been liquefied solely by the application of pressure. If the compression of SO_2 is continued, another abrupt (but small) change in volume will occur as the liquid freezes to form a solid.

Condensed phases also arise when the temperature of a gas is reduced at constant pressure. If steam (water vapor) is cooled at 1 atm pressure, it condenses to liquid water at 100°C and freezes to solid ice at 0°C. Liquids and solids form at low temperatures once the attractive forces between molecules become strong enough to overcome the kinetic energy of random thermal motion.

Six **phase transitions** occur among the three states of matter (Fig. 10.20). Solids typically melt to give liquids when they are heated, and liquids boil to give gases. Boiling is an extension of evaporation, in which the vapor escapes from the surface only. In boiling, gas bubbles form actively throughout the body of a liquid, and then rise to escape at the surface. Only when the vapor pressure of a liquid exceeds the external pressure can the liquid start to boil. The **boiling point** is the temperature at which the vapor pressure of a liquid equals the

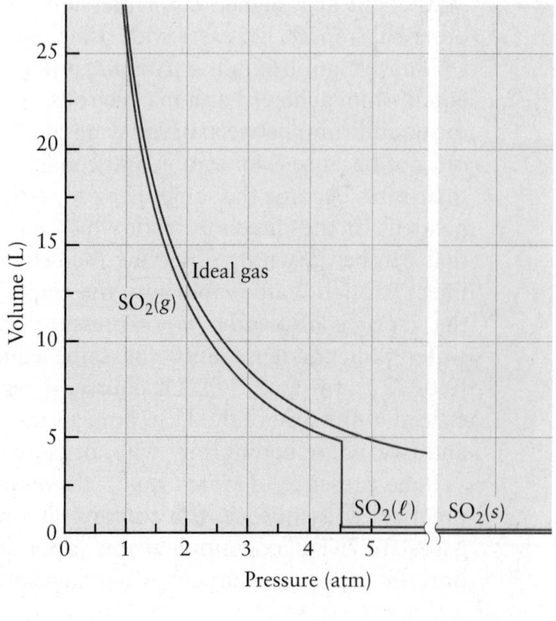

FIGURE 10.19 As 1 mol SO_2 is compressed at a constant temperature of 30°C, the volume at first falls somewhat below its ideal gas value. Then, at 4.52 atm, the volume decreases abruptly as the gas condenses to a liquid. At a much higher pressure, a further transition to the solid occurs.

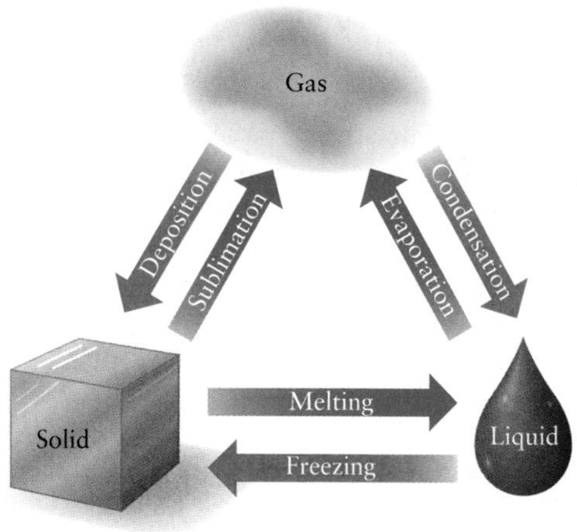

FIGURE 10.20 Direct transitions among all three states of matter not only are possible but are observed in everyday life.

external pressure. The external pressure influences boiling points quite strongly; water boils at 25°C if the external pressure is reduced below 0.03126 atm (recall that the vapor pressure of water at 25°C is just this number) but requires a temperature of 121°C to boil under an external pressure of 2.0 atm. At high elevations, the pressure of the atmosphere is lower than 1 atm; thus, water boils at a temperature less than 100°C and food cooks more slowly in boiling water than it would at a lower elevation. In contrast, the use of a pressure cooker increases the boiling temperature of water and speeds the rate at which food is cooked. The **normal boiling point** is defined as the temperature at which the vapor pressure of the liquid equals 1 atm. Figure 10.12 shows that, in general, a lower normal boiling point implies a higher vapor pressure at any fixed temperature (and, therefore, a more volatile liquid).

Melting is the conversion of a solid to the liquid state. The **normal melting point** of a solid is the temperature at which solid and liquid are in equilibrium under a pressure of 1 atm. The normal melting point of ice is 0.00°C, thus liquid water and ice coexist indefinitely (are in equilibrium) at this temperature at a pressure of 1 atm. If the temperature is reduced by even a small amount, then all the water eventually freezes; if the temperature is raised infinitesimally, all the ice eventually melts. The qualifying term *normal* is often omitted in talking about melting points because they depend only weakly on pressure.

It is sometimes possible to overshoot a phase transition, with the new phase appearing only after some delay. An example is the **superheating** of a liquid. Liquid water can reach a temperature somewhat above 100°C if heated rapidly. When vaporization of a superheated liquid does occur, it can be quite violent, with liquid thrown out of the container. Boiling chips (pieces of porous fired clay) may be added to the liquid to avoid this superheating in the laboratory. They help initiate boiling as soon as the normal boiling point is reached, by providing sites where gas bubbles can form. Heating water in a microwave oven in a very clean container can also lead to superheating, and there have been reports of violent boiling resulting in injury as the container is removed from the oven. You should carefully monitor heating times when using a microwave oven to heat water for making coffee or tea. **Supercooling** of liquids below their freezing points is also possible. In careful experiments, supercooled liquid water has been studied at temperatures below −30°C (at atmospheric pressure).

Many materials react chemically when heated, before they have a chance to melt or boil. Substances whose chemical identities change before their physical state changes do not have normal melting or boiling points. For example, sucrose (table sugar) melts but quickly begins to darken and eventually chars (Fig. 10.21). Temperatures high enough to overcome the intermolecular attractions in sugar are also sufficient to break apart the sugar molecules themselves.

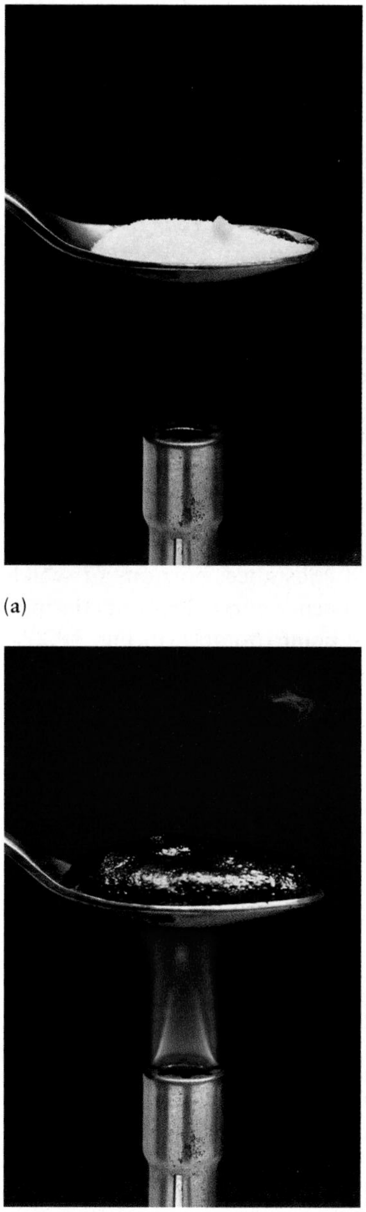

(a)

(b)

FIGURE 10.21 When sugar (a) is heated, it melts and simultaneously decomposes to a dark-colored caramelized mixture (b).

Intermolecular forces exert strong influences on phase transitions. Data presented in Section 10.3 illustrate the trend that the normal boiling point in a series of liquids increases as the strength of intermolecular forces in the liquids increases. The stronger the intermolecular attractions in a liquid, the lower its vapor pressure at any temperature and the higher its temperature must be raised to produce a vapor pressure equal to 1 atm. Melting points depend more strongly on molecular shapes and on the details of the molecular interactions than do boiling points. Consequently, their variation with the strength of the attractive forces is less systematic.

10.6 Phase Diagrams

If the temperature of a substance is held constant and the applied pressure is changed, phase transitions between two phases will be observed at particular pressures. Making the same measurements at a number of different temperatures provides the data necessary to draw the **phase diagram** for that substance—a plot of pressure against temperature that shows the stable state for every pressure–temperature combination. Figure 10.22 shows a sketch of the phase diagram for water. A great deal of information can be read from such diagrams. For each substance there is a unique combination of pressure and temperature, called the **triple point** (marked "T"), at which the gas, liquid, and solid phases coexist in equilibrium. Extending from the triple point are three lines, each denoting the conditions for the coexistence of two phases at equilibrium. Along the line TA, solid and gas are in equilibrium; along TB, solid and liquid; and along TC, liquid and gas. The regions bounded by these lines represent conditions where only one phase exists.

The gas–liquid coexistence curve extends upward in temperature and pressure from the triple point. This line, stretching from T to C in the phase diagrams, is the vapor pressure curve of the liquid substance, portions of which were shown in Figure 10.18. The gas–liquid coexistence curve does not continue indefinitely, but instead terminates at the **critical point** (point C in Fig. 10.22). Along this coexistence curve there is an abrupt, discontinuous change in the

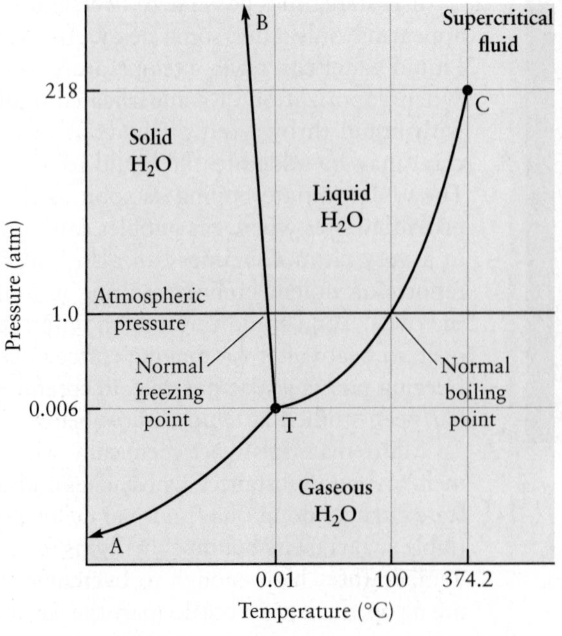

FIGURE 10.22 Phase diagram for water (the pressures and temperatures are not drawn to scale).

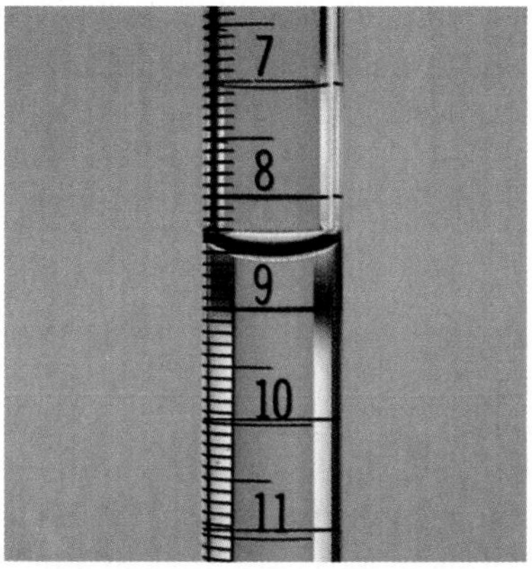

FIGURE 10.23 When a gas and liquid coexist, the interface between them is clearly visible as a meniscus. The meniscus is useful for reading the volume of a liquid in a buret. It disappears as the critical point is reached.

density and other properties from one side to the other. The differences between the properties of the liquid and the gas become smaller as the critical point is approached and disappear altogether at that point. If the substance is placed in a closed container and is gradually heated, a **meniscus** is observed at the boundary between liquid and gas (Fig. 10.23); at the critical point, this meniscus disappears. For pressures above the critical pressure (218 atm for water), it is no longer possible to identify a particular state as gas or liquid. A substance beyond its critical point is called a **supercritical fluid** because the term *fluid* includes both gases and liquids.

The liquid–solid coexistence curve does not terminate as the gas–liquid curve does at the critical point, but continues to indefinitely high pressures. In practice, such a curve is almost vertical because large changes in pressure are necessary to change the freezing temperature of a liquid. For most substances, this curve inclines slightly to the right (Figs. 10.24a, b): An increase in pressure increases the freezing point of the liquid. In other words, at constant temperature, an increase in pressure leads to the formation of a phase with higher density (smaller volume), and for most substances, the solid is denser than the liquid. Water and a few other substances are anomalous (see Fig. 10.24c); for them, the liquid–solid coexistence curve slopes up initially to the *left,* showing that an increase in pressure causes the solid to melt. This anomaly is related to the densities of the liquid and solid phases: Ice is less dense than water (which is why ice cubes float on water), so when ice is compressed at 0°C, it melts.

For most substances, including water (see Fig. 10.24c), atmospheric pressure occurs somewhere between the triple-point pressure and the critical pressure, so in our ordinary experience, all three phases—gas, liquid, and solid—are observed. For a few substances, the triple-point pressure lies *above* $P = 1$ atm, and under atmospheric conditions, there is a direct transition called **sublimation**

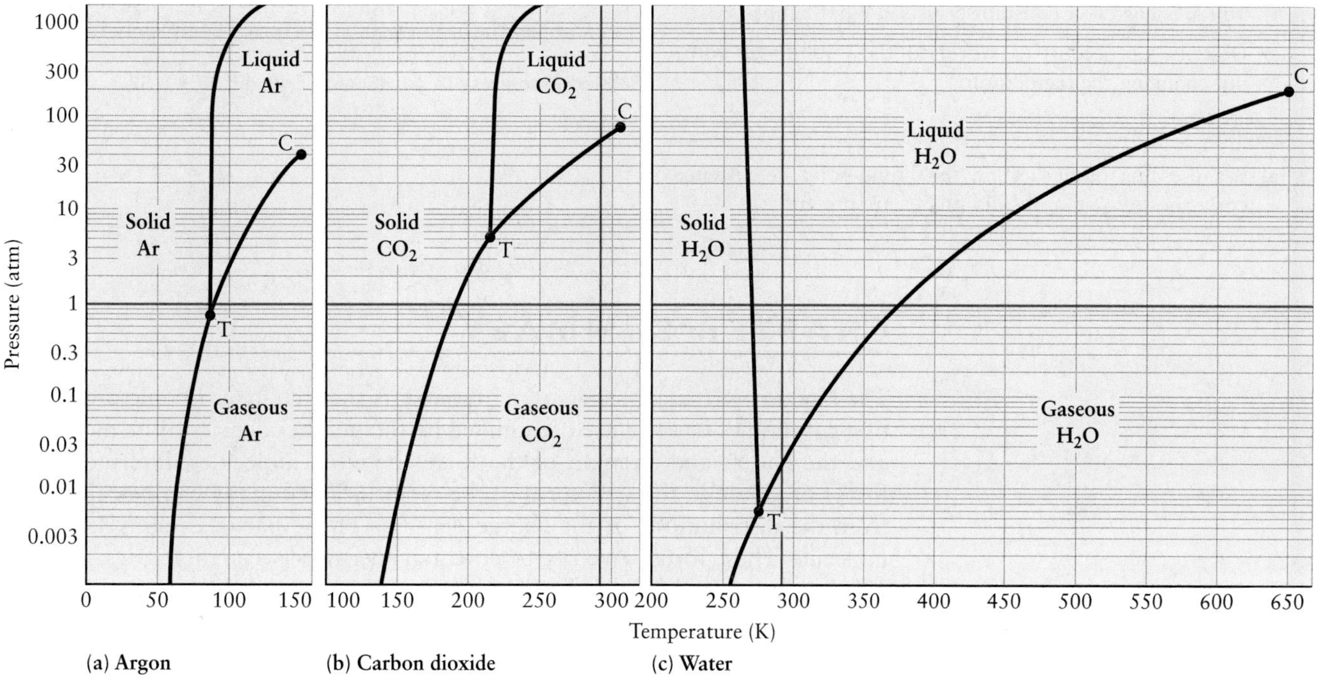

FIGURE 10.24 In these phase diagrams, the pressure increases by a factor of 10 at regular intervals along the vertical axis. This method of graphing allows large ranges of pressure to be plotted. The red horizontal and vertical lines mark a pressure of 1 atm and a temperature of 298.15 K, or 25°C. Their intersection identifies room conditions. Argon and carbon dioxide are gases at room conditions, but water is a liquid. The letter T marks the triple points of the substances, and the letter C marks their critical points. The region of stability of liquid water is larger than that of either carbon dioxide or argon.

FIGURE 10.25 Sublimation of solid carbon dioxide (dry ice). The white clouds are drops of water vapor (moisture in the air) that condense at the low temperatures near the solid surface. Gaseous carbon dioxide itself is transparent.

from solid to gas, without an intermediate liquid state. Carbon dioxide is such a substance (see Fig. 10.24b); its triple-point pressure is 5.117 atm (the triple-point temperature is −56.57°C). Solid CO_2 (dry ice) sublimes directly to gaseous CO_2 at atmospheric pressure (see Fig. 10.25). In this respect, it differs from ordinary ice, which melts before it evaporates and sublimes only at pressures below its triple-point pressure, 0.0060 atm. This fact is used in freeze-drying, a process in which foods are frozen and then put in a vacuum chamber at a pressure of less than 0.0060 atm. The ice crystals that formed on freezing then sublime, leaving a dried food that can be reconstituted by adding water.

Many substances exhibit more than one solid phase as the temperature and pressure are varied. At ordinary pressures, the most stable state of carbon is graphite, a rather soft, black solid; but at high enough pressures, the hard, transparent diamond form becomes more stable. That diamonds exist at all at atmospheric pressure is a consequence of how slowly they convert to graphite (we study the graphite–diamond case more thoroughly in our discussions of thermodynamics and spontaneous processes in Chapter 14). Below 13.2°C (and at atmospheric pressure), elemental tin undergoes a slow transformation from the metallic white form to a powdery gray form, a process referred to as "tin disease." This has caused the destruction of tin organ pipes in unheated buildings. No fewer than nine solid forms of ice are known, some of which exist only over a limited range of temperatures and pressures.

EXAMPLE 10.3

Consider a sample of argon held at $P = 600$ atm, $T = 100$ K in Figure 10.24a.

(a) What phase(s) is (are) present at equilibrium?

(b) Suppose the argon originally held at $P = 600$ atm, $T = 100$ K is heated at constant pressure. What happens?

(c) Describe a procedure to convert all the argon to gas without changing the temperature.

Solution

(a) Because this point lies on the liquid–solid coexistence curve, both liquid and solid phases are present.

(b) An increase in temperature at constant pressure corresponds to a movement to the right in the figure, and the solid argon melts to a pure liquid.

(c) If the pressure is reduced sufficiently at constant temperature (below the triple-point pressure of 0.75 atm, for example) the argon will be converted completely to gas.

Related Problems: 43, 44, 45, 46, 47, 48, 49, 50, 51, 52

CHAPTER SUMMARY

The kinetic molecular theory of matter asserts that the macroscopic properties of a gas, liquid, or solid are determined by the number density of its molecules and the nature and strength of the forces between molecules. Intermolecular forces originate in the structures of the molecules and many can be calculated from the distributions of the charge densities of the particles comprising the molecule. These forces give rise to potential energy between molecules, in magnitudes determined by the distance between the molecules. The repulsive and attractive forces are reflected in the potential energy functions, and their relative influence shown for each value of intermolecular separation. The same intermolecular forces that cause gas imperfection lead to the formation of liquids and solids. The three phases or states of matter can coexist in equilibrium. On the microscopic level, phase equilibrium is a dynamical balance in which each phase gains and loses molecules from the other phase at the same rate.

CONCEPTS AND SKILLS

Section 10.1 – Bulk Properties of Gases, Liquids, and Solids: Molecular Interpretation

Relate trends in values of bulk properties to the strength and range of intermolecular forces (Problems 1–12).

The bulk properties of gases, liquids, and solids are defined by their methods of measurement. Their magnitudes depend on the structure of the molecules, the forces between the molecules, and the average distance between molecules.

- Molar volume: the volume per mole of a substance
  - Ideal gases at standard temperature and pressure: $V_m = 22.4 \text{ L mol}^{-1}$
  - Liquids and solids: V_m = molar mass/density, typically 10–100 $\text{cm}^3 \text{ mol}^{-1}$
- Compressibility: fractional decrease in volume per unit increase in pressure
  - Large for gases
  - Small for liquids and solids
- Thermal expansion: fractional increase in volume per 1 K rise in temperature
  - Same value for all gases at a given temperature
  - Typical values for solids and liquids are of the order $\Delta V/V = 10^{-3}/\text{K}$
- Fluidity and rigidity: response to externally applied force (stress)
  - Fluids flow, in amount determined by coefficient of viscosity
  - Solids do not flow but show rigidity and mechanical strength
- Diffusion: the rate at which molecules of one substance move through the bulk of another substance.
  - Quite slow in solids, faster in liquids, rapid in gases
- Surface tension: the energy required to increase the surface area of a liquid. Materials with high surface tensions readily form spherical drops, which have the smallest possible surface area-to-volume ratio.

Section 10.2 – Intermolecular Forces: Origins in Molecular Structure

Relate magnitudes and distance dependence of intermolecular forces to the structure of molecules (Problems 13–20).

- Intermolecular forces are determined by the structure of the molecules involved. Except for van der Waals forces and hydrogen bonding, all intermolecular forces are *electrostatic.* They originate in the Coulomb interactions among all the charged particles in the molecules, and their magnitudes can be calculated from Coulomb's law when we take proper account of the molecular structures.
  - Ionic forces are the strongest and act over the longest distances (range). Their magnitude is easily calculated from Coulomb's law. They are the dominant force in ionic compounds, which most commonly appear as solids but may be gases or molten salts.
  - Ion–dipole forces act between ions and molecules with permanent dipole moments and are next strongest after ionic forces. They are relatively long-range and are important in ionic solutions.
  - Dipole–dipole forces act between neutral molecules that have permanent dipole moments. They are relatively weak and short-range; the potential energy between two dipoles falls off as $1/R^3$.

- Induced forces arise when the electron density of a neutral, nonpolar molecule is distorted or polarized by nearby charges to create a temporary dipole in the nonpolar molecule.
  - Induced dipole forces arise when an ion induces a dipole in a nonpolar atom or molecule and is then attracted to the opposite charge induced.
  - Induced dipole–induced dipole forces arise between neutral nonpolar atoms or molecules and are the only source of attractive forces in substances such as Ar.
- Repulsive forces arise when two atoms or molecules are so close together that their respective core electrons repel each other, overwhelming the attractive forces.

Section 10.3 – Intermolecular Forces in Liquids

Describe the effects of different kinds of intermolecular forces on properties of liquids (Problems 21–30).

- Intermolecular forces in liquids in general are described by the categories discussed above. In addition, hydrogen bonding occurs in substances where H is covalently bonded to N, O, or F.
  - Trends in the boiling points of hydrides reveal the special nature of intermolecular forces when H is bonded to N, O, or F. The attractive forces are much stronger than normal dipole–dipole forces, and the orientation of the bond is linear, with the hydrogen located between the two heavy atoms. Hydrogen bonds are 2 to 5 times stronger than dipole–dipole interactions.
- Water has special properties that arise from hydrogen bonding.
  - Boiling point is about 150 K higher than would occur without H-bonds.
  - Solid is less dense than liquid due to a symmetrical, open, H-bonded network.

Section 10.4 – Phase Equilibrium

Discuss the evidence that phase equilibrium is a dynamic process at the molecular level (Problems 31–38).

- Matter is organized into phases, which may be in equilibrium with each other.
  - Liquids, solids, and gases are the three normal phases of matter.
  - Supercritical fluids and plasmas (ionized gases) are states of matter that appear under specialized conditions and have exotic properties.
  - Two or more phases can coexist in equilibrium under specified conditions of temperature and pressure.
  - Phase equilibria are dynamic events; molecules are constantly shuttling back and forth between phases but with equal rates so that no macroscopic changes are observed.

Section 10.5 – Phase Transitions

Describe the effects of different kinds of intermolecular forces on phase transitions (Problems 39–42).

- Transitions between phases include: evaporation and condensation (liquid to gas and reverse), melting and freezing (solid to liquid and reverse) and sublimation and deposition (solid to gas and reverse).
- When pressure is high enough and temperature is low enough, molecules are close together, in the range where attractive forces are strongest, so gases condense to form liquids.

- When pressure is further increased and temperature is further decreased, the intermolecular attractions become even stronger, the molecules become closely packed, and liquids condense to form solids.

Section 10.6 – Phase Diagrams

Sketch the pressure–temperature phase diagram for a typical substance and identify the lines, areas, and singular points (Problems 43–52).

- Phase diagrams describe the phases of a substance that exist under various combinations of temperature and pressure in a graphical representation.
- Pressure is plotted on the y-axis and temperature along the x-axis.
- Pressure and temperature conditions under which two phases coexist lie along the *coexistence curves* that divide the two phases.
- The one condition under which three phases can coexist is marked by a single point, the *triple point,* T.
- Above a certain temperature and pressure, marked by the *critical point,* a single *supercritical fluid* exists; its properties are markedly different from normal gases or liquids.

CUMULATIVE EXERCISE

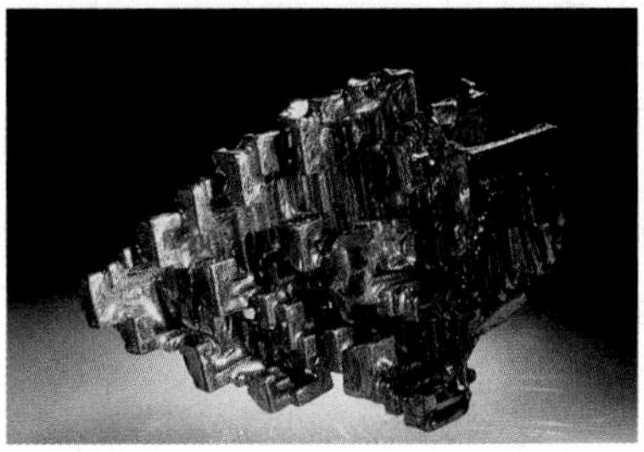

FIGURE 10.26 Elemental bismuth.

Alloys of Bismuth and their Applications

Bismuth (see Fig. 10.26) is a rather rare element in the earth's crust, but its oxides and sulfides appear at sufficient concentrations as impurities in lead and copper ores to make its recovery from these sources practical. Annual production of bismuth amounts to several million kilograms worldwide. Although elemental bismuth is a metal, its electrical conductivity is quite poor and it is relatively brittle. The major uses of bismuth arise from its low melting point (271.3°C) and the even lower melting points of its alloys, which range down to 47°C. These alloys are used as temperature sensors in fire detectors and automatic sprinkler systems because, in case of fire, they melt at a well-defined temperature, breaking an electrical connection and triggering an alarm or deluge.

(a) At its normal melting point, the density of solid bismuth is 9.73 g cm^{-3} and that of liquid bismuth is 10.05 g cm^{-3}. Does the volume of a sample of bismuth increase or decrease on melting? Does bismuth more closely resemble water or argon (see Fig. 10.24) in this regard?

(b) Since 1450 (10 years after Gutenberg), bismuth alloys have been used to cast metal type for printing. Explain why those alloys that share the melting behavior of bismuth discussed in part (a) would be especially useful for this application.

(c) A sample of solid bismuth is held at a temperature of 271.0°C and compressed. What will be observed?

(d) The vapor pressure of liquid bismuth has been measured to be 5.0 atm at a temperature of 1850°C. Does its normal boiling point lie above or below this temperature?

(e) At 1060°C, the vapor pressure of liquid bismuth is 0.013 atm. Calculate the number of bismuth atoms per cubic centimeter at equilibrium in the vapor above liquid bismuth at this temperature.

(f) The normal boiling point of liquid tin is 2270°C. Do you predict that liquid tin will be more volatile or less volatile than liquid bismuth at 1060°C?

(g) Bismuth forms two fluorides: BiF_3 and BiF_5. As is usually the case, the compound with the metal in the lower oxidation state has more ionic character, whereas that with the metal in the higher oxidation state has more covalent (molecular) character. Predict which bismuth fluoride will have the higher boiling point.

(h) Will AsF_5 have a higher or a lower normal boiling point than BiF_5?

Answers

(a) Bismuth resembles water in that its volume decreases on melting.

(b) The volume of bismuth increases on freezing; therefore, as the liquid alloy is cast, it fits tightly into its mold rather than shrinking away from the mold as most other metals do. This gives a more sharply defined metal type.

(c) The bismuth will melt.

(d) Its normal boiling point lies below this temperature.

(e) 7.2×10^{16} atoms cm^{-3}

(f) Liquid tin will be less volatile.

(g) BiF_3 will have the higher boiling point; in fact, BiF_3 boils at 900°C and BiF_5 boils at 230°C.

(h) It will have a lower normal boiling point.

PROBLEMS

Answers to problems whose numbers are red appear in Appendix G. Problems that are more challenging are indicated with asterisks.

10.1 Bulk Properties of Gases, Liquids, and Solids: Molecular Interpretation

1. A substance is nearly nonviscous and quite compressible, and it has a large coefficient of thermal expansion. Is it most likely to be a solid, a liquid, or a gas?

2. A substance is viscous, nearly incompressible, and not elastic. Is it most likely to be a solid, a liquid, or a gas?

3. A sample of volume 258 cm^3 has a mass of 2.71 kg.
   (a) Is the material gaseous or condensed?
   (b) If the molar mass of the material is 108 g mol^{-1}, calculate its molar volume.

4. A sample of volume 18.3 L has a mass of 57.9 g.
   (a) Is the material gaseous or condensed?
   (b) If the molar mass of the material is 123 g mol^{-1}, calculate its molar volume.

5. Heating a sample of matter from 20°C to 40°C at constant pressure causes its volume to increase from 546.0 to 547.6 cm^3. Classify the material as a nearly ideal gas, a nonideal gas, or condensed.

6. Cooling a sample of matter from 70°C to 10°C at constant pressure causes its volume to decrease from 873.6 to 712.6 cm^3. Classify the material as a nearly ideal gas, a nonideal gas, or condensed.

7. At 1.00 atm pressure and a temperature of 25°C, the volume of 1.0 g water is 1.0 mL. At the same pressure and a temperature of 101°C, the volume of 1.0 g water is nearly 1700 times larger. Give the reason for this large change in volume.

8. Doubling the absolute temperature of a gas essentially doubles its volume at constant pressure. Doubling the temperature of many metals, however, often increases their volumes by only a few percent. Explain.

9. Will solid sodium chloride be harder (that is, more resistant to indentation) or softer than solid carbon tetrachloride? Explain.

10. Will the surface tension of molten sodium chloride be higher than or lower than that of carbon tetrachloride? Explain.

11. Do you expect that the diffusion constant will increase or decrease as the density of a liquid is increased (by compressing it) at constant temperature? Explain. What will happen to the diffusion constant of a gas and a solid as the density increases?

12. Do you anticipate that the diffusion constant will increase as the temperature of a liquid increases at constant pressure? Why or why not? Will the diffusion constant increase with temperature for a gas and a solid? Explain.

10.2 Intermolecular Forces: Origins in Molecular Structure

13. Compare ion–dipole forces with induced dipole forces. In what ways are they similar and different? Give an example of each.

14. Compare dipole–dipole forces with dispersion forces. In what ways are they similar and different? Give an example of each.

15. Name the types of attractive forces that will contribute to the interactions among atoms, molecules, or ions in the following substances. Indicate the one(s) you expect to predominate.
   (a) KF **(b)** HI **(c)** Rn **(d)** N_2

16. Name the types of attractive forces that will contribute to the interactions among atoms, molecules, or ions in the following substances. Indicate the one(s) you expect to predominate.
(a) Ne **(b)** ClF **(c)** F_2 **(d)** $BaCl_2$

17. Predict whether a sodium ion will be most strongly attracted to a bromide ion, a molecule of hydrogen bromide, or an atom of krypton.

18. Predict whether an atom of argon will be most strongly attracted to another atom of argon, an atom of neon, or an atom of krypton.

19. **(a)** Use Figure 10.9 to estimate the length of the covalent bond in Cl_2 and the length of the ionic bond in K^+Cl^-. **Note:** The latter corresponds to the distance between the atoms in an isolated single molecule of K^+Cl^-, not in KCl(*s*) (solid potassium chloride).
(b) A book states, "The shorter the bond, the stronger the bond." What features of Figure 10.9 show that this is not always true?

20. True or false: Any two atoms held together by nonbonded attractions must be farther apart than any two atoms held together by a chemical bond. Explain.

10.3 Intermolecular Forces in Liquids

21. Under room conditions, fluorine and chlorine are gases, bromine is a liquid, and iodine is a solid. Explain the origin of this trend in the physical state of the halogens.

22. The later halogens form pentafluorides: ClF_5, BrF_5, and IF_5. At 0°C, one of these is a solid, one a liquid, and one a gas. Specify which is which, and explain your reasoning.

23. List the following substances in order of increasing normal boiling points, T_b, and explain your reasoning: NO, NH_3, Ne, RbCl.

24. List the following substances in order of increasing normal boiling points, T_b, and explain your reasoning: SO_2, He, HF, CaF_2, Ar.

25. As a vapor, methanol exists to an extent as a tetramer, $(CH_3OH)_4$, in which four CH_3OH molecules are held together by hydrogen bonds. Propose a reasonable structure for this tetramer.

26. Hypofluorous acid (HOF) is the simplest possible compound that allows comparison between fluorine and oxygen in their abilities to form hydrogen bonds. Although F attracts electrons more strongly than O, solid HOF unexpectedly contains no H · · · F hydrogen bonds! Draw a proposed structure for chains of HOF molecules in the crystalline state. The bond angle in HOF is 101 degrees.

27. Hydrazine (N_2H_4) is used as a reducing agent and in the manufacture of rocket fuels. How do you expect its boiling point to compare with that of ethylene (C_2H_4)?

28. Hydrogen peroxide (H_2O_2) is a major industrial chemical that is produced on a scale approaching 10^9 kg per year. It is widely used as a bleach and in chemical manufacturing processes. How do you expect its boiling point to compare with those of fluorine (F_2) and hydrogen sulfide (H_2S), two substances with molar masses comparable with that of hydrogen peroxide?

29. A flask contains 1.0 L (1.0 kg) of room-temperature water. Calculate the number of possible hydrogen bonds among the water molecules present in this sample. Each water molecule can accept two hydrogen bonds and also furnish the H atoms for two hydrogen bonds.

30. What is the maximum number of hydrogen bonds that can be formed in a sample containing 1.0 mol (N_A molecules) liquid HF? Compare with the maximum number that can form in 1.0 mol liquid water.

10.4 Phase Equilibrium

31. Hydrogen at a pressure of 1 atm condenses to a liquid at 20.3 K and solidifies at 14.0 K. The vapor pressure of liquid hydrogen is 0.213 atm at 16.0 K. Calculate the volume of 1.00 mol H_2 vapor under these conditions and compare it with the volume of 1.00 mol H_2 at standard temperature and pressure.

32. Helium condenses to a liquid at 4.224 K under atmospheric pressure and remains a liquid down to the absolute zero of temperature. (It is used as a coolant to reach very low temperatures.) The vapor pressure of liquid helium at 2.20 K is 0.05256 atm. Calculate the volume occupied by 1.000 mol helium vapor under these conditions and compare it with the volume of the same amount of helium at standard temperature and pressure.

33. The vapor pressure of liquid mercury at 27°C is 2.87×10^{-6} atm. Calculate the number of Hg atoms per cubic centimeter in the "empty" space above the top of the column of mercury in a barometer at 27°C.

34. The tungsten filament in an incandescent lightbulb ordinarily operates at a temperature of about 2500°C. At this temperature, the vapor pressure of solid tungsten is 7.0×10^{-9} atm. Estimate the number of gaseous tungsten atoms per cubic centimeter under these conditions.

35. Calcium carbide reacts with water to produce acetylene (C_2H_2) and calcium hydroxide. The acetylene is collected over water at 40.0°C under a total pressure of 0.9950 atm. The vapor pressure of water at this temperature is 0.0728 atm. Calculate the mass of acetylene per liter of "wet" acetylene collected in this way, assuming ideal gas behavior.

36. A metal reacts with aqueous hydrochloric acid to produce hydrogen. The hydrogen (H_2) is collected over water at 25°C under a total pressure of 0.9900 atm. The vapor pressure of water at this temperature is 0.0313 atm. Calculate the mass of hydrogen per liter of "wet" hydrogen above the water, assuming ideal gas behavior.

37. Carbon dioxide is liberated by the reaction of aqueous hydrochloric acid with calcium carbonate:

$$CaCO_3(s) + 2\,H^+(aq) \longrightarrow Ca^{2+}(aq) + CO_2(g) + H_2O(\ell)$$

A volume of 722 mL $CO_2(g)$ is collected over water at 20°C and a total pressure of 0.9963 atm. At this temperature, water has a vapor pressure of 0.0231 atm. Calculate the mass of calcium carbonate that has reacted, assuming no losses of carbon dioxide.

38. When an excess of sodium hydroxide is added to an aqueous solution of ammonium chloride, gaseous ammonia is produced:

$$NaOH(aq) + NH_4Cl(aq) \longrightarrow NaCl(aq) + NH_3(g) + H_2O(\ell)$$

Suppose 3.68 g ammonium chloride reacts in this way at 30°C and a total pressure of 0.9884 atm. At this temperature, the vapor pressure of water is 0.0419 atm. Calculate the volume of ammonia saturated with water vapor that will be produced under these conditions, assuming no leaks or other losses of gas.

10.5 Phase Transitions

39. High in the Andes, an explorer notes that the water for tea is boiling vigorously at a temperature of 90°C. Use Figure 10.18 to estimate the atmospheric pressure at the altitude of the camp. What fraction of the earth's atmosphere lies below the level of the explorer's camp?

40. The total pressure in a pressure cooker filled with water increases to 4.0 atm when it is heated, and this pressure is maintained by the periodic operation of a relief valve. Use Figure 10.24c to estimate the temperature of the water in the pressure cooker.

41. Iridium melts at a temperature of 2410°C and boils at 4130°C, whereas sodium melts at a temperature of 97.8°C and boils at 904°C. Predict which of the two molten metals has the larger surface tension at its melting point. Explain your prediction.

42. Aluminum melts at a temperature of 660°C and boils at 2470°C, whereas thallium melts at a temperature of 304°C and boils at 1460°C. Which metal will be more volatile at room temperature?

10.6 Phase Diagrams

43. At its melting point (624°C), the density of solid plutonium is 16.24 g cm^{-3}. The density of liquid plutonium is 16.66 g cm^{-3}. A small sample of liquid plutonium at 625°C is strongly compressed. Predict what phase changes, if any, will occur.

44. Phase changes occur between different solid forms, as well as from solid to liquid, liquid to gas, and solid to gas. When white tin at 1.00 atm is cooled below 13.2°C, it spontaneously changes (over a period of weeks) to gray tin. The density of gray tin is *less* than the density of white tin (5.75 g cm^{-3} vs 7.31 g cm^{-3}). Some white tin is compressed to a pressure of 2.00 atm. At this pressure, should the temperature be higher or lower than 13.2°C for the conversion to gray tin to occur? Explain your reasoning.

45. The following table gives several important points on the pressure–temperature diagram of ammonia:

	P (atm)	*T* (K)
Triple point	0.05997	195.42
Critical point	111.5	405.38
Normal boiling point	1.0	239.8
Normal melting point	1.0	195.45

Use this information to sketch the phase diagram of ammonia.

46. The following table gives several important points on the pressure–temperature diagram of nitrogen:

	P (atm)	*T* (K)
Triple point	0.123	63.15
Critical point	33.3978	126.19
Normal boiling point	1.0	77.35
Normal melting point	1.0	63.15

Use this information to sketch the phase diagram of nitrogen. The density of $N_2(s)$ is 1.03 g cm^{-3} and that of $N_2(\ell)$ is 0.808 g cm^{-3}.

47. Determine whether argon is a solid, a liquid, or a gas at each of the following combinations of temperature and pressure (use Fig. 10.24).
(a) 50 atm and 100 K
(b) 8 atm and 150 K
(c) 1.5 atm and 25 K
(d) 0.25 atm and 120 K

48. Some water starts out at a temperature of 298 K and a pressure of 1 atm. It is compressed to 500 atm at constant temperature, and then heated to 750 K at constant pressure. Next, it is decompressed at 750 K back to 1 atm and finally cooled to 400 K at constant pressure.
(a) What was the state (solid, liquid, or gas) of the water at the start of the experiment?
(b) What is the state (solid, liquid, or gas) of the water at the end of the experiment?
(c) Did any phase transitions occur during the four steps described? If so, at what temperature and pressure did they occur? (*Hint:* Trace out the various changes on the phase diagram of water [see Fig. 10.24].)

49. The vapor pressure of solid acetylene at −84.0°C is 760 torr.
(a) Does the triple-point temperature lie above or below −84.0°C? Explain.
(b) Suppose a sample of solid acetylene is held under an external pressure of 0.80 atm and heated from 10 to 300 K. What phase change(s), if any, will occur?

50. The triple point of hydrogen occurs at a temperature of 13.8 K and a pressure of 0.069 atm.
(a) What is the vapor pressure of solid hydrogen at 13.8 K?
(b) Suppose a sample of solid hydrogen is held under an external pressure of 0.030 atm and heated from 5 to 300 K. What phase change(s), if any, will occur?

51. The density of nitrogen at its critical point is 0.3131 g cm^{-3}. At a very low temperature, 0.3131 g solid nitrogen is sealed into a thick-walled glass tube with a volume of 1.000 cm^3. Describe what happens inside the tube as the tube is warmed past the critical temperature, 126.19 K.

52. At its critical point, ammonia has a density of 0.235 g cm^{-3}. You have a special thick-walled glass tube that has a 10.0-mm outside diameter, a wall thickness of 4.20 mm, and a length of 155 mm. How much ammonia must you seal into the tube if you wish to observe the disappearance of the meniscus as you heat the tube and its contents to a temperature higher than 132.23°C, the critical temperature?

ADDITIONAL PROBLEMS

53. Would you classify candle wax as a solid or a liquid? What about rubber? Discuss.

***54.** When a particle diffuses, its *mean-square displacement* in a time interval Δt is $6D\Delta t$, where D is the diffusion constant. Its *root-mean-square displacement* is the square root of this (recall the analogous root-mean-square speed from Section 9.5). Calculate the root-mean-square displacement at 25°C after 1.00 hour of (a) an oxygen molecule in air ($D = 2.1 \times 10^{-5}$ m^2 s^{-1}), (b) a molecule in liquid water ($D = 2.26 \times 10^{-9}$ m^2 s^{-1}), and (c) an atom in solid sodium ($D = 5.8 \times 10^{-13}$ m^2 s^{-1}). Note that for solids with melting points higher than sodium, the diffusion constant can be many orders of magnitude smaller.

55. Liquid hydrogen chloride will dissolve many ionic compounds. Diagram how molecules of hydrogen chloride tend to distribute themselves about a negative ion and about a positive ion in such solutions.

***56.** Section 9.6 explains that the van der Waals constant b (with units of L mol^{-1}) is related to the volume per molecule in the liquid, and thus to the sizes of the molecules. The combination of van der Waals constants a/b has units of L atm mol^{-1}. Because the liter atmosphere is a unit of energy (1 L atm = 101.325 J), a/b is proportional to the energy per mole for interacting molecules, and thus to the strength of the attractive forces between molecules, as shown in Figure 10.9. By using the van der Waals constants in Table 9.3, rank the following attractive forces from strongest to weakest: N_2, H_2, SO_2, and HCl.

57. Describe how the average kinetic and potential energies per mole change as a sample of water is heated from 10 to 1000 K at a constant pressure of 1 atm.

58. As a sample of water is heated from 0.0°C to 4.0°C, its density increases from 0.99987 to 1.00000 g cm^{-3}. What can you conclude about the coefficient of thermal expansion of water in this temperature range? Is water unusual in its behavior? Explain.

59. At 25°C, the equilibrium vapor pressure of water is 0.03126 atm. A humidifier is placed in a room of volume 110 m^3 and is operated until the air becomes saturated with water vapor. Assuming that no water vapor leaks out of the room and that initially there was no water vapor in the air, calculate the number of grams of water that have passed into the air.

60. The text states that at 1000°C, the vapor pressure of tungsten is 2×10^{-25} atm. Calculate the volume occupied per tungsten *atom* in the vapor under these conditions.

61. You boil a small quantity of water inside a 5.0-L metal can. When the can is filled with water vapor (and all air has been expelled), you quickly seal the can with a screw-on cap and remove it from the source of heat. It cools to 60°C and most of the steam condenses to liquid water. Determine the pressure inside the can. (*Hint:* Refer to Fig. 10.18.)

***62.** The air over an unknown liquid is saturated with the vapor of that liquid at 25°C and a total pressure of 0.980 atm. Suppose that a sample of 6.00 L of the saturated air is collected and the vapor of the unknown liquid is removed from that sample by cooling and condensation. The pure air remaining occupies a volume of 3.75 L at −50°C and 1.000 atm. Calculate the vapor pressure of the unknown liquid at 25°C.

63. If it is true that all solids and liquids have vapor pressures, then at sufficiently low external pressures, every substance should start to boil. In space, there is effectively zero external pressure. Explain why spacecraft do not just boil away as vapors when placed in orbit.

64. Butane-fueled cigarette lighters, which give hundreds of lights each, typically contain 4 to 5 g butane (C_4H_{10}), which is confined in a 10-mL plastic container and exerts a pressure of 3.0 atm at room temperature (25°C). The butane boils at −0.5°C under normal pressure. Butane lighters have been known to explode during use, inflicting serious injury. A person hoping to end such accidents suggests that there be less butane placed in the lighters so that the pressure inside them does not exceed 1.0 atm. Estimate how many grams of butane would be contained in such a lighter.

65. A cooling bath is prepared in a laboratory by mixing chunks of solid CO_2 with ethanol. $CO_2(s)$ sublimes at −78.5°C to $CO_2(g)$. Ethanol freezes at −114.5°C and boils at +78.4°C. State the temperature of this cooling bath and describe what will be seen when it is prepared under ordinary laboratory conditions.

66. Oxygen melts at 54.8 K and boils at 90.2 K at atmospheric pressure. At the normal boiling point, the density of the liquid is 1.14 g cm^{-3} and the vapor can be approximated as an ideal gas. The critical point is defined by $T_c = 154.6$ K, $P_c = 49.8$ atm, and (density) $d_c = 0.436$ g cm^{-3}. The triple point is defined by $T_t = 54.4$ K, $P_t = 0.0015$ atm, a liquid density equal to 1.31 g cm^{-3}, and a solid density of 1.36 g cm^{-3}. At 130 K, the vapor pressure of the liquid is 17.25 atm. Use this information to construct a phase diagram showing P versus T for oxygen. You need not draw the diagram to scale, but you should give numeric labels to as many points as possible on both axes.

67. It can be shown that if a gas obeys the van der Waals equation, its critical temperature, its critical pressure, and its molar volume at the critical point are given by the equations

$$T_c = \frac{8a}{27Rb} \qquad P_c = \frac{a}{27b^2} \qquad \left(\frac{V}{n}\right)_c = 3b$$

where a and b are the van der Waals constants of the gas. Use the van der Waals constants for oxygen, carbon dioxide, and water (from Table 9.3) to estimate the critical-point properties of these substances. Compare with the observed values given in Figure 10.24 and in Problem 66.

*68. Each increase in pressure of 100 atm decreases the melting point of ice by about 1.0°C.

(a) Estimate the temperature at which liquid water freezes under a pressure of 400 atm.

(b) One possible explanation of why a skate moves smoothly over ice is that the pressure exerted by the skater on the ice lowers its freezing point and causes it to melt. The pressure exerted by an object is the force (its mass × the acceleration of gravity, 9.8 m s^{-2}) divided by the area of contact. Calculate the change in freezing point of ice when a skater with a mass of 75 kg stands on a blade of area 8.0×10^{-5} m^2 in contact with the ice. Is this sufficient to explain the ease of skating at a temperature of, for example, −5°C (23°F)?

*69. **(a)** Sketch the phase diagram of *temperature* versus *molar volume* for carbon dioxide, indicating the region of each of the phases (gas, liquid, and solid) and the coexistence regions for two phases.

(b) Liquid water has a maximum at 4°C in its curve of density against temperature at $P = 1$ atm, and the solid is less dense than the liquid. What happens if you try to draw a phase diagram of T versus molar volume for water?

70. The critical temperature of HCl is 51°C, lower than that of HF, 188°C, and HBr, 90°C. Explain this by analyzing the nature of the intermolecular forces in each case.

71. The normal boiling points of the fluorides of the second-period elements are as follows: LiF, 1676°C; BeF_2, 1175°C; BF_3, −100°C; CF_4, −128°C; NF_3, −129°C; OF_2, −145°C; F_2, −188°C. Describe the nature of the intermolecular forces in this series of liquids, and account for the trends in boiling point.

CUMULATIVE PROBLEMS

72. At 20°C and a pressure of 1 atm, 1 mol argon gas occupies a volume of 24.0 L. Estimate the van de Waals radius for argon from the onset of the repulsive part of the argon intermolecular potential curve in Figure 9.21, and calculate the fraction of the gas volume that consists of argon atoms.

73. Other things being equal, ionic character in compounds of metals decreases with increasing oxidation number. Rank the following compounds from lowest to highest normal boiling points: AsF_5, SbF_3, SbF_5, F_2.

Chapter 11

Solutions

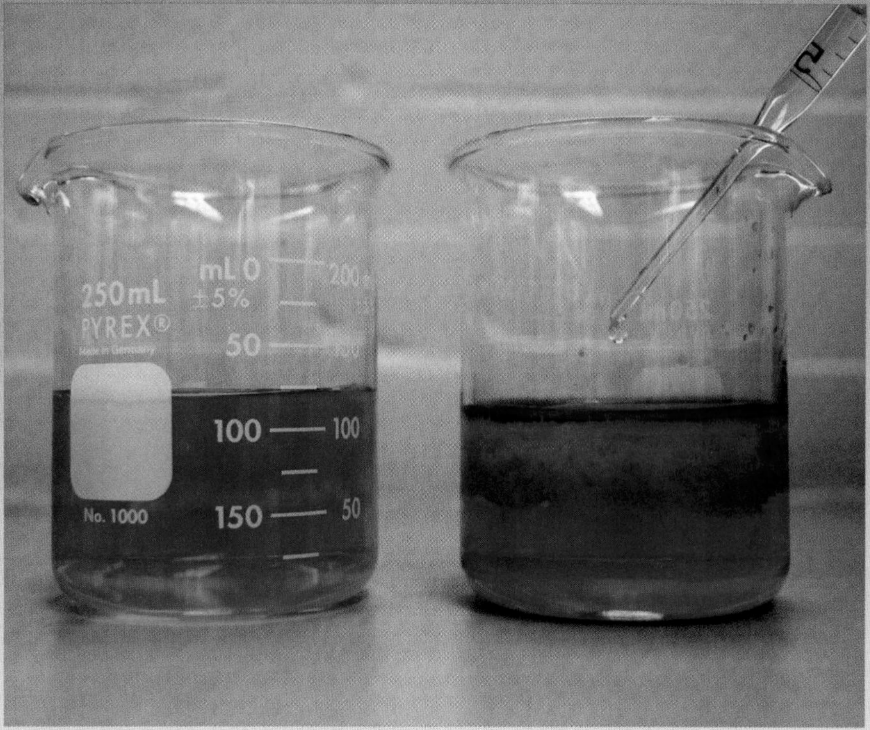

Courtesy of Ellyn Butler and Dr. Meishan Zhao, The University of Chicago

Copper ion complexes in water and ammonia solutions. The pale blue color of the copper sulfate ($CuSO_4$) solution (left) arises when the Cu^{2+} ion is complexed with six water molecules, $[Cu(H_2O)_6]^{2+}$. When ammonium hydroxide is added (right), four ammonia molecules, NH_3, replace water molecules in the complex. The $[Cu(NH_3)_4(H_2O)_2]^{2+}$ complex is dark blue in color. The oxidation number of copper is +2 in all these species.

A homogeneous system that contains two or more substances is called a **solution.** Usually, we think of a solution as a liquid that contains some dissolved substance, such as a solid or gas, and we use the term in that sense in most of this chapter. But, solutions of one solid in another are also common, one example being an alloy of gold and silver. In fact, any homogeneous system of two or more substances (liquid, solid, or gas) is a solution. The major component is usually called the **solvent,** and any minor component is called a **solute.** The solvent is regarded as a "carrier" or medium for the solute, which can participate in chemical reactions in the solution or leave the solution through precipitation or evaporation. The solvent can also participate in chemical reactions.

Description of these phenomena requires quantitative specifications of the amount of solute in the solution, or the *composition* of the solution. Solutions are formed by mixing two or more pure substances whose molecules interact directly in the mixed state. Molecules experience new intermolecular forces in moving from pure solute or solvent into the mixed state. The magnitude of these changes influences both the ease of formation and the stability of a solution.

Chemical reactions are frequently carried out in solution, and their description requires extensions of the rules of stoichiometry described in Chapter 2. We illustrate these extended rules by the important analytical techniques of titration in acid–base and oxidation–reduction reactions.

Just like pure substances, solutions can be in *phase equilibrium* with gases, solids, or other liquids. These equilibria frequently show interesting effects that depend on the molecular weight of the solute.

This chapter begins by explaining how the composition of solutions is defined and how solutions are prepared. It is important to master these concepts, because the properties and behavior of solutions are determined by their composition. With this background, we give quantitative descriptions of chemical reactions and phase equilibria in solutions, and relate these events to the nature of the species in the solution.

Throughout this chapter, it is helpful to keep in mind one guiding question: How are the properties and reactions of the pure solute modified when it is dispersed in the solvent?

11.1 Composition of Solutions

Several measures are used to specify the composition of a solution. **Mass percentage** (colloquially called weight percentage), frequently used in everyday applications, is defined as the percentage by mass of a given substance in the solution. In quantitative chemistry, the most useful measures of composition are mole fraction, molarity, and molality.

The **mole fraction** of a substance in a mixture is the number of moles of that substance divided by the total number of moles present. This term was introduced in the discussion of gas mixtures and Dalton's law (see Section 9.4). In a binary mixture containing n_1 mol of species 1 and n_2 mol of species 2, the mole fractions X_1 and X_2 are

$$X_1 = \frac{n_1}{n_1 + n_2} \qquad [11.1a]$$

$$X_2 = \frac{n_2}{n_1 + n_2} = 1 - X_1 \qquad [11.1b]$$

The mole fractions of all the species present must add up to 1. When a clear distinction can be made between solvent and solutes, the label 1 denotes the solvent, and higher numbers are given to the solutes. If comparable amounts of two liquids such as water and alcohol are mixed, the assignment of the labels 1 and 2 is arbitrary.

The **concentration** of a substance is the number of moles per unit volume. The SI units moles per cubic meter are inconveniently large for chemical work, so instead we use the **molarity,** defined as the number of moles of solute per liter of solution:

$$\text{molarity} = \frac{\text{moles solute}}{\text{liters solution}} = \text{mol L}^{-1} \qquad [11.2]$$

M is the abbreviation for "moles per liter." A 0.1 M (read "0.1 molar") solution of HCl has 0.1 mol of HCl (dissociated into ions, as explained later in this chapter) per liter of solution. Molarity is the most common way of specifying the compositions of dilute solutions. For accurate measurements it has the disadvantage of depending slightly on temperature. If a solution is heated or cooled, its volume changes, so the number of moles of solute per liter of solution also changes.

The **molality,** in contrast, involves the ratio of two masses, and so does not depend on temperature. Molality is defined as the number of moles of solute per kilogram of solvent:

$$\text{molality} = \frac{\text{moles solute}}{\text{kilograms solvent}} = \text{mol kg}^{-1} \qquad [11.3]$$

Because the density of water is 1.00 g cm^{-3} at 20°C, 1.00 L of water has mass of 1.00×10^3 g, or 1.00 kg. It follows that in a dilute aqueous solution, the number of moles of solute per liter is nearly the same as the number of moles per kilogram of water. Therefore, molarity and molality have nearly equal values. For nonaqueous solutions and concentrated aqueous solutions, this approximate equality is no longer valid.

EXAMPLE 11.1

A solution is prepared by dissolving 22.4 g of $MgCl_2$ in 0.200 L of water. Taking the density of pure water to be 1.00 g cm^{-3} and the density of the resulting solution to be 1.089 g cm^{-3}, calculate the mole fraction, molarity, and molality of $MgCl_2$ in this solution.

Solution

We are given the *mass* of the $MgCl_2$ and the *volume* of the water. The number of moles for each are

$$\text{moles } MgCl_2 = \frac{22.4 \text{ g}}{95.22 \text{ g mol}^{-1}} = 0.235 \text{ mol}$$

$$\text{moles water} = \frac{(0.200 \text{ L})(1000 \text{ cm}^3 \text{ L}^{-1})(1.00 \text{ g cm}^{-3})}{18.02 \text{ g mol}^{-1}}$$

$$= 11.1 \text{ mol}$$

$$\text{mole fraction } MgCl_2 = \frac{0.235 \text{ mol}}{(11.1 + 0.235) \text{ mol}} = 0.0207$$

To calculate the molarity, we must first determine the volume of solution. Its mass is 200 g water + 22.4 g MgCl = 222.4 g, and its density is 1.089 g cm^{-3}, so the volume is

$$\text{volume solution} = \frac{222.4 \text{ g}}{1.089 \text{ g cm}^{-3}} = 204 \text{ cm}^3 = 0.204 \text{ L}$$

$$\text{molarity } MgCl_2 = \frac{0.235 \text{ mol } MgCl_2}{0.204 \text{ L}} = 1.15 \text{ M}$$

$$\text{molality } MgCl_2 = \frac{0.235 \text{ mol } MgCl_2}{0.200 \text{ kg } H_2O} = 1.18 \text{ mol kg}^{-1}$$

Related Problems: 3, 4

EXAMPLE 11.2

A 9.386 M aqueous solution of sulfuric acid has a density of 1.5091 g cm^{-3}. Calculate the molality, the percentage by mass, and the mole fraction of sulfuric acid in this solution.

Solution

It is convenient to choose 1 L of the solution, whose mass is

$$(1000 \text{ cm}^3)(1.5091 \text{ g cm}^{-3}) = 1509.1 \text{ g} = 1.5091 \text{ kg}$$

One liter contains 9.386 mol H_2SO_4, or

$$9.386 \text{ mol } H_2SO_4 \times 98.08 \text{ g mol}^{-1} = 920.6 \text{ g } H_2SO_4$$

The mass of water in this liter of solution is then obtained by subtraction:

$$\text{Mass of water in 1 L of solution} = 1.5091 \text{ kg} - 0.9206 \text{ kg}$$

$$= 0.5885 \text{ kg}$$

(Continued)

The molality is now directly obtained as

$$\text{Molality } H_2SO_4 = \frac{9.386 \text{ mol } H_2SO_4}{0.5885 \text{ kg } H_2O} = 15.95 \text{ mol kg}^{-1}$$

and the mass percentage is

$$\text{Mass percentage } H_2SO_4 = \frac{0.9206 \text{ kg}}{1.5091 \text{ kg}} \times 100\% = 61.00\%$$

The number of moles of water is

$$\text{moles } H_2O = \frac{588.5 \text{ g}}{18.02 \text{ g mol}^{-1}} = 32.66 \text{ mol}$$

so that the mole fraction of H_2SO_4 is

$$\text{Mole fraction } H_2SO_4 = X_2 = \frac{9.386 \text{ mol}}{9.386 + 32.66 \text{ mol}} = 0.2232$$

Related Problems: 5, 6

11.1.1 Preparation of Solutions

Examples 11.1 and 11.2 show that if a known mass of solute is added to a known volume of solvent, the molarity can be calculated only if the density of the resulting solution is known. If 1 L of solvent is used, the volume of the resulting solution is less than 1 L in some cases and more in others. If a solution is to have a given molarity, it is clearly inconvenient to need to know the solution density. We avoid this problem in practice by dissolving the measured amount of solute in a smaller amount of solvent, then adding solvent continuously until the desired total volume is reached. For accurate work, solutions are prepared in a **volumetric flask,** which has a distinct ring marked around its neck to indicate a carefully calibrated volume (Fig. 11.1). Filling the flask with solvent up to this mark controls the volume of the solution.

Sometimes it may be necessary to prepare a dilute solution of specified concentration from a more concentrated solution of known concentration by adding pure solvent to the concentrated solution. Suppose that the initial concentration (molarity) is c_i and the initial solution volume is V_i. The number of moles of solute is $(c_i \text{ mol L}^{-1})(V_i \text{ L}) = c_i V_i$ mol. This number does not change on dilution to a final solution volume, V_f, because only solvent, and not solute, is being added. Thus, $c_i V_i = c_f V_f$ and the final molarity is

$$c_f = \frac{\text{moles solute}}{\text{final solution volume}} = \frac{c_i V_i}{V_f} \qquad [11.4]$$

This equation can be used to calculate the final concentration after dilution to a given final volume and also to determine what final volume should be used to obtain a given concentration.

EXAMPLE 11.3

(a) Describe how to prepare 0.500 L of a 0.100 M aqueous solution of potassium hydrogen carbonate ($KHCO_3$).

(b) Describe how to dilute this solution to a final concentration of 0.0400 M $KHCO_3$.

Solution

(a) moles solute = $(0.500 \text{ L})(0.100 \text{ mol L}^{-1}) = 0.0500$ mol

grams solute = $(0.0500 \text{ mol})(100.12 \text{ g mol}^{-1}) = 5.01$ g

because 100.12 is the molar mass of $KHCO_3$. We would, therefore, dissolve 5.01 g $KHCO_3$ in a small amount of water and dilute the solution to 0.500 L.

(b) Rearranging Equation 11.4 gives

$$V_f = \frac{c_i}{c_f} V_i$$

$$= \left(\frac{0.100 \text{ mol L}^{-1}}{0.0400 \text{ mol L}^{-1}}\right)(0.500 \text{ L}) = 1.25 \text{ L}$$

To achieve this, the solution from part (**a**) is diluted to a total volume of 1.25 L by adding water.

Related Problems: 9, 10

FIGURE 11.1 To prepare a solution of nickel chloride, $NiCl_2$, with accurately known concentration, weigh out an amount of the solid (a), transfer it to a volumetric flask (b), dissolve it in somewhat less than the required amount of water (c), and dilute to the total volume marked on the neck of the flask (d), (e). See Figure 10.23.

11.2 Nature of Dissolved Species

In the formation of a solution the attractions among the particles in the original phases (solvent-to-solvent and solute-to-solute attractions) are broken up and replaced, at least in part, by new solvent-to-solute attractions. Unlike a compound, a solution has its components present in *variable* proportions and cannot be represented by a chemical formula. Equations for simple dissolution reactions do not include the solvent as a reactant. They indicate the original state of the solute in parentheses on the left side of the equation and identify the solvent in parentheses on the right side. For example, solid (*s*) sucrose dissolves in water to give an **aqueous solution** (*aq*) solution of sucrose:

$$C_{12}H_{22}O_{11}(s) \longrightarrow C_{12}H_{22}O_{11}(aq)$$

We will see later how to describe cases in which the solvent participates in a reaction. Although the solute and solvent can be any combination of solid, liquid, and gas phases, liquid water is indisputably the best known and most important solvent. Consequently, we emphasize aqueous solutions in this chapter, but you should always remember that dissolution also occurs in many other solvents. We describe formation of aqueous solutions by considering the intermolecular forces between the solute and water molecules. Because these forces can be quite different for molecular solutes and ionic solutes, we discuss these two cases separately.

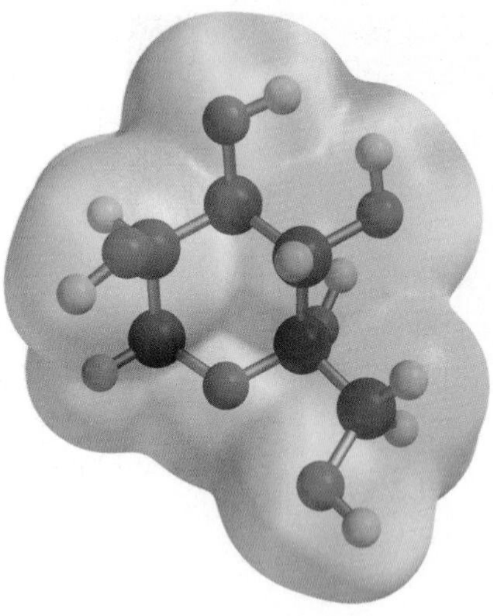

FIGURE 11.2 Electrostatic potential energy map for fructose.

11.2.1 Aqueous Solutions of Molecular Species

Molecular substances that have polar molecules are readily dissolved by water. Examples are the sugars, which have the general formula $C_m(H_2O)_n$. Specific cases are sucrose (table sugar), $C_{12}H_{22}O_{11}$; fructose (fruit sugar), $C_6H_{12}O_6$ (see Fig. 11.2); and ribose, $C_5H_{10}O_5$, a subunit in the biomolecules known as ribonucleic acids. Despite their general formula, the sugars do not contain water molecules, but they do include polar OH (hydroxyl) groups bonded to carbon atoms, which provide sites for hydrogen-bonding interactions with water molecules. These attractions replace the solute–solute interactions, and the individual aquated sugar molecules move off into the solution (Fig. 11.3). Many other molecular substances follow the same pattern, provided they are sufficiently polar. Nonpolar substances, such as carbon tetrachloride, octane, and the common oils and waxes, do not dissolve significantly in water.

11.2.2 Aqueous Solutions of Ionic Species (Electrolytes)

Potassium sulfate is an ionic solid that dissolves in water up to 120 g L^{-1} at 25°C; this maximum mass that can be dissolved in 1 L at 25°C is called the **solubility** in water. The chemical equation for this **dissolution reaction** is written as

$$K_2SO_4(s) \longrightarrow 2\ K^+(aq) + SO_4^{2-}(aq)$$

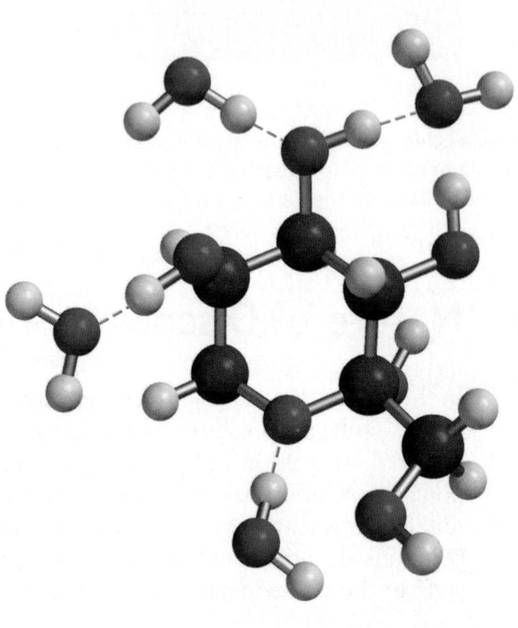

FIGURE 11.3 A molecule of fructose in aqueous solution. Note the attractions between the hydroxyl (O—H) groups of the fructose and molecules of water. The fructose molecule is aquated; the exact number and arrangement of the attached water molecules fluctuates. Also shown is one hydrogen bond between a water molecule and an oxygen atom in the fructose ring.

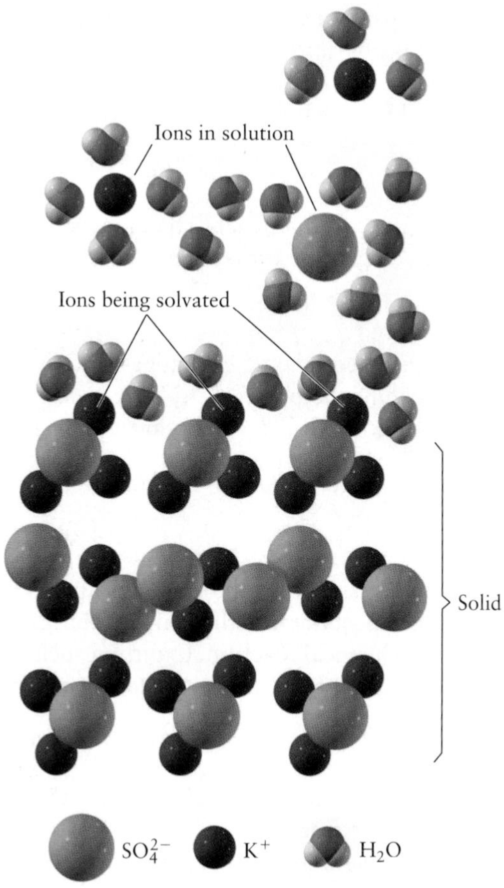

FIGURE 11.4 When an ionic solid (in this case, K_2SO_4) dissolves in water, the ions move away from their sites in the solid, where they were attracted strongly by ions of opposite electrical charge. New strong attractions replace those lost as each ion is surrounded by a group of water molecules. In a precipitation reaction, the process is reversed.

The dissolution of ionic species (Fig. 11.4) occurs through the ion–dipole forces described in Section 10.2. Each positive ion in solution is surrounded by water molecules oriented with the negative end of their dipole moments toward the positive ion. Each SO_4^{2-} anion in solution is surrounded by water molecules oriented with the positive end of their dipole moments toward the anion (hydrogen bonds may also form). When a halide such as KCl is dissolved, the anion forms a hydrogen bond with one of the H atoms in a water molecule that places the atoms O—H—Cl nearly in a straight line as described in Section 10.2.

Each ion dissolved in water and its surrounding **solvation shell** of water molecules constitute an entity held together by ion–dipole forces or by hydrogen bonds. These solvated ions can move as intact entities when an electric field is applied (Fig. 11.5). Because the resulting solution is a conductor of electricity, ionic species such as K_2SO_4 are called *electrolytes*.

In Example 11.1, it is important to note that, although the molarity of $MgCl_2$ is 1.15 M, the molarity of Cl^- ions in the solution is twice as large, or 2.30 M, because each formula unit of $MgCl_2$ dissociates to give two Cl^- ions.

Different compounds dissolve to different extents in water. Only a small amount (0.0025 g) of solid barium sulfate dissolves per liter of water at 25°C, according to the following reaction:

$$BaSO_4(s) \longrightarrow Ba^{2+}(aq) + SO_4^{2-}(aq)$$

The near-total insolubility of barium sulfate suggests that mixing a sufficiently large amount of aqueous barium ion with aqueous sulfate ion would cause the reverse reaction to occur and solid barium sulfate would appear. Of course, it is impossible to prepare a solution containing ions of one charge only. Ions of both charges must be present to maintain overall charge neutrality. But it is possible

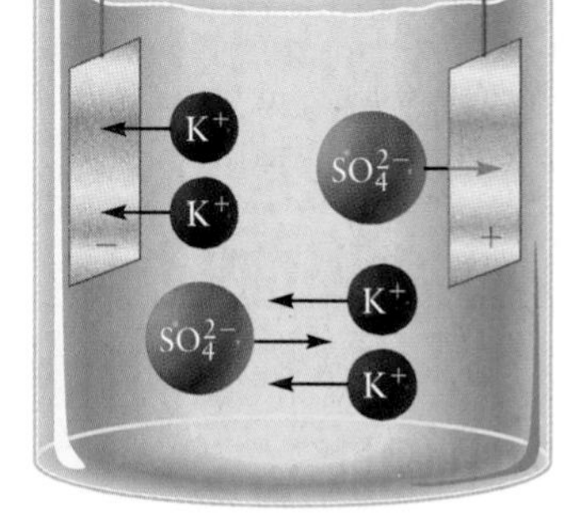

FIGURE 11.5 An aqueous solution of potassium sulfate conducts electricity. When metallic plates (electrodes) charged by a battery are put in the solution, positive ions (K^+) migrate toward the negative plate, and negative ions (SO_4^{2-}) migrate toward the positive plate.

to prepare one solution that contains a soluble barium compound in water (such as barium chloride) and a second solution that contains a soluble sulfate compound in water (such as potassium sulfate). Mixing the two solutions (Fig. 11.6) then produces solid barium sulfate through the following reaction:

$$Ba^{2+}(aq) + SO_4^{2-}(aq) \longrightarrow BaSO_4(s)$$

which is called a **precipitation reaction.**

Such precipitation reactions are sometimes written as

$$BaCl_2(aq) + K_2SO_4(aq) \longrightarrow BaSO_4(s) + 2\ KCl(aq)$$

which suggests ionic exchange—that is, the two anions exchange places. This is misleading, because $BaCl_2$, K_2SO_4, and KCl are all dissociated into ions in aqueous solution. It is more accurate to write

$$Ba^{2+}(aq) + 2\ Cl^-(aq) + 2\ K^+(aq) + SO_4^{2-}(aq) \longrightarrow BaSO_4(s) + 2\ K^+(aq) + 2\ Cl^-(aq)$$

Richard Megna/Fundamental Photographs

FIGURE 11.6 A solution of potassium sulfate is being added to one of barium chloride. A cloud of white solid barium sulfate is formed; the potassium chloride remains in solution.

The potassium and chloride ions appear on both sides of the equation. They are each a **spectator ion,** they ensure charge neutrality but do not take part directly in the chemical reaction. Omitting such spectator ions from the balanced chemical equation leads to the **net ionic equation:**

$$Ba^{2+}(aq) + SO_4^{2-}(aq) \longrightarrow BaSO_4(s)$$

A net ionic equation includes only the ions (and molecules) that actually take part in the reaction.

In dissolution and precipitation reactions, ions retain their identities, and in particular, oxidation states do not change. The ions simply exchange the positions they had in a solid (surrounded by other ions) for new positions in solution (surrounded by solvent molecules). They undergo the reverse process in precipitation.

EXAMPLE 11.4

An aqueous solution of sodium carbonate is mixed with an aqueous solution of calcium chloride, and a white precipitate immediately forms. Write a net ionic equation to account for this precipitate.

Solution

Aqueous sodium carbonate contains $Na^+(aq)$ and $CO_3^{2-}(aq)$ ions, and aqueous calcium chloride contains $Ca^{2+}(aq)$ and $Cl^-(aq)$ ions. Mixing the two solutions places $Na^+(aq)$ and $Cl^-(aq)$ ions and also $Ca^{2+}(aq)$ and $CO_3^{2-}(aq)$ ions in contact for the first time. The precipitate forms by the following reaction:

$$Ca^{2+}(aq) + CO_3^{2-}(aq) \longrightarrow CaCO_3(s)$$

because the other combination of ions leads to sodium chloride, a compound that is known to be soluble in water.

Related Problems: 13, 14

11.3 Reaction Stoichiometry in Solutions: Acid–Base Titrations

11.3.1 Reactions in Solution

Most chemical reactions that occur on the earth's surface, whether in living organisms or among inorganic substances, take place in aqueous solution. Chemical reactions carried out between substances in solution obey the requirements of stoichiometry discussed in Chapter 2, in the sense that the conservation laws embodied in balanced chemical equations are always in force. But here we must apply these requirements in a slightly different way. Instead of a conversion between masses and number of moles, using the molar mass as a conversion factor, the conversion is now between *solution volumes* and number of moles, with the concentration as the conversion factor.

For instance, consider the reaction that is used commercially to prepare elemental bromine from its salts in solution:

$$2\ Br^-(aq) + Cl_2(aq) \longrightarrow 2\ Cl^-(aq) + Br_2(aq)$$

Suppose there is 50.0 mL of a 0.0600 M solution of NaBr. What volume of a 0.0500 M solution of Cl_2 is needed to react completely with the Br^-? To answer this, find the number of moles of bromide ion present:

$$0.0500\ L \times (0.0600\ mol\ L^{-1}) = 3.00 \times 10^{-3}\ mol\ Br^-$$

Next, use the chemical conversion factor 1 mol of Cl_2 per 2 mol of Br^- to find

$$\text{moles } Cl_2 \text{ reacting} = 3.00 \times 10^{-3}\ mol\ Br^- \left(\frac{1\ mol\ Cl_2}{2\ mol\ Br^-}\right) = 1.50 \times 10^{-3}\ mol\ Cl_2$$

Finally, find the necessary volume of aqueous chlorine:

$$\frac{1.50 \times 10^{-3}\ mol}{0.0500\ mol\ L^{-1}} = 3.00 \times 10^{-2}\ L\ \text{solution}$$

The reaction requires 3.00×10^{-2} L, or 30.0 mL, of the Cl_2 solution. (In practice, an excess of Cl_2 solution would be used to ensure more nearly complete conversion of the bromide ion to bromine.)

The chloride ion concentration after completion of the reaction might also be of interest. Because each mole of bromide ion that reacts gives 1 mol of chloride ion in the products, the number of moles of Cl^- produced is 3.00×10^{-3} mol. The final volume of the solution is 0.0800 L, so the final concentration of Cl^- is

$$[Cl^-] = \frac{3.00 \times 10^{-3}\ mol}{0.0800\ L} = 0.0375\ M$$

Square brackets around a chemical symbol signify the molarity of that species.

EXAMPLE 11.5

When potassium dichromate is added to concentrated hydrochloric acid, it reacts according to the following chemical equation

$$K_2Cr_2O_7(s) + 14\ HCl(aq) \longrightarrow 2\ K^+(aq) + 2\ Cr^{3+}(aq) + 8\ Cl^-(aq) + 7\ H_2O(\ell) + 3\ Cl_2(g)$$

producing a mixed solution of chromium(III) chloride and potassium chloride and evolving gaseous chlorine. Suppose that 6.20 g of $K_2Cr_2O_7$ reacts with concentrated HCl, and that the final volume of the solution is 100.0 mL. Calculate the final concentration of $Cr^{3+}(aq)$ and the number of moles of chlorine produced.

Solution

The first step is to convert the mass of $K_2Cr_2O_7$ to moles:

$$\frac{6.20\ g\ K_2Cr_2O_7}{294.19\ g\ mol^{-1}} = 0.0211\ mol\ K_2Cr_2O_7$$

(Continued)

The balanced chemical equation states that 1 mol of $K_2Cr_2O_7$ reacts to give 2 mol of Cr^{3+} and 3 mol of Cl_2. Using these two chemical conversion factors gives

$$\text{moles } Cr^{3+} = 0.0211 \text{ mol } K_2Cr_2O_7 \left(\frac{2 \text{ mol } Cr^{3+}}{1 \text{ mol } K_2Cr_2O_7}\right)$$

$$= 0.0422 \text{ mol } Cr^{3+}$$

$$\text{moles } Cl_2 = 0.0211 \text{ mol } K_2Cr_2O_7 \left(\frac{3 \text{ mol } Cl_2}{1 \text{ mol } K_2Cr_2O_7}\right)$$

$$= 0.0633 \text{ mol } Cl_2$$

Because the final volume of the solution is 0.100 L, the concentration of $Cr^{3+}(aq)$ is

$$[Cr^{3+}] = \frac{0.0422 \text{ mol}}{0.100 \text{ L}} = 0.422 \text{ M}$$

Related Problems: 15, 16

11.3.2 Titration

FIGURE 11.7 A titration.

One of the most important techniques in analytical chemistry is **titration**—the addition of a carefully measured volume of one solution, containing substance A in known concentration, to a second solution, containing substance B in unknown concentration (see Fig. 11.7). Solution A is added through a buret, an instrument that accurately measures the volume of solution transferred as a stopcock is opened and then closed. As the solutions are mixed, A and B react quantitatively. Completion of the reaction, the **endpoint,** is signaled by a change in some physical property, such as the color of the reacting mixture. Endpoints can be detected in colorless reaction mixtures by adding a substance called an *indicator* that changes color at the endpoint. At the endpoint, the known number of moles of substance A that has been added is uniquely related to the unknown number of moles of substance B initially present by the balanced equation for the titration reaction. Titration enables chemists to determine the unknown amount of a substance present in a sample. The two most common applications of titrations involve acid–base neutralization reactions and oxidation–reduction (or redox) reactions. We describe both here briefly to illustrate the fundamental importance of solution stoichiometry calculations in titrations. Detailed discussion of acid–base and redox reactions, and the extent to which they go to completion, are presented in Chapter 15 and Chapter 17, respectively.

11.3.3 Background on Acid–Base Reactions

Table 11.1 lists the names and formulas of a number of important acids. Acids and bases have been known and characterized since ancient times. Chemical description and explanation of their properties and behavior have progressed through several stages of sophistication and generality. A broadly applicable modern treatment is presented in Chapter 15. Here, we introduce titrations using the treatment of the Swedish chemist Svante Arrhenius, who defined acids and bases by their behavior when dissolved in water.

In pure water, small but equal numbers of hydronium ions (H_3O^+) and hydroxide ions (OH^-) are present.[1] These arise from the partial ionization of water:

$$2 \, H_2O(\ell) \longrightarrow H_3O^+(aq) + OH^-(aq)$$

[1]As discussed in Chapter 15, a hydrogen ion in aqueous solution is closely held by a water molecule and is better represented as a hydronium ion, $H_3O^+(aq)$.

TABLE 11.1
Names of Common Acids

Binary Acids	Oxoacids	Organic Acids
HF, hydrofluoric acid	H_2CO_3, carbonic acid	HCOOH, formic acid
HCl, hydrochloric acid	H_3PO_3, phosphorus acid	CH_3COOH, acetic acid
HCN, hydrocyanic acid†	H_3PO_4, phosphoric acid	C_6H_5COOH, benzoic acid
H_2S, hydrosulfuric acid	HNO_2, nitrous acid	HOOCCOOH, oxalic acid
	HNO_3, nitric acid (Fig. 11.8)	
	H_2SO_3, sulfurous acid	
	H_2SO_4, sulfuric acid	
	HClO, hypochlorous acid	
	$HClO_2$, chlorous acid	
	$HClO_3$, chloric acid	
	$HClO_4$, perchloric acid	

†Contains three elements but is named as a binary acid.

FIGURE 11.8 Nitric acid, HNO_3, is manufactured from ammonia.

FIGURE 11.9 Hydrochloric acid, HCl, is used in the pickling of steel and other metals to remove oxide layers on the surface.

Following Arrhenius, we define an **acid** as a substance that when dissolved in water increases the number of hydronium ions over the number present in pure water. Gaseous hydrogen chloride reacts with water to give hydrochloric acid (Fig. 11.9):

$$H_2O(\ell) + HCl(g) \longrightarrow H_3O^+(aq) + Cl^-(aq)$$

A **base** is defined as a substance that when dissolved increases the number of hydroxide ions over the number present in pure water. Sodium hydroxide dissolves extensively in water according to the following reaction:

$$NaOH(s) \longrightarrow Na^+(aq) + OH^-(aq)$$

Ammonia is another base, as shown by the products of its reaction with water:

$$NH_3(aq) + H_2O(\ell) \longrightarrow NH_4^+(aq) + OH^-(aq)$$

When an acidic solution is mixed with a basic solution, a **neutralization reaction** occurs:

$$H_3O^+(aq) + OH^-(aq) \longrightarrow 2\ H_2O(\ell)$$

This is the reverse of the water ionization reaction shown earlier. If the spectator ions are put back into the equation, it reads

$$\underset{\text{Acid}}{HCl} + \underset{\text{Base}}{NaOH} \longrightarrow \underset{\text{Water}}{H_2O} + \underset{\text{Salt}}{NaCl}$$

showing that a salt can be defined as the product (other than water) of the reaction of an acid with a base. It is usually preferable to omit the spectator ions and to indicate explicitly only the reacting ions.

11.3.4 Acid–Base Titration

In most acid–base reactions, there is no sharp color change at the endpoint. In such cases it is necessary to add a small amount of an indicator, a dye that changes color when the reaction is complete (indicators are discussed in detail

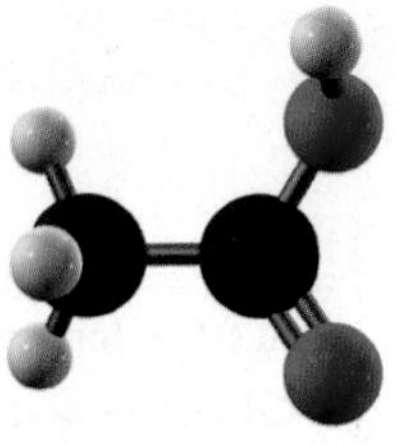

FIGURE 11.10 Acetic acid, CH_3COOH, is a common organic acid that is found in vinegar as a 3% to 5% solution by mass.

in Section 15.3). Phenolphthalein is such an indicator, changing from colorless to pink when a solution changes from acidic to basic. The concentration of acetic acid (Fig. 11.10) in an aqueous solution can be determined by adding a few drops of a phenolphthalein solution and then titrating it with a solution of sodium hydroxide of accurately known concentration. At the first permanent appearance of a pink color, the stopcock of the buret is closed. At this point, the reaction

$$CH_3COOH(aq) + OH^-(aq) \longrightarrow CH_3COO^-(aq) + H_2O(\ell)$$

has gone stoichiometrically to completion.

EXAMPLE 11.6

A sample of vinegar is to be analyzed for its acetic acid content. A volume of 50.0 mL is measured out and titrated with a solution of 1.306 M NaOH; 31.66 mL of that titrant is required to reach the phenolphthalein endpoint. Calculate the concentration of acetic acid in the vinegar (in moles per liter).

Solution

The number of moles of NaOH reacting is found by multiplying the volume of NaOH solution (31.66 mL = 0.03166 L) by its concentration (1.306 M):

$$0.03166 \text{ L} \times 1.306 \text{ mol L}^{-1} = 4.135 \times 10^{-2} \text{ mol NaOH}$$

Because 1 mol of acetic acid reacts with 1 mol of $OH^-(aq)$, the number of moles of acetic acid originally present must also have been 4.135×10^{-2} mol. Its concentration was then

$$[CH_3COOH] = \frac{4.135 \times 10^{-2} \text{ mol}}{0.0500 \text{ L}} = 0.827 \text{ M}$$

Related Problems: 25, 26

11.4 Reaction Stoichiometry in Solutions: Oxidation–Reduction Titrations

11.4.1 Background on Oxidation–Reduction (Redox) Reactions

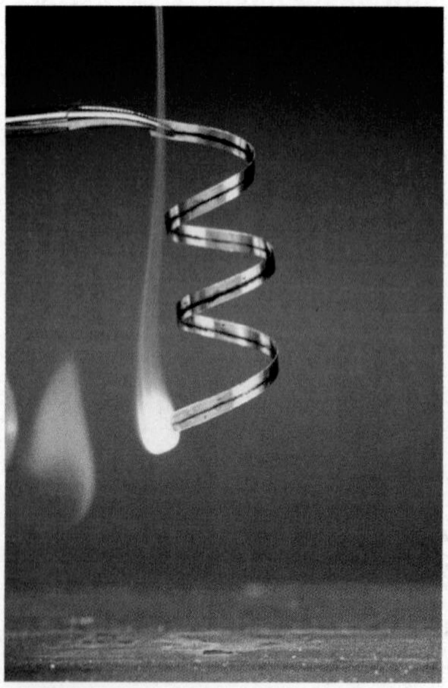

FIGURE 11.11 Magnesium burning in air gives off an extremely bright light. This characteristic led to the incorporation of magnesium into the flash powder used in early photography. Magnesium powder is still used in fireworks for the same reason.

In an **oxidation–reduction reaction** (also called a *redox reaction*), electrons are transferred between reacting species as they combine to form products. This exchange is described as a change in the oxidation number of the reactants: The oxidation number of the species giving up electrons increases, whereas that for the species accepting electrons decreases. Oxidation numbers are defined and methods for their calculation are presented in Section 3.12. A prototype redox reaction is that of magnesium (Mg) with oxygen (O) (Fig. 11.11). When this reaction is carried to completion, the product is magnesium oxide:

$$2\,Mg(s) + O_2(g) \longrightarrow 2\,MgO(s)$$

Magnesium is **oxidized** in this process; it *gives up* electrons as its oxidation number *increases* from 0 (in elemental Mg) to +2 (in MgO). Oxygen, which accepts these electrons, is said to be reduced; its oxidation number decreases from 0 to −2. The transfer of electrons (e^-) can be indicated with arrows:

$$\overset{0}{2\,Mg} + \overset{0}{O_2} \longrightarrow 2\,\overset{+2\,-2}{MgO}$$
$$\downarrow \qquad \uparrow$$
$$2 \times 2e^- = 2 \times 2e^-$$

The arrows point away from the species being oxidized (giving up electrons) and toward the species being reduced (accepting electrons). The electron "bookkeeping" beneath the equation ensures that the same number of electrons are taken up by oxygen as are given up by magnesium:

$$2 \text{ Mg atoms} \times 2 \text{ electrons per Mg atom} = 2 \text{ O atoms per formula unit} \times 2 \text{ electrons per O atom}$$

Originally, the term *oxidation* referred only to reactions with oxygen. It now is used to describe any process in which the oxidation number of a species increases, even if oxygen is not involved in the reaction. When calcium combines with chlorine to form calcium chloride,

$$\overset{0}{Ca}(s) + \overset{0}{Cl_2}(g) \longrightarrow \overset{+2\ -1}{CaCl_2}$$

$$\underset{2e^-}{\downarrow} \quad = \quad \underset{2 \times 1e^-}{\uparrow}$$

the calcium has been oxidized and the chlorine has been reduced.

Oxidation–reduction reactions are among the most important in chemistry, biochemistry, and industry. Combustion of coal, natural gas, and gasoline for heat and power are redox reactions, as are the recovery of metals such as iron and aluminum from their oxide ores and the production of chemicals such as sulfuric acid (Fig. 11.12) from sulfur, air, and water. The human body metabolizes sugars through redox reactions to obtain energy; the reaction products are liquid water and gaseous carbon dioxide.

FIGURE 11.12 Sulfuric acid, H_2SO_4, is the industrial chemical produced on the largest scale in the world, in amounts exceeding 100 million tons per year.

EXAMPLE 11.7

Determine whether the following equations represent oxidation–reduction reactions.

(a) $SnCl_2(s) + Cl_2(g) \longrightarrow SnCl_4(\ell)$

(b) $CaCO_3(s) \longrightarrow CaO(s) + CO_2(g)$

(c) $2\ H_2O_2(\ell) \longrightarrow 2\ H_2O(\ell) + O_2(g)$

Solution

(a) Tin increases its oxidation number from +2 to +4, and the oxidation number of chlorine in Cl_2 is reduced from 0 to −1. Some other chlorine atoms are unchanged in oxidation number, but this is still a redox reaction.

(b) The oxidation state of Ca remains at +2, that of O at −2, and that of C at +4. Thus, this is not a redox reaction.

(c) The oxidation number of oxygen in H_2O_2 is −1. This changes to −2 in the product H_2O and 0 in O_2. This is a redox reaction in which the same element is both oxidized and reduced.

Related Problems: 27, 28, 29, 30

11.4.2 Balancing Oxidation–Reduction Equations in Aqueous Solution

Numerous redox reactions occur in aqueous solution. For example, consider the reaction between solid copper (Cu) and an aqueous solution of silver (Ag) nitrate:

$$Cu(s) + 2\ Ag^+(aq) \longrightarrow Cu^{2+}(aq) + 2\ Ag(s)$$

The nitrate ions are spectators that do not take part in the reaction, so they are omitted in the net equation. Two electrons are transferred from each reacting Cu atom to a pair of silver ions. Copper (the electron donor) is oxidized, and silver ion (the electron acceptor) is reduced. This oxidation–reduction reaction occurs when a piece of copper is placed in an aqueous solution of silver nitrate or any other soluble silver salt (Fig. 11.13). Metallic silver immediately begins to plate out on the copper, the concentration of silver ion decreases, and blue $Cu^{2+}(aq)$ appears in solution and increases in concentration as time passes.

FIGURE 11.13 When a piece of copper wire is inserted into a solution of silver nitrate, metallic silver deposits on the wire and the solution turns blue as Cu^{2+} ions form.

It is useful to consider this chemical equation as representing the sum of an oxidation and a reduction **half-reaction,** in which electrons (e^-) appear explicitly. The oxidation of copper is written as

$$Cu(s) \longrightarrow Cu^{2+}(aq) + 2\ e^-$$

and the reduction of silver ion as

$$Ag^+(aq) + e^- \longrightarrow Ag(s)$$

In the net equation, electrons must not appear explicitly. Thus, the second equation must be multiplied by 2 before being added to the first so that the electrons cancel out on both sides. As before, this gives

$$Cu(s) + 2\ Ag^+(aq) \longrightarrow Cu^{2+}(aq) + 2\ Ag(s)$$

FIGURE 11.14 Copper(II) sulfide reacts with concentrated nitric acid to liberate nitrogen oxide and produce a solution of copper(II) sulfate, which displays the characteristic blue color of copper(II) ions in water.

Many redox reactions are too difficult to balance by the simple methods of logical reasoning described in Section 2.3. Here, we outline a systematic procedure based on half-reactions and apply it to reactions that occur in acidic or basic aqueous solution. In these reactions, water and H_3O^+ (acidic solution) or OH^- (basic solution) may take part either as reactants or as products; thus, it is necessary to *complete* the corresponding equations, as well as to balance them.

As an example, let's complete and balance the chemical equation for the dissolution of copper(II) sulfide in aqueous nitric acid (Fig. 11.14):

$$CuS(s) + NO_3^-(aq) \longrightarrow Cu^{2+}(aq) + SO_4^{2-}(aq) + NO(g)$$

Step 1 *Write two unbalanced half-equations, one for the species that is oxidized and its product and one for the species that is reduced and its product.*

Here, the unbalanced half-reaction involving CuS is

$$CuS \longrightarrow Cu^{2+} + SO_4^{2-}$$

The unbalanced half-reaction involving NO_3^- is

$$NO_3^- \longrightarrow NO$$

Step 2 *Insert coefficients to make the numbers of atoms of all elements except oxygen and hydrogen equal on the two sides of each equation.*

In this case, copper, sulfur, and nitrogen are already balanced in the two half-equations, so this step is already completed.

Step 3 *Balance oxygen by adding H_2O to one side of each half-equation.*

$$CuS + 4\ H_2O \longrightarrow Cu^{2+} + SO_4^{2-}$$

$$NO_3^- \longrightarrow NO + 2\ H_2O$$

Step 4 *Balance hydrogen. For an acidic solution, add H_3O^+ to the side of each half-equation that is "deficient" in hydrogen and add an equal amount of H_2O to the other side. For a basic solution, add H_2O to the side of each half-equation that is "deficient" in hydrogen and add an equal amount of OH^- to the other side.*

Note that this step does not disrupt the oxygen balance achieved in step 3. In this case (acidic solution), the result is

$$CuS + 12\ H_2O \longrightarrow Cu^{2+} + SO_4^{2-} + 8\ H_3O^+$$

$$NO_3^- + 4\ H_3O^+ \longrightarrow NO + 6\ H_2O$$

Step 5 *Balance charge by inserting* e^- *(electrons) as a reactant or product in each half-equation.*

$$CuS + 12\ H_2O \longrightarrow Cu^{2+} + SO_4^{2-} + 8\ H_3O^+ + 8\ e^-$$
(oxidation)

$$NO_3^- + 4\ H_3O^+ + 3\ e^- \longrightarrow NO + 6\ H_2O$$
(reduction)

Step 6 *Multiply the two half-equations by numbers chosen to make the number of electrons given off by the oxidation equal the number taken up by the reduction. Then add the two half-equations, canceling electrons. If* H_3O^+*,* OH^-*, or* H_2O *appears on both sides of the final equation, cancel out the duplications.*

Here, the oxidation half-equation must be multiplied by 3 (so that 24 electrons are produced), and the reduction half-equation by 8 (so that 24 electrons are consumed):

$$3\ CuS + 36\ H_2O \longrightarrow 3\ Cu^{2+} + 3\ SO_4^{2-} + 24\ H_3O^+ + 24\ e^-$$
$$8\ NO_3^- + 32\ H_3O^+ + 24\ e^- \longrightarrow 8\ NO + 48\ H_2O$$
$$3\ CuS + 8\ NO_3^- + 8\ H_3O^+ \longrightarrow 3\ Cu^{2+} + 3\ SO_4^{2-} + 8\ NO + 12\ H_2O$$

This procedure balances equations that are too difficult to balance by inspection. For basic solutions, remember to add H_2O and OH^-, rather than H_3O^+ and H_2O, at step 4.

EXAMPLE 11.8

Balance the following equation, which represents a reaction that occurs in basic aqueous solution:

$$Ag(s) + HS^-(aq) + CrO_4^{2-}(aq) \longrightarrow Ag_2S(s) + Cr(OH)_3(s)$$

Solution

Step 1

$$Ag + HS^- \longrightarrow Ag_2S$$
$$CrO_4^{2-} \longrightarrow Cr(OH)_3$$

Step 2

$$2\ Ag + HS^- \longrightarrow Ag_2S$$

The other half-reaction is unchanged.

Step 3 H_2O is now added to the second half-reaction to balance oxygen:

$$CrO_4^{2-} \longrightarrow Cr(OH)_3 + H_2O$$

Step 4 The right side of the silver half-reaction is deficient by 1 H. Add 1 H_2O to the right and 1 OH^- to the left:

$$2\ Ag + HS^- + OH^- \longrightarrow Ag_2S + H_2O$$

In the chromium half-reaction, the left side is deficient by 5 H atoms, so 5 H_2O is added to that side and 5 OH^- to the right side:

$$CrO_4^{2-} + 4\ H_2O \longrightarrow Cr(OH)_3 + 5\ OH^-$$

(Notice that the H_2O on the right canceled out one of the five on the left, leaving four.)

(Continued)

Step 5 Electrons are added to the right side of the silver half-reaction and to the left side of the chromium half-reaction to balance charge:

$$2\ Ag + HS^- + OH^- \longrightarrow Ag_2S + H_2O + 2\ e^-$$
(oxidation)

$$CrO_4^{2-} + 4\ H_2O + 3\ e^- \longrightarrow Cr(OH)_3 + 5\ OH^-$$
(reduction)

Step 6 The gain and loss of electrons are equalized. The first equation is multiplied by 3 so that it consumes six electrons, and the second is multiplied by 2 so that it produces six electrons.

$$6\ Ag + 3\ HS^- + 3\ OH^- \longrightarrow 3\ Ag_2S + 3\ H_2O + 6\ e^-$$
(oxidation)

$$2\ CrO_4^{2-} + 8\ H_2O + 6\ e^- \longrightarrow 2\ Cr(OH)_3 + 10\ OH^-$$
(reduction)

$$6\ Ag + 3\ HS^- + 2\ CrO_4^{2-} + 5\ H_2O \longrightarrow 3\ Ag_2S + 2\ Cr(OH)_3 + 7\ OH^-$$

Related Problems: 31, 32, 33, 34

EXAMPLE 11.9

Balance the following equation for the reaction of arsenic(III) sulfide with aqueous chloric acid:

$$As_2S_3(s) + ClO_3^-(aq) \longrightarrow H_3AsO_4(aq) + SO_4^{2-}(aq) + Cl^-(aq)$$

Solution

Step 1

$$As_2S_3 \longrightarrow H_3AsO_4 + SO_4^{2-}$$
$$ClO_3^- \longrightarrow Cl^-$$

Step 2

$$As_2S_3 \longrightarrow 2\ H_3AsO_4 + 3\ SO_4^{2-}$$

Step 3

$$As_2S_3 + 20\ H_2O \longrightarrow 2\ H_3AsO_4 + 3\ SO_4^{2-}$$
$$ClO_3^- \longrightarrow Cl^- + 3\ H_2O$$

Step 4 The chloric acid makes this an acidic solution, so H_3O^+ and H_2O are used:

$$As_2S_3 + 54\ H_2O \longrightarrow 2\ H_3AsO_4 + 3\ SO_4^{2-} + 34\ H_3O^+$$
$$ClO_3^- + 6\ H_3O^+ \longrightarrow Cl^- + 9\ H_2O$$

Step 5 Balance charge by adding electrons to each half-reaction:

$$As_2S_3 + 54\ H_2O \longrightarrow 2\ H_3AsO_4 + 3\ SO_4^{2-} + 34\ H_3O^+ + 28\ e^-$$
$$ClO_3^- + 6\ H_3O^+ + 6\ e^- \longrightarrow Cl^- + 9\ H_2O$$

Step 6 We must find the least common multiple of 28 and 6. This is 84; therefore, we need to multiply the first equation by 84/28 = 3 and the second equation by 84/6 = 14:

$$3\ As_2S_3 + 162\ H_2O \longrightarrow 6\ H_3AsO_4 + 9\ SO_4^{2-} + 102\ H_3O^+ + 84\ e^-$$
$$14\ ClO_3^- + 84\ H_3O^+ + 84\ e^- \longrightarrow 14\ Cl^- + 126\ H_2O$$

$$3\ As_2S_3 + 14\ ClO_3^- + 36\ H_2O \longrightarrow 6\ H_3AsO_4 + 9\ SO_4^{2-} + 14\ Cl^- + 18\ H_3O^+$$

11.4.3 Disproportionation

An important type of redox reaction, **disproportionation,** occurs when a single substance is both oxidized and reduced. Example 11.7, part (c), provided such a reaction:

$$2\,\overset{-1}{H_2O_2}(\ell) \longrightarrow 2\,\overset{-2}{H_2O}(\ell) + \overset{0}{O_2}(g)$$

The oxygen in hydrogen peroxide (H_2O_2) is in an intermediate oxidation state of -1; some of it is oxidized to O_2 and some is reduced to H_2O. The balancing of equations for disproportionation reactions is described in the following example.

EXAMPLE 11.10

Balance the following equation for the reaction that occurs when chlorine is dissolved in basic solution:

$$Cl_2(g) \longrightarrow ClO_3^-(aq) + Cl^-(aq)$$

Solution

Step 1 We solve this by writing the Cl_2 on the left sides of two half-equations:

$$Cl_2 \longrightarrow ClO_3^-$$
$$Cl_2 \longrightarrow Cl^-$$

Step 2

$$Cl_2 \longrightarrow 2\,ClO_3^-$$
$$Cl_2 \longrightarrow 2\,Cl^-$$

Step 3 The first half-equation becomes

$$Cl_2 + 6\,H_2O \longrightarrow 2\,ClO_3^-$$

Step 4 Now the first half-equation becomes

$$Cl_2 + 12\,OH^- \longrightarrow 2\,ClO_3^- + 6\,H_2O$$

Step 5

$$Cl_2 + 12\,OH^- \longrightarrow 2ClO_3^- + 6\,H_2O + 10\,e^- \quad \text{(oxidation)}$$

$$Cl_2 + 2\,e^- \longrightarrow 2\,Cl^- \quad \text{(reduction)}$$

Step 6 Multiply the second equation by 5 and add:

$$Cl_2 + 12\,OH^- \longrightarrow 2\,ClO_3^- + 6\,H_2O + 10\,e^-$$
$$5\,Cl_2 + 10\,e^- \longrightarrow 10\,Cl^-$$
$$6\,Cl_2 + 12\,OH^- \longrightarrow 2\,ClO_3^- + 10\,Cl^- + 6\,H_2O$$

Dividing this equation by 2 gives

$$3\,Cl_2 + 6\,OH^- \longrightarrow ClO_3^- + 5\,Cl^- + 3\,H_2O$$

Related Problems: 37, 38

11.4.4 Redox Titration

Redox titrations are illustrated by the reaction in which potassium permanganate oxidizes Fe^{2+} in acidic solution and the manganese (Mn) is reduced:

$$MnO_4^-(aq) + 5\,Fe^{2+}(aq) + 8\,H_3O^+(aq) \longrightarrow Mn^{2+}(aq) + 5\,Fe^{3+}(aq) + 12\,H_2O(\ell)$$

Redox titrations have the advantage that many involve intensely colored species that change color dramatically at the endpoint. For example, MnO_4^- is deep purple, whereas Mn^{2+} is colorless. Thus, when MnO_4^- has been added in slight excess to Fe^{2+}, the color of the solution changes permanently to purple.

The titration is begun by opening the stopcock of the buret and letting a small volume of permanganate solution run into the flask that contains the Fe^{2+} solution. A dash of purple colors the solution (Fig. 11.15a) but rapidly disappears as the permanganate ion reacts with Fe^{2+} to give the nearly colorless Mn^{2+} and Fe^{3+} products. The addition of incremental volumes of permanganate solution is continued until the Fe^{2+} is almost completely converted to Fe^{3+}. At this stage, the addition of just one drop of $KMnO_4$ imparts a pale purple color to the reaction mixture (see Fig. 11.15b) and signals the completion of the

(a) (b)

FIGURE 11.15 (a) Addition of a small amount of potassium permanganate from a buret gives a dash of purple color to the Fe^{2+} solution, which disappears quickly as the permanganate ion reacts to give Fe^{3+} and Mn^{2+}. (b) When all the Fe^{2+} ions have been consumed, additional drops of permanganate solution give a pale purple color to the solution.

reaction. The volume of the titrant $KMnO_4$ solution is calculated by subtracting the initial reading of the solution meniscus (see Fig. 10.23) in the buret from the final volume reading.

Suppose that the permanganate solution has a concentration of 0.09625 M, and 26.34 mL (0.02634 L) of it is added to reach the endpoint. The number of moles of Fe^{2+} in the original solution is calculated in two steps:

$$\text{Amount of } MnO_4^- \text{ reacting} = 0.02634 \text{ L} \times 0.09625 \text{ mol L}^{-1}$$
$$= 2.535 \times 10^{-3} \text{ mol } MnO_4^-$$

From the balanced chemical equation, each mole of MnO_4^- used causes the oxidation of 5 mol of Fe^{2+}, so

$$\text{Amount of } Fe^{2+} \text{ reacting} = 2.535 \times 10^{-3} \text{ mol } MnO_4^- \times \left(\frac{5 \text{ mol } Fe^{2+}}{1 \text{ mol } MnO_4^-}\right)$$
$$= 1.268 \times 10^{-2} \text{ mol } Fe^{2+}$$

These direct titrations form the basis of more complicated analytical procedures. Many analytical procedures are indirect and involve additional preliminary reactions of the sample before the titration can be carried out. For example, a soluble calcium salt will not take part in a redox reaction with potassium permanganate. But adding ammonium oxalate to the solution containing Ca^{2+} causes the quantitative precipitation of calcium oxalate:

$$Ca^{2+}(aq) + C_2O_4^{2-}(aq) \longrightarrow CaC_2O_4(s)$$

After the precipitate is filtered and washed, it is dissolved in sulfuric acid to form oxalic acid (Fig. 11.16):

$$CaC_2O_4(s) + 2\ H_3O^+(aq) \longrightarrow Ca^{2+}(aq) + H_2C_2O_4(aq) + 2\ H_2O$$

Finally, the oxalic acid is titrated with permanganate solution of accurately known concentration, based on the redox reaction

$$2\ MnO_4^-(aq) + 5\ H_2C_2O_4(aq) + 6\ H_3O^+(aq) \longrightarrow$$
$$2\ Mn^{2+}(aq) + 10\ CO_2(g) + 14\ H_2O(\ell)$$

In this way, the quantity of calcium can be determined indirectly by reactions that involve precipitation, acid–base, and redox steps.

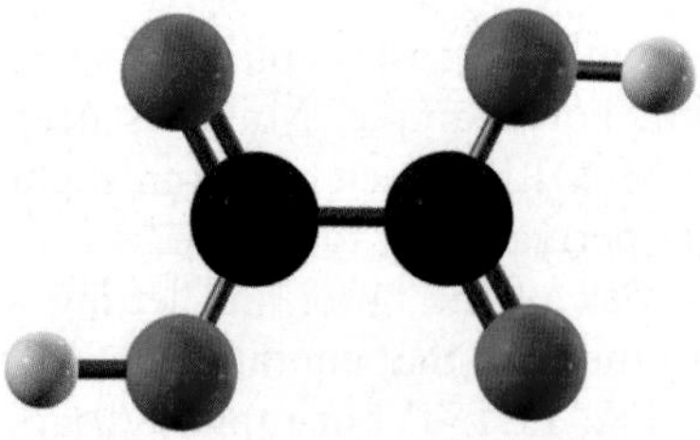

FIGURE 11.16 Oxalic acid, $H_2C_2O_4$, is an acid that can give up the two H^+ ions in aqueous solution to form the oxalate ion, $C_2O_4^{2-}$.

11.5 Phase Equilibrium in Solutions: Nonvolatile Solutes

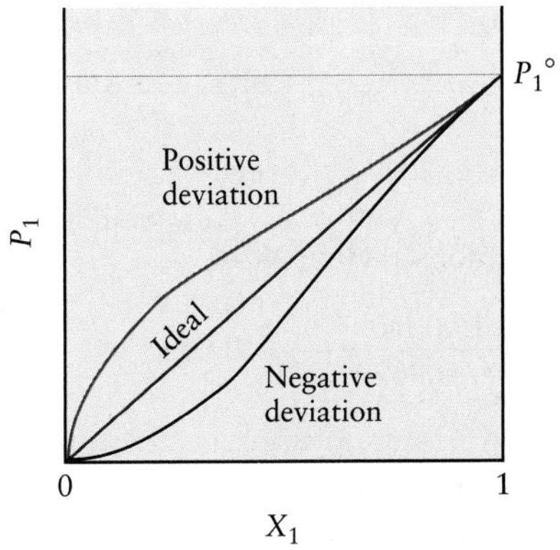

FIGURE 11.17 In an ideal solution, a graph of solvent vapor pressure P_1 versus mole fraction of solvent X_1 is a straight line. Nonideal solutions behave differently; examples of positive and negative deviations from the ideal solution are shown. The vapor pressure of pure solvent is P_1°.

We turn now from the chemical reactions of solutions to such physical properties as their vapor pressures and phase diagrams. Consider first a solution made by dissolving a nonvolatile solute in a solvent. By "nonvolatile" we mean that the vapor pressure of the solute above the solution is negligible. An example is a solution of sucrose (cane sugar) in water, in which the vapor pressure of sucrose above the solution is zero. We study the case of a volatile solute in Section 11.6.

The *solvent* vapor pressure is not zero and changes with the composition of the solution at a fixed temperature. If the mole fraction of solvent (X_1) is 1, then the vapor pressure is P_1°, the vapor pressure of pure solvent at the temperature of the experiment. When X_1 approaches 0 (giving pure solute), the vapor pressure P_1 of the solvent must go to 0 also, because solvent is no longer present. As the mole fraction X_1 changes from 1 to 0, P_1 drops from P_1° to 0. What is the shape of the curve?

The French chemist François-Marie Raoult found that for some solutions a plot of solvent vapor pressure against solvent mole fraction can be fitted closely by a straight line (Fig. 11.17). Solutions that conform to this straight-line relationship obey the following simple equation:

$$P_1 = X_1 P_1^\circ \qquad [11.5]$$

which is known as **Raoult's law**. Such a solution is called an **ideal solution**. A solution that deviates from straight-line behavior is called a **nonideal solution**. Nonideal solutions may show positive deviations (with vapor pressures higher than those predicted by Raoult's law) or negative deviations (with lower vapor pressures). On a molecular level, negative deviations arise when the solute attracts solvent molecules especially strongly, reducing their tendency to escape into the vapor phase. Positive deviations arise in the opposite case, when solvent and solute molecules are not strongly attracted to each other. Even nonideal solutions with nondissociating solutes approach Raoult's law as X_1 approaches 1, just as all real gases obey the ideal gas law at sufficiently low densities.

Raoult's law forms the basis for four properties of dilute solutions, which are called **colligative properties** (derived from Latin *colligare,* meaning "to collect together") because they depend on the collective effect of the *number* of dissolved particles rather than on the *nature* of the particular particles involved. These four properties are:

1. The lowering of the vapor pressure of a solution relative to pure solvent
2. The elevation of the boiling point of a solution relative to the pure solvent
3. The depression of the freezing point of a solution relative to the pure solvent
4. The phenomenon of osmotic pressure

11.5.1 Vapor-Pressure Lowering

Because $X_1 = 1 - X_2$ for a two-component solution, Raoult's law can be rewritten as

$$\Delta P_1 = P_1 - P_1^\circ = X_1 P_1^\circ - P_1^\circ = -X_2 P_1^\circ \qquad [11.6]$$

so that the difference in vapor pressure of the pure solvent and the solution is proportional to the mole fraction of solute. The negative sign implies **vapor-pressure lowering;** the vapor pressure is always less above a dilute solution than it is above the pure solvent.

EXAMPLE 11.11

At 25°C, the vapor pressure of pure benzene is $P_1^\circ = 0.1252$ atm. Suppose 6.40 g of naphthalene, $C_{10}H_8$ (molar mass 128.17 g mol^{-1}), is dissolved in 78.0 g of benzene (molar mass 78.0 g mol^{-1}). Calculate the vapor pressure of benzene over the solution, assuming ideal behavior.

Solution

The number of moles of solvent, n_1, is 1.00 mol in this case (because 78.0 g = 1.00 mol benzene has been used). The number of moles of solute is $n_2 = 6.40 \text{ g}/128.17 \text{ g mol}^{-1} = 0.0499$ mol $C_{10}H_8$; thus, the mole fraction X_1 is

$$X_1 = \frac{n_1}{n_1 + n_2} = \frac{1.00 \text{ mol}}{1.00 + 0.0499 \text{ mol}} = 0.9525$$

From Raoult's law, the vapor pressure of benzene above the solution is

$$P_1 = X_1 P_1^\circ = 0.9525 \times 0.1252 \text{ atm} = 0.119 \text{ atm}$$

Related Problems: 41, 42

11.5.2 Boiling-Point Elevation

The normal boiling point of a pure liquid T_b or a solution T_b' is the temperature at which the vapor pressure reaches 1 atm. Because a dissolved solute reduces the vapor pressure, the temperature of the solution must be increased to make it boil. That is, the boiling point of a solution is higher than that of the pure solvent. This phenomenon, referred to as **boiling-point elevation,** provides a method for determining molar masses.

The vapor-pressure curve of a dilute solution lies slightly below that for the pure solvent. In Figure 11.18, ΔP_1 ($= P_1 - P_1^\circ$, where $P_1 < P_1^\circ$) is the change in vapor pressure at T_b and ΔT_b is the change in temperature necessary to hold the vapor pressure at 1 atm (that is, $\Delta T_b = T_b' - T_b$ is the increase in boiling point caused by addition of solute to the pure solvent). For small concentrations of nondissociating solutes, the two curves are parallel, so

$$-\frac{\Delta P_1}{\Delta T_b} = S$$

$$\Delta T_b = -\frac{\Delta P_1}{S} = \frac{X_2 P_1^\circ}{S}$$

$$= \frac{1}{S}\left(\frac{n_2}{n_1 + n_2}\right) \quad \text{(from Raoult's law, with } P_1^\circ = 1 \text{ atm)}$$

The constant S is a property of the pure solvent only, because it is the slope of the vapor pressure curve (P_1° (at T_{high}) $- P_1^\circ$ (at T_{low}))/($T_{high} - T_{low}$) near 1 atm pressure. That is, S is independent of the solute species involved.

For very dilute solutions, $n_1 \gg n_2$; this may be simplified as follows:

$$\Delta T_b = \frac{1}{S}\frac{n_2}{n_1} = \frac{1}{S}\left(\frac{m_2/\mathcal{M}_2}{m_1/\mathcal{M}_1}\right)$$

where m_1 and m_2 are the masses of solvent and solute (in grams) and $\mathcal{M}_1$ and $\mathcal{M}_2$ are their molar masses in grams per mole. Because $\mathcal{M}_1$, like S, is a property

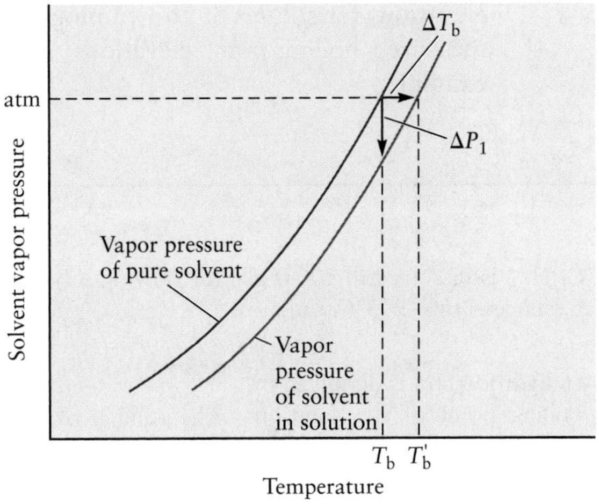

FIGURE 11.18 The vapor pressure of the solvent above a dilute solution is lower than that of the pure solvent at all temperatures. As a result, for the solution to boil (that is, for the vapor pressure to reach 1 atm), a higher temperature is required for the solution than for the pure solvent. This amounts to an elevation of the boiling point.

of the solvent only, it is convenient to combine the two and define a new constant K_b through

$$K_b = \frac{\mathcal{M}_1}{(1000 \text{ g kg}^{-1})S}$$

Then

$$\Delta T_b = K_b \left(\frac{m_2/\mathcal{M}_2}{m_1/(1000 \text{ g kg}^{-1})} \right)$$

Because m_1 is measured in grams, $m_1/(1000 \text{ g kg}^{-1})$ is the number of kilograms of solvent. Also, $m_2/\mathcal{M}_2$ is the number of moles of solute. The expression in parentheses is, therefore, the molality (m) of the solution.

$$\Delta T_b = K_b m \qquad [11.7]$$

For a given solvent, K_b is obtained by measuring the boiling-point elevations for dilute solutions of known molality (that is, containing a known amount of solute with known molar mass). Table 11.2 gives values of K_b for a number of solvents. Once K_b has been found, it can be used either to predict boiling-point

TABLE 11.2

Boiling-Point Elevation and Freezing-Point Depression Constants

Solvent	Formula	T_b (°C)	K_b (K kg mol^{-1})	T_f (°C)	K_f (K kg mol^{-1})
Acetic acid	CH_3COOH	118.1	3.07	17	3.9
Benzene	C_6H_6	80.1	2.53	5.5	4.9
Carbon tetrachloride	CCl_4	76.7	5.03	−22.9	32
Diethyl ether	$C_4H_{10}O$	34.7	2.02	−116.2	1.8
Ethanol	C_2H_5OH	78.4	1.22	−114.7	1.9
Naphthalene	$C_{10}H_8$	—	—	80.5	6.8
Water	H_2O	100.0	0.512	0.0	1.86

elevations for solutes of known molar mass or to determine molar masses from measured boiling-point elevations, as illustrated in part (b) of the following example.

EXAMPLE 11.12

(a) When 5.50 g of biphenyl ($C_{12}H_{10}$) is dissolved in 100.0 g of benzene, the boiling point increases by 0.903°C. Calculate K_b for benzene.

(b) When 6.30 g of an unknown hydrocarbon is dissolved in 150.0 g of benzene, the boiling point of the solution increases by 0.597°C. What is the molar mass of the unknown substance?

Solution

(a) Because the molar mass of biphenyl is 154.2 g mol^{-1}, 5.50 g contains 5.50 g/154.2 g mol^{-1} = 0.0357 mol. The molality, m, is

$$m = \frac{\text{mol solute}}{\text{kg solvent}} = \frac{0.0357 \text{ mol}}{0.1000 \text{ kg}} = 0.357 \text{ mol kg}^{-1}$$

$$K_b = \frac{\Delta T_b}{m} = \frac{0.903 \text{ K}}{0.357 \text{ mol kg}^{-1}}$$

$$= 2.53 \text{ K kg mol}^{-1} \text{ for benzene}$$

(b) Solving $\Delta T_b = K_b m$ for m gives

$$m = \frac{\Delta T_b}{K_b} = \frac{0.597 \text{ K}}{2.53 \text{ K kg mol}^{-1}} = 0.236 \text{ mol kg}^{-1}$$

The number of moles of solute is the product of the molality of the solution and the mass of the solvent, m_1:

$$n_2 = (0.236 \text{ mol kg}^{-1}) \times (0.1500 \text{ kg}) = 0.0354 \text{ mol}$$

Finally, the molar mass of the solute is its mass divided by its number of moles:

$$\text{molar mass of solute} = \mathcal{M}_2 = \frac{m_2}{n_2}$$

$$= \frac{6.30 \text{ g}}{0.0354 \text{ mol}} = 178 \text{ g mol}^{-1}$$

The unknown hydrocarbon might be anthracene ($C_{14}H_{10}$), which has a molar mass of 178.24 g mol^{-1}.

Related Problems: 43, 44, 45, 46

So far, only *nondissociating* solutes have been considered. Colligative properties depend on the total number of moles per liter of dissolved species present. If a solute dissociates (as sodium chloride dissolves to furnish Na^+ and Cl^- ions in aqueous solution), then the molality, m, to be used is the *total* molality. One mole of NaCl dissolves to give 2 mol of ions; thus, the total molality and the boiling-point elevation are twice as large as they would be if NaCl molecules were present in solution. One mole of $Ca(NO_3)_2$ dissolves to give 3 mol of ions (1 mol of Ca^{2+} and 2 mol of NO_3^-), giving 3 times the boiling-point elevation. The corresponding vapor-pressure lowering is greater as well. Ions behave differently than neutral molecules in solution, however, and nonideal behavior appears at lower concentrations in solutions that contain ions.

EXAMPLE 11.13

Lanthanum(III) chloride ($LaCl_3$) is a salt that completely dissociates into ions in dilute aqueous solution,

$$LaCl_3(s) \longrightarrow La^{3+}(aq) + 3\ Cl^-(aq)$$

yielding 4 mol of ions per mole of $LaCl_3$. Suppose 0.2453 g of $LaCl_3$ is dissolved in 10.00 g of H_2O. What is the boiling point of the solution at atmospheric pressure, assuming ideal solution behavior?

(Continued)

Solution

The molar mass of $LaCl_3$ is 245.3 g mol^{-1}.

$$\text{moles of LaCl}_3 = \frac{0.2453 \text{ g}}{245.3 \text{ g mol}^{-1}}$$

$$= 1.000 \times 10^{-3} \text{ mol}$$

$$\text{total molality} = m = \frac{(4)(1.000 \times 10^{-3}) \text{ mol of ions}}{0.0100 \text{ kg solvent}}$$

$$= 0.400 \text{ mol kg}^{-1}$$

This is inserted into the equation for the boiling-point elevation:

$$\Delta T_b = K_b m$$

$$= (0.512 \text{ K kg mol}^{-1})(0.400 \text{ mol kg}^{-1}) = 0.205 \text{ K}$$

$$T_b = 100.205°\text{C}$$

The actual boiling point is slightly lower than this because the solution is nonideal.

11.5.3 Freezing-Point Depression

The phenomenon of **freezing-point depression** is analogous to that of boiling-point elevation. Here, we consider only cases in which the first solid that crystallizes from solution is the pure solvent. If solute crystallizes out with solvent, the situation is more complicated.

Pure solid solvent coexists at equilibrium with its characteristic vapor pressure, determined by the temperature (Section 10.4). Solvent in solution likewise coexists with a certain vapor pressure of solvent. If solid solvent and the solvent in solution are to coexist, they must have the *same* vapor pressure. This means that the freezing temperature of a solution can be identified as the temperature at which the vapor-pressure curve of the pure solid solvent intersects that of the solution (Fig. 11.19). As solute is added to the solution, the vapor pressure of

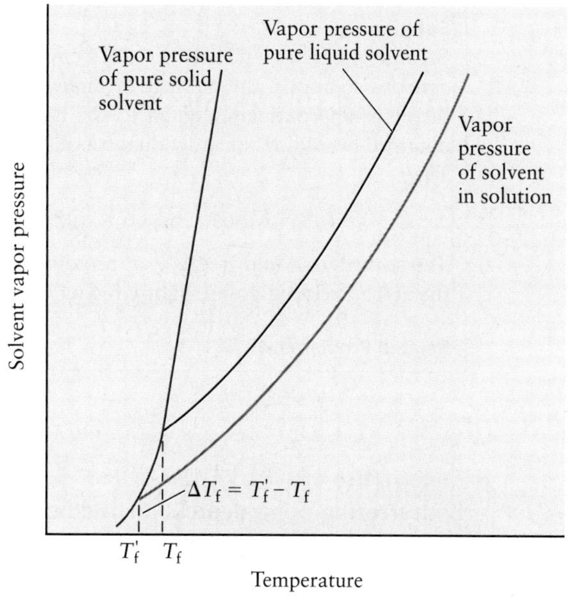

FIGURE 11.19 The vapor pressure of solvent above a dilute solution, compared with that above pure liquid and solid solvent. The depression of the freezing point from T_f to T_f' is shown.

the solvent falls and the freezing point, the temperature at which the first crystals of pure solvent begin to appear, drops. The difference $\Delta T_f = T_f' - T_f$ is therefore negative, and a freezing-point depression is observed.

The change in temperature, ΔT_f, is once again proportional to the change in vapor pressure, ΔP_1. For sufficiently small concentrations of solute, the freezing-point depression is related to the total molality, m (by analogy with the case of boiling-point elevation), through

$$\Delta T_f = T_f' - T_f = -K_f m \qquad [11.8]$$

where K_f is a positive constant that depends only on the properties of the solvent (see Table 11.2). Freezing-point depression is responsible for the fact that seawater, containing dissolved salts, has a slightly lower freezing point than fresh water. Concentrated salt solutions have still lower freezing points. Salt spread on an icy road reduces the freezing point of the ice, so the ice melts.

Measurements of the drop in the freezing point, like those of elevation of the boiling point, can be used to determine molar masses of unknown substances. If a substance dissociates in solution, the *total* molality of all species present (ionic or neutral) must be used in the calculation.

EXAMPLE 11.14

The number of moles of the major dissolved species in a 1.000-L sample of seawater are as follows. Estimate the freezing point of the seawater, assuming $K_f = 1.86$ K kg mol^{-1} for water.

Na^+	0.458 mol	Cl^-	0.533 mol
Mg^{2+}	0.052 mol	SO_4^{2-}	0.028 mol
Ca^{2+}	0.010 mol	HCO_3^-	0.002 mol
K^+	0.010 mol	Br^-	0.001 mol
Neutral species	0.001 mol		

Solution

Because water has a density of 1.00 g cm^{-3}, 1.00 L of water weighs 1.00 kg. For dilute *aqueous* solutions, the number of moles per kilogram of solvent (the molality, m) is therefore approximately equal to the number of moles per liter. The total molality, obtained by adding the individual species molalities just given, is $m = 1.095$ mol kg^{-1}. Then

$$\Delta T = -K_f m = -(1.86 \text{ K kg mol}^{-1})(1.095 \text{ mol kg}^{-1}) = -2.04 \text{ K}$$

The seawater should freeze at approximately −2°C. Nonideal solution effects make the actual freezing point slightly higher than this.

Related Problems: 47, 48

Both freezing-point depression and boiling-point elevation can be used to determine whether a species of known molar mass dissociates in solution (Fig. 11.20), as the following example shows.

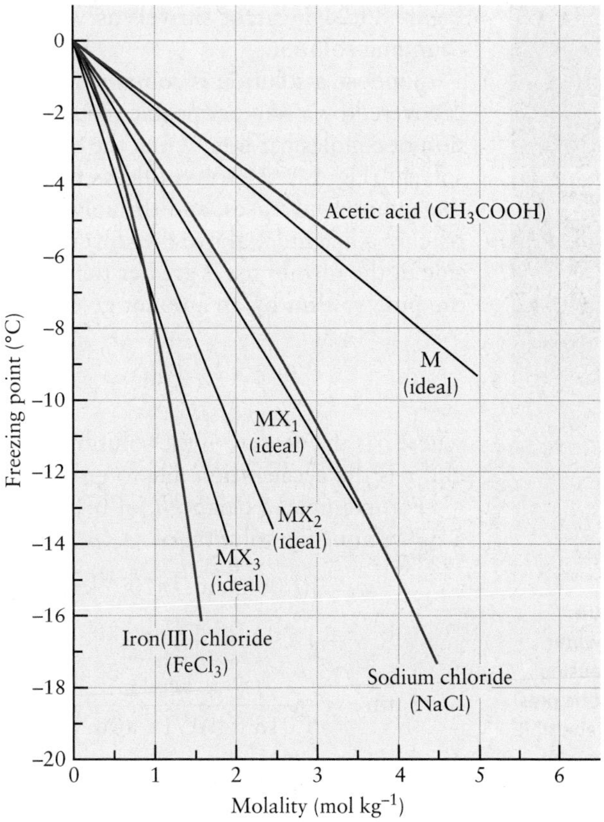

FIGURE 11.20 The heavy colored lines give the observed depression of the freezing point of water by acetic acid, NaCl, and $FeCl_3$ as the molality of the solutions increases. Straight black lines sketch the predicted ideal behavior for one through four moles of particles per mole in solution. The experimental curve for NaCl (which gives *two* moles of dissolved particles) stays close to the ideal straight line for MX; the experimental curve for $FeCl_3$ (which gives *four* moles of dissolved particles) stays fairly close to the ideal straight line for MX_3. The pattern suggests that acetic acid dissolves to give one mole of particles per mole of solute. As the molalities of the solutions increase, the observed freezing-point depressions deviate in varying ways from the straight lines.

EXAMPLE 11.15

When 0.494 g of $K_3Fe(CN)_6$ is dissolved in 100.0 g of water, the freezing point is found to be −0.093°C. How many ions are present for each formula unit of $K_3Fe(CN)_6$ dissolved?

Solution

The total molality of all species in solution is

$$m = \frac{-\Delta T_f}{K_f} = \frac{0.093 \text{ K}}{1.86 \text{ K kg mol}^{-1}} = 0.050 \text{ mol kg}^{-1}$$

Because the molar mass of $K_3Fe(CN)_6$ is 329.25 g mol^{-1}, the total molality if *no* dissociation had taken place would be

$$\frac{\left(\dfrac{0.494 \text{ g}}{329.25 \text{ g mol}^{-1}}\right)}{0.100 \text{ kg}} = 0.0150 \text{ mol kg}^{-1}$$

This is between one fourth and one third of the measured total molality in solution, so each $K_3Fe(CN)_6$ must dissociate into three to four ions. In fact, the dissociation that occurs is

$$K_3Fe(CN)_6(s) \longrightarrow 3\,K^+(aq) + [Fe(CN)_6]^{3-}(aq)$$

Deviations from ideal solution behavior have reduced the effective total molality from 0.060 to 0.050 mol kg^{-1}.

Related Problems: 51, 52

11.5.4 Osmotic Pressure

The fourth colligative property is particularly important in cellular biology because it plays a vital role in the transport of molecules across cell membranes. Such membranes are **semipermeable,** allowing small molecules such as water to pass through while blocking the passage of large molecules such as proteins and carbohydrates. A semipermeable membrane (for example, common cellophane) can be used to separate small solvent molecules from large solute molecules. In

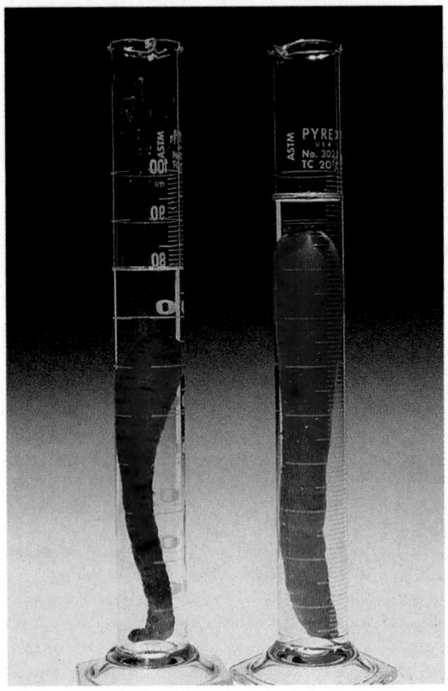

FIGURE 11.21 When a carrot is immersed in saltwater (left), water flows out into the solution, causing the carrot to shrink. The osmotic pressure outside the cells of the vegetable is greater than that inside. A carrot left in pure water (right) does not shrivel. Note that solvent molecules flow through a semi-permeable membrane from regions of low osmotic pressure to regions of high osmotic pressure. This is opposite to the direction of flow caused by differences in ordinary hydrostatic pressure.

Figure 11.21 a carrot shrivels as water in the carrot passes into the salty surrounding solution.

Suppose a solution is contained in an inverted tube, the lower end of which is covered by a semipermeable membrane. This solution has a solute concentration of c moles per liter. When the end of the tube is inserted in a beaker of pure solvent (Fig. 11.22), solvent flows from the beaker into the tube. The volume of the solution increases, and the solvent rises in the tube until, at equilibrium, it reaches a height, h, above the solvent in the beaker. The pressure on the solution side of the membrane is greater than the atmospheric pressure on the surface of the pure solvent by an amount given by the **osmotic pressure, π**:

$$\pi = \rho g h \qquad [11.9]$$

where ρ is the density of the solution (1.00 g cm^{-3} for a dilute aqueous solution) and g is the acceleration due to gravity (9.807 m s^{-2}).

For example, a height, h, of 0.17 m corresponds to an osmotic pressure for a dilute aqueous solution of

$$\pi = [(1.00 \text{ g cm}^{-3})(10^{-3} \text{ kg g}^{-1})(10^{6} \text{ cm}^{3} \text{ m}^{-3})](9.807 \text{ m s}^{-2})(0.17 \text{ m})$$
$$= 1.7 \times 10^{3} \text{ kg m}^{-1} \text{ s}^{-2} = 1.7 \times 10^{3} \text{ Pa}$$
$$\pi(\text{atm}) = \frac{1.7 \times 10^{3} \text{ Pa}}{1.013 \times 10^{5} \text{ Pa atm}^{-1}} = 0.016 \text{ atm}$$

This example illustrates how accurately very small osmotic pressures can be measured.

In 1887, Jacobus van't Hoff discovered an important relation among osmotic pressure, π, concentration, c, and absolute temperature, T:

$$\pi = cRT \qquad [11.10]$$

R is the gas constant, equal to 0.08206 L atm mol^{-1} K^{-1} if π is expressed in atmospheres and c in moles per liter. Because $c = n/V$, where n is the number of moles of solute and V is the volume of the solution, van't Hoff's equation can be rewritten as

$$\pi V = nRT$$

which bears a striking similarity to the ideal gas law. With this relation, the molar mass of a dissolved substance can be determined from the osmotic pressure of its solution.

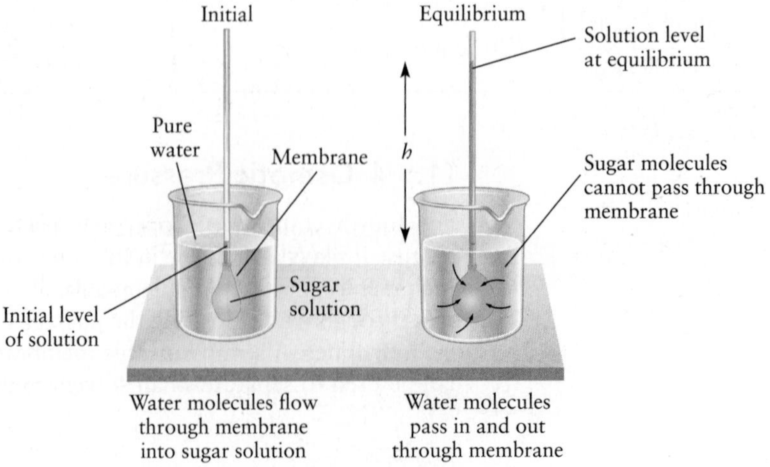

FIGURE 11.22 In this device to measure osmotic pressure, the semipermeable membrane allows solvent, but not solute, molecules to pass through. This results in a net flow of solvent into the tube until equilibrium is achieved, with the level of solution at a height, h, above the solvent in the beaker. Once this happens, the solvent molecules pass through the membrane at the same rate in both directions.

EXAMPLE 11.16

A chemist dissolves 2.00 g of a protein in 0.100 L water. The osmotic pressure is 0.021 atm at 25°C. What is the approximate molar mass of the protein?

Solution

The concentration in moles per liter is

$$c = \frac{\pi}{RT} = \frac{0.021 \text{ atm}}{(0.08206 \text{ L atm mol}^{-1} \text{ K}^{-1})(298 \text{ K})}$$

$$= 8.6 \times 10^{-4} \text{ mol L}^{-1}$$

Now 2.00 g dissolved in 0.100 L gives the same concentration as 20.0 g in 1.00 L. Therefore, 8.6×10^{-4} mol of protein must weigh 20.0 g, and the molar mass is

$$\mathcal{M} = \frac{20.0 \text{ g}}{8.6 \times 10^{-4} \text{ mol}} = 23,000 \text{ g mol}^{-1}$$

Related Problems: 53, 54

Osmotic pressure is particularly useful for measuring molar masses of large molecules such as proteins, whose solubilities may be low. In the case given in Example 11.16, the height difference h is 22 cm, an easily measured quantity. By contrast, the other three colligative properties in this example would show small effects:

$$\text{Vapor-pressure lowering} = 4.8 \times 10^{-7} \text{ atm}$$

$$\text{Boiling-point elevation} = 0.00044 \text{ K}$$

$$\text{Freezing-point depression} = 0.0016 \text{ K}$$

All these changes are too small for accurate measurement. As with the other techniques, the *total* number of moles of solute species determines the osmotic pressure if dissociation occurs.

Osmosis has other important uses. In some parts of the world, potable water is a precious commodity. It can be obtained much more economically by desalinizing brackish waters, through a process called **reverse osmosis,** than by distillation. When an ionic solution in contact with a semipermeable membrane has a pressure applied to it that exceeds its osmotic pressure, water of quite high purity passes through. Reverse osmosis is also used to control water pollution.

11.6 Phase Equilibrium in Solutions: Volatile Solutes

The preceding section described the properties of solutions of nonvolatile solutes in liquid solvents. The concept of an ideal solution can be extended to mixtures of two or more components, each of which is volatile. In this case, an ideal solution is one in which the vapor pressure of *each* species present is proportional to its mole fraction in solution over the whole range of mole fraction:

$$P_i = X_i P_i^\circ$$

where P_i° is the vapor pressure (at a given temperature) of pure substance i, X_i is its mole fraction in solution, and P_i is its partial vapor pressure over the solution. This is a generalization of Raoult's law to each component of a solution.

For an ideal mixture of two volatile substances, the vapor pressure of component 1 is

$$P_1 = X_1 P_1^\circ$$

and that of component 2 is

$$P_2 = X_2 P_2^\circ = (1 - X_1) P_2^\circ$$

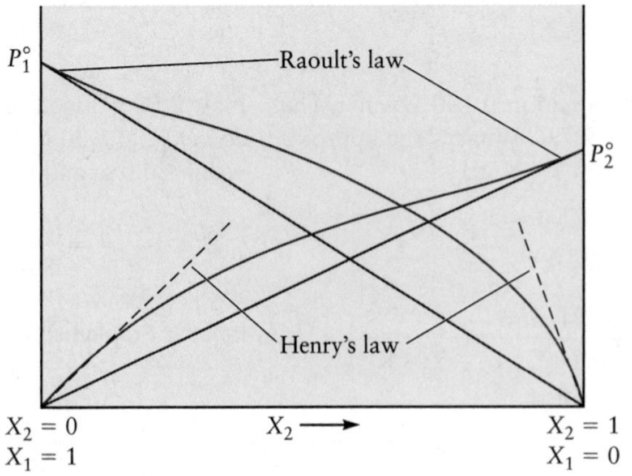

FIGURE 11.23 Vapor pressures above a mixture of two volatile liquids. Both ideal (blue lines) and nonideal behaviors (red curves) are shown. Positive deviations from ideal solution behavior are illustrated, although negative deviations are observed for other nonideal solutions. Raoult's and Henry's laws are shown as dilute solution limits for the nonideal mixture; the markers explicitly identify regions where Raoult's law represents the actual behavior of the major component. Henry's law is followed by the component with very small mole fraction, so the dashed lines represent the slope of the tangent line to the red curve as the corresponding mole fraction approaches zero.

The vapor pressures for such an ideal solution are shown in Figure 11.23, together with typical vapor pressures for a solution that shows positive deviations from ideal behavior.

11.6.1 Henry's Law

At sufficiently low mole fraction X_2, the vapor pressure of component 2 (even in a nonideal solution) is proportional to X_2:

$$P_2 = k_2 X_2 \qquad [11.11]$$

where k_2 is a constant. For X_1 small (X_2 near 1),

$$P_1 = k_1 X_1 = k_1(1 - X_2)$$

This linear vapor pressure of what is called the solute (because it is present at small mole fraction) is known as **Henry's law**: The vapor pressure of a volatile dissolved substance is proportional to the mole fraction of that substance in solution. Whenever Raoult's law is valid for a solvent, Henry's law is valid for the solute (see Fig. 11.23).

One familiar application of Henry's law is in the carbonation of beverages. If the partial pressure of CO_2 above a solution is increased, the amount dissolved in the solution increases proportionately. When the beverage can is opened, dissolved gas bubbles out of solution in response to the lower CO_2 pressure outside. Henry's law is important in biology, where gases such as oxygen dissolve in blood and other bodily fluids, and in environmental chemistry, where volatile pollutants can move between bodies of water and the atmosphere.

EXAMPLE 11.17

The Henry's law constant for oxygen dissolved in water is 4.34×10^4 atm at 25°C. If the partial pressure of oxygen in air is 0.20 atm under ordinary atmospheric conditions, calculate the concentration (in moles per liter) of dissolved oxygen in water that is in equilibrium with air at 25°C.

Solution

Henry's law is used to calculate the mole fraction of oxygen in water:

$$X_{O_2} = \frac{P_{O_2}}{k_{O_2}} = \frac{0.20 \text{ atm}}{4.34 \times 10^4 \text{ atm}} = 4.6 \times 10^{-6}$$

(Continued)

Next, the mole fraction is converted to molarity. One liter of water weighs 1000 g, so it contains

$$\frac{1000 \text{ g } H_2O}{18.02 \text{ g mol}^{-1}} = 55.5 \text{ mol water}$$

Because X_{O_2} is so small, $n_{H_2O} + n_{O_2}$ is close to n_{H_2O}, and it can be written as

$$X_{O_2} = \frac{n_{O_2}}{n_{H_2O} + n_{O_2}} \approx \frac{n_{O_2}}{n_{H_2O}}$$

$$4.6 \times 10^{-6} = \frac{n_{O_2}}{55.5 \text{ mol}}$$

Thus, the number of moles of oxygen in 1 L water is

$$n_{O_2} = (4.6 \times 10^{-6})(55.5 \text{ mol}) = 2.6 \times 10^{-4} \text{ mol}$$

and the concentration of dissolved O_2 is 2.6×10^{-4} M.

Related Problems: 57, 58

11.6.2 Distillation

The vapor pressures of the pure components of an ideal solution usually differ, and for this reason, such a solution has a composition different from that of the vapor phase with which it is in equilibrium. This can best be seen in an example.

Hexane (C_6H_{14}) and heptane (C_7H_{16}) form a nearly ideal solution over the whole range of mole fractions. At 25°C, the vapor pressure of pure hexane is $P_1^\circ = 0.198$ atm, and that of pure heptane is $P_2^\circ = 0.0600$ atm. Suppose a solution contains 4.00 mol of hexane and 6.00 mol of heptane, and that its mole fractions are, therefore, $X_1 = 0.400$ and $X_2 = 0.600$. The vapor in equilibrium with this ideal solution has partial pressures

$$P_{\text{hexane}} = P_1 = X_1 P_1^\circ = (0.400)(0.198 \text{ atm}) = 0.0792 \text{ atm}$$

$$P_{\text{heptane}} = P_2 = X_2 P_2^\circ = (0.600)(0.0600 \text{ atm}) = 0.0360 \text{ atm}$$

From Dalton's law, the total pressure is the sum of these partial pressures:

$$P_{\text{total}} = P_1 + P_2 = 0.1152 \text{ atm}$$

If X_1' and X_2' are the mole fractions in the vapor, then

$$X_1' = \frac{0.0792 \text{ atm}}{0.1152 \text{ atm}} = 0.688$$

$$X_2' = 1 - X_1' = \frac{0.0360 \text{ atm}}{0.1152 \text{ atm}} = 0.312$$

The liquid and the vapor with which it is in equilibrium have different compositions (Fig. 11.24), and the vapor is enriched in the more volatile component.

Suppose some of this vapor is removed and condensed to become liquid. The vapor in equilibrium with this new solution would be still richer in the more volatile component, and the process could be continued further (see Fig. 11.24). This progression underlies the technique of separating a mixture into its pure components by **fractional distillation,** a process in which the components are successively evaporated and recondensed. What we have described so far corresponds to a constant-temperature process, but actual distillation is conducted at constant total pressure. The vapor pressure–mole fraction plot is transformed into a boiling temperature–mole fraction plot (Fig. 11.25). Note that the component with the lower vapor pressure (component 2) has the higher boiling point, T_b^2. If the temperature of a solution of a certain composition is raised

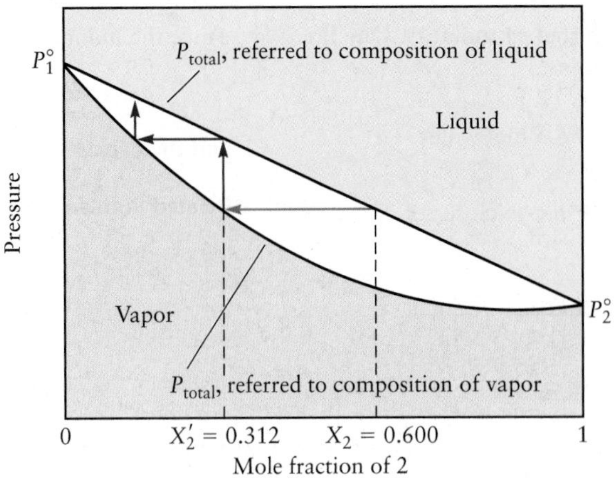

FIGURE 11.24 The composition of the vapor above a solution differs from the composition of the liquid with which it is in equilibrium. Here, the upper (straight) line is the total pressure of the vapor in equilibrium with an ideal solution having mole fraction X_2 of component 2. By moving horizontally from that line to a point of intersection with the lower curve, we can locate the mole fraction X_2' of component 2 in the vapor (red arrow). Subsequent condensations and vaporizations are shown by blue arrows.

until it touches the liquid line in the plot, the vapor in equilibrium with the solution is richer in the more volatile component 1. Its composition lies at the intersection of the horizontal constant-temperature line and the equilibrium vapor curve.

A liquid can be vaporized in different ways. It can simply be boiled until it is entirely vaporized and the final composition of the vapor is the same as that of the original liquid. It is clear that such a mixture boils over a range of temperatures, rather than at a single T_b like a pure liquid. Alternatively, if the boiling is stopped midway, the vapor fraction that has boiled off can be collected and recondensed. The resulting liquid (the condensate) will be richer in component 1 than was the original solution. By repeating the process again and again, mixtures successively richer in component 1 will be obtained. This is the principle behind the *distillation column* (Fig. 11.26). Throughout the length of the tube, such evaporations and recondensations take place, and this allows mixtures to be separated into their constituent substances. Such a process is used to separate nitrogen and oxygen in air; the air is liquefied and then distilled, with the lower boiling nitrogen ($T_b = -196°C$) vaporizing before the oxygen ($T_b = -183°C$).

Nonideal solutions may have more complicated behavior. A mixture showing large *negative* deviations from Raoult's law (one in which solute–solvent forces are strongly attractive) will show a boiling-point *maximum* (Fig. 11.27a). A solution at the maximum is called a *maximum-boiling azeotrope*; an example

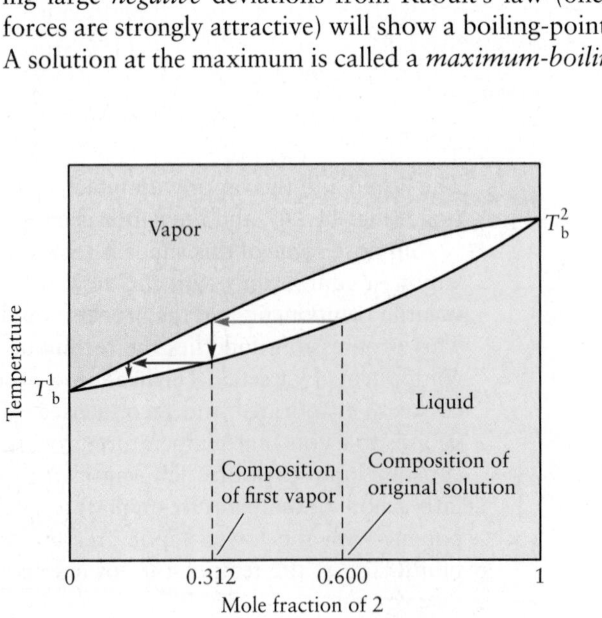

FIGURE 11.25 The boiling point of an ideal solution varies with the composition of the solution. The upper curve is the boiling temperature referred to the vapor composition, and the lower curve is the boiling temperature referred to the liquid composition. The vapors boiling off a solution that has a 0.600-mol fraction of component 2 are enriched in the more volatile component 1 to the extent that their mole fraction of component 2 is only 0.312 (red arrow). The subsequent blue arrows show the further steps used in obtaining nearly pure component 1 by fractional distillation.

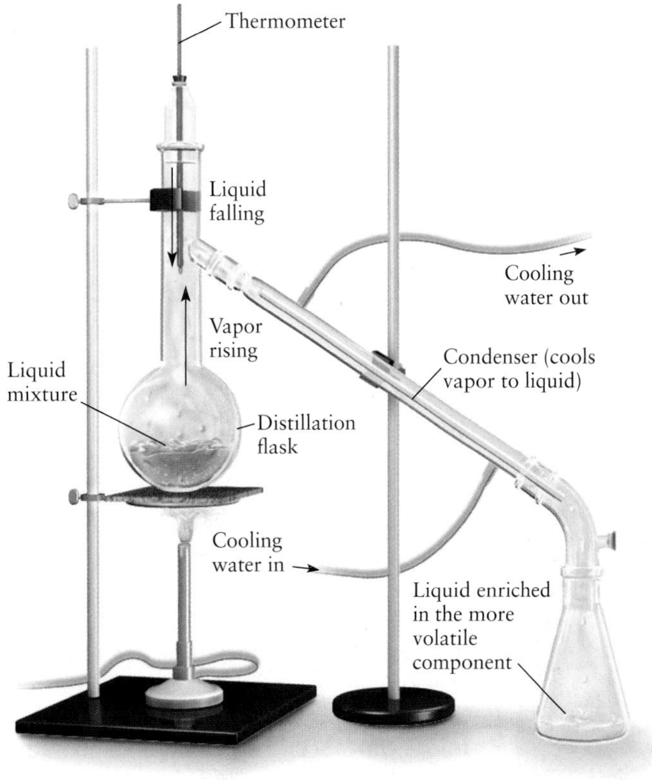

FIGURE 11.26 In a distillation column, temperature decreases with height in the column. The less volatile components condense and fall back to the flask, but the more volatile ones continue up the column into the water-cooled condenser, where they condense and are recovered in the receiver.

is that formed by the H_2O/HCl system. The boiling-point maximum occurs in this case at 108.58°C and 1 atm pressure for a composition of 20.22% HCl by mass. A mixture showing large *positive* deviations from ideal behavior may show a boiling-point *minimum* (see Fig. 11.27b) and a corresponding *minimum-boiling azeotrope.* Ethanol and water form such an azeotrope with a normal boiling point of 78.17°C and a composition of 4% water by mass. In this case, attractive forces between ethanol molecules and between water molecules are stronger than those between ethanol and water, so the solution boils at a lower temperature than either pure component. An azeotrope behaves like a single-component fluid in that it boils at a well-defined temperature and the solution and vapor have the same composition. A mixture of two substances

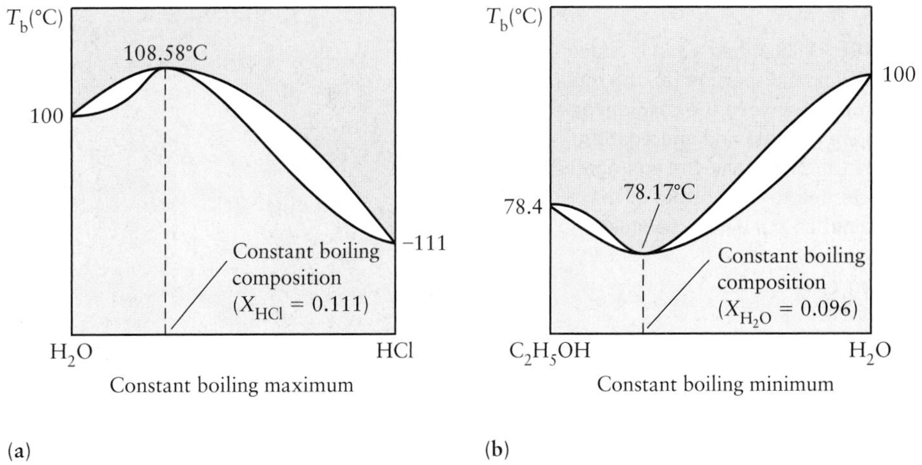

FIGURE 11.27 Dependence of boiling temperature on mole fraction for (a) maximum- and (b) minimum-boiling azeotropes. The coordinates are not to scale.

that form an azeotrope cannot be separated by fractional distillation into two pure substances, but rather into only one pure substance and a mixture with the azeotropic composition. A mixture of 50% ethanol and water, for example, can be distilled to obtain pure water and an azeotropic mixture containing 4% water and 96% ethanol. The last 4% of water cannot be removed by distillation at atmospheric pressure to obtain pure ethanol.

11.7 Colloidal Suspensions

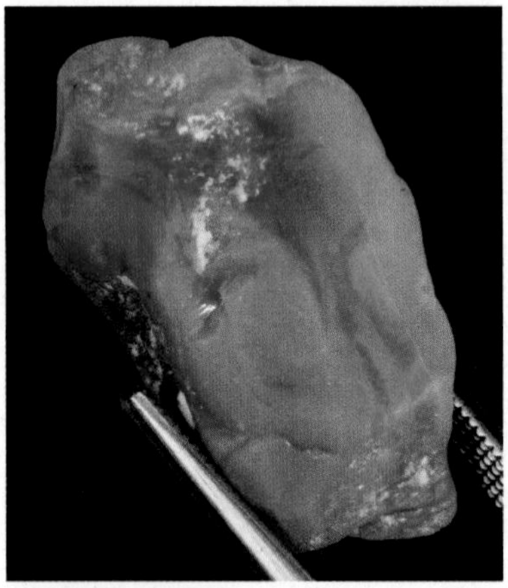

FIGURE 11.28 A natural opal.

A **colloid** is a mixture of two or more substances in which one phase is suspended as a large number of small particles in a second phase. The dispersed substance and the background medium may be any combination of gas, liquid, or solid. Examples of colloids include aerosol sprays (liquid suspended in gas), smoke (solid particles in air), milk (fat droplets and solids in water), mayonnaise (water droplets in oil), and paint (solid pigment particles in oil for oil-based paints, or pigment and oil dispersed in water for latex paints). Colloidal particles are larger than single molecules, but are too small to be seen by the eye; their dimensions typically range from 10^{-9} to 10^{-6} m in diameter. Their presence can be seen most dramatically in the way in which they scatter light; a familiar example is the passage of light from a movie projector through a suspension of small dust particles in air. The gemstone opal has remarkable optical properties that arise from colloidal water suspended in solid silicon dioxide (Fig. 11.28).

Although some colloids settle out into two separate phases if left standing long enough, others persist indefinitely; a suspension of gold particles prepared by Michael Faraday in 1857 shows no apparent settling to date. In many colloids, the particles have net positive or negative charges on their surfaces, balanced by an opposite charge of ions in solution. The settling out of such colloids is speeded by dissolving salts in the solution, a process called **flocculation**. The salts reduce the repulsive electrostatic forces between the suspended particles, causing aggregation and sedimentation (Fig. 11.29). Flocculation occurs in river deltas; when river water containing suspended clay particles meets the salt water of the ocean, the clay settles out as open, low-density sediments. Flocculating agents are deliberately added to paints so that the pigment will settle in a loosely packed sediment. When the paint is stirred, the pigment is redispersed through the medium. In the absence of such agents, the suspended particles tend to settle in compact sediments that are difficult to resuspend.

FIGURE 11.29 When a salt is added to a colloidal dispersion (a), the repulsive forces between the colloidal particles are reduced and aggregation occurs (b). Eventually, the aggregated particles fall to the bottom of the container as low-density sediment (c).

(a) (b)

FIGURE 11.30 (a) This colloidal suspension of $PbCrO_4$ appears cloudy. (b) After flocculation, the precipitate settles to the bottom.

In some cases, the formation of a colloid is not desirable, as in the precipitation of a solid from solution (see Section 11.2). Especially with metal sulfides, the solid precipitate may appear as a colloidal suspension with particles small enough to pass through ordinary filter paper (Fig. 11.30). If this happens, a precipitated solid can be separated out only by flocculation, centrifugation, or forcing the suspension through a membrane, such as cellophane, that permits passage of only the small solvent molecules.

Suspended particles are in a constant state of motion, called **Brownian motion** after Robert Brown, a Scottish botanist who used a microscope to observe the motion of pollen particles in water. Brownian motion results from the constant random buffeting of the particles by solvent molecules. In 1905, Albert Einstein showed how the motion of Brownian particles could be described on a microscopic level; his work provided one of the most striking and convincing verifications of the molecular hypothesis and of the kinetic theory of matter and led to a fairly accurate determination of Avogadro's number.

CHAPTER SUMMARY

When pure substance A is mixed with pure substance B, the resulting solution has properties different from those of the pure substances because the intermolecular forces around each molecule are now quite different. In aqueous solutions, the dissolved species are described as solute ions or molecules surrounded by solvation shells of solvent molecules, perhaps interacting with the solute molecule by hydrogen bonding or ion–dipole forces. Solutions can be in equilibrium in solid, liquid, or vapor phases, but the conditions under which phases coexist are different from those for the pure solvent. Changes in vapor pressure, freezing point, boiling point, and osmotic pressure are explained quantitatively by the number of nonvolatile solute species in the solution. When both species are volatile, the composition of the vapor phase is different from the composition of the solution, as described by Raoult's law and Henry's law, and the solution components can be separated by distillation. Many chemical reactions are carried out in liquid solutions, frequently by mixing solutions of the reactants. We write balanced chemical equations to describe the stoichiometry of reactions in solutions, and we count the number of moles of each reactant in a volume of solution through its concentration expressed in molarity. Solution stoichiometry describes many practical applications in acid–base and redox chemistry and is the basis for quantitative analytical techniques in these fields.

CONCEPTS AND SKILLS

Section 11.1 – Composition of Solutions

Express the concentration of a solute in solution in units of mass percentage, molarity, molality, and mole fraction (Problems 1–8).

- Solutions are described by their composition and method of preparation. Each definition of concentration specifies the procedure for preparing the solution.
  - Mole fraction: $X_i = n_i/n_{tot}$ where n_i is the number of moles of component i and n_{tot} is the total number of moles of all components
  - Molarity: c = moles of solute per liter of solution. The unit mol L^{-1} is given the symbol M.
  - Molality: m = moles of solute per kg of solvent
  - For *dilute, aqueous* solutions $m \approx c$

Describe how a solution of a given molarity is prepared and the effect on dilution of molarity (Problems 9–12).

- The procedure for preparing the solution is illustrated stepwise in Figure 11.1. A volume V_i of the solution at concentration c_i contains $c_i V_i$ moles of solute. Adding solvent to increase the volume to V_f reduces the concentration to $c_f = (c_i V_i)/V_f$, the moles of solute in the final solution divided by the volume of the final solution.

Section 11.2 – Nature of Dissolved Species

Describe the formation of a solution in molecular terms by comparing intermolecular forces in the pure phases and in the solution.

- Solutions are formed by breaking the bonds between molecules or ions in the solute and dispersing these species throughout the solvent.
  - The solute species are surrounded by solvent molecules, and experience very different intermolecular forces than in their pure, undissolved state.
  - Energy is required to disrupt the structure of the pure solute and solvent, and energy is released due to the attractive interactions between the solute and solvent. Formation of a solution can be endothermic or exothermic, depending on the difference between these energies.
  - Ionic solutes dissociate in solution, whereas molecular solutes remain intact when dispersed in solution.
  - When water is the solvent, the solution is called aqueous and solutes are labeled (*aq*).
  - In aqueous solutions, solute species are surrounded by solvation shells of solvent molecules held in place by intermolecular forces, primarily hydrogen bonding and ion–dipole forces.

Section 11.3 – Reaction Stoichiometry in Solutions: Acid–Base Titrations

Calculate the number of moles of substances reacting during a solution-phase reaction such as acid–base titration (Problems 15–26).

- Acid–base titration is an analytical technique for finding the concentration of acid or base in an unknown solution.
  - Stoichiometry of reactions in solutions is described by balanced equations that relate the number of moles of each reactant and product. The number of moles of each reaction species in a volume of solution is given by $n_i = c_i V$ where c_i is the concentration in mol L^{-1} of species i in the solution.

- In a neutralization reaction, if the titrant A and the molecule B in the solution of unknown concentration react in a 1:1 stoichiometric ratio, the endpoint is achieved when the number of moles of A added equals the number of moles of B initially in the unknown sample.
- In titrations a measured volume of a solution with known concentration reacts to neutralize a known volume of sample whose concentration is unknown. The endpoint, at which neutralization is achieved, is signified by an endpoint indicator, most often a color change. At the endpoint, we determine the molarity of the unknown sample using the concept in the prior bullet point.

Section 11.4 – Reaction Stoichiometry in Solutions: Oxidation–Reduction Titrations

Balance equations for redox reactions in aqueous solution, using the half-reaction method, and calculate the concentrations of substances during redox titrations (Problems 27–40).

- In oxidation–reduction (redox) reactions electrons are transferred between reacting species as they combine to form products. Reactants that lose electrons are oxidized. Reactants that gain electrons are reduced. These reactions are coupled; it is not possible to have oxidation without reduction, and vice versa.
  - Balancing the equations for redox reactions involves special procedures. The steps are:
    - Assign oxidation numbers as described in Section 3.12.
    - Write unbalanced half-equations for the species being oxidized and the species being reduced.
    - Balance all atoms except oxygen and hydrogen.
    - Determine whether the reaction takes place in acidic or basic solution.
    - In acidic solution, add one H_3O^+ to the side that needs an extra H atom and then H_2O to the other side.
    - In basic solution, add one H_2O to the side that needs an extra H atom and then OH^- to the other side.
    - Balance charge transfer in each half-equation by inserting electrons as reactants or products.
    - Multiply the two half-equations by numbers that make the number of electrons given off in oxidation the same as the number gained in reduction. Add the two half-equations together.
  - The number of moles of the reactant A added during the titration is the concentration of the titrant A in mol L^{-1} mutiplied by the volume of the titrant added in L.
  - The number of moles of the reactant B consumed during the titration is determined from the number of moles of titrant (reactant A) added and the coefficients in the balanced equation.

Section 11.5 – Phase Equilibrium in Solutions: Nonvolatile Solutes

Calculate the molar mass of a nonvolatile solute from the changes it causes in the colligative properties (vapor-pressure lowering, boiling-point elevation, freezing-point lowering, or osmotic pressure) of its dilute solution (Problems 41–56).

- Like pure substances, solutions can have solid, liquid, and gaseous phases in equilibrium with one another. The conditions where phases coexist—such as boiling points, freezing points, and vapor pressure—have different values in solution than in the pure solvent. The colligative properties of solutions describe how nonvolatile solvents cause these changes in properties of the solvent. These changes depend only on the number of solute particles, not on their nature. Vapor pressure lowering, boiling point elevation, freezing point depression, and osmotic pressure are all treated with empirical equations.

- Vapor pressure lowering: $\Delta P_1 = -X_2 P_1^{\circ}$ where X_2 is the mole fraction of the solute and P_1° is the vapor pressure of the pure solvent.
- Boiling point elevation: $\Delta T = mK_b$ (the constants have been tabulated)
- Freezing point depression: $\Delta T = -mK_f$ (the constants have been tabulated)
- A solution in contact with its pure solvent across a semi-permeable membrane experiences an increase in pressure as pure solvent flows through the membrane into the solution. This osmotic pressure can be measured quite accurately, and through the equation $\pi V = nRT$ permits determination of the molecular weight of the solute.

Section 11.6 – Phase Equilibrium in Solutions: Volatile Solutes

Discuss the meaning of Henry's law, and use it to calculate the solubilities of gases in liquids (Problems 57–60).

- Henry's law relates the solubility of a gas in a liquid to the pressure of the gas above the liquid. If the pressure of the gas is increased, more gas will be dissolved.

Relate the total pressure and composition of the vapor in equilibrium with an ideal two-component solution to the composition of the solution and the vapor pressures of its pure components (Problems 61–64).

- When the solute is volatile, the vapor pressure in equilibrium with the solution has contributions from both solute and solvent.
  - Raoult's law: vapor pressure of the solvent $P_1 = X_1 P_1^{\circ}$
  - Henry's law: vapor pressure of the solute $P_2 = X_2 k_2$

  The partial pressures of the solute and solvent in the vapor will be different from their concentrations in the solution. The vapor will be richer in the more volatile component than is the solution.

Explain how distillation is used to separate the volatile components of a binary liquid solution.

- The vapor will be richer in the more volatile component than in the solution. The process of distillation exploits this fact to separate the components of the solution by heating the solution to the temperature where the more volatile component boils, removing the vapor, and condensing it to the liquid state.

Section 11.7 – Colloidal Suspensions

Describe the physical properties of a colloidal suspension.

- A colloidal suspension consists of at least two immiscible substances (such as oil and water) and has one of the substances dispersed throughout the other as small particles, ranging in size from 1 nanometer to 1 micrometer. Particles in this size range effectively scatter incoming light and produce special visual effects.

CUMULATIVE EXERCISE

FIGURE 11.31 Tapping sap from sugar maple trees.

Manufacturing of Maple Syrup

The sap in a maple tree can be described as an approximately 3.0% (by mass) solution of sucrose ($C_{12}H_{22}O_{11}$) in water (see Fig. 11.31). Sucrose does not dissociate to any significant extent in aqueous solution.

(a) At 20°C, the density of sap is 1.010 g cm^{-3}. Calculate the molarity of sucrose in sap.

(b) A typical maple tree yields about 12 gallons of sap per year. Calculate how many grams of sucrose are contained in this volume of sap (1 gallon = 3.785 L).

(c) The rising of sap in trees is caused largely by osmosis; the concentration of dissolved sucrose in sap is higher than that of the groundwater outside the tree. Calculate the osmotic pressure of a sap solution and the height to which the sap should rise above the ground on a day when the temperature is 20°C. Approximate the groundwater as pure (although, in fact, it typically contains 0.01 to 0.03 M dissolved species). Express the answer in meters and in feet (1 m = 3.28 ft).

(d) To produce maple syrup from sap, the sap is boiled to reduce its water content. Calculate the normal boiling point of a sap solution.

(e) Maple syrup is the concentrated sap solution that results when most of the water is boiled off. The syrup has a composition of approximately 64% (by mass) sucrose and 36% water, with flavoring components present in small concentrations. If the density of the maple syrup is 1.31 g cm^{-3}, calculate the mole fraction, molarity, and molality of sucrose in maple syrup.

(f) What volume (in gallons) of maple syrup can be obtained from the sap in one typical tree?

(g) In the presence of vanadium(V) oxide, dinitrogen tetroxide oxidizes sucrose to oxalic acid ($H_2C_2O_4$) according to the following equation:

$$C_{12}H_{22}O_{11}(aq) + 9\ N_2O_4(s) \longrightarrow 6\ H_2C_2O_4(aq) + 18\ NO(g) + 5\ H_2O(\ell)$$

Calculate the mass of N_2O_4 that will react completely with 7.00 L of the sap solution from part (a), and give the concentration of oxalic acid that results.

Answers

(a) 0.089 M

(b) 1.4×10^3 g sucrose

(c) π = 2.1 atm; height = 22 m = 72 ft

(d) 100.046°C

(e) Mole fraction = 0.086; molarity = 2.4 M; molality = 5.2 mol kg^{-1}

(f) 0.56 gallon

(g) 5.2×10^2 g N_2O_4; 0.53 M

PROBLEMS

Answers to problems whose numbers are red appear in Appendix G. Problems that are more challenging are indicated with asterisks.

11.1 Composition of Solutions

1. A patient has a "cholesterol count" of 214. Like many blood-chemistry measurements, this result is measured in units of milligrams per deciliter (mg dL^{-1}).
   (a) Determine the molar concentration of cholesterol in this patient's blood, taking the molar mass of cholesterol to be 386.64 g mol^{-1}.
   (b) Estimate the molality of cholesterol in the patient's blood.
   (c) If 214 is a typical cholesterol reading among men in the United States, determine the volume of such blood required to furnish 8.10 g of cholesterol.

2. In many states, a person is legally intoxicated if his or her blood has a concentration of 0.10 g (or more) of ethyl alcohol (C_2H_5OH) per deciliter. Express this "threshold concentration" in mol L^{-1}.

3. A solution of hydrochloric acid in water is 38.00% hydrochloric acid by mass. Its density is 1.1886 g cm^{-3} at 20°C. Compute its molarity, mole fraction, and molality at this temperature.

4. A solution of acetic acid and water contains 205.0 g L^{-1} of acetic acid and 820.0 g L^{-1} of water.
   (a) Compute the density of the solution.
   (b) Compute the molarity, molality, mole fraction, and mass percentage of acetic acid in this solution.
   (c) Take the acetic acid as the solvent, and do the same for water as the solute.

5. A 6.0835 M aqueous solution of acetic acid ($C_2H_4O_2$) has a density of 1.0438 g cm^{-3}. Compute its molality.

6. A 1.241 M aqueous solution of $AgNO_3$ (used to prepare silver chloride photographic emulsions) has a density of 1.171 g cm^{-3}. Compute its molality.

7. Water is slightly soluble in liquid nitrogen. At −196°C (the boiling point of liquid nitrogen), the mole fraction of water in a saturated solution is 1.00×10^{-5}. Compute the mass of water that can dissolve in 1.00 kg of boiling liquid nitrogen.

8. Some water dissolves in liquid methane at −161°C to give a solution in which the mole fraction of water is 6.0×10^{-5}. Determine the mass of water dissolved in 1.00 L of this solution if the density of the solution is 0.78 g cm^{-3}.

9. Concentrated phosphoric acid as sold for use in the laboratory is usually 90% H_3PO_4 by mass (the rest is water). Such a solution contains 12.2 mol of H_3PO_4 per liter of solution at 25°C.
   (a) Compute the density of this solution.
   (b) What volume of this solution should be used in mixing 2.00 L of a 1.00 M phosphoric acid solution?

10. A perchloric acid solution is 60.0% $HClO_4$ by mass. It is simultaneously 9.20 M at 25°C.
   (a) Compute the density of this solution.
   (b) What volume of this solution should be used in mixing 1.00 L of a 1.00 M perchloric acid solution?

11. Suppose 25.0 g of solid NaOH is added to 1.50 L of an aqueous solution that is already 2.40 M in NaOH. Then water is added until the final volume is 4.00 L. Determine the concentration of the NaOH in the resulting solution.

12. Suppose 0.400 L of a solution of 0.0700 M nitric acid is added to 0.800 L of a solution of 0.0300 M nitric acid, giving a total volume of 1.200 L. Calculate the concentration (molarity) of nitric acid in the resulting solution.

11.2 Nature of Dissolved Species

13. Rewrite the following balanced equations as net ionic equations.
   (a) $NaCl(aq) + AgNO_3(aq) \longrightarrow AgCl(s) + NaNO_3(aq)$
   (b) $K_2CO_3(s) + 2\ HCl(aq) \longrightarrow 2\ KCl(aq) + CO_2(g) + H_2O(\ell)(s) + 2\ H_2O(\ell)$
   (c) $2\ Cs(s) + 2\ H_2O(\ell) \longrightarrow 2\ CsOH(aq) + H_2(g)$
   (d) $2\ KMnO_4(aq) + 16\ HCl(aq) \longrightarrow 5\ Cl_2(g) + 2\ MnCl_2(aq) + 2\ KCl(aq) + 8\ H_2O(\ell)$

14. Rewrite the following balanced equations as net ionic equations.
   (a) $Na_2SO_4(aq) + BaCl_2(aq) \longrightarrow BaSO_4(s) + 2\ NaCl(aq)$
   (b) $6\ NaOH(aq) + 3\ Cl_2(g) \longrightarrow NaClO_3(aq) + 5\ NaCl(aq) + 3\ H_2O(\ell)$
   (c) $Hg_2(NO_3)_2(aq) + 2\ KI(aq) \longrightarrow Hg_2I_2(s) + 2\ KNO_3(aq)$
   (d) $3\ NaOCl(aq) + KI(aq) \longrightarrow NaIO_3(aq) + 2\ NaCl(aq) + KCl(aq)$

11.3 Reaction Stoichiometry in Solutions: Acid–Base Titrations

15. When treated with acid, lead(IV) oxide is reduced to a lead(II) salt, with the liberation of oxygen:

$$2\ PbO_2(s) + 4\ HNO_3(aq) \longrightarrow 2\ Pb(NO_3)_2(aq) + 2\ H_2O(\ell) + O_2(g)$$

What volume of a 7.91 M solution of nitric acid is just sufficient to react with 15.9 g of lead(IV) oxide, according to this equation?

16. Phosphoric acid is made industrially by the reaction of fluorapatite, $Ca_5(PO_4)_3F$, in phosphate rock with sulfuric acid:

$$Ca_5(PO_4)_3F(s) + 5\ H_2SO_4(aq) + 10\ H_2O(\ell) \longrightarrow 3\ H_3PO_4(aq) + 5\ (CaSO_4{\cdot}2H_2O)(s) + HF(aq)$$

What volume of 6.3 M phosphoric acid is generated by the reaction of 2.2 metric tons (2200 kg) of fluorapatite?

17. The carbon dioxide produced (together with hydrogen) from the industrial-scale oxidation of methane in the presence of nickel is removed from the gas mixture in a scrubber containing an aqueous solution of potassium carbonate:

$$CO_2(g) + H_2O(\ell) + K_2CO_3(aq) \longrightarrow 2\ KHCO_3(aq)$$

Calculate the volume of carbon dioxide (at 50°C and 1.00 atm pressure) that will react with 187 L of a 1.36 M potassium carbonate solution.

18. Nitrogen oxide can be generated on a laboratory scale by the reaction of dilute sulfuric acid with aqueous sodium nitrite:

$$6\ NaNO_2(aq) + 3\ H_2SO_4(aq) \longrightarrow 4\ NO(g) + 2\ HNO_3(aq) + 2\ H_2O(\ell) + 3\ Na_2SO_4(aq)$$

What volume of 0.646 M aqueous $NaNO_2$ should be used in this reaction to generate 5.00 L of nitrogen oxide at a temperature of 20°C and a pressure of 0.970 atm?

19. Write a balanced equation for the acid–base reaction that leads to the production of each of the following salts. Name the acid, base, and salt.
(a) CaF_2
(b) Rb_2SO_4
(c) $Zn(NO_3)_2$
(d) KCH_3COO

20. Write a balanced equation for the acid–base reaction that leads to the production of each of the following salts. Name the acid, base, and salt.
(a) Na_2SO_3
(b) $Ca(C_6H_5COO)_2$
(c) $PbSO_4$
(d) $CuCl_2$

21. Hydrogen sulfide can be removed from natural gas by reaction with excess sodium hydroxide. Name the salt that is produced in this reaction. (**Note:** Hydrogen sulfide loses both its hydrogen atoms in the course of this reaction.)

22. During the preparation of viscose rayon, cellulose is dissolved in a bath containing sodium hydroxide and later reprecipitated as rayon using a solution of sulfuric acid. Name the salt that is a by-product of this process. Rayon production is, in fact, a significant commercial source for this salt.

23. Phosphorus trifluoride is a highly toxic gas that reacts slowly with water to give a mixture of phosphorous acid and hydrofluoric acid.
(a) Write a balanced chemical equation for this reaction.
(b) Determine the concentration (in moles per liter) of each of the acids that result from the reaction of 1.94 L of phosphorus trifluoride (measured at 25°C and 0.970 atm pressure) with water to give a solution volume of 872 mL.

24. Phosphorus pentachloride reacts violently with water to give a mixture of phosphoric acid and hydrochloric acid.
(a) Write a balanced chemical equation for this reaction.
(b) Determine the concentration (in moles per liter) of each of the acids that result from the complete reaction of 1.22 L of phosphorus pentachloride (measured at 215°C and 0.962 atm pressure) with enough water to give a solution volume of 697 mL.

25. To determine the concentration of a solution of nitric acid, a 100.0-mL sample is placed in a flask and titrated with a 0.1279 M solution of potassium hydroxide. A volume of 37.85 mL is required to reach the phenolphthalein endpoint. Calculate the concentration of nitric acid in the original sample.

26. The concentration of aqueous ammonia in a cleaning solution is determined by titration with hydrochloric acid. A volume of 23.18 mL of 0.8381 M HCl is needed to titrate a 50.0-mL sample of the ammonia solution to a methyl red endpoint. Calculate the concentration of ammonia in the cleaning solution.

11.4 Reaction Stoichiometry in Solutions: Oxidation–Reduction Titrations

27. For each of the following balanced equations, write the oxidation number above the symbol of each atom that changes oxidation state in the course of the reactions.
(a) $2\ PF_2I(\ell) + 2\ Hg(\ell) \longrightarrow P_2F_4(g) + Hg_2I_2(s)$
(b) $2\ KClO_3(s) \longrightarrow 2\ KCl(s) + 3\ O_2(g)$
(c) $4\ NH_3(g) + 5\ O_2(g) \longrightarrow 4\ NO(g) + 6\ H_2O(g)$
(d) $2\ As(s) + 6\ NaOH(\ell) \longrightarrow 2\ Na_3AsO_3(s) + 3\ H_2(g)$

28. For each of the following balanced equations, write the oxidation number above the symbol of each atom that changes oxidation state in the course of the reaction.
(a) $N_2O_4(g) + KCl(s) \longrightarrow NOCl(g) + KNO_3(s)$
(b) $H_2S(g) + 4\ O_2F_2(s) \longrightarrow SF_6(g) + 2\ HF(g) + 4\ O_2(g)$
(c) $2\ POBr_3(s) + 3\ Mg(s) \longrightarrow 2\ PO(s) + 3\ MgBr_2(s)$
(d) $4\ BCl_3(g) + 3\ SF_4(g) \longrightarrow 4\ BF_3(g) + 3\ SCl_2(\ell) + 3\ Cl_2(g)$

29. Selenic acid (H_2SeO_4) is a powerful oxidizing acid that dissolves not only silver (as does the related acid H_2SO_4) but gold, through the following reaction:

$$2\ Au(s) + 6\ H_2SeO_4(aq) \longrightarrow Au_2(SeO_4)_3(aq) + 3\ H_2SeO_3(aq) + 3\ H_2O(\ell)$$

Determine the oxidation numbers of the atoms in this equation. Which species is oxidized and which is reduced?

30. Diiodine pentaoxide oxidizes carbon monoxide to carbon dioxide under room conditions, yielding iodine as the second product:

$$I_2O_5(s) + 5\ CO(g) \longrightarrow I_2(s) + 5\ CO_2(g)$$

This can be used in an analytical method to measure the amount of carbon monoxide in a sample of air. Determine the oxidation numbers of the atoms in this equation. Which species is oxidized and which is reduced?

31. Complete and balance the following equations for reactions taking place in acidic solution.
(a) $VO_2^+(aq) + SO_2(g) \longrightarrow VO^{2+}(aq) + SO_4^{2-}(aq)$
(b) $Br_2(\ell) + SO_2(g) \longrightarrow Br^-(aq) + SO_4^{2-}(aq)$
(c) $Cr_2O_7^{2-}(aq) + Np^{4+}(aq) \longrightarrow Cr^{3+}(aq) + NpO_2^{2+}(aq)$
(d) $HCOOH(aq) + MnO_4^-(aq) \longrightarrow CO_2(g) + Mn^{2+}(aq)$
(e) $Hg_2HPO_4(s) + Au(s) + Cl^-(aq) \longrightarrow Hg(\ell) + H_2PO_4^-(aq) + AuCl_4^-(aq)$

32. Complete and balance the following equations for reactions taking place in acidic solution.
(a) $MnO_4^-(aq) + H_2S(aq) \longrightarrow Mn^{2+}(aq) + SO_4^{2-}(aq)$
(b) $Zn(s) + NO_3^-(aq) \longrightarrow Zn^{2+}(aq) + NH_4^+(aq)$
(c) $H_2O_2(aq) + MnO_4^-(aq) \longrightarrow O_2(g) + Mn^{2+}(aq)$
(d) $Sn(s) + NO_3^-(aq) \longrightarrow Sn^{4+}(aq) + N_2O(g)$
(e) $UO_2^{2+}(aq) + Te(s) \longrightarrow U^{4+}(aq) + TeO_4^{2-}(aq)$

33. Complete and balance the following equations for reactions taking place in basic solution.
(a) $Cr(OH)_3(s) + Br_2(aq) \longrightarrow CrO_4^{2-}(aq) + Br^-(aq)$
(b) $ZrO(OH)_2(s) + SO_3^{2-}(aq) \rightarrow Zr(s) + SO_4^{2-}(aq)$
(c) $HPbO_2^-(aq) + Re(s) \longrightarrow Pb(s) + ReO_4^-(aq)$
(d) $HXeO_4^-(aq) \longrightarrow XeO_6^{4-}(aq) + Xe(g)$
(e) $N_2H_4(aq) + CO_3^{2-}(aq) \longrightarrow N_2(g) + CO(g)$

34. Complete and balance the following equations for reactions taking place in basic solution.
(a) $OCl^-(aq) + I^-(aq) \longrightarrow IO_3^-(aq) + Cl^-(aq)$
(b) $SO_3^{2-}(aq) + Be(s) \longrightarrow S_2O_3^{2-}(aq) + Be_2O_3^{2-}(aq)$
(c) $H_2BO_3^-(aq) + Al(s) \longrightarrow BH_4^-(aq) + H_2AlO_3^-(aq)$
(d) $O_2(g) + Sb(s) \longrightarrow H_2O_2(aq) + SbO_2^-(aq)$
(e) $Sn(OH)_6^{2-}(aq) + Si(s) \longrightarrow HSnO_2^-(aq) + SiO_3^{2-}(aq)$

35. The following balanced equations represent reactions that occur in aqueous acid. Break them down into balanced oxidation and reduction half-equations.
(a) $2\,H_3O^+(aq) + H_2O_2(aq) + 2\,Fe^{2+}(aq) \longrightarrow 2\,Fe^{3+}(aq) + 4\,H_2O(\ell)$
(b) $H_3O^+(aq) + H_2O(\ell) + 2\,MnO_4^-(aq) + 5\,SO_2(aq) \longrightarrow 2\,Mn^{2+}(aq) + 5\,HSO_4^-(aq)$
(c) $5\,ClO_2^-(aq) + 4\,H_3O^+(aq) \longrightarrow 4\,ClO_2(g) + Cl^-(aq) + 6\,H_2O(\ell)$

36. The following balanced equations represent reactions that occur in aqueous base. Break them down into balanced oxidation and reduction half-equations.
(a) $4\,PH_3(g) + 4\,H_2O(\ell) + 4\,CrO_4^{2-}(aq) \longrightarrow P_4(s) + 4\,Cr(OH)_4^-(aq) + 4\,OH^-(aq)$
(b) $NiO_2(s) + 2\,H_2O(\ell) + Fe(s) \longrightarrow Ni(OH)_2(s) + Fe(OH)_2(s)$
(c) $CO_2(g) + 2\,NH_2OH(aq) \longrightarrow CO(g) + N_2(g) + 3\,H_2O(g)$

37. Nitrous acid (HNO_2) disproportionates in acidic solution to nitrate ion (NO_3^-) and nitrogen oxide (NO). Write a balanced equation for this reaction.

38. Thiosulfate ion ($S_2O_3^{2-}$) disproportionates in acidic solution to give solid sulfur and aqueous hydrogen sulfite ion (HSO_3^-). Write a balanced equation for this reaction.

39. Potassium dichromate in acidic solution is used to titrate a solution of iron(II) ions, with which it reacts according to

$$Cr_2O_7^{2-}(aq) + 6\,Fe^{2+}(aq) + 14\,H_3O^+(aq) \longrightarrow 2\,Cr^{3+}(aq) + 6\,Fe^{3+}(aq) + 21\,H_2O(\ell)$$

A potassium dichromate solution is prepared by dissolving 5.134 g of $K_2Cr_2O_7$ in water and diluting to a total volume of 1.000 L. A total of 34.26 mL of this solution is required to reach the endpoint in a titration of a 500.0-mL sample containing $Fe^{2+}(aq)$. Determine the concentration of Fe^{2+} in the original solution.

40. Cerium(IV) ions are strong oxidizing agents in acidic solution, oxidizing arsenious acid to arsenic acid according to the following equation:

$$2\,Ce^{4+}(aq) + H_3AsO_3(aq) + 3\,H_2O(\ell) \longrightarrow 2\,Ce^{3+}(aq) + H_3AsO_4(aq) + 2\,H^+(aq)$$

A sample of As_2O_3 weighing 0.217 g is dissolved in basic solution and then acidified to make H_3AsO_3. Its titration with a solution of acidic cerium(IV) sulfate requires 21.47 mL. Determine the original concentration of $Ce^{4+}(aq)$ in the titrating solution.

11.5 Phase Equilibrium in Solutions: Nonvolatile Solutes

41. The vapor pressure of pure acetone, $CH_3C(O)CH_3$, at 30°C is 0.3270 atm. Suppose 15.0 g of benzophenone, $C_{13}H_{10}O$, is dissolved in 50.0 g of acetone. Calculate the vapor pressure of acetone above the resulting solution.

42. The vapor pressure of diethyl ether (molar mass, 74.12 g mol^{-1}) at 30°C is 0.8517 atm. Suppose 1.800 g of maleic acid, $C_4H_4O_4$, is dissolved in 100.0 g of diethyl ether at 30°C. Calculate the vapor pressure of diethyl ether above the resulting solution.

43. Pure toluene (C_7H_8) has a normal boiling point of 110.60°C. A solution of 7.80 g of anthracene ($C_{14}H_{10}$) in 100.0 g of toluene has a boiling point of 112.06°C. Calculate K_b for toluene.

44. When 2.62 g of the nonvolatile solid anthracene, $C_{14}H_{10}$, is dissolved in 100.0 g of cyclohexane, C_6H_{12}, the boiling point of the cyclohexane is raised by 0.41°C. Calculate K_b for cyclohexane.

45. When 39.8 g of a nondissociating, nonvolatile sugar is dissolved in 200.0 g of water, the boiling point of the water is raised by 0.30°C. Estimate the molar mass of the sugar.

46. When 2.60 g of a substance that contains only indium and chlorine is dissolved in 50.0 g of tin(IV) chloride, the normal boiling point of the tin(IV) chloride is raised from 114.1°C to 116.3°C. If $K_b = 9.43$ K kg mol^{-1} for $SnCl_4$, what are the approximate molar mass and the probable molecular formula of the solute?

47. The Rast method for determining molar masses uses camphor as the solvent. Camphor melts at 178.4°C, and its large K_f (37.7 K kg mol^{-1}) makes it especially useful for accurate work. A sample of an unknown substance that weighs 0.840 g reduces the freezing point of 25.0 g of camphor to 170.8°C. What is its molar mass?

48. Barium chloride has a freezing point of 962°C and a freezing-point depression constant of 108 K kg mol^{-1}. If 12 g of an unknown substance dissolved in 562 g of barium chloride gives a solution with a freezing point of 937°C, compute the molar mass of the unknown, assuming no dissociation takes place.

49. Ice cream is made by freezing a liquid mixture that, as a first approximation, can be considered a solution of sucrose ($C_{12}H_{22}O_{11}$) in water. Estimate the temperature at which the first ice crystals begin to appear in a mix that

consists of 34% (by mass) sucrose in water. As ice crystallizes out, the remaining solution becomes more concentrated. What happens to its freezing point?

50. The solution to Problem 49 shows that to make homemade ice cream, temperatures ranging downward from -3°C are needed. Ice cubes from a freezer have a temperature of about -12°C ($+10$°F), which is cold enough, but contact with the warmer ice cream mixture causes them to melt to liquid at 0°C, which is too warm. To obtain a liquid that is cold enough, salt (NaCl) is dissolved in water, and ice is added to the saltwater. The salt lowers the freezing point of the water enough so that it can freeze the liquid inside the ice cream maker. The instructions for an ice cream maker say to add one part salt to eight parts water (by mass). What is the freezing point of this solution (in degrees Celsius and degrees Fahrenheit)? Assume that the NaCl dissociates fully into ions, and that the solution is ideal.

51. An aqueous solution is 0.8402 molal in Na_2SO_4. It has a freezing point of -4.218°C. Determine the effective number of particles arising from each Na_2SO_4 formula unit in this solution.

52. The freezing-point depression constant of pure H_2SO_4 is 6.12 K kg mol^{-1}. When 2.3 g of ethanol (C_2H_5OH) is dissolved in 1.00 kg of pure sulfuric acid, the freezing point of the solution is 0.92 K lower than the freezing point of pure sulfuric acid. Determine how many particles are formed as 1 molecule of ethanol goes into solution in sulfuric acid.

53. A 200-mg sample of a purified compound of unknown molar mass is dissolved in benzene and diluted with that solvent to a volume of 25.0 cm^3. The resulting solution is found to have an osmotic pressure of 0.0105 atm at 300 K. What is the molar mass of the unknown compound?

54. Suppose 2.37 g of a protein is dissolved in water and diluted to a total volume of 100.0 mL. The osmotic pressure of the resulting solution is 0.0319 atm at 20°C. What is the molar mass of the protein?

55. A polymer of large molar mass is dissolved in water at 15°C, and the resulting solution rises to a final height of 15.2 cm above the level of the pure water, as water molecules pass through a semipermeable membrane into the solution. If the solution contains 4.64 g polymer per liter, calculate the molar mass of the polymer.

56. Suppose 0.125 g of a protein is dissolved in 10.0 cm^3 of ethyl alcohol (C_2H_5OH), whose density at 20°C is 0.789 g cm^{-3}. The solution rises to a height of 26.3 cm in an osmometer (an apparatus for measuring osmotic pressure). What is the approximate molar mass of the protein?

11.6 Phase Equilibrium in Solutions: Volatile Solutes

57. The Henry's law constant at 25°C for carbon dioxide dissolved in water is 1.65×10^3 atm. If a carbonated beverage is bottled under a CO_2 pressure of 5.0 atm:
   (a) Calculate the number of moles of carbon dioxide dissolved per liter of water under these conditions, using 1.00 g cm^{-3} as the density of water.
   (b) Explain what happens on a microscopic level after the bottle cap is removed.

58. The Henry's law constant at 25°C for nitrogen dissolved in water is 8.57×10^4 atm, that for oxygen is 4.34×10^4 atm, and that for helium is 1.7×10^5 atm.
   (a) Calculate the number of moles of nitrogen and oxygen dissolved per liter of water in equilibrium with air at 25°C. Use Table 9.1.
   (b) Air is dissolved in blood and other bodily fluids. As a deep-sea diver descends, the pressure increases and the concentration of dissolved air in the blood increases. If the diver returns to the surface too quickly, gas bubbles out of solution within the body so rapidly that it can cause a dangerous condition called "the bends." Use Henry's law to show why divers sometimes use a combination of helium and oxygen in their breathing tanks in place of compressed air.

59. At 25°C, some water is added to a sample of gaseous methane (CH_4) at 1.00 atm pressure in a closed vessel, and the vessel is shaken until as much methane as possible dissolves. Then 1.00 kg of the solution is removed and boiled to expel the methane, yielding a volume of 3.01 L of $CH_4(g)$ at 0°C and 1.00 atm. Determine the Henry's law constant for methane in water.

60. When exactly the procedure of Problem 59 is conducted using benzene (C_6H_6) in place of water, the volume of methane that results is 0.510 L at 0°C and 1.00 atm. Determine the Henry's law constant for methane in benzene.

61. At 20°C, the vapor pressure of toluene is 0.0289 atm and the vapor pressure of benzene is 0.0987 atm. Equal numbers of moles of toluene and benzene are mixed and form an ideal solution. Compute the mole fraction of benzene in the vapor in equilibrium with this solution.

62. At 90°C, the vapor pressure of toluene is 0.534 atm and the vapor pressure of benzene is 1.34 atm. Benzene (0.400 mol) is mixed with toluene (0.900 mol) to form an ideal solution. Compute the mole fraction of benzene in the vapor in equilibrium with this solution.

63. At 40°C, the vapor pressure of pure carbon tetrachloride (CCl_4) is 0.293 atm and the vapor pressure of pure dichloroethane ($C_2H_4Cl_2$) is 0.209 atm. A nearly ideal solution is prepared by mixing 30.0 g of carbon tetrachloride with 20.0 g of dichloroethane.
   (a) Calculate the mole fraction of CCl_4 in the solution.
   (b) Calculate the total vapor pressure of the solution at 40°C.
   (c) Calculate the mole fraction of CCl_4 in the vapor in equilibrium with the solution.

64. At 300 K, the vapor pressure of pure benzene (C_6H_6) is 0.1355 atm and the vapor pressure of pure *n*-hexane (C_6H_{14}) is 0.2128 atm. Mixing 50.0 g of benzene with 50.0 g of *n*-hexane gives a solution that is nearly ideal.
   (a) Calculate the mole fraction of benzene in the solution.
   (b) Calculate the total vapor pressure of the solution at 300 K.
   (c) Calculate the mole fraction of benzene in the vapor in equilibrium with the solution.

ADDITIONAL PROBLEMS

65. Veterinarians use Donovan's solution to treat skin diseases in animals. The solution is prepared by mixing 1.00 g of $AsI_3(s)$, 1.00 g of $HgI_2(s)$, and 0.900 g of $NaHCO_3(s)$ in enough water to make a total volume of 100.0 mL.

(a) Compute the total mass of iodine per liter of Donovan's solution, in grams per liter.

(b) You need a lot of Donovan's solution to treat an outbreak of rash in an elephant herd. You have plenty of mercury(II) iodide and sodium hydrogen carbonate, but the only arsenic(III) iodide you can find is 1.50 L of a 0.100 M aqueous solution. Explain how to prepare 3.50 L of Donovan's solution starting with these materials.

66. Relative solubilities of salts in liquid ammonia can differ significantly from those in water. Thus, silver bromide is soluble in ammonia, but barium bromide is not (the reverse of the situation in water).

(a) Write a balanced equation for the reaction of an ammonia solution of barium nitrate with an ammonia solution of silver bromide. Silver nitrate is soluble in liquid ammonia.

(b) What volume of a 0.50 M solution of silver bromide will react completely with 0.215 L of a 0.076 M solution of barium nitrate in ammonia?

(c) What mass of barium bromide will precipitate from the reaction in part (b)?

***67.** A 5.0-L flask contains a mixture of ammonia and nitrogen at 27°C and a total pressure of 3.00 atm. The sample of gas is allowed to flow from the flask until the pressure in the flask has fallen to 1.00 atm. The gas that escapes is passed through 1.50 L of 0.200 M acetic acid. All the ammonia in the gas that escapes is absorbed by the solution and turns out to be just sufficient to neutralize the acetic acid present. The volume of the solution does not change significantly.

(a) Will the electrical conductivity of the aqueous solution change significantly as the gas is absorbed? Give equations for any reactions, and calculate the final concentrations of the principal ions present (if any) at the end.

(b) Calculate the percentage by mass of ammonia in the flask initially.

***68.** It was desired to neutralize a certain solution X that had been prepared by mixing solutions of potassium chloride and hydrobromic acid. Titration of 10.0 mL X with 0.100 M silver nitrate required 50.0 mL of the latter. The resulting precipitate, containing a mixture of AgCl and AgBr, was dried and found to weigh 0.762 g. How much 0.100 M sodium hydroxide should be used to neutralize 10.0 mL solution X?

***69.** Vanadic ion, V^{3+}, forms green salts and is a good reducing agent, being itself changed in neutral solutions to the nearly colorless ion $V(OH)_4^+$. Suppose that 15.0 mL of a 0.200-M solution of vanadic sulfate, $V_2(SO_4)_3$, was needed to reduce completely a 0.540-g sample of an unknown substance X. If each molecule of X accepted just one electron, what is the molecular weight of X? Suppose that each molecule of X accepted three electrons; what would be the molecular weight of X then?

***70.** A new antibiotic, A, which is an acid, can readily be oxidized by hot aqueous permanganate; the latter is reduced to manganous ion, Mn^{2+}. The following experiments have been performed with A: (a) 0.293 g A consumes just 18.3 mL of 0.080 M $KMnO_4$; (b) 0.385 g A is just neutralized by 15.7 mL of 0.490 M NaOH. What can you conclude about the molecular weight of A from (a), from (b), and from both considered together?

71. Suppose 150 mL of a 10.00% by mass solution of sodium chloride (density = 1.0726 g cm^{-3}) is acidified with sulfuric acid and then treated with an excess of $MnO_2(s)$. Under these conditions, all the chlorine is liberated as $Cl_2(g)$. The chlorine is collected without loss and reacts with excess $H_2(g)$ to form $HCl(g)$. The $HCl(g)$ is dissolved in enough water to make 250 mL of solution. Compute the molarity of this solution.

***72.** The amount of ozone in a mixture of gases can be determined by passing the mixture through an acidic aqueous solution of potassium iodide, where the ozone reacts according to

$$O_3(g) + 3\,I^-(aq) + H_2O(\ell) \longrightarrow O_2(g) + I_3^-(aq) + 2\,OH^-(aq)$$

to form the triiodide ion I_3^-. The amount of triiodide produced is then determined by titrating with thiosulfate solution:

$$I_3^-(aq) + 2\,S_2O_3^{2-}(aq) \longrightarrow 3\,I^-(aq) + S_4O_6^{2-}(aq)$$

A small amount of starch solution is added as an indicator because it forms a deep-blue complex with the triiodide solution. Disappearance of the blue color thus signals the completion of the titration. Suppose 53.2 L of a gas mixture at a temperature of 18°C and a total pressure of 0.993 atm is passed through a solution of potassium iodide until the ozone in the mixture has reacted completely. The solution requires 26.2 mL of a 0.1359-M solution of thiosulfate ion to titrate to the endpoint. Calculate the mole fraction of ozone in the original gas sample.

73. The vapor pressure of pure liquid CS_2 is 0.3914 atm at 20°C. When 40.0 g of rhombic sulfur is dissolved in 1.00 kg of CS_2, the vapor pressure of CS_2 decreases to 0.3868 atm. Determine the molecular formula for the sulfur molecules dissolved in CS_2.

74. The expressions for boiling-point elevation and freezing-point depression apply accurately to *dilute* solutions only. A saturated aqueous solution of NaI (sodium iodide) in water has a boiling point of 144°C. The mole fraction of NaI in the solution is 0.390. Compute the molality of this solution. Compare the boiling-point elevation predicted by the expression in this chapter with the elevation actually observed.

75. You take a bottle of soft drink out of your refrigerator. The contents are liquid and stay liquid, even when you shake them. Presently, you remove the cap and the liquid freezes solid. Offer a possible explanation for this observation.

76. Mercury(II) chloride ($HgCl_2$) freezes at 276.1°C and has a freezing-point depression constant K_f of 34.3 K kg mol^{-1}. When 1.36 g of solid mercury(I) chloride (empirical formula HgCl) is dissolved in 100 g of $HgCl_2$, the freezing point is reduced by 0.99°C. Calculate the molar mass of the dissolved solute species and give its molecular formula.

*77. The vapor pressure of an aqueous solution of $CaCl_2$ at 25°C is 0.02970 atm. The vapor pressure of pure water at the same temperature is 0.03126 atm. Estimate the freezing point of the solution.

78. Ethylene glycol (CH_2OHCH_2OH) is used in antifreeze because, when mixed with water, it lowers the freezing point below 0°C. What mass percentage of ethylene glycol in water must be used to reduce the freezing point of the mixture to −5.0°C, assuming ideal solution behavior?

79. A new compound has the empirical formula $GaCl_2$. This surprises some chemists who, based on the position of gallium in the periodic table, expect a chloride of gallium to have the formula $GaCl_3$ or possibly GaCl. They suggest that the "$GaCl_2$" is really $Ga[GaCl_4]$, in which the bracketed group behaves as a unit with a −1 charge. Suggest experiments to test this hypothesis.

*80. Suppose two beakers are placed in a small closed container at 25°C. One contains 400 mL of a 0.100-M aqueous solution of NaCl; the second contains 200 mL of a 0.250-M aqueous solution of NaCl. Small amounts of water evaporate from both solutions. As time passes, the volume of solution in the second beaker gradually increases, and that in the first gradually decreases. Why? If we wait long enough, what will the final volumes and concentrations be?

*81. The walls of erythrocytes (red blood cells) are permeable to water. In a salt solution, they shrivel (lose water) when the outside salt concentration is high and swell (take up water) when the outside salt concentration is low. In an experiment at 25°C, an aqueous solution of NaCl that has a freezing point of −0.046°C causes erythrocytes neither to swell nor to shrink, indicating that the osmotic pressure of their contents is equal to that of the NaCl solution. Calculate the osmotic pressure of the solution inside the erythrocytes under these conditions, assuming that its molarity and molality are equal.

82. Silver dissolves in molten lead. Compute the osmotic pressure of a 0.010 M solution of silver in lead at 423°C. Compute the height of a column of molten lead (ρ = 11.4 g cm^{-3}) to which this pressure corresponds.

83. Henry's law is important in environmental chemistry, where it predicts the distribution of pollutants between water and the atmosphere. Benzene (C_6H_6) emitted in wastewater streams, for example, can pass into the air, where it is degraded by processes induced by light from the sun. The Henry's law constant for benzene in water at 25°C is 301 atm. Calculate the partial pressure of benzene vapor in equilibrium with a solution of 2.0 g of benzene per 1000 L of water. How many benzene molecules are present in each cubic centimeter?

*84. Refer to the data of Problem 62. Calculate the mole fraction of toluene in a mixture of benzene and toluene that boils at 90°C under atmospheric pressure.

85. What is the difference between a solution and a colloidal suspension? Give examples of each, and show how, in some cases, it may be difficult to classify a mixture as one or the other.

CUMULATIVE PROBLEMS

86. A student prepares a solution by dissolving 1.000 mol of Na_2SO_4 in water. She accidentally leaves the container uncovered and comes back the next week to find only a white, solid residue. The mass of the residue is 322.2 g. Determine the chemical formula of this residue.

87. Complete combustion of 2.40 g of a compound of carbon, hydrogen, and oxygen yielded 5.46 g CO_2 and 2.23 g H_2O. When 8.69 g of the compound was dissolved in 281 g of water, the freezing point of the solution was found to be −0.97°C. What is the molecular formula of the compound?

*88. Imagine that two 1-L beakers, A and B, each containing an aqueous solution of fructose (a nonvolatile sugar with molecular weight = 180) are placed together in a box, which is then sealed. (The concentrations of the solutions are not necessarily the same.) The temperature remains constant at 26°C. Initially, there is 600 mL of solution in A and 100 mL of solution in B. As the solutions stand in the sealed box, their volumes change slowly for a while. When they stop changing, beaker A contains 400 mL and beaker B contains 300 mL. It is then determined that the solution in A is 1.5 M in fructose and has a density of 1.10 g mL^{-1}.

(a) What is the molar concentration of fructose in the solution in beaker B at the end? Explain.

(b) Calculate the concentration of fructose in the solution in A at the start.

(c) Calculate the concentration of the fructose in the solution in B at the start.

(d) The vapor pressure of pure water at 26°C is 25.2 torr. What is the pressure of water vapor in the box at the end, after the volumes have stopped changing?

Equilibrium in Chemical Reactions

© AG-PHOTOS/Shutterstock.com

Stalactites (top) and stalagmites (bottom) consist of calcium carbonate. They form when a water solution containing Ca^{2+} and HCO_3^- ions running through a cave loses water by evaporation. The concentration of ions then exceeds the equilibrium solubility of calcium carbonate. Carbon dioxide is released, and solid calcium carbonate precipitates out of the solution: $Ca^{2+}(aq) + 2\ HCO_3^-(aq) \longrightarrow CaCO_3(s) + H_2O + CO_2(g)$.

UNIT IV

In chemical reactions, bonds are broken in the reactants, and the atoms re-arrange themselves with new bonds to form the products. How far do reactions proceed toward completely converting the reactants to products? What determines the extent of their progress? Experience shows that many reactions do not go to completion, but approach instead an equilibrium state in which products and unconsumed reactants are both present in specific relative amounts. Once equilibrium has been achieved, the composition of the reaction mixture does not change further. The equilibrium composition of the mixture is related quantitatively to the equilibrium constant for the reaction. If we know the equilibrium constant, we can calculate the equilibrium composition that will result from any mixture of reactants and products present initially. This calculation is one of the most important tools available to chemists because it is used to predict and maximize the yield of reactions throughout fundamental and applied chemistry.

Temperature influences the progress of chemical reactions—driving some forward while retarding others—and is quantitatively connected to chemical equilibrium by the science of thermodynamics. Thermodynamics predicts the value of the equilibrium constant from simple physical properties of the reactants and products, and explains how that value depends on the reaction temperature. In that way, thermodynamics shows how to maximize yield of the reaction by manipulation of the temperature.

NOTE TO THE READER:
It is purely a matter of preference whether one should first study chemical equilibrium from the empirical point of view, and then study thermodynamics to provide the fundamental explanation of equilibrium, or learn thermodynamics first as essential background for the study of equilibrium. We have written this textbook to allow either approach. If your instructor prefers to cover thermodynamics before equilibrium, you should read Chapters 12, 13, and 14 straight through in the order written. If your instructor prefers to cover equilibrium from the empirical point of view before studying thermodynamics, you should skip now to Chapter 14 and omit those sections of Chapter 14 (clearly marked) that require background in thermodynamics. You should come back and read those sections later after you have studied Chapters 12 and 13.

Unit Chapters

Unit Goals

- To relate composition in the equilibrium state to the equilibrium constant
- To calculate composition in the equilibrium state from the equilibrium constant
- To describe the influence of temperature on the equilibrium constant
- To apply thermodynamics to explain these connections and maximize reaction yield

12 CHAPTER

Thermodynamic Processes and Thermochemistry

The steam locomotive operates by converting thermal energy into mechanical energy. The diesel locomotive converts chemical energy into electrical energy, then electrical energy into mechanical energy to generate motion. All these energy conversion processes are governed by thermodynamics.

Experience shows that heat is the most important factor influencing the extent of chemical reactions. Heat drives some reactions toward completion, but retards the progress of others. Therefore, it is appropriate to launch our study of chemical equilibrium by learning how to measure the heat transfer in a chemical reaction. This objective leads us into the branch of physical science called thermodynamics, which describes the meaning of heat and gives procedures for measuring heat transfer quantitatively.

Thermodynamics is a broad and general subject with applications in all branches of the physical and biological sciences and engineering; thus, we limit our discussion to those aspects necessary for chemical equilibrium. In this chapter, we demonstrate that heating a substance, the chemical system, can change the energy content of the system just as doing work on the system (e.g. pushing a ball uphill) does. Interestingly, heating a substance need not increase its temperature; when you boil water the temperature stays constant as the phase change occurs. The total amount of energy (in system and surroundings) is conserved during any such processes. The previous sentence in more concise form is the first law of thermodynamics, which is the unifying theme of this chapter. We see how to measure quantitatively the heat transfer into a system through its *heat capacity.* We see how the transfer depends on experimental conditions such as constant pressure or constant volume. Finally, we apply these methods to measure the heat transfer in chemical reactions conducted at constant pressure. Then we are ready to connect heat transfer to chemical equilibrium at the beginning of Chapter 13.

We conclude this introduction with a general overview of thermodynamics, as context for the specific studies in this and the next chapter. Thermodynamics, in which a few apparently simple laws summarize a rich variety of observed behavior, is one of the surest and most powerful branches of science. The distinctive feature of thermodynamics is the universality of its basic laws, and the beauty of the subject is the many conclusions that can be deduced from those few laws. The laws of thermodynamics cannot themselves be derived or proved. Instead, they are generalizations drawn from a great many observations of the behavior of matter. The history of thermodynamics, like that of other fields of science, has been fraught with misconceptions. As we look back on the beginnings of the discipline in the 19th century, it appears to have developed with agonizing slowness. But it *has* developed, and its laws are the pillars on which much of modern science rests. The foundations of thermodynamics are completely understood today. It is being applied in research at the forefront of science, on systems ranging from black holes in distant parts of the universe to the growth and development of the biological cell. Many new results and insights are being acquired, but the foundations are not challenged.

Thermodynamics is an *operational* science, concerned with macroscopic, measurable properties and the relations among them. Its goal is to predict what types of chemical and physical processes are possible, and under what conditions, and to calculate quantitatively the properties of the equilibrium state that ensues when a process is conducted. For example, with thermodynamics we can answer the following types of chemical questions:

1. If hydrogen and nitrogen are mixed, is it possible for them to react? If so, what will be the percentage yield of ammonia?
2. How will a particular change in temperature or pressure affect the extent of the reaction?
3. How can the conditions for the reaction be optimized to maximize its yield?

Thermodynamics is an immensely practical subject. The knowledge from thermodynamics that a particular chemical process is impossible under certain proposed conditions can prevent great loss of time and resources spent vainly trying to conduct the reaction under those conditions. Thermodynamics can also suggest ways to change conditions so that a process becomes possible.

The power of thermodynamics lies in its generality: It rests on no particular model of the structure of matter. In fact, if the entire atomic theory of matter were to be found invalid and discarded (a *very* unlikely event!), the foundations of thermodynamics would remain unshaken. Nonetheless, thermodynamics has some important limitations. Thermodynamics asserts that substances have specific measurable macroscopic properties, but it cannot explain why a particular substance has particular numerical values for these properties. Thermodynamics can determine whether a process is possible, but it cannot say how rapidly the process will occur. For example, thermodynamics predicts that diamond is an unstable substance at atmospheric pressure and will eventually become graphite, but cannot predict how long this process will take.

12.1 Systems, States, and Processes

Thermodynamics uses abstract models to represent real-world systems and processes. These processes may appear in a rich variety of situations, including controlled laboratory conditions, industrial production facilities, living systems, the environment on Earth, and space. A key step in applying the methods of thermodynamics to such diverse processes is to formulate the thermodynamic model *for each process*. This step requires precise definitions of thermodynamic terms. Students (and professors!) of thermodynamics encounter—and sometimes create—apparent contradictions that arise from careless or inaccurate use of language. Part of the difficulty is that many thermodynamic terms also have everyday meanings different from their thermodynamic usage. This section provides a brief introduction to the language of thermodynamics.

A **system** is that part of the universe of immediate interest in a particular experiment or study. The system always contains a certain amount of matter and is described by specific parameters that are controlled in the experiment. For example, the gas confined in a closed box may constitute the system, characterized by the number of moles of the gas and the fixed volume of the box. But in other experiments, it would be more appropriate to consider the gas molecules in a particular cubic centimeter of space in the middle of a room to be the system. In the first case, the boundaries are physical walls, but in the second case, the boundaries are conceptual. We explain later that the two kinds of boundaries are treated the same way mathematically. In the second example, the system is characterized by its volume, which is definite, and by the number of moles of gas within it, which may fluctuate as the system exchanges molecules with the surrounding regions.

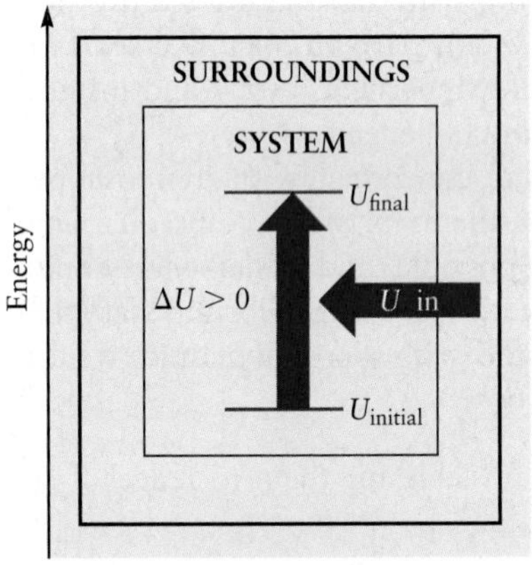

(a)

Systems are classified by the extent to which their boundaries permit exchange of matter and energy with the surrounding regions. In a **closed system**, the boundaries prevent the flow of matter into or out of it (the boundaries are **impermeable**), whereas the boundaries in an **open system** permit such flow. The amount of matter in an open system can change with time. An **isolated system** exchanges neither matter nor energy with the rest of the universe. **Rigid walls** prevent the system from gaining energy by mechanical processes such as compression and deformation; nonrigid walls permit mechanical energy transfer. **Adiabatic walls** prevent heat transfer (described in detail later) between the system and surroundings, whereas **diathermal walls** permit heat transfer. The energy content directly related to temperature is called *thermal energy*.

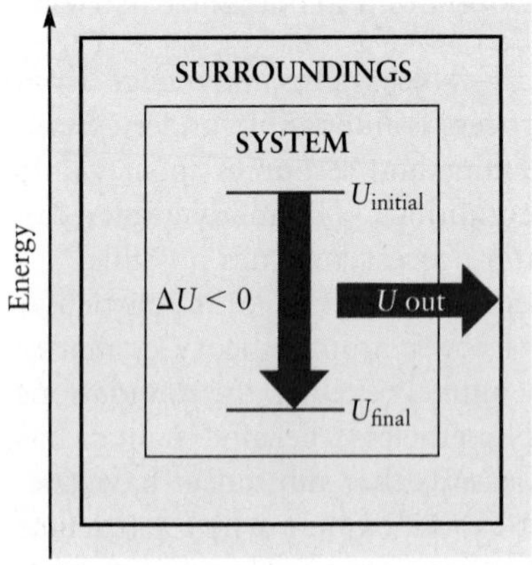

(b)

FIGURE 12.1 The thermodynamic universe. (a) The system gains energy from the surroundings. (b) The system loses energy to the surroundings.

The definition of "the system" must be tailored to the specific process under consideration. Simple physical processes such as heating or cooling a metal object are modeled as closed systems (no matter is gained or lost) with diathermal (thermal energy is transferred) and nonrigid walls (the object may expand or contract). Most chemical reactions and phase changes can be modeled as closed systems (no matter is exchanged) with diathermal (heat is transferred) and nonrigid walls. For example, the system might be a mole of water initially in the liquid phase, with a pressure of 1 atm and a temperature of 100°C. Heating the system while holding the pressure constant causes the water to evaporate. At the end of the process the system is now gaseous water at 1 atm, so it occupies a much larger volume, but no mass has been transferred to the surroundings. You will gain confidence in these classifications through experience as we examine many processes in this chapter.

The portion of the remainder of the universe that can exchange energy and matter with the system during the process of interest is called the **surroundings.** The surroundings provide the external forces that cause changes in the properties of the system during a process. The system and the surroundings together constitute the **thermodynamic universe** for that process. The thermodynamic universe for that process is isolated. Matter and energy are conserved in the thermodynamics universe while they are exchanged between the system and the surroundings during the process. Figure 12.1a shows a process in which the system gains energy U while the surroundings lose energy U.

Thermodynamics is concerned with macroscopic properties of systems and changes in these properties during processes. Such properties are of two kinds: *extensive* and *intensive.* To distinguish between them, consider the following "thought experiment." Place a thin wall through the middle of a system, dividing it into two subsystems, each of which, like the system itself, is characterized by certain properties. An **extensive property** of the system can be written as the *sum* of the corresponding property in the two subsystems. Volume, mass, and energy are typical extensive properties; the volume of a system is the sum of the volumes of its subsystems. An **intensive property** of the system is the *same* as the corresponding property of each of the subsystems. Temperature and pressure are typical intensive properties; if a system at 298 K is divided in half, the temperature of each half will still be 298 K.

A **thermodynamic state** is a macroscopic condition of a system in which the properties of the system are held at selected fixed values independent of time. The properties of the system are held constant by its boundaries and the surroundings. For example, a system comprising 2 mol helium (He) gas can be held in a piston–cylinder apparatus that maintains the system pressure at 1.5 atm, and the apparatus may be immersed in a heat bath that maintains the system temperature at 298 K. The intensive properties of pressure (P) and temperature (T) are then held constant at the values 1.5 atm and 298 K, respectively, by the surroundings. The extensive properties of the system are held constant by **constraints** at its boundaries. For example, a system comprising 2 mol helium (He) gas in a volume fixed at 5 L has constraints on its volume and the composition and mass of its contents.

After the system has been prepared by establishing a set of constraints in the surroundings, after all disturbances caused by the preparation cease and none of its properties changes with time, the system is said to have reached **equilibrium**. The same equilibrium state can be reached from different directions. The thermodynamic state of a system comprising a given quantity of a liquid or gaseous pure substance is fixed when any two of its independent properties are given. Thus, the specification of P and T for 1 mol of a pure gas fixes not merely the volume, V, but all other properties of the material, such as the internal energy, U, which is defined in Section 12.2. These relations among properties can be displayed as three-dimensional plots of any property as a function of two other properties. For 1 mol of a gas, a plot of pressure against volume and temperature is shown in Figure 12.2. The resulting surface is represented by an equation giving P as a function of V and T; this is called the *equation of state* of the substance (see Section 9.6). If we avoid regions of small V and low T where gas nonideality becomes important, the experimentally determined surface is that shown in Figure 12.2, and the equation of state is given by the ideal gas law $PV = nRT$. The points on this surface (A, B, . . .) represent experimentally measured values of P in equilibrium thermodynamic states of the system fixed by particular values of V and T. Experience shows that the values of all other macroscopic properties take on definite values at each of these states. For example, we can visualize the internal energy, U, as a similar three-dimensional plot versus T and V.

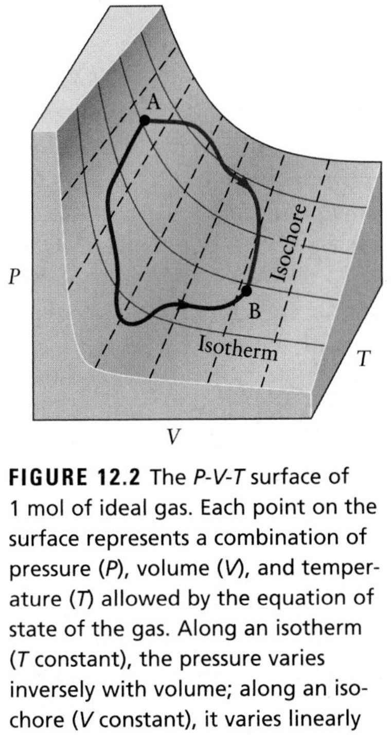

FIGURE 12.2 The *P*-*V*-*T* surface of 1 mol of ideal gas. Each point on the surface represents a combination of pressure (*P*), volume (*V*), and temperature (*T*) allowed by the equation of state of the gas. Along an isotherm (*T* constant), the pressure varies inversely with volume; along an isochore (*V* constant), it varies linearly with temperature. Two processes are shown connecting states A and B along paths that satisfy the equation of state at every point.

A **thermodynamic process** changes the thermodynamic state of a system. A process may be *physical,* such as changing the pressure of a gaseous system or boiling a liquid. A *chemical* process involves a chemical reaction (for example, the decomposition of solid $CaCO_3$, at 900 K and 1 atm pressure, to give solid CaO and gaseous CO_2 at the same temperature and pressure).

Because a process changes the state of a system, the process must start with the system in a particular equilibrium state and must also end with the system in a particular equilibrium state. Two such states A and B are indicated in

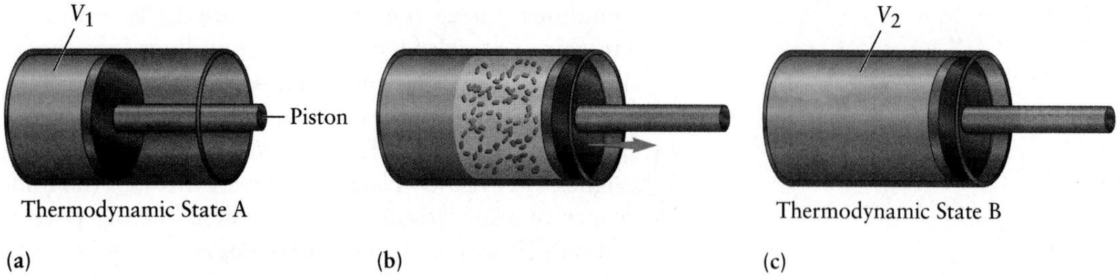

FIGURE 12.3 Stages in an irreversible expansion of a gas from an initial state (a) of volume V_1 to a final state (c) of volume V_2. In the intermediate stage shown (b), the gas is not in equilibrium; because of turbulence, pressure and temperature cannot be defined.

Figure 12.2. You might wonder whether we can sketch a *path* on the surface of equilibrium thermodynamic states to summarize the progress of the system during a process. Only special processes of the type called *reversible* can be represented in this way (see discussion in the following paragraphs).

Many conditions of a system do not correspond to any equilibrium thermodynamic state. Suppose a gas is confined by a piston in a cylinder with volume V_1 (thermodynamic state A). If the piston is abruptly pulled out to increase the volume to V_2 (Fig. 12.3), chaotic gas currents arise as the molecules begin to move into the larger volume. These intermediate stages are not thermodynamic states, because such properties as density and temperature are changing rapidly through space and in time. Eventually, the currents cease and the system approaches a new equilibrium thermodynamic state, B. States A and B are thermodynamic states, but the conditions in between cannot be described by only a few macroscopic variables, and therefore are not thermodynamic states. Such a process is called an **irreversible process.** An irreversible process cannot be represented as a path on a thermodynamic surface (as in Fig. 12.2), because the intermediate stages are not thermodynamic equilibrium states and thus do not correspond to points on the equation-of-state surface.

In contrast, a **reversible process** proceeds through a continuous series of thermodynamic states, and thus can be shown as a path on the equation-of-state surface. The term reversible is used because an infinitesimal change in external conditions suffices to reverse the direction of motion of the system. For example, if a gas is expanded by slowly pulling out a piston, only a tiny change in the force exerted from the outside is required to change the direction of motion of the piston and begin to compress the gas. Such a process is an idealization, because the final equilibrium state would be reached only after an infinite length of time; therefore, such a process could never occur in a finite time. If a real process is conducted slowly enough and in sufficiently small steps, the process can be approximated by an idealized limiting reversible process. A gas confined inside a piston–cylinder assembly will experience an irreversible compression when a kilogram of sand is suddenly dropped onto the piston. The same compression can be achieved (almost) reversibly by transferring the same kilogram of sand onto the piston one grain at the time.

An infinite number of reversible paths can be identified between any two thermodynamic states A and B. Two of them shown in Figure 12.2 could be realized by slowly changing the values of T and V in the proper sequence by manipulating the apparatus in the surroundings. We use such reversible paths throughout this textbook as a tool for calculating changes in properties caused by processes.

Certain properties of a system, called **state functions,** are uniquely determined by the thermodynamic state of the system. Volume, temperature, pressure, and the internal energy, U, are examples of state functions. The Greek letter delta (Δ) is used to indicate *changes* in the value of state functions in a

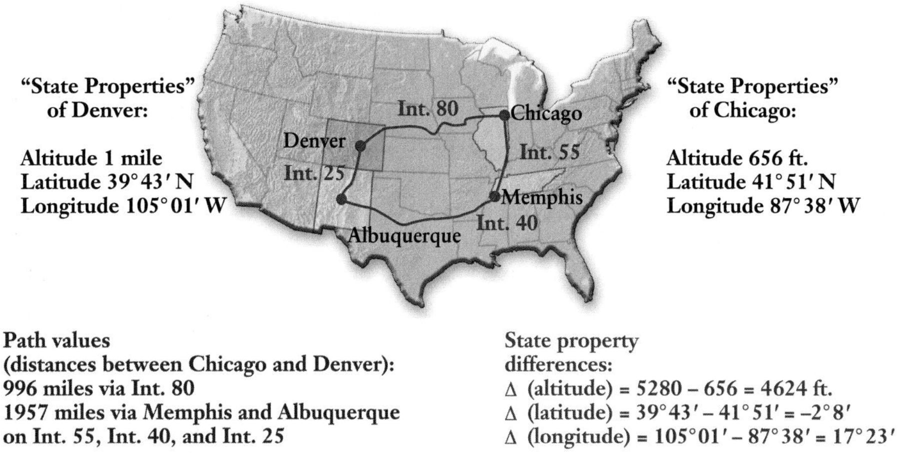

FIGURE 12.4 Differences in state properties (such as the difference in altitude between two points) are independent of the path followed. Other properties (such as the total distance traveled) depend on the particular path.

thermodynamic process. Thus, $\Delta V = V_{\text{final}} - V_{\text{initial}}$ (or $V_f - V_i$) is the change in volume between initial and final states, and $\Delta U = U_{\text{final}} - U_{\text{initial}}$ is the change in internal energy. Because U, V, and T are all state functions, the value of ΔU (or ΔV or ΔT) depends only on the initial and final states of the system, not on the path by which the process got the system from initial state to final state. The same change in any of these state functions will be measured no matter which path (reversible or irreversible) is followed between any given pair of thermodynamic states. *The change in any state function between two states is independent of path.* The converse statement is also true: If the change in a property of a system is independent of path, the property is a state function. Figure 12.4 illustrates two different paths that connect a given initial state and a common final state.

12.2 The First Law of Thermodynamics: Internal Energy, Work, and Heat

The first law of thermodynamics (which is stated at the end of this section) relates the energy change in a thermodynamic process to the amount of work done on the system and the amount of heat transferred to the system. It is first necessary to examine the ways in which amounts of heat and work are measured to understand the significance of this law. You will see that heat and work are simply different means by which energy is transferred into or out of a system.

12.2.1 Work

The mechanical definition of **work** is the product of the external force on a body times the distance through which the force acts. If a body moves in a straight line from point r_i to r_f with a constant force, F, applied along the direction of the path, the work done on the body is

$$w = F(r_f - r_i) \qquad \text{(force along direction of path)}$$

To illustrate how work can change the energy of a system, we will examine the relation between work and energy in an ordinary mechanical system.

As the first example, consider a block of mass, m, moving with initial speed v_i along a frictionless surface. We know that a force acting on an object in the direction of the initial velocity increases the speed, and therefore the kinetic energy of the object. In the following derivation, we show how the kinetic energy of an object changes when work is done on it. If a constant force, F, is exerted on it in the direction of its motion, it will experience a constant acceleration, $a = F/m$. After a time, t, the speed of the block will have increased from v_i to v_f, and its position will have changed from r_i to r_f. The work done on the block is

$$w = F(r_f - r_i) = ma(r_f - r_i)$$

The distance traveled, $r_f - r_i$, is given by the average speed, in this case $(v_i + v_f)/2$, multiplied by the elapsed time, t:

$$r_f - r_i = \left(\frac{v_i + v_f}{2}\right)t$$

When the acceleration is constant, it is equal to the change in speed, $v_f - v_i$, divided by the elapsed time:

$$a = (v_f - v_i)/t$$

Substituting both of these results into the expression for the work done gives

$$\begin{aligned} w &= m\left(\frac{v_f - v_i}{t}\right)\left(\frac{v_i + v_f}{2}\right)t \\ &= \frac{m}{2}(v_f - v_i)(v_f + v_i) \\ &= \frac{m}{2}v_f^2 - \frac{m}{2}v_i^2 \\ &= \Delta E_{kin} \end{aligned}$$

The expression on the right side of the equations is the change in kinetic energy, $\frac{1}{2}mv^2$, of the block. For this idealized example with a frictionless surface, the work done in moving the block from r_i to r_f is equal to the change in energy (in this case, kinetic) of the block.

As a second mechanical example, consider the work done in lifting an object in a gravitational field. To raise a mass, m, from an initial height, h_i, to a final height, h_f, an upward force sufficient to counteract the downward force of gravity, mg, must be exerted. The work done on the object in this case is

$$w = mg(h_f - h_i) = mg\Delta h = \Delta E_{pot}$$

This is the change in *potential* energy, mgh, of the object, showing once again that the mechanical work done in moving a body is equal to the change in energy of the body.

One important kind of mechanical work in chemistry is *pressure-volume work,* which results when a system is compressed or expanded under the influence of an outside pressure. Imagine that a gas has pressure P_i and is confined in a cylinder by a frictionless piston of cross-sectional area A and negligible mass (Fig. 12.5). The force exerted on the inside face of the piston by the gas is $F_i = P_iA$, because pressure is defined as force divided by area. If there is a gas on the outer side of the piston with pressure P_{ext} ("ext" for "external"), then if $P_{ext} = P_i$, the piston will experience no net force. If P_{ext} is increased, the gas will be compressed, and if it is decreased, the gas will expand. Consider first the case in which the external force is less than the initial force exerted by the gas, P_iA. Then the gas will expand and lift the piston from h_i to h_f. The work in this case is

$$w = -F_{ext}(h_f - h_i)$$

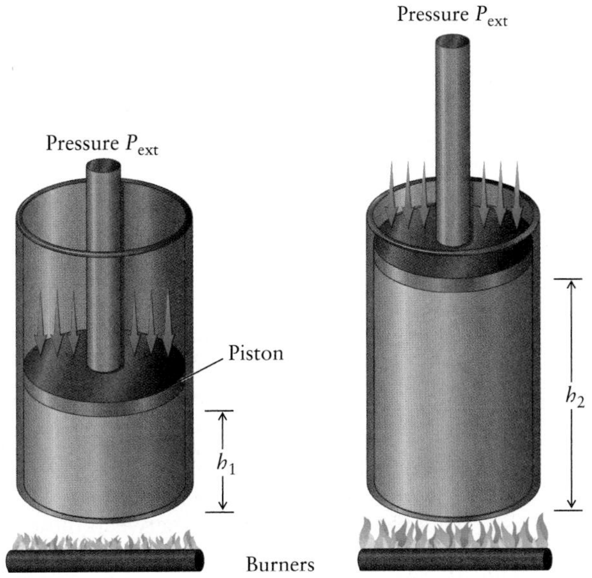

FIGURE 12.5 As the gas inside this cylinder is heated, it expands, pushing the piston against the pressure P_{ext} exerted by the gas outside. As the piston is displaced over a distance $h_f - \Delta h_i = \Delta h$, the volume of the cylinder increases by an amount $A\ \Delta h$, where A is the surface area of the piston.

The negative sign is inserted because the force from the gas outside opposes the direction of displacement of the piston during expansion of the gas inside the cylinder. This is rewritten as

$$w = -P_{ext}\, A\Delta h$$

The product $A\Delta h$ is the volume change of the system, ΔV, so the work is

$$w = -P_{ext}\ \Delta V \qquad [12.1]$$

For an expansion, $\Delta V > 0$, thus $w < 0$ and the system does work; it pushes back the surroundings. For a compression (by making $P_{ext} > P_i$), work is done *on* the system; it is pushed back by the surroundings. Again, $w = -P_{ext}\Delta V$, but now $\Delta V < 0$, so $w > 0$. If there is no volume change, $\Delta V = 0$, and no pressure–volume work is done. Finally, if there is no mechanical link to the surroundings (that is, if $P_{ext} = 0$), then once again no pressure–volume work can be performed.

If the pressure P_{ext} is expressed in pascals and the volume in cubic meters, their product is in joules (J). These are the International System of Units (SI) units for these quantities. For many purposes, it is more convenient to express pressures in atmospheres and volumes in liters; therefore, work has the unit liter-atmospheres (L atm). The two work units are related by

$$1 \text{ L atm} = (10^{-3}\text{ m}^3)(1.01325 \times 10^5 \text{ kg m}^{-1}\text{ s}^{-2}) = 101.325 \text{ J}$$

EXAMPLE 12.1

A cylinder confines 2.00 L gas under a pressure of 1.00 atm. The external pressure is also 1.00 atm. The gas is heated slowly, with the piston sliding freely outward to maintain the pressure of the gas close to 1.00 atm. Suppose the heating continues until a final volume of 3.50 L is reached. Calculate the work done on the gas and express it in joules.

Solution

This is an expansion of a system from 2.00 to 3.50 L against a constant external pressure of 1.00 atm. The work done on the system is then

$$\begin{aligned} w &= -P_{ext}\,\Delta V \\ &= -\ (1.00\text{ atm})(3.50\text{ L} - 2.00\text{ L}) \\ &= -1.50\text{ L atm} \end{aligned}$$

(Continued)

Conversion to joules gives

$$w = (-1.50 \text{ L atm})(101.325 \text{ J L}^{-1} \text{ atm}^{-1})$$
$$= -152 \text{ J}$$

Because w is negative, we see that -152 J of work was done *on* the gas. Put another way, $+152$ J of work was done *by* the gas as it expanded against atmospheric pressure.

Related Problems: 1, 2

12.2.2 Internal Energy

In the two simple mechanical cases discussed earlier, we saw how performing work on an object changes the amount of two types of energy: the kinetic energy of a moving object and the potential energy of an object in a gravitational field. In the same way, performing work can change the amount of energy in more complex cases. A third type of energy, less apparent but equally important, is **internal energy**, defined as the total energy content of a system arising from the potential energy between molecules, from the kinetic energy of molecular motions, and from chemical energy stored in chemical bonds. Potential energy between molecules includes the lattice energy of solids and the attractive and repulsive interactions between molecules in gases and liquids. Kinetic energy appears in the translation and the internal motions of individual molecules (Fig. 12.6). Gas molecules are in a constant state of motion (see Section 9.5) even when no overall gas flow is taking place in the container; the same is true of molecules in liquids and solids. In the next section we describe an experiment by James Joule and Benjamin Thompson in which they did mechanical work on water (the system) and found that the temperature of the water increased. Any mechanical or electrical work done on a system can change its internal energy (unless heat is transferred out of the system to compensate). *We conclude that P–V work is a means of changing the internal energy of a macroscopic system through purely mechanical interaction between the system and its surroundings.*

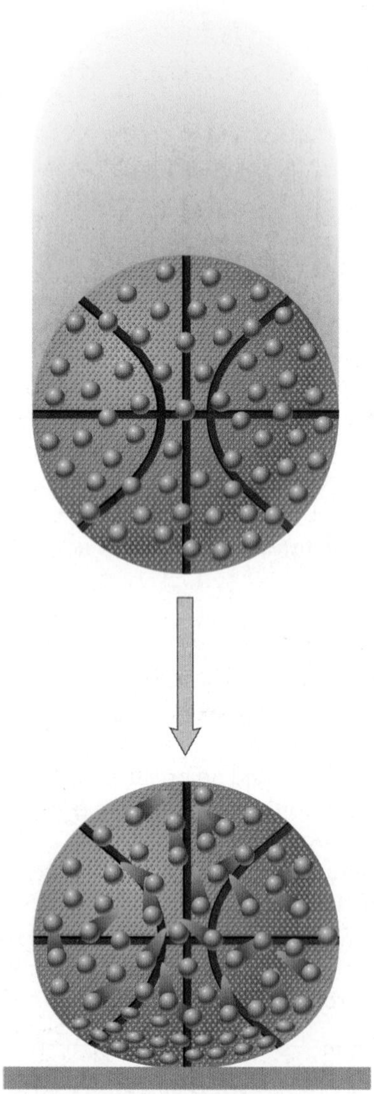

FIGURE 12.6 A ball dropped from a height increases its internal energy on impact with the ground. After impact, the molecules near the surface of the ball are pushed against one another, increasing the potential energy between the molecules. As the ball bounces, the molecules readjust their positions, after which they move a little faster. The kinetic energy of the molecules is higher than it was just before impact with the ground.

12.2.3 Heat

Now, how do we describe the role of heat in the process in Example 12.1? After all, heating the gas caused it to expand, which enabled it to move the piston and do work on the surroundings. The heater in Figure 12.5 has no mechanical "moving parts," yet it set in motion a train of events that led to a mechanical result. To explain this result, we interpret **heat** *as a means of increasing the internal energy of a system without mechanical interaction.* Justification for this interpretation is provided later in this chapter.

Heat, q, may be transferred between two objects initially at different temperatures, potentially changing the thermal energy of each. When a hot body is brought into contact with a colder body, the two temperatures change until they become equal. If a piece of hot metal is plunged into a container of water, the temperature of the water increases as its molecules begin to move faster, corresponding to an increase in the internal energy of the water. This process is sometimes described as the "flow" of heat from the hotter to the colder body. Although this picture is useful, it is based on the antiquated (and erroneous) notion that heat is a sort of fluid contained in matter.

The idea of heat flow has inspired methods for measuring the amount of energy transferred as heat; this branch of science is called **calorimetry.** One simple way is to use an **ice calorimeter** (Fig. 12.7), which consists of a bath containing ice and water, well insulated to prevent heat transfer to the surroundings.

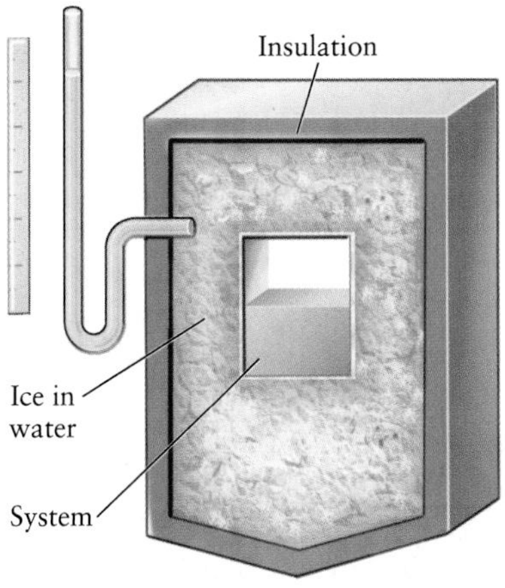

FIGURE 12.7 An ice calorimeter. As the ice melts, the volume of the ice–water mixture decreases, an effect that can be read off the scale on the left.

If heat is transferred to the bath from the system, some of the ice melts. Because a given mass of water has a smaller volume than the same mass of ice, the total volume of the ice–water mixture decreases as heat enters the bath. If twice as much heat is transferred, twice as much ice will melt and the volume change will be twice as great. The amount of heat transferred is determined from the change in volume of the contents of the calorimeter. Heat transferred from the ice bath *into* the system causes water to freeze and *increases* the total volume of the ice bath.

More contemporary versions of calorimetry use the fact that when heat is transferred to or removed from a substance in a single phase at constant pressure, the temperature changes in a reproducible way. The **specific heat capacity** of a material is the amount of heat required to increase the temperature of a 1-g mass by 1°C. If twice as much heat is transferred, the resulting temperature change will be twice as large (provided the specific heat capacity itself does not change appreciably with temperature). Thus, the temperature change of a fixed amount of a given substance is a measure of the thermal energy transferred to or from it. This is described by

$$q = mc_s\Delta T \qquad [12.2]$$

where q is the heat transferred to a body of mass m with specific heat capacity c_s to cause a temperature change of ΔT. (Be careful not to use Eq. 12.2 if a phase change or chemical reaction is occurring in the mixture being heated. Heating a pot of boiling water does not increase its temperature, it simply provides the energy for more molecules to evaporate.)

Because heat, like work, is energy in the process of being transferred, the appropriate unit for it is also the joule. Historically, however, the connections among work, heat, and energy were not appreciated until the middle of the 19th century, by which time a separate unit for heat, the calorie, was already well established. One calorie was defined as the amount of heat required to increase the temperature of 1 g water from 14.5°C to 15.5°C (or, in other words, the specific heat capacity of water, c_s, at 15°C was *defined* as 1.00 cal K^{-1} g^{-1}).

The equivalence of heat and work as means of energy transfer was suggested in 1798 by Benjamin Thompson, Count Rumford. In the course of his work as military advisor to the King of Bavaria, Thompson observed that the quantity of heat produced in boring cannons was proportional to the amount of work done in the process. Moreover, the operation could be continued indefinitely, demonstrating that heat was not a "fluid" contained in the metal of the cannon. More quantitative measurements were conducted by the German physician Julius Mayer and by the English physicist James Joule. In the 1840s, these scientists showed that the temperature of a substance could be increased by doing work on the substance, as well as by adding heat to it. Figure 12.8 shows an apparatus in which a paddle, driven by a falling weight, churns the water in a tank. Work is performed on the water, and the temperature increases. The work done on the water is $-mg\Delta h$, where Δh is the (negative) change in the height of the weight and m is its mass. The experiment is conducted in an insulated container, so no heat enters the container or leaks out to the surroundings. Because all this work goes to increase the water temperature, the specific heat capacity of the water in *joules* per gram per degree is equal to the quantity of work done divided by the product of the mass of water and its temperature increase.

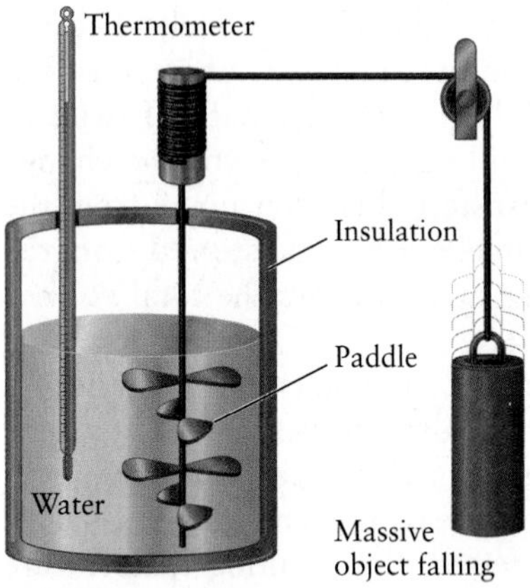

FIGURE 12.8 The falling weight turns a paddle that does work on the system (the water), causing an increase in its temperature.

EXAMPLE 12.2

Suppose a 10.00-kg mass drops through a height difference of 3.00 m, and the resulting work is used to turn a paddle in 200.0 g water, initially at 15.00°C. The final water temperature is found to be 15.35°C. Assuming that the work done is used entirely to increase the water temperature, calculate the conversion factor between joules and calories.

Solution

The total work done on the water is

$$w = -mg\Delta h = -(10.00 \text{ kg})(9.807 \text{ m s}^{-2})(-3.00 \text{ m}) = 294 \text{ J}$$

The heat (in calories) required to increase the water temperature by the same amount is

$$q = mc_s\Delta T = (200.0 \text{ g})(1.000 \text{ cal K}^{-1} \text{ g}^{-1})(0.35 \text{ K}) = 70 \text{ cal}$$

Because the work done has the same effect on the water as direct transfer of heat, these two expressions can be set equal to each other, giving

$$70 \text{ calories} = 294 \text{ joules}$$

$$1 \text{ calorie} \approx 4.2 \text{ joules}$$

Related Problems: 3, 4

These and other experiments eliminated the need for the calorie as an independent unit, and the calorie is now *defined* as

$$1 \text{ cal} = 4.184 \text{ J} \quad \text{(exactly)}$$

This book uses the joule as the primary unit for heat and energy. Because much of the chemical literature continues to use the calorie as the unit of heat, it is important to be familiar with both units.

12.2.4 The First Law of Thermodynamics

Both heat and work are forms in which energy is transferred into and out of a system; they can be thought of as energy in transit. If the energy change is caused by *mechanical* contact of the system with its surroundings, work is done; if it is caused by *thermal* contact, heat is transferred. In many processes, both heat and work cross the boundary of a system, and the change in the internal energy, U, of the system is the sum of the two contributions. We denote the internal energy by U to distinguish it from the sum of the potential and kinetic energy in a simple mechanical or electrical process, for which we use the symbol E throughout this book. This statement, called the **first law of thermodynamics,** takes the mathematical form[1]

$$\Delta U = q + w \qquad [12.3]$$

[1]Some books, especially engineering ones, define work as positive when it is done by the system. The reason is that many engineering applications focus on the work done by a particular heat engine; therefore, it is helpful to define that quantity as positive. Thus, the work is given by

$$w = P_{ext}\,\Delta V \text{ (engineering convention)}$$

and the first law reads

$$\Delta U = q - w \text{ (engineering convention)}$$

Although we do not use the engineering convention, you should check which convention is used when consulting other books. In particular, the MCAT has used the engineering convention.

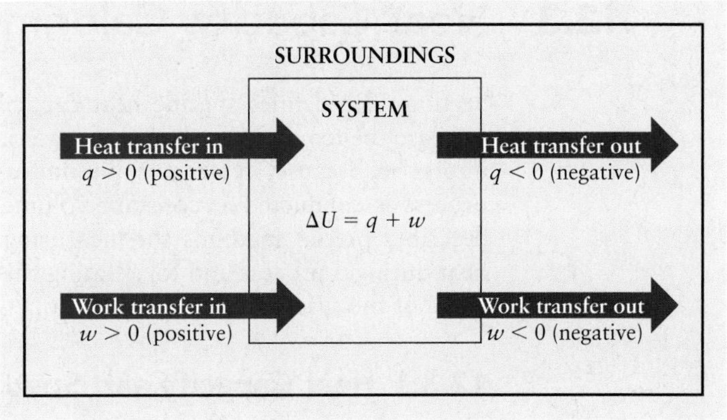

FIGURE 12.9 The first law of thermodynamics states that the change in internal energy of a system during a process is the sum of the heat transferred to the system, q, and the work done on the system, w. If heat is transferred out of the system, q is negative. Likewise, if the system does work on the surroundings, w is negative.

A system cannot be said to "contain" work or heat, because both "work" and "heat" refer not to states of the system but to *processes* that transform one state into another. In the Joule experiment shown in Figure 12.8, the work done on the water (the system) by the falling weight increased the temperature of the water. Work was performed on the system, and no heat was transferred; thus, the first law for this process takes the form $\Delta U = w$. The same change in state of the system can be achieved by transferring heat to the system without work being done; for this process, $\Delta U = q$. Because q and w depend on the particular process (or path) connecting the states, they are not state functions. But their sum, $\Delta U = q + w$, is independent of path; therefore, internal energy is a function of state (see Fig. 12.9). The fundamental physical content of the first law of thermodynamics is the following observation:

There exists a state function called internal energy represented by U whose changes $\Delta U = q + w$ are independent of path, although the values of q and w individually depend on the path of the process.

We stated earlier that the laws of thermodynamics cannot be derived or proved; they are generalizations of the results of countless experiments on a tremendous variety of substances. It is not possible even to "check" the first law by independently measuring ΔU, w, and q, because no "energy gauges" exist to determine energy changes ΔU. But what we *can* do is to measure w and q for a series of different processes connecting the same initial and final states. Every time, we find that their sum, $q + w$, is always the same.

In any process, the heat *added to* the system is *removed from* the surroundings; thus,

$$q_{sys} = -q_{surr}$$

In the same way, the work done *on* the system is done *by* the surroundings; thus,

$$w_{sys} = -w_{surr}$$

Adding these two and invoking the first law give

$$\Delta U_{sys} = -\Delta U_{surr}$$

Thus, the energy changes of system and surroundings have the same magnitude but opposite signs. The total energy change of the thermodynamic universe for a given process (system plus surroundings) is then

$$\Delta U_{univ} = \Delta U_{sys} + \Delta U_{surr} = 0 \qquad [12.4]$$

Our conclusion is that, in any process, the total energy of the thermodynamic universe remains unchanged; the total energy is conserved while it is exchanged between the system and the surroundings.

12.3 Heat Capacity, Calorimetry, and Enthalpy

Section 12.2 defines specific heat capacity as the amount of heat required to increase the temperature of 1 g of material by 1 K. That definition is somewhat imprecise, because, in fact, the amount of heat required depends on whether the process is conducted at constant volume or at constant pressure. This section describes precise methods for measuring the amount of energy transferred as heat during a process and for relating this amount to the thermodynamic properties of the system under investigation.

12.3.1 Heat Capacity and Specific Heat Capacity

The **heat capacity**, C, is defined as the amount of energy that must be added to the system to increase its temperature by 1 K. The heat capacity is a property of the system as a whole and has units of J K^{-1}.

$$q = C\Delta T$$

Now consider a case in which two gaseous systems containing identical masses of the same substance are heated to produce identical changes in temperature. During the experiment, system 1 is held at constant volume, and system 2 at constant pressure. Which system absorbed more heat in these identical temperature changes? All the energy gained by system 1 contributed to increasing the temperature of the substance, and therefore the speed of the molecules, subject to the fixed volume. But in system 2, some of the energy gained was promptly lost as the system performed expansion work against the surroundings at constant pressure. Consequently, system 2 must absorb more thermal energy from the surroundings than does system 1 to achieve the identical temperature change. Two independent heat capacity functions must be defined: C_P, the heat capacity at constant pressure, and C_V, the heat capacity at constant volume. For any system, C_P is greater than C_V. This difference can be quite large for gases. It is usually negligible for solids and liquids, because their only volume change at constant pressure is the small expansion or contraction on heating and cooling.

In thermodynamics, the *molar* heat capacities c_V and c_P (the system heat capacities C_V and C_P divided by the number of moles of substance in the system) are particularly useful: c_V is the amount of heat required to increase the temperature of 1 mol of substance by 1 K at constant volume, and c_P is the corresponding amount required at constant pressure. If the total heat transferred to n moles at constant volume is q_V, then

$$q_V = nc_V(T_2 - T_1) = nc_V\Delta T \qquad [12.5]$$

If an amount q_P is transferred at constant pressure, then

$$q_P = nc_P\Delta T \qquad [12.6]$$

provided that c_V and c_P do not change significantly between the initial and final temperatures. The *specific heat capacity* at constant V or constant P is the system heat capacity reported per gram of substance. Extensive tabulations of molar and specific heat capacities at constant volume and at constant pressure are available. The specific heat capacities of some common substances are listed in Table 12.1, and a more extensive tabulation of molar heat capacities at constant pressure is provided in Appendix D. We relate the heat capacity of a substance to its molecular structure in Section 12.5.

TABLE 12.1

Specific Heat Capacities at Constant Pressure (at 25°C)

Substance	Specific Heat Capacity (J K^{-1} g^{-1})
$Hg(\ell)$	0.140
$Cu(s)$	0.385
$Fe(s)$	0.449
$SiO_2(s)$	0.739
$CaCO_3(s)$	0.818
$O_2(g)$	0.917
$H_2O(\ell)$	4.18

The importance of the heat capacity is illustrated by placing two objects initially at different temperatures into contact. Energy in the form of heat is exchanged between them until they reach a common temperature. If the two objects are insulated from the rest of the universe during the process, the amount of heat q_2 taken up by the cooler object is equal to $-q_1$, the amount of heat given up by the hotter object. As always, the convention followed is that energy transferred to an object has a positive sign; thus, q_2 is positive when q_1 is negative. This analysis is broadly applicable; a typical example follows.

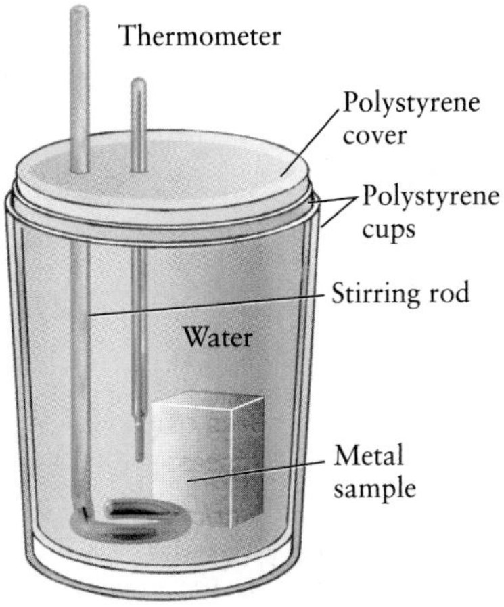

FIGURE 12.10 A Styrofoam cup calorimeter. As the piece of metal cools, it releases heat to the water. The amount of heat released can be determined from the temperature change of the water. The hot metal is the "system"; the water is the "surroundings." The Styrofoam cup wall prevents energy exchange with the remainder of the room and is the boundary of the "thermodynamic universe" for this problem.

EXAMPLE 12.3

A piece of iron weighing 72.4 g is heated to 100.0°C and plunged into 100.0 g water that is initially at 10.0°C in a Styrofoam cup calorimeter. Assume no heat is lost to the Styrofoam cup or to the environment (Fig. 12.10). Calculate the final temperature that is reached.

Solution

The "coffee cup calorimeter" operates at constant pressure determined by the atmosphere; therefore, we need specific heat data at constant pressure. Because the data involve masses, it is easier to work with specific heat capacities (see Table 12.1) than with molar heat capacities. If t_f is the final temperature (in degrees Celsius), then the equation for heat balance gives

$$m_1(c_{s1})\Delta T_1 = -m_2(c_{s2})\Delta T_2$$

$$(100.0 \text{ g H}_2\text{O})(4.18 \text{ J °C}^{-1}\text{g}^{-1})(t_f - 10.0\text{°C}) = -(72.4 \text{ g Fe})(0.449 \text{ J°C}^{-1}\text{g}^{-1})(t_f - 100.0\text{°C})$$

This is a linear equation for the unknown temperature t_f, and its solution is

$$418t_f - 4180 = -32.51t_f + 3251$$

$$t_f = 16.5\text{°C}$$

Note that specific heat capacities are numerically the same whether expressed in $\text{J K}^{-1}\text{ g}^{-1}$ or $\text{J (°C)}^{-1}\text{ g}^{-1}$ (because the degree Celsius and the kelvin have the same size). Converting 10.0°C and 100.0°C to kelvins and using specific heat capacities in units of $\text{J K}^{-1}\text{ g}^{-1}$ gives t_f = 289.7 K, which is equivalent to the previous answer.

Related Problems: 11, 12

12.3.2 Heat Transfer at Constant Volume: Bomb Calorimeters

Suppose some reacting species are sealed in a small closed container (called a *bomb*) and the container is placed in a calorimeter like the one in Figure 12.11. The reaction is initiated by a heated wire inside the bomb. As the molecules react chemically, heat is given off or taken up, and the change in temperature of the calorimetric fluid is measured. Because the container has rigid walls, its volume is constant and no pressure–volume work is done. Therefore, the change in internal energy is equal to the measured heat absorbed from the chemical reaction at constant volume:

$$\Delta U = q_V$$

Such experiments at constant volume are often inconvenient or difficult to perform. They require the use of a well-constructed reaction vessel that can resist the large pressure changes that occur in many chemical reactions.

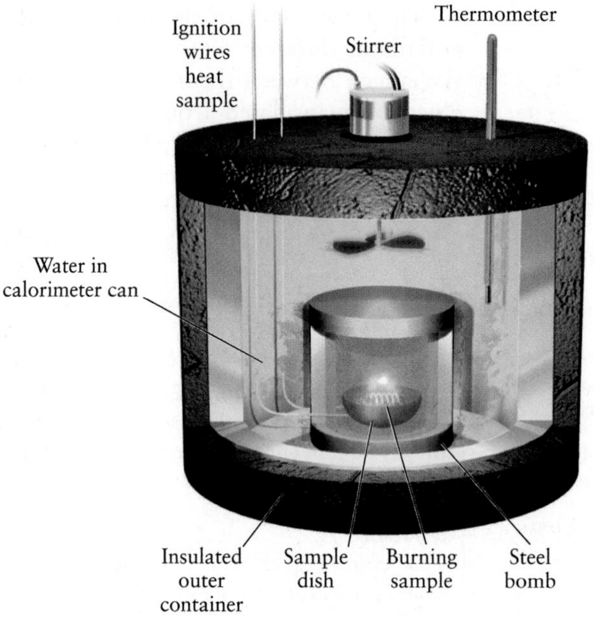

FIGURE 12.11 The combustion calorimeter is also called a "bomb calorimeter"; the combustion reaction in it is conducted at a fixed volume.

12.3.3 Heat Transfer at Constant Pressure: Enthalpy

Most chemical reactions are carried out under constant (atmospheric) pressure conditions rather than at constant volume. It is desirable to identify a state function whose change in a process is equivalent to the heat transferred at constant pressure, q_P, just as is ΔU is equivalent to q_V.

If the work done is entirely pressure–volume work, and if the external pressure is held constant, then

$$\Delta U = q_P + w = q_P - P_{ext}\Delta V$$

If the external pressure is now assumed to be equal to the internal pressure of the system P, then

$$U = q_P - P\Delta V$$

$$q_P = \Delta U + P\Delta V$$

Because P is constant, $P\Delta V = \Delta(PV)$, and this equation becomes

$$q_P = \Delta(U + PV)$$

The combination $U + PV$ appearing on the right side is now defined as the **enthalpy** H:

$$H = U + PV \qquad [12.7a]$$

thus,

$$q_P = \Delta(U + PV) = \Delta H \qquad [12.7b]$$

Because U, P, and V are state functions, H must also be a state function. Heat transfer in a process at constant pressure has therefore been related to the change in a state function.

It is important to remember that

$$\Delta H = q_P = \Delta U + P\Delta V \qquad \text{(constant pressure)}$$

holds only at constant internal pressure. If the pressure changes, the more general relationship

$$\Delta H = \Delta U + \Delta(PV)$$

must be used. Like the energy, the enthalpy change is determined by the initial and final states and is independent of the particular path along which the process is performed. This is always true for a state function.

Physical interpretation of the enthalpy function follows immediately from the equation $\Delta H = \Delta U + P\Delta V$ at constant pressure. Clearly, H has physical dimensions of energy and is, in effect, a "corrected" internal energy that reflects the consequences of changing V while thermal energy is being absorbed at constant pressure. The "correction term" $P\Delta V$ accounts precisely for the energy used by the system to do expansion work, rather than for increasing the temperature of the system. Thus, the value of q in constant pressure processes where only pressure–volume work is done is equivalent to the change in the state function enthalpy, ΔH.

12.4 The First Law and Ideal Gas Processes

We stated the first law of thermodynamics in a general form, applicable to any process that begins and ends in equilibrium states. We analyzed the heat and work terms separately and presented methods for calculating, measuring, and interpreting each. All the concepts are now in place for applying thermodynamics to the discussion of specific processes. Applications require data on certain properties of the substance being studied, such as its equation of state and its heat capacities. Thermodynamic arguments alone cannot provide the actual values of such properties; instead, thermodynamics establishes universal relations among such properties. The actual values must be obtained by methods other than thermodynamics, such as experimental measurements or theoretical calculations in statistical thermodynamics. To illustrate these points, in the next few paragraphs we obtain data on the heat capacities of ideal gases by methods outside thermodynamics. Then we use these data to apply thermodynamics to analyze particular processes carried out on ideal gases.

12.4.1 Heat Capacities of Ideal Gases

The pair of molar heat capacities c_V and c_P for an ideal monatomic gas can be calculated from the results of the kinetic theory of gases and the ideal gas equation of state. From Section 9.5, the average translational kinetic energy of n moles of an ideal gas is

$$\overline{E}_{\text{kin}} = \tfrac{3}{2}nRT$$

In a *monatomic* gas, changes in the total internal energy ΔU measured in thermodynamics can be equated to changes in the average translational kinetic energy of the atoms. If n moles of a monatomic gas is taken from a temperature T_1 to a temperature T_2, the internal energy change is

$$\Delta U = \tfrac{3}{2}nR(T_2 - T_1) = \tfrac{3}{2}nR\Delta T$$

Note that the pressure and volume do not affect U explicitly (except through temperature changes), so this result is independent of the change in pressure or volume of the gas.

Now, consider changing the temperature of an ideal gas at constant volume from the point of view of thermodynamics. Because the volume is constant (the gas is confined in a vessel with rigid, diathermal walls), the pressure–volume work, w, must be zero; therefore,

$$\Delta U = q_V = nc_V\Delta T \qquad \text{(ideal gas)}$$

where q_V is the heat transferred at constant volume. Equating this thermodynamic relation with the previous expression for ΔU (from kinetic theory) shows that

$$c_V = \tfrac{3}{2}R \qquad \text{(monatomic ideal gas)} \qquad [12.8]$$

Similarly, the molar heat capacity at constant pressure, c_P, is calculated by examining the heating of a monatomic ideal gas at constant pressure from temperature T_1 to T_2. Experimentally, such a process can be performed by placing the gas in a cylinder with a piston that moves out as the gas is heated, keeping the gas pressure equal to the outside pressure. In this case,

$$\Delta U = \tfrac{3}{2}nR\Delta T = nc_V\Delta T$$

still holds (because the energy change depends only on the temperatures for an ideal gas), but we now have

$$\Delta U = q_P + w$$

because the work is no longer zero. The work for a constant-pressure process is easily calculated from

$$w = -P\Delta V = -P(V_2 - V_1)$$

and the heat transferred is

$$q_P = nc_P\Delta T$$

Because w is negative, q_P is larger than q_V by the amount of work done by the gas as it expands. This gives

$$\Delta U = q + w$$

$$nc_V\Delta T = nc_P\Delta T - P(V_2 - V_1)$$

From the ideal gas law, $PV_1 = nRT_1$ and $PV_2 = nRT_2$; thus,

$$nc_V\Delta T = nc_P\Delta T - nR\Delta T$$

$$c_V = c_P - R$$

$$c_P = c_V + R$$

For a monatomic ideal gas, this shows that $c_P = \tfrac{5}{2}R$. It is important to use the proper units for R in these expressions for c_V and c_P. If heat is to be measured in joules, R must be expressed as

$$R = 8.315 \text{ J K}^{-1} \text{ mol}^{-1}$$

For a diatomic or polyatomic ideal gas, c_V is greater than $\tfrac{3}{2}R$, because energy can be stored in rotational and vibrational motions of the molecules; a greater amount of heat must be transferred to achieve a given temperature change. (See Section 12.5.) Even so, it is still true that

$$c_P = c_V + R \text{ (any ideal gas)} \qquad [12.9]$$

and that internal energy changes depend only on the temperature change; therefore, for a small temperature change ΔT,

$$\Delta U = nc_V\Delta T \text{ (any ideal gas)} \qquad [12.10]$$

For an ideal gas process,

$$\Delta H = \Delta U + \Delta(PV) = nc_V\Delta T + nR\Delta T$$
$$\Delta H = nc_P\Delta T \text{ (any ideal gas)} \quad [12.11]$$

because $c_P = c_V + R$. This result holds for any ideal gas process and shows that enthalpy changes, like internal energy changes, depend only on the temperature difference between initial and final states. These equations are not valid for systems other than ideal gases.

EXAMPLE 12.4

Suppose that 1.00 kJ of heat is transferred to 2.00 mol argon (at 298 K, 1 atm). What will the final temperature T_f be if the heat is transferred (**a**) at constant volume, or (**b**) at constant pressure? Calculate the energy change, ΔU, in each case.

Solution

Because argon is a monatomic, approximately ideal gas,

$$c_V = \tfrac{3}{2}R = 12.47 \text{ J K}^{-1} \text{ mol}^{-1}$$
$$c_P = \tfrac{5}{2}R = 20.79 \text{ J K}^{-1} \text{ mol}^{-1}$$

At constant volume,

$$q_V = nc_V\Delta T$$
$$1000 \text{ J} = (2.00 \text{ mol})(12.47 \text{ J K}^{-1} \text{ mol}^{-1})\Delta T$$
$$\Delta T = 40.1 \text{ K}; T_f = 298 + 40.1 = 338 \text{ K}$$
$$\Delta U = nc_V\Delta T = q_V = 1000 \text{ J}$$

At constant pressure,

$$q_P = nc_P\Delta T$$
$$1000 \text{ J} = (2.00 \text{ mol})(20.79 \text{ J K}^{-1} \text{ mol}^{-1})\Delta T$$
$$\Delta T = 24.0 \text{ K}; T_f = 298 + 24.0 = 322 \text{ K}$$
$$\Delta U = nc_V\Delta T$$
$$= (2.00 \text{ mol})(12.47 \text{ J K}^{-1} \text{ mol}^{-1})(24 \text{ K}) = 600 \text{ J}$$

Note that the expression for ΔU involves c_V even though the process is conducted at constant pressure. The difference of 400 J between the input q_P and ΔU is the work done by the gas as it expands.

Related Problems: 17, 18, 19, 20, 21, 22

12.4.2 Heat and Work for Ideal Gases

Now we have all the data needed to calculate heat and work for a variety of processes involving an ideal gas. To illustrate that q and w individually depend on the path followed, but their sum does not, consider the expansion of 1.00 mol of an ideal monatomic gas following two different paths. The system begins at state A (P_A = 2.00 atm, V_A = 10.0 L) and reaches a final state, B (P_B = 1.00 atm, V_B = 30.0 L), via either of two paths shown in Figure 12.12. Along path ACB (red arrows in Fig. 12.12), the system is first heated at constant pressure ($P_{ext} = P_A$ = 2 atm) until the volume has tripled; then it is cooled at constant volume until the pressure is halved. Along path ADB (blue arrows in Fig. 12.12), the system is cooled at constant volume until the pressure is halved, and then heated at constant pressure ($P_{ext} = P_B$ = 1 atm) until the volume has tripled.

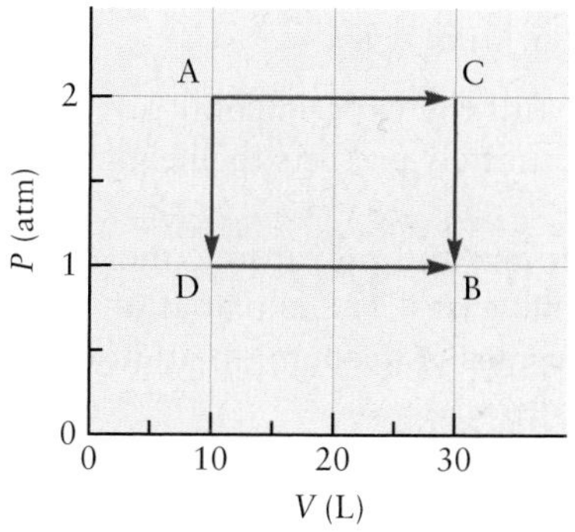

FIGURE 12.12 States A and B of a system are connected by two different ideal gas processes, one passing through state C and the other through state D.

The calculations of heat and work for each step are of the type already performed and are straightforward. Thus,

$$w_{AC} = -P_{ext}\Delta V = -P_A(V_B - V_A)$$
$$w_{CB} = 0 \text{ because } V_C = V_B$$
$$q_{AC} = q_P = nc_P\Delta T = \tfrac{5}{2}nR(T_C - T_A)$$
$$q_{CB} = q_V = nc_V\Delta T = \tfrac{3}{2}nR(T_B - T_C)$$

From the ideal gas law, $nRT_A = P_A V_A$, $nRT_B = P_B V_B$, and $nRT_C = P_C V_C = P_A V_B$ (because $P_A = P_C$ and $V_B = V_C$). Using these relations and summing over the two steps give

$$w_{ACB} = w_{AC} + w_{CB} = -P_A(V_B - V_A) = -40.0 \text{ L atm} = -4050 \text{ J}$$

$$q_{ACB} = q_{AC} + q_{CB} = \tfrac{5}{2}\, nR(T_C - T_A) + \tfrac{3}{2}\, nR(T_B - T_C)$$

$$= \tfrac{5}{2}\, P_A(V_B - V_A) + \tfrac{3}{2}\, V_B(P_B - P_A) = (100.0 - 45.0) \text{ L atm} = 5570 \text{ J}$$

The sum of these is $\Delta U = w_{ACB} + q_{ACB} = 1520$ J. (This could also have been obtained by using the ideal gas law to calculate the initial and final temperatures T_A and T_B.)

The corresponding calculation for path ADB gives

$$w_{ADB} = -2030 \text{ J and } q_{ADB} = 3550 \text{ J}$$

Even though both the work and the heat have changed, their sum is still 1520 J, illustrating that U is a state function, whereas q and w are not.

12.5 Molecular Contributions to Internal Energy and Heat Capacity

The internal energy of a system is the total of the energy involved in the motions of its molecules and the interactions between them, and so must depend on the structures of the molecules. The heat capacity at constant volume, defined as the change in internal energy per degree temperature increase, measures the ability of these mechanical motions and interactions to accommodate thermal energy.

The bridge between thermodynamics and the mechanics of molecules is provided by the kinetic theory of matter. The kinetic theory of gases (see Section 9.5) views each molecule in an ideal gas as a point particle moving independently of the others, and describes the probability of molecular speeds by the Maxwell–Boltzmann distribution. From this, the average kinetic energy per molecule is calculated to be $\bar{\varepsilon} = \left(\tfrac{3}{2}\right) k_B T$, and multiplying by the total number of molecules, N, gives the total kinetic energy in n moles of an ideal gas as $E_{kin} = \left(\tfrac{3}{2}\right) nRT$. This model calculation interprets the internal energy of a monatomic ideal gas as consisting solely of the translational kinetic energy of the atoms. Calculating the change in this internal energy as T is increased predicts the molar heat capacity at constant pressure of a monatomic ideal gas to be (see Equations 12.8 and 12.9)

$$c_P = \tfrac{3}{2} R + R = \tfrac{5}{2} R = 20.785 \text{ J mol}^{-1} \text{ K}^{-1}$$

independent of the temperature. Comparison with the experimental results in Table 12.2 shows that this model calculation accurately predicts the heat capacity for monatomic gases.

TABLE 12.2

Measured Values of c_P for Selected Gases at 298 K and 1 atm Pressure

Gas	c_P (J mol^{-1} K^{-1})
He	20.79
Ne	20.79
Ar	20.79
H_2	28.81
O_2	29.36
F_2	31.30
H_2O	33.54

The measured heat capacity values for diatomic and polyatomic molecules are significantly larger than those for monatomic gases. The explanation must lie in the existence of molecular motions, or **degrees of freedom,** in addition to translation that can accommodate thermal energy.

Figure 12.13a shows the degrees of freedom for a diatomic molecule. The molecule as a whole can undergo translational motion of its center of mass, behaving as if it were a point particle with mass equal to the total mass of the molecule, in the x-, y-, or z-directions. The molecule has three translational degrees of freedom, so the total kinetic energy is $\dfrac{1}{2}mv_x^2 + \dfrac{1}{2}mv_y^2 + \dfrac{1}{2}mv_z^2 = E_{trans} = \tfrac{1}{2}mv^2$. Each of these three degrees of freedom contributes $\tfrac{1}{2}RT$ to the

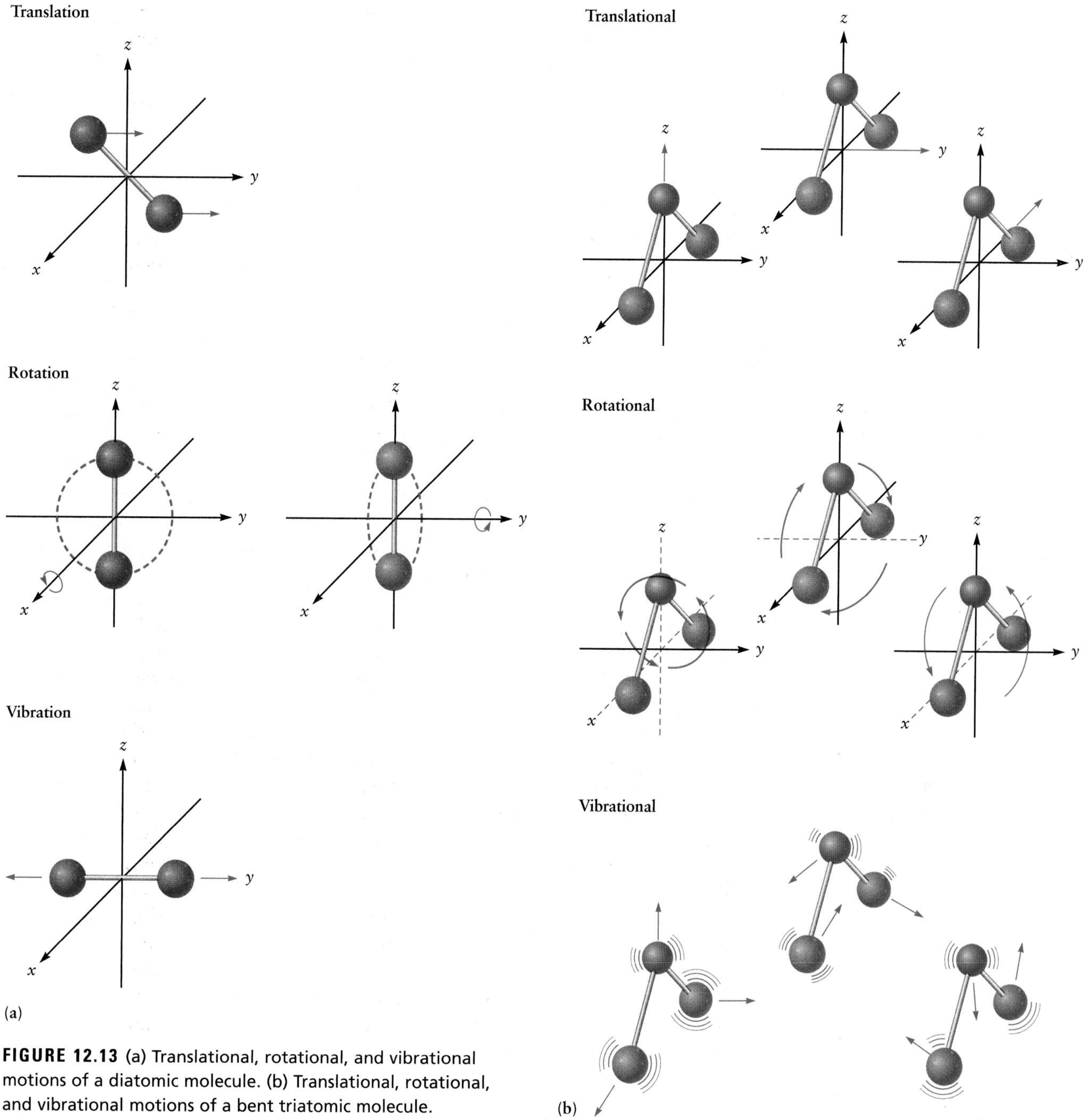

FIGURE 12.13 (a) Translational, rotational, and vibrational motions of a diatomic molecule. (b) Translational, rotational, and vibrational motions of a bent triatomic molecule.

molar heat capacity at constant volume. The molecule can rotate about its center of mass with both atoms in a plane. (See Section 3.9.) It can also tumble end-over-end, which amounts to rotation about its center of mass in a different plane. It has two rotational degrees of freedom, in each of which the kinetic energy takes the form $E_{rot} = \frac{1}{2} I\omega^2$ where I is the moment of inertia and ω is the angular speed. The molecule can stretch, or vibrate, about its equilibrium bond length R_e along the internuclear axis. (See *Connection to Instrumental Analysis* in Section 3.9) The potential energy of vibration is $PE_{vib} = \frac{1}{2} k(R_{AB} - R_e)^2$, and the kinetic energy is $KE_{vib} = \dfrac{1}{2}\mu\left(\dfrac{dR_{AB}}{dt}\right)^2$. In total, the diatomic molecule has three degrees of freedom for translational motion (kinetic energy only), two degrees of rotational freedom (kinetic energy only), and one degree of vibrational freedom (both kinetic and potential energy).

Figure 12.13b shows a similar description of the degrees of freedom of a nonlinear polyatomic molecule. Translational motion of the center of mass requires three degrees of freedom. Rotational motion can be visualized as an oblong object (such as an American football or a rugby ball) tumbling end-over-end in two separate planes, and also spiraling about its long axis. These motions require three degrees of freedom. The two planes are the same as described for the diatomic molecule. Because the spiraling motion does not occur for the diatomic molecule—it would be analogous to rotating a wire about its long axis—the diatomic molecule has one fewer degree of rotational freedom. To count the vibrational degrees of freedom, we start with the maximum total degrees of freedom that would be needed to describe the motions of all the atoms in a molecule of N atoms ($3N$), subtract away those already assigned to the center of mass (3), and those assigned to rotation (2 for a linear molecule, 3 for a molecule of general shape), and obtain $3N$-5 vibrational degrees of motion for a linear molecule and $3N$-6 for a nonlinear molecule.

To see how these modes contribute to the internal energy, let's return to the calculation of the average kinetic energy per molecule in Equations 9.12–9.14. The kinetic energy of a molecule is proportional to the square of its speed, and the second power of its speed also appears in the statistical weighting factor of the Maxwell–Boltzmann equation in Equation 9.19. Averaging the kinetic energy over this distribution produces the result $RT/2$ for each translational degree of freedom. The same result is obtained for each degree of freedom whose energy has a quadratic dependence on the speeds or coordinates, when its average is calculated over the Maxwell–Boltzmann distribution. This general conclusion is a result of the **equipartition theorem** of classical kinetic theory which depends on the mathematical form of the Maxwell–Boltzmann distribution. The result is that each translational degree of freedom contributes $RT/2$, and each rotational degree of freedom contributes $RT/2$ to the internal energy. Each vibrational degree of freedom contributes RT to the internal energy because it has one quadratic term for the kinetic energy and another for the potential energy.

We can determine the contribution of each mode to the heat capacity c_V by calculating the ratio $\Delta U/\Delta T$ as the temperature is increased at constant volume from T_1 to T_2 (for small temerature changes), sum these contributions to find the total value of c_V, and then calculate c_P by adding R to c_V, as in Equation 12.8. The results are summarized in Table 12.3 for several gases. The last two columns in Table 12.3 show that the results predicted for c_P by equipartition are significantly larger than the experimental values of c_P measured at 298 K. Figure 12.14 shows that the measured results depend on temperature, and how each mode contributes to the total heat capacity, for several gases. The translational motions contribute $(5/2)R$ at all temperatures above 0 K. The rotational contribution for H_2 increases with T, and by room temperature has reached its equipartition value of R; the total heat capacity at constant pressure for H_2 due to translational and rotational motion is $(7/2)R$ at room temperature. Heavier diatomics reach this equipartition value at even lower temperatures. Vibrational motions in H_2, N_2, and O_2 contribute significantly to the heat capacity only above room temperature. At these higher temperatures the difference between the measured values for these diatomics and $(7/2)R$ is essentially the magnitude of the vibrational contribution. At sufficiently high temperature the measured heat capacities for these diatomic gases approach the equipartition value of $(9/2)R = 37.41$ J mol^{-1} K^{-1}. Similarly the measured heat capacities for H_2O and CO_2 approach their equipartition values of $(14/2)R = 58.18$ J mol^{-1} K^{-1} and $(15/2)R = 62.37$ J mol^{-1} K^{-1}, respectively.

The fact that predictions of the equipartition theorem agree with experiment at high temperature but not at low temperature is explained by quantum

TABLE 12.3

Contributions to Heat Capacity Estimated by the Equipartition Theorem

Molecule	Degrees of Freedom				Equipartition Values (*R*)		Values of c_P (J mol^{-1} K^{-1})	
	f_{tot}	f_{tran}	f_{rot}	f_{vib}	c_V	c_P	Calculated	Measured
Ar	3	3	0	0	3/2	5/2	20.79	20.79
H_2	6	3	2	1	7/2	9/2	37.41	28.81
N_2	6	3	2	1	7/2	9/2	37.41	29.12
O_2	6	3	2	1	7/2	9/2	37.41	29.36
F_2	6	3	2	1	7/2	9/2	37.41	31.30
Cl_2	6	3	2	1	7/2	9/2	37.41	33.91
Br_2	6	3	2	1	7/2	9/2	37.41	36.02
CO	6	3	2	1	7/2	9/2	37.41	29.12
CO_2	9	3	2	4	13/2	15/2	62.37	37.13
H_2O	9	3	3	3	12/2	14/2	58.18	33.54
CH_4	15	3	3	9	24/2	26/2	108.08	35.31
C_2H_4	18	3	3	12	30/2	32/2	133.02	43.56
Si	3	0	0	3	6/2	~6/2	24.94	20.0
Al	3	0	0	3	6/2	~6/2	24.94	24.35

mechanics. The equipartition theorem is a consequence of classical mechanics, and it assumes that each mode of energy accommodation is always active. The quantum analysis starts by calculating the energy levels for each mode of motion and setting up probability for finding the molecules in excited energy states when the system is at temperature *T*. The probability function contains the factor

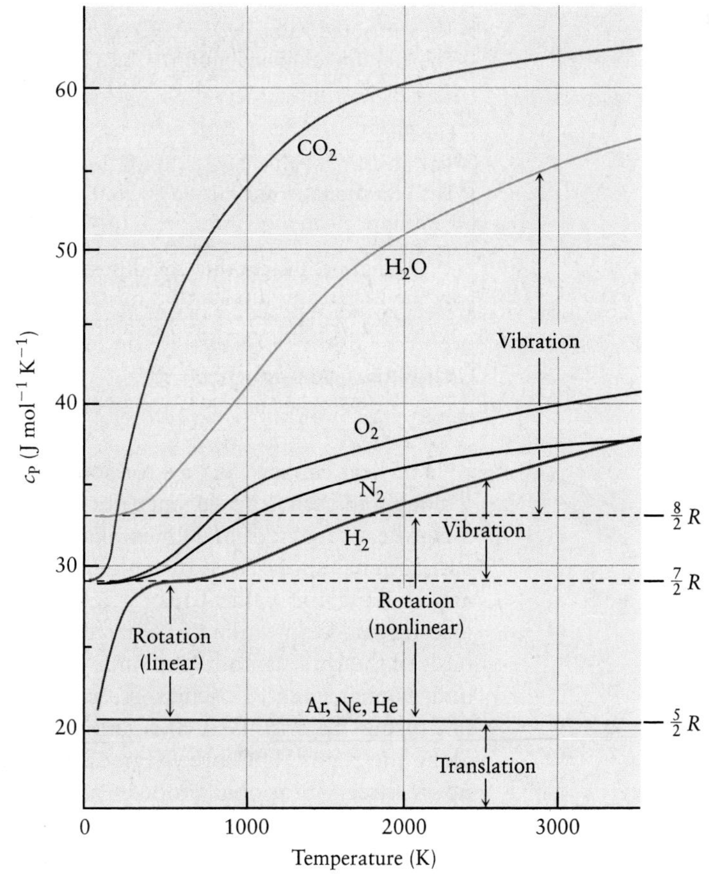

FIGURE 12.14 The temperature dependence of c_P for selected gases. Condensation to liquid and solid states at low temperatures is not included. Comparison with Table 12.3 shows that the diatomic molecules and the linear triatomic molecule CO_2 have contributions from two rotational degrees of freedom, and the non-linear triatomic H_2O has contributions from three. Vertical arrows represent the vibrational contributions for H_2 and for H_2O at specific temperatures. (Adapted from Donald R. Olander, *General Thermodynamics*, New York, CRC Press, 2007, Figure 2.5, page 57; © Cengage Learning)

$\exp(-\Delta\varepsilon/k_BT)$, where $\Delta\varepsilon$ is the gap between the ground state and a typical excited energy level. The dominant physical question is whether the temperature is sufficiently high to make $k_BT \gg \Delta\varepsilon$. If T is high enough, the quantum states will be populated and the system will behave essentially as predicted by classical mechanics. This analysis is illustrated in Example 4.8 by showing that the energy levels of He atoms confined in a three-dimensional box of macroscopic size are so closely spaced in energy that the classical prediction is a good one. We show in Chapter 20 that many rotational quantum states are populated at room temperature for the same reason. The spacing between vibrational energy levels is larger, and each case must be examined to determine whether a significant number of molecules are vibrationally excited at a particular T. The detailed procedures are demonstrated in Section 12.8, *A Deeper Look . . .* Distribution of Energy among Molecules.

EXAMPLE 12.5

Calculate the value of c_P at 298 K and 1 atm pressure predicted for CO and Br_2 by the classical equipartition theorem. Compare the predicted results with the experimental results and calculate the percent of the measured value that arises from vibrational motions.

Solution

Table 12.3 summarizes the predicted results for CO and Br_2. The total value of $c_P = (9R/2)$ because a diatomic molecule has 3 degrees of translational motion (each contributing $(R/2)$), 2 degrees of rotational motion (each contributing $(R/2)$), and 1 degree of vibrational motion contributing R for a total $(7R/2)$ to c_V. One additional term of R must be added to obtain $c_P = (9R/2)$. The numerical value is $c_P = 37.41$ J mol^{-1} K^{-1}.

Neglecting the vibrational motion predicts $c_P = (7R/2) = 29.10$ J mol^{-1} K^{-1}.

For CO:
The measured value is $c_P = 29.12$ J mol^{-1} K^{-1}.
The vibrational contribution $= 29.12 - 29.10 = 0.02$ J mol^{-1} K^{-1}.
Percent of c_P due to vibration $= (0.02/29.12) \times 100 = 0.1\%$

For Br_2:
The measured value is $c_P = 36.02$ J mol^{-1} K^{-1}.
The vibrational contribution $= 36.02 - 29.10 = 6.92$ J mol^{-1} K^{-1}.
Percent of c_P due to vibration $= (6.92/36.02) \times 100 = 19.2\ \%$

Additional insight into the different behavior of these two molecules is provided by the Boltzmann distribution for CO and Br_2 over vibrational states at room temperature in Figure 12.23 and Example 12.12 in Section 12.8.

Related Problems: 23, 24, 25, 26

The heat capacity values for solids also display quantum effects. The law of Dulong and Petit, known since about 1820, states that at room temperature the molar heat capacities of elemental solids have the value $3R$, or 24.9 J K^{-1} mol^{-1}. Research has shown that the values are substantially lower at low temperatures, approaching the value 0 as T approaches absolute 0, and increasing to the Dulong and Petit value at room temperature. See Figure 12.15a. The physical basis of the Dulong and Petit value is explained by modeling the solid as a three-dimensional stack of oscillators, in which each atom is located between a pair of "springs" along the x-axis, another pair along the y-axis, and a third pair along the z-axis. The three degrees of freedom for each atom are therefore tied up in three vibrational modes, each having a quadratic term in its potential energy and its kinetic energy. By the equipartition theorem, these modes

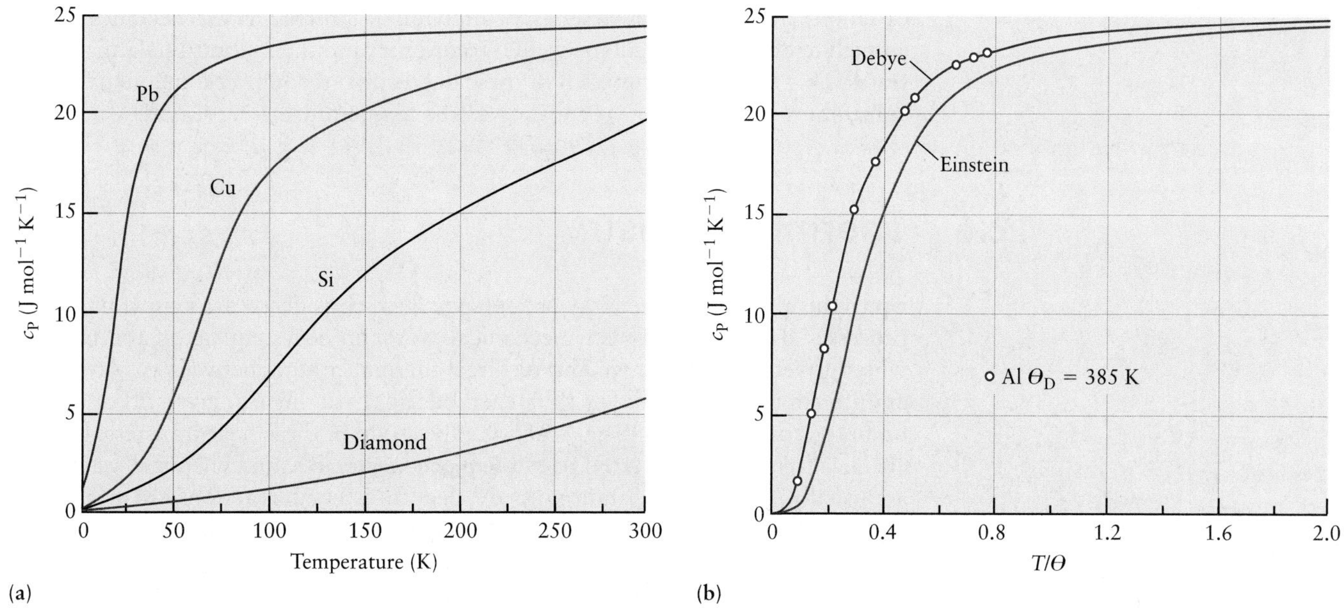

FIGURE 12.15 The temperature dependence of c_V for solids. (Note that $c_V = c_P$ because the volume of a solid does not change substantially when you exert pressure on it.) (a) Measured values for selected elemental solids. (b) Debye and Einstein models for c_v for aluminum. (Adapted from D.R. Gaskell, *Introduction to the Thermodynamics of Materials* (5th ed.), New York, Taylor and Francis, 2008, Figures 6.1, 6.2. © Cengage Learning)

contribute $6R/2 = 3R$ to c_V for the solid, as shown in Table 12.3. Einstein explained the behavior at low T in 1907 as an early application of Planck's quantum hypothesis which was introduced to explain blackbody radiation (see Section 4.2). He assumed each atom in the solid behaved as a harmonic oscillator, all moving with the same frequency ν. He calculated the average vibrational energy for these oscillators as a function of T, multiplied the result by Avogadro's number to obtain the internal energy per mole, and calculated c_V by taking the derivative with respect to T. This model leads to a characteristic temperature called the Einstein temperature $\Theta_E = h\nu/k_B$ at which the vibrational energies are excited. Einstein's calculated curve for c_V is plotted as a function of T/Θ_E in Figure 12.15b. As T increases $c_V \longrightarrow 3R$ in accord with the law of Dulong and Petit, and as $T \longrightarrow 0$, $c_V \longrightarrow 0$ in accord with experiment. Debye refined Einstein's model by allowing the atoms in the solid to vibrate about their equilibrium positions with a range of frequencies up to a maximum frequency $\nu_{max} = \nu_D$ called the Debye frequency. The Debye frequency corresponds to the shortest possible wavelength at which neighboring atoms would vibrate in opposition to each other; this is equal to twice the distance between atoms. Debye summed up Einstein's c_V contributions up to the Debye frequency, and plotted the result against the Debye temperature $\Theta_D = h\nu_D/k_B$ at which the maximum vibrational frequency becomes active. The result shown in Figure 12.15(b) reproduced the experimental data almost exactly.

We started this section with the statement that the heat capacity measures the ability of molecular motions and interactions to accommodate thermal energy. We now understand that accommodating thermal energy is fundamentally a quantum mechanical process. It is straightforward to identify modes of molecular motion and the corresponding degrees of freedom, but we must have a quantum description of the allowed energy levels for each mode in order to identify the temperature range in which each mode becomes active. If the temperature is not high enough, the modes are ineffective and might as well be absent. Figure 12.14 shows that the heat capacity of a gas plotted as a function of temperature evidences a series of individual "jumps" as kT becomes large enough to efficiently populate excited rotational, and then vibrational, quantum states. This illustrates how each of these degrees of freedom becomes active and contributes to the total heat capacity as the temperature is increased.

12.6 Thermochemistry

Up to this point, all the energy changes we have considered are simple physical processes that involve direct mechanical work on a system (as in the paddle wheel driven by a falling weight) or direct thermal contact between two systems at different temperatures. The same methods apply to the heat given off or taken up in the course of a chemical reaction. The study of these heat transfers during chemical reactions is referred to as **thermochemistry**. Because chemical reactions are usually studied at constant pressure, heat transfers in reactions are measured at constant pressure. We obtain these numbers by conducting the chemical reaction in a constant-pressure calorimeter and measuring the heat transferred as q_P. This number is the enthalpy change of the reaction, viewed as a thermodynamic process:

$$q_P = \Delta H = H_f - H_i = H_{\text{products}} - H_{\text{reactants}} = \Delta H_{\text{reaction}}$$

The result, $\Delta H_{\text{reaction}}$, is called the **reaction enthalpy** or enthalpy of reaction.

12.6.1 Enthalpies of Reaction

When carbon monoxide is burned in oxygen to produce carbon dioxide,

$$CO(g) + \tfrac{1}{2}\,O_2(g) \longrightarrow CO_2(g)$$

heat is given off. Because this energy is transferred out of the reaction vessel (the system) and into the surroundings, it has a *negative* sign. Careful calorimetric measurements show that 1.000 mol CO reacted completely with 0.500 mol O_2, at 25°C and a constant pressure of 1 atm, leads to an enthalpy change of

$$\Delta H = q_P = -2.830 \times 10^5 \text{ J} = -283.0 \text{ kJ}$$

The kilojoule (kJ), equal to 10^3 J, is used because most enthalpy changes for chemical reactions lie in the range of thousands of joules per mole.

When heat is given off by a reaction (ΔH is negative), the reaction is said to be an **exothermic reaction** (Fig. 12.16). A reaction in which heat is taken up (ΔH positive) is called **endothermic reaction** (Fig. 12.17). One example of an endothermic reaction is the preceding reaction written in the opposite direction:

$$CO_2(g) \longrightarrow CO(g) + \tfrac{1}{2}\,O_2(g) \qquad \Delta H = +283.0 \text{ kJ}$$

If the direction of a chemical reaction is reversed, the enthalpy change reverses sign. Heat is required to convert CO_2 to CO and O_2 at constant pressure. The decomposition of CO_2 into CO and O_2 is difficult to perform in the laboratory, whereas the reverse reaction is straightforward. Thermodynamics allows us to predict ΔH of the decomposition reaction with complete confidence, even if a calorimetric experiment is never actually performed for it.

Chemists have agreed on a convention for attaching reaction enthalpy values to balanced chemical equations. A reaction enthalpy written after a balanced chemical equation refers to the enthalpy change for the complete conversion of stoichiometric amounts of reactants to products; the numbers of moles of reactants and products are given by the coefficients in the equation. The preceding

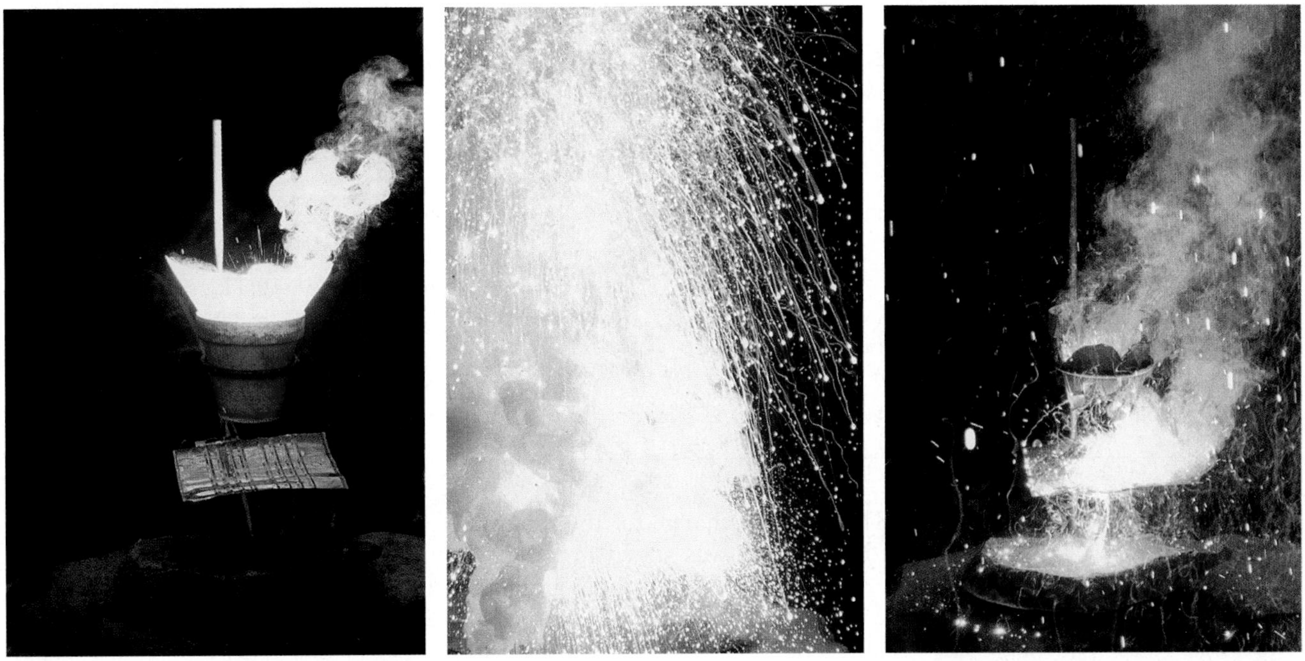

(a) (b) (c)

FIGURE 12.16 The thermite reaction, $2\ Al(s) + Fe_2O_3(s) \longrightarrow 2\ Fe(s) + Al_2O_3(s)$, is among the most exothermic of all reactions, liberating 16 kJ of heat for every gram of aluminum that reacts. (a) A piece of burning magnesium acts as a source of ignition when inserted into a pot containing a finely divided mixture of aluminum powder and iron(III) oxide. (b) After the ignition, the reaction continues on its own. (c) Enough heat is generated to produce molten iron, which can be seen flowing out of the broken pot onto the protective mat and the bottom of the stand.

equation shows the enthalpy change when 1 mol CO_2 is converted to 1 mol CO and $\frac{1}{2}$ mol O_2. If this equation is multiplied by a factor of 2, the enthalpy change must also be doubled because twice as many moles are then involved (enthalpy, like energy, is an *extensive* property).

$$2\ CO_2(g) \longrightarrow 2\ CO(g) + O_2(g) \qquad \Delta H = +566.0\ \text{kJ}$$

The molar amounts need not be integers, as the following example illustrates.

(a) (b)

FIGURE 12.17 (a) When mixed in a flask, the two solids $Ba(OH)_2{\cdot}8H_2O(s)$ and $NH_4NO_3(s)$ undergo an acid–base reaction: $Ba(OH)_2{\cdot}8\ H_2O(s) + 2\ NH_4NO_3(s) \longrightarrow Ba(NO_3)_2\ (aq) + 2\ NH_3(aq) + 10\ H_2O(\ell)$. (b) The water produced dissolves excess ammonium nitrate in an endothermic reaction. The dissolution absorbs so much heat that the water on the surface of the wet wooden block freezes to the bottom of the flask, and the block can be lifted up with the flask.

EXAMPLE 12.6

Red phosphorus reacts with liquid bromine in an exothermic reaction (Fig. 12.18):

$$2\ P(s) + 3\ Br_2(\ell) \longrightarrow 2\ PBr_3(g) \qquad \Delta H = -243\ kJ$$

Calculate the enthalpy change when 2.63 g phosphorus reacts with an excess of bromine in this way.

Solution

First, convert from grams of phosphorus to moles, using the molar mass of phosphorus, 30.97 g mol^{-1}:

$$\text{moles P} = \frac{2.63\ g\ P}{30.97\ g\ mol^{-1}} = 0.0849\ mol$$

Given that an enthalpy change of −243 kJ is associated with 2 mol P, it is readily seen that the enthalpy change associated with 0.0849 mol is

$$\Delta H = 0.0849\ mol\ P \times \left(\frac{-243\ kJ}{2\ mol\ P}\right) = -10.3\ kJ$$

This reaction is exothermic, $q_P = -10.3$ kJ, so 10.3 kJ of heat is released to the surroundings.

Related Problems: 27, 28, 29, 30

FIGURE 12.18 Red phosphorus reacts exothermically in liquid bromine. The rising gases are a mixture of the product PBr_3 and unreacted bromine that has boiled off.

The enthalpy change for the reaction of 1 mol carbon monoxide with oxygen to form CO_2 was stated to be −283.0 kJ. In a second experiment, the heat evolved when 1 mol carbon (graphite) is burned in oxygen to carbon dioxide at 25°C is readily measured to be $q_{surr} = 393.5$ kJ and $q_{surr} = -q_{sys}$ so the enthalpy of reaction is −393.5 kJ:

$$C(s,gr) + O_2(g) \longrightarrow CO_2(g) \qquad \Delta H = -393.5\ kJ$$

Now, suppose we need to know the enthalpy change for the reaction

$$C(s,gr) + \tfrac{1}{2}\ O_2(g) \longrightarrow CO(g) \qquad \Delta H = ?$$

This reaction cannot be performed simply in the laboratory. If 1 mol graphite is heated with $\frac{1}{2}$ mol oxygen, almost half the carbon burns to $CO_2(g)$, and the remainder is left as unreacted carbon. Nevertheless, thermodynamics allows us to predict the heat that *would* evolve if we could perform the reaction as written. This is possible because H is a state function, and thus ΔH for the reaction is independent of the path followed from reactants to products. We are free to select any path for which we have all the data needed for the calculation. In Figure 12.19, we illustrate the path in which 1 mol C is burned with O_2 to CO_2 (with $\Delta H = -393.5$ kJ), and to this is added the calculated enthalpy change for the (hypothetical) process in which CO_2 is converted to CO and O_2 ($\Delta H =$ +283.0 kJ). The total ΔH is the sum of the two known enthalpy changes, −393.5 kJ + 283.0 kJ = −110.5 kJ. To see this more clearly, the reactions are written out as follows:

$$\begin{array}{lll} C(s,\ gr) + O_2(g) \longrightarrow CO_2(g) & & \Delta H_1 = -393.5\ kJ \\ \quad CO_2(g) \longrightarrow CO(g) + \tfrac{1}{2}O_2(g) & & \Delta H_2 = +283.0\ kJ \\ \hline C(s,\ gr) + \tfrac{1}{2}O_2(g) \longrightarrow CO(g) & \Delta H = \Delta H_1 + \Delta H_2 & = -110.5\ kJ \end{array}$$

If two or more chemical equations are added to give another chemical equation, the corresponding enthalpies of reaction must be added.

This statement is known as **Hess's law** and derives from the fact that enthalpy is a state function. It is proper to include in the pathway any convenient reaction for which the enthalpy change is known, even if that reaction is difficult to study directly and its enthalpy change is known only from studies of the reverse reaction. The step labeled (hypothetical) earlier is a good example.

The corresponding internal energy change ΔU might be desired for this reaction. The quantity ΔU is simple to calculate because of the relation

$$\Delta H = \Delta U + \Delta(PV)$$

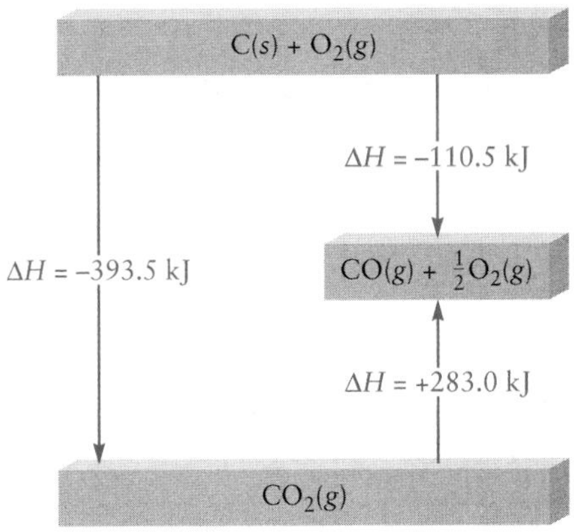

FIGURE 12.19 Because enthalpy is a state property, the enthalpy change for the reaction of carbon with oxygen to give carbon monoxide (red arrow) can be determined through measurements along a path that is less direct but easier to study (blue arrows). The enthalpy change sought is the sum of the enthalpy change to burn carbon to carbon dioxide and that to convert carbon dioxide to carbon monoxide and oxygen.

or

$$\Delta U = \Delta H - \Delta(PV)$$

The gases can be assumed to obey the ideal gas law, so

$$\Delta(PV) = \Delta(nRT) = RT\,\Delta n_g$$

because the temperature is constant at 25°C. Here, Δn_g is the change in the number of moles of *gas* in the reaction as written:

$$\begin{aligned}\Delta n_g &= \text{total moles of product gases} - \text{total moles of reactant gases}\\ &= 1\text{ mol} - \tfrac{1}{2}\text{ mol} = \tfrac{1}{2}\text{ mol}\end{aligned}$$

(Graphite is a solid, and its volume is negligible compared with the volumes of the gases.) Hence,

$$\Delta(PV) = RT\,\Delta n_g = (8.315\text{ J K}^{-1}\text{ mol}^{-1})(298\text{ K})(\tfrac{1}{2}\text{ mol}) = 1.24 \times 10^3\text{ J} = 1.24\text{ kJ}$$

Note that R must be expressed in J K^{-1} mol^{-1} to obtain the result in joules. Therefore,

$$\Delta U = -110.5\text{ kJ} - 1.24\text{ kJ} = -111.7\text{ kJ}$$

For reactions in which only liquids and solids are involved, or those in which the number of moles of gas does not change, the enthalpy and energy changes are almost equal, and their difference can be neglected.

Phase changes are not chemical reactions, but their enthalpy changes can be analyzed in the same way. Heat must be absorbed by ice to transform to water, so the phase change is endothermic, with ΔH positive:

$$H_2O(s) \longrightarrow H_2O(\ell) \qquad \Delta H_{fus} = +6.007\text{ kJ mol}^{-1}$$

Here, ΔH_{fus} is the **molar enthalpy of fusion**, the heat that must be transferred at constant pressure to melt 1 mole of substance. When a liquid freezes, the reaction is reversed and an equal amount of heat is given off to the surroundings; that is, $\Delta H_{freez} = -\Delta H_{fus}$. The vaporization of 1 mole of liquid at constant pressure and temperature requires an amount of heat called the **molar enthalpy of vaporization**, ΔH_{vap},

$$H_2O(\ell) \longrightarrow H_2O(g) \qquad \Delta H_{vap} = +40.66\text{ kJ mol}^{-1}$$

whereas the condensation of a liquid from a vapor is an exothermic process, with $\Delta H_{cond} = -\Delta H_{vap}$. Table 12.4 lists enthalpies of fusion and vaporization.

TABLE 12.4

Enthalpy Changes of Fusion and Vaporization†

Substance	ΔH_{fus} (kJ mol^{-1})	ΔH_{vap} (kJ mol^{-1})
NH_3	5.650	23.35
HCl	1.992	16.15
CO	0.836	6.04
CCl_4	2.500	30.00
H_2O	6.007	40.66
NaCl	28.800	170.00

†The enthalpy changes are measured at the normal melting point and the normal boiling point, respectively.

EXAMPLE 12.7

To vaporize 100.0 g carbon tetrachloride at its normal boiling point, 349.9 K, and $P = 1$ atm, 19.5 kJ of heat is required. Calculate ΔH_{vap} for CCl_4 and compare it with ΔU for the same process.

Solution

The molar mass of CCl_4 is 153.8 g mol^{-1}; thus, the number of moles in 100.0 g is

$$\frac{100.0 \text{ g } CCl_4}{153.8 \text{ g mol}^{-1}} = 0.6502 \text{ mol } CCl_4$$

The enthalpy change for vaporizing 1 mol CCl_4 is then

$$\left(\frac{19.5 \text{ kJ}}{0.6502 \text{ mol } CCl_4}\right) \times 1.00 \text{ mol } CCl_4 = 30.0 \text{ kJ} = \Delta H_{vap}$$

The internal energy change is then

$$\Delta U = \Delta H_{vap} - \Delta(PV) = \Delta H_{vap} - RT\,\Delta n_g$$

Inserting $T = 349.9$ K and $\Delta n_g = 1$ (because there is an increase of 1 mol of gaseous products for each mole of liquid that is vaporized) gives

$$\Delta U = 30.0 \text{ kJ} - (8.315 \text{ J K}^{-1} \text{ mol}^{-1})(349.9 \text{ K})(1.00 \text{ mol})(10^{-3} \text{ kJ J}^{-1})$$
$$= (30.0 - 2.9) \text{ kJ} = +27.1 \text{ kJ}$$

Thus, of the 30.0 kJ of energy transferred from the surroundings in the form of heat, 27.1 kJ is used to increase the internal energy of the molecules (ΔU) and 2.9 kJ is used to expand the resulting vapor, $\Delta(PV)$.

Related Problems: 31, 32, 33, 34

12.6.2 Standard-State Enthalpies

Absolute values of the enthalpy of a substance, like absolute values of the internal energy, cannot be measured or calculated. Only *changes* in enthalpy can be measured. Just as altitudes are measured relative to a standard altitude (sea level), it is necessary to adopt a reference state for the enthalpies of substances. To cope with this problem, chemists define *standard states* for chemical substances as follows:

For solids and liquids, the standard state is the thermodynamically stable state at a pressure of 1 atm and at a specified temperature.

For gases, the standard state is the gaseous phase at a pressure of 1 atm, at a specified temperature and exhibiting ideal gas behavior.

For dissolved species, the standard state is a 1-M solution at a pressure of 1 atm, at a specified temperature and exhibiting ideal solution behavior.

Standard-state values of enthalpy and other quantities are designated by attaching a superscript ° (pronounced "naught") to the symbol for the quantity and writing the specified temperature as a subscript. Any temperature may be chosen as the "specified temperature." The most common choice is 298.15 K (25°C exactly); if the temperature of a standard state is not explicitly indicated, 298.15 K should be assumed to be the value.

Once standard states have been defined, the zero of the enthalpy scale is defined by arbitrarily setting the enthalpies of selected reference substances to zero in their standard states. This is completely analogous to assigning zero as the altitude at sea level. Chemists have agreed to the following: *The chemical elements in their standard states at 298.15 K have zero enthalpy.* A complication immediately arises because some elements exist in various allotropic forms that differ in structure and all physical properties, including enthalpy. For example, oxygen can be prepared as $O_2(g)$ or $O_3(g)$ (ozone), and carbon exists in numerous allotropic forms, including graphite, diamond, and the fullerenes (see Section 7.5). Chemists have agreed to assign zero enthalpy to the form that is most stable at 1 atm and 298.15 K. Thus, $O_2(g)$ is assigned zero enthalpy in its standard state at 298.15 K, whereas the standard-state enthalpy of $O_3(g)$ is not zero in its standard state at 298.15 K. The most stable form of carbon at 1 atm and 298.15 K is graphite, which is assigned zero enthalpy; the standard state enthalpy values for diamond and all the fullerenes are not zero.[2]

The enthalpy change for a chemical reaction in which all reactants and products are in their standard states and at a specified temperature is called the **standard enthalpy** (written $\Delta H°$) for that reaction. The standard enthalpy is the central tool in thermochemistry because it provides a systematic means for comparing the energy changes due to bond rearrangements in different reactions. Standard enthalpies can be calculated from tables of reference data. For this purpose, we need one additional concept. The **standard enthalpy of formation** $\Delta H_f°$ of a compound is defined to be the enthalpy change for the reaction that produces 1 mol of the compound from its elements in their stable states, all at 25°C and 1 atm pressure. For example, the standard enthalpy of formation of liquid water is the enthalpy change for the reaction

$$H_2(g) + \tfrac{1}{2}\,O_2(g) \longrightarrow H_2O(\ell) \qquad \Delta H° = -285.83 \text{ kJ}$$

$$\Delta H_f°(H_2O(\ell)) = -285.83 \text{ kJ mol}^{-1}$$

Here, the superscript ° indicates standard-state conditions (e.g., the partial pressure of each ideal gas is 1 atm), and the subscript f stands for *formation.*

The $\Delta H_f°$ for an *element* that is already in its standard state is clearly zero, because no further change is needed to bring it to standard-state conditions. But the standard enthalpy of formation of a mole of *atoms* of an element is often a large positive quantity. That is, the reaction to generate them is endothermic:

$$\tfrac{1}{2}\,H_2(g) \longrightarrow H(g) \qquad \Delta H° = +217.96 \text{ kJ}$$

$$\Delta H_f°(H(g)) = 217.96 \text{ kJ mol}^{-1}$$

For dissolved species the standard state is defined as an ideal solution with a concentration of 1 M (this is obtained in practice by extrapolating the dilute solution behavior up to this concentration). A special comment is in order on

[2]There is one exception to this choice of standard state. The standard state of phosphorus is taken to be white phosphorus, rather than the more stable red or black form, because the latter are less well characterized.

the standard enthalpies of formation of ions. When a strong electrolyte dissolves in water, both positive and negative ions form; it is impossible to produce one without the other. It is therefore also impossible to measure the enthalpy change of formation of ions of only one charge. Only the sum of the enthalpies of formation of the positive and negative ions is accessible to calorimetric experiments. Therefore, chemists have agreed that ΔH_f° of $H^+(aq)$ is set to zero to establish a reference point for the enthalpies of formation of cations and anions.

Tables of ΔH_f° for compounds are the most important data source for thermochemistry. From them it is easy to calculate ΔH° for reactions of the compounds, and thereby systematically compare the energy changes due to bond rearrangements in different reactions. Appendix D gives a short table of standard enthalpies of formation at 25°C. The following example shows how they can be used to determine enthalpy changes for reactions performed at 25°C and 1 atm pressure.

EXAMPLE 12.8

Using Appendix D, calculate ΔH° for the following reaction at 25°C and 1 atm pressure:

$$2\,NO(g) + O_2(g) \longrightarrow 2\,NO_2(g)$$

Solution

Because enthalpy is a function of state, ΔH can be calculated along any convenient path. In particular, two steps can be chosen for which ΔH is found easily. In step 1, the reactants are decomposed into the elements in their standard states:

$$2\,NO(g) \longrightarrow N_2(g) + O_2(g) \qquad \Delta H_1 = -2\,\Delta H_f^\circ(NO)$$

The minus sign appears because the process chosen is the reverse of the formation of NO; the factor of 2 is present because 2 mol NO is involved. Because oxygen is already an element in its standard state, it does not need to be changed [equivalently, $\Delta H_f^\circ(O_2(g))$ is 0].

In step 2, the elements are combined to form products:

$$N_2(g) + 2\,O_2(g) \longrightarrow 2\,NO_2(g) \qquad \Delta H_2 = 2\,\Delta H_f^\circ(NO_2)$$

The enthalpy change of the overall reaction is then the sum of these two enthalpies:

$$\begin{aligned}\Delta H^\circ &= \Delta H_1 + \Delta H_2 = -2\,\Delta H_f^\circ(NO) + 2\,\Delta H_f^\circ(NO_2)\\ &= -(2\text{ mol})(90.25\text{ kJ mol}^{-1}) + (2\text{ mol})(33.18\text{ kJ mol}^{-1})\\ &= -114.14\text{ kJ}\end{aligned}$$

Related Problems: 39, 40, 41, 42

The general pattern should be clear from this example. The change ΔH° for a reaction at atmospheric pressure and 25°C is the sum of the ΔH_f° for the *products* (multiplied by their coefficients in the balanced chemical equation) minus the sum of the ΔH_f° for the *reactants* (also multiplied by their coefficients). For a general reaction of the form

$$aA + bB \longrightarrow cC + dD$$

the standard enthalpy change is

$$\Delta H^\circ = c\,\Delta H_f^\circ(C) + d\,\Delta H_f^\circ(D) - a\,\Delta H_f^\circ(A) - b\,\Delta H_f^\circ(B)$$

This equation can be extended to calculate the standard-state enthalpy change for any chemical reaction by adding up the standard-state enthalpy of formation for all the products (each multiplied by its stoichiometric coefficient in the balanced chemical equation) and subtracting off the total for all the reactants (each multiplied by its stoichiometric coefficient in the balanced chemical equation). In mathematical form, this procedure is represented by the equation

$$\Delta H^\circ = \sum_{i=1}^{prod} n_i \Delta H_i^\circ - \sum_{j=1}^{react} n_j \Delta H_j^\circ \qquad [12.12]$$

12.6.3 Bond Enthalpies

Chemical reactions between molecules require existing bonds to break and new ones to form in a new arrangement of the atoms. Chemists have developed methods to study highly reactive intermediate species in which one or more bonds have been broken, and to find the energy required to break a particular bond. For example, a hydrogen atom can be removed from a methane molecule,

$$CH_4(g) \longrightarrow CH_3(g) + H(g)$$

leaving two fragments, neither of which has a stable valence electron structure in the Lewis electron dot picture. Both will go on to react rapidly with other molecules or fragments and eventually form the stable products of that reaction. Nonetheless, we can measure many of the properties of these reactive species during the short time they are present.

One such important measurable quantity is the enthalpy change when a bond is broken in the gas phase, called the **bond enthalpy.** This change is invariably positive because atoms bonded together have lower energy than when separated (see Fig. 6.1). For example, the bond enthalpy of a C—H bond in methane is 438 kJ mol^{-1}, measured as the standard enthalpy change for the reaction

$$CH_4(g) \longrightarrow CH_3(g) + H(g) \qquad \Delta H° = +438 \text{ kJ}$$

in which 1 mol of C—H bonds is broken, one for each molecule of methane. Bond enthalpies are fairly constant from one compound to another. Each of the following gas-phase reactions involves the breaking of a C—H bond:

$$C_2H_6(g) \longrightarrow C_2H_5(g) + H(g) \qquad \Delta H° = +410 \text{ kJ}$$
$$CHF_3(g) \longrightarrow CF_3(g) + H(g) \qquad \Delta H° = +429 \text{ kJ}$$
$$CHCl_3(g) \longrightarrow CCl_3(g) + H(g) \qquad \Delta H° = +380 \text{ kJ}$$
$$CHBr_3(g) \longrightarrow CBr_3(g) + H(g) \qquad \Delta H° = +377 \text{ kJ}$$

The approximate constancy of the measured enthalpy changes (all lie within 8% of their average value) suggests that the C—H bonds in all five molecules are similar. Because such bond enthalpies are constant from one molecule to another, it is useful to tabulate *average* bond enthalpies from measurements on a series of compounds (Table 12.5). Any given bond enthalpy will differ

TABLE 12.5

Average Bond Enthalpies

	Molar Enthalpy of Atomization (kJ mol^{-1})‡	Bond Enthalpy (kJ mol^{-1})†								
		H—	C—	C=	C≡	N—	N=	N≡	O—	O=
H	218.0	436	413			391			463	
C	716.7	413	348	615	812	292	615	891	351	728
N	472.7	391	292	615	891	161	418	945		
O	249.2	463	351	728					139	498
S	278.8	339	259	477						
F	79.0	563	441			270			185	
Cl	121.7	432	328			200			203	
Br	111.9	366	276							
I	106.8	299	240							

†From Appendix D.
‡Data from L. Pauling. *The Nature of the Chemical Bond,* 3rd ed. Ithaca, NY: Cornell University Press, 1960.

somewhat from those shown, but in most cases, the deviations are small. The constant of bond energies in a series of molecules was introduced in Section 3.9, and representative values were listed in Table 3.4. Bond energy values are related to bond enthalpy values by the relation $\Delta H = \Delta U + \Delta(PV)$, described in the paragraphs following Example 12.6.

The bond enthalpies in Table 12.5 can be used, together with enthalpies of atomization of the elements from the same table, to estimate standard enthalpies of formation ΔH_f° for molecules in the gas phase and enthalpy changes ΔH° for gas-phase reactions. This is illustrated by the following example.

EXAMPLE 12.9

Estimate the standard enthalpy of formation of dichlorodifluoromethane, $CCl_2F_2(g)$ (Fig. 12.20). This compound is also known as Freon-12 and was used as a refrigerant because of its low reactivity and high volatility. It and other related chlorofluorocarbons (CFCs) have been phased out because of their role in depleting the ozone layer in the outer atmosphere, as discussed in Section 20.6.

Solution

The standard enthalpy of formation of $CCl_2F_2(g)$ is the enthalpy change for the process in which it is formed from the elements in their standard states at 25°C:

$$C(s,gr) + Cl_2(g) + F_2(g) \longrightarrow CCl_2F_2(g)$$

This reaction can be replaced by a hypothetical two-step process: All the species appearing on the left are atomized, and then the atoms are combined to make CCl_2F_2.

$$C(s, gr) + Cl_2(g) + F_2(g) \longrightarrow C(g) + 2\ Cl(g) + 2\ F(g)$$

$$C(g) + 2\ Cl(g) + 2\ F(g) \longrightarrow CCl_2F_2(g)$$

The enthalpy change ΔH_1 for the first step is the sum of the atomization energies:

$$\Delta H_1 = \Delta H_f^\circ(C(g)) + 2\ \Delta H_f^\circ(Cl(g)) + 2\ \Delta H_f^\circ(F(g))$$

$$= 716.7 + 2(121.7) + 2(79.0) = 1118 \text{ kJ}$$

The enthalpy change ΔH_2 for the second step can be estimated from the bond enthalpies of Table 12.5. This step involves the formation (with release of heat and, therefore, negative enthalpy change) of two C—Cl and two C—F bonds per molecule. The net ΔH for this step is then

$$\Delta H_2 \approx -[2(328) + 2(441)] = -1538 \text{ kJ}$$

$$\Delta H_1 + \Delta H_2 = -1538 + 1118 = -420 \text{ kJ}$$

This ΔH_f°, -420 kJ mol^{-1}, compares fairly well with the experimental value, -477 kJ mol^{-1}. In general, much better agreement than this is not to be expected, because tabulated bond enthalpies are only average values.

Related Problems: 49, 50, 51, 52

FIGURE 12.20 Dichlorodifluoromethane, CCl_2F_2, also known as Freon-12.

12.7 Reversible Processes in Ideal Gases

Most thermodynamic processes conducted in laboratory work are irreversible, in the sense of Section 12.1. Except in the initial and final states, the system is not at equilibrium, and the equation of state relationship between observable properties does not exist. Consequently, changes in thermodynamic quantities during an irreversible process cannot in general be calculated. Nonetheless, the changes in those quantities that are state functions are well defined, as long as the initial and final equilibrium states are known. Because these changes are independent of the detailed path of the process, they can be evaluated for any known process that connects these initial and final states. Changes can be directly calculated along *reversible* paths, during which the system proceeds through a sequence of equilibrium states in which observable properties are related by the equation of state.

This section demonstrates calculations of changes in macroscopic properties caused during several specific reversible processes in ideal gases. These will serve as auxiliary calculation pathways for evaluating changes in state functions during irreversible processes. We use this procedure extensively in Chapter 13 on spontaneous processes and the second law of thermodynamics.

Recall from Section 12.1 that a true reversible process is an idealization; it is a process in which the system proceeds with infinitesimal speed through a series of equilibrium states. The external pressure P_{ext}, therefore, can never differ by more than an infinitesimal amount from the pressure, P, of the gas itself. The heat, work, energy, and enthalpy changes for ideal gases at constant volume (called an **isochoric process**) and at constant pressure (called an **isobaric process**) have already been considered. This section examines *isothermal* (constant temperature) and *adiabatic* ($q = 0$) processes.

12.7.1 Isothermal Processes

An **isothermal process** is one conducted at constant temperature. This is accomplished by placing the system in a large reservoir (bath) at fixed temperature and allowing heat to be transferred as required between system and reservoir. The reservoir is large enough that its temperature is almost unchanged by this heat transfer. In Section 12.4, U for an ideal gas was shown to depend only on temperature; therefore, $\Delta U = 0$ for any isothermal ideal gas process. (Of course the change in energy will not be zero if a liquid evaporates to an ideal gas at constant temperature, or if a chemical reaction between ideal gases occurs.) From the first law it follows that

$$w = -q \text{ (isothermal process, ideal gas)}$$

In a *reversible* process, $P_{ext} = P_{gas} \equiv P$. But the relation $w = -P_{ext}\,\Delta V$ from Section 12.2 cannot be used to calculate the work, because that expression applies only if the external pressure remains constant as the volume changes. In the reversible isothermal expansion of an ideal gas,

$$P_{ext} = P = \frac{nRT}{V} \qquad [12.13]$$

By Boyle's law, the pressure falls as the volume is increased from V_1 to V_2, as shown by the solid line in Figure 12.21. In this case, the work is calculated by approximating the process as a *series* of expansions by small amounts ΔV,

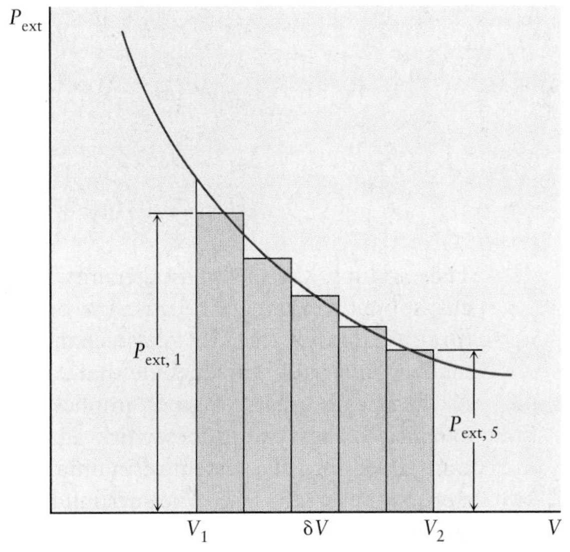

FIGURE 12.21 The area under the graph of external pressure against volume can be approximated as the sum of the areas of the rectangles shown.

during each of which P_{ext} is held constant at P_i, with $i = 1, 2, 3$, and so on labeling the step. The work done in step i is $-P_i\delta V$, and thus the total work done is the sum of the work in all steps:

$$w = -P_1\delta V - P_2\delta V - \cdots$$

The work in the complete process is the sum of the areas of the rectangles in the figure. As the step size δV is made smaller, it approaches the infinitesimal volume change that is written dV, and the corresponding increment of work becomes

$$dw = -P_{ext}\, dV$$

In this limit the sum of the areas of the rectangles approaches the area under the graph of P versus V. This limiting sum is the *integral* of P_{ext} from V_1 to V_2 and is written symbolically using an integral sign:

$$w = -\int_{V_1}^{V_2} P\, dV \quad \text{(reversible process)} \qquad [12.14]$$

Note that we used Equation 12.13 to replace P_{ext} with the internal pressure P in this reversible process. This is a compact way of saying that the work, w, for a reversible expansion process is (with a minus sign) the area under the graph of P plotted against V from V_1 to V_2.

The formal definition of an integral and procedures for evaluating integrals are covered in classes on calculus. For our purposes here it is sufficient that you interpret an integral as simply the area under a curve. Appendix C provides some useful background information and analytic expressions for a few common integrals that we use in this book.

In this case, the ideal gas law is used to write this area as

$$w = -nRT\int_{V_1}^{V_2} \frac{1}{V}\, dV \qquad [12.15]$$

since the temperature is constant. This integral is the area under a curve of $1/V$ against V (a hyperbola) from V_1 to V_2. The integral of $1/x$ evaluated between the points x_1 and x_2 is equal to $\ln(x_2/x_1)$, where ln is the natural logarithm function (see Table C.2). Because we know that $\Delta U = 0$ for an isothermal process, we can find that $q = -w$ for such a process. Combining this relation with the result of evaluating the integral in Eq. [12.15] leads us to conclude that $\Delta H = 0$ for isothermal processes involving ideal gases.

$$w = -nRT \ln \frac{V_2}{V_1}$$

$$q = -w = nRT \ln \frac{V_2}{V_1}$$

$$\Delta U = 0 \text{ because } \Delta T = 0$$

$$\Delta H = \Delta U + \Delta(PV) = \Delta U + \Delta(nRT) = 0 \qquad [12.16]$$

The enthalpy, like the internal energy, depends only on temperature for an ideal gas, so neither changes because the process was isothermal (constant temperature). Again, not that if a phase change or a chemical reaction occurs isothermally, you would *not* conclude that $\Delta U = 0$.

That V_2 is greater than V_1 implies that $w < 0$ and $q > 0$; in an isothermal expansion, the system does work against the surroundings and heat must be transferred into the system to maintain T constant. In an isothermal compression, the reverse is true: The surroundings do work on the system, and the system must then lose heat to the bath to maintain T constant.

EXAMPLE 12.10

Calculate the heat and the work associated with a process in which 5.00 mol of gas expands reversibly at constant temperature $T = 298$ K from a pressure of 10.00 to 1.00 atm.

Solution

At constant T and n, q and w are given by the equations just above Equation 12.16. Inserting the data for this example gives

$$\frac{V_2}{V_1} = \frac{P_1}{P_2} = \frac{10.0 \text{ atm}}{1.00 \text{ atm}} = 10.0$$

Thus,

$$w = -(5.00 \text{ mol})(8.315 \text{ J K}^{-1} \text{ mol}^{-1})(298 \text{ K}) \ln 10.0$$
$$= -2.85 \times 10^4 \text{ J} = -28.5 \text{ kJ}$$
$$q = -w = 28.5 \text{ kJ}$$

Related Problems: 55, 56

12.7.2 Adiabatic Processes

An **adiabatic process** is one in which there is no transfer of heat into or out of the system. This is accomplished by placing an adiabatic wall (thermal insulation) around the system to prevent heat flow.

$$q = 0$$
$$\Delta U = w$$

Consider a small adiabatic change. The volume changes by an amount dV and the temperature by an amount dT. Now U depends only on temperature for an ideal gas (no phase changes or chemical reactions), so

$$dU = nc_V \, dT$$

As always, the work is given by $-P_{ext} \, dV$. Setting these equal gives

$$nc_V \, dT = -P_{ext} \, dV$$

In other words, the temperature change dT is related to the volume change dV in such a process.

If the process is *reversible,* as well as adiabatic, so that $P_{ext} \approx P$, the ideal gas law can be used to write

$$nc_V \, dT = -P \, dV = -\frac{nRT}{V} \, dV$$

The equation is simplified by dividing both sides through by nT, making the left side depend only on T and the right side only on V:

$$\frac{c_V}{T} \, dT = -\frac{R}{V} \, dV$$

Suppose now that the change is not infinitesimal but large. How are temperature and volume related in this case? If a series of such infinitesimal changes is added together, the result is an integral of both sides of the equation from the initial state (specified by T_1 and V_1) to the final state (specified by T_2 and V_2):

$$c_V \int_{T_1}^{T_2} \frac{1}{T} \, dT = -R \int_{V_1}^{V_2} \frac{1}{V} \, dV$$

Here, c_V has been assumed to be approximately independent of T over the range from T_1 to T_2. Evaluating the integrals gives

$$c_V \ln \frac{T_2}{T_1} = -R \ln \frac{V_2}{V_1} = R \ln \frac{V_1}{V_2}$$

A more useful form results from taking antilogarithms of both sides:

$$\left(\frac{T_2}{T_1}\right)^{c_V} = \left(\frac{V_1}{V_2}\right)^{R} = \left(\frac{V_1}{V_2}\right)^{c_P - c_V}$$

The last step used the fact that $R = c_P - c_V$. Thus,

$$\left(\frac{T_2}{T_1}\right) = \left(\frac{V_1}{V_2}\right)^{c_P/c_V - 1} = \left(\frac{V_1}{V_2}\right)^{\gamma - 1}$$

where $\gamma = c_P/c_V$ is the ratio of specific heats. This can be rearranged to give

$$T_1 V_1^{\gamma-1} = T_2 V_2^{\gamma-1} \qquad [12.17]$$

In many situations, the initial thermodynamic state and, therefore, T_1 and V_1 are known. If the final volume V_2 is known, T_2 can be calculated; if T_2 is known, V_2 can be calculated.

In some cases, only the final pressure, P_2, of an adiabatic process is known. In this case, the ideal gas law gives

$$\frac{P_1 V_1}{T_1} = \frac{P_2 V_2}{T_2}$$

Multiplying this by Equation 12.17 gives

$$P_1 V_1^{\gamma} = P_2 V_2^{\gamma} \qquad [12.18]$$

This can be used to calculate V_2 from a known P_2.

Once the pressure, temperature, and volume of the final state are known, the energy and enthalpy changes and the work done are straightforward to calculate:

$$\Delta U = nc_V(T_2 - T_1) = w \text{ (reversible adiabatic process for ideal gas)}$$

$$\Delta H = \Delta U + \Delta(PV) = \Delta U + (P_2V_2 - P_1V_1)$$

or more simply,

$$\Delta H = nc_P \Delta T$$

EXAMPLE 12.11

Suppose 5.00 mol of an ideal monatomic gas at an initial temperature of 298 K and pressure of 10.0 atm is expanded adiabatically and reversibly until the pressure has decreased to 1.00 atm. Calculate the final volume and temperature, the energy and enthalpy changes, and the work done.

Solution

The initial volume is

$$V_1 = \frac{nRT_1}{P_1}$$

$$= \frac{(5.00 \text{ mol})(0.08206 \text{ L atm K}^{-1} \text{ mol}^{-1})(298 \text{ K})}{(10.0 \text{ atm})} = 12.2 \text{ L}$$

and the heat capacity ratio for a monatomic gas is

$$\gamma = \frac{c_P}{c_V} = \frac{\frac{5}{2}R}{\frac{3}{2}R} = \frac{5}{3}$$

For a reversible adiabatic process,

$$\frac{P_1}{P_2} V_1^{\gamma} = V_2^{\gamma}$$

$$(10.0)(12.2 \text{ L})^{5/3} = V_2^{5/3}$$

$$V_2 = (12.2 \text{ L})(10.0)^{3/5} = 48.7 \text{ L}$$

(Continued)

The final temperature can now be calculated from the ideal gas law:

$$T_2 = \frac{P_2V_2}{nR} = 119 \text{ K}$$

From this the work done and the energy change can be found,

$$w = \Delta U = nc_V\Delta T$$
$$= (5.00 \text{ mol})(\tfrac{3}{2} \times 8.315 \text{ J K}^{-1} \text{ mol}^{-1})(119 \text{ K} - 298 \text{ K})$$
$$= -11,200 \text{ J}$$

as well as the enthalpy change:

$$\Delta H = nc_P\Delta T$$
$$= (5.00 \text{ mol})(\tfrac{5}{2} \times 8.315 \text{ J K}^{-1} \text{ mol}^{-1})(119 \text{ K} - 298 \text{ K})$$
$$= -18,600 \text{ J}$$

Related Problems: 57, 58

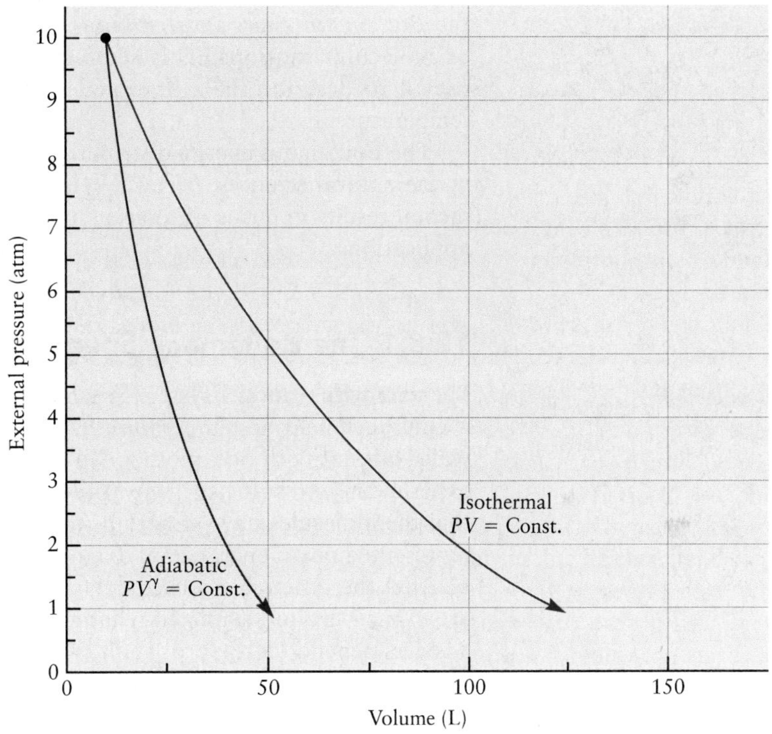

FIGURE 12.22 A comparison of reversible isothermal and adiabatic expansions. Using the technique shown in Figure 12.21, the adiabatic work is 40% of the isothermal work.

Figure 12.22 compares the adiabatic expansion of Example 12.11 to the isothermal expansion of Example 12.10. Note that the initial states were the same in the two cases, as were the final pressures. However, the final volume is larger by more than a factor of 2 for the isothermal expansion, and the work output for the adiabatic case is only 40% of the output from the isothermal expansion. Because $\gamma > 1$, the adiabatic line falls off more rapidly with increasing volume than the isothermal line. Because no heat is transferred in an adiabatic expansion, the work comes from the internal energy of the gas; thus, the temperature declines.

12.8 A DEEPER LOOK Distribution of Energy among Molecules

The kinetic molecular theory of gases (Section 9.5) relates the macroscopic properties of a gas to the structure of the constituent molecules, the forces between them, and their motions. Because the number of molecules in a sample of gas is so incredibly large—28 g nitrogen contains 6.02×10^{23} molecules—we give up the idea of following the detailed motions of any one molecule and rely on a statistical description that gives the *probability* of finding a molecule in the gas at a certain position, with a certain speed, with a certain value of energy, and so on. Treating the molecules as point masses obeying classical mechanics and using simple statistical arguments, the kinetic theory shows that the temperature of the gas is proportional to the average kinetic energy per molecule. This relation not only provides a microscopic interpretation of the concept of temperature, but it also indicates the typical values of molecular kinetic energy that occur in a gas at a particular temperature.

Now we want to determine the relation between temperature and the energy involved in other kinds of molecular motions that depend on molecular structure, not just the translation of the molecule. This relation is provided by the *Boltzmann energy distribution*, which relies on the quantum description of molecular motions. This section defines the Boltzmann distribution and uses it to describe the vibrational energy of diatomic molecules in a gas at temperature T.

The Boltzmann energy distribution is one of the most widely used relations in the natural sciences, because it provides a reliable way to interpret experimental results in terms of molecular behavior. You should become skilled in its applications.

12.8.1 The Boltzmann Energy Distribution

We start with a model system in which gaseous molecules move around inside a container held at temperature T. The molecules collide with the container walls, but not with one another. We can achieve this condition by setting up the experiment with sufficiently low pressure in the system. But this time we assume that the molecules have quantum states described by a quantum number n and represented on an energy level diagram where the energy of each state is labeled ε_n. After the system has settled down to equilibrium, how many of the molecules are in their ground state? To what extent are the excited states populated? The answers depend on the probability that a molecule in the gas is in the quantum state n, which is given by the **Boltzmann energy distribution:**

$$P(n) = C \exp(-\varepsilon_n/k_B T) \qquad [12.19]$$

where C is a normalization factor and k_B is Boltzmann's constant. This equation was derived for classical systems by Ludwig Boltzmann even before quantum mechanics had been invented. Max Planck used a version of the Boltzmann distribution in formulating his theory of blackbody radiation (see Section 4.2) to obtain the probability that his quantized oscillators would radiate energy when the blackbody was at temperature T. We do not derive the distribution, but illustrate its application and interpretation.

12.8.2 Vibrational Energy Distribution

We apply the Boltzmann distribution to describe the probability of finding molecules in each of the vibrational states in a sample of CO held at temperature T. We describe the vibrational motions using the harmonic oscillator model, for which the allowed energy levels are

$$\varepsilon_n = \left(n + \frac{1}{2}\right)h\nu$$

where $n = 0, 1, 2, 3, \ldots$ and the vibrational frequency is related to the force constant k by

$$\nu = \frac{1}{2\pi}\sqrt{\frac{k}{\mu}}$$

and μ is the reduced mass. These equations define the energy level diagram, which has uniformly spaced levels separated by

$$h\nu = \frac{h}{2\pi}\sqrt{\frac{k}{\mu}} \qquad [12.20]$$

We calculate the reduced mass of $^{12}C^{16}O$ using the isotopic masses in Table 19.1 to be

$$\mu = \frac{m_C m_O}{m_C + m_O}$$

$$= \frac{(12.00)(15.99)\ \text{amu}}{27.99}\left(\frac{1\ \text{g}}{6.02 \times 10^{23}\ \text{amu}}\right)\left(\frac{1\ \text{kg}}{10^3\ \text{g}}\right)$$

$$\mu = 1.14 \times 10^{-26}\,\text{kg}$$

The value of the force constant for CO is 1902 N m^{-1}, as measured in vibrational spectroscopy (see Section 20.3). The value of the energy level separation is then

$$h\nu = \left(\frac{6.63 \times 10^{-34}\ \text{J s}}{2\pi}\right)\left(\frac{1.902 \times 10^{3}\,\text{N m}^{-1}}{1.14 \times 10^{-26}\ \text{kg}}\right)^{1/2}$$

$$h\nu = 4.52 \times 10^{-20}\ \text{J}$$

The relative probability of finding molecules in the excited state n and in the ground state $n = 0$ is given by

$$\frac{P(n)}{P(0)} = \frac{C\exp(-\varepsilon_n/k_BT)}{C\exp(-\varepsilon_0/k_BT)} = \exp(-[\varepsilon_n - \varepsilon_0]/k_BT) \qquad [12.21]$$

Inserting the energy level expression for the harmonic oscillator gives

$$\frac{P(n)}{P(0)} = \exp(-[(n + \tfrac{1}{2})h\nu - \tfrac{1}{2}h\nu]/k_BT) = \exp(-nh\nu/k_BT) \qquad [12.22]$$

The relative populations of the first excited state $n = 1$ and the ground state are determined by the ratio $h\nu/k_BT$. We know (Chapter 3, page 94) that $h\nu$ is the quantum of vibrational energy needed to put a CO molecule in its first excited state, and we have calculated that value to be $h\nu = 4.52 \times 10^{-20}$ J. From Section 9.5 we know that the average kinetic energy of a molecule in the gas is (3/2) k_BT, which is (1/2)k_BT for each of the x, y, and z directions of motion. Therefore, we interpret k_BT as a measure of the average energy available to each molecule in a gas at temperature T. So, the ratio $h\nu/k_BT$ determines whether there is sufficient energy in the gas to put the molecules into excited states. At 300 K, the value of k_BT is 4.14×10^{-21} J, which is a factor of 10 smaller than the vibrational quantum of CO. Inserting these numbers into Equation 12.22 gives the relative probability as 3.03×10^{-5}. This means that only 3 molecules in a group of 100,000 are in the first excited state at 300 K. (More precisely, the ratio between the number of molecules in the first excited state to the number of molecules in the ground vibrational state is 3/100,000.) At 1000 K, the value of k_BT is 1.38×10^{-20} J, which is closer to the value of the CO vibrational quantum and gives a relative population of 4.41×10^{-2}.

This case study shows that CO molecules do not have significant vibrational energy unless the temperature is quite high. This happens because CO has a triple bond and, therefore, a large force constant ($k = 1902$ N m^{-1}). The correlation between force constant and bond order in diatomic molecules is explained by molecular orbital theory, and is summarized in Figure 6.21. Other diatomic molecules will have quantitatively different distributions, as determined by their structure.

EXAMPLE 12.12

Calculate the population of the first and second vibrational excited states, relative to the ground state, for Br_2 at $T = 300$ K and 1000 K. For Br_2, the measured vibrational frequency is 9.68×10^{12} s^{-1}. Interpret your results in relation to the chemical bond in Br_2.

Solution

Use Equation 12.22 with $n = 1$ to obtain N_1/N_0 and $n = 2$ to obtain N_2/N_0. It is convenient to evaluate the important quantities before substituting into the equation.

Evaluate the vibrational energy of Br_2:

$$h\nu = (6.63 \times 10^{-34}\text{ J s})(9.68 \times 10^{12}\text{ s}^{-1}) = 6.42 \times 10^{-21}$$

Evaluate k_BT at 300 K and 1000 K:

$$k_BT = (300\text{ K})(1.380 \times 10^{-23}\text{ J K}^{-1}) = 4.14 \times 10^{-21}\text{ J}$$

$$k_BT = (1000\text{ K})(1.380 \times 10^{-23}\text{ J K}^{-1}) = 1.38 \times 10^{-20}\text{ J}$$

The population ratios at 300 K are:

$$\frac{P(1)}{P(0)} = \exp\left[-\frac{6.42 \times 10^{-21}\text{ J}}{4.14 \times 10^{-21}\text{ J}}\right] = 0.212$$

$$\frac{P(2)}{P(0)} = \exp\left[-\frac{2(6.42 \times 10^{-21}\text{ J})}{4.14 \times 10^{-21}\text{ J}}\right] = 0.045$$

The population ratios at 1000 K are:

$$\frac{P(1)}{P(0)} = \exp\left[-\frac{6.42 \times 10^{-21}\text{ J}}{1.38 \times 10^{-20}\text{ J}}\right] = 0.628$$

$$\frac{P(2)}{P(0)} = \exp\left[-\frac{2(6.42 \times 10^{-21}\text{ J})}{1.38 \times 10^{-20}\text{ J}}\right] = 0.395$$

The distribution for the first six states at 300 K and 1000 K is shown in Figure 12.23.

The quantized energy of vibration is much less for Br_2 than for CO for two reasons: the single bond in Br_2 has a much smaller force constant than the triple bond in CO, and the Br atoms are much more massive than C and O atoms.

Related Problems: 59, 60, 61, 62, 63, 64

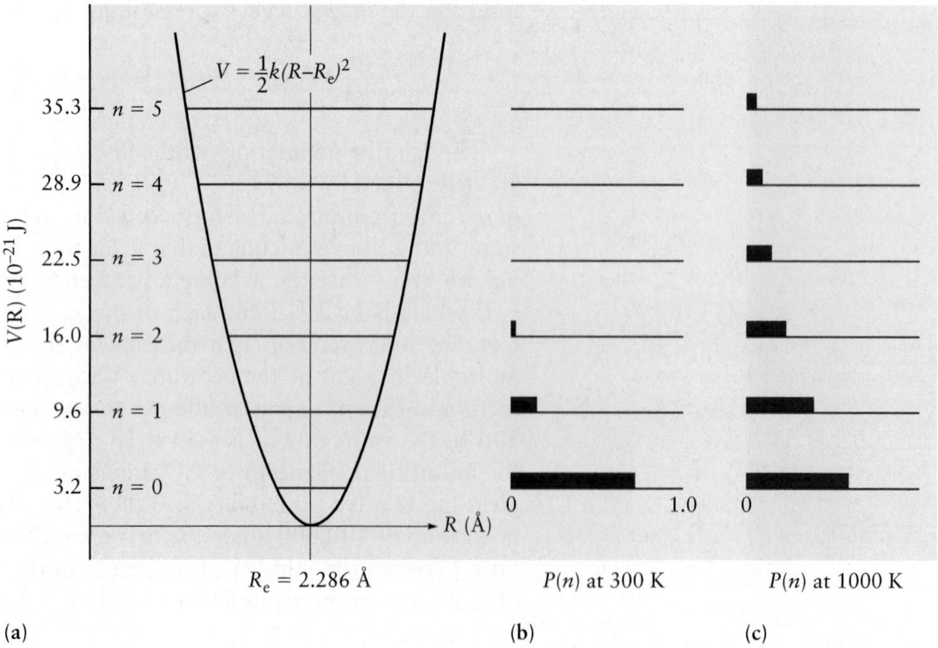

FIGURE 12.23 The potential energy curve and energy levels for the harmonic oscillator model for Br_2.

CHAPTER SUMMARY

The central goal of this chapter is to define and measure the heat liberated or absorbed during a chemical reaction. We achieved that goal in Section 12.6 by introducing the enthalpy change of a reaction, which is measured by performing the reaction in a constant-pressure calorimeter. Tabulating data for reactions in which both products and reactants are in their standard states gives the standard enthalpy of reaction, which enables systematic comparison of the energy changes due to bond rearrangements in different reactions. The standard enthalpy change for a reaction is readily calculated from tables of the standard enthalpy of formation for the products and reactants. These basic tools of thermochemistry completely describe the energy transfers in chemical reactions.

The discussion of basic thermodynamics in Sections 12.1 through 12.4 can be viewed as background necessary for achieving the central goal. Sections 12.5 and 12.7 provide the molecular interpretation of these relations. Even so, this material is important in its own right and will be used repeatedly throughout the book. Similarly, the discussion of reversible isothermal and adiabatic processes in Section 12.6 provides background needed later in the book to calculate changes in state functions for irreversible processes.

CONCEPTS AND SKILLS

Section 12.1 – Systems, States, and Processes

Give precise definitions for the terms thermodynamic system, open system, closed system, thermodynamic state, reversible process, irreversible process.

- The system is that part of the universe of interest, for example, a chemical reaction, an engine, a human being.
- The surroundings are that part of the universe that exchanges matter and energy with the system during a process.
- The thermodynamic universe is the combination of the system and the surroundings for a particular process of interest; it is assumed to be closed and isolated.
- A closed system is one in which no exchange of matter between system and surroundings is permitted.
- An adiabatic system is one in which no exchange of heat between system and surroundings is permitted.
- A thermodynamic state is a condition in which all macroscopic properties of a system are determined by the external conditions imposed on the system (for example, n, T, and P).
- A thermodynamic system is in thermodynamic equilibrium if none of its macroscopic properties is changing over time. (**Note:** Some states that appear not to be changing may not be true equilibrium states because the changes are too slow to be observed, for example, diamond turning into graphite.)
- Properties of a system may either be independent of the amount of material present (intensive, like T and P) or proportional to the quantity of material present (extensive, like internal energy and heat capacity).
- Processes can be either irreversible (a small change in the external condition will not reverse the course of the process) or reversible (a small change in the external condition will reverse the course of the process). The latter are idealizations, often called quasi-static, and would take an infinite amount of time to occur in a real sense.

Define and give examples of properties that are state functions of a system.

- A state function is a property whose value depends only on the current state of the system and not on the path by which that state was reached (examples include T and P).
- A path-dependent quantity is one in which the value does depend on the details of the path taken, work and heat being the most common examples.

Section 12.2 – The First Law of Thermodynamics: Internal Energy, Work, and Heat

Calculate the work done on an ideal gas when it is compressed reversibly (Problems 1–2).

- Work (w) is force times displacement $w = Fd$. Perhaps the most important type of work in chemistry is pressure–volume work, in which a system either expands against or is compressed by the external pressure.
- We chose the sign convention so that if the work $w = -\int_{\text{i,path}}^{\text{f}} P_{\text{ext}}\, dV$ is positive, work is done on the system.

Give a physical interpretation to the concept of heat, and calculate the change in temperature of a given quantity of a substance from its heat capacity and the amount of heat transferred to it (Problems 3–8).

- Heat transfer is measured by calorimetry. $q_{\text{cal}} = mc_{\text{s}}\Delta T$ (if there are no phase changes or chemical reactions), where m is the mass of the heat-absorbing substance in the calorimeter (usually water) and c_{s} is the specific heat of the substance. The heat added to the system, q, $= -q_{\text{cal}}$ (treating the water in the calorimeter as the surroundings and using $q_{\text{sys}} = -q_{\text{surr}}$).

State the first law of thermodynamics and give a physical interpretation (Problems 9–10).

- $\Delta U = q + w$. The internal energy of a system is a state function. Although q and w are functions of the path, their sum is a state function. Heat transferred and work done must leave the energy of the thermodynamic universe unchanged.

$$q_{sys} = -q_{surr}$$

$$w_{sys} = -w_{surr}$$

$$\Delta U_{sys} = -\Delta U_{surr}$$

Section 12.3 – Heat Capacity, Calorimetry, and Enthalpy

Calculate the final temperature reached when two substances of different mass, heat capacity, and temperature are placed in thermal contact (Problems 11–16).

- The heat q_2 taken up by the cooler body is $-q_1$, the heat given up by the warmer body. This fact gives the equation

$$m_1 c_{s1}(T_{\text{f}} - T_1) = -m_2 c_{s2}(T_{\text{f}} - T_2)$$

which is solved for T_{f}.

Section 12.4 – The First Law and Ideal Gas Processes

Calculate the amounts of heat and work and the change in the internal energy of an ideal gas during expansion and compression (Problems 17–22).

- Heat capacities of ideal monatomic gases

$$c_{\text{V}}\ (\text{ideal gas}) = (3/2)R$$

$$c_{\text{P}} = c_{\text{V}} + R = (5/2)R$$

- Heat and work for ideal gases: The calculations in section 12.4.2 associated with Figure 12.12 demonstrate that the values of q and w depend on the path over which a process occurs.

Section 12.5 – Molecular Contributions to Internal Energy and Heat Capacity

Estimate the heat capacity for a gas based on its molecular structure (Problems 23–26).

- From the known structure of the gaseous molecule, we determine the number of degrees of freedom for rotational and vibrational motions, and add these to the three degrees of freedom for translation of the molecule. Assuming ideal gas behavior, we calculate c_V by adding up the contributions from all these degrees of freedom using the classical equipartition theorem. Each translation degree of freedom contributes $R/2$, each rotational degree of freedom contributes $R/2$, and each vibrational degree of freedom contributes R. We add one additional term R to convert c_V to c_P. The results are summarized and illustrated in Table 12.3.

Section 12.6 – Thermochemistry

Calculate the energy and enthalpy changes for chemical reactions from the standard molar enthalpies of formation of reactants and products (Problems 39–46).

- The most useful thermochemical data are tables of the standard enthalpy of formation ΔH_f° for compounds, defined as the enthalpy of formation of a compound in its standard state from the elements in their standard states at 1 atm and 298.15 K.
- The change in standard state enthalpy for any reaction can be calculated from the standard state enthalpy of formation of its products and reactants as

$$\Delta H^\circ = \sum_{i=1}^{prod} n_i \Delta H_i^\circ - \sum_{j=1}^{react} n_j \Delta H_j^\circ$$

Use bond enthalpies to estimate enthalpies of formation of gaseous compounds (Problems 49–52).

- Bond enthalpy is the enthalpy change associated with breaking a chemical bond. It is based on the idea that bond enthalpy, like bond energy (see Chapter 3), is approximately independent of the rest of the molecule. Average bond enthalpies are tabulated from measurements over a broad range of molecules in which the same bond appears.

Section 12.7 – Reversible Processes in Ideal Gases

Calculate the heat absorbed and work done by an ideal gas when it expands reversibly and either isothermally or adiabatically (Problems 55–58).

- **Isothermal processes:** In this case, heat will flow (in either direction) to offset the cost of PV work, whereas T remains constant.

$$dw = -PdV$$

$$w = -\int_{V_1}^{V_2} P\,dV$$

- For an ideal gas, $P = nRT/V$. Because T is constant it comes outside the integral to give

$$w = -nRT\int_{V_1}^{V_2} (1/V)\,dV$$

$$w = -nRT\ln(V_2/V_1)$$

$$q = +nRT\ln(V_2/V_1)$$

$$\Delta U = 0$$

$$\Delta H = 0$$

- **Adiabatic processes:** No heat flows, so all of the energy comes from or goes into the internal energy of the system. The key steps are:

$$q = 0, \text{ so } \Delta U = w$$

$$nc_V\,dT = -P\,dV = -\frac{nRT}{V}\,dV \text{ rearranging, integrating, and "simplifying"}$$

gives $P_1V_1^{\gamma} = P_2V_2^{\gamma}$, where $\gamma = c_P/c_V = 5/3$ for ideal gases.

- The important result is that the pressure decreases faster in an adiabatic expansion than in an isothermal expansion because there is no heat to enable the expansion work and keep the pressure higher.

Section 12.8 – A Deeper Look . . . Distribution of Energy among Molecules

Use the Boltzmann distribution to determine the relative population of two quantum states in a gas at temperature T (Problems 59–64).

- The relative probability of finding molecules in the excited state n and in the ground state $n = 0$ is given by

$$\frac{P(n)}{P(0)} = \frac{C\exp(-\varepsilon_n/k_BT)}{C\exp(-\varepsilon_0/k_BT)} = \exp(-[\varepsilon_n - \varepsilon_0]/k_BT$$

CUMULATIVE EXERCISES

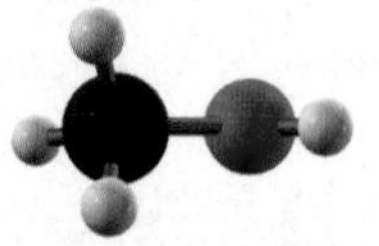

FIGURE 12.24 Methanol, CH_3OH.

Vanessa Vick/Science Source

FIGURE 12.25 A methanol-powered bus.

Methanol as a Gasoline Substitute

Methanol (CH_3OH) (Fig. 12.24) is used as a substitute for gasoline in certain high-performance vehicles (Fig. 12.25). To design engines that will run on methanol, we must understand its thermochemistry.

(a) The methanol in an automobile engine must be in the gas phase before it can react. Calculate the heat (in kilojoules) that must be added to 1.00 kg liquid methanol to increase its temperature from 25.0°C to its normal boiling point, 65.0°C. The molar heat capacity of liquid methanol is 81.6 J K^{-1} mol^{-1}.

(b) Once the methanol has reached its boiling point, it must be vaporized. The molar enthalpy of vaporization for methanol is 38 kJ mol^{-1}. How much heat must be added to vaporize 1.00 kg methanol?

(c) Once it is in the vapor phase, the methanol can react with oxygen in the air according to

$$CH_3OH(g) + \tfrac{3}{2}O_2(g) \longrightarrow CO_2(g) + 2\,H_2O(g)$$

Use average bond enthalpies to estimate the enthalpy change in this reaction, for 1 mol of methanol reacting.

(d) Use data from Appendix D to calculate the actual enthalpy change in this reaction, assuming it to be the same at 65°C as at 25°C.

(e) Calculate the heat released when 1.00 kg gaseous methanol is burned in air at constant pressure. Use the more accurate result of part (d), rather than that of part (c).

(f) Calculate the *difference* between the change in enthalpy and the change in internal energy when 1.00 kg gaseous methanol is oxidized to gaseous CO_2 and H_2O at 65°C.

(g) Suppose now that the methanol is burned inside the cylinder of an automobile. Taking the radius of the cylinder to be 4.0 cm and the distance moved by the piston during one stroke to be 12 cm, calculate the work done on the gas per stroke as it expands against an external pressure of 1.00 atm. Express your answer in liter-atmospheres and in joules.

Answers

(a) 102 kJ

(b) 1.2×10^3 kJ

(c) −508 kJ

(d) −676.49 kJ

(e) 2.11×10^4 kJ

(f) $\Delta H - \Delta U = 43.9$ kJ

(g) −0.60 L atm = −61 J

PROBLEMS

Answers to problems whose numbers are red appear in Appendix G. Problems that are more challenging are indicated with asterisks.

12.2 The First Law of Thermodynamics: Internal Energy, Work, and Heat

1. Some nitrogen for use in synthesizing ammonia is heated slowly, maintaining the external pressure close to the internal pressure of 50.0 atm, until its volume has increased from 542 to 974 L. Calculate the work done on the nitrogen as it is heated, and express it in joules.

2. The gas mixture inside one of the cylinders of an automobile engine expands against a constant external pressure of 0.98 atm, from an initial volume of 150 mL (at the end of the compression stroke) to a final volume of 800 mL. Calculate the work done on the gas mixture during this process, and express it in joules.

3. When a ball of mass m is dropped through a height difference Δh, its potential energy changes by the amount $mg\,\Delta h$, where g is the acceleration of gravity, equal to 9.81 m s^{-2}. Suppose that when the ball hits the ground, all that energy is converted to heat, increasing the temperature of the ball. If the specific heat capacity of the material in the ball is 0.850 J K^{-1} g^{-1}, calculate the height from which the ball must be dropped to increase the temperature of the ball by 1.00°C.

4. During his honeymoon in Switzerland, James Joule is said to have used a thermometer to measure the temperature difference between the water at the top and at the bottom of a waterfall. Take the height of the waterfall to be Δh and the acceleration of gravity, g, to be 9.81 m s^{-2}. Assuming that all the potential energy change $mg\,\Delta h$ of a mass m of water is used to heat that water by the time it reaches the bottom, calculate the temperature difference between the top and the bottom of a waterfall 100 meters high. Take the specific heat capacity of water to be 4.18 J K^{-1} g^{-1}.

5. The specific heat capacities of Li(*s*), Na(*s*), K(*s*), Rb(*s*), and Cs(*s*) at 25°C are 3.57, 1.23, 0.756, 0.363, and 0.242 J K^{-1} g^{-1}, respectively. Compute the molar heat capacities of these elements and identify any periodic trend. If there is a trend, use it to predict the molar heat capacity of francium, Fr(*s*).

6. The specific heat capacities of $F_2(g)$, $Cl_2(g)$, $Br_2(g)$, and $I_2(g)$ are 0.824, 0.478, 0.225, and 0.145 J K^{-1} g^{-1}, respectively. Compute the molar heat capacities of these elements and identify any periodic trend. If there is a trend, use it to predict the molar heat capacity of astatine, $At_2(g)$.

7. The specific heat capacities of the metals nickel, zinc, rhodium, tungsten, gold, and uranium at 25°C are 0.444, 0.388, 0.243, 0.132, 0.129, and 0.116 J K^{-1} g^{-1}, respectively. Calculate the molar heat capacities of these six metals. Note how closely the molar heat capacities for these metals, which were selected at random, cluster about a value of 25 J K^{-1} mol^{-1}. The rule of Dulong and Petit states that the molar heat capacities of the metallic elements are approximately 25 J K^{-1} mol^{-1}.

8. Use the empirical rule of Dulong and Petit stated in Problem 7 to estimate the specific heat capacities of vanadium, gallium, and silver.

9. A chemical system is sealed in a strong, rigid container at room temperature, and then heated vigorously.
   (a) State whether ΔU, q, and w of the system are positive, negative, or zero during the heating process.
   (b) Next, the container is cooled to its original temperature. Determine the signs of ΔU, q, and w for the cooling process.
   (c) Designate heating as step 1 and cooling as step 2. Determine the signs of $(\Delta U_1 + \Delta U_2)$, $(q_1 + q_2)$, and $(w_1 + w_2)$, if possible.

10. A battery harnesses a chemical reaction to extract energy in the form of useful electrical work.
   (a) A certain battery runs a toy truck and becomes partially discharged. In the process, it performs a total of 117.0 J of work on its immediate surroundings. It also gives off 3.0 J of heat, which the surroundings absorb. No other work or heat is exchanged with the surroundings. Compute q, w, and ΔU of the battery, making sure each quantity has the proper sign.
   (b) The same battery is now recharged exactly to its original condition. This requires 210.0 J of electrical work from an outside generator. Determine q for the battery in this process. Explain why q has the sign that it does.

12.3 Heat Capacity, Calorimetry, and Enthalpy

11. Suppose 61.0 g hot metal, which is initially at 120.0°C, is plunged into 100.0 g water that is initially at 20.00°C. The metal cools down and the water heats up until they reach a common temperature of 26.39°C. Calculate the specific heat capacity of the metal, using 4.18 J K^{-1} g^{-1} as the specific heat capacity of the water.

12. A piece of zinc at 20.0°C that weighs 60.0 g is dropped into 200.0 g water at 100.0°C. The specific heat capacity of zinc is 0.389 J K^{-1} g^{-1}, and that of water near 100°C is 4.22 J K^{-1} g^{-1}. Calculate the final temperature reached by the zinc and the water.

13. Very early in the study of the nature of heat it was observed that if two bodies of equal mass but different temperatures are placed in thermal contact, their specific heat capacities depend inversely on the change in temperature each undergoes on reaching its final temperature. Write a mathematical equation in modern notation to express this fact.

14. Iron pellets with total mass 17.0 g at a temperature of 92.0°C are mixed in an insulated container with 17.0 g water at a temperature of 20.0°C. The specific heat capacity of water is 10 times greater than that of iron. What is the final temperature inside the container?

15. In their *Memoir on Heat,* published in 1783, Lavoisier and Laplace reported, "The heat necessary to melt ice is equal to three quarters of the heat that can raise the same mass of water from the temperature of the melting ice to that of boiling water" (English translation). Use this 18th-century observation to compute the amount of heat (in joules) needed to melt 1.00 g ice. Assume that heating 1.00 g water requires 4.18 J of heat for each 1.00°C throughout the range from 0°C to 100°C.

16. Galen, the great physician of antiquity, suggested scaling temperature from a reference point defined by mixing equal masses of ice and boiling water in an insulated container. Imagine that this is done with the ice at 0.00°C and the water at 100.0°C. Assume that the heat capacity of the container is negligible, and that it takes 333.4 J of heat to melt 1.000 g ice at 0.00°C to water at 0.00°C. Compute Galen's reference temperature in degrees Celsius.

12.4 The First Law and Ideal Gas Processes

17. If 0.500 mol neon at 1.00 atm and 273 K expands against a constant external pressure of 0.100 atm until the gas pressure reaches 0.200 atm and the temperature reaches 210 K, calculate the work done on the gas, the internal energy change, and the heat absorbed by the gas.

18. Hydrogen behaves as an ideal gas at temperatures greater than 200 K and at pressures less than 50 atm. Suppose 6.00 mol hydrogen is initially contained in a 100-L vessel at a pressure of 2.00 atm. The average molar heat capacity of hydrogen at constant pressure, c_P, is 29.3 J K^{-1} mol^{-1} in the temperature range of this problem. The gas is cooled reversibly at constant pressure from its initial state to a volume of 50.0 L. Calculate the following quantities for this process.
   (a) Temperature of the gas in the final state, T_2
   (b) Work done on the gas, w, in joules
   (c) Internal energy change of the gas, ΔU, in joules
   (d) Heat absorbed by the gas, q, in joules

19. Suppose 2.00 mol of an ideal, monatomic gas is initially at a pressure of 3.00 atm and a temperature $T = 350$ K. It is expanded *irreversibly* and *adiabatically* ($q = 0$) against a constant external pressure of 1.00 atm until the volume has doubled.
(a) Calculate the final volume.
(b) Calculate w, q, and ΔU for this process, in joules.
(c) Calculate the final temperature of the gas.

20. Consider the free, isothermal (constant T) expansion of an ideal gas. "Free" means that the external force is zero, perhaps because a stopcock has been opened and the gas is allowed to expand into a vacuum. Calculate ΔU for this irreversible process. Show that $q = 0$, so that the expansion is also adiabatic ($q = 0$) for an ideal gas. This is analogous to a classic experiment first performed by Joule.

21. If 6.00 mol argon in a 100-L vessel initially at 300 K is compressed adiabatically ($q = 0$) and irreversibly until a final temperature of 450 K is reached, calculate the energy change of the gas, the heat added to the gas, and the work done on the gas.

22. A gas expands against a constant external pressure of 2.00 atm until its volume has increased from 6.00 to 10.00 L. During this process, it absorbs 500 J of heat from the surroundings.
(a) Calculate the energy change of the gas, ΔU.
(b) Calculate the work, w, done on the gas in an irreversible adiabatic ($q = 0$) process connecting the same initial and final states.

12.5 Molecular Contributions to Internal Energy and Heat Capacity

23. Calculate the value of c_P at 298 K and 1 atm pressure predicted for $I_2(g)$ and $HI(g)$ by the classical equipartition theorem. Compare the predicted results with the experimental results (see Appendix D) and calculate the percent of the measured value that arises from vibrational motions.

24. Calculate the value of c_P at 298 K and 1 atm pressure predicted for $CH_4(g)$ and $C_2H_4(g)$ by the classical equipartition theorem. Compare the predicted results with the experimental results (see Appendix D) and calculate the percent of the measured value that arises from vibrational motions.

25. **(a)** Calculate the change in enthalpy when 2.00 moles of argon is heated from 298 K to 573 K at constant pressure of 1 atmosphere.
(b) Calculate the change in enthalpy when 2.00 moles of ethylene C_2H_4 is taken through the same process. In both cases assume the heat capacity values predicted by equipartition are valid through the temperature range stated.

26. **(a)** Calculate the change in enthalpy when 20.0 grams of aluminum metal is heated from 298 K to 573 K at constant pressure of 1 atm.
(b) Calculate the change in enthalpy when 20.0 grams of metallic lead is taken through the same process. In both cases assume the heat capacity values predicted by equipartition are valid through the temperature range stated.

12.6 Thermochemistry

27. For each of the following reactions, the enthalpy change written is that measured when the numbers of moles of reactants and products taking part in the reaction are as given by their coefficients in the equation. Calculate the enthalpy change when 1.00 *gram* of the underlined substance is consumed or produced.
(a) $4\ Na(s) + O_2(g) \longrightarrow \underline{2\ Na_2O(s)}$ $\quad \Delta H = -828$ kJ
(b) $CaMg(CO_3)_2(s) \longrightarrow CaO(s) + \underline{MgO(s)} + 2\ CO_2(g)$ $\quad \Delta H = +302$ kJ
(c) $H_2(g) + \underline{2\ CO(g)} \longrightarrow H_2O_2(\ell) + 2\ C(s)$ $\quad \Delta H = +33.3$ kJ

28. For each of the following reactions, the enthalpy change given is that measured when the numbers of moles of reactants and products taking part in the reaction are as given by their coefficients in the equation. Calculate the enthalpy change when 1.00 *gram* of the underlined substance is consumed or produced.
(a) $Ca(s) + \underline{Br_2(\ell)} \longrightarrow CaBr_2(s)$ $\quad \Delta H = -683$ kJ
(b) $6\ Fe_2O_3(s) \longrightarrow \underline{4\ Fe_3O_4(s)} + O_2(g)$ $\quad \Delta H = +472$ kJ
(c) $\underline{2\ NaHSO_4(s)} \longrightarrow 2\ NaOH(s) + 2\ SO_2(g) + O_2(g)$ $\quad \Delta H = +806$ kJ

29. Liquid bromine dissolves readily in aqueous NaOH:

$$Br_2(\ell) + 2\ NaOH(aq) \longrightarrow NaBr(aq) + NaOBr(aq) + H_2O(\ell)$$

Suppose 2.88×10^{-3} mol of $Br_2(\ell)$ is sealed in a glass capsule that is then immersed in a solution containing excess $NaOH(aq)$. The capsule is broken, the mixture is stirred, and a measured 121.3 J of heat evolves. In a separate experiment, simply breaking an empty capsule and stirring the solution in the same way evolves 2.34 J of heat. Compute the heat evolved as 1.00 mol $Br_2(\ell)$ dissolves in excess $NaOH(aq)$.

30. A chemist mixes 1.00 g $CuCl_2$ with an excess of $(NH_4)_2HPO_4$ in dilute aqueous solution. He measures the evolution of 670 J of heat as the two substances react to give $Cu_3(PO_4)_2(s)$. Compute the ΔH that would result from the reaction of 1.00 mol $CuCl_2$ with an excess of $(NH_4)_2HPO_4$.

31. Calculate the enthalpy change when 2.38 g carbon monoxide (CO) vaporizes at its normal boiling point. Use data from Table 12.2.

32. Molten sodium chloride is used for making elemental sodium and chlorine. Suppose the electrical power to a vat containing 56.2 kg molten sodium chloride is cut off and the salt crystallizes (without changing its temperature). Calculate the enthalpy change, using data from Table 12.4.

33. Suppose an ice cube weighing 36.0 g at a temperature of −10°C is placed in 360 g water at a temperature of 20°C. Calculate the temperature after thermal equilibrium is reached, assuming no heat loss to the surroundings. The enthalpy of fusion of ice is $\Delta H_{fus} = 6.007$ kJ mol^{-1}, and the molar heat capacities c_P of ice and water are 38 and 75 J K^{-1} mol^{-1}, respectively.

34. You have a supply of ice at 0.0°C and a glass containing 150 g water at 25°C. The enthalpy of fusion for ice is $\Delta H_{fus} = 333$ J g^{-1}, and the specific heat capacity of water is 4.18 J K^{-1} g^{-1}. How many grams of ice must be added to the glass (and melted) to reduce the temperature of the water to 0°C?

35. The measured enthalpy change for burning ketene (CH_2CO)

$$CH_2CO(g) + 2\ O_2(g) \longrightarrow 2\ CO_2(g) + H_2O(g)$$

is $\Delta H_1 = -981.1$ kJ at 25°C. The enthalpy change for burning methane

$$CH_4(g) + 2\ O_2(g) \longrightarrow CO_2(g) + 2\ H_2O(g)$$

is $\Delta H_2 = -802.3$ kJ at 25°C. Calculate the enthalpy change at 25°C for the reaction

$$2\ CH_4(g) + 2\ O_2(g) \longrightarrow CH_2CO(g) + 3\ H_2O(g)$$

36. Given the following two reactions and corresponding enthalpy changes,

$$CO(g) + SiO_2(s) \longrightarrow SiO(g) + CO_2(g) \qquad \Delta H = +520.9 \text{ kJ}$$

$$8\ CO_2(g) + Si_3N_4(s) \longrightarrow 3\ SiO_2(s) + 2\ N_2O(g) + 8\ CO(g) \qquad \Delta H = +461.05 \text{ kJ}$$

compute the ΔH of the reaction

$$5\ CO_2(g) + Si_3N_4(s) \longrightarrow 3\ SiO(g) + 2\ N_2O(g) + 5\ CO(g)$$

37. The enthalpy change to make diamond from graphite is 1.88 kJ mol^{-1}. Which gives off more heat when burned—a pound of diamonds or a pound of graphite? Explain.

38. The enthalpy change of combustion of monoclinic sulfur to $SO_2(g)$ is −9.376 kJ g^{-1}. Under the same conditions, the rhombic form of sulfur has an enthalpy change of combustion to $SO_2(g)$ of −9.293 kJ g^{-1}. Compute the ΔH of the reaction

$$\text{S(monoclinic)} \longrightarrow \text{S(rhombic)}$$

per gram of sulfur reacting.

39. Calculate the standard enthalpy change $\Delta H°$ at 25°C for the reaction

$$N_2H_4(\ell) + 3\ O_2(g) \longrightarrow 2\ NO_2(g) + 2\ H_2O(\ell)$$

using the standard enthalpies of formation ($\Delta H_f°$) of reactants and products at 25°C from Appendix D.

40. Using the data in Appendix D, calculate $\Delta H°$ for each of the following processes:

(a) $2\ NO(g) + O_2(g) \longrightarrow 2\ NO_2(g)$
(b) $C(s) + CO_2(g) \longrightarrow 2\ CO(g)$
(c) $2\ NH_3(g) + \frac{7}{2}\ O_2(g) \longrightarrow 2\ NO_2(g) + 3\ H_2O(g)$
(d) $C(s) + H_2O(g) \longrightarrow CO(g) + H_2(g)$

41. Zinc is commonly found in nature in the form of the mineral sphalerite (ZnS). A step in the smelting of zinc is the roasting of sphalerite with oxygen to produce zinc oxide:

$$2\ ZnS(s) + 3\ O_2(g) \longrightarrow 2\ ZnO(s) + 2\ SO_2(g)$$

(a) Calculate the standard enthalpy change $\Delta H°$ for this reaction, using data from Appendix D.
(b) Calculate the heat absorbed when 3.00 metric tons (1 metric ton = 10^3 kg) of sphalerite is roasted under constant-pressure conditions.

42. The thermite process (see Fig. 12.16) is used for welding railway track together. In this reaction, aluminum reduces iron(III) oxide to metallic iron:

$$2\ Al(s) + Fe_2O_3(s) \longrightarrow 2\ Fe(s) + Al_2O_3(s)$$

Igniting a small charge of barium peroxide mixed with aluminum triggers the reaction of a mixture of aluminum powder and iron(III) oxide; the molten iron produced flows into the space between the steel rails that are to be joined.

(a) Calculate the standard enthalpy change $\Delta H°$ for this reaction, using data from Appendix D.
(b) Calculate the heat given off when 3.21 g iron(III) oxide is reduced by aluminum at constant pressure.

43. The dissolution of calcium chloride in water

$$CaCl_2(s) \longrightarrow Ca^{2+}(aq) + 2\ Cl^-(aq)$$

is used in first-aid hot packs. In these packs, an inner pouch containing the salt is broken, allowing the salt to dissolve in the surrounding water.

(a) Calculate the standard enthalpy change $\Delta H°$ for this reaction, using data from Appendix D.
(b) Suppose 20.0 g $CaCl_2$ is dissolved in 0.100 L water at 20.0°C. Calculate the temperature reached by the solution, assuming it to be an ideal solution with a heat capacity close to that of 100 g pure water (418 J K^{-1}).

44. Ammonium nitrate dissolves in water according to the reaction

$$NH_4NO_3(s) \longrightarrow NH_4^+(aq) + NO_3^-(aq)$$

(a) Calculate the standard enthalpy change $\Delta H°$ for this reaction, using data from Appendix D.
(b) Suppose 15.0 g NH_4NO_3 is dissolved in 0.100 L water at 20.0°C. Calculate the temperature reached by the solution, assuming it to be an ideal solution with a heat capacity close to that of 100 g pure water (418 J K^{-1}).
(c) From a comparison with the results of Problem 43, can you suggest a practical application of this dissolution reaction?

45. The standard enthalpy change of combustion [to $CO_2(g)$ and $H_2O(\ell)$] at 25°C of the organic liquid cyclohexane, $C_6H_{12}(\ell)$, is −3923.7 kJ mol^{-1}. Determine the ΔH_f° of $C_6H_{12}(\ell)$. Use data from Appendix D.

46. The standard enthalpy change of combustion [to $CO_2(g)$ and $H_2O(\ell)$] at 25°C of the organic liquid cyclohexane, $C_6H_{10}(\ell)$, is −3731.7 kJ mol^{-1}. Determine the ΔH_f° of $C_6H_{10}(\ell)$.

47. A sample of pure solid naphthalene ($C_{10}H_8$) weighing 0.6410 g is burned completely with oxygen to $CO_2(g)$ and $H_2O(\ell)$ in a constant-volume calorimeter at 25°C. The amount of heat evolved is observed to be 25.79 kJ.

(a) Write and balance the chemical equation for the combustion reaction.

(b) Calculate the standard change in internal energy ($\Delta U°$) for the combustion of 1.000 mol naphthalene to $CO_2(g)$ and $H_2O(\ell)$.

(c) Calculate the standard enthalpy change ($\Delta H°$) for the same reaction as in part (b).

(d) Calculate the standard enthalpy of formation per mole of naphthalene, using data for the standard enthalpies of formation of $CO_2(g)$ and $H_2O(\ell)$ from Appendix D.

48. A sample of solid benzoic acid (C_6H_5COOH) that weighs 0.800 g is burned in an excess of oxygen to $CO_2(g)$ and $H_2O(\ell)$ in a constant-volume calorimeter at 25°C. The temperature increase is observed to be 2.15°C. The heat capacity of the calorimeter and its contents is known to be 9382 J K^{-1}.

(a) Write and balance the equation for the combustion of benzoic acid.

(b) Calculate the standard change in internal energy ($\Delta U°$) for the combustion of 1.000 mol benzoic acid to $CO_2(g)$ and $H_2O(\ell)$ at 25°C.

(c) Calculate the standard enthalpy change ($\Delta H°$) for the same reaction as in part (b).

(d) Calculate the standard enthalpy of formation per mole of benzoic acid, using data for the standard enthalpies of formation of $CO_2(g)$ and $H_2O(\ell)$ from Appendix D.

49. A second CFC used as a refrigerant and in aerosols (besides that discussed in Example 12.9) is CCl_3F. Use the atomization enthalpies and average bond enthalpies from Table 12.5 to estimate the standard enthalpy of formation (ΔH_f°) of this compound in the gas phase.

50. The compound CF_3CHCl_2 (with a C—C bond) has been proposed as a substitute for CCl_3F and CCl_2F_2 because it decomposes more quickly in the atmosphere and is much less liable to reduce the concentration of ozone in the stratosphere. Use the atomization enthalpies and average bond enthalpies from Table 12.5 to estimate the standard enthalpy of formation (ΔH_f°) of CF_3CHCl_2 in the gas phase.

51. Propane has the structure $H_3C{-}CH_2{-}CH_3$. Use average bond enthalpies from Table 12.5 to estimate the change in enthalpy $\Delta H°$ for the reaction

$$C_3H_8(g) + 5\ O_2(g) \longrightarrow 3\ CO_2(g) + 4\ H_2O(g)$$

52. Use average bond enthalpies from Table 12.5 to estimate the change in enthalpy $\Delta H°$ for the reaction

$$C_2H_4(g) + H_2(g) \longrightarrow C_2H_6(g)$$

Refer to the molecular structures on page 109.

53. The following reaction

$$BBr_3(g) + BCl_3(g) \longrightarrow BBr_2Cl(g) + BCl_2Br(g)$$

has a ΔH close to zero. Sketch the Lewis structures of the four compounds, and explain why ΔH is so small.

54. At 381 K, the following reaction takes place:

$$Hg_2Cl_4(g) + Al_2Cl_6(g) \longrightarrow 2\ HgAlCl_5(g) \quad \Delta H = +10\ \text{kJ}$$

(a) Offer an explanation for the very small ΔH for this reaction for the known structures of the compounds

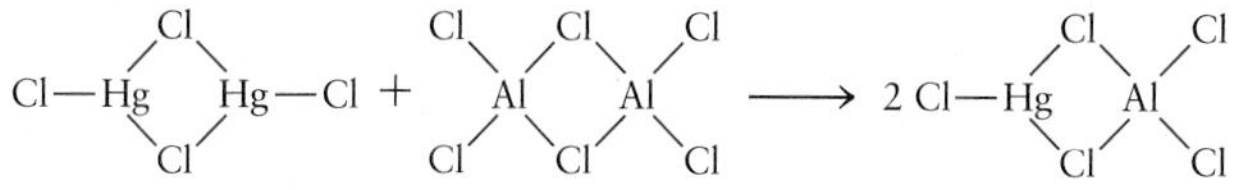

(b) Explain why the small ΔH in this reaction is evidence against

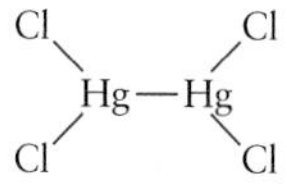

as the structure of $Hg_2Cl_4(g)$.

12.7 Reversible Processes in Ideal Gases

55. If 2.00 mol of an ideal gas at 25°C expands isothermally and reversibly from 9.00 to 36.00 L, calculate the work done on the gas and the heat absorbed by the gas in the process. What are the changes in energy (ΔU) and in enthalpy (ΔH) of the gas in the process?

56. If 54.0 g argon at 400 K is compressed isothermally and reversibly from a pressure of 1.50 to 4.00 atm, calculate the work done on the gas and the heat absorbed by gas in the process. What are the changes in energy (ΔU) and in enthalpy (ΔH) of the gas?

57. Suppose 2.00 mol of a monatomic ideal gas ($c_V = \frac{3}{2}R$) is expanded adiabatically and reversibly from a temperature T = 300 K, where the volume of the system is 20.0 L, to a volume of 60.0 L. Calculate the final temperature of the gas, the work done on the gas, and the energy and enthalpy changes.

58. Suppose 2.00 mol of an ideal gas is contained in a heat-insulated cylinder with a moveable frictionless piston. Initially, the gas is at 1.00 atm and 0°C. The gas is compressed reversibly to 2.00 atm. The molar heat capacity at constant pressure, c_P, equals 29.3 J K^{-1} mol^{-1}. Calculate the final temperature of the gas, the change in its internal energy, ΔU, and the work done on the gas.

12.8 A Deeper Look . . . Distribution of Energy among Molecules

59. Calculate the relative populations of two quantum states separated by an energy of 0.4×10^{-21} J if the temperature is 25°C.

60. Calculate the relative populations of two quantum states separated by an energy of 40×10^{-21} J if the temperature is 25°C.

61. Estimate the ratio of the number of molecules in the first excited vibrational state of the molecule N_2 to the number in the ground state, at a temperature of 450 K. The vibrational frequency of N_2 is 7.07×10^{13} s^{-1}.

62. The vibrational frequency of the ICl molecule is 1.15×10^{13} s^{-1}. For every million (1.00×10^{6}) molecules in the ground vibrational state, how many will be in the first excited vibrational state at a temperature of 300 K?

63. The force constant for HF is 966 N m^{-1}. Using the harmonic oscillator model, calculate the relative population of the first excited state and the ground state at 300 K.

64. The force constant for HBr is 412 N m^{-1}. Using the harmonic oscillator model, calculate the relative population of the first excited state and the ground state at 300 K.

ADDITIONAL PROBLEMS

65. At one time it was thought that the molar mass of indium was near 76 g mol^{-1}. By referring to the law of Dulong and Petit (see Problem 7), show how the measured specific heat of metallic indium, 0.233 J K^{-1} g^{-1}, makes this value unlikely.

66. The following table shows how the specific heat at constant pressure of liquid helium changes with temperature. Note the sharp increase over this temperature range:

Temperature (K):

1.80	1.85	1.90	1.95	2.00	2.05	2.10	2.15

c_s (J K^{-1} g^{-1}):

2.81	3.26	3.79	4.42	5.18	6.16	7.51	9.35

Estimate how much heat it takes at constant pressure to increase the temperature of 1.00 g He(ℓ) from 1.8 to 2.15 K. (*Hint:* For each temperature interval of 0.05 K, take the average, c_s, as the sum of the values at the ends of the interval divided by 2.)

67. Imagine that 2.00 mol argon, confined by a moveable, frictionless piston in a cylinder at a pressure of 1.00 atm and a temperature of 398 K, is cooled to 298 K. Argon gas may be considered ideal, and its molar heat capacity at constant pressure is $c_P = (5/2)R$, where $R = 8.315$ J K^{-1} mol^{-1}. Calculate:
   - **(a)** The work done on the system, w
   - **(b)** The heat absorbed by the system, q
   - **(c)** The energy change of the system, ΔU
   - **(d)** The enthalpy change of the system, ΔH

68. Suppose 1.00 mol ice at −30°C is heated at atmospheric pressure until it is converted to steam at 140°C. Calculate q, w, ΔH, and ΔU for this process. For ice, water, and steam, c_P is 38, 75, and 36 J K^{-1} mol^{-1}, respectively, and can be taken to be approximately independent of temperature. ΔH_{fus} for ice is 6.007 kJ mol^{-1}, and ΔH_{vap} for water is 40.66 kJ mol^{-1}. Use the ideal gas law for steam, and assume that the volume of 1 mol ice or water is negligible relative to that of 1 mol steam.

69. The gas inside a cylinder expands against a constant external pressure of 1.00 atm from a volume of 5.00 L to a volume of 13.00 L. In doing so, it turns a paddle immersed in 1.00 L water. Calculate the temperature increase of the water, assuming no loss of heat to the surroundings or frictional losses in the mechanism. Take the density of water to be 1.00 g cm^{-3} and its specific heat to be 4.18 J K^{-1} g^{-1}.

70. Suppose 1.000 mol argon (assumed to be an ideal gas) is confined in a strong, rigid container of volume 22.41 L at 273.15 K. The system is heated until 3.000 kJ (3000 J) of heat has been added. The molar heat capacity of the gas does not change during the heating and equals 12.47 J K^{-1} mol^{-1}.
   - **(a)** Calculate the original pressure inside the vessel (in atmospheres).
   - **(b)** Determine q for the system during the heating process.
   - **(c)** Determine w for the system during the heating process.
   - **(d)** Compute the temperature of the gas after the heating, in degrees Celsius. Assume the container has zero heat capacity.
   - **(e)** Compute the pressure (in atmospheres) inside the vessel after the heating.
   - **(f)** Compute ΔU of the gas during the heating process.
   - **(g)** Compute ΔH of the gas during the heating process.
   - **(h)** The correct answer to part (g) exceeds 3.000 kJ. The increase in enthalpy (which at one time was misleadingly called the "heat content") in this system exceeds the amount of heat actually added. Why is this not a violation of the law of conservation of energy?

71. When glucose, a sugar, reacts fully with oxygen, carbon dioxide and water are produced:

$$C_6H_{12}O_6(s) + 6\ O_2(g) \longrightarrow 6\ CO_2(g) + 6\ H_2O(\ell) \qquad \Delta H° = -2820 \text{ kJ}$$

Suppose a person weighing 50 kg (mostly water, with specific heat capacity 4.18 J K^{-1} g^{-1}) eats a candy bar containing 14.3 g glucose. If all the glucose reacted with oxygen and the heat produced were used entirely to increase the person's body temperature, what temperature increase would result? (In fact, most of the heat produced is lost to the surroundings before such a temperature increase occurs.)

72. In walking 1 km, you use about 100 kJ of energy. This energy comes from the oxidation of foods, which is about 30% efficient. How much energy do you save by walking 1 km instead of driving a car that gets 8.0 km L^{-1} gasoline (19 miles/gal)? The density of gasoline is 0.68 g cm^{-3} and its enthalpy of combustion is -48 kJ g^{-1}.

73. Liquid helium and liquid nitrogen are both used as coolants; $He(\ell)$ boils at 4.21 K, and $N_2(\ell)$ boils at 77.35 K. The specific heat of liquid helium near its boiling point is 4.25 J K^{-1} g^{-1}, and the specific heat of liquid nitrogen near *its* boiling point is 1.95 J K^{-1} g^{-1}. The enthalpy of vaporization of $He(\ell)$ is 25.1 J g^{-1}, and the enthalpy of vaporization of $N_2(\ell)$ is 200.3 J g^{-1} (these data are calculated from the values in Appendix F). Discuss which liquid is the better coolant (on a per-gram basis) *near* its boiling point and which is better *at* its boiling point.

74. When 1.00 g potassium chlorate ($KClO_3$) is dissolved in 50.0 g water in a Styrofoam calorimeter of negligible heat capacity, the temperature decreases from 25.00°C to 23.36°C. Calculate q for the water and $\Delta H°$ for the process.

$$KClO_3(s) \longrightarrow K^+(aq) + ClO_3^-(aq)$$

The specific heat of water is 4.184 J K^{-1} g^{-1}.

75. The enthalpy of combustion and the standard enthalpy of formation of a fuel can be determined by measuring the temperature change in a calorimeter when a weighed amount of the fuel is burned in oxygen.

(a) Write a balanced chemical equation for the combustion of isooctane, $C_8H_{18}(\ell)$, to $CO_2(g)$ and $H_2O(\ell)$. Isooctane is a component of gasoline and is used as a reference standard in determining the "octane rating" of a fuel mixture.

(b) Suppose 0.542 g isooctane is placed in a fixed-volume (bomb) calorimeter, which contains 750 g water, initially at 20.450°C, surrounding the reaction compartment. The heat capacity of the calorimeter itself (excluding the water) has been measured to be 48 J K^{-1} in a separate calibration. After the combustion of the isooctane is complete, the water temperature is measured to be 28.670°C. Taking the specific heat of water to be 4.184 J K^{-1} g^{-1}, calculate ΔU for the combustion of 0.542 g isooctane.

(c) Calculate ΔU for the combustion of 1 mol isooctane.

(d) Calculate ΔH for the combustion of 1 mol isooctane.

(e) Calculate $\Delta H°_f$ for the isooctane.

76. The enthalpy change to form 1 mol $Hg_2Br_2(s)$ from the elements at 25°C is -206.77 kJ mol^{-1}, and that of $HgBr(g)$ is 96.23 kJ mol^{-1}. Compute the enthalpy change for the decomposition of 1 mol $Hg_2Br_2(s)$ to 2 mol $HgBr(g)$:

$$Hg_2Br_2(s) \longrightarrow 2\ HgBr(g)$$

***77.** The gas most commonly used in welding is acetylene, $C_2H_2(g)$. When acetylene is burned in oxygen, the reaction is

$$C_2H_2(g) + \tfrac{5}{2}\ O_2(g) \longrightarrow 2\ CO_2(g) + H_2O(g)$$

(a) Using data from Appendix D, calculate $\Delta H°$ for this reaction.

(b) Calculate the total heat capacity of 2.00 mol $CO_2(g)$ and 1.00 mol $H_2O(g)$, using $c_P(CO_2(g)) = 37$ J K^{-1} mol^{-1} and $c_P(H_2O(g)) = 36$ J K^{-1} mol^{-1}.

(c) When this reaction is performed in an open flame, almost all the heat produced in part (a) goes to increase the temperature of the products. Calculate the maximum flame temperature that is attainable in an open flame burning acetylene in oxygen. The actual flame temperature would be lower than this because heat is lost to the surroundings.

***78.** The enthalpy of reaction changes somewhat with temperature. Suppose we wish to calculate ΔH for a reaction at a temperature T that is different from 298 K. To do this, we can replace the direct reaction at T with a three-step process. In the first step, the temperature of the reactants is changed from T to 298 K. ΔH for this step can be calculated from the molar heat capacities of the reactants, which are assumed to be independent of temperature. In the second step, the reaction is conducted at 298 K with an enthalpy change $\Delta H°$. In the third step, the temperature of the products is changed from 298 K to T. The sum of these three enthalpy changes is ΔH for the reaction at temperature T.

An important process contributing to air pollution is the following chemical reaction

$$SO_2(g) + \tfrac{1}{2}\ O_2(g) \longrightarrow SO_3(g)$$

For $SO_2(g)$, the heat capacity c_P is 39.9, for $O_2(g)$ it is 29.4, and for $SO_3(g)$ it is 50.7 J K^{-1} mol^{-1}. Calculate ΔH for the preceding reaction at 500 K, using the enthalpies of formation at 298.15 K from Appendix D.

79. At the top of the compression stroke in one of the cylinders of an automobile engine (that is, at the minimum gas volume), the volume of the gas-air mixture is 150 mL, the temperature is 600 K, and the pressure is 12.0 atm. The ratio of the number of moles of octane vapor to the number of moles of air in the combustion mixture is 1.00:80.0. What is the maximum temperature attained in the gas if octane burns explosively before the power stroke of the piston (gas expansion) begins? The gases may be considered to be ideal, and their heat capacities at constant pressure (assumed to be temperature-independent) are

$$c_P(C_8H_{18}(g)) = 327\ \text{J K}^{-1}\ \text{mol}^{-1}$$

$$c_P(O_2(g)) = 35.2\ \text{J K}^{-1}\ \text{mol}^{-1}$$

$$c_P(N_2(g)) = 29.8\ \text{J K}^{-1}\ \text{mol}^{-1}$$

$$c_P(CO_2(g)) = 45.5\ \text{J K}^{-1}\ \text{mol}^{-1}$$

$$c_P(H_2O(g)) = 38.9\ \text{J K}^{-1}\ \text{mol}^{-1}$$

The enthalpy of formation of $C_8H_{18}(g)$ at 600 K is -57.4 kJ mol^{-1}.

80. Initially, 46.0 g oxygen is at a pressure of 1.00 atm and a temperature of 400 K. It expands adiabatically and reversibly until the pressure is reduced to 0.60 atm, and it is then compressed isothermally and reversibly until the volume returns to its original value. Calculate the final pressure and temperature of the oxygen, the work done and heat added to the oxygen in this process, and the energy change ΔU. Take $c_P(O_2) = 29.4 \text{ J K}^{-1} \text{ mol}^{-1}$.

81. A young chemist buys a "one-lung" motorcycle but, before learning how to drive it, wants to understand the processes that occur in its engine. The manual says the cylinder has a radius of 5.00 cm, a piston stroke of 12.00 cm, and a (volume) compression ratio of 8:1. If a mixture of gasoline vapor (taken to be C_8H_{18}) and air in mole ratio 1:62.5 is drawn into the cylinder at 80°C and 1.00 atm, calculate:

(a) The temperature of the compressed gases just before the spark plug ignites them. (Assume that the gases are ideal, that the compression is adiabatic, and that the average heat capacity of the mixture of gasoline vapor and air is $c_P = 35 \text{ J K}^{-1} \text{ mol}^{-1}$.)
(b) The volume of the compressed gases just before ignition.
(c) The pressure of the compressed gases just before ignition.
(d) The maximum temperature of the combustion products, assuming combustion is completed before the piston begins its downstroke. Take $\Delta H_f^\circ(C_8H_{18}) = -57.4 \text{ kJ mol}^{-1}$.
(e) The temperature of the exhaust gases, assuming the expansion stroke to be adiabatic.

82. Nitromethane, CH_3NO_2, is a good fuel. It is a liquid at ordinary temperatures. When the liquid is burned, the reaction involved is chiefly

$$2\ CH_3NO_2(\ell) + \tfrac{3}{2}\ O_2(g) \longrightarrow 2\ CO_2(g) + N_2(g) + 3\ H_2O(g)$$

The standard enthalpy of formation of liquid nitromethane at 25°C is -112 kJ mol^{-1}; other relevant values can be found in Appendix D.

(a) Calculate the enthalpy change in the burning of 1 mol liquid nitromethane to form gaseous products at 25°C. State explicitly whether the reaction is endothermic or exothermic.
(b) Would more or less heat be evolved if *gaseous* nitromethane were burned under the same conditions? Indicate what additional information (if any) you would need to calculate the exact amount of heat, and show just how you would use this information.

83. Dry air containing a small amount of CO was passed through a tube containing a catalyst for the oxidation of CO to CO_2. Because of the heat evolved in this oxidation, the temperature of the air increased by 3.2 K. Calculate the weight percentage of CO in the sample of air. Assume that the specific heat at constant pressure for air is $1.01 \text{ J K}^{-1} \text{ g}^{-1}$.

84. When 1 mol isobutene, a gas with formula C_4H_8, is burned at 25°C and 1 atm to form CO_2 and gaseous water, the enthalpy change is −2528 kJ.

(a) Calculate, with the aid of any information needed from Table D-4 in Appendix D, the standard enthalpy of formation of isobutene.
(b) Suppose that 0.50 mol isobutene is burned adiabatically at constant pressure in the presence of an excess of oxygen, with 5.0 mol oxygen left at the end of the reaction. The heat capacity of the reaction vessel is 700 J K^{-1}, and pertinent molar heat capacities (in joules per kelvin per mole) are $CO_2(g)$, 37; $H_2O(g)$, 34; $O_2(g)$, 29. What is the approximate final temperature of this system (including the reaction vessel)?

85. Find the maximum possible temperature that may be reached when 0.050 mol $Ca(OH)_2(s)$ is allowed to react with 1.0 L of a 1.0-M HCl solution, both initially at 25°C. Assume that the final volume of the solution is 1.0 L, and that the specific heat at constant pressure of the solution is constant and equal to that of water, $4.18 \text{ J K}^{-1} \text{ g}^{-1}$.

CUMULATIVE PROBLEMS

86. Suppose 32.1 g $ClF_3(g)$ and 17.3 g $Li(s)$ are mixed and allowed to react at atmospheric pressure and 25°C until one of the reactants is used up, producing $LiCl(s)$ and $LiF(s)$. Calculate the amount of heat evolved.

87. Calculate ΔH_f° and ΔU_f° for the formation of silane, $SiH_4(g)$, from its elements at 298 K, if 250 cm³ of the gaseous compound at $T = 298$ K and $P = 0.658$ atm is burned in a constant-volume gas calorimeter in an excess of oxygen and causes the evolution of 9.757 kJ of heat. The combustion reaction is

$$SiH_4(g) + 2\ O_2(g) \longrightarrow SiO_2(s, \text{quartz}) + 2\ H_2O(\ell)$$

and the formation of silane from its elements is

$$Si(s) + 2\ H_2(g) \longrightarrow SiH_4(g)$$

88. **(a)** Draw Lewis diagrams for O_2, CO_2, H_2O, CH_4, C_8H_{18}, and C_2H_5OH. In C_8H_{18}, the eight carbon atoms form a chain with single bonds; in C_2H_5OH, the two carbon atoms are bonded to one another. Using average bond enthalpies from Table 12.5, compute the enthalpy change in each of the following reactions, if 1 mol of each carbon compound is burned, and all reactants and products are in the gas phase.
(b) $CH_4 + 2\ O_2 \longrightarrow CO_2 + 2\ H_2O$ (burning methane, or natural gas)
(c) $C_8H_{18} + \frac{25}{2}\ O_2 \longrightarrow 8\ CO_2 + 9\ H_2O$ (burning octane, in gasoline)
(d) $C_2H_5OH + 3\ O_2 \longrightarrow 2\ CO_2 + 3\ H_2O$ (burning ethanol, in gasohol)

89. By considering the nature of the intermolecular forces in each case, rank the following substances from smallest to largest enthalpy of vaporization: KBr, Ar, NH_3, and He. Explain your reasoning.

90. A supersonic nozzle is a cone-shaped object with a small hole in the end through which a gas is forced. As it moves through the nozzle opening, the gas expands in a manner that can be approximated as reversible and adiabatic. Such nozzles are used in molecular beams (see Section 18.5) and in supersonic aircraft engines to provide thrust, because as the gas cools, its random thermal energy is converted into directed motion of the molecules with average velocity v. Little thermodynamic work is done because the external pressure is low; thus, the net effect is to convert thermal energy to net translational motion of the gas molecules. Suppose the gas in the nozzle is helium; its pressure is 50 atm and its temperature is 400 K before it begins its expansion.

(a) What are the average speed and the average velocity of the molecules *before* the expansion?

(b) What will be the temperature of the gas after its pressure has decreased to 1.0 atm in the expansion?

(c) What is the average velocity of the molecules at this point in the expansion?

91. **(a)** Draw a Lewis diagram for carbonic acid, H_2CO_3, with a central carbon atom bonded to the three oxygen atoms.

(b) Carbonic acid is unstable in aqueous solution and converts to dissolved carbon dioxide. Use bond enthalpies to estimate the enthalpy change for the following reaction:

$$H_2CO_3 \longrightarrow H_2O + CO_2$$

13 CHAPTER

Spontaneous Processes and Thermodynamic Equilibrium

The reaction between solid sodium and gaseous chlorine proceeds imperceptibly, if at all, until the addition of a drop of water sets it off.

In both fundamental research and practical applications of chemistry, chemical reactions are carried out by mixing the reactants and regulating external conditions such as temperature and pressure. Two questions arise immediately:

1. Is it possible for the reaction to occur at the selected conditions?
2. If the reaction is possible, what determines the ratio of products and reactants at equilibrium?

Predicting the equilibrium composition for a chemical reaction is the central goal of Unit 4, and in this chapter, we develop the conceptual basis for answering these two key questions.

Thermodynamics can tell us whether a proposed reaction is possible under particular conditions even before we attempt the reaction. We can find out whether a proposed reaction is possible by determining whether it is a spontaneous thermodynamic process. In this context, "spontaneous" has a precise technical meaning (see later for definition) that should not be confused with its conversational meaning, such as describing the spontaneous behavior of people in social situations. If the reaction is spontaneous, thermodynamics can also predict the ratio of concentrations of products and reactants at equilibrium. This chapter develops the thermodynamic methods for predicting whether a reaction is spontaneous, and Chapter 14 uses these results to determine the equilibrium ratio of products and reactants.

But, we cannot use thermodynamics to predict the rate of a spontaneous reaction or how long it will take to reach equilibrium. These questions are the subject of chemical kinetics. To obtain a large amount of product from a spontaneous reaction in a short time, we need a reaction that is spontaneous and fast. Chapter 18 discusses the rates of chemical reactions. Manipulating conditions to optimize the yield of chemical reactions in practical applications requires the concepts from thermodynamics and kinetics from all three chapters.

The criteria for predicting spontaneity of physical and chemical processes are provided by the second law of thermodynamics, which is a brilliant abstraction and generalization from the observed facts of directionality in processes involving heat transfer. This is accomplished by introducing a new state function called *entropy*, which is denoted by S. We define the entropy function in such a way that the total entropy change of a thermodynamic universe (system and surroundings) is positive in the direction of spontaneous processes: $\Delta S_{tot} > 0$. Sections 13.1 through 13.5 develop the second law and demonstrate methods for calculating ΔS_{univ} and predicting spontaneity.

When processes are conducted at constant T and P, the criteria for spontaneity and for equilibrium are stated more conveniently in terms of another state function called the *Gibbs free energy* (denoted by G), which is derived from S. Because chemical reactions are usually conducted at constant T and constant P, their thermodynamic description is based on ΔG rather than ΔS. This chapter concludes by restating the criteria for spontaneity of chemical reactions in terms of ΔG. Chapter 14 shows how to identify the equilibrium state of a reaction, and calculate the equilibrium constant from the standard Gibbs free energy change $\Delta G°$.

For all these reasons, ΔG is the most important thermodynamic concept in the entire field of chemical equilibrium. Your goals in this chapter should be to understand the meaning of the state function G, to become skilled in calculating its changes, ΔG, and to interpret both the magnitude and the sign of these changes.

13.1 The Nature of Spontaneous Processes

In preparation for setting up the second law of thermodynamics, and stating precisely the criteria for spontaneity, we examine several familiar examples of spontaneous processes and describe their features in general terms. A spontaneous change is one that *will* occur by itself without outside intervention, once conditions have been established for its initiation. The change may be fast or slow, and we may have to wait a significant period to determine whether it *does* occur.

One of the most striking features of spontaneous change is that it follows a specific direction when starting from a particular initial condition. A dramatic example is shown in the series of photographs in Figure 13.1, which shows the freezing of supercooled distilled water poured from a water bottle. The liquid

(a) (b) (c)

FIGURE 13.1 Supercooled water freezes spontaneously, set off when any small particle nucleates crystallization. (a) Distilled water is cooled to −8°C, but is still liquid. (b) As the liquid water falls into a plastic bowl, ice crystals form spontaneously. (c) The spontaneous process is, in this case, rapid; the supercooled liquid water freezes during pouring, making a column of ice.

water, at a temperature of −8°C, freezes instantly as it hits a particle of ice (the particle nucleates the reaction; so you see why water must be distilled before trying to get it that cold). The freezing of supercooled water at this temperature happens spontaneously in the forward direction, but the reverse reaction is impossible (ice held at −8°C, 1 atm does not melt). One of the goals of thermodynamics is to account for this *directionality* of spontaneous change. The first law of thermodynamics provides no guidance in predicting or explaining directionality. Energy is conserved both in a forward process and in its reverse; nothing in the first law indicates a preference for one direction of change over the other.

Spontaneous processes familiar in chemical laboratories also follow specific directions:

1. We measure *heat transfer* from a hot body to a cold one when they are brought into thermal contact, but we never detect heat flowing spontaneously in the opposite direction.
2. We observe a gas *expanding* into a region of lower pressure, but we never see the reverse process, a gas compressing itself spontaneously into a small part of its container.
3. We place a drop of red ink into a beaker of water and watch the color spread by *diffusion* of the ink particles until the water is uniformly pink, but we never see the ink spontaneously reorganize as a small red drop in a volume of otherwise colorless water.
4. We place 10 g sucrose (ordinary table sugar) in a beaker and add 100 mL water at 80°C. The sucrose *dissolves* to form a uniform solution. We never observe the spontaneous reappearance of a mound of sucrose at the bottom of a beaker of water.
5. We open a container of acetone on the laboratory bench. We detect the aroma of acetone because some of the molecules have *evaporated* from the liquid and then diffused through the atmosphere to our position. We never observe the molecules to retreat spontaneously into the container.

All the previous examples are physical processes that are both spontaneous and rapid. We initiate the process and see the result quickly thereafter, with no further intervention.

Like the simple physical processes described earlier, spontaneous chemical reactions are inherently directional, though the reaction may be quite slow. Ordinary experience shows that a mixture of hydrogen and oxygen gases exists indefinitely at room temperature. Yet, if the mixture is exposed to a small amount of powdered platinum metal, or an electric spark, the gases react explosively to produce water. The reverse reaction is not spontaneous; we never observe the spontaneous decomposition of water into gaseous hydrogen and oxygen. Figure 1.2 shows decomposition of water into hydrogen and oxygen gases by electrolysis in an electrochemical cell, where electrical energy is continually provided by the external circuit to drive this nonspontaneous reaction. Another example is shown in the photograph at the beginning of this chapter, where the reaction between sodium metal and chlorine gas occurs explosively after a drop of water is added. We never observe the spontaneous decomposition of sodium chloride into sodium metal and gaseous chlorine. A third example, less dramatic than the previous two, is the spontaneous reaction of copper metal with oxygen at room temperature. This is seen in many older municipal buildings, especially in Europe, where copper metal was used as roofing material. With time, these roofs develop the blue–green patina of copper oxides. We never observe the spontaneous reappearance of shiny metallic copper on these old roofs.

These three examples illustrate that the actual outcome of a spontaneous chemical reaction depends on the reaction rate. The possibility of reaction between hydrogen and oxygen was there all along, but the rate was too slow to be observed until the powdered metal or the electrical spark accelerated the reaction. The possibility of reaction between metallic copper and oxygen was there all along, and the rate was large enough to be observed, if not dramatically fast. Thermodynamics determines whether a reaction is possible, whereas chemical kinetics determines whether it is practical. At the end of this chapter, you will be able to predict whether a chemical reaction is spontaneous, and by the end of the next chapter, you will be able to predict its equilibrium state. But, you must wait until Chapter 18 to see whether a spontaneous reaction can be carried out at a useful rate.

The direction of each of these spontaneous processes is readily apparent by observing the initial and final states, regardless of their paths. This suggests the existence of a new *state function* that indicates the directionality of spontaneous processes. That state function will turn out to be **entropy,** and it will be defined so that the sign of its change indicates the direction in which a proposed process will be spontaneous. Entropy has the interesting property that we cannot predict spontaneity by considering the system alone; we must also consider the entropy changes in the surroundings during the process.

To develop the entropy function, we must first describe spontaneous processes in the language of thermodynamics summarized in Section 12.1. You should revisit each of the examples just discussed to see how it specifically fits this language. Initially, two objects in different thermodynamic states are brought into contact; one will be called "system" and the other "surroundings." Barriers (constraints) in place between them prevent their interaction. When the constraints are removed, a spontaneous process may occur in which the system and surroundings exchange energy and matter and in which the volume of both system and surroundings may change. Because the two together constitute a *thermodynamic universe,* the total amount of energy, volume, and matter shared between them is fixed. During the process, some or all of these quantities are redistributed between the system and surroundings.

EXAMPLE 13.1

In the following two spontaneous processes, identify the system, the surroundings, and the constraint(s) removed to enable the processes to occur: (a) a piece of hot metal is cooled by immersion into a barrel of cold water; (b) a teaspoon of table sugar is dissolved in a cup of hot tea.

Solution

(a) The hot metal is the system, and the barrel of water is the surroundings. Initially we suppose they are brought into contact with an adiabatic wall between them to prevent heat flow so $q_{sys} = -q_{surr} = 0$. We initiate the cooling process by removing the adiabatic wall, whereupon thermal energy is transferred from the metal to the water. The thermal energy initially contained just in the system is now dispersed throughout the combined system and surroundings. Together the system and the surroundings are enclosed within adiabatic walls, so there is no interaction with the rest of the physical universe. The spontaneous process involves only energy transfer between system and surroundings.

(b) The sugar is the system, and the hot tea is the surroundings. Initially we imagine that the spoonful of sugar is placed in contact with the hot tea, and is surrounded by an adiabatic wall that prevents heat flow, a rigid wall that defines the volume of the sugar, and an impermeable wall that keeps the sugar separated from the tea. We initiate the dissolution process by removing the adiabatic wall, the rigid wall, and the impermeable wall. After this process the sugar molecules and the thermal energy are dispersed throughout the volume of the combined system and surroundings. The system and surroundings together are enclosed within adiabatic, rigid, and impermeable walls so there is no interaction with the rest of the physical universe. The spontaneous process involves transfer of energy, sugar molecules, and volume between the system and the surroundings.

Related Problems: 1, 2

Spontaneous processes are particular examples of irreversible processes defined in Section 12.1. In stark contrast with reversible processes, they do not proceed through a sequence of equilibrium states, and their direction cannot be reversed by an infinitesimal change in the direction of some externally applied force. Spontaneous processes cannot be represented as paths on the equation-of-state surface in Figure 12.2, but their initial and final equilibrium states can be represented as points on that surface.

What determines whether a process under consideration will be spontaneous? Where does a spontaneous process end? How are energy, volume, and matter partitioned between the system and surroundings at equilibrium? What is the nature of the final equilibrium state? These questions cannot be answered by the first law. Their answers require the second law and properties of the entropy, and a few developments are necessary before we can address these questions. We define entropy in terms of molecular motions in Section 13.2 and in terms of macroscopic process variables in Section 13.3. Finally, we present the methods for calculating entropy changes and for predicting spontaneity in Sections 13.4 and 13.5.

13.2 Entropy and Spontaneity: A Molecular Statistical Interpretation

What is entropy and why should it be related to the spontaneity of processes in nature? These are deep questions that we can only begin to answer here. To do so, we step outside the confines of classical thermodynamics, which is concerned only with macroscopic properties, and examine the microscopic molecular basis for the second law. Such an approach, called **statistical thermodynamics**, shows that spontaneous change in nature can be understood by using probability theory to predict and explain the behavior of the many atoms and molecules that comprise a macroscopic sample of matter. Statistical thermodynamics also provides theoretical procedures for calculating the entropy of a system from molecular properties.

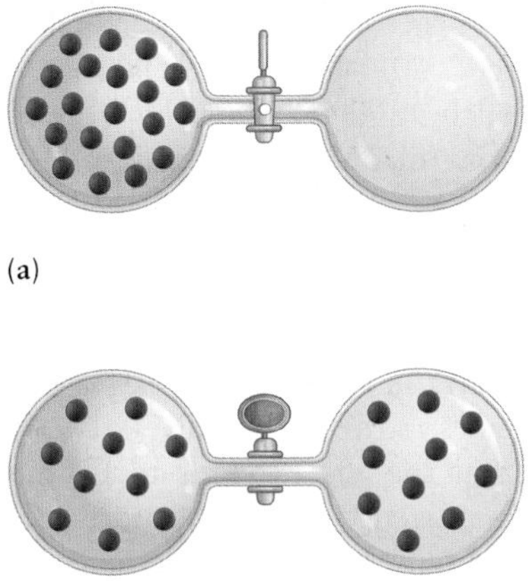

(a)

(b)

FIGURE 13.2 The free expansion of a gas into a vacuum. (a) The gas is initially confined to the bulb on the left, with the stopcock (the constraint) closed. (b) Half of the gas is found in each bulb, at equilibrium, after the stopcock is opened.

FIGURE 13.3 The 16 possible microstates of a system of four molecules that may occupy either side of a container. In only one of these are all four molecules on the left side.

13.2.1 Spontaneity and Molecular Motions

Consider a particularly simple spontaneous process: the free *adiabatic* expansion of 1 mol of an ideal gas into a vacuum (Fig. 13.2). The gas is initially held in the left bulb in volume $V/2$, whereas the right bulb is evacuated. Before it is opened, the stopcock is a constraint that holds all the molecules in the left bulb. After the stopcock is opened, the gas expands to fill the entire volume, V.

Now examine the same free expansion from a microscopic point of view. Imagine that the path of one particular tagged molecule can be followed during the expansion and for some period of time after final equilibrium has been established, perhaps through a series of time-lapse snapshots showing the locations of all molecules in the gas. From its starting position on the left side at the beginning of the experiment, the tagged molecule will cross to the right, then back to the left, and so forth. If enough time elapses, the molecule will eventually spend equal amounts of time on the two sides, because there is no physical reason for it to prefer one side to the other. As the adiabatic, free ($P_{ext} = 0$) process occurs, $q = 0$ and $w = 0$, so $\Delta U = 0$. Recall from Section 9.5 that the molecules in an ideal gas have an average internal energy that is proportional to T, so because U does not change the temperature does not change. The energy of the molecules remains the same at all locations, and the two sides of the container are identical. No experimental way exists to track the progress of one specific molecule in the gas; therefore, the best we can ask for is the *probability* that the molecule is on the left side at any given instant. (See Appendix C.6 for a review of probability concepts and methods.) The physical reasoning just summarized justifies that the probability that the molecule is on the left side will be $\frac{1}{2}$, the same as the probability that it is on the right side. Probability provides the key to understanding the direction of spontaneous change because it enables us to compare the likelihood of finding all the molecules on the left side—after the constraint has been removed—with the likelihood of finding the molecules uniformly distributed over the combined volume of both sides.

Just how unlikely is a spontaneous compression of 1 mol gas from the combined volume back to the left side? The probability that one particular molecule is on the left side at a given time is $\frac{1}{2}$. A second specific molecule may be on either the left or the right, so the probability that both are on the left is $\frac{1}{2} \times \frac{1}{2} = \frac{1}{4}$. As shown in Figure 13.3, in a gas containing a total of four molecules, all four molecules will be on the left only $\frac{1}{2} \times \frac{1}{2} \times \frac{1}{2} \times \frac{1}{2} = \frac{1}{16}$ of the time. Continuing this argument for all $N_A = 6.0 \times 10^{23}$ molecules leads to the probability that all N_A molecules are on the left:

$$\frac{1}{2} \times \frac{1}{2} \times \cdots \times \frac{1}{2} = \left(\frac{1}{2}\right)^{6.0\times10^{23}}$$

To evaluate this number, it is helpful to rewrite it in scientific notation as 1 divided by 10 raised to the power a (or, equivalently, as 10^{-a}):

$$\frac{1}{10^a} = \left(\frac{1}{2}\right)^{6.0\times10^{23}} = \frac{1}{2^{6.0\times10^{23}}}$$

A calculator can handle numbers this large (or this small) only if logarithms are used. Taking the base-10 logarithms of both denominators gives

$$\begin{aligned} a &= \log 2^{(6.0\times10^{23})} = 6.0 \times 10^{23} \log 2 \\ &= (6.0 \times 10^{23})(0.30) \\ &= 1.8 \times 10^{23} \end{aligned}$$

The probability that all the molecules will be on the left is 1 in $10^{1.8\times10^{23}}$.

This is a vanishingly small probability, because $10^{1.8 \times 10^{23}}$ is an unimaginably large number. It is vastly larger than the number 1.8×10^{23} (which is a large number already). To realize this, consider how such numbers would be written. The number 1.8×10^{23} is fairly straightforward to write out. It contains 22 zeros:

$$180{,}000{,}000{,}000{,}000{,}000{,}000{,}000$$

The number $10^{1.8 \times 10^{23}}$ is 1 followed by 1.8×10^{23} zeros. Written out, such a number would more than fill all the books in the world. Put in other terms, 1.8×10^{23} corresponds to the number of molecules in about 5 cm^3 water, but $10^{1.8 \times 10^{23}}$ is far larger than the number of molecules in the entire universe!

The statistical molecular picture explains that the gas expands to fill the whole available volume when the constraint is removed because this more uniform configuration of the molecules is overwhelmingly more probable than the initial configuration with all the molecules on the left side. The same explanation shows that the gas is never observed to compress spontaneously into a smaller volume because this nonuniform configuration of the molecules is overwhelmingly improbable in the absence of the constraint. Nothing in the laws of mechanics prevents a gas from compressing spontaneously. But this event is never seen because it is so improbable.

Thus, spontaneity in nature results from the random, statistical behavior of large numbers of molecules. *The directionality of spontaneous change is a consequence of the large numbers of molecules in the macroscopic systems treated by thermodynamics.* In systems containing fewer molecules, the situation can be quite different because the uniform configuration of molecules, although still the most probable, is no longer so overwhelmingly the most probable configuration. For example, if there were only 6 molecules, instead of 6×10^{23}, it would not be surprising to find them all on the left side at a given time. In fact, the probability that 6 molecules are on the left side is $\left(\frac{1}{2}\right)^6 = \frac{1}{64}$, so there is 1 chance in 64 that this will occur. Such small systems exhibit *statistical fluctuations* of the molecular configuration. Some of the fluctuations correspond to the initial nonuniform configuration maintained by the constraint, despite the absence of the constraint.

EXAMPLE 13.2

Calculate the probability of a spontaneous compression of 1.00 mol gas by 0.01%—that is, the probability that all the molecules will be found in a volume $V' = 0.9999V$ at a certain time.

Solution

In this case, the probability that a given molecule is in V' is not $\frac{1}{2}$ but 0.9999 (the probability that it is in the remainder of V is 0.0001). The probability that all N_A molecules are in V' is

$$(0.9999)^{N_A} = 1/10^a$$

$$a = -N_A \log(0.9999) = (6.0 \times 10^{23})(4.3 \times 10^{-5})$$

$$= 2.6 \times 10^{19}$$

The chance of such a compression occurring spontaneously is 1 in $10^{2.6 \times 10^{19}}$, which is still vanishingly small. Thus, a spontaneous compression of even a fraction of a percent will not be seen.

Related Problems: 7, 8

13.2.2 Entropy and Molecular Motions

We want to define a new function called entropy that will increase when the system undergoes a spontaneous process. How can we relate entropy to the properties of the system and formulate a definition to achieve this goal? Spontaneous processes occur when constraints are removed from a system; the molecules respond by moving to explore the suddenly increased range of motions now available to them. (Here we are assuming that the gas in the system is isolated from the surroundings; otherwise we must consider the entropy change of both the system and the surroundings to determine what process happens spontaneously.) In the free expansion of a gas, the molecules are initially confined in one part of the container. After the constraint is removed (the stopcock is opened), they are free to stay where they were, but they also are free to move throughout the larger combined volume of the two regions. Qualitatively, the numerical value of the entropy of a macroscopic system held in a particular thermodynamic state should depend on the *range of possible motions* (that is, the range of possible positions and momenta) available to the molecules while the system is held in that particular thermodynamic state. Any change in the macroscopic properties that enables the molecules to move out into larger regions of space or that increases the range of molecular speeds should increase the entropy of the system.

In preparation for defining entropy, we need a precise way to describe the "range of possible motions" of the molecules. This is accomplished by counting the number of microscopic, mechanical states, or **microstates**, available to molecules of the system. This number, denoted by Ω, counts all the possible combinations of positions and momenta available to the N molecules in the system when the system has internal energy U and volume V. For the simple model in Figure 13.3, $\Omega = 16$. In general, Ω is a function of U, V, and N, denoted as $\Omega(U, V, N)$. If the system is a monatomic ideal gas, the atoms do not interact. The position of each atom ranges freely throughout the entire volume, V. The internal energy consists of the total kinetic energy of the atoms, given as

$$U = \sum_{i=1}^{N} \varepsilon_i = \sum_{i=1}^{N} \frac{[p_{xi}^2 + p_{yi}^2 + p_{zi}^2]}{2m}$$

FIGURE 13.4 The fundamental relation between entropy (S) and the number of microstates (W) was derived by Ludwig Boltzmann in 1868. On his tombstone in Vienna is carved the equation he obtained, $S = k \log W$. We would write "ln" instead of "log" for the natural logarithm and Ω instead of "W" for the number of microstates.

thus, the momenta of the atoms range through all values that satisfy the condition

$$2mU = \sum_{i=1}^{N} [p_{xi}^2 + p_{yi}^2 + p_{zi}^2]$$

Although we do not provide the details of the calculation, the value of Ω for a monatomic ideal gas is given by

$$\Omega(U, V, N) = g\, V^N U^{(3N/2)}$$

where g is a collection of constants. Now imagine the walls defining the system are manipulated to change the values of U and V. The range of positions and momenta available to the molecules will increase or decrease accordingly. Because N is a very large number—of order 10^{23}—the value of Ω will increase or decrease dramatically when the volume of the system is increased or decreased in an expansion or compression, respectively. It will increase or decrease dramatically when the internal energy of the system is increased or decreased by heating or cooling, respectively.

The equation connecting entropy S and the number of available microstates Ω is

$$S = k_B \ln \Omega \qquad [13.1]$$

which was originally discovered by the Austrian physicist Ludwig Boltzmann in the late 19th century (Fig. 13.4). **Boltzmann's constant** k_B (see Eq. 9.14) is identified as R/N_A, the ratio of the universal gas constant R to Avogadro's number N_A. Thus, entropy has the physical dimensions J K^{-1}. It is impossible to overstate the importance of Boltzmann's relation, because it provides the link between the microscopic world of atoms and molecules and the macroscopic world of bulk matter and thermodynamics. Although this equation holds quite generally, it is difficult to apply because calculating Ω is a daunting theoretical task except for the simplest ideal systems. Other equations are used for practical applications in statistical thermodynamics. For our purposes here, the equation provides qualitative insight into the physical meaning of entropy and qualitative interpretation of the magnitude and sign of entropy changes caused by specific thermodynamic processes. The following example illustrates these insights in a simple case in which only the volume of the system changes in the process.

EXAMPLE 13.3

Consider the free expansion of 1 mol gas from $V/2$ to V (see Fig. 13.2). Use Boltzmann's relation to estimate the change in entropy for this process.

Solution

Consider the entire apparatus consisting of the filled and the evacuated bulbs to be a thermodynamic universe, so any exchange of thermal energy occurs solely between them; the pair of bulbs is taken to be thermally insulated from their surroundings. Then examine the effects of doubling V on Ω. If the volume is doubled, the number of positions available to a given molecule is doubled also. Therefore, the number of states available to the molecule should be proportional to the volume, V:

$$\text{number of states available per molecule} = cV$$

where c is a proportionality constant. The state of a *two*-molecule system is given jointly by the states of the molecules in it, so the number of microscopic states available is just the product of the number of states for each molecule, $(cV) \times (cV) = (cV)^2$. For an N-molecule system,

$$\text{microscopic states available} = \Omega = (cV)^N$$

Inserting this expression into Boltzmann's relation gives the entropy change for the free expansion of 1 mol (N_A atoms) gas from a volume $V/2$ to V:

$$\Delta S(\text{microscopic}) = N_A k_B \ln(cV) - N_A k_B \ln(cV/2)$$
$$= N_A k_B \ln\left(\frac{cV}{cV/2}\right) = N_A k_B \ln 2$$

Note that the constant c has dropped out. The calculated change in entropy on expansion of the gas is clearly positive, which is consistent with the increase in the number of available microstates on expansion.

In the next section, the entropy change for this process from the macroscopic view will be calculated to be

$$\Delta S\ (\text{thermodynamic}) = R \ln 2$$

illustrating explicitly that Boltzmann's microscopic description accurately explains the measured macroscopic results.

Related Problems: 3, 4, 5, 6, 7, 8, 9, 10

This example illustrates why Boltzmann's relation must involve the logarithmic function. Because entropy is an extensive variable, its value is proportional to N. But Ω depends on N through the *power* to which cV is raised. Therefore, doubling N doubles S but leads to Ω being squared.

Examples of other processes in which the entropy of the system increases include phase transitions such as the melting of a solid. In a solid, cohesive forces hold the atoms or molecules near their equilibrium positions in the crystal lattice, whereas in the liquid they can move far away from these fixed positions. More microstates are available to the molecules in the liquid, corresponding to a larger value of entropy; for melting of a solid, ΔS_{sys} is positive. In the same way, the entropy increases for evaporation of a liquid because the number of microstates available to the molecules increases enormously. Whereas the molecules in a liquid remain at the bottom of their container, those in a gas move throughout the container, so the number of microstates increases upon vaporization. In some solid-solid phase transformations it is difficult to predict qualitatively which phase has higher entropy. Nonetheless, calculating the relative number of microstates available correctly predicts the sign of the entropy change in a transformation from one such phase to the other.

So far, we have considered microstates that involve only the positions of the molecules in a system. The distribution of energies can also contribute to the number of microstates. For example, a gas in which all the molecules have the same speed has fewer microstates available to it than does a gas with a distribution of molecular speeds; a spontaneous process is observed experimentally in which a system initially constrained to a single molecular speed moves toward the Maxwell–Boltzmann speed distribution, because this distribution maximizes the number of microstates available for a given total energy. This textbook gives only qualitative microscopic interpretations of these more complex examples. In later courses you will see more thorough interpretations based on the more advanced equations that replace Equation 13.1.

13.3 Entropy and Heat: Macroscopic Basis of the Second Law of Thermodynamics

The second law of thermodynamics, which is stated as an abstract generalization of engineering observations on the efficiency of heat engines, has two important consequences for chemistry. It defines the entropy function in terms of measurable macroscopic quantities, providing a means to calculate the changes in entropy caused by specific processes. And, it defines the criterion for predicting whether a particular process will be spontaneous. It is thus useful to re-state the second law in a form directly applicable to chemistry: Part One defines the entropy function, and Part Two shows how to predict spontaneity. We start the discussion with a nonmathematical, qualitative summary of the arguments on efficiency, from which we state the second law in engineering terms and then define the entropy function (Part One of the second law). Section 13.4 applies the definition to calculate entropy changes for a wide variety of processes. Section 13.5 uses these calculations to predict spontaneity of processes (Part Two of the second law).

Section 13.8 presents a mathematical route to the same results developed qualitatively here. Either Section 13.3 or Section 13.8 provides adequate background for the calculations in Sections 13.4 and 13.5, which are essential for chemistry.

13.3.1 Background of the Second Law of Thermodynamics

The second law of thermodynamics originated in practical concerns over the efficiency of steam engines at the dawn of the Industrial Age, late in the 18th century, and required about a century for its complete development as an engineering tool. The central issues in that development are summarized as follows. In each stroke of a steam engine, a quantity of hot steam at high pressure is injected into a piston-cylinder assembly, where it immediately expands and pushes the piston outward against an external load, doing useful work by moving the load. The external mechanical structure to which the piston is attached includes a reciprocating mechanism that returns the piston to its original position at the end of the stroke so that a new quantity of steam can be injected to start the next stroke. The engine operates as a cyclic process, returning to the same state at the beginning of each stroke. In each stroke, the expansion process is highly irreversible because the steam is cooled and exhausted from the cylinder at the end of the stroke. In essence, the engine extracts thermal energy from a hot reservoir, uses some of this energy to accomplish useful work, then discards the remainder into a cooler reservoir (the environment). See Figure 13.5 for a schematic representation of a heat engine. The energy lost to the environment cannot be recovered by the engine. The internal combustion engine operates in a similar manner, with injection of hot steam replaced by *in situ* ignition

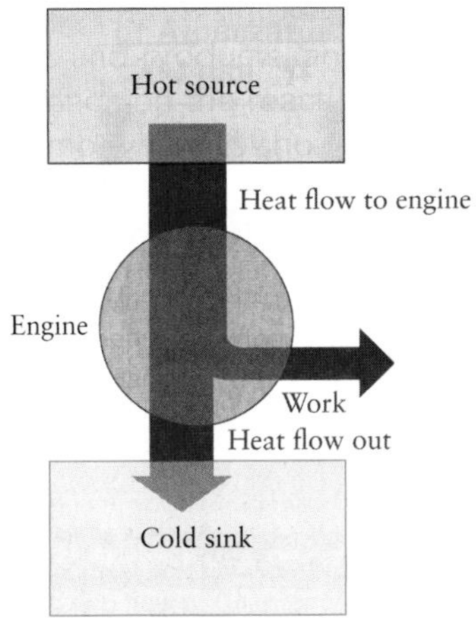

FIGURE 13.5 Schematic representation of a heat engine.

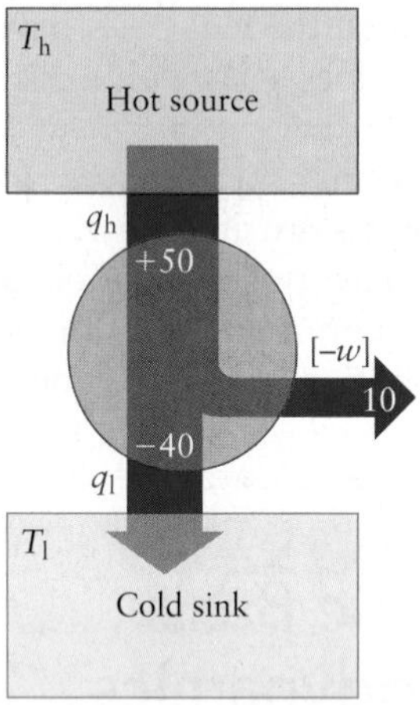

FIGURE 13.6 Operation of a heat engine. In this example the engine withdraws 50 kJ of thermal energy from the hot reservoir, performs 10 kJ of work on its surroundings, and discharges 40 kJ of thermal energy to the lower temperature reservoir.

of combustible fuel that burns in highly exothermic reactions. Its expansion process is similarly irreversible, and some of the energy expended by the hot gases is unrecoverable.

In both engines the efficiency, defined as the ratio of work accomplished *by* the engine in a cycle ($-w$ in Figure 13.6) to the heat invested to drive that cycle, q_h in Figure 13.6, can be improved by reducing the unrecoverable losses to the environment in each cycle. Seeking to maximize efficiency, Sadi Carnot, an officer in Napoleon's French Army Corps of Engineers, modeled operation of the engine with an idealized cyclic process now known as the *Carnot cycle.* He concluded that unrecoverable losses of energy to the environment cannot be completely eliminated, no matter how carefully the engine is designed. Even if the engine is operated as a reversible process (in which case, displacement of the external load is too slow to be of practical interest), its efficiency cannot exceed a fundamental limit known as the *thermodynamic efficiency.* Thus, an engine with 100% efficiency cannot be constructed.

Carnot's conclusion has been restated in more general terms by the English physicist Lord Kelvin in the following form:

> *There is no device that can transform heat withdrawn from a reservoir completely into work with no other effect.*

and by the German physicist Rudolf Clausius in the following form:

> *There is no device that can transfer heat from a colder to a warmer reservoir without net expenditure of work.*

These statements are consistent with ordinary experience that (1) heat always flows spontaneously from a hotter body to a colder body, and that (2) work is always required to refrigerate a body. With confidence based on experience, Clausius, Kelvin, and later scientists and engineers have assumed these statements to be valid for *all* heat transfer processes and labeled them as equivalent formulations of the second law of thermodynamics.

13.3.2 Definition of Entropy

How do we apply these general statements to chemical processes such as the examples described in Section 13.1, which at first glance bear no resemblance to heat engines? Highlights of the argument are summarized here, and a more detailed development is presented in Section 13.8.

Carnot's analysis of efficiency for a heat engine operating *reversibly* showed that in each cycle q/T at the high temperature reservoir and q/T at the low temperature environment summed to zero:

$$\frac{q_h}{T_h} + \frac{q_l}{T_l} = 0$$

This result suggests that q/T is a state function in a reversible process because the sum of its changes in a cyclic process is zero. Clausius extended this result to show that the quantity $\int(1/T)\,dq_{rev}$ is independent of path in *any* reversible process and is, therefore, a state function. Clausius then *defined* the entropy change $\Delta S = S_f - S_i$ of a system in a process starting in state i and ending in state f by the equation

$$\Delta S = S_f - S_i = \int_i^f \frac{dq_{rev}}{T} \qquad [13.2]$$

The definition in Equation 13.2 constitutes Part One of the second law in the form directly related to chemistry. Entropy is therefore a state function and has physical dimensions J K^{-1}. We calculate $\Delta S = S_f - S_i$ for a specific process by (1) identifying its initial and final equilibrium states i and f as points on the equation-of-state surface in Figure 12.2, (2) selecting *any convenient reversible path* between them along which dq_{rev} and T are known, and (3) evaluating this integral along the selected path. It does not matter that the actual process of interest may be irreversible. Because entropy is a state function, its change depends only on the initial and final states, and not at all on the path. *Therefore, we are free to choose any reversible path that connects the initial and final states, purely on grounds of convenience, for calculating* ΔS.

This is a beautiful consequence of the fact that S is a state function. Section 13.4 provides detailed procedures for calculating ΔS for numerous types of systems and processes. The calculated values of ΔS are then used in Section 13.5 in Part Two of the second law to predict whether a particular contemplated process will be spontaneous.

The preceding discussion emphasizes that the second law, like all other laws of science, is a bold extrapolation of the results of a great deal of direct experimental observation under controlled conditions. Based on his understanding of the Carnot cycle, Clausius assumed, or postulated, that Equation 13.2 is valid for all thermodynamic systems and all processes that begin and end in equilibrium states. For every system there exists a state function called entropy whose changes during a process depend only on the initial and final states, and the magnitude of the changes can be calculated along any convenient reversible path connecting these particular states. In Section 13.5 we see how these calculated entropy changes predict whether a process is spontaneous or not. The second law is not proved to be true by a single definitive experiment, it is not derived as a consequence of some more general theory, and it is not handed down by some higher authority. Rather, it is invoked as one of the "starting points" of thermodynamics, and conclusions drawn from its application to a great variety of processes (not necessarily involving heat engines) are compared with the results obtained in experimental studies. To date, no disagreements have been found between predictions properly made from the second law and the results of properly designed experiments.

13.4 Entropy Changes in Reversible Processes

This section outlines procedures for calculating entropy changes that occur in the system and in the surroundings during several types of processes.

13.4.1 ΔS_{sys} for Isothermal Processes

If the reversible process selected as the pathway connecting the initial and final states is isothermal, the calculation simplifies immediately. Because T is constant, it comes outside the integral:

$$\Delta S = \int_i^f \frac{dq_{rev}}{T} = \frac{1}{T}\int_i^f dq_{rev} = \frac{q_{rev}}{T} \quad [13.3]$$

Here, q_{rev} is the *finite* amount of heat absorbed by the system during the entire reversible isothermal process.

Compression/Expansion of an Ideal Gas

Consider an ideal gas enclosed in a piston-cylinder arrangement that is maintained at constant temperature in a heat bath. The gas can be compressed (or expanded) reversibly by changing the position of the piston to accomplish a specified change in volume. In Section 12.7, the heat transferred between system and bath when the gas is expanded (or compressed) isothermally and reversibly from volume V_1 to V_2 is shown to be

$$q_{rev} = nRT \ln\left(\frac{V_2}{V_1}\right)$$

The resulting change in entropy, therefore, is

$$\Delta S = nR \ln\left(\frac{V_2}{V_1}\right) \quad \text{(constant } T\text{)} \qquad [13.4]$$

From Equation 13.4 we see that the entropy of a gas increases during an isothermal expansion ($V_2 > V_1$) and decreases during a compression ($V_2 < V_1$). Boltzmann's relation (see Eq. 13.1) provides the molecular interpretation of these results. The number of microstates available to the system, Ω, increases as the volume of the system increases and decreases as volume decreases, and the entropy of the system increases or decreases accordingly.

Phase Transitions

Another type of constant-temperature process is a phase transition such as the melting of a solid at constant pressure. This occurs reversibly at the fusion temperature, T_f, because an infinitesimal change in external conditions, such as reducing the temperature, can reverse the melting process. The reversible heat absorbed by the system when 1 mol of substance melts is $q_{rev} = \Delta H_{fus}$, so

$$\Delta S_{fus} = \frac{q_{rev}}{T_f} = \frac{\Delta H_{fus}}{T_f} \qquad [13.5]$$

The entropy increases when a solid melts or a liquid vaporizes, and it decreases when the phase transition occurs in the opposite direction. Again, Boltzmann's relation provides the molecular interpretation. When a solid melts or a liquid vaporizes, the number of accessible microstates Ω increases, and thus the entropy increases.

EXAMPLE 13.4

Calculate the entropy change when 3.00 mol benzene vaporizes reversibly at its normal boiling point of 80.1°C. The molar enthalpy of vaporization of benzene at this temperature is 30.8 kJ mol^{-1}. (As the vaporization occurs, the pressure of benzene above the liquid is maintained at 1 atm by increasing the volume available to the gas.)

Solution

The entropy change when 1 mol benzene is vaporized at 80.1°C (=353.25 K, 1 atm), is

$$\Delta S_{vap} = \frac{\Delta H_{vap}}{T_b} = \frac{30,800 \text{ J mol}^{-1}}{353.25 \text{ K}} = +87.2 \text{ J K}^{-1}$$

When 3.00 mol is vaporized, the entropy change is three times as great:

$$\Delta S = (3.00 \text{ mol})(+87.2 \text{ J K}^{-1} \text{ mol}^{-1}) = +262 \text{ J K}^{-1}$$

Related Problems: 11, 12

Remarkably, most liquids have similar values for the molar entropy of vaporization. **Trouton's rule** summarizes this observation:

$$\Delta S_{vap} = 88 \pm 5 \text{ J K}^{-1} \text{ mol}^{-1} \quad [13.6]$$

Note that the ΔS_{vap} of benzene, calculated in Example 13.4, is within this range. The constancy of ΔS_{vap} means that ΔH_{vap} and T_b, which vary widely from substance to substance, must do so in the same proportion. Trouton's rule allows enthalpies of vaporization to be estimated from boiling temperatures. However, exceptions exist; the molar entropy of vaporization for water is 109 J K^{-1} mol^{-1}. The value for water is unusually high because hydrogen bonding in liquid water means that there are many fewer allowed configurations (lower entropy) at a given temperature than in other liquids; thus, water shows a much greater increase in the number of microstates on vaporization.

13.4.2 ΔS_{sys} for Processes with Changing Temperature

Now consider a reversible process in which the temperature changes. In this case, Equation 13.2 must be used:

$$\Delta S = \int_i^f \frac{1}{T} dq_{rev}$$

In the integral, i and f represent, respectively, the initial and final equilibrium states for the process. The calculation must be conducted along a reversible path connecting i and f. For a reversible *adiabatic* process, $q = 0$ and, therefore, $\Delta S = 0$. Such a process is also called an **isentropic process** (that is, the entropy is constant).

In a reversible *isochoric* process, the volume is held constant and the system is heated or cooled by contact with a reservoir whose temperature differs from that of the system by an infinitesimal amount, dT. As long as there are no phase changes or chemical reactions, the heat transferred in this case is

$$dq_{rev} = nc_V \, dT$$

and the entropy change of the system as it is heated from T_1 to T_2 is

$$\Delta S = \int_{T_1}^{T_2} \frac{1}{T} \, dq_{rev} = \int_{T_1}^{T_2} \frac{nc_V}{T} dT$$

If c_V is independent of T over the temperature range of interest, it can be removed from the integral, giving the result

$$\Delta S = nc_V \int_{T_1}^{T_2} \frac{1}{T} dT = nc_V \ln\left(\frac{T_2}{T_1}\right) \quad \text{(constant } V\text{)} \quad [13.7]$$

The analogous result for the entropy change of the system in a reversible *isobaric* process (constant *pressure*) is

$$\Delta S = \int_{T_1}^{T_2} \frac{nc_P}{T} dT = nc_P \ln\left(\frac{T_2}{T_1}\right) \quad \text{(constant } P\text{)} \quad [13.8]$$

Entropy always increases with increasing temperature. From the kinetic theory of ideal gases in Chapter 9, it is clear that increasing the temperature of the gas increases the magnitude of the average kinetic energy per molecule and, therefore, the range of momenta available to molecules. This, in turn, increases Ω for the gas and, by Boltzmann's relation, the entropy of the gas.

Now consider an experiment in which identical samples of a gas are taken through identical temperature increases, one sample at constant V and the other at constant P. Let's compare the entropy changes in the two processes. From the previous discussion it follows that $\Delta S_P > \Delta S_V$ because $c_P > c_V$. The molecular interpretation is based on the discussion of c_P and c_V in Section 12.3. The gas heated at constant P increases in volume, as well as in temperature; its molecules therefore gain access to a greater range of positions, as well as a greater range of momenta. Consequently, the gas heated at constant P experiences a *greater* increase in Ω than does the gas heated at constant V and, therefore, a greater increase in S.

The following example illustrates that the entropy is a state function, for which changes are independent of the path followed.

EXAMPLE 13.5

(a) Calculate the entropy change for the process described in Example 12.10: 5.00 mol argon expands reversibly at a constant temperature of 298 K from a pressure of 10.0 to 1.00 atm.

(b) Calculate the entropy change for the same initial and final states as in part (a) but along a different path. First, the 5.00 mol argon expands reversibly and *adiabatically* between the same two pressures. This is the path followed in Example 12.11; it causes the temperature to decrease to 118.6 K. Then the gas is heated at constant pressure back to 298 K.

Solution

(a) At constant temperature, the entropy change is

$$\Delta S = nR \ln\left(\frac{V_2}{V_1}\right) = nR \ln\left(\frac{P_1}{P_2}\right)$$

$$= (5.00 \text{ mol})(8.315 \text{ J K}^{-1} \text{ mol}^{-1}) \ln 10.0$$

$$= +95.7 \text{ J K}^{-1}$$

(b) For the adiabatic part of this path, the entropy change is zero. When the gas is then heated reversibly at constant pressure from 118.6 to 298 K, the entropy change is

$$\Delta S = nc_P \ln\left(\frac{T_2}{T_1}\right)$$

$$= (5.00 \text{ mol})\left(\frac{5}{2} \times 8.315 \text{ J K}^{-1} \text{ mol}^{-1}\right) \ln \frac{298 \text{ K}}{118.6 \text{ K}}$$

$$= +95.7 \text{ J K}^{-1}$$

This is the same as the result from part (a), an illustration of the fact that the entropy is a state function. By contrast, the amounts of heat for the two paths are different: 28.5 and 18.6 kJ.

Related Problems: 15, 16

13.4.3 ΔS for Surroundings

Usually, the surroundings can be treated as a large *heat bath* that transfers heat to or from the system at the fixed temperature of the bath. In such cases, the heat capacity of the surroundings (heat bath) must be so large that the heat transferred during the process does not change the temperature of the bath. The heat gained by the surroundings during a process is the heat lost from the system. If the process occurs at constant P, then

$$q_{surr} = -\Delta H_{sys}$$

and the entropy change of the surroundings is

$$\Delta S_{surr} = \frac{-\Delta H_{sys}}{T_{surr}} \quad [13.9]$$

Here the surroundings must be large so that the actual heat transferred to it does not bring it out of equilibrium, even though an irreversible process may be happening to the system. If the process occurring in the system is exothermic, the surroundings gain heat and the entropy change of the surroundings is positive. Similarly, an endothermic process in the system is accompanied by a negative entropy change in the surroundings, because the surroundings give up heat during the process to keep the system at the temperature of the heat bath.

If the surroundings lack sufficient heat capacity to maintain constant temperature during the process, then entropy changes for the surroundings must be calculated by the methods demonstrated above for the system, taking explicit account of the temperature change and heat capacity of the surroundings. Examples of both cases are included in the problems at the end of this chapter.

If the process on the system is not at constant P, you can still use $q_{surr} = -q_{sys}$ (the actual q_{sys}, not the q_{sys} for the reversible path). Then the entropy change of the surroundings equals $q_{surr}/T_{surr} = -q_{sys}/T_{surr}$.

13.5 Entropy Changes and Spontaneity

Part Two of the second law states that a process can occur spontaneously if the total entropy change for the thermodynamic universe of the process is positive:

$$\Delta S_{tot} = \Delta S_{sys} + \Delta S_{surr} > 0$$

We illustrate the calculation first for spontaneous cooling of a hot body and then for irreversible expansion of an ideal gas. Finally we discuss the so-called inequality of Clausius, which provides the basis for Part Two of the second law.

Spontaneous Cooling of a Hot Body

Consider a spontaneous process in which a sample of hot metal is cooled by sudden immersion in a cold bath. Heat flows from the metal into the bath until they arrive at the same temperature. This spontaneous process is accompanied by an increase in the total entropy for the thermodynamic universe of the process, as illustrated by the following example.

EXAMPLE 13.6

A well-insulated ice-water bath at 0.0°C contains 20 g ice. Throughout this experiment, the bath is maintained at the constant pressure of 1 atm. When a piece of nickel at 100°C is dropped into the bath, 10.0 g of the ice melts. Calculate the total entropy change for the thermodynamic universe of this process. (Specific heats at constant P: nickel, 0.46 J K^{-1} g^{-1}; water, 4.18 J K^{-1} g^{-1}; ice, 2.09 J K^{-1} g^{-1}. Enthalpy of fusion of ice, 334 J g^{-1}.)

Solution

Consider the nickel to be the *system* and the ice-water bath to be the *surroundings* in this experiment. Heat flows from the nickel into the bath and melts some of the ice. Consequently, the entropy of the nickel decreases and the entropy of the bath increases. The final equilibrium temperature of both system and bath is 0.0°C, as indicated by the presence of some ice in the bath at equilibrium.

Before calculating ΔS_{Ni} it is necessary to calculate the mass of the nickel from the calorimetry equation as follows:

$$\text{heat lost by Ni} = \text{heat gained by ice bath} = \text{heat used in melting ice}$$

$$-m(0.46 \text{ J K}^{-1} \text{ g}^{-1})(273.15 \text{ K} - 373.15 \text{ K}) = (10.0 \text{ g})(334 \text{ J g}^{-1})$$

$$m = 73 \text{ g}$$

(Continued)

Because the nickel is cooled at constant *P*, the entropy change for the nickel is calculated as

$$\Delta S_{Ni} = (73 \text{ g})(0.46 \text{ J g}^{-1} \text{ K}^{-1}) \ln(0.73) = -10 \text{ J K}^{-1}$$

Because the ice bath has remained at constant *T* throughout the experiment, it can be treated as a "large heat bath," and its entropy change is calculated as

$$\Delta S_{bath} = \frac{-\Delta H_{sys}}{T_{bath}} = \frac{-(-334 \text{ J g}^{-1})(10.0 \text{ g})}{273.15 \text{ K}} = 12 \text{ J K}^{-1}$$

Now,

$$\Delta S_{tot} = \Delta S_{Ni} + \Delta S_{bath} = -10 + 12 = +2 \text{ J K}^{-1}$$

Thus, the process is spontaneous, driven by the fact that the entropy gain of the melting ice exceeds the entropy loss of the cooling metal.

Related Problems: 17, 18, 19, 20

Irreversible Expansion of an Ideal Gas

Consider a gas confined within a piston-cylinder arrangement and held at constant temperature in a heat bath. Suppose the external pressure is abruptly reduced and held constant at the new lower value. The gas immediately expands against the piston until its internal pressure declines to match the new external pressure. The total entropy of system plus surroundings will increase during this expansion. In preparation for a quantitative example, a general comparison of irreversible and reversible processes connecting the same initial and final states provides insight into why the total entropy increases in a spontaneous process.

The work performed *by* a system ($-w$) as it undergoes an irreversible isothermal expansion is always less than when the expansion is conducted reversibly. To see this, return to the definition of work done *on* a system:

$$w = -\int P_{ext}\, dV$$

During an expansion, P_{ext} must be less than *P*, the pressure of the gas. For a reversible expansion, P_{ext} is only infinitesimally smaller (so the system is always close to equilibrium); but for an irreversible expansion, P_{ext} is measurably smaller. Therefore, the area under a graph of P_{ext} plotted against *V* is less than that of a graph of *P* against *V* (Fig. 13.7), so

$$-w_{irrev} = \int P_{ext}\, dV < \int P\, dV = -w_{rev}$$

and the work performed by the system, $-w_{irrev}$, is less than $-w_{rev}$. If the system is viewed as an "engine" for performing useful work on the surroundings, a reversible process is always more efficient than an irreversible one.

If the reversible and irreversible processes have the same initial and final states, then ΔU is the same for both.

$$\Delta U = w_{irrev} + q_{irrev} = w_{rev} + q_{rev}$$

But because

$$-w_{irrev} < -w_{rev}$$

we must have

$$w_{irrev} > w_{rev}$$

and

$$q_{irrev} < q_{rev}$$

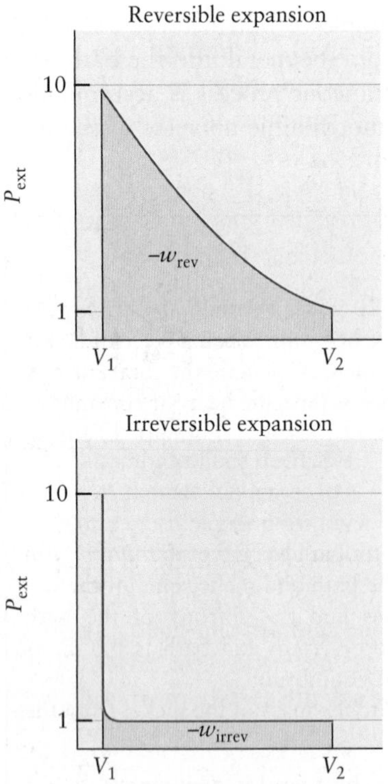

FIGURE 13.7 Work done by a system in reversible and irreversible expansions between the same initial and final states. The work performed is greater for the reversible process.

The heat absorbed is a *maximum* when the process is conducted reversibly.

The following example illustrates these inequalities.

EXAMPLE 13.7

Calculate the heat absorbed and the work done on a system of 5.00 mol of an ideal gas as it expands irreversibly at constant temperature $T = 298$ K from a pressure of 10.0 to 1.00 atm. The external pressure is held constant at 1.00 atm.

Solution

The initial volume V_1 is

$$V_1 = \frac{nRT}{P_1}$$

$$= \frac{(5.00 \text{ mol})(0.08206 \text{ L atm K}^{-1} \text{ mol}^{-1})(298 \text{ K})}{10.0 \text{ atm}} = 12.2 \text{ L}$$

The final volume is 10 times this, or

$$V_2 = 122 \text{ L}$$

For a constant external pressure,

$$w_{\text{irrev}} = -P_{\text{ext}}\Delta V$$

$$= -(1.00 \text{ atm})(122 \text{ L} - 12.2 \text{ L}) = -110 \text{ L atm}$$

$$= -11.1 \text{ kJ}$$

At constant T, $\Delta U = 0$, however, so

$$q_{\text{irrev}} = -w_{\text{irrev}} = 11.1 \text{ kJ}$$

In Example 12.10, a reversible expansion between the same two states was carried out, with the result that

$$w_{\text{rev}} = -28.5 \text{ kJ}$$

This demonstrates that

$$-w_{\text{irrev}} < -w_{\text{rev}}$$

and

$$q_{\text{irrev}} < q_{\text{rev}}$$

For reversible and irreversible *isothermal* processes connecting the same pair of initial and final states, it is always true that

$$q_{\text{rev}} > q_{\text{irrev}}$$

Dividing this expression by T, the temperature at which the heat is transferred, gives

$$\frac{q_{\text{rev}}}{T} > \frac{q_{\text{irrev}}}{T}$$

The left side is the entropy change ΔS_{sys},

$$\Delta S_{\text{sys}} = \frac{q_{\text{rev}}}{T}$$

so

$$\Delta S_{\text{sys}} > \frac{q_{\text{irrev}}}{T}$$

The last two equations can be combined as

$$\Delta S_{\text{sys}} \geq \frac{q}{T}$$

where the equality applies only to a reversible process. This expression, called the *inequality of Clausius*, states that in any spontaneous process the heat absorbed by the system from surroundings at the same temperature is always less than $T\Delta S_{\text{sys}}$. In a reversible process, the heat absorbed is equal to $T\Delta S_{\text{sys}}$.

Now let's apply Clausius's inequality to processes occurring within an *isolated* system. In this case, there is no transfer of heat into or out of the system, and $q = 0$. Therefore, for spontaneous processes within an isolated system, $\Delta S > 0$.

The thermodynamic universe of a process (that is, a system plus its surroundings) is clearly an isolated system to which Clausius's inequality can be applied. It follows that

1. *In a* reversible *process the total entropy of a system plus its surroundings is unchanged.*
2. *In an* irreversible *process the total entropy of a system plus its surroundings must increase.*
3. *A process for which* $\Delta S_{univ} < 0$ *is not spontaneous.*

These statements constitute the heart of the second law, because they justify its predictive power, which we stated as a postulate at the beginning of Section 13.5.

EXAMPLE 13.8

Calculate $\Delta S_{tot} = \Delta S_{sys} + \Delta S_{surr}$ for the reversible and irreversible isothermal expansions of Examples 12.10 and 13.7.

Solution

For the reversible expansion,

$$q_{rev} = 28.5 \text{ kJ}$$

$$\Delta S_{sys} = \frac{q_{rev}}{T} = \frac{28{,}500 \text{ J}}{298 \text{ K}} = +95.7 \text{ J K}^{-1}$$

The surroundings give up the same amount of heat at the same temperature. Hence,

$$\Delta S_{surr} = \frac{-28{,}500 \text{ J}}{298 \text{ K}} = -95.7 \text{ J K}^{-1}$$

or $\Delta S_{tot} = 95.7 - 95.7 = 0$ for the reversible process. For the irreversible expansion, it is still true that

$$\Delta S_{sys} = +95.7 \text{ J K}^{-1}$$

because S is a function of state, and the initial and final states are the same as for the reversible expansion. From Example 13.7, only 11.1 kJ of heat is given up by the surroundings in this case.[1] Hence,

$$\Delta S_{surr} = \frac{-11{,}100 \text{ J}}{298 \text{ K}} = -37.2 \text{ J K}^{-1}$$

and $\Delta S_{tot} = 95.7 - 37.2 = 58.5 \text{ J K}^{-1} > 0$ for the irreversible process.

13.6 The Third Law of Thermodynamics

In thermodynamic processes, only *changes* in entropy, ΔS, are measured, just as only changes in internal energy, ΔU, or enthalpy, ΔH, are measured. It is nevertheless useful to define absolute values of entropy relative to some reference state. An important experimental observation that simplifies the choice of reference state is:

In any thermodynamic process involving only pure phases in their equilibrium states, the entropy change ΔS *approaches zero as* T *approaches* 0 K.

This observation is the **Nernst heat theorem**, named after its discoverer, the German physicist Walther Nernst. It immediately suggests a choice of reference state: The entropy of any pure element in its equilibrium state is defined to approach zero as T approaches 0 K. From the Nernst theorem, the entropy change for any chemical reaction, including one in which elements react to give

[1]How can the heat from Example 13.7, which is irreversible from the perspective of the system, be reversible from the perspective of the surroundings? This can be accomplished by enclosing the gas in a material (such as a metal) that can efficiently transfer heat to and from the surroundings, and thus remain close to equilibrium, at the same time that the gas itself is far from equilibrium due to the gas currents that occur during the irreversible expansion.

a pure compound, approaches zero at 0 K. The most general form of this statement is the **third law of thermodynamics:**

> *The entropy of any pure substance (element or compound) in its equilibrium state approaches zero at the absolute zero of temperature.*

Absolute zero can never actually be reached; therefore, a small extrapolation is needed to make use of this result.

The third law, like the two laws that precede it, is a macroscopic law based on experimental measurements. It is consistent with the microscopic interpretation of the entropy presented in Section 13.2. From quantum mechanics and statistical thermodynamics, we know that the number of microstates available to a substance at equilibrium falls rapidly toward one as the temperature approaches absolute zero. Therefore, the absolute entropy defined as $k_B \ln \Omega$ should approach zero. The third law states that the entropy of a substance in its equilibrium state approaches zero at 0 K. In practice, equilibrium may be difficult to achieve at low temperatures, because particle motion becomes very slow. In solid CO, molecules remain randomly oriented (CO or OC) as the crystal is cooled, even though in the equilibrium state at low temperatures, each molecule would have a definite orientation. Because a molecule reorients slowly at low temperatures, such a crystal may not reach its equilibrium state in a measurable period. A nonzero entropy measured at low temperatures indicates that the system is not in equilibrium.

13.6.1 Standard-State Entropies

Because the entropy of any substance in its equilibrium state is zero at absolute zero, its entropy at any other temperature T is given by the entropy increase as it is heated from 0 K to T. If heat is added at constant pressure,

$$\Delta S = n\int_{T_1}^{T_2} \frac{c_P}{T}\, dT$$

Thus, S_T, the absolute entropy of 1 mol of substance at temperature T, is given by

$$S_T = \int_0^T \frac{c_P}{T}\, dT$$

It is necessary merely to measure c_P as a function of temperature and determine the area under a plot of c_P/T versus T from 0 K to any desired temperature. If a substance melts, boils, or undergoes some other phase change before reaching the temperature T, the entropy change for that process must be added to $\int (c_P/T)\, dT$.

To calculate entropy changes for chemical reactions, we find it convenient to use the same standard state already selected for enthalpy calculations in Section 12.3. For this purpose, we define the **standard molar entropy** to be the absolute molar entropy $S°$ at 298.15 K and 1 atm pressure (Fig. 13.8):

$$S° = \int_0^{298.15} \frac{c_P}{T} dT + \Delta S \text{ (phase changes between 0 and 298.15 K)} \qquad [13.10]$$

Standard molar entropies $S°$ are tabulated for a number of elements and compounds in Appendix D. If c_P is measured in J K^{-1} mol^{-1}, then the entropy $S°$ will have the same units. For dissolved ions, the arbitrary convention $S°$ ($H^+(aq)$) = 0 is applied (just as for the standard enthalpy of formation of H^+ discussed in Section 12.3). For this reason, some $S°$ values are negative for aqueous ions—an impossibility for substances.

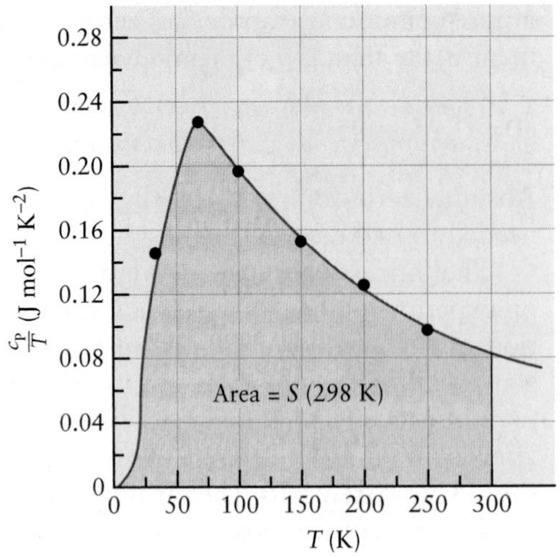

FIGURE 13.8 A graph of c_P/T versus T for platinum. The black dots represent experimental measurements. The area up to any temperature (here, 298 K) is the molar entropy at that temperature.

Tabulated standard molar entropies are used to calculate entropy changes in chemical reactions at 25°C and 1 atm, just as standard enthalpies of formation are combined to obtain enthalpies of reaction according to Hess's law (see Section 12.5).

EXAMPLE 13.9

Using the table of standard molar entropies in Appendix D, calculate $\Delta S°$ for the chemical reaction

$$N_2(g) + 2\ O_2(g) \longrightarrow 2\ NO_2(g)$$

with reactants and products at a temperature of 25°C and a pressure of 1 atm.

Solution

From the table,

$$S°(N_2(g)) = 191.50 \text{ J K}^{-1} \text{ mol}^{-1}$$

$$S°(O_2(g)) = 205.03 \text{ J K}^{-1} \text{ mol}^{-1}$$

$$S°(NO_2(g)) = 239.95 \text{ J K}^{-1} \text{ mol}^{-1}$$

The entropy change for the reaction is the sum of the entropies of the products, minus the sum of entropies of the reactants, each multiplied by its coefficient in the balanced chemical equation:

$$\begin{aligned}\Delta S° &= 2S°(NO_2(g)) - S°(N_2(g)) - 2S°(O_2(g)) \\ &= (2 \text{ mol})(239.95 \text{ J K}^{-1} \text{ mol}^{-1}) - (1 \text{ mol})(191.50 \text{ J K}^{-1} \text{ mol}^{-1}) - \\ &\quad (2 \text{ mol})(205.03 \text{ J K}^{-1} \text{ mol}^{-1}) \\ &= -121.66 \text{ J K}^{-1}\end{aligned}$$

The factors of 2 multiply $S°$ for NO_2 and O_2 because 2 mol of each appears in the chemical equation. Note that standard molar entropies, unlike standard molar enthalpies of formation ΔH_f°, are not zero for elements at 25°C. The negative $\Delta S°$ results because this is the entropy change of the system only. The surroundings must undergo a positive entropy change in such a way that $\Delta S_{tot} \geq 0$.

Related Problems: 21, 22, 23, 24, 25, 26, 27, 28

13.7 The Gibbs Free Energy

In Section 13.5, we showed that the change in entropy of a system plus its surroundings (that is, the total change of entropy, ΔS_{tot}) provides a criterion for deciding whether a process is spontaneous, reversible, or non-spontaneous:

$$\Delta S_{tot} > 0 \quad \text{spontaneous} \qquad [13.11a]$$

$$\Delta S_{tot} = 0 \quad \text{reversible} \qquad [13.11b]$$

$$\Delta S_{tot} < 0 \quad \text{nonspontaneous} \qquad [13.11c]$$

Although the algebraic sign of ΔS_{tot} is a completely general criterion for determining the spontaneity of a process, it requires calculating the entropy change for the surroundings, as well as for the system. It would be much more convenient to have a state function that predicts the feasibility of a process in the system without explicit calculations for the surroundings.

For processes that occur at constant temperature and pressure, which is the most important set of conditions for chemical applications, such a state function exists. It is called the *Gibbs free energy* and is denoted by G. After a qualitative discussion of spontaneous laboratory processes at fixed T and P, we define G and develop its properties. Finally, we apply ΔG to identify conditions for spontaneity in phase transitions and chemical reactions.

13.7.1 The Nature of Spontaneous Processes at Fixed *T* and *P*

Consider a system enclosed in a piston-cylinder assembly, which maintains pressure at the value P. The assembly is immersed in a heat bath, which maintains temperature at the value T. Experience shows that spontaneous processes under these conditions consist of spontaneous flow of molecules across a boundary completely internal to the system, separating different regions (called *phases*) of the system (Fig. 13.9).

Under these conditions, we visualize starting a spontaneous process by bringing phases A and B in Figure 13.9 into contact—both already prepared at T and P—but separated by an impermeable membrane (a constraint) that prevents exchange of matter between the phases. Removing the constraint allows spontaneous flow of molecules across the interface between phases. The system does not exchange matter with the surroundings, and the distribution of energy and volume between system and surroundings is not described explicitly. The only function of the surroundings is to maintain T and P constant throughout the experiment. Consequently, as we show in the next subsection, spontaneity of the process is determined by the change in Gibbs free energy of the *system only* while T and P remain constant.

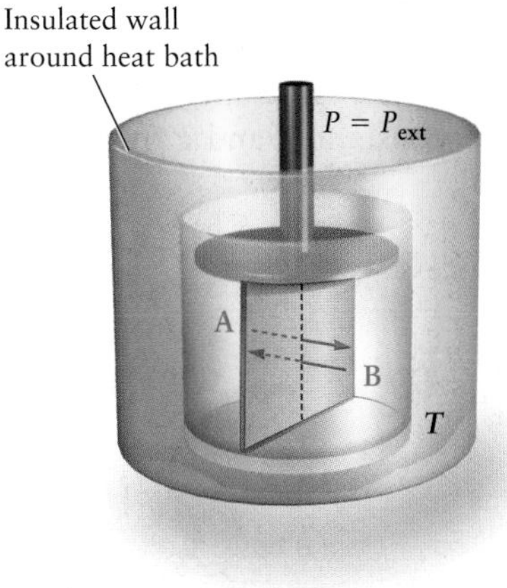

FIGURE 13.9 After the constraint between phases A and B is removed, matter can flow spontaneously between phases inside a system held at constant T (temperature) and P (pressure) by its surroundings.

These processes lead to changes in the structure or composition of the phases. Solutes are redistributed between immiscible solvents. Phase transitions occur between the solid, liquid, and gaseous states. Reactants become products. The system may gain or lose heat from the large heat reservoir while these rearrangements occur in its phases, but T remains constant. For example, the latent heat of fusion released during freezing of a liquid in the system is absorbed by the bath. Endothermic processes in the system will absorb heat from the bath, and exothermic processes will give heat to the bath. P–V work may be done on or by the system at constant P, depending on whether its density increases or decreases through the rearrangements of its phases.

Whether molecules flow spontaneously from phase A to B, or vice versa, is determined by the associated change in Gibbs free energy, as we show in the following subsection.

13.7.2 Gibbs Free Energy and Its Properties

During any process conducted at constant T and P as described earlier, the heat gained by the system is $\Delta H_{sys} = q_P$ and the heat transferred to the surroundings is $-q_P = -\Delta H_{sys}$. Because the surroundings remain at constant temperature during the process, the transfer of process heat must have the same effect on them as would a reversible transfer of the same amount of heat. Their entropy change is then

$$\Delta S_{surr} = \frac{-\Delta H_{sys}}{T_{surr}}$$

The total entropy change is

$$\Delta S_{tot} = \Delta S_{sys} + \Delta S_{surr} = \Delta S_{sys} - \frac{\Delta H_{sys}}{T_{surr}}$$

$$= \frac{-(\Delta H_{sys} - T_{surr}\,\Delta S_{sys})}{T_{surr}}$$

Because the temperature, T, is the same for both the system and the surroundings, we can rewrite this as

$$\Delta S_{tot} = \frac{-\Delta(H_{sys} - TS_{sys})}{T} \qquad [13.12]$$

We define the **Gibbs free energy** G as

$$G = H - TS \qquad [13.13]$$

therefore, Equation 13.12 becomes

$$\Delta S_{tot} = \frac{-\Delta G_{sys}}{T} \qquad [13.14]$$

Because the absolute temperature T is always positive, ΔS_{tot} and ΔG_{sys} must have the opposite sign for processes occurring at constant T and P. It follows that

$$\Delta G_{sys} < 0 \quad \text{spontaneous processes} \qquad [13.15a]$$

$$\Delta G_{sys} = 0 \quad \text{reversible processes} \qquad [13.15b]$$

$$\Delta G_{sys} > 0 \quad \text{nonspontaneous processes} \qquad [13.15c]$$

for processes conducted at constant temperature and pressure. If $\Delta G_{sys} > 0$ for a proposed process, then $\Delta G_{sys} < 0$ for the *reverse* of the proposed process, and that reverse process can occur spontaneously. The connection of these relations to experiment is shown in Figure 13.9. Initially we prepare the system at chosen values of T and P, then release the appropriate constraint to allow the process to occur. The resulting "flow" of matter between phases A and B goes in the direction that reduces the value of G for the system, so that $\Delta G < 0$ at the selected values of T and P.

13.7.3 Gibbs Free Energy and Phase Transitions

As a simple application of the Gibbs free energy, consider the freezing of 1 mol liquid water to form ice:

$$H_2O(\ell) \longrightarrow H_2O(s)$$

First, let's examine what thermodynamics predicts when this process occurs at the ordinary freezing point of water under atmospheric pressure, 273.15 K. The measured enthalpy change (the heat absorbed at constant pressure) is

$$\Delta H_{273} = q_P = -6007 \text{ J mol}^{-1}$$

At $T_f = 273.15$ K, water freezes reversibly—the system remains close to equilibrium as it freezes. Therefore, the entropy change is

$$\Delta S_{273} = \frac{q_{rev}}{T_f} = \frac{-6007 \text{ J mol}^{-1}}{273.15 \text{ K}} = -21.99 \text{ J K}^{-1} \text{ mol}^{-1}$$

The Gibbs free energy change for freezing is $\Delta G_{freezing} = G_{ice} - G_{water}$ and its value at 273 K is

$$\Delta G_{273} = \Delta H_{273} - T\,\Delta S_{273} = -6007 \text{ J mol}^{-1} - (273.15 \text{ K})(-21.99 \text{ J K}^{-1} \text{ mol}^{-1}) = 0$$

At the normal freezing point, the Gibbs free energy change is zero because the freezing of water under these conditions is an equilibrium, reversible process.

Now, let's see what thermodynamics predicts as the water is cooled below 273.15 to 263.15 K (−10.00°C). Let's calculate the change in Gibbs free energy as water freezes at this lower temperature.

Assume that ΔH and ΔS for the freezing process do not depend on temperature. Then we can write

$$\Delta G_{263} = -6007 \text{ J mol}^{-1} - (263.15 \text{ K})(-21.99 \text{ J K}^{-1} \text{ mol}^{-1}) = -220 \text{ J mol}^{-1}$$

An exact calculation takes into account that ΔH and ΔS *do* depend slightly on temperature, and it leads to

$$\Delta G_{263} = -213 \text{ J mol}^{-1}$$

for the process. Because $\Delta G < 0$, thermodynamics predicts the undercooled water will freeze spontaneously at 263.15 K. At a temperature *higher* than T_f, ΔG is greater than 0, predicting that the liquid would not freeze. This agrees with our experience in nature. Water does not freeze at atmospheric pressure if the temperature is held greater than 273.15 K; instead, the reverse process occurs, and ice melts spontaneously.

Writing the Gibbs free energy change as $\Delta G = \Delta H - T\,\Delta S$ shows that a negative value of ΔG—and therefore a spontaneous process—is favored by a negative value of ΔH and a positive value of ΔS. For freezing a liquid, ΔH is negative, but ΔS for freezing is also negative. Whether a liquid freezes depends on the competition between two factors: an enthalpy change that favors freezing and an entropy change that disfavors freezing (Fig. 13.10). At temperatures less than T_f, the former dominates and the liquid freezes spontaneously, but at temperatures greater than T_f, the latter dominates and freezing does not occur. At T_f, the Gibbs free energies of the two phases are equal ($\Delta G = 0$), and the phases coexist at equilibrium. Similar types of analysis apply to other phase transitions, such as condensing a gas to a liquid.

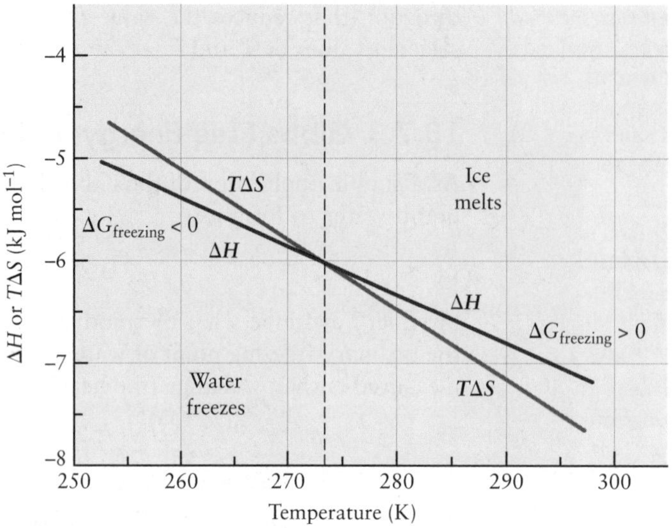

FIGURE 13.10 Plots of ΔH and $T\,\Delta S$ versus temperature for the freezing of water. At 273.15 K, the two curves cross, meaning that at this temperature, $\Delta G_{freezing} = 0$ and ice and water coexist. Below this temperature, $\Delta G_{freezing} < 0$ and the freezing of water to ice is spontaneous. Above this temperature $\Delta G_{freezing} > 0$ and the reverse process, the melting of ice to water, is spontaneous.

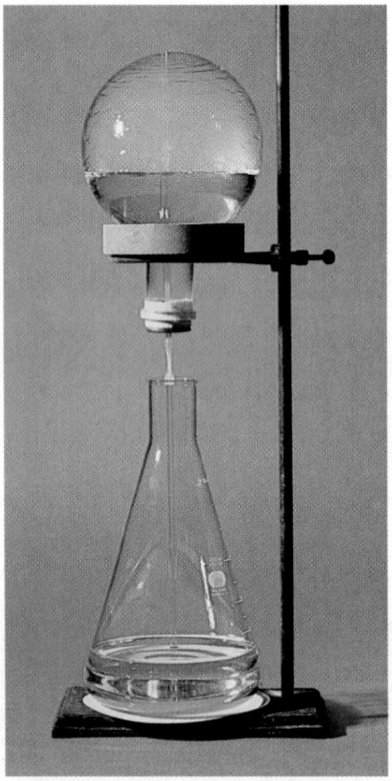

FIGURE 13.11 The dissolution of hydrogen chloride in water, $HCl(g) \rightarrow HCl(aq)$, is a spontaneous process. In this demonstration, the upper flask is filled with gaseous hydrogen chloride, and a small amount of water is injected into it. As the hydrogen chloride dissolves spontaneously, its pressure declines. The resulting air pressure difference draws water up the tube from the lower flask, allowing more hydrogen chloride to dissolve. The change is so fast that a vigorous fountain of water sprays into the upper flask. The free energy change of the process appears as work, raising the water.

13.7.4 Gibbs Free Energy and Chemical Reactions

The change in the Gibbs free energy provides a criterion for the spontaneity of any process occurring at constant temperature and pressure (Fig. 13.11). To predict whether a chemical reaction is spontaneous at given values of T and P, it is necessary to determine only the sign of ΔG for the reaction at these same conditions. From experience with other state functions, we would expect to calculate ΔG for reactions by consulting appropriate tabulations of free energy data. Because we cannot know the absolute value of the Gibbs free energy of a substance (just as we cannot know the absolute value of its internal energy U), it is convenient to define a **standard molar Gibbs free energy of formation**, ΔG_f°, analogous to the standard molar enthalpy of formation ΔH_f° introduced in Section 12.6. From tables of ΔG_f° we can calculate ΔG° for a wide range of chemical reactions, just as we used Hess's law to calculate ΔH° for a reaction from tables of ΔH_f° for products and reactants. The next few paragraphs show how tables of ΔG_f° are generated and how the data are used to determine spontaneity of reactions.

Standard-State Free Energies

The change in the Gibbs free energy for a chemical reaction performed at constant temperature is

$$\Delta G = \Delta H - T\,\Delta S \qquad [13.16]$$

where ΔH is the enthalpy change in the reaction (considered in Section 12.6) and ΔS is the entropy change in the reaction (see Section 13.4). The standard molar Gibbs free energy of formation ΔG_f° of a compound is the change in Gibbs free energy for the reaction in which 1 mol of the compound in its standard state is formed from its elements in their standard states. (You should review the definition of standard states in Section 12.6.) For example, ΔG_f° for $CO_2(g)$ is given by the Gibbs free energy change for the reaction

$$C(s) + O_2(g) \longrightarrow CO_2(g) \qquad \Delta G_f^\circ = ?$$

in which the partial pressures of all the gases is 1 atm. The value of ΔG_f° can be constructed from Equation 13.16 by using ΔH_f° and ΔS° values for this reaction and setting $T = 298.15$ K. For this reaction, $\Delta H°$ is simply $\Delta H_f°$ for $CO_2(g)$, because graphite and oxygen are elements in their standard states,

$$\Delta H° = \Delta H_f°(CO_2) = -393.51 \text{ kJ}$$

The value of $\Delta S°$ can be obtained from the absolute entropies of the substances involved at 25°C and 1 atm pressure (both elements and compounds, because the absolute entropy $S°$ of an element is not zero in its standard state).

$$\Delta S° = S°(CO_2) - S°(C) - S°(O_2) = 213.63 - 5.74 - 205.03 \text{ J K}^{-1}$$
$$= +2.86 \text{ J K}^{-1}$$

The $\Delta G_f°$ for CO_2 is then

$$\Delta G_f° = \Delta H_f° - T\,\Delta S° = -393.51 \text{ kJ} - (298.15 \text{ K})(2.86 \text{ J K}^{-1})(10^{-3} \text{ kJ J}^{-1})$$
$$= -394.36 \text{ kJ}$$

Appendix D includes a table of $\Delta G_f°$ values for numerous substances, all obtained by the method just illustrated. Note that our definition makes $\Delta G_f° = 0$ for an *element* that is already in its standard state.

Because G is a state function, chemical equations can be added together—with their $\Delta G_f°$ values combined as in Hess's law for changes in enthalpy—to calculate Gibbs free energy changes for chemical reactions under standard-state conditions.

EXAMPLE 13.10

Calculate $\Delta G°$ for the following reaction, using tabulated values for $\Delta G_f°$ from Appendix D.

$$3\ NO(g) \longrightarrow N_2O(g) + NO_2(g)$$

Solution

$$\Delta G° = \Delta G_f°(N_2O) + \Delta G_f°(NO_2) - 3\ \Delta G_f°(NO)$$
$$= (1 \text{ mol})(104.18 \text{ kJ mol}^{-1}) + (1 \text{ mol})(51.29 \text{ kJ mol}^{-1}) - (3 \text{ mol})(86.55 \text{ kJ mol}^{-1})$$
$$= -104.18 \text{ kJ}$$

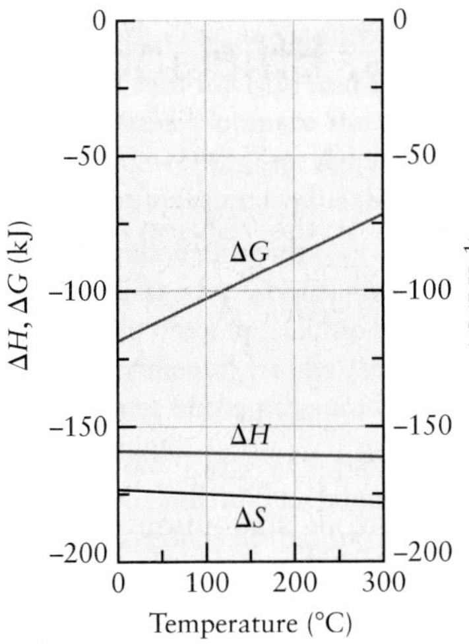

FIGURE 13.12 The entropy change of the reaction $3\ NO(g) \rightarrow N_2O(g) + NO_2(g)$ varies less than 5% between 0°C and 300°C; the enthalpy of reaction is even closer to constancy. The free energy change in the reaction shifts greatly over the temperature range, however, as the magnitude of $T\,\Delta S$ increases. Note that the units are very different on the left- (energy) and right-hand vertical axes (entropy).

Effects of Temperature on ΔG

Values of $\Delta G°$ calculated from the data in Appendix D are accurate only at $T = 298.15$ K. Values of $\Delta G°$ can be estimated for reactions at other temperatures and at $P = 1$ atm using the equation

$$\Delta G° = \Delta H° - T\,\Delta S° \qquad [13.17]$$

and tables of standard entropies and standard enthalpies of formation. The estimates will be close to the true value if $\Delta H°$ and $\Delta S°$ are not strongly dependent on T, which is usually the case (Fig. 13.12).

The value and the sign of $\Delta G°$ can depend strongly on T, even when the values of $\Delta H°$ and $\Delta S°$ do not, because of the competition between $\Delta H°$ and $T\,\Delta S°$ in Equation 13.17. If $\Delta H°$ is negative and $\Delta S°$ is positive, then the reaction is spontaneous at all temperatures when reactants and products are at

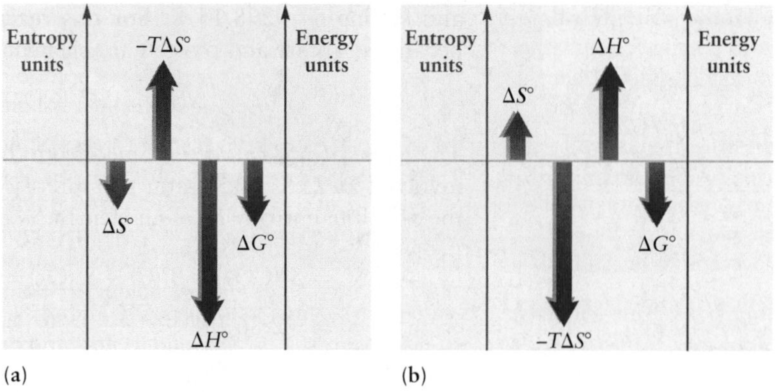

FIGURE 13.13 The competition between $\Delta H°$ and $\Delta S°$ determines the temperature range in which a reaction is spontaneous. (a) If both $\Delta H°$ and $\Delta S°$ are negative, the reaction is spontaneous at temperatures less than $T^* = \Delta H°/\Delta S°$. (b) If both $\Delta H°$ and $\Delta S°$ are positive, the reaction is spontaneous at temperatures greater than $T^* = \Delta H°/\Delta S°$. Note that the reactants and products must be in their standard states to use $\Delta G°$ rather than ΔG to determine spontaneity.

atmospheric pressure. If $\Delta H°$ is positive and $\Delta S°$ is negative, the reaction is never spontaneous. For the other possible combinations, there exists a special temperature T^*, defined by

$$T^* = \frac{\Delta H°}{\Delta S°} \quad [13.18]$$

at which $\Delta G°$ equals zero. If both $\Delta H°$ and $\Delta S°$ are positive, the reaction will be spontaneous at temperatures greater than T^*. If both are negative, the reaction will be spontaneous at temperatures less than T^* (Fig. 13.13). This discussion demonstrates an important result from chemical thermodynamics with practical importance: With knowledge of $\Delta H°$ and $\Delta S°$, we can change the temperature (holding the partial pressures of the gases involved at 1 atm each) to make a reaction spontaneous.

13.8 A DEEPER LOOK Carnot Cycles, Efficiency, and Entropy

This section provides a mathematical development of the relation between entropy and heat already presented qualitatively in Section 13.3. No additional results are obtained, but considerably greater insight is provided.

13.8.1 The Carnot Cycle

A heat engine and its operation are shown schematically in Section 13.3.1. The engine is a thermodynamic system which interacts with its surroundings in the course of a cyclic process. The engine withdraws (gains) energy in the form of heat from the high-temperature reservoir, and uses part of that energy to perform work on its surroundings. The remainder of the thermal energy is discharged (lost by the engine) to the lower temperature reservoir. In a Carnot cycle (Fig. 13.14), the engine (system) traverses two isothermal and two adiabatic paths to return to its original state. Each path is carried out reversibly (that is, in thermal equilibrium, with internal and external forces nearly balanced at every step). As the system proceeds from state A to C through state B, the system performs work (Fig. 13.14a):

$$w_{ABC} = -\int_{V_A}^{V_C} P\,dV$$

(Path ABC)

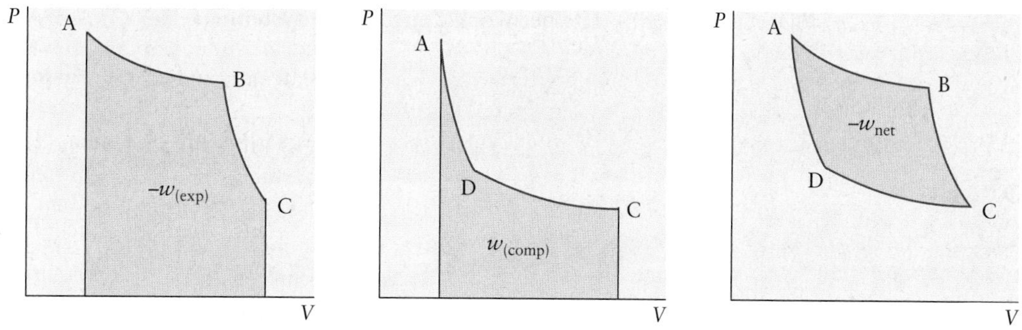

(a) Expansion (b) Compression (c) Complete cycle

FIGURE 13.14 Stages of the Carnot cycle. The work done by the system in expansion (a) and on the system by compression (b) is shown by the shaded areas. (c) The net work done per cycle is the area enclosed by the curve ABCDA.

and it is clearly negative for this (expansion) process. If the system then returns from state C to A through state D,

$$w_{CDA} = -\int_{V_C}^{V_A} P\,dV$$

(Path CDA)

which is now a positive quantity for this (compression) process; work is performed on the system (see Fig. 13.14b). In the course of the cycle, the work performed by the system is the area under curve ABC, whereas that performed on the system is the (smaller) area under curve ADC. The net result of the whole cycle is that work is performed by the system, and the amount of this work is the difference between the two areas, which is the area enclosed by the cycle (see Fig. 13.14c). As in any cyclic process, the overall energy change ΔU is zero, thus the net work equals the negative of the total heat added to the system.

We have not yet specified the material contained in the system. Assume initially that it is an ideal gas, for which the results from Section 12.7 apply directly, with T_h defined to be the higher temperature in the cycle and T_l the lower temperature.

Path AB: Isothermal Expansion (temperature T_h)

$$w_{AB} = -q_{AB} = -nRT_h \ln\left(\frac{V_B}{V_A}\right)$$

Path BC: Adiabatic Expansion

$$q_{BC} = 0$$

$$w_{BC} = nc_V(T_l - T_h) = -nc_V(T_h - T_l)$$

Path CD: Isothermal Compression (temperature T_l)

$$w_{CD} = -q_{CD} = -nRT_l \ln\left(\frac{V_D}{V_C}\right)$$

$$= nRT_l \ln\left(\frac{V_C}{V_D}\right)$$

Path DA: Adiabatic Compression

$$q_{DA} = 0$$

$$w_{DA} = nc_V(T_h - T_l)$$

The net work done on the system is

$$
\begin{aligned}
w_{net} &= w_{AB} + w_{BC} + w_{CD} + w_{DA} \\
&= -nRT_h \ln\left(\frac{V_B}{V_A}\right) - nc_V(T_h - T_l) \\
&\quad + nRT_l \ln\left(\frac{V_C}{V_D}\right) + nc_V(T_h - T_l) \\
&= -nRT_h \ln\left(\frac{V_B}{V_A}\right) + nRT_l \ln\left(\frac{V_C}{V_D}\right)
\end{aligned}
$$

This can be simplified by noting that V_B and V_C lie on one adiabatic path, and V_A and V_D lie on another. In Section 12.7, the relation for a reversible adiabatic process was found:

$$\frac{T_2}{T_1} = \left(\frac{V_1}{V_2}\right)^{\gamma-1}$$

Hence,

$$\frac{T_h}{T_l} = \left(\frac{V_C}{V_B}\right)^{\gamma-1} \quad \text{for path BC}$$

and

$$\frac{T_h}{T_l} = \left(\frac{V_D}{V_A}\right)^{\gamma-1} \quad \text{for path DA}$$

Equating these expressions gives

$$\left(\frac{V_C}{V_B}\right)^{\gamma-1} = \left(\frac{V_D}{V_A}\right)^{\gamma-1}$$

or

$$\frac{V_C}{V_B} = \frac{V_D}{V_A} \quad \text{and} \quad \frac{V_B}{V_A} = \frac{V_C}{V_D}$$

Hence, the net work done in one passage around the Carnot cycle is

$$w_{net} = -nR(T_h - T_l)\ln\frac{V_B}{V_A} \qquad [13.19]$$

13.8.2 Heat Engines

The Carnot cycle is an idealized model for a heat engine. When a certain amount of heat, q_{AB}, is added to the system at the higher temperature, T_h, a net amount of work, $-w_{net}$, is obtained from the system. In addition, some heat q_{CD} is discharged at the lower temperature, but this energy is "degraded" and is no longer available for use in the engine. The **efficiency**, ϵ, of such an engine is the ratio of the negative of the net work done on the system, $-w_{net}$, to the heat added along the high-temperature isothermal path:

$$\epsilon = \frac{-w_{net}}{q_{AB}}$$

It is the net work that is available for the performance of useful mechanical tasks such as turning electrical generators or dynamos, but it is the heat, q_{AB}, absorbed at the higher temperature, T_h, that must be "paid for" in terms of coal or oil consumed to supply it. The efficiency, ϵ, must be maximized to get out the most work possible for the lowest cost.

For the ideal gas Carnot cycle, the efficiency is easily calculated:

$$\epsilon = \frac{-w_{net}}{q_{AB}} = \frac{nR(T_h - T_l)\ln(V_B/V_A)}{nRT_h \ln(V_B/V_A)}$$

$$\epsilon = \frac{T_h - T_l}{T_h} = 1 - \frac{T_l}{T_h} \qquad [13.20]$$

This result, called the Carnot efficiency or the **thermodynamic efficiency**, places a fundamental limit on the efficiency with which heat can be converted to mechanical work. Only if the high temperature, T_h, were infinite or the low temperature, T_l, were zero would it be possible to have a heat engine operate with 100% efficiency. To maximize efficiency, the greatest possible temperature difference should be used. Although we derived this result specifically for the ideal gas, we will show later in this section that it applies to *any* reversible engine operating between two temperatures. For a *real* engine, which must operate irreversibly, the actual efficiency must be lower than the thermodynamic efficiency.

EXAMPLE 13.11

Suppose a heat engine absorbs 10.0 kJ of heat from a high-temperature source at $T_h = 450$ K and discards heat to a low-temperature reservoir at $T_l = 350$ K. Calculate the thermodynamic efficiency, ϵ, of conversion of heat to work; the amount of work performed, $-w_{net}$; and the amount of heat discharged at T_l, q_{CD}.

Solution

$$\epsilon = \frac{T_h - T_l}{T_h} = \frac{450\text{ K} - 350\text{ K}}{450\text{ K}} = 0.222$$

Therefore, the engine can be, at most, 22.2% efficient. Because $\epsilon = -w_{net}/q_{AB}$,

$$w_{net} = -\epsilon q_{AB} = -(0.222)(10.0\text{ kJ}) = -2.22\text{ kJ}$$

Because ΔU for the whole cycle is 0,

$$\Delta U = 0 = q_{AB} + q_{CD} + w_{net}$$

$$q_{CD} = -q_{AB} - w_{net} = -10.0 + 2.22 = -7.8\text{ kJ}$$

Therefore, 7.8 kJ is discharged at 350 K. This heat must be removed from the vicinity of the engine by a cooling system; otherwise, it will cause T_l to increase and reduce the efficiency of the engine.

Related Problems: 39, 40

13.8.3 Efficiency of General Carnot Engines

This subsection shows that all Carnot engines operating reversibly between two temperatures, T_h and T_l, have the same efficiency:

$$\epsilon = \frac{-w_{net}}{q_h} = \frac{T_h - T_l}{T_h} \qquad [13.21]$$

That is, the efficiency calculated for an ideal gas applies equally to any other working fluid. To demonstrate this, we assume the *contrary* true and show that assumption leads to a contradiction with experience. The assumption is, therefore, deemed false.

Assume that there *are* two reversible machines operating between the same two temperatures, T_h and T_l, one of which has an efficiency, ϵ_1, that is greater than the efficiency, ϵ_2, of the other. The two machines are adjusted so that the total work output is the same for both. The more efficient machine is run as a heat engine so that it produces mechanical work $-w_1$. This work is used to operate the other machine in the opposite sense (as a heat pump) so that $w_2 = -w_1$. The net work input to the *combined* machines is then zero, because the work produced by the first machine is used to run the second.

Now, let's examine what happens to heat in this situation. Because engine 1 is more efficient than engine 2 and because the work is the same for both, engine 1 must withdraw less heat from the hot reservoir at T_h than is discharged into the same reservoir by engine 2; there is a net transfer of heat *into* the high-temperature reservoir. For the combined engines, $\Delta U_{tot} = w_{tot} = 0$; so q_{tot} must also be zero, and a net transfer of heat *out of* the low-temperature reservoir must therefore occur. By this reasoning we have devised an apparatus that can transfer heat from a low-temperature to a high-temperature reservoir with no net expenditure of work.

But that is impossible. All our experience shows we cannot make a device that transfers heat from a cold body to a hot body without doing work. In fact, exactly the opposite is seen: Heat flows spontaneously from hotter to colder bodies. Our experience is summarized and generalized in the following statement:

> *It is impossible to construct a device that will transfer heat from a cold reservoir to a hot reservoir in a continuous cycle with no net expenditure of work.*

This is one form of the second law of thermodynamics, as stated by Rudolf Clausius.

13.8.4 Further Discussion of Efficiency

Our assumption led to a conclusion that contradicts experience. Therefore, the original assumption must have been wrong, and there *cannot* be two reversible engines operating between the same two temperatures with different efficiencies. That is, all Carnot engines must have the same efficiency, which is

$$\epsilon = \frac{T_h - T_l}{T_h}$$

for the ideal gas. For any substance, ideal or not, undergoing a Carnot cycle,

$$\epsilon = \frac{-w_{net}}{q_h} = \frac{q_h + q_l}{q_h} = \frac{T_h - T_l}{T_h}$$

The Carnot cycle forms the basis for a *thermodynamic* scale of temperature. Because $\epsilon = 1 - (T_l/T_h)$, the Carnot efficiencies determine temperature ratios and thereby establish a temperature scale. The difficulty of operating real engines close to the reversible limit makes this definition of temperature scale impractical. Instead, real gases at low pressures are used to define and determine temperatures (see Section 9.2).

The last two terms of the preceding equation can be rewritten as

$$1 + \frac{q_l}{q_h} = 1 - \frac{T_l}{T_h}$$

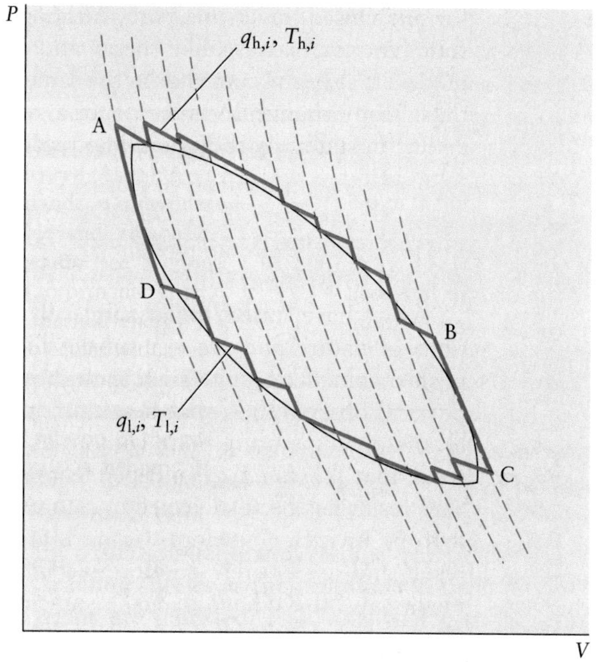

FIGURE 13.15 A general cyclic process (ABCDA) can be approximated to arbitrary accuracy by the sum of a series of Carnot cycles.

and simplified to

$$\frac{q_h}{T_h} + \frac{q_l}{T_l} = 0 \qquad [13.22]$$

This simple equation has profound importance because it contains the essence of the second law of thermodynamics, namely, that q_{rev}/T is a state function. To see this, consider a *general* reversible cyclic process and draw a series of closely spaced adiabats, as shown in Figure 13.15. (An adiabat is a curve on the PV diagram showing those thermodynamic states connected by a particular reversible adiabatic process.) Now replace each segment along the given cycle (ABCDA) with a series of alternating isothermal and adiabatic segments. Clearly, we can construct a path that is arbitrarily close to the desired curve by taking more and more closely spaced adiabats.

Now follow the evolution of $\Sigma_i(q_i/T_i)$ along this curve. The key observation is that the contributions q_i/T_i appear in pairs: For each $q_{h,i}/T_{h,i}$ from an isothermal segment along the ABC path, there is a $q_{l,i}/T_{l,i}$ along the CDA path. Any given pair forms two sides of a Carnot cycle, with the other two sides determined by adiabats; thus,

$$\frac{q_{h,i}}{T_{h,i}} + \frac{q_{l,i}}{T_{l,i}} = 0$$

Summing over i shows that

$$\sum_i \left(\frac{q_{h,i}}{T_{h,i}} + \frac{q_{l,i}}{T_{l,i}} \right) = 0$$

Because this summation follows the original reversible path to an arbitrary accuracy (if the lengths of the segments are made short enough), it follows that

$$\oint \frac{1}{T}\, dq_{rev} = 0$$

for *any* closed, reversible path. Although q is not a state function, q_{rev}/T is a state function because, like energy and enthalpy, its total change is zero for any process that begins and ends in the same state. From this result, Clausius *defined* the entropy change $\Delta S = S_f - S_i$ of a system in a process starting in state i and ending in state f by the equation

$$\Delta S = \int_i^f \frac{dq_{rev}}{T} \quad [13.23]$$

As we have stated, the principles of thermodynamics are based on observations of nature and are not subject to mathematical proof. We have accomplished something quite significant, however. From the assumption—based on physical observation—that heat cannot be transferred from a low-temperature to a high-temperature body without expenditure of work, we have derived the result that $\int(1/T)\,dq_{rev}$ is independent of path and, therefore, is a state function. This result has been subjected to rigorous testing, and no proper test has found it to be invalid. Therefore, we have great confidence in the generality of this result and are entirely comfortable in calling it a scientific law. Some critics might say, "If we have to make some assumption anyway, why don't we just assume that $\int(1/T)\,dq_{rev}$ is path independent to start with?" This is certainly possible and is practiced in many presentations of thermodynamics. Our approach is different; we prefer to base our assumptions directly on physical observation, not on abstract mathematical axioms.

CHAPTER SUMMARY

This chapter opened with the quest for methods of predicting whether a chemical reaction can occur spontaneously under a given set of conditions. The second law provides the answer: Any process is spontaneous under conditions where the total entropy of the *system and its surroundings* can increase during the process. For the particular case of constant temperature (T) and pressure (P)—the conditions most widely used for chemical reaction—the second law asserts that any process is spontaneous when the Gibbs free energy of the *system alone* can decrease during the process. The temperature dependence of the Gibbs free energy change shows that with knowledge of $\Delta H°$ and $\Delta S°$, we can identify the temperature range in which a given reaction is spontaneous under standard conditions. From the point of view of chemistry, predicting this temperature range is the most important result from the second law. All other material in this chapter can be viewed as preliminary background for arriving at this one crucial result.

The Gibbs free energy is the thermodynamic state function most naturally suited to describing the progress of chemical reactions at constant T and P. It provides the basis for predicting the equilibrium composition of the reaction mixture in Chapter 14.

CONCEPTS AND SKILLS

Section 13.1 – The Nature of Spontaneous Processes

Identify the system and surroundings involved in a spontaneous process and identify the constraint that was removed to enable the process to occur (Problems 1–2).

- Initially the system is separated from the surroundings by special walls (constraints) that prevent interaction between them. Adiabatic walls prevent flow of thermal energy (heat), rigid walls keep their volumes fixed, and impermeable walls prevent flow of matter. Removing one or more of the constraints allows the system and surroundings to exchange energy, matter, volume, or some combination of these. Then, a spontaneous process can occur if that process would increase the total entropy of the combined system and surroundings. Whether a spontaneous process does occur perceptibly depends on its rate. Thermodynamics cannot describe or predict the rates of processes.

Section 13.2 – Entropy and Spontaneity: A Molecular Statistical Interpretation

Provide a statistical interpretation of the change in entropy that occurs when a gas undergoes a change in volume (Problems 3–10).

- Increasing the volume increases the range of possible positions available to the molecules, thereby increasing Ω, the number of microstates available to the system. The entropy will increase according to Boltzmann's equation $S = k_B \ln\Omega$. Doubling the volume will change the entropy by the amount $\Delta S = N_A k_B \ln 2$. Similarly, decreasing the volume will decrease the entropy. Reducing the volume to half of the initial volume will change the entropy by the amount $\Delta S = N_A k_B \ln(0.5)$.

Section 13.3 – Entropy and Heat: Macroscopic Basis of the Second Law of Thermodynamics

Summarize the justification that entropy is a state function.

- By analyzing the Carnot cycle description of macroscopic energy transfer processes, Clausius demonstrated that the quantity $\int(1/T)dq_{rev}$ is a state function, because its change in value for any reversible process is independent of the path. Based on this result, Clausius defined the procedure for calculating the entropy change $\Delta S = S_f - S_i$ for a system between any thermodynamic states i and f $\Delta S = \int_i^f (1/T)\, dq_{rev}$. The integral can be evaluated along any reversible path between i and f so long as T and dq_{rev} are known along the path. Because S is a state function, this procedure is valid for any process—even an irreversible one—that connects states i and f.

Sections 13.4 and 13.5 – Entropy Changes in Reversible Processes and Entropy Changes and Spontaneity

Calculate the entropy change for the system and the surroundings for reversible and irreversible processes (Problems 11–20).

- Because entropy is a state function, changes in entropy are always calculated along a reversible path connecting the specified initial and final states. The details of the calculation depend on the conditions of the reversible path selected (constant T, constant P, or constant V). Specific cases are summarized as follows and in the list of Key Equations.
    - It is easy to calculate entropy changes for isothermal processes, because T is constant and comes outside the integral to give $\Delta S = q_{rev}/T$. A specific example is the isothermal compression or expansion of an ideal gas, for which $\Delta S = nR\ln(V_f/V_i)$. A second example is any phase transition at constant pressure and temperature where the phases are in equilibrium; for this case $q_{rev} = \Delta H_{trans}$. The entropy change is then $\Delta S_{trans} = \Delta H_{trans}/T_{trans}$.
    - When the temperature changes during a process that does not include a phase transition or a chemical reaction, we account for the variation of T along the process path by writing $dq_{rev} = nc_X dT$, where X represents V or P for a constant volume or a constant pressure process, respectively. When c_X is constant, the integral gives $\Delta S = nc_X \ln(T_f/T_i)$.

- Usually the surroundings are sufficiently large that they can be considered a constant temperature and constant pressure heat bath during the process. The heat lost by the system during the process is gained by the surroundings, so the entropy change for the surroundings is $\Delta S_{surr} = -\Delta H_{sys}/T_{surr}$ if the internal pressure of the system was constant. If the surroundings are not large enough to be treated in this way, ΔS_{surr} is calculated by the same procedures as ΔS_{sys}.
- To determine whether a process is spontaneous, we must calculate the total entropy change, $\Delta S_{tot} = \Delta S_{sys} + \Delta S_{surr}$. If $\Delta S_{tot} > 0$, the process is spontaneous. If $\Delta S_{tot} < 0$, the process cannot occur spontaneously. If $\Delta S_{tot} = 0$, the system is at equilibrium, and no process will occur.

Section 13.6 – The Third Law of Thermodynamics

Describe the measurements of absolute entropy, and calculate standard-state entropy changes for chemical reactions (Problems 21–28).

- The third law of thermodynamics states that the entropy of any pure substance in equilibrium approaches 0 as the temperature approaches absolute zero. Therefore the absolute entropy at any value of T can be obtained by measuring c_P and integrating the ratio c_P/T over this temperature range. The entropy changes for any phase transitions that occur in the temperature range must be included. Standard state values at $T = 298.15$ K are tabulated from results based on the equation

$$S^\circ = \int_0^{298.15} (c_P/T)\,dT + \Delta S \text{ (for all phase changes between 0 and 298.15 K)} \qquad [13.10]$$

- The standard state entropy change for a chemical reaction is calculated by summing the standard state entropy values for the products (multiplied by stoichiometric coefficients from the balanced equation for the reaction) and subtracting the standard state entropy values for the reactants (multiplied by stoichiometric coefficients).

Section 13.7 – The Gibbs Free Energy

Define the Gibbs free energy function and state the criterion it provides for the spontaneity of a process.

- When processes are conducted at fixed temperature and pressure, spontaneity is determined by changes in the Gibbs free energy, $G = H - TS$, for the system with no consideration of changes in the surroundings. If $\Delta G < 0$, the process is spontaneous. If $\Delta G > 0$, the process cannot occur spontaneously. If $\Delta G = 0$, the system is at equilibrium, and no process will occur.

Calculate the change in Gibbs free energy for reversible and spontaneous phase transition (Problems 29–32).

- This change is calculated using $\Delta G = \Delta H - T_{tr}\Delta S$ at the temperature T_{tr} of the transition. ΔH_{tr} is the measured enthalpy change of the transition occurring at constant pressure. The entropy change is calculated as $\Delta S_{tr} = \Delta H_{tr}/T_{tr}$ only if the temperature and pressure are such that the two phases are at equilibrium with each other. If they are not, then you must calculate ΔS from the sum of a few steps (since entropy is a state function). These terms are summed to obtain ΔG_{tr} for the transition.

Calculate the change in Gibbs free energy for chemical reactions and identify temperature ranges in which a particular reaction is spontaneous (Problems 33–38).

- Tabulations of the standard Gibbs free energy of formation for each substance in its standard state have been prepared by combining absolute entropy values with standard enthalpy of formation values. We can determine whether any chemical reaction is spontaneous by calculating its value of ΔG from the tabulated standard Gibbs free energy for its reactants and products. The algebraic sign of ΔG tells us whether the reaction is spontaneous. When the reactants and products are in their standard state partial pressures or concentrations, whether a reaction is spontaneous depends on the relation between $\Delta H°$ and $\Delta S°$ at the temperature of the reaction. If $\Delta H° < 0$ and $\Delta S° > 0$, the reaction is spontaneous at all temperatures. If $\Delta H° > 0$ and $\Delta S° < 0$, the reaction is never spontaneous. For other combinations, there is a special temperature $T^* = \Delta H°/\Delta S°$ at which $\Delta G° = 0$. If both $\Delta H°$ and $\Delta S°$ are positive, the reaction is spontaneous for temperatures greater than T^*. If both $\Delta H°$ and $\Delta S°$ are negative, the reaction is spontaneous for temperatures less than T^*.

KEY EQUATIONS

Sections 13.4 and 13.5—Entropy Changes in Reversible Processes. Entropy Changes and Spontaneity.

$$\Delta S = \int_{\mathrm{i}}^{\mathrm{f}} \frac{dq_{\mathrm{rev}}}{T} \qquad \text{(for all processes)}$$

$$\Delta S = \int_{\mathrm{i}}^{\mathrm{f}} \frac{dq_{\mathrm{rev}}}{T} = \frac{1}{T}\int_{\mathrm{i}}^{\mathrm{f}} dq_{\mathrm{rev}} = \frac{q_{\mathrm{rev}}}{T} \qquad \text{(constant } T\text{)}$$

$$\Delta S_{\mathrm{trans}} = \frac{q_{\mathrm{rev}}}{T_{\mathrm{trans}}} = \frac{\Delta H_{\mathrm{trans}}}{T_{\mathrm{trans}}} \qquad \text{(reversible phase transitions at constant } T \text{ and } P\text{)}$$

$$\Delta S = nR \ln\left(\frac{V_2}{V_1}\right) \qquad \text{(ideal gas, change of } V \text{ at constant } T\text{)}$$

$$\Delta S = \int_{T_1}^{T_2} \frac{nc_{\mathrm{V}}}{T}\,dT = nc_{\mathrm{V}} \ln\left(\frac{T_2}{T_1}\right) \qquad \text{(constant } V \text{ and constant } c_{\mathrm{V}}\text{) with no phase changes or chemical reactions}$$

$$\Delta S = \int_{T_1}^{T_2} \frac{nc_{\mathrm{P}}}{T}\,dT = nc_{\mathrm{P}} \ln\left(\frac{T_2}{T_1}\right) \qquad \text{(constant } P \text{ and constant } c_{\mathrm{P}}\text{) with no phase changes or chemical reactions}$$

$$\Delta S_{\mathrm{surr}} = \frac{-\Delta H_{\mathrm{sys}}}{T_{\mathrm{surr}}} \qquad \text{(surroundings are a large “heat bath” and pressure of the system is constant)}$$

$$\Delta S_{\mathrm{tot}} = \Delta S_{\mathrm{sys}} + \Delta S_{\mathrm{surr}} > 0 \qquad \text{(spontaneous process)}$$

CUMULATIVE EXERCISE

Purifying Nickel from Its Ores

Impure nickel, obtained from the smelting of its sulfide ores in a blast furnace, can be converted to metal of 99.90% to 99.99% purity by the Mond process, which relies on the equilibrium

$$\mathrm{Ni}(s) + 4\ \mathrm{CO}(g) \rightleftharpoons \mathrm{Ni(CO)_4}(g)$$

The standard enthalpy of formation of nickel tetracarbonyl, $Ni(CO)_4(g)$, is -602.9 kJ mol^{-1}, and its absolute entropy $S°$ is 410.6 J K^{-1} mol^{-1}.

(a) Predict (without referring to a table) whether the entropy change of the system (the reacting atoms and molecules) is positive or negative in this process.

(b) At a temperature where this reaction is spontaneous, predict whether the entropy change of the surroundings is positive or negative.

(c) Use the data in Appendix D to calculate $\Delta H°$ and $\Delta S°$ for this reaction.

(d) At what temperature is $\Delta G° = 0$ for this reaction?

(e) The first step in the Mond process is the equilibration of impure nickel with CO and $Ni(CO)_4$ at about 50°C. In this step, the goal is to draw as much nickel as possible into the vapor-phase complex. Calculate $\Delta G°$ for the preceding reaction at 50°C.

(f) In the second step of the Mond process, the gases are removed from the reaction chamber and heated to about 230°C. At high enough temperatures, the sign of $\Delta G°$ is reversed and the reaction occurs in the opposite direction, depositing pure nickel. In this step, the goal is to deposit as much nickel as possible from the vapor-phase complex. Calculate $\Delta G°$ for the preceding reaction at 230°C.

(g) The Mond process relies on the volatility of $Ni(CO)_4$ for its success. Under room conditions, this compound is a liquid, but it boils at 42.2°C with an enthalpy of vaporization of 29.0 kJ mol^{-1}. Calculate the entropy of vaporization of $Ni(CO)_4$, and compare it with that predicted by Trouton's rule.

(h) A recently developed variation of the Mond process carries out the first step at higher pressures and at a temperature of 150°C. Estimate the maximum pressure of $Ni(CO)_4(g)$ that can be attained before the gas will liquefy at this temperature (that is, calculate the vapor pressure of $Ni(CO)_4(\ell)$ at 150°C).

Answers

(a) Negative

(b) Positive

(c) $\Delta H° = -160.8$ kJ; $\Delta S° = -409.5$ J K^{-1}

(d) 392.7 K = 119.5°C

(e) $\Delta G° = -28.4$ kJ

(f) $\Delta G° = +46.0$ kJ

(g) 92.0 J K^{-1} mol^{-1}, close to the Trouton's rule value of 88 J K^{-1} mol^{-1}

(h) 16.7 atm

PROBLEMS

Answers to problems whose numbers are red appear in Appendix G. Problems that are more challenging are indicated with asterisks.

13.1 The Nature of Spontaneous Processes

1. For each of the following processes, identify the system and the surroundings. Identify those processes that are spontaneous. For each spontaneous process, identify the constraint that has been removed to enable the process to occur:
   (a) Ammonium nitrate dissolves in water.
   (b) Hydrogen and oxygen explode in a closed bomb.
   (c) A rubber band is rapidly extended by a hanging weight.
   (d) The gas in a chamber is slowly compressed by a weighted piston.
   (e) A glass shatters on the floor.

2. For each of the following processes, identify the system and the surroundings. Identify those processes that are spontaneous. For each spontaneous process, identify the constraint that has been removed to enable the process to occur:
   (a) A solution of hydrochloric acid is titrated with a solution of sodium hydroxide.
   (b) Zinc pellets dissolve in aqueous hydrochloric acid.
   (c) A rubber band is slowly extended by a hanging weight.
   (d) The gas in a chamber is rapidly compressed by a weighted piston.
   (e) A tray of water freezes in the freezing compartment of an electric refrigerator.

13.2 Entropy and Spontaneity: A Molecular Statistical Interpretation

3. (a) How many "microstates" are there for the numbers that come up on a pair of dice?
   (b) What is the probability that a roll of a pair of dice will show two sixes?

4. (a) Suppose a volume is divided into three equal parts. How many microstates can be written for all possible ways of distributing four molecules among the three parts?
   (b) What is the probability that all four molecules are in the leftmost third of the volume at the same time?

5. When $H_2O(\ell)$ and $D_2O(\ell)$ are mixed, the following reaction occurs spontaneously:

$$H_2O(\ell) + D_2O(\ell) \longrightarrow 2\ HOD(\ell)$$

There is little difference between the enthalpy of an O—H bond and that of an O—D bond. What is the main driving force for this reaction?

6. The two gases $BF_3(g)$ and $BCl_3(g)$ are mixed in equal molar amounts. All B—F bonds have about the same bond enthalpy, as do all B—Cl bonds. Explain why the mixture tends to react to form $BF_2Cl(g)$ and $BCl_2F(g)$.

7. Two large glass bulbs of identical volume are connected by means of a stopcock. One bulb initially contains 1.00 mol H_2; the other contains 1.00 mol helium (He). The stopcock is opened and the gases are allowed to mix and reach equilibrium. What is the probability that all the H_2 in the first bulb will diffuse into the second bulb and all the He gas in the second bulb will diffuse into the first bulb?

8. A mixture of 2.00 mol nitrogen and 1.00 mol oxygen is in thermal equilibrium in a 100-L container at 25°C. Calculate the probability that at a given time all the nitrogen will be found in the left half of the container and all the oxygen in the right half.

9. Predict the sign of the system's entropy change in each of the following processes.
   (a) Sodium chloride melts.
   (b) A building is demolished.
   (c) A volume of air is divided into three separate volumes of nitrogen, oxygen, and argon, each at the same pressure and temperature as the original air.

10. Predict the sign of the system's entropy change in each of the following processes.
   (a) A computer is constructed from iron, copper, carbon, silicon, gallium, and arsenic.
   (b) A container holding a compressed gas develops a leak and the gas enters the atmosphere.
   (c) Solid carbon dioxide (dry ice) sublimes to gaseous carbon dioxide.

13.4 Entropy Changes in Reversible Processes

11. Tungsten melts at 3410°C and has an enthalpy change of fusion of 35.4 kJ mol^{-1}. Calculate the entropy of fusion of tungsten.

12. Tetraphenylgermane, $(C_6H_5)_4Ge$, has a melting point of 232.5°C, and its enthalpy increases by 106.7 J g^{-1} during fusion. Calculate the molar enthalpy of fusion and molar entropy of fusion of tetraphenylgermane.

13. The normal boiling point of acetone is 56.2°C. Use Trouton's rule to estimate its molar enthalpy of vaporization.

14. The molar enthalpy of vaporization of liquid hydrogen chloride is 16.15 kJ mol^{-1}. Use Trouton's rule to estimate its normal boiling point.

15. If 4.00 mol hydrogen (c_P = 28.8 J K^{-1} mol^{-1}) is expanded reversibly and isothermally at 400 K from an initial volume of 12.0 L to a final volume of 30.0 L, calculate ΔU, q, w, ΔH, and ΔS for the gas.

16. Suppose 60.0 g hydrogen bromide, HBr(g), is heated reversibly from 300 to 500 K at a constant volume of 50.0 L, and then allowed to expand isothermally and reversibly until the original pressure is reached. Using $c_P(HBr(g))$ = 29.1 J K^{-1} mol^{-1}, calculate ΔU, q, w, ΔH, and ΔS for this process. Assume that HBr is an ideal gas under these conditions.

13.5 Entropy Changes and Spontaneity

17. Exactly 1 mol ice is heated reversibly at atmospheric pressure from −20°C to 0°C, melted reversibly at 0°C, and then heated reversibly at atmospheric pressure to 20°C. ΔH_{fus} = 6007 J mol^{-1}; c_P(ice) = 38 J K^{-1} mol^{-1}; and c_P(water) = 75 J K^{-1} mol^{-1}. Calculate ΔS for the system, the surroundings, and the thermodynamic universe for this process.

18. Suppose 1.00 mol water at 25°C is flash-evaporated by allowing it to fall into an iron crucible maintained at 150°C. Calculate ΔS for the water, ΔS for the iron crucible, and ΔS_{tot}, if $c_P(H_2O(\ell))$ = 75.4 J K^{-1} mol^{-1} and $c_P(H_2O(g))$ = 36.0 J K^{-1} mol^{-1}. Take ΔH_{vap} = 40.68 kJ mol^{-1} for water at its boiling point of 100°C.

19. In Example 12.3, a process was considered in which 72.4 g iron initially at 100.0°C was added to 100.0 g water initially at 10.0°C, and an equilibrium temperature of 16.5°C was reached. Take c_P(Fe) to be 25.1 J K^{-1} mol^{-1} and $c_P(H_2O)$ to be 75.3 J K^{-1} mol^{-1}, independent of temperature. Calculate ΔS for the iron, ΔS for the water, and ΔS_{tot} in this process.

20. Iron has a heat capacity of 25.1 J K^{-1} mol^{-1}, approximately independent of temperature between 0°C and 100°C.
   (a) Calculate the enthalpy and entropy change of 1.00 mol iron as it is cooled at atmospheric pressure from 100°C to 0°C.
   (b) A piece of iron weighing 55.85 g and at 100°C is placed in a large reservoir of water held at 0°C. It cools irreversibly until its temperature equals that of the water. Assuming the water reservoir is large enough that its temperature remains close to 0°C, calculate the entropy changes for the iron and the water and the total entropy change in this process.

13.6 The Third Law of Thermodynamics

21. **(a)** Use data from Appendix D to calculate the standard entropy change at 25°C for the reaction

$$N_2H_4(\ell) + 3\ O_2(g) \longrightarrow 2\ NO_2(g) + 2\ H_2O(\ell)$$

(b) Suppose the hydrazine (N_2H_4) is in the gaseous, rather than liquid, state. Will the entropy change for its reaction with oxygen be higher or lower than that calculated in part (a)? (*Hint:* Entropies of reaction can be added when chemical equations are added, in the same way that Hess's law allows enthalpies to be added.)

22. **(a)** Use data from Appendix D to calculate the standard entropy change at 25°C for the reaction

$$CH_3COOH(g) + NH_3(g) \longrightarrow CH_3NH_2(g) + CO_2(g) + H_2(g)$$

(b) Suppose that 1.00 mol each of solid acetamide, $CH_3CONH_2(s)$, and water, $H_2O(\ell)$, react to give the same products. Will the standard entropy change be larger or smaller than that calculated for the reaction in part (a)?

23. The alkali metals react with chlorine to give salts:

$$2\ Li(s) + Cl_2(g) \longrightarrow 2\ LiCl(s)$$
$$2\ Na(s) + Cl_2(g) \longrightarrow 2\ NaCl(s)$$
$$2\ K(s) + Cl_2(g) \longrightarrow 2\ KCl(s)$$
$$2\ Rb(s) + Cl_2(g) \longrightarrow 2\ RbCl(s)$$
$$2\ Cs(s) + Cl_2(g) \longrightarrow 2\ CsCl(s)$$

Using the data in Appendix D, compute $\Delta S°$ of each reaction and identify a periodic trend, if any.

24. All of the halogens react directly with $H_2(g)$ to give binary compounds. The reactions are

$$F_2(g) + H_2(g) \longrightarrow 2\ HF(g)$$
$$Cl_2(g) + H_2(g) \longrightarrow 2\ HCl(g)$$
$$Br_2(g) + H_2(g) \longrightarrow 2\ HBr(g)$$
$$I_2(g) + H_2(g) \longrightarrow 2\ HI(g)$$

Using the data in Appendix D, compute $\Delta S°$ of each reaction and identify a periodic trend, if any.

25. The dissolution of calcium chloride in water

$$CaCl_2(s) \longrightarrow Ca^{2+}(aq) + 2\ Cl^-(aq)$$

is a spontaneous process at 25°C, even though the standard entropy change of the preceding reaction is negative ($\Delta S° = -44.7$ J K^{-1}). What conclusion can you draw about the change in entropy of the surroundings in this process?

26. Quartz, $SiO_2(s)$, does not spontaneously decompose to silicon and oxygen at 25°C, 1.0 atm in the reaction

$$SiO_2(s) \longrightarrow Si(s) + O_2(g)$$

even though the standard entropy change of the reaction is large and positive ($\Delta S° = +182.02$ J K^{-1}). Explain.

27. Use the microscopic interpretation of entropy from Section 13.2 to explain why the entropy change of the system in Problem 26 is positive.

28. **(a)** Why is the entropy change of the system negative for the reaction in Problem 25, when the ions become dispersed through a large volume of solution? (*Hint:* Think about the role of the solvent, water.)
(b) Use Appendix D to calculate $\Delta S°$ for the corresponding dissolution of $CaF_2(s)$. Explain why this value is even more negative than that given in Problem 25.

13.7 The Gibbs Free Energy

29. The molar enthalpy of fusion of solid ammonia is 5.65 kJ mol^{-1}, and the molar entropy of fusion is 28.9 J K^{-1} mol^{-1}.
(a) Calculate the Gibbs free energy change for the melting of 1.00 mol ammonia at 170 K.
(b) Calculate the Gibbs free energy change for the conversion of 3.60 mol solid ammonia to liquid ammonia at 170 K.
(c) Will ammonia melt spontaneously at 170 K?
(d) At what temperature are solid and liquid ammonia in equilibrium at a pressure of 1 atm?

30. Solid tin exists in two forms: white and gray. For the transformation

$$Sn(s,\ white) \longrightarrow Sn(s,\ gray)$$

the enthalpy change is −2.1 kJ and the entropy change is −7.4 J K^{-1}.
(a) Calculate the Gibbs free energy change for the conversion of 1.00 mol white tin to gray tin at −30°C.
(b) Calculate the Gibbs free energy change for the conversion of 2.50 mol white tin to gray tin at −30°C.
(c) Will white tin convert spontaneously to gray tin at −30°C?
(d) At what temperature are white and gray tin in equilibrium at a pressure of 1 atm?

31. Ethanol's enthalpy of vaporization is 38.7 kJ mol^{-1} at its normal boiling point, 78°C. Calculate q, w, ΔU, ΔS_{sys}, and ΔG when 1.00 mol ethanol is vaporized reversibly at 78°C and 1 atm. Assume that the vapor is an ideal gas and neglect the volume of liquid ethanol relative to that of its vapor.

32. Suppose 1.00 mol superheated ice melts to liquid water at 25°C. Assume the specific heats of ice and liquid water have the same value and are independent of temperature. The enthalpy change for the melting of ice at 0°C is 6007 J mol^{-1}. Calculate ΔH, ΔS_{sys}, and ΔG for this process.

33. At 1200°C, the reduction of iron oxide to elemental iron and oxygen is not spontaneous:

$$2\ Fe_2O_3(s) \longrightarrow 4\ Fe(s) + 3\ O_2(g) \qquad \Delta G = +840\ kJ$$

Show how this process can be made to proceed if all the oxygen generated reacts with carbon:

$$C(s) + O_2(g) \longrightarrow CO_2(g) \quad \Delta G = -400\ kJ$$

This observation is the basis for the smelting of iron ore with coke to extract metallic iron.

34. The primary medium for free energy storage in living cells is adenosine triphosphate (ATP). Its formation from adenosine diphosphate (ADP) is not spontaneous:

$$ADP^{3-}(aq) + HPO_4^{2-}(aq) + H^+(aq) \longrightarrow ATP^{4-}(aq) + H_2O(\ell) \qquad \Delta G = +34.5 \text{ kJ}$$

Cells couple ATP production with the metabolism of glucose (a sugar):

$$C_6H_{12}O_6(aq) + 6\ O_2(g) \longrightarrow 6\ CO_2(g) + 6\ H_2O(\ell) \qquad \Delta G = -2872 \text{ kJ}$$

The reaction of 1 molecule of glucose leads to the formation of 38 molecules of ATP from ADP. Show how the coupling makes this reaction spontaneous. What fraction of the free energy released in the oxidation of glucose is stored in the ATP?

35. A process at constant T and P can be described as spontaneous if $\Delta G < 0$ and nonspontaneous if $\Delta G > 0$. Over what range of temperatures is each of the following processes spontaneous? Assume that all gases are at a pressure of 1 atm. (*Hint:* Use Appendix D to calculate ΔH and ΔS [assumed independent of temperature and equal to $\Delta H°$ and $\Delta S°$, respectively], and then use the definition of ΔG.)

(a) The rusting of iron, a complex reaction that can be approximated as

$$4\ Fe(s) + 3\ O_2(g) \longrightarrow 2\ Fe_2O_3(s)$$

(b) The preparation of $SO_3(g)$ from $SO_2(g)$, a step in the manufacture of sulfuric acid:

$$SO_2(g) + \tfrac{1}{2}\ O_2(g) \longrightarrow SO_3(g)$$

(c) The production of the anesthetic dinitrogen oxide through the decomposition of ammonium nitrate:

$$NH_4NO_3(s) \longrightarrow N_2O(g) + 2\ H_2O(g)$$

36. Follow the same procedure used in Problem 35 to determine the range of temperatures over which each of the following processes is spontaneous.

(a) The preparation of the poisonous gas phosgene:

$$CO(g) + Cl_2(g) \longrightarrow COCl_2(g)$$

(b) The laboratory-scale production of oxygen from the decomposition of potassium chlorate:

$$2\ KClO_3(s) \longrightarrow 2\ KCl(s) + 3\ O_2(g)$$

(c) The reduction of iron(II) oxide (wüstite) by coke (carbon), a step in the production of iron in a blast furnace:

$$FeO(s) + C(s,\ gr) \longrightarrow Fe(s) + CO(g)$$

37. Explain how it is possible to reduce tungsten(VI) oxide (WO_3) to metal with hydrogen at an elevated temperature. Over what temperature range is this reaction spontaneous? Use the data of Appendix D.

38. Tungsten(VI) oxide can also be reduced to tungsten by heating it with carbon in an electric furnace:

$$2\ WO_3(s) + 3\ C(s) \longrightarrow 2\ W(s) + 3\ CO_2(g)$$

(a) Calculate the standard free energy change ($\Delta G°$) for this reaction, and comment on the feasibility of the process at room conditions.

(b) What must be done to make the process thermodynamically feasible, assuming ΔH and ΔS are nearly independent of temperature?

13.8 A Deeper Look . . . Carnot Cycles, Efficiency, and Entropy

39. A thermodynamic engine operates cyclically and reversibly between two temperature reservoirs, absorbing heat from the high-temperature bath at 450 K and discharging heat to the low-temperature bath at 300 K.

(a) What is the thermodynamic efficiency of the engine?

(b) How much heat is discarded to the low-temperature bath if 1500 J of heat is absorbed from the high-temperature bath during each cycle?

(c) How much work does the engine perform in one cycle of operation?

40. In each cycle of its operation, a thermal engine absorbs 1000 J of heat from a large heat reservoir at 400 K and discharges heat to another large heat sink at 300 K. Calculate:

(a) The thermodynamic efficiency of the heat engine, operated reversibly

(b) The quantity of heat discharged to the low-temperature sink each cycle

(c) The maximum amount of work the engine can perform each cycle

ADDITIONAL PROBLEMS

41. Ethanol (CH_3CH_2OH) has a normal boiling point of 78.4°C and a molar enthalpy of vaporization of 38.74 kJ mol^{-1}. Calculate the molar entropy of vaporization of ethanol and compare it with the prediction of Trouton's rule.

42. A quantity of ice is mixed with a quantity of hot water in a sealed, rigid, insulated container. The insulation prevents heat exchange between the ice–water mixture and the surroundings. The contents of the container soon reach equilibrium. State whether the total *internal energy* of the contents decreases, remains the same, or increases in this process. Make a similar statement about the total *entropy* of the contents. Explain your answers.

43. **(a)** If 2.60 mol $O_2(g)$ (c_P = 29.4 J K^{-1} mol^{-1}) is compressed reversibly and adiabatically from an initial pressure of 1.00 atm and 300 K to a final pressure of 8.00 atm, calculate ΔS for the gas.

(b) Suppose a different path from that in part (a) is used. The gas is first heated at constant pressure to the same final temperature, and then compressed reversibly and isothermally to the same final pressure. Calculate ΔS for this path and show that it is equal to that found in part (a).

44. One mole of a monatomic ideal gas begins in a state with P = 1.00 atm and T = 300 K. It is expanded reversibly and adiabatically until the volume has doubled; then it is expanded irreversibly and isothermally into a vacuum until the volume has doubled again; and then it is heated reversibly at constant volume to 400 K. Finally, it is compressed reversibly and isothermally until a final state with P = 1.00 atm and T = 400 K is reached. Calculate ΔS_{sys} for this process. (*Hint:* There are two ways to solve this problem—an easy way and a hard way.)

***45.** The motion of air masses through the atmosphere can be approximated as adiabatic (because air is a poor conductor of heat) and reversible (because pressure differences in the atmosphere are small). To a good approximation, air can be treated as an ideal gas with average molar mass 29 g mol^{-1} and average heat capacity 29 J K^{-1} mol^{-1}.

(a) Show that the displacement of the air masses occurs at constant entropy ($\Delta S = 0$).

(b) Suppose the average atmospheric pressure near the earth's surface is P_0 and the temperature is T_0. The air is displaced upward until its temperature is T and its pressure is P. Determine the relation between P and T. (*Hint:* Consider the process as occurring in two steps: first a cooling from T_0 to T at constant pressure, and then an expansion from P_0 to P at constant temperature. Equate the sum of the two entropy changes to $\Delta S_{tot} = 0$.)

(c) In the lower atmosphere, the dependence of pressure on height, h, above the earth's surface can be approximated as

$$\ln (P/P_0) = -\mathcal{M}gh/RT$$

where $\mathcal{M}$ is the molar mass (kg mol^{-1}), g the acceleration due to gravity (9.8 m s^{-2}), and R the gas constant. If the air temperature at sea level near the equator is 38°C (~100°F), calculate the air temperature at the summit of Mount Kilimanjaro, 5.9 km above sea level. (For further discussion of this problem, see L. K. Nash, *J. Chem. Educ.* 61:23, 1984.)

46. Calculate the entropy change that results from mixing 54.0 g water at 273 K with 27.0 g water at 373 K in a vessel whose walls are perfectly insulated from the surroundings. Consider the specific heat of water to be constant over the temperature range from 273 to 373 K and to have the value 4.18 J K^{-1} g^{-1}.

47. Problem 20 asked for the entropy change when a piece of iron is cooled by immersion in a reservoir of water at 0°C.

(a) Repeat Problem 20(b), supposing that the iron is cooled to 50°C in a large water reservoir held at that temperature before being placed in the 0°C reservoir.

(b) Repeat the calculation supposing that four water reservoirs at 75°C, 50°C, 25°C, and 0°C are used.

(c) As more reservoirs are used, what happens to ΔS for the iron, for the water, and for the universe? How would you attempt to conduct a reversible cooling of the iron?

48. Problem 42 in Chapter 9 described an optical atomic trap. In one experiment, a gas of 500 sodium atoms is confined in a volume of 1000 μm^3. The temperature of the system is 0.00024 K. Compute the probability that, by chance, these 500 slowly moving sodium atoms will all congregate in the left half of the available volume. Express your answer in scientific notation.

***49.** Suppose we have several different ideal gases, i = 1, 2, 3,..., N, each occupying its own volume V_i, all at the same pressure and temperature. The boundaries between the volumes are removed so that the gases mix at constant temperature in the total volume $V = \Sigma_i V_i$.

(a) Using the microscopic interpretation of entropy, show that

$$\Delta S = -nR\sum_i X_i \ln X_i$$

for this process, where n is the total number of moles of gas and X_i is the mole fraction of gas i.

(b) Calculate the entropy change when 50 g each of $O_2(g)$, $N_2(g)$, and Ar(g) are mixed at 1 atm and 0°C.

(c) Using Table 9.1, calculate the entropy change when 100 L of air (assumed to be a mixture of ideal gases) at 1 atm and 25°C is separated into its component gases at the same pressure and temperature.

***50.** The N_2O molecule has the structure N—N—O. In an ordered crystal of N_2O, the molecules are lined up in a regular fashion, with the orientation of each determined by its position in the crystal. In a random crystal (formed on rapid freezing), each molecule has two equally likely orientations.

(a) Calculate the number of microstates available to a random crystal of N_A (Avogadro's number) of molecules.

(b) Calculate the entropy change when 1.00 mol of a random crystal is converted to an ordered crystal.

51. By examining the following graphs, predict which element—copper or gold—has the higher absolute entropy at a temperature of 200 K.

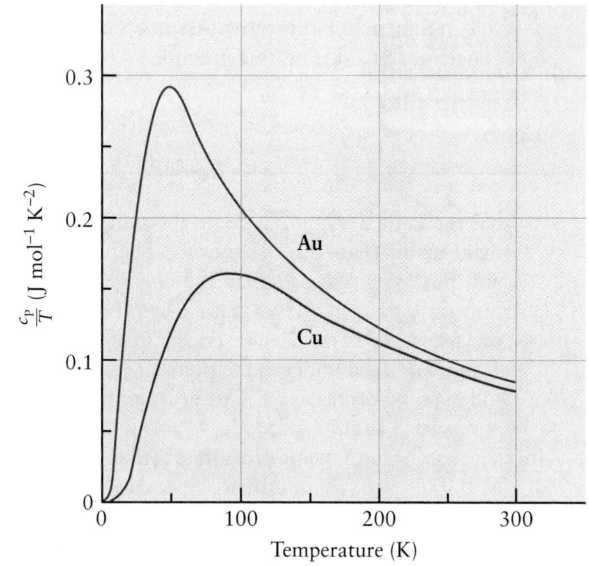

***52.** Consider the process described in Problem 19 in Chapter 12. Use the results from that problem to do the following.

(a) Calculate ΔS for the system, the surroundings, and the universe.

(b) If the absolute entropy per mole of the gas *before* the expansion is 158.2 J K^{-1} mol^{-1}, calculate ΔG_{sys} for the process.

53. Two different crystalline forms of sulfur are the rhombic form and the monoclinic form. At atmospheric pressure, rhombic sulfur undergoes a transition to monoclinic when it is heated above 368.5 K:

$$S(s, \text{rhombic}) \longrightarrow S(s, \text{monoclinic})$$

(a) What is the sign of the entropy change (ΔS) for this transition?

(b) $|\Delta H|$ for this transition is 400 J mol^{-1}. Calculate ΔS for the transition.

***54.** Use data from Appendix D to estimate the temperature at which $I_2(g)$ and $I_2(s)$ are in equilibrium at a pressure of 1 atm. Can this equilibrium actually be achieved? Refer to Appendix F for data on iodine.

55. The molar enthalpy of fusion of ice at 0°C is 6.02 kJ mol^{-1}; the molar heat capacity of undercooled water is 75.3 J mol^{-1} K^{-1}. (a) One mole of undercooled water at −10°C is induced to crystallize in a heat-insulated vessel. The result is a mixture of ice and water at 0°C. What fraction of this mixture is ice? (b) Calculate ΔS for the system.

56. A certain substance consists of two modifications A and B; $\Delta G°$ for the transition from A to B is positive. The two modifications produce the same vapor. Which has the higher vapor pressure? Which is the more soluble in a solvent common to both?

57. From the values in Appendix D, calculate the values of $\Delta G°$ and $\Delta H°$ for the reaction.

$$3\ Fe_2O_3(s) \longrightarrow 2\ Fe_3O_4(s) + \tfrac{1}{2}\ O_2(g)$$

at 25°C. Which of the two oxides is more stable at 25°C and $P_{O_2} = 1$ atm?

58. The strongest known chemical bond is that in carbon monoxide, CO, with bond enthalpy of 1.05×10^3 kJ mol^{-1}. Furthermore, the entropy increase in a gaseous dissociation of the kind $AB \rightleftharpoons A + B$ is about 110 J mol^{-1} K^{-1}. These factors establish a temperature above which there is essentially no chemistry of molecules. Show why this is so, and find the temperature.

59. The ΔG_f° of $Si_3N_4(s)$ is −642.6 kJ mol^{-1}. Use this fact and the data in Appendix D to compute $\Delta G°$ of the reaction.

$$3\ CO_2(g) + Si_3N_4(s) \longrightarrow 3\ SiO_2(\text{quartz}) + 2\ N_2(g) + 3\ C(s, \text{gr})$$

60. The compound $Pt(NH_3)_2I_2$ comes in two forms, the cis and the trans, which differ in their molecular structure. The following data are available:

	ΔH_f° (kJ mol^{-1})	ΔG_f° (kJ mol^{-1})
cis	−286.56	−130.25
trans	−316.94	−161.50

Combine these data with data from Appendix D to compute the standard entropies ($S°$) of both of these compounds at 25°C.

61. **(a)** Use data from Appendix D to calculate $\Delta H°$ and $\Delta S°$ at 25°C for the reaction

$$2\ CuCl_2(s) \rightleftharpoons 2\ CuCl(s) + Cl_2(g)$$

(b) Calculate $\Delta G°$ at 590 K, assuming $\Delta H°$ and $\Delta S°$ are independent of temperature.

(c) Careful high-temperature measurements show that when this reaction is performed at 590 K, ΔH_{590}° is 158.36 kJ and ΔS_{590}° is 177.74 J K^{-1}. Use these facts to compute an improved value of ΔG_{590}° for this reaction. Determine the percentage error in ΔG_{590}° that comes from using the 298-K values in place of 590-K values in this case.

62. **(a)** The normal boiling point of carbon tetrachloride (CCl_4) is 76.5°C. A student looks up the standard enthalpies of formation of $CCl_4(\ell)$ and of $CCl_4(g)$ in Appendix D. They are listed as −135.44 and −102.9 kJ mol^{-1}, respectively. By subtracting the first from the second, she computes $\Delta H°$ for the vaporization of CCl_4 to be 32.5 kJ mol^{-1}. But Table 12.2 states that the ΔH_{vap} of CCl_4 is 30.0 kJ mol^{-1}. Explain the discrepancy.

(b) Calculate the molar entropy change of vaporization (ΔS_{vap}) of CCl_4 at 76.5°C.

63. The typical potassium ion concentration in the fluid outside a cell is 0.0050 M, whereas that inside a muscle cell is 0.15 M.

(a) What is the spontaneous direction of motion of ions through the cell wall?

(b) In *active transport,* cells use free energy stored in ATP (see Problem 34) to move ions in the direction opposite their spontaneous direction of flow. Calculate the cost in free energy to move 1.00 mol K^+ through the cell wall by active transport. Assume no change in K^+ concentrations during this process.

CUMULATIVE PROBLEMS

64. When a gas undergoes a reversible adiabatic expansion, its entropy remains constant even though the volume increases. Explain how this can be consistent with the microscopic interpretation of entropy developed in Section 13.2. (*Hint:* Consider what happens to the distribution of velocities in the gas.)

65. The normal boiling point of liquid ammonia is 240 K; the enthalpy of vaporization at that temperature is 23.4 kJ mol^{-1}. The heat capacity of gaseous ammonia at constant pressure is 38 J mol^{-1} K^{-1}.

(a) Calculate q, w, ΔH, and ΔU for the following change in state:

$$2.00 \text{ mol } NH_3(\ell, 1 \text{ atm}, 240 \text{ K}) \longrightarrow 2.00 \text{ mol } NH_3(g, 1 \text{ atm}, 298 \text{ K})$$

Assume that the gas behaves ideally and that the volume occupied by the liquid is negligible.

(b) Calculate the entropy of vaporization of NH_3 at 240 K.

66. (a) Using the standard molar entropies for gaseous and liquid ethanol in Appendix D, calculate ΔH and ΔS for the reaction

$$C_2H_5OH(g, 1 \text{ atm}, 25°C) \longrightarrow C_2H_5OH(\ell, 1 \text{ atm}, @\ 25°C)$$

Then, treating the gas as ideal, consider whether the reaction

$$C_2H_5OH(g, 0.05 \text{ atm}, 25°C) \longrightarrow C_2H_5OH(\ell, 0.05 \text{ atm}, 25°C)$$

has the same ΔH and ΔS. To do this, you might wish to make an alternate three-step process between the initial and final states to calculate the change in state functions at 0.05 atm. (Try a path where you first change the pressure of the gas reversibly from 0.05 atm to 1 atm, and then undergo the phase transition at 1 atm, followed by decreasing the pressure of the liquid from 1 atm back to 0.05 atm.)

(b) The equilibrium vapor pressure of ethanol at 298.15 K is 0.078 atm. Will ethanol gas at 0.05 atm and 298.15 K spontaneously condense to liquid ethanol? First answer this by thinking about it qualitatively. Then support it with a calculation. Use the change in entropy of the system you calculated above, and calculate the change in entropy of the surroundings, and the change in entropy of the universe for the condensation of one mole of ethanol gas at 0.05 atm and 298.15 K to the liquid at 0.05 atm and 298.15 K. Give all three numerical answers and state whether the second law of thermodynamics predicts that the gas will condense spontaneously. Assume that for the liquid, but not the gas, that neither S nor H change appreciably as you change the pressure of the liquid (within this pressure range) while holding the temperature constant.

14 CHAPTER

Chemical Equilibrium

Gaseous ammonia, NH_3, and gaseous hydrogen chloride, HCl, react to form solid NH_4Cl, the white "smoke." In the reverse reaction, solid NH_4Cl decomposes when heated to form gaseous NH_3 and HCl.

Every time we carry out a chemical reaction—from fundamental research studies to practical industrial applications—the *yield* of the reaction is extremely important. Did we obtain all the product we could expect? Chapter 2 shows how to calculate the amount of product expected when we start a reaction with particular amounts of the reactants. This calculation assumes that the reaction goes to completion—that is, all of the limiting reagent is consumed. The resulting number, called the theoretical yield, represents the maximum amount of product that could be obtained from that reaction.

In practice, many reactions do not go to completion but rather approach a state or position of **equilibrium**. This equilibrium position, at which the reaction apparently comes to an end, is a mixture of products and unconsumed reactants present in fixed relative amounts. Once equilibrium has been achieved, there is no further net conversion of reactants to products unless the experimental conditions of the reaction (temperature and pressure) are changed. The equilibrium state is characterized by the **equilibrium constant**, which has a unique value for each reaction. Knowing the equilibrium constant and the initial amounts of reactants and products, we can calculate the composition of the equilibrium reaction mixture. Knowing the equilibrium constant and its dependence on experimental conditions, we can manipulate conditions to maximize the practical yield of that reaction. Calculating the equilibrium composition for a particular reaction and its dependence on experimental conditions is therefore a practical skill of enormous importance in chemistry.

In this chapter we define the equilibrium constant, its dependence on conditions, and its role in manipulating the yield of reactions. We illustrate these general principles with applications to reactions in the gas phase and to heterogeneous reactions. Detailed applications to reactions in aqueous solutions and electrochemical reactions are presented in the three following chapters.

The fact that reactions go to the equilibrium position was discovered empirically, and the equilibrium constant was first defined empirically. All the aforementioned applications can be accomplished with empirically determined equilibrium constants. Nonetheless, the empirical approach leaves unanswered several important fundamental questions: Why should the equilibrium state exist? Why does the equilibrium constant take its particular mathematical form? These and related questions are answered by recognizing that the chemical equilibrium position is the *thermodynamic equilibrium state* of the reaction mixture. Once we have made that connection, thermodynamics explains the existence and the mathematical form of the equilibrium constant. Thermodynamics also gives procedures for calculating the value of the equilibrium constant from the thermochemical properties of the pure reactants and products, as well as procedures for predicting its dependence on experimental conditions.

Some instructors prefer to introduce equilibrium from the empirical viewpoint and later use thermodynamics to explain the empirical developments. Others prefer to develop the background of thermodynamics first and then apply it to chemical equilibrium. We have organized this chapter to allow either approach. After an introductory section on the general nature of chemical equilibrium, we define the equilibrium constant empirically in Section 14.2, and then give a thermodynamic description in Section 14.3. Similarly, we discuss the direction of change in chemical reactions empirically in Section 14.6 and give a thermodynamic treatment in Section 14.7. Readers who have studied thermodynamics before starting this chapter should read the sections in the order presented. Readers who have not yet studied thermodynamics should skip over Sections 14.3 and 14.7 and return to them after studying Chapters 12 and 13 on thermodynamics. We provide signposts for both sets of readers at the end of each section.

14.1 The Nature of Chemical Equilibrium

14.1.1 Approach to Equilibrium

Most chemical reactions are carried out by mixing the selected reactants in a vessel and adjusting T and P until the desired products appear. Once reaction conditions have been identified, the greatest concern is to maximize the yield of the reaction.

Let's start by observing a particular reaction that illustrates all the aspects of equilibrium we need to understand. We have selected the reactions of cobalt(II) ions in aqueous solutions in which chloride ion is also present because the progress and outcome of the reactions are directly visible. The cobalt(II) ions can form

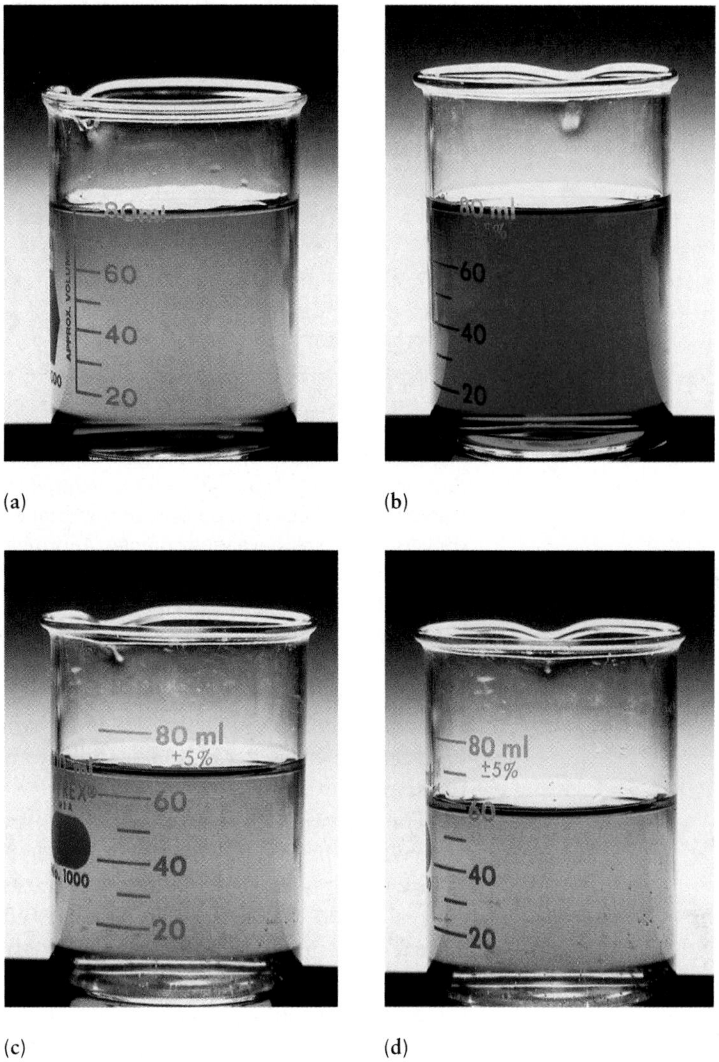

FIGURE 14.1 Chemical equilibrium in the cobalt chloride–HCl system. (a) The pink color is due to the hexaaqua complex ion $[Co(H_2O)_6]^{2+}$. (b) The blue color is due to the tetrachloro complex ion $[CoCl_4]^{2-}$. (c) Adding HCl to the pink solution in (a) converts some of the Co(II) to the tetrachloro complex. The lavender color of a sample of the resulting solution shown in (c) is produced by the combination of pink hexaaqua species and blue tetrachloro species. (d) Adding water to the blue solution in (b) converts some of the Co(II) to the hexaaqua species. The combination of the two gives the lavender color. The same equilibrium state is reached by running the reaction from the left (c) and from the right (d).

various different complex ions, depending on the amount of chloride present. For example, if $CoCl_2{\cdot}6H_2O$ is dissolved in pure water to the concentration 0.08 M, the resulting solution is pale pink in color due to the hexaaquacobalt(II) complex ion $[Co(H_2O)_6]^{2+}$ (Fig. 14.1a). If $CoCl_2{\cdot}6H_2O$ is dissolved in 10 M HCl to the concentration 0.08 M, the solution is deep blue due to the tetrachlorocobalt(II) complex ion $[CoCl_4]^{2-}$ (see Fig. 14.1b). Solutions containing a mixture of both Co(II) species are lavender, as shown in Figure 14.1c, d.

The hexaaqua complex can be converted into the tetrachloro complex by reaction with chloride ion:

$$\underset{\text{A}}{[Co(H_2O)_6]^{2+}} + \underset{\text{B}}{4\ Cl^-} \rightleftharpoons \underset{\text{C}}{[CoCl_4]^{2-}} + \underset{\text{D}}{6\ H_2O}$$

If we add concentrated HCl (reactant B, on the left side of the equation) to the pink solution in Figure 14.1a until the Co(II) concentration is 0.044 M and the HCl concentration is 5.5 M, the result is a lavender-colored solution, *a sample of which* is shown in Figure 14.1c. Optical absorption spectroscopy measurements to be described in Chapter 20 confirm the presence of both Co(II) species in the solution in Figure 14.1c. Ninety-eight percent of the Co is found in the pink hexaaqua complex, and the remaining two percent is in the blue

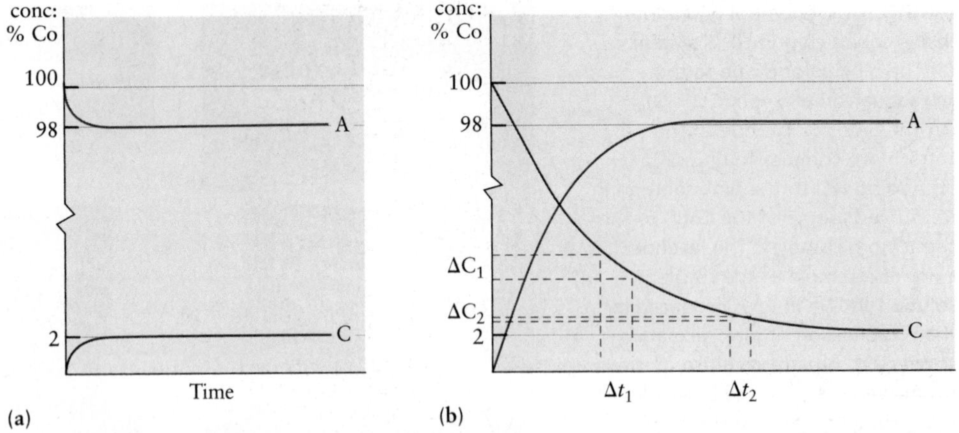

FIGURE 14.2 Sketch of the change with time of the concentrations of products and reactants in the spontaneous reactions illustrated in Figure 14.1. For ease of display, concentrations are expressed as percent of the total Co(II) present in each species. (a) Partial conversion of pink hexaaqua complex A into blue tetrachloro complex C. (b) Partial conversion of blue tetrachloro complex C into pink hexaaqua complex A. After changes in the slope of each species, concentrations become imperceptibly small, and we say the reaction has arrived at chemical equilibrium.

tetrachloro complex.[1] If the lavender solution is allowed to stand at constant temperature for several hours and the measurements repeated, the results are the same.

The reaction can also be carried out from "right to left" as written. If we add pure water (product D, on the right side of the equation) to the blue solution in Figure 14.1b until the cobalt concentration is 0.044 M and the HCl concentration is 5.5 M, the result is a lavender-colored solution, *a sample of which* is shown in Figure 14.1d. Just as we saw after running the reaction "left to right," optical absorption spectroscopy confirms that 98% of the Co(II) is present in the pink hexaaqua complex and 2% is present in the blue tetrachloro complex. Again, we find no further change in composition of the mixture, even after a long wait.

These data show that the reaction has not gone to completion but has apparently halted at an intermediate state containing products as well as unconsumed reactants. Moreover, the same final state can be achieved from either direction. This result is typical of most chemical reactions, even though it is not generally as visually apparent as here.

The questions raised in the first paragraph require quantitative investigations of the reaction mixture, which we carry out as follows. In the first experiment, we start the reaction by mixing initial concentrations of A and B, denoted as $[A]_0$ and $[B]_0$. As the reaction proceeds, we periodically sample the reaction mixture. For each sample, we measure the concentration of A, B, and C and plot concentration of each species versus time. The results of the first experiment are represented schematically in Figure 14.2a, which shows the consumption of A and the production of C. Similarly, we start the second experiment with the initial concentration $[C]_0$, and add water. The results are represented schematically in Figure 14.2b, which shows the consumption of C and the production of A.

As the reaction proceeds, we see that the concentration of each species changes progressively more slowly. This fact is indicated by the decreasing

[1]It may appear surprising to find such imbalance in the concentration of the Co(II) species when the color intensities of the pink and lavender solutions in Figure 14.1a and Figure 14.1c appear quite similar. This difference is explained by the fact that the blue tetrachloro complex absorbs light much more efficiently than the pink hexaaqua complex.

absolute values (decreasing in magnitude) of the slope [$m = \Delta(\text{conc.})/\Delta t$] sketched on the concentration curves during progressively later time intervals. Eventually, the slopes become close enough to zero to show that the concentration of each species has become constant in time. When this condition has been achieved, we say the reaction is at *chemical equilibrium*, and the reaction mixture is at the *equilibrium composition*.[2] At later times, there is no further change in composition of the reaction mixture; the concentration of each of the species X = A, B, C remains at its equilibrium value denoted by $[X]_{eq}$.

14.1.2 Characteristics of the Equilibrium State

We can gain insight into the equilibrium state by comparing chemical reactions to the familiar phase equilibrium between liquid water and water vapor (see Section 10.4 and Fig. 10.17). Let's represent the transfer of water molecules in this phase equilibrium as a chemical equation:

$$H_2O(\ell) \rightleftharpoons H_2O(g)$$

The double arrows ($\rightleftharpoons$) emphasize the dynamic nature of phase equilibrium: Liquid water evaporates to form water vapor, and at the same time vapor condenses to give liquid. An analogous dynamic description applies to a chemical equilibrium, in which bonds are broken or formed as atoms move back and forth between reactant and product molecules. When the initial concentrations of the reactants are high, collisions between their molecules cause product molecules to form. Once the concentrations of the products have increased sufficiently, the reverse reaction (forming "reactants" from "products") begins to occur. As the equilibrium state is approached, the forward and backward rates of reaction become equal and there is no further net change in reactant or product concentrations. Just as the equilibrium between liquid water and water vapor is a dynamic process on the molecular scale, with evaporation and condensation taking place simultaneously, the chemical equilibrium between reactants and products also occurs through the continuous formation of molecules of product from reactant molecules and their reaction back into reactant molecules with equal rates. Chemical equilibrium is not a static condition, although macroscopic properties such as concentrations do stop changing when equilibrium is attained. *Chemical equilibrium is the consequence of a dynamic balance between forward and backward reactions.*

The experimental results shown in Figures 14.1 and 14.2 demonstrate that the same equilibrium state is reached whether one starts with the reactants or with the products. This fact can be used to test whether a system is truly in equilibrium or whether the reaction is just so slow that changes in concentration are unmeasurably small, even though the system is far from equilibrium. If the same state is reached starting from either reactants or products, that state is a true equilibrium state.

Equilibrium states have four fundamental characteristics:

1. They display no macroscopic evidence of change.
2. They are reached through spontaneous (in the sense of Chapter 13) processes.
3. They show a dynamic balance of forward and reverse processes.
4. They are the same regardless of direction of approach.

We frequently also encounter so-called *steady states* in which the macroscopic concentrations of species are not changing with time, even though the

[2]You must exercise judgment in deciding when the slope is "sufficiently close to zero" and the concentrations are "effectively constant." There is no one instant at which equilibrium is achieved.

system is not at equilibrium. Steady states are maintained not by a dynamic balance between forward and reverse processes, but rather by the competition between a process that supplies the species to the system and a process that removes the species from the system. Many chemical reactions occur in living systems in steady states and do not represent an equilibrium between reactants and products. You must be certain that a reaction is at equilibrium and not in steady state before applying the methods of this chapter to explain the relative concentrations of reactants and products.

14.2 The Empirical Law of Mass Action

Extensive studies of the type summarized in Figure 14.2 for broad classes of reactions represented generally as

$$aA + bB \rightleftharpoons cC + dD$$

have demonstrated a most remarkable result. No matter what initial concentrations of reactants are selected at the beginning of the experiment, the value of the ratio

$$\frac{[C]_{eq}^{c}[D]_{eq}^{d}}{[A]_{eq}^{a}[B]_{eq}^{b}}$$

measured *at equilibrium* is always the same. Even if the experiment is started with an arbitrary initial mixture of reactants and products, the reaction will consume some species and produce others until it achieves this same value of this ratio at equilibrium. This ratio is called the **empirical equilibrium constant** for the reaction and denoted as K_C. The results of these studies are summarized in the following equation called the **law of mass action**, first stated in approximate form in 1864 by two Norwegians, C. M. Guldberg (a mathematician) and his brother-in-law P. Waage (a chemist):

$$\frac{[C]_{eq}^{c}[D]_{eq}^{d}}{[A]_{eq}^{a}[B]_{eq}^{b}} = K_C \qquad [14.1a]$$

The subscript C denotes that the reaction is carried out in solution and that the empirical equilibrium constant K_C is evaluated by directly measuring the concentration of each species in the equilibrium state of the reaction. In general, K_C has dimensions (concentration)$^{c+d-a-b}$; it will be dimensionless only for those reactions for which $a + b = c + d$.

Similar results have been obtained for reactions carried out in the gas phase, where the amount of each reactant and product in the reaction mixture is measured by its partial pressure P_X. For gas-phase reactions, the empirical law of mass action takes the form

$$\frac{(P_C)_{eq}^{c}(P_D)_{eq}^{d}}{(P_A)_{eq}^{a}(P_B)_{eq}^{b}} = K_P \qquad [14.1b]$$

In general, K_P has dimensions (pressure)$^{c+d-a-b}$; it will be dimensionless only for those reactions for which $a + b = c + d$.

The significance of the empirical law of mass action is twofold. First, the numerical value of K_C or K_P is an inherent property of the chemical reaction itself and does not depend on the specific initial concentrations of reactants and products selected. Second, the magnitude of K_P or K_C gives direct information about the nature of the equilibrium state or position of the reaction. If the equilibrium constant is very large, then at equilibrium the concentration or partial pressures of products are large compared with those of the reactants. In this case, stoichiometry can be used to estimate the number of moles or the masses of product formed because the reaction is near completion. If the equilibrium constant is very small, the concentration or partial pressures of reactants are large compared with those for products, and the extent of reaction is very limited. If the equilibrium constant has a value close to 1, both reactants and products are present in significant proportions at equilibrium.

The law of mass action is the basis for equilibrium calculations, which pervade the science of chemistry. All follow the same general pattern: Suppose we know the numerical value of K_C or K_P for a reaction of interest, and suppose we run an experiment by mixing selected initial concentrations of reactants. We can then use Equation 14.1 to calculate the concentration of the products and reactants at equilibrium, without doing the laborious measurements described in Figure 14.2. In this and succeeding chapters in Unit 4, we demonstrate these calculations for reactions in the gas phase, for acid–base reactions involving ionic species in aqueous solutions, and for heterogeneous reactions occurring at the interface between different states of matter. In preparation for these calculations, we now show how to write the empirical law of mass action for these classes of reactions.

14.2.1 Law of Mass Action for Gas-Phase Reactions

Many reactions are conveniently carried out entirely in the gas phase. Molecules in the gas phase are highly mobile, and the collisions necessary for chemical reactions occur frequently. Two examples are the key steps in the production of hydrogen from the methane in natural gas. The first is the **reforming reaction:**

$$CH_4(g) + H_2O(g) \longrightarrow CO(g) + 3H_2(g)$$

The second step is called the **water gas shift reaction:**

$$CO(g) + H_2O(g) \longrightarrow CO_2(g) + H_2(g)$$

This process is currently the main industrial method for the preparation of hydrogen. Its success relies critically on application of the equilibrium principles developed in this chapter.

To write the mass action law, a first glance at Equation 14.1b suggests we need only examine the balanced equation for the reaction and insert the equilibrium partial pressure of each reactant and product into Equation 14.1b and raise it to a power equivalent to its stoichiometric coefficient in the balanced equation. A deeper study of equilibrium shows that instead of inserting just the partial pressure for each reactant or product, we must insert the value of the partial pressure *relative to* a specified reference pressure P_{ref}. The result is the following expression, denoted by K with no subscript:

$$\left[\frac{(P_C/P_{ref})^c(P_D/P_{ref})^d}{(P_A/P_{ref})^a(P_B/P_{ref})^b}\right]_{eq} = K$$

Note that K is dimensionless. Collecting the terms involving P_{ref} on the right side of the equation gives

$$\left[\frac{P_C^c P_D^d}{P_A^a P_B^b}\right]_{eq} = K(P_{ref})^{(c+d-a-b)}$$

which must be equal to K_P. It is customary to choose P_{ref} as atmospheric pressure, which can be expressed as $P_{ref} = 1$ atm, $P_{ref} = 760$ torr, or $P_{ref} = 101{,}325$ Pa. If all pressures are expressed in atmospheres, then $P_{ref} = 1$ atm, and the right side shows that K_P has the same *numerical value* as the dimensionless quantity K, with P_{ref} factors serving only to make the equation dimensionally correct. If some other unit is chosen for pressures, the P_{ref} factors no longer have a numerical value of unity and must be inserted explicitly into the equilibrium expression.

The dimensionless quantity K is the *thermodynamic equilibrium constant*, which can be calculated from tabulated data on the products and reactants, even if the empirical equilibrium constant defined in Equation 14.1b is not known (see Section 14.3). Therefore, K is the preferred tool for analyzing reaction equilibria in general. The informal procedure by which we replaced K_P with K is made rigorous by the thermodynamic treatment of equilibrium in Section 14.3. Meantime, we can freely use the result in advance of formal justification.

The convention we follow in this book is always to describe chemical equilibrium in terms of the thermodynamic equilibrium constant K, even when analyzing reactions empirically. We express all partial pressures in units of atmospheres, and choose $P_{ref} = 1$ atm for gas phase reactions. This choice allows us to use the dimensionless thermodynamic equilibrium constant because P_{ref} raised to any power equals one. The P_{ref} factors will not be explicitly included because their value is unity with these choices of pressure unit and reference pressure. Thus, we write the mass action law for a general reaction involving ideal gases as

$$\frac{(P_C)_{eq}^c (P_D)_{eq}^d}{(P_A)_{eq}^a (P_B)_{eq}^b} = K \qquad [14.2a]$$

with K dimensionless. The following example illustrates these practices.

EXAMPLE 14.1

Write equilibrium expressions for the following gas-phase chemical equilibria.

(a) $2\ NOCl(g) \rightleftharpoons 2\ NO(g) + Cl_2(g)$

(b) $CO(g) + \frac{1}{2} O_2(g) \rightleftharpoons CO_2(g)$

Solution

(a) $$\frac{(P_{NO})_{eq}^2 (P_{Cl_2})_{eq}}{(P_{NOCl})_{eq}^2} = K$$

The powers of 2 come from the factors of 2 in the balanced equation.

(b) $$\frac{(P_{CO_2})_{eq}}{(P_{CO})_{eq} (P_{O_2})_{eq}^{1/2}} = K$$

Fractional powers appear in the equilibrium expression whenever they are present in the balanced equation.

Related Problems: 1, 2, 3, 4

14.2.2 Law of Mass Action for Reactions in Solution

A great many reactions are carried out in a convenient solvent for reactants and products. Dissolved reactants can be rapidly mixed, and the reaction process is easily handled. Water is a specially favored solvent because its polar structure allows a broad range of polar and ionic species to be dissolved. Water itself is partially ionized in solution, liberating H^+ and OH^- ions that can participate in reactions with the dissolved species. This leads to the important subject of acid–base equilibria in aqueous solutions (see Chapter 15), which is based on the equilibrium principles developed in this chapter. We limit the discussion in this subsection to cases in which the solvent does not participate in the reaction.

The procedure for reactions in solution is the same as that for gas-phase reactions. A deeper study of equilibrium shows that we must insert the value of each concentration at equilibrium *relative to* a specified reference concentration c_{ref} in Equation 14.1a. The result is the following expression, denoted by K with no subscript:

$$\left[\frac{([C]/c_{ref})^c([D]/c_{ref})^d}{([A]/c_{ref})^a([B]/c_{ref})^b}\right]_{eq} = K$$

The square bracket [X] represents the concentration of species X in units of mol L^{-1}. If all factors containing c_{ref} are collected on the right side and the reference state for each reactant and product is defined to be an ideal solution with a concentration c_{ref} = 1 M, the dimensionless thermodynamic equilibrium constant K is numerically equal to K_C.

Just as with gaseous reactions, the convention we follow in this book is to describe solution equilibria in terms of the thermodynamic equilibrium constant K. Thus, we express solution concentrations in units of mol L^{-1} with the reference state as c_{ref} = 1 M, and we state values of K without dimensions. Thus, the mass action law for solution reactions becomes

$$\frac{[C]^c_{eq}[D]^d_{eq}}{[A]^a_{eq}[B]^b_{eq}} = K \qquad [14.2b]$$

The following example illustrates these practices.

EXAMPLE 14.2

Household laundry bleach is a solution of sodium hypochlorite (NaOCl) prepared by adding gaseous Cl_2 to a solution of sodium hydroxide:

$$Cl_2(aq) + 2\ OH^-(aq) \rightleftharpoons ClO^-(aq) + Cl^-(aq) + H_2O(\ell)$$

The active bleaching agent is the hypochlorite ion, which can decompose to chloride and chlorate ions in a side reaction that competes with bleaching:

$$3\ ClO^-(aq) \rightleftharpoons 2\ Cl^-(aq) + ClO_3^-(aq)$$

Write the equilibrium expression for the decomposition reaction.

Solution

$$\frac{[Cl^-]^2_{eq}[ClO_3^-]_{eq}}{[ClO^-]^3_{eq}} = K$$

The exponents come from the coefficients in the balanced chemical equation.

Related Problems: 7, 8

14.2.3 Law of Mass Action for Reactions Involving Pure Substances and Multiple Phases

A variety of equilibria occur in heterogeneous systems that involve pure solids and liquids as well as gases and dissolved species. Molecular species cross the interfaces between phases in order to participate in reactions. A whole class of examples (Chapter 16) is based on the dissolution of slightly soluble salts, where the dissolved ions are in equilibrium with the pure solid. Another class includes the reaction of pure metals with acids to produce hydrogen gas:

$$Zn(s) + 2\ H_3O^+(aq) \rightleftharpoons Zn^{2+}(aq) + H_2(g) + 2\ H_2O(\ell)$$

Yet another example shows how iron sulfide residues from mining operations introduce iron and sulfur as pollutants in water streams:

$$4\ FeS_2(s) + 15\ O_2(g) + 6\ H_2O(\ell) \rightleftharpoons 4\ Fe(OH)^{2+}(aq) + 8\ HSO_4^-(aq)$$

Reactions in aqueous solution sometimes involve water as a direct participant. In addition to familiar aqueous acid–base chemistry, many organic reactions fall into this class. One example is the hydrolysis of ethyl acetate to produce acetic acid and ethanol:

$$CH_3COOC_2H_5(aq) + H_2O(\ell) \rightleftharpoons CH_3COOH(aq) + C_2H_5OH(aq)$$

Several examples illustrate the procedures for writing the mass action law for heterogeneous reactions and for reactions involving pure solids or liquids.

1. Recall the phase equilibrium between liquid water and water vapor from Section 10.4:

$$H_2O(\ell) \rightleftharpoons H_2O(g) \qquad (P_{H_2O})_{eq} = K$$

Experiments show that as long as *some* liquid water is in the container, the pressure of water vapor at 25°C is 0.03126 atm. The position of this equilibrium is not affected by the amount of liquid water present, and therefore liquid water should not appear in the mass action law. Recall that for a gas or solute, a ratio of pressures or concentrations appears in the law of mass action. This ratio is equal to 1 when the gas or solute is in its reference state (1 atm or 1 M). For a pure liquid appearing in an equilibrium chemical equation, the convention is to take that pure liquid as the reference state, so the liquid water contributes only a factor of 1 to the equilibrium expression and can thus be omitted. We postpone justification of this rule to Section 14.3.

2. An analogous situation occurs in the equilibrium between solid iodine and iodine dissolved in aqueous solution:

$$I_2(s) \rightleftharpoons I_2(aq) \qquad [I_2(aq)]_{eq} = K$$

Experiment shows that the position of the equilibrium (given by the concentration of I_2 dissolved at a given temperature) is independent of the amount of solid present, as long as there is some. The pure solid iodine does not appear in the mass action law.

3. Another example involves the decomposition of calcium carbonate:

$$CaCO_3(s) \rightleftharpoons CaO(s) + CO_2(g) \qquad (P_{CO_2})_{eq} = K$$

If calcium carbonate is heated, it decomposes into calcium oxide (lime) and carbon dioxide; the reverse reaction is favored at sufficiently high pressures of carbon dioxide. The equilibrium can be studied experimentally, and it is found that at any given temperature the equilibrium pressure of $CO_2(g)$ is constant, independent of the amounts of $CaCO_3(s)$ and $CaO(s)$, as long as some of each is present (Fig. 14.3). The two pure solids do not appear in the mass action law, which reduces to the partial pressure of CO_2.

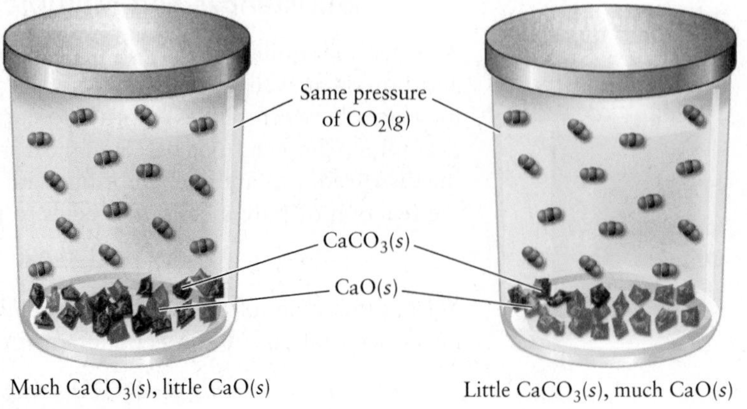

FIGURE 14.3 As long as both $CaO(s)$ and $CaCO_3(s)$ are present at equilibrium in a closed container, the partial pressure of $CO_2(g)$ at a fixed temperature does not depend on the amounts of the two solids present.

The general procedure for writing the mass action law for these more complex reactions is:

1. Gases enter the equilibrium expression as partial pressures, measured in atmospheres.
2. Dissolved species enter as concentrations, in moles per liter.
3. Pure solids and pure liquids do not appear in equilibrium expressions; neither does a solvent that participates in a chemical reaction, provided the solution is dilute.
4. Partial pressures and concentrations of products appear in the numerator, and those of reactants in the denominator; each is raised to a power equal to its coefficient in the balanced chemical equation for the reaction.

This procedure gives the dimensionless thermodynamic equilibrium constant *K* and the resulting expressions will be justified by the thermodynamic treatment of equilibrium in Section 14.3.

The following example illustrates these practices.

EXAMPLE 14.3

Hypochlorous acid (HOCl) is produced by bubbling chlorine through an agitated suspension of mercury(II) oxide in water. The chemical equation for this process is

$$2\ Cl_2(g) + 2\ HgO(s) + H_2O(\ell) \rightleftharpoons HgO \cdot HgCl_2(s) + 2\ HOCl(aq)$$

Write the equilibrium expression for this reaction.

Solution

$$\frac{[HOCl]_{eq}^2}{(P_{Cl_2}^2)_{eq}} = K$$

The HgO and $HgO \cdot HgCl_2$ do not appear because they are solids, and water does not appear because it is a pure liquid. Chlorine, as a gas, enters as its equilibrium partial pressure in atmospheres. The HOCl appears as its equilibrium concentration, in moles per liter. Both the concentration of HOCl and the partial pressure of Cl_2 are raised to the second power because their coefficients in the balanced chemical equation are 2.

Related Problems: 9, 10, 11, 12

The preceding discussion has specified the procedures for setting up the mass action law for broad classes of chemical reactions and are ready to be used in equilibrium calculations.

The empirical procedures leave unanswered numerous fundamental questions about chemical equilibrium. Why should the law of mass action exist in the first place, and why should it take the particular mathematical form shown here? Why should the equilibrium constant take a unique value for each individual chemical reaction? What factors determine that value? Why does the value of the equilibrium constant change slightly when studied over broad ranges of concentration? Why should the equilibrium constant depend on temperature? Is there a quantitative explanation for the temperature dependence?

All these questions are answered by the thermodynamic description of the equilibrium constant, provided in the next section. Readers who have already studied thermodynamics should continue to Section 14.3.

Readers who have not yet studied thermodynamics should go to Sections 14.4 and 14.5, which give detailed applications of equilibrium calculations based on the mass action law procedures just described. Sections 14.4 and 14.5 do not require background in thermodynamics.

14.3 Thermodynamic Description of the Equilibrium State

Thermodynamics views a chemical reaction as a process in which atoms "flow" from reactants to products. If the reaction is spontaneous and is carried out at constant T and P, thermodynamics requires that $\Delta G < 0$ for the process (see Section 13.7). Consequently, G always *decreases* during a spontaneous chemical reaction. When a chemical reaction has reached equilibrium, $\Delta G = 0$; that is, there is no further tendency for the reaction to occur in either the forward or the reverse direction. We will use the condition $\Delta G = 0$ in the following three subsections to develop the mass action law and the thermodynamic equilibrium constant for gaseous, solution, and heterogeneous reactions.

14.3.1 Reactions among Ideal Gases

As background, it is necessary to know how the Gibbs free energy changes with pressure at constant temperature, because in chemical equilibria, the partial pressures of gases can differ from 1 atm.

Dependence of Gibbs Free Energy of an Ideal Gas on Pressure

If the pressure of an ideal gas is changed from P_1 to P_2 with the temperature held constant, the free energy change is

$$\Delta G = \Delta(H - TS) = \Delta H - T\Delta S = -T\Delta S$$

The last equality is true because $\Delta H = 0$ when the pressure of an ideal gas is changed at constant temperature. The entropy change for an ideal gas in an isothermal process was calculated in Section 13.5:

$$\Delta S = nR \ln\left(\frac{V_2}{V_1}\right) = nR \ln\left(\frac{P_1}{P_2}\right) = -nR \ln\left(\frac{P_2}{P_1}\right)$$

so

$$\Delta G = nRT \ln\left(\frac{P_2}{P_1}\right) \qquad [14.3a]$$

Let us choose the value of P_1 to be 1 atm, which has already been defined as the standard state of gaseous substances for measurements both of enthalpy of formation (see Section 12.6) and Gibbs free energy of formation (see Section 13.7). Then, Equation 14.3a relates the Gibbs free energy of the gas at any pressure P_2 to its value at 1 atm. If we call 1 atm the *reference state* for the gas, then the change in Gibbs free energy in taking the gas isothermally from the reference state to any pressure P is given by

$$\Delta G = nRT \ln\left(\frac{P}{P_{\text{ref}}}\right) = nRT \ln P \qquad [14.3b]$$

Equation 14.3b is a shorthand version that can be used only when the pressure P is expressed in atm. The presence of P_{ref} in the denominator makes the argument of the natural logarithm function dimensionless. Choosing $P_{ref} = 1$ atm gives P_{ref} the numerical value 1, which for convenience we do not write explicitly. Nonetheless, you should always remember this (invisible) P_{ref} is required to make the equation dimensionally correct when the general pressure P in the equation is expressed in atm. If some unit of pressure other than atm is selected, P_{ref} no longer has value 1 and the P_{ref} selected must be carried explicitly in the equations.

The Equilibrium Expression for Reactions in the Gas Phase

Consider now a mixture of gases that react chemically, such as the NO, N_2O, and NO_2 given in Example 13.10:

$$3\ NO(g) \rightleftharpoons N_2O(g) + NO_2(g)$$

If all of the partial pressures are 1 atm, then ΔG for this reaction is just $\Delta G°$ at 25°C. If the pressures differ from 1 atm, ΔG must be calculated from a three-step process (Fig. 14.4). In step 1 the partial pressure of the reactant (in this case, 3 mol of NO) is changed from its initial value, P_{NO}, to the reference pressure $P_{ref} = 1$ atm:

$$\Delta G_1 = 3RT\ \ln\left(\frac{P_{ref}}{P_{NO}}\right) = RT\ \ln\left(\frac{P_{ref}}{P_{NO}}\right)^3$$

In step 2 the reaction is carried out with all reactants and products at partial pressures of $P_{ref} = 1$ atm:

$$\Delta G_2 = \Delta G°$$

In step 3 the partial pressures of the products (in this case, 1 mol of N_2O and 1 mol of NO_2) are changed from $P_{ref} = 1$ atm to P_{N_2O} and P_{NO_2}:

$$\Delta G_3 = RT\ \ln\left(\frac{P_{N_2O}}{P_{ref}}\right) + RT\ \ln\left(\frac{P_{NO_2}}{P_{ref}}\right) = RT\ \ln\left[\left(\frac{P_{N_2O}}{P_{ref}}\right)\left(\frac{P_{NO_2}}{P_{ref}}\right)\right]$$

The overall Gibbs free energy change ΔG for the reaction is the sum of the free energy changes for the three steps in the path:

$$\begin{aligned}\Delta G &= \Delta G_1 + \Delta G_2 + \Delta G_3 \\ &= \Delta G° + RT\ \ln\left[\frac{(P_{N_2O}/P_{ref})(P_{NO_2}/P_{ref})}{(P_{NO}/P_{ref})^3}\right]\end{aligned}$$

The ratio of partial pressures in square brackets above is called the reaction quotient Q; in it are the instantaneous partial pressures of the species involved

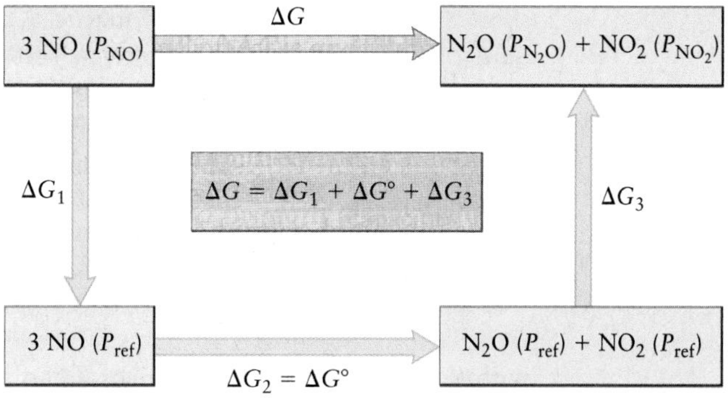

FIGURE 14.4 A three-step process (tan arrows) to calculate ΔG of a reaction (gray arrow) for which reactants and products are not in their standard states of 1 atm. This figure suggests that the reaction is run at pressure higher than 1 atm, which is true for most industrial processes. But, the procedure is valid at any pressure.

as the reaction runs spontaneously in the forward or the reverse direction. Thus, you can re-write the equation above as $\Delta G = \Delta G° + RT \ln Q$.

When a chemical reaction has reached equilibrium, $\Delta G = 0$ and Q is now equal to K. Under these conditions setting $\Delta G = 0$ and subtracting $\Delta G°$ from both sides of the preceding equation (where the subscript eq reminds you that the actual partial pressures have now reached equilibrium) gives

$$-\Delta G° = RT \ln\left[\frac{(P_{N_2O}/P_{ref})_{eq}(P_{NO_2}/P_{ref})_{eq}}{(P_{NO}/P_{ref})^3_{eq}}\right]$$

Because $\Delta G°$ depends only on temperature, the quantity $\Delta G°/RT$ must be a constant at each value of T. Therefore, in the last equation, the ratio of partial pressures inside the natural logarithm function must also be constant *at equilibrium* at each value of T. Consequently, this ratio of partial pressures is denoted by $K(T)$ and is called the *thermodynamic equilibrium constant* for the reaction. Finally, we have

$$-\Delta G° = RT \ln K(T) \qquad [14.4]$$

For the general reaction

$$aA + bB \longrightarrow cC + dD$$

the result, obtained in the same way, is

$$-\Delta G° = RT \ln\left[\frac{(P_C/P_{ref})^c_{eq}(P_D/P_{ref})^d_{eq}}{(P_A/P_{ref})^a_{eq}(P_B/P_{ref})^b_{eq}}\right] = RT \ln K(T) \qquad \text{(at equilibrium)}$$

The thermodynamic equilibrium constant for reactions involving ideal gases,

$$\left[\frac{(P_C/P_{ref})^c_{eq}(P_D/P_{ref})^d_{eq}}{(P_A/P_{ref})^a_{eq}(P_B/P_{ref})^b_{eq}}\right] = K$$

has the same form as the empirical law of mass action with K_P for gaseous reactions introduced in Section 14.2. It is common to omit the subscript eq in this expression since the context indicates whether the ratio is K or, if not at equilibrium, Q. The thermodynamic expression provides deeper understanding of that empirical result in three important ways. First, the law of mass action is seen to be a consequence of the reaction system being in thermodynamic equilibrium. Second, the thermodynamic equilibrium constant can be calculated from $\Delta G°$ (that is, from $\Delta H°$ and $\Delta S°$), so the *extent* of any equilibrium chemical reaction can be deduced from calorimetric data alone. Third, unlike K_P, the thermodynamic equilibrium constant K is always a dimensionless quantity because the pressure of a reactant or product always appears as a ratio to the reference pressure P_{ref}. The algebraic rearrangements leading to Equation 14.2a have already shown that K is numerically equal to K_P if we express all partial pressures in units of atm and select the reference state to be $P_{ref} = 1$ atm. Other choices of pressure units or reference states require the P_{ref} factors to be carried explicitly.

The convention we follow in this book, already stated in Section 14.2, is to describe chemical equilibrium in terms of the thermodynamic equilibrium constant K. The mass action law for a general reaction involving ideal gases is written as

$$\frac{(P_C)^c_{eq}(P_D)^d_{eq}}{(P_A)^a_{eq}(P_B)^b_{eq}} = K$$

with K dimensionless. The value of K can be calculated from Equation 14.4 using tables of data in Appendix D. The following example illustrates these practices.

EXAMPLE 14.4

The $\Delta G°$ of the chemical reaction

$$3\ NO(g) \longrightarrow N_2O(g) + NO_2(g)$$

was calculated in Example 13.10. Now calculate the equilibrium constant of this reaction at 25°C.

Solution

The standard free energy change for the conversion of 3 mol of NO to 1 mol of N_2O and 1 mol of NO_2 was found to be −104.18 kJ. To keep the units of the calculation correct, the $\Delta G°$ is rewritten as -104.18 kJ mol^{-1}, where "per mole" signifies "per mole of the reaction as it is written," that is, per 3 mol of NO, 1 mol of N_2O, and 1 mol of NO_2, the number of moles in the balanced equation. Substitution gives

$$\ln K = \frac{-\Delta G°}{RT}$$

$$= \frac{-(-104{,}180\ \text{J mol}^{-1})}{(8.315\ \text{J K}^{-1}\text{mol}^{-1})(298.15\ \text{K})} = 42.03$$

$$K = e^{42.03} = 1.8 \times 10^{18}$$

Related Problems: 13, 14

The conversion of $NO(g)$ to $N_2O(g)$ plus $NO_2(g)$ is spontaneous under standard conditions. The forward reaction under these conditions is scarcely observed because its rate is so slow. Nonetheless, its equilibrium constant can be calculated! Such calculations often have enormous impact in evaluating proposed solutions to practical problems. For example, the calculation shows that this reaction could be used to reduce the amount of NO in cooled exhaust gases from automobiles. The fundamental reaction tendency is there, but successful application requires finding a route to increasing the reaction rate at standard conditions. Had the equilibrium constant calculated from thermodynamics been small, this proposed application would be doomed at the outset and investment in it would not be justified.

14.3.2 Reactions in Ideal Solutions

The thermodynamic description of reactions in solution parallels that for ideal gas reactions. Although the result is not derived here, the Gibbs free energy change for n mol of a solute, as an ideal (dilute) solution changes in concentration from c_1 to c_2 mol L^{-1}, is

$$\Delta G = nRT \ln\left(\frac{c_2}{c_1}\right) \qquad \textbf{[14.5a]}$$

If the reference state for the solute is defined to be an ideal solution with a concentration $c_{\text{ref}} = 1$ M, the change in Gibbs free energy for taking the solution from the reference state to the concentration c is given by

$$\Delta G = nRT \ln\left(\frac{c}{c_{\text{ref}}}\right) = nRT \ln c \qquad \textbf{[14.5b]}$$

The overall change in Gibbs free energy for a reaction in solution is

$$\Delta G = \Delta G° \ + RT \ln\left[\frac{([C]/c_{\text{ref}})^c([D]/c_{\text{ref}})^d}{([A]/c_{\text{ref}})^a([B]/c_{\text{ref}})^b}\right]$$

where the square bracket [X] represents the concentration of species X in units of mol L^{-1}. Again you can re-write the equation above as $\Delta G = \Delta G° + RT \ln Q$, where Q is the ratio of the actual instantaneous concentrations of the species involved as the reaction runs spontaneously in the forward or the reverse direction. When the reaction arrives at equilibrium, $\Delta G = 0$, and Q is now equal to K.

Setting $\Delta G = 0$ and subtracting $\Delta G°$ from both sides gives an expression relating the equilibrium constant K to $\Delta G°$:

$$-\Delta G° = RT \ln K \qquad [14.6]$$

for reactions involving dissolved species.

In this book we describe solution equilibria in terms of the thermodynamic equilibrium constant K rather than the empirical K_C introduced in Section 14.2. Thus, we give solution concentrations in units of mol L^{-1} with the reference state as $c_{ref} = 1$ M, and state values of K as dimensionless quantities. The mass action law for solution reactions becomes

$$\left[\frac{[C]^c[D]^d}{[A]^a[B]^b}\right]_{eq} = K$$

and is numerically equivalent to K_C. (As we are now in equilibrium, each concentration is understood to be an equilibrium concentration, even though we will often omit the subscript eq in later Chapters.) When working with this expression, you must keep in mind the role of the (invisible) c_{ref} factors in making the equation dimensionally correct. This expression is the foundation for the discussions of acid–base equilibria in aqueous solutions in the next chapter.

Solution-phase equilibrium constants can be calculated from tables of standard free energies for solutes in aqueous solution at 25°C (see Appendix D). The procedure is demonstrated by the following example.

EXAMPLE 14.5

Calculate $\Delta G°$ and the equilibrium constant at 25°C for the chemical reaction

$$3\ ClO^-(aq) \rightleftharpoons 2\ Cl^-(aq) + ClO_3^-(aq)$$

whose equilibrium expression we developed in Example 14.2.

Solution

We calculate the change in standard Gibbs free energy using the ΔG_f° values for these aqueous ions in their standard states tabulated in Appendix D:

$$\begin{aligned}\Delta G° &= 2\ \Delta G_f^\circ(Cl^-(aq)) + \Delta G_f^\circ(ClO_3^-(aq)) - 3\ \Delta G_f^\circ(ClO^-(aq)) \\ &= (2\text{ mol})(-131.23\text{ kJ mol}^{-1}) + (1\text{ mol})(-7.95\text{ kJ mol}^{-1}) - \\ &\quad (3\text{ mol})(-36.8\text{ kJ mol}^{-1}) \\ &= -160.01\text{ kJ}\end{aligned}$$

As in Example 14.4, we calculate K by inserting the value for $\Delta G°$ into the equation

$$\ln K = \frac{-\Delta G°}{RT} = \frac{-(-160{,}010\text{ J mol}^{-1})}{(8.315\text{ J K}^{-1}\text{mol}^{-1})(298.15\text{ K})} = 64.54$$

$$K = e^{64.54} = 1.1 \times 10^{28}$$

The decomposition of hypochlorite ion in bleach is thus a spontaneous process at even very small concentrations of reactants and large concentrations of products, with a very large equilibrium constant at 25°C. But, this reaction is very slow at room temperature, so the laundry bleach solution remains stable. If the temperature is raised to about 75°C, the reaction occurs rapidly and the bleach solution decomposes.

14.3.3 Reactions Involving Pure Solids and Liquids and Multiple Phases: The Concept of Activity

The mass action law for homogeneous reactions in ideal gases and ideal solutions is written in Section 14.2 by straightforward inspection of the balanced equation for the reaction under study. If one or more of the reactants or products is a solid or liquid in its pure state, the procedure is less obvious, because "concentration" has no meaning for a pure species. This apparent difficulty is resolved by the concept of **activity**, which is a convenient means for comparing the properties of a substance in a general thermodynamic state with its properties in a specially selected reference state. A full treatment of equilibrium in terms of activity requires thermodynamic results beyond the scope of this book. Here we sketch the essential ideas leading to the more general form of the mass action law and merely state the range of validity of the idealized expressions presented.

The activity concept arises from the dependence of the Gibbs free energy on the pressure of a pure substance or on the composition of a solution, regardless of the phase of the system. The discussion just before Equation 14.3 shows that the change in Gibbs free energy when an ideal gas is taken from a reference state P_{ref} to any pressure P is given by

$$\Delta G = nRT \ln\left(\frac{P}{P_{ref}}\right) = nRT \ln P$$

The last form is used only when pressure P is expressed in atmospheres and $P_{ref} = 1$ atm. A similar equation can be developed for more complex systems—and keep the same simple mathematical form—if we *define* the activity a by the equation

$$\Delta G = nRT \ln a \qquad [14.7]$$

Of course if a system is already in its reference state, then $\Delta G = 0$ and the activity in this state is 1. The change in Gibbs free energy in taking a system from the reference state to any general thermodynamic state determines the activity a in the general state.

The activity is connected to pressure or to concentration by the **activity coefficient**. The *activity coefficient* γ_i of a nonideal gaseous species at pressure P_i is defined by the equation

$$a_i = \frac{\gamma_i P_i}{P_{ref}} \qquad [14.8a]$$

The reference state is chosen to be an ideal gas at one atmosphere. Comparing Equations 14.7 and 14.8a with Equation 14.3 shows that γ_i takes the value 1 for ideal gases. The activity of an ideal gas is the ratio of its pressure to the selected reference pressure. If pressures are given in units of atmospheres, then the activity of an ideal gas is numerically equal to its pressure. Similarly, the activity coefficient γ_i for solute i in a solution at concentration c_i is defined by the equation

$$a_i = \frac{\gamma_i c_i}{c_{ref}} \qquad [14.8b]$$

The activity coefficient γ_i equals 1 in the reference state, selected as an ideal solution with convenient concentration. For solute species in the dilute solutions considered in this book, concentration is most conveniently expressed in molarity, and the reference state is selected to be an ideal solution at concentration $c_{ref} = 1$ M.

The reference states for pure solids and liquids are chosen to be those forms stable at 1 atm, just as in the definition of standard states for enthalpy of formation (see Chapter 12) and Gibbs free energy of formation (see Chapter 13). Pure substances in their reference states are assigned activity of value 1.

Once reference states have been defined, the activity coefficients γ_i can be determined from experimental *P-V-T* and calorimetric data by procedures that are not described in this book. Then, Equation 14.7 can be written explicitly in terms of pressure, temperature, and concentrations when needed for specific calculations. In its present version, Equation 14.7 is especially well suited for general discussions, because it summarizes much complicated information about nonideal systems in a simple and compact form.

Starting from Equation 14.7, arguments similar to those preceding Equation 14.4 lead to the thermodynamic equilibrium constant *K* regardless of the phase of each product or reactant:

$$\left[\frac{a_C^c \cdot a_D^d}{a_A^a \cdot a_B^b}\right]_{eq} = K \qquad [14.9]$$

Substituting the appropriate ideal expression for the activity of gaseous or dissolved species from Equation 14.8a or 14.8b leads to the forms of the mass action law and the equilibrium constant *K* already derived earlier in Section 14.3 for reactions in ideal gases or in ideal solutions. We write the mass action law for reactions involving pure solids and liquids and multiple phases by substituting unity for the activity of pure liquids or solids and the appropriate ideal expression for the activity of each gaseous or dissolved species into Equation 14.9. Once a proper reference state and concentration units have been identified for *each* reactant and product, we use tabulated free energies based on these reference states to calculate the equilibrium constant.

The following example illustrates the method.

EXAMPLE 14.6

The compound urea, important in biochemistry, can be prepared in aqueous solution by the following reaction:

$$CO_2(g) + 2\ NH_3(g) \rightleftharpoons CO(NH_2)_2(aq) + H_2O(\ell)$$

(a) Write the mass action law for this reaction. **(b)** Calculate $\Delta G°$ for this reaction at 25°C. **(c)** Calculate *K* for this reaction at 25°C.

Solution

(a) In terms of activities for each species, the equilibrium expression is

$$\left[\frac{a_{urea} \cdot a_{H_2O}}{a_{CO_2} \cdot a_{NH_3}^2}\right]_{eq} = K$$

Choose the reference state for the gases to be $P_{ref} = 1$ atm, for the urea to be $c_{ref} = 1$ M, and for water to be the pure liquid with unit activity. Then at low pressures and concentrations, the limiting idealized mass action law is

$$\left[\frac{[urea]}{P_{CO_2} \cdot P_{NH_3}^2}\right]_{eq} = K$$

with [urea] in units of mol L^{-1} and the partial pressures in atm. The (invisible) P_{ref} and c_{ref} factors must always be kept in mind, because they make *K* dimensionless in this expression. This expression can be extended to higher pressures and concentrations, where nonideality becomes important, by inserting the appropriate activity coefficients.

(b) The procedure is the same as in Examples 14.4 and 14.5. The ΔG_f° values for gaseous CO_2 and NH_3 and liquid H_2O with the reference states just specified are obtained directly

(Continued)

from Appendix D. Because the data for urea in Appendix D are based on a *solid* reference state, we consult an alternate source[3] to obtain ΔG_f° (urea(aq)) = −203.84 kJ mol^{-1} with the reference state an ideal solution with c = 1 M. From these data we obtain

$$\Delta G^\circ = -203.84 - 237.18 + 394.36 + 2(16.48)$$

$$= -13.70 \text{ kJ mol}^{-1}$$

(c) Following the procedure in Example 14.5 we find K = 251.1.

This example illustrates a very important point: Before using ΔG_f° values to estimate equilibrium constants, *be certain* you know the reference state for the tabulated values.

Related Problems: 15, 16

Equilibrium expressions written in this way are accurate to about 5% when the pressures of gases do not exceed several atmospheres and the concentrations of solutes do not exceed 0.1 M. These cases cover all the applications discussed in this book, so we do not need correction factors. For concentrated ionic solutions, the corrections can become very large. In accurate studies of solution equilibria, especially in biochemical applications, activities must be used in place of partial pressures or concentrations. You should consult more advanced books when you need these techniques.

14.4 The Law of Mass Action for Related and Simultaneous Equilibria

Each time we write the mass action law, it is based on a specific balanced chemical equation in which the reaction is carried out as written "left to right." Chemical reactions can be carried out in various alternative ways, including "in reverse" and in concert with other reactions. These variations lead to relationships among equilibrium expressions, which are best illustrated in a series of examples.

14.4.1 Relationships among Equilibrium Expressions

(1) Suppose a reaction is written in two opposing directions:

$$2\ H_2(g) + O_2(g) \rightleftharpoons 2\ H_2O(g) \qquad \left[\frac{P_{H_2O}^2}{P_{H_2}^2 P_{O_2}}\right]_{eq} = K_1$$

$$2\ H_2O(g) \rightleftharpoons 2\ H_2(g) + O_2(g) \qquad \left[\frac{P_{H_2}^2 P_{O_2}}{P_{H_2O}^2}\right]_{eq} = K_2$$

In the first reaction, hydrogen and oxygen combine to form water vapor, whereas in the second, water vapor dissociates into hydrogen and oxygen. The equilibrium constants K_1 and K_2 are clearly the inverse of each other, so their product K_1K_2 = 1. This is true quite generally: The equilibrium constant for a reverse reaction is the reciprocal of the equilibrium constant for the corresponding forward reaction.

(2) What happens if a balanced chemical equation is multiplied by a constant? In the preceding example, multiplying the first equation by $\frac{1}{2}$ gives

$$H_2(g) + \tfrac{1}{2}\ O_2(g) \rightleftharpoons H_2O(g)$$

[3]Thermodynamic data for many biological molecules in solution are given in F. H. Carpenter, *J. Am. Chem. Soc., 82,* 1120 (1960).

This is a perfectly satisfactory way to write the equation for the chemical reaction; it says that 1 mol of hydrogen reacts with $\frac{1}{2}$ mol of oxygen to yield 1 mol of water vapor. The corresponding equilibrium expression is

$$\left[\frac{P_{H_2O}}{P_{H_2}P_{O_2}^{1/2}}\right]_{eq} = K_3$$

Comparison with the expression for K_1 shows that

$$K_3 = K_1^{1/2}$$

When a balanced chemical equation is multiplied by a constant factor, the corresponding equilibrium constant is raised to a power equal to that factor.

(3) A further variation is to add two equations to give a third. In this case the equilibrium constant for the third equation is the product of the equilibrium constants for the first two. For example, at 25°C,

$$2\ BrCl(g) \rightleftharpoons Cl_2(g) + Br_2(g) \qquad \left[\frac{P_{Cl_2}P_{Br_2}}{P_{BrCl}^2}\right]_{eq} = K_1 = 0.45$$

$$Br_2(g) + I_2(g) \rightleftharpoons 2\ IBr(g) \qquad \left[\frac{P_{IBr}^2}{P_{Br_2}P_{I_2}}\right]_{eq} = K_2 = 0.051$$

Adding the two chemical equations gives

$$2\ BrCl(g) + Br_2(g) + I_2(g) \rightleftharpoons 2\ IBr(g) + Cl_2(g) + Br_2(g)$$

Removing $Br_2(g)$ from both sides leaves

$$2\ BrCl(g) + I_2(g) \rightleftharpoons 2\ IBr(g) + Cl_2(g) \qquad \left[\frac{P_{IBr}^2 P_{Cl_2}}{P_{BrCl}^2 P_{I_2}}\right]_{eq} = K_3$$

and, by inspection, $K_3 = K_1K_2 = (0.45)(0.051) = 0.023$.

(4) If a second equation is *subtracted* from the first, the resulting equilibrium constant is that of the first *divided* by that of the second (subtracting a reaction is the same as adding the reverse reaction). The operations of addition and subtraction applied to chemical equations transform into multiplication and division of the equilibrium expressions and equilibrium constants.

EXAMPLE 14.7

The concentrations of the oxides of nitrogen are monitored in air-pollution reports. At 25°C, the equilibrium constant for the reaction

$$NO(g) + \tfrac{1}{2}O_2(g) \rightleftharpoons NO_2(g)$$

is

$$\left[\frac{P_{NO_2}}{P_{NO}P_{O_2}^{1/2}}\right]_{eq} = K_1 = 1.3 \times 10^6$$

and that for

$$\tfrac{1}{2}N_2(g) + \tfrac{1}{2}O_2(g) \longrightarrow NO(g)$$

is

$$\left[\frac{P_{NO}}{P_{N_2}^{1/2}P_{O_2}^{1/2}}\right]_{eq} = K_2 = 6.5 \times 10^{-16}$$

Find the equilibrium constant K_3 for the reaction

$$N_2(g) + 2\ O_2(g) \rightleftharpoons 2\ NO_2(g)$$

Solution

Adding the chemical equations for the first two reactions gives

$$\tfrac{1}{2}N_2(g) + O_2(g) \rightleftharpoons NO_2(g)$$

The equilibrium constant for this reaction, K_3', is just the product of K_1 and K_2, or K_1K_2. The constant sought, K_3, is defined by a chemical equation that is twice this, so K_3' must be raised to the power 2 (that is, squared) to give K_3:

$$K_3 = (K_3')^2 = (K_1K_2)^2 = 7.1 \times 10^{-19}$$

Related Problems: 17, 18, 19, 20

14.4.2 Consecutive Equilibria

Many real-world applications of chemistry and biochemistry involve fairly complex sets of reactions occurring in sequence and/or in parallel. Each of these individual reactions is governed by its own equilibrium constant. To describe the overall progress of the entire coupled set of reactions we write all the involved equilibrium expressions and treat them as a set of *simultaneous algebraic equations,* because the concentrations of various chemical species appear in several expressions in the set. Examination of relative values of equilibrium constants shows that some reactions dominate the overall coupled set of reactions, and this chemical insight enables mathematical simplifications in the simultaneous equations. We study coupled equilibria in considerable detail in Chapter 15 on acid–base equilibrium. In Section 14.6 we provide a brief introduction to this topic in the context of an important biological process.

14.5 Equilibrium Calculations for Gas-Phase and Heterogeneous Reactions

Equilibrium calculations, used throughout the science of chemistry in both basic and applied work, involve specific procedures. The present section presents these problem-solving techniques in the context of gas-phase and heterogeneous reactions, but they are applicable in *all* equilibrium calculations. The strategy comprises three basic steps:

1. Always start by writing the balanced equation for the reaction being studied.
2. Visualize the reaction as proceeding through three steps:
   a. the reactants are brought together at their *initial partial pressures,* but not yet allowed to react;
   b. the reaction is initiated and causes *changes in partial pressures* of all products and reactants;
   c. the reaction reaches equilibrium, and all products and reactants are present at their *equilibrium partial pressures.* (Note the similarity to the procedures described in Chapter 13 for initiating spontaneous processes by manipulating thermodynamic constraints.)
3. Develop approximation schemes when possible by neglecting a very small quantity when it is added to or subtracted from a much larger quantity.

We illustrate these calculation methods with several examples that fall into two broad classes: evaluating the equilibrium constant from reaction data, and calculating the amounts of products and reactants present at equilibrium when the equilibrium constant is known.

14.5.1 Evaluating Equilibrium Constants from Reaction Data

In Section 14.3 we showed how to evaluate K from calorimetric data on the pure reactants and products. Occasionally, these thermodynamic data may not be available for a specific reaction, or a quick estimate of the value of K may suffice. In these cases we can evaluate the equilibrium constant from measurements made directly on the reaction mixture. If we can measure the equilibrium partial pressures of all the reactants and products, we can evaluate the equilibrium constant by substituting these measured values (in units of atmospheres) into the equilibrium expression. In many cases it is not practical to measure directly the equilibrium partial pressure of each separate reactant and product. Nonetheless, the equilibrium constant can usually be derived from other available data, although the determination is less direct. We illustrate the method in the following two examples.

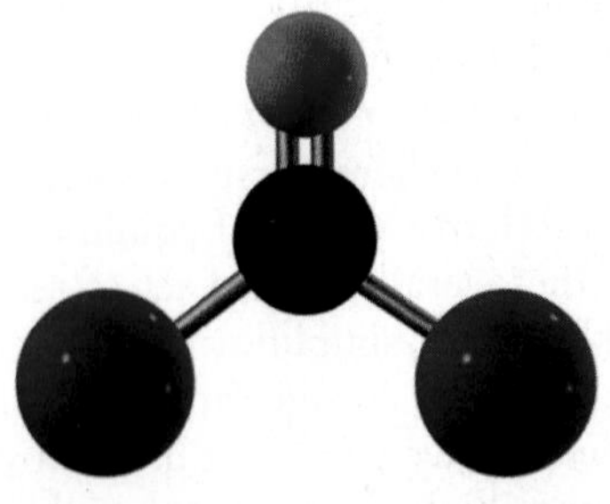

FIGURE 14.5 Phosgene, $COCl_2$, is a chemical intermediate used in making polyurethanes for foams and surface coatings.

EXAMPLE 14.8

Phosgene, $COCl_2$ (Fig. 14.5), forms from CO and Cl_2 according to the equilibrium

$$CO(g) + Cl_2(g) \rightleftharpoons COCl_2(g)$$

At 600°C, a gas mixture of CO and Cl_2 is prepared that has initial partial pressures (before reaction) of 0.60 atm for CO and 1.10 atm for Cl_2. After the reaction mixture has reached equilibrium, the partial pressure of $COCl_2(g)$ at this temperature is measured to be 0.10 atm. Calculate the equilibrium constant for this reaction. The reaction is carried out in a vessel of fixed volume.

Solution

Only the equilibrium partial pressure of $COCl_2(g)$ and the *initial* partial pressures of the other two gases are given. To find the equilibrium constant, it is necessary to determine the equilibrium partial pressures of CO and Cl_2. To do this, we set up a simple table:

	$CO(g)$	+	$Cl_2(g)$	$\rightleftharpoons$	$COCl_2(g)$
Initial partial pressure (atm)	0.60		1.10		0
Change in partial pressure (atm)	?		?		+0.10
Equilibrium partial pressure (atm)	?		?		0.10

Note that the first two lines must add to give the third. We use the relationships built into the balanced chemical equation to fill in the blanks in the table. Because this is the only reaction taking place, every mole of $COCl_2$ produced consumes exactly 1 mol of CO and exactly 1 mol of Cl_2. According to the ideal gas equation, the partial pressures of gases are proportional to the number of moles of each gas present as long as the volume and temperature are held fixed. Therefore, the change in partial pressure of each gas must be proportional to the change in its number of moles as the mixture goes toward equilibrium. If the partial pressure of $COCl_2$ *increases* by 0.10 atm through reaction, the partial pressures of CO and Cl_2 must each *decrease* by 0.10 atm. Inserting these values into the table gives

	$CO(g)$	+	$Cl_2(g)$	$\rightleftharpoons$	$COCl_2(g)$
Initial partial pressure (atm)	0.60		1.10		0
Change in partial pressure (atm)	−0.10		−0.10		+0.10
Equilibrium partial pressure (atm)	0.50		1.00		0.10

where the last line was obtained by adding the first two.

We now insert the equilibrium partial pressures into the equilibrium expression to calculate the equilibrium constant:

$$K = \left[\frac{P_{COCl_2}}{(P_{CO})(P_{Cl_2})}\right]_{eq} = \frac{(0.10)}{(0.50)(1.00)} = 0.20$$

Related Problems: 21, 22, 23, 24

For a reaction in which some of the coefficients in the balanced equation are not equal to 1, deriving expressions for the changes in the partial pressures of the products and reactants requires care. Consider the combustion of ethane at constant volume:

$$2\ C_2H_6(g) + 7\ O_2(g) \rightleftharpoons 4\ CO_2(g) + 6\ H_2O(g)$$

If the initial partial pressures are all 1.00 atm and there is a net reaction from left to right in this equation, the table is

	$2\ C_2H_6(g)$	+ $7\ O_2(g)$	$\rightleftarrows$ $4\ CO_2(g)$	+ $6\ H_2O(g)$
Initial partial pressure	1.00	1.00	1.00	1.00
Change in partial pressure	$-2y$	$-7y$	$+4y$	$+6y$
Equilibrium partial pressure	$1.00 - 2y$	$1.00 - 7y$	$1.00 + 4y$	$1.00 + 6y$

If the C_2H_6 partial pressure decreases by $2y$ atm, that of O_2 must decrease by $7y$, because the coefficients in the balanced equation are 2 and 7. The changes in partial pressures of products will have the opposite sign (positive instead of negative), because as reactants disappear, products appear. Their magnitudes are determined according to the coefficients in the balanced equation.

EXAMPLE 14.9

Graphite (a form of solid carbon) is added to a vessel that contains $CO_2(g)$ at a pressure of 0.824 atm at a certain high temperature. The pressure rises due to a reaction that produces $CO(g)$. The total pressure reaches an equilibrium value of 1.366 atm. **(a)** Write a balanced equation for the process. **(b)** Calculate the equilibrium constant.

Solution

(a) The reaction can only be the oxidation of C by CO_2 during which the CO_2 is itself reduced to CO. The reaction and its equilibrium expression are written as

$$C(s) + CO_2(g) \rightleftarrows 2\ CO(g) \qquad \left[\frac{(P_{CO})^2}{P_{CO_2}}\right]_{eq} = K$$

(b) To determine the equilibrium constant, we set up the standard table to describe reaction progress:

	$C(s)$	+	$CO_2(g)$	$\rightleftarrows$	$2\ CO(g)$
Initial partial pressure (atm)			0.824		0
Change in partial pressure (atm)			$-x$		$+2x$
Equilibrium partial pressure (atm)			$0.824 - x$		$2x$

The total pressure at equilibrium is

$$P_{tot} = 0.824 \text{ atm} - x + 2x = 0.824 + x = 1.366 \text{ atm}$$

Solving for x gives

$$x = 1.366 - 0.824 = 0.542 \text{ atm}$$

The equilibrium partial pressures of the two gases are

$$P_{CO} = 2x = 1.084 \text{ atm}$$

$$P_{CO_2} = 0.824 - 0.542 = 0.282 \text{ atm}$$

The equilibrium constant for the reaction is therefore

$$K = \frac{(1.084)^2}{0.282} = 4.17$$

Related Problems: 25, 26

14.5.2 Calculating Equilibrium Compositions When *K* Is Known

The law of mass action relates the equilibrium composition to the equilibrium constant. We can use these expressions to predict the outcomes of chemical reactions. How to approach the problem depends on the type of experimental data available.

Suppose you are asked to determine the partial pressures of reactants and products at equilibrium, given initial partial pressures of the reactants. This is illustrated by the following example.

EXAMPLE 14.10

Suppose $H_2(g)$ and $I_2(g)$ are sealed in a flask at $T = 400$ K with partial pressures $P_{H_2} = 1.320$ atm and $P_{I_2} = 1.140$ atm. At this temperature H_2 and I_2 do not react rapidly to form HI(*g*), although after a long enough time they would produce HI(*g*) at its equilibrium partial pressure. Suppose, instead, that the gases are heated in the sealed flask to 600 K, a temperature at which they quickly reach equilibrium:

$$H_2(g) + I_2(g) \rightleftharpoons 2\ HI(g)$$

The equilibrium constant for the reaction is 92.6 at 600 K:

$$\left[\frac{P_{HI}^2}{P_{H_2}P_{I_2}}\right]_{eq} = 92.6$$

(a) What are the equilibrium values of P_{H_2}, P_{I_2}, and P_{HI} at 600 K?

(b) What percentage of the I_2 originally present has reacted when equilibrium is reached?

Solution

(a) Suppose H_2 and I_2 did *not* react at 600 K. From the ideal gas law at constant volume, their partial pressures would be

$$P_{H_2}^{\circ} = 1.320\text{ atm} \times \left(\frac{600\text{ K}}{400\text{ K}}\right) = 1.980\text{ atm}$$

$$P_{I_2}^{\circ} = 1.140\text{ atm} \times \left(\frac{600\text{ K}}{400\text{ K}}\right) = 1.710\text{ atm}$$

Of course, these gases *do* react, and the extent of the reaction can be calculated. To do this, set up a table as in Example 14.8:

	$H_2(g)$	+	$I_2(g)$	$\rightleftharpoons$	2 HI(*g*)
Initial partial pressure (atm)	1.980		1.710		0
Change in partial pressure (atm)	$-x$		$-x$		$+2x$
Equilibrium partial pressure (atm)	$1.980 - x$		$1.710 - x$		$2x$

If the partial pressure of H_2 decreases by x atm as the reaction proceeds, then the partial pressure of I_2 must also decrease by x atm because each mole of H_2 reacts with 1 mol of I_2. By similar reasoning, the partial pressure of HI increases by $2x$ atm: 2 mol of HI forms from each mole of H_2. Inserting the equilibrium partial pressures into the equilibrium expression results in the equation

$$\frac{(2x)^2}{(1.980 - x)(1.710 - x)} = 92.6$$

(Continued)

Multiplying and collecting terms gives

$$88.6x^2 - 341.694x + 313.525 = 0$$

Solving for x using the quadratic formula (see Appendix C) gives

$$x = \frac{-(-341.694) \pm \sqrt{(341.694)^2 - 4(88.6)(313.525)}}{2(88.6)}$$

$$= 1.5044 \text{ atm or } 2.3522 \text{ atm}$$

The second root is physically impossible because it leads to negative answers for the equilibrium partial pressures of the $H_2(g)$ and $I_2(g)$. Discarding it leaves

$$P_{HI} = 2 \times 1.5044 \text{ atm} = 3.0088 \text{ atm}$$

$$P_{H_2} = 1.980 \text{ atm} - 1.5044 \text{ atm} = 0.4756 \text{ atm}$$

$$P_{I_2} = 1.710 \text{ atm} - 1.5044 \text{ atm} = 0.2056 \text{ atm}$$

It is a good idea to check such results by inserting the calculated equilibrium partial pressures back into the equilibrium expression to make sure that the known value of K comes out.

$$\text{Check:} \quad \frac{(3.0088)^2}{(0.4756)(0.2056)} = 92.6$$

As the final step, round off each answer to the correct number of significant digits: $P_{HI} = 3.01$ atm, $P_{H_2} = 0.48$ atm, $P_{I_2} = 0.21$ atm. Rounding off sooner makes the K calculated in the check differ somewhat from 92.6.

(b) The fraction of I_2 that has *not* reacted is the ratio of the number of moles of I_2 present at the end to that present at the beginning. Because neither volume nor temperature changes during the reaction, this equals the ratio of the final partial pressure of I_2 (0.2056 atm) to its initial partial pressure (1.710 atm):

$$\text{percentage unreacted} = \frac{0.2056 \text{ atm}}{1.710 \text{ atm}} \times 100\% = 12\%$$

The percentage that *has* reacted is then 88%.

Related Problems: 27, 28, 29, 30, 31, 32

The concentrations of any products initially present must also be included in equilibrium calculations, as illustrated by the following example.

EXAMPLE 14.11

Hydrogen is made from natural gas (methane) for immediate consumption in industrial processes, such as ammonia production. The first step is called the "steam reforming of methane":

$$CH_4(g) + H_2O(g) \rightleftharpoons CO(g) + 3\,H_2(g)$$

The equilibrium constant for this reaction is 1.8×10^{-7} at 600 K. Gaseous CH_4, H_2O, and CO are introduced into an evacuated container at 600 K, and their initial partial pressures (before reaction) are 1.40 atm, 2.30 atm, and 1.60 atm, respectively. Determine the partial pressure of $H_2(g)$ that will result at equilibrium.

(Continued)

Solution

The equilibrium expression is

$$\left[\frac{(P_{CO})(P_{H_2})^3}{(P_{CH_4})(P_{H_2O})}\right]_{eq} = K$$

Set up the table of partial pressures:

	$CH_4(g)$	+	$H_2O(g)$	$\rightleftharpoons$	$CO(g)$	+	$3\ H_2(g)$
Initial partial pressure	1.40		2.30		1.60		0
Change in partial pressure	$-y$		$-y$		$+y$		$+3y$
Equilibrium partial pressure	$1.40 - y$		$2.30 - y$		$1.60 + y$		$3y$

Insert the equilibrium partial pressures into the equilibrium expression:

$$\frac{(1.60 + y)(3y)^3}{(1.40 - y)(2.30 - y)} = 1.8 \times 10^{-7}$$

If we expand this equation by multiplying through by the denominator, a polynomial equation of fourth order in y would result, for which the quadratic formula from the preceding problem would be useless. How can we solve the equation?

The equilibrium constant in this case is quite small, so the extent of reaction will also be small. This suggests that y will be a small number relative to the partial pressures of the gases present initially. Let's try the approximation that y can be ignored where it is added to a number that is close to one; that is, let's replace $1.60 + y$ with 1.60 in the preceding equation, and make the same approximation for the two terms in the denominator. When y *multiplies* something, as in the $(3y)^3$ term, of course we cannot set it equal to zero. The result of these steps is the approximate equation

$$\frac{(1.60)(3y)^3}{(1.40)(2.30)} = 1.8 \times 10^{-7}$$

$$y^3 = 1.34 \times 10^{-8}$$

The cube roots of both sides give

$$y = 2.38 \times 10^{-3}$$

This value is indeed small compared with 1.60, 1.40, and 2.30, so our approximation of neglecting y relative to these numbers is justified. Finally, at equilibrium we have

$$P_{H_2} = 3y = 7.1 \times 10^{-3} \text{ atm}$$

Related Problems: 33, 34, 35, 36

Suppose the data for a gas-phase equilibrium are given in terms of concentrations rather than partial pressures. In such cases we convert all concentrations to partial pressures before starting the calculations. Or, we can rewrite the equilibrium expression in terms of concentration variables. The concentration [A] of an ideal gas is related to its partial pressure P_A through

$$[A] = \frac{n_A}{V} = \frac{P_A}{RT}$$

which can be written

$$P_A = RT[A]$$

We can substitute such a relation for each species appearing in the equilibrium expression. It is best to put the factors of $P_{ref} = 1$ atm back into the equilibrium expression to examine the units of the resulting equations. For the general gas-phase reaction

$$aA(g) + bB(g) \rightleftharpoons cC(g) + dD(g)$$

the equilibrium expression is

$$\left[\frac{(RT[C]/P_{ref})^c (RT[D]/P_{ref})^d}{(RT[A]/P_{ref})^a (RT[B]/P_{ref})^b}\right]_{eq} = K$$

Rearranging gives

$$\left[\frac{[C]^c[D]^d}{[A]^a[B]^b}\right]_{eq} = K(RT/P_{ref})^{+a+b-c-d}$$

This expression relates concentrations of gas-phase species at equilibrium. If $a + b - c - d = 0$ (that is, if there is no change in the total number of moles of gases in the reaction mixture), the right side of the equilibrium expression reduces to K.

EXAMPLE 14.12

At elevated temperatures, PCl_5 dissociates extensively according to

$$PCl_5(g) \rightleftharpoons PCl_3(g) + Cl_2(g)$$

At 300°C, the equilibrium constant for this reaction is $K = 11.5$. The concentrations of PCl_3 and Cl_2 at equilibrium in a container at 300°C are both 0.0100 mol L^{-1}. Calculate $[PCl_5]$.

Solution

Two moles of gases are produced for each mole of gas consumed, so RT/P_{ref} must be raised to the power $1 - 2 = -1$. Hence,

$$\left[\frac{[PCl_3][Cl_2]}{[PCl_5]}\right]_{eq} = K\left(\frac{RT}{P_{ref}}\right)^{-1} = K\left(\frac{P_{ref}}{RT}\right)$$

$$= 11.5 \times \frac{1 \text{ atm}}{(0.08206 \text{ L atm mol}^{-1}\text{ K}^{-1})(573 \text{ K})} = 0.245\,\frac{\text{mol}}{\text{L}}$$

Solving this equation for $[PCl_5]$ gives

$$[PCl_5] = \frac{[PCl_3][Cl_2]}{0.245 \text{ mol L}^{-1}} = \frac{(0.0100 \text{ mol L}^{-1})(0.0100 \text{ mol L}^{-1})}{0.245 \text{ mol L}^{-1}}$$

$$= 4.08 \times 10^{-4} \text{ mol L}^{-1}$$

Related Problems: 31, 32

14.6 The Direction of Change in Chemical Reactions: Empirical Description

The specific examples in Section 14.5 illustrate how the law of mass action gives information about the nature of the equilibrium state. The law of mass action also explains and predicts the direction in which a reaction will proceed spontaneously when reactants and products are initially mixed together with arbitrary partial pressures or compositions. This requires a new concept, the

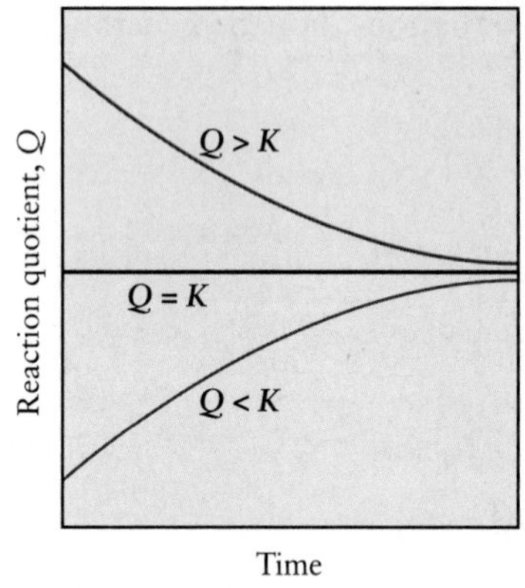

(a)

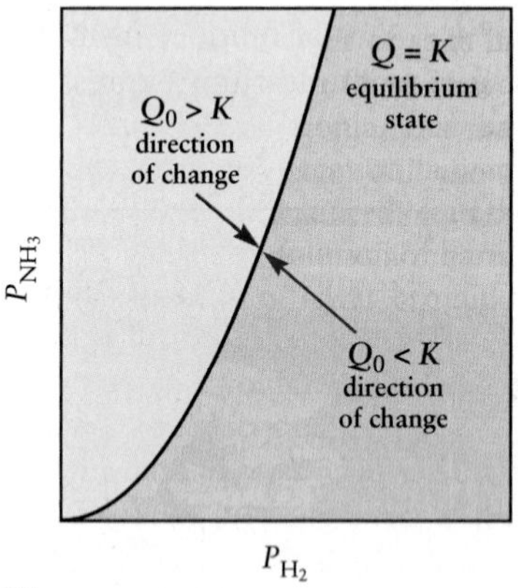

(b)

FIGURE 14.6 If nitrogen and hydrogen are mixed in 1:3 proportions together with some ammonia, they react according to the chemical equation

$$N_2(g) + 3\ H_2(g) \rightleftharpoons 2\ NH_3(g)$$

(a) If the initial reaction quotient Q_0 is less than K, it increases with time; if it is greater than K, it decreases. (b) With three moles of H_2 available for each mole of N_2, a parabolic curve represents the partial pressures of ammonia and of hydrogen that coexist at equilibrium. From initial nonequilibrium conditions on either side, the partial pressures approach equilibrium along lines with slope −2/3, because three moles of H_2 are consumed to produce two moles of NH_3.

reaction quotient Q, which is related to the equilibrium constant. Through the *principle of Le Châtelier* (described below), the mass action law also explains how a reaction in equilibrium responds to an external perturbation.

14.6.1 The Reaction Quotient

The **reaction quotient** Q for a general gas-phase reaction is defined as

$$Q = \frac{(P_C)^c (P_D)^d}{(P_A)^a (P_B)^b}$$

where the partial pressures are the actual values measured at any point during the reaction, not just at equilibrium. The distinction between Q and K is crucial. The equilibrium constant K is determined by the partial pressures of reactants and products at equilibrium, and it is a constant, dependent only on the temperature. The reaction quotient Q depends on the actual instantaneous partial pressures, whatever they may be; thus, Q changes with time. As the reaction approaches equilibrium, Q approaches K. The initial partial pressures P_A°, P_B°, P_C°, and P_D° give an initial reaction quotient Q_0, whose magnitude relative to K determines the direction in which the reaction will proceed spontaneously toward equilibrium. If Q_0 is *less* than K, then Q must *increase* as time goes on. To increase Q the reaction must proceed spontaneously in the forward direction. To see that it will, in Section 14.7 we simply rewrite $\Delta G = \Delta G° + RT \ln Q$ using equation 14.6. This gives $\Delta G = -RT \ln K + RT \ln Q = RT \ln (Q/K)$. Thus, ΔG will be negative if $Q < K$ and the reaction will proceed spontaneously in the forward direction, increasing Q. There is an increase in the product partial pressures and a decrease in reactant partial pressures as the reaction proceeds from left to right. If Q_0 is *greater* than K, similar reasoning shows that the reaction will proceed from right to left, with Q *decreasing* with time until it becomes equal to K (Fig. 14.6).

EXAMPLE 14.13

The reaction between nitrogen and hydrogen to produce ammonia

$$N_2(g) + 3\ H_2(g) \rightleftharpoons 2\ NH_3(g)$$

is essential in making nitrogen-containing fertilizers. This reaction has an equilibrium constant equal to 1.9×10^{-4} at 400°C. Suppose that 0.10 mol of N_2, 0.040 mol of H_2, and 0.020 mol of NH_3 are sealed in a 1.00-L vessel at 400°C. In which direction will the reaction proceed?

Solution

The initial pressures $P_i = n_iRT/V$ are readily calculated to be

$$P_{N_2} = 5.5 \text{ atm} \qquad P_{H_2} = 2.2 \text{ atm} \qquad P_{NH_3} = 1.1 \text{ atm}$$

The initial numerical value of Q is therefore

$$Q_0 = \frac{(P^\circ_{NH_3})^2}{P^\circ_{N_2}(P^\circ_{H_2})^3} = \frac{(1.1)^2}{(5.5)(2.2)^3} = 2.1 \times 10^{-2}$$

Because $Q_0 > K$, the reaction will proceed from right to left and ammonia will dissociate until equilibrium is reached.

Related Problems: 45, 46, 47, 48

CONNECTION TO BIOLOGY

Hemoglobin and Oxygen Transport

Humans and animals require oxygen for the aerobic metabolic processes that provide the energy necessary to sustain life. Food (a fuel) is oxidized in a series of enzyme-mediated reactions to produce energy; the products of the reactions being CO_2 and H_2O. Oxygen is carried from the lungs to tissues bound to the protein molecule hemoglobin (Hb), and it is stored in muscle by a related protein called myoglobin (Mb). Hemoglobin and myoglobin are both oxygen transport proteins in which O_2 is covalently bound to iron in an octahedral coordination complex built around the porphyrin molecule heme (see Section 8.3). Iron is coordinated to four nitrogen atoms in the planar porphyrin structure, leaving a fifth site available that binds the heme to the protein via the amino acid histidine, and a sixth site available to reversibly bind oxygen. The solubility of oxygen in whole blood can be as high as 10^{-2} mol L^{-1} only because it is chemically bound to hemoglobin. The solubility of O_2 in blood plasma, which contains no hemoglobin, is only about 10^{-4} mol L^{-1}, a value that is very close to its solubility in pure water (as determined by its Henry's Law constant, see Section 11.6). The partial pressure of O_2 in the lungs is about 100 torr, which is sufficiently high to saturate all of the available oxygen binding sites in hemoglobin. Highly oxygenated blood from the lungs is delivered throughout the body via the circulatory system, eventually reaching the capillaries, where the O_2 partial pressure is only about 40 torr. Hemoglobin releases a significant fraction of the bound oxygen under these low-pressure conditions; it can either be taken up and stored by myoglobin in muscle or transported directly to cells for their use in metabolism.

Why is oxygen bound so strongly by hemoglobin in the lungs and then released so readily to myoglobin through capillaries? The answer requires an understanding of the differences between the simple heterogeneous equilibrium established between oxygen and myoglobin and the more complicated set of simultaneous equilibria established between oxygen and hemoglobin. The differences arise as a consequence of the very different molecular structures of the two proteins. Myoglobin is a monomeric globular protein that contains 154 amino acids in a helical configuration (represented by the green ribbon structure) and a single heme group shown in red (Fig. C14.1).

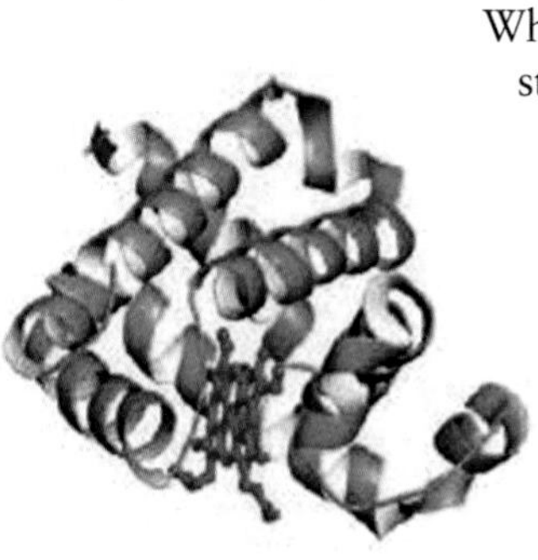

FIGURE C14.1 Myoglobin (Mb) (Adapted from Reginald H. Garrett and Charles M. Grisham, fig. 15.21 (top), p. 468, *Biochemistry*. Boston: Brooks/Cole, Cengage Learning, 2010; Cengage Learning)

Hemoglobin, on the other hand, is a tetramer with two identical α subunits that each contains 141 amino acids and a single heme group, and two identical β subunits that each contains 146 amino acids and a single heme group. Each member of an identical pair of monomers is labeled by a number, and all four monomers are colored differently, for clarity (Fig. C14.2).

The uptake of oxygen by myoglobin follows the hyperbolic Michelis–Menten equation (see Section 18.8) as shown by the blue line in Fig. C14.3, which is a plot of the **fractional saturation** versus the oxygen partial pressure. This behavior is characteristic of a simple enzyme/substrate equilibrium as represented by the following equation and equilibrium constant.

$$Mb(aq) + O_2(g) \rightleftharpoons Mb(O_2)(aq)$$

$$K = \frac{[Mb(O_2)]}{[Mb]P_{O_2}} = 271$$

The graph shows that binding saturates at very low oxygen partial pressures; half of the available sites are filled at $P_{O_2} = 1$ torr.

The experimental binding curve for hemoglobin is quite different, however. The S-shaped red curve shows that relatively few oxygen molecules are bound at low partial pressures but that the fractional saturation increases rapidly with increasing P_{O_2}; half saturation

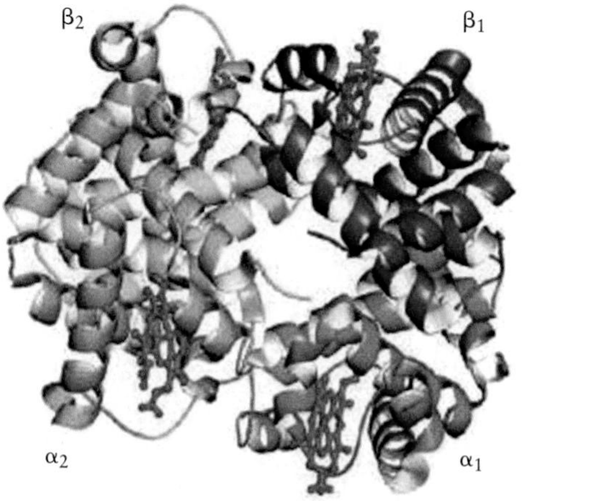

FIGURE C14.2 Hemoglobin (Hb) (Adapted from Reginald H. Garrett and Charles M. Grisham, fig. 15.21 (bottom), p. 468, *Biochemistry*. Boston: Brooks/Cole, Cengage Learning, 2010; © Cengage Learning)

occurs at about 30 torr, a much higher pressure than for myoglobin. Curves of this type are characteristic of cooperative behavior. Binding the first O_2 molecule to one heme group in hemoglobin causes a conformational change that makes it easier to bind the second O_2 molecule to a different heme group and so forth until all four sites are occupied. Evidence for this conclusion is provided by comparing the values of the consecutive equilibrium constants, as shown below.

$$Hb(aq) + O_2(g) \rightleftharpoons Hb(O_2)(aq)$$

$$K_1 = \left[\frac{[Hb(O_2)]}{[Hb]P_{O_2}}\right]_{eq} = 4.88$$

$$Hb(O_2)(aq) + O_2(g) \rightleftharpoons Hb(O_2)_2(aq)$$

$$K_2 = \left[\frac{[Hb(O_2)_2]}{[HbO_2]P_{O_2}}\right]_{eq} = 15.4$$

$$Hb(O_2)_2(aq) + O_2(g) \rightleftharpoons Hb(O_2)_3(aq)$$

$$K_3 = \left[\frac{[Hb(O_2)_3]}{[Hb(O_2)_2]P_{O_2}}\right]_{eq} = 6.49$$

$$Hb(O_2)_3(aq) + O_2(g) \rightleftharpoons Hb(O_2)_4(aq)$$

$$K_4 = \left[\frac{[Hb(O_2)_4]}{[Hb(O_2)_3]P_{O_2}}\right]_{eq} = 1750$$

The four equilibrium constants are not equal to each other, and K_4 is much larger than the other three. Reducing P_{O_2} clearly reduces the fraction of bound oxygen (Le Châtelier's principle) for both proteins but the two molecules respond quite differently to decreasing oxygen partial pressures. Hemoglobin and myoglobin are both nearly saturated at partial pressures of 100 torr (in the lungs), but the fraction of oxygen bound to hemoglobin has dropped to 55% at 30 torr (in capillaries) while the fraction bound to myoglobin in muscle remains high at 90%. These different responses to changes in the oxygen partial pressure are responsible for transfer of O_2 from hemoglobin (in blood) to myoglobin (in muscle). See chart below.

Archibald Hill derived a simple model to account for the S-shaped curve observed that provides some physical insight into the nature of the cooperativity involved. Suppose, for example, that binding was an all or nothing event. The overall reaction and resulting equilibrium constant expression would be

$$Hb + 4O_2 \rightleftharpoons Hb(O_2)_4 \qquad K = \left[\frac{[Hb](P_{O_2})^4}{[Hb(O_2)_4]}\right]_{eq}$$

from which we express the fractional saturation as follows,

$$X = \frac{[Hb(O_2)_4]}{[Hb(O_2)_4] + [Hb]} \qquad X = \frac{(P_{O_2})^4}{(P_{O_2})^4 + K}$$

which is plotted in black. The most general form proposed by Hill replaces the exponent 4 with a number n that is chosen to fit particular situations. $n = 1$ corresponds to the myoglobin case with no cooperativity, and a fit to the hemoglobin curve produces $n = 2.8$, which demonstrates a high degree of, but not perfect, cooperativity.

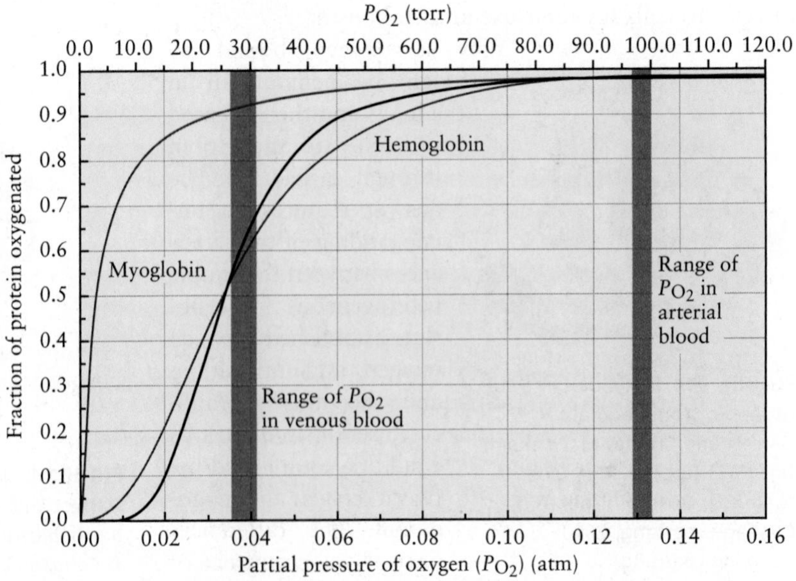

FIGURE C14.3 A plot of the fraction of the binding sites of hemoglobin (red) and myoglobin (blue) that are occupied at equilibrium as a function of the partial pressure of O_2. (Adapted from Reginald H. Garrett and Charles M. Grisham, *Biochemistry.* Boston: Brooks/Cole, Cengage Learning, 2010; © Cengage Learning)

EXAMPLE 14.14

Solid ammonium chloride is in equilibrium with ammonia and hydrogen chloride gases:

$$NH_4Cl(s) \rightleftharpoons NH_3(g) + HCl(g)$$

The equilibrium constant at 275°C is 1.04×10^{-2}.

We place 0.980 g of solid NH_4Cl into a closed vessel with volume 1.000 L and heat to 275°C. (a) In what direction does the reaction proceed? (b) What is the partial pressure of each gas at equilibrium? (c) What is the mass of solid NH_4Cl at equilibrium?

Solution

(a) We evaluate the reaction quotient

$$Q = P_{NH_3}P_{HCl}$$

Initially, $Q_0 = 0$ because neither gas is present. By comparison, $K = 1.04 \times 10^{-2}$. Because $Q < K$, the reaction will proceed spontaneously from left to right. Some of the solid NH_4Cl will decompose, and some gaseous NH_3 and HCl will appear in the vessel.

(b) We set up the standard table for the equilibrium calculation:

	$NH_4Cl(s)$	$\rightleftharpoons$	$NH_3(g)$	+	$HCl(g)$
Initial partial pressure (atm)			0		0
Change in partial pressure (atm)			$+x$		$+x$
Equilibrium partial pressure (atm)			$+x$		$+x$

Because NH_3 and HCl are formed in equimolar amounts, they will have the same partial pressure at equilibrium. The equilibrium expression is

$$(P_{NH_3}P_{HCl})_{eq} = K = 1.04 \times 10^{-2}$$

$$x^2 = 1.04 \times 10^{-2}$$

$$x = 0.102$$

At equilibrium, the partial pressures are

$$P_{NH_3} = P_{HCl} = 0.102 \text{ atm}$$

(c) The number of moles of NH_4Cl that decomposed is equal to the number of moles of each gas produced. We calculate this number, treating the gases as ideal

$$n_{NH_3} = n_{HCl} = \frac{(1.02 \times 10^{-1} \text{ atm})(1.000 \text{ L})}{(0.08206 \text{ L atm mol}^{-1} \text{ K}^{-1})(548.2 \text{ K})} = 2.268 \times 10^{-3} \text{ mol}$$

The mass of NH_4Cl consumed is $(2.268 \times 10^{-3}$ mol$)(53.49$ g mol$^{-1}) = 0.121$ g. The remaining mass is 0.859 g. The percentage decomposition of the original sample is 12.4%.

Related Problems: 49, 50

14.6.2 External Effects on *K*: Principle of Le Châtelier

Suppose a system at equilibrium is perturbed by some external stress such as a change in volume or temperature or a change in the partial pressure or concentration of one of the reactants or products. How will the system respond? The qualitative answer is embodied in a principle stated by Henri Le Châtelier in 1884:

> *A system in equilibrium that is subjected to a stress will react in a way that tends to counteract the stress.*

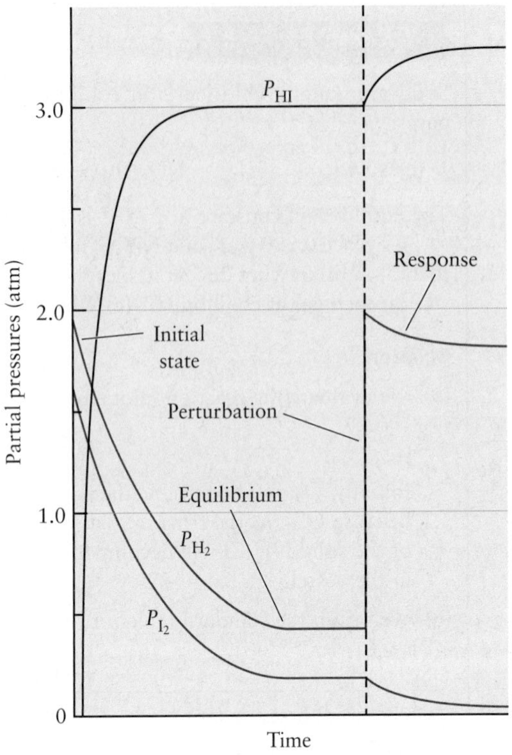

FIGURE 14.7 Partial pressures versus time for the equilibrium

$$H_2(g) + I_2(g) \rightleftharpoons 2\ HI(g)$$

The part of the graph to the left of the dashed vertical line shows the approach to equilibrium starting from the initial conditions specified in Example 14.10. Then the equilibrium state is abruptly perturbed by an increase in the partial pressure of H_2 to 2.000 atm. In accordance with Le Châtelier's principle, the system responds (Example 14.15) in such a way as to decrease the partial pressure of H_2—that is, to counteract the perturbation that moved it away from equilibrium in the first place.

Le Châtelier's principle provides a way to predict qualitatively the direction of change of a system under an external perturbation. It relies heavily on Q as a predictive tool.

Effects of Changing the Concentration of a Reactant or Product

As a simple example, consider what happens when a small quantity of a reactant is added to an equilibrium mixture. The addition of reactant lowers the reaction quotient Q below K and a net reaction takes place in the forward direction, partially converting reactants to products, until Q again equals K. The system partially counteracts the stress (the increase in the quantity of one of the reactants) and attains a new equilibrium state. If one of the *products* is added to an equilibrium mixture, Q temporarily becomes *greater* than K and a net *back* reaction occurs, partially counteracting the imposed stress by reducing the concentration of products (Fig. 14.7).

EXAMPLE 14.15

An equilibrium gas mixture of $H_2(g)$, $I_2(g)$, and $HI(g)$ at 600 K has

$$P_{H_2} = 0.4756 \text{ atm} \qquad P_{I_2} = 0.2056 \text{ atm} \qquad P_{HI} = 3.009 \text{ atm}$$

This is essentially the final equilibrium state of Example 14.10. Enough H_2 is added to increase its partial pressure to 2.000 atm at 600 K before any reaction takes place. The mixture then once again reaches equilibrium at 600 K. What are the final partial pressures of the three gases?

(Continued)

Solution

Set up the usual table, in which "initial" now means the moment after the addition of the new H_2 but before it reacts further.

	$H_2(g)$ +	$I_2(g)$ $\rightleftharpoons$	$2\ HI(g)$
Initial partial pressure (atm)	2.000	0.2056	3.009
Change in partial pressure (atm)	$-x$	$-x$	$+2x$
Equilibrium partial pressure (atm)	$2.000 - x$	$0.2056 - x$	$3.009 + 2x$

From Le Châtelier's principle it follows that net reaction to consume H_2 will occur after addition of H_2, and this fact has been used in assigning a negative sign to the change in the partial pressure of H_2 in the table.

Substitution of the equilibrium partial pressures into the equilibrium law gives

$$\frac{(3.009 + 2x)^2}{(2.000 - x)(0.2056 - x)} = 92.6$$

Expansion of this expression results in the quadratic equation

$$88.60x^2 - 216.275x + 29.023 = 0$$

which can be solved to give

$$x = 0.1425 \text{ or } 2.299$$

The second root would lead to negative partial pressures of H_2 and I_2 and is therefore physically impossible. Substitution of the first root into the expressions from the table gives

$$P_{H_2} = 2.000 - 0.1425 = 1.86 \text{ atm}$$

$$P_{I_2} = 0.2056 = -0.1425 = 0.063 \text{ atm}$$

$$P_{HI} = 3.009 + 2(0.1425) = 3.29 \text{ atm}$$

$$\text{Check:} \quad \frac{(3.29)^2}{(1.86)(0.063)} = 92.4$$

HI is synthesized industrially by reacting elemental iodine with H_2. Because iodine is so much more expensive than hydrogen, the reaction conditions are chosen to maximize the conversion of iodine to product. This objective is accomplished by keeping hydrogen greatly in excess to drive the equilibrium to the right. The product is removed continuously to provide an additional thermodynamic driving force for conversion to product. Most industrial operations are designed in such a way that products can be removed continuously to achieve high overall yields, even for reactions with small equilibrium constants.

Effects of Changing the Volume

Le Châtelier's principle also predicts the effect of a change in volume on gas-phase equilibrium. We find it easier to think about changes in volume in terms of the changes in pressure they would produce and to see how the system can respond to these changes in pressure. Decreasing the volume of a gaseous system increases its total pressure, and the system responds, if possible, to reduce the total pressure. For example, in the equilibrium

$$2\ P_2(g) \rightleftharpoons P_4(g)$$

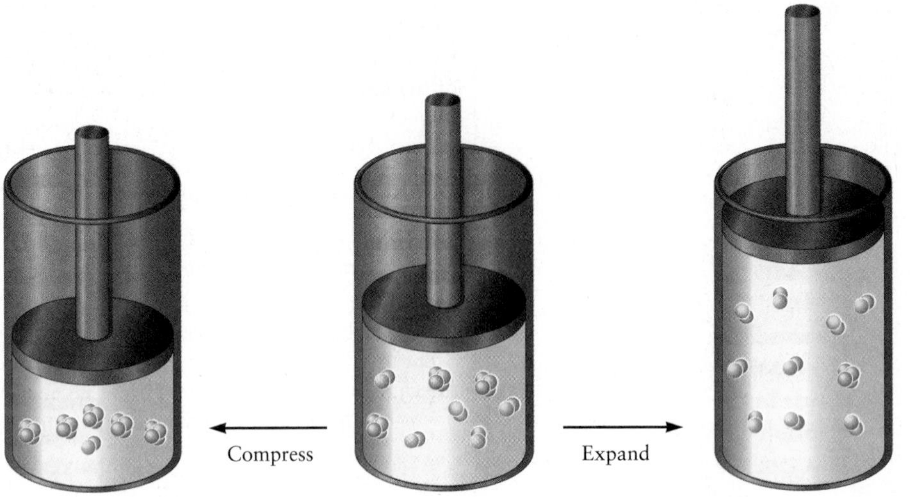

FIGURE 14.8 An equilibrium mixture of P_2 and P_4 (center) is compressed (left). Some P_2 molecules combine to give P_4 molecules, to reduce the total number of molecules and thus the total pressure. If the volume is increased (right), some P_4 molecules dissociate to pairs of P_2 molecules to increase the total number of molecules and the pressure exerted by those molecules.

the reaction shifts in the forward direction when the volume is decreased. This occurs because every two molecules of P_2 consumed produce only one molecule of P_4, thus reducing the total pressure and partially compensating for the external stress caused by the change in volume. In contrast, an *increase* in volume favors reactants over products in this system, and some P_4 dissociates to form P_2 (Fig. 14.8). If there is no difference in the total numbers of gas-phase molecules on the two sides of the equation, then a change in volume has no effect on the equilibrium.

This effect of changing the volume of an equilibrium reacting mixture can also be understood by using the reaction quotient. For the phosphorus equilibrium just described, the reaction quotient is

$$Q = \frac{P_{P_4}}{(P_{P_2})^2}$$

Initially, Q_0 equals K. Suppose the volume is then decreased by a factor of 2; because the temperature is unchanged, this initially increases each partial pressure by a factor of 2. Because there are two powers of the pressure in the denominator and only one in the numerator, this decreases Q by a factor of 2, making it lower than K. Reaction must then occur in the forward direction until Q again equals K.

When the volume of a system is decreased, its total pressure increases. Another way to increase the total pressure is to add an inert gas such as argon to the reaction mixture without changing the total volume. In this case the effect on the equilibrium is entirely different. Because the partial pressures of the reactant and product gases are unchanged by an inert gas, adding argon at constant volume has no effect on the position of the equilibrium.

Effects of Changing the Temperature

Chemical reactions can be classified as either an **endothermic reaction** (taking up heat from the surroundings) or an **exothermic reaction** (giving off heat) (see Section 12.6). Raising the temperature of an equilibrium mixture by adding heat causes reactions to occur in such a way as to absorb some of the added heat. The equilibrium in an endothermic reaction shifts from left to right, while that in an exothermic reaction shifts from right to left, with "products" reacting to give "reactants." Equivalently, we can describe the shifts in terms of the effect of temperature on equilibrium constants. The equilibrium constant for an endothermic reaction increases with increasing temperature, while that for an exothermic reaction decreases with increasing temperature.

FIGURE 14.9 The equilibrium between N_2O_4 and NO_2 depends on temperature. The tube on the right, held in an ice bath at 0°C, contains mostly N_2O_4. Its color is pale because only NO_2 is colored. The deeper color in the tube on the left, which is held at 50°C, reflects the increased NO_2 present in equilibrium at the higher temperature. The tubes contain the same masses of substance, distributed in different ways between NO_2 and N_2O_4.

This effect is illustrated by the equilibrium between nitrogen dioxide (NO_2) and its dimer, dinitrogen tetraoxide (N_2O_4), briefly considered in Example 9.6, expressed by the chemical equation

$$2\,NO_2(g) \rightleftharpoons N_2O_4(g)$$

Because NO_2 is a brown gas but N_2O_4 is colorless, the equilibrium between them can be studied by observing the color of a tube containing the two gases. At high temperatures, NO_2 predominates and a brown color results; as the temperature is lowered, the partial pressure of N_2O_4 increases and the color fades (Fig. 14.9).

The equilibrium expression for the N_2O_4–NO_2 equilibrium is

$$\left[\frac{P_{N_2O_4}}{(P_{NO_2})^2}\right]_{eq} = K$$

K has the numerical value 8.8 at $T = 25°C$, provided the partial pressures of N_2O_4 and NO_2 are expressed in atmospheres. This reaction is exothermic ($\Delta H = -58.02$ kJ mol^{-1} at 298 K) because energy must be liberated when dimers are formed. Consequently, K decreases as the temperature T increases, so the amount of N_2O_4 present for a given partial pressure of NO_2 falls with increasing temperature as the dimer dissociates at elevated temperatures.

Maximizing the Yield of a Reaction

As an application of Le Châtelier's principle, consider the reaction

$$N_2(g) + 3\,H_2(g) \rightleftharpoons 2\,NH_3(g)$$

which is the basis of the industrial synthesis of ammonia. Because this reaction is exothermic, the yield of ammonia is increased by working at as low a temperature as possible (Fig. 14.10a). At too low a temperature, the reaction is very slow, so a compromise temperature near 500°C is typically used. Because the number of moles of gas decreases as the reaction occurs, the yield of product is enhanced by decreasing the volume of the reaction vessel. Typically, total pressures of 150 to 300 atm are used (see Fig. 14.10b), although some plants work at up to 900 atm of pressure. Even at high pressures, the yield of ammonia is usually only 15% to

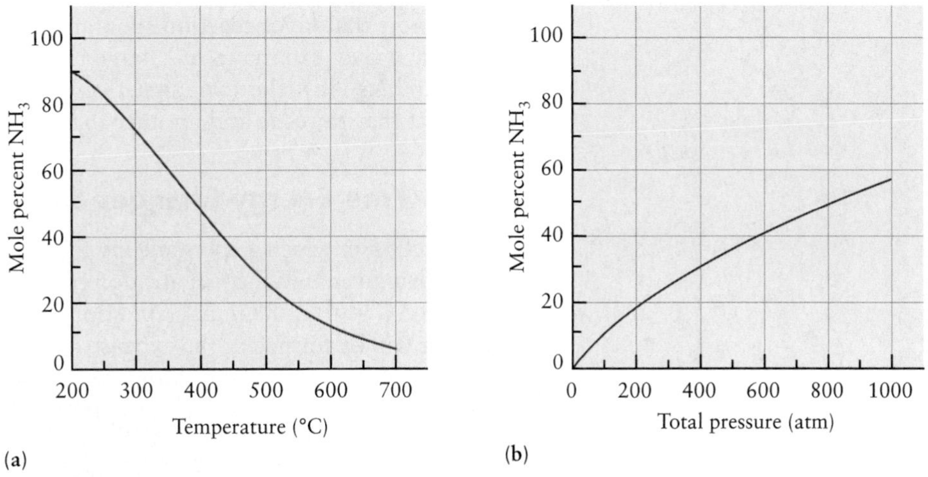

FIGURE 14.10 (a) The equilibrium mole percentage of ammonia in a 1:3 mixture of N_2 and H_2 varies with temperature; low temperatures favor high yields of NH_3. The data shown correspond to a fixed total pressure of 300 atm. (b) At a fixed temperature (here, 500°C), the yield of NH_3 increases with increasing total pressure.

20% because the equilibrium constant is so small. To overcome this, ammonia plants operate in a cyclic process in which the gas mixture is cooled after ammonia is produced so that the ammonia liquefies (its boiling point is much higher than those of nitrogen and hydrogen) and is removed from the reaction vessel. Continuous removal of products helps drive the reaction to completion.

14.7 The Direction of Change in Chemical Reactions: Thermodynamic Explanation

The specific examples in Section 14.5 demonstrate that when $K \gg 1$ the reaction has progressed far toward products, and when $K \ll 1$ the reaction has remained near reactants. The empirical discussion in Section 14.6 shows how the reaction quotient Q and Le Châtelier's principle can predict the direction of spontaneous reaction and the response of an equilibrium state to an external perturbation. Here, we use the thermodynamic description of K from Section 14.3 to provide the thermodynamic basis for these results obtained empirically in Sections 14.5 and 14.6. We identify those thermodynamic factors that determine the magnitude of K. We also provide a thermodynamic criterion for predicting the direction in which a reaction proceeds from a given initial condition.

14.7.1 The Magnitude of the Equilibrium Constant

The expression connecting the standard Gibbs free energy change and the equilibrium constant can be rewritten as

$$\ln K = \frac{-\Delta G^\circ}{RT} = \frac{\Delta S^\circ}{R} - \frac{\Delta H^\circ}{RT}$$

so that

$$K = \exp\left[\frac{-\Delta G^\circ}{RT}\right] = \exp\left[\frac{\Delta S^\circ}{R}\right] \exp\left[\frac{-\Delta H^\circ}{RT}\right]$$

Here K is large (favoring the products) if ΔS° is positive and large and ΔH° is negative and large. In other words, an increase in the number of microstates ($\Delta S^\circ > 0$) and a decrease in enthalpy ($\Delta H^\circ < 0$) both favor a large K. Thus, the same factors that favor reaction spontaneity by making ΔG° negative also favor a large K if they can make ΔG° large in magnitude as well as negative in sign. If ΔH° and ΔS° have the *same* sign, the value of K will be a compromise between one effect that raises K and another that lowers it.

14.7.2 Free Energy Changes and the Reaction Quotient

The direction in which a spontaneous chemical reaction proceeds after it is initiated with a given initial concentration of products and reactants is the direction in which $\Delta G < 0$. If the initial condition is "to the left" of the equilibrium state, products will be formed at the expense of reactants; if the initial condition is "to the right" of the equilibrium state, products will be converted back to reactants. This criterion can be made quantitative and expressed in terms of the initial concentrations as follows.

Proceeding as in Section 14.3, we find ΔG for the general gas-phase reaction

$$a\text{A} + b\text{B} \longrightarrow c\text{C} + d\text{D}$$

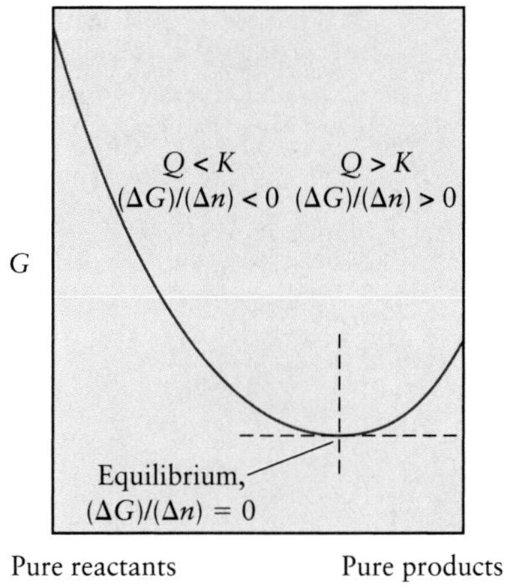

FIGURE 14.11 The free energy of a reaction system is plotted against its progress from pure reactants (left) to pure products (right). Equilibrium comes at the minimum of the curve. To the reactant side of equilibrium, $\Delta G < 0$ and $Q < K$. A reaction mixture with initial condition in this range will spontaneously move toward equilibrium by converting more reactants into products. To the product side of equilibrium, $\Delta G > 0$ and $Q > K$. A reaction mixture prepared in this range will spontaneously move toward equilibrium by converting products back into reactants.

to be

$$\Delta G = \Delta G^\circ + RT \ln \left[\frac{(P_C/P_{ref})^c (P_D/P_{ref})^d}{(P_A/P_{ref})^a (P_B/P_{ref})^b} \right]$$

At equilibrium, where $\Delta G = 0$, the combination of partial pressures appearing inside the brackets becomes the equilibrium constant, K. Away from equilibrium, this combination of partial pressures is the reaction quotient Q, introduced in Section 14.6:

$$\Delta G = \Delta G^\circ \; + RT \ln Q$$

The equilibrium constant can be substituted for ΔG° in this equation to obtain a very useful relation between ΔG, K, and Q as follows:

$$\Delta G = \Delta G^\circ + RT \ln Q = -RT \ln K + RT \ln Q = RT \ln (Q/K) \qquad [14.10]$$

If the reaction quotient Q is *less* than K, $\Delta G < 0$ and the reaction will proceed spontaneously as written, from left to right. If $Q > K$, then $\Delta G > 0$ and the *reverse* reaction (right to left) will occur spontaneously until equilibrium is reached. These conditions are represented schematically in Figure 14.11. The second law of thermodynamics thus provides a very useful criterion for the direction of reaction in terms of the initial value of the reaction quotient.

There exists a deep relationship between Figure 14.6, which represents actual events occurring in the laboratory, and Figure 14.11, which represents the thermodynamic driving force (that is, the Gibbs free energy) governing these events. The function shown in Figure 14.11 is a thermodynamic potential function analogous to the mechanical potential functions we introduced in Chapter 3. The slope of this function at every point on the curve is the thermodynamic driving force for the direction the reaction will take when the system has the composition described by that point.

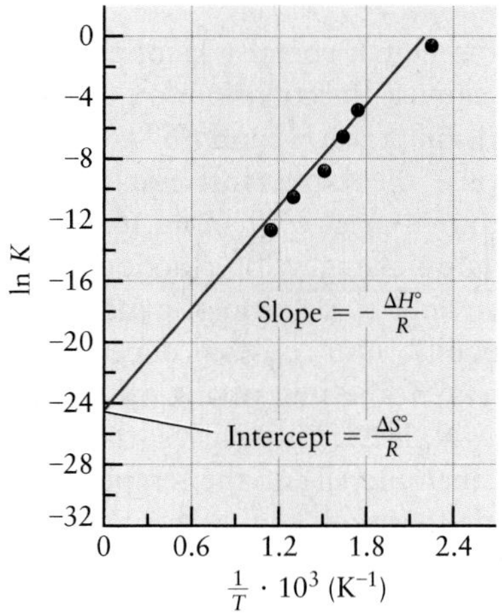

FIGURE 14.12 The temperature dependence of the equilibrium constant for the reaction

$$N_2(g) + 3\,H_2(g) \rightleftharpoons 2\,NH_3(g)$$

Experimental data are shown by points.

14.7.3 Temperature Dependence of Equilibrium Constants

Le Châtelier's principle is a qualitative way of describing the *stability* of equilibrium states against sudden perturbations in concentration, pressure, and temperature. The responses of the system to all three effects can be described quantitatively by thermodynamics. Here we describe the effect of temperature, which is the most useful of these quantitative descriptions.

The temperature dependence of the equilibrium constant is determined by the equation

$$-RT \ln K = \Delta G^\circ = \Delta H^\circ - T\Delta S^\circ$$

If ΔH° and ΔS° are independent of temperature, then all the temperature dependence of K lies in the factor of T and the equation can be used to relate the values of K at two different temperatures, as follows. At least over a limited temperature range, ΔH° and ΔS° do not vary much with temperature. To the extent that their temperature dependence may be neglected, it is evident that ln K is a linear function of $1/T$ as shown in Figure 14.12.

$$\ln K = -\frac{\Delta G^\circ}{RT} = -\frac{\Delta H^\circ}{RT} + \frac{\Delta S^\circ}{R} \qquad [14.11]$$

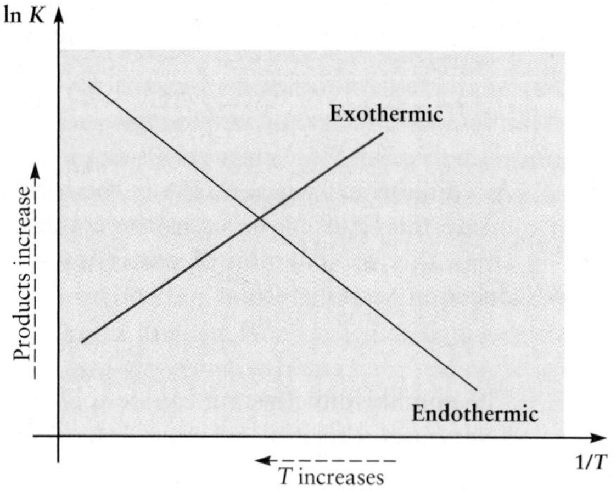

FIGURE 14.13 Sketch of ln K against $1/T$ for an exothermic reaction and for an endothermic reaction, as predicted by thermodynamics. Temperature increases to the left on this diagram. As T increases, K for the endothermic reaction increases and K for the exothermic reaction decreases in accordance with the principle of Le Châtelier.

A graph of ln K against $1/T$ is approximately a straight line with slope $-\Delta H°/R$ and intercept $\Delta S°/R$ (see Fig. 14.12). If the value of K is known for one temperature and $\Delta H°$ is also known, K can be calculated for other temperatures.

In addition to the graphical method, an equation can be obtained to connect the values of K at two different temperatures. Let K_1 and K_2 be the equilibrium constants for a reaction at temperatures T_1 and T_2, respectively. Then

$$\ln K_2 = -\frac{\Delta H°}{RT_2} + \frac{\Delta S°}{R}$$

$$\ln K_1 = -\frac{\Delta H°}{RT_1} + \frac{\Delta S°}{R}$$

Subtracting the second equation from the first gives

$$\ln\left(\frac{K_2}{K_1}\right) = -\frac{\Delta H°}{R}\left[\frac{1}{T_2} - \frac{1}{T_1}\right] \qquad [14.12]$$

This is known as the **van't Hoff equation** after the Dutch chemist Jacobus van't Hoff. Given $\Delta H°$ and K at one temperature, we can use the equation to calculate K at another temperature, within the approximation that $\Delta H°$ and $\Delta S°$ are independent of temperature. Alternatively, we can use the van't Hoff equation to determine $\Delta H°$ for reactions for which K is known at two or more temperatures. This approach is useful for finding enthalpy changes for reactions that might be difficult or inconvenient to measure directly by calorimetric methods.

The effect of a temperature change on the equilibrium constant depends on the sign of $\Delta H°$. If $\Delta H°$ is negative (the reaction is exothermic, giving off energy as heat), then increasing the temperature *reduces* K. If $\Delta H°$ is positive (the reaction is endothermic, taking up energy as heat), then increasing the temperature *increases* K. These observations obtained from thermodynamics provide the quantitative basis for Le Châtelier's principle (Fig. 14.13).

EXAMPLE 14.16

Calculate K for the equilibrium of Example 14.4 at $T = 400$ K, assuming $\Delta H°$ to be approximately independent of temperature over the range from 298 to 400 K.

Solution

The first step is to calculate $\Delta H°$ for the reaction. Appendix D provides data to calculate

$$\Delta H° = -155.52 \text{ kJ}$$

(Continued)

From the van't Hoff equation,

$$\ln\left(\frac{K_{400}}{K_{298}}\right) = -\frac{\Delta H^\circ}{R}\left[\frac{1}{400\text{ K}} - \frac{1}{298\text{ K}}\right]$$

$$= \frac{155{,}520\text{ J mol}^{-1}}{8.315\text{ J K}^{-1}\text{ mol}^{-1}}\left[\frac{1}{400\text{ K}} - \frac{1}{298\text{ K}}\right] = -16.01$$

$$\frac{K_{400}}{K_{298}} = e^{-16.01} = 1.1\times 10^{-7}$$

Taking K_{298} to be 1.8×10^{18} (from Example 14.4) gives

$$K_{400} = (1.8\times 10^{18})(1.1\times 10^{-7}) = 2.0\times 10^{11}$$

Because the reaction is exothermic, an increase in temperature reduces the equilibrium constant.

An alternative way to do this calculation would be to determine both ΔH° and ΔS° and from them to calculate ΔG at 400 K.

Related Problems: 61, 62, 63, 64, 65, 66

14.7.4 Temperature Dependence of Vapor Pressure

Suppose pure liquid water is in equilibrium with its vapor at temperature T:

$$H_2O(\ell) \rightleftharpoons H_2O(g) \qquad P_{H_2O(g)} = K$$

The temperature dependence of K (and therefore of the vapor pressure $P_{H_2O(g)}$) can be described by the van't Hoff equation in the same way as chemical reactions. If ΔH_{vap} and ΔS_{vap} are approximately independent of temperature, then from the van't Hoff equation,

$$\ln\left(\frac{K_2}{K_1}\right) = \ln\left(\frac{P_2}{P_1}\right) = -\frac{\Delta H_{vap}}{R}\left[\frac{1}{T_2} - \frac{1}{T_1}\right]$$

where P_2 and P_1 are the vapor pressures at temperatures T_2 and T_1. From this equation, the vapor pressure at any given temperature can be estimated if its value at some other temperature is known and if the enthalpy of vaporization is also known.

At the normal boiling point of a substance, T_b, the equilibrium vapor pressure is 1 atm. If T_1 is taken to correspond to T_b and T_2 to some other temperature T, the van't Hoff equation is

$$\ln P = -\frac{\Delta H_{vap}}{R}\left[\frac{1}{T} - \frac{1}{T_b}\right] \qquad [14.13]$$

where P is the equilibrium vapor pressure at temperature T, expressed in atmospheres.

EXAMPLE 14.17

The ΔH_{vap} for water is 40.66 kJ mol^{-1} at the normal boiling point, T_b = 373 K. Assuming ΔH_{vap} and ΔS_{vap} are approximately independent of temperature from 50°C to 100°C, estimate the vapor pressure of water at 50°C (323 K).

Solution

$$\ln P_{323} = \frac{-40660\text{ J mol}^{-1}}{8.315\text{ J K}^{-1}\text{ mol}^{-1}}\left[\frac{1}{323\text{ K}} - \frac{1}{373\text{ K}}\right] = -2.03$$

$$P_{323} = 0.13\text{ atm}$$

This differs slightly from the experimental value, 0.1217 atm, because ΔH_{vap} *does* change with temperature, an effect that was neglected in the approximate calculation.

Related Problems: 67, 68

14.8 Distribution of a Single Species between Immiscible Phases: Extraction and Separation Processes

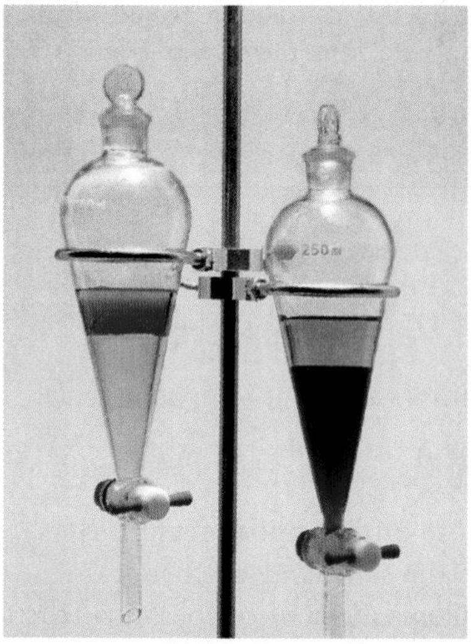

FIGURE 14.14 Iodine is dissolved in water and poured on top of carbon tetrachloride in a separatory funnel (left). After the funnel is shaken (right), the iodine reaches a partition equilibrium between the upper (aqueous) phase and the lower (CCl_4) phase. The deeper color in the lower phase indicates that iodine dissolves preferentially in the denser CCl_4 phase.

An important type of heterogeneous equilibrium involves partitioning a solute species between two immiscible solvent phases. Such equilibria are used in many separation processes in chemical research and in industry.

Suppose two immiscible liquids, such as water and carbon tetrachloride, are put in a container. "Immiscible" means mutually insoluble; these liquids separate into two phases with the less dense liquid, in this case water, lying on top of the other liquid. A visible boundary, the meniscus, separates the two phases. If a solute such as iodine is added to the mixture and the vessel is shaken to distribute the iodine through the container (Fig. 14.14), the iodine is partitioned between the two phases at equilibrium with a characteristic concentration ratio, the **partition coefficient** K. This is the equilibrium constant for the process

$$I_2(aq) \rightleftharpoons I_2(CCl_4)$$

and can be written as

$$\left[\frac{[I_2]_{CCl_4}}{[I_2]_{aq}}\right]_{eq} = K$$

in which $[I_2]_{CCl_4}$ and $[I_2]_{aq}$ are the concentrations (in moles per liter) of I_2 in the CCl_4 and aqueous phases, respectively. At 25°C, K has the value 85 for this equilibrium. The fact that K is greater than 1 shows that iodine is more soluble in CCl_4 than in water. If iodide ion (I^-) is dissolved in the water, then it can react with iodine to form the triiodide ion I_3^-:

$$I_2(aq) + I^-(aq) \longrightarrow I_3^-(aq)$$

This consumes $I_2(aq)$ and, by Le Châtelier's principle, causes more I_2 in the first equilibrium to move from the CCl_4 phase to the aqueous phase.

14.8.1 Extraction Processes

Extraction takes advantage of the partitioning of a solute between two immiscible solvents to remove that solute from one solvent into another. Suppose iodine is present as a contaminant in water that also contains other solutes that are insoluble in carbon tetrachloride. In such a case, most of the iodine could be removed by shaking the aqueous solution with CCl_4, allowing the two phases to separate, and then pouring off the water layer from the heavier layer of carbon tetrachloride. The greater the equilibrium constant for the partition of a solute from the original solvent into the extracting solvent, the more complete such a separation will be.

EXAMPLE 14.18

An aqueous solution has an iodine concentration of 2.00×10^{-3} M. Calculate the percentage of iodine remaining in the aqueous phase after extraction of 0.100 L of this aqueous solution with 0.050 L of CCl_4 at 25°C.

Solution

The number of moles of I_2 present is

$$(2.00 \times 10^{-3} \text{ mol L}^{-1})(0.100 \text{ L}) = 2.00 \times 10^{-4} \text{ mol}$$

(Continued)

Suppose that y mol remains in the aqueous phase and $(2.00 \times 10^{-4} - y)$ mol passes into the CCl_4 phase. Then

$$\left[\frac{[I_2]_{CCl_4}}{[I_2]_{aq}}\right]_{eq} = K = 85$$

$$= \frac{(2.00 \times 10^{-4} - y)/0.050}{y/0.100}$$

$$= \frac{2(2.00 \times 10^{-4} - y)}{y}$$

Note that the volumes of the two solvents used are unchanged because they are immiscible. Solving for y gives

$$y = 4.6 \times 10^{-6} \text{ mol}$$

The fraction remaining in the aqueous phase is $(4.6 \times 10^{-6})/(2.0 \times 10^{-4}) = 0.023$, or 2.3%. Additional extractions could be carried out to remove more of the I_2 from the aqueous phase.

Related Problems: 71, 72, 73, 74

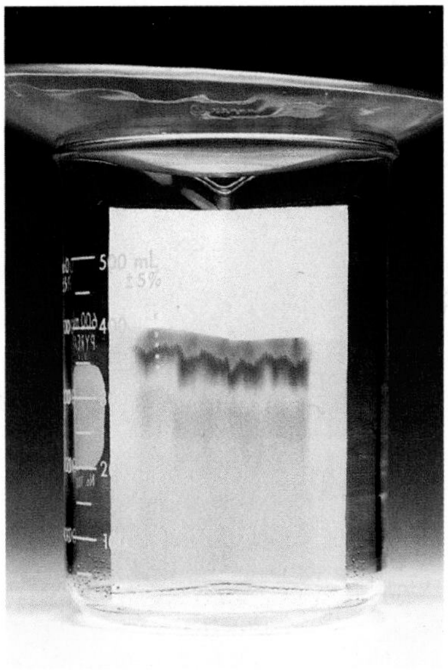

© Cengage Learning/Charles D. Winters

FIGURE 14.15 Paper chromatography separates a line of ink, drawn across the bottom of the paper, into its component colors. As the water rises through the paper, the different components of the ink are attracted differently to the water and the paper and are separated.

One extraction process used industrially on a large scale is the purification of sodium hydroxide for use in the manufacture of rayon. The sodium hydroxide produced by electrolysis typically contains 1% sodium chloride and 0.1% sodium chlorate as impurities. If a concentrated aqueous solution of sodium hydroxide is extracted with liquid ammonia, the NaCl and $NaClO_3$ are partitioned into the ammonia phase in preference over the aqueous phase. The heavier aqueous phase is added to the top of an extraction vessel filled with ammonia, and equilibrium is reached as droplets of it settle through the ammonia phase to the bottom. This procedure reduces impurity concentrations in the sodium hydroxide solution to about 0.08% NaCl and 0.0002% $NaClO_3$.

14.8.2 Chromatographic Separations

Partition equilibria are the basis of an important class of separation techniques called **chromatography.** This word comes from the Greek root *chroma,* meaning "color," and was chosen because the original chromatographic separations involved colored substances. The technique can be applied to a variety of mixtures of substances (Fig. 14.15).

Chromatography is a continuous extraction process in which solute species are exchanged between two phases. One, the mobile phase, moves with respect to the other, stationary phase. The partition ratio K of a solute A between the stationary and mobile phases is

$$\left[\frac{[A]_{stationary}}{[A]_{mobile}}\right]_{eq} = K$$

As the mobile phase containing solute passes over the stationary phase, the solute molecules move between the two phases. True equilibrium is never fully established because the motion of the fluid phase continually brings fresh solvent into contact with the stationary phase. Nevertheless, the partition coefficient K provides a guide to the behavior of a particular solute. The greater K is, the more time the solute spends in the stationary phase and therefore the slower its progress through the separation system. Solutes with different values of K are separated by their different rates of travel.

Different types of chromatography use different mobile and stationary phases; Table 14.1 lists some of the most important. *Column chromatography* (Fig. 14.16) uses a tube packed with a porous material, often a silica gel on which water has been adsorbed. Water is therefore the stationary liquid phase in this case. Other solvents such as pyridine or benzene are used in the mobile phase; in some cases it is most efficient to use different solvents in succession to separate the components of a solute mixture. As the solute fractions reach the bottom of the column, they are separated for analysis or use. Column chromatography is

TABLE 14.1

Chromatographic Separation Techniques[†]

Name	Mobile Phase	Stationary Phase
Gas–liquid	Gas	Liquid adsorbed on a porous solid in a tube
Gas–solid	Gas	Porous solid in a tube
Column	Liquid	Liquid adsorbed on a porous solid in a tubular column
Paper	Liquid	Liquid held in the pores of a thick paper
Thin layer	Liquid	Liquid or solid; solid is held on glass plate and liquid may be adsorbed on it
Ion exchange	Liquid	Solid (finely divided ion-exchange resin) in a tubular column

[†]Adapted from D. A. Skoog and D. M. West, *Analytical Chemistry* (Saunders College Publishing, Philadelphia, 1980), table 18-1.

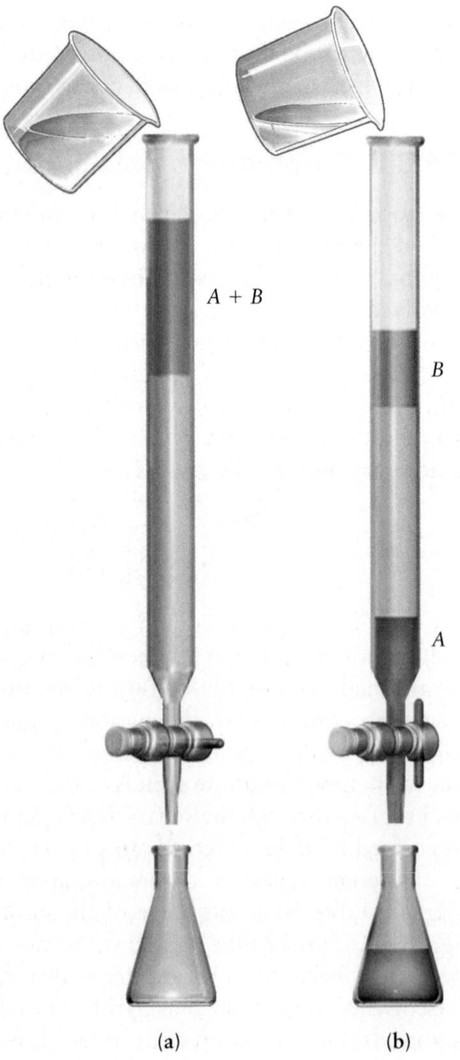

FIGURE 14.16 (a) In a column chromatograph, the top of the column is loaded with a mixture of solutes to be separated (green). (b) Upon addition of solvent, the different solutes travel at different rates, giving rise to bands. The separate fractions can be collected in different flasks for use or analysis.

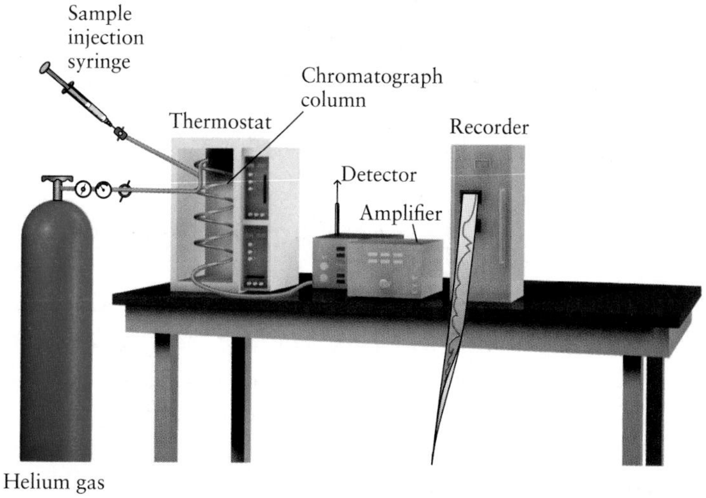

FIGURE 14.17 In a gas–liquid chromatograph, the sample is vaporized and passes through a column, carried in a stream of an inert gas such as helium or nitrogen. The residence time of any substance on the column depends on its partition coefficient from the vapor to the liquid in the column. A species leaving the column at a given time can be detected by a variety of techniques. The result is a gas chromatogram, with a peak corresponding to each substance in the mixture.

important in industry because it is easily increased from laboratory to production scale.

Gas–liquid chromatography (Fig. 14.17) is one of the most important separation techniques for modern chemical research. The stationary phase is again a liquid adsorbed on a porous solid, but the mobile phase is now gaseous. The sample is vaporized and passes through the column, carried in a stream of an inert gas such as helium or nitrogen. The residence time in the column depends on the partition coefficient of the solute species, allowing an efficient separation of mixtures. The solute leaving the column at a given time can be detected by a variety of techniques that produce a *gas chromatogram* with a peak corresponding to each solute species in the mixture. Gas–liquid chromatography is widely used for separating the products of organic reactions. It can also be used to determine the purity of substances, because even very small amounts of impurities appear clearly as separate peaks in the chromatogram. The technique is important in the separation and identification of trace amounts of possibly toxic substances in environmental and biological samples. Amounts on the order of parts per trillion (10^{-12} g in a 1-g sample) can be detected and identified.

CHAPTER SUMMARY

Most chemical reactions do not go to completion. They arrive at the equilibrium state, after which there is no further net change in the amount of products or reactants. At the microscopic level, the reaction continues in both forward and reverse directions, at equal rates in these opposing directions. So, chemical equilibrium—which we measure at the macroscopic level—is characterized by a dynamical balance of events on the microscopic level. At the equilibrium state, concentrations of products and reactants always satisfy the mass action law, and so are related by the equilibrium constant. The equilibrium constant is a unique property of each chemical reaction. Knowledge of the equilibrium constant enables us to calculate the equilibrium concentrations of products and reactants that result from any set of initial concentrations for that reaction. Equilibrium calculations have enormous predictive power for interpreting and optimizing the outcome of chemical reactions. Thermodynamics explains all the empirical observations of chemical equilibrium and provides means for quantitative predictions. Thermodynamics explains the form of the mass action law, shows how to calculate the equilibrium constant from tabulations of Gibbs free energy for products and reactants, explains the temperature dependence of the equilibrium constant, and predicts the direction of change in response to any disturbance of the equilibrium state.

CONCEPTS AND SKILLS

Section 14.1 – The Nature of Chemical Equilibrium

Describe the nature of the equilibrium state in chemical reactions.

- The equilibrium state shows no macroscopic evidence of change. On the microscopic level it is a dynamic balance between forward and reverse processes. The equilibrium state is reached through a spontaneous process. It is the same regardless of whether it is approached from the reactant side of the chemical equation or from the product side.

Section 14.2 – The Empirical Law of Mass Action

Set up the equilibrium expression for homogeneous reactions in the gas phase (Problems 1–6).

- Starting from the balanced equation for the reaction, place the partial pressure of each product in the numerator of the equilibrium expression and the partial pressure of each reactant in the denominator. Raise each of these to a power equal to its stoichiometric coefficient in the balanced reaction equation. The result is

$$\left[\frac{P_C^c P_D^d}{P_A^a P_B^b}\right]_{eq} = K$$

- Each partial pressure is expressed in units of atm, and is understood to be relative to a reference state with $P_{ref} = 1_{atm}$. Thus K is a dimensionless quantity.

Set up the equilibrium expression for homogeneous reactions in solution (Problems 7–8).

- Proceed as for the gas phase example. The result is

$$\left[\frac{[C]^c[D]^d}{[A]^a[B]^c}\right]_{eq} = K$$

- Each concentration is expressed in units of moles per liter, and is understood to be relative to a reference state with $c_{ref} = 1 \text{ mol L}^{-1}$. Thus K is a dimensionless quantity.

Set up the equilibrium expression for heterogeneous reactions (Problems 9–12).

- Proceed as in the previous cases, but representing the concentration of each product or reactant as its activity. The result is

$$\left[\frac{a_C^c a_D^d}{a_A^a a_B^b}\right]_{eq} = K$$

- Each activity is expressed relative to the proper standard state ($a = 1$) for that species, so K is a dimensionless quantity. For practical calculations involving dilute solutions and gases at low pressure, replace a for a gas with its limiting form of partial pressure in atm, a for a solute with its concentration in moles per liter, and a for a pure solid or liquid with its reference state value $a = 1$.

Section 14.3 – Thermodynamic Description of the Equilibrium State

Relate the equilibrium constant of a reaction to its change in standard Gibbs free energy (Problems 13–16).

- Calculate ΔG the change in Gibbs free energy for the reaction by a three-step process: ΔG_1 is the change for taking reactants from their initial conditions to standard state conditions. ΔG_2 is the change for converting reactants to products at standard state conditions. ΔG_3 is the change for taking products from standard state conditions to their final conditions. The equilibrium condition $\Delta G = 0$ leads to the relation $-\Delta G° = RT \ln K(T)$ from which we calculate $K(T)$ by first calculating the change in standard Gibbs free energy. Pay careful attention to the reference state for each standard free energy value you use in these calculations and note that all of these steps must be at the same temperature.

Section 14.4 – The Law of Mass Action for Related and Simultaneous Equilibria

Combine the equilibrium constants for individual reactions to obtain net equilibrium constants for combined reactions (Problems 17–20).

- $K_{\text{reverse}} = 1/K_{\text{forward}}$
- K (equation multiplied by n) $= K^n$
- $K_3 = K_1 K_2 \ldots$ when a series of reactions is added to give an overall reaction.

Section 14.5 – Equilibrium Calculations for Gas-Phase and Heterogeneous Reactions

Calculate equilibrium constants from experimental measurements of partial and total pressures (Problems 21–26).

- Substitute experimental results measured at equilibrium into the equilibrium expression and evaluate K. If the equilibrium partial pressures are not known directly, they can be calculated from other experimental data and then substituted into the equilibrium expression to determine K. The procedures are illustrated in Examples 14.9 and 14.10.

Calculate the equilibrium partial pressures of all species involved in a gas-phase or gas-solid reaction from the initial pressure(s) of the reactants (Problems 27–36).

- Assume that partial pressure of one of the species decreases by y atm while reaching the equilibrium state. Use the coefficients in the balanced reaction to write an equation for the partial pressure of each species at equilibrium in terms of y. Substitute these equilibrium partial pressures into the equilibrium relation, solve for y, and evaluate the partial pressure of each species at equilibrium.

Relate concentrations to partial pressures in equilibrium calculations (Problems 37–38).

- Define the concentration for each species A as the number of moles of A per unit volume, $[A] = n_A/V$, and relate this to the partial pressure of A through the ideal gas law $P_A = RT[A]$. Substitute the partial pressure for each species, divided by P_{ref}, into the equilibrium expression, insert the value of K, and solve to obtain the equilibrium concentration of each species.

Section 14.6 – The Direction of Change in Chemical Reactions: Empirical Description

Determine the direction in which a chemical reaction will proceed spontaneously by calculating its reaction quotient (Problems 45–46).

- Calculate the reaction quotient from the experimental data at a particular experimental condition. Compare the magnitude of Q at that condition to K and predict the direction of the reaction from the following guidelines:
  - $Q = \frac{P_C^c P_D^d}{P_A^a P_B^b}$
  - $Q < K \Rightarrow$ reaction moves to the right.
  - $Q = K \Rightarrow$ system at equilibrium, no reaction.
  - $Q > K \Rightarrow$ reaction moves to the left.

State Le Châtelier's principle and give several applications (Problems 47–58).

- Le Châtelier's Principle—A system in equilibrium that is subjected to a stress will react in a way to partially counteract the stress. Several specific cases are:
  - Stress 1: Increase concentration or pressure of species A.
  - Response 1: Reaction will move in appropriate direction to decrease A.
  - Stress 2: Increase pressure.
  - Response 2: Reaction will move in the direction that produces fewer gaseous molecules in order to decrease the pressure.
  - Stress 3: Decrease volume (same as increase pressure).
  - Response 3: Reaction moves in the direction that produces fewer gaseous molecules in order to decrease the pressure.
  - Stress 4: Increase temperature.
  - Response 4: Reaction moves in appropriate direction to absorb heat and decrease the temperature.

Section 14.7 – The Direction of Change in Chemical Reactions: Thermodynamics Explanation

Relate the temperature dependence of the equilibrium constant for a reaction to the standard enthalpy change for the reaction (Problems 59–70).

- Over temperature ranges where the standard enthalpy change of the reaction and the standard entropy change do not depend on temperature, the values of K at two different temperatures are related by the van't Hoff equation:

$$\ln\left(\frac{K_2}{K_1}\right) = -\frac{\Delta H^\circ}{R}\left[\frac{1}{T_2} - \frac{1}{T_1}\right]$$

- This equation shows that K increases with temperature for endothermic reactions and decreases with T for exothermic reactions, as predicted by the principle of Le Châtelier.

Section 14.8 – Distribution of a Single Species between Immiscible Phases: Extraction and Separation Processes

Use the law of mass action to explain the distribution of a solute between two immiscible solvents (Problems 71–74).

- Distribution of a solute between immiscible phases is a concrete illustration of the spontaneous transfer of molecules across a boundary internal to the system, at fixed T and P, as described in Figure 13.9. The law of mass action characterizes the equilibrium state after such processes, giving the relative amounts of solute in the two phases. The partition coefficient, analogous to the chemical equilibrium constant, determines the distribution between phases.

Outline the basis for separation of compounds by partition chromatography.

- Partition chromatography relies on a mobile phase containing the solute flowing past a fresh solvent into which the solute is extracted. The effectiveness of the fresh solvent in extracting the solute is determined by the partition coefficient for the solute between the two immiscible solvents chosen for the process.

CUMULATIVE EXERCISE

Gunter Marx Photography/CORBIS

FIGURE 14.18 A stockpile of sulfur near chemical plants in Los Angeles.

This cumulative exercise is divided into two parts. Readers who have not yet studied thermodynamics should stop after Part 1; those who have studied thermodynamics should continue with Part 2.

Production of Sulfuric Acid—Part 1

Sulfuric acid is produced in larger volume than any other chemical and has a tremendous number of applications, ranging from fertilizer manufacture to metal treatment and chemical synthesis. Sulfur in its pure form is shown in Figure 14.18.

The modern industrial production of sulfuric acid involves three steps, for which the balanced chemical equations are:

1. $S(s) + O_2(g) \rightleftharpoons SO_2(g)$
2. $SO_2(g) + \frac{1}{2}O_2(g) \rightleftharpoons SO_3(g)$
3. $SO_3(g) + H_2O(\ell) \rightleftharpoons H_2SO_4(\ell)$

(a) Write an equilibrium expression for each of these steps, with equilibrium constants K_1, K_2, and K_3.

(b) If these reactions could be carried out at 25°C, the equilibrium constants would be 3.9×10^{52}, 2.6×10^{12}, and 2.6×10^{14}. Write a balanced equation for the overall reaction and calculate its equilibrium constant at 25°C.

(c) Although the products of all three equilibria are strongly favored at 25°C [See the data given in part (b)], reactions 1 and 2 occur too slowly to be practical; they must be carried out at elevated temperatures. At 700°C, the partial pressures of SO_2, O_2, and SO_3 in an equilibrium mixture are measured to be 2.23 atm, 1.14 atm, and 6.26 atm, respectively. Calculate K_2 at 700°C.

(d) At 300°C, K_2 has the value 1.3×10^4. Suppose some SO_3 is introduced into an evacuated vessel at an initial partial pressure of 0.89 atm. Calculate the partial pressure of SO_2 that will be reached at equilibrium, assuming that only reaction 2 takes place under these conditions. (*Hint:* K_2 is large enough that you can assume the fraction of SO_3 dissociated is very small.)

(e) Some SO_2 is placed in a flask and heated with oxygen to 600°C, at which point $K_2 = 9.5$. At equilibrium, 62% of it has reacted to give SO_3. Calculate the partial pressure of oxygen at equilibrium in this reaction mixture.

(f) Equal numbers of moles of SO_2, O_2, and SO_3 are mixed and heated to 600°C, where their total pressure before reaction is 0.090 atm. Will reaction 2 occur from right to left or from left to right? Will the total pressure increase or decrease during the course of the reaction?

(g) Reactions 1 and 3 are both exothermic. State the effects on equilibria 1 and 3 of increasing the temperature and of decreasing the volume. (*Note:* A change in volume has little effect on liquids and solids taking part in a reaction.)

Answers—Part 1

(a) $\left[\dfrac{P_{SO_2}}{P_{O_2}}\right]_{eq} = K_1 \qquad \left[\dfrac{P_{SO_3}}{P_{SO_2}P_{O_2}^{1/2}}\right]_{eq} = K_2 \qquad \left[\dfrac{1}{P_{SO_3}}\right]_{eq} = K_3$

(b) $S(s) + \frac{3}{2}O_2(g) + H_2O(\ell) \rightleftharpoons H_2SO_4(\ell)$; $K = K_1K_2K_3 = 2.2 \times 10^{79}$

(c) K_2 is 2.63.

(d) The SO_2 partial pressure is 2.1×10^{-3} atm.

(e) The O_2 partial pressure is 0.029 atm.

(f) Left to right; pressure will decrease.

(g) Increasing the temperature will shift both equilibria to the left. Decreasing the volume will not affect equilibrium 1 and will shift equilibrium 3 to the right.

Production of Sulfuric Acid—Part 2

We continue with examination of reactions 1, 2, and 3 in the production of sulfuric acid.

(h) Calculate $\Delta G°$ for each reaction at 25°C. (Standard state of sulfur is rhombic.)

(i) Write a balanced equation for the overall reaction, and calculate its value of $\Delta G°$ at 25°C.

(j) Part (h) shows all three reactions are spontaneous at 25°C if the partial pressures of each species involved is 1 atm. Nonetheless, reactions 1 and 2 occur too slowly at 25°C to be practical; they must be carried out at higher temperatures. Calculate $\Delta G°$ for 1, 2, and 3 at 700°C.

(k) Determine the highest temperature at which all three reactions are spontaneous.

Answers—Part 2

(h) At 25°C: $\Delta G_1^\circ = -300.19$ kJ mol^{-1}; $\Delta G_2^\circ = -70.86$ kJ mol^{-1}; $\Delta G_3^\circ = -81.64$ kJ mol^{-1}

(i) $S(s) + \frac{3}{2}O_2(g) + H_2O(\ell) \rightleftharpoons H_2SO_4(\ell)$; $\Delta G_{net}^\circ = -452.89$ kJ mol^{-1}

(j) At 700° C: $\Delta G_1^\circ = -307.81$ kJ mol^{-1}; $\Delta G_2^\circ = -7.47$ kJ mol^{-1}; $\Delta G_3^\circ = +32.64$ kJ mol^{-1}

(k) Because reaction 1 has $\Delta H° < 0$ and $\Delta S° > 0$, it is spontaneous at all temperatures if the partial pressures of each species is 1 atm. (If the gaseous species are not each at 1 atm, the $RT \ln Q$ term can change the sign of ΔG to be different than $\Delta G°$, so be careful not to make conclusions about spontaneity from $\Delta G°$ alone.) Because both reaction 2 and reaction 3 have $\Delta H°$ < 0 and $\Delta S° < 0$, each is spontaneous at temperatures below the temperature T^* at which $\Delta G = 0$. With the higher ratio of $\Delta S°$ to $\Delta H°$, reaction 3 is the first to become nonspontaneous, at $T^* = 508°C$ (under conditions where every partial pressure is 1 atm).

PROBLEMS

Answers to problems whose numbers are red appear in Appendix G. Problems that are more challenging are indicated with asterisks.

14.2 The Empirical Law of Mass Action

1. Write equilibrium expressions for the following gas-phase reactions.
   (a) $2\ H_2(g) + O_2(g) \rightleftharpoons 2\ H_2O(g)$
   (b) $Xe(g) + 3\ F_2(g) \rightleftharpoons XeF_6(g)$
   (c) $2\ C_6H_6(g) + 15\ O_2(g) \rightleftharpoons 12\ CO_2(g) + 6\ H_2O(g)$
2. Write equilibrium expressions for the following gas-phase reactions.
   (a) $2\ Cl_2(g) + O_2(g) \rightleftharpoons 2\ Cl_2O(g)$
   (b) $N_2(g) + O_2(g) + Br_2(g) \rightleftharpoons 2\ NOBr(g)$
   (c) $C_3H_8(g) + 5\ O_2(g) \rightleftharpoons 3\ CO_2(g) + 4\ H_2O(g)$
3. At a moderately elevated temperature, phosphoryl chloride ($POCl_3$) can be produced in the vapor phase from the gaseous elements. Write a balanced chemical equation and an equilibrium expression for this system. Note that gaseous phosphorus consists of P_4 molecules at moderate temperatures.
4. If confined at high temperature, ammonia and oxygen quickly react and come to equilibrium with their products, water vapor and nitrogen oxide. Write a balanced chemical equation and an equilibrium expression for this system.
5. An important step in the industrial production of hydrogen is the reaction of carbon monoxide with water:

$$CO(g) + H_2O(g) \rightleftharpoons CO_2(g) + H_2(g)$$

   (a) Use the law of mass action to write the equilibrium expression for this reaction.
   (b) At 500°C, the equilibrium constant for this reaction is 3.9. Suppose that the equilibrium partial pressures of CO and H_2O are both 0.10 atm and that of CO_2 is 0.70 atm. Calculate the equilibrium partial pressure of $H_2(g)$.
6. Phosgene ($COCl_2$) is an important intermediate in the manufacture of certain plastics. It is produced by the reaction

$$CO(g) + Cl_2(g) \rightleftharpoons COCl_2(g)$$

   (a) Use the law of mass action to write the equilibrium expression for this reaction.
   (b) At 600°C, the equilibrium constant for this reaction is 0.20. Calculate the partial pressure of phosgene in equilibrium with a mixture of CO (at 0.0020 atm) and Cl_2 (at 0.00030 atm).
7. N_2O_4 is soluble in the solvent cyclohexane; however, dissolution does not prevent N_2O_4 from breaking down to give NO_2 according to the equation

$$N_2O_{4(\text{cyclohexane})} \rightleftharpoons 2\ NO_{2(\text{cyclohexane})}$$

   An effort to compare this solution equilibrium with the similar equilibrium in the gas gave the following actual experimental data at 20°C:

$[N_2O_4]$ (mol L^{-1})	$[NO_2]$ (mol L^{-1})
0.190×10^{-3}	2.80×10^{-3}
0.686×10^{-3}	5.20×10^{-3}
1.54×10^{-3}	7.26×10^{-3}
2.55×10^{-3}	10.4×10^{-3}
3.75×10^{-3}	11.7×10^{-3}
7.86×10^{-3}	17.3×10^{-3}
11.9×10^{-3}	21.0×10^{-3}

   (a) Graph the *square* of the concentration of NO_2 versus the concentration of N_2O_4.
   (b) Compute the average equilibrium constant of this reaction.
8. NO_2 is soluble in carbon tetrachloride (CCl_4). As it dissolves, it dimerizes to give N_2O_4 according to the equation

$$2\ NO_{2(CCl_4)} \rightleftharpoons N_2O_{4(CCl_4)}$$

   A study of this equilibrium gave the following experimental data at 20°C:

$[N_2O_4]$ (mol L^{-1})	$[NO_2]$ (mol L^{-1})
0.192×10^{-3}	2.68×10^{-3}
0.721×10^{-3}	4.96×10^{-3}
1.61×10^{-3}	7.39×10^{-3}
2.67×10^{-3}	10.2×10^{-3}
3.95×10^{-3}	11.0×10^{-3}
7.90×10^{-3}	16.6×10^{-3}
11.9×10^{-3}	21.4×10^{-3}

   (a) Graph the concentration of N_2O_4 versus the *square* of the concentration of NO_2.
   (b) Compute the average equilibrium constant of this reaction.
9. Using the law of mass action, write the equilibrium expression for each of the following reactions.
   (a) $8\ H_2(g) + S_8(s) \rightleftharpoons 8\ H_2S(g)$
   (b) $C(s) + H_2O(\ell) + Cl_2(g) \rightleftharpoons COCl_2(g) + H_2(g)$
   (c) $CaCO_3(s) \rightleftharpoons CaO(s) + CO_2(g)$
   (d) $3\ C_2H_2(g) \rightleftharpoons C_6H_6(\ell)$
10. Using the law of mass action, write the equilibrium expression for each of the following reactions.
   (a) $3\ C_2H_2(g) + 3\ H_2(g) \rightleftharpoons C_6H_{12}(\ell)$
   (b) $CO_2(g) + C(s) \rightleftharpoons 2\ CO(g)$
   (c) $CF_4(g) + 2\ H_2O(\ell) \rightleftharpoons CO_2(g) + 4\ HF(g)$
   (d) $K_2NiF_6(s) + TiF_4(s) \rightleftharpoons K_2TiF_6(s) + NiF_2(s) + F_2(g)$
11. Using the law of mass action, write the equilibrium expression for each of the following reactions.
   (a) $Zn(s) + 2\ Ag^+(aq) \rightleftharpoons Zn^{2+}(aq) + 2\ Ag(s)$
   (b) $VO_4^{3-}(aq) + H_2O(\ell) \rightleftharpoons VO_3(OH)^{2-}(aq) + OH^-(aq)$
   (c) $2\ As(OH)_6^{3-}(aq) + 6\ CO_2(g) \rightleftharpoons As_2O_3(s) + 6\ HCO_3^-(aq) + 3\ H_2O(\ell)$

12. Using the law of mass action, write the equilibrium expression for each of the following reactions.
(a) $6\ I^-(aq) + 2MnO_4^-(aq) + 4\ H_2O(\ell) \rightleftharpoons 3\ I_2(aq) + 2\ MnO_2(s) + 8\ OH^-(aq)$
(b) $2\ Cu^{2+}(aq) + 4\ I^-(aq) \rightleftharpoons 2\ CuI(s) + I_2(aq)$
(c) $\frac{1}{2}O_2(g) + Sn^{2+}(aq) + 3\ H_2O(\ell) \rightleftharpoons SnO_2(s) + 2\ H_3O^+(aq)$

14.3 Thermodynamic Description of the Equilibrium State

13. Calculate $\Delta G°$ and the equilibrium constant K at 25°C for the reaction

$$2\ NH_3(g) + \tfrac{7}{2}O_2(g) \rightleftharpoons 2\ NO_2(g) + 3\ H_2O(g)$$

using data in Appendix D.

14. Write a reaction for the dehydrogenation of gaseous ethane (C_2H_6) to acetylene (C_2H_2). Calculate $\Delta G°$ and the equilibrium constant for this reaction at 25°C, using data from Appendix D.

15. Use the thermodynamic data from Appendix D to calculate the equilibrium constant at 25°C for the following reactions:
(a) $SO_2(g) + \frac{1}{2}O_2(g) \rightleftharpoons SO_3(g)$
(b) $3\ Fe_2O_3(s) \rightleftharpoons 2\ Fe_3O_4(s) + \frac{1}{2}O_2(g)$
(c) $CuCl_2(s) \rightleftharpoons Cu^{2+}(aq) + 2\ Cl^-(aq)$

Write the equilibrium expression for each reaction.

16. Use the thermodynamic data from Appendix D to calculate equilibrium constants at 25°C for the following reactions.
(a) $H_2(g) + N_2(g) + 2\ O_2(g) \rightleftharpoons 2\ HNO_2(g)$
(b) $Ca(OH)_2(s) \rightleftharpoons CaO(s) + H_2O(g)$
(c) $Zn^{2+}(aq) + 4\ NH_3(aq) \rightleftharpoons Zn(NH_3)_4^{2+}(aq)$

Write the equilibrium expression for each reaction.

14.4 The Law of Mass Action for Related and Simultaneous Equilibria

17. At a certain temperature, the value of the equilibrium constant for the reaction

$$CS_2(g) + 3\ O_2(g) \rightleftharpoons CO_2(g) + 2\ SO_2(g)$$

is K_1. How is K_1 related to the equilibrium constant K_2 for the related equilibrium

$$\tfrac{1}{3}CS_2(g) + O_2(g) \rightleftharpoons \tfrac{1}{3}CO_2(g) + \tfrac{2}{3}SO_2(g)$$

at the same temperature?

18. At 25°C, the equilibrium constant for the reaction

$$6\ ClO_3F(g) \rightleftharpoons 2\ ClF(g) + 4\ ClO(g) + 7\ O_2(g) + 2\ F_2(g)$$

is 32.6. Calculate the equilibrium constant at 25°C for the reaction

$$\tfrac{1}{3}ClF(g) + \tfrac{2}{3}ClO(g) + \tfrac{7}{6}O_2(g) + \tfrac{1}{3}F_2(g) \rightleftharpoons ClO_3F(g)$$

19. Suppose that K_1 and K_2 are the respective equilibrium constants for the two reactions

$$XeF_6(g) + H_2O(g) \rightleftharpoons XeOF_4(g) + 2\ HF(g)$$

$$XeO_4(g) + XeF_6(g) \rightleftharpoons XeOF_4(g) + XeO_3F_2(g)$$

Give the equilibrium constant for the reaction

$$XeO_4(g) + 2\ HF(g) \rightleftharpoons XeO_3F_2(g) + H_2O(g)$$

in terms of K_1 and K_2.

20. At 1330 K, germanium(II) oxide (GeO) and tungsten(VI) oxide (W_2O_6) are both gases. The following two equilibria are established simultaneously:

$$2\ GeO(g) + W_2O_6(g) \rightleftharpoons 2\ GeWO_4(g)$$

$$GeO(g) + W_2O_6(g) \rightleftharpoons GeW_2O_7(g)$$

The equilibrium constants for the two are respectively 7.0×10^3 and 38×10^3. Compute K for the reaction

$$GeO(g) + GeW_2O_7(g) \rightleftharpoons 2\ GeWO_4(g)$$

14.5 Equilibrium Calculations for Gas-Phase and Heterogeneous Reactions

21. At 454 K, $Al_2Cl_6(g)$ reacts to form $Al_3Cl_9(g)$ according to the equation

$$3\ Al_2Cl_6(g) \rightleftharpoons 2\ Al_3Cl_9(g)$$

In an experiment at this temperature, the equilibrium partial pressure of $Al_2Cl_6(g)$ is 1.00 atm and the equilibrium partial pressure of $Al_3Cl_9(g)$ is 1.02×10^{-2} atm. Compute the equilibrium constant of the preceding reaction at 454 K.

22. At 298 K, $F_3SSF(g)$ decomposes partially to $SF_2(g)$. At equilibrium, the partial pressure of $SF_2(g)$ is 1.1×10^{-4} atm and the partial pressure of F_3SSF is 0.0484 atm.
(a) Write a balanced equilibrium equation to represent this reaction.
(b) Compute the equilibrium constant corresponding to the equation you wrote.

23. The compound 1,3-di-*t*-butylcyclohexane exists in two forms that are known as the chair and boat conformations because their molecular structures resemble those objects. Equilibrium exists between the two forms, represented by the equation

$$\text{chair} \rightleftharpoons \text{boat}$$

At 580 K, 6.42% of the molecules are in the chair form. Calculate the equilibrium constant for the preceding reaction as written.

24. At 248°C and a total pressure of 1.000 atm, the fractional dissociation of $SbCl_5$ is 0.718 for the reaction

$$SbCl_5(g) \rightleftharpoons SbCl_3(g) + Cl_2(g)$$

This means that 718 of every 1000 molecules of $SbCl_5$ originally present have dissociated. Calculate the equilibrium constant.

25. Sulfuryl chloride (SO_2Cl_2) is a colorless liquid that boils at 69°C. Above this temperature, the vapors dissociate into sulfur dioxide and chlorine:

$$SO_2Cl_2(g) \rightleftharpoons SO_2(g) + Cl_2(g)$$

This reaction is slow at 100°C, but it is accelerated by the presence of some $FeCl_3$ (which does not affect the final position of the equilibrium). In an experiment, 3.174 g of $SO_2Cl_2(\ell)$ and a small amount of solid $FeCl_3$ are put into an evacuated 1.000-L flask, which is then sealed and heated to 100°C. The total pressure in the flask at that temperature is found to be 1.30 atm.
(a) Calculate the partial pressure of each of the three gases present.
(b) Calculate the equilibrium constant at this temperature.

26. A certain amount of NOBr(*g*) is sealed in a flask, and the temperature is raised to 350 K. The following equilibrium is established:

$$NOBr(g) \rightleftharpoons NO(g) + \tfrac{1}{2}\,Br_2(g)$$

The total pressure in the flask when equilibrium is reached at this temperature is 0.675 atm, and the vapor density is 2.219 g L^{-1}.
(a) Calculate the partial pressure of each species.
(b) Calculate the equilibrium constant at this temperature.

27. The dehydrogenation of benzyl alcohol to make the flavoring agent benzaldehyde is an equilibrium process described by the equation

$$C_6H_5CH_2OH(g) \rightleftharpoons C_6H_5CHO(g) + H_2(g)$$

At 523 K, the value of its equilibrium constant is K = 0.558.
(a) Suppose 1.20 g of benzyl alcohol is placed in a 2.00-L vessel and heated to 523 K. What is the partial pressure of benzaldehyde when equilibrium is attained?
(b) What fraction of benzyl alcohol is dissociated into products at equilibrium?

28. Isopropyl alcohol can dissociate into acetone and hydrogen:

$$(CH_3)_2CHOH(g) \rightleftharpoons (CH_3)_2CO(g) + H_2(g)$$

At 179°C, the equilibrium constant for this dehydrogenation reaction is 0.444.
(a) If 10.00 g of isopropyl alcohol is placed in a 10.00-L vessel and heated to 179°C, what is the partial pressure of acetone when equilibrium is attained?
(b) What fraction of isopropyl alcohol is dissociated at equilibrium?

29. A weighed quantity of $PCl_5(s)$ is sealed in a 100.0-cm^3 glass bulb to which a pressure gauge is attached. The bulb is heated to 250°C, and the gauge shows that the pressure in the bulb rises to 0.895 atm. At this temperature, the solid PCl_5 is all vaporized and also partially dissociated into $Cl_2(g)$ and $PCl_3(g)$ according to the equation

$$PCl_5(g) \rightleftharpoons PCl_3(g) + Cl_2(g)$$

At 250°C, K = 2.15 for this reaction. Assume that the contents of the bulb are at equilibrium and calculate the partial pressure of the three different chemical species in the vessel.

30. Suppose 93.0 g of HI(*g*) is placed in a glass vessel and heated to 1107 K. At this temperature, equilibrium is quickly established between HI(*g*) and its decomposition products, $H_2(g)$ and $I_2(g)$:

$$2\ HI(g) \rightleftharpoons H_2(g) + I_2(g)$$

The equilibrium constant at 1107 K is 0.0259, and the total pressure at equilibrium is observed to equal 6.45 atm. Calculate the equilibrium partial pressures of HI(*g*), $H_2(g)$, and $I_2(g)$.

31. The equilibrium constant at 350 K for the reaction

$$Br_2(g) + I_2(g) \rightleftharpoons 2\ IBr(g)$$

has a value of 322. Bromine at an initial partial pressure of 0.0500 atm is mixed with iodine at an initial partial pressure of 0.0400 atm and held at 350 K until equilibrium is reached. Calculate the equilibrium partial pressure of each of the gases.

32. The equilibrium constant for the reaction of fluorine and oxygen to form oxygen difluoride (OF_2) is 40.1 at 298 K:

$$F_2(g) + \tfrac{1}{2}\,O_2(g) \rightleftharpoons OF_2(g)$$

Suppose some OF_2 is introduced into an evacuated container at 298 K and allowed to dissociate until its partial pressure reaches an equilibrium value of 1.00 atm. Calculate the equilibrium partial pressures of F_2 and O_2 in the container.

33. At 25°C, the equilibrium constant for the reaction

$$N_2(g) + O_2(g) \rightleftharpoons 2\ NO(g)$$

is 4.2×10^{-31}. Suppose a container is filled with nitrogen (at an initial partial pressure of 0.41 atm), oxygen (at an initial partial pressure of 0.59 atm), and nitrogen oxide (at an initial partial pressure of 0.22 atm). Calculate the partial pressures of all three gases after equilibrium is reached at this temperature.

34. At 25°C, the equilibrium constant for the reaction

$$2\ NO_2(g) \rightleftharpoons 2\ NO(g) + O_2(g)$$

is 5.9×10^{-13}. Suppose a container is filled with nitrogen dioxide at an initial partial pressure of 0.89 atm. Calculate the partial pressures of all three gases after equilibrium is reached at this temperature.

*35. The equilibrium constant for the synthesis of ammonia

$$N_2(g) + 3\ H_2(g) \rightleftharpoons 2\ NH_3(g)$$

has the value $K = 6.78 \times 10^5$ at 25°C. Calculate the equilibrium partial pressures of $N_2(g)$, $H_2(g)$, and $NH_3(g)$ at 25°C if the total pressure is 1.00 atm and the H:N atom ratio in the system is 3:1. (*Hint:* Try the approximation that P_{N_2} and $P_{H_2} \ll P_{NH_3}$ and see if the resulting equations are simplified.)

*36. At 400°C, $K = 3.19 \times 10^{-4}$ for the reaction in problem 35. Repeat the calculation for P_{N_2}, P_{H_2}, and P_{NH_3} assuming the same total pressure and composition. (*Hint:* Try the approximation that $P_{NH_3} \ll P_{N_2}$ and P_{H_2} and see if the resulting equations are simplified.)

37. Calculate the concentration of phosgene ($COCl_2$) that will be present at 600°C in equilibrium with carbon monoxide (at a concentration of 2.3×10^{-4} mol L^{-1}) and chlorine (at a concentration of 1.7×10^{-2} mol L^{-1}). (Use the data of problem 6.)

38. The reaction

$$SO_2Cl_2(g) \rightleftharpoons SO_2(g) + Cl_2(g)$$

has an equilibrium constant at 100°C of 2.40. Calculate the concentration of SO_2 that will be present at 100°C in equilibrium with SO_2Cl_2 (at a concentration of 3.6×10^{-4} mol L^{-1}) and chlorine (at a concentration of 6.9×10^{-3} mol L^{-1}).

39. At 298 K, the equilibrium constant for the reaction

$$Fe_2O_3(s) + 3\ H_2(g) \rightleftharpoons 2\ Fe(s) + 3\ H_2O(\ell)$$

is 4.0×10^{-6}, and that for

$$CO_2(g) + H_2(g) \rightleftharpoons CO(g) + H_2O(\ell)$$

is 3.2×10^{-4}. Suppose some solid Fe_2O_3, solid Fe, and liquid H_2O are brought into equilibrium with CO(*g*) and $CO_2(g)$ in a closed container at 298 K. Calculate the ratio of the partial pressure of CO(*g*) to that of $CO_2(g)$ at equilibrium.

40. A sample of ammonium carbamate placed in a glass vessel at 25°C undergoes the reaction

$$NH_4OCONH_2(s) \rightleftharpoons 2\ NH_3(g) + CO_2(g)$$

The total pressure of gases in equilibrium with the solid is found to be 0.115 atm.
(a) Calculate the partial pressures of NH_3 and CO_2.
(b) Calculate the equilibrium constant at 25°C.

41. The equilibrium constant for the reaction

$$NH_3(g) + HCl(g) \rightleftharpoons NH_4Cl(s)$$

at 340°C is $K = 4.0$.
(a) If the partial pressure of ammonia is $P_{NH_3} = 0.80$ atm and solid ammonium chloride is present, what is the equilibrium partial pressure of hydrogen chloride at 340°C?
(b) An excess of solid NH_4Cl is added to a container filled with ammonia at 340°C and a pressure of 1.50 atm. Calculate the pressures of $NH_3(g)$ and HCl(*g*) reached at equilibrium.

42. The equilibrium constant for the reaction

$$H_2(g) + I_2(s) \rightleftharpoons 2\ HI(g)$$

at 25°C is $K = 0.345$.
(a) If the partial pressure of hydrogen is $P_{H_2} = 1.00$ atm and solid iodine is present, what is the equilibrium partial pressure of hydrogen iodide, P_{HI}, at 25°C?
(b) An excess of solid I_2 is added to a container filled with hydrogen at 25°C and a pressure of 4.00 atm. Calculate the pressures of $H_2(g)$ and HI(*g*) reached at equilibrium.

43. Pure solid NH_4HSe is placed in an evacuated container at 24.8°C. Eventually, the pressure above the solid reaches the equilibrium pressure 0.0184 atm due to the reaction

$$NH_4HSe(s) \rightleftharpoons NH_3(g) + H_2Se(g)$$

(a) Calculate the equilibrium constant of this reaction at 24.8°C.
(b) In a different container, the partial pressure of $NH_3(g)$ in equilibrium with $NH_4HSe(s)$ at 24.8°C is 0.0252 atm. What is the partial pressure of $H_2Se(g)$?

44. The total pressure of the gases in equilibrium with solid sodium hydrogen carbonate at 110°C is 1.648 atm, corresponding to the reaction

$$2\ NaHCO_3(s) \rightleftharpoons Na_2CO_3(s) + H_2O(g) + CO_2(g)$$

($NaHCO_3$ is used in dry chemical fire extinguishers because the products of this decomposition reaction smother the fire.)
(a) Calculate the equilibrium constant at 110°C.
(b) What is the partial pressure of water vapor in equilibrium with $NaHCO_3(s)$ at 110°C if the partial pressure of $CO_2(g)$ is 0.800 atm?

14.6 The Direction of Change in Chemical Reactions: Empirical Description

45. Some Al_2Cl_6 (at a partial pressure of 0.473 atm) is placed in a closed container at 454 K with some Al_3Cl_9 (at a partial pressure of 1.02×10^{-2} atm). Enough argon is added to raise the total pressure to 1.00 atm.
(a) Calculate the initial reaction quotient for the reaction

$$3\ Al_2Cl_6(g) \rightleftharpoons 2\ Al_3Cl_9(g)$$

(b) As the gas mixture reaches equilibrium, will there be net production or consumption of Al_3Cl_9? (Use the data given in problem 21.)

46. Some SF_2 (at a partial pressure of 2.3×10^{-4} atm) is placed in a closed container at 298 K with some F_3SSF (at a partial pressure of 0.0484 atm). Enough argon is added to raise the total pressure to 1.000 atm.
(a) Calculate the initial reaction quotient for the decomposition of F_3SSF to SF_2.
(b) As the gas mixture reaches equilibrium, will there be net formation or dissociation of F_3SSF? (Use the data given in problem 22.)

47. The progress of the reaction

$$H_2(g) + Br_2(g) \rightleftharpoons 2\ HBr(g)$$

can be monitored visually by following changes in the color of the reaction mixture (Br_2 is reddish brown, and H_2 and HBr are colorless). A gas mixture is prepared at 700 K, in which 0.40 atm is the initial partial pressure of both H_2 and Br_2 and 0.90 atm is the initial partial pressure of HBr. The color of this mixture then fades as the reaction progresses toward equilibrium. Give a condition that must be satisfied by the equilibrium constant *K* (for example, it must be greater than or smaller than a given number).

48. Recall from our discussion of the NO_2–N_2O_4 equilibrium that NO_2 has a brownish color. At elevated temperatures, NO_2 reacts with CO according to

$$NO_2(g) + CO(g) \rightleftharpoons NO(g) + CO_2(g)$$

The other three gases taking part in this reaction are colorless. When a gas mixture is prepared at 500 K, in which 3.4 atm is the initial partial pressure of both NO_2 and CO, and 1.4 atm is the partial pressure of both NO and CO_2, the brown color of the mixture is observed to fade as the reaction progresses toward equilibrium. Give a condition that must be satisfied by the equilibrium constant K (for example, it must be greater than or smaller than a given number).

49. The equilibrium constant for the "water gas" reaction

$$C(s) + H_2O(g) \rightleftharpoons CO(g) + H_2(g)$$

is $K = 2.6$ at a temperature of 1000 K. Calculate the reaction quotient Q for each of the following conditions, and state which direction the reaction shifts in coming to equilibrium.

(a) $P_{H_2O} = 0.600$ atm; $P_{CO} = 1.525$ atm; $P_{H_2} = 0.805$ atm
(b) $P_{H_2O} = 0.724$ atm; $P_{CO} = 1.714$ atm; $P_{H_2} = 1.383$ atm

50. The equilibrium constant for the reaction

$$H_2S(g) + I_2(g) \rightleftharpoons 2\ HI(g) + S(s)$$

at 110°C is equal to 0.0023. Calculate the reaction quotient Q for each of the following conditions and determine whether solid sulfur is consumed or produced as the reaction comes to equilibrium.

(a) $P_{I_2} = 0.461$ atm; $P_{H_2S} = 0.050$ atm; $P_{HI} = 0.0$ atm
(b) $P_{I_2} = 0.461$ atm; $P_{H_2S} = 0.050$ atm; $P_{HI} = 9.0$ atm

51. At $T = 1200$°C the reaction

$$P_4(g) \rightleftharpoons 2\ P_2(g)$$

has an equilibrium constant $K = 0.612$.

(a) Suppose the initial partial pressure of P_4 is 5.00 atm and that of P_2 is 2.00 atm. Calculate the reaction quotient Q and state whether the reaction proceeds to the right or to the left as equilibrium is approached.
(b) Calculate the partial pressures at equilibrium.
(c) If the volume of the system is then increased, will there be net formation or net dissociation of P_4?

52. At $T = 100$°C the reaction

$$SO_2Cl_2(g) \rightleftharpoons SO_2(g) + Cl_2(g)$$

has an equilibrium constant $K = 2.4$.

(a) Suppose the initial partial pressure of SO_2Cl_2 is 1.20 atm, and $P_{SO_2} = P_{Cl_2} = 0$. Calculate the reaction quotient Q and state whether the reaction proceeds to the right or to the left as equilibrium is approached.
(b) Calculate the partial pressures at equilibrium.
(c) If the volume of the system is then decreased, will there be net formation or net dissociation of SO_2Cl_2?

53. Explain the effect of each of the following stresses on the position of the following equilibrium:

$$3\ NO(g) \rightleftharpoons N_2O(g) + NO_2(g)$$

The reaction as written is exothermic.

(a) $N_2O(g)$ is added to the equilibrium mixture without change of volume or temperature.
(b) The volume of the equilibrium mixture is reduced at constant temperature.
(c) The equilibrium mixture is cooled.
(d) Gaseous argon (which does not react) is added to the equilibrium mixture while both the total gas pressure and the temperature are kept constant.
(e) Gaseous argon is added to the equilibrium mixture without changing the volume.

54. Explain the effect of each of the following stresses on the position of the equilibrium

$$SO_3(g) \rightleftharpoons SO_2(g) + \tfrac{1}{2}\ O_2(g)$$

The reaction as written is endothermic.

(a) $O_2(g)$ is added to the equilibrium mixture without changing volume or temperature.
(b) The mixture is compressed at constant temperature.
(c) The equilibrium mixture is cooled.
(d) An inert gas is pumped into the equilibrium mixture while the total gas pressure and the temperature are kept constant.
(e) An inert gas is added to the equilibrium mixture without changing the volume.

55. In a gas-phase reaction, it is observed that the equilibrium yield of products is increased by lowering the temperature and by reducing the volume.

(a) Is the reaction exothermic or endothermic?
(b) Is there a net increase or a net decrease in the number of gas molecules in the reaction?

56. The equilibrium constant of a gas-phase reaction increases as temperature is increased. When the nonreacting gas neon is admitted to a mixture of reacting gases (holding the temperature and the total pressure fixed and increasing the volume of the reaction vessel), the product yield is observed to decrease.

(a) Is the reaction exothermic or endothermic?
(b) Is there a net increase or a net decrease in the number of gas molecules in the reaction?

57. The most extensively used organic compound in the chemical industry is ethene (C_2H_4). The two equations

$$C_2H_4(g) + Cl_2(g) \rightleftharpoons C_2H_4Cl_2(g)$$

$$C_2H_4Cl_2(g) \rightleftharpoons C_2H_3Cl(g) + HCl(g)$$

represent the way in which vinyl chloride (C_2H_3Cl) is synthesized for eventual use in polymeric plastics (polyvinyl chloride, PVC). The byproduct of the reaction, HCl, is now most cheaply made by this and similar reactions, rather than by direct combination of H_2 and Cl_2. Heat is given off in the first reaction and taken up in the second. Describe how you would design an industrial process to maximize the yield of vinyl chloride.

58. Methanol is made via the exothermic reaction

$$CO(g) + 2\,H_2(g) \longrightarrow CH_3OH(g)$$

Describe how you would control the temperature and pressure to maximize the yield of methanol.

14.7 The Direction of Change in Chemical Reactions: Thermodynamic Explanation

59. One way to manufacture ethanol is by the reaction

$$C_2H_4(g) + H_2O(g) \rightleftharpoons C_2H_5OH(g)$$

The ΔH_f° of $C_2H_4(g)$ is 52.3 kJ mol^{-1}; of $H_2O(g)$, −241.8 kJ mol^{-1}; and of $C_2H_5OH(g)$, −235.3 kJ mol^{-1}. Without doing detailed calculations, suggest the conditions of pressure and temperature that will maximize the yield of ethanol at equilibrium.

60. Dimethyl ether (CH_3OCH_3) is a good substitute for environmentally harmful propellants in aerosol spray cans. It is produced by the dehydration of methanol:

$$2\,CH_3OH(g) \rightleftharpoons CH_3OCH_3(g) + H_2O(g)$$

Describe reaction conditions that favor the equilibrium production of this valuable chemical. As a basis for your answer, compute ΔH° and ΔS° of the reaction from the data in Appendix D.

61. The equilibrium constant at 25°C for the reaction

$$2\,NO_2(g) \rightleftharpoons N_2O_4(g)$$

is 6.8. At 200°C the equilibrium constant is 1.21×10^{-3}. Calculate the enthalpy change (ΔH) for this reaction, assuming that ΔH° and ΔS° of the reaction are constant over the temperature range from 25°C to 200°C.

62. Stearic acid dimerizes when dissolved in hexane:

$$2\,C_{17}H_{35}COOH(\textit{hexane}) \rightleftharpoons (C_{17}H_{35}COOH)_2(\textit{hexane})$$

The equilibrium constant for this reaction is 2900 at 28°C, but it drops to 40 at 48°C. Estimate ΔH° and ΔS° for the reaction.

63. The equilibrium constant for the reaction

$$\tfrac{1}{2}Cl_2(g) + \tfrac{1}{2}F_2(g) \rightleftharpoons ClF(g)$$

is measured to be 9.3×10^9 at 298 K and 3.3×10^7 at 398 K.

(a) Calculate ΔG° at 298 K for the reaction.

(b) Calculate ΔH° and ΔS°, assuming the enthalpy and entropy changes to be independent of temperature between 298 and 398 K.

64. Stearic acid also dimerizes when dissolved in carbon tetrachloride:

$$2\,C_{17}H_{35}COOH(CCl_4) \rightleftharpoons (C_{17}H_{35}COOH)_2(CCl_4)$$

The equilibrium constant for this reaction is 2780 at 22°C, but it drops to 93.1 at 42°C. Estimate ΔH° and ΔS° for the reaction.

65. For the synthesis of ammonia from its elements,

$$3\,H_2(g) + N_2(g) \rightleftharpoons 2\,NH_3(g)$$

the equilibrium constant $K = 5.9 \times 10^5$ at 298 K, and $\Delta H^\circ = -92.2$ kJ mol^{-1}. Calculate the equilibrium constant for the reaction at 600 K, assuming no change in ΔH° and ΔS° between 298 K and 600 K.

66. The cumulative exercise at the end of Chapter 14 explored reaction steps in the manufacture of sulfuric acid, including the oxidation of sulfur dioxide to sulfur trioxide:

$$SO_2(g) + \tfrac{1}{2}O_2(g) \rightleftharpoons SO_3(g)$$

At 25°C the equilibrium constant for this reaction is 2.6×10^{12}, but the reaction occurs very slowly. Calculate K for this reaction at 550°C, assuming ΔH° and ΔS° are independent of temperature in the range from 25°C to 550°C.

67. The vapor pressure of ammonia at −50°C is 0.4034 atm; at 0°C, it is 4.2380 atm.

(a) Calculate the molar enthalpy of vaporization (ΔH_{vap}) of ammonia.

(b) Calculate the normal boiling temperature of $NH_3(\ell)$.

68. The vapor pressure of butyl alcohol (C_4H_9OH) at 70.1°C is 0.1316 atm; at 100.8°C, it is 0.5263 atm.

(a) Calculate the molar enthalpy of vaporization (ΔH_{vap}) of butyl alcohol.

(b) Calculate the normal boiling point of butyl alcohol.

69. Although iodine is not very soluble in pure water, it dissolves readily in water that contains $I^-(aq)$ ion, thanks to the reaction

$$I_2(aq) + I^-(aq) \rightleftharpoons I_3^-(aq)$$

The equilibrium constant of this reaction was measured as a function of temperature with these results:

T:	3.8°C	15.3°C	25.0°C	35.0°C	50.2°C
K:	1160	841	689	533	409

(a) Plot ln K on the y axis as a function of $1/T$, the reciprocal of the absolute temperature.

(b) Estimate the ΔH° of this reaction.

70. Barium nitride vaporizes slightly at high temperature as it undergoes the dissociation

$$Ba_3N_2(s) \rightleftharpoons 3\,Ba(g) + N_2(g)$$

At 1000 K the equilibrium constant is 4.5×10^{-19}. At 1200 K the equilibrium constant is 6.2×10^{-12}.

(a) Estimate ΔH° for this reaction.

(b) The equation is rewritten as

$$2\,Ba_3N_2(s) \rightleftharpoons 6\,Ba(g) + 2\,N_2(g)$$

Now the equilibrium constant is 2.0×10^{-37} at 1000 K and 3.8×10^{-23} at 1200 K. Estimate ΔH° of *this* reaction.

14.8 Distribution of a Single Species between Immiscible Phases: Extraction and Separation Processes

71. An aqueous solution, initially 1.00×10^{-2} M in iodine (I_2), is shaken with an equal volume of an immiscible organic solvent, CCl_4. The iodine distributes itself between the aqueous and CCl_4 layers, and when equilibrium is reached

at 27°C, the concentration of I_2 in the aqueous layer is 1.30×10^{-4} M. Calculate the partition coefficient K at 27°C for the reaction

$$I_2(aq) \rightleftharpoons I_2(CCl_4)$$

72. An aqueous solution, initially 2.50×10^{-2} M in iodine (I_2), is shaken with an equal volume of an immiscible organic solvent, CS_2. The iodine distributes itself between the aqueous and CS_2 layers, and when equilibrium is reached at 25°C, the concentration of I_2 in the aqueous layer is 4.16×10^{-5} M. Calculate the partition coefficient K at 25°C for the reaction

$$I_2(aq) \rightleftharpoons I_2(CS_2)$$

73. Benzoic acid (C_6H_5COOH) dissolves in water to the extent of 2.00 g L^{-1} at 15°C and in diethyl ether to the extent of 6.6×10^2 g L^{-1} at the same temperature.

(a) Calculate the equilibrium constants at 15°C for the two reactions

$$C_6H_5COOH(s) \rightleftharpoons C_6H_5COOH(aq)$$

and

$$C_6H_5COOH(s) \rightleftharpoons C_6H_5COOH(ether)$$

(b) From your answers to part (a), calculate the partition coefficient K for the reaction

$$C_6H_5COOH(aq) \rightleftharpoons C_6H_5COOH(ether)$$

74. Citric acid ($C_6H_8O_7$) dissolves in water to the extent of 1300 g L^{-1} at 15°C and in diethyl ether to the extent of 22 g L^{-1} at the same temperature.

(a) Calculate the equilibrium constants at 15°C for the two reactions

$$C_6H_8O_7(s) \rightleftharpoons C_6H_8O_7(aq)$$

and

$$C_6H_8O_7(s) \rightleftharpoons C_6H_8O_7(ether)$$

(b) From your answers to part (a), calculate the partition coefficient K for the reaction

$$C_6H_8O_7(aq) \rightleftharpoons C_6H_8O_7(ether)$$

ADDITIONAL PROBLEMS

75. At 298 K, unequal amounts of $BCl_3(g)$ and $BF_3(g)$ were mixed in a container. The gases reacted to form $BFCl_2(g)$ and $BClF_2(g)$. When equilibrium was finally reached, the four gases were present in these relative chemical amounts: BCl_3(90), BF_3(470), $BClF_2$(200), $BFCl_2$(45).

(a) Determine the equilibrium constants at 298 K of the two reactions

$$2\ BCl_3(g) + BF_3(g) \rightleftharpoons 3\ BFCl_2(g)$$

$$BCl_3(g) + 2\ BF_3(g) \rightleftharpoons 3\ BClF_2(g)$$

(b) Determine the equilibrium constant of the reaction

$$BCl_3(g) + BF_3(g) \rightleftharpoons BFCl_2 + BClF_2(g)$$

and explain why knowing this equilibrium constant really adds nothing to what you knew in part (a).

76. Methanol can be synthesized by means of the equilibrium reaction

$$CO(g) + 2\ H_2(g) \rightleftharpoons CH_3OH(g)$$

for which the equilibrium constant at 225°C is 6.08×10^{-3}. Assume that the ratio of the pressures of $CO(g)$ and $H_2(g)$ is 1:2. What values should they have if the partial pressure of methanol is to be 0.500 atm?

77. At equilibrium at 425.6°C, a sample of *cis*-1-methyl-2-ethylcyclopropane is 73.6% converted into the *trans* form:

$$cis \rightleftharpoons trans$$

(a) Compute the equilibrium constant K for this reaction.

(b) Suppose that 0.525 mol of the *cis* compound is placed in a 15.00-L vessel and heated to 425.6°C. Compute the equilibrium partial pressure of the *trans* compound.

78. The equilibrium constant for the reaction

$$(CH_3)_3COH(g) \rightleftharpoons (CH_3)_2CCH_2(g) + H_2O(g)$$

is 2.42 at 450 K.

(a) A pure sample of the reactant, which is named "*t*-butanol," is confined in a container of fixed volume at a temperature of 450 K and at an original pressure of 0.100 atm. Calculate the fraction of this starting material that is converted to products at equilibrium.

(b) A second sample of the reactant is confined, this time at an original pressure of 5.00 atm. Again, calculate the fraction of the starting material that is converted to products at equilibrium.

79. At 627°C and 1 atm, SO_3 is partly dissociated into SO_2 and O_2:

$$SO_3(g) \rightleftharpoons SO_2(g) + \tfrac{1}{2}O_2(g)$$

The density of the equilibrium mixture is 0.925 g L^{-1}. What is the degree of dissociation of SO_3 under these circumstances?

80. Acetic acid in the vapor phase consists of both monomeric and dimeric forms in equilibrium:

$$2\ CH_3COOH(g) \rightleftharpoons (CH_3COOH)_2(g)$$

At 110°C the equilibrium constant for this reaction is 3.72.

(a) Calculate the partial pressure of the dimer when the total pressure is 0.725 atm at equilibrium.

(b) What percentage of the acetic acid is dimerized under these conditions?

***81.** Repeat the calculation of problem 36 with a total pressure of 100 atm. (*Note:* Here it is necessary to solve the equation by successive approximations. See Appendix C.)

82. At 900 K the equilibrium constant for the reaction

$$\tfrac{1}{2}O_2(g) + SO_2(g) \rightleftharpoons SO_3(g)$$

has the value $K = 0.587$. What will be the equilibrium partial pressure of $O_2(g)$ if a sample of SO_3 weighing 0.800 g is heated to 900 K in a quartz-glass vessel whose volume is 100.0 cm^3? (*Note:* Here it is necessary to solve the equation by successive approximations. See Appendix C.)

***83.** At 298 K chlorine is only slightly soluble in water. Thus, under a pressure of 1.00 atm of $Cl_2(g)$, 1.00 L of water at equilibrium dissolves just 0.091 mol of Cl_2.

$$Cl_2(g) \rightleftharpoons Cl_2(aq)$$

In such solutions the $Cl_2(aq)$ concentration is 0.061 M and the concentrations of $Cl^-(aq)$ and $HOCl(aq)$ are both 0.030 M. These two additional species are formed by the equilibrium

$$Cl_2(aq) + H_2O(\ell) \rightleftharpoons H^+(aq) + Cl^-(aq) + HOCl(aq)$$

There are no other Cl-containing species. Compute the equilibrium constants K_1 and K_2 for the two reactions.

84. At 400°C the reaction

$$BaO_2(s) + 4\ HCl(g) \rightleftharpoons BaCl_2(s) + 2\ H_2O(g) + Cl_2(g)$$

has an equilibrium constant equal to K_1. How is K_1 related to the equilibrium constant K_2 of the reaction

$$2\ Cl_2(g) + 4\ H_2O(g) + 2\ BaCl_2(s) \rightleftharpoons 8\ HCl(g) + 2\ BaO_2(s)$$

at 400°C?

85. Ammonium hydrogen sulfide, a solid, decomposes to give $NH_3(g)$ and $H_2S(g)$. At 25°C, some $NH_4HS(s)$ is placed in an evacuated container. A portion of it decomposes, and the total pressure at equilibrium is 0.659 atm. Extra $NH_3(g)$ is then injected into the container, and when equilibrium is reestablished, the partial pressure of $NH_3(g)$ is 0.750 atm.

(a) Compute the equilibrium constant for the decomposition of ammonium hydrogen sulfide.
(b) Determine the final partial pressure of $H_2S(g)$ in the container.

86. The equilibrium constant for the reaction

$$KOH(s) + CO_2(g) \rightleftharpoons KHCO_3(s)$$

is 6×10^{15} at 25°C. Suppose 7.32 g of KOH and 9.41 g of $KHCO_3$ are placed in a closed evacuated container and allowed to reach equilibrium. Calculate the pressure of $CO_2(g)$ at equilibrium.

87. The equilibrium constant for the reduction of nickel(II) oxide to nickel at 754°C is 255.4, corresponding to the reaction

$$NiO(s) + CO(g) \rightleftharpoons Ni(s) + CO_2(g)$$

If the total pressure of the system at 754°C is 2.50 atm, calculate the partial pressures of $CO(g)$ and $CO_2(g)$.

88. Both glucose (corn sugar) and fructose (fruit sugar) taste sweet, but fructose tastes sweeter. Each year in the United States, tons of corn syrup destined to sweeten food are treated to convert glucose as fully as possible to the sweeter fructose. The reaction is an equilibrium:

$$\text{glucose} \rightleftharpoons \text{fructose}$$

(a) A 0.2564 M solution of pure glucose is treated at 25°C with an enzyme (catalyst) that causes the preceding equilibrium to be reached quickly. The final concentration of fructose is 0.1175 M. In another experiment at the same temperature, a 0.2666 M solution of pure fructose is treated with the same enzyme and the final concentration of glucose is 0.1415 M. Compute an average equilibrium constant for the preceding reaction.
(b) At equilibrium under these conditions, what percentage of glucose is converted to fructose?

89. At 300°C the equilibrium constant for the reaction

$$PCl_5(g) \rightleftharpoons PCl_3(g) + Cl_2(g)$$

is $K = 11.5$.

(a) Calculate the reaction quotient Q if initially $P_{PCl_3} =$ 2.0 atm, $P_{Cl_2} = 6.0$ atm, and $P_{PCl_5} = 0.10$ atm. State whether the reaction proceeds to the right or to the left as equilibrium is approached.
(b) Calculate P_{PCl_3}, P_{Cl_2}, and P_{PCl_5} at equilibrium.
(c) If the volume of the system is then increased, will the amount of PCl_5 present increase or decrease?

90. Although the process of dissolution of helium gas in water is favored in terms of energy, helium is only very slightly soluble in water. What keeps this gas from dissolving in great quantities in water?

91. The hydrogenation of pyridine to piperidine

$$C_5H_5N(g) + 3\ H_2(g) \rightleftharpoons C_5H_{11}N(g)$$

is an equilibrium process whose equilibrium constant is given by the equation

$$\log_{10} K = -20.281 + \frac{10.560\ \text{K}}{T}$$

(a) Calculate the value of K at $T = 500$ K.
(b) If the partial pressure of hydrogen is 1.00 atm, what fraction of the nitrogen is in the form of pyridine molecules at $T = 500$ K?

92. The breaking of the O—O bond in peroxydisulfuryl difluoride (FO_2SOOSO_2F) gives FO_2SO:

$$(FO_2SO)_2 \rightleftharpoons 2\ FO_2SO$$

The compound on the left of this equation is a colorless liquid that boils at 67.1°C. Its vapor, when heated to about 100°C, turns brown as the product of the reaction forms. Suppose that, in a sample of the vapor, the intensity of the brown color doubles between 100°C and 110°C and that the total pressure increases only by the 2.7% predicted for an ideal gas. Estimate $\Delta H°$ for the preceding reaction.

93. Polychlorinated biphenyls (PCBs) are a major environmental problem. These oily substances have many uses, but they resist breakdown by bacterial action when spilled in the environment and, being fat-soluble, can accumulate to dangerous concentrations in the fatty tissues of fish and animals. One little-appreciated complication in controlling the problem is that there are *209* different PCBs, all now in the environment. They are generally similar, but their solubilities in fats differ considerably. The best measure of this is K_{ow}, the equilibrium constant for the partition of a PCB between the fat-like solvent octanol and water.

$$\mathrm{PCB}(aq) \rightleftharpoons \mathrm{PCB}(octanol)$$

An equimolar mixture of PCB-2 and PCB-11 in water is treated with an equal volume of octanol. Determine the ratio between the amounts of PCB-2 and PCB-11 in the water at equilibrium. At room temperature, K_{ow} is 3.98×10^4 for PCB-2 and 1.26×10^5 for PCB-11.

94. Refer to the data in problems 73 and 74. Suppose 2.00 g of a solid consisting of 50.0% benzoic acid and 50.0% citric acid by mass is added to 100.0 mL of water and 100.0 mL of diethyl ether and the whole assemblage is shaken. When the immiscible layers are separated and the solvents are removed by evaporation, two solids result. Calculate the percentage (by mass) of the major component in each solid.

95. At 25°C the partition coefficient for the equilibrium

$$I_2(aq) \rightleftharpoons I_2(CCl_4)$$

has the value $K = 85$. To 0.100 L of an aqueous solution, which is initially 2×10^{-3} M in I_2, we add 0.025 L of CCl_4. The mixture is shaken in a separatory funnel and allowed to separate into two phases, and the CCl_4 phase is withdrawn.

(a) Calculate the fraction of the I_2 remaining in the aqueous phase.

(b) Suppose the remaining aqueous phase is shaken with another 0.025 L of CCl_4 and again separated. What fraction of the I_2 from the original aqueous solution is *now* in the aqueous phase?

(c) Compare your answer with that of Example 14.18, in which the same total amount of CCl_4 (0.050 L) was used in a *single* extraction. For a given total amount of extracting solvent, which is the more efficient way to remove iodine from water?

96. From the data in Appendix D calculate $\Delta H°$, $\Delta G°$, and K, for the following reaction at 298 K:

$$6\ CH_4(g) + \tfrac{9}{2}O_2(g) \rightleftharpoons C_6H_6(\ell) + 9\ H_2O(\ell)$$

CUMULATIVE PROBLEMS

97. The reaction

$$P_4(g) \rightleftharpoons 2\ P_2(g)$$

is endothermic and begins to occur at moderate temperatures.

(a) In which direction do you expect deviations to occur from Boyle's law, $P \propto 1/V$ (constant T), for gaseous P_4?

(b) In which direction do you expect deviations to occur from Charles's law, $V \propto T$ (constant P), for gaseous P_4?

98. A 4.72-g mass of methanol, CH_3OH, is placed in an evacuated 1.00-L flask and heated to 250°C. It vaporizes and then reaches the following equilibrium:

$$CH_3OH(g) \rightleftharpoons CO(g) + 2\ H_2(g)$$

A tiny hole forms in the side of the container, allowing a small amount of gas to effuse out. Analysis of the escaping gases shows that the rate of escape of the hydrogen is 33 times the rate of escape of the methanol. Calculate the equilibrium constant for the preceding reaction at 250°C.

99. The triple bond in the N_2 molecule is very strong, but at high enough temperatures even it breaks down. At 5000 K, when the total pressure exerted by a sample of nitrogen is 1.00 atm, $N_2(g)$ is 0.65% dissociated at equilibrium:

$$N_2(g) \rightleftharpoons 2\ N(g)$$

At 6000 K with the same total pressure, the proportion of $N_2(g)$ dissociated at equilibrium rises to 11.6%. Use the van't Hoff equation to estimate the ΔH of this reaction.

100. At 25°C the equilibrium constant for the reaction

$$CaSO_4(s) + 2\ H_2O(g) \rightleftharpoons CaSO_4{\cdot}2H_2O(s)$$

is 1.6×10^3. Over what range of relative humidities do you expect $CaSO_4(s)$ to be converted to $CaSO_4{\cdot}2H_2O$? (*Note:* The relative humidity is the partial pressure of water vapor divided by its equilibrium vapor pressure and multiplied by 100%. Use Table 10.3.)

101. Calculate the equilibrium pressure (in atmospheres) of $O_2(g)$ over a sample of pure $NiO(s)$ in contact with pure $Ni(s)$ at 25°C. The $NiO(s)$ decomposes according to the equation

$$NiO(s) \longrightarrow Ni(s) + \tfrac{1}{2}O_2(g)$$

Use data from Appendix D.

102. **(a)** From the values in Appendix D, find the enthalpy change and the Gibbs free energy change when one mole of benzene C_6H_6 is vaporized at 25°C.

(b) Calculate the vapor pressure of benzene at 25°C.

(c) Assuming that the enthalpy and entropy of vaporization are constant, estimate the normal boiling point of benzene.

103. Snow and ice sublime spontaneously when the partial pressure of water vapor is below the equilibrium vapor pressure of ice. At 0°C the vapor pressure of ice is 0.0060 atm (the triple-point pressure of water). Taking the enthalpy of sublimation of ice to be 50.0 kJ mol^{-1}, calculate the partial pressure of water vapor below which ice will sublime spontaneously at −15°C.

104. The sublimation pressure of solid NbI_5 is the pressure of gaseous NbI_5 present in equilibrium with the solid. It is given by the empirical equation

$$\log P = -6762/T + 8.566$$

The vapor pressure of liquid NbI_5, on the other hand, is given by

$$\log P = -4653/T + 5.43$$

In these two equations, T is the absolute temperature in kelvins and P is the pressure in atmospheres.

(a) Determine the enthalpy and entropy of sublimation of $NbI_5(s)$.

(b) Determine the enthalpy and entropy of vaporization of $NbI_5(\ell)$.

(c) Calculate the normal boiling point of $NbI_5(\ell)$.

(d) Calculate the triple-point temperature and pressure of NbI_5. (*Hint:* At the triple point of a substance, the liquid and solid are in equilibrium and must have the same vapor pressure. If they did not, vapor would continually escape from the phase with the higher vapor pressure and collect in the phase with the lower vapor pressure.)

105. In an extraction process, a solute species is partitioned between two immiscible solvents. Suppose the two solvents are water and carbon tetrachloride. State which phase will have the higher concentration of each of the following solutes: (a) CH_3OH, (b) C_2Cl_6, (c) Br_2, (d) NaCl. Explain your reasoning.

106. The gaseous compounds allene and propyne are isomers with formula C_3H_4. Calculate the equilibrium constant and the standard enthalpy change at 25°C for the isomerization reaction

$$\text{allene}(g) \rightleftharpoons \text{propyne}(g)$$

from the following data, all of which apply to 298 K:

	ΔH_f° (kJ mol^{-1})	ΔG_f° (kJ mol^{-1})
Allene	192	202
Propyne	185	194

107. **(a)** Calculate the standard free-energy change and the equilibrium constant for the dimerization of NO_2 to N_2O_4 at 25°C (see Appendix D).

(b) Calculate ΔG for this reaction at 25°C when the pressures of NO_2 and N_2O_4 are each held at 0.010 atm. Which way will the reaction tend to proceed?

108. There are two isomeric hydrocarbons with formula C_4H_{10}, butane and isobutane, which we denote here B and I. The standard enthalpies of formation for the gaseous species are −124.7 kJ mol^{-1} for B, −131.3 kJ mol^{-1} for I; the standard free energies of formation are −15.9 kJ mol^{-1} for B, −18.0 kJ mol^{-1} for I.

(a) Which is the more stable under standard conditions, and which has the higher entropy?

(b) The reaction B $\rightleftharpoons$ I can occur in the presence of a catalyst. Calculate the equilibrium constant at 298 K for the conversion of B to I, and calculate the percentage of B in the equilibrium mixture.

109. At 3500 K the equilibrium constant for the reaction $CO_2(g) + H_2(g) \rightleftharpoons CO(g) + H_2O(g)$ is 8.28.

(a) What is $\Delta G°(3500)$ for this reaction?

(b) What is ΔG at 3500 K for transforming 1 mol CO_2 and 1 mol H_2, both held at 0.1 atm, to 1 mol CO and 1 mol H_2O, both held at 2 atm?

(c) In which direction would this last reaction run spontaneously?

110. At 1200 K in the presence of solid carbon, an equilibrium mixture of CO and CO_2 (called "producer gas") contains 98.3 mol percent CO and 1.69 mol percent of CO_2 when the total pressure is 1 atm. The reaction is

$$CO_2(g) + C(\textit{graphite}) \rightleftharpoons 2\ CO(g)$$

(a) Calculate P_{CO} and P_{CO_2}.

(b) Calculate the equilibrium constant.

(c) Calculate $\Delta G°$ for this reaction.

111. **(a)** Formulate the equilibrium expression for the endothermic reaction

$$AgCl\cdot NH_3(s) \rightleftharpoons AgCl(s) + NH_3(g)$$

(b) What is the effect on P_{NH_3} at equilibrium if additional $AgCl(s)$ is added?

(c) What is the effect on P_{NH_3} at equilibrium if additional $NH_3(g)$ is pumped into or out of the system, provided that neither of the two solid phases shown in the chemical equation is completely used up?

(d) What is the effect on P_{NH_3} of lowering the temperature?

112. Solid ammonium carbonate decomposes according to the equation

$$(NH_4)_2CO_3(s) \rightleftharpoons 2\ NH_3(g) + CO_2(g) + H_2O(g)$$

At a certain elevated temperature the total pressure of the gases NH_3, CO_2, and H_2O generated by the decomposition of, and at equilibrium with, pure solid ammonium carbonate is 0.400 atm. Calculate the equilibrium constant for the reaction considered. What would happen to P_{NH_3} and P_{CO_2} if P_{H_2O} were adjusted by external means to be 0.200 atm without changing the relative amounts of $NH_3(g)$ and $CO_2(g)$ and with $(NH_4)_2CO_3(s)$ still being present?

15

CHAPTER

Acid–Base Equilibria

© Hans Reinhard/Okapia/Science Source

Many naturally occurring dyes change color as the acidity of their surroundings changes. The compound cyanidin is blue in the basic sap of the cornflower and red in the acidic sap of the poppy. Such dyes can be used as indicators of the degree of acidity in a medium.

According to Section 11.3, an acid is a substance that upon dissolving in water increases the concentration of hydronium (H_3O^+) ions above the value found in pure water, and a base is a substance that increases the concentration of hydroxide (OH^-) ions above its value in pure water. Despite this carefully phrased definition, it is commonplace to describe acids as substances that dissociate to give protons (which upon hydration become hydronium ions) and bases as substances that dissociate to give hydroxide ions. If the dissociation is complete, we can easily calculate the concentration of hydronium and hydroxide ions in the solution and then calculate the yield of acid–base neutralization reactions, and acid–base titrations, by the methods of stoichiometry in solution. But experience shows that many acid–base reactions do not go to completion. So, to predict the amount (or concentration) of their products, we have to use the ideas and methods of chemical equilibrium developed in Chapter 14. The present chapter deals with this fundamental aspect of acid–base chemistry which is key to its applications in the physical and biological sciences, in engineering, and in medicine: How far do acid–base reactions proceed before reaching equilibrium?

The reactions of acids and bases lie at the heart of nearly all branches of chemistry. Many acid–base reactions occur in the gaseous and solid states and in nonaqueous solutions. This chapter focuses on acid–base reactions in aqueous solutions, which play important roles in everyday life. Vinegar, orange juice, and battery fluid are familiar acidic aqueous solutions. Basic solutions are produced when such common products as borax, baking soda, and antacids are dissolved in water. Acid–base chemistry will appear many times throughout the rest of this book. We consider the effects of acidity on the dissolution of solids in Chapter 16, on redox reactions in electrochemical cells in Chapter 17, and on the rates of reaction in Chapter 18. At the end of this chapter we relate acid strength to molecular structure in organic acids, and in Chapter 23 we point out the central roles of amino acids and nucleic acids in biochemistry.

15.1 Classifications of Acids and Bases

15.1.1 Arrhenius Acids and Bases

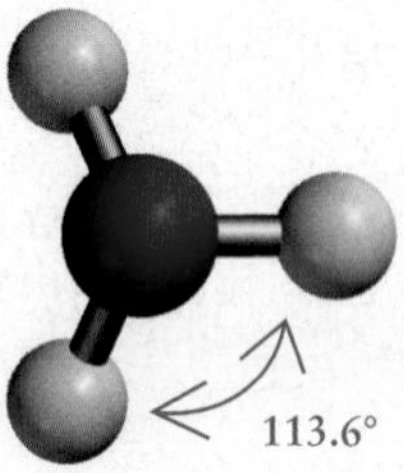

FIGURE 15.1 The structure of the hydronium ion (H_3O^+) in the gas phase.

In Section 11.3 we described the Arrhenius definition of acids and bases. An **acid** is a substance that when dissolved in water increases the concentration of hydronium ion (H_3O^+) above the value it takes in pure water. A **base** increases the concentration of hydroxide ion (OH^-). Arrhenius acids and bases typically release ions into the solution. Throughout this chapter, we will use the hydronium ion (Fig. 15.1) to better represent the true nature of hydrogen ions in water. The hydronium ion may have as many as four molecules of water attached to it by hydrogen bonding. (See Section 10.3.) Depending upon the specific context we refer to this species as the hydrogen ion, the proton, or the hydronium ion in this chapter.

The Arrhenius model views acid–base neutralization as an exchange reaction in which the H^+ from the acid and the OH^- from the base separate from their original bonding partners and combine to form the neutral water molecule, and the original bonding partners combine to form a salt. (See Section 11.3.)

The Arrhenius model also describes other familiar reactions of the acids. Three broad types summarize the results of a very large number of reactions:

1. Acids react with carbonates and hydrogen carbonates such as Na_2CO_3 or $NaHCO_3$ to liberate gaseous CO_2 and produce a salt and water. An example reaction is

$$NaHCO_3(s) + HCl(aq) \longrightarrow NaCl(aq) + H_2O(\ell) + CO_2(g) \qquad \text{(overall)}$$

$$NaHCO_3(s) + H_3O^+(aq) \longrightarrow Na^+(aq) + 2H_2O(\ell) + CO_2(g) \qquad \text{(net ionic)}$$

Notice that in the *net ionic* version of the reaction above, it is recognized that because HCl is a strong acid, $HCl(aq)$ is really $H_3O^+(aq) + Cl^-(aq)$, and the Cl^- is canceled from both sides to write the net ionic reaction of sodium carbonate with hydrochloric acid.

2. Acids react with oxides of metals to form salts and water. An example reaction is

$$CuO(s) + H_2SO_4(aq) \longrightarrow CuSO_4(aq) + H_2O(\ell) \qquad \text{(overall)}$$

$$CuO(s) + 2H_3O^+(aq) \longrightarrow Cu^{2+}(aq) + 3H_2O(\ell) \qquad \text{(net ionic)}$$

FIGURE 15.2 As zinc dissolves in dilute hydrochloric acid, bubbles of hydrogen appear, and zinc chloride forms in solution.

3. Acids react with zinc, iron, and many metal elements to generate gaseous H_2 and form salts (Fig. 15.2). An example reaction is

$$Zn(s) + 2HCl(aq) \longrightarrow ZnCl_2(aq) + H_2(g) \qquad \text{(overall)}$$

$$Zn(s) + 2H_3O^+(aq) \longrightarrow Zn^{2+}(aq) + H_2(g) + 2H_2O(\ell) \qquad \text{(net ionic)}$$

In the first example, hydrogen ions in the solution react with the carbonate or the hydrogen carbonate ions to form carbonic acid (H_2CO_3), which decomposes to produce water and carbon dioxide. The CO_2 has very limited solubility in water and escapes as gas bubbles.

In the second example, hydrogen ions in the solution are attracted to the negative charge on the oxide ions at the surface of the solid, and react with these ions to produce molecules of H_2O. The metal ions are freed from the solid into the solution.

In the third example, hydrogen ions are displaced from the solution by so-called "active metals." This reaction will be discussed in detail in Chapter 17.

The first example illustrates a key aspect of acid–base reactions: A proton is transferred from a molecule defined to be the acid, H_3O^+, to an acceptor molecule HCO_3^-. Proton transfer reactions in which the acceptor is some species other than OH^- are common throughout organic chemistry, inorganic chemistry, and biochemistry. Proton transfer reactions may occur in solvents other than water, and even in the gaseous and solid states. All these diverse reactions can be organized and discussed systematically by using generalized definitions of acids and bases, which include the Arrhenius definition as a limiting case. We describe these more general definitions as background for discussing acid–base equilibria for very broad classes of compounds. The first of these models was proposed independently by Johannes Brønsted and Thomas Lowry in 1923, and the second was introduced by G. N. Lewis in the same year.

15.1.2 Brønsted–Lowry Acids and Bases

A **Brønsted–Lowry acid** is defined as a substance that can *donate* a proton, and a **Brønsted–Lowry base** is a substance that can *accept* a proton. In a Brønsted–Lowry acid–base reaction, a proton is transferred from the acid to the base. For example, when acetic acid is dissolved in water, an acid–base reaction occurs in which protons are transferred from acetic acid molecules to water molecules:

$$\underset{\text{Acid}_1}{CH_3COOH(aq)} + \underset{\text{Base}_2}{H_2O(\ell)} \rightleftharpoons \underset{\text{Base}_1}{CH_3COO^-(aq)} + \underset{\text{Acid}_2}{H_3O^+(aq)}$$

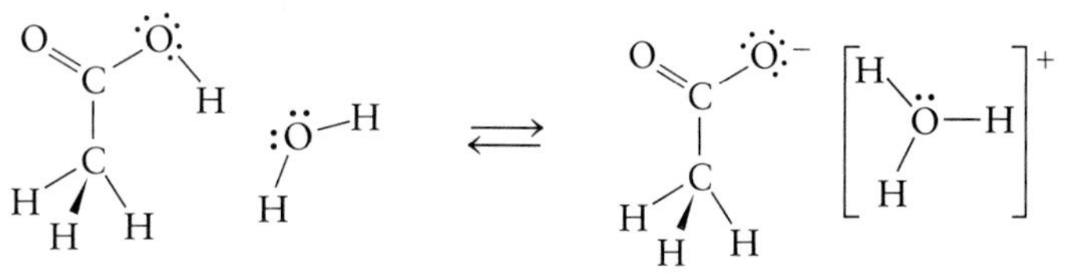

The net result is the transfer of a proton from acetic acid to water to form the hydronium ion $H_3O^+(aq)$ (Fig. 15.1).

Examination of ball-and-stick models and Lewis electron diagrams for the molecules gives some sense of how this reaction occurs. Acetic acid molecules and water molecules experience numerous collisions while diffusing through solution. In some fraction of these collisions the molecules happen to be positioned so the O—H bond on the acetic acid and the O atom on the water molecule are nearly in a straight line. Some fraction of these specially aligned collisions occur with sufficient kinetic energy to push these three atoms and all their electrons quite close together, distort the structure of the molecules, and increase their potential energy. This highly energetic configuration of the two molecules can relax by rearranging the electron density around the nuclei to form a new stable state with lower potential energy, in which the bonds in the molecules have been rearranged. It appears that a proton (H^+) is released from the hydroxyl group (O—H) of acetic acid and attaches itself to a lone pair on the water molecule, forming the new covalent bond in the product hydronium ion and freeing the bonding pair in the hydroxyl group to form the acetate ion. The net result of these proton transfer reactions is the conversion of a bonding electron pair in acetic acid to a nonbonding pair in the acetate ion and a conversion of one of the nonbonding pairs of electrons in the reactant water molecule to a covalently shared pair in the new O—H bond in the hydronium ion (H_3O^+)(*aq*). In fact we cannot claim to track the motions of any specific electron pairs during this reaction; we know from Chapters 4 through 6 that quantum mechanics does not allow us to know that level of detail. Nonetheless, the description of this reaction as the result of the redistribution of two electron pairs that breaks the O—H bond in acetic acid and forms a new O—H bond in the hydronium ion is a convenient, visual summary of the events that occur in the reaction.

Acids and bases occur as **conjugate acid–base pairs,** which are related by transfer of a proton. In an acid–base reaction, the acid donates a proton to the base and in the process is converted to its conjugate base. The base accepts the proton and is converted to its conjugate acid. In the preceding example, acetic acid CH_3COOH and the acetate anion CH_3COO^- form such a pair, in which CH_3COO^- is the conjugate base of CH_3COOH (equivalently, CH_3COOH is the conjugate acid of CH_3COO^-). In the same reaction, H_3O^+ and H_2O form a conjugate acid–base pair, where H_2O is the base and H_3O^+ is its conjugate acid. The equilibrium state of an acid–base reaction will include the acid and its conjugate base, as well as the base and its conjugate acid. It is useful to picture this equilibrium as the competition between two bases for hydrogen ions. For example, when ammonia is dissolved in water,

$$\underset{\text{Acid}_1}{H_2O(\ell)} + \underset{\text{Base}_2}{NH_3(aq)} \rightleftharpoons \underset{\text{Acid}_2}{NH_4^+(aq)} + \underset{\text{Base}_1}{OH^-(aq)}$$

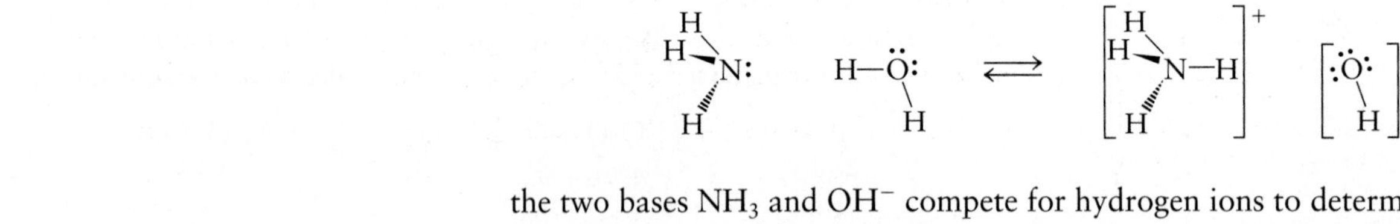

the two bases NH_3 and OH^- compete for hydrogen ions to determine the equilibrium position of the reaction.

Unlike the Arrhenius definition, the Brønsted–Lowry definition has the advantage of not being limited to aqueous solutions, and is used to describe proton transfer reactions in nonaqueous solvents. An example with liquid ammonia as the solvent is

$$\underset{\text{Acid}_1}{HCl(\text{in } NH_3)} + \underset{\text{Base}_2}{NH_3(\ell)} \rightleftharpoons \underset{\text{Acid}_2}{NH_4^+(\text{in } NH_3)} + \underset{\text{Base}_1}{Cl^-(\text{in } NH_3)}$$

The NH_3 acts as both base and solvent, just as H_2O acts as both base and solvent in the proton transfer reaction with acetic acid discussed above.

Some molecules and ions function either as acids or bases depending on reaction conditions and are called **amphoteric.** The most common example is water itself. Water acts as an acid when donating a proton to NH_3 (the conjugate base of water being OH^-) and as a base in accepting a hydrogen ion from CH_3COOH (the conjugate acid of water being H_3O^+). In the same way, the hydrogen carbonate ion can act as an acid

$$HCO_3^-(aq) + H_2O(\ell) \rightleftharpoons H_3O^+(aq) + CO_3^{2-}(aq)$$

or as a base:

$$HCO_3^-(aq) + H_2O(\ell) \rightleftharpoons H_2CO_3(aq) + OH^-(aq)$$

Many examples of proton transfer reactions found in organic chemistry and biochemistry are described by the Brønsted–Lowry model. The various organic functional groups defined in Chapter 7 can participate in these reactions. Ethane (CH_3CH_3) is an extremely weak acid (see later) that can donate a proton to form its conjugate base $CH_3CH_2^-$, an example of a carbanion, a species that plays an important role in organic reaction mechanisms. Ethanol (CH_3CH_2OH) can donate a proton to form its conjugate base $CH_3CH_2O^-$, but in this anion the extra electron is localized primarily on the O atom, not a C atom. The carboxylic acids, of which acetic acid is an example, can give up the proton that is bonded to O (the so-called acidic proton) to form their anionic conjugated bases. Methylamine (CH_3NH_2) can accept a proton to form its conjugate acid the methylammonium ion ($CH_3NH_3^+$). The acid–base behavior of organic molecules is discussed in some detail in Section 15.8.

EXAMPLE 15.1

The following two reactions involve proton transfer. In each reaction, identify the acid and the base. Identify the conjugate base of the acid, and the conjugate acid of the base. Draw Lewis electron dot diagrams for each molecule, and describe the rearrangement of bonding in the reaction as nonbonding electron pairs become covalently shared pairs and vice versa.

(a) $H_2O(\ell) + NH_4^+(aq) \longrightarrow H_3O^+(aq) + NH_3(aq)$

(b) $CH_3CH_2OH(aq) + NH_2^-(aq) \longrightarrow CH_3CH_2O^-(aq) + NH_3(aq)$

Solution

(a) The ammonium ion is the acid, and water is the base. A proton (H^+) is released from the ammonium ion and attaches itself to a lone pair on the water molecule to form the new covalent bond in the species H_3O^+, the conjugate acid of water (the base). The reaction converts one of the original N—H bonding pairs into a lone (nonbonding) pair forming NH_3, the conjugate base of NH_4^+ (the acid).

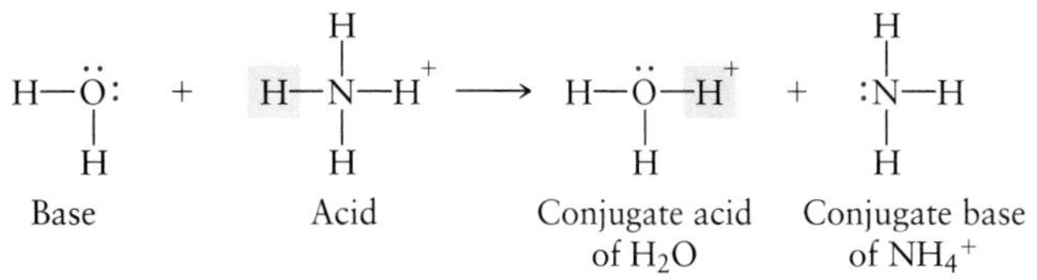

(b) Ethanol is the acid and amide ion is the base. A proton (H^+) is released from ethanol and attaches itself to a lone pair on the amide ion, producing ammonia (the conjugate acid of the amide ion) and the ethoxide ion (the conjugate base of ethanol). An electron pair that was originally a covalently shared pair in ethanol becomes a nonbonding electron pair in the conjugate base of ethanol as the H^+ moves over to accept (bond to) the nonbonding electron pair in NH_2^-. We use this language of *accepting an electron pair* for the H^+ to motivate the Lewis definition of acids and bases in the next section.

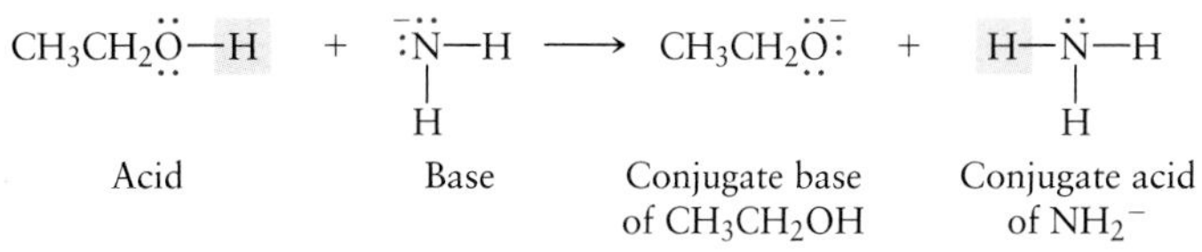

Related Problems: 1, 2, 3, 4

15.1.3 Lewis Acids and Bases

The Lewis model (Chapter 3), which focuses on the electron pair bond, provides a more general definition of acid–base behavior, of which the Arrhenius and Brønsted–Lowry definitions are special cases. Examination of the Lewis electron dot diagrams for molecules in a typical Brønsted–Lowry acid–base reaction shows that two bases compete for a proton by offering electron pairs that can be accepted by the proton to form new bonds.

$$\underset{\text{Acid}_1}{HF(aq)} + \underset{\text{Base}_2}{H_2O(\ell)} \rightleftharpoons \underset{\text{Acid}_2}{H_3O^+(aq)} + \underset{\text{Base}_1}{F^-(aq)}$$

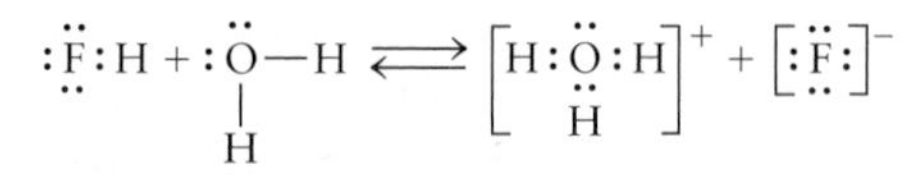

A **Lewis base** is any species that donates lone-pair electrons, and a **Lewis acid** is any species that accepts such electron pairs. The Brønsted–Lowry acids and bases considered so far fit this description (with the Lewis acid, H^+, acting as an acceptor toward various Lewis bases such as NH_3 and OH^-, the electron pair donors). The Lewis acid–base model is much more general than the Brønsted–Lowry model because it applies to other reactions that do *not* involve hydrogen ion transfers. An example is the reaction between electron-deficient BF_3 and electron-rich NH_3.

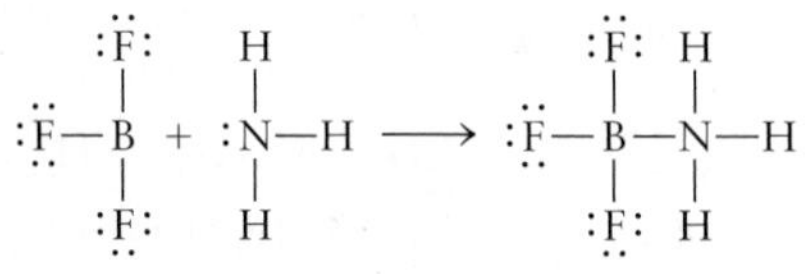

Here ammonia, the Lewis base, donates lone-pair electrons to BF_3, the Lewis acid or electron acceptor. The pair of electrons donated by the Lewis base is not removed from its valence shell but instead is shared with the Lewis acid to form a new covalent bond. The bond that forms is called a **coordinate covalent bond** (or a **dative bond**, see Section 8.2) in which both electrons are provided by a lone pair on the Lewis base.

Octet-deficient compounds involving elements of Group III, such as boron and aluminum, are often strong Lewis acids because Group III atoms (electron pair acceptors) can achieve octet configurations by forming coordinate covalent bonds with electron pair donors such as atoms and ions from Groups V through VII. Compounds of main-group elements from the later periods can also act as Lewis acids because they can accept (or share) additional electrons through valence shell expansion (see Section 3.9). In such reactions the central atom accepts a share in additional lone pairs beyond the eight electrons needed to satisfy the octet rule. For example, $SnCl_4$ is a Lewis acid that accepts electron pairs from two chloride ions to form $SnCl_6^{2-}$ as follows:

$$SnCl_4(\ell) + 2Cl^-(aq) \longrightarrow [SnCl_6]^{2-}(aq)$$

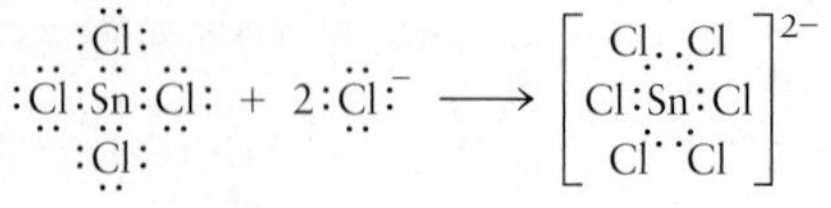

After the reaction, each tin atom is surrounded by 12 rather than 8 valence electrons.

EXAMPLE 15.2

In the following reactions, identify the Lewis acid and the Lewis base.

(a) $AlCl_3 + Cl^- \longrightarrow AlCl_4^-$

(b) $CH_3COOH(aq) + NH_3(aq) \longrightarrow CH_3COO^-(aq) + NH_4^+(aq)$

(c) $Co^{3+}(aq) + 6F^-(aq) \longrightarrow [CoF_6]^{3-}(aq)$

Solution

(a) In $AlCl_3$ the Al atom has only six valence electrons. Thus $AlCl_3$ acts as a Lewis acid to accept electrons from the lone pairs on the Lewis base Cl^-.

(b) The Lewis base NH_3 donates a lone pair of electrons to form a coordinate covalent bond with the proton from acetic acid. The proton is the Lewis acid, not the acetic acid.

(c) The transition metal ion Co^{3+} is the Lewis acid, accepting electron pairs from the Lewis bases, the F^- ions, to form coordinate covalent bonds.

Related Problems: 5, 6, 7, 8

The Lewis definition systematizes the chemistry of a great many binary oxides, which can be considered to be anhydrides of acids or bases. An **acid anhydride** is obtained by removing water from an oxoacid (see Section 11.3) until only the oxide remains; thus, CO_2 is the anhydride of carbonic acid (H_2CO_3).

EXAMPLE 15.3

What is the acid anhydride of phosphoric acid (H_3PO_4)?

Solution

If the formula unit H_3PO_4 (which contains an odd number of hydrogen atoms) is doubled, $H_6P_2O_8$ is obtained. Subtraction of 3 H_2O from this gives P_2O_5, which is the empirical formula of tetraphosphorus decaoxide (P_4O_{10}). This compound is the acid anhydride of phosphoric acid.

Related Problems: 9, 10

The oxides of most of the nonmetals are acid anhydrides, which react with an excess of water to form acidic solutions. Examples include

$$N_2O_5(s) + H_2O(\ell) \longrightarrow 2HNO_3(aq) \longrightarrow 2H^+(aq) + 2\ NO_3^-(aq)$$

$$SO_3(g) + H_2O(\ell) \longrightarrow H_2SO_4(aq) \longrightarrow H^+(aq) + HSO_4^-(aq)$$

Silica (SiO_2) is the acid anhydride of the very weak silicic acid H_2SiO_3, a gelatinous material that is insoluble in water but readily dissolves in strongly basic aqueous solutions according to

$$H_2SiO_3(s) + 2OH^-(aq) \longrightarrow SiO_3^{2-}(aq) + 2H_2O(\ell)$$

An oxide of Group I and II metals is a **base anhydride,** obtained by removing water from the corresponding hydroxides. Calcium oxide, CaO, is the base anhydride of calcium hydroxide, $Ca(OH)_2$. The removal of water from $Ca(OH)_2$ is the reverse of the addition of water to the oxide:

$$CaO(s) + H_2O(\ell) \longrightarrow Ca(OH)_2(s)$$

$$Ca(OH)_2(s) \longrightarrow CaO(s) + H_2O(\ell)$$

The base anhydride of NaOH is Na_2O. Oxides of metals in the middle groups of the periodic table (III through V) lie on the border between ionic and covalent compounds and are frequently amphoteric. An example is aluminum

FIGURE 15.3 Among the oxides of the main-group elements, acidity tends to increase from left to right and from bottom to top in the periodic table. Oxygen difluoride is only weakly acidic, however. Oxides shown in light blue are derived from metals, those in dark blue from metalloids, and those in dark red from nonmetals.

Increasing acidity →

I	II	III	IV	V	VI	VII
Li_2O	BeO	B_2O_3	CO_2	N_2O_5	(O_2)	OF_2
Na_2O	MgO	Al_2O_3	SiO_2	P_4O_{10}	SO_3	Cl_2O_7
K_2O	CaO	Ga_2O_3	GeO_2	As_2O_5	SeO_3	Br_2O_7
Rb_2O	SrO	In_2O_3	SnO_2	Sb_2O_5	TeO_3	I_2O_7
Cs_2O	BaO	Tl_2O_3	PbO_2	Bi_2O_5	PoO_3	At_2O_7

← Increasing basicity

Increasing basicity ↓ (left side); Increasing acidity ↑ (right side)

oxide (Al_2O_3), which dissolves to only a limited extent in water but much more readily in either acids or bases:

$$\text{Acting as a base: } Al_2O_3(s) + 6H_3O^+(aq) \longrightarrow 2Al^{3+}(aq) + 9H_2O(\ell)$$

$$\text{Acting as an acid: } Al_2O_3(s) + 2OH^-(aq) + 3H_2O(\ell) \longrightarrow 2Al(OH^-)_4(aq)$$

Figure 15.3 summarizes the acid–base character of the main-group oxides in the first periods.

Although oxoacids and hydroxides are Arrhenius acids and bases (they release $H_3O^+(aq)$ or $OH^-(aq)$ into aqueous solution), acid and base anhydrides do not fall into this classification because they contain neither H^+ nor OH^- ions. Acid anhydrides are acids in the Lewis sense, (they accept electron pairs), and base anhydrides are bases in the Lewis sense, (their O^{2-} ions donate electron pairs). The reaction between an acid anhydride and a base anhydride is then a Lewis acid–base reaction. An example of such a reaction is

$$CaO(s) + CO_2(g) \longrightarrow CaCO_3(s)$$

Here, the Lewis base CaO donates an electron pair (one of the lone pairs of the oxygen atom) to the Lewis acid (CO_2) to form a coordinate covalent bond in the resulting ion. Similar Lewis acid–base reactions can be written for other acid–base anhydride pairs. Sulfur trioxide, for example, reacts with metal oxides to form sulfates:

$$MgO(s) + SO_3(g) \longrightarrow MgSO_4(s)$$

Note that these reactions are not redox reactions (oxidation numbers do not change), nor are they acid–base reactions in the Arrhenius or Brønsted–Lowry sense. But they are usefully classified as acid–base reactions in the Lewis sense.

15.1.4 Comparison of Arrhenius, Brønsted–Lowry, and Lewis Definitions

The neutralization reaction between HCl and NaOH,

$$\underset{\text{Acid}}{HCl} + \underset{\text{Base}}{NaOH} \longrightarrow \underset{\text{Water}}{H_2O} + \underset{\text{Salt}}{NaCl}$$

introduced in Section 11.3 shows the progressive generality in these definitions. By the Arrhenius definition, HCl is the acid and NaOH is the base. By the Brønsted–Lowry definition, H_3O^+ is the acid and OH^- is the base. According to the Lewis model, H^+ is the acid and OH^- is the base, because the proton accepts the lone pair donated by hydroxide ion in the reaction

$$H^+(aq) + OH^-(aq) \longrightarrow H_2O(\ell)$$

15.2 Properties of Acids and Bases in Aqueous Solutions: The Brønsted–Lowry Scheme

Chapters 10 and 11 describe the special properties of liquid water. Because of its substantial dipole moment, water is especially effective as a solvent, stabilizing both polar and ionic solutes. In acid–base reactions water is not only the solvent, but also participates as a reactant. Water plays an integral role in virtually all biochemical reactions essential to the survival of living organisms; these reactions involve acids, bases, and ionic species. In view of the wide-ranging importance of these reactions, we devote the remainder of this chapter to acid–base behavior and related ionic reactions in aqueous solution. The Brønsted–Lowry definition of acids and bases is especially well suited to describe these reactions.

15.2.1 Autoionization of Water

Water can act as both acid and base in the same reaction. The resulting equilibrium is

$$\underset{\text{Acid}_1}{H_2O(\ell)} + \underset{\text{Base}_2}{H_2O(\ell)} \rightleftharpoons \underset{\text{Acid}_2}{H_3O^+(aq)} + \underset{\text{Base}_1}{OH^-(aq)}$$

or

$$2\ H_2O(\ell) \rightleftharpoons H_3O^+(aq) + OH^-(aq)$$

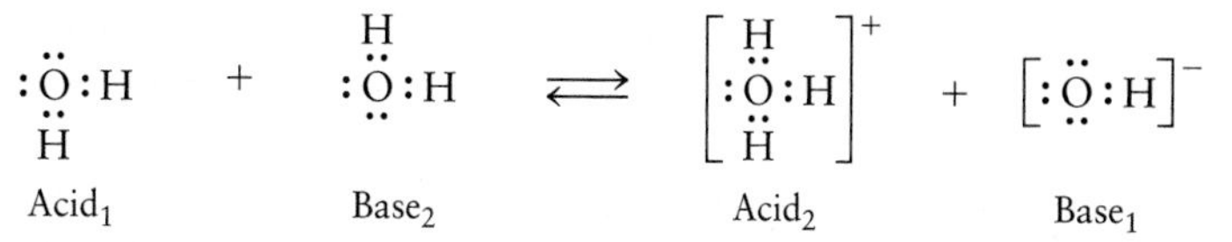

This reaction is called the **autoionization of water**, which leads to small but measurable concentrations of hydronium and hydroxide ions at equilibrium. The equilibrium expression for this reaction is

$$[H_3O^+][OH^-] = K_w \qquad [15.1]$$

Note that we have omitted the subscript eq for simplicity; the concentrations in this and all equilibrium constant expressions are equilibrium ones. The equilibrium constant for this particular reaction has a special symbol: K_w, and a special name, the **ion product constant for water**; its value is 1.0×10^{-14} at 25°C. Because the liquid water appears in this equilibrium reaction equation as a pure substance, it is considered already to be in its reference state, and therefore contributes only the factor 1 to the mass action law equilibrium expression. The reasons for this are discussed more fully in Sections 14.2 and 14.3. The temperature dependence of K_w is given in Table 15.1; all problems in this chapter are assumed to refer to 25°C unless otherwise stated.

TABLE 15.1

Temperature Dependence of K_w

T(°C)	K_w	pH of Water
0	0.114×10^{-14}	7.47
10	0.292×10^{-14}	7.27
20	0.681×10^{-14}	7.08
25	1.01×10^{-14}	7.00
30	1.47×10^{-14}	6.92
40	2.92×10^{-14}	6.77
50	5.47×10^{-14}	6.63
60	9.61×10^{-14}	6.51

Pure water contains no ions other than H_3O^+ and OH^-, and to maintain overall electrical neutrality, an equal number of ions of each type must be present. Putting these facts into the equilibrium expression Equation 15.1 gives

$$[H_3O^+] = [OH^-] = y$$
$$y^2 = 1.0 \times 10^{-14}$$
$$y = 1.0 \times 10^{-7}$$

so that in pure water at 25°C the concentrations of both H_3O^+ and OH^- are 1.0×10^{-7} M.

15.2.2 Strong Acids and Bases

An aqueous acidic solution contains an excess of H_3O^+ over OH^- ions. A **strong acid** is one that ionizes almost completely in aqueous solution. When the strong acid HCl (hydrochloric acid) is put in water, the reaction

$$\underset{\text{Base}_2}{H_2O(\ell)} + \underset{\text{Acid}_1}{HCl(aq)} \longrightarrow \underset{\text{Acid}_2}{H_3O^+(aq)} + \underset{\text{Base}_1}{Cl^-(aq)}$$

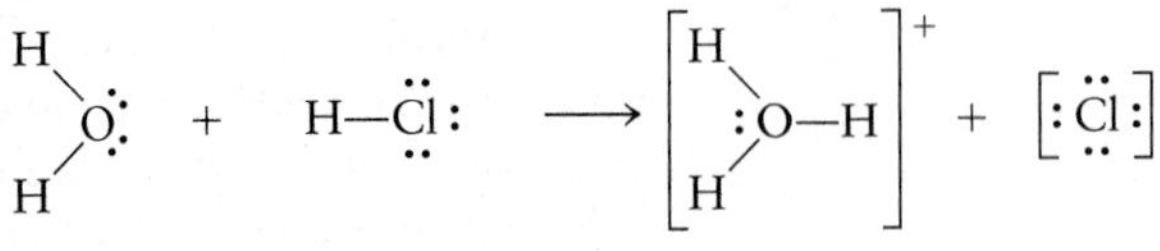

occurs. A single rather than a double arrow indicates that the reaction is essentially complete. (*Note:* Although we use this convention in Sections 15.2, 15.3, and 15.4, we often simply use single arrows in the rest of this chapter and subsequent chapters even when the equilibrium constant favors reactants.) Another strong acid is perchloric acid ($HClO_4$). (See Table 11.1.) If 0.10 mol of either of these acids is dissolved in enough water to make 1.0 L of solution, 0.10 M concentration of $H_3O^+(aq)$ results. Because the acid–base properties of solutions are determined by their concentrations of $H_3O^+(aq)$, these two strong acids have the same effect in water despite differences we shall see shortly in their intrinsic abilities to donate hydrogen ions. Water is said to have a **leveling effect** on a certain group of acids (HCl, HBr, HI, H_2SO_4, HNO_3, and $HClO_4$) because they all behave as strong acids when water is the solvent. The reactions of these acids with water all lie so far to the right at equilibrium that the differences between the acids are negligible. The concentration of H_3O^+ in a 0.10 M solution of *any* strong acid that donates one hydrogen ion per molecule is simply 0.10 M. We use this result in Equation 15.1 to obtain the OH^- concentration

$$[OH^-] = \frac{K_w}{[H_3O^+]} = \frac{1.0 \times 10^{-14}}{0.10} = 1.0 \times 10^{-13} \text{ M}$$

In the same fashion, we define a **strong base** as one that reacts essentially completely to give $OH^-(aq)$ ion when put in water. The amide ion (NH_2^-) and the hydride ion (H^-) are both strong bases. For every mole per liter of either of these ions that is added to water, one mole per liter of $OH^-(aq)$ forms. Note that the other products in these reactions are aqueous NH_3 and H_2, respectively:

$$\underset{\text{Acid}_1}{H_2O(\ell)} + \underset{\text{Base}_2}{NH_2^-(aq)} \longrightarrow \underset{\text{Base}_1}{OH^-(aq)} + \underset{\text{Acid}_2}{NH_3(aq)}$$

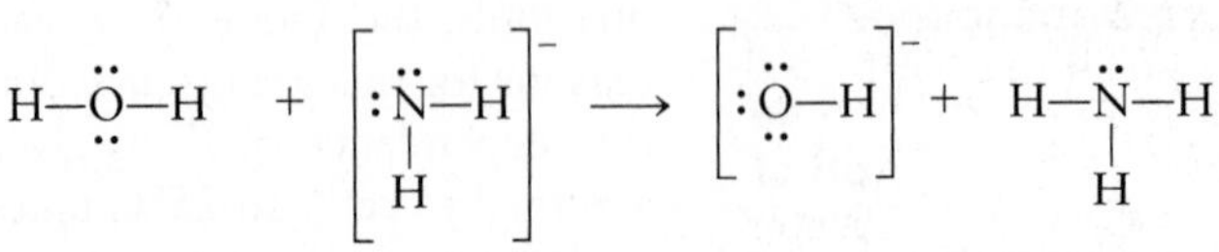

$$\underset{\text{Acid}_1}{H_2O(\ell)} + \underset{\text{Base}_2}{H^-(aq)} \longrightarrow \underset{\text{Base}_1}{OH^-(aq)} + \underset{\text{Acid}_2}{H_2(aq)}$$

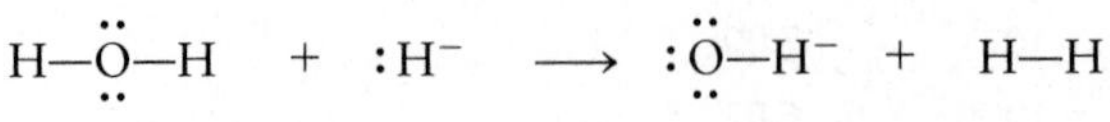

The important base sodium hydroxide, an ionic solid, increases the OH^- concentration in water when it dissolves:

$$NaOH(s) \longrightarrow Na^+(aq) + OH^-(aq)$$

For every mole of NaOH that dissolves in water, one mole of $OH^-(aq)$ forms, so NaOH is a strong base. Strong bases are leveled in aqueous solution in the same way that strong acids are leveled. If 0.10 mol of NaOH or NH_2^- or H^- is put into enough water to make 1.0 L of solution, then in every case

$$[OH^-] = 0.10 \text{ M}$$

$$[H_3O^+] = \frac{1.0 \times 10^{-14}}{0.10} = 1.0 \times 10^{-13} \text{ M}$$

The OH^- contribution from the autoionization of water is negligible here, as was the contribution of H_3O^+ from autoionization in the 0.10 M HCl and $HClO_4$ solutions. When only a very small amount of strong acid or base is added to pure water (for example, 10^{-7} mol L^{-1}), we have to include the autoionization of water to describe the concentration of hydronium and hydroxide ions accurately.

FIGURE 15.4 A simple pH meter with a digital readout. Most soft drinks are quite acidic due to the reaction of dissolved CO_2 with water to make carbonic acid (H_2CO_3).

15.2.3 The pH Function

In aqueous solution the concentration of hydronium ion can range from 10 M to 10^{-15} M. It is convenient to compress this enormous range by introducing a logarithmic acidity scale, called **pH** and defined by

$$pH = -\log_{10} [H_3O^+] \quad [15.2]$$

Pure water at 25°C has $[H_3O^+] = 1.0 \times 10^{-7}$ M, so

$$pH = -\log_{10} (1.0 \times 10^{-7}) = -(-7.00) = 7.00$$

A 0.10 M solution of HCl has $[H_3O^+] = 0.10$ M, so

$$pH = -\log_{10} (0.10) = -\log_{10} (1.0 \times 10^{-1}) = -(-1.00) = 1.00$$

and at 25°C a 0.10 M solution of NaOH has

$$pH = -\log_{10}\left(\frac{1.0 \times 10^{-14}}{0.10}\right) = -\log_{10}(1.0 \times 10^{-13}) = -(-13.00) = 13.00$$

As these examples show, calculating the pH is especially easy when the concentration of H_3O^+ is exactly a power of 10, because the logarithm is then just the power to which 10 is raised. In other cases we need a calculator. When we know the pH, we calculate the concentration of H_3O^+ by raising 10 to the power (−pH).

Because the most commonly encountered H_3O^+ concentrations are less than 1 M, the negative sign is put in the definition of the pH function to give a positive value in most cases. A *high* pH means a *low* concentration of H_3O^+ and vice versa. At 25°C,

pH < 7	Acidic solution	[15.3a]
pH = 7	Neutral solution	[15.3b]
pH > 7	Basic solution	[15.3c]

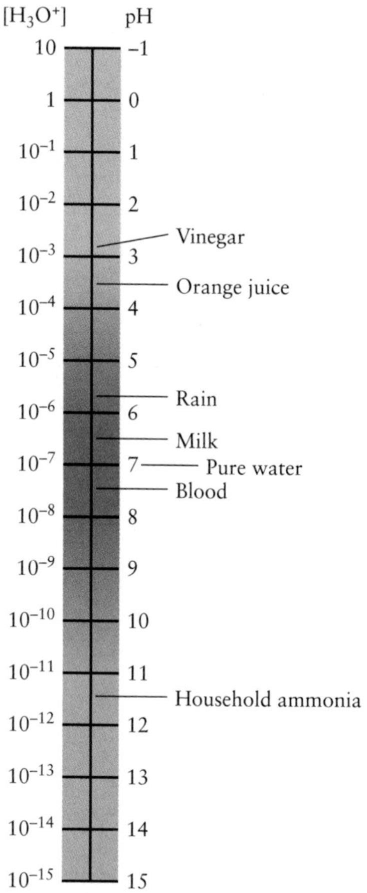

FIGURE 15.5 Many everyday materials are acidic or basic aqueous solutions with a wide range of pH values.

At other temperatures the pH of water differs from 7.00 (see Table 15.1). A change of one pH unit implies that the concentrations of H_3O^+ and OH^- change by a factor of 10 (that is, one order of magnitude). The pH is most directly measured with a **pH meter** (Fig. 15.4). The mechanism by which pH meters operate is described in Chapter 17. Figure 15.5 shows the pH values for several common fluids.

EXAMPLE 15.4

(a) A solution is prepared by dissolving 0.23 mol of NaH(*s*) in enough water to form 2.8 L of solution. Calculate its pH.

(b) The pH of some orange juice at 25°C is 2.85. Calculate $[H_3O^+]$ and $[OH^-]$.

Solution

(a) Because H^- is a strong base, it reacts to give 0.23 mol of OH^-:

$$NaH(s) + H_2O(\ell) \longrightarrow Na^+(aq) + OH^-(aq) + H_2(g)$$

The concentration is

$$[OH^-] = \frac{0.23 \text{ mol}}{2.8 \text{ L}} = 8.2 \times 10^{-2} \text{ M}$$

$$[H_3O^+] = \frac{1.0 \times 10^{-14}}{8.2 \times 10^{-2}} = 1.22 \times 10^{-13} \text{ M}$$

The pH is then

$$pH = -\log_{10}(1.22 \times 10^{-13}) = 12.91$$

(b)

$$pH = 2.85 = -\log_{10}[H_3O^+]$$

$$[H_3O^+] = 10^{-2.85}$$

This can be evaluated by using a calculator to give

$$[H_3O^+] = 1.4 \times 10^{-3} \text{ M}$$

$$[OH^-] = \frac{1.0 \times 10^{-14}}{1.4 \times 10^{-3}} = 7.1 \times 10^{-12}$$

Related Problems: 13, 14, 15, 16

15.3 Acid and Base Strength

Acids are classified as strong or weak, depending on the extent to which they are ionized in solution. In a **weak acid** the transfer of hydrogen ions to water does not proceed to completion. A weak acid such as acetic acid is thus also a **weak electrolyte**; its aqueous solutions do not conduct electricity as well as a strong acid of the same concentration because fewer ions are present. A weak acid shows smaller values for colligative properties than a strong acid (recall the effect of dissolved acetic acid on the freezing point of water in Fig. 11.20).

The Brønsted–Lowry theory helps to establish a quantitative scale for acid strength. The ionization of an acid (symbolized by "HA") in aqueous solution can be written as

$$HA(aq) + H_2O(\ell) \rightleftharpoons H_3O^+(aq) + A^-(aq)$$

where A^- is the conjugate base of HA. The equilibrium expression for this chemical reaction (see Section 14.2) is

$$\frac{[H_3O^+][A^-]}{[HA]} = K_a \qquad [15.4]$$

where the subscript "a" stands for "acid."[1] For example, if the symbol A^- refers to the cyanide ion (CN^-), we write

$$\frac{[H_3O^+][CN^-]}{[HCN]} = K_a$$

FIGURE 15.6 Hydrogen cyanide, HCN, is a highly toxic gas that dissolves in water to form equally toxic solutions of hydrocyanic acid.

where K_a is the **acid ionization constant** for hydrogen cyanide (Fig. 15.6) in water and has a numerical value of 6.17×10^{-10} at 25°C. Table 15.2 gives values of K_a, and the useful quantity $pK_a = -\log_{10} K_a$, for a number of important acids. The acid ionization constant is a quantitative measure of the strength of the acid in a particular solvent (in the present case, water). Because a weak acid does not

[1] $H_2O(\ell)$ is in its reference state, so it contributes the factor 1 (see Sections 14.2 and 14.3).

TABLE 15.2

Ionization Constants of Acids in Water at 25°C

Acid	HA	A^-	K_a	pK_a
Hydroiodic	HI	I^-	$\sim 10^{11}$	~ -11
Hydrobromic	HBr	Br^-	$\sim 10^{9}$	~ -9
Perchloric	$HClO_4$	ClO_4^-	$\sim 10^{7}$	~ -7
Hydrochloric	HCl	Cl^-	$\sim 10^{7}$	~ -7
Chloric	$HClO_3$	ClO_3^-	$\sim 10^{3}$	~ -3
Sulfuric (1)	H_2SO_4	HSO_4^-	$\sim 10^{2}$	~ -2
Nitric	HNO_3	NO_3^-	~ 20	~ -1.3
Hydronium ion	H_3O^+	H_2O	1	0.0
Iodic	HIO_3	IO_3^-	1.6×10^{-1}	0.80
Oxalic (1)	$H_2C_2O_4$	$HC_2O_4^-$	5.9×10^{-2}	1.23
Sulfurous (1)	H_2SO_3	HSO_3^-	1.54×10^{-2}	1.81
Sulfuric (2)	HSO_4^-	SO_4^{2-}	1.2×10^{-2}	1.92
Chlorous	$HClO_2$	ClO_2^-	1.1×10^{-2}	1.96
Phosphoric (1)	H_3PO_4	$H_2PO_4^-$	7.52×10^{-3}	2.12
Arsenic (1)	H_3AsO_4	$H_2AsO_4^-$	5.0×10^{-3}	2.30
Chloroacetic	$CH_2ClCOOH$	CH_2ClCOO^-	1.4×10^{-3}	2.85
Hydrofluoric	HF	F^-	6.6×10^{-4}	3.18
Nitrous	HNO_2	NO_2^-	4.6×10^{-4}	3.34
Formic	HCOOH	$HCOO^-$	1.77×10^{-4}	3.75
Benzoic	C_6H_5COOH	$C_6H_5COO^-$	6.46×10^{-5}	4.19
Oxalic (2)	$HC_2O_4^-$	$C_2O_4^{2-}$	6.4×10^{-5}	4.19
Hydrazoic	HN_3	N_3^-	1.9×10^{-5}	4.72
Acetic	CH_3COOH	CH_3COO^-	1.76×10^{-5}	4.75
Propionic	CH_3CH_2COOH	$CH_3CH_2COO^-$	1.34×10^{-5}	4.87
Pyridinium ion	$HC_5H_5N^+$	C_5H_5N	5.6×10^{-6}	5.25
Carbonic (1)	H_2CO_3	HCO_3^-	4.3×10^{-7}	6.37
Sulfurous (2)	HSO_3^-	SO_3^{2-}	1.02×10^{-7}	6.91
Arsenic (2)	$H_2AsO_4^-$	$HAsO_4^{2-}$	9.3×10^{-8}	7.03
Hydrosulfuric	H_2S	HS^-	9.1×10^{-8}	7.04
Phosphoric (2)	$H_2PO_4^-$	HPO_4^{2-}	6.23×10^{-8}	7.21
Hypochlorous	HClO	ClO^-	3.0×10^{-8}	7.53
Hydrocyanic	HCN	CN^-	6.17×10^{-10}	9.21
Ammonium ion	NH_4^+	NH_3	5.6×10^{-10}	9.25
Carbonic (2)	HCO_3^-	CO_3^{2-}	4.8×10^{-11}	10.32
Arsenic (3)	$HAsO_4^{2-}$	AsO_4^{3-}	3.0×10^{-12}	11.53
Hydrogen peroxide	H_2O_2	HO_2^-	2.4×10^{-12}	11.62
Phosphoric (3)	HPO_4^{2-}	PO_4^{3-}	2.2×10^{-13}	12.67
Water	H_2O	OH^-	1.0×10^{-14}	14.00

(1), (2), and (3) indicate the first, second, and third ionization constants, respectively.

react completely with the solvent, the concentrations of the products are low at equilibrium, and K_a is small. A strong acid reacts nearly completely so the concentration of HA in the denominator is nearly 0, and K_a is large. Determining the value of K_a for strong acids is difficult because the concentration of HA is too small to be measured accurately. Relative acid strengths for such strong acids are estimated by using a solvent that is a weaker base than water. If the solvent is a sufficiently weak base (i.e., has significantly less tendency to accept H^+ than does H_2O), a relatively strong acid may not dissociate completely. For example, in the solvent diethyl ether ($C_2H_5OC_2H_5$), HCl is a weaker proton donor than $HClO_4$. Therefore HCl is a weaker acid than $HClO_4$.

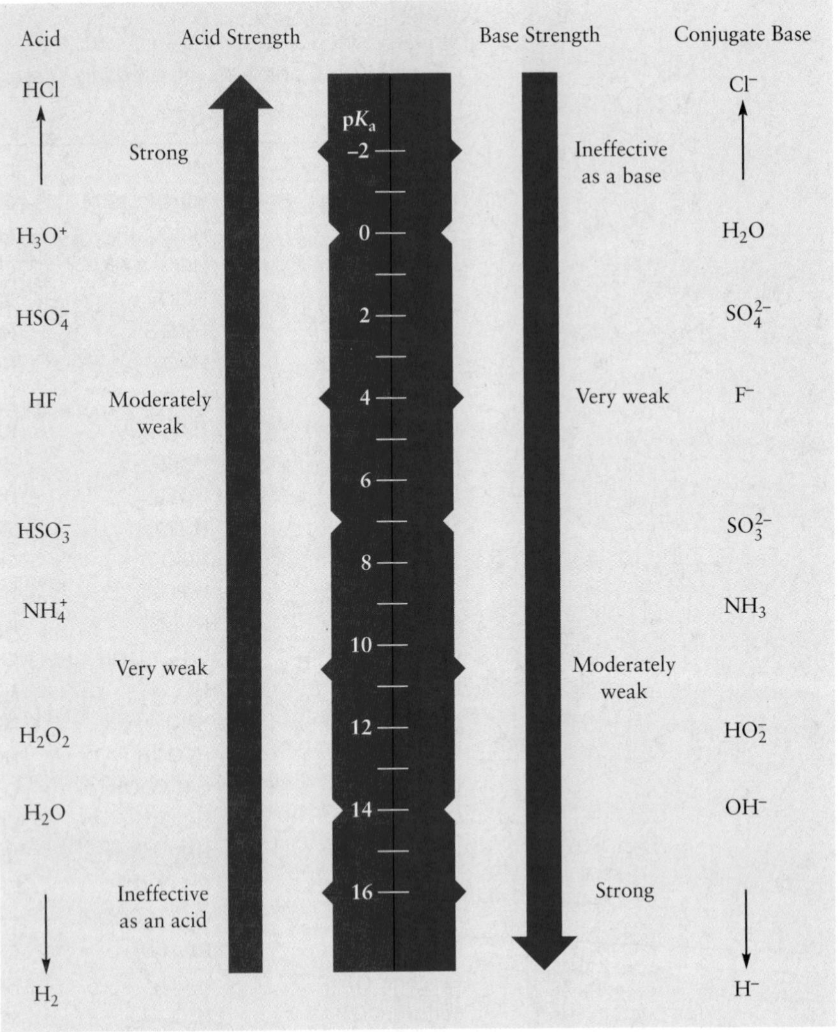

FIGURE 15.7 The relative strengths of some acids and their conjugate bases.

The strength of a base is inversely related to the strength of its conjugate acid; the weaker the acid, the stronger its conjugate base, and vice versa. This fact is demonstrated by the values in Figure 15.7. We can see the origin of this fact by writing the equation for the ionization of a base such as ammonia in water in the form

$$\underset{\text{Acid}_1}{H_2O(\ell)} + \underset{\text{Base}_2}{NH_3(aq)} \rightleftarrows \underset{\text{Acid}_2}{NH_4^+(aq)} + \underset{\text{Base}_1}{OH^-(aq)}$$

which gives an equilibrium expression of the form

$$\frac{[NH_4^+][OH^-]}{[NH_3]} = K_b$$

where the subscript "b" on K_b stands for "base." Because $[OH^-]$ and $[H_3O^+]$ are related through the water autoionization equilibrium expression

$$[OH^-][H_3O^+] = K_w$$

the K_b expression can be written as

$$K_b = \frac{[NH_4^+]K_w}{[NH_3][H_3O^+]} = \frac{K_w}{K_a}$$

where K_a is the acid ionization constant for NH_4^+, the conjugate acid of the base NH_3. This general relationship between the K_b of a base and the K_a of its conjugate acid shows that K_b need not be tabulated separately from K_a, because the two are related through

$$K_w = K_a K_b \qquad [15.5]$$

Equation 15.5 is the basis of the inverse relationship between the strength of a base and its conjugate acid illustrated in Figure 15.7. Taking the logarithm of both sides of the equation and multiplying through by -1 shows that

$$pK_a + pK_b = pK_w$$

for a conjugate acid–base pair. This equation shows the inverse relationship in terms of pK values: the weaker the acid (pK_a large, K_a small), the stronger the conjugate base (pK_b small, K_b large). When values are needed for K_b or pK_b, they can be easily calculated from the tabulated values for K_a and pK_a. For example, we obtain the value of K_b for ammonia (NH_3) from the K_a value for the ammonium ion (NH_4^+) in Table 15.2:

$$K_{b,NH_3} = \frac{K_w}{K_{a,NH_4^+}} = \frac{1.0 \times 10^{-14}}{5.6 \times 10^{-10}} = 1.8 \times 10^{-5}$$

The ammonium ion is a very weak acid. Its conjugate base ammonia is a moderately weak base.

When two bases compete for protons, the stronger base dominates the equilibrium state in the sense that it holds the larger fraction of protons and is therefore present in a lesser amount in its basic form at equilibrium. The stronger acid donates hydrogen ions to the stronger base, producing a weaker acid and a weaker base at equilibrium. To see this consider the equilibrium

$$\underset{\text{Acid}_1}{HF(aq)} + \underset{\text{Base}_2}{CN^-(aq)} \rightleftharpoons \underset{\text{Acid}_2}{HCN(aq)} + \underset{\text{Base}_1}{F^-(aq)}$$

with equilibrium constant

$$\frac{[HCN][F^-]}{[HF][CN^-]} = K$$

The two bases F^- and CN^- compete for hydrogen ions. We can construct this net reaction by starting with one acid ionization reaction,

$$HF(aq) + H_2O(\ell) \rightleftharpoons H_3O^+(aq) + F^-(aq)$$

$$\frac{[H_3O^+][F^-]}{[HF]} = K_a = 6.6 \times 10^{-4}$$

and *subtracting* from it a second acid ionization reaction:

$$HCN(aq) + H_2O(\ell) \rightleftharpoons H_3O^+(aq) + CN^-(aq)$$

$$\frac{[H_3O^+][CN^-]}{[HCN]} = K'_a = 6.17 \times 10^{-10}$$

When the net reaction is the difference of two reactions, the equilibrium constant for the net reaction is the ratio of those for the separate reactions. (Section 14.4) The numerical value of K is

$$K = \frac{K_a}{K'_a} = \frac{6.6 \times 10^{-4}}{6.17 \times 10^{-10}} = 1.1 \times 10^6$$

FIGURE 15.8 Hydrogen fluoride, HF, is a colorless liquid that boils at 19.5°C. It dissolves in water to give solutions of hydrofluoric acid, a weak acid.

Because HCN is a weaker acid than HF (Fig. 15.8), K'_a is smaller than K_a, and K is larger than 1. The equilibrium described by K lies strongly to the right. Most of the hydrogen ions end up associated with $CN^-(aq)$, the stronger base, rather than with $F^-(aq)$, the weaker base. The dissolved species present in greatest amount at equilibrium are HCN(*aq*), which is a weaker acid than HF(*aq*), and $F^-(aq)$ which is a weaker base than $CN^-(aq)$.

Similar manipulations of chemical reaction equations and their associated equilibrium expressions predict the direction and extent of proton transfer in all Brønsted–Lowry acid–base reactions.

EXAMPLE 15.5

Predict the direction of proton transfer in the reaction of acetic acid and ammonia in aqueous solution. Describe qualitatively the extent of reaction and the composition of the equilibrium state.

$$CH_3COOH(aq) + NH_3(aq) \longrightarrow CH_3COO^-(aq) + NH_4^+(aq)$$

Solution

The equilibrium expression for this reaction is

$$\frac{[CH_3COO^-][NH_4^+]}{[CH_3COOH][NH_3]} = K$$

This reaction can be constructed by subtracting the second acid ionization from the first acid ionization below:

$$CH_3COOH(aq) + H_2O(aq) \longrightarrow CH_3COO^-(aq) + H_3O(aq)$$

$$\frac{[H_3O^+][CH_3COO^-]}{[CH_3COOH]} = K_{a1}$$

$$NH_4^+(aq) + H_2O(aq) \longrightarrow H_3O^+(aq) + NH_3(aq)$$

$$\frac{[H_3O^+][NH_3]}{[NH_4^+]} = K_{a2}$$

The equilibrium constant for the net reaction is the ratio of the equilibrium constants for the separate reactions. Inserting values from Table 15.2 gives the result

$$K = \frac{K_{a1}}{K_{a2}} = \frac{1.76 \times 10^{-5}}{5.6 \times 10^{-10}} = 3.1 \times 10^4$$

The equilibrium described by K lies strongly to the right. The net result is donation of H^+ by the stronger acid $CH_3COOH(aq)$ to the stronger base $NH_3(aq)$ to produce the weaker acid $NH_4^+(aq)$ and the weaker base $CH_3COO^-(aq)$. Most of the H^+ ends up associated with the $NH_3(aq)$ rather than the $CH_3COO^-(aq)$. If the reaction begins with the same amounts of acetic acid and ammonia, the ratio of products to reactants in the equilibrium state is 30,000:1.

Related Problems: 23, 24

15.3.1 Molecular Structure and Acid Strength

The strength of an acid is determined by its tendency to lose a proton, converting it to its conjugate base, which is an anion. The more stable the anion, the further the reaction can proceed from left to right. Acid strength, therefore, depends on the ability of the anion to accommodate its negative charge, which increases with the electronegativity of the central atom in the anion.

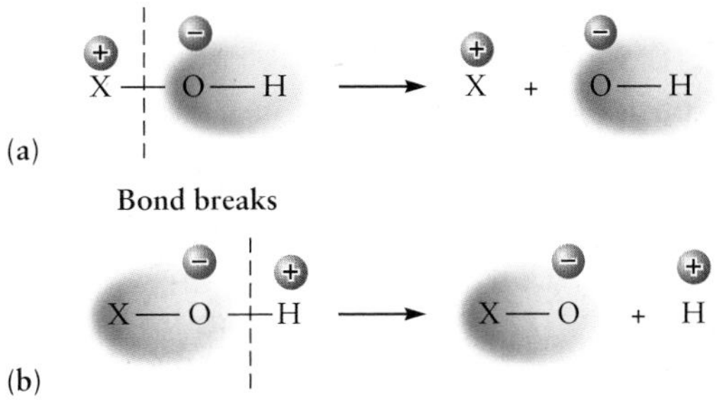

FIGURE 15.9 In part (a) the atom X is electropositive, so extra electron density (blue areas) accumulates on the OH group. The X—O bond then breaks easily, making the compound a base. In part (b) X is electronegative, so electron density is drawn from the H atom to the X—O bond. Now it is the O—H bond that breaks easily, and the compound is an acid.

Trends in the relative strength of oxoacids are explained by the influence of electronegativity and bond polarity on the ease of donating a proton. The proton donated by an **oxoacid** in aqueous solution was previously bonded to an oxygen atom in the acid molecule. Examples include sulfuric acid (H_2SO_4), nitric acid (HNO_3), and phosphoric acid (H_3PO_4). If the central atom is designated X, then oxoacids have the structure

$$-\text{X}-\text{O}-\text{H}$$

where X can be bonded to additional —OH groups, to oxygen atoms, or to hydrogen atoms. How does the strength of the oxoacid change as the electronegativity of X changes? Consider first the extreme case in which X is a highly electropositive element, such as an alkali metal. Of course, NaOH is not an acid at all, but a base. The sodium atom in Na—O—H gives up a full electron to make Na^+ and OH^- ions. Because the X—O bond here is almost completely ionic, the OH^- group has a net negative charge that holds the H^+ tightly to the oxygen and prevents formation of H^+ ions. The less electropositive alkaline-earth elements behave similarly. They form hydroxides, such as $Mg(OH)_2$, that are somewhat weaker bases than NaOH but in no way act as acids.

Now suppose the central atom X becomes more electronegative, reaching values between 2 and 3, as in the oxoacids of the elements B, C, P, As, S, Se, Br, and I. As X becomes more effective at withdrawing electron density from the oxygen atom, the X—O bond becomes more covalent. This leaves less negative charge on the oxygen atom, and consequently the oxoacid releases H^+ more readily (Fig. 15.9). Other things being equal, acid strength should increase with increasing electronegativity of the central atom. This trend is observed among the oxoacids listed in Table 15.2.

The strength of oxoacids with a given central element X increases with the number of lone oxygen atoms attached to the central atom. If the formula of these acids is written as $XO_n(OH)_m$, the corresponding acid strengths fall into distinct classes according to the value of n, the number of lone oxygen atoms (see Table 15.3). Each increase of 1 in n increases the acid ionization constant K_a by a factor of about 10^5. Another way to describe this effect is to focus on the stability of the conjugate base, $XO_{n+1}(OH)^-_{m-1}$, of the oxoacid. The greater the number of lone oxygen atoms attached to the central atom, the more easily the net negative charge can be spread out over the ion, and therefore the more stable the base. This leads to a larger K_a.

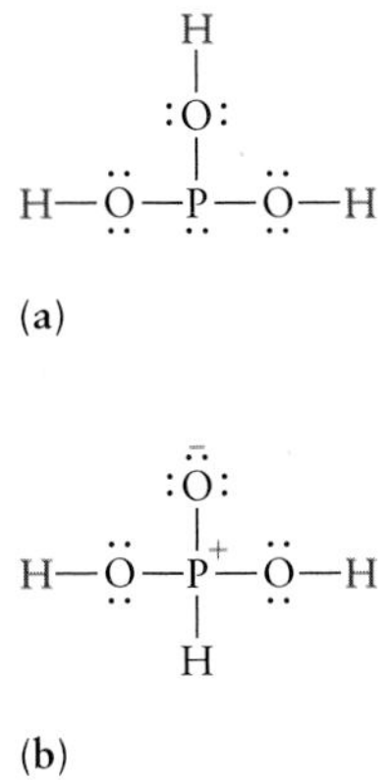

FIGURE 15.10 (a) The simplest Lewis diagram that can be drawn for H_3PO_3 gives an incorrect structure. This acid would be triprotic, like H_3PO_4. (b) The observed structure of H_3PO_3 requires assigning formal charge to the P atom and the lone O atom. The hydrogen atom attached to the P is not released into acid solution, so the acid is diprotic.

An unusual and interesting structural result can be obtained from Table 15.3. Figure 15.10a shows the simplest Lewis diagram for phosphorous acid (H_3PO_3) in which each atom achieves an octet configuration. Such a diagram could also be written $P(OH)_3$ and would be analogous to $As(OH)_3$, which has no lone oxygen atoms bonded to the central atom ($n = 0$). On the basis of this analogy, we would expect the value of K_a for $P(OH)_3$ to be on the order of 10^{-9} (a very weak acid). But in fact, H_3PO_3 is only a moderately weak acid ($K_a = 1 \times 10^{-2}$) and fits better into the class of acids with one lone oxygen atom bonded to the central atom. X-ray diffraction measurements support this structural conclusion inferred from

TABLE 15.3

Acid Ionization Constants for Oxoacids of the Nonmetals

$X(OH)_m$ Very Weak	K_a	$XO(OH)_m$ Weak	K_a	$XO_2(OH)_m$ Strong	K_a	$XO_3(OH)_m$ Very Strong	K_a
$Cl(OH)$	3×10^{-8}	$H_2PO(OH)$	8×10^{-2}	$SeO_2(OH)_2$	10^3	$ClO_3(OH)$	2×10^7
$Te(OH)_6$	2×10^{-8}	$IO(OH)_5$	2×10^{-2}	$ClO_2(OH)$	5×10^2		
$Br(OH)$	2×10^{-9}	$SO(OH)_2$	2×10^{-2}	$SO_2(OH)_2$	1×10^2		
$As(OH)_3$	6×10^{-10}	$ClO(OH)$	1×10^{-2}	$NO_2(OH)$	2×10^1		
$B(OH)_3$	6×10^{-10}	$HPO(OH)_2$	1×10^{-2}	$IO_2(OH)$	1.6×10^1		
$Ge(OH)_4$	4×10^{-10}	$PO(OH)_3$	8×10^{-3}				
$Si(OH)_4$	2×10^{-10}	$AsO(OH)_3$	5×10^{-3}				
$I(OH)$	4×10^{-11}	$SeO(OH)_2$	3×10^{-3}				
		$TeO(OH)_2$	3×10^{-3}				
		$NO(OH)$	5×10^{-4}				

chemical behavior. So, the structure of H_3PO_3 is best represented by Figure 15.10b and corresponds either to a Lewis diagram with more than eight electrons around the central phosphorus atom or to one with formal charges on the central phosphorus and lone oxygen atoms. The formula of this acid is written as $HPO(OH)_2$ in Table 15.3. Unlike phosphoric acid (H_3PO_4), which is a triprotic acid, H_3PO_3 is a diprotic acid. The third hydrogen atom, the one directly bonded to the phosphorus atom, does not ionize even in strongly basic aqueous solution.

15.3.2 Indicators

An **acid–base indicator** is a soluble dye that changes color noticeably over a fairly narrow range of pH. The typical indicator is a weak organic acid that has a different color from its conjugate base (Fig. 15.11). Litmus changes from red to blue as its acid form is converted to base. Good indicators have such intense colors that only a few drops of a dilute indicator solution must be added to the solution being studied. The very low concentration of indicator molecules has almost no effect on the pH of solution. The color changes of the indicator reflect the effects of the *other* acids and bases present in the solution.

If the acid form of a given indicator is represented as HIn and the conjugate base form as In^-, their acid–base equilibrium is

$$HIn(aq) + H_2O(\ell) \rightleftharpoons H_3O^+(aq) + In^-(aq)$$

$$\frac{[H_3O^+][In^-]}{[HIn]} = K_a$$

where K_a is the acid ionization constant for the indicator. This expression can be rearranged to give

$$\frac{[H_3O^+]}{K_a} = \frac{[HIn]}{[In^-]} \qquad [15.6]$$

FIGURE 15.11 Color differences in four indicators: bromophenol red, thymolphthalein, phenolphthalein, and bromocresol green. In each case the acidic form is on the left and the basic form is on the right.

If the concentration of hydronium ion $[H_3O^+]$ is large relative to K_a, this ratio is large, and [HIn] is large compared with $[In^-]$. The solution has the color of the acid form of the indicator because most of the indicator molecules are in the acid form. Litmus, for example, has a K_a near 10^{-7}. If the pH is 5, then

$$\frac{[H_3O^+]}{K_a} = \frac{10^{-5}}{10^{-7}} = 100$$

Thus, approximately 100 times as many indicator molecules are in the acid form as in the base form, and the solution is red.

As the concentration of hydronium ion is reduced, more molecules of acid indicator ionize to give the base form. When $[H_3O^+]$ is near K_a, almost equal amounts of the two forms are present, and the color is a mixture of the colors of the two indicator states (violet for litmus). A further decrease in $[H_3O^+]$ to a value much smaller than K_a then leads to a predominance of the base form, with the corresponding color being observed.

Different indicators have different values for K_a and thus show color changes at different pH values (Fig. 15.12). The weaker an indicator is as an acid, the higher the pH at which the color change takes place. Such color changes occur over a range of 1 to 2 pH units. Methyl red, for example, is red when the pH is below 4.8 and yellow above 6.0; shades of orange are seen at intermediate pH values. This limits the accuracy to which the pH can be

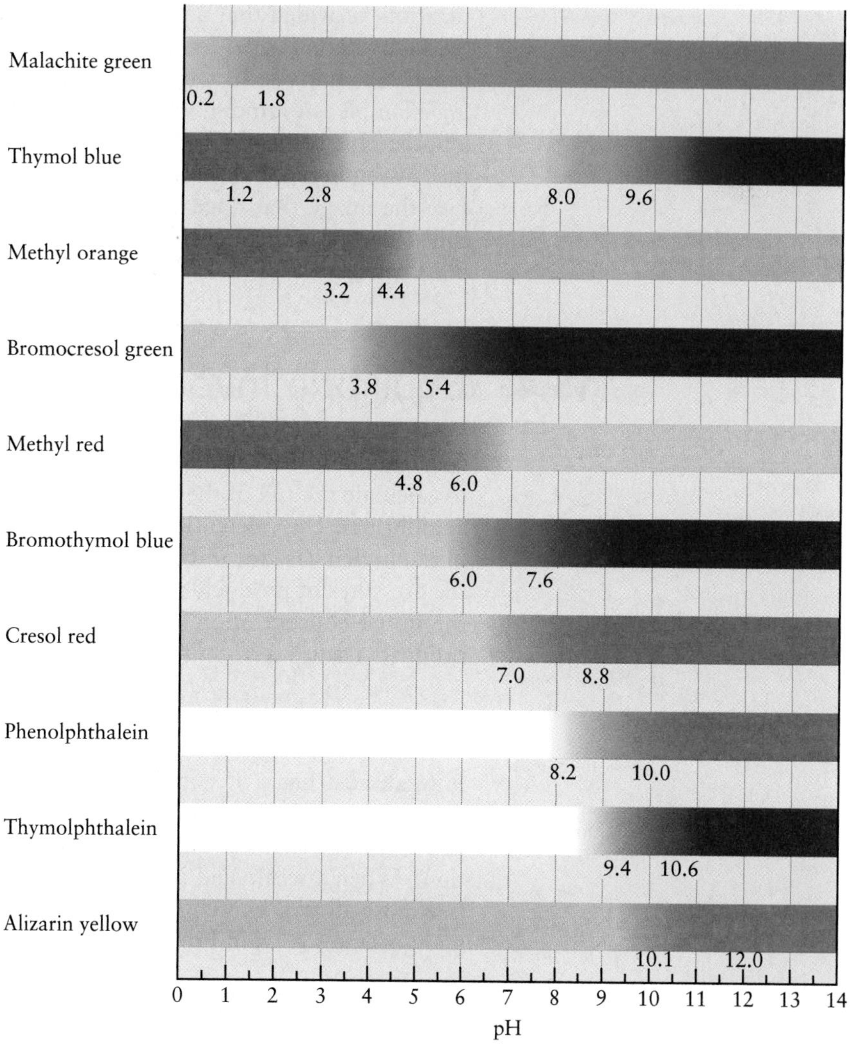

FIGURE 15.12 Indicators change their colors at very different pH values, so the best choice of indicator depends on the particular experimental conditions.

FIGURE 15.13 Red cabbage extract is a natural pH indicator. When the solution is highly acidic, the extract gives the solution a red color. As the solution becomes less and less acidic (more basic), the color changes from red to violet to yellow.

determined through the use of indicators. Section 15.6 shows that this fact does not affect the analytical determination of acid or base concentrations through titration, provided that an appropriate indicator is used.

Many natural dyes found in fruits, vegetables, and flowers act as pH indicators by changing color with changes in acidity (Fig. 15.13). A particularly striking example is cyanidin, which is responsible both for the red color of poppies and the blue color of cornflowers. The sap of the poppy is sufficiently acidic to turn cyanidin red, but the sap of the cornflower is basic and makes the dye blue. (See the image that appears at the start of this chapter.) Related natural dyes called anthocyanins contribute to the colors of raspberries, strawberries, and blackberries.

15.4 Equilibria Involving Weak Acids and Bases

The reactions of weak acids and bases with water do not go to completion. So, to calculate the pH of their solutions, we use K_a or K_b and the laws of chemical equilibrium. The calculations follow the pattern of Example 14.10 for gas equilibria. In that case the initial gas-phase pressures $P°$ are known, and we calculate the pressures of products resulting from the incomplete reaction. Here we know the initial concentration of acid or base, and calculate the concentrations of products resulting from its partial reaction with water.

15.4.1 Weak Acids

A weak acid has a K_a smaller than 1. Values of the pK_a start at zero for the strongest weak acid and range upward for progressively weaker acids. (If the pK_a is greater than 14, the compound is ineffective as an acid in aqueous solution.) When a weak acid is dissolved in water, the original concentration of HA is almost always known, but partial reaction with water consumes some HA and generates A^- and H_3O^+:

$$HA(aq) + H_2O(\ell) \rightleftharpoons H_3O^+(aq) + A^-(aq)$$

To calculate the amounts of H_3O^+, A^-, and HA at equilibrium, we use the methods of Chapter 14, with partial pressures replaced by concentrations. A new feature here is that one of the products (H_3O^+) can also come from a second source, the autoionization of the solvent, water. In most of the applications we study, this second effect is small and can be neglected in the equations. Even so, it is a good idea to verify at the end of each calculation that the $[H_3O^+]$ from the acid ionization alone exceeds 10^{-7} M by at least one order of magnitude. Otherwise, we have to use the more complete method of analysis given in Section 15.9.

EXAMPLE 15.6

Acetic acid (CH_3COOH) has a K_a of 1.76×10^{-5} at 25°C. Suppose 1.000 mol is dissolved in enough water to give 1.000 L of solution. Calculate the pH and the fraction of acetic acid ionized at equilibrium.

Solution

The initial concentration of acetic acid is 1.000 M. If y mol L^{-1} ionizes, then

	$CH_3COOH(aq)$ + $H_2O(\ell)$ $\rightleftharpoons$	$H_3O^+(aq)$ +	$CH_3COO^-(aq)$
Initial concentration (M)	1.000	≈0	0
Change in concentration (M)	$-y$	$+y$	$+y$
Equilibrium concentration (M)	$1.000 - y$	y	y

Note that we ignored the H_3O^+ initially present from the ionization of water because we expect it to be smaller than y for all but the weakest acids or the most dilute solutions. Then, the equilibrium expression states that

$$\frac{[H_3O^+][CH_3COO^-]}{[CH_3COOH]} = \frac{y^2}{1.000 - y} = K_a = 1.76 \times 10^{-5}$$

This equation could be solved using the quadratic formula, as we did in Chapter 14, or by using a programmable calculator. But it is important to develop approximation methods that enable quick solutions and also develop chemical insight and intuition into the equilibrium process. These approximations are based on the fact that acetic acid is a weak acid, and therefore only a small fraction is ionized at equilibrium. Therefore it is reasonable to assume that y is small relative to 1.000 (that is, the final equilibrium concentration of CH_3COOH is close to its initial concentration). This gives

$$\frac{y^2}{1.000 - y} \approx \frac{y^2}{1.000} = 1.76 \times 10^{-5}$$

$$y = 4.20 \times 10^{-3}$$

so

$$[H_3O^+] = y = 4.20 \times 10^{-3}\ \text{M}$$

Now we have to check whether the approximations are indeed valid. First, y is indeed much smaller than the original concentration of 1.000 M (by a factor of 200), so neglecting it in the denominator is justified. Second, the concentration of H_3O^+ from acetic acid (4.2×10^{-3} M) is large compared with 10^{-7} M, so neglecting the water ionization is also justified. Therefore,

$$\text{pH} = -\log_{10}(4.2 \times 10^{-3}) = 2.38$$

(Continued)

The fraction ionized is the ratio of the concentration of CH_3COO^- present at equilibrium to the concentration of CH_3COOH present in the first place:

$$\frac{y}{1.000} = y = 4.2 \times 10^{-3}$$

The percentage of the acetic acid that is ionized is 0.42%. Fewer than one in a hundred of the molecules of this typical weak acid dissociate in this solution.

Related Problems: 27, 28, 29, 30

Approximation methods of the type used in this example are also used in Example 14.10 and are discussed more extensively in Appendix C. As the solution of a weak acid becomes more dilute, a greater fraction of it ionizes, as shown by the following example.

EXAMPLE 15.7

Suppose that 0.00100 mol of acetic acid is used instead of the 1.000 mol in the preceding example. Calculate the pH and the percentage of acetic acid ionized.

Solution

Once again we assume that the contribution of the ionization of water to $[H_3O^+]$ is negligible, and check the validity of the assumption at the end of the calculation. Hence,

$$[CH_3COOH] = 0.00100 - y$$

$$[CH_3COO^-] = [H_3O^+] = y$$

$$\frac{y^2}{0.00100 - y} = 1.76 \times 10^{-5}$$

In this case, y is *not* small relative to 0.00100 because a substantial fraction of the acetic acid molecules ionizes, so y cannot be neglected in the denominator. There are two alternatives for solving the problem in this case.

The first method is to solve the quadratic equation obtained by multiplying out the equilibrium expression:

$$y^2 + (1.76 \times 10^{-5})y - (1.76 \times 10^{-8}) = 0$$

Using the quadratic formula gives

$$y = 1.24 \times 10^{-4}\ \text{M} = [H_3O^+] = [CH_3COO^-]$$

Because $y \gg 10^{-7}$ M, neglecting water ionization in setting up the problem is justified.

$$\text{pH} = -\log_{10}(1.24 \times 10^{-4}) = 3.91$$

$$\text{percentage ionized} = \frac{1.24 \times 10^{-4}\ \text{M}}{0.00100\ \text{M}} \times 100\% = 12.4\%$$

The other approach to solving the problem is to use successive approximations (see Appendix C). Begin by neglecting y in the denominator (relative to 0.00100), which gives

$$\frac{y^2}{0.00100} = 1.76 \times 10^{-5}$$

$$y = 1.33 \times 10^{-4}\ \text{M}$$

This is just what we do in Example 15.6. Now, continue by reinserting this *approximate* value of y into the *denominator* and then recalculate y:

$$\frac{y^2}{0.00100 - 0.00133} = 1.76 \times 10^{-5}$$

$$y^2 = 1.53 \times 10^{-8}$$

$$y = 1.33 \times 10^{-4}\ \text{M}$$

If we insert this new value of y into the denominator and iterate the calculation once again, we obtain 1.24×10^{-4} M, the same value produced by the quadratic formula. We terminate the iteration process when two successive results for y agree to the desired number of significant figures.

Related Problems: 31, 32

The method of successive approximations is often faster to apply than the quadratic formula. Keep in mind that the accuracy of a result is limited both by the accuracy of the input data (values of K_a and initial concentrations) and by the fact that solutions are not ideal. It is pointless to calculate equilibrium concentrations to any degree of accuracy higher than 1% to 3%.

15.4.2 Weak Bases

There is a perfect parallel between the behavior of weak bases and weak acids. The definition and description of weak acids, the existence of K_a, and the production of $H_3O^+(aq)$ ion when they are dissolved in water can be applied directly to weak bases, the existence of K_b, and the production of $OH^-(aq)$ ion. A **weak base** such as ammonia reacts only partially with water to produce $OH^-(aq)$:

$$H_2O(\ell) + NH_3(aq) \rightleftharpoons NH_4^+(aq) + OH^-(aq)$$

$$\frac{[NH_4^+][OH^-]}{[NH_3]} = K_b = 1.8 \times 10^{-5}$$

The K_b of a weak base is smaller than 1, and the weaker the base, the smaller the K_b. If the K_b of a compound is smaller than 1×10^{-14}, that compound is ineffective as a base in aqueous solution.

The analogy to weak acids continues in the calculation of the aqueous equilibria of weak bases, as the following example shows.

EXAMPLE 15.8

Calculate the pH of a solution made by dissolving 0.0100 mol of NH_3 in enough water to give 1.000 L of solution at 25°C. The K_b for ammonia is 1.8×10^{-5}.

Solution

Set up the table of the changes in concentrations that occur as the reaction goes to equilibrium, neglecting the small contribution to $[OH^-]$ from the autoionization of water:

	$H_2O(\ell) + NH_3(aq)$	$\rightleftharpoons NH_4^+(aq)$	$+ OH^-(aq)$
Initial concentration (M)	0.0100	0	≈0
Change in concentration (M)	$-y$	$+y$	$+y$
Equilibrium concentration (M)	$0.0100 - y$	y	y

Substitution into the equilibrium expression gives

$$\frac{y^2}{0.0100 - y} = K_b = 1.8 \times 10^{-5}$$

which can be solved for y by either the quadratic formula or the method of successive approximations to obtain:

$$y = 4.15 \times 10^{-4} \text{ M} = [OH^-]$$

The product of $[H_3O^+]$ and $[OH^-]$ is always K_w. Hence,

$$[H_3O^+] = \frac{K_w}{[OH^-]} = \frac{1.0 \times 10^{-14}}{4.15 \times 10^{-14}} = 2.4 \times 10^{-11} \text{ M}$$

$$\text{pH} = -\log_{10}(2.4 \times 10^{-11}) = 10.62$$

The pH is greater than 7, as expected for a solution of a base.

Related Problems: 35, 36

15.4.3 Hydrolysis

Most of the acids considered up to now have been uncharged species with the general formula HA. In the Brønsted–Lowry picture there is no reason why the acid should be an electrically neutral molecule. When NH_4Cl, a salt, dissolves

(a) (b) (c)

FIGURE 15.14 Proof that ammonium chloride is an acid. (a) A solution of sodium hydroxide, with the pink color of phenolphthalein indicating its basic character. (b and c) As aqueous ammonium chloride is added, a neutralization reaction takes place and the solution turns colorless from the bottom up.

in water, NH_4^+ ions are present. Some of these ionize by transferring hydrogen ions to water in a straightforward Brønsted–Lowry acid–base reaction:

$$\underset{\text{Acid}_1}{NH_4^+(aq)} + \underset{\text{Base}_2}{H_2O(\ell)} \rightleftharpoons \underset{\text{Acid}_2}{H_3O^+(aq)} + \underset{\text{Base}_1}{NH_3(aq)}$$

$$\frac{[H_3O^+][NH_3]}{NH_4^+} = K_a = 5.6 \times 10^{-10}$$

NH_4^+ acts as an acid here, just as acetic acid did in Examples 15.6 and 15.7. Because K_a is much smaller than 1, only a small amount of H_3O^+ is generated. NH_4^+ is a weak acid, but it is nonetheless an acid, and a solution of ammonium chloride has a pH below 7 (Fig. 15.14).

Hydrolysis is the label for the reaction of a substance with water. This term is applied especially to a reaction in which the pH changes from 7 upon dissolving a salt (in this case, NH_4Cl) in water. There is no need for any special description of hydrolysis. The hydrolysis reaction that takes place when NH_4Cl dissolves in water is completely described as a Brønsted–Lowry reaction in which water acts as a base and NH_4^+ acts as an acid to give a pH below 7. In parallel fashion, dissolving a salt whose anion is a weak base produces a basic solution. This too is a case of hydrolysis; it is simply another Brønsted–Lowry acid–base reaction, with water acting now as an acid (a hydrogen ion donor).

Two salts that give basic solutions are sodium acetate and sodium fluoride. When these salts dissolve in water, they furnish acetate (CH_3COO^-) and fluoride (F^-) ions, respectively, both of which act as Brønsted–Lowry bases,

$$H_2O(\ell) + CH_3COO^-(aq) \rightleftharpoons CH_3COOH(aq) + OH^-(aq)$$

$$H_2O(\ell) + F^-(aq) \rightleftharpoons HF(aq) + OH^-(aq)$$

causing the OH^- concentration to increase and give a pH above 7.

EXAMPLE 15.9

Suppose 0.100 mol of $NaCH_3COO$ is dissolved in enough water to make 1.00 L of solution. What is the pH of the solution?

(Continued)

Solution

The K_a for CH_3COOH, the conjugate acid of CH_3COO^-, is 1.76×10^{-5}. Therefore, K_b for the acetate ion is

$$\frac{[CH_3COOH][OH^-]}{[CH_3COO^-]} = K_b = \frac{K_w}{K_a} = 5.7 \times 10^{-10}$$

The equilibrium here is

	$H_2O(\ell) + CH_3COO^-(aq)$	$\rightleftarrows$	$CH_3COOH(aq)$	$+ OH^-(aq)$
Initial concentration (M)	0.100		0	≈0
Change in concentration (M)	$-y$		$+y$	$+y$
Equilibrium concentration (M)	$0.100 - y$		y	y

Substitution into the equilibrium expression gives

$$\frac{y^2}{0.100 - y} = K_b = 5.7 \times 10^{-10}$$

$$y = 7.5 \times 10^{-6}\ \text{M} = [OH^-]$$

Note that y is small relative to 0.100 and fairly large relative to 10^{-7}.

$$[H_3O^+] = \frac{K_w}{[OH^-]} = \frac{1.0 \times 10^{-14}}{7.5 \times 10^{-6}} = 1.3 \times 10^{-9}\ \text{M}$$

$$\text{pH} = 8.89$$

Related Problems: 41, 42

Hydrolysis does not occur with all ions, only with those that are conjugate acids of weak bases or conjugate bases of weak acids. Chloride ion is the conjugate base of the strong acid HCl and consequently is ineffective as a base, unlike $F^-(aq)$ and $CH_3COO^-(aq)$. Its interaction with water would therefore scarcely change the $OH^-(aq)$ concentration. So, a solution of NaCl is neutral, whereas a solution of NaF is slightly basic.

15.5 Buffer Solutions

A **buffer solution** maintains relatively constant pH upon small additions of acid or base. Typically, a buffer solution contains a weak acid and its conjugate weak base in approximately equal concentrations. Buffer solutions play important roles in controlling the solubility of ionic compounds in solution (see Chapter 16) and in maintaining the pH in biochemical and physiological processes. Many life processes are sensitive to pH and require regulation within a small range of H_3O^+ and OH^- concentrations. Organisms have built-in buffers to protect them against large changes in pH. Human blood has a pH near 7.4, which is maintained by a combination of carbonate, phosphate, and protein buffer systems. Conditions that drive blood pH below 7.0 or above 7.8 lead quickly to death.

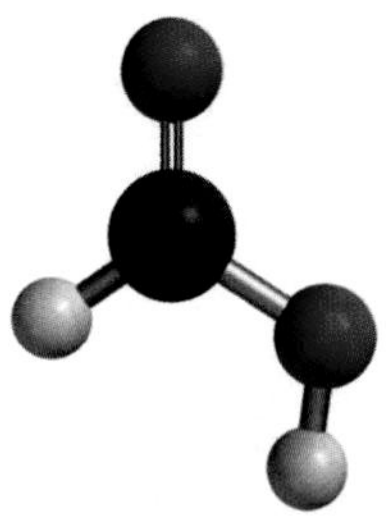

FIGURE 15.15 Formic acid, HCOOH, is the simplest carboxylic acid, with only a hydrogen atom attached to the —COOH group.

15.5.1 Calculations of Buffer Action

Consider a typical weak acid, formic acid (HCOOH) (Fig. 15.15), and its conjugate base, formate ion ($HCOO^-$). The latter can be obtained by dissolving a salt such as sodium formate (NaHCOO) in water. The acid–base equilibrium established between these species is given by

$$HCOOH(aq) + H_2O(\ell) \longrightarrow H_3O^+(aq) + HCOO^-(aq)$$

with an acid ionization constant

$$\frac{[H_3O^+][HCOO^-]}{[HCOOH]} = K_a = 1.77 \times 10^{-4}$$

Section 15.2 describes the pH of a solution containing only a weak acid (such as HCOOH) or only a weak base (such as $HCOO^-$). Suppose now that the weak acid and its conjugate base are *both* present initially. The resulting calculations resemble closely the calculation of Example 14.15, in which equilibrium was established from an initial mixture containing both reactants and products.

EXAMPLE 15.10

Suppose 1.00 mol of HCOOH and 0.500 mol of NaHCOO are added to water and diluted to 1.00 L. Calculate the pH of the solution.

Solution

	$HCOOH(aq)$ + $H_2O(\ell)$	$\rightleftharpoons$ $H_3O^+(aq)$	+ $HCOO^-(aq)$
Initial concentration (M)	1.00	≈0	0.500
Change in concentration (M)	$-y$	$+y$	$+y$
Equilibrium concentration (M)	$1.00 - y$	y	$0.500 + y$

The equilibrium expression is

$$\frac{y(0.500 + y)}{1.00 - y} = K_a = 1.77 \times 10^{-4}$$

Because y is likely to be small relative to 1.00 and to 0.500, we write the approximate expression

$$\frac{y(0.500)}{1.00} \approx 1.77 \times 10^{-4}$$

$$y \approx 3.54 \times 10^{-4} \text{ M} = [H_3O^+]$$

A glance verifies that y is indeed small relative to 1.00 and 0.500. Then

$$\text{pH} = -\log_{10}(3.54 \times 10^{-4}) = 3.45$$

Related Problems: 43, 44

To understand how buffer solutions work, let's write the equilibrium expression for the ionization of a weak acid HA in the form

$$[H_3O^+] = K_a \frac{[HA]}{[A^-]}$$

The concentration of hydronium ion depends on the ratio of the concentration of the weak acid to the concentration of its conjugate base. The key to effective buffer action is to keep these concentrations nearly equal and fairly large. Adding a small amount of base to an effective buffer takes away only a few percent of the HA molecules by converting them into A^- ions and adds only a few percent to the amount of A^- originally present. The ratio $[HA]/[A^-]$ decreases, but only very slightly. Added acid consumes a small fraction of the base A^- to generate a bit more HA. The ratio $[HA]/[A^-]$ now increases, but again the change is only slight. Because the concentration of H_3O^+ is tied directly to this ratio, it changes only slightly. The following example illustrates buffer action quantitatively.

EXAMPLE 15.11

Suppose 0.10 mol of a strong acid such as HCl is added to the solution in Example 15.10. Calculate the pH of the resulting solution.

Solution

The strong acid HCl ionizes essentially completely in dilute aqueous solution. Initially assume that *all* the hydrogen ions from HCl are taken up by formate ions to produce formic acid, some of which ionizes back to formate ion and H_3O^+. This is simply a way of describing one possible route by which equilibrium is approached and does not define the sequence of reactions that actually occurs. The position of the final equilibrium does not depend on the route by which it is attained.

Because 0.10 mol of HCl reacts with an equal number of moles of $HCOO^-$, the concentrations of $HCOO^-$ and HCOOH *before* ionization are

$$[HCOO^-]_0 = 0.50 - 0.10 = 0.40 \text{ M}$$

$$[HCOOH]_0 = 1.00 + 0.10 = 1.10 \text{ M}$$

The table to calculate the concentrations at equilibrium is

	$HCOOH(aq)$ + $H_2O(\ell)$ ⇌	$H_3O^+(aq)$ +	$HCOO^-(aq)$
Initial concentration (M)	1.10	≈0	0.40
Change in concentration (M)	$-y$	$+y$	$+y$
Equilibrium concentration (M)	$1.10 - y$	y	$0.40 + y$

The equilibrium expression then becomes

$$\frac{y(0.40 + y)}{1.10 - y} = 1.77 \times 10^{-4}$$

Because y is again likely to be small relative to both 0.40 and 1.10,

$$y \approx \left(\frac{1.10}{0.40}\right)(1.77 \times 10^{-4}) = 4.9 \times 10^{-4}$$

$$\text{pH} = 3.31$$

Even though 0.10 mol of a strong acid was added, the pH changed only slightly, from 3.45 to 3.31. In contrast, the same amount of acid added to a liter of pure water would change the pH from 7 to 1.

Related Problems: 45, 46

Note that we solved this problem by first performing a stoichiometric (limiting reactant) calculation and then an equilibrium calculation. A similar strategy works if a strong base such as OH^- is added instead of a strong acid. The base reacts with formic acid to produce formate ions. Adding 0.10 mol of OH^- to the $HCOOH/HCOO^-$ buffer of Example 15.10 increases the pH only to 3.58. In the absence of the buffer system, the same base would raise the pH to 13.00.

In any buffer there is competition between the tendency of the acid to donate hydrogen ions to water (increasing the acidity) and the tendency of the base to accept hydrogen ions from water (increasing the basicity). The resulting pH depends on the magnitude of K_a. If K_a is large relative to 10^{-7}, the acid ionization will win out and acidity will increase, as in the HCOOH/$HCOO^-$ buffer. A basic buffer (with pH > 7) can be prepared by working with an acid–base pair with K_a smaller than 10^{-7}. In this case the net reaction produces OH^-, and we must use K_b to determine the equilibrium state. A typical example is the NH_4^+/NH_3 buffer prepared by mixing ammonium chloride with ammonia.

EXAMPLE 15.12

Calculate the pH of a solution made by adding 0.100 mol of NH_4Cl and 0.200 mol of NH_3 to water and diluting to 1.000 L. K_a for NH_4^+ is 5.6×10^{-10}.

Solution

Because $K_a \ll 10^{-7}$ for NH_4^+ (equivalently, $K_b \gg 10^{-7}$ for NH_3), the net reaction is production of OH^- ions. Therefore, the equilibrium is written to show the net transfer of hydrogen ions from water to NH_3:

	$H_2O(\ell)$ + $NH_3(aq)$	$\rightleftharpoons$ $NH_4^+(aq)$	+ $OH^-(aq)$
Initial concentration (M)	0.200	0.100	0
Change in concentration (M)	$-y$	$+y$	$+y$
Equilibrium concentration (M)	$0.200 - y$	$0.100 + y$	y

The equilibrium expression is

$$\frac{y(0.100 + y)}{0.200 - y} = K_b = \frac{K_w}{K_a} = 1.8 \times 10^{-5}$$

If y is small relative to the original concentrations of both NH_3 and NH_4^+, then

$$\frac{y(0.100)}{0.200} \approx 1.8 \times 10^{-5}$$

$$y \approx \left(\frac{0.200}{0.100}\right)(1.8 \times 10^{-5}) = 3.6 \times 10^{-5} \ll 0.100,\ 0.200$$

$$[H_3O^+] = \frac{1.0 \times 10^{-14}}{[OH^-]} = \frac{1.0 \times 10^{-14}}{3.6 \times 10^{-5}} = 2.8 \times 10^{-10}\ \text{M}$$

$$\text{pH} = -\log_{10}(2.8 \times 10^{-10}) = 9.55$$

15.5.2 Designing Buffers

Control of pH is vital in synthetic and analytical chemistry, just as it is in living organisms. Procedures that work well at a pH of 5 may fail when the concentration of hydronium ion in the solution is raised tenfold to make the pH 4. Fortunately, it is possible to prepare buffer solutions that maintain the pH close to any desired value by the proper choice of a weak acid and the ratio of its concentration to that of its conjugate base. Let's see how to choose the best conjugate acid–base system and how to calculate the required acid–base ratio.

In Examples 15.10, 15.11, and 15.12, the equilibrium concentrations of acid and base in the buffer systems were close to the initial concentrations. When this is the case, the calculation of pH is simplified greatly, because for either an acidic *or* a basic buffer,

$$K_a = \frac{[H_3O^+][A^-]}{[HA]} \approx \frac{[H_3O^+][A^-]_0}{[HA]_0}$$

$$[H_3O^+] \approx \frac{[HA]_0}{[A^-]_0} K_a$$

and the pH is given by

$$pH \approx pK_a - \log_{10} \frac{[HA]_0}{[A^-]_0} \qquad [15.7]$$

We obtained Equation 15.7 by taking the logarithm and changing sign. It is easy to verify that this simple equation gives the correct result for the three preceding examples. However, it must be used with some care because it is only approximate. It is valid only when *both* $[H_3O^+]$ and $[OH^-]$ are small relative to $[HA]_0$ and $[A^-]_0$; this means the extent of ionization must be small.

We can use this expression relating pH to pK_a to design buffers with a specific value of pH. In an optimal buffer the acid and its conjugate base are purposely very nearly equal in concentration; if the difference in concentrations is too great, the buffer is less resistant to the effects of adding acid or base. To select a buffer system, we choose an acid with a pK_a as close as possible to the desired pH. Then we adjust the concentrations of acid and conjugate base to give exactly the desired pH.

EXAMPLE 15.13

Design a buffer system with pH 4.60.

Solution

From Table 15.2, the pK_a for acetic acid is 4.75, so the CH_3COOH/CH_3COO^- buffer is a suitable one. The concentrations required to give the desired pH are related by

$$pH = 4.60 = pK_a - \log_{10} \frac{[CH_3COOH]_0}{[CH_3COO^-]_0}$$

$$\log_{10} \frac{[CH_3COOH]_0}{[CH_3CHOO^-]_0} = pK_a - pH = 4.75 - 4.60 = 0.15$$

$$\frac{[CH_3COOH]_0}{[CH_3COO^-]_0} = 10^{0.15} = 1.4$$

We can establish this ratio by dissolving 0.100 mol of sodium acetate and 0.140 mol of acetic acid in water and diluting to 1.00 L, or 0.200 mol of $NaCH_3COO$ and 0.280 mol of CH_3COOH in the same volume, and so on. As long as the ratio of the concentrations is 1.4, the solution will be buffered at approximately pH 4.60.

Related Problems: 47, 48, 49, 50

The preceding example shows that the *absolute* concentrations of acid and conjugate base in a buffer are much less important than is their ratio in determining the pH. Nonetheless, the absolute concentrations do affect the capacity of the solution to resist changes in pH when acid or base is added. The higher the concentrations of buffering species, the smaller the change in pH when a fixed amount of a strong acid or base is added. In Example 15.10 a change in buffer concentrations from 1.00 M and 0.500 M to 0.500 M and 0.250 M does not alter the original pH of 3.45 because the ratio of acid to base concentrations is unchanged. The pH after 0.100 mol of HCl is added does change to the value 3.15 rather than 3.31. The buffer at lower concentration is less resistant to pH

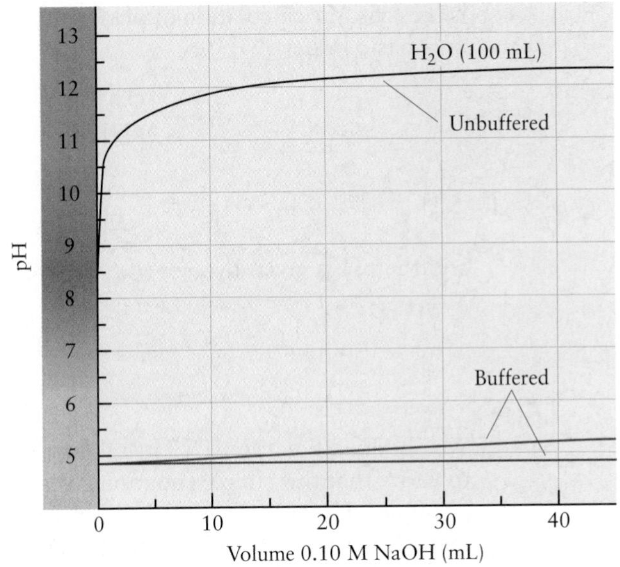

FIGURE 15.16 Addition of a given volume of base to buffered and unbuffered solutions causes a much greater change in the pH of the unbuffered solution. Of the two buffered solutions, the one with higher buffer concentration resists pH changes more effectively. The red line represents 100 mL of a buffer that is 0.1 M in both CH_3COOH and CH_3COO^-; the blue line represents the same volume of a buffer that is 1.0 M in both components.

change (Fig. 15.16). The buffering capacity of *any* buffer solution is exhausted if enough strong acid (or strong base) is added to use up the original amount of weak base (or weak acid) through chemical reaction.

15.6 Acid–Base Titration Curves

Section 11.3 describes an acid–base titration as the addition of carefully metered volumes of a basic solution of known concentration to an acidic solution of unknown concentration—or the addition of acid to base—to reach an *endpoint* at which the unknown solution has been completely neutralized. The endpoint is signaled by the color change of an indicator or by a sudden rise or fall in pH. The pH of the reaction mixture changes continuously over the course of an acid–base titration, but changes abruptly only near the endpoint. A graph of the pH versus the volume V of titrating solution added is a **titration curve**. Its shape depends on the value of K_a and on the concentrations of the acid and base reacting. The concepts and tools of acid–base equilibria predict the exact shapes of titration curves when these quantities are all known. The same concepts allow K_a and the concentration of the unknown solution to be calculated from an experimental titration curve. Here we examine three categories of titration and determine the titration curve for each: strong acid reacting with strong base, weak acid reacting with strong base, and strong acid reacting with weak base. Titrations of a weak acid with a weak base (and the reverse) are not useful for analytical purposes.

15.6.1 Titration of a Strong Acid with a Strong Base

The addition of a strong base to a strong acid (or the reverse) is the simplest type of titration. The chemical reaction is the neutralization:

$$H_3O^+(aq) + OH^-(aq) \longrightarrow 2\ H_2O(\ell)$$

Suppose a solution of 100.0 mL (0.1000 L) of 0.1000 M HCl is titrated with 0.1000 M NaOH. What does the titration curve look like? The curve can be measured experimentally, and the result is shown in Figure 15.17. We can also

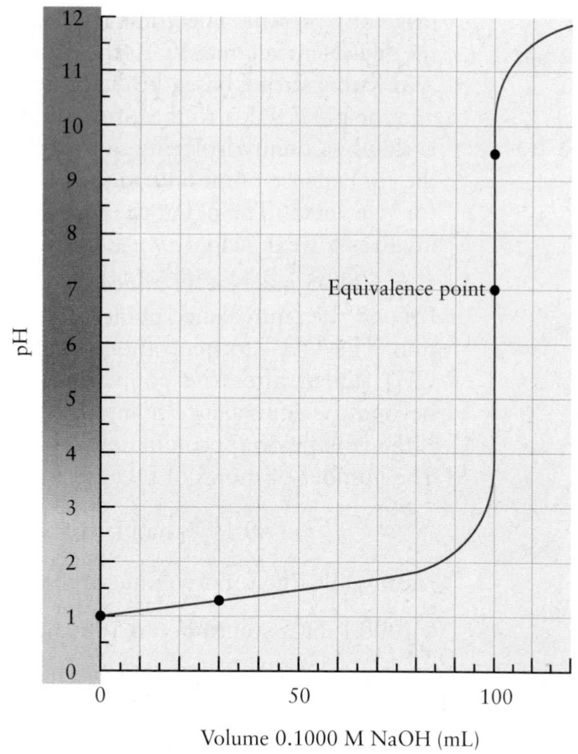

FIGURE 15.17 A titration curve for the titration of a strong acid by a strong base. The curve shown is for 100.0 mL of 0.1000 M HCl titrated with 0.1000 M NaOH.

construct the curve theoretically by calculating the pH of the reaction mixture at many different points during the addition of the NaOH solution and plotting the results. The following example illustrates the theoretical procedure.

1. **V = 0 mL NaOH added**
   Initially, $[H_3O^+] = 0.1000$ M, so the pH is 1.000. The number of moles of H_3O^+ present initially is

$$n_{H_3O^+} = [H_3O^+](\text{volume}) = (0.1000 \text{ mol L}^{-1})(0.1000 \text{ L})$$
$$= 1.000 \times 10^{-2} \text{ mol}$$

2. **V = 30.00 mL NaOH added**
   30.00 mL of 0.1000 M NaOH solution contains

$$(0.1000 \text{ mol L}^{-1})(0.03000 \text{ L}) = 3.000 \times 10^{-3} \text{ mol OH}^-$$

   This reacts with (and neutralizes) an equal number of moles of the H_3O^+ ion present initially and reduces $n_{H_3O^+}$ to

$$n_{H_3O^+} = (1.000 \times 10^{-2} - 3.000 \times 10^{-3}) \text{ mol} = 7.00 \times 10^{-3} \text{ mol}$$

   In addition—and this is very important to remember—the volume of the titration mixture has increased from 100.0 to 130.0 mL (that is, from 0.1000 to 0.1300 L). The concentration of H_3O^+ at this point in the titration is

$$[H_3O^+] = \frac{n_{H_3O^+}}{V_{tot}} = \frac{7.00 \times 10^{-3} \text{ mol}}{0.1300 \text{ L}} = 0.0538 \text{ M}$$
$$\text{pH} = 1.27$$

3. **V = 100.00 mL NaOH added**
   This is called the **equivalence point**, that point in the titration at which the number of moles of base added equals the number of moles of acid

originally present. The volume of base added up to the equivalence point is the **equivalent volume**, V_e. At the equivalence point in the titration of a strong acid with a strong base, the concentrations of OH^- and H_3O^+ must be equal and the pH 7.0 due to the autoionization of water. At this point the solution is simply a nonhydrolyzing salt (in this case, NaCl) in water. The pH is 7 at the equivalence point only in the titration of a strong acid with a strong base (or vice versa). The pH at the equivalence point differs from 7 if the titration involves a weak acid or weak base.

4. **V = 100.05 mL NaOH added**

Beyond the equivalence point, OH^- is added to a neutral unbuffered solution. The OH^- concentration can be found from the number of moles of OH^- added after the equivalence point has been reached. The volume beyond the equivalence point at this stage is 0.05 mL, or 5×10^{-5} L. (This is the volume of approximately one drop of solution added from the buret.) The number of moles of OH^- in this volume is

$$(0.1000 \text{ mol L}^{-1})(5 \times 10^{-5} \text{ L}) = 5 \times 10^{-6} \text{ mol}$$

Meanwhile, the total volume of the titration mixture is

$$0.1000 \text{ L HCl solution} + 0.10005 \text{ L NaOH solution} = 0.20005 \text{ L solution}$$

Therefore, the concentration of OH^- is

$$[OH^-] = \frac{\text{moles } OH^-}{\text{total volume}} = \frac{5 \times 10^{-5} \text{ mol}}{0.2005 \text{ L}} = 2.5 \times 10^{-5} \text{ M}$$

$$[H_3O^+] = 4 \times 10^{-10} \text{ M; pH} = 9.4$$

As the titration curve shows (see Fig. 15.17), the pH increases dramatically in the immediate vicinity of the equivalence point: $[H_3O^+]$ changes by four orders of magnitude between 99.98 mL and 100.02 mL NaOH! Any indicator whose color changes between pH = 5.0 and pH = 9.0 therefore signals the endpoint of the titration to an accuracy of ±0.02 mL in 100.0 mL, or ±0.02%. The titration **endpoint**, the experimentally measured volume at which the indicator changes color, is then almost identical to the equivalence point, the theoretical volume at which the chemical amount of added base equals that of acid present originally.

The titration of a strong base by a strong acid is entirely parallel. In this case the pH starts at a higher value and *drops* through a pH of 7 at the equivalence point. The roles of acid and base, and of H_3O^+ and OH^-, are reversed in the equations already given.

15.6.2 Titration of a Weak Acid with a Strong Base

We turn now to the titration of a weak acid with a strong base (the titration of a weak base with a strong acid is analogous). The equivalence point has the same meaning as for a strong acid titration. At the equivalence point, the number of moles of base added (in volume V_e) is equal to the number of moles of acid originally present (in volume V_0). So, once again

$$c_0 V_0 = c_t V_e$$

where c_0 is the original weak acid concentration and c_t is the OH^- concentration in the titrating solution.

The calculation of the titration curve differs from the strong acid–strong base case in that now equilibrium (reflected in the K_a of the weak acid) enters

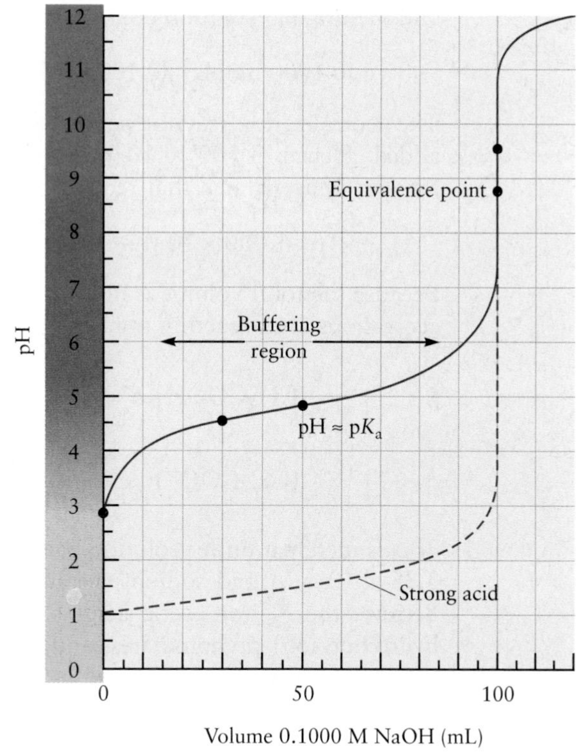

FIGURE 15.18 A titration curve for the titration of a weak acid by a strong base. The curve shown in red is for 100.0 mL of 0.1000 M CH_3COOH titrated with 0.1000 M NaOH. For comparison, the dashed blue line shows the titration curve for a strong acid of the same amount and concentration as presented in Figure 15.17.

the picture. As an example, consider the titration of 100.0 mL of a 0.1000 M solution of acetic acid (CH_3COOH) with 0.1000 M NaOH. For this titration,

$$V_e = \frac{c_0}{c_t} V_0 = \left(\frac{0.1000\text{ M}}{0.1000\text{ M}}\right)(100.0\text{ mL})$$
$$= 100.0\text{ mL}$$

There are four distinct ranges in the titration; each corresponds to a type of calculation that we have illustrated already. Figure 15.18 shows pH versus volume for this titration, and the following paragraphs outline the calculation of four typical points on the curve.

1. **$V = 0$ mL NaOH added**
   At the beginning of the titration—before any NaOH is added—the problem is simply the ionization of a weak acid, considered in Section 15.4. A calculation analogous to those of Examples 15.6 and 15.7 gives a pH of 2.88.
2. **$0 < V < V_e$**
   In this range the acid has been partially neutralized by added NaOH solution. Because OH^- is a stronger base than acetate ion, it reacts almost completely with the acid originally present:

$$CH_3COOH(aq) + OH^-(aq) \rightleftharpoons H_2O(\ell) + CH_3COO^-(aq)$$
$$K = \frac{1}{K_b} = \frac{K_a}{K_w} = 2 \times 10^9 \gg 1$$

As a specific example, suppose that 30.00 mL of 0.1000 M NaOH has been added. The 30.00 mL NaOH contains

$$(0.1000\text{ mol L}^{-1})(0.03000\text{ L}) = 3.000 \times 10^{-3}\text{ mol OH}^-$$

and the original solution contained

$$(0.1000 \text{ mol L}^{-1})(0.1000 \text{ L}) = 1.000 \times 10^{-2} \text{ mol } CH_3COOH$$

The neutralization reaction generates one CH_3COO^- ion for every OH^- ion added. Hence, 3.000×10^{-3} mol of CH_3COO^- ions is generated. The amount of acetic acid that remains unreacted is

$$1.000 \times 10^{-2} - 3.000 \times 10^{-3} = 7.00 \times 10^{-3} \text{ mol}$$

Because the total volume is now 130.0 mL (or 0.1300 L), the nominal concentrations after reaction are

$$[CH_3COOH] \approx \frac{7.00 \times 10^{-3} \text{ mol}}{0.1300 \text{ L}} = 5.38 \times 10^{-2} \text{ M}$$

$$[CH_3COO^-] \approx \frac{3.00 \times 10^{-3} \text{ mol}}{0.1300 \text{ L}} = 2.31 \times 10^{-2} \text{ M}$$

This is merely a buffer solution containing acetic acid at a concentration of 5.38×10^{-2} M and sodium acetate at a concentration of 2.31×10^{-2} M. Because the K_a for acetic acid is larger than 10^{-7}, hydronium ion (not hydroxide ion) predominates, and this is an acidic buffer. The pH can be found from the procedure used in Example 15.10 or, more approximately from Equation 15.7:

$$\text{pH} \approx \text{p}K_a - \log_{10}\frac{[CH_3COOH]_0}{[CH_3COO^-]_0} = 4.75 - \log_{10}\frac{5.38 \times 10^{-2}}{2.31 \times 10^{-2}} = 4.38$$

This region of the titration shows clearly the buffering action of a mixture of a weak acid with its conjugate base. At the half-equivalence point $V = V_e/2$, $[CH_3COOH]_0 = [CH_3COO^-]_0$, which corresponds to an equimolar buffer; at this point pH $\approx$ pK_a. On either side of this point the pH rises relatively slowly as the NaOH solution is added.

3. $V = V_e$

   At the equivalence point, c_tV_e mol of OH^- has been added to the same number of moles of CH_3COOH. An identical solution could have been prepared simply by adding $c_tV_e = 1.000 \times 10^{-2}$ mol of the base CH_3COO^- (in the form of $NaCH_3COO$) to 0.2000 L of water. The pH at the equivalence point corresponds to the hydrolysis of CH_3COO^-:

$$H_2O(\ell) + CH_3COO^-(aq) \rightleftharpoons CH_3COOH(aq) + OH^-(aq)$$

   as considered in Example 15.9. At the equivalence point the pH is 8.73.

   Note that in the titration of a weak acid by a strong base the equivalence point comes not at pH 7 but at a higher (more basic) value. By the same token, the equivalence point in the titration of a weak base by a strong acid occurs at a pH lower than 7.

4. $V > V_e$

   Beyond the equivalence point, $OH^-(aq)$ is added to a solution of the base CH_3COO^-. The $[OH^-]$ comes almost entirely from the hydroxide ion added beyond the equivalence point; very little comes from the reaction of the CH_3COO^- with water. Beyond V_e, the pH for the titration of a weak acid by a strong base is very close to that for a strong acid by a strong base.

The equivalent volume V_e is readily determined in the laboratory by using an indicator that changes color near pH 8.7, the pH at the equivalence point of the acetic acid titration. A suitable choice would be phenolphthalein, which changes from colorless to red over a pH range from 8.2 to 10.0. The slope of pH versus volume of strong base is less steep near the equivalence point for a weak acid than it is for a strong acid, so determination of the equivalent volume—and of the unknown weak acid concentration—is somewhat less accurate.

The following example shows how to determine concentrations and ionization constants for unknown acids and bases from titrations.

EXAMPLE 15.14

A volume of 50.00 mL of a weak acid of unknown concentration is titrated with a 0.1000 M solution of NaOH. The equivalence point is reached after 39.30 mL of NaOH solution has been added. At the half-equivalence point (19.65 mL) the pH is 4.85. Calculate the original concentration of the acid and its ionization constant K_a.

Solution

The number of moles of acid originally present, c_0V_0, is equal to the number of moles of base added at the equivalence point, c_tV_e, so

$$c_0 = \frac{V_e}{V_0}\,c_t = \frac{39.50\text{ mL}}{50.00\text{ mL}} \times 0.100\text{ M} = .0786\text{ M}$$

This is the original concentration of acid.

At the half-equivalence point,

$$\text{pH} = 4.85 \approx \text{p}K_a$$

$$K_a = 10^{-4.85} = 1.4 \times 10^{-5}$$

The unknown acid could be propionic acid, CH_3CH_2COOH (see Table 15.2).

Related Problems: 61, 62

15.7 Polyprotic Acids

So far, we have only considered acids that are monoprotic. A molecule of a **monoprotic acid** can donate a single hydrogen ion to an acceptor molecule. A molecule of a **polyprotic acid** can donate two or more hydrogen ions to acceptors. Sulfuric acid is a familiar and important example. It reacts in two stages; first

$$H_2SO_4(aq) + H_2O(\ell) \longrightarrow H_3O^+(aq) + HSO_4^-(aq)$$

to give the hydrogen sulfate ion, and then

$$HSO_4^-(aq) + H_2O(\ell) \longrightarrow H_3O^+(aq) + SO_4^{2-}(aq)$$

FIGURE 15.19 The indicator phenolphthalein is pink in basic solution *(left).* When dry ice (solid carbon dioxide) is placed in the bottom of the beaker *(right),* it dissolves to form carbonic acid. This neutralizes the base in solution, and causes the indicator to change to its colorless form characteristic of an acid solution.

The hydrogen sulfate ion is amphoteric, meaning that it is a base in the first reaction (with conjugate acid H_2SO_4) and an acid in the second (with conjugate base SO_4^{2-}). In its first ionization H_2SO_4 is a strong acid, but the product of that ionization (HSO_4^-) is a weak acid. So, the H_3O^+ produced in a solution of H_2SO_4 comes primarily from the first ionization, and the solution has a pH close to that of a monoprotic strong acid of the same concentration. But when this solution reacts with a strong base, its neutralizing power is twice that of a monoprotic acid of the same concentration, because each mole of sulfuric acid can react with and neutralize two moles of hydroxide ion.

15.7.1 Weak Polyprotic Acids

Weak polyprotic acids ionize in two or more stages. Examples are carbonic acid (H_2CO_3), formed from solvated CO_2 (carbonated water; Fig. 15.19), and phosphoric acid (H_3PO_4). Carbonic acid can give up one hydrogen ion to form HCO_3^- (hydrogen carbonate ion) or two hydrogen ions to form CO_3^{2-} (carbonate ion). Phosphoric acid ionizes in three stages, giving successively $H_2PO_4^-$, HPO_4^{2-}, and PO_4^{3-}.

Two simultaneous equilibria are involved in the ionization of a diprotic acid such as H_2CO_3:[2]

$$H_2CO_3(aq) + H_2O(\ell) \rightleftharpoons H_3O^+(aq) + HCO_3^-(aq)$$

$$\frac{[H_3O^+][HCO_3^-]}{[H_2CO_3]} = K_{a1} = 4.3 \times 10^{-7}$$

and

$$HCO_3^-(aq) + H_2O(\ell) \rightleftharpoons H_3O^+(aq) + CO_3^{2-}(aq)$$

$$\frac{[H_3O^+][CO_3^{2-}]}{[HCO_3^-]} = K_{a2} = 4.8 \times 10^{-11}$$

We emphasize two important points at the outset:

1. The $[H_3O^+]$ in the two ionization equilibria are one and the same.
2. K_{a2} is invariably smaller than K_{a1} because the negative charge left behind by the loss of a hydrogen ion in the first ionization causes the second hydrogen ion to be more tightly bound.

Exact calculations of simultaneous equilibria can be complicated. They simplify considerably when the original acid concentration is not too small and the ionization constants K_{a1} and K_{a2} differ substantially in magnitude (by a factor of 100 or more). The latter condition is almost always satisfied. Under such conditions, the two equilibria can be treated sequentially, as in the following example.

EXAMPLE 15.15

Calculate the equilibrium concentrations of H_2CO_3, HCO_3^-, CO_3^{2-}, and H_3O^+ in a saturated aqueous solution of CO_2, in which the original concentration of H_2CO_3 is 0.034 M.

Solution

The H_3O^+ arises both from the ionization of H_2CO_3 and from the subsequent ionization of HCO_3^-, but because $K_{a1} \gg K_{a2}$ it is reasonable to ignore the contribution of $[H_3O^+]$ from the second ionization (as well as from the autoionization of water). These approximations will be checked later in the calculation.

If y mol L^{-1} of H_2CO_3 ionizes, the following approximations apply:

	$H_2CO_3(aq)$ +	$H_2O(\ell)$ ⇌	$H_3O^+(aq)$ +	$HCO_3^-(aq)$
Initial concentration (M)	0.034		≈0	0
Change in concentration (M)	$-y$		$+y$	$+y$
Equilibrium concentration (M)	$0.034 - y$		y	y

(Continued)

[2]An accurate description of carbonic acid and solvated CO_2 is somewhat more complicated than indicated here. In fact, most of the dissolved CO_2 remains as $CO_2(aq)$, and only a small fraction actually reacts with water to give $H_2CO_3(aq)$. However, we will indicate by $[H_2CO_3]$ the total concentration of both of these species. Approximately 0.034 mol of CO_2 dissolves per liter of water at 25°C and atmospheric pressure.

where equating both $[HCO_3^-]$ and $[H_3O^+]$ to y involves the assumption that the subsequent ionization of HCO_3^- has only a small effect on its concentration. Then the first ionization equilibrium can be written as

$$\frac{[H_3O^+][HCO_3^-]}{[H_2CO_3]} = K_{a1}$$

$$\frac{y^2}{0.034 - y} = 4.3 \times 10^{-7}$$

Solving this equation for y gives

$$y = 1.2 \times 10^{-4}\ \text{M} = [H_3O^+] = [HCO_3^-]$$

$$[H_2CO_3] = 0.034 - y = 0.034\ \text{M}$$

The second equilibrium can be written as

$$\frac{[H_3O^+][CO_3^{2-}]}{[HCO_3^-]} = K_{a2} = \frac{(1.2 \times 10^{-4})[CO_3^{2-}]}{1.2 \times 10^{-4}} = 4.8 \times 10^{-11}$$

$$[CO_3^{2-}] = 4.8 \times 10^{-11}\ \text{M}$$

The concentration of the base produced in the second ionization, $[CO_3^{2-}]$, is numerically equal to K_{a2}.

Now we must check the validity of the assumptions. Because

$$[CO_3^{2-}] = 4.8 \times 10^{-11}\ \text{M} \ll 1.2 \times 10^{-4}\ \text{M} = [HCO_3^-]$$

we were justified in ignoring the effect of the second ionization on the concentrations of HCO_3^- and H_3O^+. The additional concentration of H_3O^+ furnished by HCO_3^- is only 4.8×10^{-11} M. Finally, $[H_3O^+]$ is much larger than 10^{-7} M, so neglecting the water autoionization was also justified.

Related Problems: 63, 64

FIGURE 15.20 Phosphoric acid, H_3PO_4, is a triprotic acid that is used in the manufacture of phosphate fertilizers and in the food industry.

For a triprotic acid such as H_3PO_4 (Fig. 15.20), the concentration of the base (PO_4^{3-}) resulting from the third ionization could have been calculated in a similar manner.

An analogous procedure applies to the reactions of a base that can accept two or more hydrogen ions. In a solution of sodium carbonate (Na_2CO_3), the carbonate ion reacts with water to form first HCO_3^- and then H_2CO_3:

$$H_2O(\ell) + CO_3^{2-}(aq) \rightleftharpoons HCO_3^-(aq) + OH^-(aq)$$

$$\frac{[OH^-][HCO_3^-]}{[CO_3^{2-}]} = K_{b1} = \frac{K_w}{K_{a2}} = 2.1 \times 10^{-4}$$

$$H_2O(\ell) + HCO_3^-(aq) \rightleftharpoons H_2CO_3(aq) + OH^-(aq)$$

$$\frac{[OH^-][H_2CO_3]}{[HCO_3^-]} = K_{b2} = \frac{K_w}{K_{a1}} = 2.3 \times 10^{-8}$$

In this case $K_{b1} \gg K_{b2}$, so essentially all the OH^- arises from the first reaction.

The ensuing calculation of the concentrations of species present is just like that in Example 15.15.

15.7.2 Effect of pH on Solution Composition

Changing the pH of a solution shifts the positions of all acid–base equilibria, including those involving polyprotic acids. Acid–base equilibrium expressions and equilibrium constants are used to calculate the amount of the change. For example, the two equilibria that apply to solutions containing H_2CO_3, HCO_3^-, and CO_3^{2-} can be written as

$$\frac{[HCO_3^-]}{[H_2CO_3]} = \frac{K_{a1}}{[H_3O^+]} \qquad \frac{[CO_3^{2-}]}{[HCO_3^-]} = \frac{K_{a2}}{[H_3O^+]}$$

At a given pH, the right sides are known and the relative amounts of the three carbonate species can be calculated. This is illustrated by the following example.

EXAMPLE 15.16

Calculate the fraction of carbonate present as H_2CO_3, HCO_3^-, and CO_3^{2-} at pH 10.00.

Solution

At this pH, $[H_3O^+] = 1.0 \times 10^{-10}$ M, and the values of K_{a1} and K_{a2} for the preceding equations are used to find

$$\frac{[HCO_3^-]}{[H_2CO_3]} = \frac{4.3 \times 10^{-7}}{1.0 \times 10^{-10}} = 4.3 \times 10^3$$

$$\frac{[CO_3^{2-}]}{[HCO_3^-]} = \frac{4.8 \times 10^{-11}}{1.0 \times 10^{-10}} = 0.48$$

It is most convenient to rewrite these ratios with the same species (say, HCO_3^-) in the denominator. The first equation becomes

$$\frac{[H_2CO_3]}{[HCO_3^-]} = \frac{1}{4.3 \times 10^3} = 2.3 \times 10^{-4}$$

The fraction of each species present is obtained by dividing the concentration of each species by the sum of the three concentrations. For H_2CO_3 this gives

$$\text{fraction } H_2CO_3 = \frac{[H_2CO_3]}{[H_2CO_3] + [HCO_3^-] + [CO_3^{2-}]}$$

This can be simplified by dividing numerator and denominator by $[HCO_3^-]$ and substituting the ratios that have already been calculated:

$$\text{fraction } H_2CO_3 = \frac{[H_2CO_3]/[HCO_3^-]}{([H_2CO_3]/[HCO_3^-]) + 1 + ([CO_3^{2-}]/[HCO_3^-])}$$

$$= \frac{2.3 \times 10^{-4}}{2.3 \times 10^{-4} + 1 + 0.48} = 1.6 \times 10^{-4}$$

In the same way the fractions of HCO_3^- and CO_3^{2-} present are calculated to be 0.68 and 0.32, respectively.

Related Problems: 67, 68

Repeating the calculation of Example 15.16 at a series of different pH values gives the graph shown in Figure 15.21. At high pH, CO_3^{2-} predominates, and at low pH, H_2CO_3 is the major species. At intermediate pH (near pH 8, the approximate pH of seawater), the hydrogen carbonate ion HCO_3^- is most prevalent. The variation in composition of sedimentary rocks from different locations can be traced back to the effect of pH on solution composition. Sediments containing carbonates were formed from alkaline (high-pH) lakes and oceans in which CO_2 was present mainly as CO_3^{2-} ion. Sediments deposited from waters with intermediate pH are hydrogen carbonates or mixtures between carbonates and hydrogen carbonates. An example of the latter is trona

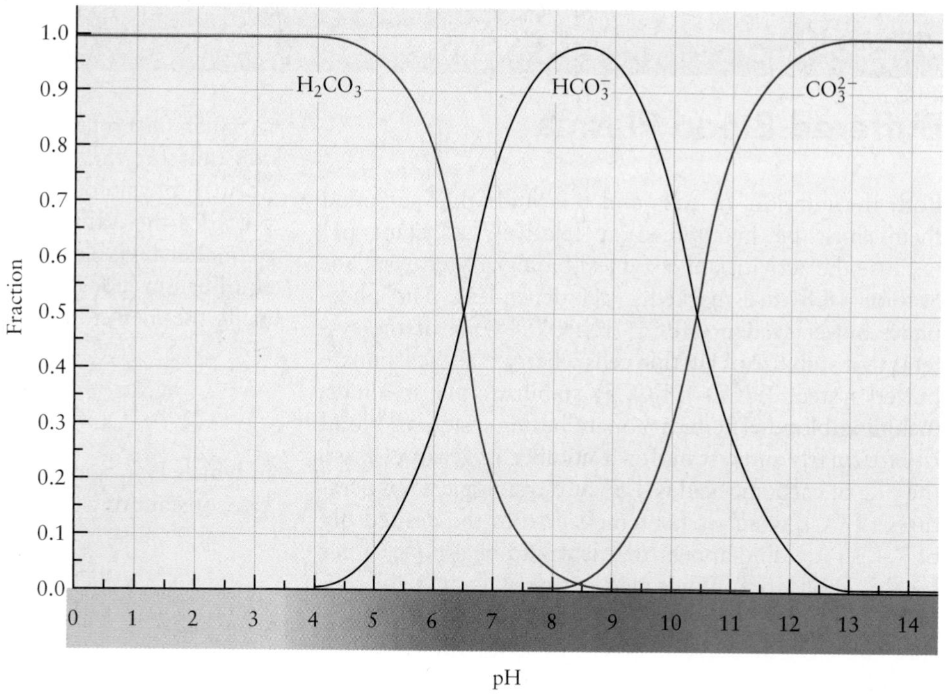

FIGURE 15.21 The equilibrium fractions of H_2CO_3, HCO_3^- and CO_3^{2-} that are present in aqueous solution at different values of the pH. (See cautionary footnote in Section 15.7.1 on page 646.)

($2Na_2CO_3 \cdot NaHCO_3 \cdot 2H_2O$), an ore from the western United States that is an important source of both carbonates of sodium. Acidic waters did not deposit carbonates, but instead released $CO_2(g)$ to the atmosphere.

15.8 Organic Acids and Bases: Structure and Reactivity

One of the goals of chemistry is to understand the relationship between *structure* and *function*. How does the molecular structure of a compound (including its attached functional groups) affect its chemical reactivity? As an example of the exploration of structure–function relationships, we consider here the trends in acid strength in organic compounds. This extends the earlier discussion of acid strength in inorganic acids (see Section 15.3) and illustrates the usefulness of chemical concepts such as electronegativity and resonance in predicting reactivity. This section depends on the thermodynamic description in Section 14.3.

Table 15.2 lists a number of organic acids with pK_a values in aqueous solution between 0 (strong acid) and 14 (weak acid). Almost all of them are carboxylic acids with the characteristic —COOH functional group, which can donate a proton to a water solvent molecule by breaking the O—H bond to form a negative ion and H^+. Acid ionization constants can be defined for many other functional groups besides carboxylic acids, because many organic reactions involve the transfer of hydrogen ions or catalysis by acids or bases. Consequently these acid ionization constants provide important information about reactivity of the compounds.

Aqueous solutions can be used to describe acidity only in compounds with values of the pK_a between about 14 and 0. Values much greater than 14 mean that there is negligible ionization in aqueous solution, and values much less than 0 mean that ionization is essentially complete. Because our goal is a scale of *relative* acidities, we can use information from nonaqueous solvents as well as from water. In a sufficiently basic solvent, even a very weak acid will show some

CONNECTION TO BIOLOGY

Buffered Blood Plasma

Both the interiors of cells and the fluids that surround them must be maintained at relatively constant pH because the activities of enzymes (biological catalysts, see Section 18.8) are markedly pH dependent. The phosphate buffer system ($HPO_4^{2-}/H_2PO_4^-$) is one of the systems that stabilize pH within cells whereas the bicarbonate buffer system (HCO_3^-/H_2CO_3) stabilizes pH in fluids, including blood. The bicarbonate buffer system in blood is particularly interesting for a number of reasons. First, the pK_a of carbonic acid is 3.57 at physiological temperatures (37°C), which is far removed from the desired pH of 7.4, so it would appear that it would be a poor choice for this application. What makes it work is the tight coupling of the buffer system with the dissolution equilibrium of CO_2 in plasma. Not only does this coupling provide an additional source of carbonic acid but it also provides a mechanism for the dynamic regulation of a number of metabolic reactions through changes in respiration rates. Finally, both pH and the concentration of dissolved CO_2 enhance and regulate the delivery of O_2 from the lungs to tissues via the capillaries. Let's examine each of these issues in turn.

We can use Equation 15.7 (often called the Henderson–Hasselbalch equation in biochemistry) to calculate the buffering capacity (the quantity of acid or base added that changes the pH by 1 unit) of the bicarbonate system in plasma

$$\text{pH} = 7.4 = 3.57 - \log \frac{[H_2CO_3]}{[HCO_3^-]}$$

so the ratio of bicarbonate to carbonic acid is about 6800 in plasma under normal physiological conditions. The concentration of bicarbonate in blood is one of several indicators of acid–base imbalance and it is generally measured as part of an electrolyte panel in routine blood tests; a typical value is 20 mM. The carbonic acid concentration under these conditions is only about 3 μM. Using the approach developed in Example 15.8, we calculate the buffering capacity of this system to be only 2.7 μM with respect to added OH^-, which is a typical OH^- concentration in plasma. The system would have essentially no buffering capacity were it not coupled to the equilibrium established between gas phase and dissolved CO_2 as represented by the following reactions,

$$CO_2(g) \rightleftharpoons CO_2(aq)$$

$$CO_2(aq) + H_2O(\ell) \overset{K_h}{\rightleftharpoons} H_2CO_3(aq)$$

$$H_2CO_3(aq) \overset{K_a}{\rightleftharpoons} H^+(aq) + HCO_3^-(aq)$$

in which (aq) refers to the plasma phase. The equilibrium constant for the second reaction is rather small under normal physiological conditions (in mammals) but it provides the additional source of carbonic acid needed to make a viable buffer system. We can combine the equilibrium constants for the simultaneous reactions using the method developed in Section 14.4 to get

$$K_a K_h = \frac{[H^+][HCO_3^-]}{[CO_2(aq)]} = \frac{[H^+][HCO_3^-]}{(P_{CO_2})/K}$$

in which P_{CO_2} and K are the partial pressure and Henry's law constant of CO_2. We can rearrange this equation to get

$$\text{pH} = \text{p}K_a \text{p}K_h + \frac{K[HCO_3^-]}{(P_{CO_2})}$$

$\text{p}K_a = 3.57$ and $\text{p}K_h = 2.52$ so the first two terms in the Henderson–Hasselbalch equation for the coupled reactions yield an initial pH = 6.1, which is still much lower than the desired value of 7.4 but within the range considered acceptable, and certainly much better than what could be accomplished using the bicarbonate buffer system alone. This coupled buffer system is an open system in which the total carbonic acid pool available for buffering is largely provided by dissolved CO_2. The Henry's law constant for CO_2 in plasma at 37°C is about 2 L atm mol^{-1} so the concentration of dissolved CO_2 at the typical partial pressure found in the lungs (40 torr) is about 1.2 mM, which provides an adequate concentration of acid to ensure sufficient buffering capacity.

Rapidly metabolizing tissues, such as muscle, produce large quantities of CO_2 and hydrogen ions, which are both heterotropic effectors that promote the release of oxygen from the oxygenated form of hemoglobin. One of the mechanisms by which this occurs is through the formation of salt bridges, ionic bonds between amino acids that stabilize certain protein structures over others. Low pH leads to the formation of salt bridges by protonating the side chain of the amino acid histidine, which binds electrostatically to the deprotonated carboxylate group of the amino acid aspartate in the same chain, resulting in a conformational change that favors the quaternary structure (see Section 7.6) of deoxyhemoglobin, thus promoting the release of O_2. CO_2 produces a similar effect, though indirectly, through the second and third reactions listed above. The regulation of O_2 binding through these mechanisms is referred to as the Bohr effect (named after Christian Bohr, the father of Neils Bohr), which has at least two important physiological consequences. First, it enhances the total fraction of bound oxygen that can be delivered by hemoglobin as it moves from $P_{O_2} \approx 100$ torr in the lungs to $P_{O_2} \approx 30$ torr in tissues. Only about 66% of the oxygen stored in hemoglobin is available to be transferred if the pH is the same in both environments. This fraction increases to 77% at pH 7.2 (characteristic of actively metabolizing tissues) with no CO_2 present, and to 88% at pH 7.2 with a CO_2 partial pressure of 40 torr. In addition to increasing the overall quantity of O_2 available, heterotropic regulation provides a mechanism to respond differentially to cells undergoing metabolism at different rates. Figure C15.1 shows that lowering the pH dramatically increases the amount of oxygen transferred. The curves all shift to the right, indicating a smaller fraction of bound oxygen at typical partial pressures found in venous blood. Increasing the partial pressure of CO_2 has the same effect as shown by the pair of curves labeled pH 7.4 (no CO_2) and pH 7.4 (40 torr CO_2).

Regulating pH is vital to all cells because many cellular processes are catalyzed by enzymes, the activities of which depend strongly on pH. Coupling of the bicarbonate buffer system in blood to CO_2 provides a mechanism for controlling pH through respiration. Increases in the plasma hydrogen ion concentration shift the equilibrium toward carbonic acid, which ultimately produces gaseous CO_2 in the reverse of the series of reactions listed earlier. pH can be restored to the desired value by increasing the rate of respiration, which removes $CO_2(g)$ and shifts the equilibria back to the right. It is possible to overdo it, however. Hyperventilation (breathing faster and/or deeper than normal) can remove such large amounts of CO_2 that the pH rises to unhealthy levels, causing symptoms such as dizziness or fainting. Sudden exposure to high altitudes, for example during skiing and mountain climbing, can also cause hyperventilation. Adaptation to high altitudes is accomplished, in part, by excreting bicarbonate through the kidneys, thus shifting the equilibria to the right and increasing pH.

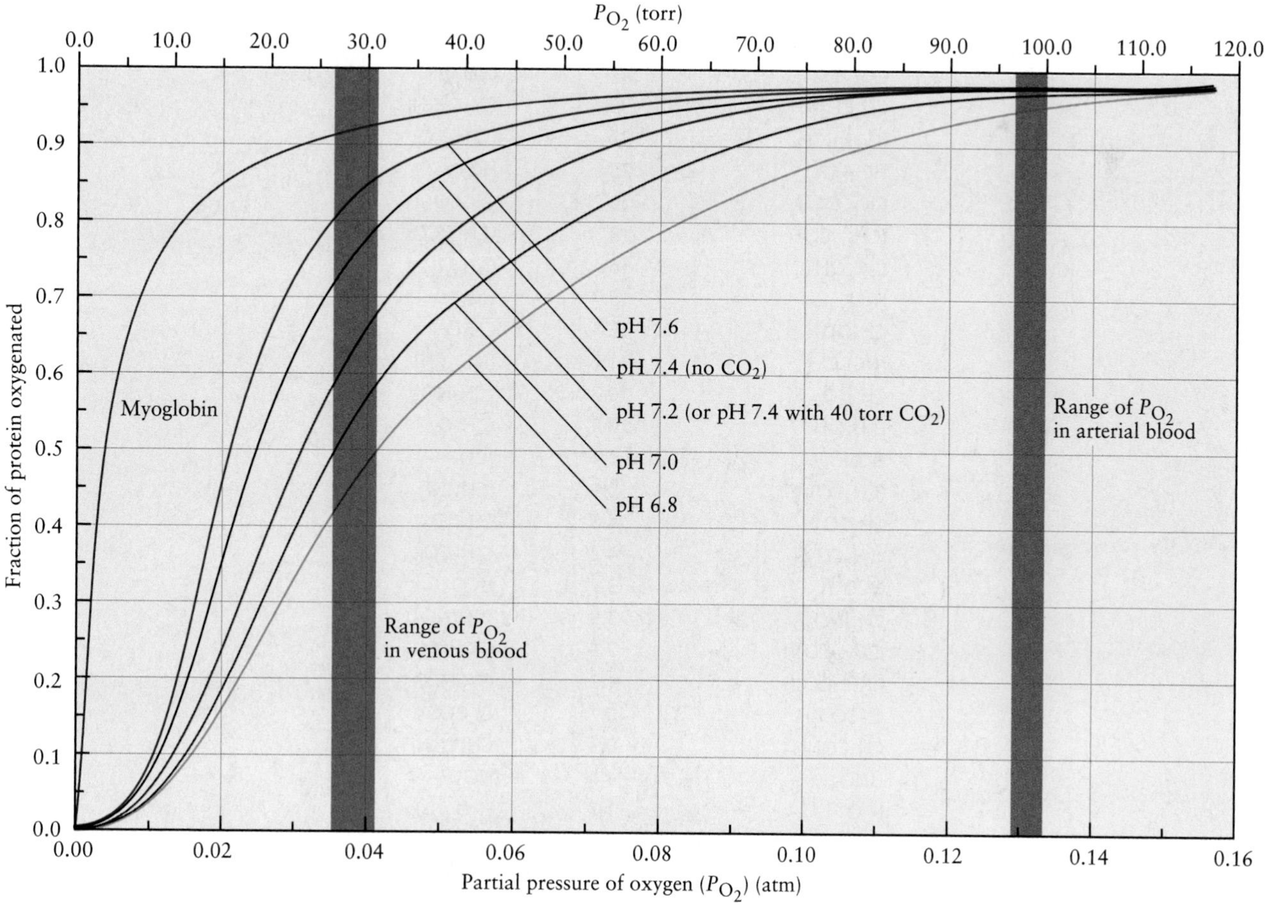

FIGURE C15.1 The effect of pH on the fraction of protein that is oxygenated, plotted as a function of the O_2 partial pressure. (Adapted from Reginald H. Garrett and Charles M. Grisham, *Biochemistry, 4e,* figure 14.5, p. 474, Boston: Brooks/Cole, Cengage Learning, 2010; © Cengage Learning)

ionization; in a sufficiently acidic solvent, even a very strong acid will be only partially ionized. This allows an extension of the pK_a scale above 14 and below 0, provided we consistently think of it as only a relative scale. We thus generalize our earlier definition of K_a, which was limited to aqueous solvents, to describe acid ionization in a general solvent:

$$\text{HA} + \text{solvent} \rightleftharpoons \text{solvent-H}^+ + \text{A}^-$$

Our assumption is that the *difference* in pK_a values for two different acids in a given solvent

$$\text{p}K_a - \text{p}K_a' = -\log_{10}(K_a/K_a')$$

depends only weakly on the particular solvent chosen, even though the actual pK_a values are very sensitive to the solvent. If this is the case, we can extend the pK_a scale beyond the range that comes from measurements of pH in aqueous solution. The result is given in Table 15.4.

TABLE 15.4

Acidities of Organic Compounds

Acid	pK_a	Conjugate Base
CH_3-CH_3	50	$CH_3-\ddot{C}H_2^-$
CH_4	49	$:CH_3^-$
$CH_2=CH_2$	44	$CH_2=\ddot{C}H^-$
C_6H_6	43	$C_6H_5^-$
$C_6H_5CH_3$	41	$C_6H_5\ddot{C}H_2^-$
$CH_3CH_2NH_3^+$	35	$CH_3CH_2NH_2$
$(C_6H_5)_3CH$	32	$(C_6H_5)_3\ddot{C}^-$
$HC\equiv CH$	25	$HC\equiv \ddot{C}^-$
CH_3COCH_3	20	$CH_3CO\ddot{C}H_2^-$
$(CH_3)_3C\ddot{O}H$	18	$(CH_3)_3C\ddot{O}{:}^-$
$C_2H_5\ddot{O}H$	16	$C_2H_5\ddot{O}{:}^-$
$H_2\ddot{O}$	15.7	$H\ddot{O}{:}^-$
CH_3OH	15	$CH_3\ddot{O}{:}^-$
$(NC)_2CH_2$	11.2	$(NC)_2\ddot{C}H^-$
CH_3NO_2	10.2	$:CH_2NO_2^-$
$C_6H_5\ddot{O}H$	10.0	$C_6H_5\ddot{O}{:}^-$
HCN	9.1	$:CN^-$
$(CH_3CO)_3CH$	5.9	$(CH_3CO)_3\ddot{C}^-$
CH_3CO_2H	4.8	$CH_3CO_2^-$
$C_6H_5CO_2H$	4.2	$C_6H_5CO_2^-$
HCO_2H	3.7	HCO_2^-
$CH_2(NO_2)_2$	3.6	$:CH(NO_2)_2^-$
$ClCH_2CO_2H$	2.9	$ClCH_2CO_2^-$
Cl_2CHCO_2H	1.3	$Cl_2CHCO_2^-$
Cl_3CCO_2H	0.7	$Cl_3CCO_2^-$
$CH_3CONH_3^+$	0.3	$CH_3CO\ddot{N}H_2$
HNO_3	−1.4	NO_3^-
H_3O^+	−1.7	H_2O
$CH_3\ddot{O}H_2^+$	−2.2	$CH_3\ddot{O}H$
$C_2H_5\ddot{O}H_2^+$	−2.4	$C_2H_5\ddot{O}H$
$(CH_3)_3C\ddot{O}H_2^+$	−3.8	$(CH_3)_3C\ddot{O}H$
$C_6H_5\ddot{O}H_2^+$	−6.7	$C_6H_5\ddot{O}H$
$(CH_3)_2C=\ddot{O}H^+$	−7.2	$(CH_3)_2C=\ddot{O}$
$R-C\equiv NH^+$	~−10	$R-C\equiv N:$

A thermodynamic analysis shows that this difference in pK_a values is related to the difference in the standard Gibbs free energy changes for the two reactions:

$$\Delta G° - \Delta G°' = -RT(\ln K_a - \ln K_a') = 2.3\ RT(pK_a - pK_a') \qquad [15.8]$$

The factor 2.3 appears when we relate the natural logarithm to the base 10 logarithm. The difference in standard Gibbs free energy changes can be separated into one contribution arising from standard enthalpy changes and one from absolute entropy changes:

$$\Delta G° - \Delta G°' = \Delta H° - \Delta H°' - T(\Delta S° - \Delta S°')$$

If we change the structure of an organic acid, what will be the effect on the pK_a? In general, modest structural changes such as the replacement of a functional group have only small effects on the entropy change of ionization, and most of the influence comes through changes in the enthalpy change of ionization. The quantitative magnitude of this effect is calculated in the following example.

EXAMPLE 15.17

How large a change in energy (or enthalpy) of reaction is needed for an acid ionization to lower the pK_a by one unit at 25°C? Assume that the entropy of ionization is unchanged. (We use unprimed variables to indicate the reaction with the lower pK_a, or higher K_a.)

Solution

The entropy of ionization is unchanged, so $\Delta S° = \Delta S°'$, and $\Delta H° - \Delta H°' = \Delta G° - \Delta G°'$. If the change in pK_a is 1.00, then the change in Gibbs free energy of reaction is

$$\begin{aligned}\Delta G° - \Delta G°' &= -RT(\ln K_a - \ln K_a') \\ &= 2.3\ RT(pK_a - pK_a') = 2.3\ RT(-1) \\ &= 2.3(8.315\ \text{J mol}^{-1}\ \text{K}^{-1})(298\ \text{K})(-1) \\ &= -5700\ \text{J mol}^{-1} = -5.7\ \text{kJ mol}^{-1}\end{aligned}$$

If we assume that this change in Gibbs free energy is entirely due to a change in enthalpy, then one unit of pK_a change corresponds to a lowering of the enthalpy of the products relative to the reactants by 5.7 kJ mol^{-1}. (Note that the lowering the energy of the products in the acid ionization reaction relative to the reactants increases the K_a and lowers the pK_a.)

Related Problems: 69, 70

The sign of the effect is clear from Example 15.17: If a change in structure of a molecule lowers its pK_a (making it a stronger acid), then the energy change for the acid ionization must be lowered. Another way to say this is that the effect of the structural change is to lower the energy of the conjugate base relative to that of the original acid, in other words, to increase the stability of the conjugate base.

An exploration of structure–function relations that lie behind the data in Table 15.4 must examine the effects of changes in molecular structure on the relative energetic stability of acid and conjugate base. Here we focus on the importance of resonance in Lewis dot structures in lowering the energy of the conjugate base.

Let's consider why the acidity of carboxylic acids is considerably greater than that of alcohols. For example, the pK_a of acetic acid, CH_3COOH, is considerably lower than that of ethanol, CH_3CH_2OH (4.8 versus 16). Some of this may be due to the "inductive effect" of substituting an electronegative O atom for two H atoms. But, inductive effects do not typically cause such large changes (more than 11 units of pH). An alternative explanation is found in the concept of resonance stabilization, which was introduced in Section 3.9 and discussed in the context of organic molecules in Section 7.4. A single Lewis diagram can be drawn for the conjugate base of ethanol:

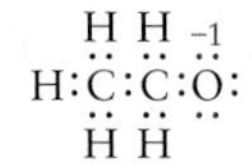

On the other hand, the conjugate base of acetic acid is represented by a resonance hybrid of two Lewis diagrams:

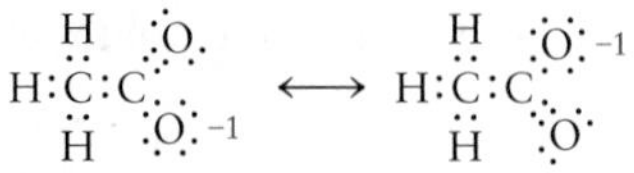

The possibility of more than one resonance form stabilizes the acetate ion and contributes to the greater acidity of acetic acid.

A second example of resonance stabilization affecting acidity arises in a comparison of phenols and alcohols. Phenol (also called carbolic acid) has a pK_a of 10, whereas the pK_a values for typical alcohols range from 16 to 18. The reason for this difference is the greater stability of the conjugate base (the phenoxide ion, $C_6H_5O^-$) due to the spreading out of the negative charge over the aromatic ring. Several different resonance forms contribute to the stability of the phenoxide ion.

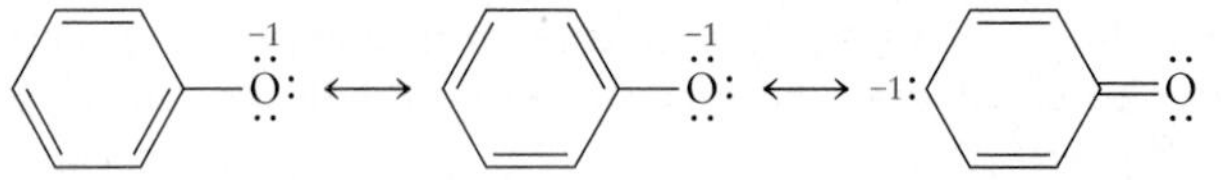

It is important to consider both the reactant and the product when you make these arguments. The reactant acid phenol has two resonance structures, so is itself stabilized by resonance. If a reactant acid is resonance stabilized, that will favor the reactants in the acid ionization equilibrium, making it less acidic. But the phenoxide ion product has three resonance structures, so shows an even larger resonance stabilization. Thus, on the whole, phenol is made more acidic by virtue of the resonance stabilization of the conjugate base product.

As you consider the effect of resonance stabilization on acid strength, or the effect of the electronegativity on oxoacid strength (see Section 15.3.1), it is important to keep in mind the caution "all other factors being equal." A beginning student might fallaciously argue that HF would be more acidic than HCl because F is more electronegative than Cl. The —X—O—H series are similar enough in the other important factors that one gets reasonable predictions based on electronegativity arguments alone.

Finally, the free energy of solvation of the ions in solution also plays a key role in acid strength. The solvation of an ion involves both a lowering in energy and a change in entropy as polar water molecules rearrange themselves around, and potentially hydrogen bond to, the newly formed anion A^-. The free energy of solvation of small ions in water (also called the free energy of hydration) ranges from −200 to −400 kJ mol^{-1} for singly charged ions.

EXAMPLE 15.18

By using resonance Lewis diagrams, predict which will be the stronger acid: H_2SO_3 or H_2SO_4 (consider the first acid ionization), all other factors being equal.

(Continued)

Solution

There are two resonance Lewis structures for the conjugate base of H_2SO_3, HSO_3^-:

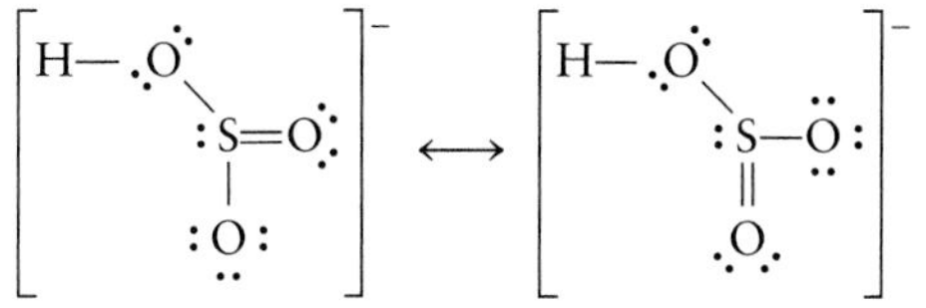

The conjugate base of H_2SO_4, HSO_4^-, has three equivalent resonance structures.

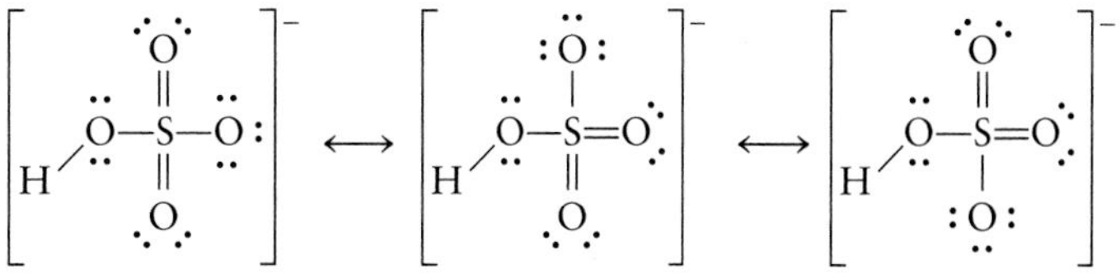

The extra stabilization, resulting from three rather than two resonance structures, of the conjugate base makes H_2SO_4 the stronger acid.

Related Problems: 69, 70

15.9 A DEEPER LOOK Exact Treatment of Acid–Base Equilibria

Sections 15.4 and 15.5 outline methods for calculating equilibria involving weak acids, bases, and buffer solutions. There we assume that the amount of hydronium ion (or hydroxide ion) resulting from the ionization of water can be neglected in comparison with that produced by the ionization of dissolved acids or bases. In this section, we replace that approximation by a treatment of acid–base equilibria that is exact, within the limits of the mass-action law. This approach leads to somewhat more complicated equations, but it serves several purposes. It has great practical importance in cases in which the previous approximations no longer hold, such as very weak acids or bases or very dilute solutions. It includes as special cases the various aspects of acid–base equilibrium considered earlier. Finally, it provides a foundation for treating amphoteric equilibrium later in this section.

Consider a general case in which the initial concentration of a weak acid HA is called c_a, and the initial concentration of its conjugate base (which for simplicity is assumed to come from the salt NaA) is c_b. In solution there will be five dissolved species: HA, A^-, Na^+, H_3O^+, and OH^-. It is necessary to write down and solve five independent equations that relate the equilibrium concentrations of these species to the initial concentrations c_a and c_b and to K_a, the acid ionization constant of HA. The first equation is simply

$$[Na^+] = c_b \qquad \text{(a)}$$

reflecting the fact that the $Na^+(aq)$ from the dissolved salt is a spectator ion that does not take part in the acid–base equilibrium. Next there are two equilibrium relations:

$$[H_3O^+][OH^-] = K_w \qquad \text{(b)}$$

$$\frac{[H_3O^+][A^-]}{[HA]} = K_a \qquad \text{(c)}$$

The fourth relation is one of stoichiometry, or conservation of "A-material":

$$c_a + c_b = [HA] + [A^-] \qquad \text{(d)}$$

The original A-material was introduced either as acid or as base, with a total concentration of $c_a + c_b$. When equilibrium is reached, some redistribution has doubtless occurred, but the *total* concentration, $[HA] + [A^-]$, must be the same. The fifth and final relation results from charge balance. The solution must be electrically neutral, so the total amount of positive charge must be equal to the total amount of negative charge:

$$[Na^+] + [H_3O^+] = [A^-] + [OH^-] \qquad \text{(e)}$$

These five independent equations completely determine the five unknown concentrations. To solve them, begin by substituting (a) into (e) and solving for $[A^-]$:

$$[A^-] = c_b + [H_3O^+] - [OH^-] \qquad \text{(e}'\text{)}$$

Next, insert (e′) into (d) and solve for [HA]:

$$[HA] = c_a + c_b - [A^-] = c_a - [H_3O^+] + [OH^-] \qquad \text{(d}'\text{)}$$

Next, substitute both (d′) and (e′) into (c) to find

$$\frac{[H_3O^+](c_b + [H_3O^+] - [OH^-])}{(c_a - [H_3O^+] + [OH^-])} = K_a \qquad \text{(c}'\text{)}$$

There are two ways to proceed with the general equation (c′). First, we can set up the exact solution for $[H_3O^+]$, a procedure that is useful for very weak acids or bases or for very dilute solutions. Or, we can reduce (c′) in various limits to cases already considered.

For the exact solution, eliminate $[OH^-]$ in (c′) by using (b). This gives

$$\frac{[H_3O^+]\left(c_b + [H_3O^+] - \dfrac{K_w}{[H_3O^+]}\right)}{\left(c_a - [H_3O^+] + \dfrac{K_w}{[H_3O^+]}\right)} = K_a$$

The numerator and denominator are multiplied by $[H_3O^+]$ and the fraction is cleared by moving the denominator to the right side:

$$[H_3O^+](c_b[H_3O^+] + [H_3O^+]^2 - K_w) = K_a(c_a[H_3O^+] - [H_3O^+]^2 + K_w)$$

This can be rewritten as

$$[H_3O^+]^3 + (c_b + K_a)[H_3O^+]^2 - (K_w + c_aK_a)[H_3O^+] - K_aK_w = 0$$

This is a cubic equation for $[H_3O^+]$, which can be solved with a calculator. The concentrations of OH^-, A^-, and HA can then be found by successive substitutions into (b), (e′), and (d′).

Alternatively, the general equation (c′) can be examined in various limits. In an acidic buffer (as in Examples 15.10 and 15.11), if it can be assumed that $[H_3O^+] \gg [OH^-]$, then (c′) simplifies to

$$\frac{[H_3O^+](c_b + [H_3O^+])}{(c_a - [H_3O^+])} = K_a$$

which is exactly the equation used in those examples. If, in addition, $c_b = 0$, the weak-acid ionization limit of Examples 15.6 and 15.7 is reached. In a basic buffer, if it can be assumed that $[OH^-] \gg [H_3O^+]$, then (c′) simplifies to

$$\frac{(K_w/[OH^-])(c_b - [OH^-])}{(c_a + [OH^-])} = K_a$$

Here (b) was used to substitute for $[H_3O^+]$ where it multiplies the whole expression. This can be rewritten as

$$\frac{[OH^-](c_a + [OH^-])}{(c_b - [OH^-])} = \frac{K_w}{K_a} = K_b$$

which is exactly the equation used in Example 15.12. If no acid is present initially ($c_a = 0$), this expression reduces to the weak-base ionization limits of Examples 15.8 and 15.9. The general approach includes all of the previous calculations as special cases.

Unless conditions require the use of the exact solution, approximate equations are preferable because they are easier to apply and provide greater physical insight. If a calculation (ignoring water autoionization) of the ionization of a weak acid gives a concentration of H_3O^+ smaller than 10^{-6} M or if a calculation of base ionization gives a concentration of OH^- smaller than 10^{-6} M, then we have to use the more exact treatment. For buffer solutions, a pH near 7 does not necessarily mean that water ionization is important, unless the acid or base concentration becomes very small.

EXAMPLE 15.19

Calculate the pH of a 1.00×10^{-5} M solution of HCN(*aq*). The K_a of HCN(*aq*) is 6.17×10^{-10}.

Solution

Suppose the autoionization of water is ignored and the method of Examples 15.6 and 15.7 is used. This gives $[H_3O^+] = 7.9 \times 10^{-8}$ M, which of course makes no sense, because it is *lower* than the concentration of hydronium ion in pure water. HCN is a very weak acid, but it is nonetheless an acid, not a base.

So, we have to use the exact cubic equation for $[H_3O^+]$, inserting into it the proper coefficients and taking $c_a = 1.00 \times 10^{-5}$ and $c_b = 0$. This gives

$$[H_3O^+]^3 + 6.17 \times 10^{-10}[H_3O^+]^2 - 1.617 \times 10^{-14}[H_3O^+] - 6.17 \times 10^{-24} = 0$$

Unfortunately, there is no method as simple as the quadratic formula to solve a cubic equation. The easiest way to solve this equation is to try a series of values for $[H_3O^+]$ on the left side, varying them to obtain a result as close as possible to 0 (see Appendix C). It is safe to assume that the final answer will be slightly larger than 1×10^{-7}, so the initial guesses should be of that magnitude. Carrying out the procedure gives

$$[H_3O^+] = 1.27 \times 10^{-7}\ \text{M} \qquad \text{pH} = 6.90$$

Related Problems: 71, 72

15.9.1 Amphoteric Equilibria

A second situation in which an exact analysis of acid–base equilibrium is useful occurs when an amphoteric species is dissolved in water. The hydrogen carbonate ion (HCO_3^-) is amphoteric because it can act as an acid in the equilibrium

$$HCO_3^-(aq) + H_2O(\ell) \rightleftharpoons H_3O^+(aq) + CO_3^{2-}(aq)$$

$$\frac{[H_3O^+][CO_3^{2-}]}{[HCO_3^-]} = K_{a2} = 4.8 \times 10^{-11}$$

or as a base in the equilibrium

$$H_2O(\ell) + HCO_3^-(aq) \rightleftharpoons H_2CO_3(aq) + OH^-(aq)$$

$$\frac{[OH^-][H_2CO_3]}{[HCO_3^-]} = K_{b2} = 2.3 \times 10^{-8}$$

If sodium hydrogen carbonate ($NaHCO_3$) is dissolved in water, there is a competition between the tendency of HCO_3^- to accept hydrogen ions and to donate them. Because $K_{b2} > K_{a2}$, there should be more production of OH^- than of H_3O^+, so the solution should be basic.

In an exact treatment of this equilibrium, there are six unknown concentrations—those of Na^+, H_2CO_3, HCO_3^-, CO_3^{2-}, OH^-, and H_3O^+. Two equilibrium equations were already presented, and a third relates $[OH^-]$ and $[H_3O^+]$ to K_w. If $[HCO_3^-]_0$ is the original concentration of $NaHCO_3$, then from stoichiometry

$$[HCO_3^-]_0 = [HCO_3^-] + [H_2CO_3] + [CO_3^{2-}]$$

because the total amount of carbonate material is conserved. Any reduction in $[HCO_3^-]$ must be compensated by a corresponding increase in either $[H_2CO_3]$ or $[CO_3^{2-}]$. Next we use the principle of conservation of charge. The positively charged species present are Na^+ and H_3O^+, and the negatively charged species are HCO_3^-, CO_3^{2-}, and OH^-. Because there is overall charge neutrality,

$$[Na^+] + [H_3O^+] = [HCO_3^-] + 2\,[CO_3^{2-}] + [OH^-]$$

where the coefficient 2 for $[CO_3^{2-}]$ arises because each carbonate ion is doubly charged. In addition, the Na^+ concentration is unchanged, so

$$[Na^+] = [HCO_3^-]_0$$

In principle, these six equations can be solved simultaneously to calculate the exact $[H_3O^+]$ for an arbitrary initial concentration of HCO_3^-. The result is complex and gives little physical insight. Instead, we give only a simpler, approximate solution, which is sufficient in the cases considered here. Subtracting the carbonate balance equation from the charge balance equation gives

$$[H_3O^+] = [CO_3^{2-}] - [H_2CO_3] + [OH^-]$$

The three equilibrium expressions are used to rewrite this as

$$[H_3O^+] = K_{a2}\frac{[HCO_3^-]}{[H_3O^+]} - \frac{[H_3O^+][HCO_3^-]}{K_{a1}} + \frac{K_w}{[H_3O^+]}$$

where $[CO_3^{2-}]$ and $[H_2CO_3]$ have been eliminated in favor of $[HCO_3^-]$.

Multiplying by $K_{a1}[H_3O^+]$ gives

$$K_{a1}[H_3O^+]^2 + [HCO_3^-][H_3O^+]^2 = K_{a1}K_{a2}[HCO_3^-] + K_{a1}K_w$$

$$[H_3O^+]^2 = \frac{K_{a1}K_{a2}[HCO_3^-] + K_{a1}K_w}{K_{a1} + [HCO_3^-]}$$

This equation still contains two unknown quantities, $[H_3O^+]$ and $[HCO_3^-]$. Because both K_{a2} and K_{b2} are small, $[HCO_3^-]$ should be close to its original value, $[HCO_3^-]_0$. If $[HCO_3^-]$ is set equal to $[HCO_3^-]_0$, this becomes

$$[H_3O^+]^2 \approx \frac{K_{a1}K_{a2}[HCO_3^-]_0 + K_{a1}K_w}{K_{a1} + [HCO_3^-]_0}$$

which can be solved for $[H_3O^+]$. In many cases of interest, $[HCO_3^-]_0 \gg K_{a1}$, and $K_{a2}\,[HCO_3^-]_0 \gg K_w$. When this is so, the expression simplifies to

$$[H_3O^+]^2 \approx K_{a1}K_{a2}$$

$$\left[H_3O^+\right] \approx \sqrt{K_{a1}K_{a2}}$$

$$\text{pH} \approx \frac{1}{2}\,(\text{p}K_{a1} + \text{p}K_{a2})$$

so the pH of such a solution is the average of the $\text{p}K_a$ values for the two ionizations.

EXAMPLE 15.20

What is the pH of a solution that is 0.100 M in $NaHCO_3$?

Solution

First, the two assumptions are checked:

$$[HCO_3^-]_0 = 0.100 \gg 4.3 \times 10^{-7} = K_{a1}$$

$$HCO_3^-]_0 K_{a2} = 4.8 \times 10^{-12} \gg 1.0 \times 10^{-14} = K_w$$

so both are satisfied. Therefore,

$$[H_3O^+] = \sqrt{K_{a1}K_{a2}} = 4.5 \times 10^{-9}\ \text{M}$$

$$\text{pH} = 8.34$$

and the solution is basic, as expected.

15.9.2 Titration of a Polyprotic Acid

A polyprotic acid has more than one equivalence point. The first equivalence point occurs when the volume V_{e1} of base added is sufficient to remove one hydrogen ion from each acid molecule, V_{e2} is the volume sufficient to remove two hydrogen ions from each, and so forth. A diprotic acid shows two equivalence points, and a triprotic acid, three. The equivalent volumes are related to each other by

$$V_{e1} = \frac{1}{2}\,V_{e2} = \frac{1}{3}\,V_{e3}$$

Figure 15.22 shows a titration curve for triprotic phosphoric acid. The three equivalence points are at 100.0 mL, 200.0 mL, and 300.0 mL. Calculating the pH as a function of the volume of added base presents no new complications beyond those already considered. The initial pH is given by a calculation analogous to that of Example 15.15, and the pH in the flat regions between equivalence points is obtained by a buffer calculation like that for a diprotic acid. For

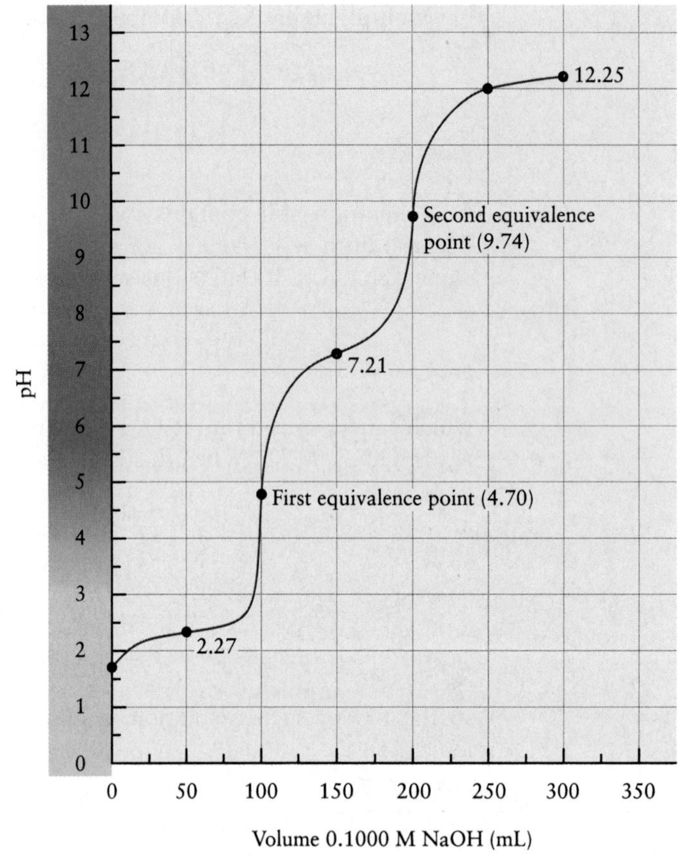

FIGURE 15.22 A titration curve for the titration of a polyprotic acid (phosphoric acid) by a strong base. The curve shown is for 100.0 mL of 0.1000 M H_3PO_4 titrated with 0.1000 M NaOH. No clear third equivalence point is seen at 300 mL because K_a for HPO_4^{2-} is not much greater than K_a for H_2O in aqueous solution.

example, the pH after addition of 50.0 mL of base is that of an equimolar $H_3PO_4/H_2PO_4^-$ buffer (subsequent ionization of $H_2PO_4^-$ can be ignored). Finally, the pH at the first equivalence point is that for a solution of NaH_2PO_4 and uses the amphoteric equilibria equations presented earlier in this section (PO_4^{3-} can be ignored in this case). The pH at the second equivalence point is an amphoteric equilibrium in which HPO_4^{2-} is in equilibrium with $H_2PO_4^-$ and with PO_4^{3-}.

CHAPTER SUMMARY

According to the Brønsted–Lowry model, acid–base chemistry encompasses a broad range of reactions in which protons are donated by species called the acid and accepted by species called the base. These reactions play key roles in situations that include industrial manufacturing processes, ordinary household substances, and the life-sustaining reactions in living systems. In aqueous solutions, all acids and bases react with the water in which they are dissolved. An acid increases the concentration of the hydronium ion H_3O^+ above its value in pure water, which is controlled by the ion product constant of water. A base increases the concentration of the hydroxide ion OH^- above the value it takes in pure water, due to the ion product constant of water. The concentration of the hydronium ion is conveniently expressed by pH, and that of the hydroxide by pOH. These quantities are related as pH + pOH = 14 by the ion product constant of water. Some acids and bases dissociate completely in solution to produce concentrations of hydronium and hydroxide equivalent to the concentration of parent acid or base with which the solutions were prepared. These are called strong acids and bases, and reactions of their solutions are completely described by the laws of stoichiometry. Most acids and bases dissociate only partially, so we must use the principles of chemical equilibrium to describe the concentration of hydronium and hydroxide in their solutions. These are the weak acids and bases, and description of their reactions requires methods of stoichiometry combined with methods of equilibrium. Acid–base neutralization reactions produce water and salts as their products. These reactions can be followed quantitatively and precisely by titration, as a means to determine the amount of acid or base in an unknown sample. The Brønsted–Lowry model does not require acids and bases to be neutral molecules. Positive and negative ions dissolved in water can react with the water as either Brønsted–Lowry acids or bases. When the salt of a weak acid is dissolved in water, hydrolysis of the anion produces the undissociated acid, and the solution is basic. Similarly, hydrolysis of the salt of a weak base produces an acidic solution. In both cases the pH is calculated by equilibrium methods. An especially interesting situation arises when a weak acid is dissolved along with one of its salts. The combination of weak acid equilibrium and hydrolysis—called a buffer solution—keeps the pH of the solution constant as small amounts of acid are added. This stabilization mechanism is extremely important in biochemical and biological situations, where success of various reactions depends critically on keeping pH constant. The extent of all these acid–base reactions can be correlated with the structures of their molecules. The extent of reaction is governed by the equilibrium constant, which in turn depends on the standard Gibbs free energy of formation of the reactants and products, which depends on their molecular structure.

CONCEPTS AND SKILLS

Section 15.1 – Classifications of Acids and Bases

Define acids and bases in the Brønsted–Lowry and Lewis models and provide several examples of their reactions with a solvent (Problems 1–12).

- In the Brønsted–Lowry model acids are proton donors, and bases are proton acceptors. Acid–base reactions are proton transfer reactions.
- In the Lewis model acids are electron pair acceptors, and bases are electron pair donors. Acid–base reactions are electron pair transfer reactions that lead to the formation of new chemical bonds.

Section 15.2 – Properties of Acids and Bases in Aqueous Solutions: The Brønsted–Lowry Scheme

Define the pH function and convert between pH and $[H_3O^+]$ (Problems 13–16).

- The autoionization of water produces free hydronium and hydroxide ions whose concentrations in liquid water obey the ion product expression for water:
    - $[H_3O^+][OH^-] = K_w = 10^{-14}$ for pure water at 25°C
    - $[H_3O^+] = [OH^-] = 10^{-7}$ M
- Strong acids and bases dissociate essentially completely in aqueous solution and increase the concentrations of hydronium and hydroxide ions, respectively.
- The pH function $pH = -\log [H_3O^+]$ is a convenient way to express the concentration of hydronium ions over a very large range of values. The pH value tells us at a glance the nature of the solution: pH < 7 is an acidic solution; pH > 7 is a basic solution; pH = 7 is a neutral solution.

Section 15.3 – Acid and Base Strength

State the relationship between the ionization constant for an acid or base and the strength of that acid or base.

- For the reaction of an acid with water $HA(aq) + H_2O(\ell) \longrightarrow H_3O^+(aq) + A^-(aq)$
    - $([H_3O^+][A^-])/[HA] = K_a$ [15.4]
    - Strong acids have $K_a > 1$, with large $[H_3O^+]$ at equilibrium.
- For the reaction of a base with water $B(aq) + H_2O(\ell) \longrightarrow BH^+(aq) + OH^-(aq)$
    - $([BH^+][OH^-])/[B] = K_b$
    - Strong bases have $K_b < 1$, with large $[OH^-]$ at equilibrium.

State the relationship between the ionization constant for an acid and that for its conjugate base (Problems 21–22).

- For a conjugate acid–base pair $K_aK_b = K_w$. Stronger conjugate acids have weaker conjugate bases and vice versa.

Describe how the strengths of oxoacids relate to their molecular structure and bonding.

- The strength of oxoacids containing the structural unit X—O—H increases as electronegativity of X increases. As X withdraws electron density from the H atom to the O—X bond, it becomes easier to break the O—H bond and release H^+.

Explain how indicators allow the pH of a solution to be estimated (Problems 25–26).

- An indicator is a weak organic acid (HIn) that has a different color from its conjugate base (In^-). As the pH of the solution changes, the ratio of [HIn] to $[In^-]$ changes according to the acid ionization equilibrium expression $K_a = ([H_3O^+]\ [In^-])/[HIn]$ and the color of the solution changes accordingly.

Section 15.4 – Equilibria Involving Weak Acids and Bases

Formulate the equilibrium expression for the ionization of a weak acid or base, and use it to determine the pH and the fraction ionized (Problems 27–36).

- For a weak acid $HA(aq) + H_2O(\ell) \longrightarrow H_3O^+(aq) + A^-(aq)$ the equilibrium expression is
    - $K_a = ([H_3O^+][A^-])/[HA]$
    - let $x = [H_3O^+] = [A^-]$
    - then $x^2/([HA] - x) = K_a$
    - If K_a is very small and the initial concentration of HA ≥ 0.05 M, use the approximation $x^2/[HA] = K_a$.
    - Otherwise, use the quadratic equation or the method of successive approximations.
- For a weak base $B(aq) + H_2O(\ell) \longrightarrow BH^+(aq) + OH^-(aq)$ the equilibrium expression is
    - $([BH^+][OH^-])/[B] = K_b$
    - Follow the same procedures as for the weak acid.
- Hydrolysis is the reaction of the salt of a weak base or the salt of a weak acid with water.
    - The cation of a weak base acts as an acid to form the conjugate base and produce an acidic solution.
    - The anion of a weak acid acts as a base to form the conjugate acid and produce a basic solution.
- For all conjugate acid–base pairs $K_b = K_w/K_a$.

Section 15.5 – Buffer Solutions

Explain the behavior of a buffer solution. Calculate its pH from the concentrations of its conjugate acid–base pair (Problems 43–46).

- A buffer solution contains either a weak acid and its salt, or a weak base and its salt, both in appreciable amounts. Adding a small amount of acid or base only slightly alters the ratio of the acid or base to its salt. Keeping this ratio nearly constant resists changes in pH when either acid or base is added.
- The pH in a buffer solution given by these approximate equations is valid when the equilibrium concentrations of the acid and its conjugate base are close to their initial concentrations:
  - $[H_3O^+] = \frac{[HA]_0}{[A^-]_0} K_a$
  - $pH \approx pK_a - \log_{10} \frac{[HA]_0}{[A^-]_0}$ **[15.7]**
- To describe the response of a buffer to an added acid or base, follow the procedures demonstrated in Example 15.11.

Design a buffer system to produce and maintain a particular pH (Problems 47–50).

- The method is based on the equation for the pH of a buffer solution. Choose an acid with a pK_a as close as possible to the desired pH. Then adjust the ratio of the concentrations of the acid and its conjugate base to produce the desired pH. The method is illustrated in Example 15.13.

Section 15.6 – Acid–Base Titration Curves

Calculate the pH at any stage in the titration of a strong acid or base by a strong base or acid (Problems 51–52).

- Titration curves for strong acids by strong bases and vice versa have four regions determined by the amount of titrant that has been added. In each of these regions the pH can be calculated by procedures of stoichiometry and definitions of solution concentration.
  - Before any titrant is added: The initial concentration of the starting acid or base determines the pH.
  - Before the equivalence point: Calculate the number of moles of acid or base remaining and divide by the total volume to calculate the concentration and determine the pH.
  - At the equivalence point: pH = 7.
  - After the equivalence point: Calculate the number of moles of excess acid or base and divide by the total volume to calculate the concentration and determine the pH.
- The choice of indicator is not critical in strong acid–strong base titrations since the slope of the transition at the equivalence point is very steep.

Calculate the pH at any stage in the titration of a weak acid or base by a strong base or acid (Problems 53–62).

- Titration curves for weak acids by strong bases and vice versa also have four regions identified by the amount of titrant that has been added. Calculating the pH in each of these requires consideration of equilibria in the solution, and each corresponds to a standard type of calculation illustrated in the chapter.
  - Before any titrant is added: This is simply the ionization of a weak acid or base, as described in Examples 15.6 and 15.7.

- Before the equivalence point: This is a buffer solution, as described in Example 15.10.
- At the equivalence point: This is a hydrolysis problem. All of the initial acid or base has been converted to a salt, which hydrolyzes back to the acid or base, so the pH $\neq$ 7. See Example 15.9.
- After the equivalence point: The problem is very close to the titration of a strong acid or base. Calculate the number of moles of excess acid or base added and divide by the total volume to obtain the concentration and the pH.

- The choice of indicator is critical in weak acid–strong base and weak base–strong acid titrations because the slope of the transition is not so steep and the pH of the equivalence point can be very different from 7. pK_a (indicator) $\approx$ pH at the equivalence point.)

Section 15.7 – Polyprotic Acids

Calculate the concentrations of all the species present in a solution of a weak polyprotic acid (Problems 63–68).

- If the initial acid concentration is not too small and the first ionization constant K_{a1} is about 100 times larger than the second ionization constant K_{a2}, the two equilibria can be treated sequentially. Set up the equilibrium expression for the first dissociation. Assume y moles of acid dissociate to produce y moles of the hydronium ion and y moles of the first conjugate base. Calculate y. Set up the equilibrium expression for the second dissociation. Assume that the equilibrium concentrations of hydronium ion and the first conjugate base take the value y determined from the first step, and calculate the equilibrium concentration of the conjugate base produced in the second ionization. The details are illustrated in Example 15.15.

Section 15.8 – Organic Acids and Bases: Structure and Reactivity

Use structure–function relations to predict effects of substitutions on relative strengths of organic acids (Problems 69, 70).

- The strength of an organic acid increases with the stability of its conjugate base anion. Any structural change that increases the stability of the conjugate base increases the strength of the organic acid. A structural change that increases the stability of the acid decreases the strength of the acid.
  - If several resonance structures are possible for the anion, its stability is increased.
  - Predicting acid strength based on simple arguments involving electronegativity or resonance should adhere to the caution "all other factors being equal."

Section 15.9 – A Deeper Look . . . Exact Treatment of Acid–Base Equilibria

Outline the procedure for the exact treatment of acid–base equilibrium and use it to find the pH of a very dilute solution of a weak acid or base (Problems 71, 72).

- For a weak acid, the dissolved species are HA, A^-, H_3O^+, and OH^-. The amount of hydronium produced by dissociation of water cannot be neglected compared to the amount produced by the dissociated acid. The concentrations of these four dissolved species must be determined from four independent equations, which are then solved simultaneously.
  - Equilibrium expression for dissociation of the acid $\frac{[H_3O^+][A^-]}{[HA]} = K_a$
  - Ion product of water $[H_3O^+][OH^-] = K_w$

- Conservation of the anion of the acid: $c_a = [HA] + [A^-]$
- Charge balance for electrical neutrality of the solution $[H_3O^+] = [A^-] + [OH^-]$

- Detailed procedures for solving these equations simultaneously are illustrated in Example 15.19.

Calculate the pH at selected points in the titration of a polyprotic acid (Problems 73, 74).

- A polyprotic acid has multiple equivalent points at which the equivalent volumes of titrant are related as $V_{e1} = \frac{1}{2}V_{e2} = \frac{1}{3}V_{e3}$.
  - The initial pH before titration begins is calculated by treating the first two equilibria sequentially as in Example 15.15.
  - Between equivalence points pH is calculated as for a buffer involving a diprotic acid.
  - The pH at each equivalence point is calculated as for an amphoteric equilibrium.
  - Detailed procedures are outlined in Figure 15.22.

CUMULATIVE EXERCISE

FIGURE 15.23 The effect of acid rain on a stand of trees in the Great Smoky Mountains of the United States.

Acid Rain

Acid rain is a major environmental problem throughout the industrialized world. One major source is the burning of fossil fuels containing sulfur (coal, oil, and natural gas). The sulfur dioxide released into the air dissolves in water or, more seriously, may be oxidized further to sulfur trioxide. The SO_3 dissolves in water to form sulfuric acid:

$$SO_3(g) + H_2O(\ell) \longrightarrow H_2SO_4(aq)$$

The net effect is to increase the acidity of the rain, which damages trees (Fig. 15.23), kills fish in lakes, dissolves stone, and corrodes metal.

(a) A sample of rainwater is tested for acidity by using two indicators. Addition of methyl orange to half of the sample gives a yellow color, and addition of methyl red to the other half gives a red color. Estimate the pH of the sample.

(b) The pH in acid rain can range down to 3 or even lower in heavily polluted areas. Calculate the concentrations of H_3O^+ and OH^- in a raindrop at pH 3.30 at 25°C.

(c) When SO_2 dissolves in water to form sulfurous acid, $H_2SO_3(aq)$, that acid can donate a hydrogen ion to water. Write a balanced chemical equation for this reaction, and identify the stronger Brønsted–Lowry acid and base in the equation.

(d) Ignore the further ionization of HSO_3^-, and calculate the pH of a solution whose initial concentration of H_2SO_3 is 4.0×10^{-4} M. (*Hint:* Use the quadratic equation in this case.)

(e) Now suppose that all the dissolved SO_2 from part (d) has been oxidized further to SO_3, so that 4.0×10^{-4} mol of H_2SO_4 is dissolved per liter. Calculate the pH in this case. (*Hint:* Because the first ionization of H_2SO_4 is that of a strong acid, the concentration of H_3O^+ can be written as 4.0×10^{-4} plus the unknown amount of dissociation from $HSO_4^-(aq)$.)

(f) Lakes have a natural buffering capacity, especially in regions where limestone gives rise to dissolved calcium carbonate. Write an equation for the effect of a small amount of acid rain containing sulfuric acid if it falls into a lake containing carbonate (CO_3^{2-}) ions. Discuss how the lake will resist further pH changes. What happens if a large excess of acid rain is deposited?

(g) A sample of 1.00 L of rainwater known to contain only sulfurous (and not sulfuric) acid is titrated with 0.0100 M NaOH. The equivalence point of the H_2SO_3/HSO_3^- titration is reached after 31.6 mL has been added. Calculate the original concentration of sulfurous acid in the sample, again ignoring any effect of SO_3^{2-} on the equilibria.

(h) Calculate the pH at the half-equivalence point, after 15.8 mL has been added. (*Hint:* Use the quadratic equation.)

Answers

(a) 4.4 to 4.8

(b) $[H_3O^+] = 5.0 \times 10^{-4}$ M; $[OH^-] = 2.0 \times 10^{-11}$ M

(c) $H_2SO_3(aq) + H_2O(\ell) \rightleftharpoons HSO_3^-(aq) + H_3O^+(aq)$. The stronger acid is H_3O^+, and the stronger base is HSO_3^-.

(d) The pH is 3.41.

(e) The pH is 3.11.

(f) The H_3O^+ in the sulfuric acid solution reacts according to $H_3O^+(aq) + CO_3^{2-}(aq) \rightleftharpoons HCO_3^-(aq) + H_2O(\ell)$. The HSO_4^- in the sulfuric acid reacts according to $HSO_4^-(aq) + CO_3^{2-}(aq) \rightleftharpoons SO_4^{2-}(aq) + HCO_3^-(aq)$. This gives rise to a HCO_3^-/CO_3^{2-} buffer that can resist further changes in pH. An excess of acid rain overwhelms the buffer and leads to the formation of H_2CO_3.

(g) 3.16×10^{-4} M

(h) The pH is 3.81.

PROBLEMS

Answers to problems whose numbers are red appear in Appendix G. Problems that are more challenging are indicated with asterisks.

15.1 Classifications of Acids and Bases

1. Which of the following can act as Brønsted–Lowry acids? Give the formula of the conjugate Brønsted–Lowry base for each of them.
   (a) Cl^- **(b)** HSO_4^- **(c)** NH_4^+
   (d) NH_3 **(e)** H_2O

2. Which of the following can act as Brønsted–Lowry bases? Give the formula of the conjugate Brønsted–Lowry acid for each of them.
   (a) F^- **(b)** SO_4^{2-} **(c)** O^{2-}
   (d) OH^- **(e)** H_2O

3. Lemon juice contains citric acid ($C_6H_8O_7$). What species serves as a base when lemon juice is mixed with baking soda (sodium hydrogen carbonate) during the preparation of some lemon cookies?

4. A treatment recommended in case of accidental swallowing of ammonia-containing cleanser is to drink large amounts of diluted vinegar. Write an equation for the chemical reaction on which this procedure depends.

5. An important step in many industrial processes is the slaking of lime, in which water is added to calcium oxide to make calcium hydroxide.
   (a) Write the balanced equation for this process.
   (b) Can this be considered a Lewis acid–base reaction? If so, what is the Lewis acid and what is the Lewis base?

6. Silica (SiO_2) is an impurity that must be removed from a metal oxide or sulfide ore when the ore is being reduced to elemental metal. To do this, lime (CaO) is added. It reacts with the silica to form a slag of calcium silicate ($CaSiO_3$), which can be separated and removed from the ore.
   (a) Write the balanced equation for this process.
   (b) Can this be considered a Lewis acid–base reaction? If so, what is the Lewis acid and what is the Lewis base?

7. Chemists working with fluorine and its compounds sometimes find it helpful to think in terms of acid–base reactions in which the fluoride ion (F^-) is donated and accepted.
   (a) Would the acid in this system be the fluoride donor or fluoride acceptor?
   (b) Identify the acid and base in each of these reactions:

$$ClF_3O_2 + BF_3 \longrightarrow ClF_2O_2 \cdot BF_4$$

$$TiF_4 + 2\ KF \longrightarrow K_2[TiF_6]$$

8. Researchers working with glasses often think of acid–base reactions in terms of oxide donors and oxide acceptors. The oxide ion is O^{2-}.
   (a) In this system, is the base the oxide donor or the oxide acceptor?
   (b) Identify the acid and base in each of these reactions:

$$2\ CaO + SiO_2 \longrightarrow Ca_2SiO_4$$

$$Ca_2SiO_4 + SiO_2 \longrightarrow 2\ CaSiO_3$$

$$Ca_2SiO_4 + CaO \longrightarrow Ca_3SiO_5$$

9. Identify each of the following oxides as an acid or base anhydride. Write the chemical formula and give the name of the acid or base formed upon reaction with water.
   (a) MgO (b) Cl_2O
   (c) SO_3 (d) Cs_2O

10. Write the chemical formula and give the name of the anhydride corresponding to each of the following acids or bases, and identify it as an acid or base anhydride.
   (a) H_3AsO_4 (b) H_2MoO_4
   (c) RbOH (d) H_2SO_3

11. Tin(II) oxide is amphoteric. Write balanced chemical equations for its reactions with an aqueous solution of hydrochloric acid and with an aqueous solution of sodium hydroxide. (*Note:* The hydroxide complex ion of tin(II) is $[Sn(OH)_3]^-$.)

12. Zinc oxide is amphoteric. Write balanced chemical equations for its reactions with an aqueous solution of hydrochloric acid and with an aqueous solution of sodium hydroxide. (*Note:* The hydroxide complex ion of zinc is $[Zn(OH)_4]^{2-}$.)

15.2 Properties of Acids and Bases in Aqueous Solutions: The Brønsted–Lowry Scheme

13. The concentration of H_3O^+ in a sample of wine is 2.0×10^{-4} M. Calculate the pH of the wine.

14. The concentration of OH^- in a solution of household bleach is 3.6×10^{-2} M. Calculate the pH of the bleach.

15. The pH of normal human urine is in the range of 5.5 to 6.5. Compute the range of the H_3O^+ concentration and the range of the OH^- concentration in normal urine.

16. The pH of normal human blood is in the range of 7.35 to 7.45. Compute the range of the concentration of H_3O^+ and the range of the OH^- concentration in normal blood.

17. The pK_w of seawater at 25°C is 13.776. This differs from the usual pK_w of 14.00 at this temperature because dissolved salts make seawater a nonideal solution. If the pH in seawater is 8.00, what are the concentrations of H_3O^+ and OH^- in seawater at 25°C?

18. At body temperature (98.6°F = 37.0°C), K_w has the value 2.4×10^{-14}. If the pH of blood is 7.4 under these conditions, what are the concentrations of H_3O^+ and OH^-?

19. When placed in water, potassium starts to react instantly and continues to react with great vigor. On the basis of this information, select the better of the following two equations to represent the reaction.

$$2\ K(s) + 2\ H_2O(\ell) \longrightarrow 2\ KOH(aq) + H_2(g)$$

$$2\ K(s) + 2\ H_3O^+(aq) \longrightarrow 2\ K^+(aq) + H_2(g) + 2H_2O(\ell)$$

State the reason for your choice.

20. Molecules of *t*-butyl chloride, $(CH_3)_3CCl$, react very slowly when mixed with water at low pH to give *t*-butyl alcohol, $(CH_3)_3COH$. When the pH is raised, the reaction takes place rapidly. Write an equation or equations to explain these facts.

15.3 Acid and Base Strength

21. Ephedrine ($C_{10}H_{15}ON$) is a base that is used in nasal sprays as a decongestant.
   (a) Write an equation for its equilibrium reaction with water.
   (b) The K_b for ephedrine is 1.4×10^{-4}. Calculate the K_a for its conjugate acid.
   (c) Is ephedrine a weaker or a stronger base than ammonia?

22. Niacin (C_5H_4NCOOH), one of the B vitamins, is an acid.
   (a) Write an equation for its equilibrium reaction with water.
   (b) The K_a for niacin is 1.5×10^{-5}. Calculate the K_b for its conjugate base.
   (c) Is the conjugate base of niacin a stronger or a weaker base than pyridine, C_5H_5N?

23. Use the data in Table 15.2 to determine the equilibrium constant for the reaction.

$$HClO_2(aq) + NO_2^-(aq) \rightleftharpoons HNO_2(aq) + ClO_2^-(aq)$$

Identify the stronger Brønsted–Lowry acid and the stronger Brønsted–Lowry base.

24. Use the data in Table 15.2 to determine the equilibrium constant for the reaction

$$HPO_4^{2-} + HCO_3^- \rightleftharpoons PO_4^{3-} + H_2CO_3$$

Identify the stronger Brønsted–Lowry acid and the stronger Brønsted–Lowry base.

25. **(a)** Which is the stronger acid—the acidic form of the indicator bromocresol green or the acidic form of methyl orange?
(b) A solution is prepared in which bromocresol green is green and methyl orange is orange. Estimate the pH of this solution.

26. **(a)** Which is the stronger base—the basic form of the indicator cresol red or the basic form of thymolphthalein?
(b) A solution is prepared in which cresol red is red and thymolphthalein is colorless. Estimate the pH of this solution.

15.4 Equilibria Involving Weak Acids and Bases

27. Aspirin is acetylsalicylic acid, $HC_9H_7O_4$, which has a K_a of 3.0×10^{-4}. Calculate the pH of a solution made by dissolving 0.65 g of acetylsalicylic acid in water and diluting to 50.0 mL.

28. Vitamin C is ascorbic acid ($HC_6H_7O_6$), for which K_a is 8.0×10^{-5}. Calculate the pH of a solution made by dissolving a 500-mg tablet of pure vitamin C in water and diluting to 100 mL.

29. **(a)** Calculate the pH of a 0.20 M solution of benzoic acid at 25°C.
(b) How many moles of acetic acid must be dissolved per liter of water to obtain the same pH as that from part (a)?

30. **(a)** Calculate the pH of a 0.35 M solution of propionic acid at 25°C.
(b) How many moles of formic acid must be dissolved per liter of water to obtain the same pH as that from part (a)?

31. Iodic acid (HIO_3) is fairly strong for a weak acid, having a K_a equal to 0.16 at 25°C. Compute the pH of a 0.100 M solution of HIO_3.

32. At 25°C, the K_a of pentafluorobenzoic acid (C_6F_5COOH) is 0.033. Suppose 0.100 mol of pentafluorobenzoic acid is dissolved in 1.00 L of water. What is the pH of this solution?

33. Papaverine hydrochloride ($papH^+Cl^-$) is a drug used as a muscle relaxant. It is a weak acid. At 25°C, a 0.205 M solution of $papH^+Cl^-$ has a pH of 3.31. Compute the K_a of the $papH^+$ ion.

34. The unstable weak acid 2-germaacetic acid (GeH_3COOH) is derived structurally from acetic acid (CH_3COOH) by having a germanium atom replace one of the carbon atoms. At 25°C, a 0.050 M solution of 2-germaacetic acid has a pH of 2.42. Compute the K_1 of 2-germaacetic acid and compare it with that of acetic acid.

35. Morphine is a weak base for which K_b is 8×10^{-7}. Calculate the pH of a solution made by dissolving 0.0400 mol of morphine in water and diluting to 600.0 mL.

36. Methylamine is a weak base for which K_b is 4.4×10^{-4}. Calculate the pH of a solution made by dissolving 0.070 mol of methylamine in water and diluting to 800.0 mL.

37. The pH at 25°C of an aqueous solution of hydrofluoric acid, HF, is 2.13. Calculate the concentration of HF in this solution, in moles per liter.

38. The pH at 25°C of an aqueous solution of sodium cyanide (NaCN) is 11.50. Calculate the concentration of CN^- in this solution, in moles per liter.

39. You have 50.00 mL of a solution that is 0.100 M in acetic acid, and you neutralize it by adding 50.00 mL of a solution that is 0.100 M in sodium hydroxide. The pH of the resulting solution is not 7.00. Explain why. Is the pH of the solution greater than or less than 7?

40. A 75.00-mL portion of a solution that is 0.0460 M in $HClO_4$ is treated with 150.00 mL of 0.0230 M KOH(*aq*). Is the pH of the resulting mixture greater than, less than, or equal to 7.0? Explain.

41. Suppose a 0.100 M solution of each of the following substances is prepared. Rank the pH of the resulting solutions from lowest to highest: NH_4Br, NaOH, KI, $NaCH_3COO$, HCl.

42. Suppose a 0.100 M solution of each of the following substances is prepared. Rank the pH of the resulting solutions from lowest to highest: KF, NH_4I, HBr, NaCl, LiOH.

15.5 Buffer Solutions

43. "Tris" is short for tris(hydroxymethyl)aminomethane. This weak base is widely used in biochemical research for the preparation of buffers. It offers low toxicity and a pK_b (5.92 at 25°C) that is convenient for the control of pH in clinical applications. A buffer is prepared by mixing 0.050 mol of tris with 0.025 mol of HCl in a volume of 2.00 L. Compute the pH of the solution.

44. "Bis" is short for bis(hydroxymethyl)aminomethane. It is a weak base that is closely related to tris (see problem 43) and has similar properties and uses. Its pK_b is 8.8 at 25°C. A buffer is prepared by mixing 0.050 mol of bis with 0.025 mol of HCl in a volume of 2.00 L (the same proportions as in the preceding problem). Compute the pH of the solution.

45. **(a)** Calculate the pH in a solution prepared by dissolving 0.050 mol of acetic acid and 0.020 mol of sodium acetate in water and adjusting the volume to 500 mL.
(b) Suppose 0.010 mol of NaOH is added to the buffer from part (a). Calculate the pH of the solution that results.

46. Sulfanilic acid ($NH_2C_6H_4SO_3H$) is used in manufacturing dyes. It ionizes in water according to the equilibrium equation

$$NH_2C_6H_4SO_3H(aq) + H_2O(\ell) \rightleftharpoons NH_2C_6H_4SO_3^-(aq) + H_3O^+(aq) \quad K_a = 5.9 \times 10^{-4}$$

A buffer is prepared by dissolving 0.20 mol of sulfanilic acid and 0.13 mol of sodium sulfanilate ($NaNH_2C_6H_4SO_3$) in water and diluting to 1.00 L.
(a) Compute the pH of the solution.
(b) Suppose 0.040 mol of HCl is added to the buffer. Calculate the pH of the solution that results.

47. A physician wishes to prepare a buffer solution at pH = 3.82 that efficiently resists changes in pH yet contains only small concentrations of the buffering agents. Determine which one of the following weak acids, together with its sodium salt, would probably be best to use: *m*-chlorobenzoic acid, $K_a = 1.04 \times 10^{-4}$; *p*-chlorocinnamic acid, $K_a = 3.89 \times 10^{-5}$; 2,5-dihydroxybenzoic acid, $K_a = 1.08 \times 10^{-3}$; or acetoacetic acid, $K_a = 2.62 \times 10^{-4}$. Explain.

48. Suppose you were designing a buffer system for imitation blood and wanted the buffer to maintain the blood at the realistic pH of 7.40. All other things being equal, which buffer system would be preferable: H_2CO_3/HCO_3^- or $H_2PO_4^-/HPO_4^{2-}$? Explain.

49. You have at your disposal an ample quantity of a solution of 0.0500 M NaOH and 500 mL of a solution of 0.100 M formic acid (HCOOH). How much of the NaOH solution should be added to the acid solution to produce a buffer of pH 4.00?

50. You have at your disposal an ample quantity of a solution of 0.100 M HCl and 400 mL of a solution of 0.0800 M NaCN. How much of the HCl solution should be added to the NaCN solution to produce a buffer of pH 9.60?

15.6 Acid–Base Titration Curves

51. Suppose 100.0 mL of a 0.3750 M solution of the strong base $Ba(OH)_2$ is titrated with a 0.4540 M solution of the strong acid $HClO_4$. The neutralization reaction is

$$Ba(OH)_2(aq) + 2\ HClO_4(aq) \longrightarrow Ba(ClO_4)_2(aq) + 2\ H_2O(\ell)$$

Compute the pH of the titration solution before any acid is added, when the titration is 1.00 mL short of the equivalence point, when the titration is at the equivalence point, and when the titration is 1.00 mL past the equivalence point. (*Note:* Each mole of $Ba(OH)_2$ gives *two* moles of OH^- in solution.)

52. A sample containing 26.38 mL of 0.1439 M HBr is titrated with a solution of NaOH having a molarity of 0.1219 M. Compute the pH of the titration solution before any base is added, when the titration is 1.00 mL short of the equivalence point, when the titration is at the equivalence point, and when the titration is 1.00 mL past the equivalence point.

53. A sample containing 50.00 mL of 0.1000 M hydrazoic acid (HN_3) is being titrated with 0.1000 M sodium hydroxide. Compute the pH before any base is added, after the addition of 25.00 mL of the base, after the addition of 50.00 mL of the base, and after the addition of 51.00 mL of the base.

54. A sample of 50.00 mL of 0.1000 M aqueous solution of chloroacetic acid, $CH_2ClCOOH$ ($K_1 = 1.4 \times 10^{-3}$), is titrated with a 0.1000 M NaOH solution. Calculate the pH at the following stages in the titration, and plot the titration curve: 0, 5.00, 25.00, 49.00, 49.90, 50.00, 50.10, and 55.00 mL NaOH.

55. The base ionization constant of ethylamine ($C_2H_5NH_2$) in aqueous solution is $K_b = 6.41 \times 10^{-4}$ at 25°C. Calculate the pH for the titration of 40.00 mL of a 0.1000 M solution of ethylamine with 0.1000 M HCl at the following volumes of added HCl: 0, 5.00, 20.00, 39.90, 40.00, 40.10, and 50.00 mL.

56. Ammonia is a weak base with a K_b of 1.8×10^{-5}. A 140.0-mL sample of a 0.175 M solution of aqueous ammonia is titrated with a 0.106 M solution of the strong acid HCl. The reaction is

$$NH_3(aq) + HCl(aq) \longrightarrow NH_4^+(aq) + Cl^-(aq)$$

Compute the pH of the titration solution before any acid is added, when the titration is at the half-equivalence point, when the titration is at the equivalence point, and when the titration is 1.00 mL past the equivalence point.

57. Sodium benzoate, the sodium salt of benzoic acid, is used as a food preservative. A sample containing solid sodium benzoate mixed with sodium chloride is dissolved in 50.0 mL of 0.500 M HCl, giving an acidic solution (benzoic acid mixed with HCl). This mixture is then titrated with 0.393 M NaOH. After the addition of 46.50 mL of the NaOH solution, the pH is found to be 8.2. At this point, the addition of one more drop (0.02 mL) of NaOH raises the pH to 9.3. Calculate the mass of sodium benzoate (NaC_6H_5COO) in the original sample. (*Hint:* At the equivalence point, the *total* number of moles of acid [here HCl] equals the *total* number of moles of base [here, both NaOH and NaC_6H_5COO].)

58. An antacid tablet (such as Tums or Rolaids) weighs 1.3259 g. The only acid-neutralizing ingredient in this brand of antacid is $CaCO_3$. When placed in 12.07 mL of 1.070 M HCl, the tablet fizzes merrily as $CO_2(g)$ is given off. After all of the CO_2 has left the solution, an indicator is added, followed by 11.74 mL of 0.5310 M NaOH. The indicator shows that at this point the solution is definitely basic. Addition of 5.12 mL of 1.070 M HCl makes the solution acidic again. Then 3.17 mL of the 0.5310 M NaOH brings the titration exactly to an endpoint, as signaled by the indicator. Compute the percentage by mass of $CaCO_3$ in the tablet.

59. What is the mass of diethylamine, $(C_2H_5)_2NH$, in 100.0 mL of an aqueous solution if it requires 15.90 mL of 0.0750 M HCl to titrate it to the equivalence point? What is the pH at the equivalence point if $K_b = 3.09 \times 10^{-4}$? What would be a suitable indicator for the titration?

60. A chemist who works in the process laboratory of the Athabasca Alkali Company makes frequent analyses of ammonia recovered from the Solvay process for making sodium carbonate. What is the pH at the equivalence point if she titrates the aqueous ammonia solution (approximately 0.10 M) with a strong acid of comparable concentration? Select an indicator that would be suitable for the titration.

61. If 50.00 mL of a 0.200 M solution of the weak base *N*-ethyl-morpholine ($C_6H_{13}NO$) is mixed with 8.00 mL of 1.00 M HCl and then diluted to a final volume of 100.0 mL with water, the result is a buffer with a pH of 7.0. Compute the K_b of *N*-ethylmorpholine.

62. The sodium salt of cacodylic acid, a weak acid, has the formula $NaO_2As(CH_3)_2 \cdot 3H_2O$. Its molar mass is 214.02 g mol^{-1}. A solution is prepared by mixing 21.40 g of this substance with enough water to make 1.000 L of solution. Then 50.00 mL of the sodium cacodylate solution is mixed with 29.55 mL of 0.100 M HCl and enough water to bring the volume to a total of 100.00 mL. The pH of the solution is 6.00. Determine the K_a of cacodylic acid.

15.7 Polyprotic Acids

63. Arsenic acid (H_3AsO_4) is a weak triprotic acid. Given the three acid ionization constants from Table 15.2 and an initial concentration of arsenic acid (before ionization) of 0.1000 M, calculate the equilibrium concentrations of H_3AsO_4, $H_2AsO_4^-$, $HAsO_4^{2-}$, AsO_4^{3-}, and H_3O^+.

64. Phthalic acid ($H_2C_8H_4O_4$, abbreviated H_2Ph) is a diprotic acid. Its ionization in water at 25°C takes place in two steps:

$$H_2Ph(aq) + H_2O(\ell) \rightleftharpoons H_3O^+(aq) + HPh^-(aq) \qquad K_{a1} = 1.26 \times 10^{-3}$$

$$HPh^-(aq) + H_2O(\ell) \rightleftharpoons H_3O^+(aq) + Ph^{2-}(aq) \qquad K_{a2} = 3.10 \times 10^{-6}$$

If 0.0100 mol of phthalic acid is dissolved per liter of water, calculate the equilibrium concentrations of H_2Ph, HPh^-, Ph^{2-}, and H_3O^+.

65. A solution as initially prepared contains 0.050 mol L^{-1} of phosphate ion (PO_4^{3-}) at 25°C. Given the three acid ionization constants from Table 15.2, calculate the equilibrium concentrations of PO_4^{3-}, HPO_4^{2-}, $H_2PO_4^-$, H_3PO_4, and OH^-.

66. Oxalic acid ionizes in two stages in aqueous solution:

$$H_2C_2O_4(aq) + H_2O(\ell) \rightleftharpoons H_3O^+(aq) + HC_2O_4^-(aq) \qquad K_{a1} = 5.9 \times 10^{-2}$$

$$HC_2O_4^-(aq) + H_2O(\ell) \rightleftharpoons H_3O^+(aq) + C_2O_4^{2-}(aq) \qquad K_{a2} = 6.4 \times 10^{-5}$$

Calculate the equilibrium concentrations of $C_2O_4^{2-}$, $HC_2O_4^-$, $H_2C_2O_4$, and OH^- in a 0.10 M solution of sodium oxalate ($Na_2C_2O_4$).

67. The pH of a normal raindrop is 5.60. Compute the concentrations of $H_2CO_3(aq)$, $HCO_3^-(aq)$, and $CO_3^{2-}(aq)$ in this raindrop if the total concentration of dissolved carbonates is 1.0×10^{-5} mol L^{-1}.

68. The pH of a drop of acid rain is 4.00. Compute the concentrations of $H_2CO_3(aq)$, $HCO_3^-(aq)$, and $CO_3^{2-}(aq)$ in the acid raindrop if the total concentration of dissolved carbonates is 3.6×10^{-5} mol L^{-1}.

15.8 Organic Acids and Bases: Structure and Reactivity

69. Which will be the stronger acid: propene ($CH_2{=}CHCH_3$) or propane (C_3H_8)? Explain by using resonance Lewis structures.

70. Predict, using resonance Lewis dot structures, which will be the stronger base, HSO_3^- or HSO_4^-. Give a thermodynamic argument for how an equilibrium constant of a reaction might be affected if the reactant, rather than the product, is resonance stabilized.

15.9 A Deeper Look . . . Exact Treatment of Acid–Base Equilibria

71. Thiamine hydrochloride (vitamin B_1 hydrochloride, HC_{12}-$H_{17}ON_4SCl_2$) is a weak acid with $K_a = 3.4 \times 10^{-7}$. Suppose 3.0×10^{-5} g of thiamine hydrochloride is dissolved in 1.00 L of water. Calculate the pH of the resulting solution. (*Hint:* This is a sufficiently dilute solution that the autoionization of water cannot be neglected.)

72. A sample of vinegar contains 40.0 g of acetic acid (CH_3COOH) per liter of solution. Suppose 1.00 mL is removed and diluted to 1.00 L, and 1.00 mL of *that* solution is removed and diluted to 1.00 L. Calculate the pH of the resulting solution. (*Hint:* This is a sufficiently dilute solution that the autoionization of water cannot be neglected.)

73. At 25°C, 50.00 mL of a 0.1000 M solution of maleic acid, a diprotic acid whose ionization constants are $K_{a1} = 1.42 \times 10^{-2}$ and $K_{a2} = 8.57 \times 10^{-7}$, is titrated with a 0.1000 M NaOH solution. Calculate the pH at the following volumes of added base: 0, 5.00, 25.00, 50.00, 75.00, 99.90, 100.00, and 105.00 mL.

74. Quinine ($C_{20}H_{24}O_2N_2$) is a water-soluble base that ionizes in two stages, with $K_{b1} = 3.31 \times 10^{-6}$ and $K_{b2} = 1.35 \times 10^{-10}$, at 25°C. Calculate the pH during the titration of an aqueous solution of 1.622 g of quinine in 100.00 mL of water as a function of the volume of added 0.1000 M HCl solution at the following volumes: 0, 25.00, 50.00, 75.00, 99.90, 100.00, and 105.00 mL.

ADDITIONAL PROBLEMS

75. Although acetic acid is normally regarded as a weak acid, it is about 34% dissociated in a 10^{-4} M solution at 25°C. It is less than 1% dissociated in 1 M solution. Discuss this variation in degree of dissociation with dilution in terms of Le Châtelier's principle, and explain how it is consistent with the supposed constancy of equilibrium constants.

76. Suppose that a 0.10 M aqueous solution of a monoprotic acid HX has just 11 times the conductivity of a 0.0010 M aqueous solution of HX. What is the approximate dissociation constant of HX? (*Hint:* In thinking about this problem, consider what the ratio of the conductivities would be if HX were a strong acid and if HX were extremely weak, as limiting cases.)

77. The ionization constant of chloroacetic acid ($ClCH_2COOH$) in water is 1.528×10^{-3} at 0°C and 1.230×10^{-3} at 40°C. Calculate the enthalpy of ionization of the acid in water, assuming that ΔH and ΔS are constant over this range of temperature.

78. The autoionization constant of water (K_w) is 1.139×10^{-15} at 0°C and 9.614×10^{-14} at 60°C.

(a) Calculate the standard enthalpy of autoionization of water.

$$2\ H_2O(\ell) \rightleftharpoons H_3O^+(aq) + OH^-(aq)$$

(b) Calculate the standard entropy of autoionization of water.

(c) At what temperature will the pH of pure water be 7.00, from these data?

79. Calculate the concentrations of H_3O^+ and OH^- at 25°C in the following:

(a) Orange juice (pH 2.8)
(b) Tomato juice (pH 3.9)
(c) Milk (pH 4.1)
(d) Borax solution (pH 8.5)
(e) Household ammonia (pH 11.9)

80. Try to choose which of the following is the pH of a 6.44×10^{-10} M $Ca(OH)_2(aq)$ solution, without doing any written calculations.

(a) 4.81
(b) 5.11
(c) 7.00
(d) 8.89
(e) 9.19

81. $Cl_2(aq)$ reacts with $H_2O(\ell)$ as follows:

$$Cl_2(aq) + 2\ H_2O(\ell) \rightleftharpoons H_3O^+(aq) + Cl^-(aq) + HOCl(aq)$$

For an experiment to succeed, $Cl_2(aq)$ must be present, but the amount of $Cl^-(aq)$ in the solution must be minimized. For this purpose, should the pH of the solution be high, low, or neutral? Explain.

82. Use the data in Table 15.2 to determine the equilibrium constant for the reaction

$$H_2PO_4^-(aq) + 2\ CO_3^{2-}(aq) \rightleftharpoons PO_4^{3-}(aq) + 2\ HCO_3^-(aq)$$

83. The first acid ionization constant of the oxoacid H_3PO_2 is 8×10^{-2}. What molecular structure do you predict for H_3PO_2? Will this acid be monoprotic, diprotic, or triprotic in aqueous solution?

84. Oxoacids can be formed that involve several central atoms of the same chemical element. An example is $H_3P_3O_9$, which can be written $P_3O_6(OH)_3$. (Sodium salts of these polyphosphoric acids are used as "builders" in detergents to improve their cleaning power.) In such a case, we would expect acid strength to correlate approximately with the *ratio* of the number of lone oxygen atoms to the number of central atoms (this ratio is 6:3 for $H_3P_3O_9$, for example). Rank the following in order of increasing acid strength: H_3PO_4, $H_3P_3O_9$, $H_4P_2O_6$, $H_4P_2O_7$, $H_5P_3O_{10}$. Assume that no hydrogen atoms are directly bonded to phosphorus in these compounds.

85. Urea (NH_2CONH_2) is a component of urine. It is a very weak base, having an estimated pK_b of 13.8 at room temperature.

(a) Write the formula of the conjugate acid of urea.

(b) Compute the equilibrium concentration of urea in a solution that contains no urea but starts out containing 0.15 mol L^{-1} of the conjugate acid of urea.

86. Exactly 1.0 L of solution of acetic acid gives the same color with methyl red as 1.0 L of a solution of hydrochloric acid. Which solution will neutralize the greater amount of 0.10 M $NaOH(aq)$? Explain.

*87. The K_a for acetic acid drops from 1.76×10^{-5} at 25°C to 1.63×10^{-5} at 50°C. Between the same two temperatures, K_w increases from 1.00×10^{-14} to 5.47×10^{-14}. At 50°C the density of a 0.10 M solution of acetic acid is 98.81% of its density at 25°C. Will the pH of a 0.10 M solution of acetic acid in water increase, decrease, or remain the same when it is heated from 25°C to 50°C? Explain.

88. Calculate the pH of a solution that is prepared by dissolving 0.23 mol of hydrofluoric acid (HF) and 0.57 mol of hypochlorous acid (HClO) in water and diluting to 3.60 L. Also, calculate the equilibrium concentrations of HF, F^-, HClO, and ClO^-. (*Hint:* The pH will be determined by the stronger acid of this pair.)

89. For each of the following compounds, indicate whether a 0.100 M aqueous solution is acidic (pH < 7), basic (pH > 7), or neutral (pH = 7): HCl, NH_4Cl, KNO_3, Na_3PO_4, $NaCH_3COO$.

*90. Calculate $[H_3O^+]$ in a solution that contains 0.100 mol of NH_4CN per liter.

$$NH_4^+(aq) + H_2O(\ell) \rightleftharpoons H_3O^+(aq) + NH_3(aq) \qquad K_a = 5.6 \times 10^{-10}$$

$$HCN(aq) + H_2O(\ell) \rightleftharpoons H_3O^+(aq) + CN^-(aq) \qquad K_a = 6.17 \times 10^{-10}$$

91. Discuss the justification for this statement: "Although one does not normally regard NH_4^+ as an acid, it is actually only slightly weaker as an acid than hydrocyanic acid, HCN, in aqueous solution."

*92. Imagine that you want to do physiological experiments at a pH of 6.0 and the organism with which you are working is sensitive to most available materials other than a certain weak acid, H_2Z, and its sodium salts. K_{a1} and K_{a2} for H_2Z are 3×10^{-1} and 5×10^{-7}. You have available 1.0 M aqueous H_2Z and 1.0 M NaOH. How much of the NaOH solution should be added to 1.0 L of the acid solution to give a buffer at pH = 6.0?

93. A buffer solution is prepared by mixing 1.00 L of 0.050 M pentafluorobenzoic acid (C_6F_5COOH) and 1.00 L of 0.060 M sodium pentafluorobenzoate (NaC_6F_5COO). The K_a of this weak acid is 0.033. Determine the pH of the buffer solution.

94. A chemist needs to prepare a buffer solution with pH = 10.00 and has both Na_2CO_3 and $NaHCO_3$ in pure crystalline form. What mass of each should be dissolved in 1.00 L of solution if the combined mass of the two salts is to be 10.0 g?

95. Which of these procedures would *not* make a pH = 4.75 buffer?
   (a) Mix 50.0 mL of 0.10 M acetic acid and 50.0 mL of 0.10 M sodium acetate.
   (b) Mix 50.0 mL of 0.20 M acetic acid and 50.0 mL of 0.10 M NaOH.
   (c) Start with 50.0 mL of 0.20 M acetic acid and add a solution of strong base until the pH equals 4.75.
   (d) Start with 50.0 mL of 0.20 M HCl and add a solution of strong base until the pH equals 4.75.
   (e) Start with 100.0 mL of 0.20 M sodium acetate and add 50.0 mL of 0.20 M HCl.

96. It takes 4.71 mL of 0.0410 M NaOH to titrate a 50.00-mL sample of flat (no CO_2) GG's Cola to a pH of 4.9. At this point the addition of one more drop (0.02 mL) of NaOH raises the pH to 6.1. The only acid in GG's Cola is phosphoric acid. Compute the concentration of phosphoric acid in this cola. Assume that the 4.71 mL of base removes only the first hydrogen from the H_3PO_4; that is, assume that the reaction is

$$H_3PO_4(aq) + OH^-(aq) \longrightarrow H_2O(\ell) + H_2PO_4^-(aq)$$

*97. Sodium carbonate exists in various crystalline forms with different amounts of water of crystallization, including Na_2CO_3, $Na_2CO_3 \cdot 10H_2O$, and others. The water of crystallization can be driven off by heating; the amount of water removed depends on the temperature and duration of heating.

A sample of $Na_2CO_3 \cdot 10H_2O$ had been heated inadvertently, and it was not known how much water had been removed. A 0.200-g sample of the solid that remained after the heating was dissolved in water, 30.0 mL of 0.100 M NaOH was added, and the CO_2 formed was removed. The solution was acidic; 6.4 mL of 0.200 M NaOH was needed to neutralize the excess acid. What fraction of the water had been driven from the $Na_2CO_3 \cdot 10H_2O$?

98. An aqueous solution of sodium carbonate, Na_2CO_3, is titrated with strong acid to a point at which two H^+ ions have reacted with each carbonate ion. (a) If 20.0 mL of the carbonate solution reacts with just 40.0 mL of 0.50 M acid, what is the molarity of the carbonate solution? (b) If the solution contains 5.0 percent by mass sodium carbonate, what is the density of the solution? (c) Suppose that you wanted to prepare a liter of an identical solution by starting with crystalline sodium carbonate decahydrate, $Na_2CO_3 \cdot 10H_2O$, rather than with solid Na_2CO_3 itself. How much of this substance would you need?

99. Three flasks, labeled *A*, *B*, and *C*, contained aqueous solutions of the same pH. It was known that one of the solutions was 1.0×10^{-3} M in nitric acid, one was 6×10^{-3} M in formic acid, and one was 4×10^{-2} M in the salt formed by the weak organic base aniline with hydrochloric acid ($C_6H_5NH_3Cl$). (Formic acid is monoprotic.) (a) Describe a procedure for identifying the solutions. (b) Compare qualitatively (on the basis of the preceding information) the strengths of nitric and formic acids with each other and with the acid strength of the anilinium ion, $C_6H_5NH_3^+$. (c) Show how the information given may be used to derive values for K_a for formic acid and K_b for aniline. Derive these values.

100. Novocain, the commonly used local anesthetic, is a weak base with $K_b = 7 \times 10^{-6}$ M. (a) If you had a 0.0200 M solution of Novocain in water, what would be the approximate concentration of OH^- and the pH? (b) Suppose that you wanted to determine the concentration of Novocain in a solution that is about 0.020 M by titration with 0.020 M HCl. Calculate the expected pH at the equivalence point.

101. A 0.1000 M solution of a weak acid, HA, requires 50.00 mL of 0.1000 M NaOH to titrate it to its equivalence point. The pH of the solution is 4.50 when only 40.00 mL of the base has been added.

(a) Calculate the ionization constant K_a of the acid.

(b) Calculate the pH of the solution at the equivalence point.

102. The chief chemist of Victory Vinegar Works, Ltd., interviews two chemists for employment. He states, "Quality control requires that our high-grade vinegar contain 5.00 ± 0.01% acetic acid by mass. How would you analyze our product to ensure that it meets this specification?"

Anne Dalton says, "I would titrate a 50.00-mL sample of the vinegar with 1.000 M NaOH, using phenolphthalein to detect the equivalence point to within ±0.02 mL of base."

Charlie Cannizzarro says, "I would use a pH meter to determine the pH to ±0.01 pH units and interface it with a computer to print out the mass percentage of acetic acid."

Which candidate did the chief chemist hire? Why?

103. Phosphonocarboxylic acid

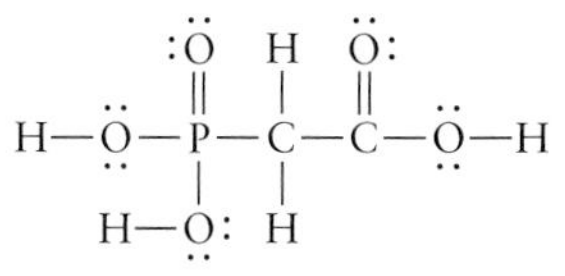

effectively inhibits the replication of the herpes virus. Structurally, it is a combination of phosphoric acid and acetic acid. It can donate three protons. The equilibrium constant values are $K_{a1} = 1.0 \times 10^{-2}$, $K_{a2} = 7.8 \times 10^{-6}$, and $K_{a3} = 2.0 \times 10^{-9}$. Enough phosphonocarboxylic acid is added to blood (pH 7.40) to make its total concentration 1.0×10^{-5} M. The pH of the blood does not change. Determine the concentrations of all four forms of the acid in this mixture.

104. Egg whites contain dissolved carbon dioxide and water, which react together to give carbonic acid (H_2CO_3). In the few days after an egg is laid, it loses carbon dioxide through its shell. Does the pH of the egg white increase or decrease during this period?

105. If you breathe too rapidly (hyperventilate), the concentration of dissolved CO_2 in your blood drops. What effect does this have on the pH of the blood?

106. A reference book states that a saturated aqueous solution of potassium hydrogen tartrate is a buffer with a pH of 3.56. Write two chemical equations that show the buffer action of this solution. (Tartaric acid is a diprotic acid with the formula $H_2C_4H_4O_6$. Potassium hydrogen tartrate is $KHC_4H_4O_6$.)

***107.** Glycine, the simplest amino acid, has both an acid group and a basic group in its structure ($H_2N—CH_2—COOH$). In aqueous solution it exists predominantly as a self-neutralized species called a zwitterion ($H_3\overset{+}{N}—CH_2—COO^-$). The zwitterion therefore behaves both as an acid and as a base, according to the equilibria at 25°C:

$$H_3\overset{+}{N}—CH_2—COO^-(aq) + H_2O(\ell) \rightleftharpoons H_2N—CH_2—COO^-(aq) + H_3O^+(aq) \qquad K_a = 1.7 \times 10^{-10}$$

$$H_3\overset{+}{N}—CH_2—COO^-(aq) + H_2O(\ell) \rightleftharpoons H_3\overset{+}{N}—CH_2—COOH(aq) + OH^-(aq) \qquad K_b = 2.2 \times 10^{-12}$$

Calculate the pH of a 0.10 M aqueous solution of glycine at 25°C. (*Hint:* You may need to take account of the autoionization of water.)

***108.** Consider the two following nitrophenol structures:

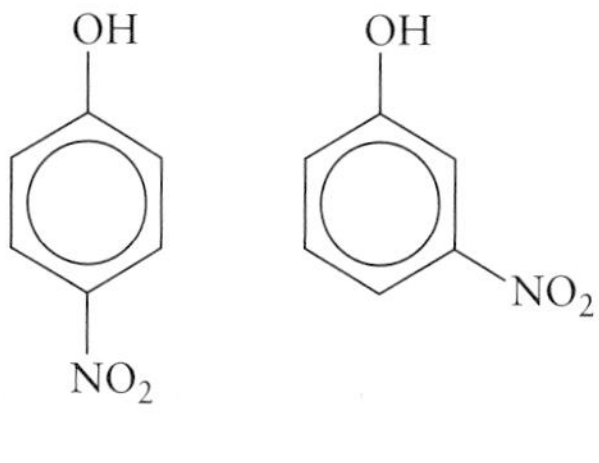

p-nitrophenol *m*-nitrophenol

Predict which will be the stronger acid. (*Hint:* Consider possible resonance structures analogous to those given in the text for phenol.)

CUMULATIVE PROBLEMS

109. Use the data from Table 15.1, together with Le Châtelier's principle, to decide whether the autoionization of water is exothermic or endothermic.

110. Baking soda (sodium hydrogen carbonate, $NaHCO_3$) is used in baking because it reacts with acids in foods to form carbonic acid (H_2CO_3), which in turn decomposes to water and carbon dioxide. In a batter the carbon dioxide appears as gas bubbles that cause the bread or cake to rise.

(a) A rule of thumb in cooking is that $\frac{1}{2}$ teaspoon baking soda is neutralized by 1 cup of sour milk. The acid component of sour milk is lactic acid ($HC_3H_5O_3$). Write an equation for the neutralization reaction.

(b) If the density of baking soda is 2.16 g cm^{-3}, calculate the concentration of lactic acid in the sour milk, in moles per liter. Take 1 cup = 236.6 mL = 48 teaspoons.

(c) Calculate the volume of carbon dioxide that is produced at 1 atm pressure and 350°F (177°C) from the reaction of $\frac{1}{2}$ teaspoon of baking soda.

111. Boric acid, $B(OH)_3$, is an acid that acts differently from the usual Brønsted–Lowry acids. It reacts with water according to

$$B(OH)_3(aq) + 2\ H_2O(\ell) \rightleftharpoons B(OH)_4^-(aq) + H_3O^+(aq) \qquad K_a = 5.8 \times 10^{-10}$$

(a) Draw Lewis structures for $B(OH)_3$ and $B(OH)_4^-$. Can these be described as Lewis acids or Lewis bases?

(b) Calculate the pH of a 0.20 M solution of $B(OH)_3(aq)$.

112. At 40°C and 1.00 atm pressure, a gaseous monoprotic acid has a density of 1.05 g L^{-1}. After 1.85 g of this gas is dissolved in water and diluted to 450 mL, the pH is measured to be 5.01. Determine the K_a of this acid and use Table 15.2 to identify it.

113. At 25°C, the Henry's law constant for carbon dioxide dissolved in water is 1.8×10^3 atm. Calculate the pH of water saturated with $CO_2(g)$ at 25°C in Denver, where the barometric pressure is 0.833 atm.

16

CHAPTER

Solubility and Precipitation Equilibria

Sodium acetate crystals ($NaC_2H_3O_2$) form quickly in a supersaturated solution when a small speck of solute is added.

Dissolution and precipitation are chemical reactions by which solids pass into and out of solution. A brief introduction and several examples appear in Section 11.2. These reactions involve equilibria between dissolved species and species in the solid state, and so are described by the general principles of chemical equilibrium in Chapter 14. These reactions rank alongside acid–base reactions in practical importance. The dissolution and reprecipitation of solids permit chemists to isolate single products from reaction mixtures and to purify impure solid samples. Understanding the mechanisms of these reactions helps engineers prevent formation of deposits in water processing and distribution systems and helps doctors reduce the incidence of painful kidney stones. Dissolution and precipitation control the formation of mineral deposits and profoundly affect the ecologies of rivers, lakes, and oceans.

The theme of this chapter is to understand how to manipulate solubility equilibria in order to control the maximum concentration of particular ionic solids in the solution. In the first section we present the general aspects of the equilibria that govern the extent of dissolution and precipitation reactions. In the remaining sections we develop the quantitative descriptions for these equilibria, including the effects of adding more solutes, adding acids or bases, and adding ligands that can bind to dissolved metal ions to form complex ions.

16.1 The Nature of Solubility Equilibria

16.1.1 General Features of Solubility Equilibria

Solubility equilibria resemble the equilibria between volatile liquids (or solids) and their vapors in a closed container. In both cases, particles from a condensed phase tend to escape and spread through a larger, but limited, volume. In both cases, equilibrium is a dynamic compromise in which the rate of escape of particles from the condensed phase is equal to their rate of return. In a vaporization–condensation equilibrium, we assumed that the vapor above the condensed phase was an ideal gas. The analogous starting assumption for a dissolution–precipitation reaction is that the solution above the undissolved solid is an ideal solution. A solution in which sufficient solute has been dissolved to establish a dissolution–precipitation equilibrium between the solid substance and its dissolved form is called a **saturated solution.**

Le Châtelier's principle applies to these equilibria, as it does to all equilibria. One way to exert a stress on a solubility equilibrium is to change the amount of solvent. Adding solvent reduces the concentration of dissolved substance; more solid then tends to dissolve to restore the concentration of the dissolved substance to its equilibrium value. If an excess of solvent is added so that all of the solid dissolves, then obviously the solubility equilibrium ceases to exist and the solution is *unsaturated.* In a vaporization–condensation equilibrium, this corresponds to the complete evaporation of the condensed phase. Removing solvent from an already saturated solution forces additional solid to precipitate in order to maintain a constant concentration. A volatile solvent is often removed by simply letting a solution stand uncovered until the solvent evaporates. When sufficient solvent has evaporated, the solid forms as crystals on the bottom and sides of the container (Fig. 16.1).

FIGURE 16.1 The beaker contains an aqueous solution of K_2PtCl_4, a substance used as a starting material in the synthesis of anticancer drugs. It is loosely covered to keep out dust and allowed to stand. As the water evaporates, the solution becomes saturated and deposits solid K_2PtCl_4 in the form of long, needle-like red crystals.

Controlled precipitation by manipulating solubility is a widely used technique for purifying reaction products in synthetic chemistry. Side reactions can generate significant amounts of impurities; other impurities may enter with the starting materials, and catalysts (introduced to increase the reaction rates) must be removed from the final products. Running a reaction may take only hours, but the *workup* (separation of crude product) and subsequent purification may require weeks. **Recrystallization,** one of the most powerful methods for purifying solids, relies on differences between the solubilities of the desired substance and its contaminants. An impure product is dissolved and reprecipitated, repeatedly if necessary, with careful control of the factors that influence solubility. Manipulating solubility requires an understanding of the equilibria that exist between an undissolved substance and its solution.

In recrystallization a solution begins to deposit a compound when it is brought to the point of saturation with respect to that compound. In dissolution the solvent attacks the solid and solvent molecules **solvate** each new solute molecule as it is released from the dissolving solid. In precipitation the reverse occurs: Solute-to-solute attractions are reestablished as the solute leaves the solution. Often, solute-to-solvent attractions persist right through the process of precipitation, and solvent incorporates itself into the solid. When lithium sulfate

(Li_2SO_4) precipitates from water, it brings with it into the solid one molecule of water per formula unit:

$$2\,Li^+(aq) + SO_4^{2-}(aq) + H_2O(\ell) \longrightarrow Li_2SO_4 \cdot H_2O(s)$$

Such loosely bound solvent is known as solvent of crystallization (Fig. 2.1). Dissolving and then reprecipitating a compound may thus furnish material that has a different chemical formula and a different mass. Consequently, recrystallization processes for purification of reaction products must be planned carefully.

Dissolution–precipitation reactions frequently come to equilibrium slowly. Days or even weeks of shaking a solid in contact with a solvent may be required before the solution becomes saturated. Moreover, solutions sometimes become **supersaturated,** a state in which the concentration of dissolved solid exceeds its equilibrium value and the solid remains in solution rather than form a precipitate. This state is analogous to superheating a pure liquid above its boiling point or supercooling it below its freezing point. A supersaturated solution may persist for months or years and require extraordinary measures to be brought to equilibrium by precipitating the solid, although thermodynamics shows that the possibility for equilibrium is there all along. Precipitation from a supersaturated solution is a spontaneous process, but its rate may be slow. The generally sluggish approach to equilibrium in dissolution–precipitation reactions is quite the opposite of the rapid rates at which acid–base reactions reach equilibrium.

16.1.2 The Solubility of Ionic Solids

The **solubility** of a substance in a solvent is defined as the greatest amount (expressed either in grams or in moles) that will dissolve in equilibrium in a specified volume of solvent at a particular temperature. Although solvents other than water are used in many applications, aqueous solutions are the most important and are the exclusive concern here. Salts show a wide range of solubilities in water (Fig. 16.2). Silver perchlorate ($AgClO_4$) dissolves to the remarkable extent of about 5570 g (or almost 27 mol) per liter of water at 25°C, but at the same temperature only about 0.0018 g (or 1.3×10^{-5} mol) of silver chloride (AgCl) dissolves per liter of water. Many salts with even lower solubilities are known. Solubilities often depend strongly on temperature (Fig. 16.3). Most dissolution reactions for ionic solids are endothermic (heat is absorbed), so by Le Châtelier's principle the solubility increases with increasing temperature.

FIGURE 16.2 Vastly different quantities of different compounds will dissolve in 1 L of water at 20°C. Clockwise from the front are borax, potassium permanganate, lead(II) chloride, sodium phosphate decahydrate, calcium oxide, and potassium dichromate.

FIGURE 16.3 Most solubilities increase with increasing temperature, but some decrease. Note that the changes are not always smooth because different solid hydrates form over different temperature ranges.

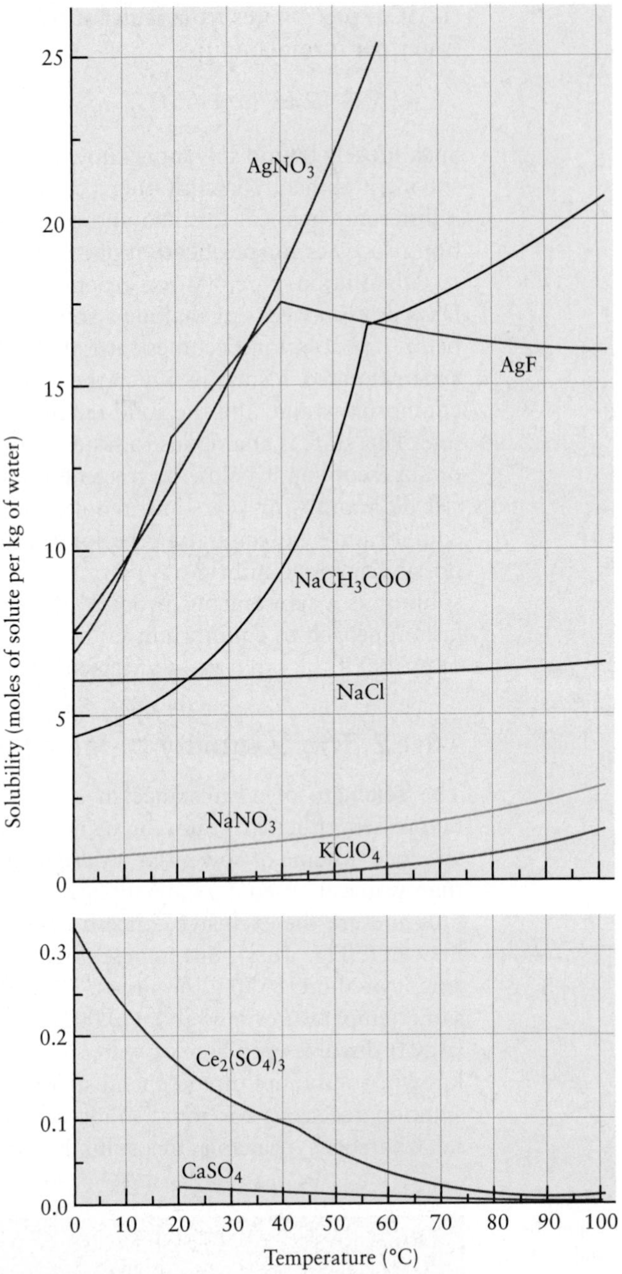

FIGURE 16.4 A yellow precipitate of lead iodide forms when a drop of KI solution is added to a solution of $Pb(NO_3)_2$.

Those dissolution reactions that are exothermic (such as for $CaSO_4$) show the opposite behavior.

Although all ionic compounds dissolve to some extent in water, those having solubilities (at 25°C) of less than 0.1 g L^{-1} are called *insoluble*. Those having solubilities of more than 10 g L^{-1} are *soluble*, and the intermediate cases (0.1 to 10 g L^{-1}) are said to be *slightly soluble*. Fortunately, it is not necessary to memorize long lists of solubility data. Table 16.1 lists some generalizations concerning groups of salts and gives enough factual data to support good predictions about precipitation or dissolution in thousands of situations of practical importance. Knowing the solubilities of ionic substances, even in these qualitative terms, provides a way to predict the courses of numerous reactions. For example, when a solution of KI is added to one of $Pb(NO_3)_2$, K^+ and NO_3^- ions are brought into contact, as are Pb^{2+} and I^- ions. From the table, KNO_3 is a soluble salt but PbI_2 is insoluble; therefore, a precipitate of PbI_2 will appear (Fig. 16.4).

TABLE 16.1

Solubilities of Ionic Compounds in Water

Anion	Soluble†	Slightly Soluble	Insoluble
NO_3^- (nitrate)	All	—	—
CH_3COO^- (acetate)	Most	—	$Be(CH_3COO)_2$
ClO_3^- (chlorate)	All	—	—
ClO_4^- (perchlorate)	Most	$KClO_4$	—
F^- (fluoride)	Group I, AgF, BeF_2	SrF_2, BaF_2, PbF_2	MgF_2, CaF_2
Cl^- (chloride)	Most	$PbCl_2$	AgCl, Hg_2Cl_2
Br^- (bromide)	Most	$PbBr_2$, $HgBr_2$	AgBr, Hg_2Br_2
I^- (iodide)	Most	—	AgI, Hg_2I_2, PbI_2, HgI_2
SO_4^{2-} (sulfate)	Most	$CaSO_4$, Ag_2SO_4, Hg_2SO_4	$SrSO_4$, $BaSO_4$, $PbSO_4$
S^{2-} (sulfide)	Groups I and II, $(NH_4)_2S$	—	Most
CO_3^{2-} (carbonate)	Group I, $(NH_4)_2CO_3$	—	Most
SO_3^{2-} (sulfite)	Group I, $(NH_4)_2SO_3$	—	Most
PO_4^{3-} (phosphate)	Group I, $(NH_4)_3PO_4$	—	Most
OH^- (hydroxide)	Group I, $Ba(OH)_2$	$Sr(OH)_2$, $Ca(OH)_2$	Most

†Soluble compounds are defined as those that dissolve to the extent of 10 or more grams per liter; slightly soluble compounds, 0.1 to 10 grams per liter; and insoluble compounds, less than 0.1 gram per liter at room temperature.

16.2 Ionic Equilibria between Solids and Solutions

When an ionic solid such as CsCl dissolves in water, it breaks up into ions that move apart from each other and become solvated by water molecules (*aquated*, or *hydrated;* Fig. 16.5). The aquated ions are shown in the chemical equation for the solubility equilibrium. Thus,

$$CsCl(s) \rightleftharpoons Cs^+(aq) + Cl^-(aq)$$

shows that the dissolved particles are ions. For a highly soluble salt (such as CsCl), the concentrations of the ions in a saturated aqueous solution are so large that the solution is nonideal. There is much association among the ions in solution, resulting in temporary pairs of oppositely charged ions and in larger clusters as well. In such cases the simple type of equilibrium expression developed in Sections 14.2 and 14.3 does not apply; therefore, we restrict our attention to sparingly soluble and "insoluble" salts, for which the concentrations in

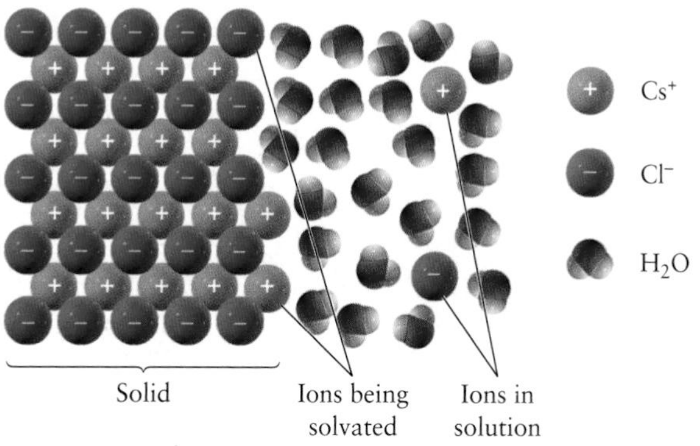

FIGURE 16.5 The dissolution of the ionic solid CsCl in water. Note the different solvation patterns of positive and negative ions. Water molecules arrange themselves so that the atoms with partial negative charges (oxygen) point toward the positively charged Cs^+ ions. The water molecules form hydrogen bonds to the Cl^- anions, so the O—H—Cl atoms lie nearly in a straight line.

a saturated solution are 0.1 mol L^{-1} or less—low enough that the interactions among the solvated ions are relatively small.

The sparingly soluble salt silver chloride, for example, establishes the following equilibrium when placed in water:

$$AgCl(s) \rightleftharpoons Ag^+(aq) + Cl^-(aq)$$

The equilibrium law for this reaction (written by following the rules for heterogeneous equilibria from Sections 14.2 and 14.3) is

$$[Ag^+][Cl^-] = K_{sp}$$

where the subscript "sp," standing for **"solubility product,"** distinguishes the K as referring to the dissolution of a slightly soluble ionic solid in water. Note that we have omitted the subscript eq for simplicity; the concentrations in this and all equilibrium constant expressions to follow are equilibrium ones. At 25°C, K_{sp} has the numerical value 1.6×10^{-10} for silver chloride. In the AgCl solubility product expression, the concentrations of the two ions produced are raised to

TABLE 16.2

Solubility Product Constants K_{sp} at 25°C

Iodates	
$AgIO_3$	$[Ag^+][IO_3^-] = 3.1 \times 10^{-8}$
$CuIO_3$	$[Cu^+][IO_3^-] = 1.4 \times 10^{-7}$
$Pb(IO_3)_2$	$[Pb^{2+}][IO_3^-]^2 = 2.6 \times 10^{-13}$
Carbonates	
Ag_2CO_3	$[Ag^+]^2[CO_3^{2-}] = 6.2 \times 10^{-12}$
$BaCO_3$	$[Ba^{2+}][CO_3^{2-}] = 8.1 \times 10^{-9}$
$CaCO_3$	$[Ca^{2+}][CO_3^{2-}] = 8.7 \times 10^{-9}$
$PbCO_3$	$[Pb^{2+}][CO_3^{2-}] = 3.3 \times 10^{-14}$
$MgCO_3$	$[Mg^{2+}][CO_3^{2-}] = 4.0 \times 10^{-5}$
$SrCO_3$	$[Sr^{2+}][CO_3^{2-}] = 1.6 \times 10^{-9}$
Chromates	
Ag_2CrO_4	$[Ag^+]^2[CrO_4^{2-}] = 1.9 \times 10^{-12}$
$BaCrO_4$	$[Ba^{2+}][CrO_4^{2-}] = 2.1 \times 10^{-10}$
$PbCrO_4$	$[Pb^{2+}][CrO_4^{2-}] = 1.8 \times 10^{-14}$
Oxalates	
CuC_2O_4	$[Cu^{2+}][C_2O_4^{2-}] = 2.9 \times 10^{-8}$
FeC_2O_4	$[Fe^{2+}][C_2O_4^{2-}] = 2.1 \times 10^{-7}$
MgC_2O_4	$[Mg^{2+}][C_2O_4^{2-}] = 8.6 \times 10^{-5}$
PbC_2O_4	$[Pb^{2+}][C_2O_4^{2-}] = 2.7 \times 10^{-11}$
SrC_2O_4	$[Sr^{2+}][C_2O_4^{2-}] = 5.6 \times 10^{-8}$
Sulfates	
$BaSO_4$	$[Ba^{2+}][SO_4^{2-}] = 1.1 \times 10^{-10}$
$CaSO_4$	$[Ca^{2+}][SO_4^{2-}] = 2.4 \times 10^{-5}$
$PbSO_4$	$[Pb^{2+}][SO_4^{2-}] = 1.1 \times 10^{-8}$
Fluorides	
BaF_2	$[Ba^{2+}][F^-]^2 = 1.7 \times 10^{-6}$
CaF_2	$[Ca^{2+}][F^-]^2 = 3.9 \times 10^{-11}$
MgF_2	$[Mg^{2+}][F^-]^2 = 6.6 \times 10^{-9}$
PbF_2	$[Pb^{2+}][F^-]^2 = 3.6 \times 10^{-8}$
SrF_2	$[Sr^{2+}][F^-]^2 = 2.8 \times 10^{-9}$
Chlorides	
$AgCl$	$[Ag^+][Cl^-] = 1.6 \times 10^{-10}$
$CuCl$	$[Cu^+][Cl^-] = 1.0 \times 10^{-6}$
Hg_2Cl_2	$[Hg_2^{2+}][Cl^-]^2 = 1.2 \times 10^{-18}$
Bromides	
$AgBr$	$[Ag^+][Br^-] = 7.7 \times 10^{-13}$
$CuBr$	$[Cu^+][Br^-] = 4.2 \times 10^{-8}$
Hg_2Br_2	$[Hg_2^{2+}][Br^-]^2 = 1.3 \times 10^{-21}$
Iodides	
AgI	$[Ag^+][I^-] = 1.5 \times 10^{-16}$
CuI	$[Cu^+][I^-] = 5.1 \times 10^{-12}$
PbI_2	$[Pb^{2+}][I^-]^2 = 8.5 \times 10^{-9}$
Hg_2I_2	$[Hg_2^{2+}][I^-]^2 = 1.2 \times 10^{-28}$
Hydroxides	
$AgOH$	$[Ag^+][OH^-] = 1.5 \times 10^{-8}$
$Al(OH)_3$	$[Al^{3+}][OH^-]^3 = 3.7 \times 10^{-15}$
$Fe(OH)_3$	$[Fe^{3+}][OH^-]^3 = 1.1 \times 10^{-36}$
$Fe(OH)_2$	$[Fe^{2+}][OH^-]^2 = 1.6 \times 10^{-14}$
$Mg(OH)_2$	$[Mg^{2+}][OH^-]^2 = 1.2 \times 10^{-11}$
$Mn(OH)_2$	$[Mn^{2+}][OH^-]^2 = 2.0 \times 10^{-13}$
$Zn(OH)_2$	$[Zn^{2+}][OH^-]^2 = 4.5 \times 10^{-17}$

the first power because their coefficients are 1 in the chemical equation. Solid AgCl does not appear in the equilibrium expression; the amount of pure solid AgCl does not affect the equilibrium as long as *some* is present. If no solid is present, the system is not in a state of equilibrium, and the product of the two ion concentrations does not obey the solubility equilibrium expression.

16.2.1 Solubility and K_{sp}

The molar solubility of a salt in water is not the same as its solubility product constant, but a simple relation often exists between them. For example, let's define S as the molar solubility of AgCl(s) in water at 25°C. Then, from stoichiometry, $[Ag^+] = [Cl^-] = S$ is the molarity of either ion at equilibrium. Hence,

$$[Ag^+][Cl^-] = S^2 = K_{sp} = 1.6 \times 10^{-10}$$

Taking the square roots of both sides of the equation gives

$$S = 1.26 \times 10^{-5} \text{ M}$$

which rounds off to 1.3×10^{-5} M. This is the molar solubility of AgCl in water. It is converted to a gram solubility by multiplying by the molar mass of AgCl:

$$(1.26 \times 10^{-5} \text{ mol L}^{-1})(143.3 \text{ g mol}^{-1}) = 1.8 \times 10^{-3} \text{ g L}^{-1}$$

Therefore, 1.8×10^{-3} g of AgCl dissolves per liter of water at 25°C.

Solubility product constants (like solubilities) can be sensitive to temperature. At 100°C the K_{sp} for silver chloride is 2.2×10^{-8}; hot water dissolves about 12 times as much silver chloride as does water at 25°C. Refer to Table 16.2 for the solubility product constants at 25°C of a number of important sparingly soluble salts. The temperature dependence of equilibrium constants is discussed in detail in Chapter 14, Section 14.7.3.

EXAMPLE 16.1

The K_{sp} of calcium fluoride is 3.9×10^{-11}. Calculate the concentrations of calcium and fluoride ions in a saturated solution of CaF_2 at 25°C, and determine the solubility of CaF_2 in grams per liter.

Solution

The solubility equilibrium is

$$CaF_2(s) \rightleftharpoons Ca^{2+}(aq) + 2\ F^-(aq)$$

and the expression for the solubility product is

$$[Ca^{2+}][F^-]^2 = K_{sp}$$

The concentration of fluoride ion is squared because it has a coefficient of 2 in the chemical equation.

If S mol of CaF_2 dissolves in 1 L, the equilibrium concentration of Ca^{2+} will be $[Ca^{2+}] = S$. The concentration of F^- will be $[F^-] = 2S$, because each mole of CaF_2 produces *two* moles of fluoride ions. Therefore,

$$[Ca^{2+}][F^-]^2 = S \times (2S)^2 = 4S^3 = K_{sp} = 3.9 \times 10^{-11}$$

Solving for S^3 gives

$$S^3 = \tfrac{1}{4}(3.9 \times 10^{-11})$$

Taking the cube roots of both sides of this equation gives

$$S = 2.1 \times 10^{-4}$$

The equilibrium concentrations are therefore

$$[Ca^{2+}] = S = 2.1 \times 10^{-4} \text{ M}$$

$$[F^-] = 2S = 4.3 \times 10^{-4} \text{ M}$$

Because the molar mass of CaF_2 is 78.1 g mol^{-1}, the gram solubility is

$$\text{gram solubility} = (2.1 \times 10^{-4} \text{ mol L}^{-1})(78.1 \text{ g mol}^{-1})$$
$$= 0.017 \text{ g L}^{-1}$$

Related Problems: 7, 8, 9, 10, 11, 12

It is possible to reverse the procedure just outlined, of course, and determine the value of K_{sp} from measured solubilities, as the following example illustrates.

EXAMPLE 16.2

Silver chromate (Ag_2CrO_4) is a red solid that dissolves in water to the extent of 0.029 g L^{-1} at 25°C. Estimate its K_{sp} and compare your estimate with the value in Table 16.2.

Solution

The molar solubility is calculated from the gram solubility and the molar mass of silver chromate (331.73 g mol^{-1}):

$$\text{molar solubility} = \frac{0.029 \text{ g L}^{-1}}{331.73 \text{ g mol}^{-1}} = 8.74 \times 10^{-5} \text{ mol L}^{-1}$$

An extra significant digit is carried in this intermediate result to avoid round-off errors. Because each mole of Ag_2CrO_4 that dissolves gives *two* moles of silver ion, the concentration of $Ag^+(aq)$ is

$$[Ag^+] = 2 \times 8.74 \times 10^{-5} \text{ M} = 1.75 \times 10^{-4} \text{ M}$$

and that of CrO_4^{2-} is simply 8.74×10^{-5} M. The solubility product constant is then

$$K_{sp} = [Ag^+]^2[CrO_4^{2-}] = (1.75 \times 10^{-4})^2 \times (8.74 \times 10^{-5})$$
$$= 2.7 \times 10^{-12}$$

This estimate is about 42% greater than the tabulated value, 1.9×10^{-12}. The text below explains this discrepancy. See also Section 16.5.

Related Problems: 13, 14, 15, 16

Computing K_{sp} from a solubility and solubility from K_{sp} is valid if the solution is ideal and if there are no side reactions that reduce the concentrations of the ions after they enter solution. If such reactions are present, they cause higher solubilities than are predicted from the K_{sp} expression. For example, the solubility computed for $PbSO_4(s)$ at 25°C from its K_{sp} is 0.032 g L^{-1}, whereas that measured experimentally is 0.0425 g L^{-1}. The difference is caused by the presence of species other than $Pb^{2+}(aq)$ in solution, such as $PbOH^+(aq)$. Further discussion of these side reactions is deferred to Section 16.5.

16.3 Precipitation and the Solubility Product

So far, we have considered only cases in which a single slightly soluble salt attains equilibrium with its component ions in water. The relative concentrations of the cations and anions in such solutions echo their relative numbers of moles in the original salt. Thus, when AgCl is dissolved, equal numbers of moles of $Ag^+(aq)$ and $Cl^-(aq)$ ions result, and when Ag_2SO_4 is dissolved, twice as many moles of $Ag^+(aq)$ ions as $SO_4^{2-}(aq)$ are produced. A solubility product relationship such as

$$[Ag^+][Cl^-] = K_{sp}$$

is more general than this, however, and continues in force even if the relative number of moles of the two ions in solution differ from those in the pure solid compound. Such a situation often results when two solutions are mixed to give a precipitate or when another salt is present that contains an ion common to the salt under consideration.

16.3.1 Precipitation from Solution

Suppose a solution is prepared by mixing one soluble salt, such as $AgNO_3$, with a solution of a second, such as NaCl. Will a precipitate of very slightly soluble

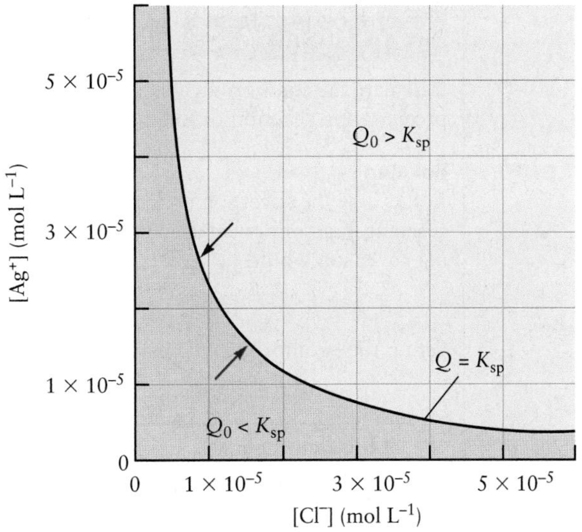

FIGURE 16.6 Some solid silver chloride is in contact with a solution containing $Ag^+(aq)$ and $Cl^-(aq)$ ions. If a solubility equilibrium exists, then the product Q of the concentrations of the ions $[Ag^+] \times [Cl^-]$ is a constant, K_{sp} (curved line). When Q exceeds K_{sp}, solid silver chloride tends to precipitate until equilibrium is attained. When Q is less than K_{sp}, additional solid tends to dissolve. If no solid is present, Q remains less than K_{sp}. The red and blue arrows represent the path toward equilibrium for a 1:1 salt like AgCl; the system moves along a path with slope = 1.

silver chloride form? The answer is found by examining the reaction quotient Q that was defined in Section 14.6. The initial reaction quotient Q_0, when the mixing of the solutions is complete but before any reaction occurs, is

$$Q_0 = [Ag^+]_0[Cl^-]_0$$

If $Q_0 < K_{sp}$, no solid silver chloride can appear. On the other hand, if $Q_0 > K_{sp}$, solid silver chloride precipitates until the reaction quotient Q reaches K_{sp} (Fig. 16.6).

EXAMPLE 16.3

An emulsion of silver chloride for photographic film is prepared by adding a soluble chloride salt to a solution of silver nitrate. Suppose 500 mL of a solution of $CaCl_2$ with a chloride ion concentration of 8.0×10^{-6} M is added to 300 mL of a 0.0040 M solution of $AgNO_3$. Will a precipitate of AgCl(s) form when equilibrium is reached?

Solution

The "initial concentrations" to be used in calculating Q_0 are those *before* reaction but *after* dilution through mixing the two solutions. The initial concentration of $Ag^+(aq)$ after dilution from 300 mL to 800 mL of solution is

$$[Ag^+]_0 = 0.00400 \text{ M} \times \left(\frac{300 \text{ mL}}{800 \text{ mL}}\right) = 0.0015 \text{ M}$$

and that of $Cl^-(aq)$ is

$$[Cl^-]_0 = 8.0 \times 10^{-6} \text{ M} \times \left(\frac{500 \text{ mL}}{800 \text{ mL}}\right) = 5.0 \times 10^{-6} \text{ M}$$

The initial reaction quotient is

$$Q_0 = [Ag^+]_0[Cl^-]_0 = (0.0015)(5.0 \times 10^{-6}) = 7.5 \times 10^{-9}$$

Because $Q_0 > K_{sp}$, a precipitate of silver chloride appears at equilibrium, although there may be too little to detect visually. Another possible precipitate, calcium nitrate, is far too soluble to form in this experiment (see Table 16.1).

Related Problems: 17, 18, 19, 20

The equilibrium concentrations of ions after the mixing of two solutions to give a precipitate are most easily calculated by supposing that the reaction first goes to completion (consuming one type of ion) and that subsequent dissolution of the solid restores some of that ionic species to solution—just the approach used in Example 15.11 for the addition of a strong acid to a buffer solution.

EXAMPLE 16.4

Calculate the equilibrium concentrations of silver and chloride ions resulting from the precipitation reaction of Example 16.3.

Solution

In this case the silver ion is clearly in excess; therefore, the chloride ion is the limiting reactant. If all of it were used up to make solid AgCl, the concentration of the remaining silver ion would be

$$[Ag^+] = 0.0015 - 5.0 \times 10^{-6} = 0.0015 \text{ M}$$

Set up the equilibrium calculation as

	$AgCl(s) \rightleftharpoons$	$Ag^+(aq)$ +	$Cl^-(aq)$
Initial concentration (M)		0.0015	0
Change in concentration (M)		$+y$	$+y$
Equilibrium concentration (M)		$0.0015 + y$	y

so that the equilibrium expression is

$$(0.0015 + y)y = K_{sp} = 1.6 \times 10^{-10}$$

This quadratic equation can be solved using the quadratic formula function on your calculator but *only* if it carries 10 significant figures. It can be solved more easily by making the approximation that y is much smaller than 0.0015; the equation therefore simplifies to

$$0.0015y \approx 1.6 \times 10^{-10}$$

$$y \approx 1.1 \times 10^{-7} \text{ M} = [Cl^-]$$

The assumption about the size of y was justified. The concentration of silver ion is

$$[Ag^+] = 0.0015 \text{ M}$$

Related Problems: 21, 22, 23, 24

16.3.2 The Common-Ion Effect

Suppose a small amount of NaCl(s) is added to a saturated solution of AgCl. What happens? Sodium chloride is quite soluble in water and dissolves to form $Na^+(aq)$ and $Cl^-(aq)$ ions, raising the concentration of chloride ion. The quantity $Q_0 = [Ag^+][Cl^-]$ then exceeds the K_{sp} of silver chloride, and silver chloride precipitates until the concentrations of $Ag^+(aq)$ and $Cl^-(aq)$ are sufficiently reduced that the solubility product expression once again is satisfied.

The same equilibrium may be approached from the other direction. The amount of AgCl(s) that can dissolve in a solution of sodium chloride is less than the amount that could dissolve in the same volume of pure water. Because $[Ag^+][Cl^-] = K_{sp}$, a graph of the equilibrium concentration of silver ion against chloride concentration has the form of a hyperbola (Fig. 16.7). The presence of excess $Cl^-(aq)$ reduces the concentration of $Ag^+(aq)$ permitted, and the solubility of AgCl(s) is reduced. In the same way, the prior presence of $Ag^+(aq)$ in the solvent (for example, when an attempt is made to dissolve AgCl in water that already contains $AgNO_3$) reduces the amount of $Cl^-(aq)$ permitted at equilibrium and also reduces the solubility of AgCl. This is referred to as the **common-ion effect:** If the solution and the solid salt to be dissolved in it have an ion in common, then the solubility of the salt is depressed.

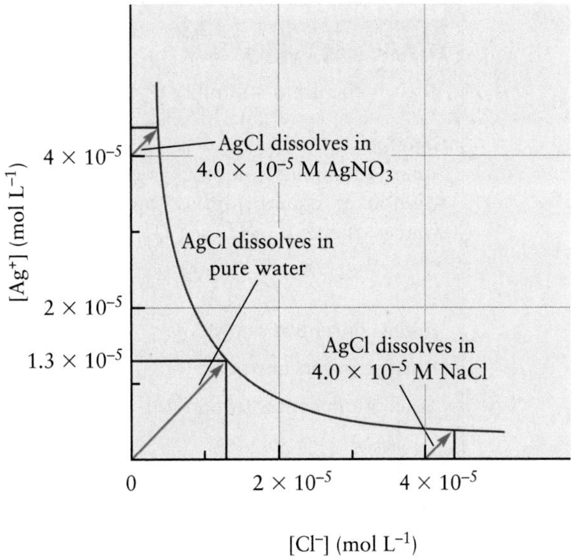

FIGURE 16.7 The presence of a dissolved common ion reduces the solubility of a salt in solution. As the AgCl dissolves, the concentrations of the ions follow the paths shown by the green arrows until they reach the red equilibrium curve. The molar solubilities are proportional to the lengths of the blue lines: 1.3×10^{-5} mol L^{-1} for AgCl in pure water, but only 0.37×10^{-5} mol L^{-1} in either 4.0×10^{-5} M $AgNO_3$ or 4.0×10^{-5} M NaCl.

Let's examine the quantitative consequences of the common-ion effect. Suppose an excess of AgCl(*s*) is added to 1.00 L of a 0.100 M NaCl solution and the solubility is again determined. If *S* mol of AgCl dissolves per liter, the concentration of $Ag^+(aq)$ will be *S* mol L^{-1} and that of $Cl^-(aq)$ will be

$$[Cl^-] = 0.100 + S$$

because the chloride ion comes from two sources: the 0.100 M NaCl and the dissolution of AgCl. The expression for the solubility product is written as

$$[Ag^+][Cl^-] = S(0.100 + S) = K_{sp} = 1.6 \times 10^{-10}$$

The solubility of AgCl(*s*) in this solution must be smaller than it is in pure water, which is much smaller than 0.100. That is,

$$S < 1.3 \times 10^{-5} \ll 0.100$$

Thus, (0.100 + *S*) can be approximated by 0.100 (as in Example 16.4), giving

$$(0.100)S \approx 1.6 \times 10^{-10}$$
$$S \approx 1.6 \times 10^{-9}$$

This is indeed much smaller than 0.100, so the approximation was a very good one. Therefore, at equilibrium,

$$[Ag^+] = S = 1.6 \times 10^{-9} \text{ M}$$
$$[Cl^-] = 0.100 \text{ M}$$

The gram solubility of AgCl in this example is

$$(1.6 \times 10^{-9} \text{ mol L}^{-1})(143.3 \text{ g mol}^{-1}) = 2.3 \times 10^{-7} \text{ g L}^{-1}$$

The solubility of AgCl in 0.100 M NaCl is lower than that in pure water by a factor of about 8000.

EXAMPLE 16.5

What is the gram solubility of $CaF_2(s)$ in a 0.100 M solution of NaF?

Solution

Again, the molar solubility is denoted by S. The only source of the $Ca^{2+}(aq)$ in the solution at equilibrium is the dissolution of CaF_2, whereas the $F^-(aq)$ has two sources, the CaF_2 and the NaF. Hence,

	$CaF_2(s) \rightleftharpoons Ca^{2+}(aq)$	+	$2\ F^-(aq)$
Initial concentration (M)	0		0.100
Change in concentration (M)	$+S$		$+2S$
Equilibrium concentration (M)	S		$0.100 + 2S$

If $0.100 + 2S$ is approximated as 0.100, then

$$[Ca^{2+}][F^-]^2 = K_{sp}$$

$$S(0.100)^2 = 3.9 \times 10^{-11}$$

$$S = 3.9 \times 10^{-9}$$

Clearly,

$$2S = 7.8 \times 10^{-9} \ll 0.100$$

so the assumption was justified. The gram solubility of CaF_2 is

$$(3.9 \times 10^{-9}\ \text{mol L}^{-1})(78.1\ \text{g mol}^{-1}) = 3.0 \times 10^{-7}\ \text{g L}^{-1}$$

and the solubility in this case is reduced by a factor of 50,000.

Related Problems: 25, 26, 27, 28

16.4 The Effects of pH on Solubility

Some solids are only weakly soluble in water but dissolve readily in acidic solutions. Copper and nickel sulfides from ores, for example, can be brought into solution with strong acids, a fact that aids greatly in the separation and recovery of these valuable metals in their elemental forms. The effect of pH on solubility is shown dramatically in the damage done to buildings and monuments by acid rain (Fig. 16.8). Both marble and limestone are made up of small crystals of calcite ($CaCO_3$), which dissolves to only a limited extent in "natural" rain (with a pH of about 5.6) but dissolves much more extensively as the rainwater becomes more acidic. The reaction

$$CaCO_3(s) + H_3O^+(aq) \longrightarrow Ca^{2+}(aq) + HCO_3^-(aq) + H_2O(\ell)$$

causes this increase. This section examines the role of pH in solubility.

16.4.1 Solubility of Hydroxides

One direct effect of pH on solubility occurs with the metal hydroxides. The concentration of OH^- appears explicitly in the expression for the solubility product of such compounds. Thus, for the dissolution of $Zn(OH)_2(s)$,

$$Zn(OH)_2(s) \rightleftharpoons Zn^{2+}(aq) + 2\ OH^-(aq)$$

FIGURE 16.8 The calcium carbonate in marble and limestone is very slightly soluble in neutral water. Its solubility is much greater in acidic water. Objects carved of these materials dissolve relatively rapidly in areas where rain, snow, or fog is acidified from air pollution. Shown here is the damage to a marble statue of George Washington between 1935 (left) and 1994 (right).

the solubility product expression is

$$[Zn^{2+}][OH^-]^2 = K_{sp} = 4.5 \times 10^{-17}$$

As the solution is made more acidic, the concentration of hydroxide ion decreases, causing an increase in the concentration of $Zn^{2+}(aq)$ ion. Zinc hydroxide is thus more soluble in acidic solution than in pure water.

EXAMPLE 16.6

Compare the solubility of $Zn(OH)_2$ in pure water with that in a solution buffered at pH 6.00.

Solution

In pure water the usual solubility product calculation applies:

$$[Zn^{2+}] = S \qquad [OH^-] = 2S$$

$$S(2S)^2 = 4S^3 = K_{sp} = 4.5 \times 10^{-17}$$

$$S = 2.2 \times 10^{-6} \text{ M} = [Zn^{2+}]$$

so the solubility is 2.2×10^{-6} mol L^{-1}, or 2.2×10^{-4} g L^{-1}. Using

$$[OH^-] = 2S = 4.5 \times 10^{-6} \text{ M}$$

the resulting solution is found to have pH 8.65.

In the second case it is assumed that the solution is buffered sufficiently that the pH remains 6.00 after dissolution of the zinc hydroxide. Then

$$[OH^-] = 1.0 \times 10^{-8} \text{ M}$$

$$[Zn^{2+}] = \frac{K_{sp}}{[OH^-]^2} = \frac{4.5 \times 10^{-17}}{(1.0 \times 10^{-8})^2} = 0.45 \text{ M}$$

so that 0.45 mol L^{-1}, or 45 g L^{-1}, should dissolve in this case. When ionic concentrations are this high, the simple form of the solubility expression will likely break down, but the qualitative conclusion is still valid: $Zn(OH)_2$ is far more soluble at pH 6.00 than in pure water.

Related Problems: 31, 32

16.4.2 Solubility of Salts of Bases

Metal hydroxides can be described as salts of a strong base, the hydroxide ion. The solubility of salts in which the anion is a different weak or strong base is also affected by pH. For example, consider a solution of a slightly soluble fluoride, such as calcium fluoride. The solubility equilibrium is

$$CaF_2(s) \rightleftharpoons Ca^{2+}(aq) + 2\,F^-(aq) \qquad K_{sp} = 3.9 \times 10^{-11}$$

As the solution is made more acidic, some of the fluoride ion reacts with hydronium ion through

$$H_3O^+(aq) + F^-(aq) \rightleftharpoons HF(aq) + H_2O(\ell)$$

Because this is just the reverse of the acid ionization of HF, its equilibrium constant is the reciprocal of K_a for HF, or $1/(3.5 \times 10^{-4}) = 2.9 \times 10^3$. As acid is added, the concentration of fluoride ion is reduced, so the calcium ion concentration must increase to maintain the solubility product equilibrium for CaF_2. As a result, the solubility of fluoride salts increases in acidic solution. The same applies to other ionic substances in which the anion is a weak or a strong base. By contrast, the solubility of a salt such as AgCl is only very slightly affected by a decrease in pH. The reason is that HCl is a strong acid, so Cl^- is ineffective as a base. The reaction

$$Cl^-(aq) + H_3O^+(aq) \longrightarrow HCl(aq) + H_2O(\ell)$$

occurs to a negligible extent in acidic solution.

16.5 Complex Ions and Solubility

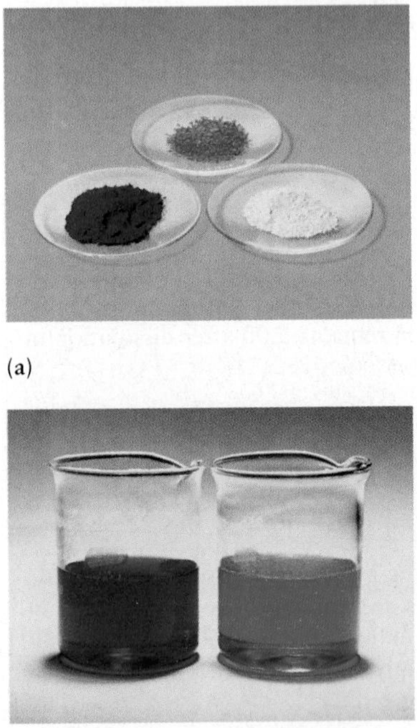

(a)

(b)

FIGURE 16.9 (a) From left to right, crystals of $Cu(NH_3)_4SO_4$, $CuSO_4{\cdot}5H_2O$, and $CuSO_4$. (b) Aqueous solutions of copper sulfate containing (*left*) and not containing (*right*) ammonia.

A transition-metal ion may form a **coordination complex** in solution or in the solid state; in the complex the metal ion is surrounded by a group of anions or neutral molecules called **ligands.** (See Section 8.2.) The interaction involves the sharing by the metal ion of a lone pair on each ligand molecule, giving a partially covalent bond with that ligand. Such complexes often have strikingly deep colors. When exposed to gaseous ammonia, greenish white crystals of copper sulfate ($CuSO_4$) give a deep blue crystalline solid with the chemical formula $[Cu(NH_3)_4]SO_4$ (Fig. 16.9). The anions in the solid are still sulfate ions (SO_4^{2-}) but the cations are now **complex ions,** or coordination complexes of the central Cu^{2+} ion with four ammonia molecules, each coordination complex is represented by the molecular formula $Cu(NH_3)_4^{2+}$. The ammonia molecules in this complex are called ammine ligands. They coordinate to the copper ion through their lone-pair electrons (Fig. 16.10), acting as Lewis bases toward the metal ion, the Lewis acid. When the solid is dissolved in water, the deep blue color remains. This is evidence that the complex persists in water, because when ordinary $CuSO_4$ (without ammonia ligands) is dissolved in water, a much paler blue color results (see Fig. 16.9b).

Here we explore the effects of the formation of complex ions on equilibria in aqueous solutions. The microscopic structure and bonding in these complexes is presented in Chapter 8.

16.5.1 Complex-Ion Equilibria

When silver ions are dissolved in an aqueous ammonia solution, doubly coordinated silver–ammonia complexes, shown in Figure 16.11, form in two stepwise reactions:

$$Ag^+(aq) + NH_3(aq) \rightleftharpoons Ag(NH_3)^+(aq)$$

$$\frac{[Ag(NH_3)^+]}{[Ag^+][NH_3]} = K_1 = 2.1 \times 10^3$$

TABLE 16.3

Formation Constants of Coordination Complexes in Aqueous Solution

	K_f	K_1	K_2	K_3	K_4	K_5	K_6
Ammines							
$Ag(NH_3)_2^+$	1.7×10^7	2.1×10^3	8.2×10^3				
$Co(NH_3)_6^{2+}$	2.5×10^4	1.0×10^2	32	8.5	4.4	1.1	0.18
$Cu(NH_3)_4^{2+}$	1.1×10^{12}	1.0×10^4	2×10^3	5×10^2	90		
$Ni(NH_3)_6^{2+}$	8×10^{-7}	5×10^2	1.3×10^2	40	12	3.3	0.8
$Zn(NH_3)_4^{2+}$	5×10^8	1.5×10^2	1.8×10^2	2×10^2	90		
Chlorides							
$AgCl_2^-$	1.8×10^5	1.7×10^3	1.0×10^2				
$FeCl_4^-$	0.14	28	4.5	0.1	1.1×10^2		
$HgCl_4^{2-}$	1.2×10^{15}	5.5×10^6	3×10^6	7	10		
$PbCl_4^{2-}$	25	40	1.5	0.8	0.5		
$SnCl_4^{2-}$	30	32	5.4	0.6	0.3		
Hydroxides							
$Co(OH)_3^-$	3×10^{10}	4×10^4	1	8×10^5			
$Cu(OH)_4^{2-}$	3×10^{18}	1×10^7	5×10^6	2×10^3	30		
$Ni(OH)_3^-$	2×10^{11}	9×10^4	4×10^3	6×10^2			
$Pb(OH)_3^-$	4×10^{14}	7×10^7	1.1×10^3	5×10^3			
$Zn(OH)_4^{2-}$	5×10^{14}	2.5×10^4	8×10^6	70	33		

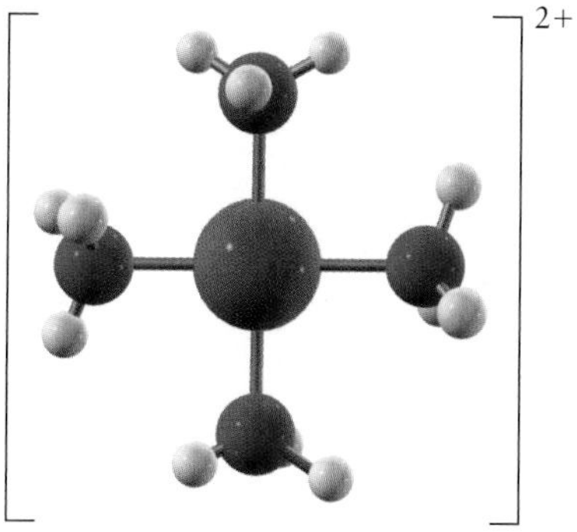

FIGURE 16.10 The structure of $Cu(NH_3)_4^{2+}$.

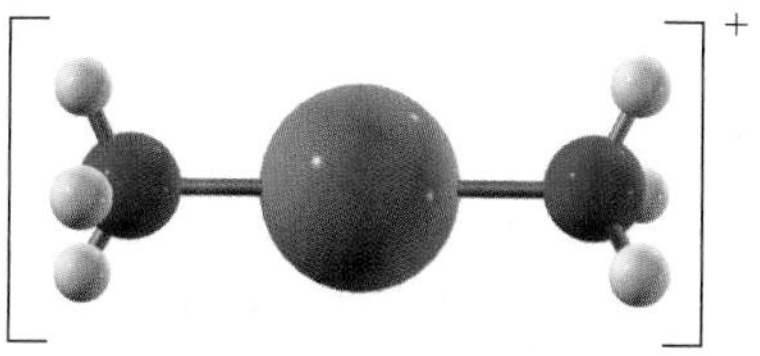

FIGURE 16.11 The structure of $Ag(NH_3)_2^+$.

$$Ag(NH_3)^+(aq) + NH_3(aq) \rightleftharpoons Ag(NH_3)_2^+(aq)$$

$$\frac{[Ag(NH_3)_2^+]}{[Ag(NH_3)^+][NH_3]} = K_2 = 8 \times 10^3$$

If these two chemical equations are added (and their corresponding equilibrium laws are multiplied), the result is

$$Ag^+(aq) + 2\ NH_3(aq) \rightleftharpoons Ag(NH_3)_2^+(aq)$$

$$\frac{[Ag(NH_3)_2^+]}{[Ag^+][NH_3]^2} = K_f = K_1K_2 = 1.7 \times 10^7$$

where K_f is the **formation constant** of the full complex ion $Ag(NH_3)_2^+$. Table 16.3 lists formation constants for a representative selection of complex ions. The larger the formation constant K_f, the more stable the corresponding complex ion (for ions with the same number of ligands).

Because K_1 and K_2 of the $Ag(NH_3)_2^+$ complex ion are both large, a silver salt dissolved in water that contains a high concentration of ammonia will be primarily in the form of the complex ion $[Ag(NH_3)_2]^+$ at equilibrium.

EXAMPLE 16.7

Suppose 0.100 mol of $AgNO_3$ is dissolved in 1.00 L of a 1.00 M in solution of NH_3. Calculate the concentrations of the Ag^+ and $Ag(NH_3)^+$ ions present at equilibrium.

Solution

Suppose that most of the Ag^+ is present as $Ag(NH_3)_2^+$ (this will be checked later). Then

$$[Ag(NH_3)_2^+]_0 = 0.100\ M$$

$$[NH_3]_0 = 1.00\ M - (2 \times 0.100)\ M = 0.80\ M$$

(Continued)

after each silver ion has become complexed with two ammonia molecules. The two stages of the dissociation of the $Ag(NH_3)_2^+$ ion are the reverse reaction of the complexation, so their equilibrium constants are the reciprocals of K_2 and K_1, respectively:

$$Ag(NH_3)_2^+(aq) \rightleftharpoons Ag(NH_3)^+(aq) + NH_3(aq)$$

$$\frac{[Ag(NH_3)^+][NH_3]}{[Ag(NH_3)_2^+]} = \frac{1}{K_2} = \frac{1}{8.2 \times 10^3}$$

$$Ag(NH_3)^+(aq) \rightleftharpoons Ag^+(aq) + NH_3(aq)$$

$$\frac{[Ag^+][NH_3]}{[Ag(NH_3)^+]} = \frac{1}{K_1} = \frac{1}{2.1 \times 10^3}$$

If y mol L^{-1} of $Ag(NH_3)_2^+$ dissociates at equilibrium according to the first equation (and not many of the $Ag(NH_3)^+$ dissociate via the second equation).

	$Ag(NH_3)_2^+(aq)$ $\rightleftharpoons$	$Ag(NH_3)^+(aq)$ +	$NH_3(aq)$
Initial concentration (M)	0.100	0	0.80
Change in concentration (M)	$-y$	$+y$	$+y$
Equilibrium concentration (M)	$0.100 - y$	y	$0.80 + y$

then the first equilibrium expression becomes

$$\frac{y(0.80 + y)}{0.10 - y} = \frac{1}{K_2} = \frac{1}{8.2 \times 10^3}$$

$$y = 1.5 \times 10^{-5}\ M = [Ag(NH_3)^+]$$

We can then calculate the concentration of free Ag^+ ions from the equilibrium law for the second step of the dissociation of the complex ion:

$$\frac{[Ag^+][NH_3]}{[Ag(NH_3)^+]} = \frac{1}{K_1}$$

$$\frac{[Ag^+](0.80)}{1.5 \times 10^{-5}} = \frac{1}{2.1 \times 10^3}$$

$$[Ag^+] = 9 \times 10^{-9}\ M$$

It is clear that the original assumption was correct, and most of the silver present is tied up in the $Ag(NH_3)_2^+$ complex.

Related Problems: 35, 36

There is a close similarity between the working of Example 16.7 and an acid–base calculation. The first step (the assumption that the reaction goes to completion and is followed by a small amount of back dissociation) is analogous to the procedure for dealing with the addition of a small amount of a strong acid to a solution of a weak base. The subsequent calculation of the successive dissociation steps resembles the calculation of polyprotic acid equilibria in Example 15.15. The only difference is that in complex-ion equilibria it is conventional to work with formation constants, which are the inverse of the dissociation constants used in acid–base equilibria.

The formation of coordination complexes can have a large effect on the solubility of a compound in water. Silver bromide is only very weakly soluble in water,

$$AgBr(s) \rightleftharpoons Ag^+(aq) + Br^-(aq) \qquad K_{sp} = 7.7 \times 10^{-13}$$

but addition of thiosulfate ion ($S_2O_3^{2-}$) (Figure 16.12) to the solution allows the complex ion $Ag(S_2O_3)_2^{3-}$ to form:

$$AgBr(s) + 2\ S_2O_3^{2-}(aq) \rightleftharpoons Ag(S_2O_3)_2^{3-}(aq) + Br^-(aq)$$

FIGURE 16.12 The thiosulfate ion (top) looks structurally similar to the sulfate ion SO_4^{2-} (bottom), but has one oxygen atom replaced by a sulfur atom. It is prepared, however, by the reaction of elemental sulfur with the sulfite ion (SO_3^{2-}).

FIGURE 16.13 An illustration of the effect of complex ion formation on solubility. Each test tube contains 2.0 g AgBr, but the one on the left also contains dissolved thiosulfate ion, which forms a complex ion with Ag^+. Almost none of the white solid AgBr has dissolved in pure water, but almost all of it has dissolved in the solution containing thiosulfate.

This greatly increases the solubility of the silver bromide (Fig. 16.13). The formation of this complex ion is an important step in the development of photographic images; thiosulfate ion is a component of the fixer that brings silver bromide into solution from the unexposed portion of the film.

EXAMPLE 16.8

Calculate the solubility of AgBr in a 1.00 M aqueous solution of ammonia.

Solution

We tentatively assume that almost all the silver that dissolves is complexed as $Ag(NH_3)_2^+$ (this will be checked later). The overall reaction is then

$$\text{AgBr}(s) + 2\ \text{NH}_3(aq) \rightleftharpoons \text{Ag(NH}_3)_2^+(aq) + \text{Br}^-(aq)$$

Note that this is the sum of the two reactions

$$\text{AgBr}(s) \rightleftharpoons \text{Ag}^+(aq) + \text{Br}^-(aq) \qquad K_{sp} = 7.7 \times 10^{-13}$$

$$\text{Ag}^+(aq) + 2\ \text{NH}_3(aq) \rightleftharpoons \text{Ag(NH}_3)_2^+(aq) \qquad K_f = 1.7 \times 10^7$$

so its equilibrium constant is the product $K_{sp}K_f = 1.3 \times 10^{-5}$.

If S mol L^{-1} of AgBr dissolves, then

$$S = [\text{Br}^-] \approx [\text{Ag(NH}_3)_2^+]$$

$$[\text{NH}_3] = 1.00 - 2S$$

because 2 mol of NH_3 is used up for each mole of complex formed. The equilibrium expression is

$$\frac{S^2}{(1.00 - 2S)^2} = K_{sp}K_f = 1.3 \times 10^{-5}$$

$$S = 3.6 \times 10^{-3}\ \text{M} = [\text{Ag(NH}_3)_2^+] = [\text{Br}^-]$$

To check the original assumption, calculate the concentration of free silver ion:

$$[\text{Ag}^+] = \frac{K_{sp}}{[\text{Br}^-]} = \frac{K_{sp}}{S} = 2.1 \times 10^{-10} \ll [\text{Ag(NH}_3)_2^+]$$

verifying that almost all the silver is complexed. The solubility is therefore 3.6×10^{-3} mol L^{-1}, significantly greater than the solubility in pure water:

$$\text{solubility in pure water} = \sqrt{K_{sp}} = 8.8 \times 10^{-7}\ \text{mol L}^{-1}$$

Related Problems: 39, 40

Another interesting effect of complex ions on solubilities is illustrated by the addition of iodide ion to a solution containing mercury(II) ion. After a moderate amount of iodide ion has been added, an orange precipitate forms (Fig. 16.14) through the reaction

$$\text{Hg}^{2+}(aq) + 2\ \text{I}^-(aq) \rightleftharpoons \text{HgI}_2(s)$$

With further addition of iodide ion, however, the orange solid redissolves because complex ions form:

$$\text{HgI}_2(s) + \text{I}^-(aq) \rightleftharpoons \text{HgI}_3^-(aq)$$

$$\text{HgI}_3^-(aq) + \text{I}^-(aq) \rightleftharpoons \text{HgI}_4^{2-}(aq)$$

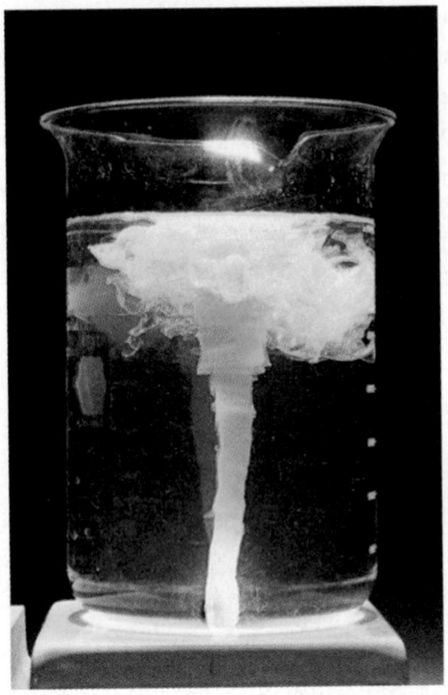

FIGURE 16.14 The "orange tornado" is a striking demonstration of the effects of complex ions on solubility. A solution is prepared with an excess of $I^-(aq)$ over $Hg^{2+}(aq)$ so that the Hg^{2+} is complexed as HgI_3^- and HgI_4^{2-}. A magnetic stirrer is used to create a vortex in the solution. Addition of a solution containing Hg^{2+} down the center of the vortex then causes the orange solid HgI_2 to form in a layer at the edges of the vortex, giving the tornado effect.

In the same way, silver chloride will dissolve in a concentrated solution of sodium chloride by forming soluble $AgCl_2^-$ complex ions. Complex ion formation affects solubility in the opposite direction from the common-ion effect of Section 16.3.

16.5.2 Acidity and Amphoterism of Complex Ions

When dissolved in water, many metal ions increase the acidity of the solution. The iron(III) ion is an example: Each dissolved Fe^{3+} ion is strongly solvated by six water molecules, leading to a complex ion $Fe(H_2O)_6^{3+}$. This complex ion can act as a Brønsted–Lowry acid, donating hydrogen ions to the solvent, water:

$$\underset{\text{Acid}_1}{Fe(H_2O)_6^{3+}(aq)} + \underset{\text{Base}_2}{H_2O(\ell)} \rightleftharpoons \underset{\text{Acid}_2}{H_3O^+(aq)} + \underset{\text{Base}_1}{Fe(H_2O)_5OH^{2+}(aq)}$$

$$\frac{[H_3O^+][Fe(H_2O)_5OH^{2+}]}{[Fe(H_2O)_6^{3+}]} = K_a = 7.7 \times 10^{-3}$$

Metal ion hydrolysis fits into the general scheme of the Brønsted–Lowry acid–base reaction.

> **EXAMPLE 16.9**
>
> Calculate the pH of a solution that is 0.100 M in $Fe(NO_3)_3$.
>
> **Solution**
>
> The iron(III) is present as $Fe(H_2O)_6^{3+}$, which reacts as a weak acid:
>
> $$Fe(H_2O)_6^{3+}(aq) + H_2O(\ell) \rightleftharpoons Fe(H_2O)_5OH^{2+}(aq) + H_3O^+(aq)$$
>
> with K_a equal to 7.7×10^{-3}. If y mol L^{-1} of $[Fe(H_2O)_6]^{3+}$ reacts, then (neglecting the ionization of water itself)
>
> $$[H_3O^+] = [Fe(H_2O)_5OH^{2+}] = y$$
>
> $$[Fe(H_2O)_6^{3+}] = 0.100 - y$$
>
> The equilibrium expression has the form
>
> $$\frac{y^2}{0.100 - y} = 7.7 \times 10^{-3}$$
>
> $$y = 2.4 \times 10^{-2}\ M = [H_3O^+]$$
>
> so the pH is 1.62. Solutions of iron(III) salts are strongly acidic.
>
> **Related Problems: 43, 44**

TABLE 16.4

pH of 0.1 M Aqueous Metal Nitrate Solutions at 25°C

Metal Nitrate	pH
$Fe(NO_3)_3$	1.6
$Pb(NO_3)_2$	3.6
$Cu(NO_3)_2$	4.0
$Zn(NO_3)_2$	5.3
$Ca(NO_3)_2$	6.7
$NaNO_3$	7.0

Another acceptable way to write the reaction that makes iron(III) solutions acidic is

$$Fe^{3+}(aq) + 2\ H_2O(\ell) \rightleftharpoons H_3O^+(aq) + FeOH^{2+}(aq)$$

in which the specific mention of the six waters of hydration is now omitted. The $FeOH^{2+}$ complex ions are brown, but Fe^{3+} ions are almost colorless. This reaction occurs to a sufficient extent to make a solution of $Fe(NO_3)_3$ in water pale brown. When strong acid is added, the equilibrium is driven back to the left and the color fades. Table 16.4 gives values of the pH for 0.1 M solutions of several metal ions. Those that form strong complexes with hydroxide ion have low pH

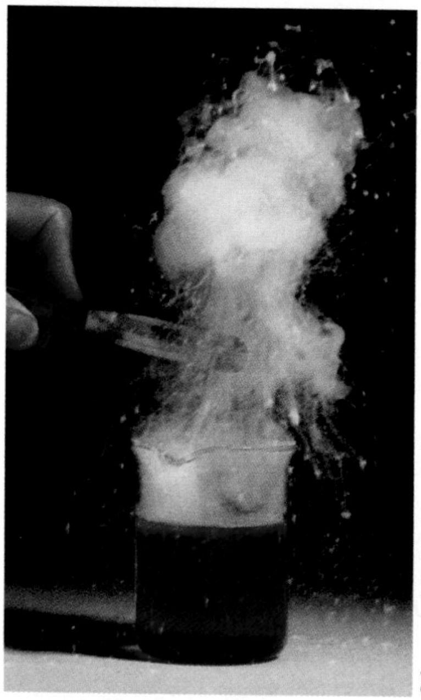

FIGURE 16.15 The acidity of metal ions is illustrated by the vigorous reaction of anhydrous $AlCl_3$ with water to generate hydrated aluminum oxides, HCl(*aq*), and heat. The HCl turns the indicator, bromocresol red, to its red acid form.

(Fig. 16.15), whereas those that do not form such complexes give neutral solutions (pH 7).

Different cations behave differently as water ligands are replaced by hydroxide ions in an increasingly basic solution. A particularly interesting example is Zn^{2+}. It forms a series of hydroxo complex ions:

$$Zn^{2+}(aq) + OH^-(aq) \rightleftharpoons ZnOH^+(aq)$$

$$ZnOH^+(aq) + OH^-(aq) \rightleftharpoons Zn(OH)_2(s)$$

$$Zn(OH)_2(s) + OH^-(aq) \rightleftharpoons Zn(OH)_3^-(aq)$$

$$Zn(OH)_3^-(aq) + OH^-(aq) \rightleftharpoons Zn(OH)_4^{2-}(aq)$$

In the Brønsted–Lowry theory, a polyprotic acid, $Zn(H_2O)_4^{2+}(aq)$, donates hydrogen ions in succession to make all the product ions. The second product, $Zn(OH)_2$, is amphoteric; it can react as either acid or base. It is only slightly soluble in pure water (its K_{sp} is only 1.9×10^{-17}). If enough acid is added to solid $Zn(OH)_2$, the OH^- ligands are removed, forming the soluble Zn^{2+} ion; if enough base is added, OH^- ligands attach to form the soluble $Zn(OH)_4^{2+}$ (zincate) ion. Thus, $Zn(OH)_2$ is soluble in strongly acidic *or* strongly basic solutions but is only slightly soluble at intermediate pH values (Fig. 16.16). This amphoterism can be used to separate Zn^{2+} from other cations that do not share the property. For example, Mg^{2+} adds a maximum of two OH^- ions to form $Mg(OH)_2$, a sparingly soluble hydroxide. Further addition of OH^- does not lead to the formation of new complex ions. If a mixture of Mg^{2+} and Zn^{2+} ions is made sufficiently basic, the Mg^{2+} precipitates as $Mg(OH)_2$ but the zinc remains in solution as $Zn(OH)_4^{2-}$, allowing the two to be separated. In the same way, aluminum is separated from iron industrially by dissolving solid $Al(OH)_3$ in strong base as $Al(OH)_4^-(aq)$ while $Fe(OH)_3$ remains as a precipitate.

16.6 A DEEPER LOOK Selective Precipitation of Ions

FIGURE 16.16 Zinc hydroxide is insoluble in water (*center*) but dissolves readily in acid (*left*) and base (*right*). The indicator used is bromocresol red, which turns from red to yellow in acidic solution.

One way to analyze a mixture of ions in solution is to separate the mixture into its components by exploiting the differences in the solubilities of compounds containing the ions. To separate silver ions from lead ions, for example, a search is made for compounds of these elements that (1) have a common anion and (2) have widely different solubilities. The chlorides AgCl and $PbCl_2$ are two such compounds, for which the solubility equilibria are

$$AgCl(s) \rightleftharpoons Ag^+(aq) + Cl^-(aq) \qquad K_{sp} = 1.6 \times 10^{-10}$$

$$PbCl_2(s) \rightleftharpoons Pb^{2+}(aq) + 2\,Cl^-(aq) \qquad K_{sp} = 2.4 \times 10^{-4}$$

Lead chloride is far more soluble in water than is silver chloride. Consider a solution that is 0.10 M in both Ag^+ and Pb^{2+}. Is it possible to add enough Cl^- to precipitate almost all the Ag^+ ions but leave all the Pb^{2+} ions in solution? If so, a quantitative separation of the two species can be achieved.

For Pb^{2+} to remain in solution, its reaction quotient must remain smaller than K_{sp}: $Q = [Pb^{2+}][Cl^-]^2 < K_{sp}$. Inserting K_{sp} and the concentration of Pb^{2+} gives

$$[Cl^-]^2 < \frac{K_{sp}}{[Pb^{2+}]} = \frac{2.4 \times 10^{-4}}{0.10} = 2.4 \times 10^{-3}$$

The square root of this is

$$[Cl^-] < 4.9 \times 10^{-2}\ M$$

Thus, as long as the chloride ion concentration remains smaller than 0.049 M, no $PbCl_2$ should precipitate. To reduce the *silver* ion concentration in solution as far as possible (that is, to precipitate out as much silver chloride as possible), the chloride ion concentration should be kept as high as possible without exceeding 0.049 M. If exactly this concentration of $Cl^-(aq)$ is chosen, then at equilibrium,

$$[Ag^+] = \frac{K_{sp}}{[Cl^-]} = \frac{1.6 \times 10^{-10}}{0.049} = 3.3 \times 10^{-9}\ M$$

At that concentration of Cl^-, the concentration of Ag^+ has been reduced to 3.3×10^{-9} from the original concentration of 0.10 M. In other words, only about three Ag^+ ions in 10^8 remain in solution, but all the Pb^{2+} ions are left in solution. A nearly perfect separation of the two ionic species has been achieved.

This calculation gave the optimal theoretical separation factor for the two ions. In practice it is necessary to keep the chloride concentration lower. If $[Cl^-]$ is ten times smaller, or 0.0049 M, about three Ag^+ ions in 10^7 remain in solution with the Pb^{2+}. This is 10 times more Ag^+ than if $[Cl^-] = 0.049$ M, but the separation of Ag^+ from Pb^{2+} is still very good.

Figure 16.17 shows graphically how ions can be separated based on solubility. The relation between the concentration of Pb^{2+} and Cl^- ions in contact with solid $PbCl_2$

$$[Pb^{2+}] = \frac{K_{sp}}{[Cl^-]^2}$$

can be rewritten in a useful form by taking the common logarithms of both sides:

$$\log_{10}[Pb^{2+}] = -2\ \log_{10}[Cl^-] + \log_{10}K_{sp}$$

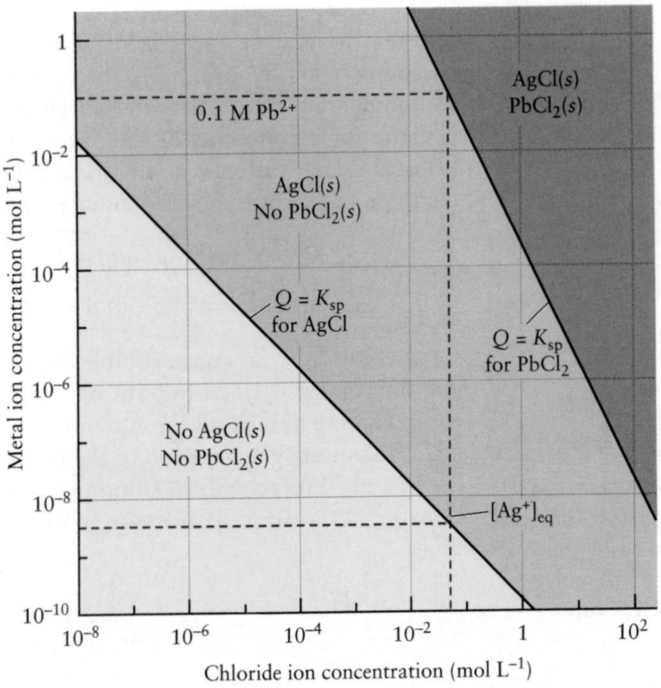

FIGURE 16.17 To separate a mixture of Ag^+ and Pb^{2+} ions, a chloride ion concentration is selected that gives a Pb^{2+} concentration below the equilibrium curve for $PbCl_2$ (so all Pb^{2+} remains in solution) but well above the equilibrium curve for Ag^+. As a result, nearly all the Ag^+ precipitates as AgCl. If $[Pb^{2+}]_0 = [Ag^+]_0 = 0.1$ M, then the maximum $[Cl^-]$ is found by tracing the horizontal red line and then dropping down the vertical red line to find $[Cl^-] = 0.049$ M. The concentration of Ag^+ still in solution is found by tracing the horizontal blue line from the intersection of the $[Cl^-]$ line of the AgCl equilibrium curve back to the vertical axis where $[Ag^+] = 3.3 \times 10^{-9}$ M.

That is, a graph of $\log_{10} [Pb^{2+}]$ against $\log_{10} [Cl^-]$ is a straight line with slope −2. The corresponding graph for the AgCl solubility equilibrium has slope −1.

$$\log_{10}[Ag^+] = -\log_{10}[Cl^-] + \log_{10}K_{sp}$$

If the concentration of Cl^- and the initial concentrations of the metal ions correspond to a point that lies between the two lines in Figure 16.17, then AgCl precipitates but $PbCl_2$ does not.

16.6.1 Metal Sulfides

Controlling the solubility of metal sulfides has important applications. According to Table 16.2, most metal sulfides are very slightly soluble in water; only a very small amount of a compound such as ZnS(*s*) will dissolve in water. Although it is tempting to write the resulting equilibrium as

$$ZnS(s) \rightleftharpoons Zn^{2+}(aq) + S^{2-}(aq) \qquad K_{sp} = ?$$

in analogy with the equilibria for other weakly soluble salts, this is misleading, because S^{2-}, like O^{2-}, is a very strong base (stronger than OH^-) and reacts almost quantitatively with water.

$$S^{2-}(aq) + H_2O(\ell) \longrightarrow HS^-(aq) + OH^-(aq)$$

The value of K_b for this reaction is on the order of 10^5; this means essentially no S^{2-} is present in aqueous solution. The net dissolution reaction is found by adding the two preceding equations,

$$ZnS(s) + H_2O(\ell) \rightleftharpoons Zn^{2+}(aq) + OH^-(aq) + HS^-(aq)$$

for which the equilibrium constant is

$$[Zn^{2+}][OH^-][HS^-] = K \approx 2 \times 10^{-25}$$

TABLE 16.5

Equilibrium Constants for Metal Sulfide Dissolution at 25°C

Metal Sulfide	K†
CuS	5×10^{-37}
PbS	3×10^{-28}
CdS	7×10^{-28}
SnS	9×10^{-27}
ZnS	2×10^{-25}
FeS	5×10^{-19}
MnS	3×10^{-14}

†K is the equilibrium constant for the reaction $MS(s) + H_2O(\ell) \rightleftharpoons M^{2+}(aq) + OH^-(aq) + HS^-(aq)$.

Table 16.5 gives values of the equilibrium constants for comparable reactions of other metal sulfides. As the pH decreases, the concentration of OH^- decreases. At the same time, the concentration of HS^- also decreases as the equilibrium

$$HS^-(aq) + H_3O^+(aq) \rightleftharpoons H_2S(aq) + H_2O(\ell)$$

shifts to the right upon addition of H_3O^+. If both $[OH^-]$ and $[HS^-]$ decrease, then $[Zn^{2+}]$ must increase in order to maintain a constant value for the product of the three concentrations. As a result, the solubility of ZnS(*s*) increases as the pH of the solution decreases. Other metal sulfides behave the same way, becoming more soluble in acidic solution.

A quantitative calculation of metal sulfide solubility requires treating several simultaneous equilibria, as the following example illustrates.

FIGURE 16.18 Hydrogen sulfide, H_2S, is a poisonous, foul-smelling gas. When dissolved in water, it gives a weak acid, hydrosulfuric acid.

EXAMPLE 16.10

In a solution that is saturated with H_2S (see Fig. 16.18), $[H_2S]$ is fixed at 0.1 M. Calculate the molar solubility of FeS(*s*) in such a solution if it is buffered at pH 3.0.

Solution

If the pH is 3.0, then

$$[OH^-] = 1 \times 10^{-11} \text{ M}$$

(Continued)

In addition, the acid ionization of H_2S must be considered:

$$H_2S(aq) + H_2O(\ell) \rightleftharpoons H_3O^+(aq) + HS^-(aq)$$

Substitution of $[H_2S] = 0.1$ M for a saturated solution and $[H_3O^+] = 1 \times 10^{-3}$ M (at pH 3.0) into the equilibrium expression for this reaction gives

$$\frac{[H_3O^+][HS^-]}{[H_2S]} = \frac{[1 \times 10^{-3}][HS^-]}{0.1} = K_a = 9.1 \times 10^{-8}$$

$$[HS^-] = 9 \times 10^{-6} \text{ M}$$

where K_a came from Table 15.2. For the reaction

$$FeS(s) + H_2O(\ell) \rightleftharpoons Fe^{2+}(aq) + HS^-(aq) + OH^-(aq)$$

the equilibrium constant from Table 16.3 is

$$[Fe^{2+}][HS^-][OH^-] = 5 \times 10^{-19}$$

Substituting the values of $[HS^-]$ and $[OH^-]$ and solving for $[Fe^{2+}]$ give

$$[Fe^{2+}](9 \times 10^{-6})(1 \times 10^{-11}) = 5 \times 10^{-19}$$

$$[Fe^{2+}] = 6 \times 10^{-3} \text{ M}$$

Hence, 6×10^{-3} mol of FeS dissolves per liter under these conditions.

Related Problems: 55, 56

By adjusting the pH through appropriate choice of buffers, as in Example 16.10, conditions can be selected so that metal ions of one element remain entirely in solution, whereas those of a second element in the mixture precipitate almost entirely as solid metal sulfide (Fig. 16.19). Such a procedure is important for separating metal ions in qualitative analysis.

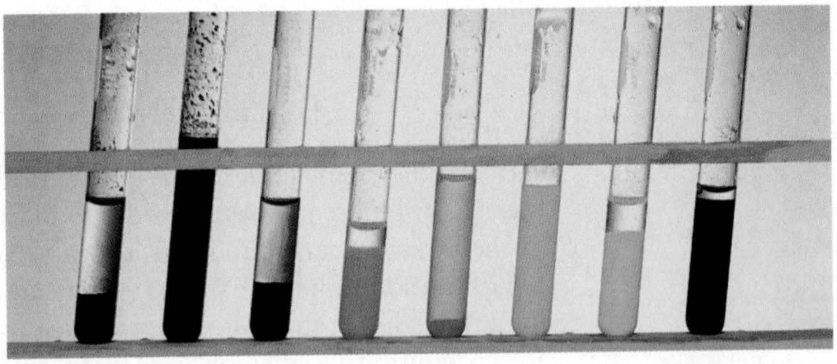

FIGURE 16.19 These sulfides are insoluble at pH 1, so they can be separated out of a mixture containing other, more soluble sulfides. From left to right, they are PbS, Bi_2S_3, CuS, CdS, Sb_2S_3, SnS_2, As_2S_3, and HgS. The precipitates occupy the lower part of the test tubes; in the 2nd and 6th from the left, the precipitate appears as a dispersion of fine particles. The colored bands observed at each meniscus are refracted images of the colored solids at the bottom of each tube.

CHAPTER SUMMARY

Dissolution–precipitation reactions involve equilibrium between a substance in its solid form and molecules or ions of that same substance dissolved in solution. This equilibrium is described by the mass action law, so knowledge of the equilibrium constant permits manipulation of the concentrations in solution through the principles described in Chapter 14. Numerous separation and purification procedures are based on such manipulations. The significance of these dissolution–precipitation equilibria in practical applications is comparable to that of acid–base reactions. Solutions of slightly soluble solids are said to be saturated when the dissolved concentration is at equilibrium with the solid. In accordance with Le Châtelier's principle, if the solution goes to a state of supersaturation, it will return to equilibrium by precipitating solid out of solution. Similarly, according to Le Châtelier, solubility can be increased by increasing temperature for an endothermic dissolution reaction, or by decreasing temperature for an exothermic dissolution reaction. Solubility is greatly influenced by the presence of other species to the solution, through the way they shift the equilibrium concentration of the dissolved solid species. This general principle explains the common ion effect and the influence of pH on solubility.

CONCEPTS AND SKILLS

Section 16.1 – The Nature of Solubility Equilibria

Discuss the dynamical processes that lead to solubility equilibria (Problems 1–4).

- When a solid substance is in contact with a solution in which this same substance is dissolved, molecules of the substance pass back and forth between the solid and dissolved states, just as molecules move between a liquid and its vapor. When the rate of passage from the solid into the solution is equal to the rate of the reverse passage from solution into the solid, a state of dissolution–precipitation equilibrium exists. At equilibrium, the solution is saturated and cannot accommodate more solute. The solubility S of a substance is the greatest amount that will dissolve in a specified volume of solvent at a specified temperature, and is expressed in g L^{-1} or mol L^{-1}. The solubility depends on temperature. In accordance with Le Châtelier's principle, solubility will increase with temperature if the dissolution reaction is endothermic, and decrease as temperature increases if the reaction is exothermic. A nonequilibrium supersaturated state can exist temporarily, in which the concentration of dissolved species exceeds the equilibrium concentration. The solution can return to equilibrium by precipitation of the excess dissolved species.

Section 16.2 – Ionic Equilibria between Solids and Solutions

Relate the solubilities of sparingly soluble salts in water to their solubility product constants (Problems 5–16).

- Slightly soluble ionic solids dissociate in solution to establish equilibrium with the dissolved cations and ions. The equilibrium expression is written in accordance with the stoichiometry of the compound. The solid does not appear in the equilibrium expression because it is in its standard state which has activity $a = 1$.
  - $AB(s) \rightleftharpoons A^+(aq) + B^-(aq)$
  - $K_{sp} = [A^+][B^-] = S^2$ if S is the molar solubility and the solid is dissolved in pure water.
  - $AB_2(s) \rightleftharpoons A^+(aq) + 2B^-(aq)$
  - $K_{sp} = [A^+][B^-]^2 = 4S^3$, and so on, for different stoichiometric ratios if the solid is dissolved in pure water.

Section 16.3 – Precipitation and the Solubility Product

Use the reaction quotient to predict whether a precipitate will form when two solutions are mixed, and then calculate the equilibrium concentrations that result (Problems 17–24).

- At any point in the reaction for the dissolution of the $AB(s)$ above $Q = [A^+][B^-]$. When $Q > K_{sp}$, a precipitate will form. Assume the reaction goes to completion and consumes all of one ion (the limiting reagent). Subsequent dissolution restores some of that ion to establish equilibrium between the precipitate and the solution. See Examples 16.3 and 16.4.

Calculate the solubility of a sparingly soluble salt in a solution that contains a given concentration of a common ion (Problems 25–30).

- The common ion effect describes the equilibrium of a solid with a solution that already contains some ions of the same type as one of those in the solid. No matter what the source of the ions, the total concentration of each in the solution must obey the equilibrium expression. The concentration of the "common ion" at equilibrium is the sum of its concentration initially present and the additional amount produced in the dissolution of the solid. Substitute this total concentration in the equilibrium expression and calculate the solubility at equilibrium. See Example 16.5.

Section 16.4 – The Effects of pH on Solubility

Determine the dependence on pH of the solubility of a salt of a weak base (Problems 31–34).

- Solubility of metal hydroxides increases at lower pH because the acid present ties up hydroxide ions in solution and so shifts the dissolution equilibrium to the right. Solubility of salts in which the anion is a weak base increases at lower pH because the hydronium ion in solution consumes the anion and so shifts the dissolution equilibrium to the right.

Section 16.5 – Complex Ions and Solubility

Calculate the concentrations of molecular and ionic species in equilibrium with complex ions (Problems 35–38).

- The equilibrium expression relates the concentration of the complex ion and the concentration of the species that participate in its formation to the formation constant K_f for the ion. Set up the equilibrium expression by writing the total concentration of each species as the sum of its concentration initially present and the additional amount produced in formation of the complex ion, and solve for the resulting equilibrium concentration. See Example 16.7.

Determine the effect of complex ion formation on the solubility of sparingly soluble salts containing a common cation (Problems 39–40).

- Suppose a sparingly soluble salt such as AgCl is already in equilibrium with a solution. Adding a new reactant that forms a soluble complex with Ag^+ ions will consume the free Ag^+ ions in the solution. By Le Châtelier's principle this will drive the dissolution equilibrium to the right and increase the solubility of AgCl above its value in pure water. The resulting equilibrium is the sum of the dissolution–precipitation reaction and the complex formation reaction, and the resulting equilibrium constant is the product $K_{sp}K_f$. See Example 16.8.

Calculate the pH of aqueous solutions containing metal cations (Problems 41–46).

- Many metal cations are highly solvated in aqueous solutions to form complex ions that function as Brønsted–Lowry acids by donating protons to the solution. If the acid dissociation constant K_a is known for the complex ion, the equilibrium concentration of H_3O^+ and the pH are readily calculated. See Example 16.9.

Section 16.6 – A Deeper Look . . . Selective Precipitation of Ions

Specify the optimal conditions for separating two elements based on the differing solubilities of their ionic compounds (Problems 49–56.)

- Identify compounds of the two elements that have a common anion and widely different solubilities. Add enough of the common anion to precipitate all of the less soluble compound and leave the more soluble compound in solution. The maximum amount that can be added is determined by keeping the reaction quotient Q for the more soluble compound below its K_{sp} value. The quantitative degree of separation depends on the relative magnitudes of the K_{sp} values. This method is illustrated in Figure 16.17.

CUMULATIVE EXERCISE

Carbonate Minerals in Fresh Water and Seawater

The carbonates (see Fig. 16.20) are among the most abundant and important minerals in the earth's crust. When these minerals come into contact with fresh water or seawater, solubility equilibria are established that greatly affect the chemistry of the natural waters. Calcium carbonate ($CaCO_3$), the most important natural carbonate, makes up limestone and other forms of rock such as marble. Other carbonate minerals include dolomite, $CaMg(CO_3)_2$, and magnesite, $MgCO_3$. These compounds are sufficiently soluble that their solutions are nonideal, so calculations based on solubility product expressions are only approximate.

(a) The rare mineral nesquehonite contains $MgCO_3$ together with water of hydration. A sample containing 21.7 g of nesquehonite is acidified and heated, and the volume of $CO_2(g)$ produced is measured to be 3.51 L at 0°C and $P = 1$ atm. Assuming all the carbonate has reacted to form CO_2, give the chemical formula for nesquehonite.

(b) Write a chemical equation and a solubility product expression for the dissolution of dolomite in water.

(c) In a sufficiently basic solution, the carbonate ion does not react significantly with water to form hydrogen carbonate ion. Calculate the solubility (in grams per liter) of limestone (calcium carbonate) in a 0.10 M solution of sodium hydroxide. Use the K_{sp} from Table 16.2.

(d) In a strongly basic 0.10 M solution of Na_2CO_3, the concentration of CO_3^{2-} is 0.10 M. What is the gram solubility of limestone in this solution? Compare your answer with that for part (c).

(e) In a mountain lake having a pH of 8.1, the total concentration of carbonate species, $[CO_3^{2-}] + [HCO_3^-]$, is measured to be 9.6×10^{-4} M, whereas the concentration of Ca^{2+} is 3.8×10^{-4} M. Calculate the concentration of CO_3^{2-} in this lake, using $K_a = 4.8 \times 10^{-11}$ for the acid ionization of HCO_3^- to CO_3^{2-}. Is the lake unsaturated, saturated, or supersaturated with respect to $CaCO_3$?

(f) Will acid rainfall into the lake increase or decrease the solubility of limestone rocks in the lake's bed?

(g) Seawater contains a high concentration of Cl^- ions, which form weak complexes with calcium, such as the ion pair $CaCl^+$. Does the presence of such complexes increase or decrease the equilibrium solubility of $CaCO_3$ in seawater?

FIGURE 16.20 Carbonate minerals. Calcite (*left*) and aragonite (*middle*) are both $CaCO_3$, and smithsonite (*right*) is $ZnCO_3$.

Answers

(a) $MgCO_3 \cdot 3H_2O$

(b) $CaMg(CO_3)_2(s) \rightleftharpoons Ca^{2+}(aq) + Mg^{2+}(aq) + 2\ CO_3^{2-}(aq)$

$$[Ca^{2+}][Mg^{2+}][CO_3^{2-}]^2 = K_{sp}$$

(c) 9.3×10^{-3} g L^{-1}

(d) 8.7×10^{-6} g L^{-1}, smaller than in part (c) because of the common-ion effect

(e) 5.8×10^{-6} M. $Q = 2.2 \times 10^{-9} < K_{sp} = 8.7 \times 10^{-9}$, so the lake is slightly less than saturated

(f) Increase

(g) Increase

PROBLEMS

Answers to problems whose numbers are red appear in Appendix G. Problems that are more challenging are indicated with asterisks.

16.1 The Nature of Solubility Equilibria

1. Gypsum has the formula $CaSO_4 \cdot 2H_2O$. Plaster of Paris has the chemical formula $CaSO_4 \cdot \frac{1}{2}\ H_2O$. In making wall plaster, water is added to plaster of Paris and the mixture then hardens into solid gypsum. How much water (in liters, at a density of 1.00 kg L^{-1}) should be added to 25.0 kg of plaster of Paris to turn it into gypsum, assuming no loss from evaporation?

2. A 1.00-g sample of magnesium sulfate is dissolved in water, and the water is then evaporated away until the residue is bone dry. If the temperature of the water is kept between 48°C and 69°C, the solid that remains weighs 1.898 g. If the experiment is repeated with the temperature held between 69°C and 100°C, however, the solid has a mass of 1.150 g. Determine how many waters of crystallization per $MgSO_4$ there are in each of these two solids.

3. The following graph shows the solubility of KBr in water in units of grams of KBr per 100 g of H_2O. If 80 g of KBr is added to 100 g of water at 10°C and the mixture is heated slowly, at what temperature will the last KBr dissolve?

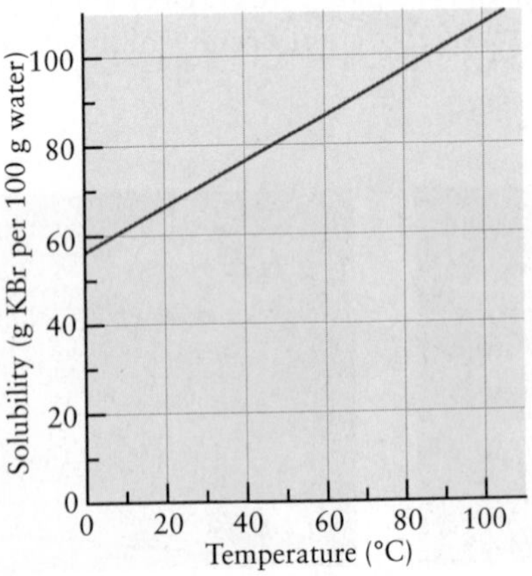

4. Figure 16.3 shows the solubility of $AgNO_3$ in water in units of moles of $AgNO_3$ per kilogram of H_2O. If 255 g of $AgNO_3$ is added to 100 g of water at 95°C and cooled slowly, at what temperature will the solution become saturated?

16.2 Ionic Equilibria between Solids and Solutions

5. Iron(III) sulfate, $Fe_2(SO_4)_3$, is a yellow compound that is used as a coagulant in water treatment. Write a balanced chemical equation and a solubility product expression for its dissolution in water.

6. Lead antimonate, $Pb_3(SbO_4)_2$, is used as an orange pigment in oil-based paints and in glazes. Write a balanced chemical equation and a solubility product expression for its dissolution in water.

7. Thallium(I) iodate ($TlIO_3$) is only slightly soluble in water. Its K_{sp} at 25°C is 3.07×10^{-6}. Estimate the solubility of thallium(I) iodate in water in units of grams per 100.0 mL of water.

8. Thallium thiocyanate (TlSCN) is only slightly soluble in water. Its K_{sp} at 25°C is 1.82×10^{-4}. Estimate the solubility of thallium thiocyanate in units of grams per 100.0 mL of water.

9. Potassium perchlorate, $KClO_4$, has a K_{sp} at 25°C of 1.07×10^{-2}. Compute its solubility in grams per liter of solution.

10. Ammonium hexachloroplatinate(IV), $(NH_4)_2(PtCl_6)$, is one of the few sparingly soluble ammonium salts. Its K_{sp} at 20°C is 5.6×10^{-6}. Compute its solubility in grams per liter of solution.

11. The solubility product constant of mercury(I) iodide is 1.2×10^{-28} at 25°C. Estimate the concentration of Hg_2^{2+} and I^- in equilibrium with solid Hg_2I_2.

12. The solubility product constant of Hg_2Cl_2 is 2×10^{-18} at 25°C. Estimate the concentration of Hg_2^{2+} and Cl^- in equilibrium with solid Hg_2Cl_2 at 25°C.

13. The solubility of silver chromate (Ag_2CrO_4) in 500 mL of water at 25°C is 0.0129 g. Calculate its solubility product constant.

14. At 25°C, 400 mL of water can dissolve 0.00896 g of lead iodate, $Pb(IO_3)_2$. Calculate K_{sp} for lead iodate.

15. At 100°C, water dissolves 1.8×10^{-2} g of AgCl per liter. Compute the K_{sp} of AgCl at this temperature.

16. A mass of 0.017 g of silver dichromate ($Ag_2Cr_2O_7$) will dissolve in 300 mL of water at 25°C. Calculate the solubility product constant K_{sp} of silver dichromate.

16.3 Precipitation and the Solubility Product

17. A solution of barium chromate ($BaCrO_4$) is prepared by dissolving 6.3×10^{-3} g of this yellow solid in 1.00 L of hot water. Will solid barium chromate precipitate upon cooling to 25°C, according to the solubility product expression? Explain.

18. A solution is prepared by dissolving 0.090 g of PbI_2 in 1.00 L of hot water and cooling the solution to 25°C. Will solid precipitate result from this process, according to the solubility product expression? Explain.

19. A solution is prepared by mixing 250.0 mL of 2.0×10^{-3} M $Ce(NO_3)_3$ and 150.0 mL of 10×10^{-2} M KIO_3 at 25°C. Determine whether $Ce(IO_3)_3(s)$ ($K_{sp} = 1.9 \times 10^{-10}$) tends to precipitate from this mixture.

20. Suppose 100.0 mL of a 0.0010 M $CaCl_2$ solution is added to 50.0 mL of a 6.0×10^{-5} M NaF solution at 25°C. Determine whether $CaF_2(s)$ ($K_{sp} = 3.9 \times 10^{-11}$) tends to precipitate from this mixture.

21. Suppose 50.0 mL of a 0.0500 M solution of $Pb(NO_3)_2$ is mixed with 40.0 mL of a 0.200 M solution of $NaIO_3$ at 25°C. Calculate the $[Pb^{2+}]$ and $[IO_3^-]$ when the mixture comes to equilibrium. At this temperature, K_{sp} for $Pb(IO_3)_2$ is 2.6×10^{-13}.

22. Silver iodide (AgI) is used in place of silver chloride for the fastest photographic film because it is more sensitive to light and can therefore form an image in a very short exposure time. A silver iodide emulsion is prepared by adding 6.60 L of 0.10 M NaI solution to 1.50 L of 0.080 M $AgNO_3$ solution at 25°C. Calculate the concentration of silver ion remaining in solution when the mixture comes to equilibrium and its chemical amount relative to the amount present initially.

23. When 50.0 mL of 0.100 M $AgNO_3$ and 30.0 mL of 0.0600 M Na_2CrO_4 are mixed, a precipitate of silver chromate (Ag_2CrO_4) is formed. The solubility product K_{sp} of silver chromate in water at 25°C is 1.9×10^{-12}. Calculate the $[Ag^+]$ and $[CrO_4^{2-}]$ remaining in solution at equilibrium.

24. When 40.0 mL of 0.0800 M $Sr(NO_3)_2$ and 80.0 mL of 0.0500 M KF are mixed, a precipitate of strontium fluoride (SrF_2) is formed. The solubility product K_{sp} of strontium fluoride in water at 25°C is 2.8×10^{-9}. Calculate the $[Sr^{2+}]$ and $[F^-]$ remaining in solution at equilibrium.

25. Calculate the solubility (in mol L^{-1}) of $CaF_2(s)$ at 25°C in a 0.040 M aqueous solution of NaF.

26. Calculate the mass of AgCl that can dissolve in 100 mL of 0.150 M NaCl solution.

27. The solubility product of nickel(II) hydroxide, $Ni(OH)_2$, at 25°C is $K_{sp} = 1.6 \times 10^{-16}$.
(a) Calculate the molar solubility of $Ni(OH)_2$ in pure water at 25°C.
(b) Calculate the molar solubility of $Ni(OH)_2$ in 0.100 M NaOH.

28. Silver arsenate (Ag_3AsO_4) is a slightly soluble salt having a solubility product of $K_{sp} = 1.0 \times 10^{-22}$ at 25°C for the equilibrium

$$Ag_3AsO_4(s) \rightleftharpoons 3\ Ag^+(aq) + AsO_4^{3-}(aq)$$

(a) Calculate the molar solubility of silver arsenate in pure water at 25°C.
(b) Calculate the molar solubility of silver arsenate in 0.10 M $AgNO_3$.

29. A saturated solution of $Mg(OH)_2$ at 25°C is prepared by equilibrating solid $Mg(OH)_2$ with water. Concentrated NaOH is then added until the solubility of $Mg(OH)_2$ is 0.0010 times that in H_2O alone. (Ignore the change in volume resulting from the addition of NaOH.) The solubility product K_{sp} of $Mg(OH)_2$ is 1.2×10^{-11} at 25°C. Calculate the concentration of hydroxide ion in the solution after the addition of the NaOH.

30. A saturated solution of BaF_2 at 25°C is prepared by equilibrating solid BaF_2 with water. Powdered NaF is then dissolved in the solution until the solubility of BaF_2 is 1.0% of that in H_2O alone. The solubility product K_{sp} of BaF_2 is 1.7×10^{-6} at 25°C. Calculate the concentration of fluoride ion in the solution after addition of the powdered NaF.

16.4 The Effects of pH on Solubility

31. Compare the molar solubility of AgOH in pure water with that in a solution buffered at pH 7.00. Note the difference between the two: When AgOH is dissolved in pure water, the pH does not remain at 7.

32. Compare the molar solubility of $Mg(OH)_2$ in pure water with that in a solution buffered at pH 9.00.

33. For each of the following ionic compounds, state whether the solubility will increase, decrease, or remain unchanged as a solution at pH 7 is made acidic.
(a) PbI_2 (b) AgOH (c) $Ca_3(PO_4)_2$

34. For each of the following ionic compounds, state whether the solubility will increase, decrease, or remain unchanged as a solution at pH 7 is made acidic.
(a) $SrCO_3$ (b) Hg_2Br_2 (c) MnS

16.5 Complex Ions and Solubility

35. Suppose 0.10 mol of $Cu(NO_3)_2$ and 1.50 mol of NH_3 are dissolved in water and diluted to a total volume of 1.00 L. Calculate the concentrations of $Cu(NH_3)_4^{2+}$ and of Cu^{2+} at equilibrium.

36. The formation constant of the $TlCl_4^-$ complex ion is 1×10^{18}. Suppose 0.15 mol of $Tl(NO_3)_3$ is dissolved in 1.00 L of a 0.50 M solution of NaCl. Calculate the concentration at equilibrium of $TlCl_4^-$ and of Tl^{3+}.

37. The organic compound "18-crown-6" binds alkali metals in aqueous solution by wrapping around and enfolding the ion. It presents a niche that nicely accommodates the K^+ ion but is too small for the Rb^+ ion and too large for the Na^+ ion. The values of the equilibrium constants show this:

$$Na^+(aq) + 18\text{-crown-6}(aq) \rightleftharpoons \text{Na-crown}^+(aq) \qquad K = 6.6$$

$$K^+(aq) + 18\text{-crown-6}(aq) \rightleftharpoons \text{K-crown}^+(aq) \qquad K = 111.6$$

$$Rb^+(aq) + 18\text{-crown-6}(aq) \rightleftharpoons \text{Rb-crown}^+(aq) \qquad K = 36$$

An aqueous solution is initially 0.0080 M in 18-crown-6(*aq*) and also 0.0080 M in $K^+(aq)$. Compute the equilibrium concentration of free K^+. ("Free" means not tied up with the 18-crown-6.) Compute the concentration of free Na^+ if the solution contains 0.0080 M $Na^+(aq)$ instead of $K^+(aq)$.

38. The organic compound 18-crown-6 (see preceding problem) also binds strongly with the alkali metal ions in methanol.

$$K^+ + 18\text{-crown-6} \rightleftharpoons [\text{complex}]^+$$

In methanol solution the equilibrium constant is 1.41×10^6. A similar reaction with Cs^+ has an equilibrium constant of only 2.75×10^4. A solution is made (in methanol) containing 0.020 mol L^{-1} each of K^+ and Cs^+. It also contains 0.30 mol L^{-1} of 18-crown-6. Compute the equilibrium concentrations of both the uncomplexed K^+ and the uncomplexed Cs^+.

39. Will silver chloride dissolve to a significantly greater extent in a 1.00 M NaCl solution than in pure water due to the possible formation of $AgCl_2^-$ ions? Use data from Tables 16.2 and 16.3 to provide a quantitative answer to this question. What will happen in a 0.100 M NaCl solution?

40. Calculate how many grams of silver chloride will dissolve in 1.0 L of a 1.0 M NH_3 solution through formation of the complex ion $Ag(NH_3)_2^+$.

41. The pH of a 0.2 M solution of $CuSO_4$ is 4.0. Write chemical equations to explain why a solution of this salt is neither basic [from the reaction of $SO_4^{2-}(aq)$ with water] nor neutral, but acidic.

42. Will a 0.05 M solution of $FeCl_3$ be acidic, basic, or neutral? Explain your answer by writing chemical equations to describe any reactions taking place.

43. The acid ionization constant for $Co(H_2O)_6^{2+}(aq)$ is 3×10^{-10}. Calculate the pH of a 0.10 M solution of $Co(NO_3)_2$.

44. The acid ionization constant for $Fe(H_2O)_6^{2+}(aq)$ is 3×10^{-6}. Calculate the pH of a 0.10 M solution of $Fe(NO_3)_2$, and compare it with the pH of the corresponding iron(III) nitrate solution from Example 16.9.

45. A 0.15 M aqueous solution of the chloride salt of the complex ion $Pt(NH_3)_4^{2+}$ is found to be weakly acidic with a pH of 4.92. This is initially puzzling because the Cl^- ion in water is not acidic and NH_3 in water is *basic*, not acidic. Finally, it is suggested that the $Pt(NH_3)_4^{2+}$ ion as a group donates hydrogen ions. Compute the K_a of this acid, assuming that just one hydrogen ion is donated.

46. The pH of a 0.10 M solution of $Ni(NO_3)_2$ is 5.0. Calculate the acid ionization constant of $Ni(H_2O)_6^{2+}(aq)$.

47. K_{sp} for $Pb(OH)_2$ is 4.2×10^{-15}, and K_f for $Pb(OH)_3^-$ is 4×10^{14}. Suppose a solution whose initial concentration of $Pb^{2+}(aq)$ is 1.00 M is brought to pH 13.0 by addition of solid NaOH. Will solid $Pb(OH)_2$ precipitate, or will the lead be dissolved as $Pb(OH)_3^-(aq)$? What will be $[Pb^{2+}]$ and $[Pb(OH)_3^-]$ at equilibrium? Repeat the calculation for an initial Pb^{2+} concentration of 0.050 M. (*Hint:* One way to solve this problem is to *assume* that $Pb(OH)_2(s)$ is present and calculate $[Pb^{2+}]$ and $[Pb(OH)_3^-]$ that would be in equilibrium with the solid. If the sum of these is less than the original $[Pb^{2+}]$, the remainder can be assumed to have precipitated. If not, there is a contradiction and we must assume that *no* $Pb(OH)_2(s)$ is present. In this case we can calculate $[Pb^{2+}]$ and $[Pb(OH)_3^-]$ directly from K_f.)

48. K_{sp} from $Zn(OH)_2$ is 4.5×10^{-17}, and K_f for $Zn(OH)_4^{2-}$ is 5×10^{14}. Suppose a solution whose initial concentration of $Zn^{2+}(aq)$ is 0.010 M is brought to pH 14.0 by addition of solid NaOH. Will solid $Zn(OH)_2$ precipitate, or will the zinc be dissolved as $Zn(OH)_4^{2-}(aq)$? What will be $[Zn^{2+}]$ and $[Zn(OH)_4^{2-}]$ at equilibrium? Repeat the calculation at pH 13 for an initial Zn^{2+} concentration of 0.10 M. See the hint in problem 47.

16.6 A Deeper Look . . . Selective Precipitation of Ions

49. An aqueous solution at 25°C is 0.10 M in both Mg^{2+} and Pb^{2+} ions. We wish to separate the two kinds of metal ions by taking advantage of the different solubilities of their oxalates, MgC_2O_4 and PbC_2O_4.
   (a) What is the highest possible oxalate ion concentration that allows only one solid oxalate salt to be present at equilibrium? Which ion is present in the solid—Mg^{2+} or Pb^{2+}?
   (b) What fraction of the less soluble ion still remains in solution under the conditions of part (a)?

50. An aqueous solution at 25°C is 0.10 M in Ba^{2+} and 0.50 M in Ca^{2+} ions. We wish to separate the two by taking advantage of the different solubilities of their fluorides, BaF_2 and CaF_2.
   (a) What is the highest possible fluoride ion concentration that allows only one solid fluoride salt to be present at equilibrium? Which ion is present in the solid—Ba^{2+} or Ca^{2+}?
   (b) What fraction of the less soluble ion still remains in solution under the conditions of part (a)?

51. The cations in an aqueous solution that contains 0.100 M $Hg_2(NO_3)_2$ and 0.0500 M $Pb(NO_3)_2$ are to be separated by taking advantage of the difference in the solubilities of their iodides. The K_{sp} of PbI_2 and of Hg_2I_2 are given in Table 16.2. What should be the concentration of iodide ion for the best separation? In the "best" separation, one of the cations should remain entirely in solution and the other should precipitate as fully as possible.

52. The cations in an aqueous solution that contains 0.150 M $Ba(NO_3)_2$ and 0.0800 M $Ca(NO_3)_2$ are to be separated by taking advantage of the difference in the solubilities of their sulfates. $K_{sp}(BaSO_4) = 1.1 \times 10^{-10}$ and $K_{sp}(CaSO_4) = 2.4 \times 10^{-5}$. What should be the concentration of sulfate ion for the best separation?

53. Calculate the $[Zn^{2+}]$ in a solution that is in equilibrium with ZnS(*s*) and in which $[H_3O^+] = 1.0 \times 10^{-5}$ M and $[H_2S] = 0.10$ M.

54. Calculate the $[Cd^{2+}]$ in a solution that is in equilibrium with CdS(*s*) and in which $[H_3O^+] = 1.0 \times 10^{-3}$ M and $[H_2S] = 0.10$ M.

55. What is the highest pH at which 0.10 M Fe^{2+} will remain entirely in a solution that is saturated with H_2S at a concentration of $[H_2S] = 0.10$ M? At this pH, what would be the concentration of Pb^{2+} in equilibrium with solid PbS in this solution?

56. What is the highest pH at which 0.050 M Mn^{2+} will remain entirely in a solution that is saturated with H_2S at a concentration of $[H_2S] = 0.10$ M? At this pH, what would be the concentration of Cd^{2+} in equilibrium with solid CdS in this solution?

ADDITIONAL PROBLEMS

57. Write a chemical equation for the dissolution of mercury(I) chloride in water, and give its solubility product expression.

*58. Magnesium ammonium phosphate has the formula $MgNH_4PO_4 \cdot 6H_2O$. It is only slightly soluble in water (its K_{sp} is 2.3×10^{-13}). Write a chemical equation and the corresponding equilibrium law for the dissolution of this compound in water.

59. Soluble barium compounds are poisonous, but barium sulfate is routinely ingested as a suspended solid in a "barium cocktail" to improve the contrast in X-ray images. Calculate the concentration of dissolved barium per liter of water in equilibrium with solid barium sulfate.

60. A saturated aqueous solution of silver perchlorate ($AgClO_4$) contains 84.8% by mass $AgClO_4$, but a saturated solution of $AgClO_4$ in 60% aqueous perchloric acid contains only 5.63% by mass $AgClO_4$. Explain this large difference using chemical equations.

61. Suppose 140 mL of 0.0010 M $Sr(NO_3)_2$ is mixed with enough 0.0050 M NaF to make 1.00 L of solution. Will $SrF_2(s)$ ($K_{sp} = 2.8 \times 10^{-9}$) precipitate at equilibrium? Explain.

62. The concentration of calcium ion in a town's supply of drinking water is 0.0020 M. (This water is referred to as hard water because it contains such a large concentration of Ca^{2+}.) Suppose the water is to be fluoridated by the addition of NaF for the purpose of reducing tooth decay. What is the maximum concentration of fluoride ion that can be achieved in the water before precipitation of CaF_2 begins? Will the water supply attain the level of fluoride ion recommended by the U.S. Public Health Service, about 5×10^{-5} M (1 mg fluorine per liter)?

63. Suppose that 150 mL of 0.200 M K_2CO_3 and 100 mL of 0.400 M $Ca(NO_3)_2$ are mixed together. Assume that the volumes are additive, that $CaCO_3$ is completely insoluble, and that all other substances that might be formed are soluble. Calculate the mass of $CaCO_3$ precipitated, and calculate the concentrations in the final solution of the four ions that were present initially.

64. The solubility of $CaCO_3$ in water is about 7 mg L^{-1}. Show how one can calculate the solubility product of $BaCO_3$ from this information and from the fact that when sodium carbonate solution is added slowly to a solution containing equimolar concentrations of Ca^{2+} and Ba^{2+}, no $CaCO_3$ is formed until about 90% of the Ba^{2+} has been precipitated as $BaCO_3$.

65. It is sometimes asserted that carbonates are soluble in strong acids because a gas is formed that escapes (CO_2). Suppose that CO_2 were extremely soluble in water (as, for example, ammonia is) and therefore it did not leave the site of the reaction, but that otherwise, its chemistry was unchanged. Would calcium carbonate be soluble in strong acids? Explain.

66. The solubility products of $Fe(OH)_3$ and $Ni(OH)_2$ are about 10^{-36} and 6×10^{-18}, respectively. Find the approximate pH range suitable for the separation of Fe^{3+} and Ni^{2+} by precipitation of $Fe(OH)_3$ from a solution initially 0.01 M in each ion, as follows: (a) Calculate the lowest pH at which all but 0.1% of the Fe^{3+} will be precipitated as $Fe(OH)_3$; (b) calculate the highest pH possible without precipitation of $Ni(OH)_2$.

67. The two solids CuBr(*s*) and AgBr(*s*) are only very slightly soluble in water: $K_{sp}(CuBr) = 4.2 \times 10^{-8}$ and $K_{sp}(AgBr) = 7.7 \times 10^{-13}$. Some CuBr(*s*) and AgBr(*s*) are both mixed into a quantity of water that is then stirred until it is saturated with respect to both solutes. Next, a small amount of KBr is added and dissolves completely. Compute the ratio of $[Cu^+]$ to $[Ag^+]$ after the system reestablishes equilibrium.

*68. The two salts $BaCl_2$ and Ag_2SO_4 are both far more soluble in water than either $BaSO_4$ ($K_{sp} = 1.1 \times 10^{-10}$) or AgCl ($K_{sp} = 1.6 \times 10^{-10}$) at 25°C. Suppose 50.0 mL of 0.040 M $BaCl_2(aq)$ is added to 50.0 mL of 0.020 M $Ag_2SO_4(aq)$. Calculate the concentrations of $SO_4^{2-}(aq)$, $Cl^-(aq)$, $Ba^{2+}(aq)$, and $Ag^+(aq)$ that remain in solution at equilibrium.

69. The Mohr method is a technique for determining the amount of chloride ion in an unknown sample. It is based on the difference in solubility between silver chloride (AgCl; $K_{sp} = 1.6 \times 10^{-10}$) and silver chromate ($Ag_2CrO_4$; $K_{sp} = 1.9 \times 10^{-12}$). In using this method, one adds a small amount of chromate ion to a solution with unknown chloride concentration. By measuring the volume of $AgNO_3$

added before the appearance of the red silver chromate, one can determine the amount of Cl^- originally present. Suppose we have a solution that is 0.100 M in Cl^- and 0.00250 M in CrO_4^{2-}. If we add 0.100 M $AgNO_3$ solution drop by drop, will AgCl or Ag_2CrO_4 precipitate first? When $Ag_2CrO_4(s)$ first appears, what fraction of the Cl^- originally present remains in solution?

70. Oxide ion, like sulfide ion, is a strong base. Write an equation for the dissolution of CaO in water and give its equilibrium constant expression. Write the corresponding equation for the dissolution of CaO in an aqueous solution of a strong acid, and relate its equilibrium constant to the previous one.

71. Water that has been saturated with magnesia (MgO) at 25°C has a pH of 10.16. Write a balanced chemical equation for the equilibrium between MgO(*s*) and the ions it furnishes in aqueous solution, and calculate the equilibrium constant at 25°C. What is the solubility, in moles per liter, of MgO in water?

72. To 1.00 L of a 0.100 M $AgNO_3$ solution is added an excess of sodium chloride. Then 1.00 L of 0.500 M $NH_3(aq)$ is added. Finally, sufficient nitric acid is added until the pH of the resulting solution is 1.0. Write balanced equations for the reactions that take place (if any) at each of three steps in this process.

73. Only about 0.16 mg of AgBr(*s*) will dissolve in 1.0 L of water (this volume of solid is smaller than the head of a pin). In a solution of ammonia that contains 0.10 mol ammonia per liter of water, there are about 555 water molecules for every molecule of ammonia. However, more than 400 times as much AgBr (68 mg) will dissolve in this solution as in plain water. Explain how such a tiny change in the composition of the solution can have such a large effect on the solubility of AgBr.

***74.** **(a)** Calculate the solubility of calcium oxalate (CaC_2O_4) in 1.0 M oxalic acid ($H_2C_2O_4$) at 25°C, using the two acid ionization constants for oxalic acid from Table 15.2 and the solubility product $K_{sp} = 2.6 \times 10^{-9}$ for CaC_2O_4.
(b) Calculate the solubility of calcium oxalate in pure water at 25°C.
(c) Account for the difference between the results of (a) and (b).

***75.** When 6 M HCl is added to solid CdS, some of the solid dissolves to give the complex ion $CdCl_4^{2-}(aq)$.
(a) Write a balanced equation for the reaction that occurs.
(b) Use data from Tables 15.2 and 16.5 and the formation constant of $CdCl_4^{2-}$ ($K_f = 8 \times 10^2$) to calculate the equilibrium constant for the reaction of part (a).
(c) What is the molar solubility of CdS per liter of 6 M HCl?

***76.** Using data from Table 16.3, calculate the concentrations of $Hg^{2+}(aq)$, $HgCl^+(aq)$, and $HgCl_2(aq)$ that result when 1.00 L of a 0.100 M $Hg(NO_3)_2$ solution is mixed with an equal volume of a 0.100 M $HgCl_2$ solution. (*Hint:* Use the analogy with amphoteric equilibria discussed in Section 10.8.)

***77.** Calculate the concentration of $Cu^{2+}(aq)$ in a solution that contains 0.020 mol of $CuCl_2$ and 0.100 mol of NaCN in 1.0 L.

$$Cu^{2+}(aq) + 4\ CN^-(aq) \rightleftharpoons Cu(CN)_4^{2-}(aq)$$
$$K = 2.0 \times 10^{30}$$

(*Hint:* Do not overlook the reaction of CN^- with water to give HCN.)

78. An aqueous solution of $K_2[Pt(OH)_6]$ has a pH greater than 7. Explain this fact by writing an equation showing the $Pt(OH)_6^{2-}$ ion acting as a Brønsted–Lowry base and accepting a hydrogen ion from water.

79. In Example 16.9 we included only the first acid dissociation K_{a1} of a 0.100 M aqueous solution of $Fe(H_2O)_6^{3+}$. Subsequent dissociation can also occur, with $K_{a2} = 2.0 \times 10^{-5}$, to give $Fe(H_2O)_4(OH)_2^+$.
(a) Calculate the concentration of $Fe(H_2O)_4(OH)_2^+$ at equilibrium. Does the pH change significantly when this second dissociation is taken into account?
(b) We can describe the same reaction as the dissociation of a complex ion $Fe(OH)_2^+$ to Fe^{3+} and two OH^- ions. Calculate K_f, the formation constant for $Fe(OH)_2^+$.

CUMULATIVE PROBLEMS

80. The volume of a certain saturated solution is greater than the sum of the volumes of the water and salt from which it is made. Predict the effect of increased pressure on the solubility of this salt.

81. Codeine has the molecular formula $C_{18}H_{21}NO_3$. It is soluble in water to the extent of 1.00 g per 120 mL of water at room temperature and 1.00 g per 60 mL of water at 80°C. Compute the molal solubility (in mol kg^{-1}) at both temperatures, taking the density of water to be fixed at 1.00 g cm^{-3}. Is the dissolution of codeine in water endothermic or exothermic?

82. Suppose 1.44 L of a saturated solution of strontium carbonate ($SrCO_3$) in boiling water at 100°C is prepared. The solution is then strongly acidified and shaken to drive off all the gaseous CO_2 that forms. The volume of this gas (at a temperature of 100°C and a partial pressure of 0.972 atm) is measured to be 0.20 L (200 mL).
(a) Calculate the molar solubility of $SrCO_3$ in water at 100°C.
(b) Estimate the solubility product constant K_{sp} of $SrCO_3$ at this temperature.
(c) Explain why the actual K_{sp} at this temperature may be lower than you predicted in part (b).

83. A buffer is prepared by adding 50.0 mL of 0.15 M $HNO_3(aq)$ to 100.0 mL of 0.12 M NaHCOO(*aq*) (sodium formate). Calculate the solubility of $CaF_2(s)$ in this solution.

17

CHAPTER

Electrochemistry

A lightning strike contains about 10^{20} electrons.

Electrochemistry is the branch of chemistry concerned with the interconversion of chemical and electrical energy through oxidation–reduction reactions. Electrons are transferred between two half-reactions through external electrical circuits, in contrast to solution-phase redox reactions discussed in Section 11.4. Energy produced by spontaneous chemical reactions may be converted to electrical energy; conversely, electrical energy can be used to drive chemical reactions that are not normally spontaneous. Electrochemistry is historically a vital field, primarily for its important industrial applications. Bulk commodity chemicals, such as chlorine, and most metals are produced using large-scale electrochemical processes. In addition, considerable efforts have been directed toward preventing the corrosion (electrochemically triggered oxidation) of metals, for example, alloy-based structural materials of pervasive importance in vehicles, ships, airplanes, buildings, and bridges. Another important practical application has been the development of batteries, which produce electrical energy from spontaneous redox reactions in electrochemical cells.

The field of electrochemistry is much broader than is represented by its industrial applications alone, and it has been transformed into a molecular science over the past forty years or so. Considerable insight into the molecular nature of electrochemical reactions has been provided by new techniques that couple electrochemical approaches with spectroscopic techniques and computational methods. The oxidation and reduction of single molecules have been detected and studied using the scanning electrochemical microscope, an analog of the scanning tunneling microscope described in Chapter 1, for example. Electrochemical methods are now widely used in medical diagnostics, where they routinely measure very low concentrations of molecules of clinical significance; current detection limits are the attomole (10^{-18} mol) level, which is about a million molecules. Finally, the current national interest in alternative energy provides enormous opportunities for research and development in electrochemistry. Solar energy conversion, fuel cells, and energy storage are technologies that rely on electrochemical transformations that occur at the molecular level. Electrochemistry has already made and will undoubtedly continue to make important contributions to developing improved energy technologies. New experimental and theoretical understanding of electrochemical processes at the molecular level, along with the opportunities to apply these methods to technological problems of global importance, make it a particularly exciting time to learn about the field.

We begin our discussion of electrochemistry by reviewing redox reactions in aqueous solution, introducing the components of electrochemical cells, and distinguishing galvanic cells from electrolytic cells. We recommend that you review the procedure in Section 11.4 for balancing redox reactions before continuing on to the next section, in which we discuss electrochemical processes from both thermodynamic and molecular points of view, as well as a variety of applications.

17.1 Electrochemical Cells

Electrochemical reactions interconvert chemical and electrical energy through coupled redox reactions in which, generally, the oxidation and reduction half-reactions are separated in space in an **electrochemical cell.** Each half-reaction occurs on or near the surface of an **electrode**, and the electrons produced at the electrode where oxidation is occurring are transferred to the other electrode by a wire. The half of the electrochemical cell in which reduction occurs is also connected to the half of the electrochemical cell where oxidation occurs by an ionic conductor to allow ions to move between the solutions surrounding the electrodes, so that they remain electrically neutral overall. Each is called a **half-cell.** The Gibbs free energy made available in a spontaneous reaction in a **galvanic cell** can be converted into electrical energy, which can be used to do work. Conversely, electrical work done *on* the system by an external power supply in an **electrolytic cell** provides a source of free energy to drive redox reactions that are not normally spontaneous. Perhaps the most powerful characteristic of electrochemistry is its ability to control reversible chemical reactions by supplying or extracting electrical energy; this ability enables both very large-scale industrial chemical technologies (making chlorine from salt water or extracting metals from their ores) as well as very sensitive schemes for chemical analysis (sub-picogram sensitivity for clinical applications).

17.1.1 Galvanic Cells

Let's begin by considering the following redox reaction

$$Cu(s) + 2\ Ag^{+}(aq) \longrightarrow Cu^{2+}(aq) + 2\ Ag(s)$$

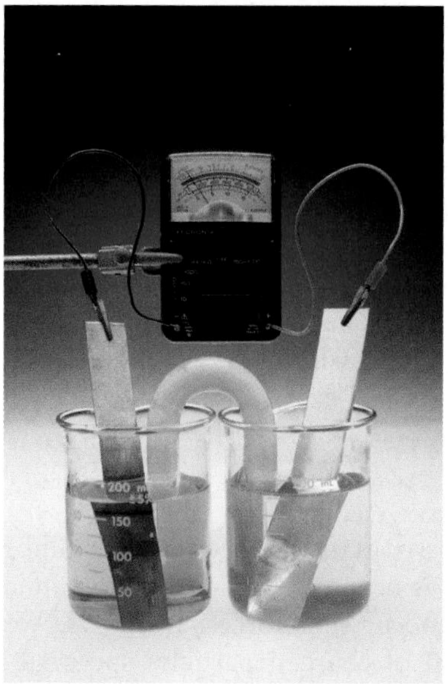

FIGURE 17.1 A metallic copper anode reacts to give a blue solution containing copper(II) ions as silver ions plate out on a silver cathode in a galvanic cell. Typically ammeters assume positive charge carriers, so point opposite to the electron flow.

which was first discussed in Section 11.4 and illustrated in Figure 11.13. It is clear from the figure in Chapter 11 that copper metal has been oxidized to form copper ions (which are responsible for the blue color), and silver ions have been reduced to silver metal, which appears in the form of the tree-like structures seen coating the wire. The reaction may be written as the sum of two half-reactions:

$$Cu(s) \longrightarrow Cu^{2+}(aq) + 2e^-$$

$$2\ Ag^+(aq) + 2e^- \longrightarrow 2\ Ag(s)$$

which makes it clear that copper metal has been oxidized (it has lost electrons) and silver ions have been reduced (they have gained electrons). Let's now run the same reaction in a galvanic cell, like the one shown in the photograph in Figure 17.1 and also illustrated schematically in Figure 17.2.

The beaker on the left contains a strip of copper metal immersed in an aqueous solution of $Cu(NO_3)_2$, while the beaker on the right contains a strip of silver metal immersed in an aqueous solution of $AgNO_3$. A wire connects the two metal electrodes, allowing electrons to flow between them, and the ionic conductor connecting the cells in this case is called a **salt bridge** (an ionic solution such as $NaNO_3$ stabilized in a gel or prevented from leaving the bridge by porous plugs on each end). Copper has still been oxidized (as evidenced by the blue color of the $Cu^{2+}(aq)$ ions in Fig. 17.1), and silver ions have still been reduced (as could be determined by weighing the silver electrode). The difference, however, is that the energy of the electrons being transferred through the wire in the galvanic cell is available to do work (run an electric motor, for example). The **ammeter** shown in the figure measures the magnitude and the direction of the electrical current that flows between the electrodes (in this case the electrons flow from the copper to the silver electrode).

Electrochemists have defined a standard terminology and notation that applies to both galvanic and electrolytic cells. The **anode** is always the electrode at which oxidation occurs, and the **cathode** is always the electrode at which reduction occurs; these names were coined by Michael Faraday, along with many other terms used in electrochemistry. The components of electrochemical cells are represented by a shorthand notation, which, for the cell just described, is written as

$$Cu|Cu^{2+}||Ag^+|Ag$$

FIGURE 17.2 A schematic of the galvanic cell shown in Figure 17.1. Electrons flow from the copper electrode to the silver electrode through an external circuit and the current is measured with an ammeter. Anions migrate toward the copper electrode and cations move toward the silver electrode in the respective solutions with sodium and nitrate ions moving through the salt bridge to maintain electrical neutrality.

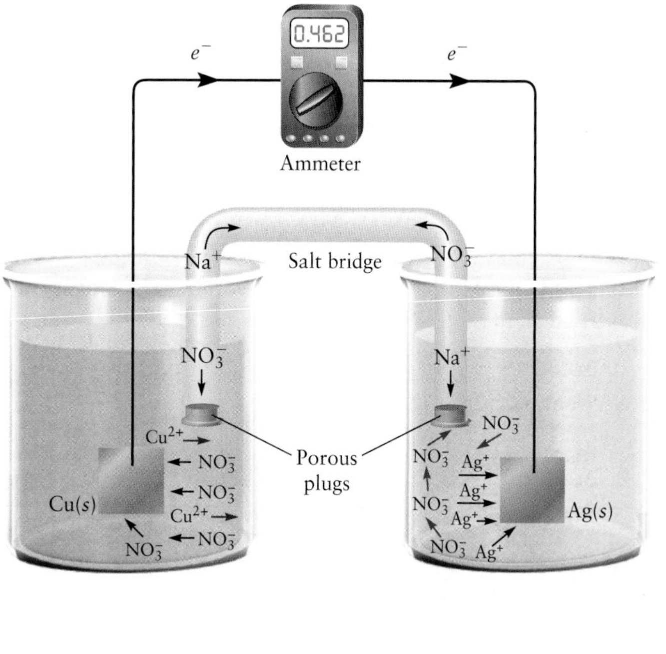

in which the single vertical lines represent phase boundaries (between the metal electrodes and the electrolyte solution in this example) and the double lines represent the salt bridge. The oxidation (anode) half-reaction is always written on the left, by convention, irrespective of whether the cell is a galvanic cell or an electrolytic cell. Species in aqueous solution are written immediately adjacent to the double line. Commas are used to indicate the presence of different components in the same phase, an aqueous solution of hydrochloric acid being represented as $|H^+, Cl^-|$, for example. Finally, unless otherwise indicated, you may assume that all reactions are carried out in aqueous solution and that the shorthand notation used for the ions H^+, Ag^+, and Cl^- implies that we mean $H^+(aq)$, $Ag^+(aq)$, and $Cl^-(aq)$, respectively.

17.1.2 Electrolytic Cells

The galvanic cell reaction described above is spontaneous because the Gibbs free energy change for the reaction as written is negative. This thermodynamic driving force can be related to an electrostatic driving force as follows. We showed in Section 3.3 that systems composed of charged particles spontaneously seek states of the lowest electrostatic potential energy just like mechanical systems spontaneously seek states of the lowest mechanical potential energy. Electrons being transferred in a galvanic cell reaction spontaneously move from a region of higher electrostatic potential energy on the anode to a region of lower electrostatic potential energy on the cathode, in the same way that water flows downhill. Electrons will flow from the anode to the cathode until the potential energies of the electrons in both half-cells are equal to one another, at which point the system has reached equilibrium; there is no longer any electrostatic driving force for the reaction.

Galvanic cell reactions occur spontaneously as written; consequently, their reverse reactions are not spontaneous. The reverse reactions can be made to occur by applying an external driving force that increases the electrostatic potential energies of electrons on the cathode side of the reaction until they are higher than those on the anode side. The external driving force is provided by a power supply in electrolytic cells. Electrons flow in the opposite direction, the species being oxidized when the cell was operating as a galvanic cell is now being reduced, and the electrode originally serving as the anode becomes the cathode and vice versa.

EXAMPLE 17.1

The final step in the production of magnesium from seawater is the electrolysis of *molten* magnesium chloride, with the overall reaction

$$Mg^{2+} + 2\,Cl^- \longrightarrow Mg(\ell) + Cl_2(g)$$

Write equations for the half-reactions occurring at the anode and at the cathode, and identify the direction in which electrons flow through the external circuit.

Solution

Oxidation, the loss of electrons, occurs at the anode, by the half-reaction

$$2\,Cl^- \longrightarrow Cl_2(g) + 2\,e^-$$

Reduction, the gain of electrons, occurs at the cathode, by the half-reaction

$$Mg^{2+} + 2\,e^- \longrightarrow Mg(\ell)$$

Electrons released at the anode, where chlorine is liberated, travel through the external circuit to the cathode, where molten magnesium is produced.

Related Problems: 1, 2

It is customary in electrochemistry to measure the difference between the **electrostatic potential** of the electrons at the cathode and the anode rather than the differences in their electrostatic potential energy. The electrostatic potential of a positively charged particle or a negatively charged particle at a particular point in space (or in your electrochemical cell) is typically given the symbol Φ, but in electrochemistry it is often given the symbol E. This is confusing because E is also often used for energy, so we will describe electrostatic potential using the symbol Φ, then introduce the change in notation afterward when you have a good understanding of the difference between electrostatic potential and electrostatic potential energy.

The SI unit for potential is the volt: 1 V = 1 J C^{-1}. The change in the potential energy of a particle of a unit positive test charge (+1 C) that results from a change in potential of +1 V is given by

$$\Delta E_P = q\Delta\Phi = (1\ \text{C})(1\ \text{V}) = 1\ \text{J}$$

where q is the charge on the particle. Of course, if the moving particle is an electron, q is negative, so the change in potential energy is negative when the electron moves through the same potential difference $\Delta\Phi$ of +1 V. Using the formula above, in MKS units, the change in potential energy of an electron of charge -1.6×10^{-19} C when it moves through a potential difference of +1 V is

$$\Delta E_P = q\Delta\Phi = -e(\Delta\Phi) = (-1.6 \times 10^{-19}\ \text{C})(+1\ \text{V}) = -1.6 \times 10^{-19}\ \text{J}$$

The energy unit eV is particularly convenient, since the above result is −1 eV. You can conclude that if you stretch a wire between the positive and negative terminals of a 9 V battery, an electron would flow from the negative terminal to the positive terminal, thus increasing its electrostatic potential, corresponding to changing its potential energy by −9 eV. For an electron, uphill in electrostatic potential is downhill in potential energy. It is customary in electrochemical applications to replace the symbol for a change in electrostatic potential $\Delta\Phi$ by ΔE, but this E is not energy, it is electrostatic potential. You can readily tell which E is which since energy would typically have units of joules, J, or electron volts, eV, and electrostatic potential is given in units of volts, V, where 1 V = 1 J C^{-1}. Thus if you use eV as your unit of energy, in this case electrostatic potential energy, and V for your unit of electrostatic potential, an electron moving through an electrostatic potential difference ΔE of +1 V has undergone a change in potential energy ΔE_P of −1 eV.

The electrostatic driving force in electrochemistry is expressed in terms of the potential, rather than the potential energy. Electrons spontaneously move from regions of higher potential energy to regions of lower potential energy, as discussed earlier, or from more negative to more positive potentials. Spontaneous processes are those for which $\Delta E_P < 0$ or $\Delta E > 0$ (the latter is the change in electrostatic potential in volts). These are equivalent statements, but you should be alert to the fact that the signs are opposites of one another and that spontaneous processes are characterized by *increases* in the potential, in contrast to what you have been used to seeing for spontaneous processes expressed in terms of changes in thermodynamic variables. We connect the two driving forces in the next section. A **voltmeter** measures the potential differences between two half-cells, the result being the **cell potential** (or voltage) as discussed later. Reversing the direction of a spontaneous process requires the application of an external potential that is greater in magnitude and opposite in sign to the potential of the spontaneous reaction. External potentials can be applied by inserting a power supply or a battery in the circuit of an electrolytic cell as shown in Figure 17.3.

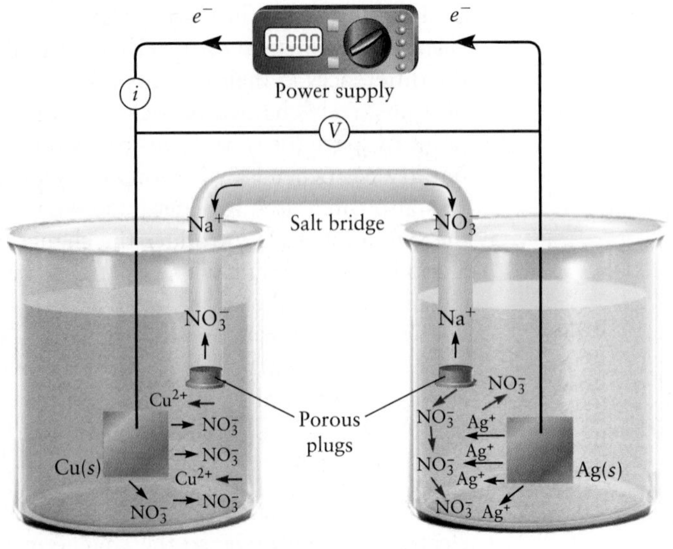

FIGURE 17.3 The spontaneous reaction in the galvanic cell shown in Figure 17.2 can be reversed by the application of an external potential from a power supply connected into the circuit as shown schematically. The power supply is in the off position in the figure; the meter would read either voltage or current during operation.

17.1.3 Faraday's Laws

Alessandro Volta discovered the galvanic cell in 1800 and constructed a "battery of cells" consisting of a number of platelets of silver and zinc that were separated from one another by porous strips of paper saturated with a salt solution. This voltaic pile, as it was then known, demonstrated that electricity could be produced by chemical reactions. By 1807, Sir Humphry Davy had prepared elemental sodium and potassium using a battery to electrolyze their respective hydroxides, demonstrating that electricity could drive chemical reactions. Michael Faraday was the first to demonstrate quantitative relationships between the amount of charge that flowed and the quantities of materials produced or consumed in electrochemical reactions, from which he asserted the fundamental duality of these processes: "the electricity which decomposes, and that which is evolved by the decomposition of, a certain quantity of matter, are alike." Faraday arranged a series of electrochemical cells through which the same amount of charge was passed and measured the quantities of the materials produced or consumed. The relationships he discovered are known as **Faraday's laws**, which we state as follows:

1. The mass of a given substance that is produced or consumed in an electrochemical reaction is proportional to the quantity of electric charge passed.
2. Equivalent masses[1] of different substances are produced or consumed in electrochemical reactions by a given quantity of electric charge passed.

These laws, which summarize the stoichiometry of electrochemical processes, were discovered by Michael Faraday in 1833, more than half a century before the electron was discovered and the nature of electricity was understood on the atomic scale. Faraday also measured, for the first time, the charge to mass ratio of the hydrogen ion, long before its chemical nature had been definitively established, by measuring the amount of charge required to generate 1 g of hydrogen gas. The value he obtained (about 1000 times smaller than e/m_e) provided a benchmark against which J. J. Thomson would later compare the charge

[1]The electron had not yet been discovered at the time of Faraday's experiments and so he was only able to assert that there were "equivalent masses" produced or consumed when a certain amount of charge was passed. The modern definition of the equivalent mass of an element or a compound in a redox reaction is its molar mass divided by the number of moles of electrons transferred per mole of substance in the corresponding half-reaction.

to mass ratio of the electron and conclude that "the charge on the electron must be very large or the mass of the electron must be very small, compared to the hydrogen ion."

Let's express Faraday's laws in modern terms and relate the changes in masses observed to the charge that flows in electrochemical experiments. The magnitude of the charge on a single electron (expressed in coulombs) has been very accurately determined to be

$$|-e| = e = 1.602176565 \times 10^{-19}\ \text{C}$$

so the magnitude of the charge of one mole of electrons is equal to

$$Q = (6.02214129 \times 10^{23})(1.602176565 \times 10^{-19}\ \text{C})$$
$$= 96{,}485.34\ \text{C}$$

which is an enormously large quantity of charge. Here we have introduced the symbol $Q = Ne$ for the magnitude of charge that flowed in the electrochemical cell. It is equal to the absolute value of the charge on an electron multiplied by the total number of electrons that were transferred during the electrochemical reaction. The magnitude of the charge per mole of electrons has been given a special name and symbol, the **Faraday constant,**

$$F = 96{,}485.34\ \text{C mol}^{-1}$$

in recognition of the central importance of the mole concept in chemistry.

We typically measure the total magnitude of charge passed in electrochemical experiments by measuring the current and the time. The **electric current** is the amount of charge that flows through a circuit per second, measured in the SI base unit, the ampere (A). The coulomb (C), a derived unit, is defined as the total magnitude of charge transferred by 1 A of current flowing for one second.

$$1\ \text{C} = (1\ \text{A})(1\ \text{s}) \qquad [17.1]$$

We can use Equation 17.1 to calculate the total magnitude, or amount, of charge that has passed when a current of I amperes has flowed for t seconds as

$$Q = It$$

or express the number of moles of electrons n transferred as

$$n = \frac{It}{96{,}485\ \text{C mol}^{-1}}$$

We can calculate the number of moles (and therefore the number of grams) of reactants consumed and products formed in electrochemical reactions if we know how many electrons are transferred in a particular reaction and how many moles of electrons have passed through the cell. Suppose, for example, that we have a zinc–silver galvanic cell in which the anode half-reaction is

$$\text{Zn}(s) \longrightarrow \text{Zn}^{2+}(aq) + 2\,e^{-}$$

and the cathode half-reaction is

$$\text{Ag}^{+}(aq) + e^{-} \longrightarrow \text{Ag}(s)$$

Each mole of electrons produced arises from the oxidation of 1/2 mol of zinc metal (because each Zn atom gives up two electrons) and causes the reduction of 1 mol of silver ions. From the molar masses of silver and zinc, we calculate that (65.38)/2 = 32.69 g of zinc is dissolved at the anode, and 107.87 g of silver is deposited at the cathode. The same relationships hold if the cell is operated as an electrolytic cell, but in that case silver is dissolved and zinc is deposited.

EXAMPLE 17.2

An electrolytic cell is constructed in which the silver ions in silver chloride are reduced to silver at the cathode and copper is oxidized to $Cu^{2+}(aq)$ at the anode. A current of 0.500 A is passed through the cell for 101 minutes. Calculate the mass of copper dissolved and the mass of silver deposited.

Solution

$$t = (101 \text{ min})(60 \text{ s min}^{-1}) = 6.06 \times 10^3 \text{ s}$$

The number of moles of electrons passed through the circuit during this time is given by

$$n = \frac{(0.500 \text{ C s}^{-1})(6.06 \times 10^3 \text{ s})}{96{,}485 \text{ C mol}^{-1}} = 3.14 \times 10^{-2} \text{ mol } e^-$$

The half-cell reactions are

$$AgCl(s) + e^- \longrightarrow Ag(s) + Cl^-(aq) \qquad \text{(cathode)}$$

$$Cu(s) \longrightarrow Cu^{2+}(aq) + 2\,e^- \qquad \text{(anode)}$$

so the masses of silver deposited and copper dissolved are

$$\text{mass Ag} = (3.14 \times 10^{-2} \text{ mol } e^-) \times \left(\frac{1 \text{ mol Ag}}{1 \text{ mol } e^-}\right) \times (107.87 \text{ g mol}^{-1})$$

$$\text{mass Ag} = 3.39 \text{ g Ag deposited}$$

$$\text{mass Cu} = (3.14 \times 10^{-2} \text{ mol } e^-) \times \left(\frac{1 \text{ mol Cu}}{2 \text{ mol } e^-}\right) \times (63.55 \text{ g mol}^{-1})$$

$$\text{mass Cu} = 0.998 \text{ g Cu dissolved}$$

Related Problems: 3, 4, 5, 6

17.2 Cell Potentials and the Gibbs Free Energy

We discussed pressure–volume work in Chapters 12 and 13 and, in particular, demonstrated that the maximum work done by the system occurs for reversible processes. We now consider a different kind of work, **electrical work**, which is fundamental in electrochemistry, and we investigate its properties. We showed earlier that the change in the potential energy ΔE_P associated with the transfer of one electron through a potential difference $\Delta\Phi$ equals $-e\Delta\Phi$. If N electrons are transferred, then $\Delta E_P = -Ne\Delta\Phi$. Using $Ne = Q$ gives

$$\Delta E_P = -Q\Delta\Phi = -nF\Delta\Phi$$

where n is the number of moles of electrons transferred, and, because the electrons are negatively charged, the charge Q on n moles of electrons equals $-nF$ (where F is Faraday's constant).

Throughout this text we will refer to the difference in the potential $\Delta\Phi$ as the cell potential difference ΔE_{cell} so the change in the potential energy is written as

$$\Delta E_P = -Q\Delta E_{cell} = -nF\Delta E_{cell}$$

where ΔE_{cell} is measured in volts and ΔE_P is measured in joules. The change in the potential energy *of the electrons* is defined as the **electrical work**, w_{elec}, done on the electrons. (In a galvanic cell running spontaneously, w_{elec} is a negative number since the cell is doing net work on the surroundings.)

$$w_{elec} = \Delta E_P = -Q\Delta E_{cell} = -nF\Delta E_{cell} \quad [17.2]$$

where $\Delta E_{cell} = E_{cathode} - E_{anode}$ (because the final location of the electrons is at the cathode, where they execute the reduction reaction, and their initial location is at the anode, where they are produced in the oxidation reaction). One would typically measure the voltage difference ΔE_{cell} before connecting the wire between the two half cells. (Some texts simply use E_{cell}, but since this is a potential difference we retain the symbol Δ.)

Work is done *on* the system when electrons move from regions of more positive potentials to regions of more negative potentials (increasing their potential energy) and work is done *by* the system when electrons move from regions of more negative potentials to regions of more positive potentials (decreasing their potential energy). The cell potential ΔE_{cell} is positive for galvanic cells, which can produce electrical work, and negative for electrolytic cells in which electrical work is done *on* the system by an external power supply.

EXAMPLE 17.3

A 6.00 V battery delivers a steady current of 1.25 A for a period of 1.50 hours. Calculate the total magnitude of charge Q, in coulombs, that passes through the circuit and the electrical work done *by* the battery.

Solution

Recalling that 1 A = 1 C s^{-1} and 1 V = 1 J C^{-1} we calculate the total magnitude of charge

$$Q = It = (1.25 \text{ C s}^{-1})(1.50 \text{ hr})(3600 \text{ s hr}^{-1})$$
$$= 6750 \text{ C}$$

and the electrical work done *on* the battery (noting that electrons have a negative charge, and flow spontaneously uphill in cell potential) is

$$w_{elec} = -Q\Delta E_{cell} = -(6750 \text{ C})(6.00 \text{ J C}^{-1})$$
$$= -4.05 \times 10^4 \text{ J}$$

This is the work done *on* the battery, so the work done *by* the battery is the negative of this quantity, or +40.5 kJ.

The change in the Gibbs free energy not only predicts the direction of spontaneous change but is also equal to the reversible work done (other than pressure–volume work) on the system at constant temperature and pressure, which we identify as the electrical work w_{elec} as follows:

$$w_{elec,rev} = \Delta G \qquad \text{(at constant } P \text{ and } T) \quad [17.3]$$

This equation shows that if the electrochemical reaction has a negative ΔG, then w_{elec} is negative and the electrochemical cell can do net work on the surroundings.

We derive this result beginning with the definition of the Gibbs free energy function G:

$$G = H - TS = U + PV - TS$$

For processes occurring at constant pressure P and constant temperature T (the usual case in electrochemical cells),

$$\Delta G = \Delta U + P\Delta V - T\Delta S$$

The first law of thermodynamics is

$$\Delta U = q + w$$

which may be rewritten as

$$\Delta U = q + w_{\text{elec}} - P\Delta V$$

to explicitly identify electrical work w_{elec} and pressure–volume work, which is $-P_{\text{ext}}\Delta V = -P\Delta V$ at constant internal pressure. Substituting for ΔU into the equation for the change in the Gibbs free energy gives

$$\Delta G = q + w_{\text{elec}} - P\Delta V + P\Delta V - T\Delta S = q + w_{\text{elec}} - T\Delta S$$

For electrochemical reactions that are run reversibly (which is rather easy to achieve in practice by controlling the current)

$$q = q_{\text{rev}} = T\Delta S$$

allowing us to show that reversible electrical work is equal to the change in the Gibbs free energy

$$w_{\text{elec,rev}} = \Delta G$$

The change in the Gibbs free energy of the system is negative for a galvanic cell, so w_{elec} is negative and there is net electrical work done *by* the system. Energy, in the form of useful electrical work, is extracted *from* the system and made available to do work *on* the surroundings. The opposite is true for electrolytic cells. Work is done *on* the system, increasing its Gibbs free energy, most commonly as an increase in the potential energy of the electrons. The maximum work done on the surroundings ($-w_{\text{elec}}$) by a galvanic cell is done when the cell is made to operate reversibly, because it is working against the maximum possible opposing force at all times, just like for reversible PV work. Cells operating irreversibly (in other words, with large currents permitted to flow) produce less electrical work than those operating reversibly.

The electrical work w_{elec} is also equal to the product of the charge transferred and the difference in the electrostatic potential, which can be calculated as follows

$$w_{\text{elec,rev}} = \Delta G = -nF\Delta E_{\text{cell}}$$

for n moles of electrons moving through a potential difference ΔE_{cell}. Galvanic cells, for which $\Delta E_{\text{cell}} > 0$, can produce electrical work, whereas electrolytic cells, for which $\Delta E_{\text{cell}} < 0$, consume electrical work.

17.2.1 Standard States and Standard Cell Potentials

Recall from Chapters 12 and 14 that the *standard states* of chemical substances are defined at 1 atm pressure and at a specified temperature. Electrochemists use *activities* (see Section 14.3) to define standard states, the standard states of gases and solutes being defined at unit activity. Recall that the standard state for solutes is defined to be a *hypothetical* state in which the solute concentration is 1 M, but the solute behaves as if it were ideal. The potential of the standard hydrogen electrode (SHE), the primary reference electrode used in electrochemistry (see Section 17.2.2), is defined under conditions of unit activity for the hydrogen ions in solution as well as unit activity for the hydrogen gas in equilibrium with the electrolyte. We specify concentrations in molarity in this chapter, but alert you that 1 M is not unit activity when reading more advanced treatments of the subject. The standard free energy change ΔG° for a reaction in which all reactants and products are in their standard states can be calculated from a table of standard free energies of formation ΔG_f° of the reactants and the products (see Appendix D). We have shown that the cell potential is related to the change in the Gibbs free energy for reversible processes, allowing us to define a **standard cell potential** ΔE°_{cell} as

$$\Delta G^\circ = -nF\Delta E^\circ_{cell} \qquad \text{(reversible)} \qquad [17.4]$$

in which all reactants and products are in their standard states (gases at 1 atm pressure, solutes at 1 M concentration, pure metals in their most stable states and at a specified temperature). It is important to note that the condition of reversibility implies that no current is flowing. Standard cell potentials are measured using voltmeters having extremely high internal resistance to ensure that this is the case; these potentials are also referred to as *open-circuit* potentials. Standard cell potentials are intrinsic electrical properties of electrochemical cells that can be calculated from the associated standard free energy changes ΔG° using Equation 17.4. Conversely, and perhaps more importantly, ΔG° (and hence equilibrium constants) for a variety of chemical reactions can be determined simply by measuring the standard cell potentials (ΔE°_{cell}) for reactions that can be carried out by electrochemical means. Many tabulated equilibrium constants have, in fact, been measured using electrochemical methods because they are more convenient and, often, more accurate than other methods. For example, the concentrations of ions in solutions of sparingly soluble salts (see Section 16.1) are quite low but can be readily determined by measuring cell potentials.

EXAMPLE 17.4

A $Zn^{2+}|Zn$ half-cell is connected to a $Cu^{2+}|Cu$ half-cell to make a galvanic cell, in which $[Zn^{2+}] = [Cu^{2+}] = 1.00$ M. The cell potential at 25°C is measured to be $\Delta E^\circ_{cell} = 1.10$ V, which corresponds to the overall cell reaction

$$Zn(s) + Cu^{2+}(aq) \longrightarrow Zn^{2+}(aq) + Cu(s)$$

Calculate ΔG° for this reaction under the given conditions.

Solution

$$\begin{aligned}\Delta G^\circ &= -nF\Delta E^\circ_{cell} \\ &= -(2.00 \text{ mol})(96{,}485 \text{ C mol}^{-1})(1.10 \text{ V}) \\ &= -2.12 \times 10^5 \text{ J} = -212 \text{ kJ}\end{aligned}$$

Related Problems: 11, 12

17.2.2 Standard Reduction Potentials

We could tabulate cell potentials for every conceivable electrochemical reaction, but the list would be very long. We can instead construct electrochemical cells in which one half-reaction has been chosen to be our reference half-reaction and then measure the cell potentials that result when this reference half-reaction is coupled to all other half-reactions of interest. This procedure is analogous to the one we used to find the changes in internal energy, enthalpy, entropy, and Gibbs free energy for chemical reactions using the standard energies and enthalpies of formation, absolute entropies, and standard Gibbs free energies of formation.

The primary reference electrode has been chosen, by convention, to be the **standard hydrogen electrode** (SHE), often called the **normal hydrogen electrode** (NHE) by electrochemists. The standard hydrogen electrode, as illustrated in Figure 17.4, consists of a platinum electrode immersed in a solution in

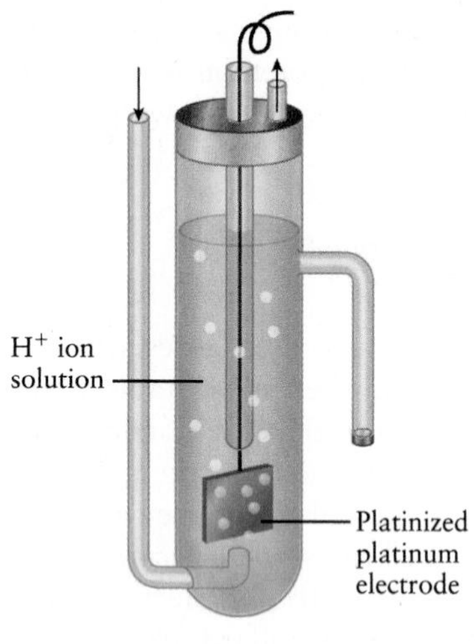

FIGURE 17.4 Schematic of a standard hydrogen electrode.

which $[H_3O^+] = 1$ M (or, more precisely, at a concentration of H_3O^+ where $a_{H_3O^+} = 1$), under a hydrogen partial pressure $P_{H_2} = 1$ atm ($a_{H_2} = 1$).[2]

Cell potentials for a number of half-reactions have been measured with reference to the SHE, which has been defined to be 0.00 V, and the results tabulated as standard *reduction* potentials by convention (see Appendix E). Half-reactions with positive potentials, with respect to SHE, proceed as reductions whereas half-reactions with negative potentials, with respect to SHE, proceed as oxidations. Standard cell potentials for any electrochemical cell can be calculated from standard reduction potentials using

$$\Delta E^{\circ}_{cell} = E^{\circ}_{cathode} - E^{\circ}_{anode} \qquad [17.5]$$

where $E^{\circ}_{cathode}$ is the standard reduction potential for the reaction occurring at the cathode (reduction) and E°_{anode} is the standard reduction potential for the reaction occurring, in reverse, at the anode (oxidation). The procedure for finding standard cell potentials for any cell is simple:

1. Identify the cathode half-reaction.
2. Identify the anode half-reaction.
3. $\Delta E^{\circ}_{cell} = E^{\circ}_{cathode} - E^{\circ}_{anode}$

If all the reactants and products are in their standard states, reactions for which $\Delta E^{\circ}_{cell} > 0$ are spontaneous as written and can be used to construct galvanic cells; those for which $\Delta E^{\circ}_{cell} < 0$ are not spontaneous as written and can be driven only by applying an external potential in electrolytic cells. The procedure we just described for finding cell potentials *always works* and can help avoid confusion with sign conventions when working electrochemical problems.

Let's illustrate this procedure for calculating a standard cell potential using the $Zn|Zn^{2+}||Cu^{2+}|Cu$ cell described in Example 17.4. The overall reaction is

$$Zn(s) + Cu^{2+}(aq) \longrightarrow Cu(s) + Zn^{2+}(aq)$$

in which $Cu^{2+}(aq)$ is reduced at the cathode and $Zn(s)$ is oxidized at the anode. The standard reduction potentials of the two half-reactions obtained from Appendix E are:

$$Zn^{2+}(aq) + 2\,e^- \longrightarrow Zn(s) \qquad E^{\circ} = -0.76 \text{ V}$$

$$Cu^{2+}(aq) + 2\,e^- \longrightarrow Cu(s) \qquad E^{\circ} = 0.34 \text{ V}$$

and so the standard cell potential for the $Zn|Zn^{2+}||Cu^{2+}|Cu$ cell is

$$\Delta E^{\circ}_{cell} = E^{\circ}_{cathode} - E^{\circ}_{anode} = 0.34 \text{ V} - (-0.76 \text{ V}) = 1.10 \text{ V}$$

in agreement with the value measured in Example 17.4. Note that it is conventional in electrochemistry to write positive potentials without explicitly including the sign.

The standard reduction potentials tabulated in Appendix E are arranged in order, with the most positive potentials at the top and the most negative potentials at the bottom. We can determine immediately by inspection whether a given reaction will be spontaneous or not; galvanic cells are those for which the

[2]Note that some sources measure the standard reduction potentials with respect to a different reference electrode than the SHE; a popular choice was the saturated calomel electrode (SCE), described later in this chapter and depicted as the reference electrode in Figure 17.5. Noting that the reduction potential for the SCE is 0.242 V with respect to the standard SHE reference cell, one need only add 0.242 V to any reduction potential measured with respect to the SCE to convert it to one measured with respect to the SHE.

reduction potential of the cathode half-reaction is more positive (higher on the list) than the reduction potential of the anode half-reaction (lower on the list), as long as both half-cells are in their standard states. (See Section 17.4 for cells not in their standard states.)

The procedure we have just outlined can be used to find cell potentials from standard reduction potentials for any reaction of interest, even those for which it is necessary to multiply one half-reaction by an integer to balance the charge transferred in the overall cell reaction. Let's calculate the cell potential for the $Cu|Cu^{2+}||Ag^{+}|Ag$ cell we introduced at the beginning of the chapter. The reaction is

$$2\ Ag^{+}(aq) + Cu(s) \longrightarrow 2\ Ag(s) + Cu^{2+}(aq)$$

which we can represent as the sum of the following two half-reactions:

$$2\ Ag^{+}(aq) + 2\ e^{-} \longrightarrow 2\ Ag(s)$$
$$Cu(s) \longrightarrow Cu^{2+}(aq) + 2\ e^{-}$$
$$\overline{2\ Ag(s) + Cu^{2+}(aq) \longrightarrow 2\ Ag^{+}(aq) + Cu(s)}$$

We identify the reduction of Ag^{+} as the cathode half-reaction and calculate the standard cell potential as before:

$$\Delta E^{\circ}_{cell} = E^{\circ}_{cathode} - E^{\circ}_{anode} = 0.799\text{ V} - 0.34\text{ V} = 0.459\text{ V}$$

Note that we did *not* multiply the standard reduction potential for the half-reaction $Ag^{+}(aq) + e^{-} \longrightarrow Ag(s)$ by 2 when calculating the standard cell potential because potentials are intensive quantities that do not depend upon the quantity of material involved. The cell potential is simply the difference in the potentials of the two half-reactions. The change in potential energy, on the other hand, or that of any other extensive thermodynamic property of interest, does depend on the quantity of material involved. The change in the Gibbs free energy of the system discussed previously, for example, is

$$\Delta G^{\circ} = -nF\Delta E^{\circ}_{cell} = (-2\text{ mol})(96{,}485\text{ C mol}^{-1})(0.459\text{ V}) = -88.6\text{ kJ}$$

The negative ΔG° indicates the reaction would proceed spontaneously (given that the solutes are at standard state concentrations). The work done *on* the system, w_{elec}, is negative, so the electrical work done *by* the cell is 88.6 kJ (if the cell were run with an opposing force so it could run close to reversibly).

Suppose we want to find the standard reduction potential for a half-reaction that is not listed in Appendix E. We could write the half-reaction of interest as the sum or difference of a pair of half-reactions tabulated in Appendix E. The number of electrons transferred in the reaction of interest, however, may be different than the numbers of electrons transferred in the standard reactions, and so we must first calculate the Gibbs free energy change for the reaction and then calculate the standard cell potential using Equation 17.4, as illustrated by the following example.

Suppose we want to find the standard reduction potential for the half-reaction

$$Cu^{2+} + e^{-} \longrightarrow Cu^{+} \qquad E^{\circ}_{3}(Cu^{2+}|Cu^{+}) = ?$$

using the following standard reduction potentials from Appendix E.

$$Cu^{2+} + 2\ e^{-} \longrightarrow Cu \qquad E^{\circ}_{1} = E^{\circ}(Cu^{2+}|Cu) = 0.340\text{ V}$$
$$Cu^{+} + e^{-} \longrightarrow Cu \qquad E^{\circ}_{2} = E^{\circ}(Cu^{+}|Cu) = 0.522\text{ V}$$

We calculate the change in the Gibbs free energy for the half-reaction as

$$\Delta G_3^\circ = \Delta G_1^\circ - \Delta G_2^\circ$$

$$-n_3 F E_3^\circ = -n_1 F E_1^\circ - (-n_2 F E_2^\circ) = -n_1 F E_1^\circ + n_2 F E_2^\circ$$

$$E_3^\circ = \frac{n_1 E_1^\circ - n_2 E_2^\circ}{n_3}$$

The standard reduction potential for the half-reaction

$$Cu^{2+} + e^- \longrightarrow Cu^+$$

therefore, is

$$E_3^\circ(Cu^{2+}|Cu^+) = \frac{(2\text{ mol})(0.340\text{ V}) - (1\text{ mol})(0.522\text{ V})}{1\text{ mol}} = 0.158\text{ V}$$

We can also use the procedure just described to calculate cell potentials for balanced chemical reactions of interest from standard reduction potentials, but the number of electrons in the numerator and the denominator will always be equal to one another and will cancel out; it is easier, therefore, to use the expression defined above:

$$\Delta E_{\text{cell}}^\circ = E_{\text{cathode}}^\circ - E_{\text{anode}}^\circ$$

EXAMPLE 17.5

An aqueous solution of potassium permanganate ($KMnO_4$) appears deep purple. The permanganate ion can be reduced to the pale-pink manganese(II) ion (Mn^{2+}) in aqueous acidic solutions. The standard reduction potential of an $MnO_4^-|Mn^{2+}$ half-cell is $E_{\text{cell}}^\circ = 1.49$ V. Suppose this half-cell is combined with a $Zn^{2+}|Zn$ half-cell in a galvanic cell, with $[Zn^{2+}] = [MnO_4^-] = [Mn^{2+}] = [H_3O^+] = 1$ M.

(a) Write equations for the reactions at the anode and the cathode.

(b) Write a balanced equation for the overall cell reaction.

(c) Calculate the standard cell potential difference, $\Delta E_{\text{cell}}^\circ$.

Solution

(a) Permanganate ions will be reduced at the cathode because $E°(MnO_4^-|Mn^{2+}) = 1.49$ V is more positive than $E°(Zn^{2+}|Zn) = -0.76$ V. The balanced half-cell reaction requires the presence of H_3O^+ ions and water, giving

$$MnO_4^- + 8\,H_3O^+ + 5\,e^- \longrightarrow Mn^{2+} + 12\,H_2O$$

The anode half-reaction for the oxidation of Zn is

$$Zn \longrightarrow Zn^{2+} + 2\,e^-$$

(b) The number of electrons delivered to the cathode must equal the number of electrons released by the anode, so the first equation must be multiplied by 2 and the second equation by 5 before adding them together to give the overall reaction:

$$2\,MnO_4^- + 16\,H_3O^+ + 5\,Zn \longrightarrow 2\,Mn^{2+} + 24\,H_2O + 5\,Zn^{2+}$$

(c) The galvanic cell potential is the difference between the standard reduction potential for permanganate (at the cathode) and that for zinc (at the anode):

$$\Delta E° = E°(MnO_4^-|Mn^{2+}) - E°(Zn^{2+}|Zn)$$

$$= 1.49 - (-0.76) = 2.25\text{ V}$$

Note that the half-cell potentials are not multiplied by their coefficients (2 and 5) before subtraction. Half-cell potentials are *intensive* properties of a galvanic cell and are therefore independent of the amount of the reacting species.

Related Problems: 13, 14

17.2.3 Oxidizing and Reducing Agents

Strong **oxidizing agents** are chemical species that are themselves easily reduced. These species are characterized by large positive reduction potentials and appear at the top of the list in Appendix E. Fluorine has the most positive reduction potential listed, and fluorine molecules are extremely eager to accept electrons to become fluoride ions, as expected from fluorine's large electron affinity and the effective solvation of $F^{-}(aq)$. Other strong oxidizing agents include hydrogen peroxide (H_2O_2) and aqueous solutions of permanganate ions (MnO_4^-). Strong **reducing agents,** on the other hand, are easily oxidized with very negative reduction potentials, and they are listed near the bottom of the table in Appendix E. The alkali and alkaline earth metals are especially good reducing agents, as expected from their low ionization potentials.

Oxygen itself is a good oxidizing agent in acidic solution at pH 0 because it has a reasonably positive standard reduction potential:

$$O_2 + 4\ H_3O^+ + 4\ e^- \longrightarrow 6\ H_2O \qquad E^\circ = 1.229\ V$$

Ozone (O_3) is an even stronger oxidizing agent than molecular oxygen, as shown by its more positive half-cell reduction potential in acidic aqueous solution:

$$O_3 + 2\ H_3O^+ + 2\ e^- \longrightarrow O_2 + 3\ H_2O \qquad E^\circ = 2.07\ V$$

The strong oxidizing power of ozone is exploited commercially as a bleach for wood pulp, and as a disinfectant and sterilizing agent for water, where it oxidizes algae and organic impurities but leaves no undesirable residue. The corresponding half-reactions in basic solutions, 1 M OH^-, are given below. They may be calculated using K_w and the methods in Section 17.3. Both oxygen and ozone are less effective oxidizing agents in basic solutions than in acidic solutions, as can be seen from their standard reduction potentials at pH 14 (standard basic conditions), which are significantly less positive than those under acidic conditions.

$$O_2 + 2\ H_2O + 4\ e^- \longrightarrow 4\ OH^- \qquad E^\circ = 0.401\ V$$

$$O_3 + H_2O + 2\ e^- \longrightarrow O_2 + 2\ OH^- \qquad E^\circ = 1.24\ V$$

Note that the O_2 and O_3 in these half reactions are gaseous species at a pressure of 1 atm in the standard state.

17.2.4 Reduction Potential Diagrams and Disproportionation

We can summarize the half-reactions of copper in a **reduction potential diagram** of the form

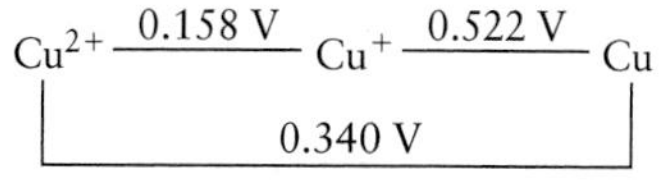

in which the species are arranged in sequence with the most oxidized form on the left and the most reduced form on the right. The standard reduction potential for each of the three half-reactions is written above the line that connects the half-reactions. Diagrams like these are very useful in helping us predict which ions are unstable with respect to *disproportionation,* a process in which a pair of ions in the same oxidation state are oxidized and reduced (see Example 11.10). A species will undergo a disproportionation reaction if the reduction potential for the half-reaction connecting the species in the center to the one on the right is more positive than the reduction potential for the half-reaction connecting the species in the center to the one on the left. The disproportionation reaction for Cu^+ is

$$2\ Cu^+ \longrightarrow Cu^{2+} + Cu \quad \Delta E^\circ = 0.552\ V - 0.158\ V = 0.324\ V$$

This reaction is spontaneous at standard conditions (and even at rather small concentrations of Cu^+) with negative $\Delta G°$ and positive $\Delta E°$. It is responsible for the negligible concentrations of Cu^+ found in aqueous solution. Interestingly enough, most transition metals, in fact, do not form stable M^+ ions in aqueous solution for this reason. The energy required to further ionize M^+ (essentially the second ionization energy, IE_2) is paid for by the large (in magnitude, as ΔH_{hyd} is negative in sign) energy of hydration of M^{2+}, making the disproportionation reaction thermodynamically favorable overall.

EXAMPLE 17.6

Hydrogen peroxide, H_2O_2, is a possible product of the reduction of oxygen in acidic solution:

$$O_2 + 2\,H_3O^+ + 2\,e^- \longrightarrow H_2O_2 + 2\,H_2O \qquad E_3^\circ = ?$$

It can then be further reduced to water:

$$H_2O_2 + 2\,H_3O^+ + 2\,e^- \longrightarrow 4\,H_2O \qquad E_2^\circ = 1.77\ \text{V}$$

(a) Use the half-cell potential just given for the reduction of H_2O_2, together with that given earlier,

$$O_2 + 4\,H_3O^+ + 4\,e^- \longrightarrow 6\,H_2O \qquad E_1^\circ = 1.229\ \text{V}$$

to calculate the standard half-cell potential for the reduction of O_2 to H_2O_2 in acidic solution.

(b) Write a reduction potential diagram for O_2, H_2O_2, and H_2O.

(c) Is H_2O_2 stable with respect to disproportionation in acidic, pH = 0, solution?

Solution

(a) The desired half-cell reaction is obtained by *subtracting* the reaction with potential E_2° from that with potential E_1°. The half-cell reduction potentials are not subtracted, however, but rather combined as described earlier in this section. Taking $n_1 = 4$, $n_2 = 2$, and $n_3 = 2$ gives

$$E_3^\circ = \frac{n_1E_1^\circ - n_2E_2^\circ}{n_3}$$

$$= \frac{(4\ \text{mol})(1.229\ \text{V}) - (2\ \text{mol})(1.77\ \text{V})}{2\ \text{mol}} = 0.69\ \text{V}$$

(b) The reduction potential diagram is obtained by omitting the electrons, water, and H_3O^+ from the corresponding half-equations:

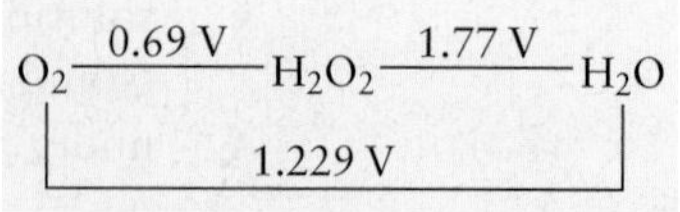

(c) H_2O_2 is thermodynamically unstable to disproportionation in acidic solution because the half-cell potential to its right (1.77 V) is higher than that to its left (0.69 V). The disproportionation of H_2O_2 is also spontaneous in neutral solution, but it is slow enough that aqueous solutions of hydrogen peroxide can be stored for a long time without deteriorating, as long as they are kept out of the light.

Related Problems: 23, 24, 25, 26

17.3 Concentration Effects and the Nernst Equation

Concentrations and pressures are rarely fixed at their standard state values in real-world applications. It is therefore necessary to understand how concentration and pressure affect cell potentials by applying the thermodynamic principles presented in Chapter 14 to electrochemical cells. In Chapter 14, we showed that the free energy change is related to the reaction quotient Q through

$$\Delta G = \Delta G° + RT \ln Q$$

Combining this equation with

$$\Delta G = -nF\Delta E_{cell}$$

and

$$\Delta G° = -nF\Delta E_{cell}^\circ$$

gives

$$-nF\Delta E_{\text{cell}} = -nF\Delta E^{\circ}_{\text{cell}} + RT \ln Q$$

from which we get

$$\Delta E_{\text{cell}} = \Delta E^{\circ}_{\text{cell}} - \frac{RT}{nF} \ln Q \qquad [17.6]$$

which is known as the **Nernst equation.**

The Nernst equation can be rewritten in terms of common (base-10) logarithms by using the fact that

$$\ln Q \approx 2.303 \log_{10} Q$$

At 25°C (298.15 K), the combination of constants 2.303 RT/F becomes

$$2.303 \frac{RT}{F} = (2.303) \frac{(8.315 \text{ J K}^{-1} \text{ mol}^{-1})(298.15 \text{ K})}{96{,}485 \text{ C mol}^{-1}}$$

$$= 0.0592 \text{ J C}^{-1} = 0.0592 \text{ V}$$

because 1 joule per coulomb is 1 volt. The Nernst equation then becomes

$$\Delta E_{\text{cell}} = \Delta E^{\circ}_{\text{cell}} - \frac{0.0592 \text{ V}}{n} \log_{10} Q \quad (\text{at } 25°\text{C}) \qquad [17.7]$$

which is its most familiar form, with n being the number of moles of electrons transferred in the overall chemical reaction as written. We remind you that these measurements are made under conditions in which no current flows and that n appears only to relate the thermodynamic and electrostatic work to one another. In a galvanic cell made from zinc, aluminum, and their ions, for example

$$3\ Zn^{2+} + 6\ e^{-} \longrightarrow 3\ Zn$$

$$2\ Al \longrightarrow 2\ Al^{3+} + 6\ e^{-}$$

$$\overline{3\ Zn^{2+} + 2\ Al \longrightarrow 2\ Al^{3+} + 3\ Zn}$$

$n = 6$ mol (per mol of stoichiometric reaction as written).

The Nernst equation is used to find the concentration dependence of half-cell potentials, as well.

$$E_{\text{hc}} = E^{\circ}_{\text{hc}} - \frac{0.0592 \text{ V}}{n_{\text{hc}}} \log_{10} Q_{\text{hc}}$$

for the potential of any half-cell reaction at 25°C, where n_{hc} is the number of moles of electrons appearing in the half-reaction and Q_{hc} is the reaction quotient for the half-cell reaction written as a reduction. Using the $Zn^{2+}|Zn$ half-cell as an example, the half-reaction written as a reduction is

$$Zn^{2+} + 2\ e^{-} \longrightarrow Zn$$

so $n_{\text{hc}} = 2$ and $Q_{\text{hc}} = 1/[Zn^{2+}]$. The cell potential for an electrochemical cell constructed from two half-cells that are not in their standard states is still calculated as before, $\Delta E_{\text{cell}} = E_{\text{cathode}} - E_{\text{anode}}$, but with the standard reduction potentials corrected to reflect the non-standard conditions using the Nernst equation.

EXAMPLE 17.7

Suppose the $Zn|Zn^{2+}||MnO_4^-|Mn^{2+}$ cell from Example 17.5 is operated at pH 2.00 with $[MnO_4^-] = 0.12$ M, $[Mn^{2+}] = 0.0010$ M, and $[Zn^{2+}] = 0.015$ M. Calculate the cell potential ΔE_{cell} at 25°C.

Solution

Recall that the overall reaction is

$$2\ MnO_4^-(aq) + 5\ Zn(s) + 16\ H_3O^+(aq) \longrightarrow 2\ Mn^{2+}(aq) + 5\ Zn^{2+}(aq) + 24\ H_2O(\ell)$$

From Example 17.5, for every 5 mol of Zn oxidized (or 2 mol of MnO_4^- reduced), 10 mol of electrons pass through the external circuit, so $n = 10$. A pH of 2.00 corresponds to a hydronium ion concentration of 0.010 M. To solve this problem, you may either use the Nernst equation, Equation 17.7, directly, with $n = 10$, or calculate the potentials for each half-reaction using the Nernst equation and then calculate the cell potential in the usual way. The Nernst equation for a half-reaction is

$$E_{hc} = E^{\circ}_{hc} - \frac{0.0592\ \text{V}}{n_{hc}} \log_{10} Q_{hc}$$

Substituting the standard reduction potentials and concentrations given conditions for each half-reaction gives

$$E(MnO_4^-|Mn^{2+}) = 1.49\ \text{V} - \frac{0.0592\ \text{V}}{5} \log_{10} \frac{(10^{-3})}{(0.12)(10^{-2})^8} = 1.33\ \text{V}$$

and

$$E(Zn|Zn^{2+}) = -0.76\ \text{V} - \frac{0.0592\ \text{V}}{2} \log_{10} \frac{1}{0.015} = -0.814\ \text{V}$$

giving

$$\Delta E_{cell} = E_{cathode} - E_{anode} = 1.33 - (-0.814) = 2.14\ \text{V}$$

Related Problems: 27, 28

17.3.1 Measuring Equilibrium Constants

Electrochemistry provides a convenient and accurate way to measure equilibrium constants for many solution-phase reactions; cell potentials are directly related to equilibrium constants as follows.

$$\Delta G^{\circ} = -nF\Delta E^{\circ}_{cell}$$

and

$$\Delta G^{\circ} = -RT \ln K$$

so

$$RT \ln K = nF\Delta E^{\circ}_{cell}$$

$$\ln K = \frac{nF}{RT} \Delta E^{\circ}_{cell}$$

and

$$\log_{10} K = \frac{n}{0.0592\ \text{V}} \Delta E^{\circ}_{cell} \quad \text{(at 25°C)} \qquad [17.8]$$

The same result can be obtained in a slightly different way, by returning to the Nernst equation, which reads (at 25°C),

$$\Delta E_{\text{cell}} = \Delta E^{\circ}_{\text{cell}} - \frac{0.0592 \text{ V}}{n} \log_{10} Q$$

Suppose all species are initially present under standard-state conditions; $Q = 1$, so the cell potential ΔE_{cell} is the standard cell potential $\Delta E^{\circ}_{\text{cell}}$. Q increases (or decreases, if $\Delta E^{\circ}_{\text{cell}} < 0$) and ΔE_{cell} decreases (or increases) as the reaction proceeds, with ΔE_{cell} approaching 0 V at equilibrium, at which point

$$\Delta E^{\circ}_{\text{cell}} = \frac{0.0592 \text{ V}}{n} \log_{10} K$$

which is the same as the relationship just obtained. This equation allows us to calculate equilibrium constants from standard cell potentials, as the following example illustrates.

EXAMPLE 17.8

Calculate the equilibrium constant for the redox reaction

$$2\ MnO_4^-(aq) + 5\ Zn(s) + 16\ H_3O^+(aq) \longrightarrow 2\ Mn^{2+}(aq) + 5\ Zn^{2+}(aq) + 24\ H_2O(\ell)$$

at 25°C using the cell potential calculated in Example 17.5.

Solution

$\Delta E^{\circ}_{\text{cell}} = 2.25$ V and $n = 10$ for this reaction as shown in Examples 17.5 and 17.7. Therefore,

$$\log_{10} K = \frac{n}{0.0592 \text{ V}} \Delta E^{\circ}_{\text{cell}} = \frac{10}{0.0592 \text{ V}} (2.25 \text{ V}) = 380$$

$$K = 10^{380}$$

This overwhelmingly large equilibrium constant reflects the strength of permanganate ion as an oxidizing agent and that of zinc as a reducing agent. No MnO_4^- ions are present at equilibrium, for all practical purposes.

Related Problems: 35, 36

The foregoing example illustrates how equilibrium constants for redox reactions can be measured electrochemically. Related procedures can also be used to measure the solubility product constants of sparingly soluble ionic compounds or the ionization constants of weak acids and bases. The solubility product constant of AgCl can be measured electrochemically as follows. An electrochemical cell is constructed in which one half-cell contains solid AgCl and Ag metal, in equilibrium with a known concentration of $Cl^-(aq)$ (established with 0.001 M NaCl, for example) so that an unknown but definite concentration of $Ag^+(aq)$ is present. A silver electrode is used so that the half-cell reaction involved is either the reduction of $Ag^+(aq)$ or the oxidation of Ag. This is, in effect, an $Ag^+|Ag$ half-cell whose potential is to be determined. The second half-cell can be any cell whose standard reduction potential is accurately known, and its choice is a matter of convenience. The standard hydrogen electrode is the reference cell in the following example.

EXAMPLE 17.9

A galvanic cell is constructed from a half-cell containing silver and silver chloride and the SHE in order to determine an unknown concentration of Ag^+ ions. The cell diagram is

$$Pt|H_2(1\ atm)|H_3O^+(1\ M)||Cl^-(1.00 \times 10^{-3}\ M) + Ag^+(?\ M)|AgCl|Ag$$

and the unknown concentration of Ag^+ is represented by $Ag^+(?\ M)$. The SHE is the anode, and the measured cell potential is $\Delta E_{cell} = 0.397$ V. Calculate the silver ion concentration in the cell and K_{sp} for AgCl at 25°C.

Solution

The half-cell reactions are

$$H_2 + 2\ H_2O \longrightarrow 2\ H_3O^+ + 2\ e^- \quad \text{(anode)}$$

$$2\ Ag^+ + 2\ e^- \longrightarrow 2\ Ag \quad \text{(cathode)}$$

$$\Delta E^\circ_{cell} = E^\circ(\text{cathode}) - E^\circ(\text{anode}) = 0.800 - 0.000\ V = 0.800\ V$$

Note that $n = 2$ for the overall cell reaction, and the reaction quotient simplifies to

$$Q = \frac{[H_3O^+]^2}{[Ag^+]^2\ P_{H_2}} = \frac{1}{[Ag^+]^2}$$

because $[H_3O^+] = 1$ M and $P_{H_2} = 1$ atm. The Nernst equation is

$$\Delta E_{cell} = \Delta E^\circ_{cell} - \frac{0.0592\ V}{n} \log_{10} Q$$

$$\log_{10} Q = \frac{n}{0.0592\ V}(\Delta E^\circ_{cell} - \Delta E_{cell}) = \frac{2}{0.0592\ V}(0.800 - 0.397\ V) = 13.6$$

$$Q = 10^{13.6} = 4 \times 10^{13} = 1/[Ag^+]^2$$

This can be solved for the silver ion concentration $[Ag^+]$ to give

$$[Ag^+] = 1.6 \times 10^{-7}\ M$$

so that

$$[Ag^+][Cl^-] = (1.6 \times 10^{-7})(1.00 \times 10^{-3})$$

$$K_{sp} = 1.6 \times 10^{-10}$$

Related Problems: 41, 42

17.3.2 pH Meters

Cell potentials are sensitive to pH if H_3O^+ is produced or consumed in either half-cell reaction. A simple cell can be constructed to measure pH as follows:

$$Pt|H_2(1\ atm)|H_3O^+(\text{aqueous, variable})||H_3O^+(1\ M)|H_2(1\ atm)|Pt$$

If the half-cell reactions are written as

$$2\ H_3O^+(1\ M) + 2\ e^- \longrightarrow H_2(1\ atm) + 2\ H_2O(\ell) \quad \text{(cathode)}$$

$$H_2(1\ atm) + 2\ H_2O(\ell) \longrightarrow 2\ H_3O^+(\text{var}) + 2\ e^- \quad \text{(anode)}$$

(In labeling the first as the cathode we have assumed that the cell with variable $[H_3O^+]$ has an H_3O^+ concentration < 1 M.)

then $n = 2$ and $Q = [H_3O^+(\text{aqueous, variable})]^2$ because the other concentrations and gas pressures are unity. The cell potential calculated using the Nernst equation is (for a cell operating at 25°C)

$$\Delta E_{\text{cell}} = \Delta E^{\circ}_{\text{cell}} - \frac{0.0592\text{ V}}{n} \log_{10} Q$$

which becomes

$$\Delta E_{\text{cell}} = -\frac{0.0592\text{ V}}{2} \log_{10} [H_3O^+]^2$$

$$= -0.0592\text{ V} \log_{10} [H_3O^+] = (0.0592\text{ V})\text{ pH}$$

because $\Delta E^{\circ}_{\text{cell}} = 0$. The measured cell potential is directly proportional to the pH.

We have just described a simple **pH meter**. It is not convenient to bubble hydrogen gas through both the unknown and reference half-cells, however, so a smaller and more portable pair of electrodes is used to replace the hydrogen half-cells. The early commercial pH meters used two electrodes that were dipped into a solution of unknown pH. One of these, called the *indicator electrode* (because it indicates the unknown pH of the analyte), was made of an AgCl-coated silver electrode in contact with an HCl solution of known concentration (i.e., 1.0 M) contained in a glass tube terminated by a thin-walled, bulb-shaped, glass membrane. A pH-dependent potential develops across this thin glass membrane when the glass electrode is immersed in a solution of different, and unknown, $[H_3O^+]$ concentration. See Figure 17.5. The reference electrode used was often the **saturated calomel electrode**, depicted on the right side of Figure 17.5. The overall cell can be represented as

$$\text{Ag}|\text{AgCl}|\text{Cl}^-(1.0\text{ M}), H_3O^+(1.0\text{ M})|\text{glass}|H_3O^+(\text{var})||\text{Cl}^-(\text{sat})|Hg_2Cl_2(s)|\text{Hg}|\text{Pt}$$

with half-reactions

$$2\text{ Ag}(s) + 2\text{ Cl}^-(1.0\text{ M}) \longrightarrow 2\text{ AgCl}(s) + 2\,e^- \qquad \text{(anode)}$$

$$H_3O^+(1.0\text{ M}) \longrightarrow H_3O^+(\text{var})$$

$$Hg_2Cl_2(s) + 2\,e^- \longrightarrow 2\text{ Hg}(\ell) + 2\text{ Cl}^-(\text{sat}) \qquad \text{(cathode)}$$

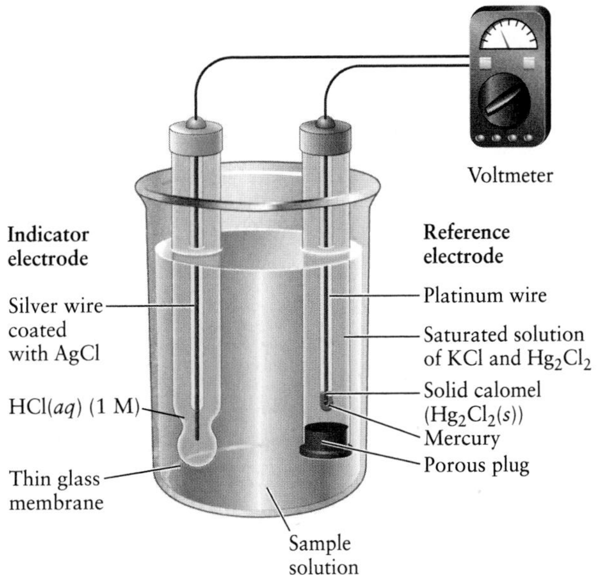

FIGURE 17.5 Schematic of an early pH meter.

The standard reduction potential of the first half-reaction can be combined with the reduction potential of the saturated calomel electrode (which is 0.242 V, with the $Cl^-(aq)$ fixed by the concentration of the KCl electrolyte, but not equal to 1 M, so not a *standard* reduction potential) to give a constant reference potential ΔE_{ref}. The second reaction is the source of the variable potential in the cell, which corresponds to the free energy of dilution of H_3O^+ from a concentration of 1.0 M to an unknown concentration. That potential appears as a junction potential across the thin glass membrane of the indicator electrode. The Nernst equation for the overall cell operated at 25°C can be written as

$$\begin{aligned}\Delta E_{cell} &= \Delta E_{ref} - \frac{0.0592\ \text{V}}{1}\log_{10}\frac{[H_3O^+(\text{var})]}{1.00} \\ &= \Delta E_{ref} + (0.0592\ \text{V})\ \text{pH}\end{aligned}$$

and the pH of the unknown solution is

$$\text{pH} = \frac{\Delta E_{cell} - \Delta E_{ref}}{0.0592\ \text{V}}$$

The glass indicator electrode has a number of advantages over other possible pH electrodes. It responds only to changes in $[H_3O^+]$, and it does so over a wide range of pH. It is unaffected by strong oxidizing agents that would make hydrogen electrodes unreliable. Highly colored solutions that would render acid–base indicators useless do not interfere with the glass indicator electrode.

Most modern pH meters use a combination probe like the one shown in Figure 17.6, in which the indicator and reference electrodes are housed in one body. You are probably familiar with probes like this one from laboratory courses. Another significant advantage of the combination probe is that it can be miniaturized to permit insertion into individual living cells and probe pH in biological samples.

A number of **ion-selective electrodes** have been developed using similar principles to selectively measure ion concentrations to levels as low as 10^{-10} M. The simplest example of such an electrode would be an Ag/AgCl reference electrode, the potential being sensitive to the $Cl^-(aq)$ concentrations in the solution of interest. Modern ion selective electrodes are constructed from polymer membranes that have been tailored to transport only the ions of interest, greatly reducing background and improving sensitivity.

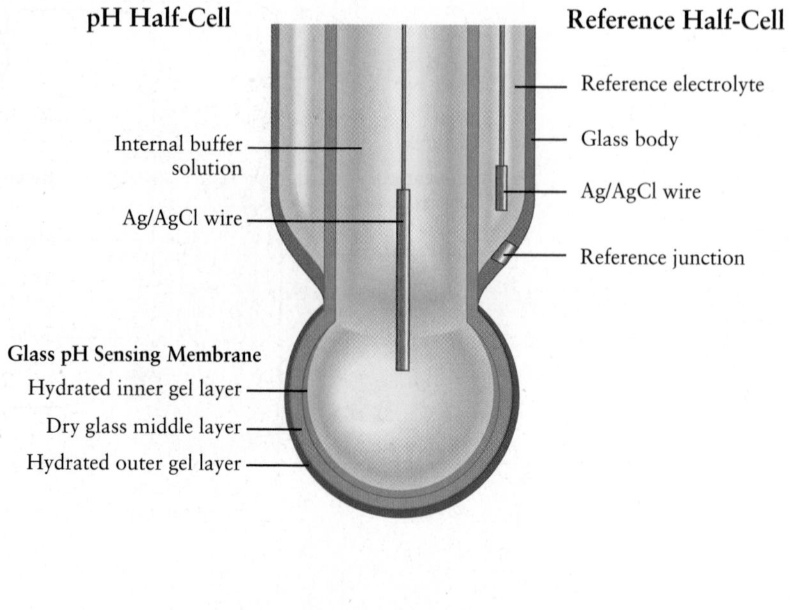

FIGURE 17.6 A combination pH probe.

17.4 Molecular Electrochemistry

The term **molecular electrochemistry** was coined recently to identify a branch of the broader field of electrochemistry that focuses on developing a molecular level understanding of electron transfer processes at electrodes and exploiting the interactions between molecules and electrode surfaces to optimize electrochemical reactions. Fundamental questions of interest include the coupling of electron and proton transfer, especially in biological systems; the way electron transfer reactions initiate bond breaking and formation; the role molecules play in catalyzing electrochemical reactions; and the way redox enzymes work. A wide range of applications is being pursued using this approach as well. The electrochemical synthesis of organic molecules, the development and application of biological sensors to problems of fundamental interest and medical diagnostics, the design of better catalysts for fuel cells, and the application of semiconductor photoelectrochemistry to problems as diverse as organic waste remediation and solar energy conversion are all active areas of research and development in this field. We discuss several of these applications in this section.

17.4.1 Electrochemical Organic Synthesis

The large-scale commercial synthesis of organic compounds uses simple starting materials derived from petroleum and converts them into the desired products in reactions that are generally carried out at high temperatures and pressures using catalysts. Many of these reactions use H_2 and O_2 to reduce and oxidize hydrocarbons to produce intermediates or the desired products. It is interesting to think about these transformations as redox reactions and to compare the driving force available in the homogeneous gas phase reactions in electrochemical terms. The driving force for the gas phase oxidations and reductions is limited by the potentials for the reduction of oxygen or the oxidation of hydrogen, which is a range of about 1V under standard conditions, which corresponds to a change in the Gibbs free energy of about 100 kJ mol^{-1} if one mole of electrons is transferred in the balanced chemical reaction. Synthetic organic reactions carried out electrochemically can be driven by potentials that range between −2.5 V to 3.5 V, providing much larger driving forces and the possibility of greater control over the thermodynamics and kinetics of reactions by controlling potential. This approach also has the advantage that it does not rely on powerful, but toxic and potentially dangerous, reducing agents like the alkali metals or oxidizing agents like Cl_2 or $Cr_2O_7^{2-}$. It has also been suggested that electrons are probably the least expensive and most widely available redox reagents on a per-mole basis.

The most famous example of a large-scale electrochemical organic synthesis is the Monsanto process for the production of adiponitrile, a precursor for hexamethylene diamine, which is one of the monomers used to make the large-volume commodity chemical nylon. The reaction mechanism is complex, but the following is one of the reactions that produces adiponitrile

$$CH_2{=}CHCN + 2\,e^- \longrightarrow 2\,CH_2{=}CHCN^-$$

$$2\,CH_2{=}CHCN^- + 2\,H^+ \longrightarrow NC(CH_2)_4CN$$

which is then reduced in hydrogen to form the hexamethylenediamine monomer. The reaction is carried out in very large electrochemical reactors with steel electrodes coated by a thin film of Cd and the reactants, products, and electrolyte carried in an aqueous solution/organic solvent emulsion. This process produces some 200,000 tons of adiponitrile each year. Electrochemical organic synthesis has become increasingly popular in recent years for the synthesis of organic molecules on much smaller scales, such as pharmaceuticals, because of its speed and selectivity.

17.4.2 Enzyme-Based Electrochemical Sensors

Approximately 5% of the world's population, and nearly 8% of the population of the United States, are diabetic; they are unable either to produce insulin, or to respond to the insulin they do produce to regulate the absorption of glucose, a major source of energy, into cells. Diabetics must keep their blood sugar levels controlled pretty tightly to avoid serious medical problems, some of which are potentially life-threatening. They do so by monitoring their blood sugar several times daily and injecting themselves with synthetic human insulin to regulate glucose uptake. In the prior section on the Nernst equation you learned how the potential difference ΔE of an electrochemical cell depends on the concentrations of the species involved. This affords a way to measure glucose concentrations in a blood sample electrochemically.

Glucose sensors monitor either the reduction of glucose by the enzyme glucose dehydrogenase GDH(FAD), or the oxidation of glucose by the enzyme glucose oxidase GOx($FADH_2$), where FAD and $FADH_2$ are the reduced and oxidized forms of cofactors (see Section 18.8) associated with the enzymes that provide the reducing or oxidizing equivalents in the form of (H^+, e^-) pairs. The first reaction is used in test strips; an example is shown in Figure 17.7.

In the particular test strip shown, reduction occurs at the Ag/AgCl reference electrode and oxidation of glucose (in the blood sample) is catalyzed by an enzyme called glucose dehydrogenase (GDH), which converts an alcohol functional group into a ketone (see Section 7.6). The overall set of reactions is summarized as follows.

$$2\,\text{AgCl}(s) + 2\,e^- \longrightarrow 2\,\text{Ag}(s) + 2\,\text{Cl}^-$$

$$\text{glucose} + \xrightarrow{\text{GDH}} \text{gluconolactone} + 2\,\text{H}^+ + 2\,e^-$$

$$\text{glucose} + 2\,\text{AgCl}(s) \xrightarrow{\text{GDH}} \text{gluconolactone} + 2\,\text{Ag}(s) + 2\,\text{Cl}^- + 2\,\text{H}^+$$

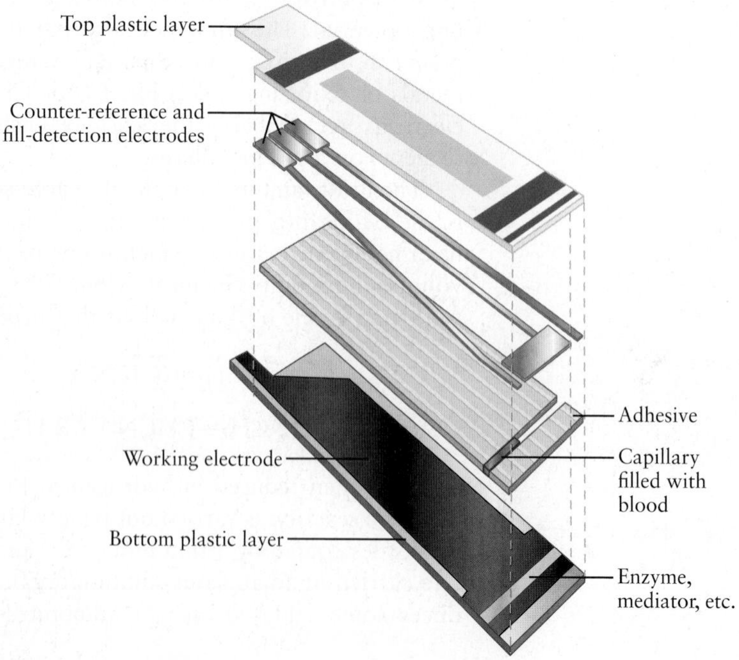

FIGURE 17.7 Glucose sensor test strip showing the capillary partially filled with blood, the carbon working electrode and the counter-reference and fill-detection electrodes. The planar electrodes are separated from one another by 50 μm.

Electron transfer from the enzyme to the electrode surface is not very efficient, however, because the enzyme's active site is buried inside the protein and there is no easy path for the electrons to follow. The problem was solved by incorporating molecules called *mediators* into the device that facilitate electron transfer over large distances by a series of efficient electron transfer reactions. The design of the mediators is one part of the clever chemistry referred to earlier.

17.4.3 Electrogenerated Chemiluminescence and Cyclic Voltammetry

Electrogenerated chemiluminescence (ECL) has become a very important analytical technique for detecting biologically important molecules, with applications in clinical diagnostics, food and water testing, the detection of biowarfare agents, and basic research. It is very sensitive (picomolar concentrations can be measured) and the response is linearly proportional to concentration. ECL generally results from charge transfer reactions between radical anions and radical cations that produce excited molecular electronic states, which then emit light. The radicals are generated electrochemically, and the series of reactions can be written as

$$A + e^- \longrightarrow A^{-\bullet}$$
$$D \longrightarrow D^{+\bullet} + e^-$$
$$A^{-\bullet} + D^{+\bullet} \longrightarrow A^* + D$$
$$A^* \longrightarrow A + h\nu$$

in which A is the oxidized form of an electron acceptor and $A^{-\bullet}$ is its radical anion formed upon reduction; D is an electron donor and $D^{+\bullet}$ is its radical cation formed upon oxidation; A^* is an electronically excited state of A and $h\nu$ is the energy of the emitted light. The electrochemical generation of the radical anion is illustrated in Figure 17.8a. Using an external voltage source, an electrode may be raised in free energy (by lowering its electric potential) until an electron can transfer from the electrode to the lowest unoccupied molecular orbital (LUMO) of an organic molecule. In this case the molecule is acting as an electron acceptor A. Figure 17.8b depicts what happens when an external voltage source is applied to lower the free energy of an electrode (by raising its electric potential). That allows an electron to transfer from the highest occupied molecular orbital (HOMO) of the molecule, now acting as an electron donor, D, to the electrode. The electron transfer reaction that produces A^*, which

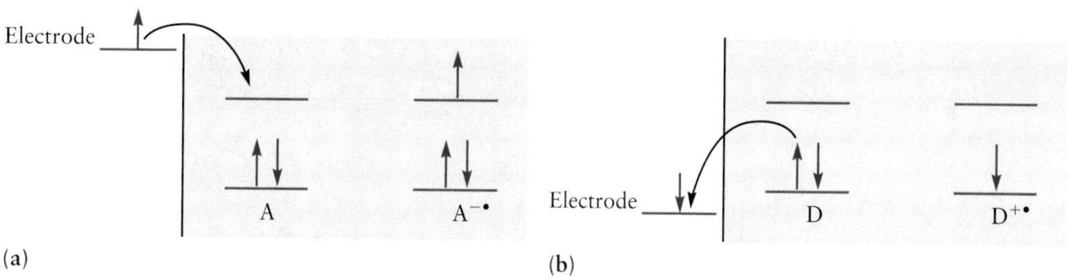

FIGURE 17.8 Generation of a radical cation and a radical anion in an ECL experiment. (a) The molecule is acting as an electron acceptor A. The radical anion of A is formed when the electrode potential is made more negative, raising its energy above the LUMO of A. (b) The same molecule is acting as an electron donor D. The radical cation of D is formed when the electrode potential is made more positive, lowering its energy below the HOMO of D.

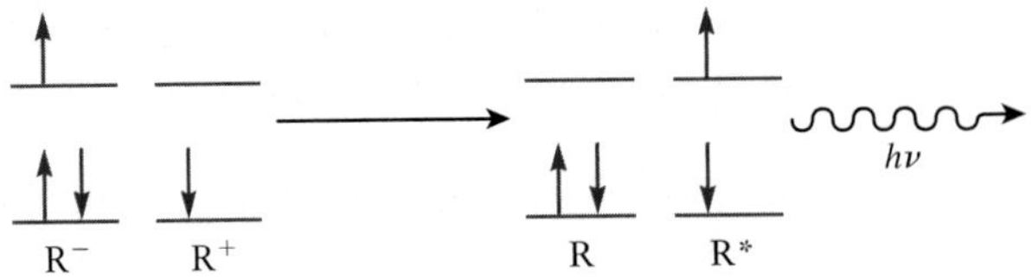

FIGURE 17.9 Reaction of R^- and R^+ to form R + R*. The latter is electronically excited, so is the species that emits light in an ECL experiment. Note that if two different molecules are the acceptor and donor, then A^- is on the left in the reactant side of this figure and A* is on the right in the product side of this figure. (One of the paired electrons in A^- was transferred to the singly occupied orbital in D^+ to form D + A*.)

subsequently returns to the ground state A by emitting a photon, is illustrated in Figure 17.9.

The ECL process is most easily illustrated using a specific example in which we identify the redox potentials of the donor and acceptor using an electrochemical technique known as cyclic voltammetry, and connect them to the absorption and emission spectra of the system. **Cyclic voltammetry** measures the current that flows as the electrode potential is scanned in the forward direction (typically toward more negative potentials) and then in the reverse direction. The cyclic nature of the scan gives the technique its name. Cyclic voltammetry is perhaps the electrochemists' most powerful method for the study of electrochemical processes. It can quickly identify redox active materials and locate the potentials at which various oxidations and reduction reactions occur. Cyclic voltammetry can also provide a great deal of information about the kinetics of electrochemical reactions, but that discussion is beyond the scope of this textbook. The application of cyclic voltammetry to ECL affords us the opportunity to introduce you to the technique at an elementary level as well as to illustrate how ECL experiments are actually carried out. Figure 17.10 shows the cyclic voltammogram of a redox-active dye molecule (PM 567) that has been tailored for ECL applications. The same molecule, in this example, is reduced to form the radical anion and oxidized to form the radical cation.

The potential is set initially to 0 V (measured with respect to SCE in Figure 17.10) and then scanned through a cycle, first toward more negative potentials, then toward more positive potentials, and finally returning to zero. As shown in the voltammogram in Figure 17.10, the current begins to rise as the potential reaches about −1.4 V, signifying the reduction of PM to $PM^{-\bullet}$ at the cathode as depicted schematically in Figure 17.8a.

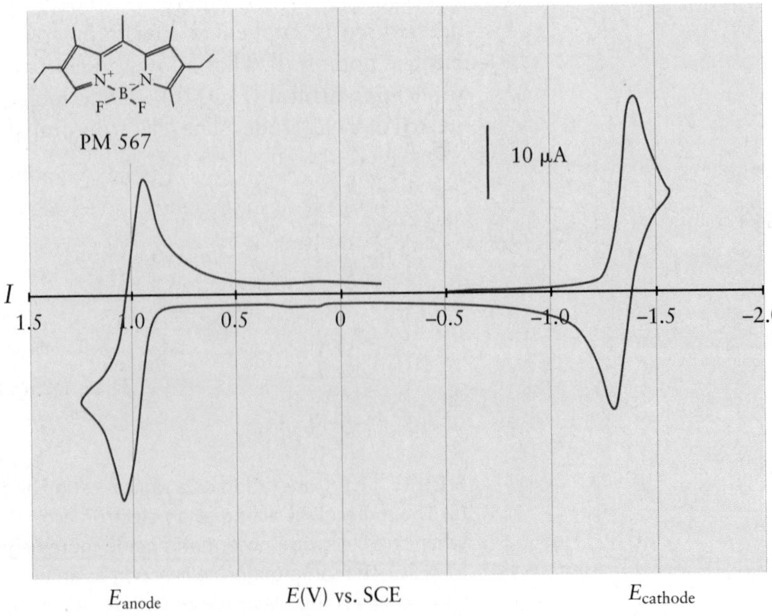

FIGURE 17.10 Cyclic voltammogram for an ECL dye.

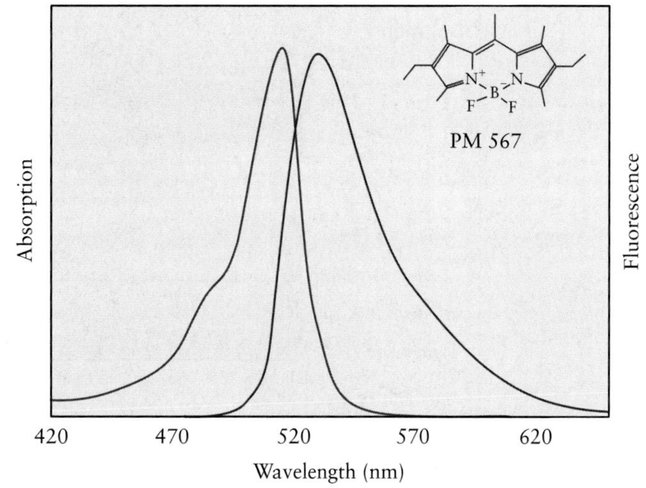

FIGURE 17.11 Absorption and fluorescence spectra of an ECL dye.

The scan is reversed by now using the external voltage source to lower the energy of the electrode (by raising its electric potential) and current begins to flow in the opposite direction as $PM^{-\bullet}$ is oxidized back to PM. When the potential is scanned through potentials more positive than about 1 V, PM is oxidized to $PM^{+\bullet}$; this is depicted schematically in Figure 17.8b. Finally, when the scan is reversed again, $PM^{+\bullet}$ may be reduced back to PM. The shapes of these curves contain a great deal of information about the kinetics of the reactions, with the current rising and then falling due to an increase and then a decrease in the rates of diffusion of reactants to the electrode surface. The redox potential for the $PM|PM^{-\bullet}$ couple is located between the two peaks in the absolute value of the wave near −1.4 V, and the redox potential for the $PM^{+\bullet}|PM$ couple is located between the two peaks in the absolute value of the wave near 1 V. We can calculate the maximum amount of energy available to create an excited state, and the energy of the emitted photon from the reaction

$$PM^{-\bullet} + PM^{+\bullet} \longrightarrow PM + PM^{*} \longrightarrow 2\ PM + h\nu$$

as follows. We have associated the potential of the $PM^{+\bullet}|PM$ couple with the HOMO (highest occupied molecular orbital) and that of the $PM|PM^{-\bullet}$ couple with the LUMO (lowest unoccupied molecular orbital), so the energy available is given by the potential difference between the two redox couples times the charge of the electron, giving $-e\Delta E = 2.4$ eV for this system. Note that this association is a qualitative one, as it neglects the contribution of entropy to the Gibbs free energy. The absorption and fluorescence spectra of PM 567 are shown in Figure 17.11. The peak in the absorption spectrum (violet curve) appears at 514 nm, which corresponds to an energy of about 2.4 eV, so the energy available from the recombination reaction is sufficient to create the excited state. The maximum in the fluorescence spectrum (blue curve) is shifted to somewhat longer wavelengths, as is typical for molecules in solution, appearing at 567 nm.

EXAMPLE 17.10

Sketch and interpret the cyclic voltammogram you expect for the $[Fe(CN)_6]^{3-}/[Fe(CN)_6]^{4-}$ redox couple measured with respect to the SHE reference electrode. The standard reduction potential for this system is 0.36 V.

(Continued)

Solution

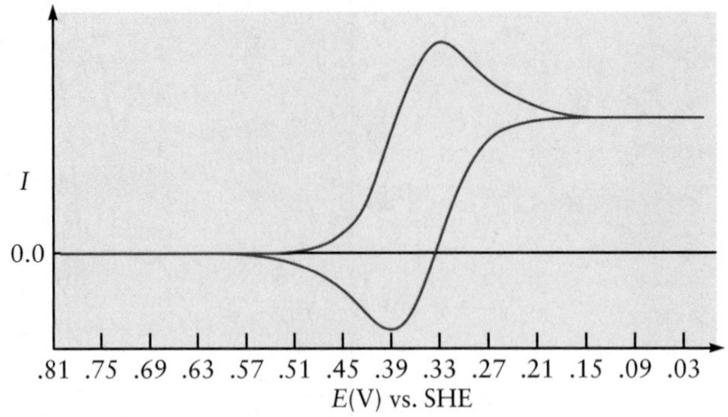

The cyclic voltammogram for this system is sketched above. The potential is set initially at 0.8 V and no current flows. A detectable current begins to flow as the potential is scanned to more negative values, the current beginning to appear at about 0.5 V. The current is due to the reduction of $[Fe(CN)_6]^{3-}$ to $[Fe(CN)_6]^{4-}$ at the cathode. The current peaks and then declines as the potential is scanned to more negative values. The direction of the potential scan is reversed at about 0 V and a current flowing in the opposite direction begins to appear at about 0.34 V, indicating the re-oxidation of $[Fe(CN)_6]^{4-}$ to $[Fe(CN)_6]^{3-}$. The current first increases and then decreases as the potential is scanned to more positive values. The redox potential is determined graphically by locating the midpoint between the two peaks, in this case 0.36 V.

Related Problems: 43, 44

EXAMPLE 17.11

Suppose that you were asked to synthesize a molecule for use in ECL that emits light at 600 nm and has a redox potential associated with the LUMO located at −1.05 V (with respect to SHE), for compatibility with other requirements in a particular analytical application. (Assume the emission spectrum is not significantly shifted from the absorption spectrum.) What is the redox potential associated with the HOMO?

Solution

The energy of a photon is related to its wavelength by $E = h\nu = hc/\lambda$, where λ is the wavelength, h is Planck's constant and c is the speed of light (see Equations 4.1 and 4.7). Solving for E gives

$$E = (6.626 \times 10^{-34}\ \text{J s})(3 \times 10^{8}\ \text{m s}^{-1})/(600 \times 10^{-9}\ \text{m})$$

$$= 3.3 \times 10^{-19}\ \text{J} = 2.06\ \text{eV}$$

which is equal to the difference in the energies of the two orbitals $\varepsilon_{\text{LUMO}} - \varepsilon_{\text{HOMO}}$. The energy of the LUMO with redox potential −1.05 V is 1.05 eV (with respect to SHE) so the energy of the HOMO must be −1.01 eV with an associated potential located at 1.01 V.

Related Problems: 45, 46

CONNECTION TO BIOLOGY

ECL in Clinical Assays

We have illustrated the principles of ECL using as an example a molecule that serves as both the electron acceptor and the electron donor. Biological applications of ECL, however, require molecules both that are stable and that show ECL in *aqueous* solution. The discovery of ECL in the water-soluble coordination complex $Ru(bpy)_3^{2+}$, shown in Figure C17.1, was a promising development, but it turned out to be difficult, if not impossible, to find a suitable electron acceptor that could be reduced electrochemically in water. The difficulty arises because potentials more negative than the potential for water reduction are needed to produce species that can reduce $Ru(bpy)_3^{3+}$ generated electrochemically by oxidation of $Ru(bpy)_3^{2+}$ (see Section 17.8, *A Deeper Look* . . . Electrolysis of Water and Aqueous Solutions). The problem was solved by finding a way to produce reducing species by *oxidation,* using molecules called coreactants that decompose upon oxidation to produce products with the required reducing power, allowing a single applied potential step to initiate both reactions.

The molecule tripropylamine, $N(CH_2CH_2CH_3)_3$ abbreviated $Pr_2NCH_2CH_2CH_3$, is a particularly good coreactant for the efficient generation of ECL from $Ru(bpy)_3^{2+}$ in aqueous solution. $Ru(bpy)_3^{2+}$ and tripropylamine are simultaneously oxidized during a single potential step to produce $Ru(bpy)_3^{3+}$ and $Pr_2N(CH_2CH_2CH_3)^{+\bullet}$ respectively. The tripropylamine radical cation then loses a proton to form the strongly reducing free radical species $Pr_2N(^{\bullet}CHCH_2CH_3)$, which reacts with $Ru(bpy)_3^{3+}$ to produce an excited state of $Ru(bpy)_3^{2+}$ that emits light. The overall sequence is summarized as follows:

$$Ru(bpy)_3^{2+} \longrightarrow Ru(bpy)_3^{3+} + e^-$$

$$Pr_2N(CH_2CH_2CH_3) \longrightarrow Pr_2N(CH_2CH_2CH_3)^{+\bullet} + e^-$$

$$Pr_2N(CH_2CH_2CH_3)^{+\bullet} \longrightarrow Pr_2N(^{\bullet}CHCH_2CH_3) + H^+$$

$$Ru(bpy)_3^{3+} + Pr_2N(^{\bullet}CHCH_2CH_3) \longrightarrow (Ru(bpy)_3^{2+})^* + Pr_2N(^{+}CH_2CH_2CH_3)$$

ECL is used in a large number of clinically important assays. $Ru(bpy)_3^{2+}$ is covalently bound, using different linking molecules, to a probe that binds to target molecules of interest. The target molecules fall into major classes that include: antigens that trigger an immune response, producing antibodies, and the nucleic acids DNA and RNA. Antibody-based assays are used to measure estrogen and testosterone levels, detect the hepatitis and HIV viruses, and measure insulin levels, for example. Antibodies to a particular antigen are covalently attached to a solid surface, and $Ru(bpy)_3^{2+}$–labeled antibodies and the sample of interest are placed in solution. Any antigen present will be recognized and bound to both the surface-bound antibodies and the solution-phase antibodies, linking them together. A number of schemes have been developed to present these samples to electrode surfaces for ECL analysis, magnetic beads that attach to electrode surfaces or miniature electrodes printed in arrays of small wells, for example. Assays based on the recognition of specific nucleic acid sequences proceed in an analogous way.

ECL has a number of significant advantages over other biological assays. First, the labels are completely stable until the potential is stepped to generate the reactive species. Second, and perhaps most importantly, there is no background emission like that found in fluorescence experiments. Light emission occurs in ECL only when the electrode is stepped through the potential region of interest. Third, it is a very sensitive technique because modern photomultipliers or other detectors can detect single photons. Detection limits for ECL have been demonstrated to be as low as 10^{-12} M, making it among the most sensitive analytical methods available.

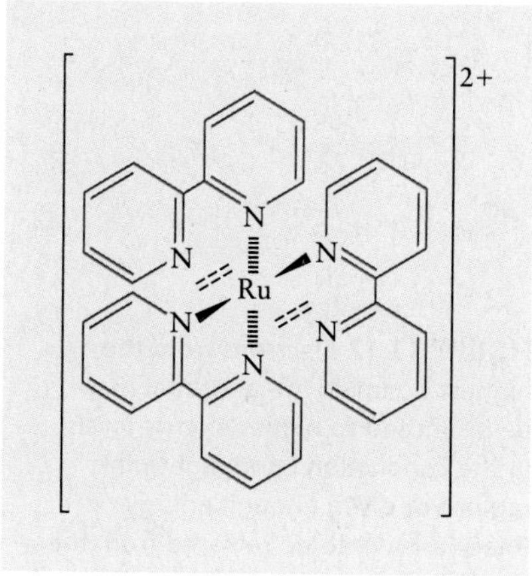

FIGURE C17.1 $Ru(bpy)_3^{2+}$.

17.4.4 Photoelectrochemistry

To understand the transfer of electrons between metals and molecular species, it is helpful to make a qualitative energy level diagram for the electronic energy levels in a metal. Figure 17.12 shows an energy level diagram for copper metal. The energies of the electronic energy levels are plotted along the vertical axis on the left, and the corresponding potentials are plotted along the vertical axis on the right. In Figure 17.12, $\varepsilon = 0$ is defined as the energy for an electron pulled far from the copper (as in the photoelectric effect in Section 4.4). Later figures in this section show energies and potentials with respect to the potential of a reference electrode (rather than with respect to a free electron's energy). The energies of the orbitals of a copper metal electrode are represented by the solid blocks of color in Figure 17.12. The energy levels of metals are spaced so closely together that they form a continuous *band* of levels (see Section 21.3). The orbitals of bulk metals are constructed from the orbitals of the individual metal atoms using LCAO MO theory, by analogy to the MOs constructed for 1,3-butadiene shown in Figure 6.53. Only half of the four π butadiene orbitals are occupied because each orbital can accommodate two electrons. The same is true for metals; half of the delocalized orbitals are occupied for univalent metals that contribute one electron per atom to form a band. The solid block of color represents the occupied copper orbitals, whereas the unoccupied orbitals are represented by the more transparent block of color. We identify the highest occupied orbital (analogous to the HOMO—highest occupied molecular orbital—in organic chemistry) as the Fermi level with energy ε_F. The energy of the Fermi level, for a metal in vacuum, is given by $\varepsilon_F = -\Phi$, where Φ is the work function of the metal, as defined in our discussion of the photoelectric effect in Section 4.4. The work function of copper is 4.7 eV, so Fig. 17.12 plots the Fermi level at −4.7 eV. Measurements of the absolute potential of SHE give 4.44 V at 25°C, corresponding to an absolute energy of −4.44 eV; this is depicted on the vertical axis on the right. It is typical, however, to measure electrostatic potentials with respect to SHE, so in subsequent figures in this chapter you will see the label 0.0 (SHE) and the potentials of the other electrodes considered will be adjusted to that scale. For example, the Fermi level of copper is at a potential of +0.26 V with respect to SHE, corresponding to being *lower* in energy than SHE by 0.26 eV.

Some metals, called semiconductors, evidence an energetic gap between the highest occupied energy levels in the valence band and the lowest unoccupied levels in the conduction band. The left-hand side of Figure 17.13 shows two energy levels of a molecule or a semiconductor that absorbs light (a **chromophore**)

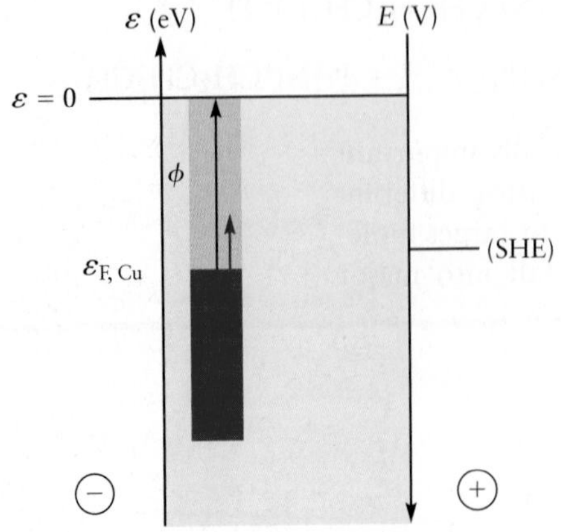

FIGURE 17.12 Electrons from the highest occupied metal orbital may be promoted to higher energy levels in the conduction band (light pink region) or given enough energy (black arrow), to be released from the metal (as in the photoelectric effect).

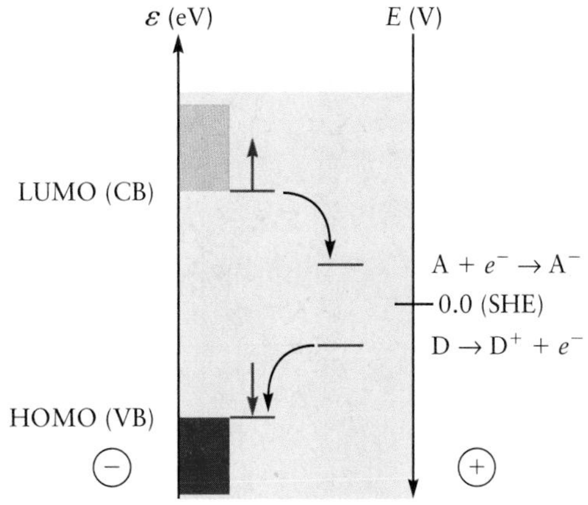

FIGURE 17.13 Photoexcitation promotes electrons to higher energies (more negative potentials), which make them stronger reducing agents, capable of reducing an acceptor A to A^-. The vacancies (holes) left behind can now accept electrons, which make them stronger oxidizing agents, capable of oxidizing a donor D to D^+. The left side of the diagram shows the energy levels in the valence band (VB) of the semiconductor in dark pink, with one representative energy level pulled out so you can see the "hole" created by promoting an electron to an energy level in the conduction band (CB).

along with the redox levels of an electron acceptor A and an electron donor D on the right-hand side of the diagram. The lower energy level in the chromophore is called the HOMO (highest occupied molecular orbital), and the higher energy level is called the LUMO (lowest unoccupied molecular orbital); the corresponding levels in the semiconductor are called the **valence band (VB)** and the **conduction band (CB)**, respectively. We treat the semiconductor bands as if they were molecular orbitals, and we do not consider how the energies of these bands bend as a function of distance from the surface, which is an important consideration in more advanced work. We refer to the energy level splitting as the bandgap for molecular systems as well as semiconductors, for convenience and consistency. Redox potentials (also represented as orbital energy levels) for two different redox reactions, $A + e^- \longrightarrow A^-$ and $D \longrightarrow D^+ + e^-$, are located on the right-hand side of the diagram; we have chosen the energy levels associated with these reactions to lie between the energy levels of the chromophore for the purpose of this discussion. Values of the potential and the potential energy, referenced to the SHE, are plotted on the two vertical axes; note the direction of increase of each.

Let's use this energy level diagram to understand the mechanism by which light absorption drives redox reactions that are not normally spontaneous. Light absorption promotes electrons from the lower level (the valence band) to the upper level (the conduction band), increasing their potential energy. The potential of these electrons is more negative than when they were in the HOMO, so these excited states have become stronger reducing agents than the ground state of the system. Vacancies (holes) left in the conduction band are now free to accept electrons from other species, making excited states more powerful oxidizing agents than ground states. Photoexcitation, therefore, has created excited states in which the electrons and the holes are simultaneously more strongly reducing and more strongly oxidizing, respectively, than the ground state molecules themselves.

Let's now consider direct photoelectrochemical water splitting by wide bandgap semiconductors such as TiO_2 or $SrTiO_3$, each with a band gap of about 3 V. When the semiconductor absorbs UV light with an energy greater than 3 eV, an electron can be promoted into the conduction band, leaving a hole in the valence band. The conduction band electron in TiO_2 can reduce an aqueous proton, as shown in the energy level diagram in Figure 17.14, while the hole accepts an electron as water oxidizes. The net reaction when four electrons are transferred in this way is the sum of two half-reactions:

$$4\,H^+(aq) + 4\,e^- \longrightarrow 2\,H_2(g)$$

$$2\,H_2O(\ell) \longrightarrow O_2(g) + 4\,H^+(aq) + 4\,e^-$$

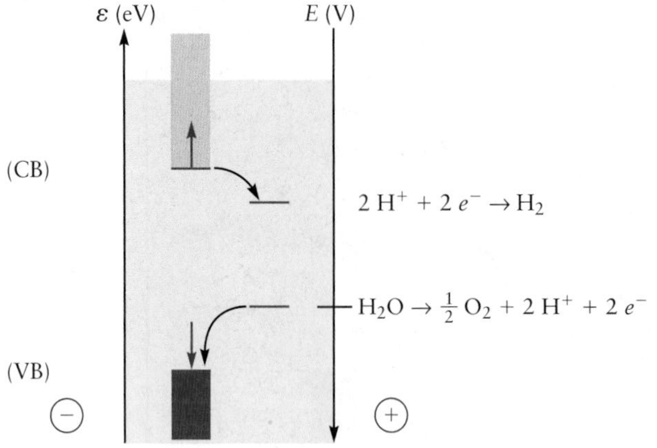

FIGURE 17.14 The potential of the conduction band of wide bandgap semiconductors is sufficiently negative to reduce hydrogen ions and that of the valence band sufficiently positive to oxidize water. The low energy edge of the conduction band of TiO_2 is 0.2 eV higher in energy (lower in electrostatic potential) than NHE.

The overall thermodynamics of the system as well as the stoichiometry of the reaction on the molecular level determines the energy required for the overall reaction. The standard Gibbs free energy of formation of water is

$$H_2(g) + \tfrac{1}{2}O_2(g) \longrightarrow H_2O(\ell) \qquad \Delta G_f^\circ = -237.2 \text{ kJ mol}^{-1}$$

so the light absorbed must provide at least 237.2 kJ mol^{-1} to decompose 1 mol of water. A mole of 3 eV photons carries 289 kJ of energy, but the energy absorbed by chromophores is partially degraded into heat, unfortunately, before ultimately being made available to drive the water splitting redox reactions. The net reaction of the two half reactions above is $2\,H_2O(\ell) \longrightarrow 2\,H_2(g) + O_2(g)$, but that requires four moles of electrons. The energy required per mole of water split is, as expected:

$$\frac{-nF\Delta E^\circ_{\text{cell}}}{2} = \frac{-(4\text{ mol})(-1.229)(96.485\text{ kJ mol}^{-1}\text{ eV}^{-1})}{2} = 237.2\text{ kJ}$$

Two electrons raised in energy through a potential difference of -1.229 V each would provide enough energy to decompose 1 molecule of water, but there is no molecular mechanism that would produce one oxygen *molecule*, as required. Water splitting, therefore, is a four-electron redox process that proceeds through a series of electron transfer reactions that involve several intermediates.

The energetic considerations discussed above have important consequences for the design of practical systems for photoelectrochemical water splitting. Colloidal semiconductors like TiO_2 (illustrated in Fig. 17.15) or $SrTiO_3$ have become the leading candidates for practical systems. They are more stable than molecular systems, and they have high surface-area-to-volume ratios that support the catalysts required to achieve acceptable reaction rates.

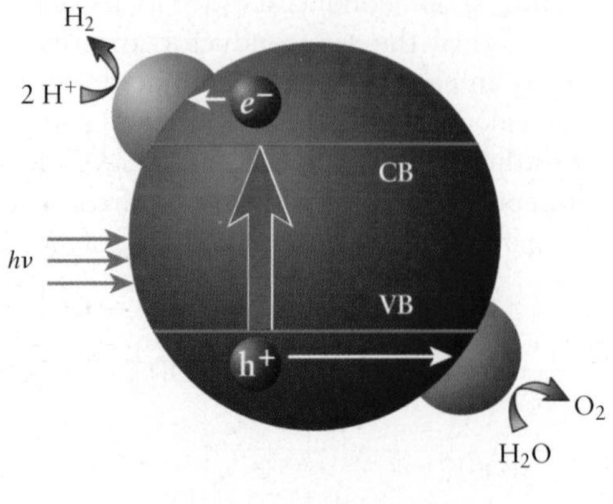

FIGURE 17.15 Direct photoelectrochemical water splitting by colloidal TiO_2, shown with hydrogen and oxygen evolution catalyst particles attached.

The water splitting reaction is thermodynamically allowed, but too slow for practical applications unless hydrogen reduction catalysts, like Pt, and oxidation catalysts, like RuO_2, are used. The wide bandgap semiconductors used for purposes of illustration are not suitable for practical applications because they absorb radiation only in the ultraviolet region of the spectrum, which accounts for less than 10% of the energy provided by sunlight. In addition, most of the energy of any ultraviolet radiation that is absorbed would be lost as heat; the fraction ultimately available is the ratio of the water oxidation potential energy to the energy of the bandgap, or about 40% (1.229 eV/3 eV). Most of the energy that is lost is due to the fact that the valence band in these wide bandgap materials is much more positive than that required to oxidize water, and so much current research in this area has been devoted to discovering new materials with smaller bandgaps and with valence bands at more negative potentials. Unfortunately, most materials with these properties decompose under illumination because the holes generated oxidize the semiconductors themselves; for example, metal sulfides and metal selenides decompose as

$$S^{2-} \xrightarrow{h\nu} S + 2\,e^- \text{ and } Se^{2-} \xrightarrow{h\nu} Se + 2\,e^-.$$

EXAMPLE 17.12

Would GaAs be a suitable candidate as a semiconductor electrode for direct photoelectrochemical water splitting? The valence band lies at about 0.7 V with respect to SHE and the conduction band lies at −0.7 V with respect to SHE; the band gap is 1.42 eV. Why or why not?

Solution

The redox potential of the conduction band is sufficiently negative to reduce H^+ to H_2 but the potential of the valence band is not sufficiently positive to oxidize water.

Related Problems: 47, 48

Improving the efficiencies of semiconductor-based solar cells is a topic of great current research interest, motivated both by a desire to more fully understand the basic physical and chemical processes involved as well as to guide the design of practical cells. Wide bandgap semiconductors like TiO_2 and $SrTiO_3$ have conduction bands with potentials sufficiently negative to reduce water but their valence bands are located at much more positive potentials than necessary to oxidize water, resulting in most of the energy absorbed being lost as heat. Photosynthesis, the process in nature by which solar energy is captured and stored, uses coupled systems to absorb light over a significant wavelength range and efficiently transfer energy to the sites where oxidation reduction reactions occur. Considerable effort is currently being devoted to develop semiconductor-based solar cells that operate using principles inspired by the natural photosynthetic systems. The accompanying *Connection to Energy:* Solar Energy Conversion introduces you to one approach that has been developed to increase the efficiency of direct photoelectrochemical water splitting as well as to generate electricity, using dye-sensitized semiconductor solar cells.

EXAMPLE 17.13

What properties should you specify for a dye to be used as a sensitizer in a dye-sensitized solar cell used for direct photoelectrochemical water splitting or for generating electricity directly in a cell like the Graetzel cell discussed in the *Connection to Energy:* Solar Energy Conversion?

Solution

The LUMO of the dye must lie at higher energy than the conduction band of the semiconductor in order for electron injection to occur; this requirement must be met for both applications.

Related Problems: 49, 50

CONNECTION TO ENERGY

Solar Energy Conversion

The sun is the ultimate source of energy for the Earth; some of that energy is stored in plants, which provide food and energy for current needs. Plants were also the precursors to fossil fuels, our main source of energy for electrical power generation, transportation, and manufacturing. The finite supply of fossil fuels, combined with a growing population and increasing energy requirements in developing countries, has stimulated tremendous interest in new sources of renewable energy such as solar, wind, and biomass. We describe two approaches to solar energy conversion and storage that are based upon photoelectrochemical reaction schemes: one to produce fuel, the other to produce electricity.

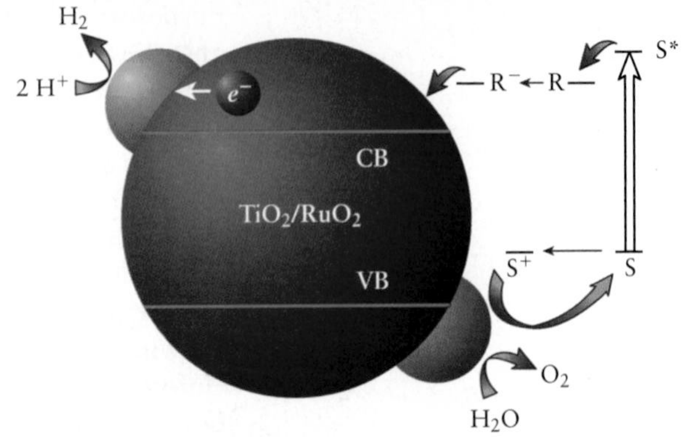

FIGURE C17.2 Photoelectrochemical water splitting with addition of a dye sensitizer S and catalysts.

Artificial Photosynthesis

Photosynthesis, which occurs in plants and some bacteria, converts solar energy into chemical potential energy that is stored in the bonds of carbohydrates. We recover this energy when metabolizing food or from the combustion of fossil fuels or biomass. Sunlight is the source of the Gibbs free energy required to drive these thermodynamically uphill reactions; absorption of a photon provides the energy to excite an electron from a HOMO to a LUMO to drive subsequent redox reactions (see Fig. 17.13). Natural photosynthetic systems are highly organized arrays of molecules that absorb much of the available sunlight and transfer the excitation energy efficiently to the photosynthetic reaction centers that catalyze the redox reactions.

The goal of *artificial photosynthesis* is to design simpler synthetic systems that perform the same functions as natural photosynthesis, the conversion and storage of solar energy as fuel, with a focus on hydrogen. Figure C17.2 shows how the addition of a dye sensitizer and a mediator to the system illustrated in Figure 17.15 enables water splitting at visible wavelengths. The basic idea is fairly straightforward but producing practical devices remains a daunting challenge. Light absorbed by sensitizers generates excited states that can inject electrons into the conduction band of the semiconductor particle, with the electrons ultimately reducing water (beginning with the $2\ H^+ \longrightarrow H_2(g) + 2\ e^-$ half reaction in Fig. 17.14 and again in Fig. C17.2 in shorthand notation not showing the electrons) to generate H_2 gas that may be stored and used as a fuel. The LUMO of the sensitizer must lie at higher energy than that of the conduction band to allow electron injection and the redox potential associated with the HOMO must be sufficiently positive to oxidize water. Redox mediators (if used) must have a wider "bandgap" than the sensitizer. Small Pt and RuO_2 particles (gold spheres) catalyze the reduction and oxidation reactions, respectively. We introduce this single-component model system because it is easy to visualize and illustrates principles that underlie one approach to develop more efficient solar cells.

Dye-Sensitized Solar Cells

The vast majority of solar cells in current use are Si-based photovoltaic devices in which light absorption excites electron-hole pairs in *p-n* junctions (see Section 22.7), generating electrons that flow through external circuits producing electrical work. These cells can be very efficient (more than 20%) but they are also very expensive due to demanding material requirements.

Considerable effort has been devoted to the development of dye-sensitized semiconductor solar cells (DSC) over the past several decades. Research in this area was pioneered in large part by the Swiss physical chemist Michael Grätzel, who has made important contributions to the development of systems for direct water splitting as well as for electricity generation. A schematic of the Grätzel cell is shown in Figure C17.3; it has many of the characteristics of the dye-sensitized semiconductor system discussed earlier. Sensitizing dyes are adsorbed on, or chemically bound to, the surfaces of wide bandgap semiconductor nanoparticles. Light absorbed by the dyes excites electrons into the LUMO from which they are injected into the conduction band of the semiconductor.

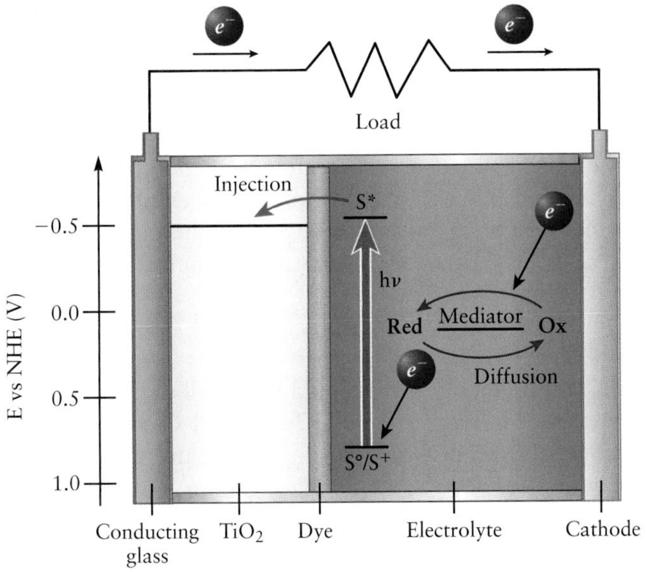

FIGURE C17.3 Schematic of a Grätzel cell.

Unlike in the previous example, however, electrons flow through an external circuit where they can do work. The electrons reenter the cell at the cathode and reduce the oxidized form of a redox mediator, which is then oxidized by the hole left behind in the sensitizer. A number of factors have been carefully considered to design cells with conversion efficiencies as high as 12% and projected lifetimes as long as 10 years. The morphology of the semiconductor nanoparticles, the electronic energy level structures of the dye molecules, the coverage of dye molecules on each nanoparticle and the way in which these nanoscopic structures are assembled have been carefully chosen to provide very large surface areas for efficient light absorption and electron injection. Dyes have been synthesized with LUMOs located slightly above the energy of the semiconductor conduction band to ensure efficient injection and with HOMOs located at the lowest possible energies to maximize the fraction of the incident light absorbed as well as to be compatible with the redox potential of the mediator. Examples of sensitizers include derivatives of $Ru(bpy)_3$ in which one or more of the bipyridine ligands have been replaced by isothiocyante (—N=C=S) ligands to extend the absorption onset to longer wavelengths, as shown in Figure C17.4. The solar spectrum is represented in the figure by the spectrum of a blackbody at the temperature of the sun to give you a sense of how much of the available radiation is absorbed by these dyes. The vertical axis is the percent conversion efficiency (incident photons to electrons).

The redox mediator transports electrons from the cathode to the radical cation created by photoexcitation and charge injection, regenerating the ground state of the sensitizer and preventing direct electron-hole pair recombination, an undesirable competing reaction. The relevant reactions are

$$S \xrightarrow{h\nu} S^+ + e^-(TiO_2^{cb}) \qquad \text{charge injection}$$

$$3\,I + e^- \longrightarrow I^- + I_2 \longrightarrow I_3^- \qquad \text{mediator reduction}$$

$$S^+ + I_3^- \longrightarrow S + I_2 + I \qquad \text{sensitizer regeneration}$$

The iodide/triodide redox couple has been the mediator of choice because it is sufficiently soluble, does not absorb significant amounts of light, has appropriate redox potentials, rapid dye regeneration kinetics and, most importantly, very slow rates for the direct recombination of the conduction band electrons with triodide, another undesirable competing reaction. The choice of redox mediator is critical; the performance of the iodide/triiodide couple is unmatched to date but there is considerable interest in finding other mediators that would increase efficiency. The standard reduction potential for the I/I_3^- couple is 0.35 V (SHE) and that of the radical cation is about 1.1 V. The energy difference is lost as heat; designing systems with more closely matched energy levels could greatly improve the efficiencies of these devices.

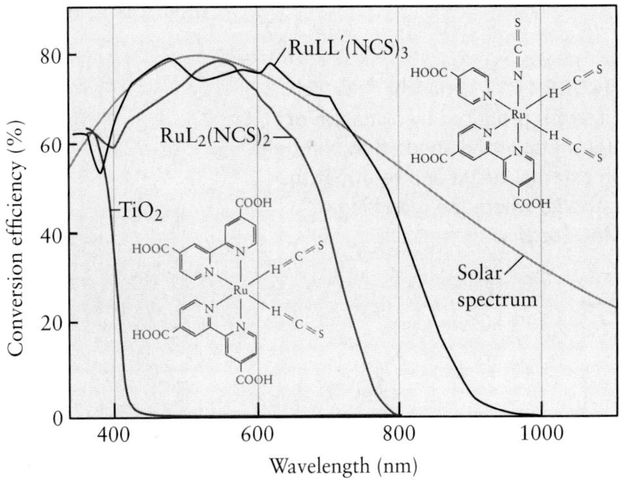

FIGURE C17.4 Conversion efficiency of molecules used as dye-sensitizers, absorbing photons in the visible and near infrared region of the solar spectrum.

17.5 Batteries and Fuel Cells

Batteries and fuel cells are examples of galvanic cells that convert chemical energy into electrical energy, the difference being that batteries carry all of their fuel with them and must be replaced or recharged, whereas fuel cells operate continuously as long as fuel and oxygen are being supplied. You are familiar with the batteries you use in cell phones, laptops, and other consumer products, but applications of batteries and fuel cells for transportation and energy storage are becoming increasingly important as well, driven primarily by energy and environmental concerns. We discuss the chemistry of several of the most common battery systems in current use and introduce you to fuel cells in this section.

17.5.1 Batteries

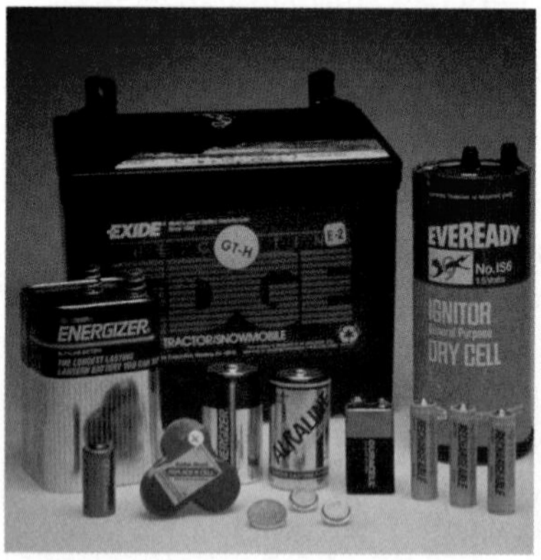

FIGURE 17.16 Batteries vary in size and chemistry. Shown here are an automobile lead-storage battery, rechargeable nickel–cadmium cells, alkaline cells, and zinc–carbon dry cells.

The origin of the **battery** is lost in history, but it could very well date back some 2000 years. A jar discovered near Baghdad in 1936 has all of the elements needed to generate a voltage, and replicas of the "Baghdad Battery" have produced about 2 V, using common electrolytes available in ancient times, like grape juice. Whether the jar was indeed used to electroplate objects with gold remains a matter of speculation. The development of the modern battery began with the invention of the voltaic pile in 1800, as mentioned earlier. The pile was a collection of galvanic cells (a battery of cells) connected in series (cathode to anode) so that the individual cell voltages added together; enough cells added together could produce a battery that could deliver quite a shock. Modern usage makes no distinction between a single cell and the voltaic pile, and the word *battery* has come to mean the cell itself (Fig. 17.16). Cells that are discarded when their electrical energy has been spent are called **primary cells**, and those that can be recharged are called **secondary cells.**

The most familiar primary cell is the **Leclanché cell** (also called a *zinc–carbon dry cell*) used for flashlights, portable radios, and a host of other purposes. More than five billion such dry cells are used worldwide each year, and estimates place the quantity of zinc consumed for this purpose at more than 30 metric tons per day. The "dry cell" is not really dry at all. Rather, its electrolyte is a moist powder containing ammonium chloride and zinc chloride. Figure 17.17 is a cutaway illustration of a dry cell, which consists of a zinc shell that serves as the anode

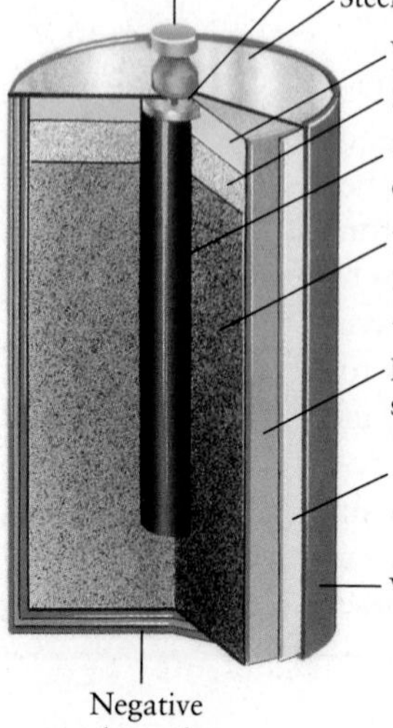

FIGURE 17.17 Leclanché "dry"cell. Electrons produced by oxidation of zinc metal at the anode flow through an external circuit and return at the cathode, where the reduction of MnO_2 occurs.

(negative terminal) and an axial graphite rod for a cathode (positive terminal), with the rod surrounded by a densely packed layer of graphite and manganese dioxide.

Each of these components performs an interesting and essential function. Oxidation occurs at the zinc anode:

$$Zn(s) \longrightarrow Zn^{2+}(aq) + 2\,e^- \text{ (anode)}$$

The moist salt mixture (like the salt bridge in Fig. 17.2) enables ion exchange between the sites of the anode and cathode half-reactions to maintain electrical neutrality as electrons flow through the external circuit. Manganese dioxide, the ultimate electron acceptor, is reduced to Mn_2O_3 by $H_2(g)$, produced by the reduction of ammonium ions in the following series of reactions:

$$2\,NH_4^+(aq) + 2\,e^- \longrightarrow H_2(g) + 2\,NH_3(aq)$$

$$H_2(g) + 2\,MnO_2(s) \longrightarrow Mn_2O_3(s) + H_2O(\ell)$$

Mixing powdered graphite with powdered MnO_2 greatly increases the effective surface area of the cathode, reduces the internal resistance of the cell, and enables currents of several amperes to flow. The overall cell reaction is

$$Zn(s) + 2\,MnO_2(s) + 2\,NH_4^+(aq) \longrightarrow Zn^{2+}(aq) + Mn_2O_3(s) + 2\,NH_3(aq) + H_2O(\ell)$$

The cell components are hermetically sealed (airtight) in a steel shell that is in contact with the zinc anode and thus acts as the negative terminal of the battery. Zinc–carbon dry cells produce 1.5 V when new.

The disadvantage of the Leclanché cell is that the concentration of protons (available as $NH_4^+(aq)$) decreases with time, and the battery's voltage falls as it is used. The zinc anode also corrodes as it oxidizes, and the electrolyte leaks out. The **alkaline dry cell** has largely replaced the Leclanché cell as the dominant primary cell for consumer applications; potassium hydroxide replaces ammonium chloride, and the half-cell reactions are

$$Zn(s) + 2\,OH^- \longrightarrow Zn(OH)_2(s) + 2\,e^- \quad \text{(anode)}$$

$$2\,MnO_2(s) + H_2O(\ell) + 2\,e^- \longrightarrow Mn_2O_3(s) + 2\,OH^- \quad \text{(cathode)}$$

The overall cell reaction is then

$$Zn(s) + 2\,MnO_2(s) + H_2O(\ell) \longrightarrow Zn(OH)_2(s) + Mn_2O_3(s)$$

All of the reactants and products in alkaline cells are pure solids or liquids whose concentrations are fixed, and so the cell voltage remains stable over time.

The third primary cell we discuss is the **zinc–mercuric oxide cell** shown in Figure 17.18. It is commonly packaged in the shape of a small button that is convenient for use in automatic cameras, hearing aids, digital calculators, and quartz–electric watches. The anode is a mixture of mercury and zinc, and the cathode is made of steel in contact with solid mercury(II) oxide (HgO). The electrolyte is a 45% KOH solution that saturates an absorbent material. The anode half-reaction is the same as that in an alkaline dry cell,

$$Zn(s) + 2\,OH^- \longrightarrow Zn(OH)_2(s) + 2\,e^- \quad \text{(anode)}$$

but the cathode half-reaction is now

$$HgO(s) + H_2O(\ell) + 2\,e^- \longrightarrow Hg(\ell) + 2\,OH^- \quad \text{(cathode)}$$

The overall reaction is

$$Zn(s) + HgO(s) + H_2O(\ell) \longrightarrow Zn(OH)_2(s) + Hg(\ell)$$

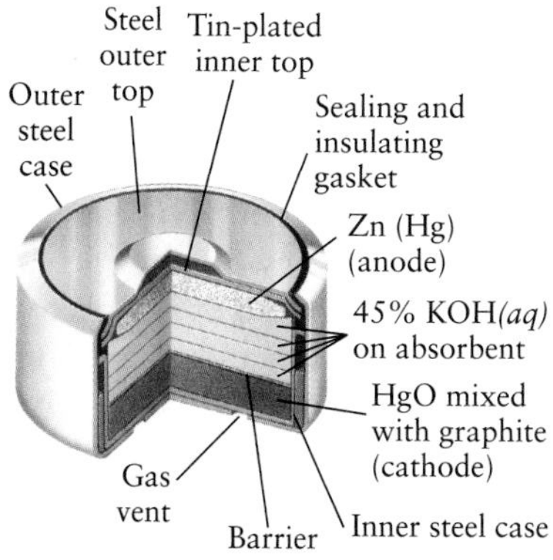

FIGURE 17.18 A zinc–mercuric oxide dry cell, used in electric watches and cameras.

The voltage (1.34 V) produced by these batteries is very stable over time, which made them especially valuable for use in communication equipment and scientific instruments.

Environmental concerns about mercury led federal regulators to ban their manufacture, however, and they have been replaced by lithium batteries (not lithium-ion batteries, discussed next) for portable electronics applications in cameras, watches, and calculators, for example. Lithium metal is the anode and MnO_2 is the cathode, with lithium perchlorate dissolved in a mixture of propylene carbonate and 1,2-dimethoxyethane as the electrolyte. The half-cell and overall cell reactions, which produce a nominal battery voltage of 3 V, are:

$$\mathrm{Li} \longrightarrow \mathrm{Li^+} + e^-$$

$$\underline{(\mathrm{Mn^{4+}})\mathrm{O_2} + \mathrm{Li^+} + e^- \longrightarrow (\mathrm{Mn^{3+}})\mathrm{O_2}\,(\mathrm{Li^+})}$$

$$\mathrm{Li} + (\mathrm{Mn^{4+}})\mathrm{O_2} \longrightarrow (\mathrm{Mn^{3+}})\mathrm{O_2}\,(\mathrm{Li^+})$$

17.5.2 Rechargeable Batteries

Rechargeable batteries have become very important in recent years for use in consumer products and transportation. The lithium-ion battery has become the dominant battery for use in laptops and cell phones. The lead acid battery has long been used in automotive applications with the nickel metal hydride (NiMH) battery having been developed as a rechargeable alternative for applications in hybrids and plug-in electric vehicles. Nickel metal hydride batteries are also popular alternatives to alkaline batteries for consumer applications because they can be recharged; they are manufactured in the same standard sizes. Batteries are recharged by applying an external potential that is greater in magnitude and opposite in sign to the galvanic cell potential, which causes the redox reactions to run in reverse, regenerating the reactants. Not all redox reactions are chemically reversible, however, which limits the choices for constructing rechargeable batteries. The Leclanché cell, for example, is not chemically reversible because the primary products are gases, hydrogen and ammonia, which are irreversibly bound up as water and $[Zn(NH_3)_2]^{2+}(aq)$ ions.

Rechargeable batteries are characterized by their voltages, the maximum current available (sustained for laptops or peak for camera flashes) and the total energy stored. The amount of charge available in a battery before it must be recharged is called its capacity, which for consumer applications has been defined as the maximum current that can be sustained for 20 hours of operation without dropping the voltage below an acceptable value. A 100 watt hour (Whr) battery, for example, will deliver 5 A for 20 hours. The battery in your laptop might have a rating of 3500 mAhr, meaning that it will provide 3.5 A for an hour or 1 A for 3.5 hours. The energy available from a battery before it must be recharged is given by the capacity times the voltage, so your laptop might provide $ItV = (3500\ \text{mAh})(10\ \text{V}) = 35\ \text{J}$ of energy before it needs to be recharged. Battery manufacturers continue to strive to develop batteries that deliver the most energy per unit volume or per-unit weight, depending upon their applications. These quantities are called the volumetric or gravimetric energy densities, measured in units Wh/L or Wh/kg, respectively. Modern Li-ion (see next page) laptop batteries achieve about 200 Wh/kg. Light metals such as lithium and magnesium are the most attractive cations for battery applications with hydroxide being among the most attractive anions. Figure 17.19 shows representative energy densities currently available in various classes of batteries.

The **lead–acid battery** is currently the most widely used rechargeable battery in vehicles. A 12-V lead storage battery consists of six 2.0-V cells (Fig. 17.20) connected in series (cathode to anode) and housed in a hard rubber or plastic case. The anodes in each cell are made of lead grids filled with spongy lead to

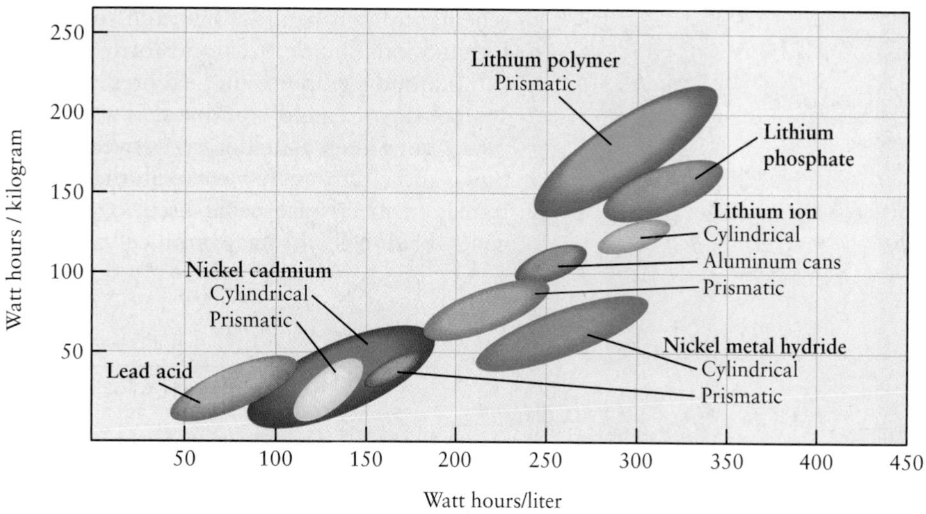

FIGURE 17.19 Ranges of typical energy densities for different classes of batteries.

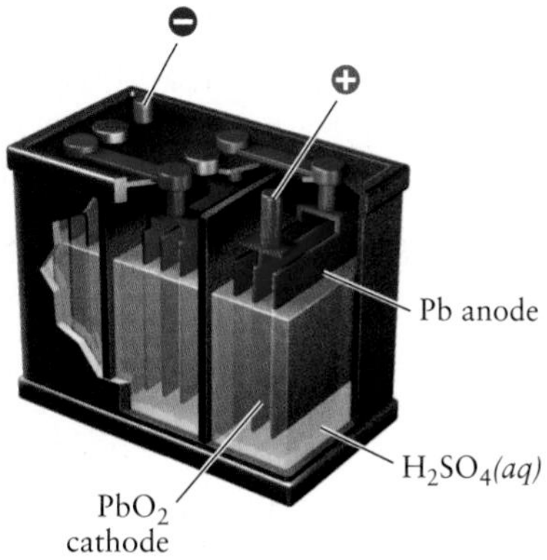

FIGURE 17.20 A lead–acid storage battery consists of several cells connected in series. The electrodes are both constructed from lead grids filled with spongy Pb (cathode) and PbO_2 (anode) and the electrolyte is sulfuric acid.

maximize the surface area in contact with the electrolyte; the cathodes are lead grids filled with PbO_2. The electrolyte is sulfuric acid solution (37% by mass).

Pb is oxidized to Pb^{2+} at the anode during discharge, while Pb^{4+} in PbO_2 is reduced to Pb^{2+} at the cathode; Pb^{2+} precipitates as lead sulfate on both electrodes. The half-cell reactions for the galvanic cells are

$$Pb(s) + SO_4^{2-} \longrightarrow PbSO_4(s) + 2\,e^- \qquad \text{(anode)}$$

$$PbO_2(s) + SO_4^{2-} + 4\,H_3O^+ + 2\,e^- \longrightarrow PbSO_4(s) + 6\,H_2O(\ell) \qquad \text{(cathode)}$$

$$Pb(s) + PbO_2(s) + 2\,SO_4^{2-} + 4\,H_3O^+ \longrightarrow 2\,PbSO_4(s) + 6\,H_2O(\ell)$$

The anode and cathode have both been largely converted to $PbSO_4(s)$ when the storage battery is fully discharged, and the concentration of sulfuric acid has fallen, because it is a reactant. Measuring the density of the electrolyte is a quick way to estimate the amount of charge available in the battery because the density of pure sulfuric acid is almost twice the density of water.

The lead–acid battery is recharged by applying a voltage greater than 12 V with the opposite polarity, which returns the battery to its initial state. One of the attractive features of lead–acid batteries is their ability to undergo thousands of charge/discharge cycles before they ultimately fail because $PbSO_4$ flakes fall off from the electrodes (preventing regeneration of Pb and PbO_2) or they develop internal short circuits. Lead–acid batteries designed for use in vehicles are not allowed to discharge completely. An alternator driven by the engine charges the battery continuously, or at least intermittently. About 1.8×10^7 J can be provided by an average automobile battery before it needs recharging, and currents as large as 100 A are available for the short time needed to start the engine.

A drawback of the lead–acid storage battery is its low energy density, the amount of energy obtainable per kilogram of battery mass. This limitation is not important when a battery is used primarily to start a gasoline-powered automobile, but it precludes the battery's use in vehicles driven by electric motors. The low energy density of the lead–acid battery limits the range of electric vehicles, and it has stimulated electrochemists to develop secondary batteries that have much higher energy densities.

Lithium-ion batteries are the most widely used rechargeable batteries for portable electronics and are used in electric cars. They have very high energy densities, due to lithium's low atomic mass, and are much safer than lithium batteries because they do not contain elemental lithium, which is very reactive. The chemistry of the lithium-ion battery is quite interesting, as shown schematically in Figure 17.21 and summarized by the reactions shown below. Lithium ions shuttle back and forth through an electrolyte between a $LiCoO_2$ cathode and a graphite anode represented as Li_yC_6 (we represent graphite by the symbol C_6 to remind you that it is a two-dimensional sheet of benzene rings).

Li^+ ions intercalate (insert) between the atomic planes of these layered structures, and electrons flow through the external circuit to balance charge. Other transition metals are being used in newer versions of this technology, so the general half-cell and the overall cell reactions may be written as:

$$Li_yC_6 \longrightarrow y\ Li^+ + y\ e^- + C_6 \qquad \text{(anode)}$$

$$Li_{1-y}MO_2 + y\ Li^+ + y\ e^- \longrightarrow LiMO_2 \qquad \text{(cathode)}$$

$$Li_yC_6 + Li_{1-y}MO_2 \longrightarrow LiMO_2 + C_6$$

The simplest way to think of this reaction is to consider what happens when single Li^+ ions shuttle back and forth. We can represent that situation by the following reactions, with M = Co.

$$LiC_6 \longrightarrow Li^+ + e^- + C_6 \qquad \text{(anode)}$$

$$CoO_2 + Li^+ + e^- \longrightarrow LiCoO_2 \qquad \text{(cathode)}$$

$$LiC_6 + CoO_2 \longrightarrow LiCoO_2 + C_6$$

in which an electron occupies an antibonding orbital of one graphite ring to form a (localized) anion for every Li^+ ion intercalated into graphite. Graphite is oxidized when lithium migrates to the cathode and electrons travel through the external circuit to reduce Co(IV) to Co(III).

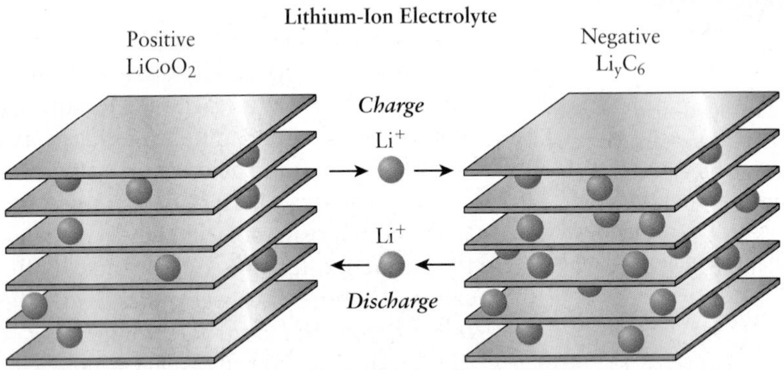

FIGURE 17.21 Lithium-ion battery schematic.

Nickel metal hydride (NiMH) batteries have replaced NiCd batteries as the rechargeable battery of choice for consumer applications like flashlights and have also begun to displace lead storage batteries from automotive applications, particularly for hybrid vehicles. The anodes in nickel metal hydride batteries are constructed from a variety of complex rare earth or nickel alloys, the hydrides of which are oxidized at the anode to form the elemental metals. The cathode is NiOOH (nickel oxyhydride), at which the nickel is reduced from Ni^{3+} to Ni^{2+}. The electrolyte is typically KOH. The half-cell reactions and overall cell reaction are:

$$MH + OH^- \longrightarrow M + H_2O + e^- \qquad \text{(anode)}$$

$$NiOOH + H_2O + e^- \longrightarrow Ni(OH)_2 + OH^- \qquad \text{(cathode)}$$

$$MH + NiOOH \longrightarrow M + Ni(OH)_2$$

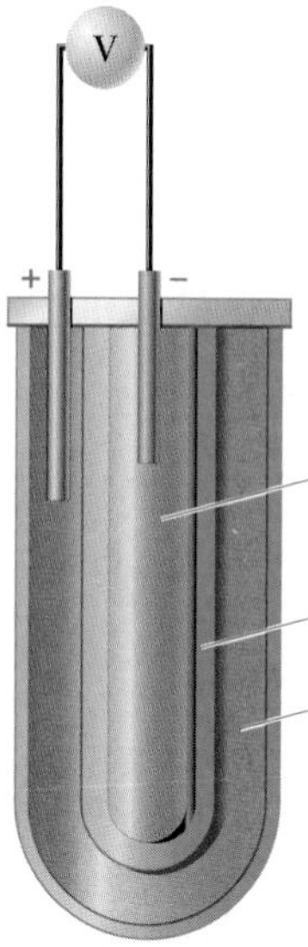

FIGURE 17.22 Schematic of a sodium–sulfur battery. Na^+ ions generated by oxidation at the anode also serve to neutralize charge as they migrate through beta-alumina to the cathode.

Another class of interesting batteries with high energy densities uses alkali metals (lithium or sodium) as anodes and sulfur as the electron acceptor. Sulfur does not conduct electricity, so graphite is used as the cathode. These batteries operate at high temperatures because the elements must be liquids for efficient electrical conduction (sulfur melts at 112°C, lithium at 186°C, and sodium at 98°C). The sodium–sulfur cell (Fig. 17.22), for example, has an optimal operating temperature of 250°C. The half-cell reactions are

$$2\ Na \longrightarrow 2\ Na^+ + 2\ e^- \qquad \text{(anode)}$$

$$S + 2\ e^- \longrightarrow S^{2-} \qquad \text{(cathode)}$$

$$2\ Na + S \longrightarrow 2\ Na^+ + S^{2-}$$

The cathode reaction shown is an oversimplification because sulfide ions react with sulfur to form polysulfides:

$$S^{2-} + nS \longrightarrow S_{n+1}^{2-}$$

but the fundamental principles of the cell operation are the same.

What makes the sodium–sulfur cell possible is a remarkable property of a compound called beta-alumina, which has the composition $NaAl_{11}O_{17}$. Beta-alumina allows sodium ions to migrate through its structure very easily, but it blocks the passage of polysulfide ions. Therefore, it functions as a semipermeable medium like the membranes used in osmosis (see Section 11.5). This ionic solid electrolyte serves to physically separate the reactants from one another while permitting the transport of the redox-active species $Na^+(aq)$; the lithium–sulfur battery operates on similar principles.

Sodium–sulfur batteries, with their high energy densities, were originally developed with transportation applications in mind. NiMH batteries have become the current favorite for automotive applications, but the sodium–sulfur battery has found an important application in large-scale energy storage, where the high operating temperatures are less problematic. The city of Presidio, Texas (population ca. 4000), has suffered intermittent power outages over the years because it has been connected to the electric power grid by a single transmission line. The city has recently installed a sodium–sulfur battery that is about the size of a house that will provide up to 4 MW of power for 8 hours, enough to power the entire town overnight. These batteries are also being used in remote locations to store energy generated from wind turbines or photovoltaic systems, making these locations independent of the electric utility grid.

17.5.3 Fuel Cells

Fuel cells convert chemical potential energy into electrical energy using coupled redox reactions, like batteries, but they operate as long as they are being supplied with fuel. Batteries carry all of their "fuel" with them and must be replaced or recharged to continue to provide energy. Fuel cells operate using the same oxidation reactions as combustion but are much more efficient, in principle, because different thermodynamic considerations apply. We discuss two classes of fuel cells: polymer electrolyte membrane (PEM) fuel cells that have been developed for transportation and solid oxide fuel cells (SOFC) that are currently the dominant technology for large-scale stationary power generation. We focus on hydrogen fuel cells because of the attractive features they offer, but fuel cells that use methane or methanol as fuels have been developed and may have advantages in certain applications, for example, in areas where natural gas is readily available. However, carbon-based fuels, of course, emit CO_2 and contribute to climate change (see Section 20.6).

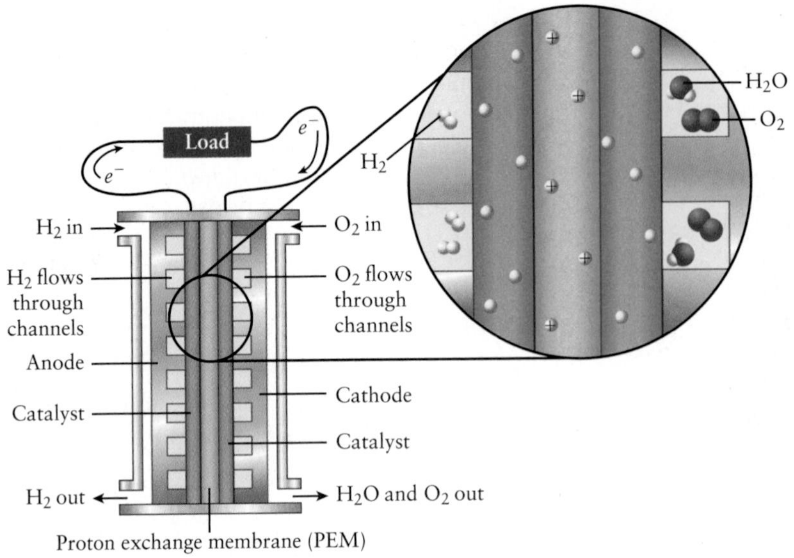

FIGURE 17.23 Polymer electrolyte membrane (PEM) fuel cell.

The basic components and operation of a PEM fuel cell are shown in Figure 17.23. The hydrogen fuel, entering from the left, is dissociated catalytically and ionized at the anode to produce protons and electrons. The electrons travel through the external circuit, producing work, while the protons move through the membrane toward the cathode. The electrons and protons react catalytically with molecular oxygen at the cathode in a four-electron reduction to produce water.

The overall cell reaction is

$$2\ H_2(g) + O_2(g) \longrightarrow 2\ H_2O(\ell)$$

with a standard cell potential $\Delta E^{\circ}_{cell} = 1.229$ V. The polymer membranes in current use are fluorocarbons (hydrocarbons in which all of the hydrogen atoms have been replaced by fluorine) with sulfonic acid $-(SO_3H)$ functional groups attached. Protons transfer by hopping between the sulfonic acid side groups, thus moving along and between the polymer chains. The operating temperature of current PEM fuel cells is limited to about 80°C because the polymer membranes must remain hydrated for this proton transfer mechanism to work. Developing alternative membrane materials that transport protons at higher temperatures is an active area of research in materials science and engineering. Increasing the activity and reducing the cost of the oxygen reduction catalysts is also another area of intense current interest. Present generation PEM fuel cells use finely divided Pt nanoparticles supported on carbon substrates. A factor of 10 improvement in oxygen reduction activity has been demonstrated recently by controlling the surface morphology and subsurface electronic structure of platinum nanocrystals that have been alloyed with other metals such as nickel. The surface layer of these particles is pure Pt, which has long been known to be effective at breaking O—O bonds and forming O—H bonds. A subsurface alloy with nominal composition $PtNi_3$ alters the electronic structure of the surface layer to make it an even more effective catalyst for oxygen reduction. Collaboration between experimental and computational materials scientists on the development of new catalysts for oxidation–reduction and other important catalytic reactions is a particularly exciting area of current research.

To better understand the attractive characteristics of fuel cells, let's compare the energy available from the hydrogen oxygen PEM fuel cell with that available from the combustion of hydrogen in an internal combustion engine. The relevant reaction for transportation is

$$H_2(g) + \tfrac{1}{2}\ O_2(g) \longrightarrow H_2O(g) \qquad \Delta H^{\circ}_f = -241.8\ \text{kJ mol}^{-1}$$

because water is produced as a high-temperature exhaust gas. We calculate the maximum work available from combustion as follows:

$$-w_{\text{max}} = \epsilon q_{\text{p}} = (1 - T_{\text{l}}/T_{\text{h}})q_{\text{p}}$$

where ϵ is the Carnot efficiency, q_{p} is the heat transferred at constant pressure (supplied to the hot reservoir by the exothermic combustion reaction), T_{l} is the temperature of the exhaust gas, and T_{h} is the operating temperature of the internal combustion engine. The Carnot efficiency of an internal combustion engine is about 45%, so the maximum theoretical work, $-w_{\text{max}}$, available from hydrogen used as a fuel in such an engine is

$$-w_{\text{max}} = (0.4)(241.8 \text{ kJ mol}^{-1}) = +96 \text{ kJ mol}^{-1}$$

Note that we have used the convention in which w is the work done on the system, so $-w$, which has a positive value above, is the work done *on* the surroundings by the engine. The maximum work available if the same reaction is carried out in a fuel cell is $-w_{\text{max}} = -\Delta G_{\text{f}}^{\text{o}} = 228.6 \text{ kJ mol}^{-1}$, which is more than twice that available, in principle, from the internal combustion engine. $\Delta G^{\text{o}} \approx \Delta H^{\text{o}}$ for combustion reactions, so finding ways to capture that energy without suffering the losses imposed by the Carnot efficiency is an important goal. Electric vehicles powered by fuel cells are also significantly more efficient than those powered by internal combustion engines, not only because of the more favorable thermodynamic considerations but also because electric motors are mechanically more efficient than internal combustion engines, the former being about 90% efficient compared with 75% efficiency for the latter. We can compare the so-called "tank-to-wheel" efficiencies of a vehicle powered by a hydrogen fuel cell to one powered by a hydrogen-fueled internal combustion engine by multiplying the work available by the mechanical efficiencies of each engine and taking the ratio.

$$\frac{\text{Overall efficiency fuel cell vehicle}}{\text{Overall efficiency internal combustion engine vehicle}} = \frac{(0.9)(228.6 \text{ kJ mol}^{-1})}{(0.75)(96 \text{ kJ mol}^{-1})} = 2.89$$

The U.S. Department of Energy has been evaluating a fleet of fuel cell vehicles for several years now and reports an average efficiency for the fuel cells in the 53–58% range, which is very close to their target of 60%. (Efficiency is defined as the ratio of the electrical power produced to the heating value of the fuel.) Honda introduced a hydrogen-powered fuel cell car called the FCX Clarity in 2013. Honda reports a tank-to-wheel efficiency of 60%, which is, indeed, about three times higher than that of vehicles powered by internal combustion engines. There are, of course, many practical issues that need to be addressed before vehicles powered by fuel cells are widely deployed, but it is a promising technology and one that clearly illustrates the advantages of capturing chemical energy directly as electrical energy, without the inevitable thermodynamic losses associated with combustion. Electric cars recharged by electricity from solar panels are also a superb option.

Solid oxide fuel cells (SOFC) appear to be the current technology of choice for large-scale stationary power generation, whether to produce backup power for critical applications such as hospitals or for generating power in remote locations that are not served by the electrical power grid. A schematic of an SOFC is shown in Figure 17.24. The electrolytes in these cells are oxide ceramics (see Section 22.2), an important example of which is zirconium oxide (ZrO_2) that has been doped with yttrium oxide, Y_2O_3. Doping replaces some of the Zr^{4+} ions with Y^{3+} ions, which creates oxide "vacancies" because only three oxide ions per Y^{3+} are required for charge neutrality. Oxide ions can move through the lattice at sufficiently high temperatures (1000 °C), which provides

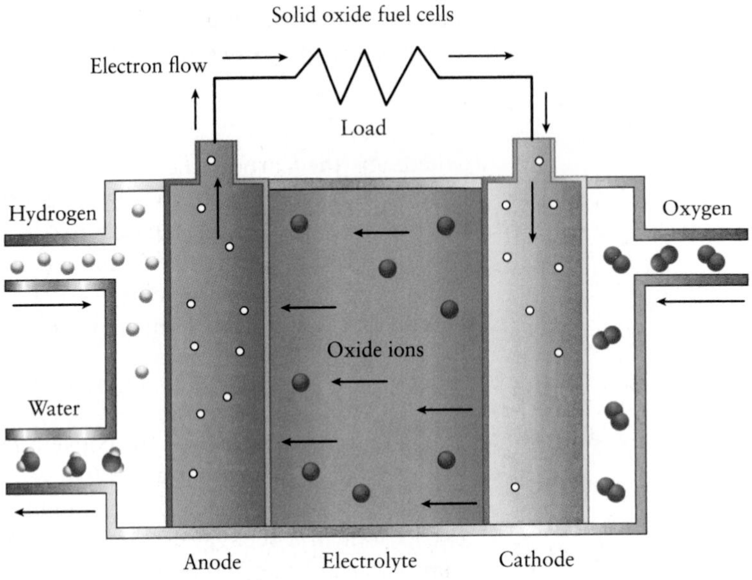

FIGURE 17.24 Solid oxide fuel cell.

a mechanism for ionic conduction. The overall cell reaction for a typical solid oxide fuel cell that uses hydrogen is

$$H_2(g) + \tfrac{1}{2}O_2(g) \longrightarrow H_2O(\ell) \qquad \Delta H_f^\circ = -285.8 \text{ kJ mol}^{-1}$$

when the product is liquid water. The same amount of energy is available from this reaction when the product is steam because the energy stored in the steam is recovered, as it cools to the boiling point, from the enthalpy of condensation, and from cooling the liquid water further to ambient temperatures. The standard cell potentials are similar for both high-temperature and low-temperature products, but the waste heat can be captured in stationary solid oxide fuel cells, unlike those that power vehicles. The intrinsic overall efficiencies of solid oxide fuel cells have been demonstrated to be around 50%, increasing if the steam is used to drive turbines to around 70%, and up to 80% or so if the waste heat is recovered. These systems can be quite large; commercial versions can provide 250 kW to 1 MW of power, which is sufficient for emergency backup at large installations like hospitals.

17.6 Corrosion and Corrosion Prevention

The **corrosion** of metals is one of the most significant problems faced by advanced industrial societies (Fig. 17.25). It has been estimated that in the United States alone, the annual cost of corrosion amounts to tens of *billions* of dollars. Effects of corrosion are both visible (the formation of rust on exposed iron surfaces) and invisible (the cracking and resulting loss of strength of metal beneath the surface). The mechanism of corrosion must be understood before processes can be developed for its prevention.

Although corrosion is a serious problem for many metals, we focus on the spontaneous electrochemical reactions of iron. Corrosion can be pictured as a "short-circuited" galvanic cell, in which some regions of the metal surface act as cathodes and others as anodes, and the electric "circuit" is completed by electrons flowing through the iron itself. These electrochemical cells form in

FIGURE 17.25 Rust.

parts of the metal where there are impurities or in regions that are subject to stress. The anode reaction is

$$Fe(s) \longrightarrow Fe^{2+}(aq) + 2\,e^-$$

Various cathode reactions are possible. In the absence of oxygen (for example, at the bottom of a lake), the corrosion reactions are

$$\begin{array}{lr} Fe(s) \longrightarrow Fe^{2+}(aq) + 2\,e^- & \text{(anode)} \\ 2\,H_2O(\ell) + 2\,e^- \longrightarrow 2\,OH^-(aq) + H_2(g) & \text{(cathode)} \\ \hline Fe(s) + 2\,H_2O(\ell) \longrightarrow Fe^{2+}(aq) + 2\,OH^-(aq) + H_2(g) & \end{array}$$

These reactions are generally slow and do not cause serious amounts of corrosion. Far more extensive corrosion takes place when the iron is in contact with both oxygen and water. In this case the cathode reaction is

$$\tfrac{1}{2}\,O_2(g) + 2\,H_3O^+(aq) + 2\,e^- \longrightarrow 3\,H_2O(\ell)$$

The Fe^{2+} ions formed simultaneously at the anode migrate to the cathode, where they are further oxidized by O_2 to the +3 oxidation state to form rust ($Fe_2O_3{\cdot}xH_2O$), a hydrated form of iron(III) oxide:

$$2\,Fe^{2+}(aq) + \tfrac{1}{2}\,O_2(g) + (6 + x)H_2O(\ell) \longrightarrow Fe_2O_3{\cdot}xH_2O + 4\,H_3O^+(aq)$$

The hydronium ions produced in this reaction allow the corrosion cycle to continue.

When a portion of the paint that protects a piece of iron or steel is chipped off (Fig. 17.26), the exposed area acts as the cathode because it is open to the atmosphere (air and water) and is therefore rich in oxygen, whereas oxygen-poor areas *under* the paint act as anodes. Rust forms on the cathode (the visible, exposed region), and pitting (loss of metal through oxidation of iron and flow of metal ions to the cathode) occurs at the anode. This pitting can lead to loss of structural strength in girders and other supports. The most serious harm done by corrosion is not the visible rusting but the damage done beneath the painted surface.

A number of factors speed corrosion. Dissolved salt provides an electrolyte that improves the flow of charge through solution; a well-known example is the rapid rusting of cars in areas where salt is spread on icy roads. Higher acidity also increases corrosion, as seen by the role of H_3O^+ as a reactant in the reduction process at the cathode. Acidity is enhanced by the presence of dissolved

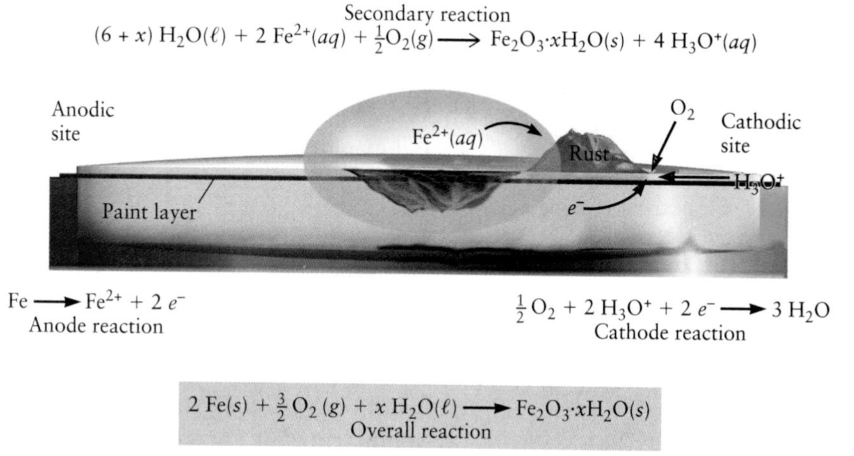

FIGURE 17.26 Schematic of iron corrosion. Pitting occurs in the anodic region, and rust appears in the cathodic region. The movement of hydronium ions through the hydrated rust pile is not shown.

CO_2 (which produces H_3O^+ and HCO_3^- ions) and by air pollution from oxides of sulfur, which leads to the formation of dissolved sulfuric acid in acid precipitation.

Corrosion of iron can be inhibited in a number of ways. Coatings of paint or plastics obviously protect the metal, but they can crack or suffer other damage, thereby localizing and accentuating the process. An important method of protecting metals arises from the phenomenon of **passivation**, in which a thin metal oxide layer forms on the surface and prevents further electrochemical reactions. Some metals become passivated spontaneously upon exposure to air; aluminum, for example, reacts with oxygen to form a thin protective layer of Al_2O_3. Special paints designed to prevent rusting contain potassium dichromate ($K_2Cr_2O_7$) and lead oxide (Pb_3O_4), which cause the superficial oxidation and passivation of iron. Stainless steel is an alloy of iron with chromium in which the chromium leads to passivation and prevents rusting.

A different way of preventing iron corrosion is to use a **sacrificial anode.** A comparison of the standard reduction potentials of iron and magnesium

$$Fe^{2+} + 2\,e^- \longrightarrow Fe(s) \qquad E° = -0.41\ V$$

$$Mg^{2+} + 2\,e^- \longrightarrow Mg(s) \qquad E° = -2.39\ V$$

shows that Mg^{2+} is much harder to reduce than Fe^{2+} or, conversely, that Mg(s) is more easily oxidized than Fe(s). A piece of magnesium in electrical contact with iron is oxidized in preference to the iron, and the iron is therefore protected. The magnesium is the sacrificial anode, and once it is consumed by oxidation it must be replaced. This method is used to protect ship hulls, bridges, and iron water pipes from corrosion. Magnesium plates are attached at regular intervals along a piece of buried pipe, and it is far easier to replace them periodically than to replace the entire pipe.

17.7 Electrometallurgy

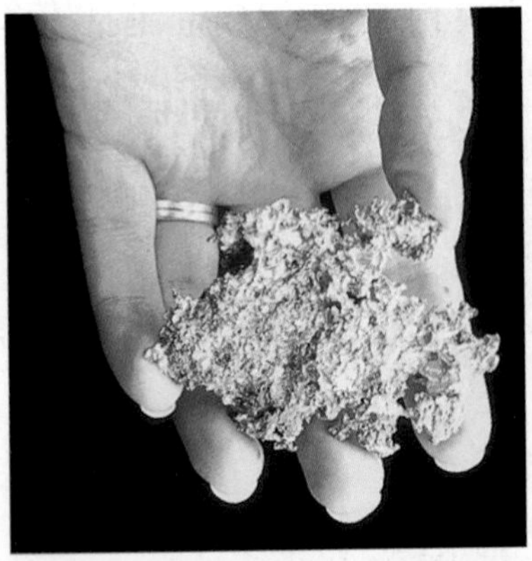

FIGURE 17.27 A specimen of native copper.

The recovery of metals from their sources in the Earth is the science of **extractive metallurgy**, a discipline that draws on chemistry, physics, and engineering for its methods. As a science it is a comparatively recent subject, but its beginnings, which were evidently in the Near East about 6000 years ago, marked the emergence of humanity from the Stone Age. The earliest known metals were undoubtedly gold, silver, and copper because they could be found in their native (elemental) states (Fig. 17.27). Gold and silver were valued for their ornamental uses, but they are too soft to have been made into tools. Iron was also found in elemental form—although rarely—in meteorites.

Most metals in nature are combined with other elements, such as oxygen and sulfur in ores, and chemical processes are required to free them. As Table 17.1 shows, the free energies of formation of most metal oxides are negative, indicating that the reverse reactions, which would yield the free metal and oxygen, have positive free energy changes. Scientists can carry out one of these reverse reactions to obtain the free metal only by coupling it with a second, spontaneous chemical reaction. The greater the cost in free energy, the more difficult the production of the free metal. Thus, silver and gold (at the bottom of Table 17.1) exist in nature as elements, and mercury can be released from its oxide or sulfide ore (cinnabar) merely by moderate heating (see Fig. 1.6). Extracting pure copper, zinc, and iron requires more stringent conditions; the ores of these metals are reduced in chemical reactions at high temperatures, collectively called **pyrometallurgy.** These reactions are carried out in huge furnaces, in which a fuel such as coke (coal from which volatile components have been

TABLE 17.1

Metal Oxides Arranged According to Ease of Reduction

Metal Oxide	Metal	$n^\dagger$	$\Delta G_f^\circ/n^\dagger$ (kJ mol^{-1})	A Method of Production of the Metal
MgO	Mg	2	−285	Electrolysis of $MgCl_2$
Al_2O_3	Al	6	−264	Electrolysis
TiO_2	Ti	4	−222	Reaction with Mg
Na_2O	Na	2	−188	Electrolysis of NaCl
Cr_2O_3	Cr	6	−176	Electrolysis, reduction by Al
ZnO	Zn	2	−159	Smelting of ZnS
SnO_2	Sn	4	−130	Smelting
Fe_2O_3	Fe	6	−124	Smelting
NiO	Ni	2	−106	Smelting of nickel sulfides
PbO	Pb	2	−94	Smelting of PbS
CuO	Cu	2	−65	Smelting of $CuFeS_2$
HgO	Hg	2	−29	Moderate heating of HgS
Ag_2O	Ag	2	−6	Found in elemental form
Au_2O_3	Au	6	>0	Found in elemental form

†The standard free energies of formation of the metal oxides in kJ mol^{-1} are adjusted for fair comparison by dividing them by n, the total decrease in oxidation state required to reduce the metal atoms contained in the oxide to oxidation states of 0. Thus, the reduction of Cr_2O_3 involves a change in the oxidation state of two chromium atoms from +3 to 0, so $n = 2 \times 3 = 6$.

expelled) serves both as the reducing agent and as the source of heat to maintain the required high temperatures. The process, called **smelting,** involves both chemical change and melting. The combustion of carbon

$$C(s) + O_2(g) \longrightarrow CO_2(g) \qquad \Delta G^\circ = -394 \text{ kJ mol}^{-1}$$

provides the driving force for the overall reaction.

Even smelting is not sufficient to recover the metals at the top of Table 17.1, which have the most negative free energies of formation for their oxides (and for their sulfides as well). These metals have particularly high values of free energy relative to their compounds found in readily available ores. **Electrometallurgy**, or electrolytic production, provides the best ways of recovering such elements from their ores. Electrochemical cells are also used to purify the metals produced by the techniques of pyrometallurgy.

17.7.1 Aluminum

Aluminum is the third most abundant element in the Earth's crust (after oxygen and silicon), accounting for 8.2% of the total mass. It occurs most commonly in association with silicon in the aluminosilicates of feldspars and micas and in clays, the products of weathering of these rocks. The most important ore for aluminum production is bauxite, a hydrated aluminum oxide that contains 50% to 60% Al_2O_3; 1% to 20% Fe_2O_3; 1% to 10% silica; minor concentrations of titanium, zirconium, vanadium, and other transition-metal oxides; and the balance (20% to 30%) water. Bauxite is purified via the **Bayer process**, which takes advantage of the fact that the amphoteric oxide alumina is soluble in strong bases but iron(III) oxide is not. Crude bauxite is dissolved in sodium hydroxide

$$Al_2O_3(s) + 2\,OH^-(aq) + 3\,H_2O(\ell) \longrightarrow 2\,Al(OH)_4^-(aq)$$

and separated from hydrated iron oxide and other insoluble impurities by filtration. Pure hydrated aluminum oxide precipitates when the solution is cooled to supersaturation and seeded with crystals of the product:

$$2\ Al(OH)_4^-(aq) \longrightarrow Al_2O_3 \cdot 3H_2O(s) + 2\ OH^-(aq)$$

The water of hydration is removed by calcining at high temperature (1200°C).

Compared with copper, iron, gold, and lead, which were known in antiquity, aluminum is a relative newcomer. Sir Humphry Davy obtained it as an alloy of iron and proved its metallic nature in 1809. It was first prepared in relatively pure form in 1825 by H. C. Oersted through reduction of aluminum chloride with an amalgam of potassium dissolved in mercury

$$AlCl_3(s) + 3\ K(Hg)_x(\ell) \longrightarrow 3\ KCl(s) + Al(Hg)_{3x}(\ell)$$

after which the mercury was removed by distillation. Aluminum remained largely a laboratory curiosity until 1886, when Charles Hall in the United States (then a 21-year-old graduate of Oberlin College) and Paul Héroult (a Frenchman of the same age) independently invented an efficient process for its production. In the 1990s the worldwide production of aluminum by the **Hall–Héroult process** was approximately 1.5×10^7 metric tons per year.

The Hall–Héroult process involves the cathodic deposition of aluminum, from molten cryolite (Na_3AlF_6) containing dissolved Al_2O_3, in electrolysis cells (Fig. 17.28). Each cell consists of a rectangular steel box some 6 m long, 2 m wide, and 1 m high, which serves as the cathode, and massive graphite anodes that extend through the roof of the cell into the molten cryolite bath. Enormous currents (50,000 to 100,000 A) are passed through the cell, and as many as one hundred such cells may be connected in series.

Molten cryolite, which is completely dissociated into Na^+ and AlF_6^{3-} ions, is an excellent solvent for aluminum oxide, giving rise to an equilibrium distribution of ions such as Al^{3+}, AlF^{2+}, AlF_2^+, . . ., AlF_6^{3-}, and O^{2-} in the electrolyte. Cryolite melts at 1000°C, but its melting point is lowered by dissolved aluminum oxide, so the operating temperature of the cell is about 950°C. Compared with the melting point of pure Al_2O_3 (2050°C), this is a low temperature, and it is the reason the Hall–Héroult process has succeeded. Molten aluminum is somewhat denser than the melt at 950°C and therefore collects at the bottom of the cell, from which it is tapped periodically. Oxygen is the primary anode product, but it reacts with the graphite electrode to produce carbon dioxide. The overall cell reaction is

$$2\ Al_2O_3 + 3\ C \longrightarrow 4\ Al + 3\ CO_2$$

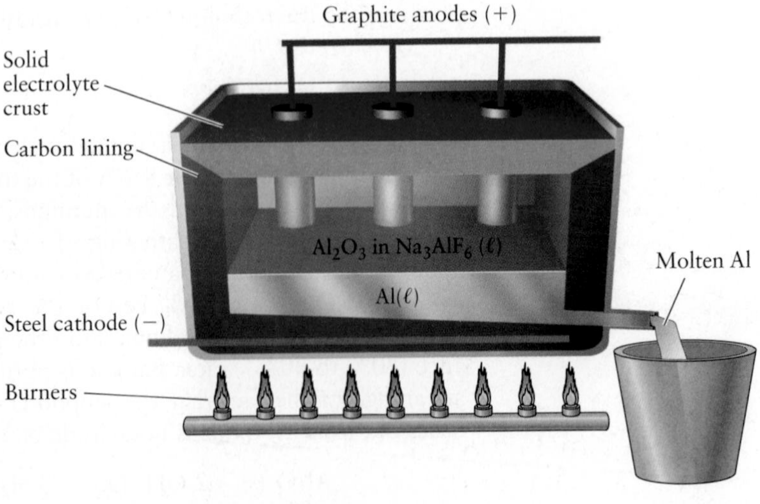

FIGURE 17.28 An electrolytic cell used in the Hall–Héroult process for the commercial production of aluminum.

TABLE 17.2

Densities of Selected Metals

Metal	Density (g cm^{-3}) at Room Conditions
Li	0.534
Na	0.971
Mg	1.738
Al	2.702
Ti	4.54
Zn	7.133
Fe	7.874
Ni	8.902
Cu	8.96
Ag	10.500
Pb	11.35
U	18.95
Au	19.32
Pt	21.45

Aluminum and its alloys have a tremendous variety of applications. Many of these make use of aluminum's low density (Table 17.2), an advantage over iron or steel when weight savings are desirable—such as in the transportation industry, which uses aluminum in vehicles from automobiles to satellites. Aluminum's high electrical conductivity and low density make it useful for electrical transmission lines. For structural and building applications, its resistance to corrosion is an important feature, as is the fact that it becomes stronger at subzero temperatures. (Steel and iron sometimes become brittle under these circumstances.) Household products that contain aluminum include foil, soft drink cans, and cooking utensils.

17.7.2 Magnesium

Like aluminum, magnesium is an abundant element on the surface of the Earth, but it is not easy to prepare in elemental form. Although ores such as dolomite [$CaMg(CO_3)_2$] and carnallite [$KMgCl_3{\cdot}6(H_2O)$] exist, the major commercial source of magnesium and its compounds is seawater. Magnesium forms the second most abundant positive ion in the sea, and scientists separate Mg^{2+} from the other cations in seawater (Na^+, Ca^{2+}, and K^+, in particular) by taking advantage of the fact that magnesium hydroxide is the least soluble hydroxide of the group. Economical recovery of magnesium requires a low-cost base to treat large volumes of seawater and efficient methods for separating the $Mg(OH)_2(s)$ that precipitates from the solution. One base that is used in this way is calcined dolomite, prepared by heating dolomite to high temperatures to drive off carbon dioxide:

$$CaMg(CO_3)_2(s) \longrightarrow CaO \cdot MgO(s) + 2\ CO_2(g)$$

The greater solubility of calcium hydroxide ($K_{sp} = 5.5 \times 10^{-6}$) relative to magnesium hydroxide ($K_{sp} = 1.2 \times 10^{-11}$) leads to the reaction

$$CaO \cdot MgO(s) + Mg^{2+}(aq) + 2\ H_2O(\ell) \longrightarrow 2\ Mg(OH)_2(s) + Ca^{2+}(aq)$$

The magnesium hydroxide that is produced in this process includes not only the magnesium from the seawater but also that from the dolomite.

An interesting alternative to dolomite as a base for magnesium production is a process used off the coast of Texas (Fig. 17.29). Oyster shells (composed largely of $CaCO_3$) are calcined to give lime (CaO), which is added to the seawater to yield magnesium hydroxide. The $Mg(OH)_2$ slurry (a suspension in water) is washed and filtered in huge nylon filters.

After purification, $Mg(OH)_2$ can be reacted with carbon dioxide to give magnesium carbonate, used for coating sodium chloride in table salt to prevent caking and for antacid remedies. Another alternative is to add hydrochloric acid to the magnesium hydroxide to neutralize it and yield hydrated magnesium chloride:

$$Mg(OH)_2(s) + 2\ HCl(aq) \longrightarrow MgCl_2(aq) + 2\ H_2O(\ell)$$

After the water is evaporated, the solid magnesium chloride is melted (m.p. 708°C) in a large steel electrolysis cell that holds as much as 10 tons of the molten salt. The steel in the cell acts as the cathode during electrolysis, with graphite anodes suspended from the top. The cell reaction is

$$MgCl_2(\ell) \longrightarrow Mg(\ell) + Cl_2(g)$$

The molten magnesium liberated at the cathode floats to the surface and is dipped out periodically, while the chlorine released at the anodes is collected and reacted with steam at high temperatures to produce hydrochloric acid. This is recycled for further reaction with magnesium hydroxide.

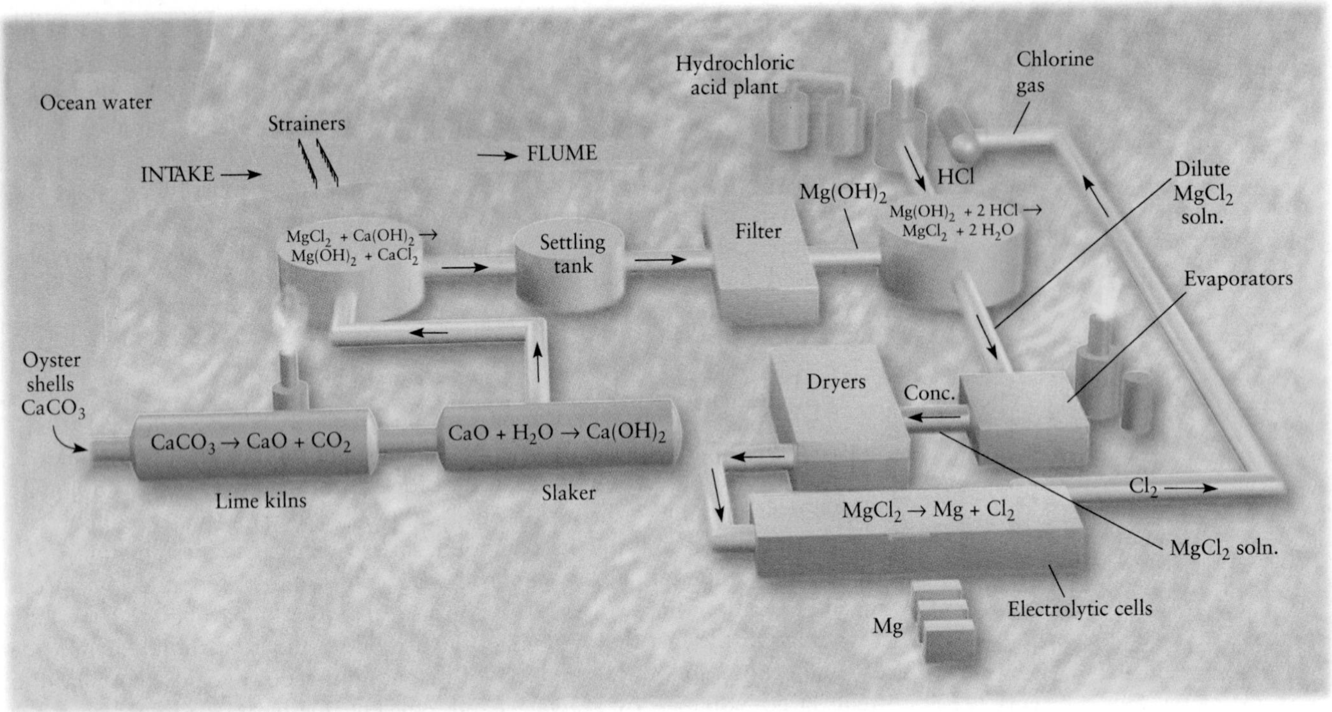

FIGURE 17.29 The production of magnesium hydroxide starts with the addition of lime (CaO) to seawater. Reaction of the magnesium hydroxide with hydrochloric acid produces magnesium chloride, which, after drying, is reduced electrolytically to produce magnesium metal.

Until 1918, elemental magnesium was used mainly in fireworks and flashbulbs, which took advantage of its great reactivity with the oxygen in air and the bright light given off in that reaction (see Fig. 11.11). Since then, many further uses for the metal and its alloys have been developed. Magnesium is even less dense than aluminum and is used in alloys with aluminum to lower density and improve resistance to corrosion under basic conditions. As discussed in Section 17.7, magnesium is used as a sacrificial anode to prevent the oxidation of another metal with which it is in contact. It is also used as a reducing agent to produce other metals such as titanium, uranium, and beryllium from their compounds.

17.7.3 Electrorefining and Electroplating

Metals that have been produced by pyrometallurgical methods, such as copper, silver, nickel, and tin, are too impure for many purposes, and **electrorefining** is used to purify them further. Crude metallic copper is cast into slabs, which are used as anodes in electrolysis cells that contain a solution of $CuSO_4$ in aqueous H_2SO_4. Thin sheets of pure copper serve as cathodes, and the copper that dissolves at the anodes is deposited in purer form on the cathodes (Fig. 17.30). Impurities that are more easily oxidized than copper, such as nickel, dissolve along with the copper but remain in solution; elements that are less easily oxidized, such as silver and gold, do not dissolve but fall away from the anode as a metallic slime. Periodically, the anode slime and the solution are removed and further processed for recovery of the elements they contain.

A related process is **electroplating**, in which electrolysis is used to plate out a thin layer of a metal on another material, often a second metal. In chrome plating, the piece of metal to be plated is placed in a hot bath of sulfuric acid and chromic acid (H_2CrO_4) and is made the cathode in an electrolytic cell. As current passes through the cell, chromium is reduced from the +6 oxidation state

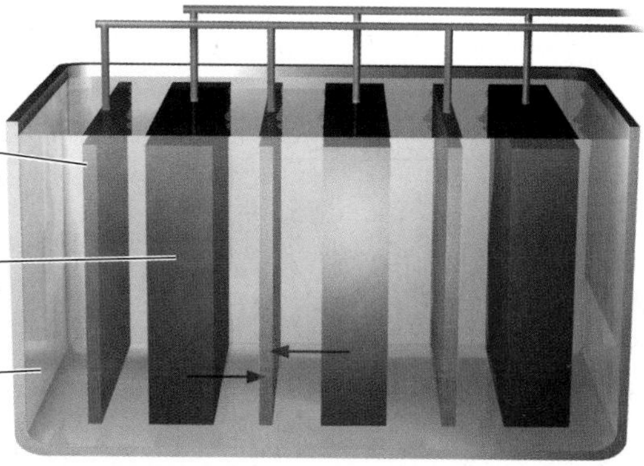

FIGURE 17.30 Electrolytic refining of copper. Many slabs of impure copper (anodes) alternating with thin sheets of pure copper (cathodes) are immersed into a dilute acidic solution of $CuSO_4$. Copper metal oxidized from the impure anodes enters the solution as Cu^{2+} ions that migrate to the cathodes where they plate out as copper metal of higher purity.

in chromic acid to elemental chromium and plates out on the cathode. A decorative chromium layer can be as thin as 2.5×10^{-5} cm (corresponding to 2 g of Cr per square meter of surface). Thicker layers ranging up to 10^{-2} cm are found in hard chromium plate, prized for its resistance to wear and used in automobile trim. Steel can be plated with cadmium to improve its resistance to corrosion in marine environments. Gold and silver are used both for decorative plating and (because they are good conductors of electricity) on electronic devices.

EXAMPLE 17.14

Suppose a layer of chromium 3.0×10^{-3} cm thick is to be plated onto an automobile bumper with a surface area of 2.0×10^3 cm^2. If a current of 250 A is used, how long must current be passed through the cell to achieve the desired thickness? The density of chromium is 7.2 g cm^{-3}.

Solution

The volume of the Cr is the product of the thickness of the layer and the surface area:

$$\text{volume} = (3.0 \times 10^{-3}\text{ cm})(2.0 \times 10^3\text{ cm}^2) = 6.0\text{ cm}^3$$

The mass of chromium is the product of this volume and the density:

$$\text{mass Cr} = (6.0\text{ cm}^3)(7.2\text{ g cm}^{-3}) = 43.2\text{ g}$$

From this, the number of moles of Cr that must be reduced is

$$\frac{43.2\text{ g}}{52.00\text{ g mol}^{-1}} = 0.831\text{ mol Cr}$$

Because Cr is being reduced from oxidation state +6 in H_2CrO_4 to 0 in the elemental form, six electrons are required for each atom of Cr deposited. The number of moles of electrons is then

$$0.831\text{ mol Cr} \times \left(\frac{6\text{ mol }e^-}{1\text{ mol Cr}}\right) = 4.98\text{ mol }e^-$$

$$\text{total magnitude of charge} = (4.98\text{ mol})(96{,}485\text{ C mol}^{-1}) = 4.81 \times 10^5\text{ C}$$

(Continued)

The required electrolysis time is the total magnitude of charge divided by the current (in amperes):

$$\text{time} = \frac{4.81 \times 10^5 \text{ C}}{250 \text{ C s}^{-1}} = 1.9 \times 10^3 \text{ s} = 32 \text{ min}$$

Related Problems: 61, 62

17.8 A DEEPER LOOK Electrolysis of Water and Aqueous Solutions

In Section 17.7 we discussed applications of electrolysis in the extraction and purification of metals from their ore sources. We examine here the electrolysis of water and aqueous solutions. Consider first the electrolysis of water using a pair of inert electrodes such as platinum, for which the half-cell reactions are

$$2\,H_3O^+ + 2\,e^- \longrightarrow H_2 + 2\,H_2O \qquad \text{(cathode)}$$
$$3\,H_2O \longrightarrow \tfrac{1}{2}\,O_2 + 2\,H_3O^+ + 2\,e^- \qquad \text{(anode)}$$
$$H_2O \longrightarrow H_2 + \tfrac{1}{2}\,O_2$$

A practical problem immediately arises. The concentration of $H_3O^+(aq)$ and $OH^-(aq)$ ions in pure water at 25°C is only 1.0×10^{-7} M, so the *rate* of electrolysis will be exceedingly small. This practical consideration is put aside for the moment because it does not alter the *thermodynamic* analysis.

The concentration of H_3O^+ in pure water at 25°C is 1.0×10^{-7} M, so we must use the Nernst equation to calculate the reduction potentials for each half-reaction as follows:

$$E_{\text{cathode}} = E^{\circ}_{\text{cathode}} - \frac{0.0592 \text{ V}}{n_{\text{hc}}} \log Q_{\text{hc}}$$
$$= 0.00 - \frac{0.0592}{2} \log \frac{P_{H_2}}{[H_3O^+]^2}$$

If $H_2(g)$ is produced at atmospheric pressure, this result simplifies to

$$E_{\text{cathode}} = 0.00 - \frac{0.0592 \text{ V}}{2} \log \frac{1}{(10^{-7})^2}$$
$$= -0.414 \text{ V}$$

The reduction potential for the anode half-reaction

$$\tfrac{1}{2}\,O_2(g) + 4\,H_3O^+(aq) + 4\,e^- \longrightarrow 6\,H_2O(\ell)$$

is calculated using the standard reduction potential for that reaction as tabulated in Appendix E:

$$E_{\text{anode}} = E^{\circ}_{\text{anode}} - \frac{0.0592 \text{ V}}{2} \log \frac{1}{(P_{O_2})^{1/2}[H_3O^+]^2}$$
$$= 1.229 \text{ V} - \frac{0.0592 \text{ V}}{2} \log \frac{1}{(10^{-7})^2}$$
$$= 0.815 \text{ V}$$

if $P_{O_2} = 1$ atm. The overall cell potential is

$$\Delta E_{cell} = E_{cathode} - E_{anode} = -0.414 - 0.815 \text{ V} = -1.229 \text{ V}$$

so the decomposition of water to produce hydrogen and oxygen does not occur spontaneously. It can be made to occur by applying a potential that is sufficiently large, called the **decomposition potential** of water, which is 1.229 V though practical applications require a larger potential, called an overpotential. Electrolysis of water then produces hydrogen bubbles at the cathode and oxygen bubbles at the anode (see Fig. 1.2).

What happens when we attempt to run electrolysis reactions in aqueous solutions? The products that are produced depend upon the nature of the solutes, their standard reduction potentials, and their concentrations. We could imagine the following reactions occurring in a 0.10 M aqueous solution of NaCl.
Cathode:

$$Na^+(0.1 \text{ M}) + e^- \longrightarrow Na$$

$$2\ H_3O^+(10^{-7}\text{ M}) + 2\ e^- \longrightarrow H_2 + 2\ H_2O$$

Anode:

$$2\ Cl^- \longrightarrow Cl_2 + 2\ e^-$$

$$6\ H_2O \longrightarrow O_2 + 4\ H_3O^+(10^{-7}\text{ M}) + 4\ e^-$$

Let's examine each of these reactions in turn. The reduction potential for the first half-reaction is

$$E(Na^+|Na) = E°(Na^+|Na) - \frac{0.0592 \text{ V}}{1} \log \frac{1}{[Na^+]}$$

$$= -2.71 - 0.06 = -2.77 \text{ V}$$

which is more negative than the reduction potential for H_3O^+ in pure water: $E^o = -0.414$ V. Hydronium ions are reduced preferentially over sodium ions under these conditions. The reduction potential for the third half-reaction is

$$E(Cl_2|Cl^-) = E°(Cl_2|Cl^-) - \frac{0.0592 \text{ V}}{1} \log \frac{[Cl^-]}{P_{Cl_2}^{1/2}}$$

$$= 1.36 + 0.06 = 1.42 \text{ V}$$

which is more positive than the reduction potential for water under these conditions: $E^o = 0.815$ V. Cl_2 is more easily reduced than O_2, so running this half reaction in reverse at the anode is more difficult than producing $O_2(g)$. Thus, $O_2(g)$ is generated at the anode, not $Cl_2(g)$. Increasing the potential above 1.229 V only *increases* the rate of the electrolysis reaction. It is not possible to generate sodium and chlorine electrolytically in aqueous solutions of NaCl.

Suppose now that 0.10 M NaI is substituted for the 0.10 M NaCl solution. Sodium ions still will not be reduced; however, $E(I_2|I^-)$ is

$$E(I_2|I^-) = E°(I_2|I^-) - \frac{0.0592 \text{ V}}{1} \log [I^-]$$

$$= 0.535 + 0.059 = 0.594 \text{ V}$$

The iodine reduction potential is more negative than the reduction potential of $O_2(g)$ in water at pH = 7, so the oxidation of 0.10 M I^- occurs in preference to the oxidation of water. The anode reaction is therefore

$$I^-(0.10 \text{ M}) \longrightarrow \tfrac{1}{2} I_2(s) + e^-$$

and the overall cell reaction is

$$H_3O^+(10^{-7}\ M) + I^-(0.10\ M) \longrightarrow \tfrac{1}{2} H_2(g) + \tfrac{1}{2} I_2(s) + H_2O(\ell)$$

The intrinsic cell voltage is

$$\Delta E_{cell} = E(\text{cathode}) - E(\text{anode}) = -0.414 - 0.594 = -1.008\ V < 0$$

$H_2(g)$ and $I_2(s)$ will be generated by applying a potential greater than the decomposition potential of the solution, which is 1.008 V. The concentration of I^- begins to decrease as the electrolysis proceeds, making the potential of the $(I_2|I^-)$ couple more positive. When the iodide ion concentration reaches about 2×10^{-5} M, $E(I_2|I^-) = 0.815$ V, the external voltage required to maintain electrolysis would have to be increased to 1.229 V. At this point, water will start to be electrolyzed, and oxygen will be produced at the anode.

Our results for the electrolysis of *neutral* aqueous solutions are summarized as follows:

1. A species can be reduced only if its reduction potential is more positive than −0.414 V.
2. A species can be oxidized only if its reduction potential is less than 0.815 V.

In solutions with pH different from 7, these results must be modified, as shown by the following example.

EXAMPLE 17.15

An aqueous 0.10 M solution of $NiCl_2$ is electrolyzed under 1 atm pressure. Determine the products formed at the anode and the cathode and the decomposition potential, if the pH is (a) 7.0; (b) 0.0.

Solution

The reduction of Ni^{2+} at the cathode has the half-cell potential

$$Ni^{2+}(aq) + 2\,e^- \longrightarrow Ni(s)$$

$$E(Ni^{2+}|Ni) = E°(Ni^{2+}|Ni) - \frac{0.0592\ V}{2} \log \frac{1}{[Ni^{2+}]}$$

$$= -0.23 - 0.03 = -0.26\ V$$

and the oxidation of the Cl^- at the anode has the *reduction* potential

$$\tfrac{1}{2} Cl_2(g) + e^- \longrightarrow Cl^-(aq)$$

$$E(Cl_2|Cl^-) = E°(Cl_2|Cl^-) - \frac{0.0592\ V}{2} \log \frac{[Cl^-]}{P_{Cl_2}^{1/2}}$$

$$= 1.36\ V - (0.0592\ V) \log 0.2 = 1.40\ V$$

(a) In neutral solution the cathode half-reaction is

$$Ni^{2+}(aq) + 2\,e^- \longrightarrow Ni(s)$$

because $E(Ni^{2+}|Ni) = -0.26\ V > -0.414$ V. The anode half-reaction is

$$3\,H_2O(\ell) \longrightarrow 2\,H_3O^+(10^{-7}\ M) + \tfrac{1}{2} O_2(g) + 2\,e^-$$

because $E(Cl_2|Cl^-) = 1.40\ V > 0.815$ V (the reduction potential for this half-reaction). The cell voltage is

$$\Delta E_{cell} = E(\text{cathode}) - E(\text{anode})$$

$$= -0.26 - 0.815\ V = -1.08\ V$$

so the decomposition potential is 1.08 V.

(b) In 1.0 M acid solution (pH = 0.0), the anode half-reaction is still

$$3\,H_2O(\ell) \longrightarrow 2\,H_3O^+(1\ M) + \tfrac{1}{2} O_2(g) + 2\,e^-$$

The reduction potential for this reaction is now $E_{hc} = E°_{hc} = 1.229$ V, which is still less than 1.40 V, the reduction potential for the competing reaction involving chlorine.

The cathode half-reaction now becomes

$$2\,H_3O^+(1\ M) + 2\,e^- \longrightarrow H_2(g) + 2\,H_2O(\ell)$$

because $E(H_3O^+|H_2) = 0.0\ V > -0.26$ V for $E(Ni^{2+}|Ni)$. We now have

$$\Delta E_{cell} = E(\text{cathode}) - E(\text{anode})$$

$$= 0.00 - 1.229 = -1.229\ V$$

so the decomposition potential of the solution is now 1.229 V, just as it is for pure water.

Related Problems: 63, 64

CHAPTER SUMMARY

Electrochemical reactions are an important class of oxidation–reduction (redox) reactions that interconvert chemical and electrical energy. The free energy released in a spontaneous chemical reaction can be used to generate electricity, or electrical energy can be provided from an external source to drive chemical reactions that are not normally spontaneous. The key to this flexibility is the separation of the oxidation and reduction parts of the reaction with the electrons being transferred through an external circuit. The cell potential ΔE_{cell} is simply related to the Gibbs free energy for an electrochemical reaction, providing a quantitative measure of the driving force for the reaction. The cell potential goes to zero for reactions at equilibrium, so the standard cell potential $\Delta E^{\circ}_{\text{cell}}$ is a direct measure of the equilibrium constant in electrochemical reactions. Standard reduction potentials can be associated with orbital energies and provide a molecular level interpretation of electrochemical reactions and processes. Electrochemical organic synthesis, sensors, and photoelectrochemistry were discussed from this point of view. Electrochemistry is enormously important in many existing technologies such as energy conversion and storage, and large-scale chemical syntheses of commodity chemicals like chlorine. It plays an important role in alternative energy sources such as fuel cells for deployment in transportation and small dedicated power plants. It allows the storage of solar energy in lightweight batteries for electric cars and as chemical energy potentially stored in hydrogen produced by photoelectrochemical water splitting.

CONCEPTS AND SKILLS

Section 17.1 – Electrochemical Cells

Sketch a typical galvanic cell, identify all of its components, specifically the anode and the cathode, write out the oxidation and reduction reactions explicitly, and use the standard shorthand notation to represent the cell (Problems 1–2).

- There are many possible choices. The sketch in Figure 17.2 could be modified to illustrate any number of cells by substituting different metals and electrolytes in each half-cell. The galvanic cell represented by the notation $Zn|Zn^{2+}||Cu^{2+}|Cu$ would have a Zn electrode immersed in a solution of $Zn(NO_3)_2$ in the left-hand beaker and a Cu electrode immersed in a solution of $Cu(NO_3)_2$ in the right-hand beaker. Cu would be the cathode and Zn would be the anode in this cell, with the reactions being

$$Zn(s) \longrightarrow Zn^{2+} + 2\,e^-$$

$$Cu^{2+} + 2\,e^- \longrightarrow Cu(s)$$

Use Faraday's laws to calculate the quantities of substances produced or consumed at the electrodes of electrochemical cells in relation to the total charge passing through the circuit (Problems 3–10).

- The number of moles of a substance that are oxidized or reduced in an electrochemical reaction is proportional to the number of moles of electrons passed through the cell. The magnitude of the charge per mole of electrons is 96,485 C mol^{-1} and it is called the Faraday constant, *F*.

Section 17.2 – Cell Potentials and the Gibbs Free Energy

Explain the relationship between the change in the Gibbs free energy and the electrical work done by the system for a galvanic cell (Problems 11–12).

- The electrical work is always $w_{\text{elec}} = \Delta G$ for a cell operated reversibly, which for a galvanic cell is a negative quantity; work is done by the system, and it is paid for by the decrease in the Gibbs free energy of the system. The change in the Gibbs free energy under standard conditions is calculated from the standard cell potential $\Delta G^\circ = -nF\Delta E^\circ_{cell}$.

Represent cell reactions as combinations of half-reactions, and calculate standard cell potentials from the standard reduction potentials of the half-reactions (Problems 13–16).

- Identify the oxidation and reduction half-reactions, and calculate the standard cell potential from the standard reduction potentials using $\Delta E^\circ_{cell} = E^\circ_{cathode} - E^\circ_{anode}$.

Find standard reduction potentials for half-reactions not listed in Appendix E by combining the tabulated half-reactions and their standard reduction potentials (Problems 23–24).

- Write the half-reaction whose potential is not known in terms of the half-reactions whose potentials are tabulated. If the net half-reaction is calculated by adding a reduction half-reaction with corresponding reduction potential E°_1 to the reverse of another reduction reaction, with reduction potential E°_2, then the resulting half-reaction has reduction potential $E^\circ_3 = (n_1E^\circ_1 - n_2E^\circ_2)/n_3$.

Use reduction potential diagrams to determine strengths of oxidizing and reducing agents and stability toward disproportionation (Problems 17–26).

- Strong oxidizing agents are those that are easily reduced, whereas strong reducing agents are those that are easily oxidized. They are listed near the top and bottom of the table in Appendix E, respectively.
- The tendency for a given species to disproportionate can be determined by a reduction potential diagram that arranges a series of species with the most oxidized form on the left and the most reduced form on the right. Disproportionation will occur when the reduction potential for the reaction that connects a species to one on its right is more positive than the reduction potential that connects it to one on its left.

Section 17.3 – Concentration Effects and the Nernst Equation

Apply the Nernst equation to calculate cell potentials for cells in which the reactants and products are not in their standard states, and to calculate equilibrium constants for chemical reactions from cell potentials (Problems 27–38).

$$\Delta E_{cell} = \Delta E^\circ_{cell} - \frac{RT}{nF}\ln Q \text{ and } \log_{10}K = \frac{nF}{RT}\Delta E^\circ_{cell}$$

Describe the principles that underlie the use of electrochemical cells as pH meters (Problems 39–40).

- The Pt|H_2|H^+ cell potential depends upon the concentration of $H^+(aq)$ in solution. A pH meter measures the half-cell potential of a solution of unknown pH against a reference that is calibrated against the SHE, with the unknown pH being calculated using $\text{pH} = \dfrac{\Delta E_{cell} - \Delta E_{ref}}{(0.0592 \text{ V})}$.

Calculate solubility product constants K_{sp} from cell potentials (Problems 41, 42).

- K_{sp} is calculated from cell potentials using the Nernst equation, using the same equation presented earlier for calculating equilibrium constants.

Section 17.4 – Molecular Electrochemistry

Discuss electrochemical processes from a molecular point of view, considering electron transfer processes between electrodes and redox couples as well as between redox couples, and be able to describe the salient features of the examples chosen (Problems 43–50).

- Electrochemical organic synthesis achieves great selectivity by using potential to control the thermodynamic driving forces and kinetics of electrochemical reactions.
- Enzyme-catalyzed reactions coupled to electrochemical cells provide extremely sensitive and selective sensors.
- Light absorbed by semiconductors produces electron-hole pairs that can be used to carry out oxidation–reduction reactions. Dye-sensitized TiO_2 has been shown to be a promising candidate for generating hydrogen from water using sunlight.

Section 17.5 – Batteries and Fuel Cells

Discuss the electrochemistry of a primary battery and contrast it with that of a secondary battery and a fuel cell (Problems 51–58).

- Primary batteries cannot be recharged, due to one or more irreversible steps during the discharge reaction, the formation of a gaseous product, or corrosion of one of its components. Reactions in secondary cells can be reversed by applying voltages opposite in sign and greater in magnitude than the galvanic cell potential.

Section 17.6 – Corrosion and Corrosion Prevention

Discuss the electrochemical corrosion of metals and describe measures that may be used to minimize it (Problems 59–62).

- Electrochemical corrosion of metals refers to the oxidation of the metal and the subsequent dissolution of the ions, resulting in the disintegration of materials. Metals may be protected by painting; by passivating layers such as the native oxides that form on aluminum or the thin layer of protective chromium oxide that forms on stainless steel; or by attaching sacrificial anodes to the material that are preferentially oxidized.

Section 17.7 – Electrometallurgy

Describe the Hall–Héroult process for the production of aluminum and the methods used to recover magnesium from seawater (Problems 63–64).
Calculate theoretical yields for redox reactions, and the current and time required to produce specified quantities of products in electrometallurgy (Problems 65–68).

- Al_2O_3 is dissolved in molten cryolite, Na_3AlF_6, and reduced electrolytically to produce molten aluminum. Magnesium is produced from seawater in a multistep process. The mineral dolomite is calcined to drive off CO_2, producing $CaO{\cdot}MgO$, which reacts with Mg^{2+} ions to produce magnesium hydroxide. The hydroxide reacts with HCl to produce the chloride, which is reduced electrolytically to produce the metal.
- Use Faraday's laws to relate current to metal deposited in electrorefining and electroplating operations (Problems 69–70).

Section 17.8 – A Deeper Look . . . Electrolysis of Water and Aqueous Solutions

Predict the products liberated at the anode and cathode of an electrolysis cell with a given aqueous electrolyte composition (Problems 71–72).

- Oxidation and reduction reactions in aqueous solution are limited to a potential window that is set by the oxidation and reduction of water at pH = 7. Applied potentials more positive than 0.814 V will oxidize $OH^-(aq)$ to O_2 and potentials more negative than −0.414 V will reduce $H^+(aq)$ to H_2.

CUMULATIVE EXERCISE

FIGURE 17.31 A pile of manganese metal.

Manganese—A Versatile Reagent and Essential Mineral

Manganese is the twelfth-most-abundant element on the Earth's surface (Fig. 17.31). Its most important ore source is pyrolusite (MnO_2). The preparation and uses of manganese and its compounds (which range up to +7 in oxidation state) are intimately bound up with electrochemistry.

(a) Elemental manganese in a state of high purity can be prepared by electrolyzing aqueous solutions of Mn^{2+}. At which electrode (anode or cathode) does the Mn appear? Electrolysis is also used to make MnO_2 in high purity from Mn^{2+} solutions. At which electrode does the MnO_2 appear?

(b) The Winkler method is an analytical procedure for determining the amount of oxygen dissolved in water. In the first step, $Mn(OH)_2(s)$ is oxidized by gaseous oxygen to $Mn(OH)_3(s)$ in basic aqueous solution. Write the oxidation and reduction half-equations for this step, and write the balanced overall equation. Then use Appendix E to calculate the standard voltage that would be measured if this reaction were carried out in an electrochemical cell.

(c) Calculate the equilibrium constant at 25°C for the reaction in part (b).

(d) In the second step of the Winkler method, the $Mn(OH)_3$ is acidified to give Mn^{3+}, and iodide ion is added. Will Mn^{3+} spontaneously oxidize I^-? Write a balanced equation for its reaction with I^-, and use data from Appendix E to calculate its equilibrium constant. Titration of the I_2 produced completes the use of the Winkler method.

(e) Manganese(IV) is an even stronger oxidizing agent than manganese(III). It oxidizes zinc to Zn^{2+} in the dry cell. Such a battery has a cell voltage of 1.5 V. Calculate the electrical work done by this battery in 1.00 hour if it produces a steady current of 0.70 A.

(f) Calculate the mass of zinc reacting in the process described in part (e).

(g) The reduction potential of permanganate ion (+7 oxidation state) in acidic aqueous solution is given by

$$MnO_4^-(aq) + 8\ H_3O^+(aq) + 5\ e^- \longrightarrow Mn^{2+}(aq) + 12\ H_2O(\ell) \qquad E^\circ = 1.491\ V$$

whereas that of the analogous fifth-period species, pertechnetate ion, is

$$TcO_4^-(aq) + 8\ H_3O^+(aq) + 5\ e^- \longrightarrow Tc^{2+}(aq) + 12\ H_2O(\ell) \qquad E^\circ = 0.500\ V$$

Which is the stronger oxidizing agent, permanganate ion or pertechnetate ion?

(h) A galvanic cell is made from two half-cells. In the first, a platinum electrode is immersed in a solution at pH 2.00 that is 0.100 M in both MnO_4^- and Mn^{2+}. In the second, a zinc electrode is immersed in a 0.0100 M solution of $Zn(NO_3)_2$. Calculate the cell voltage that will be measured.

Answers

(a) Mn appears at the cathode and MnO_2 at the anode.

(b)

$$Mn(OH)_2(s) + OH^-(aq) \longrightarrow Mn(OH)_3(s) + e^- \qquad \text{(oxidation)}$$

$$O_2(g) + 2\ H_2O(\ell) + 4\ e^- \longrightarrow 4\ OH^-(aq) \qquad \text{(reduction)}$$

$$4\ Mn(OH)_2(s) + O_2(g) + 2\ H_2O(\ell) \longrightarrow 4\ Mn(OH)_3(s)$$

$$\Delta E^\circ = 0.401 - (-0.40) = 0.80\ V$$

(c) $K = 1 \times 10^{54}$

(d) Mn^{3+} will spontaneously oxidize I^-.

$$2\ Mn^{3+}(aq) + 2\ I^-(aq) \longrightarrow 2\ Mn^{2+}(aq) + I_2(s) \qquad K = 9 \times 10^{32}$$

(e) 3.8×10^3 J

(f) 0.85 g Zn is oxidized.

(g) Permanganate ion

(h) 2.12 V

PROBLEMS

Answers to problems whose numbers are red appear in Appendix G. Problems that are more challenging are indicated with asterisks.

17.1 Electrochemical Cells

1. Diagram the following galvanic cell, indicating the direction of flow of electrons in the external circuit and the motion of ions in the salt bridge.

$$Pt(s)|Cr^{2+}(aq), Cr^{3+}(aq)||Cu^{2+}(aq)|Cu(s)$$

Write a balanced equation for the overall reaction in this cell.

2. Diagram the following galvanic cell, indicating the direction of flow of electrons in the external circuit and the motion of ions in the salt bridge.

$$Ni(s)|Ni^{2+}(aq)||HCl(aq)|H_2(g)|Pt(s)$$

Write a balanced equation for the overall reaction in this cell.

3. A quantity of electricity equal to 6.95×10^4 C passes through an electrolytic cell that contains a solution of $Sn^{4+}(aq)$ ions. Compute the maximum chemical amount, in moles, of Sn(*s*) that can be deposited at the cathode.

4. A quantity of electricity equal to 9.263×10^4 C passes through a galvanic cell that has an Ni(*s*) anode. Compute the maximum chemical amount, in moles, of $Ni^{2+}(aq)$ that can be released into solution.

5. A galvanic cell is constructed that has a zinc anode immersed in a $Zn(NO_3)_2$ solution and a platinum cathode immersed in an NaCl solution equilibrated with $Cl_2(g)$ at 1 atm and 25°C. A salt bridge connects the two half-cells.
   (a) Write a balanced equation for the cell reaction.
   (b) A steady current of 0.800 A is observed to flow for a period of 25.0 minutes. How much charge passes through the circuit during this time? How many moles of electrons is this charge equivalent to?
   (c) Calculate the change in mass of the zinc electrode.
   (d) Calculate the volume of gaseous chlorine generated or consumed as a result of the reaction.

6. A galvanic cell consists of a cadmium cathode immersed in a $CdSO_4$ solution and a zinc anode immersed in a $ZnSO_4$ solution. A salt bridge connects the two half-cells.
   (a) Write a balanced equation for the cell reaction.
   (b) A current of 1.45 A is observed to flow for a period of 2.60 hours. How much charge passes through the circuit during this time? How many moles of electrons is this charge equivalent to?
   (c) Calculate the change in mass of the zinc electrode.
   (d) Calculate the change in mass of the cadmium electrode.

7. An acidic solution containing copper ions is electrolyzed, producing gaseous oxygen (from water) at the anode and copper at the cathode. For every 16.0 g of oxygen generated, 63.5 g of copper plates out. What is the oxidation state of the copper in the solution?

8. Michael Faraday reported that passing electricity through one solution liberated 1 mass of hydrogen at the cathode and 8 masses of oxygen at the anode. The same quantity of electricity liberated 36 masses of chlorine at the anode and 58 masses of tin at the cathode from a second solution. What were the oxidation states of hydrogen, oxygen, chlorine, and tin in these solutions?

9. Liquid potassium chloride, KCl(ℓ), is decomposed in an electrolytic cell to form potassium and chlorine. Liquid KCl consists of K^+ and Cl^- ions.
   (a) Write balanced equations for the half-cell reactions at the anode and at the cathode, and for the overall cell reaction.
   (b) If a current of 2.00 A is passed through the cell for a period of 5.00 hours, calculate the mass of metal deposited and of gas liberated.

10. In the Hall–Héroult process for the electrolytic production of aluminum, Al^{3+} ions from Al_2O_3 dissolved in molten cryolite (Na_3AlF_6) are reduced to $Al(\ell)$ while carbon (graphite) is oxidized to CO_2 by reaction with oxide ions.
   (a) Write balanced equations for the half-reactions at the anode and at the cathode, and for the overall cell reaction.
   (b) If a current of 50,000 A is passed through the cell for a period of 24 hours, what mass of aluminum will be recovered?

17.2 Cell Potentials and the Gibbs Free Energy

11. A $Ni|Ni^{2+}||Ag^+|Ag$ galvanic cell is constructed in which the standard cell potential is 1.03 V. Calculate the free energy change at 25°C when 1.00 g of silver plates out, if all concentrations remain at their standard value of 1 M throughout the process. What is the maximum electrical work done by the cell on its surroundings during this experiment?

12. A $Zn|Zn^{2+}||Co^{2+}|Co$ galvanic cell is constructed in which the standard cell potential is 0.48 V. Calculate the free energy change at 25°C per gram of zinc lost at the anode, if all concentrations remain at their standard value of 1 M throughout the process. What is the maximum electrical work done by the cell on its surroundings during this experiment?

13. A galvanic cell is constructed in which a $Br_2|Br^+$ half-cell is connected to a $Co^{2+}|Co$ half-cell.
   (a) By referring to Appendix E, write balanced chemical equations for the half-reactions at the anode and the cathode and for the overall cell reaction.
   (b) Calculate the cell potential, assuming that all reactants and products are in their standard states.

14. A galvanic cell is constructed in which a $Pt|Fe^{2+}\ Fe^{3+}$ half-cell is connected to a $Cd^{2+}|Cd$ half-cell.
   (a) Referring to Appendix E, write balanced chemical equations for the half-reactions at the anode and the cathode and for the overall cell reaction.
   (b) Calculate the cell potential, assuming that all reactants and products are in their standard states.

15. In a galvanic cell, one half-cell consists of a zinc strip dipped into a 1.00 M solution of $Zn(NO_3)_2$. In the second half-cell, solid indium adsorbed on graphite is in contact with a 1.00 M solution of $In(NO_3)_3$. Indium is observed to plate out as the galvanic cell operates, and the initial cell potential is measured to be 0.425 V at 25°C.
   (a) Write balanced equations for the half-reactions at the anode and the cathode.
   (b) Calculate the standard reduction potential of an $In^{3+}|In$ half-cell. Consult Appendix E for the reduction potential of the $Zn^{2+}|Zn$ electrode.

16. In a galvanic cell, one half-cell consists of gaseous chlorine bubbled over a platinum electrode at a pressure of 1.00 atm into a 1.00 M solution of NaCl. The second half-cell has a strip of solid gallium immersed in a 1.00 M $Ga(NO_3)_3$ solution. The initial cell potential is measured to be 1.918 V at 25°C, and as the cell operates, the concentration of chloride ion is observed to increase.
   (a) Write balanced equations for the half-reactions at the anode and the cathode.
   (b) Calculate the standard reduction potential of a $Ga^{3+}|Ga$ half-cell. Consult Appendix E for the reduction potential of the $Cl_2|Cl^-$ electrode.

17. Would you expect powdered solid aluminum to act as an oxidizing agent or as a reducing agent?

18. Would you expect potassium perchlorate, $KClO_4(aq)$, in a concentrated acidic solution to act as an oxidizing agent or as a reducing agent?

19. Bromine is sometimes used in place of chlorine as a disinfectant in swimming pools. If the effectiveness of a chemical as a disinfectant depends solely on its strength as an oxidizing agent, do you expect bromine to be better or worse than chlorine as a disinfectant, at a given concentration?

20. Many bleaches, including chlorine and its oxides, oxidize dye compounds in cloth. Predict which of the following will be the strongest bleach at a given concentration and pH 0: $NaClO_3(aq)$, $NaClO(aq)$, $Cl_2(aq)$. How does the strongest chlorine-containing bleach compare in strength with ozone, $O_3(g)$?

21. Suppose you have the following reagents available at pH 0, atmospheric pressure, and 1 M concentration:

$$Co(s),\ Ag^+(aq),\ Cl^-(aq),\ Cr(s),\ BrO_3^-(aq),\ I_2(s)$$

   (a) Which is the strongest oxidizing agent?
   (b) Which is the strongest reducing agent?
   (c) Which reagent will reduce $Pb^{2+}(aq)$ while leaving $Cd^{2+}(aq)$ unreacted?

22. Suppose you have the following reagents available at pH 0, atmospheric pressure, and 1 M concentration:

$$Sc(s),\ Hg_2^{2+}(aq),\ Cr_2O_7^{2-}(aq),\ H_2O_2(aq),\ Sn^{2+}(aq),\ Ni(s)$$

   (a) Which is the strongest oxidizing agent?
   (b) Which is the strongest reducing agent?
   (c) Which reagent will oxidize $Fe(s)$ while leaving $Cu(s)$ unreacted?

23. **(a)** Use the data from Appendix E to calculate the half-cell potential $E°$ for the half-reaction

$$Mn^{3+}(aq) + 3\,e^- \longrightarrow MN(s)$$

   (b) Consider the disproportionation reaction

$$3\,Mn^{2+}(aq) \rightleftharpoons Mn(s) + 2\,Mn^{3+}(aq)$$

   Will Mn^{2+} disproportionate in aqueous solution?

24. The following standard reduction potentials have been measured in aqueous solution at 25°C:

$$Tl^{3+} + e^- \longrightarrow Tl^{2+} \qquad E° = -0.37 \text{ V}$$

$$Tl^{3+} + 2\,e^- \longrightarrow Tl^+ \qquad E° = 1.25 \text{ V}$$

(a) Calculate the half-cell potential for the half-reaction

$$Tl^{2+} + e^- \longrightarrow Tl^+$$

(b) Consider the disproportionation reaction

$$2\,Tl^{2+}(aq) \rightleftharpoons Tl^{3+}(aq) + Tl^+(aq)$$

Will Tl^{2+} disproportionate in aqueous solution?

25. The following reduction potentials are measured at pH 0:

$$BrO_3^- + 6\,H_3O^+ + 5\,e^- \longrightarrow \frac{1}{2}Br_2(\ell) + 9\,H_2O \qquad E° = 1.52 \text{ V}$$

$$Br_2(\ell) + 2\,e^- \longrightarrow 2\,Br^- \qquad E° = 1.065 \text{ V}$$

(a) Will bromine disproportionate spontaneously in acidic solution?
(b) Which is the stronger reducing agent at pH 0: $Br_2(\ell)$ or Br_2?

26. The following reduction potentials are measured at pH 14:

$$ClO^- + H_2O + 2\,e^- \longrightarrow Cl^- + 2\,OH^- \qquad E° = 0.90 \text{ V}$$

$$ClO_2^- + H_2O + 2\,e^- \longrightarrow ClO^- + 2\,OH^- \qquad E° = 0.59 \text{ V}$$

(a) Will ClO^- disproportionate spontaneously in basic solution?
(b) Which is the stronger reducing agent at pH 14: ClO^- or Cl^-?

17.3 Concentration Effects and the Nernst Equation

27. A galvanic cell is constructed that carries out the reaction

$$Pb^{2+}(aq) + 2\,Cr^{2+}(aq) \longrightarrow Pb(s) + 2\,Cr^{3+}(aq)$$

If the initial concentration of $Pb^{2+}(aq)$ is 0.15 M, that of $Cr^{2+}(aq)$ is 0.20 M, and that of $Cr^{3+}(aq)$ is 0.0030 M, calculate the initial voltage generated by the cell at 25°C.

28. A galvanic cell is constructed that carries out the reaction

$$2\,Ag(s) + Cl_2(g) \longrightarrow 2\,Ag^+(aq) + 2\,Cl^-(aq)$$

If the partial pressure of $Cl_2(g)$ is 1.00 atm, the initial concentration of $Ag^+(aq)$ is 0.25 M, and that of $Cl^-(aq)$ is 0.016 M, calculate the initial voltage generated by the cell at 25°C.

29. Calculate the reduction potential for a $Pt|Cr^{3+}, Cr^{2+}$ half-cell in which $[Cr^{3+}]$ is 0.15 M and $[Cr^{2+}]$ is 0.0019 M.

30. Calculate the reduction potential for an $I_2(s)|I^-$ half-cell in which $[I^-]$ is 1.5×10^{-6} M.

31. An $I_2(s)|I^-$ (1.00 M) half-cell is connected to an $H_3O^+|H_2$ (1 atm) half-cell in which the concentration of the hydronium ion is unknown. The measured cell potential is 0.841 V, and the $I_2|I^-$ half-cell is the cathode. What is the pH in the $H_3O^+|H_2$ half-cell?

32. A Cu^{2+}(1.00 M)$|$Cu half-cell is connected to a $Br_2(\ell)|Br^-$ half-cell in which the concentration of bromide ion is unknown. The measured cell potential is 0.963 V, and the $Cu^{2+}|Cu$ half-cell is the anode. What is the bromide ion concentration in the $Br_2(\ell)|Br^-$ half-cell?

33. The following reaction occurs in an electrochemical cell:

$$3\,HClO_2(aq) + 2\,Cr^{3+}(aq) + 12\,H_2O(\ell) \longrightarrow 3\,HClO(aq) + Cr_2O_7^{2-}(aq) + 8\,H_3O^+(aq)$$

(a) Calculate $\Delta E°$ for this cell.
(b) At pH 0, with $[Cr_2O_7^{2-}] = 0.80$ M $[HClO_2] = 0.15$ M, and $[HClO] = 0.20$ M, the cell potential is found to be 0.15 V. Calculate the concentration of $Cr^{3+}(aq)$ in the cell.

34. A galvanic cell is constructed in which the overall reaction is

$$Cr_2O_7^{2-}(aq) + 14\,H_3O^+(aq) + 6\,I^-(aq) \longrightarrow 2\,Cr^{3+}(aq) + 3\,I_2(s) + 21\,H_2O(\ell)$$

(a) Calculate $\Delta E°$ for this cell.
(b) At pH 0, with $[Cr_2O_7^{2-}] = 1.5$ M and $[I^-] = 0.40$ M, the cell potential is found to equal 0.87 V. Calculate the concentration of $Cr^{3+}(aq)$ in the cell.

35. By using the half-cell potentials in Appendix E, calculate the equilibrium constant at 25°C for the reaction in problem 33. Dichromate ion ($Cr_2O_7^{2-}$) is orange, and Cr^{3+} is light green in aqueous solution. If 2.00 L of 1.00 M $HClO_2$ solution is added to 2.00 L of 0.50 M $Cr(NO_3)_3$ solution, what color will the resulting solution have?

36. By using the half-cell potentials in Appendix E, calculate the equilibrium constant at 25°C for the reaction

$$6\,Hg^{2+}(aq) + 2\,Au(s) \rightleftharpoons 3\,Hg_2^{2+}(aq) + 2\,Au^{3+}(aq)$$

If 1.00 L of a 1.00 M $Au(NO_3)_3$ solution is added to 1.00 L of a 1.00 M $Hg_2(NO_3)_2$ solution, calculate the concentrations of Hg^{2+}, Hg_2^{2+}, and Au^{3+} at equilibrium.

37. The following standard reduction potentials have been determined for the aqueous chemistry of indium:

$$In^{3+}(aq) + 2\,e^- \longrightarrow In^+(aq) \qquad E° = -0.40 \text{ V}$$

$$In^+(aq) + e^- \longrightarrow In(s) \qquad E° = -0.21 \text{ V}$$

Calculate the equilibrium constant (K) for the disproportionation of $In^+(aq)$ at 25°C.

$$3\,In^+(aq) \rightleftharpoons 2\,In(s) + In^{3+}(aq)$$

38. Use data from Appendix E to compute the equilibrium constant for the reaction

$$Hg^{2+}(aq) + Hg(\ell) \rightleftharpoons Hg_2^{2+}(aq)$$

39. A galvanic cell consists of a $Pt|H_3O^+(1.00\ M)|H_2(g)$ cathode connected to a $Pt|H_3O^+(aq)|H_2(g)$ anode in which the concentration of H_3O^+ is unknown but is kept constant by the action of a buffer consisting of a weak acid, HA(0.10 M), mixed with its conjugate base, A^-(0.10 M). The measured cell potential is $\Delta E_{cell} = 0.150$ V at 25°C, with a hydrogen pressure of 1.00 atm at both electrodes. Calculate the pH in the buffer solution, and from it determine the K_a of the weak acid.

40. In a galvanic cell, the cathode consists of a $Ag^+(1.00\ M)|Ag$ half-cell. The anode is a platinum wire, with hydrogen bubbling over it at 1.00-atm pressure, which is immersed in a buffer solution containing benzoic acid and sodium benzoate. The concentration of benzoic acid (C_6H_5COOH) is 0.10 M, and that of benzoate ion ($C_6H_5COO^-$) is 0.050 M. The overall cell reaction is then

$$Ag^+(aq) + \frac{1}{2}H_2(g) + H_2O(\ell) \longrightarrow Ag(s) + H_3O^+(aq)$$

and the measured cell potential is 1.030 V. Calculate the pH in the buffer solution and determine the K_a of benzoic acid.

41. A galvanic cell is constructed in which the overall reaction is

$$Br_2(\ell) + H_2(g) + 2\ H_2O(\ell) \longrightarrow 2\ Br^-(aq) + 2\ H_3O^+(aq)$$

(a) Calculate $\Delta E°$ for this cell.
(b) Silver ions are added until AgBr precipitates at the cathode and $[Ag^+]$ reaches 0.060 M. The cell potential is then measured to be 1.710 V at pH = 0 and = 1.0 atm. Calculate $[Br^-]$ under these conditions.
(c) Calculate the solubility product constant K_{sp} for AgBr.

42. A galvanic cell is constructed in which the overall reaction is

$$Pb(s) + 2\ H_3O^+(aq) \longrightarrow Pb^{2+}(aq) + H_2(g) + 2\ H_2O(\ell)$$

(a) Calculate $\Delta E°$ for this cell.
(b) Chloride ions are added until $PbCl_2$ precipitates at the anode and $[Cl^-]$ reaches 0.15 M. The cell potential is then measured to be 0.22 V at pH = 0 and P_{H_2} = 1.0 atm. Calculate $[Pb^{2+}]$ under these conditions.
(c) Calculate the solubility product constant K_{sp} of $PbCl_2$.

17.4 Molecular Electrochemistry

43. Sketch and interpret the cyclic voltammogram you expect for the $[Ru(NH_3)_6]^{3+}/[Ru(NH_3)_6]^{2+}$ redox couple measured with a Ag/AgCl reference electrode. The standard reduction potential for this system is −0.215 V measured with respect to SCE and the standard reduction potential of the Ag/AgCl reference electrode with respect to SHE is 0.2223 V (see Appendix E).

44. Ferrocene is an organometallic complex that contains iron sandwiched between two five-membered cyclopentadiene rings (C_5H_5). Ferrocene methanol, a derivative, is another convenient reference redox couple used in electrochemistry. Sketch and interpret the cyclic voltammogram for the ferrocene methanol couple (Fc^+/Fc) measured with respect to SHE; the standard redox potential is 0.64 V.

45. What is the wavelength and color of light emitted in an ECL experiment in which the redox potential associated with the HOMO is 1.5 V and that associated with the LUMO is −0.9 V? The experiment is conducted in a nonaqueous solution.

46. Suppose that you were asked to synthesize a molecule for use in ECL that emits light at 520 nm and has a potential associated with the HOMO located at 0.5 V, for compatibility with other requirements in a particular analytical application. What is the redox potential associated with the LUMO?

47. Would n-CdS be a suitable semiconductor for direct photoelectrochemical water splitting? Why or why not? The conduction band lies at about −0.65 V vs NHE and the valence band lies at about 1.75 V vs NHE.

48. The p-GaP semiconductor has a bandgap of 2.3 eV with a conduction band redox potential located at −1.2 V vs SHE. Would this material be suitable for direct photoelectrochemical water splitting? Why or why not?

49. Strontium titanate ($SrTiO_3$) is another wide bandgap semiconductor that has been investigated for use in dye-sensitized solar cells. Depending on morphology and pH the conduction band of $SrTiO_3$ lies around −0.2 V vs. SHE. What energies for the HOMO and LUMO of a sensitizing dye would you specify for electron injection into the conduction band of strontium titanate and for water oxidation by the radical cation formed by photoexcitation?

50. A Grätzel cell is to be designed using strontium titanate as the semiconductor and the iodide/triode mediator discussed in the *Connection to Energy: Solar Energy Conversion.* Where should the energy levels of the sensitizer lie for maximum solar-to-electrical conversion efficiency?

17.5 Batteries and Fuel Cells

51. Calculate the potential of a lead–acid cell if all reactants and products are in their standard states. What will be the voltage if six such cells are connected in a series?

52. Calculate the standard potential of the zinc–mercuric oxide cell shown in Figure 17.18. (*Hint:* The easiest way to proceed is to calculate $\Delta G°$ for the corresponding overall reaction, and then find $\Delta E°$ from it.) Take $\Delta G_f^\circ[Zn(OH)_2(s)] = -553.5\ kJ\ mol^{-1}$.

53. **(a)** What quantity of charge (in coulombs) is a fully charged 12-V lead–acid storage battery theoretically capable of furnishing if the spongy lead available for reaction at the anodes weighs 10 kg and there is excess PbO_2?
(b) What is the theoretical maximum amount of work (in joules) that can be obtained from this battery?

54. **(a)** What quantity of charge (in coulombs) is a fully charged 1.34-V zinc–mercuric oxide watch battery theoretically capable of furnishing if the mass of HgO in the battery is 0.50 g?
(b) What is the theoretical maximum amount of work (in joules) that can be obtained from this battery?

55. The concentration of the electrolyte, sulfuric acid, in a lead–acid storage battery diminishes as the battery is discharged. Is a discharged battery recharged by replacing the dilute H_2SO_4 with fresh, concentrated H_2SO_4? Explain.

56. One cold winter morning the temperature is well below 0°F. In trying to start your car, you run the battery down completely. Several hours later, you return to replace your fouled spark plugs and find that the liquid in the battery has now frozen even though the air temperature is actually a bit higher than it was in the morning. Explain how this can happen.

57. Consider the fuel cell that accomplishes the overall reaction

$$H_2(g) + \frac{1}{2}O_2(g) \longrightarrow H_2O(\ell)$$

If the fuel cell operates with 60% efficiency, calculate the amount of electrical work generated per gram of water produced. The gas pressures are constant at 1 atm, and the temperature is 25°C.

58. Consider the fuel cell that accomplishes the overall reaction

$$CO(g) + \frac{1}{2}O_2(g) \longrightarrow CO_2(g)$$

Calculate the maximum electrical work that could be obtained from the conversion of 1.00 mol of CO(*g*) to $CO_2(g)$ in such a fuel cell operated with 100% efficiency at 25°C and with the pressure of each gas equal to 1 atm.

17.6 Corrosion and Corrosion Prevention

59. Two half-reactions proposed for the corrosion of iron in the absence of oxygen are

$$Fe(s) \longrightarrow Fe^{2+}(aq) + 2\,e^-$$

$$2\,H_2O(\ell) + 2\,e^- \longrightarrow 2\,OH^-(aq) + H_2(g)$$

Calculate the standard cell potential generated by a galvanic cell running this pair of half-reactions. Is the overall reaction spontaneous under standard conditions? As the pH falls from 14, will the reaction become spontaneous?

60. In the presence of oxygen, the cathode half-reaction written in the preceding problem is replaced by

$$\frac{1}{2}O_2(g) + 2\,H_3O^+(aq) + 2\,e^- \longrightarrow 3\,H_2O(\ell)$$

but the anode half-reaction is unchanged. Calculate the standard cell potential for *this* pair of reactions operating as a galvanic cell. Is the overall reaction spontaneous under standard conditions? As the water becomes more acidic, does the driving force for the rusting of iron increase or decrease?

61. Could sodium be used as a sacrificial anode to protect the iron hull of a ship?

62. If it is shown that titanium can be used as a sacrificial anode to protect iron, what conclusion can be drawn about the standard reduction potential of its half-reaction?

$$Ti^{3+}(aq) + 3\,e^- \longrightarrow Ti(s)$$

17.7 Electrometallurgy

63. In the Downs process, molten sodium chloride is electrolyzed to produce sodium. A valuable byproduct is chlorine. Write equations representing the processes taking place at the anode and at the cathode in the Downs process.

64. The first element to be prepared by electrolysis was potassium. In 1807, Humphry Davy, then 29 years old, passed an electric current through molten potassium hydroxide (KOH), obtaining liquid potassium at one electrode and water and oxygen at the other. Write equations to represent the processes taking place at the anode and at the cathode.

65. A current of 55,000 A is passed through a series of 100 Hall–Héroult cells for 24 hours. Calculate the maximum theoretical mass of aluminum that can be recovered.

66. A current of 75,000 A is passed through an electrolysis cell containing molten $MgCl_2$ for 7.0 days. Calculate the maximum theoretical mass of magnesium that can be recovered.

67. An important use for magnesium is to make titanium. In the Kroll process, magnesium reduces titanium(IV) chloride to elemental titanium in a sealed vessel at 800°C. Write a balanced chemical equation for this reaction. What mass of magnesium is needed, in theory, to produce 100 kg of titanium from titanium(IV) chloride?

68. Calcium is used to reduce vanadium(V) oxide to elemental vanadium in a sealed steel vessel. Vanadium is used in vanadium steel alloys for jet engines, high-quality knives, and tools. Write a balanced chemical equation for this process. What mass of calcium is needed, in theory, to produce 20.0 kg of vanadium from vanadium(V) oxide?

69. Galvanized steel consists of steel with a thin coating of zinc to reduce corrosion. The zinc can be deposited electrolytically, by making the steel object the cathode and a block of zinc the anode in an electrochemical cell containing a dissolved zinc salt. Suppose a steel garbage can is to be galvanized and requires that a total mass of 7.32 g of zinc be coated to the required thickness. How long should a current of 8.50 A be passed through the cell to achieve this?

70. In the electroplating of a silver spoon, the spoon acts as the cathode and a piece of pure silver as the anode. Both dip into a solution of silver cyanide (AgCN). Suppose that a current of 1.5 A is passed through such a cell for 22 minutes and that the spoon has a surface area of 16 cm^2. Calculate the average thickness of the silver layer deposited on the spoon, taking the density of silver to be 10.5 g cm^{-3}.

17.8 A Deeper Look . . . Electrolysis of Water and Aqueous Solutions

71. An electrolytic cell consists of a pair of inert metallic electrodes in a solution buffered to pH = 5.0 and containing nickel sulfate ($NiSO_4$) at a concentration of 1.00 M. A current of 2.00 A is passed through the cell for 10.0 hours.
(a) What product is formed at the cathode?
(b) What is the mass of this product?
(c) If the pH is changed to pH = 1.0, what product will form at the cathode?

72. A 0.100 M neutral aqueous $CaCl_2$ solution is electrolyzed using platinum electrodes. A current of 1.50 A passes through the solution for 50.0 hours.
(a) Write the half-reactions occurring at the anode and at the cathode.
(b) What is the decomposition potential?
(c) Calculate the mass, in grams, of the product formed at the cathode.

ADDITIONAL PROBLEMS

73. The drain cleaner Drano consists of aluminum turnings mixed with sodium hydroxide. When it is added to water, the sodium hydroxide dissolves and releases heat. The aluminum reacts with water to generate bubbles of hydrogen and aqueous ions. Write a balanced net ionic equation for this reaction.

74. Sulfur-containing compounds in the air tarnish silver, giving black Ag_2S. A practical method of cleaning tarnished silverware is to place the tarnished item in electrical contact with a piece of zinc and dip both into water containing a small amount of salt. Write balanced half-equations to represent what takes place.

75. A current passed through inert electrodes immersed in an aqueous solution of sodium chloride produces chlorate ion, $ClO_3^-(aq)$, at the anode and gaseous hydrogen at the cathode. Given this fact, write a balanced equation for the chemical reaction if gaseous hydrogen and aqueous sodium chlorate are mixed and allowed to react spontaneously until they reach equilibrium.

76. A galvanic cell is constructed by linking a $Co^{2+}|Co(s)$ half-cell to an $Ag^+|Ag(s)$ half-cell through a salt bridge and then connecting the cobalt and silver electrodes through an external circuit. When the circuit is closed, the cell potential is measured to be 1.08 V, and silver is seen to plate out while cobalt dissolves.
(a) Write the half-reactions that occur at the anode and at the cathode and the balanced overall cell reaction.
(b) The cobalt electrode is weighed after 150 minutes of operation and is found to have decreased in mass by 0.36 g. By what amount has the silver electrode increased in mass?
(c) What is the average current drawn from the cell during this period?

77. The galvanic cell $Zn(s)|Zn^{2+}(aq)||Ni^{2+}(aq)|Ni(s)$ is constructed using a completely immersed zinc electrode that weighs 32.68 g and a nickel electrode immersed in 575 mL of 1.00 M $Ni^{2+}(aq)$ solution. A steady current of 0.0715 A is drawn from the cell as the electrons move from the zinc electrode to the nickel electrode.
(a) Which reactant is the limiting reactant in this cell?
(b) How long does it take for the cell to be completely discharged?
(c) How much mass has the nickel electrode gained when the cell is completely discharged?
(d) What is the concentration of the $Ni^{2+}(aq)$ when the cell is completely discharged?

78. A newly discovered bacterium can reduce selenate ion, $SeO_4^{2-}(aq)$, to elemental selenium, Se(*s*), in reservoirs. This is significant because the soluble selenate ion is potentially toxic, but elemental selenium is insoluble and harmless. Assume that water is oxidized to oxygen as the selenate ion is reduced. Compute the mass of oxygen produced if all the selenate in a 1012-L reservoir contaminated with 100 mg L^{-1} of selenate ion is reduced to selenium.

79. Thomas Edison invented an electric meter that was nothing more than a simple coulometer, a device to measure the amount of electricity passing through a circuit. In this meter, a small, fixed fraction of the total current supplied to a household was passed through an electrolytic cell, plating out zinc at the cathode. Each month the cathode could then be removed and weighed to determine the amount of electricity used. If 0.25% of a household's electricity passed through such a coulometer and the cathode increased in mass by 1.83 g in a month, how many coulombs of electricity were used during that month?

80. The chief chemist of the Brite-Metal Electroplating Co. is required to certify that the rinse solutions that are discharged from the company's tin-plating process into the municipal sewer system contain no more than 10 ppm (parts per million) by mass of Sn^{2+}. The chemist devises the following analytical procedure to determine the concentration. At regular intervals, a 100-mL (100-g) sample is withdrawn from the waste stream and acidified to pH = 1.0. A starch solution and 10 mL of 0.10 M potassium iodide are added, and a 25.0-mA current is passed through the solution between platinum electrodes. Iodine appears as a product of electrolysis at the anode when the oxidation of Sn^{2+} to Sn^{4+} is practically complete and signals its presence with the deep blue color of a complex formed with starch. What is the maximum duration of electrolysis to the appearance of the blue color that ensures that the concentration of Sn^{2+} does not exceed 10 ppm?

81. Estimate the cost of the electrical energy needed to produce 1.5×10^{10} kg (a year's supply for the world) of aluminum from $Al_2O_3(s)$ if electrical energy costs 10 cents per kilowatt-hour (1 kWh = 3.6 MJ = 3.6×10^6 J) and if the cell potential is 5 V.

82. Titanium can be produced by electrolytic reduction from an anhydrous molten salt electrolyte that contains titanium(IV) chloride and a spectator salt that furnishes ions to make the electrolyte conduct electricity. The standard enthalpy of formation of $TiCl_4(\ell)$ is -750 kJ mol^{-1}, and the standard entropies of $TiCl_4(\ell)$, $Ti(s)$, and $Cl_2(g)$ are 253, 30, and 223 J K^{-1} mol^{-1}, respectively. What minimum applied voltage will be necessary at 100°C?

83. A half-cell has a graphite electrode immersed in an acidic solution (pH 0) of Mn^{2+} (concentration 1.00 M) in contact with solid MnO_2. A second half-cell has an acidic solution (pH 0) of H_2O_2 (concentration 1.00 M) in contact with a platinum electrode past which gaseous oxygen at a pressure of 1.00 atm is bubbled. The two half-cells are connected to form a galvanic cell.

(a) Referring to Appendix E, write balanced chemical equations for the half-reactions at the anode and the cathode and for the overall cell reaction.

(b) Calculate the cell potential.

84. By considering these half-reactions and their standard reduction potentials,

$$Pt^{2+} + 2\,e^- \longrightarrow Pt \qquad E^\circ = 1.2 \text{ V}$$

$$NO_3^- + 4\,H_3O + 3\,e^- \longrightarrow NO + 6\,H_2O \qquad E^\circ = 0.96 \text{ V}$$

$$PtCl_4^{2-} + 2\,e^- \longrightarrow Pt + 4\,Cl^- \qquad E^\circ = 0.73 \text{ V}$$

account for the fact that platinum will dissolve in a mixture of hydrochloric acid and nitric acid (*aqua regia*) but will not dissolve in either acid alone.

85. (a) One method to reduce the concentration of unwanted $Fe^{3+}(aq)$ in a solution of $Fe^{2+}(aq)$ is to drop a piece of metallic iron into the storage container. Write the reaction that removes the Fe^{3+}, and compute its standard cell potential.

(b) Referring to problem 23, suggest a way to remove unwanted $Mn^{3+}(aq)$ from solutions of $Mn^{2+}(aq)$.

86. (a) Based only on the standard reduction potentials for the $Cu^{2+}|Cu^+$ and $I_2(s)|I^-$ the half-reactions, would you expect $Cu^{2+}(aq)$ to be reduced to $Cu^+(aq)$ by $I^-(aq)$?

(b) The formation of solid CuI plays a role in the interaction between $Cu^{2+}(aq)$ and $I^-(aq)$.

$$Cu^{2+}(aq) + I^-(aq) + e^- \rightleftharpoons CuI(s) \qquad E^\circ = 0.86 \text{ V}$$

Taking into account this added information, do you expect Cu^{2+} to be reduced by iodide ion?

87. In some old European churches, the stained-glass windows have so darkened from corrosion and age that hardly any light comes through. Microprobe analysis showed that tiny cracks and defects on the glass surface were enriched in insoluble Mn(III) and Mn(IV) compounds. From Appendix E, suggest a reducing agent and conditions that might successfully convert these compounds to soluble Mn(II) without simultaneously reducing Fe(III) (which gives the glass its colors) to Fe(II). Take MnO_2 as representative of the insoluble Mn(III) and Mn(IV) compounds.

88. (a) Calculate the half-cell potential for the reaction

$$O_2(g) + 4\,H_3O^+(aq) + 4\,e^- \longrightarrow 6\,H_2O(\ell)$$

at pH 7 with the oxygen pressure at 1 atm.

(b) Explain why aeration of solutions of I^- leads to their decomposition. Write a balanced equation for the redox reaction that occurs.

(c) Will the same problem arise with solutions containing Br^- or Cl^-? Explain.

(d) Will decomposition be favored or opposed by increasing acidity?

89. An engineer needs to prepare a galvanic cell that uses the reaction

$$2\,Ag^+(aq) + Zn(s) \longrightarrow Zn^{2+}(aq) + 2\,Ag(s)$$

and generates an initial voltage of 1.50 V. She has 0.010 M $AgNO_3(aq)$ and 0.100 M $Zn(NO_3)_2(aq)$ solutions, as well as electrodes of metallic copper and silver, wires, containers, water, and a KNO_3 salt bridge. Sketch the cell. Clearly indicate the concentrations of all solutions.

90. Consider a galvanic cell for which the anode reaction is

$$Pb(s) \longrightarrow Pb^{2+}(1.0 \times 10^{-2} \text{ M}) + 2\,e^-$$

and the cathode reaction is

$$VO^{2+}(0.10 \text{ M}) + 2\,H_3O^+(0.10 \text{ M}) + e^- \longrightarrow V^{3+}(1.0 \times 10^{-5} \text{ M}) + 3\,H_2O(\ell)$$

The measured cell potential is 0.640 V.

(a) Calculate $E°$ for the $VO^{2+}|V^{3+}$ half-reaction, using $E°(Pb^{2+}|Pb)$ from Appendix E.
(b) Calculate the equilibrium constant (K) at 25°C for the reaction

$$Pb(s) + 2\ VO^{2+}(aq) + 4\ H_3O^+(aq) \rightleftharpoons Pb^{2+}(aq) + 2\ V^{3+}(aq) + 6\ H_2O(\ell)$$

91. Suppose we construct a pressure cell in which the gas pressures differ in the two half-cells. Suppose such a cell consists of a $Cl_2(0.010\ atm)|Cl^-(1\ M)$ half-cell connected to a $Cl_2(0.50\ atm)|Cl^-(1\ M)$ half-cell. Determine which half-cell will be the anode, write the overall equation for the reaction, and calculate the cell potential.

92. A student decides to measure the solubility of lead sulfate in water and sets up the electrochemical cell

$$Pb|PbSO_4|SO_4^{2-}(aq, 0.0500\ M)||Cl^-(aq, 1.00\ M)|AgCl|Ag$$

At 25°C the student finds the cell potential to be 0.546 V, and from Appendix E the student finds

$$AgCl(s) + e^- \rightleftharpoons Ag(s) + Cl^-(aq) \qquad E° = 0.222\ V$$

What does he find for the K_{sp} of $PbSO_4$?

93. A wire is fastened across the terminals of the Leclanché cell in Figure 17.17. Indicate the direction of electron flow in the wire.

94. Overcharging a lead–acid storage battery can generate hydrogen. Write a balanced equation to represent the reaction taking place.

95. An ambitious chemist discovers an alloy electrode that is capable of catalytically converting ethanol reversibly to carbon dioxide at 25°C according to the half-reaction

$$C_2H_5OH(\ell) + 15\ H_2O(\ell) \longrightarrow 2\ CO_2(g) + 12\ H_3O^+(aq) + 12\ e^-$$

Believing that this discovery is financially important, the chemist patents its composition and designs a fuel cell that may be represented as

$$Alloy|C_2H_5OH(\ell)|CO_2(g) + H_3O^+(1\ M)||H_3O^+(1\ M)|O_2|Ni$$

(a) Write the half-reaction occurring at the cathode.
(b) Using data from Appendix D, calculate $\Delta E°$ for the cell at 25°C.
(c) What is the $E°$ value for the ethanol half-cell?

96. Iron or steel is often covered by a thin layer of a second metal to prevent rusting: Tin cans consist of steel covered with tin, and galvanized iron is made by coating iron with a layer of zinc. If the protective layer is broken, however, iron will rust more readily in a tin can than in galvanized iron. Explain this observation by comparing the half-cell potentials of iron, tin, and zinc.

97. An electrolysis cell contains a solution of 0.10 M $NiSO_4$. The anode and cathode are both strips of Pt foil. Another electrolysis cell contains the same solution, but the electrodes are strips of Ni foil. In each case a current of 0.10 A flows through the cell for 10 hours.
(a) Write a balanced equation for the chemical reaction that occurs at the anode in each cell.
(b) Calculate the mass, in grams, of the product formed at the anode in each cell. (The product may be a gas, a solid, or an ionic species in solution.)

98. A potential difference of 2.0 V is impressed across a pair of inert electrodes (e.g., platinum) that are immersed in a 0.050 M aqueous KBr solution. What are the products that form at the anode and the cathode?

99. An aqueous solution is simultaneously 0.10 M in $SnCl_2$ and in $CoCl_2$.
(a) If the solution is electrolyzed, which metal will appear first?
(b) At what decomposition potential will that metal first appear?
(c) As the electrolysis proceeds, the concentration of the metal being reduced will drop and the potential will change. How complete a separation of the metals using electrolysis is theoretically possible? In other words, at the point where the second metal begins to form, what fraction of the first metal is left in solution?

100. A 55.5-kg slab of crude copper from a smelter has a copper content of 98.3%. Estimate the time required to purify it electrochemically if it is used as the anode in a cell that has acidic copper(II) sulfate as its electrolyte and a current of 2.00×10^3 A is passed through the cell.

101. Sheet iron can be galvanized by passing a direct current through a cell containing a solution of zinc sulfate between a graphite anode and the iron sheet. Zinc plates out on the iron. The process can be made continuous if the iron sheet is a coil that unwinds as it passes through the electrolysis cell and coils up again after it emerges from a rinse bath. Calculate the cost of the electricity required to deposit a 0.250-mm-thick layer of zinc on both sides of an iron sheet that is 1.00 m wide and 100 m long, if a current of 25 A at a voltage of 3.5 V is used and the energy efficiency of the process is 90%. The cost of electricity is 10 cents per kilowatt–hour (1 kWh = 3.6 MJ). Consult Appendix F for data on zinc.

CUMULATIVE PROBLEMS

102. A 1.0 M solution of NaOH is electrolyzed, generating $O_2(g)$ at the anode. A current of 0.15 A is passed through the cell for 75 minutes. Calculate the volume of (wet) oxygen generated in this period if the temperature is held at 25°C and the total pressure is 0.985 atm. (*Hint:* Use the vapor pressure of water at this temperature from Table 10.3.)

103. Use standard entropies from Appendix D to predict whether the standard potential of the $Cu|Cu^{2+}||Ag^+|Ag$ cell (diagrammed in Fig. 17.2) will increase or decrease if the temperature is raised above 25°C.

104. About 50,000 kJ of electrical energy is required to produce 1.0 kg of Al from its $Al(OH)_3$ ore. The major energy cost in recycling aluminum cans is the melting of the aluminum. The enthalpy of fusion of Al(*s*) is 10.7 kJ mol^{-1}. Compare the energy cost for making new aluminum with that for recycling.

105. **(a)** Use the following half-reactions and their reduction potentials to calculate the K_{sp} of AgBr:

$$Ag^+ + e^- \longrightarrow Ag(s) \qquad E° = 0.7996 \text{ V}$$

$$AgBr(s) + e^- \longrightarrow Ag(s) + Br^- \qquad E° = 0.0713 \text{ V}$$

(b) Estimate the solubility of AgBr in 0.10 M NaBr(*aq*).

106. Amounts of iodine dissolved in aqueous solution, $I_2(aq)$, can be determined by titration with thiosulfate ion ($S_2O_3^{2-}$). The thiosulfate ion is oxidized to $S_4O_6^{2-}$ while the iodine is reduced to iodide ion. Starch is used as an indicator because it has a strong blue color in the presence of dissolved iodine.

(a) Write a balanced equation for this reaction.

(b) If 56.40 mL of 0.100 M $S_2O_3^{2-}$ solution is used to reach the endpoint of a titration of an unknown amount of iodine, calculate the number of moles of iodine originally present.

(c) Combine the appropriate half-cell potentials from Appendix E with thermodynamic data from Appendix D for the equilibrium

$$I_2(s) \rightleftharpoons I_2(aq)$$

to calculate the equilibrium constant at 25°C for the reaction in part (a).

Rates of Chemical and Physical Processes

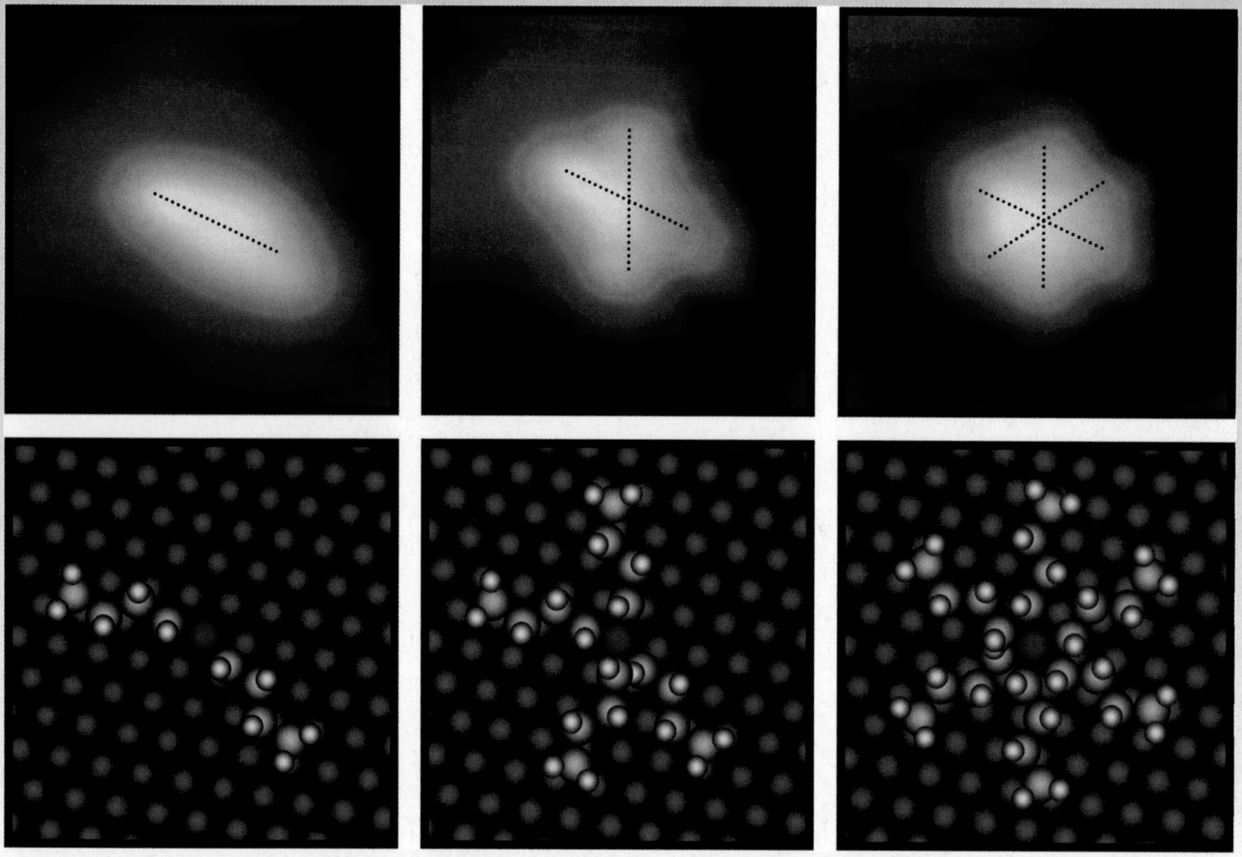

J. Phys. Chem. C, 2009, 113(25), pp 10913–10920

Scanning tunneling microscope images of dibutyl sulfide adsorbed on a single crystal gold surface, measured at 7 K, 13 K, and 25 K (top row, left to right), with corresponding ball-and-stick models shown in the bottom row. The molecule occupies a single unique site at 7 K, then samples the two lowest energy, nearly equivalent sites at 13 K, and finally rotates freely at 25 K. These "molecular rotors" are being studied to understand how more complex "molecular machines" operate in biological processes and potential applications in nanotechnology are also being explored. This series of images was chosen to illustrate the role of temperature in activating molecular motions and to anticipate the central role it plays in controlling the rates of chemical reactions.

UNIT V

Thermodynamics explains *why* chemical reactions occur; minimizing the Gibbs free energy is the driving force toward chemical equilibrium. Thermodynamics provides deep insight into the nature of chemical equilibrium, but it gives no answer to the crucial question of how rapidly that equilibrium is achieved.

Chemical kinetics explains *how* reactions occur by studying their rates and mechanisms. Chemical kinetics explains how the speeds of different chemical reactions vary from explosive rapidity to glacial sluggishness and how slow reactions can be accelerated by materials called catalysts. Chemical kinetics has enormous practical importance because it provides the basis for optimizing conditions to carry out chemical reactions at reasonable speed, under proper control.

The central goal in chemical kinetics is to find the relationship between the rate of a reaction and the amount of reactants present. Once this connection is established, the influence of external conditions—principally the temperature—can be explored. Modern experiments probe how the reaction rate and product branching depend on the vibrational motions and the relative velocity between the colliding molecules.

Nuclear chemistry represents a particularly simple limiting form of kinetics in which unstable nuclei decay with a constant probability during any time interval. Its richness arises from the multiplicity of decay paths that are possible, which arise from the mass-energy relationships that determine nuclear stability.

The interactions between molecules and electromagnetic radiation form the basis of molecular spectroscopy, allowing us to determine the structure of molecules and investigate their vibrational motion and excited electronic states. The absorption of visible and ultraviolet light can result in chemical reactions that are different than those initiated thermally. The role of photo-initiated reactions in the atmosphere and in photosynthesis are discussed.

Unit Chapters

Unit Goals

- To relate the rate of a chemical reaction to the instantaneous concentration(s) of reactants by determining the rate law and the rate constant for the reaction
- To describe the influence of temperature on the reaction rate by measuring the activation energy for the reaction
- To explain the mechanism of a complex reaction by identifying the separate elementary reaction steps through which it proceeds
- To explain the role of catalysts in manipulating reaction rates
- To introduce experiments that probe energy transfer and reaction dynamics at the level of individual collisions between molecules
- To develop an elementary description of the rates of nuclear reactions, emphasizing the half-life of radioactive species
- To survey the applications and consequences of nuclear reactions in medicine, biology, energy production, and the environment
- To introduce and discuss methods and applications of molecular spectroscopy to determining molecular structure, probing the vibrational motion and excited states of molecules, and investigating the importance of reactions initiated by the absorption of light in atmospheric chemistry and in photosynthesis

18 CHAPTER

Chemical Kinetics

Powdered chalk (mostly calcium carbonate $CaCO_3$) reacts rapidly with dilute hydrochloric acid because it has a large total surface area. A stick of chalk has a much smaller surface area, so it reacts much more slowly.

Why do some chemical reactions proceed with lightning speed when others require days, months, or even years to produce detectable amounts of products? How do catalysts increase the rates of chemical reactions? Why do small changes in temperature often have such large effects on the cooking rate of food? How does a study of the rate of a chemical reaction inform us about the way in which molecules combine to form products? All of these questions involve studies of reaction rates, which is the subject of chemical kinetics.

Chemical kinetics is a complex subject, the details of which are not currently as well understood as in chemical thermodynamics. For many reactions the equilibrium constants are known accurately, but the rates and detailed reaction pathways remain poorly understood. This is particularly true of reactions in which many species participate in the overall process connecting reactants to products. One good example is the reaction

$$5\,Fe^{2+}(aq) + MnO_4^-(aq) + 8\,H_3O^+(aq) \longrightarrow 5\,Fe^{3+}(aq) + Mn^{2+}(aq) + 12\,H_2O(\ell)$$

We can measure the equilibrium constant for this reaction easily from the voltage of a galvanic cell, and then calculate from it the equilibrium concentrations that will result from arbitrary initial conditions. It is considerably harder to determine the exact pathway by which the reaction goes from reactants to products. This path certainly does *not* involve the simultaneous collision of five Fe^{2+} ions and one MnO_4^- ion with eight H_3O^+ ions, because such a collision would be exceedingly rare. Instead, the path proceeds through a series of elementary steps involving two or at most three ions, such as

$$Fe^{2+}(aq) + MnO_4^-(aq) \longrightarrow Fe^{3+}(aq) + MnO_4^{2-}(aq)$$

$$MnO_4^{2-}(aq) + H_3O^+(aq) \longrightarrow HMnO_4^-(aq) + H_2O(\ell)$$

$$HMnO_4^-(aq) + Fe^{2+}(aq) \longrightarrow HMnO_4^{2-}(aq) + Fe^{3+}(aq)$$

Other postulated steps take the process to its final products. Some of these steps are slow, and others fast; taken together they constitute the **reaction mechanism**.

The primary goal of chemical kinetics is to deduce the mechanism of a reaction from experimental studies of its rate. For this, we have to measure how the rate depends on the concentrations of the reacting species. This chapter lays out the methods and concepts for measuring and interpreting reaction rates and for identifying the mechanism. We then introduce modern methods to probe the dynamics of the reactions at the molecular level.

18.1 Rates of Chemical Reactions

The speed of a reaction depends on many factors. Concentrations of reacting species certainly play a major role in speeding up or slowing down a particular reaction (Fig. 18.1). As we see in Section 18.5, many reaction rates are extremely

FIGURE 18.1 The rate of reaction of zinc with aqueous sulfuric acid depends on the concentration of the acid. The dilute solution reacts slowly (left), and the more concentrated solution reacts rapidly (right).

FIGURE 18.2 Steel wool burning in oxygen.

sensitive to temperature. This means that careful control of temperature is critical for quantitative measurements in chemical kinetics. Finally, the rate often depends crucially on the physical forms of the reactants, as shown in the chapter opening photograph. An iron nail oxidizes only very slowly in dry air to iron oxide, but steel wool burns spectacularly in oxygen (Fig. 18.2). Because the quantitative study of heterogeneous reactions—those involving two or more phases, such as a solid and a gas—is difficult, we begin with homogeneous reactions, which take place entirely within the gas phase or solution. In Section 18.8 we turn briefly to some important aspects of heterogeneous reactions.

18.1.1 Measuring Reaction Rates

A kinetics experiment measures the rate of change of the concentration of a substance participating in a chemical reaction. How can we experimentally monitor a changing concentration? If the reaction is slow enough, we can let it run for a measured time and then abruptly "quench" (effectively stop) it by rapidly cooling the reaction mixture sufficiently. At that low temperature the composition of the reaction mixture remains constant, so we have time to analyze the mixture for some particular reactant or product. This procedure is not useful for rapid reactions, especially those involving gas mixtures, because they are difficult to cool quickly. An alternative is to probe the concentrations by the absorption of light. Chapter 20 shows that different molecules absorb at different wavelengths. If a wavelength is absorbed by only one particular reactant or product, measuring the amount of light absorbed by the reaction mixture at that wavelength determines the concentration of the absorbing species. A series of such measurements at different times reveals the rate of change of the concentration. Often, a flash of light can also be used to initiate a very fast reaction, whose rate is then tracked by measuring absorption at a particular wavelength.

The average rate of a reaction is analogous to the average speed of a car. If the average position of a car is recorded at two different times, then

$$\text{average speed} = \frac{\text{distance traveled}}{\text{time elapsed}} = \frac{\text{change in location}}{\text{change in time}}$$

In the same way, the **average reaction rate** is obtained by dividing the change in concentration of a product (or the negative of the change in concentration of a reactant) by the time interval over which that change occurs:

$$\text{average reaction rate} = \frac{\text{change in concentration of product}}{\text{change in time}}$$

If concentration is measured in mol L^{-1} and time in seconds, then the rate of a reaction has units of mol L^{-1} s^{-1}.

Consider a specific example. In the gas-phase reaction

$$NO_2(g) + CO(g) \longrightarrow NO(g) + CO_2(g)$$

NO_2 and CO are consumed as NO and CO_2 are produced. If a probe can measure the NO concentration, the average rate of reaction can be estimated from the ratio of the change in NO concentration $\Delta[NO]$ to the time interval Δt:

$$\text{average rate} = \frac{\Delta[NO]}{\Delta t} = \frac{[NO]_f - [NO]_i}{t_f - t_i}$$

This estimate depends on the time interval Δt that is selected, because the rate at which NO is produced changes with time. Average rates for this reaction can be calculated for various time periods (see Fig. 18.3), the average rate during the first 50 s being given by

$$\text{average rate} = \frac{\Delta[\text{NO}]}{\Delta t} = \frac{(0.0160 - 0)\ \text{mol L}^{-1}}{(50 - 0)\ \text{s}} = 3.2 \times 10^{-4}\ \text{mol L}^{-1}\ \text{s}^{-1}$$

During the second 50 s, the average rate is 1.6×10^{-4} mol L^{-1} s^{-1}, and during the third 50 s it is 9.6×10^{-5} mol L^{-1} s^{-1}. Clearly, this reaction slows as it progresses, and its average rate indeed depends on the time interval chosen. Figure 18.3 shows a graphical method for determining average rates. The average rate is the slope of the straight line connecting the concentrations at the initial and final points of a time interval.

The **instantaneous rate** of a reaction is obtained by considering smaller and smaller time increments Δt (with correspondingly smaller values of Δ[NO]). As Δt approaches 0, the rate becomes the slope of the line tangent to the curve at time t (see Fig. 18.3). This slope is written as the derivative of [NO] with respect to time:

$$\text{instantaneous rate} = \text{limit}_{\Delta t \longrightarrow 0} \frac{[\text{NO}]_{t+\Delta t} - [\text{NO}]_t}{\Delta t} = \frac{d[\text{NO}]}{dt}$$

Throughout the rest of this book, we refer to the instantaneous rate simply as the *rate*. The instantaneous rate of a reaction at the moment that it begins (at $t = 0$) is the **initial rate** of that reaction.

The rate of this sample reaction could just as well have been measured by monitoring changes in the concentration of CO_2, NO_2, or CO instead of NO. Because every molecule of NO produced is accompanied by one molecule of CO_2, the rate of increase of CO_2 concentration is the same as that of NO. The concentrations of the two reactants, NO_2 and CO, *decrease* at the same rate that

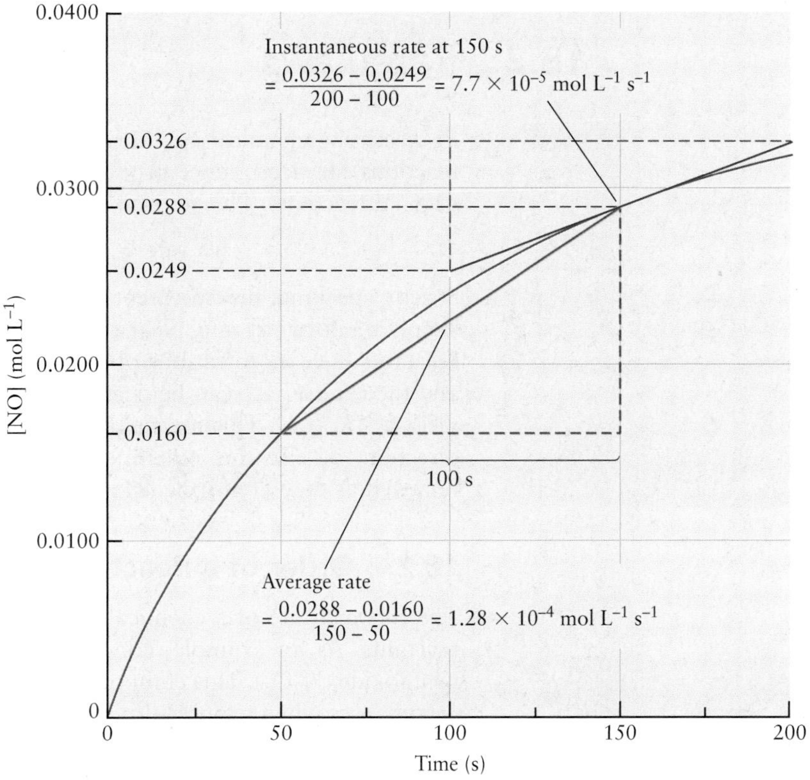

FIGURE 18.3 A graph of the concentration of NO against time in the reaction $NO_2 + CO \longrightarrow NO + CO_2$. The average rate during the time interval from 50 to 150 seconds is obtained by dividing the change in NO concentration by the duration of the interval (green box). Note that the average rate (green line) underestimates the true production rate over the time interval. The instantaneous rate 150 seconds after the start of the reaction is found by calculating the slope of the line tangent to the curve at that point (red box).

the concentrations of the products increase, because the coefficients in the balanced equation are also both equal to 1. This is summarized as

$$\text{rate} = -\frac{d[\mathrm{NO_2}]}{dt} = -\frac{d[\mathrm{CO}]}{dt} = \frac{d[\mathrm{NO}]}{dt} = \frac{d[\mathrm{CO_2}]}{dt}$$

Another gas-phase reaction is

$$2\ \mathrm{NO_2}(g) + \mathrm{F_2}(g) \longrightarrow 2\ \mathrm{NO_2F}(g)$$

This equation states that two molecules of NO_2 disappear and two molecules of NO_2F appear for each molecule of F_2 that reacts. Thus, the NO_2 concentration changes twice as fast as the F_2 concentration; the NO_2F concentration also changes twice as fast and has the opposite sign. We write the rate in this case as

$$\text{rate} = -\frac{1}{2}\frac{d[\mathrm{NO_2}]}{dt} = -\frac{d[\mathrm{F_2}]}{dt} = \frac{1}{2}\frac{d[\mathrm{NO_2F}]}{dt}$$

The rate of change of concentration of each species is divided by its coefficient in the balanced chemical equation. Rates of change of reactants appear with negative signs and those of products with positive signs. For the general reaction

$$a\mathrm{A} + b\mathrm{B} \longrightarrow c\mathrm{C} + d\mathrm{D}$$

the rate is

$$\text{rate} = -\frac{1}{a}\frac{d[\mathrm{A}]}{dt} = -\frac{1}{b}\frac{d[\mathrm{B}]}{dt} = \frac{1}{c}\frac{d[\mathrm{C}]}{dt} = \frac{1}{d}\frac{d[\mathrm{D}]}{dt} \qquad [18.1]$$

These relations hold true provided there are no transient intermediate species or, if there are intermediates, their concentrations are independent of time for most of the reaction period.

18.2 Rate Laws

In discussing chemical equilibrium we stressed that both forward and reverse reactions can occur; once products are formed, they can react back to give the original reactants. The net rate is the difference:

$$\text{net rate} = \text{forward rate} - \text{reverse rate}$$

Strictly speaking, measurements of concentration give the net rate rather than simply the forward rate. Near the beginning of a reaction that starts from pure reactants the concentrations of reactants are far higher than those of products, and the reverse rate can be neglected. In addition, many reactions go to "completion" ($K \gg 1$). This means they have a measurable rate only in the forward direction, or else the experiment can be arranged so that the products are removed as they are formed. This section focuses on forward rates exclusively.

18.2.1 Order of a Reaction

The forward rate of a chemical reaction depends on the concentrations of the reactants. As an example, consider the decomposition of gaseous dinitrogen pentaoxide (N_2O_5). This compound is a white solid that is stable below 0°C but decomposes when vaporized:

$$\mathrm{N_2O_5}(g) \longrightarrow 2\ \mathrm{NO_2}(g) + \tfrac{1}{2}\mathrm{O_2}(g)$$

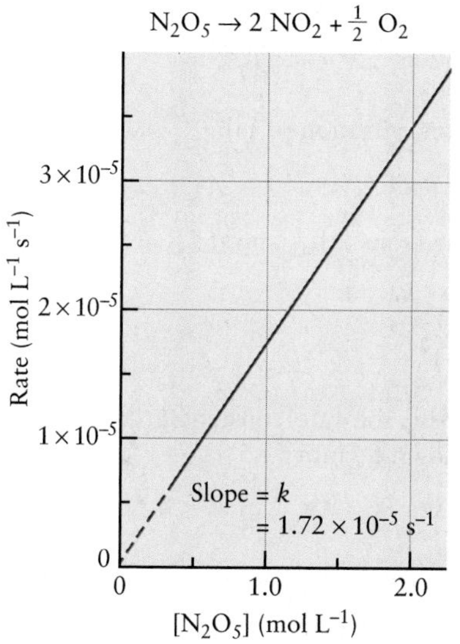

FIGURE 18.4 The rate of decomposition of $N_2O_5(g)$ at 25°C is proportional to its concentration. The slope of this line is equal to the rate constant k for the reaction.

The rate of the reaction depends on the concentration of $N_2O_5(g)$. Figure 18.4 shows the graph of rate versus concentration to be a straight line that can be extrapolated to pass through the origin. So, the rate can be written

$$\text{rate} = k[N_2O_5]$$

This relation between the rate of a reaction and concentration is called an **empirical rate expression** or *rate law*, and the proportionality constant k is called the **rate constant** for the reaction. Like an equilibrium constant, a rate constant is independent of concentration but depends on temperature, as we describe in Section 18.5.

For many (but not all) reactions with a single reactant, the rate is proportional to the concentration of that reactant raised to a power. That is, the rate expression for

$$a\text{A} \longrightarrow \text{products}$$

frequently has the form

$$\text{rate} = k[\text{A}]^n$$

It is important to note that the power n in the rate expression has no direct relation to the coefficient a in the balanced chemical equation. This number has to be determined experimentally for each rate law. For the decomposition of ethane at high temperatures and low pressures,

$$C_2H_6(g) \longrightarrow 2\ CH_3(g)$$

the rate expression has the form

$$\text{rate} = k[C_2H_6]^2$$

Therefore, $n = 2$ even though the coefficient in the chemical equation is 1.

The power to which the concentration is raised is called the **order** of the reaction with respect to that reactant. Thus, the decomposition of N_2O_5 is a **first-order reaction**, whereas that of C_2H_6 is a **second-order reaction**. Some processes are **zeroth-order reactions** over a range of concentrations. Because $[\text{A}]^0 = 1$, such reactions have rates that are independent of concentration:

$$\text{rate} = k \qquad \text{(for zeroth-order kinetics)}$$

The order of a reaction does not have to be an integer; fractional powers are sometimes found. At 450 K, the decomposition of acetaldehyde (CH_3CHO) is described by the rate expression

$$\text{rate} = k[CH_3CHO]^{3/2}$$

These examples demonstrate that reaction order is an experimentally determined property that cannot be predicted from the form of the chemical equation.

The following example illustrates how the order of a reaction can be deduced from experimental data.

EXAMPLE 18.1

At elevated temperatures, HI reacts according to the chemical equation

$$2\ HI(g) \longrightarrow H_2(g) + I_2(g)$$

At 443°C, the rate of the reaction increases with concentration of HI as follows:

[HI] (mol L^{-1})	0.0050	0.010	0.020
Rate (mol L^{-1} s^{-1})	7.5×10^{-4}	3.0×10^{-3}	1.2×10^{-2}

(Continued)

(a) Determine the order of the reaction and write the rate expression.

(b) Calculate the rate constant, and give its units.

(c) Calculate the reaction rate for a 0.0020 M concentration of HI.

Solution

(a) The rate expressions at two different concentrations $[HI]_1$ and $[HI]_2$ are

$$\text{rate}_1 = k([HI]_1)^n$$

$$\text{rate}_2 = k([HI]_2)^n$$

After dividing the second equation by the first, the rate constant k drops out, leaving the reaction order n as the only unknown quantity.

$$\frac{\text{rate}_2}{\text{rate}_1} = \left(\frac{[HI]_2}{[HI]_1}\right)^n$$

We can now substitute any two sets of data into this equation and solve for n. Taking the first two sets with $[HI]_1 = 0.0050$ M and $[HI]_2 = 0.010$ M gives

$$\frac{3.0 \times 10^{-3}}{7.5 \times 10^{-4}} = \left(\frac{0.010}{0.0050}\right)^n$$

which simplifies to

$$4 = (2)^n$$

By inspection, $n = 2$, so the reaction is second order in HI. When the solution of the equation is less obvious, we can take the logarithms of both sides, giving (in this case)

$$\log_{10} 4 = n \log_{10} 2$$

$$n = \frac{\log_{10} 4}{\log_{10} 2} = \frac{0.602}{0.301} = 2$$

The rate expression has the form

$$\text{rate} = k[HI]^2$$

(b) The rate constant k is calculated by inserting any of the sets of data into the rate expression. Taking the first set gives

$$7.5 \times 10^{-4} \text{ mol L}^{-1} \text{ s}^{-1} = k(0.0050 \text{ mol L}^{-1})^2$$

Solving for k gives

$$k = 30 \text{ L mol}^{-1} \text{ s}^{-1}$$

(c) Finally, the rate is calculated for $[HI] = 0.0020$ M:

$$\text{rate} = k[HI]^2 = (30 \text{ L mol}^{-1} \text{ s}^{-1})(0.0020 \text{ mol L}^{-1})^2$$

$$= 1.2 \times 10^{-4} \text{ mol L}^{-1} \text{ s}^{-1}$$

So far, each example reaction rate has depended only on a single concentration. In reality, many rates depend on the concentrations of two or more different chemical species, and the rate expression is written in a form such as

$$\text{rate} = -\frac{1}{a}\frac{d[A]}{dt} = k[A]^m[B]^n$$

Again the exponents m and n do not derive from the coefficients in the balanced equation for the reaction; they must be determined experimentally and are usually integers or half-integers.

The exponents m, n, . . . give the order of the reaction, just as in the simpler case where only one concentration appeared in the rate expression. The preceding reaction is said to be mth order in A, meaning that a change in the concentration of A by a certain factor leads to a change in the rate by that factor raised to the mth power. The reaction is nth order in B, and the **overall reaction order** is $m + n$. For the reaction

$$H_2PO_2^-(aq) + OH^-(aq) \longrightarrow HPO_3^{2-}(aq) + H_2(g)$$

the experimentally determined rate expression is

$$\text{rate} = k[H_2PO_2^-][OH^-]^2$$

so the reaction is said to be first order in $H_2PO_2^-(aq)$ and second order in $OH^-(aq)$, with an overall reaction order of 3. The units of k depend on the reaction order. If all concentrations are expressed in mol L^{-1} and if $p = m + n + \cdots$ is the overall reaction order, then k has units of $\text{mol}^{-(p-1)}\ L^{p-1}\ s^{-1}$.

EXAMPLE 18.2

Use the preceding rate expression to determine the effect of the following changes on the rate of decomposition of $H_2PO_2^-(aq)$:

(a) Tripling the concentration of $H_2PO_2^-(aq)$ at constant pH

(b) Changing the pH from 13 to 14 at a constant concentration of $H_2PO_2^-(aq)$

Solution

(a) Because the reaction is first order in $H_2PO_2^-(aq)$, tripling this concentration will triple the reaction rate.

(b) A change in pH from 13 to 14 corresponds to an increase in the $OH^-(aq)$ concentration by a factor of 10. Because the reaction is second order in $OH^-(aq)$ (that is, this term is squared in the rate expression), this will increase the reaction rate by a factor of 10^2, or 100.

Related Problems: 5, 6

Rate expressions that depend on more than one concentration are more difficult to obtain experimentally than those that depend just on one. One method is to find the instantaneous initial rates of reaction for several values of one of the concentrations, holding the other initial concentrations fixed from one run to the next. The experiment can then be repeated, changing one of the other concentrations. The following example illustrates this procedure.

EXAMPLE 18.3

The reaction of $NO(g)$ with $O_2(g)$ gives $NO_2(g)$:

$$2\ NO(g) + O_2(g) \longrightarrow 2\ NO_2(g)$$

From the dependence of the initial rate $(-\frac{1}{2}\,d[NO]/dt)$ on the initial concentrations of NO and O_2, determine the rate expression and the value of the rate constant.

[NO] (mol L^{-1})	[O_2] (mol L^{-1})	Initial Rate (mol L^{-1} s^{-1})
1.0×10^{-4}	1.0×10^{-4}	2.8×10^{-6}
1.0×10^{-4}	3.0×10^{-4}	8.4×10^{-6}
2.0×10^{-4}	3.0×10^{-4}	3.4×10^{-5}

(Continued)

Solution

When $[O_2]$ is multiplied by 3 (with [NO] constant), the rate is also multiplied by 3 (from 2.8×10^{-6} to 8.4×10^{-6}), so the reaction is first order in O_2. When [NO] is multiplied by 2 (with $[O_2]$ constant), the rate is multiplied by

$$\frac{3.4 \times 10^{-5}}{8.4 \times 10^{-6}} \approx 4 = 2^2$$

so the reaction is second order in NO. Thus, the form of the rate expression is

$$\text{rate} = k[O_2][NO]^2$$

To evaluate k, we insert any set of data into the equation. From the first set,

$$2.8 \times 10^{-6} \text{ mol L}^{-1} \text{ s}^{-1} = k(1.0 \times 10^{-4} \text{ mol L}^{-1})(1.0 \times 10^{-4} \text{ mol L}^{-1})^2$$

$$k = 2.8 \times 10^{6} \text{ L}^2 \text{ mol}^{-2} \text{ s}^{-1}$$

Related Problems: 7, 8

18.2.2 Integrated Rate Laws

Measuring an initial rate involves determining small changes in concentration $\Delta[A]$ that occur during a short time interval Δt. Sometimes it can be difficult to obtain sufficiently precise experimental data for these small changes. An alternative is to use an **integrated rate law**, which expresses the concentration of a species directly as a function of the time. For any simple rate expression, a corresponding integrated rate law can be obtained.

First-Order Reactions

Consider again the reaction

$$N_2O_5(g) \longrightarrow 2\ NO_2(g) + \tfrac{1}{2}\ O_2(g)$$

whose rate law has been determined experimentally to be

$$\text{rate} = -\frac{d[N_2O_5]}{dt} = k[N_2O_5]$$

This is a first-order reaction. If we let $[N_2O_5] = c$, a function of time, we have

$$\frac{dc}{dt} = -kc$$

We seek a function whose slope at every time is proportional to the value of the function itself. This function can be found through the use of calculus. Separating the variables (with concentration c on the left and time t on the right) gives

$$\frac{1}{c}\,dc = -k\,dt$$

Integrating from an initial concentration c_0 at time $t = 0$ to a concentration c at time t (see Appendix C, Section C.5) gives

$$\int_{c_0}^{c} \frac{1}{c}\,dc = -k\int_0^t dt$$

$$\ln c - \ln c_0 = -kt$$

$$\ln (c/c_0) = -kt$$

$$c = c_0\, e^{-kt} \qquad [18.2]$$

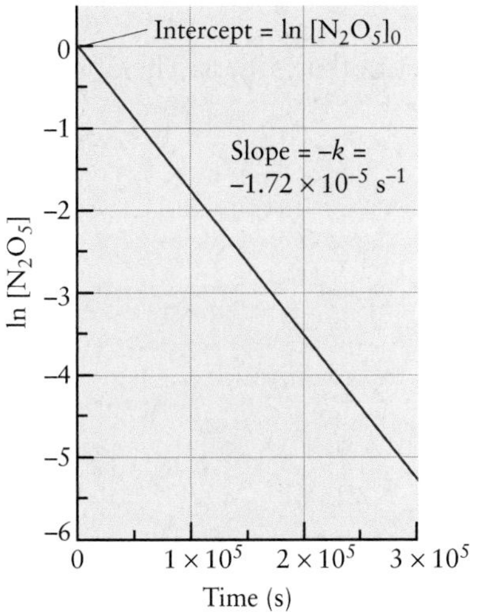

FIGURE 18.5 In a first-order reaction such as the decomposition of N_2O_5, a graph of the natural logarithm of the concentration against time is a straight line, whose negative slope gives the rate constant for the reaction.

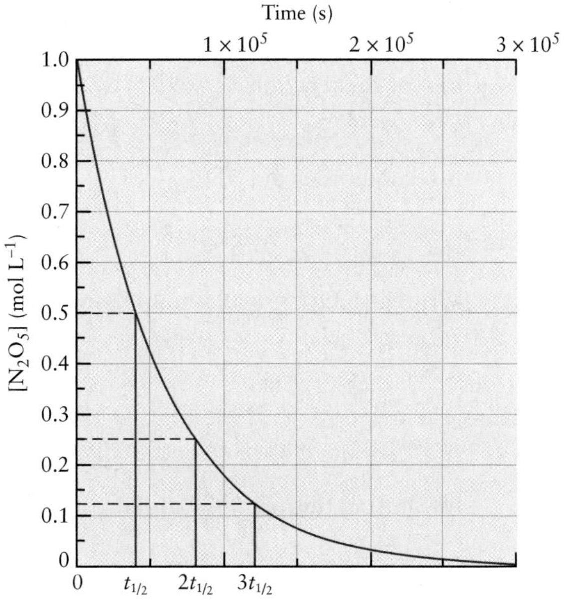

FIGURE 18.6 The same data as in Figure 18.5 are graphed in a concentration-versus-time picture. The half-life $t_{1/2}$ is the time it takes for the concentration to be reduced to half its initial value. In two half-lives, the concentration falls to one quarter of its initial value.

The concentration falls off exponentially with time. For a first-order reaction, a plot of ln c against t is a straight line with slope $-k$ (Fig. 18.5).

A useful concept in discussions of first-order reactions is the **half-life** $t_{1/2}$ which is defined as the time it takes for the original concentration c_0 to be reduced to half its value, $c_0/2$. Setting $c = c_0/2$ at time $t = t_{1/2}$ in Equation 18.2 gives

$$\ln\left(\frac{c_0/2}{c_0}\right) = -\ln 2 = -kt_{1/2}$$

$$t_{1/2} = \frac{\ln 2}{k} = \frac{0.6931}{k} \qquad [18.3]$$

If k has units of s^{-1}, (ln 2)/k is the half-life in seconds. During each half-life, the concentration of A falls to half its value (at the start of that time interval) again (Fig. 18.6).

EXAMPLE 18.4

(a) What is the rate constant k for the first-order decomposition of $N_2O_5(g)$ at 25°C if the half-life of $N_2O_5(g)$ at that temperature is 4.03×10^4 s?

(b) What percentage of the N_2O_5 molecules will *not* have reacted after one day?

Solution

(a)

$$t_{1/2} = \frac{\ln 2}{k} = 4.03 \times 10^4 \text{ s}$$

Solving for the rate constant k gives

$$k = \frac{\ln 2}{t_{1/2}} = \frac{0.6931}{4.03 \times 10^4 \text{ s}} = 1.72 \times 10^{-5} \text{ s}^{-1}$$

(b) From the integrated rate law for a first-order reaction,

$$\frac{c}{c_0} = e^{-kt}$$

Putting in the value for k and setting t to 1 day = 8.64×10^4 s gives

$$\frac{c}{c_0} = \exp\left[-(1.72 \times 10^{-5} \text{ s}^{-1})(8.64 \times 10^4 \text{ s})\right]$$

$$= e^{-1.49} = 0.226$$

Therefore, 22.6% of the molecules will not yet have reacted after one day.

Related Problems: 11, 12

Second-Order Reactions

Integrated rate laws can be obtained for reactions of other orders. The observed rate of the reaction

$$2\ NO_2(g) \longrightarrow 2\ NO(g) + O_2(g)$$

is second order in $[NO_2]$:

$$\text{rate} = -\frac{1}{2}\frac{d[NO_2]}{dt} = k[NO_2]^2$$

Writing $[NO_2] = c$ and multiplying both sides of the equation by -1 gives

$$\frac{dc}{dt} = -2kc^2$$

$$\frac{1}{c^2}\,dc = -2k\,dt$$

Integrating this from the initial concentration c_0 at time 0 to c at time t gives

$$\int_{c_0}^{c}\frac{1}{c^2}\,dc = -2k\int_0^t dt$$

$$-\frac{1}{c} + \frac{1}{c_0} = -2kt$$

$$\frac{1}{c} = \frac{1}{c_0} + 2kt \qquad [18.4]$$

For such a second-order reaction, a plot of $1/c$ against t is linear (Fig. 18.7). The factor 2 multiplying kt in this expression arises from the stoichiometric coefficient 2 for NO_2 in the balanced equation for the specific example reaction. For other second-order reactions with different stoichiometric coefficients for the reactant (see the thermal decomposition of ethane described on page 759), we must modify the integrated rate law accordingly.

The concept of half-life has little use for second-order reactions. Setting $[NO_2]$ equal to $[NO_2]_0/2$ in the preceding equation and solving for t gives

$$\frac{2}{[NO_2]_0} = 2kt_{1/2} + \frac{1}{[NO_2]_0}$$

$$t_{1/2} = \frac{1}{2k[NO_2]_0}$$

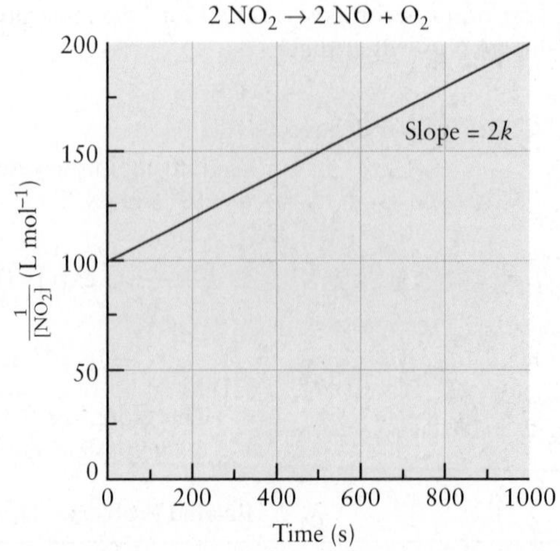

FIGURE 18.7 For a second-order reaction such as $2\ NO_2 \longrightarrow 2\ NO + O_2$, a graph of the reciprocal of the concentration against time is a straight line with slope $2k$.

For second-order reactions, the half-life is not a constant; it depends on the initial concentration.

EXAMPLE 18.5

The dimerization of tetrafluoroethene (C_2F_4) to octafluorocyclobutane (C_4F_8) is second order in the reactant C_2F_4, and at 450 K its rate constant is $k = 0.0448$ L mol^{-1} s^{-1}. If the initial concentration of C_2F_4 is 0.100 mol L^{-1}, what will its concentration be after 205 s?

Solution

For this second-order reaction,

$$\frac{1}{c} - \frac{1}{c_0} = 2kt$$

Solving for the concentration c after a time $t = 205$ s gives

$$\frac{1}{c} = (2)(0.0448 \text{ L mol}^{-1} \text{ s}^{-1})(205 \text{ s}) + \frac{1}{0.100 \text{ mol L}^{-1}} = 28.4 \text{ L mol}^{-1}$$

$$c = 3.52 \times 10^{-2} \text{ mol L}^{-1}$$

Related Problems: 15, 16

Empirical rate laws are established in two ways: by measuring the dependence of initial rates on the concentrations of each reactant and by plotting the concentration as a function of time, as illustrated earlier. It is often the case that rate laws cannot be definitively established by one concentration–time plot alone and is helpful to plot the data in several ways to choose the best fit, as shown in Figure 18.8. You may use this approach in one of your undergraduate laboratories. Determining the order of a reaction in more complicated cases requires sophisticated data analysis using specialized computer codes designed for chemical kinetics.

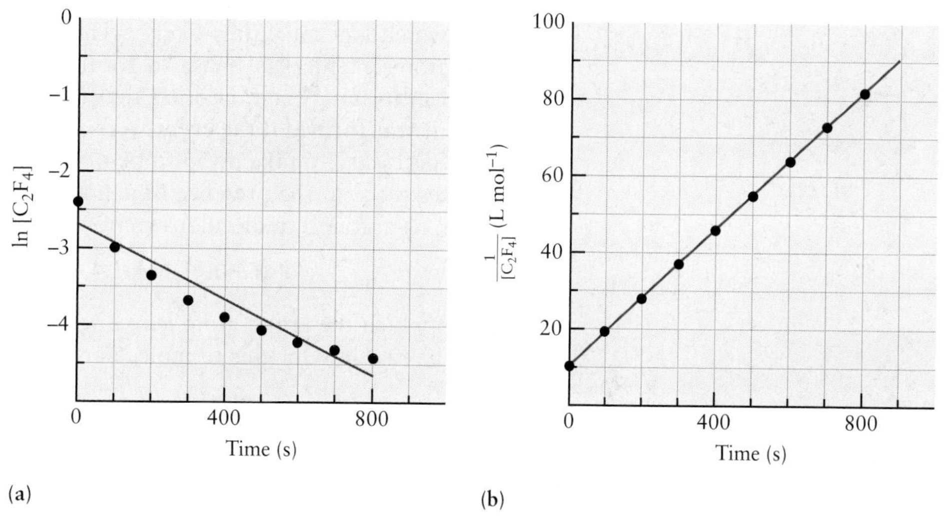

FIGURE 18.8 For the reaction in Example 18.5, (a) plotting the logarithm of the concentration of C_2F_4 against time tests for first-order kinetics and (b) plotting the reciprocal of the concentration of C_2F_4 against time tests for second-order kinetics. It is clear that the assumption of first-order kinetics does not fit the data as well; no straight line will pass through the data.

18.3 Reaction Mechanisms

Many reactions do not occur in a single step, but rather proceed through a sequence of steps to arrive at the products. Each step is called an **elementary reaction** and occurs through the collisions of atoms, ions, or molecules, as discussed in Section 9.7. The rate expression for an overall reaction cannot be derived from the stoichiometry of the balanced equation, and must be determined experimentally. But the rate of an *elementary* reaction *is* directly proportional to the product of the concentrations of the reacting species, each raised to a power equal to its coefficient in the balanced elementary equation.

18.3.1 Elementary Reactions

A **unimolecular reaction** is an elementary reaction that involves only a single reactant molecule. An example is the dissociation of energized N_2O_5 molecules in the gas phase:

$$N_2O_5^*(g) \longrightarrow NO_2(g) + NO_3(g)$$

The asterisk indicates that the N_2O_5 molecules have far more than average thermal energy. This step is unimolecular and has the rate expression

$$\text{rate} = k[N_2O_5^*]$$

An important class of unimolecular reactions is the decay of radioactive nuclei, considered in Chapter 19.

The most common type of elementary reaction involves the collision of two atoms, ions, or molecules and is called a **bimolecular reaction**. An example is the reaction

$$NO(g) + O_3(g) \longrightarrow NO_2(g) + O_2(g)$$

The frequency at which a given NO molecule collides with ozone molecules is proportional to the concentration of ozone: If there are twice as many ozone molecules per unit volume, each NO molecule will undergo twice as many collisions per unit time with ozone as it moves through space, and NO will react twice as rapidly. The rate of collisions of *all* the NO molecules in the container is proportional to the concentration of NO as well, so the rate law of a bimolecular reaction like this one has the form

$$\text{rate} = k[NO][O_3]$$

A **termolecular reaction** step involves the simultaneous collision of three molecules, which is a much less likely event. An example is the recombination of iodine atoms in the gas phase to form iodine molecules. So much energy is released in forming the I—I bond that the molecule would simply fly apart as soon as it was formed if the event were a binary collision. A third atom or molecule is necessary to take away some of the excess energy. If iodine recombination takes place in the presence of a sufficiently high concentration of an inert gas such as argon, termolecular reactions

$$I(g) + I(g) + Ar(g) \longrightarrow I_2(g) + Ar(g)$$

occur in which the argon atom leaves with more kinetic energy than it had initially. The rate law for this termolecular reaction is

$$\text{rate} = k[I]^2[Ar]$$

Elementary reactions involving collisions of four or more molecules are not observed, and even termolecular collisions are rare if other pathways are possible.

Elementary reactions in liquid solvents involve encounters of solute species with one another. If the solution is ideal, the rates of these processes are proportional to the product of the concentrations of the solute species involved. Solvent molecules are always present and may affect the reaction, even though they do not appear in the rate expression because the solvent concentration cannot be varied appreciably. A reaction such as the recombination of iodine atoms occurs readily in a liquid. It appears to be second order with rate law

$$\text{rate} = k[\text{I}]^2$$

only because the third bodies involved are solvent molecules, with constant concentrations. In the same way, a reaction between a solvent molecule and a solute molecule appears to be unimolecular, and only the concentration of solute molecules enters the rate expression for that step.

18.3.2 Reaction Mechanisms

A **reaction mechanism** is a detailed sequence of elementary reactions, with their rates, that are combined to yield the overall reaction. It is often possible to write several reaction mechanisms, each of which is consistent with a given overall reaction. One of the goals of chemical kinetics is to use the observed rate of a reaction to choose among various conceivable reaction mechanisms.

The gas-phase reaction of nitrogen dioxide with carbon monoxide provides a good example of a reaction mechanism. The generally accepted mechanism at low temperatures has two steps, both bimolecular:

$$\text{NO}_2(g) + \text{NO}_2(g) \longrightarrow \text{NO}_3(g) + \text{NO}(g) \qquad \text{(slow)}$$

$$\text{NO}_3(g) + \text{CO}(g) \longrightarrow \text{NO}_2(g) + \text{CO}_2(g) \qquad \text{(fast)}$$

For any reaction mechanism, combining the steps must give the overall reaction. When each elementary step occurs the same number of times in the course of the reaction, the chemical equations can simply be added. (If one step occurs twice as often as the others, it must be multiplied by 2 before the elementary reactions are added.) In this case, we add the two chemical equations to give

$$2\ \text{NO}_2(g) + \text{NO}_3(g) + \text{CO}(g) \longrightarrow \text{NO}_3(g) + \text{NO}(g) + \text{NO}_2(g) + \text{CO}_2(g)$$

Canceling out the NO_3 and one molecule of NO_2 from each side leads to

$$\text{NO}_2(g) + \text{CO}(g) \longrightarrow \text{NO}(g) + \text{CO}_2(g)$$

A **reaction intermediate** (here, NO_3) is a chemical species that is formed and consumed in the reaction but does not appear in the overall balanced chemical equation. One of the major challenges in chemical kinetics is to identify intermediates, which are often so short-lived that they are difficult to detect directly. Modern experimental methods to study highly reactive intermediates are discussed in Section 18.6.4.

EXAMPLE 18.6

Consider the following reaction mechanism:

$$\text{Cl}_2(g) \longrightarrow 2\ \text{Cl}(g)$$

$$\text{Cl}(g) + \text{CHCl}_3(g) \longrightarrow \text{HCl}(g) + \text{CCl}_3(g)$$

$$\text{CCl}_3(g) + \text{Cl}(g) \longrightarrow \text{CCl}_4(g)$$

(Continued)

(a) What is the molecularity of each elementary step?

(b) Write the overall equation for the reaction.

(c) Identify the reaction intermediate(s).

Solution

(a) The first step is unimolecular, and the other two are bimolecular. (Note the CCl_4 product of the third reaction would have to undergo collisions with an inert molecule to lose some of its energy; it is assumed that those inert collision partners are in high concentrations.)

(b) Adding the three steps gives

$$Cl_2(g) + 2\ Cl(g) + CHCl_3(g) + CCl_3(g) \longrightarrow 2\ Cl(g) + HCl(g) + CCl_3(g) + CCl_4(g)$$

The two species that appear in equal amounts on both sides cancel out to leave

$$Cl_2(g) + CHCl_3(g) \longrightarrow HCl(g) + CCl_4(g)$$

(c) The two reaction intermediates are Cl and CCl_3.

Related Problems: 21, 22

18.3.3 Kinetics and Chemical Equilibrium

There is a direct connection between the rates for the elementary steps in a chemical reaction mechanism and the overall equilibrium constant K. To see this connection, consider the reaction

$$2\ NO(g) + 2\ H_2(g) \rightleftharpoons N_2(g) + 2\ H_2O(g)$$

This reaction is believed to occur in a three-step process involving N_2O_2 and N_2O as intermediates:

$$NO(g) + NO(g) \underset{k_{-1}}{\overset{k_1}{\rightleftharpoons}} N_2O_2(g)$$

$$N_2O_2(g) + H_2(g) \underset{k_{-2}}{\overset{k_2}{\rightleftharpoons}} N_2O(g) + H_2O(g)$$

$$N_2O(g) + H_2(g) \underset{k_{-3}}{\overset{k_3}{\rightleftharpoons}} N_2(g) + H_2O(g)$$

These elementary reactions are shown as equilibria, so the reverse reactions (from products to reactants) are included here as well; k_1, k_2, and k_3 are the rate constants for the forward elementary steps, and k_{-1}, k_{-2}, and k_{-3} are the rate constants for the corresponding reverse reactions.

We now invoke the **principle of detailed balance,** which states that at equilibrium the rate of *each* elementary process is balanced by (equal to) the rate of its reverse process. For the preceding mechanism we conclude that

$$k_1[NO]^2_{eq} = k_{-1}[N_2O_2]_{eq}$$

$$k_2[N_2O_2]_{eq}[H_2]_{eq} = k_{-2}[N_2O]_{eq}[H_2O]_{eq}$$

$$k_3[N_2O]_{eq}[H_2]_{eq} = k_{-3}[N_2]_{eq}[H_2O]_{eq}$$

The equilibrium constants[1] K_1, K_2, and K_3 for the elementary reactions are equal to the ratio of the forward and reverse reaction rate constants:

$$K_1 = \frac{[N_2O_2]_{eq}}{[NO]_{eq}^2} = \frac{k_1}{k_{-1}}$$

$$K_2 = \frac{[N_2O]_{eq}[H_2O]_{eq}}{[N_2O_2]_{eq}[H_2]_{eq}} = \frac{k_2}{k_{-2}}$$

$$K_3 = \frac{[N_2]_{eq}[H_2O]_{eq}}{[N_2O]_{eq}[H_2]_{eq}} = \frac{k_3}{k_{-3}}$$

The steps of the mechanism are now added together to obtain the overall reaction. Recall from Section 14.4 that when reactions are added, their equilibrium constants are multiplied. Therefore, the overall equilibrium constant K is

$$K = K_1K_2K_3 = \frac{k_1k_2k_3}{k_{-1}k_{-2}k_{-3}} = \frac{[N_2O_2]_{eq}[N_2O]_{eq}[H_2O]_{eq}[N_2]_{eq}[H_2O]_{eq}}{[NO]_{eq}^2[N_2O_2]_{eq}[H_2]_{eq}[N_2O]_{eq}[H_2]_{eq}}$$

$$= \frac{[H_2O]_{eq}^2[N_2]_{eq}}{[NO]_{eq}^2[H_2]_{eq}^2}$$

The concentrations of the intermediates N_2O_2 and N_2O cancel out, giving the usual expression of the mass-action law.

This result can be generalized to any reaction mechanism. The product of the forward rate constants for the elementary reactions divided by the product of the reverse rate constants is always equal to the equilibrium constant of the overall reaction. If there are several possible mechanisms for a given reaction (which might involve intermediates other than N_2O_2 and N_2O), their forward and reverse rate constants will all be consistent in this way with the equilibrium constant of the overall reaction.

18.4 Reaction Mechanisms and Rate

In many reaction mechanisms, one step is significantly slower than all the others; this step is called the **rate-determining step**. Because an overall reaction can occur only as fast as its slowest step, that step is crucial in determining the rate of the reaction. This is analogous to the flow of automobile traffic on a highway which has a slowdown at some point. The rate at which cars can complete a trip down the full length of the highway (in cars per minute) is approximately equal to the rate at which they pass through the bottleneck.

If the rate-determining step is the first one, the analysis is particularly simple. An example is the reaction

$$2\ NO_2(g) + F_2(g) \longrightarrow 2\ NO_2F(g)$$

[1]Thermodynamic equilibrium constants are dimensionless because they are expressed in terms of activities rather than partial pressure or concentration. The convention in chemical kinetics is to use concentrations rather than activities, even for gaseous species. Therefore, the equilibrium constants K_1, K_2, and K_3 introduced here are the empirical equilibrium constants K_c described briefly in Section 14.2. These constants are not dimensionless and must be multiplied by the concentration of the reference state, $c_{ref} = RT/P_{ref}$, raised to the appropriate power to be made equal to the thermodynamic equilibrium constant. Nevertheless, to maintain consistency with the conventions of chemical kinetics, such constants as K_1, K_2, and K_3 are referred to as equilibrium constants in this section and are written without the subscript c.

for which the experimental rate law is

$$\text{rate} = k_{\text{obs}}[\text{NO}_2][\text{F}_2]$$

A possible mechanism for the reaction is

$$\text{NO}_2(g) + \text{F}_2(g) \xrightarrow{k_1} \text{NO}_2\text{F}(g) + \text{F}(g) \qquad \text{(slow)}$$

$$\text{NO}_2(g) + \text{F}(g) \xrightarrow{k_2} \text{NO}_2\text{F}(g) \qquad \text{(fast)}$$

The first step is slow and determines the rate, $k_1[\text{NO}_2][\text{F}_2]$, in agreement with the observed rate expression. The subsequent fast step does not affect the reaction rate because fluorine atoms react with NO_2 almost as soon as they are produced.

Mechanisms in which the rate-determining step occurs after one or more fast steps are often signaled by a reaction order greater than 2, by a nonintegral reaction order, or by an inverse concentration dependence on one of the species taking part in the reaction. An example is the reaction

$$2\ \text{NO}(g) + \text{O}_2(g) \longrightarrow 2\ \text{NO}_2(g)$$

for which the experimental rate law is

$$\text{rate} = k_{\text{obs}}[\text{NO}]^2[\text{O}_2]$$

One possible mechanism would be a single-step termolecular reaction of two NO molecules with one O_2 molecule. This would be consistent with the form of the rate expression, but termolecular collisions are rare, and if there is an alternative pathway it is usually followed.

One such alternative is the two-step mechanism

$$\text{NO}(g) + \text{NO}(g) \underset{}{\overset{k_1}{\rightleftharpoons}} \text{N}_2\text{O}_2(g) \qquad \text{(fast equilibrium)}$$

$$\text{N}_2\text{O}_2(g) + \text{O}_2(g) \overset{k_2}{\rightleftharpoons} 2\ \text{NO}_2(g) \qquad \text{(slow)}$$

Because the slow step determines the overall rate, we can write

$$\text{rate} = k_2[\text{N}_2\text{O}_2][\text{O}_2]$$

(Note that the slow step produces two NO_2 molecules as does the net reaction, so the overall rate is correctly expressed.) The concentration of a reactive intermediate such as N_2O_2 cannot be varied at will. Because the N_2O_2 reacts only slowly with O_2, the reverse reaction (to 2 NO) is possible and must be taken into account. In fact, it is reasonable to assume that all of the elementary reactions that occur *before* the rate-determining step are in equilibrium, with the forward and reverse reactions occurring at the same rate. In this case, we have

$$\frac{[\text{N}_2\text{O}_2]}{[\text{NO}]^2} = \frac{k_1}{k_{-1}} = K_1$$

$$[\text{N}_2\text{O}_2] = K_1[\text{NO}]^2$$

$$\text{rate} = k_2K_1[\text{NO}]^2[\text{O}_2]$$

This result is consistent with the observed reaction order, with $k_2K_1 = k_{\text{obs}}$.

EXAMPLE 18.7

In basic aqueous solution the reaction

$$\text{I}^- + \text{OCl}^- \longrightarrow \text{Cl}^- + \text{OI}^-$$

(Continued)

follows a rate law that is consistent with the following mechanism:

$$OCl^-(aq) + H_2O(\ell) \underset{k_{-1}}{\overset{k_1}{\rightleftharpoons}} HOCl(aq) + OH^-(aq) \qquad \text{(fast equilibrium)}$$

$$I^-(aq) + HOCl(aq) \xrightarrow{k_2} HOI(aq) + Cl^-(aq) \qquad \text{(slow)}$$

$$OH^-(aq) + HOI(aq) \xrightarrow{k_3} H_2O(\ell) + OI^-(aq) \qquad \text{(fast)}$$

What rate law is predicted by this mechanism?

Solution

The rate is determined by the slowest elementary step, the second one:

$$\text{rate} = k_2[I^-][HOCl]$$

But, the HOCl is in equilibrium with OCl^- and OH^- due to the first step:

$$\frac{[HOCl][OH^-]}{[OCl^-]} = K_1 = \frac{k_1}{k_{-1}}$$

Solving this for [HOCl] and inserting it into the previous expression gives the prediction

$$\text{rate} = k_2K_1\frac{[I^-][OCl^-]}{[OH^-]}$$

which is, in fact, the experimentally observed rate law.

Related Problems: 25, 26, 27, 28, 29, 30

The rate law of the foregoing example depends on the inverse of the concentration of OH^- ion. Such a form is often a clue that a rapid equilibrium occurs in the first steps of a reaction, preceding the rate-determining step. Fractional orders of reaction provide a similar clue, as in the reaction of H_2 with Br_2 to form HBr,

$$H_2 + Br_2 \longrightarrow 2\ HBr$$

for which the initial reaction rate (before very much HBr builds up) is

$$\text{rate} = k_{\text{obs}}[H_2][Br_2]^{1/2}$$

How can such a fractional power appear? One reaction mechanism that predicts this rate law is

$$Br_2 + M \underset{k_{-1}}{\overset{k_1}{\rightleftharpoons}} Br + Br + M \qquad \text{(fast equilibrium)}$$

$$Br + H_2 \xrightarrow{k_2} HBr + H \qquad \text{(slow)}$$

$$H + Br_2 \xrightarrow{k_3} HBr + Br \qquad \text{(fast)}$$

Here M stands for a second molecule that does not react but that supplies the energy to break up the bromine molecules. For such a mechanism the reaction rate is determined by the slow step:

$$\text{rate} = k_2[Br][H_2]$$

However, [Br] is fixed by the establishment of equilibrium in the first reaction,

$$\frac{[Br]^2}{[Br_2]} = K_1 = \frac{k_1}{k_{-1}}$$

so

$$[Br] = K_1^{1/2}[Br_2]^{1/2}$$

The rate expression predicted by this mechanism is thus

$$\text{rate} = k_2K_1^{1/2}[H_2][Br_2]^{1/2}$$

This is in accord with the observed fractional power in the rate law. On the other hand, the simple bimolecular mechanism

$$H_2 + Br_2 \xrightarrow{k_1} 2\ HBr \qquad \text{(slow)}$$

predicts a rate law:

$$\text{rate} = k_1[H_2][Br_2]$$

This disagrees with the observed rate law, so it can be ruled out as the major contributor to the measured rate.

This discussion shows that deducing a rate law from a proposed mechanism is relatively straightforward, but doing the reverse is much harder. In fact, several competing mechanisms often give rise to the same rate law, and only some independent type of measurement can distinguish between them. A proposed reaction mechanism cannot be proven to be correct if its predictions agree with an experimental rate law, but it can be proven wrong if its predictions disagree with the experimental results.

A classic example is the reaction

$$H_2 + I_2 \longrightarrow 2\ HI$$

for which the observed rate law is

$$\text{rate} = k_{obs}[H_2][I_2]$$

(Contrast this with the rate law already given for the analogous reaction of H_2 with Br_2.) This is one of the earliest and most extensively studied reactions in chemical kinetics, and until 1967 it was widely believed to occur as a one-step elementary reaction. At that time J. H. Sullivan investigated the effect of illuminating the reacting sample with light, which splits some of the I_2 molecules into iodine atoms. If the mechanism we proposed for H_2 and Br_2 is correct here as well, the effect of the light on the reaction should be small because it leads only to a small decrease in the I_2 concentration.

Instead, Sullivan observed a dramatic *increase* in the rate of reaction under illumination, which could be explained only by the participation of iodine *atoms* in the reaction mechanism. One such mechanism is

$$I_2 + M \underset{k_{-1}}{\overset{k_1}{\rightleftharpoons}} I + I + M \qquad \text{(fast equilibrium)}$$

$$H_2 + I + I \xrightarrow{k_2} 2\ HI \qquad \text{(slow)}$$

for which the rate law is

$$\text{rate} = k_2[H_2][I]^2 = k_2K_1[H_2][I_2]$$

This mechanism gives the same rate law that is observed experimentally, and it is consistent with the effect of light on the reaction. The other reaction mechanism also appears to contribute significantly to the overall rate.

This example illustrates the hazards of trying to determine reaction mechanisms from rate laws: several mechanisms can fit any given empirical rate law, and it is always possible that a new piece of information suggesting a different mechanism will be found. A number of experimental methods have been developed to investigate reaction mechanisms. Isotopic labelling has been used for nearly a century to trace the fates of individual atoms as they are converted from reactants into products. Molecular beam methods, discussed in Section 18.6, generate radical intermediates in a vacuum environment and collide them with a chosen reactant to selectively study the intermediate elementary steps in overall reaction mechanisms. Time-resolved spectroscopic methods developed over the past 40 years or so have allowed us to identify transient intermediates present in extremely low concentrations with lifetimes as short as a femtosecond (10^{-15} s). These methods provide great insight into the mechanisms of chemical reactions occurring in environments that include low-density gases (atmospheric chemistry and the chemistry of interstellar space), solutions, the solid state (including geochemistry), and the biology of life processes.

18.4.1 The Steady-State Approximation

In some reaction mechanisms there is no single step that is much slower than the others, so the methods discussed so far cannot predict the rate law. In such cases we use the **steady-state approximation**, which states that the concentrations of reactive intermediates remain nearly constant through most of the reaction.

To illustrate this approximation, let's examine the mechanism proposed by F. A. Lindemann for the dissociation of molecules in the gas phase. A molecule such as N_2O_5 undergoes collisions with neighboring molecules M, where M can stand for another N_2O_5 molecule or for an inert gas such as argon. Through such collisions the N_2O_5 molecule can become excited (or activated) to a state indicated by $N_2O_5^*$:

$$N_2O_5(g) + M(g) \underset{k_{-1}}{\overset{k_1}{\rightleftharpoons}} N_2O_5^*(g) + M(g)$$

The reverse process, with rate constant k_{-1}, is also indicated because the activated molecule can be deactivated by collisions with other molecules. The second step is the unimolecular decomposition of $N_2O_5^*$:

$$N_2O_5^*(g) \xrightarrow{k_2} NO_3(g) + NO_2(g)$$

Subsequent reaction steps to form O_2 and NO_2 from NO_3 occur rapidly and do not affect the measured rate:

$$NO_3(g) + NO_2(g) \xrightarrow{k_3} NO(g) + NO_2(g) + O_2(g) \qquad \text{(fast)}$$

$$NO_3(g) + NO(g) \xrightarrow{k_4} 2\ NO_2(g) \qquad \text{(fast)}$$

The $N_2O_5^*$ is a reactive intermediate; it is produced at a rate $k_1[N_2O_5][M]$ from collisions of N_2O_5 molecules with other molecules and is lost at a rate $k_{-1}[N_2O_5^*][M]$ due to deactivation and at a rate $k_{-2}[N_2O_5^*][M]$ due to dissociation. The net rate of change of $[N_2O_5^*]$ is then

$$\frac{d[N_2O_5^*]}{dt} = k_1[N_2O_5][M] - k_{-1}[N_2O_5^*][M] - k_2[N_2O_5^*]$$

At the beginning of the reaction, $[N_2O_5^*] = 0$, but this concentration builds up after a short time to a small value. The steady-state approximation consists of the assumption that after this short time the rates of production and loss of $N_2O_5^*$ become equal, and

$$\frac{d[N_2O_5^*]}{dt} = 0$$

The steady-state concentration of $[N_2O_5^*]$ persists practically unchanged throughout most of the course of the reaction.

Setting the net rate of change of the $N_2O_5^*$ concentration to 0 gives

$$\frac{d[N_2O_5^*]}{dt} = 0 = k_1[N_2O_5][M] - k_{-1}[N_2O_5^*][M] - k_2[N_2O_5^*]$$

Solving for $[N_2O_5^*]$ gives

$$[N_2O_5^*](k_2 + k_{-1}[M]) = k_1[N_2O_5][M]$$

$$[N_2O_5^*] = \frac{k_1[N_2O_5][M]}{k_2 + k_{-1}[M]}$$

The rate of the overall reaction $N_2O_5 \longrightarrow 2\ NO_2 + \frac{1}{2}O_2$ is

$$\text{rate} = \frac{1}{2}\frac{d[NO_2]}{dt} = k_2[N_2O_5^*] = \frac{k_1k_2[N_2O_5][M]}{k_2 + k_{-1}[M]}$$

This expression has two limiting cases:

1. *Low pressure* When [M] is small enough, $k_2 \gg k_{-1}[M]$ and we can use the approximation

$$\text{rate} = k_1[N_2O_5][M] \qquad \text{(second order)}$$

   This same result would be found by assuming the first step to be rate-determining.

2. *High pressure* When [M] is large enough, $k_{-1}[M] \gg k_2$ and we can use the approximation

$$\text{rate} = \left(\frac{k_1}{k_{-1}}\right) k_2[N_2O_5] = K_1k_1[N_2O_5] \qquad \text{(first order)}$$

   This same result would be found by assuming the second step to be rate-determining.

The steady-state approximation is a more general approach than those considered earlier and produces identical results under the limiting conditions defined above. The steady-state approximation is invoked in many areas of chemical kinetics, an example of which is the kinetics of enzyme-catalyzed reactions discussed in Section 18.8.

18.4.2 Chain Reactions

A **chain reaction** proceeds through a series of elementary steps, some of which are repeated many times. Chain reactions have three stages: (1) **initiation**, in which two or more reactive intermediates are generated; (2) **propagation**, in which products are formed but reactive intermediates are continuously regenerated; and (3) **termination**, in which two intermediates combine to give a stable product.

An example of a chain reaction is the reaction of methane with fluorine to give CH_3F and HF:

$$CH_4(g) + F_2(g) \longrightarrow CH_3F(g) + HF(g)$$

Although in principle this reaction could occur through a one-step bimolecular process, that route turns out to be too slow to contribute significantly under normal reaction conditions. Instead, the mechanism involves a chain reaction of the following type:

$$CH_4(g) + F_2(g) \longrightarrow CH_3(g) + HF(g) + F(g) \qquad \text{(initiation)}$$

$$CH_3(g) + F_2(g) \longrightarrow CH_3F(g) + F(g) \qquad \text{(propagation)}$$

$$CH_4(g) + F(g) \longrightarrow CH_3(g) + HF(g) \qquad \text{(propagation)}$$

$$CH_3(g) + F(g) + M(g) \longrightarrow CH_3F(g) + M(g) \qquad \text{(termination)}$$

In the initiation step, two reactive intermediates (CH_3 and F) are produced. During the propagation steps, these intermediates are not used up while reactants (CH_4 and F_2) are being converted to products (CH_3F and HF). The propagation steps can be repeated again and again, until eventually two reactive intermediates come together in a termination step. As we see in Chapter 23, chain reactions are important in building up long-chain molecules called polymers.

The chain reaction just considered proceeds at a constant rate, because each propagation step both uses up and produces a reactive intermediate. (Note that the concentrations of the CH_4 and F_2 reactants are large compared to the concentrations of the reactive intermediates CH_3 and F.) The concentrations of the reactive intermediates remain approximately constant and are determined by the rates of chain initiation and termination. Another type of chain reaction is possible in which the number of reactive intermediates increases during one or more propagation steps. This is called a **branching chain reaction.** An example is the reaction of oxygen with hydrogen. The mechanism is complex and can be initiated in various ways, leading to the formation of several reactive intermediates such as O, H, and OH. Some propagation steps are of the type already seen for CH_4 and F_2, such as

$$OH(g) + H_2(g) \longrightarrow H_2O(g) + H(g)$$

in which one reactive intermediate (OH) is used up and one (H) is produced. Other propagation steps are branching:

$$H(g) + O_2(g) \longrightarrow OH(g) + O(g)$$

$$O(g) + H_2(g) \longrightarrow OH(g) + H(g)$$

In these steps, each reactive intermediate used up causes the generation of two others. This leads to rapid growth in the number of reactive species, speeding the rate further and possibly causing an explosion. Branching chain reactions are critical in the fission of uranium (see Chapter 19).

18.5 Effect of Temperature on Reaction Rates

The first four sections of this chapter describe the experimental determination of rate laws and their relation to assumed mechanisms for chemical reactions. Now we have to find out what determines the actual magnitudes of rate constants (either for elementary reactions or for overall rates of multistep reactions), and how temperature affects reaction rates. To consider these matters, it is necessary to connect molecular collision rates to the rates of chemical reactions. We limit the discussion to gas-phase reactions, for which the kinetic theory of Chapter 9 is applicable.

18.5.1 Gas-Phase Reaction Rate Constants

In Section 9.7, we applied the kinetic theory of gases to estimate the frequency of collisions between a particular molecule and other molecules in a gas. In

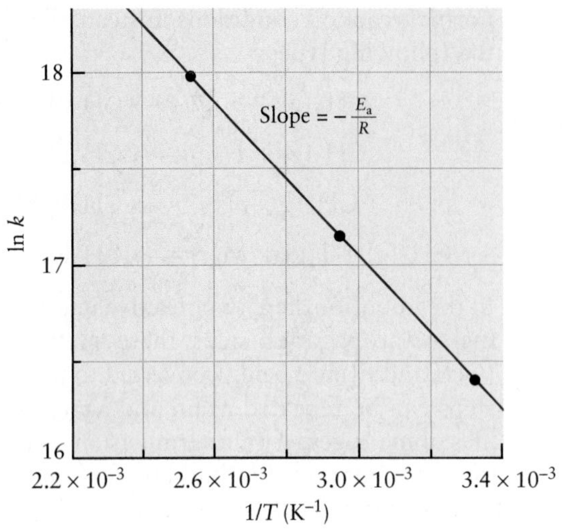

FIGURE 18.9 An Arrhenius plot of ln *k* against 1/*T* for the reaction of benzene vapor with oxygen atoms. An extrapolation to $1/T = 0$ gives the constant ln *A* from the intercept of this line.

Example 9.12, we calculated this frequency to be $4.1 \times 10^9\ s^{-1}$ under room conditions for a typical small molecule such as oxygen. If every collision led to reaction, the reaction would be practically complete in about 10^{-9} s. Some reactions do proceed at rates almost this high. An example is the bimolecular reaction between two CH_3 radicals to give ethane, C_2H_6,

$$2\ CH_3 \longrightarrow C_2H_6$$

for which the observed rate constant is $1 \times 10^{10}\ L\ mol^{-1}\ s^{-1}$. For initial pressure of CH_3 near 1 atm at 25°C, the concentration initially is about 0.04 M. The second-order integrated rate law from Section 18.2 predicts that the concentration would drop to 0.02 M after a period of 10^{-9} s. But, reaction rates that are much lower—by factors of 10^{12} or more—are common. The naive idea that "to collide is to react" clearly must be modified if we are to understand these lower rates.

We find a clue in the observed temperature dependence of reaction rate constants. The rates of many reactions increase extremely rapidly as temperature increases; typically a 10°C rise in temperature may double the rate. In 1889 Svante Arrhenius suggested that rate constants vary exponentially with inverse temperature,

$$k = Ae^{-E_a/RT} \qquad [18.5]$$

where E_a is a constant with dimensions of energy and A is a constant with the same dimensions as k. Taking the natural logarithm of this equation gives

$$\ln k = \ln A - \frac{E_a}{RT} \qquad [18.6]$$

So a plot of ln k against $1/T$ should be a straight line with slope $-E_a/R$ and intercept ln A. Many rate constants do show just this kind of temperature dependence (Fig. 18.9).

EXAMPLE 18.8

The decomposition of hydroxylamine (NH_2OH) in the presence of oxygen follows the rate law

$$-\frac{d[NH_2OH]}{dt} = k_{obs}[NH_2OH][O_2]$$

where k_{obs} is 0.237×10^{-4} L mol^{-1} s^{-1} at 0°C and 2.64×10^{-4} L mol^{-1} s^{-1} at 25°C. Calculate E_a and the factor A for this reaction.

Solution

Let us write the Arrhenius equation at two different temperatures T_1 and T_2:

$$\ln k_1 = \ln A - \frac{E_a}{RT_1} \quad \text{and} \quad \ln k_2 = \ln A - \frac{E_a}{RT_2}$$

If the first equation is subtracted from the second, the term ln A cancels out, leaving

$$\ln k_2 - \ln k_1 = \ln \frac{k_2}{k_1} = -\frac{E_a}{R}\left(\frac{1}{T_2} - \frac{1}{T_1}\right)$$

which can be solved for E_a. In the present case, T_1 = 273 K and T_2 = 298 K; therefore,

$$\ln \frac{2.64 \times 10^{-4}}{0.237 \times 10^{-4}} = \frac{-E_a}{8.315 \text{ J K}^{-1} \text{ mol}^{-1}}\left(\frac{1}{298 \text{ K}} - \frac{1}{273 \text{ K}}\right)$$

$$2.410 = \frac{E_a}{8.315 \text{ J K}^{-1} \text{ mol}^{-1}}(3.07 \times 10^{-4} \text{ K}^{-1})$$

$$E_a = 6.52 \times 10^4 \text{ J mol}^{-1} = 65.2 \text{ kJ mol}^{-1}$$

Now that E_a is known, the constant A can be calculated by using data at either temperature. At 273 K,

$$\ln A = \ln k_1 + \frac{E_a}{RT}$$

$$= \ln(0.237 \times 10^{-4}) + \frac{6.52 \times 10^4 \text{ J mol}^{-1}}{(8.315 \text{ J K}^{-1} \text{ mol}^{-1})(273 \text{ K})}$$

$$= -10.65 + 28.73 = 18.08$$

$$A = e^{18.08} = 7.1 \times 10^7 \text{ L mol}^{-1} \text{ s}^{-1}$$

We could determine E_a and A more accurately from measurements at a series of temperatures and a least-squares fit to a plot such as that in Figure 18.9.

Related Problems: 35, 36

Arrhenius believed that for molecules to react upon collision they must become "activated," so the parameter E_a came to be known as the **activation energy.** His ideas were refined by later scientists. In 1915 A. Marcelin pointed out that, while molecules make many collisions, not all collisions are reactive. Only those collisions for which the collision energy (i.e., the relative translational kinetic energy, $(1/2)\mu v_{rel}^2$, of the colliding molecules) exceeds some critical energy result in reaction. Thus, Marcelin gave a dynamic interpretation for the activation energy inferred from reaction rates.

The strong temperature dependence of rate constants, described by the Arrhenius law, is explained by the Maxwell–Boltzmann distribution of relative

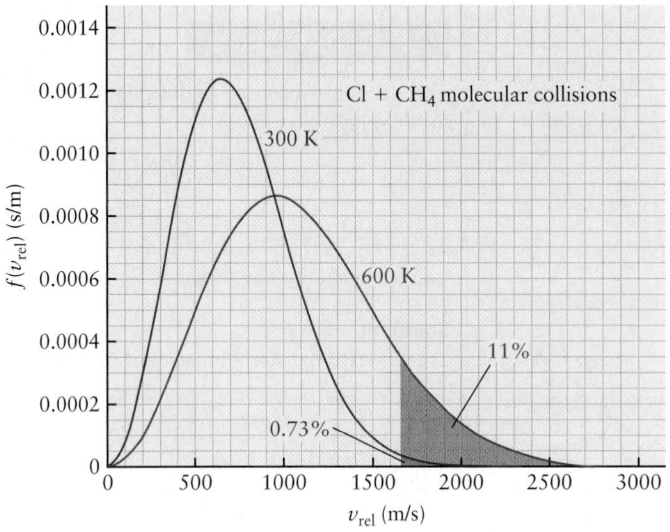

FIGURE 18.10 The Maxwell–Boltzmann distribution of relative speeds for $^{35}Cl + CH_4$ collisions at 300 K and 600 K. The fraction of collisions that have a relative kinetic energy greater than E_a is shown by the shaded areas. Doubling the temperature increases the fraction of collisions with the needed relative kinetic energy by a factor of 15, from 0.73% to 11%. This figure shows only translational energy; internal vibrational and rotational energy also promote reactions.

molecular speeds, v_{rel} (see Section 18.6). If E_a is roughly the critical relative collision energy required for a pair of molecules to react, only a small fraction of the collisions between molecules will have at least this much energy at moderate temperature. Figure 18.10 plots the Maxwell–Boltzmann distribution of the *relative* speeds between ^{35}Cl atoms and methane molecules at two temperatures. The activation energy E_a of the $Cl + CH_4 \longrightarrow HCl + CH_3$ reaction is about 15 kJ mol^{-1}. For collisions with $v_{rel} > (2\,E_a/\mu)^{1/2} = 1.65 \times 10^3$ m s^{-1} the relative kinetic energy is larger than 15 kJ mol^{-1}. The fraction of collisions with large enough v_{rel} thus corresponds to the area under the Maxwell–Boltzmann distribution curve between $v_{rel} = 1.65 \times 10^3$ m s^{-1} and ∞. As the temperature increases, the distribution function spreads out to include higher energies. The fraction of collisions having more than the critical energy E_a increases in agreement with the increase in rate calculated with Arrhenius's law and measured in experiments. The reaction rate is then proportional to $\exp(-E_a/RT)$. So, both the strong temperature dependence and the order of magnitude of the experimental rate constants are explained by the kinetic theory of gases.

18.5.2 The Reaction Coordinate and the Activated Complex

Why should there be a critical collision energy for reaction to occur between two molecules, and why is this energy similar to E_a, the activation energy that you calculate from a plot of ln k versus $1/T$? To understand the first of these questions, let's consider the physical analogy of marbles rolling on a hilly surface. As a marble rolls up a hill, its potential energy increases and its kinetic energy decreases; it slows down as it climbs the hill. If it can reach the top of the hill, it will fall down the other side, whereupon its kinetic energy will increase and its potential energy will decrease. Not every marble will make it over the hill. If its initial speed, and therefore kinetic energy, is too small, a marble will roll only part way up and then fall back down. Only those marbles with initial kinetic energy higher than some critical threshold will pass over the hill.

We can use this physical model to describe molecular collisions and reactions. As two reactant molecules, atoms, or ions approach each other along a **reaction path,** their potential energy increases as the bonds within them distort. At some maximum potential energy the collision partners become connected in an unstable entity called the **activated complex** or *transition state*. The activated complex is the cross-over stage where the smooth ascent in potential energy as the reactants come together becomes a smooth descent as the product molecules separate. As in the case of the marbles, not all pairs of colliding bodies react.

Only those pairs with sufficient relative kinetic energy can stretch bonds and rearrange atoms enough to become the transition state through which reactants become products. If the energy ε_B (the height of the potential energy hill) of the barrier to reach the transition state is too high, almost all colliding pairs of reactant molecules separate from each other without reacting. The height ε_B of the barrier is close to the measured activation energy for the reaction.

Figure 18.11 shows a graph of the potential energy versus position along the reaction path for the reaction

$$NO_2(g) + CO(g) \longrightarrow NO(g) + CO_2(g)$$

Two barrier energies are shown in Figure 18.11: $\varepsilon_{B,f}$ is the barrier energy for the forward reaction, and is approximately equal to activation energy $E_{a,f}$ for the forward reaction, and $\varepsilon_{B,r}$ is the barrier for the reverse reaction, in which NO takes an oxygen atom from CO_2 to form NO_2 and CO. It is approximately equal to the activation energy $E_{a,r}$ for the reverse reaction. The difference between the two is ΔU, the change in internal energy of the chemical reaction:

$$\Delta U = \varepsilon_{B,f} - \varepsilon_{B,r} \approx E_{a,f} - E_{a,r}$$

Whereas ΔU is a thermodynamic quantity that can be obtained from calorimetric measurements or calculated with the electronic structure methods described in Chapter 6, $E_{a,f}$ and $E_{a,r}$ must be found from the temperature dependence of the rate constants for the forward and reverse reactions. In this reaction the forward and reverse activation energies are roughly 132 and 358 kJ mol^{-1}, respectively, and ΔU from thermodynamics is -226 kJ mol^{-1}.

The activation energy for an elementary reaction in the gas phase is always positive—although in some cases it can be quite small—because there is always some energy barrier to surmount. Rates of elementary reactions therefore increase with increasing temperature. This is not necessarily true for rates of overall reactions consisting of more than one elementary reaction. These sometimes have "negative activation energies," which means that the overall reaction rate is slower at higher temperature. How can this be? Let's examine a specific example: the reaction of NO with oxygen

$$2\ NO(g) + O_2(g) \longrightarrow 2\ NO_2(g)$$

has the observed rate law

$$\text{rate} = k_{obs}[NO]^2[O_2]$$

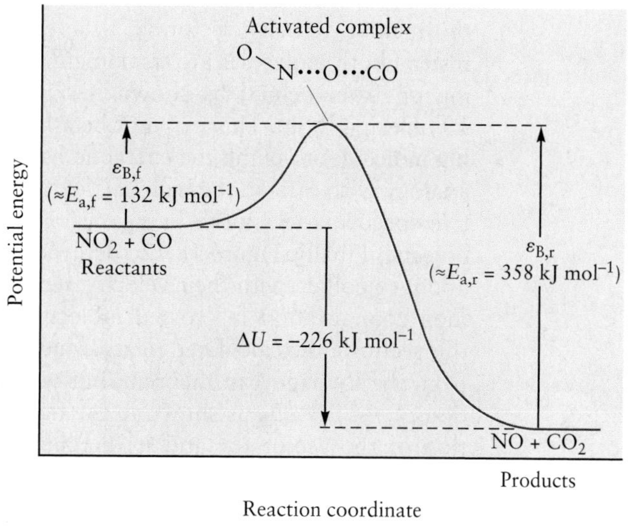

FIGURE 18.11 The energy profile along the reaction coordinate for the reaction $NO_2 + CO \longrightarrow NO + CO_2$. This direct reaction dominates the kinetics at high temperatures (above about 500 K).

where k_{obs} *decreases* with increasing temperature. In Section 18.4 we accounted for this rate expression with a two-step mechanism. The first step is a rapid equilibrium (with equilibrium constant K_1) between two NO molecules and their dimer, N_2O_2. The second step is the slow reaction (with rate constant k_2) of N_2O_2 with O_2 to form products. The overall rate constant is therefore the product of k_2 and K_1. Whereas k_2 is the rate constant for an elementary reaction, and so increases with increasing temperature, K_1 is an *equilibrium* constant and may decrease as temperature increases. Provided the reaction is sufficiently exothermic (as it is in this case), K_1 will decrease so rapidly with increasing temperature that the product k_2K_1 will decrease as well. This combination of effects explains the observation of "negative activation energies" in some overall chemical reactions.

18.6 Molecular Theories of Elementary Reactions

The kinetic theory of gases, introduced in Section 9.5, is the starting point for the development of microscopic theories of reaction rates, a field of study that is called **reaction dynamics**. We introduce you to several different theories and experimental approaches to the study of reaction dynamics in this section to provide you with insight into the nature of chemical reactions at the molecular level. These include collision theory, built upon the kinetic theory of gases; transition state theory, which is based in statistical mechanics; and the experimental technique that uses molecular beams to probe the details of reactive collisions between single pairs of molecules.

18.6.1 Collision Theory

Collision theory provides a method for calculating rate constants for bimolecular reactions based upon the properties of the reacting molecules that influence the probabilities of reactive encounters; these include size, relative kinetic energy, and relative orientation.

Let's begin by giving you a feel for why it is the relative kinetic energy, $\varepsilon_{rel} = \frac{1}{2}\mu v_{rel}^2$, of two colliding molecules, calculated from their relative speed v_{rel}, that matters in a collision. Figure 18.12 plots three examples of collisions. In all of them the speed of molecule 2 (blue) is 40 miles per hour and the speed of molecule 1 (red) is 30 miles per hour, but the relative speed is very different in each. In Figure 18.12a, molecule 2 is traveling in the $+x$ direction and molecule 1 is traveling in the $-x$ direction, so the relative speed (the magnitude of the relative velocity vector $\vec{v}_{rel}$ where $\vec{v}_{rel} = \vec{v}_2 - \vec{v}_1$) is 70 miles per hour. If instead both molecules are traveling in the $+x$ direction, with molecule 2 catching up from behind as shown in Figure 18.12b, the relative speed is only 10 miles per hour. This may not be a large enough relative speed for the colliding molecules to climb the energetic barrier en route to the transition state. (An analogy with cars on the highway is frightening, but makes it clear that the relative speed matters more in a collision than the speeds of the two individual objects.) Finally, Figure 18.12c depicts a situation where the two molecules are about to collide with their velocity vectors perpendicular to each other (a common configuration in crossed molecular beam experiments discussed later in this section); the calculated relative speed is 50 miles per hour (this is easy to see from the Pythagorean theorem, but you should calculate this using the definition $\vec{v}_{rel} = \vec{v}_2 - \vec{v}_1$, as shown in the figure caption so you get used to the direction of the vector $\vec{v}_{rel}$ and its magnitude, the relative speed v_{rel}). Thus, in a gaseous mixture of two types of molecules that collide with each other, and potentially react with each collision, the collisions with the higher relative

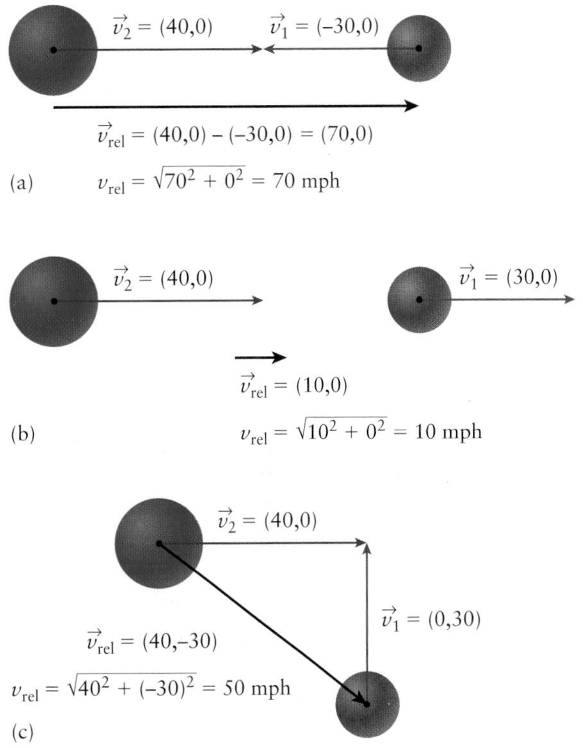

FIGURE 18.12 Calculating the relative velocity and the relative speed for three collisions. We use $\vec{v}_{rel} = \vec{v}_2 - \vec{v}_1$ to calculate the relative velocity $\vec{v}_{rel}$ (see Appendix C.7 for help with vectors), and $v_{rel} = \sqrt{(v_{rel,x})^2 + (v_{rel,y})^2}$, by the Pythagorean theorem, to calculate the relative speed. You may determine $\vec{v}_{rel}$ by graphically subtracting $\vec{v}_1$ from $\vec{v}_2$.

speeds that are more likely to result in a reaction where bonds break and other bonds form. This is the reason why in Figure 18.10 we plotted the Maxwell-Boltzmann distribution of *relative* speeds between the colliding Cl atoms and CH_4 molecules. The Maxwell-Boltzmann distribution of relative speeds, in a bulk mixture of the two gases at temperature T, is

$$f(v_{rel}) = 4\pi\left(\frac{\mu}{2\pi k_B T}\right)^{\frac{3}{2}} v_{rel}^2 \exp(-\mu v_{rel}^2/2k_B T) \qquad [18.7]$$

This looks very similar to the Maxwell-Boltzmann speed distribution in Equation 9.19 for an individual molecule, but the mass m of the molecule is replaced in Equation 18.7 by the reduced mass μ of the colliding molecules, where $\mu = m_1m_2/(m_1 + m_2)$ and the speed v is now the relative speed v_{rel}. Because of this similarity, the average relative speed $\overline{v_{rel}}$ is given by Equation 9.20, but with m replaced by the reduced mass μ of the colliding molecules.

EXAMPLE 18.9

A ^{35}Cl atom with velocity $\vec{v}_1 = (v_{1x}, v_{1y}) = (300\ \text{m s}^{-1}, -400\ \text{m s}^{-1})$ collides with a methane molecule with velocity $\vec{v}_2 = (v_{2x}, v_{2y}) = (-500\ \text{m s}^{-1}, 300\ \text{m s}^{-1})$.

(a) What is the speed of the Cl atom and the speed of the CH_4 molecule?

(b) What is the relative speed of the two colliding molecules?

(c) What is the relative kinetic energy in this collision?

(d) Would this relative kinetic energy alone be enough to surmount the barrier to the reaction Cl + $CH_4 \longrightarrow$ HCl + CH_3 if that barrier were 15 kJ mol^{-1}? (Assume that the collision is not a glancing one, so that all of the relative kinetic energy could be converted to potential energy.)

Solution

(a) The speed of the Cl atom = $\sqrt{300^2 + (-400)^2}$ = 500 m s^{-1} and the speed of the methane molecule = $\sqrt{(-500)^2 + 300^2}$ = 583 m s^{-1}.

(b) The relative speed is the magnitude of the relative velocity vector. $\vec{v}_{rel} = \vec{v}_2 - \vec{v}_1 = (-500,300) - (300,-400) = (-800,700)$ in units of m s^{-1}. This gives a relative speed of $v_{rel} = \sqrt{(-800)^2 + 700^2}$ = 1063 m s^{-1}.

(Continued)

(c) The relative kinetic energy is

$$\varepsilon_{rel} = \frac{1}{2}\mu v_{rel}^2$$

$$= \frac{1}{2}\left(\frac{(35)(16)}{51} \text{ g mol}^{-1}\right)\left(\frac{\text{kg}}{1000 \text{ g}}\right)(1063 \text{ m s}^{-1})^2$$

$$= \frac{1}{2}(0.011 \text{ kg mol}^{-1})(1063 \text{ m s}^{-1})^2$$

$$= 6203 \text{ J mol}^{-1} = 6.2 \text{ kJ mol}^{-1}$$

(d) No, this does not provide enough energy to surmount the barrier, even in the best case of it not being a glancing collision.

Related Problems: 43, 44

Now let's consider a very simple type of reaction, one in which there is no barrier, so all collisions have enough relative kinetic energy to surmount the barrier to the reaction. (For example, the reaction between two radicals is often nearly barrier-less.) One would conclude that if the two molecules collide, they react, so the reaction rate should be equal to the collision rate that we discussed in Section 9.7 and is depicted again in Figure 18.13. Specifically, the number of collisions per second per m^3 of volume between molecules of type 1 (say Cl atoms) and molecules of type 2 (say CH_4 molecules) is:

$$\frac{\text{collisions of type 1 molecules with type 2}}{\text{m}^3 \text{ s}} = \pi d^2 \overline{v_{rel}} \frac{N_1}{V}\frac{N_2}{V} \quad [18.8]$$

where d is the sum of the radii of the two colliding molecules, or the average of their diameters, as shown in Figure 18.13. The expression looks very similar to the expression for the rate of the A + B $\longrightarrow$ C + D bimolecular elementary reaction

$$\text{rate (in mol L}^{-1}\text{ s}^{-1}) = \frac{d[\text{C}]}{dt} = k[\text{A}][\text{B}] \quad [18.9]$$

with the time t in seconds, except the concentrations in Equation 18.8 are expressed in molecules per m^3 and the concentrations in Equation 18.9 are expressed in moles per liter. The rate calculated in Equation 18.8 is called the *gas kinetic collision rate*. If a reaction proceeds with that fast rate, one concludes that the barrier to the reaction is small and that there are no other restrictions, such as the relative orientation of the two colliding molecules, that would prevent a collision from resulting in a reaction. Thus, to within-unit conversions, we can immediately deduce the rate constant for a reaction with no barrier that

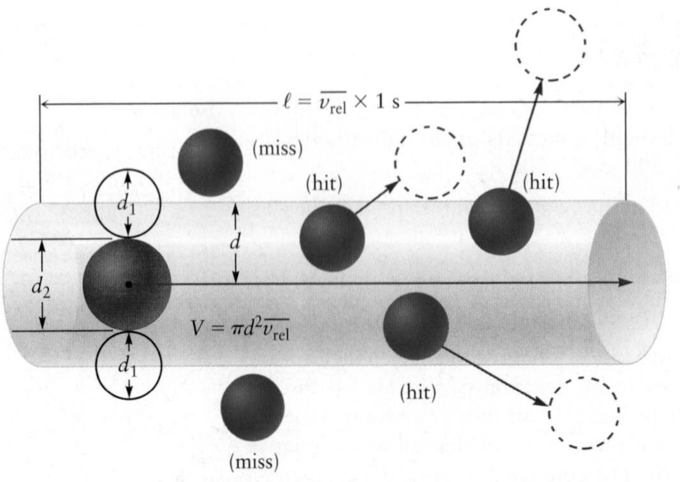

FIGURE 18.13 Molecule A (red) sweeps out a cylinder of volume $\pi d^2 \overline{v_{rel}}$ per second, where $d = \frac{(d_1 + d_2)}{2}$. It will collide with all of the B molecules (blue) whose centers lie within the cylinder.

evidences a rate consistent with the gas kinetic collision rate. It is, comparing Equations 18.8 and 18.9, $k = \pi d^2 \overline{v_{rel}}$. Figure 18.13 shows that the effective cross-sectional area that the two colliding molecules present to each other is πd^2. The symbol σ_c is used for this hard-sphere **collision cross section**, so $\sigma_c = \pi d^2$. Thus, for a reaction without a barrier, the hard sphere collision rate constant $k = \sigma_c \overline{v_{rel}}$.

EXAMPLE 18.10

The measured rate constant for the reaction of ^{35}Cl atoms with methane at 300 K is 1.0×10^{-13} cm^3 s^{-1}.

(a) If the concentration of ^{35}Cl atoms were 3×10^5 cm^{-3} (think of this as molecules per cm^3, but don't carry the molecules with your units) and the concentration of methane were 9×10^5 cm^{-3}, calculate the rate of the reaction in mol L^{-1} s^{-1}.

(b) Compare your result with the gas kinetic collision rate of ^{35}Cl atoms with methane at this temperature, taking the hard sphere diameter of a methane molecule to be the Lennard-Jones parameter $\sigma = 3.82$ Å in Table 9.4, and approximating the hard sphere radius of a Cl atom by its van der Waals radius of 175 pm. (This is larger, as expected, than the atomic radius for Cl in Appendix F, as the latter is calculated by halving the equilibrium bond length R_e for two covalently bound Cl atoms.)

(c) Calculate the gas kinetic rate constant in L mol^{-1} s^{-1} and compare it to the measured rate constant. Discuss the difference.

Solution

(a) rate $= k[\text{Cl}][\text{CH}_4]$

$$= (1 \times 10^{-13}\ \text{cm}^3\ \text{s}^{-1})\left(\frac{\text{L}}{1000\ \text{cm}^3}\right)\left(\frac{6.022 \times 10^{23}}{\text{mol}}\right)[\text{Cl}][\text{CH}_4]$$

$$= (6.022 \times 10^7\ \text{L mol}^{-1}\ \text{s}^{-1})[\text{Cl}][\text{CH}_4]$$

$$= (6.022 \times 10^7\ \text{L mol}^{-1}\ \text{s}^{-1})\left(\frac{3 \times 10^8\ \text{L}^{-1}}{6.022 \times 10^{23}\ \text{mol}^{-1}}\right)\left(\frac{9 \times 10^8\ \text{L}^{-1}}{6.022 \times 10^{23}\ \text{mol}^{-1}}\right)$$

$$= 4.5 \times 10^{-23}\ \text{mol L}^{-1}\ \text{s}^{-1}$$

(b) To calculate the gas kinetic collision rate, first calculate the average relative speed between a Cl atom and a CH_4 molecule at 300 K.

$$\overline{v_{rel}} = \sqrt{\frac{8k_BT}{\pi\mu}}$$

$$= \sqrt{\frac{8(1.38 \times 10^{-23}\ \text{J K}^{-1})(300\ \text{K})}{\pi(0.011\ \text{kg mol}^{-1})\left(\frac{\text{mol}}{6.022 \times 10^{23}}\right)}}$$

$$= 759.7\ \text{m s}^{-1}$$

Using Equation 18.8, the number of collisions per second per m^3 between Cl and methane is

$$\frac{\text{collisions of Cl with CH}_4}{\text{m}^3\ \text{s}} = \pi(3.66 \times 10^{-10}\text{m}^3)^2(759.7\ \text{m s}^{-1})(3 \times 10^{11}\ \text{m}^{-3})(9 \times 10^{11}\ \text{m}^{-3})$$

$$= 8.63 \times 10^7\ \text{m}^{-3}\ \text{s}^{-1}$$

(Continued)

This gives a gas kinetic collision rate, in units of mol L^{-1} s^{-1} equal to

$$(8.63 \times 10^7 \text{ m}^{-3} \text{ s}^{-1})\left(\frac{\text{m}^3}{1000 \text{ L}}\right)\left(\frac{\text{mol}}{6.022 \times 10^{23}}\right) = 1.43 \times 10^{-19} \text{ mol L}^{-1} \text{ s}^{-1}$$

(c) The gas kinetic collision rate is roughly a factor of 3000 larger than the measured rate of reaction. There must be a barrier to this reaction. We further show in the following derivation that if the collision is a glancing one, the relative kinetic energy must be *much larger than* the barrier height to convert enough of the relative kinetic energy to potential energy on impact.

So how do we get an expression for the rate constant when there is a barrier to the reaction? We can make a simple model similar to that above, but incorporate one important new feature—that if a collision is a glancing one not all of the relative kinetic energy can be converted to potential energy when the reactant molecules collide. If you imagine one of the collisions depicted in Figure 18.13 where the red molecule barely collides with one of the blue molecules near the edge of the collision volume, such a collision will not be able to convert much of the relative kinetic energy to potential energy on impact. We consider this carefully in the next few pages. Depending on the level of your introductory chemistry class, your instructor may suggest you skip to the paragraph immediately after Equation 18.12 and study the concept of an impact parameter for a collision presented in Figure 18.14. Thus, you can easily understand the quantitative expressions in Equations 18.13-18.18, then skip again to the final result in Equation 18.19 and the applications presented after that. Key steps in the derivation include:

1. Defining an energy-dependent **reactive cross section** $\sigma_r(\varepsilon_{rel})$ to account for the fact that only those collisions that meet a specific energy criterion will lead to reaction. Reactive cross sections are generally smaller than hard sphere collision cross sections because not every collision is effective for reaction.
2. Asserting that there is an energy threshold that determines whether a particular collision leads to reaction or not; those collisions with relative energies above this threshold are reactive and those with relative energies below this threshold are not. We find the threshold energy and determine the functional dependence of $\sigma_r(\varepsilon_{rel})$ on ε_{rel} by analyzing the portion of the relative kinetic energy from the component of $\vec{v}_{rel}$ along the line of centers when the colliding molecules, modeled as hard spheres, impact each other.

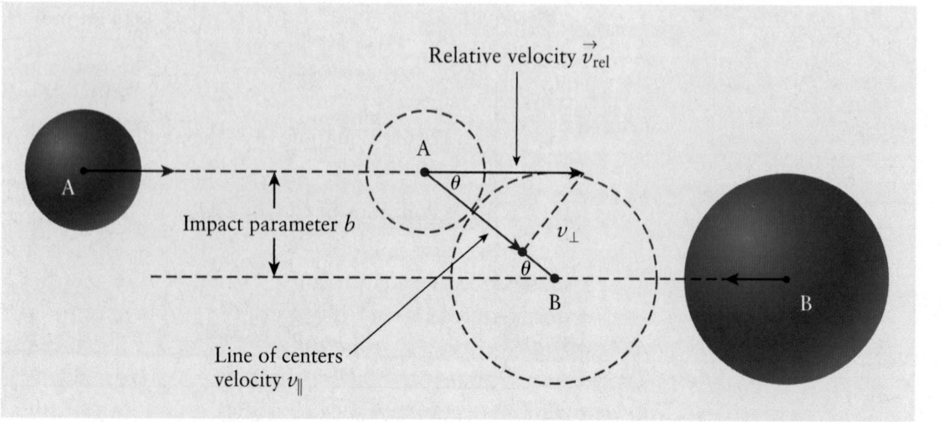

FIGURE 18.14 Molecule A is moving towards molecule B with relative velocity vector v_{rel} shown. The distance between the parallel lines is called the impact parameter b. (**Note:** In the diagram, the molecular mass of reactant A is assumed to be half of the molecular mass of reactant B.)

3. Recognizing that there is a distribution of relative kinetic energies given by the Maxwell–Boltzmann distribution $f(\varepsilon_{rel})$ and integrating the energy-dependent reaction probability over all possible relative kinetic energies weighted by that distribution.

We begin with the bimolecular rate equation introduced above and indicate our intention to integrate over the energy-dependent factors as follows. Showing that the rate is a function of ε_{rel} gives

$$\text{rate}(\varepsilon_{rel}) = k_r\left(\frac{N_1}{V}\right)\left(\frac{N_2}{V}\right) = \sigma_r(\varepsilon_{rel})v_{rel}\left(\frac{N_1}{V}\right)\left(\frac{N_2}{V}\right) \quad [18.10]$$

$$= \sigma_r(\varepsilon_{rel})\left(\frac{2\varepsilon_{rel}}{\mu}\right)^{1/2}\left(\frac{N_1}{V}\right)\left(\frac{N_2}{V}\right)$$

$$\text{total rate} = \left(\int_0^\infty \sigma_r(\varepsilon_{rel})\left(\frac{2\varepsilon_{rel}}{\mu}\right)^{1/2} f(\varepsilon_{rel})\,d\varepsilon_{rel}\right)\left(\frac{N_1}{V}\right)\left(\frac{N_2}{V}\right)$$

from which we see that the rate constant is given by

$$k_r = \int_0^\infty \sigma_r(\varepsilon_{rel})\left(\frac{2\varepsilon_{rel}}{\mu}\right)^{1/2} f(\varepsilon_{rel})\,d\varepsilon_{rel} \quad [18.11]$$

The next step is to relate $f(\varepsilon_{rel})d\varepsilon_{rel}$ to $f(v_{rel})dv_{rel}$ where $f(v_{rel})$ is the Maxwell–Boltzmann distribution of relative speeds of the two colliding reactant molecules given in Equation 18.7. We make the following substitutions into Equation 18.7

$$\varepsilon_{rel} = \tfrac{1}{2}\mu v_{rel}^2,\quad v_{rel}^2 = \left(\frac{2\varepsilon_{rel}}{\mu}\right),\quad dv_{rel} = d\varepsilon_{rel}/(2\mu\varepsilon_{rel})^{1/2}$$

which allows us to relate $f(v_{rel})dv_{rel}$, and $f(\varepsilon_{rel})d\varepsilon_{rel}$ as follows.

$$f(v_{rel})dv_{rel} = 4\pi\left(\frac{\mu}{2\pi k_B T}\right)^{3/2}\frac{2\varepsilon_{rel}}{\mu}\exp(-\varepsilon_{rel}/k_B T)\frac{d\varepsilon_{rel}}{(2\mu\varepsilon_{rel})^{1/2}}$$

$$= 2\pi\left(\frac{1}{\pi k_B T}\right)^{3/2}\varepsilon_{rel}^{1/2}\exp(-\varepsilon_{rel}/k_B T)d\varepsilon_{rel} = f(\varepsilon_{rel})d\varepsilon_{rel}$$

Substituting this result into the integral in Equation 18.11 gives

$$k_r = 2\pi\left(\frac{1}{\pi k_B T}\right)^{3/2}\int_0^\infty \sigma_r(\varepsilon_{rel})\left(\frac{2\varepsilon_{rel}}{\mu}\right)^{1/2}\varepsilon_{rel}^{1/2}\exp(-\varepsilon_{rel}/k_B T)\,d\varepsilon_{rel}$$

$$= \left(\frac{8}{\pi\mu k_B T}\right)^{1/2}\left(\frac{1}{k_B T}\right)\int_0^\infty \varepsilon_{rel}\sigma_r(\varepsilon_{rel})\exp(-\varepsilon_{rel}/k_B T)\,d\varepsilon_{rel} \quad [18.12]$$

We develop a model to determine the functional dependence of σ_r on ε_{rel} by asserting that only the component of the relative velocity directed along the line of centers $v_{\parallel}$ is effective in activating a reaction. Figure 18.14 shows the molecules A and B represented as hard spheres separated by the distance d when they collide. The velocity of molecule A, $\vec{v}_2$, relative to the velocity of molecule B, $\vec{v}_1$, is the vector $\vec{v}_{rel}$. We draw a line parallel to $\vec{v}_{rel}$ that goes through the center of B; the distance b between the two parallel lines is called the **impact parameter**. We find the component of the velocity directed along the line of centers, $v_{\parallel}$, using the geometric construction illustrated in Figure 18.14. $v_{\parallel} = v_{rel}\cos\theta$. The angle θ is calculated using the particular collision's impact parameter b and the distance d, where $\sin\theta = b/d$.

Thus, the total relative kinetic energy ε_{rel} may be written as the sum of two parts:

$$\varepsilon_{rel} = (1/2)\mu v_{rel}^2 = (1/2)\mu v_{\|}^2 + (1/2)\mu v_{\perp}^2 \quad [18.13]$$

We are modeling the colliding molecules as hard spheres, so when they collide at a distance d between them $v_{\|}$ must go to zero; the kinetic energy from this component of the velocity, $(1/2)\mu v_{\|}^2$, is converted to potential energy to surmount the barrier to the chemical reaction. Thus a collision with impact parameter $b = 0$ can convert all of the relative kinetic energy to potential energy on impact, while a collision with a larger impact parameter b (a glancing collision) converts a smaller fraction of the relative kinetic energy to potential energy on impact. Using Equation 18.13 and letting $(1/2)\mu v_{\|}^2$ be converted to potential energy on impact, $V_{max}(d)$, gives

$$(1/2)\mu v_{rel}^2 = V_{max}(d) + (1/2)\mu v_{\perp}^2 \quad [18.14]$$

Substituting $v_{\perp} = v_{rel}\sin\theta = v_{rel}(b/d)$, then solving for the maximum potential energy on impact gives

$$V_{max}(d) = (1/2)\mu v_{rel}^2(1 - (b/d)^2) = \varepsilon_{rel}(1 - (b/d)^2) \quad (\text{for } 0 \le b \le d) \quad [18.15]$$

The fraction of the relative kinetic energy that is useful to surmount the barrier to the reaction thus decreases as b increases (and when $b > d$ there is no collision at all). For each ε_{rel} then, we need to find the critical impact parameter b_o where the potential energy on impact, $V_{max}(d)$, is just large enough to surmount the barrier to the chemical reaction ε_B. Using Equation 18.14 with $V_{max}(d) = \varepsilon_B$ and $b = b_o$ gives

$$\varepsilon_B = \varepsilon_{rel}(1 - (b_o/d)^2) \quad (\text{for } 0 \le b_o \le d) \quad [18.16]$$

and solving for the critical impact parameter b_o gives

$$b_o = d(1 - (\varepsilon_B/\varepsilon_{rel}))^{1/2} \quad (\text{for } 0 \le b_o \le d) \quad [18.17]$$

Since any collision with $b < b_o$ results in reaction, the reaction cross section σ_r at a given ε_{rel} is πb_o^2, and it is smaller than the collision cross section πd^2 because a collision with an impact parameter greater than b_o will not convert enough relative kinetic energy on impact to potential energy. This explains why the rate for a reaction with a barrier is less than the gas kinetic rate, as we saw in Example 18.10. Note that the reaction cross section is now dependent on the relative kinetic energy, only equaling πd^2 at large relative kinetic energy. Figure 18.15 shows this graphically. We thus have a rate constant $k(\varepsilon_{rel}) = \sigma_r v_{rel}$ for the reaction that depends on ε_{rel} because both v_{rel} and the reaction cross section depend on ε_{rel}. Using Equation 18.15 the reaction cross section is

$$\sigma_r = \pi b_o^2 = \pi d^2(1 - (\varepsilon_B/\varepsilon_{rel})) \quad (\text{for } 0 \le b_o \le d) \quad [18.18]$$

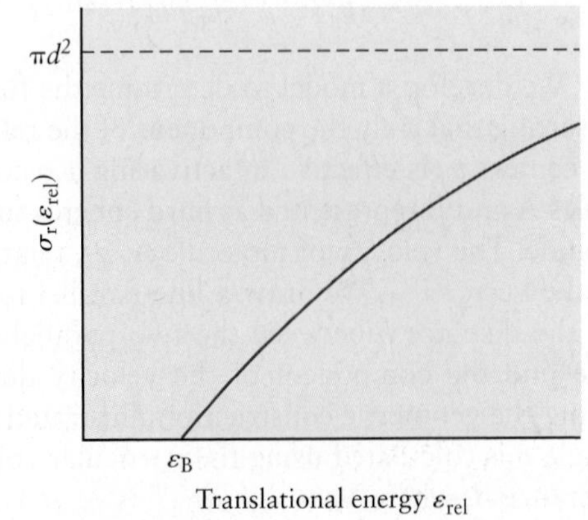

FIGURE 18.15 Dependence of the reactive cross section $\sigma_r(\varepsilon_{rel})$ on the relative translational energy ε_{rel}.

We are now in a position to derive an expression for the rate constant by substituting Equation 18.18 into the integral in Equation 18.12 and integrating to get (with the lower limit set to ε_B as ε_{rel} must be $\geq \varepsilon_B$ for reaction to occur)

$$\int_0^\infty \varepsilon_{rel}\sigma_r(\varepsilon_{rel}) \exp(-\varepsilon_{rel}/k_BT)\, d\varepsilon_{rel} = \sigma_c \int_{\varepsilon_B}^\infty \varepsilon_{rel}\left(1 - \frac{\varepsilon_B}{\varepsilon_{rel}}\right) \exp(-\varepsilon_{rel}/k_BT)\, d\varepsilon_{rel}$$

$$= (k_BT)^2\ \sigma_c\ \exp(-\varepsilon_B/k_BT)$$

with $\sigma_c = \pi d^2$. We arrive at our final goal by substituting this result into Equation 18.12, giving

$$k_r = \left(\frac{8}{\pi\mu k_BT}\right)^{1/2}\left(\frac{1}{k_BT}\right)(k_BT)^2\ \sigma_c\ \exp(-\varepsilon_B/k_BT)$$

$$= \sigma_c\left(\frac{8k_BT}{\pi\mu}\right)^{1/2} \exp(-\varepsilon_B/k_BT) \qquad [18.19]$$

which expresses the temperature-dependent rate constant in terms of molecular parameters. The rate constant depends upon the rate of collisions (which depends upon both the collision cross section and the relative speed), and the probability that a particular collision has sufficient energy directed along the line of centers to cause reaction. Comparing this expression with the empirical Arrhenius equation provides a physical rationale for the form of the Arrhenius equation and an interpretation for each of its factors. ε_B is a threshold energy, below which no reaction occurs and the derivation showed that for glancing collisions the relative kinetic energy must be larger than ε_B. As temperature increases, the relative energy of the colliding molecules increases, more collisions overcome the energy threshold, and the reaction rate increases. Finally, since the activation energy $E_a/k_B = -d \ln k_r/d(1/T)$, if we treat the factors in front of the exponential as roughly constant with temperature we would conclude that the barrier height ε_B equals E_a. Then you would identify the Arrhenius pre-exponential factor A with the first two factors in Equation 18.19, suggesting that A can be interpreted as a collision rate. It also shows that a more realistic form of the Arrhenius equation should have a temperature-dependent pre-exponential factor that reflects the $\sqrt{T}$ dependence of molecular speeds on temperature. This rather weak temperature dependence is generally overwhelmed by the much stronger temperature dependence of the exponential factor, but it is observed experimentally and must be accounted for in accurate kinetic studies.

How well does this simple theory agree with experiment? We obtain experimental activation energies and pre-exponential factors by fitting gas-phase elementary reaction rate data to the Arrhenius form. The value of A can be compared with theory, once we estimate the molecular diameter to determine the hard sphere cross section. For the elementary reaction

$$2NOCl(g) \longrightarrow 2\ NO(g) + Cl_2(g)$$

the measured rate constant is 0.16 times the calculated rate constant. This result indicates that not all collisions lead to reaction, even if the molecules have sufficient relative kinetic energy. What other effect have we overlooked? The relative orientations of the colliding molecules should certainly play a role in determining whether a particular collision results in a reaction. It seems obvious from the models shown in Figure 18.16 that the most favorable orientation is the one in which the chlorine ends of the two NOCl molecules collide head on. We improve on the predictions of collision theory by multiplying by a **steric factor** P (0.16 for this reaction) to account for the fact that only a fraction of the collisions occur with the proper orientation to lead to reaction. The steric factor

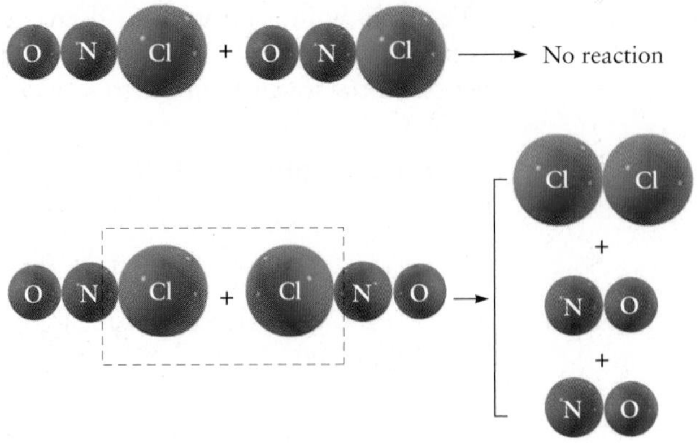

FIGURE 18.16 The steric effect on the probability of a reaction. The two NOCl molecules must approach each other in such a way that the two chlorine atoms are close together, if the encounter is to produce $Cl_2(g)$ and $NO(g)$.

TABLE 18.1

Steric Factors for Gas-Phase Reactions

Reaction	Steric Factor P
$2\ NOCl \rightarrow 2\ NO + Cl_2$	0.16
$2\ NO_2 \rightarrow 2\ NO + O_2$	5.0×10^{-2}
$2\ ClO \rightarrow Cl_2 + O_2$	2.5×10^{-3}
$H_2 + C_2H_4 \rightarrow C_2H_6$	1.7×10^{-6}
$K + Br_2 \rightarrow KBr + Br$	4.8

Adapted from P. W. Atkins and J. de Paula, *Physical Chemistry*, 9th ed. New York: W. H. Freeman, 2010, Table 22.1, p. 942.

is an empirical correction that has to be identified by comparing results of the simple theory with experimental data. It can be predicted in more advanced theories, but only for especially simple reactions.

Steric factors for a number of bimolecular reactions are given in Table 18.1; they are simply the ratios of the measured rate constants to those predicted using the hard-sphere collision rates for each reaction. These data show some interesting features. P generally decreases down the series as the molecules become progressively more complex. There are a larger number of possible relative orientations for more complex molecules, which means that a smaller fraction of those relative orientations will be optimal. The unusually large cross section for the reaction $K + Br_2 \longrightarrow KBr + Br$ is quite interesting; it appears that the molecules react *before* they collide, because the cross section is *larger* than the hard-sphere cross section. This result is explained by the harpoon mechanism mentioned in Section 3.8. The highly electronegative Br atoms in Br_2 pull an electron away from K, which has a very low ionization energy. The strong Coulomb attraction between the ions pulls them together to initiate the reaction. The electron is the harpoon, being fired from the potassium atom to draw in the bromine molecule.

18.6.2 Transition State Theory

The goal of transition state theory (also called activated complex theory) is to provide an approach to the calculation of rate constants from molecular properties that is more sophisticated than collision theory but not intractable. A full discussion of the theory is beyond the scope of this textbook, but an outline of the general approach used and the thermodynamic viewpoint developed provides additional insight into the molecular details of chemical reaction dynamics. Interpreting thermodynamic driving forces in terms of molecular structure and motion has greatly enhanced our understanding of the rates and mechanisms of chemical reactions of interest in chemistry and biology. We illustrate this point of view in our discussion of the mechanisms of enzyme catalysis in Section 18.8.

Transition state theory postulates the existence of an **activated complex**, as shown in Figure 18.11, for example, which is the configuration of atoms at the highest point on the potential energy function that describes the passage from reactants to products. The terms *transition state* and *activated complex* are often used interchangeably but we find it helpful to refer to the geometric configuration of the atoms as the activated complex and the location of that configuration on a potential energy diagram as the transition state. The activated complex is assumed to exist as if it were in equilibrium with the reactants and the theory focuses on calculating the rate at which the activated complex passes through the transition state to form products. The equilibrium constant is calculated using the methods of statistical thermodynamics in which all of the motions of the molecules are considered separately and explicitly, particularly the rotational and vibrational motions. Motion though the transition state is imagined to proceed along one set of displacements associated with a particular molecular vibration. The simplest situation to imagine is the reaction A + BC → products, in which the relevant mode of the activated complex is one in which the A—B and B—C bonds vibrate out of phase with respect to one another (an antisymmetric stretch, see Section 20.4). We imagine the frequency of that mode decreasing as the reaction proceeds with the vibrational energy stored in the B—C bond being transformed into the translational kinetic energy of the separated AB + C products. Reaction rates predicted using transition state theory are often expressed by the Eyring formula:

$$k_r = \kappa \frac{k_B T}{h} K^{\dagger}$$

in which κ(the transmission coefficient) measures the probability that the system will proceed through the transition state and $k_B T/h$ is the rate at which the activated complex dissociates to form products. $k_B T/h$ is of the order 5×10^{12} s^{-1} at 300 K, which is a useful number to remember when estimating reaction rates using transition state theory.

The thermodynamic formulation of transition state theory is more useful than the method just described because it is very difficult to calculate equilibrium constants involving activated complexes. We substitute $\exp(-\Delta G^{\dagger}/RT)$ for $K^{\dagger}$ in the preceding expression to get

$$\begin{aligned} k_r &= \kappa \frac{k_B T}{h} \exp(-\Delta G^{\dagger}/RT) \\ &= \kappa \frac{k_B T}{h} \exp(\Delta S^{\dagger}/R) \exp(-\Delta H^{\dagger}/RT) \end{aligned}$$

which we compare to the Arrhenius equation

$$k_r = A \exp(-E_a/RT)$$

Transition state theory provides some additional insight into the factors that govern activation. The Arrhenius activation energy E_a may now be expressed as $\Delta H^{\dagger}$, an activation enthalpy, and the product of $\exp(\Delta S^{\dagger})$ and $\kappa(k_B T/h)$ may be associated with the pre-exponential factor A. Although it is possible in principle, and sometimes in practice, to calculate enthalpies and entropies of activation, the real utility of this approach is that it provides us with a conceptual framework from which to analyze the driving forces for particular chemical reactions from a molecular point of view. This approach is extremely useful in helping to understand patterns of reactivity in complex systems like enzymes, for example, and it has been adopted widely for that purpose, as we illustrate briefly in the *Connection to Biology* at the end of this chapter.

18.6.3 Isotope Effects in Chemical Kinetics

Isotopes are widely used as tracers in kinetics to help determine reaction mechanisms, as discussed in Section 18.4. Isotopic substitution of the lighter elements in molecules also leads to rather dramatic effects on the rates of chemical reactions that have important ramifications in a number of fields of science. Some plants, for example, preferentially utilize $^{12}CO_2$ in photosynthesis, and the ratio of the various carbon isotopes measured in materials derived from plants can be used to identify their sources. The ratio of $^{12}CO_2$ to the heavier isotopes in atmospheric CO_2 identifies fossil fuels as the source of the recent increases in the concentrations of this greenhouse gas. Kinetic isotope effects have a purely quantum mechanical origin. They arise because the zero-point energy of the

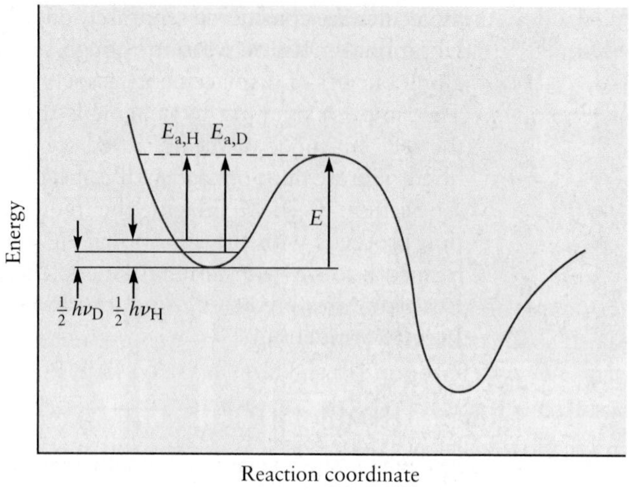

FIGURE 18.17 Energy profiles along a C—H bond reaction coordinate showing the origin of the primary kinetic isotope effect. The zero-point vibrational energy of the C—D bond is less than the zero-point vibrational energy of the C—H bond, which leads to an increase in the activation energy E_a for reactions that involve breaking C—D bonds.

heavier isotopes is less than the zero-point energy of the lighter isotopes, producing an increase in activation energies and a decrease in reaction rates.

The physical origin of kinetic isotope effects is most easily understood by examining a potential energy diagram like that shown in Figure 18.17. The largest effects observed occur for the high-frequency stretching motions of X—H bonds, where X is typically one of the main group elements important in organic chemistry, like carbon, nitrogen, and oxygen. We have drawn this diagram in a slightly different way than those in earlier figures to emphasize the fact that both the reactants and the products have potential energy diagrams, like those introduced in Chapter 3, that describe their bond lengths, dissociation energies, and vibrational frequencies (see Section 20.3). Kinetic isotope effects are considered *primary* when the isotope involved is one of the atoms in a bond and *secondary* when it is in an adjacent bond; the former are generally much larger than the latter, which are nevertheless useful in mechanistic studies. We typically calculate the *maximum* primary kinetic isotope effect by assuming that the zero-point energy of the relevant vibration in the activated complex is identical for both isotopes and it is only the difference in the zero-point energies in the reactants that matters. Let's use the C—H stretching vibration as an example and calculate the kinetic isotope effect that results from deuterium substitution using the Arrhenius equation. We label the rate constants and activation energies associated with the C—H bond and the C—D bond as $k_H(T)$, $E_{a,H}$ and $k_D(T)$, $E_{a,D}$, respectively, identifying the energy difference between the minimum of the potential energy well for the reactants and the maximum in the potential energy curve for the activated complex simply as E. The activation energies are then given by $E_{a,H} = E - \frac{1}{2}h\nu_{CH}$ and $E_{a,D} = E - \frac{1}{2}h\nu_{CD}$. Substituting these values into the Arrhenius equation and taking the ratios of the rate constants gives

$$\frac{k_H(T)}{k_D(T)} = \frac{A\exp(-E_{a,H}/RT)}{A\exp(-E_{a,D}/RT)} = \frac{\exp[(-E + \frac{1}{2}h\nu_{CH})/RT]}{\exp[(-E + \frac{1}{2}h\nu_{CD})/RT]}$$

$$= \exp[\tfrac{1}{2}(h\nu_{CH} - h\nu_{CD})]/RT$$

We can calculate the reduction in rate expected when deuterium is substituted for hydrogen in a C—H bond using vibrational frequencies measured by infrared spectroscopy (see Section 20.3) and converting them into vibrational energies using $E = h\nu$. Typical energies for C—H stretching vibrations are about 36 kJ mol^{-1} and those for C—D vibrations are about 25 kJ mol^{-1}, respectively, giving a ratio $k(D)/k(H) = 0.12$ at 300 K. The rate decreases by a factor of 8.

Measuring the effect of deuterium substitution on the rates of organic reactions provides insight into reaction mechanisms. Experimental kinetic isotope effects in the range 5 to 10 near room temperature or above strongly imply C—H bond breaking at the labeled carbon atom. Secondary kinetic isotope effects involving C—H bond breaking are smaller, but measurable (around 1.5) and point to the role of the increased reduced mass of a functional group adjacent to the carbon atom undergoing reaction.

18.6.4 Molecular Beams

The kinetics experiments described thus far are all carried out by varying the macroscopic properties of the system, such as reactant concentrations and temperature. This approach necessarily averages over the details of individual reactive encounters that depend on particular molecular details such as the distribution of energy among the various modes of the reactants (see Section 12.8) and their relative orientations when they collide with one another. Studying reactions in crossed molecular beams provides a powerful alternative that allows us to investigate the details of chemical reactions at the molecular level. A crossed molecular beam apparatus comprises an ultrahigh vacuum chamber, collimated sources of molecules with well-defined kinetic energies, internal energies, and orientations in space that intersect and react, and one or more detectors that identify the reaction products and characterize their energy distributions as a function of position in three dimensions (see Fig. 18.18).

This apparatus allows the reactant molecules to be prepared in highly selective states. A velocity selector located between the beam source and the collision region (see Fig. 9.17) introduces only those molecules whose velocities fall within a small range. This method allows for much greater control over relative energies than can be achieved simply by controlling the temperature in a macroscopic kinetics experiment. Electric and magnetic fields or laser radiation can be applied to select reactants according to how rapidly they rotate and vibrate. The velocity of products can be measured, their angular distribution relative to the directions of the initial beams can be determined, and their vibration and rotation energies can be determined. All this information allows a much more detailed examination of the way in which molecules collide and react. For example, if the activated complex rotates many times before it finally breaks up into products, the angular distribution of products should be uniform as shown in Figure 18.19a. But if it breaks up before it has a chance to rotate, the angular distribution of products should be very nonuniform and should depend on the directions of the original beams as shown in Figure 18.19b. This type of measurement gives information on the lifetime of an activated complex.

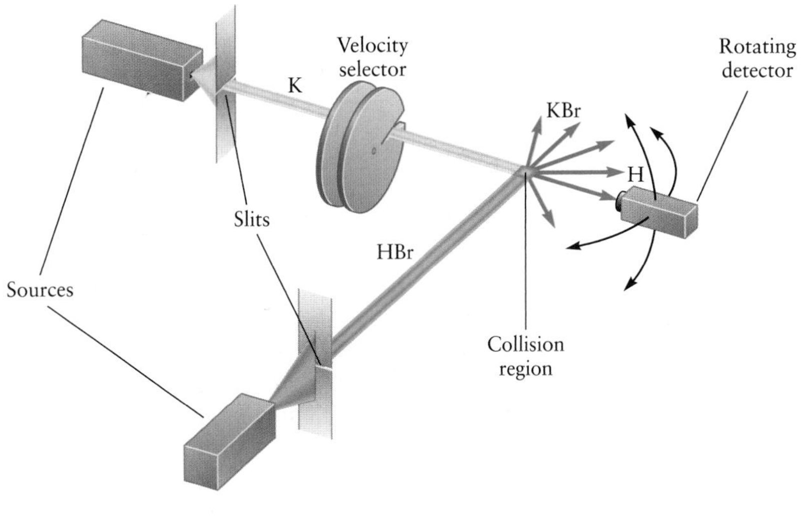

FIGURE 18.18 Diagram of a crossed molecular beam experiment. The reaction being studied is K + HBr → KBr + H. The curved arrows on the right side of the figure represent rotation of the detector in the vertical and horizontal planes.

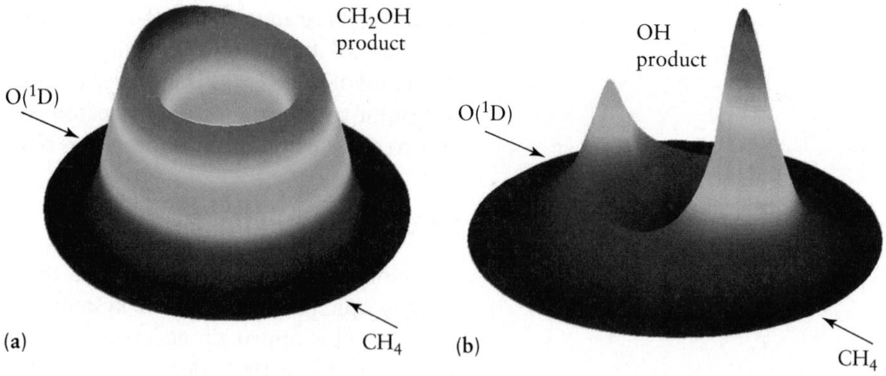

FIGURE 18.19 Angular distributions in the center-of-mass reference frame of the products of the O(^{1}D) + CH_4 reaction in a crossed molecular beams experiment. (a) The long lifetime of the complex en route to H + CH_2OH gives the complex time to rotate before breaking up into products. (b) The CH_3 + OH product channel proceeds through a short-lived activated complex. (Adapted from Xueming Yang, "Multiple channel dynamics in the O(1D) reaction with alkanes" Phys. Chem. Chem. Phys., 2006, 8, 205–215, Figures 6 and 9.)

Molecular beams are limited to reactions that are carried out in a vacuum, where well-defined beams of reactant molecules can be prepared. This limits their application to gas-phase reactions and to reactions of gaseous molecules with solid surfaces. Molecular beam methods cannot be used to study kinetics in liquid solvents, though they can be used to study reactions at the liquid-gas interface. The detailed information they provide for gas–gas and gas–surface reactions allows precise testing of models and theories for the dynamics of these classes of reactions.

EXAMPLE 18.11

In a crossed molecular beams experiment on the F_2 + C_2H_4 reaction, the fluorine molecular beam crossed the ethene molecular beam at right angles, similar to the experiment shown in Figure 18.18. Using velocity selectors, the F_2 reactant molecules selected had a speed of 700 m s^{-1} in the $+x$ direction and the ethene molecules selected had a speed of 1713 m s^{-1} in the $+y$ direction. What is the relative kinetic energy of such a collision? Using a hard sphere model, what is the largest impact parameter that has a chance to result in the reactants surmounting the 23 kJ mol^{-1} potential energy barrier and going on to products? Assume the hard sphere radius of F_2 = 1.8 Å and of C_2H_4 = 2.3 Å.

Solution

You may choose either reactant as #1; here we choose the F_2 molecule. $\vec{v}_{rel} = \vec{v}_2 - \vec{v}_1 = (0,1713) - (700,0) = (-700,1713)$ and the magnitude of the relative velocity is 1851 m s^{-1}. The relative kinetic energy is

$$\varepsilon_{rel} = \frac{1}{2}\mu v_{rel}^2 = \frac{1}{2}\left(16.12\frac{\text{g}}{\text{mol}}\right)\left(\frac{1\ \text{kg}}{1000\ \text{g}}\right)(1851\ \text{m s}^{-1})^2 = 27.6\ \text{kJ mol}^{-1}.$$

Using Equation 18.17 $b_o = 4.1\ \text{Å}\left(1 - \dfrac{23\ \text{kJ mol}^{-1}}{27.6\ \text{kJ mol}^{-1}}\right)^{\frac{1}{2}} = 1.7\ \text{Å}.$

18.7 Reactions in Solution

Reactions in solution are much more complicated to describe than their gas-phase counterparts because the transport mechanism in solutions is much different. Molecules in the gas phase travel in straight lines between collisions, whereas solute molecules in solution are continually being bounced around by collisions with solvent molecules that lead to transport by diffusion. Collisions between reactants in solution occur much less frequently than for those in the gas phase, but the reactants may be bound together by a "cage" of solvent molecules for sufficiently long periods that they can acquire enough energy through collisions with the solvent to react. We can use the steady-state approximation to help understand the process.

Consider the two-step sequence

$$A + B \underset{k_{-1}}{\overset{k_1}{\rightleftharpoons}} AB$$

$$AB \xrightarrow{k_2} \text{products}$$

We invoke the steady-state approximation to calculate the concentration of the intermediate AB, from which we can calculate the rate of the overall reaction as follows.

$$\text{rate} = k_2[\text{AB}]$$

$$\frac{d[\text{AB}]}{dt} = k_1[\text{A}][\text{B}] - k_{-1}[\text{AB}] - k_2[\text{AB}] = 0$$

$$[\text{AB}] = \frac{k_1[\text{A}][\text{B}]}{k_{-1} + k_2}$$

$$\text{rate} = k_2[\text{AB}] = \frac{k_1 k_2[\text{A}][\text{B}]}{k_{-1} + k_2}$$

Let's investigate the limits $k_2 >> k_{-1}$ and $k_2 < k_{-1}$, respectively. The overall rate in the former case is given by

$$\text{rate} = k_1[\text{A}][\text{B}]$$

and if k_1 is limited by the rate at which the reactants encounter one another by diffusion, we call this a **diffusion-controlled reaction**, with a rate constant conventionally called k_d. If instead the rate constant for the second step is slower than that of the first step and equilibrium is established in the first step, then the rate expression becomes

$$\text{rate} = k_2 K_1[\text{A}][\text{B}]$$

where K_1 is the equilibrium constant for the first reaction in the series. A slow second step implies a large activation energy, so reactions of this type are often called *activation-energy controlled.*

A simple model for diffusion-controlled reactions was constructed by M. V. Smoluchowski in 1916, shortly after Einstein presented his model of Brownian motion. The idea is that molecules diffuse toward one another with mean-square displacements given by the formula

$$\overline{(\Delta r)^2} = 6Dt$$

where D is the diffusion constant in units $m^2\ s^{-1}$. Typical values for diffusion constants of small molecules in solution are of the order of $10^{-9}\ m^2\ s^{-1}$, which means that a small molecule will travel about 10 μm in a second, in contrast to molecular speeds in gases that are the order of a few hundred meters per second near room temperature. The molecule does not travel far because its motion is that depicted in Figure 9.25; it keeps changing direction as it becomes trapped in local solvent cages and then escapes to nearby cages, but in ever-changing directions. Smoluchowski's model assumes that a number of B molecules, for example, are diffusing toward a stationary molecule A and solves the radial diffusion equation to find the rate at which B molecules arrive at A. Reaction is assumed to occur when the molecules (represented as hard spheres of radii r_A

and r_B) collide, being separated by a distance $R^* = (r_A + r_B)$. There is no reason to prefer B over A for the species that is diffusing, so an effective diffusion constant $D = D_A + D_B$ is used instead. The final result is

$$k_d = 4\pi R^* D N_A$$

where N_A is Avogadro's number, not the number density of A molecules. Bimolecular rate constants have units L mol^{-1} s^{-1}, from which we can extract a typical value for R^* by substituting 10^{-9} m^2 s^{-1} for a typical diffusion constant in liquids into the above expression and the measured diffusion controlled rate constant. Diffusion sets an upper limit for the rates of many solution-phase reactions, including enzyme-catalyzed reactions (see Section 18.8). Enzymes with the greatest catalytic efficiencies cannot convert substrates into products any faster than they encounter them. Enzymes whose rates are diffusion-limited are said to have reached "catalytic perfection."

18.8 Catalysis

A **catalyst** is a substance that takes part in a chemical reaction and speeds up the rate but undergoes no permanent chemical change itself. Catalysts therefore do not appear in the overall balanced chemical equation. But their presence very much affects the rate law, modifying and speeding existing pathways or, more commonly, providing completely new pathways by which a reaction can occur (Fig. 18.20). Catalysts have significant effects on reaction rates even when they are present in very small amounts. Industrial chemistry devotes great effort to finding catalysts to accelerate particular desired reactions without increasing the generation of undesired products.

Catalysts increase reaction rates by providing alternate pathways for reactions to occur, pathways that lower the activation energy E_a, as shown in Figure 18.21. Catalysts lower the activation energies for both the forward *and* the reverse reaction by the same amount, however, which means that they increase the rate of approach to equilibrium as shown in Figure 14.8, for example. Catalysts do not change the thermodynamics of chemical reactions because Gibbs free energies, and therefore equilibrium constants, are state functions and changes in the values of state functions are independent of the path. Reactions that are not allowed thermodynamically will not proceed even under the influence of a catalyst. Catalysts are vitally important in the chemistry of life

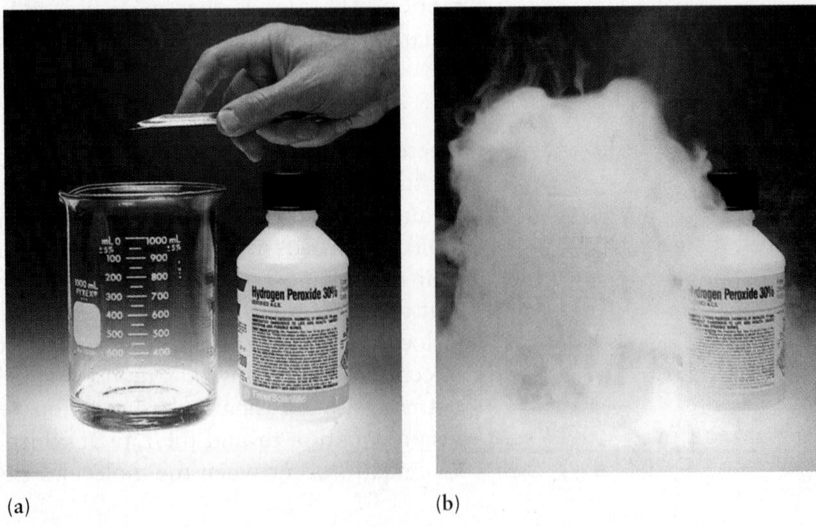

FIGURE 18.20 (a) The decomposition of hydrogen peroxide, H_2O_2, to water and oxygen is catalyzed by adding a very small amount of transition metal oxide. (b) The water evolves as steam because of the heat given off in the reaction.

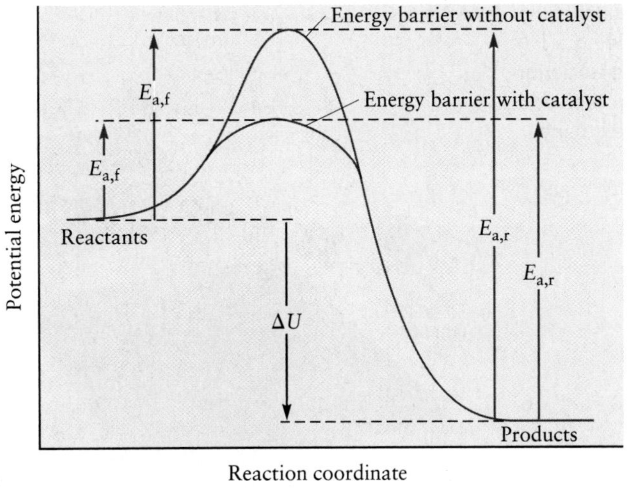

FIGURE 18.21 The most important way in which catalysts speed reactions is by reducing the activation energy. Both the uncatalyzed (red) and catalyzed (blue) reaction coordinates are shown. Here we roughly associate the empirical activation energy with the energy barrier (a risky thing to do, especially in solution).

processes as well as in industrial chemistry. Virtually all biochemical reactions are catalyzed by enzymes, and catalysts are used in virtually all industrial reactions. Products are often removed as they are formed in industrial processes so that equilibrium is never reached and very high conversion efficiencies are realized. An **inhibitor**, in contrast, slows down the rate of a chemical reaction, often by increasing an activation energy or by physically blocking a reactive site. Inhibitors are important in industrial chemistry because they reduce the rates of undesirable side reactions, allowing the desired products to form in greater yield. They are also important in biology, where they influence the rates of enzyme-catalyzed reactions by blocking the active sites of enzymes.

Catalysis are classified into two types: homogeneous and heterogeneous. In **homogeneous catalysis** the catalyst is present in the same phase as the reactants, as when a gas-phase catalyst speeds up a gas-phase reaction, or a species dissolved in solution speeds up a reaction in solution. Chlorofluorocarbons and oxides of nitrogen are homogeneous catalysts responsible for the destruction of ozone in the stratosphere. These reactions are examined in more detail in Section 20.6. A second example is the catalysis of the oxidation–reduction reaction

$$Tl^+(aq) + 2\ Ce^{4+}(aq) \longrightarrow Tl^{3+}(aq) + 2\ Ce^{3+}(aq)$$

by silver ions in solution. The direct reaction of Tl^+ with a single Ce^{4+} ion to give Tl^{2+} as an intermediate is slow. The reaction can be speeded up by adding Ag^+ ions, which take part in a reaction mechanism of the form

$$Ag^+ + Ce^{4+} \underset{k_{-1}}{\overset{k_1}{\rightleftharpoons}} Ag^{2+} + Ce^{3+} \qquad \text{(fast)}$$

$$Tl^+ + Ag^{2+} \xrightarrow{k_2} Tl^{2+} + Ag^+ \qquad \text{(slow)}$$

$$Tl^{2+} + Ce^{4+} \xrightarrow{k_3} Tl^{3+} + Ce^{3+} \qquad \text{(fast)}$$

The Ag^+ ions are not permanently transformed by this reaction because those used up in the first step are regenerated in the second; they play the role of catalyst in significantly speeding the rate of the overall reaction.

In **heterogeneous catalysis** the catalyst is present as a phase distinct from the reaction mixture. The most important case is the catalytic action of certain solid surfaces on gas-phase and solution-phase reactions. A critical step in the production of sulfuric acid relies on a solid oxide of vanadium (V_2O_5) as catalyst. Many other solid catalysts are used in industrial processes. One of the best studied is the addition of hydrogen to ethylene to form ethane:

$$C_2H_4(g) + H_2(g) \longrightarrow C_2H_6(g)$$

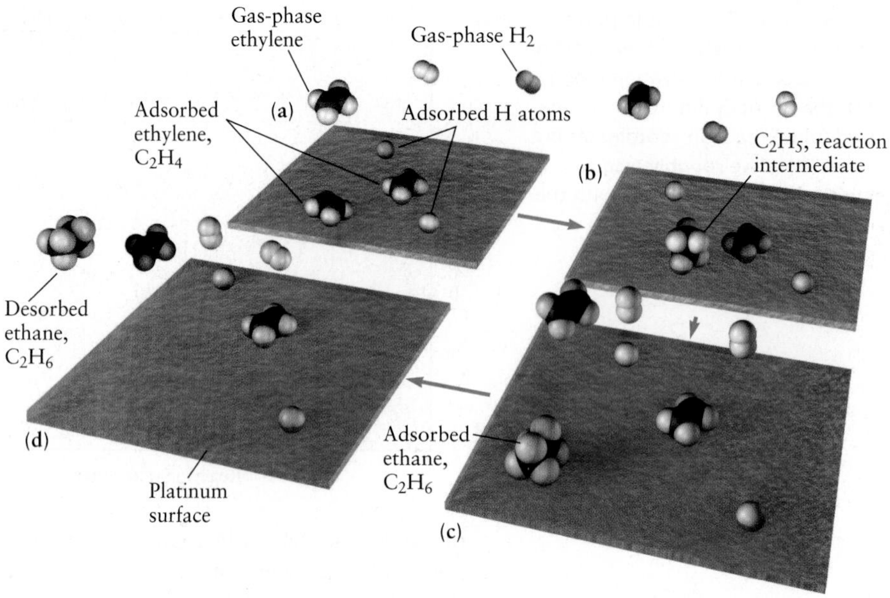

FIGURE 18.22 Platinum catalyzes the reaction $H_2 + C_2H_4$ by providing a surface that promotes the dissociation of H_2 to H atoms, which can then add to the C_2H_4 stepwise to give ethane, C_2H_6.

The process occurs extremely slowly in the gas phase but is catalyzed by a platinum surface (Fig. 18.22).

Another example is the solid catalyst used to reduce the emission of pollutants such as unburned hydrocarbons, carbon monoxide, and nitrogen oxides in the exhaust streams of automobile engines (Fig. 18.23). A **catalytic converter** is designed to simultaneously *oxidize* hydrocarbons and CO through the reactions

$$CO, C_xH_y, O_2 \xrightarrow{\text{Catalyst}} CO_2, H_2O$$

and *reduce* nitrogen oxides through the reactions

$$NO, NO_2 \xrightarrow{\text{Catalyst}} N_2, O_2$$

Clearly, the best catalyst for the reduction reactions may not be the best for the oxidation reactions, so two catalysts are combined. The noble metals, although expensive, are particularly useful. Typically, platinum and rhodium are deposited on a fine honeycomb mesh of alumina (Al_2O_3) to give a large surface area that increases the contact time of the exhaust gas with the catalysts. The platinum serves primarily as an oxidation catalyst and the rhodium as a reduction catalyst. Catalytic converters can be poisoned with certain metals that block their active

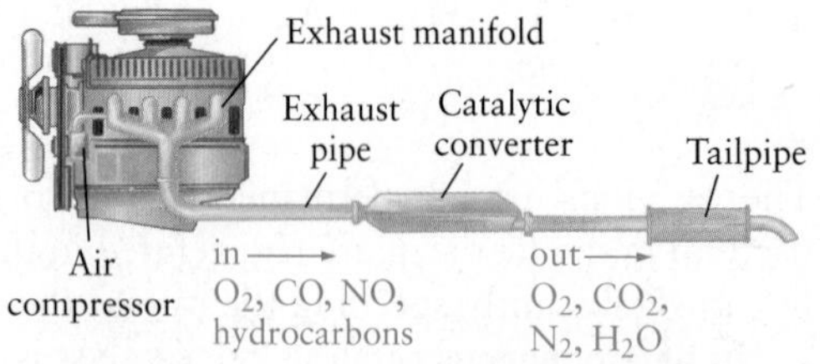

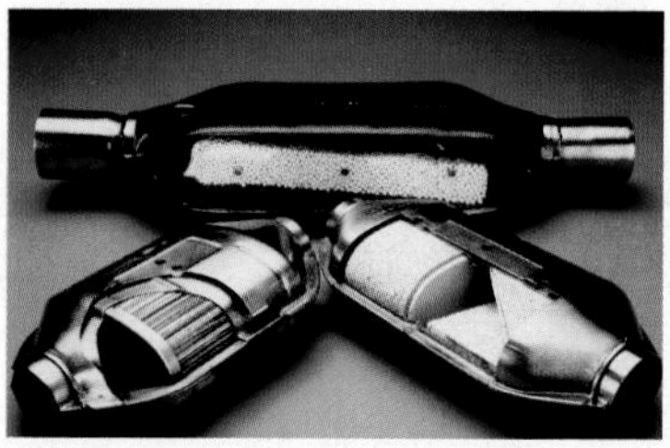

Raymond Reuter/Sygma/Corbis

FIGURE 18.23 (a) The arrangement of a catalytic converter used to reduce automobile pollution. (b) Cutaway views of several catalytic converters showing different structures for organizing metal catalysts, platinum, palladium, and rhodium on different substrates and supports. A steel-alloy heating element raises the temperature to 400°C in seconds, activating the catalysts and reducing the pollution emitted in the first minutes after the car is started.

sites and reduce their effectiveness. Because lead is one of the most serious such poisons, automobiles with catalytic converters must use unleaded fuel.

18.8.1 Enzyme Catalysis

The vast majority of chemical reactions in living organisms are catalyzed by **enzymes**, proteins that have evolved to enhance reaction rates by many orders of magnitude with exquisite selectivity. Classes of enzyme-catalyzed reactions include oxidation–reduction, hydrolysis, isomerization, and the transfer of functional groups, for example. The reactants (**substrates**) may be either small organic molecules or specific regions of much larger molecules such as proteins and nucleic acids. The oxidation of glucose to carbon dioxide and water in the series of metabolic reactions called glycolysis is an excellent example; glucose, a sugar, is perfectly stable on the shelf but is rapidly metabolized in cells to provide energy. As another example, consider the hydrolysis of urea by the enzyme urease, as represented by the following reaction.

$$H_3O^+(aq) + (NH_2)_2CO(aq) + H_2O(\ell) \xrightarrow{\text{Urease}} 2NH_4^+(aq) + HCO_3^-(aq)$$

The ratio of the rate constants for the catalyzed and uncatalyzed reactions is called the catalytic power, which for urease is an astonishing factor of 10^{14}. In the subsections that follow we introduce you to the general characteristics of enzymes, present a discussion of the kinetics of enzyme-catalyzed reactions, and introduce you to some mechanisms of enzyme-catalyzed reactions.

18.8.2 Enzyme Kinetics

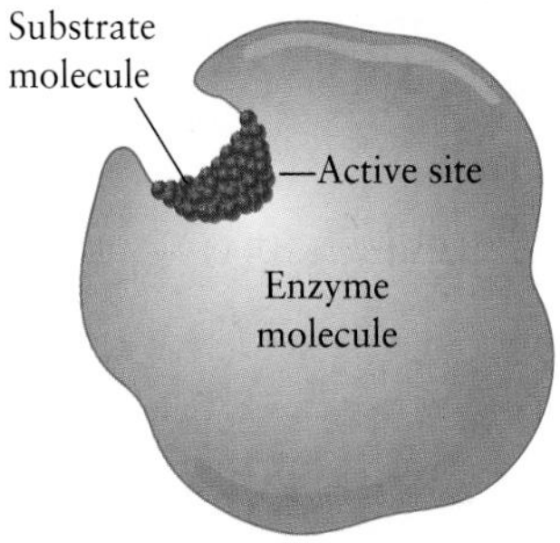

FIGURE 18.24 A sketch of an enzyme showing the active site where the enzyme binds the substrate.

Enzymes are proteins (polypeptides with molar masses of the order 10^4 g mol^{-1}) that selectively bind substrates in a small region called the active site, as shown in Figure 18.24. A particular enzyme is selective with respect both to substrate and to the reaction it catalyzes. Each individual step in a multistep metabolic pathway, for example, must produce actual yields of products in excess of 90% of the theoretical yield to avoid accumulation of undesirable side products.

Enzymes are identified using both common and systematic names, as in other areas of chemistry. Many common names were derived by adding the suffix *-ase* to the name of the substrate: urease catalyzes the hydrolysis of urea, and phosphatases catalyze the hydrolysis of phosphate ester bonds ($RO{-}PO_3R'$), for example. Other common names provide no clue whatsoever as to the nature of the substrate or the reaction. Catalase enhances the rate of hydrogen peroxide decomposition, and proteases are a class of enzymes that hydrolyze peptide bonds ($RO{-}NHR'$). The International Commission on Enzymes established a systematic classification of enzymes that divides them into six classes, based upon the general nature of the reactions they catalyze, and several subclasses within each class. The major classes and the reactions they catalyze include (1) *oxidoreductases*—redox reactions; (2) *transferases*—functional group transfer reactions, for example, methyl group transfer; (3) *hydrolases*—hydrolysis reactions; (4) *lyases*—addition to carbon–carbon double bonds; (5) *isomerases*—isomerization reactions; and (6) *ligases*—bond-forming reactions. Many enzymes require **cofactors** to carry out their catalytic functions. These are often metal ions, such as Fe^{2+}/Fe^{3+} in the cytochromes, which catalyze electron-transfer reactions, or they may be small organic molecules called **coenzymes,** examples of which include vitamins.

Enzymes work by finding alternate pathways with lower activation energies, as represented by the graph in Figure 18.21. It is customary in enzyme kinetics to refer to the activation energy as a free energy of activation, symbolized as $\Delta G^{\ddagger}$, and to identify the driving forces for enzyme-catalyzed reactions in terms of the enthalpies and entropies of activation, $\Delta H^{\ddagger}$ and $\Delta S^{\ddagger}$, as discussed in Section 18.6.

The kinetics of enzyme-catalyzed reactions are generally treated using the steady-state approximation in a form that has become known as Michaelis–Menten kinetics. The overall reaction is summarized by the following mechanism

$$E + S \underset{k_{-1}}{\overset{k_1}{\rightleftharpoons}} ES$$

$$ES \xrightarrow{k_2} E + P$$

in which E represents the free enzyme, S is the substrate, ES represents the complex formed when the substrate binds to the active site, and P is the final product. The rate constants k_1 and k_{-1} are the forward and reverse rate constants of the first step, in which an equilibrium may be established, and k_2 is the rate constant of the second step, which is assumed to be irreversible. We find the initial rate (before any back reaction between P and E might occur) by invoking the steady-state approximation to find the concentration of the enzyme–substrate complex, [ES], as discussed in Section 18.4. The concentration of [ES] reaches a steady state when the rate of formation equals the rate of decay.

$$\frac{d[\mathrm{ES}]}{dt} = 0 = k_1[\mathrm{E}][\mathrm{S}] - k_{-1}[\mathrm{ES}] - k_2[\mathrm{ES}]$$

We could solve this equation to find [ES] in terms of [E], but it is not straightforward to measure the concentration of the free enzyme directly, so an alternative approach has been developed. The total concentration $[E_T]$ of enzyme present is the sum of the concentrations of the free enzyme [E] and the enzyme–substrate complex [ES], a quantity that is readily accessible experimentally. Setting $[E_T] = [E] + [ES]$ allows us to make the substitution $[E] = [E_T] - [ES]$ in the preceding steady-state equation to get

$$\frac{d[\mathrm{ES}]}{dt} = 0 = k_1[\mathrm{E_T}][\mathrm{S}] - k_1[\mathrm{ES}][\mathrm{S}] - k_{-1}[\mathrm{ES}] - k_2[\mathrm{ES}]$$

and solve for [ES] to give

$$[\mathrm{ES}] = \frac{k_1[\mathrm{E_T}][\mathrm{S}]}{(k_{-1} + k_2) + k_1[\mathrm{S}]}$$

We define the *Michaelis constant* K_m (in units of *molarity*) as

$$K_m = \frac{k_{-1} + k_2}{k_1}$$

which allows us to express the steady-state concentration of [ES] as

$$[\mathrm{ES}] = \frac{[\mathrm{E_T}][\mathrm{S}]}{K_m + [\mathrm{S}]}$$

The rate of formation of product is then given by

$$\frac{d[\mathrm{P}]}{dt} = k_2[\mathrm{ES}] = \frac{k_2[\mathrm{E_T}][\mathrm{S}]}{K_m + [\mathrm{S}]} \qquad [18.20]$$

which is called the **Michaelis–Menten equation.** Figure 18.25 shows a plot of the overall rate as a function of substrate concentration [S].

The rate is linear (first order) at low substrate concentrations but rolls over to become independent of [S] at higher concentrations, saturating at a maximum rate V_{max}. The physical origin of this saturation behavior is easy to understand.

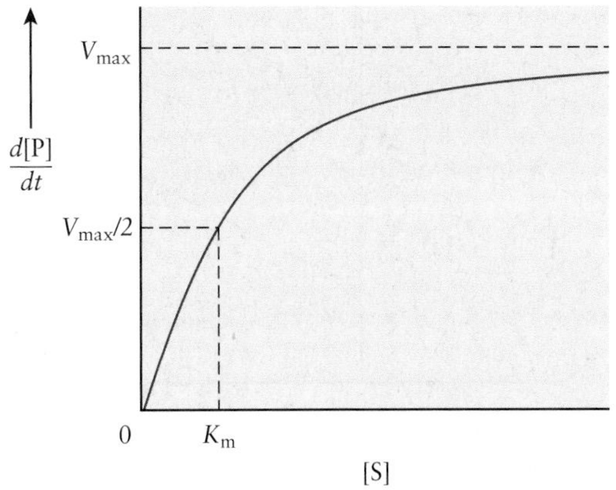

FIGURE 18.25 Michaelis–Menten plot of the dependence of the rate of an enzyme-catalyzed reaction on the concentration of the substrate, showing saturation behavior.

The first step in the overall reaction is first order in E, first order in S, and second order overall. The rate increases with increasing [S], for a fixed concentration of enzyme, until all of the available actives sites have bound substrate, after which point no further increase in rate with increasing [S] occurs. We can calculate V_{max} from Equation 18.20 by taking the limit in which $[S] >> K_m$ and setting $K_m = 0$ in the denominator to get $V_{max} = k_2[E_T]$. The Michaelis–Menten equation is often written in the following alternative form to emphasize the central role of V_{max} in enzyme kinetics.

$$\frac{d[P]}{dt} = k_2[ES] = \frac{V_{max}[S]}{K_m + [S]}$$

Rearranging this equation into the form

$$K_m = [S]\left[\left(\frac{V_{max}}{dP/dt}\right) - 1\right]$$

allows us to find K_m experimentally. $K_m = [S]$ when $dP/dt = V_{max}/2$, so we simply measure the substrate concentration for which the rate is exactly half the maximum rate, as indicated on the graph in Figure 18.25. This procedure is only approximate, however, because it may be very difficult to arrange conditions that permit the accurate determination of V_{max}.

The **turnover number** k_{cat} is an important measure of catalytic activity for enzyme-catalyzed reactions. It is defined as the number of substrate molecules converted into product per enzyme molecule per second (turnover numbers are defined in analogous ways for other kinds of catalysts as well) and is easily calculated by measuring the maximum rate for a known enzyme concentration. The maximum rate for a given total enzyme concentration is obtained from the alternative form of the Michaelis-Menten equation by making $[S] >> K_m$ and recognizing that $[ES] = [E_T]$ at saturation, from which we get

$$k_{cat} \equiv k_2 = \frac{V_{max}}{E_T}$$

Turnover numbers for different enzymes vary over an enormously wide range; each catalase molecule can decompose 40 million molecules of hydrogen peroxide per second, whereas the turnover number of chymotrypsin (see the *Connection to Biology* in Chapter 7) is about 100 and that of lysozyme is about 0.5.

Enzymes rarely operate near saturation under physiological conditions, so it is interesting to consider another measure of catalytic efficiency for low substrate concentrations. Recall from Section 18.4 that reactions for which $k_2 >> k_{-1}[M]$ show second-order kinetics; that is, the first step is rate-limiting. In the present

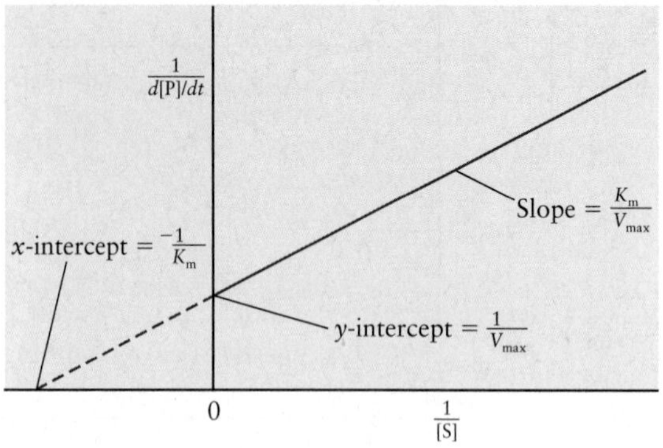

FIGURE 18.26 A graph of the linear form of the Michaelis–Menten equation that shows how the parameters K_m and V_{max} are determined.

case we can begin with the alternative form of the Michaelis–Menten equation, substitute $V_{max} = k_{cat}[E_T]$ and set $[S] = K_m$ to get

$$\frac{d[P]}{dt} = \left(\frac{k_{cat}}{K_m}\right)[E][S]$$

where [E] is the concentration of the free enzyme. The maximum rate for this first step is the rate at which E and S encounter one another; diffusion-limited rate constants are typically about 10^9 L mol^{-1} s^{-1} for small substrates (such as H_2O_2), and about an order of magnitude smaller for high molecular weight substrates, such as proteins and nucleic acids (see Section 18.6). Comparing experimental (k_{cat}/K_m) ratios to the diffusion-limited rate constants provides an important measure of the catalytic efficiency of enzymes. Finally, a linear version of the Michaelis–Menten equation makes it convenient to obtain both V_{max} and K_m graphically. We rewrite the equation in the following form

$$\frac{1}{d[P]/dt} = \left(\frac{K_m}{V_{max}}\right)\left(\frac{1}{[S]}\right) + \frac{1}{V_{max}}$$

and plot the inverse of the rate versus 1/[S], as shown in Figure 18.26. The slope is K_m/V_{max} and the y-intercept is $1/V_{max}$, allowing us to extract the values of both constants from one plot.

The kinetics of enzyme-catalyzed reactions are temperature-dependent, like most chemical reactions, but the rates begin to fall off gradually with temperature above 50–60°C or so, as shown in Figure 18.27. The rates increase with temperature, typically doubling every 10°C like many other reactions, but eventually beginning to decline as the three-dimensional structures of proteins, which are essential for their catalytic function, unfold and degrade.

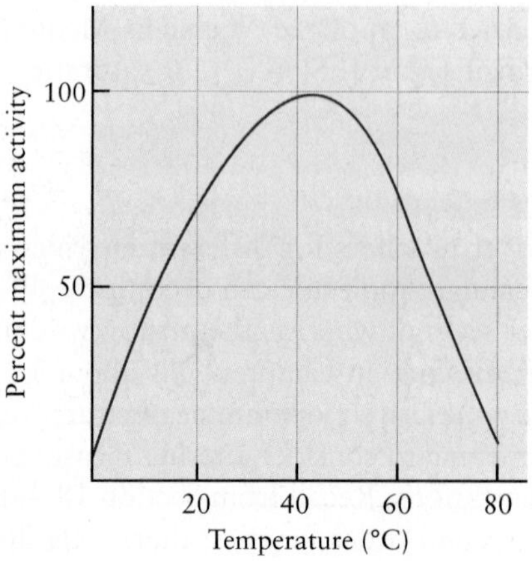

FIGURE 18.27 Plot of the activity of an enzyme versus temperature. The increase in rate at low temperatures is described by the Arrhenius equation whereas the falloff in rate at higher temperatures is due to loss of structure necessary for function.

CONNECTION TO BIOLOGY

Mechanisms of Enzyme-Catalyzed Reactions

The high degree of specificity observed for enzyme-catalyzed reactions was first attributed to a "lock-and-key" mechanism, in which binding of substrates (the key) was effected through molecular recognition by the active site of enzymes (the lock), as suggested by Figure C18.1.

Enzymes are considerably more flexible than suggested by the lock-and-key model, however, and the realization that enzyme–substrate interactions were more dynamic than suggested by that model led to the development of the **induced fit** hypothesis, in which the binding of substrate by active sites is a cooperative endeavor. The conformation of the enzyme and the substrate change dynamically during binding, catalysis, and product release to destabilize the ES and EP (enzyme-product) complexes and stabilize the transition state (or create a new one). The specificity of this process is illustrated schematically in Figure C18.2 for the enzyme hexokinase, which adds a phosphate group to six-carbon sugars but not to smaller molecules, even those of similar size and shape like glycerol ($C_3H_5(OH)_3$) shown in the inset. The protein consists of two subunits (domains) with a cleft in between, in which the residues of the active site reside. Binding of glucose causes a conformational change in the protein that brings the two domains into closer contact, thus creating the active site. This induced-fit mechanism explains how enzymes have evolved to confer such great selectivity; catalysis depends upon the properties of both the enzyme and the substrate, a very restrictive set of requirements.

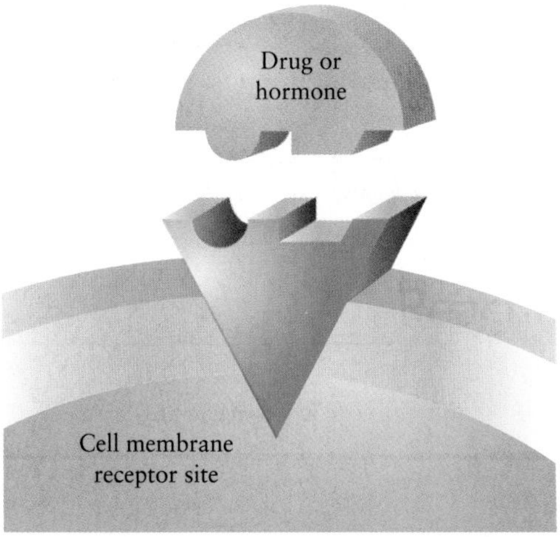

FIGURE C18.1 Lock and key mechanism.

Plotting the free-energy system as a function of reaction coordinate helps us understand how enzymes catalyze reactions from the point of view of chemical thermodynamics. Molecular models (see later) help us visualize the mechanisms by which they operate. Figure C18.3 shows a plot of the Gibbs free energy for the uncatalyzed reaction (a) as well as for the enzyme-catalyzed reaction (b). The activated complex for the uncatalyzed reaction is represented by the symbol $X^{\dagger}$, and the enzyme transition-state complex in the enzyme-catalyzed reaction is represented by the symbol $EX^{\dagger}$. The energy of the $EX^{\dagger}$complex is certainly lower than the energy of the $X^{\dagger}$complex for the uncatalyzed reaction, which suggests a mechanism for accelerating the rate by lowering the activation energy barrier. Notice, however, that there is a shallow well that corresponds to the binding energy of the ES complex. Enzymes only catalyze reactions when they destabilize (lower the energy of) the transition state complex more than they stabilize the ES complex. Let's see how this comes about by separately analyzing the enthalpic and entropic contributions to the Gibbs free energy associated with binding of substrate.

Attractive intermolecular interactions (see Section 10.2) ensure that the binding of substrate is exothermic, which favors spontaneity. Strong binding of substrate,

FIGURE C18.2 The enzyme hexokinase showing (a) the active site before binding substrate and (b) binding of glucose accompanied by global structural changes that brings the two domains together and closes off the active site. The inset shows the relative sizes of glucose, glycerol and water.

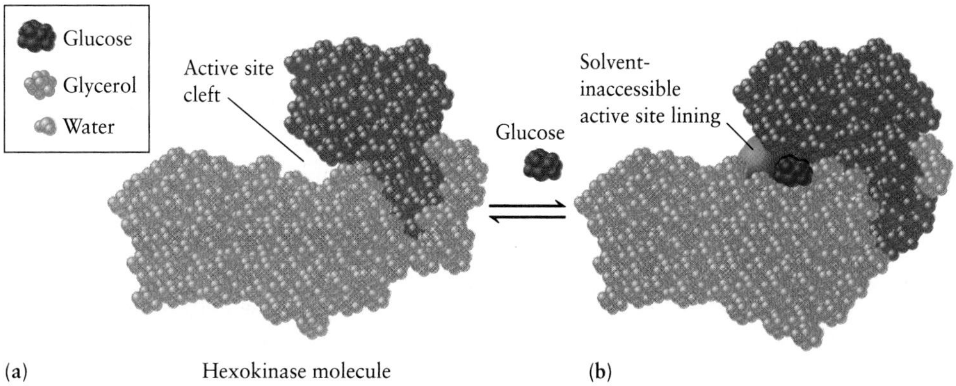

however, is undesirable, for it would create a permanent ES complex, making an enzyme unavailable for catalysis. The key to efficient catalytic activity is the fine tuning of the ES stabilization energy that allows for selective molecular recognition and binding without rendering the enzyme inactive. Two effects counteract the strong driving force provided by enthalpy. First, there is a loss of entropy that accompanies the formation of the ES complex from E and S, particularly due to the reduction in entropy of small molecule substrates as they effectively leave solution. Second, the ES complex is destabilized to some extent by strain, distortion, and loss of solvent. Figure C18.3 shows how the combination of transition state stabilization and ES complex destabilization leads to a reduction in the activation energy.

What are the mechanisms by which enzymes actually work to catalyze reactions? Structures of the active sites of enzymes, with and without bound substrates, have been investigated in great detail by X-ray crystallography in the solid state and by a number of spectroscopic techniques, such as NMR (see Section 20.4), in solution. These studies, supported by molecular dynamics simulations, have led to the conclusion that the "preorganization" of the active site prepares it for its catalytic role in which it directs the substrate to adopt the conformation most favorable for effecting the reaction of interest. We have already shown, in the *Connection to Biology* in Chapter 7, how the three amino acid residues that comprise the active site of the enzyme chymotrypsin are organized in space to optimally carry out the hydrolysis of peptide bonds. Protein motion is essential to catalysis: to configure the active site to recognize and bind substrates in optimal conformations, to locate catalytic groups near where they are needed, facilitate bond breaking and bond formation, and, finally, to assist in the conversion to products and their release.

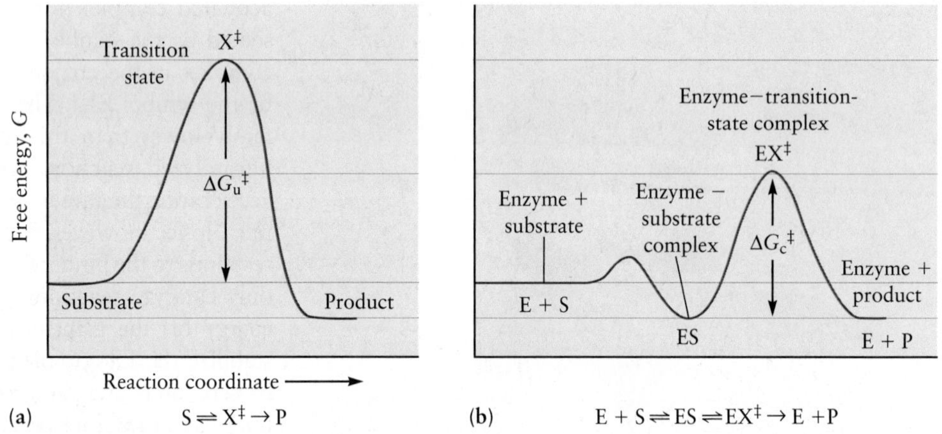

FIGURE C18.3 (a) Free energy profile for the uncatalyzed conversion of substrate to product. (b) Energy profile for the enzyme-catalyzed reaction that shows the preferential stabilization of the enzyme transition-state complex with respect to the enzyme substrate complex.

CHAPTER SUMMARY

Chemical kinetics is the study of the rates of chemical reactions and the factors that control those rates. The rates of chemical reactions depend on the concentrations of all of the reactants and products as well as the temperature. The concentration dependence of reaction rates is given by empirical rate laws in which the rate depends on the concentration of each reactant raised to some power that is not, in general, related to the stoichiometric coefficients in the balanced equation for the reaction. Rate laws may be determined by measuring the initial rate of a reaction as a function of initial concentration or by inspecting the concentration-time profile over the course of the reaction. Concentrations decrease exponentially with time for first-order reactions and the half-life is a characteristic time scale for the reaction. Most chemical reactions occur as a series of elementary reactions, which constitute the reaction mechanism. The steady-state approximation provides a simple framework to analyze the kinetics of multistep reactions. This approximation, which is widely used in many areas of chemistry and biochemistry, shows how a reaction may display either first- or second-order kinetics, depending on the concentration of inert collision partners. It also accounts for the rate-limiting step, which determines the overall rate of the reaction. Spectroscopic detection or chemical trapping of reaction intermediates is a powerful way to establish a reaction mechanism. The rates of elementary reactions increase with increasing temperature. As temperatures rise, more molecules have enough energy to overcome the potential energy barrier to the reaction and proceed to form products. Microscopic theories have been developed that provide molecular-level understanding of the factors that govern the rates of chemical reactions. These theories are based on the collision theory, which asserts that the rates are proportional to the rates of collisions times the probability of particular collisions leading to reaction. The probabilities are expressed in terms of the reactive cross section, which is interpreted as the effective size two colliding molecules present to each other that ensures reaction upon every collision. Catalysts increase the rates of chemical reactions without being either consumed or produced. Synthetic catalysts are widely used in industrial processes, and natural catalysts called enzymes catalyze and regulate the rates of virtually all biological processes. Catalysts work by lowering the activation barrier for chemical reactions.

CONCEPTS AND SKILLS

Section 18.1 – Rates of Chemical Reactions

Describe experimental methods for measuring average and instantaneous rates (Problems 1–2).

- The rates of chemical reactions are measured by determining the concentrations of reactants and products as a function of time. Concentrations may be measured by a variety of chemical or instrumental methods. The average rate is the change in concentration measured over a period of time divided by the time. The instantaneous rate is the slope of a line that is tangent to the concentration-time plot at the time of interest.

Section 18.2 – Rate Laws

Deduce rate laws and reaction orders from experimental measurements of the dependence of reaction rates on concentrations (Problems 5–8).

- Rate laws and reaction orders are deduced from the concentration dependence of the reaction rate as follows. The reaction is run under two conditions with the concentration of only one reactant being changed; if the rate doubles when the concentration of that reactant doubles, then the reaction is first order with respect to that reactant; if it quadruples, then it is second order. The process is repeated, changing the concentrations of each of the reactants in a pair of experiments while keeping all of the other concentrations and the other conditions the same. The rate constant is determined from any single experiment, once the dependence on concentrations has been determined, by dividing the rate by the product of the concentrations raised to their respective orders.

Use the integrated rate laws for first- and second-order reactions to calculate the concentrations remaining after a certain elapsed time (Problems 9–18).

- Concentrations at a given elapsed time are calculated by solving the rate equations, inserting the initial conditions and the rate constant as known values.
  - First-order reactions $c = c_0\, e^{-kt}$
  - Second-order reactions $\dfrac{1}{c} = \dfrac{1}{c_0} + 2kt$

18.3 – Reaction Mechanisms

Describe the relationship between the equilibrium constant for an elementary reaction and the corresponding forward and reverse rate constants (Problems 23–24).

- The equilibrium constant is the ratio of rate constants for the forward and reverse reactions $K_{eq} = k_f/k_r$ as required by the principle of detailed balance for elementary processes.

18.4 – Reaction Mechanisms and Rate

Deduce the rate law from a mechanism characterized by a single rate-determining step (Problems 25–30).

- The rate law for a mechanism characterized by a single rate-limiting step is given by the rate law for that elementary process, which is determined by inspection.

Use the steady-state approximation to deduce rate laws when no single rate-determining step exists (Problems 31–34).

- The rate law for the dissociation of a molecule in the Lindemann mechanism is $\text{rate} = \dfrac{k_1 k_2[A][M]}{k_2 + k_{-1}[M]}$ with limiting forms: rate $= k_1[A][M]$ at low pressure and rate $= (k_1/k_{-1})k_2[A]$ at high pressure.
- Rate laws for other reactions may be deduced using the approximation that the concentration of each reactive intermediate is roughly constant in time.

18.5 – Effect of Temperature on Reaction Rates

Calculate Arrhenius factors and activation energies from measurements of the temperature dependence of rate constants (Problems 35–40).

- Arrhenius factors are calculated by plotting ln k versus $1/T$, the slope of which is $-E_a/R$.

Discuss the connection between activation energy and the energy distribution of molecules, and relate the forward and reverse activation energies to each other through thermodynamics (Problems 41–42).

- The activation energy is roughly equal to the barrier energy, which is the minimum amount of energy necessary to initiate a chemical reaction. The fraction of collisions having a relative kinetic energy that is greater than or equal to the barrier is determined by the area under the curve of the Maxwell–Boltzmann energy distribution of relative speeds, shown in Fig. 18.10 (and given later in Eq. 18.7), between E_a and infinity. The fraction of collisions with energy greater than E_a increases with increasing temperature. The difference between the barrier heights for the forward and reverse reactions is the 0 K enthalpy change of the reaction.

18.6 – Molecular Theories of Elementary Reactions

Outline the quantitative calculation of rate constants, using the collision theory of gases (Problems 45–46).

- Collision theory calculates the rates of collisions between molecules using the kinetic theory of gases, and estimates the rates of reactive collisions by asserting that reactions occur only when the relative kinetic energy of the colliding molecules along the line of centers on impact exceeds a minimum energy ε_B.

18.7 – Reactions in Solution

Identify reactions as diffusion-controlled or activation-energy controlled (Problems 47–48).

- Diffusion-controlled reactions are second-order reactions with rate-limiting steps determined by an effective diffusion constant that is the sum of the diffusion constants of the reactants. Activation-energy controlled reactions are ones for which the rate of diffusion for reactants A and B to encounter each other is much faster than the rate at which the AB complex forms products.

18.8 – Catalysis

Describe several types of catalysts and their effects on chemical reactions.

- Catalysts increase the rates of chemical reactions by lowering the activation barrier, typically by providing alternate paths.
- Homogeneous catalysts are present in the same phase as the reaction being catalyzed, whereas heterogeneous catalysts are present in a different phase.
- Enzymes are biological catalysts that bind substrates with exquisite selectivity, position reactants at optimal locations, and stabilize transition states, all of which leads to lower activation barriers than those in the uncatalyzed reaction.

Relate the rate of an enzyme-catalyzed reaction to the concentrations of substrate and enzyme in the reaction mixture (Problems 49–50).

- Enzyme-catalyzed reactions often follow the Michaelis–Menten equation, which shows saturation of the rate at high substrate concentrations.

$$\frac{d[P]}{dt} = k_2[ES] = \frac{k_2[E_T][S]}{K_m + [S]} \qquad [18.20]$$

CUMULATIVE EXERCISE

FIGURE 18.28 Air pollution is damaging the white marble of one of the world's most famous monuments, the Taj Mahal. Particulate matter and acid rain both contribute to the problem.

Sulfite and Sulfate Kinetics in Atmospheric Chemistry

Sulfur dioxide dissolves in water droplets (fog, clouds, and rain) in the atmosphere (Fig. 18.28) and reacts according to the equation

$$SO_2(aq) + 2\ H_2O(\ell) \longrightarrow HSO_3^-(aq) + H_3O^+(aq)$$

$HSO_3^-(aq)$ (hydrogen sulfite ion) is then slowly oxidized by oxygen that is also dissolved in the droplets:

$$2\ HSO_3^-(aq) + O_2(aq) + 2\ H_2O(\ell) \longrightarrow 2\ SO_4^{2-}(aq) + 2\ H_3O^+(aq)$$

Although the second reaction has been studied for many years, only recently was it shown to proceed by the steps

$$2\ HSO_3^-(aq) + O_2(aq) \longrightarrow S_2O_7^{2-}(aq) + H_2O(\ell) \qquad \text{(fast)}$$

$$S_2O_7^{2-}(aq) + 3\ H_2O(\ell) \longrightarrow 2\ SO_4^{2-}(aq) + 2\ H_3O^+(aq) \qquad \text{(slow)}$$

The reactive intermediate, $S_2O_7^{2-}(aq)$, which had not been detected in earlier studies of this reaction, is well known in other reactions. It is the *disulfate* ion.

In an experiment at 25°C, a solution was mixed with the realistic initial concentrations of 0.270 M $HSO_3^-(aq)$ and 0.0135 M $O_2(aq)$. The initial pH was 3.90. The following table tells what happened in the solution, beginning at the moment of mixing.

Time (s)	$[HSO_3^-]$ (M)	$[O_2]$ (M)	$[S_2O_7^{2-}]$ (M)	$[HSO_4^-] + [SO_4^{2-}]$ (M)
0.000	0.270	0.0135	0.000	0.000
0.010	0.243	0.000	13.5×10^{-3}	0.000
10.0	0.243	0.000	11.8×10^{-3}	3.40×10^{-3}
45.0	0.243	0.000	7.42×10^{-3}	12.2×10^{-3}
90.0	0.243	0.000	4.08×10^{-3}	18.8×10^{-3}
150.0	0.243	0.000	1.84×10^{-3}	23.3×10^{-3}
450.0	0.243	0.000	0.034×10^{-3}	26.9×10^{-3}
600.0	0.243	0.000	0.005×10^{-3}	27.0×10^{-3}

(a) Determine the average rate of increase of the total of the concentrations of the sulfate plus hydrogen sulfate ions during the first 10 s of the experiment.

(b) Determine the average rate of disappearance of hydrogen sulfite ion during the first 0.010 s of the experiment.

(c) Explain why the hydrogen sulfite ion stops disappearing after 0.010 s.

(d) Plot the concentration of disulfate ion versus time, and use the curve to graphically estimate the instantaneous rate of disappearance of disulfate ion 90.0 s after the reaction starts.

(e) Determine the order with respect to the disulfate ion of the second step of the conversion, and the rate constant of that step.

(f) Determine the half-life of the second step of the conversion process.

(g) At 15°C the rate constant of the second step of the conversion is only 62% of its value at 25°C. Compute the activation energy of the second step.

(h) The first step of the conversion occurs much faster when 1.0×10^{-6} M $Fe^{2+}(aq)$ ion is added (but the rate of the second step is unaffected). What role does Fe^{2+} play?

(i) Write a balanced equation for the overall reaction that gives sulfuric acid from SO_2 dissolved in water droplets in the air.

Answers

(a) 3.40×10^{-4} mol L^{-1} s^{-1}

(b) 2.7 mol L^{-1} s^{-1}

(c) All of the oxygen is consumed, so the first step of the process is over.

(d) 5.43×10^{-5} mol L^{-1} s^{-1}

(e) First order; $k = 0.0133$ s^{-1}

(f) 52 s

(g) 34 kJ mol^{-1}

(h) Fe^{2+} acts as a catalyst.

(i) $2\ SO_2(aq) + O_2(aq) + 6\ H_2O(\ell) \longrightarrow SO_4^{2-}(aq) + 4\ H_3O^+(aq)$

PROBLEMS

Answers to problems whose numbers are red appear in Appendix G. Problems that are more challenging are indicated with asterisks.

18.1 Rates of Chemical Reactions

1. Use Figure 18.3 to estimate graphically the instantaneous rate of production of NO at $t = 200$ s.

2. Use Figure 18.3 to estimate graphically the instantaneous rate of production of NO at $t = 100$ s.

3. Give three related expressions for the rate of the reaction

$$N_2(g) + 3\ H_2(g) \longrightarrow 2\ NH_3(g)$$

assuming that the concentrations of any intermediates are constant and that the volume of the reaction vessel does not change.

4. Give four related expressions for the rate of the reaction

$$2\ H_2CO(g) + O_2(g) \longrightarrow 2\ CO(g) + 2\ H_2O(g)$$

assuming that the concentrations of any intermediates are constant and that the volume of the reaction vessel does not change.

18.2 Rate Laws

5. Nitrogen oxide reacts with hydrogen at elevated temperatures according to the following chemical equation:

$$2\ NO(g) + 2\ H_2(g) \longrightarrow N_2(g) + 2\ H_2O(g)$$

It is observed that, when the concentration of H_2 is cut in half, the rate of the reaction is also cut in half. When the concentration of NO is multiplied by 10, the rate of the reaction increases by a factor of 100.

(a) Write the rate expression for this reaction, and give the units of the rate constant k.

(b) If [NO] were multiplied by 3 and [H_2] by 2, what change in the rate would be observed?

6. In the presence of vanadium oxide, $SO_2(g)$ reacts with an excess of oxygen to give $SO_3(g)$:

$$SO_2(g) + \tfrac{1}{2}\ O_2(g) \xrightarrow{V_2O_5} SO_3(g)$$

This reaction is an important step in the manufacture of sulfuric acid. It is observed that tripling the SO_2 concentration increases the rate by a factor of 3, but tripling the SO_3 concentration decreases the rate by a factor of $1.7 \approx \sqrt{3}$. The rate is insensitive to the O_2 concentration as long as an excess of oxygen is present.

(a) Write the rate expression for this reaction, and give the units of the rate constant *k*.

(b) If $[SO_2]$ is multiplied by 2 and $[SO_3]$ by 4 but all other conditions are unchanged, what change in the rate will be observed?

7. In a study of the reaction of pyridine (C_5H_5N) with methyl iodide (CH_3I) in a benzene solution, the following set of initial reaction rates was measured at 25°C for different initial concentrations of the two reactants:

$[C_5H_5N]$ (mol L^{-1})	$[CH_3I]$ (mol L^{-1})	Rate (mol L^{-1} s^{-1})
1.00×10^{-4}	1.00×10^{-4}	7.5×10^{-7}
2.00×10^{-4}	2.00×10^{-4}	3.0×10^{-6}
2.00×10^{-4}	4.00×10^{-4}	6.0×10^{-6}

(a) Write the rate expression for this reaction.

(b) Calculate the rate constant *k* and give its units.

(c) Predict the initial reaction rate for a solution in which $[C_5H_5N]$ is 5.0×10^{-5} M and $[CH_3I]$ is 2.0×10^{-5} M.

8. The rate for the oxidation of iron(II) by cerium(IV)

$$Ce^{4+}(aq) + Fe^{2+}(aq) \longrightarrow Ce^{3+}(aq) + Fe^{3+}(aq)$$

is measured at several different initial concentrations of the two reactants:

$[Ce^{4+}]$ (mol L^{-1})	$[Fe^{2+}]$ (mol L^{-1})	Rate (mol L^{-1} s^{-1})
1.1×10^{-5}	1.8×10^{-5}	2.0×10^{-7}
1.1×10^{-5}	2.8×10^{-5}	3.1×10^{-7}
3.4×10^{-5}	2.8×10^{-5}	9.5×10^{-7}

(a) Write the rate expression for this reaction.

(b) Calculate the rate constant *k* and give its units.

(c) Predict the initial reaction rate for a solution in which $[Ce^{4+}]$ is 2.6×10^{-5} M and $[Fe^{2+}]$ is 1.3×10^{-5} M.

9. The reaction $SO_2Cl_2(g) \longrightarrow SO_2(g) + Cl_2(g)$ is first order, with a rate constant of 2.2×10^{-5} s^{-1} at 320°C. The partial pressure of $SO_2Cl_2(g)$ in a sealed vessel at 320°C is 1.0 atm. How long will it take for the partial pressure of $SO_2Cl_2(g)$ to fall to 0.50 atm?

10. The reaction $FClO_2(g) \longrightarrow FClO(g) + O(g)$ is first order with a rate constant of 6.76×10^{-4} s^{-1} at 322°C.

(a) Calculate the half-life of the reaction at 322°C.

(b) If the initial partial pressure of $FClO_2$ in a container at 322°C is 0.040 atm, how long will it take to fall to 0.010 atm?

11. The decomposition of benzene diazonium chloride

$$C_6H_5N_2Cl \longrightarrow C_6H_5Cl + N_2$$

follows first-order kinetics with a rate constant of 4.3×10^{-5} s^{-1} at 20°C. If the initial partial pressure of $C_6H_5N_2Cl$ is 0.0088 atm, calculate its partial pressure after 10.0 hours.

12. At 600 K, the rate constant for the first-order decomposition of nitroethane

$$CH_3CH_2NO_2(g) \longrightarrow C_2H_4(g) + HNO_2(g)$$

is 1.9×10^{-4} s^{-1}. A sample of $CH_3CH_2NO_2$ is heated to 600 K, at which point its initial partial pressure is measured to be 0.078 atm. Calculate its partial pressure after 3.0 hours.

13. Chloroethane decomposes at elevated temperatures according to the reaction

$$C_2H_5Cl(g) \longrightarrow C_2H_4(g) + HCl(g)$$

This reaction obeys first-order kinetics. After 340 s at 800 K, a measurement shows that the concentration of C_2H_5Cl has decreased from 0.0098 mol L^{-1} to 0.0016 mol L^{-1}. Calculate the rate constant *k* at 800 K.

14. The isomerization reaction

$$CH_3NC \longrightarrow CH_3CN$$

obeys the first-order rate law

$$\text{rate} = -k[CH_3NC]$$

in the presence of an excess of argon. Measurements at 500 K reveal that in 520 s the concentration of CH_3NC decreases to 71% of its original value. Calculate the rate constant *k* of the reaction at 500 K.

15. At 25°C in CCl_4 solvent, the reaction

$$I + I \longrightarrow I_2$$

is second order in the concentration of the iodine atoms. The rate constant *k* has been measured as 8.2×10^9 L mol^{-1} s^{-1}. Suppose the initial concentration of I atoms is 1.00×10^{-4} M. Calculate their concentration after 2.0×10^{-6} s.

16. HO_2 is a highly reactive chemical species that plays a role in atmospheric chemistry. The rate of the gas-phase reaction

$$HO_2(g) + HO_2(g) \longrightarrow H_2O_2(g) + O_2(g)$$

is second order in $[HO_2]$, with a rate constant at 25°C of 1.4×10^9 L mol^{-1} s^{-1}. Suppose some HO_2 with an initial concentration of 2.0×10^{-8} M could be confined at 25°C. Calculate the concentration that would remain after 1.0 s, assuming no other reactions take place.

17. The rate for the reaction

$$OH^-(aq) + NH_4^+(aq) \longrightarrow H_2O(\ell) + NH_3(aq)$$

is first order in both OH^- and NH_4^+ concentrations, and the rate constant *k* at 20°C is 3.4×10^{10} L mol^{-1} s^{-1}. Suppose 1.00 L of a 0.0010 M NaOH solution is rapidly mixed with the same volume of 0.0010 M NH_4Cl solution. Calculate the time (in seconds) required for the OH^- concentration to decrease to a value of 1.0×10^{-5} M.

18. The rate for the reaction

$$OH^-(aq) + HCN(aq) \longrightarrow H_2O(\ell) + CN^-(aq)$$

is first order in both OH^- and HCN concentrations, and the rate constant k at 25°C is 3.7×10^9 L mol^{-1} s^{-1}. Suppose 0.500 L of a 0.0020 M NaOH solution is rapidly mixed with the same volume of a 0.0020 M HCN solution. Calculate the time (in seconds) required for the OH^- concentration to decrease to a value of 1.0×10^{-4} M.

18.3 Reaction Mechanisms

19. Identify each of the following elementary reactions as unimolecular, bimolecular, or termolecular, and write the rate expression.
(a) $HCO + O_2 \longrightarrow HO_2 + CO$
(b) $CH_3 + O_2 + N_2 \longrightarrow CH_3O_2 + N_2$
(c) $HO_2NO_2 \longrightarrow HO_2 + NO_2$

20. Identify each of the following elementary reactions as unimolecular, bimolecular, or termolecular, and write the rate expression.
(a) $BrONO_2 \longrightarrow BrO + NO_2$
(b) $HO + NO_2 + Ar \longrightarrow HNO_3 + Ar$
(c) $O + H_2S \longrightarrow HO + HS$

21. Consider the following reaction mechanism:

$$H_2O_2 \longrightarrow H_2O + O$$

$$O + CF_2Cl_2 \longrightarrow ClO + CF_2Cl$$

$$ClO + O_3 \longrightarrow Cl + 2\ O_2$$

$$Cl + CF_2Cl \longrightarrow CF_2Cl_2$$

(a) What is the molecularity of each elementary step?
(b) Write the overall equation for the reaction.
(c) Identify the reaction intermediate(s).

22. Consider the following reaction mechanism:

$$NO_2Cl \longrightarrow NO_2 + Cl$$

$$Cl + H_2O \longrightarrow HCl + OH$$

$$OH + NO_2 + N_2 \longrightarrow HNO_3 + N_2$$

(a) What is the molecularity of each elementary step?
(b) Write the overall equation for the reaction.
(c) Identify the reaction intermediate(s).

23. The rate constant of the elementary reaction

$$BrO(g) + NO(g) \longrightarrow Br(g) + NO_2(g)$$

is 1.3×10^{10} L mol^{-1} s^{-1} at 25°C, and its equilibrium constant is 5.0×10^{10} at this temperature. Calculate the rate constant at 25°C of the elementary reaction

$$Br(g) + NO_2(g) \longrightarrow BrO(g) + NO(g)$$

24. The compound $IrH_3(CO)(P(C_6H_5)_3)_2$ exists in two forms: the meridional ("mer") and facial ("fac"). At 25°C in a nonaqueous solvent, the reaction mer $\longrightarrow$ fac has a rate constant of 2.33 s^{-1}, and the reaction fac $\longrightarrow$ mer has a rate constant of 2.10 s^{-1}. What is the equilibrium constant of the mer-to-fac reaction at 25°C?

18.4 Reaction Mechanisms and Rate

25. Write the overall reaction and rate laws that correspond to the following reaction mechanisms. Be sure to eliminate intermediates from the answers.

(a) $A + B \underset{k_{-1}}{\overset{k_1}{\rightleftharpoons}} C + D$ (fast equilibrium)

$C + E \overset{k_2}{\longrightarrow} F$ (slow)

(b) $A \underset{k_{-1}}{\overset{k_1}{\rightleftharpoons}} B + C$ (fast equilibrium)

$C + D \underset{k_{-2}}{\overset{k_2}{\rightleftharpoons}} E$ (fast equilibrium)

$E \overset{k_3}{\longrightarrow} F$ (slow)

26. Write the overall reaction and the rate laws that correspond to the following reaction mechanisms. Be sure to eliminate intermediates from the answers.

(a) $2\ A + B \underset{k_{-1}}{\overset{k_1}{\rightleftharpoons}} D$ (fast equilibrium)

$D + B \overset{k_2}{\longrightarrow} E + F$ (slow)

$F \overset{k_3}{\longrightarrow} G$ (fast)

(b) $A + B \underset{k_{-1}}{\overset{k_1}{\rightleftharpoons}} C$ (fast equilibrium)

$C + D \underset{k_{-2}}{\overset{k_2}{\rightleftharpoons}} F$ (fast equilibrium)

$F \overset{k_3}{\longrightarrow} G$ (slow)

27. HCl reacts with propene (CH_3CHCH_2) in the gas phase according to the overall reaction

$$HCl + CH_3CHCH_2 \longrightarrow CH_3CHClCH_3$$

The experimental rate expression is

$$\text{rate} = k[HCl]^3[CH_3CHCH_2]$$

Which, if any, of the following mechanisms are consistent with the observed rate expression?

(a) $HCl + HCl \rightleftharpoons H + HCl_2$ (fast equilibrium)
$H + CH_3CHCH_2 \longrightarrow CH_3CHCH_3$ (slow)
$HCl_2 + CH_3CHCH_3 \longrightarrow CH_3CHClCH_3 + HCl$ (fast)

(b) $HCl + HCl \rightleftharpoons H_2Cl_2$ (fast equilibrium)
$HCl + CH_3CHCH_2 \rightleftharpoons CH_3CHClCH_3^*$ (fast equilibrium)
$CH_3CHClCH_3^* + H_2Cl_2 \longrightarrow CH_3CHClCH_3 + 2\ HCl$ (slow)

(c) $HCl + CH_3CHCH_2 \rightleftharpoons H + CH_3CHClCH_2$ (fast equilibrium)
$H + HCl \rightleftharpoons H_2Cl$ (fast equilibrium)
$H_2Cl + CH_3CHClCH_2 \longrightarrow HCl + CH_3CHClCH_3$ (slow)

28. Chlorine reacts with hydrogen sulfide in aqueous solution

$$Cl_2(aq) + H_2S(aq) \longrightarrow S(s) + 2\,H^+(aq) + 2\,Cl^-(aq)$$

in a second-order reaction that follows the rate expression

$$\text{rate} = k[Cl_2][H_2S]$$

Which, if any, of the following mechanisms are consistent with the observed rate expression?

(a) $Cl_2 + H_2S \longrightarrow H^+ + Cl^- + Cl^+ + HS^-$ (slow)
$Cl^+ + HS^- \longrightarrow H^+ + Cl^- + S$ (fast)

(b) $H_2S \rightleftharpoons HS^- + H^+$ (fast equilibrium)
$HS^- + Cl_2 \longrightarrow 2\,Cl^- + S + H^+$ (slow)

(c) $H_2S \rightleftharpoons HS^- + H^+$ (fast equilibrium)
$H^+ + Cl_2 \rightleftharpoons H^+ + Cl^- + Cl^+$ (fast equilibrium)
$Cl^+ + HS^- \longrightarrow H^+ + Cl^- + S$ (slow)

29. Nitryl chloride is a reactive gas with a normal boiling point of −16°C. Its decomposition to nitrogen dioxide and chlorine is described by the equation

$$2\,NO_2Cl \longrightarrow 2\,NO_2 + Cl_2$$

The rate expression for this reaction has the form

$$\text{rate} = k[NO_2Cl]$$

Which, if any, of the following mechanisms are consistent with the observed rate expression?

(a) $NO_2Cl \longrightarrow NO_2 + Cl$ (slow)
$Cl + NO_2Cl \longrightarrow NO_2 + Cl_2$ (fast)

(b) $2\,NO_2Cl \rightleftharpoons N_2O_4 + Cl_2$ (fast equilibrium)
$N_2O_4 \longrightarrow 2\,NO_2$ (slow)

(c) $2\,NO_2Cl \rightleftharpoons ClO_2 + N_2O + ClO$ (fast equilibrium)
$N_2O + ClO_2 \rightleftharpoons NO_2 + NOCl$ (fast equilibrium)
$NOCl + ClO \longrightarrow NO_2 + Cl_2$ (slow)

30. Ozone in the upper atmosphere is decomposed by nitrogen oxide through the reaction

$$O_3 + NO \longrightarrow O_2 + NO_2$$

The experimental rate expression for this reaction is

$$\text{rate} = k[O_3][NO]$$

Which, if any, of the following mechanisms are consistent with the observed rate expression?

(a) $O_3 + NO \longrightarrow O + NO_3$ (slow)
$O + O_3 \longrightarrow 2\,O_2$ (fast)
$NO_3 + NO \longrightarrow 2\,NO_2$ (fast)

(b) $O_3 + NO \longrightarrow O_2\ NO_2$ (slow)

(c) $NO + NO \rightleftharpoons N_2O_2$ (fast equilibrium)
$N_2O_2 + O_3 \longrightarrow NO_2 + 2\,O_2$ (slow)

31. Consider the mechanism of problem 25(a). Suppose *no* assumptions are made about the relative rates of the steps. By making a steady-state approximation for the concentration of the intermediate (C), express the rate of production of the product (F) in terms of the concentrations of A, B, D, and E. In what limit does this reduce to the result of problem 25(a)?

32. Consider the mechanism of problem 25(b). Suppose *no* assumptions are made about the relative rates of the steps. By making a steady-state approximation for the concentrations of the intermediates (C and E), express the rate of production of the product (F) in terms of the concentrations of A, B, and D. In what limit does this reduce to the result of problem 25(b)?

33. The mechanism for the decomposition of NO_2Cl is

$$NO_2Cl \underset{k_{-1}}{\overset{k_1}{\rightleftharpoons}} NO_2 + Cl$$

$$NO_2Cl + Cl \xrightarrow{k_2} NO_2 + Cl_2$$

By making a steady-state approximation for [Cl], express the rate of appearance of Cl_2 in terms of the concentrations of NO_2Cl and NO_2.

34. A key step in the formation of sulfuric acid from dissolved SO_2 in acid precipitation is the oxidation of hydrogen sulfite ion by hydrogen peroxide:

$$HSO_3^-(aq) + H_2O_2(aq) \longrightarrow HSO_4^-(aq) + H_2O(\ell)$$

The mechanism involves peroxymonosulfurous acid, SO_2OOH^-:

$$HSO_3^-(aq) + H_2O_2(aq) \underset{k_{-1}}{\overset{k_1}{\rightleftharpoons}} SO_2OOH^-(aq) + H_2O(\ell)$$

$$SO_2OOH^-(aq) + H_3O^+(aq) \xrightarrow{k_2} HSO_4^-(aq) + H_3O^+(aq)$$

By making a steady-state approximation for the reactive intermediate concentration, $[SO_2OOH^-(aq)]$, express the rate of formation of $HSO_4^-(aq)$ in terms of the concentrations of $HSO_3^-(aq)$, $H_2O_2(aq)$, and $H_3O^+(aq)$.

18.5 Effect of Temperature on Reaction Rates

35. The rate of the elementary reaction

$$Ar + O_2 \longrightarrow Ar + O + O$$

has been studied as a function of temperature between 5000 and 18,000 K. The following data were obtained for the rate constant *k*:

Temperature (K)	k (L mol^{-1} s^{-1})
5,000	5.49×10^6
10,000	9.86×10^8
15,000	5.09×10^9
18,000	8.60×10^9

(a) Calculate the activation energy of this reaction.
(b) Calculate the factor *A* in the Arrhenius equation for the temperature dependence of the rate constant.

36. The gas-phase reaction

$$H + D_2 \longrightarrow HD + D$$

is the exchange of isotopes of hydrogen of atomic mass 1 (H) and 2 (D, deuterium). The following data were obtained for the rate constant *k* of this reaction:

Temperature (K)	k (L mol^{-1} s^{-1})
299	1.56×10^4
327	3.77×10^4
346	7.6×10^4
440	10^6
549	1.07×10^6
745	8.7×10^7

(a) Calculate the activation energy of this reaction.
(b) Calculate the factor A in the Arrhenius equation for the temperature dependence of the rate constant.

37. The rate constant of the elementary reaction

$$BH_4^-(aq) + NH_4^+(aq) \longrightarrow BH_3NH_3(aq) + H_2(g)$$

is $k = 1.94 \times 10^{-4}$ L mol^{-1} s^{-1} at 30.0°C, and the reaction has an activation energy of 161 kJ mol^{-1}.
(a) Compute the rate constant of the reaction at a temperature of 40.0°C.
(b) After equal concentrations of $BH_4^-(aq)$ and $NH_4^+(aq)$ are mixed at 30.0°C, 1.00×10^4 s is required for half of them to be consumed. How long will it take to consume half of the reactants if an identical experiment is performed at 40.0°C?

38. Dinitrogen tetraoxide (N_2O_4) decomposes spontaneously at room temperature in the gas phase:

$$N_2O_4(g) \longrightarrow 2\ NO_2(g)$$

The rate law governing the disappearance of N_2O_4 with time is

$$-\frac{d[N_2O_4]}{dt} = k[N_2O_4]$$

At 30°C, $k = 5.1 \times 10^6$ s^{-1} and the activation energy for the reaction is 54.0 kJ mol^{-1}.
(a) Calculate the time (in seconds) required for the partial pressure of $N_2O_4(g)$ to decrease from 0.10 atm to 0.010 atm at 30°C.
(b) Repeat the calculation of part (a) at 300°C.

39. The activation energy for the isomerization reaction of CH_3NC in Problem 14 is 161 kJ mol^{-1}, and the reaction rate constant at 600 K is 0.41 s^{-1}.
(a) Calculate the Arrhenius factor A for this reaction.
(b) Calculate the rate constant for this reaction at 1000 K.

40. Cyclopropane isomerizes to propylene according to a first-order reaction:

$$\text{cyclopropane} \longrightarrow \text{propylene}$$

The activation energy is $E_a = 272$ kJ mol^{-1}. At 500°C, the reaction rate constant is 6.1×10^{-4} s^{-1}.
(a) Calculate the Arrhenius factor A for this reaction.
(b) Calculate the rate constant for this reaction at 25°C.

41. The activation energy of the gas-phase reaction

$$OH(g) + HCl(g) \longrightarrow H_2O(g) + Cl(g)$$

is 3.5 kJ mol^{-1}, and the change in the internal energy in the reaction is $\Delta U = -66.8$ kJ mol^{-1}. Calculate the activation energy of the reaction

$$H_2O(g) + Cl(g) \longrightarrow OH(g) + HCl(g)$$

42. The compound HOCl is known, but the related compound HClO, with a different order for the atoms in the molecule, is not known. Calculations suggest that the activation energy for the conversion HOCl ⟶ HClO is 311 kJ mol^{-1} and that for the conversion HClO ⟶ HOCl is 31 kJ mol^{-1}. Estimate ΔU for the reaction HOCl ⟶ HClO.

18.6 Molecular Theories of Elementary Reactions

43. The reaction of OH with hydrocarbons initiates the formation of atmospheric aerosols, and is an important elementary step in combustion. The rate constants for the reaction of OH with ethane measured between 822 and 1367 K are well fit to an Arrhenius expression with $k = (1.1 \times 10^{-10}\ \text{cm}^3\ \text{s}^{-1})\exp(-2537\ \text{K}/T)$. In contrast, the rate constant for the reaction of OH with ethene is nearly constant with temperature at $k = 6 \times 10^{-12}$ cm^3 s^{-1}. What can you conclude about the probable barrier heights for each reaction? For the OH + ethane reaction, would an OH radical with a speed equal to twice the average speed of OH at 1000 K have a chance of having a reactive collision with a ethane molecule that also had a speed equal to twice the average speed in the Maxwell-Boltzmann distribution for ethane molecules? Explain.

44. Calculate the maximum potential energy on impact that the collision described in problem 43 could achieve if the OH molecule were instead catching up to the ethane molecule from behind.

45. Use collision theory to estimate the preexponential factor in the rate constant for the elementary reaction

$$NOCl + NOCl \longrightarrow 2\ NO + Cl_2$$

at 25°C. Take the average diameter of an NOCl molecule to be 3.0×10^{-10} m and use the steric factor P from Table 18.1.

46. Use collision theory to estimate the preexponential factor in the rate constant for the elementary reaction

$$NO_2 + NO_2 \longrightarrow 2\ NO + O_2$$

at 500 K. Take the average diameter of an NO_2 molecule to be 2.6×10^{-10} m and use the steric factor P from Table 18.1.

18.7 Reactions in Solution

47. The rate constant for the reaction $CH_3Br + Cl^- \longrightarrow CH_3Cl + Br^-$ in acetone is 5.9×10^{-3} L mol^{-1} s^{-1}. Is this reaction diffusion-controlled or limited by a large activation energy?

48. The rate constant for the reaction $H^+ + HS^- \longrightarrow H_2S$ in aqueous solution is 7.5×10^{10} L mol^{-1} s^{-1}. Is this reaction diffusion-controlled or limited by a large activation energy?

18.8 Catalysis

49. Certain bacteria use the enzyme penicillinase to decompose penicillin and render it inactive. The Michaelis–Menten constants for this enzyme and substrate are $K_m = 5 \times 10^{-5}$ mol L^{-1} and $k_2 = 2 \times 10^3$ s^{-1}.
(a) What is the maximum rate of decomposition of penicillin if the enzyme concentration is 6×10^{-7} M?
(b) At what substrate concentration will the rate of decomposition be half that calculated in part (a)?

50. The conversion of dissolved carbon dioxide in blood to HCO_3^- and H_3O^+ is catalyzed by the enzyme carbonic anhydrase. The Michaelis–Menten constants for this enzyme and substrate are $K_m = 8 \times 10^{-5}$ mol L^{-1} and $k_2 = 6 \times 10^5$ s^{-1}.
(a) What is the maximum rate of reaction of carbon dioxide if the enzyme concentration is 5×10^{-6} M?
(b) At what CO_2 concentration will the rate of decomposition be 30% of that calculated in part (a)?

ADDITIONAL PROBLEMS

51. Hemoglobin molecules in blood bind oxygen and carry it to cells, where it takes part in metabolism. The binding of oxygen

$$\text{hemoglobin}(aq) + O_2(aq) \longrightarrow \text{hemoglobin} \cdot O_2(aq)$$

is first order in hemoglobin and first order in dissolved oxygen, with a rate constant of 4×10^7 L mol^{-1} s^{-1}. Calculate the initial rate at which oxygen will be bound to hemoglobin if the concentration of hemoglobin is 2×10^{-9} M and that of oxygen is 5×10^{-5} M.

***52.** Suppose 1.00 L of 9.95×10^{-3} M $S_2O_3^{2-}$ is mixed with 1.00 L of 2.52×10^{-3} M H_2O_2 at a pH of 7.0 and a temperature of 25°C. These species react by two competing pathways, represented by the balanced equations

$$S_2O_3^{2-} + 4\ H_2O_2 \longrightarrow 2\ SO_4^{2-} + H_2O + 2\ H_3O^+$$

$$2\ S_2O_3^{2-} + H_2O_2 + 2\ H_3O^+ \longrightarrow S_4O_6^{2-} + 4\ H_2O$$

At the instant of mixing, the thiosulfate ion ($S_2O_3^{2-}$) is observed to be disappearing at the rate of 7.9×10^{-7} mol L^{-1} s^{-1}. At the same moment, the H_2O_2 is disappearing at the rate of 8.8×10^{-7} mol L^{-1} s^{-1}.

(a) Compute the percentage of the $S_2O_3^{2-}$ that is, at that moment, reacting according to the first equation.

(b) It is observed that the hydronium ion concentration drops. Use the data and answer from part (a) to compute how many milliliters per minute of 0.100 M H_3O^+ must be added to keep the pH equal to 7.0.

53. 8.23×10^{-3} mol of InCl(*s*) is placed in 1.00 L of 0.010 M HCl(*aq*) at 75°C. The InCl(*s*) dissolves quite quickly, and then the following reaction occurs:

$$3\ In^+(aq) \longrightarrow 2\ In(s) + In^{3+}(aq)$$

As this disproportionation proceeds, the solution is analyzed at intervals to determine the concentration of $In^+(aq)$ that remains.

Time (s)	$[In^+]$ (mol L^{-1})
0	8.23×10^{-3}
240	6.41×10^{-3}
480	5.00×10^{-3}
720	3.89×10^{-3}
1000	3.03×10^{-3}
1200	3.03×10^{-3}
10,000	3.03×10^{-3}

(a) Plot ln $[In^+]$ versus time, and determine the apparent rate constant for this first-order reaction.

(b) Determine the half-life of this reaction.

(c) Determine the equilibrium constant *K* for the reaction under the experimental conditions.

***54.** A compound called di-*t*-butyl peroxide [abbreviation DTBP, formula $(CH_3)_3COOC(CH_3)_3$] decomposes to give acetone [$(CH_3)_2CO$] and ethane (C_2H_6):

$$(CH_3)_3COOC(CH_3)_3\,(g) \longrightarrow 2\ (CH_3)_2CO(g) + C_2H_6(g)$$

The *total* pressure of the reaction mixture changes with time, as shown by the following data at 147.2°C:

Time (min)	P_{tot} (atm)	Time (min)	P_{tot} (atm)
0	0.2362	26	0.3322
2	0.2466	30	0.3449
6	0.2613	34	0.3570
10	0.2770	38	0.3687
14	0.2911	40	0.3749
18	0.3051	42	0.3801
20	0.3122	46	0.3909
22	0.3188		

(a) Calculate the partial pressure of DTBP at each time from these data. Assume that at time 0, DTBP is the only gas present.

(b) Are the data better described by a first-order or a second-order rate expression with respect to DTBP concentration?

55. The reaction of OH^- with HCN in aqueous solution at 25°C has a forward rate constant k_f of 3.7×10^9 L mol^{-1} s^{-1}. Using this information and the measured acid ionization constant of HCN (see Table 15.2), calculate the rate constant k_r in the first-order rate law rate $= k_r[CN^-]$ for the transfer of hydrogen ions to CN^- from surrounding water molecules:

$$H_2O(\ell) + CN^-(aq) \longrightarrow OH^-(aq) + HCN(aq)$$

56. Carbon dioxide reacts with ammonia to give ammonium carbamate, a solid. The reverse reaction also occurs:

$$CO_2(g) + 2\ NH_3(g) \rightleftharpoons NH_4OCONH_2(s)$$

The forward reaction is first order in $CO_2(g)$ and second order in $NH_3(g)$. Its rate constant is 0.238 atm^{-2} s^{-1} at 0.0°C (expressed in terms of partial pressures rather than concentrations). The reaction in the reverse direction is zero order, and its rate constant, at the same temperature, is 1.60×10^{-7} atm s^{-1}. Experimental studies show that, at all stages in the progress of this reaction, the net rate is equal to the forward rate minus the reverse rate. Compute the equilibrium constant of this reaction at 0.0°C.

57. For the reactions

$$I + I + M \longrightarrow I_2 + M$$

$$Br + Br + M \longrightarrow Br_2 + M$$

the rate laws are

$$-\frac{d[I]}{dt} = k_I[I]^2[M]$$

$$-\frac{d[Br]}{dt} = k_{Br}[Br]^2[M]$$

The ratio k_I/k_{Br} at 500°C is 3.0 when M is an Ar molecule. Initially, $[I]_0 = 2[Br]_0$, while [M] is the same for both reactions and is much greater than $[I]_0$. Calculate the ratio of the time required for [I] to decrease to half its initial value to the same time for [Br] at 500°C.

58. In some reactions there is a competition between kinetic control and thermodynamic control over product yields. Suppose compound A can undergo two elementary reactions to stable products:

$$A \underset{k_{-1}}{\overset{k_1}{\rightleftharpoons}} B \quad \text{or} \quad A \underset{k_{-2}}{\overset{k_2}{\rightleftharpoons}} C$$

For simplicity we assume first-order kinetics for both forward and reverse reactions. We take the numerical values $k_1 = 1 \times 10^8\ s^{-1}$, $k_{-1} = 1 \times 10^2\ s^{-1}$, $k_2 = 1 \times 10^9\ s^{-1}$, and $k_{-2} = 1 \times 10^4\ s^{-1}$.

(a) Calculate the equilibrium constant for the equilibrium

$$B \rightleftharpoons C$$

From this value, give the ratio of the concentration of B to that of C at equilibrium. This is an example of thermodynamic control.

(b) In the case of kinetic control, the products are isolated (or undergo additional reaction) before the back reactions can take place. Suppose the back reactions in the preceding example (k_{-1} and k_{-2}) can be ignored. Calculate the concentration ratio of B to C reached in this case.

59. Compare and contrast the mechanisms for the two gas-phase reactions

$$H_2 + Br_2 \longrightarrow 2\ HBr$$

$$H_2 + I_2 \longrightarrow 2\ HI$$

60. In Section 18.4 the steady-state approximation was used to derive a rate expression for the decomposition of $N_2O_5(g)$:

$$\text{rate} = \frac{k_1k_2[M][N_2O_5]}{k_2 + k_{-1}[M]} = k_{eff}[N_2O_5]$$

At 300 K, with an excess of nitrogen present, the following values of k_{eff} as a function of total pressure were found:

P (atm)	k_{eff} (s^{-1})	*P* (atm)	k_{eff} (s^{-1})
9.21	0.265	0.625	0.116
5.13	0.247	0.579	0.108
3.16	0.248	0.526	0.104
3.03	0.223	0.439	0.092
		0.395	0.086

Use the data to estimate the value of k_{eff} at very high total pressure and the value of k_1 in L mol^{-1} s^{-1}.

61. The decomposition of ozone by light can be described by the mechanism

$$O_3 + \text{light} \xrightarrow{k_1} O_2 + O$$

$$O + O_2 + M \xrightarrow{k_2} O_3 + M$$

$$O + O_3 \xrightarrow{k_3} 2\ O_2$$

with the overall reaction being

$$2\ O_3 + \text{light} \longrightarrow 3\ O_2$$

The rate constant k_1 depends on the light intensity and the type of light source used. By making a steady-state approximation for the concentration of oxygen atoms, express the rate of formation of O_2 in terms of the O_2, O_3, and M concentrations and the elementary rate constants. Show that only the ratio k_3/k_2, and not the individual values of k_2 and k_3, affects the rate.

***62.** In Section 18.4 we considered the following mechanism for the reaction of Br_2 with H_2:

$$Br_2 + M \underset{k_{-1}}{\overset{k_1}{\rightleftharpoons}} Br + Br + M$$

$$Br + H_2 \xrightarrow{k_2} HBr + H$$

$$H + Br_2 \xrightarrow{k_3} HBr + Br$$

Although this is adequate for calculating the *initial* rate of reaction, before product HBr builds up, there is an additional process that can participate as the reaction continues:

$$HBr + H \xrightarrow{k_4} H_2 + Br$$

(a) Write an expression for the rate of change of [H].

(b) Write an expression for the rate of change of [Br].

(c) As hydrogen and bromine atoms are both short-lived species, we can make the steady-state approximation and set the rates from parts (a) and (b) to 0. Express the steady-state concentrations [H] and [Br] in terms of concentrations of H_2, Br_2, HBr, and M. [*Hint:* Try adding the rate for part (a) to that for part (b).]

(d) Express the rate of production of HBr in terms of concentrations of H_2, Br_2, HBr, and M.

63. The following observations have been made about a certain reacting system: (i) When A, B, and C are mixed at about equal concentrations in neutral solution, two different products are formed, D and E, with the amount of D about 10 times as great as the amount of E. (ii) If everything is done as in (i) except that a trace of acid is added to the reaction mixture, the same products are formed, except that now the amount of D produced is much smaller than (about 1% of) the amount of E. The acid is not consumed in the reaction. The following mechanism has been proposed to account for some of these observations and others about the order of the reactions:

(1) $A + B \underset{k_{-1}}{\overset{k_2}{\rightleftharpoons}} F$ (rapid equilibrium)

(2) $C + F \xrightarrow{k_2} D$ (negligible reverse rate)

(3) $C + F \xrightarrow{k_3} E$ (negligible reverse rate)

(a) Explain what this proposed scheme of reactions implies about the dependence (if any) of the rate of formation of D on the concentrations of A, of B, and of C. What about the dependence (if any) of the rate of formation of E on these same concentrations?

(b) What can you say about the relative magnitudes of k_2 and k_3?

(c) What explanation can you give for observation (ii) in view of your answer to (b)?

64. Iron(II) ion is oxidized by chlorine in aqueous solution, the overall equation being

$$2\ Fe^{2+} + Cl_2 \longrightarrow 2\ Fe^{3+} + 2\ Cl^-$$

It is found experimentally that the rate of the overall reaction is decreased when either the iron(III) ion or the chloride-ion concentration is increased. Which of the following possible mechanisms is consistent with the experimental observations?

(a) (1) $Fe^{2+} + Cl_2 \underset{k_{-1}}{\overset{k_1}{\rightleftharpoons}} Fe^{3+} + Cl^- + Cl$ (rapid equilibrium)

(2) $Fe^{2+} + Cl \xrightarrow{k_2} Fe^{3+} + Cl^-$ (negligible reverse rate)

(b) (3) $Fe^{2+} + Cl_2 \underset{k_{-3}}{\overset{k_3}{\rightleftharpoons}} Fe(IV) + 2\ Cl^-$ (rapid equilibrium)

(4) $Fe(IV) + Fe^{2+} \xrightarrow{k_4} 2\ Fe^{3+}$ (negligible reverse rate)

where Fe(IV) is Fe in the (+IV) oxidation state.

65. Manfred Eigen, a German physical chemist working during the 1970s and 1980s, earned a Nobel Prize for developing the "temperature-jump" method for studying kinetics of very rapid reactions in solution, such as proton transfer. Eigen and his co-workers found that the specific rate of proton transfer from a water molecule to an ammonia molecule in a dilute aqueous solution is $k = 2 \times 10^5\ s^{-1}$. The equilibrium constant K_b for the reaction of ammonia with water is 1.8×10^{-5}. What, if anything, can be deduced from this information about the rate of transfer of a proton from NH_4^+ to a hydroxide ion? Write equations for any reactions you mention, making it clear to which reaction(s) any quoted constant(s) apply.

66. Consider the reaction

$$A + B \rightleftharpoons C + D$$

with all reactants and products gaseous (for simplicity) and an equilibrium constant K. (a) Assume that the elementary steps in the reaction are those indicated by the stoichiometric equation (in each direction), with specific rate constants for the forward reaction and the reverse reaction, respectively, k_f and k_r. Derive the relation between k_f, k_r, and K. Comment on the general validity of the assumptions made about the relation of elementary steps and the stoichiometric equation and also on the general validity of K. (b) Assume that the reaction as written is exothermic. Explain what this implies about the change of K with temperature. Explain also what it implies about the relation of the activation energies of the forward and reverse reactions and how this relation is consistent with your statement about the variation of K with temperature.

67. The gas-phase decomposition of acetaldehyde can be represented by the overall chemical equation

$$CH_3CHO \longrightarrow CH_4 + CO$$

It is thought to occur through the sequence of reactions

$$CH_3CHO \longrightarrow CH_3 + CHO$$

$$CH_3 + CH_3CHO \longrightarrow CH_4 + CH_2CHO$$

$$CH_2CHO \longrightarrow CO + CH_3$$

$$CH_3 + CH_3 \longrightarrow CH_3CH_3$$

Show that this reaction mechanism corresponds to a chain reaction, and identify the initiation, propagation, and termination steps.

68. Lanthanum(III) phosphate crystallizes as a hemihydrate, $LaPO_4 \cdot \frac{1}{2}\ H_2O$. When it is heated, it loses water to give anhydrous lanthanum(III) phosphate:

$$2(LaPO_4 \cdot \tfrac{1}{2}\ H_2O(s)) \longrightarrow 2\ LaPO_4(s) + H_2O(g)$$

This reaction is first order in the chemical amount of $LaPO_4 \cdot \frac{1}{2}\ H_2O$. The rate constant varies with temperature as follows:

Temperature (°C)	k (s^{-1})
205	2.3×10^{-4}
219	3.69×10^{-4}
246	7.75×10^{-4}
260	12.3×10^{-4}

Compute the activation energy of this reaction.

69. The water in a pressure cooker boils at a temperature greater than 100°C because it is under pressure. At this higher temperature, the chemical reactions associated with the cooking of food take place at a greater rate.

(a) Some food cooks fully in 5 min in a pressure cooker at 112°C and in 10 minutes in an open pot at 100°C. Calculate the average activation energy for the reactions associated with the cooking of this food.

(b) How long will the same food take to cook in an open pot of boiling water in Denver, where the average atmospheric pressure is 0.818 atm and the boiling point of water is 94.4°C?

70. **(a)** A certain first-order reaction has an activation energy of 53 kJ mol^{-1}. It is run twice, first at 298 K and then at 308 K (10°C higher). All other conditions are identical. Show that, in the second run, the reaction occurs at double its rate in the first run.

(b) The same reaction is run twice more at 398 K and 408 K. Show that the reaction goes 1.5 times as fast at 408 K as it does at 398 K.

*71. The gas-phase reaction between hydrogen and iodine

$$H_2(g) + I_2(g) \underset{k_r}{\overset{k_f}{\rightleftharpoons}} 2\ HI(g)$$

proceeds with a forward rate constant at 1000 K of k_f = 240 L mol^{-1} s^{-1} and an activation energy of 165 kJ mol^{-1}. By using this information and data from Appendix D, calculate the activation energy for the reverse reaction and the value of k_r at 1000 K. Assume that ΔH and ΔS for the reaction are independent of temperature between 298 and 1000 K.

72. The following reaction mechanism has been proposed for a chemical reaction:

$$A_2 \underset{k_{-1}}{\overset{k_1}{\rightleftharpoons}} A + A \quad \text{(fast equilibrium)}$$

$$A + B \underset{k_{-2}}{\overset{k_2}{\rightleftharpoons}} AB \quad \text{(fast equilibrium)}$$

$$AB + CD \xrightarrow{k_3} AC + BD \quad \text{(slow)}$$

(a) Write a balanced equation for the overall reaction.
(b) Write the rate expression that corresponds to the preceding mechanism. Express the rate in terms of concentrations of reactants only (A_2, B, CD).
(c) Suppose that the first two steps in the preceding mechanism are endothermic and the third one is exothermic. Will an increase in temperature increase the reaction rate constant, decrease it, or cause no change? Explain.

73. How would you describe the role of the CF_2Cl_2 in the reaction mechanism of Problem 21?

74. In Section 18.8 we wrote a mechanism in which silver ions catalyze the reaction of Tl^+ with Ce^{4+}. Determine the rate law for this mechanism by making a steady-state approximation for the concentration of the reactive intermediate Ag^{2+}.

75. The rates of enzyme catalysis can be lowered by the presence of inhibitor molecules I, which bind to the active site of the enzyme. This adds the following additional step to the reaction mechanism considered in Section 18.7:

$$E + I \underset{k_{-3}}{\overset{k_3}{\rightleftharpoons}} EI \quad \text{(fast equilibrium)}$$

Determine the effect of the presence of inhibitor at total concentration $[I]_0 = [I] + [EI]$ on the rate expression for formation of products derived at the end of this chapter.

76. The enzyme lysozyme kills certain bacteria by attacking a sugar called *N*-acetylglucosamine (NAG) in their cell walls. At an enzyme concentration of 2×10^{-6} M, the maximum rate for substrate (NAG) reaction, found at high substrate concentration, is 1×10^{-6} mol L^{-1} s^{-1}. The rate is reduced by a factor of 2 when the substrate concentration is reduced to 6×10^{-6} M. Determine the Michaelis–Menten constants K_m and k_2 for lysozyme.

CUMULATIVE PROBLEMS

77. The rate of the gas-phase reaction

$$H_2 + I_2 \longrightarrow 2\ HI$$

is given by

$$\text{rate} = -\frac{d[I_2]}{dt} = k[H_2][I_2]$$

with k = 0.0242 L mol^{-1} s^{-1} at 400°C. If the initial concentration of H_2 is 0.081 mol L^{-1} and that of I_2 is 0.036 mol L^{-1}, calculate the initial rate at which heat is absorbed or emitted during the reaction. Assume that the enthalpy change at 400°C is the same as that at 25°C.

78. The rate of the reaction

$$2\ ClO_2(aq) + 2\ OH^-(aq) \longrightarrow ClO_3^-(aq) + ClO_2^-(aq) + H_2O(\ell)$$

is given by

$$\text{rate} = k[ClO_2]^2[OH^-]$$

with k = 230 L^2 mol^{-2} s^{-1} at 25°C. A solution is prepared that has initial concentrations $[ClO_2]$ = 0.020 M, [HCN] = 0.095 M, and $[CN^-]$ = 0.17 M. Calculate the initial rate of the reaction.

79. A gas mixture was prepared at 500 K with total pressure 3.26 atm and a mole fraction of 0.00057 of NO and 0.00026 of O_3. The elementary reaction

$$NO + O_3 \longrightarrow NO_2 + O_2$$

has a second-order rate constant of 7.6×10^7 L mol^{-1} s^{-1} at this temperature. Calculate the initial rate of the reaction under these conditions.

80. The activation energy for the reaction

$$2\ NO_2 \longrightarrow 2\ NO + O_2$$

is E_a = 111 kJ mol^{-1}. Calculate the root-mean-square velocity of an NO_2 molecule at 400 K and compare it to the velocity of an NO_2 molecule with kinetic energy E_a/N_A.

19 CHAPTER

Nuclear Chemistry

Tim Wright/Encyclopedia/Corbis

A spent fuel rod assembly being lowered in a cooling pond for long-term storage. The blue glow is Cerenkov radiation, light emitted by high-energy beta particles slowing to the speed of light in water.

In the preceding chapters, we discuss chemical reactions essentially as rearrangements of atoms from one set of "bonding partners" to another set. We turn now to a different kind of reaction in which the elemental identity of the atomic nucleus changes in the course of the reaction. This is a fascinating subject that arose early in the 20th century from experimental studies of the physical structure of the atom (see Section 1.4).

A fundamentally different new feature distinguishes this field from the material in earlier chapters. The separate laws of conservation of matter and conservation of energy must be generalized. Nuclear reactions occur at energies sufficiently high that energy and matter are interconverted. This is the origin of the enormous amounts of energy that can be released during nuclear reactions. Mass and energy are related through Einstein's relation $E = mc^2$, and their *sum* is conserved during nuclear reactions.

19.1 Mass–Energy Relationships in Nuclei

Recall from Section 1.4 that almost all the mass of an atom is concentrated in a very small volume in the nucleus. The small size of the nucleus (which occupies less than one *trillionth* of the space in the atom) and the strong forces between the protons and neutrons that make it up largely isolate its behavior from the outside world of electrons and other nuclei. This greatly simplifies our analysis of nuclear chemistry, allowing us to examine single nuclei without concern for the atoms, ions, or molecules in which they may be found.

A nucleus (or **nuclide**) is built from two kinds of **nucleons**—protons and neutrons. A nucleus is characterized by its number of protons Z (also called the atomic number) and its mass number, A. The mass number A of a nucleus is the sum of the number of protons Z and the number of neutrons N; so $A = Z + N$ is an integer closest to the relative atomic mass of the nucleus. A nuclide or an elementary particle such as an electron is designated with the symbol ${}^{A}_{Z}X$, where X is the chemical symbol for the particle. The chemical symbol of an element is determined by the number of protons in the nucleus, each isotope having a different number of neutrons. The letter Z in the symbol determines the charge, $+Ze$, on a nuclide, where it equals the number of protons, and on particles such as electrons and positrons (described later), which are assigned $Z = -1$ and $+1$, respectively, to reflect their charge only. (See Table 19.1.)

The absolute masses of nuclides and of elementary particles such as the proton, neutron, and electron are far too small to express in kilograms or grams without the constant use of 10 raised to some large negative exponent. A unit of mass that is better sized for measuring on the submicroscopic scale is needed. An **atomic mass unit** (abbreviated u) is defined as exactly $\frac{1}{12}$ the mass of a single atom of ^{12}C. Because 1 mol of ^{12}C atoms weighs exactly 12 g, one atom weighs $12/N_A$ grams, and

$$1\ \text{u} = \frac{1}{12}\left(\frac{12\ \text{g mol}^{-1}}{6.0221420 \times 10^{23}\ \text{mol}^{-1}}\right)$$

$$= 1.6605387 \times 10^{-24}\ \text{g} = 1.6605387 \times 10^{-27}\ \text{kg}$$

The conversion factor between atomic mass units and grams is numerically equal to the inverse of Avogadro's number N_A, and the mass of a single atom in atomic mass units is numerically equal to the mass of one mole of atoms in grams. Thus, one atom of ^{1}H has a mass of 1.007825 u because 1 mol of ^{1}H has a mass of 1.007825 g. The *dalton* is a mass unit that is equivalent to the atomic mass unit and is used frequently in biochemistry.

Table 19.1 lists the masses of elementary particles and selected neutral atoms, obtained from mass spectrometry (see Fig. C1.1). Note that each atomic mass includes the contribution from the surrounding Z electrons as well as that from the nucleus.

19.1.1 The Einstein Mass–Energy Relationship

Some nuclides are stable and can exist indefinitely, but others are unstable (radioactive) and decay spontaneously to form other nuclides in processes to be considered in Section 19.2. Figure 19.1 shows the distribution of the number of neutrons N ($= A - Z$) against atomic number Z (protons in the nucleus) for the known nuclides of the elements. For the light elements, $N/Z \approx 1$ for stable nuclides, meaning nearly equal numbers of protons and neutrons are found in the nucleus. For the heavier elements, $N/Z > 1$ and there are progressively more neutrons than protons in the nucleus. One of the goals of nuclear chemistry is

TABLE 19.1

Masses of Selected Elementary Particles and Atoms

Elementary Particle	Symbol	Mass (u)	Mass (kg)
Electron	${}_{-1}^{0}e^{-}$	0.000548579911	$9.1093819 \times 10^{-31}$
Positron	${}_{1}^{0}e^{+}$	0.000548579911	$9.1093819 \times 10^{-31}$
Proton	${}_{1}^{1}p^{+}$	1.0072764669	$1.6726216 \times 10^{-27}$
Neutron	${}_{0}^{1}n$	1.0086649158	$1.6749272 \times 10^{-27}$

Atom	Mass (u)	Atom	Mass (u)
${}_{1}^{1}H$	1.007825032	${}_{11}^{23}Na$	22.9897697
${}_{1}^{2}H$	2.014101778	${}_{12}^{24}Mg$	23.9850419
${}_{1}^{3}H$	3.016049268	${}_{14}^{30}Si$	29.97377022
${}_{2}^{3}He$	3.016029310	${}_{15}^{30}P$	29.9783138
${}_{2}^{4}He$	4.002603250	${}_{16}^{32}S$	31.9720707
${}_{3}^{7}Li$	7.0160040	${}_{17}^{35}Cl$	34.96885271
${}_{4}^{7}Be$	7.01692983	${}_{20}^{40}Ca$	39.9625912
${}_{4}^{8}Be$	8.00530509	${}_{22}^{49}Ti$	48.947871
${}_{4}^{9}Be$	9.0121821	${}_{35}^{81}Br$	80.916291
${}_{4}^{10}Be$	10.0135337	${}_{37}^{87}Rb$	86.909183
${}_{5}^{8}B$	8.024607	${}_{38}^{87}Sr$	86.908879
${}_{5}^{10}B$	10.0129370	${}_{53}^{127}I$	126.904468
${}_{5}^{11}B$	11.0093055	${}_{88}^{226}Ra$	226.025403
${}_{6}^{11}C$	11.011433	${}_{88}^{228}Ra$	228.031064
${}_{6}^{12}C$	12 exactly	${}_{89}^{228}Ac$	228.031015
${}_{6}^{13}C$	13.003354838	${}_{90}^{232}Th$	232.038050
${}_{6}^{14}C$	14.003241988	${}_{90}^{234}Th$	234.043595
${}_{7}^{14}N$	14.003074005	${}_{91}^{231}Pa$	231.035879
${}_{8}^{16}O$	15.994914622	${}_{92}^{231}U$	231.036289
${}_{8}^{17}O$	16.9991315	${}_{92}^{234}U$	234.040945
${}_{8}^{18}O$	17.999160	${}_{92}^{235}U$	235.043923
${}_{9}^{19}F$	18.9984032	${}_{92}^{238}U$	238.050783
${}_{11}^{21}Na$	20.99764		

Values are taken from National Institute of Standards and Technology Special Publication 961, Jan. 2001

to understand the differing stabilities of the isotopes. Nuclear mass provides the key to this understanding.

Note from Table 19.1 that the neutron has a slightly greater mass than the proton. The neutron is stable inside a nucleus, but in free space it is unstable, decaying into a proton and an electron with a half-life of about 10 minutes:

$$ {}_{0}^{1}n \longrightarrow {}_{1}^{1}p^{+} + {}_{-1}^{0}e^{-} + \bar{\nu} $$

Here ${}_{0}^{1}n$ and ${}_{1}^{1}p^{+}$ represent the neutron and the proton, respectively. We have written this as a **balanced nuclear equation**. To check such an equation for balance, first verify that the total mass numbers (the superscripts) are equal on the two sides. Then verify that the nuclear charges are balanced by checking the sum of the subscripts. The third product is an antineutrino (see Section 19.2.1).

Mass is not conserved in the reaction just written. By subtracting the mass of the reactant from the masses of the products (using Table 19.1; the antineutrino's mass is negligible), the change in mass is calculated to be

$$ \Delta m = -1.3947 \times 10^{-30} \text{ kg} = -8.398690 \times 10^{-4} \text{ u} $$

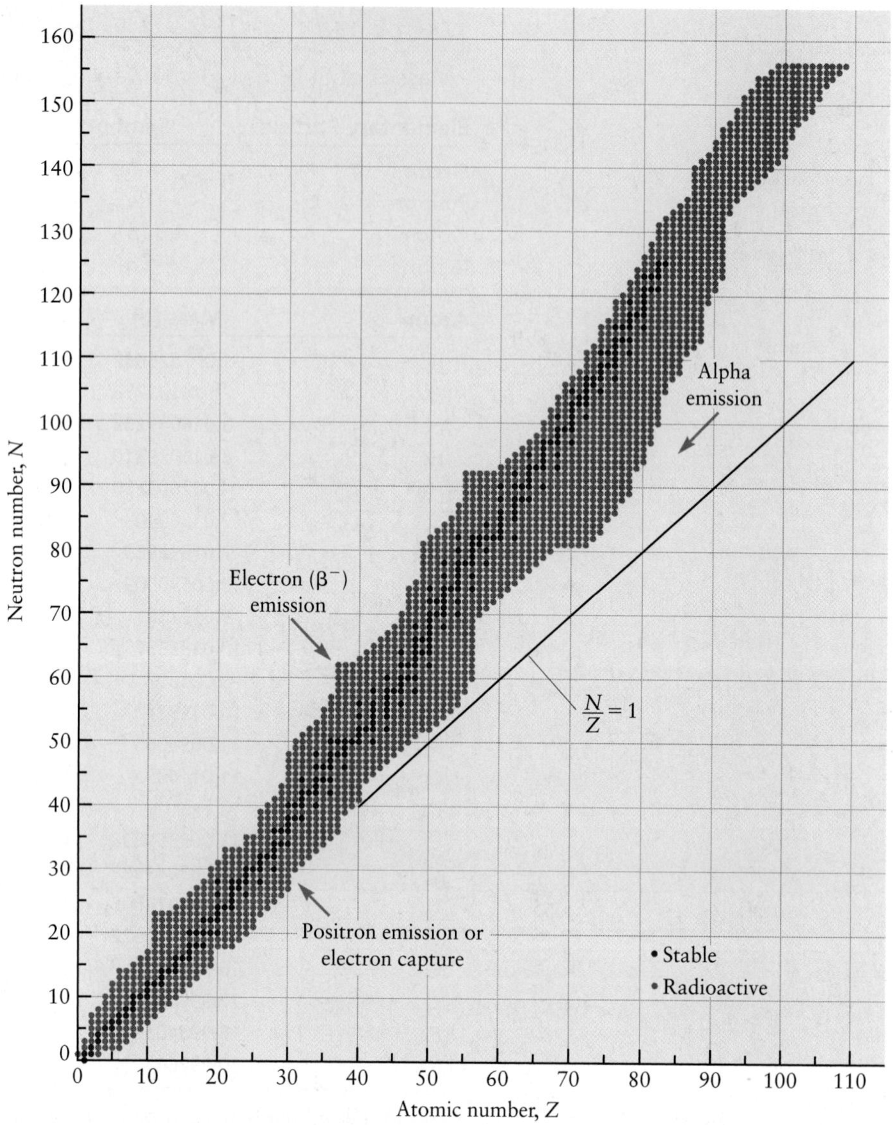

FIGURE 19.1 A plot of N versus Z for nuclides with the stable nuclides represented by blue dots and the known radioactive nuclides represented by red dots. The stable nuclides lie along a "line of stability" with slope $N/Z = 1$ (shown) for the lighter elements, increasing to 1.5 for the heavier elements. Arrows show the directions of decay of unstable nuclides via alpha emission, beta emission, positron emission, or electron capture.

What has happened to the "lost" mass? It has been converted to energy—specifically, kinetic energy carried off by the electron and proton. According to Einstein's special theory of relativity, a change in mass always accompanies a change in energy:

$$E = mc^2$$

or, better,

$$\Delta E = c^2 \, \Delta m \qquad [19.1]$$

where c is the speed of light in a vacuum and Δm is the change in mass (mass of products minus mass of reactants). To express ΔE in joules, c must have units of meters per second, and Δm, units of kilograms. In the decay of one neutron, for example,

$$\Delta E = c^2 \, \Delta m = (2.9979 \times 10^8 \text{ m s}^{-1})^2(-1.3947 \times 10^{-30} \text{ kg}) = -1.2535 \times 10^{-13} \text{ J}$$

In ordinary chemical reactions, the change in mass is negligibly small. In the combustion of 1 mol of carbon to CO_2, the energy change is about 400 kJ, corresponding to a change in mass of only 4×10^{-9} g.

Changes in energy in nuclear reactions are nearly always expressed in more convenient energy units than the joule—namely, the **electron volt** (eV) and the **megaelectron volt** (MeV). The electron volt is defined as the energy given to an electron when it is accelerated through a potential difference of exactly 1 V:

$$1 \text{ eV} = 1.60217646 \times 10^{-19} \text{ C} \times 1 \text{ V} = 1.60217646 \times 10^{-19} \text{ J}$$

The megaelectron volt is 1 million times larger:

$$1 \text{ MeV} = 1.60217646 \times 10^{-13} \text{ J}$$

A change in mass of 1 u = $1.6605387 \times 10^{-27}$ kg corresponds to

$$\Delta E = c^2 \Delta m = (2.9979246 \times 10^8 \text{ m s}^{-1})^2 (1.6605387 \times 10^{-27} \text{ kg})$$
$$= 1.4924178 \times 10^{-10} \text{ J}$$

Converting to MeV,

$$\frac{1.4924178 \times 10^{-10} \text{ J}}{1.60217646 \times 10^{-13} \text{ J MeV}^{-1}} = 931.494 \text{ MeV}$$

We say that the **energy equivalent** of 1 u is 931.494 MeV. In the decay of a neutron,

$$\Delta E = \left(-8.39869 \times 10^{-4} \text{ u}\right)\left(931.494 \text{ MeV u}^{-1}\right) = -0.782 \text{ MeV}$$

19.1.2 Binding Energies of Nuclei

The **binding energy** E_B of a nucleus is defined as the negative of the energy change ΔE that would occur if that nucleus were formed from its component protons and neutrons. For the ${}^4_2\text{He}$ nucleus, for example,

$$2\ {}^1_1p^+ + 2\ {}^1_0n \longrightarrow {}^4_2\text{He}^{2+} \qquad \Delta E = -E_B$$

Binding energies of nuclei are calculated with Einstein's relationship and accurate data from a mass spectrometer. Most measurements determine the mass of the atom with most or all of its electrons, not the mass of the bare nucleus (recall that it is the ^{12}C *atom*, not the ^{12}C nucleus, that is defined to have atomic mass 12). Therefore, we take the nuclear binding energy E_B of the helium nucleus to be the negative of the energy change ΔE for the formation of an *atom* of helium from hydrogen *atoms* and neutrons.

$$2\ {}^1_1\text{H} + 2\ {}^1_0n \longrightarrow {}^4_2\text{He}$$

The correction for the differences in binding energies of electrons (which of course are present both in the hydrogen atoms and in the atom being formed) is a very small one that need not concern us here.

EXAMPLE 19.1

Calculate the binding energy of ^{4}He from the data in Table 19.1, and express it both in joules and in megaelectron volts (MeV).

Solution

The mass change is

$$\Delta m = m[{}^4_2\text{He}] - 2m[{}^1_1\text{H}] - 2m[{}^1_0n]$$
$$= 4.00260325 \text{ u} - 2(1.00782503 \text{ u}) - 2(1.00866492 \text{ u}) = -0.03037665 \text{ u}$$

(Continued)

Einstein's relation then gives

$$\Delta E = (-0.03037665 \text{ u})(1.6605387 \times 10^{-27} \text{ kg u}^{-1})(2.9979246 \times 10^{8} \text{ m s}^{-1})^2$$

$$= -4.533465 \times 10^{-12} \text{ J}$$

$$E_B = 4.533465 \times 10^{-12} \text{ J}$$

If 1 mol of $^4_2$He atoms were formed in this way, the energy change would be greater by a factor of Avogadro's number N_A, giving $\Delta E = -2.73 \times 10^{12}$ J mol^{-1}. This is an enormous quantity, seven orders of magnitude greater than typical energy changes in chemical reactions.

(Note that we have used the mass of the neutron given in Table 19.1 rather than in the Physical Constants, updated in 2010, given in the back of this text. The atomic masses in Table 19.1 are determined by mass differences, so one should use the masses of the elementary particles consistent with those in nuclear chemistry calculations. We also use the conversion factor 1 u = $1.6605387 \times 10^{-27}$ kg given at the start of Section 19.1, although the last digit was revised to a 9 in 2010.)

The energy change, in MeV, accompanying the formation of a $^4_2$He atom is

$$\Delta E = (-0.03037665 \text{ u})(931.494 \text{ MeV u}^{-1}) = -28.2957 \text{ MeV}$$

Related Problems: 3, 4

There are four nucleons in the $^4_2$He nucleus, so the binding energy *per nucleon* is

$$\frac{E_B}{4} = \frac{28.2957 \text{ MeV}}{4} = 7.07392 \text{ MeV}$$

The binding energy per nucleon is a direct measure of the stability of the nucleus. It varies with atomic number among the stable elements, increasing to a maximum of about 8.8 MeV for iron and nickel, as Figure 19.2 illustrates.

19.2 Nuclear Decay Processes

Why are some nuclei **radioactive,** decaying spontaneously, when others are stable? Thermodynamics gives the criterion $\Delta G < 0$ for a process to be spontaneous at constant T and P. The energy change ΔE in a nuclear decay process is so great that the free energy change is essentially equal to it, and the thermodynamic criterion for spontaneous nuclear reaction consequently simplifies to

$$\Delta E < 0 \qquad \text{or} \qquad \Delta m < 0$$

The equivalence of these criteria follows from Einstein's mass–energy relationship. Spontaneous transformations of one nucleus into others can occur only if the combined mass of products is less than the mass of the original nuclide.

Before we address the details of nuclear decay, it is necessary to mention **antiparticles.** Each subatomic particle is thought to have an antiparticle of the same mass but opposite charge. Thus, the antiparticle of an electron is a **positron** (e^+), and the antiparticle of a proton is a negatively charged particle called an **antiproton.** Antiparticles have only transient existences because matter and antimatter annihilate one another when they come together, emitting radiation carrying an equivalent amount of energy. In the case of an electron and a positron, two oppositely directed **gamma rays** (γ-rays; high-energy photons), called the **annihilation radiation,** are emitted. Another important particle–antiparticle pair is the neutrino (symbol ν) and antineutrino ($\bar{\nu}$), both of which are uncharged and almost massless. (They differ only in parity, a quantum-mechanical symmetry

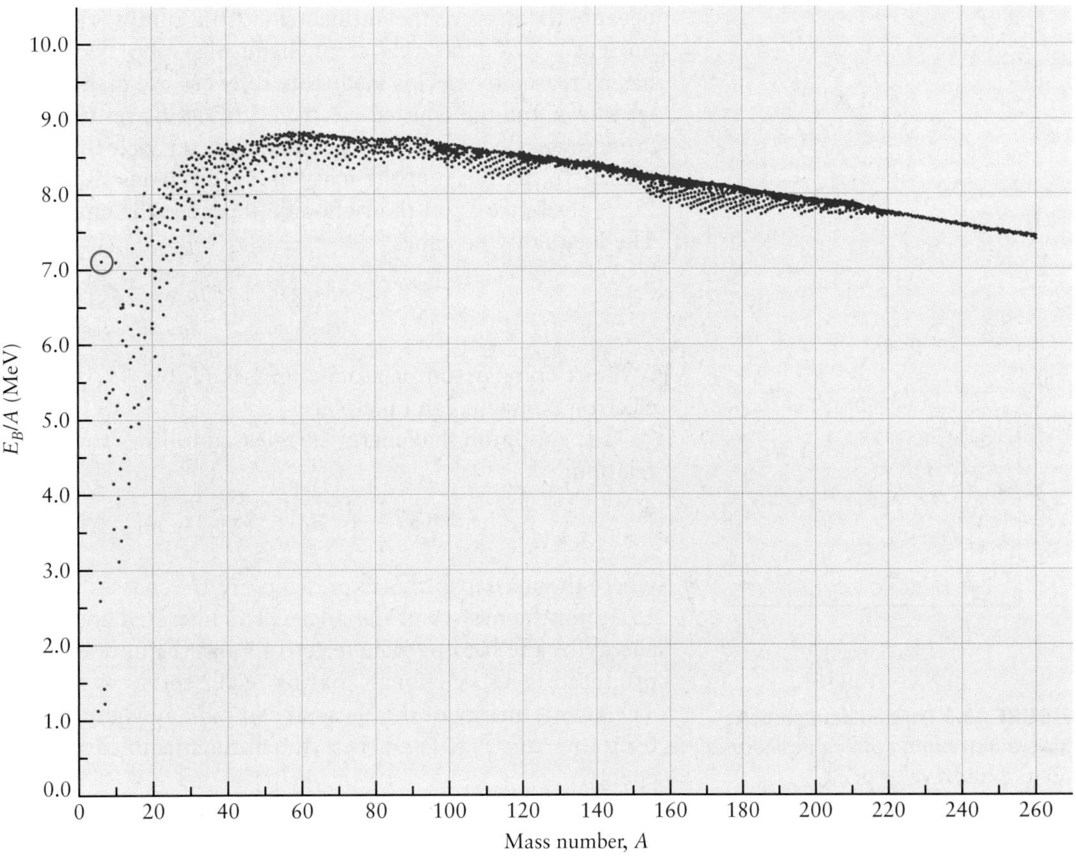

FIGURE 19.2 The variation of the binding energy per nucleon with mass number. A total of 2212 nuclides are shown, of which fewer than 270 are stable. The greatest nuclear stability is seen in the vicinity of ^{56}Fe. Note that ^{4}He (red circle) is unusually stable for its mass number.

upon reflection.) Neutrinos interact so weakly with matter that elaborate detectors are required to record the extremely rare events induced by them in selected nuclides.

19.2.1 Beta Decay by Electron Emission (β^- decay)

If an unstable nuclide contains fewer protons than do stable isotopes of the same mass number, it is called "proton deficient." Such a nucleus can decay by transforming one of its neutrons into a proton and emitting a high-energy electron $_{-1}^{0}e^-$, also called a **beta particle,** and an antineutrino ($\overline{\nu}$). The superscript 0 indicates a mass number of 0 because an electron contains neither protons nor neutrons and has a much smaller mass than a nucleus. The subscript -1 indicates the negative charge on the particle. It is not a true "atomic number," but writing the symbol in this way is helpful in balancing nuclear equations. The nuclide that results has the same mass number A, but its atomic number Z_{da} is one larger than the parent nucleus's ($Z_{\mathrm{da}} = Z_{\mathrm{pa}} + 1$) because a neutron has been transformed into a proton. Examples of **electron emission** from unstable nuclei are

$$^{14}_{6}\mathrm{C} \longrightarrow {}^{14}_{7}\mathrm{N} + {}_{-1}^{0}e^- + \overline{\nu} \qquad {}^{24}_{11}\mathrm{Na} \longrightarrow {}^{24}_{12}\mathrm{Mg} + {}_{-1}^{0}e^- + \overline{\nu}$$

The criterion $\Delta m < 0$ for spontaneous β^- decay requires that

$$\underset{\text{(Parent nucleus)}}{m[{}^{A}_{Z_{\mathrm{pa}}}\mathrm{X}_{\mathrm{pa}}]} > \underset{\text{(Daughter nucleus)}}{m[{}^{A}_{Z_{\mathrm{pa}}+1}\mathrm{X}_{\mathrm{da}}]} + m[{}_{-1}^{0}e^-]$$

because the mass of the antineutrino ($\bar{\nu}$) is almost zero, and where the daughter nucleus ${}_{Z_{pa}+1}^{A}X_{da}$ has a different chemical symbol because the number of protons has increased by 1. This inequality refers to the masses of the parent and daughter *nuclei* and the emitted electron. It must be reformulated in terms of neutral *atom* masses because the mass spectrometer yields masses of atoms rather than nuclei. To do this, add the mass of Z_{pa} electrons to both sides, giving a total of Z_{pa} + 1 electrons on the right side (because the emitted ${}_{-1}^{0}e^-$ is also counted). The inequality becomes

$$\underset{\text{(Parent atom)}}{m[{}_{Z_{pa}}^{A}X_{pa}]} > \underset{\text{(Daughter atom)}}{m[{}_{Z_{pa}+1}^{A}X_{da}]}$$

A direct comparison of atomic masses (Table 19.1) lets us determine whether electron emission can take place.

The condition that energy be released follows from the Einstein mass–energy relation:

$$\Delta E = c^2 \left\{ m[{}_{Z_{pa}+1}^{A}X_{da}] - m[{}_{Z_{pa}}^{A}X_{pa}] \right\} < 0$$

where the masses in this expression are the masses of the atoms given in Table 19.1, not the masses of the nuclei. The liberated energy is carried off as kinetic energy by the beta particle (electron) and the antineutrino, because the nucleus produced is heavy enough that its recoil energy is small and can be neglected. The kinetic energy of the electron can fall anywhere in a continuous range from 0 up to $-\Delta E$ (Fig. 19.3), with the antineutrino carrying off the balance of the energy.

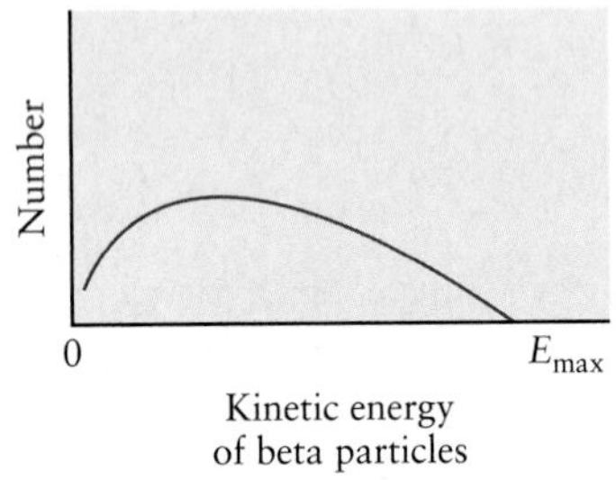

FIGURE 19.3 Emitted beta particles have a distribution of kinetic energies up to a cutoff value of E_{max}.

EXAMPLE 19.2

Calculate the maximum kinetic energy of the electron in the decay

$${}_{6}^{14}C \longrightarrow {}_{7}^{14}N + {}_{-1}^{0}e^- + \bar{\nu}$$

Solution

Table 19.1 gives the masses of the relevant atoms. From them we calculate

$$\Delta m = m[{}_{7}^{14}N] - m[{}_{6}^{14}C] = 14.0030740 \text{ u} - 14.0032420 \text{ u} = -0.0001680 \text{ u}$$

$$\Delta E = (-1.68 \times 10^{-4} \text{ u})(931.5 \text{ MeV u}^{-1}) = -0.156 \text{ MeV}$$

The ΔE of the nuclear reaction is negative, so some of the energy released to the surroundings may be carried off by the electron. The maximum kinetic energy of the electron is 0.156 MeV.

Related Problem: 8

19.2.2 Positron Emission (β^+ decay)

When a nucleus has too many protons for stability relative to the number of neutrons it contains, it may decay by emitting a positron. In this event a proton is converted to a neutron, and a high-energy positron (symbolized ${}_{1}^{0}e^+$) and a neutrino (ν) are emitted. As in electron emission, the mass number of the nuclide remains the same, but now the atomic number of the daughter nucleus Z_{da} is 1 less than the parent nucleus's ($Z_{da} = Z_{pa} - 1$). Examples of **positron emission** are

$${}_{6}^{11}C \longrightarrow {}_{5}^{11}B + {}_{1}^{0}e^+ + \nu \qquad {}_{10}^{19}Ne \longrightarrow {}_{9}^{19}F + {}_{1}^{0}e^+ + \nu$$

The positron has the same mass as an electron but has a positive charge. The mass criterion $\Delta m < 0$ is different from that for electron emission. For the bare nuclei the criterion is

$$m[{}_{Z_{pa}}^{A}X_{pa}] > m[{}_{Z_{pa}-1}^{A}X_{da}] + m[{}_{1}^{0}e^{+}]$$

(Parent nucleus) (Daughter nucleus)

Adding the mass of Z_{pa} electrons to both sides gives the mass of the neutral atom ${}_{Z_{pa}}^{A}X_{pa}$ on the left,[1] but only $Z_{pa} - 1$ electrons are needed on the right side to form a neutral atom ${}_{Z_{pa}-1}^{A}X_{da}$. This leaves additional electron and positron masses on the right. These masses are the same, so the criterion for spontaneous positron emission is

$$m[{}_{Z_{pa}}^{A}X_{pa}] > m[{}_{Z_{pa}-1}^{A}X_{da}] + 2m[{}_{1}^{0}e^{+}]$$

(Parent atom) (Daughter atom)

The energy change in positron emission is

$$\Delta E = c^2\left\{m[{}_{Z_{pa}-1}^{A}X_{da}] + 2m[{}_{1}^{0}e^{+}] - m[{}_{Z_{pa}}^{A}X_{pa}]\right\} < 0$$

where the first and last masses in this expression are each neutral atom's mass from Table 19.1. The kinetic energy $(-\Delta E)$ is distributed between the positron and the neutrino.

EXAMPLE 19.3

Calculate the maximum kinetic energy of the positron emitted in the decay

$${}_{6}^{11}C \longrightarrow {}_{5}^{11}B + {}_{1}^{0}e^{+} + \nu$$

Solution

The change in mass is

$$\begin{aligned}\Delta m &= m[{}_{5}^{11}B] + 2m[{}_{1}^{0}e^{+}] - m[{}_{6}^{11}C] \\ &= 11.0093055 \text{ u} + 2(0.00054858 \text{ u}) - 11.011433 \text{ u} \\ &= -0.00130 \text{ u}\end{aligned}$$

The energy change is then

$$\Delta E = (-0.00130 \text{ u})(931.5 \text{ MeV u}^{-1}) = -0.960 \text{ MeV}$$

The maximum kinetic energy of the positron is +0.960 MeV.

Related Problem: 7

19.2.3 Electron Capture

If $m[{}_{Z_{pa}}^{A}X_{pa}]$ exceeds $m[{}_{Z_{pa}-1}^{A}X_{da}]$ (both referring to neutral atoms) but by less than 1.0972×10^{-3} u (equivalent to $2m[{}_{1}^{0}e^{+}]$), positron emission cannot occur. An alternative process that *is* allowed is **electron capture**, in which an electron outside the nucleus of the parent atom is captured by the nucleus and a proton is converted to a neutron. As in positron emission, the mass number is unchanged and the atomic number decreases by 1. In this case, however, the only particle emitted is a neutrino. An example is

$${}_{92}^{231}U + {}_{-1}^{0}e^{-} \longrightarrow {}_{91}^{231}Pa + \nu$$

[1] Of course this statement is not true, as when Z electrons are bound to a nucleus, the total mass of the atom is not the mass of the bare nucleus added to the mass of Z free electrons. There are mass changes associated with this reaction, just as there are with all others. The *difference* between atomic masses, however, is very close to accurate in this calculation because it is only in error by the binding energy of one electron.

The mass criterion for electron capture is simply that

$$m[{}^{A}_{Z_{pa}}X_{pa}] > m[{}_{Z_{pa}-1}^{A}X_{da}]$$
$$\text{(Parent atom)} \qquad \text{(Daughter atom)}$$

so the energy change is

$$\Delta E = c^2 \left\{ m[{}_{Z_{pa}-1}^{A}X_{da}] - m[{}^{A}_{Z_{pa}}X_{pa}] \right\} < 0$$

where the masses in this expression are the masses of the atoms given in Table 19.1, not the masses of the nuclei. Electron capture is common for heavier neutron-deficient nuclei, for which the mass change with atomic number is too small for positron emission to be possible. In the preceding example,

$$\Delta m = m[{}^{231}_{91}\text{Pa}] - m[{}^{231}_{92}\text{U}] = 231.035879 \text{ u} - 231.036289 \text{ u} = -0.00041 \text{ u}$$

Because $\Delta m < 0$, this process can and does occur by electron capture, but because 0.00041 u is less than $2m[{}^{0}_{1}e^{+}]$, the analogous process could not occur through positron emission.

19.2.4 Alpha Decay

The three forms of decay discussed so far all lead to changes in the atomic number Z but not in the mass number A. Another process, **alpha decay**, involves the emission of an alpha particle (${}^{4}_{2}\text{He}^{2+}$ ion) and a decrease in mass number A by 4 (the atomic number Z and the neutron number N each decrease by 2). An example is

$${}^{238}_{92}\text{U} \longrightarrow {}^{234}_{90}\text{Th} + {}^{4}_{2}\text{He}$$

Such alpha decay occurs chiefly for elements in the unstable region beyond bismuth ($Z = 83$) in the periodic table.

The criterion for alpha decay is once again $\Delta m < 0$. In the preceding example,

$$\Delta m = m[{}^{234}_{90}\text{Th}] + m[{}^{4}_{2}\text{He}] - m[{}^{238}_{92}\text{U}]$$
$$\text{(Atom)} \qquad \text{(Atom)} \qquad \text{(Atom)}$$

$$= 234.043595 \text{ u} + 4.002603 \text{ u} - 238.050783 \text{ u} = -0.004585 \text{ u} < 0$$

$$\Delta E = (-0.004585 \text{ u})(931.5 \text{ MeV u}^{-1}) = -4.27 \text{ MeV}$$

Most of the energy is carried away by the lighter helium atom, with a small fraction appearing as recoil energy of the heavy thorium atom. The energy released into recoil kinetic energy, $1/2\ \mu v_{rel}^2$, a maximum of 4.27 MeV, is partitioned between the two products, with the kinetic energy of the He nucleus being partitioned 234/238 of the relative kinetic energy and the kinetic energy of the Th nucleus being partitioned 4/238 of the relative kinetic energy.

19.2.5 Other Modes of Decay

The arrows in Figure 19.1 show the directions of the types of decay discussed so far. Nuclei whose values of N/Z are too great move toward the line of stability by electron emission, whereas those whose values are too small undergo positron emission or electron capture. Nuclei that are simply too massive can move to lower values of both N and Z by alpha emission. The range of stable nuclides is limited by the interplay of attractive nuclear forces and repulsive Coulomb forces between protons in the nucleus. Scientists believe that it is unlikely that additional stable elements will be found, although long-lived radioactive nuclides may well exist at still higher Z.

When nuclei are very proton-deficient or very neutron-deficient, an excess particle may "boil off," that is, be ejected directly from the nucleus. These decay modes are called **neutron emission** and **proton emission**, respectively, and move nuclides down or to the left in Figure 19.1. Finally, certain unstable nuclei undergo spontaneous **fission**, in which they split into two nuclei of roughly equal size. Nuclear fission will be discussed in more detail in Section 19.5.

19.2.6 Detecting and Measuring Radioactivity

Many methods have been developed to detect, identify, and quantitatively measure the products of nuclear reactions. Some are quite simple, but others require complex electronic instrumentation. Perhaps the simplest radiation detector is the *photographic emulsion*, first used by Henri Becquerel, the discoverer of radioactivity. In 1896 Becquerel reported his observation that potassium uranyl sulfate ($K_2UO_2(SO_4)_2 \cdot 2H_2O$) could expose a photographic plate even in the dark. Such detectors are still used today in the form of film badges (radiation dosimeters) that are worn to monitor exposure to penetrating radiation. The degree of darkening of the film is proportional to the quantity of radiation received.

Working in Rutherford's laboratory, Hans Geiger and Ernest Marsden used a screen coated with zinc sulfide to detect the arrival of alpha particles by the pinpoint scintillations of light they produce. That simple device has been developed into the modern **scintillation counter.** Instead of a ZnS screen, the modern scintillation counter uses a crystal, such as sodium iodide, in which a small fraction of the Na^+ ions have been replaced by thallium (Tl^+) ions. The crystal emits a pulse of light when it absorbs a beta particle or a gamma ray, and a photomultiplier tube converts the light pulses to electrons and multiplies that electronic pulse proportional to the light pulse intensity.

The **Geiger-Müller counter** (Fig. 19.4) consists of a cylindrical tube, usually of glass, coated internally with metal to provide a negative electrode and with a wire down the center for a positive electrode. The tube is filled to a total pressure of about 0.1 atm with a mixture of a noble gas and a quench gas, an organic vapor or a halogen, e.g., 90% argon and 10% ethyl alcohol vapor, and a potential difference of about 1000 V is applied across the electrodes. When a high-energy electron (beta particle) enters the tube, it produces positive ions and electrons. The light electrons are quickly accelerated toward the positively charged wire. As they advance, they encounter and ionize other neutral atoms. An avalanche of electrons builds up, and a large electron current flows into the central wire. This causes a drop in the potential difference, which is recorded, and the multiplicative electron discharge is quenched by the alcohol molecules. In this way single beta particles produce electrical pulses that can be amplified

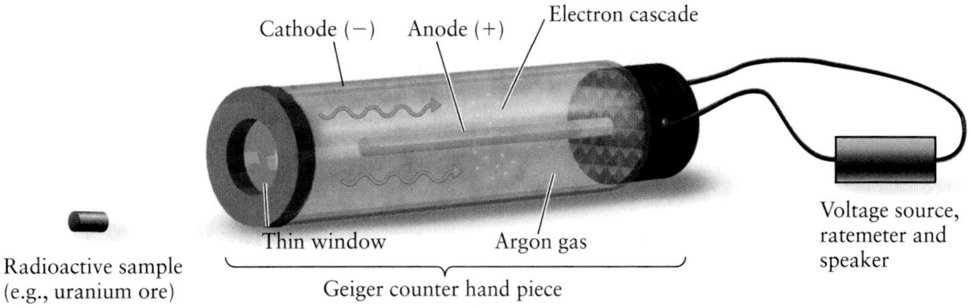

FIGURE 19.4 In a Geiger-Müller tube, radiation ionizes gas in the tube, freeing electrons that are accelerated to the anode wire in a cascade. Their arrival creates an electrical pulse, which is detected by a ratemeter. The ratemeter displays the accumulated pulses as the number of ionization events per minute.

and counted. Portable Geiger-Müller counters are widely used in uranium prospecting and to measure radiation in workplaces. Various tube sizes are used to detect different levels of radiation, and modifications must be made to detect γ rays more efficiently. **Proportional counters** are similar to Geiger-Müller counters but with the ability to measure energies like scintillation counters. They are filled with inert gases (typically He or Ar) along with a small partial pressure of a polyatomic molecule like CH_4 that improves the instrument response time. The magnitude of each current pulse in these counters is proportional to the energy of the incident particles, hence the name. Proportional counters can be modified to detect neutrons by using $^{10}BF_3$ as the fill gas, which efficiently captures neutrons to form an unstable nucleus that decays by emitting α particles, which are readily detected. Finally, many modern detectors are solid-state semiconductor devices in which the ionizing radiation creates a number of electron-hole pairs in a semiconductor junction (see Section 22.7) producing a current pulse that is proportional to the intensity of the incident radiation.

19.3 Kinetics of Radioactive Decay

The decay of any given unstable nucleus is a random event and is independent of the number of surrounding nuclei that have decayed. When the number of nuclei is large, we can be confident that during any given period a definite fraction of the original number of nuclei will have undergone a transformation into another nuclear species. In other words, the rate of decay of a collection of nuclei is proportional to the number of nuclei present, showing that nuclear decay follows a first-order rate equation of the type discussed in Chapter 18. All the results developed in that chapter apply to the present situation; for example, the integrated rate law has the form

$$N = N_i \, e^{-kt}$$

where N_i is the number of nuclei originally present at $t = 0$. The decay constant k (or a Greek lambda, λ, as used by nuclear scientists) is related to a half-life $t_{1/2}$ through

$$t_{1/2} = \frac{\ln 2}{k} = \frac{0.6931}{k}$$

just as in the first-order gas-phase chemical kinetics of Section 18.2. The half-life is the time required for the nuclei in a sample to decay to one-half their initial number, and it can range from less than 10^{-21} s to more than 10^{25} years for unstable nuclei. Table 19.2 lists the half-lives and decay modes of some unstable nuclides.

There is one important practical difference between chemical kinetics and nuclear kinetics. In chemical kinetics the concentration of a reactant or product is monitored over time, and the rate of a reaction is then found from the rate of change of that concentration. In nuclear kinetics the rate of occurrence of decay events, $-dN/dt$, is measured directly with a Geiger counter or other radiation detector. This decay rate—the average disintegration rate in numbers of nuclei per unit time—is called the **activity *A*.**

$$A = -\frac{dN}{dt} = kN \qquad [19.2]$$

TABLE 19.2

Decay Characteristics of Some Radioactive Nuclei

Nuclide	$t_{1/2}$	Decay Mode†	Daughter
$^{3}_{1}H$ (tritium)	12.26 years	e^-	$^{3}_{2}He$
$^{8}_{4}Be$	$\sim 1 \times 10^{-16}$ s	α	$^{4}_{2}He$
$^{14}_{6}C$	5730 years	e^-	$^{14}_{7}N$
$^{22}_{11}Na$	2.601 years	e^+	$^{22}_{10}Ne$
$^{24}_{11}Na$	15.02 hours	e^-	$^{24}_{12}Mg$
$^{32}_{15}P$	14.28 days	e^-	$^{32}_{16}S$
$^{35}_{16}S$	87.2 days	e^-	$^{35}_{17}Cl$
$^{36}_{17}Cl$	3.01×10^5 years	e^-	$^{36}_{18}Ar$
$^{40}_{19}K$	1.28×10^9 years	e^- (89.3%)	$^{40}_{20}Ca$
		E.C. (10.7%)	$^{40}_{18}Ar$
$^{59}_{26}Fe$	44.6 days	e^-	$^{59}_{27}Co$
$^{60}_{27}Co$	5.27 years	e^-	$^{60}_{28}Ni$
$^{90}_{38}Sr$	29 years	e^-	$^{90}_{39}Y$
$^{109}_{48}Cd$	453 days	E.C.	$^{109}_{47}Ag$
$^{125}_{53}I$	59.7 days	E.C.	$^{125}_{52}Te$
$^{131}_{53}I$	8.041 days	e^-	$^{131}_{54}Xe$
$^{127}_{54}Xe$	36.41 days	E.C.	$^{127}_{53}I$
$^{137}_{57}La$	$\sim 6 \times 10^4$ years	E.C.	$^{137}_{56}Ba$
$^{222}_{86}Rn$	3.824 days	α	$^{218}_{84}Po$
$^{226}_{88}Ra$	1600 years	α	$^{222}_{86}Rn$
$^{232}_{90}Th$	1.40×10^{10} years	α	$^{228}_{88}Ra$
$^{235}_{92}U$	7.04×10^8 years	α	$^{231}_{90}Th$
$^{238}_{92}U$	4.468×10^9 years	α	$^{234}_{90}Th$
$^{239}_{93}Np$	2.350 days	e^-	$^{239}_{94}Pu$
$^{239}_{94}Pu$	2.411×10^4 years	α	$^{235}_{92}U$

†E.C. stands for electron capture; e^+ for positron emission; e^- for electron emission; α for alpha emission.

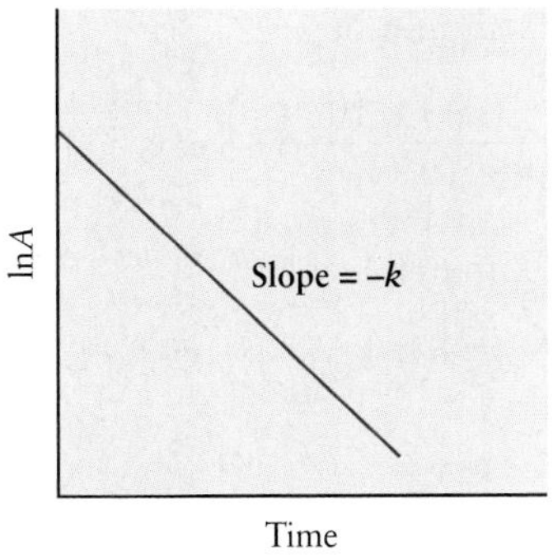

FIGURE 19.5 A graph of the logarithm of the activity of a radioactive nuclide against time is a straight line with slope $-k = -(\ln 2)/t_{1/2}$. The decay rate constant k can also be measured as the ratio of the activity to the number of atoms of the radioisotope present at time t if the latter is known.

Because the activity is proportional to the number of nuclei N, it also decays exponentially with time:

$$A = A_i e^{-kt} \qquad [19.3]$$

A plot of ln A against time t is linear with slope $-k = -(\ln 2)/t_{1/2}$, as Figure 19.5 shows. The activity A is reduced to half its initial value in a time $t_{1/2}$. Once A and k are known, the number of nuclei N at that time can be calculated from

$$N = \frac{A}{k} = \frac{At_{1/2}}{\ln 2} = \frac{At_{1/2}}{0.6931}$$

The S.I. unit of activity is the *becquerel* (Bq), defined as 1 radioactive disintegration per second. An older and much larger unit of activity is the *curie* (abbreviated Ci), which is defined as 3.7×10^{10} disintegrations per second. The activity of 1 g of radium is 1 Ci.

EXAMPLE 19.4

Tritium (^3H) decays by beta emission to ^3He with a half-life of 12.26 years. A sample of a tritiated compound has an initial activity of 0.833 Bq. Calculate the number N_i of tritium nuclei in the sample initially, the decay constant k, and the activity after 2.50 years.

Solution

Convert the half-life to seconds:

$$t_{1/2} = (12.26 \text{ yr})(60 \times 60 \times 24 \times 365.25 \text{ s yr}^{-1}) = 3.869 \times 10^8 \text{ s}$$

The number of nuclei originally present was

$$N_i = \frac{A_i t_{1/2}}{\ln 2} = \frac{(0.833 \text{ s}^{-1})(3.869 \times 10^8 \text{ s})}{0.6931} = 4.65 \times 10^8 \ ^3\text{H nuclei}$$

The decay constant k is calculated directly from the half-life:

$$k = \frac{\ln 2}{t_{1/2}} = \frac{0.6931}{3.869 \times 10^8 \text{ s}} = 1.791 \times 10^{-9} \text{ s}^{-1}$$

To find the activity after 2.50 years, convert this time to seconds (7.889×10^7 s) and use

$$A = A_i\, e^{-kt} = (0.833 \text{ Bq}) \exp[-(1.791 \times 10^{-9} \text{ s}^{-1})(7.889 \times 10^7 \text{ s})] = 0.723 \text{ Bq}$$

Related Problems: 21, 22

19.3.1 Radioactive Dating

The decay of radioactive nuclides with known half-lives enables geochemists to measure the ages of rocks from their isotopic compositions. Suppose that a uranium-bearing mineral was deposited some 2 billion (2.00×10^9) years ago and has remained geologically unaltered to the present time. The ^{238}U in the mineral decays with a half-life of 4.468×10^9 years to form a series of shorter-lived intermediates, ending in the stable lead isotope ^{206}Pb (Fig. 19.6). The fraction of uranium remaining after 2.00×10^9 years should be

$$\frac{N}{N_i} = e^{-kt} = e^{-0.6931\, t/t_{1/2}} = \exp\left(\frac{-0.6931 \times 2.00 \times 10^9 \text{ yr}}{4.468 \times 10^9 \text{ yr}}\right) = 0.733$$

and the number of ^{206}Pb atoms should be approximately

$$(1 - 0.733)\, N_i = 0.267\, N_i(^{238}\text{U})$$

The ratio of abundances

$$\frac{N(^{206}\text{Pb})}{N(^{238}\text{U})} = \frac{0.267}{0.733} = 0.364$$

therefore is determined by the time elapsed since the deposit was originally formed—in this case, 2.00×10^9 years. Of course, to calculate the age of the mineral, we work backward from the measured $N(^{206}\text{Pb})/N(^{238}\text{U})$ ratio.

To use the method, it is necessary to be certain that the stable nuclide generated (^{206}Pb in this case) arises only from the parent species (^{238}U here) and that neither lead nor uranium has left or entered the rock over the course of geologic time. If possible, it is desirable to measure the ratios of several different isotopically paired species in the same rock sample. For example, ^{87}Rb decays to ^{87}Sr

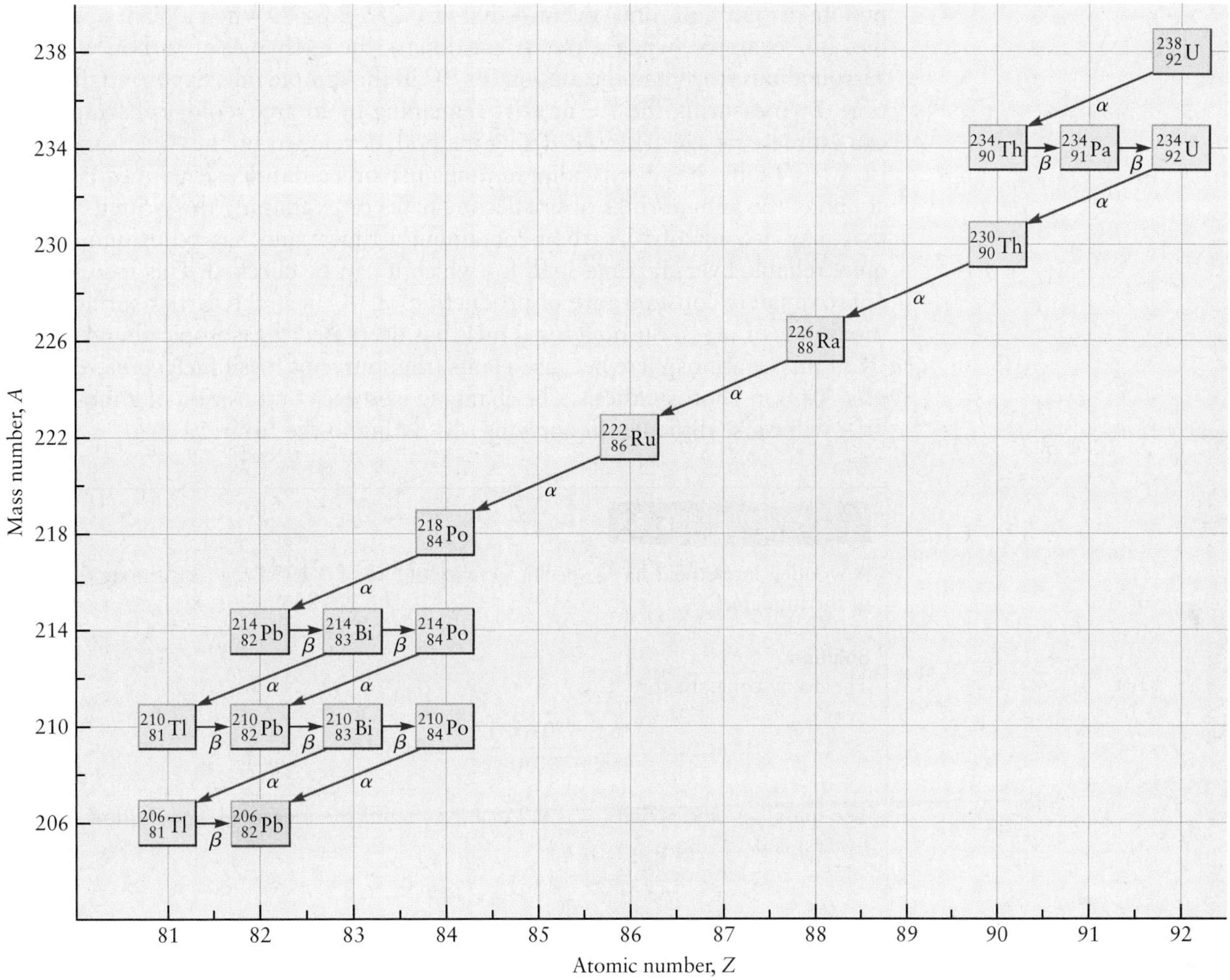

FIGURE 19.6 The radioactive nuclide ^{238}U decays via a series of alpha and beta emissions to the stable nuclide ^{206}Pb.

with a half-life of 4.9×10^{10} years, and one mode of ^{40}K decay (with a half-life of 1.28×10^{9} years) is to ^{40}Ar. Each pair should ideally yield the same age. Detailed analysis of a large number of samples suggests that the oldest surface rocks on earth are about 3.8 billion years old. An estimate of 4.5 billion years for the age of the earth and solar system comes from indirect evidence involving an isotopic analysis of meteorites, which are believed to have formed at the same time.

A somewhat different type of dating uses measurements of ^{14}C decay, which covers the range of human history and prehistory back to about 60,000 years ago. This unstable species (with a half-life of 5730 years) is produced continuously in the atmosphere. Cosmic rays of very high energy cause nuclear reactions that produce neutrons. These can collide with $^{14}_{7}N$ nuclei to produce $^{14}_{6}C$ by the reaction

$$^{14}_{7}N + ^{1}_{0}n \longrightarrow ^{14}_{6}C + ^{1}_{1}H$$

The resulting ^{14}C enters the carbon reservoir on the Earth's surface, mixing with stable ^{12}C as dissolved $H^{14}CO_3^-$ in the oceans, as $^{14}CO_2$ in the atmosphere, and in the tissues of plants and animals. This mixing, which is believed to have occurred at a fairly constant rate over the past 50,000 years, means that the ^{14}C in a living organism has a specific activity of close to 15.3 disintegrations per minute per gram of total carbon—that is, 0.255 Bq g^{-1}. When a plant or animal dies (for example, when a tree is cut down), the exchange of carbon with the surroundings stops and the amount of ^{14}C in the sample falls exponentially with time. By measuring the ^{14}C activity remaining in an archaeological sample, we can estimate its age. This ^{14}C dating method, developed by the American chemist W. F. Libby, has been calibrated against other dating techniques (such as counting the annual rings of bristlecone pines or examining the written records that may accompany a carbon-containing artifact) and has been found to be quite reliable over the time span for which it can be checked. This indicates an approximately constant rate of production of ^{14}C near the earth's surface over thousands of years. Burning fossil fuels has decreased the isotopic abundance of $^{14}CO_2$ in the atmosphere because plants (the source of fossil fuels) preferentially use $^{12}CO_2$ in photosynthesis. The changing isotopic composition of atmospheric CO_2 will cause difficulty in applying ^{14}C dating in the future.

EXAMPLE 19.5

A wooden implement has a specific activity of ^{14}C of 0.195 Bq g^{-1}. Estimate the age of the implement.

Solution

The decay constant for ^{14}C is

$$k = \frac{0.6931}{5730 \text{ yr}} = 1.21 \times 10^{-4} \text{ yr}^{-1}$$

The initial specific activity was 0.255 Bq g^{-1}, and the measured activity now (after t years) is 0.195 Bq g^{-1}, so

$$A = A_i e^{-kt}$$

$$0.195 \text{ Bq g}^{-1} = 0.255 \text{ Bq g}^{-1} e^{-(1.21\times10^{-4})t}$$

$$\ln\left(\frac{0.195}{0.255}\right) = -(1.21 \times 10^{-4} \text{ yr}^{-1})t$$

$$t = 2200 \text{ yr}$$

The implement comes from a tree cut down approximately 2200 years ago.

Related Problems: 25, 26

19.4 Radiation in Biology and Medicine

Radiation has both harmful and beneficial effects for living organisms. All forms of radiation cause damage in direct proportion to the amount of energy they deposit in cells and tissues. The damage takes the form of chemical changes in cellular molecules, which alter their functions and lead either to uncontrolled multiplication and growth of cells or to their death. Alpha particles lose their kinetic energy over very short distances in matter (typically 10 cm in air or 0.05 cm in water or tissues), producing intense ionization in their wakes until they accept electrons and are neutralized to harmless helium atoms. Radium, for example, is an alpha emitter that substitutes for calcium in bone tissue and destroys its capacity to produce both red and white blood cells. Beta particles, gamma rays, and x-rays have greater penetrating power than alpha particles and so present a radiation hazard even when their source is well outside an organism.

The amount of damage produced in tissue by any of these kinds of radiation is proportional to the number of particles or photons and to their energy. A given activity of tritium causes less damage than the same activity of ^{14}C, because the beta particles from tritium have a maximum kinetic energy of 0.0179 MeV, whereas those from ^{14}C have an energy of 0.156 MeV. What is important is the amount of ionization produced or the quantity of energy deposited by radiation. For this purpose, several units are used. The *rad* (*r*adiation *a*bsorbed *d*ose) is defined as the amount of radiation that deposits 10^{-2} J of energy per kilogram of tissue. The damage produced in human tissue depends on still other factors, such as the nature of the tissue, the kind of radiation, the total radiation dose, and the dose rate. To take all these into account, the *rem* (*r*oentgen *e*quivalent in *m*an) has been defined to measure the effective dosages of radiation received by humans. A physical dose of 1 rad of beta or gamma radiation translates into a human dose of 1 rem. Alpha radiation is more toxic; a physical dose of 1 rad of alpha radiation equals about 10 rems. The SI unit for absorbed radiation, analogous to the rad, is the gray (Gy); 1 Gy = 1 joule per kg of material. So, 1 Gy = 100 rad. The SI unit for effective dosage of radiation is the sievert (Sv). The sievert is defined in the same way as the rem, except the delivered dose is expressed in Gy instead of rad; consequently, 1 Sv = 100 rem.

Exposure to radiation is unavoidable. The average person in the United States receives about 100 millirems (mrem) or 1 millisievert (mSv) annually from natural sources that include cosmic radiation and radioactive nuclides such as ^{40}K and ^{222}Rn. Another 50 to 100 mrem or 0.5 to 1.0 mSv (variable) comes from human activities (including dental and medical x-ray examinations and airplane flights, which increase exposure to cosmic rays higher in the atmosphere). Permissible exposure levels are difficult to establish because there is no general agreement about the appropriate dose–response model for radiation damage among scientists and policy makers. The U.S. Environmental Protection Agency (EPA) uses a linear dose–response model, which asserts that damage is proportional to exposure, down to levels where control groups show no effect. It is well known, however, that there are biological repair mechanisms that protect organisms from damage at low exposures, and it might be more reasonable to establish a higher "no observable adverse effect level" or NOAEL. At the time of this writing the EPA appears to continue to embrace the more conservative linear dose–response model. The problem is made even more complex by the necessity to distinguish between tissue damage in an exposed individual and genetic damage, which may not become apparent for several generations. The adverse effects of single exposures are much easier to quantify. A single dose less than 0.25 Sv will likely cause no effect, a 1-Sv dose will cause severe illness, and a 5-Sv dose is fatal, unless the exposed individual received prompt medical treatment. (The treatment of exposed workers following the Chernobyl nuclear accident in 1986 demonstrated that immediate treatment is effective.)

EXAMPLE 19.6

The beta decay of ^{40}K that is a natural part of the body makes all human beings slightly radioactive. An adult weighing 70.0 kg contains about 170 g of potassium. The relative natural abundance of ^{40}K is 0.0118%, its half-life is 1.28×10^9 years, and its beta particles have an average kinetic energy of 0.55 MeV.

(a) Calculate the total activity of ^{40}K in this person.

(b) Determine (in Gy per year) the annual radiation absorbed dose arising from this internal ^{40}K.

(Continued)

Solution

(a) First calculate the decay constant of ^{40}K in s^{-1}:

$$k = \frac{0.693}{t_{1/2}} = \frac{0.693}{(1.28 \times 10^9 \text{ yr})(365 \times 24 \times 60 \times 60 \text{ s yr}^{-1})}$$

$$= 1.72 \times 10^{-17} \text{ s}^{-1}$$

$$\text{number of } ^{40}\text{K atoms} = \frac{170 \text{ g}}{40.0 \text{ g mol}^{-1}}(1.18 \times 10^{-4})(6.02 \times 10^{23} \text{ mol}^{-1})$$

$$= 3.02 \times 10^{20}$$

$$A = -\frac{dN}{dt} = kN = (1.72 \times 10^{-17} \text{ s}^{-1})(3.02 \times 10^{20}) = 5.19 \times 10^3 \text{ s}^{-1}$$

(b) Each disintegration of ^{40}K emits an average of 0.55 MeV of energy, and we assume that all of this energy is deposited within the body. From part (a), 5.19×10^3 disintegrations occur per second, and we know how many seconds are in a year. The total energy deposited per year is then

$$5.19 \times 10^3 \text{ s}^{-1} \times (60 \times 60 \times 24 \times 365 \text{ s yr}^{-1}) \times 0.55 \text{ MeV} = 9.0 \times 10^{10} \text{ MeV yr}^{-1}$$

Next, because a Gy is 1 J per kilogram of tissue, we express this answer in joules per year:

$$(9.0 \times 10^{10} \text{ MeV yr}^{-1})(1.602 \times 10^{-13} \text{ J MeV}^{-1}) = 0.0144 \text{ J yr}^{-1}$$

Each kilogram of body tissue receives 1/70.0 of this amount of energy per year, because the person weighs 70.0 kg. The dose is thus 21×10^{-5} J kg^{-1} yr^{-1}, which is equivalent to 0.21 Gy yr^{-1} or 21 mrad yr^{-1}. This is about a fifth of the annual background dosage received by a person.

Related Problems: 35, 36

Although radiation can do great harm, it confers great benefits in medical applications. The diagnostic importance of X-ray imaging hardly needs mention. Computerized axial tomography (CAT) provides high-resolution three-dimensional images of bones, soft tissues, and tumors; it has become an indispensable tool in modern medicine. Ionizing radiation of all types (X-rays, gamma rays, and particles) are used both externally and internally to treat cancers. The beta-emitting ^{131}I nuclide finds use in the treatment of cancer of the thyroid because iodine is taken up preferentially by the thyroid gland. Precise control of X-ray and particle beam cross-section shapes, coupled with precise alignment of the beams using CAT scans enable radiation oncologists to irradiate tumors selectively, eliminating damage to adjacent tissues as close as 1 mm to the perimeter of the beams.

Positron emission tomography (PET) is an important diagnostic technique using radiation (see *Connection to Medicine*). The vast majority of PET scans are used in oncology. Patients are injected with ^{18}F-labelled glucose that is taken up preferentially by tumors because cancer cells metabolize glucose at greater rates than normal cells. Not only do PET scans provide images that may be more detailed than those provided by MRI or CT scans, they also monitor the metabolic activity of tumors, allowing oncologists to assess whether they are malignant or benign and to follow their response to treatment.

CONNECTION TO MEDICINE

Isotopes and Nuclear Medicine

Radiotracers

The discovery of artificial radioactivity in 1934, and the invention of the cyclotron (a particle accelerator) shortly thereafter, made it possible to study the mechanisms of chemical reactions at a level of detail not previously possible. Molecules were labeled in specific positions with radioisotopes (**radiotracers**) that were incorporated using standard synthetic methods, and the disposition of the tracers among the various products was determined by measuring their activity. The method was quickly adopted to study the mechanisms of biochemical and biological processes because it is extremely sensitive, and also because radioactive isotopes of the biologically important elements were easily prepared using cyclotrons. Early studies focused on phosphorus and iodine metabolism using ^{32}P and ^{131}I, respectively. Considerable insight into the mechanism of photosynthesis was provided by pioneering studies that used ^{18}O (a heavy stable isotope of oxygen)–labeled water to show that it was the source of the O_2 produced and not CO_2. $^{14}CO_2$ was also used to trace the location of carbon in the various carbohydrate products produced and establish the photosynthetic pathways involved.

Several dozen radioisotopes are now produced routinely for a variety of applications in medical diagnostics and radiation therapy. Neutron-deficient isotopes are generally made in medical cyclotrons, whereas proton-deficient isotopes are generally made in nuclear reactors. ^{99m}Tc, an isomer of ^{99}Tc, a γ emitter, is the most widely used radioisotope because its 6-hour half-life is particularly convenient; it is long enough so that it can be transported from the source to the imaging center, but short enough that it emits γ-rays at a sufficiently high rate to ensure high resolution. Compounds containing ^{99m}Tc have been specifically designed to incorporate the tracer into tissues and structures of interest in order to image organs, bones, and tumors, as well as to monitor blood flow through the circulatory system, in particular the heart, brain, and lungs. Other radioisotopes are used to monitor thyroid function, and to image tumors and measure their metabolic activity. Positron emission tomography (PET), using ^{18}F, has become one of the most powerful of the radiotracer-based imaging techniques in recent years, with about 90% of PET scans being used to image tumors.

Positron Emission Tomography

Positron emission tomography detects both members of a pair of 0.511 MeV photons called *annihilation radiation* that are emitted when positrons and electrons

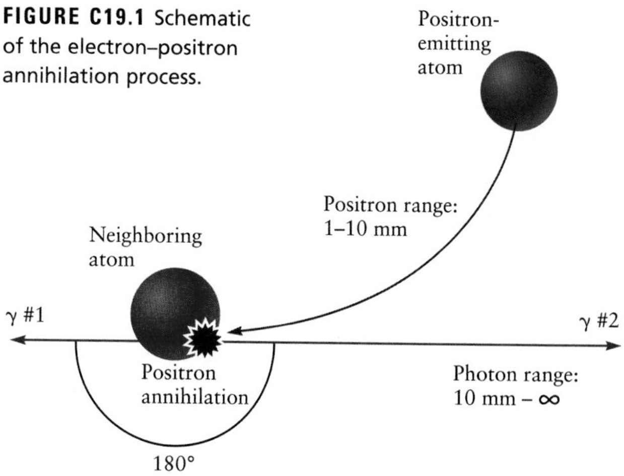

FIGURE C19.1 Schematic of the electron–positron annihilation process.

annihilate one another in a collision, the location of the radiotracer being determined by working backward from the positions of the photons measured by a large cylindrical detector. The reaction is written as follows, and a schematic of the electron–positron annihilation process is shown in Figure C19.1.

$$^{0}_{1}e + {}^{0}_{-1}e \longrightarrow 2\,\gamma\ (511\ \text{MeV})$$

The most important radioisotopes used for PET imaging include ^{11}C, ^{13}N, ^{15}O, and ^{18}F, with half-lives of approximately 20, 10, 2, and 110 minutes respectively. They are synthesized by bombarding a target with high-energy (few MeV) particles in a cyclotron or a linear accelerator, as shown by the following nuclear reactions.

$$^{1}_{1}p + {}^{14}_{7}N \longrightarrow {}^{11}_{6}C + {}^{4}_{2}He$$

$$^{1}_{1}p + {}^{13}_{6}C \longrightarrow {}^{13}_{7}N + {}^{1}_{0}n$$

$$^{2}_{1}H + {}^{14}_{7}N \longrightarrow {}^{15}_{8}O + {}^{1}_{0}n$$

$$^{1}_{1}p + {}^{18}_{8}O \longrightarrow {}^{18}_{9}F + {}^{1}_{0}n$$

The schematic of a cyclotron (Fig. C19.2) identifies its key features. Ions are injected into the center of the cyclotron, which is an evacuated chamber that contains a pair of D-shaped electrodes (dees) separated by a gap and a magnet that establishes a constant field perpendicular to the plane of the dees. A very high frequency-alternating voltage is applied to the dees, which causes the ions to accelerate each time they cross the gap. The magnetic field forces the ions to follow circular orbits of increasing radii with increasing kinetic energy, resulting in a spiral trajectory. They are eventually extracted and bombard the target as shown.

The radioisotopes must be incorporated into molecules very quickly after they are made; rapid organic

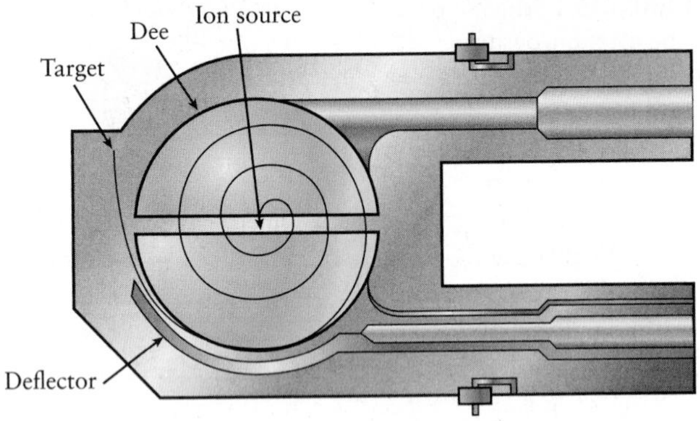

FIGURE C19.2 Schematic of a cyclotron.

synthetic methods developed at Brookhaven National Laboratories were key to the widespread implementation of PET technology. The synthesis of 2-deoxy-2-(^{18}F)fluoro-D-glucose, (FDG), in particular, was significant because it enabled the relatively long-lived ^{18}F isotope to be incorporated into a sugar that is metabolized by all cells. ^{18}FDG is the dominant isotope used in cancer studies because it is preferentially taken up in cancer cells due to their high metabolic rates. One of the earliest applications of PET used ^{11}C to label drugs and investigate their mechanisms of action by imaging the brain. $^{15}O_2$ and $^{15}OH_2$ are used to measure blood flow in the heart and brain, as is $^{13}NH_3$, and the early pioneering studies of the interactions of drugs that cross the blood–brain barrier were conducted using drugs labeled with ^{11}C.

The combination of positron emission tomography with computerized tomography (CT) or magnetic resonance imaging (MRI) allows clinicians to acquire functional (metabolic) and structural (anatomical) images sequentially, enabling them to locate tumors with very high precision. Figure C19.3 shows a PET/CT scanner in which the patient is transferred between the instruments on a gurney that preserves the patient's general alignment. Extremely small gold seeds (less than 1 mm) that are clearly visible in the CT images are implanted at several locations in patients to serve as accurate reference markers for position.

The recent development of 4D PET/CT takes the process one step further, allowing radiation oncologists to image patients in real time and monitor the movement of tumors as patients breathe. The method is so sophisticated that radiation from a pulsed linear accelerator can be delivered to the tumor at a particular time in the breathing cycle, minimizing irradiation of normal tissue and the radiation damage that it causes. The PET/CT scans in Figure C19.4 illustrate the power of the method. The tumor is clearly visible in Figure C19.4a. Figure C19.4b shows that it has largely been destroyed after a course of radiation. The field of nuclear medicine continues to evolve to provide us with diagnostic and therapeutic tools of unprecedented power and scope.

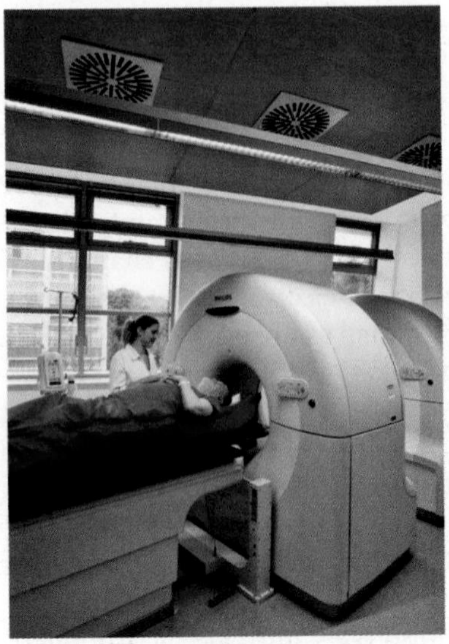

Rinecker Proton Therapy Center

FIGURE C19.3 A PET/CT scanner.

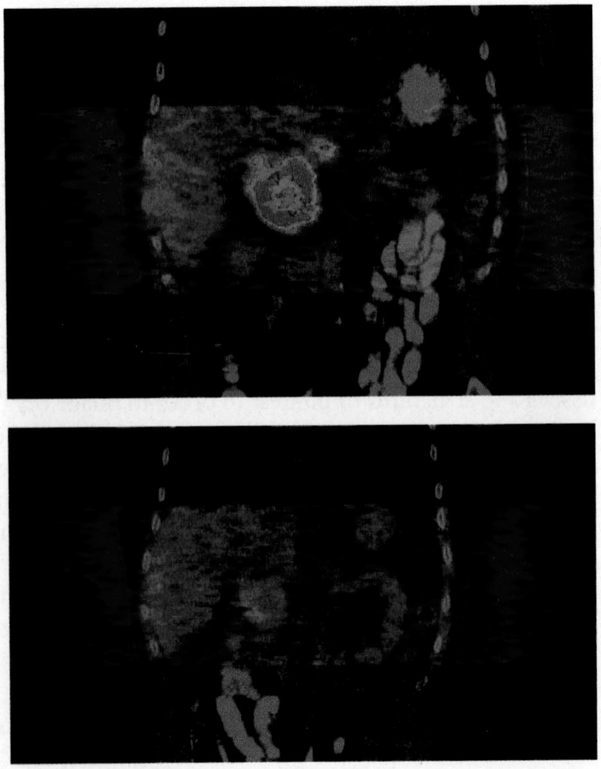

(a) Dwight E. Heron, MD, FACRO

(b) Dwight E. Heron, MD, FACRO

FIGURE C19.4 Scans from a 4D PET/CT scanner. (a) A tumor is visualized (orange and yellow areas). (b) Tumor area shown after treatment with radiation.

Less direct benefits come from other applications. An example is the study of the mechanism of photosynthesis, in which carbon dioxide and water are combined to form glucose in the green leaves of plants.

$$6\ CO_2(g) + 6\ H_2O(\ell) \longrightarrow C_6H_{12}O_6(s) + 6\ O_2(g)$$

Radioactive tracers have been invaluable in establishing the mechanisms of organic and biochemical reactions. ^{14}C, for example, can be introduced into starting re-agents using standard synthetic methods and its position in the products is used to distinguish among several possible mechanisms. Radioactive tracers are also widely used in medical diagnosis. The radioimmunoassay technique, invented by Nobel laureate Rosalyn Yalow, determines the levels of drugs and hormones in body fluids. Antigens (the drug or hormone of interest) are labeled with a radioactive tracer and bound to a known amount of antibody. Samples of blood containing unknown amounts of the antigen are added to the complex, and the unlabeled antigen displaces the labeled antigen, whose concentrations are determined by measuring its activity in solution after rinsing.

19.5 Nuclear Fission

By the 1920s, physicists and chemists were using particle accelerators to bombard samples with high-energy particles to *induce* nuclear reactions. The English physicist James Chadwick (yet another illustrious Rutherford student) identified the neutron in 1932, a dozen years after Rutherford predicted its existence, as a product of the reaction between alpha particles from a radioactive source and ^9Be:

$$^{4}_{2}He + ^{9}_{4}Be \longrightarrow ^{1}_{0}n + ^{12}_{6}C$$

Shortly after Chadwick's discovery, a group of physicists in Rome, led by Enrico Fermi, began to study the interaction of neutrons with the nuclei of various elements. The experiments produced a number of radioactive species, and it was evident that the absorption of a neutron increased the $N:Z$ ratio in target nuclei above the stability line (see Fig. 19.1). One of the targets used was uranium, the heaviest naturally occurring element. Several radioactive products resulted, none of which had chemical properties characteristic of the elements between $Z = 86$ (radon) and $Z = 92$ (uranium). It appeared to the Italian scientists in 1934 that several new transuranic elements ($Z > 92$) had been synthesized, and an active period of investigation followed.

In 1938 in Berlin, Otto Hahn and Fritz Strassmann sought to characterize the supposed transuranic elements. To their bewilderment, they found instead that barium ($Z = 56$) appeared among the products of the bombardment of uranium by neutrons. Hahn informed his former colleague Lise Meitner, and she conjectured that the products of the bombardment of uranium by neutrons were not transuranic elements but fragments of uranium atoms resulting from a process she termed **fission**. The implication of this phenomenon—the possible release of enormous amounts of energy—was immediately evident. When the outbreak of World War II appeared imminent in the summer of 1939, Albert Einstein signed a letter drafted by the top nuclear scientists to President Franklin Roosevelt to inform him of the possible military uses of fission and of his concern that Germany might develop a nuclear explosive. As a result, in 1942 President Roosevelt authorized the Manhattan District Project, an intense, coordinated effort by a large number of physicists, chemists, and engineers to make a fission bomb of unprecedented destructive power.

The operation of the first atomic bomb hinged on the fission of uranium in a chain reaction induced by absorption of neutrons. The two most abundant isotopes of uranium are ^{235}U and ^{238}U, whose natural relative abundances are

0.720% and 99.275%, respectively. Both undergo fission upon absorbing a neutron, the latter only with "fast" neutrons and the former with both "fast" and "slow" neutrons. In the early days of neutron research, it was not known that the absorption of neutrons by nuclei depends strongly upon neutron velocity. By accident, Fermi and his colleagues discovered that experiments conducted on a wooden table led to a much higher yield of radioactive products than those performed on a marble-topped table. Fermi then repeated the irradiation experiments with a block of paraffin wax interposed between the radium–beryllium neutron source and the target sample, with the startling result that the induced level of radioactivity was greatly enhanced. Within hours Fermi had found the explanation: The high-energy neutrons emitted from the radium–beryllium source were reduced to thermal energy by collision with the low-mass nuclei of the paraffin molecules. Because of their lower energies, their probability of reaction with ^{235}U was greater and a higher yield was achieved. Hydrogen nuclei and the nuclei of other light elements such as ^{12}C (in graphite) are very effective in reducing the energies of high-velocity neutrons and are called **moderators.**

The fission of ^{235}U follows many different patterns, and some 40 elements have been identified among the fission products. In a typical fission event two particular nuclides are produced together with two or three secondary neutrons; collectively, they carry away about 170 MeV of kinetic energy. Usually, the daughter nuclei have different Z and A numbers, so the fission process is asymmetric. Three of the many pathways are

$$
{}^{1}_{0}n + {}^{235}_{92}\text{U} \longrightarrow \begin{cases} {}^{72}_{30}\text{Zn} + {}^{162}_{62}\text{Sm} + 2\ {}^{1}_{0}n \\ {}^{80}_{38}\text{Sr} + {}^{153}_{54}\text{Xe} + 3\ {}^{1}_{0}n \\ {}^{94}_{36}\text{Kr} + {}^{139}_{56}\text{Ba} + 3\ {}^{1}_{0}n \end{cases}
$$

Figure 19.7 shows the distribution of the nuclides produced. On average, more than one neutron is emitted per neutron absorbed in the fission process, which

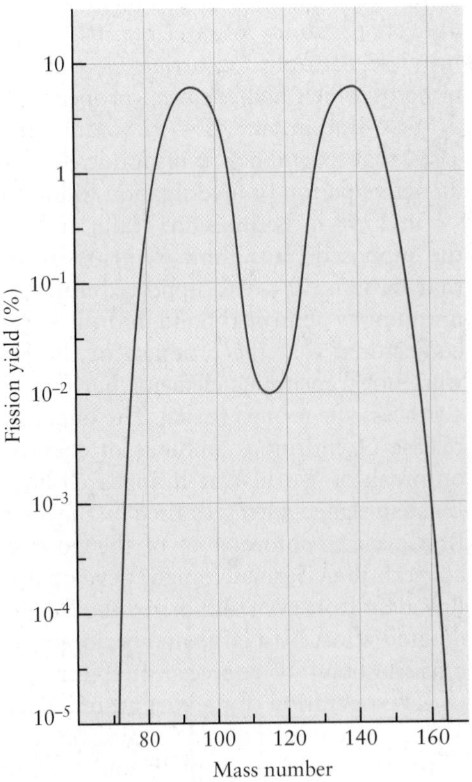

FIGURE 19.7 The distribution of nuclides produced in the fission of ^{235}U after capturing a neutron has two peaks. Nuclei having mass numbers in the vicinity of $A = 95$ and $A = 139$ are formed with the highest yield; those with $A \approx 117$ are produced with lower probability.

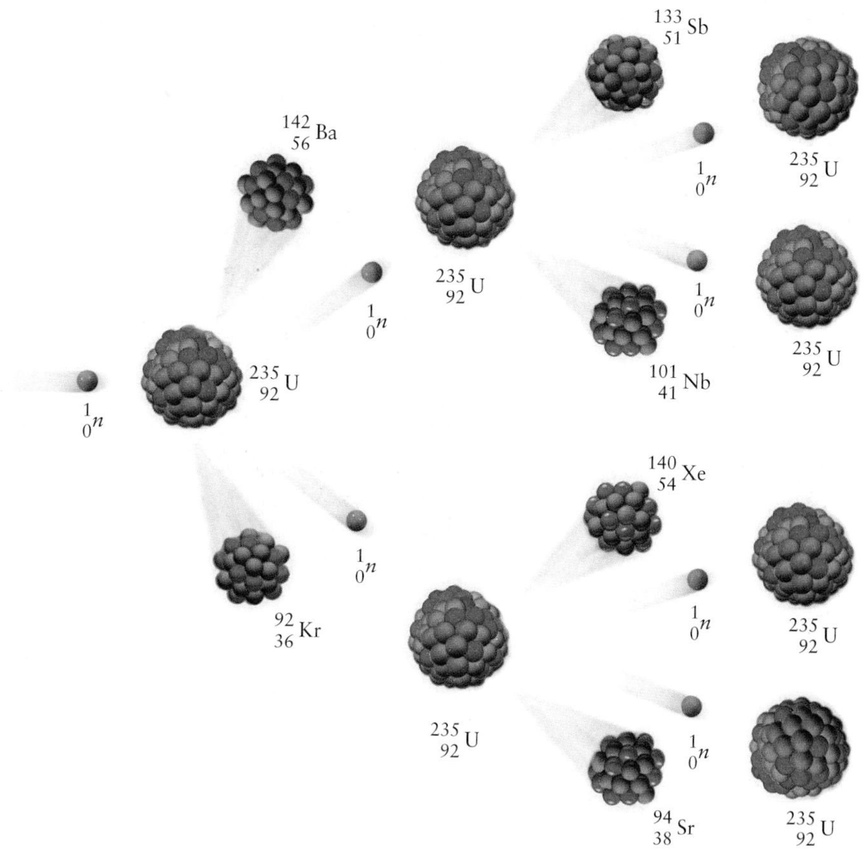

FIGURE 19.8 The number of free or ejected neutrons increases exponentially in a chain reaction. Not all of the emitted neutrons are shown.

leads to a branching chain reaction (see Section 18.4) with the number of neutrons growing exponentially with time (Fig. 19.8). Permitted to proceed unchecked, this reaction would quickly lead to the release of enormous quantities of energy. Neutrons can be lost by various processes, so it is not obvious that a self-propagating reaction will occur. On December 2, 1942, Fermi and his associates demonstrated that a self-sustaining neutron chain reaction occurred in a uranium "pile" with a graphite moderator. They limited its power output to $\frac{1}{2}$ J s^{-1} by inserting cadmium control rods to absorb neutrons, thereby balancing the rates of neutron generation and loss.

Fermi's work made two developments possible: (1) the exploitation of nuclear fission for the controlled generation of energy in nuclear reactors and (2) the production of ^{239}Pu, a slow- and fast-neutron fissionable isotope of plutonium, as an alternative to ^{235}U for the construction of atomic bombs. For the sudden release of energy required in an explosive, it was necessary to obtain ^{235}U or ^{239}Pu in a state free of neutron-absorbing impurities. Both alternatives were pursued simultaneously. The first required enriching ^{235}U from its relative abundance of 0.72% in natural uranium. This was accomplished through gaseous diffusion (see Section 9.7). For the second alternative, ^{239}Pu was recovered from the partially spent uranium fuel of large nuclear reactors by means of redox reactions, precipitation, and solvent extraction.

Although Fermi and his associates were the first scientists to demonstrate a self-sustaining nuclear chain reaction, a natural uranium fission reactor "went critical" about 1.8 billion years ago in a place now called Oklo, in the Gabon Republic of equatorial Africa. In 1972 French scientists discovered that the ^{235}U content of ore from a site in the open-pit mine at Oklo was only 0.7171%; the normal content of ore from other areas of the mine was 0.7207%. Although this deviation was not large, it was significant, and an investigation revealed that other elements were also present in the ore, in the exact proportions expected

FIGURE 19.9 A nuclear power plant. The large structure on the left is a cooling tower; the containment building is the smaller building on the right with the domed top.

after nuclear fission. This discovery established that a self-sustaining nuclear reaction had occurred at Oklo. The geological age of the ore body was found to be about 1.8×10^9 years, and the original ^{235}U concentration is calculated to have been about 3%. From the size of the active ore mass and the depletion of ^{235}U, it is estimated that the reactor generated about 15,000 megawatt-years (5×10^{17} J) of energy over about 100,000 years.

19.5.1 Nuclear Power Reactors

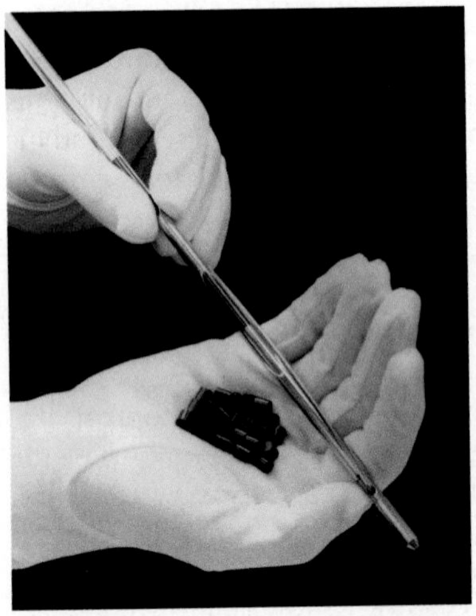

FIGURE 19.10 UO_2 pellets and fuel rods.

Most **nuclear power reactors** in the United States (Fig. 19.9) use pellets of UO_2 (Fig. 19.10) that have been sintered to form hard ceramics, which are then inserted into tubes made from a special steel alloy and bundled to form a fuel rod assembly. The uranium is primarily ^{238}U, but the amount of ^{235}U is enriched above natural abundance to a level of about 3%. The moderator used to slow the neutrons (to increase the efficiency of the fission) is ordinary water in most cases, so these reactors are called "light-water" reactors. The controlled release of energy by nuclear fission in power reactors demands a delicate balance between neutron generation and neutron loss. As mentioned earlier, this is accomplished by means of steel control rods containing ^{113}Cd or ^{10}B, isotopes that have a very large neutron-capture probability, and which effectively control the neutron flux. These rods are automatically inserted into or withdrawn from the fissioning system in response to a change in the neutron flux. As the nuclear reaction proceeds, the moderator (water) is heated and transfers its heat to a steam generator. The high-pressure steam then drives turbines that generate electricity (Fig. 19.11). The nonnuclear part of a nuclear power plant is essentially identical to that of a conventional fossil-fuel–fired power plant, the difference being the source of energy to heat the water.

The power reactors discussed so far rely on the fission of ^{235}U, an isotope in limited supply. An alternative is to convert the much more abundant ^{238}U to fissionable plutonium (^{239}Pu) by neutron bombardment:

$$^{238}_{92}\text{U} + {}^{1}_{0}n \longrightarrow {}^{239}_{93}\text{Np} + {}^{0}_{-1}e^{-} \longrightarrow {}^{239}_{94}\text{Pu} + 2\,{}^{0}_{-1}e^{-}$$

Fissionable ^{233}U can also be made from thorium:

$$^{232}_{90}\text{Th} + {}^{1}_{0}n \longrightarrow {}^{233}_{91}\text{Pa} + {}^{0}_{-1}e^{-} \longrightarrow {}^{233}_{92}\text{U} + 2\,{}^{0}_{-1}e^{-}$$

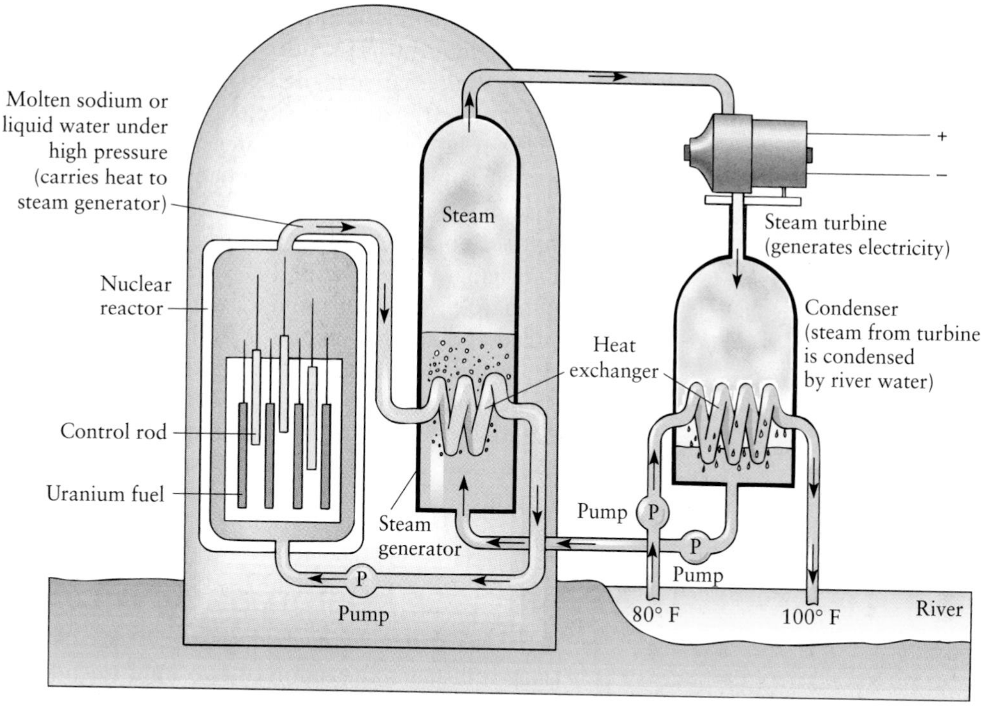

FIGURE 19.11 A schematic diagram of a pressurized-water nuclear power reactor.

In a *breeder reactor*, in addition to heat being generated by fission, neutrons are absorbed in a blanket of uranium or thorium. This causes the preceding reactions to occur and generates additional fuel for the reactor to use. An advanced technology for breeder reactors uses liquid sodium instead of water as the coolant, allowing the use of faster neutrons than in water-cooled reactors. The faster neutrons cause more complete consumption of radioactive fuels, increasing efficiency and greatly reducing radioactive waste.

The risks associated with the operation of nuclear reactors are small but not negligible, as demonstrated by the failure of the Three Mile Island reactor in the United States in 1979, the disaster at Chernobyl in the former Soviet Union in 1987, and the tsunami that led to the meltdown of the Fukushima Daiichi nuclear power plant in Japan in 2011. If a reactor has to be shut down quickly, there is danger of a meltdown, in which the heat from the continuing fission processes melts the uranium fuel. Coolant must be circulated until heat from the decay of short-lived isotopes has been dissipated. The Three Mile Island accident resulted in a partial meltdown because some water coolant pumps were inoperative and others were shut down too soon, causing damage to the core and a slight release of radioactivity into the environment. The Chernobyl disaster was caused by a failure of the water-cooling system and a meltdown. The rapid and uncontrolled nuclear reaction that took place set the graphite moderator on fire and caused the reactor building to rupture, spreading radioactive nuclides with an activity estimated at 2×10^{20} Bq into the atmosphere. A major part of the problem was that the reactor at Chernobyl, unlike those in the United States, was not in a massive containment building. The Fukushima nuclear disaster occurred on March 11, 2011. Three of the nuclear power plant's six reactors suffered a meltdown of the reactor core. This incident occurred as a result of destruction caused by an earthquake-generated tsunami. The incident resulted in a large amount of radioactive water being released into the ocean and surrounding environment.

The safe disposal of the radioactive wastes from nuclear reactors is an important and controversial matter. A variety of proposals have been made, including the burial of radioactive waste in deep mines on either a recoverable or a permanent basis, burial at sea, and launching the waste into outer space. The first alternative is the only one that appears credible. The essential requirement is that the disposal site(s) be stable with respect to possible earthquakes or invasion by underground water. Spent nuclear fuel can be encased in blocks of borosilicate glass, packed in metal containers, and buried in stable rock formations. For a nuclide such as ^{239}Pu, whose half-life is 24,000 years, a storage site that is stable over 240,000 years is needed before the activity drops to 0.1% of its original value. The Berkeley physicist Richard Muller has suggested that we compare the level of radioactivity of the waste to that which has already been mined and accept some probability greater than zero that there will be a leak. His calculations suggest that a facility that would provide a 99% level of confidence that there would be no leak would reduce the time needed to store the material to 300 years, by comparing the radioactivity resulting from such a leak not to zero but to the natural background levels found in the surrounding terrain.

19.6 Nuclear Fusion and Nucleosynthesis

Nuclear fusion is the union of two light nuclides to form a heavier nuclide with the release of energy. Fusion processes are often called *thermonuclear reactions* because they require that the colliding particles possess very high kinetic energies, corresponding to temperatures of millions of degrees, before they are initiated. They are the processes that occur in the sun and other stars. In 1939 the German physicists Hans Bethe (and, independently, Carl von Weizsäcker) proposed that in normal stars (main sequence) the following reactions take place:

$$^{1}_{1}H + ^{1}_{1}H \longrightarrow ^{2}_{1}H + ^{0}_{1}e^{+} + \nu$$

$$^{2}_{1}H + ^{1}_{1}H \longrightarrow ^{3}_{2}He + \gamma$$

$$^{3}_{2}He + ^{3}_{2}He \longrightarrow ^{4}_{2}He + 2\,^{1}_{1}H$$

In the first reaction, two high-velocity protons fuse to form a deuteron, with the emission of a positron and a neutrino that carry away (as kinetic energy) the additional 0.415 MeV of energy released. In the second reaction, a high-energy deuteron combines with a high-velocity proton to form a helium nucleus of mass 3 and a gamma ray. The third reaction completes the cycle with the formation of a normal helium nucleus ($^{4}_{2}He$) and the regeneration of two protons. Each of these reactions is exothermic, but up to 1.25 MeV is required to overcome the repulsive barrier between the positively charged nuclei. The overall result of the cycle is to convert hydrogen nuclei to helium nuclei, and the process is called **hydrogen burning.**

As such a star ages and accumulates helium, it begins to contract under the influence of its immense gravity. As it contracts, its helium core heats up; when it reaches a temperature of about 10^8 K, a stage of **helium burning** begins. The first reaction that occurs is

$$2\,^{4}_{2}He \rightleftharpoons ^{8}_{4}Be$$

This reaction is written as an equilibrium because the ^{8}Be with a half-life of only 6.7×10^{-17} s quickly reverts to helium nuclei. Even with this short half-life, the ^{8}Be nuclei are believed to occasionally react with alpha particles to form stable ^{12}C:

$$^{8}_{4}Be + ^{4}_{2}He \longrightarrow ^{12}_{6}C$$

The overall effect of the helium-burning phase of a star's life is to convert three helium nuclei to a carbon nucleus, just as helium was formed from four hydrogen nuclei in the hydrogen-burning phase. The density of the core of a star that is burning helium is on the order of 10^5 g cm^{-3}.

This process of **nucleosynthesis** continues beyond the formation of ^{12}C to produce ^{13}N, ^{13}C, ^{14}N, ^{15}O, ^{15}N, and ^{16}O. The stars in this stage, classified as red giants, are near the end of their lives and expand as the temperature of the core rises. Similar cycles occur until the temperature of a star core is about 4×10^9 K, its density is about 3×10^6 g cm^{-3}, and the nuclei are ^{56}Fe, ^{59}Co, and ^{60}Ni. These are the nuclei that have the maximum binding energy per nucleon (see Fig. 19.2). It is thought that the synthesis of still heavier nuclei occurs in the immense explosions of supernovae.

Heavy elements can also be produced in particle accelerators, which accelerate ions to high speeds, causing collisions that generate the new elements. Technetium, for example, is not found in nature but was first produced in 1937 when high-energy deuterons were directed at a molybdenum target:

$$^{96}_{42}\text{Mo} + {}^{2}_{1}\text{H} \longrightarrow {}^{97}_{43}\text{Tc} + {}^{1}_{0}n$$

The first **transuranic element** was produced in 1940. Neptunium ($Z = 93$) results from the capture of a neutron by ^{238}U, followed by beta decay. Subsequent work by the American chemist Glenn Seaborg and others led to the production of plutonium ($Z = 94$) and heavier elements. In recent years, nuclides with Z as high as 118 have been made, but only in tiny quantities. These nuclides have very short half-lives.

Controlled nuclear fusion is an attractive alternative to nuclear fission that could produce virtually unlimited quantities of energy from readily available sources of fuel with no adverse environmental impact. Two major efforts are underway to harness the power of the fusion of deuterium with tritium to form helium in a controlled way. There is virtually a limitless supply of deuterium that can be extracted from seawater, and tritium can be produced as a byproduct of the fusion reaction as discussed below. The reaction

$$^{2}_{1}\text{H} + {}^{3}_{1}\text{H} \longrightarrow {}^{4}_{2}\text{He} + {}^{1}_{0}n$$

produces 17.6 MeV *per event,* which corresponds to about 340,000,000 kJ g^{-1} of fuel burned. To put this into perspective, a conventional power plant produces about 1 GW (10^9 W) of power, consuming about 2.5 million tons of coal per year. The same amount of energy produced by nuclear fusion would consume less than 10 kg of the hydrogen isotopes, so it is little wonder that nuclear fusion research and development is being so vigorously pursued.

Initiating controlled fusion is difficult because the two nuclei must come together with sufficient energy to overcome the repulsive Coulomb potential energy barrier, which is about 1 MeV for these hydrogen isotopes. This potential barrier can be thought of as an activation energy for the reaction; temperatures of the order of 100 million K are required for it to proceed at an acceptable rate (see Section 18.5). Two very different approaches have been developed over many years in attempts to achieve these extraordinarily high temperatures; they are called magnetic confinement and inertial confinement, respectively. Magnetic confinement fusion uses magnetic fields as "walls" to contain very high temperature plasmas (ionized gases) because there are no material walls that could survive under these conditions. The fuel is heated, in part, by the absorption of electromagnetic radiation in the same way that microwave ovens heat our food. Inertial confinement fusion, on the other hand, uses short bursts of very high-power laser radiation to compress and heat the fuel to very high temperatures and densities. The main difference between the two approaches is that magnetic confinement heats the fuel at somewhat lower pressures but for longer times, whereas inertial confinement both heats and compresses the fuel almost instantaneously.

There are currently two large-scale efforts underway to demonstrate that controlled nuclear fusion can generate more energy than that required for initiation, one focused on magnetic confinement and the other on inertial confinement. ITER (originally the International Thermonuclear Fusion Reactor but now said to be derived from the Latin word for "journey" for political reasons) is an international collaboration that is constructing a large facility in the south of France with the specific goal of producing 10 times as much energy as that required for initiation. The ITER tokamak is built around a donut-shaped vacuum chamber that contains a plasma (a gas of ionized particles, see Sec. 1.4). A large number of superconducting magnets generate a magnetic field some 200,000 times stronger than that of the earth, which confines the plasma to the interior of the vacuum chamber and prevents it from heating the walls. Figure 19.12 shows a schematic of the ITER tokamak. The design calls for an operating temperature of 150 million K and power output of 500 MW. The ITER tokamak has also been designed to produce tritium from the reaction of neutrons with lithium in the "blanket" that covers the interior surfaces of the vacuum vessel, the balanced reaction being

$$^{6}_{3}\text{Li} + ^{1}_{0}n \longrightarrow ^{3}_{1}\text{H} + ^{4}_{2}\text{He}$$

A demonstration power plant (DEMO) is intended to follow, based upon the anticipated success of the ITER experiment. The goal of that project is to demonstrate continuous fusion that produces sustained power in the 2 to 4 GW range, which is comparable to the largest existing conventional power plants.

The National Ignition Facility (NIF) is a project of the United States Department of Energy designed to demonstrate net energy production by inertial confinement fusion. The world's largest laser, with the greatest amount of energy per pulse, has been constructed and tested. A total of 192 amplified laser beams are focused onto a tiny, hollow beryllium sphere, about the size of a peppercorn, which contains a fraction of a milligram of the target mixture of deuterium and tritium in the form of a frozen solid held at 18 K (see Fig. 19.13).

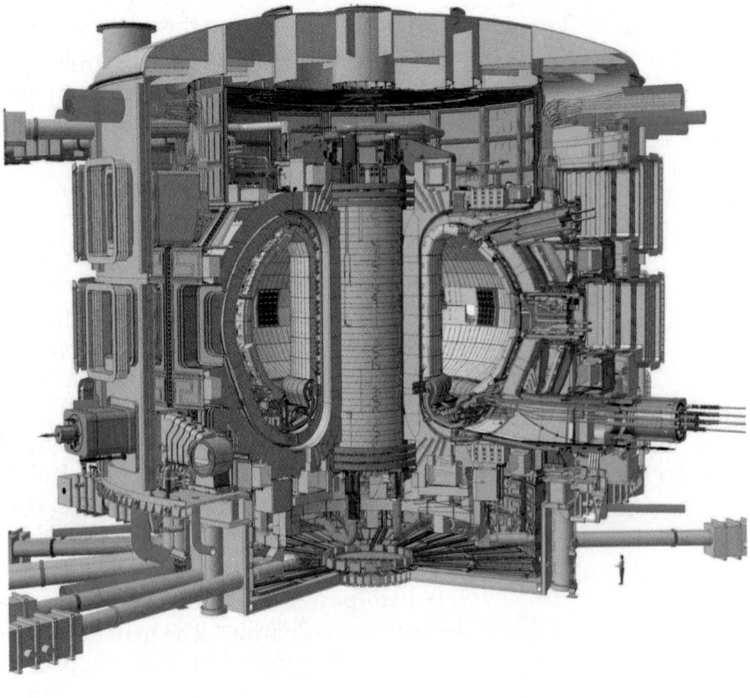

FIGURE 19.12 Cutaway of the ITER tokamak, showing the torus-shaped vacuum vessel (yellow) that is about 60 ft in diameter along with the external magnets. The entire machine will weigh some 23,000 tons.

(a)

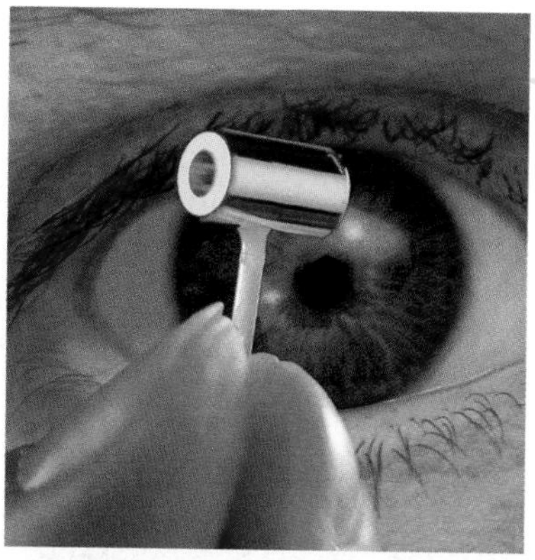
(b)

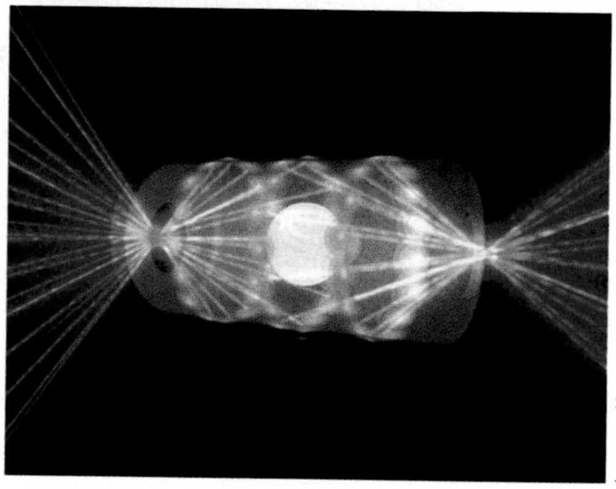
(c)

Lawrence Livermore National Laboratory

FIGURE 19.13 (a) National Ignition Facility target chamber for inertial confinement fusion experiments. 192 laser beams enter through the ports shown and are focused onto the target, which is held at the tip of the positioner seen in the foreground. (b) Target holder. (c) Laser beams focused on sample holder generate X-rays that cause the fuel pellet to implode, initiating fusion.

The laser beam generates more power than the entire U.S. electrical grid during its short pulse. The extremely high power compresses the fuel and heats it to the 100 million K required to initiate the fusion reaction. The system passed a critical milestone in February 2014 when researchers for the first time initiated a controlled nuclear fusion event that produced more energy than was used to initiate the reaction. There are significant challenges that must be met before inertial confinement fusion becomes an important practical source of energy. The initial goal of the NIF is to produce 20 MJ of energy per pulse with the laser system currently capable of generating only about two pulses per day. Scaling this approach to provide power at levels that approach those provided by conventional power plants will require lasers capable of delivering perhaps 10 pulses per second that, in turn, would consume about 850,000 fuel pellets per day, which is a daunting challenge indeed.

CHAPTER SUMMARY

The identities of the elements are not preserved in nuclear reactions—elements decay into lighter daughter elements in fission reactions and heavier elements are synthesized from lighter elements in fusion reactions. Mass changes in nuclear reactions are relatively small, but the accompanying energy changes are enormous; they are related by Einstein's famous formula $E = mc^2$. The isotopes of the lighter elements ($Z < 20$ or so) are stable when the ratio of the number of neutrons to the number of protons (N/Z) is approximately equal to 1. For these low Z elements, isotopes with $N/Z < 1$ will decay via positron emission or electron capture to decrease the number of protons in the nucleus, whereas those with $N/Z > 1$ will decay via electron emission to increase the number of protons in the nucleus. A fourth decay channel, alpha particle emission, becomes important for heavier nuclei. Radioactive decay follows first order kinetics, and the half-life $t_{1/2}$ is a convenient measure of the timescale of the reaction. Half-lives range from 10^{-16} s to 10^{20} years, which is an incredibly wide range of timescales. The half-life of ^{238}U has been used to calibrate geological timescales, and ^{14}C dating is a well-established method for dating human artifacts in anthropology. X-rays, gamma rays, electrons, positrons, and alpha particles are all used in medical diagnostics and therapy with increasing efficacy and fewer side effects. Nuclear fusion, which led to the formation of the heavier elements from hydrogen in the process called nucleosynthesis, continues to hold promise as a source of clean power for the future.

CONCEPTS AND SKILLS

Section 19.1 – Mass–Energy Relationships in Nuclei

Identify and characterize patterns of nuclear stability and the nuclear decay processes.

- Stable nuclei are those for which $N/Z = 1$ for $Z < 20$ and $N/Z \approx 1.5$ for $Z > 40$.
- Alpha decay is the spontaneous emission of α particles ($^{4}_{2}He$ nuclei) from heavy nuclei ($Z > 83$); the parent nucleus loses two protons and two neutrons.

Describe mass energy relationships in nuclear reactions (Problems 1–6).

- Einstein's relation $E = mc^2$ establishes the relationship between mass and energy in nuclear reactions; the conversion of very small quantities of mass produces tremendous quantities of energy. Spontaneous nuclear reactions are those for which $\Delta m < 0$.

Section 19.2 – Nuclear Decay Processes

Summarize the characteristics of the three kinds of ionizing radiation identified by Rutherford.

- α particles are $^{4}_{2}He$ nuclei that are the least penetrating form of ionizing radiation, β rays are electrons that penetrate a few millimeters into matter, and γ rays are very high energy electromagnetic radiation that penetrate deeply.
- Electron emission and positron emission (called β^- and β^+ decay, respectively) and electron capture (EC) are processes that convert protons into neutrons (and vice versa) to bring the N/Z ratio closer to the line of stability.

Describe the various kinds of radiation detectors.

- Photographic film darkens when exposed to radiation, the degree of darkening being proportional to the dose. Geiger-Müller counters are gas-filled tubes across which a large voltage is applied. Radiation ionizes gas molecules, and the high-energy primary electrons produced are converted into a large number of low-energy secondary electrons. Geiger-Müller counters record the resulting large current pulses. Proportional counters are gas-filled tubes, the gases being ionized by radiation. Proportional counters measure the activity by counting the number of pulses per second, and they measure the energy of the emitted particles or rays by measuring the pulse height current or number of electrons per pulse.

Write balanced nuclear equations for electron emission (one form of beta decay), positron emission, electron capture, and alpha decay, and calculate the maximum kinetic energies of the particles emitted (Problems 7–18).

- Charge, mass number, and atomic number are conserved in a balanced nuclear reaction.

Section 19.3 – Kinetics of Radioactive Decay

Solve problems involving the half-life or decay constant of a radioactive sample and its activity (Problems 19–24).

- Nuclear decay follows first-order kinetics characterized by a half-life $t_{1/2} = \ln 2/k = 0.693/k$ where k is the first-order rate constant.

Apply the kinetics of nuclear decay to the dating of rocks or artifacts (Problems 25–30).

- The ages of rocks are determined by comparing the ratio of the numbers of daughter to parent nuclei. Radiocarbon dating using ^{14}C assumes that living things have $^{14}C/^{12}C$ ratios that are in equilibrium with the atmosphere that persists until death, after which time the ^{14}C decays with its characteristic 5730-year half-life.

Section 19.4 – Radiation in Biology and Medicine

Discuss the interactions of radiation with various kinds of matter and the measurement of radiation dosage (Problems 31–36).

- The radiation dose is the energy deposited per kg of tissue; it is measured in rads or grays (SI unit). The relative biological damage is given by the dose multiplied by a factor that takes into account the damage caused by different kinds of radiation. Alpha particles and high-energy neutrons cause about 20 times as much damage as β or γ rays.

Section 19.5 – Nuclear Fission

Describe the processes of nuclear fission and calculate the amounts of energy released when it occurs (Problems 37–41).

- Nuclear fission is the spontaneous splitting of a nucleus into two smaller fragments, whereas nuclear fusion is the synthesis of a heavier nucleus from a pair of lighter nuclei.
- Nuclear chain reactions occur when more than one neutron is produced per fission event, allowing self-propagating reactions to occur.

Explain the benefits and risks associated with nuclear power.

- Nuclear power has the potential to provide essentially unlimited energy without contributing to global climate change. The major risks include high-level waste disposal and the possibility of diverting nuclear fuel to make nuclear weapons.

Section 19.6 – Nuclear Fusion and Nucleosynthesis

Explain the potential benefits of nuclear fusion.

- Fusion of hydrogen isotopes in normal stars produces He, which reacts further to produce most of the heavier elements in a sequence of fusion reactions called nucleosynthesis.

FIGURE 19.14 Radon most commonly enters houses through the foundation or basement walls.

CUMULATIVE EXERCISE

Radon in the Environment

Radioactive ^{222}Rn and ^{220}Rn form constantly from the decay of uranium and thorium in rocks and soil and, being gaseous, seep out of the ground. The radon isotopes decay fairly quickly, but their products, which are also radioactive, are then in the air and attach themselves to dust particles. Thus, airborne radioactivity can accumulate to worrisome levels in poorly ventilated basements (Fig. 19.14) in ground that is rich in uranium and thorium.

(a) Describe the composition of an atom of ^{222}Rn and compare it with that of an atom of ^{220}Rn.

(b) Although ^{222}Rn is a decay product of ^{238}U, ^{220}Rn comes from ^{232}Th. How many alpha particles are emitted in the formation of these radon isotopes from their uranium or thorium starting points? (*Hint:* Alpha decay changes the mass number *A*, but other decay processes do not.)

(c) Can alpha decay alone explain the formation of these radon isotopes from ^{238}U and ^{232}Th? If not, state what other types of decay must occur.

(d) Can $^{222}_{86}Rn$ and $^{220}_{86}Rn$ decay by alpha particle emission? Write balanced nuclear equations for these two decay processes, and calculate the changes in mass that would result. The masses of ^{222}Rn and ^{220}Rn atoms are 222.01757 and 220.01140 u, respectively; those of ^{218}Po and ^{216}Po are 218.0089 and 216.00192 u, respectively.

(e) Calculate the energy change in the alpha decay of one ^{220}Rn nucleus, in megaelectron volts and in joules.

(f) The half-life of ^{222}Rn is 3.82 days. Calculate the initial activity of 2.00×10^{-8} g of ^{222}Rn, in disintegrations per second.

(g) What will be the activity of the ^{222}Rn from part (f) after 14 days?

(h) The half-life of ^{220}Rn is 54 s. Are the health risks of exposure to a given amount of radon for a given short length of time greater or smaller for ^{220}Rn than for ^{222}Rn?

Answers

(a) An atom of ^{222}Rn has 86 electrons outside the nucleus. Inside the nucleus are 86 protons and $222 - 86 = 136$ neutrons. An atom of ^{220}Rn has the same number of electrons and protons, but only 134 neutrons in its nucleus.

(b) Four alpha particles are produced to make ^{222}Rn from ^{238}U; three are produced to make ^{220}Rn from ^{232}Th.

(c) If $^{238}_{92}U$ were to lose four alpha particles, $^{222}_{84}Po$ would result instead of $^{222}_{84}Rn$. Two $^{0}_{-1}e^{-}$ beta particles must be ejected from the nucleus along the way to raise the atomic number to $Z = 86$. The same is true of the production of ^{220}Rn from ^{232}Th.

(d) $^{222}_{86}Rn \longrightarrow {}^{218}_{84}Po + {}^{4}_{2}He$; $\Delta m = -0.00617 \text{ u} < 0$; allowed

$^{220}_{86}Rn \longrightarrow {}^{216}_{84}Po + {}^{4}_{2}He$; $\Delta m = -0.0069 \text{ u} < 0$; allowed

(e) $\Delta E = -6.4 \text{ MeV} = -1.03 \times 10^{-12} \text{ J}$

(f) $A = 1.14 \times 10^{8} \text{ s}^{-1}$

(g) $A = 9.0 \times 10^{6} \text{ s}^{-1}$

(h) Greater

PROBLEMS

Answers to problems whose numbers are red appear in Appendix G. Problems that are more challenging are indicated with asterisks.

19.1 Mass–Energy Relationships in Nuclei

1. Complete and balance the following equations for nuclear reactions that are thought to take place in stars:
   (a) $2\ {}^{12}_{6}C \longrightarrow ? + {}^{1}_{0}n$
   (b) $? + {}^{1}_{1}H \longrightarrow {}^{12}_{6}C + {}^{4}_{2}He$
   (c) $2\ {}^{3}_{2}He \longrightarrow ? + 2\ {}^{1}_{1}H$

2. Complete and balance the following equations for nuclear reactions that are used in particle accelerators to make elements beyond uranium:
   (a) ${}^{4}_{2}He + {}^{253}_{99}Es \longrightarrow ? + 2\ {}^{1}_{0}n$
   (b) ${}^{249}_{98}Cf + ? \longrightarrow {}^{257}_{103}Lr + 2\ {}^{1}_{0}n$
   (c) ${}^{238}_{92}U + {}^{12}_{6}C \longrightarrow {}^{244}_{98}Cf + ?$

3. Calculate the total binding energy, in both kJ per mole and MeV per atom, and the binding energy per nucleon of the following nuclides, using the data from Table 19.1.
   (a) ${}^{40}_{20}Ca$ (b) ${}^{87}_{37}Rb$ (c) ${}^{238}_{92}U$

4. Calculate the total binding energy, in both kilojoules per mole and MeV per atom, and the binding energy per nucleon of the following nuclides, using the data from Table 19.1.
   (a) ${}^{10}_{4}Be$ (b) ${}^{35}_{17}Cl$ (c) ${}^{49}_{22}Ti$

5. Use the data from Table 19.1 to predict which is more stable: four protons, four neutrons, and four electrons organized as two ^{4}He atoms or as one ^{8}Be atom. What is the mass difference?

6. Use the data from Table 19.1 to predict which is more stable: 16 protons, 16 neutrons, and 16 electrons organized as two ^{16}O atoms or as one ^{32}S atom. What is the mass difference?

19.2 Nuclear Decay Processes

7. The nuclide ${}^{8}_{5}B$ decays by positron emission to ${}^{8}_{4}Be$. What is the energy released (in MeV)?

8. The nuclide ${}^{10}_{4}Be$ undergoes spontaneous radioactive decay to ${}^{10}_{5}B$ with emission of a beta particle. Calculate the maximum kinetic energy of the emitted beta particle.

9. Write balanced equations that represent the following nuclear reactions.
   (a) Electron emission by ${}^{39}_{17}Cl$
   (b) Positron emission by ${}^{22}_{11}Na$
   (c) Alpha emission by ${}^{224}_{88}Ra$
   (d) Electron capture by ${}^{82}_{38}Sr$

10. Write balanced equations that represent the following nuclear reactions.
   (a) Alpha emission by ${}^{155}_{70}Yb$
   (b) Positron emission by ${}^{26}_{14}Si$
   (c) Electron capture by ${}^{65}_{30}Zn$
   (d) Electron emission by ${}^{100}_{41}Nb$

11. The stable isotopes of neon are ^{20}Ne, ^{21}Ne, and ^{22}Ne. Predict the nuclides formed when ^{19}Ne and ^{23}Ne decay.

12. The two stable isotopes of carbon are ^{12}C and ^{13}C. Predict the nuclides formed when ^{11}C and ^{14}C decay. Is alpha emission by ^{14}C possible?

13. The free neutron is an unstable particle that decays into a proton. What other particle or particles is formed in neutron decay, and what is the maximum kinetic energy (in MeV) that it can possess?

14. The radionuclide ${}^{210}_{84}Po$ decays by alpha emission to a daughter nuclide. The atomic mass of ${}^{210}_{84}Po$ is 209.9829 u, and that of its daughter is 205.9745 u.
   (a) Identify the daughter, and write the nuclear equation for the radioactive decay process.
   (b) Calculate the total energy released per disintegration (in MeV).
   (c) Calculate the kinetic energy of the emitted alpha particle.

15. The natural abundance of ^{30}Si is 3.1%. Upon irradiation with neutrons, this isotope is converted to ^{31}Si, which decays to the stable isotope ^{31}P. This provides a way of introducing trace amounts of phosphorus into silicon in a much more uniform fashion than is possible by ordinary mixing of silicon and phosphorus and gives semiconductor devices the capability of handling much higher levels of power. Write balanced nuclear equations for the two steps in the preparation of ^{31}P from ^{30}Si.

16. The most convenient way to prepare the element polonium is to expose bismuth (which is 100% ^{209}Bi) to neutrons. Write balanced nuclear equations for the two steps in the preparation of polonium.

17. One convenient source of neutrons is the reaction of an alpha particle from an emitter such as polonium (^{210}Po) with an atom of beryllium (^{9}Be). Write nuclear equations for the reactions that occur.

18. Three atoms of element 111 (roentgenium) were produced in 1994 by bombarding ^{209}Bi with ^{64}Ni.
   (a) Write a balanced equation for this nuclear reaction. What other species is produced?
   (b) Write a balanced equation for the alpha decay process of this nuclide of roentgenium.

19.3 Kinetics of Radioactive Decay

19. How many radioactive disintegrations occur per minute in a 0.0010-g sample of ^{209}Po that has been freshly separated from its decay products? The half-life of ^{209}Po is 103 years.

20. How many alpha particles are emitted per minute by a 0.0010-g sample of ^{238}U that has been freshly separated from its decay products? Assume that each decay emits one alpha particle. The half-life of ^{238}U is 4.47×10^9 years.

21. The nuclide ^{19}O, prepared by neutron irradiation of ^{19}F, has a half-life of 29 s.

(a) How many ^{19}O atoms are in a freshly prepared sample if its decay rate is $2.5 \times 10^4\ s^{-1}$?

(b) After 2.00 min, how many ^{19}O atoms remain?

22. The nuclide ^{35}S decays by beta emission with a half-life of 87.1 days.

(a) How many grams of ^{35}S are in a sample that has a decay rate from that nuclide of $3.70 \times 10^2\ s^{-1}$?

(b) After 365 days, how many grams of ^{35}S remain?

23. Astatine is the rarest naturally occurring element, with ^{219}At appearing as the product of a very minor side branch in the decay of ^{235}U (itself not a very abundant isotope). It is estimated that the mass of all the naturally occurring ^{219}At in the upper kilometer of the earth's surface has a steady-state value of only 44 mg. Calculate the total activity (in disintegrations per second) caused by all the naturally occurring astatine in this part of the earth. The half-life of ^{219}At is 54 s, and its atomic mass is 219.01 u.

24. Technetium has not been found in nature. It can be obtained readily as a product of uranium fission in nuclear power plants, however, and is now produced in quantities of many kilograms per year. One medical use relies on the tendency of ^{99m}Tc (an excited nuclear state of ^{99}Tc) to concentrate in abnormal heart tissue. Calculate the total activity (in disintegrations per second) caused by the decay of 1.0 μg of ^{99m}Tc, which has a half-life of 6.0 hours.

25. The specific activity of ^{14}C in the biosphere is 0.255 Bq g^{-1}. What is the age of a piece of papyrus from an Egyptian tomb if its beta counting rate is 0.153 Bq g^{-1}? The half-life of ^{14}C is 5730 years.

26. The specific activity of a wood article found in the Lascaux Caves in France is 0.0375 Bq g^{-1} due to ^{14}C. Calculate the age of the article.

27. Over geological time, an atom of ^{238}U decays to a stable ^{206}Pb atom in a series of eight alpha emissions, each of which leads to the formation of one helium atom. A geochemist analyzes a rock and finds that it contains 9.0×10^{-5} cm^3 of helium (at 0°C and atmospheric pressure) per gram and 2.0×10^{-7} g of ^{238}U per gram. Estimate the age of the mineral, assuming that all of the He is retained, and given that $t_{1/2}$ of ^{238}U is 4.47×10^9 years.

28. The isotope ^{232}Th decays to ^{208}Pb by the emission of six alpha particles, with a half-life of 1.39×10^{10} years. Analysis of 1.00 kg of ocean sediment shows it to contain 7.4 mg of ^{232}Th and 4.9×10^{-3} cm^3 of gaseous helium at 0°C and atmospheric pressure. Estimate the age of the sediment, assuming no loss or gain of thorium or helium from the sediment since its formation and assuming that the helium arose entirely from the decay of thorium.

29. The half-lives of ^{235}U and ^{238}U are 7.04×10^8 years and 4.47×10^9 years, respectively, and the present abundance ratio is $^{238}U/^{235}U = 137.7$. It is thought that their abundance ratio was 1 at some time *before* our earth and solar system were formed about 4.5×10^9 years ago. Estimate how long ago the supernova occurred that supposedly produced all the uranium isotopes in equal abundance, including the two longest lived isotopes, ^{238}U and ^{235}U.

30. Using the result of problem 29 and the accepted age of the earth, 4.5×10^9 yr, calculate the $^{238}U/^{235}U$ ratio at the time the earth was formed.

19.4 Radiation in Biology and Medicine

31. Write balanced equations for the decays of ^{11}C and ^{15}O, both of which are used in positron emission tomography to scan the uptake of glucose in the body.

32. Write balanced equations for the decays of ^{13}N and ^{18}F, two other radioisotopes that are used in positron emission tomography. What is the ultimate fate of the positrons?

33. The positrons emitted by ^{11}C have a maximum kinetic energy of 0.99 MeV, and those emitted by ^{15}O have a maximum kinetic energy of 1.72 MeV. Calculate the ratio of the number of millisieverts of radiation exposure caused by ingesting a given fixed chemical amount (equal numbers of atoms) of each of these radioisotopes.

34. Compare the relative health risks of contact with a given amount of ^{226}Ra, which has a half-life of 1622 years and emits 4.78-MeV alpha particles, with contact with the same chemical amount of ^{14}C, which has a half-life of 5730 years and emits beta particles with energies of up to 0.155 MeV.

35. The nuclide ^{131}I undergoes beta decay with a half-life of 8.041 days. Large quantities of this nuclide were released into the environment in the Chernobyl accident. A victim of radiation poisoning has absorbed 5.0×10^{-6} g (5.0 μg) of ^{131}I.

(a) Compute the activity, in becquerels, of the ^{131}I in this person, taking the atomic mass of the nuclide to equal 131 g mol^{-1}.

(b) Compute the radiation absorbed dose, in milligrays, caused by this nuclide during the first *second* after its ingestion. Assume that beta particles emitted by ^{131}I have an average kinetic energy of 0.40 MeV, that all of this energy is deposited within the victim's body, and that the victim weighs 60 kg.

(c) Is this dose likely to be lethal? Remember that the activity of the ^{131}I diminishes as it decays.

36. The nuclide ^{239}Pu undergoes alpha decay with a half-life of 2.411×10^4 years. An atomic energy worker breathes in 5.0×10^{-6} g (5.0 μg) of ^{239}Pu, which lodges permanently in a lung.

(a) Compute the activity, in becquerels, of the ^{239}Pu ingested, taking the atomic mass of the nuclide to be 239 g mol^{-1}.

(b) Determine the radiation absorbed dose, in milligrays, during the first *year* after its ingestion. Assume that alpha particles emitted by ^{239}Pu have an average kinetic energy of 5.24 MeV, that all of this energy is deposited within the worker's body, and that the worker weighs 60 kg.

(c) Is this dose likely to be lethal?

19.5 Nuclear Fission

37. Strontium-90 is one of the most hazardous products of atomic weapons testing because of its long half-life ($t_{1/2}$ = 28.1 years) and its tendency to accumulate in bone.

(a) Write nuclear equations for the decay of ^{90}Sr via the successive emission of two electrons (beta particles).

(b) The atomic mass of ^{90}Sr is 89.9073 u and that of ^{90}Zr is 89.9043 u. Calculate the energy released per ^{90}Sr atom, in MeV, in decaying to ^{90}Zr.

(c) What will be the initial activity of 1.00 g of ^{90}Sr released into the environment, in disintegrations per second?

(d) What activity will the material from part (c) show after 100 years?

38. Plutonium-239 is the fissionable isotope produced in breeder reactors; it is also produced in ordinary nuclear plants and in weapons tests. It is an extremely poisonous substance with a half-life of 24,100 years.

(a) Write an equation for the decay of ^{239}Pu via alpha emission.

(b) The atomic mass of ^{239}Pu is 239.05216 u and that of ^{235}U is 235.04393 u. Calculate the energy released per ^{239}Pu atom, in MeV, in decaying via alpha emission.

(c) What will be the initial activity, in disintegrations per second, of 1.00 g of ^{239}Pu buried in a disposal site for radioactive wastes?

(d) What activity will the material from part (c) show after 100,000 years?

39. The three naturally occurring isotopes of uranium are ^{234}U (half-life 2.5×10^5 years), ^{235}U (half-life 7.0×10^8 years), and ^{238}U (half-life 4.5×10^9 years). As time passes, will the average atomic mass of the uranium in a sample taken from nature increase, decrease, or remain constant?

40. Natural lithium consists of 7.42% ^{6}Li and 92.58% ^{7}Li. Much of the tritium ($^3_1$H) used in experiments with fusion reactions is made by the capture of neutrons by ^{6}Li atoms.

(a) Write a balanced nuclear equation for the process. What is the other particle produced?

(b) After ^{6}Li is removed from natural lithium, the remainder is sold for other uses. Is the molar mass of the leftover lithium greater or smaller than that of natural lithium?

41. Calculate the amount of energy released, in kilojoules per *gram* of uranium, in the fission reaction

$$^{235}_{92}\text{U} + {}^1_0n \longrightarrow {}^{94}_{36}\text{Kr} + {}^{130}_{56}\text{Ba} + 3\,{}^1_0n$$

Use the atomic masses in Table 19.1. The atomic mass of ^{94}Kr is 93.919 u and that of ^{139}Ba is 138.909 u.

19.6 Nuclear Fusion and Nucleosynthesis

42. Calculate the amount of energy released, in kilojoules per *gram* of deuterium (^{2}H), for the fusion reaction

$$^2_1\text{H} + {}^2_1\text{H} \longrightarrow {}^4_2\text{He}$$

Use the atomic masses in Table 19.1. Compare your answer with that from the preceding problem.

ADDITIONAL PROBLEMS

43. When an electron and a positron meet, they are replaced by two gamma rays, called the "annihilation radiation." Calculate the energies of these radiations, assuming that the kinetic energies of the incoming particles are 0.

44. The nuclide $^{231}_{92}$U converts spontaneously to $^{231}_{91}$Pa.

(a) Write two balanced nuclear equations for this conversion, one if it proceeds by electron capture and the other if it proceeds by positron emission.

(b) Using the nuclidic masses in Table 19.1, calculate the change in mass for each process. Explain why electron capture can occur spontaneously in this case but positron emission cannot.

45. The radioactive nuclide $^{64}_{29}Cu$ decays by beta emission to $^{64}_{30}Zn$ or by positron emission to $^{64}_{28}Ni$. The maximum kinetic energy of the beta particles is 0.58 MeV, and that of the positrons is 0.65 MeV. The mass of the neutral $^{64}_{29}Cu$ atom is 63.92976 u.

(a) Calculate the mass, in atomic mass units, of the neutral $^{64}_{30}Zn$ atom.

(b) Calculate the mass, in atomic mass units, of the neutral $^{64}_{28}Ni$ atom.

46. A puzzling observation that led to the discovery of isotopes was the fact that lead obtained from uranium-containing ores had an atomic mass lower by two full atomic mass units than lead obtained from thorium-containing ores. Explain this result, using the fact that decay of radioactive uranium and thorium to stable lead occurs via alpha and electron emission.

47. By 1913, the elements radium, actinium, thorium, and uranium had all been discovered, but element 91, between thorium and uranium in the periodic table, was not yet known. The approach used by Meitner and Hahn was to look for the parent that decays to form actinium. Alpha and beta emission are the most important decay pathways among the heavy radioactive elements. What elements would decay to actinium by each of these two pathways? If radium salts show no sign of actinium, what does this suggest about the parent of actinium? What is the origin of the name of element 91, discovered by Meitner and Hahn in 1918?

48. Working in Rutherford's laboratory in 1932, Cockcroft and Walton bombarded a lithium target with 700-keV protons and found that the following reaction occurred:

$$^{7}_{3}\text{Li} + ^{1}_{1}\text{H} \longrightarrow 2\ ^{4}_{2}\text{He}$$

Each of the alpha particles was found to have a kinetic energy of 8.5 MeV. This research provided the first experimental test of Einstein's $\Delta E = c^2\Delta m$ relationship. Discuss. Using the atomic masses from Table 19.1, calculate the value of c needed to account for this result.

49. **(a)** Calculate the binding energy per nucleon in $^{30}_{15}P$.

(b) The radioactive decay of the $^{30}_{15}P$ occurs through positron emission. Calculate the maximum kinetic energy carried off by the positron.

(c) The half-life for this decay is 150 s. Calculate the rate constant k and the fraction remaining after 450 s.

50. Selenium-82 undergoes *double* beta decay:

$$^{82}_{34}\text{Se} \longrightarrow ^{82}_{36}\text{Kr} + 2\ ^{0}_{-1}e^- + 2\bar{\nu}$$

This low-probability process occurs with a half-life of 3.5×10^{27} s, one of the longest half-lives ever measured. Estimate the activity in an 82.0-g (1.00 mol) sample of this isotope. How many ^{82}Se nuclei decay in a day?

51. Gallium citrate, which contains the radioactive nuclide ^{67}Ga, is used in medicine as a tumor-seeking agent. Gallium-67 decays with a half-life of 77.9 hours. How much time is required for it to decay to 5.0% of its initial activity?

52. The nuclide ^{241}Am is used in smoke detectors. As it decays (with a half-life of 458 years), the emitted alpha particles ionize the air. When combustion products enter the detector, the number of ions changes and with it the conductivity of the air, setting off an alarm. If the activity of ^{241}Am in the detector is 3×10^4 Bq, calculate the mass of ^{241}Am present.

53. The half-life of ^{14}C is $t_{1/2}$ = 5730 years, and 1.00 g of modern wood charcoal has an activity of 0.255 Bq.

(a) Calculate the number of ^{14}C atoms per gram of carbon in modern wood charcoal.

(b) Calculate the fraction of carbon atoms in the biosphere that are ^{14}C.

54. Carbon-14 is produced in the upper atmosphere by the reaction

$$^{14}_{7}\text{N} + ^{1}_{0}n \longrightarrow ^{14}_{6}\text{C} + ^{1}_{1}\text{H}$$

where the neutrons come from nuclear processes induced by cosmic rays. It is estimated that the steady-state ^{14}C activity in the biosphere is 1.1×10^{19} Bq.

(a) Estimate the total mass of carbon in the biosphere, using the data in problem 53.

(b) The earth's crust has an average carbon content of 250 parts per million by mass, and the total crustal mass is 2.9×10^{25} g. Estimate the fraction of the carbon in the earth's crust that is part of the biosphere. Speculate on the whereabouts of the rest of the carbon in the earth's crust.

55. Analysis of a rock sample shows that it contains 0.42 mg of ^{40}Ar for every 1.00 mg of ^{40}K. Assuming that all the argon resulted from decay of the potassium and that neither element has left or entered the rock since its formation, estimate the age of the rock. (*Hint:* Use data from Table 19.2.) Note that not all the ^{40}K decays to ^{40}Ar.

***56.** Cobalt-60 and iodine-131 are used in treatments for some types of cancer. Cobalt-60 decays with a half-life of 5.27 years, emitting beta particles with a maximum energy of 0.32 MeV. Iodine-131 decays with a half-life of 8.04 days, emitting beta particles with a maximum energy of 0.60 MeV.

(a) Suppose a fixed small number of moles of each of these isotopes were to be ingested and remain in the body indefinitely. What is the *ratio* of the number of millisieverts of total lifetime radiation exposure that would be caused by the two radioisotopes?

(b) Now suppose that the contact with each of these isotopes is for a fixed short period, such as 1 hour. What is the ratio of millisieverts of radiation exposure for the two in this case?

57. Boron is used in control rods in nuclear power reactors because it is a good neutron absorber. When the isotope ^{10}B captures a neutron, an alpha particle (helium nucleus) is emitted. What other atom is formed? Write a balanced equation.

*58. The average energy released in the fission of a ^{235}U nucleus is about 200 MeV. Suppose the conversion of this energy to electrical energy is 40% efficient. What mass of ^{235}U is converted to its fission products in a year's operation of a 1000-megawatt nuclear power station? Recall that 1 W is 1 J s^{-1}.

59. The energy released by a bomb is sometimes expressed in tons of TNT (trinitrotoluene). When one ton of TNT explodes, 4×10^9 J of energy is released. The fission of 1 mol of uranium releases approximately 2×10^{13} J of energy. Calculate the energy released by the fission of 1.2 kg of uranium in a small atomic bomb. Express your answer in tons of TNT.

60. The solar system abundances of the elements Li, Be, and B are four to seven orders of magnitude lower than those of the elements that immediately follow them: C, N, and O. Explain.

*61. The sun's distance from earth is approximately 1.50×10^8 km, and the earth's radius is 6371 km. The earth receives radiant energy from hydrogen burning in the sun at a rate of 0.135 J s^{-1} cm^{-2}. Using the data of Table 19.1, calculate the mass of hydrogen converted per second in the sun.

CUMULATIVE PROBLEMS

62. In 1951 wood from two sequoia trees was dated by the ^{14}C method. In one tree, clean borings located between the growth rings associated with the years A.D. 1057 and 1087 (that is, wood known to have grown 880 ± 15 year prior to the date of measurement) had a ^{14}C activity about 0.892 of that of wood growing in 1951. A sample from a second tree had an activity about 0.838 of that of new wood, and its age was established as 1377 ± 4 year by tree-ring counting.
   (a) What ages does carbon dating associate with the wood samples?
   (b) What values of $t_{1/2}$ can be deduced if the tree-ring dates given are used as the starting point?
   (c) Discuss assumptions underlying the calculations in (a) and (b), and indicate in what direction failures of these assumptions might affect the calculations.

63. A typical electrical generating plant has a capacity of 500 megawatt (MW; 1 MW = 10^6 J s^{-1}) and an overall efficiency of about 25%. (a) The combustion of 1 kg of bituminous coal releases about 3.2×10^4 kJ and leaves an ash residue of 100 g. What weight of coal must be used to operate a 500-MW generating plant for 1 year, and what weight of ash must be disposed of? (b) Enriched fuel for nuclear reactors contains about 4% ^{235}U, fission of which gives 1.9×10^{10} kJ per mole ^{235}U. What weight of ^{235}U is needed to operate a 500-MW power plant, assumed to have 25% efficiency, for 1 year, and what weight of fuel must be reprocessed to remove radioactive wastes? (c) The radiation from the sun striking the earth's surface on a sunny day corresponds to a power of 1.5 kW m^{-2}. How large must the collection surface be for a 500-MW solar-generating plant? (Assume that there are 6 hours of bright sun each day and that storage facilities continue to produce power at other times. The efficiency for solar-power generation would be about 25%.)

64. Examine the ratio of atomic mass to atomic number for the elements with *even* atomic number through calcium. This ratio is approximately the ratio of the average mass number to the atomic number.
   (a) Which two elements stand out as different in this set of ten?
   (b) What would be the "expected" atomic mass of argon, based on the correlation considered here?
   (c) Show how the anomaly in the ordering of natural atomic masses of argon and potassium can be accounted for by the formation of "extra" ^{40}Ar via decay of ^{40}K atoms.

65. Hydrazine, $N_2H_4(\ell)$, reacts with oxygen in a rocket engine to form nitrogen and water vapor:

$$N_2H_4(\ell) + O_2(g) \longrightarrow N_2(g) + 2\,H_2O(g)$$

   (a) Calculate $\Delta H°$ for this highly exothermic reaction at 25°C, using data from Appendix D.
   (b) Calculate $\Delta U°$ of this reaction at 25°C.
   (c) Calculate the total change in mass, in grams, during the reaction of 1.00 mol of hydrazine.

66. The long-lived isotope of radium, ^{226}Ra, decays by alpha particle emission to its daughter radon, ^{222}Rn, with a half-life of 1622 years. The energy of the alpha particle is 4.79 MeV. Suppose 1.00 g of ^{226}Ra, freed of all its radioactive progeny, were placed in a calorimeter that contained 10.0 g of water, initially at 25°C. Neglecting the heat capacity of the calorimeter and heat loss to the surroundings, calculate the temperature the water would reach after 1.00 hour. Take the specific heat of water to be 4.18 J K^{-1} g^{-1}.

67. The radioactive nuclide $^{232}_{90}Th$ has a half-life of 1.39×10^{10} years. It decays by a series of consecutive steps, the first two of which involve $^{228}_{88}Ra$ (half-life 6.7 years) and $^{228}_{89}Ac$ (half-life 6.13 hours).

(a) Write balanced equations for the first two steps in the decay of ^{232}Th, indicating all decay products. Calculate the total kinetic energy carried off by the decay products.

(b) After a short initial time, the rate of formation of ^{228}Ra becomes equal to its rate of decay. Express the number of ^{228}Ra nuclei in terms of the number of ^{232}Th nuclei, using the steady-state approximation from Section 18.4.

68. Zirconium is used in the fuel rods of most nuclear power plants. The following half-cell reduction potential applies to aqueous acidic solution:

$$ZrO_2(s) + 4\ H_3O^+(aq) + 4e^- \longrightarrow Zr(s) + 6\ H_2O^+(\ell)$$
$$E° = -1.43\ V$$

(a) Predict whether zirconium can reduce water to hydrogen. Write a balanced equation for the overall reaction.

(b) Calculate $\Delta E°$ and K for the reaction in part (a).

(c) Can your answer to part (b) explain the release of hydrogen in the Three Mile Island accident and the much greater release of hydrogen (which subsequently exploded) at Chernobyl?

20

CHAPTER

Molecular Spectroscopy and Photochemistry

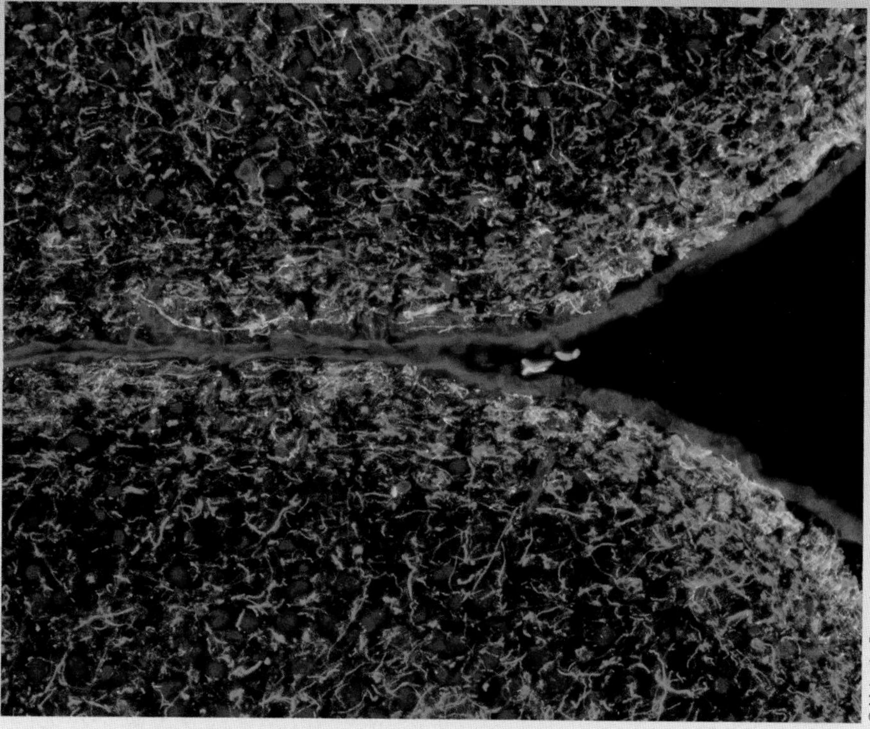

Fluorescence microscope image of the mouse cerebral cortex. Three different dyes were used to selectively image structural proteins called neurofilaments, a small protein called GFAP that is a component of intermediate filaments, and cellular nuclei.

Spectroscopy, broadly defined, refers to the field of science that is devoted to understanding the interactions between electromagnetic radiation and matter, and exploiting that understanding to develop and apply new experimental techniques to probe the structures and dynamics of material systems. The line spectra of atoms and the blackbody radiation spectrum played central roles in the development of quantum mechanics, and spectroscopic measurements have either stimulated or confirmed many of the key developments in theoretical physics of the 20th century. Molecular spectroscopy enables us to determine the three-dimensional structures of molecules, to detect and identify single molecules, and to measure the rates and mechanisms of physical processes and chemical reactions on the femtosecond (10^{-15} s) timescale. Spectroscopic imaging techniques have given us unprecedented insight into biological structure and function, and the development of medical imaging techniques over the past 40 years has revolutionized the practice of medicine. Our goal for this chapter is to introduce you to the basic concepts and methods of molecular spectroscopy and photochemistry to give you an appreciation for how they have contributed to our understanding of the nature of the chemical bond and to the development of the bonding theories discussed in Chapter 6. We also provide an introduction to the analytical applications of these methods that you will encounter in more advanced chemistry courses. Finally, we introduce you to molecular photophysics and photochemistry and illustrate the importance of these processes with examples drawn from atmospheric photochemistry and photosynthesis.

20.1 Introduction to Molecular Spectroscopy

Molecular spectroscopy experiments measure the intensity of radiation absorbed, emitted, or scattered by molecules as a function of frequency. Examples of atomic absorption and emission spectra are shown and discussed in Section 4.2. Peaks or bands (broad peaks) appear in molecular absorption and emission spectra at frequencies that satisfy the resonance condition, $|\Delta E| = h\nu$, where ΔE is the energy difference between two levels and ν is the frequency of the radiation. $\Delta E = E_f - E_i$ is positive if the molecule absorbs the radiation (a photon) and negative if the molecule emits radiation, so we have noted the absolute value of ΔE by $|\Delta E|$. Transitions between molecular energy levels can also be caused by **Raman scattering**, a form of inelastic light scattering. Photons can undergo *inelastic* collisions with molecules; energy conservation requires that $\Delta E = h\nu_i - h\nu_s$, where ν_i and ν_s are the frequencies of the incident and scattered radiation, respectively.

20.1.1 Molecular Energy Levels and Diatomic Potential Energy Curves

The Born-Oppenheimer approximation allows us to calculate the electronic wave functions for molecules as a function of the positions of the nuclei, and to generate a potential energy function for each electronic state, which governs the nuclear motion for that electronic state (see Section 6.1). Figure 20.1a shows the potential energy function for the lowest energy (ground) electronic state of HCl as well as that of one of its excited electronic states. As you learned in Chapter 3, the minimum occurs at the equilibrium internuclear separation for each state, which for the ground electronic state is R_e, the equilibrium bond length. You can see that the excited state of HCl plotted in Figure 20.1a has a larger equilibrium bond length than that in the ground electronic state, a smaller **force constant** k (fitting the potential energy near the minimum in each electronic state to $V(R) = \frac{1}{2}k(R - R_e)^2$), and a smaller dissociation energy. A few vibrational levels of each electronic state are shown; the black arrow shown in Figure 20.1a represents a transition between the ground vibrational level in the ground electronic state to an excited vibrational level in the excited electronic state. Such transitions, which result in a change in the electronic configuration of a molecule, occur as a result of the absorption of ultraviolet and, for molecules other than HCl, visible radiation. They are discussed in more detail in Section 20.5. Figure 20.1b shows the ro-vibrational energy levels of a diatomic molecule, labeled by two quantum numbers v and J. Figure 20.1b allows you to see that the typical energy changes involved in promoting a molecule from one rotational state to another without a change in vibrational quantum number (this is called a "pure rotational transition") are much smaller energy changes than those for transitions that simultaneously change the rotational and vibrational state, which is represented by the longer arrow. Rotational and vibrational transitions are discussed in detail in Section 20.3.

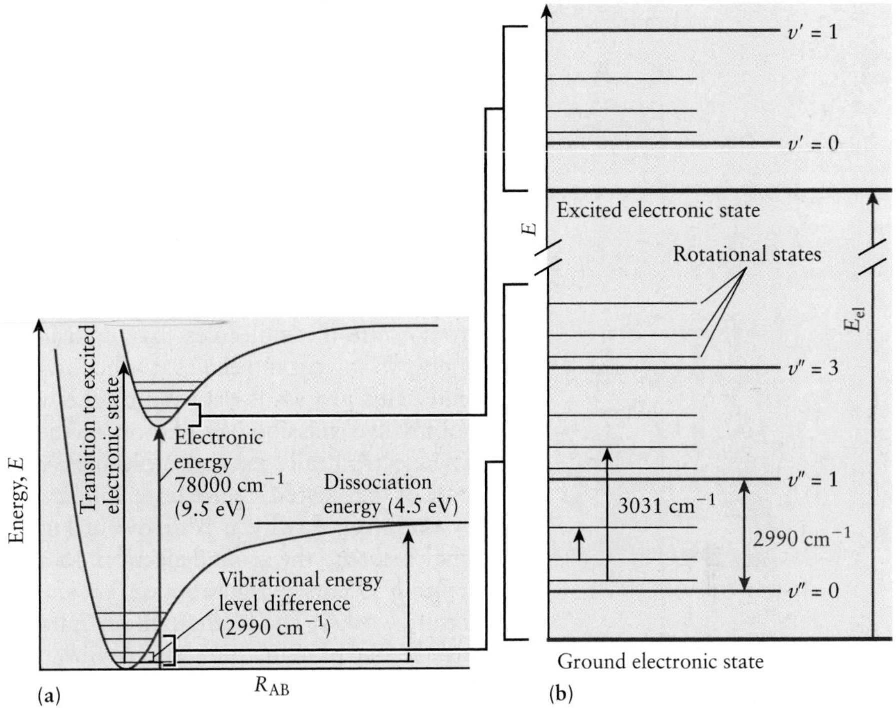

FIGURE 20.1 (a) Absorption from the ground electronic state of HCl to an excited electronic state. The vibrational energy levels in each are shown by a blue line. (a, Adapted from Peter Atkins, Julio De Paula, and Ronald Friedman, *Quanta, Matter and Change*, W. H. Freeman and Company, New York, p. 300.) (b) Vibrational and rotational energy levels in the ground state (double primed) and the excited electronic state (single primed). The rotational level energy differences are shown exaggerated for clarity. The small unlabeled black arrow shows a pure rotational transition from the $J = 2$ quantum state to the $J = 3$ quantum state; the actual energy difference (in units of hc) is 62 cm^{-1}. The longer black arrow labeled 3031 cm^{-1} shows an infrared absorption from $(v, J) = (0,1)$ to $(1,2)$ in the ground electronic state; this transition changes both the vibrational and the rotational quantum number. E_{el} is measured from the minimum of the ground state curve to the minimum of the excited state (red arrow).

We showed in Section 12.5 that molecular motions may be separated classically into translations, rotations, and vibrations (see Fig. 12.13 in particular). The couplings between these motions are sufficiently small that we may approximate the total energy as the sum of these independent contributions.

$$E_{tot} = E_{trans} + E_{rot} + E_{vib} + E_{el} \qquad [20.1]$$

Absorption or emission of light in the microwave region of the optical spectrum can cause the rotational state of a molecule to change; for diatomics the rotational state is labeled by two quantum numbers J and M_J. If a molecule absorbs light in the infrared region of the spectrum the photon provides enough energy to change the vibrational quantum state at the same time. Figure 20.1b shows the energies of different vibrational states of HCl in blue lines, labeled by the quantum number v. (In Section 12.8 we introduced the vibrational energy levels of a diatomic molecule but used n as the quantum number. The label v is the more common one used in molecular spectroscopy, so will be used consistently in this chapter.) A photon does not carry enough linear momentum to substantially change the velocity of the center of mass of a molecule, so E_{trans} may be considered a constant when a photon is absorbed. You can see in Figure 20.1b that the change in electronic energy E_{el} when you promote the molecule to an excited electronic state can be quite large; this is because in an electronic transition typically an electron in the molecule is promoted from a lower energy molecular orbital to a higher energy molecular orbital.

20.1.2 The Fate of Excited Electronic States

When a photon promotes a molecule to an excited electronic state, several processes can contribute to the spectroscopic experiment. The fate of an electronically excited molecule depends on whether the excited electronic state can support bound vibrational quantum states or not. The excited state of HCl depicted in Figure 20.1 can support bound vibrational quantum states, but a lower energy excited electronic state of HCl is like the repulsive $1\sigma_u^*$ state of

H_2^+ depicted in Figure 6.6; it will quickly dissociate (typically in less than a picosecond). If the excited electronic state can support bound vibrational quantum states, like the excited *singlet* state shown in Figure 20.2, then the electronically excited molecule can spontaneously emit a photon, returning to one of the ro-vibrational levels in the ground electronic state. (A molecule in a singlet state has half of its electrons with $m_s = +1/2$ and the other half with $m_s = -1/2$.) Detecting these emitted photons in a spectroscopic experiment can measure the energies of the quantized states of the molecule both in the excited electronic state and in the ground electronic state, as we will see in Section 20.2. Figure 20.2 assumes that the molecules are in solution, so while in the excited electronic state the molecules may collisionally relax (by transferring vibrational energy to solvent molecules) to the lowest vibrational state of the excited singlet state. This process is depicted by the wavy black arrow in Figure 20.2; it does not involve emission of a photon, so the process is called a *nonradiative* process. The electronically excited molecules, now in the lowest vibrational quantum state of the excited singlet state, might then spontaneously emit a photon shown by the green downwards arrow in Figure 20.2 to transition from the excited singlet state to the ground electronic state. The process is called **fluorescence** and typically occurs within about 10^{-9} s. (Now you know how a fluorescent light fixture works!) The final quantum state of the molecule after emitting a photon may be one of many different ro-vibrational quantum states in the ground electronic state. Figure 20.2 also depicts a process called **phosphorescence**; it occurs on much longer time scales than fluorescence, so the yield of photons from this process is typically small if it competes with fluorescence.

To understand electronic transitions in molecules, consider exciting an electron from the highest energy occupied molecule orbital of ethene, a π orbital, to the lowest unoccupied orbital, a π^* orbital. The constituent atoms are all from the first and second row of the periodic table; for such molecules the electron is unlikely to change its spin when it is promoted to the π^* orbital. Thus, the excited state reached when the molecule absorbs a photon, called a singlet state, has the electronic configuration, $(\pi)^1(\pi^*)^1$ (we have omitted the occupied σ_{C-H} and σ_{C-C} bonding molecular orbitals here) with one electron spin up ($m_s = 1/2$)

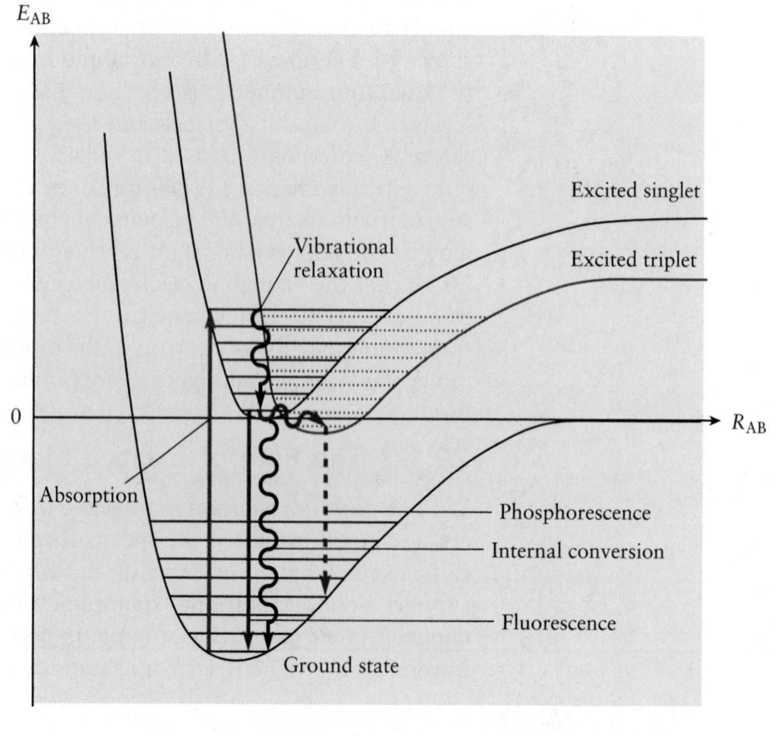

FIGURE 20.2 Absorption, emission, and nonradiative relaxation processes.

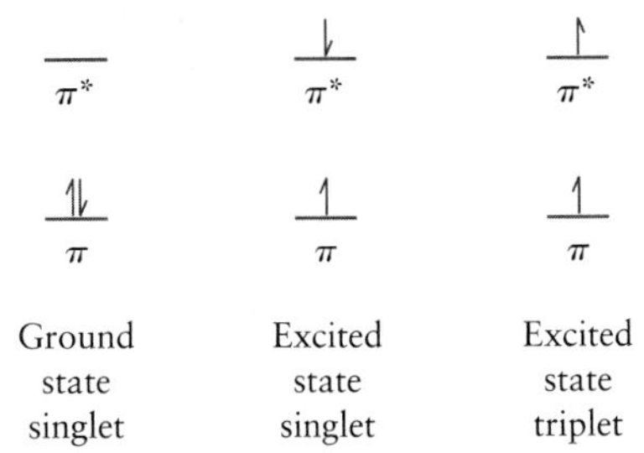

FIGURE 20.3 The ground electronic state of ethene has two electrons in the highest occupied molecular orbital, one spin up and one spin down, so the projection of the total spin on the z-axis in zero, $M_S = \sum_{i=1}^{N} m_{s,i} = 0$. The quantum number S for this multielectron molecule is 0, so there is only one (S, M_S) spin state that characterizes the ground electronic state of ethene. The ground electronic state thus is called a singlet state. If one promotes one of the electrons to a π^* orbital without changing its spin, then you reach the lowest excited singlet state of ethene. If you allow an electron to change its spin, then you can reach the lowest excited triplet state of ethene. This state is characterized by three possible values of M_S: $M_S = -1$ (both electrons spin down), $M_S = 0$, one electron spin up and one electron spin down, and $M_S = +1$ (both electrons spin up, as shown in the figure). Thus, this excited state is called a triplet state, as there are three distinct spin states represented by the three values of M_S.

and the other electron spin-down ($m_s = -1/2$), just as they had when they occupied the same orbital, as depicted in Figure 20.3. From the excited singlet state, a process called intersystem crossing can occur; intersystem crossing changes the spin state of the molecule to a triplet state. Figure 20.3 shows one of three possible spin configurations for a triplet state; one electron has flipped its spin so now the electron in the π orbital has a parallel spin to the electron in the π^* orbital. Intersystem crossing is a nonradiative process; it occurs without the absorption or emission of a photon (so the molecule initially has the same total energy). The triplet electronic state of the ethene molecule shown by the blue curve in Figure 20.2 has a slightly lower electronic energy than the corresponding singlet state as shown in Figure 20.2. Collisions of the solvent molecules with the triplet excited state molecule can transfer energy out of the vibrational motion, so again Figure 20.2 shows a slightly downward tilting red arrow to depict this. Finally, the excited triplet state molecules may emit a photon to transition back to one of the vibrational levels of the ground electronic state, as shown by the red arrow pointing straight downward in Figure 20.2. The light emitted during this transition, called phosphorescence, is not emitted as promptly as fluorescence (roughly, because it requires the unlikely event of a spin flip), so phosphorescence can appear microseconds or even a full second after the molecule is initially excited to an excited electronic state unless a competing process depletes the excited state molecules. This is so long that collisions with the solvent can relax the excited triplet state molecule to the high vibrational energy levels in the ground electronic state via a nonradiative process that competes with phosphorescence, so the yield of photons from phosphorescence can be orders of magnitude smaller than the yield of photons from fluorescence.

20.1.3 Overview of Spectroscopic Methods

In the sections that follow we discuss how we measure the following properties of molecules using the various spectroscopic techniques described:

1. Bond lengths and bond angles, from microwave, infrared, and Raman spectroscopy
2. Bond force constants and effective reduced masses from vibrational spectroscopy
3. Identification of functional groups and their relative locations in molecules, from vibrational spectroscopy, electronic spectroscopy, and nuclear magnetic resonance spectroscopy

Spectroscopic techniques have historically been classified in terms of the relevant frequencies of the electromagnetic radiation involved. Table 20.1 shows a number of important spectroscopic techniques, the relevant region of the electromagnetic spectrum, the nature of the energy levels involved, and the kinds of information obtained about the properties of molecules. We list frequencies, wavelengths, and energies in several sets of units that are used in practice, for convenience. Radiofrequency and microwave spectra are generally reported as a function of frequency (Hz), infrared and visible spectra in terms of frequency or wavelength, and ultraviolet and X-ray spectra in terms of wavelength, or energy in eV, for example.

Energies and frequencies are often reported in units of *wave numbers* (typically cm^{-1}) where $\tilde{\nu} = (1/\lambda)$ and λ is measured in cm. This is a convenient unit that is proportional to energy because $E = h\nu = hc/\lambda = (hc)(1/\lambda)$. Microwave spectra, for example, are presented either as a function of frequency (GHz) or energy (cm^{-1}), infrared and Raman spectra most often use wave numbers, and ultraviolet and visible spectra are presented as functions of wavelength (nm) or energy (eV).

TABLE 20.1

Spectroscopic Methods

Spectral Region	Frequency	Wavelength	Transition Energies	Excitations	Information Obtained
Radio waves	10^7–10^9 Hz	30–0.3 m	0.0033–0.033 cm^{-1}	Nuclear spin	Bonding environment, molecular structure
Microwave, far infrared	10^9–10^{12} Hz	0.3 m–300 μm	0.033–33 cm^{-1}	Rotational	Bond lengths and bond angles
Near infrared	10^{12}–10^{14} Hz	300 μm–3 μm	33–3300 cm^{-1}	Vibrational	Bond force constants and effective reduced mass
Visible, ultraviolet	10^{14}–10^{17} Hz	3000 nm–3 nm	3300–330,000 cm^{-1}	Valence electrons	Electronic energy levels and molecular environment
X-ray	10^{17}–10^{19} Hz	3 nm–0.3 Å	413–41,300 eV	Core electrons	Ionization energies and atomic environment

20.2 Experimental Methods in Molecular Spectroscopy

Let's consider two simple experiments in optical molecular spectroscopy. The first measures the absorption of light, tuned to different wavelengths (and, correspondingly, energies), by a sample of molecules; the resulting data is called an optical absorption spectrum. The second experiment uses photons of a specific energy to promote molecules to an excited electronic state and then detects either all the photons emitted from the excited molecules or resolves the emitted photons by their energies. An example of the second is laser-induced fluorescence.

20.2.1 Traditional Optical Absorption Spectroscopy

Figure 20.4 shows how one might measure an optical absorption spectrum. Figure 20.4a shows an idealized experiment in which the incoming light beam is attenuated when its wavelength is such that the molecules of interest may absorb the light and undergo a transition from a lower energy quantum state to a higher energy quantum state. Thus, the ideal sample cell contains only the molecules of interest and none of the incoming light beam is scattered from the sample cell windows. Typically the incoming light is first filtered through a monochromator that contains a diffraction grating, so all the photons impinging on the sample have close to the same wavelength. (A diffraction grating consists of a series of closely spaced metal lines fabricated on a glass substrate that diffract light of different wavelengths into different directions, so you can change the wavelength of the light that passes out of the exit slit on the monochromator by tilting its diffraction grating; the principle is similar to that for electron diffraction depicted in Figure 4.32). This beam of light with intensity I_0 is directed at the sample of molecules. (The intensity might be measured in photons per cross sectional area of the beam per second passing the entrance to the sample cell, photons $cm^{-2}\ s^{-1}$, or in J $cm^{-2}\ s^{-1}$ if you multiply by the photon energy.) If the wavelength of the light is such that the molecules of interest in the sample cell absorb some of the incoming photons, the light that exits the sample cell, the transmitted light, has a lower intensity, I_t. The higher the concentration N/V of light-absorbing molecules, the weaker the intensity of the transmitted beam. The beam of light travels a distance ℓ through the sample, so if ℓ is longer the intensity I_t of the transmitted beam is decreased. Finally, each molecule has a wavelength dependent absorption cross section σ that the

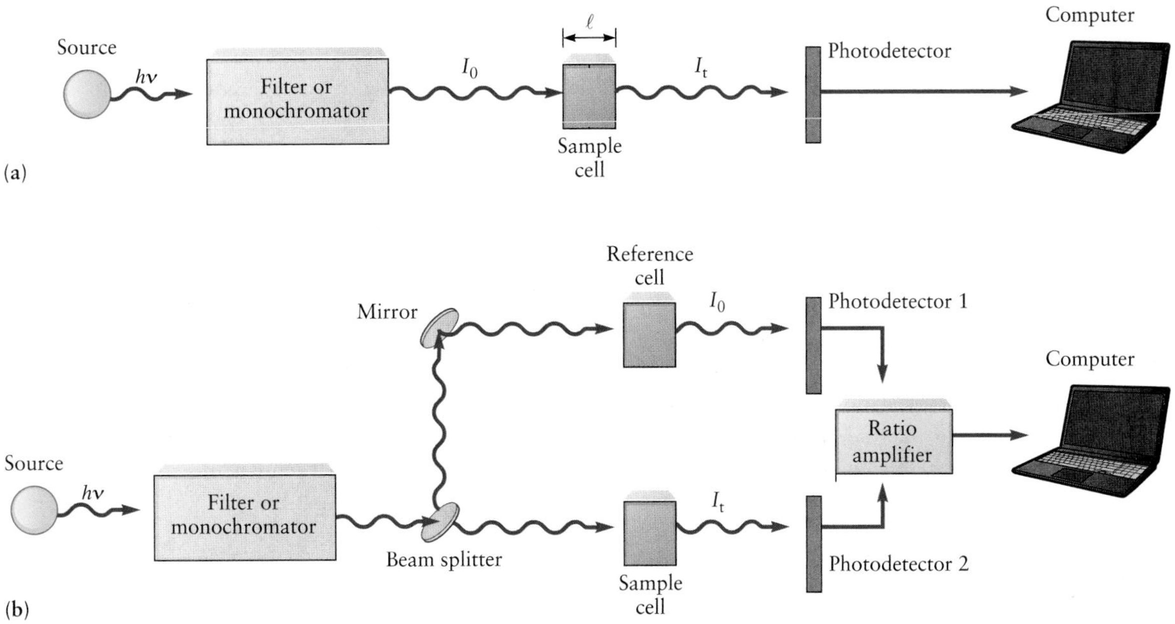

FIGURE 20.4 (a) Ideal absorption experiment; only the molecules of interest in the sample cell absorb the incoming light, and no light is scattered. (b) Schematic of double-beam spectrophotometer. The beam splitter directs the second beam through a reference cell that is identical to the sample cell but contains only the solvent, not the solute molecules of interest; the sample cell contains both solvent and the solute molecules of interest.

experiment is designed to determine. The larger the absorption cross-section, the more the light will be attenuated. All these factors are accounted for in the *Beer-Lambert law.*

$$\frac{I_t}{I_0} = e^{-\frac{N}{V}\sigma\ell} \qquad [20.2]$$

(If the sample is a mixture or if the concentration of the sample is unknown, this law is simply written as $I_t = I_0 \exp[-\alpha\ell]$ and α is called the absorption coefficient of the sample.)

The Beer–Lambert law expresses the fact that a constant fraction of the light is absorbed per unit path length and that fraction is independent of the intensity of the incident radiation. We can also think of absorption, or Raman scattering (introduced in Section 20.1), in terms of collisions between photons and molecules, using the same ideas developed for the collision theory of gases in Section 9.5, where σ is the cross-section for the process of interest. The intensity of the radiation that is absorbed or scattered is then given by $I_{a,s} = I_0 - I_t = I_0\,[1 - \exp[-(N/V)\sigma\ell]]$. Expressing absorption, emission, and scattering intensities in terms of cross sections allows us to compare the results of these different kinds of experiments from a common point of view.

EXAMPLE 20.1

Benzene has several absorption bands in the ultraviolet region of the electromagnetic spectrum, the weakest of which appears at 260 nm with an absorption cross section of 2.15×10^{-18} cm^2 molecule^{-1}. Calculate the fraction of the incident radiation absorbed by a sample of benzene confined in a cell with a 1 cm path length at 0.001 atm and 300 K.

(Continued)

Solution

Use the Beer–Lambert law in the form $I_t/I_0 = \exp[-(N/V)\sigma\ell]$ and calculate N/V using the ideal gas law. Substituting the number density, cross section, and path length into the Beer–Lambert law gives

$$I_t/I_0 = \exp[-(2.45 \times 10^{16} \text{ molecules cm}^{-3})(2.15 \times 10^{-18} \text{ cm}^{-2} \text{ molecules}^{-1})(1 \text{ cm})]$$
$$= 0.95$$

Only about 5% of the incident radiation is absorbed under these conditions.

Related Problems: 1, 2, 3, 4, 5, 6

Chemists often need to measure the absorption spectrum of a solute in solution. In this case one often uses a different form of the Beer–Lambert law, called **Beer's law**, in which concentrations are expressed in molarity, not as number densities, and the expression uses common (base 10) logarithms instead of natural (base *e*) logarithms. The integrated form of Beer's law is

$$I_t = I_0 10^{-\varepsilon c \ell}$$

by analogy to the Beer–Lambert law, but it is usually expressed in its logarithmic form as

$$A = \log\left(\frac{I_0}{I_t}\right) = \varepsilon c \ell \qquad [20.3]$$

where A is the **absorbance** (sometimes called the optical density), c is the concentration in mol L^{-1}, ε is the **molar absorption coefficient** (molar extinction coefficient in some older textbooks) expressed in units M^{-1} cm^{-1}, and ℓ is the path length in cm, as before. Chemists chose these units for convenience because they generally measure concentrations in mol L^{-1} and because 1 cm is the path length through the sample cell in a typical commercial spectrophotometer. The ratio of intensities of the incident light, I_0, and the transmitted light, I_t, is measured experimentally and the results are reported either as the absorbance A or transmittance T of the sample, which are defined and related as follows

$$A = \log\left(\frac{I_0}{I_t}\right) = -\log\left(\frac{I_t}{I_0}\right) = -\log T$$

The transmittance is often converted to percent transmission by multiplying T by 100, as shown in Figure 20.10 in Section 20.3, for example. Ultraviolet and visible absorption spectra generally graph absorbance as a function of wavelength or wave number, whereas infrared and microwave spectra generally graph percent transmission as a function of wave number.

A molar absorption coefficient ε can be related to the corresponding absorption cross section σ by comparing the integrated form of Beer's law to the Beer–Lambert law.

$$I_t/I_0 = 10^{-\varepsilon c \ell} = e^{-(N/V)\sigma\ell}$$

Taking natural logarithms and rearranging gives

$$(\ln 10)(-\varepsilon c \ell) = -(N/V)\sigma\ell$$

which can be solved, with the substitution that $N = nN_A$ (where N_A is Avogadro's number) to get

$$\varepsilon = (nN_A/V)\sigma/2.303c = N_A\sigma/(2.303)(10^3) = 2.6 \times 10^{20}\ \sigma$$

where the factor of 10^3 arises because the cross section σ and the path length ℓ have units of cm^2 (per molecule) and cm, respectively, giving nN_A/V in units of molecules cm^{-3} while the molarity c is expressed in mol L^{-1}.

Thus far we have discussed the idealized experiment depicted in Figure 20.4a. In real absorption measurements, however, the solvent may also absorb some of the incident light. A dual-beam spectrophotometer, shown in Figure 20.4b, is commonly used to measure an I_t/I_0 ratio that reflects only the absorption of light by the solute molecules of interest.[1] The beam splitter directs half of the incident light through the sample cell and the other half of the incident light through a reference cell filled with solvent alone.

The dual beam spectrometer shown in Figure 20.4b is used primarily for absorption experiments in the ultraviolet and visible regions of the spectrum. Modern instruments often use spectrographs, analogous to those illustrated in Figure 4.15 with a charge-coupled device (CCD) detector, like that in your cell phone or digital camera, replacing the photographic film. This kind of detection scheme is much faster than scanning because the intensity at all wavelengths of interest is measured simultaneously. *Fourier transform* (FT) methods have largely replaced incident frequency scanning methods in infrared (FTIR) and nuclear magnetic resonance (FT-NMR) spectroscopy, but a detailed description of those methods is not presented in this text.

20.2.2 Laser-Induced Fluorescence

A direct absorption experiment such as that described above becomes very difficult if the molecule you are interested in has a very small concentration (perhaps it is a transient radical intermediate in a kinetics experiment), or if it has a small absorption cross section. Spectroscopy was revolutionized by the discovery of *light amplification by the process of stimulated emission of radiation*; the devices that use this process to generate high intensity beams of light are called *lasers*. A modern dye laser includes a diffraction grating in the optical cavity, so the laser light consists of photons with a narrow energy distribution (typically $< 1\ \text{cm}^{-1}$), or bandwidth, centered at a chosen wavelength. This center wavelength may be tuned over a wide range by tilting the diffraction grating. The narrow-bandwidth laser light is used to promote the molecules in the sample to an excited electronic state, but instead of measuring the direct absorption of the light as the laser beam is passed through a sample, a *laser-induced fluorescence* (LIF) experiment detects the small yield of photons spontaneously emitted from the excited-state molecules. A typical detector might be a photomultiplier tube, which outputs a pulse of electrons when the emitted photons strike the photocathode inside. A schematic of the LIF experiment is shown in Figure 20.5. (A similar schematic is applicable to a Raman scattering experiment.) Samples are illuminated by laser light, photons with a narrow energy distribution (or other source of nearly monochromatic radiation). When the energy of the laser photons is tuned to an energy $h\nu_{\text{laser}} = E_f - E_i$, the energy needed to promote a molecule in the sample from its initial quantum state with energy E_i to a final quantum state with energy E_f, the molecules can absorb the laser light. The emitted radiation, produced as the excited molecules undergo a spontaneous

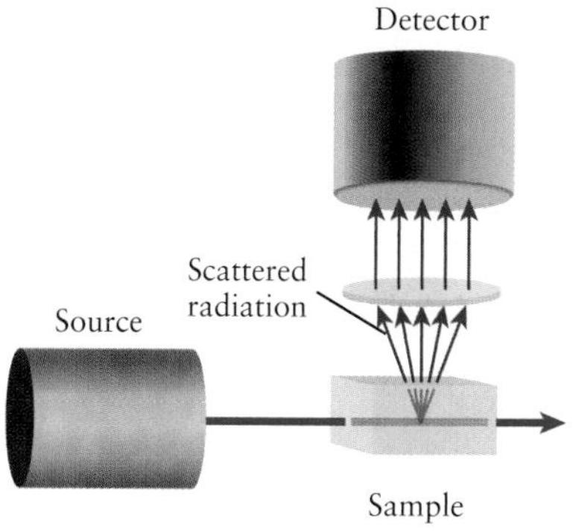

FIGURE 20.5 Schematic of an emission or scattering experiment. (Adapted from Peter Atkins, Julio de Paula and Ronald Friedman, *Quanta, Matter and Change*, W. H. Freeman and Company, New York, 2009, p. 339.)

[1] A simple derivation reveals why Figure 20.4b shows I_0 not as the intensity of the incident light, but as the intensity of the light transmitted through the reference cell. To analyze this experiment, use N_{sol}/V for the number of solvent molecules per unit volume and σ_{sol} for the wavelength dependent absorption cross section of the solvent molecules. If in Figure 20.4b you designate the intensity of the light incident on both cells by I_{in}, then by the Beer-Lambert law the light *transmitted* through the reference cell would have intensity $I_0 = I_{\text{in}} \exp[-(\sigma_{\text{sol}}(N_{\text{sol}}/V)\ell]$ while the light transmitted through the sample cell would have intensity $I_t = I_{\text{in}} \exp[-(\sigma_{\text{sol}}(N_{\text{sol}}/V)+\sigma(N/V))\ell]$. Expressing the second equation as a product of two exponentials $I_t = I_{\text{in}} \exp[-(\sigma_{\text{sol}}(N_{\text{sol}}/V)\ell] \exp[-\sigma(N/V))\ell]$ and substituting in the first expression for I_0 gives the familiar expression $I_t = I_0 \exp[-\sigma(N/V))\ell]$. Thus in the dual-beam spectrophotometer the ideal experiment is realized by taking the ratio, $I_t/I_0 = \exp[-\sigma(N/V))\ell]$ of intensities of the two *transmitted* beams; this ratio accurately reproduces the ratio I_t/I_0 of the transmitted beam to the incident beam in the idealized absorption experiment in Figure 20.4a.

transition back to a lower energy quantum state, is collected by a lens along a path at right angles to the incident beam. The intensity of the emitted photons is measured as a function of the wavelength of the incident radiation. Thus, when you detect emitted radiation you know you have tuned the wavelength of the incident laser beam to an energy $h\nu_{laser} = hc(1/\lambda_{laser})$ such that $h\nu_{laser} + E_i = E_f$, where E_f is the energy of the quantum state reached upon absorption of the photon. The key advantage is that the experiment is nearly background free; no emitted light is detected unless the molecule absorbs the laser light, and you are measuring the appearance of emitted photons rather than the tiny reduction of the intensity of the laser beam. In a laser-induced fluorescence experiment one can also measure the wavelengths of the emitted photons. As long as the excited state molecule emits a photon before undergoing collisions with other molecules in the sample, a simple conservation of energy expression allows you to derive the energies of the ro-vibrational levels in the ground electronic state from such a measurement.

Figure 20.6 shows the electronic absorption spectra of anthracene and the resulting fluorescence and phosphorescence spectra. For each absorption or emission transition shown, the vibrational quantum number in a C—C stretching

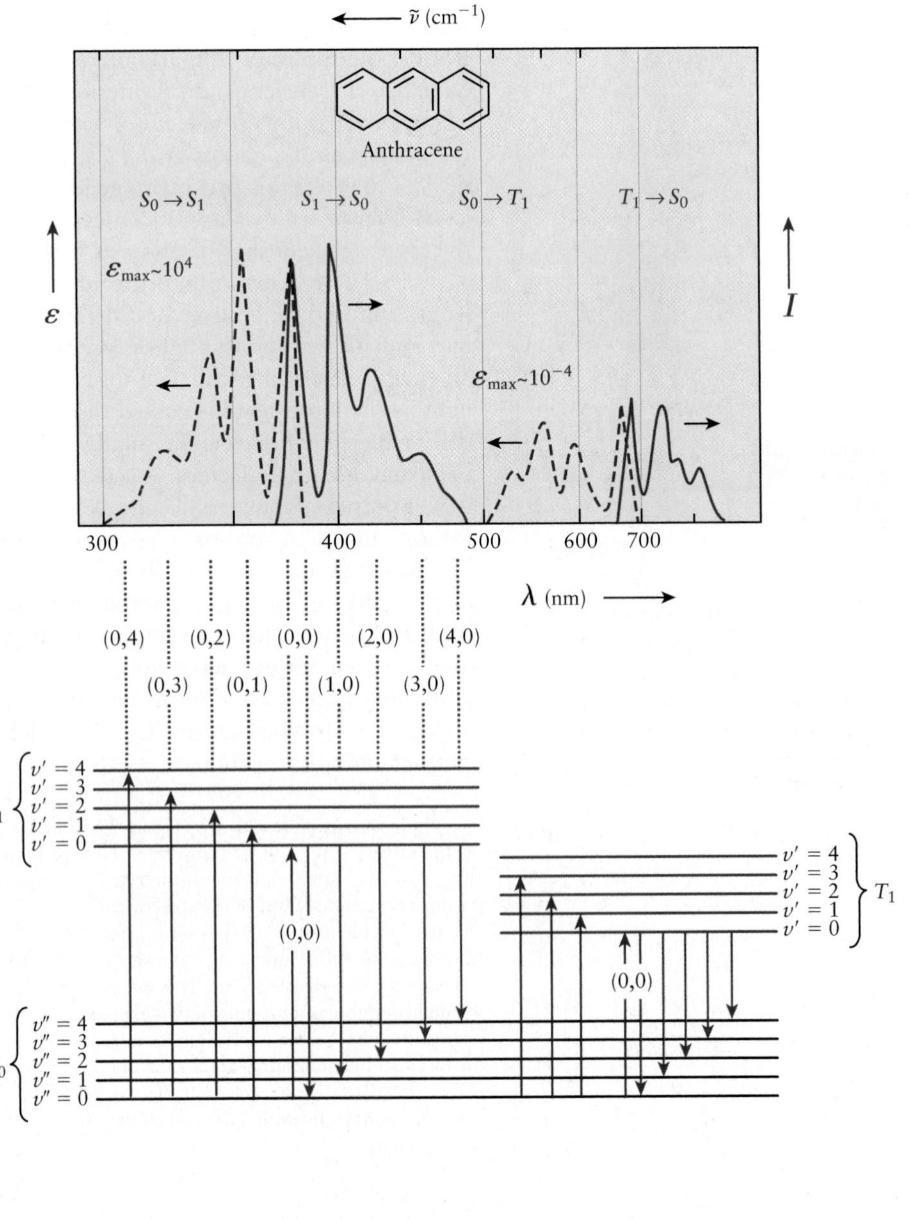

FIGURE 20.6 Absorption and fluorescence spectra of anthracene showing approximate mirror image symmetry between the shapes of the envelopes in the two spectra. (Adapted from Nicholas J. Turro, V. Ramamurthy, and J. C. Scaiano, *Principles of Modem Photochemistry*, University Science Books, Sausalito, 2009, p. 206.)

mode in the higher energy singlet (S_1) or triplet (T_1) electronic states is labeled by v' while the vibrational quantum number of the same mode in the lower energy electronic state S_0 is labeled with the vibrational quantum number v''. In a LIF experiment on anthracene, if you tuned your laser to produce photons with an energy that could promote $v'' = 0$ molecules in the ground electronic state to the lowest vibrational state in the S_1 excited state (the transition labeled (0,0) in the figure, the molecules can fluoresce, producing emitted photons with slightly smaller energies than the laser photons used to excite the anthracene. This is because the final state of the anthracene molecules in S_0, after emitting a photon, may have $v'' = 0, 1, 2, 3$ or 4 in a C—C stretching mode. The energies of these emitted photons are shown by the downward arrows in Figure 20.6 labeled (v'', v')= (0,0), (1,0), (2,0), (3,0) and (4,0). You can see that the (4,0) transition leaves the anthracene molecule in a final state in S_0 with four quanta in the C—C stretching vibrational mode.

EXAMPLE 20.2

While the spectra in Figure 20.6 were taken in solution, anthracene molecules can be cooled and isolated from collisions with other gases by using a molecular beam (see Section 18.6). The photon energy needed for the (0,0) transition is then measured to be 27,695 cm^{-1}. What is the wavelength (measured in vacuum) and the photon energy (in joules per photon) of the laser light that would be needed to excite anthracene from the $v'' = 0$ level of the ground electronic state, S_0, to the $v' = 0$ level of the S_1 excited electronic state? If you then separate the visible photons emitted from molecules in this S_1 $v' = 0$ quantum state using a diffraction grating, what is the energy of the emitted photon when the final state of anthracene, after emitting the photon, in S_0 is $v'' = 2$? The vibrational frequency of the C-C stretching mode in the ground electronic state is 1408 cm^{-1}. (The final vibrational eigenstates in the emission spectrum of anthracene are more complicated than a simple progression in this stretching vibration; we follow an early simpler interpretation of the spectrum.)

Solution

The laser photon energy needed to excite the (0,0) transition is 27,695 cm^{-1} and the wave number $\tilde{\nu} = (1/\lambda)$, so the wavelength λ of the photon is 1/(27,695 cm^{-1}) = 3.61076×10^{-5} cm = 361.076 nm. Multiply 27,695 cm^{-1} by hc (with c in cm s^{-1}) to get the laser photon energy in joules. The result is 5.5014×10^{-19} J.

The emitted photon has a smaller energy. If the molecule emitted a 27,695 cm^{-1} photon the final vibrational state in S_0 would be $v'' = 0$, so if the final state is instead $v'' = 2$ then the emitted photon energy is smaller, 27695 cm^{-1} − 2(1408 cm^{-1}) = 24879 cm^{-1}.

Related Problems: 7, 8

20.2.3 Thermal Occupation of Molecular Energy Levels

For a sample of molecules to absorb or emit photons of a particular energy and thus undergo a transition from an initial quantum state to a final quantum state, you must of course have molecules in the initial quantum state. If your sample is at thermal equilibrium, the fraction of the total number of molecules at a particular energy level i with energy E_i is given by the **Boltzmann energy distribution**

$$N_i/N = C\, g_i \exp(-E_i/k_B T)$$

where N_i is the number of molecules with energy E_i, N is the total number of molecules in the system, g_i is the number of distinct quantum states with energy E_i (g_i is called the *degeneracy* of the energy level), k_B is Boltzmann's constant (1.38×10^{-23} J K^{-1}), and C is a normalization factor. If energies per mole are used, k_B is replaced by the gas constant R. The exponential falloff means that at thermal equilibrium few molecules have energies that are large compared with $k_B T$.

If you wish to calculate the *ratio* of the number of molecules with energy E_i to the number of molecules with energy E_j, the proportionality constant in the prior equation Equation 20.4 cancels from the numerator and denominator to give

$$\frac{N_i}{N_j} = \frac{g_i e^{-E_i/k_B T}}{g_j e^{-E_j/k_B T}} = \frac{g_i}{g_j} e^{-\frac{E_i - E_j}{k_B T}} \quad [20.4]$$

In Section 12.8 we introduced the Boltzmann distribution and its application to vibrational energy levels of diatomic molecules. When the vibrational energy level spacing is large, like for HCl shown in Figure 20.1, most of the HCl molecules in a sample at 300 K are in the lowest vibrational energy level, $v'' = 0$, in the ground electronic state. In contrast, in a thermal population at 300 K of Br_2 molecules there is substantial fraction of molecules in $v'' = 1$ because the vibrational energy level spacing in Br_2 is 323 cm^{-1}, much smaller than that for HCl. Example 12.12 showed that for Br_2 $N_{v''=1}/N_{v''=0} = 0.212$, so although there are fewer Br_2 molecules in $v'' = 1$ than in $v'' = 0$, the population in $v'' = 1$ is only a factor of 5 smaller than that in $v'' = 0$. It is important to note that the ratios of degeneracies can make the population in a higher energy level larger than the population in a lower energy level. A single quantum state at a high energy will have fewer molecules populating it than a single quantum state at a lower energy, but if there are, for example, three quantum states with the same energy E_i (so $g_i = 3$) and only one quantum state with the energy E_j (so $g_j = 1$) then the population with energy E_i can be larger even when E_i is the higher of the two energies. This is because you are summing the population in all three quantum states at energy E_i. This is illustrated in Example 20.3 below.

EXAMPLE 20.3

Calculate the ratio of the populations in the first two rotational energy levels of carbon monoxide, the lowest $J = 0$ energy level and the higher $J = 1$ energy level, at 300 K if the energy difference between the levels is 3.8 cm^{-1} and the degeneracies g_J of the two levels are $g_0 = 1$ and $g_1 = 3$, respectively. (You will see in Section 20.3 that there are $2J + 1$ rotational quantum states at each energy level E_J.)

Solution

Substituting the values provided into the Equation 20.4 and expressing k_B in wave numbers ($k_B = 0.695$ cm^{-1} K^{-1}) gives

$$\frac{N_{J=1}}{N_{J=0}} = \frac{g_{J=1}}{g_{J=0}} e^{-\frac{E_{J=1} - E_{J=0}}{k_B T}}$$

$$= \frac{3}{1} e^{-\frac{3.8\ \text{cm}^{-1} - 0\ \text{cm}^{-1}}{(0.695\text{cm}^{-1}\ \text{K}^{-1})(300\ \text{K})}}$$

$$= 2.95$$

The rotational level spacings are so much smaller than $k_B T$, and the degeneracy of the upper level a factor of 3 greater than that of the lower level, that there are more molecules in the $J = 1$ level than in the $J = 0$ level.

Related Problems: 9, 10

20.3 Rotational and Vibrational Spectroscopy

We introduced you to the properties of the chemical bond and to the shapes of molecules in Chapter 3, and showed how molecular orbital theory explained the trends in bond order, bond lengths, bond dissociation energies, and bond force constants observed for homonuclear diatomic molecules and ions (see Fig. 6.20). Where did these data come from? How do we measure bond lengths and force constants and how can we use these data to estimate bond dissociation energies? We begin this section with a discussion of the spectroscopy of diatomic molecules to show how the properties of individual bonds are measured and then proceed to discuss the spectroscopy of polyatomic molecules from which we learn much about their three-dimensional structures.

20.3.1 Diatomic Molecules

The Born-Oppenheimer approximation, introduced in Section 6.1, allows us to find the electronic wave functions for diatomic molecules with the nuclei at fixed positions. The variation in the electronic energy as a function of internuclear separation generates the potential energy function that governs nuclear motion. We can write the Schrödinger equation to solve for the nuclear wave functions and energy levels if we know the form of this potential energy function. The nuclear Schrödinger equation can be solved exactly for several model potential energy functions to get wave functions and energy levels associated with rotational and vibrational motion. These include: 1) The rigid rotor model, in which the bond length is fixed and the potential energy term is constant, used to interpret rotational spectroscopy experiments; 2) The harmonic oscillator model, which represents the chemical bond as a spring with a potential given by Hooke's law, used to interpret vibrational spectroscopy experiments; and 3) The Morse potential, which is used to interpret the vibrational spectra of molecules in which the potential energy function is not harmonic, as well as to provide a method for estimating bond dissociation energies from these spectra. These methods are used for polyatomic molecules as well but we discuss them in detail only for diatomic molecules to illustrate their important general characteristics.

Rotational Spectroscopy

We treat the rotations of diatomic molecules using the quantum mechanical linear rigid rotor model, in which the bond length is fixed and the molecule rotates about an axis that is perpendicular to the internuclear axis and passes through its center of mass, as illustrated in Figure 20.7. The rotational motion is described by a single moment of inertia, which is defined as

$$I = \mu R_e^2$$

where R_e is the equilibrium bond length and $\mu = m_1 m_2/(m_1 + m_2)$ is the reduced mass.

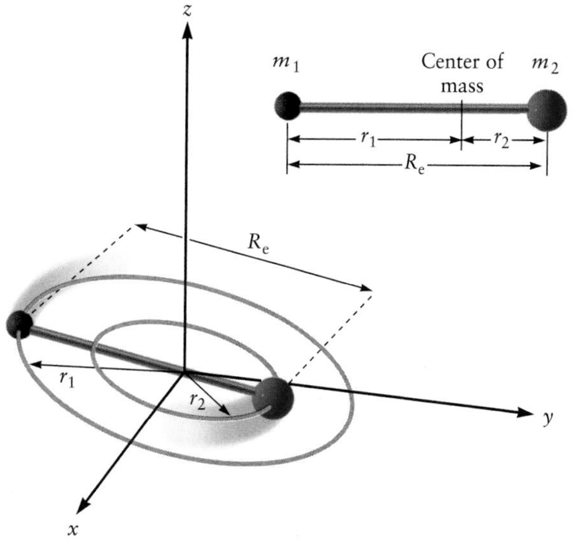

FIGURE 20.7 A diatomic molecule rotates about its center of mass, which for a diatomic molecule is closer to the atom with the larger mass

$$r_1 = \left(\frac{m_2}{m_1 + m_2}\right) R_e \text{ and } r_2 = \left(\frac{m_1}{m_1 + m_2}\right) R_e$$

The wave functions for the rigid rotor are identical to the angular parts of the hydrogen atom wave functions, except of course now the θ and ϕ angles represent the orientation of the internuclear axis of the diatomic. The rotational angular momentum is quantized in the same way. The square of the angular momentum and its projection along the laboratory z-axis can only take on the following values:

Square of angular momentum:

$$J(J+1)(h/2\pi)^2 \qquad J = 0, 1, 2, 3... \qquad [20.5a]$$

Projection of angular momentum on z-axis:

$$M_J(h/2\pi) \qquad M_J = -J, -J+1, ...0..., J-1, J \qquad [20.5b]$$

Equation 20.5b shows that the degeneracy of each level is given by $g_J = 2J + 1$. Interestingly, the $J = 0$ state has a spherically symmetric angular distribution similar to the s-orbitals of one electron atoms in Chapter 5, so you are just as likely to find the internuclear axis pointing along the x-axis, y-axis, z-axis or any θ,ϕ angle in between. Thus molecules in $J = 0$ pack like spheres, perhaps the most famous example being *para*-H_2 with paired nuclear spins (see Section 20.4).

The energy levels for the linear rigid rotor are given by

$$E_J = \frac{h^2}{8\pi^2 I} J(J+1) = hBJ(J+1)\; J = 0, 1, 2, ... \qquad [20.6a]$$

or

$$E_J = hc\tilde{B}J(J+1) \qquad [20.6b]$$

where $B = (h/8\pi^2 I)$ and $\tilde{B} = (h/8\pi^2 cI)$ are rotational constants, the former measured in Hz and the latter measured in cm^{-1}. Both units are widely used in rotational spectroscopy, and conversion between them is straightforward: $1\ cm^{-1} = 30 \times 10^9$ Hz (30 GHz). A rotational absorption spectrum using microwave radiation, presented next, allows you to measure the rotational constant. Since $\tilde{B} = (h/8\pi^2 cI)$ this gives you the moment of inertia I, and using $I = \mu R_e^2$ you can derive the internuclear distance R_e! This is how the equilibrium internuclear distance of HCl, and many other heteronuclear diatomic molecules, was first determined; decades later the measured R_e's were confirmed by solving the Schrödinger equation for the potential energy curve depicted in Figure 20.1.

Microwave absorption spectroscopy has historically been one of our most important experimental methods for accurately measuring bond lengths in heteronuclear diatomic molecules. Microwave emission spectra are used to identify molecules in interstellar space as well as to monitor the concentrations of gases of importance in the atmospheric chemistry of the Earth. Absorption and emission of microwave radiation can be interpreted classically by considering the change in direction of the dipole moment of a heteronuclear diatomic molecule as it rotates. If you imagine that the diatomic molecule in Figure 20.8 is HCl, the H atom has a partial positive charge $\delta+$ and the Cl atom has a partial negative charge $\delta-$, so as the HCl molecule rotates the dipole moment vector, pointing from the Cl atom to the H atom along the internuclear axis, rotates with it. This change in dipole moment direction is what allows the molecule to interact with the oscillating electric field of the microwave light, and absorb a photon. Thus molecules must have permanent dipole moments in order to absorb or emit microwave or far-infrared radiation as a result of rotational motion. Quantum mechanics imposes two additional requirements: conservation of energy and conservation of angular momentum. Conservation of energy is enforced by the familiar Bohr frequency rule $|\Delta E| = h\nu$ that relates the photon frequency to the energy difference between the levels. Conservation of angular momentum

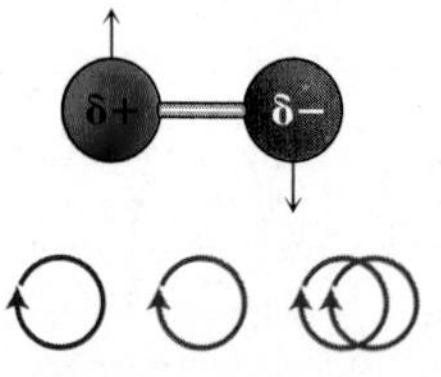

FIGURE 20.8 The angular momentum of a photon (blue circular arrow) is transferred to a molecule (red circular arrows) upon absorption of microwave radiation. The straight black arrows show the instantaneous tangential velocities of the atoms in the molecule as it rotates. (Adapted from Peter Atkins and Julio de Paula, *Physical Chemistry*, W. H. Freeman and Company, New York, 2010, p. 457.)

requires that $\Delta J = \pm 1$ for microwave absorption and emission, respectively, because the photon carries one unit of angular momentum. (Only J is changing, so the energy of the molecule increases when $\Delta J = +1$; a photon is absorbed. We'll see in the next section on infrared spectroscopy that $\Delta J = +1$ and $\Delta J = -1$ are both possible when a molecule absorbs an infrared photon, as the change in energy is primarily determined by the change in the vibrational quantum number. The photon's unit of angular momentum may add to or subtract from the molecule's when the photon is absorbed.) Figure 20.8 shows one unit of angular momentum being transferred to a molecule upon absorption. The angular momentum of a molecule decreases by one unit when a photon is emitted, the angular momentum being carried away by the photon. This **selection rule,** based upon the conservation of angular momentum, is rigorous. There are other kinds of useful selection rules in spectroscopy that are somewhat weaker because they are based upon approximations, such as the harmonic oscillator approximation introduced later.

We can use Equation 20.6b to calculate the frequencies of the allowed transitions for microwave absorption in wave numbers, by letting J_i and J_f represent the rotational quantum numbers of the initial and final states and imposing the $\Delta J = +1$ selection rule. Using the energy in rotation $E_{rot} = hc\tilde{B}(J)(J+1)$ gives

$$\begin{aligned} E_{photon} &= hc\tilde{\nu} = E_{rot,f} - E_{rot,i} \\ \tilde{\nu} &= \tilde{B}\left[J_f(J_f+1) - J_i(J_i+1)\right] \\ &= \tilde{B}\left[(J_i+1)(J_i+2) - J_i(J_i+1)\right] \\ &= 2\tilde{B}(J_i+1) \end{aligned} \qquad [20.7]$$

Equation 20.7 predicts that the microwave absorption spectrum of a heteronuclear diatomic molecule consists of a series of equally spaced lines separated by $2\tilde{B}$. Figure 20.9 shows the energy level diagram for a rigid rotor, with the allowed transitions represented by vertical arrows and the spectrum represented schematically by the stick diagram at the bottom of the figure. The microwave

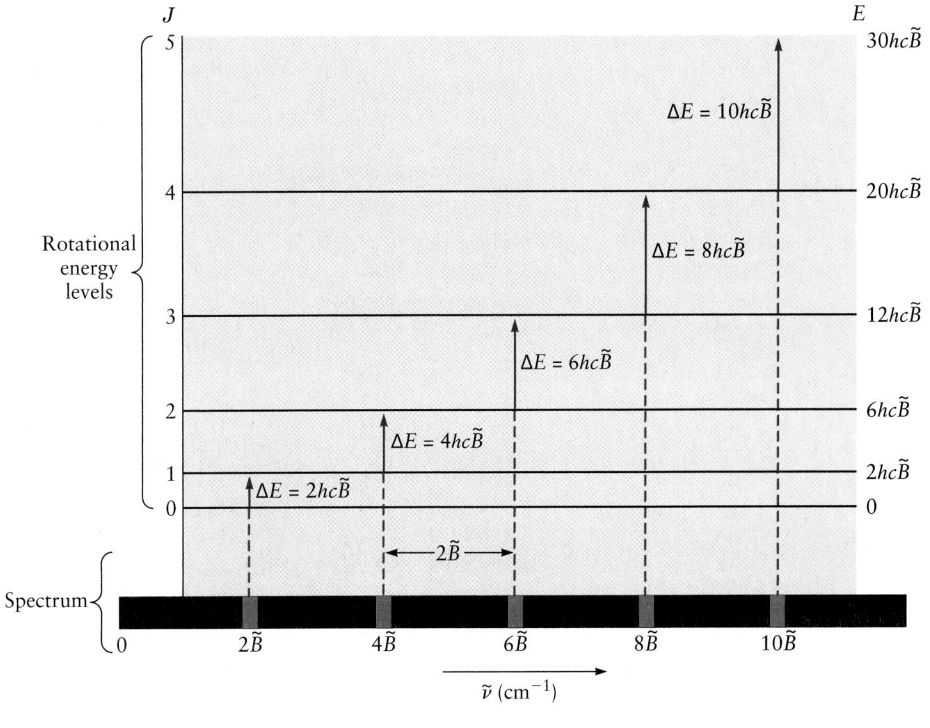

FIGURE 20.9 Rotational energy levels and allowed transitions for a heteronuclear diatomic molecule predicted by the rigid rotor model. The energy levels are given by $E_J = hc\tilde{B}J(J+1)$ and the selection rule $\Delta J = +1$ for absorption predicts a series of spectral lines equally spaced by $2\tilde{B}$, where $\tilde{B}$ is the rotational constant in cm^{-1}.

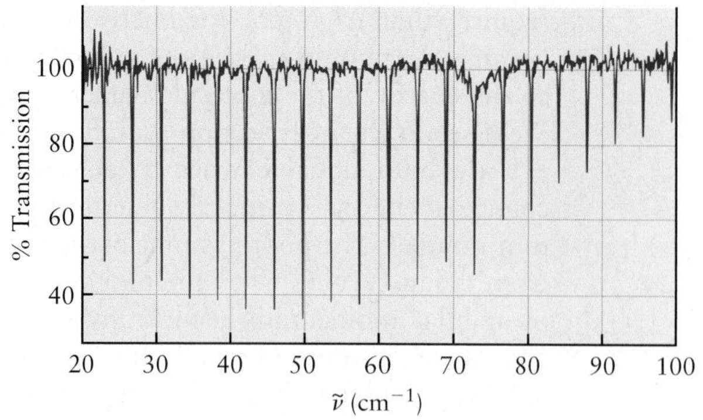

FIGURE 20.10 Microwave absorption spectrum of gas-phase carbon monoxide. The spacing between adjacent lines is 3.8 cm^{-1}, from which the rotational constant $\tilde{B} = 1.9$ cm^{-1} is determined. (Adapted from G. W. Chantry, ed., *Modern Aspects of Microwave Spectroscopy*, Academic Press, 1979.)

absorption spectrum of carbon monoxide is shown in Figure 20.10. The separation between each pair of lines is 3.8 cm^{-1} from which we determine the rotational constant $\tilde{B} = 1.9$ cm^{-1}. From this you can calculate the R_e of CO (see Example 20.4 below). The relative intensities of the lines observed in the experimental spectrum reflects both the increasing degeneracy of each energy level with increasing J as well as the exponentially decreasing populations of each (J,M_J) quantum state as determined by the Boltzmann distribution.

Bond lengths of heteronuclear diatomic molecules are simply calculated from the experimentally measured rotational constants and the atomic masses, as illustrated in Example 20.4.

EXAMPLE 20.4

The microwave absorption spectrum of gaseous NaH (isotope: $^{23}Na^{1}H$) has been measured experimentally; microwaves with wavelengths of 1.02 mm excite the transition from $J = 0$ to $J = 1$. Calculate the bond length of the NaH molecule. Use isotopic masses provided in Table 19.1.

Solution

The reduced mass is

$$\mu = \frac{m_{Na}m_H}{m_{Na} + m_H} = \frac{(22.9898 \text{ u})(1.0078 \text{ u})}{22.9898 + 1.0078 \text{ u}}$$

$$= 0.9655 \text{ u} = 1.603 \times 10^{-27} \text{ kg}$$

The energy change is

$$\Delta E = h\nu = \frac{hc}{\lambda}$$

which in this case is

$$\Delta E_{rot} = \frac{(6.626 \times 10^{-34} \text{ J s})(2.998 \times 10^{8} \text{ m s}^{-1})}{1.02 \times 10^{-3} \text{ m}} = 1.95 \times 10^{-22} \text{ J}$$

$$= \frac{h^2}{8\pi^2 I}[(1)(2) - (0)(1)] = \frac{h^2}{4\pi^2 I}$$

Solving for the moment of inertia I gives

$$I = \frac{h^2}{4\pi^2 \Delta E_{rot}} = \frac{(6.626 \times 10^{-34} \text{ J s})^2}{4\pi^2(1.95 \times 10^{-22} \text{ J})} = 5.70 \times 10^{-47} \text{ kg m}^2$$

(Continued)

The moment of inertia I is related to the bond length R_e by $I = \mu R_e^2$ so

$$R_e^2 = \frac{I}{\mu} = \frac{5.70 \times 10^{-47}\ \text{kg m}^2}{1.603 \times 10^{-27}\ \text{kg}} = 3.56 \times 10^{-20}\ \text{m}^2$$

$$R_e = 1.89 \times 10^{-10}\ \text{m} = 1.89\ \text{Å}$$

Related Problems: 11, 12, 13, 14

Microwave absorption cannot be used to measure bond lengths in homonuclear diatomic molecules because they have no permanent dipole moments. Rotational Raman spectroscopy, however, can be used to measure bond lengths in both heteronuclear and homonuclear diatomic molecules because it depends upon a different kind of interaction. Electromagnetic radiation from a polarized laser can induce a dipole moment in a molecule by displacing its electron density; the induced dipole moment is proportional to the component of the molecular **polarizability** that is oriented parallel to the electric field of the laser radiation. Polarizability (see Section 10.2) is a measure of the degree to which the electron distributions of atoms and molecules distort in the presence of electric fields; larger atoms, with more loosely bound electrons, have larger polarizabilities than smaller atoms, and molecules with delocalized electrons have larger polarizabilities than those that don't. Figure 20.11 shows that the polarizability of a diatomic molecule is generally larger when the molecule is aligned with the electric field than when it is oriented perpendicular to the field. The polarizability of a homonuclear diatomic molecule oscillates at twice the rotational frequency, producing an oscillating dipole moment that emits radiation at a frequency shifted from that of the incident radiation by that amount. We interpret Raman scattering using quantum mechanics as arising from the absorption of radiation from rotational levels of the ground electronic state to a set of "virtual" electronic states followed by emission to different rotational (or vibrational) levels of the ground electronic state, as shown in Figure 20.12.

We determine $\Delta E_{rot}/hc$ for Raman transitions by measuring the differences in frequencies of the incident (laser) radiation and the observed scattered radiation $\tilde{\nu} = \tilde{\nu}_i - \tilde{\nu}_s$ where the negative of $\tilde{\nu}$ is called the Raman shift and $\tilde{\nu}_i$ and $\tilde{\nu}_s$ are the frequencies of the incident and scattered radiation, respectively. The scattered radiation may appear either at higher or lower frequencies than the incident radiation, depending upon whether the final rotational level lies lower or higher in energy than the initial rotational level, as depicted in Figure 20.12. The selection rule for rotational Raman spectroscopy is $\Delta J = 0, \pm 2$; transitions

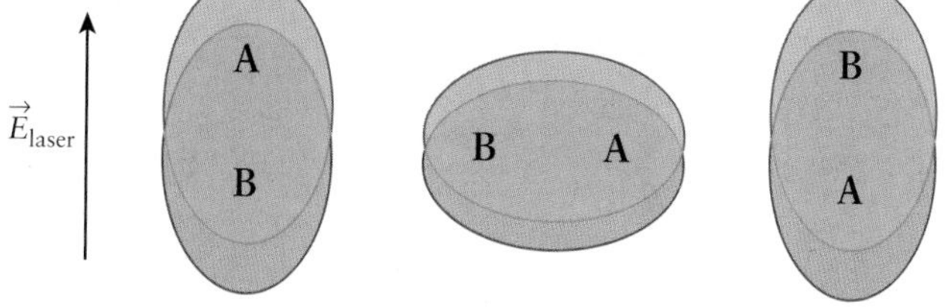

FIGURE 20.11 Schematic of the polarization induced in a diatomic molecule by polarized laser light with the electric field of the laser, $\vec{E}_{laser}$, pointing in the direction shown. The component of the molecular polarizability along the bond axis is, in general, larger than that perpendicular to the axis so the dipole moment that results oscillates at twice the rotational frequency. (We have chosen to represent the polarizability by an ellipse in which the magnitude is proportional to the lengths of the major and minor axes, which is more intuitive than the conventional representation.)

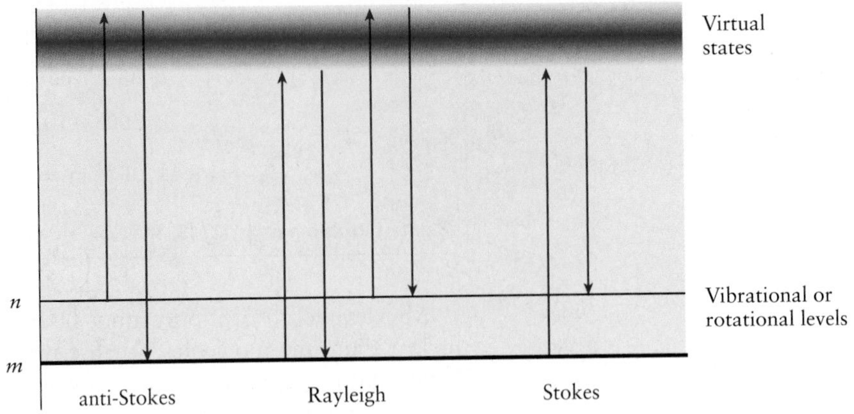

FIGURE 20.12 Quantum mechanical picture of Raman scattering arising from absorption and emission by virtual electronic states.

for which $\Delta J = 0$ produce Raman scattered radiation at the incident frequency, those for which $\Delta J = +2$ produce Raman scattered radiation at lower frequencies (Stokes lines), and transitions for which $\Delta J = -2$ produce Raman scattered radiation at higher frequencies (anti-Stokes lines). The selection rule for rotational Raman scattering is most easily understood by enforcing the conservation of angular momentum. Unlike absorption and emission, with only one photon involved, Raman scattering is a two-photon process that allows for several different combinations of the angular momenta of the incident and scattered photons. The projection of the angular momentum along the propagation direction of the light can have one of two values, +1 or −1 (in units of $h/2\pi$). If the incident and scattered photons have parallel angular momentum vectors, then the change in the J quantum number of the molecule is zero, leading to the $\Delta J = 0$ selection rule. If the incident and scattered photons have antiparallel angular momentum (think of two vectors, one the negative of the other), then two units of angular momentum must have been transferred to or from the molecule, leading to the $\Delta J = \pm 2$ selection rule.

Figure 20.13 is a figure of some historical significance. The rotational Raman spectrum of O_2 is shown on the left and that of N_2 is shown on the right. These spectra were taken in 1929, just a year after the discovery of the Raman effect by the Indian physicist Sir C. V. Raman, for which he was awarded the

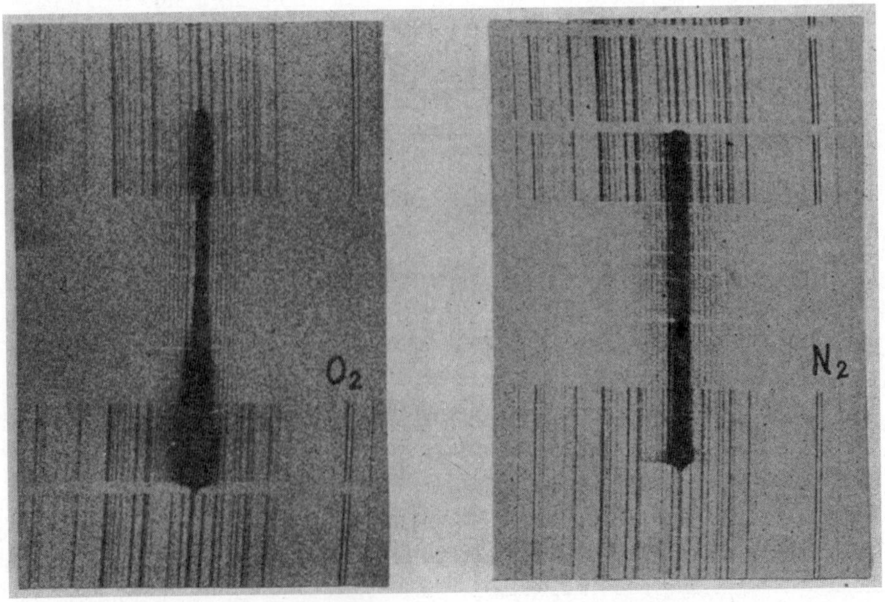

FIGURE 20.13 First published rotational Raman spectra of O_2 and N_2. The spectrum of N_2 provided the first measurement of the bond length of that molecule. (Rasetti, F., *Proceedings of the National Academy of Sciences of the United States of America,* vol. 15, no. 6 (June 15, 1929), pp. 515–519.)

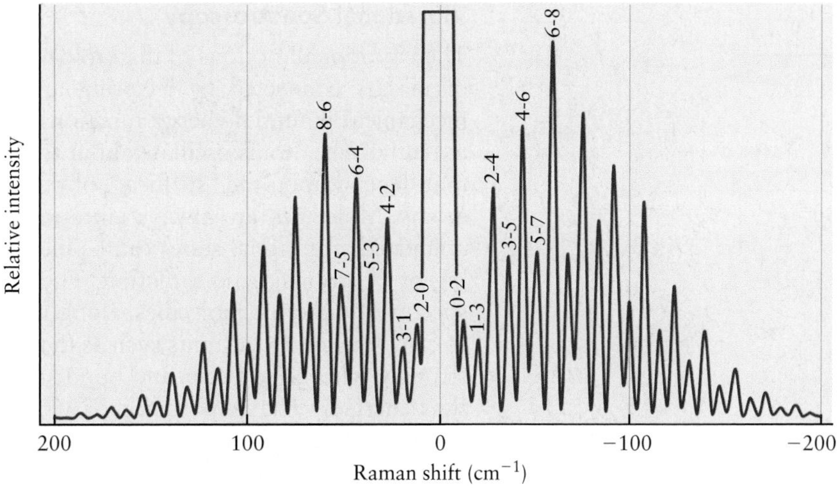

Courtesy of Bryar Broocks, Department of Chemistry, University of Idaho

FIGURE 20.14 The pure rotational Raman spectrum of N_2. The scattered photon has a smaller energy than the incident photon for the features on the right side, so the Raman shift $\nu_s - \nu_i$ is negative, ΔE_{rot} is positive, and $\Delta J = +2$.

Nobel Prize. (The discovery of lasers decades later made these spectra much easier to obtain.) The spectra consist of a series of lines, analogous to those in the microwave spectrum of CO, from which the rotational constants, moments of inertia, and bond lengths of these molecules were determined. The rotational Raman spectrum of N_2 provided the very first measurement of the equilibrium bond length (R_e) of that molecule. Rotational Raman spectroscopy provided much, if not most, of the spectroscopic data shown in Figure 6.20 for the second period homonuclear diatomic molecules, which was of pivotal importance in developing an understanding of the nature of the chemical bond.

Figure 20.14 shows the pure rotational Raman spectrum of N_2 recorded using a modern spectrometer. The Stokes lines appear as the envelope on the right, with negative Raman shifts, and the anti-Stokes lines appear as the envelope on the left, with positive Raman shifts. The overall shape of the envelopes is determined by the same factors as those for microwave absorption spectra, the degeneracy of the levels, and the Boltzmann factor. Note the interesting alternating intensities of the lines in the spectrum, which results from the effects of nuclear spin (see Section 20.4) on the effective symmetry of the molecule. Different patterns are observed for different nuclear spins: the pattern seen in the figure arises because each nitrogen atom in N_2, ^{14}N, has a nuclear spin of one. A different pattern appears for H_2 because each H atom has a nuclear spin of $\frac{1}{2}$, and if you look closely at the spectrum of O_2 shown in Figure 20.13 you will see that every other line is missing, because the spin of the ^{16}O nucleus is zero.

EXAMPLE 20.5

The rotational constant $\tilde{B}$ for H_2 is 60.86 cm^{-1}. Predict the rotational Raman spectrum of H_2 and calculate the bond length of the molecule.

Solution

The rotational Raman spectrum of H_2 consists of two envelopes of lines, a series of Stokes lines at lower frequencies than that of the excitation source and a series of anti-Stokes lines at higher frequencies. The separation between adjacent lines is $4\tilde{B} = 243.4$ cm^{-1}, from which the rotational constant was determined. The intensity of the lines in each of the envelopes peaks at low values of J, because the rotational energies of H_2 are comparable to thermal energies at 300 K and only the first few rotational levels are populated to a significant extent. The bond length is calculated from the moment of inertia, determined from the rotational constant, and the reduced mass by rearranging the equation for $\tilde{B}$ below Equation 20.6b to get

$$I = (h/8\pi^2 c\tilde{B}) = 4.6 \times 10^{-46} \text{ kg m}^2$$

The reduced mass of H_2 is $\mu = 8.3 \times 10^{-28}$ kg from which we calculate the bond length using $I = \mu R_e^2$ to get $R_e = 0.74$ Å.

Related Problems: 15, 16, 51, 52

Vibrational Spectroscopy

The vibrational motions of diatomic molecules are analogous to those of a pair of masses connected by a spring and they are analyzed using two simple mechanical potential energy functions, the harmonic oscillator and the Morse potential. The atoms oscillate about the equilibrium bond length at frequencies that depend upon the "stiffness" of the spring as well as the masses of the two atoms. Molecules are always vibrating, even at 0 K, but transitions between quantized vibrational states can be induced by absorption, emission, or scattering of electromagnetic radiation. The frequencies observed in the vibrational spectra of diatomic molecules, along with the known reduced masses, are used to calculate force constants such as those plotted in Figure 6.20, which correlate strongly with bond order and bond strength, as determined by bond dissociation energies. The importance of vibrational spectroscopy in developing an understanding of chemical bonding can hardly be overstated.

Figure 20.15 shows a more accurate potential energy function for a diatomic molecule, one called the Morse potential, along with the potential energy function for the harmonic oscillator model. Let's review the characteristics of potential energy functions before introducing the harmonic oscillator model. The potential energy of the atoms separated at infinity is often taken to be zero, but when one is studying molecular spectroscopy it is usual to shift the zero of potential energy to the minimum of the ground electronic state potential energy curve so that the expression for E_{vib} that we introduced in Section 12.8 may be used. If you start with the atoms far apart, large R, and move them toward each other, the potential energy decreases as the bond is formed, reaching its minimum value at the equilibrium bond length, R_e. The two atoms feel an attractive force between them in this region of the potential, between $R = R_e$ and longer internuclear distances. At smaller internuclear distances than $R = R_e$, the two atoms feel a repulsive force between them. The potential energy curve near the minimum closely resembles a parabola, which is the potential energy function for the harmonic oscillator model, one of the most important mechanical models in physics. We discuss the classical and quantum harmonic oscillator models to provide a basis for interpreting the vibrational spectra of molecules. We introduced you to the harmonic oscillator model (without calling it that) in our discussion of blackbody radiation in Section 4.2. Our model for that problem was a mass connected to a wall by a spring. Hooke's law, which is empirical, states that the restoring force acting on that mass is directly proportional to the

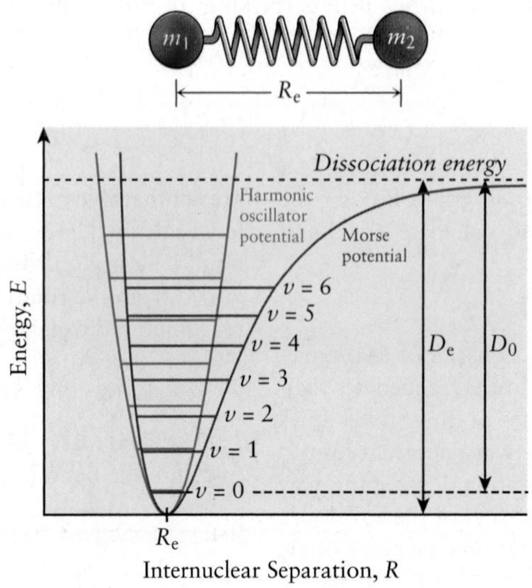

FIGURE 20.15 Potential energy curves for a diatomic molecule (represented by a ball and spring model) approximated by a Morse potential (blue) and a harmonic oscillator potential (green) that have the same equilibrium bond length R_e, and similar curvature near the minimum. The Morse potential leads to dissociation at long internuclear distances. The bond dissociation energy D_0 and the energy difference D_e between the minimum of the potential energy curve and the dissociation limit are shown. They differ by ½ $h\nu$.

displacement of the mass from its equilibrium position but acting in the opposite direction. Let's now consider two masses connected by a spring, the masses representing the atoms of a diatomic molecule and the spring representing the chemical bond. For the case of two masses connected by a spring we simply replace the single mass used in our first model by the reduced mass of the system, which is given by

$$\mu = m_1 m_2/(m_1 + m_2)$$

Let's apply Hooke's law to a diatomic molecule with reduced mass μ and equilibrium bond length R_e. The restoring force is directly proportional to the change in bond length for small displacements from equilibrium

$$F(R) = -k(R - R_e) \qquad [20.8]$$

where the **force constant** k is a measure of the stiffness of the bond, expressed in N m^{-1} and $(R - R_e)$ is the displacement from the equilibrium bond length. The force constant k will be large for stiff bonds like the triple bond of CO, and smaller for single bonds. The potential energy function $V(R)$ that results from this restoring force is obtained by integrating using Equation 20.8 with $F(R) = -dV/dR$ to get

$$V(R) = \tfrac{1}{2}k(R - R_e)^2 \qquad [20.9]$$

which is shown as the green parabolic curve in Figure 20.15. You can see that the harmonic oscillator potential approximates the real potential energy function in the region around the minimum closely, and also gives a good prediction for the energies of the lowest vibrational quantum states. This close agreement justifies the use of the harmonic oscillator approximation in vibrational spectroscopy of the low energy vibrational states of a molecule. If the two masses in the system just described are displaced from their equilibrium separation, they will begin to oscillate back and forth about that position with a characteristic frequency (in round-trip cycles per second) given by

$$v = \frac{1}{2\pi}\sqrt{\frac{k}{\mu}} \qquad [20.10]$$

in which μ is the reduced mass and k is the force constant of the harmonic oscillator. Classical harmonic oscillators vibrate at a natural frequency that is determined by Equation 20.10; they can vibrate with any amplitude (and energy, therefore) but the vibrational energies, E_{vib}, of quantum harmonic oscillators are restricted to fixed values as we now show.

The energies and wave functions for the quantum harmonic oscillator model are obtained by solving the one-dimensional Schrödinger equation with the harmonic oscillator potential energy term added to the kinetic energy term.

$$-\frac{h^2}{8\pi^2 m}\frac{d^2\psi(x)}{dx^2} + \frac{1}{2}kx^2\psi(x) = E_{vib}\psi(x) \qquad [20.11]$$

The x in this equation represents $R - R_e$ for a diatomic molecule, and the mass m would be replaced by the reduced mass μ. The method by which this equation is solved is beyond the scope of this textbook, but the wave functions and energy levels are easy to write down, graph, and interpret. You may verify the solutions by substituting them into the Schrödinger equation above. The first four wave functions for the quantum harmonic oscillator are listed in Table 20.2 and plotted in Figure 20.16. In this table and figure, $x - x_0 = x - 0 = R - R_e$ for a diatomic, so if you plot the harmonic potential as a function of R instead of x, then the minimum of the harmonic potential is at $R = R_e$, as shown in Figure 20.15.

TABLE 20.2

The First Four Harmonic Oscillator Wave Functions

$$\psi_3(x) = \left(\frac{\alpha^3}{9\pi}\right)^{1/4}\left(2\alpha x^3 - 3x\right)e^{-\alpha x^2/2}$$

$$\psi_2(x) = \left(\frac{\alpha}{4\pi}\right)^{1/4}\left(2\alpha x^2 - 1\right)e^{-\alpha x^2/2}$$

$$\psi_1(x) = \left(\frac{4\alpha^3}{\pi}\right)^{1/4} x e^{-\alpha x^2/2}$$

$$\psi_0(x) = \left(\frac{\alpha}{\pi}\right)^{1/4} e^{-\alpha x^2/2}$$

with $\alpha = \frac{2\pi}{h}(k\mu)^{1/2}$

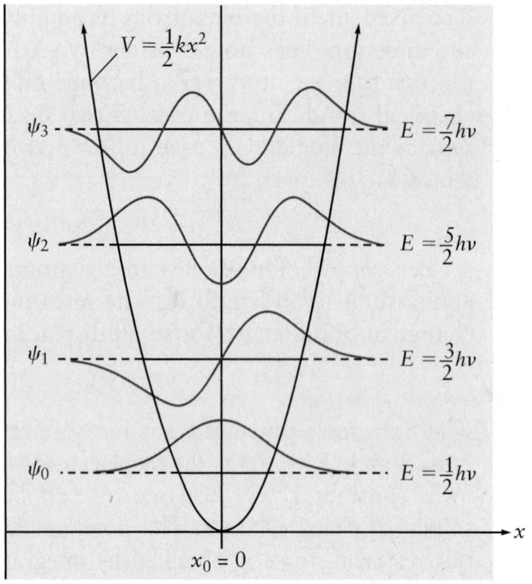

FIGURE 20.16 Energy levels and wave functions for the lower energy states of the harmonic oscillator.

The frequency of a quantum harmonic oscillator is calculated using the same formula as that for the classical harmonic oscillator, Equation 20.10. The quantized energy levels of the harmonic oscillator are given by

$$E_v = h\nu\left(v + \frac{1}{2}\right) \qquad [20.12]$$

where ν is its fundamental vibrational frequency and v is a quantum number that can have the values 0, 1, 2, ... The energy levels are equally spaced, like the steps of a ladder, with adjacent levels being separated by $\Delta E = h\nu$. The higher energy vibrations may be thought of as harmonics of the fundamental, by analogy to the one-dimensional particle-in-a-box model, discussed in Section 4.6. Unlike the particle-in-a-box, however, the $v = 0$ state of the harmonic oscillator is allowed because the term $E = \frac{1}{2}h\nu$ ensures that the total energy of the oscillator never equals zero, which would violate the uncertainty principle. In the ground vibrational state all molecules have $\frac{1}{2}h\nu$ of vibrational energy in each vibrational mode, even at 0 K; this motion is called **zero-point motion** and the energy is called the **zero-point energy.** Differences in zero-point energy account for very important isotope effects on reaction rates in chemical kinetics, as discussed in Section 18.6. Vibrational spectra are usually presented as a function of frequency expressed in wave numbers; the highest-frequency molecular vibration is that of H_2, at 4400 cm^{-1}, and the low-frequency end of the vibrational spectral region for molecules is generally around 100 cm^{-1}, frequencies that are characteristic of metal–metal stretching vibrations, for example.

Transitions between vibrational energy levels can occur by absorption or emission of infrared radiation, or by Raman scattering. Molecules must have dipole moments that change as they vibrate in order for them to absorb or emit infrared radiation, by the same classical argument presented for rotation earlier. Vibrational Raman scattering, on the other hand, requires a change in the molecular polarizability during vibration. The angular momentum selection rule introduced in Section 20.1 applies rigorously to vibrational spectroscopy as well, but it is a little harder to visualize than for rotational spectroscopy. The harmonic oscillator approximation introduces an additional, but weaker, quantum mechanical selection rule $\Delta v = \pm 1$ for both infrared absorption and emission and for Raman scattering. That selection rule is relaxed for anharmonic potential energy functions, like the Morse potential for a diatomic molecule shown in Figure 20.15.

EXAMPLE 20.6

The infrared spectrum of gaseous NaH (isotope: $^{23}Na^{1}H$) was determined via absorption spectroscopy. A single absorption band was observed at $\lambda = 8.53 \mu m$. Calculate the vibrational force constant of the NaH molecule using the harmonic oscillator model using the masses of the isotopes provided in Table 19.1.

Solution
The vibrational frequency is

$$\nu = \frac{c}{\lambda} = \frac{2.998 \times 10^{8} \text{ m s}^{-1}}{8.53 \times 10^{-6} \text{ m}} = 3.515 \times 10^{13} \text{ s}^{-1} = \frac{1}{2\pi}\sqrt{\frac{k}{\mu}}$$

Solving for the constant k (using the reduced mass from Example 20.4)

$$k = \mu(2\pi\nu)^2$$
$$= (1.603 \times 10^{-27} \text{ kg})(2\pi)^2(3.515 \times 10^{13} \text{ s}^{-1})^2$$
$$= 78.2 \text{ kg s}^{-2} = 78.2 \text{ J m}^{-2}$$

Related Problems: 17, 18, 19, 20

Force constants are a measure of the "stiffness" of chemical bonds. If you calculate the potential energy change from $V = 0$ at equilibrium geometry to the V when R is 0.1 Å larger than R_e, you will see that the potential energy V increases by a larger amount for bonds with a larger force constant k; it requires more energy to stretch bond with a large force constant (the spring is stiffer) than a bond with a smaller k. Figure 6.20 shows a strong correlation between force constants and other properties such as bond lengths (if the valence electrons involved in bonding are from orbitals with the same principle quantum number in the comparison group) and bond dissociation energies. Example 20.7 shows how the force constants plotted in Figure 6.20 were determined from vibrational spectra.

EXAMPLE 20.7

Homonuclear diatomic molecules do not absorb infrared radiation but their vibrational spectra are readily obtained using Raman spectroscopy, as shown for H_2, N_2 and O_2 in the *Connection to Instrumental Analysis* in Chapter 3. The measured vibrational frequencies for the second period diatomic molecules are: $^{7}Li_2$ (351 cm^{-1}); $^{11}B_2$ (1051 cm^{-1}); $^{12}C_2$ (1855 cm^{-1}); $^{14}N_2$ (2358 cm^{-1}); $^{16}O_2$ (1580 cm^{-1}); $^{19}F_2$ (919 cm^{-1}). Calculate the force constants for each molecule, using the vibrational frequencies given and the atomic masses for the appropriate isotope provided in Table 19.1.

Solution
Let's calculate the force constant for N_2 as an example, leaving the rest of the molecules for you to work out on your own. We modify the formula given in Example 20.6 to account for frequencies expressed in wave numbers and substitute the experimental values and the reduced mass for N_2 to solve for the force constant.

$$\tilde{\nu} = \frac{\nu}{c} = \frac{1}{2\pi c}\sqrt{\frac{k}{\mu}}$$

$$k = \mu(2\pi c\tilde{\nu})^2$$
$$= \left[1.16 \times 10^{-26} \text{kg}\right]\left[(2\pi)(3 \times 10^{10} \text{cm s}^{-1})(2358 \text{ cm}^{-1})\right]^2$$
$$= 2292 \text{ N m}^{-1}$$

which is in good agreement with the value plotted in Figure 6.20.

Related Problems: 17, 18, 19, 20

The differences in the vibrational frequencies observed for the series C_2—F_2 are dominated by the differences in force constants, the reduced masses varying by less than a factor of two. This allows us to determine the bond orders of individual bonds in polyatomic molecules simply by inspection; single bonds between the second period elements give rise to bands in the 1000 cm^{-1} region of the spectrum, double bonds around 1600 cm^{-1}, and triple bonds around 2300 cm^{-1}. These characteristic frequencies are very helpful in the analysis of the vibrational spectra of polyatomic molecules, as we show later. Examining trends in the vibrational frequencies of a series of diatomic hydrides reveals the importance of the reduced mass in determining vibrational frequencies and also allows us to establish the typical frequencies of C—H and O—H vibrations in an IR absorption spectrum. The vibrational frequencies of the hydrogen halides are: $^{1}H^{19}F$ (4139 cm^{-1}); $^{1}H^{35}Cl$ (2991 cm^{-1}); $^{1}H^{81}Br$ (2449 cm^{-1}); and $^{1}H^{127}I$ (2308 cm^{-1}). The reduced masses in each of these cases is essentially the mass of the hydrogen atom, which is why the vibrational frequencies are so high. The motion is that of the hydrogen atom vibrating against the more massive halogen atoms, which can be thought of as stationary. The vibrational frequencies of the gas-phase diatomic hydrides $^{12}C^{1}H$ (2859 cm^{-1}) and $^{16}O^{1}H$ (3765 cm^{-1}) are also typical for these bonds in polyatomic molecules; frequencies that are characteristic of particular bonds are extremely useful in the analysis of the vibrational spectra of polyatomic molecules.

While the harmonic oscillator model is very useful for predicting the energies of the low-lying vibrational energy levels, $v = 0$, 1 or 2, it does not predict the energies of the higher vibrational levels very well. The higher energy levels are much better reproduced using an anharmonic potential energy function like the **Morse potential energy function** shown in blue in Figure 20.15. The Morse potential (as it is often called) is a simple analytical expression that is widely used to describe the anharmonic potential energy function of a diatomic molecule, as it gives a better prediction for the actual vibrational energy levels observed spectroscopically. It is

$$V(R) = D_e[1 - \exp(-a(R - R_e)]^2 \quad [20.13]$$

where $V(R)$ is the potential energy curve, R is the distance between the nuclei, R_e is the equilibrium bond length, D_e is the "depth" of the potential energy well, and a is a constant that determines the "width" of the potential energy well. (Note that the Morse potential energy function gives a good representation of a diatomic molecule's potential energy in the bound region, but not at very short internuclear distances.) If you solve the Schrödinger equation for the vibrational energy levels in a Morse potential, the vibrational states become closer in energy as the energy in vibration is increased, as shown in Figure 20.15. This is represented mathematically by an anharmonic correction term to the equation for the vibrational energy levels. At the dissociation energy D_e the Morse potential predicts that the energy difference between two neighboring vibrational energy levels is reduced to zero. $E_{v+1} - E_v$ is proportional to the vibrational quantum number v (decreasing as v increases) so a plot of the experimentally measured energy differences as a function of $v + 1/2$, called a *Birge–Sponer plot*, intercepts the horizontal axis when the vibrational spacing is zero at the dissociation limit. Calculating the area under the curve gives the sum of the energy differences between all the vibrational levels up to the dissociation limit, giving a very good estimate of the bond dissociation energy. This method is illustrated in Figure 20.17 and Example 20.8.

EXAMPLE 20.8

Estimate the bond dissociation energy for H_2^+ using Birge–Sponer extrapolation. The measured frequencies of the successive transitions (0→1, 1→2, etc.) in cm^{-1} are 2191, 2064, 1941, 1821, 1705, 1591, 1479, 1368, 1257, 1145, 1033, 918, 800, 677, 548, and 411. The 0→1 vibrational transition is plotted at 0 + 1/2 in the Birge–Sponer plot so that the area under the curve gives the bond dissociation energy.

Solution

Plot the transition frequencies versus the lower quantum number + 1/2, as illustrated in Figure 20.17. Fit to a straight line using the linear least squares function on your calculator or spreadsheet to determine $y = -116.05x + 2237.7$. The area under the curve (a triangle) is 1/2(19.28)(2237.7 cm^{-1}) = 2.16×10^4 cm^{-1}, very close to the actual D_0.

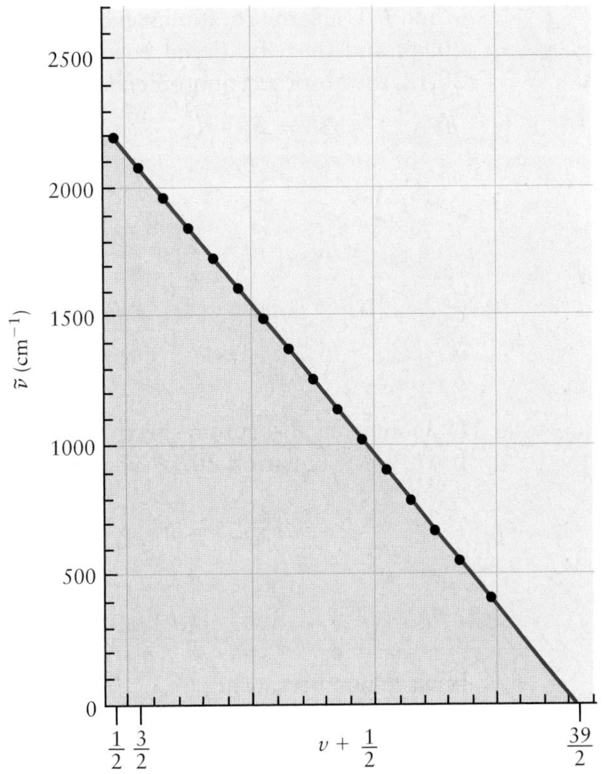

FIGURE 20.17 Birge–Sponer plot used to determine the bond dissociation energy of H_2^+. (Adapted from Peter Atkins and Julio de Paula, *Physical Chemistry*, W. H. Freeman and Company, New York, 2010, p. 467.)

So far we have discussed low-resolution infrared absorption spectra, where the wavelength distribution of the light source used to excite the molecular vibration was not very narrow. If you use an infrared laser to excite the molecule, or take a Fourier transform infrared absorption spectrum you find that what appeared to be one absorption band peaked near the energy $h\nu_{\text{vib}}$ actually consists of many sharp peaks. We have added the subscript "vib" so that we can distinguish the frequency of the photon absorbed, ν_{photon}, and its energy $h\nu_{\text{photon}}$, from the frequency of the harmonic oscillator vibration, ν_{vib}. They are different because the rotational quantum state of the molecule must also change when it absorbs a photon in order to obey the selection rule $\Delta J = \pm 1$ for single photon absorption. (The same selection rule, $\Delta J = \pm 1$, applies to infrared emission spectra, but then of course the vibrational quantum number v decreases by 1.) Figure 20.18 shows a high-resolution infrared absorption spectrum of CO.

It is easy to calculate the energy of the absorbed photon if the infrared light promotes the molecule from an initial ro-vibrational quantum state labeled by the quantum numbers v_{i} and J_{i} to a final state labeled by the quantum numbers

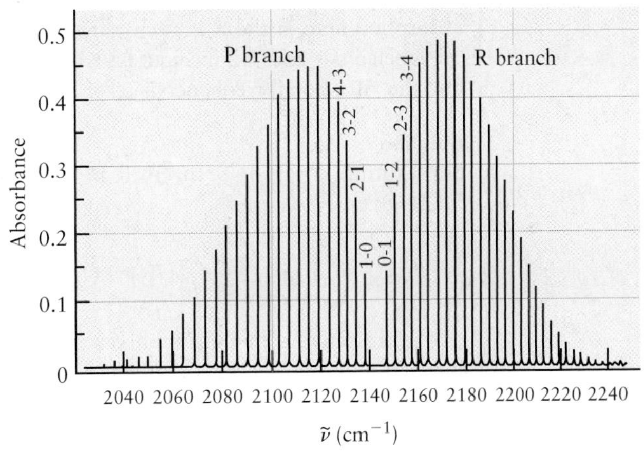

FIGURE 20.18 Vibration–rotation spectrum of carbon monoxide in the gas phase, measured using infrared absorption spectroscopy. Absorbance (see Eq. 20.3) is plotted as a function of the frequency in wave numbers of the incident photon. (Adapted from the work of Husheng Yang, Department of Chemistry, University of Idaho.)

v_f and J_f. Using the rotational energy in the rigid-rotor approximation, Equation 20.6a, and the vibrational energy in the harmonic approximation, Equation 20.12, the absorbed photon energy would be

$$h\nu_{\text{photon}} = \Delta E = E_f - E_i$$

$$h\nu_{\text{photon}} = (E_{\text{vib,f}} + E_{\text{rot,f}}) - (E_{\text{vib,i}} + E_{\text{rot,i}})$$

$$h\nu_{\text{photon}} = h\nu_{\text{vib}}\left(v_f + \frac{1}{2}\right) + \frac{h^2}{8\pi^2 I}J_f(J_f+1) - h\nu_{\text{vib}}\left(v_i + \frac{1}{2}\right) - \frac{h^2}{8\pi^2 I}J_i(J_i+1)$$

$$h\nu_{\text{photon}} = h\nu_{\text{vib}}(v_f - v_i) + \frac{h^2}{8\pi^2 I}\left[J_f(J_f+1) - J_i(J_i+1)\right] \quad [20.14]$$

In an infrared absorption spectrum where $\Delta v = +1$, ΔJ can be either $+1$ or -1. If $\Delta J = -1$, Equation 20.14 gives the energy of the absorbed photon to be

$$h\nu_{\text{photon}} = h\nu_{\text{vib}} + \frac{h^2}{8\pi^2 I}\left[(J_i - 1)(J_i) - J_i(J_i+1)\right]$$

$$= h\nu_{\text{vib}} - \frac{h^2}{4\pi^2 I}J_i \quad [20.15]$$

In wave numbers, using $h\nu_{\text{photon}} = hc\tilde{\nu}_{\text{photon}}$ and $h\nu_{\text{vib}} = hc\tilde{\nu}_{\text{vib}}$ this becomes

$$\tilde{\nu}_{\text{photon}} = \tilde{\nu}_{\text{vib}} - 2\tilde{B}J_i \quad [20.16]$$

These photon energies excite transitions in what is called the P branch ($\Delta J = -1$) of the infrared absorption spectrum. But of course, you can't have a transition where $\Delta J = -1$ and $J_i = 0$, because that would give $J_f = -1$ and no such quantum state exists! Also, an infrared absorption with $\Delta J = 0$ is not observed as this does not satisfy angular momentum conservation when a single photon is absorbed (unless there is a change in electronic angular momentum, as in NO.) Thus, you can easily see why the spectrum of carbon monoxide in Figure 20.18 appears to be missing an absorption line right in the middle of the spectrum. The missing line would have been at the energy where $h\nu_{\text{photon}} = h\nu_{\text{vib}}$. Though the low-resolution infrared absorption spectrum shows a band centered roughly at $h\nu_{\text{photon}} = h\nu_{\text{vib}}$ there is actually no absorption of light at this precise energy.

EXAMPLE 20.9

Beginning with Equation 20.14, derive an expression for the energies of the transitions in the high energy half of the infrared absorption spectrum of CO; this portion of the spectrum is called the R branch and the transitions in this region follow the selection rules $\Delta v = +1$, $\Delta J = +1$. Then calculate the energy of the absorbed photon for two absorption lines in the R branch, the one with $J_i = 0$ and the one with $J_i = 1$. (These are labeled 0-1 and 1-2 in Figure 20.18.) The rotational constant of CO is 1.92 cm^{-1} and the vibrational frequency is 2143 cm^{-1}.

Solution

Substituting $J_f = J_i + 1$ (for an R branch transition) and $v_f = v_i + 1$ into Equation 20.14 gives

$$h\nu_{\text{photo}} = h\nu_{\text{vib}} + \frac{h^2}{8\pi^2 I}\left[(J_i+1)(J_i+2) - J_i(J_i+1)\right] = h\nu_{\text{vib}} + \frac{h^2}{4\pi^2 I}(J_i+1).$$

Expressing this in wave numbers gives $\tilde{\nu}_{\text{photon}} = \tilde{\nu}_{\text{vib}} + 2\tilde{B}(J_i + 1)$. When $J_i = 0$, an infrared photon of energy 2143 cm^{-1} + 2(1.92 cm^{-1}) = 2147 cm^{-1} can be absorbed

(Continued)

(adding a quantum of vibrational energy to the CO and changing its rotational state to $J_f = 1$). When $J_i = 1$, an infrared photon of energy 2143 cm^{-1} + 4(1.92 cm^{-1}) = 2151 cm^{-1} can be absorbed. Note that the spacing between the two adjacent absorption lines is $2\tilde{B}$, the same as that in the pure rotational absorption spectrum of CO, so this infrared spectrum can also be used to determine the equilibrium bond length of CO.

Related Problems: 19, 20

20.3.2 Polyatomic Molecules

Rotational Spectroscopy

Pure rotational spectra of polyatomic molecules are considerably more complicated than those of diatomic molecules because they may have up to three different moments of inertia that result from rotations about three different molecular axes (see Section 12.5 and Figure 12.13). The symmetry of the molecule determines the number of distinct values for the moments of inertia. Linear molecules have one unique axis and a single moment of inertia for rotation about any axis perpendicular to that axis. The analysis of pure rotational spectra for *linear* triatomic molecules is only slightly more complicated than that for diatomic molecules because there are two bond lengths and only one rotational constant (for the two rotational degrees of freedom). We can determine both bond lengths by measuring the rotational constants of at least two molecules in which one or more isotopic substitutions have been made and solving two equations for the two unknowns. The calculation is straightforward, but tedious, and examples can be found in standard physical chemistry textbooks. For a nonlinear polyatomic molecule there are three rotational degrees of freedom, and three mutually perpendicular axes about which the molecule's rotation can be described. All axes are equivalent for spherically symmetric molecules such as CH_4 and SF_6, with tetrahedral and octahedral symmetry, respectively; these spherical rotors have three moments of inertia that are equal to one another. Molecules such as NH_3 and CH_3Cl have one unique axis and two different moments of inertia, one for rotation about the unique axis and one for rotations about any axis perpendicular to that axis. These molecules are called symmetric top rotors. Finally, molecules with three different moments of inertia, such as H_2O, are called asymmetric rotors. The analysis of rotational spectra for nonlinear polyatomic molecules is significantly more complex than that for linear triatomic molecules, but it is carried out in the same way, and has enabled scientists to determine the geometries of many small polyatomic molecules, like those shown in Chapters 3, 6, 7 and 8, very accurately.

Vibrational Spectroscopy

The number and frequencies of the bands observed in vibrational spectra are used to identify molecules, to characterize bonding, and to probe their local environments in complex systems such as those found in biology, geology, or materials science, for example. Molecules containing N atoms have $3N$ degrees of freedom in total, corresponding to the three translational degrees of freedom for each atom if they were not bound together. Three of these degrees of freedom correspond to translations of a molecule through space, and there are either 2 (linear molecule) or 3 (nonlinear molecule) rotational degrees of freedom. Subtracting these motions from the total leaves $3N - 5$ vibrational degrees of freedom for linear molecules and $3N - 6$ vibrational degrees of freedom for nonlinear molecules. The vibrational motions of relatively small molecules, especially rather symmetric ones, can be represented as a superposition of a number of "normal modes" in which the motions of the atoms are arranged in uniquely different ways. The normal modes of water include a symmetric H—O—H stretch (both bonds stretching and compressing in phase), an antisymmetric H—O—H stretch (both bonds stretching and compressing out of phase) and an H—O—H bending motion, for example. The energy in each vibrational mode is quantized (as it was for the single vibrational mode, the stretch, of a diatomic molecule). In the harmonic approximation, the total energy in vibration is then the sum of the vibrational energy in each normal mode

$$E_{\text{vib}} = h\nu_1\left(v_1 + \frac{1}{2}\right) = h\nu_2\left(v_2 + \frac{1}{2}\right) = h\nu_3\left(v_3 + \frac{1}{2}\right)$$

where ν_1, ν_2 and ν_3 are the vibrational frequencies of the symmetric H—O—H stretch, H—O—H bending motion, and antisymmetric H—O—H stretch respectively and the numbers v_1, v_2, and v_3 are the number of quanta of vibration in each of these normal modes. Thus, to identify the vibrational quantum state of H_2O you specify the values of three quantum numbers (v_1, v_2, v_3).

It is instructive to examine the vibrations of a small polyatomic molecule, CO_2, to get a feel for some characteristic vibrational motions, their frequencies, and their spectroscopic activity before turning our attention to the spectroscopy of more complicated polyatomic molecules. CO_2 is a linear triatomic molecule with $3N - 5 = 4$ different "normal" modes of vibration, as shown in Figure 20.19. There are two stretching modes and two bending modes (only one of the two bending modes is shown in Figure 20.19. the other is identical except the motion is perpendicular to the plane of the page). Stretching modes are those in which the bond length changes without changing the bond angles, and bending modes are those in which the bond angles change without changing the bond lengths. The two oxygen atoms move in different directions (along the bond axis) during one cycle of the symmetric stretching vibration (a) whereas they move in the same direction (both to the right, then both to the left) during one cycle of the antisymmetric stretch (b). During the antisymmetric stretch motion the center C atom moves in the opposite direction of the O atoms to keep the center-of-mass of the molecule stationary. The bond angle changes during the two bending vibrations (c), which are degenerate. The two bending modes are distinct because the motions occur in two mutually perpendicular planes. If the dipole moment or polarizability of the molecule changes during each of these vibrations they may be observed in infrared or Raman spectra, respectively.

Let's examine how the dipole moment changes during each of the vibrations, beginning with the degenerate bending modes. The permanent dipole moment of CO_2 is zero by symmetry, but there is clearly an oscillating dipole moment associated with the bending mode so this mode is infrared active. The dipole moment of CO_2 does not change during the symmetric stretching motion, so that mode cannot be excited by infrared absorption. The dipole moment of the molecule clearly changes during the antisymmetric stretching vibration, so that mode can be excited by infrared absorption. The infrared spectrum of CO_2

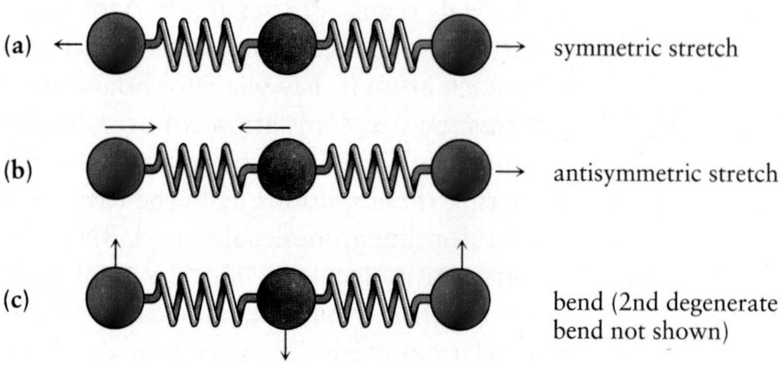

FIGURE 20.19 The normal modes of vibration of CO_2. The fourth mode, a second bending mode, is not shown as it is identical to the bend shown, except that the motion is perpendicular to the plane of the page.

TABLE 20.3

CO_2 Vibrational Modes, Frequencies, and Spectroscopic Activity

Mode	Frequency	Activity
Symmetric stretch	1354 cm^{-1}	Raman
Bend (doubly degenerate)	673 cm^{-1}	IR
Antisymmetric stretch	2396 cm^{-1}	IR

consists of two bands, one at 673 cm^{-1}, which is assigned to the doubly degenerate bending mode and one at 2396 cm^{-1}, which is assigned to the antisymmetric stretch.

Visualizing the change in polarizability during the vibrations is a little more difficult than visualizing the change in dipole moment. It is helpful to think about the polarizability of each of the bonds separately to see how the polarizability of the molecule changes during vibrations. We need to consider only the component of the polarizability that is parallel to the bond axis to understand the stretching modes. The polarizability increases as the bonds stretch and decreases as the bonds are compressed. The two C—O bonds oscillate in phase during the symmetric stretch, so the changes in the bond polarizability are additive. The bond polarizabilities oscillate out of phase during the antisymmetric stretch, and there is negligible change in the overall polarizability during that vibration. The symmetric stretch is Raman-active and the antisymmetric stretch is not. Understanding the Raman activity of the bending mode is not so easy to explain visually, so we just state the result; that mode is not Raman-active. The frequencies and IR or Raman activities of the three modes are summarized in Table 20.3. Note that none of the vibrational modes of CO_2 is both infrared active and Raman active; this is a consequence of inversion symmetry (see Section 6.1). There is a mutual exclusion rule for all centrosymmetric molecules: Modes that are infrared-active are not Raman-active, and vice versa.

Infrared and Raman spectroscopy tend to be complementary methods even for molecules that lack inversion symmetry, because the nature of the interactions between the molecules and electromagnetic radiation are different in the two cases. IR spectroscopy tends to emphasize vibrations involving polar bonds. Raman spectroscopy, on the other hand, is more sensitive to the motions of highly polarizable covalent bonds, such as those involving π bonds or conjugated systems, and rather less sensitive to the vibrations of very polar or ionic bonds, especially those involving the less polarizable lighter elements. The complementarity of the methods is illustrated by comparing the IR and Raman spectra of the amino acid cystine, as shown in Figure 20.20. Cystine is a dimer of the amino acid cysteine; the disulfide bond formed between two cysteine residues located in different regions of a protein cross-links the polymer chains and is a contributing factor to their tertiary structures. The S—S stretch observed near 500 cm^{-1} is barely visible in the infrared spectrum whereas it dominates the Raman spectrum.

Vibrational spectra of polyatomic molecules with more than three atoms are considerably more complex than those of diatomic molecules or CO_2. Fortunately, the vibrational modes of functional groups containing a few atoms are often independent of the rest of the molecule to which they are bound, in the same way that their chemical properties are quite similar in different molecules, as discussed in Section 7.6. Even the simplest amino acid glycine (H_2NCH_2COOH), for example, has 24 normal modes, but its spectrum can be analyzed by looking for vibrations that are characteristic of the carboxylic acid (—COOH) and amino (—NH_2) groups. These complex spectra provide "fingerprints" that are

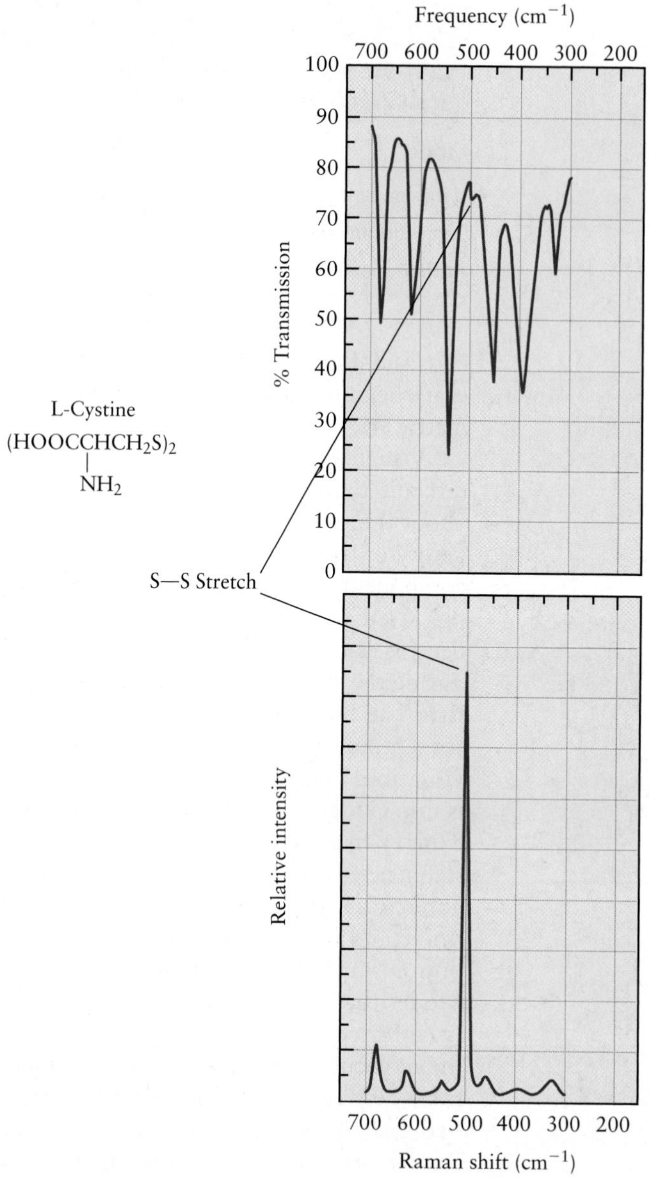

FIGURE 20.20 Infrared and Raman spectrum of the amino acid L-cystine, showing the complementary nature of the techniques. (Adapted from Dana W. Mayo, Foil A. Miller, and Robert W. Hannah, *Course Notes on the Interpretation of Infrared and Raman Spectra*, John Wiley & Sons, Inc., Hoboken, New Jersey, 2004, p. 367.)

used to identify and characterize molecules. The vibrational spectra of larger molecules are generally analyzed by looking for characteristic frequencies of functional groups, which are largely independent of other features of the molecule of which they are a part. The characteristic frequencies of these functional groups can be interpreted as a set of "normal" or "local" modes that involve only the atoms of a particular group. The stretching vibrations of methyl ($-CH_3$) or methylene ($-CH_2$) groups can be thought of as "normal" modes of those groups, comprising symmetric and antisymmetric combinations of the individual C—H bond stretches. C—X stretching vibrations, where X is a halogen, the O—H stretch of an alcohol or the C≡N stretch of the nitrile functional group, on the other hand, can be thought of as "local" modes. Table 20.4 is an abbreviated version of a table of characteristic frequencies used to analyze infrared and Raman spectra and Example 20.10 shows how to use these characteristic group frequencies to analyze the infrared spectrum of a polyatomic molecule. Note that if the molecule is in solution, the O—H and N—H stretching frequencies may be dramatically changed by hydrogen bonding with solvent molecules.

TABLE 20.4

Characteristic Vibrational Frequencies and Infrared Absorption Intensities of Selected Vibrations and Functional Groups

Frequency (cm^{-1})	Bond or Group	Vibration	Relative Intensity
3650–3200	O—H	Stretching	Weak to strong
3550–3100	N—H	Stretching	Medium
3300–2700	C—H	Stretching	Weak to medium
2250–2100	C≡C	Stretching	Weak
1820–1630	C=O	Stretching	Strong
1680–1600	C=C	Stretching	Weak to medium
1430–1390	C—N	Stretching	Strong
1250–1000	C—O	Stretching	Strong

EXAMPLE 20.10

Identify some characteristic group frequencies in the spectrum of an organic molecule shown below and identify the class of molecules to which it belongs. Propose a structure if you can.

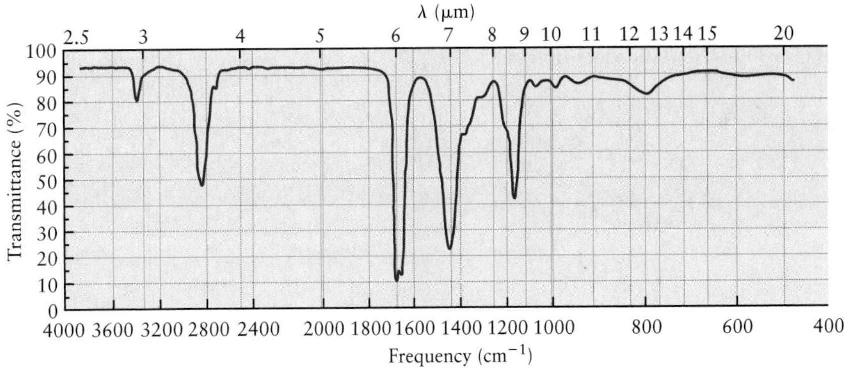

Solution

The highest frequency mode shown appears around 3350 cm^{-1}, a region characteristic of N—H stretching vibrations, the stronger absorption appearing around 2850 cm^{-1} is characteristic of C—H stretching vibrations in alkanes, and the very strong absorption band appearing near 1650 cm^{-1} is characteristic of the carbonyl (C=O) stretching vibration. The absorption near 1450 cm^{-1} is associated with a C—N stretching vibration and the one near 1175 cm^{-1} is characteristic of a C—C single bond stretch. Taken together these observations suggest that the molecule is an amide, with the RC(O)N—functional group. The presence of the C—H stretching vibrations characteristic of an alkane as well as the C—C single bond stretch suggest that it is an alkyl amide. The molecule is $H_3CCH_2C(O)NHCH_3$.

Related Problems: 31, 32

20.4 Nuclear Magnetic Resonance Spectroscopy

Just as electrons have spin, so do some nuclei have spins, and the energies of their spin states are split apart by a magnetic field. Hydrogen (^1H) nuclei have "spin up" and "spin down" states in which the proton behaves as if it were a tiny bar magnet. The energies of these states are degenerate in the absence of a magnetic field, but when an external magnetic field is applied the protons with spin up align parallel to the magnetic field (lower energy configurations) and those with spin down align antiparallel to the field (higher energy configurations). The analogy with bar magnets suggests the difference in energy between these two states will be determined by the magnitude of the applied magnetic field, B, as shown in Figure 20.21. The magnitude of the energy splitting is $\Delta E = g_N \mu_N B$ where g_N is the "nuclear g factor" for protons and μ_N is a constant called the **nuclear magneton.** Figure 20.21 shows that both the spin up and the spin down nuclei have energy E_0 when the magnitude B of the applied magnetic field, plotted on the horizontal axis, is 0, but when the magnitude of the applied magnetic field is increased to B_0, the energies split, so $E_{\text{spin-down}} - E_{\text{spin-up}} = \Delta E = g_N \mu_N B_0$. These nuclei can change spin state by absorbing or emitting photons with energy $h\nu = g_N \mu_N B$, which fall in the FM radio frequency (rf) region of the electromagnetic spectrum. In nuclear magnetic resonance (NMR) spectroscopy, low energy radio waves (photons with frequencies typically between 300 and 600 MHz) "tickle" the nuclei in a molecule, and cause spin transitions.

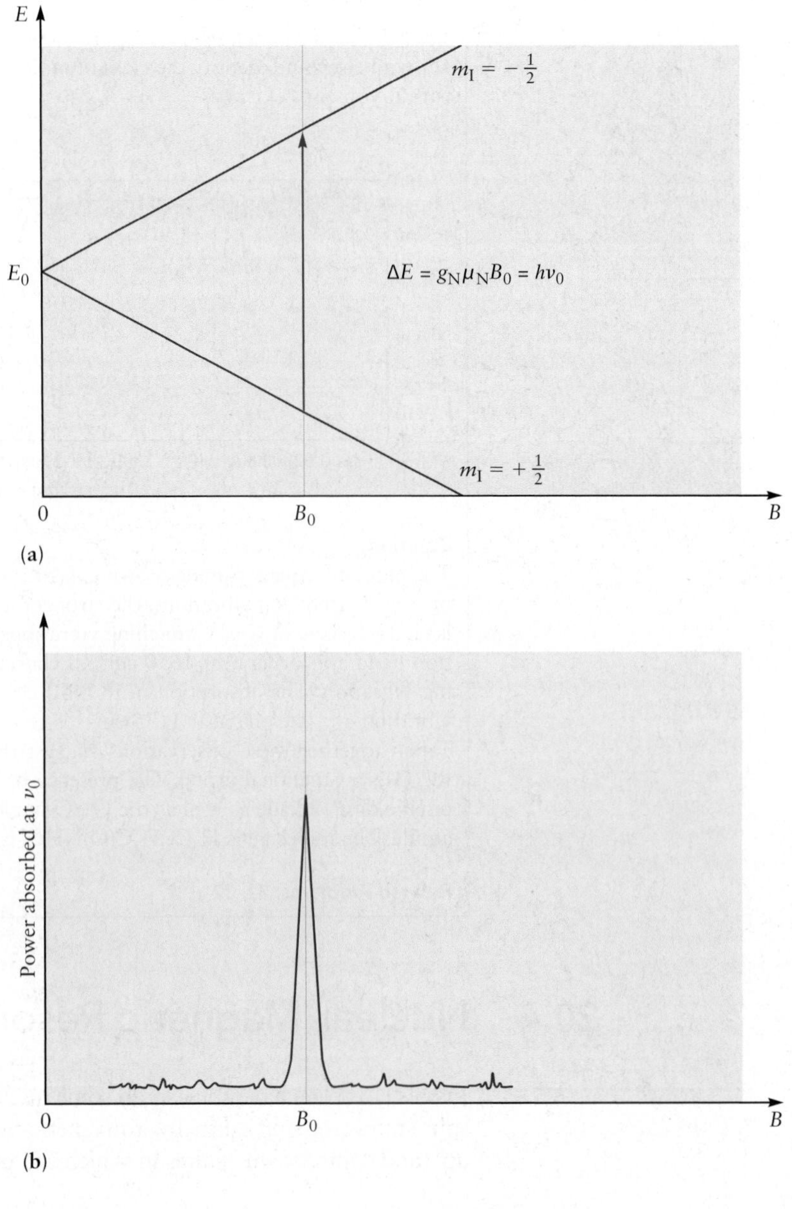

FIGURE 20.21 (a) Energy level splitting diagram for a proton in an external magnetic field. The vertical line shows the absorption transition for a 7.05 T field. (b) Proton NMR spectrum obtained by scanning the magnetic field in a 300-MHz NMR spectrometer.

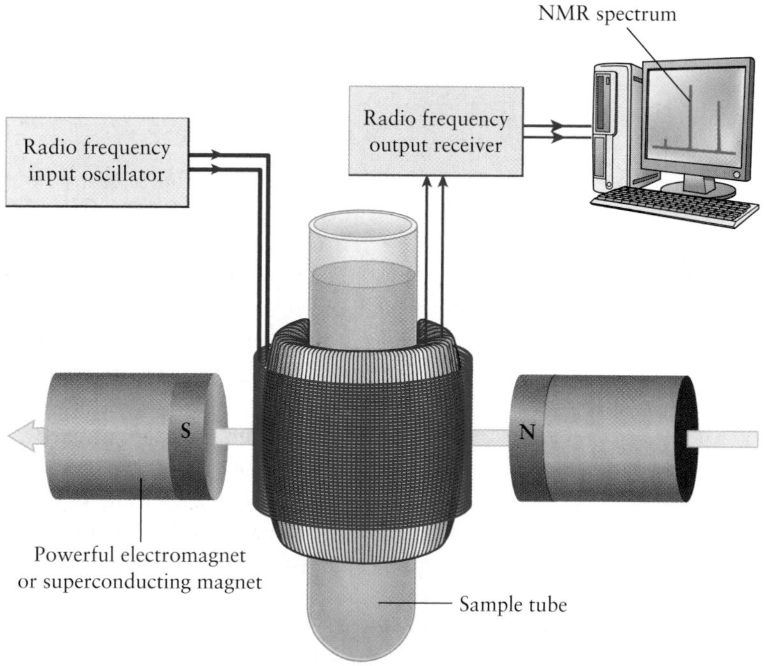

FIGURE 20.22 Schematic of an FT NMR spectrometer.

The NMR spectrum can be recorded in various ways. The earliest commercial NMR spectrometers operated in continuous wave mode, in which the sample is irradiated at constant frequency ν_0 while the magnitude of the external magnetic field B is swept through a range of values. The rf power absorbed by the sample is recorded at each value of B. When the magnitude of the magnetic field is scanned to the value $B_0 = h\nu_0/g_N\mu_N$, satisfying the resonance condition, suddenly some of the rf power is absorbed by the sample, as shown in Figure 20.21b, and a peak appears in the spectrum. Newer instruments rely on Fourier Transform NMR spectroscopy (FT NMR) in which the magnetic field is fixed and the frequency of the radio waves is scanned; a schematic is shown in Figure 20.22. The sample is contained in a small (5 mm diameter) tube that is placed in a magnetic field created by a superconducting magnet. Radiofrequency radiation is transmitted to the sample by a set of coils placed around the sample, and a second set of coils detects the signal. The sample, held in a fixed magnetic field, is irradiated with a short, intense burst of rf power spanning a frequency range broad enough to stimulate many transitions. The specific frequencies of all the nuclear spin transitions in the polyatomic sample are measured simultaneously. A complete spectrum can be acquired within a few seconds. The measurements can be repeated many times and the results averaged in computer memory to reduce the effects of random background noise. FT NMR spectrometers detect numerous nuclei besides ^{1}H (^{13}C holds special interest in organic chemistry and biochemistry), but our discussion is limited to ^{1}H.

Nuclear spins are sensitive to the chemical environment of a nucleus. Electrons with non-zero angular momentum that are near the nucleus establish an internal magnetic field that modifies the local effective field felt by each proton to a value different from that of the externally applied field. The resulting **chemical shift** causes protons within different structural units of the molecule to show NMR peaks at different values of the applied magnetic field. All protons in chemically equivalent environments will contribute to a single absorption peak in the spectrum. The relative area under each absorption peak is proportional to the number of protons within each equivalent group. In order to standardize procedures, chemical shift values are reported relative to the selected reference compound tetramethylsilane (TMS) by adding a very small amount of

TMS to the sample. The chemical shift for each proton in the spectrum is then defined in parts per million (ppm) as

$$\delta = \frac{B_s - B_r}{B_r} \times 10^6 \qquad [20.17]$$

where B_s is the value of the magnetic field at the peak in the sample spectrum and B_r is the value at the TMS peak. Extensive tables of chemical shift values relative to TMS aid the interpretation of spectra. For example, these considerations predict that the NMR spectrum of ethanol, CH_3CH_2OH, should have three peaks of relative area 3:2:1. This is exactly the result obtained when the spectrum is acquired with low resolution (Figure 20.23a).

Nuclear spins are also sensitive to other nearby nuclear spins. Because each spin behaves as if it were a small bar magnet, each influences the local value of the magnetic field felt by others nearby. The result is **spin–spin splitting,** which breaks the broad peak of each equivalent group of protons in to a multiplet of narrower, sharper peaks. So long as the chemical shift between chemically non-equivalent groups of protons is large, the multiplet patterns can be interpreted by qualitative rules that we do not describe here. For example, when applied to ethanol, these rules predict splitting of the CH_3 protons into a triplet and splitting of the CH_2 protons into a quartet, as seen in high resolution experimental spectra (see Figure 20.23b). Through **chemical shifts** and **spin–spin coupling,** NMR spectroscopy offers a way to identify the bonding groups in a molecule and interpret molecular structure.

NMR spectroscopy is one of the most important techniques available for investigating structure and dynamics in many areas of chemistry, biology, and medicine. Synthetic chemists use NMR, along with mass spectrometry, to verify the identities of new products as well as those of intermediates formed in complex multistep reactions. Biologists use NMR to study the three-dimensional structures and reactions of proteins, enzymes, and nucleic acids (DNA and RNA). The method is complementary to X-ray diffraction (see Section 21.1); it does not require crystals to be grown, the molecules are studied in their native

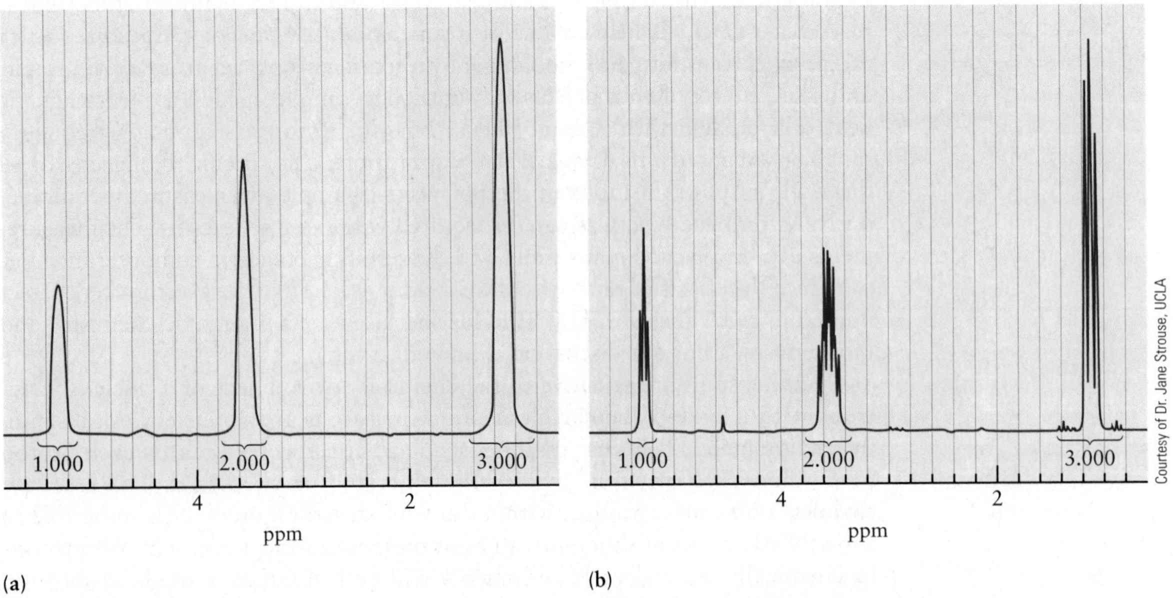

FIGURE 20.23 Proton NMR spectrum for ethanol. (a) The low-resolution spectrum shows a single broad peak for each chemically equivalent group of protons. (b) In high resolution, spin–spin splitting separates the peak for each chemically equivalent group of protons into a multiplet. The relative area under each peak, which is proportional to the number of protons responsible for the peak, is given by the number under the brackets below each peak.

environments, and it enables binding and other reactions to be studied that would be very difficult to accomplish using X-ray diffraction. An adaptation of NMR used in medical diagnosis is called magnetic resonance imaging (MRI). Images are obtained with millimeter spatial resolution, and specialized techniques have been developed for different applications that include imaging the major organs for function, identifying tumors, and monitoring blood flow to look for evidence of possible cardiovascular disease, for example.

20.5 Electronic Spectroscopy and Excited State Relaxation Processes

Figure 20.1 shows an excited electronic state of HCl that can support bound vibrational energy levels; the vibrational frequency in the excited state is typically different from that in the ground electronic state, as shown in Figure 20.1b. The total internal energy in a molecule may be represented as the sum of its rotational energy E_{rot}, its vibrational energy E_{vib}, and its electronic energy E_{el}. These are depicted in Figure 20.1b. The energetic separations between electronic states are relatively large (a few eV) so transitions between these states occur as a result of absorption or emission of electromagnetic radiation in the ultraviolet and visible regions of the spectrum. Electronic absorption spectroscopy is also widely used as an analytical method to measure the concentrations of species in solution because the measurements are easily quantified. Laser-induced fluorescence, described in Section 20.2, is a very sensitive analytical technique (detection limit ~1 molecule); it is used for applications ranging from biological imaging to following the dynamics of energy and electron transfer processes on timescales as short as a few femtoseconds (1 fs = 10^{-15} s).

In Section 20.1 we also briefly presented the radiative and nonradiative processes by which molecules in excited electronic states can lose energy. Figure 20.2 showed an electronically excited molecule returning to the ground electronic state by emitting fluorescence or phosphorescence, and through several nonradiative processes that dissipate the energy as heat. Energy and electron transfer to neighboring molecules are important processes that play central roles in a variety of natural phenomena and underlie a number of important technologies; examples of the former include photosynthesis (see Section 20.7) and examples of the latter include artificial photosynthesis (see Section 17.6 and the accompanying *Connection to Energy*). Understanding the factors that govern the rates of radiative and nonradiative processes is essential for the design and operation of lasers (see Section 20.8). Photoexcitation may also lead to photochemical reactions such as *cis–trans* isomerization, the first step in the vision process.

Molecular orbital (MO) theory, introduced in Chapter 6, provides the framework for understanding the electronic states of molecules and the nature of the transitions among them. We restrict our discussion in this section to the spectroscopy and photochemistry of a few representative organic molecules, having introduced you to the spectroscopy of inorganic compounds in Chapter 8. We use a simple model for electronic excitation, describing it in terms of an excitation of an electron in a molecule from a low energy orbital to a higher energy orbital; in reality the electronic configuration of an excited state molecule may be a linear combination of many such excited electronic configurations, and the dominant electronic configuration characterizing the excited state may change as a function of internuclear distance.

When light with a wavelength of 163 nm is absorbed by ethene, an electron is promoted from the π bonding molecular orbital to the higher energy π^* orbital.

These orbitals are depicted in Figure 6.16. Two excited states are possible, the lowest excited singlet state S_1, often labeled $^1\pi\pi^*$ for ethene, and the corresponding triplet state T_1, labeled $^3\pi\pi^*$ for ethene. (See Fig. 20.3 for the singlet and triplet spin states.) E_{el} for the triplet state is typically somewhat smaller than for the singlet state, similar to the $\pi\pi^*$ excited singlet and triplet states of anthracene shown in Figure 20.6. Exciting an electron from a π to a π^* orbital changes the bond order of the carbon-carbon bond to 1 (it was a double bond in the ground electronic state S_0), so the vibrational frequency of the C—C stretching vibration is smaller in the excited state than in the ground state. The equilibrium C—C internuclear distance is also longer in the excited state (typical C—C bond distances for single, double and triple bonds are given in Table 3.5.) Typically the lowest excited electronic state of a molecule is reached when an electron is promoted from the **highest occupied molecular orbital (HOMO)** (for ethene, the bonding π orbital) to the **lowest unoccupied molecular orbital (LUMO)** (for ethene, the anti-bonding π^* orbital). These orbitals are important in spectroscopy, photochemistry, and in many organic chemical reaction mechanisms because changes in the occupancy of these orbitals affect bond order and reactivity. The simple MO diagram of ethene suggests the possibility of other transitions at higher energies, for example a $\sigma \rightarrow \sigma^*$ transition at very high energies, some of which have been observed experimentally.

The molar absorption coefficient for the strongest absorption band in ethene, which appears at $\lambda_{max} = 163$ nm is 1.5×10^4 L mol^{-1} cm^{-1}; this value is characteristic of strong $\pi \rightarrow \pi^*$ transitions. To give you a feeling for what strong absorption means let's calculate what fraction of a beam of 163 nm radiation is absorbed by ethene gas contained in a 1 cm path length cell at 0.001 atm and 300 K. Converting to molarity and using Beer's law, Equation 20.3, we find that $A = 0.61$, and that this dilute, but strongly absorbing, sample transmits only about 25% of the incident radiation.

Let's continue to focus on the π electrons and examine the energy level diagrams and transitions between singlet states in some conjugated polyenes and aromatic hydrocarbons to which you were introduced in Section 6.11. Conjugated polyenes are hydrocarbon chains with alternating single and double bonds. Resonance structures suggest, and simple MO theory confirms, that the π electrons in these systems are delocalized over the entire molecule. Figure 6.53 shows the four π orbitals of *trans*-1,3-butadiene, the simplest conjugated polyene. While the lowest π orbital of 1,3-butadiene, delocalized over all three carbon-carbon bonds, is lower in energy than the π orbital of ethene, the next higher π orbital is the HOMO. The LUMO of 1,3 butadiene is the lower of the two π^* molecular orbitals in Figure 6.53. Thus promoting an electron from the HOMO to the LUMO of 1,3-butadiene requires less energy than it did in ethene, and the absorption spectrum is shifted to correspondingly lower energies. Table 20.5 gives λ_{max}

TABLE 20.5

Absorption of Light by Molecules with Conjugated π Electron Systems

Molecule	Number of C=C Bonds	Wavelength of Maximum Absorption (nm)
C_2H_4	1	162
C_4H_6	2	217
C_6H_8	3	251
C_8H_{10}	4	304

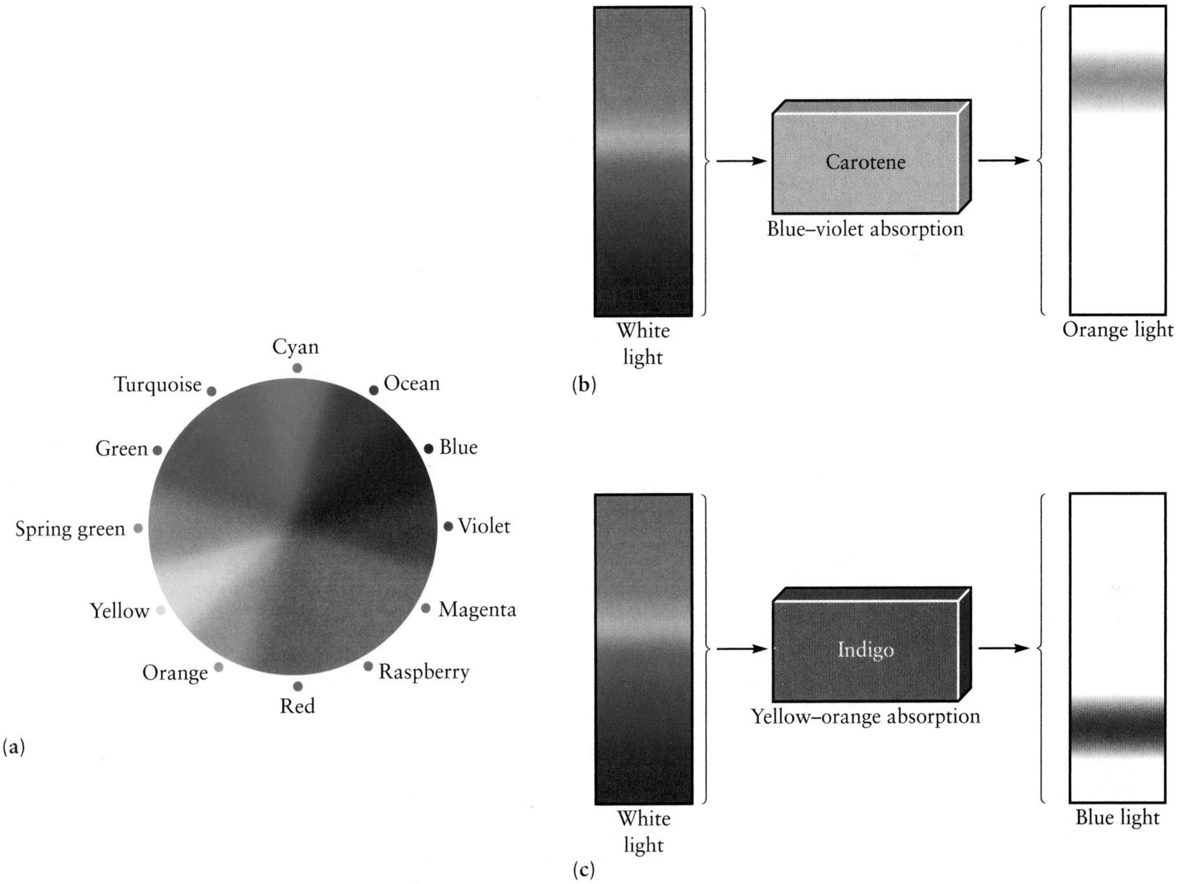

FIGURE 20.24 (a) Color wheel showing the colors of visible light and the corresponding complementary color at 180° from each color. (b) Schematic of the absorption of visible light by a solution of carotene. (c) Schematic of the absorption of visible light by a solution of indigo.

for several unsaturated hydrocarbon molecules with conjugated π electrons. Though all these molecules absorb in the ultraviolet, so appear colorless, the longest wavelength absorption bands of conjugated polyenes with eight or more double bonds appear in the visible region of the spectrum.

When a molecule absorbs light in the visible region of the electromagnetic spectrum (see Figure 4.3), the color we perceive is related to the absorption spectrum of the material, but only indirectly. What we see is the light transmitted through or reflected from the material, not the light absorbed. So, the color perceived is considered to be a **complementary color** to the color most strongly absorbed by the molecule, as shown in Figure 20.24. The color wheel shows the three primary colors, red, green, and blue (RGB), and their complementary colors, cyan, magenta, and yellow (CMY). The RGB system is used in computer displays and in television, and the CMYK system (K is black) is used in color printing, like that in this textbook. Figure 20.24b shows the spectrum of the transmitted light and the perceived color for carotene that absorbs in the blue–violet region of the spectrum, and Figure 20.24c shows the spectrum of the transmitted light and the perceived color (blue) for indigo, which absorbs in the yellow–orange region. Indigo is used to dye blue jeans; beta-carotene is responsible for the orange color of carrots, the colors of some processed foods, and the yellow and orange in certain bird feathers. The absorption spectra of these two compounds are shown in Figure 20.25.

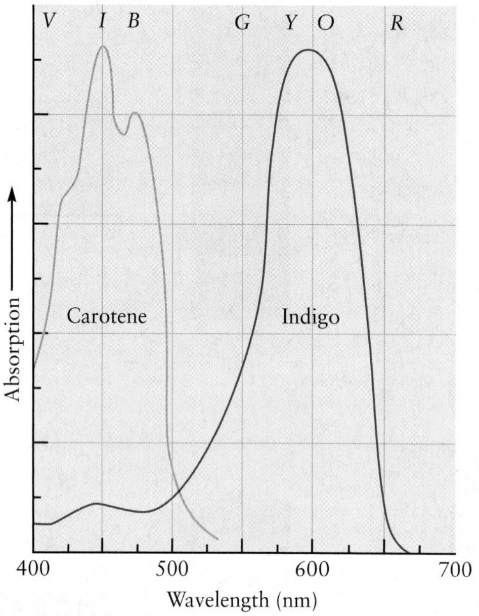

FIGURE 20.25 Absorption spectra for the dyes indigo (dark blue) and carotene (orange). The familiar mnemonic for remembering the colors in the visible region of the electromagnetic spectrum for the absorbed light is written on the top of the spectra.

EXAMPLE 20.11

Suppose you set out to design a new green dye. Over what range of wavelengths would you want your trial compound to absorb light?

Solution

A good green dye must transmit green light and absorb all of the other colors to a significant extent so you should select a dye that absorbs in the violet–blue *and* orange–red regions of the spectrum (see Fig. 20.24).

The naturally occurring substance chlorophyll, which is responsible for the green colors of grass and leaves, absorbs light over just these wavelength ranges, converting solar energy to chemical energy stored as carbohydrates in plants. Chlorophyll is also used commercially as a green dye.

Related Problems: 37, 38

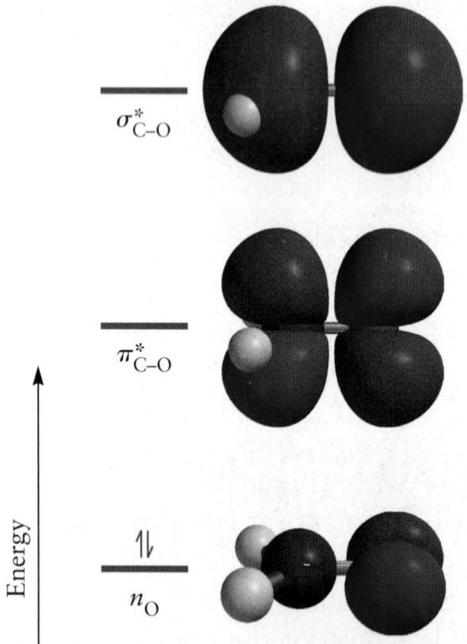

FIGURE 20.26 Approximate molecular orbitals and energy level diagram for formaldehyde. The O 2s, σ_{C-H} orbitals and the σ^*_{C-H} orbitals are omitted, as they are unimportant for the UV spectroscopy of formaldehyde.

We now consider the electronic structure and spectra of formaldehyde, a prototype for molecules in which transitions involving nonbonding electrons are important. The Lewis dot model for formaldehyde shows 12 valence electrons, two C—H bonding pairs in σ orbitals at low energies, the four electrons in the C=O double bond (one pair in a low energy σ orbital and the other pair in the $\pi_{C=O}$ orbital), and two lone pairs on the oxygen atom. One of the oxygen lone pairs resides in the O 2*s* orbital, which is at sufficiently low energy that we do not consider it when forming the MOs that are relevant to the spectroscopy and photochemistry of formaldehyde. The second lone pair on the oxygen atom resides in an O $2p_y$ orbital that is oriented perpendicular to both the carbon-oxygen internuclear axis and the O $2p_z$ orbital that we use to form π orbitals. Some of the orbitals and the associated energy level diagram for formaldehyde are shown in Figure 20.26.

The HOMO of formaldehyde is the nonbonding O $2p_y$ orbital, n_O, and the LUMO is the $\pi^*_{C=O}$ orbital, so the lowest energy transition is now an $n \rightarrow \pi^*$ transition. The energy difference between the n_O and the $\pi^*_{C=O}$ orbitals is not very large, however, so that $n \rightarrow \pi^*$ and $\pi \rightarrow \pi^*$ transitions are both important in the spectroscopy and photochemistry of formaldehyde. These transitions lead to $(\sigma)^2(\pi)^2(n)^1(\pi^*)^1$ and $(\sigma)^2(\pi)^1(n)^2(\pi^*)^1$ excited state configurations, respectively. The excited state configuration that results from an $n \rightarrow \pi^*$ transition is labeled simply the $n\pi^*$ configuration and that resulting from a $\pi \rightarrow \pi^*$ transition is labeled the $\pi\pi^*$ configuration.

Photoelectron spectroscopy, discussed for atoms in Chapter 5 and for diatomic molecules in Chapter 6, provides important clues about the nature of these molecular orbitals; the photoelectron spectrum of formaldehyde is shown in Figure 20.27a. The three sets of peaks observed arise from photoionization in which the electron originated in three distinct molecular orbitals. The **vibrational fine structure** observed as a series of closely spaced lines in each band allows us to identify the orbitals involved and to confirm the validity of our energy level diagram. Energy in excess of the minimum necessary to ionize a molecule can excite vibrations in the positive ion left behind, and the nature of those excitations can be used to identify the orbital from which the electron was removed. Removing an electron from a bonding MO weakens the corresponding bond and lowers that bond's vibrational frequency in the ion from that in the parent molecule. The equilibrium bond length in the ion increases from that in the neutral, so the ion is formed in vibrational eigenstates with many quanta in that bond's stretching motion. (This is a result of what is called the **Franck-Condon principle**—the amplitude of the ion's vibrational wave function at the inner turning point of its C—O vibrational motion has internuclear distances similar to that of the neutral molecule's equilibrium geometry.) Removing an electron from an antibonding orbital strengthens the bond (increases its bond order), so the resulting ion has a larger vibrational frequency in the bond on which the antibonding orbital was localized. Photoelectron spectra from removing an electron from a nonbonding orbital does not affect bonding very much; the vibrational frequencies of the ion are very close

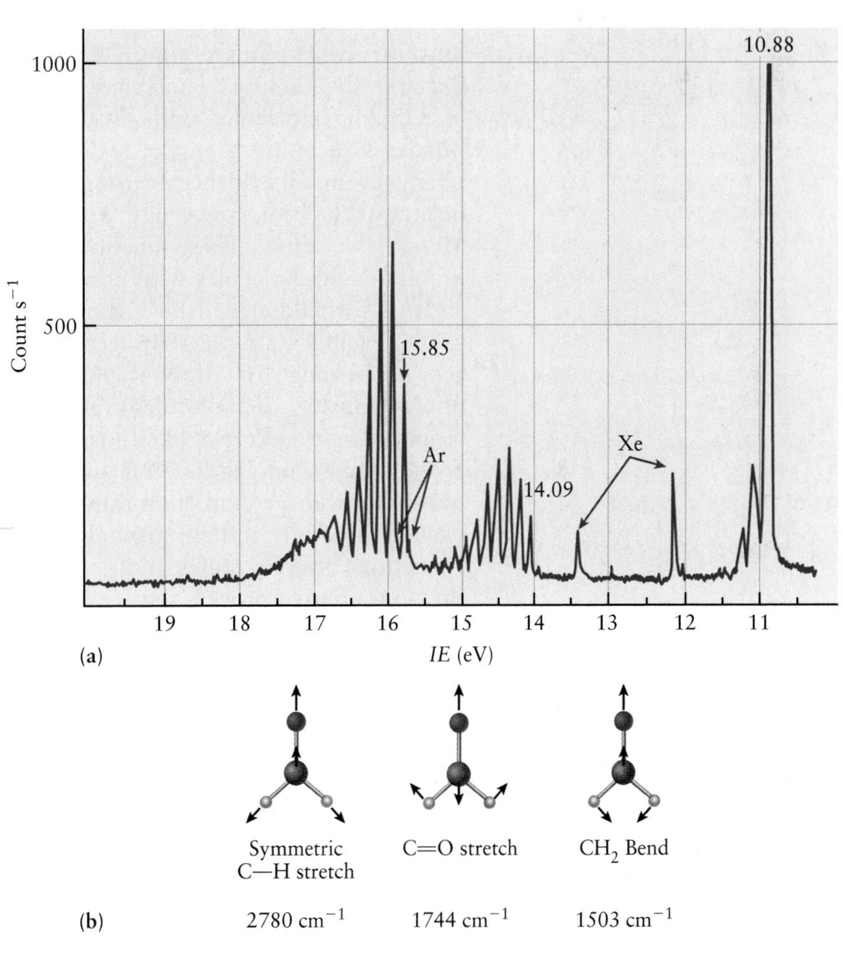

FIGURE 20.27 (a) Photoelectron spectrum of formaldehyde. The lines labeled Ar and Xe are due to photoemission from residual rare gases in the sample chamber. (b) Formaldehyde vibrational modes and frequencies in the ground electronic state. (Adapted from D. W. Turner, C. Baker, A. D. Baker, and C. R. Brundle, *Molecular Photoelectron Spectroscopy*, John Wiley & Sons, London, 1970, p. 141.)

to those in the parent molecule and the equilibrium bond lengths in the neutral molecule and ion are similar, so the strongest transition does not involve vibrational excitation of the ion. These general characteristics are summarized in the *Connection to Instrumental Analysis* in Chapter 6. Let's use this approach to interpret the photoelectron spectrum of formaldehyde and identify the orbitals associated with each set of bands. The peak that appears at 10.88 eV arises from electrons that have the lowest binding energies and it is assigned, therefore, to the HOMO. There is very little vibrational fine structure, with most of the intensity appearing in the 10.88 eV peak. Two small peaks appear next to the 10.88 eV peak; they correspond to forming ions with a quantum in vibration in one of its vibrational modes, and the shifts in cm^{-1} from the 10.8 eV peak give the vibrational frequencies of the formaldehyde cation. The ion's vibrational frequencies observed agree closely with the C—H stretching, C=O stretching, and H—C—H bending vibrations of formaldehyde in its ground electronic state shown in Figure 20.27b. This allows us to confirm our assignment of the HOMO as the nonbonding orbital. Let's now examine the set of peaks between 14 and 15 eV. The frequency observed in the vibrational progression corresponds to multiple excitations of a single mode with a frequency of about 1200 cm^{-1} and deuterium isotope substitution has no effect on the frequency observed. We thus assign this mode to the C=O stretch in the ion and identify the orbital from which the electron was removed as the bonding π orbital on the basis of the decrease in vibrational frequency (the ground-state frequency is 1744 cm^{-1}). The lower vibrational frequency is consistent with a reduction in the CO bond order that results from the removal of an electron from a bonding orbital localized on the CO bond. The vibrational frequency measured in the progression around 16 eV is lower than the frequency of the corresponding vibration in the ground state of formaldehyde. It is further reduced on deuterium substitution by almost exactly $1/\sqrt{2}$, so the vibration must involve H atom motion. These results suggest that the photoelectron produced in the 15.85-eV band came from one of the bonding CH σ orbitals.

Organic molecules with C=C double bonds, conjugated C=C double bonds, a carbonyl group (C=O), and aromatic functional groups like the phenyl group all evidence a strong $\pi \rightarrow \pi^*$ transitions (molar absorption coefficients at the peak wavelength λ_{max} for the most intense peaks lie in the range 10^3–10^5 M^{-1} cm^{-1}. These functional groups are thus often referred to as chromophores. In molecules with nonbonding electron pairs on O, N or S atoms, such as formaldehyde, there is also an $n \rightarrow \pi^*$ transition to a lower lying singlet excited state; these are generally weaker than $\pi \rightarrow \pi^*$ transitions with ε_{max} in the range 10–10^3 M^{-1} cm^{-1}. They are central to the spectroscopy and photochemistry of molecules containing carbonyl and heteroaromatic chromophores, for example, like those found in amino acids and nucleic acids. In Section 20.1 and Figure 20.2 we presented the radiative (fluorescence and phosphorescence) and non-radiative **photophysical processes** that return excited molecules to their ground states. The nonradiative process that causes transitions between states of the same spin is called **internal conversion** whereas the nonradiative process that causes transitions between different spin states is called **intersystem crossing**. Molecules formed in excited vibrational levels after internal conversion or intersystem crossing can lose vibrational energy through collisions with solvent molecules in condensed phases, a process generally referred to as vibrational relaxation. This returns the molecules to a thermally equilibrated distribution, favoring the lower vibrational levels of that state.

When a molecule absorbs a high energy photon, the energy can also result in fission of a bond in the molecule. This is a **photochemical process** called **photodissociation**. Imagine exciting H_2^+ to the repulsive excited electronic state

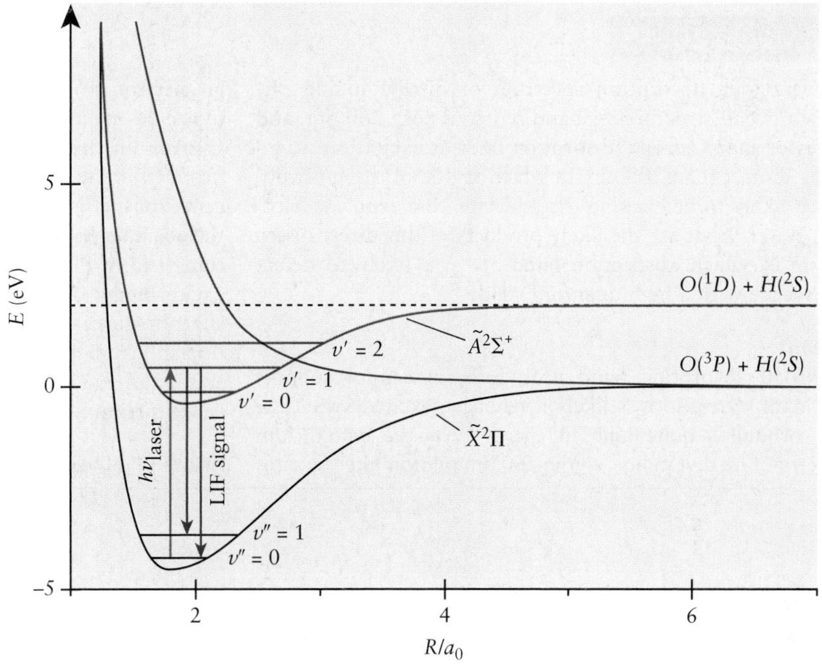

FIGURE 20.28 Three electronic states of OH: the ground electronic state ($\tilde{X}$), the lowest excited electronic state ($\tilde{A}$), and a repulsive excited state that results in the OH molecules in the $\tilde{A}$ state predissociating to H + O.

labeled $1\sigma_u^*$ in Figure 6.6. The force on the excited state pushes the two atoms away from each other, and the molecule dissociates to H + H^+. The excited state does not have a bound region of the potential, so it cannot support any bound vibrational states. Molecules may have both bound excited electronic states and repulsive ones. If you promote HCl to the excited state depicted in Figure 20.1, the absorption spectrum in this region has sharp features corresponding to exciting to different vibrational levels in the excited electronic state, but if you excite HCl at photon energies near 65,500 cm^{-1} the absorption spectrum is broad and structureless. The excited state reached at these energies is directly repulsive (you have promoted an electron to the antibonding σ^* orbital of the H—Cl bond) and the molecule dissociates to H + Cl on this repulsive excited electronic state in less than a few fsec (it does not have time to undergo internal conversion to the ground electronic state). The excited state of OH reached at 32,477 cm^{-1} (called the Ã state) shown in Figure 20.28 is formed from a bound excited state coupled to a repulsive excited state. The OH molecules in the higher rotational levels of $v' = 1$ and all of the rotational states in $v' = 2$ in the Ã state can couple to this repulsive excited state and undergo a process called predissociation. The typically sharp absorption features appear broadened for vibrational levels in the Ã state that are above the energy where the bound excited state crosses the repulsive excited state. Finally, even if the excited electronic state is a bound one, evidencing sharp peaks in the absorption spectrum spaced by the $h\nu_{vib}$ for one of the excited state vibrational modes, the molecule can undergo internal conversion to high vibrational levels of the ground electronic state with enough energy to break a bond. A competition between vibrational relaxation and dissociation of the highly vibrationally excited molecules after internal conversion to S_0 ensues if the vibrational energy is sufficient to break one of the bonds in the molecule. For example, when $^1n\pi^*$ excited formaldehyde undergoes internal conversion to S_0, both dissociation to H_2 + CO and to H + HCO compete with vibrational relaxation. Spectroscopists describe this as predissociation via internal conversion to the high vibrational levels in the ground electronic state.

EXAMPLE 20.12

The ultraviolet absorption spectrum of methyl iodide evidences a broad structureless band centered near 260 nm, and a series of sharp, strong absorption lines at excitation wavelengths shorter than 200 nm. In which region of the spectrum are you likely to be exciting the molecule to a repulsive electronic state? What are the likely products of this direct dissociation? In which absorption band are you likely to detect fluorescence with a high quantum yield?

Solution

The broad absorption band near 260 nm cannot support vibrational states, so it is likely a repulsive excited state. The σ^*_{C-H} orbital is quite high in energy near the equilibrium geometry of methyl iodide, so the 260 nm photon likely excites an electron to a σ^* orbital localized on the C—I bond. The molecules would feel a repulsive force between the I and the C atom on this excited state, so the likely photodissociation products are methyl (CH_3) radicals and iodine atoms. Fluorescence must compete with this very fast dissociation process, so the **quantum yield** of I atoms would be close to 1 and the quantum yield of fluorescence would be very small. If you instead excite methyl iodide with 200 nm photons into one of the sharp features in the higher energy region of the UV absorption spectrum, this excited state can support bound vibrational states and you should be able to observe a substantial fluorescence quantum yield.

Related Problems: 43, 44

20.6 Introduction to Atmospheric Chemistry

A great deal of attention has been focused on three areas of atmospheric science in the last forty years or so: increased air pollution in the troposphere, the depletion of ozone in the stratosphere, and climate change. Oxides of nitrogen (collectively called NO_x) and oxides of sulfur (collectively called SO_x) come from the combustion of fuels, the former mostly from vehicles and the latter mostly from power plants. They lead to the production of ground-level ozone as well as to acid rain, both of which are serious environmental pollutants. Stratospheric ozone depletion is considered to be a contributing factor to the significant increases in the incidence of skin cancer observed over the past 30 years, based upon both epidemiology and animal model studies. Ozone depletion has been traced to a class of refrigerants and propellants called chlorofluorocarbons (CFCs), which have been banned by international agreement. Climate change has generally been attributed to increases in the concentrations of CO_2, CH_4, N_2O, and other atmospheric "greenhouse gases" over the past 150 years; these increases in concentration have been attributed to human activities.

Evidence for stratospheric ozone depletion began to appear globally in the early 1970s, and the existence of a dramatic "ozone hole" over Antarctica was confirmed by the mid 1980s. A mechanism for the catalytic destruction of ozone by chlorine radicals produced by photolysis of CFCs was first proposed in the early 1970s, and by 1987 the evidence was so convincing that the international community banned the production of these chemicals worldwide. In this section we discuss the chemistry and photochemistry central to stratospheric ozone depletion because the chemical mechanisms are now quite well established and those scientific advances led to an international solution to the problem.

20.6.1 The Atmosphere

Although the chemical composition of the atmosphere summarized in Table 9.1 describes the average makeup of the portion of the atmosphere closest to the Earth's surface, it masks the variation in chemical composition with height, the dramatic role of local fluctuations in the concentrations of trace gases, and the dynamics underlying the observed average concentrations. The atmosphere is a complex chemical system that is far from equilibrium. Its properties are determined by an intricate combination of thermodynamic and kinetic factors. It is a multilayered structure, bathed in radiation from the sun, which interacts with the oceans and landmasses at the Earth's surface. We generally identify four layers of the atmosphere, on the basis of characteristic temperatures, temperature profiles, and densities (Fig. 20.29). In the outer two layers (the **thermosphere** and the **mesosphere**, above 50 km), the density is low and the gases have been extensively ionized by intense, high-energy solar radiation. The third layer, the **stratosphere**, occupies the region from 12 km to 50 km (approximately) above the Earth's surface. The **troposphere** is the lowest region, reaching up to about 12 km from the Earth's surface. In the troposphere, warmer air lies beneath cooler air. This is a dynamically unstable situation because warmer air is less dense than cooler air and tends to rise so convection takes place, mixing the gases in the troposphere and determining the weather. In the stratosphere the temperature increases with height and there is little vertical mixing from convection. Mixing across the borders between the layers is also slow, so most chemical processes in each layer can be described separately.

The study of atmospheric chemistry dates back to the 18th century. Cavendish, Priestley, Lavoisier, and Ramsay were the first scientists to study the composition of the atmosphere. Recent advances in atmospheric science have come from the further development and integration of a number of scientific disciplines. The sensitivity of chemical analysis has greatly improved; concentrations of trace gases in the atmosphere are routinely measured at the part-per-trillion (ppt) level. We are now able to generate high-spatial-resolution maps of the distributions of chemical species across the globe daily using a combination of land-based, airborne, and satellite sensors. NASA's Aura satellite, for example, has four spectrometers that measure microwave, infrared, visible, and ultraviolet spectra of some 30 atmospheric constituents each day to map out the spatial distributions of these species in three dimensions. These maps provide important data that atmospheric chemists and meteorologists use to test and refine their climate models. Laboratory studies of the relevant reactions provide rate constants, activation energies, and mechanisms that must be included in these models. The role of heterogeneous chemical reactions in the atmosphere began

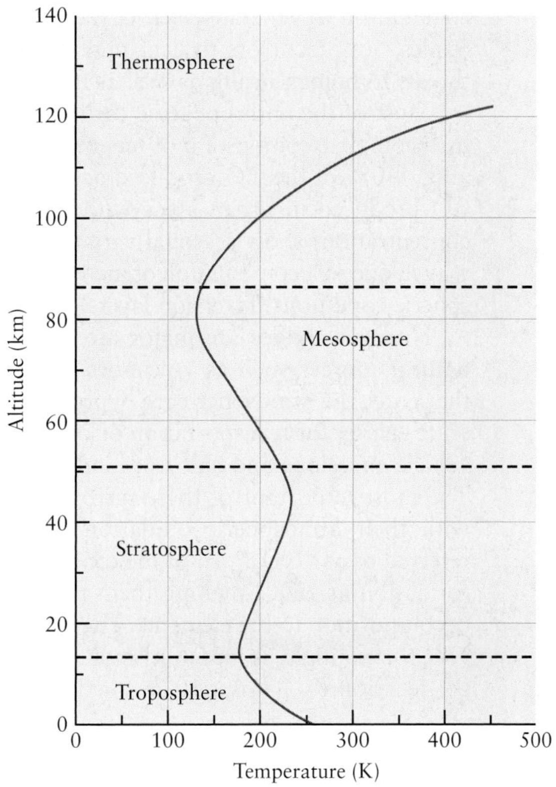

FIGURE 20.29 Variation of temperature with altitude in the atmosphere, showing its layered structure.

to be appreciated about 15 years ago, and we have learned a great deal about the reactions of molecules like $ClONO_2$ that occur on, and can be catalyzed by, particulates like ice crystals and sulfate aerosols, for example; these reactions are part of a series of reactions that lead to the formation of the ozone hole. Finally, large-scale computer simulations are now able to tackle the very difficult problem of analyzing systems of coupled chemical reactions including transport. Dynamic simulations of the reactions within and diffusion between the layers of the atmosphere have contributed a great deal to our understanding of atmospheric chemical processes.

20.6.2 Tropospheric Chemistry

The troposphere is the part of the atmosphere in contact with the Earth's surface. It is therefore most directly and immediately influenced by human activities, especially by the gases or small particles emitted into the air by motor vehicles, power plants, and factories. Some pollutants have long lifetimes and are spread fairly evenly over the Earth's surface; others are localized around cities or in regions where they are produced and where local weather patterns keep them from dispersing. Six pollutants are of the greatest concern: carbon monoxide, nitrogen oxides (NO_x), ozone, sulfur oxides (SO_x), particulate matter, and lead. We do not discuss particulate matter or lead, except to point out their effects on health, because the chemistry involved is unrelated to the chemistry of the other four pollutants. We begin by discussing the atmospheric chemistry and photochemistry of carbon monoxide, the nitrogen oxides, and ozone, because they arise mostly from vehicle emissions and because their reactions are coupled. Carbon monoxide, a product of incomplete combustion of fuels, is a poison at sufficiently high concentrations. Nitrogen oxides, generated at high temperatures in engines, lead to the production of both ground-level ozone and photochemical smog. Ozone is considered a secondary pollutant because it is generated by subsequent reactions of the primary pollutant, the nitrogen oxides. Finally, we discuss the atmospheric chemistry of sulfur oxides, which are produced primarily by the combustion of coal and oil to generate electricity. Sulfur oxides, and to a lesser extent nitrogen oxides, are precursors to acid rain. All of these criteria pollutants pose serious threats to human health as well as to the health of plants and animals.

Most of the anthropogenic carbon monoxide in the atmosphere is produced by incomplete combustion of fuel in motor vehicles; this source is responsible for about 60% of the CO produced across the United States, but up to 95% of the CO produced in urban areas with high densities of motor vehicle traffic. CO concentrations show seasonal variations with the highest levels occurring in the winter, due to a combination of increased combustion for heating as well as atmospheric conditions (inversion layers) that tend to trap CO and other pollutants.

Nitrogen oxides are major air pollutants. Although they are produced by natural sources such as volcanoes and lightning, most of the nitrogen oxides that enter the atmosphere are byproducts of combustion reactions used to generate energy for transportation or other purposes. About half come from motor vehicle emissions and half from industrial, commercial, and residential sources. The most important of these nitrogen oxides are NO and NO_2, which, along with their atmospheric oxidation products such as HNO_3, are collectively referred to as "NO_x." All of the oxides of nitrogen have positive free energies of formation at 25°C, making them thermodynamically unstable with respect to decomposition to the elements. The standard free energy of formation (ΔG_f°) for NO, for example, is +86.6 kJ mol^{-1}, which implies an equilibrium constant for the reaction $K_{298} = 6 \times 10^{-16}$ for the reaction

$$\tfrac{1}{2}\,N_2(g) + \tfrac{1}{2}\,O_2(g) \longrightarrow NO(g),$$

giving equilibrium concentrations of NO of approximately 10^{-10} ppm at room temperature. Nitrogen oxides form to a greater extent when air is heated to higher temperatures in industrial processes or in automobile engines; the equilibrium concentration of NO rises to about 100 ppm at 800°C, the operating temperature of an internal combustion engine, and to about 1000 ppm at 1500°C, the operating temperature of a gas turbine. Nitrogen oxides accumulate to concentrations much higher than their equilibrium concentrations at ambient temperatures because they decompose slowly. The long residence times of these oxides in the atmosphere demonstrate the greater importance of kinetics, relative to thermodynamics, in atmospheric chemistry.

NO_2 is the key player in the tropospheric chemistry of the nitrogen oxides. NO_2 is the reddish-brown gas that obscures visibility on smoggy days in urban areas such as Los Angeles and Houston. The photolysis of NO_2 is responsible for the generation of ground-level ozone, as well as for initiating a series of reactions that lead to photochemical smog, beginning with the following reactions.

$$NO_2 + h\nu \longrightarrow NO + O$$

$$O_2 + O + M \longrightarrow O_3 + M$$

where M is a third body that carries away some of the excess vibrational energy of the products (see Section 18.4). Ground-level ozone concentrations are controlled to some extent by the following reaction in which NO acts as a scavenger.

$$O_3 + NO \longrightarrow O_2 + NO_2$$

Ozone concentrations in metropolitan areas begin to increase in the mornings, as commuters head off to work, and then decrease during the evenings and overnight. The NO scavenging reaction reduces ozone concentrations, and the NO_2 that is produced is not photolyzed during those hours. The NO_2 and O_3 concentrations predicted by the preceding mechanism are much lower than observed, so other mechanisms for producing these species must contribute. The most important of these involve the hydroxyl radical, which is produced in the atmosphere by photolysis of ozone and subsequent reaction of the photogenerated oxygen atom with water as follows. (The single dot represents a free radical; we add this indicator below only to OH and organic radicals, not other radicals such as NO and NO_2)

$$O_3 + h\nu \longrightarrow O_2 + O$$

$$O + H_2O \longrightarrow 2\ \bullet OH$$

The hydroxyl radical is an extremely powerful oxidizing agent not only because it has an unfilled valence shell but also because of the extraordinary thermodynamic stability of the products created: CO_2 from CO oxidation and H_2O from H atom abstraction reactions. The hydroxyl radical is sometimes called "the detergent of the atmosphere" because it cleans up harmful pollutants like CO and hydrocarbons. It also prevents hydrochlorofluorocarbons from leaving the troposphere and entering the stratosphere, where they have the potential to destroy ozone. Unfortunately, however, the hydroxyl radical also plays an important role in the generation of photochemical smog and in the reactions that lead to acid rain through the following series of reactions.

$$CO + \bullet OOH \longrightarrow CO_2 + \bullet OH$$

$$RH + \bullet OH \longrightarrow \bullet R + H_2O$$

$$\bullet R + O_2 + M \longrightarrow \bullet ROO + M$$

where RH is a hydrocarbon and •R is the radical produced by hydrogen abstraction. The hydroperoxyl radical •OOH and any number of peroxyl radicals •ROO formed from various hydrocarbons oxidize NO to NO_2 as follows

$$NO + \bullet OOH \longrightarrow NO_2 + \bullet OH$$

$$NO + \bullet ROO \longrightarrow NO_2 + \bullet RO$$

thus providing the source of the "extra" NO_2 to account for the levels of ozone produced from its photolysis and subsequent reactions.

The last set of reactions we consider are those between peroxyl radicals (•ROO), peroxyacyl radicals (•RC(O)OO), and NO_2 to form peroxyalkylnitrates and peroxyacylnitrates as follows

$$\bullet ROO + NO_2 \longrightarrow ROONO_2$$

$$\bullet RC(O)OO + NO_2 \longrightarrow RC(O)OONO_2$$

The latter class of compounds, the peroxyacylnitrates or **PANs**, are among the most abundant nitrogen-containing compounds in the troposphere, and they are distinct signatures of photochemical smog in polluted urban areas. The oxides of sulfur create pollution problems on regional and even global scales because they have longer lifetimes in the atmosphere than the nitrogen oxides. Some of the SO_2 and SO_3 in the air comes from biological processes and volcanoes, but about 70% of SO_2 emissions have anthropogenic origins, the most important of which is the combustion of fossil fuels (coal in particular) to generate electricity in power plants. If the sulfur is not removed from the fuel or the exhaust gas, SO_2 enters the atmosphere as a stable but reactive pollutant. Direct oxidation of SO_2 by atmospheric oxygen is allowed thermodynamically, but the kinetics are so slow in the absence of a catalyst that the reaction does not contribute significantly to tropospheric SO_3 concentrations. Further oxidation of SO_2 by atmospheric radicals leads to sulfur trioxide by the following series of reactions, and the reaction of the atomic oxygen produced by the decomposition of the hydroperoxy (•OOH) radical can react with molecular oxygen in the third step to provide another source of ground-level ozone.

$$SO_2 + \bullet OH \longrightarrow \bullet SO_2OH$$

$$SO_2OH + \bullet O_2 \longrightarrow SO_3 + \bullet OOH$$

$$\bullet OOH \longrightarrow \bullet O + \bullet OH$$

NO_2 and SO_3 are both acid anhydrides that react with hydroxyl radicals and water vapor in the air to form nitric (HNO_3) and sulfuric (H_2SO_4) acids, respectively, which may deposit on the Earth in both wet and dry forms.

$$NO_2 + \bullet OH \longrightarrow HNO_3$$

$$SO_3 + H_2O \longrightarrow H_2SO_4$$

Acid deposition is the most general term for this set of processes; acid precipitation refers to the wet processes that include rain, snow, sleet, or hail, with **acid rain** being the process most familiar to the public. Even "pure" rainwater is slightly acidic due to the reaction of atmospheric CO_2 with H_2O to form carbonic acid, H_2CO_3, resulting in a pH in the range 5.3–5.6 (the presence of other acidic species contributes to this variability). Acid rain, on the other hand, is characterized by pH values in the pH = 4–5 range, which is deleterious to both health and the environment. The pH of acid rain varies considerably across the United States, ranging from 5.7 in the West to less than 4.7 in parts of the Northeast in 2012. The acid rain problem in the Northeast is due to the large number of coal-fired power plants, steel mills, and other industries that rely on coal and emit SO_2 in neighboring states such as Ohio and Pennsylvania. In the years from 2002 to 2012 the pH of acid rain in the Northeast improved substantially (in 2002, it was more acidic, near 4.3). The environmental consequences are also more serious, particularly for aquatic species, because lakes in the Northeast tend to have granite beds, which do not provide any buffering capacity like lakes in which limestone ($CaCO_3$) is the lakebed.

The oxides of nitrogen and sulfur are respiratory irritants that are especially problematic for those in "sensitive groups" that include children, the elderly, and individuals with respiratory problems. Health effects can be serious and even catastrophic. More than 4,000 people died in London during a five-day period in 1952. Cold weather necessitated burning large quantities of sulfur-rich coal, the emissions from which reacted with the typical London fog to form the deadly acid aerosol. As recently as 1982, the pH of the air in the Los Angeles basin fell to 1.5, in some areas leaving residents breathing nearly 0.1 M nitric acid. The Clean Air Act has substantially improved this situation.

20.6.3 Stratospheric Chemistry

Light emitted from the sun resembles that of a 5777 K black body (see Section 4.2). Although the intensity of the radiation drops rapidly toward shorter wavelengths in the UV region of the spectrum, a considerable number of very high-energy UV photons strikes the Earth's atmosphere every second. The energy of a 200 nm photon (at the border between the ultraviolet and vacuum ultraviolet) is more than 6 eV, or about 600 kJ mol^{-1}, which is sufficient to break all but the strongest chemical bonds. Radiation at shorter wavelengths causes ionization, which is even more damaging to molecules and biological tissues. Fortunately, there is a substantial layer of ozone in the stratosphere (see Fig. 20.30) that absorbs most of the harmful ultraviolet radiation not absorbed by O_2, affording us substantial protection. In the mid-1970s, evidence began to accumulate that suggested that the concentration of stratospheric ozone was decreasing dramatically and that CFCs transported from the troposphere to the stratosphere were responsible. We describe the natural cycle for the production and destruction of O_3 in the stratosphere, the mechanism by which it absorbs UV radiation, the role of Cl radicals in ozone depletion, the factors that led to the dramatic seasonal variation in the ozone concentrations over the Antarctic (and more recently, the Arctic) and, finally, the corrective steps that have been taken and the results observed to date.

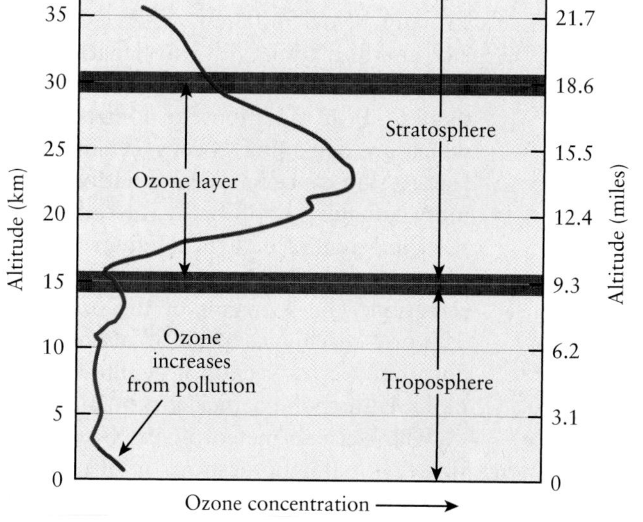

FIGURE 20.30 Variation in ozone concentrations with altitude in the atmosphere showing the stratospheric ozone layer and accumulation in the troposphere. (Adapted from Lucy Pryde Eubanks, Catherine H. Middlecamp, Carl E. Heltzel, and Steven W. Keller, *Chemistry in Context*, 6th ed., McGraw-Hill, New York, 2009, p. 58.)

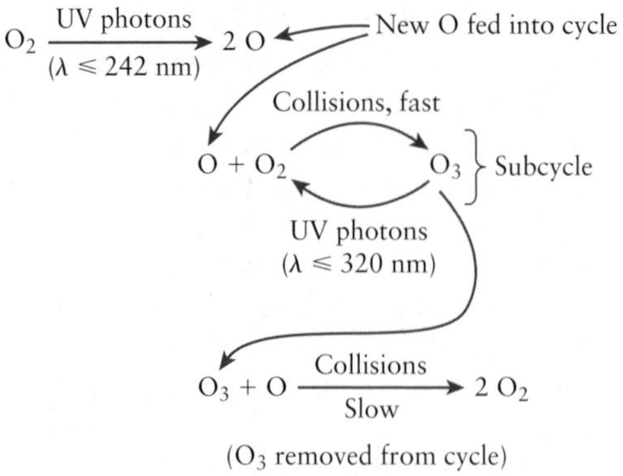

FIGURE 20.31 The Chapman cycle showing the key reactions for ozone formation and destruction in the stratosphere. (Adapted from Lucy Pryde Eubanks, Catherine H. Middlecamp, Carl E. Heltzel, and Steven W. Keller, *Chemistry in Context*, 6th ed., McGraw-Hill, New York, 2009, p. 74.)

The steady-state concentration of O_3 in the stratosphere is reasonably well accounted for by the Chapman cycle (Fig. 20.31), named after the English physicist, Sydney Chapman. Chapman proposed the following series of coupled photochemical and thermal reactions: Photolysis of O_2 by UV radiation ($\lambda < 242$ nm) produces oxygen atoms, which then react with O_2 in the second step of the cycle to produce O_3. The second reaction may be reversed photochemically by the absorption of longer wavelength ($\lambda < 340$ nm) UV radiation, or the ozone produced may react thermally with oxygen atoms to regenerate O_2. The *steady state* concentration of O_3 depends upon: (1) the O_2 concentration at the altitude of interest and (2) the flux and wavelength distribution of UV radiation at that altitude. The average ozone concentration in the atmosphere is plotted as a function of altitude in Figure 20.30; note the accumulation of ozone over a rather broad region of the stratosphere. The Chapman cycle actually overestimates stratospheric ozone concentrations because it does not consider other naturally occurring pathways for ozone depletion, the most important being reactions with •H and •OH radicals formed by the photolysis of stratospheric water vapor, but also reactions catalyzed by NO and NO_2, natural sources of which include oxidation of N_2 in forest fires, by lightning or volcanic eruptions.

UV radiation has been classified into three types based upon its absorption by a series of chosen glasses, rather like the three kinds of ionizing radiation were classified by their ability to penetrate three different kinds of matter. Ozone does not absorb appreciably at wavelengths longer than 340 nm, so the solar flux at the surface of the Earth over the region 400 − 340 nm is about the same as at the "top" of the atmosphere. Radiation from this region is called UV-A, and it is the least damaging to biological tissues. O_3 begins to absorb strongly below 340 nm, whereas O_2 begins to absorb strongly only below 242 nm. Radiation with wavelengths between 340 and 280 nm (where proteins begin to absorb) is called UV-B radiation, and it is more damaging to biological tissues. UV-B radiation can ionize molecules and form reactive free radicals, which are damaging to cells. It can also cause specific mutations in DNA that lead to skin cancers. UV-C radiation, with wavelengths less than 280 nm, is the most damaging of all but it is effectively absorbed by atmospheric O_2.

The extent of biological damage caused by UV radiation is a strong function of wavelength; 280-nm radiation is a million times more damaging than 340-nm radiation. The strongest of the ozone UV absorption bands begins at about 350 nm, reaching a peak at about 280 nm, as shown in Figure 20.32. The absorption cross section is significant, $\sigma \approx 10^{-17}$ cm^2, which corresponds to a molar extinction coefficient $\varepsilon \approx 10^3$.

The German meteorologist Paul Crutzen at Mainz was the first to demonstrate the catalytic destruction of ozone by a free radical, NO, which was produced by natural processes. This discovery led Harold Johnston at Berkeley to propose that the NO produced by supersonic transports could pose a serious

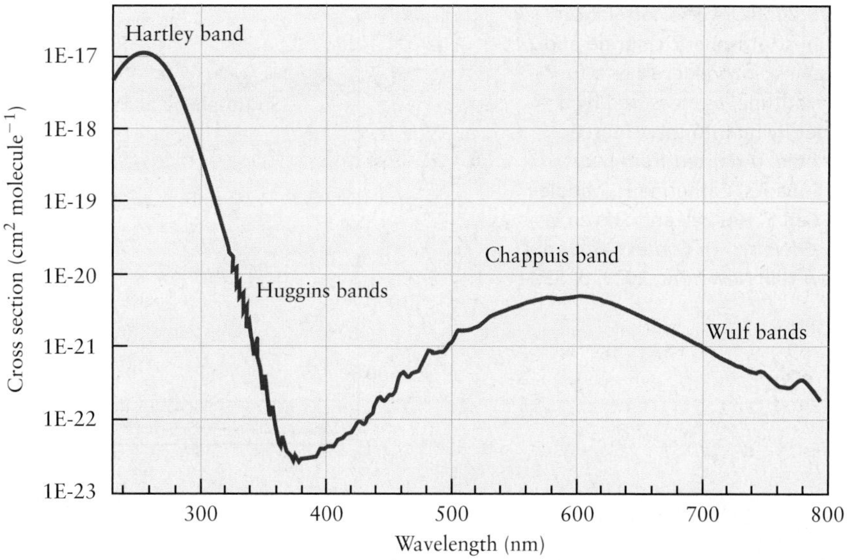

FIGURE 20.32 Ozone absorption in the visible and ultraviolet. (Adapted from J. Orphal, A critical review of the absorption cross-sections of O_3 and NO_2 in the ultraviolet and visible, *Journal of Photochemistry and Photobiology A: Chemistry* 157 (2003), 185–209.)

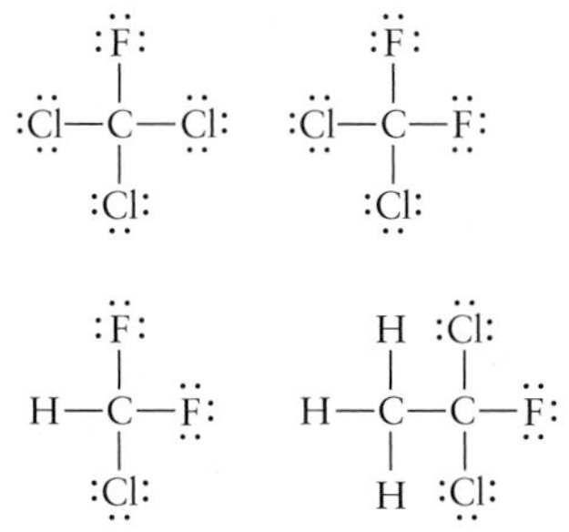

FIGURE 20.33 Lewis dot structures of four chlorofluorocarbons (CFCs).

threat to the ozone layer, and scientists began to consider other potential ozone-depleting molecules. The American chemists Mario Molina and Sherwood Rowland, both then at the University of California at Irvine (Rowland went to high school, college, and graduate school at The University of Chicago), proposed that the photolysis of CFCs (see Figure 20.33) could produce reactive chlorine-containing free radicals that could also catalyze ozone destruction, by the mechanism proposed below. The Freons CFC-11 and CFC-12 shown are two of the most important CFCs developed commercially in the 1930s for use as refrigerants and propellants.

Photolysis of CFCs with UV light with wavelengths shorter than 220 nm produces chlorine atoms and radical fragments as shown below, for example

$$CFCl_3 + h\nu \longrightarrow \bullet CFCl_2 + \bullet Cl$$

$$CF_2Cl_2 + h\nu \longrightarrow \bullet CF_2Cl + \bullet Cl$$

The chlorine atom is a very reactive free radical that is thought to destroy ozone by the catalytic mechanism shown below. The net reaction is the conversion of two O_3 molecules into three O_2 molecules.

$$2\,[Cl\bullet + O_3 \longrightarrow ClO\bullet + O_2]$$

$$2\,ClO\bullet \longrightarrow ClOOCl$$

$$ClOOCl + h\nu \longrightarrow ClOO\bullet + Cl\bullet$$

$$ClOO\bullet \longrightarrow Cl\bullet + O_2$$

$$2\,O_3 \longrightarrow 3\,O_2$$

How were these reactions, which were studied in the laboratory, linked to the stratospheric ozone problem, and how did scientists rule out naturally occurring sources of Cl radicals as the active catalyst? Several lines of evidence supported the conclusion that the photolysis of CFCs was the source of the reactive Cl. First, an inverse correlation between ClO• and ozone concentrations was established in an experiment in which a NASA jet measured the concentrations of these gases as it flew from Chile to Antarctica, entering the ozone hole as shown in Figure 20.34. (The seasonal depletion of the ozone layer over Antarctica is particularly dramatic, so this region was chosen for this experiment.) Although correlation does not prove causality, the level of detail shown in the mirror images was considered extremely good evidence that CFCs were

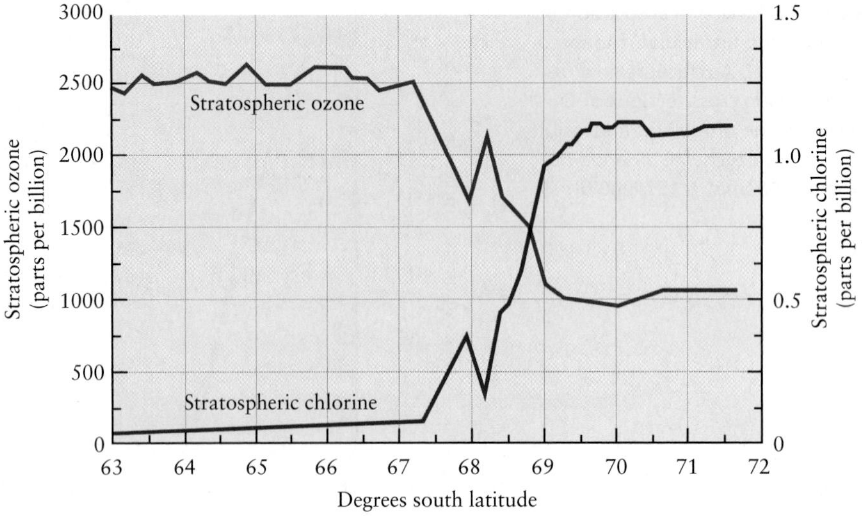

FIGURE 20.34 Inverse correlation between stratospheric chlorine and stratospheric ozone levels as a function of latitude, as measured by a NASA jet flying into the Antarctic ozone hole. (Adapted from Lucy Pryde Eubanks, Catherine H. Middlecamp, Carl E. Heltzel, and Steven W. Keller, *Chemistry in Context*, 6th ed., McGraw-Hill, New York, 2009, p. 58.)

(a)

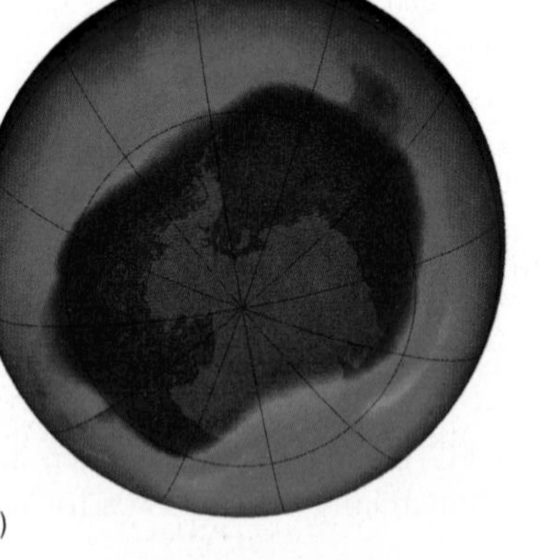

(b)

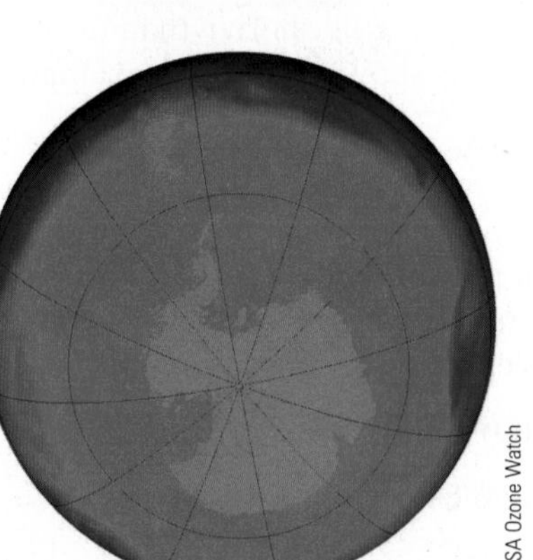

NASA Ozone Watch

FIGURE 20.35 (a) Image of the Antarctic ozone hole taken by the Ozone Monitoring Instrument on September 25, 2006. The hole (one of the largest ever recorded) covered an area greater than North America and the ozone levels were below 150 ppb, compared with the global annual average of about 350 ppb. (b) Ozone levels above Antarctica on December 25, 2006 showing substantial recovery.

responsible for ozone depletion. Second, satellites began to observe that the concentrations of halogen-containing compounds in the stratosphere were appreciable. Finally, the appearance of HF and other fluorine-containing compounds in the stratosphere and the correlation of their concentrations with the amounts and compositions of the CFCs provided the definitive proof. There are no naturally occurring sources of fluorine that could account for the HF appearing in the stratosphere. Detailed climate models, using accurate kinetic data measured in the laboratory, provided additional supporting evidence for the proposed mechanism. Molina, Rowland, and Crutzen were awarded the 1995 Nobel Prize in Chemistry for their pioneering work on this problem.

The ozone concentrations above Antarctica show dramatic seasonal variations in the ozone concentrations and depletion much greater than that averaged over the entire atmosphere. Figure 20.35 shows the ozone "hole" over Antarctica on September 25, 2006 and a corresponding image taken on December 25, 2006. The purple in the false color scale represents the lowest ozone concentrations, green is the average, and red the highest. A Dobson Unit is about 1 part per billion (ppb). Comparing the two images, we see that the ozone concentration over Antarctica during the Antarctic spring has dropped to less than a third of its average value. What mechanism causes such a dramatic variation? Several factors, acting in concert, are responsible. The Antarctic land mass is huge, and the region above the South Pole is the coldest spot on Earth. Circulating air masses prevent warmer air from entering the region during the Antarctic winter (July through September), creating conditions favorable for the formation of polar stratospheric clouds (PSCs). These PSCs are thought to catalyze the conversion of less reactive molecules like $ClONO_2$ and HCl into more reactive species like HOCl and Cl_2, which are then stored on the surfaces of these clouds during the winter months, effectively "pre-concentrating" them. Sunlight returns in spring and photolyzes the accumulated store of chlorine-containing precursors, releasing relatively large amounts of Cl• in a relatively short period of time. The hole appears quickly and dramatically. As spring turns into summer the clouds evaporate, the weather patterns change, and air from other regions of the atmosphere flows into the region, refilling the ozone hole. The situation is not as dramatic in the Arctic, presumably due to the smaller land mass, warmer temperatures, and lack of circulating winds, but ozone depletion over the North Pole is clearly evident and being closely monitored.

The international community quickly came together to address this problem globally, once the underlying science had been established and accepted. As early as the late 1970s, just a few years after Rowland and Molina's pioneering work, the U.S. banned the use of CFCs as propellants in spray cans and, a decade later, as foaming agents. In 1987, the Montreal Protocol was signed by nearly 100 countries. It called for the reduction of global CFC production to half the 1986 levels by 1998; underdeveloped countries were given longer to comply. Replacing CFCs with hydrochlorofluorocarbons (HCFCs) caused the trend in the stratospheric chlorine concentrations to reverse, but even HCFCs can lead to ozone depletion. The 1993 Copenhagen amendment to the Montreal protocol requires discontinuing their production by 2030, so HCFCs are currently being replaced by hydrofluorcarbons (HFCs).

20.7 Photosynthesis

About 1% of the energy that reaches the Earth's surface is converted into chemical energy by **photosynthesis**—a remarkably efficient sequence of photophysical processes and oxidation–reduction reactions that leads to the formation of a six-carbon sugar molecule plus molecular oxygen as follows

$$6\ CO_2 + 12\ H_2O \longrightarrow C_6H_{12}O_6 + 6\ O_2 + 6\ H_2O \qquad \Delta G_f^\circ = +2872\ \text{kJ mol}^{-1}$$

This reaction is the reverse of the spontaneous oxidation of glucose to CO_2 and H_2O by combustion or metabolism, and it requires an external driving force to occur as written. The energy of the light absorbed in the photosynthetic process increases the Gibbs free energy of the reactants and provides the necessary driving force. The end result of this carbon dioxide fixation reaction is the conversion of solar energy into chemical potential energy. We focus, in this section, on the elementary physical and chemical photosynthetic processes and do not discuss in any detail the structures and organization of the components of the photosynthetic membranes that carry out the reactions. These components include a number of membrane-bound enzyme complexes that are highly organized spatially to catalyze the various light-driven and dark-driven reactions of photosynthesis. The biochemistry required to understand the details of these enzyme-catalyzed reactions is beyond the scope of this textbook. Our primary objective is to show how the concepts we introduced at the beginning of this chapter, along with the molecular level picture of redox processes developed in Chapter 17, can be used to explain the initial steps of energy transduction in photosynthesis.

Photosynthesis occurs both in bacteria and in green plants by similar mechanisms, the differences being in the nature of their respective photosystems and reducing agents. Photosynthesis consists of a series of light reactions, discussed below, and dark reactions leading to CO_2 fixation that we just touch upon. The light reactions and dark reactions have been shown to be independent of one another. Irradiating photosynthetic systems in the absence of CO_2 generates O_2; transferring a system prepared in this way into the dark and supplying it with CO_2 results in CO_2 fixation. The most general reaction for CO_2 fixation can be written as

$$CO_2 + 2\ H_2A \longrightarrow (CH_2O) + 2A + H_2O$$

where H_2A is a hydrogen donor that is the reducing agent, and CH_2O is the carbohydrate group, the building block for sugars, starches, and cellulose in plants. The hydrogen donor in bacterial photosynthesis is often H_2S, and it is H_2O in plants. Isotope labeling experiments showed that the O_2 produced in plant photosynthesis came from the water and not the CO_2, and we write the general photosynthetic reaction as shown so that we can track the source of the oxygen atoms through the process. The energy provided by light is ultimately transduced in the form of the high-energy molecule ATP and a strong reducing agent called NADPH by the coupled reactions.

$$2\,H_2O + 2\,NADP^+ + xADP + xP_1 \xrightarrow{nh\nu} O_2 + 2\,NADPH + 2\,H^+ + xATP + xH_2O$$

Hydrolysis of ATP to ADP is coupled to many otherwise nonspontaneous reactions in biochemistry because it provides a driving force (ΔG) of about 50 kJ mol^{-1}. ATP is synthesized as part of the photosynthetic process by an enzyme powered by a transmembrane proton gradient that is set up by a series of light-driven electron transfer reactions. It is used as a source of free energy that couples into the dark reactions leading to CO_2 fixation. The reducing agent NADPH, the other primary product of photosynthesis, provides reducing equivalents for CO_2 reduction to CH_2O.

The initial photosynthetic events are very similar in both bacteria and in plants. Light is absorbed by a number of pigments, the most important of which are the chlorophylls, which are shown along with their absorption spectra in Figure 20.36.

Chlorophylls are coordination complexes of magnesium and various substituted tetrapyrroles that are built around the large planar ring structure shown. The basic structure of chlorophyll is similar to that of heme, the iron-bearing component in myoglobin and hemoglobin (see the *Connection to Biology* in Chapter 8). The two strong absorption bands observed near 650 nm and 450 nm arise from $\pi \rightarrow \pi^*$ transitions in the large aromatic ring, and they are responsible

FIGURE 20.36 Chlorophyll structures (a) and spectra (b).

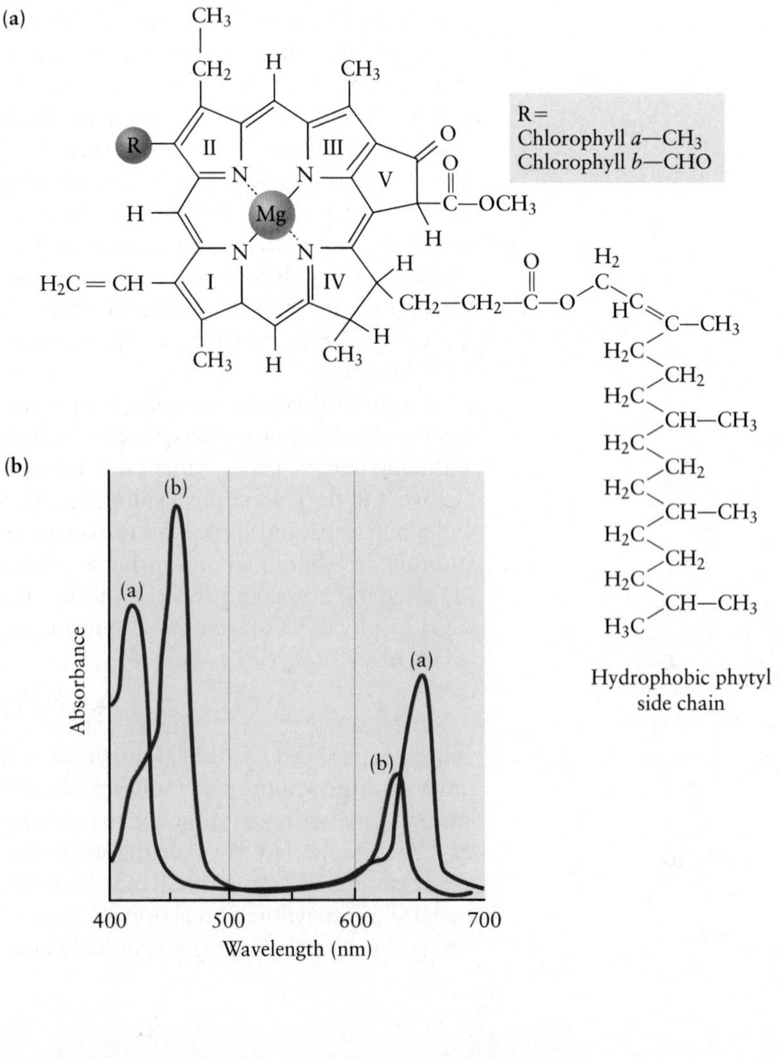

FIGURE 20.37 Structures of the accessory pigments (a) β-carotene and (b) phycocyanobilin.

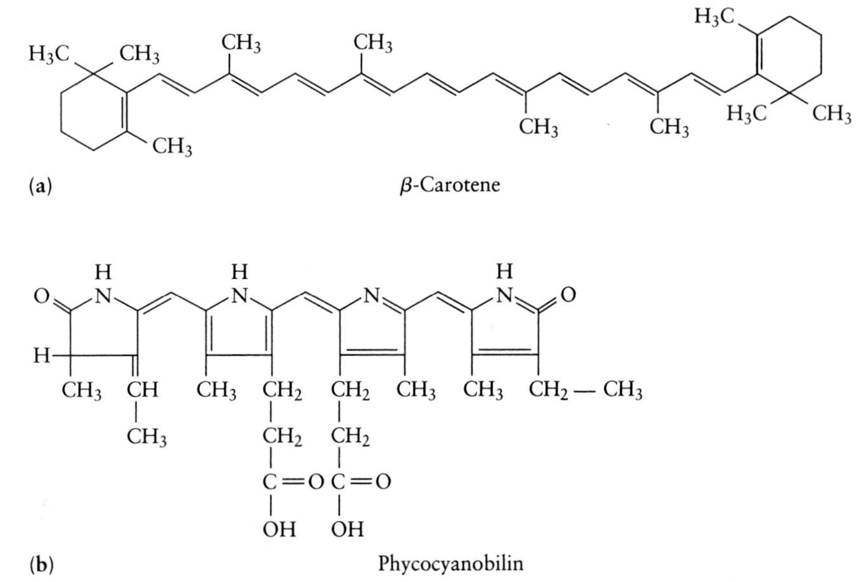

for the green color of plants. The structures of accessory pigments like carotene (which gives carrots their color) and phycocyanobilin are shown in Figure 20.37; their function is to absorb light that is not absorbed by chlorophyll, thereby utilizing the solar spectrum more efficiently.

We discussed various pathways for excited state relaxation in Sections 20.1 and 20.5 and alluded to the importance of energy transfer and electron transfer processes in photosynthesis. A photosynthetic unit comprises hundreds of chlorophyll molecules and accessory pigments whose purpose is to absorb large quantities of light and efficiently funnel the energy to a special pair of chlorophyll molecules called the reaction center. Most of the chlorophyll in the photosynthetic unit acts like a large antenna that captures light. Energy is transferred between the molecules that comprise the array by a dipole–dipole or Förster energy transfer mechanism, a nonradiative process by which excited molecules transfer energy to neighboring ground-state molecules without either emission by the donor or absorption by the acceptor. The mechanism is analogous to the coupling of two tuning forks by the air. A vibrating tuning fork will transfer energy to a quiet tuning fork if the two are placed close enough together. In a similar way the energy stored in the excited states of donor molecules is transferred to the ground states of acceptor molecules if they are near one another. The rate of Förster energy transfer falls off as the inverse sixth power of the distance between donors and acceptors (the same dependence as van der Waals forces) with a critical transfer distance R_0 that gives a rough measure of the length over which energy is transferred efficiently. R_0 is typically of the order of a few nm, which explains why the antenna chlorophyll molecules must be packed together pretty tightly for efficient energy transfer. Even though the process is not radiative, the rate of energy transfer depends on the overlap of the emission spectrum of the donor and the absorption spectrum of the acceptor. Figure 20.38 shows the energy transfer process and the overlap of the emission and absorption spectra.

Energy transferred to the special pair in the reaction center creates excited states that initiate a series of electron transfer reactions ultimately leading to oxygen evolution and/or CO_2 fixation. The structure of the reaction center in the bacterium *Rhodopseudomonas viridis* is shown in Figure 20.39. The crystal structure shown was determined by X-ray diffraction (see Section 21.1) in 1984 by three German biochemists, Hartmut Michel, Johann Deisenhofer, and Robert Huber, for which they received the Nobel Prize in Chemistry. Not only did they locate the bacteriochlorophyll molecules that form the special pair, but they also

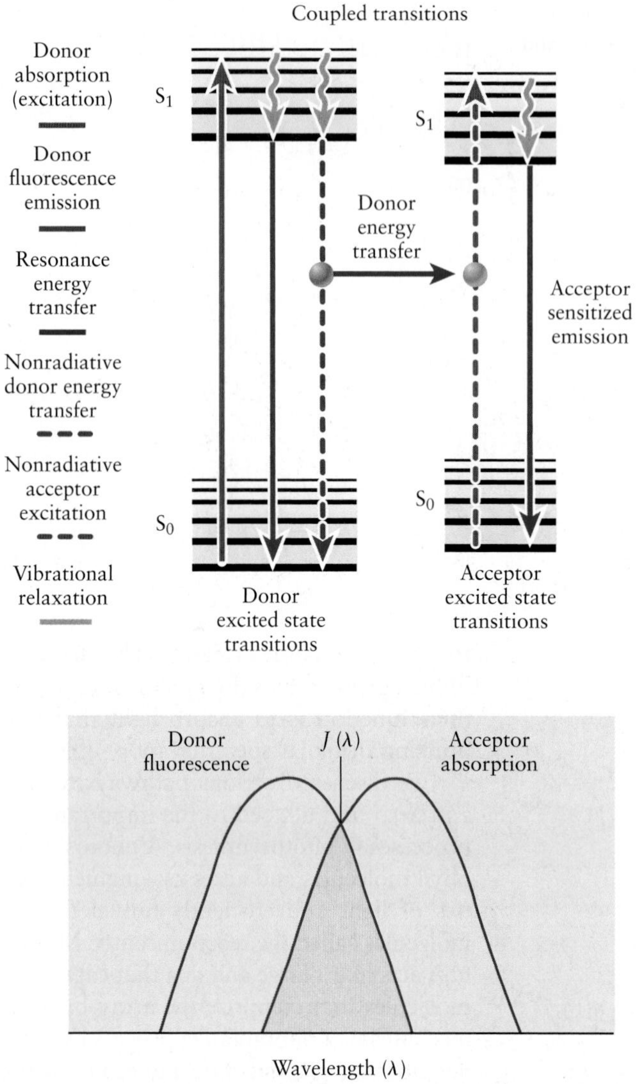

FIGURE 20.38 Schematic of Förster energy transfer mechanism and spectral overlap.

identified the positions of the molecules involved in the subsequent steps in photosynthesis.

Figure 20.40 shows how photoexcitation of the special pair creates excited states of bacteriochlorophyll that are both reducing equivalents (the electron promoted to the LUMO) and oxidizing equivalents (the "hole" left behind in the HOMO). The energies of the species are plotted as electrode potentials, referenced to the standard hydrogen electrode (SHE). Recall that the electrical potential is the potential energy per unit charge (see Section 17.3), and "more negative potentials" in electrochemistry means "higher energy" in spectroscopy because the charge on the electron is negative. Electrons excited to the LUMO undergo a series of enzyme-catalyzed electron transfer reactions that ultimately reduce $NADP^+$ to NADPH. Holes left behind in the HOMO oxidize a hydrogen donor, H_2S, in this example. The special pair in bacterial photosynthesis is called P780 because it absorbs light at 780 nm.

Returning to Figure 20.39 we see how the bacteriochlorophyll and related molecules are organized spatially to carry out these energy transfer reactions efficiently. A quinone is ultimately reduced to a hydroquinone that diffuses to the cytochrome shown, where it is oxidized back to quinone and the protons released generate a proton gradient that drives ATP synthesis as discussed.

Bacterial systems, in general, do not produce holes that are sufficiently oxidizing to oxidize water to O_2, so nature has created a second system that is used by

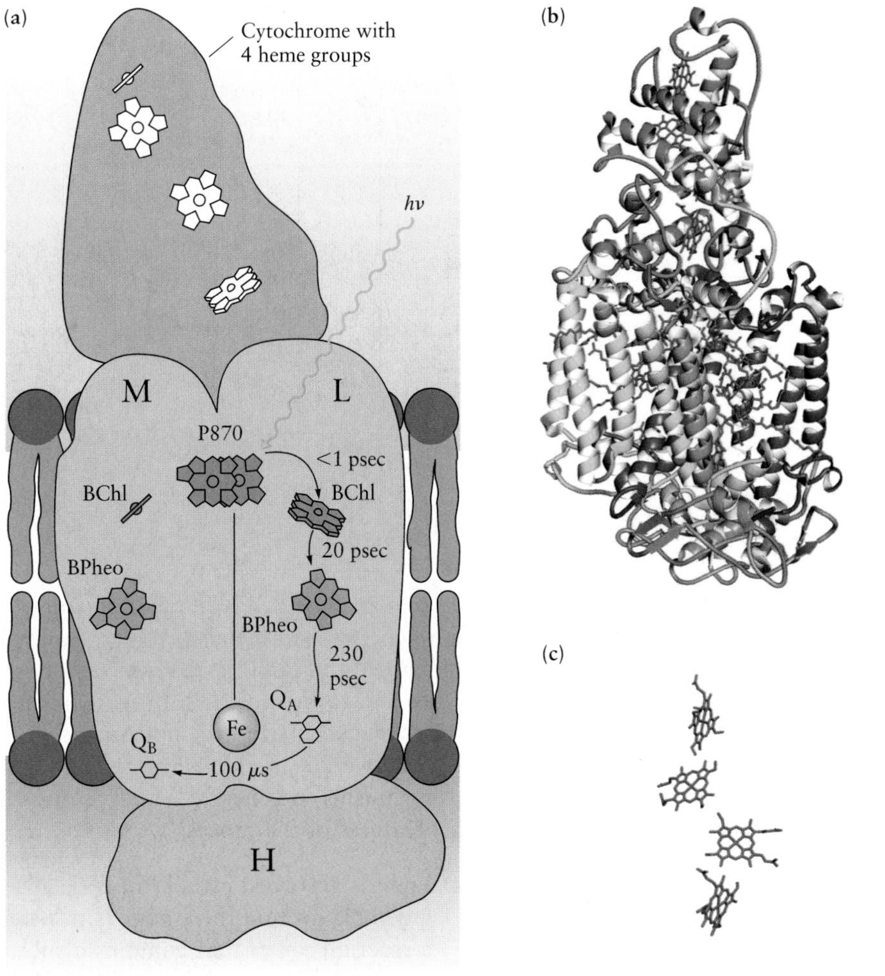

Note: The cytochrome subunit is membrane associated via a diacylglycerol moiety on its N-terminal Cys residue:

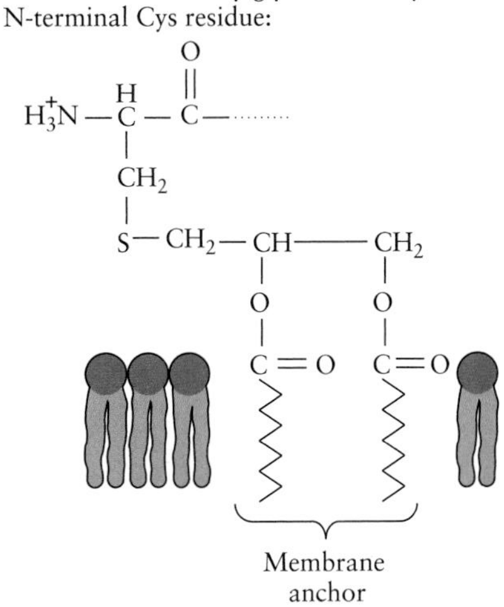

FIGURE 20.39 Structure of the photosynthetic reaction center. (a) Artist's rendition of reaction center components and their organization in the cell membrane. (b) Crystal structure of the reaction center and associated cytochrome. (c) Structure of the reaction center with the proteins eliminated to show the relationships among the chlorophyll, heme, and quinone groups.

green plants for this purpose. The *Z* scheme of photosynthesis is illustrated schematically in Figure 20.41; it consists of two coupled photosystems called PSI and PSII that produce reducing equivalents sufficiently negative to reduce $NADP^+$ to NADPH and oxidizing equivalents sufficiently positive to oxidize water. The oxygen evolving and the reducing functions of the photosynthetic system are separated and account for the existence of the light and dark reactions. The pigment in PSI absorbs light at 680 nm and is called P680, whereas that in PSII absorbs at

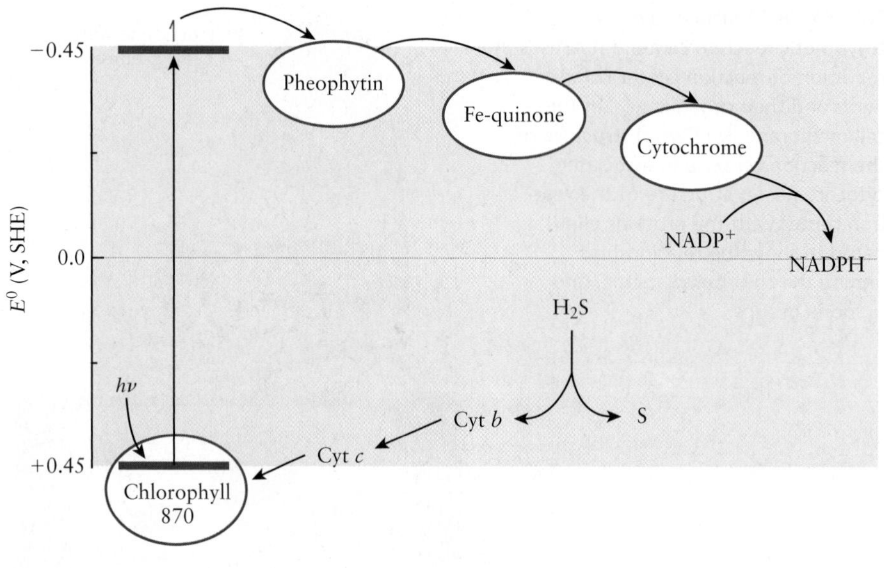

FIGURE 20.40 Energy levels (shown as reduction potentials vs. SHE) and electron transfer processes in bacterial photosynthesis.

700 nm and is called P700. Not only does the absorption of two photons provide more energy than the absorption of one photon in bacterial photosynthesis, but the HOMO in P680 is at lower energy than either of the HOMOs in P780 or P700, and so the hole is sufficiently oxidizing to oxidize water. As illustrated in electrochemical terms, the potential is sufficiently positive.

The overall stoichiometry of photosynthesis depends on the pathways used by organisms to generate ATP from ADP when driven by light. The important key features of the process we emphasize can be summarized as follows:

1. Light is absorbed by chlorophylls and accessory pigments in bacteria and in green plants, and the energy is transferred in a few picoseconds (10^{-12} s) to a reaction center that contains a special pair of chlorophyll molecules.
2. Light absorbed by the special pair creates excited states that are both more strongly reducing and strongly oxidizing than the ground state. The basic mechanism of energy transduction is the creation of these oxidizing and reducing equivalents that initiate different series of redox reactions that ultimately lead to CO_2 fixation and oxygen evolution.

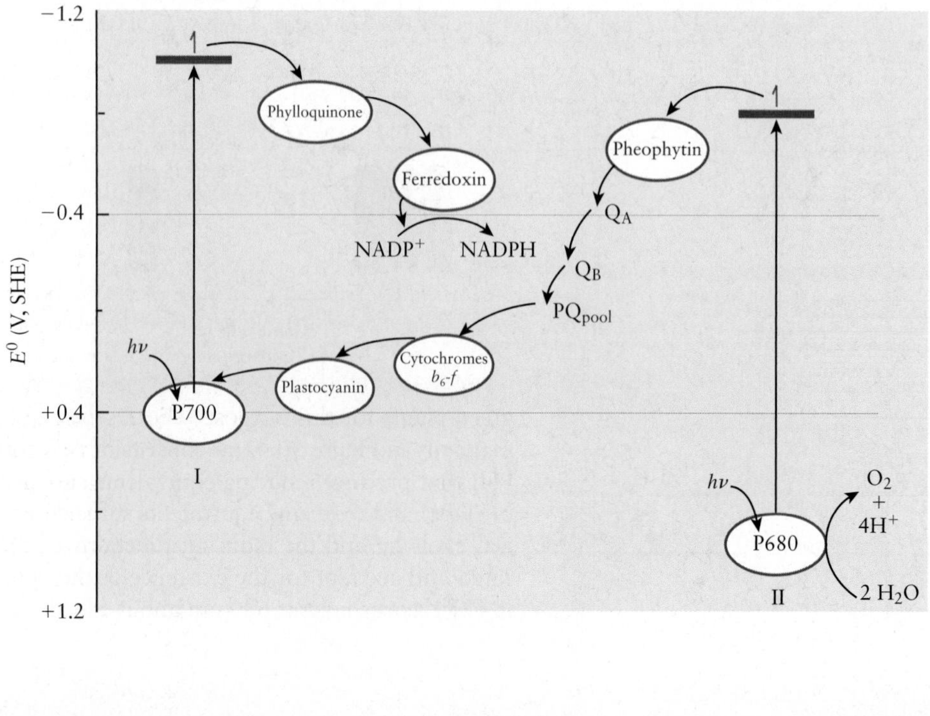

FIGURE 20.41 The *Z* scheme of photosynthesis in green plants.

3. Bacteria have a single photosynthetic system that absorbs longer wavelengths than plants and does not produce holes sufficiently oxidizing to oxidize water.
4. Green plants have two photosystems that operate in tandem to generate more energy through the absorption of two photons and, more importantly, produce holes that can oxidize water to O_2.

20.8 A DEEPER LOOK Lasers

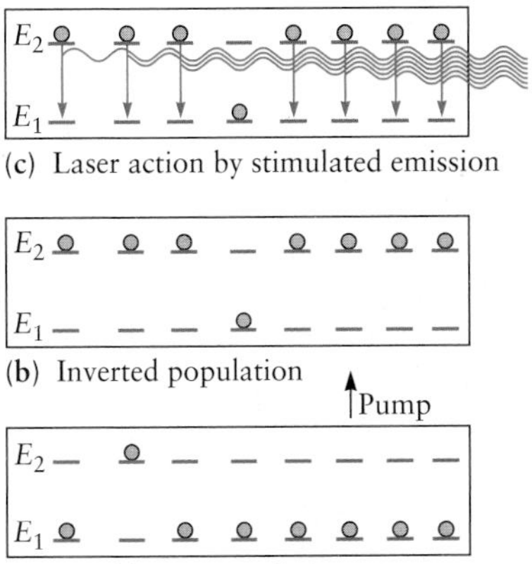

FIGURE 20.42 Operation of a laser device requires a pump source to change the equilibrium population (a) to one with a population inversion (b) and then stimulated emission (c), shown by the red downward arrows from the inverted population. (Adapted from P. W. Atkins, *Physical Chemistry*, 3rd ed., Oxford University Press, 1986.)

Lasers are devices that produce very intense beams of photons with a high degree of spatial coherence; they are the key elements in a number of important technologies with applications as diverse as machining, welding, communication, data storage, and surgery, to name a few. Fiber optic networks transmit huge volumes of digitized voice, video, and data across and between continents at nearly the speed of light. Data stored on CDs and DVDs are written by "burning" submicron size holes in the disks using a solid-state laser and read by measuring the reflectivity of the surface to locate the holes and smooth surfaces, which represent "ones" and "zeros", respectively. There is enough energy delivered by the 192 beams of the National Ignition Facility laser (see Section 19.7) to heat a mixture of hydrogen isotopes to a few million degrees and initiate fusion. Although Einstein almost certainly did not foresee this enormously diverse range of applications, the possibility of light amplification through stimulated emission of radiation (giving rise to the acronym *laser*) was implicit in his treatment of the matter-radiation interaction and he is reported to have suggested that excess energy could be released as light, under certain circumstances. The key process of stimulated emission is depicted in Figure 20.42c. Once you have managed to get more molecules in the upper energy level than the lower energy level, then electromagnetic radiation in the laser cavity, perhaps selected to be of a certain wavelength by a diffraction grating, stimulates other excited molecules to emit a photon of exactly the same frequency.

Lasers have been constructed from many different materials and have a variety of different characteristics that make them suitable for particular applications. The lasing medium may be a gas, liquid, solid, or plasma, producing radiation from the far infrared to the X-ray region of the electromagnetic spectrum. Lasers may operate continuously, like laser pointers used in the classroom, or they may be pulsed for durations as short as 12 femtoseconds (12×10^{-15} s).

Figure 20.43 shows the principle of operation of a Nd:YAG laser, a solid-state laser that produces light with a wavelength of 1064 nm. The principles of operation are similar to most other lasers. Before the laser is powered, the population of the excited state N_2 is always less than that of the ground state N_1, as required by the Boltzmann distribution. When an excitation or "pump" source (electrically powered flashlamps, in the case of a Nd:YAG laser) is powered, the population N_2 in the upper energy level shown in Figure 20.43 quickly increases—a situation in which $N_2 > N_1$ is known as a *population inversion* (see also Figure 20.42b). While photons can be spontaneously emitted (as in fluorescence) to transfer molecules from the upper energy state to the lower energy state, a process called stimulated emission (see Figure 20.42c) in the optical cavity of the laser can produce a beam of radiation very different from that of ordinary spontaneous emission; it is highly collimated and the photons are travelling in phase with each other. (Explanations for the last two assertions are beyond the scope of this textbook, but are important characteristics of many lasers.) The trick, of course, is how to prepare a system that is not in thermal equilibrium. There are two general approaches. The first is to excite the system with a very short burst of intense radiation, intense enough to create the population

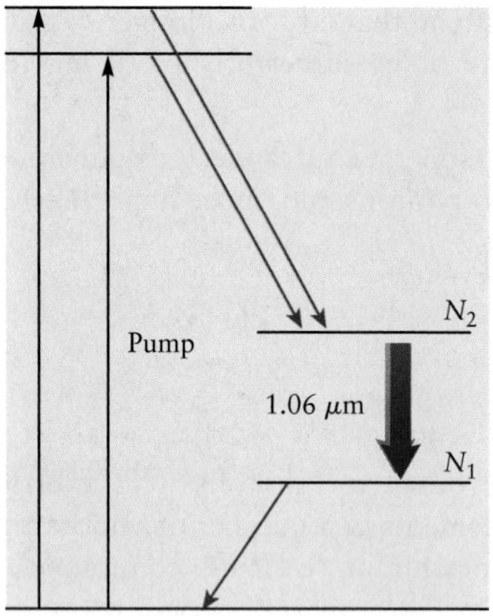

FIGURE 20.43 Nd:YAG laser operation. (Adapted from Peter Atkins and Julio de Paula, *Physical Chemistry*, W. H. Freeman and Company, New York, 2010, p. 733.)

inversion but short enough to preclude significant spontaneous emission. The second is to use a medium with three or four states that provide a bottleneck that allows population to accumulate in the upper state. The latter is the approach used in a Nd:YAG laser. (In diode and excimer lasers, a population inversion is easily achieved because the ground electronic state of the molecules involved does not have a bound potential well.) Two other conditions are required for sustaining laser emission. The radiation must be confined in such a way to ensure that emitted photons stimulate emission from other excited atoms, ions, or molecules. This is generally accomplished by housing the lasing medium in a resonant cavity, terminated by a pair of highly reflective mirrors that keep the radiation circulating in the cavity, as well as to ensure that all of the photons are travelling in the same direction. A practical laser, however, must extract some of the radiation from the cavity to be useful; a partially transmitting mirror is generally used for this purpose, with a portion of the radiation being reflected back into the cavity to sustain the coherence in stimulated emission while allowing some of the radiation to escape as a laser beam. Some of the more common lasers with which you may be familiar include the continuous red HeNe gas laser, used in supermarket scanners, or the solid-state Nd:YVO_4 laser used in green laser pointers. Both of these lasers use multiple energy levels to achieve and maintain population inversions in the continuous mode of operation; the Nd:YAG laser can operate either in continuous or pulsed modes, the latter being used to generate the high powers necessary to initiate nuclear fusion.

Figure 20.43 shows an excitation source, the "pump" beam. Lasers are pumped by other lasers or by broadband emission from flash lamps or laser diodes; the lasing medium should have a large absorption coefficient at the pump wavelength(s) to create a significant population of excited states. Relaxation back to the ground state via emission or nonradiative processes must be slow. Non-radiative relaxation or energy transfer to the upper level of a second set of states, the level labeled with population N_2, must be fast, relaxation to the lower level, the level labeled with population N_1, must be slow, and, after a photon is produced from the stimulated emission process, transferring the system to the level labeled with population N_1, the relaxation back to the initial state (the initial state that the pump beam needs) must be fast; these conditions lead to the creation of a population inversion between the two levels on the right-hand side of the energy level diagram. Figure 20.43 shows the energy level diagram for the Nd:YAG laser in which the lasing medium consists of a particular yttrium aluminum garnet, $Y_3Al_5O_{12}$ in which about 1% of the Y^{3+} ions have been replaced by Nd^{3+} ions. Two excited states lying relatively close in energy are pumped either by diode lasers at the appropriate wavelengths or by broadband diode lasers or flash-lamps. Rapid nonradiative relaxation populates the N_2 level (with 3 unpaired electrons) creating the population inversion that results in lasing at 1.06 μm. Relaxation from the N_1 level back to the ground state of the system is rapid, as required to sustain the population inversion. Nd:YAG lasers are used in many of the applications cited earlier with the fundamental frequency (1.064 μm) multiplied in special crystals to produce radiation at shorter wavelengths. Frequency doubling produces the 532 nm green light of laser pointers whereas frequency tripling produces the 355 nm UV radiation used in the National Ignition Facility. A variant, the Nd^{3+}:YVO_4 laser, produces the green light of laser pointers upon a frequency doubling.

CHAPTER SUMMARY

Spectroscopy—understanding and exploiting the interactions between radiation and matter—has provided us with a suite of techniques with which to study the structures and reactions of molecules in environments that range from dilute gases in interstellar space to those in living cells. Each of the different kinds of motions of a molecule is quantized, and each can be selectively excited by the absorption of electromagnetic radiation of the appropriate frequency. The energies of the transitions increase in the order: nuclear spin flips, molecular rotations, molecular vibrations, electronic excitations. They are excited by radiofrequency, microwave, infrared, and visible and ultraviolet radiation, respectively. We obtain bond lengths and angles, bond dissociation energies, and the energies of molecular orbitals from the spectra of molecules. Time-resolved methods allow us to follow molecular dynamics and chemical reactions on time scales as short as 10^{-14} seconds and imaging techniques provide molecularly specific information on length scales as short as 100 nm. Absorption of ultraviolet and visible radiation produces electronically excited states of molecules, the first step in a number of important processes in spectroscopy and photochemistry. Fluorescence spectroscopy is one of our most sensitive tools for detecting and imaging molecules. Energy and electron transfer are key steps in solar energy conversion processes, including photosynthesis. Understanding atmospheric photochemistry in greater detail will enable us to address problems of truly global scope—air pollution, ozone destruction, and climate change.

CONCEPTS AND SKILLS

Section 20.1 – Introduction to Molecular Spectroscopy

Identify the different kinds of spectroscopic techniques based upon the information provided and the region of the electromagnetic spectrum involved. See Table 20.1.

Identify and characterize the major radiative and nonradiative relaxation processes in molecules.

- Radiative processes include fluorescence, the emission of light resulting from transitions between states of the same spin, and phosphorescence, the emission of light resulting from transitions between states of different spin. Nonradiative processes include internal conversion (relaxation between states of the same spin without emission) and intersystem crossing (relaxation between states of different spin). No light is absorbed or emitted during a nonradiative process.

Section 20.2 – Experimental Methods in Molecular Spectroscopy

State the Beer–Lambert law and use it to calculate the intensity of absorption, emission, or scattering (Problems 1–6).

- The Beer–Lambert law is $I_t = I_0 \exp[-(N/V)\sigma\ell]$ where (N/V) is the molecular number density, σ is a cross section measured in cm^2 molecule^{-1} and ℓ is a path length measured in cm.

Describe how laser induced fluorescence can give you information on the vibrational frequencies of molecules in excited electronic states, and how dispersing the emission by wavelength can reveal the vibrational frequencies of the molecule in the ground electronic state (Problems 7, 8).

- Starting with molecules in the $v = 0$ level of the ground electronic state, you can tune the energy of the laser photon to excite the molecule to different vibrational levels in the excited electronic state. When the laser photon has the right energy to promote the molecule to a vibrational level in the excited electronic state, you see emitted photons coming from the sample. If you then leave the excitation laser tuned to excite one vibrational state in the excited state and disperse the emitted photons by their energy, you can, by subtracting the energy of the emitted photon from the energy of the absorbed photon, measure the vibrational energy levels in the ground electronic state.

Identify the essential components of a molecular spectroscopy experiment.

- Spectroscopy experiments require a source of electromagnetic radiation, an element that disperses the radiation spatially to select wavelengths, and a detector.

Use the Boltzmann distribution to calculate the population differences between two levels as a function of their energy difference and the temperature (Problems 9, 10, 23, 24).

- The Boltzmann distribution is $N_i/N \propto g_i \exp(-E_i/k_BT)$ from which we calculate population ratios using Equation 20.4.

Section 20.3 – Rotational and Vibrational Spectroscopy

Calculate the energy levels and the separation between energy levels of a linear rigid rotor and predict the appearance of pure rotational spectra (Problems 11–14).

- The energy levels of a linear rigid rotor are given by $E_J = hc\tilde{B}J(J+1)$ where $\tilde{B}$ is a rotational constant in cm^{-1}. Only transitions for which $\Delta J = 1$ are allowed in absorption so the spectrum consists of a series of lines separated by $2\tilde{B}$ with relative intensities determined by the degeneracies of the two levels involved and their populations as given by the Boltzmann distribution.

Calculate bond lengths from rotational spectral data or predict the energies of the photons absorbed in a pure rotational spectrum given the bond length of a diatomic molecule (Problems 11–14).

- The rotational constant, which is determined from the spacing between adjacent levels, is related to the moment of inertia in Equation 20.6. The bond length is determined by substituting the reduced mass into the definition of the moment of inertia. See Example 20.4.

Describe rotational Raman spectroscopy and the appearance of a rotational Raman spectrum (Problems 15–16).

- Only those transitions for which $\Delta J = 0, \pm 2$ are allowed in rotational Raman spectra of diatomic molecules. The appearance of the spectra depends upon the nuclear spins for homonuclear diatomic molecules as well as the relative populations of the levels that are determined by the degeneracies of the levels and the Boltzmann distribution.

Describe the harmonic oscillator model, calculate force constants from vibrational frequencies and reduced masses for diatomic molecules, and interpret vibrational frequencies in terms of force constants and reduced masses (Problems 17–20, 25–28).

- The harmonic oscillator model assumes that the potential energy function, $V(x) = \frac{1}{2}kx^2$, is given by Hooke's law, where $x = R - R_e$ for a diatomic molecule.
- Solving the Schrödinger equation for this model problem gives a set of wave functions and a formula for calculating the quantized energy levels, which is $E_v = h\nu\left(v + \frac{1}{2}\right)$ where v is the vibrational quantum number and ν is the vibrational frequency of the oscillator. v can be $0, 1, 2, 3\ldots$ and the term $\frac{1}{2}h\nu$ is the zero-point energy of the oscillator.
- The bond force constant is related to the vibrational frequency by $k = \mu(2\pi\nu)^2$.
- Vibrational frequencies are proportional to the square root of the force constant and inversely proportional to the square root of the reduced mass; they increase with increasing bond order, for a series of molecules with the same reduced mass, such as C—C single, double and triple bonds, and decrease with increasing reduced mass for a series of molecules with similar force constants.

Describe the anharmonic oscillator model, the effects of anharmonicity on the appearance of infrared spectra, and how bond dissociation energies can be calculated from the spectra.

- The energy levels of an anharmonic oscillator are not equally spaced, they decrease with increasing v, and the harmonic oscillator selection rule $\Delta v = \pm 1$ is relaxed. A plot of the energy differences between adjacent levels as a function of the vibrational quantum number $v + 1/2$ is approximately a straight line; extrapolating that straight line to the x intercept and taking the area under the curve gives the bond dissociation energy.

Describe the characteristics of a high-resolution infrared absorption spectrum of a heteronuclear diatomic molecule (Problems 21–22).

- When you excite a molecule with a laser photon characterized by a precise energy, what appeared a lower resolution to be a band centered at $h\nu_{\text{photon}} = h\nu_{\text{vib}}$ is actually the envelope of two sets of sharp absorption features. The P branch of the absorption spectrum results from $\Delta J = -1$ transitions to the excited vibrational state, while the R branch of the absorption spectrum results from $\Delta J = +1$ transitions to the excited vibrational state.

Describe the vibrational spectra of polyatomic molecules (Problems 29–32).

- A polyatomic molecule with N atoms has $3N - 5$ vibrational modes if the molecule is linear and $3N - 6$ vibrational modes if it is not. The vibrational frequencies of functional groups tend to be similar in many molecules and can be used for qualitative analysis.

20.4 – Nuclear Magnetic Resonance Spectroscopy

Describe the physical basis for NMR spectroscopy.

- Many nuclei have intrinsic magnetic moments that arise from their nuclear spin. Acting as tiny bar magnets, the nuclear spins can align parallel or antiparallel to an external magnetic field. The two nuclear spin states of a ^{1}H atom nucleus have the same energy in the absence of an external magnetic field, but split in energy in the presence of an external magnetic field. The energy splitting is proportional to the strength of the applied magnetic field.

Discuss what can be learned from NMR spectra (Problems 33–34).

- Integrating the peaks in a low-resolution ^{1}H NMR spectrum allows you to determine how many H atoms are in each functional group (e.g. 3 atoms in a CH_3 group, one in an OH group). The chemical shift depends upon the charge density at the magnetic nucleus of interest as well as local fields induced by the interaction of the magnetic field with neighboring nuclei; it is characteristic of particular functional groups. The spin–spin splitting pattern identifies the number of protons on carbon atoms adjacent to the carbon atom to which the particular proton of interest is bound. Taken together, the chemical shifts and spin–spin splitting patterns allow chemists to locate functional groups within molecules and to determine molecular structures.

20.5 – Electronic Spectroscopy and Excited State Relaxation Processes

Be able to identify the molecular orbitals of common chromophores, discuss the various kinds of transitions between them, and predict how the equilibrium bond length and the vibrational frequency of a bond might change if you promote an electron from a nonbonding orbital to a π^* orbital localized on that bond, or from a π orbital localized on that bond to a π^* orbital localized on that bond (Problems 35–36, 39–40, 58–59, 61).

- The highest occupied and lowest unoccupied orbitals of organic chromophores may include the σ, σ^*, π, π^*, and n orbitals. $\sigma \rightarrow \sigma^*$ transitions typically occur in the UV while the lowest energy $\pi \rightarrow \pi^*$ transitions for linear polyenes begin in the UV and shift toward the visible as the number of carbon atoms increases.

Describe how molecular photoelectron spectroscopy provides information about the bonding characteristics of molecular orbitals.

- The number and frequencies of the vibrational bands observed in the spectra provide information about the nature of the molecular ion, from which we can deduce the characteristics of the orbital from which the electron was removed. Few bands with vibrational frequencies close to that of the parent molecule indicate that the electron was ionized from a nonbonding orbital, many bands with vibrational frequencies lower than that of the parent molecule indicate that the electron came from a bonding orbital, and an intermediate number of bands with frequencies somewhat larger than that of the parent molecule indicate that the orbital was antibonding.

Define the Franck–Condon principle and provide a physical interpretation.

- The Franck–Condon principle asserts that the most intense absorption band in a series that involves simultaneous electronic and vibrational excitation is the one in which the overlap of the vibrational wave functions in the ground and excited electronic states is the greatest. The principle is a consequence of the fact that nuclei move much more slowly than electrons and can be considered to be stationary during an electronic transition.

20.6 – Introduction to Atmospheric Chemistry

Identify the different regions of the atmosphere, current areas of concern and the physical processes involved (Problems 5 and 45–48, but 45–48 focus only on the bonding of atmospherically relevant molecules).

- We are mostly concerned with the chemistry of the troposphere, the region of the atmosphere that extends from the surface of the Earth to an altitude of about 15 km, and the stratosphere, the region between 15 and 30 km. Air pollution is the chief area of concern in the troposphere, with the dominant sources being vehicle emissions and emissions from power plants and manufacturing. Oxides of nitrogen and sulfur are respiratory irritants and precursors to acid rain, which is damaging both to human health and physical structures like buildings, roads, and bridges. CFCs produced for use as refrigerants and propellants have depleted stratospheric ozone through a series of photochemical reactions that result in catalysis of the reaction $2O_3 \rightarrow 3O_2$. Anthropogenic greenhouse gases, such as CO_2, CH_4, and N_2O are thought to be largely responsible for the increase in the temperature of the Earth that has been observed since the Industrial Revolution.

20.7 – Photosynthesis

Explain the mechanism by which photosynthetic bacteria and green plants convert light energy to chemical energy (Problems 49–50).

- Visible light absorbed by chlorophyll molecules and accessory pigments is transferred to the photosynthetic reaction center where a series of light-driven and dark reactions produce carbohydrates and O_2 from CO_2 and H_2O.

20.8 – A Deeper Look . . . Lasers

Explain what is meant by a population inversion and describe how stimulated emission produces an intense beam of laser photons, all with the same wavelength, from excited molecules that would normally fluoresce by emitting photons of disparate wavelengths.

CUMULATIVE EXERCISE

Bromine

Elemental bromine is a brownish-red liquid that was first isolated in 1826. It is currently produced by oxidation of bromide ion in natural brines with elemental chlorine.

(a) What is the ground-state configuration of the valence electrons of bromine molecules (Br_2)? Is bromine paramagnetic or diamagnetic?

(b) What is the electron configuration of the Br_2^+ molecular ion? Is its bond stronger or weaker than that in Br_2? What is its bond order?

(c) Bromine compounds have been known and used for centuries. The deep purple color that symbolized imperial power in ancient Rome originated with the compound dibromoindigo, which was extracted in tiny quantities from purple snails (about 8000 snails per gram of compound). What color and maximum wavelength of *absorbed* light would give a deep purple (violet) color?

(d) What excited electronic state is responsible for the brownish-red color of bromine? Refer to Figure 6.18, but note the different row of the periodic table for Br.

(e) The two naturally occurring isotopes of bromine are ^{79}Br and ^{81}Br, with masses of 78.918 and 80.916 u, respectively. The wavelength of the $J = 0$ to $J = 1$ rotational transition in $^{79}Br^{81}Br$ is measured to be 6.18 cm. Use this information to calculate the bond length in the Br_2 molecule, and compare the result with that listed in Table 3.3.

(f) The wavelength of the vibrational transition in the $^{79}Br^{81}Br$ molecule is 3.09×10^{-5} m. Calculate the force constant for the bond in this molecule.

(g) The action of light on bromine compounds released into the air (such as by leaded gasoline) causes the formation of the BrO radical. Give the bond order of this species by comparing it with the related radical OF.

(h) There is concern that synthetic bromine-containing compounds, in addition to chlorofluorocarbons, are helping to destroy ozone in the stratosphere. The BrO [see part (g)] can take part with ClO in the following catalytic cycle:

$$Cl + O_3 \longrightarrow ClO + O_2$$

$$Br + O_3 \longrightarrow BrO + O_2$$

$$ClO + BrO \longrightarrow Cl + Br + O_2$$

Write the overall equation for this cycle.

Answers

(a) $(\sigma_{g4s})^2(\sigma_{u4s}^*)^2(\sigma_{g4p_z})^2(\pi_{u4p})^4(\pi_{g4p}^*)^4$; diamagnetic

(b) $(\sigma_{g4s})^2(\sigma_{u4s}^*)^2(\sigma_{g4p_z})^2(\pi_{u4p})^4(\pi_{g4p}^*)^3$; stronger; bond order is $\frac{3}{2}$ versus 1

(c) Yellow light, near 530 nm (see Figure 4.3)

(d) The lowest energy excited state, which arises from the excitation of an electron from the filled π_{g4p}^* orbital to the unfilled $\sigma_{u4p_z}^*$ orbital

(e) 2.28 Å (from Table 3.3: 2.286 Å)

(f) 247 J Nm^{-1}

(g) $\frac{3}{2}$ order

(h) Overall: $2\ O_3 \rightarrow 3\ O_2$

PROBLEMS

Answers to problems whose numbers are red appear in Appendix G. Problems that are more challenging are indicated with asterisks.

20.2 Experimental Methods in Molecular Spectroscopy

1. The percentage transmittance of light at 250 nm through a certain aqueous solution is 20.0% at 25°C. The experimental cell length is 1.0 cm, and the concentration of the solution is 5×10^{-4} mol L^{-1}. Calculate the absorbance. Calculate the molar absorption coefficient.

2. Beer's law is used to measure the concentration of species in solutions, once a "standardization curve" has been prepared for that species. In one such experiment, percent transmission was measured for a series of solutions with known concentrations and the results were as follows:

Concentration μg mL^{-1}	1.0	2.0	3.0	4.0	5.0
Transmission, percent	66.8	44.7	29.2	19.9	13.3

Plot these results to obtain the standardization curve. An unknown concentration of the same species, measured in the same transmission cell, transmitted 35% of the incoming light. Calculate the concentration of the unknown solution.

3. Beer's law can be used to determine the concentration of two substances A and B in solution, provided they do not react or interact, so they absorb radiation independently. The following data were obtained for A and B in three different solutions:

	[A] mol L^{-1}	[B] mol L^{-1}	% Transmittance at λ = 400 nm	% Transmittance at λ = 500 nm
Solution 1	0.0010	0	10.0	60.0
Solution 2	0	0.0050	80.0	20.0
Solution 3	?	?	40.0	50.0

Calculate the concentrations of A and B in Solution 3.

4. The absorption of ultraviolet light by proteins at wavelength 280 nm is caused mostly by the amino acids tyrosine and tryptophan along the protein molecular chains. The molecular absorption coefficients for these two amino acids are:

$$\varepsilon_{tryp}^{280} = 5690 \text{ L cm}^{-1} \text{ mol}^{-1}$$

$$\varepsilon_{tyro}^{280} = 1280 \text{ L cm}^{-1} \text{ mol}^{-1}$$

Experiments are carried out on a protein with molecular weight 26,000, which contains two units of tryptophan and six units of tyrosine along the chain. The absorption is measured in a cell 1 cm long, and the protein concentration is 1.0 mg mL^{-1}. Calculate the absorbance and the percent transmission.

5. Stratospheric ozone absorbs ultraviolet radiation (UVB) that is known to cause cancer in humans. The onset of the UVB region of the spectrum is 320 nm and the absorption cross section for ozone (O_3) that wavelength is about 5×10^{-20} cm^2 $molecule^{-1}$ (see Fig. 20.32). The average ozone number density in the stratosphere is about 5×10^{12} molecules cm^{-3}. What thickness of the ozone layer is required to absorb 60% of the UV radiation at 320 nm?

6. The absorption cross section for fluorescein, a dye that is used as a fluorescent probe for biological imaging (see the Chapter Opener image), is 9.25×10^{-16} cm^2 $molecule^{-1}$ at 500 nm. How much light is transmitted through a 1-cm path length by a standard 10^{-6} M solution of fluorescein used to label biological samples?

7. Laser-induced fluorescence is the technique of choice for detecting OH radicals in the atmosphere because the technique is so sensitive; it can detect OH radicals with concentrations of 10^6 molecules cm^{-3}. (LIF instruments are mounted in aircraft!) The difference in energy between the minimum of the ground electronic state potential energy curve of OH and the minimum of the excited electronic $\tilde{A}$ state is 32,684 cm^{-1}. (This minimum to minimum energy difference is given the label E_{el} in Figure 20.1 and Equation 20.1, but in spectroscopic literature it is usually given by the symbol T_e.) Calculate the energy of the laser photon needed to excite an OH radical that is initially in the $v'' = 0$ level of the ground electronic state, S_0, to the $v' = 1$ level of the $\tilde{A}$ state, using a harmonic approximation for the OH vibration. The vibrational frequency of the OH radical is 3738 cm^{-1} in the ground electronic state and 3179 cm^{-1} in the $\tilde{A}$ state.

8. If after laser exciting the OH radical to the $v' = 1$ level of the $\tilde{A}$ state as described in the prior problem, what energy photon would be emitted if the final vibrational state of the OH in the ground electronic state were $v'' = 3$?

9. The vibrational frequency ν for H_2 is 4400 cm^{-1} and the energy difference between its two lowest rotational energy levels, $J = 0$ and $J = 1$, is 121.7 cm^{-1}. Calculate the relative populations of the $v = 1$ and $v = 0$ vibrational energy levels and the relative populations of the two lowest rotational energy levels for H_2 at the operating temperature of the space shuttle's combustion chamber, which is 3300°C. Comment on your results.

10. The vibrational frequency ν for Br_2 is 323 cm^{-1} and the energy difference between its two lowest rotational energy levels, $J = 0$ and $J = 1$, is 0.164 cm^{-1}. Calculate the relative populations of the $v = 1$ and $v = 0$ vibrational energy levels and the relative populations of the two lowest rotational energy levels for Br_2 at 300 K. Comment on your results.

20.3 Rotational and Vibrational Spectroscopy

11. Use data from Tables 19.1 and 3.3 to predict the energy spacing between the ground state and the first excited rotational state of the $^{14}N^{16}O$ molecule.

12. Use data from Tables 19.1 and 3.3 to predict the energy spacing between the ground state and the first excited rotational state of the $^{1}H^{19}F$ molecule.

13. The first three absorption lines in the pure rotational spectrum of gaseous $^{12}C^{16}O$ are found to have the frequencies 1.15×10^{11}, 2.30×10^{11}, and 3.46×10^{11} s^{-1}. Calculate:
(a) The moment of inertia I of CO (in kg m^2)
(b) The energies of the $J = 1, J = 2$, and $J = 3$ rotational levels of CO, measured from the $J = 0$ state (in joules)
(c) The C—O bond length (in angstroms)

14. Four consecutive absorption lines in the pure rotational spectrum of gaseous $^{1}H^{35}Cl$ are found to have the frequencies 2.50×10^{12}, 3.12×10^{12}, 3.74×10^{12}, and 4.37×10^{12} s^{-1}. Calculate:
(a) The moment of inertia I of HCl (in kg m^2)
(b) The energies of the $J = 1, J = 2$, and $J = 3$ rotational levels of HCl, measured from the $J = 0$ state (in joules)
(c) The H—Cl bond length (in angstroms)
(d) The initial and final J states for the observed absorption lines

15. The rotational constant $\tilde{B} = 1.45$ cm^{-1} for O_2. Calculate the O_2 bond length and predict the appearance of its rotational Raman spectrum, recalling that transitions between states with even J are forbidden due to nuclear spin statistics considerations.

16. The rotational constant $\tilde{B}$ of $^{7}Li_2$ is 0.6727 cm^{-1}. Calculate the $^{7}Li_2$ bond length and predict the appearance of its rotational Raman spectrum without the effect of nuclear spin statistics on the relative intensities of the lines. Compare your results with those of Example 20.5.

17. The Li_2 molecule (^{7}Li isotope) shows a very weak infrared line in its vibrational spectrum at a wavelength of 2.85×10^{-5} m. Calculate the force constant for the Li_2 molecule.

18. The Na_2 molecule (^{23}Na isotope) shows a very weak infrared line in its vibrational spectrum at a wavelength of 6.28×10^{-5} m. Calculate the force constant for the Na_2 molecule, and compare your result with that of problem 17. Give a reason for any difference.

19. The "signature" infrared absorption that indicates the presence of a C—H stretching motion in a molecule occurs at wavelengths near 3.4×10^{-6} m. Use this information to estimate the force constant of the C—H stretch. Take the reduced mass in this motion to be approximately equal to the mass of the hydrogen atom (a good approximation when the H atom is attached to a heavy atom).

20. Repeat the calculation of the problem 19 for the N—H stretch, where absorption occurs near 2.9×10^{-6} m. Which bond is stiffer: N—H or C—H?

21. An HCl molecule initially in the $(v, J) = (0, 2)$ quantum state can absorb infrared photons. What are the two most likely final quantum states after absorbing an infrared photon? Calculate the energy of the infrared photon absorbed for each of these transitions. Can it also absorb a microwave photon? If yes, calculate that photon energy also. The rotational constant of HCl is 10 cm^{-1} and its vibrational frequency is 2990 cm^{-1}.

22. A vibrationally excited CO molecule initially in the $(v, J) = (1, 2)$ quantum state can emit a microwave photon, changing its rotational state, or it can emit infrared photons of two different energies. What is the final quantum state reached in these transitions? Calculate the energy of microwave photon and the infrared photon emitted in these three transitions. Example 20.9 has information on the rotational constant of CO and its vibrational frequency.

23. Estimate the ratio of the number of molecules in the first excited vibrational state of the molecule N_2 to the number in the ground state, at a temperature of 450 K. The vibrational frequency of N_2 is 7.07×10^{13} s^{-1}.

24. The vibrational frequency of the ICl molecule is 1.15×10^{13} s^{-1}. For every million (1.00×10^{6}) molecules in the ground vibrational state, how many will be in the first excited vibrational state at a temperature of 300 K?

25. The measured vibrational frequencies for the hydrogen halides given in Section 20.3 are: $^{1}H^{19}F$ (4139 cm^{-1}); $^{1}H^{35}Cl$ (2991 cm^{-1}); $^{1}H^{81}Br$ (2449 cm^{-1}); and $^{1}H^{127}I$ (2308 cm^{-1}). Calculate the force constants for each molecule, using the vibrational frequencies given and the atomic masses provided in Table 19.1. Which is the more important factor in determining the vibrational frequencies in this series of molecules—the force constants or the reduced masses?

26. The vibrational frequencies for the mixture of gases shown in the Chapter 3 *Connection to Instrumental Analysis* include: H_2 (4160 cm^{-1}) N_2 (2331 cm^{-1}) CO (2143 cm^{-1}), and O_2 (1550 cm^{-1}). What can you conclude about the bond orders by inspection? Calculate the force constants for each molecule to check the accuracy of your estimate.

27. Substitution of hydrogen by deuterium is a convenient way to label organic molecules at specific sites for the purpose of investigating reaction mechanisms. At what frequency would the C—D stretch of a phenyl ring appear if the frequency of the C—H stretch is 3062 cm^{-1}?

28. If the spacings between adjacent vibrational features in a photoelectron spectroscopy experiment decrease, changing by a factor of $1/\sqrt{2}$, when you deuterate the molecule being ionized, is it more likely that the orbital from which you are ejecting an electron is from a nonbonding orbital, a π orbital localized on a C—C or C—O bond, or a σ orbital localized on a C—H bond? Explain.

29. How many normal vibrational modes does each of the following molecules have?
(a) NH_3
(b) C_2H_4
(c) CCl_2F_2
(d) CH_3CH_2OH

30. Consider the vibrations of the molecules in the previous problem from a localized point of view, as the vibrations of individual bonds. Infrared absorption is proportional to the permanent dipole moment of each individual bond. Which of the molecules will absorb IR radiation most strongly and why?

31. The IR spectra of nonane (C_9H_{20}) and 1-hexanol ($C_6H_{13}OH$) are shown below. Assign each spectrum to the correct compound and identify the frequencies and the functional groups used to support your assignment.

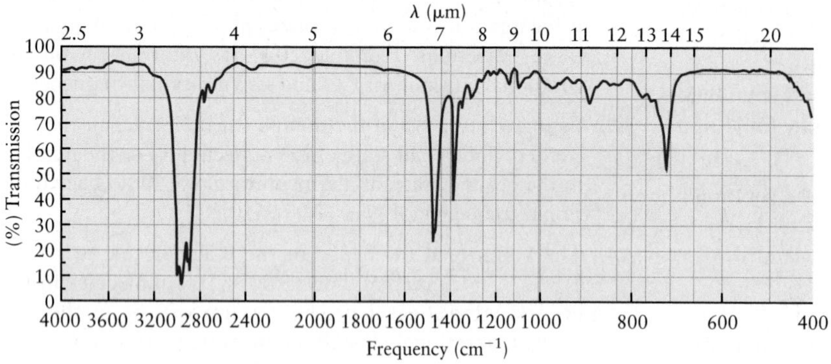

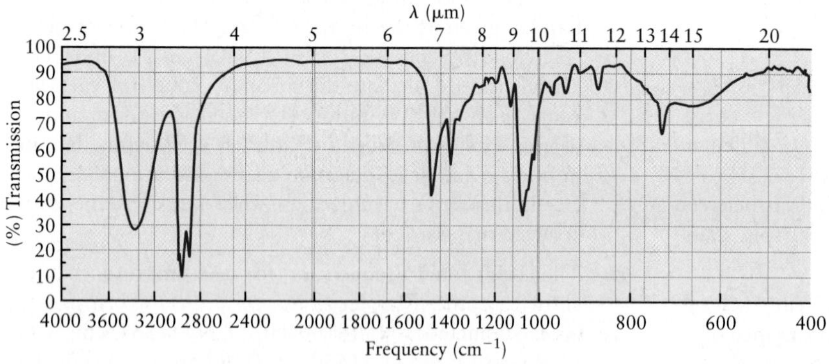

32. The IR spectrum of 2-methyl-1-butanol ($H_3CCH_2CH(CH_3)CH_2OH$) and *tert*-butyl methyl ether [$(CH_3)_3COCH_3$] are shown below. Assign each spectrum to the correct compound and identify the frequencies and the functional groups used to support your assignment.

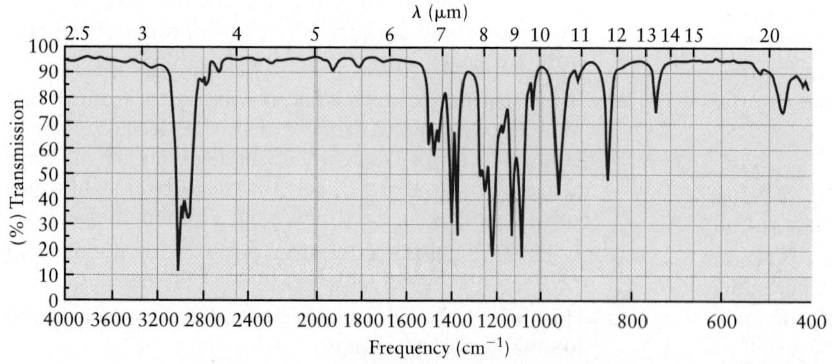

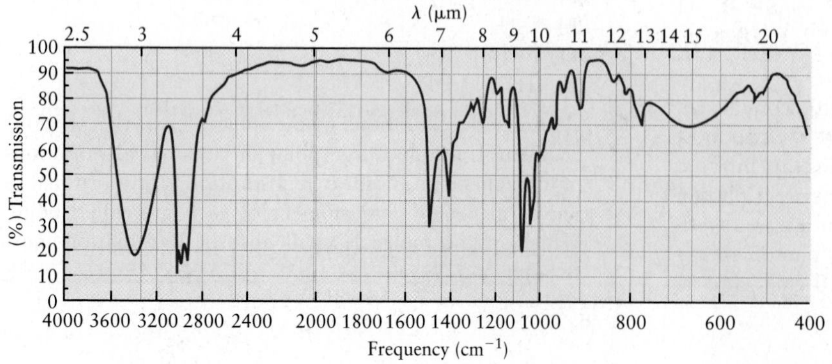

20.4 Nuclear Magnetic Resonance Spectroscopy

33. Give the number of peaks and the relative peak areas that should be observed in the low-resolution proton magnetic resonance spectra of the following molecules:

$$CH_3CH_2CH_2CH_3,\ CH_3OCH_3,\ CH_3NHCH_3.$$

34. The organic compound 1,4-dimethylbenzene (also known as *p*-xylene) has the formula $(CH_3)_2C_6H_4$. Its structure has two CH_3 (methyl) groups substituted at opposite positions on the benzene (C_6H_6) ring. Predict the number of peaks in the low-resolution proton NMR spectrum of this compound and the relative areas of the peaks.

20.5 Electronic Spectroscopy and Excited State Relaxation Processes

35. Suppose that an ethene molecule gains an additional electron to give the $C_2H_4^-$ ion. Will the bond order of the carbon–carbon bond increase or decrease? Explain.

36. Suppose that an ethene molecule is ionized by a photon to give the $C_2H_4^+$ ion. Will the bond order of the carbon–carbon bond increase or decrease? Explain.

37. The color of the dye "indanthrene brilliant orange" is evident from its name. In what wavelength range would you expect to find the maximum in the absorption spectrum of this molecule? Refer to the color spectrum in Figure 4.3.

38. In what wavelength range would you expect to find the maximum in the absorption spectrum of the dye "crystal violet"?

39. The structure of the molecule cyclohexene is shown below:

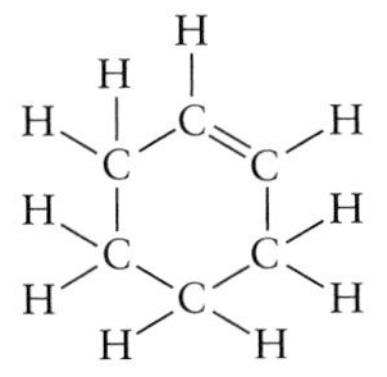

Does the absorption of ultraviolet light by cyclohexene occur at shorter wavelengths than in benzene? Explain.

40. The naphthalene molecule has a structure that corresponds to two benzene molecules fused together:

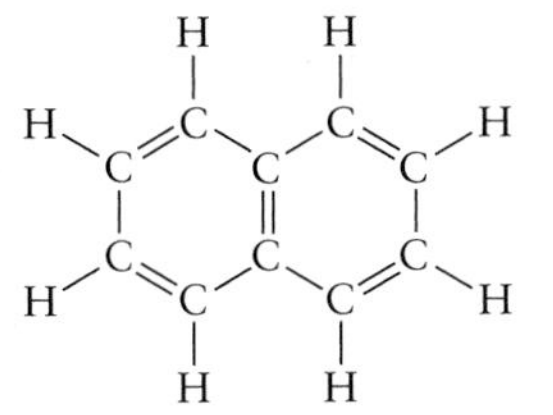

The π-electrons in this molecule are delocalized over the entire molecule. The wavelength of maximum absorption in the UV-visible part of the spectrum in benzene is 255 nm. Is the corresponding wavelength shorter or longer than 255 nm for naphthalene?

41. Use data from Table 3.3 to give an upper bound on the wavelengths of light that are capable of dissociating a molecule of ClF.

42. Use data from Table 3.3 to give an upper bound on the wavelengths of light that are capable of dissociating a molecule of ICl.

43. Plot a potential energy curve for the ground electronic state of methyl iodide and the excited electronic state reached when it absorbs a 260 nm photon, putting the potential energy V on the vertical axis and the distance between the carbon and iodine atom on the horizontal axis. The equilibrium internuclear distance of the C—I bond is about 200 pm and the C—I bond energy is about 220 kJ mol^{-1}. Assume the excited state potential energy curve at $R_{C-I} = 200$ pm is reached by the 260 nm photon, in rough agreement with the Franck-Condon principle. The absorption spectrum is broad and structureless.

44. Plot a potential energy curve for the ground state of HCl and the excited state reached in the broad structureless absorption band peaking near 65,500 cm^{-1}. The equilibrium bond length of HCl and the bond energy is given in Table 3.3.

20.6 Introduction to Atmospheric Chemistry

45. The bond dissociation energy of a typical C—F bond in a chlorofluorocarbon is approximately 440 kJ mol^{-1}. Calculate the maximum wavelength of light that can photodissociate a molecule of CCl_2F_2, breaking such a C—F bond.

46. The bond dissociation energy of a typical C—Cl bond in a chlorofluorocarbon is approximately 330 kJ mol^{-1}. Calculate the maximum wavelength of light that can photodissociate a molecule of CCl_2F_2, breaking such a C—Cl bond.

47. Draw a Lewis diagram(s) for the ozone molecule (O_3). Determine the steric number and hybridization of the central oxygen atom, and identify the molecular geometry. Describe the nature of the π bonds and give the bond order of the O—O bonds in ozone.

48. The compounds carbon dioxide (CO_2) and sulfur dioxide (SO_2) are formed by the burning of coal. Their apparently similar formulas mask underlying differences in molecular structure. Determine the shapes of these two types of molecules, identify the hybridization at the central atom of each, and compare the natures of their π bonds.

20.7 Photosynthesis

49. One way in which photosynthetic bacteria store chemical energy is through the conversion of a compound called adenosine diphosphate (ADP), together with hydrogen phosphate ion, to adenosine triphosphate (ATP):

$$ADP^{3-} + HPO_4^{2-} + H_3O^+ \longrightarrow ATP^{4-} + 2\,H_2O$$
$$\Delta G = +34.5 \text{ kJ (pH 7)}$$

Suppose some chlorophyll molecules absorb 1.00 mol of photons of blue light with wavelength 430 nm. If *all* this energy could be used to convert ADP to ATP at room conditions and pH 7, how many molecules of ATP would be produced per photon absorbed? (The actual number is smaller because the conversion is not 100% efficient.)

50. Repeat the calculation of problem 49 for red light with wavelength 700 nm.

ADDITIONAL PROBLEMS

51. In Example 20.4, we determined that the moment of inertia of the NaH molecule is 5.70×10^{-47} kg m^2.

(a) Calculate the relative population of the $J = 5$ level and the ground state at 25°C.

(b) Calculate the relative population of the $J = 15$ level and the ground state at 25°C.

(c) Calculate the relative population of the $J = 25$ level and the ground state at 25°C.

52. The rotational constant for $^{14}N_2$, measured for the first time using rotational Raman spectroscopy (see Fig. 20.9, is 1.99 cm^{-1}. Calculate (a) the moment of inertia I of N_2 in kg m^2, (b) the energies of the first three excited rotational levels, relative to that of the ground level, in wave numbers, and (c) the frequency of the $J = 0 \rightarrow J = 1$ transition in GHz.

53. What are the moments of inertia of $^1H^{19}F$ and $^1H^{81}Br$, expressed in kg m^2? Compute the spacings $\nu = \Delta E/h$ of the rotational states, in s^{-1}, between $J = 0$ and 1 and between $J = 1$ and 2. Explain, in one sentence, why the large change in mass from 19 to 81 causes only a small change in rotational energy differences.

54. The average bond length of a molecule can change slightly with vibrational state. In $^{23}Na^{35}Cl$, the frequency of light absorbed in a change from the $J = 1$ to the $J = 2$ rotational state in the ground vibrational state ($n = 0$) was measured to be $\nu = 2.60511 \times 10^{10}$ s^{-1}, and that for a change from $J = 1$ to $J = 2$ in the first excited vibrational state ($n = 1$) was $\nu = 2.58576 \times 10^{10}$ s^{-1}. Calculate the average bond lengths of NaCl in these two vibrational states, taking the relative atomic mass of ^{23}Na to be 22.9898 and that of ^{35}Cl to be 34.9689.

55. The vibrational frequencies of $^{23}Na^1H$, $^{23}Na^{35}Cl$, and $^{23}Na^{127}I$ are 3.51×10^{13} s^{-1}, 1.10×10^{13} s^{-1}, and 0.773×10^{13} s^{-1}, respectively. Their bond lengths are 1.89 Å, 2.36 Å, and 2.71 Å. What are their reduced masses? What are their force constants? If NaH and NaD have the same force constant, what is the vibrational frequency of NaD? D is ^2H.

56. Recall that nuclear spin states in nuclear magnetic resonance are typically separated by energies of 2×10^{-5} kJ mol^{-1} to 2×10^{-4} kJ mol^{-1}. What are the ratios of occupation probability between a pair of such levels at thermal equilibrium and a temperature of 25°C?

57. The vibrational temperature of a molecule prepared in a supersonic jet can be estimated from the observed populations of its vibrational levels, assuming a Boltzmann distribution. The vibrational frequency of HgBr is 5.58×10^{12} s^{-1}, and the ratio of the number of molecules in the $n = 1$ state to the number in the $n = 0$ state is 0.127. Estimate the vibrational temperature under these conditions.

58. An electron in the π orbital of ethylene (C_2H_4) is excited by a photon to the π^* orbital. Do you expect the equilibrium bond length in the excited ethene molecule to be greater or less than that in ground-state ethene? Will the vibrational frequency in the excited state be higher or lower than in the ground state? Explain your reasoning.

59. One isomer of retinal is converted to a second isomer by the absorption of a photon:

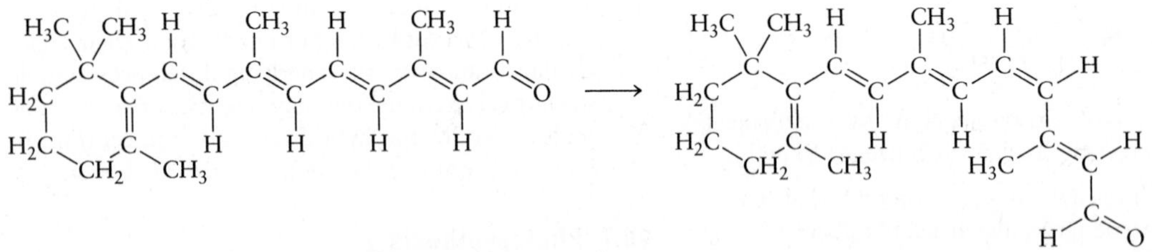

This process is a key step in the chemistry of vision. Although free retinal (in the form shown to the left of the arrow) has an absorption maximum at 376 nm, in the ultraviolet region of the spectrum this absorption shifts into the visible range when the retinal is bound in a protein, as it is in the eye.

(a) How many of the C=C bonds are *cis* and how many are *trans* in each of the preceding structures? (When assigning labels, consider the relative positions of the two largest groups attached at each double bond.) Describe the motion that takes place upon absorption of a photon.

(b) If the ring and the —CHO group in retinal were replaced by —CH_3 groups, would the absorption maximum in the molecule shift to longer or shorter wavelengths?

***60.** The ground-state electron configuration of the H_2^+ molecular ion is $(\sigma_{g1s})^1$.

(a) A molecule of H_2^+ absorbs a photon, promoting the electron to the σ^*_{u1s} molecular orbital. Predict what happens to the molecule.

(b) Another molecule of H_2^+ absorbs even more energy promoting the electron to the σ_{g2s} molecular orbital. Predict what happens to this molecule.

***61. (a)** Draw a Lewis diagram for formaldehyde (H_2CO), and decide the hybridization of the central carbon atom.

(b) Formulate the molecular orbitals for the molecule.

(c) A strong absorption is observed in the ultraviolet region of the spectrum and is attributed to a $\pi \rightarrow \pi^*$ transition. Another, weaker transition is observed at lower frequencies. What electronic excitation causes the weaker transition?

62. Write balanced chemical equations that describe the formation of nitric acid and sulfuric acid in rain, starting with the sulfur in coal and the oxygen, nitrogen, and water vapor in the atmosphere.

63. Compare and contrast the roles of ozone (O_3) and nitrogen dioxide (NO_2) in the stratosphere and in the troposphere.

64. Draw a schematic diagram of the steps in bacterial photosynthesis, numbering them in sequence and showing the approximate spatial relations of the involved molecules.

65. Do you expect the energy of the special pair of bacteriochlorophyll molecules to be higher or lower than the energy of an isolated bacteriochlorophyll? (*Hint:* Think about the analogy between the mixing of atomic orbitals to make molecular orbitals and the mixing of molecular orbitals on two nearby molecules.)

CUMULATIVE PROBLEMS

66. At thermal equilibrium, is the rate as a molecule is excited from $n = 0$ to the $n = 1$ level greater than or less than the rate for the reverse process? What is the ratio of the rate constants? (*Hint:* Think of the analogy with the chemical equilibrium between two species.)

***67.** It is important to know the dissociation constant of an indicator in order to use it properly in acid–base titrations. Spectrophotometry can be used to measure the concentration of these intensely colored species in acidic versus basic solutions, and from these data the equilibrium between the acidic and basic forms can be calculated. In one such study on the indicator *m*-nitrophenol, a 6.36×10^{-4} M solution was examined by spectrophotometry at 390 nm and 25°C in the following experiments. In highly acidic solution, where essentially all the indicator was in the form HIn, the absorbance was 0.142. In highly basic solution, where essentially all of the indicator was in the form In^-, the absorbance was 0.943. In a further series of experiments, the pH was adjusted using a buffer solution of ionic strength *I*, and absorbance was measured at each pH value. The following results were obtained:

pH	*I*	*A*
8.321	0.10	0.527
8.302	0.08	0.518
8.280	0.06	0.505
8.251	0.04	0.493
8.207	0.02	0.470

Calculate pK_a for the indicator at each value of ionic strength.

68. The hydroxyl radical has been referred to as the "chief clean-up agent in the troposphere." Its concentration is approximately zero at night and becomes as high as 1×10^7 molecules per cm^3 in highly polluted air.

(a) Calculate the maximum mole fraction and partial pressure of OH in polluted air at 25°C and atmospheric pressure.

(b) Write an equation for the reaction of HO with NO_2 in the atmosphere. How does the oxidation state of nitrogen change in this reaction? What is the ultimate fate of the product of this reaction?

69. In unpolluted air at 300 K, the hydroxyl radical OH reacts with CO with a bimolecular rate constant of 1.6×10^{11} L mol^{-1} s^{-1} and with CH_4 with a rate constant of 3.8×10^9 L mol^{-1} s^{-1}. Take the partial pressure of CO in air to be constant at 1.0×10^{-7} atm and that of CH_4 to be 1.7×10^{-6} atm, and assume that these are the primary mechanisms by which OH is consumed in the atmosphere. Calculate the half-life of OH under these conditions.

Materials

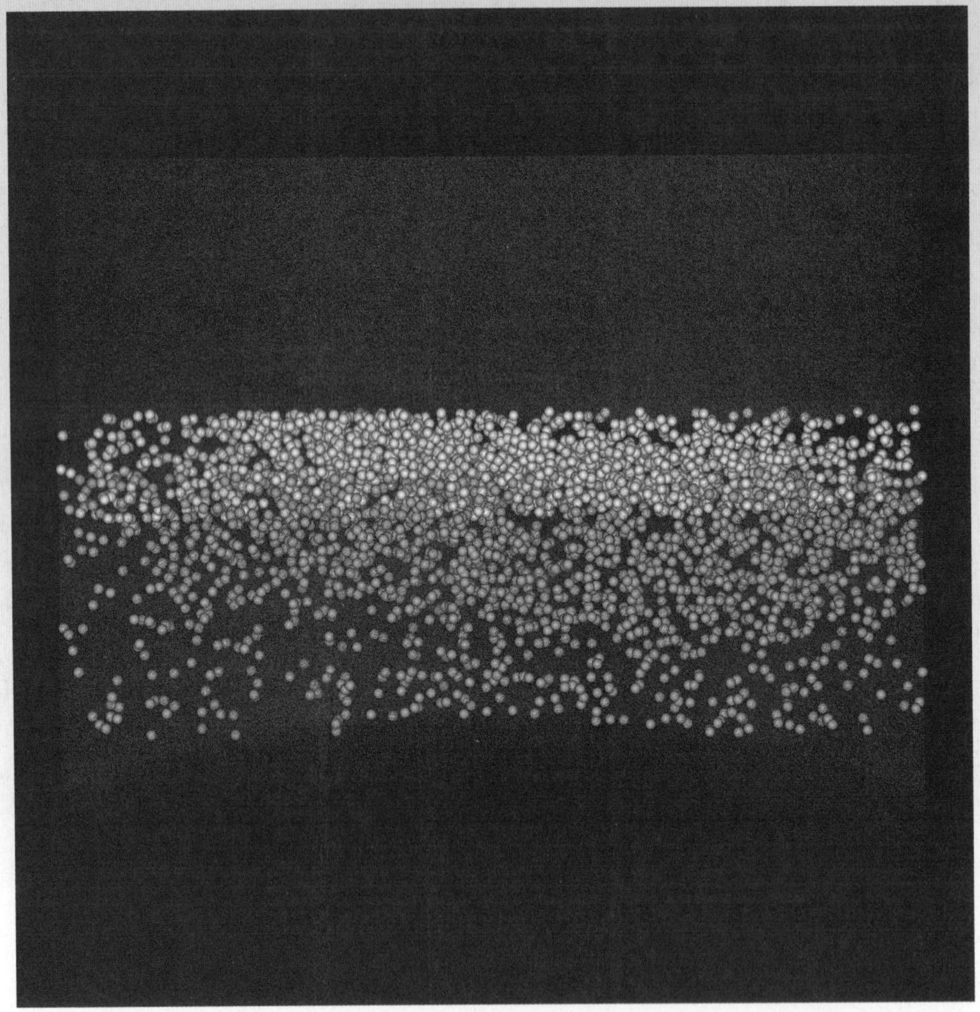
Courtesy of Cameca Instruments Inc.

Cross-sectional view of a three-dimensional map of dopant atoms (light blue spheres) implanted into a typical silicon transistor structure. Red dots represent the silicon atoms (only 2% are shown for clarity), and the gray spheres represent a native silicon dioxide layer located at the interface between the crystalline silicon substrate and layer of deposited polycrystalline silicon.

UNIT VI

Throughout history, the discovery of new materials from which to fashion the structures, machines, and devices of everyday life has set off great change in human affairs. Modern science and engineering—with chemistry in the central role—provide routes for modifying properties of materials to meet specific applications and for synthesizing and processing new materials designed from the beginning to have specific properties. The mechanical, thermal, electrical, and optical properties of a material depend on the extended nano-structural arrangement of chemical bonds within it. This arrangement can be created, modified, and tailored through chemical reactions. The contemporary disciplines of materials chemistry, solid state chemistry, and materials science and engineering—among the most active branches of chemistry today—all rely on the "properties↔structure↔reactions" correlation to make the leap from chemical bonding in isolated molecules (Unit II) to engineering applications.

Unit Chapters

Unit Goals

- To study the relationship between crystal symmetry and atomic-level structure as revealed by X-ray diffraction experiments
- To survey chemical bonding in classes of solids and correlate bonding with the properties of solids
- To explore three essential classes of materials—ceramics, optical and electronic materials, and polymers
- To illustrate the role of modern chemistry in measuring properties and identifying applications for both natural and manufactured materials

CHAPTER 21

Structure and Bonding in Solids

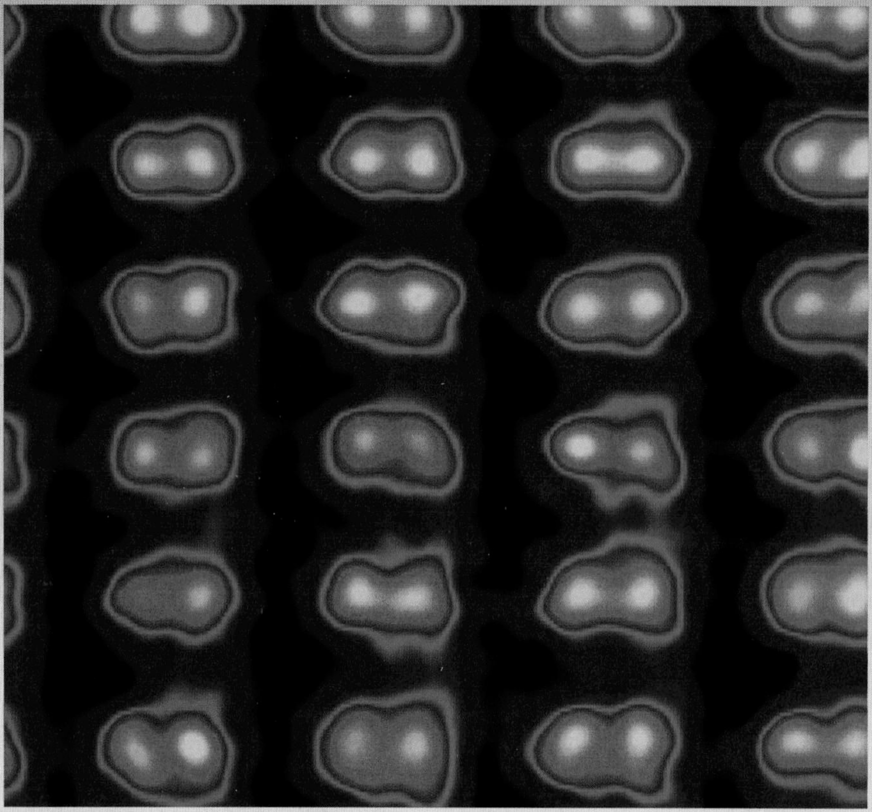

Pairs of silicon atoms separated by only 0.78 Å are clearly resolved in this ultrahigh-resolution electron microscope image.

In this chapter we begin a radical departure from the emphasis on the single molecule that permeates the previous 20 chapters. Unit II describes the chemical bonding and structure of single, isolated molecules. Unit III explains the macroscopic properties of gases and liquids through intermolecular forces that originate in the structures of individual molecules. While the behavior of many molecules contributes to these properties, the mechanism of their contribution is essentially that a few molecules come close enough to a "target" molecule to experience the intermolecular forces it sets up. Units IV and V deal with the equilibrium and rate aspects of chemical reactions as consequences of collisions between individual molecules.

The behavior of solids is a different story altogether. Chapter 10 identifies *rigidity* as the unique characteristic of solids, in dramatic contrast to the fluidity of gases and liquids. A rigid material retains its shape when an external mechanical force, called stress, is applied. A rigid material shows structural strength by not flowing under stress. All classes of solids behave this way because—with one exception—they are not collections of molecules held together by intermolecular forces. Rather, solids are extended arrays of strong chemical bonds between atoms, almost as if they were "super-molecules." These arrays extend over macroscopic distances, and it is the *collective behavior* of this set of chemical bonds that imparts strength and rigidity to solids.

A solid whose structure is highly ordered and symmetrical over macroscopic distances is called a **crystal**. Although the symmetry and beauty of crystals have always excited curiosity and wonder, the science of **crystallography** began only in the latter part of the 18th century. In those closing days of the Age of Enlightenment—while Lavoisier led the modern approach to chemistry—another brilliant French thinker established the fundamental laws of crystallography. René-Just Haüy was struck by the observation that when he accidentally dropped a crystal of calcite (a form of calcium carbonate), it fractured into smaller crystals with the same interfacial angles between their planar surfaces as in the original crystal. The statement that constant interfacial angles are observed when crystals are cleaved is now known as **Haüy's law.** Haüy concluded that the outward symmetry of crystals (Fig. 21.1) implies a highly regular internal structure and the existence of a smallest crystal unit. His inferences were correct. What distinguishes the crystalline state from the gaseous and liquid states is the nearly perfect positional order of the atoms, ions, or molecules in crystals. This crystalline order has been confirmed experimentally by X-ray diffraction and explained theoretically by the quantum theory of solids.

FIGURE 21.1 Calcite.

We begin this chapter with a look at the microscopic structure of a perfect crystal, and establish the methods for determining structure and the language for describing it. We then examine the types of chemical bonding in solids, identifying the forces that hold together different kinds of solids. The perfect crystal is the idealized model for investigations in solid state science, just as the ideal gas is the starting point for studies of fluid behavior. We use it as the point of reference for describing less ordered condensed phases of matter—defective crystals and amorphous solids—in terms of their deviations from perfect order and the consequent changes in properties. We end the chapter with a brief introduction to diffusion in solids, the mechanism by which a free atom migrates through an extended solid state structure. Diffusion has great influence on the rate and equilibrium of chemical reactions in the solid state and on tailoring the properties of the solids through carefully controlled incorporation of impurities.

21.1 Crystal Symmetry and the Unit Cell

The unifying aspect of crystal structure is the repetition, over long distances, of the same basic structural features in the arrangement of the atoms. The most fundamental way to characterize and classify these structures is based on the numbers and kinds of their **symmetry elements.** When the result of rotation, reflection, or inversion of an object can be exactly superimposed on the original object—that is, matched point for point to the original object—the structure is said to contain the corresponding symmetry element. Examples include an axis of rotation, a plane of reflection (mirror plane), or a central point (inversion center), as shown in Figure 21.2. These symmetry operations can be applied to geometrical shapes, to physical objects, and to molecular structures.

Consider a cube as an example. Suppose the center of the cube is placed at the origin of its coordinate system and the symmetry operations that transform it into identity with itself are counted (Fig. 21.3). The x, y, and z coordinate axes are 4-fold axes of rotational symmetry, denoted by C_4, because a cube that is rotated through a multiple of 90° (= 360°/4) about any one of these axes is

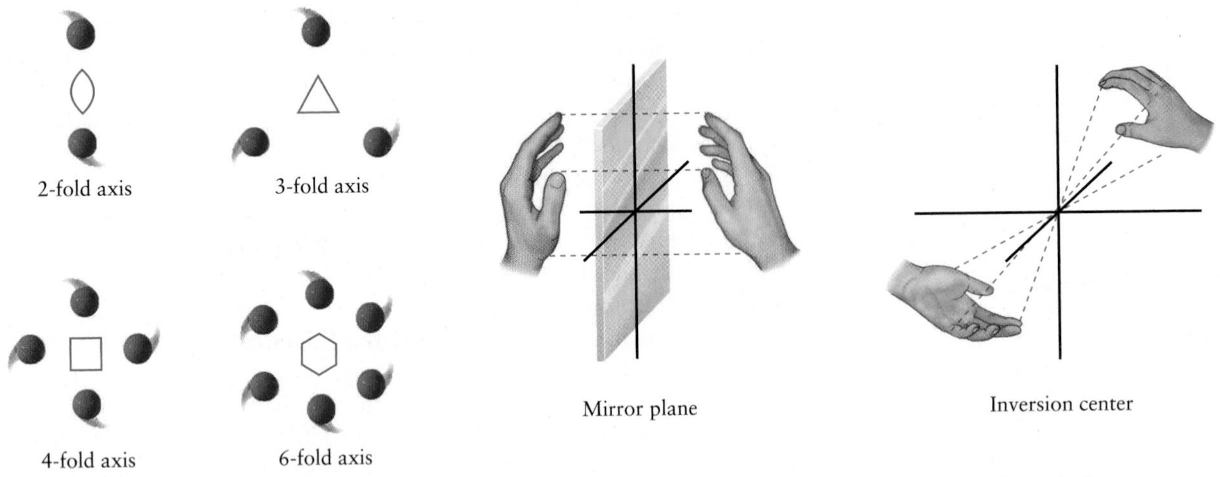

FIGURE 21.2 Three types of symmetry operation: rotation about an *n*-fold axis (with each axis going through the center of the structure and perpendicular to the plane of the page), reflection in a plane, and inversion through a point.

indistinguishable from the original cube. Similarly, a cube has four 3-fold axes of rotational symmetry, designated C_3, that are the body diagonals of the cube, connecting opposite vertices. In addition, a cube has six 2-fold rotational axes of symmetry, defined by the six axes that pass through the centers of edges and through the coordinate origin. Next, the cube has nine mirror planes of symmetry (designated by the symbol m), which reflect any point in one half of the cube into an equivalent point in the other half. Finally, a cube has a center of inversion (inversion, designated i, is reflection through a point located at the center of the cube).

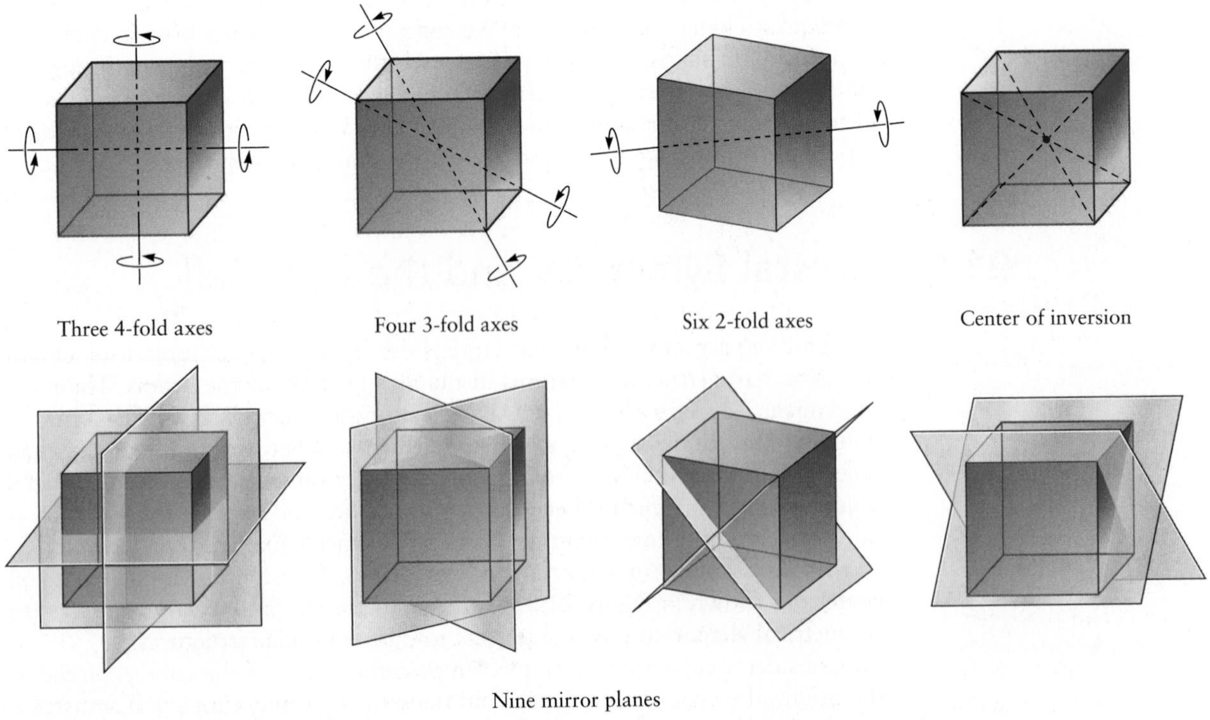

FIGURE 21.3 Symmetry operations on a cube. There are three 4-fold axes, only two of which are shown, four 3-fold axes, only two of which are shown, and six 2-fold axes, only one of which is shown. The rest can be determined by connecting equivalent pairs of points by inspection. The nine planes of symmetry are also shown, as is the center of inversion.

EXAMPLE 21.1

Identify the symmetry elements of the ammonia molecule (NH_3).

Solution

If the ammonia molecule is drawn as a pyramid with the nitrogen atom at the top (Fig. 21.4), then the only axis of rotational symmetry is a 3-fold axis passing downward through the N atom. Three mirror planes intersect at this 3-fold axis.

Related Problems: 3, 4

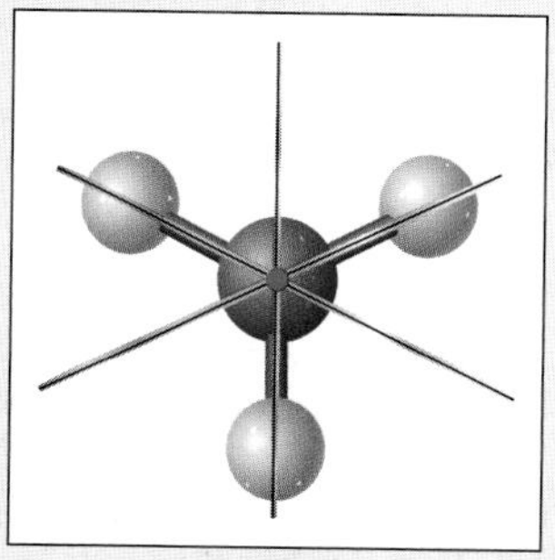

FIGURE 21.4 An ammonia molecule has a 3-fold axis rotation axis and three mirror planes. The rotation axis is represented by the red dot, which indicates that the rotation occurs around an axis perpendicular to the plane of the page. The three mirror planes are also oriented perpendicular to the plane of the page.

21.1.1 Unit Cells in Crystals

The symmetry operations just described can be applied to crystals as well as to individual molecules or shapes. Identical sites within a crystal recur regularly because of long-range order in the organization of the atoms. The three-dimensional array made up of all the points within a crystal that have the same environment in the same orientation is a **crystal lattice**. Such a lattice is an abstraction "lifted away" from a real crystal, embodying the scheme of repetition at work in that crystal. The lattice of highest possible symmetry is that of the **cubic system**. This lattice is obtained by filling space with a series of identical cubes, each of which is called a **unit cell** of the system. A single unit cell contains all structural information about its crystal, because in principle the crystal could be constructed by making a great many copies of a single original unit cell and stacking them in a three-dimensional array. Unit cells fill space.

Other crystal systems besides cubic can be defined by their own unique unit cells. Constraints of symmetry permit only seven types of three-dimensional lattices. Each type has a unit cell with the shape of a parallelepiped (Fig. 21.5), whose size and shape are fully described by three edge lengths (a, b, and c) and the three angles between those edges (α, β, and γ). These lengths and angles are the **cell constants**. The symmetry that defines each of the seven crystal systems imposes conditions on the shape of the unit cell summarized in relations among the cell constants (Table 21.1, Fig. 21.6). The unit cell chosen is the smallest unit that has all the symmetry elements of the crystal lattice; there is no benefit in using large cells once all the symmetry elements have been included. A unit cell of the minimum size is **primitive** and shares each of the eight *lattice points* at its corners with seven other unit cells, giving one lattice point per unit cell.

Other possible unit cells with the same volume (an infinite number, in fact) could be constructed, and each could generate the macroscopic crystal by repeated elementary translations, but only those shown in Figure 21.6 possess the symmetry elements of their crystal systems. Figure 21.7 illustrates a few of the infinite number of cells that can be constructed for a two-dimensional rectangular lattice. Only the rectangular cell B in the figure has three 2-fold rotation axes and two mirror planes. Although the other cells all have the same area, each of them has only one 2-fold axis and no mirror planes; they are therefore not acceptable unit cells.

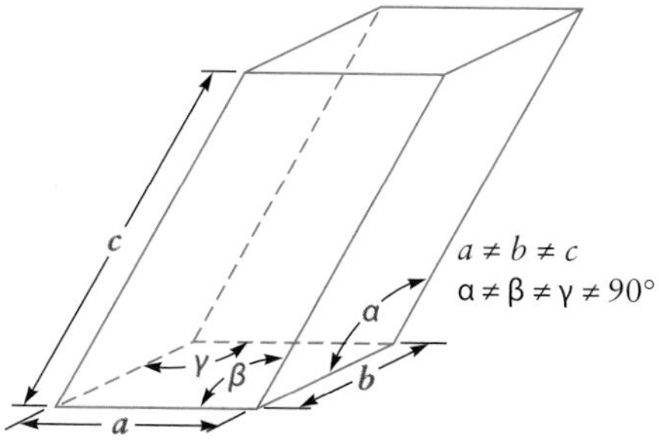

Triclinic

FIGURE 21.5 Unit cells always have three pairs of mutually parallel faces. Only six pieces of information are required to construct a scale model of a unit cell: the three cell edges (a, b, and c) and the three angles between the edges (α, β, and γ). By convention, γ is the angle between edges a and b, α the angle between b and c, and β the angle between a and c.

TABLE 21.1

The Seven Crystal Systems

Crystal System	Minimum Essential Symmetry	Conditions on Unit-Cell Edges and Angles
Hexagonal	One 6-fold rotation	$a = b$; $\alpha = \beta = 90°$, $\gamma = 120°$
Cubic	Four independent 3-fold rotations†	$a = b = c$; $\alpha = \beta = \gamma = 90°$
Tetragonal	One 4-fold rotation	$a = b$; $\alpha = \beta = \gamma = 90°$
Trigonal	One 3-fold rotation	$a = b = c$; $\alpha = \beta = \gamma \neq 90°$
Orthorhombic	Three mutually perpendicular 2-fold rotations	$\alpha = \beta = \gamma = 90°$
Monoclinic	One 2-fold rotation	$\alpha = \beta = 90°$
Triclinic	No symmetry required	None

†Each of these axes makes 70.53° angles with the other three.

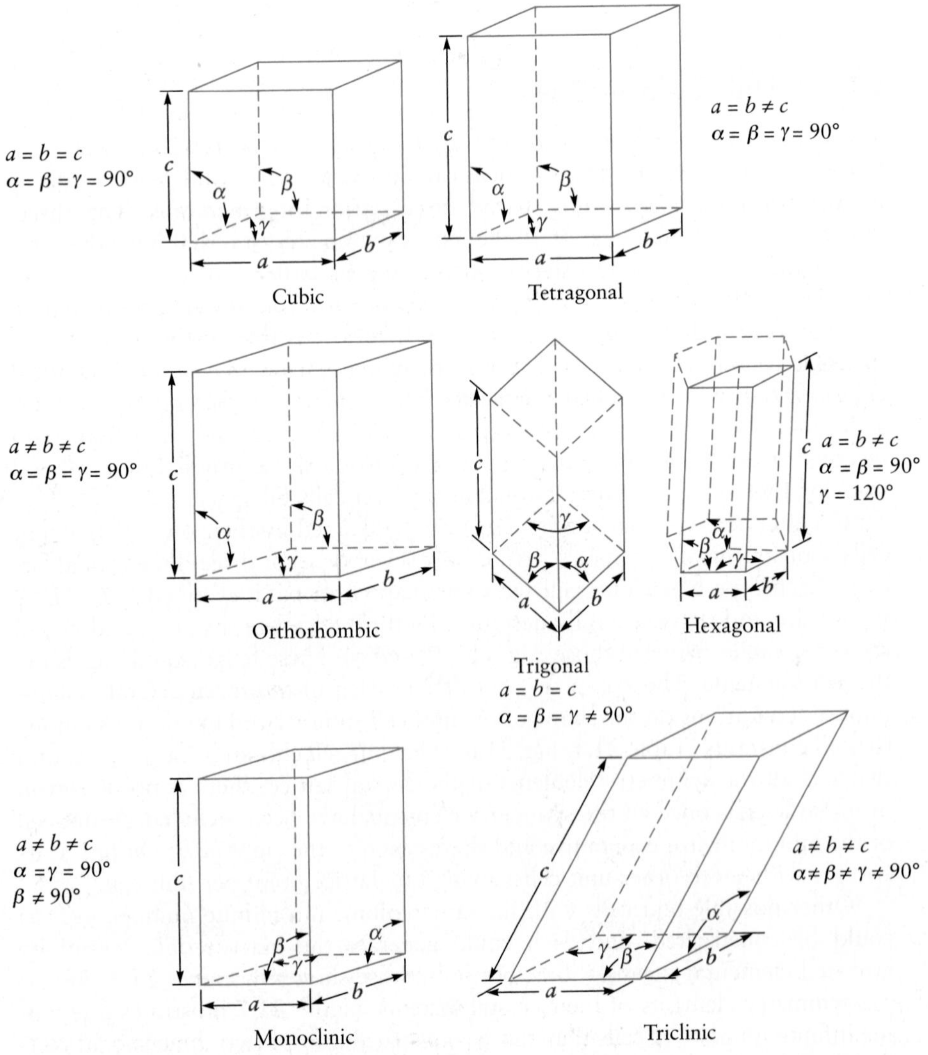

FIGURE 21.6 Shapes of the unit cells in the seven crystal systems. The more symmetric crystal systems have more symmetric cells.

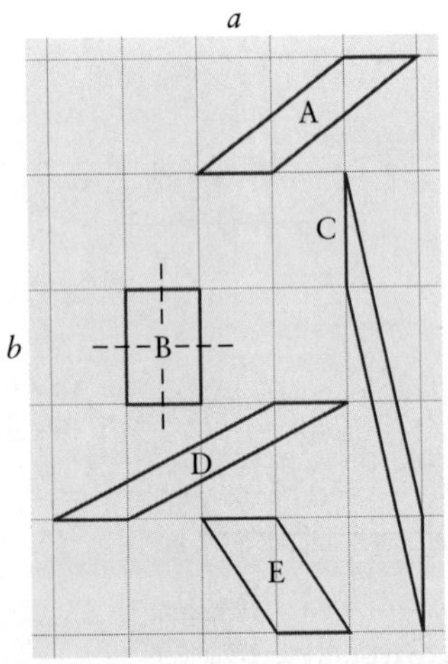

FIGURE 21.7 Cells with and without mirror planes in a two-dimensional lattice. Each cell has the same area, but only cell B possesses mirror planes in addition to the 2-fold rotation axes required by the lattice.

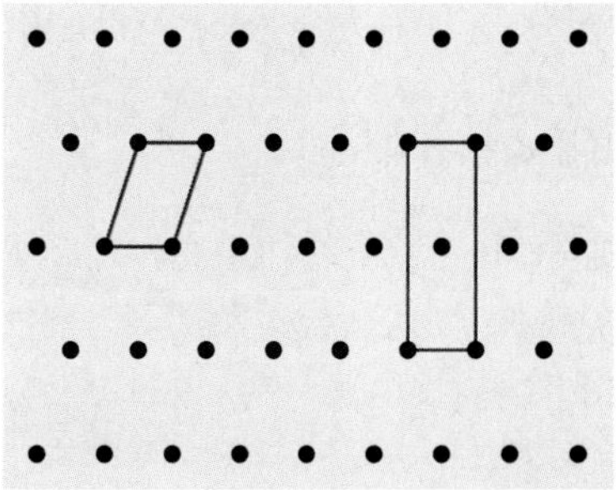

FIGURE 21.8 In this two-dimensional lattice, every lattice point is at the intersection of a horizontal mirror line and a vertical mirror line. It is possible to draw a primitive unit cell (red), but the larger, centered unit cell (blue) is preferred because it also has two mirror lines, the full symmetry of the lattice. A similar argument applies to the choice of unit cells on three-dimensional lattices.

Sometimes the smallest, or primitive, unit cell does not have the full symmetry of the crystal lattice. If so, a larger *nonprimitive* unit cell that does have the characteristic symmetry is deliberately chosen (Fig. 21.8). Only three types of nonprimitive cells are commonly used in the description of crystals: **body-centered, face-centered**, and **side-centered**. They are shown in Figure 21.9 for a cubic unit cell.

21.1.2 Scattering of X-Rays by Crystals

In the 19th century, crystallographers could classify crystals into the seven crystal systems only on the basis of their external symmetries. They could not measure the dimensions of unit cells or the positions of atoms within them. Several developments by German physicists changed this situation at the turn of the century. Wilhelm Roentgen's discovery of X-rays in 1895 provided a tool of enormous power for determining the structures of crystals. Max von Laue suggested that crystals might serve as three-dimensional gratings for the diffraction of electromagnetic radiation with a wavelength comparable to the distance between planes of atoms. Friedrich and Knipping demonstrated experimentally in 1912 that this was indeed the case, and von Laue was awarded the Nobel Prize in physics in 1914 for his theory of the diffraction of X-rays by crystals. At about the same time, W. H. Bragg and W. L. Bragg (father and son) at Cambridge University in England also demonstrated the diffraction of X-rays by crystals and shared the Nobel Prize in physics the following year. (W. L. Bragg was 22 years old and still a student at Cambridge when he discovered the diffraction law.) The formulation of the diffraction law proposed by the Braggs is equivalent to von Laue's suggestion and somewhat simpler to visualize. So we will follow an approach similar to theirs.

When electromagnetic radiation passes through matter, it interacts with the electrons in atoms, and some of it is scattered as spherical waves going out from the atoms in the solid. Suppose that X-radiation strikes two neighboring scattering centers. The expanding spheres of scattered waves soon encounter each other and interfere. In some directions, the waves are in phase and reinforce

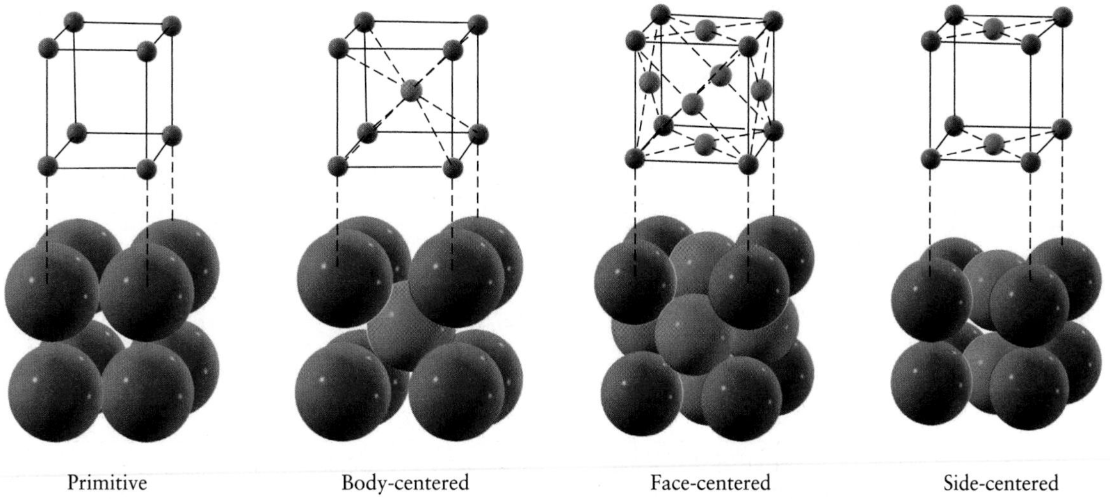

FIGURE 21.9 Centered lattices, like all lattices, have lattice points at the eight corners of the unit cell. A body-centered lattice has an additional lattice point at the center of the cell, a face-centered lattice has additional points at the centers of the six faces, and a side-centered lattice has points at the centers of two parallel sides of the unit cell. (**Note:** The colored dots in the lattice diagrams represent lattice points, not atoms.)

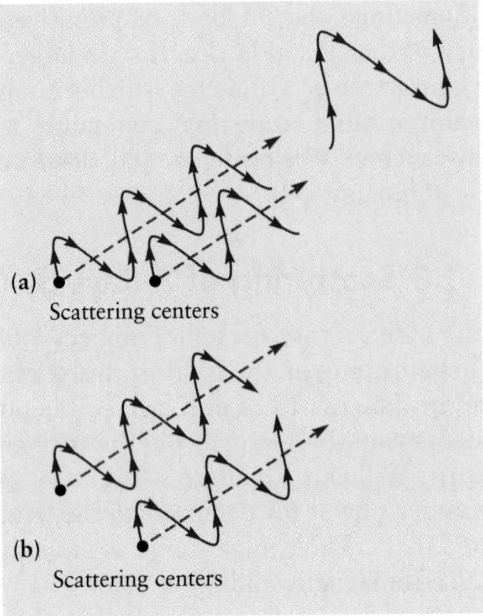

FIGURE 21.10 A beam of X-rays (not shown) is striking two scattering centers, which emit scattered radiation. The difference in the lengths of the paths followed by the scattered waves determines whether they interfere (a) constructively or (b) destructively. This path difference depends on both the distance between the centers and the direction in which the scattered waves are moving. If the incoming waves travel a different distance to each scattering center, that must be accounted for as well (see Fig. 21.11).

each other, or interfere *constructively* (Fig. 21.10a); in others they are out of phase and cancel each other out, or interfere *destructively* (see Fig. 21.10b). Constructive interference occurs when the paths traversed by two waves differ in length by a whole number of wavelengths. The amplitudes of waves that interfere constructively add to one another, and the intensity of the scattered radiation in that direction is proportional to the square of the total amplitude.

Figure 21.11 illustrates the constructive interference of X-rays scattered by the electrons in atoms in equally spaced planes separated by the distance d. A parallel bundle of coherent X-rays of a single known wavelength is allowed to fall on the surface of a crystal, making an angle θ with a set of parallel planes of atoms in the crystal. The scattering angle 2θ is then varied by rotating the crystal about an axis perpendicular to the plane of the figure. Line AD in Figure 21.11 represents a wave front of waves that are in phase as they approach the crystal. The wave that is scattered at B follows the path ABC, and the one that is scattered at F follows the path DFH. The second wave travels a greater distance than the first, and the difference in path length is the sum of the two segments EF and FG. To achieve constructive interference in the scattered waves (that is, for the phases to be the same along the wave front CH), this additional

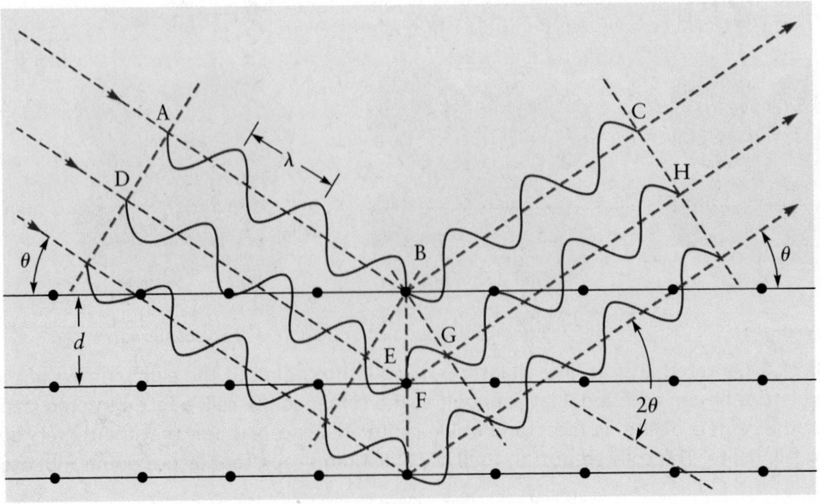

FIGURE 21.11 Constructive interference of X-rays scattered by atoms in lattice planes. Three beams of X-rays, scattered by atoms in three successive layers of a simple cubic crystal, are shown. Note that the phases of the waves are the same along the line CH, indicating constructive interference at this scattering angle 2θ.

distance traveled by the second wave must be an integral multiple of the X-ray wavelength λ:

$$EF + FG = n\lambda \qquad n = 1, 2, 3, \ldots$$

From trigonometry, the lengths of these two segments are equal to each other and to $d \sin \theta$, where d is the interplanar spacing. Therefore, constructive interference occurs only when

$$n\lambda = 2d \sin \theta \qquad n = 1, 2, 3, \ldots \qquad [21.1]$$

It is easy to verify that for angles that meet this condition, the waves scattered from the third and subsequent planes are also in phase with the waves scattered from the first two planes.

The preceding condition on allowed wavelengths is called the **Bragg law**, and the corresponding angles are called *Bragg angles* for that particular set of parallel planes of atoms. It appears as though the beam of X-rays has been reflected symmetrically from those crystal planes, and we often speak colloquially of the "Bragg reflection" of X-rays. The X-rays have not been reflected, however, but have undergone constructive interference, more commonly called diffraction. The case $n = 1$ is called first-order Bragg diffraction, $n = 2$ is second-order, and so forth.

We now possess a tool of immense value for determining the interplanar spacings of crystals. If a crystal is turned through different directions, other parallel sets of planes with different separations are brought into the Bragg condition. The symmetry of the resulting diffraction pattern identifies the crystal system, and the Bragg angles determine the cell constants. Moreover, the *intensities* of the diffracted beams permit the locations of the atoms in the unit cell to be determined.

Analogous scattering techniques use beams of neutrons. In that case the scattering interaction is between the magnetic moments of the incident neutrons and nuclei in the solid, but the principles are the same as for X-ray diffraction. Of course, it is the wave character of neutrons (in particular, their de Broglie wavelength; see Chapter 4) that is responsible for neutron diffraction. Recall that the de Broglie wavelength and neutron momentum are related by $\lambda = h/p$.

EXAMPLE 21.2

A diffraction pattern of aluminum is obtained by using X-rays with wavelength $\lambda = 0.709$ Å. The second-order Bragg diffraction from the parallel faces of the cubic unit cells is observed at the angle $2\theta = 20.2°$. Calculate the lattice parameter a.

Solution

From the Bragg condition for $n = 2$,

$$2\lambda = 2d \sin \theta$$

the spacing between planes, which is the lattice parameter, is

$$d = \frac{\lambda}{\sin \theta} = \frac{0.709 \text{Å}}{\sin(10.1°)} = 4.04 \text{ Å} = a$$

Related Problems: 5, 6

21.2 Crystal Structure

The crystal lattice is an abstract construction whose points of intersection describe the underlying symmetry of a crystal. To flesh out the description of a particular solid state structure, we must identify some structural elements that are "pinned" to the lattice points. These structural elements can be atoms, ions, or even groups of atoms as we see in this and the next chapter. We begin with some illustrative simple cases. Some of the chemical elements crystallize in particularly simple solid structures, in which a single atom is situated at each point of the lattice.

Polonium is the only element known to crystallize in the **simple cubic lattice,** with its atoms at the intersections of three sets of equally spaced planes that meet at right angles. Each unit cell contains one Po atom, separated from each of its six nearest neighbors by 3.35 Å.

The alkali metals crystallize in the **body-centered cubic (bcc) structure** at atmospheric pressure (Fig. 21.12). A unit cell of this structure contains two lattice points, one at the center of the cube and the other at any one of the eight corners. A single alkali-metal atom is associated with each lattice point. An alternative way to visualize this is to realize that each of the eight atoms that lie at the corners of a bcc unit cell is shared by the eight unit cells that meet at those corners. The contribution of the atoms to one unit cell is therefore $8 \times \frac{1}{8} = 1$ atom, to which is added the atom that lies wholly within that cell at its center.

The metals aluminum, nickel, copper, and silver, among others, crystallize in the **face-centered cubic (fcc) structure** shown in Figure 21.13. This unit cell contains four lattice points, with a single atom associated with each point. No atom

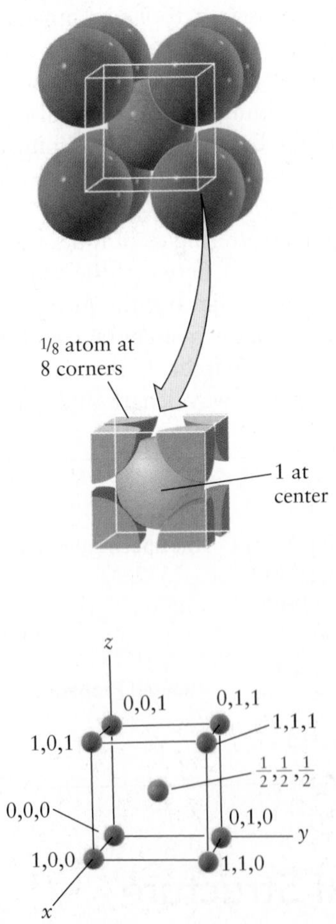

FIGURE 21.12 The bcc structure. An atom is located at the center of each cubic cell (orange) as well as at each corner of the cube (blue). The atoms are reduced slightly in size to make positions clear.

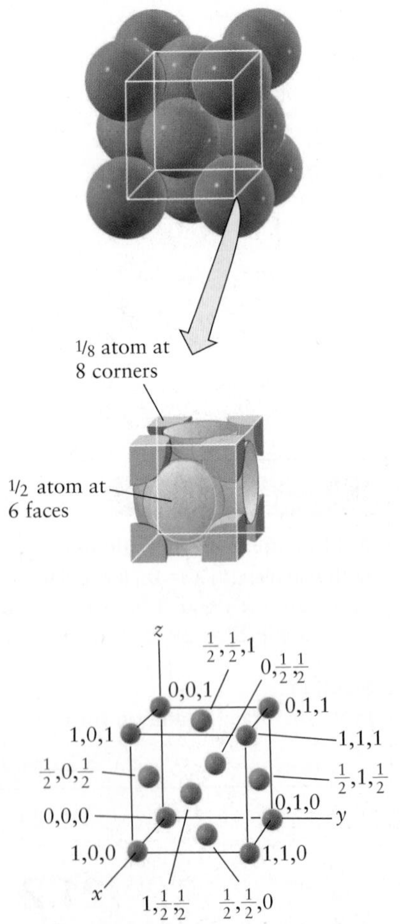

FIGURE 21.13 The fcc structure. Atoms are located at the centers of the faces (orange) as well as at the corners of the cube (blue). The atoms are reduced slightly in size to make positions clear.

lies wholly within the unit cell; there are atoms at the centers of its six faces, each of which is shared with another cell (contributing $6 \times \frac{1}{2} = 3$ atoms), and an atom at each corner of the cell (contributing $8 \times \frac{1}{8} = 1$ atom), for a total of four atoms per unit cell.

The volume of a unit cell is given by the formula

$$V_c = abc\sqrt{1 - \cos^2\alpha - \cos^2\beta - \cos^2\gamma + 2\cos\alpha\cos\beta\cos\gamma} \quad [21.2]$$

When the angles are all 90° (so that their cosines are 0), this formula reduces to the simple result $V = abc$ for the volume of a rectangular box. If the mass of the unit cell contents is known, the theoretical cell density can be computed. This density must come close to the measured density of the crystal, a quantity that can be determined by entirely independent experiments. For an *element* whose crystal contains n_c atoms per unit cell, the calculated cell density is

$$\text{density} = \rho = \frac{\text{mass}}{\text{volume}} = \frac{n_c\left(\frac{\mathcal{M}}{N_A}\right)}{V_c} = \frac{n_c\mathcal{M}}{N_A V_c}$$

where we have used the fact that the molar mass $\mathcal{M}$ of the element divided by Avogadro's number N_A is the mass of a single atom. This equation can also be used to calculate Avogadro's number from the measured density and cell constants, as the following example illustrates.

EXAMPLE 21.3

Sodium has a density of $\rho = 0.9700$ g cm^{-3} at 20°C, and its lattice parameter is $a = 4.2856$ Å. What is the value of Avogadro's number, given that the molar mass of sodium is 22.9898 g mol^{-1}?

Solution

Sodium has two atoms per unit cell because its structure is bcc, and the volume of the unit cell is a^3. Solve the foregoing equation for Avogadro's number:

$$N_A = \frac{n_c\mathcal{M}}{\rho a^3}$$

$$= \frac{(2)(22.9898 \text{ g mol}^{-1})}{(0.9700 \text{ g cm}^{-3})(4.2856 \times 10^{-8} \text{ cm})^3}$$

$$= 6.022 \times 10^{23} \text{ mol}^{-1}$$

Related Problems: 15, 16

It is useful to define the locations of atoms in the unit cell with a set of three numbers measured in units of the lattice parameter(s). For this purpose, one corner of the unit cell is taken to be at the origin of the coordinate axes appropriate to the crystal system, and an atom at that lattice point has the coordinates (0, 0, 0). Equivalent atoms at the seven remaining corners of the cell then have the coordinates (1, 0, 0), (0, 1, 0), (0, 0, 1), (1, 0, 1), (1, 1, 0), (0, 1, 1), and (1, 1, 1). These atom positions are generated from an atom at (0, 0, 0) by successive translations through unit distances along the three axes. An atom in a body-centered site has coordinates $\left(\frac{1}{2}, \frac{1}{2}, \frac{1}{2}\right)$ and the prescription for locating it is to proceed from the coordinate origin at (0, 0, 0) a distance $a/2$ along a, then a distance $b/2$ along b, and finally $c/2$ along c. In the same way, an atom in a face-centered site has coordinates such as $\left(\frac{1}{2}, \frac{1}{2}, 0\right)$, $\left(\frac{1}{2}, 0, \frac{1}{2}\right)$, or $\left(0, \frac{1}{2}, \frac{1}{2}\right)$.

So far, we have considered only cubic metals in which one atom corresponds to each lattice point. More complicated structures also occur, even for the elements, in which atoms occupy positions that are not lattice points. Diamond has an fcc structure with eight atoms (not four) per unit cell (see Fig. 21.22). Boron has a tetragonal structure with a very complex unit cell that contains 50 atoms. In molecular crystals the number of atoms per unit cell can be still greater; in a protein with a molecular mass of 10^5, there may be tens of thousands of atoms per unit cell. The computer-generated model of the protein myoglobin shown in Figure 23.21 is representative of the kinds of complex structures that are now routinely determined using modern X-ray crystallography.

21.2.1 Atomic Packing in Crystals

As we begin to add structural features like atoms onto crystal lattice sites, we have to pay attention to the size of the features. How efficiently can we pack atoms onto a lattice? We have already discussed two measures of the "size" of an atom or molecule. In Section 9.6, the van der Waals parameter b was related to the volume excluded per mole of molecules, so b/N_A is one measure of molecular size. In Section 5.5, we defined an approximate radius of an atom as the distance at which the electron density had fallen off to a particular value, or as the radius of a sphere containing a certain fraction of the total electron density. A third related measure of atomic size is based on the interatomic separations in a crystal. The radius of a noble-gas or metallic atom can be approximated as half the distance between the center of an atom and the center of its nearest neighbor in the crystal. We picture crystal structures as resulting from packing spheres in which nearest neighbors are in contact.

The nearest neighbor separation in a simple cubic crystal is equal to the lattice parameter a, so the atomic radius in that case is $a/2$. For the bcc lattice, the central atom in the unit cell "touches" each of the eight atoms at the corners of the cube, but those at the corners do not touch one another, as Figure 21.12 shows. The nearest-neighbor-distance is calculated from the Cartesian coordinates of the atom at the origin (0, 0, 0) and that at the cell center ($a/2$, $a/2$, $a/2$). By the Pythagorean theorem, the distance between these points is $\sqrt{(a/2)^2 + (a/2)^2 + (a/2)^2} = a\sqrt{3}/2$, so the atomic radius is $a\sqrt{3}/4$. Figure 21.13 shows that in an fcc crystal the atom at the center of a face [such as at (0, $a/2$, $a/2$)] touches each of the neighboring corner atoms [such as at (0, 0, 0)], so the nearest-neighbor distance is $\sqrt{(a/2)^2 + (a/2)^2} = a\sqrt{2}/2$ and the atomic radius is $a\sqrt{2}/4$. Table 21.2 summarizes the results for cubic lattices.

What is the most dense crystal packing that can be achieved? To answer this question, construct a crystal by first putting down a plane of atoms with the highest possible density, shown in Figure 21.14a. Each sphere is in contact with six other spheres in the plane. Then put down a second close-packed plane on top of the first one (see Fig. 21.14b) in such a way that each sphere in the second plane is in contact with three spheres in the plane below it; that is, each sphere in the second plane forms a tetrahedron with three spheres beneath. When the third plane is laid down, there are two possibilities. In Figure 21.14c, the atoms

TABLE 21.2

Structural Properties of Cubic Lattices

	Simple Cubic	Body-Centered Cubic	Face-Centered Cubic
Lattice points per cell	1	2	4
Number of nearest neighbors	6	8	12
Nearest-neighbor distance	A	$a\sqrt{3}/2 = 0.866a$	$a\sqrt{2}/2 = 0.707a$
Atomic radius	$a/2$	$a\sqrt{3}/4 = 0.433a$	$a\sqrt{2}/4 = 0.354a$
Packing fraction	$\frac{\pi}{6} = 0.524$	$\frac{\sqrt{3}\pi}{8} = 0.680$	$\frac{\sqrt{2}\pi}{6} = 0.740$

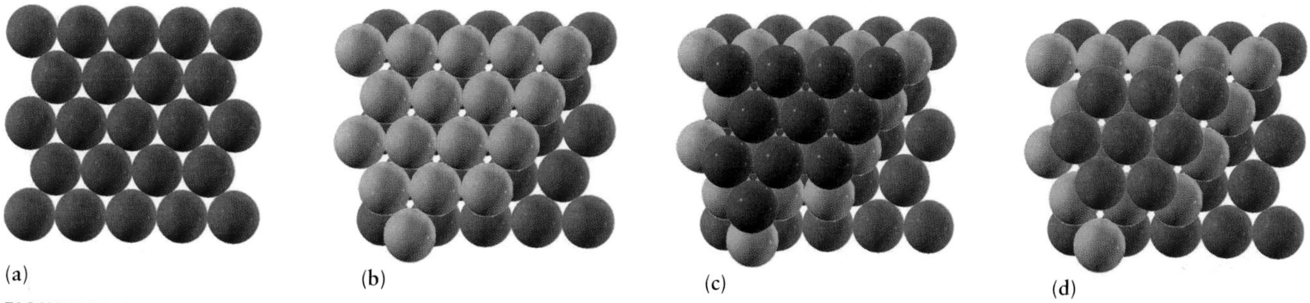

FIGURE 21.14 Close-packing of spheres. (a) One layer, with each atom surrounded by six nearest neighbors. (b) Two layers, with the atoms of the second layer centered on top of indentations in the layer below. (c) The third layer can be placed on sites that are not directly above the sites in the first layer (note that the red spheres show through). If this pattern is repeated as *abcabc* . . . it gives cubic close-packing. (d) Alternatively, the third layer can be placed directly over the sites in the first layer (note that the white triangular spaces show through). The repeated pattern *ababab* . . . gives hexagonal close-packing.

in the third plane lie on sites not directly over those in the first layer, whereas in Figure 21.14d the third-plane atoms are directly over the first-plane atoms.

Clearly, there are two choices for placing each plane, and an infinite number of crystal structures can be generated that have the same atomic packing density. The two simplest such structures correspond to the periodic layer sequences *abcabcab* . . . and *abababa*. . . . The first of these is the fcc structure already discussed, and the second is a close-packed structure in the hexagonal crystal system termed **hexagonal close-packed (hcp) structure.** In each of these simple structures, atoms occupy 74.0% of the unit cell volume, as the following example shows. (Atoms that crystallize in the bcc structure occupy only 68.0% of the crystal volume, and the packing fraction for a simple cubic array is only 52.4%.)

EXAMPLE 21.4

Calculate (a) the atomic radius of an aluminum atom and (b) the fraction of the volume of aluminum that is occupied by its atoms.

Solution

(a) Aluminum crystallizes in the fcc crystal system, and its unit cell therefore contains four atoms. Because a face-centered atom touches each of the atoms at the corners of its face, the atomic radius r_1 (see the triangle in Fig. 21.15) can be expressed by

$$4r_1 = a\sqrt{2}$$

Using the value $a = 4.04$ Å derived from X-ray diffraction (see Example 21.2) and solving for r_1 gives

$$r_1 = 1.43 \text{ Å}$$

(b) The fraction of the volume of an aluminum single crystal that is occupied by its atoms is

$$f = \frac{4\left[\frac{4}{3}\pi r_1^3\right]}{a^3} = \frac{4\frac{4}{3}\pi\left[\frac{a\sqrt{2}}{4}\right]^3}{a^3} = 0.740$$

Related Problem: 23

21.2.2 Interstitial Sites

The ways in which the empty volume is distributed in a crystal are both interesting and important. For the close-packed fcc structure, two types of **interstitial sites**, upon which the free volume in the unit cell is centered, are identifiable. An *octahedral site* is surrounded at equal distances by six nearest-neighbor atoms. Figure 21.15 shows that such sites lie at the midpoints of the edges of the fcc unit cell. A cell has twelve edges, each of which is shared by four unit cells, so the edges contribute three octahedral interstitial sites per cell. In addition, the site at the center of the unit cell is also octahedral, so the total number of

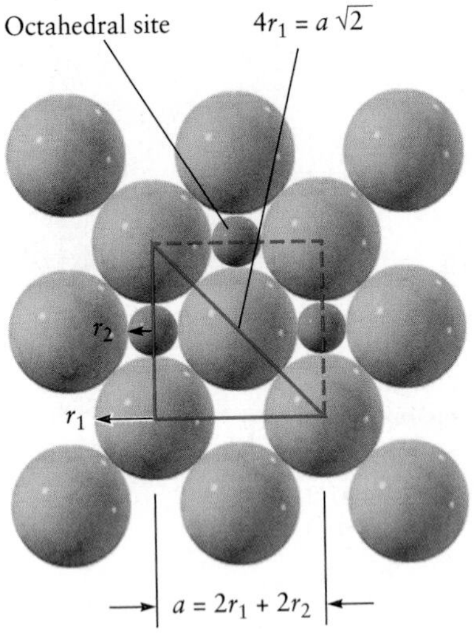

FIGURE 21.15 Octahedral sites in an fcc lattice. The geometric procedure for relating the site radius r_2 to the atom radius r_1 is shown.

octahedral sites per fcc unit cell is four, the same as the number of atoms in the unit cell.

With a bit of simple geometry, we can calculate the size of an octahedral site in an fcc structure or, more precisely, the radius r_2 of a smaller atom that would fit in the site without overlapping its neighboring atoms. Figure 21.15 represents a cell face in which the length of the diagonal is $4r_1$ and the length of the cell edge is $2r_1 + 2r_2$, where r_1 is the radius of the host atoms and r_2 is the radius of the octahedral site. From the figure,

$$r_1 = a\frac{\sqrt{2}}{4}$$

$$a = 2r_2 + 2r_1 = 2r_2 + 2a\frac{\sqrt{2}}{4}$$

$$r_2 = \frac{a}{2} - a\frac{\sqrt{2}}{4} = 0.146a$$

and the ratio of the octahedral-site radius to the host-atom radius is

$$\frac{r_2}{r_1} = \frac{0.146a}{a\sqrt{2}/4} = 0.414$$

The second type of interstitial site, known as a *tetrahedral site*, lies at the center of the space defined by four touching spheres. In an fcc cell, a tetrahedral site occurs in the volume between a corner atom and the three face-centered atoms nearest to it. Geometrical reasoning like that used for the octahedral site gives the following ratio of the radius of a tetrahedral site to that of a host atom:

$$\frac{r_2}{r_1} = 0.225$$

The fcc unit cell contains eight tetrahedral sites, twice the number of atoms in the cell.

Interstitial sites are important when a crystal contains atoms of several kinds with considerably different radii. We will return to this shortly when we consider the structures of ionic crystals.

21.3 Cohesion in Solids

In addition to symmetry, the nature of the bonding forces between atoms provides a useful way to classify solids. This classification does indeed lead to an understanding of the remarkable differences in the chemical and physical properties of different materials. We now consider crystals held together through ionic, metallic, or covalent bonding interactions, and the one class of solids held together by intermolecular forces.

21.3.1 Ionic Crystals

Compounds formed by atoms with significantly different electronegativities are largely ionic, and to a first approximation the ions can be treated as hard, charged spheres that occupy positions on the crystal lattice (see the ionic radii in Appendix F). All the elements of Groups I and II of the periodic table react with Group VI and VII elements to form ionic compounds, the great majority of which crystallize in the cubic system. The alkali-metal halides (except for the cesium halides), the ammonium halides, and the oxides and sulfides of the alkaline-earth metals all crystallize in the *rock-salt*, or **sodium chloride structure**

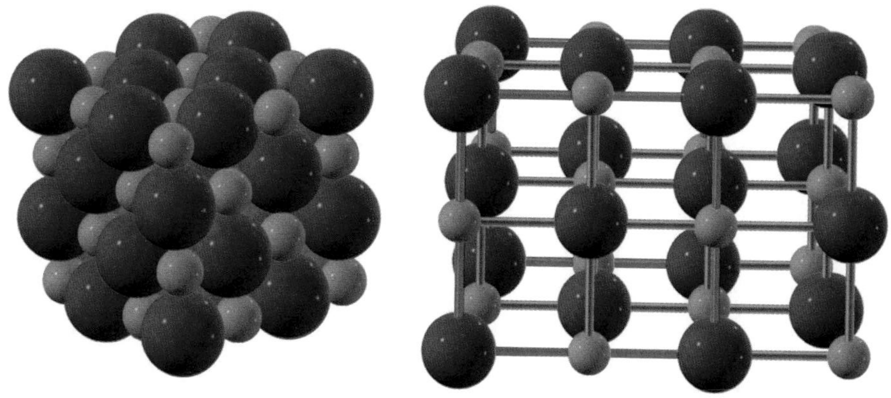

FIGURE 21.16 The sodium chloride, or rock-salt, structure. On the left, the sizes of the Na^+ ions (purple) and the Cl^- ions (green) are drawn to scale. On the right, the ions are reduced in size to allow a unit cell (shown by red lines) to be outlined clearly.

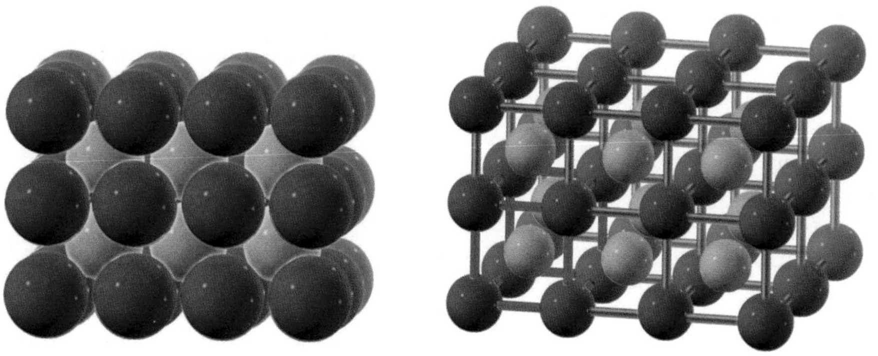

FIGURE 21.17 The structure of cesium chloride. On the left, the sizes of the Cs^+ ions (purplish pink) and the Cl^- ions (green) are drawn to scale. On the right, they are reduced in size to allow a unit cell (shown by red lines) to be outlined clearly. Note that the lattice in this structure is simple cubic, with one Cs^+ ion and one Cl^- ion per unit cell.

of ionic crystals shown in Figure 21.16. It may be viewed as an fcc lattice of anions whose octahedral sites are all occupied by cations or, equivalently, as an fcc lattice of cations whose octahedral sites are all occupied by anions. Either way, each ion is surrounded by six equidistant ions of the opposite charge. The rock-salt structure is a stable crystal structure when the cation-anion radius ratio lies between 0.414 and 0.732, if cations and anions are assumed to behave as incompressible charged spheres.

When the hard-sphere cation-anion radius ratio exceeds 0.732, as it does for the cesium halides, a different crystal structure called the **cesium chloride structure of ionic crystals**, is more stable. It may be viewed as two interpenetrating simple cubic lattices, one of anions and the other of cations, as shown in Figure 21.17. When the cation–anion radius ratio is less than 0.414, the *zinc blende*, or *sphalerite*, *structure* (named after the structure of ZnS) results. This crystal consists of an fcc lattice of S^{2-} ions, with Zn^{2+} ions occupying half of the available tetrahedral sites in alternation, as Figure 21.18a illustrates. Fluorite (CaF_2) has yet another structure; the unit cell is based on an fcc lattice of Ca^{2+} ions. The F^- ions occupy all eight of the tetrahedral sites, so the unit cell contains four Ca^{2+} and eight F^- ions (see Fig. 21.18b). The radius ratios (0.414 and 0.732) at which crossovers from one type of crystal to another occur are not accidental numbers. Recall from Section 21.2 that 0.414 is the ratio of the octahedral-site radius to the host-atom radius for an fcc lattice; only when this size ratio is exceeded does the ion inserted into that site come into contact with ions of the opposite sign in the rock-salt structure. The number 0.732 comes from a corresponding calculation of the radius ratio of the interstitial site at the center of a simple cubic unit cell (see problem 25). It is important to realize that the radius-ratio criterion for the stability limits of the structures of binary ionic compounds assumes that the ions are incompressible and that the wave functions do not overlap. The criterion fails when these approximations are not met.

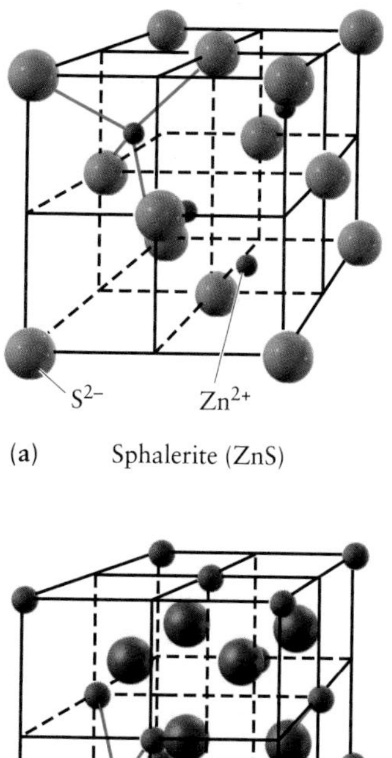

(a) Sphalerite (ZnS)

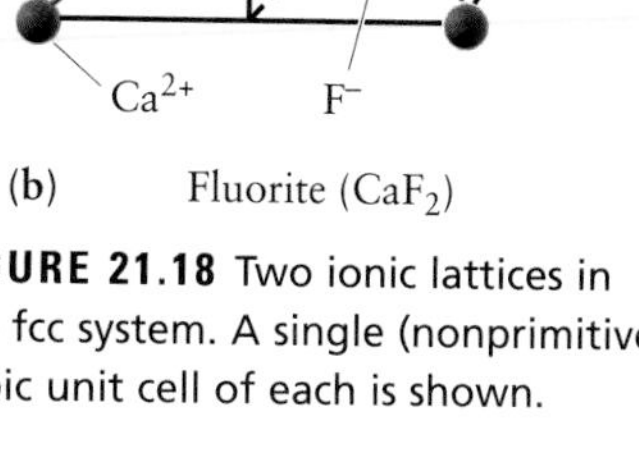

(b) Fluorite (CaF_2)

FIGURE 21.18 Two ionic lattices in the fcc system. A single (nonprimitive) cubic unit cell of each is shown.

The strength and range of the electrostatic attractions make ionic crystals hard, high-melting, brittle solids that are electrical insulators. Melting an ionic crystal, however, disrupts the lattice and sets the ions free to move, so ionic liquids are good electrical conductors.

21.3.2 Metallic Crystals

The type of bonding found in metals is quite different from that in other crystals. As we compare the various main group and transition metals in the periodic table we see only small differences in electronegativity. So, there is little tendency for ionic bonding in metals. The electronic configurations of metal atoms, even in the transition metals, do not have nearly-filled subshells, so there is little tendency to form covalent bonds by sharing electrons to achieve a stable octet. The familiar classical models of chemical bonding (see Chapter 3) do not extend to metals.

Prior to quantum mechanics, bonding in metals was described by the Drude model, named for the German physicist Paul Drude. The solid was viewed as a fixed array of positively charged metal ions, each localized at a site of the crystal lattice. These fixed ions were surrounded by a sea of mobile electrons, one contributed by each of the atoms in the solid. The number density of electrons was equal to the number density of positively charged ions, so the metal was electrically neutral. The sea of delocalized electrons would interact with the stable ions to give a strong cohesive force keeping the metal bound together. The Drude model accounts for the malleability (deformation in response to mechanical force, like hammering) and ductility (ease of drawing into a fine wire) of metals. As some ions move to new positions in response to these mechanical disturbances, delocalized electrons can rapidly adjust to maintain metallic bonding in the deformed or drawn solid. The ease with which metals conduct electricity is explained because these delocalized electrons can respond to any applied electric fields.

The delocalized sea of electrons picture of metal bonding survives in the quantum theory of solids, which is an extension of the molecular orbital description of molecular bonding (see Chapter 6). The valence electrons in a metal are delocalized in huge molecular orbitals that extend over the entire crystal and provide the "glue" that holds together the positively charged ion cores of the metal atoms. To understand the origin of these molecular orbitals, suppose just two sodium atoms are brought together, each in its electronic ground state with the configuration $1s^22s^22p^63s^1$. As the atoms approach each other, the wave functions of their 3*s* electrons combine to form two molecular orbitals—one in which their phases are symmetric (σ_{3s}) and another in which they are antisymmetric (σ^*_{3s}). Solving the Schrödinger equation yields two energy states, one above and the other below the energy of the atomic 3*s* levels, analogous to the formation of a hydrogen molecule from two hydrogen atoms described in Chapter 6. If both valence electrons are put into the level of lower energy with spins opposed, the result is a Na_2 molecule. If a third sodium atom is added, the 3*s* atomic levels of the atoms split into three sublevels (Fig. 21.19). Two electrons occupy the lowest level with their spins opposed, and the third electron occupies the middle level. The three energy levels and the three electrons belong collectively to the three sodium atoms. A fourth sodium atom could be added so that there would be four closely spaced energy sublevels, and this process could be carried on without limit.

The foregoing is not a mere "thought experiment." Sodium vapor contains about 17% Na_2 molecules at its normal boiling point. Larger sodium clusters have been produced in molecular beam experiments, and mass spectrometry

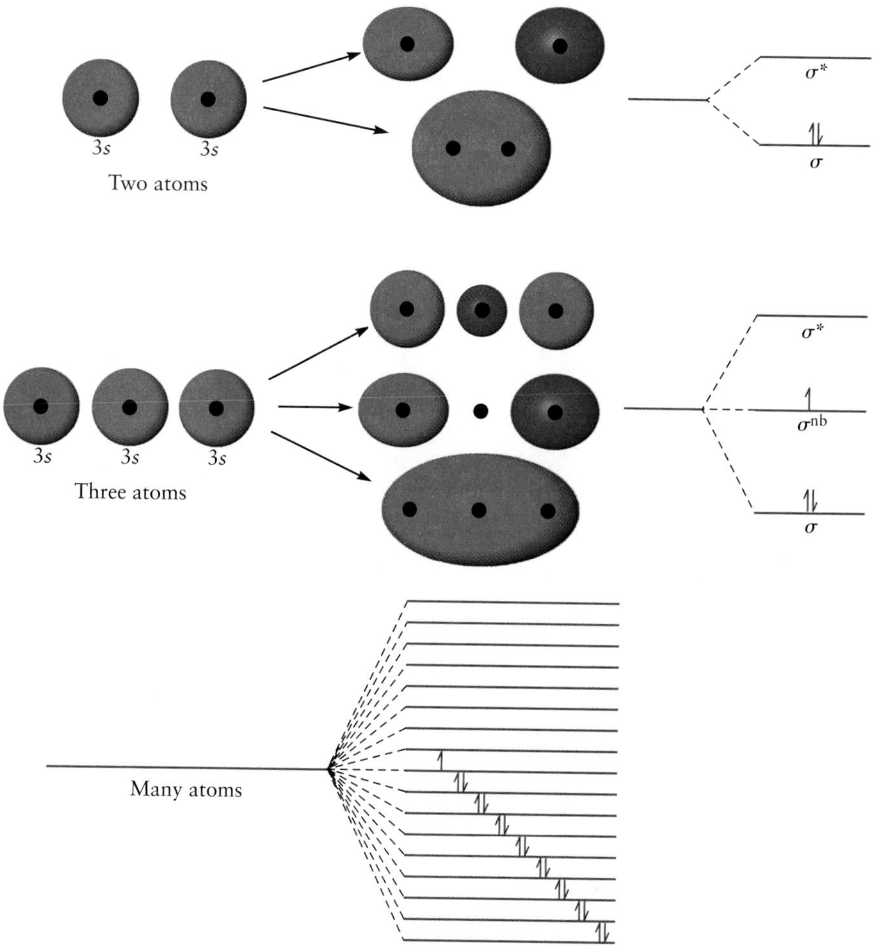

FIGURE 21.19 As sodium atoms are brought together, the molecular orbitals formed from their 3*s* atomic orbitals spread out into a band of levels, half occupied by electrons.

shows that such clusters can contain any desired number of atoms. For each added Na atom, another energy sublevel is added. Because the sublevels are so very closely spaced in a solid (with, say, 10^{23} atoms), the collection of sublevels can be regarded as an *energy band*. Figure 21.19 depicts the formation of bands of sublevels that broaden (become delocalized) as the spacing between the nuclear centers decreases, and the 3*s* electrons go into these bands. The electrons that belong to the 1*s*, 2*s*, and 2*p* atomic levels of sodium are only very slightly broadened at the equilibrium internuclear separation of the crystal, so they retain their distinct, localized character as the core levels of the ions at the lattice sites. Chapter 22 explores the electrical properties of metals, which arise from this band structure.

Most metals have crystal structures of high symmetry and crystallize in bcc, fcc, or hcp lattices (Fig. 21.20). Relatively few metals (Ga, In, Sn, Sb, Bi, and Hg) have more complex crystal structures. Many metals undergo phase transitions to other structures when the temperature or pressure is changed. Both liquid and crystal phases can be metals; in fact, the conductivity usually drops by only a small amount when a metal melts. The electron sea provides very strong binding in most metals, as shown by their high boiling points. Metals have a very large range of melting points. Gallium melts at 29.78°C (Fig. 21.21), and mercury stays liquid at temperatures that freeze water. Many transition metals require temperatures in excess of 1000°C to melt, and tungsten, the highest melting elemental metal, melts at 3410°C (see Section 8.1).

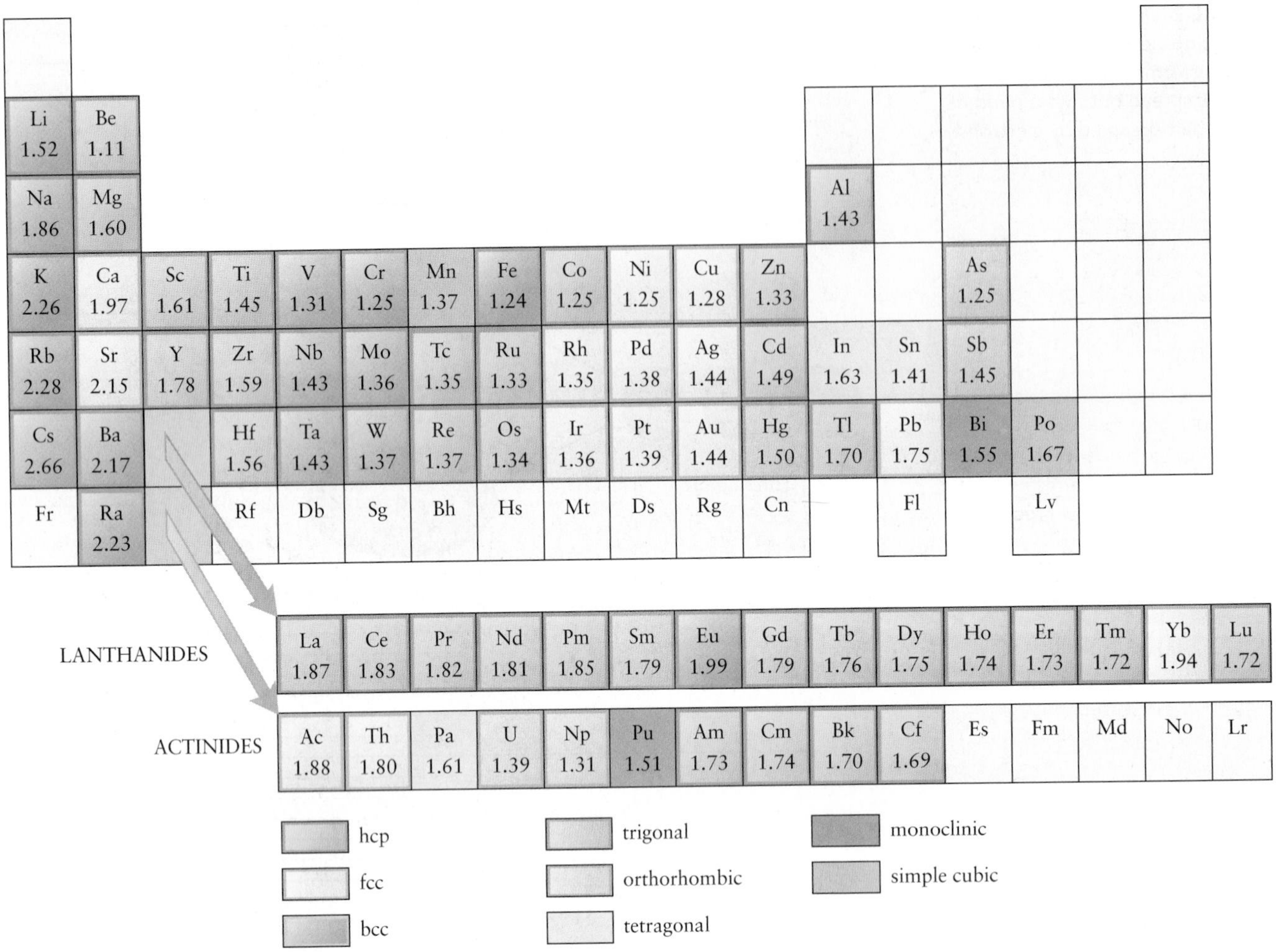

FIGURE 21.20 Crystal structures of the metallic elements at 25°C and 1 atm pressure. Atomic radii (Å) are calculated as one half the closest atom–atom distance in each structure; in most cases this is the same radius as calculated using the hard sphere contact model of Example 21.4. There are no known crystal structures for those elements for which atomic radii are not listed.

FIGURE 21.21 Solid gallium has a low melting point, low enough to melt from the heat of the body.

21.3.3 Covalent Crystals

We turn finally to a class of crystalline solids whose atoms are linked by covalent bonds rather than by the electrostatic attractions of ions or the valence electron "glue" in a metal. The archetype of the covalent crystal is diamond, which belongs to the cubic system. The ground-state electron configuration of a carbon atom is $1s^22s^22p^2$, and as shown in Section 6.9, its bonding is described by four hybrid sp^3 orbitals directed to the four corners of a regular tetrahedron. Each of the equivalent hybrid orbitals contains one electron that can spin-pair with the electron in one of the sp^3 orbitals of another carbon atom. Each carbon atom can thus link covalently to four others to yield the space-filling network shown in Figure 21.22. Covalent crystals are also called "network crystals," for obvious reasons. In a sense, every atom in a covalent crystal is part of one giant molecule that is the crystal itself. These crystals have very high melting points because of the strong attractions between covalently bound atoms. They are hard and brittle. Chapter 22 describes the electrical properties of covalent crystals.

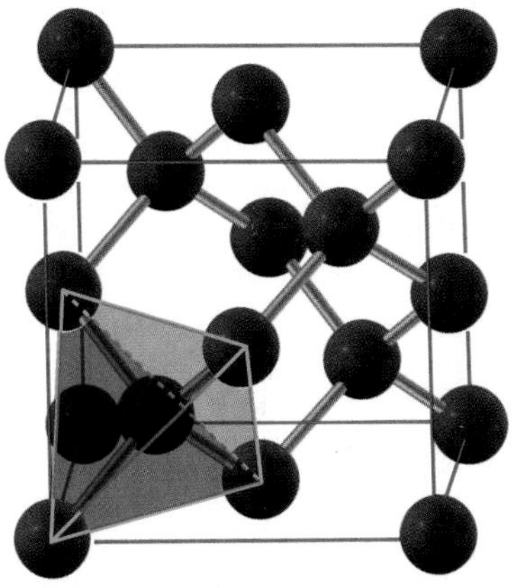

FIGURE 21.22 The structure of diamond. Each carbon atom has four nearest neighbors surrounding it at the corners of a tetrahedron.

21.3.4 Molecular Crystals

Molecular crystals include the noble gases; oxygen; nitrogen; the halogens; compounds such as carbon dioxide; metal halides of low ionicity such as Al_2Cl_6, $FeCl_3$, and $BiCl_3$; and the vast majority of organic compounds. All these molecules are held in their lattice sites by the intermolecular forces discussed in Sections 9.6 and 10.2. The trade-off between attractive and repulsive forces among even small molecules in a molecular crystal is complex because so many atoms are involved. A useful simplification is to picture a molecule as a set of fused spheres centered at each nucleus. The radius of each sphere is the van der Waals radius of the element involved. In molecular crystals, such shapes pack together so that no molecules overlap but empty space is minimized. Figure 21.23 depicts such a "space-filling model" of cyanuric triazide (C_3N_{12}), showing how nature solves the problem of efficiently packing many copies of the rather complicated molecular shape of C_3N_{12} in a single layer. In the three-dimensional molecular crystal of C_3N_{12}, many such layers stack up with a slight offset that minimizes unfilled space between layers.

Van der Waals forces are much weaker than the forces that operate in ionic, metallic, and covalent crystals. Consequently, molecular crystals typically have low melting points and are soft and easily deformed. Although at atmospheric pressure the noble-gas elements crystallize in the highly symmetric fcc lattice shown in Figure 21.13, molecules (especially those with complex geometries) more often form crystals of low symmetry in the monoclinic or triclinic systems. Molecular crystals are of great scientific value. If proteins and other macromolecules are obtained in the crystalline state, their structures can be determined by X-ray diffraction. Knowing the three-dimensional structures of biological molecules is the starting point for understanding their functions.

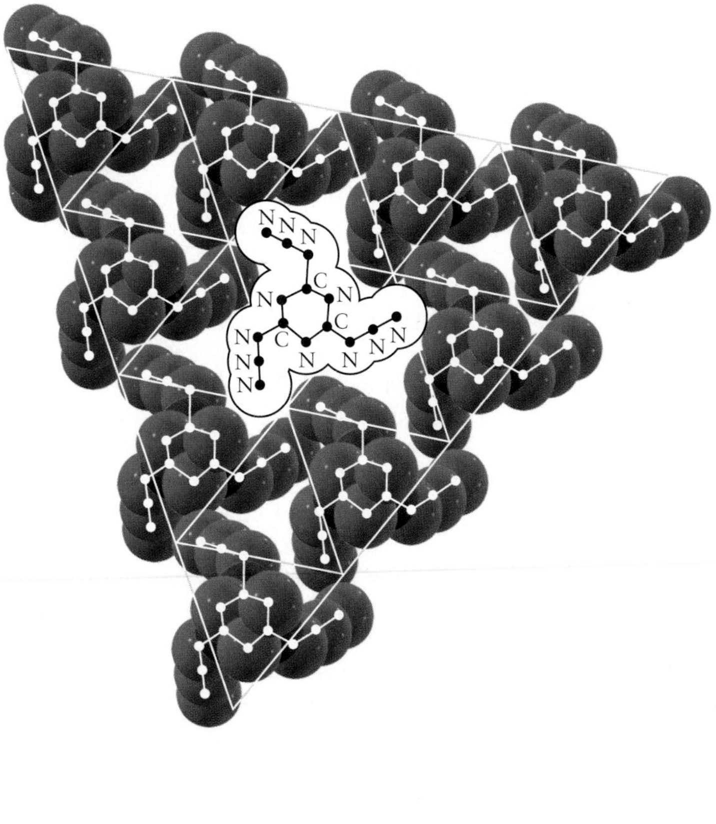

FIGURE 21.23 The van der Waals radii of the carbon and nitrogen atoms superimposed on an outline of the molecular structure of cyanuric triazide, C_3N_{12}, to show the volume of space from which each molecule excludes the others. Van der Waals forces in the molecular crystal hold the molecules in contact in a pattern that minimizes empty space. The thin white lines emphasize the 3-fold symmetry of the pattern.

21.3.5 Crystal Structures of the Elements

The chemical elements provide examples of three of the four classes of crystalline solids described in this section. Only ionic solids are excluded, because a single element cannot have the two types of atoms of different electronegativities needed to form an ionic material. We have already discussed some of the structures formed by metallic elements, which are sufficiently electropositive that their atoms readily give up electrons to form the electron sea of metallic bonding. The nonmetallic elements are more complex in their structures, reflecting a competition between intermolecular and intramolecular bonding and producing molecular or covalent solids with varied properties.

Each halogen atom has seven valence electrons and can react with one other halogen atom to form a diatomic molecule. Once this single bond forms, there is no further bonding capacity; the halogen diatomic molecules interact with one another only through relatively weak van der Waals forces and form molecular solids with low melting and boiling points.

The Group VI elements oxygen, sulfur, and selenium display dissimilar structures in the solid state. Each oxygen atom (with six valence electrons) can form one double or two single bonds. Except in ozone, its high-free-energy form, oxygen uses up all its bonding capacity with an intramolecular double bond, forming a molecular liquid and a molecular solid that are only weakly bound. In contrast, diatomic sulfur molecules (S=S) are relatively rare, being encountered only in high-temperature vapors. The favored forms of sulfur involve the bonding of every atom to two other sulfur atoms. This leads to either rings or chains, and both are observed. The stable form of sulfur at room temperature consists of S_8 molecules, with eight sulfur atoms arranged in a puckered ring (Fig. 21.24). The weak interactions between S_8 molecules make elemental sulfur a rather soft molecular solid. Above 160°C the rings in molten sulfur break open and relink to form long, tangled chains, producing a highly viscous liquid. An unstable ring form of selenium (Se_8) is known, but the thermodynamically stable form of this element is a gray crystal of metallic appearance that consists of very long spiral chains with weak interchain interaction. Crystalline tellurium has a similar structure. The Group VI elements thus show a trend (moving down the periodic table) from the formation of multiple bonds toward the chains and rings characteristic of atoms that each form two single bonds.

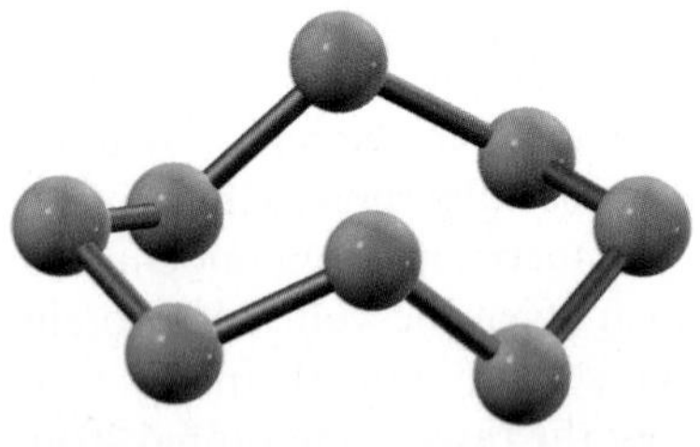

FIGURE 21.24 The structure of the S_8 sulfur molecule. The orthorhombic unit cell of rhombic sulfur, the most stable form of elemental sulfur at room temperature, is large and contains 16 of these S_8 molecules for a total of 128 atoms of sulfur.

A similar trend is evident in Group V. Only nitrogen forms diatomic molecules with triple bonds, in which all the bonding capacity is used between pairs of atoms. Elemental phosphorus exists in three forms, in all of which each phosphorus atom forms three single bonds rather than one triple bond. White phosphorus (Fig. 21.25a) consists of tetrahedral P_4 molecules, which interact with each other through weak van der Waals forces. Black phosphorus and red phosphorus (see Fig. 21.25b, c) are higher melting network solids in which the three bonds formed by each atom connect it directly or indirectly with all the other atoms in the sample. Unstable solid forms of arsenic and antimony that consist of As_4 or Sb_4 tetrahedra like those in white phosphorus can be prepared by rapid cooling of the vapor. The stable forms of these elements have structures related to that of black phosphorus.

The elements considered so far lie on the border between covalent and molecular solids. Other elements, those of intermediate electronegativity, exist as solids on the border between metallic and covalent; these are called *metalloids*. Antimony has a metallic luster, for example, but is a rather poor conductor of electricity and heat. Silicon and germanium are each an example of a **semiconductor**, with electrical conductivities far lower than those of metals but still significantly higher than those of true insulators such as diamond. Section 22.7 examines the special properties of these materials more closely.

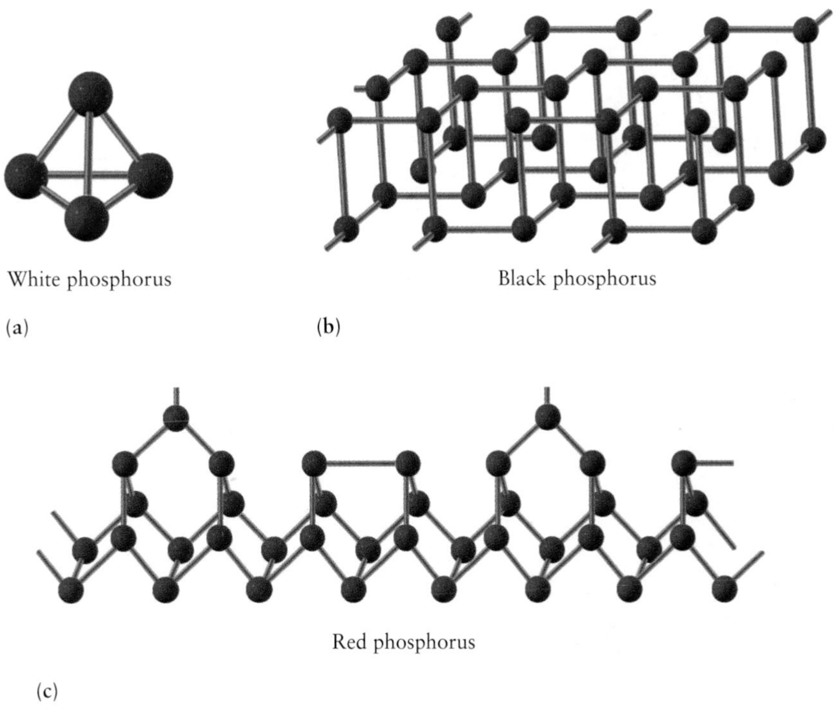

FIGURE 21.25 Structures of elemental phosphorus.

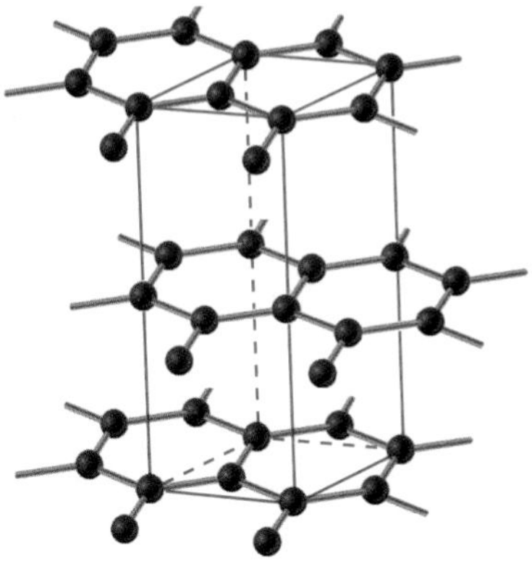

FIGURE 21.26 The structure of graphite.

Some elements of intermediate electronegativity exist in two crystalline forms with very different properties. White tin has a tetragonal crystal structure and is a metallic conductor. Below 13°C it crumbles slowly to form a powder of gray tin (with the diamond structure) that is a poor conductor. Its formation at low temperature is known as the "tin disease" and can be prevented by the addition of small amounts of bismuth or antimony. The thermodynamically stable form of carbon at room conditions is not the insulator diamond, but graphite. Graphite consists of sheets of fused hexagonal rings with only rather weak interactions between layers (Fig. 21.26). Each carbon atom shows sp^2 hybridization, with its remaining p orbital (perpendicular to the graphite layers) taking part in extended π-bonding interactions over the whole plane. Graphite can be pictured as a series of interlocked benzene rings, with π-electron delocalization contributing significantly to its stability. The delocalized electrons give graphite a significant value for conductivity in the planes of fused hexagons approaching that of the metallic elements. The conductivity and relative chemical inertness of graphite make it useful for electrodes in electrochemistry.

21.4 Defects and Amorphous Solids

Although real crystals display beautiful symmetries to the eye, they are not perfect. As a practical matter, it is impossible to rid a crystal of all impurities or to ensure that it contains perfect periodic ordering. So, we describe real crystals as "perfect crystals with defects," and define means to characterize these defects. If so many defects are present that crystalline order is destroyed, we describe the material as an **amorphous solid**.

21.4.1 Point Defects

Point defects in a pure crystalline substance include *vacancies*, in which atoms are missing from lattice sites, and **interstitials**, in which atoms are inserted in

Schottky defects in a metal or noble gas crystal

(a)

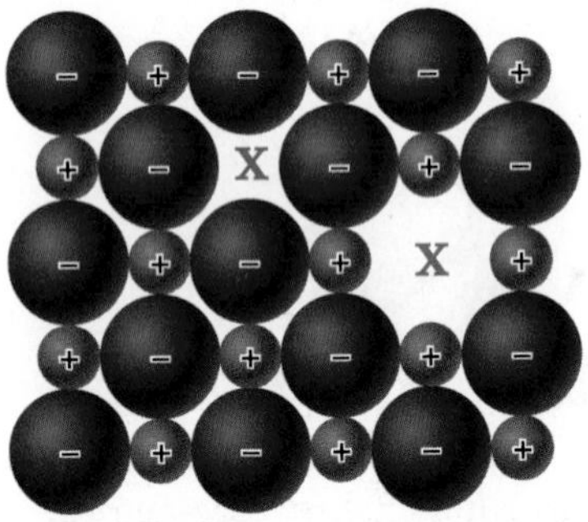

Schottky defects in an ionic crystal

(b)

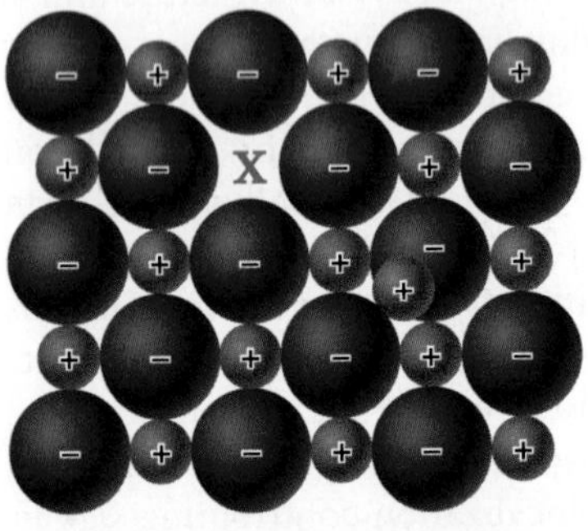

Frenkel defect in an ionic crystal

(c)

FIGURE 21.27 Point imperfections in a lattice. The red X's denote vacancies.

sites different from their normal sites. In real crystals, a small fraction of the normal atom sites remain unoccupied. Such a vacancy is called a **Schottky defect,** and their concentration depends on temperature:

$$N = N_s \exp(-\Delta G/RT)$$

where N is the number of lattice vacancies per unit volume, N_s is the number of atom sites per unit volume, and ΔG is the molar free energy of formation of vacancies. Figure 21.27a illustrates Schottky defects in the crystal structure of a metal or noble gas. Schottky defects also occur in ionic crystals but with the restriction that the imperfect crystals remain electrically neutral. Thus, in sodium chloride, for every missing Na^+ ion there must also be a missing Cl^- ion (see Fig. 21.27b).

In certain kinds of crystals, atoms or ions are displaced from their regular lattice sites to interstitial sites, and the crystal defect consists of the lattice vacancy plus the interstitial atom or ion. Figure 21.27c illustrates this type of lattice imperfection, known as a **Frenkel defect**. The silver halides (AgCl, AgBr, AgI) are examples of crystals in which Frenkel disorder is extreme. The crystal structures of these compounds are established primarily by the anion lattice, and the silver ions occupy highly disordered, almost random, sites. The rate of diffusion of silver ions in these solids is exceptionally high, as studies using radioactive isotopes of silver have shown. Both Frenkel and Schottky defects in crystals are mobile, jumping from one lattice site to a neighboring site with frequencies that depend on the temperature and the strengths of the atomic forces. Diffusion in crystalline solids is due largely to the presence and mobility of point defects; it is a thermally activated process, just like the rates of chemical reactions considered in Chapter 18. The coefficient of self-diffusion has the form

$$D = D_0 \exp[-E_a/RT] \qquad [21.3]$$

where E_a is the activation energy. The rates of diffusive motion in crystalline solids vary enormously from one substance to another. In a crystal of a low-melting metal such as sodium, an average atom undergoes about 10^8 diffusive jumps per second at 50°C, whereas in a metal such as tungsten that melts at 3410°C, an average atom jumps to another lattice site less than once per year at 1000°C!

If an alkali halide crystal such as NaCl is irradiated with X-rays, ultraviolet radiation, or high-energy electrons, some Cl^- ions may lose an electron:

$$Cl^- + h\nu \longrightarrow Cl + e^-$$

The resulting Cl atom, being uncharged and much smaller than a Cl^- ion, is no longer strongly bound in the crystal and can diffuse to the surface and escape. The electron can migrate through the crystal quite freely until it encounters an anion vacancy and is trapped in the Coulomb field of the surrounding cations (Fig. 21.28). This crystal defect is called an **F-center** (from the German word *Farbenzentrum,* meaning *color center*). It is the simplest of a family of electronic crystal defects. As the name suggests, it imparts a color to ionic crystals (Fig. 21.29).

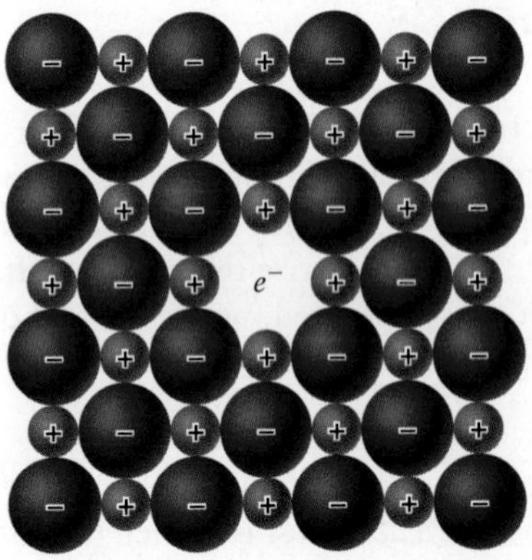

FIGURE 21.28 An F-center in a crystal.

21.4.2 Nonstoichiometric Compounds

As Chapter 1 emphasizes, the law of definite proportions was one of the principal pieces of evidence that led to the acceptance of Dalton's atomic theory. It is now recognized that a great many solid-state binary compounds do *not* have fixed and unvarying compositions but exist over a range of compositions in a

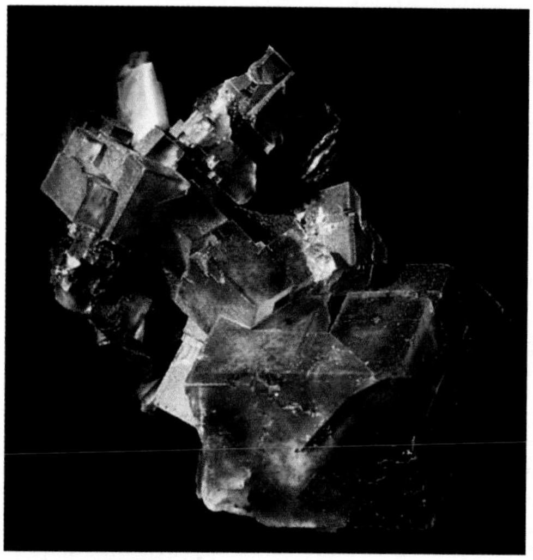

FIGURE 21.29 Pure calcium fluoride, CaF_2, is white but the natural sample of calcium fluoride (fluorite) shown here is a rich purple because F-centers are present. These are lattice sites where the F^- anion is replaced by an electron only.

single phase. Thus, FeO (wüstite) has the composition range $Fe_{0.85}O_{1.00}$ to $Fe_{0.95}O_{1.00}$ and is never found with its nominal 1:1 composition. The compounds NiO and Cu_2S also deviate considerably from their nominal stoichiometries.

The explanation depends on the existence of more than one oxidation state for the metal. In wüstite, iron can exist in either the +2 or the +3 oxidation state. Suppose a solid were to begin at the hypothetical composition of $Fe_{1.00}O_{1.00}$, with iron entirely in the +2 oxidation state. For every two Fe^{3+} ions introduced, three Fe^{2+} ions must be removed to maintain overall charge neutrality. The total number of moles of iron is then less than that in the ideal FeO stoichiometry. The departure from the nominal stoichiometry can be far more extreme than that found in wüstite. The composition of "TiO" ranges from $Ti_{0.75}O$ to $Ti_{1.45}O$. Nickel oxide varies only from $Ni_{0.97}O$ to NiO in composition, but the variation is accompanied by a dramatic change in properties. When the compound is prepared in the 1:1 composition, it is pale green and is an electrical insulator. When it is prepared in an excess of oxygen, it is black and conducts electricity fairly well. In the black material, a small fraction of Ni^{2+} ions are replaced by Ni^{3+} ions, and compensating vacancies occur at some nickel atom sites in the crystal.

EXAMPLE 21.5

The composition of a sample of wüstite is $Fe_{0.930}O_{1.00}$. What percentage of the iron is in the form of iron(III)?

Solution

For every 1.00 mol of oxygen atoms in this sample, there is 0.930 mol of iron atoms. Suppose y mol of the iron is in the +3 oxidation state and $0.930 - y$ is in the +2 oxidation state. Then the total positive charge from the iron (in moles of electron charge) is

$$+3y + 2(0.930 - y)$$

This positive charge must exactly balance the 2 mol of negative charge carried by the mole of oxygen atoms (recall that each oxygen atom has oxidation number −2). We conclude that

$$3y + 2(0.930 - y) = +2$$

Solving this equation for y gives

$$y = 0.140$$

The percentage of iron in the form of Fe^{3+} is then the ratio of this number to the total number of moles of iron, 0.930, multiplied by 100%:

$$\%\ \text{iron in form of } Fe^{3+} = \frac{0.140}{0.930} \times 100\% = 15.1\%$$

Related Problems: 37, 38

21.4.3 Alloys

The nonstoichiometric compounds just described are ionic materials with compositional disorder. A related type of disorder is exhibited by an **alloy**, a mixture of elements that displays metallic properties.

There are two types of alloys. In a **substitutional alloy**, some of the metal atoms in a crystal lattice are replaced by other atoms (usually of comparable size). Examples are brass, in which approximately one third of the atoms in a copper crystal are replaced by zinc atoms, and pewter, an alloy of tin that contains 7% copper, 6% bismuth, and 2% antimony. In an **interstitial alloy**, atoms of one or more additional elements enter the interstitial sites of the host metal lattice. An example is steel, in which carbon atoms occupy interstitial sites of an iron crystal, making the material stronger and harder than pure iron. Mild steel contains less than 0.2% C and is used for nails, whereas high-carbon steels can contain up to 1.5% C and are used in specialty applications such as tools and springs. *Alloy steels* are both substitutional and interstitial; atoms from metals such as chromium and vanadium substitute for iron atoms, with carbon remaining in interstitial sites. Alloy steels have a variety of specialized purposes, ranging from cutlery to bicycle frames.

21.4.4 Amorphous Solids and Glasses

The arrangements of atoms, ions, or molecules in crystalline solids exhibit high degrees of spatial order. Now let us briefly consider solids that lack this characteristic. An **amorphous solid**, commonly called a **glass**, resembles a crystalline solid in many respects. They may have chemical compositions, mechanical properties such as hardness and elasticity, and electrical and magnetic properties that are similar to those of crystals. Like crystals, glasses may have molecular, ionic, covalent, or metallic bonding. On an atomic scale, however, amorphous solids lack the regular periodic structure of crystals. They are states of matter in which so many defects are present that crystalline order is destroyed.

Some substances have a strong tendency to solidify as glasses. The best example is the material used in common window panes, with the approximate chemical formula $Na_2O{\cdot}CaO{\cdot}(SiO_2)_6$. This is a partly ionic, partly covalent material with Na^+ and Ca^{2+} ions distributed through a covalently bonded Si—O network. Glass-forming ability is not restricted to a few special materials, however. If a substance can be liquefied, it can almost certainly be prepared in an amorphous state. Even metals, which are known primarily in the crystalline state, have been made into amorphous solids. The trick is to bypass crystallization by cooling molten material very fast. One technique involves shooting a jet of liquid metal at a rapidly rotating cold cylinder, which produces a continuous ribbon of amorphous metal at a rate up to 2 km per minute.

FIGURE 21.30 Optical fibers (extremely thin glass fibers of specialized composition) carry information in the form of light waves.

On the molecular level, a strong tendency to form glasses is associated with the presence of long or irregularly shaped molecules that can easily become tangled and disordered. Even slowly cooling a liquid assembly of such molecules may not afford enough time for them to organize into a crystalline lattice before solidification. Instead of a sharp liquid-to-crystal transition, such glass-formers transform continuously, over a range of temperature, into amorphous solids. They lend themselves to fabrication into articles of every conceivable shape, because the flow properties of the work piece can be managed by controlling its temperature. This plasticity is the reason that glass has played an indispensable role in science, industry, and the arts.

One of the most exciting uses of a glass is the transmission of voice messages, television images, and data as light pulses. Tens of thousands of audio messages can be transmitted simultaneously through glass fibers no greater in diameter than a human hair. This is done by encoding the audio signal into electronic impulses that modulate light from a laser source. The light then passes down the glass fiber as though it were a tube. Chemical control of the glass composition reduces light loss and permits messages to travel many kilometers without amplification (Fig. 21.30).

21.5 A DEEPER LOOK Lattice Energies of Crystals

The **lattice energy** of a crystal is the energy required to separate the crystal into its component atoms, molecules, or ions at 0 K. In this section we examine the calculation and measurement of lattice energies for molecular and ionic crystals.

21.5.1 Lattice Energy of a Molecular Crystal

The lattice energy of a molecular crystal can be estimated by using the simple Lennard–Jones potential of Section 9.6:

$$V_{\mathrm{LJ}}(R) = 4\varepsilon\left[\left(\frac{\sigma}{R}\right)^{12} - \left(\frac{\sigma}{R}\right)^{6}\right]$$

Table 9.4 lists the values of ε and σ for various atoms and molecules. Note that this potential is most useful when the species in the crystal are noble gas atoms or nonpolar molecules like methane that interact due to van der Waals forces. To obtain the total potential energy for 1 mol, sum over all pairs of atoms or molecules:

$$V_{\mathrm{tot}} = \tfrac{1}{2}\sum_{i=1}^{N_{\mathrm{A}}}\sum_{j=1}^{N_{\mathrm{A}}} V_{\mathrm{LJ}}(R_{ij})$$

where R_{ij} is the distance between atom i and atom j. The factor $\frac{1}{2}$ arises because each interaction between a pair of atoms should be counted only once, not twice. For a crystal of macroscopic size, this can be rewritten as

$$V_{\mathrm{tot}} = \frac{N_{\mathrm{A}}}{2}\sum_{j=1}^{N_{\mathrm{A}}} V_{\mathrm{LJ}}(R_{ij})$$

where i is taken to be some atom in the middle of the crystal. Taking the nearest neighbor distance to be R_0, we define a ratio of distances $p_{ij} = R_{ij}/R_0$ and rewrite V_{tot} for the Lennard–Jones potential as

$$V_{\mathrm{tot}} = \frac{N_{\mathrm{A}}}{2}(4\varepsilon)\left[\sum_j\left(\frac{\sigma}{p_{ij}R_0}\right)^{12} - \sum_j\left(\frac{\sigma}{p_{ij}R_0}\right)^{6}\right]$$

$$= 2\varepsilon N_{\mathrm{A}}\left[\left(\frac{\sigma}{R_0}\right)^{12}\sum_j (p_{ij})^{-12} - \left(\frac{\sigma}{R_0}\right)^{6}\sum_j (p_{ij})^{-6}\right]$$

The two summations are dimensionless properties of the lattice structure, and accurate values can be obtained by summing over the first few sets of nearest neighbors (Table 21.3). The resulting total energy for the fcc lattice is

$$V_{\mathrm{tot}} = 2\varepsilon N_{\mathrm{A}}\left[12.132\left(\frac{\sigma}{R_0}\right)^{12} - 14.454\left(\frac{\sigma}{R_0}\right)^{6}\right]$$

The equilibrium atomic spacing at $T = 0$ K should be close to the one that gives a minimum in V_{tot}, which can be calculated by differentiating the preceding expression with respect to R_0 and setting the derivative to 0. The result is

$$R_0 = 1.09\sigma$$

TABLE 21.3

Lattice Sums for Molecular Crystals (fcc Structure)

	Number, n	p_{ij}	$n(p_{ij})^{-12}$	$n(p_{ij})^{-6}$
Nearest neighbors	12	1	12	12
Second nearest neighbors	6	$\sqrt{2}$	0.0938	0.750
Third nearest neighbors	24	$\sqrt{3}$	0.0329	0.889
Fourth nearest neighbors	12	2	0.0029	0.188
Fifth nearest neighbors	24	$\sqrt{5}$	0.0015	0.192
	⋮	⋮	⋮	⋮
Total			12.132	14.454

TABLE 21.4

Properties of Noble-Gas Crystals†

	R_0 (Å)		Lattice Energy (kJ mol^{-1})	
	Predicted	Observed	Predicted	Observed
Ne	3.00	3.13	1.83	1.88
Ar	3.71	3.76	7.72	7.74
Kr	3.92	4.01	11.50	11.20
Xe	4.47	4.35	15.20	16.00

†All data are extrapolated to 0 K and zero pressure.

and the value of V_{tot} at this value of R_0 is

$$V_{\text{tot}} = -8.61\varepsilon N_{\text{A}}$$

The corresponding potential energy when the atoms or molecules are completely separated from one another is zero. The lattice energy is the difference between these quantities and is a positive number:

$$\text{lattice energy} = -V_{\text{tot}} = 8.61\varepsilon N_{\text{A}}$$

This overestimates the true lattice energy because of the quantum effect of zero-point energy (see Section 4.6). When a quantum correction is applied, the binding energy is reduced by 28%, 10%, 6%, and 4% for Ne, Ar, Kr, and Xe, respectively. Table 21.4 shows the resulting predictions for crystal lattice energies and nearest neighbor distances. The agreement with experiment is quite reasonable, considering the approximations inherent in the use of a Lennard-Jones potential derived entirely from gas-phase data. For helium the amplitude of zero-point motion is so great that if a crystal did form, it would immediately melt. Consequently, helium remains liquid down to absolute zero at atmospheric pressure.

21.5.2 Lattice Energy of an Ionic Crystal

In Section 3.8, we calculated the potential energy of a gaseous diatomic ionic molecule relative to the separated ions by means of Coulomb's law:

$$V = \frac{q_1 q_2}{4\pi\epsilon_0 R_0}$$

where R_0 is the equilibrium internuclear separation. Coulomb's law can also be used to calculate the lattice energies of ionic compounds in the crystalline state.

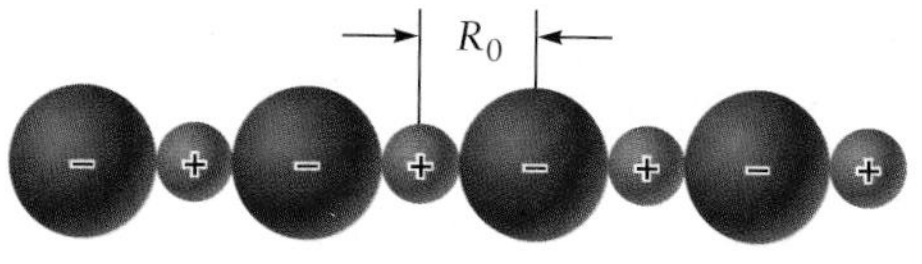

FIGURE 21.31 Lattice energy for a one-dimensional ionic crystal.

For simplicity, consider a hypothetical one-dimensional crystal (Fig. 21.31), in which ions of charge $+e$ and $-e$ alternate with an internuclear separation of R_0. One ion, selected to occupy an arbitrary origin, will interact attractively with all ions of opposite sign to make the following contribution to the crystal energy:

$$V_{\text{attraction}} = -\frac{e^2}{4\pi\epsilon_0 R_0}\left[2(1) + 2(\tfrac{1}{3}) + 2(\tfrac{1}{5}) + \ldots\right]$$

Here the factors of 2 come from the fact that there are *two* ions of opposite sign at a distance R_0 from a given ion, two at a distance $3R_0$, two at $5R_0$, and so forth. The negative sign occurs because the ions that occupy odd-numbered sites have a charge opposite that of the ion at the origin, and their interaction with the ion at the origin is attractive. The ion at the origin also interacts repulsively with all ions of the same sign to make the following contribution to the crystal energy:

$$V_{\text{repulsion}} = +\frac{e^2}{4\pi\epsilon_0 R_0}\left[2(\tfrac{1}{2}) + 2(\tfrac{1}{4}) + 2(\tfrac{1}{6}) + \ldots\right]$$

The net interaction of N_A such ions of each sign with one another is:

$$V_{\text{net}} = -\frac{N_A e^2}{4\pi\epsilon_0 R_0}\left[2 - \frac{2}{2} + \frac{2}{3} - \frac{2}{4} + \frac{2}{5} - \frac{2}{6} + \cdots\right]$$

We must be very careful with factors of 2. To obtain the potential energy for the interaction of N_A positive ions with N_A negative ions, it is necessary to multiply the total potential energy of a given ion due to all others by $2N_A$ and then divide by 2 to avoid counting the interaction of a given pair of ions twice. This gives the preceding result.

If such a calculation is carried out for a real three-dimensional crystal, the result is a series (such as that just given in brackets) whose value sums to a dimensionless number that depends upon the crystal structure. That number is called the **Madelung constant**, M, and its value is independent of the unit-cell dimensions. Table 21.5 lists the values of the Madelung constant for several crystal structures. The lattice energy is again the opposite of the total potential energy. Expressed in terms of the Madelung constant, it is

$$\text{lattice energy} = \frac{N_A e^2 M}{4\pi\epsilon_0 R_0} \qquad [21.4]$$

TABLE 21.5

Madelung Constants

Lattice	M
Rock salt	1.7476
CsCl	1.7627
Zinc blende	1.6381
Fluorite	2.5194

EXAMPLE 21.6

Calculate the electrostatic part of the lattice energy of sodium chloride, given that the internuclear separation between Na^+ and Cl^- ions is 2.82 Å.

Solution

Using the Madelung constant of 1.7476 for this structure gives

$$\frac{(6.02 \times 10^{23}\ \text{mol}^{-1})(1.602 \times 10^{-19}\ \text{C})^2(1.7476)}{(4\pi)(8.854 \times 10^{-12}\ \text{C}^2\ \text{J}^{-1}\ \text{m}^{-1})(2.82 \times 10^{-10}\ \text{m})} = 8.61 \times 10^5\ \text{J mol}^{-1} = 861\ \text{kJ mol}^{-1}$$

Related Problems: 39, 40

Ionic lattice energies are measured experimentally by means of a thermodynamic cycle developed by Max Born and Fritz Haber. The **Born–Haber cycle** is an application of Hess's law (the first law of thermodynamics). It is illustrated by a determination of the lattice energy of sodium chloride, which is ΔU for the reaction

$$NaCl(s) \longrightarrow Na^+(g) + Cl^-(g) \qquad \Delta U = ?$$

This reaction can be represented as a series of steps, each with a measurable energy or enthalpy change. In the first step, the ionic solid is converted to the elements in their standard states:

$$NaCl(s) \longrightarrow Na(s) + \tfrac{1}{2}Cl_2(g)$$

$$\Delta U_1 \approx \Delta H = -\Delta H_f^\circ(NaCl(s)) = +411.2 \text{ kJ}$$

In the second step, the elements are transformed into gas-phase atoms:

$$Na(s) \longrightarrow Na(g) \qquad \Delta U \approx \Delta H = \Delta H_f^\circ(Na(g)) = +107.3 \text{ kJ}$$

$$\tfrac{1}{2}\,Cl_2(g) \longrightarrow Cl(g) \qquad \Delta U \approx \Delta H = \Delta H_f^\circ(Cl(g)) = +121.7 \text{ kJ}$$

$$Na(s) + \tfrac{1}{2}\,Cl_2(g) \longrightarrow Na(g) + Cl(g) \qquad \Delta U_2 = +229.0 \text{ kJ}$$

Finally, in the third step, electrons are transferred from the sodium atoms to the chlorine atoms to give ions:

$$Na(g) \longrightarrow Na^+(g) + e^- \qquad \Delta U = IE_1(Na) = 496 \text{ kJ}$$

$$Cl(g) + e^- \longrightarrow Cl^-(g) \qquad \Delta U = -EA(Cl) = -349 \text{ kJ}$$

$$Na(g) + Cl(g) \longrightarrow Na^+(g) + Cl^-(g) \qquad \Delta U_3 = +147 \text{ kJ}$$

Here $EA(Cl)$ is the electron affinity of Cl, and $IE_1(Na)$ is the first ionization energy of Na. The total energy change is

$$\Delta U = \Delta U_1 + \Delta U_2 + \Delta U_3 = 411 + 229 + 147 = +787 \text{ kJ}$$

The small differences between ΔU and ΔH were neglected in this calculation. If their difference is taken into account (using $\Delta H = \Delta U + RT\Delta n_g$, where Δn_g is the change in the number of moles of gas molecules in each step of the reaction), then ΔU_2 is decreased by $\frac{3}{2}RT$ and ΔU_1 by $\frac{1}{2}RT$, giving a net decrease of $2RT$ and changing ΔU to $+782 \text{ kJ mol}^{-1}$. Comparing this experimental lattice energy with the energy calculated in Example 21.6 shows that the latter is approximately 10% greater, presumably because short-range repulsive interactions and zero-point energy were not taken into account in the lattice energy calculation.

CHAPTER SUMMARY

This chapter describes the distinctive properties of solids as the consequences of the collective behavior of an extended array of chemical bonds. The chemical bonds involved are already familiar in the context of isolated molecules. What is new here is the extended array of these bonds in the solid state. We can determine the array experimentally by X-ray diffraction. To describe the array we introduce the concepts of local symmetry and repetitive, long-range order; together, these define the unit cell and the crystal lattice. By placing structural elements at the points of the lattice, we complete the description of the structure of a perfect crystal. The number of atoms per unit cell depends on packing density of the atoms in the lattice. The properties of different classes of solids depend on the type of bonding involved: ionic, covalent, metallic, or molecular.

The perfect crystal is the fundamental starting point for investigations in the solid state. Practical materials can be described and categorized by the nature and extent of their deviations from perfect crystal structure.

CONCEPTS AND SKILLS

Section 21.1 – Crystal Symmetry and the Unit Cell

Relate the general properties of solids to the nature of the bonds between atoms and to the spatial arrangement of atoms in a solid.

- Solids differ from gases and liquids because they are rigid objects that retain their shapes even when external forces are applied. The rigidity and mechanical strength of solids are due to strong, directional bonds between atoms.
- Solids whose structures are highly ordered and symmetrical over macroscopic distances are called crystals with structures that are defined mathematically and measured experimentally.
- Crystal structure is defined in terms of the crystal lattice, a mathematical abstraction that represents the ordered and repetitive nature of the structure. A crystal lattice is constructed from a set of specific displacements along different directions. Each individual displacement generates a new lattice point; repeating this process many times generates the crystal lattice.

Identify the symmetry elements of different crystal systems (Problems 1–4).

- A symmetry element is an operation of rotation, reflection, or inversion which produces a new orientation of an object that can be exactly superimposed on the original orientation.
- The lattice embodies all the symmetry in the structure. Lattice points are identified by the fundamental symmetry operations of rotation, reflection, and inversion.
- The unit cell is the smallest region of a crystal lattice that contains all the structural information about the crystal. Each unit cell has characteristic lengths and angles. There are seven types of crystal structures (the seven crystal systems—see Table 21.1), each defined by the properties of its unit cell: hexagonal, cubic, tetragonal, trigonal, orthorhombic, monoclinic, and triclinic. The crystal lattice may be visualized as a three-dimensional stack of unit cells.

Section 21.2 – Crystal Structure

Explain how X-rays and neutrons are diffracted by crystals, and use information from such experiments to calculate lattice spacings (Problems 5–10).

- X-rays and neutrons have wavelengths comparable to the distance d between lattice planes in solids. Waves scattered from atoms in the solid experience constructive and destructive interference to produce a diffraction pattern in which peaks in the diffracted intensity appear at angles θ that satisfy the diffraction condition.

$$n\lambda = 2d \sin \theta \qquad n = 1, 2, 3, \ldots$$

- The distance d between planes in a crystal can be measured experimentally by X-ray diffraction once the diffraction angle (and n, the order of diffraction, are known.

Describe the packing of atoms in simple crystal lattices (Problems 11–24).

- Once the crystal lattice has been identified, description of the structure is completed by specifying the structural elements that are located at lattice points. The resulting structures are named by characteristics of the unit cell. Three examples are displayed by the elemental metals.

Simple cubic (one atom per unit cell)	Po
Body-centered cubic (two atoms per unit cell)	alkali metals
Face-centered cubic (four atoms per unit cell)	Al, Ni, Cu, Ag

- The volume of a unit cell is given by

$$V_c = abc\sqrt{1 - \cos^2\alpha - \cos^2\beta - \cos^2\gamma + 2\cos\alpha\cos\beta\cos\gamma} \qquad [21.2]$$

The efficiency of packing atoms onto lattice sites is determined by the size of the atom and the dimensions of the unit cell. Table 21.2 lists the fractional volume occupied by the atoms (packing fraction) for cubic lattices.

Section 21.3 – Cohesion in Solids

Compare the natures of the forces that hold atoms or molecules in their lattice sites in ionic, metallic, covalent, and molecular crystals (Problems 27–32).

- Different types of chemical bonding appear in solids and are responsible for the differences in mechanical and structural properties of different types of solids.
- Because electrostatic forces are strong and operate over large distances, ionic crystals are hard, brittle solids that have high melting points and are poor conductors of heat and electricity. They crystallize in structures determined primarily by atomic packing density.
  - Zinc blende structure for cation–anion radius ratio smaller than 0.414
  - Rock-salt structure for cation–anion radius ratio between 0.414 and 0.732
  - Cesium chloride structure for cation–anion radius ratio greater than 0.732
- Bonding in metallic crystals is explained as a sea of delocalized electrons around positively charged ions located at the lattice sites. The number density of electrons is equal to the number density of positive ions, so the metal is electrically neutral. The bonds are quite strong, evidenced by the high melting and boiling points of metals. Metals are malleable and ductile because the highly mobile electrons can rapidly adjust when lattice ions are pushed to new locations by external mechanical forces. Metals are good conductors of heat and electricity because the delocalized electrons respond easily to applied external fields.
- Covalent crystals are held together by strong, highly directional bonds usually described by the valence bond hybrid orbital method. Because of the nature of their bonds, covalent crystals have very high melting points and are hard and brittle.
- Molecular crystals are held together by van der Waals forces, the same as the intermolecular forces in gases and liquids. Because these are much weaker than ionic, metallic, and covalent bonds, the molecular crystals are usually soft, easily deformed, and have low melting points.

Section 21.4 – Defects and Amorphous Solids

Describe the kinds of equilibrium defects that are present in crystalline solids and the properties of amorphous solids (Problems 35–36). Real-world crystals do not have perfect symmetry and order. It is useful to characterize practical materials by the ways in which they deviate from the structure of the perfect crystal models described earlier.

- Point defects where atoms are missing from lattice sites are called vacancies, or Schottky defects. Their number density depends on the temperature and on the Gibbs free energy of formation of defects.
- Point defects where atoms are located between lattice sites are called interstitials. If a lattice atom is displaced to an interstitial site, the combination of the vacancy and the interstitial is called a Frenkel defect.
- Diffusion in solids occurs as Schottky or Frenkel defects hop from one lattice site to another by a thermally activated process akin to chemical reactions. The diffusion constant is given by

$$D = D_0 \exp[-E_a/RT] \qquad [21.3]$$

- Amorphous solids, or glasses, may have chemical compositions and mechanical properties similar to crystalline materials, and they may have ionic, metallic, covalent, or molecular bonding. But at the microscopic level they lack crystalline order. This arises from kinetic effects during solidification that reduce the mobility of atoms or molecules and prevent them from achieving ordered structures. Long or irregularly shaped molecules that are easily entangled lead to glass formation.

Determine the amount of different oxidation states present in nonstoichiometric solids (Problems 37–38).

- Determine the empirical formula for a nonstoichiometric metal oxide. Assuming the mixed oxide was initially 1:1, let y be the number of moles of metal ions in the +2 state that must be removed in order to accommodate each metal ion of the other oxidation state; the remaining metal ions must be in the +2 state. Enforce charge balance between the two states of metal and the total negative charge in the oxide anions to evaluate y. The method is illustrated in Example 21.5.

Section 21.5 – A Deeper Look . . . Lattice Energies of Crystals

Calculate lattice energies of molecular and ionic crystals (Problems 39–42).

- The lattice energy of a crystal is the energy required to separate the crystal into its component ions, atoms, or molecules at 0 K. The lattice energy is the negative of the total potential energy in the lattice. The total potential energy in an ionic crystal is calculated by adding up the pairwise Coulomb interactions among a mole of positive and negative ions separated by distance R_0 throughout the crystal. The geometry of the lattice is accounted for by the Madelung constant M, which multiples the pairwise interaction.

$$\text{lattice energy} = \frac{N_A e^2 M}{4\pi\epsilon_0 R_0} \qquad [21.4]$$

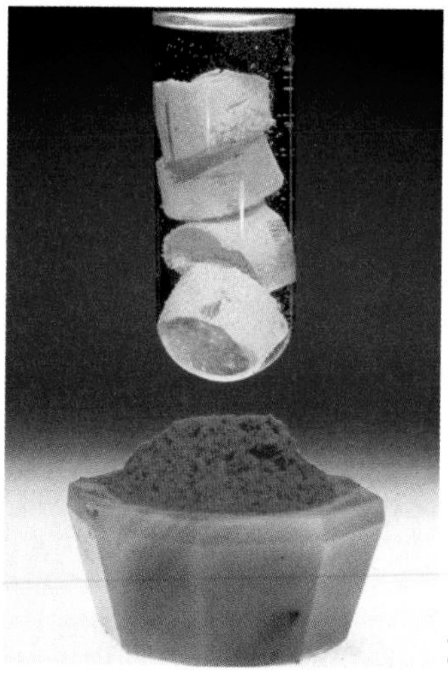

FIGURE 21.32 Two forms of elemental phosphorus: white and red.

CUMULATIVE EXERCISE

The Many States of Phosphorus

Solid elemental phosphorus appears in a rich variety of forms (Fig. 21.32), with crystals in all seven crystal systems reported under various conditions of temperature, pressure, and sample preparation.

(a) The thermodynamically stable form of phosphorus under room conditions is black phosphorus. Its unit cell is orthorhombic with edges of lengths

3.314, 4.376, and 10.48 Å. Calculate the volume of one unit cell, and determine the number of phosphorus atoms per unit cell, if the density of this form of phosphorus is 2.69 g cm^{-3}.

(b) The form of phosphorus that is easiest to prepare from the liquid or gaseous state is white phosphorus, which consists of P_4 molecules in a cubic lattice. When X-rays of wavelength 2.29 Å are scattered from the parallel faces of its unit cells, the first-order Bragg diffraction is observed at an angle 2θ of 7.10°. Calculate the length of the unit-cell edge for white phosphorus. At what angle will third-order Bragg diffraction be seen?

(c) Amorphous red phosphorus has been reported to convert to monoclinic, triclinic, tetragonal, and cubic red forms with different heat treatments. Identify the changes in the shape of the unit cell as a cubic lattice is converted first to tetragonal, then monoclinic, then triclinic.

(d) A monoclinic form of red phosphorus has been studied that has cell edge lengths 9.21, 9.15, and 22.60 Å, with an angle β of 106.1°. Each unit cell contains 84 atoms of phosphorus. Estimate the density of this form of phosphorus.

(e) Phosphorus forms many compounds with other elements. Describe the nature of the bonding in the solids white elemental phosphorus (P_4), black elemental phosphorus, sodium phosphate (Na_3PO_4), and phosphorus trichloride (PCl_3).

Answers

(a) Volume is 152.0 $Å^3$; eight atoms per unit cell

(b) 18.5 Å; angle $2\theta = 21.4°$

(c) Cubic to tetragonal: One cell edge is stretched or shrunk. Tetragonal to monoclinic: A second cell edge is stretched or shrunk, and the angles between two adjacent faces (and their opposite faces) are changed from 90°. Monoclinic to triclinic: The remaining two angles between faces are deformed from 90°.

(d) 2.36 g cm^{-3}

(e) P_4(white) and PCl_3 are molecular solids; P(black) is covalent; Na_3PO_4 is ionic.

PROBLEMS

Answers to problems whose numbers are red appear in Appendix G. Problems that are more challenging are indicated with asterisks.

21.1 Crystal Symmetry and the Unit Cell

1. Which of the following has 3-fold rotational symmetry? Explain.
   (a) An isosceles triangle
   (b) An equilateral triangle
   (c) A tetrahedron
   (d) A cube

2. Which of the following has 4-fold rotational symmetry? Explain.
   (a) A cereal box (exclusive of the writing on the sides)
   (b) A stop sign (not counting the writing)
   (c) A tetrahedron
   (d) A cube

3. Identify the symmetry elements of the CCl_2F_2 molecule (see Fig. 12.20).

4. Identify the symmetry elements of the PF_5 molecule (see Fig. 3.37a).

5. The second-order Bragg diffraction of X-rays with λ = 1.660 Å from a set of parallel planes in copper occurs at an angle 2θ = 54.70°. Calculate the distance between the scattering planes in the crystal.

6. The second-order Bragg diffraction of X-rays with λ = 1.237 Å from a set of parallel planes in aluminum occurs at an angle 2θ = 35.58°. Calculate the distance between the scattering planes in the crystal.

7. The distance between members of a set of equally spaced planes of atoms in crystalline lead is 4.950 Å. If X-rays with λ = 1.936 Å are diffracted by this set of parallel planes, calculate the angle 2θ at which fourth-order Bragg diffraction will be observed.

8. The distance between members of a set of equally spaced planes of atoms in crystalline sodium is 4.28 Å. If X-rays with λ = 1.539 Å are diffracted by this set of parallel planes, calculate the angle 2θ at which second-order Bragg diffraction will be observed.

9. The members of a series of equally spaced parallel planes of ions in crystalline LiCl are separated by 2.570 Å. Calculate all the angles 2θ at which diffracted beams of various orders may be seen, if the X-ray wavelength used is 2.167 Å.

10. The members of a series of equally spaced parallel planes in crystalline vitamin B_{12} are separated by 16.02 Å. Calculate all the angles 2θ at which diffracted beams of various orders may be seen, if the X-ray wavelength used is 2.294 Å.

21.2 Crystal Structure

11. A crucial protein at the photosynthetic reaction center of the purple bacterium *Rhodopseudomonas viridis* (see Section 20.7) has been separated from the organism, crystallized, and studied by X-ray diffraction. This substance crystallizes with a primitive unit cell in the tetragonal system. The cell dimensions are $a = b$ = 223.5 Å and c = 113.6 Å.
   (a) Determine the volume, in cubic angstroms, of this cell.
   (b) One of the crystals in this experiment was box-shaped, with dimensions 1 × 1 × 3 mm. Compute the number of unit cells in this crystal.

12. Compute the volume (in cubic angstroms) of the unit cell of potassium hexacyanoferrate(III) ($K_3Fe(CN)_6$), a substance that crystallizes in the monoclinic system with a = 8.40 Å, b = 10.44 Å, and c = 7.04 Å and with β = 107.5°.

13. The compound $Pb_4In_3B_{17}S_{18}$ crystallizes in the monoclinic system with a unit cell having a = 21.021 Å, b = 4.014 Å, c = 18.898 Å, and the only non-90° angle equal to 97.07°. There are two molecules in every unit cell. Compute the density of this substance.

14. Strontium chloride hexahydrate ($SrCl_2 \cdot 6H_2O$) crystallizes in the trigonal system in a unit cell with a = 8.9649 Å and α = 100.576°. The unit cell contains three formula units. Compute the density of this substance.

15. At room temperature, the edge length of the cubic unit cell in elemental silicon is 5.431 Å, and the density of silicon at the same temperature is 2.328 g cm^{-3}. Each cubic unit cell contains eight silicon atoms. Using only these facts, perform the following operations.
   (a) Calculate the volume (in cubic centimeters) of one unit cell.
   (b) Calculate the mass (in grams) of silicon present in a unit cell.
   (c) Calculate the mass (in grams) of an atom of silicon.
   (d) The mass of an atom of silicon is 28.0855 u. Estimate Avogadro's number to four significant figures.

16. One form of crystalline iron has a bcc lattice with an iron atom at every lattice point. Its density at 25°C is 7.86 g cm^{-3}. The length of the edge of the cubic unit cell is 2.87 Å. Use these facts to estimate Avogadro's number.

17. Sodium sulfate (Na_2SO_4) crystallizes in the orthorhombic system in a unit cell with a = 5.863 Å, b = 12.304 Å, and c = 9.821 Å. The density of these crystals is 2.663 g cm^{-3}. Determine how many Na_2SO_4 formula units are present in the unit cell.

18. The density of turquoise, $CuAl_6(PO_4)_4(OH)_8(H_2O)_4$, is 2.927 g cm^{-3}. This gemstone crystallizes in the triclinic system with cell constants a = 7.424 Å, b = 7.629 Å, c = 9.910 Å, α = 68.61°, β = 69.71°, and γ = 65.08°. Calculate the volume of the unit cell, and determine how many copper atoms are present in each unit cell of turquoise.

19. An oxide of rhenium has a structure with a Re atom at each corner of the cubic unit cell and an O atom at the center of each edge of the cell. What is the chemical formula of this compound?

20. The mineral perovskite has a calcium atom at each corner of the unit cell, a titanium atom at the center of the unit cell, and an oxygen atom at the center of each face. What is the chemical formula of this compound?

21. Iron has a body-centered cubic structure with a density of 7.86 g cm^{-3}.
   (a) Calculate the nearest neighbor distance in crystalline iron.
   (b) What is the lattice parameter for the cubic unit cell of iron?
   (c) What is the atomic radius of iron?

22. The structure of aluminum is fcc and its density is ρ = 2.70 g cm^{-3}.
   (a) How many Al atoms belong to a unit cell?
   (b) Calculate a, the lattice parameter, and d, the nearest neighbor distance.

23. Sodium has the body-centered cubic structure, and its lattice parameter is 4.28 Å.
   (a) How many Na atoms does a unit cell contain?
   (b) What fraction of the volume of the unit cell is occupied by Na atoms, if they are represented by spheres in contact with one another?

24. Nickel has an fcc structure with a density of 8.90 g cm^{-3}.
   (a) Calculate the nearest neighbor distance in crystalline nickel.
   (b) What is the atomic radius of nickel?
   (c) What is the radius of the largest atom that could fit into the interstices of a nickel lattice, approximating the atoms as spheres?

25. Calculate the ratio of the maximum radius of an interstitial atom at the center of a simple cubic unit cell to the radius of the host atom.

26. Calculate the ratio of the maximum radius of an interstitial atom at the center of each face of a bcc unit cell to the radius of the host atom.

21.3 Cohesion in Solids

27. Classify each of the following solids as molecular, ionic, metallic, or covalent.
   (a) $BaCl_2$ (b) SiC
   (c) CO (d) Co

28. Classify each of the following solids as molecular, ionic, metallic, or covalent.
   (a) Rb (b) C_5H_{12}
   (c) B (d) Na_2HPO_4

29. The melting point of cobalt is 1495°C, and that of barium chloride is 963°C. Rank the four substances in problem 27 from lowest to highest in melting point.

30. The boiling point of pentane (C_5H_{12}) is slightly less than the melting point of rubidium. Rank the four substances in problem 28 from lowest to highest in melting point.

31. Explain the relationship between the number of bonds that can be formed by a typical atom in a crystal and the possibility of forming linear, two-dimensional, and three-dimensional network structures.

32. Although large crystals of sugar (rock candy) and large crystals of salt (rock salt) have different geometric shapes, they look much the same to the untrained observer. What physical tests other than taste might be performed to distinguish between these two crystalline substances?

33. By examining Figure 21.17, determine the number of nearest neighbors, second nearest neighbors, and third nearest neighbors of a Cs^+ ion in crystalline CsCl. The nearest neighbors of the Cs^+ ion are Cl^- ions, and the second nearest neighbors are Cs^+ ions.

34. Repeat the determinations of the preceding problem for the NaCl crystal, referring to Figure 21.16.

21.4 Defects and Amorphous Solids

35. Will the presence of Frenkel defects change the measured density of a crystal?

36. What effect will the (unavoidable) presence of Schottky defects have on the determination of Avogadro's number via the method described in problems 15 and 16?

37. Iron(II) oxide is nonstoichiometric. A particular sample was found to contain 76.55% iron and 23.45% oxygen by mass.
   (a) Calculate the empirical formula of the compound (four significant figures).
   (b) What percentage of the iron in this sample is in the +3 oxidation state?

38. A sample of nickel oxide contains 78.23% Ni by mass.
   (a) What is the empirical formula of the nickel oxide to four significant figures?
   (b) What fraction of the nickel in this sample is in the +3 oxidation state?

21.5 A Deeper Look . . . Lattice Energies of Crystals

39. Calculate the energy needed to dissociate 1.00 mol of crystalline RbCl into its gaseous ions if the Madelung constant for its structure is 1.7476 and the radii of Rb^+ and Cl^- are 1.48 Å and 1.81 Å, respectively. Assume that the repulsive energy reduces the lattice energy by 10% from the pure Coulomb energy.

40. Repeat the calculation of problem 39 for CsCl, taking the Madelung constant from Table 21.5 and taking the radii of Cs^+ and Cl^- to be 1.67 Å and 1.81 Å.

41. (a) Use the Born–Haber cycle, with data from Appendices D and F, to calculate the lattice energy of LiF.
   (b) Compare the result of part (a) with the Coulomb energy calculated by using an Li—F separation of 2.014 Å in the LiF crystal, which has the rock-salt structure.

42. Repeat the calculations of problem 41 for crystalline KBr, which has the rock-salt structure with a K—Br separation of 3.298 Å.

ADDITIONAL PROBLEMS

43. Some water waves with a wavelength of 3.0 m are diffracted by an array of evenly spaced posts in the water. If the rows of posts are separated by 5.0 m, calculate the angle 2θ at which the first-order "Bragg diffraction" of these water waves will be seen.

44. A crystal scatters X-rays of wavelength $\lambda = 1.54$ Å at an angle 2θ of 32.15°. Calculate the wavelength of the X-rays in another experiment if this same diffracted beam from the same crystal is observed at an angle 2θ of 34.46°.

45. The number of beams diffracted by a single crystal depends on the wavelength λ of the X-rays used and on the volume associated with one lattice point in the crystal—that is, on the volume V_p of a primitive unit cell. An approximate formula is

$$\text{number of diffracted beams} = \frac{4}{3}\pi\left(\frac{2}{\lambda}\right)^3 V_p$$

(a) Compute the volume of the conventional unit cell of crystalline sodium chloride. This cell is cubic and has an edge length of 5.6402 Å.

(b) The NaCl unit cell contains four lattice points. Compute the volume of a primitive unit cell for NaCl.

(c) Use the formula given in this problem to estimate the number of diffracted rays that will be observed if NaCl is irradiated with X-rays of wavelength 2.2896 Å.

(d) Use the formula to estimate the number of diffracted rays that will be observed if NaCl is irradiated with X-rays having the shorter wavelength 0.7093 Å.

46. If the wavelength λ of the X-rays is too large relative to the spacing of planes in the crystal, no Bragg diffraction will be seen because $\sin\theta$ would be larger than 1 in the Bragg equation, even for $n = 1$. Calculate the longest wavelength of X-rays that can give Bragg diffraction from a set of planes separated by 4.20 Å.

47. The crystal structure of diamond is fcc, and the atom coordinates in the unit cell are (0, 0, 0), $\left(\frac{1}{2}, \frac{1}{2}, 0\right)$, $\left(\frac{1}{2}, 0, \frac{1}{2}\right)$, $\left(0, \frac{1}{2}, \mathrm{v}\right)$, $\left(\frac{1}{4}, \frac{1}{4}, \frac{1}{4}\right)$, $\left(\frac{3}{4}, \frac{1}{4}, \frac{3}{4}\right)$, $\left(\frac{3}{4}, \frac{3}{4}, \frac{1}{4}\right)$, and $\left(\frac{1}{4}, \frac{3}{4}, \frac{3}{4}\right)$. The lattice parameter is $a = 3.57$ Å. What is the C—C bond distance in diamond?

48. Polonium is the only element known to crystallize in the simple cubic lattice.

(a) What is the distance between nearest neighbor polonium atoms if the first-order diffraction of X-rays with $\lambda = 1.785$ Å from the parallel faces of its unit cells appears at an angle of $2\theta = 30.96°$ from these planes?

(b) What is the density of polonium in this crystal (in g cm^{-3})?

49. At room temperature, monoclinic sulfur has the unit-cell dimensions $a = 11.04$ Å, $b = 10.98$ Å, $c = 10.92$ Å, and $\beta = 96.73°$. Each cell contains 48 atoms of sulfur.

(a) Explain why it is not necessary to give the values of the angles α and γ in this cell.

(b) Compute the density of monoclinic sulfur (in g cm^{-3}).

50. A compound contains three elements: sodium, oxygen, and chlorine. It crystallizes in a cubic lattice. The oxygen atoms are at the corners of the unit cells, the chlorine atoms are at the centers of the unit cells, and the sodium atoms are at the centers of the faces of the unit cells. What is the formula of the compound?

***51.** Show that the radius of the largest sphere that can be placed in a tetrahedral interstitial site in an fcc lattice is $0.225r_1$, where r_1 is the radius of the atoms making up the lattice. (*Hint:* Consider a cube with the centers of four spheres placed at alternate corners, and visualize the tetrahedral site at the center of the cube. What is the relationship between r_1 and the length of a diagonal of a face? The length of a body diagonal?)

52. What is the closest packing arrangement possible for a set of thin circular discs lying in a plane? What fraction of the area of the plane is occupied by the discs? Show how the same reasoning can be applied to the packing of infinitely long, straight cylindrical fibers.

53. Name two elements that form molecular crystals, two that form metallic crystals, and two that form covalent crystals. What generalizations can you make about the portions of the periodic table where each type is found?

54. The nearest-neighbor distance in crystalline LiCl (rock-salt structure) is 2.570 Å; the bond length in a gaseous LiCl molecule is significantly shorter, 2.027 Å. Explain.

55. **(a)** Using the data of Table 9.4, estimate the lattice energy and intermolecular separation of nitrogen in its solid state, assuming an fcc structure for the solid lattice.

(b) The density of cubic nitrogen is 1.026 g cm^{-3}. Calculate the lattice parameter a and the nearest neighbor distance. Compare your answer with that from part (a).

***56.** Solid CuI_2 is unstable relative to CuI at room temperature, but $CuBr_2$, $CuCl_2$, and CuF_2 are all stable relative to the copper(I) halides. Explain by considering the steps in the Born–Haber cycle for these compounds.

57. A crystal of sodium chloride has a density of 2.165 g cm^{-3} in the absence of defects. Suppose a crystal of NaCl is grown in which 0.15% of the sodium ions and 0.15% of the chloride ions are missing. What is the density in this case?

58. The activation energy for the diffusion of sodium atoms in the crystalline state is 42.22 kJ mol^{-1}, and $D_0 = 0.145$ cm^2 s^{-1}.

(a) Calculate the diffusion constant $D = D_0 \exp(-E_a/RT)$ of sodium in the solid at its melting point (97.8°C).

(b) What is the root-mean-square displacement of an average sodium atom from an arbitrary origin after the lapse of 1.0 hour at $t = 97.8°C$? (*Hint:* Use Equation 9.32 in Chapter 9.)

59. A compound of titanium and oxygen contains 28.31% oxygen by mass.

(a) If the compound's empirical formula is Ti_xO, calculate x to four significant figures.

(b) The nonstoichiometric compounds Ti_xO can be described as having a Ti^{2+}—O^{2-} lattice in which certain Ti^{2+} ions are missing or are replaced by Ti^{3+} ions. Calculate the fraction of Ti^{2+} sites in the nonstoichiometric compound that are vacant and the fraction that are occupied by Ti^{3+} ions.

60. Classify the bonding in the following amorphous solids as molecular, ionic, metallic, or covalent.

(a) Amorphous silicon, used in photocells to collect light energy from the sun

(b) Polyvinyl chloride, a plastic of long-chain molecules composed of $-CH_2CHCl-$ repeating units, used in pipes and siding

(c) Soda-lime-silica glass, used in windows

(d) Copper-zirconium glass, an alloy of the two elements with approximate formula Cu_3Zr_2, used for its high strength and good conductivity

CUMULATIVE PROBLEMS

61. Sodium hydride (NaH) crystallizes in the rock-salt structure, with four formula units of NaH per cubic unit cell. A beam of monoenergetic neutrons, selected to have a velocity of 2.639×10^3 m s^{-1}, is scattered in second order through an angle of $2\theta = 36.26°$ by the parallel faces of the unit cells of a sodium hydride crystal.

(a) Calculate the wavelength of the neutrons.

(b) Calculate the edge length of the cubic unit cell.

(c) Calculate the distance from the center of an Na^+ ion to the center of a neighboring H^- ion.

(d) If the radius of an Na^+ ion is 0.98 Å, what is the radius of an H^- ion, assuming the two ions are in contact?

62. Chromium(III) oxide has a structure in which chromium ions occupy two-thirds of the octahedral interstitial sites in a hexagonal close-packed lattice of oxygen ions. What is the *d*-electron configuration on the chromium ion?

63. A useful rule of thumb is that in crystalline compounds every nonhydrogen atom occupies 18 Å^3, and the volume occupied by hydrogen atoms can be neglected. Using this rule, estimate the density of ice (in g cm^{-3}). Explain why the answer is so different from the observed density of ice.

64. Estimate, for the F-centers in CaF_2, the wavelength of maximum absorption in the visible region of the spectrum that will give rise to the color shown in Figure 21.29.

22 CHAPTER

Inorganic Materials

Azurite is a basic copper carbonate with chemical formula $Cu_3(CO_3)_2(OH)_2$.

Having laid the conceptual foundation for relating properties of solids to chemical bonding in Chapter 21, we turn now to applications of these concepts to three important classes of materials: ceramics, electronic materials, and optical materials. All of these are synthetic materials fashioned from inorganic, nonmetallic substances by chemical methods of synthesis and processing.

Ceramics are one of the oldest classes of materials prepared by humankind. New discoveries in ceramics are occurring at a startling rate, and major new technologic advances will certainly come from these discoveries. Ceramics have value both as structural materials—the role emphasized in this chapter—and for their wide range of electronic and optical properties.

High-speed computing, fast communication, and rapid display of information were major technologic achievements in the second half of the 20th century. These developments will grow even more rapidly in the 21st century, when every home will be digitally connected to the Internet and every cell phone will display images requiring broadband transmission. These developments were made possible by the "microelectronics revolution," beginning with the first integrated circuits fabricated on microchips by Jack Kilby at Texas Instruments and Robert Noyce at Fairchild in 1958 to 1959.[1] The speed of computers increased dramatically through advances in transistor design; advances in the solid-state laser enabled high-speed communication via fiber optics. All these advances in device design relied critically on equally dramatic accomplishments in the growth and processing of materials, which in every case involved making and breaking chemical bonds in solid-state materials. Our goals in this chapter are to introduce the optical and electronic properties of materials and to show how they depend on chemical structure.

Electronic properties describe the movement of charged particles in a material in response to an applied electric field. If the charges are free to move throughout the material, the process is *electrical conduction,* measured by the *electrical conductivity* of the material. Differences in the magnitude of the conductivity distinguish metals, semiconductors, and insulators. If the charges can move only limited distances and are then halted by opposing binding forces, separation of positive and negative charges leads to *electric polarization* of the material, measured by its *dielectric constant.* Conduction involves dissipation of energy as heat, whereas polarization involves storage of potential energy in the material.

Optical properties describe the response of a material to electromagnetic radiation, particularly visible light. The list of optical properties is long, including reflection of light from a surface, refraction (bending the direction) of light as it passes from one medium into another, absorption, and transmission. We limit the discussion here to the generation and detection of light in solid materials, as extensions of the molecular processes of emission and excitation already described in Chapter 20. The absorption of light creates the bright colors of inorganic pigments and the conversion of solar energy into electrical energy in solar cells.

In this survey of mechanical, electrical, and optical properties, keep two questions in mind: (1) How does a material respond on an atomic level to applied mechanical stress and to electrical and optical fields? (2) If the chemical structure is modified, how does this change influence the response of the material to these forces?

22.1 Minerals: Naturally Occurring Inorganic Materials

This chapter begins with a survey of the naturally occurring inorganic, nonmetallic minerals that are the starting materials for synthesis and processing of inorganic materials.

22.1.1 Silicates

Silicon and oxygen make up most of the earth's crust, with oxygen accounting for 47% and silicon for 28% of its mass. The silicon–oxygen bond is strong and partially ionic. It forms the basis for a class of minerals called **silicates,** which make up the bulk of the rocks, clays, sand, and soils in the earth's crust. From time

[1]Kilby was awarded the Nobel Prize in Physics in 2000, but Noyce, who had died, did not share the award. The Nobel Prize is not awarded posthumously.

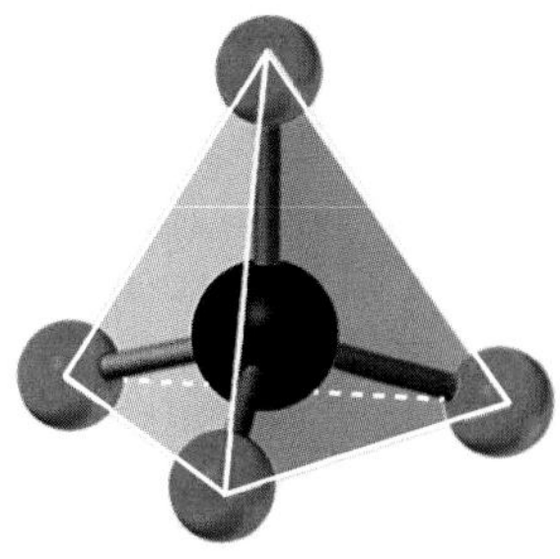

FIGURE 22.1 In silicates, four O atoms (red) are arranged in a tetrahedron around an Si atom. The stoichiometry is changed when one or more O atoms is shared with neighboring tetrahedra.

TABLE 22.1

Silicate Structures

Structure	Figure Number	Corners Shared at Each Si Atom	Repeat Unit	Si:O Ratio	Example
Tetrahedra	22.2a	0	SiO_4^{4-}	1:4	Olivines
Pairs of tetrahedra	22.2b	1	$Si_2O_7^{6-}$	$1:3\frac{1}{2}$	Thortveitite
Closed rings	22.2c	2	SiO_3^{2-}	1:3	Beryl
Infinite single chains	22.2d	2	SiO_3^{2-}	1:3	Pyroxenes
Infinite double chains	22.2e	$2\frac{1}{2}$	$Si_4O_{11}^{6-}$	$1:2\frac{3}{4}$	Amphiboles
Infinite sheets	22.2f	3	$Si_2O_5^{2-}$	$1:2\frac{1}{2}$	Talc
Infinite network	22.2g	4	SiO_2	1:2	Quartz

immemorial, silicates have provided the ingredients for building materials such as bricks, cement, concrete, and glass (which are considered later in this chapter).

The structure-building properties of silicates (Table 22.1 and Fig. 22.1) originate in the tetrahedral orthosilicate anion (SiO_4^{4-}), in which the negative charge of the silicate ion is balanced by the compensating charge of one or more cations. The simplest silicates consist of individual SiO_4^{4-} anions (Fig. 22.2a), with cations arranged around them on a regular crystalline lattice. Such silicates are properly called *orthosilicates*. Examples are forsterite (Mg_2SiO_4) and fayalite (Fe_2SiO_4),

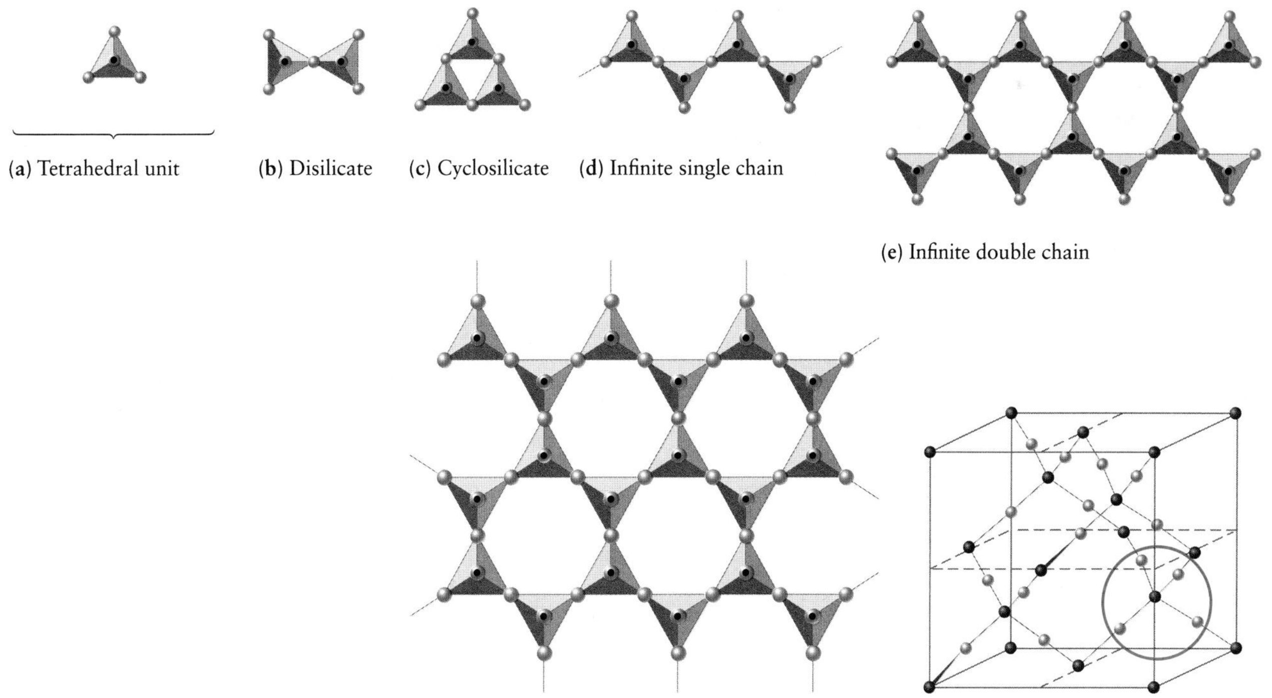

(a) Tetrahedral unit (b) Disilicate (c) Cyclosilicate (d) Infinite single chain (e) Infinite double chain (f) Infinite sheet (g) Infinite network (cristobalite, SiO_2)

FIGURE 22.2 Classes of silicate structures. (a) Symbol used for the SiO_4^{4-} tetrahedron. This top view of the symbol shows a red circle to represent the fourth oxygen atom at the top of the tetrahedron. The black dot represents the silicon atom at the center of the tetrahedron. Bear in mind that all of these structures are actually three-dimensional. Planar projections are used here for convenience of representation. (b) Disilicate. (c) Cyclosilicate. (d) Infinite single chain. (e) Infinite double chain. (f) Infinite sheet. (g) Infinite network (cristobalite, SiO_2) with an example of the tetrahedral arrangement of O atoms about a silicon atom circled.

which are the extreme members of a class of minerals called **olivines**, $[Mg,Fe]_2SiO_4$. There is a continuous range of proportions of magnesium and iron in the olivines.

Other silicate structures form when two or more SiO_4^{4-} tetrahedra link and share oxygen vertices. The simplest such minerals are the **disilicates**, such as thortveitite, $Sc_2(Si_2O_7)$, in which two tetrahedra are linked (see Fig. 22.2b). Additional linkages of tetrahedra create the ring, chain, sheet, and network structures shown in Figure 22.2 and listed in Table 22.1. In each of these, the fundamental tetrahedron is readily identified, but the Si : O ratio is no longer 4 because oxygen (O) atoms are shared at the linkages.

EXAMPLE 22.1

By referring to Table 22.1, predict the structural class in which the mineral Egyptian blue ($CaCuSi_4O_{10}$) belongs. Give the oxidation state of each of its atoms.

Solution

Because the silicon/oxygen (Si : O) ratio is 4 : 10, or $1:2\frac{1}{2}$, this mineral should have an infinite sheet structure with the repeating unit $Si_2O_5^{2-}$. The oxidation states of Si and O are +4 and −2, as usual, and that of calcium (Ca) is +2. For the total oxidation number per formula unit to sum to 0, the oxidation state of copper (Cu) must be +2.

Related Problems: 3, 4

FIGURE 22.3 The mineral quartz is one form of silica, SiO_2.

The physical properties of the silicates correlate closely with their structures. Talc, $Mg_3(Si_4O_{10})(OH)_2$, is an example of an infinite layered structure (see Fig. 22.2f). In talc, all of the bonding interactions among the atoms occur in a single layer. Layers of talc sheets are attracted to one another only by van der Waals interactions, which (being weak) permit one layer to slip easily across another. This accounts for the slippery feel of talc (called talcum powder). When all four vertices of each tetrahedron are linked to other tetrahedra, three-dimensional network structures such as cristobalite (see Fig. 22.2g) or quartz (Fig. 22.3) result. Note that the quartz network carries no charge; consequently, there are no cations in its structure. Three-dimensional network silicates such as quartz are much stiffer and harder than the linear and layered silicates, and they resist deformation well.

Asbestos is a generic term for a group of naturally occurring, hydrated silicates that can be processed mechanically into long fibers (Fig. 22.4). Some of these silicates, such as tremolite, $Ca_2Mg_5(Si_4O_{11})_2(OH)_2$, show the infinite double-chain structure of Figure 22.2e. Another kind of asbestos mineral is chrysotile, $Mg_3(Si_2O_5)(OH)_4$. As the formula indicates, this mineral has a sheet structure (see Fig. 22.2f), but the sheets are rolled into long tubes. Asbestos minerals are fibrous because the bonds along the strandlike tubes are stronger than those that hold different tubes together. Asbestos is an excellent thermal insulator that does not burn, resists acids, and is strong. For many years, it was used in cement for pipes and ducts and woven into fabric to make fire-resistant roofing paper and floor tiles. Its use has decreased significantly in recent years because inhalation of its small fibers can cause the lung disease asbestosis. The risk comes with breathing asbestos dust that is raised during mining and manufacturing processes or that is released in buildings in which asbestos-containing materials are fraying, crumbling, or being removed.

FIGURE 22.4 The fibrous structure of asbestos is apparent in this sample.

22.1.2 Aluminosilicates

An important class of minerals called **aluminosilicates** results from the replacement of some of the Si atoms in silicates with aluminum (Al) atoms. Aluminum is the third most abundant element in the earth's crust (8% by mass), where it occurs largely in the form of aluminosilicates. Aluminum in minerals can be a

FIGURE 22.5 Naturally occurring muscovite mica. The mechanical properties of crystals of mica are quite anisotropic. Thin sheets can be peeled off a crystal of mica by hand, but the sheets resist stresses in other directions more strongly. Transparent, thin sheets of mica, sometimes called isinglass, have been used for heat-resistant windows in stoves or in place of window glass.

simple cation (Al^{3+}), or it can replace silicon in tetrahedral coordination. When it replaces silicon, it contributes only three electrons to the bonding framework in place of the four electrons of Si atoms. The additional required electron is supplied by the ionization of a metal atom such as sodium (Na) or potassium (K); the resulting alkali-metal ions occupy nearby sites in the aluminosilicate structure.

The most abundant and important of the aluminosilicate minerals in the earth's surface are the *feldspars*, which result from the substitution of aluminum for silicon in three-dimensional silicate networks such as quartz. The Al ions must be accompanied by other cations such as sodium, potassium, or calcium to maintain overall charge neutrality. Albite is a feldspar with the chemical formula $NaAlSi_3O_8$. In the high-temperature form of this mineral, the Al and Si atoms are distributed at random (in 1:3 proportion) over the tetrahedral sites available to them. At lower temperatures, other crystal structures become thermodynamically stable, with partial ordering of the Al and Si sites.

If one of the four Si atoms in the structural unit of talc, $Mg_3(Si_4O_{10})(OH)_2$, is replaced by an Al atom and a K atom is furnished to supply the fourth electron needed for bonding in the tetrahedral silicate framework, the result is the composition $KMg_3(AlSi_3O_{10})(OH)_2$, which belongs to the family of **micas** (Fig. 22.5). Mica is harder than talc, and its layers slide less readily over one another, although the crystals still cleave easily into sheets. The cations occupy sites between the infinite sheets, and the van der Waals bonding that holds adjacent sheets together in talc is augmented by an ionic contribution. The further replacement of the three Mg^{2+} ions in $KMg_3(AlSi_3O_{10})(OH)_2$ with two Al^{3+} ions gives the mineral muscovite, $KAl_2(AlSi_3O_{10})(OH)_2$. Writing its formula in this way indicates that there are Al atoms in two kinds of sites in the structure: One Al atom per formula unit occupies a tetrahedral site, substituting for one Si atom, and the other two Al atoms are between the two adjacent layers. The formulas that mineralogists and crystallographers use convey more information than the usual empirical chemical formula of a compound.

22.1.3 Clay Minerals

Clays are minerals produced by the weathering action of water and heat on primary minerals. Their compositions can vary widely as a result of the replacement of one element with another. Invariably, they are microcrystalline or powdered in form and are usually hydrated. Often, they are used as supports for catalysts, as fillers in paint, and as ion-exchange vehicles. The clays that readily absorb water and swell are used as lubricants and bore-hole sealers in the drilling of oil wells.

The derivation of clays from talcs and micas provides a direct way to understand the structures of the clays. The infinite-sheet mica pyrophyllite, $Al_2(Si_4O_{10})(OH)_2$, serves as an example. If one of six Al^{3+} ions in the pyrophyllite structure is replaced by one Mg^{2+} ion and one Na^+ ion (which together carry the same charge), a type of clay called montmorillonite, $MgNaAl_5(Si_4O_{10})_3(OH)_6$, results. This clay readily absorbs water, which infiltrates between the infinite sheets and hydrates the Mg^{2+} and Na^+ ions there, causing the montmorillonite to swell (Fig. 22.6).

A different clay derives from the layered mineral talc, $Mg_3(Si_4O_{10})(OH)_2$. If iron(II) and aluminum replace magnesium and silicon in varying proportions and water molecules are allowed to take up positions between the layers, the swelling clay vermiculite results. When heated, vermiculite pops like popcorn, as the steam generated by the vaporization of water between the layers puffs the flakes up into a light, fluffy material with air inclusions. Because of its porous structure, vermiculite is used for thermal insulation or as an additive to loosen soils.

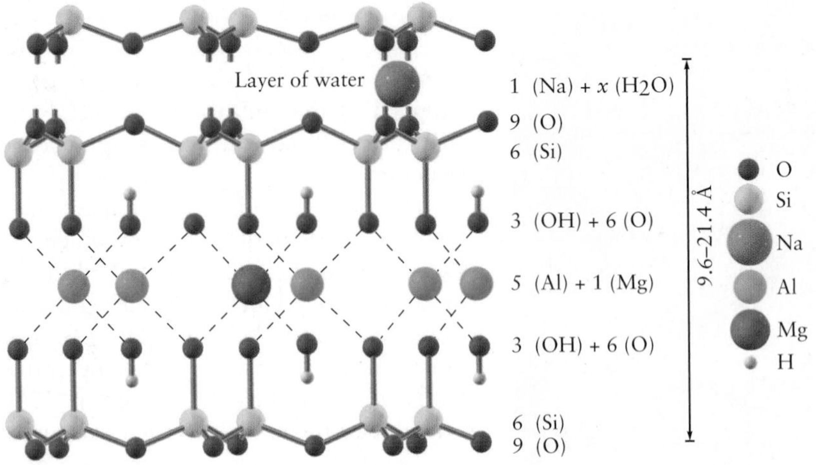

FIGURE 22.6 Structure of the clay mineral montmorillonite. Insertion of variable amounts of water causes the distance between layers to swell from 9.6 Å to more than 20 Å. When 1 Al^{3+} ion is replaced by an Mg^{2+} ion, an additional ion such as Na^+ is introduced into the water layers to maintain overall charge neutrality.

22.1.4 Zeolites

Zeolites are a class of three-dimensional aluminosilicates. Like the feldspars, they carry a negative charge on the aluminosilicate framework that is compensated by neighboring alkali-metal or alkaline-earth cations. Zeolites differ from feldspars in having much more open structures that consist of polyhedral cavities connected by tunnels (Fig. 22.7). Many zeolites are found in nature, but they can also be synthesized under conditions controlled to favor cavities of uniform size and shape. Most zeolites accommodate water molecules in their cavities, where they provide a mobile phase for the migration of the charge-compensating cations. This enables zeolites to serve as ion-exchange materials (in which one kind of positive ion can be readily exchanged for another) and is the key to their ability to soften water. Water "hardness" arises from soluble calcium and magnesium salts such as $Ca(HCO_3)_2$ and $Mg(HCO_3)_2$. Such salts are converted to insoluble carbonates (boiler scale) when the water is heated and form objectionable precipitates (bathtub ring) with soaps. When hard water is passed through a column packed with a zeolite that has Na ions in its structure, the Ca and Mg ions exchange with the Na ions and are removed from the water phase:

$$2\,\mathrm{NaZ}(s) + \mathrm{Ca}^{2+}(aq) \rightleftharpoons \mathrm{CaZ}_2(s) + 2\,\mathrm{Na}^+(aq)$$

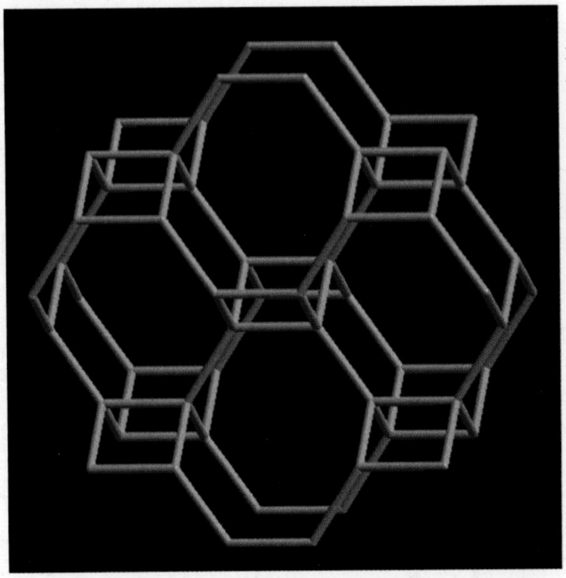

C. Baerlocher, L.B. McCusker, Atlas of Zeolite Structure Types. http://www.iza-sc.ethz.ch/IZA-SC/Atlas/AtlasHome.html

FIGURE 22.7 Structure of the synthetic zeolite Li-ABW $[Li_4(Si_4Al_4O_{16})\cdot 4H_2O]$.

When the ion-exchange capacity of the zeolite is exhausted, this reaction can be reversed by passing a concentrated solution of sodium chloride through the zeolite to regenerate it in the sodium form.

A second use of zeolites derives from the ease with which they adsorb small molecules. Their spongelike affinity for water makes them useful as drying agents; they are put between the panes of double-pane glass windows to prevent moisture from condensing on the inner surfaces. The pore size of zeolites can be selected to allow molecules that are smaller than a certain size to pass through but hold back larger molecules. Such zeolites serve as "molecular sieves" for gas purification. Ca^{+} exchange zeolites can also preferentially trap nitrogen, with its large quadrupole moment, allowing oxygen molecules in an air stream to pass through.

Perhaps the most exciting use of zeolites is as catalysts. Molecules of varying sizes and shapes have different rates of diffusion through a zeolite; this feature enables chemists to enhance the rates and yields of desired reactions and suppress unwanted reactions. The most extensive applications of zeolites currently are in the catalytic cracking of crude oil, a process that involves breaking down long-chain hydrocarbons and re-forming them into branched-chain molecules of lower molecular mass for use in high-octane unleaded gasoline. A relatively new process uses "shape-selective" zeolite catalysts to convert methanol (CH_3OH) to high-quality gasoline. Plants have been built to make gasoline by this process, using methanol derived from coal or from natural gas.

22.2 Properties of Ceramics

A **ceramic** is a synthetic material that has as its essential components inorganic, nonmetallic materials. This broad definition includes cement, concrete, and glass, in addition to the more traditional fired clay products such as bricks, roof tiles, pottery, and porcelain. The use of ceramics predates recorded history; the emergence of civilization from a primitive state is chronicled in fragments of pottery. No one knows when small vessels were first shaped by human hands from moist clay and left to harden in the heat of the sun. Such containers held nuts, grains, and berries well, but they lost their shape and slumped into formless mud when water was poured into them. Then someone discovered (perhaps by accident) that if clay was placed in the glowing embers of a fire, it became as hard as rock and withstood water well. Molded figures (found in what is now the Czech Republic) that were made 24,000 years ago are the earliest fired ceramic objects discovered so far, and fired clay vessels from the Near East date from 8000 B.C. With the action of fire on clay, the art and science of ceramics began.

Ceramics offer stiffness, hardness, resistance to wear, and resistance to corrosion (particularly by oxygen and water), even at high temperature. They are less dense than most metals, which makes them desirable metal substitutes when weight is a factor. Most are good electrical insulators at ordinary temperatures, a property that is exploited in electronics and power transmission. Ceramics retain their strength well at high temperatures. Several important structural metals soften or melt at temperatures 1000°C below the melting points of their chemical compounds in ceramics. Aluminum, for example, melts at 660°C, whereas aluminum oxide (Al_2O_3), an important compound in many ceramics, does not melt until a temperature of 2051°C is reached.

Against these advantages must be listed some serious disadvantages. Ceramics are generally brittle and low in tensile strength. They tend to have high thermal expansion but low thermal conductivity, making them subject to **thermal shock,** in which sudden local temperature change causes cracking or shattering. Metals and plastics dent or deform under stress, but ceramics cannot

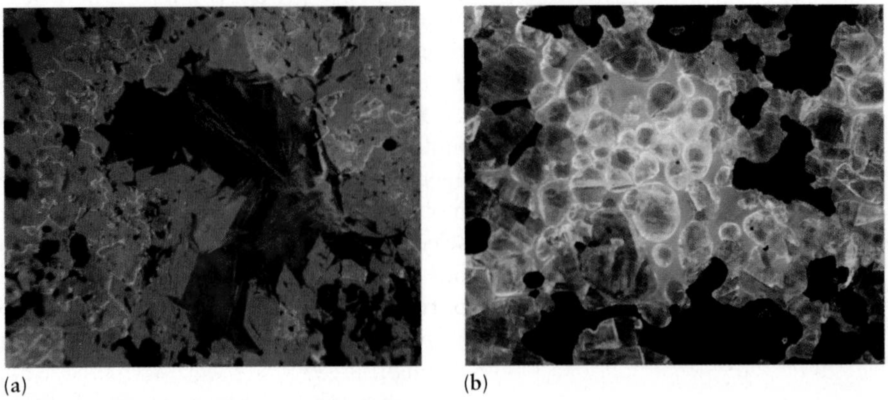

FIGURE 22.8 Microstructures of aluminosilicate ceramics, viewed by the different colors of light emitted after bombardment by electrons. (a) Forsterite (red), spinel (green), and periclase (dark brown) grains. (b) Periclase (blue) and oldhamite (yellow) grains.

absorb stress in this way: instead, they break. A major drawback of ceramics as structural materials is their tendency to fail unpredictably and catastrophically in use. Moreover, some ceramics lose mechanical strength as they age, an insidious and serious problem.

22.2.1 Composition and Structure of Ceramics

Ceramics use a variety of chemical compounds, and useful ceramic bodies are nearly always mixtures of several compounds. **Silicate ceramics**, which include the commonplace pots, dishes, and bricks, are made from aluminosilicate clay minerals. All contain the tetrahedral SiO_4 grouping discussed in Section 22.1. In **oxide ceramics**, silicon is a minor or nonexistent component. Instead, a number of metals combine with oxygen to give compounds such as alumina (Al_2O_3), magnesia (MgO), or yttria (Y_2O_3). *Nonoxide ceramics* contain compounds that are free of oxygen as principal components. Some important compounds in nonoxide ceramics are silicon nitride (Si_3N_4), silicon carbide (SiC), and boron carbide (approximate composition B_4C).

One important property of ceramics is their porosity. Porous ceramics have small openings into which fluids (typically air or water) can infiltrate. Fully dense ceramics have no channels of this sort. Two ceramic pieces can have the same chemical composition but quite different densities if the first is porous and the second is not.

A **ceramic phase** is any portion of the whole body that is physically homogeneous and bounded by a surface that separates it from other parts. Distinct phases are visible at a glance in coarse-grained ceramic pieces; in a fine-grained piece, phases can be seen with a microscope. When examined on a still finer scale, most ceramics, like metals, are microcrystalline, consisting of small crystalline grains cemented together (Fig. 22.8). The **microstructure** of such objects includes the sizes and shapes of the grains, the sizes and distribution of voids (openings between grains) and cracks, the identity and distribution of impurity grains, and the presence of stresses within the structure. Microstructural variations have enormous importance in ceramics because slight changes at this level strongly influence the properties of individual ceramic pieces. This is less true for plastic and metallic objects.

FIGURE 22.9 Silicon carbide ball bearings are harder, stiffer, and have smoother surfaces than steel bearings, resulting in improved performance due to reduced friction and less heat. Their high melting points make them the bearings of choice for demanding, high temperature applications.

The microstructure of a ceramic body depends markedly on the details of its fabrication. The techniques of forming and firing a ceramic piece are as important as its chemical composition in determining ultimate behavior because they confer a unique microstructure. This fact calls attention to the biggest problem with ceramics as structural materials: inconsistent quality. Ceramic engineers can produce parts that are harder than steel (Fig. 22.9), but not reliably so because of the difficulties of monitoring and controlling microstructure. Gas turbine engines fabricated of silicon nitride, for example, run well at 1370°C, which is hot enough to soften or melt most metals. The higher operating temperature increases engine efficiency, and the ceramic turbines weigh less, which further boosts fuel economy. Despite these advantages, there is no commercial ceramic gas turbine. Acceptable ceramic turbines have to be built from selected, pretested components. The testing costs and rejection rates are so high that economical mass production has been impossible so far.

22.2.2 Making Ceramics

The manufacture of most ceramics involves four steps: (1) the preparation of the raw material; (2) the forming of the desired shape, often achieved by mixing a powder with water or other binder and molding the resulting plastic mass; (3) the drying and firing of the piece, also called its **densification**, because pores (voids) in the dried ceramic fill in; and (4) the finishing of the piece by sawing, grooving, grinding, or polishing.

The raw materials for traditional ceramics are natural clays that come from the earth as powders or thick pastes and become plastic enough after adjustment of their water content to be formed freehand or on a potter's wheel. Special ceramics (both oxide and nonoxide) require chemically pure raw materials that are produced synthetically. Close control of the purity of the starting materials for these ceramics is essential to produce finished pieces with the desired properties. In addition to being formed by hand or in open molds, ceramic pieces are shaped by the squeezing (compacting) of the dry or semidry powders in a strong, closed mold of the desired shape, at either ordinary or elevated temperatures (hot pressing).

Firing a ceramic causes **sintering** to occur. In sintering, the fine particles of the ceramic start to merge together by diffusion at high temperatures. The density of the material increases as the voids between grains are partially filled. Sintering occurs below the melting point of the material and shrinks the ceramic body. In addition to the merging of the grains of the ceramic, firing causes partial melting, chemical reactions among different phases, reactions with gases in the atmosphere of the firing chamber, and recrystallization of compounds with an accompanying growth in crystal size. All of these changes influence the microstructure of a piece and must be understood and controlled. Firing accelerates physical and chemical changes, of course, but thermodynamic equilibrium in a fired ceramic piece is rarely reached. Kinetic factors—including the rate of heating, the length of time at which each temperature is held, and the rate of cooling—influence microstructure. As a result, the use of microwave radiation (as in microwave ovens) rather than kilns to fire ceramics is under development in ceramic factories, because it promises more exact control of the heating rate, and thereby more reliable quality.

22.3 Silicate Ceramics

The silicate ceramics include materials that vary widely in composition, structure, and use. They range from simple earthenware bricks and pottery to cement, fine porcelain, and glass. Their structural strength is based on the same linking of silicate ion tetrahedra that gives structure to silicate minerals in nature.

22.3.1 Pottery and Structural Clay Products

Aluminosilicate clays are products of the weathering of primary minerals. When water is added to such clays in moderate amount, a thick paste results that is easily molded into different shapes. Clays expand as water invades the space between adjacent aluminosilicate sheets of the mineral, but they release most of this water to a dry atmosphere and shrink. A small fraction of the water or hydroxide ions remains rather tightly bound by ion–dipole forces to cations between the aluminosilicate sheets and is lost only when the clay is heated to a high temperature. The firing of aluminosilicate clays simultaneously causes irreversible chemical changes to occur. The clay kaolinite ($Al_2Si_2O_5(OH)_4$) undergoes the following reaction:

$$\underset{\text{(kaolinite)}}{3\ Al_2Si_2O_5(OH)_4} \longrightarrow \underset{\text{(mullite)}}{Al_6Si_2O_{13}(s)} + \underset{\text{(silica)}}{4\ SiO_2(s)} + \underset{\text{(water)}}{6\ H_2O(g)}$$

The fired ceramic body is a mixture of two phases: mullite and silica. Mullite, a rare mineral in nature, takes the form of needlelike crystals that interpenetrate and confer strength on the ceramic. When the temperature is above 1470°C, the silica phase forms as minute grains of cristobalite, one of the several crystalline forms of SiO_2.

If chemically pure kaolinite is fired, the finished ceramic object is white. Such purified clay minerals are the raw material for fine china. As they occur in nature, clays contain impurities, such as transition-metal oxides, that affect the color of both the unfired clay and the fired ceramic object if they are not removed. The colors of the metal oxides arise from their absorption of light at visible wavelengths, as explained by crystal field theory (see Section 8.4). Common colors for ceramics are yellow or greenish yellow, brown, and red. Bricks are red when the clay used to make them has high iron content.

Before a clay is fired in a kiln, it must first be freed of moisture by slowly heating to about 500°C. If a clay body dried at room temperature were to be placed directly into a hot kiln, it would literally explode from the sudden, uncontrolled expulsion of water. A fired ceramic shrinks somewhat as it cools, causing cracks to form. These imperfections limit the strength of the fired object and are undesirable. The occurrence of imperfections can be reduced by coating the surface of a partially fired clay object with a **glaze**, a thin layer that minimizes crack formation in the underlying ceramic by holding it in a state of tension as it cools. Glazes, as their name implies, are glasses that have no sharp melting temperature, but rather harden and develop resistance to shear stresses increasingly as the temperature of the high-fired clay object is gradually reduced. Glazes generally are aluminosilicates that have high aluminum content to raise their viscosity, and thereby reduce the tendency to run off the surface during firing. They also provide the means of coloring the surfaces of fired clays and imparting decorative designs to them. Transition-metal oxides (particularly those of titanium [Ti], vanadium [V], chromium [Cr], manganese [Mn], iron [Fe], cobalt [Co], nickel [Ni], and Cu) are responsible for the colors. The oxidation state of the transition metal in the glaze is critical in determining the color produced and is controlled by regulating the composition of the atmosphere in the kiln; atmospheres rich in oxygen give high oxidation states, and those poor or lacking in oxygen give low oxidation states.

22.3.2 Glass

Glassmaking probably originated in the Near East about 3500 years ago. It is one of the oldest domestic arts, but its beginnings, like those of metallurgy, are obscure. Both required high-temperature, charcoal-fueled ovens and vessels made of materials that did not easily melt to initiate and contain the necessary chemical reactions. In the early period of glassmaking, desired shapes were fabricated by sculpting them from solid chunks of glass. At a later date, molten glass was poured in successive layers over a core of sand. A great advance was the invention of glassblowing, which probably occurred in the first century B.C. A long iron tube was dipped into molten glass and a rough ball of viscous

© Charles D. Winters/Science Source

FIGURE 22.10 Handcrafted glassware is trimmed after being blown into the desired shape.

material was caused to accumulate on its end by rotating the tube. Blowing into the iron tube forced the soft glass to take the form of a hollow ball (Fig. 22.10) that could be further shaped into a vessel and severed from the blowing tube with a blade. Artisans in the Roman Empire developed glassblowing to a high degree, but with the decline of that civilization, the skill of glassmaking in Europe deteriorated until the Venetians redeveloped the lost techniques a thousand years later.

Glasses are amorphous solids of widely varying composition (see Section 21.4 for a discussion of the physical properties of glass). In this chapter, the term *glass* is used in the restricted and familiar sense to refer to materials formed from silica, usually in combination with metal oxides. The absence of long-range order in glasses has the consequence that they are **isotropic**—that is, their physical properties are the same in all directions. This has advantages in technology, including that glasses expand uniformly in all directions with an increase in temperature. The mechanical strength of glass is intrinsically high, exceeding the tensile strength of steel, provided the surface is free of scratches and other imperfections. Flaws in the surface provide sites where fractures can start when the glass is stressed. When a glass object of intricate shape and nonuniform thickness is suddenly cooled, internal stresses are locked in; these stresses may be relieved catastrophically when the object is heated or struck (Fig. 22.11). Slowly heating a strained object to a temperature somewhat below its softening point and holding it there for a while before allowing it to cool slowly is called **annealing**; it gives short-range diffusion of atoms a chance to occur and to eliminate internal stresses.

The softening and annealing temperatures of a glass and other properties, such as density, depend on its chemical composition (Table 22.2). Silica itself (SiO_2) forms a glass if it is heated above its melting point and then cooled

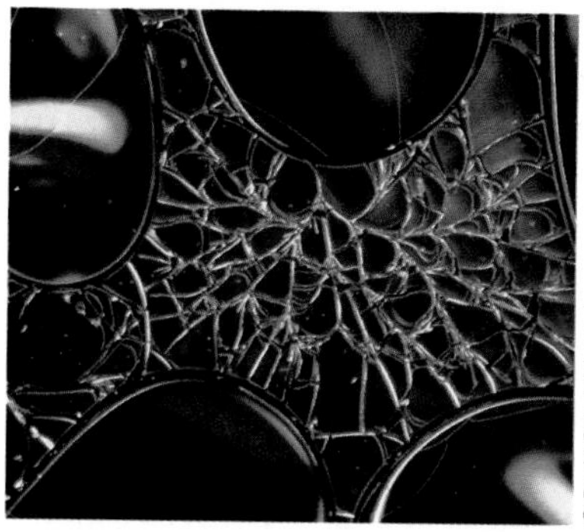

© David Guizak/Phototake

FIGURE 22.11 The strains associated with internal stresses in rapidly cooled glass can be made visible by viewing the glass in polarized light. They appear as colored regions.

TABLE 22.2

Composition and Properties of Various Glasses

Silica Glass	Soda-lime Glass	Borosilicate Glass	Aluminosilicate Glass	Leaded Glass
Composition				
SiO_2, 99.9% H_2O, 0.1%	SiO_2, 73% Na_2O, 17% CaO, 5% MgO, 4% Al_2O_3, 1%	SiO_2, 81% B_2O_3, 13% Na_2O, 4% Al_2O_3, 2% B_2O_3, 5%	SiO_2, 63% Al_2O_3, 17% CaO, 8% MgO, 7% Al_2O_3, 2%	SiO_2, 56% PbO, 29% K_2O, 9% Na_2O, 4%
Coefficient of Linear Thermal Expansion (°C^{-1} × 10^7)†				
5.5	93	33	42	89
Softening Point (°C)				
1580	695	820	915	630
Annealing Point (°C)				
1050	510	565	715	435
Density (g cm^{-3})				
2.20	2.47	2.23	2.52	3.05
Refractive Index‡ at λ = 589 nm				
1.459	1.512	1.474	1.530	1.560

†The coefficient of linear thermal expansion is defined as the fractional increase in length of a body when its temperature is increased by 1°C.

‡The refractive index is a vital property of glass for optical applications. It is defined by $n = \sin\theta_i/\sin\theta_r$, where θ_i is the angle of incidence of a ray of light on the surface of the glass and θ_r is the angle of refraction of the ray of light in the glass.

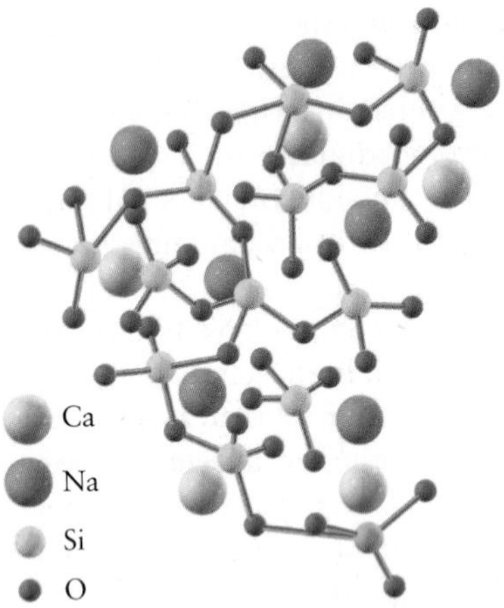

FIGURE 22.12 Structure of a soda-lime glass. Note the tetrahedral coordination of oxygen atoms around each silicon atom.

rapidly to avoid crystallization. The resulting vitreous (glassy) silica has limited use because the high temperatures required to shape it make it quite expensive. Sodium silicate glasses are formed in the high-temperature reaction of silica sand with anhydrous sodium carbonate (soda ash, Na_2CO_3):

$$Na_2CO_3(s) + n\ SiO_2(s) \longrightarrow Na_2O\cdot(SiO_2)_n(s) + CO_2(g)$$

The melting point of the nonvolatile product is about 900°C, and the glassy state results if cooling through that temperature is rapid. The product, called "water glass," is water-soluble; thus, it is unsuitable for making vessels. Its aqueous solutions, however, are used in some detergents and as adhesives for sealing cardboard boxes.

An insoluble glass with useful structural properties results if lime (CaO) is added to the sodium carbonate–silica starting materials. **Soda-lime glass** is the resulting product, with the approximate composition $Na_2O\cdot CaO\cdot(SiO_2)_6$. Soda-lime glass is easy to melt and shape and is used in applications ranging from bottles to window glass. It accounts for more than 90% of all the glass manufactured today. The structure of this ionic glass is shown schematically in Figure 22.12. It is a three-dimensional network of the type discussed in Section 22.1, but with random coordination of the silicate tetrahedra. The network is a giant "polyanion," with Na^+ and Ca^{2+} ions distributed in the void spaces to compensate for the negative charge on the network.

Replacing lime and some of the silica in a glass by other oxides (Al_2O_3, B_2O_3, K_2O, or PbO) modifies its properties noticeably. For example, the thermal conductivity of ordinary (soda-lime) glass is quite low, and its coefficient of thermal expansion is high. This means that internal stresses are created when its surface is subjected locally to extreme heat or cold, and it may shatter. The coefficient of thermal expansion is appreciably lower in certain borosilicate glasses, in which many of the silicon sites are occupied by boron. Pyrex, the most familiar of these glasses, has a coefficient of linear expansion about one-third that of ordinary soda-lime glass and is the preferred material for laboratory glassware and household ovenware. Vycor has an even smaller coefficient of expansion (approximating that of fused silica) and is made by chemical treatment of a borosilicate glass to leach out its sodium. This leaves a porous structure that is densified by increasing the temperature and shrinking the glass to its final volume.

22.3.3 Cements

Hydraulic cement was first developed by the ancient Romans, who found that a mixture of lime (CaO) and dry volcanic ash reacts slowly with water, even at low temperatures, to form a durable solid. They used this knowledge to build the Pantheon in Rome, a circular building whose concrete dome, spanning 143 feet without internal support, still stands nearly 2000 years after its construction! The knowledge of cement making was lost for centuries after the fall of the Roman Empire. It was rediscovered in 1824 by an English bricklayer, Joseph Aspdin, who patented a process for calcining a mixture of limestone and clay. He called the product **Portland cement** because, when mixed with water, it hardened to a material that resembled a kind of limestone found on the Isle of Portland. Portland cement is now manufactured in every major country, and annual worldwide production is currently about 800 million metric tons, exceeding the production of all other materials. Portland cement opened up a new age in the methods of constructing highways and buildings: Rock could be crushed and then molded in cement, rather than shaped with cutting tools.

Portland cement is a finely ground, powdered mixture of compounds produced by the high-temperature reaction of lime, silica, alumina, and iron oxide. The lime (CaO) may come from limestone or chalk deposits, and the silica (SiO_2) and alumina (Al_2O_3) are often obtained in clays or slags. The blast

TABLE 22.3

Composition of Portland Cement

Oxide	Percentage by Mass
Lime (CaO)	61–69
Silica (SiO_2)	18–24
Alumina (Al_2O_3)	4–8
Iron(III) oxide (Fe_2O_3)	1–8
Minor oxides (MgO, Na_2O, K_2O, SO_3)	2–4

furnaces of steel mills are a common source of slag, which is a byproduct of the smelting of iron ore. The composition of slag varies, but it can be represented as a calcium aluminum silicate of approximate formula $CaO{\cdot}Al_2O_3{\cdot}(SiO_2)_2$. Molten slag solidifies into "blast furnace clinkers" on quenching in water. This material is crushed and ground to a fine powder, blended with lime in the correct proportion, and burned again in a horizontal rotary kiln at temperatures up to 1500°C to produce "cement clinker." A final stage of grinding and the addition of about 5% gypsum ($CaSO_4{\cdot}2\,H_2O$) to lengthen the setting time completes the process of manufacture. Table 22.3 presents the composition of a typical Portland cement and gives percentages of the separate oxides; these simple materials, which are the "elements" of cement making, combine in the cement in more complex compounds such as tricalcium silicate, $(CaO)_3{\cdot}SiO_2$, and tricalcium aluminate, $(CaO)_3{\cdot}Al_2O_3$.

Cement *sets* when the semiliquid slurry first formed by the addition of water to the powder becomes a solid of low strength. Subsequently, it gains strength in a slower *hardening* process. Setting and hardening involve a complex group of exothermic reactions in which several hydrated compounds form. Portland cement is a **hydraulic cement**, because it hardens not by loss of admixed water, but by chemical reactions that incorporate water into the final body. The main reaction during setting is the hydration of the tricalcium aluminate, which can be approximated by the following equation:

$$(CaO)_3{\cdot}Al_2O_3(s) + 3\,(CaSO_4{\cdot}2H_2O)(s) + 26\,H_2O(\ell) \longrightarrow (CaO)_3{\cdot}Al_2O_3{\cdot}(CaSO_4)_3{\cdot}32H_2O(s)$$

The product forms after 5 or 6 hours as a microscopic forest of long crystalline needles that lock together to solidify the cement. Later, the calcium silicates react with water to harden the cement. For example:

$$6\,(CaO)_3{\cdot}SiO_2(s) + 18\,H_2O(\ell) \longrightarrow (CaO)_5{\cdot}(SiO_2)_6{\cdot}5H_2O(s) + 13\,Ca(OH)_2(s)$$

The hydrated calcium silicates develop as strong tendrils that coat and enclose unreacted grains of cement, each other, and other particles that may be present, binding them in a robust network. Most of the strength of cement comes from these entangled networks, which, in turn, depend ultimately for strength on chains of O—Si—O—Si silicate bonds. Hardening is slower than setting; it may take as long as 1 year for the final strength of a cement to be attained.

Portland cement is rarely used alone. Generally, it is combined with sand, water, and lime to make **mortar**, which is applied with a trowel to bond bricks or stone together in an assembled structure. When Portland cement is mixed with sand and aggregate (crushed stone or pebbles) in the proportions of 1:3.75:5 by volume, the mixture is called **concrete**. Concrete is outstanding in its resistance to compressive forces and is therefore the primary material in use for the foundations of buildings and the construction of dams, in which the compressive loads are enormous. The stiffness (resistance to bending) of concrete is high, but its fracture toughness (resistance to impact) is substantially lower and its tensile strength is relatively poor. For this reason, concrete is usually reinforced with steel rods when it is used in structural elements such as beams that are subject to transverse or tensile stresses.

As excess water evaporates from cement during hardening, pores form that typically comprise 25% to 30% of the volume of the solid. This porosity weakens concrete, and recent research has shown that the fracture strength is related inversely to the size of the largest pores in the cement. A new material, called "macro defect–free" (MDF) cement, has been developed in which the size of the pores is reduced from about a millimeter to a few micrometers by the addition of water-soluble polymers that make a doughlike "liquid" cement that is moldable with the use of far less water. Unset MDF cement is mechanically kneaded and extruded into the desired shape. The final result possesses substantially increased bending resistance and fracture toughness. MDF cement can even be molded into springs and shaped on a conventional lathe. When it is reinforced with organic fibers, its toughness is further increased. The development of MDF cement is a good illustration of the way in which chemistry and engineering collaborate to furnish new materials.

22.4 Nonsilicate Ceramics

Many useful ceramics exist that are *not* based on the Si—O bond and the SiO_4 tetrahedron. They have important uses in electronics, optics, and the chemical industry. Some of these materials are oxides, but others contain neither silicon nor oxygen.

22.4.1 Oxide Ceramics

Oxide ceramics are materials that contain oxygen in combination with any of a number of metals. These materials are named by adding an *-ia* ending to the stem of the name of the metallic element. Thus, if the main chemical component of an oxide ceramic is Be_2O_3, it is a *beryllia* ceramic; if the main component is Y_2O_3, it is an *yttria* ceramic; and if it is MgO, it is a *magnesia* ceramic. As Table 22.4 shows, the melting points of these and other oxides are substantially higher than the melting points of the elements themselves. Such high temperatures are hard to achieve and maintain, and the molten oxides corrode most container materials. Oxide ceramic bodies are therefore not shaped by melting the appropriate oxide and pouring it into a mold. Instead, these ceramics are fabricated by sintering, like the silicate ceramics.

Alumina (Al_2O_3) is the most important nonsilicate ceramic material. It melts at a temperature of 2051°C and retains strength even at temperatures of 1500°C to 1700°C. Alumina has a large electrical resistivity and withstands both thermal shock and corrosion well. These properties make it a good material for spark plug insulators, and most spark plugs now use a ceramic that is 94% alumina.

High-density alumina is fabricated in such a way that open pores between the grains are nearly completely eliminated; the grains are small, with an average diameter as low as 1.5 μm. Unlike most others, this ceramic has good mechanical strength against impact, which has led to its use in armor plating. The ceramic absorbs the energy of an impacting projectile by breaking; thus, penetration does not occur. High-density alumina is also used in high-speed cutting tools for machining metals. The temperature resistance of the ceramic allows much faster cutting speeds, and a ceramic cutting edge has no tendency to weld to the metallic work piece, as metallic tools do. These properties make

TABLE 22.4

Melting Points of Some Metals and Their Oxides

Metal	Melting Point (°C)	Oxide	Melting Point (°C)
Be	1287	BeO	2570
Mg	651	MgO	2800
Al	660	Al_2O_3	2051
Si	1410	SiO_2	1723
Ca	865	CaO	2572
Y	1852	Y_2O_3	2690

alumina cutting tools superior to metallic tools, as long as they do not break too easily. High-density alumina is also used in artificial joints (Fig. 22.13).

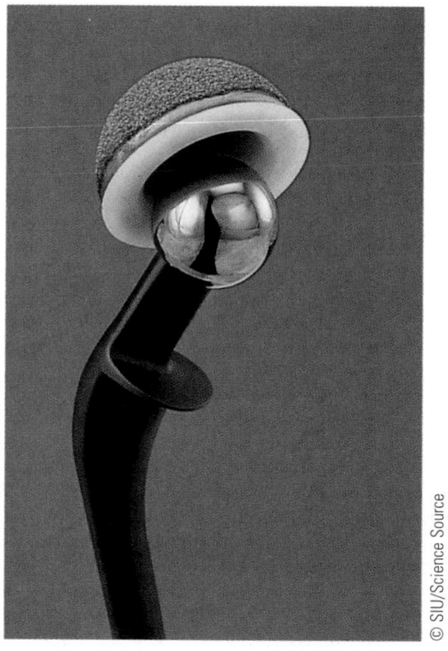

FIGURE 22.13 The socket in this artificial hip is made of high-density alumina.

If Al_2O_3 doped with a small percentage of MgO is fired in a vacuum or a hydrogen atmosphere (instead of air) at a temperature of 1800°C to 1900°C, even very small pores, which scatter light and make the material white, are removed. The resulting ceramic is translucent. This material is used to contain the sodium in high-intensity sodium discharge lamps. With envelopes of high-density alumina, these lamps can be operated at temperatures of 1500°C to give a whiter and more intense light. Old-style sodium-vapor lamps with glass envelopes were limited to a temperature of 600°C because the sodium vapor reacted with the glass at higher temperatures. At low temperatures, the light from a sodium-vapor lamp has an undesirable yellow color.

Magnesia (MgO) is mainly used as a **refractory**—a ceramic material that withstands a temperature of more than 1500°C without melting (MgO melts at 2800°C). A major use of magnesia is as insulation in electrical heating devices, because it combines high thermal conductivity with excellent electrical resistance. Magnesia is prepared from magnesite ores, which consist of $MgCO_3$ and a variety of impurities. When purified magnesite is heated to 800°C to 900°C, carbon dioxide is driven off to form MgO(*s*) in fine grains:

$$MgCO_3(s) \longrightarrow MgO(s) + CO_2(g)$$

After cooling, fine-grained MgO reacts vigorously with water to form magnesium hydroxide:

$$MgO(s) + H_2O(\ell) \longrightarrow Mg(OH)_2(s)$$

Heating fine-grained MgO(*s*) to 1700°C causes the MgO grains to sinter, giving a "dead-burned magnesia" that consists of large crystals and does not react with water. The $\Delta G°$ of the reaction between MgO and H_2O does not change when magnesia is dead burned. The altered microstructure (larger crystals vs. small) makes the reaction with water exceedingly slow, however.

22.4.2 Superconducting Ceramics

The oxide ceramics discussed so far in this chapter all consisted of single chemical compounds, except for minor additives. A natural idea for new ceramics is to make materials that contain two (or more) oxides in equal or nearly equal molar amounts. Thus, if $BaCO_3$ and TiO_2 are mixed and heated to high temperature, they react to give the ceramic barium titanate:

$$BaCO_3(s) + TiO_2(s) \rightarrow BaTiO_3(s) + CO_2(g)$$

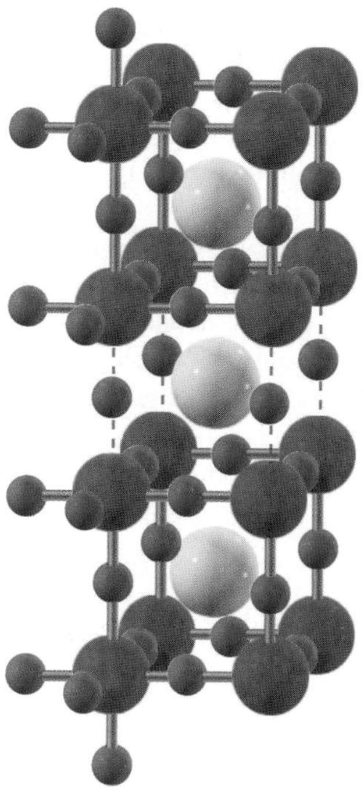

FIGURE 22.14 Structure of perovskite ($CaTiO_3$). A stack of three unit cells is shown, with some additional O atoms (red) from neighboring cells. Each Ti atom (gray) is surrounded by six O atoms; each Ca atom (white) has eight O atoms as nearest neighbors.

Barium titanate, which has many novel properties, is a **mixed oxide ceramic**. It has the same structure as the mineral *perovskite*, $CaTiO_3$ (Fig. 22.14), except, of course, that Ba replaces Ca. Perovskites typically have two metal atoms for every three O atoms, giving them the general formula ABO_3, where A stands for a metal atom at the center of the unit cube and B stands for an atom of a different metal at the cube corners.

Research interest in perovskite ceramic compositions has grown explosively after the discovery that some of them are able to act as a **superconductor** at relatively high temperatures. A superconducting material offers no resistance whatsoever to the flow of an electric current. The phenomenon was discovered by the Dutch physicist Heike Kamerlingh-Onnes in 1911, when he cooled mercury below its superconducting transition temperature of 4 K. Such low temperatures are difficult to achieve and maintain, but if practical superconductors could be made to work at higher temperatures (or even at room temperature!), then power transmission, electronics, transportation, medicine, and many other aspects of human life would be transformed. More than 60 years of research

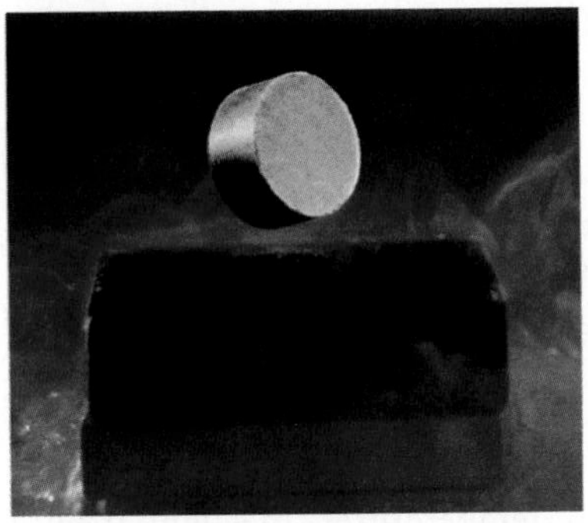

FIGURE 22.15 Levitation of a small magnet above a disk of superconducting material. A superconducting substance cannot be penetrated by an external magnetic field. At room temperature, the magnet rests on the ceramic disk. When the disk is cooled with liquid nitrogen, it becomes superconducting and excludes the magnet's field, forcing the magnet into the air.

with metallic systems culminated in 1973 with the discovery of a niobium–tin alloy with a world-record superconducting transition temperature of 23.3 K. Progress toward higher temperature superconductors then stalled until 1986, when K. Alex Müller and J. Georg Bednorz, who had had the inspired notion to look for higher transition temperatures among perovskite ceramics, found a Ba—La—Cu—O perovskite phase having a transition temperature of 35 K. This result motivated other scientists, who soon discovered another rare-earth–containing perovskite ceramic that became a superconductor at 90 K. This result was particularly exciting because 90 K exceeds the boiling point of liquid nitrogen (77 K), a relatively cheap refrigerant (Fig. 22.15). This 1-2-3 compound (so called because its formula, $YBa_2Cu_3O_{(9-x)}$, has one Y, two Ba, and three Cu atoms per formula unit) is not an ideal perovskite because it has fewer than nine O atoms in combination with its six metal atoms. The deficiency makes x in the formula somewhat greater than 2, depending on the exact method of preparation. The structure of this nonstoichiometric solid is shown in Figure 22.16. More recently, the maximum superconducting transition temperature has increased to 125 K in another class of ceramics that does not contain rare-earth elements.

A crucial concern in the application of superconducting ceramics is to devise ways to fabricate the new materials in desired shapes such as wires. This will be quite a challenge because these superconductors are ceramics and have the brittleness and fragility typical of ceramic materials.

22.4.3 Nonoxide Ceramics

In nonoxide ceramics, nitrogen (N) or carbon (C) takes the place of oxygen in combination with silicon or boron. Specific substances are boron nitride (BN), boron carbide (B_4C), the silicon borides (SiB_4 and SiB_6), silicon nitride (Si_3N_4), and silicon carbide (SiC). All of these compounds possess strong, short covalent bonds. They are hard and strong, but brittle. Table 22.5 lists the enthalpies of the chemical bonds in these compounds.

Much research has aimed at making gas-turbine and other engines from ceramics. Of the oxide ceramics, only alumina and zirconia (ZrO_2) are strong enough, but both resist thermal shock too poorly for this application. Attention has therefore turned to the nonoxide **silicon nitride** (Si_3N_4). In this **network solid** (Fig. 22.17), every Si atom bonds to four N atoms that surround it at the corners of a tetrahedron; these tetrahedra link into a three-dimensional network by sharing corners. The Si—N bond is covalent and strong (the bond enthalpy is 439 kJ mol^{-1}). The similarity to the joining of SiO_4 units in silicate minerals (see Section 22.1) is clear.

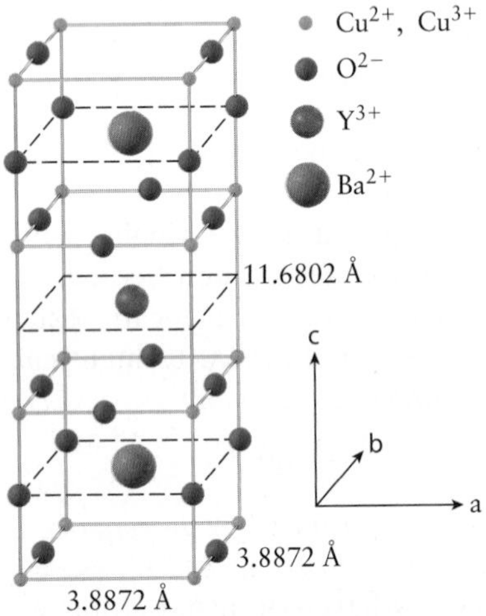

FIGURE 22.16 The structure of $YBa_2Cu_3O_{9-x}$ is a layered perovskite. Each Cu atom (orange) is bonded to O atoms (red). The layers differ because one-third of them contain Y atoms (blue-gray), whereas two-thirds contain Ba atoms (green). This structure is a variation of the perovskite structure in Figure 22.14. In a hypothetical "$BaCuO_3$" structure, every third Ba has been replaced by a Y, and the O atoms in the layer containing the Y have been removed.

TABLE 22.5

Bond Enthalpies in Nonoxide Ceramics

Bond	Bond Enthalpy (kJ mol^{-1})
B—N	389
B—C	448
Si—N	439
Si—C	435
B—Si	289
C—C	350

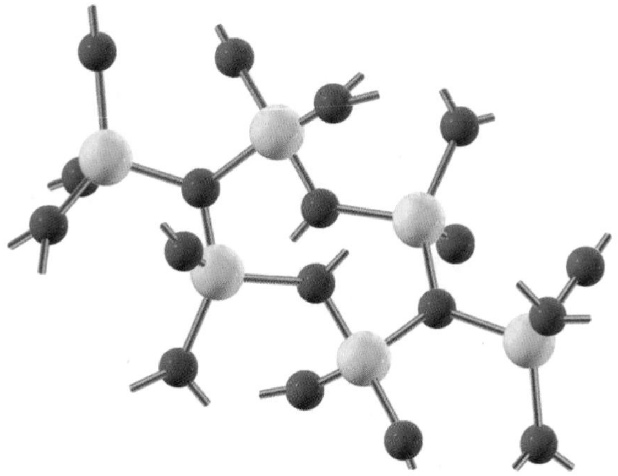

FIGURE 22.17 Structure of silicon nitride (Si_3N_4). Each Si atom is bonded to four N atoms, and each N atom is bonded to three Si atoms. The result is a strong network.

At first, silicon nitride appears chemically unpromising as a high-temperature structural material. It is unstable in contact with water because the following reaction has a negative $\Delta G°$:

$$Si_3N_4(s) + 6\ H_2O(\ell) \longrightarrow 3\ SiO_2(s) + 4\ NH_3(g)$$

In fact, this reaction causes finely ground Si_3N_4 powder to give off an odor of ammonia in moist air at room temperature. Silicon nitride is also thermodynamically unstable in air, reacting spontaneously with oxygen:

$$Si_3N_4(s) + 3\ O_2(g) \longrightarrow 3\ SiO_3(s) + 2\ N_2(g)$$

The reaction has a $\Delta G°$ of -1927 kJ mol^{-1}. In practice, neither reaction occurs at a perceptible rate when Si_3N_4 is in bulk form. Initial contact of oxygen or water with $Si_3N_4(s)$ forms a surface film of $SiO_2(s)$ that protects the bulk of the $Si_3N_4(s)$ from further attack. When *strongly* heated in air (to about 1900°C), Si_3N_4 does decompose, violently, but until that temperature is reached, it resists attack.

Fully dense silicon nitride parts are stronger than metallic alloys at high temperatures. Ball bearings made of dense silicon nitride work well without lubrication at temperatures up to 700°C; for example, they last longer than steel ball bearings. Because the strength of silicon nitride increases with the density attained in the production process, the trick is to form a dense piece of silicon nitride in the desired shape. One method for making useful shapes of silicon nitride is "reaction bonding." Powdered silicon is compacted in molds, removed, and then fired under an atmosphere of nitrogen at 1250°C to 1450°C. The following reaction forms the ceramic:

$$3\ Si(s) + 2\ N_2(g) \longrightarrow Si_3N_4(s)$$

The parts neither swell nor shrink significantly during the chemical conversion from Si to Si_3N_4, making it possible to fabricate complex shapes reliably. Unfortunately, reaction-bonded Si_3N_4 is still somewhat porous and is not strong enough for many applications. In the "hot-pressed" forming process, Si_3N_4 powder is prepared in the form of exceedingly small particles by reaction of silicon tetrachloride with ammonia:

$$3\ SiCl_4(g) + 4\ NH_3(g) \longrightarrow Si_3N_4(s) + 12\ HCl(g)$$

The solid Si_3N_4 forms as a smoke. It is captured as a powder, mixed with a carefully controlled amount of MgO additive, placed in an enclosed mold, and sintered at 1850°C under a pressure of 230 atm. The resulting ceramic shrinks to nearly full density (no pores). Because the material does not flow well (to fill a complex mold completely), only simple shapes are possible. Hot-pressed silicon nitride is impressively tough and can be machined only with great difficulty and with diamond tools.

In the silicate minerals of Section 22.1, AlO_4^{5-} units routinely substitute for SiO_4^{4-} tetrahedra as long as positive ions of some type are present to balance the electric charge. This fact suggests that in silicon nitride some Si^{4+} ions, which lie at the centers of tetrahedra of N atoms, could be replaced by Al^{3+} if a compensating replacement of O^{2-} for N^{3-} were simultaneously made. Experiments show that ceramic alloying of this type works well, giving many new ceramics with great potential called **sialons** (named for the four elements Si—Al—O—N). These ceramics illustrate the way a structural theme in a naturally occurring material guided the search for new materials.

Boron has one fewer valence electron than carbon, and nitrogen has one more valence electron. *Boron nitride* (BN) is therefore isoelectronic with C_2, and it is not surprising that it has two structural modifications that resemble the structures of graphite and diamond. In hexagonal boron nitride, the B and N atoms take alternate places in an extended "chicken-wire" sheet in which the B—N distance is 1.45 Å. The sheets stack in such a way that each B atom has a N atom directly above it and directly below it, and vice versa. The cubic form of boron nitride has the diamond structure, is comparable in hardness to diamond, and resists oxidation better. Boron nitride is often prepared by **chemical vapor deposition**, a method used in fabricating several other ceramics as well. In this method, a controlled chemical reaction of gases on a contoured, heated surface gives a solid product of the desired shape. If a cup made of BN is needed, a cup-shaped mold is heated to a temperature exceeding 1000°C and a mixture of $BCl_3(g)$ and $NH_3(g)$ is passed over its surface. The reaction

$$BCl_3(g) + NH_3(g) \longrightarrow BN(s) + 3\ HCl(g)$$

deposits a cup-shaped layer of BN(*s*). Boron nitride cups and tubes are used to contain and evaporate molten metals.

Silicon carbide (SiC) is diamond in which half of the C atoms are replaced by Si atoms. Also known by its trade name of Carborundum, silicon carbide was developed originally as an abrasive, but it is now used primarily as a refractory and as an additive in steel manufacture. It is formed and densified by methods similar to those used with silicon nitride. Silicon carbide is often produced in the form of small plates or whiskers to reinforce other ceramics. Fired silicon carbide whiskers are quite small (0.5 μm in diameter and 50 μm long) but are strong. They are mixed with a second ceramic material before that material is formed. Firing then gives a **composite ceramic.** Such composites are stronger and tougher than unreinforced bodies of the same primary material. Whiskers serve to reinforce the main material by stopping cracks; they either deflect advancing cracks or soak up their energy, and a widening crack must dislodge them to proceed (Fig. 22.18). Recently, SiC has been used in high-power, high-temperature semiconductor devices (Section 22.7).

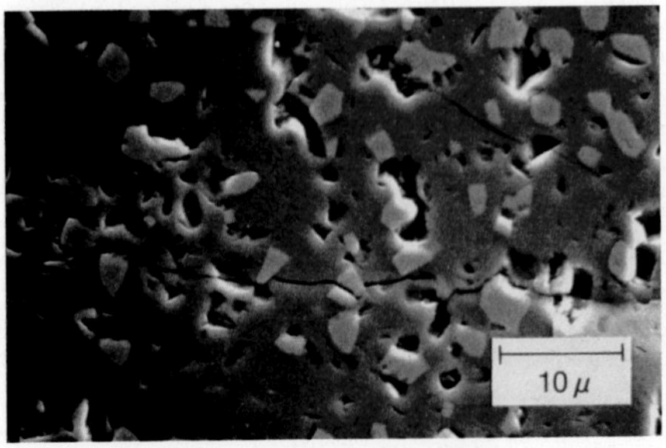

Carborundum Company

FIGURE 22.18 Microstructure of a composite ceramic. The light-colored areas are TiB_2, the gray areas are SiC, and the dark areas are voids. The TiB_2 acts to toughen the SiC matrix. Note that the crack shown passing through from left to right is forced to deflect around the TiB_2 particles.

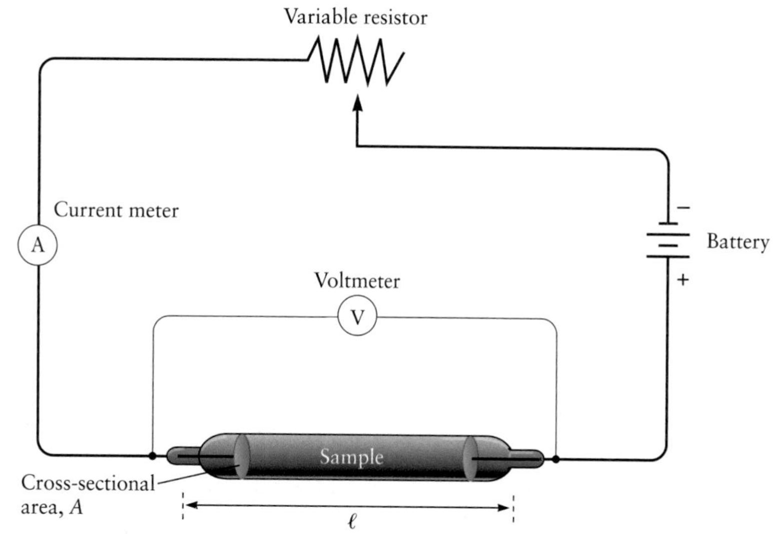

FIGURE 22.19 Test circuit for measuring conductivity.

22.5 Electrical Conduction in Materials

Electronic properties describe the movement of charged particles in a material in response to an applied electric field. If the charges are free to move throughout the material, the process is *electrical conduction,* measured by the *electrical conductivity* of the material. Differences in the magnitude of the conductivity distinguish metals, semiconductors, and insulators. If the charges can move only limited distances and are then halted by opposing binding forces, separation of positive and negative charges leads to *electric polarization* of the material, measured by its *dielectric constant.* Conduction involves dissipation of energy as heat, whereas polarization involves storage of potential energy in the material.

An electric field applied to a material that has free charged particles causes these particles to flow through the material and into the external circuit. How do we define and measure the conductivity of a material? On what macroscopic properties does it depend? How does it relate to the detailed chemical structure of the material?

22.5.1 Measurement of Conductivity

The electrical conductivity of a material is measured by placing a cylindrical sample of cross-sectional area, A, and length, ℓ, in a simple electrical circuit as a resistor in series with a power supply and ammeter; a voltmeter measures the actual voltage across the sample (Fig. 22.19). If the voltage, V, is varied and the resulting current, I, is measured at each voltage, a plot of I versus V is a straight line (Fig. 22.20):

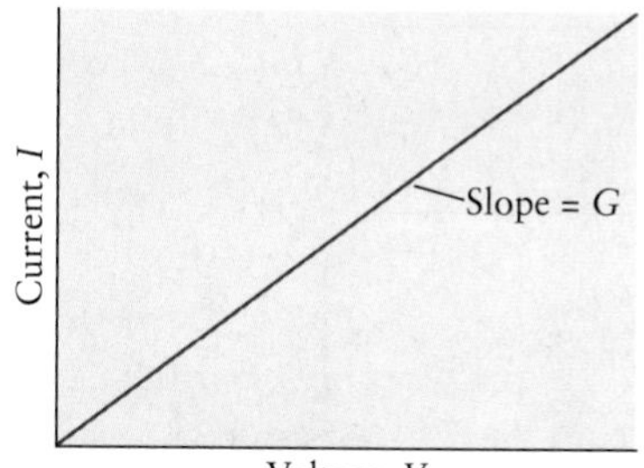

FIGURE 22.20 Plot of current against voltage is a straight line whose slope is the conductance G.

$$I = GV \qquad [22.1]$$

The resulting slope is called the **conductance** and is denoted by G. If V is measured in volts and I in amperes (A), then G has units of siemens (1 siemens = 1 A/V). If the value of G is constant, the material follows Ohm's law

$$V = IR$$

and G is the reciprocal of the **resistance** R, which is measured in ohms (Ω; 1 Ω = 1 V/A). So far, the measurement depends on the size and shape of the material sample, in addition to its composition. To remove these geometric effects, we note from experiment that R increases as ℓ increases, and it decreases as A increases. Therefore, we define the material property **resistivity**, denoted by ρ, as the proportionality constant that summarizes these two effects:

$$R = \frac{\ell}{A}\rho \qquad [22.2]$$

From this definition, ρ has dimensions of Ω m. Finally, we define the material property **conductivity**, σ, as the reciprocal of the resistivity:

$$\sigma = \frac{1}{\rho} = \frac{\ell}{RA} \qquad [22.3]$$

from which σ has units of $(\Omega\ \text{m})^{-1}$ or S m^{-1}. We now rewrite Ohm's law in a form independent of sample geometry. First, define the current density $J(\text{A m}^{-2})$ through the sample as I/A and define the electric field $E(\text{V m}^{-1})$ through the sample as V/ℓ. Inserting these definitions and the definition of σ from Equation 22.3 into Equation 22.1 gives:

$$J = \sigma E \qquad [22.4]$$

According to Equation 22.4, the current density flowing through a material sample is proportional to the electric field applied to the sample, and the proportionality constant is the conductivity of the material of which the sample is made. This is the equation we use to relate the conductivity of a material to its microstructural properties. Table 22.6 lists conductivities for several common metals at room temperature.[2]

TABLE 22.6

Electrical Conductivity of Selected Metals at Room Temperature

Metal	Conductivity $[(\Omega\ \text{m})^{-1}]$
Silver	6.8×10^7
Copper	6.0×10^7
Gold	4.3×10^7
Aluminum	3.8×10^7
Iron	1.0×10^7
Platinum	0.94×10^7
Stainless steel	0.2×10^7

22.5.2 Microscopic Origins of the Conductivity

Insight into the conductivity is provided by measuring the electrical conductivity of aqueous ionic solutions (Fig. 22.21; this topic is referred to in Chapters 11 and 15). The conductivity of pure water, multiply distilled to remove all impurities, is about 0.043×10^{-6} $(\Omega\ \text{cm})^{-1}$. Exposed to the air, pure water dissolves CO_2, which forms carbonic acid, H_2CO_3; dissociation produces H_3O^+ and HCO_3^-, which increase the conductivity to about 1×10^{-6} $(\Omega\ \text{cm})^{-1}$. As ionic

[2]Because the meter is inconveniently large for measuring the dimensions of most samples in material studies, you will frequently see ρ expressed in units of Ω cm and σ in units of $(\Omega\ \text{cm})^{-1}$. In base International System of Units (SI), resistivity has units expressed as kg $\text{m}^3\ \text{s}^{-3}\ \text{A}^{-2}$.

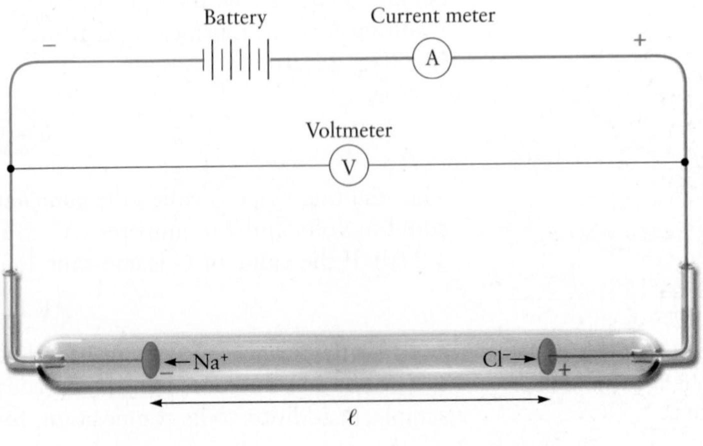

FIGURE 22.21 Apparatus for measuring the conductivity of an aqueous solution of NaCl.

solutes are added to water, the conductivity increases rapidly; a 1.0-M solution in NaOH has conductivity of about 0.180 $(\Omega\ \text{cm})^{-1}$ at 25°C. The conductivity depends strongly on both concentration and ionic species. The concentration dependence is summarized by the **molar conductivity**, defined by $\Lambda_m = \sigma/c$, where c is the concentration of the ion measured in moles per liter. To gain fundamental understanding of the conduction mechanism, we relate Λ_m to the details of the ionic motion in dilute solutions in response to the electric field. This leads to the concept of *mobility* of the ion, denoted by μ, which shows how its surroundings influence the response of an ion to an applied field.

As the electric field is turned on, each ion is accelerated by the field and experiences a force whose magnitude and direction are given by the vector relation

$$\vec{F} = ze\vec{E}$$

where z is the charge on the ion in units of the elementary charge, e, and the expression is valid for either cations or anions (the force points in opposite directions in the two cases). As the ion is accelerated by the field, however, its forward motion is retarded by friction from the surrounding solvent molecules. This retarding force $\vec{F}'$ opposes the direction of motion of the ion and is proportional to its velocity, $\vec{v}$:

$$\vec{F}' = -f\vec{v}$$

where f is the frictional drag coefficient. In due course, these forces balance and the ion achieves its steady drift velocity $\vec{v}_d$. This force balance condition provides the following relation:

$$\vec{v}_d = ze\frac{\vec{E}}{f}$$

This equation shows that the electric field increases the drift velocity, whereas the drag coefficient reduces it. We define the proportionality constant between the *drift speed* (the magnitude of the drift velocity) and the field to be the **mobility**, μ, of the ion (a positive number), giving the following relation:

$$|\vec{v}_d| = \mu|\vec{E}| \qquad [22.5]$$

Mobility has physical units of $\text{m}^2\ \text{V}^{-1}\ \text{s}^{-1}$. For ions in solution, the mobility is given by

$$\mu = \frac{|z|e}{f}$$

In more advanced work, it is possible to estimate the drag coefficient, f, and provide a theoretical prediction of the mobility. Experimentally, the mobility for ions is obtained from the molar conductivity through the following equation, which we do not justify:

$$\Lambda_m = |z|\mu F \qquad [22.6]$$

where F is the Faraday constant, 96,485.34 C mol^{-1} (the Faraday constant is introduced in Chapter 17).

Mobilities are shown in Table 22.7 for several ions in aqueous solution. One interesting feature is that the smaller cations have lower mobilities than the larger cations; for example, μ for Na^+ is smaller than that for K^+, and μ for Mg^{2+} is smaller than that for Ca^{2+}. One might expect that the smaller ions would have larger mobility, because their smaller size should encounter less

TABLE 22.7

Mobilities of Selected Ions in Aqueous Solution at 25°C

Ion	Mobility ($\text{cm}^2\ \text{V}^{-1}\ \text{s}^{-1}$)
Li^+	4.01×10^{-4}
Na^+	5.19×10^{-4}
K^+	7.62×10^{-4}
Mg^{2+}	5.50×10^{-4}
Ca^{2+}	6.17×10^{-4}
Ba^{2+}	6.60×10^{-4}
Cl^-	7.91×10^{-4}
Br^-	8.10×10^{-4}
NO_3^-	7.40×10^{-4}
ClO_4^-	6.98×10^{-4}
CH_3COO^-	4.24×10^{-4}

frictional drag when moving through the solvent. However, the greater charge density on the smaller cations attracts a larger solvation shell than occurs on the larger cations (see Section 11.2 and Fig. 11.5). The entity moving through the solution in response to the electric field is not the "bare" cation but the cation "dressed" with its solvation shell. The larger solvation shell on the smaller ions causes more frictional drag and lower mobility.

It can be shown that the conductivity of an ion is related to its mobility by the following equation:

$$\sigma_{\text{ion}} = |z| e n_{\text{ion}} \mu_{\text{ion}} \qquad [22.7]$$

where n_{ion} is the number density of ions present, expressed in number per cubic meter. The conductivity depends on two fundamental microscopic variables: the number density of carriers and the carrier mobility. Both positive and negative ions can respond to the electric field in the solution, so the total conductivity must include a contribution from each:

$$\sigma_{\text{tot}} = |z_{\text{cation}}| e n_{\text{cation}} \mu_{\text{cation}} + |z_{\text{cation}}| e n_{\text{anion}} \mu_{\text{anion}} \qquad [22.8]$$

Equations 22.7 and 22.8 are valid for non-superconducting electrical conduction processes: motion of positive and negative ions in solution and in ionic solids, and motion of electrons in solids. These equations are used in subsequent sections to discuss conduction in a variety of materials. These equations should remind you that the conductivity in any material depends on two separate microscopic parameters: the number of charge carriers present and their mobilities.

EXAMPLE 22.2

The molar conductivity of Na^+ ions in aqueous solution at 25°C has been determined to be $5.01 \times 10^{-2}\ (\Omega\ \text{cm})^{-1}\ \text{mol}^{-1}\ \text{L}$. Assume an electric field of $1.0 \times 10^2\ \text{V cm}^{-1}$ is applied to the solution. Calculate the mobility and the drift velocity of Na^+ ions.

Solution

From Equation 22.6, the mobility and molar conductivity are related by

$$\mu = \frac{\Lambda_m}{|z|F}$$
$$= \frac{5.01 \times 10^{-2}\ \Omega^{-1}\ \text{cm}^{-1}\ \text{mol}^{-1}\ \text{L}}{(1)(96{,}485\ \text{C mol}^{-1})}$$
$$= 5.19 \times 10^{-4}\ \text{cm}^2\ \text{V}^{-1}\ \text{s}^{-1}$$

We used Ohm's law $V = IR$ or $V = \text{C s}^{-1}\ \Omega$ to simplify the units, and we took $1\ \text{L} = 10^3\ \text{cm}^3$.

From Equation 22.5, the drift velocity is given by

$$v_d = \mu E$$
$$= (5.19 \times 10^{-4}\ \text{cm}^2\ \text{V}^{-1}\ \text{s}^{-1})(1.0 \times 10^2\ \text{V cm}^{-1})$$
$$= 5.19 \times 10^{-2}\ \text{cm s}^{-1}$$

Related Problems: 21, 22, 23, 24, 25, 26

22.5.3 Illustration: Conductivity in Metals

Before the quantum theory of solids (see description in Chapter 21), microscopic descriptions of metals were based on the Drude model, named for the German physicist Paul Drude. The solid was viewed as a fixed array of positively charged metal ions, each localized to a site on the solid lattice. These fixed ions were surrounded by a sea of mobile electrons, one contributed by each of the atoms in the solid. The number density of the electrons, n_{el}, is then equal to the number density of atoms in the solid. As the electrons move through the ions in response to an applied electric field, they can be scattered away from their straight-line motions by collisions with the fixed ions; this influences the mobility of the electrons. As temperature increases, the electrons move more rapidly and the number of their collisions with the ions increases; therefore, the mobility of the electrons decreases as temperature increases. Equation 22.7 applied to the electrons in the Drude model gives

$$\sigma_{el} = e n_{el} \mu_{el} \qquad [22.9]$$

which predicts that the electrical conductivity of a metal will decrease as temperature increases, because the electron mobility decreases with temperature, whereas the electron number density is independent of temperature. This simple model prediction agrees with the experimental fact that the resistivity of metals increases as temperature increases.

22.6 Band Theory of Conduction

The characteristic property of metals is their good ability to conduct electricity and heat. Both phenomena are due to the ease with which valence electrons move. Electrical conduction is a result of the flow of electrons from regions of high potential energy to those of low potential energy. Collisions of high-kinetic-energy electrons and low-kinetic-energy electrons between high- and low-temperature regions of the metal, as well as the collisional excitation of lattice vibrations, facilitate heat conduction. Why are electrons so mobile in a metal but so tightly bound to atoms in an insulating solid, such as diamond or sodium chloride?

Figure 21.19 shows how the quantum energy levels of Na atoms in a crystal are spread out into a continuous band of states, which is half-occupied by electrons. Now, suppose a small electric potential difference is applied across a sodium crystal. The spin-paired electrons that lie deep in the band cannot be accelerated by a weak electric field because occupied levels exist just above them. They have no place to go (recall the Pauli principle, which states that an energy level contains at most two electrons). At the top of the "sea" of occupied levels, however, there is an uppermost electron-occupied or half-occupied level called the **Fermi level.** Electrons that lie near that level have the highest kinetic energy of all the valence electrons in the crystal and can be accelerated by the electric field to occupy the levels above. They are free to migrate in response to the electric field so that they conduct an electric current. These same electrons at the Fermi level are responsible for the high thermal conductivities of metals. They are also the electrons freed by the photoelectric effect when a photon gives them sufficient kinetic energy to escape from the metal (see Section 4.4).

A natural question arises: Why are the alkaline-earth elements metals, given the argument just presented? A metal such as magnesium contains two $3s$ electrons, and one might expect the band derived from the broadened $3s$ level to be completely filled. The answer to this question is that the energies of the $3s$ orbitals and $3p$ orbitals for magnesium are not greatly different. When the internuclear separation becomes small enough, the $3p$ band overlaps the $3s$ band, and many unoccupied sublevels are then available above the highest filled level.

22.6.1 Band Picture of Bonding in Silicon

An analogous procedure to that shown in Figure 21.19 can be used to construct a band picture for silicon. In this case, the $4N_A$ valence atomic orbitals (both $3s$ and $3p$) from 1 mol (N_A atoms) silicon split into *two* bands in the silicon crystal, each containing $2N_A$ closely spaced levels (Fig. 22.22). The lower band is called the **valence band** and the upper one the **conduction band.** Between the top of the valence band and the bottom of the conduction band is an energy region that is

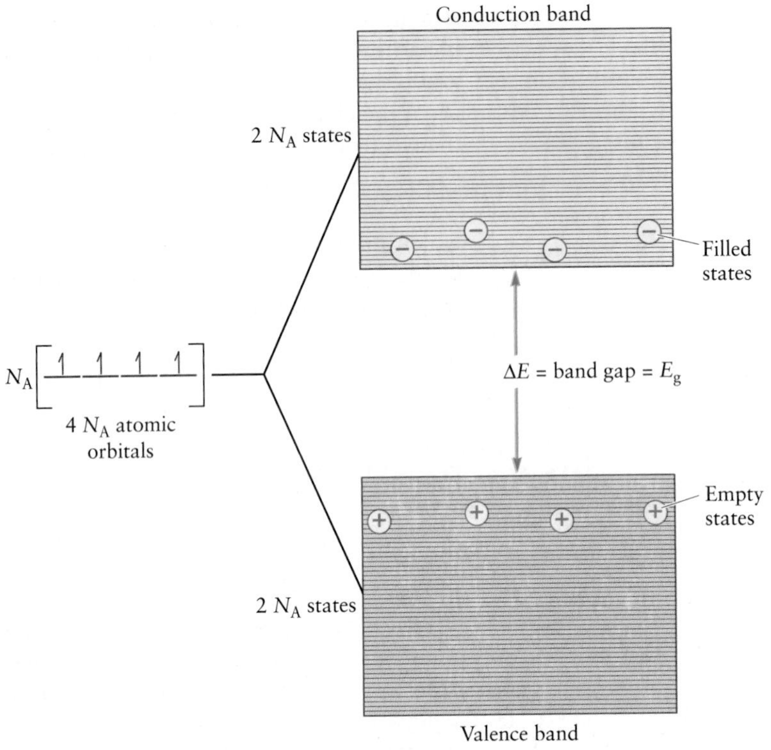

FIGURE 22.22 The valence orbitals of the silicon atom combine in crystalline silicon to give two bands of closely spaced levels. The valence band is almost completely filled, and the conduction band is almost empty.

forbidden to electrons. The magnitude of the separation in energy between the valence band and the lowest level of the conduction band is called the **band gap,** E_g, which for pure silicon is 1.94×10^{-19} J. This is the amount of energy that an electron must gain to be excited from the top of the valence band to the bottom of the conduction band. For 1 mol of electrons to be excited in this way, the energy is larger by a factor of Avogadro's number N_A, giving 117 kJ mol^{-1}. (Another unit used for band gaps is the *electron volt,* defined in Section 3.3 as 1.60218×10^{-19} J. In this unit, the band gap in Si is 1.21 eV.)

Each Si atom in the crystal contributes four valence electrons to the bands of orbitals in Figure 22.22, for a total of $4N_A$ per mole. This is a sufficient number to place two electrons in each level of the valence band (with opposing spins) and leave the conduction band empty. There are no low-lying energy levels for those electrons at the top of the valence band to enter if given a small increment in their energy. The band gap means that an electron at the top of the filled valence band must acquire an energy of at least 1.94×10^{-19} J, equivalent to 117 kJ mol^{-1}, to jump to the lowest empty level of the conduction band. This is a large amount of energy. If it were to be supplied by a thermal source, the temperature of the source would have to be on the order of

$$T = \frac{\Delta E}{R} = \frac{117{,}000 \text{ J mol}^{-1}}{8.315 \text{ J K}^{-1} \text{ mol}^{-1}} = 14{,}000 \text{ K}$$

which is far above the temperature at which a crystalline sample could exist. At room temperature, only a few electrons per mole in the extreme tail of the Boltzmann distribution have enough energy to jump the gap; therefore, the conduction band in pure silicon is sparsely populated with electrons. The result is that silicon is not a good conductor of electricity; good electrical conductivity requires a net motion of many electrons under the impetus of a small electric potential difference. Silicon has an electrical conductivity that is 11 orders of magnitude smaller than that of copper at room temperature. These are called

semiconductors, because their electrical conductivity, although smaller than that of a metal, is far greater than that of an **insulator** such as diamond, which has a larger band gap. The conductivity of a semiconductor is increased by increasing the temperature, which excites more electrons into levels in the conduction band. Another way to increase the conductivity of a semiconductor is to irradiate it with a beam of electromagnetic radiation with a frequency high enough to excite electrons from the valence band to the conduction band. This process resembles the photoelectric effect described in Section 4.4, with the difference that now the electrons are not removed from the material but only moved into the conduction band, ready to conduct a current if a potential difference is imposed.

EXAMPLE 22.3

Calculate the longest wavelength of light that can excite electrons from the valence to the conduction band in silicon. In what region of the spectrum does this wavelength fall?

Solution

The energy carried by a photon is $h\nu = hc/\lambda$, where h is Planck's constant, ν the photon frequency, c the speed of light, and λ the photon wavelength. For a photon to just excite an electron across the band gap, this energy must be equal to that band gap energy, $E_g = 1.94 \times 10^{-19}$ J, where

$$\frac{hc}{\lambda} = E_g$$

Solving for the wavelength gives

$$\lambda = \frac{hc}{E_g} = \frac{(6.626 \times 10^{-34} \text{ J s}) \times (2.998 \times 10^{8} \text{ m s}^{-1})}{1.94 \times 10^{-19} \text{ J}}$$

$$= 1.02 \times 10^{-6} \text{ m} = 1020 \text{ nm}$$

This wavelength falls in the infrared region of the spectrum. Photons with shorter wavelengths (for example, visible light) carry more than enough energy to excite electrons to the conduction band in silicon.

Related Problems: 27, 28

22.7 Semiconductors

Silicon in very high purity is said to display its **intrinsic properties**. When certain other elements are added to pure silicon in a process called **doping**, it acquires interesting electronic properties. For example, if atoms of a Group V element such as arsenic or antimony are diffused into silicon, they substitute for Si atoms in the network. Such atoms have five valence electrons, so each introduces one more electron into the silicon crystal than is needed for bonding. The extra electrons occupy energy levels just below the lowest level of the conduction band. Little energy is required to promote electrons from such a **donor impurity level** into the conduction band, and the electrical conductivity of the silicon crystal is increased without the necessity of increasing the temperature. Silicon doped with atoms of a Group V element is called an ***n*-type semiconductor** to indicate that the charge carrier is negative.

However, if a Group III element such as gallium is used as a dopant, there is one fewer electron in the valence band per dopant atom because Group III elements have only three valence electrons, not four. This situation corresponds to creating one *hole* in the valence band, with an effective charge of +1, for each Group III atom added. If a voltage difference is impressed across a crystal that is doped in this way, it causes the positively charged holes to move toward the negative source of potential. Equivalently, a valence band electron next to the (positive) hole will move in the opposite direction (that is, toward the positive source of potential). Whether we think of holes or electrons as the mobile charge carrier in the valence band, the result is the same; we are simply using different words to describe the same physical phenomenon (Fig. 22.23). Silicon that has been doped with a Group III element is called a ***p*-type semiconductor** to indicate that the carrier has an effective positive charge.

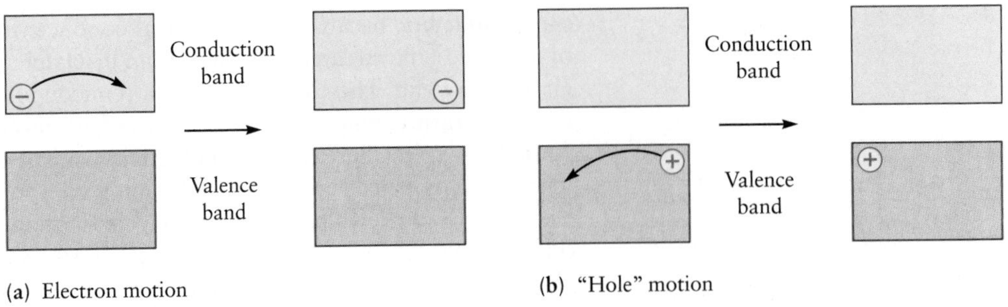

FIGURE 22.23 (a) In an *n*-type semiconductor, a small number of electrons occupy levels in the conduction band. When an electric field is imposed, each electron moves (black arrows) into one of the numerous nearby vacant energy levels in the conduction band. (b) In a *p*-type semiconductor, a small number of levels in the valence band are unoccupied. Conduction occurs as electrons in the numerous occupied levels of the valence band jump into the sparsely distributed unoccupied levels. The arrow shows the motion of an electron to occupy a previously empty level. The process can also be described as the motion of positively charged holes in the opposite direction.

A different class of semiconductors is based not on silicon, but on equimolar compounds of Group III with Group V elements. Gallium arsenide, for example, is isoelectronic to the Group IV semiconductor germanium. When GaAs is doped with the Group VI element tellurium, an *n*-type semiconductor is produced; doping with zinc, which has one *fewer* valence electron than gallium, gives a *p*-type semiconductor. Other III–V combinations have different band gaps and are useful in particular applications. Indium antimonide (InSb), for example, has a small enough band gap that absorption of infrared radiation causes electrons to be excited from the valence to the conduction band, and an electric current then flows when a small potential difference is applied. This compound is therefore used as a detector of infrared radiation. Still other compounds formed between the zinc group (zinc, cadmium, and mercury) and Group VI elements such as sulfur also have the same average number of valence electrons per atom as silicon and make useful semiconductors.

22.7.1 *p–n* Junctions and Device Performance

Of what value is it to have a semiconductor that can conduct an electric current by the flow of electrons if it is *n*-type or by the flow of holes if it is *p*-type? Many electronic functions can be fulfilled by semiconductors that possess these properties, but the simplest is **rectification**—the conversion of alternating current into direct current. Suppose thin crystals of *n*- and *p*-type silicon are placed in contact with one another and connected to a battery. In Figure 22.24a, the positive pole of the battery is connected to the *n*-type silicon, and the *p*-type silicon is connected to the negative pole. Only a small transient current can flow through the circuit in this case, because when the electrons in the conduction band of the *n*-type silicon have flowed out to the positive pole of the battery, there is none to take their place and current ceases to flow. However, if the negative pole of the battery is connected to the *n*-type silicon and the positive pole to the *p*-type silicon (see Fig. 22.24b), a steady current flows because electrons and holes move in opposite directions and recombine at the *n–p* junction. In effect, electrons flow toward the *n–p* junction in the *n*-type material, holes flow toward the *n–p* junction in the *p*-type material, and the junction is a "sink" where electrons fill the holes in the valence band and neutralize one another. If, instead of a galvanic cell, an alternating current source was connected to the *n–p* rectifier, current would flow in one direction only, creating pulsed direct current (see Fig. 22.24c).

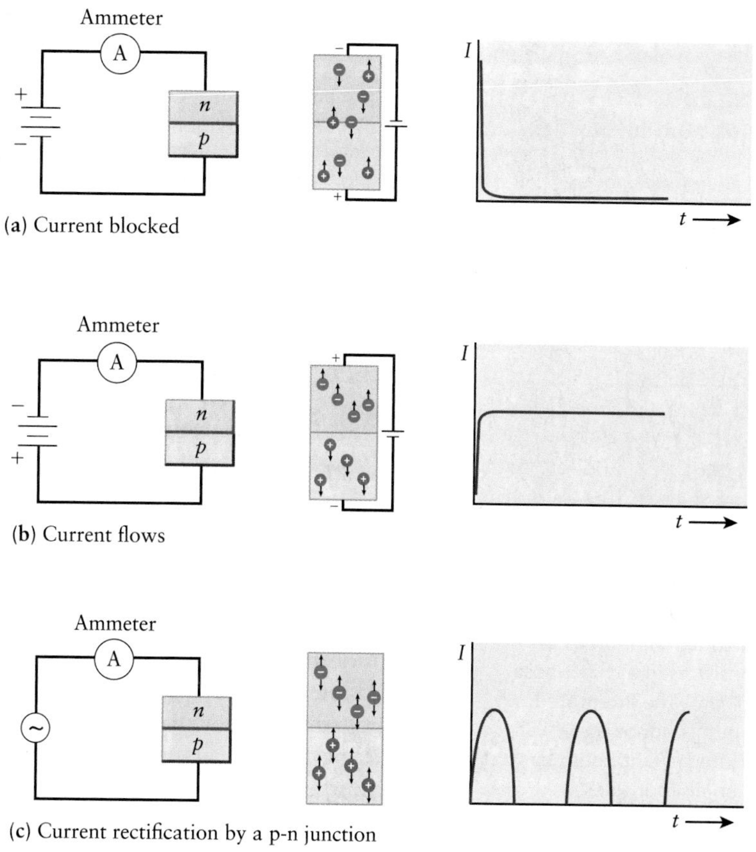

FIGURE 22.24 Current rectification by a *p–n* junction.

Gallium arsenide and other semiconductors also provide materials for making solid-state lasers, which have applications ranging from reading compact discs to performing delicate eye surgery. When an electric current is passed through a material containing *n–p* junctions, electrons from the *n* regions and holes from the *p* regions flow toward the junctions, where they recombine and emit light. The light moves through the material, stimulating additional recombinations at other junctions and the emission of additional photons (Fig. 22.25). Critical to the operation of a laser is that these photons are coherent (in phase with one another),

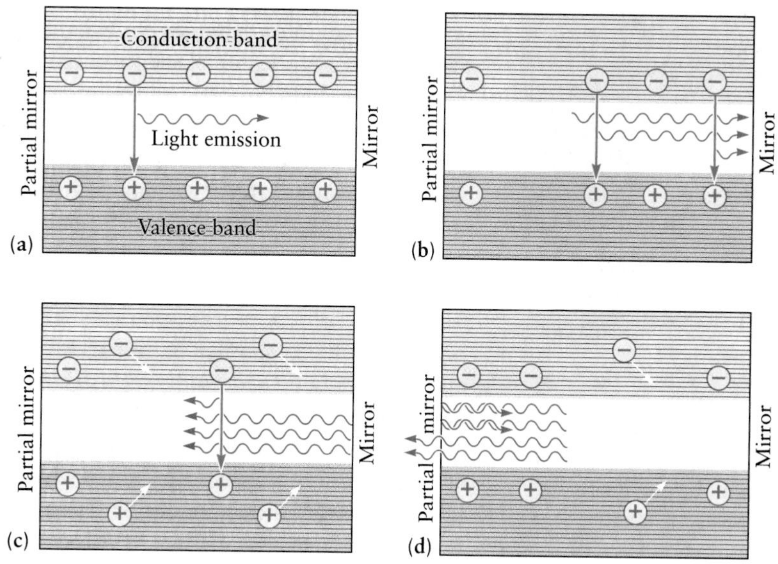

FIGURE 22.25 (a) In this solid-state laser, photons are emitted as electrons and holes recombine. (b) An emitted photon can stimulate the emission of additional photons. (c) Then reflection by a mirror on the right side sends coherent waves back through the laser medium and further amplification occurs by stimulated emission. (d) Some of the waves pass through a partially reflecting mirror on the left side.

Andre Kudyusov/Photodisc/Getty Images

FIGURE 22.26 An integrated circuit, like the microprocessor in a laptop computer, can contain hundreds of millions of transistors in a chip whose area is about 1 cm^2. The integrated circuit shown in this figure is connected to a printed circuit board by rows of wires on all four sides.

© Tommaso Guicciardini/Science Photo Library/Science Source

FIGURE 22.27 Silicon solar collectors are used on a large scale to harvest energy from the sun in this photovoltaic power plant located on one of the Tremiti Islands in Italy.

so the corresponding electric fields add constructively to create a plane wave. The light is reflected by a mirror at one end of the material and sets up a standing wave inside the semiconductor. At the other end, a partially reflecting mirror allows an intense beam of coherent light with a fixed wavelength to leave.

Semiconductors perform a wide range of electronic functions that formerly required the use of vacuum tubes. Vacuum tubes occupy much more space, generate large amounts of heat, and require considerably more energy to operate than a **transistor** does, their semiconductor counterparts. More important, semiconductors can be built into integrated circuits (Fig. 22.26) and made to store information and process it at great speeds.

Solar cells based on silicon or gallium arsenide provide a way to convert the radiant energy of the sun directly into electrical work by a technology that is virtually nonpolluting (Fig. 22.27). The high capital costs of solar cells make them uncompetitive with conventional fossil fuel sources of energy at this time, but as reserves of fossil fuels dwindle, solar energy will become an important option.

22.8 Pigments and Phosphors: Optical Displays

© Kurt Nassau/Phototake

FIGURE 22.28 Mixed crystals of two semiconductors with different band gaps, CdS (yellow) and CdSe (black), show a range of colors, illustrating a decrease in the band gap energy as the composition of the mixture becomes richer in Se.

The band gap of an insulator or semiconductor has a significant effect on its color. Pure diamond has a large band gap, so even blue light does not have enough energy to excite electrons from the valence band to the conduction band. As a result, light in the visible region of the electromagnetic spectrum (see Fig. 4.3) passes through diamonds without being absorbed and the diamonds are colorless. Cadmium sulfide (CdS; Fig. 22.28) has a band gap of 3.8×10^{-19} J, which corresponds to a wavelength of 510 nm in the visible region of the spectrum. The absorption cross section of CdS increases as the incident photon wavelength is decreased to 450 nm. Cadmium sulfide thus primarily absorbs violet and blue light but strongly transmits yellow, giving it a deep yellow color. Cadmium sulfide is the pigment called cadmium yellow. Cinnabar (HgS, Fig. 22.29) has a smaller band gap of 3.2×10^{-19} J and absorbs all light except red. It has a deep red color and is the pigment vermilion. Semiconductors with band gaps of less than 2.8×10^{-19} J absorb all wavelengths of visible light and appear black. These include silicon (see Example 22.2), germanium, and gallium arsenide.

Doping silicon brings donor levels close enough to the conduction band or acceptor levels close enough to the valence band that thermal excitation can

FIGURE 22.29 Crystalline cinnabar, HgS.

cause electrons to move into a conducting state. The corresponding doping of *insulators* or wide band-gap semiconductors can bring donor or acceptor states into positions where *visible light* can be absorbed or emitted. This changes the colors and optical properties of the materials. Nitrogen doped in diamond gives a donor impurity level in the band gap. Transitions to this level can absorb some blue light, giving the diamond an undesirable yellowish color. In contrast, boron doped into diamond gives an acceptor level that absorbs red light most strongly and gives the highly prized and rare "blue diamond."

Phosphors are wide band-gap materials with dopants selected to create new levels such that particular colors of light are emitted. Electrons in these materials are excited by light of other wavelengths or by electrons hitting their surfaces, and light is then emitted as they return to lower energy states. A fluorescent lamp, for example, is a mercury-vapor lamp in which the inside of the tube has been coated with phosphors. The phosphors absorb the violet and ultraviolet light emitted by mercury vapor and emit at lower energies and longer wavelengths, giving a nearly white light that is more desirable than the bluish light that comes from a mercury-vapor lamp without the phosphors.

Phosphors are also used in television screens. The picture is formed by scanning a beam of electrons (from an electron gun) over the screen. The electrons strike the phosphors coating the screen, exciting their electrons and causing them to emit light. In a black-and-white television tube, the phosphors are a mixture of silver doped into ZnS, which gives blue light, and silver doped into $Zn_xCd_{1-x}S$, which gives yellow light. The combination of the two provides a reasonable approximation of white. A color television uses three different electron guns, with three corresponding types of phosphor on the screen. Silver doped in ZnS gives blue, manganese doped in Zn_2SiO_4 is used for green, and europium doped in YVO_4 gives red light. Masks are used to ensure that each electron beam encounters only the phosphors corresponding to the desired color.

CHAPTER SUMMARY

The properties of solid materials are determined by their microscopic structure, which in turn depends on the nature of the chemical bonds created during synthesis and processing. Mechanical and structural properties originate in strong ionic and covalent bonding. Electrical conductivity measures the movement of charged particles throughout the material. Metals, semiconductors, and insulators are distinguished by differences in conductivity values, as explained by the differences in their band gaps. In effect, this difference measures the extent to which some valence electrons from the atoms comprising the solid are delocalized while the remainder are involved in formation of localized bonds. Optical properties measure the response of the solid to visible light. The magnitude of the band gap determines what wavelength of light is absorbed, and therefore the color of the material.

CONCEPTS AND SKILLS

Section 22.1 – Minerals: Naturally Occurring Inorganic Materials

Show how the fundamental silicate tetrahedral unit (SiO_4^{4-}) *links to other silicate tetrahedra to form rings, chains, double chains, sheets, and space-filling crystalline networks (Problems 1–4).*

- Two or more tetrahedra may link together by sharing oxygen vertices. Several examples are shown in Figure 22.2 and Table 22.1. In these linked structures the Si:O ratio is no longer 1:4 because oxygen atoms are shared at each of the linked points.

Describe the chemical compositions and structures of aluminosilicates, clays, and zeolites (Problems 5–6).

- Aluminosilicates are produced when Al atoms replace some of the Si atoms in silicates. Because the Al has only three electrons, one additional electron is provided by an alkali metal. Both the Al and alkali ions occupy tetrahedral sites in the aluminosilicate structure.
- Zeolites are aluminosilicate materials with extended three-dimensional structures that enclose polyhedral cavities. The very large surface area of these cavities makes these very effective as ion exchange materials for water purification, absorbing small molecules, and as catalysts.
- Clays are produced from primary materials by the weathering action of water and heat. The structures and properties of clays can be understood from the structure of the primary materials from which they derive, such as talcs and micas.

Section 22.2 – Properties of Ceramics

Describe the structure of ceramic materials and the ways in which they are formed.

- Ceramics are synthetic materials fabricated from inorganic, nonmetallic materials. Ceramics are mixtures that include silicates, oxides of elements other than silicon, and non-oxide materials such as carbides and nitrides. The microstructure of ceramics includes grains, voids, and cracks of various sizes and orientations. The microstructure can be controlled by sintering and densification to reduce the size of pores and voids.
- The mechanical properties of solid materials are determined by their internal structure, which in turn depends on the nature of the chemical bonds created during synthesis and processing.

Section 22.3 – Silicate Ceramics

Outline the properties of pottery, glass, and cement and the chemical reactions that give them structural strength (Problems 7–14).

- Silicate ceramics are well suited for structural applications because of their strength, which originates in the partially ionic, strong silicon–oxygen bonds in the tetrahedral orthosilicate anion. This structural unit appears in naturally occurring minerals and clays, which are fashioned into ceramic pieces through sintering and densification processes.

Section 22.4 – Nonsilicate Ceramics

List several important oxide and mixed oxide ceramics and give some of their uses (Problems 15–16).

- Nonsilicate ceramics derive comparable properties from other inorganic structural units.
- Oxide ceramics are made from oxides of numerous metals including beryllium, aluminum, calcium, and yttrium; these metal-oxide bonds are essentially ionic.

Discuss the special properties of nonoxide ceramics and the kinetic and thermodynamic factors that make them useful (Problems 17–20).

- Nonoxide ceramics are based on the nitrogen and carbon compounds of silicon and boron. These compounds have short, strong, highly directional covalent bonds, so the materials have great structural strength but are brittle.

Section 22.5 – Electrical Conduction in Materials

Explain how the conductivity of a material is measured, and relate it to the number and mobility of charge carriers (Problems 21–26).

- The electrical properties measure the movement of charged particles in a material in response to an applied electric field. If the particles are free to move throughout the materials, electrical conduction is the result, and the magnitude is characterized as the electrical conductivity of the material.
- The resistance of a cylindrical sample of material is measured as in Ohm's law. Geometrical factors are included to define the resistivity ρ, and the conductivity σ is defined as the reciprocal of the resistivity.

$$\sigma = \frac{1}{\rho} = \frac{\ell}{RA} \qquad [22.3]$$

- Ohm's law can be rewritten to state that the current density flowing through a sample is proportional to the applied electrical field, and the proportionality constant is the electrical conductivity. This expression is used to relate the conductivity of a material to its microstructural properties

$$J = \sigma E \qquad [22.4]$$

Section 22.6 – Band Theory of Conduction

Use the band model to describe the conductivity of metals, semiconductors, and insulators (Problems 27–30).

- Electrical conductivity depends on the product of two separate microscopic parameters of a material: the number density of charge carriers present and the mobility of the carriers.
- The band theory of solids explains the three broad classes of electronic conductivity seen in nature in terms of the number density of charge carriers available in classes of solids.
- Metals have high conductivity values because the number density of free, mobile electrons is quite high—at least one per atom in the solid is in the conduction band. The "electron sea" is delocalized throughout the solid, and the free electrons respond easily to applied electric fields.
- Insulating materials, such as the ceramics, have very low electrical conductivity because they have essentially no free electrons to carry current. There are no electrons in the conduction band, and the band gap is too large for electrons to be promoted from the valence band to the conduction band.
- Semiconductors have conductivity values intermediate between metals and insulators because their bandgaps are small enough that electrons can be promoted from the valence band to the conduction band with modest thermal or optical excitation.

Section 22.7 – Semiconductors

Describe the mechanism of action of intrinsic, n-type, and p-type semiconductors (Problems 31–34).

- Doping a semiconductor by adding an electron donor impurity creates new states very near the bottom of the conduction band, so electrons can be promoted from the donor into the conduction band without increasing the temperature. These promoted electrons are now free to respond to an applied electric field. This doping process leads to an n-type semiconductor, in which the charge carrier is an electron.

- Doping a semiconductor by adding an electron acceptor impurity creates new states very near the top of the valence band. An electron can move to this impurity state from the valence band, leaving a positively charged hole in the valence band. The hole can move in response to an applied electric field. This doping process leads to a *p*-type semiconductor, in which the charge carrier is a hole.
- When a *p*-type semiconductor is placed in contact with an *n*-type semiconductor and the resulting *p–n* junction is placed in an electrical circuit, the junction can either pass or block DC current, or rectify AC current depending on details of the circuit. The device can emit light due to recombination of holes and electrons at the junction.

Section 22.8 – Pigments and Phosphors: Optical Displays

Relate the band gap of a semiconductor or phosphor to the frequencies of electromagnetic radiation absorbed or emitted when electrons make transitions between the valence and conduction bands (Problems 35–36).

- Optical properties describe the response of a material to light in the visible range of the electromagnetic spectrum.
- The band gap of an insulator or semiconductor determines the wavelength of light absorbed by the materials. Because the remaining portions of white light are transmitted through the materials, the color of the materials is complementary to the wavelength of the light absorbed.
- Many insulators are not colored because their band gaps are so large that no visible light is absorbed. Adding dopants can introduce donor or acceptor states that enable absorption of visible light, so the doped materials are colored.
- Phosphors are wide band gap materials doped to emit light at specific wavelengths upon excitation by electron impact or incident light at other wavelengths. Phosphors are used to generate visible light in fluorescent fixtures, and to display signals on video monitors.

PROBLEMS

Answers to problems whose numbers are red appear in Appendix G. Problems that are more challenging are indicated with asterisks.

22.1 Minerals: Naturally Occurring Inorganic Materials

1. Draw a Lewis electron-dot diagram for the disilicate ion ($Si_2O_7^{6-}$). What changes in this structure would be necessary to produce the structure of the pyrophosphate ion ($P_2O_7^{4-}$) and the pyrosulfate ion ($S_2O_7^{2-}$)? What is the analogous compound of chlorine?
2. Draw a Lewis electron-dot diagram for the cyclosilicate ion ($Si_6O_{18}^{12-}$), which forms part of the structures of beryl and emerald.
3. Using Table 22.1, predict the structure of each of the following silicate minerals (network, sheets, double chains, and so forth). Give the oxidation state of each atom.
   (a) Andradite, $Ca_3Fe_2(SiO_4)_3$
   (b) Vlasovite, $Na_2ZrSi_4O_{10}$
   (c) Hardystonite, $Ca_2ZnSi_2O_7$
   (d) Chrysotile, $Mg_3Si_2O_5(OH)_4$
4. Using Table 22.1, predict the structure of each of the following silicate minerals (network, sheets, double chains, and so forth). Give the oxidation state of each atom.
   (a) Tremolite, $Ca_2Mg_5(Si_4O_{11})_2(OH)_2$
   (b) Gillespite, $BaFeSi_4O_{10}$
   (c) Uvarovite, $Ca_3Cr_2(SiO_4)_3$
   (d) Barysilate, $MnPb_8(Si_2O_7)_3$

5. Using Table 22.1, predict the structure of each of the following aluminosilicate minerals (network, sheets, double chains, and so forth). In each case, the Al atoms grouped with the Si and O in the formula substitute for Si in tetrahedral sites. Give the oxidation state of each atom.
   (a) Keatite, $Li(AlSi_2O_6)$
   (b) Muscovite, $KAl_2(AlSi_3O_{10})(OH)_2$
   (c) Cordierite, $Al_3Mg_2(AlSi_5O_{18})$

6. Using Table 22.1, predict the structure of each of the following aluminosilicate minerals (network, sheets, double chains, and so forth). In each case, the Al atoms grouped with the Si and O in the formula substitute for Si in tetrahedral sites. Give the oxidation state of each atom.
   (a) Amesite, $Mg_2Al(AlSiO_5)(OH)_4$
   (b) Phlogopite, $KMg_3(AlSi_3O_{10})(OH)_2$
   (c) Thomsonite, $NaCa_2(Al_5Si_5O_{20}) \cdot 6\ H_2O$

22.3 Silicate Ceramics

7. A ceramic that has been much used by artisans and craftsmen for the carving of small figurines is based on the mineral steatite (commonly known as soapstone). Steatite is a hydrated magnesium silicate that has the composition $Mg_3Si_4O_{10}(OH)_2$. It is remarkably soft—a fingernail can scratch it. When heated in a furnace to about 1000°C, chemical reaction transforms it into a hard, two-phase composite of magnesium silicate ($MgSiO_3$) and quartz in much the same way that clay minerals are converted into mullite ($Al_6Si_2O_{13}$) and cristobalite (SiO_2) on firing. Write a balanced chemical equation for this reaction.

8. A clay mineral that is frequently used together with or in place of kaolinite is pyrophyllite ($Al_2Si_4O_{10}(OH)_2$). Write a balanced chemical equation for the production of mullite and cristobalite on the firing of pyrophyllite.

9. Calculate the volume of carbon dioxide produced at standard temperature and pressure when a sheet of ordinary glass of mass 2.50 kg is made from its starting materials—sodium carbonate, calcium carbonate, and silica. Take the composition of the glass to be $Na_2O \cdot CaO \cdot (SiO_2)_6$.

10. Calculate the volume of steam produced when a 4.0-kg brick made from pure kaolinite is completely dehydrated at 600°C and a pressure of 1.00 atm.

11. A sample of soda-lime glass for tableware is analyzed and found to contain the following percentages by mass of oxides: SiO_2, 72.4%; Na_2O, 18.1%; CaO, 8.1%; Al_2O_3, 1.0%; MgO, 0.2%; BaO, 0.2%. (The elements are not actually present as binary oxides, but this is the way compositions are usually given.) Calculate the chemical amounts of Si, Na, Ca, Al, Mg, and Ba atoms per mole of O atoms in this sample.

12. A sample of Portland cement is analyzed and found to contain the following percentages by mass of oxides: CaO, 64.3%; SiO_2, 21.2%; Al_2O_3, 5.9%; Fe_2O_3, 2.9%; MgO, 2.5%; SO_3, 1.8%; Na_2O, 1.4%. Calculate the chemical amounts of Ca, Si, Al, Fe, Mg, S, and Na atoms per mole of O atoms in this sample.

13. The most important contributor to the strength of hardened Portland cement is tricalcium silicate, $(CaO)_3 \cdot SiO_2$, for which the measured standard enthalpy of formation is $-2929.2\ kJ\ mol^{-1}$. Calculate the standard enthalpy change for the production of 1.00 mol tricalcium silicate from quartz and lime.

14. One of the simplest of the heat-generating reactions that occur when water is added to cement is the production of calcium hydroxide (slaked lime) from lime. Write a balanced chemical equation for this reaction, and use data from Appendix D to calculate the amount of heat generated by the reaction of 1.00 kg lime with water at room conditions.

22.4 Nonsilicate Ceramics

15. Calculate the average oxidation number of the copper in $YBa_2Cu_3O_{9-x}$ if $x = 2$. Assume that the rare-earth element yttrium is in its usual +3 oxidation state.

16. The mixed oxide ceramic $Tl_2Ca_2Ba_2Cu_3O_{10+x}$ has zero electrical resistance at 125 K. Calculate the average oxidation number of the copper in this compound if $x = 0.50$ and thallium is in the +3 oxidation state.

17. Silicon carbide (SiC) is made by the high-temperature reaction of silica sand (quartz) with coke; the byproduct is carbon monoxide.
   (a) Write a balanced chemical equation for this reaction.
   (b) Calculate the standard enthalpy change per mole of SiC produced.
   (c) Predict (qualitatively) the following physical properties of silicon carbide: conductivity, melting point, and hardness.

18. Boron nitride (BN) is made by the reaction of boron trichloride with ammonia.
   (a) Write a balanced chemical equation for this reaction.
   (b) Calculate the standard enthalpy change per mole of BN produced, given that the standard molar enthalpy of formation of BN(*s*) is $254.4\ kJ\ mol^{-1}$.
   (c) Predict (qualitatively) the following physical properties of boron nitride: conductivity, melting point, and hardness.

19. The standard free energy of formation of cubic silicon carbide (SiC) is $62.8\ kJ\ mol^{-1}$. Determine the standard free energy change when 1.00 mol SiC reacts with oxygen to form SiO_2 (*s*, quartz) and $CO_2(g)$. Is silicon carbide thermodynamically stable in the air at room conditions?

20. The standard free energy of formation of boron carbide (B_4C) is $71\ kJ\ mol^{-1}$. Determine the standard free energy change when 1.00 mol B_4C reacts with oxygen to form $B_2O_3(s)$ and $CO_2(g)$. Is boron carbide thermodynamically stable in the air at room conditions?

22.5 Electrical Conduction in Materials

21. A cylindrical sample of solid germanium has length 55.0 mm and diameter 5.0 mm. In a test circuit, 0.150 A of current flowed through this sample when the voltage applied between its ends was 17.5 V. What is the electrical conductivity of this sample?

22. A gold wire 4.0 mm in diameter and 1.5 m in length is to be used in a test circuit. (a) Calculate the resistance of the wire. (b) Calculate the current density in the wire when the voltage applied between its ends is 0.070 V. (c) Calculate the electric field in the wire.

23. The mobilities for Na^+ and Cl^- in aqueous solution are given in Table 22.7. Calculate the conductivity of a 0.10-M solution of NaCl in water at 25°C.

24. Explain why the ionic mobility of CH_3COO^- is smaller than that for Cl^-.

25. The electrical conductivity for copper is given in Table 22.6. The electron mobility in copper at room temperature is 3.0×10^{-3} m^2 V^{-1} s^{-1}. Using the Drude model for metallic conductivity, calculate the number of free electrons per Cu atom. The density of copper is 8.9 g cm^{-3}.

26. A variety of useful metallic alloys can be prepared by dissolving Ni in Cu. The room temperature resistivity of pure copper is 1.6×10^{-8} Ω m. As nickel is dissolved in copper up to 50% mass, the resistivity increases in a nearly linear fashion to the value 47.0×10^{-8} Ω m. Explain this increase qualitatively.

22.6 Band Theory of Conduction

27. Electrons in a semiconductor can be excited from the valence band to the conduction band through the absorption of photons with energies exceeding the band gap. At room temperature, indium phosphide (InP) is a semiconductor that absorbs light only at wavelengths less than 920 nm. Calculate the band gap in InP.

28. Both GaAs and CdS are semiconductors that are being studied for possible use in solar cells to generate electric current from sunlight. Their band gaps are 2.29×10^{-19} J and 3.88×10^{-19} J, respectively, at room temperature. Calculate the longest wavelength of light that is capable of exciting electrons across the band gap in each of these substances. In which region of the electromagnetic spectrum do these wavelengths fall? Use this result to explain why CdS-based sensors are used in some cameras to estimate the proper exposure conditions.

29. The number of electrons excited to the conduction band per cubic centimeter in a semiconductor can be estimated from the following equation:

$$n_e = (4.8 \times 10^{15}\ \text{cm}^{-3}\ \text{K}^{-3/2})\, T^{3/2}\, e^{-E_g/(2RT)}$$

where T is the temperature in kelvins and E_g the band gap in joules *per mole*. The band gap of diamond at 300 K is 8.7×10^{-19} J. How many electrons are thermally excited to the conduction band at this temperature in a 1.00-cm^3 diamond crystal?

30. The band gap of pure crystalline germanium is 1.1×10^{-19} J at 300 K. How many electrons are excited from the valence band to the conduction band in a 1.00-cm^3 crystal of germanium at 300 K? Use the equation given in the preceding problem.

22.7 Semiconductors

31. Describe the nature of electrical conduction in (a) silicon doped with phosphorus and (b) indium antimonide doped with zinc.

32. Describe the nature of electrical conduction in (a) germanium doped with indium and (b) cadmium sulfide doped with arsenic.

33. In a light-emitting diode (LED), which is used in displays on electronic equipment, watches, and clocks, a voltage is imposed across an n–p semiconductor junction. The electrons on the n side combine with the holes on the p side and emit light at the frequency of the band gap. This process can also be described as the emission of light as electrons fall from levels in the conduction band to empty levels in the valence band. It is the reverse of the production of electric current by illumination of a semiconductor.

Many LEDs are made from semiconductors that have the general composition $GaAs_{1-x}P_x$. When x is varied between 0 and 1, the band gap changes and, with it, the color of light emitted by the diode. When $x = 0.4$, the band gap is 2.9×10^{-19} J. Determine the wavelength and color of the light emitted by this LED.

34. When the LED described in Problem 33 has the composition $GaAs_{0.14}P_{0.86}$ (i.e., $x = 0.86$), the band gap has increased to 3.4×10^{-19} J. Determine the wavelength and color of the light emitted by this LED.

22.8 Pigments and Phosphors: Optical Displays

35. The pigment zinc white (ZnO) turns bright yellow when heated, but the white color returns when the sample is cooled. Does the band gap increase or decrease when the sample is heated?

36. Mercury(II) sulfide (HgS) exists in two different crystalline forms. In cinnabar, the band gap is 3.2×10^{-19} J; in metacinnabar, it is 2.6×10^{-19} J. In some old paintings with improperly formulated paints, the pigment vermilion (cinnabar) has transformed to metacinnabar on exposure to light. Describe the color change that results.

ADDITIONAL PROBLEMS

37. Predict the structure of each of the following silicate minerals (network, sheets, double chains, and so forth). Give the oxidation state of each atom.

(a) Apophyllite, $KCa_4(Si_8O_{20})F \cdot 8\ H_2O$

(b) Rhodonite, $CaMn_4(Si_5O_{15})$

(c) Margarite, $CaAl_2(Al_2Si_2O_{10})(OH)_2$

38. Using Table 22.1, predict the kind of structure formed by manganpyrosmalite, a silicate mineral with chemical formula $Mn_{12}FeMg_3(Si_{12}O_{30})(OH)_{10}Cl_{10}$. Give the oxidation state of each atom in this formula unit.

39. A reference book lists the chemical formula of one form of vermiculite as

$$[(Mg_{2.36}Fe_{0.48}Al_{0.16})(Si_{2.72}Al_{1.28})O_{10}(OH)_2][Mg_{0.32}(H_2O)_{4.32}]$$

Determine the oxidation state of the iron in this mineral.

40. The most common feldspars are those that contain potassium, sodium, and calcium cations. They are called, respectively, orthoclase ($KAlSi_3O_8$), albite ($NaAlSi_3O_8$), and anorthite ($CaAl_2Si_2O_8$). The solid solubility of orthoclase in albite is limited, and its solubility in anorthite is almost negligible. Albite and anorthite, however, are completely miscible at high temperatures and show complete solid solution. Offer an explanation for these observations, based on the tabulated radii of the K^+, Na^+, and Ca^{2+} ions from Appendix F.

41. The clay mineral kaolinite ($Al_2Si_2O_5(OH)_4$) is formed by the weathering action of water containing dissolved carbon dioxide on the feldspar mineral anorthite ($CaAl_2Si_2O_8$). Write a balanced chemical equation for the reaction that occurs. The CO_2 forms H_2CO_3 as it dissolves. As the pH is lowered, will the weathering occur to a greater or a lesser extent?

42. Certain kinds of zeolite have the general formula $M_2O{\cdot}Al_2O_3{\cdot}ySiO_2{\cdot}wH_2O$, where M is an alkali metal such as sodium or potassium, y is 2 or more, and w is any integer. Compute the mass percentage of aluminum in a zeolite that has M = K, $y = 4$, and $w = 6$.

43. **(a)** Use data from Tables 15.2 and 16.2 to calculate the solubility of $CaCO_3$ in water at pH 7.

(b) Will the solubility increase or decrease if the pH is lowered and the water becomes more acidic?

(c) Calculate the maximum amount of limestone (primarily calcium carbonate) that could dissolve per year in a river at pH 7 with an average flow rate of 1.0×10^6 m³/h.

44. Silica (SiO_2) exists in several forms, including quartz (molar volume 22.69 $cm^3\ mol^{-1}$) and cristobalite (molar volume 25.74 $cm^3\ mol^{-1}$).

(a) Use data from Appendix D to calculate $\Delta H°$, $\Delta S°$, and ΔG at 25°C.

(b) Which form is thermodynamically stable at 25°C?

(c) Which form is stable at very high temperatures, provided that melting does not occur first?

45. Talc, $Mg_3Si_4O_{10}(OH)_2$, reacts with forsterite (Mg_2SiO_4) to form enstatite ($MgSiO_3$) and water vapor.

(a) Write a balanced chemical equation for this reaction.

(b) If the water pressure is equal to the total pressure, will formation of products be favored or disfavored with increasing total pressure?

(c) The entropy change for this reaction is positive. Will the slope of the coexistence curve (pressure plotted against temperature) be positive or negative?

46. In what ways does soda-lime glass resemble and in what ways does it differ from a pot made from the firing of kaolinite? Include the following aspects in your discussion: composition, structure, physical properties, and method of preparation.

47. Iron oxides are red when the average oxidation state of iron is high and black when it is low. To impart each of these colors to a pot made from clay that contains iron oxides, would you use an air-rich or a smoky atmosphere in the kiln? Explain.

48. Refractories can be classified as acidic or basic, depending on the properties of the oxides in question. A basic refractory must not be used in contact with acid, and an acidic refractory must not be used in contact with a base. Classify magnesia and silica as acidic or basic refractories.

49. Dolomite bricks are used in the linings of furnaces in the cement and steel industries. Pure dolomite contains 45.7% $MgCO_3$ and 54.3% $CaCO_3$ by mass. Determine the empirical formula of dolomite.

50. Beryllia (BeO) ceramics have some use but show only poor resistance to strong acids and bases. Write likely chemical equations for the reaction of BeO with a strong acid and with a strong base.

51. Silicon nitride resists all acids except hydrofluoric, with which it reacts to give silicon tetrafluoride and ammonia. Write a balanced chemical equation for this reaction.

52. Compare oxide ceramics such as alumina (Al_2O_3) and magnesia (MgO), which have significant ionic character, with covalently bonded nonoxide ceramics such as silicon carbide (SiC) and boron carbide (B_4C; see Problems 19 and 20) with respect to thermodynamic stability at ordinary conditions.

53. Compare the hybridization of Si atoms in Si(*s*) with that of C atoms in graphite (see Fig. 21.26). If silicon were to adopt the graphite structure, would its electrical conductivity be high or low?

54. Describe how the band gap varies from a metal to a semiconductor to an insulator.

55. Suppose some people are sitting in a row at a movie theater, with a single empty seat on the left end of the row. Every 5 minutes, a person moves into a seat on his or her left if it is empty. In what direction and with what speed does the empty seat "move" along the row? Comment on the connection with hole motion in *p*-type semiconductors.

56. A sample of silicon doped with antimony is an *n*-type semiconductor. Suppose a small amount of gallium is added to such a semiconductor. Describe how the conduction properties of the solid will vary with the amount of gallium added.

23 CHAPTER

Polymeric Materials and Soft Condensed Matter

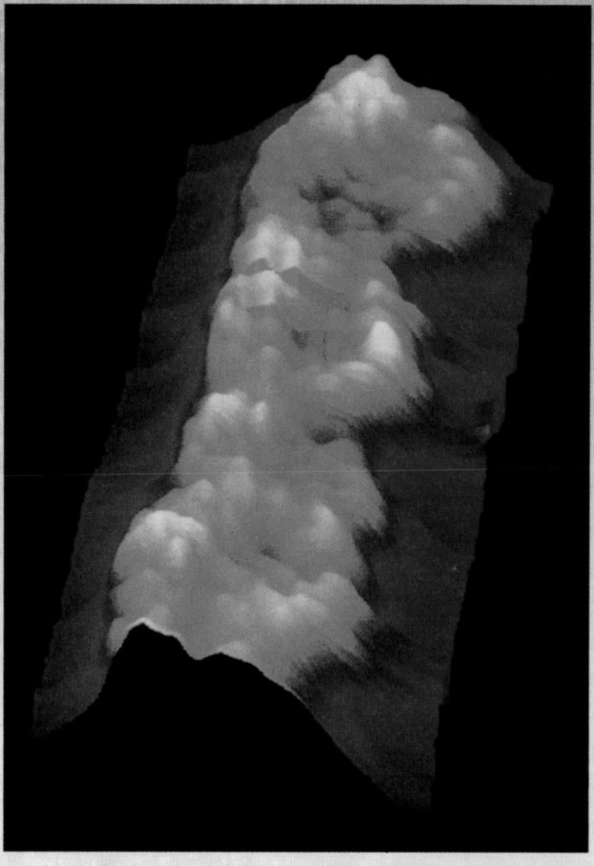

A false-color scanning tunneling micrograph (STM) of a DNA double-helix molecule adsorbed on a graphite substrate.

The organic compounds discussed in Chapter 7 were relatively small molecules, ranging from four or five atoms (such as methane or formaldehyde) to long-chain hydrocarbons up to 30 carbon atoms with relative molecular masses of several hundred. In addition to these smaller molecules, carbon atoms string together in stable chains of essentially unlimited length. Such chains provide the backbones of truly huge molecules that may contain hundreds of thousands or even millions of atoms. Such a compound, called a **polymer,** is formed by linking numerous separate small **monomer units** in strands and webs.

Although many polymers are based on the ability of carbon to form stable long-chain molecules with various functional groups attached, carbon is not unique in this ability. Recall from Chapter 22 the chains, sheets, and networks found in natural silicates, in which the elements silicon and oxygen join together to form extended structures. This chapter focuses on organic polymers, whose chemical and physical properties depend on the bonding and functional group chemistry discussed in Section 7.6. We examine both synthetic polymers, which are built largely from the hydrocarbon raw materials discussed in Section 7.1, and naturally occurring biopolymers such as starch, proteins, and nucleic acids, which are built from products of biological synthesis.

23.1 Polymerization Reactions for Synthetic Polymers

To construct a polymer, very many monomers must add to a growing polymer molecule, and the reaction must not falter after the first few molecules have reacted. This is achieved by having the polymer molecule retain highly reactive functional groups at all times during its synthesis. The two major types of polymer growth are addition polymerization and condensation polymerization.

In **addition polymerization**, monomers react to form a polymer chain without net loss of atoms. The most common type of addition polymerization involves the free-radical chain reaction of molecules that have C=C bonds. As in the chain reactions considered in Section 18.4, the overall process consists of three steps: initiation, propagation (repeated many times to build up a long chain), and termination. As an example, consider the polymerization of vinyl chloride (chloroethene, $CH_2{=}CHCl$) to polyvinyl chloride (Fig. 23.1). This process can be initiated by a small concentration of molecules that have bonds weak enough to be broken by the action of light or heat, giving radicals. An example of such an **initiator** is a peroxide, which can be represented as R—O—O—R′, where R and R′ represent alkyl groups. The weak O—O bonds break

$$R{-}\ddot{\underset{..}{O}}{-}\ddot{\underset{..}{O}}{-}R' \longrightarrow R{-}\ddot{\underset{..}{O}}\cdot + \cdot\ddot{\underset{..}{O}}{-}R' \quad \text{(initiation)}$$

to give radicals, whose oxygen valence shells are incomplete. The radicals remedy this by reacting readily with vinyl chloride, accepting electrons from the C=C bonds to reestablish a closed-shell electron configuration on the oxygen atoms:

$$R{-}\ddot{\underset{..}{O}}\cdot + CH_2{=}CHCl \longrightarrow R{-}\ddot{\underset{..}{O}}{-}CH_2{-}\underset{\underset{Cl}{|}}{\overset{\overset{H}{|}}{C}}\cdot \quad \text{(propagation)}$$

One of the two π electrons in the vinyl chloride double bond has been used to form a single bond with the R—O · radical. The other remains on the second carbon atom, leaving it as a seven–valence-electron atom that will react with another vinyl chloride molecule:

$$R{-}O{-}CH_2{-}\underset{\underset{Cl}{|}}{\overset{\overset{H}{|}}{C}}\cdot + CH_2{=}\underset{\underset{Cl}{|}}{\overset{\overset{H}{|}}{C}} \longrightarrow R{-}O{-}CH_2{-}\underset{\underset{Cl}{|}}{\overset{\overset{H}{|}}{C}}{-}CH_2{-}\underset{\underset{Cl}{|}}{\overset{\overset{H}{|}}{C}}\cdot \quad \text{(propagation)}$$

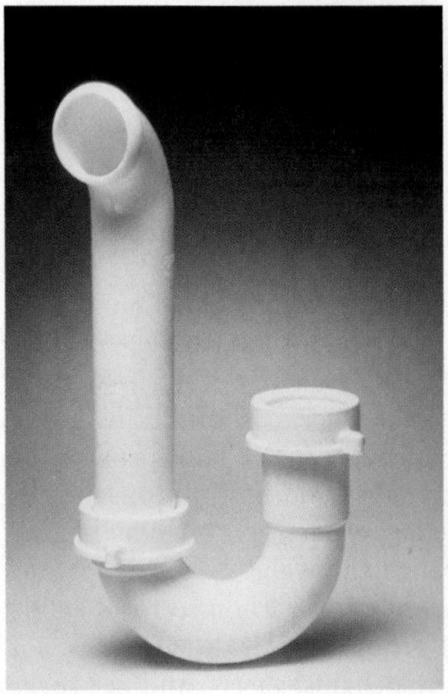

FIGURE 23.1 A pipefitting of polyvinyl chloride.

At each stage, the end group of the lengthening chain is one electron short of a valence octet and remains quite reactive. The reaction can continue, building up long-chain molecules of high molecular mass. The vinyl chloride monomers always attach to the growing chain with their CH_2 group because the odd electron is more stable on a CHCl end group. This gives the polymer a regular alternation of $-CH_2-$ and $-CHCl-$ groups. Its chemical formula is $(-CH_2CHCl-)_n$.

Termination occurs when the radical end groups on two different chains encounter each other and the two chains couple to give a longer chain:

$$\mathrm{R{-}O{-}(CH_2{-}CHCl)_{\mathit{m}}{-}CH_2{-}\underset{Cl}{\overset{H}{C}}{\cdot}} + \mathrm{{\cdot}\underset{Cl}{\overset{H}{C}}{-}CH_2{-}(CHCl{-}CH_2)_{\mathit{n}}{-}O{-}R'} \longrightarrow$$

$$\mathrm{R{-}O{-}(CH_2{-}CHCl)_{\mathit{m}}{-}CH_2{-}\underset{Cl}{\overset{H}{C}}{-}\underset{Cl}{\overset{H}{C}}{-}CH_2{-}(CHCl{-}CH_2)_{\mathit{n}}{-}O{-}R'}$$

Alternatively, a hydrogen atom may transfer from one end group to the other:

$$\mathrm{R{-}O{-}(CH_2{-}CHCl)_{\mathit{m}}{-}CH_2{-}\underset{H}{\overset{H}{C}}{-}\underset{Cl}{\overset{H}{C}}{\cdot}} + \mathrm{{\cdot}\underset{Cl}{\overset{H}{C}}{-}CH_2{-}(CHCl{-}CH_2)_{\mathit{n}}{-}O{-}R'} \longrightarrow$$

$$\mathrm{R{-}O{-}(CH_2{-}CHCl)_{\mathit{m}}{-}CH{=}CHCl} + \mathrm{CH_2Cl{-}CH_2{-}(CHCl{-}CH_2)_{\mathit{n}}{-}O{-}R'}$$

(termination)

The latter termination step leaves a double bond on one chain end and a $-CH_2Cl$ group on the other. When the polymer molecules are long, the exact natures of the end groups have little effect on the physical and chemical properties of the material. A different type of hydrogen transfer step often has a much greater effect on the properties of the resulting polymer. Suppose that a hydrogen atom transfers not from the monomer unit on the *end* of a second chain but from a monomer unit in the middle of that chain (Fig. 23.2). Then the first chain stops growing, but the radical site moves to the middle of the second chain, and growth resumes from that point, forming a *branched* polymeric chain with very different properties.

Addition polymerization can be initiated by ions as well as by free radicals. An example is the polymerization of acrylonitrile:

$$n\ \mathrm{CH_2{=}\underset{C\equiv N}{CH}} \longrightarrow \left[\mathrm{CH_2\underset{C\equiv N}{CH}}\right]_n$$

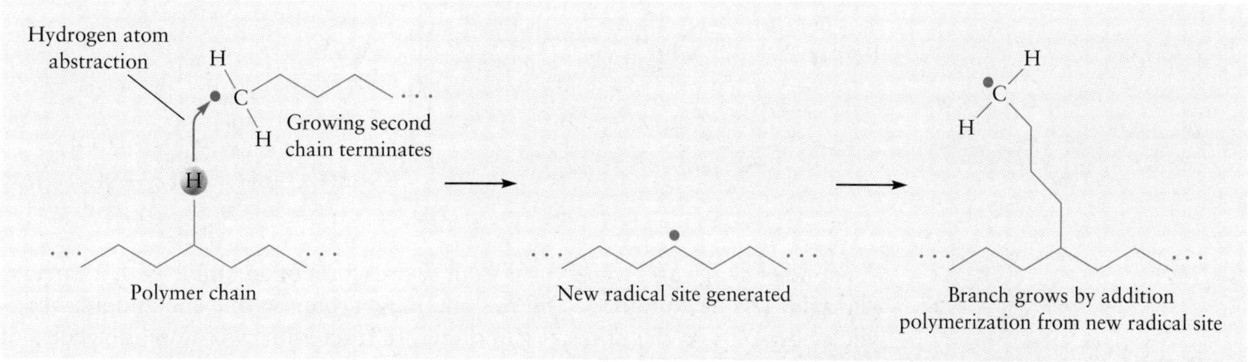

FIGURE 23.2 Chain branching can occur when a hydrogen atom is transferred (abstracted) from the middle of one chain to the free radical end of a second chain, thus terminating the growth of the second chain. The newly generated free radical in the middle of the polymer chain provides a site for the growth of a branched chain via continued monomer addition polymerization.

A suitable initiator for this process is butyl lithium, $(CH_3CH_2CH_2CH_2)^-Li^+$. The butyl anion (abbreviated Bu^-) reacts with the end carbon atom in a molecule of acrylonitrile to give a new anion:

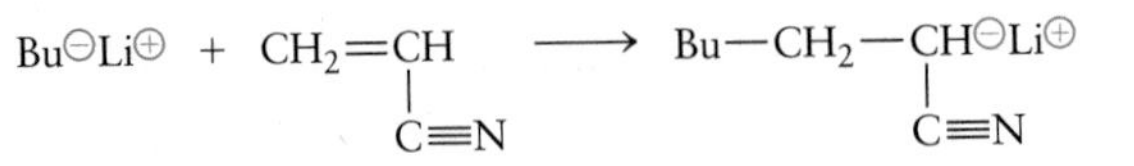

The new anion then reacts with an additional molecule of acrylonitrile:

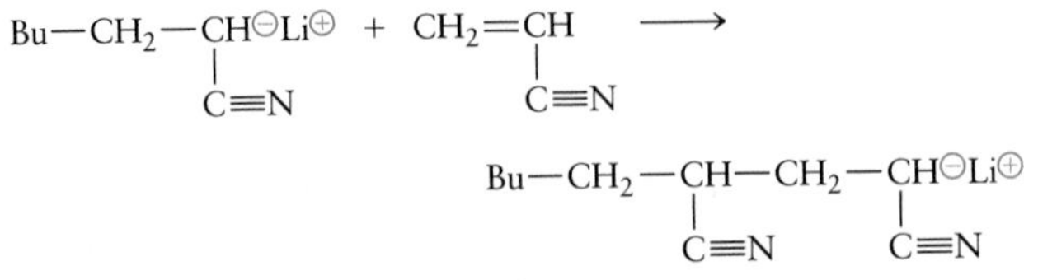

The process continues, building up a long-chain polymer.

Ionic polymerization differs from free-radical polymerization because the negatively charged end groups repel one another, ruling out termination by the coupling of two chains. The ionic group at the end of the growing polymer is stable at each stage. Once the supply of monomer has been used up, the polymer can exist indefinitely with its ionic end group, in contrast with the free-radical case, in which some reaction must take place to terminate the process. Ion-initiated polymers are called "living" polymers because, when additional monomer is added (even months later), they resume growth and increase in molecular mass. Termination can be achieved by adding water to replace the Li^+ with a hydrogen ion:

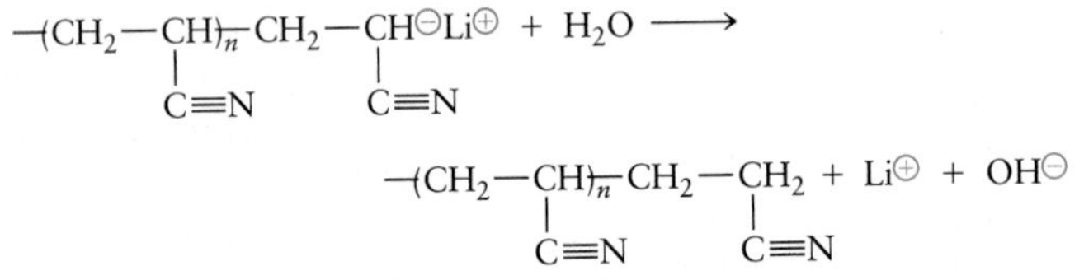

A second important mechanism of polymerization is **condensation polymerization,** in which a small molecule (frequently water) is split off as each monomer unit is attached to the growing polymer.[1] An example is the polymerization of 6-aminohexanoic acid. The first two molecules react upon heating according to

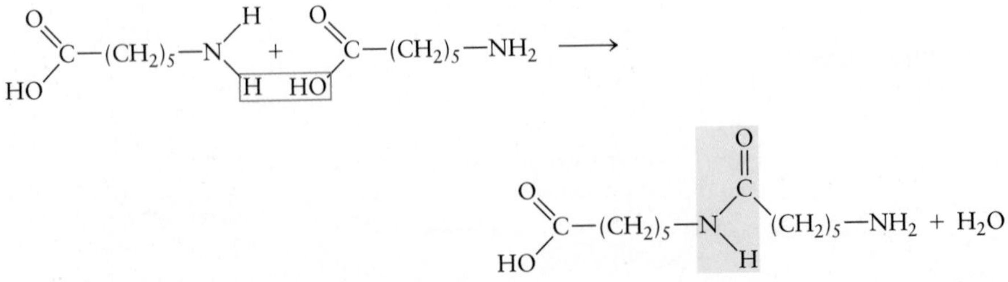

An amide linkage and water form from the reaction of an amine with a carboxylic acid. The new molecule still has an amine group on one end and a carboxylic acid group on the other; it can therefore react with two more molecules of 6-aminohexanoic acid. The process repeats to build up a long-chain molecule.

[1]Condensation reactions have appeared several times outside the context of polymer synthesis in this text. For example, two molecules of H_2SO_4 condense to form disulfuric acid ($H_2S_2O_7$), and a carboxylic acid condenses with an alcohol to form an ester (see Section 7.6).

For each monomer unit added, one molecule of water is split off. The final polymer in this case is called nylon 6 and is used in fiber-belted radial tires and in carpets.

Both addition and condensation polymerization can be carried out with mixtures of two or more types of monomers present in the reaction mixture. The result is a **random copolymer** that incorporates both types of monomers in an irregular sequence along the chain. For example, a 1:6 molar ratio of styrene to butadiene monomers is used to make styrene–butadiene rubber (SBR) for automobile tires, and a 2:1 ratio gives a copolymer that is an ingredient in latex paints.

23.1.1 Cross-Linking: Nonlinear Synthetic Polymers

If every monomer forming a polymer has only two reactive sites, then only chains and rings can be made. The 6-aminohexanoic acid used in making nylon 6 has one amine group and one carboxylic acid group per molecule. When both functional groups react, one link is forged in the polymer chain, but that link cannot react further. If some or all of the monomers in a polymer have three or more reactive sites, however, then cross-linking to form sheets or networks is possible.

One important example of cross-linking involves phenol–formaldehyde copolymers (Fig. 23.3). When these two compounds are mixed (with the phenol in excess in the presence of an acid catalyst), straight-chain polymers form. The first step is the addition of formaldehyde to phenol to give methylolphenol:

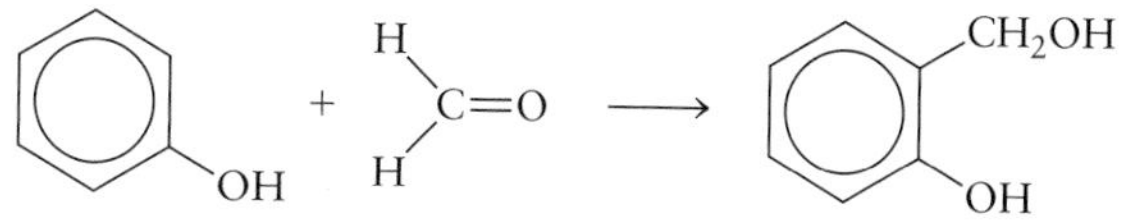

Molecules of methylolphenol then undergo condensation reactions (releasing water) to form a linear polymer called novalac:

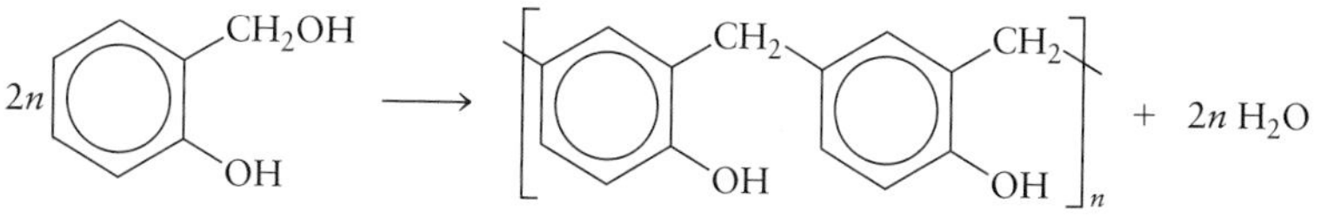

FIGURE 23.3 When a mixture of phenol (C_6H_5OH) and formaldehyde (CH_2O) dissolved in acetic acid is treated with concentrated hydrochloric acid, a phenol–formaldehyde polymer grows.

If, on the other hand, the reaction is carried out with an excess of formaldehyde, dimethylolphenols and trimethylolphenols form:

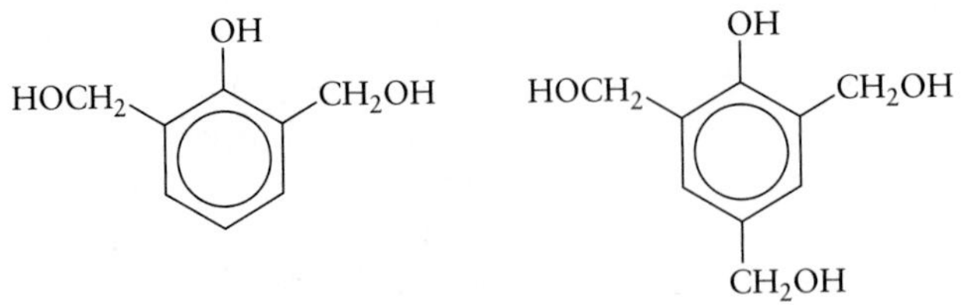

Each of these monomers has more than two reactive sites and can react with up to three others to form a cross-linked polymer that is much stronger and more impact-resistant than the linear polymer. The very first synthetic plastic, Bakelite, was made in 1907 from cross-linked phenol and formaldehyde. Modern phenol–formaldehyde polymers are used as adhesives for plywood; more than a billion kilograms are produced per year in the United States.

Cross-linking is often desirable because it leads to a stronger material. Sometimes cross-linking agents are added deliberately to form additional bonds between polymer chains. Polybutadiene contains double bonds that can be linked upon addition of appropriate oxidizing agents. One especially important kind of cross-linking occurs through sulfur chains in rubber, as we will see.

23.2 Applications for Synthetic Polymers

The three largest uses for polymers are in fibers, plastics, and elastomers (rubbers). We can distinguish between these three types of materials on the basis of their physical properties, especially their resistance to stretching. A typical fiber strongly resists stretching and can be elongated by less than 10% before breaking. Plastics are intermediate in their resistance to stretching and elongate 20% to 100% before breaking. Finally, elastomers stretch readily, with elongations of 100% to 1000% (that is, some types of rubber can be stretched by a factor of 10 without breaking). A fourth important class is the more recently developed electrically conducting polymers, which combine the optical and electronic properties of inorganic semiconductors with the processibility of conventional polymers. This section examines the major kinds of synthetic polymers and their uses.

23.2.1 Fibers

Many important fibers, including cotton and wool, are naturally occurring polymers. The first commercially successful synthetic polymers were made not by polymerization reactions but through the chemical regeneration of the natural polymer cellulose, a condensation polymer of the sugar glucose that is made by plants:

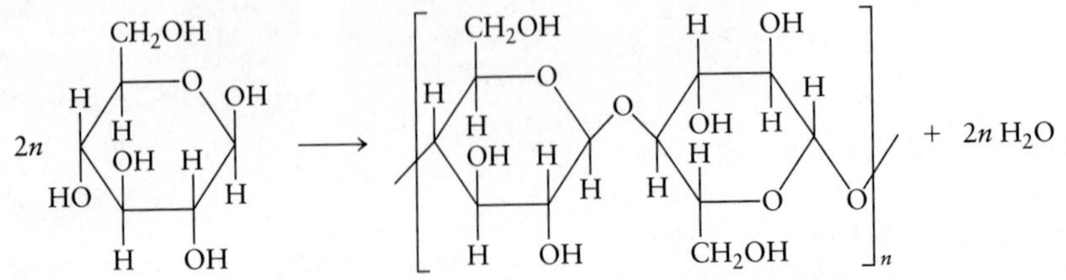

In the viscose rayon process, still used today, cellulose is digested in a concentrated solution of NaOH to convert the $-OH$ groups to $-O^-$ Na^+ ionic

FIGURE 23.4 Filter paper (cellulose) will dissolve in a concentrated ammonia solution containing $[Cu(NH_3)_4]^{2+}$ ions. When the solution is extruded into aqueous sulfuric acid, a dark blue thread of rayon (regenerated cellulose) precipitates.

groups. Reaction with CS_2 leads to the formation of about one "xanthate" group for every two glucose monomer units:

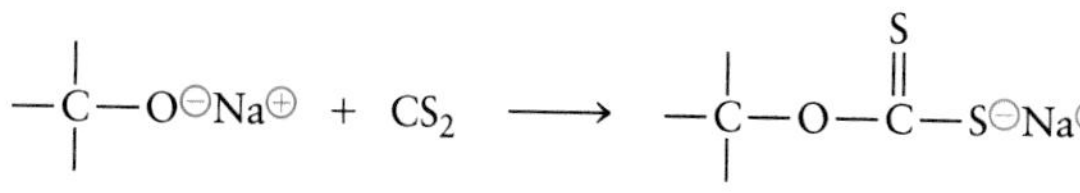

Xanthate

Such substitutions reduce the hydrogen-bond forces holding polymer chains together. In the ripening step, some of these xanthate groups are removed with regeneration of CS_2, and others migrate to the $-CH_2OH$ groups from the ring $-OH$ groups. Afterward, sulfuric acid is added to neutralize the NaOH and to remove the remaining xanthate groups. At the same time, the viscose rayon is spun out to form fibers (Fig. 23.4) while new hydrogen bonds form.

Rayon is a "semisynthetic" fiber because it is prepared from a natural polymeric starting material. The first truly synthetic polymeric fiber was nylon, developed in the 1930s by the American chemist Wallace Carothers at DuPont Company. He knew of the condensation of an amine with a carboxylic acid to form an amide linkage (see Section 7.6) and noted that, if each molecule had *two* amine or carboxylic acid functional groups, long-chain polymers could form. The specific starting materials upon which Carothers settled, after numerous attempts, were adipic acid and hexamethylenediamine:

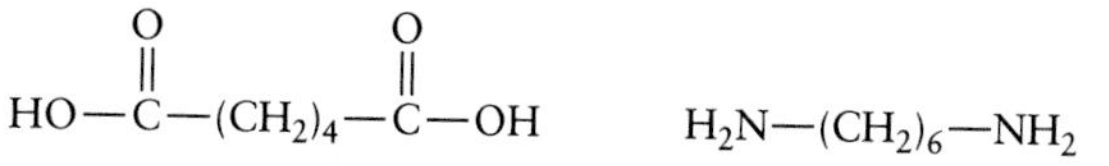

Adipic acid　　Hexamethylenediamine

The two react with loss of water, according to the equation

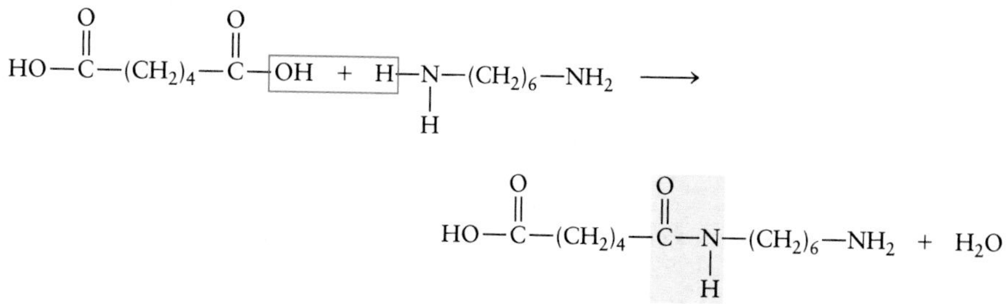

The resulting molecule has a carboxylic acid group on one end (which can react with another molecule of hexamethylenediamine) and an amine group on the other end (which can react with another molecule of adipic acid). The process can continue indefinitely, leading to a polymer with the formula

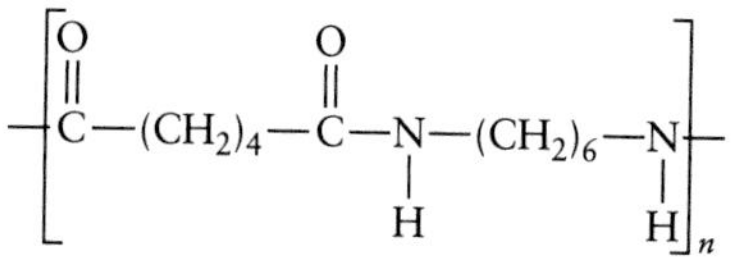

called nylon 66 (Fig. 23.5). The nylon is extruded as a thread or spun as a fiber from the melt. The combination of well-aligned polymer molecules and N—H···O hydrogen bonds between chains makes nylon one of the strongest materials known. The designation "66" indicates that this nylon has six carbon atoms on the starting carboxylic acid and six on the diamine. Other nylons can be made with different numbers of carbon atoms.

Just as a carboxylic acid reacts with an amine to give an amide, it also reacts with an alcohol to give an ester. This suggests the possible reaction of a

FIGURE 23.5 Hexamethylenediamine is dissolved in water (lower layer), and adipyl chloride, a derivative of adipic acid, is dissolved in hexane (upper layer). At the interface between the layers, nylon forms and is drawn out onto the stirring bar.

TABLE 23.1

Fibers

Name	Structural Units	Properties	Sample Uses
Rayon	Regenerated cellulose	Absorbent, soft, easy to dye, poor wash and wear	Dresses, suits, coats, curtains, blankets
Acetate	Acetylated cellulose	Fast drying, supple, shrink-resistant	Dresses, shirts, draperies, upholstery
Nylon	Polyamide	Strong, lustrous, easy to wash, smooth, resilient	Carpeting, upholstery, tents, sails, hosiery, stretch fabrics, rope
Dacron	Polyester	Strong, easy to dye, shrink-resistant	Permanent-press fabrics, rope, sails, thread
Acrylic (Orlon)	$\text{—}(CH_2\text{—}CH(C{\equiv}N))_n\text{—}$	Warm, lightweight, resilient, quick-drying	Carpeting, sweaters, baby clothes, socks

Adapted from P. J. Chenier, *Survey of Industrial Chemistry.* New York: John Wiley & Sons, 1986, Table 18.4.

dicarboxylic acid and a glycol (dialcohol) to form a polymer. The polymer produced most extensively in this way is polyethylene terephthalate, which is built up from terephthalic acid (a benzene ring with —COOH groups on both ends) and ethylene glycol. The first two molecules react according to

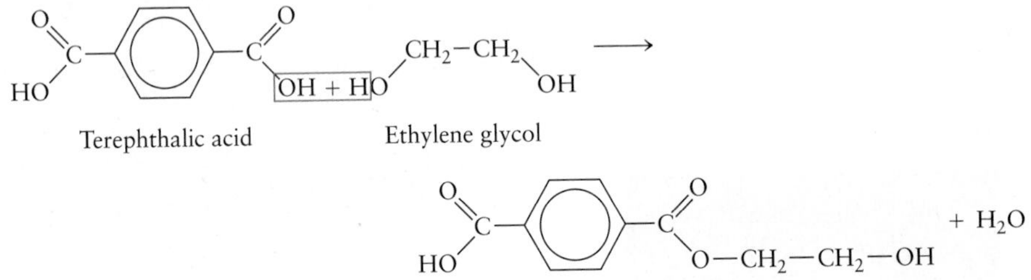

Further reaction then builds up the polymer, which is called polyester and sold under trade names such as Dacron. The planar benzene rings in this polymer make it stiffer than nylon, which has no aromatic groups in its backbone, and help make polyester fabrics crush-resistant. The same polymer formed in a thin sheet rather than a fiber becomes Mylar, a very strong film used for audio and video tapes.

Table 23.1 summarizes the structures, properties, and uses of some important fibers.

23.2.2 Plastics

Plastics are loosely defined as polymeric materials that can be molded or extruded into desired shapes and that harden upon cooling or solvent evaporation. Rather than being spun into threads in which their molecules are aligned, as in fibers, plastics are cast into three-dimensional forms or spread into films for packaging applications. Although celluloid articles were fabricated by plastic processing by the late 1800s, the first important synthetic plastic was Bakelite, the phenol–formaldehyde resin whose cross-linking was discussed earlier in this section. Table 23.2 lists some of the most important plastics and their properties.

Ethene ($CH_2{=}CH_2$) is the simplest monomer that will polymerize. (In industrial settings, ethene is typically referred to by its common name ethylene. Thus

TABLE 23.2

Plastics

Name	Structural Units	Properties	Sample Uses
Polyethylene	$-(CH_2-CH_2)_n-$	High density: hard, strong, stiff	Molded containers, lids, toys, pipe
		Low density: soft, flexible, clear	Packaging, trash bags, squeeze bottles
Polypropylene	$-(CH_2-CH(CH_3))_n-$	Stiffer, harder than high-density polyethylene, higher melting point	Containers, lids, carpeting, luggage, rope
Polyvinyl chloride	$-(CH_2-CH(Cl))_n-$	Nonflammable, resistant to chemicals	Water pipes, roofing, credit cards, records
Polystyrene	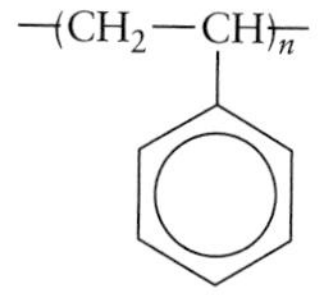	Brittle, flammable, not resistant to chemicals, easy to process and dye	Furniture, toys, refrigerator linings, insulation
Phenolics	Phenol–formaldehyde copolymer	Resistant to heat, water, chemicals	Plywood adhesive, Fiberglass binder, circuit boards

Adapted from P. J. Chenier, *Survey of Industrial Chemistry.* New York: John Wiley & Sons, 1986, pp. 252–264.

the common name for a polymer of ethene is polyethylene. The IUPAC name polyethene is not yet in widespread use, so we will refer to such polymers by their common names herein.) Through free-radical–initiated addition polymerization at high pressures (1000 atm to 3000 atm) and temperatures (300°C to 500°C), it forms polyethylene:

$$n\ CH_2{=}CH_2 \longrightarrow -(CH_2OCH_2)_n-$$

The polyethylene formed in this way is not the perfect linear chain implied by this simple equation. Free radicals frequently abstract hydrogen from the middles of chains in this synthesis, so the polyethylene is heavily branched with hydrocarbon side chains of varying length. It is called *low-density polyethylene* (LDPE) because the difficulty of packing the irregular side chains gives it a lower density (<0.94 g cm^{-3}) than that of perfectly linear polyethylene. This irregularity also makes it relatively soft, so its primary uses are in coatings, plastic packaging, trash bags, and squeeze bottles in which softness is an advantage, not a drawback.

A major breakthrough occurred in 1954, when the German chemist Karl Ziegler showed that ethene could also be polymerized with a catalyst consisting of $TiCl_4$ and an organoaluminum compound [for example, $Al(C_2H_5)_3$]. The addition of ethene takes place at each stage within the coordination sphere of the titanium atom, so monomers can add only at the end of the growing chain. The result is linear polyethylene, also called *high-density polyethylene* (HDPE) because of its density (0.96 g cm^{-3}). Because its linear chains are regular, HDPE contains large crystalline regions, which make it much harder than LDPE and thus suitable for molding into plastic bowls, lids, and toys.

A third kind of polyethylene introduced in the late 1970s is called *linear low-density polyethylene* (LLDPE). It is made by the same metal-catalyzed reactions as HDPE, but it is a deliberate copolymer with other 1-alkenes such as 1-butene. It has some side groups (which reduce the crystallinity and density), but they have a controlled short length instead of the irregular, long side branches in LDPE. LLDPE is stronger and more rigid than LDPE; it is also less expensive because lower pressures and temperatures are used in its manufacture.

If one of the hydrogen atoms of the ethene monomer unit is replaced with a different type of atom or functional group, the plastics that form upon polymerization have different properties. Substitution of a methyl group (that is, the use of propylene as monomer) leads to polypropylene:

$$\left[CH_2 - \underset{\displaystyle CH_3}{\underset{|}{CH}} \right]_n$$

This reaction cannot be carried out successfully by free-radical polymerization. It was first achieved in 1953–1954 by Ziegler and the Italian chemist Giulio Natta, who used the Ziegler catalyst later employed in making HDPE. In polypropylene, the methyl groups attached to the carbon backbone can be arranged in different conformations (Fig. 23.6). In the **isotactic form**, all the methyl groups are arranged on the same side, whereas in the **syndiotactic form** they alternate in a regular fashion. The **atactic form** shows a random positioning of methyl groups. Natta showed that the Ziegler catalyst led to isotactic polypropylene, and he developed another catalyst, using VCl_4, that gave the syndiotactic form. Polypropylene plastic is stiffer and harder than HDPE and has a higher melting point, so it is particularly useful in applications requiring high temperatures (such as the sterilization of medical instruments).

In polystyrene, a benzene ring replaces one hydrogen atom of each ethene monomer unit. Because such a ring is bulky, atactic polystyrene does not crystallize to any significant extent. The most familiar application of this polymer is in the polystyrene foam used in disposable containers for food and drinks and as insulation. A volatile liquid or a compound that dissociates to gaseous products on heating is added to the molten polystyrene. It forms bubbles that remain as the polymer is cooled and molded. The gas-filled pockets in the final product make it a good thermal insulator.

Synthetic polymers with other elements beyond carbon and hydrogen offer many additional possibilities for making plastics. Polyvinylchloride was already discussed in Section 23.1. Another well-known plastic is the solid perfluorocarbon

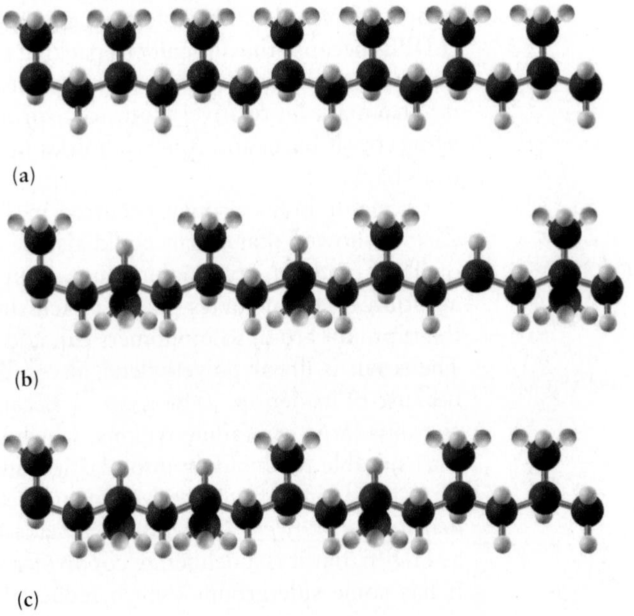

FIGURE 23.6 The structures of (a) isotactic, (b) syndiotactic, and (c) atactic polypropylene.

FIGURE 23.7 A model of the molecular structure of Teflon shows that the fluorine atoms shield the carbon chain very effectively. Note that the chain must twist to accommodate the bulk of the fluorine atoms, completing a full spiral every 26 C atoms along the chain.

called polytetrafluoroethylene (Teflon), formulated as $(-CF_2-CF_{2-})_n$, where n is a large number. Chemically, this compound is nearly completely inert, resisting attack from boiling sulfuric acid, molten potassium hydroxide, gaseous fluorine, and other aggressive chemicals. Physically, it has excellent heat stability (a working temperature up to 260°C), is a very good electrical insulator, and has a low coefficient of friction that makes it useful for bearing surfaces in machines as well as coating frying pans ("non-stick" Teflon). In Teflon the carbon atoms lie in a long chain that is encased by tightly bound fluorine atoms (Fig. 23.7). Even reactants with a strong innate ability to disrupt C—C bonds (such as fluorine itself) fail to attack Teflon at observable rates because there is no way to get past the surrounding fluorine atoms and their tightly held electrons.

23.2.3 Rubber

An **elastomer** is a polymer that can be deformed to a great extent and still recover its original form when the deforming stress is removed. The term *rubber* was introduced by Joseph Priestley, who observed that such materials can be used to rub out pencil marks. Natural rubber is a polymer of isoprene (2-methylbutadiene). The isoprene molecule contains two double bonds of which polymerization removes only one; natural rubber is therefore unsaturated, containing one double bond per isoprene unit. In polymeric isoprene, the geometry at each double bond can be either *cis* or *trans* (Fig. 23.8). Natural rubber is all-*cis* polyisoprene. The all-*trans* form also occurs in nature in the sap of certain trees and is called gutta-percha. This material is used to cover golf balls because it is particularly tough. Isoprene can be polymerized by free-radical addition polymerization, but the resulting polymer contains a mixture of *cis* and *trans* double bonds and is useless as an elastomer.

Even pure natural rubber has limited utility because it melts, is soft, and does not fully spring back to its original form after being stretched. In 1839 the American inventor Charles Goodyear discovered that if sulfur is added to rubber and the mixture is heated, the rubber hardens, becomes more resilient, and does not melt. This process is referred to as **vulcanization** and involves the formation of sulfur bridges between the methyl side groups on different chains. Small amounts of sulfur (<5%) yield an elastic material in which sulfur links between chains remain after stretching and enable the rubber to regain its original form when the external force is removed. Large amounts of sulfur give the very hard, nonelastic material ebonite.

Research on synthetic substitutes for natural rubber began in the United States and Europe before World War II. Attention focused on copolymers of butadiene with styrene (now called SBR rubber) and with acrylonitrile (NBR rubber). The Japanese occupation of the rubber-producing countries of Southeast Asia sharply curtailed the supply of natural rubber to the Allied nations, and rapid steps were taken to increase production of synthetic rubber.

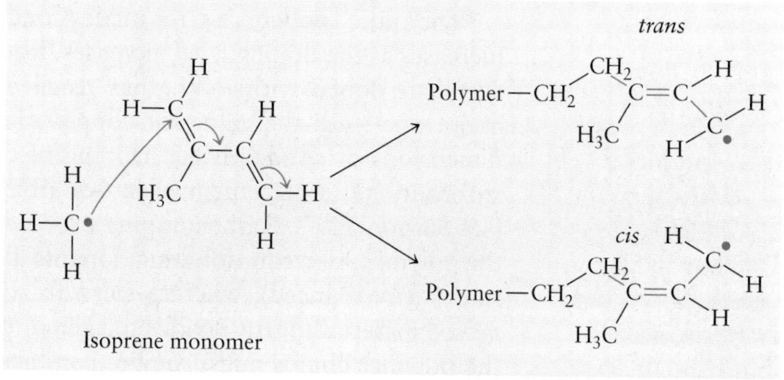

FIGURE 23.8 In the polymerization of isoprene, a *cis* or *trans* configuration can form at each double bond in the polymer. The blue arrows show the redistribution of the electrons upon bond formation.

The initial production goal was 40,000 tons per year of SBR. By 1945, U.S. production had reached an incredible total of more than 600,000 tons per year. During those few years, many advances were made in production techniques, quantitative analysis, and basic understanding of rubber elasticity. Styrene–butadiene rubber production continued after the war, and in 1950 SBR exceeded natural rubber in overall production volume for the first time. More recently, several factors have favored natural rubber: the increasing cost of the hydrocarbon feed stock for synthetic rubber, gains in productivity of natural rubber, and the growing preference for belted radial tires, which use more natural rubber.

The development of the Ziegler–Natta catalysts has affected rubber production as well. First, it facilitated the synthesis of all-*cis* polyisoprene and the demonstration that its properties were nearly identical to those of natural rubber. (A small amount of "synthetic natural rubber" is produced today.) Second, a new kind of synthetic rubber was developed: all-*cis* polybutadiene. It now ranks second in production after styrene–butadiene rubber.

23.2.4 Electrically Conducting Polymers

Electrically conducting polymers, sometimes called *synthetic metals,* have a backbone that is a π-conjugated system, with alternating double and single bonds. This system is formed by overlap of carbon $2p_z$-orbitals, as in Figure 7.12. The polymers are named after the monomer units on which their structures are based. The simplest conducting polymer is polyacetylene, which is a continuation of the 1,3-butadiene structure in Figure 6.53 to much longer chain lengths. Other conducting polymers include ring structures in the conjugated backbone. The monomeric units are shown below for *trans*-polyacetylene, polythiophene, poly(*para*-phenylene), and poly(*para*-pyridine).

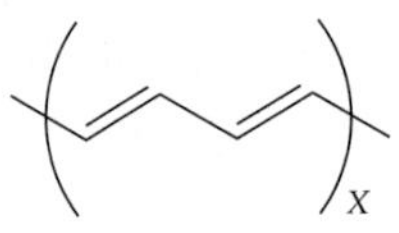

trans-Polyacetylene

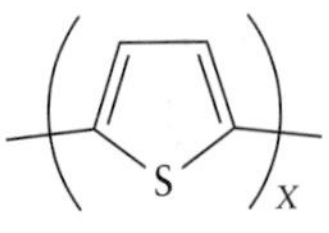

Polythiophene

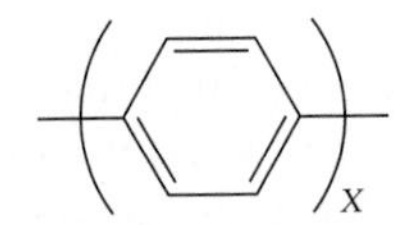

Poly(*para*-phenylene)

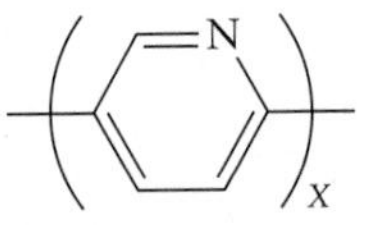

Poly(*para*-pyridine)

As the chain length increases, the energy levels shown in Figure 6.53 for 1,3-butadiene increase in number and coalesce into bands. Thus, the conjugated electronic structure for the individual linear polymer molecule is described by bands, which previously we have seen only for extended three-dimensional solids (see Figs. 21.20 and 22.22). The ground state for the polymer chain is that of an insulator, with an energy gap between occupied and empty levels.

The pure polymers are made conductive by doping; the conductivity increases as the doping level increases. Room temperature conductivity for polyacetylene doped with iodine has reached values of 5×10^4 S cm^{-1}, which is about one-tenth the value for copper. (See Section 23.1 for the definition and dimensions of conductivity.) Doping of conductive polymers does not involve substitutional replacement of lattice atoms as in the inorganic semiconductors (see Section 22.7). Rather, doping proceeds by partial oxidation or reduction of the polymer. Electron-donating dopants like Na, K, and Li produce *n*-type material (partly reduced), whereas electron acceptors like I_2, PF_6, and BF_4 produce *p*-type material (partly oxidized). The dopant ions appear interstitially between the polymer chains and promote conductivity by exchanging charges with the conjugated polymer backbones. A wide variety of interesting structural arrangements of polymer chains and dopants can be produced. The details of the conduction process depend strongly on structure and on the degree of ordering of the polymer chains. Research in this area is a fascinating interplay between concepts of solid state physics and synthetic organic chemistry.

Applications of conductive polymers rely on their combination of electrical and optical properties (comparable to metallic conductors and inorganic semiconductors) with the mechanical flexibility and the chemical processibility of organic polymers. Applications have already appeared in packaging materials for items that are sensitive to electrostatic discharges, in flexible materials for shielding against electromagnetic interference (previously achieved only with rigid metal enclosures), and in rechargeable batteries. Applications are envisioned for electrochemical drug delivery in medicine. Very recent applications include light-emitting diodes, transistors, and memory cells. This field holds rich opportunity for cross-disciplinary developments in chemistry, physics, materials science and engineering, and electrical engineering.

23.3 Liquid Crystals

A **liquid crystal** constitutes an interesting state of matter with properties intermediate between those of true liquids and those of crystals. Unlike glasses, liquid-crystal states are thermodynamically stable. Many organic materials do not show a single solid-to-liquid transition but rather a cascade of transitions involving new intermediate phases. In recent years, liquid crystals have been used in a variety of practical applications, ranging from temperature sensors to displays on calculators and other electronic devices.

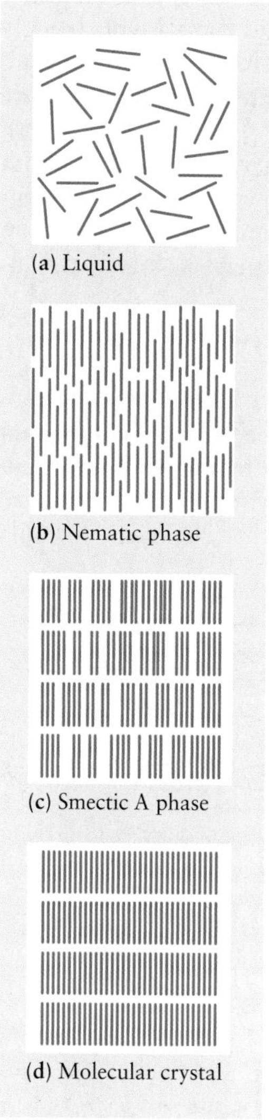

FIGURE 23.9 Different states of structural order for rod-shaped molecules. The figure is only schematic; in a real sample, the lining up of the molecules would not be so nearly perfect.

23.3.1 The Structure of Liquid Crystals

Substances that form liquid crystals are usually characterized by molecules with elongated, rod-like shapes. An example is terephthal-bis-(4-*n*-butylaniline), called TBBA, whose molecular structure can be represented as

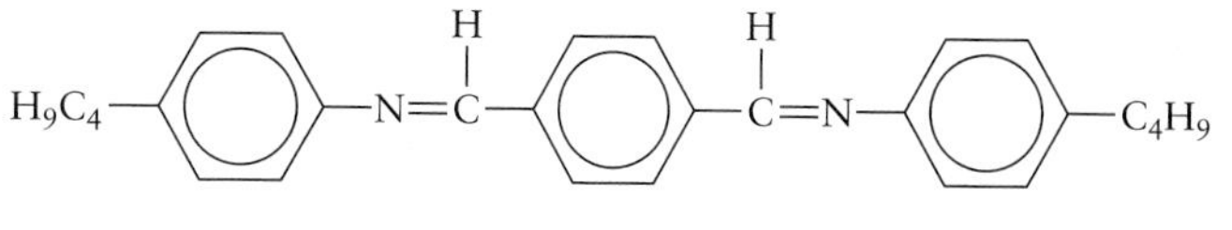

with hydrocarbon groups at the ends separated by a relatively rigid backbone of benzene rings and N=C bonds. Such rod-like molecules tend to line up even in the liquid phase, as Figure 23.9a shows. Ordering in this phase persists only over small distances, however, and on average a given molecule is equally likely to take any orientation.

The simplest type of liquid-crystal phase is the **nematic phase** (see Fig. 23.9b); TBBA undergoes a transition from liquid to nematic at 237°C. In a nematic liquid crystal, the molecules display a preferred orientation in a particular direction, but their centers are distributed at random, as they would be in an ordinary liquid. Although liquid-crystal phases are characterized by a net orientation of molecules over large distances, not all the molecules point in exactly the same direction. There are fluctuations in the orientation of each molecule, and only *on average* do the molecules have a greater probability of pointing in a particular direction.

Some liquid crystals form one or more **smectic phases.** These display a variety of microscopic structures that are indicated by the letters A, B, C, and so forth. Figure 23.9c shows one of them, the smectic A structure; the molecules

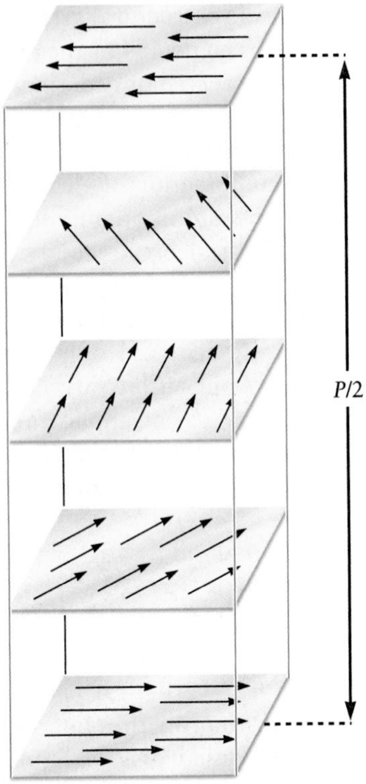

FIGURE 23.10 Several cuts through a cholesteric liquid crystal, showing how the molecular orientation changes with position. The pitch *P* is the distance over which the orientation repeats itself; here, one half of this distance is shown.

continue to display net orientational ordering, but now, unlike in the nematic phase, the centers of the molecules also tend to lie in layers. Within each layer, however, these centers are distributed at random as in an ordinary liquid. TBBA enters the smectic A phase at 200°C, before undergoing transitions to two other more ordered smectic phases at lower temperatures.

At low enough temperatures (below 113°C for TBBA), a liquid crystal freezes into a crystalline solid (see Fig. 23.9d) in which the molecules' orientations are ordered and their centers lie on a regular three-dimensional lattice. The progression of structures in Figure 23.9 illustrates the meaning of the term *liquid crystal*. Liquid crystals are solid-like in showing orientational ordering but liquid-like in the random distribution of the centers of their molecules.

A third type of liquid crystal is called **cholesteric**. The name stems from the fact that many of these liquid crystals involve derivatives of the cholesterol molecule. The structure of a cholesteric liquid crystal is shown schematically in Figure 23.10. In each plane the molecules show a nematic type of ordering, but the orientation of the molecules changes by a regular amount from plane to plane, leading to a helical structure. The distance between planes with the same orientations is referred to as the *pitch P*, which can be quite large (on the order of hundreds of nanometers or longer). A cholesteric liquid crystal will strongly diffract light with wavelengths λ comparable to the pitch. As the temperature changes, the pitch changes as well; the color of the diffracted light can therefore be used as a simple temperature sensor.

The particular orientation taken by a liquid crystal is very sensitive to both the nature of the surfaces with which it is in contact and small electric or magnetic fields. This sensitivity is the basis for the use of nematic liquid crystals in electronic display devices such as digital watches and calculators (Fig. 23.11), as well as in large-screen liquid-crystal displays.

23.3.2 Micelles and Membranes

The liquid crystals we have discussed so far have all been single-component systems, but an interesting second type can be formed from two-component mixtures. One component of the mixture is frequently water, and a typical second component is sodium stearate, which has the chemical formula

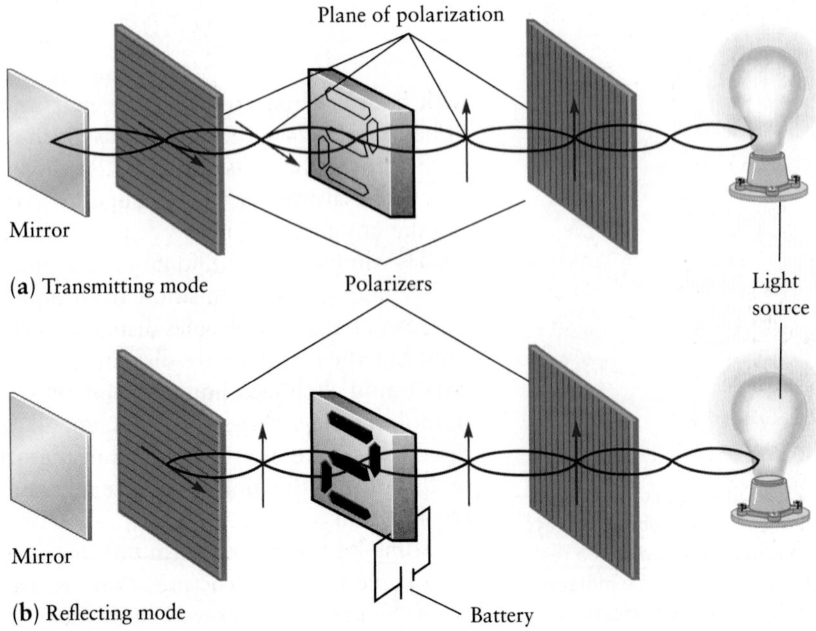

FIGURE 23.11 The mode of operation of a liquid-crystal display device. (a) The light has a polarization that permits it to pass through the second polarizing filter and strike the mirror, giving a bright display. (b) Imposition of a potential difference across some portion of the display causes the liquid-crystal molecules to rotate, creating a different polarization of light. Because the "rotated" light is blocked by the second filter, it does not reach the mirror, and that part of the display appears black.

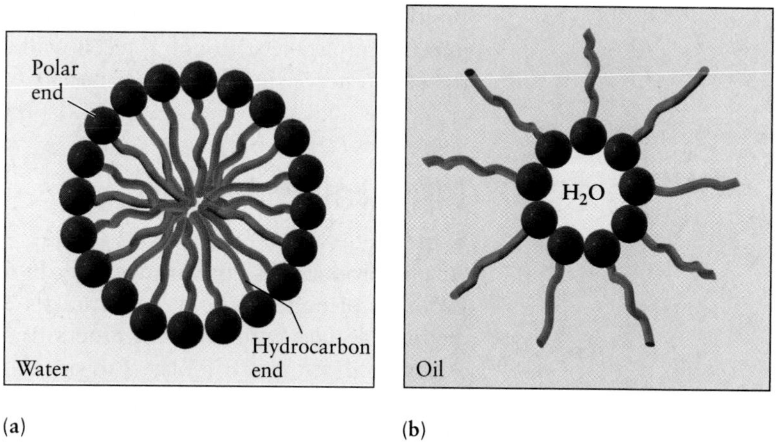

FIGURE 23.12 The structure of a micelle (a) and a reverse micelle (b).

$CH_3(CH_2)_{16}COO^-Na^+$. Preparation of this soap from animal fat was described in Section 7.6. It is a salt analogous to sodium acetate, and its special properties arise from the different natures of the two ends of the molecules. The long hydrocarbon tail is **hydrophobic** ("water fearing") because hydrocarbons do not dissolve in water and avoid contact with it. The ionic carboxylate group ($-COO^-$), on the other hand, is **hydrophilic** and dissolves readily in water both because of its ionic nature and because it can participate in hydrogen bonds. Such molecules are called **amphiphiles**.

If a very small amount of an amphiphile is dissolved in water, it will separate into solvated individual molecules; as soon as a critical concentration is exceeded, however, the molecules organize into a structure called a **micelle**, which contains 40 to 100 molecules (Fig. 23.12a). Micelles are small, nearly spherical clusters of molecules whose hydrocarbon tails are in the nonpolar interior and whose ionic groups are exposed to the water. This organization requires a decrease in entropy but leads to a significant lowering of the energy because the hydrophobic chains are removed from direct contact with water. If, on the other hand, a hydrocarbon solvent is used, *reverse micelles* can form, in which the hydrocarbon tails of the long-chain ions make contact with the solvent and small amounts of water are collected in the polar interior of the micelle (see Fig. 23.12b).

Micelle formation is critical to the action of soaps and detergents. Grease and fat are oily substances that are more soluble in hydrocarbons than in pure water. The function of a soap such as sodium stearate in the cleaning of fabrics is to detach the grease and associated materials (dirt) from the surfaces to which it has adhered and to form a suspension of oil drops surrounded by amphiphile molecules, which can then be rinsed off. The main disadvantage of natural soaps is that their salts with ions such as Ca^{2+} and Mg^{2+} (present in dirt or hard water) are not soluble and precipitate, leaving a scum or residue in the objects being washed. To prevent this, a number of analogs to natural soaps have been developed whose calcium and magnesium salts are more soluble in water. Such amphiphilic synthetic agents are called **detergents**.

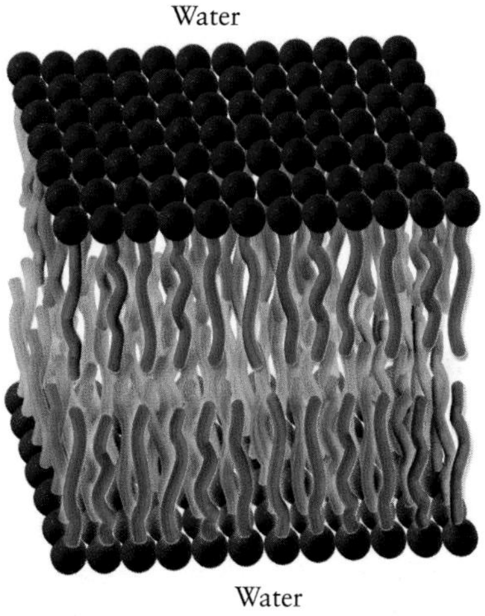

FIGURE 23.13 A bilayer membrane.

Micelles are not the only structures that can form when molecules with hydrophilic and hydrophobic sections are dissolved in water. At higher concentrations, flat **bilayer membranes** form (Fig. 23.13) and can stack into a layered or **lamellar phase** that resembles smectic liquid crystals in their macroscopic properties. A bilayer membrane consists of two planar layers of molecules, with the hydrophilic portions in contact with water and with the hydrophobic portions of one layer in contact with the corresponding hydrophobic portions of a second layer. Such membranes can be made artificially with detergent solutions and serve as models for biological membranes that enclose living cells.

Biological membranes contain embedded proteins that control the passage of ions and molecules through the cell wall. In this way they affect the response of the cell to nerve signals and hormones.

23.4 Natural Polymers

All the products of human ingenuity in the design of polymers pale beside the products of nature. Plants and animals employ a tremendous variety of long-chain molecules with different functions: some for structural strength, others to act as catalysts, and still others to provide instructions for the synthesis of vital components of the cell. In this section we discuss these three important classes of natural polymers: polysaccharides, proteins, and nucleic acids.

23.4.1 Carbohydrates and Polysaccharides

Carbohydrates form a class of compounds of carbon with hydrogen and oxygen. The name comes from the chemical formulas of these compounds, which can be written $C_n(H_2O)_m$, suggesting a "hydrate" of carbon. Simple **sugars**, or **monosaccharides**, are carbohydrates with the chemical formula $C_nH_{2n}O_n$. Sugars with three, four, five, and six carbon atoms are called trioses, tetroses, pentoses, and hexoses, respectively.

Glucose is a hexose sugar that exists in several forms in solution (Fig. 23.14). There is a rapid equilibrium between a straight-chain form (a six-carbon molecule with five —OH groups and one aldehyde —CHO group) and a cyclic form, in which the ring is composed of five carbon atoms and one oxygen, with four —OH side groups and one $-CH_2OH$ side group. In the straight-chain form, four of the carbon atoms (those numbered 2 through 5) are chiral centers, with four different groups bonded to them. As discussed in Section 7.2 (see Fig. 7.9), each such carbon atom can exist in two configurations, each labeled L- for *levo* or D- for *dextro* (Latin for *left* and *right*, respectively). These configurations give rise to $2^4 = 16$ distinct hexose sugars. The glucose formed in plant photosynthesis always has the chirality shown in Figure 23.14b. Of the 15 other straight-chain hexose sugars, the only ones found in nature are D-galactose (in the milk sugar lactose) and D-mannose (a plant sugar).

Figure 23.14 shows that glucose actually has two different ring forms, depending on whether the —OH group created from the aldehyde by the

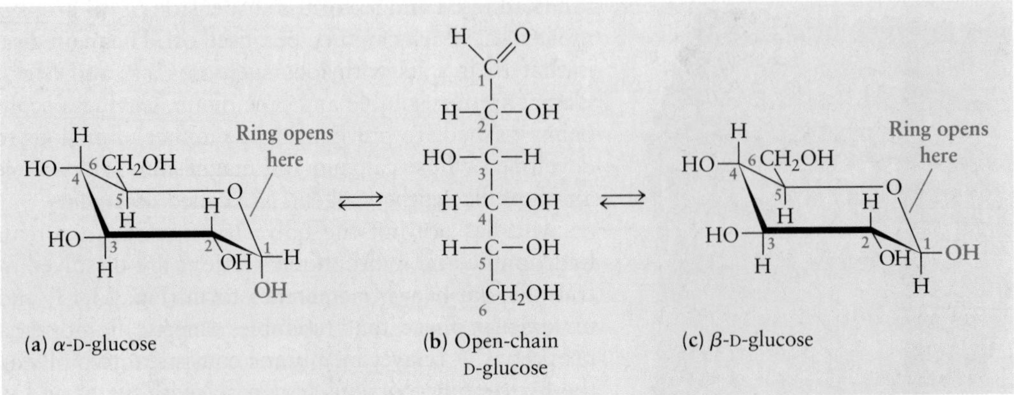

FIGURE 23.14 D-Glucose exists in two ring forms in solution (a and c), which interconvert via an open-chain form (b). The two rings differ in the placement of the —OH and —H groups on carbon atom 1.

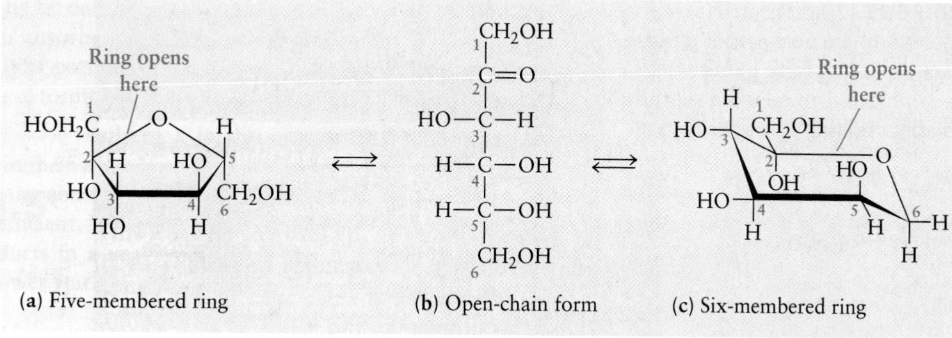

FIGURE 23.15 In aqueous solutions of the sugar D-fructose, an equilibrium exists among a five-atom ring, an open chain, and a six-atom ring. In addition to the β isomers shown here, both ring forms have α isomers, in which the $-CH_2OH$ and $-OH$ on carbon 2 are exchanged.

closing of the ring lies above or below the plane of the ring. Another way to see this is to note that closing the ring creates a fifth chiral carbon atom. The two ring forms of D-glucose are called α-D-glucose (Fig. 23.14a) and β-D-glucose (see Fig. 23.14c). In aqueous solution, these two forms interconvert rapidly via the open-chain glucose form and cannot be separated. They can be isolated separately in crystalline form, however. D-fructose, a common sugar found in fruit and honey, has the same molecular formula as D-glucose but is a member of a class of hexose sugars that are ketones rather than aldehydes. In their straight-chain forms, these sugars have the C=O bond at carbon atom 2 rather than carbon atom 1 (Fig. 23.15).

Many plant cells do not stop the synthesis process with simple sugars such as glucose, but rather continue by linking sugars together to form more complex carbohydrates. **Disaccharides** are composed of two simple sugars linked together by a condensation reaction with the elimination of water. Examples shown in Figure 23.16 are the milk sugar lactose and the plant sugar sucrose (ordinary table sugar, extracted from sugarcane and sugar beets). Further linkages of sugar units lead to polymers called **polysaccharides**. The position of the oxygen atom linking the monomer units has a fundamental effect on the properties and functions of the polymers that result. Starch (Fig. 23.17a) is a polymer of α-D-glucose and is metabolized by humans and animals. Cellulose (see

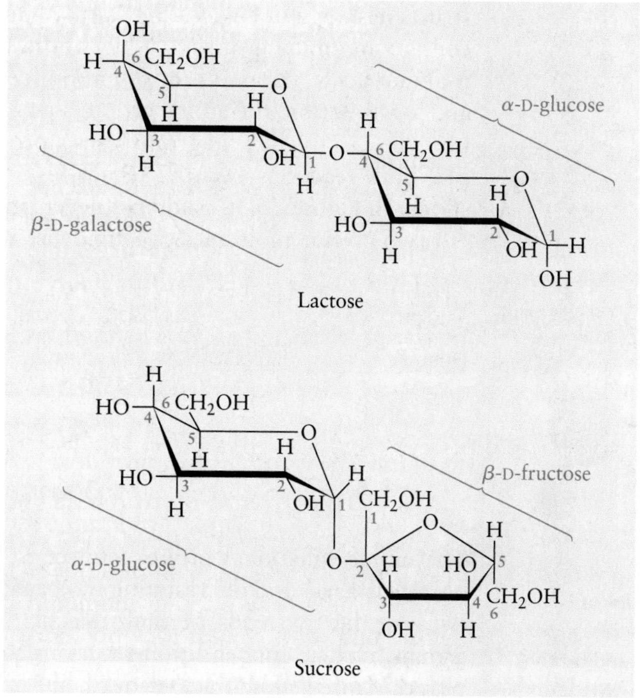

FIGURE 23.16 Two disaccharides. Their derivations from monosaccharide building blocks are shown.

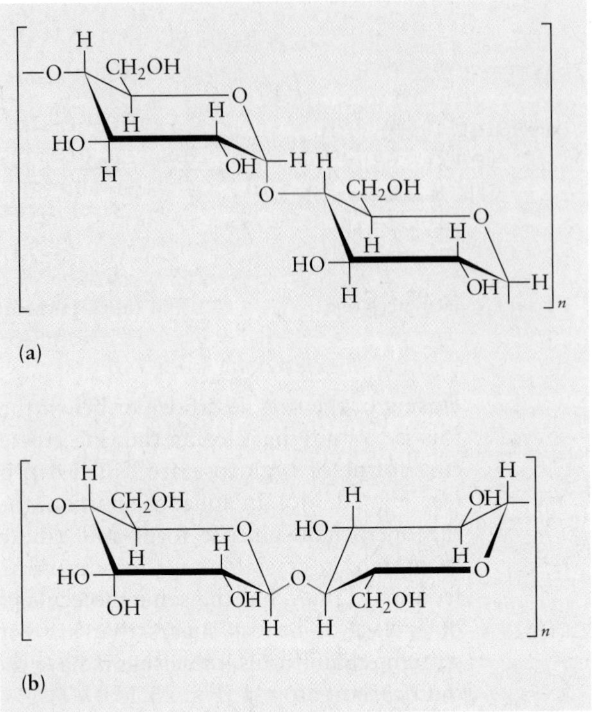

FIGURE 23.17 Both starch (a) and cellulose (b) are polymers of glucose. In starch, all the cyclic glucose units are α-D-glucose. In cellulose, all the monomer units are β-D-glucose.

Fig. 23.17b), a polymer of β-D-glucose, cannot be digested except by certain bacteria that live in the digestive tracts of goats, cows, and other ruminants and in some insects, such as termites. It forms the structural fiber of trees and plants and is present in linen, cotton, and paper. It is the most abundant organic compound on earth.

23.4.2 Amino Acids and Proteins

The monomeric building blocks of the biopolymers, called proteins, are the α-amino acids. The simplest amino acid is glycine, which has the molecular structure shown in Figure 23.18. An amino acid, as indicated by the name, must contain an amine group ($-NH_2$) and a carboxylic acid group ($-COOH$). In α-amino acids, the two groups are bonded to the same carbon atom. In acidic aqueous solution, the amine group is protonated to form $-NH_3^+$; in basic solution, the carboxylic acid group loses a proton to form $-COO^-$. At intermediate pH, both reactions occur. The net result is that the simple amino acid form shown in Figure 23.18 is almost never present in aqueous solution.

Two glycine molecules can condense with loss of water to form an amide:

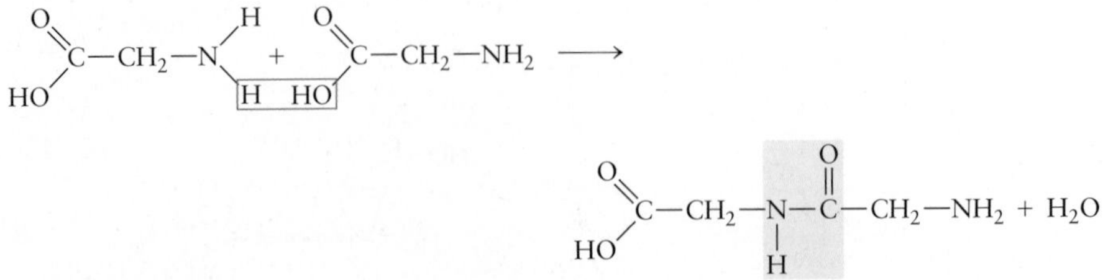

FIGURE 23.18 The structure of glycine. On the left side is the carboxylic acid group ($-COOH$), and on the right is the amine group ($-NH_2$).

The amide functional group connecting two amino acids is referred to as a **peptide linkage**, and the resulting molecule is a *dipeptide*—in this case, diglycine. Because the two ends of the molecule still have carboxylic acid and amine groups, further condensation reactions to form a **polypeptide**, a polymer comprised of many amino acid groups, are possible. If glycine were the only amino acid available, the result would be polyglycine, a rather uninteresting protein. There is a close similarity between this naturally occurring condensation polymer and the synthetic polyamide nylon. Polyglycine could be called "nylon 2," a simple polyamide in which each repeating unit contains two carbon atoms.

Nature does not stop with glycine as a monomer unit. Instead, any of 20 different α-amino acids are found in most natural polypeptides. In each of these, one of the hydrogen atoms on the central carbon atom of glycine is replaced by another side group. Alanine is the next simplest α-amino acid after glycine; it has a $-CH_3$ group in place of an $-H$ atom. This substitution has a profound consequence. In alanine, four different groups are attached to a central carbon: $-COOH$, $-NH_2$, $-CH_3$, and $-H$. There are two ways in which four different groups can be arranged in a tetrahedral structure about a central atom (see Fig. 7.9). The two optical isomers of alanine are designated by the prefixes L- and D- for *levo* and *dextro* (Latin for "left" and "right," respectively).

If a mixture of L- and D-alanine were caused to polymerize, nearly all the polymer molecules would have different structures because their sequences of D-alanine and L-alanine monomer units would differ. To create polymers with definite structures for particular roles, there is only one recourse: to build all polypeptides from one of the optical isomers so that the properties will be reproducible from molecule to molecule. Nearly all naturally occurring α-amino acids are of the L form, and most earthly organisms have no use for D-α-amino acids in making polypeptides. Terrestrial life could presumably have begun equally well using mainly D-amino acids (all biomolecules would be mirror images of their present forms). The mechanism by which the established preference was initially selected is not known.

The $-H$ group of glycine and $-CH_3$ group of alanine give just the first two amino acid building blocks. Table 23.3 shows all 20 important α-amino acids, arranged by side group. Note the variety in their chemical and physical properties. Some side groups contain basic groups; others are acidic. Some are compact; others are bulky. Some can take part in hydrogen bonds; others can complex readily with metal ions to form coordination complexes.

This variety in properties of the α-amino acids leads to even more variety in the polymers derived from them, called **proteins**. The term *protein* is usually applied to polymers with more than about 50 amino acid groups; large proteins may contain many thousand such groups. Given the fact that any one of 20 α-amino acids may appear at each point in the chain, the number of possible sequences of amino acids in even small proteins is staggering. Moreover, the amino acid sequence describes only one aspect of the molecular structure of a protein. It contains no information about the three-dimensional conformation adopted by the protein. The carbonyl group and the amine group in each amino acid along the protein chain are potential sites for hydrogen bonds, which may also involve functional groups on the amino acid side chains. Also, the cysteine side groups ($-CH_2-SH$) can react with one another, with loss of hydrogen, to form $-CH_2-S-S-CH_2-$ disulfide bridges between different cysteine groups in a single chain or between neighboring chains (the same kind of cross-linking by sulfur occurs in the vulcanization of rubber). As a result of these strong intrachain interactions, the molecules of a given protein have a rather well-defined conformation even in solution, as compared with the much more varied range of conformations available to a simple alkane chain (see Fig. 7.3). The three-dimensional structures of many proteins have been determined by X-ray diffraction.

There are two primary categories of proteins: fibrous and globular. A **fibrous protein** is usually a structural material consisting of polymer chains linked in sheets or twisted in long fibers. Silk is a fibrous protein in which the monomer units are primarily glycine and alanine, with smaller amounts of serine and tyrosine. The protein chains are cross-linked by hydrogen bonds to form

TABLE 23.3

α-Amino Acid Side Groups

Hydrogen "Side Group"		
Glycine	Gly	$-H$
Alkyl Side Groups		
Alanine	Ala	$-CH_3$
Valine	Val	$-CH-CH_3$ $\vert$ CH_3
Leucine	Leu	$-CH_2-CH-CH_3$ $\vert$ CH_3
Isoleucine	Ile	$-CH-CH_2-CH_3$ $\vert$ CH_3
Proline	Pro (structure of entire amino acid)	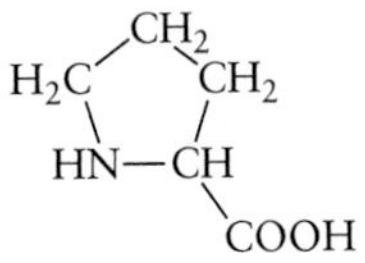
Aromatic Side Groups		
Phenylalanine	Phe	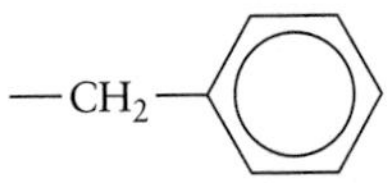
Tyrosine	Tyr	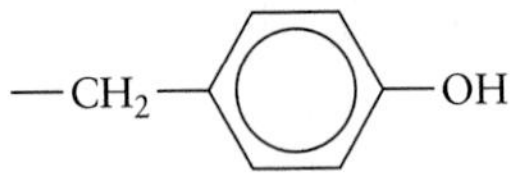
Tryptophan	Trp	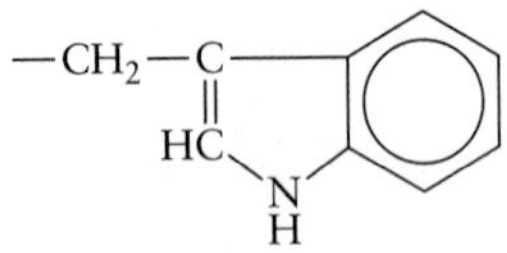
Alcohol-Containing Side Groups		
Serine	Ser	$-CH_2OH$
Threonine	Thr	OH $\vert$ $-CH$ $\vert$ CH_3
Basic Side Groups		
Lysine	Lys	$-CH_2CH_2CH_2CH_2NH_2$
Arginine	Arg	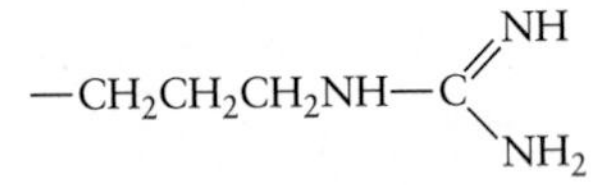
Histidine	His	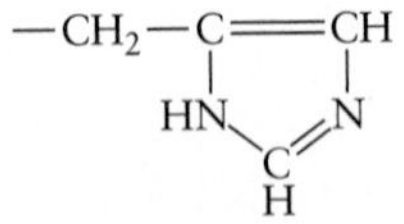
Acidic Side Groups		
Aspartic acid	Asp	$-CH_2COOH$
Glutamic acid	Glu	$-CH_2CH_2COOH$
Amide-Containing Side Groups		
Asparagine	Asn	O $\Vert$ $-CH_2C-NH_2$
Glutamine	Gln	O $\Vert$ $-CH_2CH_2C-NH_2$
Sulfur-Containing Side Groups		
Cysteine	Cys	$-CH_2-SH$
Methionine	Met	$-CH_2CH_2-S-CH_3$

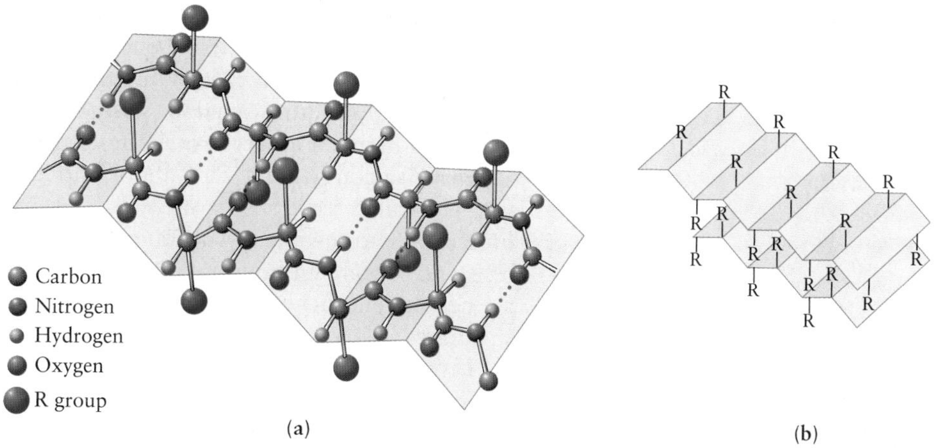

FIGURE 23.19 (a) View of a β-pleated sheet that emphasizes the interchain hydrogen bonds responsible for the structure. (b) View of two stacked β-pleated sheets that form part of a microcrystalline structure responsible for the great strength of silk.

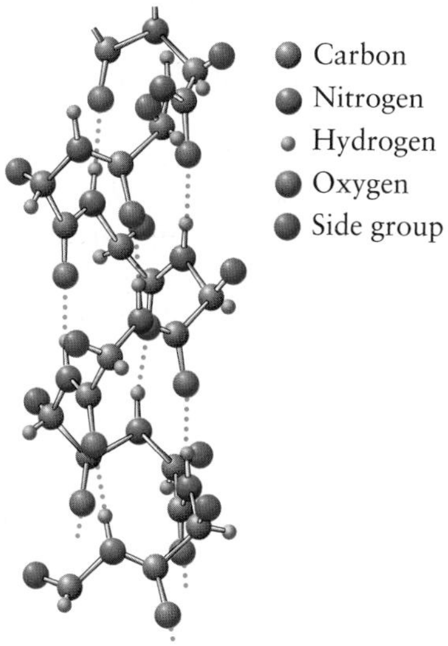

FIGURE 23.20 α-helix protein secondary structure, with interchain hydrogen bonds shown.

sheet-like structures (Fig. 23.19) that are arranged so that the nonhydrogen side groups all lie on one side of the sheet; the sheets then stack in layers. The relatively weak forces between sheets give silk its characteristic smooth feel. The amino acids in wool and hair have side chains that are larger, bulkier, and less regularly distributed than those in silk; therefore, sheet structures do not form. Instead, the protein molecules twist into a right-handed coil called an *α-helix* (Fig. 23.20). In this structure, each carbonyl group is hydrogen-bonded to the amine group of the fourth amino acid farther along the chain; the bulky side groups jut out from the helix and do not interfere with one another.

The second type of protein is the **globular protein.** Globular proteins include the carriers of oxygen in the blood (hemoglobin) and in cells (myoglobin). They have irregular folded structures (Fig. 23.21) and typically consist of 100 to 1000 amino acid groups in one or more chains. Globular proteins frequently have parts of their structures in α-helices and sheets, with other portions in more disordered forms. Hydrocarbon side groups tend to cluster in regions that exclude water, whereas charged and polar side groups tend to remain in close contact with water. The sequences of amino acid units for many such proteins have been worked out by cleaving them into smaller pieces and analyzing the

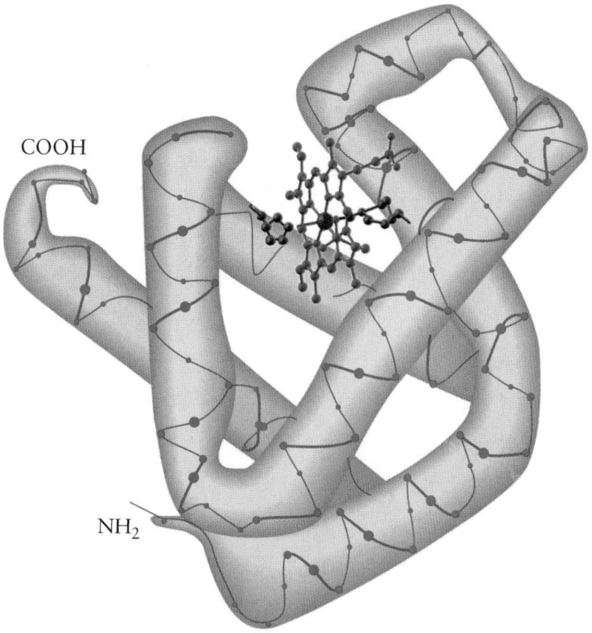

FIGURE 23.21 A computer-generated model of the structure of myoglobin. The central heme group (red) is shown in greater detail in Chapter 8—*Connection to Biology:* Coordination Complexes in Heme Proteins; two histidine groups (green) extend toward the central iron atom. Much of the protein (blue tube) is coiled in α-helices.

structure of the fragments. It took Frederick Sanger 10 years to complete the first such determination of sequence for the 51 amino acids in bovine insulin, an accomplishment that earned him the Nobel Prize in chemistry in 1958. Now, automated procedures enable scientists to rapidly determine amino acid sequences in much longer protein molecules.

Enzymes constitute a very important class of globular proteins. They catalyze particular reactions in the cell, such as the synthesis and breakdown of proteins, the transport of substances across cell walls, and the recognition and resistance of foreign bodies. Enzymes act by lowering the activation barrier for a reaction, and they must be selective so as to act only on a restricted group of substrates.

Let's examine the enzyme carboxypeptidase A, whose structure has been determined by X-ray diffraction. It removes amino acids one at a time from the carboxylic acid end of a polypeptide. Figure 23.22 shows the structure of the active site (with a peptide chain in place, ready to be cleaved). A special feature of this enzyme is the role played by the zinc ion, which is coordinated to two histidine residues in the enzyme and to a carboxylate group on a nearby glutamic acid residue. The zinc ion helps remove electrons from the carbonyl group of the peptide linkage, making it more positive and thereby more susceptible to attack by water or by the carboxylate group of a second glutamic acid residue. The side chain on the outer amino acid of the peptide being cleaved is positioned in a hydrophobic cavity, which favors large aromatic or branched side chains (such as that in tyrosine) over smaller hydrophilic side chains (such as that in aspartic acid). Carboxypeptidase A is thus selective about the sites at which it cleaves peptide chains.

The molecular "engineering" that lies behind nature's design of carboxypeptidase A and other enzymes is truly remarkable. The amino acid residues that form the active site and determine its catalytic properties are *not* adjacent to one another in the protein chain. As indicated by the numbers after the residues in Figure 23.22, the two glutamic acid residues are the 72nd and 270th amino acids along the chain. The enzyme adopts a conformation in which the key residues, distant from one another in terms of chain position, are nonetheless quite close in three-dimensional space, allowing the enzyme to carry out its specialized function.

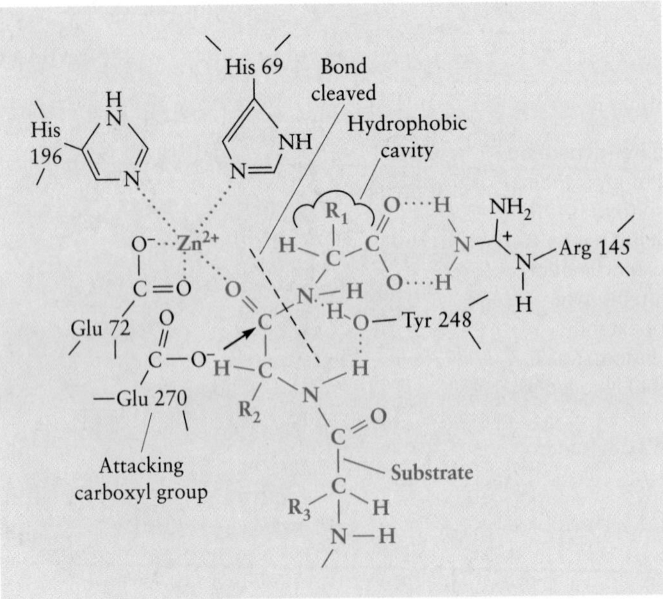

FIGURE 23.22 The active site of carboxypeptidase A. Shown in red is a substrate polypeptide that is being cleaved by the enzyme. Green is used to show the role of Zn^{2+} as a complexing ion, and blue is used to show the hydrogen bonds that maintain the geometry.

23.4.3 Nucleotides and Nucleic Acids

We have seen that proteins are copolymers made up typically of 20 types of monomer units. Simply mixing the amino acids and letting them dehydrate to form polymer chains at random would never lead to the particular structures needed by living cells. How does the cell preserve information about the amino acid sequences that make up its proteins, and how does it transmit this information to daughter cells through the reproductive process? These questions lie in the field of molecular genetics, an area in which chemistry plays the central role.

The primary genetic material is deoxyribonucleic acid (DNA). This biopolymer is made up of four types of monomer units called **nucleotides**. Each nucleotide is composed of three parts:

1. One molecule of a pyrimidine or purine base. The four bases are thymine, cytosine, adenine, and guanine (Fig. 23.23a).
2. One molecule of the sugar D-deoxyribose ($C_5H_{10}O_4$). D-Ribose is a pentose sugar with a five-membered ring.
3. One molecule of phosphoric acid (H_3PO_4).

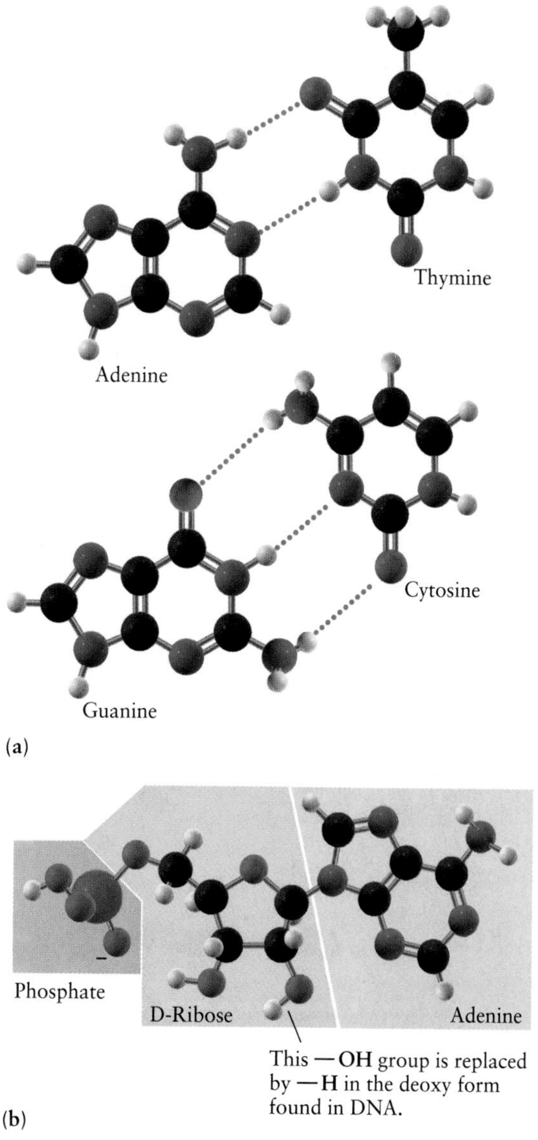

FIGURE 23.23 (a) The structures of the purine and pyrimidine bases. Hydrogen-bonding between pairs of bases is indicated by red dots. (b) The structure of the nucleotide adenosine monophosphate (AMP).

The cyclic sugar molecule links the base to the phosphate group, undergoing two condensation reactions, with loss of water, to form the nucleotide (see Fig. 23.23b). The first key to discovering the structure of DNA was the following observation: Although the proportions of the four bases in DNA from different organisms are quite variable, the chemical amount of cytosine (C) is always approximately equal to that of guanine (G) and the chemical amount of adenine (A) is always approximately equal to that of thymine (T). This suggested some type of base pairing in DNA that could lead to association of C with G and of A with T. The second crucial observation was an X-ray diffraction study by Rosalind Franklin and Maurice Wilkins that suggested the presence of helical structures of more than one chain in DNA.

James Watson and Francis Crick put together these two pieces of information in their famous 1953 proposal of a double-helix structure for DNA. They concluded that DNA consists of two interacting helical strands of nucleic acid polymer (Fig. 23.24), with each cytosine on one strand linked through hydrogen bonds to a guanine on the other and each adenine to a thymine. This accounted for the observed molar ratios of the bases, and it also provided a model for the replication of the molecule, which is crucial for passing on information during

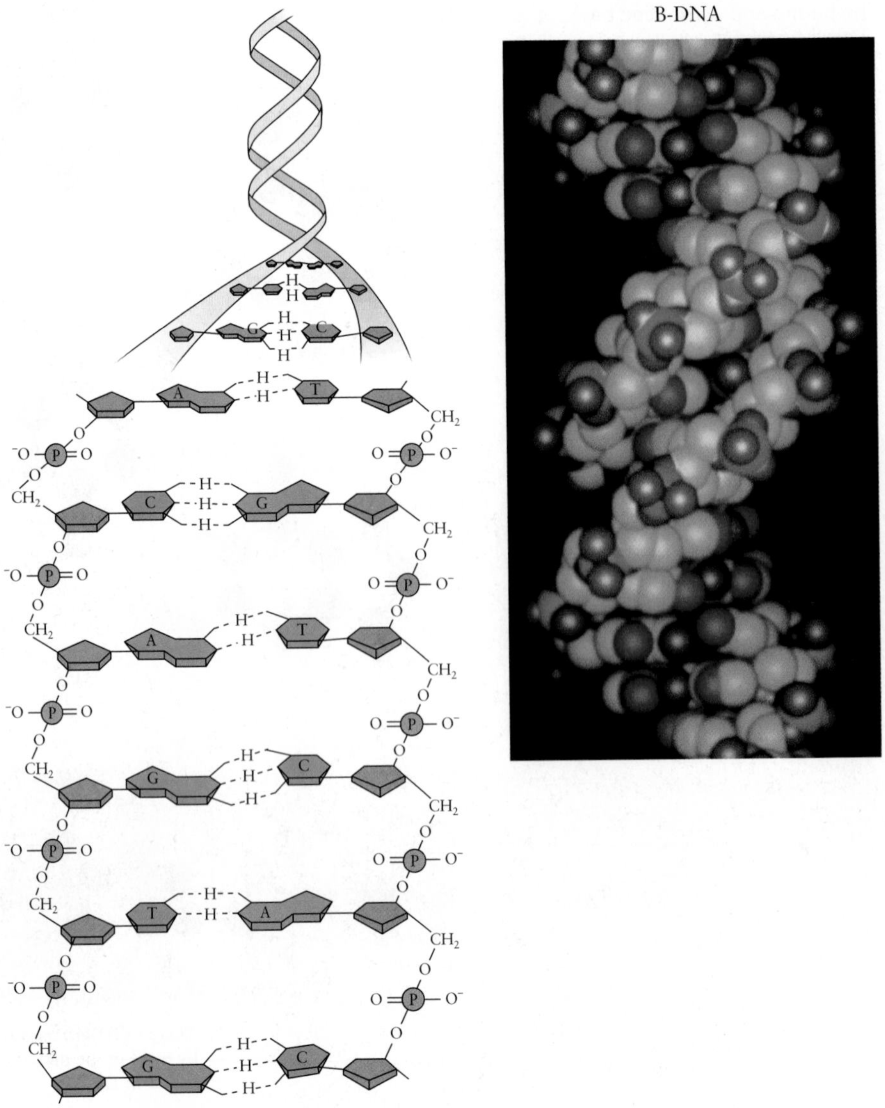

FIGURE 23.24 The double-helix structure of DNA.

the reproductive process. One DNA strand serves as a template upon which a second DNA strand is synthesized. A DNA molecule reproduces by starting to unwind at one end. As it does so, new nucleotides are guided into position opposite the proper bases on each of the two strands. If the nucleotide does not fit the template, it cannot link to the polymeric strand under construction. The result of the polymer synthesis is two double-helix molecules, each containing one strand from the original and one new strand that is identical to the original in every respect.

Information is encoded in DNA in the sequence of the base pairs. Subsequent research has broken this genetic code and established the connection between the base sequence in a segment of DNA and the amino acid sequence of the protein synthesized according to the directions in that segment. The code in a nucleic acid is read as consecutive, nonoverlapping triplets of bases, with each triplet standing for a particular amino acid. Thus, a nucleic acid strand consisting of pure cytosine gives a polypeptide of pure proline, meaning that the triplet CCC codes for proline. The nucleic acid strand AGAGAGAG . . . is read as the alternating triplets AGA and GAG and gives a polypeptide consisting of alternating arginine (coded by AGA) and glutamic acid (coded by GAG) monomer units. There are 64 (4^3) possible triplets, so typically more than one code exists for a particular amino acid. Some triplets serve as signals to terminate a polypeptide chain. Remarkably, the genetic code appears to be universal, independent of the species of plant or animal, a finding that suggests a common origin for all terrestrial life.

A single change in a base pair in DNA causes a change in one amino acid of the protein that is coded for by that DNA. Such a change may seem small, but it may have dramatic (even fatal) effects for the organism in question. For example, the triplets GAA and GAG both code for glutamic acid (an acidic residue), whereas the triplets GTA and GTG code for valine (a nonpolar residue). A single change in the central A to T in the DNA thus changes an amino acid in a protein produced by the organism. This might seem like a small effect, but it can change the structure and mode of action of the protein. This particular change is responsible for the presence of hemoglobin S (instead of hemoglobin A) in the blood of people who have sickle cell anemia. In two of the four hemoglobin S chains a glutamic acid amino acid is changed to a valine, and the result is a decrease in the solubility of S relative to A by a factor of 25. This leads to polymerization of the hemoglobin to form insoluble structures that bend red blood cells into sickle shapes, a change that can lead to early death. One of the challenges of modern genetic engineering is to use chemistry to modify coding molecules in living species to eliminate fatal or disease-causing mutants.

CHAPTER SUMMARY

Mankind has long used naturally occurring polymers, obtained from plants and animals, as useful materials. These are familiar in cotton, wool, and silk fabrics, and in structural applications of wood and rubber. Inspired by these natural products, chemists have invented synthetic routes to these and similar materials, often with modified or improved properties. Curiosity about life processes in living systems has inspired research on natural polymers, such as polysaccharides, proteins, enzymes, and DNA, to discern the relation between their structure and function. The structures of these solid synthetic and natural polymeric materials range from purely amorphous to highly crystalline. The route to controlling their molecular and crystalline structures lies through understanding their local bonding.

CONCEPTS AND SKILLS

Section 23.1 – Polymerization Reactions for Synthetic Polymers

Contrast the methods of addition and condensation polymerization (Problems 1–6).

- Polymer formation requires that many monomers must be attached to a growing polymer molecule. This requires that highly reactive functional groups must be available at each growth step. This is achieved by two main mechanisms.
  - Addition polymerization requires monomers to join the polymer without net loss of atoms. This usually involves free radical reaction of molecules that have C=C double bonds, and proceeds through three steps: initiation, propagation, and termination.
  - Condensation polymerization requires that a small molecule such as water is split off as each monomer is added to the polymer.
- Both addition and condensation polymerization can occur with mixtures of monomers to produce a random copolymer of the two monomers.
- If monomers have three or more reactive sites, cross-linking reactions to form sheets or networks is possible. Cross-linking is often brought about deliberately to obtain stronger materials.

Section 23.2 – Applications for Synthetic Polymers

Give several examples of fibers, plastics, and rubbers and describe how they are made and used (Problems 7–10).

- Synthetic polymers are selected for four major classes of applications based on their physical properties.
  - Fibers for use in fabrics are spun into threads in which their molecules are aligned. They must resist stretching, and usually break after only 10% elongation. The first purely synthetic fiber was nylon, developed by condensation polymerization.
  - Plastics are molded or extruded into desired shapes that harden upon cooling or solvent evaporation. They typically elongate 20% to 100% before breaking. Many of these are formed by addition polymerization of ethene or its derivatives. Examples include polyethylene, polystyrene, and Teflon. By using proper catalysts and manipulating the size of side group substituents, it is possible to overcome the tendency of addition polymerization to form highly branched chains. These tools make it possible to control the degree of crystallinity in polymer materials.
  - Rubbers, or elastomers, stretch readily to elongate by a factor of 10 before breaking. Natural rubber, obtained from the sap of certain trees, can be hardened and toughened by addition of sulfur in the vulcanization process. Synthetic rubber is produced by addition copolymerization of butadiene and styrene.
  - Electrically conducting polymers combine the optical and electronic properties of inorganic semiconductors with the processing ease of conventional polymers. Their structures are continuations of the 1,3-butadiene structure to greater lengths, and the electronic structure for the individual molecule is described by bands. These polymers are made electrically conducting by doping.

Section 23.3 – Liquid Crystals

Explain what a liquid crystal is, and state how nematic and smectic phases differ from ordinary liquids and crystalline solids (Problems 11–12).

- Liquid crystals have properties intermediate between those of true liquids and crystals. Unlike glasses, they are thermodynamically stable.
- Based on microscopic structural details, liquid crystals form three separate ordered phases: nematic, smectic, and cholesteric.
- Orientation of liquid crystals depends sensitively on small electric and magnetic fields. This is the basis of liquid crystal displays for digital information.

Describe the formation of ordered structures such as micelles and membranes in surfactant solutions (Problems 13–14).

- Amphiphiles are molecules with a water soluble ionic structure on one end (hydrophilic) and a long hydrocarbon tail on the other end (hydrophobic). When amphiphiles are dissolved in water above a critical concentration, they form ordered clusters called micelles in which the organic tails of the molecules are enclosed in a shell of hydrophilic structures that shield the organic phase from interaction with the water. This structure is stable because it eliminates the repulsive interaction between the organic tails and water. Amphiphiles can form reverse micelles in organic solvents by shielding the hydrophilic structures in a shell of hydrocarbon tails that are dissolved in the organic solvent. At still higher concentrations, amphiphiles can be organized into two-dimensional sheets rather than spherical clusters. The hydrophobic sides of two sheets interact to form a stable bilayer membrane in which the two facing hydrophobic sheets are sandwiched between two hydrophilic layers.

Section 23.4 – Natural Polymers

Describe the formation of polysaccharides from sugars, proteins from amino acids, and DNA from nucleotides, and the roles of these biopolymers in living cells (Problems 15–22).

- Naturally occurring polymers are synthesized by plants and animals to support a variety of life processes.
- Polysaccharides are grown by linkage of simple sugar units (carbohydrates). One example is starch, which is readily metabolized by humans and animals. Another is cellulose, which forms the structural fibers of plants and trees and appears in linen and cotton fabrics and in paper.
- The biological polymers called proteins are built up from the amino acid units, ranging in number from 50 to several thousand per polymer. Since 20 amino acids are available, a very large number of sequences can be realized. Fibrous proteins consist of polymers linked in sheets or twisted in long fibers; they are structural materials. Globular proteins include hemoglobin and enzymes; they generally control chemical reactions in living systems.
- The primary genetic material DNA is a biopolymer made up of four types of monomer units called nucleotides. DNA serves as a template during reproduction to enable a cell to preserve information about the amino acid sequences that make up its proteins, and transfer this information to daughter cells through the genetic code.

PROBLEMS

Answers to problems whose numbers are red appear in Appendix G. Problems that are more challenging are indicated with asterisks.

23.1 Polymerization Reactions for Synthetic Polymers

1. Write a balanced chemical equation to represent the addition polymerization of 1,1-dichloroethene. The product of this reaction is Saran, used as a plastic wrap.

2. Write a balanced chemical equation to represent the addition polymerization of tetrafluoroethene. The product of this reaction is Teflon.

3. A polymer produced by addition polymerization consists of ($-CH_2-O-$) groups joined in a long chain. What was the starting monomer?

4. The polymer polymethyl methacrylate is used to make Plexiglas. It has the formula

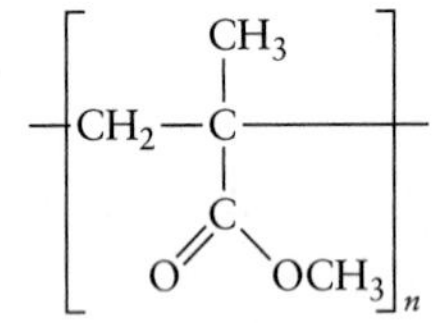

Draw the structural formula of the starting monomer.

5. The monomer glycine (NH_2-CH_2-COOH) can undergo condensation polymerization to form polyglycine, in which the structural units are joined by amide linkages.
   (a) What molecule is split off in the formation of polyglycine?
   (b) Draw the structure of the repeat unit in polyglycine.

6. The polymer forms $\text{+}NH-CH(CH_3)-\overset{\overset{O}{\|}}{C}\text{+}_n$ upon condensation polymerization with loss of water. Draw the structure of the starting monomer.

23.2 Applications for Synthetic Polymers

7. Determine the mass of adipic acid and the mass of hexamethylenediamine needed to make 1.00×10^3 kg of nylon 66 fiber.

8. Determine the mass of terephthalic acid and the mass of ethylene glycol needed to make 10.0 kg of polyester fiber.

9. In a recent year, 4.37 billion kilograms of low-density polyethylene was produced in the United States. What volume of gaseous ethene at 0°C and 1.00 atm would give this amount?

10. In a recent year, 2.84 billion kilograms of polystyrene was produced in the United States. Polystyrene is the addition polymer formed from the styrene monomer, $C_6H_5CH{=}CH_2$. How many styrene monomer units were incorporated in that 2.84 billion kilograms of polymer?

23.3 Liquid Crystals

11. Compare the natures and extents of order in the smectic liquid-crystal and isotropic liquid phases of a substance. Which has the higher entropy? Which has the higher enthalpy?

12. Nematic liquid crystals form when a liquid of long rod-like molecules is cooled. What *additional* types of intermolecular interactions would you expect to favor the formation of a *smectic* phase?

13. Consider a ternary (three-component) system of amphiphile, hydrocarbon, and water. What structure do you expect to form if small amounts of the first two components are mixed with a large amount of water?

14. In addition to spherical micelles and lamellar phases, a binary amphiphile-water mixture can also form extended cylindrical rolls, with the hydrophilic groups pointing out and the hydrophobic chains on the interior. Over what composition range are such cylinders most likely to be found, relative to spherical micelles and planar layers?

23.4 Natural Polymers

15. By referring to Figure 23.16a, draw the structure of the ring form of β-D-galactose. How many asymmetric carbon atoms (chiral centers) are there in the molecule?

16. By referring to Figure 23.23b, draw the structure of the ring form of D-ribose. How many asymmetric carbon atoms (chiral centers) are there in the molecule?

17. How many tripeptides can be synthesized using just three different species of α-amino acids?

18. How many different polypeptides, each containing ten amino acids, can be made from the amino acids listed in Table 23.3? How many different polypeptides, each containing 100 amino acids, can be made?

19. Draw the structure of the pentapeptide alanine—leucine—phenylalanine—glycine—isoleucine. Assume that the free $-NH_2$ group is at the alanine end of the peptide chain. Would this compound be more likely to dissolve in water or in octane? Explain.

20. Draw the structure of the pentapeptide aspartic acid—serine—lysine—glutamic acid—tyrosine. Assume that the free $-NH_2$ group is at the aspartic acid end of the peptide chain. Would this compound be more likely to dissolve in water or in octane? Explain.

21. Suppose a long-chain polypeptide is constructed entirely from phenylalanine monomer units. What is its empirical formula? How many amino acids does it contain if its molar mass is 17,500 g mol^{-1}?

22. A typical bacterial DNA has a molar mass of 4×10^9 g mol^{-1}. Approximately how many nucleotides does it contain?

ADDITIONAL PROBLEMS

23. In the addition polymerization of acrylonitrile, a very small amount of butyl lithium causes a reaction that can consume hundreds of pounds of the monomer; however, the butyl lithium is called an initiator, not a catalyst. Explain why.

24. Based on the facts that the free-radical polymerization of ethene is spontaneous and that polymer molecules are less disorganized than the starting monomers, decide whether the polymerization reaction is exothermic or endothermic. Explain.

25. According to a trade journal, approximately 950 million lb of 1,2-dichloroethane was exported from the United States in a recent year. (The common name of this substance is ethylene dichloride.) The article states that "between 500 million and 550 million pounds of PVC could have been made from that ethylene dichloride." Compute the range of percentage yields of PVC from 1,2-dichloroethane that is implied by these figures.

26. The complete hydrogenation of natural rubber (the addition of H_2 to all double bonds) gives a product that is indistinguishable from the product of the complete hydrogenation of gutta-percha. Explain how this strengthens the conclusion that these two substances are isomers of each other.

27. A reducing solution breaks S—S bonds in proteins, whereas an oxidizing solution allows them to re-form. Discuss how such solutions might be used to carry out the curling of hair.

28. L-Sucrose tastes sweet, but it is not metabolized. It has been suggested as a potential nonnutritive sweetener. Draw the molecular structure of L-sucrose, using Figure 23.16b as a starting point.

29. Polypeptides are synthesized from a 50:50 mixture of L-alanine and D-alanine. How many different isomeric molecules containing 22 monomer units are possible?

30. An osmotic pressure measurement taken on a solution containing hemoglobin shows that the molar mass of that protein is approximately 65,000 g mol^{-1}. A chemical analysis shows it to contain 0.344% of iron by mass. How many iron atoms does each hemoglobin molecule contain?

***31.** At very low pH, alanine is a diprotic acid that can be represented as $H_3N^+{-}CH(CH_3){-}COOH$. The pK_a of the carboxyl group is 2.3, and the pK_a of the $-NH_3^+$ group is 9.7.

(a) At pH 7, what fraction of the amino acid molecules dissolved in an aqueous solution will have the form $H_3N^+{-}CH(CH_3){-}COO^-$?

(b) What fraction of the molecules at this pH will have the form $H_2N{-}CH(CH_3){-}COOH$?

32. The sequence of bases in one strand of DNA reads ACTTGACCG. Write the sequence of bases in the complementary strand.

33. Nucleic acids are diesters of phosphoric acid. Esters are not usually acidic. Why are nucleic acids acidic, or is this name inappropriate?

34. The codons in the genetic code are sequences of three bases. Explain why sequences of only two bases could not be used to code for the 20 different amino acids commonly found in proteins.

35. The average distance between base pairs measured parallel to the axis of a double-helical DNA molecule is 3.4 angstroms. The average molecular weight of a pair of nucleotides is about 650 g mol^{-1}. What is the approximate length in millimeters of a single DNA molecule of molecular weight 2.8×10^9 g mol^{-1} (a value typical for the DNA of some bacteria)? About how many base pairs does this DNA contain?

Appendices

A crystal of elemental bismuth.

APPENDIX

A Scientific Notation and Experimental Error

A.1 Scientific Notation

Very large and very small numbers are common in chemistry. Repeatedly writing such numbers in the ordinary way (for example, the important number 602,214,100,000,000,000,000,000) would be tedious and would engender errors. *Scientific notation* offers a better way. A number in scientific notation is expressed as a number from 1 to 10 multiplied by 10 raised to some power. Any number can be represented in this way, as the following examples show.

$$643.8 = 6.438 \times 10^2$$

$$-19{,}000{,}000 = -1.9 \times 10^7$$

$$0.0236 = 2.36 \times 10^{-2}$$

$$602{,}214{,}100{,}000{,}000{,}000{,}000{,}000 = 6.022141 \times 10^{23}$$

A simple rule of thumb is that the power to which 10 is raised is n if the decimal point is moved n places to the left and is $-n$ if the decimal is moved n places to the right.

When two or more numbers written in scientific notation are to be added or subtracted, they should first be expressed as multiples of the *same* power of 10:

$$\begin{array}{rcr} 6.431 \times 10^4 & \longrightarrow & 6.431 \times 10^4 \\ +2.1 \times 10^2 & \longrightarrow & +0.021 \times 10^4 \\ +3.67 \times 10^3 & \longrightarrow & +0.367 \times 10^4 \\ \hline ? & & 6.819 \times 10^4 \end{array}$$

When two numbers in scientific notation are multiplied, the coefficients are multiplied and then the powers of 10 are multiplied (by adding the exponents):

$$1.38 \times 10^{-16} \times 8.80 \times 10^3 = (1.38 \times 8.80) \times 10^{(-16+3)}$$

$$= 12.1 \times 10^{-13} = 1.21 \times 10^{-12}$$

We divide one number by a second by dividing the coefficients and then multiplying by 10 raised to the first exponent minus the second (exponents are subtracted):

$$\frac{6.63 \times 10^{-27}}{2.34 \times 10^{-16}} = \frac{6.63}{2.34} \times \frac{10^{-27}}{10^{-16}}$$

$$= 2.83 \times 10^{[-27-(-16)]} = 2.83 \times 10^{-11}$$

Any calculator or computer equipped to perform scientific and engineering calculations can accept and display numbers in scientific notation. It cannot determine whether the input has an error, however, or whether the answer makes sense. That is your responsibility! Develop the habit of mentally estimating the order of magnitude of the answer as a rough check on your calculator's result.

A.2 Experimental Error

Chemistry is an experimental science in which every quantitative measurement is subject to some degree of error. We can seek to reduce error by carrying out additional measurements or by changing our experimental apparatus, but we can never eliminate error altogether. It is important, therefore, to be able to assess the results of an experiment quantitatively to establish the limits of the experiment's validity. Errors are of two types: random (lack of precision) and systematic (lack of accuracy).

A.2.1 Precision and Random Errors

Precision refers to the degree of agreement in a collection of experimental results and is estimated by repeating the measurement under conditions as nearly identical as possible. If the conditions are truly identical, then differences among the trials are due to random error. As a specific example, consider some actual results of an early, important experiment by American physicist Robert Millikan in 1909, to measure the magnitude of the charge e on the electron. The experiment (discussed in greater detail in Chapter 1) involved a study of the motion of charged oil drops suspended in air in an electric field. Millikan made hundreds of measurements on many different oil drops, but we shall consider only a set of results for e found for one particular drop (Table A.1). The values he found ranged from 4.894 to 4.941×10^{-10} esu. What do we choose to report as the best estimate for e? The proper procedure is to first examine the data to see whether any of the results are especially far from the rest (a value above 5×10^{-10} esu would fall into this category). Such values are likely to result from some mistake in carrying out or reporting that particular measurement and therefore are excluded from further consideration (although there have been cases in science where just such exceptional results have led to significant breakthroughs). In Millikan's data, no such points should be excluded. To obtain our best estimate for e, we calculate the **average value** (or *mean*) by adding up the values found and dividing by the number of measurements. We can write the average value of any property after a series of N measurements $x_1, x_2, \ldots, x_N$ as

$$\overline{x} = \frac{1}{N}(x_1 + x_2 + \cdots + x_N) = \frac{1}{N}\sum_{i=1}^{N} x_i$$

where a capital Greek sigma (Σ) is introduced to indicate a summation of x_i over values of i from 1 to N. In the present case, this gives an average for e of 4.917×10^{-10} esu.

This average by itself does not convey any estimate of uncertainty. If all of the measurements had given results between 4.91×10^{-10} and 4.92×10^{-10} esu, the uncertainty would be less than if the results had ranged from 4×10^{-10} to 6×10^{-10} esu. Furthermore, an average of 100 measurements should have less uncertainty than an average of 5. How are these ideas made quantitative?

TABLE A.1

Measurement Number	1	2	3	4	5	6	7	8	9	10	11	12	13
e (10^{-10} esu)	4.915	4.920	4.937	4.923	4.931	4.936	4.941	4.902	4.927	4.900	4.904	4.897	4.894

From R. A. Millikan, *Phys. Rev.* 32:349, 1911. [1 esu = 3.3356×10^{-10} C]

A statistical measure of the spread of data, called the **standard deviation** σ, is useful in this regard. It is given by the formula

$$\sigma = \sqrt{\frac{(x_1 - \bar{x})^2 + (x_2 - \bar{x})^2 + \cdots + (x_N - \bar{x})^2}{N - 1}}$$

$$= \sqrt{\frac{1}{N-1}\sum_{i=1}^{N}(x_i - \bar{x})^2}$$

The standard deviation is found by adding up the squares of the deviations of the individual data points from the average value $\bar{x}$, dividing by $N - 1$, and taking the square root. Table A.2 shows how σ is used quantitatively. A *confidence limit* is defined as

$$\text{confidence limit} = \pm\frac{t\sigma}{\sqrt{N}}$$

The table gives the factor t for various numbers of measurements, N, and for various levels of confidence.

For Millikan's data, $N = 13$ and $\sigma = 0.017 \times 10^{-10}$. For 95% confidence with 13 measurements, the table shows $t = 2.18$ and the confidence limit is

$$\text{confidence limit} = \pm\frac{(2.18)(0.017 \times 10^{-10})}{\sqrt{13}} = \pm 0.010 \times 10^{-10} \text{ esu}$$

Thus, a 95% probability exists that the *true* average (obtained by repeating the experiment under the same conditions an infinite number of times) will lie within $\pm 0.010 \times 10^{-10}$ esu of the average 4.917×10^{-10} esu. Within this 95% confidence level, our best estimate for e is written as

$$(4.917 \pm 0.010) \times 10^{-10} \text{ esu}$$

For other confidence levels and other numbers of measurements, the factor t and therefore the confidence limit change.

TABLE A.2

N	Factor *t* for Confidence Interval of			
(Number of Observations)	80%	90%	95%	99%
2	3.08	6.31	12.7	63.7
3	1.89	2.92	4.30	9.92
4	1.64	2.35	3.18	5.84
5	1.53	2.13	2.78	4.60
6	1.48	2.02	2.57	4.03
7	1.44	1.94	2.45	3.71
8	1.42	1.90	2.36	3.50
9	1.40	1.86	2.31	3.36
10	1.38	1.83	2.26	3.25
11	1.37	1.81	2.23	3.17
12	1.36	1.80	2.20	3.11
13	1.36	1.78	2.18	3.06
14	1.35	1.77	2.16	3.01
15	1.34	1.76	2.14	2.98
∞	1.29	1.64	1.96	2.58

A.2.2 Accuracy and Systematic Errors

The magnitude of the charge e on the electron has been measured by several different techniques since Millikan's day. The current best estimate for e is

$$e = (4.80320452 \pm 0.0000011) \times 10^{-10} \text{ esu}$$
$$= (1.60217657 \pm 0.00000035) \times 10^{-19} \text{ C}$$

This value lies outside the range of uncertainty we estimated from Millikan's original data. In fact, it lies well below the smallest of the 13 measurements of e. Why?

To understand this discrepancy, we need to remember that there is a second source of error in any experiment: *systematic* error that causes a shift in the measured values from the true value and reduces the **accuracy of the result.** By making more measurements, we can reduce the uncertainty due to *random* errors and improve the *precision* of our result; however, if systematic errors are present, the average value will continue to deviate from the true value. Such systematic errors may result from a miscalibration of the experimental apparatus or from a fundamental inadequacy in the technique for measuring a property. In the case of Millikan's experiment, the then-accepted value for the viscosity of air (used in calculating the charge e) was subsequently found to be wrong. This caused his results to be systematically too high.

Error thus arises from two sources. Lack of precision (random errors) can be estimated by a statistical analysis of a series of measurements. Lack of accuracy (systematic errors) is much more problematic. If a systematic error is known to be present, we should do our best to correct for it before reporting the result. (For example, if our apparatus has not been calibrated correctly, it should be recalibrated.) The problem is that systematic errors of which we have no knowledge may be present. In this case the experiment should be repeated with different apparatus to eliminate the systematic error caused by a particular piece of equipment; better still, a different and independent way to measure the property might be devised. Only after enough independent experimental data are available can we be convinced of the accuracy of a result—that is, how closely it approximates the true result.

A.3 Significant Figures

The number of *significant figures* is the number of digits used to express a measured or calculated quantity, excluding zeros that may precede the first nonzero digit. Suppose the mass of a sample of sodium chloride is measured to be 8.241 g and the uncertainty is estimated to be ±0.001 g. The mass is said to be given to four significant figures because we are confident of the first three digits (8, 2, 4) and the uncertainty appears in the fourth (1), which nevertheless is still significant. Writing additional digits beyond the 1 would not be justified, however, unless the accuracy of the weighing could be improved. When we record a volume as 22.4 L, we imply that the uncertainty in the measurement is in the last digit written ($V = 22.4 \pm 0.3$ L, for example). A volume written as 22.43 L, on the other hand, implies that the uncertainty is far less and appears only in the fourth significant figure.

In the same way, writing 20.000 m is quite different from writing 20.0 m. The second measurement (with three significant figures) could easily be made with a common meterstick. The first (with five significant figures) would require a more precise method. We should avoid reporting results such as "700 m," however, because the two trailing zeros may or may not be significant. The uncertainty in the measurement could be of order ±1 m or ±10 m or perhaps ±100 m; it is impossible to tell which without further information. To avoid this ambiguity, we can write such measurements using the scientific notation described in Section A.1. The measurement "700 m" translates into any of the following:

7.00×10^2 m	Three significant figures
7.0×10^2 m	Two significant figures
7×10^2 m	One significant figure

Frequently, it is necessary to combine several different experimental measurements to obtain a final result. Some operations involve addition or subtraction, and others entail multiplication or division. These operations affect the number of significant figures that should be retained in the calculated result. Suppose, for example, that a weighed sample of 8.241 g of sodium chloride is dissolved in 160.1 g of water. What will be the mass of the solution that results? It is tempting to simply write 160.1 + 8.241 = 168.341 g, but this is *not* correct. In saying that the mass of water is 160.1 g, we imply that there is some uncertainty about the number of tenths of a gram measured. This uncertainty must also apply to the sum of the masses, so the last two digits in the sum are not significant and should be *rounded off*, leaving 168.3 as the final result.

Following addition or subtraction, round off the result to the leftmost decimal place that contained an uncertain digit in the original numbers.

Rounding off is a straightforward operation. It consists of first discarding the digits that are not significant and then adjusting the last remaining digit. If the first discarded digit is less than 5, the remaining digits are left as they are (for example, 168.341 is rounded down to 168.3 because the first discarded digit, 4, is less than 5). If the first discarded digit is greater than 5, or if it is equal to 5 and is followed by one or more nonzero digits, then the last digit is increased by 1 (for example, 168.364 and 168.3503 both become 168.4 when rounded off to four digits). Finally, if the first digit discarded is 5 and all subsequent digits are zeros, the last digit remaining is rounded to the nearest even digit (for example, both 168.35 and 168.45 would be rounded to 168.4). This last rule is chosen so that, on the average, as many numbers are rounded up as down. Other conventions are sometimes used.

In multiplication or division it is not the number of decimal places that matters (as in addition or subtraction) but the number of significant figures in the least precisely known quantity. Suppose, for example, the measured volume of a sample is 4.34 cm^3 and its mass is 8.241 g. The density, found by dividing the mass by the volume on a calculator, for example, is

$$\frac{8.241 \text{ g}}{4.34 \text{ cm}^3} = 1.89884 \ldots \text{ g cm}^{-3}$$

How many significant figures should we report? Because the volume is the less precisely known quantity (three significant figures as opposed to four for the mass), it controls the precision that may properly be reported in the answer. Only three significant figures are justified, so the result is rounded to 1.90 g cm^{-3}.

The number of significant figures in the result of a multiplication or division is the smallest of the numbers of significant figures used as input.

It is best to carry out the arithmetical operations and *then* round the final answer to the correct number of significant figures, rather than round off the input data first. The difference is usually small, but this recommendation is nevertheless worth following. For example, the correct way to add the three distances 15 m, 6.6 m, and 12.6 m is

$$\begin{array}{r} 15\ \ \text{ m} \\ +\ 6.6 \text{ m} \\ +12.6 \text{ m} \\ \hline 34.2 \text{ m} \end{array} \longrightarrow 34 \text{ m} \qquad \text{rather than} \qquad \begin{array}{r} 15 \text{ m} \longrightarrow 15 \text{ m} \\ 6.6 \text{ m} \longrightarrow \ \ 7 \text{ m} \\ 12.6 \text{ m} \longrightarrow 13 \text{ m} \\ \hline 35 \text{ m} \end{array}$$

For the same reason, we frequently carry extra digits through the intermediate steps of a worked example and round off only for the final answer. If a calculation is done entirely on a scientific calculator or a computer, several extra digits are usually carried along automatically. Before the final answer is reported, however, it is important to round off to the proper number of significant figures.

Sometimes pure constants appear in expressions. In this case the accuracy of the result is determined by the accuracy of the other factors. The uncertainty in the volume of a sphere, $\frac{4}{3}\pi r^3$, depends only on the uncertainty in the radius r; 4 and 3 are pure constants (4.000 . . . and 3.000 . . . , respectively), and π can be given to as many significant figures (3.14159265 . . .) as are warranted by the radius.

PROBLEMS

Answers to problems whose numbers are red appear in Appendix G.

A.1 Scientific Notation

1. Express the following in scientific notation.
   - (a) 0.0000582
   - (b) 402
   - (c) 7.93
   - (d) −6593.00
   - (e) 0.002530
   - (f) 1.47

2. Express the following in scientific notation.
   - (a) 4579
   - (b) −0.05020
   - (c) 2134.560
   - (d) 3.825
   - (e) 0.0000450
   - (f) 9.814

3. Convert the following from scientific notation to decimal form.
   - (a) 5.37×10^{-4}
   - (b) 9.390×10^{6}
   - (c) -2.47×10^{-3}
   - (d) 6.020×10^{-3}
   - (e) 2×10^{4}

4. Convert the following from scientific notation to decimal form.
   - (a) 3.333×10^{-3}
   - (b) -1.20×10^{7}
   - (c) 2.79×10^{-5}
   - (d) 3×10^{1}
   - (e) 6.700×10^{-2}

5. A certain chemical plant produces 7.46×10^{8} kg of polyethylene in one year. Express this amount in decimal form.

6. A microorganism contains 0.0000046 g of vanadium. Express this amount in scientific notation.

A.2 Experimental Error

7. A group of students took turns using a laboratory balance to weigh the water contained in a beaker. The results they reported were 111.42 g, 111.67 g, 111.21 g, 135.64 g, 111.02 g, 111.29 g, and 111.42 g.
   - (a) Should any of the data be excluded before the average is calculated?
   - (b) From the remaining measurements, calculate the average value of the mass of the water in the beaker.
   - (c) Calculate the standard deviation σ and, from it, the 95% confidence limit.

8. By measuring the sides of a small box, a group of students made the following estimates for its volume: 544 cm^3, 590 cm^3, 523 cm^3, 560 cm^3, 519 cm^3, 570 cm^3, and 578 cm^3.
   - (a) Should any of the data be excluded before the average is calculated?
   - (b) Calculate the average value of the volume of the box from the remaining measurements.
   - (c) Calculate the standard deviation σ and, from it, the 90% confidence limit.

9. Of the measurements in problems 7 and 8, which is more precise?

10. A more accurate determination of the mass in problem 7 (using a better balance) gives the value 104.67 g, and an accurate determination of the volume in problem 8 gives the value 553 cm^3. Which of the two measurements in problems 7 and 8 is more accurate, in the sense of having the smaller systematic error relative to the actual value?

A.3 Significant Figures

11. State the number of significant figures in each of the following measurements.
(a) 13.604 L
(b) −0.00345°C
(c) 340 lb
(d) 3.40×10^{2} miles
(e) 6.248×10^{-27} J

12. State the number of significant figures in each of the following measurements.
(a) −0.0025 in
(b) 7000 g
(c) 143.7902 s
(d) 2.670×10^{7} Pa
(e) 2.05×10^{-19} J

13. Round off each of the measurements in problem 11 to two significant figures.

14. Round off each of the measurements in problem 12 to two significant figures.

15. Round off the measured number 2,997,215.548 to nine significant digits.

16. Round off the measured number in problem 15 to eight, seven, six, five, four, three, two, and one significant digits.

17. Express the results of the following additions and subtractions to the proper number of significant figures. All of the numbers are measured quantities.
(a) $67.314 + 8.63 - 243.198 =$
(b) $4.31 + 64 + 7.19 =$
(c) $3.1256 \times 10^{15} - 4.631 \times 10^{13} =$
(d) $2.41 \times 10^{-26} - 7.83 \times 10^{-25} =$

18. Express the results of the following additions and subtractions to the proper number of significant figures. All of the numbers are measured quantities.
(a) $245.876 + 4.65 + 0.3678 =$
(b) $798.36 - 1005.7 + 129.652 =$
(c) $7.98 \times 10^{17} + 6.472 \times 10^{19} =$
(d) $(4.32 \times 10^{-15}) + (6.257 \times 10^{-14}) - (2.136 \times 10^{-13}) =$

19. Express the results of the following multiplications and divisions to the proper number of significant figures. All of the numbers are measured quantities.
(a) $\dfrac{-72.415}{8.62} =$
(b) $52.814 \times 0.00279 =$
(c) $(7.023 \times 10^{14}) \times (4.62 \times 10^{-27}) =$
(d) $\dfrac{4.3 \times 10^{-12}}{9.632 \times 10^{-26}} =$

20. Express the results of the following multiplications and divisions to the proper number of significant figures. All of the numbers are measured quantities.
(a) $129.587 \times 32.33 =$
(b) $\dfrac{4.7791}{3.21 \times 5.793} =$
(c) $\dfrac{10566.9}{3.584 \times 10^{29}} =$
(d) $(5.247 \times 10^{13}) \times (1.3 \times 10^{-17}) =$

21. Compute the area of a triangle (according to the formula $A = \frac{1}{2}ba$) if its base and altitude are measured to equal 42.07 cm and 16.0 cm, respectively. Justify the number of significant figures in the answer.

22. An inch is defined as exactly 2.54 cm. The length of a table is measured as 505.16 cm. Compute the length of the table in inches. Justify the number of significant figures in the answer.

APPENDIX B

SI Units, Unit Conversions, and Physics for General Chemistry

This appendix reviews essential concepts of physics, as well as systems of units of measure, essential for work in general chemistry.

B.1 SI Units and Unit Conversions

Scientific work requires measurement of quantities or properties observed in the laboratory. Results are expressed not as pure numbers but rather as **dimensions** that reflect the nature of the property under study. For example, mass, time, length, area, volume, energy, and temperature are fundamentally distinct quantities, each of which is characterized by its unique dimension. The magnitude of each dimensioned quantity can be expressed in various *units* (for example, feet or centimeters for length). Several systems of units are available for use, and facility at conversion among them is essential for scientific work. Over the course of history, different countries evolved different sets of units to express length, mass, and many other physical dimensions. Gradually, these diverse sets of units are being replaced by international standards that facilitate comparison of measurements made in different localities and that help avoid complications and confusion. The unified system of units recommended by international agreement is called SI, which stands for "Système International d'Unités," or the International System of Units. In this section we outline the use of SI units and discuss interconversions with other systems of units.

The SI uses seven *base units*, which are listed in Table B.1. All other units can be written as combinations of the base units. In writing the units for a measurement, we abbreviate them (see Table B.1), and we use exponential notation to denote the power to which a unit is raised; a minus sign appears in the exponent when the unit is in the denominator. For example, velocity is a quantity with dimensions of length divided by time, so in SI it is expressed in meters per

TABLE B.1

Base Units in the International System of Units

Quantity	Unit	Symbol
Length	meter	m
Mass	kilogram	kg
Time	second	s
Temperature	kelvin	K
Number of moles (of substance)	mole	mol
Electric current	ampere	A
Luminous intensity	candela	cd

TABLE B.2

Derived Units in SI

Quantity	Unit	Symbol	Definition
Energy	joule	J	$kg\ m^2\ s^{-2}$
Force	newton	N	$kg\ m\ s^{-2} = J\ m^{-1}$
Power	watt	W	$kg\ m^2\ s^{-3} = J\ s^{-1}$
Pressure	pascal	Pa	$kg\ m^{-1}\ s^{-2} = N\ m^{-2}$
Electric charge	coulomb	C	A s
Electric potential difference	volt	V	$kg\ m^2\ s^{-3}\ A^{-1} = J\ C^{-1}$

second, or $m\ s^{-1}$. Some derived units that are used frequently have special names. Energy, for example, is the product of mass and the square of the velocity. Therefore, it is measured in units of kilogram square meters per square seconds ($kg\ m^2\ s^{-2}$), and 1 $kg\ m^2\ s^{-2}$ is called a *joule*. Other derived units, such as the pascal for measuring pressure, appear in Table B.2. Although these names provide a useful shorthand, it is important to remember their meanings in terms of the base units.

Because scientists work on scales ranging from the microscopic to the astronomical, there is a tremendous range in the magnitudes of measured quantities. Consequently, a set of **prefixes** has been incorporated into the International System of Units to simplify the description of small and large quantities (Table B.3). The prefixes specify various powers of 10 times the base and derived units. Some of them are quite familiar in everyday use: the kilometer, for example, is 10^3 m. Others may sound less familiar—for instance, the femtosecond (1 fs = 10^{-15} s) or the gigapascal (1 GPa = 10^9 Pa).

In addition to base and derived SI units, several units that are not officially sanctioned are used in this book. The first is the liter (abbreviated L), a very convenient size for volume measurements in chemistry; a liter is $10^{-3}\ m^3$, or 1 cubic decimeter (dm^3):

$$1\ L = 1\ dm^3 = 10^{-3}\ m^3 = 10^3\ cm^3$$

Second, we use the *angstrom* (abbreviated Å) as a unit of length for atoms and molecules:

$$1\ \text{Å} = 10^{-10}\ m = 100\ pm = 0.1\ nm$$

TABLE B.3

Prefixes in SI

Fraction	Prefix	Symbol	Factor	Prefix	Symbol
10^{-1}	*deci-*	d	10	*deca-*	da
10^{-2}	*centi-*	c	10^2	*hecto-*	h
10^{-3}	*milli-*	m	10^3	*kilo-*	k
10^{-6}	*micro-*	μ	10^6	*mega-*	M
10^{-9}	*nano-*	n	10^9	*giga-*	G
10^{-12}	*pico-*	p	10^{12}	*tera-*	T
10^{-15}	*femto-*	f			
10^{-18}	*atto-*	a			

This unit is used simply because most atomic sizes and chemical bond lengths fall in the range of one to several angstroms, and the use of either picometers or nanometers is slightly awkward. Next, we use the *atmosphere* (abbreviated atm) as a unit of pressure. It is not a simple power of 10 times the SI unit of the pascal, but rather is defined as follows:

$$1 \text{ atm} = 101{,}325 \text{ Pa} = 0.101325 \text{ MPa}$$

This unit is used because most chemistry procedures are carried out at pressures near atmospheric pressure, for which the pascal is too small a unit to be convenient. In addition, expressions for equilibrium constants (see Chapter 14) are simplified when pressures are expressed in atmospheres.

The non-SI temperature units require special mention. The two most important temperature scales in the United States are the Fahrenheit scale and the Celsius scale, which employ the Fahrenheit degree (°F) and the Celsius degree (°C), respectively. The *size* of the Celsius degree is the same as that of the SI temperature unit, the kelvin, but the two scales are shifted relative to each other by 273.15°C:

$$T_K = \frac{1 \text{ K}}{1°\text{C}}(t°\text{C}) + 273.15 \text{ K}$$

The size of the degree Fahrenheit is $\frac{5}{9}$ the size of the Celsius degree. The two scales are related by

$$t°\text{C} = \frac{5°\text{C}}{9°\text{F}}(t°\text{F} - 32°\text{F}) \qquad \text{or} \qquad t°\text{F} = \frac{9°\text{F}}{5°\text{C}}(t°\text{C}) + 32°\text{F}$$

The advantage of a unified system of units is that if all the quantities in a calculation are expressed in SI units, the final result must come out in SI units. Nevertheless, it is important to become familiar with the ways in which units are interconverted because units other than SI base units often appear in calculations. The *unit conversion method* provides a systematic approach to this problem.

As a simple example, suppose the mass of a sample is measured to be 64.3 g. If this is to be used in a formula involving SI units, it should be converted to kilograms (the SI base unit of mass). To do this, we use the fact that 1 kg = 1000 g and write

$$\frac{64.3 \text{ g}}{1000 \text{ g kg}^{-1}} = 0.0643 \text{ kg}$$

Note that this is, in effect, division by 1; because 1000 g = 1 kg, 1000 g kg^{-1} = 1, and we cancel units between numerator and denominator to obtain the final result. This unit conversion could also be written as

$$64.3 \text{ g}\left(\frac{1 \text{ kg}}{1000 \text{ g}}\right) = 0.0643 \text{ kg}$$

In this book we use the more compact first version of the unit conversion. Instead of *dividing* by 1000 g kg^{-1}, we can equally well *multiply* by 1 = 10^{-3} kg g^{-1}:

$$(64.3 \text{ g})(10^{-3} \text{ kg g}^{-1}) = 0.0643 \text{ kg}$$

Other unit conversions may involve more than just powers of 10, but they are equally easy to carry out. For example, to express 16.4 inches in meters, we use the fact that 1 inch = 0.0254 m (or 1 = 0.0254 m inch^{-1}), so

$$(16.4 \text{ inches})(0.0254 \text{ m inch}^{-1}) = 0.417 \text{ m}$$

More complicated combinations are possible. For example, to convert from liter-atmospheres to joules (the SI unit of energy), two separate unit conversions are used:

$$(1 \text{ L atm})(10^{-3} \text{ m}^3 \text{ L}^{-1})(101{,}325 \text{ Pa atm}^{-1}) = 101.325 \text{ kg m}^2 \text{ s}^{-2}$$
$$= 101.325 \text{ Pa m}^3 = 101.325 \text{ J}$$

When doing chemical calculations, it is very important to write out units explicitly and cancel units in intermediate steps to obtain the correct units for the final result. This practice is a way of checking to make sure that units have not been incorrectly mixed without unit conversions or that an incorrect formula has not been used. If a result that is supposed to be a temperature comes out with units of $\text{m}^3 \text{ s}^{-2}$, then a mistake has been made!

B.2 Background in Physics

Although physics and chemistry are distinct sciences with distinct objectives and methods of investigation, concepts from one can aid investigations in the other. Physical reasoning aids chemical understanding in cases where applied forces move chemical systems to new positions—or change their sizes and shapes—and change their energy. These energy changes can have chemical consequences, as shown in two specific examples. First, the total energy content of a system increases when the system is compressed by externally applied pressure (if the system is thermally isolated). Second, when significant forces exist between molecules, the mutual energy of a pair of molecules changes as the molecules are pushed closer together. Intermolecular forces profoundly influence the organization of matter into solid, liquid, and gaseous states, as well as the effectiveness of molecular collisions in causing chemical reactions. Specific illustrations appear at many places in this book. This section reviews general aspects of motion, forces, and energy as background for these specific applications. For simplicity, we limit the discussion to a *point mass*, which is an object characterized only by its total mass m. We do not inquire into the internal structure of the object. In various contexts, we represent planets, projectiles, molecules, atoms, nuclei, or electrons by this model to predict their response to applied forces.

B.2.1 Describing the Motion of an Object: Speed, Velocity, Momentum, and Acceleration

If you are not already familiar with taking the derivative df/dx of a function $f(x)$, read Appendix C.4 before reading on in this section. Likewise, if you are not familiar with a vector $\vec{v}$ and how to write it in terms of its components v_x and v_y, read Appendix C.7 before reading on in this section.

Let's do some examples of the motion of a ball to familiarize you with a bit of physics, and how vectors (Appendix C.7) and derivatives of functions (Appendix C.4) are used to solve simple physical problems. We'll start with motion in a straight line. You are surely familiar with speed = (distance traveled)/(time spent traveling). Imagine standing at $x = 2$ ft of a Cartesian coordinate system and rolling a ball towards your friend, who is standing at $x = 17.4$ ft (a positive coordinate, so your friend is in the $+x$ direction). You could easily calculate the speed you'd have to roll the ball in order for it to reach your friend in 3.5 seconds (if it did not slow down en route). That would be speed $= (x_f - x_i)/\Delta t =$ (17.4 ft − 2.0 ft)/(3.5 s) = (15.4 ft)/(3.5 s) = 4.4 ft s^{-1} = 3 miles per hour. If your friend were standing at the position $x = -13.4$ ft (at a position on the negative side of the x-axis), you could do the same calculation, but you would not write speed $= (x_f - x_i)/\Delta t = (-13.4 \text{ ft} - 2 \text{ ft})/(3.5 \text{ s})$ or you would get a negative speed! So the distance traveled is not $x_f - x_i$; that is called the *displacement* of the ball, or its change in position on the x-axis, and it is negative. The distance traveled is the magnitude of $(x_f - x_i)$, or the absolute value of $x_f - x_i$, $|x_f - x_i|$. But you have lost some information in this calculation, as you know you have to roll the ball in the $-x$ direction if you are standing at $x = 2$ ft and your friend is standing at $x = -13.4$ ft. In an instance like this, when the quantity you are interested in has both magnitude (the speed of the ball) and direction (you must roll it in the $-x$ direction), vectors become useful! The next paragraph describes velocity vectors.

To describe the direction of travel, let's write the **velocity** of the ball as $\vec{v} = (v_x, v_y)$, where v_x is called the x-component of the velocity and v_y is called the y-component of the velocity, and show the speed is the length of the velocity vector. If you point the y-axis of your coordinate system straight upwards, then of course the y-component of the velocity is 0 since you are rolling the ball on a level field in the $-x$ direction. The x component of the velocity $v_x = (x_f - x_i)/\Delta t = (-13.4 \text{ ft} - 2 \text{ ft})/(3.5 \text{ s}) = -4.4 \text{ ft s}^{-1} = -3$ miles per hour. Instead of writing $\vec{v} = (v_x, v_y) = (-4.4 \text{ ft s}^{-1}, 0)$, you could alternatively write $\vec{v} = v_x\hat{i} + v_y\hat{j} = (-4.4 \text{ ft s}^{-1})\hat{i} + 0\hat{j} = (-4.4 \text{ ft s}^{-1})\hat{i}$, where $\hat{i}$ is the unit vector (1,0) in Cartesian coordinate space. Either way of writing the velocity $\vec{v}$ shows a negative value for v_x, indicating that it is traveling in the negative direction along the x-axis, towards your friend at $x = -13.4$ ft, and the speed is simply $\|\vec{v}\|$ = length of the velocity vector $\vec{v}$. (Note: In this text we use single bars $|\vec{v}|$ to indicate the length of a vector, and the absolute value of a scalar.) By the Pythagorean theorem,

$$\text{speed} = \text{length of the velocity vector } \vec{v} = |\vec{v}| = \sqrt{(v_x)^2 + (v_y)^2}$$

Now, if the ball is slowing down en route, you have to distinguish between the instantaneous speed and the **average speed**, and the instantaneous velocity and the **average velocity**. They are equal only if the ball maintains the same speed throughout, and does not change direction. Generalizing the above example to the possibility that either the speed or the direction of travel might change, the *x-component of the velocity* and the *y-component of the velocity* are, more generally,

$$v_x = \frac{dx}{dt} \quad \text{and} \quad v_y = \frac{dx}{dt}$$

If you start your ball at $x_i = 2$ ft and impart it with an x-component of the velocity of -4.4 ft s^{-1}, you can calculate the final position of the ball after 3.5 seconds by integrating the above expressions:

$$\int_{x_i}^{x_f} dx = \int_{t=0}^{t=3.5\text{ s}} v_x dt \quad \text{and} \quad \int_{y_i}^{y_f} dy = \int_{t=0}^{t=3.5\text{ s}} v_y dt$$

$$x_f - x_i = \int_{t=0}^{t=3.5\text{ s}} v_x dt \quad \text{and} \quad y_f - y_i = \int_{t=0}^{t=3.5\text{ s}} v_y dt$$

Using $v_x = -4.4 \text{ ft s}^{-1}$ and the initial position of the ball at $x_i = 2$ ft gives $x_f = 2 \text{ ft} + \int_{t=0}^{t=3.5\text{ s}} (-4.4 \text{ ft s}^{-1})dt = 2 \text{ ft} + (-4.4 \text{ ft s}^{-1})(3.5 \text{ s}) = -13.4$ ft, so the ball reaches your friend standing at $x = -13.4$ in 3.5 seconds. If v_x were changing en route, you'd have to insert the functional form of $v_x(t)$ in the integral expression above to get the correct answer.

You are already familiar with accelerating or decelerating your car, that is, changing the speed of your car. But in chemistry and physics problems we have to be a bit more careful and talk about a change in velocity, not just a change in speed. Thus we define the **acceleration** $\vec{a}$ as a vector, because you can change a velocity vector either by changing its length (the speed) or by changing its direction. Just like the velocity vector, the acceleration vector has components, so you can write $\vec{a} = (a_x, a_y)$ where

$$a_x = \frac{dv_x}{dt} \qquad \text{and} \qquad a_y = \frac{dv_y}{dt}$$

Finally, we define the **linear momentum** of an object of mass m as $\vec{p} = m\vec{v}$, so the momentum has components $p_x = mv_x$ and $p_y = mv_y$. You are already familiar with conservation of momentum. Imagine a 100-lb female ice skater (call her particle A) pushing away from a 200-lb male ice skater (call him particle B) from a stationary initial position. If she is imparted an x-component of velocity $v_{x,\text{A}} =$ 6 mph, his speed would be slower, 3 miles per hour, and he would be pushed in the opposite direction, so his $v_{x,\text{B}} = -3$ mph. This reflects the law of *conservation of momentum*. Because initially neither were moving $p_{x,\text{initial}} = 0$ and at the end, after they have pushed back from one another, the sum of their two momenta is $p_{x,\text{final}} = p_{x,\text{final,A}} + p_{x,\text{final,B}} = m_\text{A}v_{x,\text{final,A}} + m_\text{B}v_{x,\text{final,B}} = (100 \text{ lb})(6 \text{ mph}) + (200 \text{ lb})(-3 \text{ mph})$, which sums to 0, so the total momentum is unchanged. As long as there are no external forces (see the next section) on these two ice skaters, the vector sum (adding by components, see Appendix C.7) of their momenta will be conserved.

Generalizing the definition of velocity $\vec{v}$, acceleration $\vec{a}$, and linear momentum $\vec{p}$ to three dimensions x, y, and z is easy. For example, in three dimensions the velocity vector $\vec{v} = (v_x, v_y, v_z)$, and its length, the speed, $= \sqrt{(v_x)^2 + (v_y)^2 + (v_y)^2}$.

B.2.2 Forces Change the Motion of an Object

Force is defined as the agent that changes the motion of an object; a force (push or pull on an object) has both magnitude and direction, so is given the vector symbol $\vec{F}$. This restatement of our daily experience that pushing or pulling an object causes it to move is the basis of Sir Isaac Newton's first law of motion:

> *Every object persists in a state of rest or of uniform motion in a straight line unless compelled to change that state by forces impressed upon it.*

We determine the properties of a force experimentally in the laboratory by measuring the consequences of applying the force. Suppose we have arbitrarily selected a standard test object of known mass m_1. We apply a force $\vec{F}_1$ and, by making the distance and time measurements described earlier, determine the acceleration $\vec{a}_1$ imparted to the object by the force. Experience shows that a different force imparts different acceleration to the test object. Forces can be ranked by the magnitude of the acceleration they impart to the standard test object. Now suppose we choose a second test object of mass m_2. Applying the original force $\vec{F}_1$ to this object, so the force on object 2, $\vec{F}_2$, is equal to the force you had previously exerted on object 1, $\vec{F}_1$, produces an acceleration of object 2, $\vec{a}_2$, which is different in magnitude from the acceleration $\vec{a}_1$ that it gave to the first test object. The less massive object has a larger magnitude acceleration. The results of such experiments are summarized in Newton's second law of motion:

> *The acceleration imparted to an object by an applied force has magnitude proportional to the magnitude of the force, is parallel to the direction of the force, and is inversely proportional to the mass of the object.*

In mathematical form this statement becomes

$$\vec{F} = m\vec{a} \quad \text{or} \quad \vec{a} = \vec{F}/m$$

This equation demonstrates that force must have dimensions of M L t^{-2} (mass × distance × time^{-2}). In SI units, force is expressed in newtons (N) where 1 N = 1 kg m s^{-2}. If you exert the same magnitude and direction of force on a child ice skater, of small mass, and on your heavy spouse, the child will experience a larger magnitude acceleration.

One familiar example is the force caused by gravity at the surface of the earth. Measurements show this force produces a downward acceleration (toward the center of the earth) of constant magnitude 9.80665 m s^{-2}, conventionally denoted by g, the gravitational constant. The magnitude of gravitational force exerted on a body of mass m is (writing the symbol F for the magnitude of the force $\vec{F}$)

$$F = mg$$

Another familiar example is the restoring force exerted on an object by a spring. In chemistry this is a useful model of the binding forces that keep atoms together in a chemical bond or near their "home" positions in a solid crystal. Imagine an object of mass m located on a smooth tabletop. The object is connected to one end of a coiled metallic spring; the other end of the spring is anchored to a post in the tabletop. At rest, the object is located at position x_0. Make your x-axis have its origin at the post and be directed toward the mass when it is at its rest position, so x_0 is a positive number. When $x > x_0$, the spring is stretched. When $x < x_0$, the spring is compressed. Now suppose the object is pulled away from the post in a straight line to a new position x beyond x_0 so the spring is stretched. Measurements show that at the stretched position x the magnitude of the x-component of the force exerted on the object by the spring is directly proportional to the magnitude of the displacement from the rest position:

$$|F_x| = k|(x - x_0)|$$

where the vertical bars around F_x and $(x - x_0)$ represent the magnitude or absolute value of each. In the preceding equation, k is a positive constant whose magnitude must be determined empirically in each case studied. Clearly, this force is directed back toward the rest position, and when the object is released, the recoiling spring accelerates the object back toward the rest position. When the object is displaced in the $+x$ direction from the rest position, so $x - x_0$ is positive, the restoring force points in the $-x$ direction; the x-component of the force is negative. Thus there is a minus sign in the equation for F_x, the x-component of the force

$$F_x = -k(x - x_0)$$

You can see that when the position of the mass is closer to the post (at the origin) than the rest position, $x < x_0$, the equation above is the product of a negative number, $-k$, and another negative number, $x - x_0$, so F_x has a positive value and the mass is pushed in the $+x$ direction. This is again towards its rest position, as the present position of the mass is closer to the origin than its rest position. When studying the dependence of force on position, we must pay careful attention to the coordinate system. The force and the acceleration at any point are parallel to each other, but the position of the particle (defined relative to some convenient origin of coordinates in each particular case) may point in a different direction. In applications concerned with the height of an object above the earth, the y-component of gravitational force is usually written as

$$F_y = -mg$$

to emphasize that the vertical displacement y is positive and pointing away from the surface of the earth, while the gravitational force is clearly directed toward the earth.

Once a force has been determined, the motion of the object under that force can be predicted from Newton's second law. If the x-component of the force, F_x, is constant then the x-component of the acceleration will also be constant. If the object was initially at rest at time $t = t_1$ (so $v_x(t_1) = 0$) and then is accelerated for a period of time starting at t_1 and ending at t_2, the x-component of the velocity at t_2 is given by

$$v_x(t_2) = a_x\Delta t + v_x(t_1) = a_x\Delta t$$

where $\Delta t = t_2 - t_1$. The position at t_2 is given by

$$x(t_2) = x(t_1) + \tfrac{1}{2}a_x(t_2 - t_1)^2$$

if the particle is initially at rest.

If the force depends on position, predicting the motion is slightly more complicated. One must then consider the instantaneous values of each component of the acceleration. For the x-component of the acceleration,

$$a_x = \frac{dv_x}{dt} \text{ which gives } v_x(t_2) = v_x(t_1) + \int_{t_1}^{t_2} a_x dt$$

You can see that you recover the simpler expressions above if a_x is a constant (because F_x is constant) and if $v_x(t_1) = 0$. However, when the acceleration is not constant, such as in the case of a mass attached to a spring, this gives a different result. Using $F_x = -k(x - x_0)$ gives a motion in which the object oscillates about the rest position x_0 with a period τ given by

$$\frac{1}{\tau} = \frac{1}{2\pi}\left(\frac{k}{m}\right)^{1/2}$$

The frequency ν of the oscillation (the number of cycles per unit time) is given by $\nu = \tau^{-1}$. If we represent the motion in terms of the angular frequency $\omega = 2\pi\nu$, which has dimensions radians s^{-1}, the position of the oscillating object is given by

$$x(t) - x_0 = A\cos(\omega t + \delta)$$

where the amplitude A and the phase factor δ are determined by the initial position and velocity of the object.

B.2.3 Forms of Energy

The concept of **energy** originated in the science of mechanics and was first defined as the capacity to perform work, that is, to move an object from one position to another. It is now understood that energy appears in many different forms, each of which can cause particular kinds of physical and chemical changes. From daily experience, we recognize the kinetic energy due to the speed of an onrushing automobile. To stop the automobile, its *kinetic energy* must be overcome by work performed by its brakes. Otherwise, the automobile will crash into other objects and expend its kinetic energy by deforming these objects, as well as itself. The *potential energy* of a mass of snow on a ski slope becomes the kinetic energy of an avalanche. The *electrical energy* stored in a battery can move objects by driving motors, or warm objects through electric heaters. The chemical energy stored in gasoline can move objects by powering an internal combustion engine. The *thermal energy* of hot steam can move objects by driving a steam engine.

Each of these forms of energy plays a role in chemistry, and each is described at the appropriate point in this textbook. Here we concentrate on the nature of potential and kinetic energy and their interconversion. The understanding we gain here is essential background for understanding these other forms of energy.

The **kinetic energy** of a moving object is defined by

$$\mathcal{T} = \frac{1}{2}mv^2$$

where m is the mass of the object and v is its speed. A stationary object has no kinetic energy. In the SI systems of units (see Appendix B.1), energy is expressed in joules (J). Thus, 0.5 joule is the kinetic energy of an object with mass 1 kg moving at a speed of 1 m s^{-1}. Experience shows that applying a force to a moving object changes the kinetic energy of the object. If the force is applied in the same direction as the velocity of the object, the kinetic energy is increased; if the force is opposed to the velocity, the kinetic energy is decreased.

Potential energy is the energy stored in an object due to its location relative to a specified reference position. Potential energy therefore depends explicitly on the position x of the object and is expressed as a mathematical function $V(x)$. The change in potential energy when an object is moved from position x_1 to position x_2 by a constant force is equal to the work done in moving the object:

$$\text{change in } V(x) = \text{force} \times \text{displacement} = \text{force} \times (x_2 - x_1)$$

To lift an object of mass m from the surface of the earth to the height h, some agency must perform work in the amount mgh against the downward-directed force of gravity, which has constant magnitude acceleration denoted by g. The change in potential energy is mgh. Potential energy is expressed in joules in the SI system of units.

B.2.4 Conservation of Energy

The science of mechanics deals with idealized motions of objects in which friction does not occur. The motions of an object interconvert its potential and kinetic energy subject to the restriction that their sum always remains constant. For example, consider a soccer ball rolling down the side of a steep gully at the edge of the playing field, and assume there is no friction between the ball and the surface on which it rolls. The ball rolls down one side, goes across the bottom, climbs partway up the opposite side, then stops and reverses direction. This pattern is repeated many times as the ball continues to oscillate back and forth across the bottom of the gully. On each downward leg of its journey, the ball loses potential energy and gains kinetic energy, but their sum remains constant. On each upward leg, the ball loses kinetic energy and gains potential energy, but their sum remains constant.

The description in the previous paragraph is clearly an idealization, because eventually the ball comes to rest at the bottom of the gully. In reality, the ball loses some of its kinetic energy through friction with the surface of the gully on each upward leg and on each downward leg of its journey. Both the ball and the gully surface become slightly warmer as a result. The energy lost from the purely mechanical motion is added to the *internal energy* of the ball and the gully surface. The total amount of energy has not changed; we have neither created nor destroyed energy in this process. Rather, we have identified a new mode of energy storage (called internal energy) to interpret the new effects beyond those explained by basic mechanics. Internal energy, heat, and friction are discussed in detail in Chapter 12 as part of the science of thermodynamics.

Similar arguments have extended the law of conservation of energy from the idealized motions of mechanics to include a broad variety of phenomena in which several different forms of energy are involved. This law is one of the securest building blocks in scientific reasoning. It provides the starting point for interpreting and relating a great variety of superficially different processes. In chemistry this law provides the foundation for studying complex processes in which kinetic, potential, electrical, chemical, and thermal energy are interconverted without net loss or gain.

B.2.5 Representing Energy Conservation by Potential Energy Curves

Consider again the soccer ball rolling down the walls of the gully. This process is shown schematically in Figure B.1, where the curve $V(x)$ suggests a "cross-sectional" sketch of the gully. The curve $V(x)$ actually represents the potential energy of the ball relative to its value at the bottom of the gully. The x coordinate locates the distance of the ball from the bottom of the gully to a position along its side. Suppose the ball is held in place at the position x_1. It possesses only potential energy so the total energy E, normally the sum of the kinetic energy and the potential energy $E = 1/2\ mv^2 + V(x)$ at the point x, is simply $E = V(x_1)$. At the point $x = x_1$ the speed of the ball is 0 so the kinetic energy is 0. If the ball is released, it falls down the slope and passes across the bottom, where its potential energy is zero and its kinetic energy is E. It then climbs the opposite side until it rises to position $-x_1$, where its kinetic energy is zero. The ball promptly reverses direction and retraces its path back to position x_1, where its total energy E is all potential energy.

B.2.6 Potential Energy Curves, Force, and Stability

One very important application of the potential energy function in this textbook is to provide a way to qualitatively predict the motion of an object without solving Newton's second law. The method determines the direction of the force applied to an object from knowledge of the potential energy curve as shown in Figure B.1. In preparation for this method, let's examine the potential energy curves for some familiar forces.

Let's consider a one-dimensional problem with just one particle here. A more detailed analysis is given in Section 3.3. The value of the potential energy of a particle located at the point x is always stated relative to its value at a specially selected reference point x_0. The value $V(x_0)$ is a constant and is usually assigned the value zero, and the point x_0 is frequently selected as the origin of coordinates for specifying the location of the object.

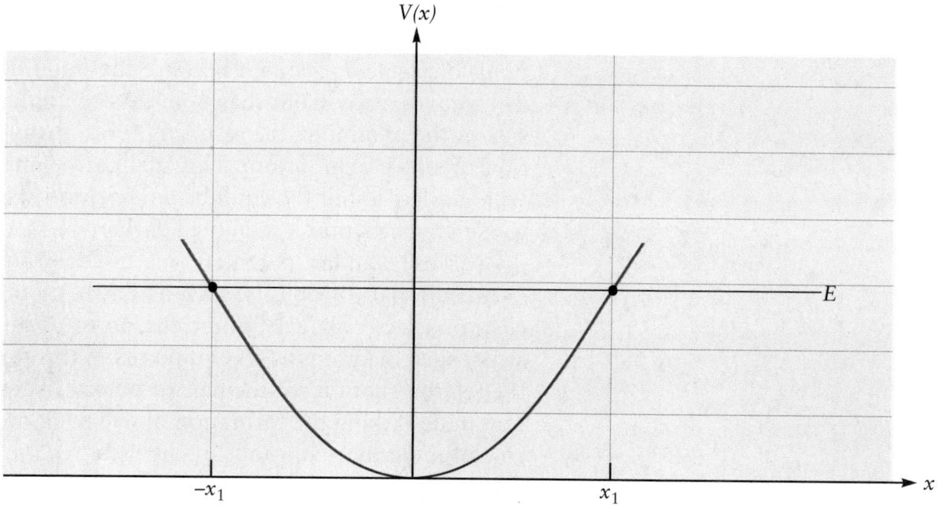

FIGURE B.1 Sketch of the potential energy of a soccer ball rolling down the sides of a gully. The ball is released at position x_1 with total energy E. In idealized motion, it continues to oscillate back and forth across the bottom of the gully. The net force on the ball is always in the direction in which the potential energy is decreasing. In real systems there is sufficient friction between the ball and the gully to bring the ball to rest at the bottom of the gully.

In the presence of a particular x-component of the force F_x, the potential energy of the object at x is defined as the work required to move the object to x from the reference position x_0. Since the work done by a constant force in moving an object is defined as (force) × (displacement), the definition of potential energy in mathematical terms is

$$V(x) - V(x_0) = (F)(x - x_0)$$

Note carefully that the displacement referred to in the definition of potential energy starts at x_0 and ends at x. For example, the potential energy of an object of mass m at height h above the surface of the earth is the work done by a force $+mg$ opposing gravity in lifting the object from h_0 to h:

$$\begin{aligned} V(h) - V(h_0) &= mg(h - h_0) \\ &= mgh - mgh_0 \end{aligned}$$

For a variable force, the definition of potential energy becomes slightly more complicated and requires explicit use of calculus. The potential energy for the linear restoring force $F_x = -k(x - x_0)$ is

$$V(x) = \tfrac{1}{2}kx^2$$

where the value $V(x_0)$ has been set to 0 and x_0 has been selected as the origin of coordinates.

It is instructive to plot these two potential energy functions as graphs (see Appendix C). The gravitational potential energy function is a straight line from the origin with slope mg, while the potential energy function for the linear restoring force is a parabola. For the parabola the potential energy of the object increases as the distance from the reference point x_0 increases.

Knowledge of the potential energy curve as a function of position lets us predict the direction of the net force on the object at each position. For example in Figure B.1 the x-component of the force is a function that depends on position, $F_x(x)$, and the force is directed toward $x_0 = 0$ whether x is negative or positive. From the definition of potential energy stated earlier, but using F now to represent the gravitational or restoring force, not the opposing force, we get

$$F_x = -dV(x)/dx$$

The net force drives the object toward the position where the potential energy is a minimum and its slope is zero. For both the gravitational force and the linear restoring force, the object experiences forces that attract it back toward the center of force.

The minimum of the potential energy curve shown in Figure B.1 is called a point of **stable equilibrium** because the potential energy of the object increases as it moves away from that point. As the ball tries to climb the wall on either side of the minimum, the restoring force always drives it back toward this position of stable equilibrium. This qualitative interpretation predicts that the ball will oscillate about the equilibrium position, as predicated by the exact solution to Newton's second law quoted earlier.

We use similar potential energy diagrams to represent the interaction between a pair of objects, such as Earth–moon, Earth–Mars, electron–nucleus, electron–electron, nucleus–nucleus, atom–atom, or molecule–molecule. We construct such diagrams at several points in this textbook and use them to interpret the relative motions of the pair of objects. These methods are extremely important in describing the formation of chemical bonds, the states of matter, and the role of molecular collisions in chemical reactions.

B.2.7 Electrical Forces

The concepts summarized so far in this section also are used to describe the mutual interactions and the motions of electrically charged particles. The only new features are to identify the force that represents electrical interactions and to obtain the corresponding potential energy function. Positive and negative charges and the electrical forces between them were first quantified by Charles Coulomb late in the 18th century. In his honor the unit of charge in SI units is the **coulomb** (C). Electrical charge is fundamentally quantized in units of the magnitude of charge carried by a single electron e, which is equal to 1.60218×10^{-19} C. The coulomb is thus an inconveniently large unit for chemical reasoning. Nonetheless, for consistency and for quantitative accuracy, physical equations involving charge use SI units.

Suppose a charge of magnitude q_1 is held at the origin of coordinates and another charge of magnitude q_2 is brought near it, at the distance r from the origin. The magnitude of the acceleration imparted to the charge q_2 by the charge q_1 fixed at the origin can be determined as described earlier for uncharged objects. From such measurements, Coulomb determined that the magnitude of the force was directly proportional to the product of the two charges and inversely proportional to the square of the distance between them:

$$|F| = \frac{|q_1 q_2|}{4\pi\epsilon_0 r^2}$$

The force is expressed in newtons (N) and charge in coulombs (C) and the constant ϵ_0, called the permittivity of the vacuum, is 8.854×10^{-12} $C^2\ J^{-1}\ m^{-1}$.

Section 3.3 describes the direction of the Coulomb force in detail, so we do not repeat that discussion here. If the two particles are either both positively charged or both negatively charged, the particles feel a **repulsive force**, so the force on particle 2 is directed away from particle 1's location and the force on particle 1 is directed away from particle 2's location. If one particle has a positive charge and the other particle has a negative charge (e.g., $q_1 > 0$ and $q_2 < 0$), the particles feel an **attractive force**, so the force on particle 2 is directed towards particle 1's location and the force on particle 1 is directed towards particle 2's location.

The **Coulomb potential energy** corresponding to this force is

$$V(r) = \frac{q_1 q_2}{4\pi\epsilon_0 r}$$

Note that the force varies as r^{-2} while the potential energy varies as r^{-1}. Note also that if the charges have the same sign, dV/dr is negative and the force between the particles is repulsive; if the charges have opposite sign, dV/dr is positive and the force between the particles is attractive. If we know the potential energy curve between two charged particles, we can predict the forces between them and their resulting motion.

The Coulomb potential energy function holds great importance in chemistry for examining the structure of atoms and molecules. In 1912, Ernest Rutherford proposed that an atom of atomic number Z comprises a dense, central nucleus of positive charge $+Ze$ surrounded by a total of Z individual electrons, each with charge $-e$, moving around the nucleus. Thus, each individual electron has a potential energy of interaction with the nucleus given by

$$V(r) = -\frac{Ze^2}{4\pi\epsilon_0 r}$$

and the force between each electron and the nucleus is attractive. Rutherford's model of the atom, firmly based on experimental results, was completely at odds with the physical theories of the day. Attempts to reconcile these results with theory led to the development of the new theory called quantum mechanics.

B.2.8 Circular Motion and Angular Momentum

An object executing uniform circular motion about a point (for example, a ball on the end of a rope being swung in circular motion) is described by position, velocity, speed, momentum, and force, just as defined earlier for objects in linear motion. It is most convenient to describe this motion in polar coordinates r and θ, which represent, respectively, the distance of the object from the center and its angular displacement from the x-axis in ordinary Cartesian coordinates. Because r is constant for circular motion, the motion variables depend directly on θ. The magnitude of the angular velocity during the time interval from t_1 to t_2 is given by

$$\omega = \left| \frac{\theta_2 - \theta_1}{t_2 - t_1} \right|$$

for small changes in the angle θ. The angular acceleration is given by the rate of change of the angular velocity $\vec{\omega}$. In linear motion, we are concerned with the momentum $p_x = mv_x$ of an object as it heads toward a particular point; the linear momentum measures the impact that the object can transfer in a collision as it arrives at the point. To extend this concept to circular motion, we define the magnitude of the **angular momentum** of an object as it revolves around a point as $L = mvr$ in the special case of uniform circular motion where the vector $\vec{r}$, with length r, is perpendicular to the velocity vector $\vec{v}$, with constant speed v (see Fig. 3.5 in Chapter 3). The angular momentum of an electron around a nucleus is a crucial feature of atomic structure, which is discussed in Chapter 5.

PROBLEMS

Answers to problems whose numbers are red appear in Appendix G.

B.1 SI Units and Unit Conversions

1. Rewrite the following in scientific notation, using only the base units of Table B.1, without prefixes.
   (a) 65.2 nanograms
   (b) 88 picoseconds
   (c) 5.4 terawatts
   (d) 17 kilovolts

2. Rewrite the following in scientific notation, using only the base units of Table B.1, without prefixes.
   (a) 66 μK
   (b) 15.9 MJ
   (c) 0.13 mg
   (d) 62 GPa

3. Express the following temperatures in degrees Celsius.
   (a) 9001°F
   (b) 98.6°F (the normal body temperature of human beings)
   (c) 20°F above the boiling point of water at 1 atm pressure
   (d) −40°F

4. Express the following temperatures in degrees Fahrenheit.
   (a) 5000°C
   (b) 40.0°C
   (c) 212°C
   (d) −40°C

5. Express the temperatures given in problem 3 in kelvins.

6. Express the temperatures given in problem 4 in kelvins.

7. Express the following in SI units, either base or derived.
   (a) 55.0 miles per hour (1 mile = 1609.344 m)
   (b) 1.15 g cm^{-3}
   (c) 1.6×10^{-19} C Å
   (d) 0.15 mol L^{-1}
   (e) 5.7×10^{3} L atm day^{-1}

8. Express the following in SI units, either base or derived.
   (a) 67.3 atm
   (b) 1.0×10^{4} V cm^{-1}
   (c) 7.4 Å $year^{-1}$
   (d) 22.4 L mol^{-1}
   (e) 14.7 lb $inch^{-2}$ (1 inch = 2.54 cm; 1 lb = 453.59 g)

9. The kilowatt-hour (kWh) is a common unit in measurements of the consumption of electricity. What is the conversion factor between the kilowatt-hour and the joule? Express 15.3 kWh in joules.

10. A car's rate of fuel consumption is often measured in miles per gallon (mpg). Determine the conversion factor between miles per gallon and the SI unit of meters per cubic decimeter (1 gallon = 3.785 dm^3, and 1 mile = 1609.344 m). Express 30.0 mpg in SI units.

11. A certain V-8 engine has a displacement of 404 in^3. Express this volume in cubic centimeters (cm^3) and in liters.

12. Light travels in a vacuum at a speed of 3.00×10^{8} m s^{-1}.
   (a) Convert this speed to miles per second.
   (b) Express this speed in furlongs per fortnight, a little-used unit of speed. (A furlong, a distance used in horse racing, is 660 ft; a fortnight is exactly 2 weeks.)

B.2 Background in Physics

13. After being spiked, a volleyball travels with speed near 100 miles per hour. Calculate the kinetic energy of the volleyball. The mass of a volleyball is 0.270 kg.

14. The fastball of a famous pitcher in the National League has been clocked in excess of 95 mph. Calculate the work done by the pitcher in accelerating the ball to that speed. The mass of a baseball is 0.145 kg.

15. A tennis ball weighs approximately 2 ounces on a postage scale. A student practices his serve against the wall of the chemistry building, and the ball achieves the speed of 98 miles per hour. Calculate the kinetic energy of the ball after the serve.

Mathematics for General Chemistry

C

APPENDIX

Mathematics is an essential tool in chemistry. This appendix reviews some of the most important mathematical techniques for general chemistry.

C.1 Using Graphs

In many situations in science, we are interested in how one quantity (measured or predicted) depends on another quantity. The position of a moving car depends on the time, for example, or the pressure of a gas depends on the volume of the gas at a given temperature. A very useful way to show such a relation is through a *graph*, in which one quantity is plotted against another.

The usual convention in drawing graphs is to use the horizontal axis for the variable over which we have control and use the vertical axis for the measured or calculated quantity. After a series of measurements, the points on the graph frequently lie along a recognizable curve, and that curve can be drawn through the points. Because any experimental measurement involves some degree of uncertainty, there will be some scatter in the points, so there is no purpose in drawing a curve that passes precisely through every measured point. If there is a relation between the quantities measured, however, the points will display a systematic trend and a curve can be drawn that represents that trend.

The curves plotted on graphs can have many shapes. The simplest and most important shape is a straight line. A straight line is a graph of a relation such as

$$y = 4x + 7$$

or, more generally,

$$y = mx + b$$

where the variable y is plotted along the vertical axis and the variable x along the horizontal axis (Fig. C.1 shows a plot of $y = 2x - 1$). The quantity m is the *slope* of the line that is plotted, and b is the *intercept*—the point at which the line crosses the y axis. This can be demonstrated by setting x equal to 0 and noting that y is then equal to b. The slope is a measure of the steepness of the line; the greater the value of m, the steeper the line. If the line goes up and to the right, the slope is positive; if it goes down and to the right, the slope is negative.

The slope of a line can be determined from the coordinates of two points on the line. Suppose, for example, that when $x = 3$, $y = 5$, and when $x = 4$, $y = 7$. These two points can be written in shorthand notation as (3, 5) and (4, 7). The slope of the line is then defined as the "rise over the run"—the change in the y coordinate divided by the change in the x coordinate:

$$\text{slope} = m = \frac{\Delta y}{\Delta x} = \frac{7-5}{4-3} = \frac{2}{1} = 2$$

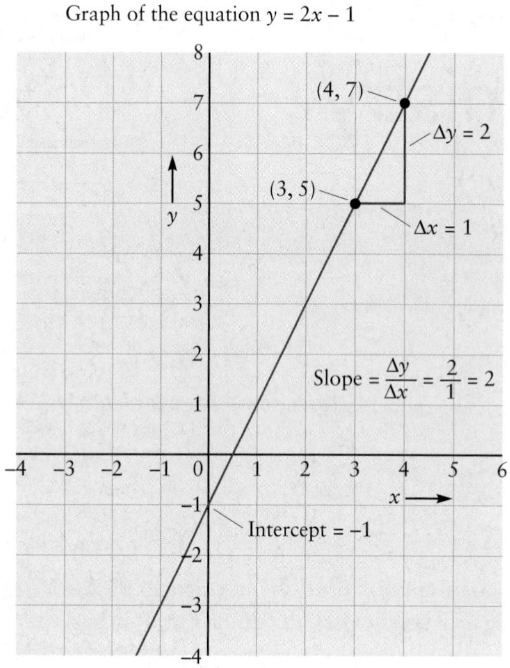

FIGURE C.1 A straight-line, or linear, relationship between two experimental quantities is a very desirable result because it is easy to graph and easy to represent mathematically. The equation of this straight line ($y = 2x - 1$) fits the general form $y = mx + b$. The line's y intercept is -1 and its slope is 2.

The symbol Δ (capital Greek delta) indicates the change in a quantity—the final value minus the initial value. In chemistry, if the two quantities being graphed have dimensions, the slope has dimensions as well. If a graph of distance traveled (in meters) against time (in seconds) is a straight line, its slope has dimensions of meters per second (m s^{-1}).

C.2 Solution of Algebraic Equations

In chemistry it is often necessary to solve an algebraic equation for an unknown quantity, such as a concentration or a partial pressure in an equilibrium-constant expression. Let us represent the unknown quantity with the symbol x. If the equation is linear, the method of solution is straightforward:

$$5x + 9 = 0$$

$$5x = -9$$

$$x = -\frac{9}{5}$$

or, more generally, if $ax + b = 0$, then $x = -b/a$.

Nonlinear equations are of many kinds. One of the most common in chemistry is the quadratic equation, which can always be rearranged into the form

$$ax^2 + bx + c = 0$$

where each of the constants (a, b, and c) may be positive, negative, or 0. The two solutions to a quadratic equation are given by the quadratic formula:

$$x = \frac{-b \pm \sqrt{b^2 - 4ac}}{2a}$$

As an example, suppose that the equation

$$x = 3 + \frac{7}{x}$$

arises in a chemistry problem. Multiplying by x and rearranging the terms gives

$$x^2 - 3x - 7 = 0$$

Inserting $a = 1$, $b = -3$, and $c = -7$ into the quadratic formula gives

$$x = \frac{-(-3) \pm \sqrt{(-3)^2 - 4(1)(-7)}}{2} = \frac{3 \pm \sqrt{9 + 28}}{2} = \frac{3 \pm \sqrt{37}}{2}$$

The two roots of the equation are

$$x = 4.5414 \qquad \text{and} \qquad x = -1.5414$$

In a chemistry problem, the choice of the proper root can frequently be made on physical grounds. If x corresponds to a concentration, for example, the negative root is unphysical and can be discarded.

The solution of cubic or higher order algebraic equations (or more complicated equations involving sines, cosines, logarithms, or exponentials) becomes more difficult, and approximate or numerical methods must be used. As an illustration, consider the equation

$$x^2\left(\frac{2.00 + x}{3.00 - x}\right) = 1.00 \times 10^{-6}$$

If x is assumed to be small relative to both 2.00 and 3.00, we obtain the simpler approximate equation

$$x^2\left(\frac{2.00}{3.00}\right) \approx 1.00 \times 10^{-6}$$

which leads to the roots $x = \pm 0.00122$. We immediately confirm that the solutions obtained in this way *are* small compared with 2.00 (and 3.00) and that our approximation was a good one. When a quantity (x in this case) is added to or subtracted from a larger quantity in a complicated equation, it is usually worthwhile, in solving the equation, to simply neglect the occurrence of the small quantity. Note that this tactic works only for addition and subtraction, never for multiplication or division.

Suppose now that the equation is changed to

$$x^2\left(\frac{2.00 + x}{3.00 - x}\right) = 1.00 \times 10^{-2}$$

In this case the equation that comes from neglecting x compared with 2.00 and 3.00 is

$$x^2\left(\frac{2.00}{3.00}\right) \approx 1.00 \times 10^{-2}$$

which is solved by $x \approx \pm 0.122$. The number 0.122 is smaller than 2.00 and 3.00, but not so small that it can be ignored. In this case more precise results can be obtained by *iteration*. Let us simply add and subtract the approximate positive root $x \approx +0.122$ as specified inside the parentheses and solve again for x:

$$x^2\left(\frac{2.00 + 0.122}{3.00 - 0.122}\right) = 1.00 \times 10^{-2}$$

$$x = 0.116$$

This new value can again be inserted into the original equation and the process repeated:

$$x^2\left(\frac{2.00 + 0.116}{3.00 - 0.116}\right) = 1.00 \times 10^{-2}$$

$$x = 0.117$$

Once the successive values of x agree to within the desired accuracy, the iteration can be stopped. Another root of the equation is obtained in a similar fashion if the iterative procedure starts with $x = -0.122$.

In some cases iteration fails. Suppose, for example, we have the equation

$$x^2\left(\frac{2.00 + x}{3.00 - x}\right) = 10.0$$

There is no particular reason to believe that x should be small compared with 2.00 or 3.00, but if we nevertheless assume that it is, we find $x = \pm\sqrt{15} = \pm 3.873$. Putting $x = +3.873$ back in for an iterative cycle leads to the equation

$$x^2 = -1.49$$

which has no real solutions. Starting with $x = -3.873$ succeeds no better; it gives

$$x^2 = -36.7$$

One way to overcome these difficulties is to solve the original equation graphically. We plot the left side of the equation against x and see at which values it becomes equal to 10 (the right side). We might initially calculate

x	$x^2\left(\frac{2.00 + x}{3.00 - x}\right)$
0	0
1	1.5
2	16

We observe that for $x = 1$ the left side is less than 10, whereas for $x = 2$ it is greater than 10. Somewhere in between there must be an x for which the left side is *equal to* 10. We can pinpoint it by selecting values of x between 1 and 2; if the left side is less than 10, x should be increased, and if it is greater than 10, x should be decreased.

x	$x^2\left(\frac{2.00 + x}{3.00 - x}\right)$
1.5	5.25
Increase to 1.8	10.26
Decrease to 1.75	9.19
Increase to 1.79	10.04

Thus, 1.79 is quite close to a solution of the equation. Improved values are easily obtained by further adjustments of this type.

C.3 Powers and Logarithms

Raising a number to a power and the inverse operation, taking a logarithm, are important in many chemical problems. Although the ready availability of electronic calculators makes the mechanical execution of these operations quite routine, it remains important to understand what is involved in such "special functions."

The mathematical expression 10^4 implies multiplying 10 by itself 4 times to give 10,000. Ten, or any other number, when raised to the power 0 always gives 1:

$$10^0 = 1$$

Negative powers of 10 give numbers less than 1 and are equivalent to raising 10 to the corresponding *positive* power and then taking the reciprocal:

$$10^{-3} = 1/10^3 = 0.001$$

We can extend the idea of raising to a power to include powers that are not whole numbers. For example, raising to the power $\frac{1}{2}$ (or 0.5) is the same as taking the square root:

$$10^{0.5} = \sqrt{10} = 3.1623\ldots$$

Scientific calculators have a 10^x (or INV LOG) key that can be used for calculating powers of 10 in cases where the power is not a whole number.

Numbers other than 10 can be raised to powers, as well; these numbers are referred to as *bases*. Many calculators have a y^x key that lets any positive number y be raised to any power x. One of the most important bases in scientific problems is the transcendental number called e (2.7182818 . . .). The e^x (or INV LN) key on a calculator is used to raise e to any power x. The quantity e^x also denoted as $\exp(x)$, is called the *exponential* of x. A key property of powers is that a base raised to the sum of two powers is equivalent to the product of the base raised separately to these powers. Thus, we can write

$$10^{21+6} = 10^{21} \times 10^6 = 10^{27}$$

The same type of relationship holds for any base, including e.

Logarithms also occur frequently in chemistry problems. The logarithm of a number is the exponent to which some base has to be raised to obtain the number. The base is almost always either 10 or the transcendental number e. Thus,

$$B^a = n \qquad \text{and} \qquad \log_B n = a$$

where a is the logarithm, B is the base, and n is the number.

Common logarithms are base-10 logarithms; that is, they are powers to which 10 has to be raised in order to give the number. For example, $10^3 = 1000$, so $\log_{10} 1000 = 3$. We shall frequently omit the 10 when showing common logarithms and write this equation as $\log 1000 = 3$. Only the logarithms of 1, 10, 100, 1000, and so on are whole numbers; the logarithms of other numbers are decimal fractions. The decimal point in a logarithm divides it into two parts. To the left of the decimal point is the *characteristic;* to the right is the *mantissa.* Thus, the logarithm in the equation

$$\log\,(7.310 \times 10^3) = 3.8639$$

has a characteristic of 3 and a mantissa of 0.8639. As may be verified with a calculator, the base-10 logarithm of the much larger (but closely related) number 7.310×10^{23} is 23.8639. As this case illustrates, the characteristic is determined solely by the location of the decimal point in the number and not by the number's precision, so it is *not* included when counting significant figures. The mantissa should be written with as many significant figures as the original number.

A logarithm is truly an exponent and as such follows the same rules of multiplication and division as other exponents. In multiplication and division we have

$$\log\,(n \times m) = \log n + \log m$$

$$\log\left(\frac{n}{m}\right) = \log n - \log m$$

Furthermore,

$$\log n^m = m \log n$$

so the logarithm of $3^5 = 243$ is

$$\log 3^5 = 5 \log 3 = 5 \times 0.47712 = 2.3856$$

There is no such thing as the logarithm of a negative number, because there is no power to which 10 (or any other base) can be raised to give a negative number.

A frequently used base for logarithms is the number e ($e = 2.7182818. . . .$). The logarithm to the base e is called the **natural logarithm** and is indicated by $\log_e$ or ln. Base-e logarithms are related to base-10 logarithms by the formula

$$\ln n = 2.3025851 \log n$$

As already stated, calculations of logarithms and powers are inverse operations. Thus, if we want to find the number for which 3.8639 is the common logarithm, we simply calculate

$$10^{3.8639} = 7.310 \times 10^3$$

If we need the number for which the natural logarithm is 2.108, we calculate

$$e^{2.108} = 8.23$$

As before, the number of significant digits in the answer should correspond to the number of digits in the *mantissa* of the logarithm.

C.4 Slopes of Curves and Derivatives

Very frequently in science, one measured quantity depends on a second one. If the property y depends on a second property x, we can write $y = f(x)$, where f is a function that expresses the dependence of y on x. Often, we are interested in the effect of a small change Δx on the dependent variable y. If x changes to $x + \Delta x$, then y will change to $y + \Delta y$. How is Δy related to Δx? Suppose we have the simple linear relation between y and x

$$y = mx + b$$

where m and b are constants. If we substitute $y + \Delta y$ and $x + \Delta x$, we find

$$y + \Delta y = m(x + \Delta x) + b$$

Subtracting the first equation from the second leaves

$$\Delta y = m\Delta x$$

or

$$\frac{\Delta y}{\Delta x} = m$$

The change in y is proportional to the change in x, with a proportionality constant equal to the slope of the line in the graph of y against x.

Suppose now we have a slightly more complicated relationship such as

$$y = ax^2$$

where a is a constant. If we substitute $y = y + \Delta y$ and $x = x + \Delta x$ here, we find

$$\begin{aligned} y + \Delta y &= a(x + \Delta x)^2 \\ &= ax^2 + 2ax\,\Delta x + (\Delta x)^2 \end{aligned}$$

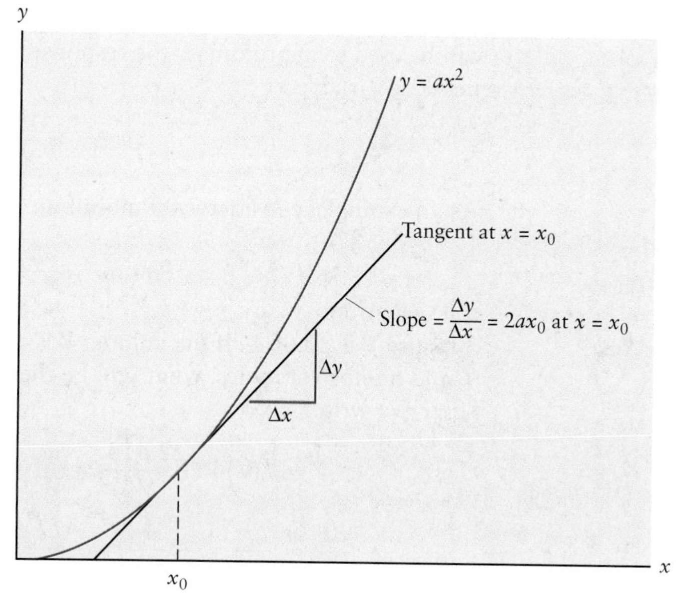

FIGURE C.2 The slope of a curve at a point.

Subtracting as before leaves

$$\Delta y = 2ax\ \Delta x + (\Delta x)^2$$

Here we have a more complicated, nonlinear relationship between Δy and Δx. If Δx is small enough, however, the term $(\Delta x)^2$ will be small relative to the term proportional to Δx, and we may write

$$\Delta y \approx 2ax\ \Delta x \qquad (\Delta x \text{ small})$$

$$\frac{\Delta y}{\Delta x} \approx 2ax \qquad (\Delta x \text{ small})$$

How do we indicate this graphically? The graph of y against x is no longer a straight line, so we need to generalize the concept of slope. We define a **tangent line** at the point x_0 as the line that touches the graph of $f(x)$ at $x = x_0$ without crossing it.[1] If Δy and Δx are small (Fig. C.2), the slope of the tangent line is

$$\frac{\Delta y}{\Delta x} = 2ax_0 \quad \text{at} \quad x = x_0$$

for the example just discussed. Clearly, the slope is not constant but changes with x_0. If we define the slope at each point on the curve by the slope of the corresponding tangent line, we obtain a *new* function that gives the slope of the curve $f(x)$ at each point x. We call this new function the *derivative* of $f(x)$ and represent it with df/dx. We have already calculated the derivatives of two functions:

$$f(x) = mx + b \Rightarrow \frac{df}{dx} = m$$

$$f(x) = ax^2 \Rightarrow \frac{df}{dx} = 2ax$$

Table C.1 shows derivatives of several other functions that are important in basic chemistry.

TABLE C.1

Derivatives of Simple Functions

Function $f(x)$	Derivative $\frac{df}{dx}$
$mx + b$	m
ax^2	$2ax$
$\frac{a}{x} = ax^{-1}$	$-\frac{a}{x^2} = -ax^{-2}$
ax^n	nax^{n-1}
e^{ax}	ae^{ax}
$\ln ax$	$\frac{1}{x}$
$\sin ax$	$a \cos ax$
$\cos ax$	$-a \sin ax$

The constants a, b, m, and n may be positive or negative.

[1]This is an intuitive, rather than a rigorously mathematical, definition of a tangent line.

The derivative gives the slope of the tangent line to $f(x)$ at each point x and can be used to approximate the response to a small perturbation Δx in the independent variable x:

$$\Delta y \approx \frac{df}{dx}\Delta x$$

As an example, consider 1.00 mol of an ideal gas at 0°C that obeys the law

$$P(\text{atm}) = \frac{22.414 \text{ L atm}}{V(\text{L})}$$

Suppose V is 20.00 L. If the volume V is changed by a small amount ΔV at fixed T and number of moles, what will be the corresponding change ΔP in the pressure? We write

$$P = f(V) = \frac{22.414}{V} = \frac{a}{V}$$

$$\frac{dP}{dV} = \frac{df}{dV} = -\frac{a}{V^2} = -\frac{22.414}{V^2} \qquad \text{(from Table C.1)}$$

$$\Delta P \approx \frac{dP}{dV}\Delta V = -\frac{22.414}{V^2}(\Delta V)$$

If $V = 20.00$ L, then

$$\Delta P \approx -\frac{22.414 \text{ L atm}}{(200.00 \text{ L})^2}\Delta V = -(0.0560 \text{ atm L}^{-1})\,\Delta V$$

C.5 Areas under Curves and Integrals

Another mathematical operation that arises frequently in science is the calculation of the area under a curve. Some areas are those of simple geometric shapes and are easy to calculate. If the function $f(x)$ is a constant,

$$f(x) = a$$

then the area under a graph of $f(x)$ between the two points x_1 and x_2 is that of a rectangle (Fig. C.3a) and is easily calculated as

$$\text{area} = \text{height} \times \text{base} = a(x_2 - x_1) = [ax_2] - [ax_1]$$

If $f(x)$ is a straight line that is neither horizontal nor vertical,

$$f(x) = mx + b$$

then the area is that of a trapezoid (Fig. C.3b) and is equal to

$$\begin{aligned}\text{area} &= \text{average height} \times \text{base}\\ &= \tfrac{1}{2}(mx_1 + b + mx_2 + b)(x_2 - x_1)\\ &= [\tfrac{1}{2}mx_2^2 + bx_2] - [\tfrac{1}{2}mx_1^2 + bx_1]\end{aligned}$$

Suppose now that $f(x)$ is a more complicated function, such as that shown in Figure C.3c. We can estimate the area under this graph by approximating it with a series of small rectangles of width Δx and varying heights. If the height of the ith rectangle is $y_i = f(x_i)$, then we have, approximately,

$$\begin{aligned}\text{area} &\approx f(x_1)\,\Delta x + f(x_2)\,\Delta x + \ldots\\ &= \sum_i f(x_i)\,\Delta x\end{aligned}$$

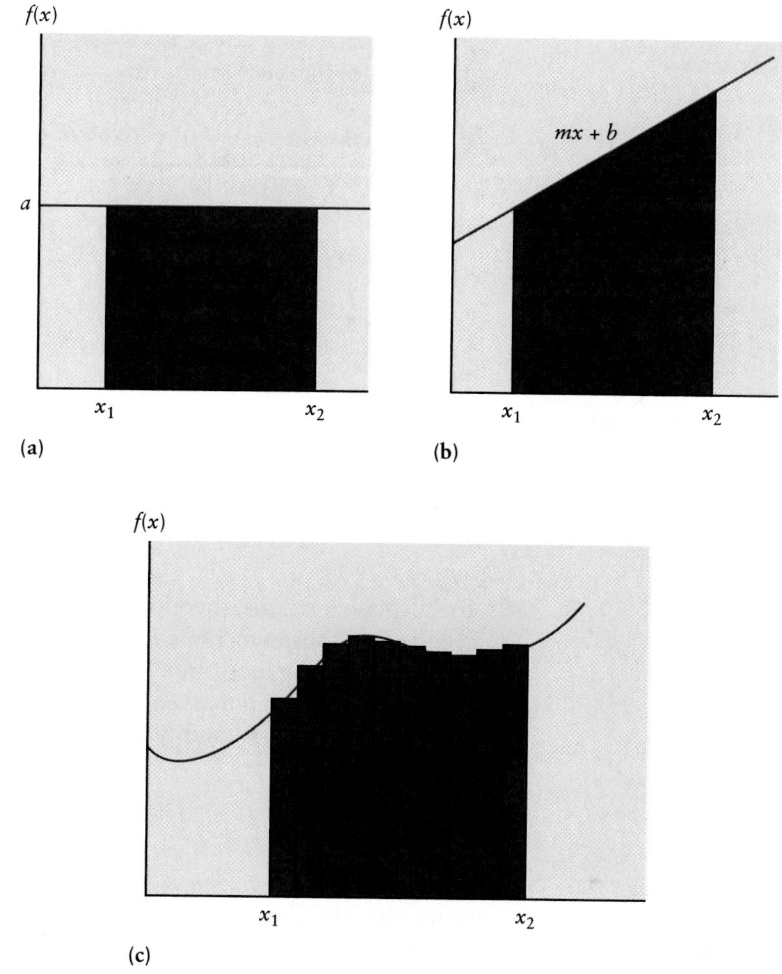

FIGURE C.3 The integral as the area under a curve.

As the widths of the rectangles Δx become small, this becomes a better approximation. We *define* the area under a curve $f(x)$ between two points as the limiting value of this sum as Δx approaches 0, called the *definite integral:*

$$\text{area} = \int_{x_1}^{x_2} f(x)\,dx$$

We have already worked out two examples of such integrals:

$$f(x) = a \Rightarrow \int_{x_1}^{x_2} a\,dx = [ax_2] - [ax_1]$$

$$f(x) = ax + b \Rightarrow \int_{x_1}^{x_2} (ax + b)\,dx = [\tfrac{1}{2}\,ax_2^2 + bx_2] - [\tfrac{1}{2}\,ax_1^2 + bx_1]$$

These integrals have an interesting form, which can be generalized to other cases as well: some function F evaluated at the upper limit (x_2) minus that function F evaluated at the lower limit (x_1).

$$\int_{x_1}^{x_2} f(x)\,dx = F(x_2) - F(x_1)$$

F is called the **antiderivative** of f because, for reasons we shall not go into here, it is obtained by the inverse of a derivative operation. In other words, if F is the antiderivative of f, then f is the derivative of F:

$$\frac{dF}{dx} = f(x)$$

TABLE C.2

Integrals of Simple Functions

Function $f(x)$	Antiderivative $F(x)$	Integral $\int_{x_1}^{x_2} f(x)\,dx$
a	ax	$ax_2 - ax_1$
$mx + b$	$\frac{1}{2}mx^2 + bx$	$[\frac{1}{2}mx_2^2 + bx_2] - [\frac{1}{2}mx_1^2 + bx_1]$
x^n $(n \neq -1)$	$\frac{1}{n+1}x^{n+1}$	$\frac{1}{n+1}(x_2^{n+1} - x_1^{n+1})$
$\frac{1}{x} = x^{-1}$ $(x > 0)$	$\ln x$	$(\ln x_2 - \ln x_1) = \ln \frac{x_2}{x_1}$
$\frac{1}{x^2} = x^{-2}$	$-\frac{1}{x}$	$-\left(\frac{1}{x^2} - \frac{1}{x_1}\right)$
e^{ax}	$\frac{e^{ax}}{a}$	$\frac{1}{a}(e^{ax_2} - e^{ax_1})$

The constants a, b, m, and n may be positive or negative.

To calculate integrals, therefore, we invert the results of Table C.1, because we need to find those functions $F(x)$ whose derivatives are equal to $f(x)$. Table C.2 lists several of the most important integrals for basic chemistry.

Several additional mathematical properties of integrals are important at this point. If a function is multiplied by a constant c, then the integral is multiplied by the same constant:

$$\int_{x_1}^{x_2} cf(x)\,dx = c\int_{x_1}^{x_2} f(x)\,dx$$

The reason is self-evident: if a function is multiplied everywhere by a constant factor, then the area under its graph must be increased by the same factor. Second, the integral of the sum of two functions is the sum of the separate integrals:

$$\int_{x_1}^{x_2} [f(x) + g(x)]\,dx = \int_{x_1}^{x_2} f(x)\,dx + \int_{x_1}^{x_2} g(x)\,dx$$

Finally, if the upper and lower limits of integration are reversed, the sign of the integral changes as well. This is easily seen from the antiderivative form:

$$\int_{x_2}^{x_1} f(x)\,dx = F(x_1) - F(x_2) = -[F(x_2) - F(x_1)] = -\int_{x_1}^{x_2} f(x)\,dx$$

C.6 Probability

Many applications in chemistry require us to interpret—and even predict—the results of measurements where we have only limited information about the system and the process involved. In such cases the best we can do is identify the possible outcomes of the experiment and assign a *probability* to each of them. Two examples illustrate the issues we face. In discussions of atomic structure, we would like to know the position of an electron relative to the nucleus. The principles of quantum mechanics tell us we can never know the exact location or trajectory of an electron; the most information we can have is the probability of finding an electron at each point in space around the nucleus. In discussing the behavior of a macroscopic amount of helium gas confined at a particular volume, pressure, and temperature we would like to know the speed with which an atom is moving in the container. We do not have experimental means to "tag" a particular atom, track its motions in the container, and measure its speed. The best we can do is estimate the probability that some typical atom is moving with each possible speed.

Everyone is familiar with the common sense concept of probability as a way to assess the likelihood of a desirable outcome in a game of chance. The purpose of this section is to give a brief introduction to probability in a form suited for scientific work.

C.6.1 Random Variable

The first step in setting up a probability model of a statistical experiment is to define the *random variable* X, the measurable quantity whose values *fluctuate,* or change, as we carry out many repetitions of the experiment. While defining the random variable, we also identify the *outcomes* of the statistical experiment, the possible values that X can take as we carry out many repetitions. If the experiment consists of flipping a coin, then X is simply the label on the side of the coin facing up, and the only possible outcomes are H for "heads" and T for "tails." If the experiment consists of rolling a die, X is the number of dots on the side facing up, with integral values from 1 to 6. If the experiment is rolling a pair of dice, X is the sum of the number of dots on the two sides facing up, with integral values from $1 + 1 = 2$ to $6 + 6 = 12$. If the experiment is to find the position of an electron in an atom, X is then r, the distance from the nucleus to the electron, and it can range from 0 to infinity. If the experiment is to find the speed of an atom in a sample of gas, X is then v, the speed of the atom, which can range from 0 to some large value.

C.6.2 Probability and Probability Functions

It is convenient to set up a graphical representation of the probability model. We represent the random variable X along the horizontal axis of an ordinary Cartesian graph. The possible outcomes of measuring X are shown as points along the horizontal axis. We see from the earlier examples that these can be either discrete or continuous values, depending on the nature of X. Along the vertical axis we want to plot $P(X)$, the probability of observing the random variable to have the value X, for each of the possible outcomes.

The **probability function** $P(X)$ is generated by making multiple measurements of the random variable X and recording the results in a histogram. The performance of a chemistry class with 50 students on a quiz with maximum score 20 is a good example of a statistical experiment. Here the random variable X is the score a student achieves, and the possible outcomes are the integers from 0 to 20. The set of graded papers constitute 50 repetitions of the statistical experiment, and the results are summarized in Figure C.4. We define the probability

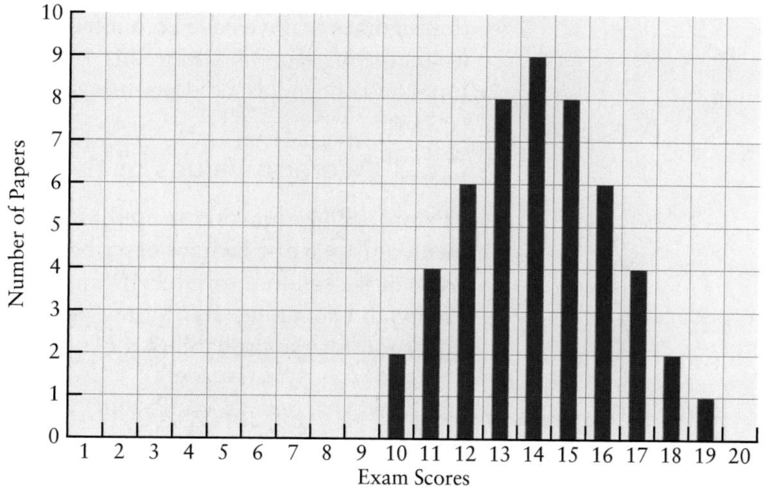

FIGURE C.4 Distribution of exam scores in a class of 50 students.

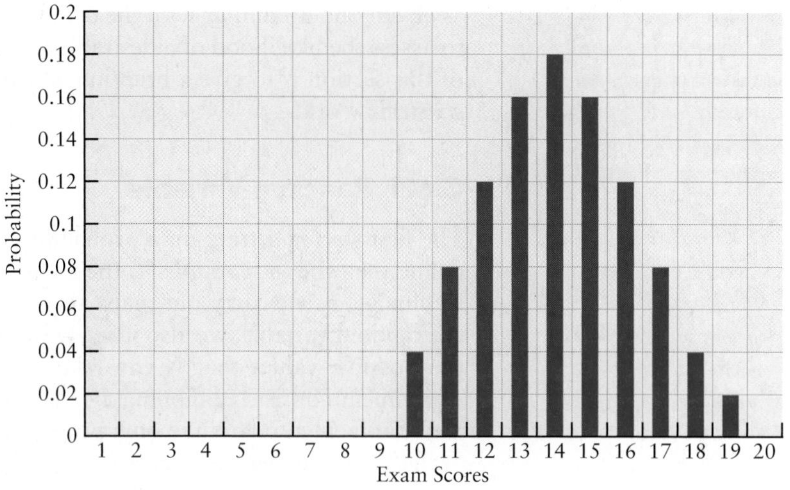

FIGURE C.5 Probability function for the exam scores in Figure C.4.

that a particular score was achieved on the exam to be the fraction of the papers with that score:

$$P(X_i) = \frac{n_i(X_i)}{N}$$

where $n_i(X_i)$ is the number of papers with score X_i and N is the total number of papers. Figure C.5 shows the probability function generated from the data in Figure C.4. The function has a maximum at $X = 14$, which is labeled the *most probable value* X_{mp}. This is the value that appears most often in the experimental data. Probability is a pure number; it does not have physical dimensions. Probability cannot be determined from a single measurement; many repetitions of the statistical experiment are required for the definition of probability.

The probability function has the interesting property that its sum over all the possible outcomes is 1:

$$P_0 + P_1 + \ldots + P_{19} + P_{20} = \frac{1}{50}(n_0 + n_1 + \ldots + n_{19} + n_{20}) = \frac{50}{50} = 1$$

We represent this long sum in the following compact way where the uppercase Greek sigma means "add up all the terms," and the running index *i* indicates the term for each possible outcome from 0 to 20:

$$\sum_{i=0}^{20} P_i = \frac{1}{50}\sum_{i=0}^{20} n_i = \frac{50}{50} = 1$$

A probability function with this property is said to be **normalized.** Normalization signifies that we have accounted for all the possible outcomes: the probability that *some* score is achieved is 1. We should always check that a probability function is normalized before using it in further calculations.

C.6.3 Average Values of the Random Variable

The probability function is used in numerous statistical analyses of the experiment, only one of which we describe here. We calculate $\overline{X}$, the *mean,* or average, value of the random variable, by multiplying each possible outcome by its probability and summing over all the possible outcomes. In our compact form this operation for the probability function in Figure C.5 is

$$\overline{X} = \sum_{i=0}^{20} X_i P(X_i) = \frac{1}{50}\sum_{i=0}^{20} X_i n_i = 14.7$$

The mean value is one way to convey the essence of the probability function in a single number: Is the overall class performance on the exam good, fair, or poor? Another way is to state the most probable value X_{mp}. In this example the mean value is slightly larger than the most probable value because the distribution slightly favors the high end of the range, courtesy of the student who scored 19 of 20! The mean and most probable values coincide only when the distribution is perfectly symmetrical about the maximum. The complete probability function gives much more insight into the results of the experiment than the mean or most probable values standing alone.

C.6.4 Probability Density: Continuous Random Variables and Probability Functions

When the random variable is continuous, as in the case of the position of the electron and speed of the helium atom referred to earlier, we must consider every point of the X axis—not just the integers—as a possible outcome. Therefore, we cannot define the probability of observing a specific value, which we call X_0. Rather, we define the probability that X falls within a narrow, infinitesimal interval between X_0 and $X_0 + dX$. When the interval dX is sufficiently small, the probability will be proportional to the width of the interval, and we define the probability as follows:

$$[\text{Probability of } X \text{ falling in the interval between } X_0 \text{ and } X_0 + dX] = \mathcal{P}(X_0)dX$$

Because this analysis is required at each point X, we drop the subscript 0, and discuss the properties of $\mathcal{P}(X)$ in general. The function $\mathcal{P}(X)$, called the **probability density function** for X, is plotted as a continuous function above the X-axis. It is defined for all points in an interval (a, b) whose end points depend on the nature of X. In some cases the end points can include $+\infty, -\infty$, or both. Note that $\mathcal{P}(X)$ has physical dimensions of X^{-1}, since the product $\mathcal{P}(X)dX$ must be dimensionless.

The probability density plays the same role for continuous variables as does $P(X)$ for discrete random variables. The normalization condition becomes

$$\int_a^b \mathcal{P}(X)\, dX = 1$$

and the mean value of X is calculated as

$$\overline{X} = \int_a^b X \cdot \mathcal{P}(X)\, dX$$

The symmetry of the probability density function determines whether the most probable and mean values of X coincide. We use a special symbol here to distinguish probability density from probability. In the main portion of the textbook we do not always make that distinction in symbols. But, it is always clear from context whether we are discussing a probability function or a probability density function.

C.6.5 Experimental Tests of Theoretical Probability

Theoretical probability identifies the possible outcomes of a statistical experiment, and uses theoretical arguments to predict the probability of each. Many applications in chemistry take this form. In atomic and molecular structure problems, the general principles of quantum mechanics predict the probability functions. In other cases the theoretical predictions are based on assumptions about the chemical or physical behavior of a system. In all cases, the validity of these predictions must be tested by comparison with laboratory measurements of the behavior of the same random variable. A full determination of experimental probability, and the mean values that come from it, must be obtained and compared with the theoretical predictions. A theoretical prediction of probability can never be tested or interpreted with a single measurement. A large number of repeated measurements is necessary to reveal the true statistical behavior.

C.7 Vectors

In chemistry, many quantities of interest have both magnitude and direction. For example, a strong (large-magnitude) force on a molecule might push it, but you need to know in what direction it is pushed. In the kinetic theory of gases, the distribution of molecular speeds is important, where the speed is the magnitude of the velocity, but when the molecule hits the wall of a container it changes its direction of travel, not its speed. If you wish to understand how an HCl molecule might reorient itself near a cation, you need to know not only how large the dipole moment of HCl is, the magnitude of the dipole moment, but also whether the positive partial charge is near the H or the Cl, as the positively charged side of the diatomic will feel a repulsive force from the cation. If you have already learned a bit about vectors in high school, you may wish to skip to the paragraph below that begins with "A very useful concept is the unit vector. . . ."

A force, velocity, or dipole moment may be represented by a **vector.** We'll restrict our discussion to vectors in a Cartesian coordinate system. In school you learned to label a location in a plane by its x and y coordinates. Figure C.6a shows the point $(x,y) = (-3,4)$. If you draw an arrow from the origin of the coordinate system to the point $(-3,4)$, this is a geometric representation of the vector $\vec{r}$. One

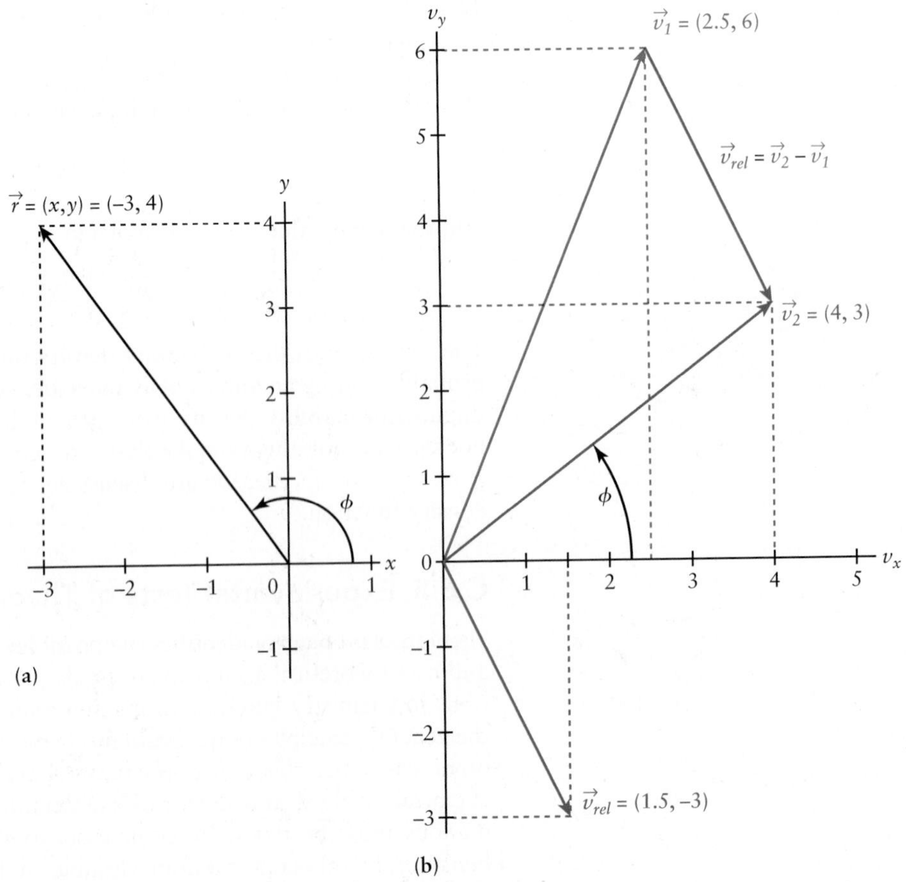

FIGURE C.6 Vectors representing position and velocity. (a) The position vector $\vec{r} = (r_x, r_y) = (x, y) = (-3, 4)$. In a physical problem you need to specify units, perhaps miles or angstroms. (b) Velocity vectors for two particles, each with components v_x and v_y. Note that the horizontal axis is labeled v_x now, as it is marked off in increments of v_x where $v_x = dx/dt$, and the vertical axis is labeled v_y now, where $v_y = dy/dt$. The relative velocity vector $\vec{v}_{rel}$ (of particle #2 with respect to particle #1) is shown in two positions, one to show its components easily and the other to show that $\vec{v}_1 + \vec{v}_{rel} = \vec{v}_2$, or $\vec{v}_{rel} = \vec{v}_2 - \vec{v}_1$.

can easily determine each *component* of the vector $\vec{r}$. The x-component, typically given the symbol r_x or simply x, is -3, and the y-component, typically given the symbol r_y or simply y, is 4. Thus, the vector $\vec{r} = (r_x, r_y) = (x, y) = (-3, 4)$. The length of the vector, also called the *magnitude* of the vector, is, by the Pythagorean theorem, $\sqrt{(r_x)^2 + (r_y)^2} = \sqrt{(-3)^2 + (4)^2} = 5$. If the distance units were meters, you would say that the point $(-3,4)$ is a distance of 5 m from the origin of the coordinate system.

The two components of the vector $\vec{r}$ also uniquely determine the direction of the vector $\vec{r}$ since the angle ϕ in Figure C.6a may be calculated from x and y. If you multiply a vector by a number c like $c = -2$ or $c = \pi$ (these numbers are called scalars), the result is simply $c\vec{r} = c(x, y) = (cx, cy)$. Thus multiplying a vector by a positive number changes the length of the vector, but not its direction, and multiplying a vector by a negative number reverses its direction.

For instance, if you wished to represent the location of your car with respect to your house, you might imagine the house at the origin of the coordinate system and direct the x-axis east and the y-axis north. If the units with which you measure distance are miles, the vector $\vec{r}$ in Figure C.6a would tell you that your car is located 5 miles from your house (the distance being the magnitude of the vector) in the northwest direction. (Of course, this distance is "as the crow flies" as you typically can't drive from one point to another by following a single straight line!) In this textbook we typically carry the units along with the numerical values to make it easy for you to keep track of them, so we might write the vector $\vec{r} = (r_x, r_y) = (x, y) = (-3 \text{ miles}, 4 \text{ miles})$. Also note that the magnitude of a vector, in this case the distance, is often denoted in mathematics texts as $\|r\|$, and those texts typically do not put an arrow over the vector $\vec{r}$. In this text we will put an arrow over the symbol used for a vector, and to simplify the notation we will indicate the magnitude of the vector $\vec{r}$ simply by r without an arrow over it (we will not put "$\| \ \|$" in the notation.)

A very useful concept is the unit vector in the same direction as $\vec{r}$. This unit vector is denoted by $\hat{r}$ where $\hat{r} = \vec{r}/r$, so the unit vector $\hat{r}$ is the vector $\vec{r}$ divided by its magnitude. Carrying the units with the components of the vector $\vec{r}$ in Figure C.6a, if the distances are measured in miles, you would calculate

$$\hat{r} = \frac{\vec{r}}{r} = \frac{(-3 \text{ miles}, 4 \text{ miles})}{5 \text{ miles}} = (-0.6, 0.8)$$

Note that even if you carry your units with the values of a vector's components, a unit vector does not have units of measure associated with it (called "dimensions" in Appendix B.1). If you calculate the magnitude of any unit vector, you will get a value of 1. (e.g., the magnitude of the $\hat{r}$ above $= \sqrt{(-0.6)^2 + (0.8)^2} = 1$). In Cartesian coordinates it is common to denote the unit vector parallel to the x-axis as $\hat{i} = (1, 0)$ and the unit vector parallel to the y-axis as $\hat{j} = (0, 1)$. (Each unit vector points in the direction of increasing x or y, respectively.) If you require a third axis, maybe because you are representing the position of a car that is northwest of your house, but also part way up a hill, the unit vectors in the same direction as the Cartesian coordinate axes are $\hat{i} = (1, 0, 0)$, $\hat{j} = (0, 1, 0)$, and $\hat{k} = (0, 0, 1)$. The magnitude of these unit vectors are, of course, 1. They are particularly useful in any Cartesian coordinate system because you can write any vector as a linear combination of them. For example, the vector $\vec{r}$ in figure C.6a $= -3(1,0) + 4(0,1) = -3\hat{i} + 4\hat{j}$.

Thus far, we have been discussing vectors whose magnitudes represent distance. Let's now consider a problem where we wish to use the length of a vector to represent the speed of an object, and the direction of the vector to indicate the direction of travel. Let $(x(t), y(t))$ be the position of the object at time t. To represent a velocity vector, you must label your Cartesian axes not by x and y, but rather by $v_x = dx/dt$ and $v_y = dy/dt$, as shown in Figure C.6b (called "velocity space"). If your unit of measure is miles per hour, then the red vector labeled $\vec{v}_1 = (v_{1x}, v_{1y}) = (2.5, 6)$ in Figure C.6b has magnitude $v_1 = \sqrt{(v_{1x})^2 + (v_{1y})^2} =$ 6.5 miles per hour. This might be the speed of car #1, and it is traveling in the northeast direction (if again moving to larger values of x is in the eastward direction and moving to larger values of y is in the northward direction). The velocity of car #2 is represented by the blue vector labeled $\vec{v}_2$, and you would calculate the speed v_2 of car #2 to be $\sqrt{(v_{2x})^2 + (v_{2y})^2} = 5$ miles per hour. Both cars are travelling eastward, (because both v_{1x} and v_{2x} are positive), but the absolute value of v_{2x} is larger than the absolute value of v_{1x}, so car #2 is traveling eastward faster than car #1. You may wish to calculate the relative velocity of car #2 with respect to car #1. This is the green vector labeled $\vec{v}_{\text{rel}}$ in Figure C.6b.

$$\vec{v}_{\text{rel}} = \vec{v}_2 - \vec{v}_1 = (4, 3) - (2.5, 6) = (1.5, -3)$$

From this calculation you see that car #2 is indeed traveling faster eastward than car #1 (the component of the relative velocity vector in the eastward direction is positive), but car #2 is traveling slower in the northward direction than car #1. Try constructing vectors such that car #2 is traveling northeast at 40 miles per hour and car #1 is traveling southwest at 40 miles per hour, so they might be speeding right toward each other. Then calculate v_{rel}, the magnitude of $\vec{v}_{\text{rel}}$. You will see that the relative speed is a scary 80 miles per hour. Relative velocity is discussed in Chapter 18 in the context of molecular collisions; a large magnitude of the relative velocity helps to surmount the energy barrier to a chemical reaction.

When we discuss attractive and repulsive forces between two molecules or particles, we want to distinguish between the vector that points from particle #1 to particle number #2; let's call it $\vec{r}_{12}$, and the vector that points from particle #2 to particle #1, $\vec{r}_{21}$. The vector $\vec{r}_{12} = \vec{r}_2 - \vec{r}_1$ and the vector $\vec{r}_{21} = \vec{r}_1 - \vec{r}_2$, so it is easy to show they have the same magnitude, $r_{12} = r_{21} = r$, but point in opposite directions $\vec{r}_{12} = -\vec{r}_{21}$. The force on particle #2 (exerted by particle #1) is $\vec{F}_2 = -\dfrac{dV}{dr}\hat{r}_{12}$ (where $V(r)$ is the potential energy as a function of the separation r) and the force on particle #1 (exerted by particle 2) is $\vec{F}_1 = -\dfrac{dV}{dr}\hat{r}_{21}$, so one immediately recovers Newton's second law, that $\vec{F}_1 = -\vec{F}_2$ (as long as there are no external forces, only the forces that each of these particles exerts on the other). We use these results in Chapter 3.

PROBLEMS

Answers to problems whose numbers are red appear in Appendix G.

C.1 Using Graphs

1. A particular plot of distance traveled against time elapsed is found to be a straight line. After an elapsed time of 1.5 hours, the distance traveled was 75 miles, and after an elapsed time of 3.0 hours, the distance traveled was 150 miles. Calculate the slope of the plot of distance against time, and give its units.

2. The pressure of a gas in a rigid container is measured at several different temperatures, and it is found that a plot of pressure against temperature is a straight line. At 20.0°C the pressure is 4.30 atm, and at 100.0°C the pressure is 5.47 atm. Calculate the slope of the plot of pressure against temperature and give its units.

3. Rewrite each of the following linear equations in the form $y = mx + b$, and give the slope and intercept of the corresponding plot. Then draw the graph.
   (a) $y = 4x - 7$
   (b) $7x - 2y = 5$
   (c) $3y + 6x - 4 = 0$

4. Rewrite each of the following linear equations in the form $y = mx + b$, and give the slope and intercept of the corresponding plot. Then draw the graph.
   (a) $y = -2x - 8$
   (b) $-3x + 4y = 7$
   (c) $7y - 16x + 53 = 0$

5. Graph the relation

$$y = 2x^3 - 3x^2 + 6x - 5$$

from $y = -2$ to $y = +2$. Is the plotted curve a straight line?

6. Graph the relation

$$y = \frac{8 - 10x - 3x^2}{2 - 3x}$$

from $x = -3$ to $x = +3$. Is the plotted curve a straight line?

C.2 Solution of Algebraic Equations

7. Solve the following linear equations for x.
   (a) $7x + 5 = 0$
   (b) $-4x + 3 = 0$
   (c) $-3x = -2$

8. Solve the following linear equations for x.
   (a) $6 - 8x = 0$
   (b) $-2x - 5 = 0$
   (c) $4x = -8$

9. Solve the following quadratic equations for x.
   (a) $4x^2 + 7x - 5 = 0$
   (b) $2x^2 = -3 - 6x$
   (c) $2x + \frac{3}{x} = 6$

10. Solve the following quadratic equations for x.
   (a) $6x^2 + 15x + 2 = 0$
   (b) $4x = 5x^2 - 3$
   (c) $\frac{1}{2 - x} + 3x = 4$

11. Solve each of the following equations for x using the approximation of small x, iteration, or graphical solution, as appropriate.
   (a) $x(2.00 + x)^2 = 2.6 \times 10^{-6}$
   (b) $x(3.00 - 7x)(2.00 + 2x) = 0.230$
   (c) $2x^3 + 3x^2 + 12x = -16$

12. Solve each of the following equations for x using the approximation of small x, iteration, or graphical solution, as appropriate.
   (a) $x(2.00 + x)(3.00 - x)(5.00 + 2x) = 1.58 \times 10^{-15}$
   (b) $x\frac{(3.00 + x)(1.00 - x)}{2.00 - x} = 0.122$
   (c) $12x^3 - 4x^2 + 35x = 10$

C.3 Powers and Logarithms

13. Calculate each of the following expressions, giving your answers the proper numbers of significant figures.
   (a) $\log(3.56 \times 10^4)$
   (b) $e^{-15.69}$
   (c) $10^{8.41}$
   (d) $\ln(6.893 \times 10^{-22})$

14. Calculate each of the following expressions, giving your answers the proper numbers of significant figures.
   (a) $10^{-16.528}$
   (b) $\ln(4.30 \times 10^{13})$
   (c) $e^{14.21}$
   (d) $\log(4.983 \times 10^{-11})$

15. What number has a common logarithm of 0.4793?

16. What number has a natural logarithm of -15.824?

17. Determine the common logarithm of 3.00×10^{121}. It is quite likely that your calculator will *not* give a correct answer. Explain why.

18. Compute the value of $10^{-107.8}$. It is quite likely that your calculator will *not* give the correct answer. Explain why.

19. The common logarithm of 5.64 is 0.751. Without using a calculator, determine the common logarithm of 5.64×10^7 and of 5.64×10^{-3}.

20. The common logarithm of 2.68 is 0.428. Without using a calculator, determine the common logarithm of 2.68×10^{192} and of 2.68×10^{-289}.

21. Use the graphical method and a calculator to solve the equation

$$\log \ln x = -x$$

for x. Give x to four significant figures.

22. Use a calculator to find a number that is equal to the reciprocal of its own natural logarithm. Report the answer to four significant figures.

C.4 Slopes of Curves and Derivatives

23. Calculate the derivatives of the following functions.
   (a) $y = 4x^2 + 4$
   (b) $y = \sin 3x + 4 \cos 2x$
   (c) $y = 3x + 2$
   (d) $y = \ln 7x$

24. Calculate the derivatives of the following functions.
   (a) $y = 6x^{19}$
   (b) $y = 7x^2 + 6x + 2$
   (c) $y = e^{-6x}$
   (d) $y = \cos 2x + \frac{7}{x}$

C.5 Areas under Curves and Integrals

25. Calculate the following integrals.

(a) $\int_2^4 (3x + 1)\, dx$

(b) $\int_0^5 x^6 dx$

(c) $\int_1^4 e^{-2x} dx$

26. Calculate the following integrals.

(a) $\int_{-1}^3 4\, dx$

(b) $\int_2^{100} \frac{1}{x}\, dx$

(c) $\int_2^4 \frac{5}{x^2}\, dx$

APPENDIX

D Standard Chemical Thermodynamic Properties

This table lists standard enthalpies of formation ΔH_f°, standard third-law entropies S°, standard free energies of formation ΔG_f°, and molar heat capacities at constant pressure, c_P, for a variety of substances, all at 25°C (298.15 K) and 1 atm. The table proceeds from the left side to the right side of the periodic table. Binary compounds are listed under the element that occurs to the left in the periodic table, except that binary oxides and hydrides are listed with the other element. Thus, KCl is listed with potassium and its compounds, but ClO_2 is listed with chlorine and its compounds.

Note that the *solution-phase* entropies are not absolute entropies but are measured relative to the arbitrary standard $S^\circ(H^+(aq)) = 0$. Consequently, some of them (as well as some of the heat capacities) are negative.

Most of the thermodynamic data in these tables were taken from the *NBS Tables of Chemical Thermodynamic Properties* (1982) and changed, where necessary, from a standard pressure of 0.1 MPa to 1 atm. The data for organic compounds $C_nH_m (n > 2)$ were taken from the *Handbook of Chemistry and Physics* (1981).

Substance	ΔH_f° (25°C) kJ mol^{-1}	S° (25°C) J K^{-1} mol^{-1}	ΔG_f° (25°C) kJ mol^{-1}	c_P (25°C) J K^{-1} mol^{-1}
H(*g*)	217.96	114.60	203.26	20.78
H_2(*g*)	0	130.57	0	28.82
H^+(*aq*)	0	0	0	0
H_3O^+(*aq*)	−285.83	69.91	−237.18	75.29
Li(*s*)	0	29.12	0	24.77
Li(*g*)	159.37	138.66	126.69	20.79
Li^+(*aq*)	−278.49	13.4	−293.31	68.6
LiH(*s*)	−90.54	20.01	−68.37	27.87
Li_2O(*s*)	−597.94	37.57	−561.20	54.10
LiF(*s*)	−615.97	35.65	−587.73	41.59
LiCl(*s*)	−408.61	59.33	−384.39	47.99
LiBr(*s*)	−351.21	74.27	−342.00	—
LiI(*s*)	−270.41	86.78	−270.29	51.04
Na(*s*)	0	51.21	0	28.24
Na(*g*)	107.32	153.60	76.79	20.79
Na^+(*aq*)	−240.12	59.0	−261.90	46.4
Na_2O(*s*)	−414.22	75.06	−375.48	69.12
NaOH(*s*)	−425.61	64.46	−379.53	59.54
NaF(*s*)	−573.65	51.46	−543.51	48.86
NaCl(*s*)	−411.15	72.13	−384.15	50.50
NaBr(*s*)	−361.06	86.82	−348.98	51.38
NaI(*s*)	−287.78	98.53	−286.06	52.09
$NaNO_3$(*s*)	−467.85	116.52	−367.07	92.88
Na_2S(*s*)	−364.8	83.7	−349.8	—
Na_2SO_4(*s*)	−1387.08	149.58	−1270.23	128.20

continued

	Substance	ΔH_f° (25°C) kJ mol^{-1}	S° (25°C) J K^{-1} mol^{-1}	ΔG_f° (25°C) kJ mol^{-1}	c_P (25°C) J K^{-1} mol^{-1}
	$NaHSO_4(s)$	−1125.5	113.0	−992.9	—
	$Na_2CO_3(s)$	−1130.68	134.98	−1044.49	112.30
	$NaHCO_3(s)$	−950.81	101.7	−851.1	87.61
	$K(s)$	0	64.18	0	29.58
	$K(g)$	89.24	160.23	60.62	20.79
	$K^+(aq)$	−252.38	102.5	−283.27	21.8
	$KO_2(s)$	−284.93	116.7	−239.4	77.53
	$K_2O_2(s)$	−494.1	102.1	−425.1	—
	$KOH(s)$	−424.76	78.9	−379.11	64.9
	$KF(s)$	−567.27	66.57	−537.77	49.04
	$KCl(s)$	−436.75	82.59	−409.16	51.30
	$KClO_3(s)$	−397.73	143.1	−296.25	100.25
	$KBr(s)$	−393.80	95.90	−380.66	52.30
	$KI(s)$	−327.90	106.32	−324.89	52.93
	$KMnO_4(s)$	−837.2	171.71	−737.7	117.57
	$K_2CrO_4(s)$	−1403.7	200.12	−1295.8	145.98
	$K_2Cr_2O_7(s)$	−2061.5	291.2	−1881.9	219.24
	$Rb(s)$	0	76.78	0	31.06
	$Rb(g)$	80.88	169.98	53.09	20.79
	$Rb^+(aq)$	−251.17	121.50	−283.98	—
	$RbCl(s)$	−435.35	95.90	−407.82	52.38
	$RbBr(s)$	−394.59	109.96	−381.79	52.84
	$RbI(s)$	−333.80	118.41	−328.86	53.18
	$Cs(s)$	0	85.23	0	32.17
	$Cs(g)$	76.06	175.49	49.15	70.79
	$Cs^+(aq)$	−258.28	133.05	−292.02	−10.5
	$CsF(s)$	−553.5	92.80	−525.5	51.09
	$CsCl(s)$	−443.04	101.17	−414.55	52.47
	$CsBr(s)$	−405.81	113.05	−391.41	52.93
	$CsI(s)$	−346.60	123.05	−340.58	52.80
II	$Be(s)$	0	9.50	0	16.44
	$Be(g)$	324.3	136.16	286.6	20.79
	$BeO(s)$	−609.6	14.14	−580.3	25.52
	$Mg(s)$	0	32.68	0	24.89
	$Mg(g)$	147.70	148.54	113.13	20.79
	$Mg^{2+}(aq)$	−466.85	−138.1	−454.8	—
	$MgO(s)$	−601.70	26.94	−569.45	37.15
	$MgCl_2(s)$	−641.32	89.62	−591.82	71.38
	$MgSO_4(s)$	−1284.9	91.6	−1170.7	96.48
	$Ca(s)$	0	41.42	0	25.31
	$Ca(g)$	178.2	154.77	144.33	20.79
	$Ca^{2+}(aq)$	−542.83	−53.1	−553.58	—
	$CaH_2(s)$	−186.2	42	−147.2	—
	$CaO(s)$	−635.09	39.75	−604.05	42.80
	$CaS(s)$	−482.4	56.5	−477.4	47.40
	$Ca(OH)_2(s)$	−986.09	83.39	−898.56	87.49
	$CaF_2(s)$	−1219.6	68.87	−1167.3	67.03
	$CaCl_2(s)$	−795.8	104.6	−748.1	72.59
	$CaBr_2(s)$	−682.8	130	−663.6	—
	$CaI_2(s)$	−533.5	142	−528.9	—
	$Ca(NO_3)_2(s)$	−938.39	193.3	−743.20	149.37
	$CaC_2(s)$	−59.8	69.96	−64.9	62.72
	$CaCO_3$(s, calcite)	−1206.92	92.9	−1128.84	81.88
	$CaCO_3$(s, aragonite)	−1207.13	88.7	−1127.80	81.25
	$CaSO_4(s)$	−1434.11	106.9	−1321.86	99.66
	$CaSiO_3(s)$	−1634.94	81.92	−1549.66	85.27

Substance	ΔH_f° (25°C) kJ mol^{-1}	S° (25°C) J K^{-1} mol^{-1}	ΔG_f° (25°C) kJ mol^{-1}	c_P (25°C) J K^{-1} mol^{-1}
$CaMg(CO_3)_2$ (*s*, dolomite)	−2326.3	155.18	−2163.4	157.53
Sr(*s*)	0	52.3	0	26.4
Sr(*g*)	164.4	164.51	130.9	20.79
Sr^{2+}(*aq*)	−545.80	−32.6	−559.48	—
$SrCl_2$(*s*)	−828.9	114.85	−781.1	75.60
$SrCO_3$(*s*)	−1220.0	97.1	−1140.1	81.42
Ba(*s*)	0	62.8	0	28.07
Ba(*g*)	180	170.24	146	20.79
Ba^{2+}(*aq*)	−537.64	9.6	−560.77	—
$BaCl_2$(*s*)	−858.6	123.68	−810.4	75.14
$BaCO_3$(*s*)	−1216.3	112.1	−1137.6	85.35
$BaSO_4$(*s*)	−1473.2	132.2	−1362.3	101.75
Sc(*s*)	0	34.64	0	25.52
Sc(*g*)	377.8	174.68	336.06	22.09
Sc^{3+}(*aq*)	−614.2	−255	−586.6	—
Ti(*s*)	0	30.63	0	25.02
Ti(*g*)	469.9	180.19	425.1	24.43
TiO_2(*s*, rutile)	−944.7	50.33	−889.5	55.02
$TiCl_4$(*ℓ*)	−804.2	252.3	−737.2	145.18
$TiCl_4$(*g*)	−763.2	354.8	−726.8	95.4
Cr(*s*)	0	23.77	0	23.35
Cr(*g*)	396.6	174.39	351.8	20.79
Cr_2O_3(*s*)	−1139.7	81.2	−1058.1	118.74
CrO_4^{2-}(*aq*)	−881.15	50.21	−727.75	—
$Cr_2O_7^{2-}$(*aq*)	−1490.3	261.9	−1301.1	—
W(*s*)	0	32.64	0	24.27
W(*g*)	849.4	173.84	807.1	21.31
WO_2(*s*)	−589.69	50.54	−533.92	56.11
WO_3(*s*)	−842.87	75.90	−764.08	73.76
Mn(*s*)	0	32.01	0	26.32
Mn(*g*)	280.7	238.5	173.59	20.79
Mn^{2+}(*aq*)	−220.75	−73.6	−228.1	50
MnO(*s*)	−385.22	59.71	−362.92	45.44
MnO_2(*s*)	−520.03	53.05	−465.17	54.14
MnO_4^-(*s*)	−541.4	191.2	−447.2	−82.0
Fe(*s*)	0	27.28	0	25.10
Fe(*g*)	416.3	180.38	370.7	25.68
Fe^{2+}(*aq*)	−89.1	−137.7	−78.9	—
Fe^{3+}(*aq*)	−48.5	−315.9	−4.7	—
$Fe_{0.947}O$(*s*, wüstite)	−266.27	57.49	−245.12	48.12
Fe_2O_3(*s*, hematite)	−824.2	87.40	−742.2	103.85
Fe_3O_4(*s*, magnetite)	−1118.4	146.4	−1015.5	143.43
$Fe(OH)_3$(*s*)	−823.0	106.7	−696.5	—
FeS(*s*)	−100.0	60.29	−100.4	50.54
$FeCO_3$(*s*)	−740.57	93.1	−666.72	82.13
$Fe(CN)_6^{3-}$(*aq*)	561.9	270.3	729.4	—
$Fe(CN)_6^{4-}$(*aq*)	455.6	95.0	695.1	—
Co(*s*)	0	30.04	0	24.81
Co(*g*)	424.7	179.41	380.3	23.02
Co^{2+}(*aq*)	−58.2	−113	−54.4	—
Co^{3+}(*aq*)	92	−305	134	—
CoO(*s*)	−237.94	52.97	−214.22	55.23

continued

	Substance	ΔH_f° (25°C) kJ mol^{-1}	S° (25°C) J K^{-1} mol^{-1}	ΔG_f° (25°C) kJ mol^{-1}	c_P (25°C) J K^{-1} mol^{-1}
	$CoCl_2(s)$	−312.5	109.16	−269.8	78.49
	$Ni(s)$	0	29.87	0	26.07
	$Ni(g)$	429.7	182.08	384.5	25.36
	$Ni^{2+}(aq)$	−54.0	−128.9	−45.6	—
	$NiO(s)$	−239.7	37.99	−211.7	44.31
	$Pt(s)$	0	41.63	0	25.86
	$Pt(g)$	565.3	192.30	520.5	25.53
	$PtCl_6^{2-}(aq)$	−668.2	219.7	−482.7	—
	$Cu(s)$	0	33.15	0	24.44
	$Cu(g)$	338.32	166.27	298.61	20.79
	$Cu^{+}(aq)$	71.67	40.6	49.98	—
	$Cu^{2+}(aq)$	64.77	−99.6	65.49	—
	$CuO(s)$	−157.3	42.63	−129.7	42.30
	$Cu_2O(s)$	−168.6	93.14	−146.0	63.64
	$CuCl(s)$	−137.2	86.2	−119.88	48.5
	$CuCl_2(s)$	−220.1	108.07	−175.7	71.88
	$CuSO_4(s)$	−771.36	109	−661.9	100.0
	$Cu(NH_3)_4^{2+}(aq)$	−348.5	273.6	−111.07	—
	$Ag(s)$	0	42.55	0	25.35
	$Ag(g)$	284.55	172.89	245.68	20.79
	$Ag^{+}(aq)$	105.58	72.68	77.11	21.8
	$AgCl(s)$	−127.07	96.2	−109.81	50.79
	$AgNO_3(s)$	−124.39	140.92	−33.48	93.05
	$Ag(NH_3)_2^{+}(aq)$	−111.29	245.2	−17.12	—
	$Au(s)$	0	47.40	0	25.42
	$Au(g)$	366.1	180.39	326.3	20.79
	$Zn(s)$	0	41.63	0	25.40
	$Zn(g)$	130.73	160.87	95.18	20.79
	$Zn^{2+}(aq)$	−153.89	−112.1	−147.06	46
	$ZnO(s)$	−348.28	43.64	−318.32	40.25
	ZnS(*s*, sphalerite)	−205.98	57.7	−201.29	46.0
	$ZnCl_2(s)$	−415.05	111.46	−369.43	71.34
	$ZnSO_4(s)$	−982.8	110.5	−871.5	99.2
	$Zn(NH_3)_4^{2+}(aq)$	−533.5	301	−301.9	—
	$Hg(\ell)$	0	76.02	0	27.98
	$Hg(g)$	61.32	174.85	31.85	20.79
	$HgO(s)$	−90.83	70.29	−58.56	44.06
	$HgCl_2(s)$	−224.3	146.0	−178.6	—
	$Hg_2Cl_2(s)$	−265.22	192.5	−210.78	—
III	$B(s)$	0	5.86	0	11.09
	$B(g)$	562.7	153.34	518.8	20.80
	$B_2H_6(g)$	35.6	232.00	86.6	56.90
	$B_5H_9(g)$	73.2	275.81	174.9	96.78
	$B_2O_3(s)$	−1272.77	53.97	−1193.70	62.93
	$H_3BO_3(s)$	−1094.33	88.83	−969.02	81.38
	$BF_3(g)$	−1137.00	254.01	−1120.35	50.46
	$BF_4^{-}(aq)$	−1574.9	180	−1486.9	—
	$BCl_3(g)$	−403.76	289.99	−388.74	62.72
	$BBr_3(g)$	−205.64	324.13	−232.47	67.78
	$Al(s)$	0	28.33	0	24.35
	$Al(g)$	326.4	164.43	285.7	21.38
	$Al^{3+}(aq)$	−531	−321.7	−485	—
	$Al_2O_3(s)$	−1675.7	50.92	−1582.3	79.04
	$AlCl_3(s)$	−704.2	110.67	−628.8	91.84
	$Ga(s)$	0	40.88	0	25.86

	Substance	ΔH_f° (25°C) kJ mol^{-1}	S° (25°C) J K^{-1} mol^{-1}	ΔG_f° (25°C) kJ mol^{-1}	c_P (25°C) J K^{-1} mol^{-1}
	Ga(*g*)	277.0	168.95	238.9	25.36
	Tl(*s*)	0	64.18	0	26.32
	Tl(*g*)	182.21	180.85	147.44	20.79
IV	C(*s*, graphite)	0	5.74	0	8.53
	C(*s*, diamond)	1.895	2.377	2.900	6.11
	C(*g*)	716.682	157.99	671.29	20.84
	CH_4(*g*)	−74.81	186.15	−50.75	35.31
	C_2H_2(*g*)	226.73	200.83	209.20	43.93
	C_2H_4(*g*)	52.26	219.45	68.12	43.56
	C_2H_6(*g*)	−84.68	229.49	−32.89	52.63
	C_3H_8(*g*)	−103.85	269.91	−23.49	73.0
	n-C_4H_{10}(*g*)	−124.73	310.03	−15.71	97.5
	C_4H_{10}(*g*, isobutane)	−131.60	294.64	−17.97	96.8
	n-C_5H_{12}(*g*)	−146.44	348.40	−8.20	120
	C_6H_6(*g*)	82.93	269.2	129.66	81.6
	C_6H_6(ℓ)	49.03	172.8	124.50	136
	CO(*g*)	−110.52	197.56	−137.15	29.14
	CO_2(*g*)	−393.51	213.63	−394.36	37.11
	CO_2(*aq*)	−413.80	117.6	−385.98	—
	CS_2(ℓ)	89.70	151.34	65.27	75.7
	CS_2(*g*)	117.36	237.73	67.15	45.40
	H_2CO_3(*aq*)	−699.65	187.4	−623.08	—
	HCO_3^-(*aq*)	−691.99	91.2	−586.77	—
	CO_3^{2-}(*aq*)	−677.14	−56.9	−527.81	—
	HCOOH(ℓ)	−424.72	128.95	−361.42	99.04
	HCOOH(*aq*)	−425.43	163	−372.3	—
	$COOH^-$(*aq*)	−425.55	92	−351.0	−87.9
	CH_2O(*g*)	−108.57	218.66	−102.55	35.40
	CH_3OH(ℓ)	−238.66	126.8	−166.35	81.6
	CH_3OH(*g*)	−200.66	239.70	−162.01	43.89
	CH_3OH(*aq*)	−245.93	133.1	−175.31	—
	$H_2C_2O_4$(*s*)	−827.2	—	—	117
	$HC_2O_4^-$(*aq*)	−818.4	149.4	−698.34	—
	$C_2O_4^{2-}$(*aq*)	−825.1	45.6	−673.9	—
	CH_3COOH(ℓ)	−484.5	159.8	−390.0	124.3
	CH_3COOH(*g*)	−432.25	282.4	−374.1	66.5
	CH_3COOH(*aq*)	−485.76	178.7	−396.46	—
	CH_3COO^-(*aq*)	−486.01	86.6	−369.31	−6.3
	CH_3CHO(ℓ)	−192.30	160.2	−128.12	—
	C_2H_5OH(ℓ)	−277.69	160.7	−174.89	111.46
	C_2H_5OH(*g*)	−235.10	282.59	−168.57	65.44
	C_2H_5OH(*aq*)	−288.3	148.5	−181.64	—
	CH_3OCH_3(*g*)	−184.05	266.27	−112.67	64.39
	CF_4(*g*)	−925	261.50	−879	61.09
	CCl_4(ℓ)	−135.44	216.40	−65.28	131.75
	CCl_4(*g*)	−102.9	309.74	−60.62	83.30
	$CHCl_3$(*g*)	−103.14	295.60	−70.37	65.69
	$COCl_2$(*g*)	−218.8	283.53	−204.6	57.66
	CH_2Cl_2(*g*)	−92.47	270.12	−65.90	50.96
	CH_3Cl(*g*)	−80.83	234.47	−57.40	40.75
	CBr_4(*s*)	18.8	212.5	47.7	144.3
	CH_3I(ℓ)	−15.5	163.2	13.4	126
	HCN(*g*)	135.1	201.67	124.7	35.86
	HCN(*aq*)	107.1	124.7	119.7	—
	CN^-(*aq*)	150.6	94.1	172.4	—
	CH_3NH_2(*g*)	−22.97	243.30	32.09	53.1

continued

	Substance	ΔH_f° (25°C) kJ mol^{-1}	S° (25°C) J K^{-1} mol^{-1}	ΔG_f° (25°C) kJ mol^{-1}	c_P (25°C) J K^{-1} mol^{-1}
	$CO(NH_2)_2(s)$	−333.51	104.49	−197.44	93.14
	$Si(s)$	0	18.83	0	20.00
	$Si(g)$	455.6	167.86	411.3	22.25
	$SiC(s)$	−65.3	16.61	−62.8	26.86
	$SiO_2(s$, quartz)	−910.94	41.84	−856.67	44.43
	$SiO_2(s$, cristobalite)	−909.48	42.68	−855.43	44.18
	$Ge(s)$	0	31.09	0	23.35
	$Ge(g)$	376.6	335.9	167.79	30.73
	$Sn(s$, white)	0	51.55	0	26.99
	$Sn(s$, gray)	−2.09	44.14	0.13	25.77
	$Sn(g)$	302.1	168.38	267.3	21.26
	$SnO(s)$	−285.8	56.5	−256.9	44.31
	$SnO_2(s)$	−580.7	52.3	−519.6	52.59
	$Sn(OH)_2(s)$	−561.1	155	−491.7	—
	$Pb(s)$	0	64.81	0	26.44
	$Pb(g)$	195.0	161.9	175.26	20.79
	$Pb^{2+}(aq)$	−1.7	10.5	−24.43	—
	$PbO(s$, yellow)	−217.32	68.70	−187.91	45.77
	$PbO(s$, red)	−218.99	66.5	−188.95	45.81
	$PbO_2(s)$	−277.4	68.6	−217.36	64.64
	$PbS(s)$	−100.4	91.2	−98.7	49.50
	$PbI_2(s)$	−175.48	174.85	−173.64	77.36
	$PbSO_4(s)$	−919.94	148.57	−813.21	103.21
V	$N_2(g)$	0	191.50	0	29.12
	$N(g)$	472.70	153.19	455.58	20.79
	$NH_3(g)$	−46.11	192.34	−16.48	35.06
	$NH_3(aq)$	−80.29	111.3	−26.50	—
	$NH_4^+(aq)$	−132.51	113.4	−79.31	79.9
	$N_2H_4(\ell)$	50.63	121.21	149.24	98.87
	$N_2H_4(aq)$	34.31	138	128.1	—
	$NO(g)$	90.25	210.65	86.55	29.84
	$NO_2(g)$	33.18	239.95	51.29	37.20
	$NO_2^-(aq)$	−104.6	123.0	−32.2	−97.5
	$NO_3^-(aq)$	−205.0	146.4	−108.74	−86.6
	$N_2O(g)$	82.05	219.74	104.18	38.45
	$N_2O_4(g)$	9.16	304.18	97.82	77.28
	$N_2O_5(s)$	−43.1	178.2	113.8	143.1
	$HNO_2(g)$	−79.5	254.0	−46.0	45.6
	$HNO_3(\ell)$	−174.10	155.49	−80.76	109.87
	$NH_4NO_3(s)$	−365.56	151.08	−184.02	139.3
	$NH_4Cl(s)$	−314.43	94.6	−202.97	84.1
	$(NH_4)_2SO_4(s)$	−1180.85	220.1	−901.90	187.49
	$P(s$, white)	0	41.09	0	23.84
	$P(s$, red)	−17.6	22.80	−12.1	21.21
	$P(g)$	314.64	163.08	278.28	20.79
	$P_2(g)$	144.3	218.02	103.7	32.05
	$P_4(g)$	58.91	279.87	24.47	67.15
	$PH_3(g)$	5.4	210.12	13.4	37.11
	$H_3PO_4(s)$	−1279.0	110.50	−1119.2	106.06
	$H_3PO_4(aq)$	−1288.34	158.2	−1142.54	—
	$H_2PO_4^-(aq)$	−1296.29	90.4	−1130.28	—
	$HPO_4^{2-}(aq)$	−1292.14	−33.5	−1089.15	—
	$PO_4^{3-}(aq)$	−1277.4	−222	−1018.7	—
	$PCl_3(g)$	−287.0	311.67	−267.8	71.84
	$PCl_5(g)$	−374.9	364.47	−305.0	112.80
	$As(s$, gray)	0	35.1	0	24.64

	Substance	ΔH_f° (25°C) kJ mol^{-1}	S° (25°C) J K^{-1} mol^{-1}	ΔG_f° (25°C) kJ mol^{-1}	c_P (25°C) J K^{-1} mol^{-1}
	$As(g)$	302.5	174.10	261.0	20.79
	$As_2(g)$	222.2	239.3	171.9	35.00
	$As_4(g)$	143.9	314	92.4	—
	$AsH_3(g)$	66.44	222.67	68.91	38.07
	$As_4O_6(s)$	−1313.94	214.2	−1152.53	191.29
	$Sb(s)$	0	45.69	0	25.33
	$Sb(g)$	262.3	180.16	222.1	20.79
	$Bi(s)$	0	56.74	0	25.52
	$Bi(g)$	207.1	186.90	168.2	20.79
VI	$O_2(g)$	0	205.03	0	29.36
	$O(g)$	249.17	160.95	231.76	21.91
	$O_3(g)$	142.7	238.82	163.2	39.20
	$OH^-(aq)$	−229.99	−10.75	−157.24	−148.5
	$H_2O(\ell)$	−285.83	69.91	−237.18	75.29
	$H_2O(g)$	−241.82	188.72	−228.59	35.58
	$H_2O_2(\ell)$	−187.78	109.6	−120.42	89.1
	$H_2O_2(aq)$	−191.17	143.9	−134.03	—
	$S(s$, rhombic)	0	31.80	0	22.64
	$S(s$, monoclinic)	0.30	32.6	0.096	—
	$S(g)$	278.80	167.71	238.28	23.67
	$S_8(g)$	102.30	430.87	49.66	156.44
	$H_2S(g)$	−20.63	205.68	−33.56	34.23
	$H_2S(aq)$	−39.7	121	−27.83	—
	$HS^-(aq)$	−17.6	62.8	12.08	—
	$SO(g)$	6.26	221.84	−19.87	30.17
	$SO_2(g)$	−296.83	248.11	−300.19	39.87
	$SO_3(g)$	−395.72	256.65	−371.08	50.67
	$H_2SO_3(aq)$	−608.81	232.2	−537.81	—
	$HSO_3^-(aq)$	−626.22	139.7	−527.73	—
	$SO_3^{2-}(aq)$	−635.5	−29	−486.5	—
	$H_2SO_4(\ell)$	−813.99	156.90	−690.10	138.91
	$HSO_4^-(aq)$	−887.34	131.8	−755.91	−84
	$SO_4^{2-}(aq)$	−909.27	20.1	−744.53	−293
	$SF_6(g)$	−1209	291.71	−1105.4	97.28
	$Se(s$, black)	0	42.44	0	25.36
	$Se(g)$	227.07	176.61	187.06	20.82
VII	$F_2(g)$	0	202.67	0	31.30
	$F(g)$	78.99	158.64	61.94	22.74
	$F^-(aq)$	−332.63	−13.8	−278.79	−106.7
	$HF(g)$	−271.1	173.67	−273.2	29.13
	$HF(aq)$	−320.08	88.7	−296.82	—
	$XeF_4(s)$	−261.5	—	—	—
	$Cl_2(g)$	0	222.96	0	33.91
	$Cl(g)$	121.68	165.09	105.71	21.84
	$Cl^-(aq)$	−167.16	56.5	−131.23	−136.4
	$HCl(g)$	−92.31	186.80	−95.30	29.12
	$ClO^-(aq)$	−107.1	42	−36.8	—
	$ClO_2(g)$	102.5	256.73	120.5	41.97
	$ClO_2^-(aq)$	−66.5	101.3	17.2	—
	$ClO_3^-(aq)$	−103.97	162.3	−7.95	—
	$ClO_4^-(aq)$	−129.33	182.0	−8.52	—
	$Cl_2O(g)$	80.3	266.10	97.9	45.40
	$HClO(aq)$	−120.9	142	−79.9	—
	$ClF_3(g)$	−163.2	281.50	−123.0	63.85
	$Br_2(\ell)$	0	152.23	0	75.69
	$Br_2(g)$	30.91	245.35	3.14	36.02

continued

	Substance	ΔH_f° (25°C) kJ mol^{-1}	S° (25°C) J K^{-1} mol^{-1}	ΔG_f° (25°C) kJ mol^{-1}	c_P (25°C) J K^{-1} mol^{-1}
	$Br_2(aq)$	−2.59	130.5	3.93	—
	$Br(g)$	111.88	174.91	82.41	20.79
	$Br^-(aq)$	−121.55	82.4	−103.96	−141. 8
	$HBr(g)$	−36.40	198.59	−53.43	29.14
	$BrO_3^-(aq)$	−67.07	161.71	18.60	—
	$I_2(s)$	0	116.14	0	54.44
	$I_2(g)$	62.44	260.58	19.36	36.90
	$I_2(aq)$	22.6	137.2	16.40	—
	$I(g)$	106.84	180.68	70.28	20.79
	$I^-(aq)$	−55.19	111.3	−51.57	−142.3
	$I_3^-(aq)$	−51.5	239.3	−51.4	—
	$HI(g)$	26.48	206.48	1.72	29.16
	$ICl(g)$	17.78	247.44	−5.44	35.56
	$IBr(g)$	40.84	258.66	3.71	36.44
VIII	$He(g)$	0	126.04	0	20.79
	$Ne(g)$	0	146.22	0	20.79
	$Ar(g)$	0	154.73	0	20.79
	$Kr(g)$	0	163.97	0	20.79
	$Xe(g)$	0	169.57	0	20.79

E

APPENDIX

Standard Reduction Potentials at 25°C

Half-Reaction	$E°$ (volts)
$F_2(g) + 2\,e^- \longrightarrow 2\,F^-$	2.87
$H_2O_2 + 2\,H_3O^+ + 2\,e^- \longrightarrow 4\,H_2O$	1.776
$PbO_2(s) + SO_4^{2-} + 4\,H_3O^+ + 2\,e^- \longrightarrow PbSO_4(s) + 6\,H_2O$	1.685
$Au^+ + e^- \longrightarrow Au(s)$	1.68
$MnO_4^- + 4\,H_3O^+ + 3\,e^- \longrightarrow MnO_2(s) + 6\,H_2O$	1.679
$HClO_2 + 2\,H_3O^+ + 2\,e^- \longrightarrow HClO + 3\,H_2O$	1.64
$HClO + H_3O^+ + e^- \longrightarrow Cl_2(g) + 2\,H_2O$	1.63
$Ce^{4+} + e^- \longrightarrow Ce^{3+}$ (1 M HNO_3 solution)	1.61
$2\,NO(g) + 2\,H_3O^+ + 2\,e^- \longrightarrow N_2O(g) + 3\,H_2O$	1.59
$BrO_3^- + 6\,H_3O^+ + 5\,e^- \longrightarrow Br_2(\ell) + 9\,H_2O$	1.52
$Mn^{3+} + e^- \longrightarrow Mn^{2+}$	1.51
$MnO_4^- + 8\,H_3O^+ + 5\,e^- \longrightarrow Mn^{2+} + 12\,H_2O$	1.491
$ClO_3^- + 6\,H_3O^+ + 5\,e^- \longrightarrow Cl_2(g) + 9\,H_2O$	1.47
$PbO_2(s) + 4\,H_3O^+ + 2\,e^- \longrightarrow Pb^{2+} + 6\,H_2O$	1.46
$Au^{3+} + 3\,e^- \longrightarrow Au(s)$	1.42
$Cl_2(g) + 2\,e^- \longrightarrow 2\,Cl^-$	1.3583
$Cr_2O_7^{2-} + 14\,H_3O^+ + 6\,e^- \longrightarrow 2\,Cr^{3+} + 21\,H_2O$	1.33
$O_3(g) + H_2O + 2\,e^- \longrightarrow O_2 + 2\,OH^-$	1.24
$O_2(g) + 4\,H_3O^+ + 4\,e^- \longrightarrow 6\,H_2O$	1.229
$MnO_2(s) + 4\,H_3O^+ + 2\,e^- \longrightarrow Mn^{2+} + 6\,H_2O$	1.208
$ClO_4^- + 2\,H_3O^+ + 2\,e^- \longrightarrow ClO_3 + 3\,H_2O$	1.19
$Br_2(\ell) + 2\,e^- \longrightarrow 2\,Br^-$	1.065
$NO_3^- + 4\,H_3O^+ + 3\,e^- \longrightarrow NO(g) + 6\,H_2O$	0.96
$2\,Hg^{2+} + 2\,e^- \longrightarrow Hg_2^{2+}$	0.905
$Ag^+ + e^- \longrightarrow Ag(s)$	0.7996
$Hg_2^{2+} + 2\,e^- \longrightarrow 2\,Hg(\ell)$	0.7961
$Fe^{3+} + e^- \longrightarrow Fe^{2+}$	0.770
$O_2(g) + 2\,H_3O^+ + 2\,e^- \longrightarrow H_2O_2 + 2\,H_2O$	0.682
$BrO_3^- + 3\,H_2O + 6\,e^- \longrightarrow Br^- + 6\,OH^-$	0.61
$MnO_4^- + 2\,H_2O + 3\,e^- \longrightarrow MnO_2(s) + 4\,OH^-$	0.588
$I_2(s) + 2\,e^- \longrightarrow 2\,I^-$	0.535
$Cu^+ + e^- \longrightarrow Cu(s)$	0.522
$O_2(g) + 2\,H_2O + 4\,e^- \longrightarrow 4\,OH^-$	0.401
$Cu^{2+} + 2\,e^- \longrightarrow Cu(s)$	0.3402
$PbO_2(s) + H_2O + 2\,e^- \longrightarrow PbO(s) + 2\,OH^-$	0.28
$Hg_2Cl_2(s) + 2\,e^- \longrightarrow 2\,Hg(\ell) + 2\,Cl^-$	0.2682
$AgCl(s) + e^- \longrightarrow Ag(s) + Cl^-$	0.2223
$SO_4^{2-} + 4\,H_3O^+ + 2\,e^- \longrightarrow H_2SO_3 + 5\,H_2O$	0.20

continued

Half-Reaction	E° (volts)
$Cu^{2+} + e^- \rightarrow Cu^+$	0.158
$S_4O_6^{2-} + 2\ e^- \rightarrow 2\ S_2O_3^{2-}$	0.0895
$NO_3^- + H_2O + 2\ e^- \rightarrow NO_2^- + 2\ OH^-$	0.01
$2\ H_3O^+ + 2\ e^- \rightarrow H_2(g) + 2\ H_2O(\ell)$	0.000 exactly
$Pb^{2+} + 2\ e^- \rightarrow Pb(s)$	−0.1263
$Sn^{2+} + 2\ e^- \rightarrow Sn(s)$	−0.1364
$Ni^{2+} + 2\ e^- \rightarrow Ni(s)$	−0.23
$Co^{2+} + 2\ e^- \rightarrow Co(s)$	−0.28
$PbSO_4(s) + 2\ e^- \rightarrow Pb(s) + SO_4^{2-}$	−0.356
$Mn(OH)_3(s) + e^- \rightarrow Mn(OH)_2(s) + OH^-$	−0.40
$Cd^{2+} + 2\ e^- \rightarrow Cd(s)$	−0.4026
$Fe^{2+} + 2\ e^- \rightarrow Fe(s)$	−0.409
$Cr^{3+} + e^- \rightarrow Cr^{2+}$	−0.424
$Fe(OH)_3(s) + e^- \rightarrow Fe(OH)_2(s) + OH^-$	−0.56
$PbO(s) + H_2O + 2\ e^- \rightarrow Pb(s) + 2\ OH^-$	−0.576
$2\ SO_3^{2-} + 3\ H_2O + 4\ e^- \rightarrow S_2O_3^{2-} + 6\ OH^-$	−0.58
$Ni(OH)_2(s) + 2\ e^- \rightarrow Ni(s) + 2\ OH^-$	−0.66
$Co(OH)_2(s) + 2\ e^- \rightarrow Co(s) + 2\ OH^-$	−0.73
$Cr^{3+} + 3\ e^- \rightarrow Cr(s)$	−0.74
$Zn^{2+} + 2\ e^- \rightarrow Zn(s)$	−0.7628
$2\ H_2O + 2\ e^- \rightarrow H_2(g) + 2\ OH^-$	−0.8277
$Cr^{2+} + 2\ e^- \rightarrow Cr(s)$	−0.905
$SO_4^{2-} + H_2O + 2\ e^- \rightarrow SO_3^{2-} + 2\ OH^-$	−0.92
$Mn^{2+} + 2\ e^- \rightarrow Mn(s)$	−1.029
$Mn(OH)_2(s) + 2\ e^- \rightarrow Mn(s) + 2\ OH^-$	−1.47
$Al^{3+} + 3\ e^- \rightarrow Al(s)$	−1.706
$Sc^{3+} + 3\ e^- \rightarrow Sc(s)$	−2.08
$Ce^{3+} + 3\ e^- \rightarrow Ce(s)$	−2.335
$La^{3+} + 3\ e^- \rightarrow La(s)$	−2.37
$Mg^{2+} + 2\ e^- \rightarrow Mg(s)$	−2.375
$Mg(OH)_2(s) + 2\ e^- \rightarrow Mg(s) + 2\ OH^-$	−2.69
$Na^+ + e^- \rightarrow Na(s)$	−2.7109
$Ca^{2+} + 2\ e^- \rightarrow Ca(s)$	−2.76
$Ba^{2+} + 2\ e^- \rightarrow Ba(s)$	−2.90
$K^+ + e^- \rightarrow K(s)$	−2.925
$Li^+ + e^- \rightarrow Li(s)$	−3.045

All voltages are standard reduction potentials (relative to the standard hydrogen electrode) at 25°C and 1 atm pressure. All species are in aqueous solution unless otherwise indicated.

F

APPENDIX

Physical Properties of the Elements

Hydrogen and the Alkali Metals (Group I Elements)

	Hydrogen	Lithium	Sodium	Potassium	Rubidium	Cesium	Francium
Atomic number	1	3	11	19	37	55	87
Atomic mass	1.00794	6.941	22.98976928	39.0983	85.4678	132.9054519	(223.0197)
Melting point (°C)	−259.14	180.54	97.81	63.65	38.89	28.40	25
Boiling point (°C)	−252.87	1347	903.8	774	688	678.4	677
Density at 25°C (g cm^{-3})	0.070 (−253°C)	0.534	0.971	0.862	1.532	1.878	
Color	Colorless	Silver	Silver	Silver	Silver	Silver	
Ground-state electron configuration	$1s^1$	[He]$2s^1$	[Ne]$3s^1$	[Ar]$4s^1$	[Kr]$5s^1$	[Xe]$6s^1$	[Rn]$7s^1$
Ionization energy†	1312.0	520.2	495.8	418.8	403.0	375.7	≈400
Electron affinity†	72.770	59.63	52.867	48.384	46.884	45.505	est. 44
Electronegativity	2.20	0.98	0.93	0.82	0.82	0.79	0.70
Ionic radius (Å)	1.46(H^-)	0.68	0.98	1.33	1.48	1.67	≈1.8
Atomic radius (Å)	0.37	1.52	1.86	2.27	2.47	2.65	≈2.7
Enthalpy of fusion†	0.1172	3.000	2.602	2.335	2.351	2.09	
Enthalpy of vaporization†	0.4522	147.1	97.42	89.6	76.9	67.8	
Bond enthalpy of M_2†	436	102.8	72.6	54.8	51.0	44.8	
Standard reduction potential (volts)	0 H^+/H_2	−3.045 Li^+/Li	−2.7109 Na^+/Na	−2.924 K^+/K	−2.925 Rb^+/Rb	−2.923 Cs^+/Cs	≈2.9 Fr^+/Fr

†In kilojoules per mole.

The Alkaline-Earth Metals (Group II Elements)

	Beryllium	Magnesium	Calcium	Strontium	Barium	Radium
Atomic number	4	12	20	38	56	88
Atomic mass	9.012182	24.3050	40.078	87.62	137.327	(226.0254)
Melting point (°C)	1283	648.8	839	769	725	700
Boiling point (°C)	2484	1105	1484	1384	1640	
Density at 25°C (g cm^{-3})	1.848	1.738	1.55	2.54	3.51	5
Color	Gray	Silver	Silver	Silver	Silver-yellow	Silver
Ground-state electron configuration	[He]2s^2	[Ne]3s^2	[Ar]4s^2	[Kr]5s^2	[Xe]6s^2	[Rn]7s^2
Ionization energy†	899.4	737.7	589.8	549.5	502.9	509.3
Electron affinity†	<0	<0	2.0	4.6	13.95	>0
Electronegativity	1.57	1.31	1.00	0.95	0.89	0.90
Ionic radius (Å)	0.31	0.66	0.99	1.13	1.35	1.43
Atomic radius (Å)	1.13	1.60	1.97	2.15	2.17	2.23
Enthalpy of fusion†	11.6	8.95	8.95	9.62	7.66	7.15
Enthalpy of vaporization†	297.6	127.6	154.7	154.4	150.9	136.7
Bond enthalpy of M_2†	9.46					
Standard reduction potential (volts)	−1.70 Be^{2+}/Be	−2.375 Mg^{2+}/Mg	−2.76 Ca^{2+}/Ca	−2.89 Sr^{2+}/Sr	−2.90 Ba^{2+}/Ba	−2.916 Ra^{2+}/Ra

†In kilojoules per mole.

Group III Elements

	Boron	Aluminum	Gallium	Indium	Thallium
Atomic number	5	13	31	49	81
Atomic mass	10.811	26.9815386	69.723	114.818	204.3833
Melting point (°C)	2300	660.37	29.78	156.61	303.5
Boiling point (°C)	3658	2467	2403	2080	1457
Density at 25°C (g cm^{-3})	2.34	2.702	5.904	7.30	11.85
Color	Yellow	Silver	Silver	Silver	Blue-white
Ground-state electron configuration	[He]2$s^2$2p^1	[Ne]3$s^2$3p^1	[Ar]3d^{10}4$s^2$4p^1	[Kr]4d^{10}5$s^2$5p^1	[Xe]4f^{14}5d^{10}6$s^2$6p^1
Ionization energy†	800.6	577.6	578.8	558.3	589.3
Electron affinity†	26.7	42.6	29	29	≈20
Electronegativity	2.04	1.61	1.81	1.78	1.83
Ionic radius (Å)	0.23 (+3)	0.51 (+3)	0.62 (+3)	0.81 (+3)	0.95 (+3)
Atomic radius (Å)	0.88	1.43	1.22	1.63	1.70
Enthalpy of fusion†	22.6	10.75	5.59	3.26	4.08
Enthalpy of vaporization†	508	291	272	243	182
Bond enthalpy of M_2†	295	167	116	106	≈63
Standard reduction potential (volts)	−0.890 $B(OH)_3$/B	−1.706 Al^{3+}/Al	−0.560 Ga^{3+}/Ga	−0.338 In^{3+}/In	0.719 Tl^{3+}/Tl

†In kilojoules per mole.

Group IV Elements

	Carbon	Silicon	Germanium	Tin	Lead
Atomic number	6	14	32	50	82
Atomic mass	12.0107	28.0855	72.64	118.710	207.2
Melting point (°C)	3550	1410	937.4	231.9681	327.502
Boiling point (°C)	4827	2355	2830	2270	1740
Density at 25°C (g cm^{-3})	2.25 (gr) 3.51 (dia)	2.33	5.323	5.75 (gray) 7.31 (white)	11.35
Color	Black (gr) Colorless (dia)	Gray	Gray-white	Silver	Blue-white
Ground-state electron configuration	$[He]2s^22p^2$	$[Ne]3s^23p^2$	$[Ar]3d^{10}4s^24p^2$	$[Kr]4d^{10}5s^25p^2$	$[Xe]4f^{14}5d^{10}6s^26p^2$
Ionization energy†	1086.4	786.4	762.2	708.6	715.5
Electron affinity†	121.85	133.6	≈120	≈120	35.1
Electronegativity	2.55	1.90	2.01	1.88	2.10
Ionic radius (Å)	0.15 (+4) 2.60 (−4)	0.42 (+4) 2.71 (−4)	0.53 (+4) 0.73 (+2) 2.72 (−4)	0.71 (+4) 0.93 (+2)	0.84 (+4) 1.20 (+2)
Atomic radius (Å)	0.77	1.17	1.22	1.40	1.75
Enthalpy of fusion†	105.0	50.2	34.7	6.99	4.774
Enthalpy of vaporization†	718.9	359	328	302	195.6
Bond enthalpy of M_2†	178	317	280	192	61
Standard reduction potential (volts)			−0.13 H_2GeO_3,H^+/Ge	−0.1364 Sn^{2+}/Sn	−0.1263 Pb^{2+}/Pb

†In kilojoules per mole.

Group V Elements

	Nitrogen	Phosphorus	Arsenic	Antimony	Bismuth
Atomic number	7	15	33	51	83
Atomic mass	14.00674	30.973762	74.92160	121.760	208.98040
Melting point (°C)	−209.86	44.1	817 (28 atm.)	630.74	271.3
Boiling point (°C)	−195.8	280	613 (subl.)	1750	1560
Density at 25°C (g cm^{-3})	0.808 (−196°C)	1.82 (white) 2.20 (red) 2.69 (black)	5.727	6.691	9.747
Color	Colorless		Gray	Blue-white	White
Ground-state electron configuration	$[He]2s^22p^3$	$[Ne]3s^23p^3$	$[Ar]3d^{10}4s^24p^3$	$[Kr]4d^{10}5s^25p^3$	$[Xe]4f^{14}5d^{10}6s^26p^3$
Ionization energy†	1402.3	1011.7	947	833.7	703.3
Electron affinity†	−7	72.03	≈80	103	91.3
Electronegativity	3.04	2.19	2.18	2.05	2.02
Ionic radius (Å)	1.71 (−3)	0.44 (+3) 2.12 (−3)	0.46 (+5) 0.58 (+3) 2.22 (−3)	0.62 (+5) 0.76 (+3) 2.45 (−3)	0.96 (+3)
Atomic radius (Å)	0.70	1.10	1.21	1.41	1.55
Enthalpy of fusion†	0.720	6.587	27.72	20.91	10.88
Enthalpy of vaporization†	5.608	59.03	334	262.5	184.6
Bond enthalpy of M_2†	945	485	383	289	194
Standard reduction potential (volts)	0.96 NO_3^-,H^+/NO	−0.276 H_3PO_4/H_3PO_3	0.234 As_2O_3,H^+/As	0.1445 Sb_2O_3,H^+/Sb	−0.46 Bi_2O_3,OH^-/Bi

†In kilojoules per mole.

The Chalcogens (Group VI Elements)

	Oxygen	Sulfur	Selenium	Tellurium	Polonium
Atomic number	8	16	34	52	84
Atomic mass	15.9994	32.065	78.96	127.60	(208.9824)
Melting point (°C)	−218.4	119.0 (mon.) 112.8 (rhom.)	217	449.5	254
Boiling point (°C)	−182.962	444.674	684.9	989.8	962
Density at 25°C (g cm^{-3})	1.14 (−183°C)	1.957 (mon.) 2.07 (rhom.)	4.79	6.24	9.32
Color	Pale blue (ℓ)	Yellow	Gray	Silver	Silver-gray
Ground-state electron configuration	$[He]2s^22p^4$	$[Ne]3s^23p^4$	$[Ar]3d^{10}4s^24p^4$	$[Kr]4d^{10}5s^25p^4$	$[Xe]4f^{14}5d^{10}6s^26p^4$
Ionization energy†	1313.9	999.6	940.9	869.3	812
Electron affinity†	140.97676	200.4116	194.967	190.15	≈180
Electronegativity	3.44	2.58	2.55	2.10	2.00
Ionic radius (Å)	1.40 (−2)	0.29 (+6) 1.84 (−2)	0.42 (+6) 1.98 (−2)	0.56 (+6) 2.21 (−2)	0.67 (+6) 2.30 (−2)
Atomic radius (Å)	0.66	1.04	1.17	1.43	1.67
Enthalpy of fusion†	0.4187	1.411	5.443	17.50	10
Enthalpy of vaporization†	6.819	238	207	195	90
Bond enthalpy of M_2†	498	429	308	225	
Standard reduction potential (volts)	1.229 $O_2,H^+/H_2O$	−0.508 S/S^{2-}	−0.78 Se/Se^{2-}	−0.92 Te/Te^{2-}	≈−1.4 Po/Po^{2-}

†In kilojoules per mole.

The Halogens (Group VII Elements)

	Fluorine	Chlorine	Bromine	Iodine	Astatine
Atomic number	9	17	35	53	85
Atomic mass	18.9984032	35.453	79.904	126.90447	(209.9871)
Melting point (°C)	−219.62	−100.98	−7.25	113.5	302
Boiling point (°C)	−188.14	−34.6	58.78	184.35	337
Density at 25°C (g cm^{-3})	1.108 (−189°C)	1.367 (−34.6°C)	3.119	4.93	
Color	Yellow	Yellow-green	Deep red	Violet-black	
Ground-state electron configuration	$[He]2s^22p^5$	$[Ne]3s^23p^5$	$[Ar]3d^{10}4s^24p^5$	$[Kr]4d^{10}5s^25p^5$	$[Xe]4f^{14}5d^{10}6s^26p^5$
Ionization energy†	1681.0	1251.1	1139.9	1008.4	≈930
Electron affinity†	328.0	349.0	324.7	295.2	≈270
Electronegativity	3.98	3.16	2.96	2.66	2.20
Ionic radius (Å)	1.33	1.81	1.96	2.20	≈2.27
Atomic radius (Å)	0.64	0.99	1.14	1.33	1.40
Enthalpy of fusion†	0.511	6.410	10.55	15.78	23.9
Enthalpy of vaporization†	6.531	20.347	29.56	41.950	
Bond enthalpy of M_2†	158	243	193	151	110
Standard reduction potential (volts)	2.87 F_2/F^-	1.358 Cl_2/Cl^-	1.065 Br_2/Br^-	0.535 I_2/I^-	≈0.2 At_2/At^-

†In kilojoules per mole.

The Noble Gases (Group VIII Elements)

	Helium	Neon	Argon	Krypton	Xenon	Radon
Atomic number	2	10	18	36	54	86
Atomic mass	4.002602	20.1797	39.948	83.798	131.293	(222.0176)
Melting point (°C)	−272.2 (26 atm)	−248.67	−189.2	−156.6	−111.9	−71
Boiling point (°C)	−268.934	−246.048	−185.7	−152.30	−107.1	−61.8
Density at 25°C (g cm^{-3})	0.147 (−270.8°C)	1.207 (−246.1°C)	1.40 (−186°C)	2.155 (−152.9°C)	3.52 (−109°C)	4.4 (−52°C)
Color	Colorless	Colorless	Colorless	Colorless	Colorless	Colorless
Ground-state electron configuration	$1s^2$	[He]$2s^22p^6$	[Ne]$3s^23p^6$	[Ar]$3d^{10}4s^24p^6$	[Kr]$4d^{10}5s^25p^6$	[Xe]$4f^{14}5d^{10}6s^26p^6$
Ionization energy†	2372.3	2080.6	1520.5	1350.7	1170.4	1037.0
Electron affinity†	<0	<0	<0	<0	<0	<0
Atomic radius (Å)	0.32	0.69	0.97	1.10	1.30	1.45
Enthalpy of fusion†	0.02093	0.3345	1.176	1.637	2.299	2.9
Enthalpy of vaporization†	0.1005	1.741	6.288	9.187	12.643	18.4

†In kilojoules per mole.

The Transition Elements

	Scandium	Yttrium	Lutetium	Titanium	Zirconium	Hafnium
Atomic number	21	39	71	22	40	72
Atomic mass	44.955912	88.90585	174.9668	47.867	91.224	178.49
Melting point (°C)	1541	1522	1656	1660	1852	2227
Boiling point (°C)	2831	3338	3315	3287	4504	4602
Density at 25°C (g cm^{-3})	2.989	4.469	9.840	4.54	6.506	13.31
Color	Silver	Silver	Silver	Silver	Gray-white	Silver
Ground-state electron configuration	[Ar]$3d^14s^2$	[Kr]$4d^15s^2$	[Xe]$4f^{14}5d^16s^2$	[Ar]$3d^24s^2$	[Kr]$4d^25s^2$	[Xe]$4f^{14}5d^26s^2$
Ionization energy†	631	616	523.5	658	660	654
Electron affinity†	18.1	29.6	≈50	7.6	41.1	≈0
Electronegativity	1.36	1.22	1.27	1.54	1.33	1.30
Ionic radius (Å)	0.81	0.93	0.848 (+3)	0.68	0.80	0.78
Atomic radius (Å)	1.61	1.78	1.72	1.45	1.59	1.56
Enthalpy of fusion†	11.4	11.4	19.2	18.62	20.9	25.5
Enthalpy of vaporization†	328	425	247	426	590	571
Standard reduction potential (volts)	−2.08 Sc^{3+}/Sc	−2.37 Y^{3+}/Y	−2.30 Lu^{3+}/Lu	−0.86 $TiO_2,H^+/Ti$	−1.43 $ZrO_2,H^+/Zr$	−1.57 $HfO_2,H^+/Hf$

†In kilojoules per mole.

continued

The Transition Elements (cont.)

	Vanadium	Niobium	Tantalum	Chromium	Molybdenum	Tungsten
Atomic number	23	41	73	24	42	74
Atomic mass	50.9415	92.90638	180.94788	51.9961	95.96	183.84
Melting point (°C)	1890	2468	2996	1857	2617	3410
Boiling point (°C)	3380	4742	5425	2672	4612	5660
Density at 25°C (g cm^{-3})	6.11	8.57	16.654	7.18	10.22	19.3
Color	Silver-white	Gray-white	Steel gray	Silver	Silver	Steel gray
Ground-state electron configuration	[Ar]$3d^34s^2$	[Kr]$4d^45s^1$	[Xe]$4f^{14}5d^36s^2$	[Ar]$3d^54s^1$	[Kr]$4d^55s^1$	[Xe]$4f^{14}5d^46s^2$
Ionization energy†	650	664	761	652.8	684.9	770
Electron affinity†	50.7	86.2	31.1	64.3	72.0	78.6
Electronegativity	1.63	1.60	1.50	1.66	2.16	2.36
Ionic radius (Å)	0.59 (+5) 0.63 (+4) 0.74 (+3) 0.88 (+2)	0.69 (+5) 0.74 (+4)	0.68 (+5)	0.63 (+3) 0.89 (+2)	0.62 (+6) 0.70 (+4)	0.62 (+6) 0.70 (+4)
Atomic radius (Å)	1.31	1.43	1.43	1.25	1.36	1.37
Enthalpy of fusion†	21.1	26.4		20.9	27.8	35.4
Enthalpy of vaporization†	512	722	781	394.7	589.2	819.3
Standard reduction potential (volts)	−1.2 V^{2+}/V	−0.62 $Nb_2O_5,H^+/Nb$	−0.71 $Ta_2O_5,H^+/Ta$	−0.74 Cr^{3+}/Cr	0.0 $H_2MoO_4,H^+/Mo$	−0.09 $WO_3,H^+/W$

	Manganese	Technetium	Rhenium	Iron	Ruthenium	Osmium
Atomic number	25	43	75	26	44	76
Atomic mass	54.938045	(97.9064)	186.207	55.845	101.07	190.23
Melting point (°C)	1244	2172	3180	1535	2310	3045
Boiling point (°C)	1962	4877	5627	2750	3900	5027
Density at 25°C (g cm^{-3})	7.21	11.50	21.02	7.874	12.41	22.57
Color	Gray-white	Silver-gray	Silver	Gray	White	Blue-white
Ground-state electron configuration	[Ar]$3d^54s^2$	[Kr]$4d^55s^2$	[Xe]$4f^{14}5d^56s^2$	[Ar]$3d^64s^2$	[Kr]$4d^75s^1$	[Xe]$4f^{14}5d^66s^2$
Ionization energy†	717.4	702	760	759.3	711	840
Electron affinity†	<0	≈ 53	≈ 14	15.7	≈ 100	≈ 106
Electronegativity	1.55	1.90	1.90	1.90	2.2	2.20
Ionic radius (Å)	0.80 (+2)		0.56 (+7) 0.27 (+4)	0.60 (+3) 0.72 (+2)	0.67 (+4)	0.69 (+6) 0.88 (+4)
Atomic radius (Å)	1.37	1.35	1.34	1.24	1.32	1.34
Enthalpy of fusion†	14.6	23.8	33.1	15.19	26.0	31.8
Enthalpy of vaporization†	279	585	778	414	649	678
Standard reduction potential (volts)	−0.183 Mn^{3+}/Mn	0.738 $TcO_4^-,H^+/TcO_2$	0.3 Re^{3+}/Re	−0.036 Fe^{3+}/Fe	0.49 Ru^{4+}/Ru^{3+}	0.85 $OsO_4,H^+/Os$

†In kilojoules per mole.

The Transition Elements (cont.)

	Cobalt	Rhodium	Iridium	Nickel	Palladium	Platinum
Atomic number	27	45	77	28	46	78
Atomic mass	58.933195	102.90550	192.217	58.6934	106.42	195.084
Melting point (°C)	1459	1966	2410	1453	1552	1772
Boiling point (°C)	2870	3727	4130	2732	3140	3827
Density at 25°C (g cm^{-3})	8.9	12.41	22.42	8.902	12.02	21.45
Color	Steel gray	Silver	Silver	Silver	Steel white	Silver
Ground-state electron configuration	[Ar]$3d^7 4s^2$	[Kr]$4d^8 5s^1$	[Xe]$4f^{14} 5d^7 6s^2$	[Ar]$3d^8 4s^2$	[Kr]$4d^{10}$	[Xe]$4f^{14} 5d^9 6s^1$
Ionization energy†	758	720	880	736.7	805	868
Electron affinity†	63.8	110	151	111.5	51.8	205.1
Electronegativity	1.88	2.28	2.20	1.91	2.20	2.28
Ionic radius (Å)	0.63 (+3) 0.72 (+2)	0.68 (+3)	0.68 (+4)	0.69 (+2)	0.65 (+4) 0.80 (+2)	0.65 (+4) 0.80 (+2)
Atomic radius (Å)	1.25	1.34	1.36	1.25	1.38	1.37
Enthalpy of fusion†	16.2	21.5	26.4	17.6	17.6	19.7
Enthalpy of vaporization†	373	557	669	428	353	564
Standard reduction potential (volts)	−0.28 Co^{2+}/Co	1.43 Rh^{4+}/Rh^{3+}	0.1 $Ir_2O_3/Ir,OH^-$	−0.23 Ni^{2+}/Ni	0.83 Pd^{2+}/Pd	1.2 Pt^{2+}/Pt

	Copper	Silver	Gold	Zinc	Cadmium	Mercury
Atomic number	29	47	79	30	48	80
Atomic mass	63.546	107.8682	196.966569	65.38	112.411	200.59
Melting point (°C)	1083.4	961.93	1064.43	419.58	320.9	−38.87
Boiling point (°C)	2567	2212	2807	907	765	356.58
Density at 25°C (g cm^{-3})	8.96	10.50	19.32	7.133	8.65	13.546
Color	Red	Silver	Yellow	Blue-white	Blue-white	Silver
Ground-state electron configuration	[Ar]$3d^{10} 4s^1$	[Kr]$4d^{10} 5s^1$	[Xe]$4f^{14} 5d^{10} 6s^1$	[Ar]$3d^{10} 4s^2$	[Kr]$4d^{10} 5s^2$	[Xe]$4f^{14} 5d^{10} 6s^2$
Ionization energy†	745.4	731.0	890.1	906.4	867.7	1007.0
Electron affinity†	118.5	125.6	222.749	<0	<0	<0
Electronegativity	1.90	1.93	2.54	1.65	1.69	2.00
Ionic radius (Å)	0.72 (+2) 0.96 (+1)	0.89 (+2) 1.26 (+1)	0.85 (+2) 1.37 (+1)	0.74 (+2)	0.97 (+2) 1.14 (+1)	1.10 (+2) 1.27 (+1)
Atomic radius (Å)	1.28	1.44	1.44	1.34	1.49	1.50
Enthalpy of fusion†	13.3	11.95	12.36	7.39	6.11	2.300
Enthalpy of vaporization†	304	285	365	131	112	59.1
Standard reduction potential (volts)	0.340 Cu^{2+}/Cu	0.800 Ag^+/Ag	1.42 Au^{3+}/Au	−0.763 Zn^{2+}/Zn	−0.403 Cd^{2+}/Cd	0.796 Hg_2^{2+}/Hg

† In kilojoules per mole.

The Lanthanide Elements

	Lanthanum	Cerium	Praseodymium	Neodymium	Promethium	Samarium	Europium
Atomic number	57	58	59	60	61	62	63
Atomic mass	138.90547	140.116	140.90765	144.242	(144.9127)	150.36	151.964
Melting point (°C)	921	798	931	1010	≈1080	1072	822
Boiling point (°C)	3457	3257	3212	3127	≈2400	1778	1597
Density at 25°C (g cm^{-3})	6.145	6.657	6.773	6.80	7.22	7.520	5.243
Color	Silver	Gray	Silver	Silver		Silver	Silver
Ground-state electron configuration	[Xe]$5d^16s^2$	[Xe]$4f^15d^16s^2$	[Xe]$4f^36s^2$	[Xe]$4f^46s^2$	[Xe]$4f^56s^2$	[Xe]$4f^66s^2$	[Xe]$4f^76s^2$
Ionization energy†	538.1	528	523	530	536	543	547
Electron affinity†	50			est. 50			
Electronegativity	1.10	1.12	1.13	1.14		1.17	
Ionic radius (Å)	1.15	0.92 (+4) 1.034 (+3)	0.90 (+4) 1.013 (+3)	0.995 (+3)	0.979 (+3)	0.964 (+3)	0.950 (+3) 1.09 (+2)
Atomic radius (Å)	1.87	1.82	1.82	1.81	1.81	1.80	2.00
Enthalpy of fusion†	5.40	5.18	6.18	7.13	12.6	8.91	(10.5)
Enthalpy of vaporization†	419	389	329	324		207	172
Standard reduction potential (volts)	−2.37 La^{3+}/La	−2.335 Ce^{3+}/Ce	−2.35 Pr^{3+}/Pr	−2.32 Nd^{3+}/Nd	−2.29 Pm^{3+}/Pm	−2.30 Sm^{3+}/Sm	−1.99 Eu^{3+}/Eu

	Gadolinium	Terbium	Dysprosium	Holmium	Erbium	Thulium	Ytterbium
Atomic number	64	65	66	67	68	69	70
Atomic mass	157.25	158.92535	162.500	164.93032	167.259	168.93421	173.054
Melting point (°C)	1311	1360	1409	1470	1522	1545	824
Boiling point (°C)	3233	3041	2335	2720	2510	1727	1193
Density at 25°C (g cm^{-3})	7.900	8.229	8.550	8.795	9.066	9.321	6.965
Color	Silver	Silver-gray	Silver	Silver	Silver	Silver	Silver
Ground-state electron configuration	[Xe]$4f^75d^16s^2$	[Xe]$4f^96s^2$	[Xe]$4f^{10}6s^2$	[Xe]$4f^{11}6s^2$	[Xe]$4f^{12}6s^2$	[Xe]$4f^{13}6s^2$	[Xe]$4f^{14}6s^2$
Ionization energy†	592	564	572	581	589	596.7	603.4
Electron affinity†				est. 50			
Electronegativity	1.20		1.22	1.23	1.24	1.25	
Ionic radius (Å)	0.938 (+3)	0.84 (+4) 0.923 (+3)	0.908 (+3)	0.894 (+3)	0.881 (+3)	0.869 (+3)	0.858 (+3) 0.93 (+2)
Atomic radius (Å)	1.79	1.76	1.75	1.74	1.73	1.72	1.94
Enthalpy of fusion†	15.5	16.3	17.2	17.2	17.2	18.2	9.2
Enthalpy of vaporization†	301	293	165	285	280	240	165
Standard reduction potential (volts)	−2.28 Gd^{3+}/Gd	−2.31 Tb^{3+}/Tb	−2.29 Dy^{3+}/Dy	−2.33 Ho^{3+}/Ho	−2.32 Er^{3+}/Er	−2.32 Tm^{3+}/Tm	−2.22 Yb^{3+}/Yb

†In kilojoules per mole.

The Actinide Elements

	Actinium	Thorium	Protactinium	Uranium	Neptunium	Plutonium	Americium
Atomic number	89	90	91	92	93	94	95
Atomic mass	(227.0277)	232.0381	231.0359	238.0289	(237.0482)	(244.0642)	(243.0614)
Melting point (°C)	1050	1750	1600	1132.3	640	624	994
Boiling point (°C)	3200	4790		3818	2732	3232	2607
Density at 25°C (g cm^{-3})	10.07	11.72	15.37	18.95	20.25	19.84	13.67
Color	Silver	Silver	Silver	Silver	Silver	Silver	Silver
Ground-state electron configuration	[Rn]$6d^{1}7s^{2}$	[Rn]$6d^{2}7s^{2}$	[Rn]$5f^{2}6d^{1}7s^{2}$	[Rn]$5f^{3}6d^{1}7s^{2}$	[Rn]$5f^{4}6d^{1}7s^{2}$	[Rn]$5f^{6}7s^{2}$	[Rn]$5f^{7}7s^{2}$
Ionization energy†	499	587	568	587	597	585	578
Electronegativity	1.1	1.3	1.5	1.38	1.36	1.28	1.3
Ionic radius (Å)	1.11 (+3)	0.99 (+4)	0.89 (+5) 0.96 (+4) 1.05 (+3)	0.80 (+6) 0.93 (+4) 1.03 (+3)	0.71 (+7) 0.92 (+4) 1.01 (+3)	0.90 (+4) 1.00 (+3)	0.89 (+4) 0.99 (+3)
Atomic radius (Å)	1.88	1.80	1.61	1.38	1.30	1.51	1.84
Enthalpy of fusion†	14.2	18.8	16.7	12.9	9.46	3.93	14.4
Enthalpy of vaporization†	293	575	481	536	337	348	238
Standard reduction potential (volts)	−2.6 Ac^{3+}/Ac	−1.90 Th^{4+}/Th	−1.0 $PaO_2^{+},H^{+}/Pa$	−1.8 U^{3+}/U	−1.9 Np^{3+}/U	−2.03 Pu^{3+}/Pu	−2.32 Am^{3+}/Am

	Curium	Berkelium	Californium	Einsteinium	Fermium	Mendelevium	Nobelium
Atomic number	96	97	98	99	100	101	102
Atomic mass	(247.0703)	(247.0703)	(251.0796)	(252.0830)	(257.0951)	(258.0984)	(259.1011)
Melting point (°C)	1340						
Boiling point (°C)							
Density at 25°C (g cm^{-3})	13.51	14					
Color	Silver	Silver	Silver	Silver			
Ground-state electron configuration	[Rn]$5f^{7}6d^{1}7s^{2}$	[Rn]$5f^{9}7s^{2}$	[Rn]$5f^{10}7s^{2}$	[Rn]$5f^{11}7s^{2}$	[Rn]$5f^{12}7s^{2}$	[Rn]$5f^{13}7s^{2}$	[Rn]$5f^{14}7s^{2}$
Ionization energy†	581	601	608	619	627	635	642
Electronegativity	1.3	1.3	1.3	1.3	1.3	1.3	1.3
Ionic radius (Å)	0.88 (+4) 1.01 (+3) 1.19 (+2)	0.87 (+4) 1.00 (+3) 1.18 (+2)	0.86 (+4) 0.99 (+3) 1.17 (+2)	0.85 (+4) 0.98 (+3) 1.16 (+2)	0.84 (+4) 0.97 (+3) 1.15 (+2)	0.84 (+4) 0.96 (+3) 1.14 (+2)	0.83 (+4) 0.95 (+3) 1.13 (+2)
Standard reduction potential (volts)	−2.06 Cm^{3+}/Cm	−1.05 Bk^{3+}/Bk	−1.93 Cf^{3+}/Cf	−2.0 Es^{3+}/Es	−1.96 Fm^{3+}/Fm	−1.7 Md^{3+}/Md	−1.2 No^{3+}/No

†In kilojoules per mole.

The Transactinide Elements†

	Lawrencium	Rutherfordium	Dubnium	Seaborgium	Bohrium	Hassium	Meitnerium
Atomic number	103	104	105	106	107	108	109
Atomic mass	(262)	(265)	(268)	(271)	(270)	(277)	(276)
Melting point (°C)	1600						
Ground-state electron configuration	[Rn]$5f^{14}7s^{2}7p^{1}$	[Rn]$5f^{14}6d^{2}7s^{2}$	[Rn]$5f^{14}6d^{3}7s^{2}$	[Rn]$5f^{14}6d^{4}7s^{2}$	[Rn]$5f^{14}6d^{5}7s^{2}$	[Rn]$5f^{14}6d^{6}7s^{2}$	[Rn]$5f^{14}6d^{7}7s^{2}$
Ionization energy		490	640	730	660	750	840

†All missing data are unknown.

Answers to Odd-Numbered Problems

G

APPENDIX

CHAPTER 1

1. Mercury is an element; water and sodium chloride are compounds. The other materials are mixtures: seawater and air are homogeneous; table salt and wood are heterogeneous. Mayonnaise appears homogeneous to the naked eye, but under magnification shows itself as water droplets suspended in oil.
3. Substances
5. 16.9 g
7. **(a)** 2.005 and 1.504 g
   (b) 2.005/1.504 = 1.333 = 4/3; SiN (or a multiple)
9. 2, 3, 4, 5
11. **(a)** HO (or any multiple, such as H_2O_2)
    (b) All would give H_2 and O_2 in 1:1 ratio.
13. 2.0 L N_2O, 3.0 L O_2
15. 0.052 m
17. 28.086
19. 11.01
21. **(a)** 145/94 = 1.54
    (b) 94 electrons
23. 95 protons, 146 neutrons, 95 electrons
25. 2.107298×10^{-22} g
27. **(a)** 283.89 **(b)** 115.36 **(c)** 164.09 **(d)** 158.03
    (e) 132.13
29. The total count of 2.52×10^9 atoms of gold has a mass of only 8.3×10^{-13} g, which is far too small to detect with a balance.
31. 1.041 mol
33. 2540 cm^3 = 2.54 L
35. 7.03×10^{23} atoms

CHAPTER 2

1. Pt: 47.06%; F: 36.67%; Cl: 8.553%; O: 7.720%
3. N_4H_6, H_2O, LiH, $C_{12}H_{26}$
5. 0.225%
7. $Zn_3P_2O_8$
9. Fe_3Si_7
11. BaN, Ba_3N_2
13. **(a)** 0.923 g C, 0.077 g H
    (b) No
    (c) 92.3% C, 7.7% H
    (d) CH
15. C_4F_8
17. **(a)** 62.1
    (b) 6
    (c) 56
    (d) Si (atomic mass 28.1), N (atomic mass 14.0)
    (e) Si_2H_6
19. **(a)** $3\,H_2 + N_2 \longrightarrow 2NH_3$
    (b) $2\,K + O_2 \longrightarrow K_2O_2$
    (c) $PbO_2 + Pb + 2\,H_2SO_4 \longrightarrow 2\,PbSO_4 + 2\,H_2O$
    (d) $2\,BF_3 + 3\,H_2O \longrightarrow B_2O_3 + 6\,HF$
    (e) $2\,KClO_3 \longrightarrow 2\,KCl + 3\,O_2$
    (f) $CH_3COOH + 2\,O_2 \longrightarrow 2\,CO_2 + 2\,H_2O$
    (g) $2\,K_2O_2 + 2\,H_2O \longrightarrow 4\,KOH + O_2$
    (h) $3\,PCl_5 + 5\,AsF_3 \longrightarrow 3\,PF_5 + 5\,AsCl_3$
21. **(a)** 12.06 g **(b)** 1.258 g **(c)** 4.692 g
23. 7.83 g $K_2Zn_3[Fe(CN)_6]_2$
25. 0.134 g SiO_2
27. 1.18×10^3 g
29. 418 g KCl; 199 g Cl_2
31. **(a)** 58.8 **(b)** Probably nickel (Ni)
33. 42.49% NaCl, 57.51% KCl
35. 14.7 g NH_4Cl, 5.3 g NH_3
37. 303.0 g Fe; 83.93%

CHAPTER 3

1. Melting point 1250°C (obs. 1541°C), boiling point 2386°C (obs. 2831°C), density 3.02 g cm^{-3} (obs. 2.99)
3. SbH_3, HBr, SnH_4, H_2Se
5. **(a)** $F_x = 0$, $F_y = -5.77 \times 10^{-9}$ N
   (b) $V = 1.15 \times 10^{-18}$ J = 7.20 eV
7. **(a)** At 2.12 Å separation, the magnitude of the force is 5.13×10^{-9} N. At 0.529 Å separation, the magnitude of the force is 8.24×10^{-8} N.
   (b) −20.4 eV
   (c) 2.68×10^6 m s^{-1}

9. **(a)** Sr **(b)** Rn **(c)** Xe **(d)** Sr

11. Using data for Be from Table 3.l, calculate $\log(IE_n)$ for $n = 1, 2, 3$, and 4. The graph of $\log(IE_n)$ versus n shows a dramatic increase between $n = 2$ and $n = 3$, suggesting two easily removed electrons outside a stable helium-like inner shell containing two electrons.

13. **(a)** Cs **(b)** F **(c)** K **(d)** At

15. K < Si < S < O < F

17.

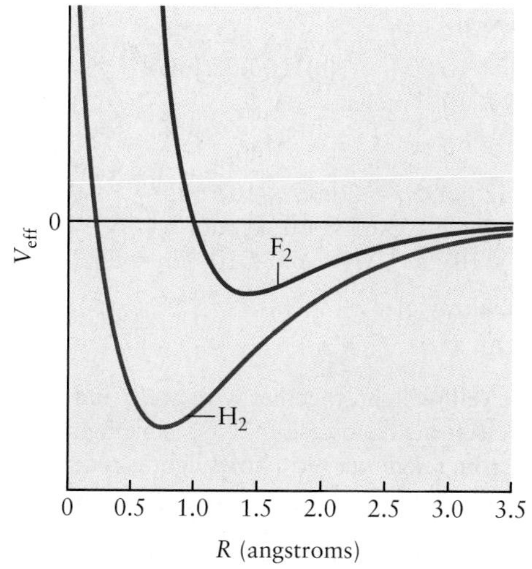

19.

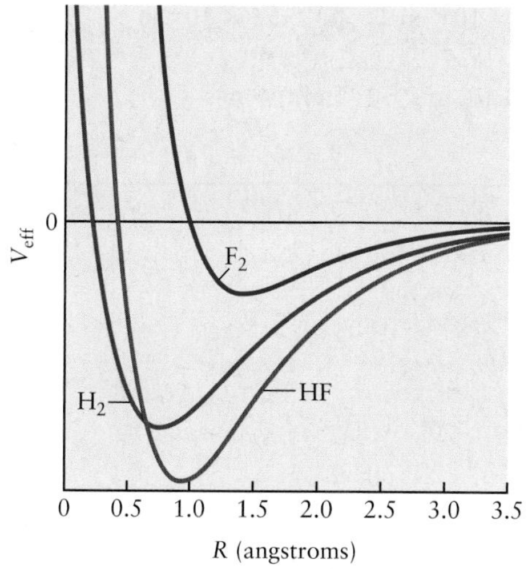

21. **(a)** 86 e^- (8 valence, 78 core)
(b) 37 e^- (1 valence, 36 core)
(c) 36 e^- (8 valence, 28 core)
(d) 52 e^- (6 valence, 46 core)

23. $K(g) + Cl(g) \longrightarrow K^+(g) + Cl^-(g)\ \Delta E = 70\ \text{kJ mol}^{-1}$

$K(g) + Cl(g) \longrightarrow K^2(g) + Cl^1(g)\ \Delta E = 1203\ \text{kJ mol}^{-1}$

25. 450 kJ mol^{-1}

27. The As–H bond length will lie between 1.42 and 1.71 Å (observed: 1.52 Å). SbH_3 will have the weakest bond.

29. Bond lengths are often shorter than the sum of atomic radii in polar molecules because of the electrostatic attraction between the oppositely charged ends of the dipole. Such shortening is slight in HI, indicating that it is not very polar.

31. Most polar: N—P > C—N > N—O > N—N: least polar

33. **(a)** CI_4 **(b)** OF_2 **(c)** SiH_4

35. ClO, 16%; KI, 74%; TlCl, 38%; InCl, 33%

37. HF, 40%; HCl, 19%; HBr, 14%; HI, 8%; CsF, 87%

39. **(a)**

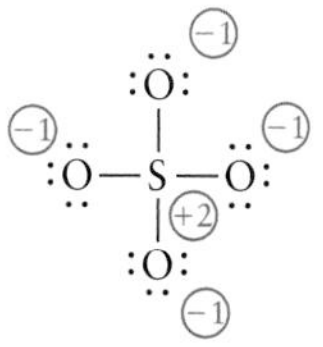

(b)

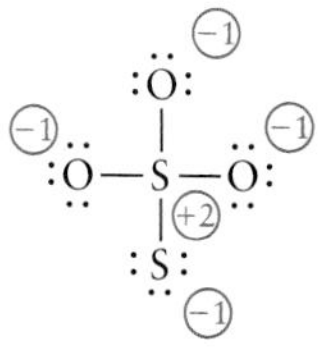

(c)

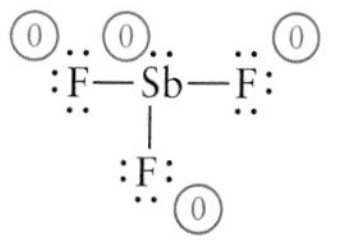

(d)

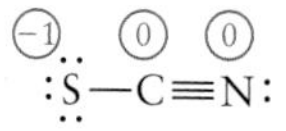

41. The first is favored.

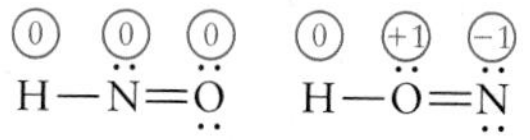

43. **(a)** Group IV, CO_2 **(b)** Group VII, Cl_2O_7
(c) Group V, V, NO_2^- **(d)** Group VI, VI, HSO_4^-

45. **(a)**

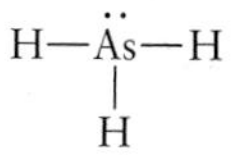

(b)

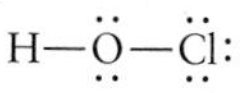

(c)

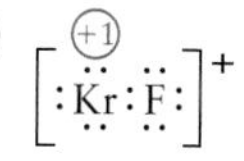

(d)

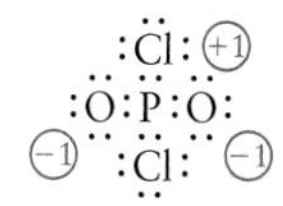

47.

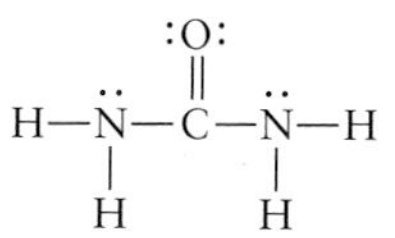

Bond lengths: N—H 1.01×10^{-10} m, N—C 1.47×10^{-10} m, Cu = O 1.20×10^{-10} m

49.

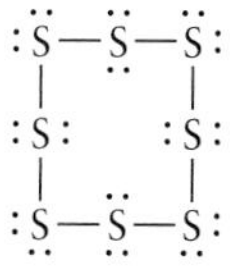

51. **(a)**

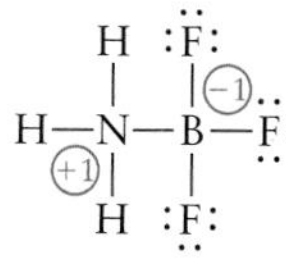

(b)

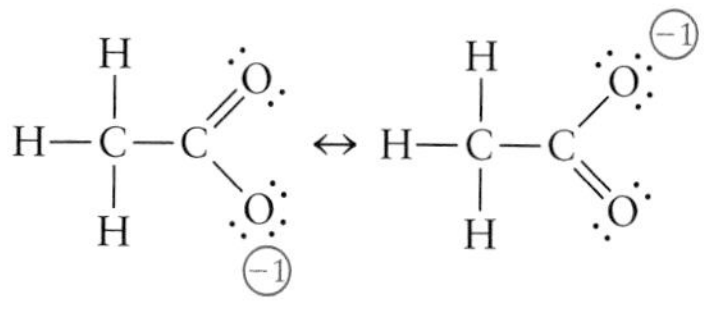

(c)

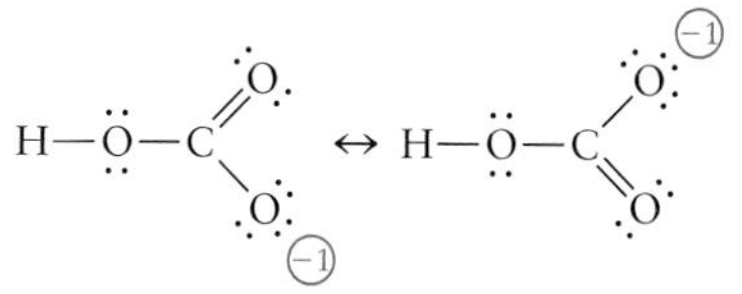

53.

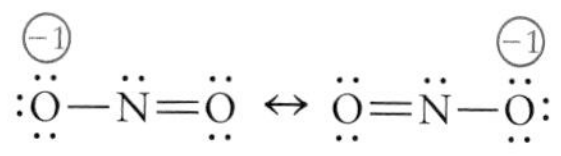

Between 1.18×10^{-10} m and 1.43×10^{-10} m.

55.

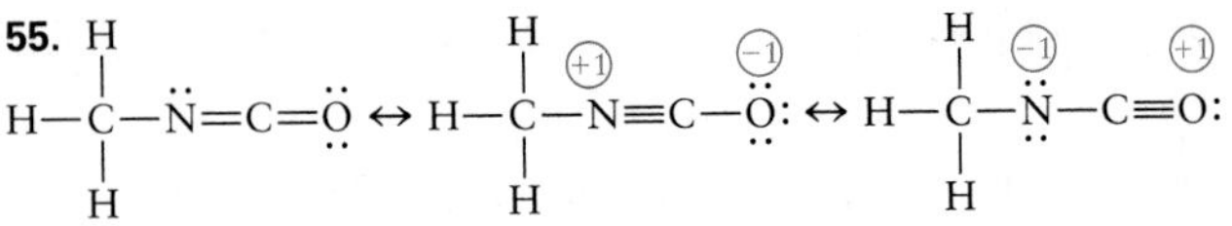

57. **(a)**

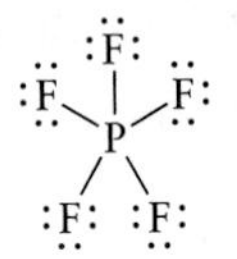

(b)

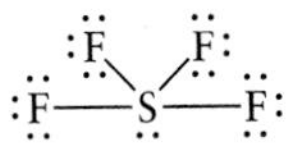

(c)

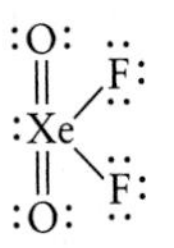

59. **(a)** SN = 4, tetrahedral **(b)** SN = 3, trigonal planar
(c) SN = 6, octahedral **(d)** SN = 4, pyramidal
(e) SN = 5, distorted T-shape

61. **(a)** SN = 6, square planar
(b) SN = 4, bent, angle < 109.5°
(c) SN = 4, pyramidal, angle < 109.5°
(d) SN = 2, linear

63. **(a)** SO_3 **(b)** NF_3 **(c)** NO_2^- **(d)** CO_3^{2-}

65. Only **(d)** and **(e)** are polar.

67. No, because VSEPR theory predicts a steric number of 3 and a bent molecule in both cases.

69. **(a)** Linear **(b)** The N end

71.

+2−1	+2−2+1	−4+1	+2+4−2	+6−2
$SrBr_2$	$Zn(OH)_4^{2-}$	SiH_4	$CaSiO_3$	$Cr_2O_7^{2-}$
+2+5−2−1	$+1-\frac{1}{2}$	+1−1		
$Ca_5(PO_4)_3F$	KO_2	CsH		

73. **(a)** CsCl, cesium chloride Cs· + :C̤̈l· → (Cs^+)(:C̤̈l:$^-$)
(b) $CaAt_2$, calcium astatide Ċa· + 2:Ȧ̤t· → (Ca^{2+})(:Ä̤t:$^-$)$_2$
(c) Al_2S_3, aluminum sulfide
2·Ȧl· + 3 ·Ṡ̤· → $(Al^{3+})_2$(:S̤̈:$^{2-}$)$_3$
(d) K_2Te, potassium telluride 2 K· + ·Ṫ̤e· → $(K^+)_2$(:T̤̈e:$^{2-}$)

75. **(a)** Aluminum oxide **(b)** Rubidium selenide
(c) Ammonium sulfide **(d)** Calcium nitrate
(e) Cesium sulfate
(f) Potassium hydrogen carbonate

77. **(a)** AgCN **(b)** $Ca(OCl)_2$ **(c)** K_2CrO_4
(d) Ga_2O_3 **(e)** KO_2 **(f)** $Ba(HCO_3)_2$

79. Na_3PO_4, sodium phosphate

81. **(a)** SiO_2 **(b)** $(NH_4)_2CO_3$ **(c)** PbO_2
(d) P_2O_5 **(e)** CaI_2 **(f)** $Fe(NO_3)_3$

83. **(a)** Copper(I) sulfide and copper(II) sulfide
(b) Sodium sulfate
(c) Tetraarsenic hexaoxide
(d) Zirconium(IV) chloride
(e) Dichlorine heptaoxide
(f) Gallium(I) oxide

CHAPTER 4

1. 0.66 m s^{-1}

3. 3.04 m

5. **(a)** 5.00×10^5 s^{-1} **(b)** 4.4 min

7. λ = 1.313 m; time = 0.0873 s

9. 2.7 K

11. Red

13. 550 nm, green

15. **(a)** 3.371×10^{-19} J **(b)** 203.0 kJ mol^{-1}
(c) 4.926×10^{-3} mol s^{-1}

17. 5.895×10^{-7} m

19. r_3 = 0.0952 nm, $E_3 = 26.06 \times 10^{-18}$ J,
energy per mole = 3.65×10^3 kJ mol^{-1},
$\nu = 1.14 \times 10^{16}$ s^{-1}, $\lambda = 2.63 \times 10^{-8}$ m = 26.3 nm

21. 72.90 nm, ultraviolet

23. Blue

25. Part of the yellow light, together with green and blue light, will eject electrons from cesium. No visible light will eject electrons from selenium; ultraviolet light is required.

27. **(a)** 7.4×10^{-20} J **(b)** 4.0×10^5 m s^{-1}

29. **(a)** 100 cm, 33 cm **(b)** two nodes

31. **(a)** 7.27×10^{-7} m **(b)** 3.96×10^{10} m
(c) 2.2×10^{-34} m

35. **(a)** 5.8×10^4 m s^{-1} **(b)** 7.9 m s^{-1}

37. **(a)**

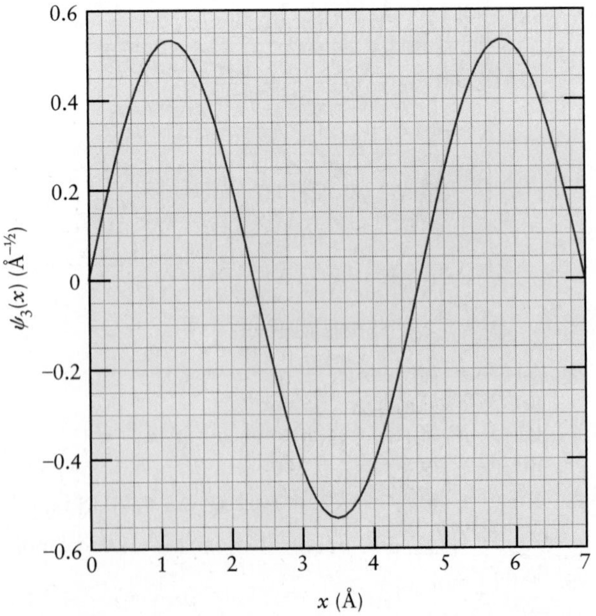

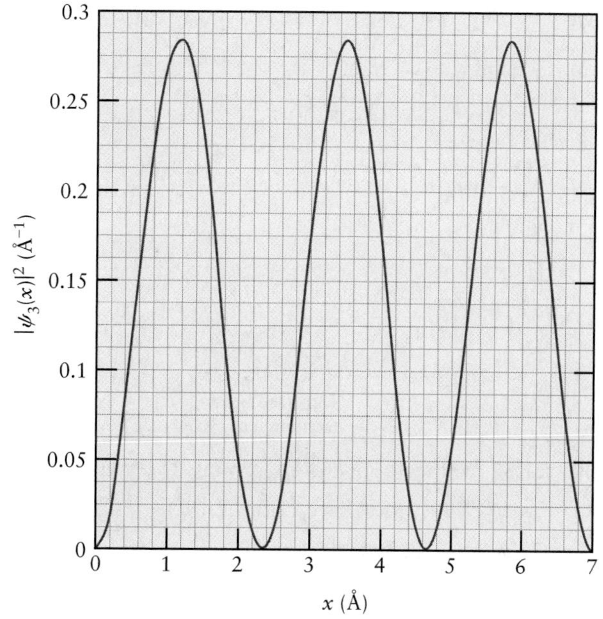

The probability of finding this particle between $x_1 =$ 3 Å and $x_2 = 4$ Å is

$$\int_{3\text{ Å}}^{4\text{ Å}} \frac{2}{7\text{Å}} \sin^2\left(\frac{3\pi x}{7\text{ Å}}\right) dx = 0.25 = 25\%$$

If you are not yet comfortable with integrals, you can estimate the shaded area under the curve as 120 rectangles multiplied by the area per rectangle in the graph of $|\psi|^2$ above, (0.01 Å^{-1})(0.2 Å) = 0.002. This gives a probability of 0.24 = 24%, so the estimated count of rectangles was close to correct.

(b) $$\mathcal{H}\psi(x) = -\frac{h^2}{8\pi^2 m}\frac{d^2}{dx^2}\sqrt{\frac{2}{7\text{ Å}}}\sin\left(\frac{3\pi x}{7\text{ Å}}\right)$$

$$= -\frac{h^2}{8\pi^2 m}\left(\frac{3\pi}{7\text{ Å}}\right)\sqrt{\frac{2}{7\text{ Å}}}\frac{d}{dx}\cos\left(\frac{3\pi x}{7\text{ Å}}\right)$$

$$= \frac{h^2}{8\pi^2 m}\left(\frac{3\pi}{7\text{ Å}}\right)^2\sqrt{\frac{2}{7\text{ Å}}}\sin\left(\frac{3\pi x}{7\text{ Å}}\right)$$

This is indeed a constant, E, multiplied by the original $\psi = \sqrt{\frac{2}{7\text{ Å}}}\sin\left(\frac{3\pi x}{7\text{ Å}}\right)$

So $E = \frac{h^2}{8\pi^2 m}\left(\frac{3\pi}{7\text{ Å}}\right)^2 = \frac{3^3 h^2}{8m(7\text{ Å})^2}$, in agreement with Equation 4.37

39. $E_1 = 3.36 \times 10^{-18}$ J; $E_2 = 1.34 \times 10^{-17}$ J; $E_3 = 3.02 \times 10^{-17}$ J; $\lambda = 1.97 \times 10^8$ m

41. **(a)** $\tilde{\Psi}_{21} = \tilde{\Psi}_{12}$ because the x- and y-axes are equivalent in a square box.
(b) Exchanging x and y corresponds to a 90-degree rotation.
(c) Exchanging labels cannot change the energy of the particle, a physically observable quantity.

CHAPTER 5

1. Only **(b)** is allowed.

3. **(a)** $4p$ **(b)** $2s$ **(c)** $6f$

5. 2.0×10^{-7} and 0

7. **(a)** 2 radial, 1 angular **(b)** 1 radial, 0 angular **(c)** 2 radial, 3 angular

9. $\psi_{p_z}^2 \propto \cos^2\theta = 0$ for $\theta = \pi/2$; d_{xz} nodal planes are the y–z and x–y planes; $d_{x^2-y^2}$ nodal planes are two planes containing the z-axis at 45° from the x- and y-axes.

11. 3.17 Å, 2.64 Å

13. $\varepsilon_{2s} = -0.41$ Ry $= -538$ kJ mol^{-1} and $\bar{r} = 3.9a_0$.

15. -1 Ry exactly, -0.397 Ry, -0.376 Ry

17. **(a)** $1s^2 2s^2 2p^2$ **(b)** [Ar]$3d^{10}4s^2 4p^4$ **(c)** [Ar]$3d^6 4s^2$

19. Be^+: $1s^2 2s^1$; C^-: $1s^2 2s^2 2p^3$; Ne^{2+}: $1s^2 2s^2 2p^4$; Mg^+: [Ne]$3s^1$; P^{2+}: [Ne]$3s^2 3p^1$; Cl^-: [Ne]$3s^2 3p^6$; As^+: [Ar]$3d^{10}4s^2 4p^2$; I^-: [Kr]$4d^{10}5s^2 5p^6$. All except Cl^- and I^- are paramagnetic.

21. **(a)** In **(b)** S^{2-} **(c)** Mn^{4+}

23. 117

25. First "noble gases" at $Z = 1, 5, 9$

27. 9.52 eV or 1.52×10^{-18} J

31. **(a)** $Z_{eff,1s} = 7.12$ **(b)** $Z_{eff,2s}$ approx equal 3.2 **(c)** $Z_{eff,2p}$ approx equal 1.9

33. **(a)** K **(b)** Cs **(c)** Kr **(d)** K **(e)** Cl^-

35. **(a)** S^{2-} **(b)** Ti^{2+} **(c)** Mn^{2+} **(d)** Sr^{2+}

37. **(a)** For definition see Section 3.4. Helium has two electrons in the 1s orbital, which has the smallest radius. Moreover, each electron only partially screens the other from the nucleus, which has charge +2. Therefore, removal of the first electron requires considerable energy.
(b) Li, because Li^+ is essentially like He and explanation in **(a)** applies.
(c) 50.4 nm

39. **(a)** Start with Ca^{2+} and Ar, which are isoelectronic; that is, they have the same number of electrons. Ca^{2+} has the higher nuclear charge and is therefore smaller. Mg is above Ca in Group 2, so its 2+ ion is smaller than that of Ca. Br^- is larger than Cl^- in Group 7, which is, in turn, larger than Ar. Therefore, $Mg^{2+} < Ca^{2+} <$ Ar $< Br^2$.
(b) Na has one s electron, well shielded from the nucleus; Ne and Na^+ are isoelectronic, but Na^+ has a net positive charge; O has an unfilled shell, but no special stability relative to Ne. Therefore, Na < O < Ne < Na^+.
(c) Al is metallic, and thus electropositive; electronegativity sequence of others follows their relative horizontal positions in the periodic table. Therefore, Al < H < O < F.

41. 318.4 nm, near ultraviolet

CHAPTER 6

1. 0, 1, 2, 3, 2, 3 (6 σ orbitals in order of increasing energy)

3. All circles.

5. The $1\sigma_g$ MO

7. State **(b)** has higher energy because it has two electrons in the anti-bonding MO.

9. H_2

11. He_2^+

15. Smaller bond energy; larger bond length

17. **(a)** F_2: $(\sigma_{g2s})^2(\sigma_{u2s}^*)^2(\sigma_{g2p})^2(\pi_{u2p})^4(\pi_{g2p}^*)^4$;
F_2^+: $(\sigma_{g2s})^2(\sigma_{u2s}^*)^2(\sigma_{g2p})^2(\pi_{u2p})^4(\pi_{g2p}^*)^3$
(b) F_2: 1, F_2^+ : $\frac{3}{2}$
(c) F_2^+ should be paramagnetic.
(d) $F_2 < F_2^+$

19. $(\sigma_{g3s})^2(\sigma_{u3s}^*)^2(\sigma_{g3p_z})^2(\pi_{u3p})^4(\pi_{g3p}^*)^2$ bond order = 2; paramagnetic

21. **(a)** F, 1 **(b)** N, $2\frac{1}{2}$ **(c)** O, $1\frac{1}{2}$

23. **(a)** is diamagnetic; **(b)** and **(c)** are paramagnetic.

25. Bond order $2\frac{1}{2}$, paramagnetic

27. The outermost electron in CF is in a π_{2p}^* molecular orbital. Removing it gives a stronger bond.

29. $(\sigma_{1s})^2(\sigma_{1s}^*)^2$ bond order = 0. It should be unstable.

31. Ground-state electronic configurations

CF: $(\sigma_{2s})^2(\sigma_{2s}^*)^2(\pi_{2p_x},\pi_{2p_y})^4(\sigma_{2p_z})^2(\pi_{2p_x}^*,\pi_{2p_y}^*)^1$

CH: $(\sigma^{nb})^2\sigma^1(\pi_x^{nb},\pi_y^{nb})^2$

CH^+: $(\sigma^{nb})^2\sigma^1(\pi_x^{nb},\pi_y^{nb})^1$

CN^-: $(\sigma_{2s})^2(\sigma_{2s}^*)^2(\pi_{2p_x},\pi_{2p_y})^4(\sigma_{2p_z})^2$

CF, CH, and CH^+ have unpaired electrons.

33. Ground-state electronic configurations and bond orders:

CF^+: $(\sigma_{2s})^2(\sigma_{2s}^*)^2(\pi_{2p_x},\pi_{2p_y})^4(\sigma_{2p_z})^2$
BO = (0.5)(8 − 2) = 3

CF^-: $(\sigma_{2s})^2(\sigma_{2s}^*)^2(\pi_{2p_x},\pi_{2p_y})^4(\sigma_{2p_z})^2(\pi_{2p_x}^*,\pi_{2p_y}^*)^2$
BO = (0.5)(8 − 4) = 2

NO: $(\sigma_{2s})^2(\sigma_{2s}^*)^2(\pi_{2p_x},\pi_{2p_y})^4(\sigma_{2p_z})^2(\pi_{2p_x}^*,\pi_{2p_y}^*)^1$
BO = (0.5)(8 − 3) = 2.5

Bond dissociation energies are ranked $CF^+ > NO > CF^-$ because bond dissociation energy correlates with bond order.

35. **(a)** IE = 15.6 eV, from σ_{g2p_z}, which is highest occupied level.
(b) IE = 16.7 eV, from $\pi_{u2p_{xy}}$, which is second-highest occupied level.

37.

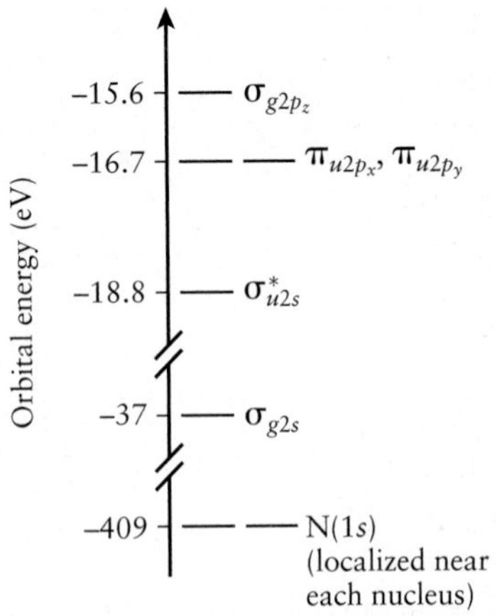

39. 11.88 eV: the $4p_x$ or the $4p_y$ orbital; 15.2 eV: the σ orbital

41. Li_2: $\psi_\sigma^{\text{bond}}(1, 2; R_{AB}) = c_1[2s^A(1)2s^B(2) + 2s^A(2)2s^B(1)]$
C_2: $\psi_\pi^{\text{bond}}(1, 2, 3, 4; R_{AB}) =$
$c_1R_{AB}[2p_y^A(1)\,2p_y^B(2)][2p_x^A(3)\,2p_x^B(4)] +$
$c_1R_A\,[2p_y^A(2)\,2p_y^B(1)][2p_x^A(4)\,2p_x^B(3)]$

43. Because the ground state of Be has no unpaired electrons, the simple VB model predicts that a bond will not form between two Be atoms. The same result is predicted by the LCAO approach. Two of the valence electrons in Be_2 occupy a bonding molecular orbital, but the other two must occupy an antibonding molecular orbital, leading to bond order zero.

45. $\psi_\sigma^{\text{bond}}(1, 2; R_{BH}) = C_1[1s^H(1)2p_z^B(2)] + C_2[1s^H(2)2p_z^B(1)]$

The simple VB model predicts a diatomic molecule BH. This is incorrect. The correct prediction is BH_3.

47. N has one unpaired electron in each of its $2p$ orbitals. Each of these can overlap with H 1s to form an s bond whose wave function has the form $\psi_\sigma = C_1[2p_a^N(1)1s^H(2)] + C_2[2p_a^N(2)1s^H(1)]$ where a is x, y, or z.

Because the $2p$ orbitals are mutually perpendicular, the simple VB model predicts a trigonal pyramid with angles of 90 degrees.

49. sp^3 hybridization of central N^- bent molecular ion

51. **(a)** sp^3 on C, tetrahedral
(b) sp on C, linear
(c) sp^3 on O, bent
(d) sp^3 on C, pyramidal
(e) sp on Be, linear

53. sp^2 hybrid orbitals. ClO_3^+ is trigonal planar, ClO_2^+ is bent.

55. sp^3 hybrid orbitals, tetrahedral

59. Sixteen electrons, so molecule is linear; sp hybridization of central N gives two σ bonds, with $2p_z$ orbitals on outer nitrogen atoms (four electrons). Lone pairs on both $2s$ orbitals on outer nitrogen atoms (four atoms). π system as in Figure 6.22: $(\pi)^4(\pi^{nb})^4$ with eight electrons. Total bond order = 4; bond order 2 per N^- bond. N_3 and N_3^+ should be bound. N_3 and N_3^+ are paramagnetic.

61. $\{\ddot{\text{O}}=\ddot{\text{N}}-\ddot{\text{O}}: \leftrightarrow :\ddot{\text{O}}-\ddot{\text{N}}=\ddot{\text{O}}\}^-$

$SN = 3$, sp^2 hybridization, bent molecule. The $2p_z$ orbitals perpendicular to the plane of the molecule can be combined into a π molecular orbital containing one pair of electrons. This orbital adds bond order 1/2 to each N—O bond, for a total bond order of 3/2 per bond. There is also a π_z^{nb} (non-bonding) MO occupied by two electrons, but that does not affect the bond order.

63. They are similar to the MOs of benzene, Figure 6.54, with a few important differences. N is more electronegative than C, so the lowest π MO has increased electron density near the N atom, as shown at http://csi.chemie.tu-darmstadt.de/ak/immel/tutorials/orbitals/molecular/pyridine.html.

The coefficient multiplying the N $2p_z$ orbital is larger for the linear combination of AOs that is used to construct the lowest π MO. Because the square of the coefficient of this atomic orbital summed over all the π MOs must equal 1, it contributes less to the highest π^* MO.

Finally, the presence of the N atom breaks the degeneracy in the middle two sets of π MOs, with the orbitals that have contributions from a $2p_z$ orbital on the N atom lowering with respect to the other in each set. Thus the π orbitals of pyridine have six different energies, and the lowest three in energy are filled.

CHAPTER 7

1. Yes, if the amount of knocking is less than that of iso-octane. Examples are the BTX compounds.

3. Ethane; $2\ C_2H_6(g) + \frac{7}{2}\ O_2(g) \longrightarrow 2\ CO_2(g) + 3\ H_2O(\ell)$

5. **(a)** $C_{10}H_{22} \longrightarrow C_5H_{10} + C_5H_{12}$
(b) Two: 1-pentene and 2-pentene

7. **(a)**

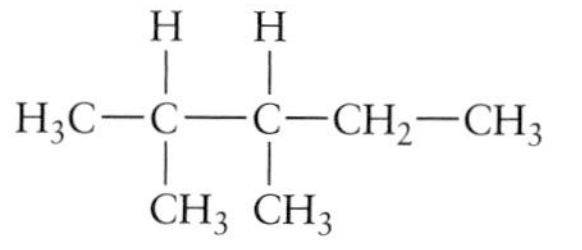

(b)

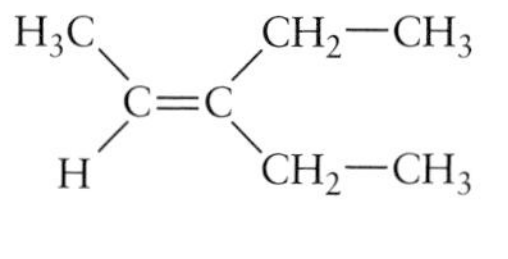

(c)

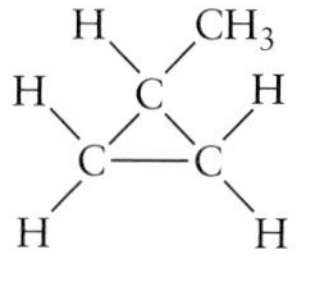

(d)

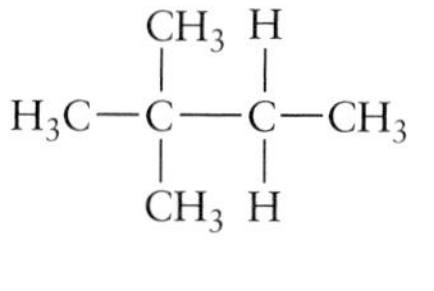

(e)

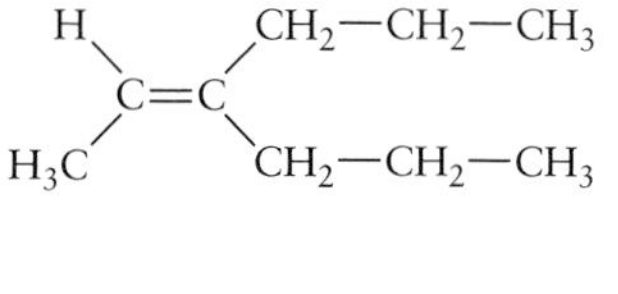

(f)

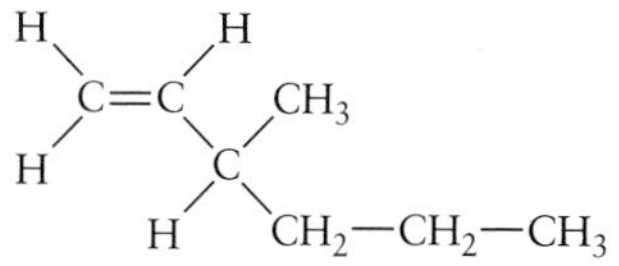

(g)

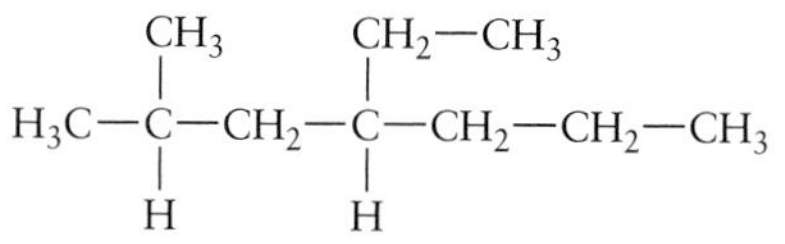

(h)

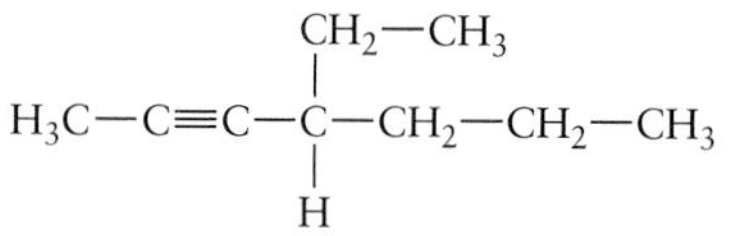

9.

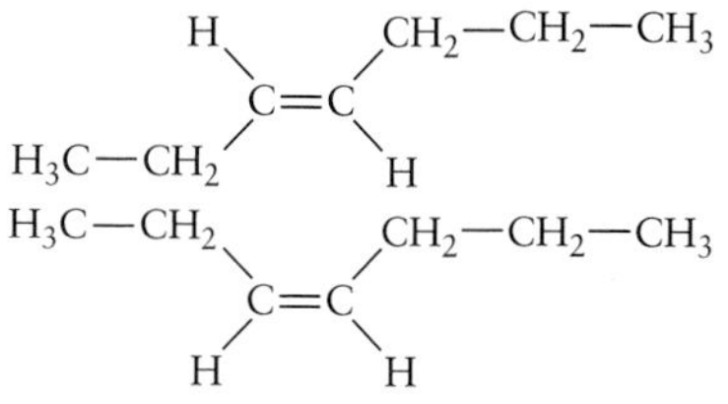

11. **(a)** 1,2-Hexadiene
(b) 1,3,5-Hexatriene
(c) 2-Methyl-1-hexene
(d) 3-Hexyne

13. **(a)** sp^2, sp, sp^2, sp^3, sp^3, sp^3
(b) All sp^2
(c) sp^2, sp^2; all others sp^3
(d) sp^3, sp^3, sp, sp, sp^3, sp^3

15. 30 double bonds; on the bonds shared by two hexagonal faces

17. 11.2%; 4.49×10^9 kg

19. **(a)** $CH_3CH_2CH_2CH_2OH + CH_3COOH \longrightarrow CH_3COOCH_2CH_2CH_2CH_3 + H_2O$
(b) $NH_4CH_3COO \longrightarrow CH_3CONH_2 + H_2O$
(c) $CH_3CH_2CH_2OH \longrightarrow CH_3CH_2CHO$ (propionaldehyde) $+ H_2$
(d) $CH_3(CH_2)_5CH_3 + 11\ O_2 \longrightarrow 7\ CO_2 + 8\ H_2O$

21. **(a)** $CH_2{=}CH_2 + Br_2 \longrightarrow CH_2BrCH_2Br$; $CH_2BrCH_2Br \longrightarrow CH_2 = CHBr + HBr$
(b) $CH_3CH_2CH = CH_2 + H_2O \longrightarrow CH_3CH_2CH(OH)CH_3$ (using H_2SO_4)
(c) $CH_3CH = CH_2 + H_2O \longrightarrow CH_3CH(OH)CH_3$ (using H_2SO_4); $CH_3CH(OH)CH_3 \longrightarrow CH_3COCH_3 + H_2$ (copper or zinc oxide catalyst)

23.

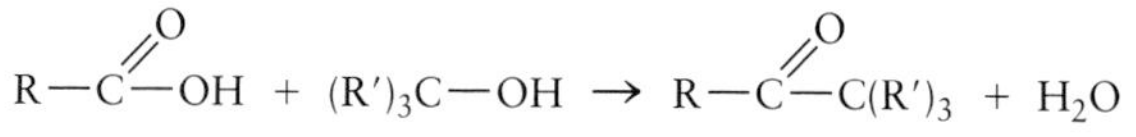

25. 79.9 L

27. $-CH_3$ carbon sp^3, other carbon sp^2. A π orbital with two electrons bonds the second carbon atom with the oxygen atom. The three groups around the second carbon form an approximately trigonal planar structure, with bond angles near 120 degrees. The geometry about the first carbon atom is approximately tetrahedral, with angles near 109.5 degrees.

29. The Lewis diagrams for HCOOH and $HCOO^-$ are

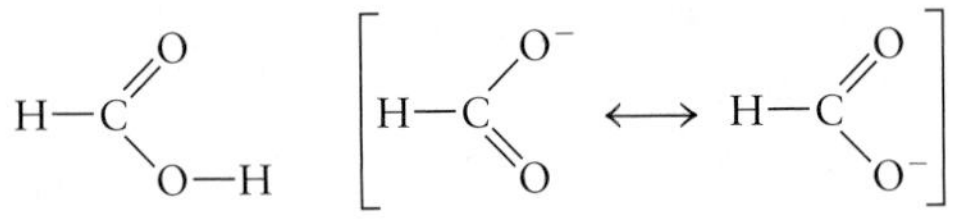

One resonance form is given for HCOOH, but two are given for the formate anion $HCOO^-$. In formic acid, one oxygen atom is doubly bonded to the carbon atom, and the other is singly bonded. In the anion, there is some double-bond character in both C—O bonds. The carbon atom in HCOOH is sp^2 hybridized (*SN* 3), and the OH oxygen atom is sp^3 hybridized (*SN* 4). The immediate surroundings of the carbon atom have trigonal planar geometry, and the C—O—H group is bent. In the $HCOO^-$, the carbon atom is sp^2 hybridized (*SN* 3) and both O atoms have a *p* orbital perpendicular to the OCO plane that can participate in π bonding. These two O atom *p* orbitals and the remaining *p* orbital on the C atom result in a three-center four-electron π system. In HCOOH, π overlap occurs between out-of-plane *p* orbitals on the carbon atom and only one oxygen atom. Both the C— to —O bond lengths in the formate ion should lie somewhere between the value for the single bond (1.36 Å) and the value for the double bond (1.23 Å).

31. **(a)** Alcohol: $CH_3CH(OH)CH_3$, isopropyl alcohol; carboxylic acid: $CH_3C(CH_3)(OCH_3)(CH_2)_3CH(CH_3) = CH_2CH = CHC(CH_3) = CHCOOH$
(b) 3,7,11-Trimethyl-2,4-dodecadiene

33. **(a)** $C_9H_8O_4$ **(b)** 1.80×10^{-3} mol

35. Dehydrogenate to make C—O bond from C—OH bond on first ring; move C—C bond from first to second ring; insert C=O group on third ring; remove hydrocarbon side chain on fourth ring together with hydrogen atom and replace it with an —OH group and a $—COCH_2OH$ group.

CHAPTER 8

1. **(a)** PtF_4 **(b)** PtF_6

3. $V_{10}O_{28}^{6-} + 16\ H_3O^+ \longrightarrow 10\ VO_2^+ + 24\ H_2O + 5$ state, V_2O_5

5. $2\ TiO_2 + H_2 \longrightarrow Ti_2O_3 + H_2O$ titanium(III) oxide

7. +2, +4, +3, +7, mixed +2 and +3, +7

9. Small ligands with large charge densities, such as F^-, can stabilize the higher oxidation states of a given transition metal.

11. Al^{3+} is a hard acid that prefers to pair with the hard base O^{2-}, whereas Ni^{2+} and Cu^{2+} are borderline acids that prefer to pair with the soft base S^{2-}.

13. Melting and boiling points. $TiCl_4$ is predicted to be a covalent compound because Ti^{4+} is a high oxidation state cation that can accept electrons from Cl^- ions to form covalent (dative) bonds.

15. Monodentate, at the N-atom lone pair

17. +2, +2, +2, 0

19. **(a)** $Na_2[Zn(OH)_4]$ **(b)** $[Co(en)_2Cl_2]NO_3$
(c) $[Pt(H_2O)_3Br]Cl$ **(d)** $[Pt(NH_3)_4(NO_2)_2]Br_2$

21. **(a)** Ammonium diamminetetraisothiocyanatochromate(III)
(b) Pentacarbonyltechnetium(I) iodide
(c) Potassium pentacyanomanganate(IV)
(d) Tetra-ammineaquachlorocobalt(III) bromide

23. $[Cu(NH_3)_2Cl_2] < KNO_3 < Na_2[PtCl_6] < [Co(NH_3)_6]Cl_3$

25. **(a)** $[Pt(NH_3)_2BrCl]$ has two isomers (*cis* and *trans*). Neither is optically active.
(b) $[Co(CN)_3(H_2O)_2Cl]^-$ has three possible isomers. None of the three isomers is optically active.
(c) $[V(C_2O_4)_3]^{3-}$ is enantiomeric with two possible optical isomers.

27. Three isomeric $[Fe(en)_2Cl_2]^+$ complexes exist.

29. **(a)** Strong: 1, weak: 5 **(b)** Strong or weak: 0
(c) Strong or weak: 3 **(d)** Strong: 2, weak: 4
(e) Strong: 0, weak: 4

31. $[Fe(CN)_6]^{3-}$: 1 unpaired, CFSE $= -2\ \Delta_o$; $[Fe(H_2O)_6]^{3+}$: = unpaired, CFSE = 0

33. d^3 gives half-filled and d^8 filled and half-filled shells for metal ions in an octahedral field. d^5 will be half-filled and stable for high spin (small Δ_o), and d^6 will be a filled subshell and stable for low spin (large Δ_o).

35. This ion does not absorb any significant amount of light in the visible range.

37. $\lambda \approx 480$ nm; $\Delta_o \approx 250$ kJ mol^{-1}

39. -500 kJ mol^{-1}

41. **(a)** Orange–yellow
(b) Approximately 600 nm (actual: 575 nm)
(c) Decrease, because CN^- is a strong-field ligand and will increase Δ_o.

43. **(a)** Pale, because F^- is an even weaker field ligand than H_2O, so it should be high-spin d^5. This is a half-filled shell and the complex will absorb only weakly.
(b) Colorless, because Hg(II) is a d^{10} filled-subshell species.

45. $[AuBr_4]^-$ is a low-spin, d^8 square planar complex, the square planar geometry being favored for larger cations. $[NiBr_4]^{2-}$ is high-spin, d^8 tetrahedral complex, the tetrahedral geometry being favored for smaller cations.

CHAPTER 9

1. $NH_4HS(s) \longrightarrow NH_3(g) + H_2S(g)$

3. $NH_4Br(s) + NaOH(aq) \longrightarrow NH_3(g) + H_2O(\ell) + NaBr(aq)$

5. 10.3 m

7. 1.40×10^4 ft

9. 1697.5 atm, 1.7200×10^3 bar

11. 0.857 atm

13. 8.00 L

15. 14.3 gill

17. 134 L

19. 35.2 psi

21. **(a)** 19.8 atm **(b)** 23.0 atm

23. The mass of a gas in a given volume changes proportionately to the absolute temperature. This statement will be true only at −19.8°C.

25. **(a)** $2\ Na(s) + 2\ HCl(g) \longrightarrow 2\ NaCl(s) + H_2(g)$
(b) 4.23 L

27. 3.0×10^6 L

29. 24.2 L

31. **(a)** 932 L H_2S **(b)** 1.33 kg, 466 L SO_2

33. $X_{SO_3} = 0.135$; $P_{SO_3} = 0.128$ atm

35. $X_{N_2} = 0.027$; $P_{N_2} = 1.6 \times 10^{-4}$ atm

37. **(a)** $X_{CO} = 0.444$ **(b)** $X_{CO} = 0.33$

39. **(a)** 5.8×10^{17} **(b)** 520 L

41. **(a)** 1.93×10^3 m s^{-1} = 1.93 km s^{-1} **(b)** 226 m s^{-1}

43. 6100 m s^{-1} (6000 K), 790 m s^{-1} (100 K)

45. 1.82%

47. Greater. As T increases, the Maxwell–Boltzmann distribution shifts to higher speeds.

49. 162 atm = 2.38×10^3 psi

51. **(a)** 27.8 atm **(b)** 24.6 atm; attractive forces dominate.

53. 6.1×10^{-9} m

55. 92.6 g mol^{-1}

57. 1830 stages

59. 7.4×10^{-7} atm; 20 m^2 s^{-1}

CHAPTER 10

1. Gas

3. **(a)** Condensed **(b)** 10.3 cm^3 mol^{-1}

5. Condensed

7. The liquid water has vaporized to steam.

9. Harder, because forces binding the particles in NaCl are stronger and resist deformation better.

11. In all three phases, the diffusion constant should decrease as its density is increased. At higher densities, molecules are closer to each other. In gases, they will collide more often and travel shorter distances between collisions. In liquids and solids, there will be less space for molecules to move around each other.

13. Both involve the interaction of an ion with a dipole. In the first case, the dipole is permanent (preexisting), whereas in the second, it is induced by the approach of the ion. Induced dipole forces are weaker than ion–dipole forces. Examples: Na^+ with HCl (ion–dipole); Na^+ with Cl_2 (induced dipole).

15. **(a)** *Ion–ion*, dispersion **(b)** *Dipole–dipole*, dispersion
(c) Dispersion **(d)** Dispersion

17. Bromide ion

19. **(a)** 2.0×10^{-10} m; 2.5×10^{-10} m
(b) KCl has a longer bond yet lower potential energy (greater bond strength).

21. Heavier gases have stronger attractive forces, favoring the liquid and solid states.

23. Ne < NO < NH_3 < RbCl; nonpolar < polar < hydrogen-bonded < ionic

25. An eight-membered ring of alternating H and O atoms is reasonable. The ring is probably not planar.

27. The two have comparable molar masses, but the hydrogen-bonding in hydrazine should give it a higher boiling point.

29. 6.7×10^{25}

31. 6.16 L; several times smaller than the volume of 1 mol at standard temperature and pressure (22.4 L mol^{-1})

33. 7.02×10^{13} atoms cm^{-3}

35. 0.9345 g L^{-1}

37. 2.92 g $CaCO_3$

39. 0.69 atm; 31% lies below

41. Iridium. Its higher melting and boiling points indicate that the intermolecular forces in iridium are stronger than those in sodium.

43. No phase change will occur.

47. **(a)** Liquid **(b)** Gas **(c)** Solid **(d)** Gas

49. **(a)** Above. If gas and solid coexist at −84.0°C, their coexistence must extend upward in temperature to the triple point.
(b) The solid will sublime at some temperature below −84.0°C.

51. The meniscus between gas and liquid phases will disappear at the critical temperature, 126.19 K.

CHAPTER 11

1. **(a)** 5.53×10^{-3} M **(b)** 5.5×10^{-3} molal
(c) 3.79 L

3. Molarity = 12.39 M in HCl; mole fraction of HCl = 0.2324; molality = 16.81 molal in HCl

5. 8.9665 molal

7. 0.00643 g H_2O

9. **(a)** 1.33 g mL^{-1} **(b)** 164 mL

11. 1.06 M in NaOH

13. **(a)** $Ag^+(aq) + Cl^-(aq) \longrightarrow AgCl(s)$
(b) $K_2CO_3(s) + 2\ H^+(aq) \longrightarrow 2\ K^+(aq) + CO_2(g) + H_2O(\ell)$
(c) $2\ Cs(s) + 2\ H_2O(\ell) \longrightarrow 2\ Cs^+(aq) + 2\ OH^-(aq) + H_2(g)$
(d) $2\ MnO_4^-(aq) + 16\ H^+(aq) + 10\ Cl^-(aq) \longrightarrow 5\ Cl_2(g) + 2\ Mn^{2+}(aq) + 8\ H_2O(\ell)$

15. 16.8 mL of 7.91 M HNO_3

17. 6.74×10^3 L $CO_2(g)$

19. **(a)** $Ca(OH)_2(aq) + 2\ HF(aq) \longrightarrow CaF_2(aq) + 2\ H_2O(\ell)$
Hydrofluoric acid, calcium hydroxide, calcium fluoride

(b) $2\ RbOH(aq) + H_2SO_4(aq) \longrightarrow Rb_2SO_4(aq) + 2\ H_2O(\ell)$
Sulfuric acid, rubidium hydroxide, rubidium sulfate

(c) $Zn(OH)_2(s) + 2\ HNO_3(aq) \longrightarrow Zn(NO_3)_2(aq) + 2\ H_2O(\ell)$
Nitric acid, zinc hydroxide, zinc nitrate

(d) $KOH(aq) + CH_3COOH(aq) \longrightarrow KCH_3COO(aq) + H_2O(\ell)$
Acetic acid, potassium hydroxide, potassium acetate

21. Sodium sulfide

23. (a) $PF_3 + 3\ H_2O \longrightarrow H_3PO_3 + 3\ HF$
(b) $[H_3PO_3] = 0.0882$ M; $[HF] = 0.265$ M

25. 0.04841 M HNO_3

27. (a) $2\ \overset{+3}{P}F_2I(\ell) + 2\ \overset{0}{Hg}(\ell) \rightarrow \overset{+2}{P_2}F_4(g) + \overset{+1}{Hg_2}I_2(s)$
(b) $2\ K\overset{+5}{Cl}\overset{-2}{O_3}(s) \rightarrow 2\ K\overset{-1}{Cl}(s) + 3\ \overset{0}{O_2}(g)$
(c) $4\ \overset{-3}{N}H_3(g) + 5\ \overset{0}{O_2}(g) \longrightarrow 4\ \overset{+2}{N}\overset{-2}{O}(g) + 6\ H_2\overset{-2}{O}(g)$
(d) $2\ \overset{0}{As}(s) + 6\ Na\overset{+1}{O}H(\ell) \longrightarrow 2\ Na_3\overset{+3}{As}O_3(s) + 3\ \overset{0}{H_2}(g)$

29. $2\ \overset{0}{Au}(s) + 6\ H_2\overset{+6}{Se}O_4 \longrightarrow \overset{+3}{Au_2}(\overset{+6}{Se}O_4)_3(aq) + 3\ H_2\overset{+4}{Se}O_3(aq) + 3\ H_2O_6(\ell)$
Au is oxidized
H_2SeO_4 is reduced (half of it)

31. (a) $2\ VO_2^+(aq) + SO_2(g) \longrightarrow 2\ VO^2(aq) + SO_4^{2-}(aq)$
(b) $Br_2(\ell) + SO_2(g) + 6\ H_2O(\ell) \longrightarrow 2\ Br^-(aq) + SO_4^{2-}(aq) + 4\ H_3O^+(aq)$
(c) $Cr_2O_7^{2-}(aq) + 3\ Np^{4+}(aq) + 2\ H_3O^+(aq) \longrightarrow 2\ Cr^{3+}(aq) + 3\ NpO_2^{2+}(aq) + 3\ H_2O(\ell)$
(d) $5\ HCOOH(aq) + 2\ MnO_4^-(aq) + 6\ H_3O^+(aq) \longrightarrow 5\ CO_2(g) + 2\ Mn^{2+}(aq) + 14\ H_2O(\ell)$
(e) $3\ Hg_2HPO_4(s) + 2\ Au(s) + 8\ Cl^-(aq) + 3\ H_3O^+(aq) \longrightarrow 6\ Hg(\ell) + 3\ H_2PO_4^-(aq) + 2\ AuCl_4^-(aq) + 3\ H_2O(\ell)$

33. (a) $2\ Cr(OH)_3(s) + 3\ Br_2(aq) + 10\ OH^-(aq) \longrightarrow 2\ CrO_4^{2-}(aq) + 6\ Br^-(aq) + 8\ H_2O(\ell)$
(b) $ZrO(OH)_2(s) + 2\ SO_3^{2-}(aq) \longrightarrow Zr(s) + 2\ SO_4^{2-}(aq) + H_2O(\ell)$
(c) $7\ HPbO_2^-(aq) + 2\ Re(s) \longrightarrow 7\ Pb(s) + 2\ ReO_4^-(aq) + H_2O(\ell) + 5\ OH^-(aq)$
(d) $4\ HXeO_4^-(aq) + 8\ OH^-(aq) \longrightarrow 3\ XeO_6^{4-}(aq) + Xe(g) + 6\ H_2O(\ell)$
(e) $N_2H_4(aq) + 2\ CO_3^{2-}(aq) \longrightarrow N_2(g) + 2\ CO(g) + 4\ OH^-(aq)$

35. (a) $Fe^{2+}(aq) \longrightarrow Fe^{3+}(aq) + e^-$ oxidation
$H_2O_2(aq) + 2\ H_3O^+(aq) + 2\ e^- \longrightarrow 4\ H_2O(\ell)$ reduction
(b) $5\ H_2O(\ell) + SO_2(aq) \longrightarrow HSO_4^-(aq) + 3\ H_3O^+(aq) + 2\ e^-$ oxidation
$Mn_4^-(aq) + 8\ H_3O^+(aq) + 5\ e^- \longrightarrow Mn^{2+}(aq) + 12\ H_2O(\ell)$ reduction
(c) $ClO_2^-(aq) \longrightarrow ClO_2(g)\ 1\ e^-$ oxidation
$ClO_2^-(aq) + 4\ H_3O^+(aq) + 4\ e^- \longrightarrow Cl^-(aq) + 6\ H_2O(\ell)$ reduction

37. $3\ HNO_2(aq) \longrightarrow NO_3^-(aq) + 2\ NO(g) + H_3O^+(aq)$

39. 7.175×10^{-3} M $Fe^{2+}(aq)$

41. 0.2985 atm

43. 3.34 K kg mol^{-1}

45. 340 g mol^{-1}

47. 1.7×10^2 g mol^{-1}

49. -2.8°C. As the solution becomes more concentrated, its freezing point decreases further.

51. 2.70 particles (complete dissociation of Na_2SO_4 gives 3 particles)

53. 1.88×10^4 g mol^{-1}

55. 7.46×10^3 g mol^{-1}

57. (a) 0.17 mol CO_2
(b) Because the partial pressure of CO_2 in the atmosphere is much less than 1 atm, the excess CO_2 will bubble out from the solution and escape when the cap is removed.

59. 4.13×10^2 atm

61. 0.774

63. (a) 0.491 **(b)** 0.250 atm **(c)** 0.575

CHAPTER 12

1. -2.16×10^4 L atm $= -2.19 \times 10^6$ J

3. 86.6 m

5. 24.8, 28.3, 29.6, 31.0, and 32.2 J K^{-1} mol^{-1}, respectively; extrapolating gives about 33.5 J K^{-1} mol^{-1} for Fr.

7. 26.1, 25.4, 25.0, 24.3, 25.4, and 27.6 J K^{-1} mol^{-1}

9. (a) w is zero, q is positive, and ΔU is positive.
(b) w is zero, q is negative, and ΔU is negative.
(c) $(w_1 + w_2)$ is zero. $(\Delta U_1 + \Delta U_2) = (q_1 + q_2)$. The latter two sums could be any of three possibilities: both positive, both negative, or both zero.

11. 0.468 J K^{-1} g^{-1}

13. $q_1 = Mc_{s1}\Delta T_1 = -q_2 = -Mc_{s2}\Delta T_2 \qquad \frac{C_{s1}}{C_{s2}} = -\frac{\Delta T_2}{\Delta T_1}$

15. 314 J g^{-1} (modern value is 333 J g^{-1})

17. $w = -323$ J; $\Delta U = -393$ J; $q = -70$ J

19. (a) 38.3 L
(b) $w = -1.94 \times 10^3$ J; $q = 0$; $\Delta U = -1.94 \times 10^3$ J
(c) 272 K

21. $\Delta U = +11.2$ kJ; $q = 0$; $w = +11.2$ kJ

23. For HI:

The measured value is $c_P = 29.16$ J mol^{-1} K^{-1}. (Appendix D)

The vibrational contribution = 29.16 − 29.10 = 0.06 J mol^{-1} K^{-1}.

Percent of c_P for HI due to vibration = (0.06/29.16) × 100 = 0.2%

For I_2:

The measured value is $c_P = 36.90$ J mol^{-1} K^{-1}. (Appendix D)

The vibrational contribution = 36.02 − 29.10 = 7.80 J mol^{-1} K^{-1}.

Percent of c_P for I_2 due to vibration = (7.80/36.90) × 100 = 21.1%

25. **(a)** for Argon ΔH = 11.4 kJ
(b) for C_2H_4 ΔH = 73.2 kJ

27. **(a)** −6.68 kJ **(b)** +7.49 kJ **(c)** +0.594 kJ

29. 41.3 kJ

31. +513 J

33. 10.4°C

35. −623.5 kJ

37. A pound of diamonds. (This is recommended as a source of heat only as a last resort, however!)

39. −555.93 kJ

41. **(a)** −878.26 kJ **(b)** -1.35×10^7 kJ of heat absorbed

43. **(a)** −81.4 kJ **(b)** 55.1°C

45. $\Delta H_f^\circ = -152.3$ kJ mol^{-1}

47. **(a)** $C_{10}H_8(s) + 12\ O_2(g) \longrightarrow 10\ CO_2(g) + 4\ H_2O(\ell)$
(b) −5157 kJ **(c)** −5162 kJ **(d)** +84 kJ mol^{-1}

49. −264 kJ mol^{-1}

51. -1.58×10^3 kJ

53. Each side of the equation has three B—Br and three B—Cl bonds.

55. $w = -6.87$ kJ; $q = +6.87$ kJ; $\Delta U = \Delta H = 0$

57. $T_f = 144$ K; $w = \Delta U = 23.89$ kJ; $\Delta H = 26.48$ kJ

59. $N_{high}/N_{low} = 0.91$

61. 5.3×10^{-4}

63. 2.42×10^{-9}

CHAPTER 13

1. **(a)** The system is all the matter participating in the reaction $NH_4NO_3(s) \longrightarrow NH_4^+(aq) + NO_3^-(aq)$. This includes solid ammonium nitrate, the water in which it dissolves, and the aquated ions that are the products of the dissolution process. The surroundings include the flask or beaker in which the system is held, the air above the system, and other neighboring materials. The dissolution of ammonium nitrate is spontaneous after any physical separation (such as a glass wall or a space of air) between the water and the ammonium nitrate has been removed.
(b) The system is all the matter participating in the reaction $H_2(g) + O_2(g) \longrightarrow$ products. The surroundings are the walls of the bomb and other portions of its environment that might deliver heat or work or absorb heat or work. The reaction of hydrogen with oxygen is spontaneous. Once hydrogen and oxygen are mixed in a closed bomb, no constraint exists to prevent their reaction. It is found experimentally that this system gives products quite slowly at room temperature (no immediate explosion). It explodes instantly at higher temperatures.
(c) The system is the rubber band. The surroundings consist of the weight (visualized as attached to the lower end of the rubber band), a hanger at the top of the rubber band, and the air in contact with the rubber band. The change is spontaneous once a constraint such as a stand or support underneath the weight is removed.
(d) The system is the gas contained in the chamber. The surroundings are the walls of the chamber and the moveable piston head. The process is spontaneous if the force exerted by the weight on the piston exceeds the force exerted by the collisions of the molecules of the gas on the bottom of the piston. (The forces due to the mass of the piston itself and friction between the piston and the walls within which it moves are neglected.)
(e) The system is the drinking glass in the process glass ⟶ fragments. The surroundings are the floor, the air, and the other materials in the room. The change is spontaneous. It occurs when the constraint, which is whatever portion of the surroundings holds the glass above the floor, is removed.

3. **(a)** $6 \times 6 = 36$ **(b)** 1 in 36

5. The tendency for entropy to increase

7. $10^{-3.62 \times 10^{23}}$ (that is, 1 part in $10^{+3.62 \times 10^{23}}$)

9. **(a)** $\Delta S > 0$ **(b)** $\Delta S > 0$ **(c)** $\Delta S < 0$

11. 9.61 J K^{-1} mol^{-1}

13. 29 kJ mol^{-1}

15. $\Delta U = \Delta H = 0$; $w = -1.22 \times 10^4$ J; $q = +1.22 \times 10^4$ J; $\Delta S = +30.5$ J K^{-1}

17. $\Delta S_{sys} = +30.2$ J K^{-1}; $\Delta S_{surr} = -30.2$ J K^{-1}; $\Delta S_{univ} = 0$

19. $\Delta S_{Fe} = -8.24$ J K^{-1}; $\Delta S_{H_2O} = +9.49$ J K^{-1}; $\Delta S_{tot} = +1.25$ J K^{-1}

21. **(a)** −116.58 J K^{-1} **(b)** Lower (more negative)

23. ΔS° = −162.54, −181.12, −186.14, −184.72, −191.08 J K^{-1}. The entropy changes in these reactions become increasingly negative with increasing atomic mass, except that the rubidium reaction is out of line.

25. ΔS_{surr} must be positive and greater in magnitude than 44.7 J K^{-1}.

27. $\Delta S_{sys} > 0$ because the gas produced (oxygen) has many possible microstates.

29. **(a)** +740 J **(b)** 2.65 kJ **(c)** No **(d)** 196 K

31. $q = +38.7$ kJ; $w = -2.92$ kJ; $\Delta U = +35.8$ kJ; $\Delta S_{sys} = +110$ J K^{-1}; $\Delta G = 0$

33. Overall reaction: $2\ Fe_2O_3(s) + 3\ C(s) \longrightarrow 4\ Fe(s) + 3\ CO_2(g)$ $\Delta G = +840\ J + 3(-400)\ J = -360\ J < 0$

35. **(a)** $0 < T < 3000$ K
(b) $0 < T < 1050$ K
(c) Spontaneous at all temperatures

37. $WO_3(s) + 3\ H_2(g) \longrightarrow W(s) + 3\ H_2O(g)$

$\Delta H^\circ = 117.41$ kJ; $\Delta S^\circ = 131.19$ J K^{-1}; $\Delta G < 0$ for $T > \Delta H^\circ/\Delta S^\circ = 895$ K

39. **(a)** 0.333 **(b)** $q = -1000$ J **(c)** $w = -500$ J

CHAPTER 14

1. (a) $\left[\frac{P_{H_2O}^2}{P_{H_2}^2 P_{O_2}}\right]_{eq} = K$ (b) $\left[\frac{P_{XeF_6}}{P_{Xe}P_{F_2}^3}\right]_{eq} = K$

(c) $\left[\frac{P_{CO_2}^{12} P_{H_2O}^6}{P_{C_6H_6}^2 P_{O_2}^{15}}\right]_{eq} = K$

3. $P_4(g) + 6\ Cl_2(g) + 2\ O_2(g) \rightleftharpoons 4\ POCl_3(g)$ $\left[\frac{P_{POCl_3}^4}{P_{P_4} P_{Cl_2}^6 P_{O_2}^2}\right]_{eq} = K$

5. (a) $\left[\frac{P_{CO_2} P_{H_2}}{P_{CO} P_{H_2O}}\right]_{eq} = K$ (b) 0.056 atm

7. (a) The graph is a straight line passing through the origin.
(b) The experimental K's range from 3.42×10^{-2} to 4.24×10^{-2}, with a mean of 3.84×10^{-2}.

9. (a) $\left[\frac{(P_{H_2S})^8}{(P_{H_2})^8}\right]_{eq} = K$ (b) $\left[\frac{(P_{COCl_2})(P_{H_2})}{(P_{Cl_2})}\right]_{eq} = K$

(c) $[P_{CO_2}]_{eq} = K$ (d) $\left[\frac{1}{(P_{C_2H_2})^3}\right]_{eq} = K$

11. (a) $\left[\frac{[Zn^{2+}]}{[Ag^+]^2}\right]_{eq} = K$

(b) $\left[\frac{[VO_3(OH)^{2-}][OH^-]}{[VO_4^{3-}]}\right]_{eq} = K$

(c) $\left[\frac{[HCO_3^-]^6}{[As(OH)_6^{3-}]^2 P_{CO_2}^6}\right]_{eq} = K$

13. $\Delta G° = -550.23$ kJ; $K = 2.5 \times 10^{96}$

15. (a) 2.6×10^{12} $\left[\frac{P_{SO_3}}{(P_{SO_2})(P_{O_2})^{1/2}}\right]_{eq} = K$

(b) 5.4×10^{-35} $[(P_{O_2})^{1/2}]_{eq} = K$
(c) 5.3×10^3 $[[Cu^{2+}][Cl^-]^2]_{eq} = K$

17. $K_1 = (K_2)^3$

19. K_2/K_1

21. 1.04×10^{-4}

23. 14.6

25. (a) $P_{SO_2} = P_{Cl_2} = 0.58$ atm; $P_{SO_2Cl_2} = 0.14$ atm (b) 2.4

27. (a) 0.180 atm (b) 0.756

29. $P_{PCl_5} = 0.078$ atm; $P_{PCl_3} = P_{Cl_2} = 0.409$ atm

31. $P_{Br_2} = 0.0116$ atm; $P_{I_2} = 0.0016$ atm; $P_{IBr} = 0.0768$ atm

33. $P_{N_2} = 0.52$ atm; $P_{O_2} = 0.70$ atm; $P_{NO} = 3.9 \times 10^{-16}$ atm

35. $P_{N_2} = 0.0148$ atm; $P_{H_2} = 0.0445$ atm; $P_{NH_3} = 0.941$ atm

37. 5.6×10^{-5} mol L^{-1}

39. 2.0×10^{-2}

41. (a) 0.31 atm (b) 1.65 atm, 0.15 atm

43. (a) 8.46×10^{-5} (b) 0.00336 atm

45. (a) 9.83×10^{-4} (b) Net consumption

47. $K > 5.1$

49. (a) $Q = 2.05$; reaction shifts to right
(b) $Q = 3.27$; reaction shifts to left

51. (a) 0.800, left
(b) $P_{P_4} = 5.12$ atm, $P_{P_2} = 1.77$ atm
(c) Dissociation

53. (a) Shifts left
(b) Shifts right
(c) Shifts right
(d) The volume must have been increased to keep the total pressure constant; shift is to left.
(e) No effect

55. (a) Exothermic (b) Decrease

57. Run the first step at low temperature and high pressure, and the second step at high temperature and low pressure.

59. Low temperature and high pressure

61. -58 kJ

63. (a) -56.9 kJ
(b) $\Delta H° = 255.6$ kJ, $\Delta S° = +4.2$ J K^{-1}

65. 4.3×10^{-3}

67. (a) 23.8 kJ mol^{-1}
(b) $T_b = 240$ K

69. (a)

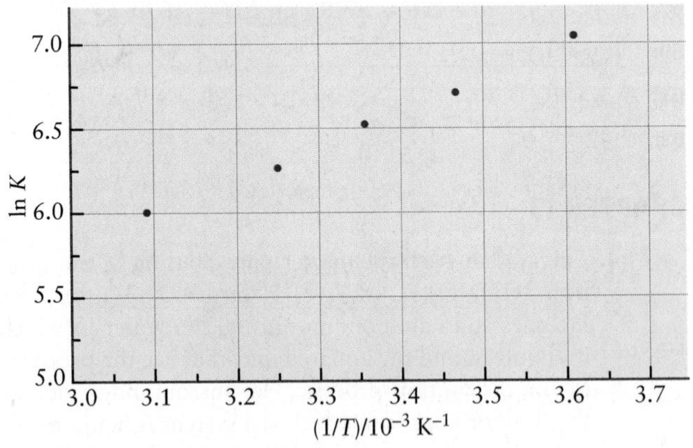

(b) -16.9×10^3 J mol^{-1}

71. 76

73. (a) $K_1 = 1.64 \times 10^{-2}$; $K_2 = 5.4$ (b) 330

CHAPTER 15

1. (a) Cl^- cannot act as a Brønsted–Lowry acid.
(b) SO_4^{2-}
(c) NH_3
(d) NH_2^- (but NH_3 put in water acts as a base, not an acid)
(e) OH^-

3. HCO_3^- serves as the base.

5. (a) $CaO(s) + H_2O(\ell) \longrightarrow Ca(OH)_2(s)$
(b) The CaO acts as a Lewis base, donating a pair of electrons (located on the oxide ion) to one of the hydrogen ions (the Lewis acid) of the water molecule.

7. (a) Fluoride acceptor
(b) Acids: BF_3, TiF_4; bases: ClF_3O_2, KF

9. (a) Base, $Mg(OH)_2$ (b) Acid, HOCl
(c) Acid, H_2SO_4 (d) Base, CsOH

11. $SnO(s) + 2\,HCl(aq) \longrightarrow Sn^{2+}(aq) + 2\,Cl^-(aq) + H_2O(\ell)$;
$SnO(s) + NaOH(aq) + H_2O(\ell) \longrightarrow Sn(OH)_3^-(aq) + Na^+(aq)$

13. 3.70

15. 3×10^{-7} M < $[H_3O^+]$ < 3×10^{-6} M;
3×10^{-9} M < $[OH^-]$ < 3×10^{-8} M

17. $[H_3O^+] = 1.0 \times 10^{-8}$ M; $[OH^-] = 1.7 \times 10^{-6}$ M

19. The first reaction is more likely to be correct. In the second case, the reactant H_3O^+ would be present at very low concentration (10^{-7} M) and would give neither a fast nor a vigorous reaction.

21. (a) $C_{10}H_{15}ON(aq) + H_2O(\ell) \rightleftharpoons C_{10}H_{15}ONH^+(aq) + OH^-(aq)$
(b) 7.1×10^{-11}
(c) Stronger

23. $K = 24$; $HClO_2$ is the stronger acid and NO_2^- the stronger base.

25. (a) Methyl orange (b) 3.8 to 4.4

27. 2.35

29. (a) 2.45 (b) 0.72 mol

31. 1.2

33. 1.2×10^{-6}

35. 10.4

37. 0.083 M

39. The reaction gives a base of moderate strength, the acetate ion, in solution, so the pH > 7.

41. HCl < NH_4Br < KI < $NaCH_3COO$ < NaOH

43. 8.08

45. (a) 4.36 (b) 4.63

47. *m*-Chlorobenzoic acid

49. 639 mL

51. 13.88, 11.24, 7.00, 2.77

53. 2.86, 4.72, 8.71, 11.00

55. 11.89, 11.52, 10.79, 8.20, 6.05, 3.90, 1.95

57. 0.97 g

59. 0.0872 g, pH = 6.23 if no approximations are made, bromothymol blue

61. 4×10^{-7}

63. $[H_3AsO_4] = 8.0 \times 10^{-2}$ M;
$[H_2AsO_4^-] = [H_3O^+] = 2.0 \times 10^{-2}$ M;
$[HAsO_4^{2-}] = 9.3 \times 10^{-8}$ M; $[AsO_4^{3-}] = 1.4 \times 10^{-17}$ M

65. $[PO_4^{3-}] = 0.020$ M; $[HPO_4^{2-}] = [OH^-] = 0.030$ M;
$[H_2PO_4^-] = 1.61 \times 10^{-7}$ M; $[H_3PO_4] = 7.1 \times 10^{-18}$ M

67. $[H_2CO_3] = 8.5 \times 10^{-6}$ M; $[HCO_3^-] = 1.5 \times 10^{-6}$ M;
$[CO_3^{2-}] = 2.8 \times 10^{-11}$ M

69. Propene

71. 46 kJ mol^{-1}

73. pK_{a1} should be smaller than 4.9; pK_{a2} should be larger.

CHAPTER 16

1. 4.65 L

3. About 48°C

5. $Fe_2(SO_4)_3(s) \rightleftharpoons 2\,Fe^{3+}(aq) + 3\,SO_4^{2-}(aq)$
$[Fe^{3+}]^2[SO_4^{2-}]^3 = K_{sp}$

7. 0.0665 g per 100 mL water

9. 14.3 g L^{-1}

11. $[I^-] = 6.2 \times 10^{-10}$ M, $[Hg_2^{2+}] = 3.1 \times 10^{-10}$ M

13. 1.9×10^{-12}

15. 1.6×10^{-8}

17. Yes. The initial reaction quotient is $6.2 \times 10^{-10} > K_{sp}$.

19. Yes

21. $[Pb^{2+}] = 2.3 \times 10^{-10}$ M; $[IO_3^-] = 0.033$ M

23. $[Ag^+] = 1.8 \times 10^{-2}$ M; $[CrO_4^{2-}] = 6.2 \times 10^{-9}$ M

25. 2.4×10^{-8} mol L^{-1}

27. (a) 3.4×10^{-6} M (b) 1.6×10^{-14} M

29. 9.1×10^{-3} M

31. In pure water: 1.2×10^{-4} M; at pH 7: 0.15 M

33. (a) Unchanged (b) Increase (c) Increase

35. $[Cu(NH_3)_4^{2+}] = 0.10$ M; $[Cu^{2+}] = 7 \times 10^{-14}$ M

37. $[K^+] = 0.0051$ M and $[Na^+] = 0.0076$ M

39. More will dissolve in 1 M NaCl (3×10^{-5} versus 1.3×10^{-5} mol L^{-1}). Less will dissolve in 0.100 M NaCl (3×10^{-6} versus 1.3×10^{-5} mol L^{-1}).

41. $Cu^{2+}(aq) + 2\,H_2O(\ell) \rightleftharpoons CuOH^+(aq) + H_3O^+(aq)$ or
$Cu(H_2O)_4^{2+}(aq) + H_2O(\ell) \rightleftharpoons Cu(H_2O)_3OH^+(aq) + H_3O^+(aq)$

43. 5.3

45. $K_a = 9.6 \times 10^{-10}$

47. In the first case, solid $Pb(OH)_2$ will precipitate; $[Pb^{2+}] = 4.2 \times 10^{-13}$ M, $[Pb(OH)_3^-] = 0.17$ M. In the second case, solid $Pb(OH)_2$ will not precipitate; $[Pb^{2+}] = 1 \times 10^{-13}$ M, $[Pb(OH)_3^-] = 0.050$ M.

49. (a) 8.6×10^{-4} M, Pb^{2+} in solid (b) 3.1×10^{-7}

51. $[I^-] = 4.1 \times 10^{-4}$ M

53. 2×10^{-13} M

55. pH = 2.4; $[Pb^{2+}] = 6 \times 10^{-11}$ M

CHAPTER 17

1. Electrons flow from the left electrode to the right as Cr(II) is oxidized to Cr(III). In the salt bridge, negative ions flow from right to left and positive ions from left to right. The overall reaction is represented as follows:

$$2\,Cr^{2+}(aq) + Cu^{2+}(s) \longrightarrow 2\,Cr^{3+}(aq) + Cu(s)$$

3. 0.180 mol

5. **(a)** $Zn(s) + Cl_2(g) \longrightarrow Zn^{2+}(aq) + 2\,Cl^-(aq)$
(b) 1.20×10^3 C; 1.24×10^{-2} mol e^-
(c) Decreases by 0.407 g **(d)** 0.152 L Cl_2 consumed

7. +2

9. **(a)** Anode: $Cl^- \longrightarrow \frac{1}{2}\,Cl_2(g) + e^-$;
cathode: $K^+ + e^- \longrightarrow K(\ell)$
total: $Cl^- + K^+ \longrightarrow K(\ell) + \frac{1}{2}\,Cl_2(g)$
(b) Mass K = 14.6 g; mass Cl_2 = 13.2 g

11. $\Delta G° = -921$ J; max work done *by* cell = +921 J

13. **(a)** Anode: $Co \longrightarrow Co^{2+} + 2\,e^-$;
cathode: $Br_2 + 2\,e^- \longrightarrow 2\,Br^2$;
total: $Co + Br_2 \longrightarrow Co^{2+} + 2\,Br^-$
(b) 1.34 V

15. **(a)** Anode: $Zn \longrightarrow Zn^{2+} + 2\,e^-$;
cathode: $In^{3+} + 3\,e^- \longrightarrow In$
(b) −0.338 V

17. Reducing agent

19. Br_2 less effective than Cl_2

21. **(a)** BrO_3^- **(b)** Cr **(c)** Co

23. **(a)** $E° = -0.183$ V **(b)** It will not disproportionate.

25. **(a)** No **(b)** Br^-

27. 0.381 V

29. −0.31 V

31. 5.17

33. **(a)** 0.31 V **(b)** 1×10^{-8} M

35. $K = 3 \times 10^{31}$, orange

37. 3×10^6

39. pH = 2.53; $K_a = 0.0029$

41. **(a)** 1.065 V **(b)** 1.3×10^{-11} M **(c)** 7.6×10^{-13}

43. The midpoint between the waves appears at −0.195 V, reflecting the difference between the reduction potentials of the SCE and Ag/AgCl reference electrodes.

45. $E = E_{HOMO} - E_{LUMO} = 2.4$ eV, $\lambda = 517$ nm, green

47. Yes, the valence band potential is sufficiently positive to oxidize water and the conduction band potential is sufficiently negative to reduce H^+.

49. The redox potential of the HOMO must be more negative than the conduction band of strontium titanate (−0.2 V) in order to inject electrons and that of the LUMO must be more positive than the water oxidation potential (1.229 V) so the minimum HOMO − LUMO gap must be around 1.5 eV.

51. 2.041 V; 12.25 V

53. **(a)** 9.3×10^6 C **(b)** 1.1×10^8 J

55. No, because $H_2SO_4(aq)$ is not the only substance whose amount changes during discharge. The accumulated $PbSO_4$ must also be removed and replaced by Pb and PbO_2.

57. 7900 J g^{-1}

59. $\Delta E = -0.419$ V, not spontaneous under standard conditions (pH = 14). If $[OH^-]$ is small enough, the equilibrium will shift to the right and the reaction will become spontaneous.

61. According to its reduction potential, yes. In practice, however, the sodium would react instantly and explosively with the water and would therefore be useless for this purpose.

63. $2\,Cl^- \longrightarrow Cl_2(g) + 2\,e^-$ anode
$Na^+ + e^- \longrightarrow Na(\ell)$ cathode

65. 4.4×10^4 kg

67. $2\,Mg + TiCl_4 \longrightarrow Ti + 2\,MgCl_2$; 102 kg

69. 42.4 min

71. **(a)** Ni
(b) 21.9 g (provided the electrolyte volume is very large)
(c) H_2

CHAPTER 18

1. 5.3×10^{-5} mol L^{-1} s^{-1}

3. $\text{rate} = -\dfrac{\Delta[N_2]}{\Delta t} = -\dfrac{1}{3}\dfrac{\Delta[H_2]}{\Delta t} = \dfrac{1}{2}\dfrac{\Delta[NH_3]}{\Delta t}$

5. **(a)** Rate = $k[NO]^2[H_2]$; k has the units L^2 mol^{-2} s^{-1}.
(b) An increase by a factor of 18

7. **(a)** Rate = $k[C_5H_5N][CH_3I]$ **(b)** $k = 75$ L mol^{-1} s^{-1}
(c) 7.5×10^{-8} mol L^{-1} s^{-1}

9. 3.2×10^4 s

11. 0.0019 atm

13. 5.3×10^{-3} s^{-1}

15. 2.3×10^{-5} M

17. 2.9×10^{-6} s

19. **(a)** Bimolecular, rate = $k[HCO][O_2]$
(b) Termolecular, rate = $k[CH_3][O_2][N_2]$
(c) Unimolecular, rate = $k[HO_2NO_2]$

21. **(a)** The first step is unimolecular; the others are bimolecular.
(b) $H_2O_2 + O_3 \longrightarrow H_2O + 2\,O_2$
(c) O, ClO, CF_2Cl, Cl

23. 0.26 L mol^{-1} s^{-1}

25. **(a)** $A + B + E \longrightarrow D + F$ $\quad \text{rate} = \dfrac{k_1k_2[A][B][E]}{k_{-1}[D]}$

(b) $A + D \longrightarrow B + F$ $\quad \text{rate} = \dfrac{k_1k_2k_3}{k_{-1}k_{-2}}\dfrac{[A][D]}{[B]}$

27. Only mechanism **(b)**

29. Only mechanism **(a)**

31. $\text{rate} = \dfrac{k_2k_1[A][B][E]}{k_2[E] + k_{-1}[D]}$ When $k_2[E] \ll k_{-1}[D]$

33. $\dfrac{d[Cl_2]}{dt} = \dfrac{k_1k_2[NO_2Cl]^2}{k_{-1}[NO_2] + k_2[NO_2Cl]}$

35. **(a)** 4.25×10^5 J **(b)** 1.54×10^{11} L mol^{-1} s^{-1}

37. **(a)** 1.49×10^{-3} L mol^{-1} s^{-1} **(b)** 1.30×10^3 s

39. **(a)** 4.3×10^{-3} s^{-1} **(b)** 1.7×10^5 s^{-1}

41. 70.3 kJ mol^{-1}

43. For OH + ethane, $\varepsilon_B \approx 21.1$ kJ mol^{-1}. For OH + ethene, $\varepsilon_B \approx 0$ kJ mol^{-1}. $E_{rel} = 83$ kJ mol$^{-1} > 21$ kJ mol^{-1} so yes, if not a glancing collision.

45. 6.0×10^9 L mol^{-1} s^{-1}

47. The rate constant is much smaller than that expected for a diffusion-controlled reaction so it is limited by a large activation energy.

49. **(a)** 1×10^{-3} mol L^{-1} s^{-1} **(b)** 5×10^{-5} M

CHAPTER 19

1. **(a)** $2\,^{12}_{6}C \longrightarrow\, ^{23}_{12}Mg + ^{1}_{0}n$
(b) $^{15}_{7}N + ^{1}_{1}H \longrightarrow\, ^{12}_{6}C + ^{4}_{2}He$
(c) $2\,^{3}_{2}He \longrightarrow\, ^{4}_{2}He + 2\,^{1}_{1}H$

3. **(a)** 3.300×10^{10} kJ mol^{-1} = 342.1 MeV total; 8.551 MeV per nucleon
(b) 7.312×10^{10} kJ mol^{-1} = 757.9 MeV total; 8.711 MeV per nucleon
(c) 1.738×10^{11} kJ mol^{-1} = 1801.7 MeV total; 7.570 MeV per nucleon

5. The two ^{4}He atoms are more stable, with a mass lower than that of one ^{8}Be atom by 9.85×10^{-5} u.

7. 16.96 MeV

9. **(a)** $^{39}_{17}Cl \longrightarrow\, ^{39}_{18}Ar + ^{0}_{-1}e^- + \bar{\nu}$
(b) $^{22}_{11}Na \longrightarrow\, ^{22}_{10}Ne + ^{0}_{-1}e^+ + \nu$
(c) $^{224}_{88}Ra \longrightarrow\, ^{220}_{86}Rn + ^{4}_{2}He$
(d) $^{82}_{38}Sr + ^{0}_{-1}e^- \longrightarrow\, ^{82}_{37}Rb + \nu$

11. ^{19}Ne decays to ^{19}F by positron emission; ^{23}Ne decays to ^{23}Na by β^- decay (electron emission).

13. An electron, 0.78 MeV

15. $^{30}_{14}Si + ^{1}_{0}n \longrightarrow\, ^{31}_{14}Si \longrightarrow\, ^{31}_{15}P + ^{0}_{-1}e^- + \bar{\nu}$

17. $^{210}_{84}Po \longrightarrow\, ^{206}_{82}Pb + ^{4}_{2}He$ $\quad ^{4}_{2}He + ^{9}_{4}Be \longrightarrow\, ^{12}_{6}C + ^{1}_{0}n$

19. 3.7×10^{10} min^{-1}

21. **(a)** 1.0×10^6 **(b)** 5.9×10^4

23. 1.6×10^{18} disintegrations s^{-1}

25. 4200 years

27. 3.0×10^9 years

29. 5.9×10^9 years

31. $^{11}_{6}C \longrightarrow\, ^{11}_{5}B + ^{0}_{1}e^+ + n$; $^{15}_{8}O \longrightarrow\, ^{15}_{7}N + ^{0}_{1}e^+ + n$

33. Exposure from ^{15}O is greater by a factor of 1.72/0.99 = 1.74.

35. **(a)** 2.3×10^{10} Bq
(b) 0.025 mGy
(c) Yes. A dose of 5 Gy has a 50% chance of being lethal; this dose is greater than 8.5 Gy in the first 8 days.

37. **(a)** $^{90}_{38}Sr \longrightarrow\, ^{90}_{40}Zr + 2\,^{0}_{-1}e^- + 2\,\bar{\nu}$
(b) 2.8 MeV
(c) 5.23×10^{12} disintegrations s^{-1}
(d) 4.44×10^{11} disintegrations s^{-1}

39. Increase (assuming no light isotopes of U are products of the decay)

41. 7.59×10^7 kJ g^{-1}

CHAPTER 20

1. **(a)** $A = 0.699$ **(b)** $\varepsilon = 1 \times 10^3$ L mol^{-1} cm^{-1}

3. $c_A = 0.000368$ mol L^{-1}

$c_B = 0.00157$ mol L^{-1}

5. Use the Beer-Lambert law in the form $I_t = I_0 \exp[-(N/V)\sigma\ell]$ and rearrange it to solve for ℓ to get

$$\begin{aligned}\ell &= -\sigma^{-1}(N/V)^{-1}\ln(I_t/I_0)\\ &= -(5 \times 10^{-20}\text{ cm}^2\text{ molecule}^{-1})^{-1} \times\\ &\quad (5 \times 10^{12}\text{ molecules cm}^{-3})^{-1}\ln(0.4)\\ &= 3.7 \times 10^6\text{ cm} = 37\text{ km}\end{aligned}$$

7. 35583.5 cm^{-1}

9. $N_{v=1}/N_{v=0} = \exp(-\Delta\varepsilon/k_BT) = 0.17$;

$N_{J=1}/N_{J=0} = 3\exp(-\Delta\varepsilon/k_BT) = 2.86$

11. 6.736×10^{-23} J = 40.56 J mol^{-1}

13. **(a)** 1.45×10^{-46} kg m^2
(b) $E_1 = 7.65 \times 10^{-23}$ J; $E_2 = 2.30 \times 10^{-22}$ J; $E_3 = 4.59 \times 10^{-22}$ J
(c) 1.13 Å

15. $I = h/(8\pi^2 c\tilde{B}) = 1.929 \times 10^{-46}$ kg m^2

17. 25.4 kg s^{-2}

19. 5.1×10^2 kg s^{-2}

21. $(v_f, J_f) = (1,3)$ and $(1,1)$. The transition from (0,2) to (1,1) requires 2950 cm^{-1}. The transition from (0,2) to (1,3) requires a 3050 cm^{-1} photon. It also can undergo a pure rotational transition from $J_i = 2$ to $J_f = 3$, by absorbing a 60 cm^{-1} photon.

23. 5.3×10^{-4}

25. Rearranging Equation 20.10 to solve for k, with vibrational frequencies given in cm^{-1}, and substituting the values given for each molecule we find that k = 966.0 N m^{-1} (^{1}H^{19}F), 516.3 N m^{-1} (^{1}H^{35}Cl), 351.7 N m^{-1} (^{1}H^{81}Br), 313.7 N m^{-1} (^{1}H^{127}I). $\mu \approx 1$ for all of these molecules so the vibrational frequency is most sensitive to differences in the force constants.

27. $(\nu_{CD})/(\nu_{CH}) \approx \sqrt{\mu_{CH}/\mu_{CD}}$ so $\nu_{CD} \approx 2165$ cm^{-1}

29. NH_3 (6); C_2H_4 (12); CCl_2F_2 (9); CH_3CH_2OH (21)

31. The top spectrum is that of nonane and the bottom spectrum is that of 1-hexanol. Prominent bands in the nonane spectrum include those due to aliphatic C—H stretching modes around 2,900 cm^{-1}, $-CH_2$ and $-CH_3$ bending modes near 1,450 and 1,375 cm^{-1}, respectively. Prominent bands in the 1-hexanol spectrum include the broad O—H stretch around 3,400 cm^{-1} characteristic of hydrogen-bonded alcohols, $-CH_2$ and $-CH_3$ bending modes near 1,450 and 1,375 cm^{-1} similar to those in nonane, and the C—O stretch around 1,075 cm^{-1}, also characteristic of alcohols.

33. 2, 3:2; 1; 2, 6:1

35. Decrease, because the electron will enter a π^* antibonding orbital

37. In the blue range, around 450 nm

39. Shorter; the π bonding is localized in cyclohexene.

41. 474 nm

43.

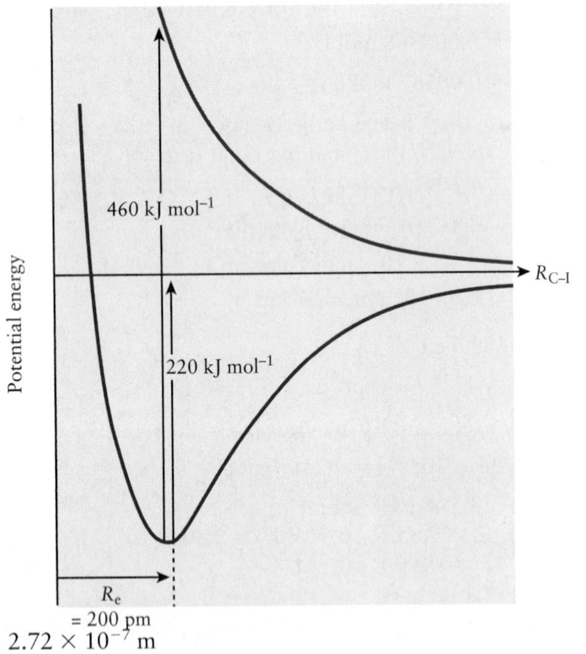

45. 2.72×10^{-7} m

47.

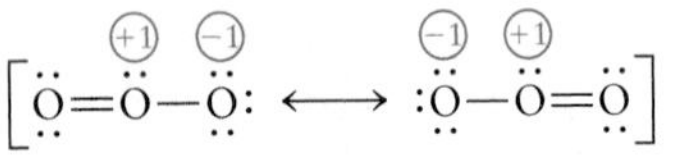

$SN = 3$, sp^2 hybridization, bent molecule

Two electrons in a π orbital formed from the three $2p_z$ orbitals perpendicular to the molecular plane, total bond order of $\frac{3}{2}$ for each O—O bond.

49. 8 molecules

CHAPTER 21

1. **(b)**, **(c)**, and **(d)**

3. Two mirror planes, one 2-fold rotation axis

5. 3.613 Å

7. 102.9°

9. ±49.87° and ±115.0°

11. **(a)** 5.675×10^6 Å^3 **(b)** 5×10^{14}

13. 4.059 g cm^{-3}

15. **(a)** 1.602×10^{-22} cm^3
(b) 3.729×10^{-22} g
(c) 4.662×10^{-23} g
(d) 6.025×10^{23} (in error by 0.05%)

17. 8

19. ReO_3

21. **(a)** 2.48 Å **(b)** 2.87 Å **(c)** 1.24 Å

23. **(a)** 2 **(b)** 0.680

25. 0.732

27. **(a)** Ionic **(b)** Covalent
(c) Molecular **(d)** Metallic

29. $CO < BaCl_2 < Co < SiC$

33. 8, 6, 12

35. Not by a significant amount because each vacancy is accompanied by an interstitial. In large numbers, such defects could cause a small bulging of the crystal and a decrease in its density.

37. **(a)** $Fe_{0.9352}O$ **(b)** 13.86% of the iron

39. 664 kJ mol^{-1}

41. **(a)** 1041 kJ mol^{-1} **(b)** 1205 kJ mol^{-1}

CHAPTER 22

1. The structures of $P_2O_7^{4-}$ and $S_2O_7^{2-}$ have the same number of electrons and are found by writing "P" or "S" in place of "Si" and adjusting the formal charges. The chlorine compound is Cl_2O_7, dichlorine heptaoxide.

3. **(a)** Tetrahedra; Ca: +2, Fe: +3, Si: +4, O: −2
(b) Infinite sheets; Na: +1, Zr: +2, Si: +4, O: −2
(c) Pairs of tetrahedra; Ca: −2, Zn: +2, Si: +4, O: −2
(d) Infinite sheets; Mg: +2, Si: +4, O: −2, H: +1

5. **(a)** Infinite network; Li: +1, Al: +3, Si: +4, O: −2
(b) Infinite sheets; K: 11, Al: +3, Si: +4, O: −2, H: +1
(c) Closed rings or infinite single chains; Al: +3, Mg: +2, Si: +4, O: −2

7. $Mg_3Si_4O_{10}(OH)_2(s) \longrightarrow 3\ MgSiO_3(s) + SiO_2(s) + H_2O(g)$

9. 234 L

11. 0.418 mol Si, 0.203 mol Na, 0.050 mol Ca, 0.0068 mol Al, 0.002 mol Mg, 0.0005 mol Ba

13. −113.0 kJ

15. $\frac{7}{3}$

17. **(a)** $SiO_2(s) + 3\ C(s) \longrightarrow SiC(s) + 2\ CO(g)$
(b) +624.6 kJ mol^{-1}
(c) Low conductivity, high melting point, very hard

19. −1188.2 kJ, unstable

21. 24 (Ω m)$^{-1}$ or 24 S m^{-1}

23. 1.26 C V^{-1} s^{-1} m^{-1} or 1.26 (Ω m)$^{-1}$ or 1.26 S m^{-1}

25. 1.5 electrons per Cu atom

27. 2.16×10^{-19} J

29. 6.1×10^{-27} electrons in a cm^3, a negligible number density

31. **(a)** Electron movement (*n*-type)
(b) Hole movement (*p*-type)

33. 680 nm, red

35. Decrease

CHAPTER 23

1. $n\ CCl_2{=}CH_2 \longrightarrow -(CCl_2{-}CH_2{-})_n$

3. Formaldehyde, $H_2C{=}O$

5. **(a)** H_2O **(b)** $-(NH{-}CH_2{-}CO{-})$

7. 646 kg adipic acid and 513 kg hexamethylenediamine

9. 3.49×10^{12} L = 3.49×10^{9} m^3 = 3.49 km^3

11. The isotropic phase has higher entropy and higher enthalpy.

13. A micelle (containing hydrocarbon in the interior and water on the outside)

15. 5 chiral centers

17. 27

19. In octane, because the side groups are all hydrocarbon

21. C_9H_9NO; 119 units

APPENDIX A

1. **(a)** 5.82×10^{25} **(b)** 1.402×10^{3} **(c)** 7.93 **(d)** -6.59300×10^{3} **(e)** 2.530×10^{-3} **(f)** 1.47

3. **(a)** 0.000537 **(b)** 9,390,000 **(c)** −0.00247 **(d)** 0.00620 **(e)** 20,000

5. 746,000,000 kg

7. **(a)** The value is 135.6 **(b)** 111.34 g **(c)** 0.22 g, 0.23 g

9. That of the mass in problem 7

11. **(a)** 5 **(b)** 3 **(c)** Either two or three **(d)** 3 **(e)** 4

13. **(a)** 14 L **(b)** −0.0034°C **(c)** 340 lb = 3.4×10^{2} lb **(d)** 3.4×10^{2} miles **(e)** 6.2×10^{-27} J

15. 2,997,215.55

17. **(a)** −167.25 **(b)** 76 **(c)** 3.1693×10^{15} **(d)** -7.59×10^{-25}

19. **(a)** −8.40 **(b)** 0.147 **(c)** 3.24×10^{-12} **(d)** 4.5×10^{13}

21. $A = 337$ cm^2

APPENDIX B

1. **(a)** 6.52×10^{-11} kg **(b)** 8.8×10^{-11} s **(c)** 5.4×10^{12} kg m^2 s^{-3} **(d)** 1.7×10^{4} kg m^2 s^{-3} A^{-1}

3. **(a)** 4983°C **(b)** 37.0°C **(c)** 111°C **(d)** −40°C

5. **(a)** 5256 K **(b)** 310.2 K **(c)** 384 K **(d)** 233 K

7. **(a)** 24.6 m s^{-1} **(b)** 1.51×103 kg m^{-3} **(c)** 1.6×10^{-29} A s m **(d)** 1.5×10^{2} mol m^{-3} **(e)** 6.7 kg m^2 s^{-3} = 6.7 W

9. 1 kW-hr = 3.6×10^{6} J; 15.3 kW-hr = 5.51×10^{7} J

11. 6620 cm^3 or 6.62 L

13. 3×10^{2} J

15. 55 J

APPENDIX C

1. 50 miles per hour

3. **(a)** Slope = 4, intercept = −7

(b) Slope $\frac{7}{2} = 3.5$, intercept $= -\frac{5}{2} = -2.5$

(c) Slope = −2, intercept $= \frac{4}{3} = 1.3333\ldots$

5. The graph is not a straight line.

7. **(a)** $-\frac{5}{7} = -0.7142857\ldots$ **(b)** $\frac{3}{4} = 0.75$

(c) $\frac{2}{3} = 0.6666\ldots$

9. **(a)** 0.5447 and −2.295 **(b)** −0.6340 and −2.366 **(c)** 2.366 and 0.6340

11. **(a)** 6.5×10^{-7} (also two complex roots) **(b)** 4.07×10^{-2} (also 0.399 and 21.011) **(c)** −1.3732 (also two complex roots)

13. **(a)** 4.551 **(b)** 1.53×10^{-7} **(c)** 2.6×10^{8} **(d)** −48.7264

15. 3.015

17. 121.477

19. 7.751 and −2.249

21. $x = 1.086$

23. **(a)** $8x$ **(b)** $3\cos 3x - 8\sin 2x$ **(c)** 3 **(d)** $\frac{1}{x}$

25. **(a)** 20 **(b)** $\frac{78125}{7}$ **(c)** 0.0675

INDEX/GLOSSARY

Page numbers followed by "f" denote figures; those followed by "t" denote tables. Page numbers preceded by "A" indicate Appendix page numbers.

Locations of Some Important Tables of Data